2017 IEEE 44th Photovoltaic Specialist Conference (PVSC 2017)

Washington, DC, USA
25-30 June 2017

Pages 2128-2832

IEEE Catalog Number: CFP17PSC-POD
ISBN: 978-1-5090-5606-4

Copyright © 2017 by the Institute of Electrical and Electronics Engineers, Inc.
All Rights Reserved

Copyright and Reprint Permissions: Abstracting is permitted with credit to the source. Libraries are permitted to photocopy beyond the limit of U.S. copyright law for private use of patrons those articles in this volume that carry a code at the bottom of the first page, provided the per-copy fee indicated in the code is paid through Copyright Clearance Center, 222 Rosewood Drive, Danvers, MA 01923.

For other copying, reprint or republication permission, write to IEEE Copyrights Manager, IEEE Service Center, 445 Hoes Lane, Piscataway, NJ 08854. All rights reserved.

*** *This is a print representation of what appears in the IEEE Digital Library. Some format issues inherent in the e-media version may also appear in this print version.*

IEEE Catalog Number: CFP17PSC-POD
ISBN (Print-On-Demand):   978-1-5090-5606-4
ISBN (Online):   978-1-5090-5605-7
ISSN: 0160-8371

Additional Copies of This Publication Are Available From:

Curran Associates, Inc
57 Morehouse Lane
Red Hook, NY 12571 USA
Phone: (845) 758-0400
Fax: (845) 758-2633
E-mail: curran@proceedings.com
Web: www.proceedings.com

2017 IEEE 44th Photovoltaic Specialist Conference (PVSC 2017)

Washington, DC, USA
25-30 June 2017

Pages 2128-2832

IEEE Catalog Number: CFP17PSC-POD
ISBN: 978-1-5090-5606-4

TABLE OF CONTENTS

OPEN CIRCUIT VOLTAGE CALCULATION USING TEMPERATURE AND IRRADIANCE 1
Andrew Melvin

EFFECT OF CL-DOPING IN ZNTEO ON PHOTOLUMINESCENCE AND PHOTOVOLTAIC
PROPERTIES OF ZNTEO-BASED INTERMEDIATE BAND SOLAR CELLS 3
T. Tanaka ; S. Tsutsumi ; Y. Okano ; K. Matsuo ; K. Saito ; Q. Guo ; M. Nishio ; T. Tayagaki ; K. M. Yu ; W. Walukiewicz

TOWARD LEAD HALIDE PEROVSKITE-BASED INTERMEDIATE BAND ABSORBERS 6
Matthew D. Sampson ; Ji-Sang Park ; Richard D. Schaller ; Maria K. Y. Chan ; Alex B. F. Martinson

TYPE-II QUANTUM DOTS FOR APPLICATION TO PHOTON RATCHET INTERMEDIATE
BAND SOLAR CELLS 10
Ryo Tamaki ; Yasushi Shoji ; Yoshitaka Okada

AN INVESTIGATION OF THE ROLE OF RECOMBINATION PROCESSES IN THE
OPERATION OF INAS/GAASL-XSBX QUANTUM DOT SOLAR CELLS 14
Y. Cheng ; A. J. Meleco ; A. J. Roeth ; V. R. Whiteside ; M. C. Debnath ; M. B. Santos ; T. D. Mishima ; S. Hatch ; H.Y. Liu ; I. R. Sellers

TEMPERATURE AND VOLTAGE-BIAS DEPENDENT TWO-STEP PHOTON ABSORPTION IN
INAS/GAASL AL0.3GAAS QUANTUM DOT IN A WELL SOLAR CELLS 18
Yushuai Dai ; Brittany L. Smith ; Michael A. Slocum ; Zachary S. Bittner ; Hyun Kum ; Julia D'Rozario ; Seth M. Hubbard

INCREASING CURRENT GENERATION BY PHOTON UP-CONVERSION IN A SINGLE-
JUNCTION SOLAR CELL WITH A HETERO-INTERFACE 23
Shigeo Asahi ; Kazuki Kusaki ; Toshiyuki Kaizu ; Takashi Kita

RTP-ASSISTED EX-SITU ANALYSIS OF (AG,CU)(IN,GA)SE2FORMATION USING
SELENIZATION 26
Sina Soltanmohammad ; William N. Shafarman

ROLE OF EV+0.98 EV TRAP IN LIGHT SOAKING-INDUCED SHORT CIRCUIT CURRENT
INSTABILITY IN CIGS SOLAR CELLS 30
P. K. Paul ; T. Jarmar ; L. Stolt ; A. Rockett ; A. R. Arehart

STUDY OF DEFECT PROPERTIES IN CUGASE2THIN-FILM SOLAR-CELLS USING
ADMITTANCE SPECTROSCOPY 33
Muhammad Monirul Islam ; Shogo Ishizuka ; Hajime Shibata ; Shigeru Niki ; Katsuhiro Akimoto ; Takeaki Sakurai

TRANSMISSIVE SPECTRUM-SPLITTING CONCENTRATOR PHOTOVOLTAIC CELLS AND
MODULES 37
Yaping Ji ; Qi Xu ; Brian Riggs ; John Robertson ; Kazi Islam ; Vince Romanin ; Dimitri D. Krut ; Jim H. Ermer ; Matthew D. Escarra

ALGAINP/GAAS TANDEM SOLAR CELLS FOR POWER CONVERSION AT 400 C AND 1000X
CONCENTRATION 42
Myles A. Steiner ; Emmett E. Perl ; John Simon ; Daniel J. Friedman ; Nikhil Jain ; Paul Sharps ; Claiborne Mcpheeters ; Minjoo L. Lee

GALNASP SOLAR CELLS GROWN BY HYDRIDE VAPOR PHASE EPITAXY FOR ONE-SUN &
LOW-CONCENTRATION III-V/SI PHOTOVOLTAICS 46
Nikhil Jain ; John Simon ; Kevin L. Schulte ; David R. Diercks ; Corinne E. Packard ; David Young ; Aaron J. Ptak

PHOTO-ELECTROCHEMICAL HYDROGEN GENERATION FROM INVERTED
METAMORPHIC MULTIJUNCTION III-VS 47
Todd G. Deutsch ; James L. Young ; Myles A. Steiner ; Henning Döscher ; Ryan M. France ; John A. Turner

ADVANCED SILICON THIN FILMS FOR HIGH-EFFICIENCY SILICON HETEROJUNCTION-
BASED SOLAR CELLS 50
A. Descoeudres ; C. Allebe ; N. Badel ; L. Barraud ; J. Champliaud ; G. Christmann ; L. Curvat ; F. Debrot ; A. Faes ; J. Geissbühler ; J. Horzel ; A. Lachowicz ; J. Levrat ; S. Martin De Nicolas ; S. Nicolay ; B. Paviet-Salomon ; L.-L. Senaud ; A. Tomasi ; C. Ballif ; M. Despeisse

MOOXAND WOXBASED HOLE-SELECTIVE CONTACTS FOR WAFER-BASED SI SOLAR
CELLS 55
Stephanie Essig ; Julie Dréon ; Jérémie Werner ; Philipp Löper ; Stefaan De Wolf ; Mathieu Boccard ; Christophe Ballif

METAL NANOPARTICLE HOLE CONTACTS FOR SILICON SOLAR CELLS 59
James Bullock ; Zhaoran Xu ; Mark Hettick ; Yimao Wan ; Ali Javey

NEAR-FIELD TRANSPORT IMAGING APPLICATION OF PHOTOVOLTAIC MATERIALS..........62
Chuanxiao Xiao ; Chun-Sheng Jiang ; John Moseley ; John Simon ; Kevin Schulte ; Aaron J. Ptak ; Steve Johnston ; Brian Gorman ; Mowafak Al-Jassim ; Nancy M. Haegel ; Helio Moutinho

APPLICATIONS OF DMD-BASED INHOMOGENEOUS ILLUMINATION PHOTOLUMINESCENCE IMAGING FOR SILICON WAFERS AND SOLAR CELLS..........66
Yan Zhu ; Mattias Klaus Juhl ; Ziv Hameiri ; Thorsten Trupke

NUMERICAL MODEL TO EXTRACT MATERIALS PROPERTIES MAP FROM SPECTRALLY RESOLVED LUMINESCENCE IMAGES..........70
Nicolas Paul ; Vincent Le Guen ; Daniel Ory ; Laurent Lombez

NON-DESTRUCTIVE CONTACT RESISTIVITY MEASUREMENTS ON SOLAR CELLS USING THE CIRCULAR TRANSMISSION LINE METHOD..........74
Geoffrey Gregory ; Andrew M. Gabor ; Andrew Anselmo ; Rob Janoch ; Zhihao Yang ; Kristopher O. Davis

RADIATION RESISTANCE OF LOW COST HIGH EFFICIENCY TRIPLE JUNCTION SOLAR CELLS..........76
Roberta Campesato ; Erminio Greco ; Giuseppe Gabetta ; Mariacristina Casale ; Gabriele Gori ; M. Sankaran ; Suresh E. Puthanveettil ; B. R. Uma ; M. Ravindra ; Sheeja Krishnan

AMORPHOUS SILICON CARBIDE REAR-SIDE PASSIVATION AND REFLECTOR LAYER STACKS FOR MULTI-JUNCTION SPACE SOLAR CELLS BASED ON GERMANIUM SUBSTRATES..........83
Stefan Janz ; Charlotte Weiss ; Christian Mohr ; Rufi Kurstjens ; Bruno Boizot ; Bianca Fuhrmann ; Victor Khorenko

HOT CARRIER TRANSPORTATION DYNAMICS IN INAS/GAAS QUANTUM DOT SOLAR CELL..........85
Tomah Sogabe ; Kohdai Nii ; Katsuyoshi Sakamoto ; Koichi Yarnaquchi ; Yoshitaka Okada

INTEGRATION OF CRACK-TOLERANT COMPOSITE GRIDLINES ON TRIPLE JUNCTION PHOTOVOLTAIC CELLS..........88
Omar K. Abudayyeh ; Geoffrey K. Bradshaw ; Steven Whipple ; David M. Wilt ; Sang M. Han

SUBCELL LIGHT CURRENT- VOLTAGE CHARACTERIZATION OF IRRADIATED MULTIJUNCTION SOLAR CELL..........93
Don Walker ; John Nocerino ; Yao Yue ; Colin J. Mann ; Simon H. Liu

ANALYTICAL METHOD FOR PREDICTING SPACECRAFT POWER GENERATION ON PARTIALLY SHADED SOLAR PANELS..........96
Gordon Wu ; Bao Hoang

EVALUATING THE EMISSIVITY OF PSEUDOMORPHIC GLASS (PMG)..........102
Ryan D. Beauchemin ; David M. Wilt ; Paul E. Hausgen

CHARACTERIZING THE IMPACT OF SOLAR SPECTRAL IRRADIANCE ON PV MODULE OUTPUT..........107
M. Schweiger ; W. Herrmann

USE OF MEASURED AEROSOL OPTICAL DEPTH AND PRECIPITABLE WATER TO MODEL CLEAR SKY IRRADIANCE..........110
Mark M. Mikofski ; Clifford W. Hansen ; William F. Holmgren ; Gregory M. Kimbal

RECENT ADVANCEMENTS IN THE NUMERICAL SIMULATION OF SURFACE IRRADIANCE FOR SOLAR ENERGY APPLICATIONS..........116
Yu Xie ; Manajit Sengupta ; Chris Deline

OPTIMAL IRRADIANCE SENSOR PLACEMENT FOR PHOTOVOLTAIC SYSTEMS USING MUTUAL INFORMATION BASED GREEDY ALGORITHM IN GAUSSIAN PROCESS..........120
Lian Lian Jiang ; R. Srivatsan ; Douglas L. Maskell

EVALUATING DIFFERENT UPSCALING APPROACHES TO DERIVE THE ACTUAL POWER OF DISTRIBUTED PV SYSTEMS..........126
Sven Killinger ; Björn Müller ; Bernhard Wille-Haussmann ; Russell Mckenna

ADVANCES IN LONG-TERM SOLAR ENERGY PREDICTION AND PROJECT RISK ASSESSMENT METHODOLOGY..........132
Alemu Tadesse ; Adam Kankiewicz ; Alex Kubiniec ; Richard Perez ; John Dise ; Thomas Hoff

DECOUPLING THIN FILM CDTE GROWTH FROM PACKAGING: TOWARD RECORD SPECIFIC POWER IN LOW COST POLYCRYSTALLINE PV..........138
D. Clayton-Warwick ; M.D. Kempe ; M. S. Dabney ; T. M. Barnes ; C. A. Wolden ; M. O. Reese

JUNCTION ACTIVATION OF CDTE/CDS SOLAR CELL USING MGCL2..........142
G. Angeles-Ordóñez ; E. Regalado-Pérez ; M.G. Reyes-Banda ; N. R. Mathews ; X. Mathew

VARIATION OF CU CONTENT OF SPRAYED CU(IN, GA)(S,SE)2SOLAR CELLS BASED ON A THIOL-AMINE SOLVENT MIXTURE..........146
Panagiota Arnou ; Sona Ulicná ; Alexander Eeles ; Mustafa Togay ; Lewis D. Wright ; Andrei V. Malkov ; John M. Walls ; Jake W. Bowers

CUINSE2 ABSORBER LAYER GROWN UNDER COPPER EXCESS WITH A COPPER POOR SURFACE FORMED BY A KF POST DEPOSITION TREATMENT 151

Finn Babbe ; Hossam Elanzeery ; Michele Melchiorre ; Susanne Siebentritt

CU2ZNSNSE4SOLAR CELLS ONTO POLYIMIDE SUBSTRATES FABRICATED AT LOW TEMPERATURE 155

Ignacio Becerril-Romero ; Simón Lopez-Marino ; Moisés Espíndola-Rodríguez ; Laura Acebo ; Markus Neuschitzer ; Yudania Sánchez ; Edgardo Saucedo ; Paul Pistor

AN OPTIMIZED PHOTOLITHOGRAPHY RECIPE FOR CU(IN1-X,GAX)(SY,SE1-Y)2(CIGSSE) SOLAR CELLS 160

Xia Hao ; Shenghao Wang ; Katsuhiro Akimoto ; Takuya Kato ; Hiroki Sugimoto ; Takeaki Sakurai

EFFECTS OF CDCL2PASSIVATION ON THIN CDTE ABSORBERS FABRICATED BY CLOSE-SPACE SUBLIMATION 164

Anna Wojtowicz ; Alexandra M. Huss ; Jennifer A. Drayton ; James R. Sites

CDS1-XSEXWINDOW LAYER FOR CDTE PREPARED BY THE EXCHANGE OF S WITH SE IN CDS FILMS 170

Geethika K. Liyanage ; Adam B. Phillips ; Zhaoning Song ; Suneth C. Watthage ; Ramez H. Ahanzhamejhad ; Michael J. Heben

EFFECT OF ILLUMINATION ON THERMAL CDCL2TREATMENT OF CDTE 175

Sudhajit Misra ; Carina E. Hahn ; Vasilios Palekis ; Christos Ferekides ; Michael A. Scarpulla

CHALLENGES IN THE INDUSTRIAL PRODUCTION OF CZTS MONOGRAIN SOLAR CELLS 178

Gerhard Peharz ; Valentin Satzinger ; Sandra Pötz ; Gernot Oreski ; Theodoros Dimopoulos ; Stefan Edinger ; Wolfeanz Hackl ; Hannes Starkl ; Parichehr Esfandiari ; Peter Krabb ; Stefan Gahr ; Lukas Plessing ; Dieter Meissner

UNDERSTANDING INSTABILITIES AND DEGRADATION DUE TO MOISTURE INGRESS IN CU(IN, GA)SE2SOLAR CELLS 182

Grace Rajan ; Shankar Karki ; Isaac Butt ; Krishna Aryal ; Tyler J. Grassman ; Angus Rockett ; Sylvain Marsillac

CONTROL OF MOSE2 FORMATION IN HYDRAZINE-FREE SOLUTION-PROCESSED CIS/CIGS THIN FILM SOLAR CELLS 186

Sona Ulicná ; Panagiota Arnou ; Alexander Eeles ; Mustafa Togay ; Lewis D. Wright ; Ali Abbas ; Andrei V. Malkov ; John M. Walls ; Jake W. Bowers

GROWTH AND PROPERTIES OF EPITAXIAL CU(IN, GA)SE2THIN FILMS DEPOSITED BY THE THREE-STAGE PROCESS FOR SOLAR CELLS 192

Takeru Yamagami ; Yuta Ando ; Ishwor Khatri ; Mutsumi Sugiyama ; Tokio Nakada

IMPROVEMENT OF CIS SOLAR CELLS WITH KF POSTDEPOSITION FOLLOWING A SIMPLE TWO-STEP SELENIZATION PROCESS 195

Yang Zhang ; Robert E. Bartolo ; Sang Jik Kwon ; Mario Dagenais

THE TWINS STRUCTURE, ELECTRICAL PROPERTIES AND CELL PERFORMANCE OF MAGNETRON SPUTTERING DEPOSITED CHLORINE DOPED CDTE 198

Ziyao Zhu ; Fu-Kuo Chiang ; Zhongming Du ; Yufeng Zhang ; Xiangxin Liu

INVESTIGATION AND MITIGATION OF SHUNTS FOR HIGHER EFFICIENCY EPITAXIAL GASB/GASB AND GASB/GAAS SOLAR CELLS 202

George T. Nelson ; Bor-Chau Juang ; Steve Johnston ; Michael A. Slocum ; Zachary S. Bittner ; Ramesh B. Lagumavarapu ; Diana Huffaker ; Seth M. Hubbard

DEVELOPMENT OF GASB SOLAR CELLS ON GAAS BY MOVPE VIA INTERFACE MISFIT TECHNIQUE 206

Michael A. Slocum ; Alessandro Giussani ; Emily Kessler ; Phil Ahrenkiel ; George T. Nelson ; Seth M. Hubbard

FABRICATION OF INGAASP SOLAR CELLS FOR CONCENTRATOR APPLICATIONS 210

Mitchell F. Bennett ; Matthew P. Lumb ; Kenneth J. Schmieder ; Brent Fisher ; Eric A. Armour ; Robert J. Walters

DETAILED CHARACTERIZATION FOR TCAD SIMULATIONS OF GAAS0.76P0.24/SI1-YGEY/SI SINGLE JUNCTION SOLAR CELLS 213

Sabina Abdul Hadi ; Timothy Milakovich ; Eugene A. Fitzgerald ; Ammar Nayfeh

COMPARATIVE STUDY OF >2 EV LATTICE-MATCHED AND METAMORPHIC (AL)GAINP MATERIALS AND SOLAR CELLS GROWN BY MOCVD 215

Daniel J. Chmielewski ; Christine Jackson ; Jacob Boyer ; Daniel Lepkowski ; John A. Carlin ; Aaron R. Arehart ; Tyler J. Grassman ; Steven A. Ringel

PERFORMANCE OF GASB PHOTOVOLTAICS WITH GRAPHENE COATING 219

Benjamin P. Conlon ; Daniel J. Herrera ; Shaimaa A. Abdallah ; Jonathan O. Okafor ; Luke F. Lester

HIGH EFFICIENCY SINGLE-JUNCTION INGAP PHOTOVOLTAIC DEVICES UNDER LOW INTENSITY LIGHT ILLUMINATION 222

Yushuai Dai ; Hyun Kum ; Michael A. Slocum ; George T. Nelson ; Seth M. Hubbard

RADIATION RESISTANT OF UPRIGHT METAMORPHIC GAINP/GAINAS/GE TRIPLE JUNCTION SOLAR CELLS FOR SPACE USE 226

Liang Fang ; Abuduwayiti Aierken ; Zhen Pan ; Qiming Zhang ; Zhanhang Li ; Heini Maliya ; Wei Gao ; Hui Gao ; Ronghua Wan ; Bao Zhang ; He Wang ; Qi Guo

HIGH EFFICIENCY GLASS WAVEGUIDING SOLAR CONCENTRATOR 229

Chehao Hu ; Yusuf Dogan ; Matthew Morrison ; A. Nanda ; D. Ma ; R. Atkins ; C. K. Madsen

GAINASP/GAINAS TANDEM SOLAR CELL WITH 32.6% ONE-SUN EFFICIENCY 232

Nikhil Jain ; Kevin L. Schulte ; John F. Geisz ; Ryan M. France ; Myles A. Steiner

EVALUATION OF TANDEM EFFICIENCIES: DILUTE NITRIDE P-I-N (BULK OR MQWS) IN CONJUNCTION WITH PRACTICAL SI SOLAR CELLS 236

Khim Kharel ; Alexandre Freundlich

GALLIUM PHOSPHIDE NANOSTRUCTURE ON SILICON BY SILICA NANOSPHERES LITHOGRAPHY AND METAL ASSISTED CHEMICAL ETCHING 240

Sangpyeong Kim ; Chaomin Zhang ; Som Dahal ; Stuart Bowden ; Christiana B. Honsberg

EFFICIENCY ENHANCEMENT OF INGAP/INGAAS/GE SOLAR CELLS WITH GRADUALLY DOPED P-N JUNCTION ACTIVE LAYERS 244

Youngjo Kim ; Sang Hyun Jung ; Chang Zoo Kim ; Kangho Kim ; Hyun-Beom Shin ; Kyung Ho Park ; Won-Kyu Park ; Jaejin Lee ; Ho Kwan Kang

ANALYSIS OF INGAP OXIDE GROWTH RATE AT HIGH TEMPERATURES AND AMBIENT CONDITIONS FOR TERRESTRIAL PHOTOVOLTAIC APPLICATIONS 247

Nicole A. Kotulak ; Matthew P. Lumb ; Raymond Hoheisel ; Erin Cleveland ; Mitchell Bennett ; Phillip P. Jenkins ; Robert J. Walters

GRAIN BOUNDARIES IN THIN-FILM POLYCRYSTALLINE GAAS SOLAR CELLS: A SIMULATION STUDY 251

Khushboo Kumari ; Sushobhan Avasthi

TIME-RESOLVED PL MEASUREMENTS IN THE GROWTH OF HIGH VOLTAGE (AL)GAINP/GAAS SOLAR CELLS 255

Xinyi Li ; Wei Zhang ; Hongbo Lu

LOW-RESISTANCE AND HIGHLY-TRANSPARENT GASB-BASED TUNNEL JUNCTIONS 259

Matthew P. Lumb ; Shawn Mack ; Maria Gonzalez ; Kenneth J. Schmieder ; Mitchell F. Bennett ; Chaffra A. Affouda ; James E. Moore ; Robert J. Walters

MODULATED PHOTOCURRENT MEASUREMENTS IN DOUBLE JUNCTION SOLAR CELLS 263

Nicolás Márquez Peraca ; Behrang H. Hamadani

EFFECT OF ATMOSPHERIC ABSORPTION BANDS ON THE OPTIMAL DESIGN OF MULTIJUNCTION SOLAR CELLS 268

William E. Mcmahon ; Daniel J. Friedman ; John F. Geisz

EFFECTS OF CONTACT CONFIGURATION AND PERIMETER RECOMBINATION ON OPTIMAL CELL SIZE FOR HIGH CONCENTRATION PHOTOVOLTAICS 272

James E. Moore ; Matthew P. Lumb ; Kenneth J. Schmieder ; Robert J. Walters ; Brent Fisher ; Matt Meitl ; Scott Burroughs

NUMERICAL SIMULATION OF DEFECTS IN III-V PV CELLS: THE EFFECT OF VOLTAGE BIAS AND DOPING CONCENTRATION 276

Vasiliki Paraskeva ; Constantinos Lazarou ; Andreas Livera ; Venizelos Venizelou ; Maria Hadjipanayi ; George E. Georghiou

IMPROVEMENT OF OPEN-CIRCUIT VOLTAGE IN METAMORPHIC GASB CELLS GROWN ON GAAS SUBSTRATES BY USING AN INTERFACIAL MISFIT ARRAY AND AN ALSB BLOCKING LAYER 281

E. J. Renteria ; S. J. Addamane ; D. M. Shima ; A. Mansoori ; A. L. Soudachanh ; G. Balakrishnan

ENERGY YIELD EVALUATION FOR FIELD OPERATION OF SOLAR CELLS IN SINGAPORE: GAAS/GAAS TANDEM VS. GAAS SINGLE-JUNCTION SOLAR CELLS 284

Maung Thway ; Zekun Ren ; Kevin Nay Yaung ; Haohui Liu ; Zhe Liu ; Samuel Raj ; Soo Jin Chua ; Armin G. Aberle ; Tonio Buonassisi ; Ian Marius Peters ; Fen Lin

SIMULATION OF THE PERFORMANCES OF MULTIJUNCTION SOLAR CELLS WITH IMPROVED VOLTAGE BY TRANSFER AND SCATTERING MATRIX METHODS 290

Gianluca Timò ; Lucio Andreani

OPTIMIZED DESIGN OF BACK-CONTACT THIN-FILM GAAS SOLAR CELLS 294

Jia-Ling Tsai ; Chung-Yu Hong ; Tien-Chien Zhan ; Yuh-Renn Wu ; Albert Lin ; Peichen Yu

DESIGN CONSIDERATIONS ON GAINNAS SOLAR CELLS WITH BACK SURFACE REFLECTORS 297

Antti Tukiainen ; Arto Aho ; Timo Aho ; Ville Polojärvi ; Mircea Guina

QUANTITATIVE ELECTROLUMINESCENCE ANALYSIS OF TRIPLE JUNCTION SOLAR CELLS TO DETERMINE SUBCELL VOLTAGE-TEMPERATURE COEFFICIENTS 301

Kevin Tyler ; Geoffrey K. Bradshaw ; Sam Wilt ; David M. Wilt ; Richard R. King

PROGRESS TOWARDS DOUBLE-JUNCTION INGAN SOLAR CELL 305

Ehsan Vadiee ; Evan A. Clinton ; Heather Mcfavilen ; Alec M. Fischer ; Yi Fang ; Joshua J. Williams ; Christiana B. Honsberg ; William A. Doolittle ; Stephen M. Goodnick

A PHYSICS-BASED SIMULATION TOOL FOR LEAKAGE CURRENTS IN C-SI PV MODULES 309

John M. Waddle ; Saroj Dahal ; Marco Nardone

BROADBAND TA2O5 MOTH-EYE ANTIREFLECTION COATINGS FOR TANDEM SOLAR CELLS ON SI 315

Bo Yuan ; Brian Thibeault ; David Payne ; James Mutitu ; Ivan Perez-Wurfl ; Kevin Dobson ; Brianna Conrad ; Allen Barnett ; Robert L. Opila

CARRIER TRANSPORT IN POLYCRYSTALLINE SILICON AT HIGH OPTICAL INJECTION: TRANSIENT PHOTOCONDUCTANCE VS. NUMERICAL MODELING 319

Uchechi Anyanwu ; Christian Harris ; Andrey Semichaevsky

IMPROVING SILICON SURFACE PASSIVATION WITH A SILICON OXIDE LAYER GROWN VIA OZONATED DEIONIZED WATER 322

Sara Bakhshi ; Ngwe Zin ; Kristopher O. Davis ; Marshall Wilson ; Ismail Kashkoush ; Winston V. Schoenfeld

DEPOSITION OF SIOC BY PLASMA-FREE ULTRA-LOW-TEMPERATURE ALD (ULT-ALD) AND ITS PASSIVATION ON P-TYPE SILICON 326

Meixi Chen ; Naoto Noda ; Raphael Rochat ; Abhishek Iyer ; James H. Hack ; Changhee Ko ; Christian Dussarrat ; Robert L. Opila

A METHOD FOR QUANTITATIVELY INVESTIGATING THE REAR-SIDE PASSIVATION PERFORMANCE OF PERC CELLS 329

Tsung-Cheng Chen ; Yung-Sheng Lin ; Chen-Hao Ku ; Ting-Wei Kuo ; Cheng-Shun Hu ; Ching-Chang Wen

FIELD-EFFECT PASSIVATION BY NEGATIVE CHARGE ON BORON EMITTER AND BORON-DOPED SURFACES BY A NOVEL LOW-COST PLASMA CHARGE INJECTION 333

Eunhwan Cho ; Young-Woo Ok ; James Hwang ; Aditi Jain ; Vijay D. Upadhyaya ; John Keith Tate ; Ajeet Rohatgi

INDUSTRY RELEVANT RIE TEXTURING FOR MC-SI DIAMOND WIRE OR DIRECT WAFER® PRODUCT: OPTIMIZED REFLECTIVITY, UNIFORMITY, AND THROUGHPUT 337

Jose Luis Cruz-Campa ; Ray Fraser ; Rob Steeman ; John Linton

SHORT-CIRCUIT CURRENT-DENSITY ENHANCEMENT OF SILICON SOLAR CELLS USING PLASMONICS ANTIREFLECTIVE COATING AND LUMINESCENT DOWNSHIFTING 343

Sheng-Kai Feng ; Wen-Jeng Ho ; Guan-Yi Li ; Jheng-Jie Liu ; Hao-Yu Yang ; Ta-Wei Chuang

EXTREMELY LOW REFLECTIVITY NANOPOROUS BLACK SILICON SURFACE BY COPPER CATALYZED ETCHING FOR EFFICIENT SOLAR CELLS 346

K A S M Ehteshamul Haque ; Wenqi Duan ; Fatima Toor

IMPACT OF FRONT SIDE PYRAMID SIZE ON THE LIGHT TRAPPING PERFORMANCE OF WAFER BASED SILICON SOLAR CELLS AND MODULES 352

Oliver Höhn ; Nico Tucher ; Benedikt Bläsi

A STUDY OF BLISTER CONTROL OF AL2O3 THIN FILM DEPOSITED BY PLASMA-ASSISTED ATOMIC LAYER DEPOSITION AFTER FIRING PROCESS 356

Min Gu Kang ; Jeong In Lee ; Hee-Eun Song ; Myeong Sangjeong ; Kyung Taekjeong ; Hyo Sikchang

PYPVCELL: AN OPEN-SOURCE SOLAR CELL MODELING LIBRARY IN PYTHON 359

Kan-Hua Lee ; Kenji Araki ; Omar Elleuch ; Nobuaki Kojima ; Masafumi Yamaguchi

IMPROVEMENT IN SURFACE PASSIVATION OF C-SI USING GRADIENT-LAYERED A-SI:H FILM FOR HIGH EFFICIENCY SILICON HETEROJUNCTION SOLAR CELLS 363

Soonil Lee ; Leo Mathew ; Rajesh Rao ; Jae Hyun Kim ; Sanjay K. Banerjee ; Edward T. Yu

PHOTOVOLTAIC PERFORMANCE ENHANCEMENT OF TEXTURED SILICON SOLAR CELLS USING LUMINESCENT DOWN-SHIFTING METHYLAMMONIUM LEAD TRIBROMIDE PEROVSKITE NANOPHOSPHORS 367

Guan-Yi Li ; Wen-Jeng Ho ; Sheng-Kai Feng ; Hao-Yu Yang ; Ta-Wei Chuang ; Bang-Jin You ; Zong-Xian Lin ; Zong-Liang Tseng ; Lung-Chien Chen

SINX THIN FILMS WITH APPROPRIATE ANTIREFLECTION AND SHIFT-CONVERSION PROPERTIES FOR SILICON SOLAR CELLS 370

E. Men-Pérez ; J. Salazar ; A. Dutt ; J. Santoyo-Salazar ; G. Santana

NUMERICAL SIMULATION OF CRYSTALLINE SILICON SOLAR CELLS WITH FULL AREA METAL OXIDE REAR CONTACTS 373

James E. Moore ; Woojun Yoon ; Phillip P. Jenkins ; Robert J. Walters

INTERDIGITATED BACK CONTACT SILICON SOLAR CELL WITH PEROVSKITE LAYER FOR FRONT SURFACE PASSIVATION AND ULTRAVIOLET RADIATION STABILITY 377

Rahul Pandey ; Shivam Gupta ; Trijul Khatri ; Rishu Chaujar

POTENTIAL OF A-SI:H/C-SI HETEROJUNCTION SOLAR CELLS WITH VERY THIN WAFERS 381

Hitoshi Sai ; Hiroshi Umishio ; Takuya Matsui ; Shota Nunomura ; Tomoyuki Kawatsu ; Hidetaka Takato ; Koji Matsubara

MANIPULATING FIXED CHARGES IN ZRO2 BY DOPING FOR PASSIVATION AND ANTIREFLECTION ON WAFER-SI SOLAR CELLS 385

Woo Jung Shin ; Laidong Wang ; Wen-Hsi Huang ; Meng Tao

LOW TEMPERATURE ANTIREFLECTION COATING FOR SILICON SOLAR CELLS 389

O. S. Shinde ; Ej Schneller ; N. Dhere ; S. V. Ghaisas

RELATIONSHIP BETWEEN POWER LOSS AND VOLTAGE APPLIED TO SOLAR CELLS IN PID-AFFECTED SOLAR MODULES 392

Fumei Wang ; Baosong Duan ; Wenshuang He ; He Wang ; Hong Yang ; Chengfeng Su ; Bojie Su ; Xue Zhang ; Yunxue Cao ; Hui Zhao

A NEW LOW-COST AND LOW-TEMPERATURE CHEMICAL PASSIVATION PROCESS FOR LARGE AREA INDUSTRIAL SINGLE CRYSTALLINE SILICON WAFERS 396

Tarun S. Yadav ; K. Sandeep ; Ashok K. Sharma ; B. Spandana ; K.L. Narasimhan ; B.M. Arora ; Anil Kottantharayil ; Prabir K. Basu

EVALUATION OF ALD PASSIVATION LAYERS FOR INDUSTRIAL PERC PROCESS 399

Chang Youn Yoo ; Keunkee Hong ; Jisun Kim ; Eunjoo Lee ; Dong Seop Kim

QUANTITATIVE ANALYSIS OF ELECTROLUMINESCENCE AND INFRARED THERMAL IMAGES FOR AGED MONOCRYSTALLINE SILICON PHOTOVOLTAIC MODULES 402

Irene Berardone ; Juan Lopez Garcia ; Marco Paggi

GAP PASSIVATION STRUCTURE FOR SCALABLE N-TYPE INTERDIGITATED ALL BACK CONTACT SILICON HETERO-JUNCTION SOLAR CELL 408

Lei Zhang ; Ujjwal Das ; Steven Hegedus

PROPOSAL OF THE BANDGAP DESIGN USING THE SUN HEIGHT OF THE CULMINATION ON THE WINTER SOLSTICE 412

Kenji Araki ; Kan-Hua Lee ; Masafumi Yamaguchi

PHOTOEXCITED CARRIERS, PHONONS, AND THEIR SCATTERING MEASURED IN SEMICONDUCTOR JUNCTIONS BY TRANSIENT EXTREME ULTRAVIOLET SPECTROSCOPY 417

Scott K. Cushing ; Brett M. Marsh ; Mihai E. Vaida ; Lucas M. Carneiro ; Ilana J. Porter ; Angela Lee ; Stephen R. Leone

ON THE USE OF VOLTAGE MEASUREMENTS FOR DETERMINING CARRIER LIFETIME AT HIGH ILLUMINATION INTENSITY 420

Robert Dumbrell ; Mattias K. Juhl ; Thorsten Trupke ; Ziv Hameiri

HIGH RESOLUTION 3D CHEMICAL CHARACTERISATION OF A CADMIUM TELLURIDE SOLAR CELL BY DYNAMIC SIMS 424

Thomas Fiducia ; Kexue Li ; Chris Grovenor ; Kurt Barth ; Walajabad Sampath ; Michael Walls

HARSH OUTDOOR EVALUATION SETUP AND FIRST POWER PRODUCTION RESULTS FOR SI MINI-MODULES COVERED BY EU3+-BASED DOWN CONVERTERS 429

Benjamín González-Díaz ; Carlos Montes ; Joaquín Sanchiz ; Luis Ocaña ; Carlos Quinto ; Cecilio Hernández-Rodríguez ; Mari Paz Friend ; Manuel Cendagorta-Galarza ; David Cañadillas ; Ricardo Guerrero-Lemus

STUDY OF MICRO-STRUCTURAL PROPERTIES OF ZNO AND TIO2THIN FILM GROWN BY SPRAY PYROLYSIS 433

G. Gordillo ; J.M. Correa ; A.A. Ramirez ; E. A. Ramírez

NONLINEAR RESPONSE OF SILICON SOLAR CELLS 437

Behrang H. Hamadani ; Andrew Shore ; Howard W. Yoon ; Mark Campanelli

EXTENDED LINEAR INTERPOLATION/EXTRAPOLATION PROCEDURE FOR ACCURATE AND VERSATILE TRANSLATION OF THE I-V CURVES OF PV CELLS AND MODULES 441

Y. Hishikawa ; H. Ohshima ; M. Higa ; K. Yamagoe ; T. Takenouchi ; T. Doi

SEVERITY TEST WITH UNEVEN LOAD DUE TO WIND ACTION ON PHOTOVOLTAIC MODULE 445

Shu-Tsung Hsu

STANDARDIZED DURABILITY TEST FOR ORGANIC PHOTOVOLTAIC AND DYE SENSITIZED SOLAR CELL 448

Shu-Tsung Hsu ; Yean-San Long ; Teng-Chun Wu

SPATIAL THICKNESS UNIFORMITY AND STRUCTURAL EVALUATION OF RF SPUTTERED ZNO THIN FILMS FOR SOLAR CELL 451

Babar Hussain ; Taj M. Khan

LOCAL MEASUREMENTS OF SURFACE CAPACITANCE BY ELECTROSTATIC FORCE MICROSCOPY ON CU(IN, GA)SE2MATERIALS 455

Tomoaki Ishii ; Takashi Minemoto ; Takuji Takahashi

A COMPARISON OF SI-BASED CAMERAS FOR IMAGING LUMINESCENCE FROM PHOTOVOLTAIC MATERIALS AND DEVICES 459

Steve Johnston

BLISTERING OF AL2O3/A-SINX:H STACKS: ANALYSIS OF THE SUBMERGED PART OF THE ICEBERG BY COLORED PICOSECOND ACOUSTIC MICROSCOPY 464

Fabien Lebreton ; Arnaud Devos ; Etienne Drahi ; Patricia De Coux ; François Silva ; Sergej Filonovich ; Pere Roca I Cabarrocas

SELF-REFERENCE PROCEDURE TO REDUCE UNCERTAINTY IN MODULE CALIBRATION 467

D.H. Levi ; C.R. Osterwald ; S. Rummel ; L. Ottoson ; A. Anderberg

UNCERTAINTY EVALUATION OF PRIMARY REFERENCE PHOTOVOLTAIC CELL CALIBRATION UNDER OUTDOOR CONDITION IN TIBET 472

Haitao Liu ; Shiyu Sang ; Guomin Zhou ; Yonghui Zhai

REQUIREMENT OF ARTIFICIAL LIGHTING SIMULATOR FOR EVALUATION EMERGING PV PERFORMANCE RATING UNDER INDOOR ENVIRONMENT 476

Yean-San Long ; Shu-Tsung Hsu ; Teng-Chun Wu

NON-CONTACT VOLTAGE MEASUREMENT OF SOLAR CELL WITH ELECTROSTATIC VOLTMETER 480

Sakutaro Miyajima ; Kensuke Nishioka ; Yoshihiro Hishikawa

NREL'S CELL AND MODULE PERFORMANCE GROUP'S ASYMPTOTIC PMAX PROTOCOL FOR PEROVSKITE DEVICES 483

Tom Moriarty ; Dean Levi

OUTDOOR OPERATING TEMPERATURE MODELING OF PHOTOVOLTAIC MODULES INCLUDING TRANSIENT EFFECT 487

Soo-Young Oh ; Min-Soo Kim ; Won-Shup So ; Woo Kyoung Kim ; Jae Hak Jung ; Chinho Park ; Benazzouz Aboubakr ; Ikken Badr ; Naimi Zakaria ; Benlarabi Ahmed

PRIMARY REFERENCE CELL CALIBRATIONS WITH REDUCED MEASUREMENT UNCERTAINTY 490

C.R. Osterwald ; L. Ottoson ; R. Williams ; C. Mack ; T. Moriarty ; K.A. Emery ; D.H. Levi

IMPLEMENTATION OF NOVEL PIN CONNECTION AND TEST ROUTINE FOR IMPROVED ACCURACY IN I-V MEASUREMENTS 496

Samuel Raj ; Johnson Kai Chi Wong ; Mohan Krishan Bhan ; Evan Palmer ; Jian Wei Ho ; Sumukh Ramprasad ; Wang Junci ; Thomas Mueller ; Armin G. Aberle

A NEW METHOD TO QUANTIFY CONTACT RESISTANCE USING LOCALIZED-ILLUMINATION PHOTOLUMINESCENCE TECHNIQUE IN A SOLAR CELL 499

Amit Singh Rajput ; Samuel Raj ; Johnson Wong ; Armin G. Aberle

IMPROVEMENT OF THE PROPERTIES OF CZTS THIN FILMS PREPARED BY SPRAY PYROLYSIS USING DMSO IN ACETONE AS SOLVENT 503

E. A. Ramírez ; A. Ramírez ; G. Gordillo

ASSESSMENT OF CARRIER LIFETIMES AND SURFACE RECOMBINATION VELOCITY THROUGH SPECTRAL MEASUREMENTS 508

John Roller ; Behrang H. Hamadani

IMPACT OF SPACE RADIATION ENVIRONMENT ON CONCENTRATOR PHOTOVOLTAIC SYSTEMS 512

Pilar Espinet-Gonzalez ; Tatiana Vinogradova ; Michael D. Kelzenberg ; Alexander Messer ; Emily C. Warmann ; Chris Peterson ; Nina Vaidya ; Ali Naqavi ; Jing-Shun Huang ; Samuel P. Loke ; Don Walker ; Colin J. Mann ; Sergio Pellegrino ; Harry A. Atwater

EXTRACTING THE FIXED CHARGE DENSITY IN HFOX FILMS GROWN ON HIGHLY-DOPED P-SI SAMPLES 517

Alexander To ; Jie Cur ; Bram Hoex

NEAR-UNITY ULTRA-WIDEBAND THERMAL INFRARED EMISSION FOR SPACE SOLAR POWER RADIATIVE COOLING 521

Ali Naqavi ; Samuel P. Loke ; Michael D. Kelzenberg ; Emily C. Warmann ; Pilar Espinet-González ; Nina Vaidya ; Jing-Shun Huang ; Tatiana A. Roy ; Alexander J. Messer ; Tatiana G. Vinogradova ; Ali Hajimiri ; Sergio Pellegrino ; Harry A. Atwater

LINE-FOCUS AND POINT-FOCUS SPACE PHOTOVOLTAIC CONCENTRATORS USING ROBUST FRESNEL LENSES, 4-JUNCTION CELLS, & GRAPHENE RADIATORS 525

Mark O'Neill ; A.J. Mcdanal ; Michael Piszczor ; Matt Myers ; Paul Sharps ; Claiborne Mcpheeters ; Jeff Steinfedt

SIMULATION OF LIGHT TRAPPING STRUCTURES FOR ENHANCING RADIATION HARDNESS IN SPACE SOLAR CELLS 531

Nizami Z. Vagidov ; Kyle H. Montgomery ; Geoffrey K. Bradshaw ; David M. Wilt

AN ALTERNATIVE METHOD FOR SOLAR CELL INTEGRATION 537

Jessica Buckner ; Tracy Davis ; Eric Muskovin ; Bernard Carpenter

NIEL DOSE ANALYSIS ON TRIPLE JUNCTION CELLS 30% EFFICIENT AND RELATED SINGLE JUNCTIONS 541

Roberta Campesato ; Erminio Greco ; Mariacristina Casale ; Massimo Gervasi ; P.G. Rancoita ; Davide Rozza ; Mauro Tacconi ; Enos Gombia ; Aldo Kingma ; Carsten Baur

THIN AND FLEXIBLE TRIPLE JUNCTION CELLS 30% EFFICIENT: QUALIFICATION RESULTS AND FUTURE SPACE APPLICATIONS 545

Roberta Campesato ; Mariacristina Casale ; Giuseppe Gabetta ; Emilio Fernandez Lisbona ; Laurent D'Abrigeon

PRINTED ASSEMBLIES OF MICROSCALE TRIPLE-JUNCTION (3J) INVERTED METAMORPHIC (IMM) GAINP/GAAS/INGAAS SOLAR CELLS 549

Boju Gai ; John Geisz ; Daniel Friedman ; Jongseung Yoon

COMPARATIVE STUDY ON NONRADIATIVE RECOMBINATION CENTERS IN PROTON IRRADIATED INAS/GAAS QUANTUM DOT STRUCTURE BY TWO WAVELENGTH EXCITED PHOTOLUMINESCENCE 552

M. D. Haque ; N. Kamata ; S-I. Sato ; S. M. Hubbard

DESIGN AND PROTOTYPING EFFORTS FOR THE SPACE SOLAR POWER INITIATIVE 558

Michael D. Kelzenberg ; Pilar Espinct-Gonzalez ; Nina Vaidya ; Tatiana A. Roy ; Emily C. Warmann ; Ali Naqavi ; Samuel P. Loke ; Jing-Shun Huang ; Tatiana G. Vinogradova ; Alexander J. Messer ; Christophe Leclerc ; Eleftherios E. Gdoutos ; Fabien Royer ; Ali Hajimiri ; Sergio Pellegrino ; Harry A. Atwater

DEFECT CHARACTERIZATION OF III-V QUANTUM STRUCTURE SOLAR CELLS USING PHOTO-INDUCED CURRENT TRANSIENT SPECTROSCOPY 562

Shin-Ichiro Sato ; Takeyoshi Sugaya ; Tetsuya Nakamura ; Takeshi Ohshima

EFFECT OF LUMINESCENCE COUPLING BETWEEN INGAP AND GAAS SUBCELLS TO EXTERNAL QUANTUM EFFICIENCY IN TRIPLE-JUNCTION SOLAR CELLS 567

Mitsunobu Suga ; Mitsuru Imaizumi ; Tetsuya Nakamur ; Takeshi Ohshima

LIGHTWEIGHT CARBON FIBER MIRRORS FOR SOLAR CONCENTRATOR APPLICATIONS 572

Nina Vaidya ; Michael D. Kelzenberg ; Pilar Espinet-Gonzalez ; Tatiana G. Vinogradova ; Jing-Shun Huang ; Christophe Leclerc ; Ali Naqavi ; Emily C. Warmann ; Sergio Pellegrino ; Harry A. Atwater

GAAS SOLAR CELLS ON V-GROOVED SILICON VIA SELECTIVE AREA GROWTH 578

Michelle Vaisman ; Nikhil Jain ; Qiang Li ; Kei May Lau ; Adele C. Tamboli ; Emily L. Warren

HIGH TEMPERATURE ANNEALING OF IN1-XGAXN MQW SOLAR CELLS 582

Joshua J. Williams ; Heather Mcfavilen ; Steven Young ; Christiana B. Honsberg ; Stephen M. Goodnick

SOLAR PROBE PLUS ARRAY RELIABILITY 585

Anton Yanchilin ; Edward Gaddy

PHOTOVOLTAIC TEMPERATURE ESTIMATION MODEL FOR RAPID IRRADIANCE CHANGE CONDITIONS IN TROPICAL REGIONS USING HEURISTIC ALGORITHMS 589

R. Srivatsan ; Lian L. Jiang ; Douglas L. Maskell

ACCURACY OF CDTE PV ENERGY PREDICTIONS USING SPECTRAL CORRECTIONS 595

Mitchell Lee ; Kendra Passow ; Paul Wolffersdorff

PLANTPREDICT: SOLAR PERFORMANCE MODELING MADE SIMPLE 600

Kendra Passow ; Lauren Ngan ; Geoffrey Rich ; Mitch Lee ; Stephen Kaplan

INTEGRABILITY COMPARISON BETWEEN BIPV AND BAPV IN TROPICAL CONDITIONS: A BANGALORE CASE-STUDY 604

Gayathri Aaditya ; Roshan R Rao ; Monto Mani

A NEW PHOTOVOLTAIC SYSTEM TOPOLOGY THROUGH LOAD MANAGEMENT 608

Joseph A. Azzolini ; Meng Tao

FIRST STEP FOR POWER GENERATION AMOUNT ESTIMATION OF SOLAR MATCHING SYSTEM 613

Kazuya Hosokawa ; Toshiaki Yachi ; Yoichi Hirata ; Yasuyuki Watanabe

IRRADIANCE AND TEMPERATURE DISTRIBUTIONS AT HIGH LATITUDES: DESIGN IMPLICATIONS FOR PHOTOVOLTAIC SYSTEMS 619

Anne Gerdimenes ; Josefine Sclj

STEP-BY-STEP EVALUATION OF PHOTOVOLTAIC MODULE PERFORMANCE RELATED TO OUTDOOR PARAMETERS: EVALUATION OF THE UNCERTAINTY 626

Anne Migan Dubois ; Jordi Badosa ; Fausto Calderón-Obaldía ; Olivier Atlan ; Vincent Bourdin ; Marko Pavlov ; Dae Young Kim ; Yvan Bonnassieux

PERFORMANCE COMPARISONS OF A PV SYSTEM BY MONITORING SOLAR IRRADIANCE WITH DIFFERENT PYRANOMETERS 632

Yasuhiro Matsumoto ; J. Antonio Urbano ; Ramón Peña ; María De La Luz Olvera ; Nun Pitalúa ; Miguel A. Luna ; René Asomoza

FINANCIAL ANALYSIS OF A GRID-CONNECTED PHOTOVOLTAIC SYSTEM IN SOUTH FLORIDA 638

Hadis Moradi ; Amir Abtahi ; Ali Zilouchian

STUDY OF PHOTOVOLTAIC SYSTEMS MONITORING METHODS..643
E. Ortega ; G. Aranguren ; M.J. Sáenz ; R. Gutiérrez ; J.C. Jimeno

GLOBAL DESIGN ASPECTS OF PERSISTENT AND AUTONOMOUS PV POWERED SYSTEMS.................648
I. M. Peters ; S. Watson ; N. Sahraei ; T. Buonassisi

HOW TO CHOOSE THE BEST EMPIRICAL MODEL FOR OPTIMUM ENERGY YIELD
PREDICTIONS..652
Steve Ransome ; Juergen Sutterlueti

MODELING AND ANALYSIS OF PHOTOVOLTAIC ELECTROCHEMICAL SYSTEM USING
MODULE-LEVEL POWER ELECTRONICS...658
Gowri M. Sriramagiri ; Nuha Ahmed ; Kevin D. Dobson ; Steven S. Hegedus

BETAVOLTAIC GENERATION FUNCTION IN SILICON...663
A.V. Sachenko ; I.O. Sokolovskyi ; M. Evstigneev

MULTI-OBJECTIVE OPTIMIZATION FOR COLOR-TUNABILITY AND TRANSPARENCY IN
COLLOIDAL QUANTUM DOT SOLAR CELLS..667
Ebuka S. Arinze ; Botong Qiu ; Nathan Palmquist ; Yan Cheng ; Yida Lin ; Gabrielle Nyirjesy ; Gary Qian ;
Susanna M. Thon

CUBIC PHASE INXGA1-XN/GAN QUANTUM WELLS FOR THEIR APPLICATION TO
TANDEM SOLAR CELLS..670
C. A. Hernández-Gutiérrez ; Y. L. Casallas-Moreno ; Dagoberto Cardona ; Yu. Kudriavtsev ; A. Morales-Acevedo
; G. Santana-Rodríguez ; M. López-López

MODELING OF P-I-N GAASPN/GAP MQWS SOLAR CELL: TOWARDS LATTICE MATCHED
III-V/SI TANDEM..673
Khim Kharel ; Alexandre Freundlich

INP QUANTUM DOT INTERMEDIATE BAND SOLAR CELL GROWN VIA MOCVD.................677
Hyun Kum ; Yushuai Dai ; Michael Slocum ; Zachary Bittner ; Seth Hubbard

MODIFIED LIMITING EFFICIENCY FOR MULTIPLE EXCITON GENERATION SOLAR
CELLS..681
Jongwon Lee ; Christiana B. Honsberg

A SIMPLE MONTE CARLO MODEL OF A HOT CARRIER CELL.......................................685
Tor Oskar Saetre

OPTIMIZATION OF SEMICONDUCTOR QUANTUM DOTS FOR LUMINESCENT SOLAR
CONCENTRATORS: MINIMIZING REABSORPTION LOSSES..690
Anatoli I. Shkrebtii ; Anatoliy V. Sachenko ; Igor O. Sokolovskyi ; Vitaliy P. Kostylyov ; Mykola R. Kulish ; Denis
V. Khomcnko ; Mykhaylo A. Evstigneev

DEVELOPMENT OF ABSORBER AND ENERGY SELECTIVE CONTACTS FOR HOT
CARRIER SOLAR CELLS..696
Santosh Shrestha ; Simon Chung ; Yuanxun Liao ; Wenkai Cao ; Neeti Gupta ; Yi Zhang ; Xiaoming Wen ; Gavin
Conibeer

GAASBI DEVICES FOR THERMAL ENERGY CONVERSION..701
Margaret Stevens ; Abigail Licht ; Nicole Pfiester ; Emily Carlson ; Kevin Grossklaus ; Thomas E. Vandervelde

ANALYTIC JV-CHARACTERISTICS OF IDEAL IMPURITY PV-CELLS..............................706
Rune Strandberg

PHOTOLUMINESCENCE PROPERTIES OF IN-PLANE ULTRAHIGH-DENSITY INAS
QUANTUM DOTS ON GAASSB/GAAS(001) FOR SOLAR CELL APPLICATIONS.................712
Ryo Sugiyama ; Naoki Akimoto ; Tomah Sogabe ; Koichi Yamaguchi

CARRIER SELECTIVE BACK CONTACT (CSBC) SOLAR CELL USING TRANSITION METAL
OXIDES..716
Astha Tyagi ; Kunal Ghosh ; Anil Kottantharayil ; Saurabh Lodha

ANALYSIS OF OPEN-CIRCUIT VOLTAGE AND CONVERSION EFFICIENCY IN QUANTUM-
DOT SOLAR CELLS VIA DETAILED-BALANCE-LIMIT THEORY.......................................721
Lin Zhu ; Hidefumi Akiyama ; Yoshihiko Kanemitsu

ZINC SELENIDE SURFACE PASSIVATION LAYER FOR SINGLE-CRYSTALLINE CZTSE
SOLAR CELLS..726
Michael A. Lloyd ; Douglas Bishop ; Brian E. Mccandless ; Robert Birkmirc

USE OF SINGLE WALL CARBON NANOTUBE FILMS DOPED WITH TRIETHYLOXONIUM
HEXACHLORANTIMONATE AS A TRANSPARENT BACK CONTACT FOR CDTE SOLAR
CELLS..730
Fadhil K. Alfadhili ; Jacob M. Gibbs ; Geethika K. Liyanage ; Patrick W. Krantz ; Suneth C. Watthage ; Zhaoning
Song ; Adam B. Phillips ; Michael J. Heben

GRAIN AND GRAIN BOUNDARY GEOMETRICAL SHAPE CONSIDERATIONS ON SODIUM
AND POTASSIUM DIFFUSION THROUGH MOLYBDENUM FILMS.....................................735
Orlando Ayala ; Chinedum Akwari ; Tasnuva Ashrafee ; Shankar Karki ; Grace Rajan ; Sylvain Marsillac

SOLUTION-PROCESSED NICKEL-ALLOYED IRON PYRITE THIN FILM AS HOLE TRANSPORT LAYER IN CADMIUM TELLURIDE SOLAR CELLS 738

Ebin Bastola ; Khagendra P. Bhandari ; Randy J. Ellingson

USE OF CDS:O AND CDSE AS WINDOW LAYERS FOR CDTE PHOTOVOLTAICS 742

Tom Baines ; Guillaume. Zoppi ; Ken Durose ; Jonathan D. Major

APPLICATIONS OF HYBRID ORGANIC-INORGANIC METAL HALIDE PEROVSKITE THIN FILM AS A HOLE TRANSPORT LAYER IN CDTE THIN FILM SOLAR CELLS 748

Khagendra P. Bhandari ; Suneth C. Watthage ; Zhaoning Song ; Adam Phillips ; Michael J. Heben ; Randy J. Ellingson

MAGNESIUM-DOPED ZINC OXIDE AS A HIGH RESISTANCE TRANSPARENT LAYER FOR THIN FILM CDS/CDTE SOLAR CELLS 752

Francesco Bittau ; Elisa Artegiani ; Ali Abbas ; Daniele Menossi ; Alessandro Romeo ; Jake W. Bowers ; John M. Walls

INVESTIGATION OF ZNL-XMGXO:A1 FILM BY RATIO FREQUENCY MAGNETRON CO-SPUTTERING AS TRANSPARENT CONDUCTIVE OXIDE LAYER 757

Jakapan Chantana ; Yuya Ishino ; Takashi Minemoto

A NEW TCO/WINDOW-BUFFER FRONT STACK FOR CDTE SOLAR CELLS AND ITS IMPLEMENTATION 761

Alan E. Delahoy ; Xuehai Tan ; Akash Saraf ; Payal Patra ; Surya Manda ; Yunfei Chen ; Krishnakumar Velappan ; Bastian Siepchen ; Shou Peng ; Ken K. Chin

SYNTHESIS OF HIGH-QUALITY AZO POLYCRYSTALLINE FILMS VIA TARGET BIAS RADIO FREQUENCY MAGNETRON SPUTTERING 767

Zhongming Du ; Yufeng Zhang ; Xiangxin Liu

CLOSE-SPACE SUBLIMATED CDTE SOLAR CELLS WITH CO-SPUTTERED CDSXSE1-XALLOY WINDOW LAYERS 771

Corey R. Grice ; Maxwell Junda ; Alexander Archer ; Jian Li ; Yanfa Yan

EFFECTS OF GRAPHENE OXIDE BARRIER ON CU2ZNSNSXSE4-XTHIN FILM SOLAR CELLS 777

Woo-Lim Jeong ; Jung-Hong Min ; In-Young Kim ; Hae-Sun Kim ; Jin-Hyeok Kim ; Dong-Seon Lee

13% CDS/CDTE SOLAR CELL USING A NANOCOMPOSITE (CUS)X(ZNS)1-X THIN FILM HOLE TRANSPORT LAYER 781

Kamala Khanal Subedi ; Khagendra P. Bhandari ; Ebin Bastola ; Randy J. Ellingson

MOLYBDENUM OXIDE AND MOLYBDENUM NITRIDE BACK CONTACTS FOR THIN-FILM CDTE SOLAR CELLS 785

Anna Kindvall ; Jason Kephart ; Walajabad Sampath

INVESTIGATION AND OPTIMIZATION OF CD-FREE BUFFER LAYERS IN2S3 AND ZN(O, S) FOR CU2ZNSN(S, SE)4-BASED SOLAR CELLS 791

Willi Kogler ; Thomas Schnabel ; Andreas Bauer ; Stefanie Spiering ; Erik Ahlswede ; Michael Powalla

REAR CONTACT PASSIVATION FOR HIGH BANDGAP CU(IN, GA)SE2 SOLAR CELLS WITH VARYING ABSORBER THICKNESS AND FLAT GA PROFILE 796

Dorothea Ledinek ; Pedro Salome ; Carl Hägglund ; Marika Edoff

LASER ANNEALED BACK CONTACTS FOR CDTE SOLAR CELLS 802

Vasilios Palekis ; Shamara Collins ; Imran Khan ; Vamsi Evani ; Sudhajit Misra ; Michael A. Scarpulla ; Mark Lonergan ; Don Morel ; Chris Ferekides

ENHANCED ANTI-REFLECTIVE COATING FOR THIN FILM SOLAR CELLS 807

Grace Rajan ; Shankar Karki ; Robert W. Collins ; Sylvain Marsillac

INFLUENCE OF AGS LAYER INSERTION AT ABSORBER/ITO INTERFACE ON STRUCTURAL AND PHOTOVOLTAIC PROPERTIES OF ULTRATHIN CU(IN,GA)SE2 SOLAR CELLS 810

Muhammad Saifullah ; Jihye Gwak ; Kihwan Kim ; Joo Hyung Park ; Junsik Cho ; Jae Ho Yun

NOVEL, FACILE BACK SURFACE TREATMENT FOR CDTE SOLAR CELLS 815

Suneth C. Watthage ; Geethika K. Liyanage ; Zhaoning Song ; Fadhil K. Alfadhili ; Rabee B. Alkhayat ; Khagendra P. Bhandari ; Randy J. Ellingson ; Adam B. Phillips ; Michael J. Heben

OPTIMIZING CDS BUFFER LAYER FOR CIGS BASED THIN FILM SOLAR CELL 820

Weijie Zhang ; Korhan Demirkan ; Geordie Zapalac ; David Spaulding ; Jochen Titus ; Neil Mackie

INVESTIGATION OF INP DEFECT CHARACTERISTICS GROWN USING NOVEL TF-VLS TECHNIQUE 823

Abhinav Chikhalkar ; Alec Fischer ; Mark Hettick ; Ali Javey ; Richard R. King

INVESTIGATION OF FAST GROWTH GAAS-BASED SOLAR CELL ON REUSABLE SUBSTRATE BY METALORGANIC CHEMICAL VAPOR DEPOSITION 827

Chaomin Zhang ; Abhinav Chikhalkar ; Ehsan Vadiee ; Richard King ; Christiana Honsberg ; Eric Armour ; Yeongho Kim

DEVELOPMENT OF ALUMINUM EPILAYERS AS BUFFERS FOR GAINAS 831
Phil Ahrenkiel ; Nathan Smaglik ; Nikhil Pokharel ; Alessandro Giussani ; Michael A. Slocum ; Seth M. Hubbard

LASER CRYSTALLIZATION OF AMORPHOUS GERMANIUM ON TITANIUM NITRIDE-
COATED STEEL FOR LOW-COST GAAS SOLAR-CELLS 837
Saloni Chaurasia ; Srinivasan Raghavan ; Sushobhan Avasthi

HIGH QUALITY EPITAXIAL GERMANIUM ON SI (100) FOR LOW -COST III–V SOLAR-
CELLS 841
Saloni Chaurasia ; Srinivasan Raghavan ; Sushobhan Avasthi

CRYSTALLINITY CONTROL IN LOW-TEMPERATURE GROWTH OF POLY-CRYSTALLINE
GE BY ION BEAM DEPOSITION 845
S. I. Maximenko ; N. A. Mahadik ; P. P. Jenkins ; R. J. Walters ; A. Giussani ; E. L. Mcclure ; S. M. Hubbard ; C. Bailey

HIGH EFFICIENCY GAINP/GAAS DOUBLE JUNCTION SOLAR CELL ON SI SUBSTRATE
ASSISTED BY THE ELECTRON BEAM TREATMENT 849
Hyo Jin Kim ; Yong Whan Kim

ANALYSIS OF DEPOSITED RESIDUES AND ITS CLEANING PROCESS ON GAAS
SUBSTRATE AFTER EPITAXIAL LIFT-OFF 854
Tatsuya Nakata ; Kentaroh Watanabe ; Hassanet Sodabanlu ; Daiki Kimura ; Naoya Miyashita ; Yoshitaka Okada ; Yoshiaki Nakano ; Masakazu Sugiyama

ULTRATHIN SILICON-AN-INSULATOR (SOI) WAFER FOR COMPLIANT SUBSTRATE 858
Shinyoung Noh ; Anita Ho-Baillie ; Stephen Bremner ; Martin A. Green ; Xiaojing Hao

CHARACTERIZATION OF GAAS SOLAR CELLS GROWN BY HYDRIDE VAPOR PHASE
EPITAXY IN HORIZONTAL REACTOR 861
Ryuji Oshima ; Kikuo Makita ; Takeyoshi Sugaya ; Akinori Ubukata

FLEXIBLE GAAS SINGLE-JUNCTION SOLAR CELLS BASED ON SINGLE-CRYSTAL-LIKE
THIN-FILM MATERIALS DIRECTLY GROWN ON METAL TAPES 866
Sara Pouladi ; Monika Rathi ; Mojtaba Asadirad ; Pavel Dutta ; Seung Kyu Oh ; Devendra Khatiwada ; Shahab Shervin ; Yao Yao ; Venkat Selvamanickam ; Jae-Hyun Ryou

REDUCED DEFECT DENSITY IN SINGLE-CRYSTALLINE-LIKE GAAS THIN FILM ON
FLEXIBLE METAL SUBSTRATES BY USING SUPERLATTICE STRUCTURES 869
M. Rathi ; P. Dutta ; D. Khatiwada ; Y. Yao ; Y. Gao ; Y. Li ; S. Sun ; S. Pouladi ; S. Reed ; A. Khadimallah ; J. Ryou ; V. Selvamanickam ; N. Zheng ; P. Ahrenkiel

ECONOMIC ANALYSIS OF TRANSFER PRINTED III–V VIRTUAL SUBSTRATES 873
Kenneth J. Schmieder ; Matthew P. Lumb ; Michael K. Yakes ; Shawn Mack ; Mitchell F. Bennett ; Sergey I. Maximenko ; Laura B. Ruppalt ; Michael A. Meeker ; Chase T. Ellis ; Matthew Meitl ; Joseph G. Tischler ; Robert J. Walters

THIN FILMS OF ZINC-DOPED GAAS BY RF MAGNETRON SPUTTERING FOR USE IN
PHOTOVOLTAIC CELLS 876
Kirby Simon ; Kyle Cepeda ; Nishit Shetty ; Elijah Thimsen

SELF ALIGNED ALUMINUM SELECTIVE EMITTER FOR N-TYPE SI CELLS 881
San Theigi ; Robert C. Reedy ; Vincenzo Lasalvia ; Paul Stradins ; Benjamin G. Lee

HOW TO REALIZE SOLAR CELLS WITH LASER STRUCTURED PLATED NI-CU-CONTACTS
WITH EXCELLENT ADHESION AND HIGH FILL-FACTORS WITHOUT PARASITIC
PLATING 884
A. Büchler ; S. Kluska ; J. Bartsch ; B. Grübel ; A.A. Brand ; S. Gutscher ; M. Glatthaar

EXPLOITING THE POTENTIALS OF THE FRONT SURFACE FIELD (FSF) INDUSTRIAL
SILICON SOLAR CELL 888
Ahrar Ahmed Chowdhury ; Yu -Chen Hsu ; Veysel Unsur ; Abasifreke Ebong

PHOTOVOLTAIC PERFORMANCE OF SILICON SOLAR CELLS ENHANCED BY
PLASMONIC SILVER NANOPARTICLES OF VARIOUS DIMENSIONS DEPOSITING
THROUGH ANODIC ALUMINUM OXIDE TEMPLATE 893
Ta-Wei Chuang ; Wen-Jeng Ho ; Sheng-Kai Feng ; Jheng-Jie Liu ; Guan-Yi Li ; Hao-Yu Yang ; Yun-Chie Yang ; Cho-Chun Chiang ; Yao-Hui Chen

MITIGATION OF POTENTIAL-INDUCED DEGRADATION 896
Orry Faur ; Maria Faur

ELECTRODEPOSITION OF SI-LAYER THROUGH REDUCTION OF DIATOMACEOUS
EARTH FOR THE APPLICATION OF SOLAR-CELLS 900
Muhammad Monirul Islam ; Imane Abdellaoui ; Takeaki Sakurai ; Saad Hamzaoui ; Katsuhiro Akimoto

EFFECT OF SI CONTENT IN A1 PASTE ON LOCAL A1 REAR CONTACTS IN PERC CELL 904
Supawan Joonwichien ; Katsuhiko Shirasawa ; Satoshi Utsunomiya ; Hidetaka Takato

NEW SILVER PASTE METALLIZATION APPROACH ON P+ DIFFUSION ZONES OF SILICON
SOLAR CELLS 907
Yunjun Li ; Mohshi Yang ; Igor Pavlovsky ; Guoping Zeng

INFLUENCES OF ANNEALING AND DEFECT LIMITATION ON P-TYPE SILICON SOLAR CELL 911

Yu-Hsuan Lin ; Sung-Yu Chen ; Kuen-Yi Wu ; Chien-Hsun Chen ; Chen-Hsun Du ; Chun-Ming Yeh

REDUCED TEMPERATURE SILVER PASTE WITH LOW CONTACT RESISTANCE FOR ADVANCED SOLAR CELL APPLICATIONS 914

Ryan Mayberry ; Daniel Holzmann ; Gerd Schulz ; Lindsey Karpowich ; Mark Naylor ; Matthias Hoerteis

BSF ISLANDS FOR REDUCED RECOMBINATION IN IBC CELLS 917

Agnes A. Mewe ; Nicolas Guillevin ; Ilkay Cesar ; Antonius R. Burgers

THERMAL STABILITY OF HYDROGENATED BORON EMITTERS 921

Khaja H. Mohammed ; Larry C. Cousar ; Philip A. Mcmeans ; Garrett Z. Evans ; Douglas A. Hutchings ; Hameed A. Naseem ; Sergiu C. Pop

LIGHT INDUCED PLATING OF SILICON SOLAR CELLS USING BORIC ACID-FREE NICKEL CHEMISTRY 925

Krystal Munoz ; Lynne Michaelson ; Joseph Karas ; Tom Tyson ; James Rand ; Stuart Bowden

BAKING TEMPERATURE DEPENDENCE OF CU PASTE ON Al-BSF CELL PROPERTIES 931

Tomohiro Saito ; Tetsuya Fukuda ; Hoang Tri Hai ; Yuji Kurimoto ; Daisuke Ando ; Yuji Sutou ; Katsuhiko Shirasawa ; Junichi Koike

THE SILVER CONTACT AND FORMATION MECHANISM OF THE BORON EMITTER AND THE CURRENT FLOW MECHANISM OF THE SOLAR CELL ELECTRODE 935

Seunghyun Shin ; Soohyun Bae ; Sungeun Park ; Yoonmook Kang ; Hae-Seok Lee ; Donghwan Kim

LASER ANNEALING TO ENHANCE PERFORMANCE OF ALL-LASER-BASED SILICON BACK CONTACT SOLAR CELLS 937

Zeming Sun ; Mool C. Gupta

LARGE AREA N-TYPE SELECTIVE EMITTER CELLS USING LASER DOPING THROUGH BORON DOPED SCREEN PRINTED PASTE 940

Ajay D Upadhyaya ; Vijaykumar D Upadhyaya ; Brian Rounsaville ; Keeya Madani ; Ajeet Rohatgi ; Toru Hanada

METALLIZED BORON-DOPED BLACK SILICON EMITTERS FOR FRONT CONTACT SOLAR CELLS 944

Guillaume Von Gastrow ; Hele Savin ; Eric Calle ; Pablo Ortega ; Ramón Alcubilla ; Andreana Daniil ; Elias Z. Stutz ; Anna Fontcuberta I Morral ; Sebastian Husein ; Tara Nietzold ; Mariana Bertoni

CONTACT RESISTANCE MEASUREMENT FOR THERMALLY DIFFUSED POINT CONTACT BY LOCALIZED DIELECTRIC BREAKDOWN SOLAR CELLS 948

Qilin Ye ; Ned J. Western ; Anqi Liao ; Stephen P. Bremner

LOW TEMPERATURE REAR SURFACE METALLIZATION OF MULTI-CRYSTALLINE SILICON SOLAR CELLS FOR IMPROVED BULK LIFETIME 953

N. J. Western ; S. P. Bremner

INVESTIGATION OF HIGH PERFORMANCE PEROVSKITE-BASED SOLAR CELLS GROWN BY HYBRID CHEMICAL VAPOR DEPOSITION TECHNIQUE 958

Huseyin Cem Gokkaya ; Shen Qian ; Zhiwei Ren ; Annie Ng ; Charles Surya

ENHANCED PEROVSKITE SOLAR CELL PERFORMANCE USING FULL SPACE DEVICE OPTIMIZATION 963

Ahmer A.B. Baloch ; Shahzada P. Aly ; Mohammad I. Hossain ; Raka Jovanovic ; Nouar Tabet ; Fahhad H. Alharbi

MEASURING OPTICAL ABSORPTION IN ORGANIC PHOTOVOLTAICS USING MONOCHROMATED ELECTRON ENERGY-LOSS SPECTROSCOPY 966

Jessica A. Alexander ; Frank J. Scheltens ; David W. Mccomb ; Lawrence F. Drummy ; Michael F. Durstock ; James B. Gilchrist ; Sandrine Hentz

ADVANCED DEPOSITION OF PHOTO-CATALYTIC TIO2 FILM BY ATMOSPHERIC SPPS FOR DYE SENSITIZED SOLAR CELLS 970

Ifeanacho Anyadiegwu ; Dickson Kindole ; Geoffrey Kibiegon Ronoh ; Yoshimasa Noda ; Yasutaka Ando

CH3NH3PBI3-XBRXPEROVSKITE SOLAR CELLS VIA SPRAY ASSISTED TWO-STEP DEPOSITION: INFLUENCE OF BROMIDE ON THE DEVICE PERFORMANCE 976

Gaoda Chai ; Shiqiang Luo ; Shizhen Wang ; Hang Zhou

MODULATED STRUCTURE TO MAXIMIZE THE OPEN-CIRCUIT VOLTAGE WITH MODERATE BAND-GAP OF SMALL MOLECULE ORGANIC SOLAR CELLS-DFT APPROACH 980

Saravanan Chinnusamy ; Amita Munshi ; Sukanya Santhosh Kumar ; W. S. Sampath ; Milind S. Dangate

PEROVSKITE GRAIN SIZE MODULATION BY ANNEALING IN METHYL-AMINE ENVIRONMENT 986

Arun Singh Chouhan ; Naga Prathibha Jasti ; Srinivasan Raghavan ; Sushobhan Avasthi ; Shreyash Hadke

FE2O3AS AN ELECTRON TRANSPORT MATERIAL FOR ORGANO-METAL HALIDE PEROVSKITE SOLAR CELLS 989

Dallas Fisher ; Pravakar P. Rajbhandari ; Tara P. Dhakal

OPTICAL EVALUATION OF PEROVSKITE FILMS IN AND FOR SOLAR CELL DEVICE STRUCTURES 993

Kiran Ghimire ; Dewei Zhao ; Changlei Wang ; Yanfa Yan ; Nikolas J. Printraza

HYBRID ORGANIC-INORGANIC SOLAR CELLS WITH A BENZOQUINONE PASSIVATING LAYER 999

James Hack ; Abhishek Iyer ; Meixi Chen ; Nicole Kotulak ; Akirt Sridharan ; Robert Opila

PRECISE 1-V CURVE MEASUREMENT PROCEDURE FOR PEROVSKITE SOLAR CELLS: APPLICATION TO VARIOUS TYPES OF DEVICES 1003

Y. Hishikawa ; M. Yoshita ; H. Shimura ; A. Sasaki ; T. Ueda

ENHANCING THE CRYSTALLINE OF PLANAR-STRUCTURE CH3NH3PBI3PEROVSKITE SOLAR CELLS VIA SANDWICH EVAPORATION TECHNIQUE 1006

Po-Tsun Kuo ; Shang-Pang Lin ; Cheng-Shian Lin ; Ching-Fuh Lin

TOWARD HIGH PERFORMANCE ORGANIC-SILICON HYBRID SOLAR CELLS 1009

Yi Lai ; Hong-Jhang Syu ; Ching-Fuh Lin

NICKEL OXIDE THIN FILMS BY RADIO FREQUENCY SPUTTER FOR INVERTED PEROVSKITE SOLAR CELLS 1012

Hyeonseok Lee ; Yu-Ting Huang ; Shien-Ping Feng

ANOMALOUS EFFICIENCY SCALING WITH DARK CURRENT IN PEROVSKITE SOLAR CELLS 1015

Vikas Nandal ; Pradeep R. Nair

NUMERICAL SIMULATION AND PERFORMANCE OPTIMIZATION OF PEROVSKITE SOLAR CELL 1018

Sai Naga Raghuram Nanduri ; Mahbube K. Siddiki ; Ghulam M. Chaudhry ; Yahya Z. Alharthi

PERFORMANCE PREDICTION FOR LARGE AREA PEROVSKITE SOLAR CELLS 1022

Yojak Raote ; Hitarth Choubisa ; Pradeep R. Nair

PHOTOCONVERSION EFFICIENCY MODELING IN PEROVSKITE SOLAR CELLS 1025

A.V. Sachenko ; V.P. Kostylyov ; A.V. Bobyl ; V.M. Vlasiuk ; I.O. Sokolovskyi ; E.I. Terukov ; M. Evstigneev

INFLUENCE OF MONO- AND DI-VALENT METAL ADDITIVES ON MORPHOLOGY AND CHARGE CARRIER DYNAMICS OF CH3NH3PBI3PEROVSKITE 1030

Niraj Shrestha ; Suneth C. Watthage ; Zhaoning Song ; Paul J. Roland ; Adam B. Phillips ; Michael J. Heben ; Randall J. Ellingson

EFFECT OF DUAL CATHODE BUFFER LAYER ON TERNARY ORGANIC SOLAR CELL 1034

Ashish Singh ; T. Bhim Raju ; Anamika Dey ; Ritesh Kant Gupta ; Parameswar K. Iyer

COPPER PLATED TOP ELECTRODE FOR AN INVERTED ORGANIC PHOTOVOLTAIC 1037

Malia Steward ; Zhan Shi ; Kyoung- Tae Kim ; Seungkeun Choi

INTERFACE BAND GAP AND CHARGE TRAPPING IN BULK HETEROJUNCTION SOLAR CELLS 1040

Marian Tzolov ; Maxwell Mcintyre

FABRICATION OF EFFICIENT CH3NH3PBI3 SOLAR CELLS IN AMBIENT AIR 1044

Feng Wang ; Ye Zhongbiao ; Hojjatollah Sarvari ; Somin Park ; Kenneth Graham ; Yuetao Zhao ; Zhi David Chen

HIGH EFFICIENCY PEROVSKITE SOLAR CELLS BY A MODIFIED LOW-TEMPERATURE SOLUTION PROCESS INTER-DIFFUSION METHOD 1048

Yangyi Yao ; Wei-Lun Hsu ; Mario Dagenais

INTERFACIAL MODIFICATION OF SOL-GEL ZNO/AZO BILAYER AS HIGHLY EFFICIENT ELECTRON TRANSPORT LAYER FOR PEROVSKITE SOLAR CELLS 1051

Shang-Hsuan Wu ; Ming-Yi Lin ; Sheng-Hao Chang ; Wei-Chen Tu ; Chi-Wei Chu ; Via-Chung Chang

THE POTENTIAL OF BIFACIAL PHOTOVOLTAICS: A GLOBAL PERSPECTIVE 1055

Xingshu Sun ; Mohammad R. Khan ; Amir Hanna ; Muhammad M. Hussain ; Muhammad A. Alam

PERFORMANCE ASSESSMENT OF STAND ALONE BIFACIAL SOLAR PANEL UNDER REAL TIME CONDITIONS 1058

Ahmer A.B. Baloch ; Maher Armoush ; Basel Hindi ; Abdelkader Bousselham ; Nouar Tabet

OPERATION AND PERFORMANCE ASSESSMENT OF GRID-CONNECTED PV SYSTEMS IN OPERATION IN MAUI, HAWAII 1061

Severine Busquet ; Jonathan Kobayashi ; Richard E. Rocheleau

A NOVEL MULTILEVEL SOLAR PANEL SYSTEM: IMPLEMENTATION AND VERIFICATION 1067

Tanmoy Debnath ; Syed N. Imtiaz ; Syed F. Nawaz ; Abdullah Al Mahmud ; Mosaddequr Rahman

PREDICTING POWER LOSS DUE TO MODULE MISMATCH IN UTILITY-SCALE PHOTOVOLTAIC SYSTEMS 1071

Stephen Kaplan ; Kendra Passow

APPLICATION OF SHAPED REFLECTORS TO INCREASE THE ENERGY HARVEST OF BIFACIAL PV SYSTEMS - ANALYZED WITH A MINIATURIZED TEST ARRAY 1077

Hartmut Nussbaumer ; Markus Klenk ; Nico Keller ; Dominic Heller ; Remo Kaslin ; Thomas Baumann ; Franz Baumgartner

TOWARDS NEW MODULE AND SYSTEM CONCEPTS FOR LINEAR SHADING RESPONSE 1081

Kostas Sinapis ; Tom T.H. Rooijakkers ; Lenneke H. Sloof ; Lars A.G. Okel ; Mark J. Jansen ; Anna J. Carr

PARTIAL SHADING ABATEMENT THROUGH CASCADED H-BRIDGE TOPOLOGY 1086

Steven Tidwell ; Joseph Latham ; Michael Mcintyre

DATA ANALYSIS FOR EFFECTIVE MONITORING OF PARTIALLY SHADED PHOTOVOLTAIC SYSTEMS 1090

Odysseas Tsafarakis ; Kostas Sinapis ; Wilfried G.J.H.M. Van Sark

BIFACIAL PHOTOVOLTAIC MODULE ENERGY YIELD CALCULATION AND ANALYSIS 1094

Christopher E. Valdivia ; Chu Tu Li ; Annie Russell ; Joan E. Haysom ; Rui Li ; David Lekx ; Mohsen M. Sepeher ; Dan Henes ; Karin Hinzer ; Henry P. Schriemer

DESIGN AND DEVELOPMENT OF A SOLAR PHOTOVOLTAIC MODULE DETECTION CONTROL SYSTEM BASED ON PLC 1100

Yiwang Wang ; Jili Zhang ; Kanglin Liu ; Houjun Tang ; Hui Pan ; Yan Lin ; Peter Yang ; Rui Wang

DETECTING CALIBRATION DRIFT AT GROUND TRUTH STATIONS A DEMONSTRATION OF SATELLITE IRRADIANCE MODELS' ACCURACY 1104

Richard Perez ; James Schlemmer ; Adam Kankiewicz ; John Dise ; Alemu Tadese ; Thomas Hoff

PERFORMANCE OF SOLAR RESOURCE MONITORING STATIONS IN HOT CLIMATE REGIONS 1110

Yahya Z. Alharthi ; Mahbube K. Siddiki ; Ghulam M. Chaudhry ; Saad Muaddi ; Ahmed Alahmed

FIRST RESULTS OF A LOW COST ALL-SKY IMAGER FOR CLOUD TRACKING AND INTRA-HOUR IRRADIANCE FORECASTING SERVING A PV-BASED SMART GRID IN LA GRACIOSA ISLAND 1116

David Cañadillas ; Walter Richardson ; Benjamín Gonzalez-Díaz ; Les E. Shephard ; Ricardo Guerrero Lemus

STATISTICAL ANALYSIS OF PV INSOLATION DATA 1122

Abdulmunim Guwaeder ; Rama Ramakumar

A COMPARISON OF PV POWER FORECASTS USING PVLIB-PYTHON 1127

William F. Holmgren ; Antonio T. Lorenzo ; Clifford Hansen

COMPARING THE TYPICAL GHI YEAR VS TYPICAL POWER YEAR 1132

Alex Kubiniec ; Adam Kankiewicz ; Alemu Tadesse

THE HOLY GRAIL OF RESOURCE ASSESSMENT: LOW COST GROUND-BASED MEASUREMENTS WITH GOOD ACCURACY 1134

Bill Marion ; Benjamin Smith

GLOBAL COMPARISON OF THE IMPACT OF TEMPERATURE AND PRECIPITABLE WATER ON CDTE AND SILICON SOLAR CELLS 1140

I. M. Peters ; L. Haohui ; T. Reindl ; T. Buonassisi

ESTIMATION OF MEAN MONTHLY GLOBAL SOLAR RADIATION USING MODEL BASED ON SUNSHINE HOURS FOR COLOMBIA 1143

Diego J. Rodríguez ; Johan Hernández ; Adolfo Jaramillo

IMPLEMENTATION OF SOLAR DIFFUSE CIE MODEL IN RAY TRACING PROGRAM FOR IRRADIANCE CALCULATIONS 1147

Liliana Ruiz Diaz ; Pierre-Alexandre Blanche ; Robert A. Norwood

INVESTIGATION OF CITY-LEVEL SITE-PAIR CORRELATIONS OF SOLAR VARIABILITY USING EMPIRICAL SATELLITE DATA 1151

Rhythm Singh ; Rangan Banerje

ULTRA-SHORT-TERM PHOTOVOLTAIC GENERATION FORECASTING MODEL BASED ON WEATHER CLUSTERING AND MARKOV CHAIN 1158

Jin Tan ; Changhong Deng

DAILY SOLAR IRRADIANCE PROFILE CHARACTERIZATION AND RAMP RATE ANALYSIS AT DIFFERENT TIME RESOLUTIONS 1163

Spyros Theocharides ; Venizelos Venizelou ; George Makrides ; George E. Georghiou

COMPARISON AND ANALYSIS OF INSTRUMENTS MEASURING PLANE OF ARRAY IRRADIANCE FOR ONE-AXIS TRACKING PV SYSTEMS 1169

Frank Vignola ; Chun-Yu Chiu ; Josh Peterson ; Michael Dooraghi ; Manajit Sengupta

A SKY IMAGE ANALYSIS SYSTEM FOR SUB-MINUTE PV PREDICTION 1175

Rodrigo Verschac ; Li Li ; Shohei Nobuhara ; Takekazu Kato

LARGE AREA NANOSTRUCTURE INTEGRATION FOR BROAD-SPECTRUM, OMNIDIRECTIONAL ANTIREFLECTION IMPROVEMENTS ON POLYMER PACKAGED, MECHANICALLY FLEXIBLE, EPITAXIAL LIFT-OFF III-V SOLAR CELLS 1181

Gabriel Cossio ; Jihwan Lee ; Gautham Ragunathan ; Andre Wibowo ; Sudersena Rao Tatavarti ; Kimberly Sablon ; Edward T. Yu

DEVELOPMENT OF BACK SURFACE TEXTURE FOR LIGHT MANAGEMENT IN EPITAXIAL LIFT OFF (ELO) QUANTUM DOT SOLAR CELLS 1184

Brittany L. Smith ; George T. Nelson ; Yushuai Dai ; Michael A. Slocum ; Andre Wibowo ; Rao Tatavarti ; Seth M. Hubbard

ENABLING HIGH-EFFICIENCY INAS/GAAS QUANTUM DOT SOLAR CELLS BY EPITAXIAL LIFT-OFF AND LIGHT MANAGEMENT 1189

F. Cappelluti ; A. P. Cédola ; A. Khalili ; Farid Elsehrawy ; G. Bauhuis ; P. Mulder ; J. Schermer ; G. Bissels ; T. Aho ; T. Niemi ; M. Guina ; D. Kim ; J. Wu ; H. Liu

CHARACTERIZATION OF ARSENIC DOPED CDTE LAYERS AND SOLAR CELLS 1193

Sachit Grover ; Xiaoping Li ; Wei Zhang ; Ming Yu ; Gang Xiong ; Markus Gloeckler ; Roger Malik

ENHANCING P-TYPE DOPING IN POLYCRYSTALLINE CDTE FILMS 1196

Brian Mccandless ; Wayne Buchanan ; Gowri Sriramagiri ; Christopher Thompson ; Joel Duenow ; David Albin ; Soren Jensen ; John Moseley ; M. Al-Jassim ; Wyatt K. Metzger

SPECTRAL AND CONCENTRATION SENSITIVITY OF MULTIJUNCTION SOLAR CELLS AT HIGH TEMPERATURE 1201

Daniel J. Friedman ; Myles A. Steiner ; Emmett E. Perl ; John Simon

ON THE USE OF TRANSPARENT CONDUCTIVE OXIDES IN HIGH CONCENTRATOR III-V MULTIJUNCTION SOLAR CELLS 1204

Ignacio Rey-Stolle ; Yeonbae Lee ; Iván Garcia ; Luis Cifuentes ; Kin Man Yu ; Carlos Algora ; Wladek Walukiewicz

COMPONENT INTEGRATION EFFECTS IN 4-JUNCTION SOLAR CELLS WITH DILUTE NITRIDE 1EV SUBCELL 1210

I. García ; M. Ochoa ; I. Lombardero ; L. Cifuentes ; P. Caño ; M. Hinojosa ; I. Rey-Stolle ; C. Algora ; A. D. Johnson ; J. I. Davies ; K.H. Tan ; W.K. Loke ; S. Wicaksono ; S. F. Yoon

BISMUTH SURFACTANT-MEDIATED GROWTH OF GANASSB(BI) SOLAR CELLS 1215

Aymeric Maros ; Chaomin Zhang ; Jongwon Lee ; Hongfeng Wang ; Stephen Bremner ; Nikolai Faleev ; Christiana B. Honsberg ; Richard. R. King

AMORPHOUS SILICON CARBIDE FOR SILICON SURFACE PASSIVATION IN CARRIER-SELECTIVE CONTACT DEVICES 1220

Mathieu Boccard ; Christophe Ballif ; Zachary C. Holman

SURFACE PASSIVATION OF BORON DIFFUSED JUNCTIONS BY BOROSILICATE GLASS AND IN SITU GROWN SILICON DIOXIDE INTERFACE LAYER 1222

Valentin D. Mihailetchi ; Haifeng Chu ; Jan Lossen ; Radovan Kopecek

IMPROVED LIGHT INCOUPLING IN PLANAR SOLAR CELLS VIA IMPROVED TEXTURE MORPHOLOGY OF PDMS SCATTERING LAYER 1228

Salman Manzoor ; Zhengshan J. Yu ; Asad Ali ; Waqar Ali ; Zachary C. Holman

DAMAGE-FREE LASER ABLATION FOR EMITTER PATTERNING OF SILICON HETEROJUNCTION INTERDIGITATED BACK-CONTACT SOLAR CELLS 1233

Menglei Xu ; Twan Bearda ; Miha Filipic ; Hariharsudan Sivaramakrishnan Radhakrishnan ; Maarten Debucquoy ; Ivan Gordon ; Jozef Szlufcik ; Jef Poortmans

BENEFITS OF A THERMAL DRIFT DURING ATOMIC LAYER DEPOSITION OF AL2O3FOR C-SI PASSIVATION 1237

Fabien Lebreton ; Andy Zauner ; Pavel Bulkin ; Francois Silva ; Sergej Filonovich ; Pere Roca I Cabarrocas

GROWTH DIFFERENCE OF AMORPHOUS SILICON BETWEEN PLASMA ENHANCED AND CATALYTIC CVD BASED ON SILICON HETEROJUNCTION SOLAR CELLS 1241

Liping Zhang ; Renfang Chen ; Zhuopeng Wu ; Chenguang Sun ; Fanying Meng ; Zhengxin Liu

DEVELOPING AN UNDERSTANDING-BASED SELECTION OF HYBRID-PEROVSKITE COMPOUNDS AND THE CU-IN HYBRID-PEROVSKITE (CIHP) FAMILY 1245

Alex Zunger ; G. Dalpian ; Qihang Liu ; L.B Abdalla ; L.L. Kazmerski

EFFECTS OF ELECTRON AND PROTON RADIATION ON PEROVSKITE SOLAR CELLS FOR SPACE SOLAR POWER APPLICATION 1248

Jing-Shun Huang ; Michael D. Kelzenberg ; Pilar Espinet-González ; Colin Mann ; Don Walker ; Ali Naqavi ; Nina Vaidya ; Emily Warmann ; Harry A. Atwater

TOWARDS PEROVSKITE SILICON TANDEM SOLAR CELLS WITH OPTIMIZED OPTICAL PROPERTIES 1253

Jan Christoph Goldschmidt ; Alexander J. Bett ; Patricia S.C. Schulze ; Nico Tucher ; Martin Bivour ; Markus Kohlstädt ; Seunghun Lee ; Simone Mastroianni ; Laura Mundt ; Markus Mundus ; Paul Ndione ; Karl Wienands ; Kristina Winkler ; Uli Würfel ; Martin Hermle ; Stefan W. Glunz

FIRST-PRINCIPLES DENSITY FUNCTIONAL THEORY CALCULATION OF METAL-
SUBSTITUTED LEAD HALIDE PEROVSKITE 1256
Ji-Sang Park ; Matthew D. Sampson ; Alex B.F. Martinson ; Maria K.Y. Chan

ESTIMATING THE EFFECTS OF MODULE AREA ON THIN-FILM PHOTOVOLTAIC SYSTEM
COSTS 1259
Kelsey A. W. Horowitz ; Ran Fu ; Xingshu Sun ; Tim Silverman ; Michael Woodhouse ; Muhammad A. Alam

COST ANALYSIS OF TANDEM MODULES 1264
Sarah E. Sofia ; Jonathan Mailoal ; Dirk Weiss ; Tonio Buonassisi ; Ian Marius Peters

CAUSE OF CURRENT-COLLECTION FAILURE OBSERVED INISC-REDUCTION PHASE OF
PV CELLS AND MODULES EXPOSED TO ACETIC ACID 1268
Tadanori Tanahashi ; Norihiko Sakamoto ; Hajime Shibata ; Atsushi Masuda

COMPARISON OF PV MODULE PERFORMANCE BEFORE AND AFTER 11, 20, AND 25.5
YEARS OF FIELD EXPOSURE 1271
Jacob Rada ; Charles Chamberlin ; Peter Lehman ; Arne Jacobson

MARRYING QUALITY ASSURANCE WITH DESIGN ENGINEERING – A WINNING
PARTNERSHIP! BUT, A CULTURAL DIVIDE? 1275
*Sarah Kurtz ; Govind Ramu ; Robert Cornell ; Sumanth Lokanath ; Edward Hsi ; Tony Sample ; Masaaki
Yamamichi ; George Kelly ; Ted Spooner ; Jonathan Previtali ; John Wohlgemuth*

UPDATED EVALUATION OF SHOCK HAZARDS TO FIREFIGHTERS WORKING IN
PROXIMITY OF PV SYSTEMS 1280
Olga Lavrova ; Jimmy E. Quiroz ; Jack Flicker ; Renee Gooding

GROWTH AND OPTIMIZATION OF GAINP/INP NANOWIRE TUNNEL DIODE 1286
Xulu Zeng ; Gaute Otnes ; Magnus Heurlin ; Magnus T Borgström

CATHODOLUMINESCENCE MAPPING FOR THE DETERMINATION OF N-TYPE DOPING IN
SINGLE GAAS NANOWIRES 1289
*Hung-Ling Chen ; Chalermchai Himwas ; Andrea Scaccabarozzi ; Pierre Rale ; Fabrice Oehler ; Aristide
Lemaître ; Laurent Lombez ; Jean-François Guillemoles ; Maria Tchemycheva ; Jean-Christophe Harmand ;
Andrea Cattoni ; Stéphane Collin*

OPTICAL OPTIMIZATION OF PASSIVATED GAAS NANOWIRE SOLAR CELLS 1294
Kyle W. Robertson ; Ray R. Lapierre ; Jacob J. Krich

HIGH EFFICIENCY GAN NANOWIRE/SI PHOTOCATHODE FOR
PHOTOELECTROCHEMICAL WATER SPLITTING 1299
Srinivas Vanka ; Sheng Chu ; Yichen Wang ; Ishiang Shih ; Hong Guo ; Zetian Mi

ANALYTIC DESCRIPTION OF THE IMPACT OF GRAIN BOUNDARIES ON VOC 1303
Paul Haney ; Benoit Gaury

ROLE OF TELLURIUM BUFFER LAYER ON CDTE SOLAR CELLS' ABSORBER/BACK-
CONTACT INTERFACE 1308
Tao Song ; James R. Sites

SIMULTANEOUS EXAMINATION OF GRAIN-BOUNDARY POTENTIAL, RECOMBINATION,
AND PHOTOCURRENT IN CDTE SOLAR CELLS USING DIVERSE NANOMETER-SCALE
IMAGING 1312
*C.S. Jiang ; H.R. Moutinho ; J. Moseley ; A. Kanevce ; J.N. Duenow ; E. Colegrove ; C. Xiao ; W.K. Metzger ;
M.M. Al-Jassim*

NANOPARTICLE/METAL REAR REFLECTORS FOR LOW- AND HIGH-TEMPERATURE
SILICON SOLAR CELLS 1317
*Syeda Qudsia ; Farah Qazi ; Mehwish Azher Javed ; Mathieu Boccard ; Zhengshan J. Yu ; Peter Firth ; Jonathan
Bryan ; Zachary C. Holman*

ABSORPTION IN EACH LAYER OF A SILICON HETEROJUNCTION SOLAR CELL 1322
*Keith R. Mcintosh ; Malcolm D. Abbott ; Benjamin A. Sudbury ; Salman Manzoor ; Zhengshan J. Yu ; Mehdi
Leilaeioun ; Jiatiwei Shi ; Zachary C. Holman*

INVESTIGATIONS ON PLASMONIC COLOR TUNING COATING ON C-SI SOLAR CELLS 1329
*Gerhard Peharz ; Wolfgang Waldhauser ; Christine Prietl ; Bettina Großschädl ; Martin C. Schubert ; Bernhard
Michl*

INVESTIGATION OF INTERFACE AND BULK LOCALIZED STATES IN A-SI:H SOLAR
CELLS 1333
Adrien Bidiville ; Takuya Matsui ; Hitoshi Sai ; Koji Matsubara

EXPERIMENTAL AND THEORETICAL STUDY OF THE INFRARED EMISSIVITY OF
CRYSTALLINE SILICON SOLAR CELLS 1339
*Alberto Riverola ; Alexander Mellor ; Diego Alonso Alvarez ; Lourdes Ferre Llin ; Ilaria Guarracino ; Christos N.
Markides ; Douglas Paul ; Daniel Chemisana ; Ned Ekins-Daukes*

HIGH PERFORMANCE MOLECULAR DONORS FOR ORGANIC SOLAR CELLS, MATERIALS
DESIGN AND DEVICE OPTIMIZATION 1342
Paul Geraghty ; Haotian Wang ; Calvin Lee ; Jegadesan Subbiah ; David Jones

ADVANCED OPTICAL MODELLING OF MICRO-TEXTURED SOLUTION-PROCESSED
SOLAR CELLS WITH CONSIDERATION OF SMALL-AREA EFFECTS...................................1346
Benjamin Lipovšek ; Marko Jošt ; Andrej Campa ; Fei Gu ; Christoph J. Brabec ; Karen Forberich ; Janez Krc ; Marko Tonic

IDENTIFICATION OF DEGRADATION PATHWAYS OF ORGANIC SOLAR CELLS USING
INFRARED SPECTROSCOPY..1350
S. Shah ; R Biswas ; T. Koschny ; V L Dalal

A DEVICE-INDEPENDENT SCREENING TECHNIQUE FOR RAPIDLY IDENTIFYING NEXT
GENERATION OPV MATERIALS...1354
Bryon W. Larson ; Andrew J. Ferguson ; Bertrand J. Tremolet De Villers ; Ross E. Larsen

NOVEL ANTHANTHRONE AND ANTHANTHRENE CO-POLYMERS AS P-TYPE
CONJUGATED SEMICONDUCTORS FOR ORGANIC PHOTOVOLTAICS.............................1360
Suru Vivian John ; Patrick Denk ; Christoph Ulbricht ; Herwig Heilbrunner ; Jean-Benoit Giguère ; Antoine Lafleur-Lambert ; Jean-Francois Morin ; Emmanuel Iwuoha ; Daniel Ayuk Mbi Egbe

REDUCING UV INDUCED DEGRADATION LOSSES OF SOLAR MODULES WITH C-SI
SOLAR CELLS FEATURING DIELECTRIC PASSIVATION LAYERS......................................1366
Robert Witteck ; Henning Schulte-Huxel ; Boris Veith-Wolf ; Malte Ruben Vogt ; Fabian Kiefer ; Marc Kontges ; Robby Peibst ; Rolf Brendel

LARGE-AREA JUNCTION DAMAGE IN POTENTIAL-INDUCED DEGRADATION OF C-SI
SOLAR MODULES 1371
Chuanxiao Xiao ; Chun-Sheng Jiang ; Steve Johnston ; Steve P. Harvey ; Peter Hacke ; Brian Gorman ; Mowafak Al-Jassim

SEARCH FOR MICROSTRUCTURAL DEFECTS AS NUCLEI FOR PID-SHUNTS IN SILICON
SOLAR CELLS 1376
Volker Naumann ; Otwin Breitenstein ; Jan Bauer ; Christian Hagendorf

INVESTIGATING PID SHUNTING IN POLYCRYSTALLINE SILICON MODULES VIA MULTI-
SCALE, MULTI-TECHNIQUE CHARACTERIZATION 1381
Steven P. Harvey ; John Moseley ; Adam Stokes ; Andrew Norman ; Brian Gorman ; Peter Hacke ; Steve Johnston ; Mowafak Al-Jassim

POTENTIAL-INDUCED DEGRADATION OF A SI NITRIDE/CRYSTALLINE SI INTERFACE
OBSERVED THROUGH MINORITY CARRIER LIFETIME MEASUREMENT..........................1385
Naoyuki Nishikawa ; Seira Yamaguchi ; Keisuke Ohdaira

FIELD INSPECTION OF PV MODULES: QUANTIFICATION OF EVA BROWNING LEVEL
USING AN IMAGE PROCESSING TOOL..1389
Sushanth Gudla ; Govindasamy Tamizhmani

PREVENTING POTENTIAL-INDUCED DEGRADATION IN CRYSTALLINE SILICON PV
MODULES: RELATIONSHIP BETWEEN DEGRADATION AND BILL OF MATERIAL 1395
Alessandro Virtuani ; Eleonora Annigoni ; Christophe Ballif

IDENTIFYING REVERSE-BIAS BREAKDOWN SITES IN CUINXGA(1-X)SE2 1400
Steve Johnston ; Elizabeth Palmiotti ; Andreas Gerber ; Harvey Guthrey ; Lorelle Mansfield ; Timothy J. Silverman ; Mowafak Al-Jassim ; Angus Rockett

HIMAWARI-8 ENABLED REAL-TIME DISTRIBUTED PV SIMULATIONS FOR
DISTRIBUTION NETWORKS 1405
Nicholas A. Engerer ; Jamie M. Bright ; Sven Killinger

REDUCED MEASUREMENT UNCERTAINTY IN PV MODULE BATCH TESTING 1411
Blagovest Mihaylov ; Bengt Jaeckel ; Juergen Arp ; Ralph Gottschalg

CLOUD MOTION IDENTIFICATION ALGORITHMS BASED ON ALL-SKY IMAGES TO
SUPPORT SOLAR IRRADIANCE FORECAST...1415
Lydie Magnone ; Fabrizio Sossan ; Enrica Scolari ; Mario Paolone

AUTOMATIC DETECTION OF INACTIVE SOLAR CELL CRACKS IN
ELECTROLUMINESCENCE IMAGES 1421
Sergiu Spataru ; Peter Hacke ; Dezso Sera

APPLYING SPATIAL DOWNSCALING AND SMART PERSISTENCE TO PROVIDE AN
IMPROVED SOLAR FORECAST TO REDUCE COMMERCIAL DEMAND CHARGES............1427
Alex Kubiniec ; Ted Belanger ; Adam Kankiewicz ; Skip Dise ; Nate Glasgow ; Alemu Tadesse

THERMAL CHARACTERISTICS OF PID-AFFECTED MONOCRYSTALLINE SILICON SOLAR
MODULES UNDER ILLUMINATED AND DARK CONDITIONS..1430
Pan Zhao ; Shuwen Guo ; He Wang ; Hong Yang ; Dengyuan Song ; Shiyu Sang ; Bojie Su ; Xue Zhang ; Yunxue Cao ; Hui Zhao

TARGETED EVALUATION OF UTILITY-SCALE AND DISTRIBUTED SOLAR FORECASTING ...1435
Matthew Lave ; Robert J. Broderick ; Laurie Burnham

RECORD EFFICIENCIES FOR SELENIUM PHOTOVOLTAICS AND APPLICATION TO INDOOR SOLAR CELLS 1441

Douglas M. Bishop ; Teodor Todorov ; Yun Seog Lee ; Oki Gunawan ; Richard Haight

CLOSE-SPACED SUBLIMATION FOR SB2SE3SOLAR CELLS 1445

Laurie J. Phillips ; Peter Yates ; Oliver S. Hutter ; Tom Baines ; Leon Bowen ; Ken Durose ; Jonathan D. Major

FABRICATION OF COPPER ARSENIC SULFIDE THIN FILMS FROM NANOPARTICLES FOR APPLICATION IN SOLAR CELLS 1449

Scott A. Mcclary ; Joseph Andler ; Carol A. Handwerker ; Rakesh Agrawal

ORIENTATION CONTROLLED GE THIN FILMS ON GLASS BY AL-INDUCED CRYSTALLIZATION 1452

Kaveh Shervin ; Khim Kharel ; Alexandre Freundlich

IN-LINE POTASSIUM FLUORIDE TREATMENT OF CIGS ABSORBERS DEPOSITED ON FLEXIBLE SUBSTRATES IN A PRODUCTION-SCALE PROCESS TOOL 1455

Ryan Kaczynski ; Jinwoo Lee ; Jane Van Alsburg ; Baosheng Sang ; Urs Schoop ; Jeffrey Britt

LIGHT-SOAK AND DARK-HEAT INDUCED CHANGES IN CU(IN, GA)SE2 SOLAR CELLS: A MACROSCOPIC TO MICROSCOPIC STUDY 1459

Rouin Farshchi ; Benjamin Hickey ; Dmitry Poplavskyy

A NEW MODEL TO DETERMINE INSTALLED SYSTEM COST AND LCOE FOR ARPA-E'S MOSAIC MICRO-CONCENTRATOR PV PROGRAM 1463

Ran Fu ; Kelsey A.W. Horowitz ; Daniel W. Cunningham ; James Zahler

FIXED-TILT 660 × CONCENTRATING PHOTOVOLTAIC SYSTEM WITH 30% EFFICIENCY 1469

Alex J. Grede ; Jared S. Price ; Baomin Wang ; Michael V. Lipski ; Brent Fisher ; Kyu-Tae Lee ; Junwen He ; Gregory S. Brulo ; Xiaokun Ma ; Scott Burroughs ; Christopher D. Rahn ; Ralph G. Nuzzo ; John A. Rogers ; Noel C. Giebink

WAFER INTEGRATED MICRO-SCALE CONCENTRATING PHOTOVOLTAICS 1473

Tian Gu ; Duanhui Li ; Lan Li ; Bradley Jared ; Gordon Keeler ; Bill Miller ; William Sweatt ; Scott Paap ; Michael Saavedra ; Ujjwal Das ; Steve Hegedus ; Anna Tanke-Pedretti ; Juejun Hu

TOWARD STATIONARY CONCENTRATOR PHOTOVOLTAIC PANELS 1476

Peter Kozodoy ; Christopher Gladden ; Michael Pavilonis ; Tobias Wheeler ; Christopher Rhodes ; Chadwick Casper ; Kevin Schneider

CPV TECHNOLOGIES NOT RELYING ON PERFECTION OF TRACKERS 1479

Kenji Araki ; Yasuyuki Ota ; Kan-Hua Lee ; Kensuke Nishioka ; Masafumi Yamaguchi

THE GETTERING EFFECT OF DIELECTRIC FILMS FOR SILICON SOLAR CELLS 1485

A. Y. Liu ; C. Sun ; V. P. Markevich ; A. R. Peaker ; J. D. Murphy ; D. Macdonald

TABULA RASA: OXYGEN PRECIPITATE DISSOLUTION THOUGH RAPID HIGH TEMPERATURE PROCESSING IN SILICON 1491

Erin E. Looney ; Hannu S. Laine ; Mallory A. Jensen ; Amanda Youssef ; Vincenzo Lasalvia ; Paul Stradins ; Tonio Buonassisi

TOWARD EFFECTIVE GETTERING IN BORON-IMPLANTED SILICON SOLAR CELLS 1494

Hannu S. Laine ; Ville Vähänissi ; Zhengjun Liu ; Ernesto Magaña ; Ashley E. Morishige ; Jan Krügener ; Kristian Salo ; Hele Savin ; Barry Lai ; David P. Fenning

IMPACT OF THE INITIAL GROWTH INTERFACE ON THE GRAIN STRUCTURE IN HPMC-SI INGOT 1498

Giri Wahyu Alam ; Etienne Pihan ; Benoit Marie ; Nathalie Mangelinck-Noël

EFFECT OF CARBON CONCENTRATION AND GROWTH CONDITIONS ON OXYGEN PRECIPITATION BEHAVIOR IN N-TYPE CZ-SI 1504

Takuto Kojima ; Ryota Suzuki ; Kosuke Kinoshita ; Kyotaro Nakamura ; Atsushi Ogura ; Yoshio Ohshita ; Isao Masada ; Shoji Tachibana

NANO-IMAGING OF PERFORMANCE IN PHOTOVOLTAICS 1508

Elizabeth M. Tennyson ; Marina S. Leite

IMPLICATIONS OF CONDUCTIVE GRAIN BOUNDARIES IN CHLORINE-TREATED CDTE SOLAR CELLS 1511

Mohit Tuteja ; Vasilios Palekis ; Allen Hall ; Scott Maclaren ; Chris S. Ferekides ; Angus A. Rockett

IMAGING THE MULTI-TEMPORAL PHOTO-CARRIER DYNAMICS AT THE NANOMETER SCALE IN ORGANIC AND INORGANIC SOLAR CELLS 1516

Pablo A. Fernández Garrillo ; Lukasz Borowik ; Florent Caffy ; Renaud Demadrille ; Benjamin Grévin

NANOSCALE TOMOGRAPHIC CHARGE TRANSPORT IN POLYCRYSTALLINE CHALCOGENIDE ABSORBERS: CDTE VERSUS CIGS 1522

Justin L. Luria ; Andrew Moore ; Sun Yu ; Mark Aindow ; Bryan D. Huey

IMPROVING THE PV MODULE SINGLE-DIODE MODEL ACCURACY WITH TEMPERATURE DEPENDENCE OF THE SERIES RESISTANCE 1526

Kyumin Lee

CELL-TO-MODULE (CTM) ANALYSIS FOR PHOTOVOLTAIC MODULES WITH SHINGLED SOLAR CELLS 1531

Max Mittag ; Tobias Zech ; Martin Wiese ; David Blasi ; Matthieu Ebert ; Harry Wirth

A PRACTICAL IRRADIANCE MODEL FOR BIFACIAL PV MODULES 1537

Bill Marion ; Sara Macalpine ; Chris Deline ; Amir Asgharzadeh ; Fatima Toor ; Daniel Riley ; Joshua Stein ; Clifford Hansen

A DETAILED MODEL OF REAR-SIDE IRRADIANCE FOR BIFACIAL PV MODULES 1543

Clifford W. Hansen ; Renee Gooding ; Nathan Guay ; Daniel M. Riley ; Johnson Kallickal ; Donald Ellibee ; Amir Asgharzadeh ; Bill Marion ; Fatima Toor ; Joshua S. Stein

VIEW FACTOR MODEL AND VALIDATION FOR BIFACIAL PV AND DIFFUSE SHADE ON SINGLE-AXIS TRACKERS 1549

Marc Abou Anoma ; David Jacob ; Ben C. Bourne ; Jonathan A. Scholl ; Daniel M. Riley ; Clifford W. Hansen

A FAST QUASI-STATIC TIME SERIES (QSTS) SIMULATION METHOD FOR PV IMPACT STUDIES USING VOLTAGE SENSITIVITIES OF CONTROLLABLE ELEMENTS 1555

Xiaochen Zhangl ; Santiago Grijalva ; Matthew J. Reno ; Jeremiah Deboever ; Robert J. Broderick

FAST DETERMINATION OF DISTRIBUTION-CONNECTED PV IMPACTS USING A VARIABLE-TIME-STEP QUASI-STATIC TIME-SERIES APPROACH 1561

Barry Mather

SCALABILITY OF THE VECTOR QUANTIZATION APPROACH FOR FAST QSTS SIMULATION 1567

Jeremiah Deboever ; Santiago Grijalva ; Matthew J. Reno ; Xiaochen Zhang ; Robert J. Broderick

MACHINE LEARNING FOR RAPID QSTS SIMULATIONS USING NEURAL NETWORKS 1573

Matthew J. Reno ; Robert J. Broderick ; Logan Blakely

ALGORITHMIC ASPECTS OF A COMMERCIAL-GRADE DISTRIBUTION SYSTEM LOAD FLOW ENGINE 1579

Francis Therrien ; Marc Belletête ; Jean-Sébastien Lacroix ; Matthew J. Reno

RESONANT AND NON-RESONANT DIELECTRIC COATINGS FOR HIGH EFFICIENCY SOLAR CELLS 1585

Dongheon Ha ; Chen Gong ; Marina S. Leite ; Jeremy N. Munday

ENHANCED LIGHT TRAPPING IN THIN SILICON SOLAR CELLS USING EFFECTIVELY TRANSPARENT CONTACTS (ETCS) 1589

Rebecca Saive ; André Augusto ; Stuart G. Bowden ; Harry A. Atwater

ENHANCED POWER CONVERSION EFFICIENCY IN SINGLE NANOWIRE DEVICES THROUGH SYMMETRY BREAKING DESIGN 1594

Jian Zhou ; Yonggang Wu ; Zihuan Xia ; Xuefei Qin ; Zongyi Zhang

CDSE(TE)/CDS/CDSE RODS VS. CDTE/CDS/CDSE SPHERES: MORPHOLOGY-DEPENDENT CARRIER DYNAMICS FOR PHOTON UPCONVERSION 1598

Eric Y. Chen ; Zhuohui Li ; Christopher C. Milleville ; Kyle R. Lennon ; Matthew F. Doty

DRIFT-DIFFUSION INGAN/GAN SOLAR CELL SIMULATOR WITH OPTICAL MANAGEMENT 1603

Y. Fang ; D. Guo ; A. Fischer ; E. Vadiee ; C. Zhang ; J. Williams ; S. M. Goodnick ; D. Vasileska

PERFORMANCE ENHANCEMENT OF A GAAS SOLAR CELL WITH COLLOIDAL QUANTUM DOTS EMBEDDED IN TRENCHES 1606

Chia-Jhe Shu ; Yu-Ming Huang ; Shun-Chieh Hsu ; Jinn-Kong Shu ; Jia-Lin Tsai ; Pei-Chen Yu ; Yung-Jr Hung ; Chien-Chung Lin

ENHANCED PHOTORESPONSE OF INN DEVICES USING INDIUM-TIN OXIDE NANORODS 1610

Lung-Hsing Hsu ; Yuh-Jen Cheng ; Peichen Yu ; Hao-Chung Kuo ; Chien-Chung Lin

PLASMONIC SILVER STRUCTURES FOR IMPROVED PEROVSKITE PHOTOVOLTAIC PERFORMANCE 1614

Arul Varman Kesavan ; Arun D Rao ; Praveen C Ramamurthy

QUANTUM CUTTING LUMINESCENT PMMA FILMS CONTAINING CE3+ - YB3+ CODOPED YAG PHOSPHOR FOR SI CONCENTRATOR SOLAR CELLS 1619

Lu Li ; Chaogang Lou ; Huihui Cao

NUMERICAL EVALUATION ON THE NANO-ROD ARRAY ON A N-SIDE-UP THIN-FILM GAAS SOLAR CELLS 1623

Po-Ching Wu ; Yan-Zhang Lin ; Shun-Chieh Hsu ; Chia-Jhe Hsu ; Chien-Chung Lin

DOWN SHIFTED CONVERSION FOR ENHANCED HIT SOLAR CELL EFFICIENCY 1627

Albert S. Lin ; Parag Parashar ; Wei-Ming Huang ; Yi-Wen Huang ; Ding-Rung Jian ; Ming-Hsuan Kao ; Shi-Wei Chen ; Chang-Hong Shen ; Jia-Min Shieh ; Tzu-Yu Chen ; Chien-Chung Lin ; Hao-Chung Kuo

THE PLANAR THERMOPHOTOVOLTAIC SELECTIVE NEARLY-PERFECT ABSORBERS/EMITTERS 1631

Parag Parashar ; Ding-Rung Jian ; Weiming Huang ; Vi-Wen Huang ; Albert Lin

HYBRID PEDOT:PSS SILICON SOLAR CELLS WITH PENCIL ROD STRUCTURES 1635
Ruei-Ying Wu ; Liang-Chian You ; Hsin-Fei Meng ; Chun-Chi Chen ; Peichen Yu

PL STUDY OF PHOSPHORUS-DOPED CDTE EVT FILMS 1638
Shamara Collins ; Imran Khan ; Vamsi Evani ; Chih An Hsu ; Vasilios Palekis ; Don Morel ; Chris Ferekides

CHARACTERIZATION OF SINGLE-SOURCE DEPOSITED CLOSE-SPACE SUBLIMATION CDTEXSE1-XTHIN FILM SOLAR CELLS 1643
Corey R. Grice ; Jian Li ; Yanfa Yan

THE INFLUENCE OF THE CU-RICH/CU-POOR SEQUENCE ON THE PROPERTIES OF CU(IN, GA)SE2 FILMS DEPOSITED BY IN-LINE CO-EVAPORATION PROCESS 1648
He Wang ; Fang Fang Liu ; Yi Tong Yang ; Li You Yao ; Peng Gao ; Zhi Bin Xiao ; Qiang Sun

DETERMINATION AND MODELING OF INJECTION DEPENDENT SERIES RESISTANCE IN CIGS SOLAR CELLS 1651
Vito Huhn ; Bart E. Pieters ; Andreas Gerber ; Yael Augarten ; Uwe Rau

LARGE GRAIN GROWTH IN CU2ZNSNS4 THIN FILMS IN THE ABSENCE OF NA USING RAPID THERMAL ANNEALING 1656
J. L. Johnson ; A. Bhatia ; J. G. Bolke ; M. A. Scarpulla

CU2ZNSNS4THIN FILMS SYNTHESIZED BY COSPUTTERING AND RAPID THERMAL ANNEALING: EFFECTS OF COMPOSITION AND TEMPERATURE 1661
J.L. Johnson ; W.M. Hlaing Oo ; M. Karmarkar ; M.A. Scarpulla

EARTH-ABUNDANT CZTSSE THIN FILM SOLAR CELLS ON FLEXIBLE STAINLESS STEEL FOIL SUBSTRATES 1665
Hae-Sun Kim ; Woo-Lim Jeong ; Dong-Seon Lee

COMPARISON OF MGCL2AND CDCL2ACTIVATION TREATMENT FOR CDTE SOLAR CELLS: RECRYSTALLIZATION AND DEFECTS 1669
Daniele Menossi ; Elisa Artegiani ; Ivan Rimmaudo ; Alessia Le Donne ; Simona Binetti ; Juan Luis Pena ; Fabio Piccinelli ; Alessandro Romeo

CHARACTERIZATION OF CDTE PHOTOVOLTAIC DEVICES PASSIVATED USING HYDROGEN PLASMA 1674
Amit Munshi ; Piotr Kaminski ; Ali Abbas ; Shiva Tarun Chenna ; Sreeram Chandralal ; John Walls ; Walajabad Sampath

GROUP-V DOPING IMPACT ON CD-RICH CDTE SINGLE CRYSTALS GROWN BY TRAVELING-HEATER METHOD 1679
Akira Nagaoka ; Kenji Yoshino ; Yoshitaro Nose ; Darius Kuciauskas ; Michael A. Scarpulla

BAND-GAP ENGINEERING IN CU2ZNSN(S,SE)4SOLAR CELLS BY POST-SULPHURIZATION OF SELENIZED ABSORBER LAYERS 1682
Markus Neuwirth ; Elisabeth Seydel ; Heinz Kalt ; Michael Hetterich

IMPACT OF GA/III PROFILE ON VOLTAGE-DEPENDENT COLLECTION LOSSES IN CIGS SOLAR CELLS 1686
Dmitry Poplavskyy ; Jeff Bailey ; Rouin Farshchi ; David Spaulding

CL DIFFUSION IN CDTE SOLAR CELLS ACTIVATED BY GASEOUS CHCLF2ATMOSPHERE 1691
I. Rimmaudo ; R. Mis Fernandez ; V. Rejon ; A. Abbas ; F. Lisco ; J.M. Walls ; J.L. Peña

STABILITY OF CD1-XZNXTE ALLOYS UNDER CDTE PROCESSING CONDITIONS 1697
Yegor Samoilenko ; Colin A. Wolden

CIGSE ABSORBER PREPARATION: AN ALTERNATIVE TO H2SE 1701
O.S. Shinde ; E.J. Schenller ; S.R. Jadkar ; S.V Ghaisas ; N. Dhere

CHARGE CONTROLLED SEQUENTIAL ELECTRODEPOSITION FOR SYNTHESIS OF CU2ZNSNS4ON MO-COATED GLASS SUBSTRATE 1704
Ashish K. Singh ; Rajiv Dubey ; Manoj Neergat ; Kavaipatti R. Balasubramaniam

EFFECT OF DEPOSITED PRESSURE ON THE CDTE THIN FILMS BY CLOSED SPACE SUBLIMATION METHOD 1707
Yufeng Zhang ; Zhongming Du ; Xiangxin Liu

ANALYZING THE COST REDUCTION POTENTIAL OF III-V/SI HYBRID CONCENTRATOR PHOTOVOLTAIC SYSTEMS 1711
Kan-Hua Lee ; Kenji Araki ; Masafumi Yamaguchi

GENERALIZED NUMERICAL DESIGN OF AXIALLY-ASYMMETRICAL AND GRID-ARRANGED STATIC CPV ARRAY FOR MAXIMIZING ANNUAL ENERGY GENERATION 1714
Kenji Araki ; Kan-Hua Lee ; Masafumi Yamaguchi

SPECTRAL TRANSMITTANCE ANALYSIS OF LIQUIDS FOR HIGH CONCENTRATION III-V PHOTOVOLTAIC IMMERSION COOLING APPLICATIONS 1719
Xinyue Han ; Yongjie Guo

OPTICAL DESIGN FOR 2-TERMINAL III-V/SI SMAC MODULE 1724
Masaaki Baba ; Kikuo Makita ; Hidenori Mizuno ; Hidetaka Takato ; Takeyoshi Sugaya ; Noboru Yamada

DESIGN OF OPTICAL ELEMENTS FOR LOW PROFILE CPV PANEL WITH SUN TRACKING FOR ROOFTOP INSTALLATION............1728

Xinbing Liu ; Zhou Lu ; Riccardo Leto ; Carlton Brule ; Nanu Brates

MICRO CHIPLET PRINTER DEVELOPMENT FOR MOSAIC PROGRAM............1733

P.Y. Maeda ; Y. D. Wang ; S. Raychaudhuri ; J. Kalb ; D. K. Biegelsen ; R. Lujan ; Q. Wang ; Y. Wang ; J. Bert ; B. Rupp ; I. Matei ; L. Crawford ; A. Plochowietz ; E.M. Chow ; J.P. Lu ; V. Gupta

MICRO-OPTICAL TANDEM LUMINESCENT SOLAR CONCENTRATOR............1737

David R. Needell ; Zach Nett ; Ognjen Ilic ; Colton R. Bukowsky ; Junwen He ; Lu Xu ; Ralph G. Nuzzo ; Benjamin G. Lee ; John F. Geisz ; A. Paul Alivisatos ; Harry A. Atwater

INCREASE IN MAXIMUM POWER OF A-SI, C-SI AND GAAS.76P.24 SOLAR CELLS UNDER LOW CONCENTRATION............1741

Hiba Riaz ; Sabina Abdul Hadi ; Ammar Nayfeh

DESIGN AND EVALUATION OF PARTIAL CONCENTRATION III-V/SI MODULE WITH ENHANCED DIFFUSE SUNLIGHT TRANSMISSION............1743

Daisuke Sato ; Noboru Yamada ; Kan-Hua Lee ; Kenji Araki ; Masafumi Yamaguchi

CONTAMINATION CONTROL CHALLENGES ON SHJ SOLAR CELL PROCESSING............1747

G. Condorelli ; P. Rotoli ; A. Canino ; A. Battaglia ; W. Favre ; A. -S. Ozanne ; A. Moustafa ; A. Danel ; D. Muñoz ; P. -J. Ribeyron ; C. Gerardi

>23% SILICON HETEROJUNCTION SOLAR CELLS IN MEYER BURGER'S DEMO LINE: RESULTS OF PILOT PRODUCTION ON MASS PRODUCTION TOOLS............1752

J. Zhao ; M. König ; A. Wissen ; V. Breus ; D. Deckerl ; M. Fritzsche ; M. Schorch ; H. J. Nonnenmacher ; M. Leonhardt ; T. Große ; J. Hausmann ; A. Waltmger ; D. Landgraf ; S. Burkhardt ; H. Mehlich ; E. Vetter ; F. Schitthelm ; Y. Yao ; T. Söderström ; A. Richter ; D. Habermann ; S. Leu

EXPERIMENTAL AND SIMULATION STUDIES ON TIO2/SILICON HETEROJUNCTION DIODES............1755

Swasti Bhatia ; Neha Raorane ; Nimisha Sreekumar ; Pradeep R. Nair ; Aldrin Antony

A STUDY ON BLISTER FORMATION AND ELECTRICAL PROPERTIES UNDER VARIOUS ANNEALING CONDITION FOR TUNNELING OXIDE PASSIVATION LAYER............1758

Sungjin Choi ; Ka-Hyun Kim ; Min Gu Kang ; Jeong In Lee ; Donghwan Kim ; Hee-Eun Song

PROCESSING APPROACHES AND CHALLENGES OF INTERDIGITATED BACK CONTACT SI SOLAR CELLS............1761

Ujjwal Das ; Lei Zhang ; Steven Hegedus

FABRICATION OF CUI/A-SI:H/C-SI STRUCTURE FOR APPLICATION TO HOLE-SELECTIVE CONTACTS OF HETEROJUNCTION SI SOLAR CELLS............1765

Kazuhiro Gotoh ; Min Cui ; Nguyen Cong Thanh ; Koichi Koyama ; Isao Takahashi ; Yasuyoshi Kurokawa ; Hideki Matsumura ; Noritaka Usami

CHARACTERISTICS OF THIN CRYSTALLINE SILICON SOLAR CELLS WITH RIB STRUCTURE............1769

Yukimi Ichikawa ; Shuhei Yoshiba ; Masakazu Hirai ; Makoto Konagai

MEASUREMENT OF TIO2/P-SI SELECTIVE CONTACT PERFORMANCE USING A HETEROJUNCTION BIPOLAR TRANSISTOR WITH A SELECTIVE CONTACT EMITTER............1773

Janam Jhaveri ; Alexander Berg ; Sigurd Wagner ; James C. Sturm

EFFECT OF GROWTH AND POST-OXIDATION ANNEALING TEMPERATURE OF THERMALLY GROWN TUNNELING SIOX, ON THE IIMPLIED VOCOF PASSIVATED CONTACTS FOR C-SI BASED SOLAR CELLS............1777

Abhijit S. Kale ; William Nemeth ; Matthew Page ; Sumit Agarwal ; Paul Stradins

PARTIALLY CONTACTED SURFACES WITH CONTACT SIZE IN THE 1 μM RANGE FOR C-SI PERC SOLAR CELLS............1781

R. Khoury ; I. Martín ; G. López ; C. Jin ; J.M. López-González ; L. Zeyu ; P. Bulkin ; E.V. Johnson ; R. Alcubilla

ENTRANCE OF LOW COST FABRICATION OF BACK-CONTACT HETEROJUNCTION SOLAR CELLS BY USING PLASMA ION IMPLANTATION............1787

Koichi Koyama ; Keisuke Ohdaira ; Hideki Matsumura

TLM MEASUREMENTS VARYING THE INTRINSIC A-SI:H LAYER THICKNESS IN SILICON HETEROJUNCTION SOLAR CELLS............1790

Mehdi Leilaeioun ; William Weigand ; Pradyumna Muralidharan ; Mathieu Boccard ; Dragica Vasileska ; Stephen Goodnick ; Zachary Holman

SOLAR CELLS APPLICATION OF P-TYPE POLY-SI THIN FILM BY ALUMINUM INDUCED CRYSTALLIZATION............1794

Shota Masuda ; Kazuhiro Gotoh ; Isao Takahashi ; Kyotaro Nakamura ; Yoshio Ohshita ; Noritaka Usami

A SELF - CONSISTENTLY COUPLED DRIFT DIFFUSION AND MONTE CARLO SIMULATOR TO MODEL SILICON HETEROJUNCTION SOLAR CELLS............1797

Pradyumna Muralidharan ; Stuart Bowden ; Stephen M. Goodnick ; Dragica Vasileska

DOPANT PATTERNING BY PECVD AND MECHANICAL MASKING FOR PASSIVATED TUNNELING CONTACT IBC CELL ARCHITECTURES 1801

William Nemeth ; Vincenzo Lasalvia ; Benjamin G. Lee ; Abhijit Kale ; Paul Stradins

ALD ALUMINUM OXIDE AS A HOLE SELECTIVE TUNNELING CONTACT FOR CRYSTALLINE SILICON SOLAR CELLS 1804

Kortan Ögütman ; Kristopher O. Davis ; Winston V. Schoenfeld ; Michael Haslinger ; Sofie Robert ; Emanuele Cornagliotti ; Joachim John

SCREEN PRINTED, LARGE AREA BIFACIAL N-PERT CELLS WITH TUNNEL OXIDE PASSIVATED BACK CONTACT 1807

Young-Woo Ok ; Ajay D Upadhyaya ; Brian Rounsaville ; Ying-Yuan Huang ; Vijaykumar D Upadhyaya ; Ajeet Rohatgi

CORRELATION BETWEEN ELECTROLUMINESCENCE AND PHOTOCONVERSION EFFICIENCY IN A-SI:H/C-SI HETEROJUNCTION SOLAR CELLS 1811

A.V. Sachenko ; A.V. Bobyl ; V.N. Verbitskiy ; V.M. Vlasyuk ; D.M. Zhigunov ; V.P. Kostylyov ; I.O. Sokolovskyi ; E.I. Terukov ; P.A. Forsh ; M. Evstigneev

AN ISOTOPE STUDY OF HYDROGEN PASSIVATION OF POLY-SI/SIOXPASSIVATED CONTACTS FOR SI SOLAR CELLS 1817

Manuel Schnabel ; William Nemeth ; Bas W.H. Van De Loo ; Bart Macco ; Wilhelmus M.M. Kessels ; Paul Stradins ; David L. Young

ALLEVIATING HYDROGEN PLASMA DAMAGE TO AMORPHOUS/CRYSTALLINE SILICON INTERFACE PASSIVATION 1820

Jianwei Shi ; Zachary C. Holman

LARGE-AREA N-TYPE TOPCON CELLS WITH SCREEN-PRINTED CONTACT ON SELECTIVE BORON EMITTER FORMED BY WET CHEMICAL ETCH-BACK 1824

Yuguo Tao ; Felix Book ; Barbara Terheiden ; Viiaykumar Upadhvaya ; Keeya Madani ; Brian Rounsaville ; Eunhwan Cho ; Ajeet Rohatgi

HYDROGEN PLASMA POST-DEPOSITION TREATMENT FOR PASSIVATION OF A-SI/C-SI INTERFACE FOR HETEROJUNCTION SOLAR CELL BY CORRELATING OPTICAL EMISSION SPECTROSCOPY AND MINORITY CARRIER LIFETIME 1828

Anishkumar Soman ; Ugochukwu Nsofor ; Lei Zhang ; Ujjwal Das ; Tingyi Gu ; Steve Hegedus

MEASURING DIODE RESISTIVITY OF PASSIVATED CONTACTS 1832

San Theingi ; William Nemeth ; David L. Young ; Paul Stradins ; Benjamin G. Lee

ULTRA-THIN CRYSTALLINE SILICON SOLAR CELLS WITH NICKEL OXIDE INTERLAYER AS HOLE-SELECTIVE CONTACT 1835

Muyu Xue ; Raisul Islam ; Junyan Chen ; Zheng Lyu ; Yusi Chen ; Daniel Dewitt ; Albert Pleus ; Christian Tae ; Ching-Ying Lu ; Kai Zang ; Jieyang Jia ; Yijie Huo ; Ted Kamins ; Krishna Saraswat ; James Harris

CRYSTALLINE SI SOLAR CELLS WITH PASSIVATING, CARRIER-SELECTIVE NICKEL OXIDE CONTACTS 1838

Woojun Yoon ; James Moore ; David Scheiman ; Eunhwan Cho ; Young-Woo Ok ; Nicole Kotulak ; Phillip P. Jenkins ; Ajeet Rohatgi ; Robert J. Walters

GAP/SI HETEROJUNCTION SOLAR CELLS GROWN BY MOLECULAR BEAM EPITAXY 1841

Chaomin Zhang ; Ehsan Vadiee ; Richard R. King ; Christiana B. Honsberg

SPIN COATED NICKEL OXIDE AND VANADIUM OXIDE LAYERS ON SILICON FOR A CARRIER SELECTIVE CONTACT SOLAR CELL 1845

Jing Zhao ; Fa-Jun Ma, Jae-Yun ; Anita Ho-Baillie ; Stephen Bremner

QUANTIFICATION OF PV MODULE DISCOLORATION USING VISUAL IMAGE ANALYSIS 1850

Shashwata Chattopadhyay ; Chetan Singh Solanki ; Anil Kottantharayil ; K.L. Narasimhan ; Juzer Vasi ; Sai Tatapudi ; Govindasamy Tamizhmani

TEMPERATURE AND POWER STUDY OF ADHERED AND RACKED DOUBLE GLASS PHOTOVOLTAIC MODULES 1855

Volker Beutner ; Rubina Singh ; Cameron Stark

FIELD INSPECTION OF PV MODULES: QUANTITATIVE DETERMINATION OF PERFORMANCE LOSS DUE TO CELL CRACKS USING EL IMAGES 1858

Carlos A. Rodríguez Castañeda ; Shashwata Chattopadhyay ; Jaewon Oh ; Sai Tatapudi ; Govindasamy Tamizhmani ; Hailin Hu

SCALE UP DESIGNS FOR HAND-HELD LIGHT-WEIGHT TPV DC POWER SUPPLY 1863

L. M. Fraas ; J. E. Avery ; L. Minkin ; Hui She ; L. Ferguson

HIGH EFFICIENCY ANTI-REFLECTIVE COATING FOR PV MODULE GLASS 1869

Brennen M. Freiburger ; Corey S. Thompson ; Robert A. Fleming ; Douglas Hutchings ; Sergiu C. Pop

INVESTIGATION OF EFFICIENCY FOR PID-AFFECTED SOLAR MODULE AT NONSTANDARD TEST CONDITIONS 1873

Shuwen Guo ; Pan Zhao ; Weijing Huang ; Jipeng Chang ; He Wang ; Hong Yang ; Chengfeng Su ; Bojie Su ; Xue Zhang ; Yunxue Cao ; Hui Zhao

THERMAL UNIFORMITY MAPPING OF PV MODULES AND PLANTS..........1877
Ashwini Pavgi ; Jaewon Oh ; Joseph Kuitche ; Sai Tatapudi ; Govindasamy Tamizhmani

CLIMATE-SPECIFIC THERMAL MODEL COEFFICIENTS FOR C-SI AND THIN-FILM PV MODULES..........1883
Ashwini Pavgi ; Joseph Kuitche ; Jaewon Oh ; Govindasamy Tamizhmani

EFFECT OF THE THERMOPHYSICAL PROPERTIES OF A PHASE CHANGE MATERIAL ON THE ELECTRICAL OUTPUT OF A CONCENTRATED PHOTOVOLTAIC SYSTEM..........1888
Jawad Sarwar ; Ahmed E. Abbas ; Konstantinos E. Kakosimos

PASSIVE COOLING OF PHOTOVOLTAICS WITH DESICCANTS..........1893
Lin J. Simpson ; Jason Woods ; Nicolas Valderrama ; Alex Hill ; Nina Vincent ; Timothy Silverman

MODIFIED MAXIMUM POWER EXTRACTION TECHNIQUE FOR RAPIDLY CHANGING NUI AND DYNAMIC LOADS..........1898
U Aswani ; S.P. Duttagupta ; T.I. Eldho ; B.V. Rao

REAL-TIME MONITORING OF PHOTO VOLTAIC RELIABILITY ONLY USING MAXIMUM POWER POINT - THE SUNS-VMP METHOD..........1904
Xingshu Sun ; Haejun Chung ; Raghu Vamsi Krishna Chavali ; Peter Bermel ; Muhammad Ashraful Alam

PHOTOVOLTAIC MODULE DURABILITY AND RELIABILITY: ANALYSIS OF A 23-YEAR-OLD ARRAY OPERATING IN QUEBEC, CANADA..........1908
Christopher Baldus-Jeursen ; Alexandre Côté ; Naveen Goswamy ; Tanya Deer ; Yves Poissant

ARE E-W TRACKERS A BETTER OPTION FOR FUTURE INVESTMENTS IN PV SECTOR-A DETAILED TECHNO-COMMERCIAL STUDY..........1912
Rakesh Bohra ; Ramesh Rame Gowda ; Mani R. Krishnan

EXPERIMENTAL EVALUATION OF THE PERFORMANCE OF CRYSTALLINE SI PV MODULE DEGRADATION AFTER 15-YEARS OF FIELD EXPOSURE..........1917
Denio A. Cassini ; Antonia Sônia A. C. Diniz ; Marcelo Machado Viana ; Michele C. C. De Oliveira ; F. C. Lins Vanessa De ; Roberto Zilles ; Lawrence L. Kazmerski

FIELD INVESTIGATIONS OF POTENTIAL-INDUCED DEGRADATION (PID) FOR CRYSTALLINE SILICON PV PANELS IN DIFFERENT CLIMATES..........1922
Yifeng Chen ; Peter Hacke ; Yong Sheng Khoo ; Kaitlyn Vansant ; Zigang Wang ; Wei Luo ; Jing Chai ; Chris Deline ; Yan Wang ; Armin G. Aberle ; Pietro P. Altermatt ; Zhiqiang Feng ; Sarah Kurtz ; Pierre J. Verlinden

DETERMINING THE POWER RATE OF CHANGE OF 353 PLANT INVERTERS TIME-SERIES DATA ACROSS MULTIPLE CLIMATE ZONES, USING A MONTH-BY-MONTH DATA SCIENCE ANALYSIS..........1927
Alan J. Curran ; Yang Hu ; Rojiar Haddadian ; Jennifer L. Braid ; David Meakin ; Timothy J. Peshek ; Roger H. French

PHOTOVOLTAIC ARRAY DIFFERENTIAL BACKSIDE EXPOSURE CONDITIONS: BACKSHEET DEGRADATION AND SITE DESIGN..........1933
Andrew Fairbrother ; Julien Avenet ; Yadong Lyu ; Matthew Boyd ; Scott Julien ; Kai-Tak Wan ; Liang Ji ; Kenneth Boyce ; Sebastien Merzlic ; Amy Lefebvre ; Greg O'Brien ; Yu Wang ; Laura Bruckman ; Roger French ; Michael Kempe ; Brian Dougherty ; Xiaohong Gu

STUDY ON RANDOM FAILURE OF CRYSTALLINE SILICON SOLAR MODULES IN THE FIELD..........1937
Xuefang Jiang ; Fumei Wang ; Ao Wang ; Hong Yang ; He Wang ; Jie Ding ; Junjun Zhang ; Jingsheng Huang

POTENTIAL INDUCED DEGRADATION (PID) POWER LOSS CORRELATION TO LEAKAGE AND REVERSE BIAS CURRENTS..........1941
Michalis Florides ; Georgios Konstantinou ; Venizelos Venizelou ; George Makrides ; George E. Georghiou

PERFORMANCE STUDY OF VARIOUS PV MODULE TECHNOLOGIES IN DESERT CONDITIONS..........1946
Jim J John ; Ammar Elnosh ; Anwar Almheiri ; Wadhah Alzahmi ; Marco Stefancich ; Pedro Banda

HIGH-SPEED MEASUREMENTS OF GENERATED POWER AND ITS RELATIONSHIP TO WEATHER OBSERVATIONS AT YOSHINOGARI MEGA SOLAR POWER PLANT..........1950
Makoto Kasu ; Shigeomi Hara ; Takumi Uematsu

IMPACT OF MISSING DATA ON THE ESTIMATION OF PHOTOVOLTAIC SYSTEM DEGRADATION RATE..........1954
Andreas Livera ; Alexander Phinikarides ; George Makrides ; George E. Georghiou

FIELD DEGRADATION AND FAILURES OF AGED CRYSTALLINE SILICON PV MODULES IN MEXICO..........1959
D. Martínez Escobar ; P. A. Sánchez-Pérez ; Rocío De La Luz Santos Magdaleno ; José Ortega Cruz ; Sai Tatapudi ; Aarón Sánchez Juárez ; Govindasamy Tamizhmani

RAPID SHUTDOWN WITH PANEL LEVEL ELECTRONICS-A SUITABLE SAFETY MEASURE?..........1965
Adam Cordova ; Christopher Merz ; Gerd Bettenwort ; Markus Hopf ; Hannes Knopf ; Joachim Laschinski

INVESTIGATING A NEW OPERATING POINT FOR PV PANELS SEEKING MAXIMUM LIFE SPAN ... 1968

Bechara Nehme ; Nacer K. M'sirdi ; Tilda Akiki

POWER GENERATION EVALUATION OF LARGE-SCALE PHOTOVOLTAIC SYSTEMS LOCATED ON INCLINED PLANE ... 1973

Naotaka Oka ; Yasuhito Takahashi ; Koji Fujiwara ; Kazuyuki Hidaka ; Hiroshi Morita

INVESTIGATING THE IMPACT OF SOLAR CELLS PARTIAL SHADING ON PHOTOVOLTAIC MODULES BY THERMOGRAPHY ... 1979

David Pera ; José A. Silva ; Sara Costa ; João M. Serra

ANNUAL DEGRADATION RATE AND ITS LINEARITY ANALYSIS USING METERED KWH DATA ... 1984

Christopher Raupp ; Govindasamy Tamizhmani

ELECTRICAL PERFORMANCE ANALYSIS OF A 27 KW GRID-CONNECTED PV SYSTEM WITH SOILING AND SHADING IN MORELOS MEXICO ... 1990

P. A. Sánchez-Pérez ; D. Martínez Escobar ; E. O. Ángel Ruiz ; R. Santos Magdaleno ; José Ortega Cruz ; A. Sánchez Juárez

MODIFIED STC CORRECTION PROCEDURE FOR ASSESSING PV MODULE DEGRADATION IN FIELD SURVEYS ... 1995

Hemant K. Singh ; R. Dubey ; S. Zachariah ; K. L. Narasimhan ; B. M. Arora ; A. Kottantharayil ; J. Vasi

DEGRADATION MODELS OF PHOTOVOLTAIC MODULE BACKSHEETS EXPOSED TO DIVERSE REAL WORLD CONDITION ... 2000

Yu Wang ; Sebastien Merzlic ; Andrew Fairbrother ; Scott Julien ; Lucas Fridman ; Camille Loyer ; Amy L. Lefebvre ; Gregory O'Brien ; Xiaohong Gu ; Liang Ji ; Ken Boyce ; Michael Kempe ; Kai-Tak Wan ; Roger H. French ; Laura S. Bruckman

ADDRESSING HOTSPOTS IN THE PRODUCT ENVIRONMENTAL FOOTPRINT OF CDTE PHOTOVOLTAICS ... 2005

Parikhit Sinha ; Andreas Wade

PHOTOVOLTAIC SMART HOME SYSTEM - DUBAI CASE STUDY ... 2011

Ammar Natsheh ; Marwa Aljaziri ; Maitha Moosa ; Gharibah Essa ; Hassa Moosa

DIRECT DRIVE PHOTOVOLTAIC MILK CHILLING EXPERIENCE IN KENYA ... 2014

Robert Foster ; Brian Jensen ; Brian Dugdill ; Wendy Hadley ; Bruce Knight ; Abudul Faraj ; Johnson Kyalo Mwove

COST OPTIMIZATION OF DECOMMISSIONING AND RECYCLING CDTE PV POWER PLANTS ... 2019

V. Fthenakis ; Z. Zhang ; J. -K Choi

CHALLENGES FOR DECISION MAKERS WHEN FEED-IN TARIFFS OR NET METERING SCHEMES CHANGE TO INCENTIVES DEPENDENT ON A HIGH SHARE OF SELF-CONSUMED ELECTRICITY ... 2025

Mattias Gustafsson

PROCEDURES TO MAKE PROJECTS ABOUT RENEWABLE ENERGY GENERATION CONNECTED TO THE GRID IN COLOMBIA ... 2031

J. A. Hernandez ; C. A. Arredondo ; D. J. Rodriguez

A CRITICAL ANALYSIS ON THE THIN CRYSTALLINE SILICON PV MODULE OF THE LIGHTWEIGHT PV SYSTEM ... 2035

Meixi Chen ; Abhishek Iyer ; Cheng-Hao Shih ; Lado Kurdgelashvili ; Robert Opila

PHOTOVOLTAIC MODULE MANUFACTURING COSTS, AVERAGE PRICES AND INDUSTRY BALANCE 2006–2016 ... 2039

Paula Mints ; Zhengshan J Yu

SOLAR CELL AND WIND ENERGY REPLACEMENT OF POWER PLANTS GLOBALLY ... 2042

Larry Partain ; Shirley Hansen ; Dirk Bennett ; Richard Hansen ; Allan Newlands ; Lewis Fraas

ANALYSIS OF LIGHT ENVIRONMENT UNDER SOLAR PANELS AND CROP LAYOUT ... 2048

Deng Wang ; Yaojie Sun ; Yandan Lin ; Yuan Gao

INTERFACE EFFECTS OF ALKALI TREATMENT ON CU-RICH THIN FILM SOLAR CELLS ... 2054

Hossam Elanzeery ; Finn Babbe ; Anastasiya Zelenina ; Michele Melchiorre ; Susanne Siebentritt

INCREASEDVOCAND FF IN ZNO1-XSX-BUFFERED CUIN1-XGAXSE2SOLAR CELLS BY CADMIUM PARTIAL ELECTROLYTE TREATMENT ... 2058

Andreas Bauer ; Dimitrios Hariskos ; Wiltraud Wischmann

PASSIVATING AND CARRIER-SELECTIVE CONTACTS - BASIC REQUIREMENTS AND IMPLEMENTATION ... 2064

S.W. Glunz ; M. Bivour ; C. Messmer ; F. Feldmann ; R. Müller ; C. Reichel ; A. Richter ; F. Schindler ; J. Benick ; M. Hermle

FIRST-PRINCIPLES MODELING OF ALKALI METAL POST DEPOSITION TREATMENT EFFECTS IN CIGS SOLAR CELLS..2070
Maria Fedina ; Hannu-Pekka Komsa ; Ville Havu ; Martti J. Puska

EXPLORING SILICON CARBIDE- AND SILICON OXIDE-BASED LAYER STACKS FOR PASSIVATING CONTACTS TO SILICON SOLAR CELLS..2073
P. Löper ; G. Nogay ; P. Wyss ; M. Hyvl ; P. Procel ; J. Stuckelberger ; A. Ingenito ; I. Mack ; Q. Jeangros ; M. Ledinsky ; A. Fejfar ; C. Allebé ; J. Horzel ; M. Despeisse ; F. Crupi ; F.-J. Haug ; C. Ballif

EFFICIENT ELECTRON CONTACTS FORN-TYPE SILICON SOLAR CELLS USING MAGNESIUM METAL, OXIDE, AND FLUORIDE..2076
Yimao Wan ; Chris Samundsett ; James Bullock ; Di Yan ; Thomas Allen ; Jun Peng ; Jie Cui ; Mark Hettick ; Ali Javey ; Andres Cuevas

GRADED (ALZGA1-Z)XIN1-XP WINDOW-EMITTER STRUCTURES FOR IMPROVED SHORT-WAVELENGTH RESPONSE..2079
Jacob T. Boyer ; Daniel L. Lepkowski ; Daniel J. Chmielewski ; Steven A. Ringel ; Tyler J. Grassman

INTEGRATION OF QUANTUM DOTS AND QUANTUM WELLS INTO INGAAS METAMORPHIC SUBCELL FOR RADIATION HARD 3-J ELO IMM PHOTOVOLTAICS......................2084
Zachary S. Bittner ; Hyun Kum ; Michael A. Slocum ; George T. Nelson ; Rao Tatavarti ; Andre Wibowo ; Seth M. Hubbard

PROTON IRRADIATION OF 3J SOLAR CELLS AT LOW TEMPERATURE.................................2087
Seonyong Park ; Jacques C. Bourgoin ; Olivier Cavani ; Sandrine Picard ; Jérôme Bourcois ; Victor Khorenko ; Carsten Baur ; Bruno Boizot

ULTRA-THIN GAAS SOLAR CELLS: RADIATION TOLERANCE AND SPACE APPLICATIONS..............2091
Louise C. Hirstl ; Michael K. Yakes ; Jeffery. H. Warner ; Mitchell F. Bennett ; Kenneth J. Schmieder ; Stephanie Tomasulo ; Erin Cleveland ; Sergey Maximenko ; James Moore ; Robert J. Walters ; Phillip P. Jenkins

LARGE AREA MULTIJUNCTION III-V SPACE SOLAR CELLS OVER 31% EFFICIENCY..........................2094
X.Q. Liu ; C. Fetzer ; P. Chiu ; M. Haddad ; X. Zhang ; R. Cravens ; D. Law ; J. Ermer ; J. Krogen ; S. Sharma ; J. Hanley

ADVANCED-ARCHITECTURE HIGH-EFFICIENCY SOLAR CELLS FOR LOW IRRADIANCE LOW TEMPERATURE (LILT) APPLICATIONS..2099
Andreea Boca ; Jonathan Grandidier ; Claiborne Mcpheeters ; Paul Sharps ; Philip Chiu ; Xing-Quan Liu ; James Ermer

ULTRA-LIGHTWEIGHT PV MODULE DESIGN FOR BUILDING INTEGRATED PHOTOVOLTAICS..2104
Ana C. Martins ; Valentin Chapuis ; Alessandro Virtuani ; Christophe Ballif

DESIGN IT WITH LSCS; AN EXPLORATION OF APPLICATIONS FOR LUMINESCENT SOLAR CONCENTRATOR PV TECHNOLOGIES..2109
Wouter Eggink ; Angèle Reinders

INVESTIGATING PV-BATTERY 3-TERMINAL INTEGRATION CONCEPT AS A SELF-SUSTAINING POWER SOLUTION..2114
Solomon N. Agbo ; Oleksandr Astakhov ; Uwe Rau ; Tsvetelina Merdzhanova

PERFORMANCE ASSESSMENT OF A BIPV ROOFING TILE IN OUTDOOR TESTING.................2118
Cristina S. Polo Lopez ; Pierluigi Bonomo ; Francesco Frontini ; Vasco Medici ; Lorenzo Nespoli

LIFE CYCLE ASSESSMENT OF TRANSPARENT ORGANIC PHOTOVOLTAIC FOR WINDOW APPLICATIONS..2124
Annick Anctil ; Eunsang Lee ; Jack Stephan ; Anjali Munasinghe ; Christopher Traverse ; Richard R. Lunt

A REDUCED ORDER MODEL FOR A TOV STUDY IN A SOLAR PV PROJECT............................2128
Ahmad Abdullah ; Billy Yancey

CYBER SECURITY ASSESSMENT OF DISTRIBUTED ENERGY RESOURCES..............................2135
Cedric Carter ; Ifeoma Onunkwo ; Patricia Cordeiro ; Jay Johnson

EVALUATION OF FAST-FREQUENCY SUPPORT FUNCTIONS IN HIGH PENETRATION ISOLATED POWER SYSTEMS..2141
Mohamed Elkhatib ; Jason Neely ; Jay Johnson

LOSS OF UTILITY DETECTION CAPABILITIES FOR TODAY'S UTILITY INTERCONNECTED PHOTOVOLTAIC INVERTERS..2147
Sigifredo Gonzalez ; Gregory Kern ; Michael Ropp

PARAMETRIC PV GRID-SUPPORT FUNCTION CHARACTERIZATION FOR SIMULATION ENVIRONMENTS..2153
Javier Hernandez-Alvidrez ; Jay Johnson

COST ANALYSIS AND COST REDUCTION OPPORTUNITIES OF RESIDENTIAL PV SYSTEM IN THE JAPAN...2159
Izumi Kaizuka ; Haruki Yamaya ; Takashi Ohigashi ; Risa Kurihara ; Osamu Ikki

SUPPLY AND DEMAND CONSTRAINTS ON FUTURE PV POWER IN THE USA..........................2163
Paul A. Basore ; Wesley J. Cole

RESIDENTIAL PHOTOVOLTAIC ELECTRICITY GENERATION IN THE EUROPEAN UNION 2017-OPPORTUNITIES AND CHALLENGES 2167
Arnulf Jäger-Waldau ; Thomas Huld ; Sandor Szabo

INVESTIGATING NANOSCALE DETERMINANTS OF CHARGE COLLECTION IN QUASI-2D PEROVSKITE SOLAR CELLS 2170
Yanqi Luo ; Xueying Li ; Bat-El Cohen ; Barry Lai ; Lioz Etgar ; David P Penning

RECENT DEVELOPMENTS OF SOLAR PHOTOVOLTAIC SYSTEMS IN INDIA 2172
Saravanan Vasudevan ; Arumugam Murugesan

OPERANDO X-RAY DIFFRACTION FOR CHARACTERIZATION OF PHOTOVOLTAIC MATERIALS 2176
Laura T Schelhasl ; Jeffrey A. Christians ; Joseph J. Berry ; Michael F. T Oney ; Christopher J. Tassone ; Joseph M. Luther ; Kevin H. Stone

X-RAY BEAM INDUCED VOLTAGE: A NOVEL TECHNIQUE FOR ELECTRICAL NANOCHARACTERIZATION OF SOLAR CELLS 2179
Michael E. Stuckelberger ; Tara Nietzold ; Bradley M. West ; Barry Lai ; Jörg M. Maser ; Volker Rose ; Mariana I. Bertoni

ELECTRO-LUMINESCENT REFRIGERATION ENABLED BY HIGHLY EFFICIENT PHOTOVOLTAICS 2185
T. Patrick Xiao ; Kaifeng Chen ; Parthiban Santhanam ; Shanhui Fan ; Eli Yablonovitch

MULTIPLE QUANTUM WELLS AS SLOWED HOT CARRIER COOLING ABSORBERS IN HOT CARRIER CELLS 2186
Gavin Conibeer ; Yi Zhang ; Simon Chung ; Yuaxun Liao ; Stephen Bremner ; Santosh Shrestha

QUANTITATIVE OPTOELECTRONIC MEASUREMENTS OF CARRIER THERMODYNAMICS PROPERTIES IN QUANTUM WELL HOT CARRIER SOLAR CELL 2192
Dac-Trung Nguven ; Laurent Lombez ; François Gibelli ; Soline Boyer-Richard ; Alain Le Corre ; Olivier Durand ; Jean-François Guillemoles

ABSORPTION ENHANCEMENT IN INGAASP/INGAP QUANTUM WELL SOLAR CELLS 2195
Islam E.H. Sayed ; Nikhil Jain ; Myles A. Steiner ; John F. Geisz ; Salah M. Bedair

CARRIER COLLECTION MODEL AND DESIGN RULE FOR QUANTUM WELL SOLAR CELLS 2201
Kasidit Toprasertpong ; Boram Kim ; Yoshiaki Nakano ; Masakazu Sugiyama

INFLUENCE OF CONDUCTION BAND OFFSETS AT WINDOW/BUFFER AND BUFFER/ABSORBER INTERFACES ON THE ROLL-OVER OF J-V CURVES OF CIGS SOLAR CELLS 2205
Giovanna Sozzi ; Simone Di Napoli ; Roberto Menozzi ; Florian Werner ; Susanne Siebentritt ; Philip Jackson ; Wolfram Witte

OVERVIEW OF SURFACE PASSIVATION SCHEMES FOR THIN FILM SOLAR CELLS 2209
Ratan Kotipalli ; Bart Vermang

TOWARDS 10% STATE-OF-THE-ART PURE SULFIDE CU2ZNSNS4 SOLAR CELL BY MODIFYING THE INTERFACE CHEMISTRY 2213
Kaiwen Sun ; Jialiang Huang ; Steve Johnston ; Chang Yan ; Fangyang Liu ; Xiaojing Hao ; Martin Green

BAND GAP CHANGES OF THE CDS BUFFER INDUCED BY POST-ANNEALING OF CU2ZNSN(S,SE)4SOLAR CELLS 2216
Mario Lang ; Nicolas Schäfer ; Christian Huber ; Thomas Schnabe ; Heinz Kalt ; Michael Hetterich

22.61 % EFFICIENT FULLY SCREEN PRINTED PERC SOLAR CELL 2220
Weiwei Deng ; Feng Ye ; Ruimin Liu ; Yunpeng Li ; Haiyan Chen ; Zhen Xiong ; Yang Yang ; Yifeng Chen ; Yongqian Wang ; Pietro P. Altermatt ; Zhiqiang Feng ; Pierre J. Verlinden

HOW TO ACHIEVE 23% EFFICIENT LARGE-AREA CU PLATED N-PERT CELLS? 2227
Monica Aleman ; Angel Uruena ; Emanuele Cornagliotti ; Patrick Choulat ; Joachim John ; Richard Russell ; Sukvhinder Singh ; Loic Tous ; Wen-Cheng Sun ; Filip Duerinckx ; Jozef Szlufcik

MICROSTRUCTURE AND RECOMBINATION ACTIVITY OF GRAIN BOUNDARIES FROM FRONT AND REAR SIDE DURING A LID-CYCLE OF MC-PERC SOLAR CELLS 2232
Tabea Luka ; Marko Turek ; Stephan Großer ; Christian Hagendorf

THERMODYNAMIC EFFICIENCY LIMIT OF BIFACIAL SOLAR CELLS FOR VARIOUS SPECTRAL ALBEDOS 2236
Thomas C.R. Russell ; Rebecca Saive ; Harry A. Atwater

PROCESS-INDUCED DEGRADATION RESISTANT N-CZ WAFERS THROUGH TABULA RASA DEFECT ENGINEERING 2242
Vincenzo Lasalvia ; William Nemeth ; Matthew Page ; Wooseok Nam ; Youngsik Han ; Sungsun Baik ; Amanda Youssef ; Tonio Buonassisi ; Paul Stradins

DETECTION OF A SHIFTING BROMINE CONCENTRATION IN HYBRID PEROVSKITES BY X-RAY FLUORESCENCE MICROSCOPY 2245
Yanqi Luo ; Parisa Khoram ; Sarah Brittman ; Barry Lai ; Erik C. Garnett ; David P. Fenning

INFLUENCE OF GRAIN SIZE AND INTERFACES ON PHOTO-STABILITY OF PEROVSKITE SOLAR CELLS 2247

Istiaque Hossain ; Liang Zhang ; Ranjith Kottokkaran ; Mohamed El-Henawey ; Pranav Joshi ; Max Noack ; Vikram Dalal

COLD THOUGHTS ON PEROVSKITE FEVER........... 2251

Tao Xu ; Jue Gong

LBIC ANALYSIS OF PEROVSKITE BASED SOLAR CELLS STABILITY 2255

Carmen M. Ruiz ; Javier Ramos ; Richard Garuz ; Damien Barakel ; Jean Reusser ; Judikaël Le Rouzo

ASSESSING JOB GROWTH AND SUSTAINABILITY IN THE US PV INDUSTRY 2258

Brion Bob

ENSURING THE RELIABILITY OF PHOTOVOLTAIC POWER SYSTEMS USING INTERNATIONAL STANDARDS AND THE IECRE CONFORMITY ASSESSMENT SYSTEM 2263

George Kelly ; Adrian Häring ; Ted Spooner ; Greg Ball ; Sarah Kurtz ; Matthias Heinze ; Masaaki Yamamichi ; Govind Ramu

A FRAMEWORK TO CALCULATE UNCERTAINTIES FOR LIFETIME ENERGY YIELD PREDICTIONS OF PV SYSTEMS 2267

Bjorn Muller ; Peter Bostock ; Boris Farnung ; Christian Reise

INTEGRATED PV-RECYCLING-MORE EFFICIENT, MORE EFFECTIVE........... 2272

Wolfram Palitzsch ; Ulrich Loser

ANALYSIS OF GAINP SOLAR CELLS GROWN BY HYDRIDE VAPOR PHASE EPITAXY 2275

Kevin L. Schulte ; John Simon ; David L. Young ; Aaron J. Ptak

INVESTIGATION OF ADHESION FORCES BETWEEN DUST PARTICLES AND SOLAR GLASS........... 2280

H.R. Moutinho ; C.-S. Jiang ; B. To ; C. Perkins ; M. Muller ; M.M. Al-Jassim ; L. Simpson

ANTI-REFLECTIVE AND ANTI-SOILING PROPERTIES OF A KLEANBOOST™, A SUPERHYDROPHOBIC NANO-TEXTURED COATING FOR SOLAR GLASS 2285

Illya Nayshevsky ; Qianfeng Xu ; Gil Barahman ; Alan Lyons

MULTILAYER-GROWN ULTRATHIN NANOSTRUCTURED GAAS SOLAR CELLS 2291

Boju Gai ; Yukun Sun ; Minjoo Lee ; Jongseung Yoon

LABORATORY STUDIES OF PARTICLE CEMENTATION AND PV MODULE SOILING 2294

Craig L. Perkins ; Matthew Muller ; Lin Simpson

VIRTUAL SUBSTRATES FOR LOW-COST HIGH EFFICIENCY III-V PHOTOVOLTAICS 2298

Sean J. Babcock ; Marlene L. Lichty ; Shankar Karki ; Grace Rajan ; Sylvain Marsillac ; Elisabeth L. Mcclure ; Seth M. Hubbard ; Christopher G. Bailey

SEASONAL TRENDS OF SOILING ON PHOTOVOLTAIC SYSTEMS 2301

Leonardo Micheli ; Daniel Ruth ; Matthew Muller

INTERRELATIONSHIPS AMONG NON-UNIFORM SOILING DISTRIBUTIONS AND PV MODULE PERFORMANCE PARAMETERS, CLIMATE CONDITIONS, AND SOILING PARTICLE AND MODULE SURFACE PROPERTIES 2307

Lawrence L. Kazmerski ; Antonia Sonia A.C. Diniz ; Daniel Sena Braga ; Cristiana Brasil Maia ; Marcelo Machado Viana ; Suellen C. Costa ; Pedro P. Brito ; Cláudio Dias Campos ; Sergio De Morais Hanriot ; Leila R. De Oliveira Cruz

PV MODULE DURABILITY -CONNECTING FIELD RESULTS, ACCELERATED TESTING, AND MATERIALS 2312

T. John Trout ; W. Gambogi ; T. Felder ; K. R. Choudhury ; L. Garreau-Iles ; Y. Heta ; K. Stika

FEMTOSECOND VS NANOSECOND: AN ANALYSIS ON THE LASER ABLATION PROPERTIES OF DIELECTRIC LAYERS FOR SOLAR CELLS 2318

Jaffar Moideen Yacob Ali ; Vinodh Shanmugam ; Carlos D. Rodríguez-Gallegos ; Bianca Lim ; Armin Aberle ; Thomas Mueller

GROWTH OF MOS2 THIN FILMS WITH MICRODOME TEXTURE AS OMNIDIRECTIONAL LIGHT TRAP FOR SOLAR CELL APPLICATIONS........... 2324

Hussain M. Abouelkhair ; Nina A. Orlovskaya ; Robert E. Peale

STUDY OF SPATIAL DISTRIBUTION OF ELECTRICAL, OPTICAL AND STRUCTURAL PROPERTIES OF MAGNETRON SPUTTERED AZO THIN FILMS........... 2330

Mohit Agarwal ; Rajiv O Dusane

MULTIBAND FORMATION IN CR DOPED CUGAS2 THIN FILMS SYNTHESIZED BY CHEMICAL SPRAY PYROLYSIS 2334

Nazmul Ahsan ; Sivaperuman Kalainatharr ; Naoya Miyashita ; Takuya Hoshii ; Yoshitaka Okada

EFFECTS OF ANNEALING AND SUBSTRATE TEMPERATURE FOR SN-S THIN FILMS 2338

Yoji Akaki ; Kazuya Iwasaki ; Shigeyuki Nakamura ; Hideaki Araki

MOLYBDENUM OXIDE THIN FILMS FOR HETEROJUNCTION SOLAR CELLS 2342

A. Dominguez ; Ateet Dutt ; O. De Melo ; G. Santana

DUAL ION BEAM SPUTTERED TCO THIN FILMS: SPUTTER-INSTIGATED PLASMONIC FEATURES FOR ULTRATHIN PHOTOVOLTAICS............2345

Vivek Garg ; Brajendra S. Sengar ; Vishnu Awasthi ; Shailendra Kumar ; Shaibal Mukherjee

COMBINATORIAL STUDY OF SN-TI-W-O TRANSPARENT CONDUCTING OXIDE THIN FILMS FOR PHOTOVOLTAIC APPLICATIONS............2349

Michael N. Gona ; Patrick J. M. Isherwood ; Jake W. Bowers ; John M. Walls

BANDGAP AND ELECTRON AFFINITY OPTIMIZATION OF ZINC OXIDE FOR N-ZNO/P-SI SINGLE HETEROJUNCTION SOLAR CELL............2355

Babar Hussain ; Aasma Aslam

MODELING AND OPTIMIZING THE EFFICIENCY OF A ZNO/ZNTE SOLAR CELL USING SCAPS SOFTWARE............2358

Amal Kabalan ; Sam Roy ; Benjamin Chen

TERNARY PHOSPHIDE SEMICONDUCTOR INMG/ZN3P2SOLAR CELLS............2361

Ryoji Katsube ; Kenji Kazumi ; Yoshitaro Nose

NUMERICAL MODELING OF WSE2SOLAR CELLS............2364

H. Kyureghian ; M. Hilfiker ; E. Ediger ; V. Medic ; N.J. Ianno

BIAXIAL-TEXTURED TITANIUM NITRIDE THIN FILMS ON LOW-COST, FLEXIBLE METAL SUBSTRATE AS A CONDUCTIVE BUFFER LAYER FOR THIN FILM SOLAR CELLS............2368

Yongkuan Li ; Yao Yao ; Ying Gao ; Sicong Sun ; Pavel Dutta ; Monika Rathi ; Jae-Hyun Ryou ; Venkat Selvamanickam

SNS BY IONIZED JET DEPOSITION FOR PHOTOVOLTAIC APPLICATIONS............2372

Daniele Menossi ; Simone Di Mare ; Ivan Rimmaudo ; Elisa Artegiani ; Giampiero Tedeschi ; Juan Luis Pena ; Fabio Piccinelli ; Andrei Salavei ; Alessandro Romeo

EFFECT OF VALENCE BAND SPLITTING ON THE ABSORPTION SPECTRA OF MONOLAYER MOS2 IN PRESENCE OF SULPHUR VACANCIES............2376

Himani Mishra ; Sitangshu Bhattacharya

THE STUDY OF SOME MATERIALS AS BUFFER LAYER IN COPPER ANTIMONY SULPHIDE (CUSBS2) SOLAR CELL USING SCAPS 1-D............2381

Muteeu Olopade ; Adeyinka Adewoyin ; Michael Chendo ; Adewumi Bolaji

INFLUENCE OF HETERO-INTERFACES ON PHOTOVOLTAIC PERFORMANCE IN SOLAR CELLS BASED ON ZNSNP2BULK CRYSTAL............2385

Shigeru Nakatsuka ; Shunsuke Akari ; Jakapan Chantana ; Takashi Minemoto ; Yoshitaro Nose

JUNCTION BY DIFFUSION OF ELEMENTAL SODIUM ALONE INTO BRIDGMAN CU(IN, GA) SE2............2388

S. Park ; C. H. Champness ; S. Vanka ; Z. Mi ; I. Shih

OXYGEN SUBSTITUTION AND SULFUR VACANCIES IN NABIS2: A PB-FREE CANDIDATE FOR SOLUTION PROCESSABLE SOLAR CELLS............2392

Robert J Patterson ; Hongze Xia ; Long Hu ; Zhilong Zhang ; Lin Yuan ; Jianfeng Yang ; Weijian Chen ; Zihan Chen ; Yijun Gao ; Yicong Hu ; Binesh Puthen Veettil ; John A. Stride ; Gavin Conibeer ; Shujuan Huang

EFFECT OF ANNEALING ON PERFORMANCE OF SOLAR CELLS WITH NEW OXIDE ABSORBER MN2V2O7............2395

Pramod Ravindra ; Eashwer Athresh ; Rajeev Ranjan ; Srinivasan Raghavan ; Sushobhan Avasthi

ELECTRO-OPTICAL PROPERTIES OF ZN2MO3O8THIN-FILMS: A NOVEL LOW-BANDGAP SOLAR ABSORBER............2399

Pramod Ravindra ; Eashwer Athresh ; Rajeev Ranjan ; Srinivasan Raghavan ; Sushobhan Avasthi

LOW TEMPERATURE SOLUTION PROCESS FOR RANDOM HIGH ASPECT RATIO SILVER NANOWIRE AS PROMISING TRANSPARENT CONDUCTIVE LAYER............2403

Arastoo Teymouri ; Supriya Pillai ; Zi Ouyang ; Xiaojing Hao ; Martin Green

OXYGEN INCORPORATION INTO SI NANOCRYSTAL/SIC MULTILAYERS............2407

Charlotte Weiss ; Andreas Reichert ; Johannes Hofmann ; Stefan Janz

DESIGN OF CASCADED HETEROSTRUCTURED P-I-I-N CDS/CDSE LOW COST SOLAR CELL............2411

M. Zinaddinov ; S. Mil'shtein

FAST C-V METHOD TO MITIGATE EFFECTS OF DEEP LEVELS IN CIGS DOPING PROFILES............2414

P. K. Paull ; J. Bailey ; G. Zapalac ; A. R. Arehart

CRYSTAL GROWTH PHENOMENA IN POLYCRYSTALLINE (CU)ZNTE/CDTE/CDS VIA MOLECULAR DYNAMICS............2419

Rodolfo Aguirre ; Jose J. Chavez ; Xiao W. Zhou ; David Zubia

USING HIGH-RESOLUTION ANOMALOUS-SCATTERING X-RAY DIFFRACTION TO OBSERVE OFF-STOICHIOMETRIC CU2ZNSNS4CRYSTAL STRUCTURES............2423

Christopher J. Bosson ; Max T. Birch ; Douglas P. Halliday ; Chiu C. Tang ; Peter D. Hatton

SIMULATION OF ZNMGO AS THE WINDOW LAYER FORCDTESOLAR CELLS 2427
Yunfei Chen ; Shou Peng ; Xin Cao ; Alan E. Delahoy ; Ken K. Chin

MODELING EFFECT OF DEFECTS ON EFFICIENCY OF NANOWIRE CDS-CDTE SOLAR CELLS 2432
Hongmei Dang ; Esther Ososanya ; Nian Zhang ; Xiaohui Wang ; Hojjatollah Sarvari ; Vijay P. Singlr

ANALYTICAL DESCRIPTION OF CHARGED GRAIN BOUNDARY RECOMBINATION IN POLYCRYSTALLINE THIN FILM SOLAR CELLS 2438
Benoit Gaury ; Paul M. Haney

IMAGING THE EFFECT OF CDSE WINDOW LAYERS IN CDTE PHOTOVOLTAICS 2443
John M. Howard ; Elizabeth M. Tennyson ; William B. Gunnarsson ; Naba R. Paudel ; Yanfa Yan ; Marina S. Leite

INVESTIGATION OF TRAPS DENSITY AND POSITION IN ALKALI TREATED CU(IN,GA)SE2 THIN FILMS AND SOLAR CELLS 2446
Shankar Karki ; Pran K. Paul ; Grace Rajan ; Chinedum Akwari ; Angus Rockett ; Steven A Ringel ; Aaron R. Arehart ; Sylvain Marsillac

THE EFFECT OF DEPOSITION STOICHIOMETRY AND POST-DEPOSITION TREATMENTS ON DEEP DEFECTS IN CDTE 2449
Imran S. Khan ; Vamsi Evani ; Shamara Collins ; Chih An Hsu ; Vasilis Palekis ; Chris Ferekides

TESTING THE LIMITS OF MECHANICALLY-SCRIBED CIGS MICROCELLS 2453
Ombline Lafont ; Nicolas Vandamme ; Leia Ruffini ; Jia Yu ; Philip Jackson ; Jose Alvarez ; Daniel Lincot

PHOTOLUMINESCENCE IMAGING ANALYSIS OF DOPING IN THIN FILM CDS AND CDS/CDTE DEVICES 2457
C. Potamialis ; F. Lisco ; B. Maniscalco ; M. Togay ; A. Abbas ; M. Biiss ; J.W. Bowers ; J.M. Waiis ; I. Rimmaudo ; R. Mis Fernandez ; V. Rejon ; J.L. Peña

APPLICATION OF MAPPING SPECTROSCOPIC ELLIPSOMETRY FOR CDSE/CDTE SOLAR CELLS: OPTIMIZATION OF LOW-TEMPERATURE PROCESSED DEVICES WITH ALL-SPUTTERED SEMICONDUCTORS 2462
Mohammed A. Razooqi ; Adam B. Phillips ; Geethika K. Liyanage ; Fadhil K. Al-Fadhili ; Maxwell M. Junda ; Nikolas J. Podraza ; Michael J. Heben ; Robert W. Collins ; Prakash Koirala

ASSESSING THE VALIDITY AND ACCURACY OF EFFECTIVE ELECTRONIC MATERIALS: CAN 1D SIMULATIONS PREDICT POLYCRYSTALLINE DEVICE PERFORMANCE? 2467
Yubo Sun ; Allison Perna ; Sudhajit Misra ; Vasilios Palekis ; Chris Ferekides ; Jeffrey Aguiar ; Peter Bermel ; Michael A. Scarpulla

CHARACTERIZING RECOMBINATION IN CDTE-BASED SOLAR CELLS BY THE TEMPERATURE AND EXCITATION DEPENDENCE OF OPEN-CIRCUIT VOLTAGE AND PHOTOLUMINESCENCE 2473
Craig H. Swartz ; Sanjoy Paul ; Corey R. Grice ; Yanfa Yan ; Lorelle Mansfield ; Sachit Grover ; Gang Xiong ; Jian V. Li

EXPERIMENTAL EVIDENCE FOR CDS-RELATED TRANSPORT BARRIER IN THIN FILM SOLAR CELLS AND ITS IMPACT ON ADMITTANCE SPECTROSCOPY 2478
Florian Werner ; Anastasiya Zelenina ; Susanne Siebentritt

TRANSPARENT CONDUCTIVE ADHESIVES FOR TANDEM SOLAR CELLS 2482
Talysa R. Klein ; Benjamin G. Lee ; Manuel Schnabel ; Emily L. Warren ; Pauls Stradins ; Adele C. Tamboli ; Maikel F.A.M. Van Hest

MODELING THREE-TERMINAL III- V LSI TANDEM SOLAR CELLS 2488
Emily L. Warren ; Michael G. Deceglie ; Paul Stradins ; Adele C. Tamboli

WAFER BONDING APPROACHES FOR III-V ON SI MULTI-JUNCTION SOLAR CELLS 2492
Laura Vauche ; Elias Veinberg-Vidal ; Clément Weick ; Christophe Morales ; Vincent Larrey ; Christophe Lecouvey ; Mickaël Martin ; Jérémy Da Fonseca ; Christophe Jany ; Thibaut Desrues ; Céline Brughera ; Philippe Voarino ; Thierry Salvetat ; Frank Fournel ; Mathieu Baudrit ; Cécilia Dupré

DESIGN ARITHMETIC OF THE LATERAL III-V / SI HYBRID MODULE 2498
Kenji Araki ; Kyotaro Nakamura ; Kan-Hua Lee ; Takefumi Kamioka ; Yu-Cian Wang ; Nobuaki Kojima ; Yoshio Ohshita ; Masafumi Yamaguchi

GAASP NANOWIRE SOLAR CELL DEVELOPMENT TOWARDS NANOWIRE/SI TANDEM APPLICATIONS 2502
Enrique Barrigon ; Yang Chen ; Gaute Otnes ; Vilgaile Dagyte ; Nicklas Anttu ; Lars Samuelson ; Magnus Borgström

DEMONSTRATION OF GAINP2/SI VOLTAGE MATCHED TANDEM SOLAR CELLS 2506
David C. Bobela ; Kenneth J. Schmieder ; Matthew P. Lumb ; James E. Moore ; Robert J Walters ; Eric A. Armour ; Leo Matthew ; Rajesh Rao ; Angelo Mascarenhas ; Kirstin Alberi

WAFER BONDED III–V ON SILICON MULTI -JUNCTION CELL WITH EFFICIENCY BEYOND 31% 2511
Romain Cariou ; Jan Benick ; Paul Beutel ; Nico Tucher ; Martin Graf ; David Lackner ; Martin Hermle ; Stefan W. Glunz ; Andreas W. Bett ; Frank Dimroth

INTEGRATION OF THIN AL FILMS ON INO.18GAO.82AS METAMORPHIC GRADE STRUCTURES FOR LOW-COST III- V PHOTOVOLTAICS..2514
Alessandro Giussani ; Michael A. Slocum ; Seth M. Hubbard ; Nathan Smaglik ; Nikhil Pokharel ; S. Phillip Ahrenkiel

TEMPERATURE DEPENDENT CHARACTERISTICS OF GAINP/GAAS/GAINNASSB SOLAR CELL UNDER SIMULATED AM0 SPECTRA 2520
Riku Isoaho ; Arto Aho ; Antti Tukiainen ; Mircea Guina

EFFICIENCY OF GAAS P/SI TWO-JUNCTION SOLAR CELLS WITH MULTI-QUANTUM WELLS: A REALISTIC MODELING WITH CARRIER COLLECTION EFFICIENCY................2524
Boram Kim ; Kasidit Toprasertpong ; Oliver Supplie ; Agnieszka Paszuk ; Thomas Hannappel ; Yoshiaki Nakano ; Masakazu Sugiyama

INVERSE METAMORPHIC III-V/EPI-SIGE TANDEM SOLAR CELL PERFORMANCE ASSESSED BY OPTICAL AND ELECTRICAL MODELING..2528
Raphaël Lachaurne ; Martin Foldyna ; Gwénaëlle Hamon ; Nicolas Vaissiére ; Jean Decobert ; Romain Cariou ; Pere Roca I Cabarrocas ; José Alvarez ; Jean-Paul Kleider

TOWARDS MONOLITHICALLY INTEGRATED GAAS ON SI TANDEM SOLAR CELL................2532
Zhen Liu ; Zekun Ren ; Haohui Liu ; Tonio Buonassisi ; Ian Marius Peters

ZNSIP2 THIN FILM GROWTH FOR SI-BASED TANDEM PHOTOVOLTAICS 2536
Aaron D. Martinez ; Elisa M. Miller ; Andrew G. Norman ; Paul Stradins ; Eric S. Toberer ; Adele C. Tamboli

IN SITU CONTROL OVER THE SUBLATTICE ORIENTATION OF GAP/SI(100): AS VIRTUAL SUBSTRATES FOR TANDEM ABSORBERS 2538
Aznieszka Paszuk ; Oliver Supplie ; Sebastian Brückner ; Matthias M. May ; Anja Dobrich ; Andreas Nägelein ; Boram Kim ; Yoshiaki Nakano ; Masakazu Sugiyama ; Peter Kleinschmidt ; Thomas Hannappel ; Thomas Hannappel

III-V/SI TANDEM CELL TO MODULE INTERCONNECTION - COMPARISON BETWEEN DIFFERENT OPERATION MODES..2543
Henning Schulte-Huxel ; Emily L. Warren ; Manuel Schnabel ; Paul Stradins ; Daniel Friedman ; Adele C. Tamboli

INGAP/GAAS/ITO/SI HYBRID TRIPLE-JUNCTION CELLS WITH GAAS/ITO BONDING INTERFACES 2548
Naoteru Shigekawa ; Tomoya Hara ; Tomoki Ogawa ; Jianbo Liang ; Takefumi Kamioka ; Kenji Araki ; Masafumi Yamaguchi

MEASUREMENTS OF POTENTIALS AT TAP CONTACTS AND ESTIMATION OF RESISTANCE ACROSS BONDING INTERFACES IN INGAP/GAAS/SI HYBRID TRIPLE-JUNCTION CELLS 2551
Naoteru Shigekawa ; Jianbo Liang

OPTIMIZATION OF A GAASP TOP CELL FOR IMPLEMENTATION IN A III-V/SI TANDEM STRUCTURE 2554
Amber C. Silvaggio ; Daniel L. Lepkowski ; Daniel J. Chmielewski ; Jacob T. Boyer ; Steven A. Ringel ; Tyler J. Grassman

THEORETICAL DESIGN OF PEROVSKITE/CDTE FOUR-TERMINAL TANDEM SOLAR CELLS..2558
Tao Tang ; Huan Zhang ; Xingzhi Du ; Yiming Lnr ; Hang Zhou

WAFER-BONDED ALGAAS///SI DUAL-JUNCTION SOLAR CELLS 2562
Elias Veinberg-Vidal ; Laura Vauche ; Clément Weick ; Jérémy Da Fonseca ; Christophe Jany ; Christophe Morales ; Christophe Lecouvey ; Thibaut Desrues ; Philippe Voarino ; Frank Fournel ; Anne Kaminski-Cachopo ; Alejandro Datas ; Pablo Garcia-Linares ; Mathieu Baudrit ; Pierre Mur ; Cécilia Dupré

ENHANCEMENT OF SI PHOTOVOLTAIC MODULE BY INTRODUCING III-V/SI HYBRID CONFIGURATIONS AND COST EVALUATIONS UNDER VARIOUS COST RATIOS OF III-V/SI PHOTOVOLTAICS..2566
Yu-Cian Wang ; Kenii Araki ; Kyotaro Nakamura ; Kan-Hua Lee ; Takefumi Kamioka ; Nobuaki Kojima ; Yoshio Ohshita ; Masafumi Yamaguchi

NUMERICAL SIMULATION OF P-TYPE FRONT JUNCTION PERL SILICON CELL FOR III-V LSI TANDEM DEVICES 2569
Chuqi Yi ; Fa-Jun Ma ; Anita Ho-Baillie ; Stephen Bremner

EPITAXIAL GAP LAYERS GROWN ON SI SUBSTRATES USING MIGRATION ENHANCED AND MOLECULAR BEAM EPITAXY..2573
Chaomin Zhang ; Allison Boley ; Nikolai Faleev ; David J. Smith ; Christiana B. Honsberg

INVESTIGATION OF CARRIER-INDUCED DEFECT BEHAVIOR IN P-TYPE MULTICRYSTALLINE SILICON 2576
Catherine E. Chan ; Tsun H. Fung ; David N.R. Payne ; Daniel Chen ; Malcolm D. Abbott ; Alison M. Ciesla ; Ran Chen ; Brett J. Hallam ; Stuart R. Wenham

MAGNETRON SPUTTERED HYDROGENATED SILICON THIN FILMS: ASSESSMENT FOR
APPLICATION IN PHOTOVOLTAICS 2582
    Dipendra Adhikari ; Maxwell M. Junda ; Sylvain X. Marsillac ; Robert W. Collins ; Nikolas J. Podraza
HIGH QUALITY AND THIN SILICON WAFER FOR NEXT GENERATION SOLAR CELLS 2588
    Yoshio Ohshita ; Takuto Kojima ; Ryota Suzuki ; Kosuke Kinoshita ; Tomoyuki Kawatsu ; Kyotaro Nakamura ;
    Atsushi Ogura
FIRST DEMONSTRATION OF RADIAL JUNCTION SILICON NANOWIRE SOLAR MINI-
MODULES PREPARED BY PECVD AND LASER SCRIBING 2593
    Mutaz Al-Ghzaiwat ; Martin Foldyna ; Takashi Fuyuki ; Wanghua Chen ; Erik V. Johnson ; Jacques Meot ; Pere
    Roca I Cabarrocas
IMPACT OF INDUCED DEFECTS ON DEVICE PERFORMANCE IN SILICON
HETEROJUNCTION SOLAR CELLS 2596
    Pradeep Balaji ; André Augusto ; Stuart G. Bowden
LASER HYDROGENATION ON HEAVILY DISLOCATED CAST-MONO SILICON CELLS 2600
    Alison M. Ciesla ; Catherine E. Chan ; Sisi Wang ; Malcolm D. Abbott ; Cheemun Chong ; Stuart R. Wenham
PERFORMANCE OPTIMIZATION OF SEMI-TRANSPARENT THIN-FILM AMORPHOUS
SILICON SOLAR CELLS 2605
    Yuan Gao ; Fai Tong Si ; Olindo Isabella ; Rudi Santbergen ; Guangtao Yang ; Jianfei Dong ; Guoqi Zhang ;
    Miro Zeman
LOW TEMPERATURE SPALLING OF SILICON: A CRACK PROPAGATION STUDY 2610
    Pablo Guimera Coll ; Tine Uberg Nærland ; Nathan Stoddard ; Michael Stuckelberger ; Mariana Bertoni
NEW FINDINGS OF THERMAL EFFECT ON PM-SI:H SOLAR CELLS OPTOELECTRONIC
PROPERTIES 2614
    L. Hamui ; L. A. Górnez-González ; G. Santana
STUDY OF PV MODULE DEGRADATION RATE PREDICTION THROUGH CORRELATION
OF FIELD-AGED AND ACCELERATED-AGED MODULE DEGRADATION DATA 2618
    Babak T. Hamzavy ; William J. Grieco ; Brian J. Fields ; Cara S. Libby ; William B. Hobbs ; Olga Lavrova ; C.
    Birk Jones
ADVANCED ANALYSIS OF MULTI WIRE WAFERING PROCESSES 2622
    Ringo Koepgel ; Samuel Brinnig ; Felix Kaule ; Hartmut Schwabe ; Stephan Schoenfelder
CONSIDERATION ON OPEN-CIRCUIT VOLTAGE OF SI HETEROJUNCTION SOLAR CELLS
UNDER LOW CONCENTRATION CONDITION 2627
    Makoto Konagai
CHARACTERIZATION OF MICROCRYSTALLINE SILICON THIN FILM SOLAR CELLS
PREPARED BY HIGH WORKING PRESSURE PLASMA-ENHANCED CHEMICAL VAPOR
DEPOSITION 2631
    Jung-Dae Kwon ; Dong-Ho Kim ; Ji-Hoon Lee ; Myungkwan Song ; Myunghun Shin
ATOMIC-LAYER-DEPOSITEDV2O5-XFILMS AS A HIGHLY-EFFICIENT P-TYPE LAYER FOR
THIN FILM A-SI SOLAR CELLS 2634
    Ji-Hoon Lee ; Myungkwan Song ; Dong-Ho Kim ; Jung-Dae Kwon
A NOVEL DEFECT PASSIVATION METHOD FOR MULTICRYSTALLINE SI WAFER BY H2S
REACTION 2637
    Hsiang-Yu Liu ; Ujjwal K. Das ; Robert W. Birkmire
CARRIER TRANSPORTATION AT NOVEL SILVER PASTE CONTACT 2642
    Takefumi Kamioka ; Satoshi Kamevama ; Kazuo Muramatsu ; Aki Tanaka ; Naotaka Iwata ; Kyotaro Nakamura ;
    Atsushi Ogura ; Yoshio Ohshita
INFLUENCE OF DEPOSITION PARAMETERS ON SILICON THIN FILMS DEPOSITED BY
MAGNETRON SPUTTERING 2646
    Grace Rajan ; Tejaswini Miryala ; Shankar Karki ; Robert W. Collins ; Nikolas Podraza ; Sylvain Marsillac
MINORITY CARRIER LIFETIME VARIATIONS IN MULTICRYSTALLINE SILICON WAFERS
WITH TEMPERATURE AND INGOT POSITION 2651
    Sissel Tind Søndergaard ; Jan Ove Odden ; Rune Strandberg
CUO NANOWIRES-BASED RADIAL HETERO-JUNCTION THIN FILM SILICON SOLAR
CELLS WITH A HIGH OPEN-CIRCUIT VOLTAGE 2656
    Xiaolin Sun ; Jiawen Lu ; Fan Yang ; Linwei Yu ; Jun Xu ; Ling Xu ; Kunji Chen
THE EFFECT OF CHEMICAL COMPOSITION ON POROUS ETCHING FOR EPI AND LIFT-
OFF WAFER PROCESS 2660
    Teng-Yu Wang ; Peng-Wei Chen ; Han-Wen Liu
ELECTRICAL AND OPTICAL PERFORMANCE OF SILICON SOLAR CELLS USING
PLASMONICS INDIUM NANOPARTICLES LAYER EMBEDDED IN SIO2ANTIREFLECTIVE
COATING 2664
    Hao-Yu Yang ; Wen-Jeng Ho ; Sheng-Kai Feng ; Jheng-Jie Liu ; Ta-Wei Chuang ; Guan-Yi Li ; Yun-Chie Yang ;
    Cho-Chun Chiang ; Yao- Hui Chen

ELECTROLUMINESCENCE ANALYSIS FOR SEPARATION OF SERIES RESISTANCE FROM RECOMBINATION EFFECTS IN SILICON SOLAR CELLS WITH INTERDIGITATED BACK CONTACT DESIGN 2667
Nuha Ahmed ; Lei Zhang ; Ujjwal Das ; Steven Hegedus

INDOOR MEASUREMENT OF ANGLE RESOLVED LIGHT ABSORPTION BY BLACK SILICON 2672
Mekbib W. Amdemeskel ; Beniamino Iandolo ; Rasmus S. Davidsen ; Ole Hansen ; Gisele A. Dos Reis Benatto ; Nicholas Riedel ; Peter B. Poulsen ; Sune Thorsteinsson ; Anders Thorseth ; Carsten Dam-Hansen

IMPACT OF NON- FLAT PHOTOGENERATION AND CARRIER PROFILES ON THE LUMINESCENT EMISSION AND DETECTION OF SILICON SOLAR CELLS 2677
Nekane Azkona ; Federico Recart ; Pedro Rodríguez ; Vanesa Fano ; Aloña Otaegi ; Juan Carlos Jimeno

DEVELOPMENT OF OUTDOOR LUMINESCENCE IMAGING FOR DRONE-BASED PV ARRAY INSPECTION 2682
Gisele A. Dos Reis Benatto ; Nicholas Riedel ; Sune Thorsteinsson ; Peter B. Poulsen ; Anders Thorseth ; Carsten Dam-Hansen ; Claire Mantel ; Soren Forchhammer ; Kenn H. B. Frederiksen ; Jan Vedde ; Michael Petersen ; Henrik Voss ; Michael Messerschmidt ; Harsh Parikh ; Sergiu Spataru ; Dezso Sera

CLIMBING DRUM PEEL (CDP) TEST METHOD FOR CHARACTERIZING ADHESION IN FLEXIBLE PV MODULES 2688
Venkata Bheemreddy ; Kedar Hardikar

ACCURACY OF SOLAR SIMULATOR SPECTRAL DETERMINATION USING BAND-PASS FILTERING METHOD 2692
Weston Dobson ; Harrison Wilterdink ; Cassidy Sainsbury ; Adrienne Blum ; Justin Dinger ; Ronald A. Sinton ; Karsten Bothe ; David Hinken ; Martin Wolf

CORRELATION OF I-V CURVE PARAMETERS WITH MODULE-LEVEL ELECTROLUMINESCENT IMAGE DATA OVER 3000 HOURS DAMP-HEAT EXPOSURE 2697
Justin S. Fada ; Andrew J. Loach ; Alan J. Curran ; Jennifer L. Braid ; Shuying Yang ; Timothy J. Peshek ; Roger H. French

A NOVEL METHOD TO INVESTIGATE STOICHIOMETRY AND PERFORMANCE OF BURIED PASSIVATED CONTACTS UTILIZING TIME-OF-FLIGHT SIMS 2702
Steven P. Harvey ; William Nemeth ; Jeff Aguiar ; Craig Perkins ; Pauls Stradins

A COMPARISON BETWEEN QUASI-STEADY STATE AND TRANSIENT PHOTOCONDUCTANCE LIFETIMES IN SILICON INGOTS: SIMULATIONS AND MEASUREMENTS 2707
Mohsen Goodarzi ; Ronald Sinton ; Daniel Chung ; Bernhard Mitchell ; Thorsten Trupke ; Daniel Macdonald

NEW DEVELOPMENT IN GLOW DISCHARGE OPTICAL EMISSION SPECTROMETRY FOR THE CHARACTERIZATION AND THE THICKNESS MEASUREMENT OF LAYERS FOR PHOTOVOLTAIC APPLICATIONS 2711
Philippe Hunault ; Matthieu Chausseau ; Patrick Chaporr ; Sofia Gaiaschi ; Anais Loubar ; Muriel Bouttcmy ; Arnaud Etcheberry

DEEP LEVEL TRANSIENT SPECTROSCOPY MEASUREMENTS OF SILICON HETEROJUNCTION CELLS 2716
Sanchit Khatavkar ; C. V. Kannan ; Vijay Kumar ; P. R. Nair ; B. M. Arora

CHARACTERIZATION OF MODULES AND ARRAYS WITH SUNS VOC 2719
Alex Killam ; Stuart Bowden

A STUDY OF PERFORMANCE CHARACTERIZATION WITH REAR LIGHT SOURCE IN CONVENTIONAL BIFACIAL SOLAR CELLS 2723
Soo Min Kim ; Sang Hoon Jung ; Rae-Won Choi ; Yong Bae Kim ; Min Gu Kang ; Hee-Eun Sonp ; Gyu-Seok Choi

ELECTRICAL CHARACTERIZATION OF THE CARRIER TRANSPORT PROPERTIES IN ACU(IN,GA)SE2SOLAR CELL 2728
Roberto Lopez ; Sanjoy Paull ; Ingrid Repins ; Jian V. Li

SYSTEMATIC THERMALPHOTOVOLTAIC SOLAR CELL OPTIMIZATION 2732
Zheng Lyu ; Muyu Xue ; Junyan Chen ; Jieyang Jia ; Shanhui Fan ; James Harris

CHARACTERIZATION OF TELLURIUM AS A BACK CONTACT FOR CDTE SOLAR CELLS 2736
C.E. Moffett ; W.S. Sampath

ON THE DIFFERENT EXPLANATIONS OF THE RECOMBINATION CURRENTS WITH HIGH IDEALITY FACTOR IN SILICON SOLAR CELLS 2740
A. Otaegi ; V. Fano ; N. Azkona ; J. R. Gutiérrez ; J. C. Jimeno

IDENTIFICATION OF SHUNTS IN A MONOLITHIC MULTIJUNCTION GAAS/GAAS DEVICE BY SPECTROMETRIC CHARACTERIZATION 2744
Felipe Oviedo ; Liu Zhe ; Zekun Ren ; Kevin Nay Yaung ; Maung Thway ; Liu Haohui ; Tonio Buonassisi ; Ian Marius Peters

A SIMULATION STUDY ON RADIATIVE RECOMBINATION ANALYSIS IN CIGS SOLAR CELL ... 2749

Sanjoy Paul ; Roberto Lopez ; Md Dalim Mia ; Craig H. Swartz ; Jian V. Li

SIMULATION AND SPECTROSCOPY OF CARRIER RELAXATION IN GASB AND GAAS ... 2755

A.C. Scofield ; A.I. Hudson ; B.L. Liang ; B.C. Juang ; D.L. Huffaker ; S.M. Hubbard ; W.T. Lotshaw

COMPUTATIONAL DESIGN OF DOPANTS IN CDTE GRAIN BOUNDARIES FOR EFFICIENT PHOTOVOLTAICS ... 2759

Fatih G. Sen ; Tadas Paulauskas ; Ce Sun ; Moon Kim ; Robert F. Klie ; Maria K.Y. Chan

ANALYSES OF PHOTOVOLTAIC POWER PLANT PERFORMANCE ESTIMATES BASED ON DETAILED LABORATORY MODULE CHARACTERIZATIONS AND TYPICAL REAL-WORLD INPUT DATA SOURCES ... 2762

Rajeev Singh ; John L.R. Watts ; Kellen Gillispie

CRITICAL EVALUATION OF THE FOUNDATIONS OF SOLAR SIMULATOR STANDARDS ... 2765

Ronald A. Sinton ; Harrison Wilterdink ; Justin Dinger ; Adrienne L. Blum ; Weston Dobson ; Cassidy Sainsbury

IMPACT OF INFRARED OPTICAL PROPERTIES ON CRYSTALLINE SI AND THIN FILM CDTE SOLAR CELLS ... 2771

Indra Subedi ; Timothy J Silverman ; Michael Deceglie ; Nikolas J. Podraza

THE IMPACT OF IMPURITIES ON THE RELATIVE EFFICIENCIES OF SOLAR CELLS FROM DIFFERENT SILICON FEEDSTOCKS ... 2776

Muhammad Tayyib ; Aleksandr Dobroliubov ; Zekija Ramic ; Muhammad Nadeem Akarm ; Jan Ove Odden

ACCURACY EVALUATION OF ABSOLUTE ELECTROLUMINESCENCE-EFFICIENCY MEASUREMENTS OF SOLAR CELLS USING A SENSITIVITY-CALIBRATED-PHOTODETECTOR CONTACT METHOD ... 2781

Masahiro Yoshita ; Yoshihiro Hishikawa ; Yoshihiko Kanemitsu ; Hidefumi Akiyama

NANOMETER-SCALE CARRIER IMAGING OF POTENTIAL-INDUCED DEGRADATION IN C-SI SOLAR CELLS ... 2785

C.-S. Jiang ; C. Xiao ; H.R. Moutinho ; S. Johnston ; M.M. Al-Jassim ; X. Yang ; Y. Chen ; J. Ye

NREL EFFORTS TO ADDRESS SOILING ON PV MODULES ... 2789

Lin J. Simpson ; Matthew Muller ; Michael Deceglie ; Helio Moutinho ; Craig Perkins ; C. S. Jiang ; David C. Miller ; Leonardo Micheli ; Govindasamy Tamizhmani ; Sai Ravi Vasista Tatapudi ; Mowafak Al-Jassim

MODELING POTENTIAL-INDUCED DEGRADATION (PID) OF FIELD-EXPOSED CRYSTALLINE SILICON SOLAR PV MODULES: FOCUS ON A REGENERATION TERM ... 2794

Eleonora Annigoni ; Alessandro Virtuani ; Fanny Sculati-Meillaud ; Christophe Ballif

SOILING LOSS ON PV MODULES AT TWO LOCATIONS IN INDIA STUDIED USING A WATER BASED ARTIFICIAL SOILING METHOD ... 2799

Sonali Bhaduri ; Sachin Zachariah ; Lawrence L. Kazmcrski ; Balasubramaniam Kavaipatti ; Anil Kottantharayil

QUANTIFYING YEAR-TO-YEAR VARIATIONS IN SOLAR PANEL SOILING FROM PV ENERGY-PRODUCTION DATA ... 2804

Michael G. Deceglie ; Leonardo Micheli ; Matthew Muller

ACCURATELY MEASURING PV SOILING LOSSES WITH SOILING STATION EMPLOYING PV MODULE POWER MEASUREMENTS ... 2808

Michael Gostein ; Bill Stueve ; Mandy Chan

PERFORMANCE OF MONOCRYSTALLINE SILICON SOLAR CELL- INFLUENCE OF DUST ON ULTRA-VIOLET AND VISIBLE REGION DURING EARLY STAGE OF DEPOSITION ... 2811

Hemaprabha Elangovan ; Upasna Ranjan ; A K Jagdish ; Praveen C. Ramamurthy ; Kamanio Chattopadhyay

A COMPREHENSIVE STUDY OF LIGHT SOAKING EFFECT IN CDTE SOLAR CELLS ... 2816

D. Guo ; A. Moore ; D. Krasikov ; I. Sankin ; D. Vasileska

CORRECTION FOR METASTABILITY IN THE QUANTIFICATION OF PID IN THIN-FILM MODULE TESTING ... 2819

Peter Hacke ; Sergiu Spataru ; Steve Johnston

A FINE MODEL OF POWER DEGRADATION FOR CRYSTALLINE SILICON SOLAR MODULES ... 2823

Wenshuang Hea ; Baosong Duan ; Fumei Wang ; Ao Wang ; Jipeng Chang ; He Wang ; Hong Yang ; Jie Ding ; Junjun Zhang ; Jingsheng Huang

TEST METHODS FOR HYDROPHOBIC COATINGS ON SOLAR COVER GLASS ... 2827

Kenan Isbilir ; Biancamaria Maniscalco ; Ralph Gottschalg ; John Michael Walls

IMPACT OF DEGRADATION RATES ON SOLAR PV FINANCING FOR PROJECTS LOCATED IN THE UNITED STATES ... 2833

Rounak A. Kharait ; Phil Stiles ; Jarrett Carriere ; Larry Mcclung

ANALYSIS OF WIND DIRECTION AND SPEED MEASUREMENTS IN ARID REGION - A SITE EVALUATION USING DATA WITH LOW TEMPORAL RESOLUTION ... 2836

Elisabeth Klimm ; Felix Guischard ; Karl-Anders Weiss

FORECASTING ENVIRONMENTAL DEGRADATION POWER LOSS IN SOLAR PANELS WITH A PREDICTIVE CRACK OPENING TEST 2839
Jason L. Lincoln ; Andrew M. Gabor ; Eric J. Schneller ; Hubert Seigneur ; Joseph Walters ; Rob Janoch ; Andrew Anselmo ; Victor Huayamave ; Winston Schoenfeld

FLUORESCENCE IMAGING ON THE CROSS-SECTION OF PHOTOVOLTAIC LAMINATES AGED UNDER DIFFERENT UV INTENSITIES 2844
Yadong Lyu ; Jae Hyun Kim ; Xiaohong Gu

STATISTICAL ANALYSIS OF DEGRADATION DATA FOR C-SI MODULES OBSERVED IN INDIA IN 2016 2849
Chiranjibi Mahapatra ; Rajiv Dubey ; Shashwata Chattopadhyay ; Sachin Zachariah ; Sanjeev Sabnis

PROCESS INDUCED DEFLECTION AND STRESS ON ENCAPSULATED SOLAR CELLS 2854
Xiaodong Meng ; Michael Stuckelberger ; Peter Hacke ; Mariana Bertoni

A UNIFIED GLOBAL INVESTIGATION ON THE SPECTRAL EFFECTS OF SOILING LOSSES OF PV GLASS SUBSTRATES: PRELIMINARY RESULTS 2858
Leonardo Micheli ; Eduardo F. Fernández ; Greg P. Smestad ; Hameed Alrashidi ; Nabin Sarmah ; Nazmi Sellami ; Ibrahim A. I. Hassan ; Amal Kasry ; Gustavo Nofuentes ; Neeru Sood ; Bala Pesala ; S. Senthilarasu ; Florencia Almonacid ; K.S. Reddy ; Matthew Muller ; Tapas K. Mallick

REFERENCE: PROCEEDINGS OF THE IEEE PVSC CONF., 2017 THE DEVELOPMENT OF A DC BREAKDOWN VOLTAGE TEST FOR PHOTOVOLTAIC INSULATING MATERIALS 2864
David C. Miller ; Bernt Ake-Sultan ; Axel Borne ; Rene Eugen ; Bradley L. Givot ; Jürgen Jung ; Steven W. Macmaster ; Byron K. Mcdanold ; Ulf H. Nilsson ; Nancy H. Phillips ; Ian A. Tappan ; Nick S. Bosco

FIELD-EVALUATION OF ELECTRODYNAMIC SCREENS FOR MAINTAINING HIGH OPTICAL EFFICIENCY OPERATION OF SOLAR COLLECTORS 2870
Cristian Morales ; Annie Bernard ; Ryan Eriksen ; Julius Yellowhair ; Sean Garner ; Ricci La Centra ; Alecia Griffin ; Alexis Lloyd ; Yujie Gao ; Ramakrishnan Lakshmanan ; Mark Horenstein ; Malay Mazumder

EFFECT OF REVERSE BIAS VOLTAGES ON SMALL SCALE GRIDDED CIGS SOLAR CELLS 2875
Soheyl Mortazavi ; Klaas Bakker ; Jome Carolus ; Michael Daenen ; Gabriela De Amorim Soares ; Henk Steijvers ; Arthur Weeber ; Mirjam Theelen

A METHOD TO EXTRACT SOILING LOSS DATA FROM SOILING STATIONS WITH IMPERFECT CLEANING SCHEDULES 2881
Matthew Muller ; Leonardo Micheli ; Alfredo A. Martinez-Morales

ANALYTICAL (S)TEM STUDIES OF DEFECTS ASSOCIATED WITH PID IN STRESSED SI PV MODULES 2887
Andrew Norman ; Adam Stokes ; John Moseley ; Steven Harvey ; Steve Johnston ; Harvey Guthrey ; Mowafak Al-Jassim

DESIGN, DEVELOPMENT, AND EVALUATION OF ELECTRODYNAMIC SCREENS FOR SELF-CLEANING SOLAR PANELS AND CONCENTRATING MIRRORS 2891
Annie Bernard ; Cristian Morales ; Ryan S. Eriksen ; Alecia C. Griffin ; Yujie Gao ; Ramakrishnan Lakshmanan ; Ricci La Centra ; Arash Sayyah ; Julius E. Yellowhair ; Sean M. Garner ; N Mark Horenstein ; Malay K. Mazumder

EVALUATING SOLAR CELL FRACTURE AS A FUNCTION OF MODULE MECHANICAL LOADING CONDITIONS 2897
Eric J. Schneller ; Andrew M. Gabor ; Jason Lincoln ; Rob Janoch ; Andrew Anselmo ; Joseph Walters ; Hubert Seigneur

COMPUTATIONAL STUDY OF THE EFFECT OF PHOTOVOLTAIC (PV) MODULE PARAMETERS ON STRESS DEVELOPMENT IN SILICON UNDER STATIC LOADING 2902
Saurabh Sethia ; Karan Shishir Yadav ; Sudharm Rathore ; Abhishek Shubhrant ; Aparna Singh

A SIMPLE METHOD FOR MEASURING SOLAR RADIATION INTENSITY BY IMAGE ANALYSES 2906
Akiko Takahashi ; Akinori Moriki ; Nobuyuki Yamada ; Jun Imai ; Shigeyuki Funabiki

DEGRADATION OF SOLDER BONDS IN FIELD AGED PV MODULES: CORRELATION WITH SERIES RESISTANCE INCREASE 2912
Abhishiktha Tummala ; Jaewon Oh ; Sai Tatapudi ; Govindasamy Tamizhmani

PERFORMANCE OF LIGHT AND DARK CURRENT-VOLTAGE CHARACTERISTICS FOR PID-AFFECTED MONOCRYSTALLINE SILICON SOLAR MODULES 2918
He Wang ; Pan Zhao ; Shuwen Guo ; Hong Yang ; Weijing Huang ; Shiyu Sang ; Bojie Su ; Xue Zhang ; Yunxue Cao ; Hui Zhao

SOILING RATES OF PV MODULES VS. THERMOPILE PYRANOMETERS 2923
Martin Waters ; Tejas Tirumalai ; Michael Gostein ; Bill Stueve

GRID INTEGRATION OF BUILDING SYSTEMS AND 1 MW PHOTOVOLTAIC ARRAY USING VOLTTRON 2926
David Raker ; Andrew Sellers ; Roshan Kini ; Michael Green ; Thomas Stuart ; Randall Ellingson ; Raghav Khanna ; Michael Heben

INTERCONNECTION STUDY OF DISTRIBUTED PV SYSTEMS BY INTERFACING MATLAB WITH OPENDSS AND GIS ... 2931

Joseph A. Ahamioje ; Hariharan Krishnaswami

NOVEL MPPT ALGORITHM FOR ACTIVE POWER CONTROL OF MULTI-LEVEL DUAL-ACTIVE BRIDGE PV CONVERTER IMPLEMENTED IN NI MYRIO ... 2936

Shilpa Marti ; Hariharan Krishnaswami

MODELING A GRID-CONNECTED PV/BATTERY MICROGRID SYSTEM WITH MPPT CONTROLLER ... 2941

Genesis Alvarez ; Hadis Moradi ; Mathew Smith ; Ali Zilouchian

>94.5%REDUCTION IN GRID-BUY ELECTRICITY AND ELIMINATION OF AM & PM ENERGY PEAKS/SPIKES BY OPTIMIZING ENERGY USAGE AND INTEGRATION OF CUSTOMER SELF-SUPPLY ROOFTOP SOLAR PV WITH ELECTRICAL & THERMAL (HOT & COLD) STORAGE BATTERIES: A CASE STUDY FOR RESIDENTIAL HAWAII ... 2947

John Borland ; Jay Moore ; Corpuz Poncho ; Takahiro Tanaka ; Harumi Mcclure

A SINGLE-STAGECˊUK-BASED TRANSFORMERLESS INVERTER FOR 1-Φ GRID-CONNECTED PV SYSTEMS ... 2952

Phani Kumar Chamarthi ; Amit Kumar Gupta ; Madhuwanti S. Joshi ; Vivek Agarwal

A STATE SPACE AVERAGE MODEL FOR DYNAMIC MICROGRID BASED SPACE STATION SIMULATIONS ... 2957

Rachid Darbali-Zamora ; Eduardo I. Ortiz-Rivera

BUCK CONVERTER AND SEPIC BASED ELECTRONIC POWER SUPPLY DESIGN WITH MPPT AND VOLTAGE REGULATION FOR SMALL SATELLITE APPLICATIONS ... 2963

Rachid Darbali-Zamora ; Nicolás Cobo-Yepes ; John E. Salazar-Duque ; Eduardo I. Ortiz-Rivera ; Amilcar A. Rincon-Charris

VIRTUAL POWER PLANT FEEDBACK CONTROL DESIGN FOR FAST AND RELIABLE ENERGY MARKET AND CONTINGENCY RESERVE DISPATCH ... 2969

Mohamed Elkhatib ; Jay Johnson ; David Schoenwald

INTELLIGENT SAMPLING OF PERIODS FOR REDUCED COMPUTATIONAL TIME OF TIME SERIES ANALYSIS OF PV IMPACTS ON THE DISTRIBUTION SYSTEM ... 2975

Jason Galtieri ; Matthew J. Reno

A PWM SCHEME TO REALISE TWO TIMES EFFECTIVE SWITCHING FREQUENCY WITH CONSTANT COMMON MODE VOLTAGE AND REACTIVE POWER CAPABILITY IN 1- Φ GRID-TIED TRANSFORMERLESS H6 PV INVERTER ... 2981

Amit Kumar Gupta ; Madhuwanti S. Joshi ; Vivek Agarwal

A SOLAR PV RETROFIT SOLUTION FOR RESIDENTIAL BATTERY INVERTERS ... 2986

Amit Kumar Gupta ; Vaibhav Pawar ; Madhuwanti S. Joshi ; Vivek Agarwal ; Deepak Chandran

COST BENEFIT AND ALTERNATIVES ANALYSIS OF DISTRIBUTION SYSTEMS WITH ENERGY STORAGE SYSTEMS ... 2991

Tom Harris ; Adarsh Nagarajan ; Murali Baggu ; Tom Bialek

EVALUATION OF PV HOSTING CAPACITIES OF DISTRIBUTION GRIDS WITH UTILIZATION OF SOLAR-ROOF-POTENTIAL-ANALYSES ... 2996

Gerd Heilscher ; Falko Ebe ; Basem Idlbi ; Jeromie Morris ; Florian Meier

EXPERIMENTAL DISTRIBUTION CIRCUIT VOLTAGE REGULATION USING DER POWER FACTOR, VOLT-VAR, AND EXTREMUM SEEKING CONTROL METHODS ... 3002

Jay Johnson ; Sigifredo Gonzalez ; Daniel B. Arnold

DYNAMIC SETPOINT CONTROL OF ELECTRIC HOT WATER HEATER TANKS FOR INCREASED INTEGRATION OF SOLAR PHOTOVOLTAIC SYSTEMS ... 3008

C. Birk Jones ; Monte Lunacek ; Matthew Lave ; Jay Johnson ; Robert Broderick

SPATIAL ANALYSIS OF RESIDENTIAL COMBINED PHOTOVOLTAIC AND BATTERY POTENTIAL: CASE STUDY UTRECHT, THE NETHERLANDS ... 3014

Geert Litjens ; Bala Bhavya Kausika ; Ernst Worrell ; Wilfried Van Sark

POWER BALANCE REQUIREMENTS FOR SUSTAINED ISLANDING OF INVERTER BASED DISTRIBUTED GENERATION ... 3020

Gregory A. Kern ; Michael Ropp ; Sigifredo Gonzalez

FULL-SCALE DEMONSTRATION OF DISTRIBUTION SYSTEM PARAMETER ESTIMATION TO IMPROVE LOW-VOLTAGE CIRCUIT MODELS ... 3025

Matthew Lave ; Matthew J. Reno ; Robert J. Broderick ; Jouni Peppanen

CREATION AND VALUE OF SYNTHETIC HIGH-FREQUENCY SOLAR INPUTS FOR DISTRIBUTION SYSTEM QSTS SIMULATIONS ... 3031

Matthew Lave ; Matthew J. Reno ; Robert J. Broderick

A DIRECT MAXIMUM POWER POINT SEARCH USING CURRENT-VOLTAGE BASED POWER-LAW RELATION FOR PHOTOVOLTAIC SYSTEM UNDER UNIFORM IRRADIANCE 3038

Hitesh K. Mehta ; Ashish K. Panchal

PASSIVITY BASED CONTROLLER FOR PHOTOVOLTAIC MODULES USING ZETA CONVERTER 3044

Daniel A. Merced Cirino ; Rachid Darbali Zamora ; Eduardo I. Ortiz Rivera

SIC SWITCH BASED SINGLE-STAGE BUCK-BOOST TRANSFORMERLESS MINI INVERTER WITH LOW LEAKAGE CURRENT AND NEGLIGIBLE DC INJECTION 3050

Soumya Ranjan Mohapatra ; Amit Kumar Gupta ; Madhuwanti S. Joshi ; Vivek Agarwal

OPEN SOURCE TOOLS FOR HIGH PERFORMANCE QUASI-STATIC-TIME-SERIES SIMULATION USING PARALLEL PROCESSING 3055

Davis Montenegro ; Roger C. Dugan ; Matthew J. Reno

MAXIMUM POWER POINT TRACKING OF PV MODULE BASED ON NEW EXPLICIT I-V RELATION 3061

Tejeswar Nukala ; A. K. Panchal

AN AUTOCORRELATION-BASED COPULA MODEL FOR PRODUCING REALISTIC CLEAR-SKY INDEX AND PHOTOVOLTAIC POWER GENERATION TIME-SERIES 3067

Joakim Munkhammar ; Joakim Widén

DYNAMIC RESPONSE OF MAXIMUM POWER POINT TRACKING USING PERTURB AND OBSERVE ALGORITHM WITH MOMENTUM TERM 3073

Gautam A. Raiker

A FRAMEWORK FOR COMPARING THE ECONOMIC PERFORMANCE AND ASSOCIATED EMISSIONS OF GRID-CONNECTED BATTERY STORAGE SYSTEMS IN EXISTING BUILDING STOCK: A NYISO CASE STUDY 3077

Julian Do Nascimento Ricardo ; Vasilis Fthenakis

IMPROVING ANY ARBITRARY MPPT HILL CLIMBER WITH ANN ESTIMATIONS 3083

Jesse Roberts ; Indranil Bhattacharya

INCREASING SOLAR PHOTOVOLTAIC PENETRATION USING THERMAL ENERGY STORAGE 3088

Alexander F. Routhier ; Christiana Honsberg

MODEL PREDICTIVE CONTROL OF GRID CONNECTED MODULAR MULTILEVEL CONVERTER FOR INTEGRATION OF PHOTOVOLTAIC POWER SYSTEMS 3092

Amir Shahirinia ; Amin Hajizadeh

MAXIMIZATION OF SELF-SUFFICIENCY WITH GRID CONSTRAINTS: PV GENERATORS, WIND TURBINES AND STORAGE TO FEED TERTIARY SECTOR USERS 3096

Filippo Spertino ; Jawad Ahmad ; Alessandro Ciocia ; Paolo Di Leo ; Francesco Giordano

SWITCHES CONTROLLING TO IMPLEMENT ADAPTIVE MULTILEVEL INVERTER ON PV SYSTEM 3102

Hadi Suhana ; Ngapuli I Sinisuka ; Muhammad Nurdin ; Yvon Besanger ; Vincent Debusschere

DEMAND RESPONSE FOR THE PROMOTION OF PHOTOVOLTAIC PENETRATION 3107

Venizelos Venizelou ; Spyros Theocharides ; George Makrides ; Venizelos Efthymiou ; George E. Georghiou

GRIDDLER AI: NEW PARADIGM IN LUMINESCENCE IMAGE ANALYSIS USING AUTOMATED FINITE ELEMENT METHODS 3113

Johnson Wong ; Percis Teena ; Daniel Inns

INTERACTION OF O2IDIMERS WITH GA IN SI AND IMPLICATIONS FOR A COMPREHENSIVE MODEL OF LIGHT- INDUCED DEGRADATION 3119

Yu Jin ; Scott T. Dunham

NUMERICAL SIMULATION OF EBIC FOR ANALYSIS OF EXTENDED DEFECTS 3123

Marco Nardone ; John Moseley ; Saroj Dahal ; Anuja V. Parikh ; John M. Waddle

COLLOIDAL QUANTUM DOT SOLAR CELL ELECTRICAL PARAMETER IMAGING USING CAMERA-BASED HIGH-FREQUENCY HETERODYNE LOCK-IN CARRIEROGRAPHY 3129

Lilei Hu ; Mengxia Liu ; Andreas Mandelis ; Qiming Sun ; Alexander Melnikov ; Edward H. Sargent

A NEW PERSPECTIVE ON POTENTIAL-INDUCED DEGRADATION OF THE SHUNTING TYPE BY MICRO RAMAN-SPECTROSCOPY AND MICRO LIGHT-BEAM-INDUCED CURRENT 3135

A. Büchler ; H. Nagel ; M. Breitwieser ; S. Kluska ; F. D. Heinz ; M. C. Schubert ; M. Glatthaar ; S. Glunz

NANOSCALE DETECTION OF DEEP LEVELS IN CIGS USING ELECTRON ENERGY LOSS SPECTROSCOPY 3139

Julia I. Deitz ; Pran K. Paul ; Shankar Karki ; Sylvain Marsillac ; Aaron R. Arehart ; Tyler J. Grassman ; David W. Mccomb

MEASUREMENT OF CARRIER DYNAMICS IN PHOTOVOLTAIC CZTSE BY TIME-RESOLVED TERAHERTZ SPECTROSCOPY 3143
Siming Li ; Michael A. Lloyd ; Andrew A. Golembeski ; Brian E. Mccndless ; Jason B. Baxter

DECOUPLING GRAIN-BOUNDARY, GRAIN-INTERIOR, AND SURFACE RECOMBINATION WITH CATHODOLUMINESCENCE 3147
John Moseley ; Pierre Rale ; Stéphane Collin ; Ana Kanevce ; Eric Colegrove ; Joel Duenow ; Soren Jensen ; Wyatt K. Metzger ; Mowafak M. Al-Jassim

HIGH RESOLUTION THZ SCANNING FOR OPTIMIZATION OF DIELECTRIC LAYER OPENING PROCESS ON DOPED SI SURFACES 3150
P. Spinelli ; F.J.K. Danzl ; D. Deligiannls ; N. Guillevin ; A.R. Burgers ; S. Sawallich ; M. Nage ; I. Cesar

DEGRADATION ASSESSMENT OF FIELDED CIGS PHOTOVOLTAIC ARRAYS 3155
Bruce H. King ; Joshua S. Stein ; Daniel Riley ; C. Birk Jones ; Charles D. Robinson

APPLICATION OF IEC 61724 STANDARDS TO ANALYZE PV SYSTEM PERFORMANCE IN DIFFERENT CLIMATES 3161
Katherine A. Klise ; Joshua S. Stein ; Joseph Cunningham

EFFECTS OF URBAN ENVIRONMENT ON SOLAR PV PERFORMANCE 3167
Panagiotis Moraitis ; Bala Bhavya Kausika ; Wilfried G.J.H.M. Van Sark

IRRADIANCE MEASUREMENT CONSIDERATIONS FOR SYSTEM PERFORMANCE ASSESSMENT WHEN MANAGING FLEETS OF PHOTOVOLTAIC ASSETS ACROSS ASIA 3172
André M. Nobre ; Shravan Karthik ; Chenxi Liu ; Rohit Jaswal ; Rupesh Baker ; Raghav Malhotra ; Alan Khor

MACHINE LEARNING IN PV FAULT DETECTION, DIAGNOSTICS AND PROGNOSTICS: A REVIEW 3178
Sandy Rodrigues ; Helena Geirinhas Ramos ; F. Morgado-Dias

OUTDOOR FIELD PERFORMANCE FROM BIFACIAL PHOTOVOLTAIC MODULES AND SYSTEMS 3184
Joshua S. Stein ; Daniel Riley ; Matthew Lave ; Clifford Hansen ; Chris Deline ; Fatima Toor

DEFINING THRESHOLD VALUES OF ENCAPSULANT AND BACKSHEET ADHESION FOR PV MODULE RELIABILITY 3190
Nick Bosco ; Joshua Eafanti ; Sarah Kurtz ; Jared Tracy ; Reinhold Dauskardt

CHARACTERIZATIONS OF AGED GLASS/ETHYLENE VINYL ACETATE/GLASS USING FLUORESCENCE SPECTROSCOPY AND INSTRUMENTED INDENTATION 3195
Jae Hyun Kim ; Yadong Lyu ; David C. Miller ; Xiaohong Gu

ENCAPSULANT ADHESION TO SURFACE METALLIZATION ON PHOTOVOLTAIC CELLS 3200
Jared Tracy ; Nick Bosco ; Reinhold Dauskardt

IMPACT OF UV LIGHT INTENSITY ON PHOTODEGRADATION OF PV BACKSHEETS 3204
Xiaohong Gu ; Li-Chieh Yu ; Yadong Lyu ; Jae Hyun Kim ; Andrew Fairbrother ; Tinh Nguyen

SURVEY OF MECHANICAL DURABILITY OF PV BACKSHEETS 3208
Michael D. Kempe ; David C. Miller ; Allen Zielnik ; Daniel Montiel-Chicharro ; Jiang Zhu ; Ralph Gottschalg

SOLAR VARIABILITY REDUCTION USING OFF-MAXIMUM POWER POINT TRACKING AND BATTERY STORAGE 3214
Jason Galtieri ; Philip T. Krein

INTEGRATION OF ELECTROCHEMICAL CAPACITORS ON SILICON PHOTOVOLTAIC MODULES FOR RAPID-RESPONSE POWER BUFFERING 3220
Yu Jiang ; Xuanyi Shi ; Derwin Lau ; Da-Wei Wang ; Zi Ouyang ; Alison Lennon

DESIGN & EVALUATION OF A HYBRID SWITCHED CAPACITOR CIRCUIT WITH WIDE-BANDGAP DEVICES FOR COMPACT MVDC PV POWER CONVERSION 3224
J. Stewart ; J. Delhotal ; J. Richards ; J. Neely ; L. Rashkin ; J. D. Flicker ; R. Kaplar ; S. Gonzalez ; J. Lehr

SOLAR ENERGY FOR CLEAN AND AFFORDABLE WATER DESALINATION 3230
V. M. Fthenakis ; Adam A. Atia

GLOBAL RESIDENTIAL AIR-CONDITIONING SECTOR AS A DRIVER FOR PHOTOVOLTAIC INDUSTRY GROWTH DURING THE 21ST CENTURY 3236
Hannu S. Laine ; Jyri Salpakari ; Marius Peters ; Erin E. Looney ; Ashley E. Morishige ; Hele Savin ; Gregory Wilson ; Tonio Buonassisi

MEASURES TO REMOVE ECONOMIC NON-MARKET FAILURE AND INSTITUTIONAL BARRIERS THAT RESTRICT PHOTOVOLTAICS SELF-CONSUMPTION AND NET-METERING IN SPAIN 3240
Enrique Rosalcs-Ascnsio ; Juan A. Méndez ; Benjamín Gonzálcz-Díaz ; Ricardo Guerrero Lemus

COST COMPETITIVE CONCENTRATOR PHOTOVOLTAICS FOR SOLAR THERMAL APPLICATIONS 3245
Brian C. Riggs ; Richard E. Biedenham ; Chris Dougher ; Yaping Vera Ji ; Qi Xu ; Vince Romanin ; Daniel S. Codd ; James M. Zahler ; Matthew D. Escarra

PREDICTING THE EFFICIENCY OF THE SILICON BOTTOM CELL IN A TWO-TERMINAL TANDEM SOLAR CELL 3250

Zhengshan J. Yu ; Zachary C. Holman

MECHANICALLY STACKED 4-TERMINAL III-V/SI TANDEM SOLAR CELLS 3254

Stephanie Essig ; Christophe Allebe ; John F. Geisz ; Myles A. Steiner ; Loris Barraud ; Antoine Descoeudres ; J. Scott Ward ; Manuel Schnabel ; David L. Young ; Matthieu Despeisse ; Christophe Ballif ; Adele Tamboli

PEROVSKITE/SILICON TANDEM SOLAR CELLS: CHALLENGES TOWARDS HIGH-EFFICIENCY IN 4-TERMINAL AND MONOLITHIC DEVICES 3256

Jérémie Werner ; Florent Sahli ; Brett Kamino ; Davide Sacchetto ; Matthias Bräuninger ; Arnaud Walter ; Christophe Ballif ; Matthieu Despeisse ; Sylvain Nicolay ; Bjoern Niesen ; Raphaël Monnard ; Stefaan De Wolf ; Soo-Jin Moon ; Loris Barraud ; Bertrand Paviet-Salomon ; Jonas Geissbuehler ; Christophe Allebé

THE OUTCOME OF REPLACING SN COMPLETELY BY GE IN KESTERITE CU2ZNSNSE4SOLAR CELLS 3260

S. Sahayaraj ; G. Brammertz ; B. Vermang ; T. Schnabel ; E. Ahlswede ; Z. Huang ; S. Ranjbar ; M. Meuris ; J. Vleugels ; J. Poortmans

TRANSITION METAL OXIDES NANO-LAYERS AS EFFICIENT BACK ELECTRON REFLECTORS FOR CU2ZNSNSE4SOLAR CELLS 3265

Sergio Giraldo ; Moisés Espíndola-Rodríguez ; Florian Oliva ; Víctor Izquierdo-Roca ; Alejandro Pérez-Rodríguez ; Edgardo Saucedo

MIXED SULFUR AND SELENIUM ANNEALING STUDY OF COMPOUND-SPUTTERED BILAYER CU2ZNSNS4/ CU2ZNSNSE4PRECURSORS 3269

N. Ross ; S. Grini ; L. Vines ; C. Platzer-Björkman

REVEALING THE ROLE OF MN INCORPORATION IN CU2ZNSN(S, SE)4PHOTOVOLTAIC ABSORBER LAYER 3275

Stener Lie ; Joel M. R. Tan ; Wenjie Li ; Shin Woei Leow ; Oki Gunawan ; Doug Bishop ; Lydia H. Wong

NON-VACUUM SINGLE STEP SYNTHESIS OF LARGE-GRAIN SIZE CZTS PHOTO ABSORBER FOR THIN FILM SOLAR CELLS BY FLUX ASSISTED CHEMICAL SPRAY 3279

Ratheesh R. Thankalekshmi ; Navjot Kaur Sidhu ; A.C. Rastogi

RAMAN SCATTERING ASSESSMENT OF POINT DEFECTS IN KESTERITE SEMICONDUCTORS: UV RESONANT RAMAN CHARACTERIZATION FOR ADVANCED PHOTOVOLTAICS 3285

Florian Oliva ; Laia Arqués Farré ; Sergio Giraldo ; Mirjana Dimitrievska ; Paul Pistor ; Alejandro Martínez-Pérez ; Lorenzo Calvo-Barrio ; Edgardo Saucedo ; Alejandro Pérez-Rodríguez ; Victor Izquierdo-Roca

ASSESSING THE DEFECT RESPONSIBLE FOR LETID: TEMPERATURE- AND INJECTION-DEPENDENT LIFETIME SPECTROSCOPY 3290

Mallory A. Jensen ; Yan Zhu ; Erin E. Looney ; Ashley E. Morishige ; Carlos Vargas ; Ziv Hameiri ; Tonio Buonassisi

MICROSCOPIC DISTRIBUTION OF LUMINESCENCE FROM DISLOCATION CLUSTERS IN MULTICRYSTALLINE SILICON WAFERS 3295

H. T. Nguyen ; M. A. Jensen ; L. Li ; C. Samundsett ; H. C. Sio ; B. Lai ; T. Buonassisi ; D. Macdonald

DO GRAIN BOUNDARIES MATTER? ELECTRICAL AND ELEMENTAL IDENTIFICATION AT GRAIN BOUNDARIES IN LETID-AFFECTED P-TYPE MULTICRYSTALLINE SILICON 3300

Mallory A. Jensen ; Ashley E. Morishige ; Sagnik Chakraborty ; Romika Sharma ; Hang Cheong Sio ; Chang Sun ; Barry Lai ; Volker Rose ; Amanda Youssef ; Erin E. Looney ; Sarah Wieghold ; Jeremy Poindexter ; Juan-Pablo Correa-Baena ; Daniel Macdonald ; Joel B. Li ; Tonio Buonassisi

PERC SOLAR CELL PERFORMANCE PREDICTIONS FROM MULTICRYSTALLINE SILICON INGOT METROLOGY DATA 3304

Bernhard Mitchell ; Daniel Chung ; Qiuxiang He ; Hua Zhang ; Zhen Xiong ; Pietro P. Altermatt ; Peter Geelan-Small ; Thorsten Trupke

PHOTOLUMINESCENCE-IMAGING-BASED EVALUATION OF NON-UNIFORM CDTE DEGRADATION 3305

Steve Johnston ; David Albin ; Peter Hacke ; Steven P. Harvey ; Helio Moutinho ; Mowafak Al-Jassim ; Wyatt K. Metzger

MACHINE LEARNING AND CORRELATIVE MICROSCOPY: HOW 'BIG DATA' TECHNIQUES CAN BENEFIT THIN FILM SOLAR CELL CHARACTERIZATION 3309

Bradley M. West ; Michael Stuckelberger ; Tara Nietzold ; Barry Lai ; Jörg Maser ; Mariana I. Bertoni

METAL INDUCED CONTACT RECOMBINATION MEASURED BY QUASI-STEADY-STATE PHOTOLUMINESCENCE 3315

Robert Dumbrell ; Mattias K. Juhl ; Mengjie Li ; Thorsten Trupke ; Ziv Hameiri

USING TIME-OF-FLIGHT SIMS TO INVESTIGATE GROUP V DOPANT DISTRIBUTION IN CDTE 3319

Steven P. Harvey ; Eric Colegrove ; Brian Mccandless ; David Albin ; Mowafak Al-Jassim ; Wyatt K. Metzger

QUANTITATIVE ANALYSIS OF ACTIVE DOPANT DISTRIBUTION AND ESTIMATION OF EFFECTIVE DIFFUSIVITY IN PHOSPHORUS- IMPLANTED EMITTER OF SI SOLAR CELL USING SCANNING NONLINEAR DIELECTRIC MICROSCOPY 3323

Kotaro Hirose ; Katsuto Tanahashi ; Hidetaka Takato ; Yasuo Cho

SIMULATION OF DRIVE-LEVEL CAPACITANCE PROFILING TO INTERPRET MEASUREMENTS ON CU(IN, GA)SE2SCHOTTKY DEVICES 3327

Geordie Zapalac ; Jeff Bailey

ANALYSIS OF THE IMPACT OF INSTALLATION PARAMETERS AND SYSTEM SIZE ON BIFACIAL GAIN AND ENERGY YIELD OF PV SYSTEMS 3333

Amir Asgharzadeh ; Tomas Lubenow ; Joseph Sink ; Bill Marion ; Chris Deline ; Clifford Hansen ; Joshua Stein ; Fatima Toor

DEPENDENCE OF STRING POWER ON ITS HEIGHT IN THE ARRAY IN YOSHINOGARI MEGA SOLAR POWER PLANT 3339

Shigeomi Hara ; Makoto Kasu ; Yasuki Masutomi

A BOTTOM-UP ENERGY SIMULATION FRAMEWORK TO ACCURATELY COMPARE PV MODULE TOPOLOGIES UNDER NON-UNIFORM AND DYNAMIC OPERATING CONDITIONS 3343

Patrizio Manganiello ; Maro Baka ; Hans Goverde ; Tom Borgers ; Jonathan Govaerts ; Arvid Van Der Heide ; Eszter Voroshazi ; Francky Catthoor

A PERFORMANCE MODEL FOR BIFACIAL PV MODULES 3348

Daniel Riley ; Clifford Hansen ; Joshua Stein ; Matthew Lave ; Johnson Kallickal ; Bill Marion ; Fatima Toor

ACCURATE MODELING OF PARTIALLY SHADED PV ARRAYS 3354

Bennet Meyers ; Mark Mikofski

EVALUATION OF UNCERTAINTY IN PV PROJECT DESIGN: DEFINITION OF SCENARIOS AND IMPACT ON ENERGY YIELD PREDICTIONS 3360

Giorgio Belluardo ; Magnus Herz ; Ulrike Jahn ; Mauricio Richter ; David Moser

MONOCRYSTALLINE 1.7 EV MGCDTE DOUBLE-HETEROSTRUCTURE SOLAR CELL WITH 11.2% EFFICIENCY 3366

Calli M. Campbell ; Xin-Hao Zhao ; Yuan Zhao ; Mathieu Boccard ; Cheng- Ying Tsai ; Jacob J. Becker ; Zachary Holman ; Yong-Hang Zhang

MBE GROWTH OF 1.7EV AL0.2GA0.8AS AND 1.42EV GAAS SOLAR CELLS ON SI USING DISLOCATIONS FILTERS: AN ALTERNATIVE PATHWAY TOWARD III-V/ SI SOLAR CELLS ARCHITECTURES 3370

Arthur Onno ; Mingchu Tang ; Mu Wang ; Yurii Maidaniuk ; Mourad Benamara ; Yuriy I. Mazur ; Gregory J. Salamo ; Lars Oberbeck ; Jiang Wu ; Huiyun Liu

III- V/SI TANDEM CELLS UTILIZING INTERDIGITATED BACK CONTACT SI CELLS AND VARYING TERMINAL CONFIGURATIONS 3371

Manuel Schnabel ; Michael Rienacker ; Agnes Merkle ; Talysa R. Klein ; Nikhil Jain ; Stephanie Essig ; Henning Schulte-Huxel ; Emily Warren ; Maikel F.A.M. Van Hest ; John Geisz ; Jan Schmidt ; Rolf Brendel ; Robby Peibst ; Paul Stradins ; Adele Tamboli

TOWARDS HIGH-EFFICIENCY GAASP/SI TANDEM CELLS 3376

S. Fan ; M. Vaisman ; K. Nay Yaung ; E. Perl ; D. Martín-Martín ; M. Leilaeioun ; Z. C. Holman ; M. L. Lee

CHARACTERIZATION OF HETEROEPITAXIAL GAAS FILMS GROWN ON SI USING SELECTIVE AREA NUCLEATION 3381

Emily L. Warren ; Emily A. Makoutz ; Michelle Vaisman ; Benjamin F. Bachman ; William E. Mcmahon ; Jeramy D. Zimmerman ; Adele C. Tamboli

EFFICIENT PHOTON UPCONVERSION IN SEMICONDUCTOR NANOSTRUCTURES: CONSTRAINTS AND OPPORTUNITIES 3384

Matthew F. Doty ; Eric Y. Chen ; Jing Zhang ; Diane G. Sellers ; Zhuohui Li ; Christopher C. Milleville ; Kyle Lennon ; Joshua M. O. Zide

ENHANCED ULTRA-THIN A-GE:H SOLAR CELLS BY PLASMONIC NANOPARTICLES EMBEDDED IN THE OPTICAL RESONANT CAVITY 3388

Brendan Brady ; Volker Steenhoff ; Benedikt Nickel ; Martin Vehse ; Alexander G. Brolo

NATIVE-METAL-OXIDE-COATED PLASMONIC ELECTRODE METASURFACES FOR NANOPHOTONIC LIGHT TRAPPING AND EFFICIENT CHARGE COLLECTION 3393

Deirdre M. O'Carroll ; Christopher E. Petoukhoff ; Zhongkai Cheng ; Zeqing Shen ; Catrice M. Carter

IN-GA PRECURSOR ISLANDS FOR CU(IN, GA)SE2MICRO-CONCENTRATOR SOLAR CELLS 3396

Katharina Eylers ; Franziska Ringleb ; Berit Heidmann ; Sergiu Levcenco ; Thomas Unold ; Hagen W. Klemm ; Gina Peschel ; Alexander Fuhrich ; Thomas Teubner ; Thomas Schmidt ; Martina Schmid ; Torta Boeck

ADVANCES IN SILICON SURFACE TEXTURIZATION BY METAL ASSISTED CHEMICAL ETCHING FOR PHOTOVOLTAIC APPLICATIONS 3402

Sylvain Le Gall ; Raphaël Lachaume ; Encarnacion Torralba ; Mathieu Halbwax ; Vincent Magnin ; Taha El Assimi ; Marin Fouchier ; Joseph Harari ; Jean-Pierre Vilcot ; Christine Cachet-Vivier ; Stéphane Bastide

SINGLE CRYSTALLINE SUBSTRATES FOR III- V GROWTH VIA EXFOLIATION OF BULK SINGLE CRYSTALS 3406
Celeste L. Melamed ; Brenden R. Ortiz ; Aaron D. Martinez ; William E. Mcmahon ; Adele C. Tamboli ; Andrew G. Norman ; Eric S. Toberer

CUZNS HOLE CONTACTS ON MONOCRYSTALLINE CDTE SOLAR CELLS 3410
Jacob J. Becker ; Xiaojie Xu ; Rachel Woods-Robinson ; Calli M. Campbell ; Maxwell Lassise ; Joel Ager ; Yong-Hang Zhang

THE EFFECT OF THE CDCL2 HEAT TREATMENT ON CDSEXTE1-X SOLAR CELLS 3413
Chih An Hsu ; Vasilios Palekis ; Imran Khan ; Shamara Collins ; Don Morel ; Chris Ferekides

EFFECTS OF CDCL2TREATMENT ON THE LOCAL ELECTRONIC PROPERTIES OF POLYCRYSTALLINE CDTE MEASURED WITH PHOTOEMISSION ELECTRON MICROSCOPY 3417
Morgann Berg ; Jason M. Kephart ; Walajabad S. Sampath ; Taisuke Ohta ; Calvin Chan

POINT DEFECTS IN CDTE BULK SINGLE CRYSTALS GROWN IN CD-RICH CONDITIONS 3422
Tursun Ablekim ; Santosh K. Swain ; Teresa M. Barnes ; Kelvin G. Lynn

OPTICAL PROPERTIES OFCDSE1-XSXANDCDSE1-YTEYALLOYS AND THEIR APPLICATION FOR CDTE PHOTOVOLTAICS 3426
Maxwell M. Junda ; Corey R. Grice ; Prakash Koirala ; Robert W. Collins ; Yanfa Yan ; Nikolas J. Podraza

BLISTERING OF MAGNETRON SPUTTERED THIN FILM CDTE DEVICES 3430
P.M. Kaminski ; S. Yilmaz ; A. Abbas ; F. Bittau ; J.W. Bowers ; R.C. Greenhalgh ; J.M. Walls

ENERGY YIELD IN HOT & SUNNY CLIMATES: IMPACT OF SILICON SOLAR CELL ARCHITECTURE AND CELL INTERCONNECTION 3435
Jan Haschke ; Johannes P. Seif ; Yannick Riesen ; Andrea Tomasi ; Jean Cattin ; Loïc Tous ; Patrick Choulat ; Monica Aleman ; Emanuele Comagliotti ; Angel Uruena ; Richard Russell ; Filip Duerinckx ; Jonathan Champliaud ; Jacques Levrat ; Amir A. Abdallah ; Brahim Aïssa ; Nouar Tabet ; Nicolas Wyrsch ; Matthieu Despeisse ; Jozef Szlufcik ; Stefaan De Wolf ; Christophe Ballif

NOVEL REAR SIDE METALLIZATION ROUTE FOR SI SOLAR CELLS USING A TRANSPARENT CONDUCTING ADHESIVE 3439
Manuel Schnabel ; Talysa R. Klein ; Benjamin G. Lee ; William Nemeth ; Vincenzo Lasalvia ; Maikel F.A.M. Van Hest ; Paul Stradins

MULTILAYER FOIL METALLIZATION FOR ALL BACK CONTACT CELLS 3442
David H. Levy ; David E. Carlson

ELECTROLUMINESCENCE EXCITATION SPECTROSCOPY: A NOVEL APPROACH TO NON-CONTACT QUANTUM EFFICIENCY MEASUREMENTS 3448
Kristopher O. Davis ; Greg S. Horner ; Joshua B. Gallon ; Leonid A. Vasilyev ; Kyle B. Lu ; Antonius B. Dirriwachter ; Terry B. Rigdon ; Eric J. Schneller ; Kortan Ogutman ; Richard K. Ahrenkiel

ILLUMINATED OUTDOOR LUMINESCENCE IMAGING OF PHOTOVOLTAIC MODULES 3452
Timothy J Silverman ; Michael G. Deceglie ; Kaitlyn Vansant ; Steve Johnston ; Ingrid Repins

ELECTROLUMINESCENT IMAGE PROCESSING AND CELL DEGRADATION TYPE CLASSIFICATION VIA COMPUTER VISION AND STATISTICAL LEARNING METHODOLOGIES 3456
Justin S. Fada ; Mohammad A. Hossain ; Jennifer L. Braid ; Shuying Yang ; Timothy J Peshek ; Roger H. French

TOWARDS DEVELOPING A STANDARD FOR TESTING BIFACIAL PV MODULES: SINGLE-SIDE VERSUS DOUBLE-SIDE ILLUMINATION METHOD I-V MEASUREMENTS UNDER DIFFERENT IRRADIANCE AND TEMPERATURE 3462
Stefan Roest ; Witek Nawara ; Bas B. Van Aken ; Elias Garcia Goma

ELECTRICAL TRANSPORT PROPERTIES FROM LONG WAVELENGTH ELLIPSOMETRY 3468
Prakash Uprety ; Maxwell M. Junda ; Indra Subedi ; Michael A. Slocum ; David V. Forbes ; Seth M. Rubbard ; Nikolas J. Podraza

IN SITU RAMAN MONITORING OF KESTERITE CU2ZNSNS4 PHASE FORMATION FROM SULFURIZATION OF SOL-GEL OXIDE PRECURSORS 3473
Osama Awadallah ; Joseph Hernandez ; Andriy Durygin ; Zhe Cheng

PERFORMANCE OF FIELD-AGED PV MODULES IN INDIA: RESULTS FROM 2016 ALL INDIA SURVEY OF PV MODULE RELIABILITY 3478
Rajiv Dubey ; Sachin Zachariah ; Shashwata Chattopadhyay ; Vivek Kuthanazhi ; Sugguna Rambabu ; Sonali Bhaduri ; Hemant K. Singh ; Archana Sinha ; Birinchi Bora ; Rajesh Kumar ; O. S. Sastry ; Chetan S. Solanki ; Anil Kottantharayil ; Brij M. Arora ; K. L. Narasimhan ; Juzer Vasi

INFERRING THE PERFORMANCE RATIO OF PV SYSTEMS DISTRIBUTED IN AN REGION: A REAL-CASE STUDY IN SOUTH TYROL 3482
Marco Pierro ; Giorgio Belluardo ; Philip Ingenhoven ; Cristina Cornaro ; David Moser

QUANTIFY PHOTOVOLTAIC MODULE DEGRADATION USING THE LOSS FACTOR MODEL PARAMETERS 3488

C. Birk Jones ; Bruce H. King ; Joshua S. Stein ; Justin S. Fada ; Alan J. Curran ; Roger H. French ; Erdmut Schnabel ; Michael Koehl ; Olga Lavrova

SIMULATING PV SYSTEM PERFORMANCE WITH COMPONENT RELIABILITY DISTRIBUTIONS 3494

Geoffrey T. Klisel ; Janine M. Freeman ; Olga Lavrova

LIFETIME AND DEGRADATION OF PRE-DAMAGED PV-MODULES – FIELD STUDY AND LAB TESTING 3500

Claudia Buerhop ; Sven Wirsching ; Simon Gehre ; Tobias Pickel ; Thilo Winkler ; Andreas Bemrrr ; Julia Merghcim ; Christian Camus ; Jens Hauch ; Christoph J. Brabec

IMM TRIPLE-JUNCTION SOLAR CELLS AND MODULES OPTIMIZED FOR SPACE AND TERRESTRIAL CONDITIONS 3506

Tatsuya Takamoto ; Hiroyuki Juso ; Kohsuke Ueda ; Hidetoshi Washio ; Hiroshi Yamaguchi ; Mitsuru Imaizumi ; Taishi Sumita ; Tetsuya Nakamura

VERY HIGH SPECIFIC POWER ELO SOLAR CELLS (>3 KW/KG) FOR UAV, SPACE, AND PORTABLE POWER APPLICATIONS 3511

D. Cardwell ; A. Kirk ; C. Stender ; A. Wibowo ; F. Tuminello ; M. Drees ; R. Chan ; M. Osowski ; N. Pan

ENHANCED ENDURANCE OF A UNMANNED AERIAL VEHICLES USING HIGH EFFICIENCY SI AND III-V SOLAR CELLS 3514

David Scheiman ; Raymond Hoheisel ; Daniel J Edwards ; Andrew Paulsen ; Justin Lorentzen ; Steve Carruthers ; Sam Carter ; Matthew Kelly ; Phillip Jenkins ; Robert Walters

HIGH PERFORMANCE, LIGHTWEIGHT GAAS SOLAR CELLS FOR AEROSPACE AND MOBILE APPLICATIONS 3520

Aarohi Vijh ; Lori Washington ; Robert C. Parenti

THROUGH-EPITAXIAL-VIA BACK-CONTACT MULTI-JUNCTION SOLAR CELLS FABRICATED USING EPITAXIAL LIFT-OFF 3524

Rekha Reddy ; Marilyn L. Nowakowski ; David Rowell ; Christopher L. Stender ; Christopher Youtsey

DESIGN OF INGAP/GAAS/LNGAAS MULTI-JUNCTION CELLS WITH REDUCED LAYER THICKNESSES USING LIGHT-TRAPPING REAR TEXTURE 3528

Lin Zhu ; Anurag Reddy ; Kentaroh Watanabe ; Masakazu Sugiyama ; Yoshiaki Nakano ; Hidefumi Akiyama

Author Index

A Reduced Order Model for a TOV Study in a Solar PV Project

Ahmad Abdullah*and Billy Yancey[†]
*Electric Power Engineers, Inc and the Department of Electrical Power and Machines
Cairo University, Faculty of Engineering
Cairo, Egypt 12613
E-mail: ahmad.abdullah@ieee.org
[†]Electric Power Engineers, Inc
Austin, TX, USA 78738
E-mail: byancey@epeconsulting.com

Abstract—Special system studies are needed to assess the different preliminary designs of solar photovoltaic (PV) projects. One of these is a Temporary Overvoltage (TOV) study. The main purpose of a TOV study is to evaluate the capability of the surge arresters (SAs) within the substation. To assess the capability of the SAs accurately, a detailed electromagnetic transient (EMT) model of the project has to be built. With the detailed EMT model, which has a large number of inverters, the run time of the model becomes prohibitive even for a single scenario. In this paper, we propose a method to systematically reduce the order of the EMT model at the substation level thus making the model suitable for TOV studies. The response of the reduced order model is then benchmarked against the response of the full order model of an 80 MW solar PV project for various TOV scenarios. Simulation results show satisfactory agreement between the response of the detailed model and the response of the reduced order model. Additionally, the run time of the proposed reduced order model is less than the run time of the full order model by a factor of ninety six.

Index Terms—Solar power generation, Photovoltaic systems, Electromagnetic transients, Surge arresters, Transient overvoltage, Surge protection, Temporary Overvoltage

I. INTRODUCTION

With the increased penetration of renewable energy into the grid, special system studies are called upon to assess their impact on various aspects of the power system. One of these aspects is evaluating the adequacy the of surge arresters within the substation. Surge arrester MCOV and energy handling capability are generally selected on an ad hoc manner in the early design stage. Assessment of the adequacy of the SAs in the substation ensures that SAs can ride through the TOV by absorbing an amount of energy that is within their energy handling capability and that the TOV level is limited to a value determined by applicable standards. The IEEE Standard C62.82.1-2010 [1] defines TOV as "an oscillatory phase-to-ground or phase-to-phase overvoltage that is at a given location of relatively long duration (seconds, even minutes) and that is undamped or only weakly damped; resulting from operation of a switching device or fault condition".

This can occur when the PV inverter is suddenly disconnected from the grid. Because inverters act as a constant current source, hence when a circuit breaker opens the inverter terminal voltage can cause voltage fluctuations. When this occurs, inverters quickly shut down, but there can be a short period of time where some inverters can create overvoltage spikes. This is a concern for PV system owners and utilities since large voltage spikes can damage other equipment that is still connected in the vicinity.

Historically, assessment of SAs had been done under specific assumptions about the nature of TOV. For example, in conventional gas generation and if the neutral of synchronous generator is ungrounded, it is known that a single line to ground fault can cause the phase to ground voltage to increase by a factor of $\sqrt{3}$. This value can be used along with the surge arrester TOV withstand capability curve [2] to judge the adequacy of the SA. However, due to the nature of the technology used in renewable energy resources, this might not be true. Renewable energy resources are generally inverter based generation. These inverters incorporate a large number of switches and are of various topologies. The temporary overvoltage withstand capability is usable for TOVs lasting at least 10 milliseconds and due to the microprocessor controller used within these inverters, most TOV events are transient in nature and the duration of the TOV events in most cases do not exceed milliseconds. Thus, the SA overvoltage withstand capability curve is of no practical usefulness in case of TOV events due to inverter based technologies. Hence, it is not always possible to assess the capability of a SA using the project configuration, grounding scheme and the TOV withstand capability.

This necessitates a paradigm shift in performing TOV studies. Building the renewable energy project in an EMT type software becomes a must to assess the performance of SAs under various scenarios. The model must include all inverters as well as all SAs characteristics in order to accurately represent the project. Running such models in EMT software requires small solution time step and thus a long simulation run time. Moreover, performing many TOV scenarios becomes a daunting task due to the long simulation run time. Thus, it is of utmost importance to develop a method to reduce the total number of switches (inverters) in the model to reduce the simulation run time.

Most equivalencing techniques [3] treat the renewable project as one unit, i.e., the whole project starting from the main power transformer (MPT) down to the medium voltage collector system and the low voltage inverters are replaced

978-1-5090-5606-4/17 $31.00 © 2017 IEEE

by a single electrical component that accurately captures the transient performance of the project as a whole. This is done mainly for grid impact studies and specifically for dynamic simulations. Popular methods such as the one in [4] is suitable only for power flow and dynamic studies not EMT type simulations.

In this paper, we provide a way to reduce the order of the solar PV project at the substation level. Each feeder of the collector system is reduced on its own to a simple generation resource and an impedance. Thus the number of the inverters in the EMT model is drastically reduced to the number of the feeders in the collector system. It thus possible to study the performance of the SAs in the substation since they, generally, are installed at the beginning of each of these collector feeders.

The paper is organized as follows. The detailed system EMT model is described in section §II. The benchmark response of the inverter as supplied by the manufacturer is shown in III. The methodology is provided in section §IV. The TOV scenarios used for comparing the detailed and reduced order model is provided in section §V. The results of the reduced order model is shown in section §VI. Conclusions are summarized in section §VII.

II. DETAILED SYSTEM MODEL

The system under study is an 80 MW solar PV project and is shown in figure 1. The project is divided into four collector feeders and two capacitor banks each rated at 4.5 MVAr. Each feeder has different number of inverters connected to it. The configuration of each feeder has been removed from the paper for confidentiality reasons.

The number of inverters on each feeder is shown in figure 1. The project has a total of 45 inverters and each one is capable of producing 1.8 MVA. Each inverter block in figure 1 has a DC to AC stage, an LC filter and an inverter step up transformer (ISU) transformer. Each ISU transformer is rated at 1.85 MVA and connected in delta-star with the star connected to the low voltage side of the inverter and ungrounded. The low voltage is 0.42 kV while the medium voltage is 34.5 kV. A schematic of the inverter is shown in figure 2.

Surge Arresters exist at the beginning of each feeder inside the substation as shown in figure 1. All surge arresters at the medium voltage level are MOV type, have the same MCOV of 24.4 kV and have the same energy handling capability of 219 kJ. The V-I characteristics of the SAs are obtained from [5] and is shown in figure 3. The project connects to the Point of Interconnection (POI) at 138 kV thorough the MPT which has a rating of 89 MVA. The feeder circuit breakers are EMA type breakers [6]. These circuit breakers are equipped with a mechanically interlocked switch on the load side that grounds the load side within 1 cycle of opening the circuit breaker's main contacts.

III. INVERTER BENCHMARK TESTS

As has been stated in section §I, the inverter response in fundamentally different from conventional synchronous

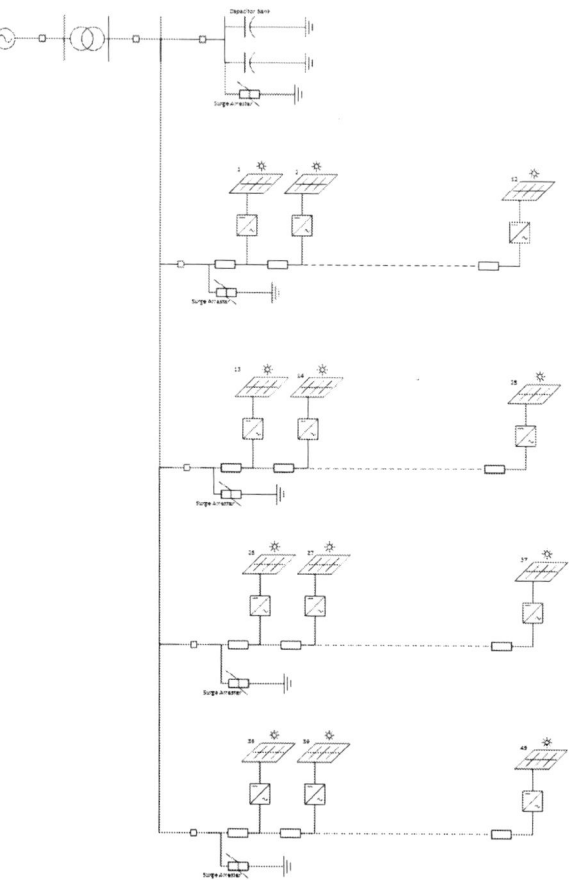

Fig. 1. Full Order EMT model

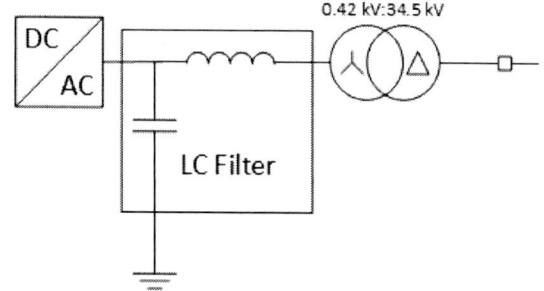

Fig. 2. Schematic of the inverter in the project

machines. To be able to successfully reduce the order of the model and design the benchmark scenarios in section §V, the response of the inverter under specific tests has to be known. The inverter manufacturer supplied two benchmark tests. The first one is a load rejection test in figure 4 and the second one is a line to line fault in figure 5. The fault is performed at the inverter terminals with the ISU transformer terminals connected to a infinite bus. It can be seen from both figure 4 and figure 5 that the load rejection test produces the worst TOV as opposed to conventional power systems where

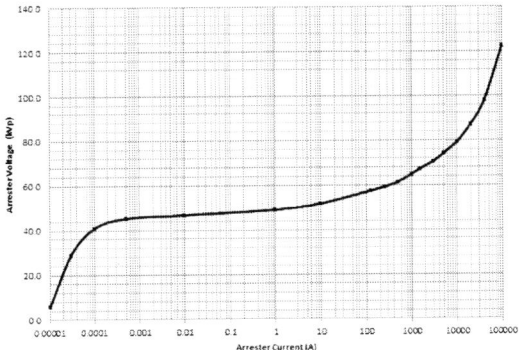

Fig. 3. Voltage-Current (V-I) Characteristics of the MV surge arrester

generally the single line to ground fault causes the highest TOV.

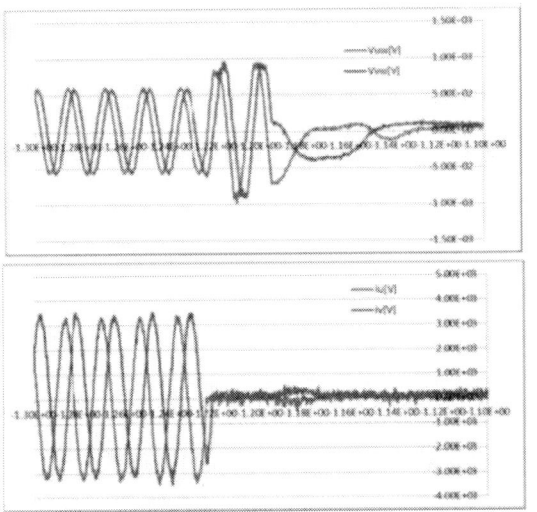

Fig. 4. Load rejection test (upper curve is voltage - lower curve is current)

IV. METHODOLOGY

Just as any electrical source can be represent by its Thevenin's or Norton's equivalent [7], [8], the inverters within the solar PV project can be modeled as such depending on the technology used within the inverter. However, most manufacturers of solar PV inverters use a technology that makes the inverters act as a current source or a voltage controlled current source. Due to that, Norton's equivalent model would be most suitable. A Norton's equivalent consists of two parts: the Norton's current source and the impedance in parallel with it.

The basic idea behind the method in this paper is to represent each feeder by a pseudo-Norton's equivalent. The pseudo-Norton's equivalent will consist of two parts: a pseudo-Norton source and an impedance in parallel. The pseudo-Norton's source will be responsible for equivalencing the low frequency response of the feeder, while the impedance

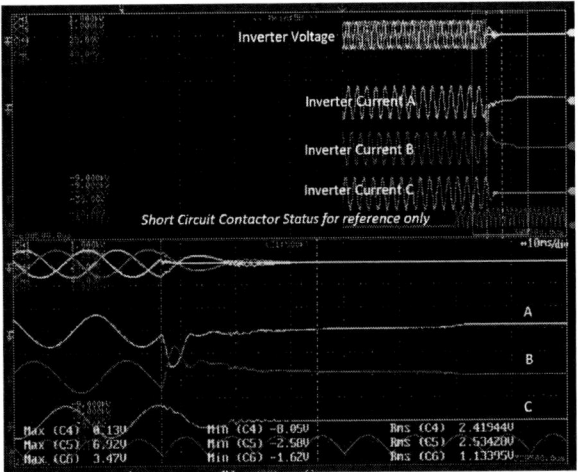

Fig. 5. Phase A to Phase B fault on the inverter terminal

in parallel will be equivalencing the high frequency response of the feeder. This effectively means that the step response of the pseudo-Norton's source should correspond to the low frequency response portion of the overall feeder frequency response. This also means that the step response of the impedance in parallel should correspond to the high frequency response portion of the overall feeder frequency response. The construction of the pseudo-Norton's source is provided in section IV-A while the construction of the impedance in parallel is provided in section §IV-B.

A. The Pseudo-Norton's Source

The pseudo-Norton's source will consist of an inverter stage with its associated controls, a filter as well a transformer. Generally speaking, any inverter must contain a filter to shape the output waveforms by rejecting the high frequency switching harmonics. Inverter manufacturers can use transformerless [9] technology, but this is outside the scope of this paper. The validity of the current methodology is yet to be investigated under transformerless technology.

To construct the pseudo-Norton's source, the inverter MVA rating must be scaled up by a factor equal to the total number of inverters on the feeder. Many vendors supply proprietary EMT models of their inverters that has the number of inverters or the MVA variable. If the user is using a custom EMT model, then the model must have the MVA rating or the number of inverters variable. The controls of the inverter are to kept the same without any change. The values of the inductance and the capacitance of LC filter are also to be scaled up by the number of inverters. Lastly, the ISU transformer MVA rating is also to be scaled up by a factor equal to the number of inverters without changing the per unit impedance of the ISU transformer.

In this paper, we used a confidential model supplied by the manufacturer. The model has proprietary control algorithms, proprietary switching topology, LC filer and an ISU transformer. We only scaled the model as described in this section.

B. The Impedance in Parallel

It is necessary that the parallel impedance represents the high frequency response of the collector system. Since it is typical in power flow studies to represent the cable sections in the collector system using pi-models, the reader is to be warned against such representation in EMT type analysis as this representation is only valid at the power frequency. The parallel impedance is nothing other than a frequency dependent impedance that captures the high frequency response of the cable sections in collector feeder. Thus the cable sections along the feeder have to be modeled by a suitable frequency dependent model. The user has two choices:

1) Perform a frequency scan on the feeder with all inverters removed from the project (ISU transformers have to be open-circuited as well as the feeder breaker). Using that frequency scan, the user can use vector fitting [10], [11] to construct a frequency dependent model. Passivity has to be enforced upon the resulting fitting by insuring that negative resistance is not a result of the fitting. Negative resistance causes instability in the EMT simulations.

2) Keep the feeder intact without performing a vector fitting. This means that the cable sections are to be kept in the model but without the inverters, LC filters or the ISU transformers.

Theoretically, both methods should represent the same impedance. The first choice above can be done very quickly in PSCAD™/EMTDC. However, PSCAD™/EMTDC does not enforce passivity on the resulting frequency dependent model. Due to that, the authors used the second choice and they will treat the first method in a separate publication.

The authors also noted that for the equivalencing to produce satisfactory results, some cable sections in feeder have to be left out from this equivalencing process. It turned out that the first cable section has to be removed from this frequency dependent impedance. The overall reduced order model is shown in figure 6.

V. TOV SCENARIOS

A total of 52 cases (8 cases per feeder) have been used to benchmark the response of the reduced order model against the full order model. The cases are grouped into three main categories: feeder energization, load rejection and faults.

Since we are only interested in the SA evaluation, the voltage waveform has been monitored at the terminal of each SA. Current waveforms will be reported upon in a different publication. Circuit breakers are given close/open signal at around t=0.15 seconds.

The feeder energization cases are designed to test the validity of the construction of the parallel frequency dependent impedance. In these cases all inverters were offline and the feeder was initially de-energized. At certain time, the

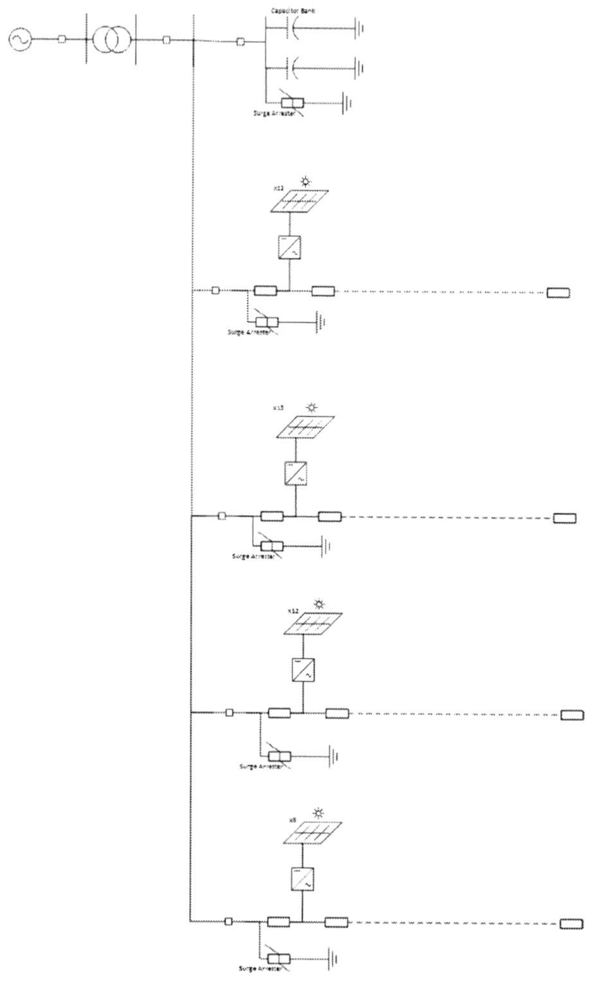

Fig. 6. Reduced Order Model

feeder breaker is closed and the response of both models are compared.

For the load rejection cases, the feeder was initially energized and then the circuit breaker was abruptly opened. Load rejection cases are the most important ones as evident from the tests supplied by the manufacturer in section §III. Load rejection cases should reveal whether scaling, grouping the inverters and placing them after the first cable section of each feeder is a valid approach.

Finally, fault cases have been performed. These are single line to ground , line-line and three phase to ground fault on the feeder. table I summarizes the cases that have been performed for each feeder.

VI. RESULTS

Due to paper length requirement, only the key scenario results of the benchmark cases are discussed as part of this study. Specifically, scenarios B2, C1, C2, and D1 we will be discussed in further detail. The voltage at the MV bus is measured and compared. It should be noted that any

TABLE I
BENCHMARK CASES

Study Scenario	Case Number	Description
Case A: Project energization	A1	Energization of a feeder with capacitor banks ON
	A2	Energization of a feeder with capacitor banks OFF
	A3	Energization of all feeders with capacitor banks ON
	A4	Energization of all feeders with capacitor banks OFF
Case B: Capacitor switching	B1	Energization of both capacitor banks at 20% project output
	B2	Energization of both capacitor banks at 100% project output
	B3	De-energization of both capacitor banks with 20% project output
	B4	De-energization of both capacitor banks with 100% project output
Case C: De-energization/ load rejection	C1	Tripping of a feeder with capacitor banks OFF
	C2	Tripping of a feeder with capacitor banks ON
	C3	Tripping of whole project with capacitor banks OFF
	C4	Tripping of whole project with capacitor banks ON
Case D: Various faults	D1	LG fault on a feeder with capacitor banks ON
	D2	LG fault on a feeder with capacitor banks OFF
	D3	LL fault on a feeder with capacitor banks ON
	D4	LL fault on a feeder with capacitor banks OFF
	D5	LLLG fault on a feeder with capacitor banks ON
	D6	LLLG fault on a feeder with capacitor banks OFF
	D7	LG fault at 138 kV HV substation bus with capacitor banks ON
	D8	LG fault at 138 kV HV substation bus with capacitor banks OFF
	D9	LLLG fault at 138 kV HV substation bus with both capacitor banks ON
	D10	LLLG fault at 138 kV HV substation bus with both capacitor banks OFF

switching operation happens around t=0.15 seconds. In all the figures below, the waveforms with dashed lines represent the reduced order model while waveforms with the continuous lines represent the full order model.

For Case A scenarios in table I, the response of the reduced order model is identical to the response of the full order model. This because in the reduced order model, the feeder has been left intact as mentioned in section §IV-B and energization occurs with the inverters offline, i.e., the inverter response is not part of the overall response and the voltage transients obtained and are due to the response of the feeders only.

For Case B scenarios, it is noted that the voltage waveforms of the reduced order model have more oscillations than the detailed EMT model. This is due to the fact that when the inverters are distributed across the feeder, cable sections of the feeder act as a filter that suppresses unwanted harmonics. However, in the reduced order model, only one cable section exists before the measuring point. Even though, we increased the inverter filter size, we do not get an exact replica of the voltage waveform of the detailed system model. The response of the reduced order model as well as the detailed model is shown in figure 7. However these higher harmonics don't affect the energy absorbed by the SAs since they are fast

transients and much below the MCOV of the SAs.

For Case C scenarios and apart from the higher frequency harmonics, the response of reduced order model of one of the phases seems to not agree very well with the response of the full order model in Case C1. However, this disparity between the two responses causes a 15 kJ difference only between the calculated absorbed energy of the SAs between the detailed and reduced order model. Given that the energy handling capability of the SAs is 219 kJ, the error is less than 7%. The response of the reduced order model as well as the detailed model is shown in figure 8. For Case C2, the reduced order model follows the detailed model except for higher frequency harmonics as in Case B scenarios. Only Case C1 shows such disparity and other Case C scenarios exhibit the same type of response as in Case B scenarios.

For Case D scenarios, the observations from Case A scenarios still hold. The response of the reduced order model as well as the detailed model is shown in figure 10.

All cases in this section has been focused on the response of Feeder 1. However, all other feeders exhibit the same response as Feeder 1. In all scenarios, the run time of the full order model is 8 hours per scenario on a workstation having 6 cores at a clock speed of 3.2 GHz and 32 GB of RAM. For the reduced order model, the simulation run time does not exceed

6 minutes for any scenario.

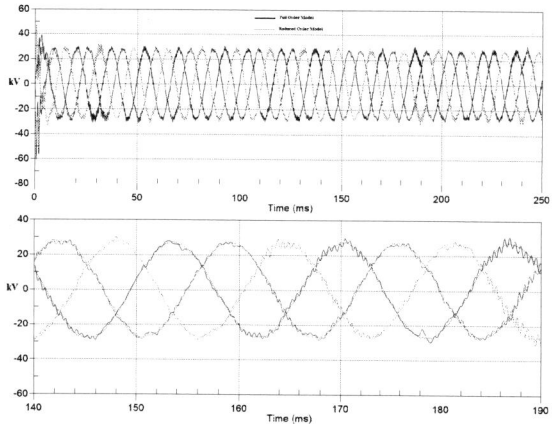

Fig. 7. 34.5 kV bus three phase voltages for Case B2 for feeder 1

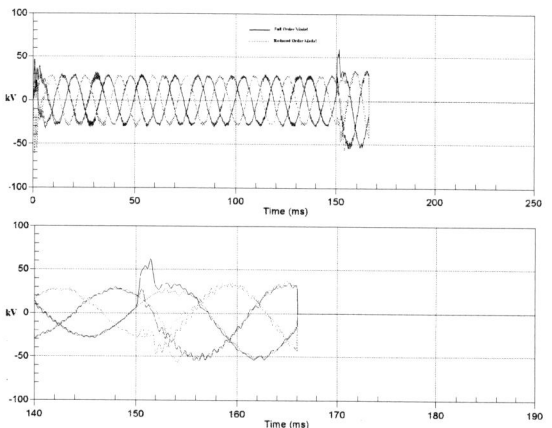

Fig. 8. 34.5 kV bus three phase voltages for Case C1 for feeder 1

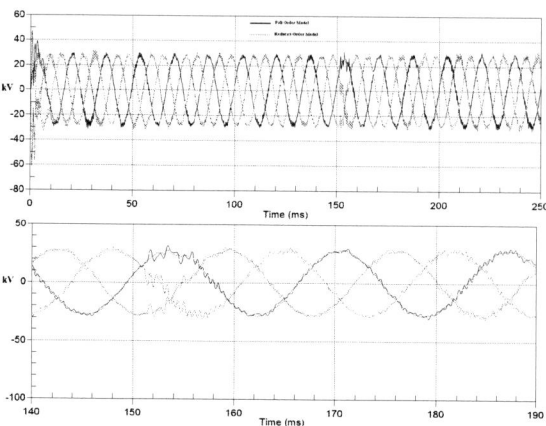

Fig. 9. 34.5 kV bus three phase voltages for Case C2 for feeder 1

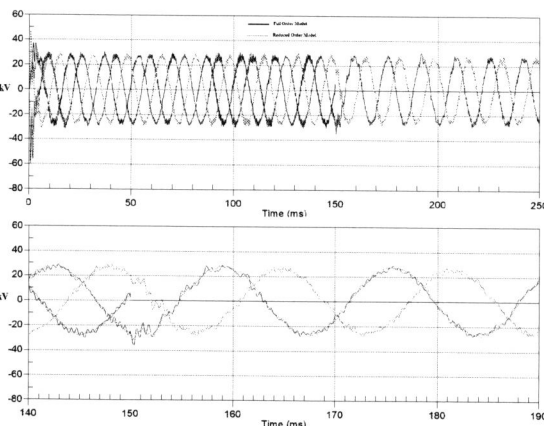

Fig. 10. 34.5 kV bus three phase voltages for Case D1 for feeder 1

VII. CONCLUSIONS

In this paper a systematic way of reducing the order of solar PV project at the substation level has been presented. The method is shown to capture the response of the collector system under various TOV scenarios to a satisfactory accuracy. The reduced order model takes much less time to run as compared to the full order model. This makes it useful in analyzing a large number of scenarios without the need for a highly capable computer.

The analysis in this paper is only limited to the SAs in the substation. If the feeder has SAs at the ends of the collector system- as usually the case- the method here will not be applicable but another method needs to be used to evaluate those SAs and will be analyzed in future research. Theoretically speaking the SAs in the substation are the ones that are under the most severe duty, and if those arresters can be shown to withstand various TOVs, then all other SAs in the project should be safe as well as long as they have the same capability.

REFERENCES

[1] IEEE Std 1312-1-1996, "IEEE Standard for Insulation Coordination - Definitions, Principles, and Rules," 1996.

[2] IEEE Std C62.22-2009, "Ieee guide for the application of metal-oxide surge arresters for alternating-current systems," pp. 1–142, July 2009.

[3] D. N. Hussein, M. Matar, and R. Iravani, "A type-4 wind power plant equivalent model for the analysis of electromagnetic transients in power systems," *IEEE Transactions on Power Systems*, vol. 28, no. 3, pp. 3096–3104, 2013.

[4] E. Muljadi, C. Butterfield, A. Ellis, J. Mechenbier, J. Hochheimer, R. Young, N. Miller, R. Delmerico, R. Zavadil, and J. Smith, "Equivalencing the collector system of a large wind power plant," in *2006 IEEE Power Engineering Society General Meeting*. IEEE, 2006, pp. 9–pp.

[5] J. Woodworth, "Arrester selection guide." [Online]. Available: http://www.arresterworks.com/arresterfacts/pdf_files/Arrester_Characteristics_for_ATPDraw_Users.xls

[6] EMA Electromechanics, "Combined vacuum substation circuit breaker and high-speed mechanically interlocked grounding switch." [Online]. Available: http://emaelectromechanics.com/vdhgsmi/

[7] N. Watson and J. Arrillaga, *Power systems electromagnetic transients simulation*. Iet, 2003, vol. 39.

[8] L. S. Bobrow, *Elementary linear circuit analysis*. Oxford Univ. Press, 1997.

[9] R. Inzunza and H. Akagi, "A 6.6-kv transformerless shunt hybrid active

filter for installation on a power distribution system," *IEEE Transactions on power electronics*, vol. 20, no. 4, pp. 893–900, 2005.

[10] B. Gustavsen and A. Semlyen, "Rational approximation of frequency domain responses by vector fitting," *IEEE Transactions on power delivery*, vol. 14, no. 3, pp. 1052–1061, 1999.

[11] A. Morched and V. Brandwajn, "Transmission network equivalents for electromagnetic transients studies," *IEEE transactions on power apparatus and systems*, no. 9, pp. 2984–2994, 1983.

Cyber Security Assessment of Distributed Energy Resources

Cedric Carter, Ifeoma Onunkwo, Patricia Cordeiro, Jay Johnson

Sandia National Laboratories, Albuquerque, New Mexico, 87185, USA

Abstract — New distributed energy resource (DER) interconnection standards require communications and interoperability to provide grid operators greater flexibility for delivering voltage and frequency support. These communication channels are designed to allow utilities, aggregators, and other grid operators the ability to enable and configure various grid-support functions. However, these capabilities expand the power system cyber security attack surface and pose a significant risk to the resilience of the electric grid if controlled in aggregate. To advise the solar industry, grid operators, and government of the current risks and provide evidence-based recommendations to the community, Sandia performed cyber security assessments of a communications-enabled PV inverter and remote grid-monitoring gateway. The team found several well-designed security features but also some weaknesses. Based on these findings, recommendations are provided to improve the security features of DER devices.

Index Terms — cyber security, distributed energy resources, PV inverters, control network security.

I. INTRODUCTION

Power system Supervisory Control and Data Acquisition (SCADA) communications are generally proprietary stovepipe systems running on dedicated communication channels with a dependence on perimeter defenses. These networks typically run between large centralized generators, substations, and other utility-owned assets and the utility management system. However, this communication network will be expanding rapidly with the addition of interoperability requirements in the forthcoming revision to the U.S. interconnection standard IEEE Std. 1547 [1]. This presents an emerging fundamental challenge in securing power systems because distributed energy resource (DER) communications run over public and poorly-secured private networks, and the addition of DER devices significantly increases the electrical grid attack surface.

In January 2017, the second installment of the Quadrennial Energy Review (QER 1.2) focused on the electricity system and found it was a strategic imperative to protect and enhance the cyber defenses of the U.S. through modernization and transformation [2]. The Department of Homeland Security (DHS) Industrial Control Systems Cyber Emergency Response Team (ICS-CERT), which coordinates control systems-related security incidents, has seen increasing numbers of incidents in recent years—with the energy sector making up a significant portion of these (15% in 2015) [3]-[4]. As an example of the risk posed by these incidents, in Dec. 2015, there were coordinated cyber-attacks on three Ukrainian Oblenergos (distribution companies), resulting in the disconnection of seven 110 kV and 2,335 kV substations, and power outages affecting approximately 225,000 customers for 3 hours [5]-[6].

Historically, DER devices were programmed statically and not designed to provide any grid-support services. However, with the increasing penetration of solar and energy storage systems, there is growing need to provide grid-support capabilities, evidenced by the recent updates to California and Hawaii interconnection rules [7]-[8] and the forthcoming full revision to IEEE 1547. Interoperability capabilities are common for most inverters deployed in the U.S. now. These devices communicate over Ethernet, Wi-Fi, Bluetooth, and serial connections using a variety of proprietary and standardized protocols (e.g., SunSpec Modbus [9]). Enphase Energy made headlines worldwide when it remotely updated 800,000 Inverters (154 MW of capacity) on the Hawaiian Islands of O'ahu, Hawai'i, Moloka'i and Lana'i in 2015 [10], [11]. While many praised the achievement as a breakthrough for reducing the costs of retrofitting power systems, others warned of the cyber security implications. If one company could remotely update the settings of 100s of megawatts of power equipment, anyone with access to that control network would be able to make malicious changes to those devices as well. Certain settings could damage equipment, cause distribution overvoltages, or initiate a blackout if the contingency reserve was not sufficient. For example, on the island of O'ahu, there will be an estimated 400 MW of installed PV capacity in 2017 but only 180 MW of contingency reserves [12]. Therefore, disconnecting or curtailing a significant portion of the solar generation on a sunny day could cause a blackout because backup power is sized for *N-1* contingencies, not a cyber-attacks.

As DER enter the Internet of Things (IoT), there have been some early cyber security warning signs. Upon gaining access to a VPN tunnel established for a DC optimizer data manager, Fred Bret-Mounet discovered 1,000 other PV devices on the same subnet. Had he desired, he could have also remotely disconnected these devices [13]-[14]. A large Distributed Denial-of-Service (DDoS) attack using a botnet of IoT devices affected many websites including Amazon, Twitter, and Netflix in Oct. 2016 [15].

Consequently, it is imperative to secure DER communications to provide grid reliability and resiliency. Many DER devices communicate via unsecured serial protocols (e.g., Modbus), so there has been an effort to develop translators that integrate with DER to take encrypted

978-1-5090-5606-4/17 $31.00 © 2017 IEEE

protocols such as OpenADR 2.0b and IEEE 2030.5 (SEP 2.0) and only unencrypt the communications within the DER [16]. This approach mitigates the security risk because the adversary needs physical access to the devices to subvert them. As part of a California Solar Initiative grant, Sandia National Laboratories (Sandia) led a team to generate cyber security recommendations for PV Inverters using SunSpec Modbus removable communications modules [17]. The team presented a number of threats, vulnerabilities and high-level recommendations for residential inverter-based DER systems covering physical security, access control, integrity, confidentiality, encryption, and policy.

Novel methods for detecting, mitigating, and recovering from cyber-attacks must also be developed to counteract rapidly evolving threats and vulnerabilities. Techniques of identifying and removing compromised/unauthorized DERs, segmenting DERs into resource pools to minimize damage in the event of successful compromise, and safeguarding the DER from mass compromise are being developed by Sandia National Laboratories and many other research institutions.

II. CYBER ASSESSMENTS

Vulnerability assessments and penetration testing are important milestones for securing complex systems exposed to unsanctioned environments. Sandia has many years of experience over a broad range of cyber assessment applications that require strict assurances. Many of the research and development efforts are designed to protect critical infrastructure for the U.S. Government and private industry.

To help protect and improve the security of DER devices in the United States and globally, the Sandia team conducted cyber assessments of interoperable DER devices. The cyber assessments were designed to better understand the risks posed by current communication practices from various vendors. The solar and DER industry can shift toward a more secure DER control infrastructure by addressing cyber security risks identified during these assessments by incorporating more security features in future hardware and software revisions. The overall goal of the cyber assessment and penetration tests was to take a snapshot of the security profile for exemplar fielded equipment in order to inform the industry of security weaknesses and provide recommendations to DER and gateway vendors. Additionally, by going through the cyber assessment process, a more formalized approach was developed for conducting these studies in the future. This procedure could be refined to create the basis of a cyber security testing protocol in the future.

Sandia's Information Design Assurance Red Team (IDART) team performs physical and cyber security assessments for government agencies using an experienced methodology [18]. *Red teaming* is defined as an authorized, adversary-based assessment for defensive purposes. The cyber assessments presented here incorporated elements of this methodology, guidelines from the NIST Guide to Industrial Control Systems (ICS) Security [19], ICS-CERT Practice Guide [20], and

collective expertise regarding networked PV inverters and PV gateways.

III. LABORATORY CONFIGURATION

In order to conduct the assessments, an isolated, controlled network was created for selected DER devices. All the experiments were performed at the Distributed Energy Technologies Laboratory (DETL) at Sandia National Laboratories. The following security tests and experiments were performed:
- Network Reconnaissance
- Packet Replay
- Man in the Middle
- Denial of Service (DoS)
- Modified Firmware Upload
- Maintained Logs per device
- Password Handling

A network was created with two residential-scale DER devices, clients, switches, and a red teaming station shown in Fig. 1. Device A used a software GUI to change settings using UDP/IP and Device B used a web-interface GUI to change settings using TCP/IP via a gateway that then issued proprietary commands to the DER. Security assessments and evaluations focused on the current version of the software installed on each device. The red team station was equipped with Kali Linux, an open-source Linux operating system equipped with security tools such as network scanners to probe for vulnerabilities and network attack tools used to identify and exploit vulnerabilities.

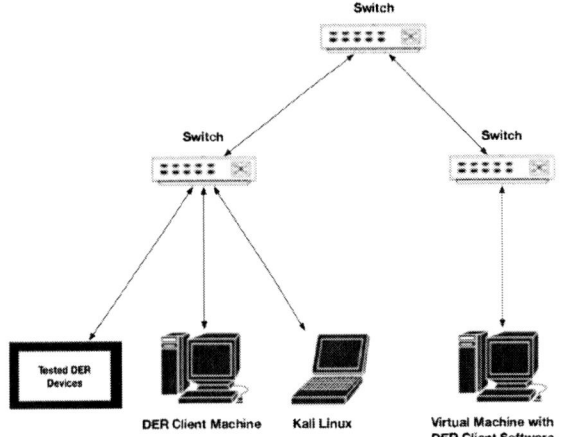

Fig. 1. Virtual and physical network diagram with hardware-in-the loop test bed for red team exploration.

This report summarizes efforts undertaken to identify and evaluate the security vulnerabilities associated with the aforementioned devices with the aim of infiltrating and compromising the system. The assessment findings led to multiple recommendations to improve the security of the system. These recommendations were provided to the DER vendors and anonymized herein to advise the greater industry.

IV. Experimental Results

A. Experiment 1 – Network Reconnaissance

Network reconnaissance is typically the first step an adversary takes to gain information about the system of interest. To gather as much information as possible about the DER devices, Nmap and OpenVAS tools running on Kali Linux were used for vulnerability scanning and device reconnaissance. The tools determined open ports and services running on the DER devices by probing and scanning all valid network ports (1- 65535), shown in Fig 2.

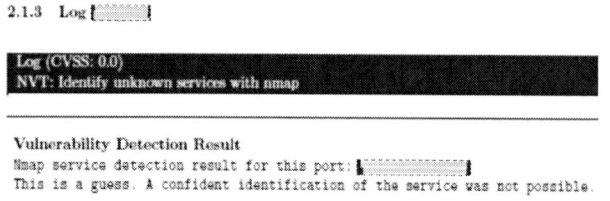

Fig. 2. OpenVAS port detection results on a DER device.

Network reconnaissance revealed the IP and MAC addresses of the DER devices, operating services, the type of system, and the open ports. The scan also identified vulnerable services running on the open ports. An additional scan identified that denial of service (DoS) attacks could potentially saturate and shut down both device A and B. For device B, multiple types of denial of service attacks were identified: two could crash the gateway and the other reported the device was susceptible to hub flood attacks. Hub flood attacks could aid an adversary in sniffing data communications when the switch turns into a learning mode. Finally, the DER scans indicated that security patches were not up to date.

B. Experiment 2 – Packet Replay

Packet replay is a network attack in which a data transmission is maliciously replayed or repeated. This test validated the authentication of data transfer from the DER's client application to the destination device. Data transmissions between the DER client applications and the devices were recorded, modified, and retransmitted as shown in Fig. 3.

By modifying the contents of the messages, an adversary could enable a range of DER actions. A configured Switched Port Analyzer (SPAN) port on a switch verified traffic generated from the DER client applications and the devices utilized both TCP/IP and UDP/IP transport protocols. Portions of DER modification requests were passed in plain text, and set point commands could be determined by inspecting the data packets of recorded traffic. To perform the packet replay attack, a DC disconnect voltage setting request from the client application was captured, modified to a new set point, and re-transmitted using an unauthenticated client script created with a UDP/IP socket library in Python. This action bypassed the DER client credentials needed to modify the DER. The falsified command was accepted by the DER and DC disconnect voltage was modified on the DER device.

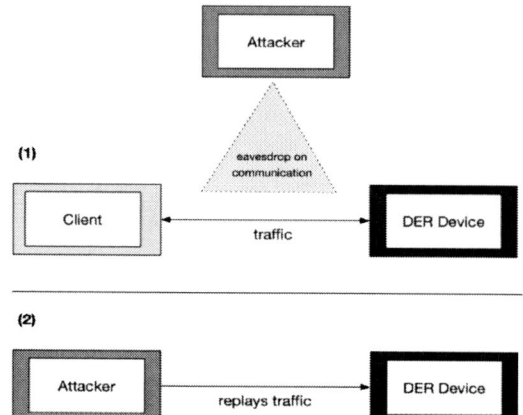

Fig. 3 Packet replay attack example.

DER applications that used TCP/IP transport are more resistant to packet replay attacks because the replay attack uses a unique session ID that the DER device generates during the initial three-way TCP handshake at the communication initialization. Session IDs are unique and pseudo-randomly generated, so they cannot be reused during a different communication session because the DER expects another unique session ID. Specifically, TCP/IP transport includes additional IP header information (e.g., sequence number) that DER applications can use to prevent packet replay attacks. Sequence number enables the server to detect and drop duplication of packets transferred to the DER device, which will prevent replay attacks because the same packet cannot be retransmitted. The number utilizes the Acknowledgement (ACK) value in sync during the transaction shown in Fig. 4. Sequence numbers also ensure that insertion of data in the data stream can be detected.

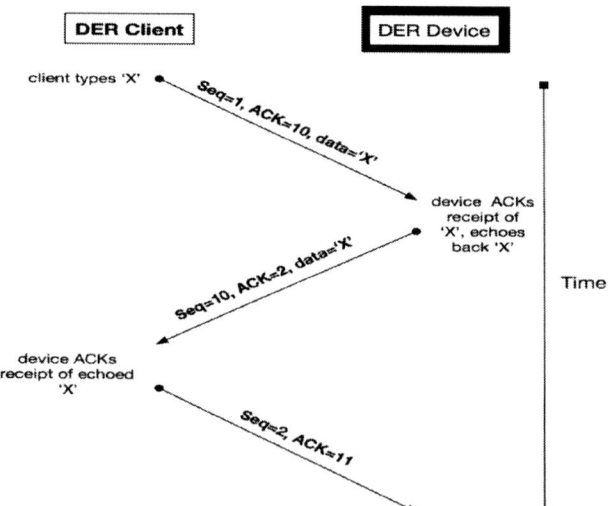

Fig. 4. Example TCP/IP transaction session.

978-1-5090-5606-4/17 $31.00 © 2017 IEEE

C. Experiment 3 – Man in the Middle

Man in the middle attacks relay altered communications between two or more parties. This experiment validates the integrity and confidentiality of data communication between the DER devices and client applications. Ettercap, a man in the middle (MiTM) testing tool, was used to eavesdrop on communications between the DER client and device, then address Resolution Protocol (ARP) poisoning and port stealing were successfully completed. ARP poisoning forced the MAC address of the adversary to be linked with the IP address of the victim. This technique enabled the interception of data in-transit. (An example of this experiment is shown in Fig. 5.) Port stealing allowed intercept and modification of data in-transit and enabled the adversary to receive, read, and modify data before it reached the destination. In port stealing, an adversary often "steals" traffic that is directed to another port of an Ethernet switch. This attack allows the adversary to receive packets that were originally directed to another computer. ARP poisoning and port stealing resulted in data interception on both devices.

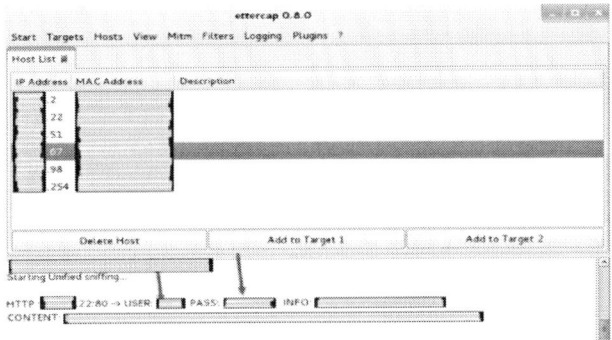

Fig. 5. MiTM attack on a DER device using Ettercap.

D. Experiment 4 – Denial of Service (DoS)

Denial of service attacks result from flooding the bandwidth or resources of the targeted system to cause severe latency, disruption of control flow, or cause intermittent connections. In a DoS attack, the adversary increases the amount of time to issue responses by submitting abnormally fast traffic to open ports on the DER device. This attack continuously flooded the DER devices with orchestrated packets. In addition, this type of attack can make the DER device unavailable for access to legitimate users as shown in Fig. 6.

This test validates the availability of data transferred between the DER client and devices. There was latency between the client application command and the DER device response. It took approximately 100 ms to submit various requests and receive responses from Device A, and 200 ms from Device B. These requests included the following:

- Passwords used to gain access
- Device name change
- Device configurations

Successful DoS attacks were demonstrated on Devices A and B, whereby the connection time to DER devices was significantly increased. This was accomplished by using a Python module and a bash script to continuously send packets every 100 ms. During the DoS attack, the legitimate user could not connect to the DER, read data points, or make modifications with the DER client. After the DoS attack, the DER client applications needed to be reconnected to make modifications to the DER.

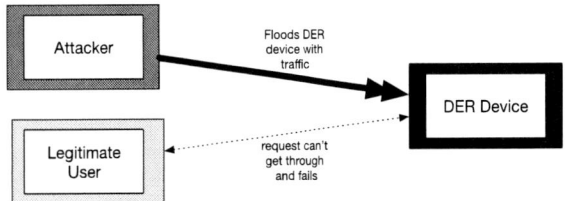

Fig. 6. Example DoS attack to a DER device.

E. Experiment 5 – Modified Firmware Upload

Modified firmware attacks surreptitiously change the overall functionality of the DER. This test validates the integrity and authentication of the firmware update process of DER embedded systems. Firmware updates can often be performed either locally or remotely, through an Ethernet, Wi-Fi, USB connection via an FTP, or telnet session. The two DER devices were tested to ascertain whether a malicious user could load unauthorized firmware.

In these experiments, firmware update files were obtained from the device manufacturers. These files were modified in a hexadecimal editor, changing either single byte values of known parameters in readable sections of the code, or randomly selected bytes in code that was not human readable. In all types of modifications, the end-devices successfully rejected the modified firmware.

Device A's rejected upgrade halted, most likely due to a cyclic redundancy check (CRC). Any intentional modification to the firmware would require the correct identification and recalculation of a valid CRC value for the modified file, as shown in Fig. 7. CRC algorithms are designed for error detection, not security, and there exist mathematical tools for reversing CRCs. Falsifying the CRC requires insight into the CRC polynomial and other parameters used, as well as discovering what segments of the file to input to the CRC computation.

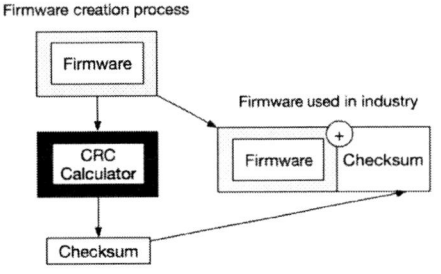

Fig 7. Example firmware creation process

The firmware file for Device B contained no human readable code. A high entropy measure of the file indicated that the data was likely encrypted. An update with a randomly modified data byte failed, necessitating a restart of the user interface. The device upgrade appeared to be secured by means of both encryption and authentication. However, the device was seen to have loaded and decrypted the file prior to rejecting the modified upgrade. Better practice would authenticate a file prior to decryption so as to prevent emplacement of malicious code on the device.

Another Device B firmware file with no readable code was randomly modified and submitted via the network as a device update. In this case, the device ceased communicating without warning and could only be recovered with a local configuration method. The modified firmware upload had failed, however the failure or protection mechanism was not conveyed to the user. Regardless of the cause, this test revealed that network access allowed an inappropriate upload to "brick" the device. Access must therefore be carefully managed on networked devices.

F. Experiment 6 – Maintained Logs per Device

Per NIST [19], "Without proper and accurate logs, it might be impossible to determine what caused a security event to occur." The two devices were found to store event logs in local memory and this evaluation inspected the devices for proper bookkeeping practices. The manufacturers' monitoring software packages were used to display the available logs of DER events. Logged events on the devices included communication device connection attempts and status, self-test results, DER parameter modifications, and grid status. Information in device A's logs were more accurate, including the date and time, the type, description, and source of the event.

However, in both devices, there was a limit to the number of events stored in local memory. It is unknown how typical operators archive the logs to maintain history for auditing. A solution to this issue can be accomplished by prompting the user to export logs once a certain memory usage is reached. In addition, stored logs should be protected from unattributed modification. Modifying stored logs can misguide auditing procedures, thus preventing an accurate case study. It is incumbent upon the user to change default passwords and control their use. A security event might not be tracked down if past events are not fully archived and attributable.

G. Experiment 7 – Password Handling

The password handling experiment determined if passwords were transferred in plaintext on the DER network. The experiment investigated the "login" and "password update/change" functions on the DER devices.

The gateway associated with Device B used unsecured web authentication mechanisms and had its passwords captured in plaintext using a network traffic analyzer tool. The login credentials of Device A were not transferred in plaintext. However, critical information for this DER, such as the serial IDs and device names, were transferred in plain text. An

adversary could potentially gain insight from this information and formulate a cyber-attack. In addition, passwords used for the "update/change" function were transferred in plaintext on both devices. Modified passwords were captured in plaintext in the "password update/change" packet while eavesdropping on the network, shown in Fig. 8.

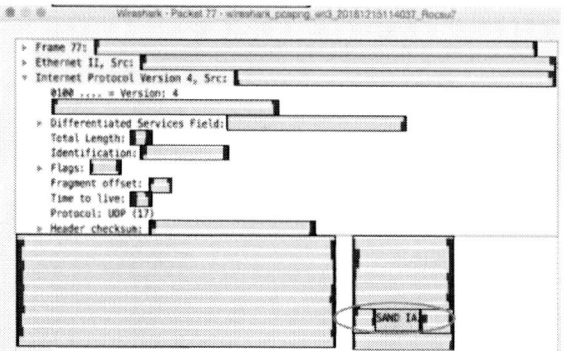

Fig. 8. Network capture of a password during the update/change function.

H. DER Assessment Summary and Recommendations

Scanning the DER network, host device ports, and services, and performing red team analyses identified cyber security vulnerabilities in the interoperable DER. The findings of this assessment are summarized in Table I.

TABLE I
DER CYBER ASSESSMENT COMPARISON

	Device A	Device B
Protocol	UDP/IP	TCP/IP
Analyzed Interface	Ethernet	Ethernet
Reconnaissance	✓	✓
Packet Replay	x	o
MiTM	x	x
DoS	x	x
Mod Firmware	o	o
Prevalent Logs	x	x
Password Handling	x	x

x = Exploits Exist, ✓= Successful, o = Incomplete

As part of the assessment, multiple recommendations were collated to assist the PV and DER community mitigate cyber weaknesses in the control networks. It is recommended to:

- Encrypt data exchanges and do not pass information in plaintext.
  - o Use of telnet for remote logins should be discontinued or upgraded to the latest version, sTELNET.
  - o Applications such as FTP should be replaced with another secure file transfer system, FTP-SSL.
- Secure password strategies and policies should be implemented and enforced for all system users. Require these credentials to be different for privileged access.
- Utilize practical firewall rules to mitigate the effects associated with Denial of Service attacks and unauthorized access to DER network.

- Lock MAC address on the network devices and on each port of a switch to prevent receiving unauthorized traffic.
- Implement the *AAA framework*: Authentication, Authorization, and Accounting.
  - Authentication: Ensure users, devices, and applications attempting to access system resources are valid and trusted.
  - Accounting: Ensure all devices and systems are accounted for cyber security best practices.
  - Authorization: Ensure users, devices, and applications attempting to access system resources are authorized for access.
- Practice *Principle of Least Privilege.*
  - Every module (such as a process, a user, or a program, depending on the subject) must be able to access only the information and resources that are necessary for its legitimate purpose.
  - If a user or resource no longer needs access to perform a legitimate task, disable their access.
  - Disable all ports that are not being used for normal operation.

V. CONCLUSION

DER devices are increasingly being connected to internet networks to exchange information with utilities, aggregators, financial institutions, and grid operators. New interconnection standards in the US will soon require DER devices to include adjustable DER grid-support functions. The confluence of these two trends exposes the grid to new cyber security risks because adversaries could change DER functions through the public internet if the DER control networks are not properly secured. Sandia National Laboratories has conducted a cyber security assessment of two residential-scale interoperable DER devices to better understand the state-of-the-art for DER communication systems. The team found multiple security weaknesses that could be exploited to gain access or control of DER devices. These findings have been shared with the device vendors to take corrective actions and with the solar industry in anonymized form to provide example concerns, best practices, and recommendations for improving the cyber security posture of the DER devices and US power system as a whole.

ACKNOWLEDGEMENT

This research was funded by the U.S. Department of Energy's SunShot National Laboratory Multiyear Partnership Program. Sandia National Laboratories is a multi-mission laboratory managed and operated by National Technology and Engineering Solutions of Sandia, LLC., a wholly owned subsidiary of Honeywell International, Inc., for the U.S.

Department of Energy's National Nuclear Security Administration under contract DE-NA-0003525.

REFERENCES

[1] IEEE 1547 Std. 1547-2008, "IEEE Standard for Interconnecting Distributed Resources with Electric Power Systems," Institute of Electrical and Electronics Engineers, Inc., New York, NY.

[2] Quadrennial Energy Review (QER), "Transforming the Nation's Electricity System." January 6, 2017.

[3] National Cybersecurity and Communications Integration Center/Industrial Control Systems Cyber Emergency Response Team, Year in Review, FY 2015.

[4] Industrial Control Systems Cyber Emergency Response Team, ICS-CERT Year in Review, 2014.

[5] K. Zetter, "Inside the Cunning, Unprecedented Hack of Ukraine's Power Grid," Wired, March 3, 2016.

[6] E-ISAC, Analysis of the Cyber Attack on the Ukrainian Power Grid: Defense Use Case, March 18, 2016.

[7] Pacific Gas and Electric Co., Electric Rule No. 21, Generating Facility Interconnections, Filed with the CPUC, Jan. 20, 2015.

[8] Hawaiian Electric Company, Inc. Rule No. 14, Service Connection and Facilities on Customer's Premises, D&O No. 33258 filed Oct. 12, 2015, effective Oct. 21, 2015.

[9] SunSpec Alliance, SunSpec Technology Overview, SunSpec Alliance Interoperability Specification, v 1.4.

[10] P. Fairley, 800,000 MicroInverters Remotely Retrofitted on Oahu—in One Day, IEEE Spectrum, 5 Feb 2015.

[11] A. Konkar, 'Something Astounding Just Happened': Enphase's Grid- Stabilizing Collaboration with Hawaiian Electric, Enphase Energy blog, 11 Mar, 2015.

[12] GE Energy Consulting, Oahu Distributed PV Grid Stability Study, Part 1: System Frequency Response to Generator Contingency Events, March 3, 2016.

[13] T. Fox-Brewster, "This Man Hacked His Own Solar Panels... And Claims 1,000 More Homes Vulnerable," Forbes, Aug. 1, 2016.

[14] F. Bret-Mounet, "All Your Solar Panels are Belong to Me," DEF CON 24, Las Vegas, Aug 4-7, 2016.

[15] K. Leswing, A massive cyberattack knocked out major websites across the internet, Business Insider, 21 Oct, 2016.

[16] B. Seal, et al., "Final Report for CSI RD&D Solicitation #4 Standard Communication Interface and Certification Test Program for Smart Inverters," June 2016.

[17] J. Henry, et al., Cyber Security Requirements and Recommendations for CSI RD&D Solicitation #4 Distributed Energy Resource Communications, Oct 2015.

[18] Sandia National Laboratories, The Information Design Assurance Red Team (IDART™), 2009. URL: http://www.idart.sandia.gov/

[19] K. Stouffer, V. Pillitteri, S. Lightman, M. Abrams, A. Hahn, "Guide to Industrial Control Systems (ICS) Security," NIST, May 2015.

[20] DHS, ICS-CERT, "Recommended Practices", May 22, 2017. URL: https://ics-cert.us-cert.gov/Recommended-Practices

Evaluation of Fast-Frequency Support Functions in High Penetration Isolated Power Systems

Mohamed Elkhatib, Jason Neely, and Jay Johnson

Sandia National Laboratories, Albuquerque, NM, 87123, USA

Abstract— Distributed Energy Resources (DER) with grid-support functions provide grid operators with essential capabilities to stabilize grid voltage and frequency. Providing ancillary services with DER is particularly attractive as renewable generation displaces traditional thermal power plants; however, optimizing operational modes and advanced function settings to provide these capabilities is not well understood. In this paper, the impact of enabling Frequency-Watt (FW) capabilities in distributed PV is evaluated for the electric grid on the Hawaiian island of Oahu to provide fast frequency reserves. The study is completed using validated bulk grid models that have been modified to include PV resources with a pointwise linear FW capability. A key contribution of this work is a detailed sensitivity analysis indicating the dependence of simulated grid dynamic performance on different grid, inverter, and FW function parameters.

Keywords—PV penetration, frequency-watt, grid support functions, island grids

I. INTRODUCTION

In the last 5 years, Distributed Energy Resources (DER) interconnection and interoperability codes and standards in the U.S. have been rapidly changing to offer grid operators additional resources to provide voltage and frequency control. The impetus to change the DER requirements arose from increasing penetrations of grid-connected, distributed, inverter-based renewable energy systems which reduced system inertia and replaced it with variable, non-dispatchable generators. Since the mid-2000s, the U.S. has relied on the IEEE 1547 series [1]-[2] of standards to harmonize the interconnection requirements across the country. However, high penetrations of photovoltaic installations—driven by state renewable portfolio standards, favorable economics, and consumer preference—have led regulators to conclude there is an eminent need to have DER grid-support capabilities for reliable power system operations. As a stop-gap measure, IEEE 1547a [3] was drafted in 2014, which permits, but does not require, DER to actively regulate voltage and frequency with agreement of the Area Electric Power System operator.

Taking the regulation process further, the Hawaiian and Californian state Public Utility Commissions changed their interconnection standards to require grid-support functions for inverter-based DER. Currently, HI Rule 14H [4] and CA Rule 21 [5] have requirements for wide frequency and voltage ride-throughs, and active and reactive power functions. IEEE 1547

is currently undergoing a full revision to add similar functions to the interconnection standard and re-harmonize the requirements across the nation, but this standard is not expected until late 2017 or 2018.

As a result of the new US interconnection requirements, and similar requirements in Europe, DER venders have added many grid interactive functions to their equipment. One of the capabilities known as Frequency-Watt (FW) or P(f) adjusts the active power injection of the DER based on the local grid frequency measurements. This function is designed to provide bulk system balancing with a droop-like controller. Typically, the requirement is defined as a parameterized function as in the German requirement in VDE-AR-N 4105 [6] and the International Electrotechnical Commission (IEC) Technical Report (TR) 61850-90-7 [7]. There is also the option to provide pointwise linear FW curves in IEC 61850-90-7 and the SunSpec Alliance Modbus information models [8], or to generate bidirectional FW functions with pointwise definitions, which are required in the Italian technical rule CEI 0-21 [9] for low voltage and CEI 0-16 for medium and high voltage [10]. Thus, there is considerable interest in determining, quantitatively, the potential impact of FW functions on grid dynamic performance and the optimal FW curves to provide this functionality. There is strong evidence that implementing FW in high PV penetration grids can greatly improve the dynamic response (higher frequency nadir, lower apex) to a contingency event [10]-[10]. However, there is a need to identify the parameters of this function that are most important to system performance. Special emphasis is placed herein on the response time of the function, which has been found to be a critical requirement for providing this capability [10].

In this paper, the bulk power system impact of distributed FW function deployment is evaluated for the Oahu grid with high PV penetration using grid modeling software. Section II reviews the FW function model and discusses its integration into the grid model. Section III provides results of a sensitivity analysis relating DER and system parameter variation to system over-frequency response. Section IV presents results of sensitivity analysis for under-frequency response. Finally, Section V presents simulation results considering the addition of synthetic inertia. Finally, conclusions are provided in Section VII.

II. FREQUENCY-WATT FUNCTIONS MODELING FOR BULK SYSTEM STUDIES

In this work, *Power System Simulation for Engineers* (PSS/E) software was used to study the impact of fast frequency response functions on bulk power system frequency response. A dynamic model of Oahu Island was obtained from Hawaiian Electric Company wherein the distributed PV penetration level was 466 MW, which represents about 40% of the peak load. DERs were aggregated at their respective transmission bus and a PSS/E user defined model (UDM) was developed to model the aggregated DER. Different fast frequency response functions were then built into the UDM and integrated into the Oahu model.

Fig. 1 shows the block diagram of the FW-enabled DER model. Therein, T_1 represents the time constant associated with frequency measurements which accounts for PLL and other related dynamics, T_3 represents the inverter's response time, P_{set} represents the power set point of the inverter in steady state, and P_{out} is the total output power of the inverter. The FW function is modeled using four frequency-power pair points as shown in Fig. 2. Other equivalent parametrized representations could be used as well.

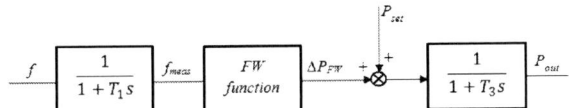

Fig. 1. Distributed PV model including FW function

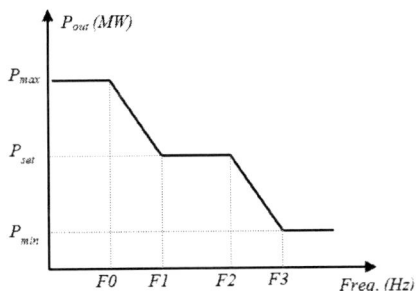

Fig. 2. Frequency-Watt (FW) function parameters

A load loss contingency was simulated to demonstrate the effectiveness of enabling the distributed PV (DPV) FW function. The contingency represents the loss of 62 MW of load. Approximately 300 MW of DPV was equipped with the FW function. Fig. 3 shows the frequency response of the system when the DPV FW function was enabled versus when the DPV FW function was disabled. When FW function was disabled, the system frequency rises to around 60.54 Hz at about 1.5 second. As a result, all legacy DPV tripped on over-

frequency[1]. Consequently, the event, which started as an over-frequency event, triggered an under-frequency event due to the loss of almost 150 MW of generation. Subsequently, the system frequency dropped to ~58.87 Hz at ~4.1 seconds resulting in a load shed of about 38 MW. Enabling the FW function resulted in significantly improved frequency response and neither generation nor load was tripped due to the contingency.

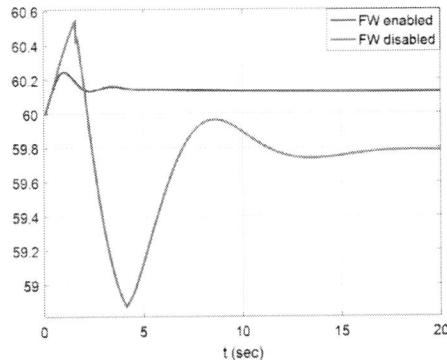

Fig. 3. Model Parameters: $T_1=T_3=0.5$ sec, $F_3=61$ Hz

III. SENSITIVITY ANALYSIS – OVERFREQUENCY EVENTS

In this section, we study the impact of FW function on over-frequency events. Sensitivity analysis is presented for the frequency response with respect to different parameters, namely: inverter's response time, FW function droop (as measured by *FW* slope), penetration of FW-enabled DERs and system inertia. A light load spring case is used in this study and the initiating contingency is the loss of 62 MW of load.

A. Sensitivity to inverter's response time

In this case, the inverter response time constant, i.e. T_3, was increased from 0.5 sec to 20 sec. As shown in Fig. 4 and 5, slowing down inverters (i.e. increasing T_3) resulted in higher magnitude of oscillations and higher frequency apex values as the controller response becomes less in-phase with the error signal. However, as shown in Fig. 5, as T_3 increases further, the frequency apex continues to increase, but the magnitude of subsequent oscillations tends to decrease again as T_3 goes above 1.5 seconds. This decrease in later oscillations is due to the added phase lag resulting in the controller once again being in phase with the error signal. In conclusion, while it is beneficial to speed up inverters to achieve better frequency response, it is important to carefully consider the possibility of unwanted oscillations when inverters act relatively fast. It is noted that the relationship between the inverter response and dynamic response can be described well in the frequency domain, which is a subject for future research.

---

[1] Note: many PV systems in HI have been remotely retrofitted with wider frequency ride-through curves (i.e., those in [4]) to avoid DPV tripping events like this [13].

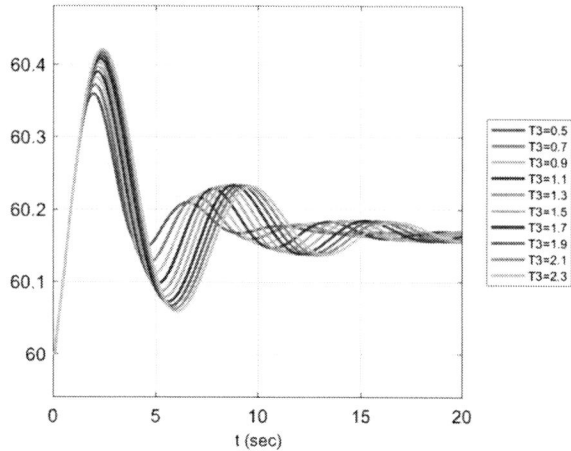

Fig. 4. Sensitivity to variation in inverter time constant T_3 with fixed model parameters: $F_3=63$ Hz, $T_1=0.5$ sec

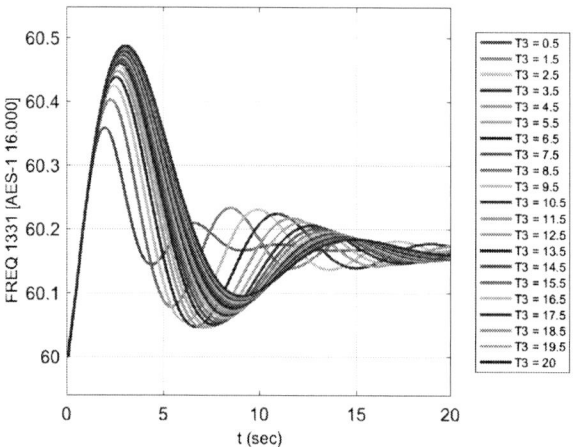

Fig. 5. Sensitivity to variation in inverter time constant T_3 with fixed model parameters: $F_3=63$ Hz, $T_1=0.5$ sec

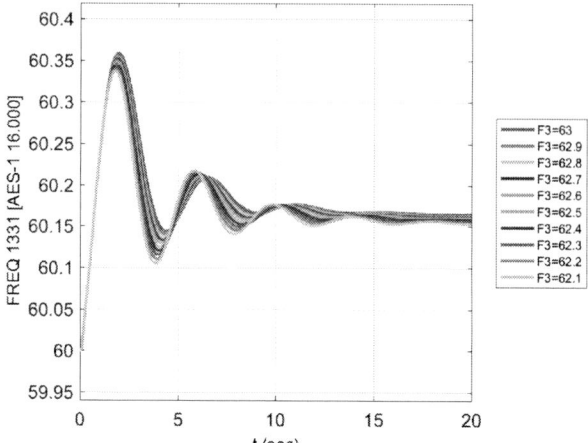

Fig. 6. Sensitivity to variation in FW curve parameter F_3 with fixed model parameters: $T_1=0.5$ sec, $T_3=0.5$ sec

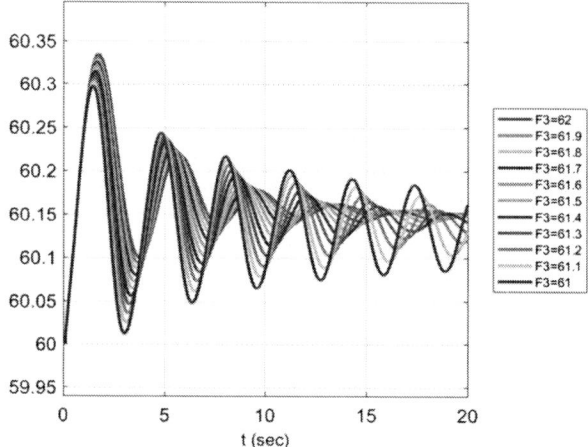

Fig. 7. Sensitivity to variation in FW curve parameter F_3 with fixed model parameters: $T_1=0.5$ sec, $T_3=0.5$ sec

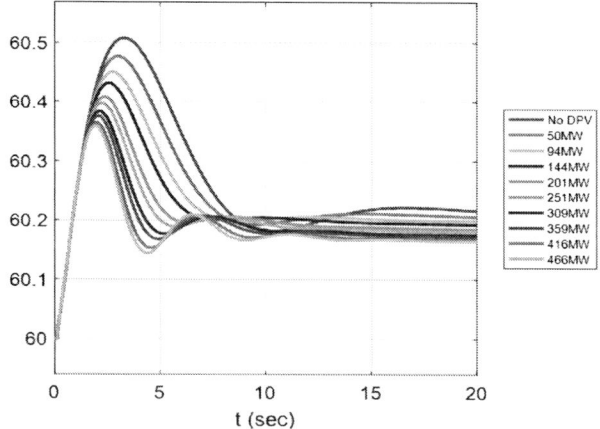

Fig. 8. Sensitivity to variation in DPV penetration level with fixed model parameters: $T_1=0.5$ sec, $T_3=0.5$ sec, $F_3=63$Hz

B. Sensitivity to FW function droop

In this case, the droop of the FW function was increased by reducing the value of F_3 which in effect makes the inverters more responsive to frequency changes. As a result, the frequency response was improved as measured by the frequency apex, shown in Fig. 6. However, steeper FW function droop may result in higher magnitude of oscillations in the frequency response as shown in Fig. 7.

C. Sensitivity to FW-enabled penetration level

Increasing the penetration level of FW-enabled DERs in the system results in better frequency response, determined by the frequency apex and settling frequency, as shown in Fig. 8. However, depending on model parameters, higher penetrations can make the system more susceptible to oscillations as shown in Fig. 9.

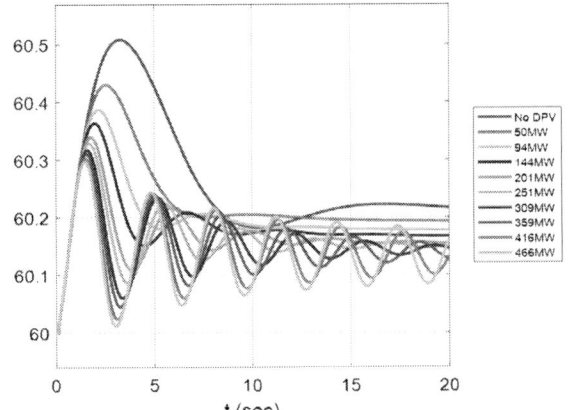

Fig. 9. Sensitivity to variation in DPV penetration level with fixed model parameters: $T_1=0.5$ sec, $T_3=0.5$ sec, $F_3=61Hz$

D. Sensitivity to system's inertia

As the penetration of inverter-based DER increases, the inertia of the system will decrease as more synchronous generators will be brought offline. Reduced system inertia may make the system more susceptible to frequency oscillations that are higher in amplitude and frequency. Fig. 10 compares the system frequency response when all synchronous machines were in service versus the case when 1 of the machines was out of service, which indicates a slightly higher oscillation amplitude and frequency.

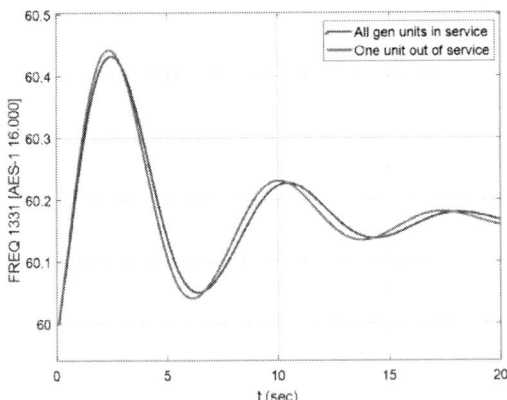

Fig. 10. Sensitivity to variation in system inertia with fixed model parameters: $T_1=0.5$ sec, $T_2=3$ sec, $F_3=63$ Hz

IV. SENSITIVITY ANALYSIS – UNDERFREQUENCY EVENTS

The sensitivity analysis presented in Section III focused on over-frequency events; however, inverter-based frequency support functions may be able to mitigate under-frequency events as well if sufficient headroom exists. Typically, PV systems operate at their maximum power point so they can only reduce their output to respond to over-frequency events. However, as the penetration of PV increases, there will be a need for PV to provide reserve for under-frequency events by curtailing their output in steady state.

To assess the impact of PV reserve level on frequency response, the FW function was enabled on 220 MW of DPV for eight base cases with different reserve levels varying from 0% to 30% by curtailing the output of DPV in steady state. A peak load summer case was used in this study and the initiating contingency was the loss of a 200 MW generator unit in the system. It is clear from Fig.11 that as the reserve level increases, the resulting frequency response of the system improves as measured by the frequency nadir. Consequently, increased PV reserve level resulted in less load shedding as shown in Fig. 12.

It is important to notice, however, that the load shedding mechanism plays a crucial rule in the ability of PV to provide reserve for under-frequency events. In the case of Figs. 11 and 12, a traditional load shedding scheme was implemented where tripping a load resulted in tripping any PV connected to the same bus. This represents typical load shedding schemes, which trips distribution feeders in whole. However, losing DPV due to load shedding has an adverse impact on frequency response due to concurrent loss of reserve.

To highlight the impact of load shedding in this case, the simulation was repeated while the transfer trip between loads and PV of the same bus was disabled. Figs. 13 and 14 show the improvement in frequency response and load shedding.

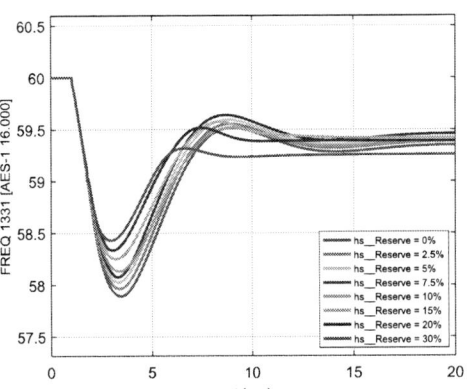

Fig. 11. Frequency response under different PV reserve levels

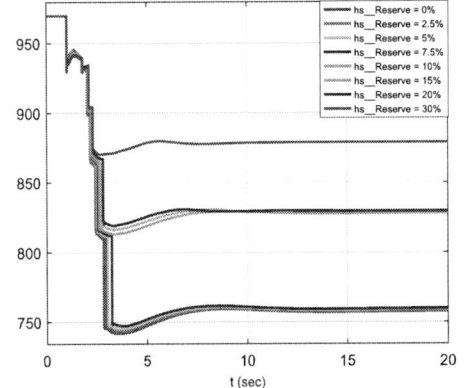

Fig. 12. Total load of the system (MW) under different PV reserve levels

978-1-5090-5606-4/17 $31.00 © 2017 IEEE

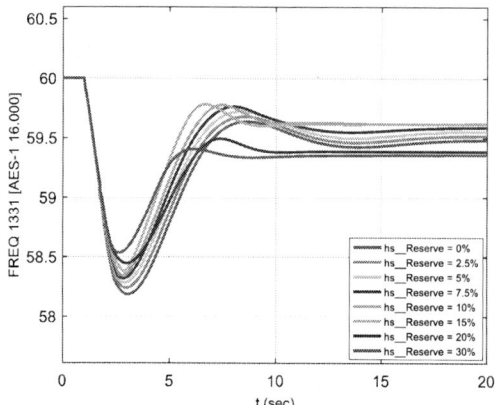

Fig. 13. Frequency response under different PV reserve levels with different load shedding scheme

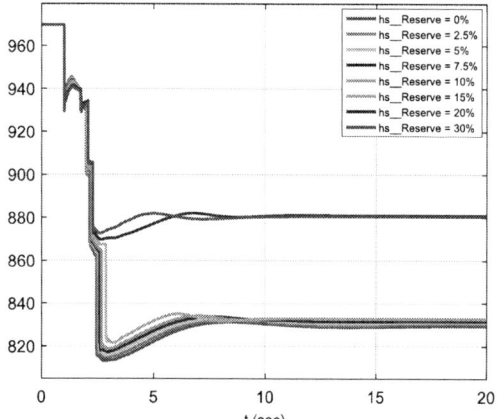

Fig. 14. Total load of the system (MW) under different PV reserve levels

V. ADDING A SYNTHETIC INERTIA COMPONENT TO THE FW FUNCTION

As discussed in Section III, to improve the frequency response using FW function, inverter response should be fast and the FW droop should be steep. However, as was shown in Section III, increasing the steepness of the FW function could result in frequency oscillations while increasing the inverter response time might potentially generate other oscillations in the system such as subsynchronous oscillations.

Basically, the output of the FW function changes proportional to the measured frequency deviation, i.e., FW is a proportional controller. Therefore, the FW function will be slow to react after the contingency and until the frequency of the system falls considerably relative to its steady state value.

One way to increase the speed of the FW function is to add a component to the output power which is proportional to the negative of the rate of change of the frequency, i.e., the FW is now a proportional-derivative controller. In essence, that component mimics the actions of the conventional inertia in opposing any change to the frequency as thus is known as

synthetic inertia. Fig. 15 shows the modified FW-enabled inverter model block diagram including the synthetic inertia component where T_2 is the synthetic inertia filler time constant and K is the synthetic inertia gain. This modified scheme will allow the FW function to respond quickly post-contingency when the rate of change of the frequency is high while frequency deviation is low and also around the frequency nadir point when the rate of change of the frequency is low but the frequency deviation is high

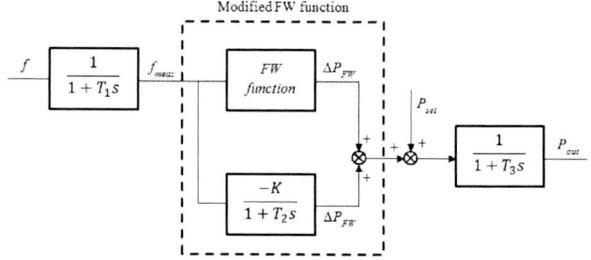

Fig. 15. Modified FW function Block Diagram.

To test the performance of the modified FW function, the base case used in section III was used under the same contingency. Fig. 16 compares the frequency response of the system with the synthetic inertia block disabled, $K=0$, versus when the synthetic inertia gain was set to 1. It is clear that the addition of a synthetic inertia component results in a significant improvement to the frequency response as measured by the reduced nadir and less oscillation. Additionally, Fig. 17 depicts the sensitivity of the frequency response to the synthetic inertia constant.

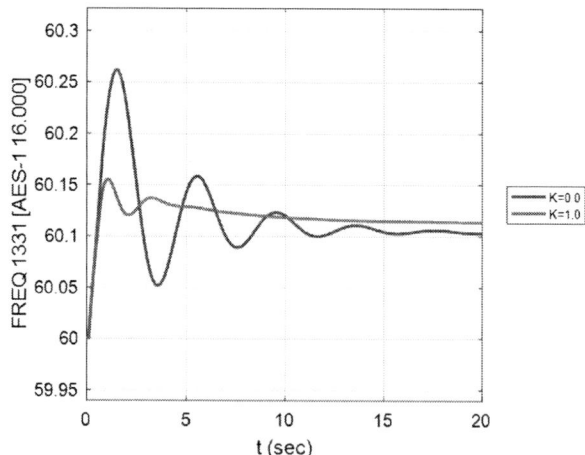

Fig. 16. Frequency response of the system with and without enabling synthetic inertia with fixed model parameters: T_1=0.5 sec, T_3=3 sec, F_3=62 Hz

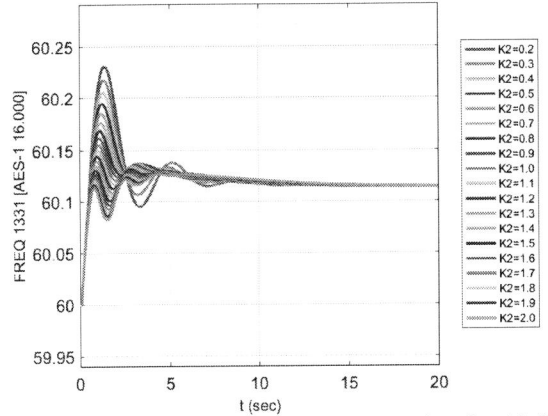

Fig. 17. Sensitivity to variation in synthetic inertia gain with fixed model parameters: $T_1=0.5$ sec, $T_3=3$ sec, $F_3=62$ Hz.

VI. CONCLUSIONS

Distributed PV systems with frequency-watt functions have the ability to greatly improve bulk system response during contingency events. Using pointwise linear FW functions with curtailment provides up-reserve headroom during risky high solar time periods when traditional generation is displaced with inertialess systems. The sensitivity study presented in this paper shows that while FW parameters can be tailored for improved response characteristics, improper tuning of the parameters could result in unwanted oscillations in the frequency response. Moreover, the results show that the addition of a derivative term to the FW function could result in significant improvement to the overall system frequency response.

While DER equipment can easily provide upward and downward active power adjustments with little delay, currently interconnection standards in the U.S. do not define an operating mode akin to Fig. 2, wherein the PV system is curtailed at nominal grid frequency and active power changes based on local grid frequency measurements. However, the new IEEE Std. 1547 language includes a "Frequency-droop (frequency/power) operation" requirement that includes a bi-directional droop curve. Coupling this function with PV curtailment would enable the fast-frequency support capabilities described in this work. As renewable energy penetrations continue to increase in bulk power systems, it becomes increasingly critical to provide ancillary services using this equipment. Therefore, it is recommended that standards development organizations consider adding pointwise linear FW functions to interconnection standards in the future.

ACKNOWLEDGMENT

The authors would like to thank Hawaiian Electric Company (HECO) for providing the dynamic model of Oahu Island and for their valuable input during the course of this study.

This work was supported by the U.S. Department of Energy through the Grid Modernization Laboratory Consortium initiative.

Sandia National Laboratories is a multi-mission laboratory managed and operated by National Technology and Engineering Solutions of Sandia, LLC., a wholly owned subsidiary of Honeywell International, Inc., for the U.S. Department of Energy's National Nuclear Security Administration under contract DE-NA0003525.

REFERENCES

[1] IEEE Standard 1547-2003, Standard for Interconnecting Distributed Resources with Electric Power Systems, 2003.

[2] IEEE Standard 1547.1-2005, Standard for Conformance Test Procedures for Equipment Interconnecting Distributed Resources with Electric Power Systems, 2005.

[3] IEEE Standard 1547a-2014, Standard for Interconnecting Distributed Resources with Electric Power Systems: Amendment 1, 2014.

[4] Hawaiian Electric Company, Inc., Rule No. 14, Service Connections and Facilities on Customer's Premises, Section H: Interconnection of distributed generating facilities with the company's distribution system, effective October 21, 2015.

[5] Pacific Gas and Electric Company, Electric Rule No. 21, Generating Facility Interconnections, Filed with the CPUC on 20 Jan, 2015.

[6] VDE Application Guide VDE-AR-N 4105: Generators in the low voltage distribution network. Application guide for generating plants' connection to and parallel operation with the low-voltage network, 1/08/2010.

[7] IEC Technical Report 61850-90-7, "Communication networks and systems for power utility automation–Part 90-7: Object models for power converters in distributed energy resources (DER) systems," Edition 1.0, Feb 2013.

[8] SunSpec Alliance Interoperability Specification, SunSpec Inverter Models, Document #12020 Version 1.5, released 4-14-2015.

[9] CEI Reference Technical Rules for the Connection of Active and Passive Consumers to the HV and MV Electrical Networks of Distribution Company, CEI 0-16 and 0-16, 2014.

[10] CEI Reference Technical Rules for the Connection of Active and Passive Users to the LV Electrical Utilities, CEI Reference 0-21, December 2013.

[11] J. Neely, J. Johnson, J. Delhotal, S. Gonzalez, M. Lave, Evaluation of PV Frequency-Watt Function for Fast Frequency Reserves, IEEE Applied Power Electronics Conference (APEC), Long Beach, CA, March , 2016

[12] J. Johnson, J. Neely, J. Delhotal, M. Lave, "Photovoltaic Frequency-Watt Curve Design for Frequency Regulation and Fast Contingency Reserves," IEEE Journal of Photovoltaics, vol. 6, no. 6, pp. 1611-1618, Nov. 2016.

[13] Peter Fairley, "800,000 Microinverters Remotely Retrofitted on Oahu—in One Day," IEEE Spectrum, 05 Feb. 2015.

Loss of Utility Detection Capabilities for Today's Utility Interconnected Photovoltaic Inverters

Sigifredo Gonzalez[1], Gregory Kern[2], Michael Ropp[3]

[1]Sandia National Laboratories, Albuquerque, New Mexico, 87123, USA
[2]SunPower Corporation, Austin, Texas, 78758, USA
[3]Northern Plains Power Technologies, Brookings, South Dakota, 57006, USA

Abstract — The level of installed photovoltaic (PV) generation has surpassed 35 GW in the United States and the solar penetration continues to increase at a high rate. Almost all installations up to now rely on utility interconnection requirements based on 2003 standards, which made distributed energy resources (DER) devices sensitive to perturbations on the utility, would quickly disconnect, and would not provide utility support when most needed. This is changing, and to minimize adverse effects on the performance of electrical power systems, PV inverters must implement voltage and frequency ride-through capabilities and provide voltage and frequency support features. These new utility support requirements have caused renewed concern of loss of utility capabilities of PV DER devices. This paper focuses on revisions to the utility interconnection requirements and investigates the impacts of these changes on the islanding detection capabilities of PV inverters.

Index Terms — anti-islanding, distributed energy resources, photovoltaics

I. INTRODUCTION

Utility scale photovoltaic (PV) plants drove the total installation for Q3 of 2016 to 4.1 GW and presently over 35 GW of PV based distributed energy resources (DER) has been installed in the United States [1]. Virtually all of the utility interconnected PV systems installed in the continental United States are systems that adhere to IEEE 1547-2003 [2] or IEEE 1547a-2014 [3] utility interconnection standard requirements. To sustain a high level of PV installations, future systems will have to minimize adverse impacts on the grid through voltage and frequency ride-through capabilities and voltage and frequency support functions.

The governing IEEE P1547 [4] utility interconnection standard and the IEEE P1547.1 [5] testing standard are undergoing full revisions, and these revisions are incorporating all of the new electrical power system (EPS) voltage and frequency ride-through capabilities and support functions. Fig. 1 shows the expansion of voltage and frequency limits that impact anti-islanding operation. These new voltage and frequency ride-through capabilities and regulating functions are needed to support the EPS. This also includes power quality, reclosure coordination, and loss of utility detection capabilities of today's DER.

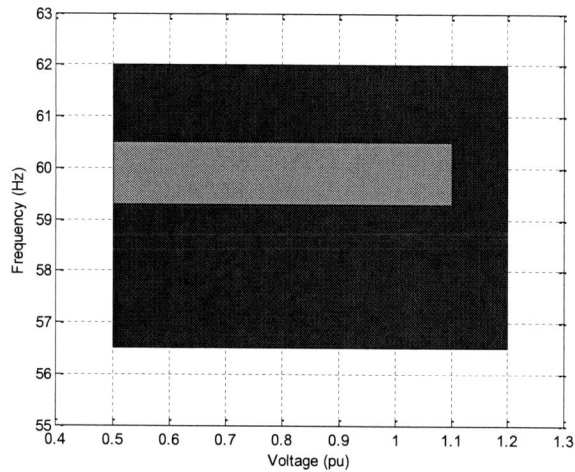

Fig. 1. Increased Voltage and Frequency Impact on Anti-islanding from IEEE 1547-2003 to P1547.

A. Methods of Anti-islanding

Some investigators are expressing concern that the new grid support functions may interfere with the detection of unintentional islands. This paper will focus on the anti-islanding detection requirement, type testing, and the aforementioned impact of grid support functions. The investigation will look into the methods [6] used to detect the loss of utility and prevent an anti-islanding event. It will also investigate how the anti-islanding test procedure of the draft IEEE P1547.1 has changed to account for the new EPS support capabilities, how these may affect the detection, and how long it takes for the DER to respond to the detection upon ceasing to energizing the utility.

B. Integrated Anti-islanding Methods

Anti-islanding techniques are commonly grouped into three categories: passive inverter-resident, active inverter-resident, and non-inverter-resident. These methods have been extensively reviewed and characterized [7-12].

• Passive inverter-resident methods utilize signal processing applied to the inverter terminal voltage, and possibly both the inverter voltage and current, to discern when an island or other abnormal grid condition has occurred. Nearly all inverters available today incorporate one or more passive methods, usually in conjunction with active methods. The primary advantages of passive methods are detection speed, low cost, general maintenance of effectiveness regardless of the number of inverters, and freedom from interference of one manufacturers' methods with another's. The primary disadvantage of passive methods is that it is very difficult to select their parameters to simultaneously achieve sensitivity and selectivity (i.e., all islands are detected, without false trips). Nearly all of these methods can be compromised if there are synchronous generators in the island. Examples of passive methods include:

o   Over/undervoltage and over/underfrequency
o   Phase jump detection or vector shift
o   Rate of Change of Frequency (RoCoF)
o   V/Hz relaying
o   Passive impedance detection based on cross-correlation
o   Detection of jumps in voltage THD, or voltage harmonic content at a specific frequency
o   Various forms of signal processing (e.g., wavelet analysis) looking for specific signatures that indicate island formation.

• Active inverter-resident methods involve the use of some deliberate perturbation of or injection into the inverter output current in order to create a condition during islanding that can be detected. Most, but not all, commercially-available inverters today use an active islanding detection method. As a group, these methods tend to be very effective with high selectivity and sensitivity in single-inverter cases, and many retain their sensitivity and selectivity in multi-inverter cases. The primary disadvantages of these methods are that they negatively impact grid stability and power quality, and their capability can degrade when more than one manufacturers' product is present in an island. Active methods are also compromised by the presence of rotating generation, although not as much as passive methods. Examples of active inverter-resident methods include:

o   Phase or frequency shift with positive feedback (e.g., Sandia Frequency Shift)
o   Positive feedback on voltage amplitude, usually triggered by a passive method (e.g., Sandia Voltage Shift)
o   Negative sequence current injection, with or without positive feedback
o   Various forms of Impedance detection, based on perturbation of current phase or magnitude, with or without positive feedback

• Non-inverter-resident methods. As the name implies, these are methods implemented outside the inverter. These methods typically involve one of two mechanisms: a) a change in circuit topology designed to disrupt

generation: load balance in an island, or b) communications between the DER site and the grid. These methods tend to be highly effective and provide very good sensitivity and selectivity. Their primary disadvantages are cost and scheduling and logistical issues associated with the installation of the additional equipment, and in addition these methods are often relatively slow to respond. Examples include:

o   Change in circuit topology
    ▪   Capacitor insertion/toggling
    ▪   Grounding or shorting switches
o   Communications
    ▪   Direct transfer trip (DTT)
    ▪   Power line carrier permissive (PLCP)
    ▪   Synchrophasor-based approaches

Table IV describes the most commonly-used methods and their susceptibility to degradation in effectiveness due to the implementation of EPS support functions.

II. ANTI-ISLANDING PROCEDURE CHANGES

The anti-islanding capabilities of inverter based DER are evaluated utilizing a circuit intended to create a condition that minimizes any voltage and frequency shift when the EPS is disconnected. For this to occur, active and reactive power from loads and from generation must be equal and set to resonate near 60 Hz. The islanding test configuration can be connected in either a wye-connection or a delta-connection. If the device under test (DUT) does not require a delta connected load, then it is suggested that the loads be connected in a wye-connected configuration. This configuration will minimize a voltage shift due to the "islanded" circuit not being referenced, which can lead to a false positive where the DUT responds correctly by ceasing to energize the utility or simulated utility in the specified response time, but it does so due to a misaligned circuit load value or a circuit that is not referenced and the voltage symmetry was insufficient to sustain the island. This type of condition may indicate that the DUT ceases to energize the utility but not because the anti-islanding algorithm was sufficient to detect the island condition.

The RLC circuit setup requires several steps to achieve an adequately tuned circuit. The existing test sequence to IEEE 1547.1 requires adjustments to the loads for a proper setting that minimize changes in voltage and frequency the moment the utility is removed. Fig. 2 shows the new RLC islanding circuit that has been introduced. This version has more detail and distinguishes between the types of systems under test and the load configurations. By default, the load configuration will be wye-connected to minimize uncertainty of unbalanced voltages causing a false positive. This test configuration circuit now has the detail needed to ensure the loads remain referenced during the opening of the utility switch (S3) which simulates the loss of the utility.

978-1-5090-5606-4/17 $31.00 © 2017 IEEE

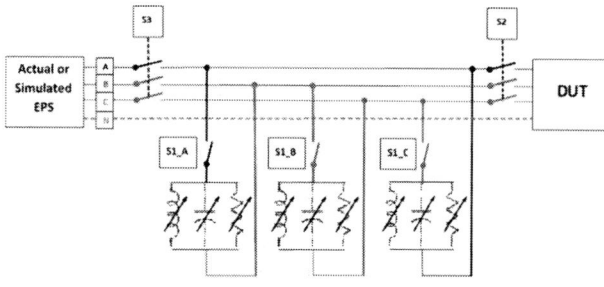

Fig. 2.   Wye-connected RLC anti-islanding test circuit.

A. Existing method of RLC Anti-islanding Test Procedure

The test procedure requires that the RLC islanding test be conducted at matched conditions, where the inductance and the capacitance are both set to resonate at 60 Hz and the amount of reactive load equals the active power of the inverter, which will equate to a Q factor close to 1. The resistive load is set to absorb as much power as the DUT is producing. The test procedure then requires up to a ±5% reactive power imbalance in 1% increments. After the 11 tests are conducted from the previous setup, the results are reviewed, and the 1% load setting that yielded the three longest trip times shall be subject to two additional iterations. If the longest run-on times are not consecutive, an additional two iterations are conducted for the nonconsecutive settings in between. For the new version of IEEE 1547.1, it is under consideration to perform the tests at two power levels (95% and 25%) instead of three (100%, 66%, and 33%). Table I shows the number of iterations per power level.

TABLE I
TEST ITERATIONS PER POWER LEVEL

Test condition	Number of tests
Matched	1
Mismatch 1% to 5%	5
Mismatch -1% to -5%	5
3-longest run on times	3
Total number to tests per power level.	14

A test matrix of DER operating conditions has been created to evaluate the DER while operating under different voltage and frequency regulating functions. Table II shows the test conditions that the anti-islanding test procedure should cover to fully assess the capabilities of the DUT. Some of the newly-mandated grid support function capabilities, such as specified power factor, volt-var, and specific reactive power, must be tested independently. Therefore a new sequence of testing is required for each of the functions.

TABLE II
TEST CONDITIONS FOR ANTI-ISLANDING TESTING

Test condition	Functions active during Anti-islanding Test
1	IEEE P1547 default settings
2	SPS, RR, FW
3	VV, RR, FW, VW
4	Watt-Var, RR, FW, VW
5	SVar, RR, FW, VW

Table legend: SPS- specified power factor, RR- ramp rate, FW-frequency-watt, VV- volt-Var, VW- volt-watt, CVar- commanded-Var

This will bring the total number of anti-islanding tests for an EPS supporting DER too approximately: 14 x 2 x 5 = 140 tests. This is a high number of tests that comes at a significant cost when the product is being developed and when the product is undergoing certification. However, there is an alternative.

B. Introduction of New Concise RLC Anti-islanding Test Procedure

Recent changes to the draft anti-islanding test procedure of IEEE 1547.1 will introduce an alternative method for evaluating the loss of utility detection of DER connected to the Area EPS. Part of the reason for the high number of tests with the traditional procedure is that the test sequence sweeps the RLC load parameters over a range in order to find the worst-case condition for the device. The alternative test procedure enables determination of this worst-case condition in fewer steps via the intermediate step of disabling the anti-islanding detection algorithm. The test utilizes a properly tuned RLC circuit that will allow the DER to intentionally island and demonstrate that the RLC circuit has been tuned correctly such that the DER, *with the anti-islanding algorithm disabled*, will have a run-on time longer than the 2 second requirement. Ideally, the DER will run-on long enough for the RLC circuit to be adjusted for a perfect 60 Hz resonance and the power generated-to-resistive load match is close enough to minimize the voltage to change when the utility is removed. The following steps show the procedure for disabling the anti-islanding algorithm:

a) Set all DUT input source parameters to the nominal operating conditions and for the DUT to operate at 95% (+/- 5%) of rated output power in maximum power tracking mode.

b) Set (or verify) all DUT parameters to the default 1547 settings. Grid support functions are set to test condition 1 shown in Table II.

c) Set the actual or simulated EPS to the DUT nominal voltage +/- 5% and frequency.

d) Record all applicable settings.

e) Close switch S1, switch S2, and switch S3, and wait until

978-1-5090-5606-4/17 $31.00 © 2017 IEEE   2149

the EUT settles at the desired power level.

f) **Disable** the anti-islanding function in the DUT.

g) Open switch S3 and verify that the test setup will support a sustained island and continue to operate for at least the required anti-islanding clearing time.

h) Repeat steps e) through g) for a total of three tests, with each test having a run-on time at least the required anti-islanded clearing time.

i) **Enable** the anti-islanding function in the DUT

j) Close switch S1, switch S2, and switch S3; re-enable operation of the EUT and wait until the EUT settles at the desired power level.

k) Open switch S3 and record the time between the opening of switch S3 and when the DUT ceases to energize and trips, thus de-energizing the test circuit.

l) Repeat steps i) and k) for a total of three tests at this power level.

This test sequence substantially reduces the number of tests while still providing a rigorous type-test of the inverter's anti-islanding method. The new anti-islanding test requirements shown in Table III provide an indication of the number of tests required for this concise anti-islanding test procedure.

TABLE III
NEW ANTI-ISLANDING TEST REQUIREMENTS

Anti-islanding algorithm status	Number of tests
disabled	3
enabled	3

The tests will be conducted at the same two power levels (95% and 25% of rated power). The tests will be conducted with the same EUT operating conditions as presented in Table II. This will bring the total number of anti-islanding tests for an EPS supporting DER to approximately 6 x 2 x 5 = 60, which is less than half the number of tests required under the present test procedure.

The following data show the test results obtained with the anti-islanding algorithm disabled and with it enabled. For each test, the test conditions are identical except for the status of the anti-islanding algorithm.

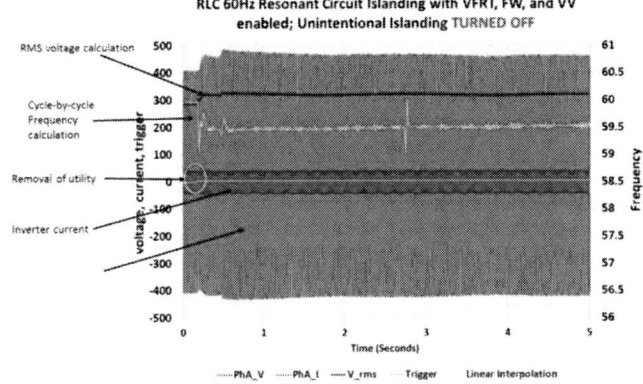

Fig. 3. UI test with algorithm "OFF" demonstrates continuous run-on.

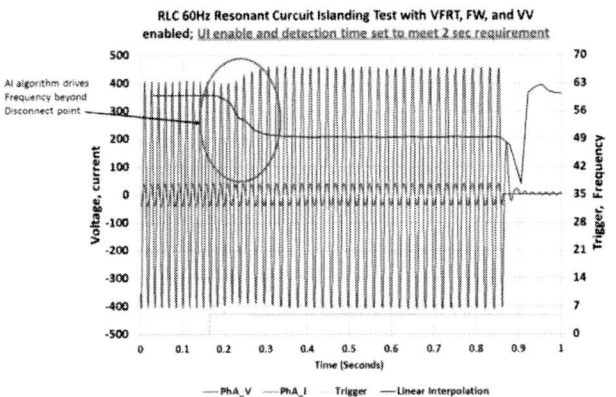

Fig. 4. UI test with algorithm "ON" demonstrates DUT detects loss of utility and ceases to energize within 2 seconds.

C. Introduction to Communication Based Anti-islanding

This anti-islanding test procedure evaluates the DUT's response to permissive signal removal. This test requires the DUT to be operating at rated power. Since EPS support functions do not adversely affect the detection and response time of the DUT to the loss of the permissive signal, there is no need to evaluate using the different operating modes in Table II and at different power levels. The test procedure is simple and only requires the interruption of the permissive signal and documenting the response of the DER. Fig. 5 shows the simplicity of the connection and that this test procedure

TABLE IV
ANTI-ISLANDING METHODS, CHARACTERISTICS, AND SUSCEPTIBILITY TO CHANGES IN PROCEDURE

Power electronic devices		
Method	**Characteristic**	**Susceptibility to EPS support functions**
RoCoF	Commonly-used passive method; trips the inverter when df/dt breaches a threshold	Will be strongly negatively impacted by new ride-through requirements
impedance	Detection with positive feedback	Minor reduction in effectiveness
Impedance	Detection without positive feedback	Minor reduction in effectiveness
Sandia Frequency Shift	implemented as phase or frequency injection	Increase in non-detection zone
perturbation	Feedback on negative sequence current	Increase in non-detection zone
Communication-Based Anti-islanding Methods		
Direct Transfer Trip (DTT)	Opening of utility breaker or isolation device Cost is obstacle unless large installation	No measurable reduction in effectiveness
Power Line Carrier Permissive (PLCP)	Opening of utility breaker or isolation device cost is issue, requires utility commitment	No measurable reduction in effectiveness
Synchrophasor	Several variants, each requiring a utility-supplied reference signal. Cost is issue, requires utility commitment	No measurable reduction in effectiveness

doesn't require any load configurations nor interrupting the actual power flow from the DUT, but the DUT is required to respond to loss of permissive signal and stop energizing the EPS within the required two seconds.

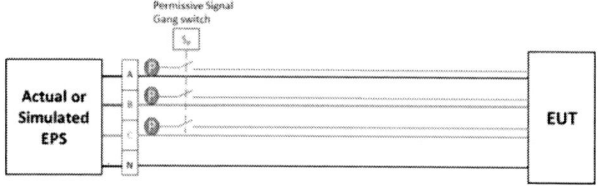

Fig. 5. Permissive signal Anti-islanding Test Configuration.

III. POWER BALANCE REQUIREMENTS

In order for a sustained island to exist, both active and reactive power must be balanced between the aggregate generation and load within the island circuit. Once the utility is disconnected from a section of an EPS, there is nowhere else that power and energy can flow except within the island. It is *after* the island has formed that the powers must be exactly balanced. In the instant just prior to the formation of the island, the degree to which the powers are balanced affects whether a sustained island is possible. In this context, a sustained island is a case where the generation continues to energize the circuit beyond the two second time limit from IEEE 1547.

PV inverters can be approximated by a constant-current model during early stages of an event and by a constant-power model over longer time periods. During the constant power period, the island voltage can be computed as:

$$V_{Island} = V_{EPS} \sqrt{P_{Gen} / P_{Load}} \qquad (1)$$

During the constant current period, the island voltage can be computed as:

$$V_{Island} = V_{EPS} \times P_{Gen} / P_{Load} \qquad (2)$$

where V_{EPS} is the EPS voltage just prior to formation of the island. P_{Gen} is the aggregate active power generation just prior to formation of the island. V_{Island} is the steady state voltage that the island will stabilize at if the DER does not trip based on voltage or frequency trips or anti-islanding protections. P_{Load} assumes that the aggregate load can be modeled as a resistance, R_{Load}, and is measured just prior to formation of the island.

$$P_{Load} = V_{EPS}^{2} / R_{Load} \qquad (3)$$

A similar equation can be used to estimate island frequency by looking at the reactive power balance.

$$F_{Island} = F_{EPS} \sqrt{(Q_L + Q_{Lder}) / (Q_C + Q_{Cder})} \quad (4)$$

where F_{EPS} is the EPS frequency just prior to formation of the island. Q_L and Q_C are the reactive powers of the inductive and capacitive loads within the island circuit, also measured just prior to formation of the island. Q_{Lder} and Q_{Cder} represent the reactive power output of the DER and presume constant reactive power operation of the DER before and after formation of the island (a presumption that is being tested at this time).

These equations do not predict the transient response of the system immediately after formation of an island, but it is expected that well behaved DER and island loads will quickly reach steady state equilibrium. These equations also do not take into account non-idealities of load components such as change in resistance as power level changes, change in capacitance or in inductance with change in voltage, or harmonic currents of magnetic components. The equations are only intended to model the fundamental frequency first order effects.

The quality factor equation from IEEE 1547.1-2005 is modified now to include the reactive power components of the DER:

$$QF = \frac{\sqrt{(Q_L + Q_{Lder}) \times (Q_C + Q_{Cder})}}{P_{Load}} \quad (5)$$

IEEE 1547-2003 voltage trips that were faster than two seconds range from 0.50 to 1.10 pu. If the island voltage exceeded these limits, the generation would trip on voltage and shutdown before the two second island protection limit. The frequency trips for small generation, less than 30 kW, were fast, 0.16 seconds, at 59.3 and 60.5 Hz. The frequency trips for larger generation could be set as wide as 57.0 to 60.5 Hz.

IEEE P1547 Draft 6 ensures a fast, 0.083 second, voltage trip at 1.20 pu, and momentary cessation at 0.50 pu. The frequency trips range from 56.5 and 62.0 Hz and are fast at 0.083 seconds.

IV. SUMMARY AND FUTURE WORK NEEDED TO COMPLETE ASSESSMENT

Much work has been conducted to group anti-islanding techniques into categories and simulations have been conducted to characterize the susceptibility of these different categories with the implementation of EPS support functions. Capabilities of DER according to the type of anti-islanding methods implemented into the device have undergone some laboratory validation experiments but more are needed to cover the multitude of combinations of categories and the matrix of operating functions.

Conducting laboratory experiments to exercise the new anti-islanding test procedures will continue and are needed to validate that the draft procedures are sufficiently robust and can assess the different types of anti-islanding categories and combinations of EPS support functions. Additional experiments will be conducted to document the variability in loads that can be tolerated and will maintain a load to generation balance that will allow the device under test to continue to operate and stay within the voltage and frequency operating ranges.

ACKNOWLEDGEMENTS

Sandia National Laboratories is a multimission laboratory managed and operated by National Technology & Engineering Solutions of Sandia, LLC, a wholly owned subsidiary of Honeywell International, Inc., for the U.S. Department of Energy's National Nuclear Security Administration under contract DE-NA0003525.

The authors would like to thank Jeff Zirzow, who provided much of the test data and Nicholas Gurule for providing the data analysis.

References

[1] Solar Energy Industries Association and GTM Research, U.S. Solar Market Insight: Year-in-Review, 2016.

[2] IEEE. *1547-2003–IEEE Standard for Interconnecting Distributed Resources with Electric Power Systems.* Approved June 12, 2003. Institute of Electrical and Electronics Engineers, Inc., New York, NY.

[3] IEEE. *1547a-2014–IEEE Standard for Interconnecting Distributed Resources with Electric Power Systems: Amendment 1.* Approved May 16, 2014. New York, NY.

[4] IEEE. *P1547/D6.0 Draft Standard for Interconnection and Interoperability of Distributed Energy Resources with Associated Electric Power Systems Interfaces.* Working Group draft not publicly released. Dec. 2016.

[5] IEEE. *P1547.1/D1.0 Draft Standard Conformance Test Procedures for Equipment Interconnecting Distributed Energy Resources with Electric Power Systems and Associated Interfaces.* Working Group draft not publicly released. Nov. 30, 2016.

[6] M. Ropp and A. Ellis, "Suggested guidelines for anti-islanding screening," Sandia National Laboratories, Albuquerque, NM, SAND2012-1365, Feb. 2012.

[7] W. Bower and M. Ropp, "Evaluation of islanding detection methods for utility interactive inverters in photovoltaic systems", Sandia National Laboratories, Albuquerque, NM, SAND2002-3591. Nov. 2002.

[8] F. de Mango, M. Liserre, A. Dell'Aquila, A. Pigazo, "Overview of anti-islanding algorithms for PV systems. Part I: passive methods", 12th Annual International Power Electronics and Motion Control Conference, pp. 1878-1883, Sept. 2006,

[9] F. de Mango, M. Liserre, A. Dell'Aquila, "Overview of anti-islanding algorithms for PV systems. Part II: active methods", 12th Annual International Power Electronics and Motion Control Conference, pp. 1884-1889, Sept. 2006.

[10] P. Mahat, Z. Chen, B. Bak-Jensen, "Review of islanding detection methods for distributed generation", Third International Conference on Electric Utility Deregulation and Restructuring and Power Technologies, pp. 2743-2748, April 2008.

[11] A. M. Massoud, K. H. Ahmed, S. J. Finney, B. W. Williams, "Harmonic distortion-based island detection technique for inverter-based distributed generation", *IET Renewable Power Generation,* vol. 3, no. 4, pp. 493-507, Sept. 2009.

[12] T. Quoc-Tran, "New methods of islanding detection for photovoltaic inverters", IEEE PES Innovative Smart Grid Technologies Conference Europe (ISGT-Europe), 5 pgs., Oct. 2016.

Parametric PV Grid-Support Function Characterization for Simulation Environments

Javier Hernandez-Alvidrez[1], and Jay Johnson[2]

[1]New Mexico State University, Las Cruces, New Mexico, 88003, USA.
[2]Sandia National Laboratories, Albuquerque, New Mexico, 87185, USA.

Abstract—Modern photovoltaic (PV) inverters and other inverter-based distributed energy resources (DER) have the ability to provide grid-support services with different advanced inverter functions. At this time, the nuanced influence of these functions on grid operations, including voltage regulation, frequency reserve deployments, protection systems, and other ancillary services is not fully understood. Researchers are increasingly turning to hardware-in-the-loop (HIL) simulations to answer renewable integration questions by coupling physical devices to power simulations. Unfortunately, it is common for these simulations to use ideal power converter models to represent the end-devices. In this paper, we perform parametric characterization of multiple PV inverter products to generate empirical equations which accurately describe the inverter behavior for volt-var, fixed power factor, and frequency-watt grid support functions over a range of operating conditions. These equations are designed for use in HIL and other power simulations to better represent the actual output of fielded equipment.

Index Terms—inverters, hardware-in-the-loop, volt-var, frequency-watt, fixed power factor, inverter model.

I. INTRODUCTION

In 2015, the California Public Utilities Commission joined Hawaii and a number of European jurisdictions requiring grid-support functionality from PV inverters [1]–[4]. A large effort is also currently underway to revise the national interconnection standard, IEEE Std. 1547 [5], to require additional distributed energy resource (DER) grid-support functions and device interoperability. The new functions required in these standards have the ability to assist grid operators to perform voltage regulation, frequency stabilization, and power system protection via different operating modes.

The combination of modes (i.e. grid support functions) and settings for specific use cases with a given circuit topology and collection of DER assets has been widely studied in the literature. Typically these studies rely on power simulations to determine the optimal settings for the given scenario. In many power simulations, advanced-grid functions are modeled as ideal functions without measurement inaccuracies, generation errors, or time delays. For instance, studies of voltage regulation [6] and hosting capacity using the volt-var function [7] assume perfect reactive power generation for a given grid voltage. However, physical devices can have significant errors in ideal behavior and response times can be lengthy [8]–[12]. The results of advanced grid function tests conducted by the Smart Grid International Research

Facility Network (SIRFN) for PV [13] and energy storage systems [14] found large errors for some devices. The latency of the frequency-watt function was found to be detrimental to providing fast contingency reserves [9], [14]; communication latencies with fixed power factor and constant reactive power were found to disrupt voltage regulation control strategies [6]; and, recently, miscalculated frequencies from inverters caused approximately 700 MW of generation to trip in the Western Interconnection [15].

In the U.S., grid-interconnected DER assets are tested to Underwriters Laboratories (UL) 1741 [16] certification standard by a Nationally Recognized Testing Laboratory (NRTL). The most recent version of UL 1741 was published in 2016 and includes a test procedure to evaluate and certify DER grid-support functions in Supplement A. The current language allows vendors to certify products that fall within the manufacturer's specified accuracy, unless specifically called out in the interconnection source requirements document (SRD)– e.g. CA Rule 21, HI Rule 14H, IEEE 1547. Most of the SRDs do not specify accuracies for the grid support functions so inverter manufacturers are free to provide generous accuracies in order to pass the tests. As a result, DER may not behave as programmed in the field and grid operators may not correctly select the DER settings or functions to provide grid services.

Researchers are increasingly employing control hardware-in-the-loop (CHIL) and power hardware-in-the-loop (PHIL) techniques to better understand how equipment will operate in the field [3]. Yet, for large HIL power system simulations, DER devices are typically modeled as ideal converters. In this work, we seek to establish a methodology for representing physical PV inverters as analytical models to accurately capture grid-support function behavior to provide recommendations on the types of models which most accurately represent the response of the equipment. We generate and compare models of two PV inverters from different vendors under a range of PV irradiance conditions with fixed power factor, volt-var, and freq-watt functions.

II. EXPERIMENTAL CONFIGURATION

To assess the DER for a range of operating conditions and functions, the SunSpec System Validation Platform (SVP) [17] was used to communicate to an Ametek RS-180 180 kVA grid simulator, a 200 kW Ametek PV Simulator, National Instruments (NI) PCIe-6259 data acquisition (DAQ)

978-1-5090-5606-4/17 $31.00 © 2017 IEEE

cards, and the Equipment Under Test (EUT). The test configuration is shown in Fig 1.

The advanced grid-support functions were configured in the EUT through SunSpec Modbus RTU or TCP commands. The grid voltage and PV irradiance were set via communications to the grid and PV simulators using the SVP scriptable interface. The IV curve of the PV simulator was programmed to provide nameplate EUT power at 1000 W/m^2 irradiance. At 1000 W/m^2 the EUT operated just under its AC nameplate power because of the device efficiency. For each setting, six cycles of the current and voltage waveforms were captured at 10 kHz via the National Instruments (NI) data acquisition system and processed for parameters of interest using the IEEE Std. 1459 [18]. Experiments were conducted for the support functions implemented in two inverters from different manufacturers labeled, hereafter, as EUT A and EUT B.

A. Volt-Var Function

For each of the inverters, the steady-state volt-var behavior of the EUTs was determined after a 2 second settling time. Grid voltages were evenly sampled 250 times (or every ~0.2 V) from V_{min} to V_{max} (the low and high voltage trip points) of the EUTs to fully characterize the curves. The curves were swept in increasing and decreasing voltage directions to capture any hysteresis at irradiance values of 100-1100 W/m^2 at 100 W/m^2 increments. Three Volt/Var curves were programmed into the devices to evaluate different VV settings.

B. Fixed Power Factor

The fixed power factor function was tested on the two EUTs at incremental values of 0.01 from PF_{min} to PF_{max}. At each power factor, the irradiance values were adjusted from 100-1100 W/m^2 at 100 W/m^2 increments.

C. Frequency-Watt Function

The freq-watt function was evaluated on the two inverters by adjusting the grid simulator frequency from 59.5 to 62.0 Hz at irradiance values of 100-1100 W/m^2 at 100 W/m^2 increments.

III. ANALYTICAL MODELS

The proposed analytical inverter model is of the form:

$$Y[P,Q] = \sum a_i \cdot f_i \tag{1}$$

where Y is the active or reactive power output of the EUT (P or Q), a_i is a boolean representing function activation, and f_i is the inverter function. In this work we investigate the EUT response by considering three functions with the values of a_i being mutually exclusive, so the output of the inverter is represented by:

$$
\begin{aligned}
Y = P, Q = &\, a_{PF} \cdot f_{PF}\left(G, PF_{settings}\right) + \\
&\, a_{VV} \cdot f_{VV}\left(G, VV_{settings}, V_{grid}\right) + \\
&\, a_{FW} \cdot f_{FW}\left(G, FW_{settings}, F_{grid}\right) +
\end{aligned} \tag{2}
$$

where the analytical model of each function is expressed with a parametric mathematical expression that depends on three parameters: EUT settings, DC power conditions (PV irradiance, G), and grid conditions (f, V). Since we consider the response from each of these functions individually, e.g.

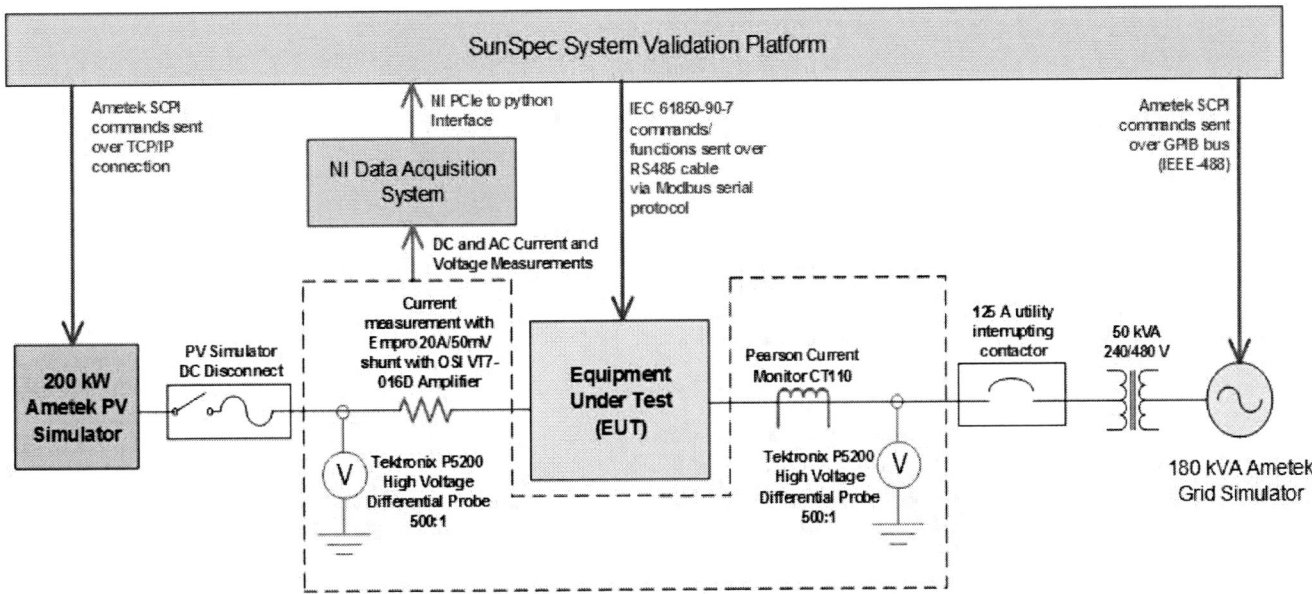

Fig. 1: Inverter Testbed at Sandia National Laboratories' Distributed Energy Technologies Laboratory (DETL).

only one a_i value can be 1 (true) at a time, then (2) can be expressed as:

$$Y = P, Q = \begin{bmatrix} a_{PF} & 0 & \\ 0 & a_{VV} & 0 \\ 0 & 0 & a_{FW} \end{bmatrix} \begin{bmatrix} f_{PF} \\ f_{VV} \\ f_{FW} \end{bmatrix} \quad (3)$$

In order to characterize the behavior of the aforementioned functions with analytical models, three different curve fitting approximation methods were evaluated: linear piecewise, least square polynomial interpolation, and Fourier approximation.

A. Linear Piecewise

In piecewise curve fitting method, the set of experimental data was approximated by a set of linear functions, where each function characterizes a particular section of the data on which the behavior can be modeled with a first order (linear) algorithm. This method proves to be quite effective in the approximation of functions with relatively long linear sections with sharp discontinuities between them, such as the ones encountered in the VV of FW functions. The general expression for this method is:

$$Y(x) = \begin{cases} A_1 \cdot x + B_1, \text{if } x_{1min} \leq x \leq x_{1max} \\ A_2 \cdot x + B_2, \text{if } x_{2min} \leq x \leq x_{2max} \\ \vdots \\ A_n \cdot x + B_n, \text{if } x_{nmin} \leq x \leq x_{nmax} \end{cases} \quad (4)$$

where n is the number of linear sections used. Coefficients A and B depend on the type of linear interpolation used to characterize the respective linear section.

B. Least Squares Polynomial Interpolation

With this method, the experimental set of data was approximated by means of an n order polynomial of the form:

$$Y(x) = a_0 + a_1 x + a_2 x^2 + ... + a_n x^n \quad (5)$$

Provided (5) passes through all ordinate values of the experimental ordered pairs: (x_0, y_0), (x_1, y_1), ...,(x_m, y_m), and m is the total number of pairs. Then, the coefficients of (5) can be obtained from:

$$\begin{bmatrix} a_0 \\ a_1 \\ \vdots \\ a_n \end{bmatrix} = \begin{bmatrix} 1 & x_0 & x_0^2 & \cdots & x_0^n \\ 1 & x_1 & x_1^2 & \cdots & x_1^n \\ \vdots & \vdots & \vdots & \ddots & \vdots \\ 1 & x_m & x_m^2 & \cdots & x_m^n \end{bmatrix}^{-1} \begin{bmatrix} 1 \\ y_0 \\ \vdots \\ y_n \end{bmatrix} \quad (6)$$

or, in a simplified way:

$$\vec{a} = \mathbf{V}(x) \cdot \vec{y} \quad (7)$$

where $\mathbf{V}(x)$ is called the Vandermonde matrix [19] and its dimension is determined by the total number of experimental ordered pair points (rows), and by the desired number of polynomial coefficients (columns). The higher the number of coefficients the higher the accuracy of the approximated polynomial at the expense of using more computational resources for coefficient calculation.

C. Fourier Approximation

In this approximation method the dynamics of the set of experimental data points is modeled as a linear combination of harmonically related sinusoids, which gives:

$$\begin{aligned} Y(x) = & A_0 + A_1 \sin\left(\omega x + \varphi_1\right) + A_2 \sin\left(2\omega x + \varphi_2\right) + ... \\ & + A_n \sin\left(n\omega x + \varphi_n\right) \end{aligned}$$
$$(8)$$

where the coefficients (A_0 to A_n) and phase angles (φ_1 to φ_n) are obtained by applying the Discrete Fourier Transform [20] to the experimental set of data. As with the case of polynomial approximation, the more terms included in the model, the more accurate the approximation will be, but at the expense of using more computational resources to compute the Fourier coefficients and phase angles.

IV. EXPERIMENTAL RESULTS

Figure 2 shows the experimental results of the Volt-Var curves implemented on EUT A and EUT B. In the figure, the programmed VV function is shown as a black solid line and each colored scatter plot represents the experimental data corresponding to a different irradiance level. From Fig. 2, it can be observed that for each irradiance level in EUT A there is a systematic offset in both the voltage and the reactive power, clearly seen in the dead band of the VV curve. The offset in the AC voltage at the Point of Common Coupling (PCC) is most likely related to measurement error of the EUT. The voltage at the PCC was independently verified to within 0.1 V by a calibrated handheld meter and the calibrated NI DAQ system. The offset in the reactive power is most likely related to the output filter of the inverter. It can also be observed that the available reactive power is curtailed at lower irradiance levels, due to the power factor constraint of EUT A. Another way of visualizing the VV behavior of the EUT is to plot the results on an active-reactive power (P-Q) plane. As shown in Fig. 3, at low irradiance, the EUT holds the active power constant as the voltage changes the reactive power. At higher irradiance levels, the EUT adjusts active power as the voltage deviates from V_{nom}. The power factor ratings of EUT A (± 0.85) and EUT B (± 0.80) are shown with red lines. EUT A VV function is constrained to the PF limits of the equipment, but the reactive power of EUT B is limited by the reactive power limits of the device ($\pm$ 1800 VAr). This must be captured in the analytical model.

Figure 4 shows the experimental results of the fixed power factor function for EUT A and EUT B. The dynamics of the function were plotted such that for each colored scatter plot the reactive power is a function of the DC irradiance level (independent variable) while keeping the power factor constant (parametric variable). From Fig. 4, it can be noticed an offset on the reactive power of EUT A for unity power factor, which directly correlates with the systematic offset in EUT A of Fig. 2. Visually, the dynamics of EUT B have a better behavior than EUT A. From Fig. 4 notice that the offset at unity power factor is very close to zero, and also the well

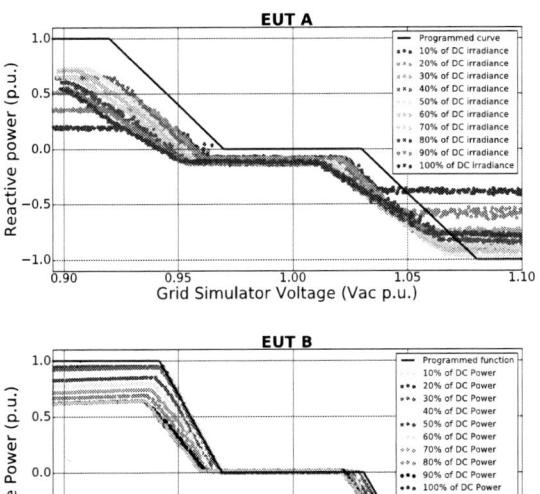

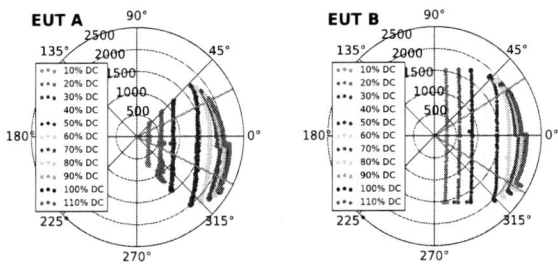

Fig. 2: Volt-Var curves.

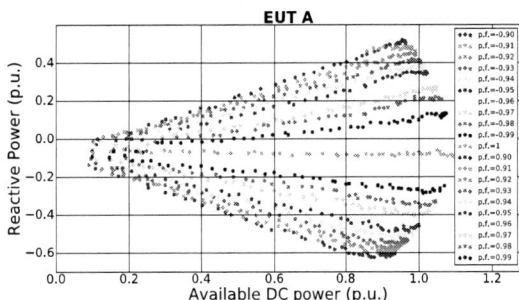

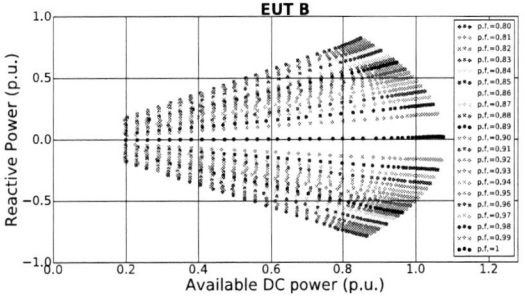

Fig. 4: Fixed power factor curves.

Fig. 3: EUT A and EUT B volt-var curves in the P-Q plane.

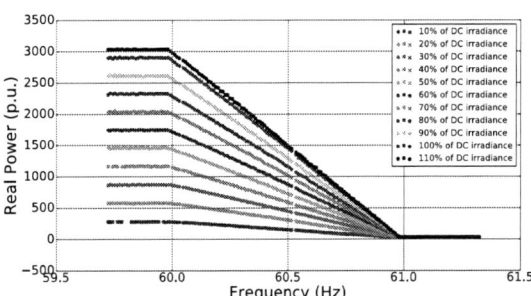

Fig. 5: Frequency-Watt curves for EUT B.

defined linearity for each power factor waveform in EUT B compared to EUT A.

Figure 5 shows the experimental results of the FW curve for EUT B, where each colored scattered waveform corresponds to a different DC irradiance level. The behavior is representative of the function, making the analytical modeling with device errors less critical.

V. ANALYTICAL MODEL IMPLEMENTATION

The three inverter functions were modeled using the aforementioned approximation methods. For each function, the implemented algorithm had three routines: (a) extracts the experimental data and identifies the independent and parametric variables according to Table I, (b) performs the approximation method, and (c) interpolates between parametric data if the desired parametric variable G is not in the data sets.

A. Volt-Var function

The three approximations of the VV function for EUT B are shown in Fig. 6. For the linear piecewise approximation, the approximation curves (green curves) model the sharp corners with a good degree of accuracy. For the polynomial interpolation, which was approximated using $n = 53$, the modeled function tracks the experimental data smoothly except for the sharp corners. In order to reduce the error to less than 1% in the sharp corners, a polynomial of at

TABLE I: Variable assignments for each function.

Inverter Function	Independent variable	Parametric variable
Volt-Var	Grid Voltage	DC irradiance
Fixed PF	DC irradiance	Power factor
Frequency-Watt	Grid Frequency	DC irradiance

least 100th order must be used, which significantly increases the computational resources needed, and may lead to ill-conditioned polynomials [19]. For the Fourier approximation, 35 harmonically related sinusoids were used. The intrinsic oscillatory nature of this method appears at the ends of the curve in the form of high frequency oscillations, although this high frequency variation could be reduced with a higher order Fourier approximation. One advantage of the Fourier method is that the sharp corners are modeled in a more accurate way compared to the polynomial interpolation.

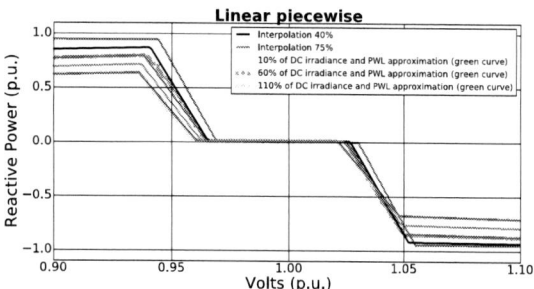

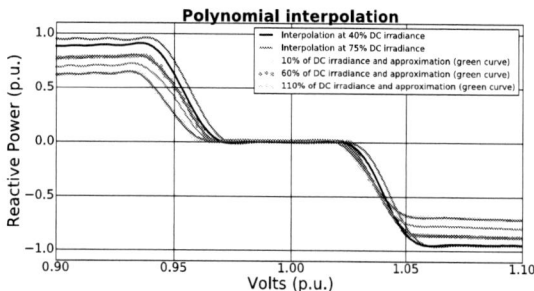

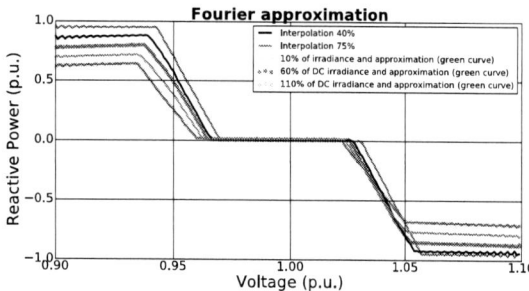

Fig. 6: Approximations of Volt-Var function for EUT B.

B. Fixed Power Factor

The results of the PF approximations are shown in Fig. 7. This function was not modeled with the Fourier approximation method due to the lack of periodicity of the experimental results. The performance of the two approximations of Fig. 7 was compared using the Pearson correlation coefficient (R^2). Based on R^2, the polynomial approximation seemed to be a better fit for this function, but, this approximation showed

evidence of overfitting, so the linear approximation may predict EUT behavior better.

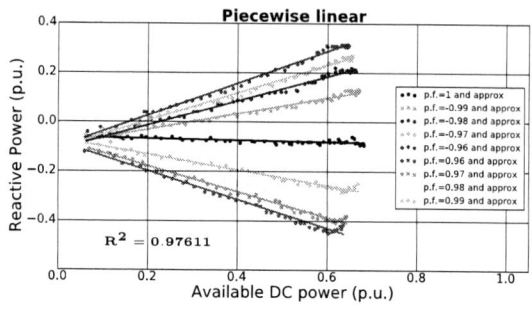

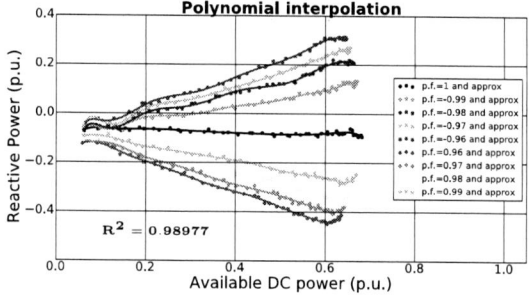

Fig. 7: Approximations of Fixed Power Factor functions for EUT A.

C. Frequency-Watt function

The results of the approximations curves for this function are shown in Fig. 8. Again, the piecewise linear method was the best option to model the sharp corners of the curve. The polynomial interpolation and Fourier approximation were given the same number of terms as with the Volt-Var function ($n = 53$ for polynomial, and 35 harmonically related terms for Fourier approximation). Since the Frequency-Watt curve has only 2 sharp corners, the accuracy of the approximation for this curve was improved at such sharp corners in comparison with the Volt-Var curve.

VI. CONCLUSIONS

In order to accurately represent inverters with grid-support functions in HIL simulations, a generalized inverter model was created. This model was represented with three different approximation methods and validated with three functions.

The piecewise linear approximation method proved to be suitable for curves with sharp corners as long as the data points of each section are linearly well-aligned. However, due to the sectionalized (non-continuous) nature of this method, interpolation between parameterized curves results were more difficult since the algorithm needs many logic statements to identify the starting and ending points of each section of the curve for every single value of the independent variable.

978-1-5090-5606-4/17 $31.00 © 2017 IEEE   2157

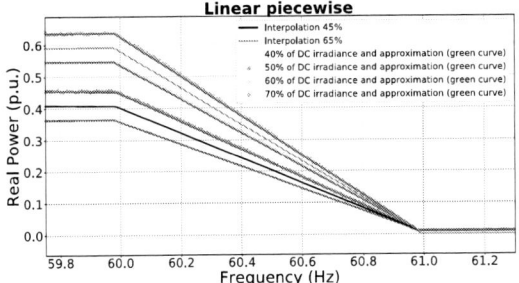

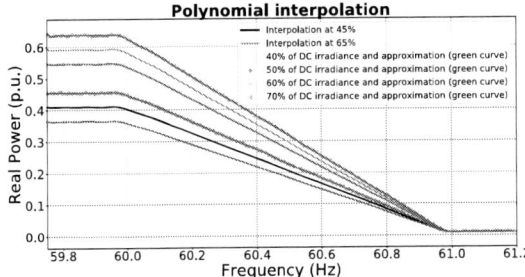

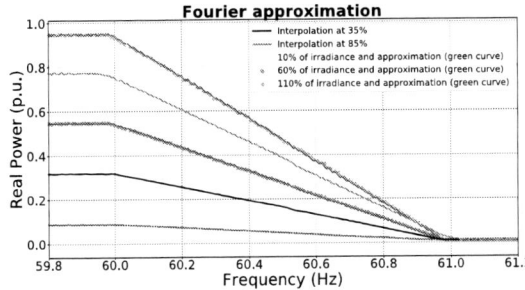

Fig. 8: Approximations of Frequency-Watt functions for EUT B.

The polynomial interpolation method proved to be effective in curves with scattered data points as well as curves with systematic offsets. This method lacked good accuracy in the neighborhood of sharp corners. But, the continuous nature of this method provided and easy way to interpolate between parameterized curves. The Fourier method proved to be effective in complex curves like VV or FW. The accuracy of this method relies on the number of harmonically related components used to model the curves. Also, this method provided a much better accuracy in the neighborhood of sharp corners when compared to polynomial interpolation. Another benefit is the ease of interpolation of this method due to its continuous nature.

VII. ACKNOWLEDGMENTS

Sandia National Laboratories is a multi-mission laboratory managed and operated by National Technology and Engineering Solutions of Sandia, LLC., a wholly owned subsidiary of Honeywell International, Inc., for the U.S. Department of Energy's National Nuclear Security Administration under contract DE-NA-0003525.

REFERENCES

[1] Pacific Gas and Electric Co, "Electric Rule No. 21, Generating Facility Interconnections, Filed with the CPUC," Jan. 20, 2015.

[2] Hawaiian Electric Company, "Inc. Rule No. 14, Service Connection and Facilities on Customers Premises, D&O No. 33258 filed Oct. 12, 2015, effective Oct 21, 2015."

[3] R. Bründlinger, T. Strasser, G. Lauss, A. Hoke, S. Chakraborty, G. Martin, B. Kroposki, J. Johnson, E. de Jong, "Lab Tests: Verifying That Smart Grid Power Converters Are Truly Smart," in *IEEE Power and Energy Magazine*, vol. 13, no. 2, March-April 2015, pp. 30–42. doi: 10.1109/MPE.2014.2 379 935.

[4] D. Rosewater, et al., "International development of energy storage interoperability test protocols for renewable energy integration," in *EU PVSEC, Hamburg, Germany*, Sept 14-18 2015.

[5] IEEE 1547 Std. 1547-2008, "IEEE Standard for Interconnecting Distributed Resources with Electric Power Systems," Institute of Electrical and Electronics Engineers, Inc. New York, NY, Standard, Sept 2008.

[6] M. J. Reno, J. E. Quiroz, O. Lavrova, R. H. Byrne, "Evaluation of communication requirements for voltage regulation control with advanced inverters," in *North American Power Symposium (NAPS)*. Denver, CO: IEEE, Sept 2016, pp. 1–6.

[7] J. Seuss, M. J. Reno, R. J. Broderick, S. Grijalva, "Improving Distribution Network PV Hosting Capacity via Smart Inverter Reactive Power Support," in *IEEE PES General Meeting*. Denver, CO: IEEE, July 2015.

[8] J. B. Ahn, J. J. Lee, J. Johnson, J. H. Bae, "Test Results for Advanced Inverter Functions Based-on IEC 61850-90-7," in *5th Asia-Pacific Forum on Renewable Energy (AFORE)*, Jeju, Korea, 4-7 November 2015.

[9] J. Neely, J. Johnson, J. Delhotal, S. Gonzalez, M. Lave, "Evaluation of PV Frequency-Watt Function for Fast Frequency Reserves," in *IEEE Applied Power Electronics Conference (APEC)*, Long Beach, CA, March 20-24 2016.

[10] S. Gonzalez, J. Johnson, J. Neely, "Electrical Power System Support-Function Capabilities of Residential and Small Commercial Inverters," in *42nd IEEE PVSC*, New Orleans, LA, June 14-18 2015.

[11] S. Gonzalez, J. Johnson, M. Reno, T. Zgonena, "Small Commercial Inverter Laboratory Evaluations of UL 1741 SA Grid-Support Function Response Times," in *43rd IEEE PVSC*, New Orleans, LA, June 5-10 2016.

[12] B. Seal, et al, "Final Report for CSI RD&D Solicitation #4 Standard Communication Interface and Certification Test Program for Smart Inverters," June 2016.

[13] J. Johnson, R. Bründlinger, C. Urrego, R. Alonso, "Collaborative Development Of Automated Advanced Interoperability Certification Test Protocols For PV Smart Grid Integration," in *EU PVSEC*, Amsterdam, Netherlands, 22-26 Sept 2014.

[14] D. Rosewater, J. Johnson, M. Verga, R. Lazzari, C. Messner, R. Brndlinger, K. Johannes, J. Hashimoto, K. Otani, "International development of energy storage interoperability test protocols for renewable energy integration," in *EU PVSEC*, Hamburg, Germany, 14-18 Sept 2015.

[15] NERC, "1200 MW Fault Induced Solar Photovoltaic Resource Interruption Disturbance Report," North American Electric Reliability Corporation, Tech. Rep., June 2017.

[16] UL-1741 Ed. 2, "Inverters, Converters, Controllers and Interconnection System Equipment for use with Distributed Energy Resources," Underwriters Laboratories, Standard, 2016.

[17] J. Johnson, B. Fox, "Automating the Sandia Advanced Interoperability Test Protocols," in *40th IEEE PVSC*, Denver, CO, 8-13 June 2014.

[18] IEEE Std 1459, "IEEE Trial-Use Standard Definitions for the Measurement of Electric Power Quantities Under Sinusoidal, Nonsinusoidal, Balanced, or Unbalanced Conditions," Institute of Electrical and Electronics Engineers, Inc. New York, NY, Standard, Approved 30 January 2000.

[19] George M. Phillips, *Interpolation and Approximation by Polynomials*. Springer, 2003.

[20] J. Alan V. Oppenheim, Ronald W. Schafer, *Discrete-Time Signal Processing*, 3rd ed. Prentice Hall, 1999.

978-1-5090-5606-4/17 $31.00 © 2017 IEEE

Cost analysis and cost reduction opportunities of residential PV system in the Japan

Izumi Kaizuka[1], Haruki Yamaya[1], Takashi Ohigashi[1], Risa Kurihara[1], and Osamu Ikki[1]

[1]RTS Corporation, 3-19-2 Hatchobori, Chuo-ku, Tokyo 104-0032, Japan
TEL: +81-3-3551-6345   E-mail: info@rts-pv.com

Abstract — Since the start of the Feed-in Tariff Program in July 2012, the PV market in Japan has been growing and annual installed capacity reached 10.8 GW in 2015. Newly installed PV capacity in 2016 is estimated to be 8.6 GW. While the majority segments of the new capacity are commercial, industrial and utility scale application, installed capacity of residential application remains relative low at around 0.9 GW in 2016. With the reduction of Feed-in tariff level, installation cost for residential PV system decreased from 472 Yen/W in the 2012 to 363 Yen/kW (USD 3.15/W) in the third quarter of 2016. However, cost is still higher level in comparison with Germany, Australia and other countries. It is observed that soft cost of residential PV system is one of the major factors for higher cost.

In this paper, reasons behind higher cost is analyzed based on residential cost breakdown based on hearings, supply chain survey. This paper also analyze the future cost reduction opportunities by the policy change and other factors.

Index Terms — FIT program, Soft cost, PV market, outlook, residential

I. CURRENT STATUS OF RESIDENTIAL PV MARKET

As shown Figure 1, the annual PV installed capacity in Japan in 2015 was 10.8 GW (DC) and 2016 installed capacity is expected to be 8.6 GW (DC). Growth of the PV market comes from increase of installed capacity of utility-scale, commercial applications. Residential PV market size of 2016 is 0.9 GW.

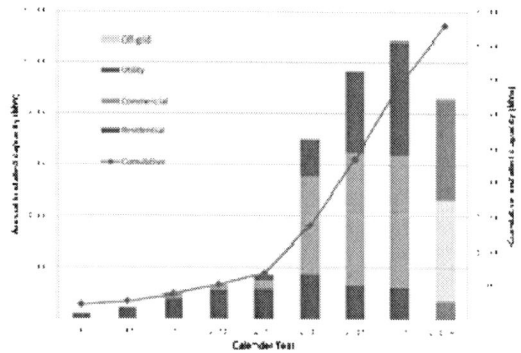

Fig. 1. Trends of installed capacity of PV system in Japan (DC).
Source: RTS Corporation

II. MARKET DRIVER OF RESIDENTIAL PV SYSTEM

Market driver of residential PV system is the Feed-in Tariff (FIT) program implemented under the Renewable Energy Law by the Ministry of Economy, Trade and Industry (METI). The owners of residential PV system can sell surplus PV power to utility companies at higher tariff (FIT) than retail electricity for 10 years. FIT level is reviewed once a fiscal year based on actual cost and assumed IRR set by the government.

Table I shows the trends of FIT level for residential application (< 10 kW application). Selling surplus power is not subject to consumption tax.

In FY 2016, the FIT levels for PV systems were set lower than those of the previous fiscal year. The tariffs and periods of purchase were set as follows: 1) 24 JPY/kWh (excl. tax) for PV systems with a capacity of 10 kW or more for the period of 20 years and 2) 33 JPY/kWh for PV systems with a capacity of below 10 kW (31 JPY/kWh for PV systems which are not required to be equipped with devices to respond to output curtailment).

Communication equipment for output curtailment is required for PV inverters installed in the service area of Hokkaido Electric Power (HEPCO), Tohoku Electric Power, Hokuriku Electric Power, Chugoku Electric Power, Shikoku Electric Power, Kyushu Electric Power or Okinawa Electric Power (OEPC). In these areas, residential PV systems are subject of output curtailment. Due to the priority of implementing curtailment, there has been few chances of curtailment for residential PV system in these electric power company's area.

TABLE I Trends of FIT level for residential application.

	Period	FIT level
	FY 2012	42 Yen/kWh
	FY 2013	38 Yen/kWh
	FY2014	37 Yen/kWh
FY 2015	with communication device	35 Yen/kWh
	without communication device	37 Yen/kWh
FY 2016	with communication device	31 Yen/kWh
	without communication device	33 Yen/kWh

Source: Ministry of Economy, Trade and Industry (METI), compiled by RTS Corporation

III. COST TRENDS OF RESIDENTIAL PV SYSTEM

According to a committee for setting the FIT level under METI, installed cost of residential PV system has been decreasing. In the third quarter of 2012 when the FIT program started was 472 Yen/W. In the third quarter of 2016 the reported cost reduced to 363 Yen/W by 23%. Latest price during January to March 2017 was 293 Yen/W.

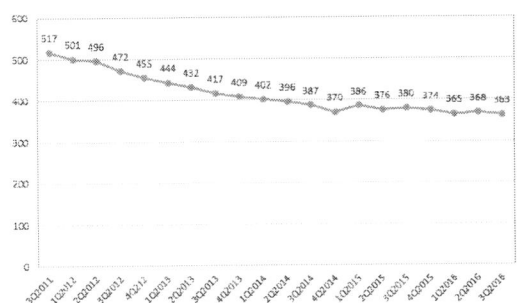

Fig. 2. Price trends of residential PV system.
Source: METI compiled by RTS Corporation

While cost reduction was observed with the reduction of the FIT level, the installed cost is relatively high in comparison with Germany or other major PV market. Figure 3 shows reported residential PV system cost by countries in 2015. Japanese install cost is the highest among 16 countries as shown in Figure 3.

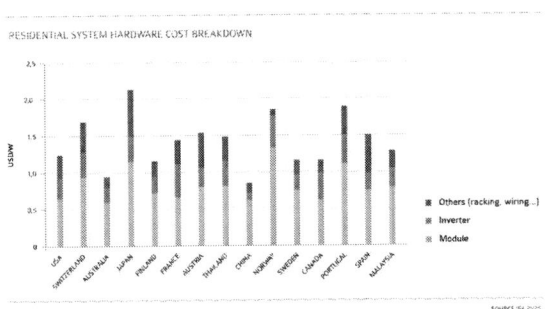

Fig. 3. Residential PV system cost by country.
Source: IEA PVPS, Trends Report 2016

IV. COST ANALYSIS OF RESIDENTIAL PV SYSTEM

Figure 4 shows the cost break down of residential PV system in the end of FY 2016 surveyed by hearings conducted by RTS Corporation. PV module cost (176 Yen/W) accounts for 58% of the total installation cost. The level of PV module

price are more than double of European level. One of the reasons of the higher priced PV modules is a supply chain for residential PV system with multiple steps. According to our survey based on hearings, PV module cost includes margins added through the supply chain and customer acquisition and other cost. Thus, as shown in Figure 5, the PV module cost for end users is significantly different from the factory gate price. It is recognized that higher cost reflects the longer supply chain between PV system suppliers (mainly Japanese PV module manufacturing companies) and end-users. It is assumed that reorganization of supply chain and optimization of PV module installation on the roofs would be the key opportunities in cost reduction in the future.

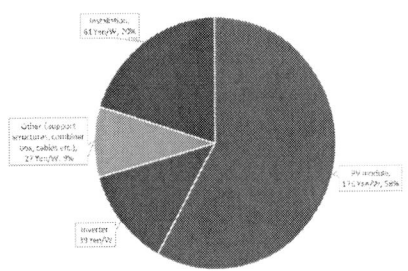

Fig. 4. **Example of cost breakdown of residential PV system.**
Source: RTS Corporation

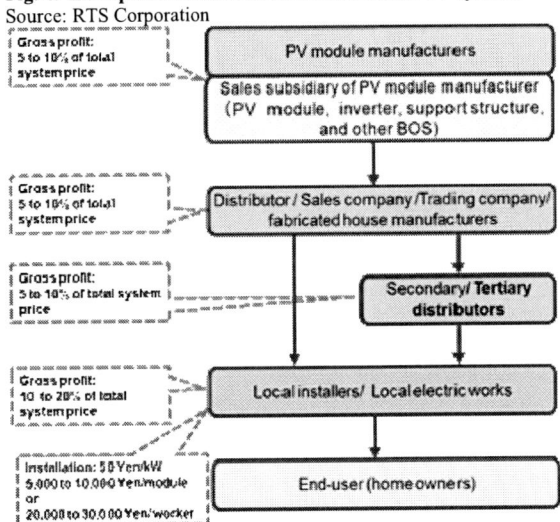

Fig. 5. Supply chain structure of Japanese residential PV system.
Source: RTS Corporation

It is also observed that sales price for end users who does not know much about market situation is set relatively higher. For example, huge price difference for the same type of the PV module by different distributors is reported as shown in Table II.

Another reason for the high price of PV module can be explained by the supplier's nationality. In residential PV market, Japanese PV module suppliers share is around 80% and the price of made-in Japan PV modules are higher than those from overseas countries such as China, South Korea and Taiwan.

TABLE II Trading price of the same PV modules by distributors

Manufacturer	Distributor	Transaction number	End user price (Yen/W)
A	A	3,706	200
	B	392	284
	C	292	288
B	A	47	186
	B	10	262
	C	85	180
	D	4	311
	E	122	196
C	A	1,019	204
	B	71	266
	C	45	321

Source: Ministry of Economy, Trade and Industry (METI), compiled by RTS Corporation

VI. COST REDUCTION OPPORTUNITIES AND OUTLOOK

While cost for residential PV system is relatively higher level in the world, it is expected that cost reduction will be advanced with several factors.

One of the factors gives significant impacts on cost reduction is the level of FIT in the future. METI revised the Renewable Energy Act in May 2016 and the revised Renewable Energy Act took effect in April 2017. With the revision, FIT level for multi-year can be set and METI announced FIT level for multiple years from FY 2017 to FY 2019 to ensure cost reduction.

For PV systems with a capacity of less than 10 kW, FITs are set at 28 Yen/kWh for FY 2017, 26 Yen/kWh for FY 2018 (target) and 24 Yen/kWh for FY 2019 (target) as shown in Table III. As for double power generation systems such as residential PV system with storage batteries, or fuel cells, FITs are set at 25 Yen/kWh for FY 2017 and FY 2018 (target) and 24 Yen/kWh for FY 2019 (target). In case, PV systems are obliged to install a device to respond to output curtailment in specific electric utility company's service areas, 2 Yen/kWh is added to above-mentioned FIT.

With scheduled reduction of FIT level towards FY 2019, it is expected that residential PV system cost reduction will be accelerated. It should be noted that FIT level of FY 2019 is below electricity retail charge and introduction of residential storage batteries is expected to increase in FY 2019.

TABLE III Trends of FIT level for residential application.

Period (Fiscal Year)		FIT: Yen/kWh	Price*: Yen/W
2016	with communication device	31	353
	W/O communication device	33	
2017	with communication device	28	336
	W/O communication device	30	
2018	with communication device	26	322
	W/O communication device	28	
2019	with communication device	24	308
	W/O communication device	26	

*: Residential system price used for basis for setting FIT level
Source: METI, compiled by RTS Corporation

The second factor assumed to have the impacts on the market growth of residential sector is growth of the market size in the future. Tighter house energy efficiency codes will be implemented in FY 2020 and national roadmap for Zero Energy House (ZEH) is supposed to be tail-wind of residential PV systems. For new houses, it is expected that residential PV system will be standard equipment for energy creation. As shown in Table IV, it is estimated that size of Japanese residential market will grow from 1.2 GW level in FY 2015 to 1.93 GW and 2.23 GW in FY 2020 in business as usual scenario and advanced scenario respectively.

Other drivers of cost reductions identified based on the analysis mentioned above are 1) rationalization of the supply chain, 2) raising the awareness of market situation for consumers, 3) Innovative channel such as websites that provides price level or compare proposals by multiple installers.

It is also expected that local government's initiatives will drive price reduction. One of the examples of such initiative is Kanagawa Solar Bank System established by Kanagawa Prefectural Government. In the initiative, consumers can contact to listed installers selected by the Prefecture based on the criteria including price level and quality of installation. Saitama Prefecture also made partnership with four residential PV system providers.

TABLE IV Residential PV system outlook in Japan (Installed capacity and number of houses)

Category		FY 2015	FY 2020 BAU* Scenario	FY 2020 Advanced Scenario
Retrofit	<10kW	454 MW (94,100)	778 MW (145,400)	884MW (16,500)
	10kW or more	245 MW (23,300)	47 MW (4,600)	51MW (5,000)
New house	<10kW	407 MW (86,100)	1,057 MW (165,300)	1,281 MW (185,200)
	10kW or more	98 MW (9,200)	48 MW (4,700)	49 MW (4,900)
Total		1,204 MW (212,700)	1,930 MW (320,000)	2,265 MW (360,000)

*: BAU represents for Business as Usual
Source: RTS Corporation, Current Status and Outlook of Residential PV Market in Japan

Based on these future trends and cost reduction opportunities, cost reduction will be advanced in the future. Figure 6 shows the future installation cost outlook of residential PV systems in Japan.

REFERENCES

[1] RTS Corporation, "PV Activities in Japan and Global PV Highlights", Vol. 23, No1, January 2017
[2] IEA PVPS, "Trends in Photovoltaic Applications 2016", October 2016
[3] Hiroshi Matsukawa, Haruki Yamaya, Takashi Ohigashi, Izumi Kaizuka and Osamu Ikki, RTS Corporation, "PV Market in Japan: Cost Trends After Implementation of The FIT Program", EUPVSEC 2014, September 2014
[4] Hiroyuki Yamada NEDO, Osamu Ikki, RTS Corporation, "PV Technology Status and Prospects", IEA PVPS Annual Report 2017, May 2017
[5] RTS Corporation, "Current Status and Outlook of Residential PV Market in Japan", October 2016

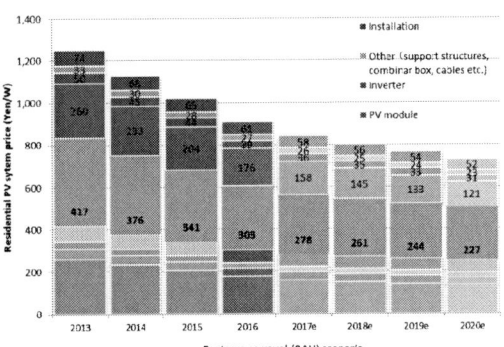

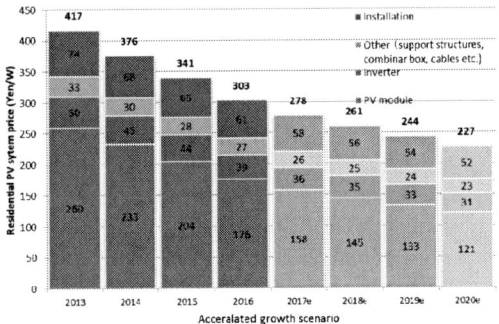

Fig. 6. Supply chain structure of Japanese residential PV system.
Source: RTS Corporation

Supply and Demand Constraints on Future PV Power in the USA

Paul A. Basore[1] and Wesley J. Cole[2]

[1]U.S. Department of Energy, Washington, DC 20585 USA
[2]National Renewable Energy Laboratory, Golden, CO 80401 USA

Abstract — **An economic analysis with a defined uncertainty range is presented that considers whether the future deployment of utility-scale PV (UPV) systems in the USA will be limited by the supply of PV modules or by market demand for these systems. An analysis of global PV module manufacturing growth and the fraction of modules produced that will be available for installation in the USA gives a plausible range of cumulative capacity for the UPV segment of 50 – 170 GW in 2030 and 60 – 500 GW in 2040. A parallel analysis of the future demand for UPV systems in the USA indicates a plausible range for cumulative capacity of 80 – 230 GW in 2030 and 150 – 530 GW in 2040. The plausible ranges for supply and demand substantially overlap in both 2030 and 2040, suggesting that neither supply nor demand is more likely to limit PV deployment in the USA. Consequently, mechanisms for enabling growth in both supply *and* demand can benefit efforts intended to increase the deployment of PV in the USA.**

Index Terms — **photovoltaics, strategy, USA, utility-scale**

I. INTRODUCTION

The adoption of photovoltaic (PV) solar power in the USA may be constrained either by the supply of appropriate products or by the market demand for those products. Supply and demand are both affected by price. Higher margins promote growth of the supply chain, while lower prices promote growth of market demand. Supply and demand are thus inextricably linked and must both be considered in strategic planning.

The analysis reported here compares the plausible range (based on our assumptions) for deployment of utility-scale photovoltaics (UPV) in the USA from a supply perspective and from a demand perspective. The PV module supply constraints are drawn from [1]. The PV system demand constraints are based on scenario analysis using the ReEDS model [2]. Further details on the models from [1] and [2] are provided in Section II and Section IV, respectively. The supply constraints arise from the high capital investment required to grow the PV

module manufacturing supply chain. The demand constraints arise from competition with other sources for electricity supply. The manufacturing supply analysis was global in scope, whereas the market demand analysis was for the contiguous USA. To compare the supply and demand of UPV using the two models, it is necessary to estimate the fraction of global module shipments that will be deployed in the USA.

Fig. 1 illustrates the logic flow used to compare the supply and demand constraints. In each block shown, there is considerable uncertainty in one or more key parameters. We attempt to bound this uncertainty by defining low and high limits of "surprise" for these key parameters. In order to capture a fairly wide range, we assign these limits to the 10th and 90th percentiles of a normal probability distribution (i.e. the edges of the 80% confidence interval). The future evolution of PV deployment predicted by this model is also probabilistic, producing a range of outcomes that can be considered "plausible," and beyond which would be "surprising."

The following sections address each of the blocks in Fig. 1, from left to right.

II. GLOBAL MODULE MANUFACTURING

The evolution of global module manufacturing capacity depends on how much capital is invested each year in supply-chain capacity and how much that new capacity costs (capex). If a given level of manufacturing capacity is to be sustained, annual investment in supply-chain capacity must be sufficient to replace existing capacity as it reaches end-of-life. Recently, the capital investment rate (CapIR) of $75M/yr for each GW/yr of production capacity ($75M/GW) is only about half of what is required to maintain manufacturing capacity, which we call the capital demand rate (CapDR, currently $150M/yr per GW/yr of production capacity) [1]. If CapIR is greater than

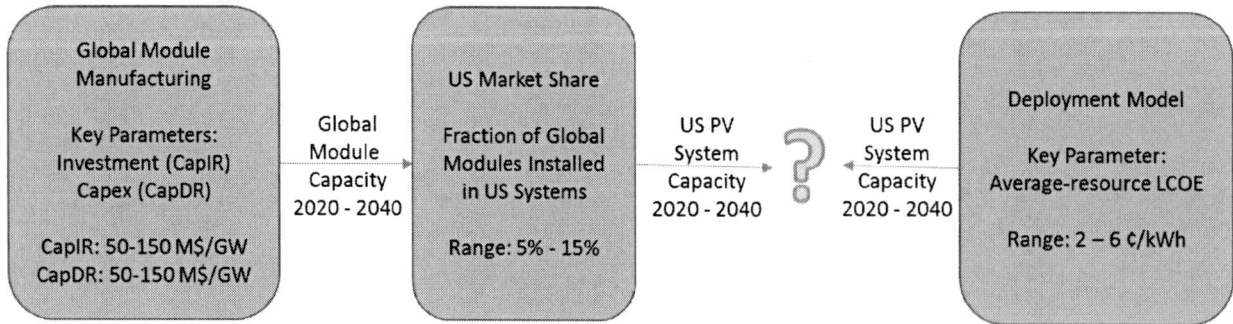

Fig. 1.   Logic flow for applying the constraints of supply (left) and demand (right). The plausible ranges shown are the limits of surprise.

CapDR (level of investment is greater than the cost of maintaining the current level of manufacturing capacity), manufacturing capacity will increase, while the opposite happens if CapIR is less than CapDR.

It would be surprising if the future investment rate dropped below $50M/GW, as that would lead to a rapid collapse of the industry. It would also be surprising if the investment rate more than doubled. It would be surprising for capital demand rate to increase, but equally surprising for it to fall below one-third of its current value. The nominal expectation is that the industry will achieve marginal sustainability, with investment rate and capital demand rate converging in the vicinity of $100M/GW.

The spreadsheet referenced in [1] uses annual PV module manufacturing capacity calculations to determine the cumulative amount of PV modules deployed in service versus time, assuming that all PV modules produced are deployed and accounting for the decommissioning of deployed PV systems at their end-of-life. Fig. 2 illustrates this calculated future global in-service PV module generation capacity (GW_{dc}) for four cases, from top to bottom:

(a) CapIR doubles ($150M/GW), CapDR drops ($50M/GW)
(b) CapIR doubles ($150M/GW), CapDR same ($150M/GW)
(c) CapIR drops ($50M/GW), CapDR drops ($50M/GW)
(d) CapIR drops ($50M/GW), CapDR same ($150M/GW)

In all cases, the investment and capital demand rates are assumed to ramp linearly to their new value over the period $2015 - 2030$ and then remain stable through 2040. The average service life of deployed modules is assumed to ramp from 25 years in 2015 to 40 years in 2030 and then remain stable through 2040. The only one of these four scenarios that yields rapid growth (top curve) combines increased investment with reduced capex. In this high-growth scenario, we constrain supply-chain capacity expansion in later years in order to maintain 80% plant utilization.

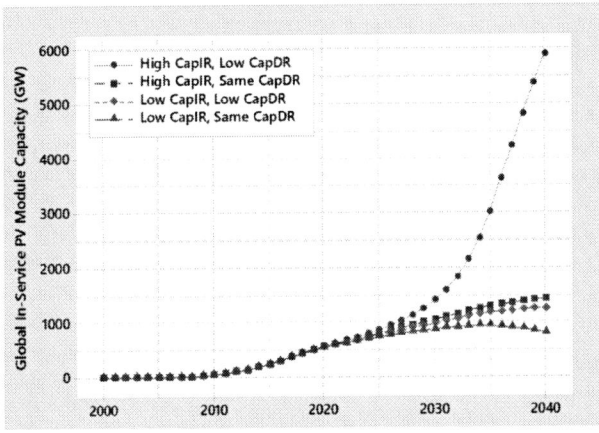

Fig. 2.   Evolution of global PV module generation capacity for four combinations of investment rate and capital demand rate.

Fig. 2 can be used to estimate the plausible limits on future global in-service PV module capacity. Assuming that CapIR and CapDR are statistically independent quantities, then the plausible range (80% confidence interval) for global in-service

capacity is as shown in Table I. The upper plausible limit is less than the top curve in Fig. 2 because that curve requires both CapIR and CapDR to achieve their limits of surprise. Such an outcome is possible, but would be "doubly surprising" and thus falls outside the surprise limit.

TABLE I
PLAUSIBLE RANGE FOR SUPPLY-LIMITED GLOBAL IN-SERVICE PV MODULE GENERATION CAPACITY FOR THE YEAR INDICATED

2030	2040
$890 - 1270$ GW_{dc}	$730 - 4400$ GW_{dc}

III. U.S. MARKET SHARE

According to the International Energy Agency [3], global cumulative PV module shipments through 2015 were 228 GW. Of this, 6.1% was installed in U.S. UPV systems. In 2015 alone, 8.2% of the global module shipment of 50.7 GW_{dc} was installed in U.S. UPV systems. In the next two decades, we specify the low and high limits of surprise for U.S. market share by assuming that the fraction of cumulative global module capacity installed in U.S. UPV systems will not drop below 5%, or exceed 15%, thus defining a plausible future range for cumulative U.S. UPV market share of $5 - 15\%$.

U.S. generation capacity is equal to global in-service module capacity times the U.S. market share. The U.S. market share is not expected to be strongly correlated to global capacity, so these two factors can be treated as statistically independent. Table II lists the plausible range (80% confidence interval) for the product of global capacity times U.S. UPV market share.

TABLE II
PLAUSIBLE RANGE FOR SUPPLY-LIMITED UPV GENERATION CAPACITY IN THE USA FOR THE YEAR INDICATED, WITHOUT CONSIDERING DEMAND CONSTRAINTS

2030	2040
$50 - 170$ GW_{dc}	$60 - 500$ GW_{dc}

IV. DEPLOYMENT MODEL

The National Renewable Energy Laboratory (NREL) has developed the Regional Energy Deployment System (ReEDS) [4] to model the influence of policies, technology, and economic developments on the evolution of the electricity generation portfolio in the contiguous USA. The model deploys new electric generating capacity, transmission capacity, and energy storage capacity as they are required to meet growth in demand, to replace power plants scheduled for retirement, and to fulfill policy requirements. The model selects new capacity from all available renewable and non-renewable options to minimize the total system cost for the nation as a whole. Regulatory constraints and incentives as they are currently enacted into law are incorporated, but no additional future constraints or incentives are assumed.

The ReEDS was run for numerous scenarios associated with what we deem to be plausible future cost reductions for PV systems. The scenarios incorporated policies that were current as of summer 2016. The output, shown in Table III, provides an indication of the market demand for UPV systems in the USA as a function of the system cost, summarized here by the levelized cost of electricity (LCOE) for a typical UPV system in a location with average U.S. solar resource (LCOE is used here as a summary metric. It not used by the model to project adoption [4].). The PV system cost was assumed to gradually decline to the value indicated in 2030 and remain at that level thereafter (in 2016 U.S. dollars). All non-solar generation technologies and energy storage technologies assumed cost trajectories from NREL's Annual Technology Baseline (ATB) [5]. The column in Table III labeled ATB is the ATB Mid-Case Scenario, which has UPV average-resource LCOE at 6¢/kWh in 2030 and 5½¢/kWh in 2040.

TABLE III
REEDS DEPLOYMENT CALCULATION FOR UPV IN THE USA FOR PLAUSIBLE VALUES OF AVERAGE-RESOURCE LCOE ACHIEVED IN 2030

Avg-Resource LCOE (/kWh)	2¢	3¢	4¢	ATB
2030 Deployment (GW_dc)	230	209	167	81
2030 U.S. Electricity Portion	10%	9%	7%	4%
2040 Deployment (GW_dc)	531	435	313	148
2040 U.S. Electricity Portion	19%	16%	12%	6%

V. COMPARING SUPPLY AND DEMAND

The plausible deployment range based on supply (Table II) and the plausible range based on demand (Table III) can be seen to substantially overlap in both 2030 and 2040. The overlap indicates deployment levels that are plausible from both the supply perspective presented here and the demand perspective presented here. The joint probability for a given level of deployment, considering both supply and demand constraints, is the product of the supply and demand probability values at that level of deployment. This is illustrated in Fig. 3 for 2030 and in Fig. 4 for 2040. For comparison, UPV capacity in the USA at mid-2017 is approximately 25 GW_dc.

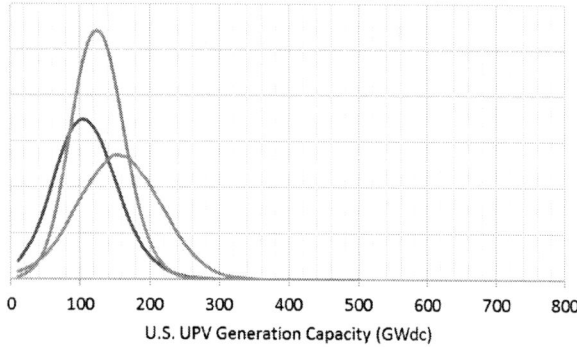

Fig. 3.   Probability distributions for supply (blue) and demand (orange) constraints on U.S. UPV deployment in 2030. The product (green) is the joint probability considering both supply and demand constraints.

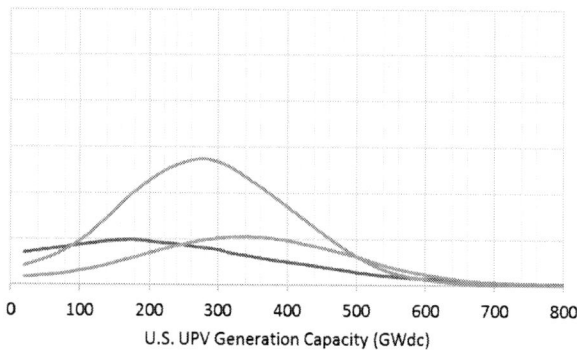

Fig. 4.   Probability distributions for supply (blue) and demand (orange) constraints on U.S. UPV deployment in 2040. The product (green) is the joint probability considering both supply and demand constraints.

The resulting plausible range for U.S. UPV deployment in 2030 and 2040, considering both supply and demand constraints, is given in Table IV. Based on the ReEDS calculation shown in Table III, this level of deployment corresponds to UPV supplying roughly 5% of U.S. electricity in 2030 and 10% in 2040.

TABLE IV
PLAUSIBLE RANGE OF FUTURE U.S. UPV GENERATION CAPACITY CONSIDERING BOTH SUPPLY AND DEMAND CONSTRAINTS

	2030	2040
Deployment (GW_dc)	70 – 170	120 – 440
U.S. Electricity Portion	3% – 7%	5% – 16%

The figures in Table IV do not include electricity generated by distributed photovoltaic systems (residential or commercial). The deployment of distributed PV (DPV) in the USA is currently similar to UPV. If that ratio is maintained in the future, then UPV and DPV together could supply roughly 10% of U.S. electricity in 2030 and 20% in 2040.

If greater PV deployment is desired, but is limited by supply of modules, a strategy for increasing PV deployment should focus on helping the module supply chain become more attractive for investment, for example by reducing capex and other manufacturing costs. On the other hand, if PV installations become limited by demand, a strategy for increasing PV deployment should focus on helping PV project developers deploy more systems, for example by reducing project costs and other barriers to deployment. Our results indicate that UPV generation capacity in the USA could plausibly be constrained by either supply or demand. Consequently, mechanisms for enabling growth in both supply *and* demand can benefit efforts intended to increase the deployment of PV in the USA.

ACKNOWLEDGEMENT

This work was supported by the U.S. Department of Energy, Solar Energy Technologies Office, a portion of which was performed under Contract No. DE-AC36-08GO28308 with the National Renewable Energy Laboratory.

REFERENCES

[1] P. A. Basore, "Paths to future growth in photovoltaics manufacturing," *Prog. Photovolt: Res. Appl.* (2016), DOI: 10.1002/pip.2761.

[2] W. Cole, T. Mai, J. Logan, D. Steinberg, J. McCall, J. Richards, B. Sigrin, and G. Porro, "2016 Standard Scenarios Report: A U.S. Electricity Sector Outlook," NREL/TP-6A20-66939, 2016.

[3] International Energy Agency, "Trends 2016 in Photovoltaic Applications," IEA PVPS T1-30:2016.

[4] K. Eurek, W. Cole, D. Bielen, N. Blair, S. Cohen, B. Frew, J. Ho, V. Krishnan, T. Mai, and D. Steinberg, "Regional Energy Deployment System (ReEDS) Model Documentation: Version 2016," NREL/TP-6A20-67067.

[5] "2016 Annual Technology Baseline," National Renewable Energy Laboratory, 2016.

Residential Photovoltaic Electricity Generation in the European Union 2017

-

Opportunities and Challenges

Arnulf Jäger-Waldau, Thomas Huld, Sandor Szabo

European Commission, Joint Research Centre, Energy Efficiency and Renewables Unit, 21027 Ispra, Italy

Abstract — Since 2000 grid-connected solar photovoltaic systems have increased their world-wide cumulative capacity about 250 times to exceed 300 GW at the end of 2016. More than another doubling is forecasted until 2020. Despite this positive development the photovoltaic market in the European Union has dramatically declined since 2012 and a major change is not in sight. The further development of the photovoltaic market in the EU is not only coupled to the question at what prices solar photovoltaic electricity can be provided and delivered to the customers, but even more to the regulations in place, which govern the installation of residential systems and the self consumption of PV electricity.

Index Terms —

I. INTRODUCTION

In 2017, the costs of direct current (DC) electricity in central Europe at the PV module level have dropped to less than 0.02 EUR/kWh and shows that it is the technology with the lowest cost for electricity generation. This cost reduction is in accordance with the drastic price reduction of international power purchase agreement (PPA) deals, which could be observed during 2016. However, the low DC generation cost is only a part of the total, as there is an additional cost component to provide the electricity to the customer where and when it is needed.

In the case of residential photovoltaic electricity generation systems, self-consumption of PV generated electricity is a possible driver for the further growth of photovoltaics in Europe after feed-in schemes already ended or are set to be phased out soon.

The roadmap of the EC initiative "New Deal for Energy Consumers: Empowering Consumers, Developing Demand Side Response; using smart technology; linking Wholesale and Retail Markets; Flanking Measures to Protect Vulnerable Customers" list as one of the options the right to self-generate and self-consume electricity.

On 30 November 2016, the European Commission presented a package of measures "Clean Energy for all Europeans" (Winter package) with the aim to keep the European Union competitive when clean energy transition is changing the global energy markets [1]. The package calls for the implementation of measures to reduce the existing barriers for self-consumption, storage and selling of electricity produced by consumers. To do so the whole electricity system should become more flexible and responsive to the way consumers produce and consume electricity.

If properly implemented, these measures could lead to a revitalization of the currently weak European PV market.

II. CURRENT SITUATION

In the EU total installed PV power was about 105 GW and about 65% of it was on rooftops – residential and commercial – at the end of 2016 [2, 3]. With current policies between this may rise to between 100 and 150 GW of Photovoltaic power installed on rooftops in the European Union by 2030. It is estimated that about 40% would be residential and 60% commercial installations.

However, the current scenarios do not take into account the rapid development of the decentralized storage markets as well as the push towards common owned and used electricity generation systems in condominium houses. These trends, together with the increasing push to smart energy home solutions including the trend to utilize an electric vehicle at least as a second car for local transport, can increase the demand for residential PV systems considerably over the next decade.

In May 2017, the weighted average price of an installed residential PV system without tax in Europe was 1.16 EUR/Wp [4]. Adding a surcharge of 0.14 EUR/Wp for fees, permitting, insurance, etc., an installed PV system costs 1 300 EUR/kWp without financing and VAT.

However, there is a wide range or residential PV prices in the European Union ranging from about EUR 1 000 per kWp to as much as EUR 2 500 per kWp. This can be explained only in parts by the degree of local competition between dealers and installers as well as differences in labour costs. Lack of competition across borders regarding installation services due to national differences in installer certification and local administrative and permitting requirements are more important.

Across Member States tax, levy and subsidy regimes for PV energy equipment purchases and electricity sales differ substantially. The same is true for permit fees or connection charges as well as other administrative costs. They are applied on a per-installation basis (and can be as high as EUR 1000), or on an installed power basis, or a combination of both.

All these different regulations result in quite heterogeneous market conditions, which finally influence the cost and economic attractiveness of PV systems.

Beside the regulation and tax differences the relative weight of the network fees in the overall electricity retail prices for the residential and commercial (industrial) sectors have a huge impact on the level and attractiveness of self-consumption. This part of the retail price affects the network operators' income by self-consumption, but on the other hand could be used for network integration, storage and self-consumption support purposes.

For many consumers the purchase of a PV system is now a 'credit card purchase', with no need for financing. However, the large differences of financing costs in the different Member States, which are based on the long term interest rates from the European Central Bank (ECB) and range between 3-11%, can still play a role. The influence of financing was already presented earlier [5].

III. RESULTS

To show the competitiveness of PV generated electricity we calculated the LCOE from PV systems using the May 2017 averaged price and compared it with the variable part of the electricity prices customers are paying in the European Union Member States (Fig. 1).

For this benchmark calculation we used operation, maintenance and repair (O&M) costs of 2%, which are higher than in other analyses. This reflects the fact that labour costs related to O&M activities have not decreased like the hardware components. Depending on the actual radiation level, the 2% O&M costs are a main cost factor besides financing costs. Adding a conservative safety margin of 0.8 EUR cent/kWh on top of the 2.0 to 2.6 EUR cent /kWh results in an electricity price of 2.8 to 3.4 EUR cent/kWh after the 20-year financial payback period, depending on the actual solar radiation.

Whether or not owners of residential photovoltaic systems still need additional financing for their system or are able to consider it as a one off 'credit card purchase' depends on their economic situation and the actual size and price of the system. Figure 1 shows the difference between a LCOE calculation without and with 3% interest rate.

As shown earlier there is a growing number of countries where electricity production from residential PV solar systems can be cheaper than the variable part of residential electricity prices, depending on the actual electricity price and the local solar radiation level. Therefore, using self-generated electricity provides a means to lower the electricity bill on one hand, and to avoid excessive penetration of PV generated electricity in

the grid network. In the case of a PV system size that generates as much electricity as the customer uses over a year, the actual consumption during the time of generation is in general about 25%-30% on residences, in commercial buildings it can be more [6,7].

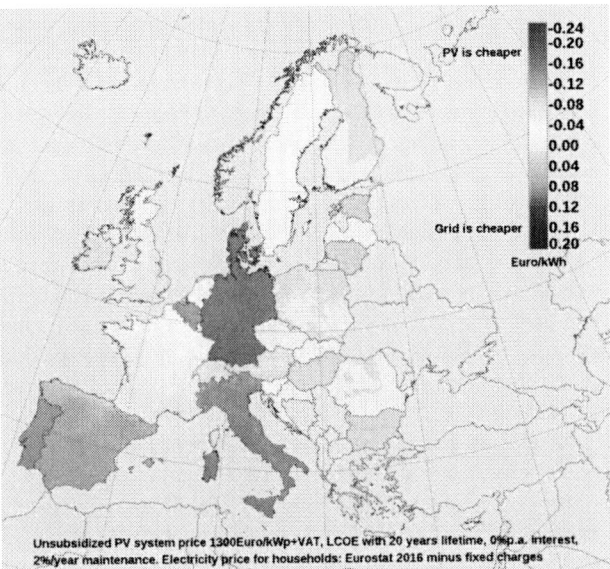

(a)

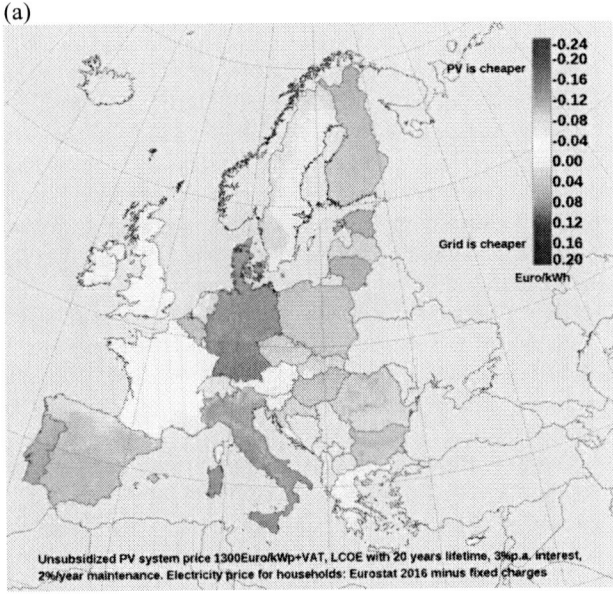

(b)

Fig. 1. Comparison of European residential electricity prices (variable part) with electricity generated by a PV solar system Without (a) and with 3% interest for the system purchase (b).

There are in principle two methods, to increase the direct consumption ("Self-Consumption") of solar electricity. One is to use intelligent control systems, which switch major loads (washing/dryer machines, heat pumps, refrigerators, air-conditioners) on when the sun is shining. The second one requires a means to store the energy, either as electricity which requires accumulators, or as "product", (heat-storage, cold-storage or pumped water), for use at night or rainy days. Storing electricity has the additional advantage of making energy offers to the network operator at times the operators chooses as being profitable.

In general, the share of self-consumed electricity from a residential or commercial system without local storage is limited and a part of the electricity has to be sold via the grid. The question is: what kind of pricing to apply for this part – contract, wholesale or day-ahead prices or what are the acceptable costs for storage?

There is a wide range of prices for PV systems as well as electricity for customers in the different Member States, which defines the attractiveness of self-consumption. The level of overall PV production affects the level of self-consumption and hence the cost of the self-consumed electricity [6]. Other parameters like the base to peak load ratio, or the composition of the electricity portfolio have a big impact on the RES levels and network costs.

As already mentioned earlier the current policies in place could result in about 100 to 150 GW of rooftop capacity in the European Union by 2030. Compared with the already installed capacity of about 65 GW this would be an increase of only 35 to 85 GW over 14 years, or between 2.5 and 6.0 GW of new installations annually. On the other side, even the conservative IEA predictions in the Energy technology Perspectives 2016 show a technical potential of 470 GW of PV system capacity on rooftops in cities of the European Union by 2030 [8].

The European Council meeting on 23-24 October 2014 has set a legally binding target for the European Union to reach 27% of its energy consumption from renewable energy sources by 2030. Already to realise this target electricity generation from Renewable Energy Resources has to reach 1,600 – 1,700 TWh in 2030, about twice the value in 2014. The largest share of this renewable energy, about 1,000 TWh will have to be supplied by solar and wind power [9]. Compared to the electricity production of about 410 TWh (approx.. 300 TWh wind and 110TWh PV) in 2016 an in increase by about 2.5 times is needed.

The 27% RES target of the European Union was decided before the Paris Agreement (COP21) was reached in December 2015. In view of the fact that the Paris Agreement went into force on 4 November 2016, a revision of the original Nationally Determined Contributions (NDC's) proposed by each country is necessary as there is a general consensus that those pledges are not sufficient to reach the 1.5 °C target by 2050.

At the moment the RES directive recast is in the parliamentary process in the European Parliament and the ITRE committee (Industry, Research and Energy) has already voiced it's opinion that at least 35% RES by 2030 are needed to stay on a feasible trajectory for the GHG reductions necessary by 20250 and to achieve the goals of the Paris

agreement. The consequences for renewable electricity implies that an additional 200 to 250 TWh, mainly from solar and wind power are needed to reach this target. Taking these developments into consideration, it is clear that Europe has a need for additional policies (higher national targets in the RE auction design, revised finance models) to increase solar photovoltaic power and especially rooftop installations to at least 350 to 400 GW by 2030.

IV. CONCLUSIONS

Already now PV generation costs can be lower than residential electricity prices for almost 80% of the European Union's population, depending on the actual electricity price and the local solar radiation level.

Increased self-consumption of PV electricity can lead to a revitalisation of the European PV market. To realise such a scenario, a single market for PV in the European Union has to be realised as well as the appropriate regulatory and legal conditions, which will enable a secure and fair financing of the necessary grid infrastructure and providing economic benefits for residential PV system operators and the society.

REFERENCES

[1] Communication from the Commission to the European Parliament, the Council, the European Economic and Social Committee, the Committee of regions and the European Investment Bank, Clean Energy For All Europeans, Brussels, 30.11.2016 COM(2016) 860 final

[2] A. Jäger-Waldau, PV Status Report 2016, October 2016, European Commission, ISBN 978-92-79-63055-2

[3] A. Jäger-Waldau, Snapshot on Photovoltaics – March 2017, Sustainability 9(5) May 2017

[4] A PVinsights, May 2017, http://pvinsights.com

[5] A. Jäger-Waldau, S. Szabo, T. Huld and H. Ossenbrink Economics of Residential PV Systems in Europe, Technical Digest of WCPEC-6, Kyoto, 2014

[6] Self-Consumption of Electricity by Households, Effects of PV System Size and Battery Storage, T. Huld, H. Ruf, G. Heilscher, Proc. 29th EUPVSC, 2014

[7] J. Moshövel, K.-P. Kairies D. Magnor, M. Leuthold ,M. Bost, S. Gährs, E. Szczechowicz, M. Cramer, D. U. Sauer, Analysis of the maximal possible grid relief from PV-peak-power impacts by using storage systems for increased self-consumption, Applied Energy, Volume 137, 1 January 2015, Pages 567–575

[8] International Energy Agency, Energy Technology Perspectives 2016, ISBN PRINT 978-92-64-25234-9 / PDF 978-92-64-25233-2

[9] R. Lacal Arantegui and A. Jäger-Waldau, Photovoltaics and Wind Status in the European Union after the Paris Agreement, Renew Sustain Energy Rev, submitted

Investigating Nanoscale Determinants of Charge Collection in Quasi-2D Perovskite Solar Cells

Yanqi Luo,[1] Xueying Li,[1] Bat-El Cohen,[2] Barry Lai,[3] Lioz Etgar,[2] and David P Fenning[*,1]

[*,1] Department of Nanoengineering, University of California San Diego, La Jolla, CA 92093
[2] The Institute of Chemistry, The Hebrew University of Jerusalem, Jerusalem, Israel, 91940
[3] Advanced Photon Source, Argonne National Laboratory, Argonne, IL 60439

Abstract — **Quasi-2D perovskite photovoltaics have been shown to have better stability than their 3D counterparts but the effects of the A-site cation on charge collection have not been fully clarified in these systems. In this study, the nanoscale carrier collection properties within quasi-2D bromide based lead perovskites (*n*=3 and 40) are found to be uncorrelated with local variations in halide stoichiometry, in contrast to the 3D perovskite. The local elemental distribution is quantified by means of synchrotron-based nanoprobe X-ray fluorescence (Nano-XRF), while the local carrier collection is simultaneous collected *via* X-ray beam induced current (XBIC).**

Index Terms — **Quasi-2D perovskites, synchrotron based characterization, nano-probe X-ray fluorescence, and X-ray beam induced current.**

I. INTRODUCTION

Organic-inorganic hybrid perovskites hold promise for use in the next-generation solar cells due to their large absorption coefficient[1] and long carrier diffusion length.[2] Limited device stability and operational lifetime remain as significant obstacles to perovskite commercialization. The most intensively studied hybrid perovskite, methylammonium lead iodide (MAPbI$_3$), shows poor stability under humid and ambient conditions.[3] Even though post-treatment with a hydrophobic surface (Teflon) passivation enhances stability in the presence of humidity,[4][4] a secondary phase of PbI$_2$ grows as the MAPbI$_3$ gradually degrades after 30 days. The electronic performance suffers as the MAPbI$_3$ absorber decays.

The significant research effort to improve perovskite stability has sparked interest in the quasi-2D perovskites as a potential path to use the dimensionality of the absorber to manipulate the chemical stability.[5] A general approach to synthesize a quasi-2D perovskite is to incorporate a larger A-site cation, such as phenylethylammonium (PEA$^+$), together with methylammonium (MA$^+$). The 3D perovskite structure is not compatible with large ionic radius of PEA because of the tolerance factor requirement; therefore, the mixture system crystallizes with PEA halide layers that break the symmetry of the 3D perovskite and forms low-dimensionality layered perovskites.[6] The molar ratio between PEA and MA is used to denote the number of expected perovskite layers, *n*, in between the spacer cation layers. The larger spacer cations can contain a long, hydrophobic chain toward improving stability in the presence of humidity.

The quasi-2D perovskites have illustrated better stability comparing to the 3D ones. Unfortunately, the optoelectronic performance of 2D perovskite is lower than their 3D counterparts and believed to be hindered by poor carrier transport in PEA regions.[7] Therefore, understanding the dimensionality effects on microscopic charge carrier collection is crucial in these layered systems.

In this study, we use synchrotron-based nanoprobe X-ray fluorescence (Nano-XRF) microscopy to identify the relationship between local nanoscale stoichiometry within the quasi-2D perovskites and charge collection. We find that the X-ray beam induced current (XBIC) maps are uncorrelated with the variations seen in local halide distribution in the quasi-2D perovskites. In contrast, the charge collection in the 3D perovskite appears to be governed by the local stoichiometry. Future work is needed to clarify the root cause determinants of charge collection variation in 2D perovskite to raise their performance toward 3D counterparts.

II. EXPERIMENTAL AND RESULTS

A series of quansi-2D bromide perovskite solar cells with different number of perovskite layers between the spacer cation layers were fabricated following a previous published method.[8] Operating solar cells were studied by synchrotron-based nanoprobe X-ray florescence at beamline 2-ID-D of the Advanced Photon Source to evaluate the relationship between nanoscale perovskite stoichiometry and the X-ray beam induced current in the absorber materials.

The elemental XRF maps reveal that the bromide distribution varies at different length scales depending on *n*. As the dimensionality increases, the perovskite film's morphology appears to change from a network-like coverage with pinholes, to what looks like a domain-separated film, and finally to randomly-distributed.

The observed variations in spatially-resolved XBIC maps are found to be independent from the local halide distribution in the quasi-2D perovskite. On the other hand, 3D bulk perovskite film suggests a relatively strong correlation with bromide distribution.

III. CONCLUSION AND OUTLOOK

The local carrier collection in quasi-2D bromide perovskites, as determined by X-ray beam induced current measurements, does not seem to be directly correlated with the halide distribution within the film. Future work is needed to investigate the dominant factor in determining the observed nanoscale variations in carrier collection to improve the efficiency of quasi-2D perovskites toward their 3D counterparts while maintaining enhanced stability.

ACKNOWLEDGEMENT

YL acknowledges the support of the UC Carbon Neutrality initiative, and DPF start up funds from the UC San Diego. LE and BEC acknowledge the Israel ministry of science and space for the financial support. This research used resources of the Advanced Photon Source, a U.S. Department of Energy (DOE) Office of Science User Facility operated for the DOE Office of Science by Argonne National Laboratory under Contract No. DE-AC02-06CH11357.

REFERENCES

[1] S. D. Stranks, G. E. Eperon, G. Grancini, C. Menelaou, M. J. P. Alcocer, T. Leijtens, L. M. Herz, A. Petrozza, H. J. Snaith, *Sci.* **2013**, *342*, 341.

[2] G. Xing, N. Mathews, S. S. Lim, Y. M. Lam, S. Mhaisalkar, T. C. Sum, *Sci.* **2013**, *342*, 344.

[3] G. E. Eperon, S. N. Habisreutinger, T. Leijtens, B. J. Bruijnaers, J. J. Van Franeker, D. W. Dequilettes, S. Pathak, R. J. Sutton, G. Grancini, D. S. Ginger, R. A. J. Janssen, A. Petrozza, H. J. Snaith, *ACS Nano* **2015**, *9*, 9380.

[4] I. Hwang, I. Jeong, J. Lee, M. J. Ko, K. Yong, *ACS Appl. Mater. Interfaces* **2015**, *7*, 17330.

[5] A. H. Slavney, R. W. Smaha, I. C. Smith, A. Jaffe, D. Umeyama, H. I. Karunadasa, *Inorg. Chem.* **2017**, *56*, 46.

[6] G. Kieslich, S. Sun, T. Cheetham, *Chem. Sci.* **2015**, *6*, 3430.

[7] D. H. Cao, C. C. Stoumpos, O. K. Farha, J. T. Hupp, M. G. Kanatzidis, *J. Am. Chem. Soc.* **2015**, *137*, 7843.

[8] B.-E. Cohen, M. Wierzbowska, L. Etgar, *Adv. Funct. Mater.* **2017**, *27*, 1604733.

Recent Developments of Solar Photovoltaic Systems in India

Saravanan Vasudevan and Arumugam Murugesan

Arunai Engineering College, Tiruvannamalai, Tamilnadu, India – 606 603

vsaranaec@yahoo.co.in, drmarumugam@yahoo.com

Abstract — **India is marching towards to achieve solar power generation target of 100 GW by the year 2022. Recent progress and developments in solar power sector of India are presented. This paper supports the prevailing progressive path for solar power development and recommends future course of actions for India.**

Index Terms — **India , initiatives, photovoltaics, solar parks.**

I. INTRODUCTION

Government of India (GoI) have recently announced a massive renewable power generation target of 1,75,000 MW by 2022, which have solar power share of 1,00,000 MW. Breakup of year wise solar power target is listed in Table I. As of 31 March 2017, 12288.83 MW of grid interactive solar photovoltaic (SPV) power was generated and 462.54 MW_{EQ} off grid SPV system was put in service [1]. However the country's estimated solar power potential is about 7, 48, 000 MW.

Table I Target (Year wise) to achieve solar power generation

Year	Rooftop (in MW)	Ground Mounted (in MW)	Total (in MW)
2015-16	200	1,800	2,000
2016-17	4,800	7,200	12,000
2017-18	5,000	10,000	15,000
2018-19	6,000	10,000	16,000
2019-20	7,000	10,000	17,000
2020-21	8,000	9,500	17,500
2021-22	9,000	8,500	17,500
Total	40,000	57,000	90,434

International Technology roadmap for photovoltaic (ITRPV) report [2] suggests that global PV capacity to grow from 303 GWp in 2016 to 1700 GWp in 2030. International Renewable Energy Agency (IRENA) predicts solar PV deployment in India will be 2, 09,000 MW by 2030. India's green bond market is expanding quickly, with more than USD 1.1 billion issued during 2015 and a total of USD 800 million in the first eight months of 2016.[3] Global investment in renewables rose from USD 50 billion in 2004 to more than USD 286 billion in 2015, especially the solar sector records USD 161 billion in 2015. In India, renewable energy's investment is increased from USD 2.7 billion (year 2004) to USD 9.7 billion (year 2016) [4] Utility-scale PV power from plants commissioned in the year 2016 typically costs between USD 0.06-0.10 per kilowatt-hour (kWh) in Europe, China, India, South Africa and United States. India stands 5th position in the world in terms of annual investments for solar photovoltaic and renewable power generation [5].

Green Climate Fund in line with India's Intended Nationally Determined Contributions (INDCs) under the UN Framework Convention on Climate Change (UNFCCC) supplements domestic resources allocation for accelerating development and deployment of renewable energy in the country. India had made substantial progress in establishing policies to promote investment in renewable energy through National Clean Energy Fund especially for solar photovoltaic systems. Creation of International Solar Alliance with various countries and organizing many investor meets/summits augments the work done in the process.[6]This paper presents the recent growth and developments of solar photovoltaic systems in India made through various studies by international agencies and initiatives by international organizations, government commitments, public-private-partnerships and investments.

II. INDIA'S ENERGY SCENARIO

As of 31 March 2017, India had 326848.53 MW of utility based installed electricity generating capacity, mostly from thermal based systems (218329.88 MW) and renewable power generation accounts for 57260.23 MW [7]. Indian Brand Equity Foundation [8] states that electricity production in India stood at 1, 107.8 billion units (BU) in the financial year (FY) 16, having a growth of around 5.64 % over the previous fiscal year. Over FY10–16, electricity production expanded at a compound annual growth rate of 6.21 %. Ministry of Power, Government of India has set a target of 1,229.4 BU of electricity to be generated in the financial year 2017-18.The annual growth rate in renewable energy generation in India has been estimated to be 27 per cent and 18 per cent for conventional energy. Around 293 global and domestic companies have committed to generate 266 GW of solar, wind, mini-hydel and biomass-based power in India over the next 5–10 years with an investment of about USD 310–350 billion.

NITI Aayog, India initiated India Energy Portal and suggested the need for global information systems based mapping of renewable energy in consultation with NREL, USA. It strongly recommends for solar power growth through action measures beyond 13th five year plan of India. [9]. As reported in the User Guide for Renewable Energy Sectors, India Energy Security Scenarios, 2047, solar PV capacity addition is expected to achieve any of the levels, where level 1corresponds to least effort scenario with a capacity addition of 37 GW and level 4 corresponds to heroic effort scenario to add 479 GW.

More impetus is shown for the growth of photovoltaic systems in India through policies such as generation based incentives, public private partnership, liberalized foreign direct investment policy, low interest funds provided from National Clean Energy Fund (NCEF) to Indian Renewable Energy Development Agency Ltd (IREDA) for on-lending to viable renewable energy projects and tax benefits. Three key regulatory instruments played a major role in the solar photovoltaic promotion in India are preferential Feed in tariffs (FiT), Renewable Purchase Obligation (RPO) and Renewable Energy Certificates (REC) over the years. The growth of solar power generation is found to be more promising as shown in Fig. 1.

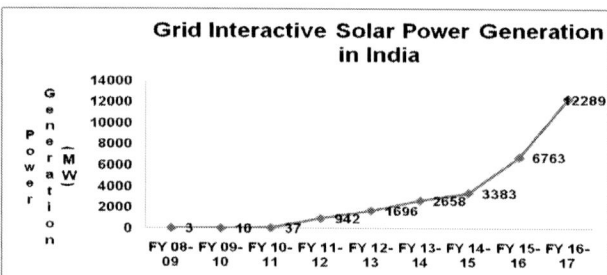

Fig. 1   Growth of grid interactive solar power generation in India

The bench mark capital cost for solar PV projects for FY 2016 – 2017 shall be INR 530.02 lakhs/MW, with the assumptions like module prices to be USD 0.48/W, PCU cost to be INR 35 lakhs/MW. The detailed breakup of utility based PV systems is given in Table II.

Table. II Bench mark cost for solar PV projects (1 MW) for FY 2016–17

Particulars	Capital Cost (INR lakhs/MW)	% of Total cost
PV modules	328.39	61.96 %
Land cost	25	4.7 %
Civil and general works	35	6.6 %
Mounting structures	35	6.6%
Power Conditioning Unit (PCU)	35	6.6%
Evacuation cost up to interconnection (Cables and Transformers)	44	8.3%
Preliminary and Pre –Operative expenses including IDC and contingency	27.63	5.21%
Total Capital Cost	530.02	100 %

Following are the terms and conditions for tariff determination obtained from Central Electricity Regulatory Commission [10], for renewable energy sources, Regulations 2017 as shown in Table. III. Under this, the solar PV plant should have capacity utilization factor (CUF) of 19%, auxiliary consumption factor shall be 0.25 % of gross generation and useful life of the system to be 25 years with total power generation of about 41.6 million units.

Table. III CERC tariff parameters for solar PV (1 MW) for FY 2016–17

CERC Tariff parameters	Values
Tariff period	25 years
Capital cost (INR lakhs)	530.02
Debt/Equity ratio	70:30
Total Debt Amount(INR lakhs)	371.02
Debt Cost (12 years of repayment period including moratorium)	12.76%
Total Equity Amount(INR lakhs)	159.01
Return on Equity(Weighted Average)	22.40%
Discount rate	10.70%
Income Tax	33.99%
Depreciation values	5.28%
O & M cost (INR lakhs)	7.00
Levelized total cost of generation, (INR/kWh)	5.68
Accelerated Depreciation (INR/kWh)	0.59
Net levelized cost of generation, (INR/kWh)	5.09

Ministry of New and Renewable Energy (MNRE), Government of India (GoI) has revised the benchmark cost for off – grid and decentralized solar PV applications programme and grid connected rooftop and small solar power plants programme for the year 2017 – 2018 and are given in Table IV and Table VI respectively

Table IV Benchmark cost for Off grid and decentralized Solar PV applications programme for the year 2017 - 2018

Category	Benchmark cost (Rs/Wp)
>1kWp to 10 kWp (with battery bank @ 7.2 VAh/Wp)	135
>1kWp to 10 kWp (with battery bank @ 3.6 VAh/Wp)	108
>1kWp to 10 kWp (with battery bank @ 1.2 VAh/Wp)	90
>10 kWp to 100 kWp (with battery bank @ 7.2 VAh/Wp)	120
>10 kWp to 100 kWp (with battery bank @ 3.6 VAh/Wp)	96
>10 kWp to 100 kWp (with battery bank @ 1.2 VAh/Wp)	80

Table V Benchmark cost for grid connected roof top and small solar power plants programme for the year 2017 – 2018

Category	Benchmark cost (Rs/Wp)
Upto 10 kWp	70
>10 – 100 kWp	65
>100 – 500 kWp	60

III. INITIATIVES AND ANNOUNCEMENTS MADE FOR SOLAR POWER DEVELOPMENT IN INDIA

MNRE has planned to set up 50 solar parks each with a capacity of 500 MW and above by 2019 - 2020, with enhancement of capacity from 20,000 MW to 40,000 MW for "Development of Solar Parks and Ultra Mega Solar Power Projects" with an estimated central financial assistance of Rs. 8100 crores under National Solar Mission. Solar Energy Corporation of India Limited, nodal agency of MNRE is committed to develop and monitor solar projects in various parts of India. [11]

Despite challenges and barriers to the adoption of solar photovoltaic systems, huge employment opportunities nearly 1, 03,000 are estimated in India (IRENA, Renewable energy & Jobs, Annual review 2016). Many recent initiatives supported by various stake holders such as international organizations, governments and public/private sectors are listed to boost solar photovoltaic power generation in India [8, 12] :

A. Indo – US Initiatives: A Joint Indo-US PACE Setter Fund has been established, with a contribution of USD 7.9 million to enhance clean energy cooperation. US Federal Agencies have committed a total of USD 4 billion for projects and equipment sourcing for the growing renewable energy sector in India.

B. GoI Initiatives: a) Ministry of Power has recommended to Goods and Service Tax council that renewable energy should be given either zero rate tax status or deemed export status to minimize further increase in power tariff in India. b) GoI has recently announced the formation of International Solar Alliance of 121 tropical countries to develop and promote solar energy. Recently, Memorandum of Understanding is signed with other countries like France, United Kingdom and Portugal to strengthen renewable energy cooperation. c)Union Cabinet of India approved 15,000 MW of grid-connected solar power projects of National Thermal Power Corp Ltd. d) Development of grid connected solar PV power plants on canal banks (50 MW) and canal tops (50 MW) by providing capital subsidies in the various states of India. e) MNRE has signed an agreement with Germany-based KfW Development Bank to fund the USD 44.7 million (Rs 300 crores) floating solar project in Maharashtra and Kerala to generate over 310 GW of green energy. f) Central Ministries have pledged to produce 5000 MW roof top solar power. Their individual break ups are: Ministry of Railways (500 MW), Department of Atomic Energy (500 MW), Department of Food and Public distribution (355.23 MW), Department of Defence Production (232.25 MW), Ministry of Steel (230MW). g) Ministry of Shipping plans to install 160.64 MW of solar and wind based power systems at all the major ports across the country by 2017 for giving a fillip to government's Green Port Initiative. h) Creation of Indo – German Solar Energy Partnership: KfW to provide concessional loans of EUR 1 billion for 5 years focusing on solar rooftops, GIZ to provide technical assistance, policy advisory services, pilot project promotion and capacity building & training, PTB to provide skill development and curriculum.

C. Financial sector Initiatives: a) World Bank Group has committed to provide USD 1 billion for India's solar energy projects and plans to work with multilateral development banks and financial institutions to develop financing instruments to support solar energy development. b) International Finance Corporation, investment arm of World Bank, plans to invest USD 125.3 million in Hero Future Energies Limited, to fund the construction of solar and wind power plants. c) Reserve Bank of India has notified to include renewable energy under priority sector lending. So, banks can provide loans up to a limit of USD 2.36 million to borrowers for renewable energy projects. d) State Bank of India has signed an agreement with World Bank for USD 626.3 million credit facility, aimed at financing grid connected rooftop solar photovoltaic projects in India.

D. Announcements and Services: a) Adani group has announced new investments exceeding INR 23, 000 crores by 2021 in solar and wind development at 8th Vibrant Gujarat Global Summit, 2017. b) Abraaj has committed to invest nearly 500 MW of utility scale solar projects with Aditya Birla Group in December 2015. c) Tamilnadu Energy Development Agency, a Government of Tamilnadu enterprise established for promoting renewable energy, invites the national /international companies for the establishment of 500 MW solar parks in Tamilnadu. [13]. d) CLP India has acquired a 49 per cent stake in SE Solar, Suzlon Group for building a 100 MW solar energy plant at Veltoor in Telangana, with USD 11.02 million (Rs 73.5 crores). e) Andhra Pradesh Government plans to establish an 'Energy University', focusing on research orientation and development of energy efficiency, energy conservation, and renewable sources. f) MNRE has developed a list of web/mobile services for the promotion of photovoltaic systems in India: i) SPIN–to enroll all empanelled agencies (private/government sector) under grid connected rooftop and small solar power plants programme. [14] ii) Solar Guidelines – for encouraging rapid development of solar power sector by facilitating information dissemination about latest solar energy projects, policy frame works. iii) Web portal of Green Energy Corridor for monitoring. g) ARUN (Atal Rooftop solar User Navigator) - mobile app for promoting and installation of solar rooftop power system in the home/consumer premises [15].

IV. RECOMMENDATIONS SUGGESTED FOR FURTHER GROWTH OF SOLAR POWER IN INDIA

The suggestions recommended for solar photovoltaic project developments in India are: i)strengthening of policy commitments, ii) incorporating latest technology insights and planning techniques, iii) integrated framework for standards and quality assurance/infrastructure, iv) solar resource planning, v) National solar energy financing mechanisms, private/public funding schemes and low cost financing, vi) Attraction of foreign direct investments, vii) no collalateral security for development of solar energy parks, viii) single window clearance for solar projects, ix) must run status for PV systems, x) support to PV manufacturing, supply chain and cross cutting power electronic technologies, xi) proactive planning through National Smart Grid Mission, xii) quicker implementation of advanced metering system, xiii) upgradation of green energy transmission corridor and grid operating protocols, xiv) availability of solar radiation data through GIS based stations, xv) organizing more consumer awareness campaigns and demonstration activities, xvi) creation of new training institutes and build institutional, technical and human capacity, xvii) support to R&D programmes with industry association and xviii) integration of solar technology with energy storage mechanisms.

V. CONCLUSIONS

India has made substantial progress in establishing policies and actions to promote investments in renewable energy especially solar photovoltaic systems. It also set targets and timeframes to achieve clean power generation through various mechanisms and collaborations with all stakeholders of the world. Besides it needs to develop and strengthen its grid infrastructure, deregulation of energy markets, rural electrification, and skilled employment opportunities. By doing so, India can ensure affordable, reliable and sustainable power.

ACKNOWLEDGEMENTS

This work was supported by the Wind Energy Division, Ministry of New & Renewable Energy, Government of India under grant (IFD Dy. No. 1429 dated 04/11/2016, Demand No. 61/69, Budget Head: 2810.00.104.04.05.31/35) and Central Power Research Institute, A Government of India Society, Ministry of Power, Bangalore through RSOP Project under grant RSOP/2015/DG/6/15122015.

REFERENCES

[1] http://mnre.gov.in/mission-and-vision2/achievements/[Accessed on 03.05.2017]

[2] http://www.itrpv.net/ International Technology Roadmap for Photovoltaic, Eighth Edition, March 2017 [Accessed on 30.03.2017]

[3] www.irena.org.IRENA (2017), REthinking Energy 2017: Accelerating the global energy transformation. International Renewable Energy Agency, Abu Dhabi. [Accessed on 23.01.2017]

[4] http://fs-unep-centre.org/publications/global-trends-renewable energy-investment-2017 [Accessed on 03.05.2017]

[5] REN21, Renewables 2016 global status report, http://www.ren21.net/status-of-renewables/global-status-report/ [Accessed on June 01, 2016]

[6] V. Saravanan, M. Arumugam, R. Venkatesan "Progress of solar photovoltaic systems in India" *32nd European Photovoltaic Solar Energy Conference and Exhibition (EUPVSEC 2016),* 2016, pp.no. 2933 – 2936.

[7] Installed Capacity, March 2017, Central Electricity Authority, Ministry of Power, Government of India, http://www.cea.nic.in/monthlyinstalledcapacity.html (Accessed on 03.05.2017)

[8] Power Sector report, April 2017.India Brand Equity Foundation, Department of Commerce, Ministry of Commerce and Industry, GoI. http://www.ibef.org/industry/power-sector-india.aspx (Accessed on 03.05.2017)

[9] http://niti.gov.in//Report_on_India's_RE_Roadmap_2030-full_report-web.pdf/ Report on India's Renewable Electricity Roadmap 2030. Toward Accelerated Renewable Electricity Deployment. NITI Aayog.

[10] http://www.cercind.gov.in/Draft_reg.html Draft CERC (Terms and Conditions for Tariff determination from Renewable Energy Sources) Regulations, 2017 (Accessed on 10.04.2017)

[11] V. Saravanan, M. Arumugam, S. Ganesh Babu "Development of Solar Parks in India" *26th International Photovoltaic Science and Engineering Conference (PVSEC-26),* 2016 (Poster)

[12] V. Saravanan, M. Arumugam," Solar photovoltaic systems in India: progress, barriers, challenges, impacts, merits, recommendations and vision for the future" *All India Power Seminar – 2016(Present Scenario in Power Sector & Future Challenges)* on the eve of Diamond Jubilee Celebrations, NLC INDIA LIMITED (Formerly Neyveli Lignite Corporation Limited), Neyveli, Tamilnadu, India. (pg.no. 2 to 8, 88 to 102,Renewable Energy Section)

[13] http://teda.in/ [Accessed on 10.04.2017]

[14] www.solarrooftop.gov.in[Accessed on 23.01.2017]

[15] https://play.google.com/store/apps/details?id=nicmnre.nicspvapplication&hl=en [launched on 24/01/2017]

Operando X-Ray Diffraction for Characterization of Photovoltaic Materials

Laura T. Schelhas[1], Jeffrey A. Christians[2], Joseph J. Berry[2], Michael F. Toney[1], Christopher J. Tassone[1], Joseph M. Luther[2], Kevin H. Stone[1]

[1]SLAC National Accelerator Laboratory, Menlo Park, CA 94025, USA
[2]National Renewable Energy Laboratory, Golden, CO 80401, USA

Abstract — **Understanding the structure-function relationship in photovoltaic materials is a key aspect in the optimization of new materials. Traditional characterization techniques often rely on measuring samples before and after synthesis or operation and then the subsequent comparison of these end points. Here we describe a methodology for X-ray diffraction of operational solar cell devices. In this work, we present, as an example, data investigating the structural tetragonal-to-cubic phase transition in $CH_3NH_3PbI_3$, organic-inorganic perovskite solar cell devices during heating and operation. We also comment on the applicability of this technique beyond hybrid perovskite photovoltaics.**

I. INTRODUCTION

Traditional structural characterization of materials is often done *ex-situ*, where the samples are measured before and after formation or operation. This approach, though insightful, misses a large amount of information on what is happening to the sample between end points. Synchrotron sources with their high flux coupled with advances in detector technology have reduced collection times for X-ray diffraction (XRD) experiments making *in-situ* data collection possible. These advances in structural characterization can provide greater insight into the formation and operation of photovoltaic materials and devices.

In recent years, organic-inorganic perovskite films have become a hot topic in photovoltaic research with photoconversion efficiencies now in excess of 22%. However, a deeper understanding of the operation of these materials has lagged behind the rapid improvement in efficiency. *In-situ* XRD techniques have begun being used to provide greater understanding of the formation mechanism in these hybrid solar perovskite materials providing key insights into reproducible synthesis of these materials [1]–[4].

Here we report methods to simultaneously study the structural properties via XRD of photovoltaic materials while also monitoring the device performance. In particular, we discuss characterization of the tetragonal-to-cubic phase transition in methylammonium lead iodide ($CH_3NH_3PbI_3$, or MAPI), during operation or *operando* as an example of this approach.

II. EXPERIMENTAL

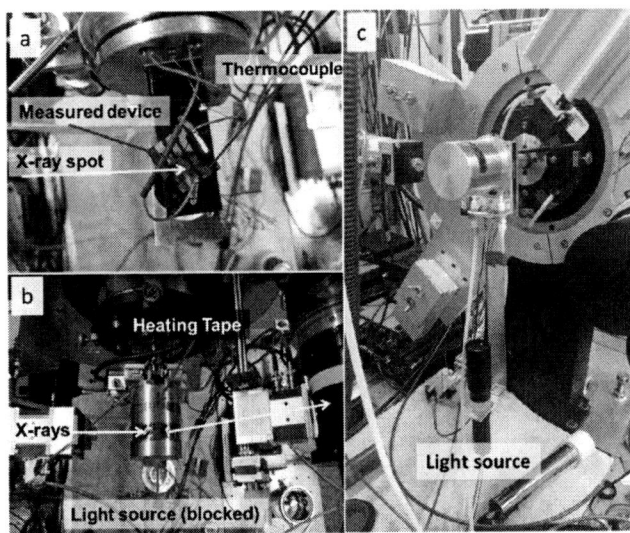

Fig. 1. *Operando* measurement chamber used for simultaneous XRD and *IV* characterization. a) chamber view with cap removed, MAPI sample mounted to the stage b) top view with cap on showing x-ray windows, and heating tape, c) wide view of chamber showing the light source mounted below the sample.

Operando measurements are carried out in a chamber build for use at the Stanford Synchrotron Radiation Lightsource (SSRL) (Fig. 1). The sample is held onto the stage with clips that also serve as electrical contacts. The device is illuminated from the bottom using a light source. Initial studies made use of a white LED source (intensity approximately 0.30 Sun). Currently, the system can be fitted with a solar simulator (HAL-320, Asahi Spectra), seen in figure 1c. The entire chamber is enclosed with a cap with X-ray transparent entrance and exit windows. Samples can be heated using kapton heating mounted to the chamber cap, a thermocouple is placed at the sample to measure the heat from the light as well as the heating tape. The atmosphere of the chamber can be controlled using gas flow through inlet/outlet valves. For example the samples can be run in inert atmospheres (helium/nitrogen/argon) to prevent degradation due to moisture. Conversely, humid environments can be used

978-1-5090-5606-4/17 $31.00 © 2017 IEEE

to explore the effects of moisture on the sample. XRD data collection can be done using a small area Pilatus 100K detector, which allows for rapid collection of a large scattering range, or a Si(111) analyzer crystal and photomultiplier tube detector for high resolution data collection. Beam damage can be a concern so minimizing exposure is recommended. [5]

Care must be taken when determining the data collection protocol when exploring these hybrid perovskite materials. For example, when exploring the structural phase transition in MAPI to avoid artifacts in the device measurements due to sample conditioning the measurements were performed at constant time intervals (figure 2). The protocol used was (i) light on, (ii) *IV* curve, (iii) XRD, (iv) light off, (v) 10 minute rest period; with constant heating during the whole cycle.

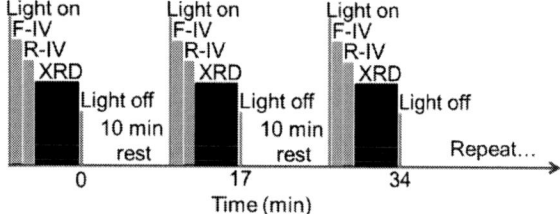

Fig. 2. Schematic representation of data collection protocol used for an *operando* heating study.[6]

Data collection is versatile in this setup allowing for customization of experiments. The heating of MAPI is just one example where this could be used. Another would be to hold a sample using pseudo max power point tracking, where the applied potential that results in the max power of the device is calculated after measuring an *IV* curve. This potential is then applied to the device and can be adjusted as needed by measuring another *IV* curve at any time. This data can then be collected under different conditions to correlate structure with device performance. Next we will summarize our first example of using this approach to study MAPI devices.

III. RESULTS AND DISCUSSION

We have used the *operando* approach described to study MAPI devices. MAPI is of particular interest due to its structural tetragonal-to-cubic phase transition near 65°C. Since the transition temperature is within the range expected in operating solar modules, it is important to understand the structural behavior at this transition and its impact on the device performance.

Previous studies have reported discontinuities in optoelectronic properties of MAPI across the tetragonal-to-cubic transition [7], [8]; however, the transition temperature varies greatly within the literature from 37 °C, reported from photoluminescence[7] to 55-57 °C reported from XRD and differential scanning calorimetry (DSC)[9]. Conversely, there are also a number of studies that indicate little optoelectronic change over the phase transition [10], [11]. Given the range of different results on this subject we employed the *operando* XRD technique to simultaneously study the structure of an operational MAPI device while measuring the device performance.

Detailed analysis of this study have previously been published.[6] But in summary we observed the disappearance of the $(211)_{tet}$ peak, a characteristic feature of the tetragonal-to-cubic phase transition which is not allowed in the cubic symmetry. Therefore by monitoring the intensity of this peak, we determined the maximum transition temperature. Figure 3, plots select XRD data of the $(211)_{tet}$ peak before and after the transition. Using this approach we were also able to simultaneously obtain *IV* curves during heating. A representative scan is shown in Figure 4.

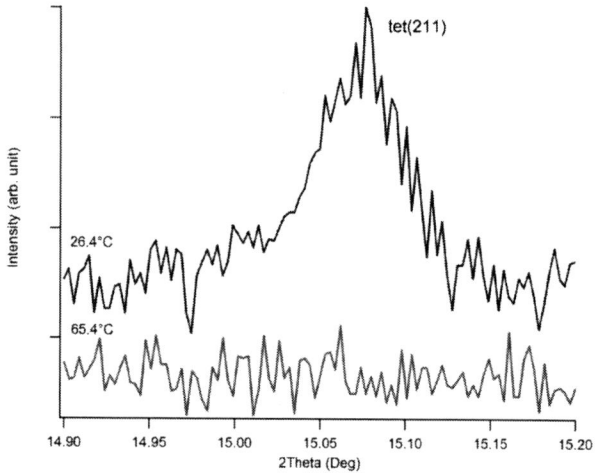

Fig. 3. Select XRD of MAPI device during heating *operando*.

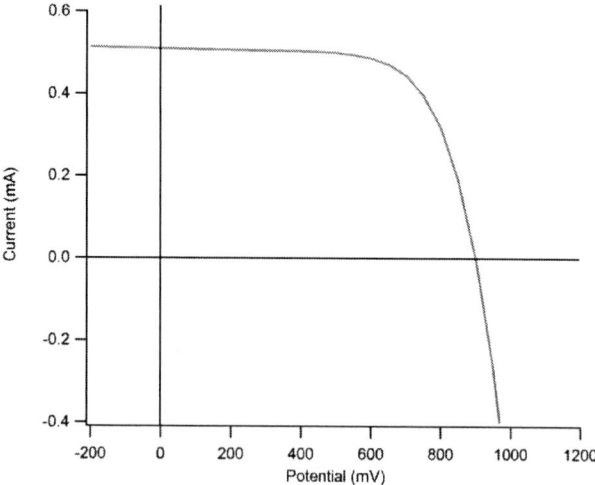

Fig. 4. *IV* curve collected at 50.5°C *operando*.

To confirm that the *operando* approach described here is valid complimentary *ex-situ* device measurements were made. MAPI devices with identical device architecture were measured as a function of temperature. Similar results were observed for both approaches, confirming the results from the *operando* studies.

IV. SUMMARY

The impact of this work is two-fold. First we confirmed that though this promising new solar cell material exhibits a structural phase change within operational temperatures it is of little concern for the ultimate efficiency of the device.[6] Second, we developed *operando* techniques for measuring solar cell materials, which will have applications far beyond just hybrid perovskite solar cells. This technique is not specific to the example given here. The chamber presented here allows for careful structural characterization under a number of different external stimuli (light, temperature, humidity, etc.). This can then be applied to studying degradation mechanisms across a broad range of photovoltaic materials, and is not just limited to the absorber itself.

REFERENCES

[1] E. L. Unger *et al.*, "Chloride in lead-chloride derived organo-metal halides for perovskite-absorber solar cells," *Chem. Mater.*, vol. 26, no. 24, pp. 7158–7165, 2014.

[2] K. W. Tan *et al.*, "Thermally Induced Structural Evolution and Performance of Mesoporous Block Copolymer-Directed Alumina Perovskite Solar Cells," *ACS Nano*, vol. 8, no. 5, pp. 4730–4739, 2014.

[3] V. L. Pool *et al.*, "Thermal engineering of FAPbI3 perovskite material via radiative thermal annealing and in situ XRD," *Nat. Commun.*, vol. 8, pp. 1–8, 2016.

[4] D. P. Nenon *et al.*, "Structural and chemical evolution of methylammonium lead halide perovskites during thermal processing from solution," *Energy Environ. Sci.*, vol. 9, pp. 2072–2082, 2016.

[5] R. L. Z. Hoye *et al.*, "Perovskite-Inspired Photovoltaic Materials: Toward Best Practices in Materials Characterization and Calculations," *Chem. Mater.*, vol. 29, no. 5, pp. 1964–1988, 2017.

[6] L. T. Schelhas *et al.*, "Monitoring a Silent Phase Transition in CH NH PbI Monitoring a Silent Phase Transition in CH3NH3PbI 3 Solar Cells via Operando X-ray Diffraction," *ACS Energy Lett.*, vol. 1, pp. 1007–1012, 2016.

[7] R. L. Milot, G. E. Eperon, H. J. Snaith, M. B. Johnston, and L. M. Herz, "Temperature-Dependent Charge-Carrier Dynamics in CH3NH3PbI3 Perovskite Thin Films," *Adv. Funct. Mater.*, vol. 25, no. 39, pp. 6218–6227, 2015.

[8] L. Cojocaru *et al.*, "Temperature Effects on the Photovoltaic Performance of Planar Structure Perovskite Solar Cells," *Chem. Lett.*, no. 110, pp. 1557–1559, 2015.

[9] T. Baikie *et al.*, "Synthesis and crystal chemistry of the hybrid perovskite (CH3NH3)PbI3 for solid-state sensitised solar cell applications," *J. Mater. Chem. A*, vol. 1, no. 18, p. 5628, 2013.

[10] C. Quarti *et al.*, "Structural and optical properties of methylammonium lead iodide across the tetragonal to cubic phase transition: implications for perovskite solar cells," *Energy Environ. Sci.*, vol. 9, pp. 155–163, 2016.

[11] T. J. Jacobsson, W. Tress, J.-P. Correa-Baena, T. Edvinsson, and A. Hagfeldt, "Room Temperature as a Goldilocks Environment for CH 3 NH 3 PbI 3 Perovskite Solar Cells: The Importance of Temperature on Device Performance," *J. Phys. Chem. C*, vol. 21, no. 120, pp. 11382–11393, 2016.

X-Ray Beam Induced Voltage: A Novel Technique for Electrical Nanocharacterization of Solar Cells

Michael E. Stuckelberger*, Tara Nietzold*, Bradley M. West*,
Barry Lai†, Jörg M. Maser†, Volker Rose†‡, and Mariana I. Bertoni*

*Defect Lab, School of Electrical, Computer and Energy Engineering, Arizona State University, Tempe, AZ, 85287, USA
†Advanced Photon Source, Argonne National Laboratory, Argonne, IL, 60439, USA
‡Center for Nanoscale Materials, Argonne National Laboratory, Argonne, IL, 60439, USA
Contact: michael.stuckelberger@alumni.ethz.ch

Abstract—The efficiency of solar cells with a polycrystalline absorber is typically limited by grain boundaries. Many questions need to be answered such as: which roles do elemental and structural inhomogeneities play with respect to electrical performance? By which mechanisms does degradation occur and how can it be mitigated? The answers to these questions lie at the nanoscale. Ideally, we would measure performance, composition, and structure *in-situ* and *operando* all at once. Correlative, synchrotron-based X-ray microscopy offers a step in this direction. Here, we present a novel technique, X-ray beam induced voltage (XBIV) measurements, which complements the set of X-ray microscopy techniques. Combining the penetration depth of visible-light microscopy with the spatial resolution of electron-beam methods, XBIV measurements shine light on recombination and absorber-layer bandgap variations at nanoscale resolution. We give experimental details and discuss the use of lock-in amplification based on first applications of XBIV measurements to $CuIn_xGa_{1-x}Se_2$ solar cells.

Index Terms—XBIV, X-ray beam induced voltage, XBIC, X-ray beam induced current, CIGS, solar cells, X-ray microscopy, lock-in

I. INTRODUCTION

Understanding the mechanism of charge collection and elemental composition at the nanoscale is of critical importance for developing high-efficiency solar cells. Although this is true for monocrystalline silicon solar cells of all cell architectures, it is much more critical for multi-crystalline silicon solar cells that dominate the photovoltaics market today [1] and for solar cells with polycrystalline thin-film absorber layers such as chalcopyrites ($CuIn_xGa_{1-x}Se_2$ (CIGS), CdTe, etc.), III/V materials deposited by methods faster than epitaxial growth, or perovskite solar cells (PSCs). In all these solar-cell types, grain boundaries play a crucial role for the overall cell performance, and although grain boundaries have been shown in some cases to have beneficial effects, they are mostly detrimental and limit the performance due to the enhanced recombination at defect states [2], [3], [4].

A tool capable of correlating composition, chemical environment, and performance at the length scale of grain boundaries could help unveil the origins of performance variation and degradation that will enable the engineering of devices with higher efficiency and longer life times. Both of which are strong levers to decrease the levelized cost of electricity (LCOE) [5].

Synchrotron-based correlative X-ray microscopy offers such a tool set. For the assessment of the electrical performance, X-ray beam induced current (XBIC) measurements have been established as a non-destructive method with a spatial resolution at the nanoscale. Its key advantage over laser- (LBIC) and electron-beam induced current (EBIC) measurements that are related to XBIC, is the combination of the penetration depth of visible-light microscopy with the resolution of electron microscopy.

XBIC measurements have first been demonstrated in 2000 [6] and have consequently raised significant interest for solar cells with multi-crystalline silicon [7], [8], CdTe [9], CIGS [9], [10], [11], [12], [13], and perovskite [14], [15] absorber layers, with feature sizes down to the sub-micrometer range. The simultaneous measurement of XBIC with elemental composition from X-ray fluorescence (XRF), structure from X-ray diffraction (XRD), or bandgap/radiative recombination from X-ray excited optical luminescence (XEOL) enables a spot-by-spot correlation.

X-ray beam induced voltage (XBIV) measurements expand the toolset of X-ray microscopy techniques further. Here, we introduce this novel technique and apply it to $CuIn_xGa_{1-x}Se_2$ solar cells. In the following, we demonstrate the utilization of lock-in amplification for XBIV and XBIC measurements, which gives access to the full current-voltage ($I(V)$) characteristics of solar cells.

II. EXPERIMENTAL: HOW TO MEASURE XBIV & XBIC?

A. Beamline settings

All measurements presented here were obtained at the beamline 2-ID-D of the Advanced Photon Source (APS) at Argonne National Laboratory [16], but we have successfully performed XBIC/XBIV measurements at APS beamlines 26-ID-C [17] as well as at the Hard X-ray Nanoprobe beamline 3-ID [18] of the National Synchrotron Light Source II at Brookhaven National Laboratory, and at the soft-X-ray beamline 10-ID-1 [19] of the Canadian Light Source.

The dwell time of 100 ms was limited by the count rate of the simultaneous XRF measurements and the speed of the scanning motor that was operated in "fly-scan" mode [20]. The incident-photon energy was 10.5 keV (above the Ga_K absorption edge), the X-ray beam spotsize was 200 nm,

978-1-5090-5606-4/17 $31.00 © 2017 IEEE

and the stepsize was 1 μm. Note that the spatial resolution of XBIC/XBIV/EBIC/electron-beam induced voltage (EBIV) measurements can be multiple times larger than the spot size due to the spread generation of electron-hole pairs in the particle shower resulting from incident high-energy particles [15].

B. Sample preparation

In contrast to electron-beam microscopy methods, XBIV and XBIC measurements do not require particular sample preparation apart from electrical connections to the operational solar cell.

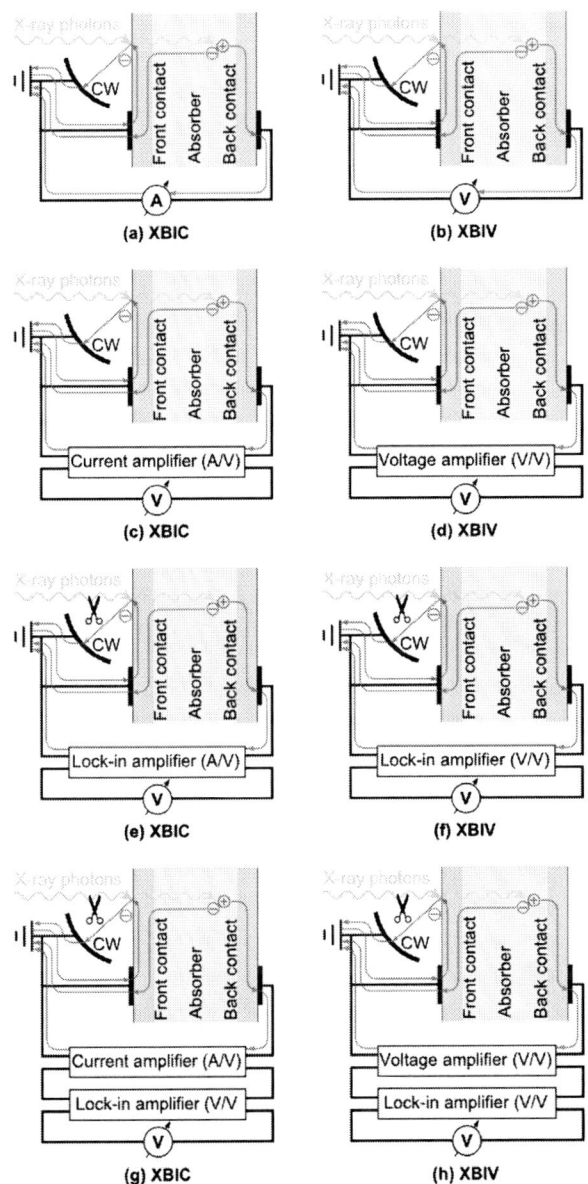

Figure 1. Electrical setup for X-ray beam induced current (XBIC, left) and voltage (XBIV, right) measurements for different amplification configurations discussed in the text and Tab. I. The abbreviation "CW" stands for Chamber Wall.

We performed the XBIC/XBIV measurements on a non-encapsulated $CuIn_xGa_{1-x}Se_2$ solar cell deposited by MiaSolé [21] on flexible steel substrates, with the X-ray beam incident at a 90° angle to the front contact surface. For subsequent temperature-dependent XBIV measurements presented with more experimental details in [22], the cell was mounted inside a heating stage [23] that we developed for *in-situ* measurements of $CuIn_xGa_{1-x}Se_2$ growth [24], [25], [26]. Here, measurements were performed at room temperature in nitrogen atmosphere.

C. Electrical connections

As we detail in [25], proper grounding of the contact that is exposed to the incident X-ray photons is critical for XBIC measurements. If the opposite side is grounded, the current of electrons replacing photoelectrons ejected from the surface is misinterpreted as XBIC signal, contributing additional error or even resulting in false measurements.

For XBIV measurements, the grounding is less critical as long as one side is grounded to avoid charging of the solar cells. We recommend to ground the front side for consistency and fast switching between XBIC and XBIV measurements as shown in Fig. 1. Due to the low signal amplitude, electrical feedthroughs (in case of a vacuum chamber) and wires should be shielded and kept as short as possible.

D. (Lock-in) signal amplification

Signal amplification is often the most challenging part of XBIC and XBIV measurements. Figure 1 shows different amplification configurations that we have tested (except for the configurations (a&h)), and summarize the findings in Tab. I. These configurations assume data acquisition through the unit that controls the scanning measurement, which avoids the synchronization of data recorded separately. Across a multitude of beamlines, we have used voltage-frequency converters with 10^5 cts/V as scalers for data acquisition. Ideally, we used two voltage-frequency converters with a split input signal and inverted polarity in one of them. This enables the measurement of both negative and positive voltages, which can be critical for small signals that include wanted or unwanted bias.

Table I
COMPARISON OF THE SIGNAL AMPLIFICATION CONFIGURATIONS SHOWN IN FIG. 1 FOR X-RAY BEAM INDUCED CURRENT (XBIC) AND VOLTAGE (XBIV) MEASUREMENTS IN TERMS OF SIGNAL/NOISE (S/N) RATIO AND BIAS SUPPRESSION. WE USED THE EMPHASIZED CONFIGURATIONS (F&G) FOR MEASUREMENTS PRESENTED HERE AND RECOMMEND THEM FOR MOST EXPERIMENTS.

Setup	XBIV/XBIC	Lock-in/DC	S/N	Bias supp.
Fig. 1(a)	XBIC	DC	− −	− −
Fig. 1(b)	XBIV	DC	− −	− −
Fig. 1(c)	XBIC	DC	−	− −
Fig. 1(d)	XBIV	DC	−	− −
Fig. 1(e)	XBIC	Lock-in	+	++
Fig. 1(f)	*XBIV*	*Lock-in*	++	++
Fig. 1(g)	*XBIC*	*Lock-in*	++	++
Fig. 1(h)	XBIV	Lock-in	++	++

Because of the use of a voltage-frequency converter, we did not test configuration (a) and used always a current-voltage converter for XBIC measurements. As current amplifier, we used a Stanford Research Systems Preamplifier SR 570. For lock-in amplification of current or voltage signals, as well as for DC amplification of the voltage signal, we used the MFLI lock-in amplifier from Zurich Instruments. When using lock-in amplification, we chopped the X-ray beam at a frequency of 318 Hz with the filter wheel MC1F10 from Thorlabs that provides a transmittance ratio of $> 10^{12}$ betweeen X-rays ON/OFF at a photon energy of 10.5 keV.

Comparing the amplification configurations, we note:

- With present photon fluxes of nanoprobe beamlines, XBIC and XBIV measurements without amplification (Fig. 1(a&b)) are out of reach, although we have reproducibly measured low counts of XBIV signal using configuration (b).

- To our knowledge, configuration (c) has been used for all previously reported XBIC measurements. Although the signal quality is often satisfying, there are intrinsic limitations apart from the obviously lacking possibility of applying bias light or voltage without lock-in amplification. Environmental changes often cause jumps in the recorded level that add not only error to the quantitative evaluation but hinder also qualitative comparisons. Configuration (d) is the analogous configuration for XBIV measurements with DC amplification.

- Configurations (e&f) are the straight-forward implementation of lock-in amplification. Especially for low-photon-flux experiments, the signal/noise ratio is boosted compared to configurations (c&d), which enables the application of bias light and voltage up to a DC signal that is several orders of magnitude higher than the AC signal, and measurement artifacts from environmental variations are strongly suppressed.

- Configurations (g&h) add a DC signal amplification prior to lock-in amplification. For XBIC measurements (g), this is our preferred setting, giving a consistently higher signal/noise ratio and less measurement artifacts than (e). The reasons are not entirely clear but it may be caused by the better current amplification of the SR 570 compared to the MFLI. However, configuration (e) should be considered if the frequency is too high for proper amplification of the AC signal. For XBIV measurements, the signal quality of configuration (f) has been high enough for our experiments such that we did not see the need for complicating the experiment by adding a voltage amplifier.

E. Data correction

For higher accuracy, we typically correct the XBIC signal pixel-by-pixel for variations of the absorber thickness:

$$\text{XBIC} \approx \frac{\text{XBIC}_{\text{raw}}}{1 - \exp\left[-\sum_i \alpha_i \cdot m_i\right]} \quad (1)$$

with XBIC_{raw} the raw XBIC measurement, $\alpha_i = \frac{\mu_i}{\rho_i} \left[\frac{\text{cm}^2}{\text{g}}\right]$ the tabulated X-ray attenuation coefficient, and $m_i \left[\frac{\text{g}}{\text{cm}^2}\right]$ the

mass density for the absorber element i that we typically measure through simultaneous XRF measurements (see [12] for XRF corrections). For the quantitative conversion of the corrected XBIC signal to the charge collection efficiency, we refer to [25].

Note that these corrections rely on the linear relationship between the XBIC signal, the number of generated electron-hole pairs, and absorbed X-ray photons. This proportionality is not given for XBIV signal (rather, XBIV depends logarithmically on the number of incident photons). Therefore, thickness variations have a smaller impact on XBIV than on XBIC signal, and the Beer-Lambert-law based correction of eq. (1) does not apply for XBIV (c.f. section V).

III. THE LOCK-IN COMPLICATION

Lock-in amplification for current measurements is standard for many PV applications such as quantum efficiency or LBIC measurements. The interpretation of such measurements assumes most often implicitly a linear relationship between the incident photon intensity and the measured current. The same is true for XBIC measurements. Comparing a good cell (or cell area) with a poor cell with different slopes of current vs. photon intensity in Fig. 2(a), the good cell generates (for perfectly linear relationships) the same additional current $\Delta I_{\text{bright}}^{\text{good}} = \Delta I_{\text{dark}}^{\text{good}}$ for a given additional photon intensity $\Delta X_{\text{bright}} = \Delta X_{\text{dark}}$, independent of any bias light that sets the solar cell into 'bright' or 'dark' condition. Similarly, $\Delta I_{\text{bright}}^{\text{poor}} = \Delta I_{\text{dark}}^{\text{poor}}$ holds. Consequently, the good solar cell outperforms the poor solar cell in all current measurements.

This is not true for voltage measurements such as in XBIV, EBIV and laser-beam induced voltage (LBIV) experiments, as the voltage depends in first approximation logarithmically on the photon intensity (see Fig. 2(b)) [27]. Therefore, depending on the absolute photon intensity, the same amount of additional photons $\Delta X_{\text{bright}} = \Delta X_{\text{dark}}$ leads in the given example to $\Delta V_{\text{dark}}^{\text{good}} > \Delta V_{\text{dark}}^{\text{poor}}$, but to $\Delta V_{\text{bright}}^{\text{good}} < \Delta V_{\text{bright}}^{\text{poor}}$.

Consequently, the application of bias light or bias current for voltage measurements of solar cells with lock-in amplification leads to ambiguous results.

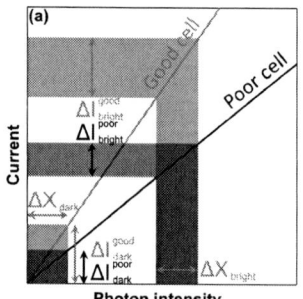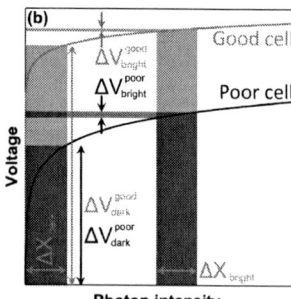

Figure 2. In experiments with lock-in amplification, the proportionality between current and photon intensity in (a) leads to a higher response of a good cell than of a poor cell $\Delta I^{\text{good}} > \Delta I^{\text{poor}}$, independent of the absolute photon intensity. In contrast, the logarithmic relationship between voltage and photon intensity in (b) does not allow for a prediction whether the good or poor cell provides a higher response ΔV.

IV. ACCESS TO THE FULL LOCAL $I(V)$ CURVE

The possibility to apply bias voltage and bias light during XBIC measurements with lock-in amplification enables the measurement of full $I(V) = \text{XBIC}(V)$ curves—notably at any given point of a nanoscale map. This opens the window to a new world of electrical nanocharacterization with the extraction of local $I(V)$ parameters under varying conditions.

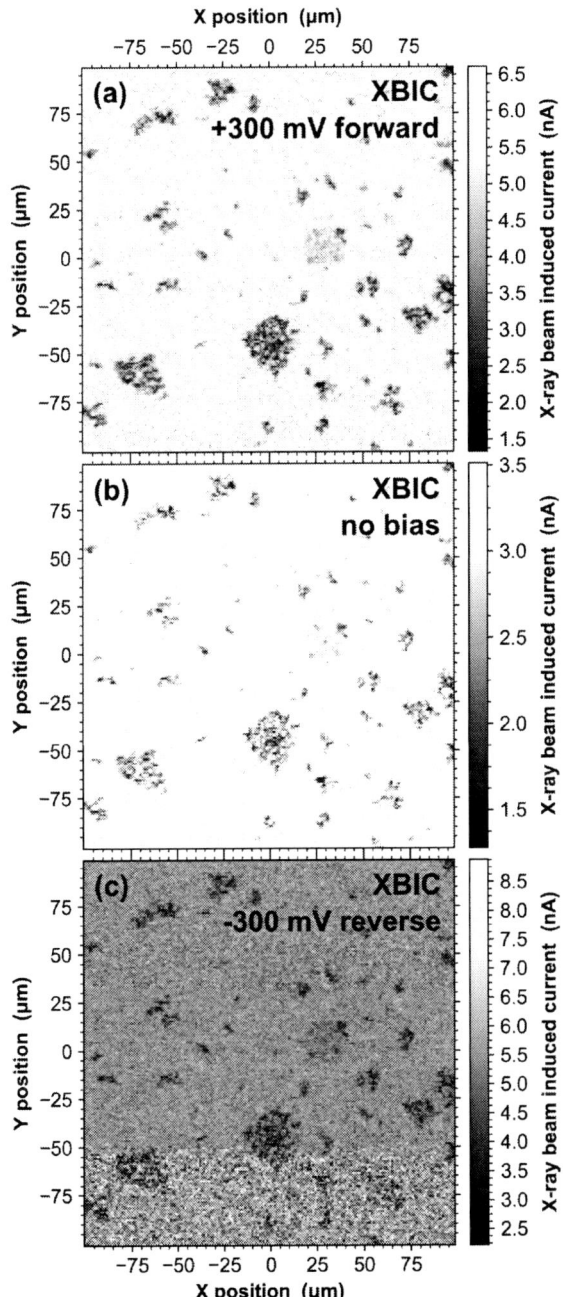

Figure 3. X-ray beam induced current map of a $\text{CuIn}_x\text{Ga}_{1-x}\text{Se}_2$ solar cell. The measurements were performed in dark with $300\,\text{mV}$ forward bias (a), $0\,\text{mV}$ bias (b), and $-300\,\text{mV}$ reverse bias (c) voltage applied. The measurement configuration corresponds to the schematics of Fig. 1(g).

Figure 3 shows such maps with different bias voltages applied from $300\,\text{mV}$ forward bias (Fig. 3(a)) to $0\,\text{mV}$ bias (Fig. 3(b)) and to $-300\,\text{mV}$ reverse bias (Fig. 3(c)).

These measurements were performed in dark. Therefore, a strong spatial gradient occurs in the quasi-Fermi-level splitting with a spot-sized area under high illumination that operates in parallel to the other cell areas. Obviously, measurements with bias light up to $1000\,\text{W/m}^2$ at AM1.5g are desirable, with the X-rays contributing locally only a negligible fraction of electron-hole pair excitation. For macroscopically large solar cells, this is currently out of reach: although we were able to take XBIC measurements with simultaneously applied bias light and bias voltage, the maximum ratio of DC/AC signal we could measure was in the order of 10^6.

Comparing the XBIC maps of Fig. 3, we note:

- Throughout the $200\,\mu\text{m} \times 200\,\mu\text{m}$ maps at all voltages, high-performing areas (bright color) can clearly be distinguished from low-performing areas that span up to hundreds of μm^2. A statistical analysis of the temperature dependence including a correlation of the performance with elemental distributions is presented in [22].

- Particularly in Fig. 3(c), the measurement is noisier at $Y < -50\,\mu\text{m}$ and $Y > +35\,\mu\text{m}$. This artifact is induced by the difficulty to lock in to the AC signal with a critically high DC/AC ratio. Note that we had to lower the amplification of the SR 570 from $100\,\text{nA/V}$ for the measurement at $0\,\text{mV}$ bias voltage to $100\,\mu\text{A/V}$ for the measurement at $\pm 300\,\text{mV}$ bias voltage to avoid saturation of the lock-in amplifier input. From the large DC current in both voltage bias directions, we conclude that shunts are at least under dark conditions not negligible in this solar cell.

- Overall, the poor areas underperform at all voltages. However, the difference between poor and good areas is smaller under reverse bias. This is expected, as the electric field supports the collection of X-ray generated charge carriers. This is particularly well visible in a squared region around $(X, Y) \approx (30\,\mu\text{m}, 10\,\mu\text{m})$, where precedent measurements induced beam damage.

V. COMPARISON OF XBIV AND XBIC

Figure 4(a) shows the XBIV measurement of the same area; it was taken just before the measurements shown in Fig. 3. The measurement conditions were the same except for the measurement mode that corresponds to Fig. 1(f).

It is not surprising that the underperforming areas in XBIC underperform also with respect to XBIV, as poor charge carrier collection translates both into low XBIC or quantum efficiency, and into low voltage. However, it is noteworthy that the area with beam damage around $(X, Y) \approx (30\,\mu\text{m}, 10\,\mu\text{m})$ only leads to lower XBIC but not to lower XBIV signal. At this moment, we do not know the reason but can exclude time effects, as the XBIV measurement was performed between the occurrence of beam damage and the XBIC measurements.

A possible explanation is based on the similarity between X-ray beam induced damage and visible-light induced damage. As discussed in [28], the studied solar cells suffer from

978-1-5090-5606-4/17 $31.00 © 2017 IEEE

performance degradation under light soaking. Depending on the mechanism that governs metastability in the cycle of light soaking/annealing, light soaking could show up as a relatively stronger reduction of the short-circuit current density (J_{sc}) than of the open-circuit voltage (V_{oc}). An alternative explanation is based on a beam-induced bandgap widening, such that an enhanced bandgap (higher V_{oc}) is compensated by the defect-induced Fermi-level splitting reduction (lower V_{oc}).

Figure 4(b) shows the pixel-by-pixel correlation between XBIV and XBIC from Fig. 3(a) as a hexbin plot. The Pearson's correlation coefficient is $r = 0.66$ with a p-value of 0.0, which indicates a statistically significant correlation between XBIV and XBIC at $300\,\text{mV}$ forward bias. The analogous correlations between XBIV and XBIC of Fig. 3(b–c) (hexbin plots not shown) yield $r = 0.59$, $p = 0.0$ and $r = 0.41$, $p = 0.0$ for 0 and $-300\,\text{mV}$ reverse bias, respectively.

From the visual comparison of the XBIV and XBIC maps, one might expect higher correlation coefficients. Although we sometimes observe correlations with $r > 0.9$, lower r

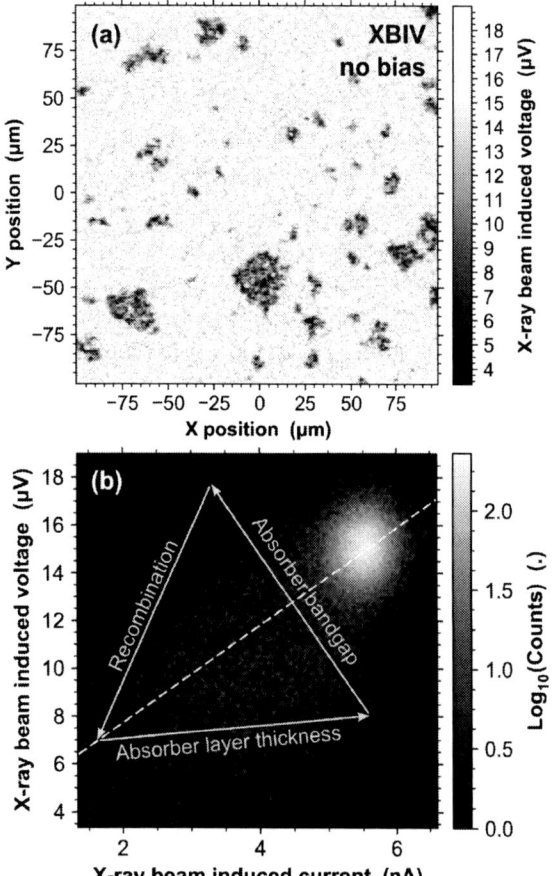

Figure 4. (a) X-ray beam induced voltage (XBIV) map of the same area as the X-ray beam induced current (XBIC) maps shown in Fig. 3. (b) Hexbin plot of the pixel-by-pixel correlation of the XBIV measurement with the XBIC measurement at $300\,\text{mV}$ forward bias shown in Fig. 3(a). Pearson's correlation coefficient $r = 0.66$ with a p-value of 0.0. The green lines indicate the qualitative dependence of XBIC and XBIV on recombination, absorber bandgap, and absorber thickness.

values do not necessarily indicate higher measurement noise. Instead, XBIC and XBIV measure two distinct properties with different dependencies on recombination, absorber bandgap, and absorber thickness as indicated in Fig. 4(b).

- *Absorber layer thickness:* Similar as for illumination of a solar cell with red light, the illumination with hard X-rays generates more electron-hole pairs for increased absorber layer thickness d. For sufficiently low X-ray absorption, the electron-hole pair generation rate G is proportional to the absorber thickness XBIC $\propto G \propto d$. In contrast, XBIV increases logarithmically with d: XBIV $\propto \log(d)$ due to the increased G. Therefore, the thickness dependence of XBIV can be neglected for small thickness variations.

- *Recombination:* The assumption of a simple equivalent circuit consisting only of a diode and a photogenerator leads to a logarithmic dependence of the open-circuit voltage on the inverse saturation current. As the saturation current varies easily over orders of magnitude, increased recombination leads to a relatively stronger decrease of the voltage than of the current, that decreases by the recombination current.

- *Absorber bandgap (E_g):* As discussed in [25], XBIC $\propto G \propto \frac{1}{E_g}$. In contrast, the approximation XBIV $\propto E_g$ holds in many cases.

These dependencies of XBIC and XBIV are crucial for further data analysis. Most importantly, their inverse dependency on the bandgap allows for a deconvolution of the collection efficiency and the bandgap for combined measurements. Particularly for complex multi-element material systems, where spatial variations of collection efficiency and bandgap are common, this deconvolution is of critical importance and provides a unique advantage over other electrical nanocharacterization techniques.

VI. CONCLUSIONS

We introduced the technique of XBIV measurements that complements the toolset of X-ray microscopy for solar cell nanocharacterization.

Utilizing for the first time lock-in amplification for both XBIC and XBIV measurements, we tested different configurations of signal amplification and found the highest signal/noise ratios for the pre-amplification of small XBIC currents prior to lock-in amplification and for the direct lock-in amplification of small XBIV voltages.

We further demonstrated XBIC measurements with positive and negative bias voltages applied to the solar cell. This gives access to the full $I(V)$ curve in dark and under illumination, to create environments close to outdoor operating conditions. In a correlative X-ray microscopy approach with XRF measurements, multidimensional maps of elemental composition and full electrical characterization at each pixel—potentially combined with a variation of the cell environment—will provide highly valuable and specific information for the improved understanding and performance of solar cells.

Beyond X-ray based solar cell characterization, the lock-in technique can be utilized for laser and electron-beam based techniques. Additionally, the combination of LBIC/LBIV and EBIC/EBIV provides similar advantages over standard LBIC or EBIC measurements.

Acknowledgments

We greatly acknowledge Rouin Farshchi, Dmitry Poplavskyy, Jeff Bailey (all MiaSolé HiTech Corp., USA) for providing $CuIn_xGa_{1-x}Se_2$ solar cells, Rupak Chakraborty, Jim Serdy, Tonio Buonassisi (all MIT, USA) for their contributions building the heating stage, David Fenning (UC San Diego, USA) for fruitful discussions, and Chris Roehrig for invaluable practical help with XBIV measurements. We acknowledge funding from the U.S. Department of Energy under contract DE-EE0005948. Work at the Advanced Photon Source was supported by the U.S. Department of Energy, Office of Science, Office of Basic Energy Sciences, under Contract No. DE-AC02-06CH11357. This material is based upon work supported in part by the National Science Foundation (NSF) and the Department of Energy (DOE) under NSF CA No. EEC-1041895. Any opinions, findings and conclusions or recommendations expressed in this material are those of the author(s) and do not necessarily reflect those of NSF or DOE.

References

[1] M. A. Green, "Commercial progress and challenges for photovoltaics," *Nature Energy*, vol. 1, pp. 1–4, 2016.

[2] Y. Yan, R. Noufi, and M. M. Al-Jassim, "Grain-boundary physics in polycrystalline CuInSe2 revisited: experiment and theory," *Physical Review Letters*, vol. 96, p. 205501, 2006.

[3] U. Rau, K. Taretto, and S. Siebentritt, "Grain boundaries in Cu(In,Ga)(Se,S)2 thin-film solar cells," *Applied Physics A*, vol. 96, pp. 221–234, 2009.

[4] M. I. Bertoni, D. P. Fenning, M. Rinio, V. Rose, M. Holt, J. Maser, and T. Buonassisi, "Nanoprobe x-ray fluorescence characterization of defects in large-area solar cells," *Energy & Environmental Science*, vol. 4, pp. 4252–4257, 2011.

[5] R. Jones-Albertus, D. Feldman, R. Fu, K. Horowitz, and M. Woodhouse, "Technology advances needed for photovoltaics to achieve widespread grid price parity," *Progress in Photovoltaics: Research and Applications*, vol. 25, no. 9, pp. 1272–1283, 2016.

[6] H. Hieslmair, A. A. Istratov, R. Sachdeva, and E. R. Weber, "New synchrotron-radiation based techniqeu to study localized defects in silicon: "EBIC" with X-ray excitation," *10th Workshop on Crystalline Silicon Solar Cell Materials and Processes*, pp. 162–165, 2010.

[7] O. F. Vyvenko, T. Buonassisi, A. A. Istratov, H. Hieslmair, A. C. Thompson, R. Schindler, and E. R. Weber, "X-ray beam induced current–a synchrotron radiation based technique for the in situ analysis of recombination properties and chemical nature of metal clusters in silicon," *Journal of Applied Physics*, vol. 91, no. 6, pp. 3614–3617, 2002.

[8] T. Buonassisi, A. A. Istratov, S. Peters, C. Ballif, J. Isenberg, S. Riepe, W. Warta, R. Schindler, G. Willeke, Z. Cai, B. Lai, and E. R. Weber, "Impact of metal silicide precipitate dissolution during rapid thermal processing of multicrystalline silicon solar cells," *Applied Physics Letters*, vol. 87, no. 12, p. 121918, 2005.

[9] M. Stuckelberger, B. West, S. Husein, H. Guthrey, M. Al-Jassim, R. Chakraborty, T. Buonassisi, J. M. Maser, B. Lai, B. Stripe, V. Rose, and M. Bertoni, "Latest developments in the x-ray based characterization of thin-film solar cells," *Proc. Photovoltaic Specialist Conference (PVSC)*, pp. 1–6, 2015.

[10] B. West, S. Husein, M. Stuckelberger, B. Lai, J. Maser, B. Stripe, V. Rose, H. Guthrey, M. Al-Jassim, and M. Bertoni, "Correlation between grain composition and charge carrier collection in Cu(In,Ga)Se2 solar cells," *Proc. Photovoltaic Specialist Conference (PVSC)*, pp. 1–6, 2015.

[11] B. West, M. Stuckelberger, H. Guthrey, L. Chen, B. Lai, J. Maser, V. Rose, J. J. Dynes, W. Shafarman, M. Al-Jassim, and M. I. Bertoni, "Synchrotron x-ray characterization of alkali elements at grain boundaries in Cu(In,Ga)Se2 solar cells," *Proc. Photovoltaic Specialist Conference (PVSC)*, pp. 31–34, 2016.

[12] B. West, M. Stuckelberger, A. Jeffries, S. Gangam, B. Lai, B. Stripe, J. Maser, V. Rose, S. Vogt, and M. Bertoni, "X-ray fluorescence at nanoscale resolution for multicomponent layered structures: a solar cell case study," *Journal of Synchrotron Radiation*, vol. 24, pp. 288–295, 2017.

[13] B. West, M. Stuckelberger, H. Guthrey, L. Chen, B. Lai, J. Maser, V. Rose, W. Shafarman, M. Al-Jassim, and M. I. Bertoni, "Grain engineering: How nanoscale inhomogeneities can control charge collection in solar cells," *Nano Energy*, vol. 32, pp. 488–493, 2017.

[14] M. Stuckelberger, T. Nietzold, G. N. Hall, B. West, J. Werner, B. Niesen, C. Ballif, V. Rose, D. P. Fenning, and M. I. Bertoni, "Elemental distribution and charge collection at the nanoscale on perovskite solar cells," *Proc. Photovoltaic Specialist Conference (PVSC)*, pp. 1191–1196, 2016.

[15] ——, "Charge collection in hybrid perovskite solar cells: relation to the nanoscale elemental distribution," *IEEE Journal of Photovoltaics*, vol. 7, no. 2, pp. 590–597, 2017.

[16] W. Yun, B. Lai, Z. Cai, J. Maser, D. Legnini, E. Gluskin, Z. Chen, A. A. Krasnoperova, Y. Vladimirsky, F. Cerrina, E. Di Fabrizio, and M. Gentili, "Nanometer focusing of hard x rays by phase zone plates," *Review of Scientific Instruments*, vol. 70, no. 5, pp. 2238–2241, 1999.

[17] R. P. Winarski, M. V. Holt, V. Rose, P. Fuesz, D. Carbaugh, C. Benson, D. Shu, D. Kline, G. B. Stephenson, I. McNulty, and J. Maser, "A hard x-ray nanoprobe beamline for nanoscale microscopy," *Journal of Synchrotron Radiation*, vol. 19, no. 6, pp. 1056–1060, 2012.

[18] E. Nazaretski, K. Lauer, H. Yan, N. Bouet, J. Zhou, R. Conley, X. Huang, W. Xu, M. Lu, K. Gofron, S. Kalbfleisch, U. Wagner, C. Rau, and Y. S. Chu, "Pushing the limits: an instrument for hard x-ray imaging below 20nm," *Journal of Synchrotron Radiation*, vol. 22, no. 2, pp. 336–341, 2015.

[19] K. V. Kaznatcheev, C. Karunakaran, U. D. Lanke, S. G. Urquhart, M. Obst, and A. P. Hitchcock, "Soft x-ray spectromicroscopy beamline at the cls: Commissioning results," *Nuclear Instruments and Methods in Physics Research A*, vol. 582, pp. 96–99, 2007.

[20] A. E. Morishige, H. S. Laine, E. E. Looney, M. A. Jensen, S. Vogt, J. B. Li, B. Lai, H. Savin, and T. Buonassisi, "Increased throughput and sensitivity of synchrotron-based characterization for photovoltaic materials," *IEEE Journal of Photovoltaics*, vol. 7, no. 3, pp. 763–771, 2017.

[21] MiaSolé, 2017, available at: http://miasole.com/, (accessed June 4, 2017).

[22] M. Stuckelberger, T. Nietzold, B. M. West, R. Farshchi, D. Poplavskyy, J. Bailey, B. Lai, J. Maser, and M. I. Bertoni, "How does CIGS performance depend on temperature at the microscale?" *Submitted for publication*, 2017.

[23] R. Chakraborty, J. Serdy, B. West, M. Stuckelberger, B. Lai, J. Maser, M. I. Bertoni, M. L. Culpepper, and T. Buonassisi, "Development of an in situ temperature stage for synchrotron x-ray spectromicroscopy," *Review of Scientific Instruments*, vol. 86, p. 113705, 2015.

[24] B. West, M. Stuckelberger, L. Chen, R. Lovelett, B. Lai, J. Maser, W. Shafarman, and M. Bertoni, "Growth of Cu(In,Ga)(S,Se)2 films: Unravelling the mysteries by in-situ x-ray imaging," *Proc. Photovoltaic Specialist Conference (PVSC)*, pp. 530–533, 2016.

[25] M. Stuckelberger, B. West, T. Nietzold, B. Lai, J. M. Maser, V. Rose, and M. I. Bertoni, "Review: Engineering solar cells based on correlative X-ray microscopy," *Journal of Materials Research*, vol. 32, no. 10, pp. 1825–1854, 2017.

[26] B. West, M. Stuckelberger, S. Wojcik, L. Chen, B. Lai, J. Maser, and M. Bertoni, "Machine learning and correlative microscopy: how 'big data' techniques can benefit thin film solar cell characterization," *Proc. Photovoltaic Specialist Conference (PVSC)*, 2017, (in press).

[27] W. Shockley, "The theory of p-n junctions in semiconductors and p-n junction transistors," *Bell System Technical Journal*, vol. 28, no. 3, pp. 435–489, 1949.

[28] R. Farshchi, B. Hickey, G. Zapalac, J. Bailey, D. Spaulding, and D. Poplavskyy, "Mechanisms for light-soaking induced carrier concentration changes in the absorber layer of Cu(In,Ga)Se2 solar cells," *Proc. Photovoltaic Specialist Conference (PVSC)*, pp. 2157–2160, 2016.

978-1-5090-5606-4/17 $31.00 © 2017 IEEE

Electro-Luminescent Refrigeration Enabled by Highly Efficient Photovoltaics

T. Patrick Xiao[1], Kaifeng Chen[2], Parthiban Santhanam[2], Shanhui Fan[2], Eli Yablonovitch[1]

[1]University of California, Berkeley, Berkeley, CA, 94720, USA
[2]Stanford University, Stanford, CA, 94305, USA

Abstract — Recognition of the need to design for luminescence extraction has led to recent records in solar cell efficiency. These design principles also create opportunities to use the photovoltaic (PV) cell and its reciprocal device, the light-emitting diode (LED), in applications beyond solar energy. As LEDs approach the limit of unity external quantum efficiency, the outgoing photons carry away more energy than supplied to the device, and the phenomenon of electroluminescent cooling becomes accessible. We investigate a refrigeration scheme that uses an LED and a PV cell – both with near-unity quantum efficiency – on the two sides of a luminescent heat engine. With the best available materials, and a device design that maximizes light extraction, electroluminescent cooling can operate with a coefficient of performance that substantially exceeds that of other solid-state cooling technologies.

Index Terms — electroluminescent cooling, light-emitting diodes, photovoltaic cells, thermophotonics

I. INTRODUCTION

In recent years, record solar conversion efficiencies have been achieved by designing the solar cell to be an efficient light emitter [1]. Progress in photovoltaics therefore runs parallel to the evolution of light-emitting diode (LED) efficiency, as both technologies must rely on the same essential physics of light extraction to reach their theoretical limits. An LED that approaches unity external quantum efficiency ($\eta_{ext} = 1$) emits one external photon for every injected electron. If the outgoing photons are more energetic than the injected electrons, entropy must be pumped out of the semiconductor lattice into the emitted radiation [2]-[4]. A significant self-cooling effect is a property exhibited only by the most efficient LEDs.

To date, LEDs have not reached the threshold of efficiency necessary to observe a net cooling effect in experiment. Nonetheless, we show that with presently available optoelectronic materials and our current knowledge of device design, LEDs and PV cells with sufficient efficiency can be realized that make electroluminescence as a cooling mechanism not only feasible, but potentially superior in performance to other solid-state cooling technologies.

II. ELECTROLUMINESCENT REFRIGERATION

Microscopically, every conversion of an electron-hole pair into a photon is a cooling event if the photon energy $\hbar\omega$ exceeds the applied bias qV to supply the carrier to the LED. In such an event, the electrons and holes extract thermal energy from the lattice of the amount $(\hbar\omega - qV)$ via the Peltier effect in order to recombine and generate luminescence. A macroscopic net cooling effect, however, is only possible to observe if the heat liberated in these successful luminescence events exceeds the heat generated by the non-radiative loss of carriers, parasitic absorption of photons, and by other dissipative mechanisms such as Joule heating. If net cooling is achieved, the relevant metrics of performance as a refrigerator are the cooling heat flux Q_c (in W/cm²) and the coefficient of performance (COP), which is the ratio of the cooling flux to the supplied work, COP = Q_c/W.

Though an LED alone can act as a refrigerator, significant improvements to the COP are possible by combining the LED with a PV cell in the configuration shown in Fig. 1, which has been called the thermophotonic heat pump [5, 6]. The two devices are separated by a macroscopic vacuum gap. The PV cell absorbs the far-field luminescent radiation from the LED and reclaims part of the optical energy as electrical power, which can be supplied back to the LED input. The addition of the PV cell greatly reduces the required amount of externally supplied work and increases the refrigerator COP:

$$\mathrm{COP} \equiv \frac{Q_c}{W} = \frac{\hbar\omega_c\Phi_c - \hbar\omega_h\Phi_h - J_cV_c - Q_{\Omega c}}{J_cV_c - J_hV_h + Q_{\Omega c} + Q_{\Omega h}} \quad (1)$$

where Φ_c, $\hbar\omega_c$, J_c, V_c, and $Q_{\Omega c}$ are, respectively, the luminescent photon flux, average photon energy, current, applied voltage, and Ohmic dissipation in the LED. The corresponding quantities for the PV cell have the subscript "h". The COP is bounded by the Carnot limit, COP_Carnot =

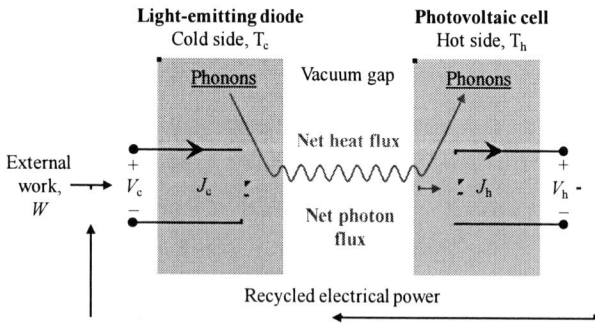

Fig. 1. Diagram of the thermophotonic configuration for electroluminescent cooling. Photons emitted by the LED transport both energy and entropy (heat), and are collected by the PV cell in the far field, which returns part of the energy to the LED electrical input. The vacuum gap suppresses conductive heat leakage.

978-1-5090-5606-4/17 $31.00 © 2017 IEEE

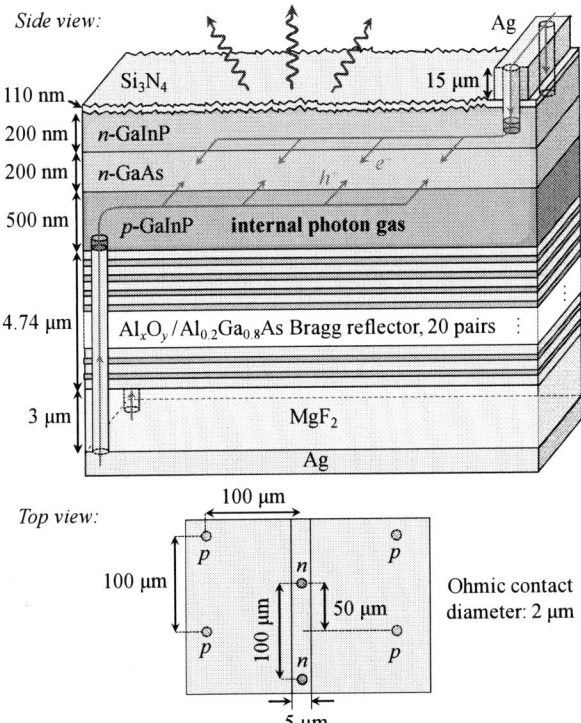

Side view:

110 nm — Si₃N₄ / Ag / 15 μm
200 nm — *n*-GaInP
200 nm — *n*-GaAs
500 nm — *p*-GaInP — internal photon gas
4.74 μm — Al$_x$O$_y$/Al$_{0.2}$Ga$_{0.8}$As Bragg reflector, 20 pairs
3 μm — MgF₂
Ag

Top view:

100 μm
100 μm
100 μm
50 μm
5 μm
Ohmic contact diameter: 2 μm

Fig. 2. Structure of the GaAs/GaInP double heterostructure LED designed for efficient external luminescence. The *n*-GaAs active region is doped to 2.5×10^{17} cm⁻³, while the GaInP cladding layers are each doped to 6×10^{17} cm⁻³. Front surface texturing and a high-contrast Bragg reflector on the rear side enhance the extraction of light out of the front surface. The structure can be repeated over a large area along the lateral dimensions, with a point-contact arrangement shown in the top view.

$T_c/(T_h - T_c)$, where T_c and T_h are the temperatures of the cold-side LED and hot-side PV cell, respectively.

The need for high LED efficiency can be understood as follows: every successful luminescence event, occurring with a probability equal to the LED external quantum efficiency $\eta_{ext,c}$, liberates $(\hbar\omega_c - qV_c)$ of lattice heat. Meanwhile, every time an injected carrier fails to leave the LED as a photon, occurring with a probability $(1 - \eta_{ext,c})$, the full input energy of qV_c is deposited back into the lattice. To produce a substantial photon flux, the voltage qV_c is necessarily a large fraction of the photon energy, so that the penalty for a failed luminescence can be many times greater than the benefit of a successful luminescence. A high quantum efficiency is required to overcome this imbalance, with a threshold for net cooling (COP ≥ 0) given by $\eta_{ext,c} \geq qV_c / \hbar\omega_c$. To obtain a significant fraction of the Carnot COP, an efficiency well above this threshold must be reached.

The net energy cost of a successful luminescence event in the LED is $(qV_c - qV_h)$, due to the recovery action of the PV cell. A high PV external quantum efficiency $\eta_{ext,h}$ maximizes the PV operating voltage V_h, improving the COP. However, this benefit is manifest only upon successful luminescence

from the LED; an efficient LED design, therefore, is of foremost importance for cooling performance.

III. LED DESIGN

The external quantum efficiency η_{ext} of an LED can be broken down into its component efficiencies:

$$\eta_{ext} = \eta_{inj} \times \eta_{int} \times C_{ext} \qquad (2)$$

where we have defined the injection efficiency η_{inj} as the probability that an injected carrier recombines in the active region, the internal quantum efficiency η_{int} is the probability that an active-region recombination event is radiative, and the photon extraction efficiency C_{ext} is the probability that an internally generated photon will escape the device.

For our analysis, we consider devices based on the *n*-GaInP/*n*-GaAs/*p*-GaInP double heterostructure, shown in Fig. 2. Exceptionally high external photoluminescence efficiencies have previously been demonstrated with these structures, including 96% [7] at room temperature and 99.5% at 100K [8], primarily due to the excellent properties of the interface. The internal quantum efficiency of the active region under low-level injection conditions ($p \ll n \approx N_D$ in the *n*-doped GaAs active region) can be given by:

$$\eta_{int} = \frac{BN_D}{A + BN_D + C_n N_D^2} \qquad (3)$$

where A is the rate of defect-mediated recombination, B is the radiative recombination coefficient, and C_n is the two-electron Auger coefficient in GaAs [9]. The defect recombination rate is given by $A = 1/\tau_{srh} + 2S/d$, where τ_{srh} is the bulk Shockley-Read-Hall (SRH) lifetime, S is the surface recombination velocity, and d is the active region thickness. Bulk lifetimes in GaAs longer than $\tau_{srh} = 21$ μs [8] and $S = 1.5$ cm/s at the GaAs/GaInP interface [10] have previously been measured at room temperature. For a 200 nm *n*-GaAs layer at the optimal doping level of $N_D = 2.5 \times 10^{17}$ cm⁻³, the internal quantum efficiency can reach $\eta_{int} = 99.79\%$ at 300K or $\eta_{int} = 99.83\%$ at 280K (for refrigeration applications), as shown by the dashed curve in Fig. 3. At voltages large enough to produce high-level injection ($n \approx p > N_D$), η_{int} falls due to increasing Auger recombination.

To calculate the injection efficiency η_{inj}, we consider (1) the leakage current arising from bulk SRH recombination in the depletion region of the *p-n* heterojunction, and (2) the leakage of hot carriers out of the active region on the opposite side. The former leakage component increases as $\exp(qV/2kT)$, and dominates the characteristics at low voltage. We calculate this leakage assuming a lifetime of $\tau_{srh} = 0.1$ μs in the *p*-GaInP layer. The latter effect is minimal except at very large biases due to the large confinement potentials of the doped heterostructure.

Once a photon has been generated internally, there are several ways in which it may be lost before exiting the device.

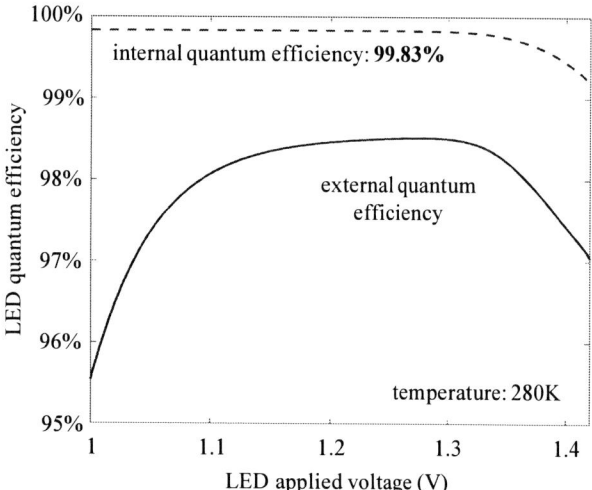

Fig. 3. Internal (dashed) and external (solid) quantum efficiency of the LED in Fig. 2 as a function of the applied voltage, evaluated at 280K. The internal efficiency under low-level injection is labeled. The external efficiency drops steeply at low bias due to SRH recombination in the depletion region and at high bias due to Auger recombination.

It can be re-absorbed in the active region, re-generating an electron-hole pair, which may then dissipate the energy by non-radiative recombination. The photon can also be lost by parasitic absorption at the rear surface, at the Ohmic contacts, and by free carriers. The remaining features of the device structure in Fig. 2 have been optimized for efficient light extraction. The rate of escape through the front surface must be maximized, while the aforementioned mechanisms of loss must be minimized.

Light extraction out of the front surface is greatly enhanced by texturing the surface of the high-index semiconductor film. In a planar film, only a very small fraction of the internal radiation ($1/4n_r^2 \approx 2\%$) is emitted into the escape cone of the front surface. The remaining photons are trapped by total internal reflection and must undergo many absorption/re-emission events (each with a possibility of loss) before entering the escape cone. A surface texture randomizes the angle of the photons with every pass through the film, greatly accelerating their extraction [11]. Since the photons no longer need to rely on re-absorption to be extracted, the active region thickness can be significantly reduced, leading to a lower rate of loss by non-radiative recombination. In addition to the surface texture, we add a Si_3N_4 ($n_r \approx 2$) anti-reflection coating to enhance the escape rate.

An excellent rear reflector is necessary for efficient light extraction. To this end, we propose to use a high-contrast, undoped $Al_{0.2}Ga_{0.8}As/Al_xO_y$ Bragg reflector, which provides near-omnidirectional reflection of photons within the luminescence spectrum. The oxide layers can be formed through the selective wet oxidation of Al-rich AlGaAs layers

grown epitaxially over the LED heterostructure [12]. Furthermore, photons that penetrate the Bragg reflector with a large angle from normal are totally internally reflected by the underlying MgF_2 low-index layer. Finally, an Ag mirror reflects the remaining photons. Averaged over the angle, energy, and polarization of the internal photons, the reflectivity of the rear mirror is $R = 99.990\%$.

Parasitic absorption by the Ohmic contacts can be minimized by using a point contact array with low surface coverage on both sides, and a lateral current-injection scheme within the semiconductor to spread the current uniformly over the device area. Carriers are supplied to the LED from a linear metal grid on the front side, and a metal backplane on the rear side. To fabricate the device, epitaxial growth and rear-side processing can be done on the growth substrate, with front-side processing following the lift-off of the epitaxial film.

The external quantum efficiency of the LED is shown in Fig. 3. An optimal choice of materials and an optimized device design lead to an external quantum efficiency $\eta_{ext} > 98\%$ at 280K for applied biases in the range ~1.1V – 1.35V. The light extraction efficiency is $C_{ext} = 98.5\%$ under low-level injection, with non-radiative recombination as the dominant photon loss mechanism.

IV. OPTIMIZED REFRIGERATION PERFORMANCE

To evaluate the room-temperature cooling properties of the thermophotonic system in Fig. 1, we choose steady-state operating temperatures of $T_c = 280K$ and $T_h = 330K$. We use GaAs as the light-emitting material on both sides. Auspiciously, the PV bandgap is ~20 meV lower than the LED bandgap due to its higher temperature, which aids the transfer of optical energy from cold to hot.

The PV cell is designed to be identical to the LED structure in Fig. 2, but with thicker GaInP cladding layers (2 μm and 10 μm for the n- and p-GaInP layers, respectively) to reduce Ohmic dissipation, though at the cost of increased free-carrier absorption. We also assume an ideal metal mesh filter with a grid spacing of ~10 μm over the PV cell front surface to fully suppress the thermal radiation heat leakage from hot to cold. This non-luminescent mode of heat transfer arises from optical phonon resonances in the III-V and dielectric layers. The Ohmic losses Q_Ω in each device are calculated with a circuit model that includes a distributed network of forward-biased diodes, reverse current sources, spreading resistance in the semiconductor, and the series resistances of the Ohmic contacts and metallization. The Ohmic dissipation in this LED design is relatively large due to the use of a lateral current injection scheme.

To find the most efficient operating point of the system at any given cooling flux, we iterate through a range of LED voltages V_c and for each value, we find the value of the PV voltage V_h that numerically optimizes the COP given in (1).

978-1-5090-5606-4/17 $31.00 © 2017 IEEE

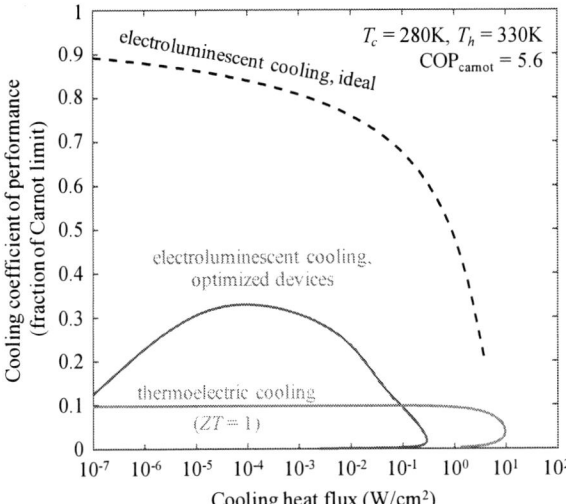

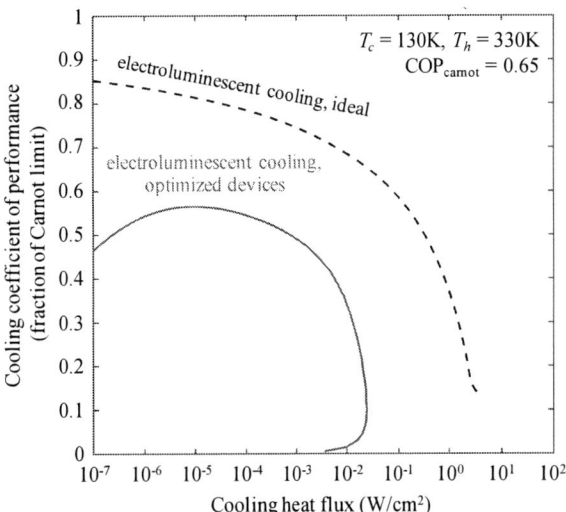

Fig. 4. The refrigeration performance, characterized by the cooling heat flux and the coefficient of the performance, is shown for the electroluminescent cooling system operating near room temperature. The performance of the LED/PV design in Fig. 2 is plotted in blue, and the ideal performance (with $\eta_{\text{ext,c}} = \eta_{\text{ext,h}} = 100\%$ and $Q_{\Omega c} = Q_{\Omega h} = 0$) is represented by the dashed curve. Electroluminescent cooling is 2-3× more efficient than state-of-the-art thermoelectric coolers (red) for fluxes of 1 to 10 mW/cm^2.

Fig. 5. The electroluminescent cooling performance is shown for a cryogenic refrigeration application, for both the optimized system (blue) and the ideal system (dashed). The performance is enhanced relative to room temperature due to an increase in the LED quantum efficiency when operated at low temperatures.

We plot the resulting relationship between the cooling heat flux Q_c and the Carnot-normalized COP in Fig. 4. We also plot the case in which both devices have perfect external luminescence efficiency and zero Ohmic losses.

The electroluminescent cooling COP in the ideal case is largest at low heat fluxes and decreases at high fluxes. This is consistent with our prediction in Section II; at low bias, the external photons carry away a larger amount of heat on average for the same quantum efficiency. For the real, optimized device (blue curve), the COP is limited at low flux by the depletion-region SRH current, which reduces $\eta_{\text{ext,c}}$. In practice, residual heat leakage may also constrain the COP at very low fluxes. At the other extreme, the COP begins to decline more rapidly above $Q_c \approx 10$ mW/cm^2 due to Ohmic dissipation. We also note from Fig. 4 that although the optimized LED design already has an operating regime with near-unity quantum efficiency, large improvements to the COP are still possible with even more ideal devices. The electroluminescent cooling application, therefore, reaps large benefits from asymptotic approaches to material and device perfection.

We compare the performance of the present cooling system with a thermoelectric cooler having a figure-of-merit $ZT = 1$ and realistic contact resistances representative of the best commercially available modules [13]. The thickness of the thermoelectric elements is varied to maximize the COP for a given cooling flux. Thermoelectric cooling can reach higher maximum cooling fluxes than the electroluminescent system, and is the most efficient choice for fluxes greater than about 0.1 W/cm^2. However, the thermoelectric COP at all fluxes is limited to ~10% of the Carnot limit for this choice of temperatures. Notably, the Carnot-normalized COP of the electroluminescent cooling system can be 3.1× what can be achieved with thermoelectrics at a moderate cooling flux of 1 mW/cm^2, and 2.3× at 10 mW/cm^2. The advantage is larger for greater temperature differences and diminishes at smaller temperature differences, and electroluminescent cooling remains the more efficient option at these fluxes down to $\Delta T \approx 15$K.

The quantum efficiency of the LED improves at low temperatures, originating mostly from an increase in η_{int}. Both the defect recombination rate A and the Auger coefficient C_n decrease at reduced temperatures [10, 14], while the radiative coefficient B increases. Taking all temperature dependences into account, η_{int} improves to 99.99% at $T_c = 130$K, resulting in an external quantum efficiency $\eta_{\text{ext}} > 99\%$ for applied biases in the range ~1.3V – 1.5V.

The cooling performance of the same devices in a cryogenic application, with $T_c = 130$K and $T_h = 330$K, is shown in Fig. 5. At these temperatures, a COP that is 49% of Carnot is possible at 1 mW/cm^2, and 34% of Carnot at 10 mW/cm^2, though the maximum heat flux is reduced because of a narrower luminescence spectrum. Thermoelectric coolers are very inefficient at these temperatures. Laser-induced optical refrigeration is also far less efficient than electroluminescent cooling, even with ideal optoelectronic devices [15]. This can be attributed to the fact that anti-Stokes fluorescence cooling is generally limited to about kT of heat extraction per photon, whereas in electroluminescent cooling the heat extraction per photon is equal to $(\hbar\omega - qV)$, which is much greater than kT except under large bias.

V. Conclusion

As LEDs approach the limit of unity external quantum efficiency, electroluminescence becomes a compelling mechanism of cooling. Using the GaAs/GaInP double heterostructure, we design an LED with the requisite efficiency to make this type of refrigeration viable: $\eta_{ext} > 98\%$ near room temperature. At moderate cooling fluxes of 1 to 10 mW/cm^2, an electroluminescent cooling system can be 2-3× more efficient than presently available thermoelectrics near room temperature, while considerably outperforming solid-state laser cooling in cryogenic applications. Nonetheless, the cooling performance can be further improved with the availability of materials having a superior luminescence efficiency.

Acknowledgements

This work was supported by the DOE 'Light-Material Interactions in Energy Conversion' Energy Frontier Research Center under grant DE-SC0001293. T. P. Xiao was supported by the National Science Foundation Graduate Research Fellowship under Grant No. DGE 1106400.

References

[1] O. D. Miller, E. Yablonovitch, and S. R. Kurtz, "Strong internal and external luminescence as solar cells approach the Shockley-Queisser limit," *IEEE Journal of Photovoltaics.*, vol. 2, no. 3, pp. 303–311, 2012.

[2] K. Lehovec, C. A. Accardo, and E. Jamgochian, "Light emission produced by current injected into a green silicon-carbide crystal," *Physical Review*, vol. 89, pp. 20-25, 1953.

[3] P. Behrdal, "Radiant refrigeration by semiconductor diodes, *Journal of Applied Physics*, vol. 58, no. 3, pp. 1369-1374, 1985.

[4] P. Santhanam, D. J. Gray, and R. J. Ram, "Thermoelectrically pumped light-emitting diodes operating above unity efficiency," *Physical Review Letters*, vol. 18, p. 097403, 2012.

[5] N.-P. Harder and M. A. Green, "Thermophotonics," *Semiconductor Science and Technology*, vol. 18, no. 5, pp. S270-278, 2003.

[6] J. Oksanen and J. Tulkki, "Thermophotonic heat pump – theoretical model and numerical simulations," *Journal of Applied Physics*, vol. 107, no. 9, 2010.

[7] H. Gauck, T. H. Gfroerer, M. J. Renn, E. A. Cornell, and K. A. Bertness, "External radiative quantum efficiency of 96% from a GaAs/GaInP heterostructure," *Applied Physics A*, vol. 64, no. 2, pp. 143-147, 1997.

[8] D. A. Bender, J. G. Cederberg, C. Wang, and M. Sheik-Bahae, "Development of high quantum efficiency GaAs/GaInP double heterostructures for laser cooling," *Applied Physics Letters*, vol. 102, no. 25, p. 252102, 2013.

[9] U. Strauss, W. W. Rühle, and K. Köhler, "Auger recombination in intrinsic GaAs," *Applied Physics Letters*, vol. 62, no. 1, pp. 55-57, 1993.

[10] B. Imangholi, M. P. Hasselbeck, M. Sheik-Bahae, R. I. Epstein, and S. Kurtz, "Effects of epitaxial lift-off on interface recombination and laser cooling in GaInP/GaAs heterostructures," *Applied Physics Letters*, vol. 86, no. 8, p. 081104, 2005.

[11] I. Schnitzer, E. Yablonovitch, C. Caneau, T. J. Gmitter, and A. Scherer, "30% external quantum efficiency from surface textured, thin-film light-emitting diodes," *Applied Physics Letters*, vol. 63, no. 16, pp. 2174-2176, 1993.

[12] M. H. MacDougal, H. Zhao, P. D. Dapkus, M. Ziari, and W. H. Steier, "Wide-bandwidth distributed Bragg reflectors using oxide/GaAs multilayers," *Electronics Letters*, vol. 30, pp. 1147-1149, 1994.

[13] D. Zhao and G. Tan, "A review of thermoelectric cooling: Materials, modeling and applications," *Applied Thermal Engineering*, vol. 66, no. 1-2, pp. 15-24, 2014.

[14] M. Takeshima, "Effect of Auger recombination on laser operation in Ga$_{1-x}$Al$_x$As," *Journal of Applied Physics*, vol. 58, no. 10, pp. 3846-3850, 1985.

[15] M. Sheik-Bahae and R. I. Epstein, "Optical refrigeration," *Nature Photonics*, vol. 1, pp. 693-699, 2007.

Multiple quantum wells as slowed hot carrier cooling absorbers in hot carrier cells

Gavin Conibeer, Yi Zhang, Simon Chung, Yuaxun Liao, Stephen Bremner, Santosh Shrestha

School of Photovoltaic and Renewable Energy Engineering, UNSW Sydney, NSW 2052, Australia

Abstract — The Hot Carrier solar cell has the potential to yield a very high efficiency, well over 50% under 1 sun. Multiple quantum wells have been shown to have significantly slower hot carrier cooling rates than bulk material [1,2] and are thus a promising candidate for hot carrier solar cell absorbers. However, the mechanism(s) by which hot carrier cooling is restricted is not clear. Presented is a systematic study of carrier cooling rates in GaAs/AlAs MQW with either varying barrier or varying well thickness. These allow a determination as to whether the mechanisms of either a reduction in hot carrier diffusion; a localisation of phonons emitted by hot carriers; or mini-gaps in the MQW phonon dispersion are responsible for reduced carrier cooling rates.

Initial devices fabricated using MQW as hot carrier absorbers indicate promising photovoltaic performance which result from collection of hot carriers.

I. INTRODUCTION

The Hot Carrier solar cell aims to tackle the carrier thermalisation loss after absorption of above band-gap photons. It has the potential to achieve very high efficiencies in a device that is essentially a single junction. Detailed balance calculations indicate limiting efficiencies as high as 65% under 1 sun and 85% under maximum concentration [1].

The key property for a hot carrier absorber is to slow the rate of carrier cooling from the picosecond timescale to at least 100s of ps, but preferably ns to be similar to the rate of radiative recombination. Hot carriers cool primarily by emission of LO phonons.

Low dimensional multiple quantum well (MQW) systems have been shown to have lower carrier cooling rates. Comparison of bulk and MQW materials has shown significantly slower carrier cooling in the latter. Bulk GaAs as compared to MQW GaAs/AlGaAs materials measured using time resolved transient absorption by Rosenwaks [2]. This shows that the carriers stay hotter for significantly longer times in the MQW samples, particularly at the higher injection levels by 1½ orders of magnitude. This is due to an enhanced 'phonon bottleneck' in the MQWs allowing the threshold intensity at which a certain ratio of LO phonon re-absorption to emission is reached which allows maintenance of a hot carrier population to be reached at a much lower illumination level. More recent work on strain balanced InGaAs/GaAsP MQWs by Hirst has also shown carrier temperatures significantly above ambient, as measured by PL [3]. Importantly increase in In content to make the wells deeper and to reduce the degree of confinement is seen to increase the effective carrier temperatures.

The mechanism for this slowed cooling in MQWs is not clear. The Hirst result of slowing cooling in wider wells indicates that slowed carrier cooling is not a direct result of electronic quantum confinement in the wells. Other possibilities include (1) the restriction of hot carrier diffusion in the direction perpendicular to the wells which locally increases the hot carrier population and leads to an onset of 'phonon bottleneck' at a lower illumination intensity; (2) reflection of phonons from the QW/barrier interfaces, effectively confining phonons in the wells and leading to an early onset of phonon bottleneck; (3) a coherent folding of the phonon band structure in the periodic QW system, resulting in mini-gaps in the phonon DOS, which for specific energies can block decay of optical to acoustic phonons and again lead to a phonon bottleneck in the optical phonon population leading to fewer opportunities for carriers to lose their energy.

In the current study, GaAs/AlAs MQW samples grown by MBE are used to comprehensively investigate these mechanisms behind the reduction of carrier cooling. A series of samples in which the well thickness is varied with constant barrier thickness is compared with a series in which the barrier thickness is varied at constant well thickness. Photoluminescence and XRD data are presented which indicate reasonably uniform material quality across the series. Time resolved photoluminescence, utilising time correlated single photon counting, is used to measure the carrier temperature with time after excitation and hence indicate carrier cooling rates. Comparison of the trends in the various series of samples is used to elucidate some insight into the nature of the reduced carrier cooling mechanisms.

The MQW samples with the highest carrier temperatures and longest carrier lifetimes have been fabricated into simple PV devices by addition of contacts and measured by illuminated I-V, indicating promising photovoltaic performance.

II. MQW GROWTH AND MATERIAL QUALITY

Multiple quantum well (MQW) samples with GaAs wells (L_W) and AlAs barriers (L_B) were grown by Molecular Beam Epitaxy (MBE) on a Veeco Gen930 system. Arsenic flux supplied as a dimer using a valved cracker source, with Al and Ga fluxes supplied by thermal sources.

The growth temperature for all layers was 620 °C as monitored by pyrometer, with the temperature calibrated by observing thermal de-oxidation of the GaAs by Reflective High Energy Electron Diffraction (RHEED). The same number of quantum wells was grown in each case, at 30 wells and 31 barrier layers, but the thickness of wells and/or barriers was varied as indicated in Table 1, hence the total thickness of each sample also varied.

TABLE 1. MBE MQWs SAMPLES, WITH INTENDED AND MEASURED SUPERLATTICE THICKNESSES.

Sample	$D = L_W + L_B$ (nm)		Difference (nm)
	Designed	XRD	
1	6+40=46	45.5	0.5
2	8+40=48	46.8	1.2
3	12+40=52	51.6	0.4
4	30+40=70	69.7	0.3
5	30+5=35	34.4	0.6
6	30+2=32	31.3	0.7

XRD, TEM and PL data all indicate that all the samples were of good quality with high crystallinity and L_W and L_B thicknesses close to those designed, with the small exception of sample 2 with $L_W=8$nm, which had about 1nm difference between XRD and intended dimensions.

III. MEASUREMENT OF CARRIER COOLING RATES WITH TIME RESOLVED PHOTOLUMINESCENCE

Absorption of high energy photons produces electron-hole pairs well above the band gap energy. These high energy carriers tend to lose their energy by thermalisation by emission of optical phonons over time. However, they also have a finite chance of spontaneously recombining at any time with an emission energy related to the carrier energy at that specific time. The distribution of these emission energies at a specific time after excitation is therefore also a snapshot of the carrier energy distribution at that time. Hence, if this is approximated to a hot thermal distribution, the carrier temperature with time and the lifetime of hot carriers at these energies can be calculated as a function of time.

Time resolved photoluminescence (tr-PL) was carried out to investigate the carrier dynamics on a nanosecond timescale, where the time resolution is 66ps and excitation wavelength is 640nm. (This excitation energy is chosen to be between the GaAs and AlAs bandgaps and hence only absorbed in the GaAs wells.) The tr-PL intensity is the spectrally integrated PL for a 20nm spectral width around the corresponding peak position.

Figure 1 shows the intensity of the integrated tr-PL signal as a function of both emitted photon wavelength and time since excitation in ns, for the series of samples with L_W increasing from 2, 8, 12 to 30nm and with common L_B of 40nm.

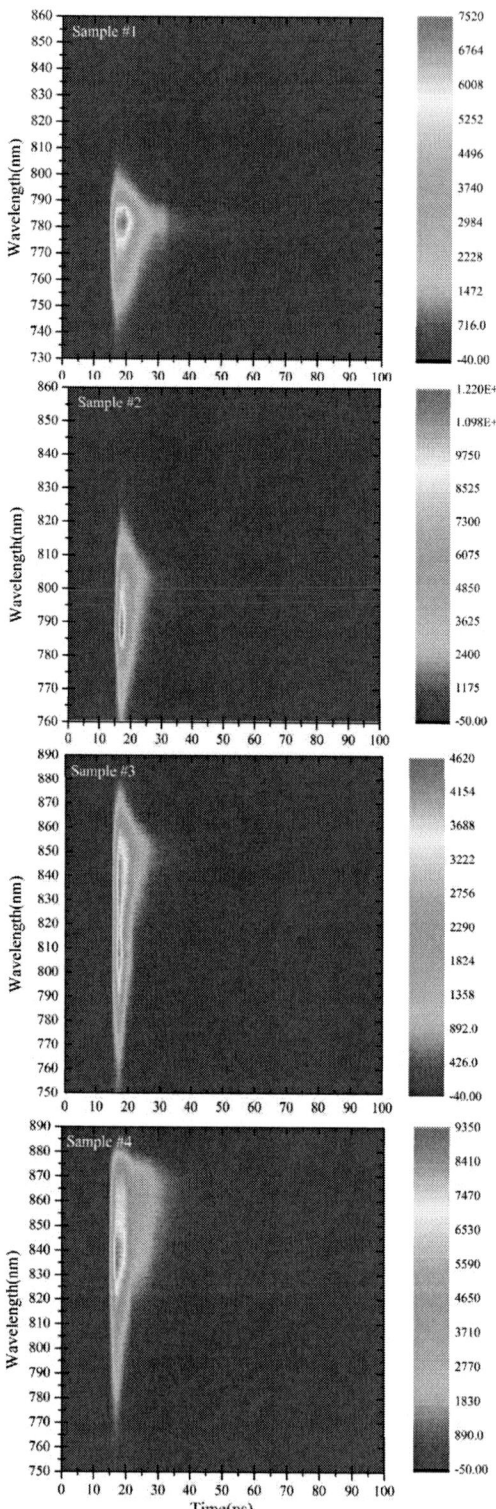

Fig. 1. tr-PL intensity as a function of wavelength and of time for samples 1 to 4 with the same L_B but increasing L_W.

Figure 2 shows tr-PL for the series of samples with L_B decreasing from 40, 5 to 2nm and with common L_W of 30nm.

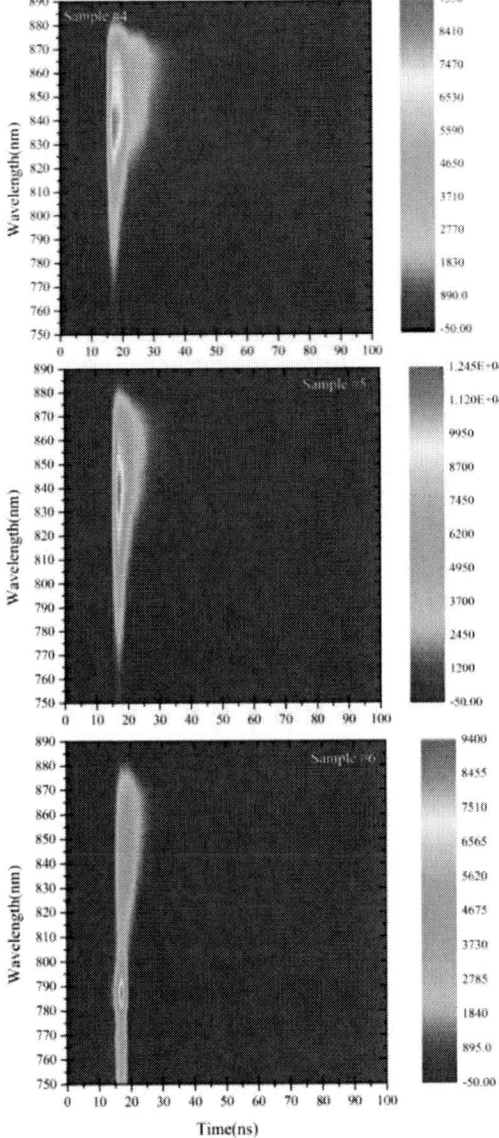

Fig. 2. tr-PL intensity as a function of wavelength and of time for samples 4,5,6 with the same L_W but decreasing L_B.

Vertical slices through the TRPL plots are used to generate full spectrum tr-PL curves at specific time delays for each sample to find the linear carrier cooling region within the high energy tail. An exponential fit to this linear portion then gives the time-dependent carrier temperature. The carrier temperatures from these 'high energy tail fittings' are plotted as a function of time delay for each of the samples with different QW thickness in Figure 3 and for those with different barrier thickness in Figure 4. A single exponential fit to each of these curves then also gives the carrier temperature and decay time constant for each QW / barrier thickness. [5]

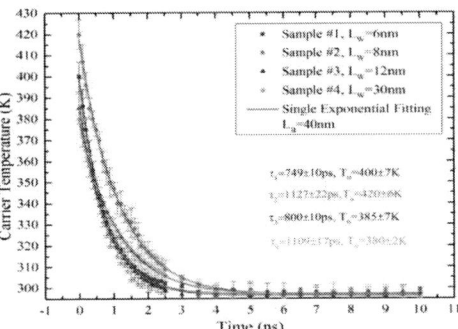

Fig. 3. Carrier temperature evolution from 'high energy tail fitting' for samples 1 to 4 with increasing QW L_W, but fixed L_B of 40nm.

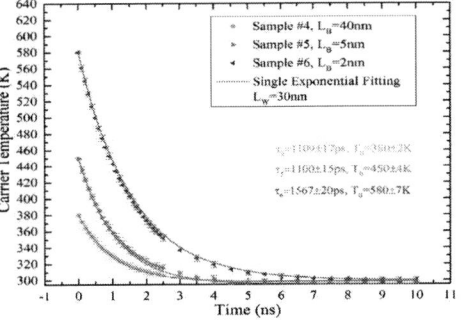

Fig. 4. Carrier temperature evolution from 'high energy tail fitting' for samples 4,5,6 with decreasing QW L_B, but fixed L_W of 30nm.

IV. COMPARISON OF CARRIER TEMPERATURE AND CARRIER LIFETIME RESULTS

The results in Figures 3 and 4 all show thermalization times of over 700 picoseconds, which is much longer than that of bulk materials. This is consistent with the long times seen in the other work on MQW carrier temperatures in [2,3]. The carrier temperature and decay time constant data from Figures 3 and 4 are plotted in Figure 5 as a function of the QW or barrier thickness.

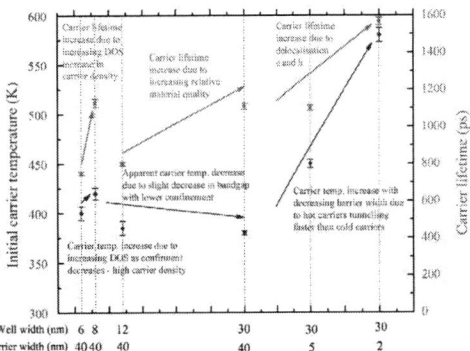

Fig. 5. Comparison of initial carrier temperatures (blue diamonds) and hot carrier lifetimes (red squares) for varying well or barrier thicknesses.

978-1-5090-5606-4/17 $31.00 © 2017 IEEE

V. HOT CARRIER COOLING IN MQWs

The XRD and TEM results are consistent with each other and indicate that these samples fabricated by MBE have relatively good crystal quality and layer thickness uniformity and that the layer thicknesses are close to the designed values. High crystal quality of the material is vital for a hot carrier absorber. This is because dislocations, crystal aggregation or strain-induced defects caused by poor crystal quality will significantly affect optical properties and make analysis of hot carrier dynamics much more difficult. CWPL is also consistent with good quality QW material and show a red shift in PL with well thickness and hence also with reducing quantum confinement. Hence these MQW samples are suitable for time resolved PL measurements and temperature fitting to indicate carrier cooling rates.

With reference to Figure 5 primarily, these initial tr-PL measurements do indeed indicate long hot carrier lifetimes in these MQWs, consistent with significantly slower hot carrier cooling rates in MQWs.

Increasing well thickness, constant barrier thickness: In considering the samples with increasing QW width and fixed barrier width of 40nm in the left half of Figure 5, there is a strong increase in both carrier temperature and hot carrier lifetime for QW widths from 6nm to 8nm. However, there is a decrease in carrier temperature for subsequent 12nm and 30nm QW samples, but a general increase in hot carrier lifetime for this same series. A potential explanation for this non-monotonic behaviour relies on the change of DOS as quantum confinement decreases for thin wells and on both a decrease in confined energy levels and an increase in effective material quality, as well thickness increases, for thicker wells.

For the increase in well thickness from 6nm to 8nm, there is a very strong decrease in quantum confinement, because this decreases exponentially with the thickness. This results in a strong increase in the DOS in the wells. As this is non-linear with well thickness it also leads to an increase in the carrier density for a given illumination intensity and consequent greater emission of optical phonons by hot carriers and hence earlier onset of a phonon bottleneck (in which optical phonons are restricted from decaying into acoustic phonons - the Klemens' mechanism - and are therefore re-absorbed by hot carriers) and hence the slowed carrier cooling rate leading to both higher carrier temperature and longer hot carrier lifetime. For thicker QWs of 12nm and 30nm there is a further reduction in quantum confinement, but again as this change is exponential, it is already quite small and so does not result in any appreciable increase in DOS in these thicker wells, although it does still have a small effect on reducing the ground state confined energy level. This results in very little change to the non-equilibrium hot carrier population but the slightly lower ground state means this population is shifted to slightly lower energies, thus giving an apparent decrease in carrier temperature.

However, the increase in QW width also increases the ratio of high quality bulk-like GaAs material to lower quality QW/barrier interface material. The radiative recombination efficiency will depend on the quality of the total GaAs thickness, so an increase in well width is likely to increase the radiative efficiency and give the increase in hot carrier radiative lifetime observed.

Decreasing barrier thickness, constant well thickness: Considering now the samples with fixed well width of 30nm and decreasing barrier width in the right half of Figure 5, there is a very strong increase in carrier temperature as barrier width decreases. However, there is no change in the hot carrier lifetime going from 40nm to 5nm barriers, but then a strong increase in this lifetime for the sample with the thinnest 2nm barriers. Because the barriers become thinner these effects are likely to be dominated by changes in the diffusion of hot carriers, differences in effective mass and hence mobility of holes and electrons and by the fact that initial excitation gives rise to the highest carrier population in the top few QWs.

More specifically, initial excitation of carriers in the QWs gives hot carrier populations isolated within each well, with the greatest excitation occurring in the top few wells dependent on the absorption coefficient of GaAs. As time progresses after the initial excitation, tunnelling diffusion of carriers occurs from the more highly populated upper wells into lower wells. This tunnelling effect will be exponentially dependent on the barrier thickness and much greater for the thinner barriers.

Because of their much smaller effective mass (0.03 cf. about 1 in GaAs), electrons will tunnel much faster than holes through a given barrier. At later measurement delay times this will tend to give a separation of charge carriers with electrons diffusing through to lower wells but holes remaining almost immobile in upper wells. This carrier separation (the Dember effect enhanced by tunnelling) delocalises electrons from holes in the depth direction and leads to a suppression of radiative recombination.[28] For an assumed constant non-radiative recombination probability, this will lead to a longer observed hot carrier or thermalisation lifetime in the thinnest barrier sample.

However, hot electrons will tunnel significantly faster than cold electrons through a given barrier because they are closer to the top of the barrier and hence their tunnelling probability is exponentially enhanced. This difference in hot/cold electron diffusion will be strongest for *thicker* barriers, such that for the 40nm barrier cold electrons will have negligible tunnelling probability but hot electrons will tunnel relatively fast into lower wells thus both cooling the population in the upper few wells and reducing the carrier density of these hot carriers as they diffuse to lower wells such that they no longer exhibit phonon bottleneck and cool relatively quickly. This therefore gives the observed relatively low carrier temperature with the thickest barriers, increasing to progressively higher temperatures as the barrier thickness decreases first to 5nm and then to 2nm.

This hot carrier diffusion effect to some extent counteracts the charge carrier separation effect on hot carrier lifetime such that there is almost no change in this lifetime in going from 40nm to 5nm barriers (even though there is the increase in carrier temperature because the reducing difference in hot/cold electron diffusion is little affected by the carrier delocalisation) but the large increase in both hot carrier lifetime and carrier temperature on going from 5nm to 2nm barriers (as the delocalisation becomes stronger but the indirectly connected differential hot/cold electron diffusion weaker).

Overall effect of MQWs on hot carrier cooling: There is clearly further work that needs to be carried out to confirm (or deny) these rather qualitative explanations of MQW hot carrier behaviour, but what is certainly very clear is that thin barriers and thick wells in the MQWs give both high carrier temperatures, up to 300K above ambient, and long hot carrier lifetimes, up to 1.5ns. This bodes very well for the incorporation of such MQW structures as hot carrier cell absorbers. The necessary long lived hot carrier properties appear to be approached in structures which will also facilitate good transport of hot carriers to the surfaces and hence to energy selective contacts for extraction of high energy and hence high voltage charge carriers. The role of the multiple interfaces in GaAs material that is close to being bulk-like would appear to be important. This would be very consistent with the very high figures of merit seen in thermoelectric MQW structures, in which electron transport is little impeded by thin low energy barriers, but phonon scattering is strongly enhanced by these same barriers which have a strong mismatch in acoustic impedance which strongly reflects and localises phonons. Work on comparing different numbers of QW interfaces in MQWs with thin widely spaced barriers will be carried out.

VI. HOT CARRIER DEVICES FROM MQWS

Initial devices have been fabricated from the sample which shows the highest carrier temperatures and lifetimes. This is sample 6 with the thinnest barriers at 2nm and 30nm QWs. Devices are made by putting mesa contacts and the front surface and contacting to the GaAs substrate. The illuminated I-V of this device at 0.8 suns is shown in Figure 6.

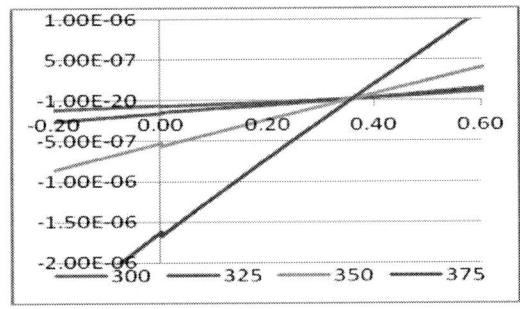

Fig. 6. Illuminated I-V at 0.8 suns for sample 6, 30nm QWs / 2nm barriers at a range of temperatures in K.

There is clear power generation in the fourth quadrant at all temperatures. With varying temperature there is a clear trend for the short circuit current to increase with temperature. This provides evidence thermal carriers are being collected as the current increases with thermally

excited carriers while maintaining the same incidence illumination. In the dark the slope does not increase as drastically showing the thermally excited carriers cannot alone lead to the large increase in power generation at higher temperatures. The fill factor is bad at about 25% for all temperatures. The behaviour with temperature is anomalous for a normal photovoltaic device and suggests behaviour similar to a thermoelectric device. The V_{OC} at 0.35V is quite reasonable but in itself does not prove hot carrier collection in a GaAs based device. (This would require V_{OC} close to or greater than the band gap.) However, with the TRPL evidence of hot carriers with long lifetimes and the unusual temperature behaviour for a photovoltaic cell, this is a very strong indication of true hot carrier collection in a hot carrier solar cell.

VII. CONCLUSIONS AND FURTHER WORK

Multiple quantum well (MQW) samples in the GaAs/AlAs system either with varying QW thickness or with varying barrier thickness, have been used to investigate the mechanisms responsible for high carrier temperatures and long hot carrier lifetimes in MQWs. XRD, TEM and PL all indicate good quality material with dimensions close to those designed.

Time resolved PL indicates that in general thicker QWs give lower carrier temperature but longer hot carrier lifetimes. But there is complexity in the onset of phonon bottleneck leading to reduced carrier cooling rate as the thickness of QWs increases. This would seem to occur at lower illumination intensities as the QW thickness increases for small thicknesses because of a rapid decrease in quantum confinement. But for thicker quantum wells quantum confinement is less important and material quality is probably dominant leading to longer hot carrier lifetimes for higher quality thicker QWs.

For variation in barrier thickness, thinner barriers are seen to give both higher carrier temperatures and longer hot carrier lifetimes with temperatures up to 300 degrees above ambient at lifetimes of approx. 1.5ns, due to differential tunneling through barriers for electrons and holes and for hot and cold carriers leading to carrier separation and suppression of recombination to give longer hot carrier lifetimes.

Promising devices showing power generation have been fabricated from these MQW samples. These will be optimized with improved contacting and selection of MQWs with better hot carrier cooling properties.

REFERENCES

[1] M.A. Green, *Third Generation Photovoltaics*, Springer-Verlag, 2003.
[2] Y. Rosenwaks, M. Hanna, D. Levi, D. Szmyd, R. Ahrenkiel, A. Nozik, *Phys Rev B*, vol. 48, pp. 14675-14678, 1993.
[3] L.C. Hirst, M. Fürher, D.J. Farrell, A. Le Bris, J.-F. Guillemoles, M.J.Y. Tayebjee, R. Clady, T.W. Schmidt, M. Sugiyama, Y. Wang, H. Fujii, N.J. Ekins-Daukes, *Proc. of SPIE*, v 8256, p 82560X, 2012.
[4] Y. Zhang, PhD thesis, UNSW 2016, "Study on III-V Materials for Hot Carrier Solar Cell Absorbers"

978-1-5090-5606-4/17 $31.00 © 2017 IEEE

Quantitative optoelectronic measurements of carrier thermodynamics properties in quantum well hot carrier solar cell

Dac-Trung Nguyen[1,2], Laurent Lombez[1,2], François Gibelli[2], Soline Boyer-Richard[1,3], Alain Le Corre[1,3], Olivier Durand[1,3] and Jean-François Guillemoles[1,2]

1 – Institut photovoltaïque d'Île-de-France (IPVF), 92160 Antony, France.

2 – Institut de Recherche et Développement sur l'Energie Photovoltaïque (IRDEP), UMR 7174 CNRS-EDF-Chimie ParisTech, 78400 Chatou, France.

3 – FOTON-OHM, UMR 6082 CNRS-INSA, 35708 Rennes, France.

Abstract —We investigated a semiconductor heterostructure based on InGaAsP multi quantum wells using optical and electrical characterizations in the scope of hot carrier solar cell device. The potential of the investigated quantum well structure to overpass the Schockley Queisser limit is discussed. Population density, temperature and quasi-Fermi level splitting of photo-generated carriers are investigated by fitting the full luminescence spectra using generalized Planck's law. A proper optical study is realized thanks to a detailed description of the absorption of excitons and free carriers in the quantum well. Optical measurements are compared to electrical measurements where the open circuit voltage electrically measured is higher than the minimum absorption threshold. To probe the hot carrier effect in such measurements we look at the changes in thermodynamic properties of carriers in the quantum well and in the barriers when changing the excitation power and the electrical bias.

Index Hot Carriers, characterization, quasi-Fermi level splitting, carrier temperature, quantum well

I. Introduction

The third generation solar cells is defined by their potential to overcome the Shockley–Queisser (SQ) limit of a single junction enabling a lower material usage and/or installation cost share. Among the proposed photovoltaic (PV) devices, hot-carrier solar cells (HCSCs), aim to reduce the two major losses in a classical PV cells (i) the non-absorption of the incoming light by reducing the absorber band gap and (ii) the thermalization process [1,4]. To this purpose, quantum well structures have shown a slower carrier cooling rate [2].
To evidence hot carriers populations, photoluminescence spectra have been analyzed and, using the generalized Planck's law, one may extract the carrier temperature when looking at the high energy tail. However this method could be inaccurate when dealing with nanostructures and high excitation fluxes. In fact, absorption is influenced by several phenomena such as the presence of exciton, free carriers and band filling. A model of the absorption allows us to fit the whole PL spectrum at any excitation power.

We therefore present photoluminescence study of InP/InGaAsP single quantum well p-i-n junction to probe the potential of this structure in the scope of HCSCs. Quantitative optical measurements indicate the presence of a hot carrier population within the quantum well (QW). This gives an extra gain of the electrochemical potential in the barrier. Such effect is analog to a Seebeck conversion process [6]. The quantitative optical measurements are then compared to the electrical characteristics. We see that the Voc overpass the absorption threshold of the QW and could potentially indicate an electrical working condition above the SQ limit. Using the method we present here the real impact of the hot carrier population within the structure can be discussed.

II. Sample and setup description

The investigated sample is based on quaternary $In_xGa_{1-x}As_yP_{1-y}$, chosen for lattice-matching with the InP substrate, and optimized from the results of our previous work [6]. It optically showed the evidence of hot carrier effects in a quantum well/barriers absorber. The multilayer wafer contains principally an intrinsic InGaAsP-based quantum well/barriers region ($In_{0.78}Ga_{0.22}As_{0.81}P_{0.19}/In_{0.8}Ga_{0.2}As_{0.435}P_{0.565}$), playing the role of absorber which generates charge carriers. This absorber is sandwiched between two gradually n- and p-doped InP layers for carriers separation and collection. The quantum well has been designed to minimize the number of energy levels in the QW and to spectrally separate absorption of the quantum well and the barriers, so as to simplify the spectral analysis.
The current-voltage characteristics under 1-sun illumination, is firstly done. Cells of several diameters from 5 to 200µm are characterized. Using an equivalent one-diode model, the ideality factor obtained from dark IV measurements is about 1.3. The results for the 11 µm diameter cell shows the best efficiency under laser excitation. The power conversion efficiencies is 11% under focused laser illumination equivalent to 15 000 Suns. Although we measure efficiency of potential HCSC, the

purpose of the design is rather to allow proper opto-electrical measurements. The good electrical behavior comes from the short-circuit current that is proportional to the laser power in the entire experimental range shown here (see Fig 1). Remarkably the open circuit voltage exceeds the lowest absorption threshold and would indicate an electrical behavior above the SQ limit. Before going to this conclusion a careful investigation is needed.

The setup we use is a hyperspectral imaging system that records quantitative luminescence spectrum (luminance). This allows accessing the thermodynamics properties of the emission that is related to the carrier thermodynamics properties. The excitation wavelength is 980nm. All measurements are carried out at room temperature.

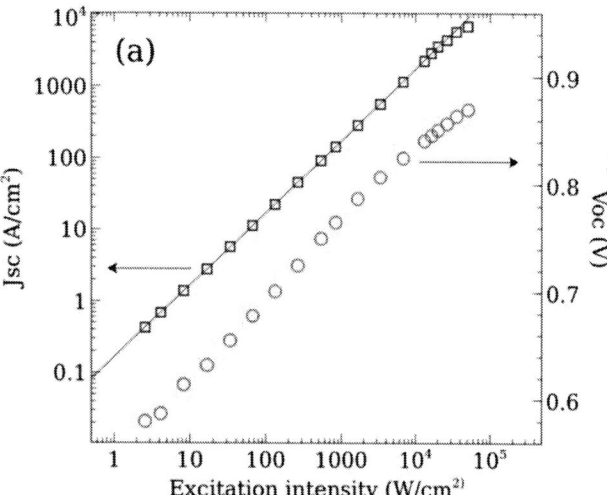

Figure 1 : Electrical properties for different laser illumination intensities. (a) : Variation of short-circuit current density (Jsc, blue squares) and open-circuit voltage (Voc, red circles) with laser power.

III. Modeling luminescence spectra

We record photoluminescence spectra and determine the carrier temperature as well as the quasi Fermi levels splitting according to the generalized Planck's law [3]:

$$\Phi(\hbar\omega, r) = A(\hbar\omega, r)\frac{1}{4\pi^2\hbar^3 c_0^2}(\hbar\omega)^2 \frac{1}{\exp\left(\dfrac{\hbar\omega - \Delta\mu\,(r)}{kT}\right) - 1}$$

Φ is the luminescence emission, $\hbar\omega$ the photon energy, A the absorptivity , $\hbar$ the reduced Planck's constant, c_0 the speed of light in vacuum, $\Delta\mu$ the quasi-Fermi level splitting and T the temperature.
Expression of the absorptivity A(E) is crucial in order to minimize errors in the evaluation of T and $\Delta\mu$. Absorptivity is calculated from absorption coefficients and thicknesses. For our well/barriers structure:

$$A = 1 - \exp\left[-(\alpha_w \cdot d_w + \alpha_b \cdot d_b)\right]$$

Where α_w and α_b are absorption coefficients of well and barriers, d_w and d_b their thicknesses. We used an absorption model taking into account excitons and free carriers in the well and free carriers in the barriers [5] :

$$\alpha_{w0}(E) = A_x \cdot \exp\left[-\frac{(E - E_x)^2}{2\Gamma_x^2}\right]$$

$$+ A_1 \cdot \frac{1}{1 + \exp\left(-\dfrac{E - E_1}{\Gamma_1}\right)} \cdot \frac{2}{1 + \exp\left(-2\pi\sqrt{\dfrac{R_y}{|E - E_c|}}\right)}$$

$$+ A_2 \cdot \frac{1}{1 + \exp\left(-\dfrac{E - E_2}{\Gamma_2}\right)}$$

for the well, and

$$\alpha_{b0}(E) = A_b \cdot \frac{1}{1 + \exp\left(-\dfrac{E - E_b}{\Gamma_b}\right)}$$

for the barriers.

Moreover, with increasing excitation intensities, electrons and holes progressively fill energy bands, resulting in a reduction of absorption coefficient (photo-bleaching). This band filling effect is also expressed by introducing occupation functions into the absorption coefficient (expression not shown here).

Here the carrier temperature and electro-chemical potential are taken separately for the well and barriers. Inclusion of the spectral dependence of the absorptivity and the band filling minimizes the uncertainty in the evaluation of T and thus $\Delta\mu$ [5].

Experimentally we record the luminescence spectra for different laser excitation power and different applied voltage. In both cases the full spectrum fit quality is good as one can see it Fig 2 for three different excitation power. .

IV. Opto-electrical measurements

Form the luminescence spectra we are then able to probe the carrier thermodynamics properties in both spectral regions as a function of the two experimental conditions. We also measure the carrier temperature as a function of the excitation power and electrical bias respectively. In both cases we observe a temperature increase above the room temperature for carrier in the QW. The power excitation dependence behavior is shown in Fig. 3.

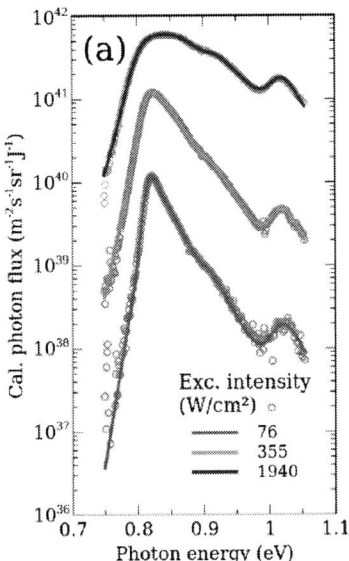

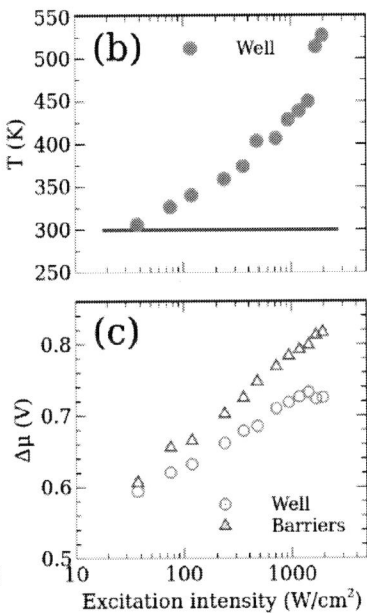

Fig. 2 : Example of 3 PL spectra and fit at low (green line), medium (orange line) and strong (red line) laser excitation.

Fig. 3 (b) and (c): Variation of carrier temperature T and electrochemical potential $\Delta\mu$ in quantum well (orange dots) and barriers (green triangles) with laser power.

The extra heat is known to be possibly converted into an extra voltage via a pseudo-Seebeck effect [6]. This is confirmed when looking at the electrochemical potential of the carriers in both spectral region as the difference between two values increase with the carrier temperature in the well. The value of $\Delta\mu$ in the barrier region is close to the lowest absorption threshold of the QW (0.8eV) approving the effect electrically observed.

The device presented here is also electrically investigated (details published elsewhere). Luminescence experiments under bias shows $\Delta\mu$ value in the barrier close to the voltage as the carriers in this spectral region are supposed to be cold (fast thermalization). We believe this experiment to be a path towards the quantification of voltage and current increase due to hot carrier effect. This should confirm the electrical behavior we observed.

V. Conclusion

We have investigated p-i-n single quantum well device by means of complementary optical and electrical measurements. A rigorous fit of the complete luminescence spectrum allows the access to carrier temperature and electrochemical potential in different spectral regions. The variation of thermodynamic properties as a function of optical excitation power or electrical bias is measured. Optical and electrical behavior would indicate a working condition above the SQ limit and needs to be confirmed.

We are working a new device generation which aims to enhance the hot carrier footprint, both optically and electrically.

References

[1] R.T. Ross, A.J. Nozik, J. Appl. Phys., 53, 3813, (1982).

[2] Y. Rosenwaks, et al. Phys. Rev.B, 48, 14675, (1993).

[3] P. Würfel, Sol. Energy Mats. and Sol. Cells., 46, (1997).

[4] J. A. R. Dimmock, et al. Prog. Photovolt: Res. Appl., 22, 151–160, (2014). J. A. R. Dimmock, et al J. Opt., 18, 074003, (2016).

[5] F. Gibelli, L. Lombez, and J.-F. Guillemoles, J. Phys.: Condens. Matter, 29, 06LT02, (2017).

[6] J. Rodière, L. Lombez, A. L. Corre, O. Durand, and J.-F. Guillemoles, Applied Physics Letters,106, 183901, (2015)

This work was carried out in the framework of a project of IPVF (Institut Photovoltaïque d'Ile-de-France). This project has been supported by the French Government in the frame of the program of investment for the future (Programme d'Investissement d'Avenir – ANR-IEED-002-01)

Absorption Enhancement in InGaAsP/InGaP Quantum Well Solar Cells

Islam E. H. Sayed[1,2], Nikhil Jain[2], Myles A. Steiner[2], John F. Geisz[2], Salah M. Bedair[1]

[1]Department of Electrical and Computer Engineering, North Carolina State University, Raleigh, NC 27695, U.S.A.

[2]National Renewable Energy Laboratory, Golden, CO 80401, U.S.A

Abstract — **InGaAsP/InGaP quantum well (QW) structure is a potential candidate for subcells in next-generation multijunction solar cells because of its tunable bandgap (1.5-1.8 eV). However, the insufficient light absorption in the QWs has previously limited the sub-bandgap quantum efficiency (QE) to less than 25%. We report on the development of InGaAsP/InGaP superlattice solar cell with improved sub-bandgap QE exceeding 75% and bandgap-voltage offset of 0.40 V. The enhancements were accomplished by reducing the background doping in the QW region, monitoring the stress evolution during growth, modifying the QW design to grow thicker wells and processing the devices with optical reflector to enhance light absorption.**

Index Terms — **bandgap engineering, InGaP, quantum wells, III-V, solar cells.**

I. INTRODUCTION

III-V multi-junction solar cells are the most efficient photovoltaic technology today with efficiencies (η) exceeding 45.0% and 30% at high solar concentration and one-sun, respectively. Lattice-matched InGaP/Ga(In)As/Ge and inverted-metamorphic (IMM) InGaP/InGaAs/GaAs triple-junction designs have reached efficiencies of 44.0% and 44.4%, respectively, under concentration [1, 2]. Increasing the number of junctions to four has increased the η to reach ~46% using wafer-bonded structures [3, 4] and IMM structures [5]. Increasing the number of subcells to five and six is a promising route to increase the η beyond 50%, motivating more design flexibility in achieving the optimal bandgaps for subcells used in these two designs.

Strain balanced Quantum wells (QWs), or superlattices, is one approach for bandgap tuning that may offer more design flexibility in developing subcells with optimal bandgaps for next generation solar cells. InGaAs/GaAsP QW structures has been employed to tune the bandgap of GaAs-based cells to 1.2 eV in triple junction InGaP/GaAs/Ge solar cells [6]. A recently proposed QW structure is InGaAsP/InGaP grown in the unintentionally doped *i* layer of $In_{0.49}Ga_{0.51}P$ p-i-n solar cell [7]. This QW structure is promising for use in five and six junction devices due to its wide and tunable bandgap (1.5-1.8 eV) and being aluminum-free. However, this structure suffers from inefficient light absorption, which has limited the external quantum efficiency (EQE) beyond the band-edge of $In_{0.49}Ga_{0.51}P$ (hereafter, sub-bandgap EQE) to less than 30%. The inefficient light absorption can be attributed to: (i) <u>strain balance limitation</u> that has put constraints on the thickness of the absorbing InGaAsP well (t_w) to be less than 30% of the total

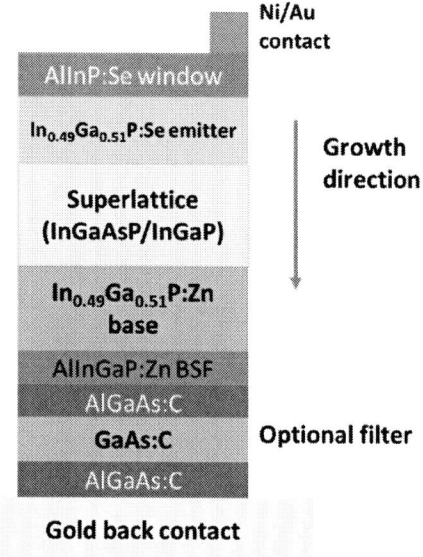

Figure 1: Schematic of InGaAsP/InGaP superlattice solar cell structure grown on GaAs substrate. Gold back metal is electroplated during processing to enhance light absorption in QWs and the substrate was removed using wet-etching. The samples are grown with optional GaAs filter to simulate the optical environment without a back reflector.

thickness of each period, and the (ii) <u>number of quantum wells is limited to 30</u> due to high carbon background doping of $4 \times 10^{15} cm^{-3}$ which corresponds to a depletion region thickness of about 0.7 μm [8, 9].

In this work, we implement four strategies to improve the light absorption in InGaAsP/InGaP QWs: (i) growing the quantum wells lattice matched rather than strain-balanced to increase the thickness of the absorbing InGaAsP quantum well relative to that of the barrier (t_b), (ii) using a back metal reflector to double the optical path length, (iii) using triethylgallium (TEGa) in InGaAsP, arsine (AsH₃), and phosphine (PH₃) to reduce carbon background doping, (iv) monitoring the stress evolution during the growth using multi-beam optical sensor (MOS) to obtain almost strain-free structures and increase the number of quantum wells that can be included in the *i* layer to one-hundred. These approaches have resulted in major enhancement in sub-bandgap EQE exceeding 75%, which is at least three-fold higher than previously strain-balanced approaches [7]. We also compare

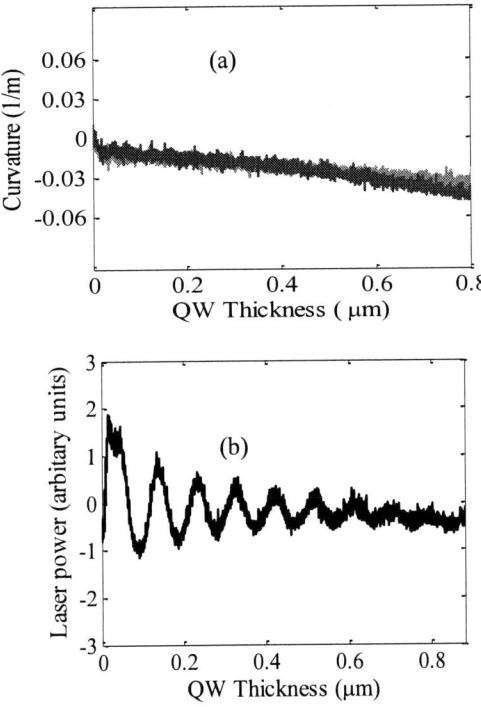

Figure 2: The two-dimensional curvature and the output laser power of 100 period QW device (MP197).

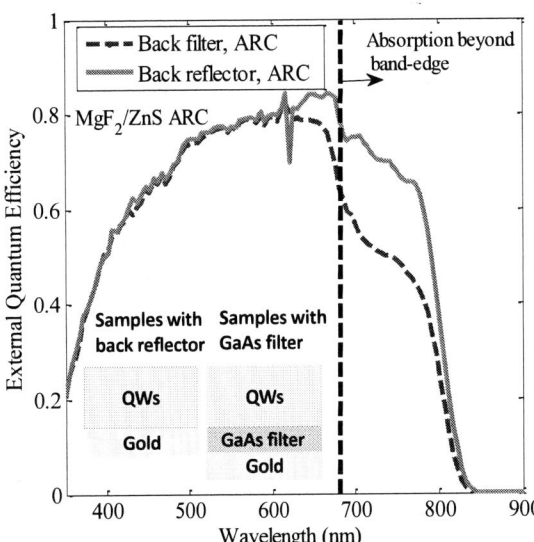

Figure 3: EQE of 100 period InGaAsP/InGaP superlattice solar cell (MP197) with back reflector and with back filter. The dotted vertical line represent the band-edge of $In_{0.49}Ga_{0.51}P$. Two pieces of the device structure were processed separately, with and without removing the optional GaAs filter as shown in the inset of Figure 3.

the carrier transport in the lattice-matched structure with the strain-balanced design through analyzing the carrier escape lifetimes for tunneling and thermionic emission.

II. EXPERIMENTAL

The QW samples were grown in a custom metal-organic vapor phase epitaxy reactor operating at atmospheric pressure. Trimethylgallium (TMGa) and TEGa precursors are used as gallium sources in InGaP and InGaAsP, respectively. The use of TEGa should produce lower carbon background doping if compared with TMGa since the decomposition process includes β-elimination and does not produce methyl-radicals. The lower background doping should help include more quantum wells and accordingly enhance the absorption in the QWs. The other precursors used in this work are: trimethyl-indium, arsine (AsH_3), phosphine (PH_3), diethylzinc, and hydrogen selenide (H_2Se/H_2). $In_{0.32}Ga_{0.68}As_{0.60}P_{0.40}/$ $In_{0.49}Ga_{0.51}P$ QWs were grown nominally lattice matched to GaAs substrates misoriented by 2° toward ($\bar{1}11$)B direction at 600 °C. The thickness of wells (t_w) and barriers (t_b) are 55Å and 25Å, respectively. All structures are grown with AlInP window and AlInGaP back surface field, as shown in Fig. 1. All devices are grown with an optional GaAs filter at the back to simulate the optical environment of a QW cell in a multijunction device.

The devices were processed with back gold reflector and then were inverted and bonded with epoxy on a supporting silicon handle. The GaAs substrate was removed using wet etching. Electroplated gold front grids were defined using standard photolithography. The final structure is shown in Fig. 1.

The curvatures of the samples were monitored during growth by multibeam optical stress (MOS) measurements whereby the stress is calculated using Stoney's equation [10][11].

III. RESULTS

The QWs grown in this work are depleted as confirmed by capacitance-voltage measurements.

Fig. 2 depicts the in-situ curvature and the laser output power (to maintain constant measured signal intensity) obtained in MOS measurements of the 100 period device (MP197) versus QW thickness. The stress calculated from the measured curvature of Fig. 2(a) is 0.087 GPa, resulting in minimal strain relaxation. The reflected laser intensity from the growing surface, Fig. 2(b), shows growth interference oscillations on a nearly flat background with no drastic increase, indicating that no significant degradation in the surface morphology has occurred during the QW growth.

Two device structures were processed separately, with and without the optional GaAs filter, as shown in the inset of Fig. 3. Fig. 3 shows the EQE of these $In_{0.32}Ga_{0.68}As_{0.60}P_{0.40}/$

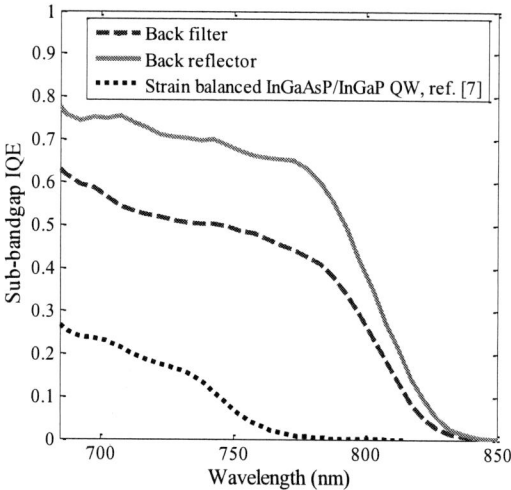

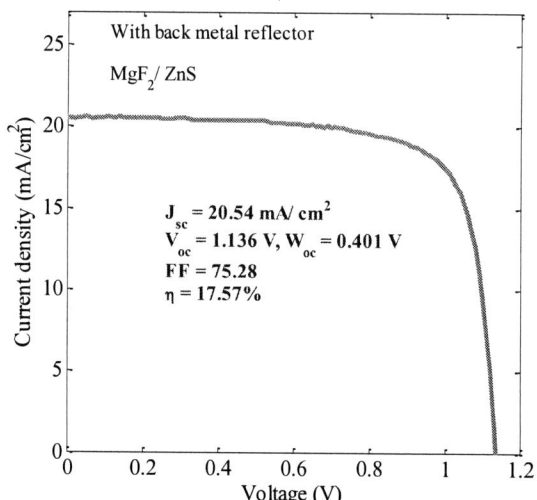

Figure 4: Comparison between the EQE of lattice-matched InGaAsP/InGaP superlattice solar cell reported here and previously reported strain balanced approaches in the wavelength range where the QWs are absorbing.

Figure 5: Current voltage characacteristics of 100 period InGaAsP/InGaP QW device.

$In_{0.49}Ga_{0.51}P$ superlattice solar cell devices after applying a ZnS/MgF_2 antireflective coating (ARC). The peak sub-bandgap EQE of the filtered device is ~61% which is at least double the peak internal quantum efficiency of the previously reported strain-balanced QW devices [7], as shown in fig. 4. The optical reflector allows photons that were not directly absorbed in the QWs in their first pass, to be absorbed in their second pass. This has improved the sub-bandgap EQE values relative to the filtered device, as shown in Figures 3-4.

Fig. 5 shows the current density-voltage curve measured at one sun AM1.5 for the 100 period QW cell (MP197) with back reflector and ARC. The short circuit current density (J_{sc}), open circuit voltage (V_{oc}), fill factor (FF), and efficiency (η) were 20.5 mA/cm^2, 1.13 V, 75.3%, and 17.6%, respectively. The bandgap-voltage offset ($W_{oc} = E_g/q - V_{oc}$), which is an indicator of quality of solar cells with different bandgaps, is 0.401 V, E_g = 1.537 eV and q is the electron charge. The bandgap was calculated from the EQE based on detailed-balance analysis same as in [12, 13]. A W_{oc} approaching or less than 0.4 V is an indicator of high internal-radiative and external-luminescence efficiency. The FF is considered low and implies that further work is needed to enhance the carrier transport across the barriers and the wells.

IV. DISCUSSION

In this work, we have grown $In_{0.32}Ga_{0.68}As_{0.60}P_{0.40}$ / $In_{0.49}Ga_{0.51}P$ QWs both lattice matched to GaAs substrates. The advantages of growing the structure lattice matched is to increase the well thickness relative to the barrier to enhance the sub-bandgap absorption. InGaAsP with compositions lattice

matched to GaAs are susceptible to spinodal decomposition and atomic ordering. The growth conditions of the InGaAsP such as growth temperature, substrate miscut, V/III ratio, and p-type doping were optimized to reduce the driving forces of spinodal-like decomposition [14, 15]. Briefly, growth temperatures less than or equal to 650 °C induce CuPt ordering which suppresses the quaternary decomposition. The thickness of the InGaP barrier is kept at 25 Å to enable carriers to tunnel across it. This has resulted in major improvements in the sub-bandgap quantum efficiency of this quantum well structure as shown in Figures 3, 4.

Previous studies have attempted to grow the quantum wells strain-balanced, where the $In_xGa_{1-x}As_zP_{1-z}$ and $In_yGa_{1-y}P$ are grown under compressive and tensile stress respectively, utilized compositions in the range of x = 60-75%, y = 30-45%, and z = 5-20% [7, 8, 16, 17]. Due to strain balance limitation, the thickness of the InGaAsP quantum well had to be thinner than $In_yGa_{1-y}P$ barrier. Beyond the $In_{0.49}Ga_{0.51}P$ bandedge (680 nm), absorption occurs only in InGaAsP well and growing the structure strain-balanced has limited the thickness of the InGaAsP to about 25% of the total period, which accordingly limited the sub-bandgap EQE to less than 25% as shown in Fig. 4 [7, 8].

We compare the carrier transport mechanisms taking place in the lattice-matched and strain-balanced quantum well structures. Quantum well solar cells requires that the generated minority carriers in the QW region to be swept across wells and barriers by an electric field in the depletion region. There are two main carrier transports that can exist in any quantum well structure as shown in Fig. 6: (i) carrier tunneling across InGaP barriers as shown in the band diagram of the lattice matched structure (Fig. 6a) and (ii) thermionic emission from InGaAsP quantum wells as shown in the band diagram of the strain

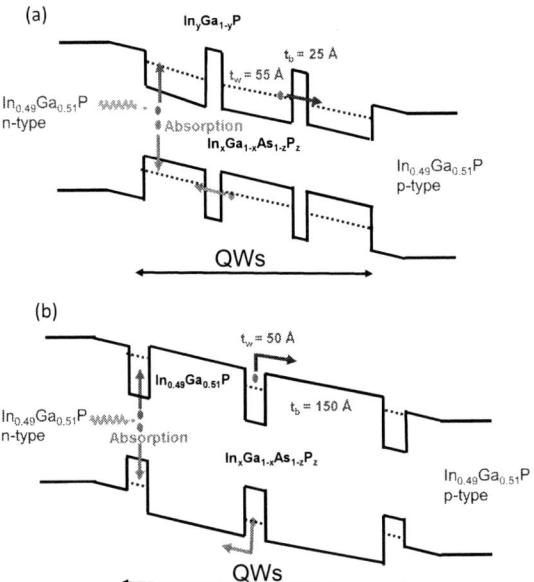

Figure 6: Band diagram of InGaAsP/InGaP quantum well with two different designs: (a) lattice-matched structure with thin InGaP barriers to allow carrier tunneling and (b) strain balanced structure with thin wells designed to promote thermionic emission.

balanced structure(Fig. 6b). Carrier tunneling is more dependent on the InGaP barrier thickness. Thermionic emission is more dependent on the barrier heights for electrons and holes. Carriers are more likely to be transported by tunneling in the lattice matched structure (Fig. 6a) due to the thin InGaP barriers (25Å). Thermionic emission is the dominant carrier transport for the strain-balanced structure (Fig, 6b) due to the fairly thin wells (50 Å) and thick barriers (150 Å). The escape lifetime ($\tau_{esc.}$) of a carrier from a QW is determined by the thermionic emission and tunneling lifetimes as follows [6]

$$\frac{1}{\tau_{esc.}} = \frac{1}{\tau_{tun}} + \frac{1}{\tau_{therm.}} \quad (1)$$

where $\tau_{tun.}$ and $\tau_{therm.}$ are the tunneling and thermionic emission lifetimes, respectively. The escape probability of minority carriers from a single quantum well can be calculated as follows

$$P = \frac{1/\tau_{esc.}}{1/\tau_{esc.} + 1/\tau_{rec.}} \quad (2)$$

where $\tau_{rec.}$ is the recombination lifetime. If we assume that the escape probabilities from all the quantum wells are the same, the total escape probability can be expressed as follows

$$P^{tot} = P^N \quad (3)$$

where N is the number of quantum wells in the intrinsic region. Eqn. (3) assumes all minority carriers will traverse N quantum wells. However, it should be mentioned that each carrier traverses different number of wells, depending on the position where the carrier was generated. Thus, Eqn. (3) assumes the worst case scenario for a carrier transport. $\tau_{tun.}$ and $\tau_{therm.}$ can be expressed as [18-20]

$$\frac{1}{\tau_{tun}} = \frac{n\pi\hbar}{2t_w^2 m_w^*} e^{-\frac{2}{\hbar}\int_0^{t_b}\sqrt{2m_b^*(E_b - E(z)z)}dz} \quad (4)$$

$$\frac{1}{\tau_{th}} = \frac{1}{t_w}\sqrt{\frac{kT}{2\pi m_w^*}} e^{-\frac{E_b}{kT}} \quad (5)$$

where m_w^* and m_b^* are the effective masses at the InGaAsP well and the InGaP barrier, respectively. E_b is the effective barrier height from the ground state (n=1). $E(z)$ is the electric field, $\hbar$ is the reduced Planck constant, and k is the Boltzmann constant. Kronig-Penney model was used to model the quantum states. The effective bandgap calculation includes strain and quantum size effects. Experimental values for the conduction and valence band offsets for InGaAsP/InGaP under the current growth conditions are scarce. The band offsets in the model are calculated using Anderson's rule based on the electron affinities and bandgap values of the InGaAsP and InGaP. The effective masses of the InGaP and InGaAsP are interpolated between the corresponding values for the binary compounds [21].

For the lattice matched superlattice structure shown in Fig. 6(a), the estimated barrier heights in the conduction and valence bands are 0.0684 and 0.3168 V, respectively. The electron effective mass in InGaAsP well and InGaP barrier are 0.0926m$_o$ and 0.1268m$_o$, respectively, where m$_o$ is the electron mass. The hole effective masses used in the well and barrier are 0.7m$_o$ and 0.1155m$_o$, respectively. For the strain balanced structure shown in Fig. 6(b), the barrier heights in the conduction and valence bands are 0.1622 and 0.2745 V, respectively. The electron effective mass in the InGaAsP well and InGaP barrier are 0.0974m$_o$ and 0.134m$_o$, respectively. The hole effective masses used in the well and barrier are 0.814m$_o$ and 0.1196m$_o$, respectively. Heavy holes and light holes masses are used for the InGaAsP and InGaP due to splitting of degeneracy because of compressive and tensile stress, respectively. The tunneling and thermionic emission escape probabilities for n =1 electron (e) and heavy-hole (hh) states are calculated using equations 1-5, assuming tunneling and thermionic emission taking place and are summarized in Table I.

For the lattice matched superlattice structure shown in Fig. 6a, the tunneling transport probability is over 98% through the 100 QWs for both n= 1 electron and heavy hole states in the conduction band and valence band. The efficient tunneling is due to using thin barriers (25 Å). The thermionic emission escape probability is 99% for electrons and ~0 for heavy holes.

Table I: Tunneling and thermionic emission escape probabilities for electrons (e's) and heavy holes (hh's) states for both the lattice matched and strain balanced structures

	Tunneling escape probability (P_{tun})		Thermionic emission escape probability $(P_{therm.})$	
	e	hh	e	hh
Lattice-matched superlattice	0.9998	0.9846	0.9991	0.0
Strain-balanced structure, ref [7]	0.0	0.0	0.9999999	0.5824

The high thermionic emission escape probability for electrons is due to the fact that only 13% of the total band offset occur in the conduction band, thus resulting in low barrier height for electrons (0.0684V). By comparing the escape probabilities for tunneling and thermionic emission shown in Table I, it is clear that both carrier transport mechanisms exist in the superlattice structure, with tunneling being more dominant.

For the strain-balanced structure shown in Fig. 6(b), the thermionic emission escape probability of n=1 electron and heavy hole states are 99% and 58%, respectively. The tunneling escape probabilities are almost zero because of the relatively thick barrier (~150 Å) for electrons and holes to tunnel. This indicates that thermionic emission is the only mechanism for carrier transport taking place in this structure. This discussion thus indicates the superiority of the lattice-matched structure over the strain balanced design and explains the sub-bandgap quantum efficiency improvements shown in Fig. 3, 4.

V. CONCLUSIONS

In this paper, we have presented an InGaAsP/InGaP superlattice solar cell with sub-bandgap exceeding 75% and bandgap-voltage offset of 0.40 V. The background doping in the QW region was reduced by using TEGa. Monitoring the stress evolution during growth using multi-beam optical sensor permitted the growth of 100 quantum wells. The QW design was improved in order to create conditions to allow for thicker absorbing wells and efficient carrier transport for both electrons and holes. The solar cells were processed with optical reflector to enhance light absorption in these QWs.

ACKNOWLEDGMENT

This work was supported by the U.S. Department of Energy Under Contract No. DE-AC36-08GO28308 with the National Renewable Energy Laboratory. S. M. Bedair acknowledges National Science Foundation for financial support. The authors would like to thank Waldo Olavarria and Michelle Young for device growth and processing, and Dr. Peter Colter for useful discussions. Islam Sayed acknowledges the support provided by all members of the III/V group at NREL, especially Daniel

Friedman, Emmett Perl, John Simon, Kevin Schulte, W. E. McMahon, and Ryan France.

REFERENCES

[1] M. A. Green, K. Emery, Y. Hishikawa, W. Warta, and E. D. Dunlop, "Solar cell efficiency tables (version 47)," *Progress in Photovoltaics: Research and Applications,* vol. 24, pp. 3-11, 2016.

[2] T. Takamoto, H. Washio, and H. Juso, "Application of InGaP/GaAs/InGaAs triple junction solar cells to space use and concentrator photovoltaic," in *Photovoltaic Specialist Conference (PVSC), 2014 IEEE 40th,* 2014, pp. 0001-0005.

[3] F. Dimroth, M. Grave, P. Beutel, U. Fiedeler, C. Karcher, T. N. Tibbits, *et al.*, "Wafer bonded four‐junction GaInP/GaAs//GaInAsP/GaInAs concentrator solar cells with 44.7% efficiency," *Progress in Photovoltaics: Research and Applications,* vol. 22, pp. 277-282, 2014.

[4] S. Philipps, A. Bett, K. Horowitz, and S. Kurtz, "Current Status of Concentrator Photovoltaic (CPV) Technology," National Renewable Energy Laboratory (NREL), Golden, CO.2015.

[5] R. M. France, J. F. Geisz, I. García, M. A. Steiner, W. E. McMahon, D. J. Friedman, *et al.*, "Design flexibility of ultrahigh efficiency four-junction inverted metamorphic solar cells," *IEEE Journal of Photovoltaics,* vol. 6, pp. 578-583, 2016.

[6] G. K. Bradshaw, C. Z. Carlin, J. P. Samberg, N. A. El-Masry, P. C. Colter, and S. M. Bedair, "Carrier transport and improved collection in thin-barrier InGaAs/GaAsP strained quantum well solar cells," *Photovoltaics, IEEE Journal of,* vol. 3, pp. 278-283, 2013.

[7] I. E. H. Sayed, C. Z. Carlin, B. G. Hagar, P. C. Colter, and S. Bedair, "Strain-Balanced InGaAsP/GaInP Multiple Quantum Well Solar Cells With a Tunable Bandgap (1.65–1.82 eV)," *IEEE Journal of Photovoltaics,* vol. 6, pp. 997-1003, 2016.

[8] I. E. Hashem, C. Z. Carlin, B. G. Hagar, P. C. Colter, and S. Bedair, "InGaP-based quantum well solar cells: Growth, structural design, and photovoltaic

properties," *Journal of Applied Physics,* vol. 119, p. 095706, 2016.

[9] I. E. H. Sayed, B. G. Hagar, C. Z. Carlin, P. C. Colter, and S. Bedair, "InGaP-based quantum well solar cells," in *Photovoltaic Specialists Conference (PVSC), 2016 IEEE 43rd,* 2016, pp. 0147-0150.

[10] J. Geisz, A. Levander, A. Norman, K. Jones, and M. Romero, "In situ stress measurement for MOVPE growth of high efficiency lattice-mismatched solar cells," *Journal of Crystal Growth,* vol. 310, pp. 2339-2344, 2008.

[11] Islam E. H. Sayed, Nikhil Jain, Myles A. Steiner , John F. Geisz, Pat Dippo, Darius Kuciauskas, Peter C. Colter, "In-situ curvature monitoring and X-ray diffraction study of InGaAsP/InGaP quantum wells," Journal of Crystal Growth (Submitted)

[12] M. Steiner, E. Perl, J. Geisz, D. Friedman, N. Jain, D. Levi, *et al.*, "Apparent bandgap shift in the internal quantum efficiency for solar cells with back reflectors," *Journal of Applied Physics,* vol. 121, p. 164501, 2017.

[13] J. F. Geisz, M. A. Steiner, I. Garcia, R. M. France, W. E. McMahon, C. R. Osterwald, *et al.*, "Generalized optoelectronic model of series-connected multijunction solar cells," *IEEE Journal of Photovoltaics,* vol. 5, pp. 1827-1839, 2015.

[14] N. Jain, R. Oshima, R. France, J. Geisz, A. Norman, P. Dippo, *et al.*, "Development of lattice-matched 1.7 eV GaInAsP solar cells grown on GaAs by MOVPE," in *Photovoltaic Specialists Conference (PVSC), 2016 IEEE 43rd,* 2016, pp. 0046-0051.

[15] R. Oshima, R. M. France, J. F. Geisz, A. G. Norman, and M. A. Steiner, "Growth of lattice-matched GaInAsP grown on vicinal GaAs (001) substrates within the miscibility gap for solar cells," *Journal of Crystal Growth,* vol. 458, pp. 1-7, 2017.

[16] I. E. H. Sayed, B. G. Hagar, C. Z. Carlin, P. C. Colter, and S. M. Bedair, "Extending the absorption threshold of InGaP solar cells to 1.60 eV using quantum wells: Experimental and modeling results," in *Photovoltaic Specialists Conference (PVSC), 2016 IEEE 43rd,* 2016, pp. 2366-2370.

[17] I. E. H. Sayed, C. Z. Carlin, B. Hagar, P. C. Colter, and S. M. Bedair, "Tunable GaInP solar cell lattice matched to GaAs," in *Photovoltaic Specialist Conference (PVSC), 2015 IEEE 42nd,* 2015, pp. 1-4.

[18] H. Schneider and K. v. Klitzing, "Thermionic emission and Gaussian transport of holes in a GaAs/Al x Ga 1− x As multiple-quantum-well structure," *Physical Review B,* vol. 38, p. 6160, 1988.

[19] D. A. Miller, G. Livescu, J. Cunningham, and W. Y. Jan, "Quantum well carrier sweep out: relation to electroabsorption and exciton saturation," *Quantum Electronics, IEEE Journal of,* vol. 27, pp. 2281-2295, 1991.

[20] J. Nelson, M. Paxman, K. Barnham, J. Roberts, and C. Button, "Steady-state carrier escape from single quantum wells," *Quantum Electronics, IEEE Journal of,* vol. 29, pp. 1460-1468, 1993.

[21] S. Adachi, "Material parameters of In1−xGaxAsyP1−y and related binaries," *Journal of Applied Physics,* vol. 53, pp. 8775-8792, 1982.

Carrier Collection Model and Design Rule for Quantum Well Solar Cells

Kasidit Toprasertpong, Boram Kim, Yoshiaki Nakano, and Masakazu Sugiyama

School of Engineering, the University of Tokyo, Bunkyo-ku, Tokyo, 113-0032, Japan

Abstract — Multiple quantum wells (MQWs) help enhance the light absorption but suffer from their poor carrier transport. Here, we propose a design rule for the cell optimization, particularly focusing on the optimal position and stack number for a given MQW structure. According to the effective-mobility model, the carrier collection from MQW is efficient when the carrier density profile is balanced and the total thickness is kept below the drift length. Based on this finding, we propose the modified detailed-balance calculation which includes the carrier collection behavior in MQW solar cells.

Index Terms — carrier collection, detailed balance, device design, multiple quantum wells, solar cells.

I. INTRODUCTION

Multiple quantum well (MQW) structures have been proposed as narrow-gap absorbers which absorb photons with energy lower than the bandgap of the host material [1]. This allows us to engineer the absorption spectrum and enhance the cell photocurrent. MQW is particularly investigated for the application on multi-junction solar cells, in which the current matching is vital to the conversion efficiency [2]. However, the alternate stacks of barriers and wells confine photogenerated carriers and obstruct the charge separation. The photocurrent enhancement due to the implementation of MQW is, therefore, less than that expected from the increased light absorption.

In spite of a number of studies on the carrier transport and the knowledge on its qualitative relation with the I-V characteristics—faster carrier escape provides better performance—[3]-[5], the quantitative and analytic relation has not been well understood. A lack of proper models hinders us from incorporating the carrier collection behavior in the performance prediction of MQW solar cells, and the configuration design, such as the optimal MQW position, was difficult. Most research so far has assumed the perfect carrier collection in the calculation model of MQW solar cells [6]-[9], which leads to inaccurate design particularly for matching the subcell photocurrent in multi-junction solar cells.

In this study, based on the experimental results and model analysis, we analyze the theoretical model for the carrier collection efficiency in MQW solar cells. Then, a design rule of the device configuration and the modification of detailed-balance model to cover the essence of MQW solar cells are discussed.

II. EXPERIMENTAL DETAILS

Three p-i-n GaAs solar cells with 27-stack $In_{0.2}Ga_{0.8}As$ (5.4 nm)/$GaAs_{0.61}P_{0.39}$ (5.7 nm) MQWs inserted in different positions in the i-regions, mentioned as MQW-top, MQW-mid, and MQW-bottom, were prepared as shown in Fig. 1. The p-type background doping in the i-regions were carefully compensated by slight n-doping so that the net background carrier concentration became below 5×10^{14} cm^{-3}. In this way, MQWs in different positions experience similar electric field and the carrier escape rates are assured to be independent of the position in the i-region.

The carrier collection efficiencies (CCE) under monochromatic illumination were evaluated from the voltage-dependent external quantum efficiency (EQE). That is, the collection efficiency of carriers excited by light with wavelength λ at voltage V is experimentally estimated by

$$CCE(V, \lambda) = \frac{EQE(V, \lambda)}{EQE(V_{reverse}, \lambda)}, \qquad (1)$$

where $V_{reverse}$ is the sufficiently-high reverse bias voltage [10]. EQEs were measured under AM1.5G bias light.

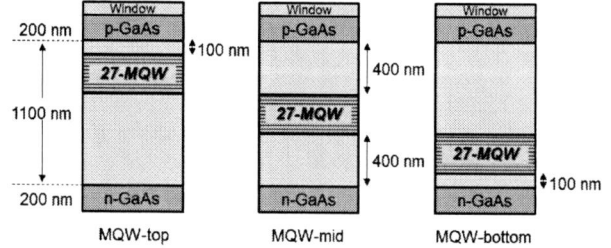

Fig. 1. MQW cell structures. 27-stack MQWs were inserted in different positions in the i-regions. The electric field in the i-region is mostly uniform due to the compensated background doping.

III. RESULTS AND DISCUSSION

A. Measurement Results

Fig. 2(a) shows CCE at 0.6 V under monochromatic excitation and AM1.5G-light bias, extracted from measured EQE using (1). Interestingly, the collection of carriers directly generated in MQW ($\lambda > 870$ nm) shows the unexpected behavior. In spite of the identical MQW and the uniform electric field, which should give the same carrier escape rate $1/\tau_{esc}$ from MQWs in all samples, MQW-top shows significantly efficient carrier collection as compared with the others. This tendency cannot be explained by the conventional model, CCE $= 1/\tau_{esc}/(1/\tau_{esc} + 1/\tau_r)$ [11], in which only the carrier escape rate $1/\tau_{esc}$ and recombination rate $1/\tau_r$ are taken into account.

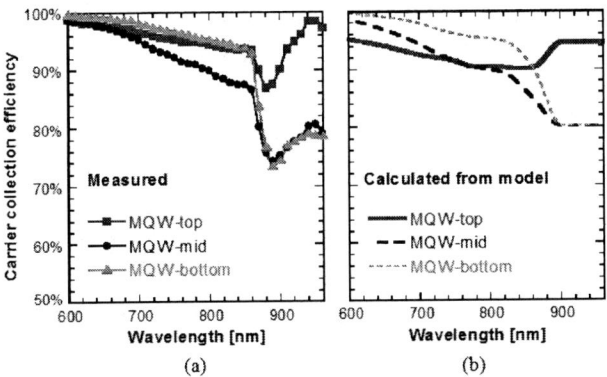

Fig. 2. CCE (a) measured and (b) calculated from (5) at 0.6 V. Collection of carriers directly generated in MQW ($\lambda > 870$ nm) is the most efficient in MQW-top.

B. Model for CCE

To understand the above results, we briefly discuss a model for carrier collection efficiency. Detailed derivation of the model has been carried out elsewhere [12], [13]. We have shown in [12], [13] that photogenerated electrons only recombine in the hole-rich region where the hole density p is much higher than the electron density n, and similarly for the recombination of photogenerated holes. By simplifying the carrier escape process as the effective mobility μ_n and μ_p, the collection efficiency of electrons generated at p-side outside the MQW, which have to move across the entire MQW, is expressed as

$$\text{CCE}_n = \exp\left[-\frac{L_{p>n}}{L_{d,n}}\right], \quad (2)$$

where $L_{p>n}$ is the hole-rich region width in MQW,

$$L_{d,n} = \mu_n \tau_{r,n} E \quad (3)$$

is the electron drift length, $\tau_{r,n}$ is the electron Shockley-Read-Hall (SRH) lifetime, and E is the electric field. Similarly, the collection efficiency of photogenerated holes moving across the entire MQW is given by

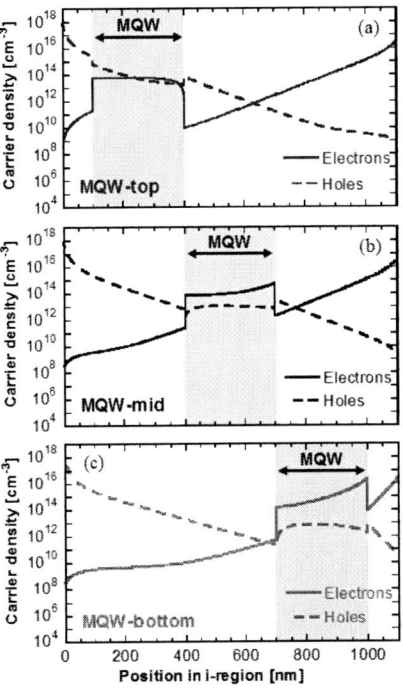

Fig. 3. Carrier distribution profiles of (a) MQW-top, (b) MQW-mid, and (c) MQW-bottom at 0.6 V under 1-sun bias light. Only the i-regions are shown. The shaded area corresponds to MQW and the non-shaded area to i-GaAs. The electron-rich and hole-rich regions in MQW are comparatively balanced in MQW-top.

$$\text{CCE}_p = \exp\left[-\frac{L_{n>p}}{L_{d,p}}\right], \quad (4)$$

where the subscript p corresponds to the property of holes.

From (2)-(4), (which discuss CCE for carriers photogenerated outside the MQW,) we derive CCE for direct photogeneration (CCE$_{\text{direct}}$). Consider MQW with the hole-rich region in $0 \leq z < L_{p>n}$ and the electron-rich region in $L_{p>n} < z \leq L_{\text{MQW}}$, where $L_{\text{MQW}} = L_{p>n} + L_{n>p}$ is the MQW total thickness. Electrons generated at $0 \leq z < L_{p>n}$ have to travel $L_{p>n} - z$ to escape from the hole-rich region and holes generated at $L_{p>n} < z \leq L_{\text{MQW}}$ have to travel $z - L_{p>n}$ in the electron-rich region. Due to gradual absorption in MQW, we can assume the uniform photogeneration. Then, the total collection efficiency becomes

$$\begin{aligned}
\text{CCE}_{\text{direct}} &= \frac{1}{L_{\text{MQW}}}\left(\int_0^{L_{p>n}} \exp\left[-(L_{p>n} - z)/L_{d,n}\right] dz \right. \\
&\left. + \int_{L_{p>n}}^{L_{\text{MQW}}} \exp\left[-(z - L_{p>n})/L_{d,p}\right] dz\right) \\
&\approx \frac{L_{p>n}}{L_{\text{MQW}}} \exp\left[-\frac{L_{p>n}}{2L_{d,n}}\right] + \frac{L_{n>p}}{L_{\text{MQW}}} \exp\left[-\frac{L_{n>p}}{2L_{d,p}}\right]. (5)
\end{aligned}$$

978-1-5090-5606-4/17 $31.00 © 2017 IEEE   2202

Thus, the CCE under any given incident spectrum is

$$\text{CCE} = \frac{G_p\,\text{CCE}_n + G_n\,\text{CCE}_p + G_\text{direct}\,\text{CCE}_\text{direct}}{G_p + G_n + G_\text{direct}}, \quad (6)$$

where G_p, G_n, and G_direct are the total absorptions at the p-side bulk, n-side bulk, and inside MQW, respectively.

Fig. 2(b) shows λ-dependent CCE simulated from (6) using $\mu_n = \mu_p = 0.28$ cm2/Vs [14], $\tau_{r,n} = 300$ ns, $\tau_{r,p} = 35$ ns, $E = 6.9$ kV/cm, G_p, G_n, and G_direct from the Lambert-Beer law, and $L_{p>n}$ and $L_{n>p}$ from Fig. 3, which depicts the carrier distribution profiles estimated from the device simulator. It can be seen that the simulation results reproduce the measured CCE (Fig. 2(a)) with a good agreement, confirming the validity of our model.

C. Cell Design

In this section, we discuss about the cell design for efficient collection of carriers generated directly in MQW (CCE_direct), which is one main challenging issue in MQW solar cells.

For MQW with a given total thickness and mobility, CCE_direct in (5) reaches the maximum at

$$\max\left[\text{CCE}_\text{direct}\right] = \exp\left[-\frac{L_\text{MQW}}{4\overline{L}_d}\right] \quad (7a)$$

when the carrier distribution satisfies

$$L_{p>n} = \frac{L_{d,n}}{L_{d,p}+L_{d,n}}L_\text{MQW}, \quad L_{n>p} = \frac{L_{d,p}}{L_{d,p}+L_{d,n}}L_\text{MQW}, \quad (7b)$$

where $\overline{L}_d = (L_{d,n}+L_{d,p})/2$ is the average drift length. This implies that in addition to the escape and the recombination, the carrier distribution is another important factor in the carrier collection process. (7b) leads to the design for the optimal MQW position for maximizing carrier collection. That is, MQW should be embedded so that the crossing point of $n(z)$ and $p(z)$ is near the center of MQW, shifted by the transport asymmetry between electrons and holes. This explains why CCE_direct (CCE at $\lambda > 870$ nm in Fig. 2) is significantly high in MQW-top, which has better balance of carrier distribution than the other two cells (Fig. 3).

It is worth noting that while the carrier distribution balancing suppresses the SRH recombination of photocarriers, it simultaneously enhances that of electrically injected carriers, increasing the dark current. While MQW-top has the highest CCE and thus EQE at $\lambda > 870$ nm, it has the lowest open-circuit voltage ($V_\text{oc} = 907\pm1$ mV in MQW-top, 922 ± 2 mV in MQW-mid, 918 ± 1 mV in MQW-bottom). The V_oc trade-off in this design should be carefully considered in single-junction cells, but can be less problematic in multi-junction cells where the current-matching is more severe to their efficiency.

Furthermore, (7a) provides the upper limit of MQW stacks. For MQW embedded in the optimal position, keeping the MQW structure thinner than the average drift length,

$$L_\text{MQW}\left(= N_\text{MQW}l\right) \leq \overline{L}_d, \quad (8)$$

is necessary to keep carrier collection efficient ($> 80\%$). N_MQW and l are the MQW stack number and period, respectively. Fig. 4 shows the allowed N_MQW for various effective mobilities $\overline{\mu} = (\mu_n + \mu_p)/2$ when $E = 5$ kV/cm, $l = 10$ nm, $\tau_{r,n} = \tau_{r,p} \equiv \tau_r$, and 1% absorption per QW are assumed based on the experimental results [15]. This limit of N_MQW consequently results in insufficient light absorption, as the corresponding one-path absorption is shown on the right axis. Fig. 4 can also be interpreted as the lower limit of MQW mobility required for cells with a given target light absorption. Together with a model for estimating μ_n and μ_p which will be discussed elsewhere, this can be used as a prescreening tool in the feasibility investigation of new MQW structures.

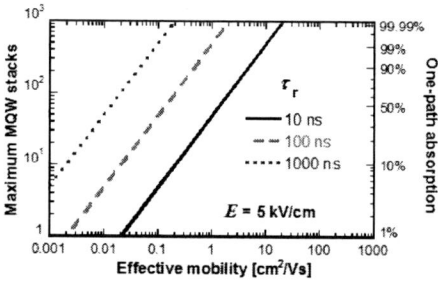

Fig. 4. Maximum stack numbers of MQW to keep carrier collection efficiency above 80%. The effective mobility is the average between that of electrons and holes. The carrier lifetimes were set to 10, 100, and 1000 ns. The corresponding one-path light absorption is shown on the right axis.

D. Modified Detailed-Balance Calculation

The detailed-balance model is modified to realistically describe MQW solar cells with imperfect carrier collection. The upper limit of MQW cell characteristics with the effective mobility $\overline{\mu}$ and the SRH lifetime $\tau_n = \tau_p = \tau_r$ is given by

$$J = J_\text{ph} - J_{01}\left(\exp\left[\frac{qV}{k_BT}\right]-1\right) - J_{02}\left(\exp\left[\frac{qV}{2k_BT}\right]-1\right), \quad (9a)$$

$$J_\text{ph} = q\,\text{CCE}\int A(\varepsilon)N_\text{illu}(\varepsilon)d\varepsilon, \quad (9b)$$

$$J_{01} = q\,\text{CCE}\int A(\varepsilon)N_\text{bb}(\varepsilon,T)d\varepsilon, \quad (9c)$$

$$J_{02} = \pi n_i k T L_\text{MQW}/2\tau_r(\phi_\text{bi}-V), \quad (9d)$$

$$\text{CCE} = \exp[-L_\text{MQW}^2/4\overline{\mu}\tau_r(\phi_\text{bi}-V)], \quad (9e)$$

$$\phi_\text{bi} = E_{g,\text{host}}, \quad (9f)$$

where $k_B T$ is the thermal energy, q the elementary charge, $A(\varepsilon)$ the absorption, $N_{\text{illu}}(\varepsilon)$ the incident photon flux spectrum, $N_{\text{bb}}(\varepsilon, T)$ the black-body photon flux spectrum, n_i the intrinsic carrier density in MQW, ϕ_{bi} the built-in potential, $E_{g,\text{host}}$ is the bandgap of the host material. Here, we assumed (i) carrier distribution in (7b), with which the maximum CCE is given by (7a), (ii) the maximum ϕ_{bi} for the non-degenerate emitter and base is given by $E_{g,\text{host}}$, (iii) the i-region entirely covered with MQW, which gives $E = (\phi_{\text{bi}} - V)/L_{\text{MQW}}$, and (iv) the dark current dominated by the recombination in MQW and the recombination in the host material is negligibly small. CCE appears in (9c) due to the non-uniform chemical potential in MQW with poor carrier transport [16], [17]. J_{02} in (9d) was derived by analytically integrating the SRH recombination. The above equation set can be used particularly for the MQW-based subcell design in multi-junction solar cells.

VI. CONCLUSION

The carrier collection behavior and the design rule for MQW solar cells were investigated. The formula for carrier collection efficiency shows that the carrier collection is the most efficient when the carrier distribution is well balanced. It was also found that MQW should not be stacked more than its drift length. The proposed model allows us to consider the non-unity carrier collection efficiency in the detailed-balance calculation, approaching the actual characteristics of MQW solar cells.

ACKNOWLEDGEMENT

This study was supported by the New Energy and Industrial Technology Development Organization (NEDO), Japan (P15003), and a Grant-in-Aid for JSPS Fellows (15J03447) from Japan Society for the Promotion of Science.

REFERENCES

[1] K. W. J. Barnham, B. Braun, J. Nelson, M. Paxman, C. Button, J. S. Roberts, and C. T. Foxon, "Short-circuit current and energy efficiency enhancement in a low-dimensional structure photovoltaic device," *Applied Physics Letters*, vol. 59, pp. 135–137, 1991.

[2] N. J. Ekins-Daukes, J. M. Barnes, K. W. J. Barnham, J. P. Connolly, M. Mazzer, J. C. Clark, R. Grey, G. Hill, M. A. Pate, and J. S. Roberts, "Strained and strain-balanced quantum well devices for high-efficiency tandem solar cells," *Solar Energy Materials and Solar Cells*, vol. 68, pp. 71–87, 2001.

[3] N. Watanabe, H. Yokoyama, N. Shigekawa, K. Sugita, and A. Yamamoto, "Barrier thickness dependence of photovoltaic characteristics of InGaN/GaN multiple quantum well solar cells," *Japanese Journal of Applied Physics*, vol. 51, p. 10ND10, 2012.

[4] M. Jo, Y. Ding, T. Noda, T. Mano, Y. Sakuma, K. Sakoda, L. Han, and H. Sakaki, "Impacts of ambipolar carrier escape on current-voltage characteristics in a type-I quantum-well solar cell," *Applied Physics Letters*, vol. 103, p. 061118, 2013.

[5] M. Sugiyama, Y. Wang, H. Fujii, H. Sodabanlu, K. Watanabe, and Y. Nakano, "A quantum-well superlattice solar cell for enhanced current output and minimized drop in open-circuit voltage under sunlight concentration," *J. Phys. D: Appl. Phys.*, vol. 46, p. 024001, 2013.

[6] S. J. Lade and Zahedi, "A revised ideal model for AlGaAs/GaAs quantum well solar cells," *Microelectronics Journal*, vol. 35, pp. 401–410, 2004.

[7] J. C. Rimada, L. Hernandez, J. P. Connolly, and K. W. J. Barnham, "Conversion efficiency enhancement of AlGaAs quantum well solar cells," *Microelectronics Journal*, vol. 38, pp. 513–518, 2007.

[8] A. G. J. Adams, B. C. Browne, I. M. Ballard, J. P. Connolly, N. L. A. Chan, A. Ioannides, W. Elder, P. N. Stavrinou, K. W. J. Barnham, and N. J. Ekins-Daukes, "Recent results for single-junction and tandem quantum well solar cells," *Progress in Photovoltaics: Research and Applications*, vol. 19, pp. 865–877, 2011.

[9] C. I. Cabrera, J. C. Rimada, J. P. Connolly, and L. Hernandez, "Modelling of GaAsP/InGaAs/GaAs strain-balanced multiple-quantum well solar cells," *Journal of Applied Physics*, vol. 113, p. 024512, 2013.

[10] H. Fujii, K. Toprasertpong, K. Wanatabe, M. Sugiyama, and Y. Nakano, "Evaluation of carrier collection efficiency in multiple quantum well solar cells," *IEEE Journal of Photovoltaics*, vol. 4, pp. 237–243, 2014.

[11] J. Nelson, M. Paxman, K. W. J. Barnham, J. S. Roberts, and C. Button, "Steady-state carrier escape from single quantum wells," *IEEE Journal of Quantum Electronics*, vol. 29, pp. 1460–1468, 1993.

[12] K. Toprasertpong, T. Inoue, A. Delamarre, K. Watanabe, M. Sugiyama, and Y. Nakano, "Photocurrent collection mechanism and role of carrier distribution in p-i-n quantum well solar cells," in *43rd IEEE Photovoltaic Specialist Conference*, 2016, p. 163.

[13] K. Toprasertpong, T. Inoue, Y. Nakano, and M. Sugiyama, "Investigation and modeling of photocurrent collection process in multiple quantum well solar cells," submitted.

[14] K. Toprasertpong, T. Tanibuchi, H. Fujii, T. Kada, S. Asahi, K. Watanabe, M. Sugiyama, T. Kita, and Y. Nakano, "Comparison of electron and hole mobilities in multiple quantum well solar cells using a time-of-flight technique," *IEEE Journal of Photovoltaics*, vol. 5, pp. 1613–1620, 2015.

[15] K. Toprasertpong, H. Fujii, T. Thomas, M. Führer, D. Alonso-Álvarez, D. J. Farrell, K. Watanabe, N. Okada, N. J. Ekins-Daukes, M. Sugiyama, and Y. Nakano, "Absorption threshold extended to 1.15 eV using InGaAs/GaAsP quantum wells for over-50%-efficient lattice-matched quad-junction solar cells," *Progress in Photovoltaics: Research and Applications*, vol. 24, pp. 533–542, 2016.

[16] U. Rau and R. Brendel, "The detailed balance principle and the reciprocity theorem between photocarrier collection and dark carrier distribution in solar cells," *Journal of Applied Physics*, vol. 84, pp. 6412–6418, 1998.

[17] K. Toprasertpong, A. Delamarre, Y. Nakano, J. -F. Guillemoles, and M. Sugiyama, to be submitted.

Influence of conduction band offsets at window/buffer and buffer/absorber interfaces on the roll-over of J-V curves of CIGS solar cells

Giovanna Sozzi[1], Simone Di Napoli[1], Roberto Menozzi[1], Florian Werner[2], Susanne Siebentritt[2], Philip Jackson[3], Wolfram Witte[3]

[1]Department of Engineering and Architecture, University of Parma, Parco Area delle Scienze 181A, 43124 Parma, Italy

[2]Laboratory for Photovoltaics, University of Luxembourg, 41, rue du Brill, L-4422 Belvaux, Luxembourg

[3]Zentrum für Sonnenenergie und Wasserstoff-Forschung Baden-Württemberg (ZSW), Meitnerstraße 1, 70563 Stuttgart, Germany

Abstract — By comparing simulated and measured current-voltage (J-V) characteristics of a high efficiency Cu(In,Ga)Se₂ (CIGS) solar cell, we investigate the effect of conduction band offsets at window/buffer and buffer/absorber interfaces on the roll-over of the J-V curves with temperature. We simulate the J-V characteristics in the temperature range 300 K - 100 K, achieving good fits with measurements by describing the transport of electrons over the barrier at the ZnO/CdS interface by the thermionic-emission theory, and including a Schottky contact at the back CIGS/Mo interface.

I. INTRODUCTION

Although thin-film Cu(In,Ga)Se₂ (CIGS) solar cells have reached efficiencies of 22.6 % [1], a complete understanding of the phenomena that may influence the device performance, such as the saturation of the cell in forward current at low temperatures (*roll-over*), is still incomplete and debated. The blocking and non-exponential behavior of the temperature-dependent J-V (J-V-T) curves known as *roll-over* was mainly attributed to a Schottky diode at the molybdenum back contact [2], or deep acceptor-like defects at the interface between the CdS buffer layer and a defect-chalcopyrite layer [3].

In this work we aim at interpreting the roll-over of measured J-V-T curves by means of numerical simulations; in particular, we will examine the way the conduction band offsets (CBOs) at hetero-interfaces affect the current and induce the roll-over. We also examine the effect of a non-ohmic contact at the CIGS/molybdenum interface.

II. METHOD

A. Sample and Measurements

The CIGS solar cell considered – fabricated at ZSW - had 21.0% efficiency measured at standard testing conditions and consists of a soda-lime glass substrate, Mo back contact, 2.4 μm thick CIGS absorber with RbF post-deposition treatment (PDT), CdS buffer, intrinsic ZnO, Al-doped ZnO window layer, and MgF₂ anti-reflective coating (ARC).

The J-V-T curves measured with a halogen lamp for illumination (adjusted to yield the expected short-circuit current) in the range 323 - 100 K, depicted in Fig. 1, exhibit a remarkable roll-over of the current at low temperature. Losses in solar cell parameters (mainly fill factor) could be a result of wire bonding and mounting in the cryostat. Even if we restrict the analysis to this one sample for a quantitative description of the roll-over, we observed similar behavior in other samples with alkali metal PDT.

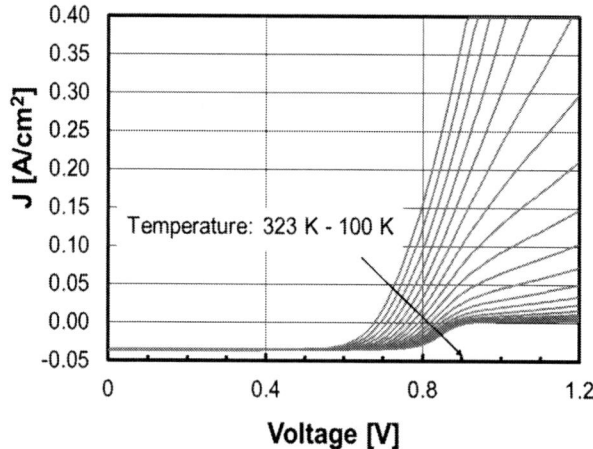

Fig. 1.   Measured J-V-T curves under illumination.

B. Simulations

We modeled the cell using the Synopsys Sentaurus-Tcad suite [4]. The cell's behavior in the dark is described by the Poisson, electron and hole continuity, and drift-diffusion equations. Thermionic emission for electrons is also accounted for at the window/buffer heterojunction when specifically indicated. Recombination via deep defects follows the Shockley – Read – Hall (SRH) model.

The measured absorber double-graded [Ga]/([Ga]+[In]) (GGI) profile across the thickness is loaded into the model of the solar cell to give the corresponding band gap grading and position-dependent optical coefficients [5].

The cell is illuminated by the AM1.5G spectrum, and light propagation is modeled by the transfer matrix method (TMM). The optical behavior is described by complex refractive indexes coming from the literature [6] or from unpublished measured data (for CIGS), that in the case of CIGS vary with the GGI ratio too [5]. The main parameters used in the simulation are displayed in Table I [5], [7]. Series resistance is optimized to get the best match on the 300 K curve.

TABLE I

MATERIAL PARAMETERS USED IN THE SIMULATIONS AT 300 K

Material	ZnO	CdS	CIGS
Eg [eV]	3.3	2.4	graded
$N_{D/A}$ (cm^{-3})	N_D: 10^{20} - AZO	N_D: $2 \cdot 10^{16}$	N_A: $5.5 \cdot 10^{15}$
	N_D: 10^{17} - i-ZnO		
m_e/m_0	0.2	0.2	0.09
m_h/m_0	1.2	0.8	0.72
χ [eV]	variable	variable	4.6
μ_e/μ_h [cm^2/(V·s)]	100/25	100/25	100/25
Bulk Traps			
N_t [cm^{-3}]	10^{16} (acc.)	$3 \cdot 10^{15}$ (acc.)	$7 \cdot 10^{13}$ (acc.)
Energy [eV]	midgap		
σ_e/σ_h [cm^2]	$10^{-15}/10^{-12}$	$10^{-15}/10^{-12}$	$4 \cdot 10^{-14}$ /$1 \cdot 10^{-15}$

In the simulations shown hereafter, we vary the CBO at different hetero-interfaces in order to discriminate between the effects due to the CBO at the buffer/window, ΔE_{BW}, and absorber/buffer, ΔE_{AB}, heterojunctions.

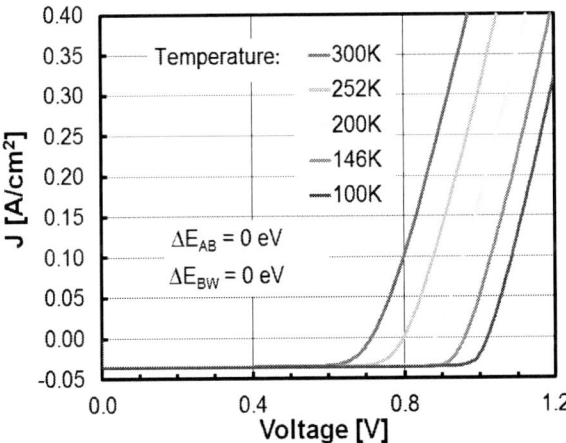

Fig. 2. Simulated J-V-T curves under illumination.

III. RESULTS

A. $\Delta E_{AB} = 0$, $\Delta E_{BW} = 0$

With no-conduction band offsets at the hetero-interfaces, the simulated J-V curves show no roll-over, Fig. 2; the curves only shift with decreasing temperature towards higher voltages, as predicted by the classical pn-junction theory.

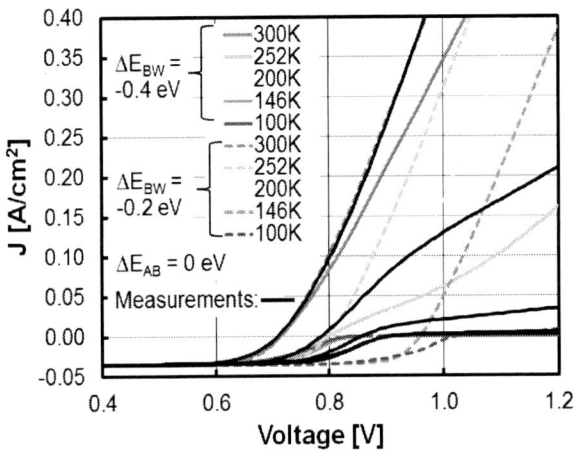

Fig. 3. Simulated (colored lines) and measured (black lines) current-voltage characteristics under illumination at different temperatures and for two values of the buffer-window CBO.

B. $\Delta E_{AB} = 0$, $\Delta E_{BW} < 0$

A negative CBO (cliff) is usually expected at the ZnO/CdS interface [8]; $\Delta E_{BW} < 0$ determines the onset of the roll-over at low temperature as shown in Fig. 3 for the two cases of ΔE_{BW} = -0.2 eV and -0.4 eV. However, for ΔE_{BW} = -0.2 eV, the roll-over only appears at the lowest temperature T = 100 K (blue dashed curve), while for ΔE_{BW} = -0.4 eV it is already

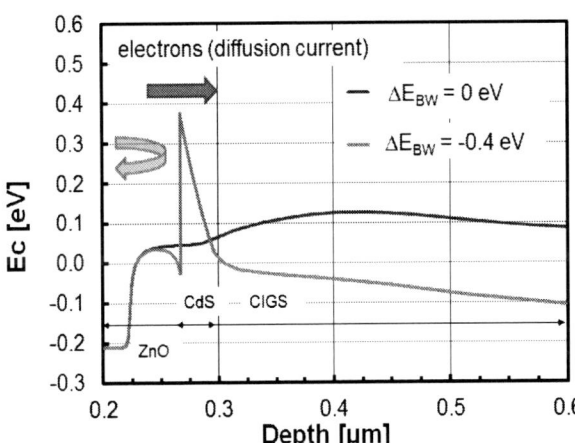

Fig. 4. Conduction band energy, E_C, versus depth at V = 1.1 V under illumination for two values of the window-buffer CBO. ΔE_{AB} = 0 eV. T = 200 K.

978-1-5090-5606-4/17 $31.00 © 2017 IEEE

remarkable at a temperature as high as T = 252 K (orange solid curve).

The way a large negative ΔE_{BW} affects the current flow can be understood looking at the conduction-band diagrams in Fig. 4: a large $\Delta E_{BW} < 0$ (red curve) increases the barrier to the flow of electrons from the window to the absorber, thus suppressing the diffusion current which dominates the J-V characteristics at voltages > V_{oc}.

The comparison with measurements (black curves in Fig. 3) shows that, in order to obtain a similar roll-over in the simulated currents, ΔE_{BW} must be larger than the expected value (i.e., ΔE_{BW} = -0.2 eV [8]), and even then the match between measurements and simulations is only qualitative.

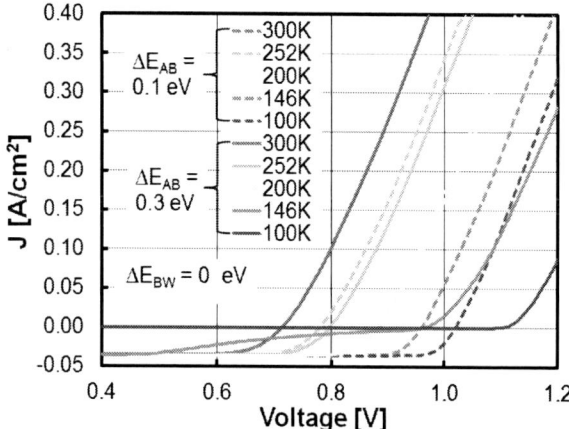

Fig. 5. Simulated current-voltage characteristics under illumination at different temperatures, for two values of the window-buffer CBO.

C. $\Delta E_{AB} > 0$, $\Delta E_{BW} = 0$

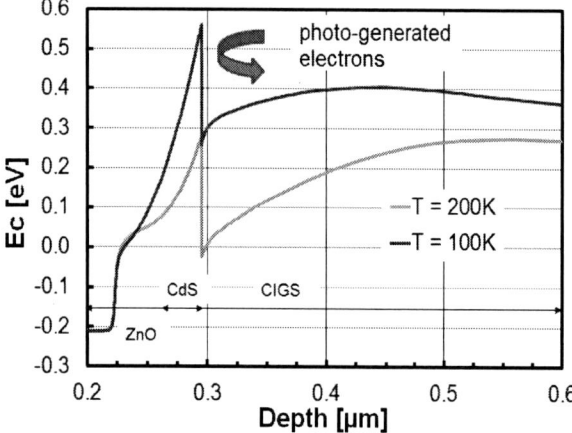

Fig. 6. Conduction band energy, E_C, versus depth at V = 0.7 V under illumination. ΔE_{AB} = 0.3 eV, ΔE_{BW} = 0 eV.

In the case of positive CBO (spike) at the CdS/CIGS interface, only the largest ΔE_{AB} = 0.3 eV (solid curves in Fig. 5), determines a change of shape in the J-V curves, mainly at V < V_{oc}, where the photo-generated current is reduced for T < 200 K, while the effect on the roll-over is negligible. The reduction of the photocurrent at low temperature is due to the larger energy barrier seen at the CdS/CIGS junction by the photo-generated electrons leaving the absorber as shown by the conduction-band diagrams in Fig. 6 (blue curve).

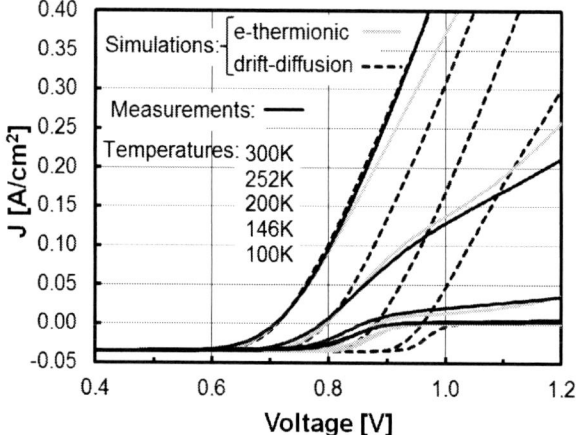

Fig. 7. Measured and simulated J-V curves under illumination at different temperatures. The simulations are performed using either a drift-diffusion or a thermionic model at the buffer-window interface. ΔE_{AB} = 0.1 eV and ΔE_{BW} = -0.2 eV.

D. $\Delta E_{AB} > 0$, $\Delta E_{BW} < 0$

When CBOs are accounted for at both hetero-interfaces, the effect of ΔE_{BW} on roll-over slightly increases (not shown here). However, assuming ΔE_{BW} = -0.2 eV [8] and ΔE_{AB} = 0.1 eV [9], does not allow to match the measured roll-over (dashed and solid black curves in Fig. 7, respectively).

Since the roll-over is mainly influenced by the barrier at the ZnO/CdS interface, we then introduced a thermionic emission model for the transport of electrons over the barrier: the simulated currents (orange curves in Fig. 7) now well reproduce the measured ones, with some deviation at high voltages.

E. Back barrier at the CIGS/molybdenum contact

A non-Ohmic back-contact acting as a counter-diode in series with the pn junction can act as a barrier for the carrier flow.

Fig. 8 illustrates the simulated current in the cases where the back contact is modeled as either ohmic (dashed black line) or Schottky (solid red line). A Schottky barrier with a 120 meV barrier for holes at the back contact causes the suppression of the forward current even for moderate values of ΔE_{AB} and ΔE_{BW}; however, the simulated current show the tendency to saturate to an almost constant value at low temperature which is not observed in the measurements (see Fig. 1 for comparison).

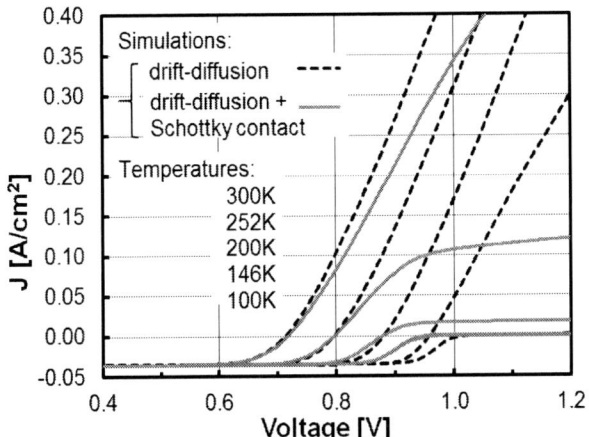

Fig. 8. Simulated current with ohmic (dashed black line) and Schottky (solid red line) back contact and drift-diffusion transport model at the interface. $\Delta E_{AB} = 0.1$ eV, $\Delta E_{BW} = -0.2$ eV.

On the other hand, if we model the back contact as a Schottky barrier for holes ($E_v - E_f = 120$ meV) together with the thermionic emission model for the transport of electrons over the barrier at the ZnO/CdS interface, the discrepancy observed between measured and modeled J-V-T curves is remarkably reduced, as shown by the best fit case in Fig. 9, where an optimized (temperature independent) value of the series resistance has been used to get the best match.

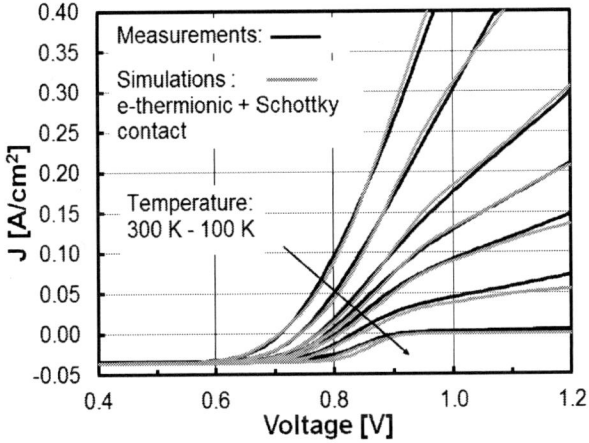

Fig. 9. Best fit of the current-voltage characteristics under illumination.

IV. CONCLUSIONS

In this work we analyzed the effect of conduction band offsets at buffer/window and buffer/absorber interfaces on the roll-over of the J-V-T curves in the 300 K - 100 K range of a

CIGS cell with 21% efficiency (with ARC). The roll-over is mainly controlled by the energy barrier at the buffer/window interface, where the transport of electrons over the barrier can be adequately described by the thermionic-emission theory. As far as the back contact is concerned, even if a back-side Schottky barrier alone cannot reproduce the experimental curves, in combination with the buffer/window interface barrier and the thermionic emission model, it yields the best match with the measured data.

ACKNOWLEDGEMENT

This project has received funding from the *European Union's Horizon 2020 research and innovation programme* under grant agreement No 641004, project Sharc25.

REFERENCES

[1] P. Jackson, R. Wuerz, D. Hariskos, E. Lotter, W. Witte, and M. Powalla, "Effects of heavy alkali elements in Cu(In,Ga)Se₂ solar cells with efficiencies up to 22.6%," *Phys. status solidi - Rapid Res. Lett.*, vol. 586, no. 8, pp. 583–586, 2016.

[2] T. Eisenbarth, T. Unold, R. Caballero, C. A. Kaufmann, and H. Schock, "Interpretation of admittance, capacitance-voltage, and current-voltage signatures in Cu(In,Ga)Se₂ thin film solar cells," *J. Appl. Phys.*, vol. 107, no. 3, p. 34509, Feb. 2010.

[3] M. Topič, F. Smole, and J. Furlan, "Examination of blocking current-voltage behaviour through defect chalcopyrite layer in ZnO/CdS/Cu(In,Ga)Se₂/Mo solar cell," *Sol. Energy Mater. Sol. Cells*, vol. 49, no. 1–4, pp. 311–317, Dec. 1997.

[4] https://www.synopsys.com/silicon/tcad.html.

[5] G. Sozzi, S. Di Napoli, R. Menozzi, R. Carron, E. Avancini, B. Bissig, S. Buecheler, and A. N. Tiwari, "Analysis of Ga grading in CIGS absorbers with different Cu content," in *43rd IEEE Photovoltaic Specialists Conference*, 2016, pp. 2279–2282, 2016.

[6] T. Hara, T. Maekawa, S. Minoura, Y. Sago, S. Niki, and H. Fujiwara, "Quantitative assessment of optical gain and loss in submicron-textured CuIn₁₋ₓGaₓSe₂ solar cells fabricated by three-stage coevaporation," *Phys. Rev. Appl.*, vol. 2, no. 3, pp. 1–17, 2014.

[7] F. Troni, F. Dodi, G. Sozzi, and R. Menozzi, "Modeling of thin-film Cu(In,Ga)Se₂ solar cells," in *IEEE International Conference on Simulation of Semiconductor Processes and Devices (SISPAD)*, 2010, pp. 33–36.

[8] T. Törndahl, C. Platzer-Björkman, J. Kessler, and M. Edoff, "Atomic layer deposition of Zn₁₋ₓMgₓO buffer layers for Cu(In,Ga)Se₂ solar cells," *Prog. Photovolt. Res. Appl.*, vol. 15, no. February 2013, pp. 225–235, 2007.

[9] H. Kashiwabara, Y. Hayase, K. Takeshita, T. Okuda, S. Niki, K. Matsubara, K. Sakurai, A. Yamada, S. Ishizuka, and N. Terada, "Study of Changes of Electronic and Structural Nature of CBD-CDS/CIGS Interface with Ga Concentration," in *2006 IEEE 4th World Conference on Photovoltaic Energy Conference*, 2006, pp. 495–498.

978-1-5090-5606-4/17 $31.00 © 2017 IEEE

Overview of Surface Passivation Schemes for Thin Film Solar Cells

Ratan Kotipalli[1] and Bart Vermang[2,3]

[1]Université catholique de Louvain (UCL), Louvain-la-Neuve, 1348, Belgium
[2]University of Hasselt – partner in Solliance, Diepenbeek, 3590, Belgium
[3]Imec – partner in Solliance, Leuven, 3001, Belgium

Abstract — **This work provides a rapid overview for the current state of surface passivation layer schemes for thin film solar cells: From its fundamentals to solar cell applications, and their perspective. It provides an overview of important literature and prospect considerations based on simulations.**

Index Terms — **Photovoltaics, Thin film solar cells, Cu(In,Ga)Se₂, Surface passivation, Front, Rear.**

I. SILICON SURFACE PASSIVATION BY USE OF A DIELECTRIC

The idea to use dielectric layers to reduce recombination at (= "passivate") interfaces stems from silicon (Si) PV. Recombination dynamics at semiconductor interfaces have been described by Shockley, Read, and Hall. Their formalism shows that Si surface passivation layers reduce electronic recombination at the interface by two key methods: (i) Chemical passivation – corresponding to a reduction in interface trap density – and (ii) field effect passivation – resulting from a fixed charge density in the passivation layer that reduces the surface minority or majority charge carrier concentration. In advanced Si solar cell design – e.g. the passivated emitter and rear solar cell (PERC), see Fig. 1 – such passivation layers are combined with micron-sized point openings that serve as electrical contacts, both at the rear and front Si surfaces [1,2].

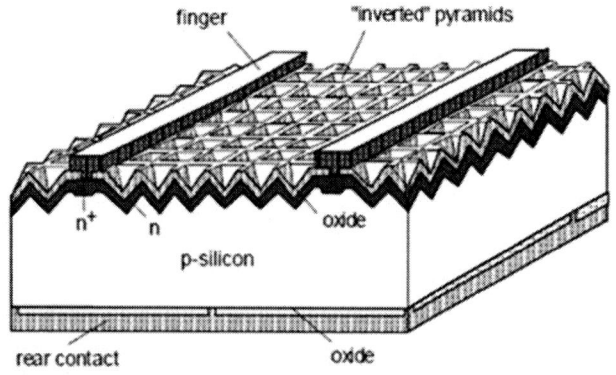

Fig. 1. Drawing of a silicon passivated emitter and rear cell (PERC), which includes SiO₂ front and rear surface passivation layers, taken from [2].

II. REAR SURFACE PASSIVATION SCHEMES FOR CIGS

Various research groups have shown that Al₂O₃ is very suitable to passivate the rear of Cu(In,Ga)Se₂ (CIGS) thin film solar cells, which can be explained by a combination of chemical and field effect passivation. Opto-electrical measurements can be used to screen interesting passivation layers, as is shown for Al₂O₃ grown on CIGS [3]. Even more, electrical measurements of similar structures show that the Al₂O₃ layers exhibit a high density of negative fixed charges (its field effect passivation) in combination with a reasonably low interface trap density (its chemical passivation) [4]. In the meanwhile, several groups successfully fabricated Al₂O₃ rear-passivated CIGS solar cells, where recombination at the Al₂O₃/CIGS interface decreased substantially. Uppsala University fabricated rear-passivated solar cells with nano-size point openings generated through the removal of nanosphere-shaped precipitates [5,6], or e-beam lithography [7]; while KIT and ZSW applied photo-lithography [8]. Fig. 2 shows a cross section image of such an Al₂O₃ rear passivated CIGS solar cell.

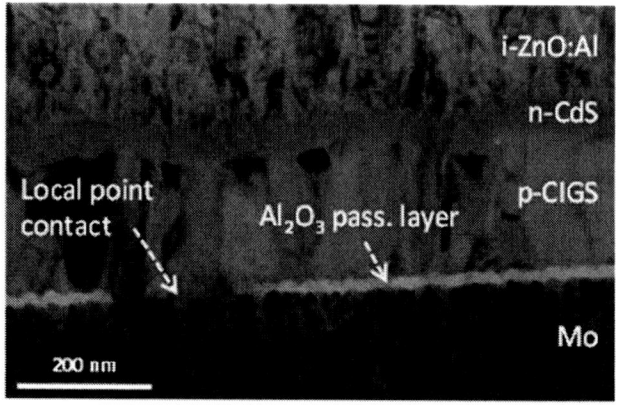

Fig. 2. Transmission electron microscopy cross-section image of an Al₂O₃ rear surface passivated cell with a well-controlled grid of nano-sized local rear point contacts, taken from [7].

Remarkable open circuit voltage (V_{OC}) results have also been achieved for ultra-thin CIGS solar cells with (= despite) a nano-structured SiO₂/CIGS rear interface [9,10], indicating that this SiO₂ layer could be an interesting alternative for Al₂O₃ passivation. M. Schmid has indeed shown reduced interface recombination at a nanostructured SiO₂/CIGS rear interface [11], very similar to previously acquired results for ultra-thin CIGS solar cells with a nano-structured Al₂O₃/CIGS rear interface [12].

III. FRONT SURFACE PASSIVATION SCHEMES FOR CIGS

First results indicate that another type of passivation layer will be required to passivate the front of CIGS solar cells. HZB, University of Parma and EMPA have used simulations to show that a positively charged surface passivation layer with nano-sized point openings (e.g. generated as in [13]) would be beneficial to passivate the CIGS/buffer front interface [14,15]. In this case, this positively charged layer causes a n-type inversion layer in the CIGS, which extends the n-type buffer layer, as is shown in Fig. 3. One might even consider to omit the buffer layer completely, and instead use a conformal but ultra-thin (to allow charge carrier tunneling) front surface passivation layer to generate an "inversion layer emitter". This approach is already applied in so-called metal-insulator-semiconductor inversion-layer (MIS/IL) Si solar cells [16]. One surface passivation layer candidate is TiO₂: grown on Si it exhibits a positive charge density [17], and it shows potential for front-passivated CIGS solar cells in [18]. Another candidate is Ga₂O₃: Imec has successfully fabricated Ga₂O₃ front surface passivated CIGS solar cells, a manuscript is in preparation.

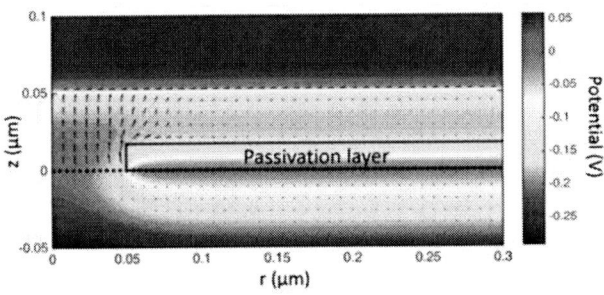

Fig. 3. Simulated potential (color scale) and electron current (arrows) for a front surface passivated CIGS solar cell, by use of a positively charged surface passivation layer, taken from [14]. Note that z = 0 µm corresponds to the CIGS front interface.

IV. POTENTIAL OF PASSIVATED EMITTER AND REAR CIGS (PERCIGS)

Integration of front and rear surface passivation layers – combined with approaches for optical confinement [19,20] – into CIGS solar cells with ever thinner absorber layers opens the door for increased cell efficiency, as compared to 'unpassivated' state-of-the-art CIGS solar cells. This is simulated in Fig. 4(a) by use of SCAPS [21], this graph shows solar cell efficiency as a function of CIGS absorber layer thickness for standard CIGS solar cells, and also for an industrially viable and an ideal case of the Passivated *Emitter*

(in this case it actually is the Front CIGS interface) and Rear CIGS (PERCIGS) solar cell design (simulation details are mentioned in the figure caption and [4]). Remarkably, for PERC-type Si solar cells a very similar trend in efficiency as a function of Si wafer thickness is seen, as is shown in Fig. 4(b). Note that this trend has been chased by Solexel Inc. who holds the world record for the thinnest Si solar cell: 21.2 % efficiency in case of a 35 µm thick Si 'absorber' layer.

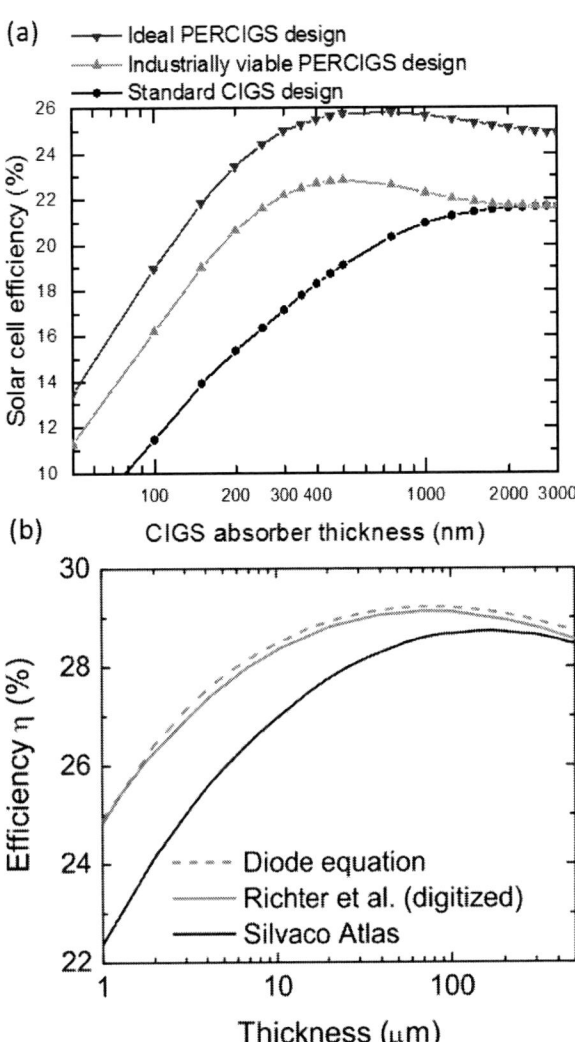

Fig. 4. (a) Solar cell efficiency simulations for state-of-the-art standard CIGS and PERCIGS (with surface passivation, but also light management) solar cells. The standard design contains low rear internal reflection (Mo/CIGS), no surface passivation layers and a high quality CIGS absorber layer (E$_{trap}$ = 0.3 eV). The industrially viable PERCIGS design contains an aluminum rear reflector (Al/Mo/(Al₂O₃/)CIGS rear), surface passivation layers as described in [4], and a high quality CIGS absorber layer. The ideal PERCIGS design contains complete light trapping, surface passivation layers as described in [4], and an excellent quality CIGS absorber layer (E$_{trap}$ = 0.1 eV). (b) A similar graph for Si solar cells, taken from [22].

V. POTENTIAL OF SURFACE PASSIVATION FOR OTHER THIN FILM

This approach of using dielectric layers to passivate thin film solar cell interfaces is also very valuable for other photovoltaic thin film materials: CdTe surfaces have been successfully passivated by Al_2O_3 films [23,24], and similar passivating layers have been applied for $Cu_2(Zn,Sn)S_4$ (CZTS) [25] and perovskite materials [26,27].

ACKNOWLEDGEMENT

This work has received funding from the European Research Council (ERC) under the European Union's Horizon 2020 research and innovation programme (grant agreement n° 715027).

REFERENCES

[1] A. Cuevas, T. Allen, J. Bullock, Y. Wan, D. Yan, and X. Zhang, "Skin care for healthy silicon solar cells," *Proc. 42nd IEEE PVSC (New-Orleans)*, pp. 1–6, 2015.

[2] M.A. Green, "The Passivated Emitter and Rear Cell (PERC): From conception to mass production," *Solar Energy Mater. Solar Cells*, vol. 143, pp. 190–197, 2015.

[3] J. Joel, B. Vermang, J. Larsen, O. Donzel-Gargand, and M. Edoff, "On the assessment of CIGS surface passivation by photoluminescence," *Phys. Status Solidi RRL*, vol. 9, no. 5, pp. 288–292, 2015.

[4] R. Kotipalli, B. Vermang, J. Joel, R. Rajkumar, M. Edoff, and D. Flandre, "Investigating the electronic properties of $Al_2O_3/Cu(In,Ga)Se_2$ interface," *AIP Advances*, vol. 1, no. 5, pp. 107101-1– 107101-6, 2015.

[5] B. Vermang, V. Fjällström, J. Pettersson, P. Salomé, and M. Edoff, "Development of rear surface passivated $Cu(In,Ga)Se_2$ thin film solar cells with nano-sized local rear point contacts," *Sol. Energ. Mat. Sol. Cells*, vol. 117, pp. 505–511, 2013.

[6] B. Vermang, J.T. Wätjen, V. Fjällström, F. Rostvall, M. Edoff, R. Kotipalli, F. Henry, and D. Flandre, "Employing Si solar cell technology to increase efficiency of ultra-thin $Cu(In,Ga)Se_2$ solar cells," *Prog. Photovoltaics: Res. Appl.*, vol. 22, no. 10, pp. 1023 – 1029, 2014.

[7] B. Vermang, J.T. Wätjen, C. Frisk, V. Fjällström, F. Rostvall, M. Edoff, P. Salomé, J. Borme, N. Nicoara, and S. Sadewasser, "Introduction of Si PERC rear contacting design to boost efficiency of $Cu(In,Ga)Se_2$ solar cells," *IEEE J. Photovoltaics*, vol. 4, no. 6, pp. 1644–1649, 2014.

[8] P. Casper, R. Hünig, G. Gomard, O. Kiowski, C. Reitz, U. Lemmer, M. Powalla, and M. Hetterich, "Optoelectrical improvement of ultra-thin $Cu(In,Ga)Se_2$ solar cells through microstructured MgF_2 and Al_2O_3 back contact passivation layer," *Phys. Status Solidi RRL*, vol. 10, no. 5, pp. 376–380, 2016.

[9] C. van Lare, G. Yin, A. Polman, and M. Schmid. "Light coupling and trapping in ultrathin $Cu(In,Ga)Se_2$ solar cells using dielectric scattering patterns," *ACS Nano*, vol. 9, no. 10, pp. 9603–9613, 2015.

[10] E. Jarzembowski, B. Fuhrmann, H. Leipner, W. Fränzel, and R. Scheer, "Ultrathin $Cu(In,Ga)Se_2$ solar cells with point-like back contact in experiment and simulation," *Thin Solid Films*, in press, DOI: 10.1016/j.tsf.2016.11.003, 2016.

[11] M Schmid, "Review on light management by nanostructures in chalcopyrite solar cells," *Semicond. Sci. Technol.*, vol. 32, no. 4, pp. 043003, 2017.

[12] B. Vermang, J.T. Wätjen, V. Fjällström, F. Rostvall, M. Edoff, R. Gunnarsson, I. Pilch, U. Helmersson, R. Kotipalli, F. Henry, and D. Flandre, "Highly reflective rear surface passivation design for ultra-thin $Cu(In,Ga)Se_2$ solar cells," *Thin Solid Films*, vol. 582, pp. 300–303, 2015.

[13] P. Reinhard, B. Bissig, F. Pianezzi, H. Hagendorfer, G. Sozzi, R. Menozzi, C. Gretener, S. Nishiwaki, S. Buecheler, and A.N. Tiwari, "Alkali-templated surface nanopatterning of chalcogenide thin films: A novel approach toward solar cells with enhanced efficiency," *Nano Lett.*, vol. 15, no. 5, pp. 3334–3340, 2015.

[14] A. Bercegol, B. Chacko, R. Klenk, I. Lauermann, M.Ch. Lux-Steiner, and M. Liero, "Point contacts at the copper-indium-gallium-selenide interface—A theoretical outlook," *J. Appl. Phys.*, vol. 119, pp. 155304-1–155304-7, 2016.

[15] G. Sozzi, S. Di Napoli, R. Menozzi, B. Bissig, S. Buecheler, and A.N. Tiwari, "Impact of front-side point contact/passivation geometry on thin-film solar cell performance," *Sol. Energ. Mat. Sol. Cells*, vol. 165, pp. 94–102, 2017.

[16] M.A. Green and R.B. Godfrey, "MIS solar cell—General theory and new experimental results for silicon," *Appl. Phys. Lett.*, vol. 29, no. 9, pp. 610–612, 1976.

[17] V.-S. Dang, H. Parala, J.H. Kim, K. Xu, N.B. Srinivasan, E. Edengeiser, M. Havenith, A.D. Wieck, T. de los Arcos, R.A. Fischer, and A. Devi, "Electrical and optical properties of TiO_2 thin films prepared by plasma-enhanced atomic layer deposition," *Phys. Status Solidi A*, vol. 211, no. 2, pp. 416–424, 2014.

[18] W. Hsu, C.M. Sutter-Fella, M. Hettick, L. Cheng, S. Chan, Y. Chen, Y. Zeng, M. Zheng, H.-P. Wang, C.-C. Chiang, and A. Javey, "Electron-selective TiO_2 contact for $Cu(In,Ga)Se_2$ solar cells," *Sci. Rep.*, vol. 5, article no. 16028, 2015.

[19] O. Poncelet, R. Kotipalli, B. Vermang, A. Macleod, L.A. Francis, and D. Flandre, "Optimisation of rear reflectance in ultra-thin CIGS solar cells towards > 20% efficiency," *Solar Energy*, vol. 146, pp. 443–452, 2017.

[20] R. Kotipalli, O. Poncelet, G. Li, Y. Zeng, L.A. Francis, B. Vermang, and D. Flandre, "Addressing the impact of rear surface passivation mechanisms on ultra-thin $Cu(In,Ga)Se_2$ solar cell performances using SCAPS 1-D model," *Solar Energy*, under review, 2017.

[21] M. Burgelman, P. Nollet, and S. Degrave, "Modelling polycrystalline semiconductor solar cells," *Thin Solid Films*, vol. 361–362, pp. 527–532, 2000.

[22] M.J. Kerr, A. Cuevas, and P. Campbell, "Limiting efficiency of crystalline silicon solar cells due to Coulomb-enhanced Auger recombination," *Prog. Photovoltaics: Res. Appl.*, vol. 11, no. 2, pp. 97–104, 2003.

[23] J. Liang, Q. Lin, H. Li, Y. Su, X. Yang, Z. Wu, J. Zheng, X. Wang, Y. Lin, and F. Pan, "Rectification and tunneling effects enabled by Al_2O_3 atomic layer deposited on back contact of CdTe solar cells," *Appl. Phys. Lett.*, vol. 107, pp. 013907-1–013907-4, 2015.

[24] B. Bissig, C. Guerra-Nunez, R. Carron, S. Nishiwaki, F. La Mattina, F. Pianezzi, P.A. Losio, E. Avancini, P. Reinhard, S.G. Haass, M. Lingg, T. Feurer, I. Utke, S. Buecheler, and A.N. Tiwari, "Surface passivation for reliable measurement of bulk electronic properties of heterojunction devices," *small*, vol. 12, no. 38, pp. 5339–5346, 2016.

[25] B. Vermang, Y. Ren, O. Donzel-Gargand, C. Frisk, J. Joel, P. Salomé, J. Borme, S. Sadewasser, C. Platzer-Björkman, and M. Edoff, "Rear surface optimization of CZTS solar cells by use of

a passivation layer with nanosized point openings," *IEEE J. Photovoltaics*, vol. 6, no. 1, pp. 332–336, 2015.

[26] Y.H. Lee, J. Luo, M.-K. Son, P. Gao, K.T. Cho, J. Seo, S.M. Zakeeruddin, M. Grätzel, and M. Khaja Nazeeruddin, "Enhanced charge collection with passivation layers in perovskite solar cells," *Adv. Mater.*, vol. 28, no. 20, pp. 3966–3972, 2016.

[27] G.W.P. Adhyaksa, L.W. Veldhuizen, Y. Kuang, S. Brittman, R.E.I. Schropp, and E.C. Garnett, "Carrier diffusion lengths in hybrid perovskites: processing, composition, aging, and surface passivation effects," *Chem. Mater.*, vol. 28, no. 15, pp. 5259–5263, 2016.

Towards 10% State-of-the-Art Pure Sulfide Cu_2ZnSnS_4 Solar Cell by modifying the Interface Chemistry

Kaiwen Sun[1], Jialiang Huang[1], Steve Johnston[2], Chang Yan[1], Fangyang Liu[1], Xiaojing Hao[1]*, Martin Green[1]

1 School of Photovoltaic and Renewable Energy Engineering, University of New South Wales (UNSW), Sydney, NSW 2052, Australia

2 National Renewable Energy Laboratory (NREL), Golden, Colorado 80401, USA

Abstract — **Microstructure and chemistry of the CZTS/ZnCdS heterointerface, playing an important role in improving CZTS device performance, is studied in this work. A favorable self-assembled ultrathin ZnS layer is obtained by controlling the absorber composition. This ZnS layer has better lattice match with the CZTS absorber and passivation effect of the charge carrier at the interface, so that it can reduce the interface defects and passivate the interface recombination, thereby increasing the Voc as well as the device performance. Ammonium hydroxide is applied as complexing agent for the SILAR process of ZnCdS, which can control the free ion concentration in the precursor solution and reduce the oxide and hydroxide byproduct during the deposition process. This new recipe leads to a dramatic boost of fill factor of the device due to the reduced electronic defects in the depletion region. A confirmed new world record 9.5% efficiency pure sulfide kesterite device is achieved by applying these techniques and over 10% efficiency is projected to be obtained by further modifying the heterojunction interface.**

I. INTRODUCTION

Cu_2ZnSnS_4 (CZTS), a Se free system, is a potential next-generation photovoltaic cell material because of the abundance and non-toxicity of its various elements and its suitable band gap. This material has been intensively studied over the past few years where gradual efficiency improvements have led to efficiencies of up to 9.5% for this pure sulfide wider band gap CZTS.[1] The main efficiency limitation for CZTS device is the open circuit voltage when comparing to the band gap of the CZTS absorber (ie., a large Voc deficit). Apart from the bulk defects and disorder in the CZTS film which cause local fluctuations of the band gap and/or the band gap edges due to locally varying electrostatic potential differences, which is believed to contribute to the V_{oc} deficit, the interface recombination is another important factor limiting the V_{oc} and thereby the device efficiency. In our recent work,[2] we found that the traditional CdS buffer forms a "cliff-like" conduction band offset (CBO) with the CZTS absorber, which is one of the key reasons that lead to severe interface recombination and V_{oc} deficit. By replacing the CdS buffer with $Zn_{1-x}Cd_xS$ (ZnCdS), where we adjust the CBO towards the optimal "spike" range (0-0.4) and improve the V_{oc} as well as the device efficiency dramatically. Nevertheless, from our device simulation results, if the interface recombination velocity is fixed to a low level and the band alignment is favorable, the V_{oc} is estimated to be above 900 mV, which is over 100 mV higher than our current champion cell. Therefore, with the more favorable CBO by applying ZnCdS buffer, there still are other interface recombination factors limiting the V_{oc}. In this study, we look into the microstructure and chemistry at the heterointerface to investigate new route other than the band alignment to improve the device performance.

II. EXPERIMENTAL

CZTS absorbers were prepared by co-sputtering Cu/ZnS/SnS precursors on Mo-coated soda lime glass using a magnetron sputtering system with a targeted Cu-poor and Zn-rich precursor composition of a (Cu/(Zn+Sn)=0.8, Zn/Sn=1.1-1.3). These CZTS precursors were then subjected to sulfurization using Rapid Thermal Processor in combined sulfur and tin sulfide vapour atmosphere at 560 ℃.

The ZnCdS buffer was prepared via successive ionic layer adsorption and reaction (SILAR) method using $ZnSO_4$ and $CdSO_4$ mixed solution as the cation precursor solution and Na_2S as anion precursor solution. After the buffer layers deposition, 60 nm thick i-ZnO and 200 nm thick ITO layers were deposited by RF magnetron sputtering. The total area of the final cells are 0.3~0.4 cm^2 defined by mechanical scribing.

The microstructure and elemental mapping across the CZTS and buffer interface were measured by JEOL JEM-ARM200F (200kV) aberration-corrected scanning transmission electron microscope (STEM) equipped with energy dispersive X-ray spectroscopy (EDAX) system. The J-V curves were performed using a solar simulator from PV Measurement (with simulated AM 1.5G illumination) calibrated with a standard Si reference and a Keithley 2400 source meter. External quantum efficiency (EQE) data were collected by a QEX10 spectral response system (PV measurements, Inc.)

III. RESULTS AND DISCUSSION

The standard absorber composition in our lab used to be Zn/Sn~1.2 and Cu/Sn~0.8. In this work, we tried three different Zn/Sn ratios from 1.3 to 1.1, while keeping Cu/Sn ratio at 0.8. The ZnCdS buffer has better epitaxial growth on all different composition absorbers than traditional CdS buffer because of the better lattice match of ZnCdS with CZTS than that of CdS. For the absorbers with higher Zn content (i.e. Zn/Sn=1.3 and 1.2), the interface between CZTS and ZnCdS shows similar microstructure and the elements distribute relatively homogeneous across the interface.

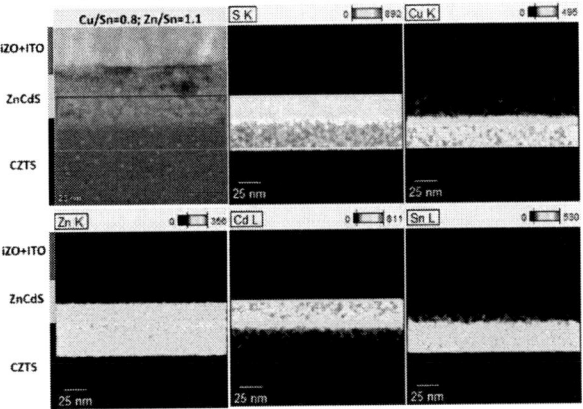

Figure 1. EDS mapping of the CZTS/ZnCdS interface region of the absorber with lower Zn content.

However, for the lower Zn ratio absorber (Zn/Sn=1.1), Zn and Cd present a distinct distribution at the surface and in the bulk region. It is indicative from **Figure 1** that Cd shows a deficient composition in the interface region while Zn demonstrates a discontinuous sufficient composition in the Cd-deficient width. This means a discontinuous Zn-rich and Cd-poor layer forms at the surface of the CZTS, which can also be confirmed by the EDS line scan of the three different interfaces. This discontinuous ZnS layer can act as a passivation layer because of its wide bandgap insulating property. A pertinent publication showed that a thin ZnS nanodot layer with suitable diameter placed between CIGS and the In_2S_3 buffer layer can serve as a passivation layer, still enabling lateral diffusion of charge carriers to the In_2S_3 contact bridge, which forms the point contact structure at the interface. The thin ZnS layer observed in low Zn ratio absorber surface in our present study shows similar structure with the nanodot structure, which is also ultrathin and has nanometer scale opening. Hence, this ZnS layer found at the interface between the absorber with low Zn content and ZnCdS buffer can not only reduce the interface defect because of its coherent crystal structure and lattice parameter with CZTS, but also passivates the charge carrier recombination due to its dielectric property.

The current density-voltage (*J-V*) curves of the CZTS solar cells from these three different composition absorbers with ZnCdS buffer are shown is **Figure 2** and the device electrical parameters are listed in **Table 1**. With the incomplete ZnS layer observed at the interface, excess charge carrier are kept away from contacts between ZnCdS and CZTS at which recombination is theoretically high. Therefore, detrimental charge carrier recombination at the heterojunction, which is one important position for performance loss in kesterite solar cells, is substantially supressed. This could explain why the

device from absorber with lower Zn content shows an increase in Voc. The fill factor of lower Zn content absorber also increases to over 62%, which can be partially related to the increasing Voc, i.e., the ZnS passivation effect.

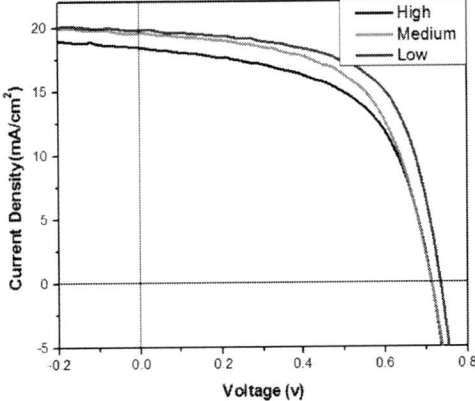

Figure 2. J-V curves of devices fabricated with absorbers of different composition (labelled with Zn/Sn ratio from high 1.3 to low 1.1).

The short circuit current density (Jsc) improvement can be confirmed from external quantum efficiency (EQE) which is not presented here. The EQE is improved at all wavelength, especially in the long-wavelength region. The EQE improvement indicates an improvement in the collection length of photo-generated carriers, which can also be attributed to passivation of the CZTS surface by the ZnS layer.

Table 1. Device characteristics of the CZTS solar cells with absorbers of different composition.

Zn/Sn ratio	Voc (mV)	Jsc (mA/cm^2)	FF (%)	Eff (%)
1.3	711.89	18.42	57.77	7.58
1.2	711.87	19.56	59.47	8.28
1.1	735.25	20.11	62.53	9.24

An issue that the ZnCdS film tends to have oxide and hydroxide impurities may be caused by the buffer layer prepared by SILAR method without complexing agents during the deposition process. This would result in electronic defects in the depletion region and impact the device performance mainly with low fill factor. In the chemical bath deposition (CBD) process, ammonium hydroxide is usually used as a complexing agent to control the free metal ion concentration and to dissolve the exceeded oxide or hydroxide byproduct during the deposition process. In our previous work,[1] no ammonium hydroxide was used in the SILAR process. That should be part of the reasons for the relative lower FF of all the CZTS devices with the ZnCdS buffer compared to the its mature CIGS counterpart.

In order to solve this problem, ammonium hydroxide is added to the cation precursor solution in our new recipe. As a result, a higher fill factor and at the same time a new word record pure sulphide kesterite solar cell is obtained, which is certified by National Renewable Energy Laboratory (NREL) as shown in **Figure 3**.[1] The fill factor is measured to be more than 68% in our lab (certified to be over 66%), which is much higher than our previously reported CZTS device with ZnCdS buffer. The effect of the ammonium hydroxide may explain the improved fill factor, which, however, still needs to be further confirmed in our coming work.

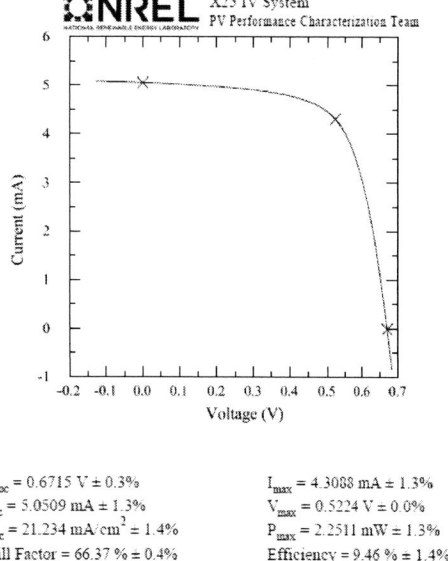

V_{oc} = 0.6715 V ± 0.3%	I_{max} = 4.3088 mA ± 1.3%
I_{sc} = 5.0509 mA ± 1.3%	V_{max} = 0.5224 V = 0.0%
J_{sc} = 21.234 mA/cm^2 = 1.4%	P_{max} = 2.2511 mW ± 1.3%
Fill Factor = 66.37 % ± 0.4%	Efficiency = 9.46 % ± 1.4%

Figure 3. Device characteristics of the CZTS solar cells with ZnCdS buffer prepared with ammonium hydroxide (certificate document from NREL)

Although the fill factor and short circuit current density (Jsc) have been enhanced substantially, the Voc of the champion cell is much lower than our normal ZnCdS buffered device. One possible reason is the composition of the obtained ZnCdS layer changes when ammonium hydroxide is added in the cation precursor solution because it would change the deposition rate of ZnS and CdS. As we found in our recent work, the band structure of the ZnCdS film varies with its composition. We also checked the Zn/Cd ratio of the ZnCdS film prepared with and without ammonium hydroxide as the complexing agent from cation precursor solution with identical initial Zn and Cd concentration. The ICP results reveals that the Zn content in the film prepared with ammonium hydroxide is lower than that of without ammonium hydroxide. This lower Zn content will drive the conduction band offset between CZTS and ZnCdS towards the unfavourable structure. To modify the composition of the ZnCdS film prepared with ammonium hydroxide complexing agent to fit the absorber is the subject of intense investigation in our current work.

The most frequently used method for determining the dominating recombination path is temperature dependent voltage analysis. In order to find out the origin of recombination limitations of the current device, the JV-T measurements are performed as shown in **Figure 4.** For comparison, the JV-T curve of our CZTS device with CdS buffer is also plotted in the figure. While the CdS reference device gives an activation energy of around 1 eV, the Voc of

the ZnCdS device extrapolates to 1.22 eV at 0 K. The improved activation energy confirms the more optimized interface by using ZnCdS buffer. Nevertheless, considering the high band gap of the CZTS absorber (~1.5 eV), the 1.22 eV activation energy is still lower than the ideal value. Therefore, it can be concluded that the CZTS devices are still limited by interface recombination. How to further modify the interface to alleviate the severe recombination occurs at the heterojunction and to promote the efficiency overpass 10% are still the main focus of our next step work.

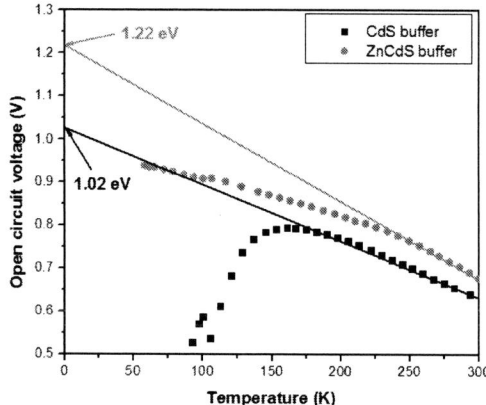

Figure 4. Temperature dependence of Voc of CZTS cells fabricating with CdS buffer (reference) and ZnCdS buffer.

IV. CONCLUSION

In this work, we investigated the microstructure and chemistry of the CZTS/ZnCdS interface. By controlling the composition of the absorber with low Zn content, we can obtain a self-assembled ultrathin ZnS layer with nanometer scale opening on the surface of absorber, which can reduce the interface defect and passivates the charge carrier recombination, thereby increasing the Voc as well as the device performance. In order to reduce the byproduct of oxide and hydroxide impurities formed during the SILAR process and improve the fill factor, ammonium hydroxide is used as complexing agent in the new recipe. As a consequence, a significant fill factor boost and a confirmed new word record efficiency of 9.5% for pure sulphide kesterite CZTS are achieved. Finally, JV-T measurement indicates that the dominant recombination mechanism for our champion CZTS device with ZnCdS buffer layer is still occurs at the interface and further modification of the heterojunction is under way.

REFERENCES

[1] M. A. Green, K. Emery, *et al.*, "Solar cell efficiency tables (version 49)," *Progress in Photovoltaics: Research and Applications,* vol. 25, pp. 3-13, 2017.

[2] K. Sun, C. Yan, F. Liu, J. Huang, X. Hao, *et al.*, "Over 9% Efficient Kesterite Cu2ZnSnS4 Solar Cell Fabricated by Using Zn$_{1-x}$Cd$_x$S Buffer Layer," *Advanced Energy Materials,* vol. 6, pp. 1600045, 2016.

Band Gap Changes of the CdS Buffer Induced by Post-Annealing of Cu₂ZnSn(S,Se)₄ Solar Cells

Mario Lang[1], Nicolas Schäfer[1], Christian Huber[1], Thomas Schnabel[2], Heinz Kalt[1], and Michael Hetterich[3]

[1]Institute of Applied Physics, Karlsruhe Institute of Technology (KIT), Wolfgang-Gaede-Str. 1, 76131 Karlsruhe, Germany

[2]Zentrum für Sonnenenergie- und Wasserstoff-Forschung Baden-Württemberg (ZSW), Meitnerstr. 1, 70563 Stuttgart, Germany

[3]Light Technology Institute, KIT, Engesserstr. 13, 76131 Karlsruhe, Germany

Abstract — Changes in band gap energy of the buffer layer of kesterite Cu₂ZnSn(S,Se)₄ solar cells were analyzed in dependence of different post-annealing temperatures and times using electroreflectance (ER). In order to measure the band gap of the buffer, the ER experiments were performed in "Brewster geometry" (angle of incidence and angle of detection equal Brewster's angle) allowing to reduce line shape distortions induced by interference effects and hence allowing reliable analysis of the ER spectra. Due to post-annealing, an irreversible reduction of the buffer layer's band gap is detected that increases for longer annealing times and higher temperatures. The decrease of band gap energy is attributed to an interdiffusion of atoms between the buffer and absorber layers.

Index Terms — CZTSSe, Electroreflectance, Buffer

I. INTRODUCTION

Post-annealing of kesterite Cu₂ZnSn(S,Se)₄ (CZTSSe) solar cells can be used to tune the band gap energy of the absorber layer [1],[2]. This property offers the possibility to bring the band gap energy closer to the optimal value without changing the composition of the material and hence can also be used to increase efficiency [3]. The variation in band gap energy is achieved by changing the degree of the so-called Cu–Zn disorder of the Cu and Zn atoms in the Cu–Zn planes [4],[5]. The degree of disorder can be changed by thermal treatments and since the band gap is directly related to the degree of disorder the thermal treatments do not only change the degree of disorder, but they also allow for changes in the band gap energy [1],[2],[5].

However, in post-annealing treatments of full solar cells the band gap energy of the absorber changes *irreversibly* for the case of too high annealing temperatures and too long processing times [1]. A possible reason for this irreversible change is the degradation of the absorber/buffer interface induced by, e.g., diffusion of Cd into the absorber [1]. Such a diffusion process should also lead to a detectable change in the properties of the buffer layer, especially the band gap.

In this work, we use electroreflectance (ER) in "Brewster geometry" (angle of incidence and angle of detection equal Brewster's angle) for investigating the changes of the buffer layer's band gap induced by thermal annealing procedures. We find that the band gap energy of the buffer layer decreases irreversibly with increasing annealing temperature and time.

II. SAMPLE FABRICATION

The standard fabrication technique for the used CZTSSe solar cells is a wet-chemical two-step approach as described by Schnabel *et al.* [6]. The precursor solution is doctor-bladed onto a Mo-coated soda-lime glass and subsequently annealed in a Se atmosphere in a rapid-thermal annealing system under N₂ pressure. The CdS buffer layer is deposited wet-chemically and an intrinsic ZnO window layer is sputtered on top. The cell is finished with an aluminum-doped ZnO front contact. Hence, the cells have the standard layer sequence of Mo/CZTSSe/CdS/i-ZnO/Al:ZnO. Typically, the [S]/([S]+[Se]) ratio is approximately 10 %.

III. ELECTROREFLECTANCE OF THE BUFFER LAYER

In this chapter, we will describe the experimental ER setup used to determine the band gap of the buffer layer in complete solar cells in more detail.

ER measurements are performed using a 0.32-m focal-length monochromator, equipped with a Xe lamp. Reflected light is detected with an electrically cooled Si photodiode in "Brewster geometry" (more information below). A reverse square wave voltage is applied using a function generator

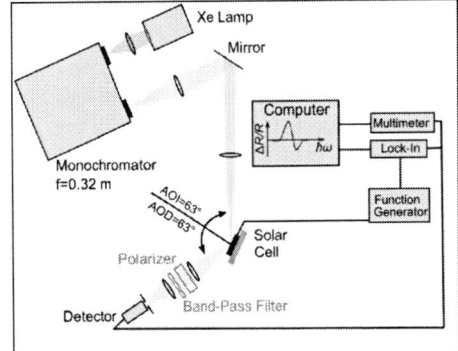

Fig. 1 Electroreflectance setup for measurements in "Brewster geometry". The angle of incidence (AOI) is chosen to match Brewster's angle of ZnO, and only p-polarized light is detected. For measurements in diffuse reflection, the AOI and the angle of detection are chosen independently.

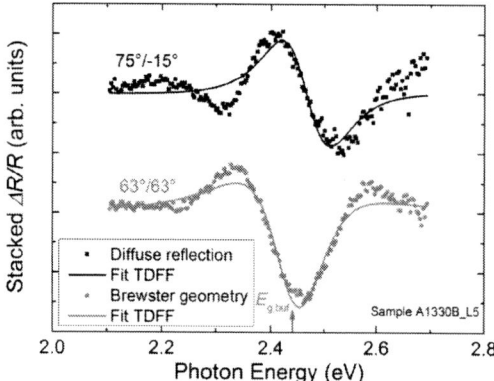

Fig. 2 ER spectra of a CdS buffer layer in a complete CZTSSe solar cell measured in diffuse reflection (black squares, angle of incidence $AOI = 75°$, angle of detection $AOD = -15°$) and in "Brewster geometry" (red circles, $AOI = AOD = 63°$). The straight lines are the TDFF fits to the data points. The red arrow indicates the band gap value determined by the fit.

with a frequency of 223 Hz to perform the electromodulation. The reflectance R was evaluated using a digital multimeter whereas the change in reflectance ΔR was evaluated by a lock-in amplifier.

ER spectra of thin-film solar cell absorbers are generally distorted due to thin-film interferences which can easily lead to misinterpretation of ER spectra [7],[8]. In order to reduce these interference distortions, diffusely reflected light can be used for ER analysis as shown by Krämmer et al. [8].

Measurements of the buffer layer's band gap are also affected by thin-film interferences since reflections at the front surface can interfere with reflections at the buffer layer. Since the photon energy of the light used for buffer measurements is significantly higher than the band gap of the absorber, there are no reflections at the backside of the absorber layer. In order to reduce the interferences in ER, diffuse reflection can also be used. However, depending on the investigated sample (e.g., its front surface roughness), the interference distortions do not vanish completely and a clear analysis of ER is problematic.

Therefore, we use ER in "Brewster geometry" to decrease interference distortions for the present analysis. In this approach, a polarizer is used in the detection path of the ER setup to detect only p-polarized light. The angle of incidence (AOI) and the angle of detection (AOD) are chosen to match Brewster's angle for ZnO ($AOI = AOD = 63°$) [9] as to prevent reflections from the front surface (air/ZnO interface). The experimental setup is schematically shown in Fig. 1.

The differences between ER spectra taken in diffuse reflection and "Brewster geometry" are shown in Fig. 2 together with the fit of the theoretically expected line shape (Third-Derivative Functional Form, TDFF) [10]. It is obvious that the spectrum measured in "Brewster geometry" is represented much clearer by the TDFF fit.

We conclude that ER in "Brewster geometry" is a powerful tool to characterize buffer layers in thin-film solar cell devices.

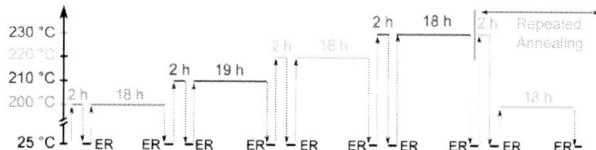

Fig. 3 The different steps of the annealing series. The same sample was heated to each temperature for 2 h or 18 h/19 h and then quickly cooled down. After each cool-down process the band gap was measured with ER.

IV. INFLUENCE OF ANNEALING ON THE BUFFER LAYER

As mentioned in chapter I, too high annealing temperatures and too long processing times lead to irreversible changes of the absorber band gap energy as observed by Krämmer et al. [1]. Interdiffusion (e.g., of Cd from the buffer into the CZTSSe absorber) is suspected to cause this irreversible shift which should also change the buffer composition and thus its properties.

In order to study the influence of thermal annealing on the CdS buffer band gap energy, the same solar cell sample underwent several annealing steps in a graphite box under N_2 pressure in a rapid thermal annealing system. After each annealing step, the band gap of the buffer was determined from the ER spectrum. The series of the different annealing steps is shown in Fig. 3. The annealing temperature was chosen to be higher than the critical temperature for the order–disorder transition, which has been determined to be $T_c = 195 \pm 5\ °C$ in this kind of CZTSSe samples [1]. Above this temperature in thermal equilibrium the sample is completely disordered meaning that the Cu and Zn atoms are randomly distributed in the Cu–Zn planes and the absorber's band gap is minimal [1],[2],[5].

At first the influence of the annealing time at one fixed temperature was investigated. To this end, a short annealing (2 h) at one fixed temperature was followed by a long

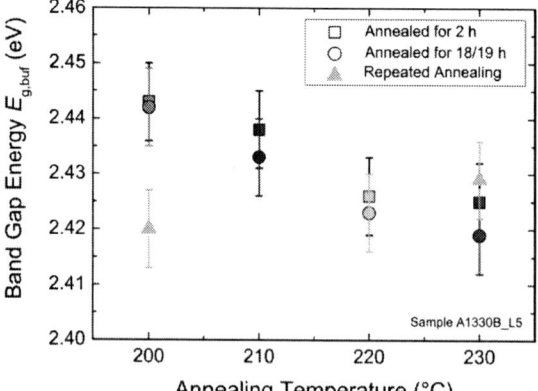

Fig. 4 Band gap energy $E_{g,buf}$ of the buffer layer for the different annealing temperatures and times (squares 2 h, circles 18/19 h). Turquois triangles show $E_{g,buf}$ after the repeated annealing steps at $T = 200\ °C$ and $T = 230\ °C$.

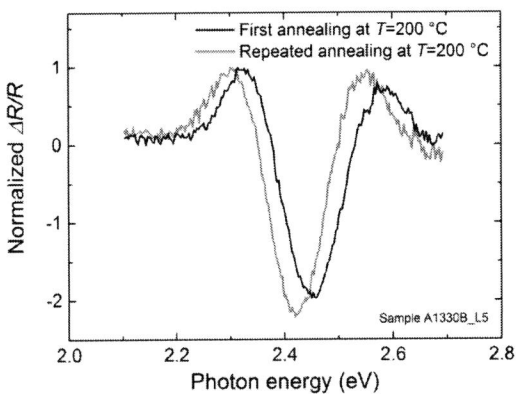

Fig. 5 ER spectra for the same annealing temperature ($T = 200$ °C) and time (18 h) at the beginning of the annealing series (black, first annealing) and after the annealing series (red, repeated annealing).

annealing (18 h) at the same temperature. In order to also investigate the influence of the annealing temperature on the band gap of the buffer layer this procedure was repeated for higher temperatures ($T = 210$ °C, $T = 220$ °C and $T = 230$ °C). Each annealing step was followed by a rapid cool-down to room-temperature in less than three minutes and a measurement of the band gap with ER.

The influence of the annealing procedures on the buffer layer's band gap is shown in Fig. 4. The data points are determined by fitting a Gaussian function to the modulus spectrum of the $\Delta R/R$ spectrum and taking the energetic position of the maximum [11]-[13]. The error bars are determined by manually estimating the upper and lower limit of an acceptable Gaussian fit. A Gaussian function was used since the line shape is not Lorentzian anymore due to inhomogeneous broadening [14].

A clear trend to a decreased band gap with increasing annealing temperature is seen, and also an increase of the annealing time at one fixed temperature tends to reduce the band gap. This is a sign of a structural/compositional change in the buffer layer if complete solar cells are post-annealed.

The observed changes should be irreversible, especially further annealing procedures should not reverse the observed trend. In order to prove the irreversibility two further annealing procedures were performed named as "repeated annealing" in Figs. 3-5. The sample was again annealed at $T = 230$ °C for 2 h and was repeatedly annealed at $T = 200$ °C for 18 h (Fig. 3). The spectrum measured after the "repeated annealing" at $T = 200$ °C is clearly shifted with respect to the one after the first annealing (Fig. 5). The difference in band gap energy is approximately $\Delta E_{g,buf} \approx 20$ meV which confirms the expected irreversibility.

Since the band gap changes are irreversible, we attribute them to interdiffusion between the buffer and the absorber, potentially involving a Cd loss in the buffer and alloying with elements from the absorber. Further structural studies into the chemistry at the interface are required to clarify this aspect in more detail.

V. CONCLUSION

In conclusion, we have demonstrated that the band gap energy of buffer layers in CZTSSe solar cells can be determined using ER spectroscopy in "Brewster geometry". The "Brewster geometry" helps to reduce line shape distortions induced by interference effects. Utilizing this approach, we have presented evidence, that annealing of complete solar cells induces an irreversible reduction of the buffer layer's band gap. This reduction increases for longer annealing times and higher temperatures. We attribute this band gap change to interdiffusion effects between buffer and absorber.

ACKNOWLEDGMENTS

We acknowledge financial support by the German Federal Ministry of Education and Research (BMBF), FKZ 03SF0530B, and the Karlsruhe School of Optics and Photonics (KSOP) at KIT.

REFERENCES

[1] C. Krämmer, C. Huber, C. Zimmermann, M. Lang, T. Schnabel, T. Abzieher, E. Ahlswede, H. Kalt, and M. Hetterich, "Reversible band gap changes in Cu$_2$ZnSn(S,Se)$_4$ via post-annealing of finished solar cells measured by electroreflectance," *Applied Physics Letters*, 105, 262104, 2014.

[2] G. Rey, A. Redinger, J. Sendler, T. P. Weiss, M. Thevenin, M. Guennou, B. El Adib, and S. Siebentritt, "The band gap of Cu$_2$ZnSnSe$_4$: Effect of order-disorder, " *Applied Physics Letters*, 105, 112106, 2014.

[3] C. Krämmer, C. Huber, T. Schnabel, C. Zimmermann, M. Lang, E. Ahlswede, H. Kalt, and M. Hetterich, "Order-disorder related band gap changes in Cu$_2$ZnSn(S,Se)$_4$: Impact on solar cell performance, " in *42nd IEEE Photovoltaic Specialist Conference*, p. 1-4, 2015.

[4] S. Schorr, "The crystal structure of kesterite type compounds: A neutron and X-ray diffraction study, " *Solar Energy Materials & Solar cells 95*, 1482-1488, 2011.

[5] J. J. S. Scragg, L. Choubrac, A. Lafond, T. Ericson, and C. Platzer-Björkman, "A low-temperature order-disorder transition in Cu$_2$ZnSnS$_4$ thin films," *Applied Physics Letters*, 104, 041911, 2014.

[6] T. Schnabel, M. Löw, and E. Ahlswede, "Vacuum-free preparation of 7.5% efficient Cu$_2$ZnSn(S,Se)$_4$ solar cells based on metal salt precursors," *Solar Energy Materials & Solar Cells*, 117, 324-328, 2013.

[7] C. Huber, C. Krämmer, D. Sperber, A. Magin, H. Kalt, and M. Hetterich, "Electroreflectance of thin-film solar cells: Simulation and experiment," *Physical Review B*, 92, 075201, 2015.

[8] C. Krämmer, C. Huber, A. Redinger, D. Sperber, G. Rey, S. Siebentritt, H. Kalt and M. Hetterich, "Diffuse electroreflectance of thin-film solar cells: Suppression of interference-related lineshape distortions," *Applied Physics Letters*, 107, 222104, 2015.

[9] Z. Qiao, C. Agashe, D. Mergel, "Dielectric modeling of transmittance spectra of thin ZnO:Al films," *Thin Solid Films* 496, 520-525, 2006.

[10] D. E. Aspnes, "Third-Derivative Modulation Spectroscopy with Low-Field Electroreflectance," *Surface Science*, 37, 418-442, 1973.

[11] T. J. C. Hosea, "Estimating Critical-Point Parameters from Kramers-Kronig Transformations of Modulated Reflectance Spectra," *Physica Status Solidi B,* 182, K43-K47, 1994.

[12] T. J. C. Hosea, "Estimating Critical-Point Parameters of Modulated Reflectance Spectra," *Physica Status Solidi B*, 189, 531-542, 1995.

[13] A. Grau, T. Passow, and M. Hetterich, "Temperature dependence of the GaAsN conduction band structure," *Applied Physics Letters,* 89, 202105, 2006.

[14] C. Krämmer, "Optoelectronic Characterization of Thin-Film Solar Cells by Electroreflectance and Luminescence Spectroscopy," *Dissertation*, Karlsruhe Institute of Technology 2015.

22.61% Efficient fully Screen Printed PERC Solar Cell

Weiwei Deng*, Feng Ye, Ruimin Liu, Yunpeng Li, Haiyan Chen, Zhen Xiong , Yang Yang, Yifeng Chen, Yongqian Wang, Pietro P. Altermatt, Zhiqiang Feng and Pierre J. Verlinden

State Key Laboratory of PV Science and Technology, Trina Solar, Changzhou, Jiangsu, China, 213031.
*weiwei.deng@trinasolar.com

Abstract —**Passivated emitter and rear solar cells (PERC) on p-type Cz Si wafers are currently being migrated to mainstream production, with ongoing improvements in recent years. We describe and characterize our recent batch of PERC cells, fabricated on 156×156 mm^2 wafers with fully screen printed technology and an industrial type process. The champion cell efficiency has reached 22.61%, with a V_{oc} of 684.4 mV and a fill factor of 81.49%, confirmed by Fraunhofer CalLab. The co-optimization of the front diffusion and Ag-Si contact increases the fill factor, a lower J_0 and higher lifetime increase V_{oc}, the two main contributors to our recent cell efficiency improvements compared to the cell we fabricated last year. The characterization and simulation results show the emitter losses dominant among all of the losses.**

Index Terms — **PERC solar cell, emitter profile, contacts, characterization, ray tracing, flash tester, device modeling**

I. INTRODUCTION

Industrial PERC cell capacity is currently 15 GW globally (compared to 84 GW standard production), and is forecasted to expand globally by about 10 GW this year [1,2]. By 2020, there may be 61 GW of PERC production (compared to 77 GW standard), see Fig. 1. The advantage of PERC cells – higher efficiencies – needs to be greater than their disadvantage – higher manufacturing costs. This paper describes how the efficiency of our champion mono PERC cells has been increased from 22.13% of last year [3] to 22.61% this year. Such advances are continuously transferred to mass production, currently having median efficiency of 21.3%.

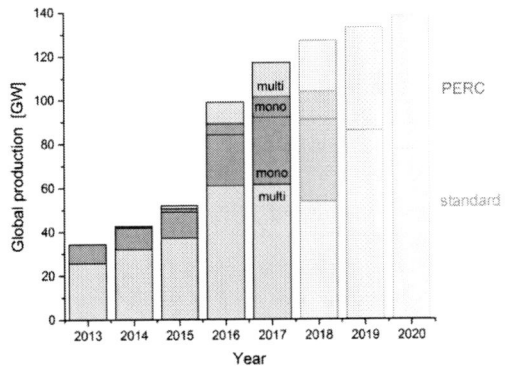

Fig. 1: Recorded and forecast [1,2] global production of PERC and standard cells on multi or mono Si wafers.

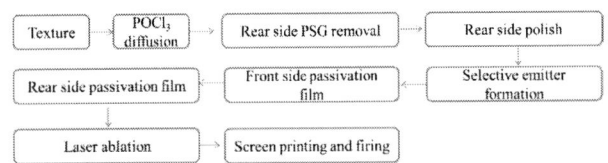

Fig. 2: Process flow for PERC solar cell fabricated at Trina.

II. CELL PROCESS AND IV CHARACTERIZATION

Our PERC cells were manufactured from commercially available 156×156 mm^2 Cz Si wafers with resistivity between 1 Ωcm and 2 Ωcm. The cells were fabricated using the industrial type process shown in Fig. 2: texturing by TMAH; cleaning with HF/HCl solution; emitter formation by POCl$_3$ diffusion in a tube furnace; etching and polishing of rear in HF and TMAH solution; selective emitter etching in HF/HNO$_3$; rear and front passivation by PECVD; picosecond laser ablation at rear; screen printing and firing.

The IV performance of the champion cell was measured at Fraunhofer ISE CalLab and is given in Table I. We repeated our in-house measurement of all the cells in the batch with the new calibration, using a Sinton FCT-450 flash tester, and reached a median efficiency of 22.55% as shown in Fig. 3. Our measured FF had to be multiplied by 1.002 to match CalLab's value. This is very little, as contacting cells by flash tester is more difficult than in a cw measurement like CalLab. Fig. 3 shows box plots, where the box height spans from the 1st to the 3rd quartile (called the interquartile range, IQR),

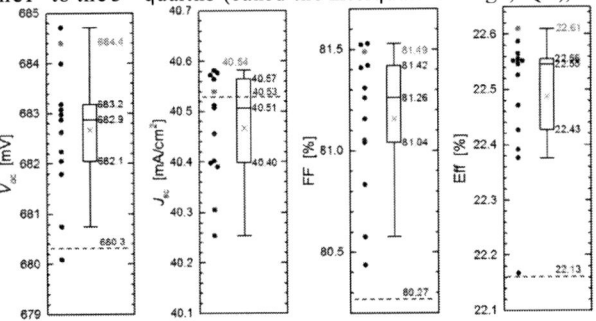

Fig. 3: In-house IV measurements of the champion batch with a Sinton FCT-450 flash tester after calibration of the champion cell (red) by CalLab, see Table I. Blue: champion cell of last year. The median line differs from the average (cross) due to asymmetric data.

978-1-5090-5606-4/17 $31.00 © 2017 IEEE

and the whiskers extend maximally by 1.5·IQR, indicating the border to outsiders. Note that the improvements from last year (indicated by the blue dashed line) are mainly due to a higher FF, followed by a higher V_{oc}. The fill factor increased 1.22%$_{abs}$, V_{oc} by 4 mV [3]. The co-optimization of the P diffusion profile and of the Ag-Si contact, as well as better controlling of the cell process ensures the high fill factor and V_{oc}, which will be discussed in the following part.

TABLE I

THE I-V PARAMETERS OF THE BEST PERC CELL, MEASURED AT CALLAB, AND MEDIAN VALUES OF THE BATCH, MEASURED IN-HOUSE.

	V_{oc}	J_{sc}	FF	Eff	V_{mpp}	J_{mpp}	Cell area
	(mV)	(mA /cm^2)	(%)	(%)	(mV)	(mA/ cm^2)	(cm^2)
Median	682.9	40.51	81.26	22.55	582.64	38.34	
Best cell	684.4	40.54	81.49	22.61	587.9	38.48	243.23

III. RECENT IMPROVEMENTS

Based on the analysis of cells we fabricated last year, one of the directions to improve the fill factor is to reduce the series resistance caused by the high contact resistance. The emitter diffusion profile and the Ag paste are the key for the contact resistance. So experiments were first designed aiming to co-optimize the diffusion profile and front metallization. We changed the emitter profile so it has a higher peak P density under the metal front contact but a lower profile in the region deeper than the well-known kink. The contact resistivity R_c was compared among cells with different emitters, using the transfer length method (TLM). To ensure good contact between the probe and the fingers, the finger was made about 200 μm wide, which is much wider than in real cells. The results are shown in Fig. 4. To compare R_c of two different emitters, we use a notched box plot [4]. The notch indicates the confidence interval of the median. If the median is taken from a larger number n of measurement points, its confidence interval shrinks according to $\pm 1.57 \cdot$IQR/$\sqrt{n}$. Hence, two experiments yield a statistically significant difference in median only if the two notches do not overlap, as is indeed the case in Fig. 4. Hence, the newly developed emitter profile enables a decreased R_c. Note that notched box plots can also be used to determine the minimal number of samples necessary to draw relevant conclusions between two experiments. The smaller the expected difference in median, the larger the number of samples must be.

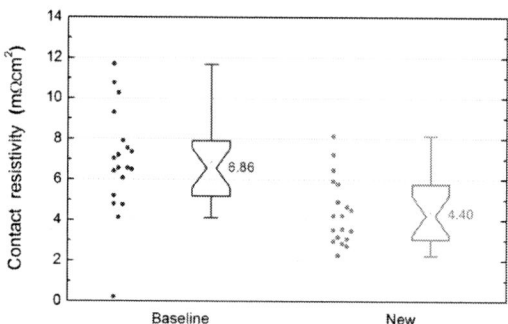

Fig. 4: The contact resistivity for cells with a different emitter profile, measured using the TLM method, and visualized with notched box plots explained in the text.

Subsequently, PERC cells were fabricated to compare the IV performance of different emitter profiles, see Table II. There is a statistically significant improvement in fill factor which is consistent with the contact resistivity measurements on the cells with different emitter profile. The minor decrease in V_{oc} is probably caused by the increased J_0 in the heavily diffused area.

TABLE II   THE I-V PARAMETERS FOR THE CELLS WITH DIFFERENT EMITTER PROFILE, MEASURED IN-HOUSE.

	V_{oc}	J_{sc}	FF	Eff	R_s
	(mV)	(mA /cm^2)	(%)	(%)	Ω cm^2
Baseline	686.6	39.85	80.44	21.99	0.574
New diffusion profile	685.7	39.93	80.78	22.12	0.532

After the diffusion profile optimization, a newly developed Ag paste 'B' was introduced on the newly developed emitter, and compared with the baseline paste A. From the TLM procedure we conclude in Fig. 5 that R_c is statistically the same for paste A and B, but the line resistiviy is significantly lower for paste B, reducing the lumped series resistance R_s of the cells and, in turn, increasing FF further. Note that by looking solely at the data points in Fig. 5(a), one would be rather unsure whether paste B yields a lower line resistivity, but the notches do clearly not overlap. Also note in Fig, 5(b) that the median values (the corner of the notches) are very similar, while the average values (crosses) are not. This is so because the average is strongly influenced by data points far away from the average, while the median is robust even to possible outsiders. This is why robust statistics is more suitable for evaluating such experiments. Also note that the number of samples in Fig. 5(b) would need to be increased considerably to have a chance of discerning any difference between the pastes, e.g. in a production line. Production lines offer the possiblity to improve cells in very small increments in a statistically save way.

978-1-5090-5606-4/17 $31.00 © 2017 IEEE

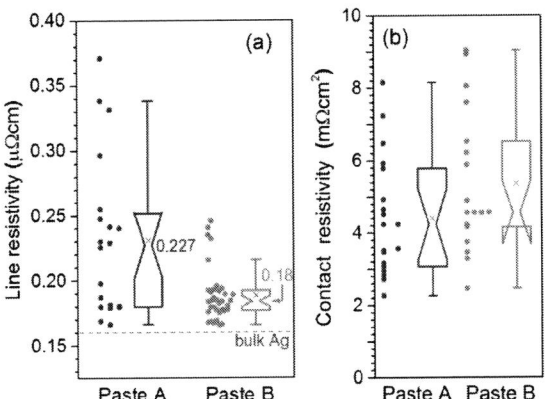

Fig. 5: Contact resistivity and line resistivity of paste A and B, measured with TLM.

IV. CELL ANALYSIS

In each batch, a set of wafers is only partially processed to serve as monitor wafers for lifetime, J_0, and dopant profile (ECV) measurements. The experiment is designed to monitor the different parameters of the cell.

The J_0 monitoring plan for the emitter is shown in Fig. 6. High resistivity Cz wafers were used, but texturing and diffusion processes were the same as in the cells because all monitoring wafers are processed simultaneously with the cells.

Fig. 7 shows the J_0 measurements on 51 or 29 lifetime samples, respectively, with either the lowly-doped or heavily-doped passivated emitter part on both textured surfaces. As shown in Ref. [7,8] by a combination of measurements and numerical modeling, the relevant J_0 value needs to be taken at the injection density yielding the highest value of J_0, and this value is still slightly lower than without measuring artifacts. Also Ref. [9] shows that care must be taken not to underestimate J_0, particularly in diffusions where J_0 is low. Note in Fig. 7 that the relevant injection density is

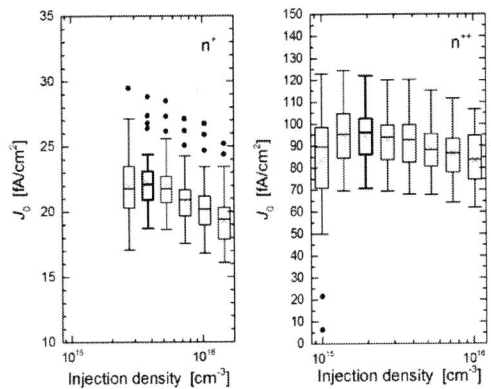

Fig. 7: Measurements of J_0 of the lowly or the heavily doped, passivated emitter part on 51 or 29 lifetimes samples, respectively.

higher if the diffusion has a higher J_0, as is explained in Ref. [8].

The quality of the rear side passivation and the J_0 of the rear side LBSF was evaluated using the monitor wafers following the process and sample structure shown in Fig. 8. The aluminum can only be deposited on one side to ensure firing as usual. To still have lifetime samples as symmetric as possible, three different kinds of monitor wafers were prepared. The first wafer was passivated on both sides, the second wafer, too, but then was Al screen printed on one side, and the third wafer obtained the final rear structure of the cells on one side and the passivation on the other side. Normally, the passivation film underneath the Al is changed during the firing process. So the J_0 of the passivation film needs to be measured also on the samples with the Al screen printing and firing. Before the lifetime measurement, the Al layer was removed in HCl and cleaned to make sure there are no residuals of Al paste.

Emitter monitoring		
40 Ωcm wafers		
Saw damage etch		
Texturing		
Cleaning		
Diffusion		
PSG removal		/
/		Etch to ~130 Ω/sq
Cleaning		
SiN$_x$ deposition double sides		
/	Print TLM	/
Firing		
$J_{0,n++}$	R_c	$J_{0,n+}$

Fig. 6: The process flow for emitter monitoring

Rear side monitoring		
High resistivity wafer (20 Ωcm)		
5 pc	10 pc	10 pc
Saw damage etch		
Cleaning		
Passivation film deposition on both sides		
/		Laser ablation
/	Rear Al	Rear Al
firing		
/	remove Al	remove Al
$J_{0,pass}$	$J_{0,pass,Al}$	$J_{0,LBSF}$

Fig. 8: The process flow for monitoring the rear side passivation quality and $J_{0,LBSF}$, and the resulting sample structure.

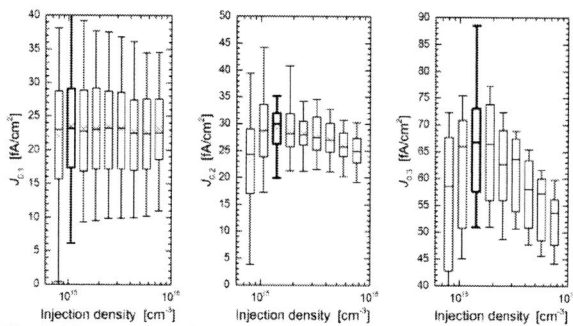

Fig. 9: Measured J_0 of the samples for the rear side monitoring

The first samples allow us to measure $J_{0.1} = 2J_{0.pass}$ and hence J_{0pass} of the passivation film. We measured each of the 5 wafers at 5 locations: one at the center, four at each corner, yielding 25 measurements. See Fig. 9. The passivation in the second samples has the same thermal history as in cell fabrication, and so they allow us to determine the relevant $J_{0.passAl}$ of the rear passivation. With $J_{02} = J_{0,pass} + J_{0,passAl}$ we obtain $J_{0,passAl} = 18.3$ fA/cm². This is for the rear passivated, Al covered surface of a 20 Ωcm wafer: However, our PERC cells have wafers near 2 Ωcm. Using $S_{eff} = (\Delta n + N_{dop})/(qn_i^2)$, there is $S_{eff} = 1.1421 \times 10^{-15} \Delta n$ cm/s. A comparison with literature data [6] yields a good agreement with a medium passivation quality. For 1.65 Ωcm PERC cells, this is $S_{eff} \approx 10$ cm/s, which is what we expect from literature.

For the determination of $J_{0,BSF}$, we use $J_{0.3} = J_{0.pass} + fJ_{0.LBSF} + (1-f)J_{0.passAl}$, where f is the fraction of LBSF area to the total area. We obtain $J_{0.LBSF} = 992$ fA/cm² for the LBSF area. However, the contacts are elliptically shaped having width W near 80 µm and depth D near 20 µm, as shown in Fig. 10. Therefore, the effective area of the contact is enlarged. This can be calculated with half the diameter of an ellipse:

$$A_{eff} = \frac{\pi}{4}\left[3\left(\frac{W}{2}+D\right) - \sqrt{\left(3\frac{W}{2}+D\right)\left(\frac{W}{2}+3D\right)}\right]$$

so for our case the effective width is 48 µm, i.e. enlarged by the factor 1.2. Hence, we need to divide the 992 fA/cm² by 1.2 to obtain the area-weighted $J_{0,LBSF}$ of 827 fA/cm². A comparison with literature data [10] shows that this value is reasonable for a 4-5 µm deep LBSF. Reproducing J0.LBSF with Sentaurus simulations shows that about 1/1000th of all possible Al-O complexes are activated during firing of the contact, which is in the usual range.

The contributions to the higher V_{oc} can be identified in a cell analysis. In each batch, a set of wafers are only partially processed to serve as samples for lifetime, J_0, and dopant profile (ECV) measurements. Fig. 6 shows the measured effective lifetime τ_{eff} of dummy wafers (boxes). The SRH lifetime τ_{SRH} can be fitted as BO complex (having $\tau_p = 10\tau_n$) self consistently with the effective surface recombination velocity S_{eff} of the passivation stack, using

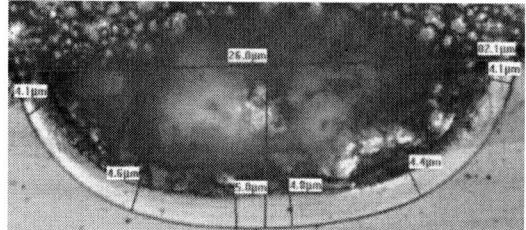

Fig. 10: Cross sectional view of the LBSF measured with 3D microscopy.

$$\frac{1}{\tau_{eff}} = \frac{1}{\tau_{bulk}} + \frac{1}{\tau_{surf}} \approx \frac{1}{\tau_{SRH}} + \frac{1}{\tau_{intr}} + 2\frac{S_{eff}}{W},$$

with wafer thickness W = 175 µm. This results in $\tau_n = 0.35$ ms, i.e. in a median bulk lifetime $\tau_{bulk} = 0.58$ ms at mpp (and 1.47 ms at open-circuit), and the self-consistent, modeled S_{eff} shown in Fig. 11. S_{eff} is modeled with the fixed charge and the density-of-states, measured by means of C-V, and with applying asymmetric capture cross-sections [5,6], although our passivation is not a Al_2O_3-SiN_x stack but similar.

The reason why the champion cell has a high V_{oc} can be found with the Sinton FCT-450 flash tester, which also measures – via a suns-V_{oc} procedure – the injection-dependent τ_{eff} of the finished cell, together with the sum of its J_0 at the front and back. Care must be taken to determine J_0 at the correct injection density, as low-injection mimics too low J_0 values. For the champion cell, we obtain $J_0 = 100$ fA/cm², while we obtain 120 fA/cm² for the area-weighted sum of our median measurements of $J_{0,n+}$, J_{0n++} and $J_{0,metal}$ at the front, and of the $J_{0,LBSF}$ and $S_{eff} = J_0(\Delta n + N_{dop})/qn_i^2$ at the rear. In order to fit the lifetime of the champion cell from the flash tester, we need to assume $\tau_n = 0.7$ ms instead of $\tau_n = 0.35$ ms. Hence, the V_{oc} is large compared to the median of that batch because J_0 at front and back is 20 fA/cm² lower, and due to a higher τ_n. A full Sentaurus model confirms this.

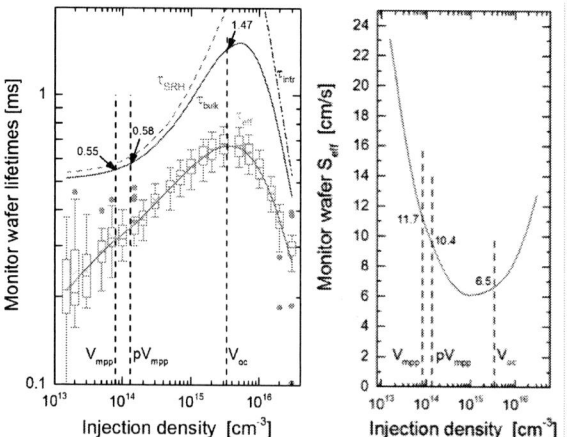

Fig. 11: Measured effective lifetime τ_{eff} of dummy wafers (boxes), and a self-consistent SRH fit of the bulk and surface recombination.

The analysis of recombination losses is routinely done with Sentaurus device simulations. Sentaurus solves the coupled set of semiconductor equations and therefore does not accept J_0 values. To incorporate the J_0 measurements into our simulations, we model the J_0 experiment with the dopant profiles measured by electrochemical capacitance-voltage (ECV), using the Profiler CVP21 from WEP and the procedures described in Ref. [11]. The only unknown – the surface SRH recombination parameter S_p – is adjusted such that the measured J_0 is obtained. S_p is then compared to previous batches and to literature values to monitor the surface passivation quality. Monitoring this quality is not trivial because a change in J_0 may not only be caused by a different passivation quality, but also by a different dopant profile, causing a different amount of Auger recombination and a different surface dopant density which, in turn, leads to a different S_p value despite of maintaining the same passivation quality. Combining Sentaurus simulations with J_0 and ECV measurements disentangles the various contributions and accelerates the improvement of emitter quality.

With the independently measured device parameters, the entire cell is routinely modeled with Sentaurus. In most cases, no adjustment of device parameters is necessary for reproducing the measured IV parameters to a precision as listed in Table III. In case the simulations reproduce the measured IV parameters not as well, it is usually best to check/repeat the J_0 measurements, as they can be rather imprecise [12]. Generally, it is insufficient to fit the simulated V_{oc} to the measured one by varying the bulk lifetime (as is sometimes done in the literature). Without a precisely known bulk lifetime, there is a risk to compensate imprecise input parameters with overestimated losses in wrong device regions. Hence, a good fit of V_{oc} is not a guarantee for a correct simulation. A wrong distribution of recombination losses among the various device regions may lead to wrong predictions for cell improvements.

Fig. 12 shows the simulated recombination losses at 1-sun. It can be expected from an optimized cell that at least two if not more device regions have similar amounts of recombination at mpp. This is so because reducing only one of them improves cell efficiency considerably less than if there is one dominating loss that is reduced. Hence, tuning multiple device regions to a similar amount of loss creates a kind of 'plateau' in efficiency improvements. This also implies that a further efficiency improvement is accomplished best when all involved device regions are improved simultaneously. In our case, it is the emitter, base and the rear contacts that dominate the losses equally strongly, while there is considerably less recombination in the LBSF's and at the rear passivated surface (Fig. 12 is plotted logarithmically). The recombination at the front metal contacts is possibly underestimated but they are not significant compared to the total losses. We have been unable to quantify them experimentally in a reliable way.

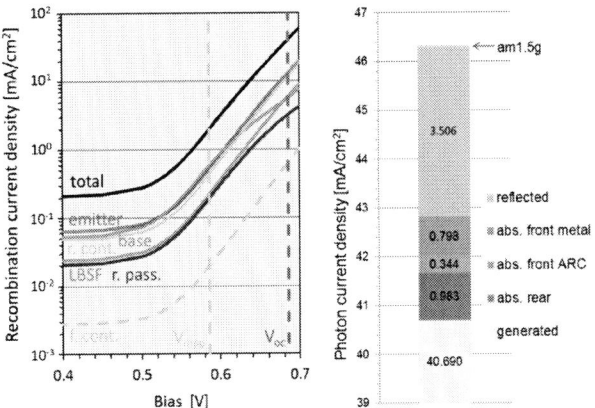

Fig. 12: The recombination losses in each device part obtained from Sentaurus device simulations at 1-sun, and the optical losses of the cell in air, obtained with the PV Lighthouse ray tracer.

Instead of quantifying the losses only at mpp, their dependence on bias reveals their dynamics. For example the saturation behavior of the base towards open-circuit is due to injection-dependent bulk lifetime and may significantly lower the fill factor (while thermal donors in n-type may significantly increase FF [13]). If changes in FF from batch to batch are solely attributed to changes in R_s, one may be led to a wrong path for further improvements. The Sinton FCT-450 flash tester is a valuable tool for measuring the injection-dependent lifetime of finished cells and hence for monitoring PERC manufacturing. There are considerably more device and process parameters in PERC cells than in standard cells, and the parameters are interlinked in a rather complex way. Hence, for improving PERC performance, a detailed and precise, yet fast and reliable monitoring e.g. with the FCT-450, is very valuable.

TABLE III   THE IV PARAMETERS OF THE CHAMPION CELL MEASURED BY CALLAB AND SIMULATED WITH SENTAURUS.

	V_{oc}	J_{sc}	FF	Eff
	(mV)	(mA /cm^2)	(%)	(%)
CalLab	684.4	40.54	81.49	22.61
Simulation	684.0	40.64	81.47	22.65

The analysis of resistive losses is routinely done by a combination of measurements and numerical device modeling. The internal resistance $R_{s\,int}$ is computed with the device model, the resistance of the metallization $R_{s\,grid}$ with a SPICE model [14] where the simulated IV curves are a direct input (without the need to fit the simulated IV curves with a 2-diode model). Fig. 13 shows the resulting lumped total series resistance R_s of the cell. It is extracted from three IV curves at 1-sun and ±10% light intensity using the method in Ref. [15].

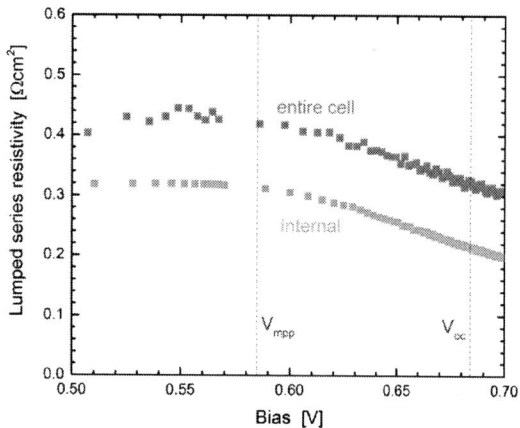

Fig. 13: Lumped series resistance, extracted from simulated IV curves.

The value at mpp (0.42 Ωcm^2) is compatible with measurements either using the LOANA tool and Ref. [12] or using the Sinton FCT-450 flash tester. It is not compatible with the value we obtained using our h.a.l.m. tester, possibly because the internal capacitance of the cell is about three times higher (28 μF/cm^2) than in standard cells and our h.a.l.m. tester may need an update. The cell's capacitance increases strongly with V_{mpp}.

$R_{s,int}$ decreases towards higher bias for two main reasons: towards open-circuit, there are less and less electrons flowing laterally along the emitter, and the base becomes more conductive towards higher injection. Such effects become particularly strong in improved emitters with a rather high sheet resistivity. $R_{s,grid}$ is rather independent of bias in most front metal grid designs. Only if narrow fingers for improving J_{sc} cause a rather high resistance, $R_{s,grid}$ may depend on bias due to distributed resistance effects.

Finally, we conducted an optical loss analysis using the online Module Ray Tracer (MRT) from PV Lighthouse to ray trace the cell, using material data extracted from ellipsometry, transmission and reflection measurements, and some refractive-index data from the PV lighthouse library where applicable. Particular care is taken for evaluating the material data at the rear of the cell, because parasitic absorption at the rear is the second largest optical loss after the front reflection, see Fig. 12. The details of our procedure are published in Ref. [16] and therefore not given here. The ray tracing yields a generation current of $J_{gen} = 40.69$ mA/cm^2 within the silicon. Inserting the optical generation profile from the MRT into Sentaurus device simulations yields a recombination loss at short-circuit of $J_{rec,sc} = 0.21$ mA/cm^2, which results in $J_{sc} = 40.48$ mA/cm^2, which is within measurement uncertainties equal to the experimental median of 40.51 mA/cm^2.

V. CONCLUSIONS

A decreased contact resistivity (from 6.8 to 4.4 mΩcm^2) and a reduced front finger line resistivity (from 0.23 to 0.18

$\mu\Omega$cm) are the main contribution to the high fill factor. J_0 reduced by 20 fAcm^2 and the bulk lifetime increased from $\tau_p = 0.35$ to 0.7 ms are the main contribution to the high V_{oc}, leading to an efficiency increase from 22.13% to 22.61%. Most of the device optimization can be transferred to the production lines because they are materialized with industry equipment and do not add significant cost. Because PERC cells have considerably more device and process parameters than standard cells, a detailed and precise monitoring and analysis is advisable.

ACKNOWLEDGEMENT

This work is supported by the Natural Science Foundation of Jiangsu Province under the Project Number of SBK2017030242.

REFERENCES

[1] http://pv.energytrend.com/research/PERC_Cells_Global _Production_Capacity_to_Reach_25GW_in_2017.html

[2] http://www.pv-tech.org/news/perc-solar-cell-migration-to -hit-25gw-in-2017-energytrend

[3] F. Ye, W. Deng, W.u Guo, R. Liu, D. Chen, Y. Chen, Y. Yang, N. Yuan, J. Ding, Z. Feng, P. Altermatt, P. Verlinden, "22.13% Efficient Industrial P-Type Mono PERC Solar Cell", in *43rd IEEE PV Specialists Conference*, 2016, p. 3360.

[4] R. McGill, J.W. Tukey, W.A. Larsen, "Variations of box plots", *The American Statistician*, vol. 32, 1978, pp. 12-16

[5] D. Schuldis, A. Richter, J. Benick, P. Saint-Cast, M. Hermle, and S. W. Glunz, "Properties of the c-Si/Al$_2$O$_3$ interface of ultrathin atomic layer deposited Al$_2$O$_3$ layers capped by SiN$_x$ for c-Si surface passivation", *Appl. Phys. Lett.* vol. 105, 2014, p. 231601.

[6] L.E. Black, T. Allen, K.R. McIntosh, A. Cuevas, "Effect of boron concentration on recombination at the p-Si–Al$_2$O$_3$ interface", *J Appl. Phys.* vol. 115 (2014), p. 093707.

[7] B. Min, A. Dastgheib-Shirazi, P.P. Altermatt, H. Kurz, "Accurate determination of the emitter saturation current density for industrial P-diffused emitters", in *29th EU PV Solar Energy Conference*, 2014, pp. 463 – 466.

[8] B. Liu, Binhui, Y. Chen, Y. Yang, D. Chen, Z Feng, P.P. Altermatt, P. Verlinden, H. Chen, "Improved evaluation of saturation currents and bulk lifetime in industrial Si solar cells by the quasi steady state photoconductance decay method", *Solar Energy Materials and Solar Cells*, vol. 149, 2016, pp. 258 – 65.

[9] A. Kimmerle, J. Greulich, A. Wolf, "Carrier-diffusion corrected J0-analysis of charge carrier lifetime measurements for increased consistency", *Solar Energy Materials and Solar Cells*, vol. 142, 2015, pp. 116–122.

[10] H. Steinkemper, M. Rauer, P. Altermatt, F.D. Heinz, C. Schmiga, M. Hermle, "Adapted parameterization of incomplete ionization in aluminum-doped silicon and impact on numerical device simulation", *J. Appl. Phys.* vol. 117, 2015, p. 074504.

[11] R. Bock, P.P. Altermatt, J. Schmidt, "Accurate extraction of doping profiles from electrochemical capacitance voltage measurements", Proc. *23rd EU PV Solar Energy Conference*, 2008, pp. 1510–1513.

[12] A.F. Thomson, Z. Hameiri, N.E. Grant, C.J. Price, Y. Di, J. Spurgin, "Uncertainty in photoconductance measurements of the emitter saturation current", *IEEE J. of PV*, vol. 3, 2013, 1200 –

1207.

[13] W. Duan, S. Yuan, Y. Sheng, W. Cai, Y. Chen, Y. Yang, P.P. Altermatt, Z. Feng, P.J. Verlinden, "A route towards high efficiency n-type PERT solar cells", Proc. *32nd EU PV Solar Energy Conference, Munich, Germany,* 2016.

[14] Y. Chen, Y. Yang, W. Deng, A. Ali, P.J. Verlinden, P.P. Altermatt, "Front metal finger inhomogeneity: its influence on optimization and on the cell efficiency distribution in production lines", *Energy Procedia* vol. 98, 2016, pp. 30 – 39.

[15] K.C. Fong, K.R. McIntosh, A.W. Blakers, "Accurate series resistance measurement of solar cells", *Prog. in PV* vol. 21, 2013, pp. 490 – 499.

[16] Y. Yang, R. Liu, K.R. McIntosh, M. Abbott, B. Sudbury, J. Holovsky, F. Ye, W. Deng, Z. Feng, P.J. Verlinden, P. P. Altermatt, "Combining ray tracing with device modeling to evaluate experiments for an optical analysis of crystalline Si solar cells and modules", *7th Int. Conf. on Si PV (SiliconPV)*, 2017 (to appear).

How to achieve 23% efficient large-area Cu plated n-PERT cells?

Monica Aleman[1], Angel Uruena[1,2], Emanuele Cornagliotti[1], Patrick Choulat[1], Joachim John[1], Richard Russell[1], Sukvhinder Singh[1], Loic Tous[1], Wen-Cheng Sun[1], Filip Duerinckx[1], Jozef Szlufcik[1]

[1]Imec, Kapeldreef 75, Leuven, 3001, Belgium
[2] now with Kaneka Belgium N.V., Nijverheidsstr. 16, Westerlo-Oevel, 2260, Belgium

Abstract — **This paper presents the different strategies which have been studied at imec in order to reduce the front surface recombination of our monofacial high-efficiency rear-junction (RJ) n-PERT diffused cells. These devices feature: a laser doped selective FSF on a lightly POCl₃ diffused front side; front contacts by Ni/Cu plating followed by immersion Ag plating; a diffused rear boron emitter passivated with an Al₂O₃/PECVD SiOₓ stack and sputtered rear AlSi contacts. Each of the three main parameters contributing to the front recombination has been tackled, starting from the reduction of the metal front recombination (by improving the laser doping process), then the reduction of the contact fraction with the implementation of busbar free cells and finally, a further reduction in the recombination of the passivated areas, by improving the thermal treatment of the dielectrics. This study shows that an average 23% efficiency is within reach for large-area double-side contacted RJ PERT cells, with voltages over 700 mV and an average FF of 80.7%.**

I. Introduction

After the introduction of the PERC technology showing a potential for efficiencies around 25% on lab-scale [1] and the development of industrial processes for the manufacturing of such cells [2], it is only recently that the industry is shifting its manufacturing from Al-BSF cells to PERC devices. With this change, companies like Solarworld, Gintech and Trina Solar keep constantly breaking their efficiency records, like the recent 22.61% efficiency for Trina Solar [3]. This information is relevant to state that changes in the PV industry typically happen slowly, and incrementally. The implementation of n-PERT solar cells is considered as the incremental step for the improvement from p-type PERC cells [4]. Its closest processing approach featuring screen-printed contacts is already under investigation in the industry, as well as some research labs [5-9]. At imec, we have focused on the use of Ni/Cu plated contacts for the front metallization, because even though the screen printing pastes have shown an exceptional improvement over the years, nickel is fundamentally a better contacting material for high efficiency devices [10] and Cu represents a more cost-effective solution than Ag for the transport of the electrical current [11].

The question presented in this paper is how to achieve 23% efficiency with RJ n-PERT cells. We've already established that the front and bulk recombination contribute to a big part of the power loss in our devices [12, 13] and that one of the main sources for the front recombination is the recombination

underneath the front metal contacts. This paper deals with 3 ways to reduce the front recombination:

1. Reduction of the metal J_0 on the front surface by improving the laser doping parameters.
2. Reducing the front contact fraction by manufacturing busbar free cells.
3. Further reducing the recombination of the passivated areas by applying a high temperature "firing step"; which integrates our experience on n-PERT cells with rear screen printed contacts into our plated high efficiency devices [14].

II. Experimental Data

The high efficiency n-PERT devices are manufactured on 170-180 µm n-Si Cz wafers with a bulk resistivity ranging from 2 to 5 Ω.cm. They feature a lightly doped FSF formed by POCl₃ diffusion (R_{sheet}~300 Ω/sq). The front contacts are plated on to laser doped lines using Light Induced Plating (LIP) of Ni, plated Cu and a very thin immersion Ag layer. Some cells feature busbars (BB) and some are BB free. The Cu plating differs for these groups: BB free cells are only plated with LIP Cu, and cells with busbars have both LIP Cu and an electroplated Cu coating.

The rear side features a diffused boron emitter R_{sheet}~100 Ω/sq. on a flat surface passivated by a stack of Al₂O₃/ PECVD SiOₓ. The rear side is metallized by a 2 um thick sputtered AlSi layer with contact openings in the dielectric stack by laser ablation. More details about the manufacturing process of these devices can be found on [15]. Figure 1 shows a cross section view of the high efficiency n-PERT cells.

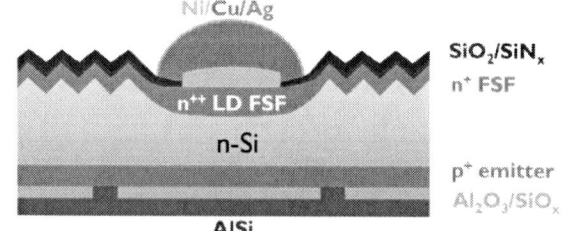

Figure 1: Cross section of the rear junction devices

Short loop experiments are performed to evaluate the impact of each new processing step through lifetime measurements.

These tests are done on symmetrical samples. Random pyramid texturing or saw damage removed (SDR) surfaces are implemented depending on the purpose (respectively FSF or emitter evaluations).

The recombination current is characterized from lifetime measurements by Quasi-Steady-State Photoconductance QSSPC, using a high-level injection approximation at an injection level $\Delta n = 1 \cdot 10^6 cm^{-3}$ and assuming an intrinsic carrier concentration n_i of $8.3 \cdot 10^9$ cm^{-3} [16]. IV curves are measured with a steady-state solar simulator from Wacom while a Grid^{TOUCH} chuck from Meyer Burger is added to the Wacom setup for the measurement of BB free cells [17]. Open-Circuit Photoluminescence (OCPL) images are obtained with a system from BTI.

a. REDUCING THE RECOMBINATION UNDERNEATH THE FRONT CONTACTS

As mentioned previously, the front contacts are formed on laser doped lines. The selective laser doping (LD) was first optimized for p-PERC cells [18]. This process was directly applied on the n-PERT platform, with a patterning speed of 5 m/s (LD1). The laser doping process for RJ PERT cells was optimized using a Design Of Experiments (DOE). The optimal selective FSF requires a scanning speed of 1 m/s (LD2). A slower speed during LD results in a deeper local doping [19].

The recombination current due to the laser doping ($J_{0, laser}$) has been characterized for these two speeds using symmetrical textured samples with a dielectrics stack of SiO$_2$/SiN$_x$ and a FSF doping (~300 Ω/sq.) corresponding to the n-PERT baseline [15]. Six squares with an area of 16 cm^2 each are patterned by LD on the test wafers while variating the front contact fraction. $J_{0, laser}$ is calculated from the slope of the curve of total recombination current vs metal fraction, considering in the computation the recombination in all the passivated areas.

The result shows a lowering of $J_{0, laser}$ from ~1-1.5 pA/cm^2 for LD1, to ~ 0.5- 0.65 pA/cm^2 for LD2. The front contact fraction is increased by the widening of the fingers during the front patterning process (from ~10 μm to 16 μm). The contact area increase is overcompensated with the reduction in $J_{0, laser}$ for LD2. So, we expect an increase of about 8 mV in open-circuit voltage (V_{oc}) by implementing the slow LD process.

When evaluating these processes in the diffused n-PERT baseline, we see an increase in V_{oc} from an average of 685-689 mV with the LD1 to 695-697 mV with the LD2. A V_{oc} as high as 700 mV has been obtained by combining this process with a lowly doped rear epitaxial emitter ($N_s < 1e^{18}$cm^3) [20,21].

The increase of the contact fraction did not show an impact on the short circuit current of these devices, both reaching values of 40.3 mA/cm^2. The total shading is mainly dominated by the plated finger width. The total plated finger width (~30 μm) is not strongly affected by the small change in the line width.

b. REDUCING THE FRONT CONTACT FRACTION: TOWARDS BUSBAR FREE CELLS

Another strong candidate for the decrease of the front recombination losses corresponds to the reduction of the contact fraction on the front side. One way to do this when using screen-printed contacts is by printing non-fire through busbars. The selective nature of the metal deposition by plating, which is one its main advantages when combined with laser processing (no alignment required), is at a loss when it comes to this path towards the contact fraction (C_f) reduction. The plated BB are only formed in the regions where the nitride is opened, thus the area is not passivated. However, as presented by ISFH, in principle it is also possible to reduce the busbar width (from 1 to 0.5 or 0.3 mm), which not only results in a lower C_f, but also less shadowing [22].

A simple calculation to estimate the impact of the use of three narrow BB (0.35 mm) in our cells, instead of the wide busbars (~1 mm), shows that the reduction of the shading fraction generated only from the plated busbars, from ~1.88% to ~0.64%, would result in a j_{sc} increase of ~ 0.5 mA/cm^2, just from the difference in the optics for wide vs narrow BB. This translates into a decrease in the total C_f of ~1.4%. A wider implementation of this approach depends on the production of interconnection tabs featuring widths of 0.3 mm.

A second option consists of lowering the number of opened lines inside the busbar area, creating "open busbars". The total reduction of the contacted area is the same as the one obtained by using 0.35 mm wide busbars. Figure 2 shows microscope images comparing the full (a), narrow (b) an open busbar (c) cases.

A third alternative corresponds to a BB free cell, where the current transport is distributed among several interconnection lines (wires or busbars). This technique was first presented in 2006 by Day4Energy [23] and is now commercialized by Meyer Burger [24]. The use of a multi busbar technology represents a similar approach [25], as commercialized by Schmidt [26]. Imec is also developing an interconnection technique based on a woven fabric which is intrinsically compatible with BB free devices [27].

Regardless of the chosen interconnection method, the removal of the busbars from the front-side of the cells will result in a reduction in the front recombination and shading losses. In theory, a current gain of at least 0.3 mA/cm^2 could be obtained by removing the narrow contacting BB. Thanks to the higher number of interconnecting points, this current gain can be further increased by reducing the total plated thickness.

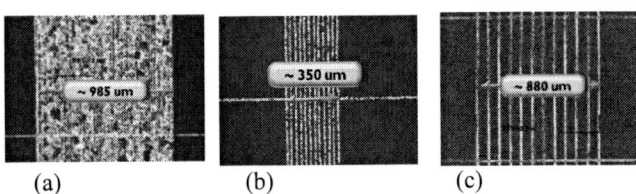

(a)   (b)   (c)

Figure 2: Microscope views of laser patterned busbars with (a) Fully ablated BB, (b) narrow BB, (c) open BB

Figure 3 shows open circuit PL images for cells with (left) narrow BB and (right) without BB. The strong recombination underneath the laser doped area along the busbars is clearly visible on the left picture, while a more homogeneous PL response is obtained on the right image.

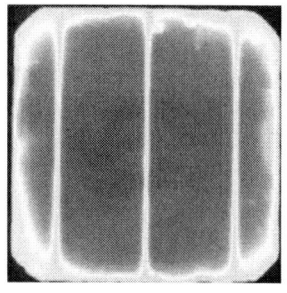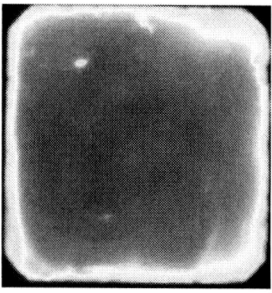

Figure 3: OCPL for a cell with narrow busbars (left) and a busbar free device (right). Both feature the same scale and illumination time.

Table I shows IV data comparing cells with narrow BB vs BB free cells. The main improvement for the BB free cells as compared to the narrow BB comes from increase in j_{sc}. V_{oc} values of 698.5 mV were measured, reaching an efficiency up to 22.8%.

TABLE I: I-V DATA FOR CELLS WITH LOWER FRONT CONTACT FRACTION

Busbars	Area [cm²]	V_{oc} [mV]	j_{sc} [mA/cm²]	FF [%]	η [%]
Narrow BB	238.9	697.0	40.0	80.0	22.3[1]
BB free	238.9	697.5	40.7	79.4	22.5[1]
BB free (best)	238.9	698.5	40.9	79.8	22.8[2]

[1] Calibrated at the FhG ISE CalLab

[2] Measured at MB by Grid[Touch], calibrated j_{sc} with a calibrated cell from FhG ISE CalLab (BB free cell from the same batch and same features)

c. FURTHER IMPROVEMENTS TO THE FRONT PASSIVATION

Our standard devices receive a belt furnace anneal (BFA) after laser processing at a temperature between 400-500°C. This process was implemented as a substitute to the forming gas anneal step, typically performed after the AlSi deposition [15]. Several observations point to the insufficiency of this step to achieve maximal passivation with the front dielectrics (SiO₂/SiNₓ:H). For example: these devices show a lower IQE response in the blue region when compared to the screen printed rear n-pert cells manufactured with the same diffusion profiles and the same front features [14].

It is well-known that a firing step (at temperatures over 700°C) contributes to the hydrogen release from the SiNₓ layer, improving the surface and bulk passivation [28]. Screen-printed cells typically follow such a high temperature firing step.

In 2014 we evaluated the impact of a firing step at high temperature to anneal the laser doping induced damage. No significant improvement was observed on the reduction of the $J_{0, laser}$ by this step as compared to the application of a low temperature BFA [29].

For this study, we concentrated on the recombination current of the passivated areas ($J_{0, pass}$). Using symmetrical lifetime test wafers we compared the BFA anneal vs a firing step. The wafers feature random pyramid texturing on a 170 um thick n-type Cz Si wafers with a base doping of ~4-5 Ω.cm and a FSF with a sheet resistance of ~300 Ω/sq. The $J_{0, pass}$ was reduced from ~16 fA/cm² for the BFA wafers down to 8.5 fA/cm² for the fired ones.

d. INTEGRATING THE FIRING STEP INTO THE FLOW

Before the addition of the firing step in the integration flow, we evaluated its impact in the recombination current of the boron emitter ($J_{0, emitter}$). A strong increase of $J_{0, emitter}$ after firing was observed for thin (6 - 8 nm) Al₂O₃ layers. The passivation quality could be maintained at a high level for Al₂O₃ layers over 10 nm thickness. This thickness dependence can be fully prevented by depositing a thin SiNₓ capping layer over the SiOₓ.

In our integration, we used a 30 nm thick SiNₓ layer over the Al₂O₃/SiOₓ stack. The rear laser ablation process was adapted to keep a constant rear contact fraction. Figure 4 shows a schema of the optimized integration flow.

Front-end: emitter and FSF formation
Thin thermal SiO₂ formation
Front SiNₓ ARC
Rear SiO₂ removal
Rear Al₂O₃ / SiOₓ / SiNₓ
Firing anneal
Rear ablation
Rear metallization AlSi PVD
Front laser doping
BFA 500°C
Front plating: LiP Ni / LiP Cu / immersion Ag

Figure 4: Schema showing the optimized integration flow

Finally, table II shows the IV data for the BB free cells including all processes presented in this paper for the optimization of the front-side. In other words, these cells feature:

- An improved LD process for the selective FSF formation
- A high resistive FSF (~350 Ω/sq.),
- A very low front contact fraction (no BB),
- An enhanced FSF passivation, by the high temperature firing of the front SiNₓ before the rear metallization.

The result is a 23.1% device and an average efficiency of 23% for 11 cells manufactured simultaneously.

TABLE II: I-V DATA FOR CELLS WITH LOWER FRONT CONTACT FRACTION

Busbars	Area [cm^2]	V_{oc} [mV]	j_{sc} [mA/cm^2]	FF [%]	η [%]
BB free (11 cells)	239	701 ± 2	40.7 ± 0.1	80.7 ± 0.3	23.0[+] ± 0.1
BB free (best)	239	699	40.7	81.2	23.1[+]

[+] internal measurement by Grid[TOUCH], calibrated J_{sc} with BB free cell calibrated at ISE CalLab

III. SUMMARY

Figure 5 shows the cumulated j_0 obtained for each configuration (decomposed as a weighted area value for the measured recombination parameters), as well as the corresponding implied and experimental V_{oc}. The implied V_{oc} at 25C is calculated from the total J_0, assuming a J_{sc} of 40 mA/cm^2. The deviation between $V_{oc, implied}$ and $V_{oc, experimental}$ can partly be attributed to the edge recombination losses, which have not been included in this analysis.

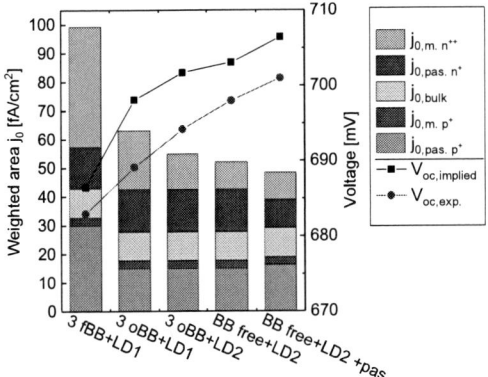

Figure 5: Evolution of the saturation current density for the different configurations of imec's RJ n-PERT diffused cells (3fBB: 3 full busbars, oBB: open BB, BB free: without BB). LD1 and LD2 correspond to the processing conditions as presented in section (a) in the text. Last data set (on the right side) corresponds the flow shown on figure 4.

As can be seen from Figure 5 and Table II, even with the diffused emitter, which features a surface doping concentration 10 times higher than the epitaxial emitters used in [20,21], V_{oc} values over 700 mV are within reach, with fill factors well above 80% and an average efficiency of 23%.

These results demonstrate the potential of combining plating and laser doping for the manufacturing of high efficiency n-PERT RJ devices. Further improvements could be achieved by evaluating the source for the recombination loss between the implied and experimental V_{oc}, potentially by a better understanding and reduction of edge recombination losses.

III CONCLUSIONS

We have presented three techniques to reduce the front recombination of our monofacial RJ high-efficiency n-PERT devices: reduction of the front $J_{0, laser}$, the reduction of the front contact fraction (with BB free cells) and the improvement of the passivation quality by implementing a high temperature firing step. With all these improvements we have achieved efficiencies up to 23.1%. Such high efficiencies are enabled thanks to the combination of selective laser doping and plating, for very narrow front lines, an excellent rear surface passivation by the Al_2O_3 on the boron emitter and the high quality rear contact formation by AlSi sputtering.

ACKNOWLEDGEMENTS

We would like to acknowledge the contribution of the people in the silicon PV department at imec Special thanks go to Izabela Kuzma-Filipek, Bartosz Zielinski, Jia Chen, Maria Recaman-Payo and Maxime Levillayer.

The authors gratefully acknowledge the financial support of imec's industrial affiliation program for Si-PV.

REFERENCES

[1] J. Zhao, A. Wang, M. A. Green, "24.5% efficiency silicon PERT cells on MCZ substrates and 24.7% efficiency PERL cells on FZ substrates", *Prog. Photovolt. Res. Appl.*, vol. 7, pp. 471-474, 1999.

[2] G. Agostinelli, J. Szlufcick, P. Choulat, G. Beaucarne, "Local contact structures for industrial PERC-type solar cells". *Proc. 20th EUPVSEC*, Barcelona, 2005.

[3] I. Clover, "Trina solar sets new 22.61% mono PERC efficiency record", *PV tech magazine*, December 19th 2016.

[4] International Technology Roadmap for Photovoltaics (ITRPV), "2015 Results including maturity reports", *7th ed.*, October 2016

[5] V. Mertens, et al., "Large Area n-Type Cz Double Side Contact Back Junction Solar Cell with 21.3% Conversion Efficiency", *Proc. 28th EUPVSEC*, Paris, 2013.

[6] W. Wang et al., "Industrial screen-printed n-type rear-junction solar cells with 20.6% efficiency", *IEEE J. Photovoltaics*, vol. 5, no. 4, pp. 1245-1249, Jul. 2015.

[7] Z-W. Peng, P-T. Hsieh, C-J. Huang, Y-J. Lin, P-K. Chang, C. Kuo, C-C. Li, "Toward 21% Efficiency nPERT Solar Cells with Selective Back Surface Field Technique", *Energy Procedia*, Vol. 92, August 2016, pp. 702-707

[8] B. Lim, T. Brendemühl, M. Berger, A. Christ, T. Dullweber, "n-PERT back junction solar cells: An option for the next industrial technology generation?", *Proc. 29th EUPVSEC*, Amsterdam, 2014.

[9] N. Wehmeier, et al., " 21.0% efficient screen-printed n-PERT back junction silicon solar cell with plasma-deposited boron diffusion source" *Solmat 158*, pp. 50-54, 2016

[10] D. K. Schroder, D. L Meier, "Solar cell contact resistance-A review", *IEEE Trans. Electron Devices*, vol. Ed.-31, no. 5, pp. 637-647, May 1984.

[12] M. Aleman, et al., "Large-area high-efficiency n-type Si rear junction cells featuring laser ablation and Cu-plated front contacts", *Proc. 28th EUPVSEC*, Paris, 2013

[13] L. Tous, et al., "Evaluation of advanced p-PERL and n-PERT large area silicon solar cells with 20.5% energy conversion efficiencies", *Prog. Photovoltaics Res. Appl., vol. 23, no. 5*, pp. 660-670, 2015.

[14] J. Chen, et al., "21.3% Large Area n-PERT Silicon Solar Cells Using Screen-Printed Aluminium with Open Circuit Voltage above 680 mV", *Proc. 32nd EUPVSEC*, Munich, 2016

[15] A. Uruena, et al., "Progress on large area n-type silicon solar cells with front laser doping and a rear emitter", *Prog. Photovolt. Res. Appl. Vol. 24*; pp.1149-1156, 2016

[16] R. Sinton, et al., "Quasi-steady state photoconductance, a new method for solar cell material and device characterization", Proc. 25th IEEE PVSC, Washington, 1996.

[17] N. Bassi, et al. "GridTouch: innovative solution for accurate IV measurement of busbarless cells in production and laboratory environments.", *Proc. of 29th EUPVSEC*, Amsterdam, 2014.

[18] B. Hallam, et al., "Hydrogen passivation of laser-induced defects for laser-doped silicon solar cells", *IEEE JPV, Vol. 4*, No 6, November 2014.

[19] B. Hallam, et al., " Deep junction laser doping for contacting buried layers in silicon solar cells", *Solmat 113*, 2013, pp. 124-134

[20] R. Hao, et al., "Kerfless Epitaxial Mono Crystalline Si Wafers With Built-In Junction and From Reused Substrates for High Efficiency PERx Cells", *IEEE JPV. Vol. 6, No 6*, November 2016

[21] I. Kuzma-Filipek et al., "22.5% n-pert solar cells on epitaxially grown silicon wafers", *Proc. 31st EUPVSEC* 2016

[22] T. Dullweber, et al., " Fine-Line printed 5 busbar PERC solar cells with conversion efficiencies beyond 21%", *Proc. 29th EU PVSEC*, 2014

[23] A. Schneider, L. Rubin, G. Rubin, "Solar Cell Efficiency Improvement by New Metallization Techniques - the Day4 Electrode", *Proc. IEEE 4th WCPEC*, 2006, pp. 1095-1098.

[24] T. Soederstroem, P. Papet, J. Ufheil, "Smart wire connection technology", *Proc. 28th EU PVSEC* Paris, 2013.

[25] S. Braun, et al., "Solar cell improvement by using a multi busbar design as front electrode", *Energy Procedia, 27* 2012 pp. 227-233.

[26] J. Walter, et al., "Multi-wire Interconnection of Busbar-free Solar Cells", *Energy Procedia*, vol. 55, pp. 380-388, 2014.

[27] T. Borgers, et al., "Multi-wire interconnection technologies weaving the way for back contact and bifacial PV modules", *Proc. 43rd IEEE PVSC*, Portland, 2016.

[28] F. Duerinckx, "Bulk and surface passivation of screen-printed multicrystalline silicon solar cells based on plasma enhanced CVD of silicon nitride", *PhD KULeuven*, 1999

[29] M. Aleman, et al., "Reducing front recombination losses to improve the efficiency of rear-junction Cu plated n-Si cells", *Proc. 29th EUPVSEC*, Paris, 2014

Microstructure and recombination activity of grain boundaries from front and rear side during a LID-cycle of mc-PERC solar cells

Tabea Luka[*1,2], Marko Turek[1], Stephan Großer[1], Christian Hagendorf[1]

[1]Fraunhofer Center for Silicon-Photovoltaics CSP, 06120 Halle (Saale), Germany
[2]Anhalt University of Applied Sciences, 06366 Köthen (Anhalt), Germany

Abstract — **Under illumination multi-crystalline PERC solar cells may suffer from a severe efficiency loss up to 20 %$_{rel}$ induced by carrier injection which is known as mc-LID (or LeTID). Lateral investigations show that grain boundaries as well as the rear side are of special interest. Target preparation and element analysis at locations with high rear recombination unveil Cu-containing particles at industrially produced mc-Si PERC solar cells. They accumulate at grain boundaries and at the rear surface where the passivation stack is damaged. The correlation of these Cu particles and mc-LID is examined.**

I. INTRODUCTION

Light exposure of multi-crystalline solar cells can lead to a severe efficiency loss referred to as mc-LID or LeTID [1, 2]. PERC cells are much stronger affected than Al-BSF cells showing an efficiency reduction of up to 20 %$_{rel}$ [1, 3]. By adapting the cell process the degradation can be reduced [2, 4, 5]. However, this implies a narrowing of the process window leaving fewer options for the optimization of cell properties. Thus, a deeper understanding of the defect is favored, allowing maximal process flexibility for an optimal efficiency output.

In previous papers, it was shown that the degradation is rather homogenous throughout the entire solar cell. However, slight deviations in the degradation of structural defects were detected [5]. It is known that grain boundaries with a high defect density and rear contacts show a reduced degradation [6, 7]. In this paper, the focus is set on the investigation of grain boundaries. Throughout degradation and regeneration, the LID extend of grain boundaries in comparison to intra grain regions is examined. Microstructural investigations are performed from the rear side with focus on grain boundaries and rear contacts after significant degradation.

The root cause of mc-LID is still unknown. However, it is widely believed that some metallic impurity causes the degradation [8, 9]. Among other defect complexes being discussed as degradation sources, Cu plays a special role due to its positive charge state, its diffusion and precipitation behavior [9-11]. Yet, until now no significant amount of Cu could be detected at mc-LID sensitive solar cells although first indications were presented in [12]. At the samples investigated in this paper, an increased Cu concentration is detected. The correlation of Cu particles and mc-LID is discussed.

II. MAIN EXPERIMENTAL RESULTS

The investigated samples are industrially produced mc-Si PERC solar cells from different cell producers. The degradation has been carried out using an LED based test setup leading to accelerated mc-LID degradation. Integral current-voltage measurements, electro luminescence (EL) and light beam induced current (LBIC) measurements are performed under standard test conditions in the solar cell analysis system LOANA.

Degraded samples were cut in smaller parts from which the Al rear-contact was removed in an ultrasonic cleaner to allow lateral high resolution investigations. After this preparation, µLBIC measurements were performed at an in-house developed LBIC system consisting of laser scanning microscope LSM700 from ZEISS extended with a LBIC acquisition system developed by point electronic GmbH. Excitation wavelengths of 555 nm and 639 nm have been used. For further microstructure analysis, scanning electron microscopy (SEM) and energy-dispersive X-ray spectroscopy (EDX) has been performed at a Hitachi SU70 SEM equipped with an EDX system. Defect specific high-resolution analysis of structure and composition has been done by target preparation of a cross-section using a focused ion beam system (FIB – ZEISS Auriga) and subsequent transmission electron microscopy (TEM - Tecnai G2 F20 and Titan G2 60-300 from FEI) and EDX mapping.

A. Lateral investigations throughout degradation and regeneration

To preclude the influence of other (non mc-LID) light induced degradation mechanisms as BO-LID or FeB-LID, a pretreatment at 25 °C under 1 sun illumination was performed for 24 h. The observed efficiency change was less than 1 %rel during this pretreatment. As the degradation is much more pronounced at elevated temperature mc-LID is the major light induced degradation mechanism in these samples.

At elevated temperature (100 °C) and varying illumination (leading to a constant carrier density of $\Delta n = 1.4 \cdot 10^{14}/cm^3$), an efficiency degradation of 20 %$_{rel}$ is detected, see Fig. 1. Subsequent to the degradation phase, regeneration is detected

---

* Corresponding author: e-mail tabea.luka@csp.fraunhofer.de, Phone: + 49 345 5589-5130, Fax: + 49 345 5589-101

which saturates at 6 %rel reduced efficiency in comparison to the initial value.

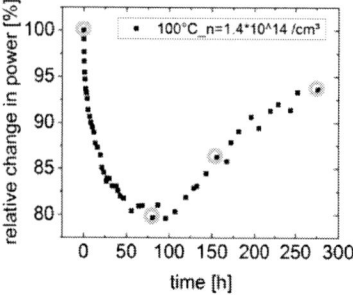

Fig. 1. Mc-LID degradation and regeneration curve of a mc-Si PERC solar cell at 100 °C with varying illumination leading to a constant carrier density of $\Delta n=1.4 \; 10^{14}/cm^3$. The gray circles mark the times where the EL images presented in Fig. 2 are taken.

Throughout degradation and regeneration, EL images are taken, see Fig. 2. The typical lateral appearance of mc-LID is observed, as grain boundaries show an enhanced LBIC signal (indicated by arrows in Fig. 2) compared to intra grain regions at maximum degradation. This holds not only for the degradation. Also grain boundaries have a higher lifetime throughout regeneration until the signal of the intra grain regions recover to the same value. This effect can be so strong that it predominates even in regions with decorated grain boundaries which are characterized by a reduced lifetime in the initial state (marked with circles in Fig. 2). Defects at decorated grain boundaries seem to disappear during degradation. However, throughout regeneration the defects reappear again. Thus, these defects do not dissolve during degradation. Instead, the surrounding (intra grains) decreased to a similar value in lifetime.

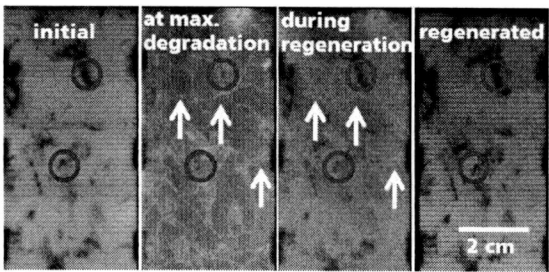

Fig. 2. EL images of a 3.8 cm x 7 cm section taken at different stages of the mc-LID degradation. The points of time at which the images were taken are marked with gray circles in Fig. 1.

B. High resolution lateral investigations

After significant degradation of 12-15 %rel, the degradation of some samples was interrupted for high resolution lateral investigation in microstructural analysis. The samples presented here were degraded at 75 °C and one sun. However, also at other degradation conditions similar results were observed.

The degraded samples are investigated from the front and rear side to get a more profound understanding of lateral variations in bulk lifetime and rear passivation quality. LBIC investigations from the front (Fig. 3a) show qualitatively the same results as EL, discussed in the previous section: Decorated grain boundaries (marked with an A) as well as grain boundaries showing reduced LID extend (marked with a B) are observed. The comparison of the spatial resolved LBIC measurements from front and rear side, Fig. 3 a) and b), reveals a qualitatively different appearance. Many grain boundaries with a higher bulk lifetime in comparison to intra grain regions (B) show an increased rear recombination, when investigated from the rear, see Fig 3b).

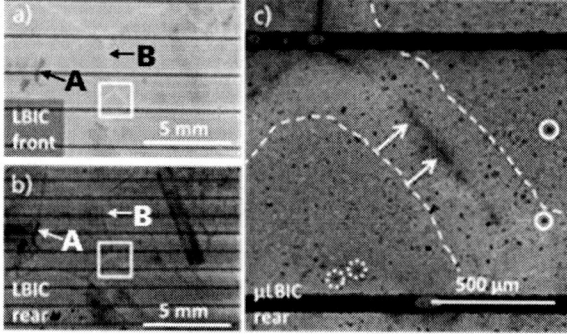

Fig. 3. LBIC (EQE) images of a degraded PERC solar cell at 960 nm from the front (a) and at 658 nm from the rear (b). (c): high resolution LBIC (μLBIC) image at 639 nm from the rear at the marked region in (a) and (b).

Conventional LBIC measurements are supplemented by higher resolution μLBIC measurements from the rear side, shown in Fig. 3 c). A relatively short excitation wavelength of 639 nm was used to localize regions with enhanced bulk lifetime and increased rear surface recombination. At grain boundaries of type B, a bright area with an enhanced LBIC signal surrounding the GB is detected (enclosed by the dashed line in Fig. 3 c). This observation is an indication for a gettering process induced by the grain boundary. Furthermore, a discontinuous increased rear side recombination at the GB (see Fig. 3 c, marked with arrows) and recombination-active intra-grain sites have been found (see Fig. 3 c, marked as dashed circles). The specific halo of the μm-sized recombination side allows a precise distinction from Al sphere residuals (marked with solid circles) which exhibit a well-defined shape without a halo.

C. Target preparation and element analysis

Microstructure information of the recombination sites is obtained by SEM and EDX analysis performed at the exposed passivated rear surface. Figure 4 a) shows a SEM image at a recombination active grain boundary which proceeds vertically in the image. An EDX mapping at this position unveils copper containing particles located at the grain boundary which corresponds to the visible particles in the SEM image (not shown here). Not only at grain boundaries, but also at intra-grain rear recombination sites particles were detected. Figure 4 b) shows an SEM image of an intra-grain recombination site as an example of a particle at the surface. An EDX-map confirmed that the passivation layer is locally absent and that at these areas Cu containing particles are located on the bare silicon. The size of the particles is in the range of around 100 nm. Furthermore, we found by EDX that the appearance of particles was restricted to areas with damaged passivation stack. However, a direct correlation was not observed. Only at approximately 50 % of the intra-grain rear recombination sites particles were detected. Even a lower percentage of the grain boundaries with enhanced rear recombination were decorated with particles.

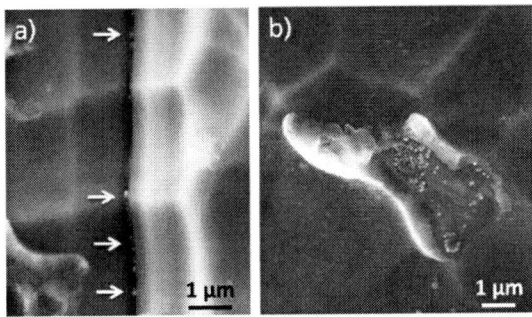

Fig. 4. a): SEM image at a grain boundary with increased rear recombination at a degraded mc PERC cell. Cu particles are marked with arrows. b): SEM image at an area with increased rear recombination located in the middle of a grain. The passivation is damaged. Cu-containing particles are found on the bare silicon.

EDX analysis at local rear contact positions also revealed Cu-containing particles on top of the eutectic. This confirms the assumption that Cu can accumulate at the rear side on positions where the diffusion barrier is locally absent (damaged passivation layer or local rear contact).

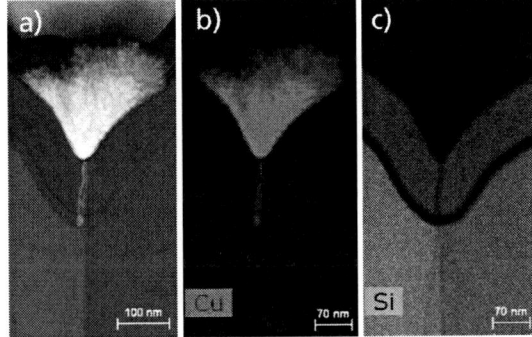

Fig. 5.: TEM of micro-particle at grain boundary; Cu and Si-mapping showing the bulk material, the passivation layer and the Cu-containing particle (color levels not on absolute scales).

A cross section has been prepared precisely at the position of a Cu-containing particle located at a grain boundary. Subsequent TEM analysis, shown in Figure 5 a), reveals a

particle on top of the passivation stack at a grain boundary. A form-fit contact of the particle to the surface topography (trench) can be found. The TEM data unveil a filled "channel" connecting the particle and the Si-substrate. Figure 5 b) and c) show EDX Cu- and Si-mappings of the cross section. It confirms that the particle as well as the channel connecting the particle with the Si contains Cu.

These investigations were done at several samples showing different mc-LID extent. Cu was detected on all samples. However, no correlation between the mc-LID extent and the Cu concentration on the sample was detected.

III. DISCUSSION

Our experiments have revealed that mc-LID degradation and regeneration occurs rather homogeneously throughout the cell. However, structural defects as grain boundaries are less affected by mc-LID during the entire degradation and regeneration process. This confirms that grain boundaries show indeed reduced LID and that the degradation is not just slowed down. If the grain boundaries would underlie a slower degradation to the same extent as the intra grain regions, a contrast inversion would occur throughout regeneration. Furthermore, it was shown that a region around the grain boundary shows this reduced recombination after degradation. This result indicates that the reduced degradation can be attributed to a metal gettering at structural defects at grain boundaries. The same effect might be the reason of the reduced degradation at local rear contacts, as there are also non-passivated structural defects which getter impurities.

An enhanced rear recombination was detected at grain boundaries. Thus, the rear side was microstructurally investigated finding Cu-containing particles. These particles were solely detected at areas with locally damaged passivation layer, as regions of μm-size inside a grain, gain boundaries, and local rear contacts. A connection to mc-LID is conceivable, since the last two are known to show less mc-LID. This connection might be attributed to a gettering at the rear that reduces the degradation extend, while in intra grain regions with intact passivation Cu gettering is inhibited as even thin amorphous oxide layers are efficient diffusion barriers [10, 13]. However, no direct correlation between Cu and mc-LID could be established and the origin of the Cu atoms at the rear side is under further investigation. Since a damaged passivation is necessary to detect Cu-particles, the Si bulk material has to be taken into account as Cu source. Nevertheless, according to the present knowledge we cannot exclude that it originates from the metallization or some contamination introduced through the ultrasonic treatment.

IV. SUMMARY

Industrially produced mc-PERC solar cells showing strong mc-LID sensitivity of more than 15 %rel were investigated.

978-1-5090-5606-4/17 $31.00 © 2017 IEEE

Lateral investigations throughout degradation and regeneration combined with high resolution LBIC showed that regions adjacent to grain boundaries with high dislocation densities are less affected by mc-LID. We concluded that this indicates metal gettering at grain boundaries and that the same effect occurs at rear contacts. This model explains the reduced degradation at grain boundaries and rear contacts in comparison to intra grain regions as well as the reduced recombination of Al-BSF cells in comparison to PERC solar cells.

Investigations of the rear surface of mc-LID sensitive cells revealed Cu-containing particles at areas with damaged or even absent passivation layer (i.e. grain boundaries, rear contacts and areas of µm-size in the middle of a grain). As the Cu-containing particles are never observed without a direct connection to the Si bulk, the Si should be considered as Cu source. Under this assumption Cu should be taken into account as possible candidate for the mc-LID defect formation. Still, the direct correlation between mc-LID and Cu is missing and the connection is to be further investigated.

REFERENCES

[1] K. Ramspeck, et al., *Proceedings of the 27th EU-PVSEC*, pp. 861-865, 2012.

[2] F. Kersten, et al., *Solar Energy Materials and Solar Cells*, 142, pp. 83-86, 2015.

[3] F. Fertig, K. Krauß, and Stefan Rein, *Phys. Status Solidi RRL*, 9, No. 1, pp. 41-46, 2015.

[4] C. E. Chan, et al., *IEEE journal of Photovoltaics*, 6, pp. 1473-1479, 2016.

[5] T. Luka et al., *Proceedings of the 31st EU PVSEC*, p. 826 – 828, 2015.

[6] T. Luka et al., Solar Energy Materials and Solar Cells, 158, pp. 43-49, 2016.

[7] M. Selinger et al., *Energy Procedia*, 92, pp. 867-872, 2016.

[8] Bredemeier et al., *AIP ADVANCES*, 6, 035119, 2016.

[9] K. Nakayashiki et al., *IEEE journal of Photovoltaics*, 6 (4), pp. 860-868, 2016.

[10] A.A. Istratov and E.R. Weber, *J. Electrochem. Soc.*, 149 (1), G21-G30, 2002.

[11] J. Lindroos and H. Savin, *Solar Energy Materials & Solar Cells*, 147, pp. 115-126, 2016.

[12] T. Luka et. al., *Physica status solidi - rapid research letters* 11, p. 1600426, 2017.

[13] M.B. Shabani et. al., *J. Electrochem. Soc.*, 143, pp. 2025-2029, 1996.

Thermodynamic Efficiency Limit of Bifacial Solar Cells for Various Spectral Albedos

Thomas C.R. Russell*, Rebecca Saive*, Harry A. Atwater

Thomas J. Watson Laboratories of Applied Physics and Material Science, California Institute of Technology,
Pasadena, CA 91125, USA
*Contributed equally to this work.

Abstract— We have adapted the Shockley-Queisser radiative flux balance model to investigate the influence of the spectral dependent albedo on the power conversion efficiency of bifacial solar cells. By taking the spectral albedo of the surroundings into account the optimal band gap becomes dependent on the type of ground-reflecting surface. Furthermore, we find that the maximum efficiency depends on the spectral albedo of the surroundings and that optimal cell performance cannot be assessed when only accounting for a spectrally-independent albedo. We predict that the power conversion efficiency for a bifacial silicon solar cell surrounded by green grass is an absolute 0.90% higher than expected from a wavelength independent albedo, and even 1.51% higher for white sand, with the spectral albedo model. Furthermore, we calculated spectral properties of antireflection coatings and we found that commonly used antireflection coatings in silicon solar cells provide good properties in order to accept back-scattered light from surfaces with different spectral albedos.

Index Terms—Bifacial Solar Cells, Spectral albedo, Thermodynamic efficiency limit, Shockley-Queisser-Limit

I. INTRODUCTION

INCREASING the power output of solar modules is a crucial step towards lowering the levelized cost of electricity of photovoltaic power plants [1, 2]. The use of bifacial solar cells is one strategy towards higher power output of solar power plants, due to the fact that the bifacial solar cells are designed to accept incident light at the front and rear of the cells. In the 1960s this concept was first introduced [3] but only recently module manufacturers have shown interested and actually started selling bifacial modules as standard technology. The gain from using a bifacial configuration often exceeds the power output of monofacial solar modules by a surprising amount and values of up to 50% have been reported [4]. The power output of a bifacial solar cell is strongly dependent on the albedo of the surrounding ground-reflecting surface as well as the geometry in which the cells are mounted [5]. Here we address the influence of the albedo on the power output of the bifacial solar cell, geometric factors affecting the performance of the bifacial solar cell have been extensively studied elsewhere [5-12]. The albedo is defined as the fraction of light that is back-reflected relative to the incident sunlight and varies strongly depending on the ground-reflecting surface properties. As examples green grass exhibits an albedo of 0.24 and snow shows an albedo of 0.85. The additional photon flux of light reflected by the ground-reflecting surface leads to effective light concentration within the cell. Therefore, both the voltages and currents of the cells increase. However, the albedo of typical reflective surfaces also exhibits a spectral dependence – which is evident from the examples of green grass and snow – that affect solar cell performance [13, 14]. Figure 1 gives an overview of the spectral and effective albedo of several ubiquitous surfaces. We show that maximum achievable power conversion efficiency may increase and the optimal thermodynamic operating point of a bifacial solar cell can deviate when taking this spectrally-dependent albedo into account.

First, the optimal thermodynamic efficiency limit for bifacial solar cells without spectral dependence of the albedo is investigated. Next, we include the spectral dependence of different materials and apply an adapted version of the model of Shockley and Queisser [15] to this case and calculate the maximum efficiency for bifacial solar cells. We find that including the spectral dependence of the surroundings can result in an increase or decrease of the maximum thermodynamic efficiency and power output for several realistic environmental conditions. In the case of green grass, the maximum power conversion efficiency increases by an absolute 0.90% and for white sand by an absolute 1.51% due to a high flux of reflected photons with energy above band gap. We use these results to study the implications for silicon bifacial solar cells.

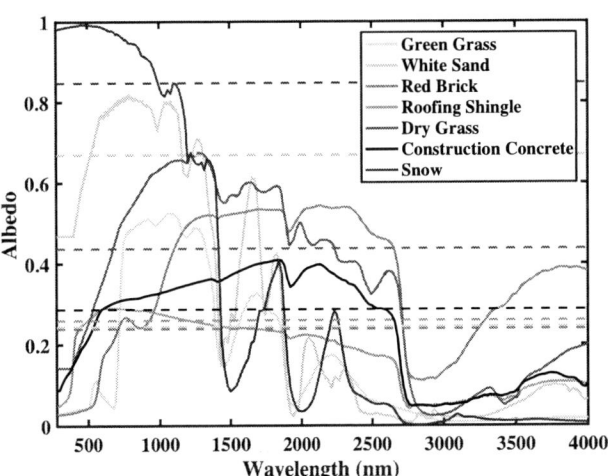

Figure 1: Spectral and effective (dashed lines) albedo of several widely-used surfaces.

More than 90% of the modules currently produced use silicon as the absorber material [16]. To date, the only bifacial modules available on the market use silicon cells, with different cell technologies [17].

II. THERMODYNAMIC EFFICIENCY LIMIT WITH SPECTRAL INDEPENDENT ALBEDO

The Shockley-Queisser radiative flux balance [15, 18] was adapted to account for power gain due to the acceptance of reflected photons by the surroundings. Both the front side AM 1.5G illumination and ground-reflected rear illumination are accepted by the bifacial solar cell and together form the total optical flux. At first, the albedo of the ground is assumed to be uniform and spectral independent, i.e. the ground has an effective albedo (R_A) [19]. Furthermore, step-function absorption, with no absorption for photons with energy below bandgap, 100% absorption for photons with energy above band gap, is assumed here. The total number of photons above bandgap, accounting for an effective albedo, is enhanced by a factor $1 + R_A$ and thereby the short-circuit current is increased linearly with the effective albedo. The open-circuit voltage (V_{oc}) is given by

$$V_{OC} = \frac{k_B T_{Cell}}{q} \ln\left(\frac{(1+R_A)\cdot I_S}{I_0} + 1\right) \qquad (1)$$

in which T_{Cell} is 300K, I_S is the sunlight generated current and I_0 is the dark saturation current. I_0 is affected by both radiative and non-radiative recombination processes. The radiative part of the dark current contains an extra factor of two in comparison with a monofacial cell, since the bifacial solar cell has no back reflector, leading to a slight decrease in V_{OC}. The power conversion efficiency of the bifacial solar cell increases logarithmically with with R_A, due to the behavior of the V_{OC} (eq. 1). This effect is shown in Fig. 2, with the power conversion efficiency of a bifacial solar cell as a function of the bandgap energy for increasing values of R_A. A single-junction bifacial solar cell is considered. The maximum conversion efficiency can be obtained via

$$\eta_{Max} = \frac{V_{OC} \cdot J_{SC} \cdot FF \cdot (1+R_A)}{P_{SUN} \cdot (1+R_A)} \qquad (2)$$

in which FF is the fill factor, J_{SC} is the short-circuit current density, V_{oc} the open-circuit voltage and P_{SUN} the solar constant of approximately 1000 W/m² total solar irradiance. In Fig. 2, the dashed black line represents the Shockley-Queisser limit for AM 1.5G spectrum, with no rear-side illumination and without back reflector. The maximum power conversion efficiency for the AM 1.5G spectrum assuming single junction, bandgap of 1.34 eV and no back reflector is around 33.1% ($R_A = 0$). The power conversion efficiency (eq. 2) is defined in such a way that the solar constant increases with R_A, i.e. the total incoming power is increased due to the ability of the back surface to intercept the scattered light. For increasing values of R_A between 0 and 1, the Shockley-Queisser limit is enhanced to a maximum of 33.7% for R_A equal to 1 and bandgap equal to 1.34 eV. The single junction monofacial solar cell with back reflector has a maximum efficiency of 33.7% for an optimal

bandgap of 1.34 eV, due to a decrease in entropy loss of the free energy of incoming photons. For this reason, either including a back reflector or assuming an effective albedo of 1, enhances the V_{oc} and thereby the efficiency to 33.7% (Fig. 2). The efficiency is the same, but the power output of the bifacial solar cell with an effective albedo of 1 is doubled with respect to the monofacial panel with back reflector.

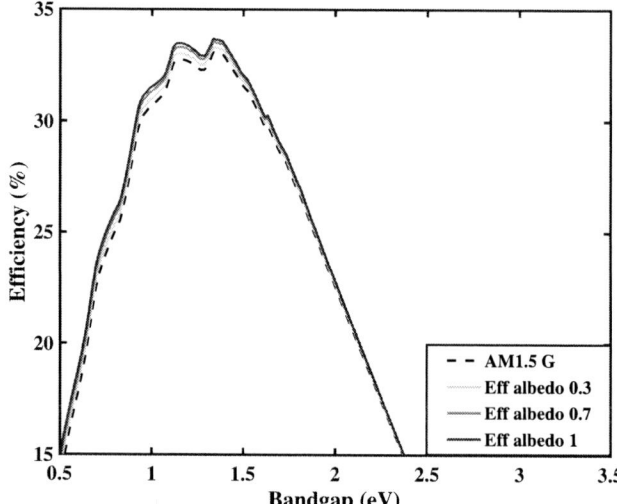

Figure 2: Maximum conversion efficiency of bifacial solar cells for increasing effective albedo.

III. THERMODYNAMIC EFFICIENCY LIMIT WITH SPECTRAL ALBEDO

In order to account for spectral effects of surroundings on the conversion efficiencies for bifacial solar cells a wavelength dependent albedo is introduced, i.e. spectral albedo. Data on spectral albedo was obtained from The ASTER spectral library version 2.0 of Jet Propulsion Laboratory [20]. Seven common surfaces both natural and artificial are chosen, including green grass, white sand, red brick, roofing shingle, dry grass, construction concrete and snow. By replacing the effective albedo with a spectral dependent albedo the incoming light on the rear side is unique for each ground-reflecting surface (Fig. 1), instead of a constant fraction of the AM 1.5G spectrum. The dashed lines in Fig. 1 show the effective albedos of the respective surfaces and are obtained via

$$R_A = \frac{\int_0^\infty R_A(E) \cdot AM\ 1.5G(E)\ dE}{\int_0^\infty AM\ 1.5G(E)\ dE} \qquad (3)$$

Both the short-circuit current density and open-circuit voltage need to be adapted to account for spectral albedos. The J_{SC} and V_{OC} as a function of the bandgap for a certain spectral albedo $(R_A(E))$ are obtained via equation 4 and 5, respectively.

$$J_{SC,alb} = q \cdot \int_{E_{gap}}^\infty \left(1 + R_A(E)\right) \cdot AM\ 1.5G(E) \cdot \frac{1}{E} dE \qquad (4)$$

$$V_{OC,alb} = \frac{k_B T_{Cell}}{q} \ln\left(\frac{q \cdot \int_{E_{gap}}^\infty (1+R_A(E)) \cdot AM\ 1.5G(E) \frac{1}{E} dE}{I_0} + 1\right)(5)$$

TABLE I
Ideal band gap, thermodynamic efficiency limit and maximum power output for different spectral and effective albedo.

	Optimal Bandgap (eV)	J_{SC} (Eff. Alb.) mA/cm²	J_{SC} (Spec. Alb.) mA/cm²	V_{OC} (Eff. Alb.) mV	V_{OC} (Spec. Alb.) mV	Fill Factor	R_A	Efficiency (Eff. Alb.) %	Efficiency (Spec. Alb.) %	Power Output (Eff. Alb.) W/m²	Power Output (Spec. Alb.) W/m²
AM1.5 G	1.34	35.18	35.18	1060.1	1060.1	0.89	0	33.08	33.08	330.96	330.96
Green Grass	1.13	54.08	55.51	871.9	872.6	0.87	0.24	33.27	33.83	409.88	421.06
White Sand	1.34	58.73	61.03	1073.4	1074.4	0.89	0.67	33.53	34.88	560.44	582.59
Red Brick	1.13	53.84	52.35	871.8	871.1	0.87	0.24	33.27	31.99	407.97	396.29
Roofing Shingle	1.34	44.31	44.75	1066.1	1066.3	0.89	0.26	33.28	33.62	415.23	423.70
Dry Grass	1.13	62.48	62.93	875.6	875.9	0.87	0.44	33.40	33.38	475.77	479.92
Construction Concrete	1.34	45.28	44.84	1066.6	1066.4	0.89	0.29	33.30	32.98	428.78	424.57
Snow	1.34	65.00	69.25	1076.0	1077.6	0.89	0.85	33.62	35.88	616.88	663.17

in which q is the electron charge, $R_A(E)$ is the albedo dependent on the photon energy and AM 1.5G is the incoming solar spectrum. Furthermore, to obtain the spectral albedo dependent power conversion efficiency, equation 2 has been adapted to the form

$$\eta_{Max,alb} = \frac{V_{OC,alb} \cdot J_{SC,alb} \cdot FF}{P_{SUN} \cdot (1+R_A)} \qquad (6)$$

At short-circuit conditions, both terms accounting for spontanuous generation due to thermal excitation at 300K and radiative recombination can be neglegted. Since the recombination rate is much smaller than the number of photons above bandgap. Figure 3 shows the short-circuit current density as a function of the bandgap for different spectral albedo surfaces.

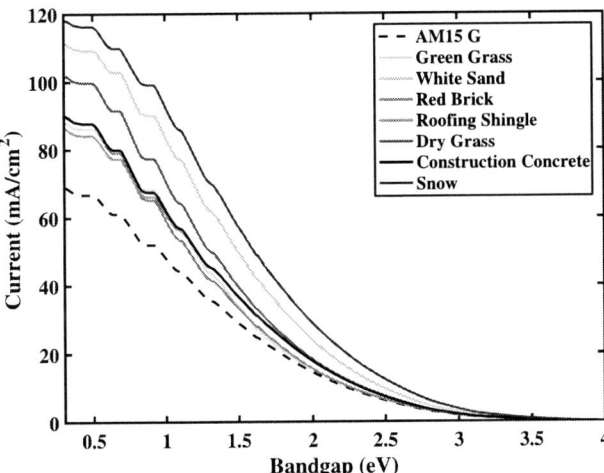

Figure 3: Ideal short-circuit current density as a function of bandgap.

The power conversion efficiency as a function of bandgap energy is shown in figure 4 for different ground-refelecting surfaces with spectral albedo taken into account. This theoretical result shows that for green grass, white sand and snow for example, the power conversion efficiency is higher than for cells with and without a mirror back reflector under AM 1.5G illumination. Therefore, it exceeds the calculated efficiency with spectral independent albedo for all bandgaps. However, red brick yields an efficiency lower than the AM 1.5G case and therefore, performs worse than expected from the spectral independent albedo. Both the maximum power conversion efficiency and optimal bandgap for a single-junction bifacial cell vary with the type of ground-reflecting surface.

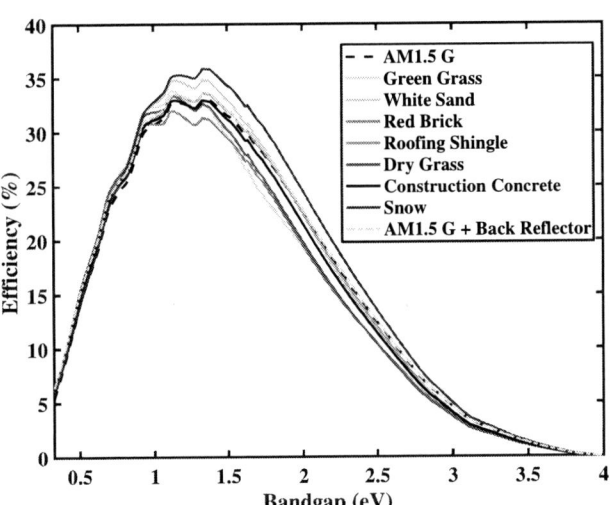

Figure 4: Thermodynamic efficiency limit for different spectral albedo.

Using snow as reflective surface shows the largest enhancement in power conversion efficiency in comparison with the AM1.5G spectrum, an absolute 2.26%. The optimal bandgap for a bifacial cell surrounded by snow is 1.34 eV, which is unchanged from the case of a monofacial solar cell. For green grass the optimal bandgap for power conversion is 1.13 eV which is closer to the bandgap of crystalline silicon. Table 1 gives an overview of the simulation result. The power output increases in each case with the additional photon flux generated

TABLE II
Thermodynamic efficiency limit and maximum power output for silicon solar cells under different spectral albedo.

	J_{SC} (Eff. Alb.) mA/cm²	J_{SC} (Spec. Alb.) mA/cm²	V_{OC} (Eff. Alb.) mV	V_{OC} (Spec. Alb.) mV	Fill Factor	R_A	Efficiency (Eff. Alb.) %	Efficiency (Spec. Alb.) %	Power Output (Eff. Alb.) W/m²	Power Output (Spec. Alb.) W/m²
AM1.5 G	43.52	43.52	865.3	865.3	0.87	0	32.69	32.69	327.07	327.07
Green Grass	54.14	55.58	871.0	871.6	0.87	0.24	32.93	33.83	409.80	421.06
White Sand	72.65	75.86	878.6	879.7	0.87	0.67	33.25	34.76	555.29	580.61
Red Brick	53.89	52.41	870.8	870.1	0.87	0.24	32.92	31.99	407.89	396.29
Roofing Shingle	54.82	55.38	871.3	871.5	0.87	0.26	32.94	33.29	415.14	419.50
Dry Grass	62.54	63.08	874.7	874.9	0.87	0.44	33.09	33.38	475.67	479.88
Construction Concrete	56.01	55.94	871.8	871.8	0.87	0.29	32.97	32.92	424.44	423.89
Snow	80.40	84.60	881.2	882.5	0.87	0.85	33.36	35.16	616.60	649.83

through reflection from the surroundings. In the case of snow the power output is increased by 7.5% when taking a spectrally dependent albedo into account and for white sand by 4.0%. However, for red brick the power output is 2.9% lower than predicted when only taking the effective albedo into account. The power output for snow, when considering a spectral albedo, is even more than double the power output under AM 1.5G illumination for a cell without back reflector, even though the effective albedo of snow is less than one. A comparison with the spectral albedos presented in Fig. 1 shows that surfaces perform best when a large portion of the reflected light spectrum exhibits photon energies above the semiconductor bandgap.

IV. PRACTICAL IMPLICATIONS FOR BIFACIAL SILICON SOLAR CELLS

We calculated the Shockley Queisser efficiency limit as well as the maximum power output for the band gap of crystalline silicon (1.125 eV). In Table 2 the simulated short-circuit current (J_{sc}), the open circuit voltage (V_{oc}), the fill factor (FF), the effective albedo (R_A), the maximum efficiency and maximum power output considering effective albedo and spectral albedo respectively are presented. We find again that white sand shows a very high increase in short circuit current to 75.86 mA/cm² compared to 72.65 mA/cm² when the effective albedo of white sand is considered. This corresponds to a 77.5% power output increase compared to AM 1.5G illumination, while one would have only expected 69.8% from the spectral independent albedo simulation. This increased short circuit current density together with an increase in open circuit voltage leads to an efficiency of 34.76% which is an absolute 1.51% higher than the efficiency expected if only taking effective albedo into account. An increase in efficiency for spectral albedo compared to effective albedo is also observed for green grass or snow. However, red brick for example shows a lower efficiency than expected from the effective albedo. For all ground-reflecting surfaces the power output is increased by collecting the back scattered light. However, for white sand and dry or green grass the actual power

output considering the spectral albedo is higher than if only considering an effective albedo while it is lower for red brick.

In a real silicon solar cell loss processes that are not covered by a pure thermodynamic analysis occur. One of the loss mechanisms is reflection of photons at the front surface. In order to mitigate reflection losses silicon solar cells feature textured surfaces coated with antireflection layers that maximize the photon flux into the crystalline silicon. The material used as an antireflection coating depends on the specific silicon technology. In crystalline homojunction silicon solar cells most commonly silicon nitride is used. In heterojunction solar cells a transparent conductive oxide (TCO) such as indium tin oxide (ITO) that simultaneously acts as charge transport layer also provides the desired antireflection properties. The optimum thickness of either the silicon nitride or the ITO depends on the spectrum of the incoming light, for AM 1.5G one obtains 79 nm of silicon nitride and 71 nm of ITO for heterojunction solar cells. In order to find out whether these antireflection coatings provide optimal conditions for rear side coatings taking spectral albedo into account we performed optical simulations using PV Lighthouse OPAL 2 [29]. As incoming spectra we weighted the AM 1.5G with the spectral albedo of the respective surfaces. We used literature optical data for the crystalline silicon [30], silicon nitride[31], amorphous silicon [32] and ITO [33]. A layer of 10 nm p-type amorphous silicon was used for the silicon heterojunction solar cell. Table 3 shows a summary of the results. The transmission into the crystalline silicon for homojunctions with silicon nitride antireflection coating and for heterojunctions with ITO antireflection coating is given as the short-circuit current density that would be obtained if all transmitted photons provided one electron-hole pair. The short current density is shown right of the respective antireflection coating. First the standard silicon nitride or ITO thickness is used to calculate the short-circuit current density, then the thickness was optimized for both cases. In all cases the value the value for the standard thickness and for the optimized thickness are very close to each with a maximum enhancement of 0.1 mA/cm² being achieved

TABLE III

Table 3: Antireflection coating thickness and respective transmission into the crystalline silicon for crystalline silicon homojunction and heterojunctions solar cells. The transmission is gives as a short-circuit current density that would be obtained if all incoming photons provided an electron-hole pair.

	Silicon Nitride (nm)	J_{SC} mA/cm^2	Silicon Nitride (nm)	J_{SC} mA/cm^2	ITO (nm)	J_{SC} mA/cm^2	ITO (nm)	J_{SC} mA/cm^2
AM 1.5 G	79	42.85	79	42.85	71	41.41	71	41.41
Green Grass	79	12.02	105	12.15	71	12.06	98	12.17
White Sand	79	32.00	84	32.01	71	31.27	75	31.28
Red Brick	79	8.97	102	9.02	71	8.96	93	9.02
Roofing Shingle	79	11.76	82	11.76	71	11.50	73	11.50
Dry Grass	79	19.49	93	19.54	71	19.29	84	19.34
Construction Concrete	79	12.34	86	12.35	71	12.12	78	12.13

for green grass. This shows that commonly used antireflection coatings already provide good conditions for rear side antireflection coatings even when spectral albedo is taken into account.

V. CONCLUSION

We show that taking into account the spectral albedo can considerably alter the power output boost of bifacial solar cells. The predicted power output when considering the spectral albedo of snow, white sand or green grass is higher than one would have expected from the effective albedo. On the other hand, red brick and construction concrete exhibit lower efficiencies. In the case of snow and white sand, the power output is relatively increased by 5.4% and 4.6%, respectively, when taking a spectrally dependent albedo into account for silicon bifacial solar cells. However, for red brick the power output is relatively lowered by 2.8% than predicted when only taking the effective albedo into account. Furthermore, we found that commonly used antireflection coating in silicon solar cells provide good properties in order to accept photons back-scattered from surfaces with different spectral albedos.

ACKNOWLEDGEMENT

This material is based upon work supported by the Engineering Research Center Program of the National Science Foundation and the Office of Energy Efficiency and Renewable Energy of the Department of Energy under NSF Cooperative Agreement No. EEC-1041895 and by the U.S. Department of Energy through the Bay Area Photovoltaic Consortium under Award Number DE-EE0004946. RS acknowledges support from the Global Climate & Energy project.

REFERENCES

[1] R. Kopecek, Y. Veschetti, E. Gerritsen, A. Schneider, C. Comparotto, V. D. Mihailetchi, et al., "Bifaciality: One small step for technology, one giant leap for kWh cost reduction," Photovoltaics International, pp. 32-45, 2014.

[2] F. Fertig, S. Nold, N. Wöhrle, J. Greulich, I. Hädrich, K. Krauß, et al., "Economic feasibility of bifacial silicon solar cells," Progress in Photovoltaics: Research and Applications, 2016.

[3] M. Hiroshi, "Radiation energy transducing device," ed: Google Patents, 1966.

[4] A. Cuevas, A. Luque, J. Eguren, and J. Del Alamo, "50 Per cent more output power from an albedo-collecting flat panel using bifacial solar cells," Solar Energy, vol. 29, pp. 419-420, 1982.

[5] B. Soria, E. Gerritsen, P. Lefillastre, and J. E. Broquin, "A study of the annual performance of bifacial photovoltaic modules in the case of vertical facade integration," Energy Science & Engineering, vol. 4, pp. 52-68, 2016.

[6] L. Kreinin, A. Karsenty, D. Grobgeld, and N. Eisenberg, "PV systems based on bifacial modules: Performance simulation vs. design factors," in Photovoltaic Specialists Conference (PVSC), 2016 IEEE 43rd, 2016, pp. 2688-2691.

[7] C. Deline, S. MacAlpine, B. Marion, F. Toor, A. Asgharzadeh, and J. S. Stein, "Evaluation and field assessment of bifacial photovoltaic module power rating methodologies," in Photovoltaic Specialists Conference (PVSC), 2016 IEEE 43rd, 2016, pp. 3698-3703.

[8] C. W. Hansen, J. S. Stein, C. Deline, S. MacAlpine, B. Marion, A. Asgharzadeh, et al., "Analysis of irradiance models for bifacial PV modules," in Photovoltaic Specialists Conference (PVSC), 2016 IEEE 43rd, 2016, pp. 0138-0143.

[9] U. A. Yusufoglu, T. M. Pletzer, L. J. Koduvelikulathu, C. Comparotto, R. Kopecek, and H. Kurz, "Analysis of the annual performance of bifacial modules and optimization methods," IEEE Journal of Photovoltaics, vol. 5, pp. 320-328, 2015.

[10] C. K. Lo, Y. S. Lim, and F. A. Rahman, "New integrated simulation tool for the optimum design of bifacial solar panel with reflectors on a specific site," Renewable Energy, vol. 81, pp. 293-307, 2015.

[11] A. Krenzinger and E. Lorenzo, "Estimation of radiation incident on bifacial albedo-collecting panels," International journal of solar energy, vol. 4, pp. 297-319, 1986.

[12] S. Guo, T. M. Walsh, and M. Peters, "Vertically mounted bifacial photovoltaic modules: A global analysis," Energy, vol. 61, pp. 447-454, 2013.

[13] M. Brennan, A. Abramase, R. W. Andrews, and J. M. Pearce, "Effects of spectral albedo on solar photovoltaic devices," Solar Energy Materials and Solar Cells, vol. 124, pp. 111-116, 2014.

[14] R. W. Andrews and J. M. Pearce, "The effect of spectral albedo on amorphous silicon and crystalline silicon solar photovoltaic device performance," *Solar Energy*, vol. 91, pp. 233-241, 2013.

[15] W. Shockley and H. J. Queisser, "Detailed balance limit of efficiency of p-n junction solar cells," *Journal of applied physics*, vol. 32, pp. 510-519, 1961.

[16] I. Fraunhofer, "Photovoltaics report," *Fraunhofer ISE, Freiburg*, 2014.

[17] F. Fertig, N. Wöhrle, J. Greulich, K. Krauß, E. Lohmüller, S. Meier, *et al.*, "Bifacial potential of single-and double-sided collecting silicon solar cells," *Progress in Photovoltaics: Research and Applications*, 2016.

[18] L. C. Hirst and N. J. Ekins-Daukes, "Fundamental losses in solar cells," *Progress in Photovoltaics: Research and Applications*, vol. 19, pp. 286-293, 2011.

[19] M. R. Khan and M. A. Alam, "Thermodynamic limit of bifacial double-junction tandem solar cells," *Applied Physics Letters*, vol. 107, p. 223502, 2015.

[20] A. Baldridge, S. Hook, C. Grove, and G. Rivera, "The ASTER spectral library version 2.0," *Remote Sensing of Environment*, vol. 113, pp. 711-715, 2009.

[21] M. A. Green, "Self-consistent optical parameters of intrinsic silicon at 300K including temperature coefficients," *Solar Energy Materials and Solar Cells*, vol. 92, pp. 1305-1310, 2008.

[22] Z. C. Holman, M. Filipič, A. Descoeudres, S. De Wolf, F. Smole, M. Topič, *et al.*, "Infrared light management in high-efficiency silicon heterojunction and rear-passivated solar cells," *Journal of Applied Physics*, vol. 113, p. 013107, 2013.

[23] Y. Han, X. Yu, D. Wang, and D. Yang, "Formation of various pyramidal structures on monocrystalline silicon surface and their influence on the solar cells," *Journal of Nanomaterials*, vol. 2013, p. 7, 2013.

[24] M. A. Green, "Radiative efficiency of state-of-the-art photovoltaic cells," *Progress in Photovoltaics: Research and Applications*, vol. 20, pp. 472-476, 2012.

[25] K. R. McIntosh and S. C. Baker-Finch, "OPAL 2: Rapid optical simulation of silicon solar cells," in *Photovoltaic Specialists Conference (PVSC), 2012 38th IEEE*, 2012, pp. 000265-000271.

[26] R. Saive, A. M. Borsuk, H. S. Emmer, C. R. Bukowsky, J. V. Lloyd, S. Yalamanchili, *et al.*, "Effectively Transparent Front Contacts for Optoelectronic Devices," *Advanced Optical Materials*, vol. 4, pp. 1470–1474, 2016.

[27] R. Saive, C. R. Bukowsky, S. Yalamanchili, M. Boccard, T. Saenz, A. M. Borsuk, *et al.*, "Effectively transparent contacts (ETCs) for solar cells," in *Photovoltaic Specialists Conference (PVSC), 2016 IEEE 43rd*, 2016, pp. 3612-3615.

[28] R. Saive, M. Boccard, T. Saenz, S. Yalamanchili, C. R. Bukowsky, P. Jahelka, *et al.*, "Silicon heterojunction solar cells with effectively transparent front contacts," *Sustainable Energy & Fuels*, vol. 1, pp. 593-598, 2017.

[29] K. R. McIntosh and S. C. Baker-Finch, "OPAL 2: Rapid optical simulation of silicon solar cells," in *Photovoltaic Specialists Conference (PVSC), 2012 38th IEEE*, 2012, pp. 000265-000271.

[30] M. A. Green, "Self-consistent optical parameters of intrinsic silicon at 300K including temperature coefficients," *Solar Energy Materials and Solar Cells*, vol. 92, pp. 1305-1310, 2008.

[31] S. C. Baker-Finch and K. R. McIntosh, "Reflection of normally incident light from silicon solar cells with pyramidal texture," *Progress in Photovoltaics: Research and Applications*, vol. 19, pp. 406-416, 2011.

[32] Z. C. Holman, A. Descoeudres, L. Barraud, F. Z. Fernandez, J. P. Seif, S. De Wolf, *et al.*, "Current losses at the front of silicon heterojunction solar cells," *Photovoltaics, IEEE Journal of*, vol. 2, pp. 7-15, 2012.

[33] Z. C. Holman, M. Filipič, A. Descoeudres, S. De Wolf, F. Smole, M. Topič, *et al.*, "Infrared light management in high-efficiency silicon heterojunction and rear-passivated solar cells," *Journal of Applied Physics*, vol. 113, p. 013107, 2013.

Process-Induced Degradation Resistant n-Cz Wafers through *Tabula Rasa* Defect Engineering

Vincenzo LaSalvia[1], William Nemeth[1], Matthew Page[1], Wooseok Nam[2], Youngsik Han[2], Sungsun Baik[2], Amanda Youssef[3], Tonio Buonassisi[3], Paul Stradins[1]

[1]*National Renewable Energy Laboratory (NREL), Golden, Colorado USA*
[2]*Woongjin Energy Co. Ltd., Daejoen City, South Korea*
[3]*Massachusetts Institute of Technology (MIT), Cambridge, Massachusetts USA*

ABSTRACT — We explore a simple annealing technique *tabula rasa* that allows monocrystalline n-type Czochralski silicon (n-Cz) wafers to avoid the oxygen precipitation that occurs during solar cell processing at high thermal budget. With this technique, the deep B emitter formation used in n-FZ based record efficiency Si cells, now becomes possible in n-Cz wafers. Wafers of both low and high O_i are subject to different *tabula rasa* processes that inject either vacancies or interstitials. As a result, we see strong variations in the carrier lifetime immediately after *tabula rasa* treatments in various ambients. However, after a thermal budget of the deep boron emitter formation, these variations disappear and we see significant lifetime improvement after *tabula rasa*, especially in wafers with high O_i content. This suggests that *tabula rasa*, an industrially relevant batch technique, could be successfully implemented for n-Cz wafers to remain oxygen precipitation-free during high efficiency n-type Cz solar cell processing and greatly improving fabrication yield.

I. INTRODUCTION

High carrier lifetime in n-Cz wafers is a requirement for production of high efficiency solar cells. This is a challenge due to a strict limitation of crystal doping variations, and grown-in impurities, oxygen, and intrinsic point defects – especially vacancies. The interaction between oxygen precipitate nuclei and vacancies may principally lead to deleterious O-precipitation in n-Cz wafers during subsequent cell processing[1]. It is during this process-induced degradation that oxygen precipitate nuclei grow as O_i accumulate, dislocations eventually form around them, and harmful metal impurities accumulate in these dislocations due to the surrounding strain field [2]. Cell processing of n-Cz wafers at temperatures necessary for diffused boron emitters result in recombination activity and severely degraded n-type cell performance.

We have reported last year on using *tabula rasa* (TR), a high temperature, wafer pre-treatment to dissolve the oxygen precipitate nuclei. *Tabula rasa* dissolves oxygen precipitate nuclei, creating a significant buffer of time before they can nucleate again during the cell process. It is desirable to utilize this effect in wafers with high oxygen content such as from the end of the ingot or crystals grown at higher rates. Other previous work[3] have suggested that TR is beneficial, but only without investigating the end result of process-induced degradation after the whole cell process, or in wafers with

varying O_i content. Here, we compare these degradation effects in TR-treated and non-TR treated wafers from different Cz crystal growths after cell processing that involved B-emitter formation. We show that despite different point defects injected into wafer by various *tabula rasa* treatments, the end result is significant resistance to process-induced degradation in any and all TR-treated wafers, especially those with high O_i content that would typically degrade considerably in carrier lifetime. Importantly, without *tabula rasa*, the n-Cz wafers cannot sustain deep B emitter formation and degrade substantially via O-precipitation. Whereas, with *tabula rasa*, n-Cz wafers successfully survive the high thermal budget of forming low surface concentration ($\sim 2 \times 10^{19}$ cm^{-3}), deep ($\sim$1.5 µm) boron emitter, thus opening a way to ultrahigh efficiency n-Cz cells similar to record cells on n-FZ [4].

II. EXPERIMENT

The experiments are carried-out on high purity, <100>, n-Cz Si wafers, doped with phosphorus at 2-4 Ω-cm. Wafers are selected from two separate Czochralski crystal growths, with two specific O_i concentrations. The wafers selected are 11.9 ppma and 16.8 ppma O_i, respectively. The wafers are 156mm pseudosquare, and are then cleaved into sections of approximately 60mm x 60mm providing 4 quarter-wafers for variable processing sequences and subsequent characterization by Sinton lifetime and photoluminescence (PL) mapping. The *tabula rasa* treatment is applied at 1100°C in a conventional quartz hot-wall quartz diffusion furnace, in either a N_2 or O_2 ambient gas flowing in the furnace at approximately 3.0 l/min. In addition to a witness quarter-wafer that does not receive any TR treatment, the three remaining quarter-wafers each receive TR treatment with O_2, N_2, or O_2 with an extended (90 minute) temperature ramp from 1100°C to 750°C in an effort to mimic typical industrial thermal cooling rates.

After the TR treatment all wafer-quarters are subjected to a thermal budget identical to that of a diffused boron emitter process, in a continuous fashion and in only a N_2 ambient environment (see Figure 2). We have previously discovered that a continuous thermal budget of both the boron shallow-diffusion and then subsequent oxidation for enhanced-diffusion identically mimics the process-induced degradation that occurs from the oxygen-precipitate nuclei (OPN) accumulating and growing throughout the cell process to the point of heightened recombination activity within the bulk of

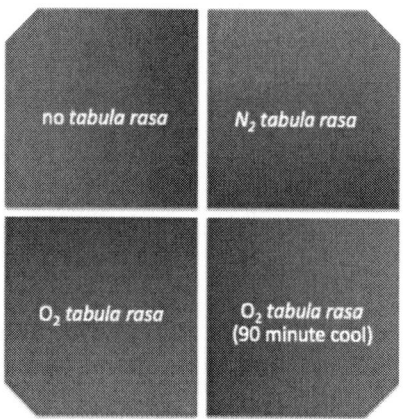

Figure 1. Quartering and assigning of the TR pre-treatments of each wafer for the experiment.

the wafer, without an added benefit from some B-induced gettering. The gettering step is then applied separately.

After the simulated boron emitter thermal budget in an inert ambient of Fig. 2, all wafers are cleaned with a full RCA procedure to remove any oxidation or nitridation that may have occurred during the thermal processing, and then passivated with ~15nm Al_2O_3 by atomic layer deposition. Following the Al_2O_3 we apply a 400˚C anneal in 10% H_2

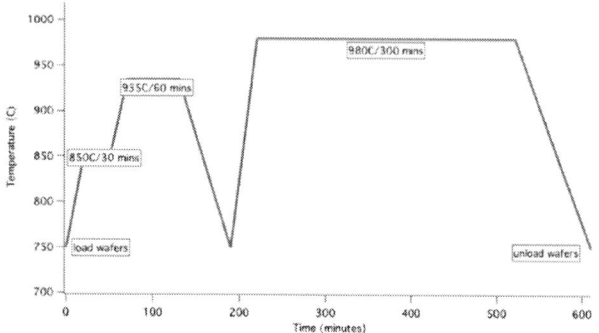

Figure 2. Thermal budget to mimic diffused-boron emitter formation in a typical n-type cell process.

forming gas in order to activate the chemically passivating component in the Al_2O_3 film, and then photoconductive decay lifetime measurements were taken on the individual wafer-quarter samples with a Sinton WCT-120 to determine bulk lifetime (Tau_{bulk}) and implied open circuit voltage (iV_{OC}). PL imaging is also taken at this time to elucidate any spatial correlations to lifetime. This step reveals the effect of *tabula rasa* treatment on B-emitter formation but without gettering.

Finally, after stripping the passivating Al_2O_3, the samples were submitted to a phosphorus gettering process utilizing a bubbled phosphorus source ($POCl_3$) in a separate quartz hot-wall furnace. After the phosphorus diffusion/gettering each sample is stripped of the resulting phosphosilicate glass films in dilute HF and then etched in a 25% KOH solution in order to remove ~5 µm of the diffused surfaces, and then once

again, each wafer-quarter was RCA cleaned and prepared for a subsequent passivation with an Al_2O_3 film via atomic layer passivation and and a forming gas anneal for activation. The carrier lifetimes of the wafer-quarters were then measured again with PCD and imaged again with PL to show comparative changes in material quality after the phosphorus gettering.

III. RESULTS

The two wafers of respectively low O_i concentration (11.8 ppma) and high O_i (15.9 ppma) are selected from controlled-growth wafer sets from Woongjin Energy Co. Ltd. We note that our *high O_i* wafers represent rather typical values of O_i in photovoltaic-grade Cz-Si, while the *low O_i* wafers have exceptionally low O_i bordering on the minimal [O_i] practically obtainable in Cz-Si [5].

The results of the first portion of this experiment are shown in Table 1 and Figure 3 and relate to the process due to the thermal budget for a diffused-boron emitter. In particular, the iV_{OC} measurements on the *low O_i* wafers subjected to the TR and cell processing thermal budgets reveal that TR in both N_2 and O_2 environments preserve relatively high iV_{oc} ~680 mV, while the wafer quarter not subject to TR degraded in bulk lifetime, resulting in iV_{OC} of ~ 40 mV less that TR treated wafers.

pre-treatment	*(15.9 ppm O_i)* iV_{OC} (mV)	*(11.8 ppm O_i)* iV_{OC} (mV)
N_2 TR	670	679
no TR/witness	575	644
O_2 TR	675	685
O_2 + 90 mins	659	652

Table 1. Resulting iV_{OC} measurements via PCD after surface passivation with Al_2O_3 on wafer-quarters after simulated thermal budget in Figure 2.

The results of the same TR treatments and thermal budgets performed on the *high O_i* wafer in Table 1 show that the samples with high O_i content degrade severely if no TR was performed, to the extent that makes the wafer unsuitable for solar cells (575 mV iV_{OC}). The PL image of this sample (not shown here) at an extended exposure reveals a swirl-ring defect pattern expected of oxygen precipitates that have nucleated and grown near the ring-like vacancy profile that sets in during the Cz crystal growth. In contrast to the no TR-treated samples, *high O_i* wafer quarters with TR show remarkable stability against process-induced degradation. Especially the wafer quarter pre-treated with TR in an O_2 ambient (an interstitial-injecting condition) shows a resulting iV_{OC} of 675 mV, just only 10 mV lower than in *low O_i* wafer. In comparing the two wafer sets in Table 1, we also conclude that generally, process-induced degradation is much less pronounced in wafers with low O_i.

978-1-5090-5606-4/17 $31.00 © 2017 IEEE

From the above results, we conclude that after the thermal budget necessary for a diffused boron emitter, the n-Cz Si without a TR pre-treatment severely degrades due to enhanced O-precipitation, the degree of degradation increasing with O_i content in the wafer.

The second portion of this experiment explores how both TR-pretreated and untreated Cz-Si react to a phosphorus gettering process typical in cell processing for back surface field (BSF) formation. It is expected that the large binding energies of transitional metal impurities to O-precipitates and O-precipitate/vacancy complexes make it extremely difficult for metallic impurities to getter out of the bulk of Cz-Si once they have bound to O-precipitates. (This process is used in the IC industry for internal impurity gettering into the wafer bulk and out of the Magic Denuded Zone). The Cz-Si surfaces are etched in KOH in order to remove any P-doped regions near the surface as to reveal a true bulk lifetime comparable to the PCD measurements of un-doped Cz-Si after the emitter thermal budget portion. Figure 4 shows the iV_{OC} of each of the wafer quarters from both after the simulated thermal budget of a diffused boron emitter, and after the phosphorus gettering step.

Figure 4 results show that the iV_{OC} of the TR pre-treated wafers almost fully recovers to their *as-received* state of ~720 mV, whereas the untreated wafer witnesses exhibit considerably diminished iV_{OC}, especially in the *high O_i* wafer. Here, even after phosphorus gettering the iV_{OC} of the no-TR wafer is still below 600 mV. In the *low O_i* wafer no-TR we observe an increase in iV_{OC} comparable to TR pre-treated wafers consistent with the lack of O-precipitation, but still less iV_{OC} than TR pre-treated wafers even with *high O_i* concentration. While more work is needed to optimize the TR process to obtain even higher carrier lifetimes after the deep B emitter formation, these results already suggest an industrial pathway to ultrahigh efficiency n-Cz Si cells with deep diffused, low surface concentration B emitter. Tabula Rasa is preformed as a batch process in a commonly used diffusion furnace, making it easy to integrate with rest of the cell's processing including B diffusion.

IV. CONCLUSION

Our study reveals that a *tabula rasa* pre-treatment improves the wafers significantly when applied as a pre-treatment to solar cell processing, especially those with high O_i. Even more importantly, this anneal can be easily integrated into industrial solar cell manufacturing as batch-processing, and more so without adding any additional capital expenditure as *tabula rasa* simply uses already existing hot-wall tube furnaces. Without TR, oxygen precipitates that typically develop during the thermal budget of n-type cell processing can lower the iV_{OC} of such wafers to 575 mV from >720 mV as-received. *Tabula rasa*, especially when used in an O_2 ambient, can boost the iV_{OC} of the same Cz-Si back up to 719 mV, offering a

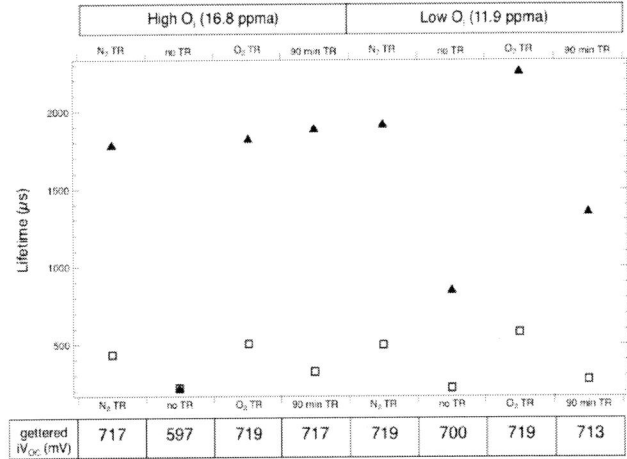

Figure 3. Bulk lifetime values for *low O_i* and *high O_i* wafers after boron emitter thermal budget (in outlined squares), and after subsequent phosphorus gettering (in closed triangles.) Respective iV_{OC} values from after phosphorus gettering processing are shown in the table just below the axis of the graph.

viable method of fabricating oxygen precipitate-free n-type Si solar cells with ultrahigh efficiencies.

Funding for this work was provided by the United Sates Department of Energy EERE contract SETP DE-EE00030301 (SuNLaMP) and under Contract No. DE-AC36-08GO28308.

REFERENCES

[1] J. D. Murphy, M. Al-Amin, K. Bothe, M. Olmo, V. V. Voronkov, and R. J. Falster, "The effect of oxide precipitates on minority carrier lifetime in n-type silicon," *Journal of Applied Physics*, vol. 118, no. 21, pp. 215706–13, Dec. 2015.

[2] P. Zhang, H. Väinölä, A. A. Istratov, and E. R. Weber, "Thermal stability of internal gettering of iron in silicon and its impact on optimization of gettering," *Appl. Phys. Lett.*, vol. 83, no. 21, pp. 4324–4326, Nov. 2003.

[3] B. Sopori, P. Basnyat, S. Devayajanam, T. Tan, A. Upadhyaya, K. Tate, A. Rohatgi, and H. Xu, "Dissolution of Oxygen Precipitate Nuclei in n-Type CZ-Si Wafers to Improve Their Material Quality: Experimental Results," *IEEE J. Photovoltaics*, vol. 7, no. 1, pp. 97–103, Dec. 2016.

[4] A. Richter, J. Benick, F. Feldmann, A. Fell, M. Hermle, and S. W. Glunz, "n-Type Si solar cells with passivating electron contact_ Identifying sources for efficiency limitations by wafer thickness and resistivity variation," *Solar Energy Materials and Solar Cells*, pp. 1–10, May 2017.

[5] K. Bothe, R. Sinton, and J. Schmidt, "Fundamental boron-oxygen-related carrier lifetime limit in mono- and multicrystalline silicon," *Prog. Photovolt: Res. Appl.*, vol. 13, no. 4, pp. 287–296, 2005.

Detection of A Shifting Bromine Concentration in Hybrid Perovskites by X-ray Fluorescence Microscopy

Yanqi Luo,[1] Parisa Khoram,[2] Sarah Brittman,[2] Barry Lai,[3] Erik C. Garnett,[2] David P. Fenning[*,1]

[*,1] Department of Nanoengineering, University of California San Diego, La Jolla, CA 92093
[2] Center for Nanophotonics, FOM Institute AMOLF, Amsterdam, the Netherlands
[3] Advanced Photon Source, Argonne National Laboratory, Argonne, IL 60439

Abstract — **Hybrid perovskite materials are promising for application as the active layers in solar cells due to their excellent optoelectronic properties. Improved understanding of the origins of hysteresis in perovskite materials must be established to overcome barriers to commercialization. In this study, we use a lateral device geometry to bias a methylammonium lead bromide thin film single crystal. Upon biasing, the halide concentration varies laterally within the crystal, as observed by means of non-destructive synchrotron-based nanoprobe X-ray fluorescence (Nano-XRF). The direct observation of a partially reversible bromine re-distribution under bias is revealed experimentally. The origins of device hysteresis and switchable effects can be partially explained by the local elemental information available from Nano-XRF.**

Index Terms — **Hybrid perovskite, halide migration, nanoproble X-ray fluorescence**

I. INTRODUCTION

The organometal halide perovskite materials, ABX_3, have risen quickly in solar cell device performance, achieving >22% efficiency.[1] Despite their excellent optoelectronic properties, hysteresis of scanning current-voltage measurement impairs the evaluation of the performance of perovskite devices and may indicate the presence of performance-limiting defects.[2]

Theoretical and experimental evidence point to ionic migration under an electric field as contributing to current-voltage hysteresis.[3] Extensive computational work particularly in the methylammonium lead iodide ($MAPbI_3$) system has shown I^- is the most mobile ion because of its low energy of migration with respect to Pb^{2+} and MA^+. [4] Analogously, in inorganic oxide perovskite materials, ABO_3, O^{2-} anions are known to migrate with low energy barriers in certain chemistries.[5] Besides halide anions, Yuan *et al.* provide evidence that MA^+ is migrates from photothermal induced resonance (PTIR) microscopy mapping under electric poling. [6]

It should be noted that many common analytical techniques used to visualize the elemental distribution of halide perovskites must be used with caution because of the sensitivity of these materials to ambient environment – especially commonly used vacuum environments [7] – in addition to the risk of optical and electrical beam damage during characterization.[8], [9] Recently, nanoprobe X-ray fluorescence mapping has been shown to be an effective, non-interfering probe for investigating the local composition in hybrid lead halide perovskite.[10], [11]

In this study, we use synchrotron-based nanoprobe X-ray fluorescence (Nano-XRF) microscopy [10] to study the bromine concentration within methylammonium lead bromide thin film single crystals during biasing. A quasi-reversible shift in the bromine distribution is revealed directly.

II. EXPERIMENTAL AND RESULTS

The perovskite single crystal thin film was spin-coated on back-contact electrodes was prepared following a previous published fabrication method.[12] The single crystal thin film was studied by synchrotron-based nanoprobe X-ray fluorescence (XRF) at beamline 2-ID-D of the Advanced Photon Source. A series of XRF maps was collected for a sequence of alternating steps of applied DC bias.

The XRF investigation illustrates that the halide concentration varies coincident with changes in the bias condition.[13] Br enrichment is found opposite to the direction of applied electric field.

III. CONCLUSION AND OUTLOOK

Systematic elemental variations are detected by nano-X-ray fluorescence in hybrid lead bromide perovskites upon biasing. Future investigation on reducing defect density within halide perovskite is needed to lower the degree of ion redistribution.

ACKNOWLEDGEMENT

YL acknowledges the support of DPF start up funds from the UC San Diego and the Sentaurus simulation from Jonathan Scharf. P.K acknowledges the ERC grant from AMOL. This research used resources of the Advanced Photon Source, a U.S. Department of Energy (DOE) Office of Science User Facility operated for the DOE Office of Science by Argonne National Laboratory under Contract No. DE-AC02-06CH11357.

REFERENCES

[1] NREL, "Best Research-Cell Efficiencies," 2017. [Online].
Available:
https://www.nrel.gov/pv/assets/images/efficiency-chart.png.

[2] H. J. Snaith, A. Abate, J. M. Ball, G. E. Eperon, T. Leijtens, N. K. Noel, S. D. Stranks, J. T. W. Wang, K.

Wojciechowski, and W. Zhang, "Anomalous hysteresis in perovskite solar cells," *J. Phys. Chem. Lett.*, vol. 5, no. 9, pp. 1511–1515, 2014.

[3] D. W. deQuilettes, W. Zhang, V. M. Burlakov, D. J. Graham, T. Leijtens, A. Osherov, V. Bulović, H. J. Snaith, D. S. Ginger, and S. D. Stranks, "Photo-induced halide redistribution in organic–inorganic perovskite films," *Nat. Commun.*, vol. 7, no. May, pp. 1–9, 2016.

[4] C. Eames, J. M. Frost, P. R. F. Barnes, B. C. O'Regan, A. Walsh, and M. S. Islam, "Ionic transport in hybrid lead iodide perovskite solar cells," *Nat. Commun.*, vol. 6, no. May, p. 7497, 2015.

[5] J. B. Goodenough, "Electronic and ionic transport properties and other physical aspects of perovskites," *Reports Prog. Phys.*, vol. 67, no. 11, pp. 1915–1993, 2004.

[6] Y. Yuan, J. Chae, Y. Shao, Q. Wang, Z. Xiao, A. Centrone, and J. Huang, "Photovoltaic Switching Mechanism in Lateral Structure Hybrid Perovskite Solar Cells," *Adv. Energy Mater.*, vol. 5, pp. 1–7, 2015.

[7] S. G. Motti, M. Gandini, A. J. Barker, J. M. Ball, A. R. Srimath Kandada, and A. Petrozza, "Photoinduced Emissive Trap States in Lead Halide Perovskite Semiconductors," *ACS Energy Lett.*, vol. 1, no. 4, pp. 726–730, 2016.

[8] N. Klein-Kedem, D. Cahen, and G. Hodes, "Effects of Light and Electron Beam Irradiation on Halide Perovskites and Their Solar Cells," *Acc. Chem. Res.*, vol. 49, no. 2, pp. 347–354, 2016.

[9] E. T. Hoke, D. J. Slotcavage, E. R. Dohner, A. R. Bowring, H. I. Karunadasa, and M. D. McGehee, "Reversible photo-induced trap formation in mixed-halide hybrid perovskites for photovoltaics," *Chem. Sci.*, vol. 6, pp. 613–617, 2015.

[10] Y. Luo, S. Gamliel, S. Nijem, S. Aharon, M. Holt, B. Stripe, V. Rose, M. I. Bertoni, L. Etgar, and D. P. Fenning, "Spatially Heterogeneous Chlorine Incorporation in Organic-Inorganic Perovskite Solar Cells," *Chem. Mater.*, vol. 28, pp. 6536–6543, 2016.

[11] M. Stuckelberger, B. West, T. Nietzold, B. Lai, J. M. Maser, V. Rose, and M. I. Bertoni, "Review: Engineering Solar Cells Based on Correlative X-Ray Microscopy," *J. Mater. Res.*, vol. 32, no. 10, pp. 1825–1854, 2017.

[12] P. Khoram, S. Brittman, W. I. Dzik, J. N. H. Reek, and E. C. Garnett, "Growth and Characterization of PDMS-Stamped Halide Perovskite Single Microcrystals," *J. Phys. Chem. C*, vol. 120, no. 12, pp. 6475–6481, 2016.

[13] Y. Luo, P. Khoram, S. Brittman, Z. Zhu, B. Lai, S. P. Ong, E. C. Garnett, and D. P. Fenning, "Direct Observation of Halide Migration and the Resulting Optoelectronic Effect within CH3NH3PbBr3 Single Crystals," *To be Submitted*.

Influence of Grain Size and Interfaces on Photo-Stability of Perovskite Solar Cells

Istiaque Hossain, Liang Zhang, Ranjith Kottokkaran, Mohamed El-Henawey, Pranav Joshi, Max Noack and Vikram Dalal

Iowa State University, Ames, Iowa 50011

Abstract — **We report on the photo-stability of perovskite solar cells prepared using solution growth and vapor phase growth. The cells prepared using solution growth had varying grain sizes achieved by using solvent annealing techniques. The cells deposited using co-evaporation were prepared at varying substrate temperatures. All the cells were exposed for 100 hours to varying degrees of intensity of light inside an environmental chamber filled with high purity nitrogen and thus was devoid of moisture and oxygen. It was found that the cells with larger grains had a significantly lower photo-degradation than cells with smaller grains. It was also found that vapor deposited cells, in general, had less photo-degradation than solution-grown cells with similar grain sizes. The photo-degradation is ascribed to the generation of ions by light during the degradation process. A measurement of ion density reveals that the short circuit current during degradation, and even before degradation, is inversely correlated to the measured ion density. The measurement of degradation at varying light intensities shows a light-induced annealing behavior similar to what is found in a-Si, leading to saturation in photo-degradation. It was also found that the interfaces have a profound influence on photo-degradation, with cells deposited on organic interfaces suffering significantly more degradation than cells deposited on inorganic interfaces.**

I. INTRODUCTION

Hybrid organic-inorganic lead-based perovskite solar cells have become a major potential new PV technology. Conversion efficiency exceeding 22% has been achieved in cells fabricated using solution growth techniques[1]. A major issue remaining in the field is the photo-induced degradation in the properties' of the solar cells and also the temperature stability of the materials. Previous work [2] has shown that light generates ions, and that these ions lead to extra recombination of photo-generated electrons and holes, leading to a decrease in conversion efficiency. An interesting question that arises is whether we can reduce the photo-degradation by reducing the rate of ion generation. Yet another question is what roles do the p and n interfaces play in photo-degradation of the cell. The degradation data also show a tendency towards saturation, a behavior very similar to what is observed in a-Si solar cells, leading to a question about the mechanism behind such saturation. In this paper, we address all these questions, by doing careful experiments and match the degradation data with measured data on ion densities and also match the reduction in short-circuit current to at various intensities to a model which includes light induced annealing of ions.

II. EXPERIMENTAL DETAILS

All the solar cells studied here were of the p-i-n superstrate geometry. The cells were deposited on glass substrates coated with ITO. Next, a p-layer was deposited. The p layer was either NiO or PTAA. Both these layers provide excellent hole transport layers with a good match to the valence band of perovskites and a Type II heterojunction to inhibit electron recombination at the p contact. The perovskite layer was deposited using either solution growth techniques [2, 3] inside a nitrogen filled glove box, or a co-evaporation technique in a vacuum system. On top of the perovskite layer, an electron conducting heterojunction layer of PCBM was deposited using spin coating, followed by deposition of Al as the back ohmic contact. Light entered the cell from the glass side. Fig. 1 illustrates a schematic diagram of the cell structure.

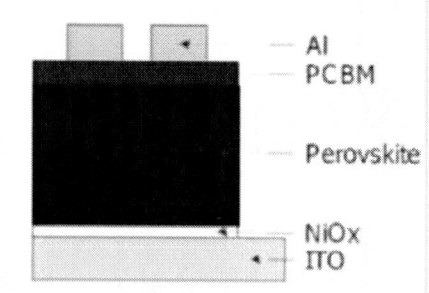

Fig. 1 Schematic diagram of perovskite cell

During the solution growth using the process developed by Xiao et al [2], some of the perovskite films were subjected to grain growth using a petri dish and DMF vapor [4], the so-called solvent annealing (SA) process. Using solvent annealing, the grain size was increased from 200-300 nm for the control sample to >1 micrometer for the SA sample. In Fig. 2, we show the grain sizes for control and SA samples.

III. RESULTS

A.I-V curves: The illuminated I-V curves, measured using an ABET full-spectrum AM1.5 simulator are shown in Fig. 3 for the control sample and the SA sample. They are very similar, with perhaps a slightly higher current for the SA sample.

978-1-5090-5606-4/17 $31.00 © 2017 IEEE

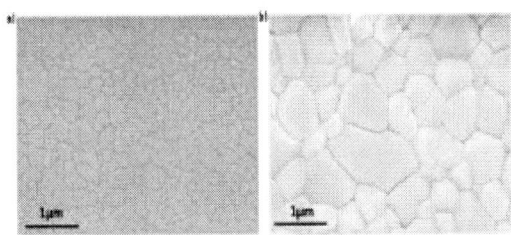

Fig. 2 Grain sizes for control(left)and solvent annealed (right) devices.

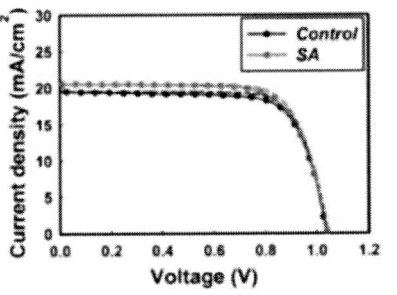

Fig. 3 I-V curves for control cell and cell with solvent anneal(SA)

B. Data on photo-induced degradation of control and SA samples:The changes in voltage and current upon light-soaking for the control and the SA cells are shown in Fig. 4.

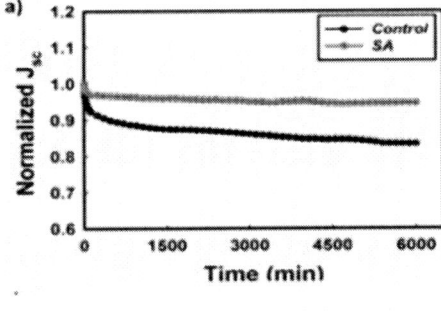

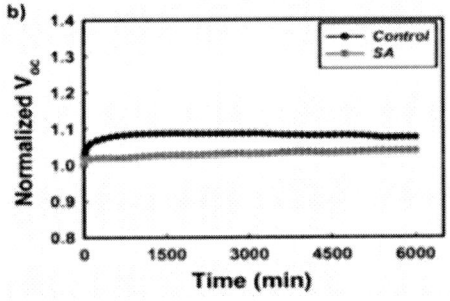

Fig. 4 Reduction in Jsc and Voc for control sample (black curve) and solvent annealed sample (red curve) upon photo-induced degradation.

C. Recovery of photo-degraded cell after thermal anneal in the dark. When the cell is allowed to sit in the dark after photo-degradation, it recovers completely after some time [5]. The recovery for the control and SA samples is shown in Fig. 5.

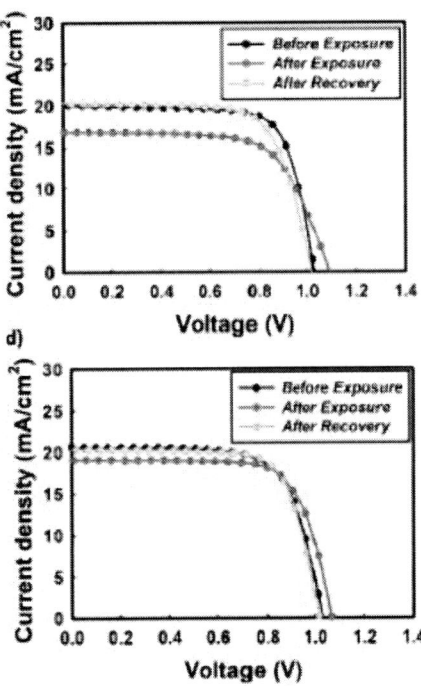

Fig. 5 Recovery of I-V curves for control sample(top) and SA sample(bottom)

D. Data on dark current measurement before and after degradation for the solvent annealed cell and the control cell:
To understand what is happening upon photo-degradation, and upon thermal anneal after degradation, we measured dark I-V curves before degradation, immediately after degradation, and during thermal anneal in dark. The data is shown in Fig. 6 for a typical cell. Note from Fig. 6 how the zero current immediately after degradation is shifted to positive voltage rather than being at zero voltage. At zero volts, there is a finite negative current. The figure also shows how the dark current recovers back to the original value with zero current at zero voltage when the cell is allowed to anneal in the dark.

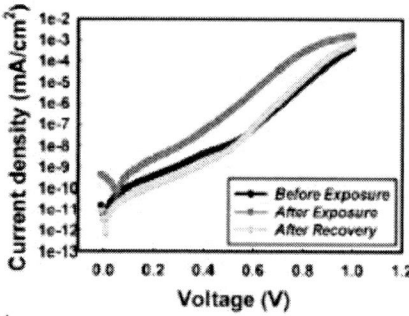

Fig. 6 Dark current-voltage curve initially (black), after degradation(red), and after recovery(green)

The negative current at zero volts in Fig. 6 suggests that positive charges in the material are moving towards the negatively charged(p)contact[5] and negative charges towards the positive(n) electrode. These ions are the light generated ions during degradation. We have suggested previously [5] that the reduction in short circuit current during degradation results from recombination with these light generated ions. To quantify this assertion, we integrated the dark current at 0V over time, giving us a measure of the total ionic charge density generated during photo-degradation. The results of this measurement, and the calculated ion density after degradation are shown in Fig. 7 (a) and (b) for the two samples before and after degradation. The figures show that the ion density in the control sample is higher than the solvent annealed sample before degradation, and that after degradation, the ion density in the control sample remains 40% higher than in the SA sample. This result strongly supports our interpretation that photo-generated ions are responsible for degradation and that samples with larger grains have fewer such ions.

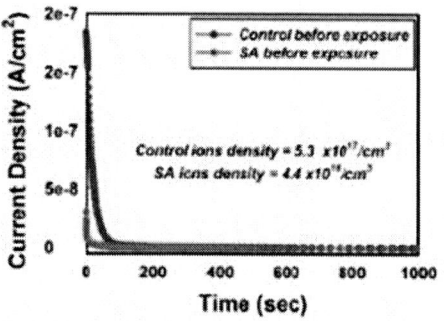

Fig. 7(a)

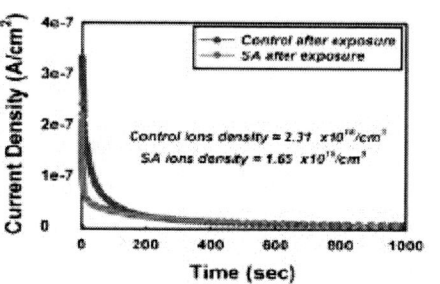

Fig. 7(b)
Fig. 7(a) and (b), showing integrated ion densities before and after degradation for the control sample and the sample SA with larger grain.

E. Saturation of degradation of current with time. A very interesting observation from Fig. 4 and Fig. 8 is that the degradation appears to saturate at all intensities. Initially, the degradation is linear with time, but as time increases, degradation slows down, implying a reverse process. This is a remarkable observation- it suggests that there is an inverse light-induced annealing process that is also occurring. This observation is very similar to the case for a-Si, where light generates defects, but over time, it also anneals these defects out through a reverse process[8].

F. Influence of contacts on photo-induced degradation:
We made several cells on both NiO and PTAA hole transport layers. The initial device efficiencies for both cells were very similar (~15% range). In Fig. 8, we show the data for degradation in current on the two p type contact materials. Very clearly, the sample deposited on PTAA degrades much more than the sample deposited on NiO. Thus, we show that the interfaces are playing a major role in the degradation of the cell.

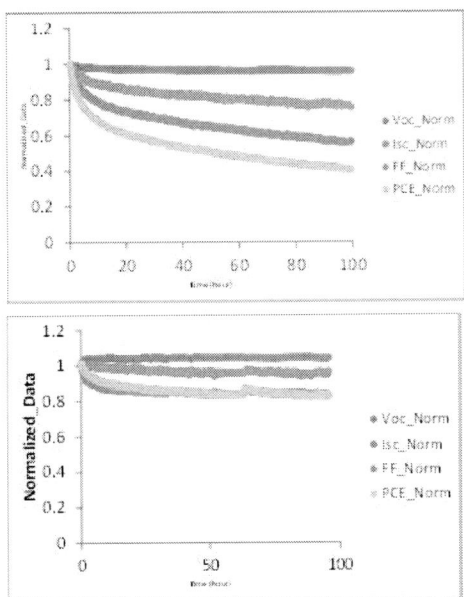

Fig. 8. Degradation of device deposited on PTAA(top and NiO(bottom).

IV. DISCUSSION

The results on measured ion densities and the reduction in short circuit current clearly show the critical role of photo-generated ions on photo-degradation of the solar cell. We have also shown that light is inducing a reverse annealing effect on ion generation, and that is why the photo-induced degradation tends to saturate. Finally, we have shown that the type of contact, whether organic (PTAA) or inorganic(NiO) also play a role in the degradation of cells with similar efficiencies. In the talk, we will also show how co-evaporated devices show a better stability than solution-grown devices.

V. ACKNOWLEDGMENTS

This work was partially supported by two NSF grants, CBET-1336134 and ECCS-1507291. Mr.El-Henewey was partially supported by a fellowship from the Government of Egypt.

VI. REFERENCES

1. W. S. Yang et al, Science, 2015, 348, 1234-1237
2. Z. Xiao et al, *Nat Mater*, 2015, **14**, 193-198.
3. H-S. Kim et Sci. Rep. 2, 591 (2012)
4. J. Liu et al, *Applied Materials & Interfaces*, 2015, **7**, 24008-24015
5. Pranav Joshi et al, AIP-Advances, **6**, 115114(2016)

Cold Thoughts on Perovskite Fever

Tao Xu, Jue Gong

Department of Chemistry and Biochemistry, Northern Illinois University, DeKalb, Illinois 60115, USA

Abstract — The latest progress in solution process-based organic-inorganic hybrid perovskites solar cells is reshaping the growth pattern of any previous photovoltaic technologies and has raised a storm of research fever. However, despite the success in boosting efficiency, it also appears high time to inject an intense dose of cold thoughts into this globally-spreading "perovskite fever", because perovskite solar cells are still facing several critical challenges. Prominently, these hybrid perovskites suffer from the materials instability in moisture, the use of environment-hazardous lead, the costly and unstable complex organics as hole transport materials, the use of precious metals as back cathode, the hysteresis in current-voltage scans, the tricky engineering of good-quality perovskite films. Moreover, there are still some missing puzzle pieces for a comprehensive basic science understanding. In this talk, I will present our work on tackling these challenges. First, that nickel, as an industrial commodity metal with work function of 5.1 eV, can replace gold as the back cathode in perovskite solar cells. Furthermore, when partial iodides in $CH_3NH_3PbI_3$ are replaced with pseudohalide ions thiocyanate (SCN-), the resulting SCN-containing perovskite films strikingly rival the conventional $CH_3NH_3PbI_3$ films in terms of moisture, and exhibit efficiency comparable to that of $CH_3NH_3PbI_3$-based cells fabricated in the same way. In addition the unusual electron-rotor interaction discovered in hybrid perovskite materials using isotope effects and the unusual optical and electronic properties of hybrid perovskite crystals under high pressure. These fundamental studies provide a new coordination in the mechanistic foundation needed for better materials by designing.

Index Terms — perovskite, solar cell, cathode, anode, pressure

I. INTRODUCTION

Hybrid organic-inorganic solid state solar cells with perovskite-structured $CH_3NH_3PbI_3$ as active layer have recently been reported with over 20% efficiency, which emerges as the most promising candidate for the next generation solar cells.[1] As a game changer in photovoltaics, perovskite type material exhibits striking excellence in both light absorption and charge transport (1069nm electron diffusion length and 1213 nm holes diffusion length). Despite its aura, the dark side of this rising star should not be ignored. Current grand challenges include the stability issue of CH3NH3PbI3,[2] the use of environment-hazardous lead,[3] the cost of complex organics as hole-blocking layer[4], and the use of expensive noble metals as back cathode.[5] On the other hand, opportunities are on the horizontal by fundamental understanding of the elementary processes of light absorption, charge separation and charge collection, which can provide valuable clues for designing better materials that overcome the drawback in current lead iodide based perovskite materials.

II. EXPERIMENTAL

$CH_3NH_3PbI_3$ precursory solution: 0.289g of PbI2 (from Aldrich) and 0.1g CH3NH3I were dissolved in 0.52 ml γ-butyrolactone t 60 oC under stirring for 3 hours. (CH3NH3)2Pb(SCN)2I2 precursory solution: 0.203g Pb(SCN)2 and 0.15g CH3NH3I were dissolved in 0.6ml dimethylformamide (DMF) under stirring at 60 oC for 3 hours. The hole transport material (HTM) layer was prepared by dissolving 92 mg 2,2',7,7'-tetrakis(N,N-di-p-methoxyphenylamine)-9,9'-spiro-bifluorene (spiro-MeOTAD), 7.2 mg Li bis-trifluoromethane sulfonimide, and 12 mg 4-tert-butylpyridine in 1 ml chlorobenzene.

Fabrication of the corresponding perovskite solar cells: the perovskite precursor solution was spin-coated at 2000 rmp for 1 min in dry air onto the photoanodes, fabricated as we described previously,[5-7] followed by annealing at 105 ℃ in air for 30 minutes. The HTM layer was spin-coated at 2500 rpm. A layer of 80 nm thick Au film was thermally coated atop the HTM layer using an Edward 306A thermal evaporator. Ni film was coated by sputtering.

Characterization: Potentiostat (Gamry Reference 600) and solar simulator (Photo Emission Inc. CA, model SS50B) were used to acquire the current density-voltage (J-V) curves. Scan electron microscope (SEM) images were taken on a Tescan SEM (model Vega II SBH). Diffusive reflectance of the devices was taken on a Filmetrics F20-EXR thin film analyser with an integrating sphere for measuring the reflectance. XRD spectra were collected on a Miniflex X-ray diffractometer.

III. RESULTS AND DISCUSSION

In order to maximize the attainable open circuit voltage (Voc), low chemical potential, namely, high work-function noble metals, such as gold, and silver,[1] are generally used as back cathode. Thermal evaporation of gold is a very costly and wasteful process because only a tiny portion of gold is eventually deposited onto the devices. Therefore, replacing gold with earth-abundant elements as the cathode in perovskite solar cells, while still retaining their high Voc and energy conversion efficiency is a pivotally critical step towards the cost-effective production of perovskite solar cells. In quest for high work-function metals, noble metals such as gold, platinum, or silver often come up as the routine choice. Herein, we break down this stereotype selection rule by

978-1-5090-5606-4/17 $31.00 © 2017 IEEE

perovskite-type solar cells using nickel as the back cathode.5 Nickel has a work-function of ~5.04 eV, very close to that of gold (~5.1 eV), but with a unit price less than 0.03% of that of gold. As configured in Figure 1, the cell contains a ~900 nm-high rutile TiO_2 nanowires (NWs) array on FTO glass as photoanode, which is filled with a layer of spinning-coated $CH_3NH_3PbI_3$, a 220 nm-thick spinning-coated spiro-MeOTAD (2,2′7,7′-tetrakis-(N,N-di-p-methoxyphenylamine)-9,9′-spirobifluorene) as hole transport material (HTM), capped by a sputtering-coated nickel film as the cathode.

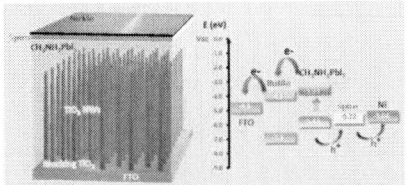

Figure 1. A schematic diagram of a perovskite solar cell consisting of a FTO anode covered with a blocking TiO_2 thin film, rutile TiO_2 nanowires, spin-coated $CH_3NH_3PbI_3$ layer and sprio-MeTAD layer, followed by a sputtered nickel cathode. The left panel shows the energy diagram of the device. [5]

Figure 2 shows the photocurrent density vs photovoltage (J-V) curves of cells using 900 nm TiO_2 nanowires as photoaondes. It can been seen that the cell using 300 nm nickel film as cathode exhibits comparable energy conversion efficiency (η=10.4%) to that of the cell using 80 nm gold film as cathode (η=11.6%). However, the cells using 150 nm nickel film as the cathodes exhibits notable drop in efficiency (η=7.7%). To troubleshoot the source of energy loss in 150 nm nickel-cathoded cells, short circuit current density (Jsc) for all cells are first compared. However, both cells exhibit almost the same Jsc, about 20 mA/cm-2. Furthermore, Voc of the cell using 150 nm nickel as cathode is 0.752V, only slightly lower than that of the cells using 80nm gold or 300 nm Ni film as cathode. Then, compared to the fill factor (FF) of 68% for the cell using gold as cathode, it is the relatively low FF (50%) found in the cell using nickel as cathode that is responsible for its low conversion efficiency. Since there is no significant difference in Voc and Jsc in comparison to the other two cells, we attribute the low FF in cell with 150 nm nickel cathode mainly to its high series resistance. Indeed, shunt resistance (Rsh) at Jsc for 150 nm nickel-backed cell is 11794Ω, which is about 80% of that for gold-backed cell; while the series resistance (Rs) at Voc for the nickel-backed cell is 264Ω, over 3 times higher than that of the gold-backed cell.

As we further increased the thickness of the nickel film to 300 nm thick, the FF of the cell increases to 0.61 and the conversion efficiency reaches 10.4%. The two-probe measurement on the edge-to-edge resistance of the 300 nm thick nickel pads is measured to be 3Ω and the four-probe sheet resistance of the 300 nm-thick nickel cathodes is measured to be 3~5Ω/square.

The stability of the photovoltaic performance of the cells using nickel cathodes are also studied. Figure 3a and 3b respectively show the J-V curves of the cells using 150nm Ni and 300 nm Ni film as the cathode. Explicitly, for the 150 nm Ni-cathoded cell, the Jsc decreased from 22.4 mA/cm-2 on day 1 to 20.4 mA/cm-2 on day 5, and FF increased from 43% on day 1 to 50% on day 5. For the 300 nm Ni-cathoded cell, Voc changed from the 0.70 V on day 1 to 0.75 V on day 5, Jsc changed slightly from 20.1 mA/cm-2 on day 1 to 20.6 mA/cm-2 on day 5, and FF increased from 50% on day 1 to 61% on day 5. We think that the decrease in Jsc for the 150 nm Ni-cathoded cell is likely due to the absorption of moisture in the $CH_3NH_3PbI_3$. The overall efficiencies increase from 6.7% on day 1 to 7.7% on day 5 for the 150 nm Ni-cathoded cell, and 7.8% on day 1 to 10.4% on day 5 for the 300 nm Ni-cathoded cell.

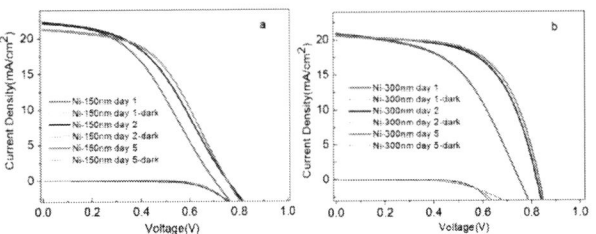

Figure 2. J-V curves for nickel-cathoded cells, (a) 150nm thick Ni and (b) 300 thick nickel, measured on 1 day 1, day 2 and day 5, respectively. Between each measurement, the devices are stored in a desiccator.[5]

Furthermore, perovskite materials decomposes in the moisture air, which is the main barrier for its future commercialization. Pseudohalides have similar chemical behaviors and properties to true halide.[16] In this study, we developed a new perovskite material $(CH_3NH_3)_2Pb(SCN)_2I_2$ for perovskite solar cells by replacing some iodides in $CH_3NH_3PbI_3$ with o thiocyanate ions. Figure 3 shows the XRD pattern of $(CH_3NH_3)_2Pb(SCN)_2I_2$ along with $CH_3NH_3PbI_3$ and CH_3NH_3I for comparison.

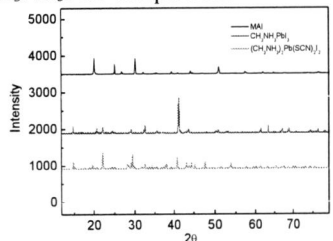

Figure 3. XRD patterns of $(CH_3NH_3)_2Pb(SCN)_2I_2$ along with $CH_3NH_3PbI_3$ and CH_3NH_3I for comparison.[7]

According to the crystal structure of $CH_3NH_3PbX_3$ perovskite material, lead ion and halides form the frame of the perovskite structure.[17] The first step of degradation in moisture involves a formation of a hydrated intermediate containing isolated PbX_6^{4-} octahedra.[13b] The formation constant of lead-halide complex is enssentially its equilibrium constant, which reflects the binding tightness between the halides and the central Pb2+. The formation constant K4 (cumulative formation constant $\beta n = K1 \cdot K2 \cdots Kn$) is calculated to be only 3.5 for $PbI_4{}^2$-,[18] which is in the range of weak interaction between I- and Pb2+. In the case of $(CH_3NH_3)_2Pb(SCN)_2I_2$, the interaction between Pb2+ and SCN- is much stronger and the formation constant K4 is up to 7 for $Pb(SCN)_4{}^{2-}$.[19] Comparing the spheric shape I- with the linear shape SCN as indicated by their Lewis structures, the lone pair of electrons from S and N in SCN- can form stronger ligand with Pb, which in turn stabilizes the frame structure of $(CH_3NH_3)_2Pb(SCN)_2I_2$. We show that direct improvement of moisture tolerance for $(CH_3NH_3)_2Pb(SCN)_2I_2$ has been observed in accelerated stability test compared with traditional $CH_3NH_3PbI_3$ perovskite material.

The accelerated moisture-tolerance tests for perovskite materials were performed by monitoring the reflection of the corresponding perovskite films in 95% relative humidity air at room temperature. Increase of reflection in Figure 3a for $CH_3NH_3PbI_3$ is due to the decomposition of perovskite structure in moisture.[8b] As can be seen, perovskite type $CH_3NH_3PbI_3$ started to decompose immediately after being exposed to moisture. After 1.5 hours, most of the black film has been decomposed to the yellowish color, originated from PbI_2, and the corresponding reflection increases from about 10% to about 20%. In the case of $(CH_3NH_3)_2Pb(SCN)_2I_2$ as shown in Figure 3b, no obvious reflection increase was observed even after 4 hours exposure to moisture. The corresponding reflection increases only 2% after 4 hours of exposure in moisture, from about 6% to 8%. Some tiny white patches emerge at the corners, which are likely due to $Pb(SCN)_2$. In addition, the black color of $(CH_3NH_3)_2Pb(SCN)_2I_2$ film can last for months in air with 40% relative humidity or below. The lower inserted images in figure 3a and 3b compare the CH3NH3PbI3 film and $CH_3NH_3Pb(SCN)_2I_2$ film after 30 days in air at 20~40% relative humidity.

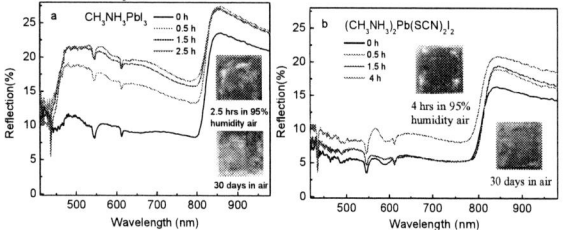

Figure 4. Accelerated stability test for perovskite films (a) CH3NH3PbI3 and (b) (CH3NH3)2Pb(SCN)2I2 in air with 95% relative humidity.[7]

We further reported the discovery of a novel electron-lattice interaction in solid-state semiconductors. We revealed the carrier–rotor coupling effect in perovskite organic–inorganic hybrid lead iodide ($CH_3NH_3PbI_3$) compounds by isotope effects. By replacing the hydrogen with deuterium in the cationic rotor in methylammonium (MA+), we synthesized

single crystals of $CH_3ND_3PbI_3$, $CD_3NH_3PbI_3$, and $CD_3ND_3PbI_3$, along with the original $CH_3NH_3PbI_3$.

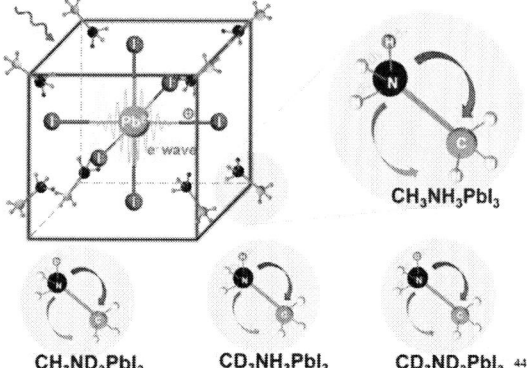

Figure 5. Four isotoped peroskite materials.[8]

Time-resolved photoluminescence (TRPL) study reveals that the carrier lifetime follows the trend of $CH_3NH_3PbI_3$ > $CH_3ND_3PbI_3$, $\approx$ $CD_3NH_3PbI_3$ > $CD_3ND_3PbI_3$, exactly in accordance with the rotational frequency of the isotope cationic rotors, i.e. $CH_3NH_3{}^+$ >$CH_3ND_3PbI_3\approx$ $CD_3NH_3PbI_3$> CD_3ND_3 as shown in Figure 6.

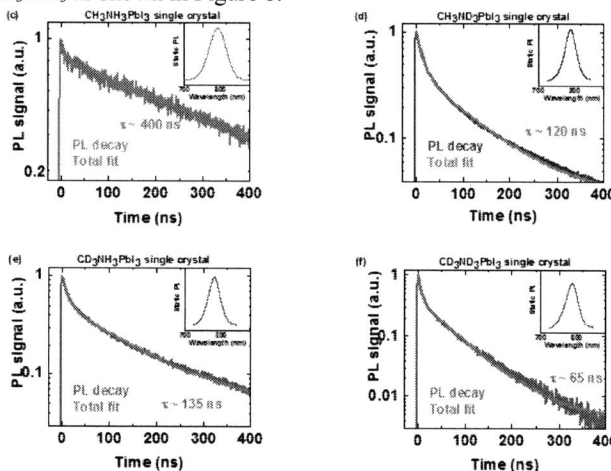

Figure 5. TRPL of the single crystals of four isotoped peroskite materials.[8]

Note that the comparison in carrier lifetime between $CD_3NH_3PbI_3$ and $CH_3ND_3PbI_3$ single crystals suggests that vibrational modes in methylammonium (MA+) have little impact on carrier lifetime. We found that polaron model suggests that there is a strong interaction between the photo-induced free electrons in the PbI_3- inorganic framework and the polarity variation of the organic rotor MA+ resulting from their rotation. The work adds a new coordination for materials-by-design in fields of photovoltaics, ferroelectrics, superconductivity, theroelectrics, piezoelectrics and so on.

V. CONCLUSIONS

Our work provided some alternative pathways to tackle the grand chanlleges in perovksite solar cells including materials cost, moisture stality and fundamental understanding of charge transport.

V. ACKNOWLEDGEMENT

This work was supported by the U. S. National Science Foundation (CBET-1150617).

VI. REFERENCES

[1] X. Liu, W. Zhao; H. Cui, Y. Xie, Y. Wang, T. Xu, F. Huang, "Organic–inorganic Halide Perovskite Based Solar Cells-Revolutionary Progress in Photovoltaics" *Inorganic Chemistry Frontiers,* , Vol. *2,* page 315-335, 2015

[2] T. Leijtens, G. E. Eperon, S. Pathak, A. Abate, M. M. Lee, H. J. Snaith, " Overcoming ultraviolet light instability of sensitized TiO_2 with meso-superstructured organometal tri-halide perovskite solar cells. " *Nat. Commun.* Vol.4, 2885. 2013.

[3] F. Hao, C. C. Stoumpos, D. H. Cao, R. P. H. Chang, M. G. Kanatzidis, "Lead-free solid-state organic-inorganic halide perovskite solar cells", *Nature Photonics* Vol. 8, page. 489-494. 2014,

[4] J. A. Christians, R. C. Fung, P. V. Kamat, "An inorganic hole conductor for organo-lead halide perovskite solar cells. Improved hole conductivity with copper iodide" *J. Am. Chem. Soc.* Vol. 136, page 758-764, 2014.

[5] Q. Jiang, S. Xia, B. Shi, X. Feng, T. Xu, "Nickel‐Cathoded Perovskite Solar Cells", *J. Phys. Chem. C.,* Vol. 118, page. 25878–25883, 2014.

[6] Q. Jiang, X. Sheng, Y. Li, X. Feng, T. Xu, "Rutile TiO2 Nanowires Perovskite Solar Cells" *Chem. Commun.* Vol. 50, page 14720-14723, 2014,

[7] Q. Jiang, D. Rebollar, J. Gong, E. L. Piacentino, C. Zheng, T. Xu, "Pseudohalide‐Induced Moisture Tolerance in Perovskite CH3NH3Pb(SCN)2I Thin Films" *Agnew. Chem. Int. Ed.,* Vol.*54,* page 7617-7620,2015,

[8] J. Gong, M. Yang, X. Ma, R. D. Schaller, G. Liu, L. Kong, Y. Yang, M. C. Beard, M. Lesslie, Y. Dai, B. Huang, K. Zhu, T. Xu, "Interaction in Organic–Inorganic Lead Iodide Perovskites Discovered by Isotope Effect J. Phys. Chem. Lett. 2016, 7, 2879–2887

LBIC analysis of perovskite based solar cells stability

Carmen M. Ruiz[1,2], Javier Ramos[2], Richard Garuz[1], Damien Barakel[1], Jean Rousset[3], Judikaël Le Rouzo[1]

[1]Aix Marseille Université, CNRS, IM2NP UMR 7334, 13397, Marseille, France

[2]Institut Photovoltaïque d'Ile de France 8, rue de la Renaissance, 92160 Antony, France

[3]Institute of Research and Development on Photovoltaic Energy (IRDEP), UMR 7174 CNRS-EDF- Chimie ParisTech, EDF R&D, Chatou, France

Abstract — One of the main obstacles to overcome on perovskite based solar cells is their inherent instability that effectively degrades de photovoltaic performance of the devices. In order to understand its mechanisms, LBIC mappings at different wavelengths have been performed at different times. The lack of homogeneity within the cell is analyzed.

Index Terms — perovskite solar cells, LBIC measurements, stability.

I. INTRODUCTION

The rapid and enormous advance in the field of perovskite solar cells (PSC) has permitted to reach power conversion efficiencies over 20% together with a great number of scientific reports during the last five years. Although some of the physical and optoelectronic properties remained unclear at the beginning, very important efforts have been done recently to elucidate and understand the outstanding behavior of these type of solar cells [1].

Nevertheless, there is still much to be done for achieving a mature competitive photovoltaic technology. In particular, severe degradation of photovoltaic performances in a rather short time are strongly limiting the potential of this material.

While some of these losses are related to atmosphere degradation and can be partially or totally limited with an appropriate encapsulation, some others seem to be inherent to the material structure, and have to be fully identified for developing techniques that would allow to reduce their importance.

Light Beam Induced Current (LBIC) is a characterization technique that permits a direct correlation of the generated photocurrent with the topography of the cell. It is specially suited to detect inhomogeneities correlated with features such as contact corrosion, bad encapsulation or absorber discontinuities, as the regions with the particular problem will present a different current value. Also, LBIC can beperformed with monochromatic light, in this way , issues related with the absorption length of the material can be detected within the solar cell.

In this work we propose the use of LBIC under both white and monochromatic illumination in order to elucidate the potential issues that would lead to current losses on perovskite based solar cells.

II. EXPERIMENTAL SECTION

A. Materials

All chemicals were commercially available. PbI_2 was acquired from TCI (>98%), $PbBr_2$ from Alfa Aesar and the organic halide salts i.e. methylammonium iodide (MAI), formamidinium iodide (FAI) and methylammonium bromide (MABr) were bought from Dyesol. TiO_2 commercial paste for mesoporous paste was obtained from Dyesol (30NR-D). Spiro-OMeTAD was purchased from Sigma Aldrich (sublimed grade). Dried solvents from Sigma Aldrich were employed without any treatment or purification.

B. Device Fabrication

Initially, FTO glass was etched using HCl (4M) and Zn powder. After that, the substrates were ultrasonicated using detergent (RBS®), rinsed with deionized water and ultrasonicated again with acetone. Then, substrates were rinsed with absolute ethanol, dried with compressed air and heated at 500°C to remove any organic residue. A TiO_2 dense blocking layer was deposited by spray pyrolysis at 450°C using a solution consisted on titanium diisopropoxide bis(acetylacetonate) diluted in acetyl acetone and pure ethanol using oxygen as carrier gas. TiO_2 nanoparticles were deposited by spin-coating emplying a solution of 30NR-D paste diluted in absolute ethanol. Later, perovskite was prepared by intermediate adduct formation according literature, both for $MAPbI_3$ [2] and mixed one [3] using spin coating to deposit it. Spiro-OMeTAD (2,2',7,7'-tetrakis(N,Ndi- p-methoxyphenyamine)-9,9-spirobifluorene) was diluted in chlorobenzene and spun coated from a solution containing FK209 colbalt salt (5% mole ratio regarding Spiro- OMeTAD) LiTFSI (50% mole ratio), tert-butyl pyridine (330% mole ratio). Finally 100nm of Au were thermallyevaporated under high vacuum.

Perovskite and Spiro-OMeTAD solutions were prepared inside a nitrogen glove box as well as their respective depositions. The active area (4x4mm) was masked to 0.09cm^2 for J-V characterization.

C. Characterization

With a careful control of the stoichiometry of the system, MAPbI$_3$ solar cells exhibiting 17.5% of power conversion efficiencies in reverse scan were obtained (15.3% in forward one, see Figure 1a). Although high voltages and photocurrents were obtained keeping fill factors high, some issues, such as hysteresis and the differences in the EQE under light biased/unbiased conditions (Figure 1b) most likely due to the presence of traps, limit the possibilities of these systems. This level of efficiency (> 17 %) is obtained in a very reproducible way though the batch and between batches.

A mixed perovskite has also been considered, using both formamidinium (FA) and methylammonium (MA) as organic cations and both bromide and iodide as halides giving a final stoichiometry of FA0.85MA0.15PbI0.85Br0.15. With this compound, the problems related to hysteresis (Figure 1c) and the differences in EQE (Figure 1d) by varying conditions problems have been reduced, so we assume the presence oftraps has been reduced. In this case, a careful control of theexcess PbI2 in the mixture of reactants must be taken intoaccount to push the efficiencies to almost 17% with negligible hysteresis.

On figure 2 we can see the corresponding LBIC mapping using white light as excitation source and an spot size of 100 microns. It appears a difference between the center of the cell and the edges of the contact, that can be related to partial overlapping of the scanning spot. It is also interesting to see the presence of inhomogeneities within the surface of the cell, that can be attributed to differences in the perovskite composition.

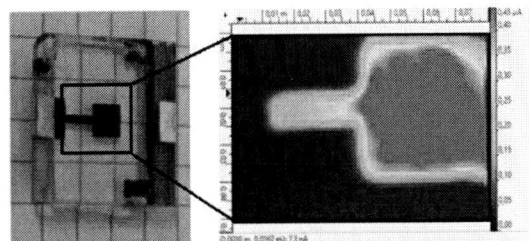

Figure 2. a) Image of a solar cell and b) Corresponding current cartography under white light, edge effects around the contact are clear, while other regions inside the cell present also different current values.

For a better understanding of this phenomenon LBIC measurements were carried out on the same cell with different monochromatic sources, as can be seen on figure 3. While the mapping obtained when a white light is used as a probe seems mostly homogeneous in the center of the cell, the study with different wavelengths gives a more complicated perspective. In particular, for larger wavelengths, the differences of collected photocurrent become more important than for the shorter ones. This can be explained because carrier collection at 700 nm can be impacted by an incomplete absorption , such as the layer thickness is not uniform, or of carrier recombination due to the presence of traps that are not uniformly distributed within the cell.

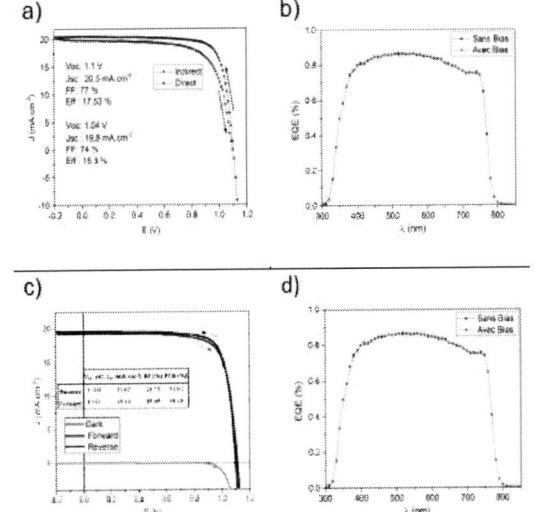

Figure 1. a) J-V characteristics of a MAPbI$_3$ solar cell b) EQE spectra of a MAPbI$_3$ based solar cell c) J-V characteristics of a mixed perovskite solar cell d) EQE spectra of a mixed perovskite based solar cell.

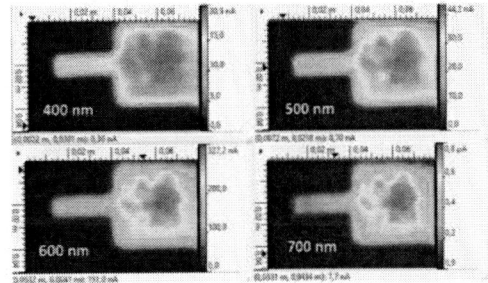

Figure 3. LBIC mappings under illumination at 400, 500, 600 and 700 nm. The inhomogeneities are more evident for larger wavelengths.

The evolution through time of these distributions of current inside the cell, tends to agree with a partial decomposition that would lead to a worst carrier collection.

Complementary capacitance measurements are being carried in order to identify these potential carrier traps.

REFERENCES

[1] J. S. Manser, J. A. Christians, and P. V. Kamat, "Intriguing Optoelectronic Properties of Metal Halide Perovskites," Chem. Rev., vol. 116, no. 21, pp. 12956–13008, 2016.

[2] N. Ahn, D.-Y. Son, I.-H. Jang, S. M. Kang, M. Choi, and N.-G. Park, "Highly Reproducible Perovskite Solar Cells with Average Efficiency of 18.3% and Best Efficiency of 19.7% Fabricated via Lewis Base Adduct of Lead(II) Iodide," J. Am. Chem. Soc., vol. 137, no. 27, pp. 8696–8699, 2015.

[3] N. J. Jeon, J. H. Noh, W. S. Yang, Y. C. Kim, S. Ryu, J. Seo, and S. Il Seok, "Compositional engineering of perovskite materials for high-performance solar cells," Nature, vol. 517, no. 7535, pp. 476–480, 2015.

Assessing Job Growth and Sustainability in the US PV Industry

Brion Bob

SunShot Initiative, Department of Energy, Washington, DC, 20585, USA

Abstract — **The US Photovoltaics industry has emerged as a remarkable job-creation engine during the past 5 years. At the same time, the price of PV systems has declined dramatically. While the number of jobs per megawatt installed has steadily declined alongside system price, the number of jobs per dollar of revenue from system sales is found to have remained largely stable. The potential implications of future system price declines and the differing behaviors of the residential, commercial and utility-scale market segments are explored and discussed.**

Index Terms — **Solar Jobs, Market Growth, Revenue Generation, Workforce Productivity, Learning Rates.**

I. INTRODUCTION AND BACKGROUND

The US solar industry has emerged over the past several years as a major source of employment and economic opportunity. According to the US Energy Employment Report published by the Department of Energy, 373,000 Americans were employed in some capacity by the US solar industry in 2016, placing the solar workforce on a similar scale as established forms of energy generation [1]. The growth in solar employment comes alongside precipitous price declines in the levelized cost of solar electricity [2-4], which has contributed to continually increasing amounts of annually installed solar capacity

The aggressive price reductions that the industry has achieved to date, and which continue at a fast rate, are important for increasing deployment rates, and within many jurisdictions, further cost reductions are still needed to motivate utilities and balancing authorities to incorporate a comparatively new technology into their generation portfolio. As a result, the SunShot 2030 vision has targeted a 3 cent/kWh levelized cost of electricity for utility scale PV systems in geographies typical of the central United States [5], which would provide low-cost electricity at rates comparable to or lower than the wholesale price of electricity from existing generators.

Due to steadily declining prices for capacity and energy, a noticeable reduction in worldwide investment in renewable energy companies, systems, and other assets was observed from 2015 to 2016 despite steadily increasing rates of deployment [6]. The eventual balance between decreasing prices and increasing deployment will determine the behavior over time of system sales revenue within the US solar industry. The following discussion represents an initial attempt to understand job growth and sustainability within the US solar industry during the past several years in terms of productivity metrics that are either industrywide or specific to an individual

sector (residential, commercial, or utility-scale) within the US solar industry.

II. RESULTS AND ANALYSIS

In the following sections, jobs are defined as workers spending greater than 50% effort on solar activities sourced from the workforce figures published in the *National Solar Jobs Census* [7] and scaled by the fraction of PV firms surveyed or the reported fraction of PV employment to provide an estimate of US PV jobs. Price and deployment forecasts for the US PV industry use data from the *Solar Market Insight* series [8] and also projections from NREL's Regional Energy Deployment System [9].

A. Productivity in the US Photovoltaics Industry

As industries scale, their workforces become more efficient at getting the job done. This leads to improvements in an industry's productivity metrics, which include the amount of PV capacity that can be installed by a single person and the amount of value that arises from that work during a given time period. Due to a steady rate of advancements in system performance, material intensity, design standardization,

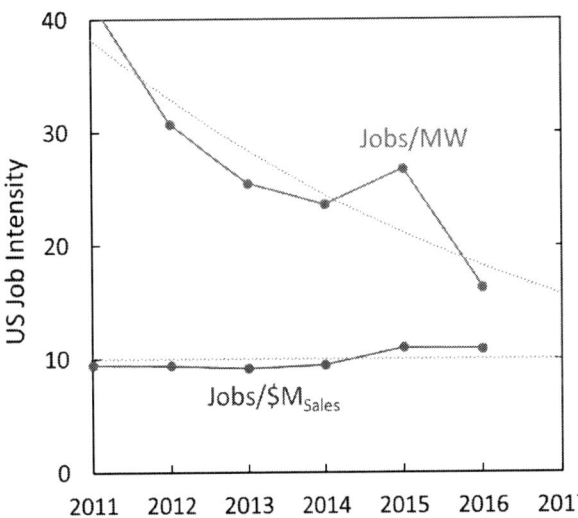

Fig. 1. The current trajectory of US employment numbers per installed megawatt of capacity (orange curve) is currently showing a rapid decline as the industry becomes more efficient and throughput increases along the entire solar value chain. US PV industry employment per unit of system sales revenue has followed a much flatter trajectory over the same timeframe.

permitting and interconnection practices, workforce skill and experience, and various other developments, the US PV industry workforce is rapidly becoming more efficient at installing systems with a given number of people, represented by the declining number of US job-years per installed MW, as shown in Figure 1.

Despite substantial changes in workforce productivity per unit capacity, the number of people that are employed per dollar of system sales has remained relatively constant during the same period. Given that the US PV industry installed roughly 7x more capacity in 2016 than it did in 2011 at less than half the price per MW of capacity, this relative stability would appear to indicate that workforce size as a function of sales revenue is linked to fundamental or slow-changing aspects of the industry's financial structure, which makes it a potentially useful tool for identifying long-term trends and projecting into the future.

The ratio of wages to overall industry revenue is in turn determined by how much value added is retained by commercial interests and how much of the industry's revenue is directed towards indirect costs that are not currently tabulated as jobs within the PV industry. These factors have either been slow to change during the past 5 years or have largely counterbalanced each other, resulting in a relatively

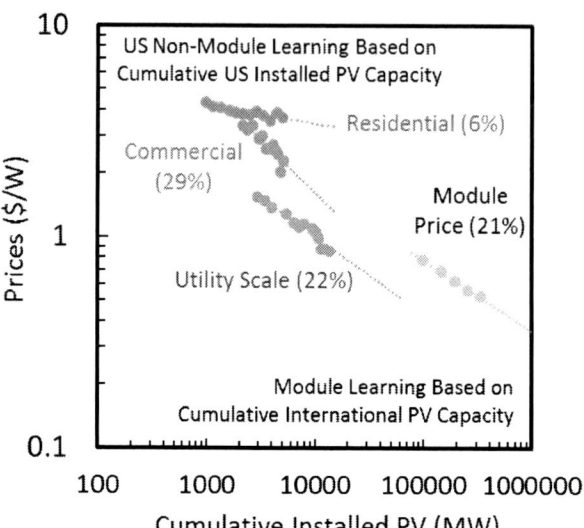

Fig. 3. Estimated learning curves for the time period 2013-2015 for non-module (blue/orange/grey) and module (yellow) prices associated with the residential, commercial, and utility scale market segments for the US PV industry.

stable value for the number of jobs that are created per dollar of system sales revenue.

A stable number of jobs per unit sales revenue in turn implies that the jobs per watt metric for the US PV industry is, for the moment, following a nearly proportional trend to annual blended average system price. The accuracy of this assumption year over year can be observed directly when the system price learning curve is plotted alongside the number of jobs required to install one MW per year, as shown in Figure 2. The nearly identical slopes of the two curves provides a direct illustration that overall industry throughput (in watts per person hour) has played a major role in reducing system costs, and that future improvements in technologies that increase the speed and ease with which PV modules and systems can be fabricated and commissioned will likely continue to occupy a central role in facilitating future cost reductions within the industry.

B. Near-Term Behavior of Jobs and Revenue

Based on the observed trends in the productivity metrics of the US PV industry during the past few years, the future balance between declining costs and increasing deployment rates, which determines the sales revenue, can be expected to largely define the future trajectory of workforce size within the PV industry. To gain some insight into near-term expectations, learning curves for the three-year period from 2013 to 2015 were prepared for international module fabrication costs and US-based non-module costs for the residential, commercial, and utility scale segments of the PV industry, as shown in

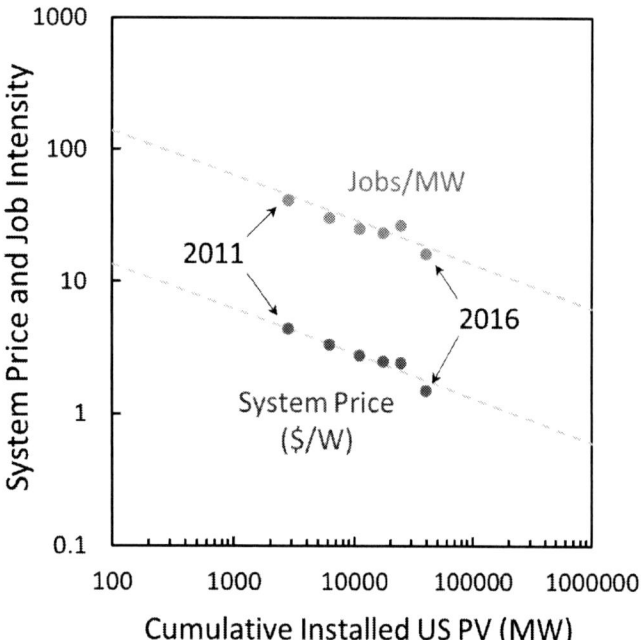

Fig. 2. Depiction of the correlation between Jobs/MW and blended average installed system price during the past 5 years. Slight fluctuations in the shape of the two curves with respect to one another represent small variations in the Job·yrs/M_{Sales}$ value for each year. The dashed lines correspond to a 21% cost reduction for each doubling in cumulative installed capacity.

Figure 3. While reported prices for the non-module aspects of residential PV systems appear to have declined relatively slowly during the past few years, reported prices of non-module aspects of the commercial and utility-scale market segments have exhibited substantial learning rates during the past 2 years of roughly 29% and 22% per doubling of installed capacity.

The module and non-module learning curves for each market segment were combined with deployment projections from GTM Research [8] to form a hypothetical near-term US PV revenue scenario. As shown in Figure 4, this combination of deployment rates and prices results in a relatively optimistic forecast for system sales within the residential market segment. As a result, overall solar industry system sales revenue continues to grow steadily despite price declines in the utility scale market segment nearly matching increases in deployment rates during the later years within this time period.

Differences between the residential, commercial, and utility scale segments also have important implications for the employment characteristics of the industry. The utility scale segment of the US PV industry bears a significant resemblance to the US wind industry, which are each capable of providing large scale energy resources at prices that are competitive with conventional wholesale generation. The rooftop segments of the PV industry allow for companies to sell directly to end users, which is vital in providing new and larger revenue streams. The PV industry has been able to emerge so quickly as a job creation engine in part due to its ability to branch out into three sectors that engage the full energy value chain and sell to a wide range of customers from utilities to businesses to individuals.

According to the data in Figure 4(a), the rooftop market segments appear to be steadily increasing in market size, while the utility scale sector is potentially closer to reaching a point where significant decreases in system pricing could begin to cause a saturation in system sales revenue. If system installation revenue begins to saturate, the addition of electricity energy storage, ancillary grid service capabilities, shared land uses, or other services that can be co-located with PV systems can help increase the value of the PV system and generate additional sales value for the industry.

The residential segment is currently deserving of additional mention as a driving force for job creation. Nearly 55% of all US solar jobs found by the most recent DOE US Energy Employment Report were in the residential PV sector [1], despite that sector accounting for closer to 33% of 2016 system sales revenue. Figure 4(b) provides the approximate job-years per million dollars for the residential and utility-scale sectors of the US PV industry from 2014-2016. The high numbers of jobs per unit of system sales revenue in the residential sector could be a reflection that installation activities, which tend to have a high domestic content, occupy a significantly higher fraction of overall system value than module fabrication activities in the residential sector. The observed differences may also be partially due to the different significance of system sales as a revenue stream in the residential vs. utility scale sectors, which is a topic for further investigation.

The near-term revenue projections shown in Figure 4 can be combined with the observed job densities of the residential, commercial, and utility scale market segments from the past few years to produce an estimate of future workforce sizes for

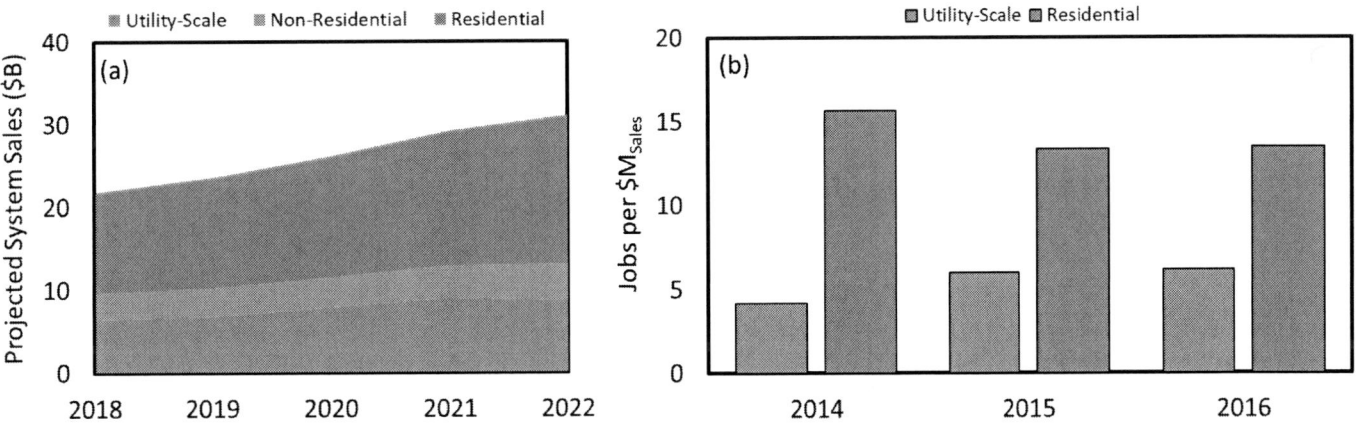

Fig. 4. (a) Projected revenue from system sales by the US PV industry for the utility-scale, commercial, and residential market segments. Future system cost has been estimated based on the apparent learning curves for module cost and for and the non-module costs of each of the system types, and deployment numbers have been sourced from the GTM Research's *Solar Market Insight* series [8]. As with all projections, future system sales numbers carry a substantial amount of uncertainty, and are subject to change as technology advances and markets respond to changes in local demand and policy. **(b)** Summary of recent employment figures per unit of reported (2014 and 2015) and modeled (2016) system sales in the utility-scale and residential sectors of the US market.

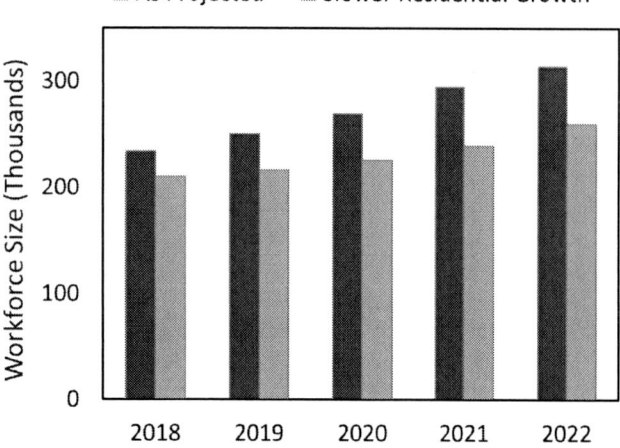

Fig. 5. Projected US PV workforce size based on near-term system deployment forecasts [8]. The two scenarios pictured represent either current learning rates in reported prices across the industry, or a reduced residential revenue base due to downward price pressure and reduced emphasis on growth and expansion.

the US PV industry. The resulting projected job numbers are shown in two scenarios in Figure 5 that examine two potential futures for the residential market segment. The larger workforce scenario includes continued growth in the residential market segment with relatively slow declines in non-module system costs, which would result in an increasingly substantial revenue base within the residential sector. The second scenario corresponds to lower system prices in the residential sector via a 20% learning rate, but without significant changes in deployed capacity. This combination of reduced prices without increased deployment would be a possible outcome of residential installers reducing sales and expansion efforts to focus on profitability rather than growth, and results in approximately a 30% reduction in residential system sales as of 2022. This slower residential growth scenario is provided to illustrate the importance of the behavior of the residential market segment in determining overall PV workforce size.

C. Projecting Farther Ahead

A sense of possible system sales trajectories during the coming decades is offered by the Regional Energy Deployment System (ReEDS) that models deployment rates for each energy generation technology geographically across the US, and the Annual Technology Baseline that tracks the likely costs associated with every major generation technology over time. Each of these models were developed and are maintained by the National Renewable Energy Laboratory [9]. Using deployment numbers from ReEDS and system costs from the Annual Technology Baseline, the hypothetical system sales revenue for the US PV industry can be projected to 2050. Two scenarios are shown in inflation adjusted 2014 dollars in

Figure 6. The mid-case scenario is a middle of the road projection scenario across all energy generation sources, while the low-cost scenario examines the potential effects of rapid price reductions for renewable energy technologies.

The difference between the near-term system sales forecasts in Figures 4 and 6 is partially due to the fact that the ReEDS model generally does not account for certain rapid market fluctuations due to events such as the ITC extension and the current reduction in module prices that together pushed US utility scale PV deployment projections well ahead of previous projections. ReEDS also optimizes regional power systems to minimize anticipated costs, which in the longer term could create a preference for the deployment of utility scale PV systems compared to rooftop systems depending on the availability of regional transmission capacity. This results in a significant projected reduction in revenue for the rooftop solar market segments and an overall reduction in industry sales revenue. Lastly, the projections in Figure 5 are reported in 2014 dollars using a 2.5% annual inflation rate, and so will make long term system sales values appear numerically lower

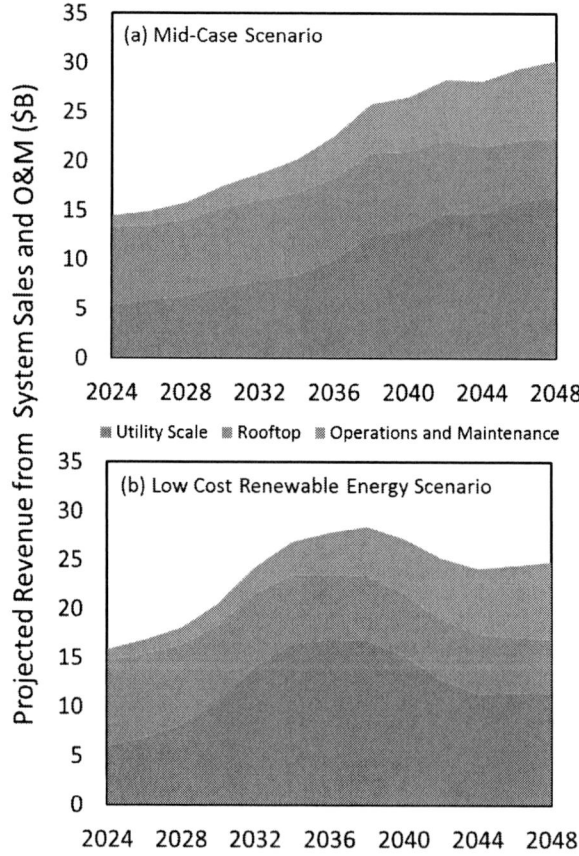

Fig. 6. Projected system sales approaching 2050 based on NREL's ReEDS model [9] for **(a)** The standard scenario for prices by energy source and **(b)** The low-cost renewables scenario, illustrating the potential effects on long term system sales of rapidly decreasing PV system prices. The data has been smoothed via a moving average in order to better illustrate long term trends.

than near-term forecasts that are more directly subjected to the lower inflation rates that have occurred since 2010.

The long-term revenue behavior between the two scenarios is illustrative of the effects of rapid declines in PV system prices on long term system sales revenue. The eventual projected size of the PV industry based on system sales and O&M revenue is slightly higher in the mid-case scenario, where moderate deployment rates are sufficient to maintain a sizeable market at the slightly higher predicted price points. Revenue from system sales in the low-cost scenario increases rapidly, but then begins to decrease and saturate as deployment rates are outpaced by declining system prices. Revenue from operations and maintenance also becomes significant during the coming decades due to the growing number of installed systems, and helps to maintain similar overall revenues in the low-cost scenario, which projects roughly 50% more PV systems being deployed by 2050 than the mid-case scenario.

Based on the data from the ReEDS projects, there is a feedback mechanism that stabilizes the long-term revenue that can be expected of the solar industry. Lower prices mean less revenue per unit of installed capacity, but also increase deployment rates. Any steps taken to increase the value of electricity, services, or other products and capabilities provided by PV systems [10] has the potential to increase deployment at a given system price point, and so increase the likelihood that high amounts of investment can be achieved and maintained over the long term.

III. SUMMARY

This discussion is intended to provide a sense of how possible trends in system deployment and technology development could affect the future size of the US photovoltaics industry workforce. During the past several years, the US PV industry has demonstrated a continuous ability to rapidly add jobs as prices have declined by roughly 60% due to a corresponding acceleration in annual system deployment. The number of jobs per unit of revenue from PV system sales is presented here as a potentially useful way to chart the job creation efficiency of the industry. Due to what appears to be relatively stable interactions between numbers of workers and the overall flow of revenue into the industry from system sales, Job·yrs/$M values have experienced only minor fluctuations even as the size of the industry increased some 7x and the number of person-hours required to install a given system size has decreased to less than half of its 2011 value. Due to the relatively constant number of jobs that are currently being supported per unit revenue within the industry, the productivity metric of Jobs/MW is currently following the same learning rate per installed capacity as the blended average system cost.

If future declines in PV system price begin to outpace increases in PV system deployment rates, the development and implementation of energy storage and other services and hardware that can increase the functionality and value of PV systems as well as engage new customer segments could play an increasingly important role in maintaining the revenue base and thus job creation capability of the PV industry as we move into a future of inexpensive and widely available renewable energy.

REFERENCES

[1] Second annual DOE Energy Employment and Jobs Report. https://energy.gov/downloads/2017-us-energy-and-employment-report, Published January 2017.

[2] On the path to Sunshot: The Role of Advances in Solar Photovoltaic Efficiency, Reliability, and Cost., M. Woodhouse, R. Jones-Albertus, D. Feldman, R. Fu, K Horowitz, D. Chung, D. Jordan, and S. Kurtz.
http://www.nrel.gov/docs/fy16osti/65872.pdf

[3] Renewable Energy and Jobs – Annual Review 2016, published by the International Renewable Energy Agency. http://www.irena.org/menu/index.aspx?CatID=141&PriMenuID=36&SubcatID=2729&mnu=Subcat

[4] SEIA 2016 Annual Report http://www.seia.org/research-resources/2016-annual-report

[5] SunShot 2030 Goals Announcement, Released November 2016, https://energy.gov/eere/sunshot/sunshot-2030

[6] Bloomberg Clean Energy Investment, End of Year 2016, https://about.bnef.com/clean-energy-investment/

[7] National Solar Jobs Census Series 2011-2016. Published by *The Solar Foundation*. http://www.thesolarfoundation.org/national/

[8] US Solar Market Insight 2016 Year in Review, GTM Research. https://www.greentechmedia.com/research/subscription/u.s.-solar-market-insight.

[9] W. Cole, T. Mai, J. Logan, D. Steinberg, J. McCall, J. Richards, B. Sigrin, and G. Porro. 2016. 2016 Standard Scenarios Report: A U.S. Electricity Sector Outlook, Golden, CO: National Renewable Energy Laboratory. NREL/TP-6A20-66939. http://www.nrel.gov/analysis/data_tech_baseline.html

[10] Solar + Storage Jobs, A Discussion Paper, Published July 21, 2016 by *The Solar Foundation*. http://www.thesolarfoundation.org/wp-content/uploads/2016/07/Solar-Storage-Jobs-A-Discussion-Paper-7.21.pdf

Ensuring the Reliability of Photovoltaic Power Systems Using International Standards and the IECRE Conformity Assessment System

George Kelly[a], Adrian Häring[b], Ted Spooner[c], Greg Ball[d], Sarah Kurtz[e],
Matthias Heinze[f], Masaaki Yamamichi[g], Govind Ramu[h]

[a]Sunset Technology, Mount Airy, MD, US; [b]SolarEdge, Niestetal, Germany; [c]UNSW, Sydney, Australia;
[d]SolarCity, San Rafael, CA, US; [e]NREL, Golden, CO, US; [f]TUV Rheinland, Pleasanton, CA, US;
[g]AIST, Tsukuba, Japan; [h]SunPower, San Jose, CA, US

Abstract — To help address the industry's needs for assuring the value and reducing the risk of investments in PV power plants; the International Electrotechnical Commission (IEC) has established a new conformity assessment system for renewable energy (IECRE). There are presently important efforts underway to define the requirements for various types of PV system certificates, the rules for mutual acceptance of these certificates, and publication of the international standards upon which these certifications will be based. This paper presents a detailed analysis of the interrelationship of these activities for initiation of IECRE PV system certifications.

Index Terms — certification, conformity assessment, PV systems, qualification, standards.

I. INTRODUCTION AND BACKGROUND

One of the key issues facing the PV industry, and renewable energy in general, is the need to assure investors of the value of their system. As one means of accomplishing this goal, the International Electrotechnical Commission (IEC) has established a new system for conformity assessments of renewable energy projects (IECRE).

Conformity assessment is an important tool used in many industries to demonstrate compliance with international standards and provide assurance of quality and reliability. In the PV industry, it has become increasingly important to provide assurance to investors that projects will produce the expected amount of energy (and revenue) specified in the design phase. A similar need exists in the wind energy and marine energy industries. As a means of providing such assurance, an effort was initiated in 2012 to establish a conformity assessment system for renewable energy projects.

The new system has been formed within the IEC, which already operates three conformity assessment systems (refer to Figure 1). The new system, authorized by a decision of the IEC Conformity Assessment Board (CAB) in June 2013, is called IECRE (RE for Renewable Energy). As shown in Figure 1, the IECRE system will consist of separate "schemes" for each industry sector. Initially, these will be Wind, Marine and Solar PV. The structure of the system allows for additional sectors (such as solar thermal or fuel cells) to be added in the future.

IECRE conformity assessments will be performed by independent third parties (certifying bodies), and based on inspection and test results provided by accredited laboratories.

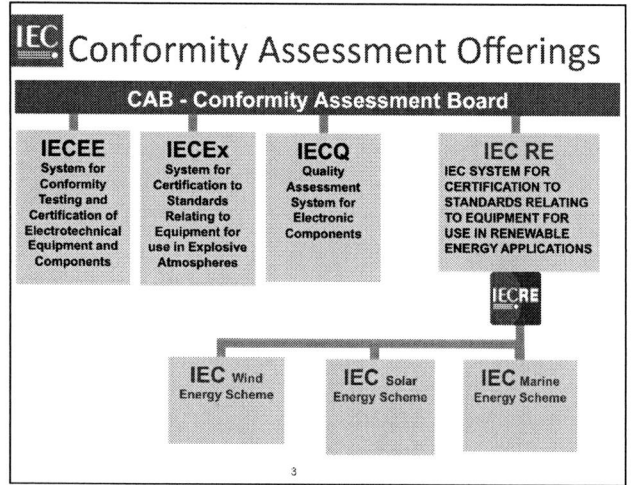

Fig. 1.   IEC conformity assessment systems

The assessments will include both "factory" and "field" aspects of PV projects (refer to Figure 2). Factory aspects include certification of hardware (modules, inverters, BOS) and auditing of the manufacturer's quality management system. Field aspects include system design, installation, commissioning, operation and maintenance.

An international team whose members were drawn from three IEC Technical Committees (TC82 Solar photovoltaic energy systems, TC88 Wind turbines, and TC114 Marine energy - Wave, tidal and other water current converters) wrote the Basic Rules for the IECRE system in the first half of 2014. Official operations of the system were kicked off in Boulder, Colorado in Sept 2014 with the first meeting of the IECRE Management Committee (REMC) and formation of an Operational Management Committee (OMC) for each RE industry sector. Rules of Procedure (RoP) have now been written for each industry sector and approved by the REMC. The second edition of the PV-RoP was approved in July 2016.

An important activity supporting the new IECRE system is the development of new or revised international standards to be used as the basis for conformity assessments. Experts in TC82 are working in several areas that are critical to PV system certification.

978-1-5090-5606-4/17 $31.00 © 2017 IEEE

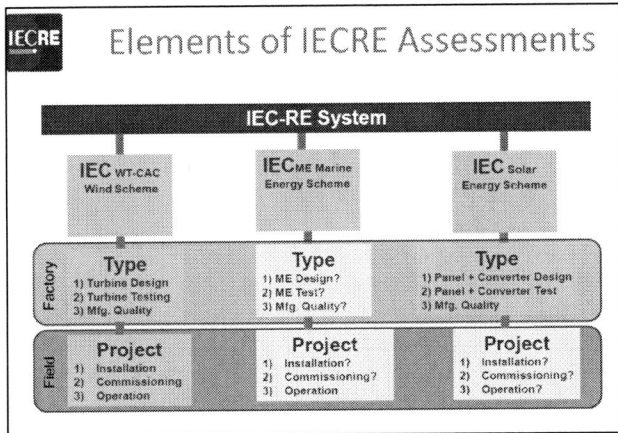

Fig. 2. Elements of IECRE conformity assessments

There is also an effort underway to define the requirements and procedures for certification bodies (CBs) and inspection bodies (IBs) to be accepted into the system and mutually recognize each other's results. This article presents a summary of IECRE activities to date, a review of activities to close gaps in the standards where needed, and the latest status of CB and IB accreditation for acceptance in the IECRE system.

II. DEFINITION OF IECRE CERTIFICATES

In order to provide meaningful system certifications, it is necessary to have published standards that define the technical requirements, as well as Operational Documents (ODs) that define the requirements for certification. Edition 2 of the PV-RoP defines several parameters for classification of systems, which in turn will affect the requirements for a particular certificate. These may be obtained at different project stages, as shown in Fig. 3.

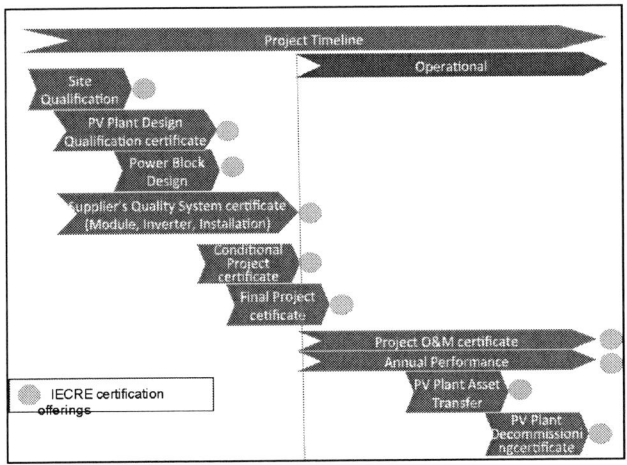

Fig. 3. Timeline View for PV System Certifications

There has been a significant amount of work within PV-OMC working group 404 (Promotion and Marketing) to identify certification offerings for which there is likely to be a demand in various PV markets. Each of the proposed certificates would be defined with an OD that identifies the requirements for issuing a valid certificate. The most recent of the ODs is shown in Table I.

TABLE I.   PV OPERATIONAL DOCUMENTS

TITLE	OD	STATUS
Conditional PV Project certificate (commissioning)	401	Published 2016
Conditional PV Project certificate (construction complete)	401-1	Draft in process
Annual PV Plant Performance certificate	402	Published 2016
PV Plant Design Qualification certificate	403	Draft in process
PV Site Qualification certificate	403-1	Draft in process
PV Power Block Design Qualification certificate	403-2	Future work
PV Asset Transfer certificate	404	Draft in process
PV Decommissioning certificate	409	Future work
PV Module Factory QMS certificate	405	Published 2016
PV System Installation QMS certificate	410	Future work
PV Inverter Factory QMS certificate	4xx	Future work

During 2016, the PV-OMC approved the first five ODs were approved by the member countries. OD-401 covers the requirements for Conditional PV Project certification, and OD-402 addresses Annual PV Plant Performance certification. The OD-405 series (3 parts) covers the requirements for quality system audits in PV module factories.

Plans are now in place to develop the remaining ODs as prioritized by the members of the PV-OMC. The most urgent need is for the PV Plant Design Qualification certificate, which will be designated OD-403. Site Qualification and Power Block Design Qualification will later become sub-parts of this document, as appropriate standards are identified for each.

The next priority is to separate the Conditional PV Project certificate into sub-parts for mechanical and electrical work, designated "construction completion" and "commissioning". This structure has been chosen to align with the existing standard practices for financing PV projects and disbursing funds to contractors.

The requirements for PV Asset Transfer and Decommissioning certificates still need to be discussed among industry stakeholders, so these have been assigned lower priority and will not be addressed until late 2017 or 2018.

III. STANDARDS DEVELOPMENT

To prepare for implementation of the new system, an effort was undertaken in 2013 to evaluate the existing international and national standards for relevance and applicability to conformity assessment, and to identify gaps in the existing standards, which needed to be addressed to ensure the success of IECRE for the PV industry.

978-1-5090-5606-4/17 $31.00 © 2017 IEEE

This study[1] helped guide the work program of TC82, focusing resources in four specific areas where critical gaps were identified:

A. Module Quality

The manufacturing consistency efforts led by the International Photovoltaic Quality Assurance Task Force (PVQAT) focused on developing a guideline for factory inspections and quality assurance during module manufacturing. The result of this effort was IEC/TS 62941[2], published in early 2016. The other major advance in module standards during the past year was the publication of new editions of the module qualification and safety standards.

B. System Installation Quality

TC82 Working Group 3 has initiated a new project to develop IEC/TS 63049 - Terrestrial photovoltaic (PV) systems - Guideline for increased confidence in PV system installation[3]. This technical specification is intended to improve the consistency of installation in a similar manner as 62941 does for module manufacturing, but there are additional challenges because the work must be performed at many different sites rather than a controlled factory setting. Therefore, the certification will concentrate on the quality system aspects that can be controlled and repeatable from project to project.

C. System Performance Measurement

The IEC 61724 series is probably the most important work in progress for implementation of IECRE certifications. IEC 61724-1[4] contains requirements for system performance monitoring, including equipment characteristics, data sampling and filtering, measured and calculated parameters, and performance metrics. IEC/TS 61724-2[5] defines a Capacity evaluation method, which would typically be used to verify initial output, as well as to compare results over time. IEC/TS 61724-3[6] describes an Energy evaluation method, using one full year's data adjusted for actual weather conditions. This evaluation will be required for the IECRE Annual PV Project certification.

D. System Commissioning and Maintenance

IEC 62446 has now been expanded to a series with the generic title "Photovoltaic (PV) systems - Requirements for testing, documentation and maintenance". The new edition of the commissioning standard, IEC 62446-1[7], was published in early 2016, and is a significant improvement to an already good standard. This standard now includes specific procedures to address different categories of PV systems, depending on size and complexity. This increased applicability of the international standard also increases its usefulness as the common basis of IECRE commissioning audits.

TC82 Working Group 3 has recently started development of a standard for system maintenance, IEC 62446-2[8]. This document will address preventative and corrective maintenance, both safety-related and performance-related, as well as troubleshooting and documentation of results. The document is expected to reach final publication later in 2017.

IV. CB/IB PARTICIPATION

The underlying principle of all IEC conformity assessments is mutual acceptance of results by all participants. The basis for establishing mutual recognition includes accreditation (assuring that an organization has competency in the relevant standards) combined with regular peer assessments to monitor performance and consistency. Figure 4 shows the general

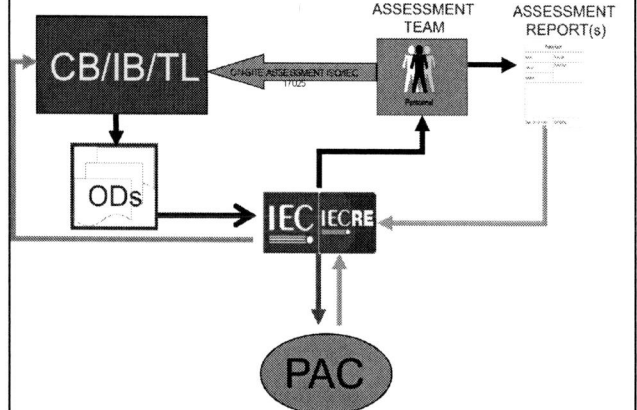

relationships involved in the peer assessment process.

Fig. 4. Peer Assessment Process Diagram

V. CONCLUSION

Conformity assessment is potentially an important tool for assuring investors of the value of PV projects and for reducing technical risks for insurance companies. The new IECRE system is a good way to accomplish this, and has significant support within the PV community. The system has begun formal operations and the first certificates (for wind turbines) have been issued in October 2016. Rules of Procedure are approved and Operational Documents are in progress for each industry sector. Certification bodies, inspection bodies and testing laboratories have applied to join the system, and are beginning their peer assessment process. It is expected that PV factory and system certifications will be issued by the end of this year.

Relevant and appropriate standards exist for most aspects of importance to PV system performance and reliability. Gaps were identified specifically in the areas of installation and operations and maintenance (O&M). IEC TC82 has been working to address these gaps, and publication of the new standards is necessary to enable various certification offerings that have an identified market demand.

Therefore, it is very important to expedite the additional standards that are presently under development and coordinate the preparation of IECRE Operational Documents to define how these standards will be used for PV system certifications.

REFERENCES

[1] G. Kelly, E. Spooner, G. Volberg, G. Ball, J. Bruckner, "Ensuring the reliability of PV systems through the selection of international standards for the IECRE conformity assessment system"; *Photovoltaic Specialist Conference (PVSC), 2014 IEEE 41st,* On page(s): 0914 - 0918.

[2] International Electrotechnical Commission, IEC/TS 62941, Guideline for increased confidence in PV module design qualification and type approval. Geneva, Switzerland: IEC Central Office, 2016.

[3] International Electrotechnical Commission, IEC/TS 63049 Terrestrial photovoltaic (PV) systems - Guideline for increased confidence in PV system installation (New Work In Process). Geneva, Switzerland: IEC Central Office, 2015.

[4] International Electrotechnical Commission, IEC 61724-1 Photovoltaic system performance - Part 1: Monitoring (New Work In Process). Geneva, Switzerland: IEC Central Office, 2015.

[5] International Electrotechnical Commission, IEC/TS 61724-2 Photovoltaic system performance - Part 2: Capacity evaluation method (New Work In Process). Geneva, Switzerland: IEC Central Office, 2015.

[6] International Electrotechnical Commission, IEC/TS 61724-3 Photovoltaic system performance - Part 3: Energy evaluation method (New Work In Process). Geneva, Switzerland: IEC Central Office, 2015.

[7] International Electrotechnical Commission, IEC 62446-1 Photovoltaic (PV) systems - Requirements for testing, documentation and maintenance - Part 1: Grid connected systems - Documentation, commissioning tests and inspection. Geneva, Switzerland: IEC Central Office, 2009.

[8] International Electrotechnical Commission, IEC 62446-2 Maintenance of PV systems (New Work In Process). Geneva, Switzerland: IEC Central Office, 2014.

A Framework to Calculate Uncertainties for Lifetime Energy Yield Predictions of PV Systems

Björn Müller[1], Peter Bostock[2], Boris Farnung[1], Christian Reise[1]

[1]Fraunhofer ISE, Fraunhofer Institute for Solar Energy Systems, 79110 Freiburg, Germany
[2]VDE Americas, San José, USA

Abstract — Today, lifetime energy yield predictions should include an assessment of the uncertainties of the predicted values as a precondition for financing large PV plants in order to properly quantify risk and return. However, the approach to perform the yield prediction with attendant uncertainty estimation is not standardized which makes it difficult to compare one investment and another. Often important sources of uncertainties are not even included in the overall uncertainty assessment. A method to calculate uncertainties and P-values for year by year and lifetime energy yield predictions of PV systems is presented. The focus is on the cumulative impact of the combination of uncertainties on delivered energy going into the project pro-forma and downside financial analysis, not on estimation of uncertainties for individual modelling steps.

The proposed approach is an effort to standardize the procedure of uncertainty calculation of predicted energy yields of PV systems in order to properly estimate financial investment risk.

I. Introduction

Up to now, there is no standardized uncertainty framework for lifetime energy yield predictions available. Furthermore, there is a lack of definitions in this field to make sure, we are talking on same things. For this reason it is difficult for financers, investors and other stakeholders to compare uncertainty assessments for yield predictions.

Furthermore, up to now often just an initial yield and the P90 of the initial yield as well as the average yield over the assumed lifetime of the system and the P90 of the average yield are given. These numbers may not be adequate to feed financial models taking effects of degradation and dimming and brightening trends including their uncertainties into account. The same applies to influences from interannual variations and the time they need to cancel out.

The aim of this paper is to develop a method to "handle" all sources of uncertainties influencing lifetime energy yield predictions and to present the information needed to feed financial models with time dependent yield estimates and P-values as desired by investors or stakeholders. This method can be seen as a part of "a common framework that can assess the impact of technical risks on the economic performance of a PV project" [1].

In this contribution, we will apply the proposed method to yield predictions of 26 PV systems described in [2] and compare the outcome against measured energy yields.

II. Definitions

It is important to be clear with all these wordings, so we will first define something. Yield is used as synonym for energy production.

A. Lifetime Energy Yield Prediction

A lifetime energy yield prediction is "an estimate of the total energy production for a PV system at a specific site" [3]. The primary aim is to predict the energy production for the Prediction Period. State of the art yield predictions can be partitioned in four main parts (see e.g. [4]):

1) Assessment of the solar resource and other meteorological quantities for the Reference Period as well as the generation or selection of the best possible time series to describe meteorological conditions for this period

2) Simulation of the PV system energy output based on modelling and parametrization of the actual PV system and using the meteorological time series from 1) (calculation of Predicted Reference Yield)

3) Estimation of long-term changes in energy yield over the Prediction Period and application of these changes to Predicted Reference Yield from 2) to determine the Predicted Yield

4) Analysis of uncertainties for the Predicted Yield

B. Predicted Reference Yield

The best estimate of the mean annual yield for the PV system under consideration within a reference period in the past. This yield is calculated using a simulation program, best estimates of the parameters for the simulation program (in other words the most likely values of all possible uncertain parameters) and the best source of meteorological data for the reference period. It is usually given as absolute value [kWh] or specific value [kWh/kWp].

C. Predicted Yield

The predicted yield is derived from predicted reference yield by applying best estimates of expected changes of the yield within the assumed lifetime of the PV system (the prediction period). These changes could be changes in system behavior (e.g. degradation) or climate changes (e.g. dimming and brightening).

Predicted Yield can be given for individual years (predicted yield in year *x*) and / or as mean annual yield over the prediction period (predicted mean annual lifetime yield).

978-1-5090-5606-4/17 $31.00 © 2017 IEEE

It can be given as absolute value [kWh], specific value [kWh/kWp] or as percentage of Predicted Reference Yield.

III. UNCERTAINTY SOURCES

According to [5], an uncertainty analysis is "a systematic process to propagate uncertainty in a model or its inputs to uncertainty in the model's output" that involves two primary steps: uncertainty quantification and uncertainty propagation. A full quantification of uncertainties includes both, the quantification of uncertainties for all parameters and all models of the modelling chain. These uncertainties are then propagated through the whole modelling chain.

However, as for the prediction of lifetime energy yields of PV systems this approach is difficult to apply (e.g. the number of models and parameters is high, each uncertainty vary with daily as well as seasonal cycles and actual site of the PV system..., see e.g. [5] for more details), it is not implemented in state of the art yield predictions. Instead for recent yield predictions the uncertainty is quantified by assigning an uncertainty distribution to the output of each modelling step to derive Predicted Yield.

A first category of uncertainty is resulting from modelling Predicted Reference Yield. In practice it is usually considered by assigning uncertainty distributions to the annual losses or gains for each modelling step calculated by the yield simulation program. For a number of modelling steps losses or gains are close to zero, while their uncertainty may be quite high. Furthermore it is physically not meaningful for some modelling steps to show positive results (e.g. it is only meaningful that soiling, shading... cause losses, not gains). In these situations it may be more appropriate to assign asymmetric uncertainty distributions to losses or gains that include only physically meaningful results.

To include uncertainties from the conversion of Predicted Reference Yield to Predicted Yield, uncertainty distributions for the estimated change rates have to be quantified as well. Note that the uncertainties of Predicted Yield for individual years will increase with time when an uncertainty for linear change rates is included in the uncertainty quantification: while the effect of e.g. an actual -0.5%/year deviation from the expected change rate for system behavior will be small for the first year of operation, the system will generate 5% less energy in year 10. For consideration of uncertainty of change rate for system behavior again an asymmetric uncertainty distribution may be more appropriate, as observations of degradation rates tend to be asymmetric [6,7] and positive degradation rates are physically not meaningful.

As a third category of uncertainty interannual variation of the Predicted Yield has to be considered. The background for the consideration of interannual variation as an uncertainty is the fact, that Predicted Reference Yield and as a follow up Predicted Yield is calculated as an annual value (it is the yield in a "typical year"). However solar radiation (as the main influencing factor for yield variations) and other meteorological quantities will vary from year to year and influence annual yields. As long as this variation is interpreted as a variation around a possible trend of meteorological quantities whose

effects on yield are considered when calculating Predicted Yields, this variation may not influence cumulated Predicted Yields to a high extend but may have a high impact on annual Predicted Yields.

IV. UNCERTAINTY PROPAGATION

After quantifying all uncertainties they are propagated to derive the final uncertainty of Predicted Yields for all individual years in the Prediction Period. At least two possible approaches for the propagation of uncertainties can be defined: the law on propagation of uncertainties (see e.g. [8]) and the propagation of probability distributions using a Monte Carlo method (see e.g. [5,9,10]). While this is a simplification, usually independence of all losses and gains is assumed for both methods. Also for law on propagation of uncertainties usually normal distributions are used.

If asymmetric uncertainty distributions are applied to some of the modelling steps, the law on propagation of uncertainties may not be applicable. There are other circumstances and conditions under which the results of a Monte Carlo approach may be preferable (see e.g. [5,9]).

P values are used to deliver information on the uncertainty distribution of the Predicted Yield.

Pxx is the value that is exceeded with the probability xx%. E.g. P50 for Predicted Yield in year 1 may be 1000 kWh/kWp, which means that 1000 kWh/kWp is the Predicted Yield that is exceeded with a probability of 50%. If the law of propagation of uncertainties is applied to propagate uncertainties, P values can be calculated based on the quantile of a normal distribution. In case a Monte Carlo approach is used they can be derived from the quantile of the empirical distribution of all Z realizations of Predicted Yield, where Z is the sample size of the Monte Carlo approach.

P values can be given as absolute value [kWh], specific value [kWh/kWp] or as percentage of Predicted Reference Yield. They can be given for individual years (PXX in Year n), for mean annual yield over the Prediction Period (PXX of Predicted Mean Annual Lifetime Yield) and/or for cumulative yield over the Prediction Period (Pxx for Predicted Lifetime Yield).

V. PRESENTATION OF RESULTS

The implementation of a Monte-Carlo simulation technique into PV energy yield simulation software means that not only one single simulation is performed, but a large number of simulations with input being randomly chosen from predefined probability distributions.

The practical implementation into the software code introduces the drawing of Z realizations from a pre-defined probability distribution of GHI, which can be denoted u_{GHI}. The probability distribution of energetic gains related to irradiation transposition can be denoted u_{GPOA}, and analog other probability distributions that are deemed necessary to be included.

Similarly, the uncertainty related to predicted PV energy yield is included by drawing Z realizations reflecting Z possible long-term rates of change in system behavior (i.e., drawing

Z realizations from predefined probability distribution of degradation rates).

Using appropriate equations for each gain or loss mechanism, the PV energy yield is calculated for each individual year of assumed PV system lifetime, with the effect of Z realizations that reflect long-term changes being considered separately for each subsequent year. Next – and independent from other uncertainty categories – different overall irradiation of each individual year can be included to reflect the effects of uncertainty related to inter-annual variation of irradiation, leading to the incorporation of all described uncertainties for each individual uncertainty category for each individual year n

throughout the assumed PV system lifetime simulated by this Monte-Carlo approach. Since this procedure can be implemented in simple loops regarding software coding, it is straightforward to numerically determine the cumulative yields of years 1 through n for all Z realizations.

Fig. 1 depicts exemplary results for this simulation procedure for an assumed PV system lifetime of 20 years. All values in Fig. 1 were normalized to the best estimate of predicted reference yield. Our assumptions of probability distributions leading to the determination of uncertainties of individual models are summarized in Table 1.

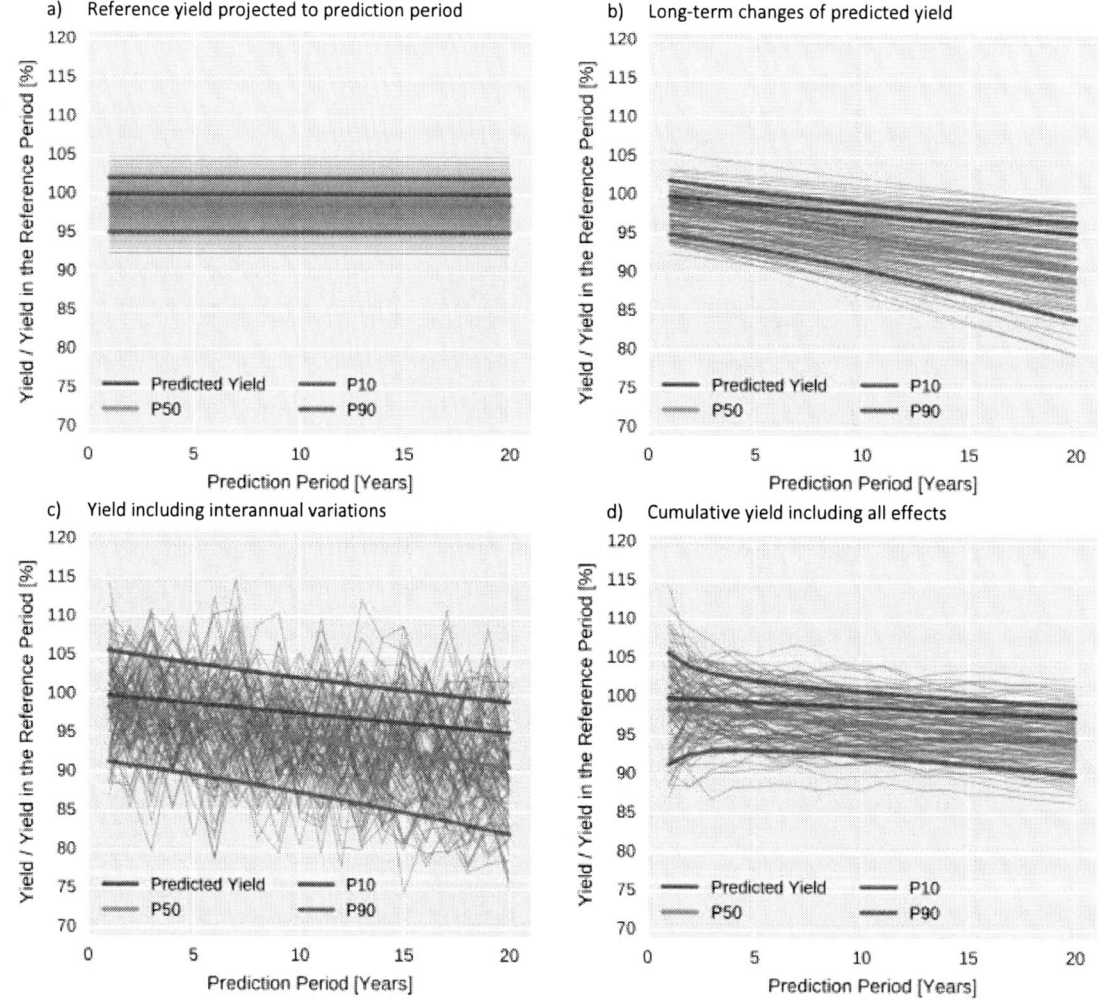

Fig 1.   Ratio of predicted yield compared to predicted reference yield for various stages in a Monte-Carlo simulation, notably a) yield in the reference projected onto prediction period (GHI and irradiation transposition); b) long-term changes (degradation and irradiation trends); c) inter-annual variations of meteorological conditions and d) the cumulative predicted PV energy yield as an expanded annual average value.

Table 1: Uncertainties of individual modelling steps used in the exemplary yield simulation. A normal (Gaussian) distribution is characterized by mean value μ and standard deviation σ, while a triangular distribution is characterized by minimum a, maximum b, and modus c.

Calculation step	Distribution	Parameter		
	normal	μ	σ	
	triangular	a	b	c
		%	%	%
Solar resource potential in the reference period				
GPOA	normal	11.4	2.5	
Yield in the reference period				
Horizon shading	triangular	-1.0	0.0	0.0
Row-shading	triangular	-5.0	0.0	-1.0
Soiling	triangular	-1.5	0.0	-0.5
Reflection	triangular	-4.1	-2.6	-3.1
STC power	normal	0.0	2.0	
Spectrum	normal	-1.0	0.5	
Irradiation level	normal	-3.9	1.9	
Temperature	normal	-2.4	1.0	
Mismatch	triangular	-1.8	0.0	-0.8
DC cabling	triangular	-2.5	-1.0	-1.5
Inverter	triangular	-5.7	0.0	-2.7
Power limitation	triangular	-1.0	0.0	0.0
Transformer	triangular	-2.0	-0.5	-1.0
Yield in the prediction period				
System behavior	triangular	-1.6	0.0	-0.6
Solar irradiation	normal	0.0	0.3	
Annual variation	normal	0.0	4.9	

Due to incorporating non-normal distributions of uncertainty, one can denote a qualitative difference between yield predicted deterministically and the P50 yield (Fig. 1a). The difference between these yields increases over time (Fig. 1b), since uncertainty associated with changing system output over time also exhibits a non-normal probability distribution. Fig. 1b also shows the general increase of uncertainty in the prediction period.

Including uncertainties of inter-annual variation of meteorological conditions leads to a stark increase of uncertainties related to individual yields (Fig. 1c). However, these uncertainties are substantially reduced for cumulative yields within only few years, as shown in Fig. 1d.

In a final step, we compared observed annual yield values to the uncertainty range as expected from the Monte-Carlo simulations. We used the performance data of 26 individual PV systems as described in [2] for this comparison. The parameters given for a single system in Table 1 were adapted to the specific systems and sites, then, a number of realizations were calculated for the set of 26 systems. Fig. 2 presents both the Monte-Carlo results and the measured values in a similar presentation as in the lower section of Fig. 1. The measured values fit quite well into the uncertainty range as expected from the Monte-Carlo simulations.

VI. CONCLUSION

A method to calculate uncertainties and P-values for year by year and life-time energy yield predictions of PV systems is presented. The method is based on a Monte-Carlo simulation which uses Gaussian or triangular distributions for individual modelling steps, including the solar resource data and long term changes of system behavior. The parameters may be adjusted to individual sites and system layouts. A comparison of expected uncertainty and observed variability showed a good agreement.

The proposed approach is an effort to standardize the procedure of uncertainty calculation of predicted energy yields of PV systems in order to properly estimate financial investment risk.

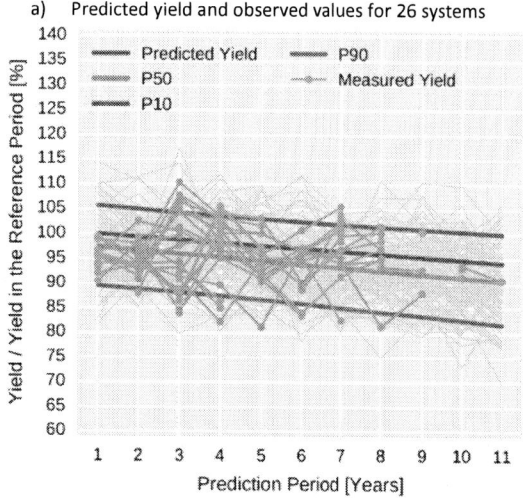 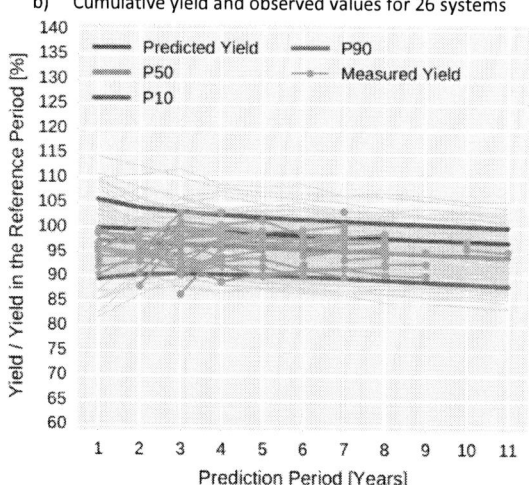

Fig 2. Predicted yield and observed annual yield values for 26 PV systems located in Germany and Spain [2]. The Monte-Carlo calculations for a single system as shown in Fig. 1c and Fig. 1d were adapted to the specific systems and sites. The measured values fit quite well into the uncertainty range as expected from the Monte-Carlo simulations.

VII. REFERENCES

1. Moser D, Del Buono M, Jahn U, Herz M, Richter M, Brabandere K de. Identification of technical risks in the photovoltaic value chain and quantification of the economic impact. *Prog. Photovolt: Res. Appl.* 2017: n/a-n/a, DOI: 10.1002/pip.2857.
2. Müller B, Hardt L, Armbruster A, Kiefer K, Reise C. Yield predictions for photovoltaic power plants: empirical validation, recent advances and remaining uncertainties. *Prog. Photovolt: Res. Appl.* 2015; **24(4)**: 570–83, DOI: 10.1002/pip.2616.
3. Huld T, Dunlop E, Beyer HG, Gottschalg R. Data sets for energy rating of photovoltaic modules. *Solar Energy* 2013; **93(0)**: 267–79, DOI: 10.1016/j.solener.2013.04.014.
4. Dirnberger D, Müller B, Reise C. PV module energy rating: opportunities and limitations. *Prog. Photovolt: Res. Appl.* 2015; **23(12)**: 1754–70, DOI: 10.1002/pip.2618.
5. Hansen CW, Martin CE. *Photovoltaic System Modeling: Uncertainty and Sensitivity Analyses*; 2015.

6. Jordan DC, Kurtz SR. Photovoltaic Degradation Rates-an Analytical Review. *Prog. Photovolt: Res. Appl.* 2013; **21(1)**: 12–29, DOI: 10.1002/pip.1182.
7. Jordan DC, Kurtz SR, VanSant K, Newmiller J. Compendium of photovoltaic degradation rates. *Progress in Photovoltaics: Research and Applications* 2016: n/a-n/a, DOI: 10.1002/pip.2744.
8. Joint Committee for Guides in Metrology (JCGM). *Evaluation of measurement data - Guide to the expression of uncertainty in measurement.* JCGM 100:2008: Bureau International des Poids et Mesures (BIPM); 2008.
9. Joint Committee for Guides in Metrology (JCGM). *Evaluation of measurement data - Supplement 1 to the "Guide to the expression of uncertainty in measurement" - Propagation of distributions using a Monte Carlo method.* JCGM 101:2008: Bureau International des Poids et Mesures (BIPM); 2008.
10. Thevenard D, Pelland S. Estimating the uncertainty in long-term photovoltaic yield predictions. *Sol Energy* 2013; **91**: 432–45, DOI: 10.1016/j.solener.2011.05.006.

Integrated PV-Recycling – More Efficient, More Effective

Wolfram Palitzsch and Ulrich Loser

Loser Chemie GmbH, 08056 Zwickau, Kopernikusstraße 38-42, Germany

Abstract — Switching to renewable energy sources, such as solar energy, is one of the key principles of the Circular Economy. Material cycling should maximise the duration of material use and profitability and minimise energy and virgin material inputs. Therefore, the first priority for cycling should be reusing materials 'as is' or with minimal modification. By combining physical and chemical technologies it is possible to follow the "zero waste concept" and to enter with PV-Recycling the Circular Economy. The step-by-step and clever separation of the individual components was crucial. Our Recycling technologies work with light, water and biodegradable auxiliaries. An optical nanotechnology helps to open the composite. This means that the glass panels do not have to be destroyed. This is very important because the front and rear glass can be sorted.

Index Terms — recycling, thin film photovoltaic, indium, gallium, glass, CIGS, CdTe, EOL, photovoltaic scrap.

I. INTRODUCTION

It is undisputed that unquestionable benefits coming up with photovoltaic systems. The emission of greenhouse-gas from the manufacturing process is insignificant compared to the lifetime greenhouse-gas savings after installation and producing electricity. This green electricity displaces our dirtiest energy sources. Therefore, it is understood that the photovoltaic industry is undergoing remarkable growth. Global solar PV installations are expected to soon reach a cumulative capacity of 200 GW [1] , which will see solar producing more energy than 30 coal or nuclear plants combined, and will help save over 100 million tons of the greenhouse-gas CO_2 annually. I

In 2050 total quantities of end-of-life photovoltaic panels are anticipated to amount to 9.57 million tones [2]. We are not there yet, but discussions on recycling have already started. We use a new method for extracting and reclaiming metals of the mostly used semiconductor layers, e.g. CIS, CIGS, GaAs or CdTe, found in thin film PV production waste. From the perspective of the high-tech industry, it is necessary that the essential materials that we need to import in some quantities, are kept in the loop.

On the other hand statutory prescriptions, as e.g. the German "Kreislaufwirtschaftsgesetz" (law encouraging closed-loop economy) are asking for a maximum quota of recycling – and a minimum use of resources (e.g. energy, raw materials). Beside this we have a few reasons for the recycling of materials in generally. Cadmium, selenium, tellurium, gallium, molybdenum, indium but also silver and silicon are some of the major elements used in these photovoltaic cells. We know about the future limits in the availability of these elements. So recycling is required as the most advisable end-of-life strategy and to save the raw materials from production wastes now.

It is well known that the rare metals combined typically only represent 1% of the mass of a photovoltaic panel, their value is significant. So for example on crystalline silicon silver is found.

However, proposals, such as the admixture in cement [3] we do not want to follow, because that means the loss of conventional and rare resources.

II. EXPERIMENTAL MATERIAL(S) AND METHODS

Solar Cell Tabbing or Interconnect Wire are primarily produced from a tin or tin alloy coated copper flat wire. Also the Solar Bus Wire is rolled from round wire, then solder coated on all four sides. Lead or lead free solders applied on all four sides for consistent bonding with tabbing ribbon (Fig. 1).

Fig. 1. Silicon scrap with interconnect wires

So the treatment with a fresh solution of aluminium chloride and water, carried out in the first part of the treatment leads to a product with a lot of impurities. The desired product poly-aluminium-hydroxide-chloride $[Al_n(OH)_mCl_{3n-m}]$ contains too much copper, tin and/or lead. Thus, it became necessary to develop a method in which the cables and connections can be separated from the silicon wafers. The process should be as simple and robust as possible. Because the connectors were soldered to the silicon cells, a temperature rise above the softening temperature of the solder should be sufficient for separation. However, the separation was too complex and residues of the solder remained on the silicon shards. Finally, we decided to dispense with small amounts of silicon and strive for a qualitatively better separation.

978-1-5090-5606-4/17 $31.00 © 2017 IEEE

For this purpose, we used the different physical properties of the brittle silicon and the plastic metal of the connectors. If the waste mixture is mechanically stressed, the brittle silicon flake, while the connectors and lines remain, will be self-connected with corresponding movement and form so-called braids, which can easily be separated, see Fig. 2.

Fig. 2. Result after mechanical treatment of Silicon/Ribbon/Bus mixture.

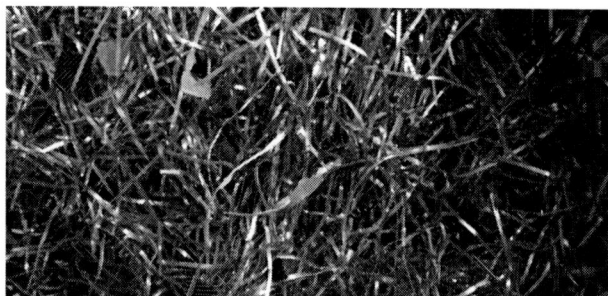

Fig. 3. Separated wires with small portion of silicon cells.

This material (Fig. 3) was analyzed to give the following composition: Cu: 89-93%, Pb: 3.8-4.1% and Ag: 0.1-1.9%. The material is so sellable, but we would like a higher value added - trials for refinement and separation are currently underway. For this process we found that the use of V-blender leads to the best results. A V-blender is made of two hollow cylindrical shells. They joined at an angle of 75° to 90°. The mixer container is mounted on trunnions to allow it to tumble. As the machine tumbles, the material continuously splits and recombines (for wires it means, the form bundles of fluff or humps), with the mixing occurring as the material free-falls randomly inside the vessel.

After the treatment of the waste, separation of the bus and ribbon wires, the material could treat as our proposed procedure: The chemical reaction between the back contact aluminium and the depleted aluminum-chloride solution is very simple and works better than the normal treatment with sodium hydroxide. Because the material was previously freed from the interfering metals copper, tin and lead, we get a cleaner product. The next step was that the remaining silver was dissolved and the silicon material was washed and dried.

The applied technology for this purpose works without the formation of nitrous gases which is expected in the application of nitric acid. We use a kind of a transport system which dissolves silver from waste with a very low concentration of Ag by the addition of a stoichiometric amount of hydrochloric acid related to the amount of silver. It is decreased by simultaneous release of the silver salt.

This fact is very important as there are also some other silicon wastes on the market. Solar cells with heterojunction contacts are able to achieve particularly high electrical voltages. This rear side contacting avoids shading losses from front contacts, and thus has the potential to achieve very high solar cell currents. It means, that this type of silicon cells do not have a rear contact from aluminium. On both sides of this cells the contacts are made from silver and ITO (indium-tin-oxide). So the treatment with aluminium chloride is not necessary. For the recycling of the mixture of Si/Ag/In/Sn it was necessary to develop a new technology. We found, that it is possible to remove both contacts in one step by treatment with methyl sulfonic acid.

Our system works at room temperature and it makes the separation of silver and indium very simple and cheap. The chemistry follows the reactions, noted as follows:

$$Ag + ITO^{[1]} + 6\ CH_3\text{-}SO_2OH \longrightarrow Ag^+ + CH_3\text{-}SO_2O^- + In^{3+} + 3\ CH_3\text{-}SO_2O^- + Sn^{2+} + 2\ CH_3\text{-}SO_2O^- + 3H_2O$$

$$Ag^+ + CH_3\text{-}SO_2O^- + In^{3+} + 3\ CH_3\text{-}SO_2O^- + Sn^{2+} + 2\ CH_3\text{-}SO_2O^- + HCl \longrightarrow$$

$$\mathbf{AgCl}_{(s)} + In^{3+} + 3\ CH_3\text{-}SO_2O^- + Sn^{2+} + 2\ CH_3\text{-}SO_2O^- + CH_3\text{-}SO_2OH$$

As can be seen, the silver can be precipitated almost stoichetrically from the resulting solution. Subsequent filtration and a washing process result in the desired clean silver salt.

The remaining solution contains methylsulfonic acid, indium mesylate and tin mesylate, and can be used further for the treatment of HJT cell scrap. The treated silicon cell scrap is available for further investigations.

---

[1] $(In_2O_3)_{0.9} \cdot (SnO_2)_{0.1}$

Now we use this method also for extracting and reclaiming metals of the mostly used semiconductor layers, e.g. CIS, CIGS, GaAs or CdTe, found in thin film PV production waste. But most of the thin film photovoltaic scrap is glass. The price therefore depends primarily on how much money you can generate for secondary raw material glass. This is also the decisive criterion for the recycling route we have to go. Usually two different grades of glass are used in one thin film photovoltaic module - front glass and rear glass. The front glass is of high quality, because it is also free of iron. This is obvious, that a pure-grade secondary material of front glass has more value as a mixture with the ferrous back glass.

Contrary to conventional technologies, like shreddering, we do not break the glass. Using technologies from the field of optical nanotechnology it has been possible to fully open the sandwich structure, without damaging the glass.

The semiconductor containing PV scrap is opened before the hydrometallurgical treatment so that the semiconductor materials to be accessible of a liquid. It does not matter how big the pieces are at the end. For single-coated, large-area material such as foils or glass plates with evaporated metals or semiconductors is diving in a bath sufficient. To resolve both the semiconductor and the silver we use a combination of methanesulfuric acid (MSA) and hydrogen peroxide.

Benefits in practical applications come, for example, from its nonoxidizing nature, the high solubility of its salts, the absence of color and odor, and the fact that it is readily biodegradable.
Consequently, MSA is becoming increasingly important in a number of applications and industries.

We developed several novel methods for dissolving and reclaiming rare metals from scrap of chalcopyrite ($CuInS_2$, $CuInSe_2$ and $Cu(InGa)Se_2$) systems, the II-VI compound semiconductor CdTe and the III-V compound semiconductor GaAs and associated photovoltaic manufacturing waste, like overspray and overlayered shields by use of MSA.

These semiconductors can't be physically separated, but our chemical approach can. By the use of alkylsulfonic acids we can extract in every case all used metals in a very short reaction time. From this recycling process of end-of-life or semi materials we obtained the result that the clean glass pieces are useable for float glass production.

Consequently, we were able to establish with our partners that the recycling into float glass is possible with such assorted raw cullet of known provenance and very low contamination. The chemical extraction operates at room temperature and the used alkylsulfonic acid is readily biodegradable (OECD 301 A).

The treatment of CIS-PV scrap with a solution of methanesulphuric acid CH_3SO_3H (MSA) and H_2O_2 results clean glass, clean plastics and a poly-metallic solution. A great benefit of this technology is that the front and back sides of the glass can be sorted directly and thus high-quality secondary raw materials are available.

After washing the glass can be crushed to a uniform size and checked for contaminants. Analytical results are presented in Table 1.

TABLE I
ANALYTICAL DATA (XRF) OF THE CLEAN GLASS AFTER TREATMENT OF CIS-PV SCRAP

Ingredient	Concentration wt.%
Al_2O_3	0,804
Fe_2O_3	0,047
CaO	9,29
MgO	3,95
Na_2O	13,47
SiO_2	71,79

III. CONCLUSION

In the long term, renewable energies will make a decisive role with regard to the quality of life on our planet. Therefore the PV industry will grow and at the same time, the efficiency will increase. But this development will also consume resources. Therefore, the industry and the policy should have in focus also the closing of cycles and support sustainability with wise laws very well. Then all we will rewarded. We can de-metalize coated and patterned silicon wafers, as well as broken solar cells or production scrap, by using a very simple method to minimize waste and to simultaneously produce marketable products. For thin film photovoltaic waste we finished piloting our new universal recycling procedure. We obtained a very high level quality of glass, usable demonstrably for float glass production.

REFERENCES

[1] calculates Bundesverband Solarwirtschaft e.V. (BSW-Solar), http://www.solarwirtschaft.de/presse/pressemeldungen/presseme ldungen-im-detail/news/photovoltaik-200-gigawatt-installiert.html [24.04.2016]

[2] V. Monier and M. Hestin, "Study on photovoltaic panels supplementing the impact assessment for a recast of the WEEE directive", Final Report, ENV.G.4/FRA/2007/0067, 14th of April 2011, p. 6.

[3] L. J. Fernandez et al., "Recycling silicon solar cell waste in cement-based systems", Solar Energy Materials and Solar Cells, 3th of March 2011.

Analysis of GaInP Solar Cells Grown by Hydride Vapor Phase Epitaxy

Kevin L. Schulte, John Simon, David L. Young, and Aaron J. Ptak

National Renewable Energy Laboratory, Golden, CO 80401 USA

Abstract — We analyzed the performance of a simple n-on-p GaInP solar cell grown without surface passivation but with a back reflector by dynamic hydride vapor phase epitaxy (D-HVPE). We employed a thin emitter because the lack of surface passivation limited collection in that layer. We obtained a maximum internal quantum efficiency of nearly 95% by using $1x10^{16}$ cm^{-3} base doping. This device exhibited an open circuit voltage (V_{OC}) of 1.27 V, short circuit current (J_{SC}) of 12.6 mA/cm^2, a fill factor (FF) of 79.3%, and 12.8% efficiency. Limits on cell performance predominantly come from V_{OC} and FF. A Suns-V_{OC} analysis of this cell indicates that 5.5% (absolute) of the FF is lost to series resistance, which stems from a high contact resistance. Another 5.5% of FF loss results from elevated dark current, leading to non-ideal diode behavior. This behavior implies that improvement in bulk material quality through reduction in concentrations of non-radiative defects will improve cell performance.

Index terms – hydride vapor phase epitaxy, GaInP, solar cells

I. INTRODUCTION

Hydride vapor phase epitaxy (HVPE) is a route to high efficiency III-V materials and solar cells with reduced deposition cost obtained through increased throughput and reduced input costs. High photovoltaic conversion efficiency is important in maximizing power to weight ratio for distributed applications such as space, unmanned aerial vehicles and portable power. High efficiency at low cost could enable the use of III-Vs for terrestrial applications such as building integrated PV due to reduced balance of systems cost [1]. The GaInP/GaAs tandem cell offers high performance in a monolithically grown two-junction solar cell, with a demonstrated efficiency of 31.6% at one-sun illumination [2]. We previously demonstrated 23.8% efficient HVPE-grown single-junction GaAs cells [3], and the development of a GaInP top cell would enable a high efficiency, all HVPE-grown GaInP/GaAs tandem with reduced cost. In this work, we analyze the performance of an HVPE-grown, unpassivated GaInP p-n junction solar cell. We demonstrate a 12.8% efficient device, and perform a fill factor analysis, identifying sources of performance loss. Pathways for improving future performance are identified.

II. EXPERIMENTAL

All materials were grown in our custom dynamic HVPE (D-HVPE) reactor detailed in ref. [4]. Growth details are detailed in ref. [5]. Metal chlorides were generated *in situ* through the reaction of HCl gas with pure Ga and In sources. PH$_3$ and AsH$_3$ were used to supply the Group V components. H$_2$Se and diethylzinc were the n- and p-type dopants, respectively. The source temperature, where the boats containing the Group III metals reside, was 800 °C, and the

deposition temperature was 650 °C. The V/III flow ratio during GaInP growth was 2 and the reactor pressure was ~1 atm for all growths. Growth rates were calibrated through selective etching and measurement of layer thicknesses by stylus profilometry.

The solar cell was grown in an inverted manner on a Si-doped, n+ (100) GaAs substrate with a 4° miscut towards the (111)B plane. Fig. 1 presents the device structure. The device lacked a front surface passivation layer, due to system constraints preventing the growth of Al-containing compounds in our specific HVPE reactor. This is not a limitation of HVPE, however [6, 7]. A 0.1 μm GaAs buffer was grown first to bury surface contamination, followed by a 0.25 μm GaInP etch stop. These layers were etched away during processing and do not appear in Fig. 1. The remaining layers are shown in the figure, grown from top to bottom. The highly doped GaInP back contact also served as a back surface field (BSF). The growth rate of the base layer was 0.4 μm/min. Base doping density was measured by standard capacitance voltage (C-V) techniques. 25 mm^2 devices were processed in an inverted configuration with a gold back reflector and with the substrate removed, as in ref. [8]. External quantum efficiency (EQE) and surface reflectivity were measured on a custom-built instrument and used to calculate internal quantum efficiency (IQE). Cell band gap was calculated from EQE as in ref. [9]. Light current density-voltage (J-V) curves were measured under a simulated AM1.5G spectrum calibrated using a GaInP reference cell. We conducted Suns-V_{OC} using a flash lamp with a slow decay, creating a variable intensity source, and V_{OC} was measured as a function of intensity. The intensity of the lamp over time was measured using a calibrated reference cell, and the data were corrected for spectral mismatch by fixing the Suns-V_{OC}

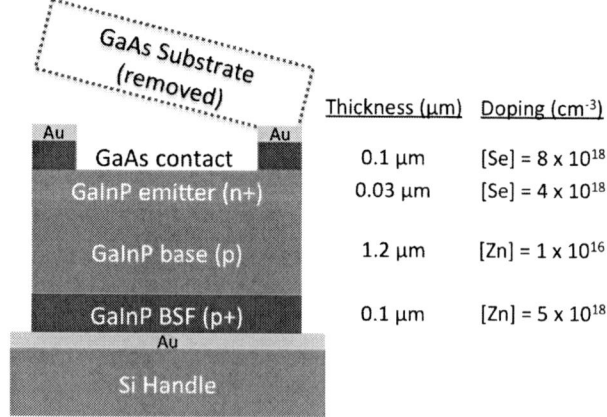

	Thickness (μm)	Doping (cm^{-3})
GaAs contact	0.1 μm	[Se] = 8 x 10^{18}
GaInP emitter (n+)	0.03 μm	[Se] = 4 x 10^{18}
GaInP base (p)	1.2 μm	[Zn] = 1 x 10^{16}
GaInP BSF (p+)	0.1 μm	[Zn] = 5 x 10^{18}

Inverted GaInP Cell

Fig. 1. Inverted GaInP solar cell device structure grown and tested in this study.

to that measured in the traditional J-V curve with the GaInP reference [10]. Since only V_{OC} is measured, and no current is drawn from the device, this method permits simulation of a J-V curve that is not affected by series resistance. Comparison to the traditional J-V curve allows direct determination of resistance losses in the cell.

III. RESULTS AND DISCUSSION

Previously we showed a strong dependence of collection efficiency on doping density in the range 1.5 x 10^{16} to 4.8 x 10^{17} cm^{-3} [5]. Long wavelength collection increased strongly as the doping density decreased, implying that at higher doping levels, the diffusion length was significantly less than the 1 μm base thickness employed [5]. Thus we targeted a doping density of 1 x 10^{16} cm^{-3} for the present cell with the structure of Fig. 1. The internal quantum efficiency for this device is shown in Fig. 2. Note that a two-layer MgF$_2$/ZnS anti-reflective coating (ARC) was evaporated onto the cell after device processing. The collection efficiency reached nearly 95% at its maximum.

We developed an optical model of the device to evaluate its performance relative to the ultimate level possible with this design (see ref. [5] for details). We calculated the absorption in the emitter and base as a function of wavelength using n and k data for disordered GaInP. The absorption in the emitter is highlighted as the shaded area in Fig. 2, while the absorption in the base is defined by the solid black curve. If we assume that the emitter contributes almost nothing to carrier collection, reasonable given the lack of passivation and high doping there, the black line indicates the maximum obtainable IQE at each wavelength for photons absorbed in the base. Comparison of the model with the experimental IQE indicates that the majority of photons that reach the base are collected. Integration of the AM1.5G spectrum with the modeled IQE (black line) and the experimental IQE yields values of 14.2 and 13.4 mA/cm^{-2}, respectively. Thus ~94% of the absorbed photons transmitted to the base are collected as current.

Fig. 3 shows the light and dark J-V curves for the cell and indicates the open circuit voltage (V_{OC}), short circuit current (J_{SC}), fill factor (FF), and cell efficiency. The measured J_{SC} differed from the integrated IQE due to remaining reflectance loss after application of the ARC. The overall efficiency was 12.8%. The band gap was 1.86 eV, as evaluated from the EQE, yielding a $W_{OC} = E_g/q - V_{OC} = 0.59$ V for a V_{OC} of 1.27. This value is elevated due to the lack of surface passivation, but also indicates that room for improvement of the bulk material quality also remains. The dark current exhibited an ideality factor near n=2, which is likely the result of non-radiative recombination at deep level defects in the depletion region (which is wide due to the asymmetric doping levels in the emitter and base). Optimal conditions for the growth of GaInP by HVPE are not well known in the literature. Thus it

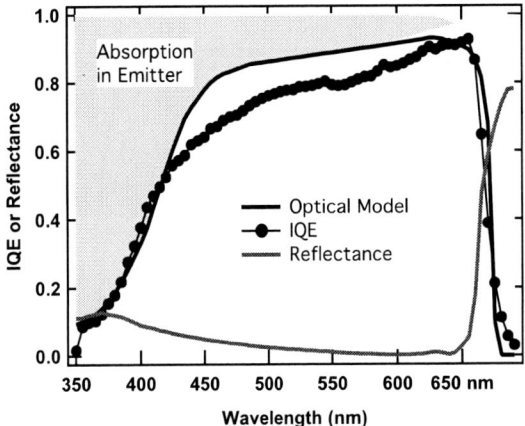

Fig. 2. (Markers): IQE for a 1.2 μm thick GaInP cell with gold back reflector and 30 nm GaInP emitter. (Black curve): Ideal IQE modeled for a solar cell with the same cell parameters as the real cell, with the emitter assumed to contribute nothing to collection.

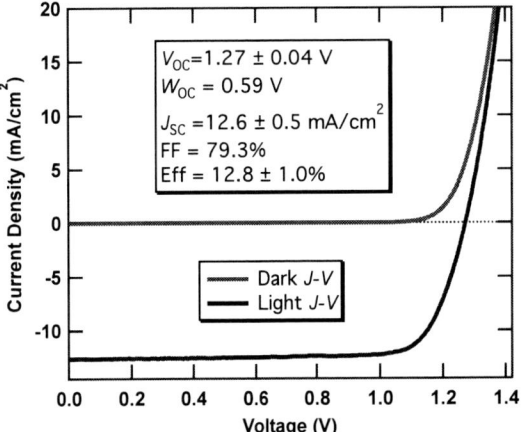

Fig. 3. Light (under AM1.5G) and dark J-V curves for an HVPE GaInP cell with the design of Fig. 1.

is likely that significant room for optimization of the bulk material quality through reduction of deep level defects remains. Reduction of intrinsic deep level defects will drive increases in V_{OC} through reduced dark current. Growth conditions such as deposition temperature and V/III ratio commonly affect deep level defect density in HVPE [11, 12] and MOVPE [13] GaAs, and likely have similar impacts on GaInP. Indeed, in a limited number of growths in our system, changes in the V/III ratio had a significant effect on device performance [5]. Deeper understanding of the relationships between growth conditions and deep level defects will allow for their mitigation, leading to development of higher efficiency devices.

Room for FF optimization exists as well, because percentages in the high 80s are common for GaInP solar cells [14]. We performed a Suns-V_{OC} measurement of the GaInP cell in order to gain more information about sources of FF

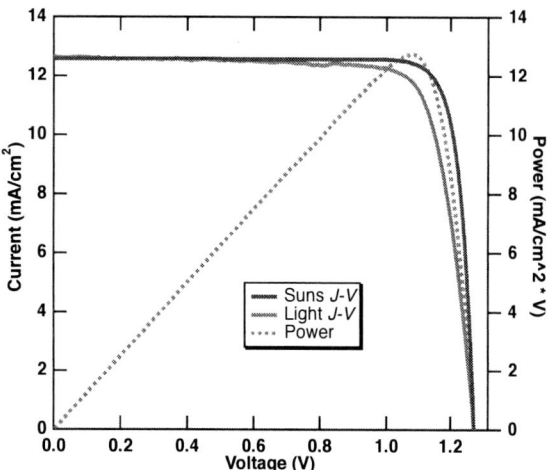

Fig. 4. Suns-V_{OC} characterization of an inverted HVPE GaInP cell.

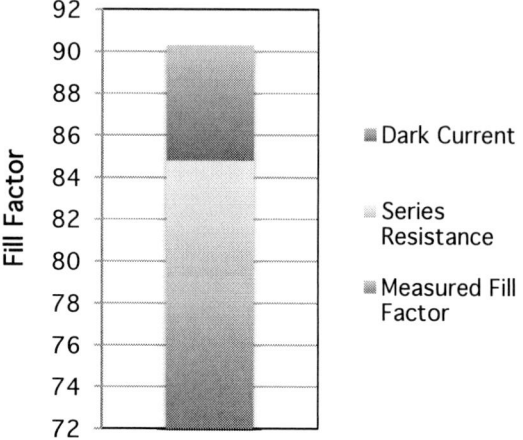

Fig. 5. Analysis of fill factor losses in the GaInP cell.

loss. In this measurement, the V_{OC} is recorded as the illumination intensity is varied, yielding a simulated light J-V curve without the effects of series resistance. Fig. 4 shows the simulated J-V curve plotted along with the J-V curve and power from the traditional light measurement. The series resistivity, r_s, is found to be 6.3 $\Omega \cdot cm^2$, from the equation:

$$r_s = \frac{V_{Suns}(J_{MP}) - V_{MP}}{J_{MP}} \quad (1)$$

where V_{MP} and J_{MP} are the voltage and current at the maximum power point (MPP) from the traditional light J-V while $V_{Suns}(J_{MP})$ is the voltage measured by Suns-V_{OC} at J_{MP} [15]. Quantification of the series resistance permits a FF loss analysis in which the FF of the cell, after correcting for losses from series and shunt resistance, is compared to the FF of an ideal diode with the same band gap. FF loss stemming from non-ideal diode behavior related to non-radiative dark current is then isolated, quantifying the degree of FF loss related to material quality [16]. The results of this analysis are presented in Fig. 5. Note that the shunt resistance, obtained by fitting the slope of the dark current near V = 0, was 5.5 x 10^5 $\Omega \cdot cm^2$, thus the resulting effects on FF were insignificant and ignored.

The series resistance led to a roughly 5.5% (absolute) loss in FF. We measured the contact and sheet resistances by the transfer length method in order to understand the limiting resistance. The sheet resistance was 176 $\Omega/\square$, which quite reasonable for a one-sun cell. The contact resistance was 0.007 Ω/cm^2, which was elevated due to insufficient doping in the contact layer. A sensitivity analysis using the model of ref. [17] indicates that the series resistance is almost completely limited by the contact resistance. Increasing the Se doping in the GaAs contact into the low to mid 10^{19} cm^{-3}

range should improve this resistance. This doping level is easily achievable in HVPE GaAs [18], and thus the contact resistance should not limit future cells.

The other major loss in fill factor is the dark current loss, which is related to elevated dark currents that result from non-radiative recombination. A diode with a high non-radiative dark current operates in an n=2 regime, rather than an n=1 regime in which the dark current is dominated by radiative recombination. The current in an n=2 diode increases more slowly with voltage, which manifests as a more rounded J-V curve, pushing the MPP to a lower voltage. At one-sun, a 5.5% absolute loss in FF (Fig. 5) results from the elevated dark current, again highlighting the importance of optimizing the bulk material quality of the HVPE-grown material. Fig. 6 shows the diode performance under low concentration, as determined from the Suns-V_{OC} measurement. Encouragingly, the V_{OC} rapidly approaches an n=1 ideality factor under low

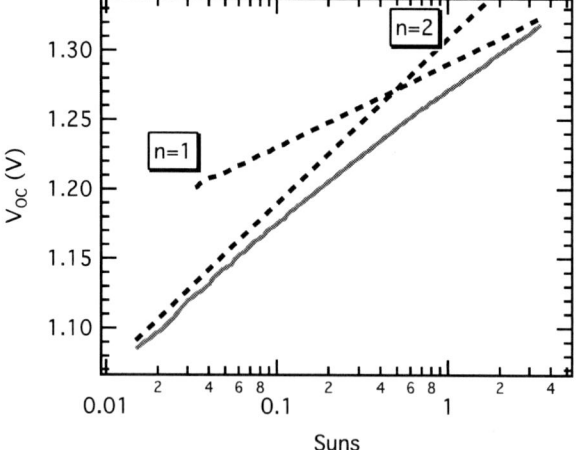

Fig. 6. V_{OC} as a function of concentration as determined from the Suns-V_{OC} measurement.

concentration, indicating that only a slight reduction in defect concentrations may be necessary to achieve n=1 performance at one-sun.

Lastly we point out that a reduction in the non-radiative defects that contribute to dark current will also increase the diffusion length, permitting the use of designs with higher base doping levels. This will increase V_{OC} through an increase in quasi-Fermi level splitting and further reduce the dark current by shrinking the width of the depletion region. Given that we have performed only limited optimization of the GaInP growth conditions, the potential for significant improvement in device performance above what has already been achieved is promising.

IV. SUMMARY

We analyzed the performance of a GaInP solar cell grown by HVPE. A device without front surface passivation but with anti-reflective coating reached an efficiency of 12.8%. Collection of photons in the base compared favorably to an optical model that assumed perfect collection. The majority of the absorption occurs in the depletion region, and dark current is likely dominated by non-radiative recombination in this region leading to an n = 2 diode ideality factor at one-sun. Sources of fill factor loss were identified through a Suns-V_{OC} analysis. Going forward, reduction of deep level defects in this material will be key to improving efficiency through increased V_{OC} and FF.

V. ACKNOWLEDGMENT

The authors would like to thank David Guiling for materials growth, Michelle Young for device processing, and Daniel J. Friedman for useful discussions. The information, data, or work presented herein was funded in part by the Advanced Research Projects Agency (ARPA-E), US Department of Energy, award #15/CJ000/07/05. This work was also supported by the U.S. Department of Energy under Contract No. DE-AC36-08GO28308 with the National Renewable Energy Laboratory. Funding provided by U.S. DOE Office of Energy Efficiency and Renewable Energy Solar Energy Technologies Program.

The U.S. Government retains and the publisher, by accepting the article for publication, acknowledges that the U.S. Government retains a nonexclusive, paid up, irrevocable, worldwide license to publish or reproduce the published form of this work, or allow others to do so, for U.S. Government purposes.

REFERENCES

[1] R. Jones‑Albertus, D. Feldman, R. Fu, K. Horowitz, and M. Woodhouse, "Technology advances needed for photovoltaics to achieve widespread grid price parity," *Progress in Photovoltaics: Research and Applications*, 2016.

[2] M. A. Green, K. Emery, Y. Hishikawa, W. Warta, E. D. Dunlop, D. H. Levi, *et al.*, "Solar cell efficiency tables (version 49)," *Progress in Photovoltaics: Research and Applications*, vol. 25, pp. 3-13, 2017.

[3] J. Simon, K. L. Schulte, N. Jain, M. Young, M. R. Young, D. L. Young, *et al.*, "Upright and inverted single junction GaAs solar cells grown by hyrdide vapor phase epitaxy," presented at the IEEE Photovoltaics Specialists' Conference, Portland, OR, 2016.

[4] D. L. Young, A. J. Ptak, T. F. Kuech, K. Schulte, and J. D. Simon, "High throughput semiconductor deposition system," ed: Google Patents, 2015.

[5] K. L. Schulte, J. Simon, J. Mangum, C. E. Packard, B. P. Gorman, N. Jain, *et al.*, "Development of GaInP Solar Cells Grown by Hydride Vapor Phase Epitaxy," *IEEE Journal of Photovoltaics*, pp. 1-6, 2017.

[6] W. Johnston Jr and W. Callahan, "High‑performance solar cell material: n‑AlAs/p‑GaAs prepared by vapor phase epitaxy," *Applied Physics Letters*, vol. 28, pp. 150-152, 1976.

[7] M. Ettenberg, A. Sigai, A. Dreeben, and S. Gilbert, "Vapor Growth and Properties of AlAs," *Journal of The Electrochemical Society*, vol. 118, pp. 1355-1358, 1971.

[8] A. Duda, J. S. Ward, and M. Young, "Inverted Metamorphic Multijunction (IMM) Cell Processing Instructions," National Renewable Energy Laboratory2012.

[9] M. A. Steiner, E. E. Perl, J. F. Geisz, D. J. Friedman, N. Jain, D. Levi, *et al.*, "Apparent bandgap shift in the internal quantum efficiency for solar cells with back reflectors," *Journal of Applied Physics*, vol. 121, p. 164501, 2017.

[10] T. Roth, J. Hohl-Ebinger, E. Schmich, W. Warta, S. W. Glunz, and R. A. Sinton, "Improving the accuracy of Suns-V OC measurements using spectral mismatch correction," in *Photovoltaic Specialists Conference, 2008. PVSC'08. 33rd IEEE*, 2008, pp. 1-5.

[11] K. Schulte and T. Kuech, "A model for arsenic anti-site incorporation in GaAs grown by hydride vapor phase epitaxy," *Journal of Applied Physics*, vol. 116, p. 243504, 2014.

[12] M. D. Miller, G. H. Olsen, and M. Ettenberg, "The effect of gas-phase stoichiometry on deep levels in vapor-grown GaAs," *Applied Physics Letters*, vol. 31, pp. 538-540, 1977.

[13] P. K. Bhattacharya, J. W. Ku, S. J. T. Owen, V. Aebi, C. B. Cooper, and R. L. Moon, "The trend of deep states in organometallic vapor phase epitaxial GaAs with varying As/Ga ratios," *Applied Physics Letters*, vol. 36, pp. 304-306, 1980.

[14] J. F. Geisz, M. A. Steiner, I. Garcia, S. R. Kurtz, and D. J. Friedman, "Enhanced external radiative efficiency for 20.8% efficient single-junction GaInP solar cells," *Applied Physics Letters*, vol. 103, p. 041118, 2013.

[15] D. Pysch, A. Mette, and S. W. Glunz, "A review and comparison of different methods to determine the series resistance of solar cells," *Solar Energy Materials and Solar Cells*, vol. 91, pp. 1698-1706, 2007.

[16] A. Khanna, T. Mueller, R. A. Stangl, B. Hoex, P. K. Basu, and A. G. Aberle, "A fill factor loss analysis method for silicon wafer solar cells," *IEEE Journal of Photovoltaics*, vol. 3, pp. 1170-1177, 2013.

[17]   D. K. Schroder and D. L. Meier, "Solar cell contact resistance—A review," *IEEE Transactions on electron devices*, vol. 31, pp. 637-647, 1984.

[18]   A. J. Ptak, J. Simon, and K. L. Schulte, "Tunnel Junction Development Using Hydride Vapor Phase Epitaxy," presented at the IEEE PVSC, Washington, DC, 2017.

Investigation of Adhesion Forces Between Dust Particles and Solar Glass

H.R. Moutinho, C.-S. Jiang, B. To, C. Perkins, M. Muller, M.M. Al-Jassim, and L. Simpson

National Renewable Energy Laboratory, Golden, Colorado, United States

Abstract — **We investigate the adhesion mechanisms that occur in the first stages of soiling by measuring adhesion forces between different kinds of particles and solar glass, using force vs. distance curves in an atomic force microscope (AFM). To simulate several degrees of roughness, we polished the samples at different conditions. We studied the effects of relative humidity by performing the measurements in a glove box and in an AFM enclosure.**

The measurements revealed that the main interactions are van der Waals and capillary forces. We show the dependence of the adhesion force on the several parameters and provide an explanation of how they operate.

Index Terms — **adhesion forces, soiling, dust particles.**

I. INTRODUCTION

When considering economic factors in solar cell technology, most of the literature has concentrated on costs, efficiency- related parameters, and reliability. However, as the deployment of solar panels has increased in the last several decades, it has become clear that the accumulation of dust particles with time—, i.e., soiling—, may significantly impact power output, and may be as important as these other factors. The influence of dust/dirt on the performance of photovoltaic (PV) modules has been the subject of many studies [1–2]. However, most studies have focused on measuring soiling rates at different locations and reporting their influence on efficiency. Although there are studies on the force mechanisms responsible for dust/glass interaction, only a few of them are applied to PV [3–4] and most are related to other fields [5–6]. In this work, our goal is to establish the basic mechanisms responsible for the first stages of soiling and to study how parameters such as relative humidity (RH) and surface roughness affect the adhesion forces.

II. EXPERIMENTAL SETUP

The main technique used in this work is force-distance curves (F-z curves) [7] obtained in an atomic force microscope (AFM), which allows us to measure the force between the particle and the surface. For measurements at zero relative humidity (RH), we used a Dimension 5000 AFM with Nanoscope V electronics (from Bruker) installed in an MBraun glove box. For measurements at other values of RH, we used a Dimension 3100 AFM with Nanoscope V electronics (from Bruker). AFM was also used to attach real dust particles to AFM tipless cantilevers.

To determine the shape and size of dust particles, we used scanning electron microscopy (SEM) performed in a field-emission Nova NanoSEM 6300 SEM (from FEI Company); and to measure the composition of dust particles, we used energy-dispersive X-ray spectroscopy (EDS), performed in a Pegasus system (from EDAX).

To emulate the surface of solar panels, we used solar glass from Saint-Gobain Company. To emulate surface roughness, caused by factors such as wind and cleaning, we polished the glass samples with lapping diamond films with different grits and obtained roughness from 0.3 to 22.2 nm.

To determine the interaction mechanisms in a controllable manner, we began the work using oxidized AFM Si tips (20-nm diameter) and SiO_2 spheres (5- and 20-μm diameters) as the particles. After this, we used Arizona road dust particles with different sizes (15 – 60-μm diameter) and shapes to measure adhesion forces. The dust particles were mounted on tipless cantilevers using the optical view of the AFM according to the following procedure: spread dust particles on a glass slide; spread glue on the glass slide on a clean area; place the end of the cantilever over the edge of the glue and move it onto the surface; retrieve the cantilever, which will have a small amount of glue on its surface; move the cantilever onto the surface on a clean area to remove excess of glue and retrieve it; locate a dust particle and position the end of the cantilever over it; move the cantilever down onto the dust particle; retrieve the cantilever with the attached dust particle. To make the process faster, we used a special glue that cures in seconds when exposed to ultraviolet light.

The adhesion-force measurements were performed with relative humidity varying from 0% to 80%. For 0% RH, we used a glove box with levels of H_2O and O_2 lower than 0.1 ppm. For other values of RH, we performed the measurements inside the enclosure of an AFM in air. To control the humidity, we used saturated solutions of salt/deionized water at room temperature and used different salts to obtain different RH values [8].

III. RESULTS AND DISCUSSION

A typical F-z curve between an AFM tip and the surface of a glass sample is shown in Fig. 1. As the piezo driver approaches the sample surface from far away, there is no interaction between the probe and the surface, and the deflection curve is horizontal (black straight line in Fig. 1). When the tip gets close enough, it is attracted down and

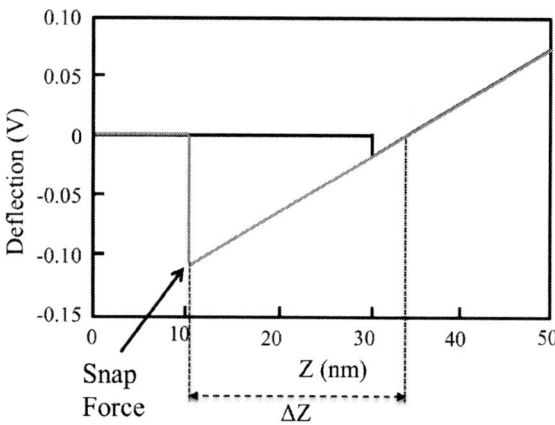

Fig. 1. F-z curve between an AFM tip and the surface of a solar glass sample.

touches the sample (small kink in the curve). As the piezo driver keeps going down, the cantilever bends up and the deflection increases linearly. At the end of its range, the piezo driver reverses its movement (red curve) and starts moving up. The deflection curve follows the inverse path, but due to the adhesion force, it departs from the F-z curve obtained during the approach. At this point, the cantilever bends down. As the piezo driver keeps moving up, the deflection and the pull-up force increase. The adhesion force is the force on the cantilever when it snaps off the surface, and it is given by the product between the known cantilever force constant (k) and its deflection ($F=k\Delta z$). It is important to mention that the horizontal part of the F-z curve and the small kink are characteristics of short-range forces, such as van der Waals forces. For long-range forces, such as electrostatic forces, the

tip usually starts feeling the interaction a few micrometers away from the surface, and the cantilever starts bending down. As the cantilever approaches the surface, the attractive interaction increases and the first part of the curve continuously bends downward until the tip touches the surface.

In Fig. 2, we show the variation of the adhesion force between oxidized AFM probes and SiO_2 particles and solar glass surfaces with different values of surface roughness at different RH values. At 0% RH, there is no water layer, and consequently, the force between the particles and surface are van der Waals forces. This is corroborated by the short-range character of the force [9]. Capillary forces appear when there is a water layer between the surface and particle, and they have two components due to the surface tension and water meniscus pressure [10, 11]. From Fig. 2, we notice that the force does not increase for RH values up to 20%–30%. After that, there is a strong increase due to capillarity. For high RH values, the interaction is a combination of van der Waals and capillary forces. However, the Hamaker constant in the van der Waals force is lower for water than for air [12], and our calculations show that the capillary forces dominate at high RH values, except for higher values of surface roughness (Fig. 2b and discussion below).

Due to the short range of capillary and van der Waals forces, surface roughness is expected to have a strong effect. This occurs because as surface roughness increases, the contact between surface and particle will occur only in a few points. In this case, a larger part of the particle is too far from the surface, and only the contact points experience the van der Waals interaction. For the case of capillary force, the rougher the sample, the more water is needed to fill the asperities. The effect of roughness on both forces can be clearly observed for the SiO_2 spheres (Fig. 2b). Notice that as the roughness increases, the force decreases. Also, the force does not change

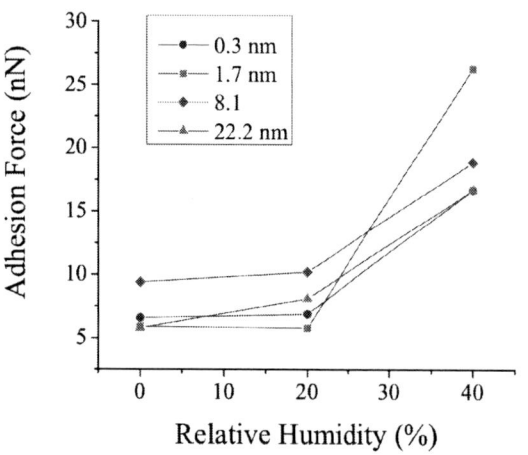

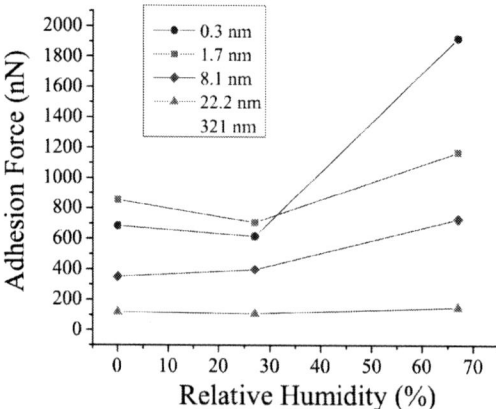

Fig. 2. Adhesion forces for AFM tip (left) and 20-μm SiO_2 sphere (right) for different values of surface roughness as a function of relative humidity.

with RH for the higher values of roughness because the asperities are too large to allow for the formation of a continuous water meniscus. However, the same does not occur for the AFM tips (Fig. 2a). Notice that there is not a clear trend on the variation of the force with surface roughness. The reason for this is the small size of the tip. By analyzing AFM linescans of the surface topography, we observed that the diameter of the tip is generally smaller than the topographic features of the solar glass surface. Because of this, the probe can penetrate the asperities, and the contact area does not depend on the roughness of the sample surface. In this case, the force is caused by the local point of contact; and because this is a local property, there is a random variation on the values obtained for different measurements, as observed in Fig. 2a.

Another characteristic of the interaction was an increase in the adhesion force with particle size, which was more pronounced for smaller values of roughness, as observed in Fig. 3. As expected, the larger the particle size, the larger is the contact area and the larger is the adhesion force. However, for similar reasons as above, as the roughness increases, the effect of probe size on the adhesion force decreases.

After the adhesion force mechanisms were established in these ideal particles, we measured the adhesion force between the surface and real dust particles. SEM analysis showed that the size of the particles varied from about 15 to 60 μm, and there was a large variation in their shape, as observed in Fig. 4. EDS analysis showed that the main composition of these particles was SiO_2. The dust particles were mounted on tipless cantilevers as described in the experimental setup section. Similar to SiO_2 spheres, the F-z curves indicated that only van der Waals and capillary forces were present. Also, in this case, there is a decrease in the adhesion forces for larger values of surface roughness, as observed in Fig. 5.

The variation of the force for particles with different sizes

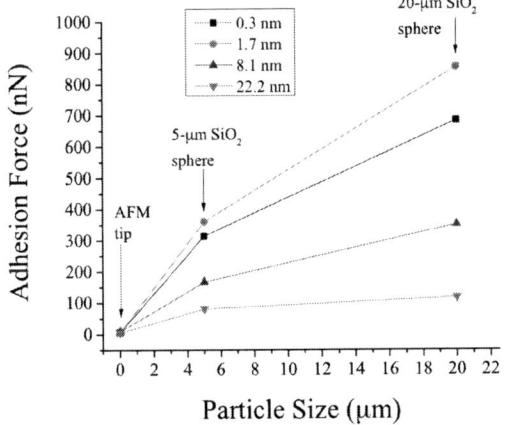

Fig. 3. Variation of the adhesion force with particle size for particles with different values of surface roughness.

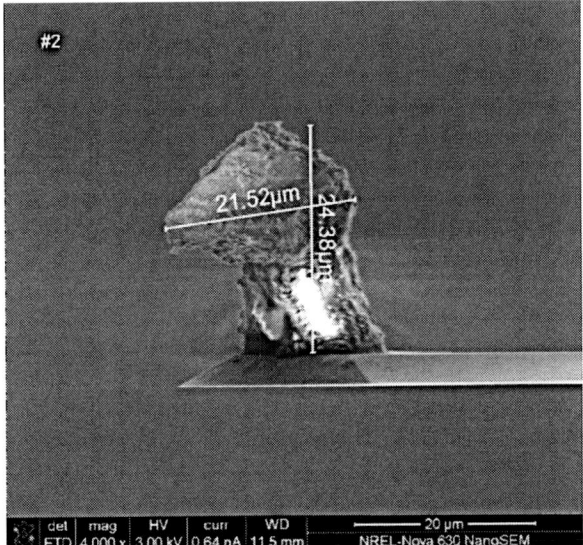

Fig. 4. SEM images of two dust particles (#2 and #4) used in force versus distance measurements.

as a function of RH is shown in Fig. 6. The increase in force with an increase in humidity is clear, and, contrary to the previous observation, there is an increase in the adhesion force for RH values as small as 20%. This was attributed to a higher hydrophilicity of the dust particles compared to AFM tips and SiO_2 spheres. Another important result from this work is that the adhesion forces are not a function of dust particle size, as observed when comparing the adhesion curves for particles #2 and #4 and the size of the particles on Fig. 4. It is clear that the smaller particle has a larger adhesion force (van der Waals and capillary) than the much larger particle. This seems to disagree with our previous experiment and with reports that show an increase in the van der Waals and capillary forces as the size of particles increases. Instead, this result shows the importance of studying real particles instead of glass spheres to determine soiling. Because of the roughness, real particles generally have a much smaller contact area than glass spheres, as we observed in several measurements. Thus, in general, the adhesion forces are larger for spheres than for dust particles of similar sizes. This is evident if we compare the values of the force in Figs. 2b and 6. Because of their short range, the van der Waals and capillary adhesion forces are determined by the local contact points, and most of the particle does not feel the attractive interaction, which explains why the force is not directly proportional to particle size in this case.

These results explain work from Cuddihy [13], who reported that winds with hurricane speeds were not able to remove particles smaller than 50 µm (smaller cross-section than large particles) from surfaces. These results may be important for companies when deciding where to deploy solar panels, concerning the type and shape of dust particles in the specific location.

Because of uncertainties in the cantilever force constant values reported by the manufacturers, the absolute values of the reported forces may not be very precise. However, this issue did not influence the reported trends or the influence of several parameters on the adhesion forces, as well as the conclusions of the work.

Finally, it is important to mention that there was no electrostatic interaction in this study between the particles and the solar glass. However, this may not be the case when the panel is deployed in the field. In that case, it is possible that the voltages present on the module during normal operation induce electrostatic forces that may be larger than both the van der Waals and capillary forces together.

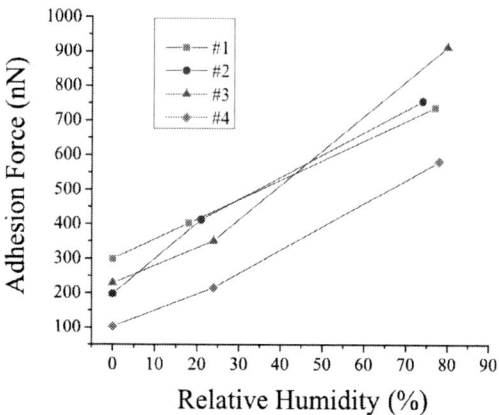

Fig. 6. Adhesion force between dust particles and glass substrate as a function of RH. The different curves are for particles of different shapes and sizes. The glass substrate has surface roughness equal to 0.3 nm. Particles are the same as the ones in Fig. 5.

IV. CONCLUSIONS

Van der Waals and capillary forces are responsible for the first interaction between particles and solar glass, and they must be considered when studying soiling on solar panels. The adhesion forces on AFM tips and SiO_2 spheres increase with relative humidity, but may be affected by surface roughness, which can decrease adhesion under certain conditions. For real dust particles, adhesion forces also increase with relative humidity and decrease with surface roughness. However, contrary to AFM tips and glass spheres, particle size is not important, which may be important information when choosing specific sites for solar panels deployment.

ACKNOWLEDGMENT

This work was supported by the U.S. Department of Energy under Contract No. DE-AC36-08GO28308 with the National Renewable Energy Laboratory. The U.S. Government retains and the publisher, by accepting the article for publication, acknowledges that the U.S. Government retains a nonexclusive, paid up, irrevocable, worldwide license to publish or reproduce the published form of this work, or allow others to do so, for U.S. Government purposes.

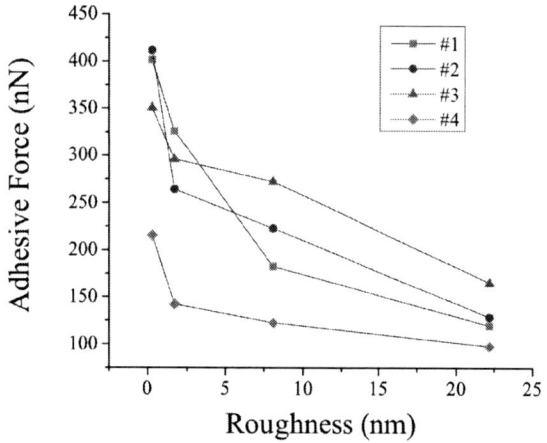

Fig. 5. Adhesion force for dust particles with different sizes and shapes as a function of glass substrate roughness. Data taken at room humidity (~20%). Particles #2 and #4 are shown in Fig. 4.

REFERENCES

[1] T. Sarver, A. Al-Qaraghuli, and L.L.Kazmerski, A comprehensive review of the impact of dust on the use of solar energy: History, investigations, results, literature, and mitigation approaches, *Ren. Sustain. Energy Rev.* **22**, 698 (2013).

[2] H.P. Garg, Effect of dirt on transparent covers in flat-plate solar energy collectors, *Solar Energy* **15**, 299 (1974).

[3] S.A.M. Said, and H.M. Walwil, Fundamental studies on dust fouling effects on PV module performance, *Solar Energy* **107**, 328 (2014).

[4] T. Sueto, Y. Ota, and K. Nishioka, Suppression of dust adhesion on a concentrator photovoltaic module using an anti-soiling photocatalytic coating, *Solar Energy* **97**, 414 (2013).

[5] J.A.S. Cleaver and J.W.G. Tyrrell, The Influence of relative humidity on particle adhesion – A review of previous work and the anomalous behavior of soda-lime glass, *KONA* **22**, 9 (2004).

[6] R. Jones, H.M. Pollock, J.A.S. Cleaver, and C.S. Hodges, Adhesion forces between glass and silicon surfaces in air studied by AFM: Effects on relative humidity, particle size, roughness, and surface treatment, *Langmuir* **18**, 8045 (2002).

[7] B. Cappella and G. Dietler, Force-distance curves by atomic force microscopy, *Surf. Sci. Reports* **34**, 1 (1999).

[8] J.F. Young, Humidity control in the laboratory using salt solutions – A Review, *J. Appl. Chem.* **17**, 241 (1967).

[9] G. Devaud, C. Haley, C. Rockwell, and A. Fischer, Surfaces that shed dust: Unraveling the mechanisms, *Proc. SPIE* **9196**, 919603-1 (2014).

[10] D.L. Malotky and M.K. Chaudhury, Investigation of capillary forces using atomic force microscopy, *Langmuir* **17**, 7823 (2001).

[11] H.-J. Butt and M. Kappl, Normal capillary forces, *Adv. Colloid Interf. Sci.* **146**, 48 (2009).

[12] L. Bergström, Hamaker constants of inorganic materials, *Adv. Colloid Interf. Sci.* **70**, 125 (1997).

[13] E. Cuddihy, Theoretical considerations of soil retention, *Sol. En. Mater.* **3**, 21, 1980.

Anti-reflective and anti-soiling properties of a KleanBoost™, a superhydrophobic nano-textured coating for solar glass

Illya Nayshevsky[1,2], QianFeng Xu[1,3], Gil Barahman[1], Alan Lyons[1,2,3]

[1] College of Staten Island, City University of New York, 2800 Victory Blvd, Staten Island, NY 10314
[2] The Graduate Center, City University of New York, 365 5th Avenue, New York, NY 10016
[3] ARL Designs LLC, 215 W 125th Street 4th Fl., New York, NY 10027

Abstract — Soiling of solar cover glass can result in a significant loss of electrical output of PV panels. Dust and other contaminants adhere strongly to the glass by known mechanisms. In contrast, anti-soiling coatings, and the mechanisms by which they function, are not well-characterized. In this paper, we examine the properties of KleanBoost™ a thin anti-reflective and anti-soiling fluoropolymer coating for glass. In particular, we examine the effect of dew and baking cycles on dust accumulation and optical transparency. A dust "herding" process was observed as the condensed water droplets reduce in diameter as they evaporate on the hydrophobic surface. Anti-soiling and cleaning mechanisms were evaluated as a function of hydrophobicity of the glass nano-coating, and the KleanBoost™ surfaces were characterized and compared to their anti-soiling and cleaning properties.

Index Terms — fluoropolymer coatings, anti-reflectivity, anti-soiling, low-iron glass, photovoltaic solar cells.

I. INTRODUCTION

The electrical output of solar PV panels is reduced by two main factors: reflection of light at the air-glass interface and scattering of light due to soiling. Reflections occur because of the abrupt change in refractive index between air (n=1) and glass (n=1.52). Anti-reflective coatings (ARCs) have been developed where a thin layer of a transparent material, with an intermediate refractive index, is applied to solar cover glass thereby reducing reflections by 50-80% [1]-[2]. Although ARCs are currently applied to ~80% of solar PV modules, ARCs do not improve soiling behavior. Soiling of panels can reduce electrical output by as much as 50%, depending upon the local environment. Although some work has been conducted to develop anti-soiling coatings for solar cover glass, commercial adoption has not occurred. There are numerous challenges to developing a commercially viable anti-soiling coating including: cost, coating life-time and anti-soiling effectiveness. Mechanisms describing how dust particles adhere to glass, such as cementation, have been published [3]. Also, it has been shown that soiling rates are highest during periods of morning dew [4]. However, the mechanisms by which anti-soiling coatings prevent dust adhesion have not been reported.

In this paper, we describe the anti-reflective and anti-soiling properties of a novel superhydrophobic fluoropolymer nano-textured coating, which we call KleanBoost™. The coating process results in a thin (~0.5 μm) layer of the fluoropolymer strongly bonded to low-iron glass. A test rig was constructed to apply a uniform layer of dust onto both uncoated and coated surfaces with and without the presence of simulated dew, mimicking the dust accumulation processes and rates in Arizona [4]. After dust deposition, samples were baked at 65°C. This process was repeated for 5 cycles before cleaning. Percent transmission through the glass was quantified as a function of dust deposition cycle as well as after cleaning. KleanBoost™ coated surfaces exhibit both anti-soiling and anti-reflective properties, especially in the presence of condensed water on the glass surface. Moreover, the KleanBoost™ coated surfaces can be cleaned using a minimal amount of water, restoring 99.7% of the light transmission before dust exposure. Water alone, without brushing, fully removes the dust indicating that the coating prevents cementation from occurring. A dust "herding" mechanism is developed to explain the anti-soiling properties.

II. METHOD

A. Materials.

KleanBoost™ coatings were fabricated in our facilities by laminating a fluoropolymer onto low iron glass substrates (3 mm thick OptiWhite from Pilkington). Nanofibrils with a diameter of ~100 nm are formed on the surface as shown in Fig. 1a. The refractive index of the coating gradually increases from 1.00 (air) to 1.35 at the bottom of the polymer layer as shown schematically in Fig. 1b. This surface morphology, combined with the intrinsically low surface energy of the fluoropolymer, creates a superhydrophobic coating immediately after fabrication with a water contact angle (WCA) of 150° as shown in Fig. 1c. In comparison, bare glass (uncoated) control samples exhibited a WCA of 10°.

B. Measurements.

Optical transmission spectra were recorded using a PerkinElmer Lambda 650 UV-Vis spectrometer in the range from 350 nm to 850 nm, with a resolution of 1 nm. A total of 3 scans were performed on each test sample. Water contact angle and sliding angle values were measured using a rame-hart model 250-F1 contact angle goniometer using 5μL droplets of DI water. Evaporation of condensate was monitored with a Nikon SMZ1500 stereo-zoom microscope. Images were analyzed with ImageJ 1.50i software (National Institute of Health).

978-1-5090-5606-4/17 $31.00 © 2017 IEEE

Fig. 1. KleanBoost™ coating on low-iron glass (a) SEM micrograph of the surface nanostructures of the fluoropolymer coating, (b) cartoon illustrating the graded index of refraction, (c) water droplet (5 µL) poised on the coated surface.

C. Artificial Soiling.

A test chamber was designed and built to deposit a uniform coating of dust onto glass coupons. The chamber was designed so that water could be condensed on the glass surfaces prior to

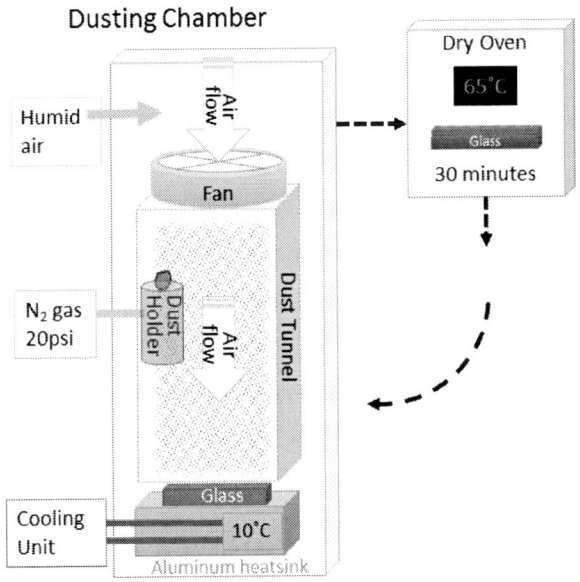

Fig. 2. Schematic of the dust deposition system illustrating the dusting chamber with: dust holder; dust tunnel with top mounted fan; chilled substrate support as well as the oven used to bake the samples after each dusting cycle.

dust deposition in order to simulate dew. A schematic of the dusting chamber is shown in Fig. 2.

Two glass test substrates (each measuring 5.0 cm x 5.0cm) is placed on a cooling plate at the bottom of the chamber. The plate is cooled to 10°C so that moisture in the chamber will condense onto the surface forming a layer of liquid water. A sonic humidifier was connected to a controller to maintain constant humidity in the chamber (set-point: 70% ±4% RH). The humidity was regulated via an on-off power switch controlled by a humidity sensor such that the humidity was maintained to within 4% of the set-point.

A coated KleanBoost™ test sample and a bare glass control sample were placed side-by-side on the cooling plate aligned to the bottom of a "dust tunnel" (10 cm x 10 cm upper opening, tapering down to 5 cm x 10 cm exit over a 20 cm length) which is designed to uniformly disperse a controlled amount of dust onto the glass surface. Standard Arizona Road Dust (A2 Fine Grade ISO 12103-1, 20 mg), loaded onto a metal mesh in the dust holder, is injected upwards into the dust tunnel by applying a pulse of dry N_2 gas (20 psi, 1 min) from under the support mesh. The dust encounters a downward stream of air from the overhead fan with a downstream flow rate of 3 m/s. The turbulent air helped to disperse the dust so that it uniformly covers the underlying glass.

D. Cleaning System.

To assess the anti-soiling properties of the KleanBoost™ coated surfaces, a dropwise cleaning system was designed and built as shown schematically in Fig. 3. A syringe is located at the top of the glass sample to be tested, and the needle of the syringe is placed 5mm above the surface. Water droplets are dispensed using a syringe pump as the glass substrate is linearly translated underneath the needle at the rate of 5 cm/min. Droplets are dispensed at a rate of 3.0 mL/min during the 1 minute of transit time (3.0 mL total volume). The volume of

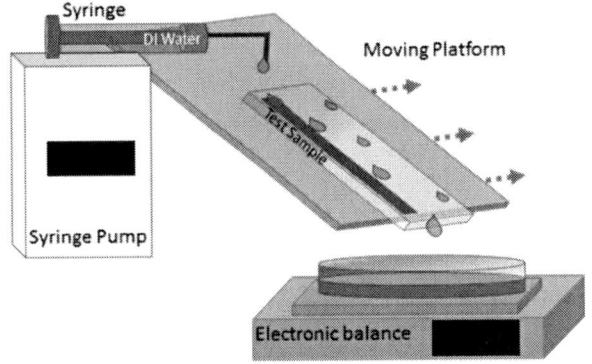

Fig. 3. Schematic of the no-touch cleaning system used to dispense water onto the dusted surfaces.

water recovered after cleaning was measured using an electronic balance.

III. RESULTS

A. Surface Properties

Surface characterization of the KleanBoost™ coated surfaces via scanning electron microscopy (SEM) showed surface of granular morphology (Fig. 4). The texture of the surface can explain the anti-reflective as well as the anti-soiling properties of the surface. The hierarchical surface roughness provides a graded refractive index as the light passes from a material of lower refractive index (n = 1.00) to that of higher refractive index (n = 1.35) [5]-[6]. These nano-scale surface features, composed of a low surface energy fluoropolymer, create a superhydrophobic surface [7] such that liquid water droplets form a high contact angle and are highly mobile compared to uncoated glass, with a sliding angle of 36.8°.

B. Anti-Reflective Properties

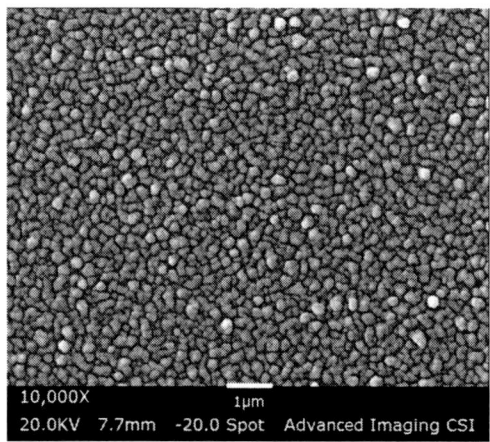

Fig. 4.   Scanning electron microscopy image of KleanBoost™ coated superhydrophobic surface, 10,000x scale bar = 1 μm.

The UV-vis transmission spectra of the uncoated and KleanBoost™ coated glass samples are shown in Fig. 5. The bare (uncoated) low-iron glass shows a maximum % transmission (%T) of 91.4% at 550 nm. Applying the anti-reflective/anti-soiling coating increases the %T by as much as 2.8% such that the maximum %T is 93.8%. The anti-reflective properties are roughly uniform over the entire visible spectrum.

C. Anti-Soiling Properties

Coating the glass substrates with dust leads to a decrease in %T for both bare glass as well as KleanBoost™ coated glass as shown in Fig. 6. For the bare glass samples, the %T decreases linearly with each successive dew-dust and bake application cycle. The average rate of transmission loss for the bare glass was 1.1 ±0.3% per dust application cycle measured over 5 dust

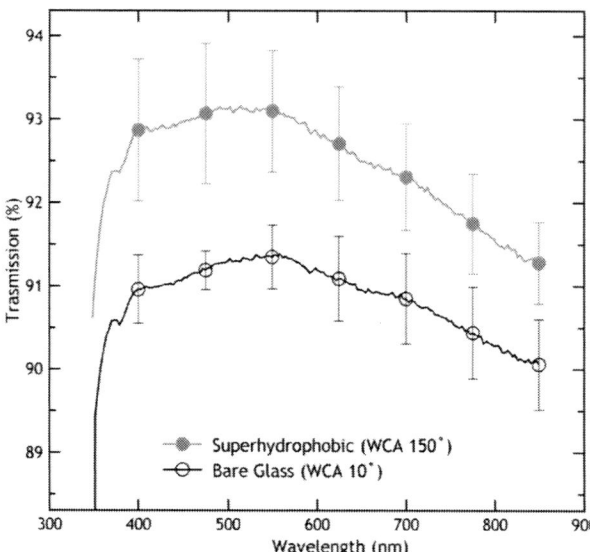

Fig. 5.   Average percent transmission (%T) spectra of low-iron glass with a superhydrophobic KleanBoost™ coating (contact angle of 150° - solid red circles) and bare, uncoated, glass (contact angle of 10° - black open circles).

cycles and averaged over 8 samples (3 measurements were taken per sample for a total of 24 measurements per application cycle). The total decrease in %T is 5.7 ±1.8% after five dew-dust and bake cycles.

In contrast, the KleanBoost™ superhydrophobic coated surfaces (WCA 150°) exhibit anti-soiling properties. On KleanBoost™ coated surfaces, the %T decrease occurs at a slower rate when exposed to the same dust deposition conditions. The average %T decrease is only 0.7% per dust application with a total %T decrease of only 3.4 ± 0.5% after 5 dew-dust and bake cycles, averaged over 4 samples (3 measurements per sample or 12 measurements per data point). The scattering of light by successive dust cycles decreases over time; the fifth dust deposition cycle resulted in only a 0.1% decrease in %T. To ensure that the same quantity of dust was applied to both bare glass and KleanBoost™ surfaces, coated and uncoated samples were placed side-by side on the cooling plate during dust deposition and the position and sequence of the coated samples was varied.

D. Cleaning with Water

The cleaning procedure using water droplets only had virtually no effect on the bare glass sample; the %T recovered only 0.2% ± 1.8% of it optical transmission to a %T value of 85.4% ± 1.8%. In contrast, washing essentially completely restores the transparency of the KleanBoost™ superhydrophobic coated glass surface, restoring 99.7% of its original %T as shown in Fig. 6. Moreover, the water recovery for KleanBoost™ is very high. Greater than >98% of the 3.0

mL of water dispensed on the surface was recovered, whereas less than 85% of the water is recovered after cleaning the uncoated glass surface.

IV. DISCUSSION

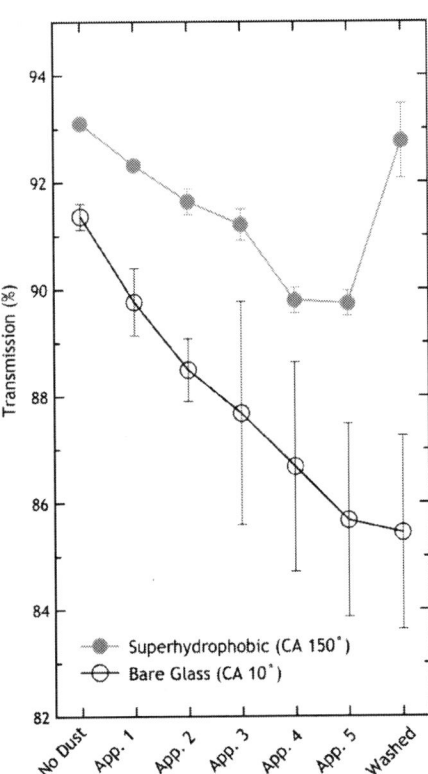

Fig. 6.   Transmission (%) at 550nm as a function of dust application cycles for low-iron glass samples with a superhydrophobic KleanBoost™ coating (contact angle of 150° - solid red circles) and uncoated glass samples (contact angle of 10° - open black circles).

KleanBoost™ superhydrophobic coatings impart both anti-reflective and anti-soiling properties to low-iron glass. The coating results in anti-reflective properties throughout the visible spectrum with a 2.4% gain in transmission at 550 nm. The superhydrophobic surface also reduces the effect of soiling in the presence of condensation and cleaning.

The anti-soiling mechanism is a direct consequence of the wetting properties of the surfaces –during both condensation as well as evaporation of water. Moisture condenses differently on bare glass compared to the superhydrophobic surface. As shown in Fig. 7a & 7b, water condenses into relatively large continuous pools of water on the hydrophilic uncoated glass surface, whereas discreet droplets, approximately 100 μm in diameter, form on the superhydrophobic surface. This filmwise vs. drop-wise condensation process on hydrophilic and

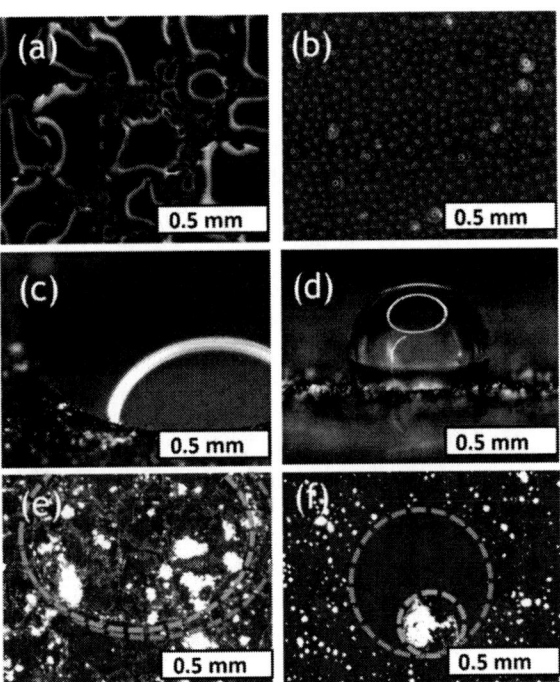

Fig. 7.   Optical micrographs of condensation and evaporation processes on dusted surfaces (a) water condensed on bare glass, (b) water condensed on KleanBoost™ glass; (c) 2μL droplet on bare glass and (d) 2μL droplet on KleanBoost™ glass. Dust remaining after 2μL droplet evaporation on (e) bare glass and (f) KleanBoost™ glass.

hydrophobic substrates is well known [8]. Once condensate forms, dust is deposited onto the surfaces where it preferentially adheres to regions of liquid water.

During drying, liquid water evaporates leaving regions of dust behind. The evaporation process, like condensation, proceeds differently on uncoated glass as compared to the superhydrophobic surface. On the hydrophilic glass the water-glass contact line is pinned due to strong interactions between H_2O and SiO_2. As the liquid evaporates, the liquid-solid contact area remains constant as the height of the droplet decreases. Evaporation from the surface of the liquid water layer facilitates convective currents that help disperse the dust over the surface area covered by water.

On the superhydrophobic surface, the liquid-solid contact line is mobile; thus, water droplets contract laterally along the surface while maintaining a relatively constant high contact angle. As the droplets evaporate, water effectively "herds" the dust particles keeping them within the ever-shrinking contact area of the drop. Eventually the droplets become sufficiently small that the droplets transition to the Wenzel state and the dust is left on a small fraction of the surface area originally covered by the condensed droplet.

Imaging the evaporation and dust-herding process is challenging due to the small size of the condensed droplets on the superhydrophobic surface. To visually confirm the dust-herding mechanism, we dispensed individual 2 µL droplets onto surfaces which were previously coated with 5 dew-dust and bake cycles as shown in Figs. 7c and 7d. The low contact angle of the droplet on hydrophilic glass (Fig. 7c) and high contact angle of the droplet on the superhydrophobic surface (Fig 7d) are readily apparent. After evaporation on the uncoated glass, dust is dispersed over the entire surface. Both regions of relatively large dust agglomerates, as well as regions of fine dust particles covering large areas of the surface can be readily seen in Fig. 7e. In contrast, essentially all of the dust captured in the droplet on the superhydrophobic surface has been herded into a centralized location leaving one large dust agglomerate. The remainder of the surface is clean – free from both agglomerates as well as regions coated with fine particles as shown in Fig. 7f.

A schematic of the dust herding process is shown in Fig. 8. On a superhydrophobic surface, corralling the dust into micron-sized piles leaves most of the surface free of dust and greatly reduces overall scattering of light. This dust herding mechanism accounts for the anti-soiling properties. The decrease in %T on the superhydrophobic is only 40% of the value on an uncoated glass surface in the presence of liquid water condensate.

Cleaning is also facilitated on a superhydrophobic surface. Water alone, without brushing, does not remove dust efficiently on a uncoated glass substrate. The water is more strongly attracted to the hydrophilic glass than the dry dust. As a result, the water forms channels, cleaning isolated regions, but avoiding the other areas of the surface. On a superhydrophobic surface, the water slides easily down the surface as individual droplets. These drops efficiently imbibe
the dust particles in its path, systematically sweeping the surface free of dust. After cleaning, 99.7% of the original %T was restored without using brushes.

V. CONCLUSIONS

The properties of KleanBoost™, an anti-reflective and anti-soiling coating on low-iron solar glass, were characterized. AR properties, comparable to commercially available ARCs, were

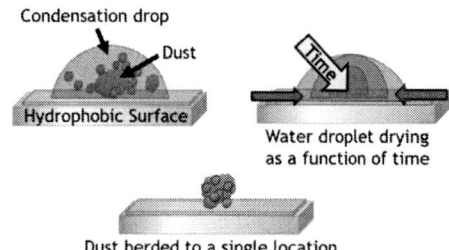

Fig. 8. Cartoon illustrating the dust herding mechanism observed on the superhydrophobic surface.

achieved by using a low index of refraction polymer combined with hierarchical roughness leading to 2.4% optical gain at 550 nm. This roughness, combined with the low surface energy of the fluoropolymer coating, results in dropwise condensation of water as well as highly mobile water droplets on the surface. As the simulated dew evaporates, the droplets contract laterally due to the high mobility of the liquid-solid contact line. This lateral motion herds the dust particles causing them to accumulate in the center of the droplets, leaving the majority of the surface dust-free. The overall effect reduces scattering and accounts for the higher %T. The low surface energy and hierarchical roughness of the superhydrophobic surface also reduces the dust-surface contact area and minimizes dust-surface chemical interactions, thereby enabling essentially complete (99.7%) cleaning of the surface with small quantities of water.

This work indicates that a superhydrophobic coating fabricated using a low index of refraction and low surface energy polymer can impart multi-functional properties: anti-reflectivity; anti-soiling; and facile cleaning, to solar cover glass. This type of coating could reduce the LCOE of solar photovoltaic energy by as much as $0.006/kWh through reduction in operation and maintenance costs. The efficient recovery of water would further increase the feasibility of cleaning in arid climates where water is scarce and expensive.

ACKNOWLEDGEMENT

The authors acknowledge support from NSF SBIR Phase II, PowerBridgeNY and the NSF Industry–University Cooperative Research Center for MetaMaterials. IN and GB acknowledge support from CSI and CUNY.

REFERENCES

[1] M. Bolen and N. Enbar, "Assessing Anti-Reflective and Anti-Soiling Coatings for Photovoltaic Modules," *Program of Technology Innovation*, Electric Power Research Institute, Sep-2016.

[2] H. K. Raut, V. A. Ganesh, A. S. Nair, and S. Ramakrishna, "Anti-reflective coatings: A critical, in-depth review," *Energy & Environmental Science*, vol. 4, no. 10, p. 3779, 2011.

[3] L. L. Kazmerski, A. S. A. C. Diniz, C. B. Maia, M. M. Viana, S. C. Costa, P. P. Brito, C. D. Campos, L. V. M. Neto, S. D. M. Hanriot, and L. R. D. O. Cruz, "Fundamental studies of the adhesion of dust to PV module chemical and physical relationships at the microscale," *2015 IEEE 42nd Photovoltaic Specialist Conference (PVSC)*, 2015.

[4] "NSRDB - TMY3: Alphabetical List by State and City," *NSRDB update - TMY3: Alphabetical List by State and City*. [Online]. Available: http://rredc.nrel.gov/solar/old_data/nsrdb/1991-

2005/tmy3/by_state_and_city_old.html. [Accessed: 19-Jan-2017].

[5] P. Buskens, M. Burghoorn, M. C. D. Mourad, and Z. Vroon, "Antireflective Coatings for Glass and Transparent Polymers," *Langmuir*, vol. 32, no. 27, pp. 6781–6793, Dec. 2016.

[6] H. K. Raut, V. A. Ganesh, A. S. Nair, and S. Ramakrishna, "Anti-reflective coatings: A critical, in-depth review," *Energy & Environmental Science*, vol. 4, no. 10, p. 3779, 2011.

[7] Feng, L., Li, S., Li, Y., Li, H., Zhang, L., Zhai, J., Song, Y., Liu, B., Jiang, L. and Zhu, D. (2002), Super-Hydrophobic Surfaces: From Natural to *Artificial. Adv. Mater.*, 14: 1857–1860. doi:10.1002/adma.200290020

[8] H. Y. Erbil, G. Mchale, and M. I. Newton, "Drop Evaporation on Solid Surfaces: Constant Contact Angle Mode," *Langmuir*, vol. 18, no. 7, pp. 2636–2641, 2002.

Multilayer-Grown Ultrathin Nanostructured GaAs Solar Cells

Boju Gai[1], Yukun Sun[3], Minjoo Lee[3], Jongseung Yoon[1,2,*]

[1]Department of Chemical Engineering and Materials Science, [2]Department of Electrical Engineering, University of Southern California, Los Angeles, California 90089, USA

[3]Department of Electrical and Computer Engineering, University of Illinois Urbana Champaign, Urbana, Illinois 61801, USA

Abstract — We present multilayer-grown ultrathin single-junction GaAs solar cells with uniform photovoltaic performance. Triple-stack ultrathin GaAs solar cells were grown by molecular beam epitaxy (MBE) using beryllium as a p-type dopant. Microscale solar cells with a vertical contact configuration exhibited excellent uniformity (<3%) of performance among top, middle, and bottom device layers due to the suppressed diffusion of p-type impurity and reduced time for epitaxial growth. Microcells implemented with hexagonally periodic TiO_2 nanoposts and metal reflector provided significantly improved performance owing to the optimized bifacial photon management, where 20.8% efficiency was obtained from 420-nm thick GaAs solar cells.

Index Terms — III-V; Photovoltaics; Multilayer epitaxy; Light trapping; Transfer printing

I. INTRODUCTION

Large-scale deployment of GaAs solar cells remains significant challenge owing to the prohibitively high materials cost [1]. While epitaxial liftoff (ELO) has been proposed to lower the cost of III-V solar cells by reusing the growth wafer multiple times, the feasibility of this method for a large number (e.g. > 20) of substrate reuses has not been fully proven yet [2]. In this regard, an alternative approach called 'multilayer epitaxial assemblies' by Yoon and co-authors has been reported to circumvent these limitations of conventional ELO, where multiple number of solar cell stacks are grown on a single growth wafer with sacrificial layers inserted between device stacks [3]. Although conceptually promising, this approach has also faced difficulties of maintaining uniform photovoltaic performance between devices layers grown in different sequences [3]-[4]. In triple-stack, n-on-p type GaAs solar cells, the performance showed systematic degradation owing to the diffusion of p-type dopant, zinc [3]. In triple-stack p-on-n type GaAs solar cells with carbon as a p-type dopant, carbon-related point defects and hydrogen passivation were responsible for the degraded performance in the middle and bottom layers [4]. Here we present novel materials platform of multilayer-grown GaAs solar cells that can circumvent above-described difficulties by exploiting ultrathin device configuration, beryllium as a p-type dopant, and nanoscale photon management enabled with periodic TiO_2 nanoposts and metallic back-side reflector [5].

II. RESULTS AND DISCUSSION

Triple-stack ultrathin GaAs solar cells were grown on a GaAs wafer by solid-source MBE system, where the individual device stack is composed of n^+-GaAs top contact (20 nm), n-AlInP window (20 nm), n-GaAs emitter (50 nm), p-GaAs base (250 nm), p^+-AlGaAs back surface field (BSF) (50 nm), and p^+-GaAs bottom contact (50 nm) (Figure 1). Details of materials specifications appears in our previous report [5]. The fabrication of microscale solar cells began with the formation of mesa (~500 x 500 μm^2)

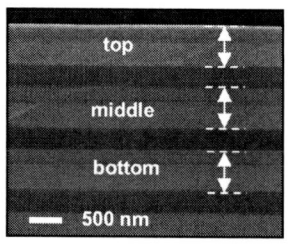

Figure 1. Cross-sectional SEM image of the triple-stack GaAs solar cells.

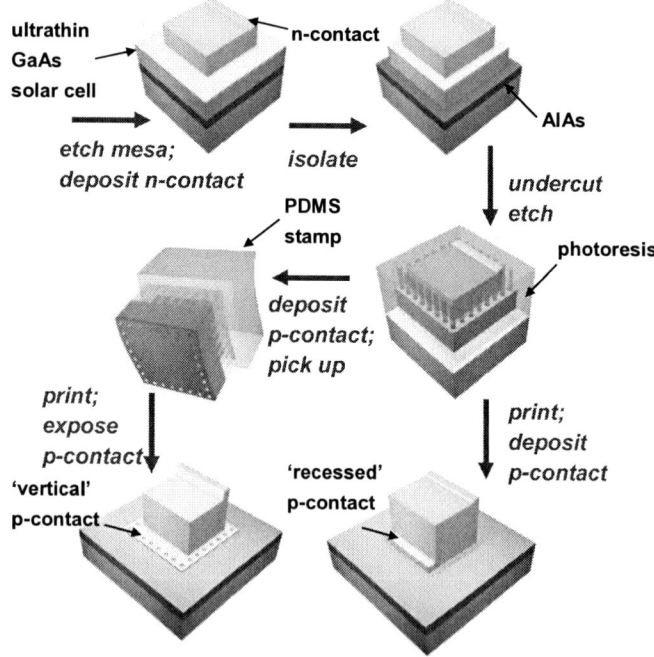

Figure 2. Schematic illustration of processing steps to fabricate triple-stack ultrathin GaAs solar cells configured with vertical- and recessed-contact configuration.

structure by photolithography and wet chemical etching using a mixture of citric acid and hydrogen peroxide (Figure 2). N-type metal contact (Pd/Ge/Au) was deposited by electron beam evaporation and liftoff, followed by the formation of second mesa structure to produce isolated arrays of GaAs microcells. After the undercut etching with a polymeric anchor structure, the arrays of microcells were released by an elastomeric stamp made of poly(dimethyl siloxane) (PDMS). In this study, both 'vertical' and 'recessed' contact designs were implemented. For the vertical contact, p-type metal contact (Pt/Ti/Pt/Au) was deposited on the entire bottom surface of lifted cells on the PDMS stamp. The lifted cells were printed on a glass substrate using a thin layer of photocurable polymer as an adhesive, followed by the exposure of p-type metal contact (for vertical contact design) through wet chemical etching or the deposition of p-type metal contact (for recessed contact design).

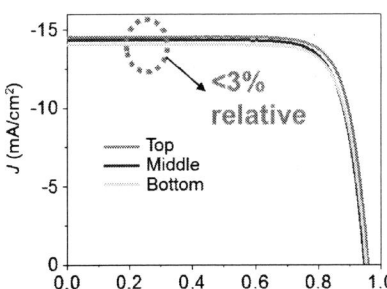

Figure 3. Representative current density (J)-voltage (V) curves of ultrathin GaAs solar cells on the source wafer, measured under simulated AM1.5G solar illumination.

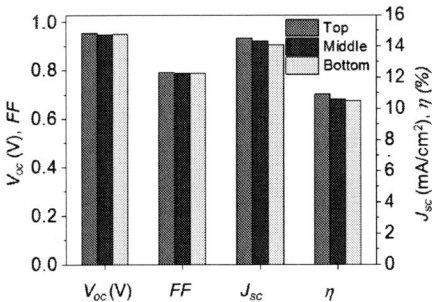

Figure 4. Corresponding photovoltaic device characteristics, extracted from Figure 3.

The photovoltaic performance of triple-stack ultrathin GaAs solar cells was characterized under AM1.5G simulated solar illumination (1000W/m^2) on wafer as depicted in Figure 3, where the average short-circuit current densities were 14.6, 14.5, 14.3 mA/cm^2, open-circuit voltages were 0.950, 0.942, and 0.941 V, fill factors were 0.78, 0.77, 0.78, and efficiencies were 10.8, 10.6, 10.5%, respectively (Figure 4). The slight variation (< 3% relative) of performance is within

the experimental error range. The internal and external quantum efficiency data also support these results.

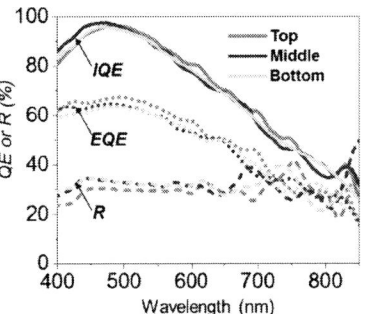

Figure 4. Internal (IQE) and external (EQE) quantum efficiencies and specular reflectance (R) spectra measured from triple-stack GaAs solar cells.

The secondary ion mass spectrometry (SIMS) data showed that the average concentration of beryllium slightly increased from top (~3.0 x 10^{17} cm^{-3}) to bottom (~7.5 x 10^{17} cm^{-3}) device layers due to the thermal diffusion (Figure 5), which

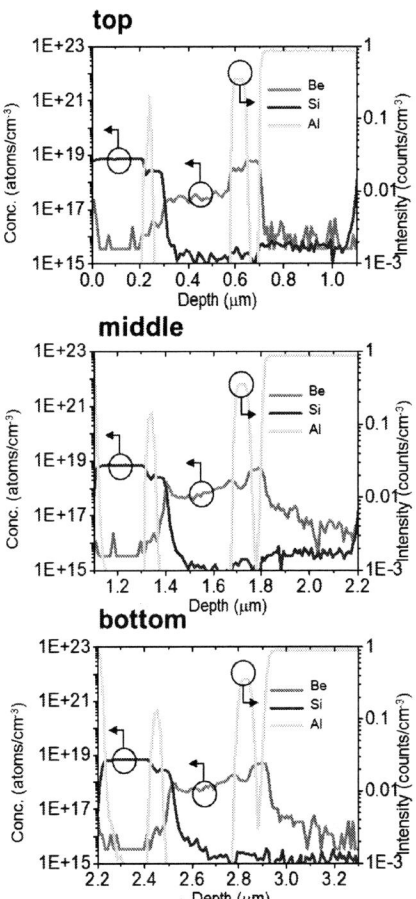

Figure 5. Atomic concentration profile of beryllium (red line) and silicon (blue line) in triple-stack GaAs solar cells, measured by SIMS.

is much smaller than in previously reported zinc-doped system and did not translate to the degraded performance.

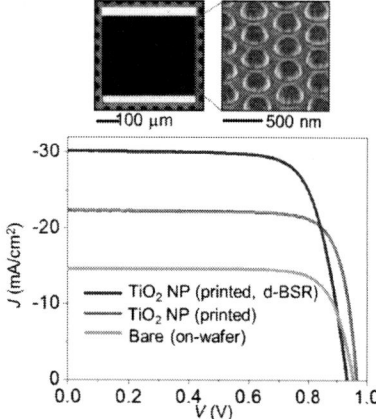

Figure 6. Representative JV curves of ultrathin GaAs solar cells with (red) and without (orange) TiO2 nanoposts (NPs) after the printing and on the growth wafer, respectively, where the p-type contact is made in a vertical configuration. The violet line shows JV curve for nanostructured GaAs solar cells with recessed contact configuration.

To compensate for the incomplete absorption of long wavelength photons, hexagonally periodic TiO_2 nanoposts (NPs) were implemented on top of the window layer by softimprint lithography and dry etching, following previously reported procedures [6]. As summarized in Figure 6, the short-circuit current density increased from 14.6 (for bare GaAs cells on the wafer) to 22.3 mA/cm^2 (for nanostructured GaAs cells after the printing) owing to the combined effects of bifacial photon management such as antireflection, diffraction, and light trapping. For the recessed contact design, the current density of printed microcells was even further increased to ~30.2 mA/cm^2, which is attributed to the effect of waveguiding and concentration for photons that are incident outside the cell area [7].

IV. CONCLUSION

We have demonstrated the growth and fabrication of triple-stack ultrathin (emitter + base = 300 nm) GaAs solar

cells having uniform (<3% relative) photovoltaic performance between device layers grown in multilayer epitaxy. Suppressed diffusion of p-type dopant (i.e. beryllium) and reduced time of epitaxial growth associated with ultrathin device configuration were responsible for the preservation of materials properties and device performance. Bifacial photon management employing TiO_2 NPs and back-side reflector together with epitaxial design to facilitate multipass light trapping collectively contributed to the significant increase of performance, where a 17.2 and 20.8% efficiencies were achieved in 420-nm-thick GaAs solar cells printed on glass with vertical and recessed contact configurations, respectively.

REFERENCES

[1] M. Woodhouse, A. Goodrich, "A Manufacturing Cost Analysis Relevant to Single- and Dual-Junction Photovoltaic Cells Fabricated with III-Vs and III-Vs Grown on Czochralski Silicon" *NREL report*, PR-6A20-60126 (2014)

[2] E. Yablonovitch, T. Gmitter, J. Harbison, J. P.; R. Bhat, "Extreme Selectivity in The Lift-Off of Epitaxial GaAs Films." *Appl. Phys. Lett.*, vol. 51, pp. 2222-2224, (1987)

[3] J. Yoon, S. Jo, I. Chun, I. Jung, H. Kim, M. Meitl, E. Menard, X. Li, J. Coleman, U. Paik, J. Rogers, "GaAs Photovoltaics and Optoelectronics Using Releasable Multilayer Epitaxial Assemblies." *Nature*, vol. 465, pp. 329-U80, (2010)

[4] D. Kang, S. Arab, S. Cronin, X. Li, J. Rogers, J. Yoon, "Carbon-doped GaAs Single Junction Solar Microcells Grown in Multilayer Epitaxial Assemblies." *Appl. Phys. Lett.*, vol. 102, pp. 253902, (2013)

[5] B. Gai, Y. Sun, H. Lim, H. Chen, J. Faucher, M. Lee, J. Yoon, "Multilayer-Grown Ultrathin Nanostructured GaAs Solar Cells as a Cost-Competitive Materials Platform for III-V Photovoltaics" *ACS Nano*, 11, pp. 992-999, (2017)

[6] S. Lee, A. Kwong, D. Jung, J. Faucher, R. Biswas, L. Shen, D. Kang, M. Lee, J. Yoon, "High Performance Ultrathin GaAs Solar Cells Enabled with Heterogeneously Integrated Dielectric Periodic Nanostructures." *ACS Nano*, vol. 9, pp. 10356-10365, (2015)

[7] S. Lee, R. Biswas, W. Li, D. Kang, L. Chan, J. Yoon, "Printable Nanostructured Silicon Solar Cells for High-Performance, Large-Area Flexible Photovoltaics." *ACS Nano* vol. 8, pp. 10507-10516, (2014)

978-1-5090-5606-4/17 $31.00 © 2017 IEEE

Laboratory Studies of Particle Cementation and PV module Soiling

Craig L. Perkins, Matthew Muller, and Lin Simpson

National Renewable Energy Laboratory, 15013 Denver West Parkway, Golden CO 80401 USA

Abstract — Soiling of photovoltaic modules by aerosols, carbonaceous materials, and dust is increasingly being recognized as an important efficiency loss mechanism. Significant soiling can involve cementation, a term that currently is only loosely defined within the PV community. In this contribution, we provide a robust and testable definition of cementation based on the ability of cemented particles to transmit shear. A novel system for experimentally probing cementation based on a quartz crystal resonator is introduced, along with initial results. Finally, we describe future efforts involving this system and how these laboratory results might be utilized to help mitigate real-world soiling losses.

I. INTRODUCTION

Efficiency losses in photovoltaic modules due to soiling by airborne particulates is estimated to increase the levelized cost of energy by up to ¢1/kW-hr, and soiling remains a significant problem despite years of study. The problem is complex. Soiling of typical glass surfaces depends on a large number of factors that include particle size, roughness, and chemical composition. It is clear that humidity and simply time-in-contact are important in particle adhesion to glass and other surfaces. Power loss due to soiling necessarily must take into account soiling layer optical properties as well as the relative rates at which particles arrive and are removed.

The difficulty of particle removal is expected to be related to the degree to which particles are bound to the surface, and for this reason a number of studies have focused on determining both theoretically and experimentally the adhesion forces between model particles and various surfaces.[1]–[3] Atomic force microscopy (AFM)-based experiments have used the AFM tip itself as a proxy for a dust particle, and typically measure forces and energies that are required to pull particles off surfaces in a direction normal to the surface. Recent work by Cho et al suggest however that particle removal by hydrodynamic drag requires energies that are about 100 times lower than the adhesion energies measured by AFM and also calculated theoretically.[4] We note here that common module cleaning techniques such as washing, air blasting, and brushing are all methods that rely on drag rather than tensile stress to remove particulates.

The results in [4] were explained on the basis the low energies required to initiate rolling of particles and overcoming the static friction forces between particles and the surfaces of interest. One relatively new method of studying friction forces between surfaces and particles relies on quartz crystal resonators, a type of sensor commonly referred to as a quartz crystal microbalance.[5], [6] This technique relies on the high sensitivity of a QCM to the shear modulus of material that is adsorbed upon the sensor. In this paper we demonstrate how a QCM can be used to probe the interactions of particles with a model glass surface and how the transition from sliding contact to adhered particles depends on humidity cycling. Based on this transition and the ability of particles to transmit shear force, we propose a robust and testable definition of cementation as it applies to dust on glass surfaces.

II. EXPERIMENTAL

Our QCM flow cell apparatus has been described previously.[7] For these experiments, the apparatus was used with two gas streams: dry nitrogen and water-saturated nitrogen produced with a home-made bubbler. A SiO_2-coated resonator was used a model for the silica-based glass and anti-reflection surfaces commonly found in PV modules. Measurements were conducted at room temperature. In order to make connections with our on-going AFM-based experiments that use SiO_2 spheres as proxies for dust particles, the set of particles that we have studied initially include 50 μm diameter glass spheres. We also report measurements on standard Arizona road dust and halite. Particles were applied to the resonator using a cotton swab and in a manner such that the resonator surface was not touched with the swab.

III. RESULTS

Figure 1 shows output from the QCM with the resonator in an open configuration that allowed sequential deposition of Arizona road dust. Frequency changes plotted against the left axis are commonly used to determine mass gain per area.

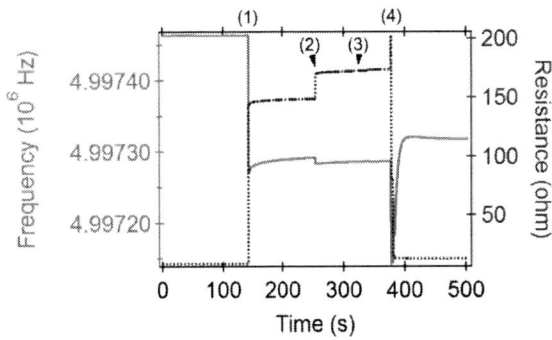

Figure 1. Changes in frequency and resistance with dust depositions in steps 1-3 and air blast (4).

978-1-5090-5606-4/17 $31.00 © 2017 IEEE

Less common is use of the crystal resistance values that are shown plotted against the right hand axis. Increases in crystal resistance are a measure of shear wave damping in adsorbed materials: for rigidly bound thin films of inorganic materials, for example as measured during CdS chemical bath deposition, resistance changes are small (<10 ohm). In contrast, a large resistance change is observed upon the initial deposition of dust (point 1). Also observed is a sharp decrease in crystal frequency that is usually associated with an increase in mass of the resonator. Subsequent dust depositions in steps (2) and (3) show smaller and smaller changes in frequency and resistance, indicating that dust accumulation under these conditions (dry nitrogen) is self-limiting. In step (4), an air blast was used to remove dust from the resonator. After the transient from the air blast, it can be seen that crystal resistance drops to near its initial value of eight ohms, indicating that shear wave damping is low. The crystal frequency remains below its initial value however. The combination of these two measurements indicates that the air blast removed loosely bound particles, leaving behind only those rigidly adhered to the resonator surface.

Some aspects of initial measurements such as the one depicted in Figure 1 warranted further examination. For example, simple conversion of the observed frequency change in Fig.1 to mass/area resulted in very low values of about 10 ng/cm^2, seemingly not consistent with a dust layer that was clearly observable by eye. In addition, the normally very stable resonator frequency is seen to drift upwards for minutes after initial dust applications. To remove uncertainties associated with the complex composition and size distribution of Arizona road dust as well as with humidity and temperature changes that were possible with the QCM in an open configuration, we examined particles with known size, shape, and composition and we used the QCM in a flow cell configuration as depicted in [7]. To allow comparison with AFM experiments that utilized SiO2 and glass spheres as probes of particle-surface interactions, we chose 50 μm diameter glass spheres.

Figure 2 shows three traces of frequency versus time for a control consisting of the clean, unloaded sensor (dark yellow), the sensor loaded with 50 μm glass spheres and sitting in static (not-flowing) humidified nitrogen (black), and the sensor loaded with glass spheres with dry-humid cycling (gray). Humidity cycling for the control and spheres was done with one minute of water-saturated nitrogen followed by five minutes of dry nitrogen. Inspection of the curves reveals several interesting phenomena. For one, comparison of the black and yellow curves at t=0 reveals an initially puzzling observation, namely that deposition of glass spheres increased rather than decreased the resonant frequency of the QCM. A review of the literature revealed the reason for this behavior: micron-sized particles that are poorly adhered to a QCM are inertially clamped in space while the crystal moves beneath them.[6] Frictional forces between the particles and the crystal

essentially increase the stiffness of the oscillator, thereby increasing its resonant frequency.

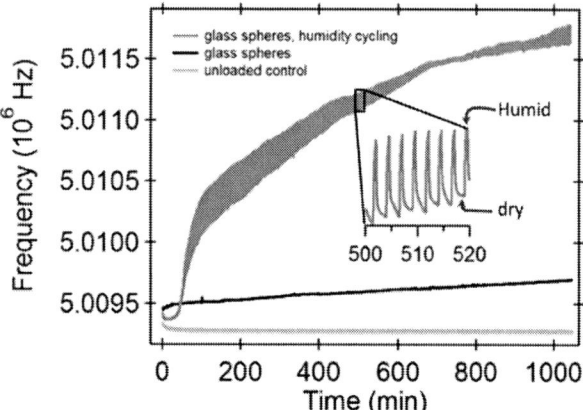

Figure 2. QCM output for a control (yellow trace), the sensor loaded with 50 micron glass spheres in non-flowing humidified nitrogen (black trace), and with glass spheres and humidity cycles (gray).

Another intriguing feature of the data in Fig. 2 is the upward drift in frequency with time observed in both data sets. Again, previous work provides a qualitative answer: capillary aging, or an increase in friction coefficient as a function of time in contact, has been observed for a wide variety of materials.[8] The process involves thermally activated formation of capillary bridges between particle asperities and a surface. It can be seen that humidity cycling (gray trace) increases the rate at which friction between particles and the resonator increases. The inset shows that the sensor is picking up the small mass changes associated with changes in the equilibrium water layer thickness during each wet-dry cycle.

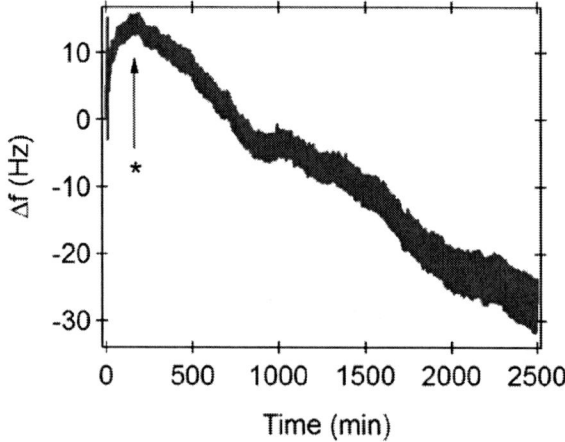

Figure 3. QCM output with halite particles and humidity cycling. The * indicates the onset of cementation.

Figure 3 shows another data set taken with halite particles placed on the QCM and with humidity cycling. Data are plotted as a change in frequency rather than absolute frequency to allow comparison between sensors that have slightly different resonant frequencies. As in the case with the inert glass spheres, frequency changes are initially positive, indicating that frictional forces between the salt particles and the sensor have increased. After about 25 humidity cycles at t=175 minutes (indicated with *), the frequency starts to decrease and continues to do so until the measurement was stopped after about two days. We interpret this change in sign in the Δf vs time curve as the onset of particle cementation, where some fraction of the salt particles become rigidly adhered such that they oscillate in phase with the quartz resonator.

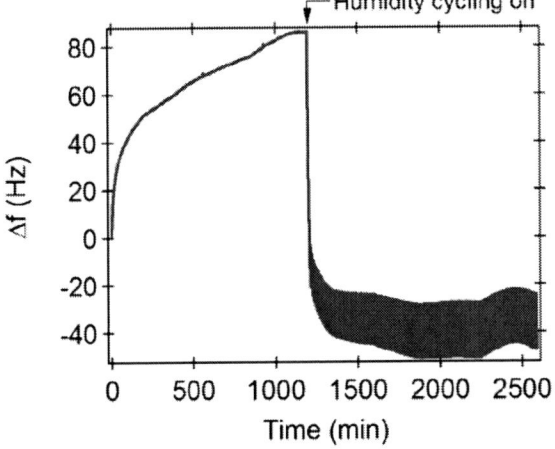

Figure 4. Palygorskite with and without humidity cycling.

Figure 4 shows the behavior of a magnesium-bearing clay mineral, palygorskite (< 45 μm B-80 bleaching clay, Jaxon Filtration) with and without humidity cycling at room temperature. Recently palygorskite has been implicated in cementation reactions occurring in the Middle East.[9] Dry palygorskite loaded onto the sensor shows the aging behavior seen with glass spheres and other materials in which friction between the quartz sensor and the clay particles increases with time. Turning on humidity cycling causes an immediate cementation reaction, faster even than in the case of the halite shown in Figure 3.

IV. DISCUSSION

It is clear that using a QCM to probe particle-surface interactions has both advantages and disadvantages. On the one hand, a typical set-up allows the measurement of a large number of particles relative to single particle AFM-based measurements. By simultaneous measurement of crystal frequency and resistance, one is able to ascertain whether or not actual cementation has taken place. In this context, we use the term cementation in the same way that cement researchers do: it is a type of phase transition commonly referred to as a percolation transition in which individual particles are transformed into a mechanically monolithic solid that is able to transmit shear forces.[10] Having a well-defined definition for cementation, one that is relevant to understanding particle removal by shear processes such as washing, air blasting, and brushing, should help future work especially in cases where it may be unclear whether or not cementation has taken place.[9]

A disadvantage to using this system is that opposing effects on frequency changes (decreases with small, strongly bound particles, and increases with large, weakly bound particles) means that quantitative extraction of mass loading and frictional coefficients is difficult for real dust having a large dispersion in particle size. Furthermore, for materials with a density of silica, this change-over in behavior occurs at several tens of microns in diameter, a size critical in module soiling. Future work with this system will implement one solution to this issue by using particles of known size dispersion and composition. With known particle size, it should be possible to assay common minerals for their propensity to undergo cementation phase transitions, anti-soiling coatings, and the dependence of cementation processes on temperature, time, humidity cycles.

V. CONCLUSIONS

We have repurposed a flow cell QCM apparatus to study the interactions of micron-sized particles with a model glass surface, silica. We find that Arizona road dust application without humidity cycling is self-limiting, and that only some of this test dust is strongly adhered. Using a simpler system of 50 μm glass spheres, we find consistent with prior work that adsorption of large weakly adhered particles causes an increase in the resonant frequency of a QCM because of frictional forces between the particles and the resonator. Also consistent with prior work, we see an increase over time in the coefficient of friction and that humidity cycling causes a large increase in the rate at which friction increases. We show that our set-up can detect the onset of cementation and provide a robust, testable definition for that process. Further work with this system should provide guidance for PV module cleaning based on dust composition, module surface coating, as well as recommendations on cleaning intervals based on the anticipated onset of cementation.

REFERENCES

[1] L. L. Kazmerski *et al.*, "Fundamental studies of the adhesion of dust to PV module chemical and physical relationships at the microscale," presented at the 2015 IEEE 42nd Photovoltaic Specialist Conference, PVSC 2015, 2015.

[2] J. a. S. Cleaver and J. W. G. Tyrrell, "The Influence of Relative Humidity on Particle Adhesion – a Review of Previous Work

and the Anomalous Behaviour of Soda-lime Glass," *KONA Powder Part. J.*, vol. 22, pp. 9–22, 2004.

[3] X. Xiao and L. Qian, "Investigation of Humidity-Dependent Capillary Force," *Langmuir*, vol. 16, no. 21, pp. 8153–8158, Oct. 2000.

[4] K. L. Cho *et al.*, "Shear-Induced Detachment of Polystyrene Beads from SAM-Coated Surfaces," *Langmuir*, vol. 31, no. 40, pp. 11105–11112, Oct. 2015.

[5] S. Berg and D. Johannsmann, "High Speed Microtribology with Quartz Crystal Resonators," *Phys. Rev. Lett.*, vol. 91, no. 14, Oct. 2003.

[6] A. Pomorska, D. Shchukin, R. Hammond, M. A. Cooper, G. Grundmeier, and D. Johannsmann, "Positive Frequency Shifts Observed Upon Adsorbing Micron-Sized Solid Objects to a Quartz Crystal Microbalance from the Liquid Phase," *Anal. Chem.*, vol. 82, no. 6, pp. 2237–2242, Mar. 2010.

[7] C. L. Perkins, "Molecular Anchors for Self-Assembled Monolayers on ZnO: A Direct Comparison of the Thiol and Phosphonic Acid Moieties," *J. Phys. Chem. C*, vol. 113, no. 42, pp. 18276–18286, Oct. 2009.

[8] L. Bocquet, E. Charlaix, S. Ciliberto, and J. Crassous, "Moisture-induced ageing in granular media and the kinetics of capillary condensation," *Nature*, vol. 396, no. 6713, pp. 735–737, Dec. 1998.

[9] A. Boumiz, C. Vernet, and F. Cohen Tenoudji, "Mechanical properties of cement pastes and mortars at early ages," *Adv. Cem. Based Mater.*, vol. 3, no. 3–4, pp. 94–106, 1996.

[10] K. Ilse, M. Werner, V. Naumann, B. W. Figgis, C. Hagendorf, and J. Bagdahn, "Microstructural analysis of the cementation process during soiling on glass surfaces in arid and semi-arid climates," *Phys. Status Solidi RRL – Rapid Res. Lett.*, vol. 10, no. 7, pp. 525–529, Jul. 2016.

Virtual Substrates for Low-Cost High Efficiency III-V Photovoltaics

Sean J. Babcock[1], Marlene L. Lichty[1], Shankar Karki[1], Grace Rajan[1],
Sylvain Marsillac[1], Elisabeth L. McClure[2], Seth M. Hubbard[2], and Christopher G. Bailey[1]

[1]Old Dominion University, Norfolk, VA, 23529, USA
[2]Rochester Institute of Technology, Rochester, NY, 14623, USA

Abstract—The use of the low-cost thin-film vapor-liquid-solid (TF-VLS) method in the manufacturing of III-V solar cell substrates has the potential to provide a lightweight, flexible, and cheaper alternative to traditional III-V substrates typical of state-of-the-art power generation technology. The TF-VLS process has recently been shown to produce high optoelectronic quality polycrystalline InP on lightweight flexible metal foils. In this work, the novel TF-VLS method is employed to grow InP and InAs using following structures: Mo(foil)/InP, Si(wafer)/Mo/InP, and Mo(foil)/InAs. As a result of InP trials, XRD measurements have identified the presence of polycrystalline InP peaks and the absence of In peaks, signifying full conversion from In to InP for both Mo(foil) and Si(wafer)/Mo(sputtered) substrates. Photoluminescence measurements showed that both samples emit near single crystal InP bandedge of 1.34 eV with FWHM values in close agreement with each other. The TF-VLS method was expanded to InAs with initial trials indicating polycrystalline InAs XRD peaks and the absence of In peaks indicating full conversion from In to InAs.

Index Terms—molybdenum, indium, phosphide, asrnide, low cost, silicon dioxide, photovoltaic cells.

I. INTRODUCTION

Today's terrestrial solar cell market is comprised predominately of silicon based devices due to silicon's abundance in nature (second only to oxygen in the earth's crust). However, the efficiency of silicon technology has plateaued with the world record cell efficiency at 25.6% [1]. Solar arrays typical of today's space missions take advantage of the high efficiency and performance that are characteristic of III-V photovoltaics (PV), but at a much higher cost compared to silicon. A III-V triple-junction device currently holds the world record cell efficiency of 46% under (AM1.5) 508 suns [1]. In order for the terrestrial solar cell market to break away from plateaued silicon technology and take advantage high-efficiency devices, innovative growth methods must be implemented to combat the primary factor influencing III-V device costs, expensive epitaxy growth processes. The work discussed here aims to replace expensive epitaxially grown substrates currently used to grow III-V devices with a low-cost, lightweight, flexible alternative using the novel thin-film vapor-liquid-solid (TF-VLS) growth method. This method has recently been used to successfully grow high optoelectronic quality polycrystalline InP on flexible metal foils and has the potential to be expanded to include a wide array of binary and ternary materials which would allow for bandgap/lattice constant tuning.

In 2014, the National Renewable Energy Laboratory (NREL) published a study on the manufacturing costs of III-V photovoltaics [2]. The study analyzed and broke down the cost (in $/W) of each layer in a single junction GaAs device into material, labor and maintenance, utility, and depreciation costs (assuming 25% device efficiency). The study showed that the substrate layer accounted for 62% of overall device cost after peeling off (epitaxially lifting) and reusing the substrate 20 times. In comparison, the GaAs base layer grown via MOVPE accounted for only 25% of device cost. With substrate costs this high, there is an obvious effort towards the development of engineering solutions that will drive down overall device cost.

In addition to cost reduction, this research aims to use the TF-VLS technique to target materials of specific lattice constants, which translate into bandgap targeting. Currently, commercially available substrates are limited to III-V binary materials. Multijunction PV architectures could benefit from PV substrates with alternative lattice constants not currently commercially available. This could be used in PV technology to target specific solar spectrums that vary geographically on Earth and even more significantly in space. For example, solar cells deployed to the Mars surface would experience a solar spectrum different from that of AM0 due to the effects of atmospheric dust filtering solar radiation [3]. A solar array could be equipped with devices with bandgaps that are optimized to absorb solar radiation specific to an environment. TF-VLS technology could also be used to grow lattice matched devices in the effort to push the boundaries of efficiency. For example, device modeling has identified a 5.8 Å lattice matched multijunction device to exceed 51% efficiency under 100-sun illumination [4]. This was demonstrated by using specific compositions of InGaAs, InGaAsP, and InAlAs to target bandgaps of 0.94, 1.39, and 1.93 eV, respectively, to optimize solar spectrum absorption. Therefore, in order to simultaneously achieve lattice-constant tuned substrates and reduce the cost of III-V multijunction architecture we will employ TF-VLS fabrication method to grow polycrystalline thin-film III-V materials on lightweight flexible metal foils.

II. EXPERIMENTAL

The TF-VLS method (first described by Kapadia, et al) is an adaptation of the VLS (vapor-liquid-solid) method traditionally used to grow nanowires. [5]. This work has

978-1-5090-5606-4/17 $31.00 © 2017 IEEE

shown that TF-VLS can be used to successfully grow poly-crystalline thin-film InP and InAs on molybdenum substrates (Mo(foil) and Si(wafer)/Mo(sputtered)). The TF-VLS based polycrystalline InP film growth process, as illustrated in Fig. 1, begins by depositing a layer of indium (0.2-2 μm) on

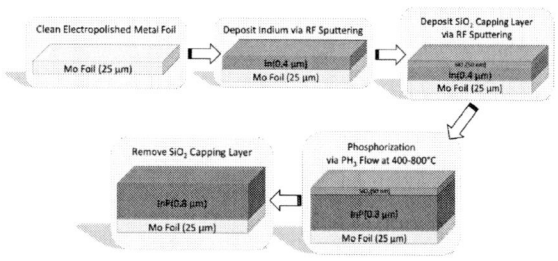

Fig. 1: TF-VLS based polycrystalline InP process steps.

electropolished molybdenum foil (~25 μm) via RF sputtering. An SiO$_2$ (50 nm) capping layer is then be applied via electron beam deposition. The capping layer allows the liquid indium to maintain a planar geometry by preventing it from dewetting during the phosphorization step. It is important to note that these experiments omitted the SiO$_2$ capping layer typically included in the TF-VLS process to simplify the experiment and to have a comprehensive baseline for future SiO$_2$-capped studies. Using a graphite sample susceptor in a laminar flow quartz tube style MOCVD reactor, the samples were then heated to 400-800°C, which is well beyond the melting temperature of indium (157°C). After temperature stabilization, phosphine gas is introduced into the chamber. Phosphorous then diffuse through the SiO$_2$ capping layer down to the molybdenum surface which allows for the precipitation of solid InP crystals. Nucleation prefers the Mo/In interface and does not occur in the bulk of the In before coming in contact with the Mo surface. As more P atoms diffuse through the capping layer, the crystals continue to grow. This continues until all of the In is converted into InP and the crystals converge into each other forming grain boundaries of a continuous polycrystalline indium phosphide film. The time it takes for phosphorization to complete is on the order of tens of minutes and depends on the thickness of the initial In layer and temperature. Finally, SiO$_2$ is then etched away using hydrogen fluoride.

III. RESULTS

Following the TF-VLS processing steps, SEM, XRD, and PL measurements were performed on the resulting Mo(foil)/InP and Si(wafer)/Mo/InP stacks. Fig. 2 shows SEM images of TF-VLS grown (at 750°C for 20 mins) InP on two types of substrates: Mo foil (a ,b) and Si(wafer)/Mo(sputtered, 700 nm) (c ,d). Again, the SiO$_2$ step has been omitted in all samples described in this narrative. The latter experiment was designed to determine the effects of the TF-VLS method on sputtered Mo with the added benefit of simplified characterization such as cross-section SEM while keeping within the spirit of the low-cost initiative. Using a rigid Si

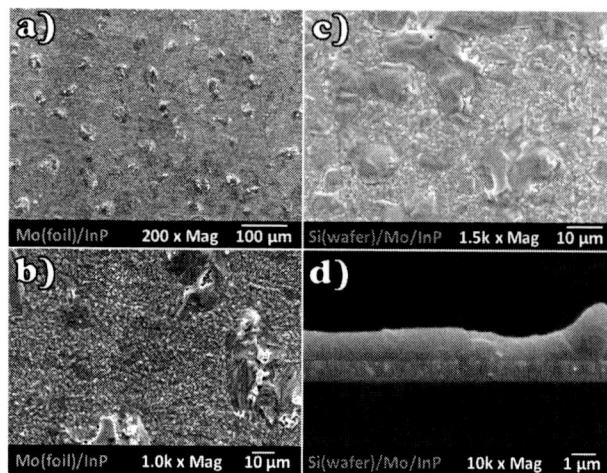

Fig. 2: SEM images of TF-VLS grown InP on Mo foil: a) 200X, and b) 1,000X, and on Si(wafer)/Mo(sputtered, 700 nm): c) 1,500X, d) 10,000X cross-section.

wafer as a substrate not only makes cleaving for cross-section measurements possible, but it also allows for more uniform heat transfer during phosphorization compared to Mo foil due to the foils non-uniform contact with the susceptor. The top image (Fig. 2a) shows uniform surface coverage of indium phosphide islands with a density of roughly 200 islands per square millimeter. A closer look (Fig. 2b) reveals that the islands are surrounded by very small indium phosphide grains fully concealing the Mo foil substrate. Fig. 2c shows that the planar surface morphology looks similar to that of the Mo foil substrate in that they both demonstrate a rough island covered surface. However, this sample does appear to have somewhat smoother and more continuous islands than the Mo foil sample but with slight Mo surface exposure. Had the SiO$_2$ capping layer been used, it would be expected that the island morphology would be that of a uniformly smooth surface, thereby eliminating the island morphology due to the capping layer's role of maintaining planar In geometry during phosphorization. To improve upon the small grain sizes seen in both samples, the use of electron beam deposition to apply the In layer has shown to provide for a smoother initial In surface (with almost indistinguishable grain sizes) when compared to sputtering. For this reason, ebeam deposited indium will be employed in future experiments. Fig. 2d shows cross-section of a mostly uniform indium phosphide layer on sputtered Mo. However, this image was captured at a non-island location where the indium phosphide fully covered the sputtered Mo surface. The image is encouraging because it demonstrates that InP crystal growth can occur in a planar format without deformations or holes throughout the layer.

Fig. 3a shows powder XRD results of TF-VLS grown InP samples shown in Fig. 2. The XRD peaks are consistent with that of polycrystalline InP with peaks corresponding to multiple crystal planes. What is most significant about this

978-1-5090-5606-4/17 $31.00 © 2017 IEEE

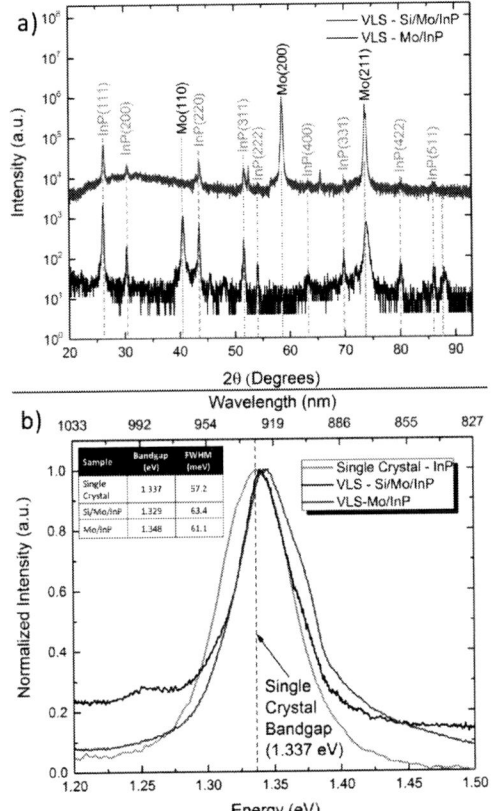

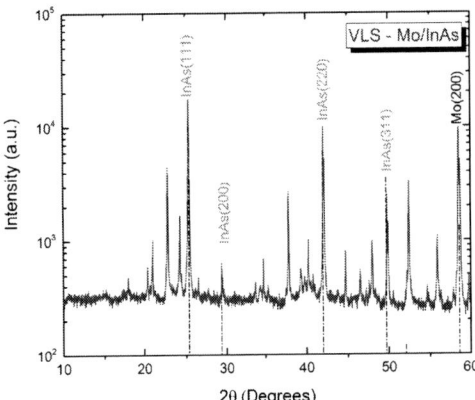

Fig. 4: Powder XRD results of TF-VLS grown InAs on Mo foil.

Fig. 3: TF-VLS grown InP on Mo(foil) (blue line) and Si(wafer)/Mo(sputtered, 700 nm) (black line) substrates a) Powder XRD, b) photoluminescence.

IV. SUMMARY

In this work, it was shown that TF-VLS InP can be achieved using Mo foil and sputtered Mo on Si wafers as substrates without the use of the SiO$_2$ capping layer. Future efforts will focus on the use of the capping layer to provide for a more planer InP surface geometry. Also shown in this work, TF-VLS InAs on Mo foil was achieved but demonstrated unidentified XRD peaks. Further InAs experiments will be performed in response to this result. Following the success of the aforementioned studies, TF-VLS grown InP (and InAs) will be used as the substrate to epitaxially grown InP (and InAs) single junction devices, thereby pushing toward the overall objective of reducing the cost of III-V photovoltaics.

ACKNOWLEDGMENT

This work was supported by the National Science Foundation's (NSF) SoLar Engineering Academic Program (SoLEAP) contract number 1355678 and the Office of Naval Research contract number ONR BAA N00173-03.

REFERENCES

[1] M. A. Green, K. Emery, Y. Hishikawa, W. Warta, and E. D. Dunlop, "Solar cell efficiency tables (version 48)," *Progress in Photovoltaics: Research and Applications*, vol. 24, no. 7, pp. 905–913, 2016.

[2] M. Woodhouse and A. Goodrich, "Manufacturing cost analysis relevant to single-and dual-junction photovoltaic cells fabricated with III-Vs and III-Vs grown on czochralski silicon (presentation)," tech. rep., National Renewable Energy Laboratory (NREL), Golden, CO., 2014.

[3] G. A. Landis and D. Hyatt, "The solar spectrum on the martian surface and its effect on photovoltaic performance," in *2006 IEEE 4th World Conference on Photovoltaic Energy Conference*, vol. 2, pp. 1979–1982, IEEE, 2006.

[4] M. S. Leite, R. L. Woo, J. N. Munday, W. D. Hong, S. Mesropian, D. C. Law, and H. A. Atwater, "Towards an optimized all lattice-matched InAlAs/InGaAsP/InGaAs multijunction solar cell with efficiency > 50%," *Applied Physics Letters*, vol. 102, no. 3, 2013.

[5] R. Kapadia, Z. Yu, H. Wang, M. Zheng, C. Battaglia, M. Hettick, D. Kiriya, K. Takei, P. Lobaccaro, J. Beeman, J. Ager, R. Maboudian, D. Chrzan, and A. Javey, "A direct thin-film path towards low-cost large-area III-V photovoltaics.," *Scientific Reports*, vol. 3, 2013.

plot is the lack of pure unbound indium peaks which would have appeared at 33, 36, and 40 degrees, thereby indicating full phosphorization of In into InP. Fig. 3b shows normalized photoluminescence plots of VLS grown InP compared to that of single crystal indium phosphide. All samples emit near the InP bandedge of 1.34 eV with FWHM values in close agreement with each other, thereby indicating the crystallinity properties of the TF-VLS grown sample are similar to that of single crystal InP.

Having achieved TF-VLS grown polycrystalline InP, the low cost method was expanded to the growth of InAs using the same initial Mo(foil)/In stack, but instead of phoshpine, arsine was flowed into the chamber. While PL measurements are not currently available, powder XRD results (Fig. 4) show clear InAs peaks and the absence of In peaks, indicating full conversion from In to InAs. However, there exists the presence of a number of unidentified peaks. The lattice constants were calculated to be unusually high (6.3-7.3 Å), possibly indicating multiple InAs phases and/or antimony contamination.

Seasonal Trends of Soiling on Photovoltaic Systems

Leonardo Micheli,[a,b] Daniel Ruth,[a] and Matthew Muller[a]

[a] National Renewable Energy Laboratory, Golden, Colorado, 80401, USA

[b] Department of Chemistry, Colorado School of Mines, Golden, Colorado, 80401, USA

Abstract — **This work investigates the seasonal variability of photovoltaic soiling losses over a 12-month period for 15 soiling stations deployed in the USA. We present a new parameter able to rank the sites according to the cumulative losses occurring over 3- and 6- month periods. The relations between soiling losses and particulate matter are briefly discussed, as well. Moving from long-term to shorter-term data increases the complexity of the analysis: monthly correlations are found to have lower accuracy than the longer-term ones presented in the literature.**

I. INTRODUCTION

Soiling—i.e., accumulation of dust, dirt, or particles on a photovoltaic (PV) surface—can cause dramatic reduction in PV performance [1]. Last year, we presented a preliminary study on the soiling losses of six soiling stations installed in the USA [2]. We showed that the concentrations of airborne particulate matter (PM) and the precipitation pattern were the best soiling predictors. These results were then confirmed in Ref. [3], where performance data of 20 soiling stations were investigated against 100 parameters. In that work, we found that, when long-term performances of different sites are compared, the correlation between soiling losses and PM has the highest coefficient of determination. Among the meteorological parameters, the average length of the dry period showed the best correlation with the soiling losses, followed by the maximum length of the dry period.

Rainfall patterns and particulate emissions follow seasonal trends that can strongly affect the soiling losses. Indeed, long dry periods can result in higher soiling than that occurring during rainy periods. The identification of seasonal patterns is therefore essential to characterize a site correctly and to determine the most adequate cleaning schedule. The present work aims to investigate, with a systematic approach, the seasonal soiling trends occurring at different sites over a 12-month period and to provide an instrument to quantify the seasonal soiling. In the second part of the paper, we present an analysis of the causes of seasonal soiling.

II. SEASONAL SOILING: DEFINITION AND CLASSIFICATION

A. Previous definitions and aim of this work

Seasonality is a concept that has been widely discussed in fields such as meteorology. Colwell [4], using the term "contingency," defined seasonality as the degree to which time determines states, or the degree to which they are statistically dependent on each other. In this work, the main aim is identifying how much soiling can vary in one year and which factors are driving these changes. Studying the repeatability of seasonality across different years is beyond the scope of this paper, which focuses only on 12-month datasets; therefore, the term "variability" has been preferred to "seasonality" to describe the seasonal trends of soiling independently of their year-to-year recurrence.

A number of works have discussed the seasonal effects of PV soiling. In 2005, Marion et al. [5] considered the effects of a seasonal soiling on the performance ratio. One year later, Kimber et al. [6] showed that PV systems from the Southwest USA had the lowest efficiencies during the summer dry season. El-Nashar [7] investigated the seasonal dust deposition in Abu Dhabi on a thermal solar collector. Caron and Littman [8] found that soiling in agricultural areas was governed by rainfall patterns and by the local seasonal tilling and harvesting activities. These previous works typically discuss the seasonal soiling in a qualitative and often site-specific manner. The goal of this paper, instead, is to provide quantitative instruments to characterize the seasonal soiling losses occurring over a 12-month period. Indeed, the identification of the seasonal soiling trends at a site—and of its atmospheric and pollution data—can help in planning the most-adequate cleaning schedule and, thus, to enhance the energy yield while minimizing the maintenance costs.

B. Seasonality Index

A number of indexes have been used in the past to quantify the seasonality of a location. In this work, we have considered the "Seasonality Index" (SI), a parameter introduced in 1981 by Walsh and Lawler [9] to describe the degree of variability in monthly rainfall through one year. The SI consists of the sum of the absolute deviations of the total monthly accumulated rain from the monthly mean, divided by the total rain accumulated in one year. The authors identified seven classes, reported and described in Table I. In this work, the SI has been adapted to describe the variability of soiling across a 12-month period and renamed as "Soiling Variability Index".

III. SOILING METRICS

A. Soiling stations

Among the sites investigated in Ref. [3], this study considers only those 15 sites with at least one full year of data available, listed in Table II each per letters A–P. Sites where data have been missing for more than two consecutive weeks have not been included. Three sites had data for more than two or three consecutive years available: each year has then been analyzed as an independent dataset and numbered under the letter classification. So the total population considered in this work has been of 19 datasets. A soiling station is installed at each site: each station consists of two identical cells, a pyranometer, and a weather station. One of the two PV cells is regularly cleaned whereas the other is allowed to naturally soil. Soiling is quantified using the daily soiling ratio (daily SRatio), determined using the same procedure described in [3]. The daily soiling ratio is calculated using the short-circuit currents recorded between 11:00 AM and 1:00 PM and when the plane-of-array irradiance is higher than 500 W/m^2. Monthly and annual SRatios can be obtained by averaging the daily values.

TABLE I
SEASONALITY INDEX (SI) CLASSIFICATION, AS DESCRIBED IN [9].

SI	Class
< 0.2	Very equable
≥ 0.2 and < 0.4	Equable, but with a definite wetter season
≥ 0.4 and < 0.6	Rather seasonal with a short drier season
≥ 0.6 and < 0.8	Seasonal
≥ 0.8 and < 1.0	Markedly seasonal, with a longer drier season
≥ 1.0 and < 1.2	Most rain in 3 months or less
≥ 1.2	Extreme, almost all rain in 1–2 months

B. Quantifying the soiling variability

The "Soiling Variability Index" (SVI) presented in this work has been adapted from the seasonality index (SI). The SI has originally been developed to consider the monthly accumulated rainfall values. So, to make it applicable to soiling, the following monthly soiling metric (S_m) has been considered:

$$S_m(m) = \sum_{d=1}^{n_d} (1 - dailySRatio(d)), \quad (1)$$

where m is the month, d is the day, and n_d is the number of days of the mth-month. S_m is not a direct measure of the energy loss, but an indicator of the potential impact of soiling on the energy yield. S_m is 0 if no soiling occurred; otherwise, it is always greater than 0. In order not to affect the S_m of months with a reduced number of daily data, a linear regression has been conducted previously to estimate any missing daily value of SRatios. The SVI for a site over a 12-month period is calculated as:

$$SVI(site) = \frac{\sum_{m=1}^{12} \left| S_m(m) - \left(S_{m\_sum}/12 \right) \right|}{S_{m\_sum}}, \quad (2)$$

where S_{m_sum} is the sum of the monthly S_m. As the original index, no correction has been made to balance the different number of days among the various months. SVI varies between zero (no variability: all the months have the same soiling) to 1.83 (maximum variability: all the soiling is accumulated in one month).

TABLE II
DESCRIPTION OF LOCATION, LAND COVER, AND CHARACTERISTIC WEATHER FOR EACH SOILING SITE USED IN THIS STUDY.

Site	Data Collection Period	County
A	01/15 – 12/15	Colfax, NM
B	01/15 – 12/15	Luna, NM
C	01/15 – 12/15	Imperial, CA
D	07/15 – 06/16	Fresno, CA
E 1	01/14 – 12/14	Los Angeles, CA
E 2	01/15 – 12/15	Los Angeles, CA
F	01/15 – 12/15	Yuma, AZ
G	08/15 – 07/16	San Luis Obispo, CA
H 1	01/14 – 12/14	Pima, AZ
H 2	01/15 – 12/15	Pima, AZ
J	06/13 – 05/14	Kern, CA
K	12/14 – 11/15	Winkler, TX
L	05/14 – 04/15	Iron, UT
M	12/13 – 11/14	Kern, CA
N	01/14 – 12/14	Polk, FL
O 1	06/13 – 05/14	Adams, CO
O 2	06/14 – 05/15	Adams, CO
O 3	06/15 – 05/16	Adams, CO
P	01/15 – 12/15	Maricopa, AZ

C. Results

The annual SRatio and the SVI calculated for each site are reported in Table III. For the purpose of operations and maintenance (O&M) decisions, it is valuable to understand if a disproportionate amount of soiling occurs in a cumulative period such as the dry season or during other marked periods such as agricultural harvesting. The SVI will provide more practicality if it can distinguish sites with such periods. Therefore, the SVI is correlated against a number of variables derived from the monthly soiling metric to investigate the ability of the SVI to determine the seasonal soiling occurring for a number of consecutive months (Table IV). As shown in Fig. 1, for the sites analyzed in this work, the SVI shows a linear relation with the relative cumulative soiling accumulated in the worst (most-soiled) month and in the worst three consecutive month periods. This result suggests that the SVI could be used to determine high soiling seasons occurring within a 12-month period.

TABLE III
ANNUAL SRATIO, SOILING VARIABILITY INDEX, AND UNCERTAINTY SVI FOR THE SITES INVESTIGATED IN THIS STUDY. UNCERTAINTY IS CALCULATED AS REPORTED IN SECTION III-D.

Site	SRatio	SVI	Uncertainty (%)
A	>0.99	0.8	18
B	0.98	0.1	36
C	0.97	0.4	4
D	0.98	0.4	4
E 1	0.99	0.6	12
E 2	0.99	0.7	6
F	>0.99	0.4	36
G	0.98	1.0	3
H 1	0.99	0.5	10
H 2	>0.99	0.2	43
J	0.92	0.5	1
K	0.99	1.1	12
L	>0.99	1.0	15
M	0.99	0.6	5
N	0.99	0.4	12
O 1	0.98	0.6	6
O 2	0.99	0.3	13
O 3	0.99	1.0	6
P	0.99	0.4	30

TABLE IV
PARAMETERS USED TO DESCRIBE SOILING VARIABILITY OF A SITE. THE PERCENTAGES REPRESENT THE R^2 OBTAINED WHEN THEY ARE RELATED TO THE SOILING VARIABILITY INDEX.

Parameter	Description	R^2 (%)
Months to 50%	Number of months needed to achieve the 50% of the total S_m	82
Max monthly S_m	Max S_m registered in one month	73
Max 3-month S_m	Max S_m when three consecutive months are considered	82
Max 6-month S_m	Max S_m when six consecutive months are considered	37

The datasets analyzed in this work fall into five of the seven categories given in Table I. The behavior of 19 datasets as divided into the SVI categories is as follows:

- SVI < 0.2: seasonal variability in soiling is not present or is negligible in these datasets. Indeed, the variation of monthly SRatio is very limited (< 1%) and the losses are equally distributed during the year (about 50% of the losses recorded in the worst 6 months).
- 0.2 < SVI < 0.4: these datasets are affected by limited seasonal soiling, with 60% to 70% of their soiling losses occurring in 6 months.

- 0.4 < SVI < 0.6: seasonal soiling can have a non-negligible impact because the value of the monthly SRatio can vary up to 15% during the year. 70% to 80% of the total losses occur in 6 months.
- 0.6 < SVI < 0.8: datasets with high variability in soiling. Most of the losses occur in 3 to 4 months and 85% to 90% of soiling is experienced in 6 months.
- SVI > 1: soiling in these datasets has extreme variability. Around 30% of soiling occurs in the worst month and almost all the losses (> 95%) occur in 6 months.

These results should be extended and refined through the analysis of more datasets.

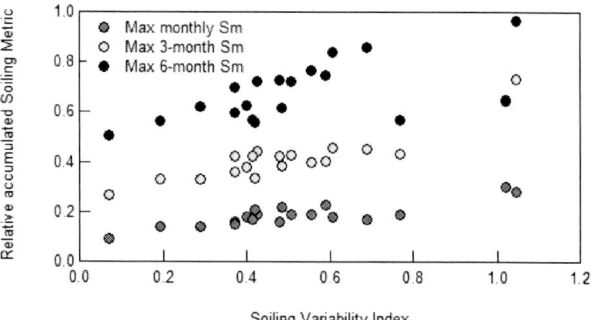

Fig. 1. The relative accumulated soiling metric in 1-month, 3-month, and 6-month period plotted against the soiling variability index of each site. The parameters are described in Table IV.

D. Soiling variability index vs. soiling ratio

The determination of O&M decisions cannot omit the absolute impact of soiling. Indeed, conducting cleaning at a site with high seasonal variation but low soiling might not be cost effective. Therefore, the analysis of site soiling should take into account both the SVI and the annualized SRatio. Indeed, despite being able to identify sites with distinct periods of high relative cumulative losses (Fig. 1), the SVI is not able to distinguish high vs. low soiling loss sites. The SVI would rate similarly a site with high soiling losses and a site with low soiling losses if both of them show the same distribution over the year. This is the case of Site C and Site P (Fig. 2), where 20% of the losses occur in the worst month and 30% to 40% occur in the three worst consecutive months. The SVI of these two sites is similar (0.40 vs 0.42), even if the annual SRatios are strongly different (0.97 vs 0.99). Another interesting case is Site A, with very low soiling losses (monthly SRatio ≥ 0.99) and a high variability, because more than 40% of the losses are concentrated in three months. Despite the high SVI, this seasonal pattern would have a very limited impact on the PV system because the absolute soiling losses are negligible at any time of the year.

A way to quantify the actual magnitude of seasonal soiling is by considering the impact of the uncertainty on the soiling measurement through a Monte Carlo computation. Indeed, the daily soiling ratios measured at the soiling stations are subject to uncertainties [10] and the uncertainty can be particularly significant for low soiling ratio sites. To estimate the uncertainty in SVI, each dataset has been modified, generating daily SRatio values equal to:

$$SRatio^*(d) = SRatio(d) \cdot (1 + x^* 0.005), \qquad (3)$$

where SRatio*(d) is the new daily soiling ratio value in input for a d day, and x is a randomly generated number, regenerated on each day, with a value between -1 and +1. SRatio*(d) is set to always be ≤ 1. In this approach, a fixed 5% uncertainty on the daily SRatio value has been considered. For each site, the dataset is regenerated 1,000 times, and each time, a new SVI is calculated. The results of this analysis (Table III) show that the uncertainty on SVI tends to increase with the soiling ratios, ranging from values of 1% to 42%.

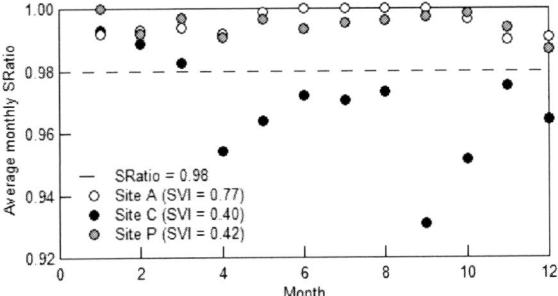

Fig. 2. Average monthly SRatios of Sites A, C, and P. Site A has high seasonality and low soiling losses. Site C has intermediate seasonality and high losses. Site P has intermediate seasonality and low losses. Months are numbered 1 to 12, from the start of the data collection to the end, so there is not necessarily any correspondence between the number and an actual month.

IV. PARTICULATE MATTER AND SEASONAL PV SOILING

A. Data sourcing

The investigation described in the previous section has focused on analyzing the performance data recorded at different stations. The results presented so far could not be used for predicting future soiling variability at a site. Seasonality is generally determined using multi-year datasets. Lacking such long soiling datasets, the prediction of seasonal soiling relies on identifying its correlation with other more widely available parameters. In our previous work [3], we showed the relations among the mean soiling ratios and a number of meteorological and environmental parameters on a multi-month time scale. In this section, we briefly discuss the relations between soiling and one of the most impactful parameters among those investigated in [3]—the particulate matter, on a short-term scale. The investigation here is limited to the PM_{10} data, calculated from the closest monitoring stations and from the stations located within 50 and 100 km from each site. Daily PM data have been downloaded from the U.S. Environmental Protection Agency databases [11]. PM

values for each site have to be determined from nearby monitoring stations by using an inverse distance weighting algorithm [12]. Daily PM data are more challenging to process than yearly ones: PM concentrations are recorded by each station at different time intervals and some of the stations have data missing for some time periods. In the present approach, the daily concentration of a site is calculated as the average of the measurements available from the nearby stations, weighted according to the distance and the spatial distribution of the stations active on that day. The rainfall data used in this analysis have been sourced from the PRISM database [13].

B. Impact of airborne particulate matter

An R^2 of 0.39 is found if the monthly accumulated daily losses, measured through the S_m, are compared against the accumulated daily PM_{10} concentrations (Fig. 3a). This value can be increased to 0.47 if monthly data are replaced with data accumulated in three consecutive months (Fig. 3b). Data occurring for dry periods longer than 90 days are marked in white because they appear to have a different trend. Indeed, without those data, the R^2 would rise to 0.63.

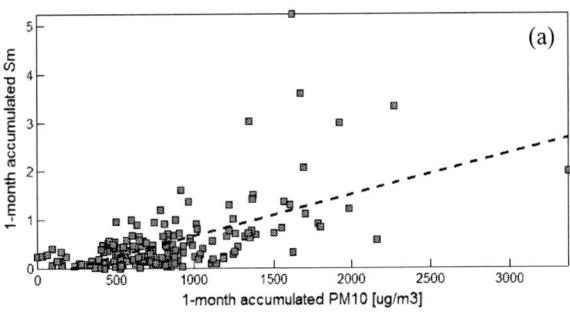

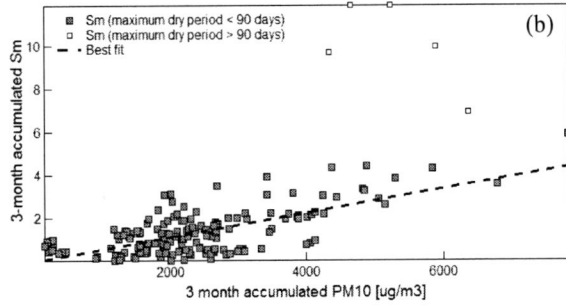

Fig. 3. (a) S_m accumulated each month plotted against the PM_{10} concentration accumulated in the same month. (b) S_m accumulated in three months plotted against the PM_{10} accumulated in the same time period. The maximum length of the dry period counts the number of days elapsed since the last rainfall and, thus, can be higher than the number of days in a month. Data occurring for dry periods longer than 90 days are marked in white.

The lower R^2 registered for PM_{10} in this investigation compared to the previous study [3] are probably due to a number of reasons. First, when short time periods are considered, the variability of a number of parameters becomes more relevant than for annual or longer time periods. Indeed, the complexity of correlating short-term PM10 concentrations and PV soiling has already been discussed in literature [14], [15]. For example, the seasonal behavior is expected to be strongly affected by the rainfall pattern and by the distribution of the dry periods, in particular. This is confirmed by a visual analysis of the soiling profile of Site D, in Fig. 4: the daily SRatios are higher when precipitations are more frequent, between November and May. Second, the discontinuity problems with daily PM10 registered by the EPA monitoring stations result in increased uncertainty: the results of this analysis could be enhanced by using alternative data process approaches. Third, the EPA monitoring stations might not be able, in some cases, to register the local seasonal PM10 trend of a site. For example, Site P is surrounded by agricultural fields outside of Denver whereas the PM stations are within the suburban Denver landscape (Fig. 5). In this case, the PM stations will not be able to record the effects of the rural activities that are affecting the site during the harvesting/tilling seasons.

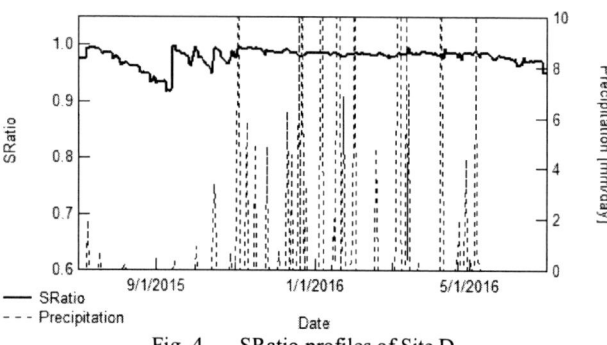

Fig. 4.   SRatio profiles of Site D.

V. CONCLUSIONS

This paper presents the initial results of an investigation on seasonal PV soiling. The daily soiling performance of 15 soiling stations installed in the USA has been analyzed and compared.

The first part of the work has focused on the definition and identification of seasonal soiling trends. A new parameter, seasonal variability index (SVI), obtained by adapting the seasonality index, has been introduced to quantify the seasonal behavior of soiling over a 12-month period. This parameter provided a means to classify the datasets investigated in this study into five groups depending on the number of months in which most of the losses are experienced. The SVI is a valid instrument for characterizing soiling variability, but it cannot distinguish sites with high or low soiling losses because it is based on the analysis of relative cumulative losses only. A Monte Carlo computational approach has been used to investigate the impact of soiling station measurement uncertainty on the SVI. The results show that stations at low soiling loss sites can have disproportionately higher

uncertainty in the SVI. In this sense, the high uncertainty in the SVI signals that the SVI has low usefulness, which is consistent with the outcome that there is low practical value in calculating an SVI when the annual soiling losses are lower than 1%.

Fig. 5.   Aerial view of Site O in Colorado. The green mark indicates the location of the site; the orange and yellow marks represent the PM_{10} and $PM_{2.5}$ monitoring stations available nearby. The green circle delimitates the area within 50 km from the site. A yellow 20-km circle is represented around the Denver metro area (centered at lat. 39.742043 and long. -104.991531): it can be seen that the monitoring stations are concentrated in the urban areas.
Source: 39.75685 & -104.62025. Google Earth, 12/30/16. 01/19/17.

In the second part of the paper, the relations among short-term soiling, particulate matter, and rainfall patterns were investigated. This kind of study is required to be able to predict the seasonal behavior of soiling at a site. The correlations among monthly soiling and pollution data have been found to have lower accuracy that those reported previously for longer-term data. These results have been enhanced when the soiling period increased from one month to three months: an R^2 as high as 0.63 was obtained between 3 month cumulative soiling and PM_{10} values. A number of factors that can impact this correlation have been listed and should be investigated in the future to enable the prediction of seasonal soiling trends.

ACKNOWLEDGEMENT

The authors are thankful to Sarah Kurtz and Paul Ndione for their useful comments.

This work was supported by the U.S. Department of Energy under Contract No. DE-AC36-08GO28308 with Alliance for Sustainable Energy, LLC, the Manager and Operator of the National Renewable Energy Laboratory.

The U.S. Government retains and the publisher, by accepting the article for publication, acknowledges that the U.S. Government retains a nonexclusive, paid-up, irrevocable, worldwide license to publish or reproduce the published form of this work, or allow others to do so, for U.S. Government purposes.

REFERENCES

[1] T. Sarver, A. Al-Qaraghuli, and L. L. Kazmerski, "A comprehensive review of the impact of dust on the use of solar energy: History, investigations, results, literature, and mitigation approaches," *Renew. Sustain. Energy Rev.*, vol. 22, pp. 698–733, 2013.

[2] L. Micheli, M. Muller, and S. Kurtz, "Determining the effects of environment and atmospheric parameters on PV field performance," in *2016 IEEE 43rd Photovoltaic Specialist Conference (PVSC)*, 2016, pp. 1724–1729.

[3] L. Micheli and M. Muller, "An investigation of the key parameters for predicting PV soiling losses," *Prog. Photovoltaics Res. Appl.*, vol. 25, no. 4, pp. 291–307, Apr. 2017.

[4] R. K. Colwell, "Predictability, constancy, and contingency of periodic phenomena," *Ecology*, vol. 55, no. 5, pp. 1148–1153, 1974.

[5] B. Marion, J. Adelstein, H. Hadyen, B. Hammond, and T. Flecther, "Performance parameters for grid-connected PV systems," in *31st IEEE Photovoltaics Specialists Conference and Exhibition*, 2005, pp. 1601–1606.

[6] A. Kimber, L. Mitchell, S. Nogradi, and H. Wenger, "The effect of soiling on large grid-connected photovoltaic systems in California and the Southwest region of the United States," in *Photovoltaic Energy Conversion, Conference Record of the 2006 IEEE 4th World Conference on*, 2006, pp. 2391–2395.

[7] A. M. El-Nashar, "Effect of dust deposition on the performance of a solar desalination plant operating in an arid desert area," *Sol. Energy*, vol. 75, no. 5, pp. 421–431, Nov. 2003.

[8] J. R. Caron and B. Littmann, "Direct monitoring of energy lost due to soiling on first solar modules in California," *IEEE J.*

Photovoltaics, vol. 3, no. 1, pp. 336–340, 2013.

[9] R. P. D. Walsh and D. M. Lawler, "Rainfall seasonality: description, spatial patterns and change through time," *Weather*, vol. 36, no. 7, pp. 201–208, 1981.

[10] M. Muller, L. Micheli, and A. A. Martinez-Morales, "A method to extract soiling loss data from soiling stations with imperfect cleaning schedules," in *2017 IEEE 44th Photovoltaic Specialist Conference (PVSC)*, 2017.

[11] U.S. Environmental Protection Agency, "Air Quality System Data Mart [internet database]." [Online]. Available: https://www.epa.gov/airdata. [Accessed: 20-Jan-2016].

[12] S. Falke, R. Husar, and B. Schichtel, "Mapping Air Pollutant Concentrations from Point Monitoring Data I: Declustering and Temporal Variance." [Online]. Available: http://capita.wustl.edu/capita/capitareports/mappingairquality/mappingaqi.pdf. [Accessed: 21-Nov-2016].

[13] PRISM Climate Group - Oregon State University, "PRISM Gridded Climate Data." [Online]. Available: http://prism.oregonstate.edu. [Accessed: 03-Jun-2017].

[14] L. Boyle, H. Flinchpaugh, and M. Hannigan, "Assessment of PM dry deposition on solar energy harvesting systems: Measurement–model comparison," *Aerosol Sci. Technol.*, vol. 50, no. 4, pp. 380–391, Apr. 2016.

[15] B. Guo, W. Javed, S. Khan, B. Figgis, and T. Mirza, "Models for prediction of soiling-caused photovoltaic power output degradation based on environmental variables in Doha, Qatar," in *ASME 2016 10th International Conference on Energy Sustainability collocated with the ASME 2016 Power Conference and the ASME 2016 14th International Conference on Fuel Cell Science, Engineering and Technology*, 2016, pp. 1–8.

Interrelationships Among Non-Uniform Soiling Distributions and PV Module Performance Parameters, Climate Conditions, and Soiling Particle and Module Surface Properties

Lawrence L. Kazmerski[a,b,*], *Fellow, IEEE*, Antonia Sonia A.C. Diniz[b], *Member, IEEE*,
Daniel Sena Braga[b], Cristiana Brasil Maia[b], Marcelo Machado Viana[c], Suellen C. Costa[b], Pedro P. Brito[b],
Cláudio Dias Campos[b], Sergio de Morais Hanriot[b], and Leila R. de Oliveira Cruz[d], *Member, IEEE*

[a]University of Colorado Boulder, Renewable & Sustainable Energy Institute (RASEI), Boulder, Colorado USA,
[b]Pontifícia Universidade Católica de Minas Gerais (PUCMinas), Belo Horizonte, Brasil,
[c]Universidade Federal de Minas Gerais (UFMG), Belo Horizonte, Brasil,
[d]Instituto Militar de Engenharia (IME), Rio de Janeiro, Brasil *Corresponding author: solarpvkaz@gmail.com*

Abstract — Soiling of PV modules is a growing concern because of the derating of energy output from solar installations, especially in the growing markets in the sun-rich areas of the world that ironically also might have the greatest issues with dust accumulation. This paper examines the relationships among soiling levels and PV module performance, the climate conditions, and the physical and chemical properties of the soiling particles and surfaces involved. Specifically, these studies focus on the commonly encountered non-uniform soiling of module surfaces. These non-uniform accumulations not only cause decrease in the power produced, but result in shading that can cause increased area heating ("hot spots") leading to module degradation. This paper evaluates the effects of non-uniform soiling patterns (categorized as edge build-up, waves, and blotches) on the J-V characteristics, documenting changes in the shape of these characteristics with the geometry and thickness distributions on the module surfaces. These studies are performed on crystalline Si framed modules and thin-film CdTe frameless modules. The non-uniform distributions are also related to the temperature distributions—with the temperature mapping evaluated using IR cameras. Hot spots with temperature increases of more than 15°C in some cases for modules operating under normal sunlight conditions. These soiling patterns, performance parameters, and temperature distributions/levels are determined for modules in several Brazil climate zones—although some results are compared to cases evaluated from the other world geographical regions. The soiling patterning is due to primarily wind, moisture. In some cases, the adhesion of the particles can be correlated with the modules surface conditions and the corresponding chemistry of the particles under the specific climate conditions.

Index Terms – Module soiling, non-uniform soiling, hot spots, reliability, I-V characteristics, temperature mapping

I. INTRODUCTION

The integrity of the first surface of interaction for the incoming solar photons remains a priority for PV module reliability and performance though this issue has been examined for more than 7 decades [1]. Soiling of PV module surface is a growing concern because of the derating of energy output from solar installation, especially in the growing markets in the sun-rich areas of the world that ironically also might have the greatest issues with dust accumulation. With the opening of the solar markets in sunbelt and desert areas of the world, areas that have not only the highest solar resource availability but also the most critical dust issues, the *understanding and mitigation* of this problem become crucial for avoiding any showstopper for rapid and widespread deployment. The soiling issue is not, of course, confined to these harsh-climate areas. Significant losses can also be encountered in areas that have tropical and temperate environments.

This paper continues to contribute within the worldwide efforts toward examining, understanding, and adding to the knowledge base of the effects and fundamental interaction of these soiling particles with the PV module surface. We build on our previous reports and observations of the soiling process, adhesion mechanisms, and performance effects, primarily focusing on our work in Brazil (and complementing with recent comparative observations and results from the other regions.

The primary focus is of these studies is on the interrelationship of non-uniform distributions on PV module with electrical, thermal, physical/chemical properties of the soiling material, and climate-zone influence. This includes: (1) Definitions of the climate zones in the major geographical region for these investigations (Brazil); (2) Observations of the common non-uniform dust patterns and some general classifications; (3) In-depth evaluation of the effects of these soiling patterns on the module I-V characteristics; (4) Correlation of these patterns with the time-dependent module temperature distributions and possible "hot-spot" formation; (5) Initial associations of the pattern formation with the climate-zone conditions (wind, moisture), the properties of the module surface, and the chemistry and physical nature of the soiling material. The purpose is to validate some critical relationships between the module performance parameters and the soiling process, important for modeling and simulations, predictions, and potential mitigations or minimization of these externally produced device-reliability issues.

II. EXPERIMENTAL FEATURES

Current-voltage characteristics were carried out on our soiling monitoring stations (Atonometrics), recording the short-circuit current, maximum power, and when required, the total IV curve. The solar irradiance was monitored using calibrated reference cells (and checked periodically against a separate reference cell and a calibrated pyranometer for consistency). The stations incorporated weather stations to record all meteorological

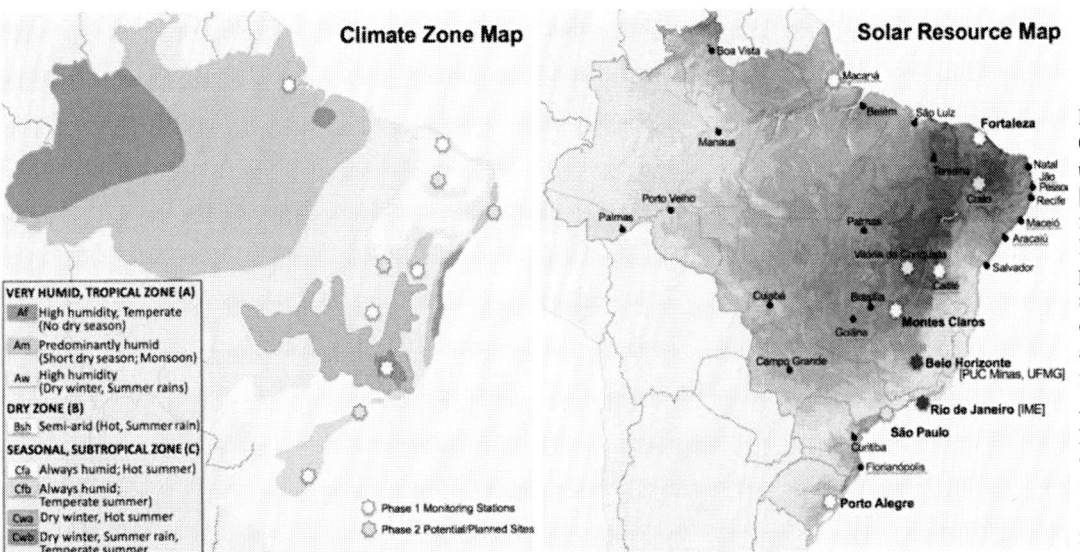

Figure 1. *Left:* Climate zones, based upon Köppen-Geiger [7,8], showing chosen first-phase monitoring locations & additional potential monitoring sites (based upon climate-zone coverage and priority PV-installation locations; *Right:* Collaborating partners (blue suns) & confirmed monitoring sites (blue and yellow suns) shown on solar resource map.

conditions for the given location (climate zone). Module temperature was recorded with an RTD on the back surface of the module. Module temperature mapping was accomplished using a relatively low-cost IR camera (Seek Compact Pro [2]) attached to either an iPhone or an iPad, with 320x240 sensor and 32-degree field-of-view. (An NECSan-ei Instrument Model TH71-2MX IR camera was also used for some of the mapping studies.) Temperatures were also independently (regularly) checked using IR detectors and thermocouples. All data are corrected for temperature and irradiance levels. Silicon (Canadian Solar-265 W and Kyocera-185 W) modules and First Solar CdTe modules were used for the *Crystalline-Si* and *Thin-Film Monitoring Stations*, respectively. These are shown in Fig. 2.

III. SUMMARY RESULTS AND DISCUSSION

Previously, we have reported fundamental studies on the adhesion of individual dust/soiling particles to the surface of module glass at the microscale, providing some initial insight into relationships between these properties and the chemistry of the particle and module glass surfaces and the ambient conditions [3,4]. This presentation extends our studies to the macroscale, examining real soiling patterns on exposed modules (climate

zones)—coupled with performance, temperature, and links to the microscale studies.

A. Climate Zone Considerations

Climate has been reported by several groups to be a critical factor for PV system reliability in general, and for soiling in particular [5-7]. The important environmental factors of temperature, humidity, rainfall, wind conditions, etc. have been classified into "climate zones", the most quoted of which were reported by Köppen (and updated/refined by several groups) [8,9]. Based upon these, we have developed the climate zone map shown in Fig. 1.

The major planned and existing installations in Brazil intersect these climate zones (Table 1). Our soiling monitoring stations have been chosen to cover 5 of these zones initially—and intersect in locations that have major planned PV plants. The studies reported in this paper are from the monitoring station locations identified in Fig. 1.

B. Soiling and IV Characteristics: Interpretations

It is common to evaluate the soiling effects on module performance by monitoring the short-circuit current. It has been

Figure 2. Monitoring stations – Left: Thin-Film Station, 110-W First Solar module (with soiled module on top having about 6% loss in power); Right: Crystalline Si Station (with 265-W Canadian Solar module). Each of the initial 4 Brazil locations has one of each of these stations installed.

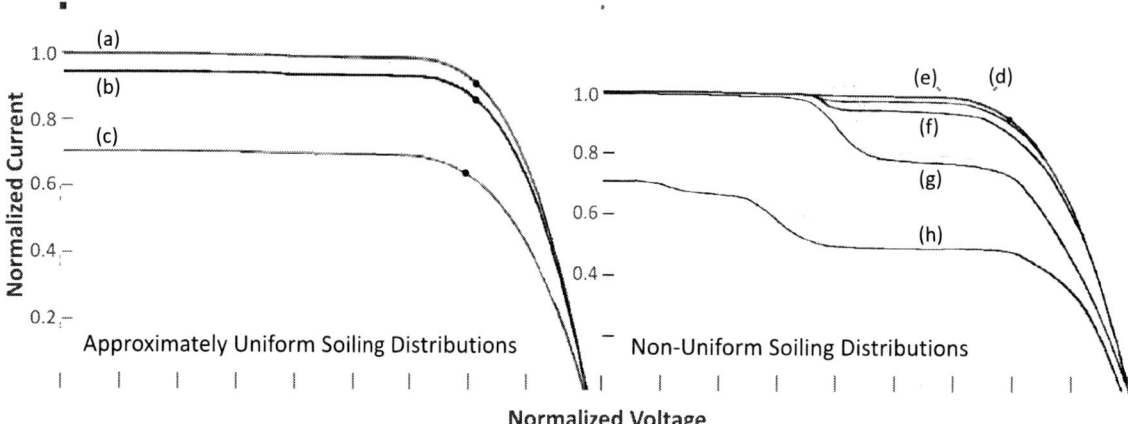

Figure 3. IV Characteristics of several soiling conditions and patterns for crystalline Si module. Left: (a) for clean module and (b) and (c) 7 gm/cm^2 and 33 gm/cm^2 approximately "uniform" soiling, respectively. Right: (d) clean module; (e) medium-level soiling around sides and bottom of frame (edge build-up); (f) higher-level soiling around sides and bottom of frame; (g) high level at sides and corners; (h) with blotch covering several cells near center of module with high-level edge build up.

shown that it is critical to monitor the maximum power output as well because non-uniform distributions can lead to misleading results [10-11]. The study of the non-uniform dust accumulation on PV modules surfaces has been shown to affect the shape of the IV characteristics and possibly generate hot spots [12-14]. We have observed these effects and have categorized the shapes of the IV characteristics for Si (framed) and thin-film CdTe (unframed) modules on several identified soiling patterns. We have initially and broadly categorized these as:

(1) *edge build-up* (accumulations at one or more of the module edges, typically more prevalent with framed modules);

(2) *waves* (undulations of soiling that can occur with wind or rain influences); and

(3) *blotches* (irregular shapes of soiling that can occur anywhere on the module surface, and in multiple places. Bloch sized can vary—and could include such irregularities as bird and animal droppings).

We use these designations to interpret the shapes and levels of the resulting module IV characteristics. Initially, this generates a database of distinctive characteristics that can then be used in the inverse fashion—to identify the major soiling patterns or combinations that are occurring. These can then be used in various modeling and simulation activities.

Our approach compares *field-observed* soiled modules with various soiling patterns and "laboratory" *simulations* using both calibrated (for transmittance and spectral characteristics) papers/organics and artificially produced dust patterns using typical module dust samples.

As an example, Fig. 3 presents several generated IV-characteristics illustrating the effects of the soiling patterns. (Extensive examples and detailed correlations (and accompanying photos of the pertinent module and condition) will be presented in the final paper for this conference.). The first set is for the case of uniform (i.e., nearly evenly distributed thickness or g/cm^2 layers) deposits over the entire module sur-

Table 1. Brazil added PV Installations by State in 2015. Color dots indicate climate zones (Fig. 1) of major PV installations.

Brasil State		Number of Projects	Installed Capacity (MW)
Bahia	●	140	3717
Ceará	●	30	754
Goiás		4	60
Mato Grosso do Sul	●	1	20
Minas Gerais	●●	43	1225
Paraíba		17	430
Pernambuco		47	1315
Piauí		79	2047
Rio Grande do Norte		68	1900
São Paulo		43	1235
Tocantins	●	21	455
Total		**493**	**13,159**

face. This results only in the loss of the short circuit current as a function of the deposit thickness and soiling particle characteristics. Nature, however, does not usually cooperate in universally providing this more easily predictive case. The more complex non-uniform distribution is the far more common.

Figure 3 d-g shows a series of IV characteristic for edge soiling edge build-up along the frame of a Si module—comparing these IV shapes to the edge plus blotch patterns (Fig 3h). The characteristic shape is affected by both the incursion of the soiling from the edge into the module and the thickness of the build-up. We have documented these effects for both field-deployed modules and extensive laboratory-produced test cases. The correlation of these provide the interpretations and the predictions. The cases in Fig. 3 are for modules examined over a function of time from our monitoring station in Belo

978-1-5090-5606-4/17 $31.00 © 2017 IEEE

Figure 4. Module with non-uniform soiling at corner and edges that results in the heating of those regions (see Fig. 5).

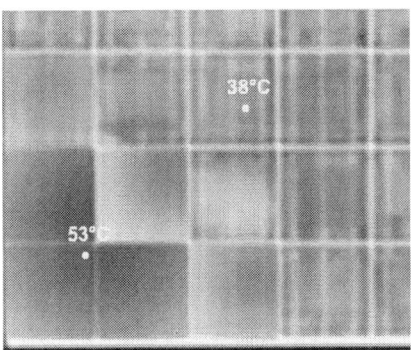

Figure 5. Example of thermal (IR) scan of module (lower right-hand corner of Si module) with edge soiling build-up on left side and bottom. 3 cells have elevated temperature where soiling is highest (corner/frame region).

Horizonte and compared to modules on the roof of Mineirão Stadium (1.4 MW array). The dust generally builds on the sides and bottom module edges along the frame. The soiling composition has been reported previously [3,4] and is fine-grained (~20μm diameter particles average) and composed of Fe-oxides (hematite and enstatite) with secondary silicates, calcite and dolomite minerals. (These edge build-ups were also simulated in laboratory conditions with almost identical results.)

C. Soiling Patterns and Temperature Mapping

For selected soiling patterns on the modules, we monitored the temperature distribution along with the IV characteristics. IR mapping of the temperature clearly shows the development of hot regions (as a function of exposure to the sun's radiation). An example of one such temperature map of a non-uniform soiling case (Fig. 4 for corner and edges) is shown in Fig. 5 for a crystalline Si module. The cells showing increased temperatures correlate directly with soil build-up regions on the module surface. Initially, the regions beneath the soiled regions are slightly cooler because they are shaded from the sun's radiation. We monitor the module glass surface temperature as a function of time—as well as the cell temperature. As time increases, the temperature of those cells shaded by the soiling increases with the increasing temperature of the affected cell. This map is taken as the module temperature increases—showing the rise in temperature along the edge cells. From these IR measurements, cell temperatures exceeding 16°C have been routinely recorded in Si modules under normal operating under operating conditions—and as high as 30°C. These hot spots are in the range that can possibility cause other failure mechanisms to be triggered as a result. These hot regions are directly associated soiling pattern types defined in Section 1B. The issues appear to be more critical in Si modules than in thin films, but some issues are also reported for thin-film modules. The thin-film modules are far less prone to such hot spot creation due to the cell patterning and to their frameless construction.

III. SUMMARY AND PAPER HIGHLIGHTS

This paper confirms and extends the results of several previous investigations of soiling effects on module performance, in additions to some first-time correlations among the performance parameters, dust patterns and properties, and the resulting modules temperature distributions. Specifically, this paper has been concerned with (major contributions):

- In-depth evaluations of the effects of several specifically identified and defined soiling patterns on the module I-V characteristics. This provides the correlation of the shape and several features of the IV-curves with the soiling patterns that have developed on the module surface. The studies are reported for both field experience and laboratory simulations. Detailed examples will be provided for all three pattern types, with specific identification of the effects on the shapes of the IV patterns;

- Correlation of these soiling patterns with the time-dependent module temperature distributions are confirmed, and possible "hot-spot" formations are observed (with temperature increases of more than 18°C-30°C-over 'normal' regions). The module temperatures are mapped using low-cost but reasonable resolution IR camera techniques. The direct effect of the rising temperature on the IV characteristics is also reported and animated in real time.

- Initial associations of the pattern formation with the climate-zone conditions (wind, moisture), the properties of the module surface, and the chemistry and physical nature of the soiling material.

The studies are carried out primarily in Brazil in several of the 5 climate zones that our monitoring stations are deployed. The climate zone maps of Brazil are reported, based upon Köppen-Geiger climate-zone designations. These results are based on media (video) segments showing real-time and on-site data acquisition (IV-parameters and temperature mapping), and animations to illustrate and explain the processes involved.

978-1-5090-5606-4/17 $31.00 © 2017 IEEE

ACKNOWLEDGEMENT

The authors gratefully acknowledge the support of CAPES for part of this work through "Pesquisa Visitante Estrangeiro," and to Ciência sem Fronteiras that supports the student and postdocs involved with our Brasil dust/soiling reliability project. We also acknowledge also the help and advice of the Solar Energy Institute for India and the United States (SERIIUS), which maintains a strong leading research and development program in soiling and dust research and mitigation. This group is funded jointly by the United States Department of Energy and the India Office of Science and Technology in the Ministry of New and Renewable Energy. We especially appreciate the expert help and guidance of Sonali Bhaduri of IIT Bombay.

REFERENCES

[1] Suellen C. Costa, Antonia Sonia A.C. Diniz, and L.L. Kazmerski, "Dust and Soiling Issues and Impacts Relating to Solar Systems: Literature Review Update for 2012-2015," Renewable & Sustainable Energy Rev. (2016).
[doi: 10.1016/j.rser.2016.04.059]

[2] See Seek Thermal (www.thermal.com).

[3] L.L. Kazmerski, Antonia Sonia A.C. Diniz, Cristiana Brasil Maia, Marcelo Machado Viana, Suellen C. Costa, Pedro P. Brito, Cláudio Dias Campos, Lauro V. Machado Neto, Sergio de Morais Hanriot, and Leila R. de Oliveira Cruz, "Fundamental studies of the adhesion of dust to PV module surfaces: chemical and physical relationships at the microscale" Proc. 42nd IEEE Photovoltaic Spec. Conf.-New Orleans (IEEE, New York; 2015); Also IEEE J. Photovoltaics 6, 719-729 (2016)

[4] L.L. Kazmerski, A.S.A.C. Diniz, C. Brasil Maia, M.M. Viana, S.C. Costa, P.P. Brito, C.D. Campos, S de Morais Hanrio, and L.R. Cruz, "Soiling particle interactions on PV modules: Surface and inter-particle adhesion and chemistry effects", Proc. 43rd IEEE Photovoltaic Spec. Conf.-Portland (IEEE, New York; 2016). DOI 10.1109/PVSC.2016.7749916

[5] Jim J. John, Sonali Warade, Govindasamy Tamizhmani, and Anil Kottantharayil, "Study of soiling loss on photovoltaic modules with artificially deposited dust of different gravimetric densities and compositions," Proc. 42nd IEEE PVSC-New Orleans (IEEE, New

York; 2015). Also, IEEE J. Photovoltaics (2016). DOI: 10.1109/JPHOTOV.2015.2495208

[6] M. Mani and R. Pillai)." Impact of dust on solar photovoltaic (PV) performance: Research status, challenges and recommendations," Renewable and Sustainable Energy Reviews 3124-3131 (2013).

[7] Julius Tanesab, David Parlevliet, Jonathan Whale, and Tania Urmee, "Dust Effect and its Economic Analysis on PV Modules Deployed in a Temperate Climate Zone," Energy Procedia 100, 65-68 (2016).

[8] Wladimir Köppen "Die Wärmezonen der Erde, nach der Dauer der heissen, gemässigten und kalten Zeit und nach der Wirkung der Wärme auf die organische Welt betrachtet" International Journal of Mechanical, Aerospace, Industrial, Mechatronic and Manufacturing Engineering Vol:5, No:10, 2011

[9] M. Kottek, J. Grieser, C. Beck, B. Rudolf, and R. Rubel, Meteorologische Zeitschrift, **15**, 259-263 (2006).

[10] Gostein, M., B. Littmann, J.R. Caron, and L. Dunn, "Comparing PV power plant soiling measurements extracted from PV module irradiance and power measurements." Proc. 39th IEEE Photovoltaic Specialists Conference (IEEE, NY; 2013) pp. 3004–9. DOI:10.1109/PVSC.2013.6745094.

[11] L. Dunn, B. Littmann, J.R. Caron, and M. Gostein, "PV module soiling measurement uncertainty analysis," Proc. 39th IEEE Photovoltaic Spec. Conf. (2013). DOI: 10.1109/PVSC.2013.6744236

[12] Shaharin A. Sulaiman, Haizatul H. Hussain, Nik Siti H. Nik Leh, and Mohd S. I. Razali. "Effects of Dust on the Performance of PV Panels," International Journal of Mechanical, Aerospace, Industrial, Mechatronic and Manufacturing Engineering 5, 2028-2033 (2011).

[13] Qasem, H., Thomas R. Betts, Harald Müllejans, Hassan AlBusairi, and Ralph Gottschalg, "Dust-induced shading on photovoltaic modules." Prog. in Photovoltaics 22, 218-226 (2014).

[14] E. Lorenzo, R. Moretón, and I. Luque, "Dust effects on PV array performance: In-field observations with non-uniform patterns," Prog. in Photovoltaics 22, 666-670 (2013).

[15] Emad Talib Hashim Tabark and Abed Hussien. "Dust Effect on the Efficiency of Silicon Mono Crystalline Solar Modules at Different Tilt Angles at Al-Jadryia Climate Conditions," J. Engineering 22, 56-73 (2016).

PV Module Durability –connecting field results, accelerated testing, and materials

T. John Trout[1], W. Gambogi[1], T. Felder[1], K. R. Choudhury[1], L . Garreau-Iles[2], Y. Heta[3], and K. Stika[1]

(1) DuPont, Wilmington, DE, 19803, USA (2) Du Pont de Nemours International S.A. Geneva, Switzerland, (3) DuPont K.K., Kanagawa Science Park, Kanagawa, Japan

Abstract-- **This presentation will discuss the framework employed at DuPont for using the analysis of modules from the field to guide development of accelerated tests and understand aging and degradation of PV modules, components and materials. Our framework starts with analysis of the stresses that are applied to modules in the field and in accelerated tests, including UV irradiation, humidity, temperature, and temperature cycling, either applied singly or in sequences. It then examines the measurements or responses that can then be evaluated or measured to determine the effects of these stresses. Measurements and indicators of aging and degradation include visual and physical degradation and changes of polymeric and other components, corrosion of metallization and bus bars, loss of power, and changes in IV characteristics and electroluminescent imaging. To correctly analyze and compare the data to the field it is important to understand the evolution and changes in materials in modules over time. Comparisons of these characteristics from the field and from accelerated tests can then be used to develop accelerated tests and determine the stress levels that match the levels of aging and degradation mechanisms found in typical field exposures. We will illustrate this work with examples from our analysis of fielded modules from around the globe and show how we have used it to understand degradation mechanisms and develop accelerated test protocols that more closely match what is seen in the field.**

Index Terms — **materials, durability, reliability, testing**

I. INTRODUCTION

Understanding the performance of PV modules and their aging and degradation is critical to designing and developing new materials and modules that maximize the return on investment for PV installations. At DuPont, we have been intensively studying the reliability and aging of PV modules, components and materials for more than a decade. This paper describes the approaches and methods that we use and illustrate our work with examples and key learnings.

Our approach begins with field degradation. DuPont has a global team evaluating and assessing fields in order to understand the degradation modes and mechanisms of modules, components, and materials. Data is collected and analyzed by component, material, mounting, age, and climate. Insights from these investigations are then used to guide and develop new accelerated tests. A core guiding principle is that accelerated tests should match the degradation seen in the field. Furthermore, any failures that are not observed in the field must not be produced by tests that are too highly-accelerated.

Our accelerated test program is actively studying a broad range of modules and materials. Data from our field experience guides and provides goals for accelerated tests. A simple formalism of Sample / Stress / Measurement is used to systematically organize and design tests. This approach is illustrated with several examples including Damp Heat and Sequential Tests.

In the final section, we discuss materials and illustrate changes that have occurred and their effect on module performance and durability.

II. FIELD ASSESMENT

The literature contains many references and several reviews on power loss and degradation of modules from the field over different times periods. [1]-[3] There are fewer studies and more limited information on the degradation of backsheets. To understand the degradation of modules and develop detailed data on backsheets, a global field program was developed at DuPont.

The goals of our field program are to study module and component degradation and relate that to the materials of construction. In the course of this work we have developed visual inspection protocols and a Failure Mode Effects Analysis (FMEA) approach. We use analytical tools in the field including thermal imaging to assess hot spots and FTIR spectroscopy to identify materials and elucidate degradation mechanisms. Selected modules from the field are then obtained for more in-depth physical and chemical analysis in the lab. [4] We actively collaborate with many partners across the globe to conduct these field assessments.

The data collected from the field and the lab include: (1) Cell defects such as corrosion, antireflection coating (ARC) delamination, and snail trails; (2) EVA defects including yellowing and delamination; and (3) Backsheet defects including front and airside delamination, yellowing, cracking, peeling, and localized burning. Power loss information is also collected and analyzed when available.

The data collected on fields and modules goes into a database that allows statistical analysis of many variables and at many levels. Our 2016 analysis more than doubled the amount of installations and modules from 2015, resulting in data from 197 installations from around the world that account for 1.9 million modules and 453 MW of installed power. The ages of panels range from recent installation to over 25 years, with the average age being 3.4 years.

Figure 1 shows the breakdown of the analysis by defect category. Overall, 22% of the panels inspected had some defects. Of these 22% defects, 11% were cell related, 7.5%

were backsheet related, 2.7% were related to encapsulant, and 0.5% were caused by other sources. The data covers a range of climates with the largest data set from temperate climates (351 MW), followed by smaller sample sizes for desert (54 MW) and tropical (47 MW) climates.

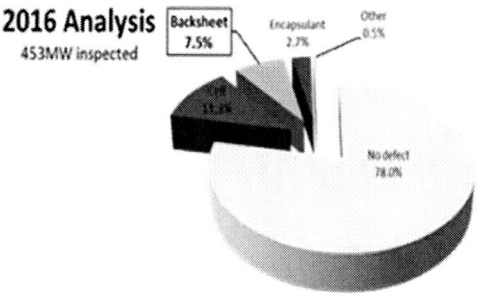

Figure 1: Defects by type for 2016 Field analysis of 453 MW and 197 installations.

The trends for defect rates by climate are shown in figure 2. There is an overall trend of defects with climate showing desert > tropical > temperate (if we note and discount a large contribution of snail trail defects in the temperate segment). Cell and metallization defects show little trend with climate. Polymer components (EVA and backsheet) show a pronounced trend with climate.

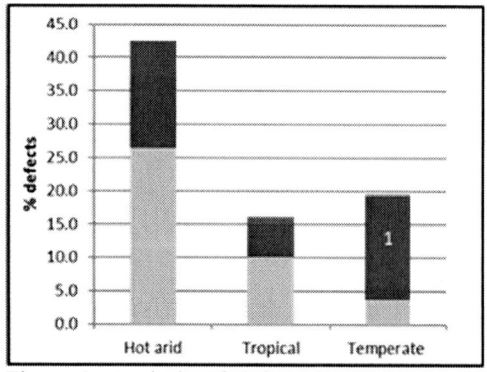

Figure 2. Analysis of defects by type and climate (Green=backsheet, violet = EVA, red = cell, blue = other) [1]Temperate cell defects had a large contribution from snail trails likely due to sampling.

Analysis of roof and ground mount is shown in figure 3. The data shows much higher total defect levels for roof mounted systems. Cell defects were similar for both mounting types while backsheet defects were 3 times higher for roof systems. It is likely that the reason for higher defect levels are the typically 15-25°C higher temperature on roof systems, accelerating the degradation.

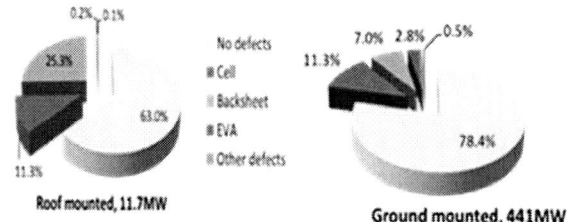

Figure 3: Comparison of defect rates for roof and ground mounted systems.

Analysis of the data by backsheet type and age shows several interesting trends. Figure 4 plots the percentage of backsheet defects for each installation by age. It is easy to see the time period over which each backsheet type has been used with Tedlar® being the dominant backsheet for fields from 15 to over 25 years. PVDF and PET backsheets appear to show an increase in defects with increasing age. For Tedlar® backsheets, the low percentage of defects does not increase with time.

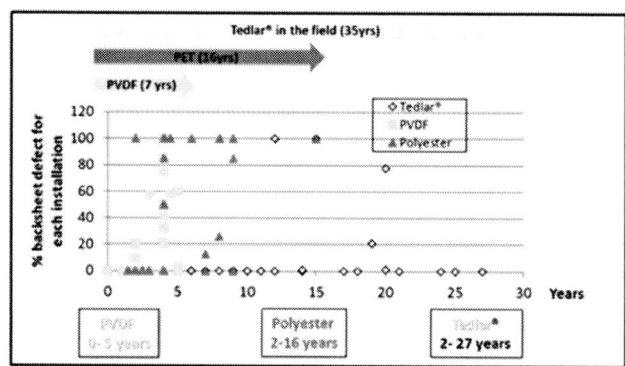

Figure 4: Defects vs years in the field for various backsheet types.

Table 1 shows the defect proportion for different backsheets over the entire time period. Table 2 displays the defect proportion by years of service with the number of installations and size, and the percentage of Tedlar® fields. It is notable that

Material (av. age in years)	% of fields with > 20% defects	Defect rate (MW basis)
Tedlar® (5.4)	11%	0.1%
PVDF (2.3)	43%	7%
Polyester(3.0)	50%	8.6%

Table 1. Defect rates by backsheet type.

the overall defect percentage declines with increasing age. This is likely due to the increase in the amount of Tedlar® PVF-based backsheets in older fields.

Years of service	Installations inspected	MW inspected	Defect rate (MW basis)	% Tedlar
0 - 5	114	396	5.7%	18%
5 - 10	42	54.6	20.7%	32%
10 - 15	14	1.0	1.1%	94%
15 - 20	10	0.1	8.8%	28%
20 - 27	15	1.0	0.5%	98%

Table 2: Defect rates vs years in the field.

The data yields some insights and offers new hypotheses on degradation in different climates. Combining this data with the assumption that the harshest climates should cause the most degradation and defects, we can rank the harshness of climates as desert > tropical > temperate. We can also evaluate the relative importance of climate stress factors in the different climate zones. By comparing the stress factors and levels in different climates with the defect data, it appears that elevated temperature is likely the most important factor followed by UV level. The higher degradation rates of roof systems is also consistent with higher temperature being the dominant factor.

The data also shows the difficulties and the uncontrolled nature of field data analysis. Even with the current sample size, we feel the need for an even broader sample of field data (greater data for tropical and desert climates, and roof mounted systems) for improved statistical analysis. Additionally, there are many hidden and uncontrolled variables involved. This is especially relevant when comparing modules of different ages. As modules get older they experience climate stresses for longer periods of time. Also, older modules are made of materials and components that can be quite different from those used in newer modules. It is important to understand and to try to disentangle these differences. In spite of these caveats there is much to learn from the study of fielded modules. "Long term outdoor exposure is the ultimate test for all module components, material quality and manufacturing quality" [5]

III. ACCELERATED TESTS

Different accelerated tests are used to simulate or predict the performance of modules in the field which experience real world stresses. Acceleration is achieved by higher stress levels (more intense light or higher humidity, e.g.) or through the use of higher temperatures to accelerate degradation. To systematically design accelerated tests, we use the stress and response formalism shown in Figure 5.

Sample	Stresses	Response [Measurement]
Module		Power loss, IV curve, EL, Visual, Insulation
Cell	Field Exposure DH, UV, TC, HF, ML,... Sequential Tests	Contact Resistance, SEM, EL Image, Visual
Backsheet		Mechanical Properties, Adhesion, Color, Molecular Weight Change, Gloss, Visual
EVA		Color, VA content, Acetic Acid concentration, Visual

Figure 5. Stress/response formalism for accelerated tests.

The design of accelerated tests start with a consideration of the sample, which can be a module or component or a material. Next, the stress should be considered. These can be single stresses such as UV exposure and Damp Heat. Multiple stresses can also be applied in sequence, simultaneously, and in complex combinations. Actual field exposure is complex with multiple variable, sequential, and simultaneous stresses. Finally, the measurements are selected that can gauge or quantify the effect of the stresses. These measurements and the resulting values are dependent on the choice of sample and the stress employed. For backsheets, visual changes, mechanical properties, adhesion, color, gloss, polymer molecular weight and chemical change are measured.

The most interesting and valuable part of this methodology is comparing accelerated tests to results from the field. This yields important information on the accelerated tests as well as the modules, components, and materials being tested. To develop and validate accelerated tests, it is critical to use materials and components that are the same or as close as possible to those from the field. Once the tests are developed and validated, new or modern materials and components can be tested and performance compared to the degradation of older materials. This can yield important information and insight on the expected field performance of new materials.

A highly sought-after goal is to correlate power generation performance of modules in the field and with those under accelerated tests. A model for power performance of modules over their lifespan has been reported by the IEA [6]. This can be combined with the composite field power loss data from the NREL study [1] to yield a useful conceptual model to compare to accelerated tests. The IEA model is shown below Figure 6).

There are two important features of the model- the baseline degradation rate (until wear-out), and the wear-out or end of life failure. Applying the NREL mean degradation rates of 0.5% - 0.8% / year to the baseline rate yields 80% retained power in 25 years and 76% power in 30 years (using 0.8%). The end of life failure rate is characterized by much higher degradation rates (> 50%). NREL also reported the degradation rate by climate: desert 1.2%, tropical 0.8%, and temperate 0.6% although they were not statistically distinguishable [1]. It is interesting to note that this climatic ordering of degradation is the same as found by us for defect rates earlier in section II.

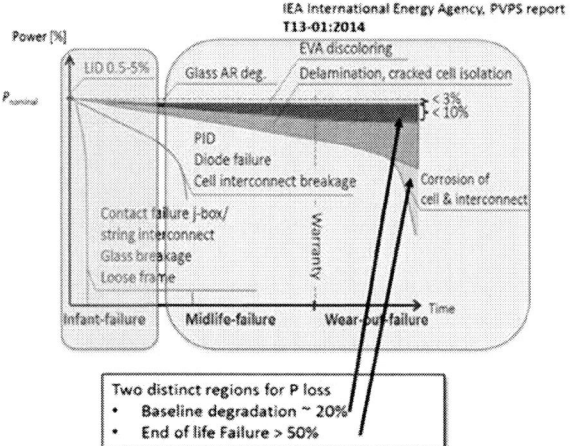

Figure 6. IEA model for module degradation with two degradation regions.

The method presented here can be used to compare, evaluate, and develop new accelerated tests. Backsheet measurements are an important response due to their importance (second highest defect in the field analysis), the multiple measures available to quantify the effect of the stress, and the high readout sensitivity. Degradation of different backsheets in different accelerated tests can be compared to field results and used to determine which tests are most effective in reproducing field aging and defects. Table 3 shows the results for three different accelerated tests on modules and backsheets, for four major types of backsheet currently in use.

Stress	PPE	KPE	PolyAmide	TPT/TPE	Comment
Field	Yellowing Mech Prop Loss Cracking	Cracking Front Side Yellowing	Yellowing Mech Prop Loss Cracking	Low defects	Effects of simultaneous and sequential stresses
Damp Heat (1000 hrs)	Slight Yellowing	No Change	Mech Prop Loss	No Change	Misses UV degradation
UV (4000 hrs)	Yellowing Mech Prop Loss	No Change	Mech Prop Loss	No Change	Misses hydrolysis and moisture
DH/UV/TC (MAST Sequential Test)	Yellowing Mech Prop Loss Cracking	Cracking Front Side Yellowing	Yellowing Mech Prop Loss Cracking	No Change	Combines key stresses Gives best correlation

Table 3. Comparison of accelerated test results to field observation.

The data shows that the Sequential Test combining Damp Heat, UV, and Thermal Cycling has the best correlation to backsheet degradation seen in the field. A single stress of UV is next best while Damp Heat is the least predictive. It is also interesting to note that the Tedlar® defect data is not useful in comparing these tests due to very low levels of change (low sensitivity).

This methodology can also be used to evaluate a single accelerated test and we will illustrate this by an analysis of Damp Heat. In this case we will use several of the sample types and measures that are available.

Samples include both modules and backsheets. The principle measures are power loss and EL imaging (for modules), acetic acid levels (for EVA), and mechanical properties (for backsheets). These measures were then compared to fielded modules. As discussed earlier, power loss before end of life is in the 20% - 24% range. Figure 7 shows the loss of power in Damp Heat with >30% steep power loss occurring at 2700 hours. The acetic acid concentration has also been measured for both fielded modules and modules from Damp Heat Testing. The average acetic acid level from measurements of fielded modules is ~ 740 ppm for 25 years exposure ranging from 300 to 1,300 ppm. In Damp Heat, 3000 hours exposure creates much higher acetic acid levels, >4000 ppm, a level far higher than measured on fielded modules.

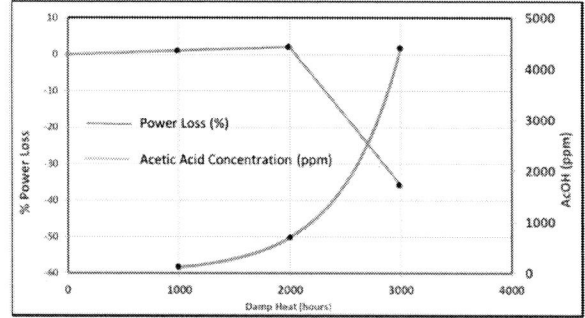

Figure 7. Damp heat exposure effects on power loss and acetic acid concentration in minimodules.

EL images from fielded modules and from Damp Heat exposure also show important differences. Typical EL images are shown in Figure 8 for three modules with different locations, backsheets, and ages. These EL images do not show a fixed signature of degradation and is typical of the hundreds of modules studied. On the right is the typical EL degradation seen in Damp Heat that starts with a darkening at the edges of the cell and progresses to darken the entire cell. This dark edge degradation signature is rarely seen in the field and implies that Damp Heat is producing a different degradation mechanism. This observation should raise concern of using this extended damp heat test in isolation to predict field performance.

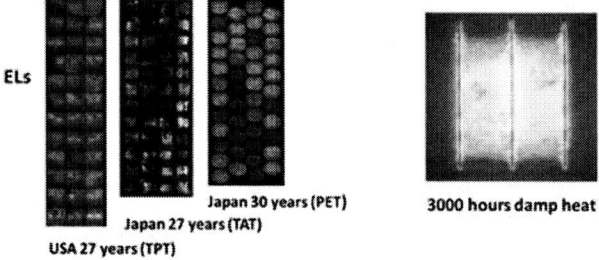

Figure 8. EL images from 3 fielded modules with 25-30 years in the field (left). EL image after extended damp heat (right).

Backsheet mechanical properties (retention of Elongation) were also measured and compared for a Damp Heat test and backsheet from the field. Modules from Miyako Island (Japan) show 80 – 100% retention of mechanical properties (elongation) for modules in service over 25-30 years in a tropical environment. In Damp Heat tests, this level of retention corresponds to less than ~1000 hours. At 3000 hours of Damp Heat there is no retention of elongation.

This analysis demonstrated that different components and materials of modules degrade differently with different mechanisms and at different rates. The difference in EL signatures questions the applicability of using Damp Heat to predict module power performance. Considering the outcome of these tests, we conclude that Damp Heat can be useful for comparative testing but should not extend beyond 1000 hours for backsheet related tests, unless test to failure is desired.

Table 3 showed that the most predictive accelerated test for backsheet degradation was the MAST Sequential Accelerated Test illustrated in figure 9 that combines Damp Heat, UV, and Thermal Cycling. [7-8] The design of the test incorporates exposure levels from dosages and analysis of field degradation. As discussed earlier, Damp Heat exposure of 1000 hours is used. The UV dosage is 4000 hours, corresponding to 24 years of backside UV field exposure. Thermal cycling is set at 600 cycles. An initial stress of Damp Heat and UV is applied, then up to three sequences of TC and UV are applied. The results of this accelerated test show degradations that are found in the field for the major backsheet types. These include backsheet yellowing, cracking, and loss of mechanical properties. All of these defects and failures are the result of polymer degradation.

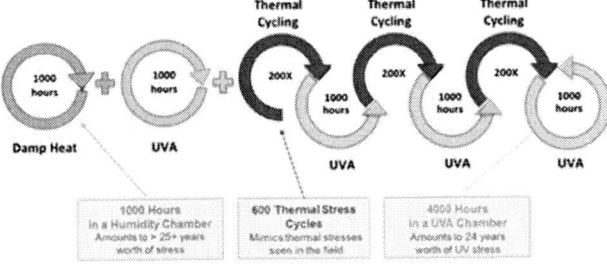

Figure 9. DuPont MAST Sequential test combining Damp Heat, UVA, and TC.

IV. MATERIALS

There are historic examples in PV of material choices and changes that have affected module performance and degradation. The initial JPL / NASA block design and buy programs tested many materials and designs and ended with Block V that designed and specified TPT backsheets. [9] The large and well-studied problem of EVA yellowing that is visible in many old modules has largely been fixed with a formulation change. Snail Trails are a more recent issue resulting from an EVA formulation change. [10] There have also been recent failures of backsheets based on new materials and designs.

Our analysis of modules from the field and from Damp Heat tests has found some interesting and important changes in power loss and EL signatures that can be traced to changes in materials – specifically metallization materials and silver paste. We find that this change occurred in about 2012.

Figure 10 shows the power loss in Damp Heat for minimodules made with cells with old (pre-2012) and new (post 2012) paste formulations. The older pastes show appreciable power loss between 2000 and 3000 hours, while with newer pastes the loss occurs in ~ 5000 hours. EL images are shown in Figure 9, with very different signatures of the degradation. The older pastes show the dark edge degradation that was discussed earlier. The newer paste in figure 11 shows degradation starting along the bus bars. This is confirmed in both Damp Heat and PCT tests.

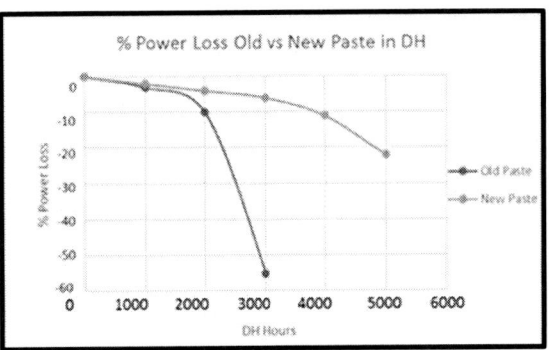

Figure 10: Power loss in damp heat with new and old pastes.

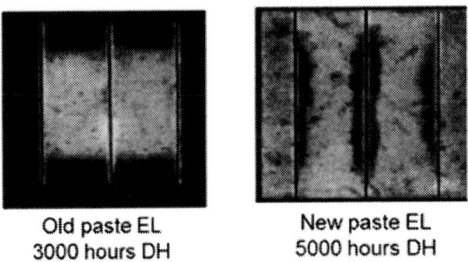

Figure 11: EL images after damp heat from cells with old and new paste.

There are several important learnings from this analysis. First "old pastes" should be used in Damp Heat tests if comparison is to be made to the field as there is a different mode of degradation with newer pastes. The data also shows that newer pastes are more resistant to degradation. If we assume the degradation is from corrosion of finger lines from acetic acid (produced by damp heat) and concomitant increase in contact resistance, then we can expect that the newer pastes are more resistant to acetic acid corrosion. This implies that newer modules with newer pastes should have a lower power degradation rate than older modules. It also means that the acetic acid corrosion model is less important for new modules going forward.

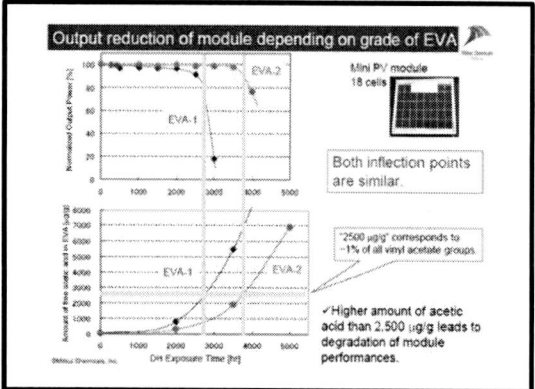

Figure 12: Power loss and acetic acid concentrations after damp heat exposure for new and old brands of EVA.

Results from Mitsui in 2012 [11] suggest that recent compositions of EVA may in addition produce less acetic acid and reduce power loss in modules. Their data also suggests that 2,500 ppm acetic acid is needed for power loss from corrosion of contacts, which is substantially higher than the 600-1,300 ppm value measured by us in fielded modules. With the deployment of EVAs that generate less acetic acid, combined with newer pastes that are more corrosion-resistant, we can expect modules with superior performance over time.

Other changes are also being made that could affect future module performance and durability. The thickness of cells and has decreased dramatically. Will we see more cell cracking and will it affect module durability? Will other changes in materials to reduce cost have impacts in the future? The methods and tests presented here can help us gain confidence that we are making the right choices going forward.

V. CONCLUSIONS

Understanding and predicting PV modules performance is a very challenging undertaking requiring the combination of analysis of field results, carefully designed accelerated tests, and thorough understanding of the materials and components making up modules. Over the last decade at DuPont we have developed and applied this methodology to advance our understanding of PV module reliability. Field results are the starting point, guide, and ultimate arbiter for this endeavor. Accelerated tests can add needed data and controls but must be carefully compared to field results. All of the components and materials that make up modules must be considered to understanding the reliability and durability of PV modules. Materials do matter, and the choices of these components and materials affect the performance and durability of the module over its service life. With the current and future changes to materials, cell and module components and structures, carefully designed approaches and tests can help us gain confidence that we are making the right choices today and for the years to come.

REFERENCES

[1] D. C. Jordan, S. R. Kurtz, K. VanSaint, and J. Newmiller, "Compendium of photovoltaic degradation rates", *Prog. Photovolt.: Res. Appl.*, wileyonlinelibrary.com DOI:10.1002/pip. 2744, 2016.

[2] D. C. Jordan and S. R. Kurtz, "Photovoltaic degradation rates- an analytical review", *Prog. Photovolt.: Res. Appl.*, wileyonlinelibrary.com DOI:10.1002/pip.1182, 2013.

[3] R. Dubey, S. Chattopadhyay, V. Kuthanazhi, J. John, F. Ansari, S. Rambabu, B. M. Arora, A. Kottantharayil, K. L. Narasimhan, J. Vasi, B. Bora, Y. K. Singh, K. Yadav, M. Banger, R. Singh and O. S. Sastry "All-India survey of photovoltaic module reliability: 2014", www.ncpre.iitb.ac.in 2016.

[4] W. Gambogi, K. Stika, A. Bradley, B. Hamzavy, R. Smith, and M. DeBergalis, "Material characterization in PV modules", Atlas/NIST PV Conference, Gaithersburg, MD, 2011.

[5] A. Skoczek, T. Sample, and E. D. Dunlop, "The results of performance measurements of field-aged crystalline silicon photovoltaic modules", *Prog. Photovolt.: Res. Appl.*, wileyonlinelibrary.com DOI:10.1002/pip.874, 2008.

[6] IEA-PVPS "Review of failures of photovoltaic modules", Report IEA-PVPS T13-01:2014.

[7] W. Gambogi, J. Kopchik, T. Felder, S MacMaster, A Bradley, B. Hamzavy, B.-L. Yu, K. Stika, L. Garreau-Iles, C.-F. Wang, Z. Pan, H. Hu, Y. Heta, and T. J. Trout "Improving accelerated test methods for PV module service life prediction", SNEC, 2015.

[8] T. Felder, W. Gambogi, K. Stika, B.-L. Yu, A. Bradley, H. Hu, L. Garreau-Iles, and T. J. Trout, "Sequential accelerated tests: improving the correlation of accelerated tests to module performance in the field", SPIE, 2016

[9] Flat Plate Solar Array Project Final Report, Volumes I and VI, October 1986, JPL Publication 86-31. "The block program approach to photovoltaic module development", Smokler, M. I.; Otth, D. H.; Ross, R. G., Jr., Photovoltaic Specialists Conference, 18th, Las Vegas, NV, October 21-25, 1985, Conference Record (A87-19826 07-44). New York, Institute of Electrical and Electronics Engineers, Inc., 1985, p. 1150-1158.

[10] J. J. Fan, Y. Daliang, X. Yao, Z. Pan, M. Terry, W. Gambogi, K. Stika, J Liu, W. Tao, Z. Liu, Y. Liu, M. Wang, Q. Wu, and T. J. Trout, "Study on snail trail formation in PV module through modeling and accelerated aging tests", *Solar Energy Materials & Solar Cells*, vol. 164, pp. 80-86, 2017.

[11] T. Shioda, "Acetic acid production rate in EVA encapsulants and it's influence on performance of PV modules", 2nd Atlas/NIST Materials Durability Workshop, Gaithersburg, MD, 2013.

Femtosecond vs Nanosecond: An Analysis on the Laser Ablation Properties of Dielectric Layers for Solar Cells

Jaffar Moideen Yacob Ali[1,2], Vinodh Shanmugam[1], Carlos D. Rodríguez-Gallegos[1,2], Bianca Lim[1],
Armin Aberle[1,2], Thomas Mueller[1]

[1]Solar Energy Research Institute of Singapore (SERIS), National University of Singapore, Singapore 117574
[2]Department of Electrical and Computer Engineering, National University of Singapore, Singapore 117583

Abstract—This paper reports on the ablation and micro-structuring properties of dielectric layers using a high average power femtosecond and nanosecond laser sources. The dielectrics investigated include SiN_X, AlO_X/SiN_X stack and thermal SiO_2/SiN_X stack deposited on planar n-type silicon wafer. Initially, single pulse ablation properties such as threshold fluence and energy penetration depth were determined for both laser sources. In femtosecond ablation, the presence of two different ablation regimes: gentle and strong ablation was identified. An analytical model has been developed to estimate the line width micro-machined at different pulse spacing. The modelled line width is in good agreement with the experimentally measured values for femtosecond ablation due to negligible debris deposition. As such, for line ablation using femtosecond laser, the reduction in the threshold fluence with respect to pulse overlap ratio has been estimated. These ablation properties are very useful in carrying out precise and low damage structuring of dielectrics for ablation intensive architectures such as interdigitated back contact (IBC) solar cells.

Index Terms—*femtosecond laser, solar cell, dielectric ablation.*

I. INTRODUCTION

Laser, as a processing tool, has gained significant attention in the PV industry due to its inherent advantages such as high speed processing, versatility and precision that are vital in the fabrication of cost effective and highly efficient solar cells [1]. Being a non-contact process, laser material processing induce far less damage to the substrate than mechanical scribing. Recently, ultrashort pulse laser technology using femtosecond laser source is being adapted into solar cell processing [2]. A femtosecond (fs) laser pulse is shorter than the timescale of photon – electron – lattice interactions and thus the laser pulse ends before the excited electrons could transfer energy to the ions [3]. Consequently, it is possible to obtain highly localized energy deposition with negligible thermal diffusion to the surrounding area. The use of lasers in solar cell processing can only be successful if it does not have detrimental effects on the solar cell performance. As such, it is vital to understand the fundamentals of laser interaction with dielectric layers and silicon. This work focuses mainly on the ablation and micro-structuring properties of three widely used dielectric layers using an industrial laser tool that consists of high average power femtosecond (fs) and nanosecond (ns) laser

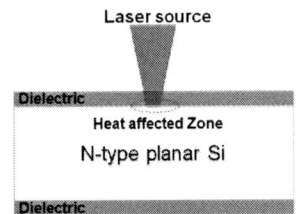

Fig. 1. Schematic of the test sample illustrating the process of indirect ablation.

TABLE I
DIELECTRIC LAYERS AND THEIR THICKNESS

Dielectric Layer	Thickness (nm)
SiN_X	70
AlO_X / SiN_X stack	10 / 70
SiO_2 / SiN_X stack	30 / 160

sources having a Gaussian pulse profile. The nanosecond laser source produces pulses at 38 ns duration and 532 nm central wavelength. The femtosecond laser source produces pulses at 480 fs duration and 1030 nm central wavelength. However, in this study, the fs laser source is used in second harmonic generation mode (SHG) at 515 nm for comparison with ns laser source. The laser beam scanning system has a maximum marking speed of 20 m/s. Furthermore, the parameters such as average output power, pulse repetition rate, marking speed and focus position are controlled via software interface.

II. EXPERIMENTAL DETAILS

The sample structure and the thickness of the dielectric layers used in this study are detailed in Fig. 1 and Table I respectively. The ablation process is an indirect one where the laser pulse is absorbed in the silicon and the dielectric is lifted off by vapor pressure.

III. RESULTS AND DISCUSSIONS

A. Single Pulse Ablation Parameters

The single pulse ablation parameters such as threshold fluence F_{TH}, spot radius ω_0 at the interaction surface and energy penetration depth z_0 are experimentally determined in

978-1-5090-5606-4/17 $31.00 © 2017 IEEE

TABLE II
SINGLE PULSE ABLATION PARAMETERS

Dielectric Layer	Ns Ablation			Fs Gentle Ablation			Fs Strong Ablation		
	F_{TH} (J/cm^2)	ω_0 (μm)	z_0 (μm)	F_{TH} (J/cm^2)	ω_0 (μm)	z_0 (μm)	F_{TH} (J/cm^2)	ω_0 (μm)	z_0 (μm)
SiN_X	3.8	23.7	1.96	0.137	18.0	81.5	1.44	35.7	264.9
AlO_X / SiN_X stack	4.3	24.4	1.87	0.299	16.5	61.2	2.33	35.6	252.7
SiO_2 / SiN_X stack	4.4	23.5	1.74	0.463	19.5	54.6	2.97	41.8	421.5

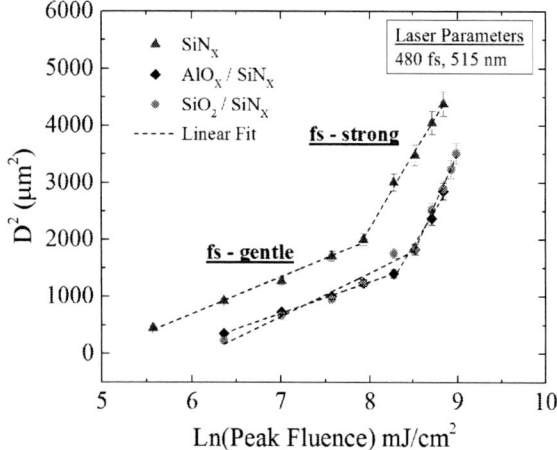

Fig. 2. D^2 vs Ln (Peak Fluence) using fs laser.

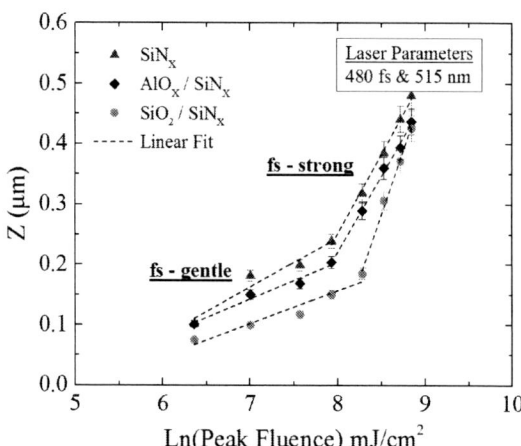

Fig. 4. Z vs Ln (Peak Fluence) using fs laser.

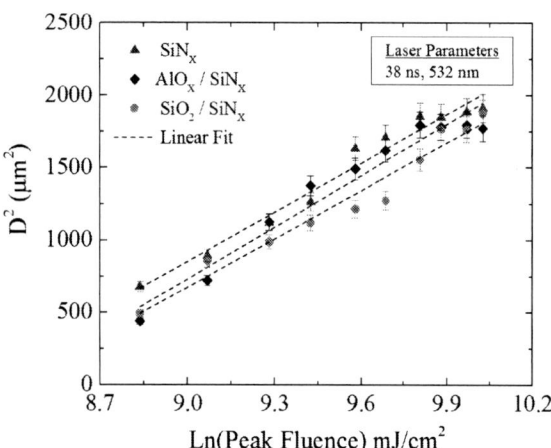

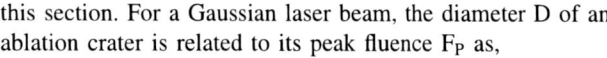
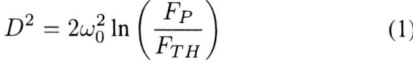

Fig. 3. D^2 vs Ln (Peak Fluence) using ns laser.

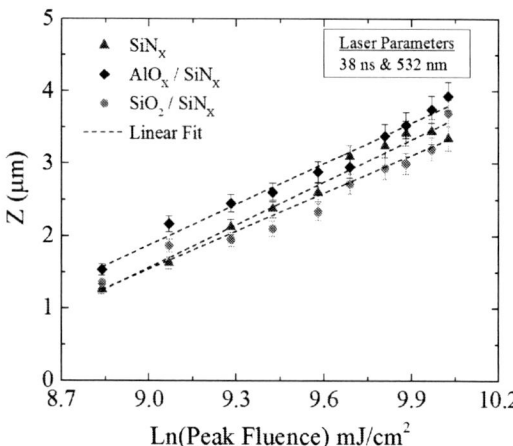

Fig. 5. Z vs Ln (Peak Fluence) using ns laser.

this section. For a Gaussian laser beam, the diameter D of an ablation crater is related to its peak fluence F_P as,

$$D^2 = 2\omega_0^2 \ln\left(\frac{F_P}{F_{TH}}\right) \quad (1)$$

where F_{TH} is defined as the minimum peak fluence required to completely remove the dielectric layer. When the peak fluence is equal to or less than the threshold fluence, we get $D^2 = 0$. Thus, plotting D^2 against Ln (F_P) and extrapolating the linear fit will give us F_{TH} [4] and ω_0 can be estimated from the slope of the linear fit. This parameter will vary for different dielectric layers. Fig. 2 and Fig. 3 show the plot of D^2 vs Ln

(Peak Fluence) for different dielectric layers ablated using fs and ns laser sources respectively.

Based on the Beer–Lambert law, the depth Z of an ablated crater is related to peak fluence as,

$$Z = z_0 \ln\left(\frac{F_P}{F_{TH}}\right) \quad (2)$$

where z_0 is the energy penetration depth per pulse. Fig. 4 and Fig. 5 show the plot of Z vs Ln (Peak Fluence) using ns and fs laser sources respectively. z_0 is estimated from the slope of linear fit. The single pulse ablation parameters extracted from the above graphs are presented in Table II.

978-1-5090-5606-4/17 $31.00 © 2017 IEEE

For ns ablation, we could see that D^2 and Z are increasing linearly with Ln (F_P). However, ablation using fs laser has two regimes that are characterized by different slopes. Commonly, these two regimes are termed in literature as gentle and strong ablation regimes [5]. In this work they are distinguished using subscripts G and S respectively. In the gentle ablation regime (at lower fluence), the density and temperature of hot electrons are lower than the ones in the strong ablation regime. Therefore, the rate of electronic heat conduction is lower. Due to lower temperature difference between the electrons and the lattice, thermal equilibrium is achieved faster. Thus, the gentle regime is characterized by reduced electronic heat conduction rate that happens in a shorter timescale and the ablation is mainly governed by optical penetration of the laser beam [6]. In the strong ablation regime (at higher fluence), the temperature and density of hot electrons are very high resulting in higher electronic heat conduction rate. Moreover, it takes longer time to achieve thermal equilibrium due to large temperature difference. Hence this regime is characterized by enhanced electronic heat conduction rate that happens for a longer duration and the ablation is governed by heat transport of thermally driven electrons [7]. We could see that the F_{TH} values increase with respect to layer thickness in indirect ablation. For fs ablation, the threshold fluence and energy penetration depth are very low in comparison with ns ablation. This is due to the extreme high intensities of fs laser that results in enhanced light absorption. It is interesting to note that the interaction spot radius in the gentle regime is almost half as that of the strong regime. As such, using fs laser in the gentle regime, that require low F_{TH} for ablation and has low ω_0 and z_0, could allow precise structuring of dielectrics without considerable damage to the underlying silicon.

B. Line Ablation Properties

Laser processing for solar cell fabrication requires precise structuring of dielectrics and underlying silicon. Therefore, an analytical model has been developed to determine the width of laser scribes using the single pulse ablation constants (i.e. F_{TH}, ω_0 and z_0). For simplicity, the re-deposition of ablated material is ignored and is assumed that it does not affect the succeeding pulses.

Let us consider a laser groove produced by an array of closely spaced pulses as shown in Fig. 6. Due to pulse overlap,

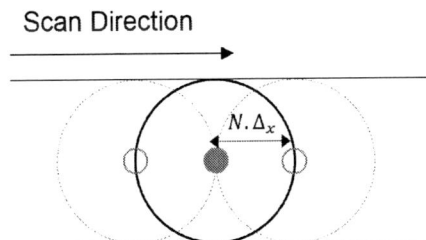

Scan Direction

Fig. 6. Pulse overlapping in line ablation. Small inner circles indicate the spot center and large outer circles indicate the fluence distribution. Shaded inner circle represents the spot that is considered for F_{equi} calculation.

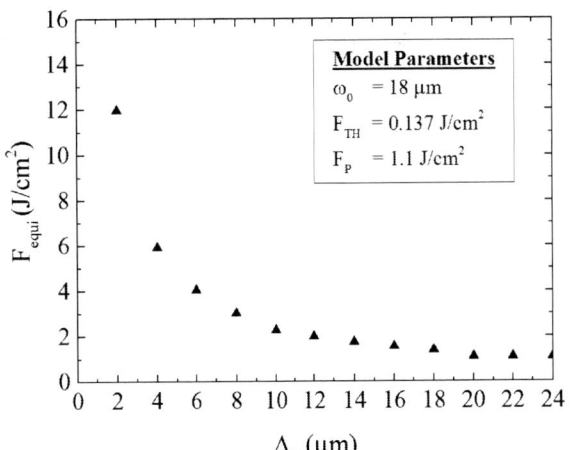

Fig. 7. Variation in F_{equi} with respect to Δ_X.

the equivalent fluence F_{equi} received by the surface is much higher than the incoming peak fluence F_P. Let the horizontal spacing between two adjacent laser pulses be represented as Δ_X and it can be varied by either the pulse repetition rate f_{rep} or the laser marking speed v according to the following relation,

$$\Delta_X = \frac{v}{f_{rep}} \tag{3}$$

The fluence of the nth pulse F (n) at the spot center that is considered along the midpoint of the groove is calculated as [8]

$$F(n) = F_P \exp\left(-\frac{2\left(n\,\Delta_X\right)^2}{\omega_0^2}\right) \tag{4}$$

where n Δ_X is the distance between the nth pulse and the spot center. The number of overlapping pulses N_{OV} is obtained by dividing the maximum spot radius (for a given F_P and F_{TH}) by pulse spacing Δ_X. N_{OV} is written as,

$$N_{OV} = \frac{\frac{\omega_0}{2}\sqrt{\ln\left(\frac{F_P}{F_{TH}}\right)}}{\Delta_X} \tag{5}$$

The equivalent fluence is then estimated by summing up the pulse fluence overlapping in both directions and can be estimated as,

$$F_{equi} = \sum_{i=-N_{OV}}^{N_{OV}} F(i) = \sum_{i=-N_{OV}}^{N_{OV}} F_P \exp\left(-\frac{2\left(i\,\Delta_X\right)^2}{\omega_0^2}\right) \tag{6}$$

Fig. 7 shows the increase in F_{equi} with respect to pulse spacing Δ_X. As per (6), the overlapping pulses have similar fluence distribution in both the directions. Hence, the resulting line diameter W_{LINE} is estimated by considering only the peak

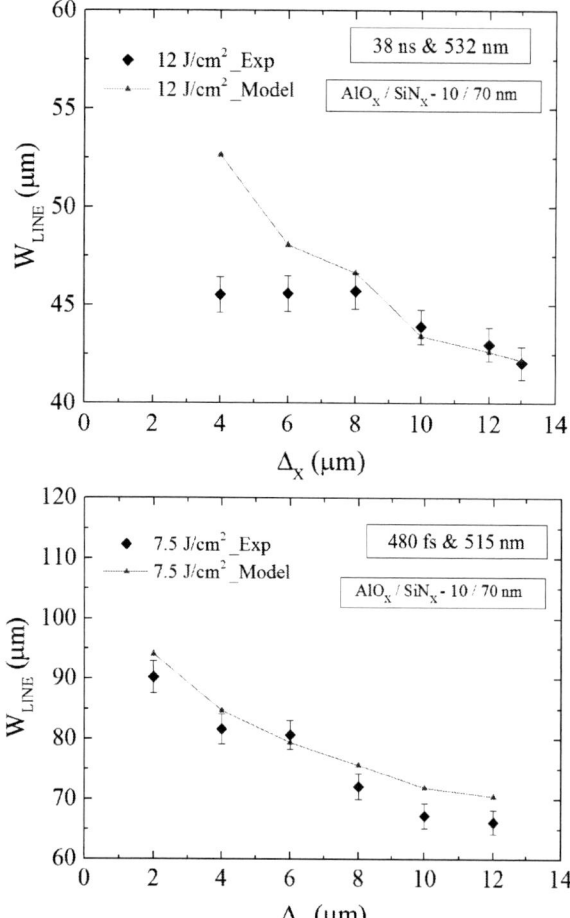

Fig. 8. W_{LINE} vs Δ_X using ns (top) and fs (bottom) laser sources.

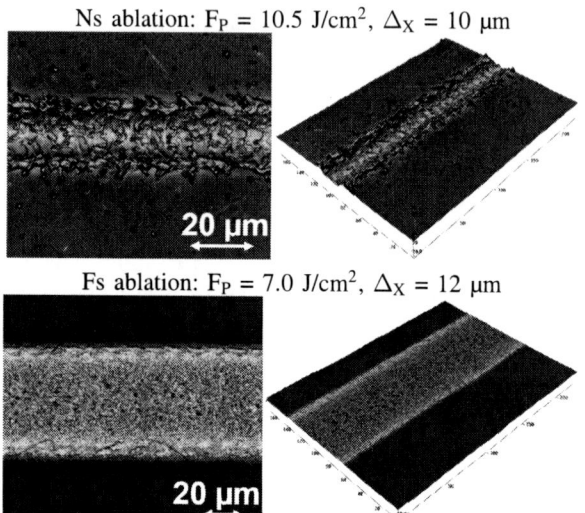

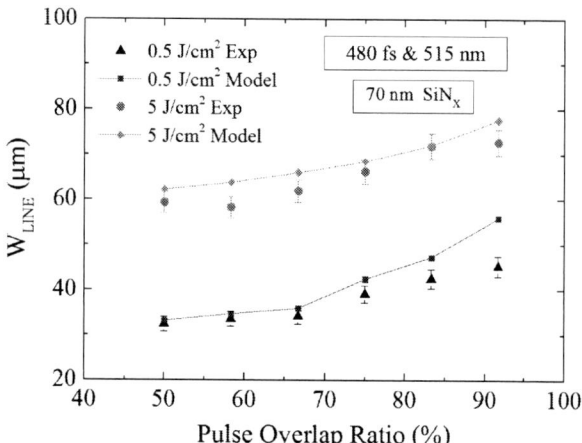

Fig. 9. 3D optical microscopic image of ns (top) and fs (bottom) ablated laser grooves.

Fig. 10. W_{LINE} vs Pulse overlap ratio for fs gentle (0.5 J/cm²) and fs strong (5 J/cm²) pulse fluences.

fluence and the fluence accumulated in one direction. W_{LINE} is given as,

$$W_{LINE} = \sqrt{2}\omega_0 \sqrt{\ln\left(\frac{F_P + \left(\frac{(F_{equi} - F_p)}{2}\right)}{F_{TH}}\right)} \quad (7)$$

Fig. 8 shows the variation of line width with respect to Δ_X for fs and ns ablation. The modelled line width is in good agreement with the experimentally measured values for fs ablation due to its inherent properties such as rapid vaporization and negligible debris deposition. Fig. 9 shows 3D optical microscopic images of fs and ns laser ablated grooves. Fs laser ablated groove has no debris deposition and melt formation even at a high pulse energy (strong ablation regime). However, for ns ablation, the edges of laser groove has considerable taper formation due to re-deposited debris and melted silicon. Hence, for $\Delta_X \leq 8$ μm (i.e. beyond 40%

pulse overlap ratio), the line diameter stays almost constant due to significant debris settlement that are difficult to remove.

Fig. 10 shows experimental and modelled line width with respect to pulse overlap ratio at both fs gentle and fs strong pulse fluence values. We can see that the modelled and measured values are quite similar. As a result, (7) could reliably be used for modelling line width in fs ablation. The pulse overlap (PO) ratio is written as,

$$PO = \left(1 - \frac{\Delta_X}{\omega_0}\right) \times 100\% = \left(1 - \frac{v}{f_{rep}\,\omega_0}\right) \times 100\% \quad (8)$$

C. Threshold Fluence vs Pulse Overlap Ratio

When performing laser ablation with multiple pulse overlaps, the ablation starts at pulse fluence far below the threshold, a phenomenon termed as the incubation effect [9]. This effect

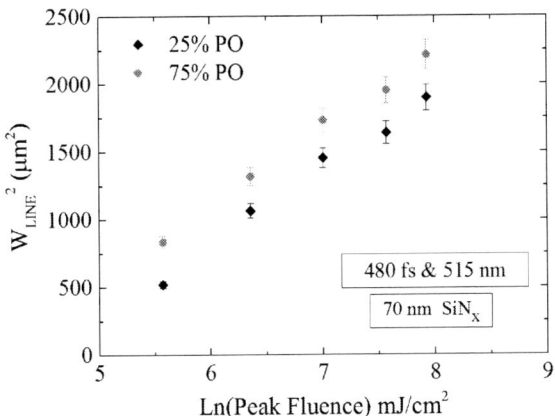

Fig. 11. W_{LINE}^2 vs Ln (Peak Fluence).

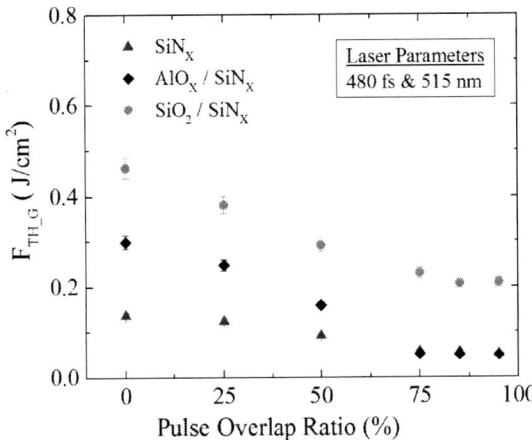

Fig. 12. F_{TH_G} vs Pulse overlap ratio for all three dielectric layers.

TABLE III
F_{TH_G} VS PULSE OVERLAP RATIO

Pulse Overlap Ratio (%)	F_{TH_G} (J/cm^2)		
	SiN_x	AlO_x / SiN_x	SiO_2 / SiN_x
0 (Single Pulse)	0.137	0.299	0.463
25	0.124	0.248	0.457
50	0.092	0.159	0.292
75	0.057	0.0507	0.23
85	0.055	0.05	0.207
95	0.051	0.049	0.21

is commonly studied for the case of single pulse ablation where the subsequent pulses impinge on the same spot. However, structures such as IBC solar cells require continuous scribing where the pulse overlap ratio is varied based on the selected scribe width and acceptable laser induced damage. Hence, in this section, the reduction in the threshold fluence while scribing a continuous line is determined for various pulse overlap ratios. The methodology is similar to the determination of F_{TH} from single pulse ablation. Lines were scribed at fixed pulse overlap ratio (25%, 50%, 75%, 85% and 95%) and by varying the pulse fluence F_P. The threshold fluence is then obtained by plotting W_{LINE}^2 vs Ln (Peak Fluence) for each pulse overlap ratio. From Fig. 8 and Fig. 9, we observed that in ns ablation, the line width does not vary much with respect to pulse spacing Δ_X due to considerable debris re-deposition. As such, the reduction in threshold fluence with respect to pulse overlap ratio is only determined for the case of fs ablation.

Fig. 11 shows the plot of W_{LINE}^2 vs Ln (Peak Fluence) for the ablation of 70 nm SiN_x at 25% and 75% pulse overlap ratios. The extracted threshold fluence at various pulse overlap ratios for all three dielectric layers is shown in Fig. 12 and the data is presented in Table III. Since the estimated fluence denotes the onset of ablation, it represents the gentle regime and is written as F_{TH_G}.

IV. SUMMARY AND FUTURE WORK

The single pulse ablation parameters such as F_{TH}, ω_0 and z_0 were experimentally determined for both fs and ns sources.

Ablation using the fs laser source has two operating regimes, gentle and strong. Operating in the gentle regime produces ablation spots that are much smaller than the focus spot radius. However, moving to the strong regime could result in much wider line widths. This has to be taken into consideration while carrying out precise ablation and patterning of dielectrics such as in the fabrication of IBC solar cells. An analytical model has been successfully developed to accurately estimate the scribe width with respect to pulse spacing while carrying out continuous ablation. It is seen that the scribe width in ns ablation stays almost constant due to significant debris settlement that are difficult to remove by subsequent laser pulses. Moreover, we have estimated the reduction in threshold fluence with respect to pulse overlap ratio for the case of fs ablation.

As a future work, the impact of gentle and strong fs ablation on the underlying silicon will be studied. In the gentle regime, the energy penetration depth is confined to <100 nm. As such, any laser induced damage is expected to be on the silicon surface allowing for an instant laser damage removal. However, in the strong regime, though the energy penetration depth is <300 nm, there is a possibility of laser induced bulk defects due to thermal diffusion of excited electrons.

Furthermore, it is important to study the impacts of fs gentle and fs strong ablation on the performance of silicon solar cells. A good starting point would be to investigate aluminum local back surface field (Al-LBSF) solar cells since they are not very sensitive to laser induced damages. It would be useful to know if such a cell structure is affected by fs strong ablation.

ACKNOWLEDGMENT

SERIS is sponsored by the National University of Singapore (NUS) and Singapore's National Research Foundation (NRF) through the Singapore Economic Development Board (EDB).This research was supported by the National Research Foundation, Prime Minister's Office, Singapore under its Clean Energy Research Programme project grant (EIRP, Award No. NRF2012EWT-EIRP001-023).

REFERENCES

[1] A. Niyibizi, B. W. Ikua, P. N. Kioni, and P. Kihato, "Laser material processing in crystalline silicon photovoltaics," 2012.

[2] J. Hermann, M. Benfarah, S. Bruneau, E. Axente, G. Coustillier, T. Itina, J. Guillemoles, and P. Alloncle, "Comparative investigation of solar cell thin film processing using nanosecond and femtosecond lasers," *Journal of Physics D: Applied Physics*, vol. 39, no. 3, p. 453, 2006.

[3] M. Shirk and P. Molian, "A review of ultrashort pulsed laser ablation of materials," *Journal of Laser Applications*, vol. 10, no. 1, pp. 18–28, 1998.

[4] J. M. Liu, "Simple technique for measurements of pulsed gaussian-beam spot sizes," *Optics Letters*, vol. 7, no. 5, pp. 196–198, 1982. [Online]. Available: http://ol.osa.org/abstract.cfm?URI=ol-7-5-196

[5] A. Borowiec and H. Haugen, "Femtosecond laser micromachining of grooves in indium phosphide," *Applied Physics A*, vol. 79, no. 3, pp. 521–529, 2004. [Online]. Available: http://dx.doi.org/10.1007/s00339-003-2377-0

[6] P. Mannion, J. Magee, E. Coyne, and G. M. O'Connor, "Ablation thresholds in ultrafast laser micromachining of common metals in air," in *Opto Ireland*. International Society for Optics and Photonics, 2003, pp. 470–478.

[7] J. Schille, L. Schneider, and U. Loeschner, "Process optimization in high-average-power ultrashort pulse laser microfabrication: how laser process parameters influence efficiency, throughput and quality," *Applied Physics A*, vol. 120, no. 3, pp. 847–855, 2015.

[8] J. Furmanski, A. Rubenchik, M. Shirk, and B. Stuart, "Deterministic processing of alumina with ultrashort laser pulses," *Journal of Applied Physics*, vol. 102, no. 7, p. 073112, 2007.

[9] J. Bonse, S. Baudach, J. Krüger, W. Kautek, and M. Lenzner, "Femtosecond laser ablation of silicon–modification thresholds and morphology," *Applied Physics A: Materials Science & Processing*, vol. 74, no. 1, pp. 19–25, 2002.

Growth of MoS₂ Thin Films with Microdome Texture as Omnidirectional Light Trap for Solar Cell Applications

Hussain M. Abouelkhair[1], Nina A. Orlovskaya[2], and Robert E. Peale[1]

[1]Department of Physics, University of Central Florida, Orlando, Florida, 32816, United States

[2]Department of Mechanical & Aerospace Engineering, University of Central Florida, Orlando, Florida, 32816, United States

Abstract — Antireflection by microdome texture on MoS2 thin films is reported. MoS2 films with uniformly distributed microdomes were grown by atmospheric pressure chemical vapor deposition, with structure and composition confirmed by Raman, x-ray diffraction, and energy dispersive x-ray analysis. The morphology and distribution of the domes were investigated by scanning electron microscopy. Finite difference time domain simulations of reflectance show that parabolic domes of 0.5 μm base diameter and 1 μm height can eliminate reflection for incidence angles up to 50°, which can improve light harvesting and efficiency of MoS2-based solar cells.

Index Terms — antireflection microdomes, chemical vapor deposition, molybdenum disulfide, Raman spectroscopy, scanning electron microscopy, solar cells, x-ray diffraction

I. INTRODUCTION

Light harvesting of incident light on solar cells is critical for boosting their efficiency. A 33% reflectance means we lose one third of the incident light energy. Minimizing this reflectance to zero would increase the efficiency by 50% if all the absorbed photons generate hole-electron pairs that are collected. Two techniques are known to reduce reflections. The first is smooth antireflection coatings, which strongly reduce reflectance at a specific wavelength and for a narrow range of light incident angles by destructive interference [1]-[3]. The second is engineered micro- or nano-structures on the front surface of solar cells [4]-[6], which advantageously feature broad spectral range and omnidirectionality in comparison with the smooth coatings [5]-[9]. For such antireflection structures, three main geometries have been studied, namely pyramids [6, 10-14], domes [5], [8], [9], [15]-[17], and pillar [18]-[22]. Nanotips, nanocones, and nanowires can be considered special cases of the basic shapes by considering elongation in one direction. The usual Fresnel reflection due to the sudden refractive index discontinuity is decreased or almost eliminated by the surface structures, which gradually change the effective refractive index between the two media [23]. Inspiration comes from the dome structures on the corneal surface of moths' eyes, which efficiently suppress reflectance for better light-harvesting and night vision [15], [17], [23]-[26]. The optimum structure geometries depend on the materials considered, while ease of fabrication depends strongly on the specific shapes. Thus, experimentally optimizing for structure shape is impractical. Simulating these structures using rigorous three-dimensional finite difference time domain (3D-FDTD) method facilitates the study and optimization of structures geometries.

MoS₂ is one of the promising materials for solar cells due to its favorable optical, electrical, chemical, and mechanical properties [27]-[29]. MoS₂ absorption coefficient ~ 10^5 cm^{-1} exceeds that of silicon [28], [29], potentially allowing thinner, lighter, and cheaper solar cells. The bandgap of MoS₂ is 1.3 eV, which matches well the solar spectrum. The predicted high mobility (410 cm² V^{-1} s^{-1}), mechanical flexibility and the chemical inertness to most acids of MoS₂ have attracted the attention [30]. On the other hand, MoS₂ has high refractive index ~ 3.5 which causes 31% reflection loss for light normally incident on its smooth surface [31]. Therefore front-surface antireflection structures are essential to maximize light harvesting and minimize losses [8].

Here we investigate different anti-reflection structure geometries on a smooth film of MoS₂ for solar cell applications. This study is relevant to MoS₂ films with microdome texture, which we grew by atmospheric pressure chemical vapor deposition. The microdomes which appear spontaneously without any lithographic processing are shown to be effective as an omnidirectional light trap. The demonstrated method of growth has significance for MoS₂-based solar cells and other optical and optoelectronic applications.

II. SIMULATION

Microdomes of MoS₂ on a smooth film of MoS₂ supported by sapphire substrate were simulated and analyzed using the 3D-FDTD method to determine the best structure geometry and arrangement that results in minimum reflectance at different angles and broad spectrum. The incident light source was a plane wave with either S or P polarization. The reflectance from unpolarized light source was calculated as follows:

$$\langle |E|^2 \rangle = \frac{1}{2} |\bar{E}_s|^2 + \frac{1}{2} |\bar{E}_p|^2 \qquad (1)$$

Where $\langle |E|^2 \rangle$ represents the time-averaged electric field intensity of an unpolarized beam light source. $|\bar{E}_s|^2$ and $|\bar{E}_p|^2$

represents the reflectance from S and P polarized light source respectively. Bloch boundaries were used vertically and perfectly matched layers (PML) boundaries were used horizontally above the domes and below the substrate. The wavelength dependence of refractive index and extinction coefficient of the domes, the film and, the substrate was considered in the simulation. We used the FDTD software (Lumerical Solutions, Inc.) to calculate total reflectance (specular plus diffuse) for different shapes and geometries.

Parameters that affect reflectance include dome base diameter, height, and spatial distribution on the surface. We first investigated the effect of different base diameters on reflectance using monochromatic incident light with 500 nm wavelength, which corresponds to the peak of the solar spectrum [8]. The best base diameter that minimized reflectance was then considered in simulations at different wavelengths from 300 to 1200 nm. The effect of height and distribution have been investigated as well.

III. EXPERIMENTAL

MoS_2 was grown by a two-step process. First, a 100 nm molybdenum thin film was electron-beam evaporated on a pre-cleaned c-plane sapphire wafer. Second, the molybdenum film was sulfurized in a two-zone atmospheric pressure chemical vapor deposition (APCVD) system, which is illustrated schematically in Fig. 1.

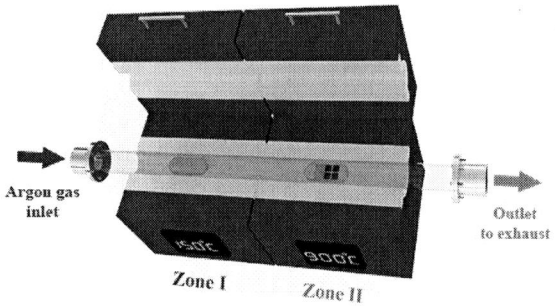

Fig. 1.   Schematic illustration of a two-zone APCVD setup.

The sulfur powder (Sigma-Aldrich purum ≥ 99.5%) was placed in the middle of the first zone (150 °C), and the Mo film was placed face-up in the middle of the second zone (900 °C). Ultrahigh pure argon carried sulfur vapor to the growth zone at a rate of 24 sccm/min. Once the second zone reaches 900 °C, it was kept for 30 minutes (sample A) or 60 minutes (sample B) at this temperature, then cooled down to room temperature at a rate of 20 °C/min.

Raman spectroscopy was carried out at room temperature using a Renishaw's inVia micro-Raman spectrometer system with 532 nm 100 mW laser excitation. The laser spot size was 1–2 μm. X-ray diffraction (XRD) was performed using a

PANalytical X'Pert[3] MRD X-ray diffractometer with a hybrid monochromator source, operating at 45 kV and 40 mA. The surface morphology of as-grown MoS_2 thin films was characterized by a Zeiss ULTRA-55 FEG scanning electron microscope (FEG-SEM). The composition was analyzed by an energy dispersive x-ray analyzer (EDX) coupled to the FEG-SEM. The total reflectance (specular plus diffuse) measurements were performed using LABSPHERE integrating sphere attached to CARY 500 spectrophotometer.

IV. RESULTS AND DISCUSSION

A. Simulation

Fig. 2 shows the total reflectance of parabolic domes with different base diameters. The square array of domes is on a 2.5-μm-thick smooth film of MoS_2 supported by a sapphire substrate. Each reflectance point in the plot represents the average reflectance over all light incident angles from 0° to 80°. This demonstrates the omnidirectionality of these structures. The domes height is half the diameter in all cases. The incident light wavelength is 500 nm.

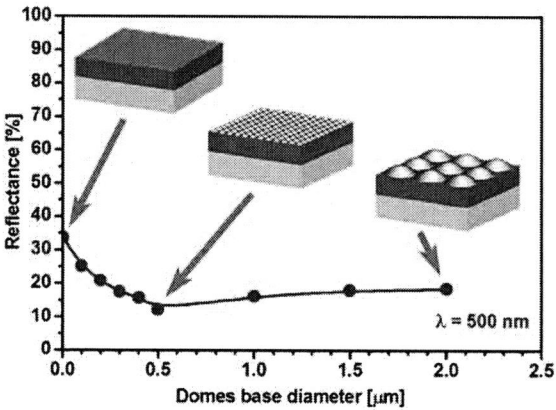

Fig. 2.   The reflectance of MoS_2 domes with different base diameters.

The reflectance decreases from 33.6% to 12.2% with increasing the base diameter from 0 to 0.5 μm. Then the reflectance slightly increases to 18.5% with increasing the diameter from 0.5 μm to 2 μm. Therefore we chose the 0.5 μm diameter for further studies over a broad range of wavelengths. As shown in Fig. 2 the most effective base diameter equals the incident wavelength, but we next demonstrate that the same base diameter is nearly equally effective at all relevant wavelengths.

Fig. 3 shows a comparison between the reflectance of a smooth film and 0.5 μm diameter domes over a broad range of wavelengths from 300 nm to 1200 nm. Each reflectance point in the plot represents the average reflectance over all light incident angles from 0° to 80°. The reflectance decreased by 51% over the whole range. This demonstrates the broad spectral range for the effectiveness of these structures. The wavelength independence of reflectance reinforces the effective medium theory [32] and shows that the antireflection effect is due the gradual change of effective refractive index rather than scattering.

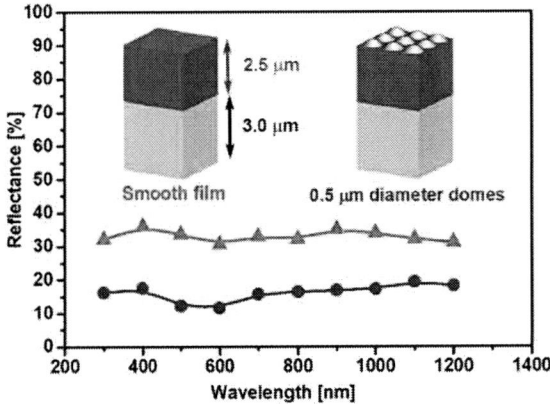

Fig. 3. The reflectance of a square array of MoS_2 domes with 0.5 µm base diameter compared to that of the smooth film.

Fig. 4 shows the reflectance of square arrays of parabolic domes with different heights but constant 0.5 µm base diameter.

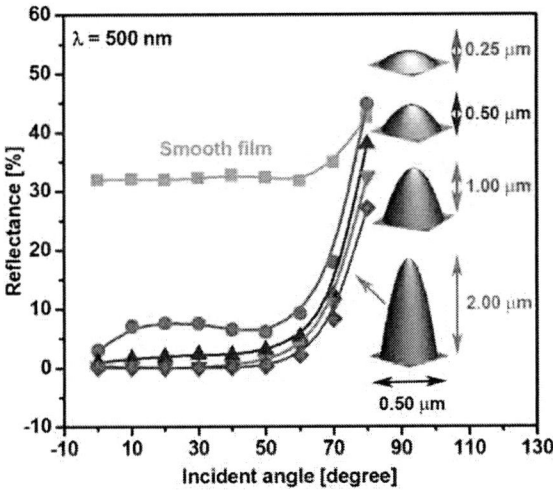

Fig. 4. The reflectance of square arrays of MoS_2 parabolic domes with different heights. The domes base diameter is 0.5 µm for all heights.

The total (specular plus diffuse) reflectance averaged over all the incident angles from 0° to 80° decreased from 12.18% to 4.22% as the height of the dome increased from 0.25 µm to 2

µm. Importantly, the average reflectance of domes with heights 1 µm and 2 µm is practically zero over all the incident light angles from 0° to 50°.

Fig. 5 shows the reflectance averaged over incident angles from 0° to 80° for different spatial distributions of domes and monochromatic incident light with 500 nm wavelength. The hexagonal array of parabolic domes with 0.5 µm base diameter and 0.25 µm height reduces the reflectance more efficiently than a square array of these domes, but this effect is less than the difference in fill factors: For the square array of parabolic domes, the fill factor is 39%, while for the hexagonal array it is 45%. In addition, we compared the former two arrays to hexagonal arrays of pyramids, cones, and hemispherical domes of the same base area and height. The average reflectance of the pyramids was the highest while the hexagonal array of parabolic domes was lowest. Cones had the second highest reflectance after the pyramids. Thus smooth structures such as domes are better than faceted structures such as pyramids.

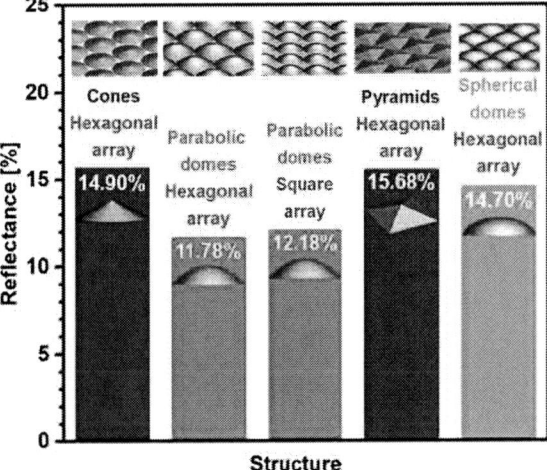

Fig. 5. The total reflectance of arrays of different structures and spatial arrangments.

B. Raman Spectroscopy

Fig. 6 compares Raman spectra of samples A and B to a reference spectrum of c-plane sapphire. The sapphire reference spectrum has five peaks located at 378, 417, 449, 576 and 750 cm^{-1}, which match published [33] values of 378, 418, 451, 578, and 751 cm^{-1}. A sixth previously reported peak at 432 cm^{-1} [33] is an unresolved shoulder in our spectrum. The MoS_2 samples A and B have Raman spectra comprising four peaks located at 283, 379, 406 and 450 cm^{-1} in agreement with earlier reports for 2H-MoS_2, which has first order Raman-active modes E_{1g}, E^1_{2g}, and A_{1g} [34]-[37]. The peaks are slightly red-shifted and broadened, as expected for polycrystalline material in comparison with single crystal [36]. It is evident from the

978-1-5090-5606-4/17 $31.00 © 2017 IEEE

Raman spectra of samples A and B that the difference in growth time does not have a great effect on the quality of the as-grown MoS₂ film.

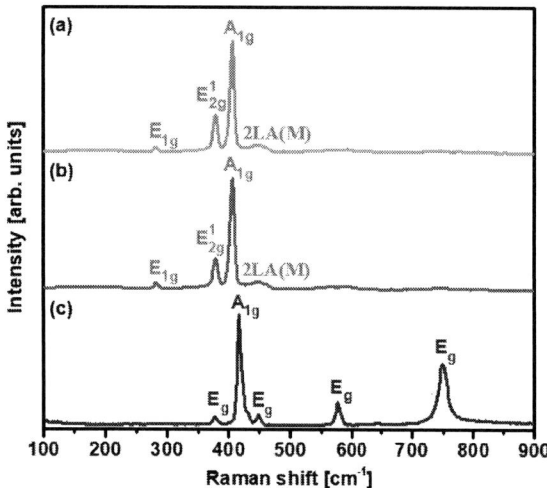

Fig. 6. Raman spectra of MoS₂-on-sapphire (a) sample A, (b) sample B, and (c) blank c-plane sapphire wafer.

C. X-ray Diffraction

Fig. 7 shows symmetric out-of-plane X-ray diffraction patterns for samples A and B of MoS₂. Six diffraction peaks of MoS₂ are identified which corresponding to the (002), (004), (100), (006), (110) and (008) planes, respectively. The location of these peaks matches well the XRD card PDF # 00-37-1492.

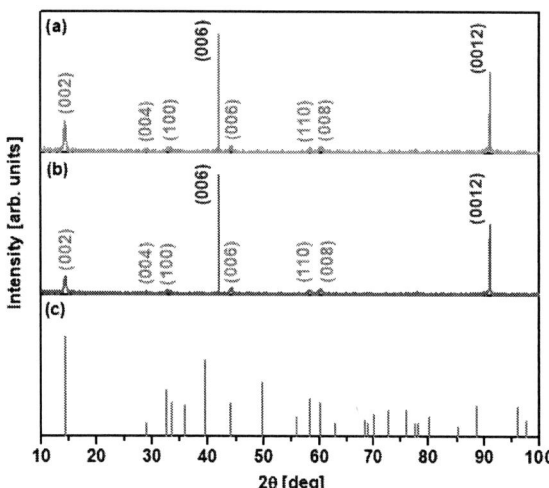

Fig. 7. Symmetric Out-of-plane XRD patterns of (a) sample A, (b) sample B and (c) reference XRD pattern of MoS₂. MoS₂ peaks are labeled in red and sapphire peaks are labeled in blue.

The presence of diffraction peaks from unparallel planes confirms the polycrystalline nature of the film. Two diffraction peaks (006) and (0 0 12) from the sapphire substrate are identified (PDF # 01-070-5679). The relative strength of the (002) peak indicates preferred orientation such that the (002) plane of MoS₂ is parallel to the (006) plane of the sapphire substrate. The polycrystalline nature according to the x-ray diffraction agrees well with Raman analysis.

D. SEM and EDX

Fig. 8a shows a SEM image of the surface of evaporated Mo thin film before sulfurization. The Mo film was smooth and dense. SEM images of the surface of sample B of MoS₂ at different magnifications are shown in Fig. 8b-d. The microdomes are randomly distributed and uniformly cover the whole surface. EDX data of sample B over an area equivalent to the field of view of Fig. 8d gives S/Mo ratio of 1.96, which indicate that the film is almost stoichiometric. A reasonable explanation of the cause of the formation of these microdomes is the small thickness of the Mo thin film and coefficient of thermal expansion mismatch between the sapphire substrate (7.3×10^{-6}/°C) and the grown MoS₂ (10.7×10^{-6}/°C) [38], [39].

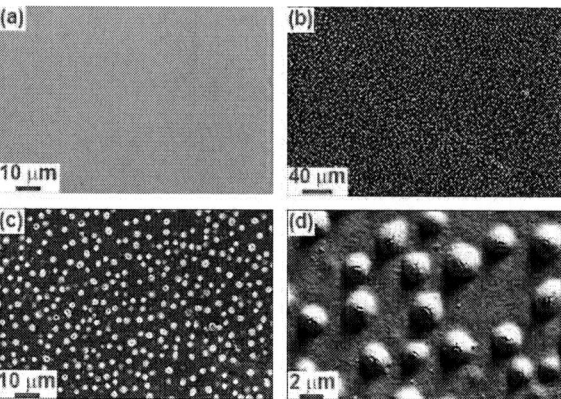

Fig. 8. SEM images of the surface of (a) Mo thin film before sulfurization (b-d) Sample B of MoS₂ at different magnifications.

E. Reflectance Measurement And Simulation

Fig. 9 compares the experimentally measured total reflectance (specular plus diffuse) spectrum of MoS₂ sample B with that simulated for a square array of parabolic domes (2 μm in diameter and 1 μm in height) with a period of 2.2 μm. The simulation of an area of a few microns requires large computing resources, therefore we chose to simulate a periodic spatial distribution with Bloch boundary condition. Although the regular distribution of simulated domes differs from the random arrangement on the actual sample, it nevertheless demonstrates a strong reduction in reflectance that agrees well with experiment. Based on experimental measurement and simulation domes with 2 μm base diameter could reduce the reflectance by 45% which consequently improves light harvesting and solar cell efficiency.

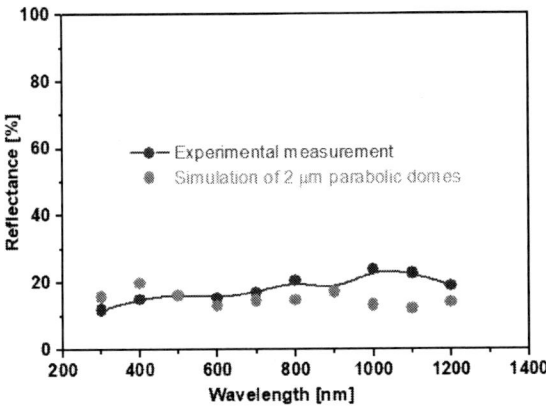

Fig. 9. The reflectance of MoS₂-on-sapphire sample B in comparison with simulation for parabolic domes with 2 μm base diameter.

V. CONCLUSION

Different structures and geometries have been studied and analyzed by FDTD method to assess their performance for antireflection applications. A comparison between cones, pyramids, parabolic and spherical domes of the same base area and height showed that parabolic domes are the best antireflection structures. A parabolic dome of 0.5μm base diameter and 2μm height can reduce reflectance to almost zero for all incident angles from 0° to 50°. These antireflections structures can boost MoS₂-based solar cells efficiency by almost 50%. MoS₂ films with microdome texture were grown by APCVD. SEM images revealed the uniform random distribution of these microdomes. Stoichiometric composition and structure of MoS₂ have been confirmed by EDX, Raman spectroscopy, and x-ray diffraction. Based on experimental measurement and simulation domes with 2 μm base diameter and 1 μm height could reduce the reflectance by 45%.

REFERENCES

[1] S. Lien, D. Wuu, W. Yeh, and J. Liu, "Tri-layer antireflection coatings (SiO₂/SiO₂–TiO₂/TiO₂) for silicon solar cells using a sol–gel technique," *Solar Energy Materials and Solar Cells*, vol. 90, pp. 2710-2719, 2006.

[2] V. M. Aroutiounian, K. Martirosyan, and P. Soukiassian, "Almost zero reflectance of a silicon oxynitride/porous silicon double layer antireflection coating for silicon photovoltaic cells," *Journal of Physics D: Applied Physics*, vol. 39, pp. 1623-1625, 2006.

[3] D. Bouhafs, "Design and simulation of antireflection coating systems for optoelectronic devices: Application to silicon solar cells," *Solar Energy Materials and Solar Cells*, vol. 52, pp. 79-93, 1998.

[4] B. W. Schneider, N. N. Lal, S. Baker-Finch, and T. P. White, "Pyramidal surface textures for light trapping and antireflection in perovskite-on-silicon tandem solar cells," *Optics Express*, vol. 22 pp. a1422-a1430, 2014.

[5] Y. C. Wang, H. Y. Cheng, Y. T. Yen, T. T. Wu, C. H. Hsu, H. W. Tsai, C. H. Shen, J. M. Shieh, and Y. L. Chueh, "Large-scale micro- and nanopatterns of Cu(In,Ga)Se₂ thin film solar cells by mold-assisted chemical-etching process," *ACS Nano*, vol. 9, pp. 3907-3916, 2015.

[6] H. P. Wang, T. Y. Lin, M. L. Tsai, W. C. Tu, M. Y. Huang, C. W. Liu, Y. L. Chueh, and J. H. He, "Toward efficient and omnidirectional n-type Si solar cells: concurrent improvement in optical and electrical characteristics by employing microscale hierarchical structures," *ACS Nano*, vol. 8, pp. 2959-69, 2014.

[7] D. Iencinella, E. Centurioni, R. Rizzoli, and F. Zignani, "An optimized texturing process for silicon solar cell substrates using TMAH," *Solar Energy Materials and Solar Cells*, vol. 87, pp. 725-732, 2005.

[8] L. Han and H. Zhao, "Simulation analysis of GaN microdomes with broadband omnidirectional antireflection for concentrator photovoltaics," *Journal of Applied Physics*, vol. 115, p. 133102, 2014.

[9] M. Nam, J. Lee, and K.-K. Lee, "Efficiency improvement of solar cells by importing microdome-shaped anti-reflective structures as a surface protection layer," *Microelectronic Engineering*, vol. 88, pp. 2314-2318, 2011.

[10] R. Dewan, I. Vasilev, V. Jovanov, and D. Knipp, "Optical enhancement and losses of pyramid textured thin-film silicon solar cells," *Journal of Applied Physics*, vol. 110, p. 013101, 2011.

[11] P. Papet, O. Nichiporuk, A. Kaminski, Y. Rozier, J. Kraiem, J. F. Lelievre, A. Chaumartin, A. Fave, and M. Lemiti, "Pyramidal texturing of silicon solar cell with TMAH chemical anisotropic etching," *Solar Energy Materials and Solar Cells*, vol. 90, pp. 2319-2328, 2006.

[12] H. H. Lin, W. H. Chen, and F. C. Hong, "Improvement of polycrystalline silicon wafer solar cell efficiency by forming nanoscale pyramids on wafer surface using a self-mask etching technique," *Journal of Vacuum Science & Technology B*, vol. 31, p. 31401, 2013.

[13] S. C. Baker-Finch and K. R. McIntosh, "Reflection of normally incident light from silicon solar cells with pyramidal texture," *Progress in Photovoltaics: Research and Applications*, vol. 19, pp. 406-416, 2011.

[14] W. H. Southwell, "Pyramid-array surface-relief structures producing antireflection index matching on optical surfaces," *Journal of the Optical Society of America A*, vol. 8, pp. 549-553, 1991.

[15] R. Dewan, S. Fischer, V. B. Meyer-Rochow, Y. Ozdemir, S. Hamraz, and D. Knipp, "Studying nanostructured nipple arrays of moth eye facets helps to design better thin film solar cells," *Bioinspiration and Biomimetics*, vol. 7, p. 016003, 2012.

[16] J. Zhu, C. M. Hsu, Z. Yu, S. Fan, and Y. Cui, "Nanodome solar cells with efficient light management and self-cleaning," *Nano Letters*, vol. 10, pp. 1979-84, 2010.

[17] L. Yang, Q. Feng, B. Ng, X. Luo, and M. Hong, "Hybrid moth-eye structures for enhanced broadband antireflection characteristics," *Applied Physics Express*, vol. 3, p. 102602, 2010.

[18] Y. Kim, N. D. Lam, K. Kim, W. K. Park, and J. Lee, "Ge nanopillar solar cells epitaxially grown by metalorganic chemical vapor deposition," *Scientific Reports*, vol. 7, p. 42693, 2017.

[19] B. D. Choudhury and S. Anand, "Rapid thermal annealing treated spin-on doped antireflective radial junction Si nanopillar solar cell," *Optics Express*, vol. 25, pp. A200-A207, 2017.

[20] J. Proust, A. L. Fehrembach, F. Bedu, I. Ozerov, and N. Bonod, "Optimized 2D array of thin silicon pillars for efficient antireflective coatings in the visible spectrum," *Scientific Reports*, vol. 6, p. 24947, 2016.

[21] P. Spinelli, M. A. Verschuuren, and A. Polman, "Broadband omnidirectional antireflection coating based on subwavelength surface Mie resonators," *Nature communications*, vol. 3, p. 692, 2012.

[22] R. Rebigan, A. Avram, F. Craciunoiu, R. Tomescu, E. Budianu, M. Purica, and M. Popescu, "Silicon plasma processing for antireflective micro-textured surfaces with applications for solar cells," presented at the CAS 2013 (International Semiconductor Conference), Sinaia, Romania, 2013.

[23] P. B. Clapham and M. C. Hutley, "Reduction of Lens Reflexion by the "Moth Eye" Principle," *Nature*, vol. 244, pp. 281-282, 1973.

[24] C. G. Bernhard and W. H. Miller, "A corneal nipple pattern in insect compound eyes," *Acta Physiologica Scandinavica*, vol. 56, pp. 385-386, 1962.

[25] P. I. Stavroulakis, S. A. Boden, T. Johnson, and D. M. Bagnall, "Suppression of backscattered diffraction from sub-wavelength 'moth-eye' arrays," *Opt Express*, vol. 21, pp. 1-11, 2013.

[26] C.-H. Sun, P. Jiang, and B. Jiang, "Broadband moth-eye antireflection coatings on silicon," *Applied Physics Letters*, vol. 92, p. 061112, 2008.

[27] M. Shanmugam, C. A. Durcan, and B. Yu, "Layered semiconductor molybdenum disulfide nanomembrane based Schottky-barrier solar cells," *Nanoscale*, vol. 4, pp. 7399-7405, 2012.

[28] A. Jäger-Waldau, M. C. Lux-Steiner, and E. Bucher, "MoS_2, $MoSe_2$, WS_2 and WSe_2 thin films for photovoltaics," *Solid State Phenomena*, vol. 37-38, pp. 479-484, 1994.

[29] L. Britnell, R. M. Ribeiro, A. Eckmann, R. Jalil, B. D. Belle, A. Mishchenko, Y. J. Kim, R. V. Gorbachev, T. Georgiou, S. V. Morozov, A. N. Grigorenko, A. K. Geim, C. Casiraghi, A. H. Castro Neto, and K. S. Novoselov, "Strong light-matter interactions in heterostructures of atomically thin films," *Science*, vol. 340, pp. 1311-1314, 2013.

[30] Z. Yu, Y. Pan, Y. Shen, Z. Wang, Z. Y. Ong, T. Xu, R. Xin, L. Pan, B. Wang, L. Sun, J. Wang, G. Zhang, Y. W. Zhang, Y. Shi, and X. Wang, "Towards intrinsic charge transport in monolayer molybdenum disulfide by defect and interface engineering," *Nature communications*, vol. 5, p. 5290, 2014.

[31] H. Zhang, Y. Ma, Y. Wan, X. Rong, Z. Xie, W. Wang, and L. Dai, "Measuring the refractive index of highly crystalline monolayer MoS_2 with high confidence," *Scientific Reports*, vol. 5, p. 8440, 2015.

[32] T. C. Choy, *Effective medium theory principles and applications*. Oxford, United Kingdom: Oxford University Press, 2016.

[33] M. C. Munisso, W. Zhu, and G. Pezzotti, "Raman tensor analysis of sapphire single crystal and its application to define crystallographic orientation in polycrystalline alumina," *Physica Status Solidi (b): Basic Solid State Physics*, vol. 246, pp. 1893-1900, 2009.

[34] H. Li, Q. Zhang, C. C. R. Yap, B. K. Tay, T. H. T. Edwin, A. Olivier, and D. Baillargeat, "From bulk to monolayer MoS_2: evolution of Raman scattering," *Advanced Functional Materials*, vol. 22, pp. 1385-1390, 2012.

[35] A. M. Stacy and D. T. HoduL, "Raman spectra of IVB and VIB transition metal disulfides using laser energies near the absorption edges," *Journal of Physics and Chemistry of Solids*, vol. 46, pp. 405-409, 1985.

[36] G. L. Frey, R. Tenne, M. J. Matthews, M. S. Dresselhaus, and G. Dresselhaus, "Raman and resonance Raman investigation of MoS_2 nanoparticles," *physical Review B: Condensed Matter*, vol. 60, pp. 2883-2892, 1999.

[37] J. M. Chen and C. S. Wang, "Second order Raman spectrum of MoS_2" *Solid State Communications*, vol. 14, pp. 857-860, 1974.

[38] W. M. Yim and R. J. Paff, "Thermal expansion of AlN, sapphire, and silicon," *Journal of Applied Physics*, vol. 45, pp. 1456-1457, 1974.

[39] E. M. Dudnik and V. K. Oganesyan, "Thermal expansion of some sulfides of the transition metals," *Soviet Powder Metallurgy and Metal Ceramics*, vol. 5, pp. 125-127, 1966.

Study of spatial distribution of electrical, optical and structural properties of magnetron sputtered AZO thin films

Mohit Agarwal[a] and Rajiv O Dusane[b]

[a]Department of Electronics and Communication Engineering, Thapar University, Patiala 147004, Punjab
[b]Department of Metallurgical Engineering & Materials Science, Indian Institute of Technology Bombay, Mumbai, 400076, India

Abstract — **Aluminum doped zinc oxide (AZO) is a commonly used transparent conducting oxide (TCO) material that is currently an integral part of thin film solar cells and silicon-based heterojunction photovoltaics. However, the large area uniformity and homogeneity of TCOs are major challenges to use this material as a front electrode in solar modules. This requires a detailed study of spatial distribution of TCOs' properties over the large area deposition. In this report, the electrical, optical and structural properties of AZO thin films deposited using RF magnetron sputtering technique have been studied spatially. It is observed from the results that the highly transparent and highly conducting films can be deposited over the substrate area which are approximately equal and above to the target area. The film crystallinity deteriorates away from the target region which strongly supports the XRD results.**

I. INTRODUCTION

TCOs are highly electrically conductive material having high transparency in the visible part of the solar spectrum, commonly deposited using CVD, RF/DC sputtering techniques [1]. These TCOs are currently being used in variety of applications such as solar cells, liquid crystal displays, light emitting diodes, low-emissivity windows and thin film transistors, etc. [3-8]. In particular, they are widely used in a variety of photovoltaic devices such as crystalline-Si heterojunction with intrinsic thin layer (HIT) cells, silicon based thin film solar cell and organic photovoltaics [9, 10]. Highly transparent and conductive TCO films are required for the photovoltaic application. Currently, many oxides have emerged as commercially important transparent conductors such as indium oxide, tin oxide, and zinc oxide [1]. The most common substitutional impurities used to increase the conductivity are Al in Zinc Oxide, Sn in indium oxide and Sb in tin oxide. The advantages of ZnO over all other TCOs are its low material cost, non-toxicity, high crystallinity and stability in hydrogen plasma as compared to SnO_2 [11]. In recent years, TCO thin films are extensively used as a front contact layer in thin film and flexible photovoltaic technology [12]. To obtain high quality of AZO thin films various deposition techniques such as pulsed laser deposition [13], ultrasonic spray pyrolysis [14-16], electrostatic spray deposition [16], chemical vapor deposition [17] and sol-gel method, etc. [18, 20] have been used worldwide. The high-quality AZO films can be prepared by reactive and non-reactive magnetron sputtering [20, 21]. The use of magnetron sputtering has many advantages such as easy control over preferred orientations, uniform films, large area deposition, good adhesion of the films on substrates, high deposition rate

and film density improvement [22-25]. However, it has been shown that plasma ion-bombardment significantly affects the electrical conductivity uniformity over the substrate in magnetron sputtering method [26]. On the other hand, some studies show that the spatial non-homogeneity in conductivity is due to the bombardment of energetic electrons on the film [27]. In agreement to these results, some groups also suggested that the bombardment of energetic oxygen atoms or oxygen ions can also affect the film properties leading to the spatial non-homogeneity of conductivity [28].

In this paper, AZO films were deposited on corning glass by RF magnetron sputtering method using 2 wt% of alumina doped zinc oxide ceramic target. The spatial distributions of electrical and crystalline properties of AZO thin films have been discussed.

II. EXPERIMENTAL DETAILS

AZO thin films were deposited on Corning glass substrates by RF magnetron sputtering method using 3-inch ceramic target. The sputtering was carried out in an argon ambiance.

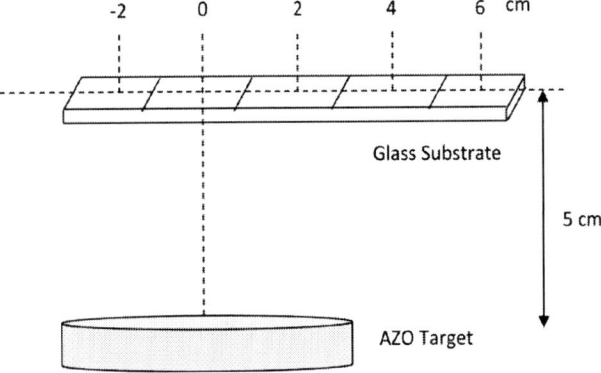

Figure 1 The schematic diagram of the substrate position relative to the center of AZO ceramic target.

The base pressure inside the chamber was 10^{-7} mbar created by turbo molecular pump. Before loading the substrate into the chamber, the target was sputtered for 5 minutes to clean the target. The distance between the substrate and target was kept constant at 50 mm and the films were deposited by applying RF power of 200 W. The substrate heating was kept constant at 200 °C during the deposition of AZO thin films. To investigate the spatial distribution of electrical, optical and structural properties of AZO thin films, 5 X 8 cm^2 area corning glass is used for the deposition. The arrangement of the substrate has been shown in Figure 1.

The crystallographic studies of the film were carried out by X-ray diffraction method (PANalytical X'PERT PRO MRD). The thickness variation of the films as a function of position with respect to the center of the target was measured using thickness profilometer. The electrical resistivity was measured at room temperature by the Van der Pauw method. The optical transmittance of AZO films was measured in the wavelength range 300 to 900 nm by the UV-Visible spectrometer. The microstructure of AZO thin films was studied by the scanning electron microscope (JEOL JSM 7600F).

III. RESULTS AND DISCUSSION

The thickness of AZO thin films varies significantly as a function of position on the substrate as reported by earlier studies [30]. Minami et. al. reported that the deposition rate decreases rapidly as the substrate to target relative position changes in either direction with respect to the center of target [25]. Figure 2 shows the characteristics of AZO thin films as a function of position on the substrate deposited at a particular condition.

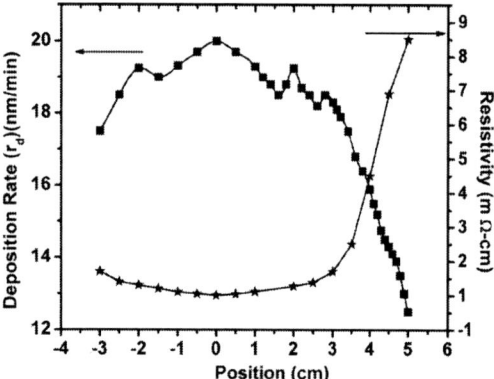

Figure 2 The deposition rate and resistivity of AZO thin films as a function of distance relative to origin of substrate corresponding to center of the target

The position 0 cm in this figure corresponds to the center of the target. The deposition rate is high (19 to 20 nm/min) in the target range of ±2 cm, and it decreases sharply as the substrate to target relative increases further. It can be seen from the Figure 2 that nearly 95% uniformity in the film thickness is found in the central substrate region of diameter ≥ ½ of target diameter. Moving 4-5 cm away on the substrate from the center of target in lateral distances, the thin film thickness decreased down to 1/2 of the maximal value. It is also observed from Figure 2 that in front of the target erosion area the deposition rate has irregular changes. The AZO thin films are having very low resistivity (5x10-4 Ω-cm) even at very high deposition rate as can be seen in Figure 2. More interestingly, the area of lower resistivity was almost equal to the available sputtering area of the target. The resistivity of AZO thin films increases up to the 8x10-3 Ω-cm as a lateral

distance increases to 5 cm from the center of the target. The same behavior of the spatial distribution of electric resistivity is observed by Minami et.al. [23]. The results can be explained due to the fact that one can expect very high power density in front of the target region compared to the region outside the target peripheral. The high-power density could lead to the large grain size.

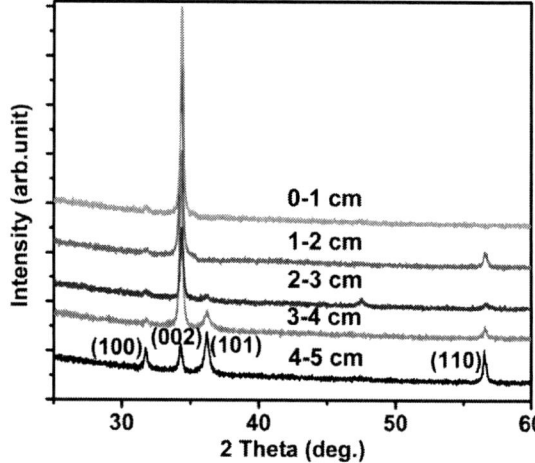

Figure 3 The XRD patterns of AZO thin films as a function of different defined positions on the substrate.

To confirm this XRD pattern of AZO thin film was taken at different places on the glass substrate. The XRD results show that the crystalline properties of the films are strongly influenced by the different lateral positions on the substrate. The 0-1 cm as shown in Figure 3 indicates the position on the substrate that corresponds to the 1 cm around to the center of the target and other positions have been defined similarly. It can be seen from XRD that major peak at 34.4 theta shows the AZO film is polycrystalline with an orientation perpendicular to the substrate surface. It is reported that AZO thin films having (002) orientations shows the better conductivity.

Further around 2 cm to the center of the target the AZO thin film shows the highly intense peak of (002) orientation. As we move further away from the defined center on the substrate some other peaks get appeared corresponding to (100), (101) and (110) orientations. Moving 4-5 cm the (101) orientation becomes more dominating than (002) orientated peak; the resistivity also increases in the same position as shown in Figure 1. The appearance of other orientation in XRD data may be explained by the variation of sputtered particle fluxes at different lateral positions. The change in the mutual ratios of Al, Zn and O may also be the reason for different orientations of AZO thin films at 4-5 cm away from the center.

The results could be compared with the resistivity curve. It is clear from Figures 2 and 3 that spatial distributions of resistivity and that of the XRD pattern of the film can be correlated each other. The FWHM of the AZO film which is calculated separately from the XRD (002) peak, varies from 0.25 to 0.28 as we move from the center to 5 cm away.

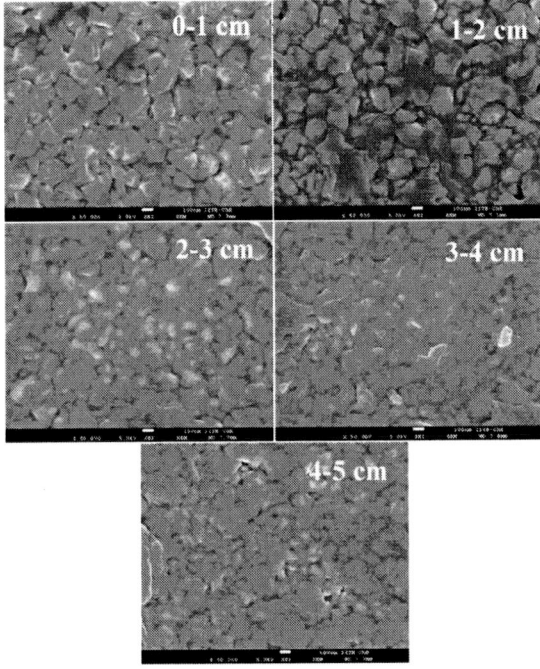

Figure 4 The SEM images of AZO thin films at the different substrate to target position.

Figure 4 shows the SEM image of the AZO films deposited on glass substrate at a different relative position to the center of the target. The SEM image also shows the strong correlation between the resistivity values at different positions. The resistivity of AZO films is low in the target range of ±3 cm since SEM images are showing the presence of highly regular shaped and good density grains. However, as we further move away from the target center, the irregular – shape grain could be seen. There are some pin holes also visible at 5 cm away from the target center. This results the increase in the resistivity of the AZO thin film. The ion bombardment of the film by energetic oxygen atoms or oxygen ions may affect the crystalline properties and hence make the non-homogeneity in the film resistivity.

Figure 5 shows the optical transmittance data of AZO thin films that were taken at different positions from the defined center. At all positions, the average transmittance is nearly 85 %. From Figure 5 it is clear that higher conductivity leads to decrease in transmittance in near infrared range. This can be attributed due to the light has been absorbed by the free carrier in the AZO films in near infrared range. It is also seen from Figure 5 that the AZO thin films deposited in the central region shows a clear blue shift in the transmittance spectra. This result was highly expectable as the high conductivity is found in the same area. The high conductivity is due to high

charge carrier concentration which leads to the widening of the optical band gap so called Burstein-Moss effect.

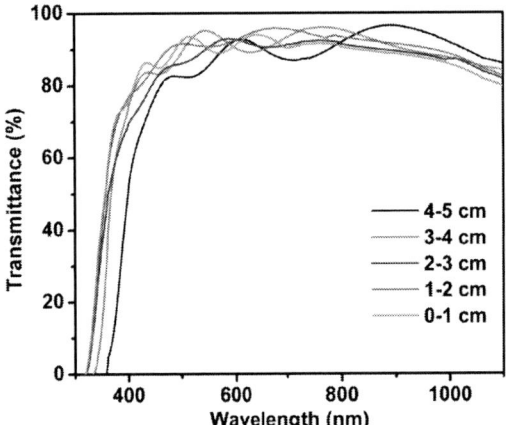

Figure 5 The optical transmittance of AZO thin films at the different substrate to target position.

III. Conclusions

Highly conductive and transparent AZO thin films were prepared by RF magnetron sputtering method. This study examines the spatial distribution of electrical, optical and structural properties of AZO thin films deposited on the glass substrate. It is found that highly conducting AZO films can be obtained within the area of ±3 cm around the center of the target. As the distance from the center of the target increases the properties of AZO, thin films start degrading. The XRD shows that the target facing area on the substrate has dominant (002) peak which is requisite for higher conductivity while other unwanted peaks (100), (101) and (110) start dominating in the remaining area. Further the SEM and transmittance data shows the strong correlation between the XRD and conductivity results.

References

[1] A. Stadler, Transparent Conducting Oxides—An Up-To-Date Overview, Materials, 5 (2012) 661-683.
[2] S.J. Tark, M.G. Kang, S. Park, S.H. Lee, C.S. Son, J.C. Lee, D. Kim, Characterization of hydrogenated Al-doped ZnO films prepared by multi-step texturing for photovoltaic applications, Current Applied Physics, 11 (2011) 362-367.
[3] G. Luka, T.A. Krajewski, B.S. Witkowski, G. Wisz, I.S. Virt, E. Guziewicz, M. Godlewski, Aluminum-doped zinc oxide films grown by atomic layer deposition for transparent electrode applications, Journal of Materials Science: Materials in Electronics, 22 (2011) 1810-1815.
[4] H. Saarenpää, T. Niemi, A. Tukiainen, H. Lemmetyinen, N. Tkachenko, Aluminum doped zinc oxide films grown by atomic layer deposition for organic photovoltaic devices, Solar Energy Materials and Solar Cells, 94 (2010) 1379-1383.

[5] Y.S. Park, M.S. Seo, J.S. Kim, J. Lee, The effect of substrate temperature on Al-doped ZnO characteristics for organic thin film transistor applications, Materials Research Bulletin, 48 (2013) 5136-5140.

[6] S.L. Ou, D.S. Wuu, S.P. Liu, Y.C. Fu, S.C. Huang, R.H. Horng, Pulsed laser deposition of ITO/AZO transparent contact layers for GaN LED applications, Optics Express, 19 (2011) 16244-16251.

[7] C.Y. Park, B.H. Choi, J.H. Lee, Electrical and optical properties of index-matched transparent conducting oxide layers for liquid crystal on si projection displays, Japanese Journal of Applied Physics, 52 (2013).

[8] F. Giovannetti, S. Föste, N. Ehrmann, G. Rockendorf, High transmittance, low emissivity glass covers for flat plate collectors: Applications and performance, Solar Energy, 104 (2014) 52-59.

[9] S.J. Tark, C.S. Son, D. Kim, Changes in interface properties of TCO/a-Si:H layer by Zn buffer layer in silicon heterojunction solar cells, Korean Journal of Materials Research, 21 (2011) 341-346.

[10] Y. Lu, X. Zhang, J. Huang, J. Li, T. Wei, P. Lan, Y. Yang, H. Xu, W. Song, Investigation on antireflection coatings for Al:ZnO in silicon thin-film solar cells, Optik, 124 (2013) 3392-3395.

[11] J.-H. Lan, J. Kanicki, A. Catalano, J. Keane, Alternative transparent conducting oxide to ITO for the a-Si:H TFT-LCD applications, Proceedings of the International Workshop on Active Matrix Liquid Crystal Displays, AMLCDs, 1995, pp. 54-57.

[12] M. Agarwal, P. Modi, R.O. Dusane, Study of electrical, optical and structural properties of Al-doped ZnO thin films on PEN substrates, Journal of Nano- and Electronic Physics, 5 (2013).

[13] P. Gondoni, M. Ghidelli, F. Di Fonzo, V. Russo, P. Bruno, J. Martí-Rujas, C.E. Bottani, A. Li Bassi, C.S. Casari, Structural and functional properties of Al:ZnO thin films grown by Pulsed Laser Deposition at room temperature, Thin Solid Films, 520 (2012) 4707-4711.

[14] A. Djelloul, M.S. Aida, J. Bougdira, Photoluminescence, FTIR and X-ray diffraction studies on undoped and Al-doped ZnO thin films grown on polycrystalline α-alumina substrates by ultrasonic spray pyrolysis, Journal of Luminescence, 130 (2010) 2113-2117.

[15] J.H. Lee, B.O. Park, Characteristics of Al-doped ZnO thin films obtained by ultrasonic spray pyrolysis: Effects of Al doping and an annealing treatment, Materials Science and Engineering B: Solid-State Materials for Advanced Technology, 106 (2004) 242-245.

[16] K. Mahmood, S.B. Park, Atmospheric pressure based electrostatic spray deposition of transparent conductive ZnO and Al-doped ZnO (AZO) thin films: Effects of Al doping and annealing treatment, Electronic Materials Letters, 9 (2013) 161-170.

[17] A. Illiberi, P.J.P.M. Simons, B. Kniknie, J. Van Deelen, M. Theelen, M. Zeman, M. Tijssen, W. Zijlmans, H.L.A.H. Steijvers, D. Habets, A.C. Janssen, E.H.A. Beckers, Growth of ZnO x:Al by high-throughput CVD at atmospheric pressure, Journal of Crystal Growth, 347 (2012) 56-61.

[18] H. Karaagac, E. Yengel, M. Saif Islam, Physical properties and heterojunction device demonstration of aluminum-doped ZnO thin films synthesized at room ambient via sol-gel method, Journal of Alloys and Compounds, 521 (2012) 155-162.

[19] M.H. Mamat, Z. Khusaimi, M.Z. Musa, M.F. Malek, M. Rusop, Fabrication of ultraviolet photoconductive sensor using a novel aluminium-doped zinc oxide nanorod-nanoflake network thin film prepared via ultrasonic-assisted sol-gel and immersion methods, Sensors and Actuators, A: Physical, 171 (2011) 241-247.

[22] H.X. Chen, J.J. Ding, X.G. Zhao, S.Y. Ma, Microstructure and optical properties of ZnO:Al films prepared by radio frequency reactive magnetron sputtering, Physica B: Condensed Matter, 405 (2010) 1339-1344.

[21] J.I. Nomoto, T. Hirano, T. Miyata, T. Minami, Preparation of Al-doped ZnO transparent electrodes suitable for thin-film solar cell applications by various types of magnetron sputtering depositions, Thin Solid Films, 520 (2011) 1400-1406.

[22] F.H. Wang, H.P. Chang, C.C. Tseng, C.C. Huang, Effects of H2 plasma treatment on properties of ZnO:Al thin films prepared by RF magnetron sputtering, Surface and Coatings Technology, 205 (2011) 5269-5277.

[23] T. Minami, K. Oohashi, S. Takata, Preparations of ZnO:Al transparent conducting films by d.c. magnetron sputtering, Thin Solid Films, 194 (1990) 721-729.

[24] Y. Nishi, K. Hirohata, N. Tsukamoto, Y. Sato, N. Oka, Y. Shigesato, High rate reactive magnetron sputter deposition of Al-doped ZnO with unipolar pulsing and impedance control system, Journal of Vacuum Science and Technology A: Vacuum, Surfaces and Films, 28 (2010) 890-894.

[25] F. Jiao, C. Liao, J.F. Han, Z. Zhou, Preparation of large area Al-ZnO thin film by DC magnetron sputtering, Guang Pu Xue Yu Guang Pu Fen Xi/Spectroscopy and Spectral Analysis, 29 (2009) 698-701.

[26] N. Ito, N. Oka, Y. Sato, Y. Shigesato, Effects of energetic ion bombardment on structural and electrical properties of Al-doped ZnO films deposited by RF-superimposed DC magnetron sputtering, Japanese Journal of Applied Physics, 49 (2010) 0711031-0711035.

[27] K. Tominaga, T. Yuasa, M. Kume, O. Tada, Influence of energetic oxygen bombardment on conductive ZnO films, Japanese Journal of Applied Physics, Part 1: Regular Papers & Short Notes, 24 (1985) 944-949.

[28] T. Minami, H. Nanto, H. Sato, S. Takata, Effect of applied external magnetic field on the relationship between the arrangement of the substrate and the resistivity of aluminium-doped ZnO thin films prepared by r.f. magnetron sputtering, Thin Solid Films, 164 (1988) 275-279.

Multiband Formation in Cr doped CuGaS₂ Thin Films Synthesized by Chemical Spray Pyrolysis

Nazmul Ahsan[1], Sivaperuman Kalainathan[2], Naoya Miyashita[1], Takuya Hoshii[3], and Yoshitaka Okada[1]

[1]Research Center for Advanced Science and Technology (RCAST), The University of Tokyo, Japan
[2]Centre for Crystal Growth, School of Advanced Sciences, VIT University, Vellore, India
[3]School of Engineering, Tokyo Institute of Technology, Tokyo, Japan

Abstract — We study the formation of multiband electronic structures for Cr doped chalcopyrite CuGaS₂ thin films synthesized by chemical spray pyrolysis. Electronic transition studies using photo-modulated reflectance (PR) spectra revealed that the bandgaps were widened when Cr was added, and were accompanied by a new transition due to intermediate bands (IB) formation within the bandgap. This is in consistent with our calculation based on density functional theory. PR transitions due to native defects were prominent for pure CuGaS₂ thin films, and their transition strengths were reduced in the Cr added films. These observations are consistent with photo-luminescence (PL) spectra measured at room temperature.

Index Terms — Intermediate band solar cell, chalcopyrite, II-VI semiconductors, spray pyrolysis, density functional theory

I. INTRODUCTION

The addition of an intermediate band (IB) can extend the functional limit posed by the conventional two-band character of semiconductor materials. For example, the concept of intermediate band solar cells (IBSC) makes use of an intermediate state or band to excite electrons by below-bandgap photons from the valence band (VB) to the conduction band (CB) by two-step photon absorption (TSPA) that accounts to a higher solar energy conversion efficiency [1]. Although the key operational demonstration of TSPA at room temperature has been presented recently, the energy conversion efficiency boost is yet very low, and much room remains for the development of multiband materials for practical application [2].

The quest for three band system have been spanned over various IB materials in recent years such as thin films of highly mismatched alloy (HMA) III-V dilute nitrides and II-VI oxides [2-5], deep impurity doped hosts [6], and nanostructures using quantum dots [6], quantum rings [7], etc. Key factors that drive IB formation process are different in each approach, and consist of host electronic structures, structural dimensionality, nature of the impurity atoms, etc. In HMA thin films, isoelectric impurities atoms with much different affinity and size in III-V and II-VI hosts modifies electronic structures remarkably with distinct three band features [8-10]. Low-dimensional structures such as QD array produce distinct mini-bands due to quantum-size effect [1]. Fabrication of large number in density and stacks of QD array can promote the needed optical absorption for IR-generated currents. In order to improve the IBSC efficiency, host materials with much wider band gap such as AlGaAs or InGaP are required [11].

Copper-based chalcopyrite semiconductors attract research interests mainly due to their higher absorption coefficients suitable for thin-film solar cell application. Of particular, the direct bandgap of CuGaS₂ or 'CGS' in short lies in the green region of the visible spectrum at room temperature [12]. It can be a suitable IB host since its bandgap of 2.4eV is relatively wide, and can accommodate impurity bands sufficiently deep to avoid thermal escape of photo-generated carriers staying at the IB bands. Doping of transition metals in CGS has been predicted to be a potential candidate for IB solar cells [13]. Earlier reports for doping of transition metals such as Fe [14], V [15], Mn [16], Cr [17], Zn, Ti [18] to the CGS hosts have been predicted for the creation of IB. The valency match and the less distortion in lattice make the transition and rare earth elements to a suitable dopant for a chalcopyrite lattice.

Recently, we have reported that chemical spray pyrolysis technique can be employed for Cr doped CGS thin film synthesis on glass substrates with single phase crystalline qualities [19]. In the present paper, at first, we study the optical properties using photo-modulated reflectance (PR) and photoluminescence (PL) characterizations at room temperature. Next, electronic band structure analyses are performed based on density functional theory (DFT). We attribute the sub-bandgap transitions in PR to the impurity band in the Cr added CGS thin films, and show that the analyses are in agreement with the DFT calculation.

II. EXPERIMENTAL

Thin films were synthesized by chemical spray pyrolysis technique on glass substrates at 250 °C. The precursor solution for the host CGS synthesis consists 0.1 M of copper acetate, gallium chloride and thiourea dissolved in deionised water. For Cr doping, the chromium chloride was varied between 1-4 weight percent (wt.%). Thin films were around 2 μm-thick. The Xray diffraction patterns were recorded using Powder X-ray Diffractometer and showed (112) oriented single phase. Details of the synthesis and physical characterization are published elsewhere [19]. Photo-modulated reflectance (PR) spectroscopy was performed at room temperature utilizing a

978-1-5090-5606-4/17 $31.00 © 2017 IEEE

mechanically chopped 405 nm line of a 30 mW solid-state laser. The applied laser intensity for PR measurements was kept constant at 33 mW/cm². Photo-luminescence (PL) spectroscopy was performed at room temperature utilizing 405 nm line a solid-state laser. In order to improve the PL signals by suppressing wave-guiding effect of the glass substrate, signal collection was made under total reflection condition monitored by reflectivity from substrate using a secondary photo detector. The electronic structures were calculated based on density functional theory (DFT) using GGA exchange-correlation functional.

III. RESULTS AND DISCUSSION

A. Photo-modulated reflectance (PR) characterization

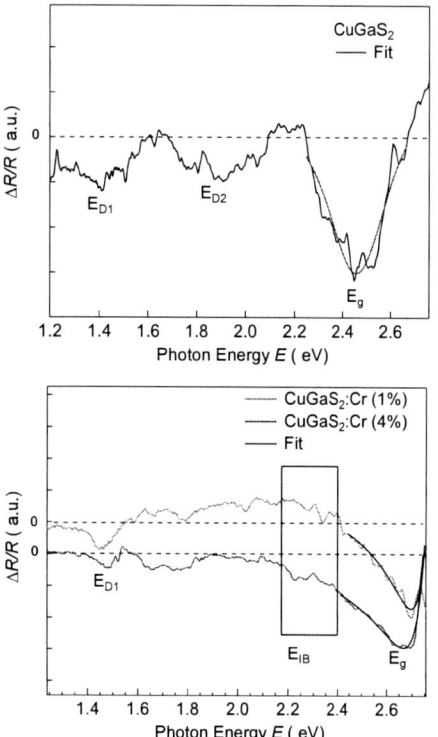

Fig 1. Room temperature PR spectra of CuGaS₂ thin films *without* and *with* Cr impurities. PR features are assigned to two native defects E_{D1} and E_{D2} at low energy region, and an intermediate bands E_{IB} and bandgaps E_g at high energy region. Solid lines are least squares fits for the bandgaps.

PR spectra of the samples are shown in Figs. 1. The structures in the high energy regions are related to the host bandgap E_g (VB→CB) transitions. The E_g values and width (Γ) of the transitions were retrieved using a third derivative low electric field model, which are 2.47 eV and 0.22 eV for

the CGS film without Cr. The E_g and Γ values are 2.87 eV and 0.03 eV for 1% Cr, and 2.87 and 0.09 eV for 4% Cr-added samples, respectively. The high Γ values suggest for the presence of compositional and spatial inhomogeneity in the thin films. On the other hand, Cr addition extend the bandgaps of the host material from 2.47 eV to 2.87 eV. The long tails below the bandgaps are commonly observed in materials when compositional inhomogeneity is prominent.

Inside the host bandgap, as shown in Fig. 1(a), there appears two additional transitions E_{D1} and E_{D2} involving defect bands of CGS native defects. With Cr addition, transition strength of both of the defect bands in Fig. 1(b) are much suppressed. This leads to an overall improvement of the PR background in the bandgap region, and followed by transitions E_{IB} below the E_g location in the Cr added films in Fig. 1(b). This transition feature (VB→IB) indicates the emergence of an intermediate band (IB) inside the gap. The energy position E_{IB} is around 2.25 eV. The widening of the bandgap accompanied by an IB formation in Cr added CGS is in consistent with our calculation based on density functional theory discussed below.

B. Photo-luminescence (PL) characterization

Figure 2 shows the PL spectra for CGS thin films with different level of Cr dopants. The sharp line at about 1.53 eV is grating-scattered light of the diode-pumped laser (405 nm or 3.06 eV). Another narrow peaks at 1.91 eV also are related to the scattered light. Although in standard case optical filters are set to cancel scattered light out, such a measure was not taken in the present experiments in order to allow PL detection from hosts' high bandgaps.

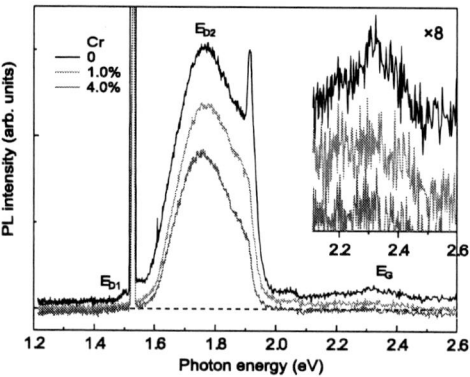

Fig.2. Room-temperature PL spectra for the CGS thin films with different Cr contents. Transitions are assigned to two native defects E_{D1} and E_{D2} at low energy region, and bandgaps E_g at high energy regions. The inset 2 shows the enlargement of the PL magnitude between 2.2 – 2.6 eV in which response from the IB bands are expected.

978-1-5090-5606-4/17 $31.00 © 2017 IEEE

At low energy region, a native defect peak overlapped at the 1.53 eV line, assigned E_{D1} in PR, and is faintly observed in the pure CGS. The broad and strong PL peak at around 1.8 eV matches well with the PR assigned peak E_{D2}. Past reports on MOVPE grown samples attributed E_{D2} to a defect band due to low Cu/Ga ratios (< 0.84). A similar band is also reported for thin films produced under S-poor conditions, and annealing under S-overpressure has been shown effective to remove the defects. PL intensities of the defect bands are observed reduced with Cr addition, and are consistent with similar level of reduction in PR transition strengths of the defects. Perhaps, the local atomic rearrangement due to Cr addition modifies the formation process of native defects related to Cu/Ga ratio.

At the high energy region in Fig. 3, weak PL signals can be observed at around 2.33 eV for the pure CGS thin film, and nearly matches with PR assigned E_g peak except a minor red-shift of the PL by around 100 meV. Such a red-shift can be ascribed to the presence of long low-energy tails observed in PR. Past report show that PL signals often are generated from tail states where strong thermal trapping and re-trapping occur, and can give rise to S-shape behaviour in temperature-dependent PL peaks. An improvement in material and spatial homogeneity can reduce the red-shift. The Cr-added samples did not produce E_g related PL at photon energy up to 2.6 eV beyond which the direct beam of excitation laser affects PL background significantly, and perhaps buries the E_g signals of Cr added samples underneath.

The inset of Fig. 2 shows the enlargement of the PL magnitude at the high energy region in which response from the IB band is expected. Weak peaks can be observed for the Cr added samples nearly at the host E_g location. Since PR analysis also suggests about the presence of the E_{IB} nearby, it is difficult to distinguish the assignment of the peaks. These signals are likely due to IB bands since both PR studies and calculation suggest about the location of the E_g at higher energies than that of the pure CGS.

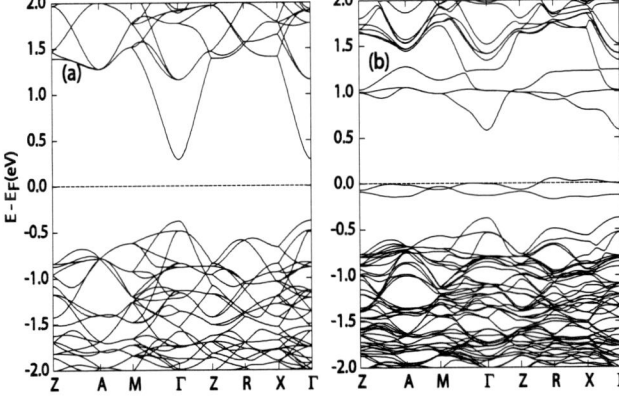

Fig 3. The bulk electronic band structure for a narrow range of energy for (a) CuGaS$_2$ and (b) CuGaCrS$_2$ in the GGA calculations

after dynamical relaxation. Fermi levels of the compounds are set at the energy zero. Structures are displayed in the main directions of the corresponding first Brillouin zone.

C. Band structure analysis

In order to analyse the electronic structure, band structures were calculated by first principles theoretical calculations based on density functional theory (DFT). The 2×2×1 super-cell considered in the present work is a 64-atom system consisting four units of conventional 16-atom chalcopyrite tetragonal cells stacked along the x- and y-axes. One Ga atom of the supercell was replaced with the Cr accounting the impurity concentration to 6.25% (or, x=0.0625 in CuGa$_{1-x}$Cr$_x$S$_2$). This value is comparable to that used in the present experiments.

For the self-consistent DFT calculations, norm-conserving, nonlocal pseudopotentials and linear combination of atomic orbitals are used. An extended size of basis set (double zeta) has been used. The exchange-correlation effects of the valence electrons were described through the generalized gradient approximation (GGA), within the Perdew-Burke-Ernzerhof (PBE) functional. In order to get accurate results, atomic coordinates were optimized by relaxation and minimization of the total energy and atomic forces. Before relaxation, CGS lattice constants were set, a=5.372 Å and c/2a=0.98.

Figure 3 shows the band structures for a narrow energy range along the high symmetrical directions of the first Brillouin zone for the pure CGS and Cr doped crystals. The direct bandgap character of the host VB-CB is nearly preserved after Cr addition except that the band-gap is widened and an intermediate band of states appears in the bandgap. It is noticeable that the Fermi level exists inside the intermediate band, and hence metallic in character. This situation of IB state occupation is suitable for TSPA, a necessary requirement for the IBSC operation [20]. The values for the direct bandgaps are 0.64 eV for the pure CGS and 0.88 eV for the Cr-doped crystals. Although experimental values of our CGS is around 2.4 eV [19], such an underestimation of the calculated bandgaps is an inherent feature of the DFT method.

These narrow IB are fairly separated from VB and CB. We notice that several location of IB compete for transitions between VB and CB. Between IB-CB, two indirect gaps close to the IB minima nearby R and X points compete with the direct gap at Γ point. In a similar manner between IB-CB, the two IB maxima nearby R and X points compete with the direct bandgap at Γ point. Within the IB, the energy difference of the minima and maxima at R and X points with respect to the Γ point ranges between 20-50 meV. This value is fairly comparable to the thermal energy at room temperature, and can promote inter-valley carrier-exchange within the IB and thus, can enhance TSPA involving VB, IB and CB.

IV. CONCLUSION

The key finding of the present study is the observation of sub-band gap optical response for Cr doped CGS thin films, and is assigned to an intermediate band. Band structure analysis based on DFT shows that the intermediate band is metallic in character which is suitable for efficient TSPA activity during an IBSC operation.

ACKNOWLEDGEMENT

This work is supported by New Energy and Industrial Technology Development Organization (NEDO), and Ministry of Economy, Trade and Industry (METI), Japan. Parts of this work is performed under the JSPS-DST bilateral program, and supported by Hirose International Scholarship Foundation, Japan.

REFERENCES

[1] Y. Okada, N. J. Ekins-Daukes, T. Kita, R. Tamaki, M. Yoshida, A. Pusch, *et al.*, "Intermediate band solar cells: Recent progress and future directions," *Applied Physics Reviews,* vol. 2, p. 021302, 2015.

[2] N. Ahsan, N. Miyashita, M. M. Islam, K. M. Yu, W. Walukiewicz, and Y. Okada, "Two-photon excitation in an intermediate band solar cell structure," *Appl. Phys. Lett.,* vol. 100, p. 172111, 2012.

[3] T. Tanaka, K. M. Yu, A. Levander, O. Dubon, L. Reichertz, N. Lopez, *et al.*, "Demonstration of ZnTe1-xOx Intermediate Band Solar Cell," *Jpn. J. Appl. Phys.,* vol. 50, p. 082304, 2011.

[4] T. Tanaka, M. Miyabara, Y. Nagao, K. Saito, Q. Guo, M. Nishio, *et al.*, "Photogenerated Current By Two-Step Photon Excitation in ZnTeO Intermediate Band Solar Cells with n-ZnO Window Layer," *IEEE J. Photovolt.,* vol. 4, p. 196, 2014.

[5] N. Lo´pez, L. A. Reichertz, K. M. Yu, K. Campman, and W. Walukiewicz, "Engineering the Electronic Band Structure for Multiband Solar Cells," *Phys. Rev. Lett.,* vol. 106, p. 028701, 2011.

[6] R. Oshima, A. Takata, and Y. Okada, "Strain-compensated InAs/GaNAs quantum dots for use in high-efficiency solar cells," *Appl. Phys. Lett.,* vol. 93, p. 083111, 2008.

[7] J. Wu, D. Shao, Z. Li, M. O. Manasreh, V. P. Kunets, Z. M. Wang, *et al.*, "Intermediate-band material based on GaAs quantum rings for solar cells," *Appl. Phys. Lett.,* vol. 95, p. 071908, 2009.

[8] W. Walukiewicz, W. Shan, K. M. Yu, J. W. A. III, E. E. Haller, I. Miotkowski, *et al.*, "Interaction of Localized Electronic States with the Conduction Band: Band Anticrossing in II-VI Semiconductor Ternaries," *Phys. Rev. Lett.,* vol. 85, p. 1552, 2000.

[9] J. Wu, W. Walukiewicz, K. M. Yu, J. W. Ager, E. E. Haller, I. Miotkowski, *et al.*, "Origin of the large band-gap bowing in highly mismatched semiconductor alloys," *Phys. Rev. B,* vol. 67, p. 035207, 2003.

[10] W. Shan, K. M. Yu, W. Walukiewicz, J. Wu, J. W. A. III, and E. E. Haller, "Band anticrossing in dilute nitrides," *J. Phys.: Cond. Mat.,* vol. 16, p. S3355, 2004.

[11] Y. Dai, M. A. Slocum, Z. Bittner, S. Hellstroem, D. V. Forbes, and S. M. Hubbard, "Optimization in wide-band-gap quantum dot solar cells," *43rd IEEE Photovoltaic Specialists Conference (PVSC),* pp. 0151 - 0154, 2016.

[12] W.-J. Jeong and G.-C. Park, "Structural and electrical properties of CuGaS2 thin films by electron beam evaporation," *Solar Energy Materials & Solar Cells,* vol. 75, pp. 93 –100, 2003.

[13] A. Martí, D. F. Marrón, and A. Luque, "Evaluation of the efficiency potential of intermediate band solar cells based on thin-film chalcopyrite materials," *J. Appl. Phys.,* vol. 103, p. 073706, 2008.

[14] K. Sato and T. Teranishi, "Effect of Delocalization of d-Electrons on the Optical Reflectivity Spectra of $CuGa_{1-x}Fe_xS_2$ and $CuAl_{1-x}Fe_xS_2$ Systems," *Jpn. J. Appl. Phys.,* vol. 19, pp. 101 - 105, 1980.

[15] P. Palacios, K. Sánchez, J. C. Conesa, J. J. Fernández, and P. Wahnón, "Theoretical modelling of intermediate band solar cell materials based on metal-doped chalcopyrite compounds," *Thin Solid Films,* vol. 515, pp. 6280 –6284, 2007.

[16] Y.-J. Zhao and A. Zunger, "Electronic structure and ferromagnetism of Mn-substituted $CuAlS_2$, $CuGaS_2$, $CuInS_2$, $CuGaSe_2$, and $CuGaTe_2$," *Phys. Rev. B,* vol. 69, p. 104422, 2004.

[17] P. Palacios, I. Aguilera, P. Wahnón, and J. C. Conesa, "Thermodynamics of the Formation of Ti- and Cr-doped $CuGaS_2$ Intermediate-band Photovoltaic Materials," *J. Phys. Chem. C,* vol. 112, pp. 9525 - 9529, 2008.

[18] Y. Seminóvski, P. Palacios, and P. Wahnón, "Intermediate band position modulated by Zn addition in Ti doped $CuGaS_2$," *Thin Solid Films,* vol. 519, pp. 7517 - 7521, 2011.

[19] S. Kalainathan, N. Ahsan, T. Hoshii, and Y. Okada, "Tailoring of Intermediate Band in CuGaS2 Thin film via Chromium doping by facile chemical Spray Pyrolysis technique," *Materials Science in Semiconductor Processing,* vol. submitted.

[20] A. Luque and A. Martí, "A Metallic Intermediate Band High Efficiency Solar Cell," *Prog. Photovolt: Res. Appl.,* vol. 9, p. 73, 2001.

Effects of Annealing and Substrate Temperature for Sn-S Thin Films

Yoji Akaki[1], Kazuya Iwasaki[1], Shigeyuki Nakamura[2], Hideaki Araki[3]

[1]National Institute of Technology, Miyakonojo College, Miyazaki, 885-8567, Japan, [2]National Institute of Technology, Tsuyama College, Okayama, 708-8509, Japan, [3]National Institute of Technology, Nagaoka College, Niigata, 940-8532, Japan

Abstract — Sn and SnS thin films were deposited by thermal evaporation. The substrate temperature was from room temperature to 300 °C. After deposition, the thin films prepared at R.T. were annealed in a H_2S atmosphere at temperature ranging from 100 to 500 °C for 1 hour. SnS thin films were obtained from SnS precursor films annealed below 150 °C. The XRD patterns of all the thin films deposited at substrate temperature ranging from 150 to 300 °C indicated such films were SnS single phase from. From cross-sectional SEM photographs, SnS precursor films deposited at 300 °C was found to be denser than that at room temperature.

Index Terms —evaporation, hydrogen sulfide, solar cell, tin, tin sulfide.

I. INTRODUCTION

Tin sulfide (SnS) which consist of low-cost, earth-abundant, non-toxic elements, may be one of the promising material for photovoltaic applications due to the band gap energy of 1.3 eV with a high optical absorption coefficient of 10^4 cm^{-1}. SnS thin films have been prepared using a variety of techniques such as sulfurization using sulfur powder [1,2], vacuum evaporation [3,4], chemical vapor deposition [5], chemical bath deposition [6], spray pyrolysis deposition [7], hot wall deposition [8], sputtering [9], and atomic layer deposition [10]. However the conversion efficiency of SnS solar cell is still low, at 4.36 % [10].

In our previous paper [11], Sn thin films have been prepared on glass substrates using a thermal evaporation method, and the films were subsequently annealed from 100 to 500 °C in H_2S atmosphere. We reported that the SnS single phase by X-ray diffraction (XRD) measurement was obtained from the thin films annealed at 250 °C in H_2S atmosphere. However, the surface roughness of the thin films was very large. In this paper, we report on the effect of substrate temperature and annealing for Sn and SnS thin films deposited by a vacuum evaporation.

II. EXPERIMENTAL PROCEDURE

Sn thin films were deposited onto glass substrates by thermal evaporation. The source material used was Sn powder with 99.99% purity and SnS with 99.9% purity. A pressure before the evaporation was below 2×10^{-3} Pa. The substrate temperature was from room temperature to 300 °C. After the deposition, the thin films prepared at R.T. were annealed in a H_2S atmosphere at temperature ranging from 100 to 500 °C for 1 hour.

The crystalline structure of the thin films was analyzed by X-ray diffraction (XRD, Rigaku SmartLab using Cu-Kα radiation), and the composition of the thin films was analyzed by energy dispersive X-ray spectroscopy (EDS, BRUKER QUANTAX FlatQUAD). The band gap of the thin films was determined from transmittance and reflectance measured by a UV-Visible-NIR Spectrophotometer (Hitachi U-4000). Resistivity of the thin films was measured by four-point prove method (Mitsubishi Chemical Loresta-EP MCP-T360). Morphological properties of the thin films were observed by a field emission scanning electron microscope (FE-SEM, Hitachi SU8020).

III. RESULTS AND DISCUSSION

Figure 1 (a) and (b) shows XRD pattern of the thin films annealed for S and SnS precursor films deposited at R.T. and the PDF card of Sn (PDF card No. 00-004-0673), SnS (PDF card No. 00-014-0620), and SnS_2 (PDF card No. 01-083-1705). From XRD patterns for Sn precursor films (Fig. 1 (a)), the as-deposited film and films annealed from 100 to 200 °C corresponded to the Sn phase. Sulfurization did not occur below 200 °C because the melting point of Sn is 230 °C. Thin films annealed from 250 to 350 °C exhibited the orthorhombic phase of SnS. However, the films show diffraction peaks corresponding to not only the SnS but also the Sn phase. Further, thin films annealed from 400 to 500 °C showed mainly the SnS_2 phase. From XRD patterns for SnS precursor films (Fig. 1 (b)), SnS single phase appeared by annealing at below 150 °C, SnS_2 single phase by annealing at above 350 °C. The XRD patterns of all the thin films deposited at substrate temperature ranging from 150 to 300 °C indicated such films were SnS single phase from.

Figure 2 (a) and (b) shows the Sn/S ratios of the thin films annealed for Sn and SnS precursor films, respectively, deposited at R.T. Sn/S ratio of the thin film annealed at 300 °C for Sn precursor film (Fig. 2 (a)) was 0.97 which was closest to SnS stoichiometry. However, the ratios of the thin films annealed above 350 °C were considerably different from that of SnS. Sn/S ratio of the thin films annealed above 400 °C was 0.60 - 0.65 which was a stoichiometry close to SnS_2. Sn/S ratio of the thin films annealed at around 100 °C for SnS precursor films (Fig. 2 (b)) was approximately 1.0 which was a stoichiometry close to SnS. However, the ratios of the thin

978-1-5090-5606-4/17 $31.00 © 2017 IEEE

films annealed above 350 °C were considerably different from that of SnS. Sn/S ratios of the thin films annealed at above 300 °C were approximately 0.57 - 0.62 which were a stoichiometry close to SnS_2. Sn/S ratio of the thin films deposited at substrate temperature ranging from 150 to 300 °C was 1.13 – 1.15.

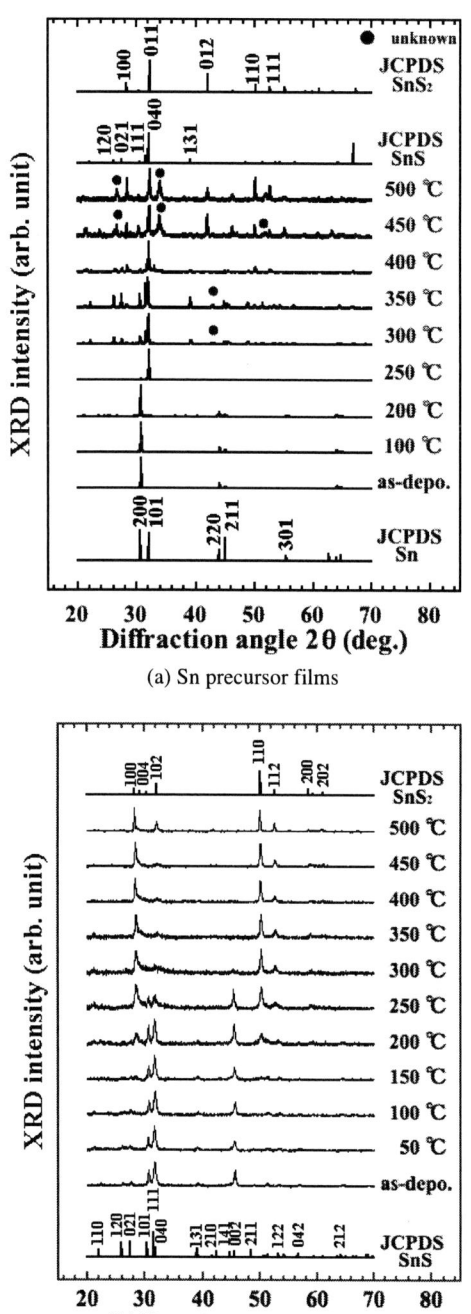

(a) Sn precursor films

(b) SnS precursor films

Fig. 1. XRD pattern of the thin films annealed for S and SnS precursor films deposited at R.T.

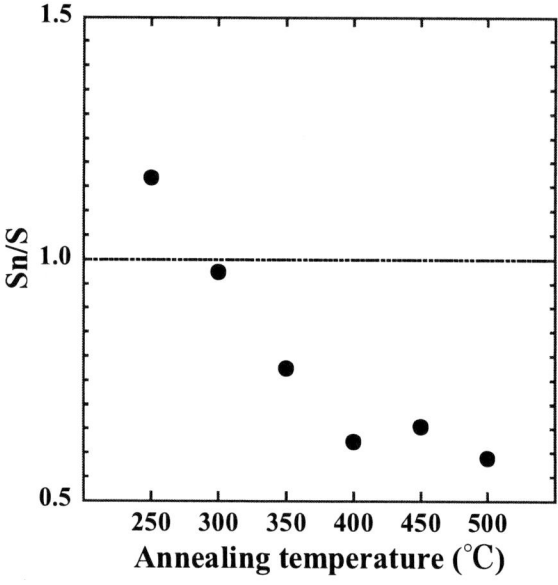

(a) Sn precursor films

(b) SnS precursor films

Fig. 2. Sn/S ratios of the thin films annealed for Sn and SnS precursor films deposited at R.T.

Figure 3 shows transmittance spectra of the thin films annealed for Sn precursor films deposited at R.T. The transmittance of as-deposited film and films annealed below 200 °C was almost 0% because of Sn metal. The transmittance of the thin films annealed from 250 to 400 °C was over 10% above 1000 nm and approximately 0% below 900 nm. The absorption coefficient of thin films annealed from 250 to 450 °C was over 10^4 cm^{-1} at photon energy above 1.3 eV. This indicates that SnS crystal exists in the films. However, the absorption coefficient of thin films annealed at 500 °C was over 10^4 cm^{-1} at photon energy above 2.4 eV. The direct band gap of the thin films annealed from 250 to 400 °C estimated from the transmittance and reflectance spectra was around 1.4-1.5 eV, which is similar to the direct band gap of SnS [1]. However, the direct band gap changed with increasing annealing temperature and the band gap of the thin films annealed from 400 to 500 °C was 2.4-2.5 eV, which is similar to the direct band gap of SnS$_2$ [12].

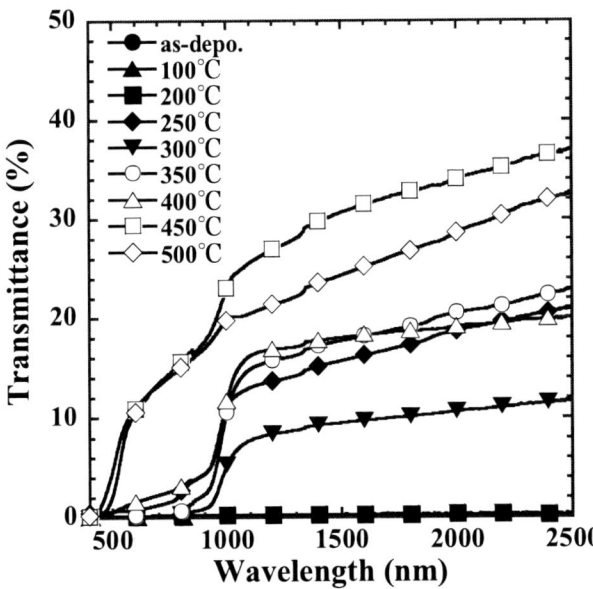

Fig. 3.   Transmittance spectra of the thin films annealed for Sn precursor films deposited at R.T.

The transmittance of as-deposited film and films annealed below 200 °C was approximately 20% above 1000 nm and approximately 0% below 800 nm. The absorption coefficient of thin films annealed below 200 °C was over 10^4 cm^{-1} at photon energy above 1.4 eV. This indicates that SnS crystals exist in the films. However, the absorption coefficient of thin films annealed above 350 °C was over 10^4 cm^{-1} at photon energy above 2.3 eV. The direct band gap of the thin films annealed below 200 °C estimated from the transmittance and reflectance spectra has at around 1.4-1.5 eV, which is similar to the direct band gap of SnS [1]. However, the direct band gap changed with increasing annealing temperature and the band gap of the thin films annealed above 350 °C was 2.3-2.4 eV, which is similar to the direct band gap of SnS$_2$ [12]. The direct band gap of the thin films deposited at substrate temperature ranging from 150 to 300 °C was 1.4-1.5 eV, which is similar to the direct band gap of SnS [1].

From cross-sectional SEM photographs of the thin films annealed for Sn precursor films deposited at R.T., many voids are confirmed in the thin films, and hence the surface is very rough. The voids in the thin films decrease with increasing the annealing temperature from cross-sectional SEM photographs

of the thin films annealed for SnS precursor films deposited at R.T. Figure 4 shows cross-sectional SEM photographs of SnS films deposited at R.T and 300 °C. The SnS thin films deposited at 300 °C was denser than that at room temperature.

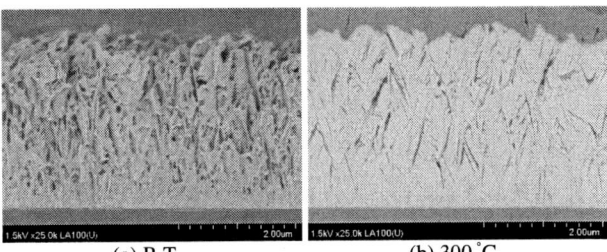

(a) R.T   (b) 300 °C

Fig. 4.   Cross-sectional SEM photographs of the thin films annealed for SnS precursor films deposited at R.T and 300 °C

IV. CONCLUSIONS

Sn and SnS thin films were deposited by thermal evaporation. The substrate temperature was varied from room temperature to 300 °C. After deposition, the thin films prepared at R.T. were annealed in a H$_2$S atmosphere at temperature ranging from 100 to 500 °C for 1 hour.

XRD patterns of the thin films annealed for Sn precursor films deposited at R.T. revealed that thin films annealed from 250 to 350 °C exhibited the orthorhombic phase of SnS with Sn phase. The thin films annealed from 400 to 500 °C showed mainly the SnS$_2$ phase. XRD patterns of the thin films annealed for SnS precursor films revealed that the single phase SnS films were fabricated by annealing below 150 °C, whereas SnS$_2$ single phase by annealing above 350 °C. The XRD patterns of all the thin films deposited at substrate temperature ranging from 150 to 300 °C indicated such films were SnS single phase from

From transmittance and reflectance spectra of the thin films annealed for Sn precursor films deposited at R.T., it is found that the direct band gap of the thin films annealed from 250 to 400 °C estimated from the transmittance and reflectance spectra has at around 1.4-1.5 eV, which is similar to the direct band gap of SnS. The thin films annealed for SnS precursor films deposited at R.T. revealed that the direct band gap of the thin films annealed below 200 °C estimated at around 1.4-1.5 eV. Furthermore, the direct band gap of the thin films deposited at substrate temperature ranging from 150 to 300 °C was 1.4-1.5 eV.

From cross-sectional SEM photographs of the thin films annealed for Sn precursor films, many voids was confirmed in the thin films, the surface roughness is very rough. The voids in the thin films decrease with increasing the annealing temperature from the thin films annealed for SnS precursor films deposited at R.T. From cross-sectional SEM photographs, SnS precursor films found to be deposited at 300 °C was denser than that at room temperature.

REFERENCES

[1] K. T. R. Reddy, P.P. Reddy, P. K. Datta, and R. W. Miles, "Formation of polycrystalline SnS layers by a two-step process," *Thin Solid Films*, vol. 403, pp. 116-119, 2002.

[2] M. Sugiyama, K. Miyauchi, T. Minemura, K. Ohtsuka, K.Noguchi, and H.Nakanishi, "Preparation of SnS Films by Sulfurization of Sn Sheet," *Jpn. J. Appl. Phys.*, vol. 47, pp. 4494-4495, 2008.

[3] A. Tanusevski, and D. Poelman, "Optical and photoconductive properties of SnS thiin films prepared by electron beam evaporation," *Sol. Energy Mater. Sol. Cells*, vol. 80, pp. 297-303, 2003.

[4] Y. Kawano, J. chantana, and T.Minemoto, "Impact of growth temperature on the properties of SnS film prepared by thermal evaporation and its photovoltaic performance," Curr. Appl. Phys., vol. 15, pp. 897-901, 2015.

[5] L. S. Price, I. P. Parkin, A. M. E. Hardy, and R. J. H. Clark, "Atmospheric Pressure Chemical Vapor Deposition of Tin Sulfides (SnS, Sn_2S_3, and SnS_2) on Glass," *Chem. Mater.*, vol. 11, pp. 1792-1799, 1999.

[6] D. Avellaneda, G. Delgado, M. T. S. Nair, and P. K. Nair, "Structural and chemical transformations in SnS thin films used in chemically deposited photovoltaic cells," *Thin Solid Films*, vol. 515, pp. 5771-5776, 2007.

[7] K. T. R. Reddy, N. K. Reddy, and R. W. Miles, "Photovoltaic properties of SnS based solar cells," *Sol. Energy Mater. Sol. Cells*, vol. 90, pp. 3041-3046, 2006.

[8] S. A. Bashkirov, v. F. Gremenok, V. A. Ivanov, V. V. lazenka, and K. Bente, "Tin sulfide thin films and Mo/p-SnS/n-CdS/ZnO heterojunctions for photovoltaic applications," *Thin Solid films*, vol. 520, pp. 5807-5810, 2012.

[9] K. Hartman, J. L. Johnson, M. I. Bertoni, D. Recht, M. J. Aziz, M. A. Scarplla, T. Buonassisi, "SnS thin-films by RF sputtering at room temperature," Thin Solid Films, vol. 519, pp. 7421-7424, 2011.

[10] P. Sinsermsuksakul, L. Sun, S. W. Lee, H. H. Park, S. B. Kim, C. Yang, and R. G. Gordon, "Overcoming Efficiency Limitations of SnS-Based Solar Cells," *Adv. Energy Mater.*, vol. 4, p. 1400496, 2014.

[11] K. Iwasaki S. Nakamura, Y. Akaki, "Properties of Sn-S thin films prepared by sulfulization," Technical digest of 6th World Conference on Photovoltaic Energy Conversion (Kyoto, 2014) p.475.

[12] B.Thangaraju, and P. Kaliannan, "Spray pyrolytic deposition and characterization of SnS and SnS_2 thin films," *J. Phys. D: Appl. Phys.*, vol. 33, p. 1054, 2000.

Molybdenum oxide thin films for heterojunction solar cells

A. Domínguez[1], Ateet Dutt[1*], O. de Melo[2] and G. Santana[1*]

[1] Instituto de Investigaciones en Materiales, Universidad Nacional Autónoma de México. A.P. 70-360, Coyoacán, C.P. 04510, México, D.F.

[2] Departamento de Física, Universidad de la Habana 27 y J. Plaza de la Revolución. La Habana. Cuba.

* Corresponding authors: Guillermo Santana, gsantana@iim.unam.mx, and Ateet Dutt, adutt@cinvestav.mx

Abstract —

In the present work, MoO_x nanostructures were obtained from Mo seed layers, grown by a sputtering technique. The X-Ray Diffraction (XRD), Atomic Force Microscopy (AFM), UV-Visible (UV-VIS) were applied to investigate the structural, morphological and optical properties of the samples. By optimizing the different deposition conditions, MoOx thin films were grown from 90 nm to 340 nm. The analysis made using the XRD Spectroscopy showed the presence of matrix MoOx (where x; could vary from 1 to 3). Especially, it was noticed that the samples have an oxidation state value (x = 2) before treatment, whereas, it changes to 3 after the treatment. Additionally, changes were found in the photoconductivity response of the thin films because of the heat treatment. Thin films grown in this work could be used for the fabrication of hole collector as well as the back contact region in the p-n type solar cells.

I. INTRODUCTION

Crystalline silicon is one of the most widely used semiconductor material in the photovoltaic industry, while in parallel much research is going on the efficiency increment of solar cells. Decreasing the cost of solar cells construction is the main objective of majority photovoltaic researches [1].

Growing n^+ emitter layer is a critical and an economical way of preparation of solar cells in bulk form [2-6]. Especially, using diluted orthophosphoric acid (H_3PO_4) in the spin coating or spray pyrolysis is one of the widest techniques to grow this kind of structures. These cells are of great importance in developing countries where sophisticated and expensive equipment are rarely available. Much research has been done to meet the requirements of PV industry, including low cost, high throughput using various phosphorus diffusion techniques.

In this report, we have developed a thin layer of MoO_x, and further, it has been inserted in a homemade orthophosphoric based solar cell. The concept of inserting a thin layer of transition metal oxides (TMO) such as MoO_x is to make compatible contacts with proper value of work function to avoid the many contact losses in the superficial of the solar cell [7-11].

A brief study of MoO_x as a hole selective layer for the silicon solar cells has already been presented by Bullock et al. [12]. It was successfully demonstrated, the growth of thin films of MoO_x

on the type p and n-type c-Si has a significant role in the development of selective contacts.

In the present work, we have investigated the formation of Mo seed layers by a sputtering technique. The thickness of the deposited thin layer of Mo was cross-checked by techniques such as SEM, AFM. Also, using XRD, we found variation in the oxidation state of MoO_x. On the other hand, we also fabricated orthophosphoric solar cell utilizing the homemade equipment. Finally, we have compared the optoelectronic properties of the final structure of the cell with and without MoO_x thin layer as back surface field. The study made in the present work shows the favorable results of passivation as well as hole selective feature of MoO_x thin layer in p-n type solar cells. Suitable applications with an additional feature of cost sustainability could promote the development of thin film proposed in the present work for the bulk production of silicon solar cells.

II. EXPERIMENTAL

Thin films were deposited on n-type [1 0 0], p-type and glass substrates by using DC Cressington sputtering unit. The pressure of 0.02 mbar was maintained using a simple mechanical pump, and then a flow of Ar (inert gas) of high purity was directed until the system pressure achieved the value of 0.1 mbar. Subsequently, the growth was started by establishing a current of 40mA. The thickness of the deposited film was monitored by using an average density of 5.58 g / cm^3.

In this work, crystalline structure and planer orientation of the MoO_x films were characterized in a Rigaku Ultima 4 operating at 40kV / 44mA in the range of 10-70 °. Topography images of the samples were obtained using an AFM Jeol brand, model JSPM-4210. The areas were analyzed in rectangular sections with dimensions of 1μm and were measured in taping mode. The images obtained were processed using the WinSPM DPS 2.0 program. UV-Vis spectrosocpy was carried out to calculate the band gap of material using Filmetrics F10-RT beam in the spectral range contemplated from 380-700 nm (wavelength). For

978-1-5090-5606-4/17 $31.00 © 2017 IEEE

optoelectrical measurements, we used a Keithley 2450 SourceMeter Meter / Source Device in both dark and illuminated conditions.

III. Results and discussion

One of the most used characterization techniques for the identification of materials as well as of the crystalline phases both in powders and in thin film form is undoubtedly XRD. In Figure 1, the diffractogram of the samples obtained in this work before the thermal treatment is shown. As in general, the presence of three different phases orthorhombic network (α - MoO_3), monoclinic network (β - MoO_2) and orthorhombic network (Γ-$Mo_4 O_{11}$) can be observed (JCPDS 00-013-0142).

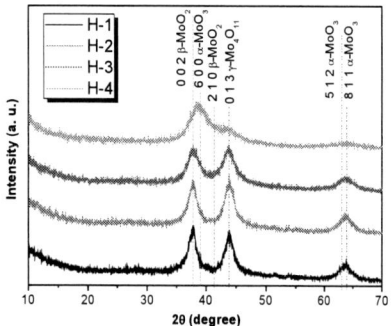

Figure1. Representative XRD spectra of as-deposited thin films

Figure 2 shows the spectra of the samples after the thermal treatment.

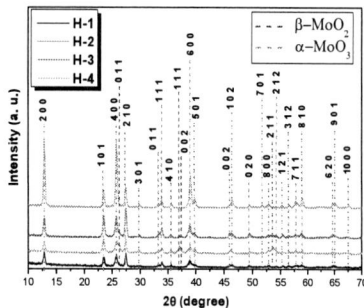

Figure2. Representative XRD spectra of annealed thin films

As can be seen, in Figure 2 narrower diffraction peaks can be observed, than those present in the films before the thermal treatment (Figure 1), thus evidencing a higher degree of crystallization in samples after being subjected to thermal treatment. The identified diffraction peaks correspond mostly to MoO_3 and a few to MoO_2, with no indication of the possible presence of Mo_4O_{11} or any other sub-phase. An important aspect, which can be observed from the diffraction pattern, is that there was no presence of MoO_2 in the thinner samples (H-1 and H-2), whereas, it was present in the sample H-4 (thickest). Another important aspect, which could be seen, is that there was an increase in the degree of crystallization of the films at the diminution of the thickness of thin films.

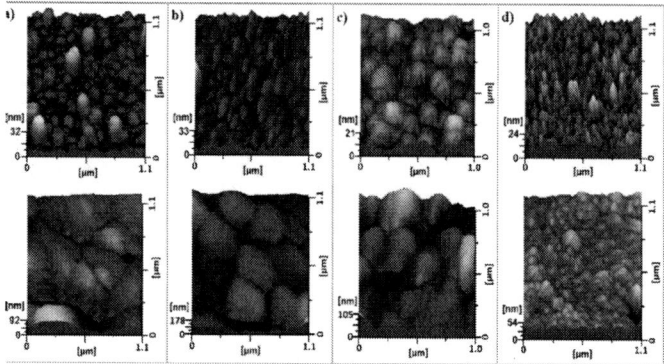

Figure3. AFM analysis of the sample before and after treatment

From Figure 3 it can be observed that the average roughness of the films was around 9 nm - 17.0 nm before the heat treatment. Even more, it can be seen that the films were thicker and they were less rough. On the other hand, it can be seen that the average roughness of the sample increased to around 41.4 nm – 80.6 nm after the heat treatment.

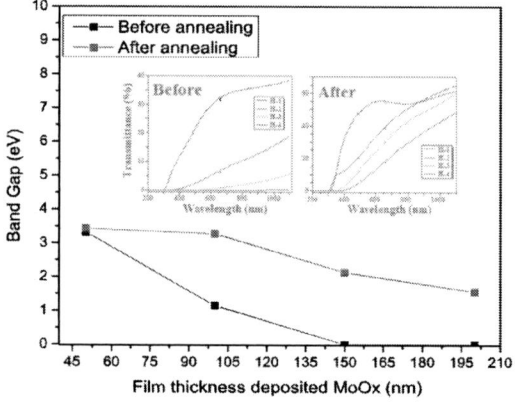

Figure 4. Film thickness variation of the sample with the band gap of the material. Inset shows the UV-Vis spectra of the sample before and after heat treatment

On the other hand Figure, 4 shows the absorption coefficients of the samples without and with thermal treatment, respectively. As shown in Figure 4, before the thermal treatment, the thicker samples have a higher absorption coefficient. However, after treatment, sample H-1 shown the highest absorption coefficient, despite having a larger gap than the rest of the films. It could be due to the presence of electronic states within the band gap, by either defects or impurities, which allow the absorption of photons. After carrying out structural and morphological studies, the optoelectronic characterization was made to find the prospective of material for device fabrication.

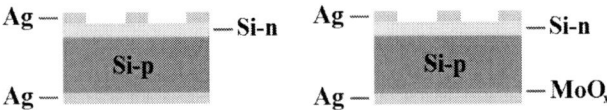

Figure 5 Schematic of the solar cell structure with and without molybdenum oxide thin film.

Furthermore, optoelectronic properties of the samples were studied under dark and illuminated conditions. It was found that the electrical conductivity of the samples was reduced by six magnitudes after the heat treatment. Especially, for the samples, H-3 and H-4, 66% and 48% increase in the electrical conductivity were observed when they were illuminated. Finally, as illustrated in Figure 5 home-made orthophosphoric cell (with and without MoO_x layer) was made by using H_3PO_4. The increase in efficiency of the cell was observed which could be attributed to the excellent passivation and charge selectivity due to thin MoO_x layer.

IV. Conclusions

In this work, we have explored different deposition conditions to obtain MoO_x thin films by using the sputtering unit. Using XRD technique, it was found that the corresponding oxidation state of the sample was X=2, whereas, X= 3 was more prominent after the heat treatment. Significant change in the band gap also with respect to heat treatment confirmed the change in the oxidation state as well as in the structure of the thin films. Finally, hike in the efficiency of the device due to the insertion of thin Mo layer confirms the suitability of the material for future fabrication of the type HIT or p-n solar cells.

V. Acknowledgements

The authors acknowledge financial support for this project from DGAPA-UNAM PAPIIT Projects IN108215 and IN107017. We want to thank Carlos Ramos Vilchis and Cain Gonzalez for their technical assistance.

VI. References

[1] K. Wakisaka, M. Taguchi, T. Sawada, M. Tanaka, T. Matsuyama, T. Matsuoka, S. Tsuda, S. Nakano, Y. Kishi, Y. Kuwano, Proceeding of the 22nd IEEE Photovolt. Spec. Conf., Las Vagas, USA, 1992, pp. 887–892.

[2] D. Bouhafs, A. Moussi, M. Boumaour, S.E.K. Abaïdiab L. Mahiou, " N+ silicon solar cells emitters realized using phosphoric acid as doping source in a spray process" Thin Solid Films vol. 510, pp. 325-328, 2006

[3] H. Nayaka, M. Nishida, Y. Takeda, S. Moriuchi, T. Tonegawa, T. Machida, T. Nunoi, "Poly crystalline silicon solar cells with Vgrooved Surface" Sol. Energy Mater. Sol. Cells vol. 34, pp 219-225, 1994

[4] A. Rohatgi, Z. Chen, P. Sana, J. Crotty, J. Salami Sol. Energy Mater. Sol. Cells vol.34, pp 227, 1994

[5] F. Duerinckx, J. Szlufcik, J. Nijs, R. Mertens, C. Gerhards, C. Marckmann, P. Fath, G. Willeke J. Schmid, H.A. Ossenbrink, P. Helm, H. Ehmann, E.D. Dunlop (Eds.), 2nd World Conference and exhibition on Photovoltaic Solar Energy Conversion, Vienna, AUSTRIA (July 6–10 1998), p. 1248

[6] R. Preu, R. Lüdeman, G. Emanuel, W. Wettling, W. Eversheim, G. Güthenke, D. Untiedt, G. Schweitzer W. Hoffmann, J.L Bal, H. Ossenbrink, W. Palz, P. Helm (Eds.), 16th European Photovoltaic Energy Conference, Glasgow, U.K. (May 1–5 2000), p. 1451

[7] S. Chen, J. R. Manders, S.-W. Tsang, et al., "Metal oxides for interface engineering in polymer solar cells" Journal of Materials Chemistry, vol.22, pp. 24202–24212, 2012.

[8] M. T. Greiner and Z.-H. Lu, "Thin-film metal oxides in organic semiconductor devices: their electronic structures, work functions and interfaces" NPG Asia Mater vol. 5, pp. e55, 2013

[9] S. Il Park, S. Jae Baik, J.-S. Im, et al., "Towards a high efficiency amorphous silicon solar cell using molybdenum oxide as a window layer instead of conventional p-type amorphous silicon carbide" Applied Physics Letters vol.99, pp.063504, 2011.

[10] L. Fang, S. J. Baik, J. W. Kim, et al., "Tunable work function of a WOx buffer layer for enhanced photocarrier collection of pin-type amorphous silicon solar cells" Journal of Applied Physics vol. 109, pp. 104501, 2011.

[11] C. Battaglia, X. Yin, M. Zheng, et al., Hole selective MoOx contact for silicon solar cells, Nano letters vol. 14 (2), pp 967–971, 2014

[12] J. Bullock, A. Cuevas, T. Allen, and C. Battaglia, "Molybdenum oxide MoOx: A versatile hole contact for silicon solar cells" App. Phys. Lett. Vol. 105, pp. 232109, 2014.

Dual ion beam sputtered TCO thin films: Sputter-instigated plasmonic features for ultrathin photovoltaics

Vivek Garg [1], Brajendra S. Sengar [1], Vishnu Awasthi[1], Shailendra Kumar[2], and Shaibal Mukherjee [1, *]

[1]Hybrid Nanodevice Research Group (HNRG), Electrical Engineering, Indian Institute of Technology (IIT) Indore, Indore-453552, India
[2]Raja Ramanna Center for Advanced Technology, Indore-452013, India

Abstract — **Electrical and optical properties of dual ion-beam sputtered (DIBS) Ga-doped ZnO (GZO) and Ga-doped MgZnO (GMZO) individual films are analyzed. Usage of secondary ion source favors excitation of sputter-instigated plasmonic feature in individual thin film. The plasmonic feature observed in GZO and GMZO thin films due to the formation of metal and metal oxide nanoclusters. The plasmon generation is verified by electron energy loss spectra obtained by ultraviolet photoelectron spectroscopy, spectroscopic Ellipsometry, and field emission scanning-electron microscopy measurements. This is promising in terms of increasing the efficiency of ultrathin solar cells by increasing optical path length in the absorbing layer.**
Index terms—DIBS, UPS, plasmons, GMZO, GZO.

I. INTRODUCTION

Cost-effective and efficient thin film solar cells (SCs) fabrication is a challenge for contemporary researchers. The leading approaches to overcome these challenges are: (1) to use cost-effective and more efficient thin film materials and (2) by using cost-effective nanoclusters in thin films to compensate for the reduction in physical layer thickness by enhancing optical layer thickness. In recent years, focused research on ZnO films has been gaining momentum and is now becoming the most established alternate as a transparent conducting oxide (TCO) material due to its higher transparency in visible region, high band gap tunability, high electro-chemical stability, and large exciton binding energy (60 meV). A number of ZnO based TCO materials are investigated such as MgZnO, GaZnO, and AlZnO etc. The incorporation of Mg in ZnO helps in the band gap tunability. Moreover, incorporation of both Ga and Al in ZnO matrix adds a flexibility to enhance the carrier concentration. [1]

In order to improve the performance of thin film SCs and reduce their manufacturing cost simultaneously, the concept of plasmonic is introduced.[2]-[4] Enhanced absorption coefficient by the excitation of plasmons greatly influences the performance of SCs with reduced physical layer. The usage of nanoclusters of noble metal such as Au, Pt and Ag is quite common as plasmonic layer.[5] Even though these nanoclusters offer higher carrier concentrations and limited chemical interactions, the realization of such noble metal nanoclusters is costly due to material cost and their incorporation in thin films SCs is a complicated procedure.

Therefore, there is a requirement of cost-effective alternatives for Plasmon excitation mechanism in ultrathin SCs. Here, we also report a straightforward and single-step procedure for formation of Ga, Zn, Mg and their oxide nanoclusters in GZO and GMZO films, which act as TCO and buffer layer in SCs. It has been reported earlier that these metallic or metal-oxide nanoclusters in semiconducting films are created due to difference in the sputtering rates of different constituent elements of semiconductors.[6]

II. EXPERIMENTAL TECHNIQUE

GZO and GMZO thin films with thicknesses of 200 nm and 100 nm are grown separately on Si substrate 200 and 500°C respectively, by Elettrorava DIBS system illustrated in Fig.1. DIBS deposition system is equipped with radio frequency (RF) ion source (deposition source) and direct current coupled

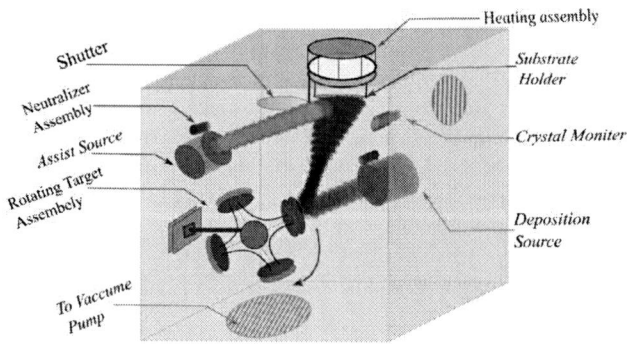

Fig.1: Schematic illustration of DIBS deposition system.

(DC) ion source (assist source). As compared to single ion beam deposition, in DIBS the film stoichiometry is enhanced by mixing oxygen into the working gas (Ar) in the assist ion source [6]-[8]. Besides, assist ion source hinders three-dimensional island formation and removes weak dangling bonds during film deposition process which improves growth uniformity and adhesion of film to the substrate. Further, it also helps in-situ substrate pre-cleaning before growth, and also favors formation of metal, metal oxide nanoclusters in TCO thin films.

978-1-5090-5606-4/17 $31.00 © 2017 IEEE

Moreover, the background pressure in the deposition chamber is $\sim 1\times 10^{-8}$ mbar. The deposition of individual thin films are carried out in Ar atmosphere with a chamber working pressure of 2.4×10^{-4} mbar and with RF power of 68 W. In addition, direct-coupled assist source is also turned on along with deposition source during growth. Moreover, the GZO and GMZO films growth is carried out using a 4N (99.99%) pure 4-in-diameter 1 at. %, and 3 at. % of Ga-doped-ZnO (99.99%), and Ga doped $Mg_{0.05}Zn_{0.95}O$ target respectively. UPS measurements are performed by utilizing the beam line of INDUS-1 synchrotron source.

SE measurement is carried out using J.A. Woolam M-2000D Ellipsometer. Further, data analysis is performed using a three-layer optical model consisting of a Si substrate, GZO or GMZO layer, and a top roughness layer. Theoretical spectra by using fitting parameters of the General Oscillator (Gen-Osc) model is in good agreement with experimental data with low value of mean square error (MSE) 5.27 (GMZO) and 9.64 (GZO) confirms the suitability and reliability of the model. [9]-[11] Tauc-Lorentz (T-L) oscillators in GenOsc model is utilized to model the complex dielectric function, which is occurring due to excitonic transitions.

III. RESULTS AND DISCUSSION

A. Electrical and Optical Properties

GZO and GMZO thin films have been studied for their electrical properties. For this, a four-probe Hall-effect measurement along with 2612A Keithley source meter is used. Hall measurement has revealed n-type conductivity in both GZO and GMZO films with carrier density of 1.06×10^{21} and 4.36×10^{19} cm^{-3}, resistivity of 4.11×10^{-3} and $7.43' 10^{-2}$ W-cm, and mobility of 14.83 and 1.93 cm^{2}V^{-1}s^{-1}, respectively, at room temperature.

The transmission spectra of GZO and GMZO thin films grown on sapphire substrate are shown in Fig. 2(a) and (b). Both the films are highly transparent. The transmittance of GZO film is better than 96% while transmittance of GMZO film is better than 94% in 400-to 800 nm wavelength range.

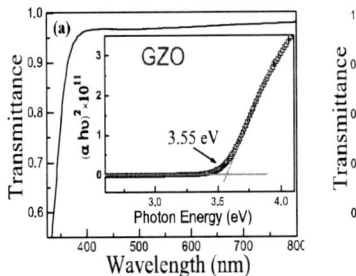 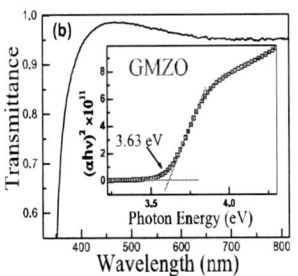

Fig.2: Transmission spectra of (a) GZO and (b) GMZO. Insets show respective Taucs plots.

Band gap evaluated from Taucs plot, obtained from transmission spectra, are 3.55 and 3.63 eV for GZO and GMZO, respectively.

B. Spectroscopic Ellipsometry (SE) Measurement

Fig. 3(a) depicts the spectral variation of the real part (ε_1) and the imaginary part (ε_2) of the complex dielectric function, $\varepsilon(E)=\varepsilon_1(E)+i\varepsilon_2(E)$, in the photon energy range of 1.5-6 eV for GZO and GMZO films. The values of ε_1 and ε_2 have significant role in designing optical and optoelectronic devices, since they are closely associated to the electronic polarizability of ions and the localized fields inside materials.[5] The origin of ε_1 peaks in GMZO is excitonic and mainly due to exciton-phonon complex transitions, as reported elsewhere [12]. Fig. 3(a) shows the maximum value of ε_1 is 4.79 (at $\sim$3.54 eV) and 5.47 (at $\sim$3.72 eV) for GZO and GMZO, respectively. It can be noted that linewidth of ε_1 of GZO is relatively broader than that of GMZO. In general, the broadening parameter of ε_1 peak is inversely proportional to corresponding excitonic lifetime.[13] Therefore, the results indicate that the excitonic lifetime is longer in GMZO as compared to that in GZO, which can be concluded, in other words, that excitonic stability is more in GMZO than in GZO.[6]

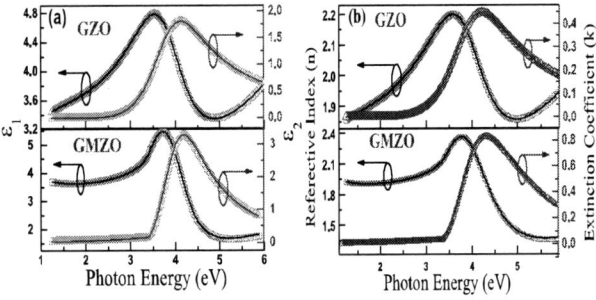

Fig.3: Variations of (a) ε_1 and ε_2, (b) n and k of GZO and GMZO thin films with incident photon energy.

Fig. 3(a) depicts that onset of ε_2, which corresponds to the absorption at the fundamental Eg, observed at 3.53 and 3.65 eV for GZO and GMZO, respectively. There is an unusual peak in ε_2 spectra of GZO and GMZO at photon energy larger than Eg (>4eV). This peak is due to additional absorption in films, which might be caused by particle plasmons produced by metal or metal oxide nanoclusters in GZO and GMZO thin films. Additionally, spectral variation in refractive index (n) and extinction coefficient (k) of GZO and GMZO films are plotted in Fig. 3(b). Maximum value of n is 2.2 and 2.37 for GZO and GMZO, respectively. The value of k is nearly zero below 2.75 and 3.4 eV for GZO and GMZO, respectively, which indicates that these films are completely transparent below the respective photon energy. The onset in absorption spectra above Eg~3.5 eV, is observed in both the films, which is caused by interband transition. The particle plasmon features can be observed in the spectra of both GZO and GMZO in the form of peak at 4.22 and 4.28 eV, respectively, which might correspond to plasmon resonance peak of nanoclusters formed in corresponding thin film.

C. Electron Energy Loss Spectra

Some broad plasmon peaks are detected in the UPS spectra of GMZO and GZO thin films. Fig. 4(a-c) shows the UPS spectra

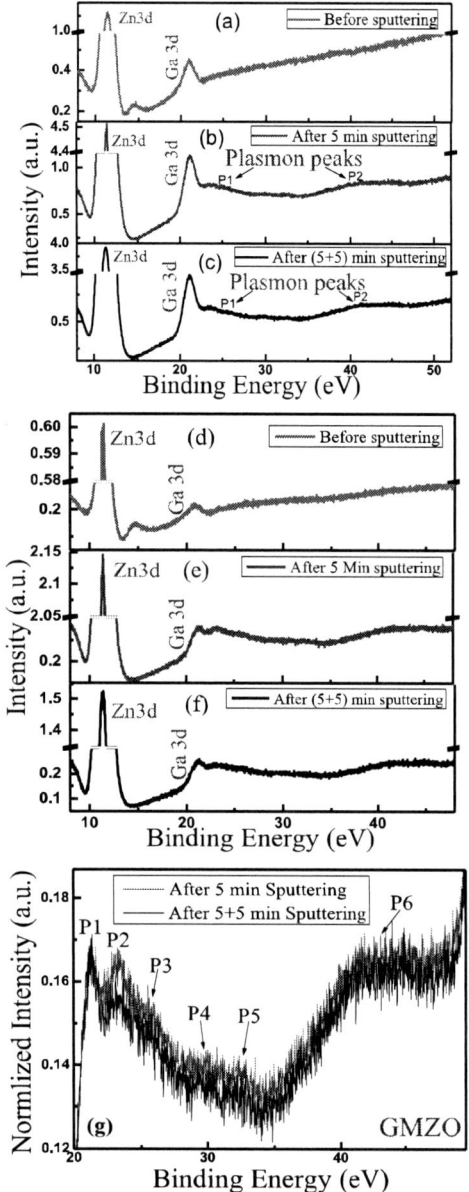

Fig.4: UPS spectra of GZO (a) before, (b) after 5 minute, and(c) after (5+5) minute, and of GMZO (d) Before sputtering, and (e) After 5 minute (f) After (5+5) minute of Ar⁺ sputtering and (g) Extended normalized spectra of 5 and (5+5) minute sputtered GMZO film.

of GZO film before and after 5 and (5+5) minute of Ar+ sputtering. Broad plasmon peaks (P1 and P2) are observed at does not exhibit existence of any such broad plasmon peaks. Plasmon peaks in sputtered film are generally formed due to kinetic energy loss of Zn-3d core photoelectrons after scattering with valence plasmons as a result of inelastic scattering mechanism.[13] These broad plasmon peaks can be the valence bulk plasmon (VBP) and valence surface plasmons (VSP) of Zn, ZnO, Ga, GaO nanoclusters embedded within GZO matrix.[5,6] Moreover, Fig. 4(d-f) shows the UPS spectra of GMZO film before and after5 and (5+5) minute of Ar+ sputtering. For GMZO, broad plasmonic peaks P1–P6 shown in Fig.4.(g), attributed to the collective contribution of particle plasmons energy of Ga, GaO, Mg, MgO, MgZnO, Zn and ZnO nanoclusters in ZnO and in air medium as tabulated in Table 1. Additionally, theoretically calculated values of the VBP, VSP and particle plasmon resonance energy (PPRE) of different nanoclusters in ZnO, and in air medium are tabulated in Table 1. These results are in good agreement with FESEM micrograph of GZO and GMZO shown in Fig. 5(a) and 5(b), respectively, and it confirms the formation of nanoclusters in GZO and GMZO thin films.

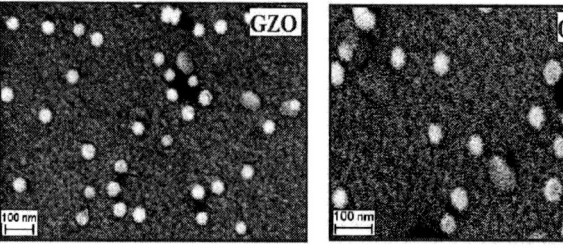

Fig.5: FESEM images of (a) GZO and (b) GMZO.

The particle plasmon resonance energy is in the range of ~1.87-10.04 eV for different nanoclusters in both ZnO and air medium. The energy range of nanoclusters fall in the ultraviolet-visible-infrared range of solar spectrum and can certainly improve the absorption cross-section by scattering mechanism, where the effective optical path length in the photoactive medium increases and hence ultimately enhance the performance of the solar cell.[14]-[16] Currently, high-efficiency and ultra-thin solar cells use precious metals like Au and Ag with particle plasmon energy of 2.8 and 4.5 eV, respectively which adds to the cost of device fabrication. Nevertheless, these precious and noble metals can be replaced by the nanoclusters generated by a short-duration Ar+ sputtering of DIBS deposited films for realizing economical and high-performance photovoltaics. Moreover, sputter-instigated plasmons can be incorporated in different layers of thin film solar cell architecture i.e. Plasmon enhanced low loss back contact, buffer-less solar cells, backscattering layers etc. for ultrathin solar cell applications. Additionally, it also motivates toward realization of all sputtered plasmon enhanced solar cells.

TABLE I:
CALCULATED ENERGY VALUES OF VBPs, VSPs AND PARTICLE
PLASMON RESONANCE IN ZnO AND AIR MEDIUM.

Thin Film	Clusters of elements and compounds	Energy of VBP (eV)	Energy of VSP (eV)	Particle plasmon resonance energy (eV) (in ZnO medium)	Particle plasmon resonance energy (eV) (in air medium)
GZO	Ga	14.53	10.27	3.82	6.49
	Zn	13.46	9.52	3.54	6.02
	GaO	23.69	16.75	6.10	10.04
	ZnO	7.31	5.17	1.87	3.04
GMZO	Ga	14.53	10.27	3.82	6.49
	Mg	10.89	7.70	2.86	4.87
	Zn	13.46	9.52	3.54	6.02
	MgO	9.57	6.77	2.42	3.84
	GaO	23.69	16.75	6.10	10.04
	ZnO	7.31	5.17	1.87	3.04
	MgZnO	21.41	15.14	5.46	8.81

IV. CONCLUSIONS

The electrical and optical properties are studied for the individual thin films of GZO and GMZO. The spectroscopic Ellipsometry reveals the favorable optical constants of both films for solar cell application and confirms plasmon generation. Moreover, the broad plasmon peaks observed in UPS spectra due to metal and metal oxide nanoclusters. The particle plasmon resonance energy in GZO and GMZO are in the range of ~1.87-10.04 eV for different nanoclusters are in both ZnO and air medium. Moreover, FESEM confirms generation of nanoclusters in the individual thin films. From this, we can conclude that, the DIBS system can be used for one-step generation of metal and metal oxide nanoclusters in TCO thin films. These nanoclusters are extremely promising to enhance the optical scattering and trapping of the incident light, which increases the optical path length in the absorber layer for achieving all sputtered cost-effective ultra-thin SCs and eventually increases its efficiency.

ACKNOWLEDGEMENT

This work is partially supported by Board of Research in Nuclear Sciences (BRNS), Department of Atomic Energy (DAE), Government of India, Clean Energy Research Initiative (CERI), Department of Science and Technology (DST), Government of India, and UGC-DAE CSR. Brajendra S. Sengar and Vivek Garg are also thankful to CSIR and UGC, respectively for the award of fellowship. Dr. Shaibal Mukherjee is thankful to DeitY YFRF, Government of India award. We are also thankful to DIBS facility equipped at Sophisticated Instrument Centre (SIC) at IIT Indore

REFERENCES

[1] W. S. Liu, W.K. Chen, and K. P. Hsueh, "Transparent conductive Ga-doped MgxZn1−xO films with high optical transmittance prepared by radio frequency magnetron sputtering", Journal of Alloys and Compounds,552, 255-263, 2013.

[2] M. A Green, and S. Pillai. "Harnessing plasmonics for solar cells." Nat Photon 6(3): 130-132, 2012.

[3] F. Beck, S. Mokkapati, A. Polman, and K.R. Catchpole "Asymmetry in photocurrent enhancement by plasmonic nanoparticle arrays located on the front or on the rear of solar cells." Applied Physics Letters 96(3): 033113, 2010.

[4] Harry. A. Atwater and Albert Polman, "Plasmonics for improved photovoltaic devices", Nature materials 9 (3), 205, 2010.

[5] V. Garg, B.S Sengar, V. Awasthi, Aaryashree, P. Sharma, C. Mukherjee, S. Kumar and S. Mukherjee, "Localized Surface Plasmon Resonance on Au Nanoparticles: Tuning and Exploitation for Performance Enhancement in Ultrathin Photovoltaics." RSC Advances, 6, 26216-26226, 2016.

[5] V. Awasthi, V. Garg, B. S. Sengar, S.K. Pandey, Aaryashree, S. Kumar, C. Mukherjee, and S. Mukherjee. "Impact of sputter-instigated plasmonic features in TCO films: for ultrathin photovoltaic applications." Applied Physics Letters, 110(10): 103903, 2017.

[6] S. K Pandey, V. Awasthi, B. S. Sengar, V. Garg, P Sharma, S. Kumar, C. Mukherjee, and S. Mukherjee. "Band alignment and photon extraction studies of Na-doped MgZnO/Ga-doped ZnO heterojunction for light-emitter applications." Journal of applied physics 118(16): 165301, 2015.

[7] R. Singh, P. Sharma, M. A. Khan, V. Garg, V. Awasthi, A. Kranti, and S. Mukherjee. "Investigation of barrier inhomogeneities and interface state density in Au/MgZnO: Ga Schottky contact." Journal of Physics D: Applied Physics 49(44): 445303, 2016.

[8] P. Sharma, R. Singh, V. Awasthi, S.K Pandey, V. Garg., S. Mukherjee "Detection of a high photoresponse at zero bias from a highly conducting ZnO: Ga based UV photodetector." RSC Advances 5(104): 85523-85529, 2015.

[9] G. Jellison, and L. Boatner "Optical functions of uniaxial ZnO determined by generalized ellipsometry." Physical Review B, 58(7): 3586, 1998.

[10] B. S. Sengar, V. Garg, V. Awasthi, Aaryashree, S. Kumar, C. Mukherjee, M Gupta, and S. Mukherjee . "Growth and characterization of dual ion beam sputtered Cu2ZnSn(S,Se)4 thin films for cost-effective photovoltaic application." Solar Energy 139: 1-12, 2016.

[11] J. N. Hilfiker, N. Singh, T. Tiwald, D. Convey, M.S. Smith, J. H. Baker and H. G. Tompkins. "Survey of methods to characterize thin absorbing films with spectroscopic ellipsometry." Thin Solid Films 516(22): 7979-7989, 2008.

[12] H. Neumann, W. Hörig, E. Reccius, H. Sobotta, B. Schumann, and G. Kühn. "Growth and optical properties of CuGaTe2 thin films." Thin Solid Films 61, no. 1, 13-22, 1979.

[13] V. Awasthi, S.K. Pandey, V. Garg, B.S. Sengar, P. Sharma, S. Kumar, C. Mukherjee and S. Mukherjee "Plasmon generation in sputtered Ga-doped MgZnO thin films for solar cell applications" Journal of applied physics 119(23): 233101, 2016.

[14] A. Calzolari, A. Ruini and A. Catellani "Transparent Conductive Oxides as Near-IR Plasmonic Materials: The Case of Al-Doped ZnO Derivatives." ACS Photonics 1(8), 703-709, 2014.

[15] V. E. Ferry, J. N. Mundey, and H. A. Atwater. "Design considerations for plasmonic photovoltaics." Advanced materials 22(43), 4794-4808, 2010.

[16] A. Calzolari, A. Ruini, and A. Catellani. "Transparent Conductive Oxides as Near-IR Plasmonic Materials: The Case of Al-Doped ZnO Derivatives." ACS Photonics 1(8), 703-709. 2014.

Combinatorial study of Sn-Ti-W-O transparent conducting oxide thin films for photovoltaic applications

Michael N. Gona, Patrick J. M. Isherwood, Jake W. Bowers, John M. Walls

Centre for Renewable Energy Systems Technology (CREST), Holywell Park, The Wolfson School of Mechanical, Electrical and Manufacturing Engineering, Loughborough University, Loughborough, Leicestershire, LE11 3TU, United Kingdom

Abstract — a combinatorial study of transparent conducting oxide thin films based on SnO_2–TiO_2–WO_3 phase space is reported. These multinary oxide films were fabricated by magnetron reactive co-sputtering of tin monoxide (SnO), titanium (Ti) and tungsten (W) targets. SnO_2–TiO_2–WO_3 film compositions with Ti/Sn ratio (0.02 – 0.12) and W/(Ti+Sn) ratio (0.02 – 0.25) were explored. The effect of oxygen partial pressure on composition, structure and optical properties was evaluated. High optical transparency above 80% across the visible spectrum was obtained for sputtered ternary SnO_2-TiO_2 oxide films for oxygen partial pressure >19.4%. A positive correlation between optical bandgap and Ti/Sn ratio was observed. However, optical properties deteriorated as Ti-content increased in the as-deposited SnO_2-TiO_2-WO_3 films. All studied as-deposited SnO_2-TiO_2-WO_3 thin films were found to be highly resistive. X-ray diffraction data indicated no long-range structural order.

Index Terms — transparent conducting oxides, amorphous thin films, magnetron reactive sputtering, combinatorial technique

I. INTRODUCTION

Transparent conducting oxides (TCOs) perform critical functions in a wide range of opto-electronic devices. For instance, thin-film solar cells have traditionally used doped wide-bandgap TCOs such as SnO_2:F, In_2O_3:Sn and ZnO:Al as transparent conducting electrodes, anti-reflection coatings and chemical barriers, etc. [1]–[6]. However, these crystalline TCOs need elevated deposition temperatures to optimize their physical, optical and charge-carrier transfer properties. This is a severe handicap for temperature-sensitive fabrication of next-generation organic and flexible photovoltaic devices. Amorphous TCOs thin films have been reported as potential alternatives to their conventional crystalline counterparts [7]–[9]. The approach follows the ability to tailor the physical, optical and charge-carrier transfer properties over a range of atomic compositions in the candidate amorphous thin films [10]–[13]. More recent studies of multicomponent oxides on InZnO [14], ZnO-In_2O_3-SnO_2 [15] and ZnSnO [16] have enunciated this concept. Potential amorphous TCO candidates include the nontoxic and environmental-friendly TiO_2-SnO_2-WO_3 thin films. TiO_2 and SnO_2 have similar crystal structures, chemical bonds, bond-lengths and band energies [17]–[19]. Both are also remarkably stable oxides. Earlier reports on magnetron sputtered WO_3 thin films have shown the possibility of modulating the physical, optical and electronic by judicial control of depositions conditions [20]–[22]. Ti-doped WO_3 [23], [24] and W-doped TiO_2 [25] studies have shown considerable possibilities that W–Ti oxide films offer in optimizing the TCO thin-film structure, physical and opto-electronic properties for high-efficiency and low-cost next-generation photovoltaic devices. Magnetron reactive cosputtering technique [26] provides opportunity of producing TiO_2-SnO_2-WO_3 thin films with controllable stoichiometry and composition. This report presented here investigates the structural and optical properties of sputtered TiO_2-SnO_2-WO_3 thin films in relation to the composition and the effect of reactive gas (O_2) pressure.

II. EXPERIMENTAL

Film synthesis - compositionally graded Sn-Ti-W-O thin films were deposited by magnetron co-sputtering of ceramic SnO-target (99.99% purity), Ti-target (99.99% purity) and W-target (99.99% purity). The films were deposited using an AJA International Orion 8HV sputter coater onto standard RCA pre-cleaned soda-lime glass (SLG) substrates (size 10cm × 10cm) at room temperature. A metallic mask was used to create a 5×5 matrix with each sample size of 1.8cm × 1.8cm. The sputtering chamber base pressure was maintained below 10^{-7} Torr. The working pressure and Ar (sputtering gas) flow rate of 1mTorr and 5SCCM respectively, were selected after preliminary investigations to determine the optimum deposition conditions. An oxygen (O_2 100% purity) supply provided the reactive gas. The O_2 flow rate was varied to control the oxygen partial pressure during depositions. Independent RF, DC and pulsed DC power sources were used to excite a three-gun assembly, shown in Fig. 1.

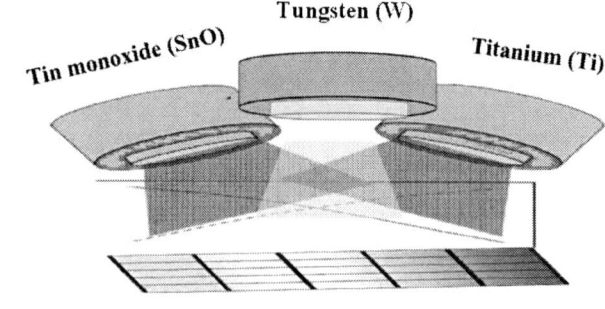

Fig. 1. The schematic of Tin monoxide (SnO), Tungsten (W) and Titanium (Ti) target arrangement in the magnetron sputtering chamber

The target power was adjusted to control the sputtered flux condensing rate on stationary substrates. The coating duration was set to give the desired film thickness. Prior to each coating, each target was presputtered for 120s (with shutter closed) to remove target surface contaminants. All Sn-Ti-W-O thin films were deposited at room temperature and analyzed as-deposited.

Film analysis - At each coordinate on the 5 × 5 matrix, the analysis focused on structure, composition and optical properties. This allowed for rapid correlation of the results. Crystal structure was measured using a Bruker D2 phaser X-ray diffractometer with a Cu-Kα X-ray source and Lynxeye™ detector. A Cary 5000 UV-VIS-NIR spectrophotometer was used to determine the optical properties. An AMBIOS XP-2 profilometer was used to determine the film thickness. ThermoFisher Scientific™ K-Alpha™ X-ray photoelectron spectrometer (XPS) was used to probe the elemental contents of the samples.

III. RESULTS AND DISCUSSION

A. RT magnetron sputtered SnO₂ and TiO₂ thin films

Optical properties - The dependence of the deposition rate, physical and optical properties of RT sputtered SnO₂ thin films on oxygen partial pressure (ppO2) was investigated. Fig. 1 shows behavior of the deposition rate, optical transmittance (in the visible spectrum) and optical bandgap with sputtering reactive gas.

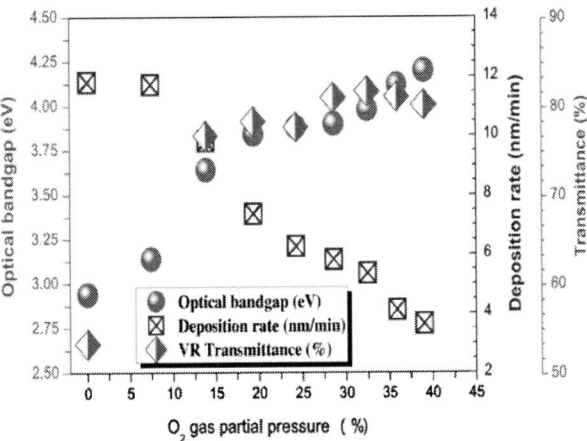

Fig. 2. Room temperature reactive sputtering of SnO₂ thin films

The SnO target power was set at 180W (RF supply). The sputtered SnO₂ thin films exhibited an average thickness of 440nm. The average deposition rate decreased from a maximum of 12.5nm/min at ppO₂ of 0% (100% Ar plasma) to less than 3nm/min for ppO₂ > 40%. The deposition rate rapidly decreases when even small amounts of O₂ are added to an

Argon-plasma. This decreasing behavior has previously been reported when a ZnO and SiO₂ targets have been sputtered using various Ar/O2 gas mixtures [27][28]. However, optical properties of these films improved with the increase in ppO₂. Optical transmittance in the visible spectrum of SnO₂ films increased from 53% at ppO₂ of 0% to 85% for a ppO₂ of 40%. This trend is mirrored by the optical bandgap behavior.

TiO₂ thin films were deposited using a metallic Ti target (180W Pulsed DC). Fig. 2 shows variation of TiO₂ film deposition rate, average VR transmittance and optical bandgap with O₂ partial pressure (ppO₂).

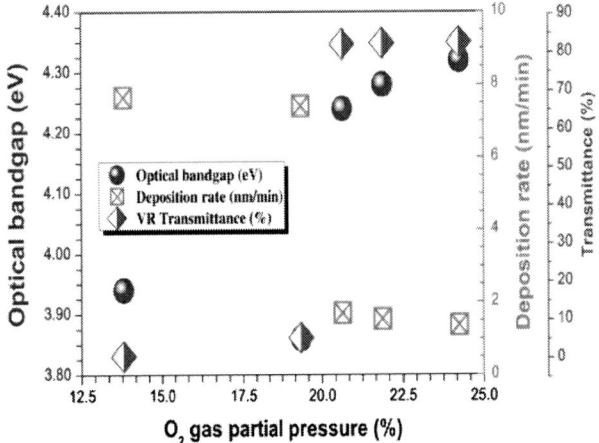

Fig. 3. Room temperature reactive sputtering of TiO₂ thin films

There is a markedly abrupt drop in the TiO₂ deposition rate at the ppO₂ of 40%. The deposition rate at averages 7.5nm/min for ppO₂ < 40% and then substantially reduces to below 2nm/nm thereafter. This sudden reduction in growth rate is most likely a result of TiO₂ formation on the Ti target surface [29]. The Ti target transitions from operating in "metallic mode" with high yield rate to "oxide mode" low yield mode. The TiO₂ thin films deposited in "metallic mode" conditions were less optically transparent (average VR transmittance below 10%) compared to films deposited in the "oxide mode" region with average VR transmittance of 80%.

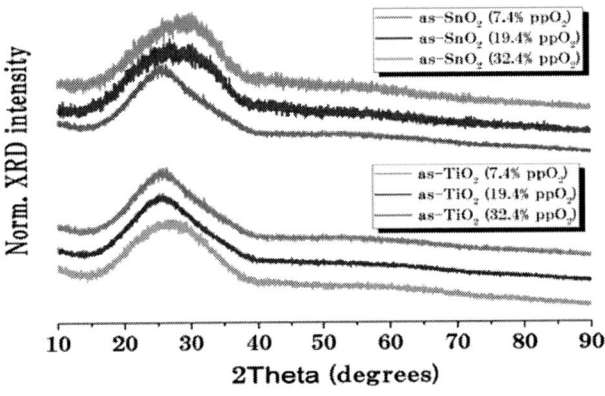

Fig. 4. XRD patterns of RT as-deposited SnO₂ and TiO₂ thin films

Film structure – X-ray patterns of as-deposited SnO$_2$ and TiO$_2$ films deposited at oxygen partial pressure (ppO$_2$) levels of 7.4%, 19.4% and 32.4% are shown in Fig. 4. Apart from the "broad-hump" shape from the glass substrate shown between 20° and 40°, no sharp diffraction peaks can be seen in all the samples. These profiles are indicative of the absence of long-range structural order in the RT sputtered films. However, it is difficult to determine whether the as-deposited SnO$_2$ and TiO$_2$ films are amorphous or composed of nano-crystallites of only few nanometers which cannot be detected by XRD, because both structural phases exhibit a broad X-ray diffraction peak.

B. RT magnetron sputtered SnO$_2$-TiO$_2$ thin films

Fig. 5 shows the colour map of Ti/Sn ratio of a 5 x 5 spatial matrix for room temperature co-sputtered Ti (60W DC power) and SnO (60W RF power). The deposition conditions were set at pressure of 1mTorr, Ar-flow of 5SCCM and oxygen partial pressure (ppO$_2$) of 24.2%. The average thickness of the sample was 270nm. XPS analysis showed Ti atomic percent with a maximum at 12.5% and high Sn atomic percent varying between 87.4% and 96.9%.

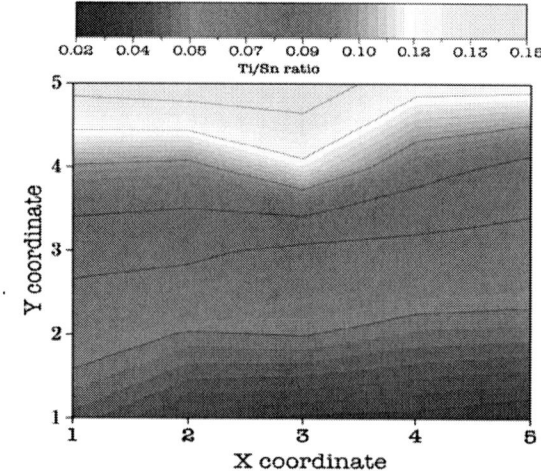

Fig. 5. Compositional mapping of the ratio Ti/Sn of a 10cm × 10cm graded Sn-Ti-O thin film; each [x,y] corresponds to one 1.8cm× 1.8cm cell

The Ti/Sn atomic ratio varies from 0.04 to 0.125 across the Sn-Ti-O film sputtered at a ppO$_2$ of 24.2%. The average Ti/Sn atomic ratio corresponds to the deposition rates of sputtered TiO$_2$ and SnO$_2$ fluxes on the substrate surface.

Chemical composition and optical bandgap

Fig. 6 (a) and Fig. 6 (b) Ti/Sn atomic ratio and optical bandgaps at selected coordinates (1,1) (1,2), (1,3) and (1,4) and (2,1), (2,4) and (2,5) respectively. The graphs show a positive correlation between the Ti/Sn atomic ratio and optical bandgap in Sn-rich compositions. This result is in good agreement with optical transmittance data obtained from spectrophotometry and the literature [19], and provides a

quick way of determining the composition of the SnO$_2$-TiO$_2$ films.

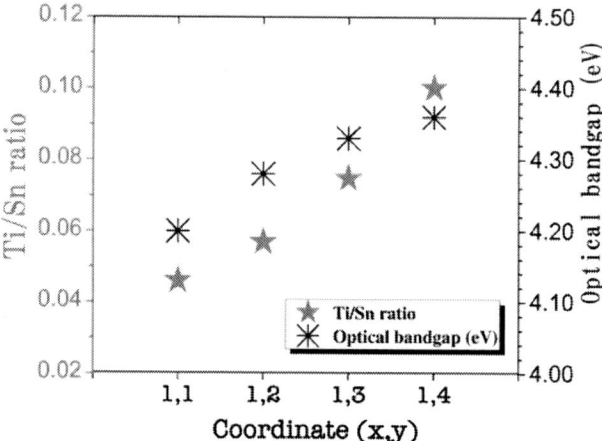

Fig. 6. Optical bandgap and Ti/Sn ratio at selected sample coordinates

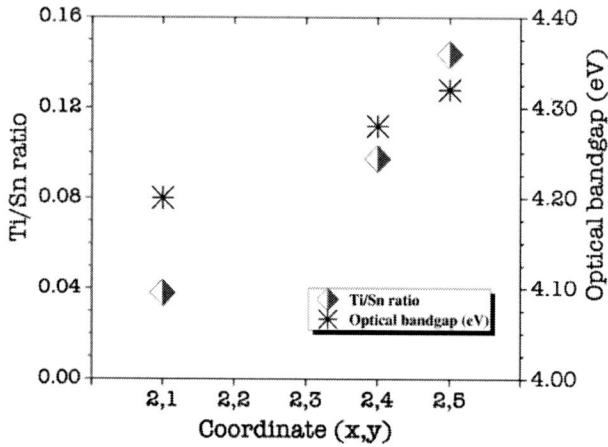

Fig. 7. Optical bandgap and Ti/Sn ratio at selected coordinates

C. RT magnetron sputtered SnO$_2$-TiO$_2$-WO$_3$ thin films

Chemical composition and optical properties

X-ray photoelectron spectroscopy was used to obtain the chemical composition and valence state of the elements present in the grown films. The XPS survey spectra of the films identified the main constituents as Sn, Ti, W, O, and C. The carbon peak seen in the XPS spectra was due to carbon from air exposure before being placed in the XPS system.

Fig. 8, Fig.9 and Fig.10 show composition colour maps of the sputtered SnO$_2$-TiO$_2$-WO$_3$ thin films deposited by co-sputtering of W-target (120W RF), Ti-target (105W Pulsed DC) and SnO-target (60W DC). The sputtering pressure was set at 1mTorr, with Ar-flow of 5SCCM with oxygen partial pressure (ppO$_2$) of 24.2%. The average thickness of the deposited film was 492nm. XPS analysis showed average Ti atomic percent of 0.46%, Sn atomic percent of 8.87% and W atomic percent of 11.54%.

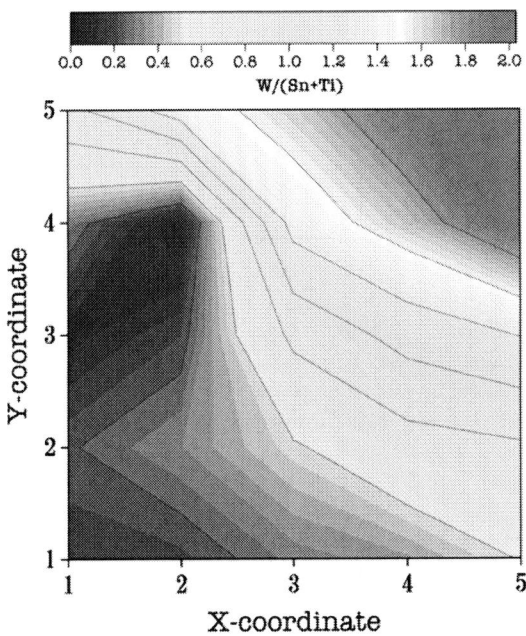

Fig. 8. Compositional mapping of the ratio W/(Sn+Ti) of a 10cm × 10cm; each [x,y] corresponds to one 1.8cm × 1.8cm cell in *SnO₂-TiO₂-WO₃* thin films

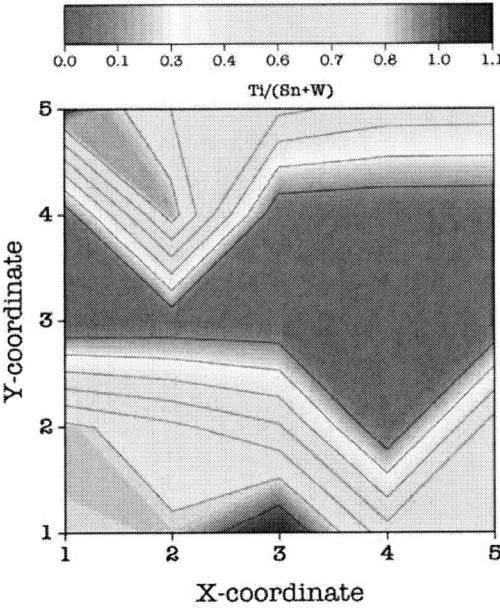

Fig. 9. Compositional mapping of the ratio Ti/(Sn+W) of a 10cm × 10cm; each [x,y] corresponds to one 1.8cm × 1.8cm cell in *SnO₂-TiO₂-WO₃* thin films

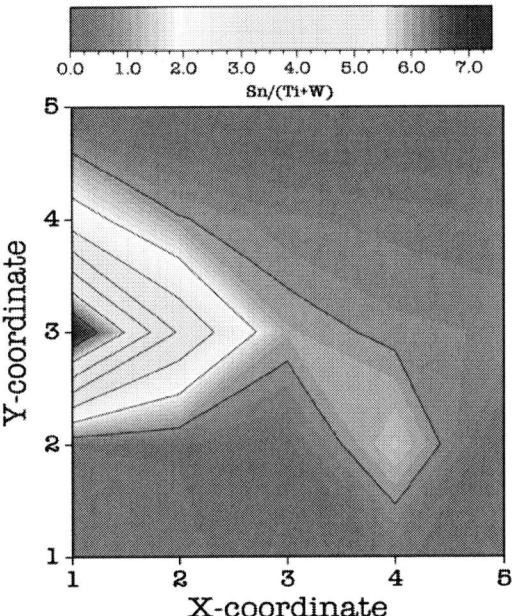

Fig. 10. Compositional mapping of the ratio Sn/(Ti+W) of a 10cm × 10cm; each [x,y] corresponds to one 1.8cm × 1.8cm cell in *SnO₂-TiO₂-WO₃* thin films

Fig. 11 and Fig. 12 show optical transmittance and reflectance curves for respective coordinates on the matrix grid of the SnO_2-TiO_2-WO_3 thin films. These films exhibited high transmittance (>80%) across the visible spectrum. No free carrier absorption effects were observed. This was reflected in high sheet resistances observed in as-deposited SnO_2-TiO_2-WO_3 films. Fig. 12 shows a marked decrease in transmittance as Ti atomic content increased.

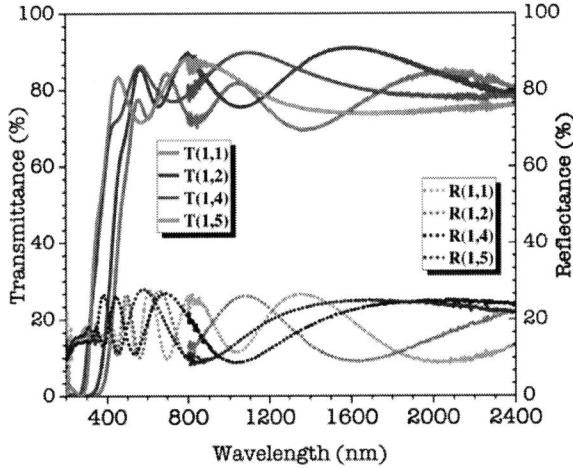

Fig. 11. Optical transmittance and reflectance plots of as-deposited *SnO₂-TiO₂-WO₃* thin films (at coordinates (1,1), (1,2), (1,4) and (1,5))

978-1-5090-5606-4/17 $31.00 © 2017 IEEE   2352

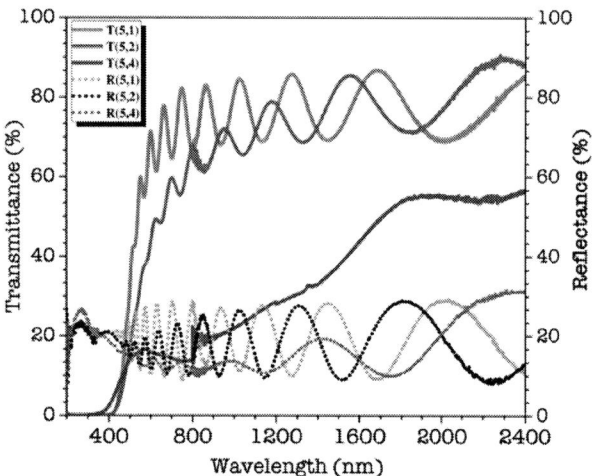

Fig. 12. Optical transmittance and reflectance plots of as-deposited SnO_2-TiO_2-WO_3 thin films (at coordinates (5,1), (5,2) and (5,4))

IV. SUMMARY

Multinary oxide SnO_2-TiO_2-WO_3 thin films were fabricated by reactive co-sputtering of SnO, Ti and W targets. The structural and optical properties of compositional ranges Ti/Sn of (0.02 – 0.12) and W/(Ti+Sn) of (0.02 - 0.25) were explored. XRD data of the as-deposited SnO_2-TiO_2-WO_3 thin films reveal the absence of long-range structural order. It was observed that film deposition rate for Ti, W and SnO targets declined with an increase in O_2 partial pressure. Sputtered films Optical transmittance in the visible spectrum of SnO_2 films increased with oxygen partial pressure and this correlated well with the optical bandgap behavior. Optical properties deteriorated as Ti-content increased in the as-deposited SnO_2-TiO_2-WO_3 films. All studied as-deposited SnO_2-TiO_2-WO_3 thin films were found to be highly resistive.

ACKNOWLEDGEMENT

The authors wish to thank Dr. Keith Yendall and Rhiannon Buckton of the Loughborough Materials Characterization Centre, Loughborough University for their assistance with XRD and XPS measurements, respectively. M. N. Gona gratefully acknowledges financial support from the Commonwealth Scholarship Commission (UK) for PhD scholarship.

REFERENCES

[1] N. K. Temizer, S. Nori, and J. Narayan, "Ga and Al doped zinc oxide thin films for transparent conducting oxide applications : Structure-property correlations Ga and Al doped zinc oxide thin films for transparent conducting oxide applications :

Structure-property correlations," vol. 23705, 2014.

[2] M. Wuttig, "Correlation between structure , stress and deposition parameters in direct current sputtered zinc oxide films," no. May, 2016.

[3] J. Herrero, "Comparison study of ITO thin films deposited by sputtering at room temperature onto polymer and glass substrates," vol. 481, pp. 129–132, 2005.

[4] T. Minami, "Present status of transparent conducting oxide thin-film development for Indium-Tin-Oxide (ITO) substitutes," vol. 516, pp. 5822–5828, 2008.

[5] H. Search, C. Journals, A. Contact, M. Iopscience, and I. P. Address, "Study on Crystallinity of Tin-Doped Indium Oxide Films Sputtering Deposited by DC Magnetron," vol. 1870, 1870.

[6] B. Stjerna, E. Olsson, C. G. Granqvist, B. Stjerna, and E. Olsson, "Optical and electrical properties of radio frequency sputtered tin oxide films doped with oxygen vacancies , F , Sb , or Mo Optical and electrical properties of radio frequency films doped with oxygen vacancies , F , Sb , or MO sputtered tin oxide," vol. 3797, no. 1994, 2009.

[7] A. Walsh, J. L. F. Da Silva, S. Wei, and C. Postal, "Interplay between Order and Disorder in the High Performance of Amorphous Transparent Conducting Oxides," no. 17, pp. 5119–5124, 2009.

[8] H. Hosono, "Ionic amorphous oxide semiconductors : Material design , carrier transport , and device application," vol. 352, pp. 851–858, 2006.

[9] A. Walsh, J. L. F. Da Silva, S.-H. Wei, and C. Postal, "Multi-component Transparent Conducting Oxides: Progress in Materials Modelling," pp. 1–12, 2011.

[10] F. Funabiki, T. Kamiya, and H. Hosono, "Doping effects in amorphous oxides," pp. 447–457, 2012.

[11] E. A. Davis and N. F. Mott, "Conduction in non-crystalline systems V . Conductivity , optical absorption and photoconductivity in amorphous semiconductors," vol. 8086, no. May, 2017.

[12] A. Takagi, K. Nomura, H. Ohta, H. Yanagi, T. Kamiya, M. Hirano, and H. Hosono, "Carrier transport and electronic structure in amorphous," vol. 486, pp. 38–41, 2005.

[13] X. Zhou, J. Xu, L. Yang, X. Tang, Q. Wei, and Z. Yu, "Amorphous In_2Ga_2ZnO_7 films with adjustable structural, electrical and optical properties deposited by magnetron sputtering," *Opt. Mater. Express*, vol. 5, no. 7, p. 1628, 2015.

[14] M. F. A. M. Van Hest, M. S. Dabney, J. D. Perkins, and D. S. Ginley, "High-mobility molybdenum doped indium oxide," vol. 496, pp. 70–74, 2006.

[15] C. A. Hoel, T. O. Mason, and K. R. Poeppelmeier, "Transparent Conducting Oxides in the ZnO-In2O3 -SnO2 System," *Chem. Mater.*, vol. 22, pp. 3569–3579, 2010.

[16] C. W. Gorrie, M. Reese, J. D. Perkins, M. F. A. M. Van Hest, J. L. Alleman, M. S. Dabney, B. To, D. S. Ginley, J. J. Berry, M. F. A. M. Van Hest, J. L. Alleman, M. S. Dabney, B. To, D. S. Ginley, J. J. Be, M. F. A. M. Van Hest, J. L. Alleman, M. S. Dabney, B. To, D. S. Ginley, and J. J. Berry, "Transparent

conducting contacts based on zinc oxide substitutionally doped with gallium," *Conf. Rec. IEEE Photovolt. Spec. Conf.*, pp. 7–9, 2008.

[17] M. Dou and C. Persson, "Comparative study of rutile and anatase SnO2 and TiO 2: Band-edge structures, dielectric functions, and polaron effects," *J. Appl. Phys.*, vol. 113, no. 8, 2013.

[18] T. Hitosugi, N. Yamada, S. Nakao, Y. Hirose, and T. Hasegawa, "Properties of TiO2-based transparent conducting oxides," *Phys. Status Solidi*, vol. 207, no. 7, pp. 1529–1537, 2010.

[19] S. Chen, J. R. Manders, S. Tsang, and F. So, "Metal oxides for interface engineering in polymer solar cells," pp. 24202–24212, 2012.

[20] M. Vargas, D. M. Lopez, N. R. Murphy, J. T. Grant, and C. V. Ramana, "Effect of W–Ti target composition on the surface chemistry and electronic structure of WO3–TiO2 films made by reactive sputtering," *Appl. Surf. Sci.*, vol. 353, pp. 728–734, 2015.

[21] M. Vargas, E. J. Rubio, A. Gutierrez, and C. V Ramana, "WO3 films made by co-sputter deposition Spectroscopic ellipsometry determination of the optical constants of titanium-doped WO 3 films made by co-sputter deposition," vol. 133511, 2014.

[22] S. F. E. Akbarnejad and A. S. Elahi, "Growth and Characterization of Tungsten Oxide Thin Films using the Reactive Magnetron Sputtering System," *J. Inorg. Organomet. Polym. Mater.*, vol. 26, no. 4, pp. 889–894, 2016.

[23] P. S. Patil, S. H. Mujawar, A. I. Inamdar, and P. S. Shinde, "Structural , electrical and optical properties of TiO 2 doped WO 3 thin films," vol. 252, pp. 1643–1650, 2005.

[24] C. V Ramana, G. Baghmar, E. J. Rubio, and M. J. Hernandez, "Optical Constants of Amorphous , Transparent Titanium-Doped Tungsten Oxide Thin Films," 2013.

[25] D. Chen, G. Xu, L. Miao, L. Chen, S. Nakao, and P. Jin, "W-doped anatase TiO 2 transparent conductive oxide films: Theory and experiment," *J. Appl. Phys.*, vol. 63707, no. 2010, pp. 2–6, 2015.

[26] W. D. Sproul, D. J. Christie, and D. C. Carter, "Control of reactive sputtering processes," vol. 491, pp. 1–17, 2005.

[27] C. R. Aita, A. J. Purdes, K. L. Lad, and P. D. Funkenbusch, "The effect of O2 on reactively sputtered zinc oxide," *J. Appl. Phys.*, vol. 51, no. 10, pp. 5533–5536, 1980.

[28] C. R. Aita and N. C. Tran, "Sputter deposition of platinum films in argon/oxygen and neon/oxygen discharges," *J. Appl. Phys.*, vol. 56, no. 4, pp. 958–963, 1984.

[29] R. Snyders, J. Dauchot, and M. Hecq, "Synthesis of Metal Oxide Thin Films by Reactive Magnetron Sputtering in Ar / O 2 Mixtures : An Experimental Study of the Chemical Mechanisms," pp. 113–126, 2007.

Bandgap and Electron Affinity Optimization of Zinc Oxide for n-ZnO/p-Si Single Heterojunction Solar Cell

Babar Hussain[1,4,5]* and Aasma Aslam[2,3]

[1]Energy Production and Infrastructure Center, Department of Electrical and Computer Engineering,
University of North Carolina at Charlotte, Charlotte, NC, 28223, USA
[2]Department of Electrical and Computer Engineering, University of Illinois at Chicago, Chicago, IL, USA
[3]Department of Information Technology, Hazara University, Mansehra, Pakistan
[4]National Institute of Lasers and Optronics, Nilore 45650, Islamabad, Pakistan
[5]Intel Corporation, Rio Rancho, NM, USA

*Corresponding Author: babar.hussain@intel.com

Abstract—This paper reports the influence of bandgap and/or electron affinity tuning of zinc oxide on the performance of n-ZnO/p-Si single heterojunction photovoltaic cell. The simulations using PC1D reveal that the open circuit voltage and fill factor can be improved significantly by optimizing valence-band and conduction-band off-sets by engineering bandgap and electron affinity of zinc oxide. The overall conversion efficiency of more than 20.3% can be achieved without additional cost or any change in device structure. It has been found out that the improvement in efficiency is mainly due to reduction in conduction band offset. Furthermore, increase in bandgap of ZnO by gallium alloying is demonstrated experimentally.

Index Terms—Zinc oxide, electron affinity, bandgap tuning, solar cells.

I. INTRODUCTION

Zinc oxide (ZnO) is an emerging material in semiconductor industry due to its abundance and being environmentally friendly. The only major drawback of ZnO is that it cannot be p-doped which hinders its use to make homojunction device. But the n-ZnO has found its applications in several optoelectronic devices such as photovoltaic cells [1]. Since the proposed use of n-ZnO as emitter layer and antireflection (AR) coating, several researchers have employed n-ZnO thin films to fabricate potentially high efficiency and low cost solar cell [2–4]. Apart from several other properties which make ZnO a unique wide bandgap material, its bandgap and electron affinity can be tuned over a large range by doping or alloying. Recently, nickel (Ni) doped ZnO thin films were prepared by spray pyrolysis and an optical bandgap decrease from 3.47 eV for the undoped ZnO film to 2.87 eV for 15% Ni doping was achieved [5]. In 2010, Mayer et al. demonstrated that the bandgap of ZnO prepared by pulsed laser deposition can be narrowed down to 2 eV by Se incorporation [6]. Later, the same research group reported effects of growth parameters on electron affinity of ZnO [7]. Also, there are various reports available demonstrating significant reduction in conduction band offset (or electron affinity) by incorporating magnesium (Mg) in ZnO. We have previously reported synthesis of ZnO thin films using metal organic chemical vapor deposition (MOCVD) with optimized parameters for the fabrication of n-ZnO/p-Si solar cell [1]. It was anticipated that the overall conversion efficiency of 19% and fill factor of 81% can be achieved using the proposed structure. Increase in effective bandgap (blueshift) of ZnO by Gallium (Ga) alloying was also demonstrated.

In this paper, we report simulations based optimization of bandgap and electron affinity of ZnO to enhance the conversion efficiency of ZnO/Si single heterojunction solar cell. The schematic of the solar cell structure is depicted in Fig. 1. The effects of valence-band and conduction-band off-set engineering on the open circuit voltage (V_{OC}), short circuit current density (J_{SC}), fill factor (FF), and overall conversion efficiency (η) have been investigated using PC1D software.

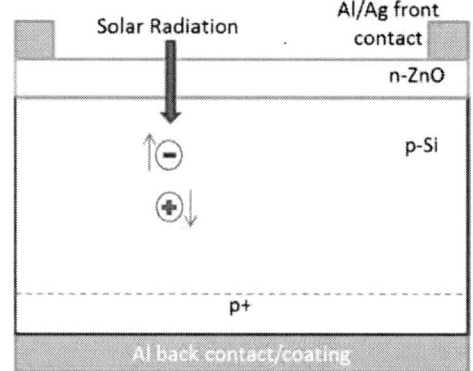

Fig. 1. Schematic of the n-ZnO/p-Si single heterojunction solar cell structure.

II. RESULTS AND ANALYSIS

We have prepared ZnO thin films by RF sputtering and have performed detailed characterization. The experimental details have been reported elsewhere [8]. The photoluminescence and absorption measurements performed in our labs showed bandgap value of 3.27 eV which was used in the simulation. The most common value of electron affinity (4.5 eV) provided in literature was used initially. The absorption spectrum of ZnO of thickness ~0.5 μm measured in our lab using Filmetrics tool was used in the simulation to investigate the effect of electron affinity. The details of the n-ZnO/p-Si solar cell modeling, structure schematic, and other optimized parameters for PC1D can be found in earlier report by Hussain et al. [1].

A. PC1D Simulations

Figure 2 illustrates improvement in efficiency with reduction in electron affinity of ZnO. It is obvious that the conversion efficiency exceeds 20% by lowering electron affinity to 4.3 eV. The simple reason behind this phenomenon is reduction in conduction band offset that leads to a decrease in the dark current. In other words, when a barrier for majority carrier electrons is formed by conduction band offset, it increases the probability of recombination via interface defects by Shottky-Read-Hall (SRH) mechanism. The maximum efficiency of 20.34% can be achieved with ZnO having bandgap of 3.27 eV and electron affinity of ~4.1 eV. Further reduction in electron affinity deteriorates the cell efficiency. This can be theoretically confirmed by band-bending diagram of ZnO/Si junction as shown in Fig. 3. Since the electron affinity of Si is ~4.05 eV, the electron affinity of ZnO below this value results in formation of a spike in the conduction band of n-ZnO region. This spike acts as a potential barrier and blocks electron flow from p-Si to n-ZnO region. Therefore, it is difficult for p-Si region to contribute in the photocurrent. This reasoning is supported by Figure 4 that depicts significant increase in V_{OC} with reduction in electron affinity. A negligible increase in J_{SC} can be attributed to the same reason.

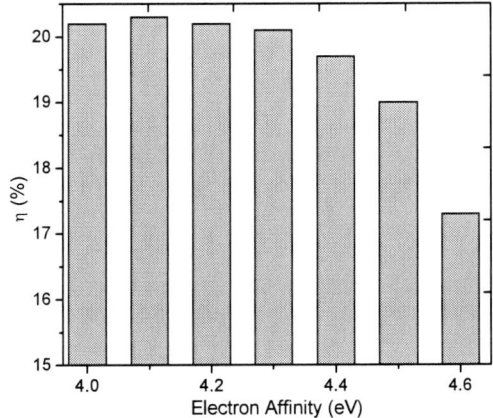

Fig. 2. Influence of electron affinity of ZnO (bandgap: 3.27 eV) on the efficiency of n-ZnO/p-Si heterojunction solar cell.

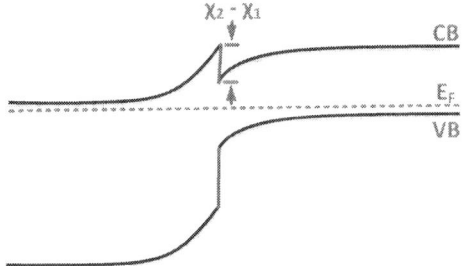

Fig. 3. Schematic of the band-bending when electron affinity of ZnO (χ_1), at left, is lower than that of Si (χ_2), at right. CB: conduction band, VB: valence band, E_F: fermi level.

The efficiency of the solar cell alters by modifying the bandgap as well. Change in conversion efficiency with bandgap value of ZnO is shown in Fig. 5 for three different values of electron affinity. The absorption spectrum was altered for values of bandgap other than 3.27 eV to get realistic results. The efficiency increases by decreasing the bandgap (or valence-band off-set). This improvement in efficiency cannot be explained by band-bending diagram based on the famous Anderson's rule which ignores the effects of chemical bonding. The chemical bonding or electrical polarization due to interface states can alter the band bending significantly. Figure 5 also illustrates that the efficiency reduces significantly below a certain bandgap value. It is predictable because a considerable part of solar spectrum gets absorbed in ZnO for such a small bandgap value. The ZnO layer is much thinner (0.5 μm) than Si (160 μm) but it has a higher absorption coefficient due to direct bandgap of ZnO.

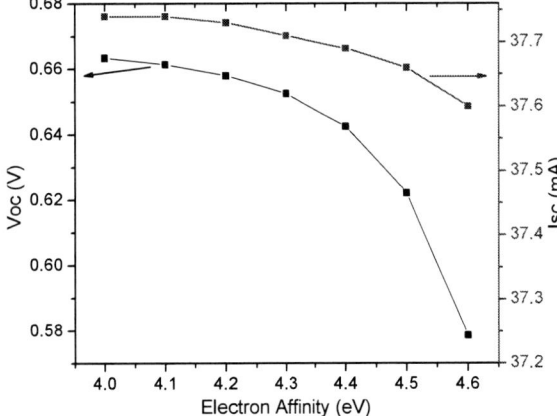

Fig. 4. Effect of electron affinity of ZnO (bandgap: 3.27 eV) on the V_{OC} and I_{SC} of n-ZnO/p-Si solar cell.

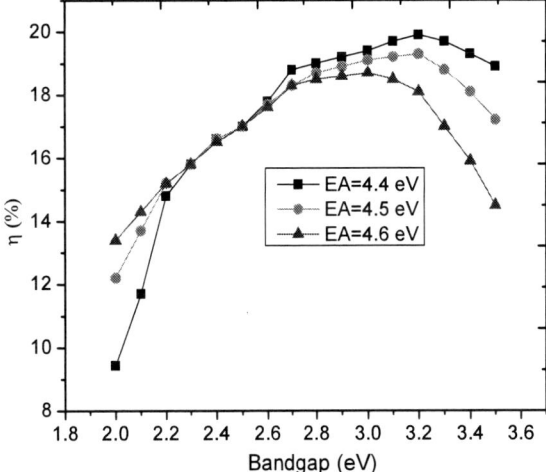

Fig. 5. Change in overall conversion efficiency of n-ZnO/p-Si solar cell with modification of bandgap value of ZnO for three different values of electron affinity (EA). Few data points have been interpolated because numerical solution was not converging for those points in PC1D.

B. Gallium Alloying in ZnO to Increase Bandgap

We have grown Ga rich ZnO:Ga films using MOCVD to examine the bandgap tuning. Trimethylgallium was used as the Ga source. The experimental details are provided elsewhere [1]. The photoluminescence (PL) spectra of both ZnO and ZnO:Ga were dominated by the near band edge (NBE) emission as depicted in Fig. 6. The spectra are normalized to their maximum values for clarity. The bandgap is blue-shifted by ~105 meV (12 nm). The molar ratio of Ga was ~50% during our growth process which lead to a bandgap around 3.35 eV (370 nm). This is in accordance with the model reported by Zhao et al. [9]. We attribute this increase in bandgap of ZnO to well-known Burstein-Moss effect in which effective bandgap of a heavily doped semiconductor is increased as the absorption edge in conduction band moves to higher energies because all states close to the conduction band edge are filled.

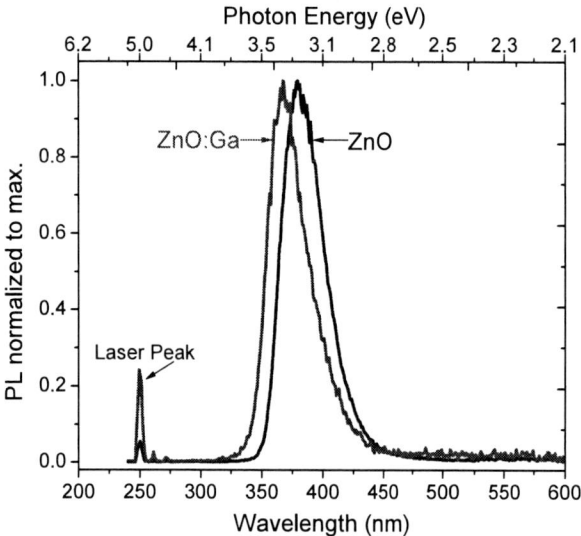

Fig. 6. PL spectra of pristine ZnO and Ga-rich ZnO:Ga showing increase in bandgap due to Burstein-Moss shift suitable for ZnO use as a front layer of n-ZnO/p-Si solar cell [1].

III. CONCLUSION

It is ascertained that the open circuit voltage of n-ZnO/p-Si single heterojunction solar cell can be significantly improved by tuning bandgap and/or electron affinity of ZnO by doping or alloying. The experimentally measured bandgap and absorption spectrum of ZnO was used in simulations using modified PC1D software. The major reason of improvement in the solar cell efficiency is reduced conduction band offset that results in decrease in the dark current. The best values achieved for open circuit voltage, short circuit current density, fill factor, and conversion efficiency are 0.662 V, 37.7 mA/cm^2, 0.815, and 20.34%, respectively, for ZnO having a bandgap of 3.27 eV and electron affinity of 4.1 eV. The development of ZnO/Si solar cell is reported somewhere else [10, 11].

ACKNOWLEDGEMENTS

The author is grateful to Dr. Yong Zhang from UNCC for useful discussions regarding this work.

REFERENCES

[1] B. Hussain, A. Ebong, and I. Ferguson, "Zinc oxide as an active n-layer and antireflection coating for silicon based heterojunction solar cell," Solar Energy Materials & Solar Cells, vol. 139, pp. 95–100, 2015.

[2] S. Baturay, Y. S. Ocak, and D. Kaya, "The effect of Gd doping on the electrical and photoelectrical properties of Gd:ZnO/p-Si heterojunctions," Journal of Alloys and Compounds, vol. 645, pp. 29–33, 2015.

[3] X. Ren et al., "Topology and texture controlled ZnO thin film electrodeposition for superior solar cell efficiency," Solar Energy Materials & Solar Cells, vol. 134, pp. 54–59, 2015.

[4] X. Zeng, X. Wen, X. Sun, W. Liao, and Y. Wen, "Boron-doped zinc oxide thin films grown by metal organic chemical vapor deposition for bifacial a-Si:H/c-Si heterojunction solar cells," Thin Solid Films, vol. 605, pp. 257–262, 2016.

[5] S. C. Das et al., "Band gap tuning in ZnO through Ni doping via spray pyrolysis," J. Phys. Chem., vol. 117, pp. 12745–12753, 2013.

[6] M. A. Mayer et al., "Band structure engineering of ZnO1-xSex alloys," Applied Physics Letters, vol. 97, pp. 022104-1–3, 2010.

[7] M. A. Mayer, K. M. Yu, E. E. Haller, and W. Walukiewicz, "Tuning structural, electrical, and optical properties of oxide alloys: ZnO1-xSex," Journal of Applied Physics, vol. 111, pp. 113505-1–6, 2012.

[8] B. Hussain, A. Ali, V. Unsur, and A. Ebong, "On structural and electrical characterization of n-ZnO/p-Si single heterojunction solar cell," 43rd IEEE PVSC, in press.

[9] J. Zhao, X. W. Sun, and S. T. Tan, "Bandgap-Engineered Ga-Rich GaZnO Thin Films for UV Transparent Electronics," IEEE Transactions on Electron Devices, vol. 56, pp. 2995–2999, 2009.

[10] B. Hussain, "Improvement in open circuit voltage of n-ZnO/p-Si solar cell by using amorphous-ZnO at the interface," Prog Photovolt Res Appl. 0:1–9, 2017.

[11] B. Hussain, "Development of n-ZnO/p-Si single heterojunction solar cell with and without interfacial layer," The University of North Carolina at Charlotte, 2017, 154; 10258481.

Modeling and Optimizing the Efficiency of a ZnO/ZnTe Solar Cell Using SCAPS Software

Amal Kabalan, Sam Roy and Benjamin Chen

Bucknell University, Lewisburg, PA 17837, United States of America

Abstract — This paper presents a systematic study on optimizing the efficiency of ZnO/ZnTe solar cells. Solar Cell Capacitance Simulator was used to perform the analysis. The cell structure under study consists of ZnO/CdS/ZnTe deposited on Indium Tin Oxide covered glass. Three main parameters were optimized: (a) the layer thickness of the absorbent film (ZnTe) and the window layer (ZnO) (b) the lattice mismatch between the metal contact and the window layer (ZnO) (c) adding a buffer layer between the absorbent and the window layers. We found the highest efficiency can be obtained with an absorbent layer greater than 15 μm, with using Aluminum as the metal contact since it has the lowest lattice mismatch with ZnO, and with adding a CdS buffer layer between ZnO and ZnTe. All the above results were verified using SCAPS software. The highest efficiency obtained was 17.25 % with a short current density (J_{sc}) 9.8 mA/cm^2, open circuit voltage (V_{oc}), 1.89 V and Fill Factor of 92.74 %.

I. INTRODUCTION

Solar cells based on II-VI semiconductors (with II = Zn, Cd, Hg and VI = S, Se, Te) are wide gap compounds that are of special interest for solar energy conversion due to their high photosensitivity, direct optical transitions and high values of absorption coefficient (10^4 cm^{-1}) in the visible region of the spectrum. The efficiency of of ZnTe-based solar cells that has been reported range from 7 – 10% with a current target of 15 - 20% range [1]. This paper will focus on three parameters that affect the efficiency of ZnTe/ZnO solar cells and will optimize them using SCAPS Software. The three parameters are:
a) Thickness of the absorbent layer and the window layer
b) The lattice mismatch between the window layer and the metal contact
c) The effect of a buffer layer between the absorbent layer and the window layer

The solar simulator *Solar Cell Capacitance Simulator* (SCAPS) will be used to run the above study. SCAPS is a one-dimensional solar cell simulation program that has been developed by researchers at the Department of Electronics and Information Systems (ELIS) of the University of Gent, Belgium.

II. SIMULATION

The simulator software was used to numerically analyze the effects of physical parameters on the performance of the cell, namely: the short-circuit current density, open-circuit voltage, fill factor and the photovoltaic conversion efficiency. SCAPS-1D software was used for all modeling purposes. We started with a solar cell structure that consists of ZnO as the window layer, ZnTe as the absorbent layer deposited on Indium Tin Oxide (ITO) covered glass. Copper was initially used as contacts. A sketch and schematic of the solar cell designed can be seen below in Fig. 2a and Fig. 2b, respectively. The input parameters of the different layers used are shown in Table 1 [2-4]. The working point values have been set as follows: Temperature set to 300 K, applied voltage set to 1 V, the frequency set to 1×10^6 Hz and the number of points set to 5. The above parameters are input values that the software requests before starting the simulation.

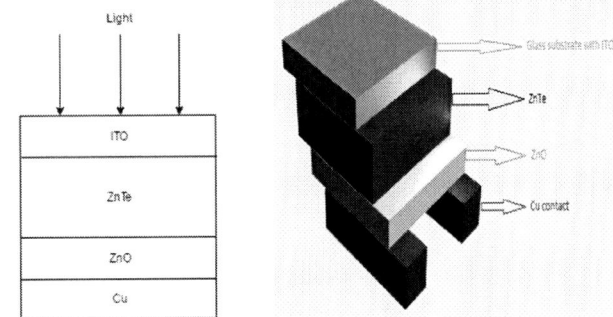

Fig. 1.   (a) Sketch (b) Schematic of a ZnO/ZnTe solar cell

TABLE I
PHYSICAL PARAMETERS OF DIFFERENT LAYERS

Parameters	ITO	p- ZnTe	n- ZnO	Cu	Parameters	p- ZnTe	n- ZnO
Surface recombination velocity (electrons) (cm/s)	1×10^7	-	-	1×10^6	VB effective density of states ($1/cm^3$)	1.166×10^{19}	1.8×10^{19}
Surface recombination velocity (holes) (cm/s)	1×10^5	-	-	1×10^6	Electron Thermal Velocity (cm/s)	3.24×10^7	2.4×10^7
Metal Work Function (eV)	-	-	-	4.7	Hole Thermal Velocity (cm/s)	1.51×10^7	1.3×10^7
Thickness (nm)	-	100	200	-	Electron Mobility (cm^2/Vs)	330	100
Bandgap (eV)	-	2.190	3.300	-	Hole Mobility (cm^2/Vs)	80	31
Electron Affinity (eV)	-	3.730	4.550	-	Shallow uniform donor density ND ($1/cm^3$)	0	10^{17}
Dielectric Permittivity (relative)	-	10.300	9.000	-	Shallow uniform donor density NA ($1/cm^3$)	2.16×10^{19}	0
CB effective density of states ($1/cm^3$)	-	1.176×10^{18}	3.1×10^{18}	-			

III. RESULTS

A. Optimization I: Comparing Lattice Mismatch

The optimization process was started by comparing the lattice constants of the metal contacts and their percentage mismatch with ZnO. The four contacts that were compared are aluminum (Al), copper (Cu), molybdenum (Mo) and platinum (Pt). The results from the comparison can be seen below in Table 3. [5]. Since Al has the lowest mismatch with ZnO, copper has now been replaced with aluminum to be used as the metal contact for our design.

TABLE II
COMPARING LATTICE MISMATCH BETWEEN CONTACTS

Element	Lattice Constant (nm)	Lattice Mismatch (%)
Al	4.0	12.66
Pt	3.920	14.41
Cu	3.610	21.18
Mo	3.150	31.22

B. Optimization II: Adding CdS as a Buffer Layer

To reduce the stress between ZnO and ZnTe we added a CdS layer between the two materials. To study the effect of CdS on the efficiency we used the software to calculate the efficiency which showed an increase from 14.86% to 15.67%. Based on the above results, CdS was added to our design as a buffer layer between ZnTe and ZnO.

C. Optimization III: Varying Layer Thickness

To get a better estimate of the impact of thickness on the efficiency, the thickness of ZnTe was varied from 0.8 μm to 25 μm and the thickness of ZnO was varied from 0.005 μm to 0.09 μm. The results of the ZnTe and ZnO thickness variation can be found below in Fig. 3 and Fig. 4 which shows that while the efficiency increases with increase in thickness of ZnTe, efficiency decreases with increase in ZnO thickness. The highest efficiency was reported at 24 μm of ZnTe and 5 nm of ZnO. Since it is difficult to fabricate a ZnO thickness of 5 nm we selected a thickness of 10 nm for the current analysis. As for ZnTe, the thickness 24 μm corresponding to the highest efficiency has been selected.

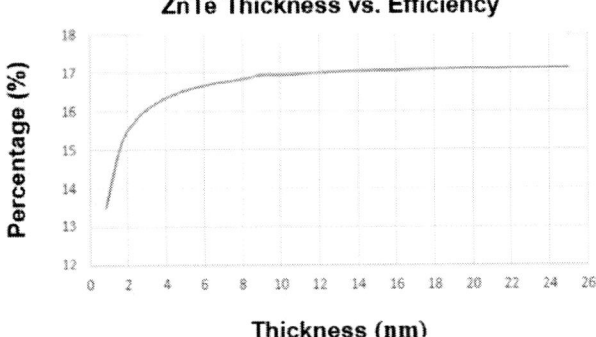

Fig. 2.   ZnTe thickness versus efficiency of the solar cell

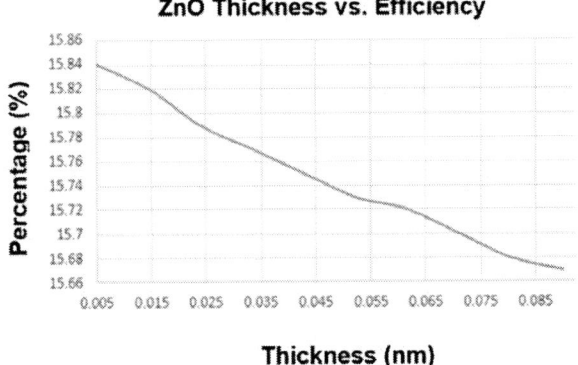

Fig. 3.   ZnO thickness versus efficiency of the solar cell

The efficiencies of the updated solar cell design were re-calculated after making the changes mentioned in optimization steps 1 through 3. The results of the simulation can be found in Table 3. Mo and Al both have the highest expected efficiency of 17.25%, but since the lattice mismatch with ZnO is the highest (31.22%) for Mo among the 4 contacts, Mo was not considered as the metal contact for our design. The highest efficiency for ZnTe/ ZnO solar cells prior to this paper was reported to be around 10% whereas after the optimization steps, we were able to model a solar cell structure with 17.25% as shown in Fig. 4.

TABLE III
COMPARING SOLAR CELLS EFFICINCIES

Element	Lattice Mismatch (%)	Efficiency
Al	12.66	17.25
Pt	14.41	17
Cu	21.18	17.21

Mo	31.22	17.25

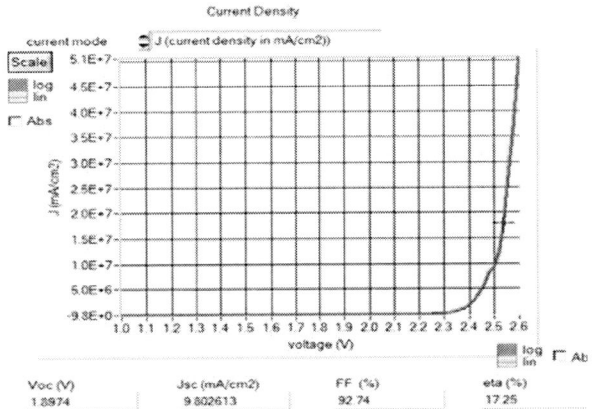

Fig. 4.   I-V of the final ZnO/ZnTe solar cell design

CONCLUSION

The highest efficiency for ZnTe/ ZnO solar cells prior to this paper was reported to be around 10%. After multiple optimization steps, we were able to model a design and simulate the efficiency to be over 17.2%. The parameters that were optimized were absorbent and window layer thickness, type of contacts and the addition of a CdS buffer layer.

REFERENCES

[1] M. A. Green, "Third generation photovoltaics: Solar cells for 2020 and beyond," *Physica E: Low-dimensional Systems and Nanostructures*, vol. 14, no. s 1–2, pp. 65–70, Apr. 2002.

[2] O. Skhouni, A. El Manouni, B. Mari, and H. Ullah, "Numerical study of the influence of ZnTe thickness on CdS/ZnTe solar cell performance," *The European Physical Journal Applied Physics*, vol. 74, no. 2, p. 24602, May 2016.

[3] S. draogo, Zougmor, J. M. Ndjaka, and H. P. Corporation, "Numerical analysis of Copper-Indium-Gallium-Diselenide-Based solar cells by SCAPS-1D," *International Journal of Photoenergy*, vol. 2013, Sep. 2013.

[4] Y. Rosenwaks, L. Burstein, Y. Shapira, and D. Huppert, "Studies of surface recombination velocity at copper/cadmium sulfide (1120) interfaces," *The Journal of Physical Chemistry*, vol. 94, no. 17, pp. 6842–6847, Aug. 1990.

[5] Networks Sandbox, "Interactive periodic table," 2000. [Online]. Available: http://www.infoplease.com/periodictable.php. Accessed: Dec. 15, 2016

Ternary phosphide semiconductor in Mg/Zn_3P_2 solar cells

Ryoji Katsube, Kenji Kazumi, and Yoshitaro Nose

Department of Materials Science and Engineering, Kyoto University, Kyoto 606-8501, Japan

Abstract — Zn_3P_2 is a promising candidate as an absorber in thin film solar cells because of the earth-abundant constituents and the optical properties. In the present study, we first report on formation of a novel Mg-P-Zn ternary phosphide semiconductor, $Mg_xZn_yP_2$, in Mg/Zn_3P_2 junction which was utilized in the champion device of Zn_3P_2-based solar cells. $Mg_xZn_yP_2$ has a certain range of Mg/Zn ratio and a different crystal structure from that of Zn_3P_2. $Mg_xZn_yP_2$ must play a key role in Zn_3P_2 solar cells since the cells show much higher efficiency compared to Zn_3P_2-based cells with other structures.

I. INTRODUCTION

Developments of thin film photovoltaics (TFPVs) based on CdTe and $Cu(In,Ga)Se_2$ have made a steady progress in the past three decades. These devices are now in mass production by companies such as First Solar and Solar Frontier. Although the productions of the above TFPVs are increasing year by year, they contain the elements in limited supply such as Te, In, and Ga. Towards terawatt scale deployment of TFPVs, it is desirable to develop solar absorbers consisting of earth-abundant elements.

Zn_3P_2 is a potential candidate as an earth-abundant alternative to CdTe because manufacturing processes of Zn_3P_2-based photovoltaics are expected to be similar to those of CdTe-based devices [1]. The highest efficiency of Zn_3P_2-based photovoltaics was achieved utilizing $Mg/p-Zn_3P_2$ junction [2]. The performance of the device was known to be improved by thermal annealing at 100 °C, and the effects of annealing on the interface structure were discussed in the previous studies. The earliest work by Bhushan and Catalano suggests that n-type inversion of $p-Zn_3P_2$ by interstitial incorporation of Mg [3,4]. However, in 2010s, Kimball et al. reported that Mg-doped Zn_3P_2 did not show n-type conductivity [5] and that the formation of Mg-Zn-P alloys at the interface contributed to the efficiency [6]. As the above, the interface structure of $Mg/p-Zn_3P_2$ solar cells is still controversial and there are no reports on microscopic observation of $Mg/p-Zn_3P_2$ interface.

In this paper, we thus reveal the nano-scale and atomic-atomic level structure of $Mg/p-Zn_3P_2$ junctions in the solar cells through scanning transmission electron microscope (STEM) observations and selected-area electron diffraction (SAED) analyses.

II. EXPERIMENTAL PROCEDURE

The Zn_3P_2 bulk crystal ingots were grown by the physical vapor transport technique as described elsewhere [7], using zinc shots (Kojundo Chemical Lab., 99.99%) and red phosphorus flakes (Kojundo Chemical Lab., 99.9999%) as starting materials. Prior to the syntheses, zinc shots were chemically-etched using 0.1 M solution of HCl for 30 sec. The grown ingots were mechanically sliced, diced, and polished to obtain the Zn_3P_2 plates with mirror surfaces. The area and the thickness were 2.2×2.2 mm^2 and 1 mm, respectively. The plates exhibit a p-type conductivity and have the resistivity of 10 Ω cm and the carrier concentration of 10^{16} cm^2 V^{-1} s^{-1}. The plates were etched in bromine water (Nacalai tesque, 35 g L^{-1}) diluted 20 times with ultra-pure water for 10 min and subsequently rinsed in ultra-pure water just before being set in the vacuum chamber. Mg electrodes were deposited on the Zn_3P_2 plates by thermal evaporation of Mg shots (Kojundo Chemical Lab., 99.9%) under a pressure of 2–3 $\times 10^{-2}$ Pa without intentional heating of the plates. The thicknesses of the films were controlled to be several μm. The Ohmic contact on the other side was formed by Ag film which was also deposited by thermal evaporation of Ag shots (Kojundo Chemical Lab., 99.99%). Annealing of the samples was carried out in sealed quartz ampules under a pressure of 10^{-2} Pa. The temperature range was from 100 to 300 °C and the annealing duration was fixed at 1 h.

The *I-V* characteristics of the samples were collected using a source meter (I-V 2400, Keithley Instruments) at room temperature. Au needle and Cu plate were used as the probes for Mg-deposited side and Ag back contact, respectively. The STEM and SAED observations were carried out with a transmission electron microscope (JEM-2100F, JEOL) for thin samples prepared by focused ion beam (SMI9200, Seiko Instruments Inc.).

III. RESULTS AND DISCUSSIONS

Fig. 1 shows the *I-V* curves of $Mg/p-Zn_3P_2/Ag$ samples before and after annealing at 100 and 300 °C. The annealings lead to the reduction of the current at both forward and reverse bias, in other words, the increase of the series resistance. The sample after annealing at 300 °C shows no rectifying behavior due to the high series resistance. The ideality factor and the barrier height of the sample before annealing were determined to be 3.91 and 0.97 eV, respectively, based on the thermionic

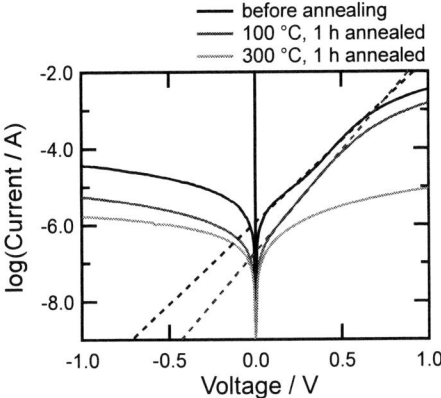

Fig. 1. *I-V* curves of Mg/p-Zn$_3$P$_2$/Ag samples before and after annealing.

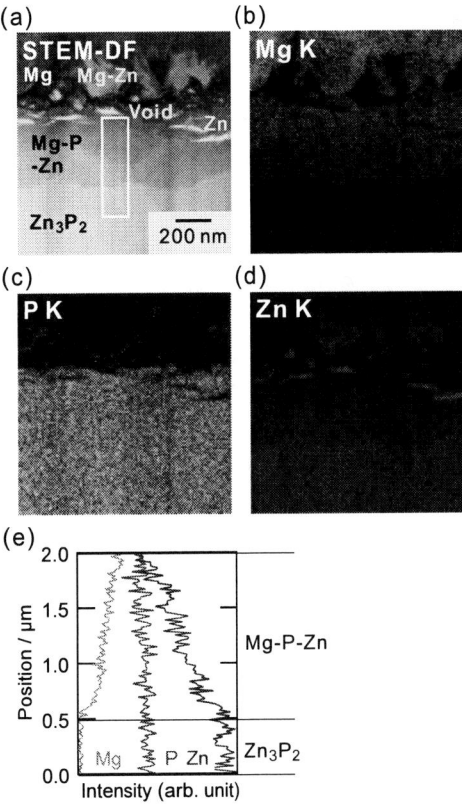

Fig. 2. (a) Cross-sectional STEM dark field image of the Mg/p-Zn$_3$P$_2$/Ag sample after annealing at 300 °C for 1 h, EDS mapping for (b) Mg K, (c) P K, and (d) Zn K, and (e) EDS line profile of the area surrounded by the yellow lines in (a).

emission theory [8,9]. The reduction in the ideality factor to 3.16 was indicated in the sample annealed at 100 °C together with a slight increase in barrier height. The observed behavior in the *I-V* curves implies the change of the carrier transport characteristics due to the structure change at the interface.

Fig. 2 shows the cross-sectional images by STEM-EDS for the Mg/p-Zn$_3$P$_2$/Ag samples after annealing at 300 °C for 1 h. The area including Mg, P, and Zn with a thickness over 1 μm is located near the interface. The compositions of Zn and Mg continuously change in the direction perpendicular to the original Mg/Zn$_3$P$_2$ interface, whereas that of phosphorus is almost constant as shown Fig. 2(e). Furthermore, according to the formation of the Mg-P-Zn area, some voids and metallic substances such as Mg-Zn intermetallics and Zn were formed at the Mg side of the interface.

SAED patterns of the Zn$_3$P$_2$ area and the Mg-P-Zn area were collected to clarify the crystal structure of the Mg-P-Zn area which should be the "Mg-Zn-P alloy" in the paper by Kimball et al. Fig. 3 shows the SAED patterns under the same direction of incident electron beam. The SAED pattern from the Zn$_3$P$_2$ area can be indexed as [100] zone axis of Zn$_3$P$_2$. On the other hand, the pattern from the Mg-P-Zn area shown in Fig. 3(b) is totally different from Fig. 3(a). Therefore, the Mg-P-Zn area is not an alloy but a novel Mg-P-Zn ternary compound semiconductor, Mg$_x$Zn$_y$P$_2$. In the Mg-P-Zn system, a compound, Mg$_{1.75}$Zn$_{1.25}$P$_2$ belonging to trigonal space group was reported [10]. The obtained SAED pattern is not inconsistent with the reported structure, although the composition is different from that of Mg$_{1.75}$Zn$_{1.25}$P$_2$.

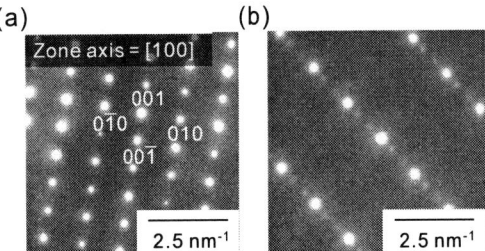

Fig. 3. SAED patterns from (a) the Zn$_3$P$_2$ area and (b) the Mg-P-Zn area of Fig. 2(a) under the same direction of incident electron beam.

IV. CONCLUSIONS

In the present study, we reported on the formation of ternary phosphide semiconductor, Mg$_x$Zn$_y$P$_2$, at the interface of the Mg/p-Zn$_3$P$_2$ junction. It was clarified that the carrier transport behavior observed in the *I-V* curves depended on the phosphide formation by annealing. Consequently, the properties of the ternary phosphide and related junction state should be investigated to achieve higher efficiency in Zn$_3$P$_2$ solar cells.

ACKNOWLEDGEMENT

This work was partly supported by JSPS-KAKENHI Grant Number 26289279. The authors are grateful to Mr. N. Sasaki (Kyoto Univ.) for his experimental supports for preparation of the samples for STEM observations.

REFERENCES

[1] J. Collier, S. Wu, and D. Apul, "Life cycle environmental impacts from CZTS (copper zinc tin sulfide) and Zn_3P_2 (zinc phosphide) thin film PV (photovoltaic) cells," *Energy*, vol. 74, no. C, pp. 314–321, 2014.

[2] A. Catalano, J. V Masi, and N. C. Wyeth, "Schottky Barrier Grid Devices on Zn_3P_2," in *2nd E. C. Photovoltaic Solar Energy Conference*, 1979, pp. 440–446.

[3] A. Catalano and M. Bhushan, "Evidence of p/n homojunction formation in Zn_3P_2," *Applied Physics Letters*, vol. 37, no. 6, pp. 567–569, 1980.

[4] M. Bhushan, "Mg diffused zinc phosphide n/p junctions," *Journal of Applied Physics*, vol. 53, no. 1, pp. 514–519, 1982.

[5] G. M. Kimball, N. S. Lewis, and H. A. Atwater, "Mg doping and alloying in Zn_3P_2 heterojunction solar cells," in *35th IEEE Photovoltaic Specialists Conference*, 2010, pp. 1039–1043.

[6] G. M. Kimball, N. S. Lewis, and H. A. Atwater, "Direct evidence of Mg-Zn-P alloy formation in Mg/Zn_3P_2 solar cells," in *37th IEEE Photovoltaic Specialists Conference*, 2011, pp. 1–4.

[7] R. Katsube and Y. Nose, *Journal of Materials Chemistry C*, accepted, doi: 10.1039/C7TC01047H.

[8] F. A. Padovani and R. Stratton, "Field and thermionic-field emission in schottky barriers," *Solid State Electronics*, vol. 9, pp. 695–707, 1966.

[9] C. R. Crowell and V. L. Rideout, "Normalized thermionic-field (T-F) emission in metal-semiconductor (Schottky) barriers," *Solid State Electronics*, vol. 12, no. 2, pp. 89–105, 1969.

[10] P. Klüfers and A. Mewis, "Zur Struktur der Verbindungen $BaZn_2P_2$ und $BaZn_2As_2$," *Zeitschrift für Naturforschung B*, vol. 33, no. 2, pp. 151–155, 1978.

NUMERICAL MODELING OF WSe₂ SOLAR CELLS

H. KYUREGHIAN, M. HILFIKER, E. EDIGER, V. MEDIC AND N.J. IANNO

Department of Electrical and Computer Engineering, University of Nebraska-Lincoln, Lincoln, Nebraska 68588, USA.

Abstract - In this work PC1D software was used to characterize a single crystal WSe₂/thin film heterojunction solar cell. The results were compared to experimentally obtained values, where the efficiency of the simulated device (7.2%) was in reasonable agreement with the measured value (7.8%) in spite of the uncertainty of the experimental details. The main reason for the relatively low experimental efficiency has been shown to be the halogen lamp illumination spectra and the choice of Au as the back contact. Simulated efficiencies in excess of 16% are shown for an AM1.5 Illumination spectra with potentially higher values for a better choice of back contact metal. In addition, simulation of thin film devices shows efficiencies in excess of 18% may be achieved.

Index Terms — chalcogenide, tungsten selenide, photovoltaic cell simulation

I. INTRODUCTION

There are several indications that WSe₂ may be a suitable photovoltaic material. Early literature reported a band gap of 1.35 eV, very close to the optimal value of 1.34 eV for a single-junction solar cell. More recently WSe2 bandgap values have computationally been determined to be

1.512 eV and 1.447 eV for the direct and indirect band gaps respectively. Also the overall material costs of thin film WSe₂ cells compare very favorably to a wide range of commonly studied thin film systems such as CIGS and CdTe; or in terms of estimated per watt raw material cost, based on single junction devices.

In addition, WSe2 has only one stable crystalline structure at room temperature - hexagonal, with unit cell parameters of $a = b = 3.327$ A° and $c = 15.069$ A° ; Hermann Mauguin space group *P* 63/*mmc*, Point group 6/mmm. The only other stable compound of W and Se, the amorphous WSe3 decomposes to WSe₂ at 220°C. Both p and n-type WSe₂ bulk crystals have been produced, leading to the possibility of homojunction devices. Single crystal mobilities are well in excess of 100 cm²/V-sec, with lifetimes in the 6-30 ns range, with an absorption coefficient in the 10^5 cm⁻¹ range over the visible spectrum, [1,2]. These properties should yield a high efficiency solar cell. Single crystal WSe₂/thin film heterojunction PV cells have been reported, with a p-type absorber and efficiency of around 7% [3] and 7.8 % [1]. A thin film based heterojunction has also been described, [4], although with a very low efficiency, < 0.0017% based on reported V_{OC} and ISC values.

In view of these favorable attributes we have employed PC1D software [5,6] to simulate the performance of WSe₂ single crystal based heterojunction photovoltaic devices reported in the literature [1]. This will provide some insight into the limitations of actual devices, provide a theoretical maximum efficiency and provide direction as to the steps needed to improve device efficiency. In addition, simulations of thin film devices will also be shown based on measured thin film properties, and the device limitations determined from the single crystal results.

Simulations were carried out with input parameters obtained through a combination of existing literature and direct measurements performed on in-house grown thin films [7].

II. Device Model and Material Properties

The device is taken from [1] and is reasonably well specified. It is an as-grown single crystal p-type absorber 2 microns thick with a mobility of 200 cm²/V-sec, a carrier concentration of 1E17/cm³ and Au back contact. We assumed a lifetime of 6ns, which is the lower end of the reported values. As described in [7], refractive index $n(\lambda)$ and extinction coefficient $k(\lambda)$ spectra for WSe₂ were obtained from ellipsometric measurements. The WSe₂ parameters are seen in Table I. The value for electron affinity, 4.0 eV, is taken from [8] (based on n-type WSe2). The dielectric constant for WSe₂ comes from [9], where the approximate average of the graph over the relevant energy spectrum was taken. Electron and hole effective masses are very important parameters, as they determine the effective density of states Nc and Nv and through them, the intrinsic carrier concentrations. Several different values have been reported including $m* = 7.597 \cdot 10{-}31 kg = 0.834 m_e$ and $m*= 0.327 m_e$.[10,11] As a first approximation, then, we use the average of those two values, 0.58me for the hole effective mass. The electron effective mass is estimated through the simplified relationship between E(k), m* and the published E vs k diagrams [12].

The heterojunction window layer is sputter deposited ZnO (thickness not specified, 50 nm employed in these simulations) with a doping concentration between 1E19 and 1E20/cm3. The electronic and optical properties were taken from the literature and seen in Table II.[13-15] The top contact grid is indium. In [1] it was clearly shown the ZnO deposition conditions strongly influenced the device efficiency. In this work will simulate the most efficient device (7.8%).

The reflectivity as a function of wavelength was calculated based on the ZnO index of refraction and thickness and was incorporated into the simulation. The grid coverage was assumed to be 5% and was also incorporated into the simulation. The ideal equilibrium band diagram as calculated from PC1D is seen in Fig. 1.

TABLE I
SUMMARY OF WSE₂ PARAMETERS

Band gap	1.39 eV
Electron affinity	4.0 eV
Dielectric constant	20.5
Electron effective mass	0.79 m_o
Hole effective mass	0.58 m_o
Intrinsic carrier density	$9.5 \cdot 106 cm^{-3}$
Auger coefficient	$1E-26 cm^6$
Mobility model parameters	
$\mu high,p$	30.435 $cm^2/V\, s$, 32.161$cm^2/V\, s$
$\mu high,n$	22.412 $cm2/V\, s$, 23.683$cm2/V\, s$
$\mu low,p$	0.127 $cm^2/V\, s$
$\mu low,n$	0.094 $cm^2/V\, s$
$Nref$	$1.197941 \cdot 1017 cm^{-3}$(p), $1.149403 \cdot 1017 cm^{-3}$(n)
α	12.46884 (p), 10 (n)
Refractive index spectrum	internal study
Absorption coefficient spectrum	internal study

TABLE II
SUMMARY OF ZNO PARAMETERS

Parameter	Value
Band Gap (eV)	3.27
Mobility Fixed (cm^2/Vs)	17.7
Electron Affinity (eV)	4.5
Dielectric Constant (rel.)	8.66
Nc/Nv Ratio	0.293
Intrinsic Concentrations	ni(200)=2.123E-23 cm^{-3}, ni(300)=2.107E-9 cm^{-3}, ni(400)=2.384e-2 cm^{-3}
N-type Doping (cm^-3)	1.00E+19 cm^{-3}
Effective Mass	m_e=.26m_o m_h=.59m_o
Absorption File	Variable
Refractive Index File	Variable
Lifetime (us)	тn=тp=1e-3

A key aspect of the reported results [1] is the spectrum employed to quantify the device. It was stated a halogen lamp of 70mW/cm^2 intensity was used for the measurements. This spectrum was assumed to be that of a quartz halogen lamp as seen in Fig 2.

The results are as follows:

Spectra: Halogen Lamp
Device: Au/WSe₂/ZnO/In grid
Simulation Results:

- Isc = 1.14 Amps
- Voc = .85 Volts
- Eff = 7.8%

Experimental Results:

- Isc = 1.7Amps
- Voc = .525 V
- Eff = 7.8%

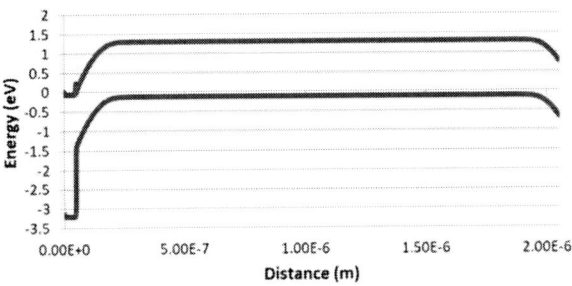

Fig. Equilibrium Band diagram of ZnO/WSe2/Au device.

While the simulated efficiency is reasonable the short circuit current and open circuit voltage are not a good match. For this simulation an Au work function of 5.1 eV was employed. It was found that the efficiency and open circuit voltage are extremely sensitive to this value. A change of as little as 0.06eV drives the open circuit voltage to 0.55V and the simulated efficiency to less than 3%. The reported range of work functions for Au is between 5.1 and 5.4 eV. [17]

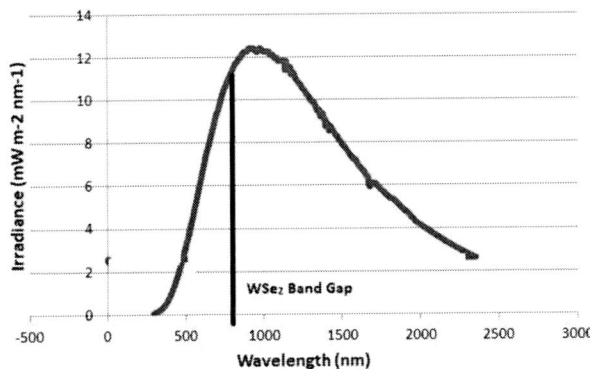

Fig. 2 Halogen spectrum. [16]

While the device has not been simulated exactly it is extremely useful to examine the potential efficiency under AM1.5 illumination conditions. In addition to changing the spectra the thickness of the ZnO was adjusted to 60 nm to enhance the anti-reflection properties of the structure. Under these conditions the following simulated results were obtained:

- Isc = -2.475 Amps
- Voc = .8827 Volts
- Eff = 16.29%
-

A brief examination of the band structure shows that Au forms a barrier contact with WSe2, therefore further improvements in efficiency can be expected for better matched back contact metal. Therefore, utilizing a material with a work function closer to the electronic affinity of WSe2 should raise the overall efficiency. By applying a back contact of Copper (Work Function ~4.65eV)[9], a simulated efficiency of 18% was

achieved, where the simulation assumes an ideal contact between WSe2 and Copper.

In summary the relatively low efficiency reported for single crystal WSe2/thin film heterojunction devices appears to be a result of the halogen lamp illumination spectrum and possibly the choice of Au as the back contact.

ACKNOWLEDGEMENT

The authors wish to acknowledge the support of the UCARE program at the University of Nebraska-Lincoln and the Nebraska Center for Energy Sciences Summer Research Program for their support.

REFERENCES

[1] M. Vogt, M. C. Lux-Steiner, P. Dolatzoglou, and E. Bucher, "Comparison between the photovoltaic performance of WSe2 Heterojunctions Perpared by ITO or ZnO magnetron sputtering", *Proceed. Of the 8th E.C. Photovolatic Energy Conference*, Florence, Italy, 1988, p. 1112

[2] A. Jakubowicz, D. Mahalu, M. Wolf, A. Wold, R. Tenne, "WSe2: Optical and electrical properties as related to surface passivation of recombination centers", *Phys. Rev. B*, vol. 40, pp. 2992–3000, 1989.

[3] M. C. Lux-Steiner, M. Vo¨gt, P. Dolatzoglou, A. Ja¨ger-Waldau, E. Bucher, "Preparation and so- lar cell performance of n-zno/p-wse2 heterojunctions", *Technical Digest of the International PVSEC-3*, Tokyo, Japan B-III-p-3, 1987, p. 687.

[4] A. Jager-Waldau, M. C. Lux-Steiner, E. Bucher, "MoS2, MoSe2, WS2 and WSe2 thin films for photovoltaics", *Solid State Phenomena*, vols.37-38, pp.479–484, 1994.

[5] P.A. Basore, "Numerical modeling of textured silicon solar cells using PC1D", *IEEE Trans. on Electron Devices*, vol. 37, pp. 337-343, 1990.

[6] C. Jiang, T.Z. Li, X. Zhang, L. Hou, "Simulation of silicon solar cell using PC1D", *Advanced Materials Research*, vols. 383-390, pp. 7032-1036, 2011.

[7] Q Qinglei Ma, Hrachya Kyureghian, Joel D. Banninga, N. J. Ianno, "Thin Film WSe2 for use as a photovoltaic absorber material", *MRS Proceedings*, (2014) 1670, mrss14-1670-e01-02 doi:10.1557/opl.2014.739.

[8] O. Lang, Y. Tomm, R. Schlaf, C. Pettenkofer, W. Jaegermann, "Single crystalline GaSe/WSe2 heterointerfaces grown by van der waals epitaxy. II. junction characterization", *Journal of Applied Physics*, vol. 75 (12), pp. 7814–7820, 1994.

[9] B. Davey, B. L. Evans, "The optical properties of MoTe2 and WSe2, *Physica Status Solidi A*, vol. 13 (2), pp. 483–491, 1972.

[10] K. K. Patel, K. D. Patel, M. Patel, C. A. Patel, V. M. Pathak, R. Srivastava, "Structural and ther- moelectric properties of tungsten diselenide crystals", *AIP Conference Proceedings 1393* (1) (2011) 251–252.

[11] G. K. Solanki, D. N. Gujarathi, M. P. Deshpande, D. Lakshminarayana, M. K. Agarwal, Trans- port property measurements in tungsten sulphoselenide single crystals grown by a CVT technique, Crystal Research and Technology 43 (2), 2008, pp.179–185.

[12] A. Jain, S. P. Ong, G. Hautier, W. Chen, W. D. Richards, S. Dacek, S. Cholia, D. Gunter, D. Skinner, G. Ceder, K. a. Persson, "The Materials Project: A materials genome approach to accelerating materials innovation", *APL Materials*, vol. 1 (1), pp. 011002, 2013

[13] B. Hussain, A. Ebong, Ian Ferguson, "Zinc oxide as an active n-layer and antireflection coating for silicon based heterojunction solar cell", *Solar Energy Materials & Solar Cells,* vol. 139, pp. 95-100, 2015.

[14] N. S. Pesika, K. J. Stebe, and P. C. Searson, "Determination of the Particle Size Distribution of Quantum Nanocrystals from Absorbance Spectra", *Advanced Materials,* vol. No. 15, 1289-1291, 2003.

[15] P. Banarjee, W. J Lee, K. R. Bae, S. B. Lee, and G. W. Rubloff, "Structural, Electrical, and Optical Properties of Atomic Layer Deposition Al-Doped ZnO films" , *Journal of Applied Physics*, vol. 108, pp. 043504-1 – 043504-7, 2010.

[16] Newport Oriel Product Training. htts://www.newport.com/medias/sys_master/images/images/hf b/hdf/8797196451870/Light-Sources.pdf

[17]. H. Michaelson, "The work function of the elements and its periodicity", *Journal of Applied Physics*, vol. 48, pp. 4729-4733, 1977.

Biaxial-textured Titanium Nitride thin films on low-cost, flexible metal substrate as a conductive buffer layer for thin film solar cells

Yongkuan Li, Yao Yao, Ying Gao, Sicong Sun, Pavel Dutta, Monika Rathi, Jae-Hyun Ryou, and Venkat Selvamanickam

Department of Mechanical Engineering, Advanced Manufacturing Institute & Texas Center for Superconductivity, University of Houston, Houston, TX 77204, USA

Abstract — **Biaxial textured TiN thin films have been grown as the first conductive buffer layer for thin film silicon and III-V photovoltaics based on low-cost, flexible Hastelloy substrates. The trends of TiN (002) peak position, out-of-plane and in-plane texture quality and resistivity of biaxially-textured TiN films grown at varying N_2 flow rates by reactive sputtering have been studied. The TiN film with N/Ti ratio of 0.8 shows the lowest resistivity and the best texture, making it suitable as the first conductive buffer layer for thin film silicon and III-V solar cells fabricated on inexpensive metal substrates.**

Index Terms — **conductive buffer layers, single crystal like materials, Titanium Nitride, biaxial-textured, thin film solar cell.**

I. INTRODUCTION

The contribution of solar energy to electricity generated in the U.S. is mainly limited by the higher cost compared to the other energy sources. [1] Thin film solar cells have been demonstrated on glass, polymer and metal substrates instead of wafer substrates to reduce the cost. However, these films are either amorphous or polycrystalline, leading to lower conversion efficiency compared with crystalline solar cells. [2] We have demonstrated a method to fabricate single-crystalline-like films for GaAs and Si photovoltaics on inexpensive, flexible substrates. In this method, an Ion Beam Assisted Deposition (IBAD) technique was used to grow biaxial-textured Magnesium Oxide (MgO) thin films over low-cost, flexible Hastelloy C-276 substrates by a continuous reel-to-reel process. [3] Buffer layers from MgO to Germanium (Ge) [4] were deposited epitaxially to transfer the biaxial texture to both Si [5] and GaAs [6] thin film solar cells. However, because of the intervening oxide buffer layers, complex etching and deposition processes are required to fabricate the bottom contacts for the devices. This scheme can lead to a higher fabrication cost, lower yield caused by less area for light absorption and possibly a relatively higher shunt current and higher series resistance. In this context, we have developed an architecture of biaxial-textured films on flexible metal substrates using conductive buffer layers so as to use the substrate as the bottom contact and thereby eliminate the afore-said problems.

Titanium Nitride (TiN) is a promising candidate replacing MgO as the first textured layer. First, TiN has high electrical conductivity close to metal [7], high thermal stability and is a good diffusion barrier to various elements. Second, IBAD biaxial-textured TiN has been grown over various amorphous substrates [8], and conductive buffer architectures based on the

TiN were reported for Yttrium Barium Copper Oxide (YBCO). [9] As the first step to build up conductive buffer layers over IBAD TiN (usually several nanometers thick), a thicker TiN layer should be homo-epitaxially grown. In order to study the electrical and structural properties and to optimize the growth parameters of the epitaxial TiN film, this layer was grown on a biaxial-textured MgO substrate by reactive magnetron sputtering. Other layers grown epitaxially on TiN, and over IBAD TiN are being experimented on and will be presented later.

II. EXPERIMENTS

The growth of biaxially-textured MgO on metal substrates was done by a reel-to-reel magnetron sputtering system; details are provided in reference 4. TiN thin films were grown using a co-sputter system equipped with four magnetron sources for four different sputtering targets, which facilitates multi-layer growth. The chamber was pumped down to a base pressure of 2×10^{-7} Torr. High purity Ar (99.999%) and N_2 (99.999%) were used. For TiN deposition, a Ti target was utilized, and the temperature, the sputtering power, Ar flow and the target-substrate distance were set to 700 °C, 250 W, 20 sccm, and 10 cm, respectively. Four different N_2 flow rates of 1.75, 3, 6 and 12 sccm were used to grow four samples. After 30 minutes of deposition, the thicknesses were measured to be around 1200, 400, 300, and 200 nm respectively. The surface morphologies were imaged by a LEO 1525 Scanning Electron Microscope (SEM). Energy-dispersive X-ray spectroscopy (EDS) data was obtained with a JEOL JSM-6330F system. High resolution X-ray diffraction (HRXRD) patterns were collected using a Rigaku Smartlab system and a Bruker 2D General Area Detector Diffraction System. Resistivities of the films were measured by a Four-probe Hall measurement (Ecopia HMS5000) at room temperature.

III. RESULTS AND DISCUSSION

The θ-2θ scans of the TiN films grown on biaxially-textured MgO on Hastelloy substrate are presented in Fig. 1 in a log scale. All TiN films show (002) peak at around 42° as the preferred orientation, indicating that the materials are out-of-plane aligned. A very weak peak A at 38.26° can possibly be identified as a strained TiN (111) peak (since the standard peak position (JCPDS, No. 01-087-0629) is at 36.646°), which grows stronger when N_2 flow is low. Peak B and Peak C are both Hastelloy peaks.

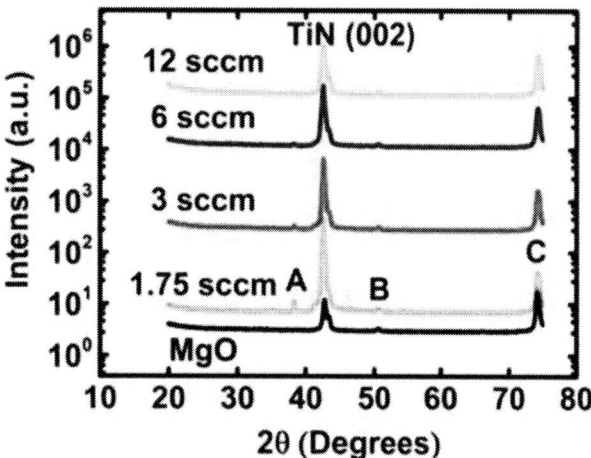

Fig. 1.   θ-2θ scans of the TiN samples grown with different N_2 flows on biaxially-textured MgO on metal substrate. Peak A is possibly a strained TiN (111) peak. Peak B and C are both peaks of Hastelloy.

The TiN (002) peak positions are derived by fitting the peaks to a Gaussian function and are listed in Tab. 1. According to Bragg's Law:

$$n \cdot \lambda = 2d_{hkl} \cdot \sin\theta_{hkl} \tag{1}$$

where n is a positive integer, λ is the wavelength of the incident beam (X-ray for XRD), d_{hkl} is the spacing of the diffraction planes *hkl*, and θ_{hkl} is the angle between the incident beam and the reflecting planes. For materials with a cubic unit cell, the spacing of *hkl* plane can be expressed in terms of the lattice constant as:

$$1/d_{hkl}^2 = (h^2+k^2+l^2)/a^2 \tag{1}$$

where a is the lattice constant of the materials. Then we can calculate the actual lattice constant from the measured XRD peak through the following equation:

$$a = a_{JCPDS} \cdot \sin\theta_{hkl(JCPDS)} \cdot \frac{1}{\sin\theta_{hkl}} \tag{3}$$

where a_{JCPDS} is the standard powder materials lattice constant form the JCPDS card, and $\theta_{hkl(JCPDS)}$ is the θ value of *hkl* planes from the JCPDS card. With a reference of TiN JCPDS card # 01-087-0629, the lattice constant and the lattice mismatch to GaAs with in plane 45° lattice rotation (JCPDS card # 00-014-0450) are calculated with equation (3) and listed in Tab. 1. Considering that the lattice constant of TiN is larger than that of MgO, the TiN lattice will be compressed in plane and be expanded out of plane, resulting in relative larger lattice constants from the θ-2θ scans compared to the values from the JCPDS card (which is 4.244 Å). But, it is seen from Table I that only the data of samples grown with 6 and 12 sccm N_2 are following this trend. That means another factor is affecting the lattice constant, which is most probably the content of N which

will be discussed later. The least lattice mismatch to GaAs is achieved in the sample grown with 3 sccm N_2 flow.

TABLE I. TiN (002) PEAK POSITIONS, LATTICE CONSTANT AND THE LATTICE MISMATCH TO GaAs AFTER 45° IN-PLANE ROTATION.

N_2 Flow (sccm)	TiN (002) peak positions (°)	Lattice constant a (Angstrom)	Lattice mismatch to GaAs after 45° in-plane rotation
1.75	42.59	4.242	5.76 %
3	42.62	4.240	5.72 %
6	42.55	4.246	5.86 %
12	42.43	4.258	6.11 %

Fig. 2 displays the pole figure of {220} peaks of the TiN films grown with 3 sccm N_2, which shows clearly the four-fold symmetry of the cubic material, indicating the in-plane alignment of the TiN thin film. All the other samples with different N_2 flow demonstrate similar four-fold symmetry. Combined with evidence of strong out-of-plane alignment, it can be concluded that the TiN films are clearly biaxially textured (single crystal like).

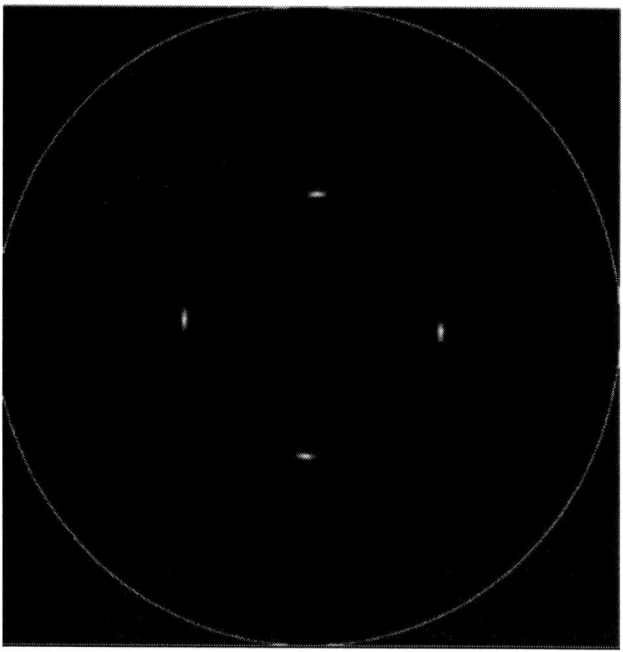

Fig. 2.   Pole figure of {220} peaks of the TiN films grown with 3 sccm N_2.

Fig. 3 (a) shows a SEM image of the TiN film grown with 3 sccm N_2, and the SEM images of the other samples with different flow are all similar. The film surface is smooth and shiny, with a uniform distribution of very fine grains with size of 50-100 nm. A photograph of the flexible TiN film on the Hastelloy substrate is displayed in Fig. 3 (b).

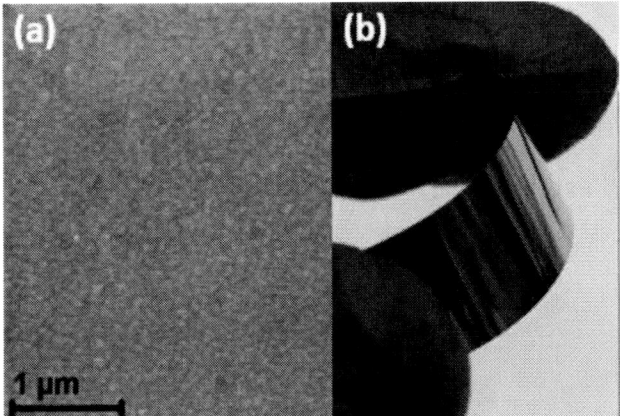

Fig. 3. (a) SEM Image of TiN film grown with 3 sccm N_2. (b) Photograph of a TiN film grown with 3 sccm N_2 over flexible Hastelloy.

Elemental concentration of all samples made with different N_2 flows and a TiN sample grown with a TiN target (sample *a*, with flow of Ar only) was measured. Since the EDS measurement result is not accurate for N, the N to Ti atomic ratio (N/Ti ratio) was first calculated and then normalized by dividing it by the N/Ti ratio of sample *a*, and the data are plotted versus the N_2 flow in Fig. 4. It is found that the normalized N/Ti ratio increases from 0.63 to 0.98 accordingly but not linearly with increasing N_2 flow. TiN is reported to crystallize as a rock salt structure with N/Ti ratio in the range of 0.6 to 1.2 [10], indicating that the error of normalization is acceptable. Lower N/Ti ratio results in a smaller lattice constant as shown in Table I, which indicates that more N vacancies could be formed. But when N/Ti ratio becomes as low as 0.63, the lattice parameter increases instead, indicating the formation of Ti interstitial atoms before the formation of Di-Titanium nitride (Ti₂N).

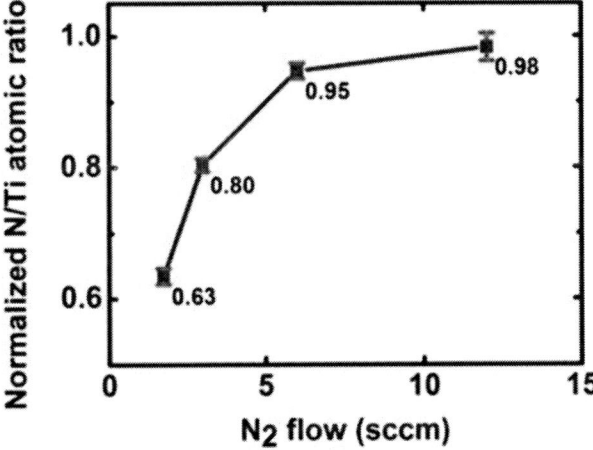

Fig. 4. Plot of the normalized N/Ti atomic ratios with standard errors versus the N_2 flow. The data has been normalized by dividing the N/Ti ratio atomic ratio by the N/Ti atomic ratio of a TiN sample deposited from a TiN target with Ar flow only.

The resistivity of the samples is plotted versus the normalized N/Ti ratio. As shown in Fig. 5, with lowering N content, the resistivity first decreases and then increases, reaching a minimum of 4.17 μΩ•cm. Yokota et. al. [11] reported the same bowl-like relationship for TiN film over Si substrate between resistivity and N/Ti ratio, but with the lowest TiN resistivity value of 30 μΩ•cm, compared to a resistivity of bulk TiN of 21.7 μΩ•cm. Karr et. al. [12] grew TiN thin film over MgO wafers and found that the lowest value for their samples was 12.4 μΩ•cm. The groups that fabricated biaxial-textured TiN over IBAD TiN did not report the resistivity data probably since it is hard to measure the TiN thin film resistivity with thickness at a nanometer scale over conductive substrates with thickness in micrometer scale.

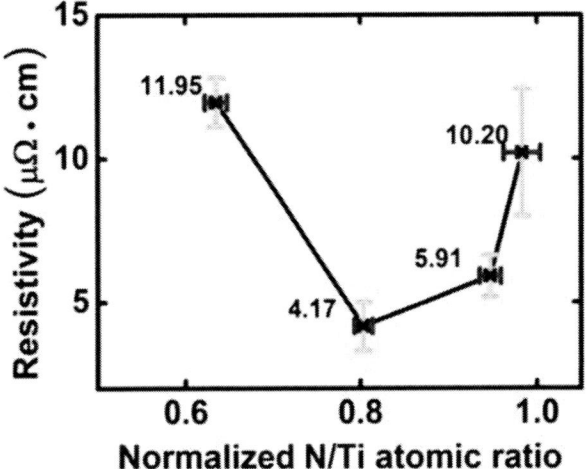

Fig. 5. Resistivity of TiN samples grown with different N_2 flows versus the normalized N/Ti ratio, with standard errors for both variables. The resistivity for each sample is marked in the plot.

The texture of the single-crystal-like thin films is qualified by two metrics: 1. the out-of-plane texture $\Delta\omega$, measured by the full-width-at-half-maximum (FWHM) of the rocking curve of the TiN (002) peak, 2. the in-plane texture $\Delta\phi$, measured as the average of the four FWHM values of the TiN {220} phi scan peaks integrated from (220) XRD pole figure. The $\Delta\omega$ is derived by fitting the rocking curve of the sample to a Gaussian function, and the $\Delta\phi$ is derived by a Gaussian fitting of the phi scan pattern integrated from the pole figure measurement as shown in Fig. (2). Both $\Delta\omega$ and $\Delta\phi$ values of the TiN films are plotted in Fig. 6 versus the normalized N/Ti ratio.

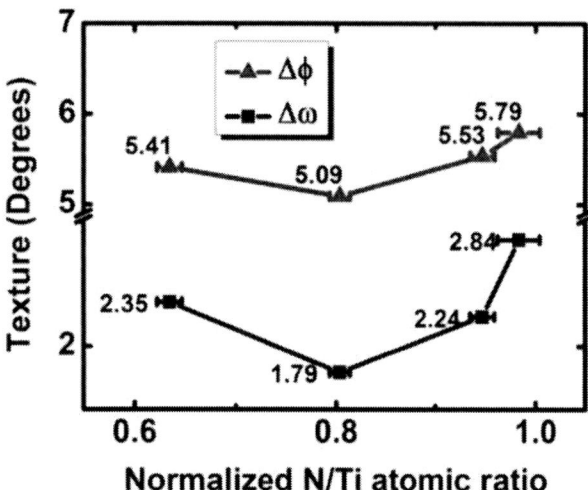

Fig. 6. Out-of-plane texture ($\Delta\omega$) and in-plane texture ($\Delta\phi$) of TiN films versus the normalized N/Ti ratio with standard errors.

As plotted in Fig. (6), the best texture is achieved at N/Ti ratio of 0.80, and both $\Delta\omega$ and $\Delta\phi$ become larger with either more N content or Ti content in the film. The TiN lattice with more N vacancies matches better with MgO lattice, so the corresponding samples can achieve better texture. Vice versa, the broader texture of the sample with N/Ti ratio of 0.63 is probably caused by Ti interstitial atoms. The samples having lower $\Delta\omega$ and $\Delta\phi$ values will consist of grains with less misalignment, and thus will have fewer defects. The defects likely contribute to the scattering of carriers, so the resistivity trend in Fig. 5 can be correlated to the texture trend in Fig. 6. Better texture, fewer defects, lower resistivity and lower lattice mismatch (to GaAs or Si) of the buffer layers are all needed for a better solar cell performance using biaxially-textured substrates. In this sense, the best sample is the TiN thin film grown with 3 sccm N_2 on the biaxially-textured substrate.

IV. CONCLUSION

TiN thin films have been grown with different N_2 flow using reactive magnetron sputtering over biaxial-textured MgO substrates. The N content in the films decreases with decreasing N_2 flow, probably causing N vacancies and then Ti interstitial atoms before Ti_2N formation. The TiN (002) peak position changes at different N_2 flows, resulting in the least lattice mismatch to GaAs in the sample with N/Ti ratio of 0.8. The resistivity and the texture of the TiN films both decrease when N/Ti ratio becomes lower likely due to formation of more N vacancies leading to better lattice match between TiN and MgO and in turn fewer defects that can act as scattering centers, while the probable formation of Ti interstitial atoms have the opposite effect. The sample with N/Ti ratio of 0.8 has the lowest resistivity and the best texture, which is suitable as the first

conductive buffer layer for both Si and GaAs solar cells based on low-cost, biaxially-textured substrates. Other layers built on TiN, and over IBAD TiN instead of IBAD MgO are underway.

REFERENCES

[1] Arvind Shah, P Torres, Reto Tscharner, N Wyrsch, and H Keppner, "Photovoltaic technology: the case for thin-film solar cells", science, vol. 285, pp. 692-98, 1999.

[2] K. L. Chopra, P. D. Paulson, and V. Dutta, "Thin-film solar cells: an overview", Progress in Photovoltaics, vol. 12, pp. 69-92, 2004.

[3] James R Groves, Paul N Arendt, Stephen R Foltyn, Quanxi Jia, Terry G Holesinger, Harriet Kung, Raymond F DePaula, Paul C Dowden, Eric J Peterson, and Liliana Stan, "Recent progress in continuously processed IBAD MgO template meters for HTS applications", Physica C: Superconductivity, vol. 382, pp. 43-47, 2002.

[4] V Selvamanickam, S Sambandam, A Sundaram, S Lee, A Rar, X Xiong, A Alemu, C Boney, and A Freundlich, "Germanium films with strong in-plane and out-of-plane texture on flexible, randomly textured metal substrates", Journal of Crystal Growth, vol. 311, pp. 4553-57, 2009.

[5] Ying Gao, Pavel Dutta, Monika Rathi, Yao Yao, Milko Iliev, Jae-Hyun Ryou, and Venkat Selvamanickam, "Heteroepitaxial silicon thin films on flexible polycrystalline metal substrates for crystalline photovoltaic solar cells: a comparison between physical vapor deposition and plasma-enhanced chemical vapor deposition", in 40th IEEE Photovoltaic Specialist Conference (PVSC), p. 1287, 2014.

[6] P Dutta, M Rathi, N Zheng, Y Gao, Y Yao, J Martinez, P Ahrenkiel, and V Selvamanickam, "High mobility single-crystalline-like GaAs thin films on inexpensive flexible metal substrates by metal-organic chemical vapor deposition", Applied Physics Letters, vol. 105, p. 092104, 2014.

[7] K. Yokota, T. Kasuya, K. Nakamura, M. Ohnishi, and F. Miyashita, "Ion beam current dependence of compositions and resistivities on titanium nitride films deposited onto silicon by an ion beam assisted deposition method", Nuclear Instruments & Methods in Physics Research Section B-Beam Interactions with Materials and Atoms, vol. 242, 390-92, 2006.

[8] R Hühne, S Fähler, and B Holzapfel, "Thin biaxially textured TiN films on amorphous substrates prepared by ion-beam assisted pulsed laser deposition", Applied physics letters, vol. 85, pp. 2744-46, 2004.

[9] R. Huhne, K. Guth, R. Gartner, M. Kidszun, F. Thoss, B. Rellinghaus, L. Schultz, and B. Holzapfel, "Application of textured IBAD-TiN buffer layers in coated conductor architectures", Superconductor Science & Technology, vol. 23, p. 014010, 2010.

[10] K. Vasu, M. G. Krishna, and K. A. Padmanabhan, "Substrate-temperature dependent structure and composition variations in rf magnetron sputtered Titanium Nitride thin films", Applied Surface Science, vol. 257, pp. 3069-74, 2011.

[11] K. Yokota, T. Kasuya, K. Nakamura, M. Ohnishi, and F. Miyashita, "Ion beam current dependence of compositions and resistivities on titanium nitride films deposited onto silicon by an ion beam assisted deposition method", Nuclear Instruments & Methods in Physics Research Section B-Beam Interactions with Materials and Atoms, vol. 242(1-2), pp. 390-392, 2006.

[12] B. W. Karr, D. G. Cahill, I. Petrov, and J. E. Greene, "Effects of high-flux low-energy ion bombardment on the low-temperature growth morphology of TiN (001) epitaxial layers", Physical Review B, vol. 61(23), pp. 16137-16143, 2000.

SnS by Ionized Jet Deposition for photovoltaic applications

Daniele Menossi, Simone Di Mare, Ivan Rimmaudo°, Elisa Artegiani, Giampiero Tedeschi[+], Juan Luis Pena°, Fabio Piccinelli*, Andrei Salavei and Alessandro Romeo

Department of Computer Science, University of Verona, Verona, 37123, Italy

° Centro de Investigación y de Estudios Avanzados del IPN Unidad Mérida, Depto. Física Aplicada, Km. 6, Antigua Carretera a Progreso, C.P., 97310 Mérida, Yucatán, Mexico.

[+]Noivion S.r.l., Rovereto, 38068, Italy

*Department of Biotechnology, University of Verona, Verona, 37123, Italy

Abstract — **Tin sulfide is an excellent candidate for the mass production of solar cells as it is composed of abundant and not toxic elements. We have prepared SnS polycrystalline films by a novel technique. This technology allows to transfer atoms from a target of SnS to the substrate due to the generation of plasma from the target, giving enough energy to have a good crystallization of the material also at low temperature. Ionized Jet deposition (IJD) is an improved Pulsed Electron Deposition technique in which a pulsed high power electron beam penetrates into the target resulting in a rapid evaporation of the material, and its transformation in plasma state. The non-equilibrium extraction (ablation) facilitates stoichiometric composition of the plasma. This is particularly advantageous in the case of complex, multicomponent materials.**

Index Terms — **SnS, Pulsed Electron deposition, thin films.**

I. INTRODUCTION

Tin sulfide is an excellent candidate for the mass production of solar cells as it is composed of abundant and not toxic elements. On the contrary materials currently used for the mass production of thin film solar panels (mainly CdTe and CIGS) suffer from limitations related to scarcity issues and/or toxicity of some elements that compose them thus representing a constraint to the spread on a large scale of the cheapest thin film technology. The research in SnS is done actively by major research centers such as MIT and Harvard University confirming the great interest in this material. However, much still needs to be done: efficiencies not greater than 5% [1] have been obtained mainly due to the difficulty of maintaining a proper stoichiometry, avoiding secondary phases and the lack of an ideal n-layer to make a good junction with SnS.

Ionized Jet deposition (IJD), which is an improved version of pulsed jet deposition, allows to control the stoichiometry and to maintain low substrate temperature. This separates the temperature effects on the crystallization (not needed anymore) from the elements interdiffusion in the junction.

The IJD method is based on the ionization of a gas stream flowing through a metallic nozzle, which serves simultaneously as the auxiliary electrode for the plasma discharge ignition. A pulse at high voltage (up to 25 kV) and a short duration (less than 1 μs) is applied to the cathode. Through a system of trigger and auxiliary electrodes, it causes strong ionization of the gas jet ablating the target surface (see Fig. 1). So a pulsed high power electron beam penetrates approximately for 1 μm into the target, resulting in a rapid evaporation of the target material and its transformation in plasma phase. The non-equilibrium extraction of the target material (ablation) facilitates stoichiometric composition of the plasma. Under optimum conditions, the target stoichiometry is thus preserved in the deposited films. The main feature of the pulsed systems is the ability to generate a high power density of about 10^8 W/cm^2 at the target surface. As a result, the thermodynamic properties of the target material, such as the melting point and specific heat, become unimportant for the evaporation process.

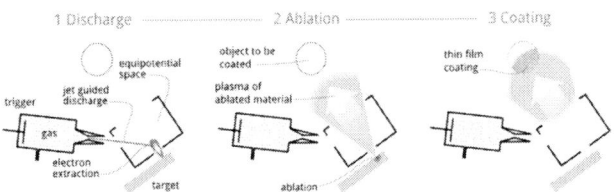

Fig. 1.   Scheme of the ionized jet deposition process.

The work here presented is based on our previous work on SnS deposition and solar cell preparation by thermal evaporation in vacuum [2] and on the promising results coming from the application of IJD on CdTe based solar cell fabrication process [3].

In our laboratories SnS solar cells with 1.5% conversion efficiencies have been fabricated [4]. SnS thin films are typically made in superstrate configuration by vacuum thermal evaporation (VTE). On a soda lime glass ITO and ZnO layers (300 nm and 100 nm respectively) are deposited by RF sputtering in a reactive atmosphere of Ar/O$_2$ at 400°C. The TCO is then annealed in vacuum at 450°C. As buffer layer, 50 nm of CdS are deposited by VTE at 100°C, then 2 μm SnS thin films are deposited on this stack by VTE at different temperatures (best efficiencies have been obtained for 270°C); 99.99 % pure SnS pieces are used as source material.

978-1-5090-5606-4/17 $31.00 © 2017 IEEE

II. EXPERIMENTAL DETAILS

The IJD-SnS layers have been deposited in a dedicated vacuum chamber. This is equipped with a IJD gun from Noivion S.r.l., and the vacuum is provided by a rotary and turbo-molecular pumping system, able to reach 10^{-7} mbar.

The distance between target and electron gun is fixed to 4 cm. During the deposition, the chamber pressure is brought between $10^{-4} - 10^{-3}$ mbar and IJD acceleration voltage is tuned in the range of 13-16 kV. Different substrate temperatures have been tested: from room temperature up to 300 – 400°C.

The SnS solar cells were prepared following the standard procedure mentioned above: on a common soda lime glass, a 300 nm ITO and a 100 nm ZnO layers are deposited and are followed by a deposition of 50 nm of CdS by vacuum thermal evaporation. Finally the back contact is made by depositing a 50 nm gold film, by VTE and with substrate at room temperature. SnS layers have been studied in terms of physical properties by atomic force microscopy (AFM), using a NT-MDT -SMENA A system, equipped with gold coated silicon tips from the same company, in semi-contact mode and by X-ray diffraction (XRD) analysis, processed with a Thermo ARL X'TRA powder diffractometer (in Bragg-Brentano geometry), equipped with a Cu-anode X-ray source (Kα, λ=1.5418 Å) and a Peltier Si (Li) cooled solid state detector). Preliminary finished devices were analyzed by current density–voltage (J-V) analyses, performed with a Keithley Source Meter 2420 at room temperature.

III. ANALYSIS AND DISCUSSION

SnS has been deposited by ionized jet deposition on the CdS/ZnO/ITO stacks. Different samples have been made by applying acceleration voltages of 4-5 kV, with a power ranging from 30 to 40 W and the repetition rates used were in the order of 80 Hz. Nevertheless with these process parameters, the ablation time resulted to be quite long. In order to reach SnS film thicknesses of about 2 micrometers, acceleration voltages between 10 to 12 kV have been applied, corresponding to a power of 200-250 W, with same ablation frequencies of 80 Hz.

At the same time different pressure values during ablation have been tested. For pressures lower than 10^{-4} mbar, plasma revealed not to be sufficiently confined and the deposition is quite slow; higher pressures, around $5*10^{-4} - 10^{-3}$ mbar are necessary to reach sufficient deposition rates, of about 8-10 nm/sec.

In these conditions different samples have been processed with different substrate temperatures. Samples at room temperature have shown delamination of the layers most probably due to the lattice mismatch between CdS and SnS. This can be reduced by increasing the substrate temperature during SnS deposition. Tests show that with temperatures exceeding 200°C the absorber layer results to be more adhesive and compact.

This behavior has been explained performing atomic force microscopy measurements. In fact AFM pictures, see Fig. 2, show a reduction of grain size by increasing the substrate temperature. At the same time these grains tend to be more dense and packed. However, if we compare the results with SnS deposited by VTE (Fig. 3) we observe a smoother surface for IJD-samples. At a substrate temperature of 400°C the grains seem to be small enough to make the surface appear quite smooth. Also macroscopically layers deposited at 400°C look reflective and very smooth.

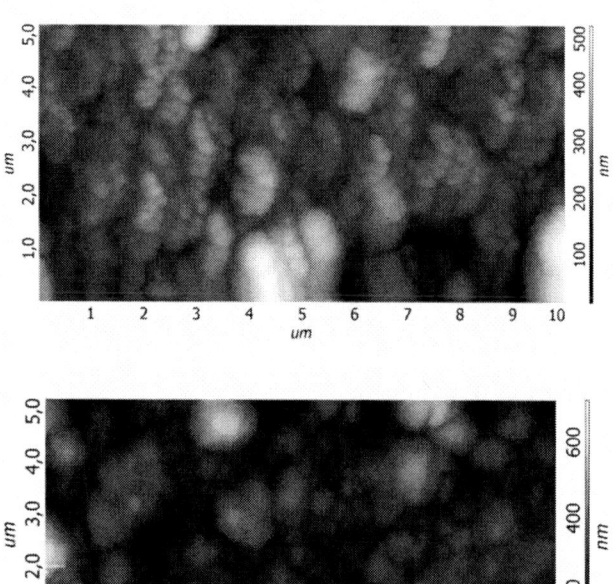

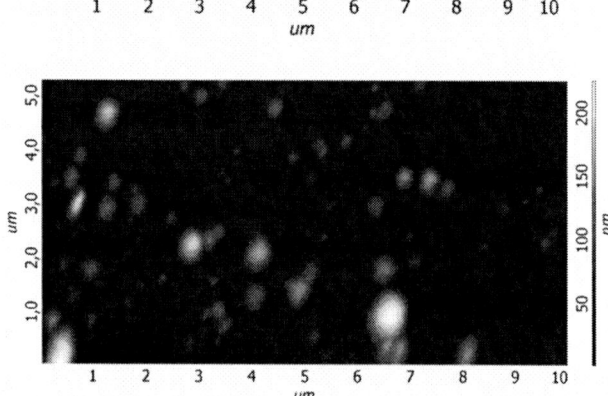

Fig. 2.   SnS layers deposited by IJD at 200°C, 300°C and 400°C (from top to bottom).

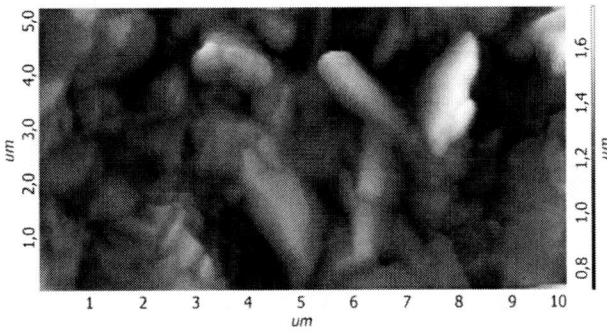

Fig. 3.   SnS layer by VE deposited at 300°C.

A similar comparison between samples has been done by XRD analysis. In Fig. 4 are shown XRD patterns of SnS deposited by IJD at different substrate temperatures. The main reflections, also detected for VTE-SnS (see Fig. 4), such as (010) and (111) are observed and very few other peaks are revealed. This might confirm the smooth and uniform low crystallized layer. In fact generally thin films show more peaks as the grains undergo large crystallization, as also attested by the post deposition treated SnS layers reported by the same authors [5]. However from XRD no secondary phases are detected and so a good quality in terms of homogeneity and uniformity can be addressed to these films, deposited between 200°C and 400°C. This confirms the potentiality of the technique for low substrate temperature deposition.

More investigations to confirm absence of secondary phases will be done by Raman spectroscopy.

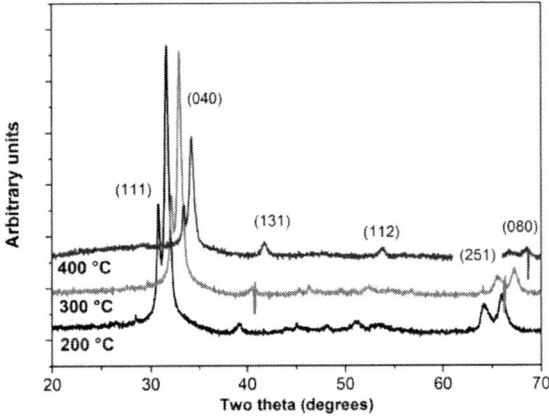

Fig. 4.   XRD spectra of SnS films deposited by IJD, at different substrate temperatures, namely (black) 200°C, (red) 300°C and (blue) 400°C.

Prototype solar cells, fabricated with IJD-SnS absorber layer have been tested, by means of current density-voltage measurements. So far have been obtained photovoltaic parameters of about J_{SC} = 2.1 mA/cm^2, V_{OC} = 0.12 V, FF = 31% and efficiencies of 0.08%.

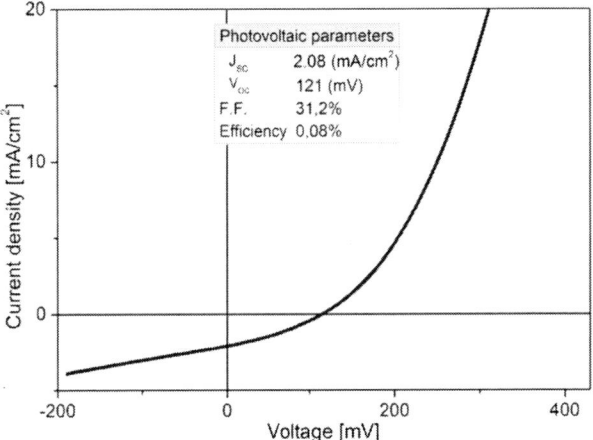

Fig. 5.   J-V behavior of a representative IJD SnS solar cell.

The low performance of these devices, Fig. 5, is might due to the too high-defective interface between CdS and SnS layers, considering to the strong energy band offset between them [6]. Other buffer layers different from CdS have been recently experimented, namely Zn(O,S), $Cd_{1-x}Zn_xS$, SnS_2, TiO_2 and a-Si [7]. So far the best results have been obtained by Buonassisi et al., using a Zn(O,S)/ZnO buffer layer structure [8]. Further work is in progress, experimenting different buffer layers than CdS, which could better match with the SnS-absorber layer and combining this with the novel IJD technique in order to fabricate more efficient solar cells.

IV. CONCLUSIONS

SnS thin film layers have been fabricated for the first time by an improved technique of electron pulsed deposition called ionized jet deposition.

Different vacuum pressures as well as different ablation powers have been tested and the deposition process has been optimized. Comparison of IJD-SnS layers deposited at different temperatures shows the good quality of layers at 300°C temperatures. The morphology of IJD-SnS at high temperature is very different from the VTE one, since its crystal size at these temperatures is smaller.

However XRD patterns show that the IJD- layers grown at any temperature between 200°C and 400°C seem not to have secondary phases and confirm the low influence of substrate temperature. These results are promising for optimizing a low-substrate temperature deposition process.

Preliminary solar cell devices show the formation o p/n junction even if efficiencies are still below 1%. With different buffer layers, in progress of study, it is expected an increase in the photovoltaic parameters of these devices.

ACKNOWLEDGEMENTS

This work has been supported by Fondazione Cassa di Risparmio di Trento e Rovereto with project TIN-Jet, by CONACYT-SENER (Mexico) and by CeMIE-Sol (Grant Nos. 207450/P25).

REFERENCES

[1] P. Sinsermsuksakul, L. Sun, S. W. Lee, H. H. Park, S. B. Kim, C. Yang, and R. G. Gordon, "Overcoming efficiency limitations of SnS-based solar cells," *Advanced Energy Materials*, vol. 4, no. 15, pp. 1400496–2, Oct-2014.

[2] S. Di Mare, A. Salavei, D. Menossi, F. Piccinelli, P. Bernardi, E. Artegiani, A. Kumar, G. Mariotto, and A. Romeo, "A study of SnS recrystallization by post deposition treatment," in *Proceedings of 43rd IEEE Photovoltaic Specialists Conference*, 2016, pp. 431–434.

[3] A. Salavei, G. Tedeschi, D. Menossi, S. Di Mare, F. Piccinelli, and A. Romeo, "CdTe thin film solar cells by pulsed electron deposition," in *Photovoltaic*

Specialist Conference (PVSC), 2016 IEEE 43rd, 2016, pp. 137–140.

[4] S. Di Mare, A. Salavei, F. Piccinelli, and A. Romeo, "Analysis of SnS Growth and Post Deposition Treatment by Congruent Physical Vapor Deposition," *Proc. EU PVSEC 2015*, pp. 1349–1352, 2015.

[5] S. Di Mare, A. Salavei, F. Piccinelli, and A. Romeo, "Analysis of SnS Growth and Post Deposition Treatment by Congruent Physical Vapor Deposition," in *31st European Photovoltaic Solar Energy Conference and Exhibition*, 2015.

[6] M. Sugiyama, K. T. R. Reddy, N. Revathi, Y. Shimamoto, and Y. Murata, *Thin Solid Films* 519(21), 7429 (2011).

[7] P. Sinsermsuksakul, K. Hartman, S. B. Kim, J. Heo, L. Sun, H. H. Park, R. Chakraborty, T. Buonassisi, *and* R. G. Gordon, "Enhancing the efficiency of SnS solar cells via band-offset engineering with a zinc oxysulfide buffer layer", *Appl. Phys. Lett.* 102, 053901 (2013).

[8] V. Steinmann , R. Jaramillo , K. Hartman , R. Chakraborty , R. E. Brandt , J. R. Poindexter , Y. S. Lee , L. Sun , A. Polizzotti , H. H. Park , R. G. Gordon , and T. Buonassisi, "3.88% Effi cient Tin Sulfi de Solar Cells using Congruent Thermal Evaporation", *Adv. Mater.* 2014, 26, pp. 7488–7492.

Effect of valence band splitting on the absorption spectra of monolayer MoS₂ in presence of sulphur vacancies.

Himani Mishra, Sitangshu Bhattacharya

Department of Microelectronics

Indian Institute of Information Technology, Allahabad,Uttar Pradesh, 211015, India

rse2016002, sitansghu@iiita.ac.in

Abstract - **Single layer Molybdenum Disulfide (MoS₂), a direct band gap transitional metal dichalcogenide (TMDC) has attracted a lot of research and study due to its excellent electro-optical integrity. Optical absorption within this monolayer MoS₂ is extremely influenced by the presence of sulphur vacancies. At reduced dimensions the interaction between these vacancy created trap centres and charge carriers becomes more prominent leading to the formation of bound excitons. Here we demonstrate the absorption spectra of a single layer MoS₂ through many body perturbation theory in the presence of sulphur vacancy sites. We use a fully relativistic approach within the GW approximation containing the non collinear core correction with full spinor wave functions. The absorption spectra calculation is achieved through the Bathe-Salpeter equation to include the excitonic excitations at room temperature. Our computations exhibit a Gaussian absorption spectra observed with double excitonic peaks A and B, unlike the step function profile without the incorporation of excited states. This double peak corresponds to valence band splitting at the K point of the brillouin zone which is most prominent at the top of the valence band and is a result of spin orbit coupling of the excitons. The absorption edge demonstrates a red shift when investigated in the presence of sulphur vacancies which can be attributed to inter excitionic interactions and the reduction in bandgap in the presence of sulphur vacancies. A change in the value of absorption coefficient is observed as a result of localization of excitons in the traps. Our outcomes clarify the vacancy and exciton material science of MoS₂ offering another course towards fitting its physical properties by defect engineering.**

KEYWORDS Excitonic peak, Monolayer Molybdenum Sulphide, Spin orbit coupling, Sulphur vacancies.

I. INTRODUCTION

As far back as the disclosure of graphene an allotrope of carbon, showing properties such as, high carrier mobility [1], miniaturized scale ballistic transport [2], strange quantum hall impact [3], 2.3% absorption of visible light [4] a considerable measure of attraction have been given to the two dimensional (2D) materials in the zone of nano-electronics and optics. In the field of photovoltaic, 2D materials are believed to speak to amazing properties. Graphene passivated silicon solar cells are found to exhibit an increase in the power conversion efficiency by 13.95 % [5] [6]. This pulls in enthusiasm of specialists to investigate other 2D materials, for example, TMDCs to comprehend their electronic and optical properties. Semiconducting nature of these materials makes them closer in properties to silicon which is the core material of solar industry and research.

Absorption of 5-10% incident light is possible in monolayer MoS₂ with a thickness of as low as 1 nm [7]. Similar results have been accounted in the case of other TMDCs namely MoSe₂ and WS₂ in this manner accomplishing one order of extent higher sunlight absorption than GaAs and Silicon. [7][9] The power conversion efficiency for 1 nm thick TMDC hetero structures are achieved upto ~1% which is three times better than the best existing ultrathin solar cells. [7] It had also appeared through first rule estimations that a monolayer TMDC with sub-nanometer thickness can absorb as much daylight as can be absorbed by a 50 nm of silicon solar cell and create electrical streams as high as 4.5 mA/cm². [7] This demonstrates the commitment of 2D monolayers can be colossal in delivering great quality solar cells and that their structure should be examined deliberately to comprehend the components of absorption and electron-hole pair generation inside these materials. Electronic structure in layered MoS₂ is an aggregate consequence of quantum control, interlayer association, and crystal symmetry. Valence band splitting is demonstrated just for few layered structure of MoS₂ after which it starts to diminish due to strong inter layer interactions. The effect of this valence band splitting on the absorption spectra is very prominent and its effect in the presence of vacancies has to be evaluated.

2D monolayer molybdenum disulfide can be extracted from bulk MoS₂ crystal by chemical or physical exfoliation and also through Chemical Vapour Deposition process. [10][11][12] Extraction through any of these processes produces defects in the monolayer because of its compound nature and high volatility which affects the absorbance peak within the material. Sulphur vacancies are the most prominent defects in these monolayers with a vacancy density of up to ~1.3×10^{13} cm⁻² [13] [14].

In this paper we will explore that how these defects affect the absorbance within the material with respect to intrinsic 2D MoS₂ monolayer and also the effect of valence band splitting on the absorption spectra. We will have a clear idea of the effect of bound excitons within the sulphur traps on the absorption spectra which will be very useful to understand sunlight absorption and electron-hole pair generation in these materials.

II. COMPUTATIONAL METHODS

To study the electronic and optical properties of the 2D monolayer MoS₂ structure we have used a three step approach where DFT+GW+BSE calculations generate the desired optical spectrum which will be compared for different cases. This three step process can be explained as:

DFT: Neutral excitations as a combination of single.

GW: Neutral excitations as a combination of single but with corrected quasiparticle bandgap.

BSE: Real two body propagator and many body problem.

Density functional theory calculations are based on the well-known Kohn sham equation which can be given as

$$\left[-\frac{1}{2}\nabla^2 + V^H + V^{xc}\right]\varphi_i = \varepsilon_i\varphi_i \qquad (1)$$

$$n = \sum_i^{occ}|\varphi_i|^2 \qquad (2)$$

where V^H and V^{xc} are the Hartree and exchange-correlation potential respectively. Here we have utilised the Perdew–Burke–Ernzerhof (PBE) exchange and correlation for performing density functional theory calculation within Quantum espresso package. [8] All the ground state calculations are performed in the norm conserving pseudo potential with nonlinear core correction for the inclusion of the core as well as valence electron charge density in the plane wave's basis set with a wave function kinetic energy cut off of 80 Hartree. Pseudo potential used for the calculation is fully relativistic in nature generated within the oncvpsp package which is necessary for the generation of the full spinor wave function otherwise the effects of spin are neglected in the calculation. A 2x2x1 2D monolayer MoS_2 super cell has been used for all the calculations with 10.83 Å of vacuum at both the ends along the z-direction for isolation in the same direction. Non-collinear spin orbit calculations within 10x10x1 Monkhorst−Pack k-point mesh have been performed to examine the effect of spin on the absorption spectra of the 2D MoS_2 monolayer. The time reversal symmetry (K and -K) and spatial symmetry is maintained throughout the calculations, as is required to be maintained in case of the material being investigated.

The presence of molybdenum vacancies is also investigated in the monolayer MoS_2. To investigate these vacancies we have performed ground state calculations for molybdenum vacancies in monolayer MoS_2 and found that the material in this case becomes metallic in nature and as a result becomes incompetent for solar cell applications.

Ground state PBE calculations were very successful in describing the ground state properties of most of the materials but does not incorporate the effects of quasi particles anywhere. To incorporate the excited state corrections the quasi particle band structure is extracted using the GW approximation (GWA). GWA is a generalisation of the Hartree–Fock method with the only exception of the dynamically screened Coulomb interaction instead of the static one. This approximation solves the Dyson equation which is the most reliable method to calculate the self-energy of most of the materials. It can be shown that the quasiparticle energies E_i can be obtained from the quasiparticle equation

$$\left[-\frac{1}{2}\nabla^2(r) + V^H(r)\right]\psi_i(r) + \int d^3r' \sum (r,r';E_i)\,\psi_i(r') = E_i\psi_i(r) \qquad (3)$$

where $\sum$ denotes the self-energy and contains the exchange and correlation. The solution of this equation is very complex even when applied to simple electron gas case and hence we need an approximation for the self-energy calculation which is given by GWA where the effects of screening are also taken into consideration with warnier approximations.

For GW calculations it is required to have an approximation for energy dependence of the GW function. Most widely used estimation of this kind is the Plasmon-pole approximation where the persistent change of charge density is displayed within a plane-wave representation of the dielectric matrix. It utilises gauss integration method for approximating the energy dependence for each matrix element at real positive energies by a simple Plasmon pole. As a result plasmon pole approximation enables exact energy integrations for the calculation of GW electron self-energy [14]. We have used the non-self-consistent GW calculation with plasmon pole approximation for the calculation of the quasi particle energies. Accuracy of our quasi particle band structure is dependent on the successful convergence of the dielectric matrix, empty states and the size of the monkhorst−Pack k-point mesh. A coulomb truncated GW calculation had been performed in order to cut off the coulomb screening in the z-direction as here we are concerned with a two dimensional monolayer. The truncation is highly dependent on the size of the supercell being investigated and should be slightly lesser in size as compared with the supercell dimensions. It is important to mention here that the incorporation of the semi-core states of d and f orbitals is vital for successful calculations.

The third and the final step of the calculation is the generation of the absorption spectra using Bethe Salpeter equation (BSE) named after Hans Bethe and Edwin Salpeter. [22] Optical absorption spectra is obtained by the imaginary part of the inverse of the dielectric function. The macroscopic dielectric function is given in equation 1

$$\varepsilon_M(\omega)$$
$$= 1 - \lim_{q \to 0} v_0(q) \sum_\lambda \frac{\left|\sum_{(n_1 n_2)}\langle n_1|e^{-iq.r}|n_2\rangle A_\lambda^{(n_1 n_2)}\right|^2}{E_\lambda^{exc} - \omega - i\eta}$$
$$(4)$$

where $A_\lambda^{n_1 n_2}$, E_λ^{exc} are the eigenvectors and eigenvalues of the (resonant) two-particle Hamiltonian.

The excitonic Hamiltonian can be given as

$$H_{(vck)(v'c'k')}^{reso} = (E_{ck} - E_{vk})\delta_{vv'}\delta_{cc'}\delta_{kk'} + 2v_{vck}^{v'c'k'} - W_{vck}^{v'c'k'} \qquad (5)$$

Optical calculations are performed with the complete Bethe-Salpeter Equation kernel having both the exchange and statically screened interactions solved using the diagonalization method. These calculations are also performed in the plasmon pole approximation and the quasi particle database is given in this calculation in order to incorporate the effects of quasi particle energy states. The direction of dipole operator has been recognized by the plane of the monolayer MoS₂ monolayer.

The convergence of the BS equation is highly dependent on the k-point sampling and hence we have used a 10x10x1 Monkhorst–Pack mesh both in the case of defect less and with defect 2D MoS₂ monolayer after carefully examining it over other mesh sizes.

BSE and GW calculation are performed within the YAMBO package and all the necessary parameters have been converged carefully in order to ascertain correct results.

III. RESULTS AND DISCUSSION

Fig. 1 represents the optimised structures of 2x2x1 supercell of monolayer MoS₂ with and without vacancy. The electron localisation function for the particular structures is also given in the figure which is the measure of the probability of finding an electron in the vicinity of a reference electron located at a given point with same spin. [15] We can see that the electron localisation function becomes negligible for the sulphur vacancy site. This is due to the fact that there is no atom present at the vacancy site and hence no electron can be localised over there. A dangling bond is formed over there which will act as trap centre for excitons. To investigate it further and have a clear idea of the properties of MoS₂ monolayer with vacancies we have evaluated the electronic band structure and optical spectrum of the material which will be discussed in detail.

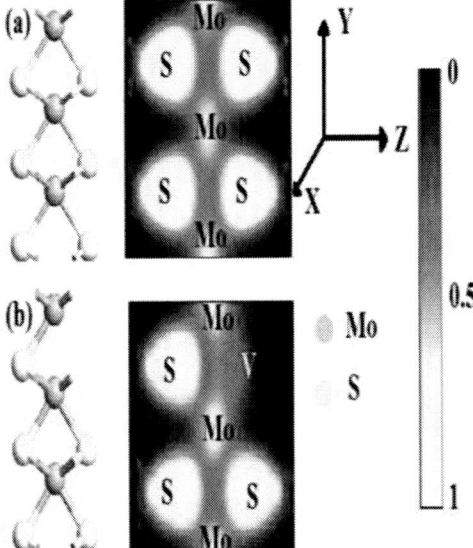

Fig. 1 Optimised structures and electron localisation functions of (a) 2D monolayer MoS₂ (b) 2D monolayer MoS₂ with sulphur vacancies.

Fig.2 represents the comparison of the band structures of 2D monolayer MoS₂ and the same with sulphur vacancies in ground state energy. The valence band maxima and conduction band minima are located on the K and K' point of the brillouin zone indicating a direct bandgap of 1.71 eV

for intrinsic monolayer MoS₂ and 0.32 eV for monolayer with defects which is in good agreement with the reported values from first principle calculations. [16] [17] Many new energy states are observed in the band structure of monolayer MoS₂ due to presence of the sulphur vacancy, because of which the bandgap of the vacancy induced MoS₂ monolayer is found to be lower than that of intrinsic one. Lack of inversion symmetry [18] and strong spin orbit coupling of the d orbitals splits the top of the valence band at K point in both the cases. This splitting is only visible when full spinor wave functions are evaluated with non collinear calculations as it is necessary to break the degeneracy of the energy states at K point [19].

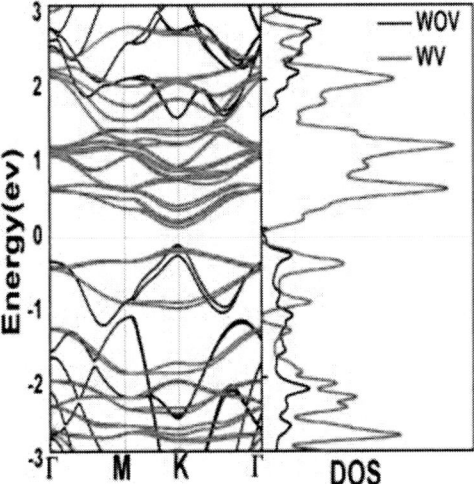

Fig. 2 Ground state band structure and density of states of 2D monolayer MoS₂ without vacancy (WOV) and with vacancy (WV).

The top of the valence band is splitted by 148 meV in case of intrinsic monolayer. The splitting in the band induced due to the presence of the vacancy which becomes the top of the valence band in this case is approximated to be 57.6 meV.

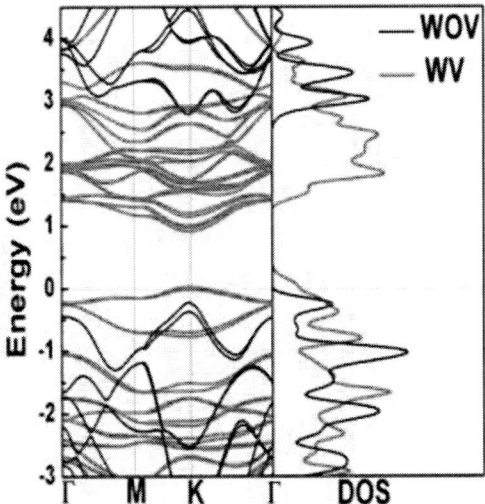

Fig. 3 Quasi particle band structure and density of states of 2d monolayer MoS2 with and without vacancy.

In fig. 3 we have demonstrated the quasi particle band structure and density of states of the MoS$_2$ monolayer with self-energy correction. The overall nature of the E-k curve remains same as that of ground state band structure but it is shifted in energy and has a slightly increased band gap. The quasi particle bandgap from GW calculation is approximated to be 2.8 eV for intrinsic monolayer MoS$_2$ and 0.82 eV for monolayer MoS$_2$ with defects. This is due to the addition of the self-energy of the excitons formed in this monolayer MoS$_2$. We have also given the density of states from the GW calculation which is having both spin-up and spin-down contributions almost equal to half. The valence band splitting contributions are almost same as that of ground state results for both the cases.

	Without defect	With defect
DFT	1.71 eV	0.32 eV
GW	2.8 eV	0.82 eV

Table 1 DFT and GW Band gap for monolayer MoS2 without and with vacancy.

In fig. 4 the absorption spectrum of 2D monolayer MoS$_2$ has been represented. The results of both RPA and BSE calculations have been evaluated at 300 K. All the calculations are done at 300 K as there is no considerable change in the nature of the absorption spectrum at both these temperatures that is 0 K and 300 K. In this fig. we have shown the calculated absorption spectra with and without excitonic effects. The excitonic calculations demonstrates a blue shift in comparison to RPA calculations which is due to the discrepancy in the RPA method. The RPA spectrum resembles to the step function profile after a threshold of 2.5 eV as in the case of independent particles.

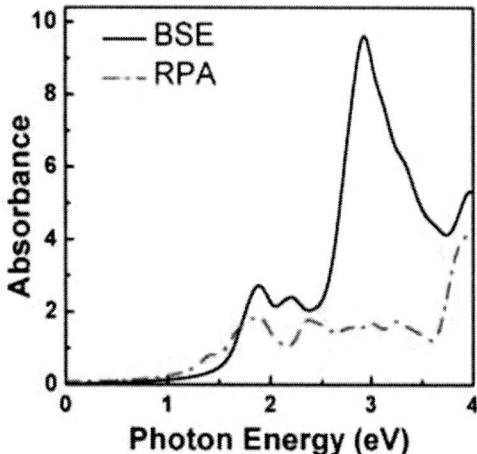

Fig. 4 BSE and RPA absorption spectra of 2D monolayer MoS2.

The excitonic optical spectrum is having three main features namely the two excitonic peaks accompanied by

an abrupt peak. The difference in the two peaks A and B at 1.9 eV and 2.1 eV which are formed due to the presence excitons is given by valence band splitting at the K point of the brillouin zone. This difference should be equal in magnitude to the valence band splitting on top of the valence band. Our absorption spectra at lower energy is very close to the experimental results observed at these energies [20]. The experimental setup is not able to measure the optical spectra above the threshold of 2.4 eV energy and hence the abrupt high peak at 3 eV can only be produced through theoretical calculations [20]. This peak at 3.0 eV, corresponds to the part of the Brillouin zone between K and Γ symmetry points where we observe a high density of states which can be seen from fig. 2 due to the parallel conduction and valence bands [21].

Fig. 5 represents the comparison of the absorption spectra of monolayer MoS$_2$ with and without sulphur vacancy. A red shift is demonstrated in the absorption spectra of the vacancy induced monolayer. The absorption coefficient is also seen to be slightly lower for the vacancy case attributed to the creation of trap centres within the structure. The first excitonic peak in the optical spectrum of defective monolayer is seen at 1.5 eV which corresponds to the first exciton or rather we can say first bright exciton. Second excitonic peak at 1.92 eV is also present in the structure. The presence of these excitons in the monolayer MoS$_2$ makes them highly suitable for application in the excitonic solar cells and also for the passivation of silicon based solar cells.

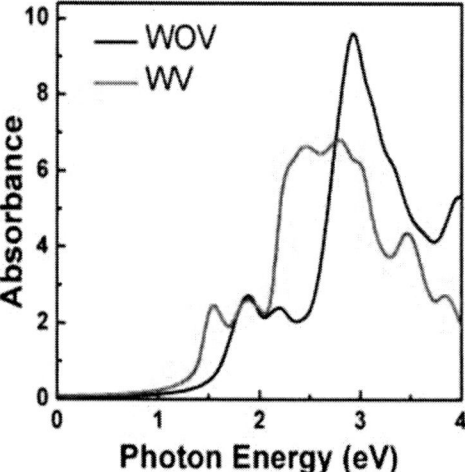

Fig. 5 Absorption spectra of 2D monolayer MoS$_2$ without vacancy (WOV) and with vacancy (WV).

IV. CONCLUSION

Here we have demonstrated the band structure, density of states and absorption curve of 2D monolayer MoS$_2$ with and without sulphur vacancies in order to understand the effect of these defects on the electronic and optical properties of the monolayer molybdenum sulfide. The sulphur vacancies being the dominant defects generated in monolayer MoS$_2$ have been studied and the results show a considerable change in both the electronic as well as optical properties of the defected and defect less monolayer MoS_2. The energy states generated in the case of monolayer MoS_2 with sulphur vacancies within the bandgap when compared with the intrinsic monolayer act as traps and provide ground for electron hole pair recombination which is a demerit for solar cells. The absorption peak is also lowered due to the presence of vacancies as bound excitons are generated in the presence of vacancies which will result in the lower absorption of the incident light. So, now we should look forward for various methods to passivate the dangling bonds generated because of these vacancies.

ACKNOWLEDGMENTS

The authors acknowledges financial support from the Department of science and technology through projects and the support from Nano Electro-Thermal Lab in Indian Institute of Information Technology, Allahabad, India.

REFERENCES

[1] Geim, Andre Konstantin. "Graphene: status and prospects." *science* 324.5934 (2009): 1530-1534.

[2] Mayorov, Alexander S., et al. "Micrometer-scale ballistic transport in encapsulated graphene at room temperature." *Nano letters* 11.6 (2011): 2396-2399.

[3] Novoselov, Konstantin S., et al. "Room-temperature quantum Hall effect in graphene." *Science* 315.5817 (2007): 1379-1379.

[4] Nair, Rahul Raveendran, et al. "Fine structure constant defines visual transparency of graphene." *Science* 320.5881 (2008): 1308-1308.

[5] Li, Xinming, et al. "Graphene-on-silicon Schottky junction solar cells." *Advanced Materials* 22.25 (2010): 2743-2748.

[6] Song, Yi, et al. "Role of interfacial oxide in high-efficiency graphene–silicon Schottky barrier solar cells." *Nano letters* 15.3 (2015): 2104-2110.

[7] Bernardi, Marco, Maurizia Palummo, and Jeffrey C. Grossman. "Extraordinary sunlight absorption and one nanometer thick photovoltaics using two-dimensional monolayer materials." *Nano letters* 13.8 (2013): 3664-3670.

[8] Giannozzi, Paolo, et al. "QUANTUM ESPRESSO: a modular and open-source software project for quantum simulations of materials." *Journal of physics: Condensed matter* 21.39 (2009): 395502 [9] Lin, Shisheng, et al. "Interface designed MoS2/GaAs heterostructure solar cell with sandwich stacked hexagonal boron nitride." *Scientific reports* 5 (2015).

[9] Li, Bo, et al. "Growth of large area few-layer or monolayer MoS2 from controllable MoO3 nanowire nuclei." *RSC Advances* 4.50 (2014): 26407-26412.

[10] McCreary, Kathleen M., et al. "Large-Area Synthesis of Continuous and Uniform MoS2 Monolayer Films on Graphene." *Advanced Functional Materials* 24.41 (2014): 6449-6454.

[11] Lan, Feifei, et al. "Epitaxial growth of single-crystalline monolayer MoS2 by two-step method." *ECS Solid State Letters* 4.3 (2015): P19-P21.

[12] Hong, Jinhua, et al. "Exploring atomic defects in molybdenum disulphide monolayers." *Naturecommunications* 6 (2015).

[13] Qiu, Hao, et al. "Hopping transport through defect-induced localized states in molybdenum disulphide." *Nature communications* 4 (2013).

[14] Engel, G. E., et al. "Calculation of the GW self-energy in semiconducting crystals." *Physical Review B* 44.24 (1991): 13356.

[15] Becke, Axel D., and Kenneth E. Edgecombe. "A simple measure of electron localization in atomic and molecular systems." *The Journal of chemical physics* 92.9 (1990): 5397-5403.

[16] Lebegue, S., and O. Eriksson. "Electronic structure of two-dimensional crystals from ab initio theory." *Physical Review B* 79.11 (2009): 115409.

[17] Kuc, Agnieszka, Nourdine Zibouche, and Thomas Heine. "Influence of quantum confinement on the electronic structure of the transition metal sulfide T S 2." *Physical Review B* 83.24 (2011): 245213.

[18] Molina-Sanchez, Alejandro, and Ludger Wirtz. "Phonons in single-layer and few-layer MoS 2 and WS 2." *Physical Review B* 84.15 (2011): 155413.

[19] Xiao, Di, et al. "Coupled spin and valley physics in monolayers of MoS 2 and other group-VI dichalcogenides." *Physical Review Letters* 108.19 (2012): 196802.

[20] Molina-Sánchez, Alejandro, et al. "Effect of spin-orbit interaction on the optical spectra of single-layer, double-layer, and bulk MoS 2." *Physical Review B* 88.4 (2013): 045412.

[21] Peter, Y. Yu. "Optical Properties I." *Fundamentals of Semiconductors*. Springer Berlin Heidelberg, 1996. 233-331.

[22] H. A. Bethe and E. Salpeter: Quantum Mechanics of One- and Two-Electron Atoms (Springer-Verlag, Berlin, 1957), sect. 61.

[23] Cheiwchanchamnangij, Tawinan, and Walter RL Lambrecht. "Quasiparticle band structure calculation of monolayer, bilayer, and bulk MoS2." *Physical Review B* 85.20 (2012): 205302.

[24] W. Greiner, J. Reinhardt (2003).Quantum Electrodynamics (3rd ed.). Springer. ISBN 978-3-540-44029-1.

[25] Aulbur, W.G. Jonsson, L. Wilkins, J.W. (2000), Solid State Physics.

The Study of Some Materials as Buffer Layer in Copper Antimony Sulphide (CuSbS₂) Solar Cell Using SCAPS 1-D

Muteeu Olopade*, Adeyinka Adewoyin, Michael Chendo, and Adewumi Bolaji

Department of Physics, University of Lagos, Akoka, Lagos, Nigeria

Abstract —This study involves the investigation of different thin film materials as buffer layer in CuSbS₂ solar cells. The materials considered were CdS, InS, ZnSe and ZnS. SCAPS-1D was used for the modeling and simulation of the device. A base model using CdS was developed, simulated and an efficiency of 3.24% was obtained. Thereafter, other buffer layer materials were introduced into the device and results show that the use of InS as buffer layer gave the best performing device with an efficiency of 3.78%. The thickness of InS layer was then optimized and a buffer layer thickness of 0.03μm gave the best efficiency of 3.88%.

Index Terms — CuSbS₂, buffer layer, ZnS and SCAPS-1D

I. INTRODUCTION

Thin film solar cells based on the chalcogenide absorbers such as CdTe and Cu(In,Ga)(S,Se)₂ have reached highest energy conversion efficiencies of 16.5% and 20.3% respectively [1]. However, the toxicity (Cd and Se) and the scarcity of raw materials (Te, In, and Ga) are major problems limiting the wide spread and utilization of these devices. Also, copper zinc tin sulfoselenide (CZTSSe) solar cell, another promising device in terms of efficiency have achieved a certified 12.6% efficiency via a hydrazine-based solution process [2]. In this device, Se is another compound that is not desirable because of its toxicity. Therefore, earth-abundant absorber materials with almost no toxicity needs to be intensively explored for thin film photovoltaics.

One promising absorber material is the Chalcostibite, copper antimony sulphide (CuSbS₂). It is a ternary layered semiconductor material suitable for sustainable and scalable photovoltaics due to its low-toxicity, low cost of the elements, optimal band gap (~1.5 eV) and high absorption coefficient of over $10^4 cm^{-1}$. These materials have been characterized by structural, electrical and photo electrochemical techniques in order to establish their potential as absorber layer materials for photovoltaic applications. Also, the ideal band gap of CuSbS₂ satisfies the Shockley–Queisser requirements for the efficient harvesting of the solar spectrum [3].

Septina et. al., reported a conversion efficiency of 3.13% has been achieved with the fabrication of a CuSbS₂- based solar cell from electrodeposited Cu/Sb stacked layers followed by sulfurization at 450°C in H₂S gas for 30 min, [1]. However, current efficiencies are still very low.

In this paper, the basic model of the CuSbS₂ thin film solar cell simulated was carried out first with a view of establishing a standard to be used for the study. Then, the thickness of the CuSbS₂ thin film layer was optimized. Finally, different buffer layer materials were introduced into the device with a view of analyzing their impacts on the performance of CuSbS₂ thin film solar cell and subsequently, the thickness of the best performing buffer layer was optimized.

II. DEVICE AND STRUCTURE

In this study, the basic structure of the device under study consist of a p-CuSbS₂ absorber layer, CdS buffer layer and a window layer made of n-ZnO:Al. The back contact of the device is Molybdenum and the entire structure is assumed to be deposited on a soda-lime glass. The schematic of the solar cell is shown in Fig. 1.

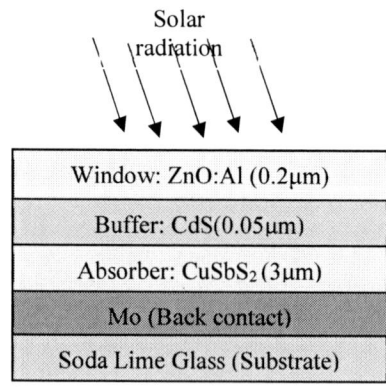

Fig.1: Schematic of 1-D CuSbS₂ substrate solar cells

TABLE I
PHYSICAL PARAMETERS USED IN THE SIMULATION

Parameters	p-CuSbS$_2$	CdS	InS	ZnSe	ZnS	ZnO:Al
Thickness(μm)	3	0.05	0.05	0.05	0.05	0.2
Band gap(eV)	1.5	2.4	2.1	2.9	3.6	3.4
Electron affinity (eV)	4.5	4.2	4.7	4.09	4.5	4.65
Dielectric permittivity (relative)	10.0	10	13.5	10	10	9.0
CB density of state (cm^{-3})	2.2e18	1.5e18	1.8e19	1.5e18	1.5e18	2.2e18
VB density of state (cm^{-3})	1.8e19	1.8e19	4.0e13	1.8e19	1.8e19	1.8e19
Electron mobility (cm^2/Vs)	100	100	400	50	50	100
Hole mobility (cm^2/Vs)	25	25	210	20	20	25
Donor density N$_D$ (cm^{-3})	0	0	10	0	0	1e17
Acceptor density(N$_A$) (cm^{-3})	2e16	1e17	1e18	5.5e7	1e17	0
Electron thermal velocity (cm/s)	1e7	1e7	1e7	1e7	1e7	1e7
Hole thermal velocity (cm/s)	1e7	1e7	1e7	1e7	1e7	1e7

The Solar Cell Capacitance Simulator (SCAPS-1D) simulation software developed by Prof. M. Burgelman et al. in the Department of Electronics and Information Systems at University of Gent, Belgium [4] was employed in our present study for the modeling and simulation of the device.

SCAPS is capable of solving the basic semiconductor equations, Poisson equation and continuity equations for electrons and holes as presented in the work of Mostefaoui et. al. [5]. The default illumination spectrum is set to the global AM 1.5 standard and the operating temperature is set to 300K. The baseline values of the physical parameters used in the study are all cited from experimental study, reasonable estimates in some cases, or literatures as presented in Table 1 [6 - 9]. The graphs of the simulation results data were plotted using the Microsoft Excel.

III. RESULTS AND DISCUSSION

A. Device simulation of baseline parameters

The simulation of the device using the baseline parameters gave an efficiency (η) of 3.24% with an open circuit voltage (Voc) of 0.48V, short circuit current (Jsc) of 13.66mA/cm^2 and a fill factor of 49.56%. A plot of the current density (J) against voltage is shown in fig. 2. This result is a close approximation to the obtain experimental result [3, 9, 10].

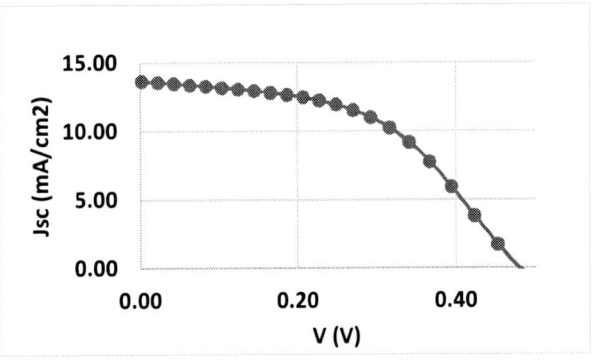

Fig. 2. J-V plot of the base model CuSbS$_2$ solar cell with CdS buffer layer.

B. Alternative buffer layers without defects

The performance characteristics of the solar cells with different buffer layer materials are as summarized in Table 2. Table 2 includes all the photovoltaic parameters (Jsc, Voc, η and FF).

Table 2. The photovoltaic parameters for the CuSbS$_2$ based solar cell with different buffer layers.

The graph of the current density against open circuit voltage for the device with different buffer material is shown in figure 3.

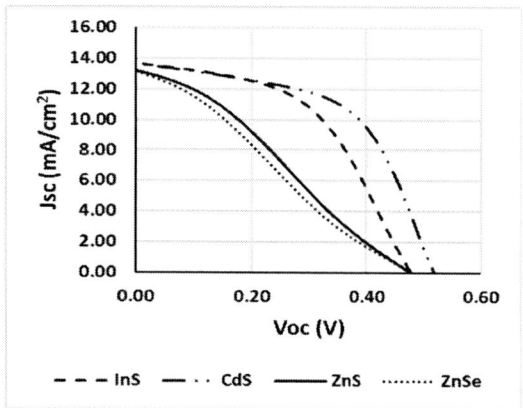

Fig. 3. J-V Characteristics CuSbS₂ of solar cell with different buffer layer.

In the simulation of our solar cells with CdS and other materials (InS, ZnSe, ZnS) as buffer layer, it was observed that the solar cell with InS had an efficiency of 3.78% which is an improvement on the reported efficiency of 3.13% [1]. Zinc selenide (ZnSe) had the least efficiency of 1.82%. The performance of these buffer layers in the copper antimony sulphides thin film solar cell is quite similar to that of CIGS as reported by Mostefoui et. al. [5]. In their report, ZnSe gave the least efficiency.

C. Effect of buffer layer thickness on device performance

The analysis of all the results shows that the best performance device was obtained with InS buffer material. To achieve the best photovoltaic parameters (Jsc, Voc, η and FF) of a solar cell, this buffer layer should be optimized. To achieve this the buffer layer thickness was varied from 0.01μm to 0.09μm. Table 3 shows the effect of the variation of the buffer layer thickness on the performance of the device. There corresponding J-V characteristic curve is shown in figure 4.

Buffer material	Voc (V)	Jsc (mA/cm²)	FF (%)	η (%)
CdS	0.48	13.66	49.56	3.24
InS	0.51	13.62	54.20	3.78
ZnS	0.48	13.22	31.69	2.00
ZnSe	0.48	13.12	28.76	1.82

Table 3. Variation of buffer layer thickness on device performance

Buffer material	Voc (V)	Jsc (mA/cm²)	FF (%)	η (%)
0.01	0.49	14.13	49.11	3.46
0.03	0.51	14.01	53.95	3.88
0.05	0.51	13.62	54.20	3.78
0.07	0.51	13.00	54.06	3.59
0.09	0.51	12.34	54.31	3.42
0.11	0.51	11.78	54.71	3.28

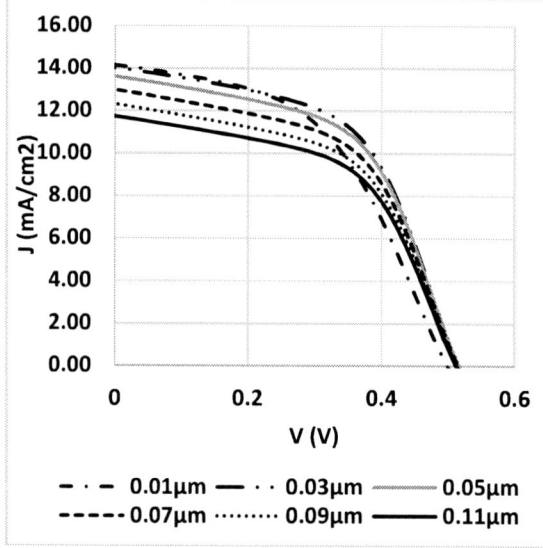

Fig. 4. J-V Characteristics CuSbS₂ of solar cell with different buffer layer thickness of InS

The results show that thinner buffer layer shows higher performance. A buffer layer thickness within the range of

$0.03\mu m$ and $0.05\mu m$ would be the most ideal for this device.

VI. CONCLUSION

The results of the simulations showed the InS material gave the best performance of the $CuSbS_2$ thin film solar cell. Also, CdS will be a close alternative to the InS buffer material, since their performances are relatively close.

ACKNOWLEDGEMENT

We acknowledge the developer of SCAPS-1D program; Prof. Burgelman and his group at the Department of Electronics and Information Systems, University of Gent, Belgium for make the software available for our use.

REFERENCES

[1] Septina, W. and S. Ikeda, Y. Iga, T. Harada, and M. Matsumura. *"Thin film solar cell based on CuSbS2 absorber fabricated from an electrochemically deposited metal stack."* Thin Solid Films, 550:700 – 704, 2014.

[2] Yang, Bo and L. Wang, J. Han, Y. Zhou, H. Song, S. Chen, J. Zhong, Lu Lv, D. Niu, and J. Tang. *"CuSbS₂ as a promising earth-abundant photovoltaic absorber material: A combined theoretical and experimental study."* Chemistry of Materials, 26(10):3135–3143, 2014.

[3] Colombara, D. and L. M. Peter, K. D. Rogers, J. D. Painter, and S. Roncallo. *"Formation of CuSbS₂ and CuSbSe₂ thin films via chalcogenisation of Sb-Cu metal precursors."* Thin Solid Films, 519(21):7438 – 7443, 2011.

[4] Burgelman M., Nollet, P., and S. Degrave, (2000). *"Modelling Polycrystalline semiconductor solar cells",* Thin Solid Films, 361-362, pages 527-532.

[5] Mostefaoui M., Mazari H., Khelifi S., Bouraiou A., Dabou R. (2015). *"Simulation of High Efficiency CIGS solar cells with SCAPS-1D software".* Energy Procedia 74 (2015) 736 – 744.

[6] Olopade M. A., Oyebola O.O., and Adeleke B. S., (2012). *"Investigation of some materials as buffer layer in copper zinc tin sulphide (Cu₂ZnSnS₄) solar cells by SCAPS-1D."* Advances in Applied Science Research, 3 (6):3396-3400 2012.

[7] Olopade M.A. and Adewoyin A., Olorode D. , Chendo M. . *"Effect of band gap grading on the performance characteristics of Cu₂ZnSnS₄ solar cell",* Photovoltaic Specialist Conference (PVSC) pg. 2394-2396, IEEE, 2014.

[8] Peijie L. and L. Lin, Y. Jinling, S. Cheng, Peimin Lu and Qiao Zheng. *"Numerical simulation of Cu₂ZnSnS₄ based solar cells with In₂S₃ buffer layers by SCAPS-*

1D". Journal of Applied Science and Engineering, *Vol. 17, No. 4, pp. 383_390 2014.*

[9] Acosta O. and R. E., Avellaneda D., Shaji S., Castillo G.A., Das Roy T.K., and B. Krishnan. *"CuSbS2 thin films by heating Sb₂S₃/Cu layers for PV applications".* Journal of Materials Science: Materials in Electronics, 25(10):4356–4362, 2014.

[10] Chan Choi Y., Yeom E.J, Ahn .T.K and Prof. Sang I.S. *"CuSbS₂-sensitized inorganic–organic heterojunction solar cells fabricated using a metal–thiourea complex solution".* Angewandte Chemie, Volume 127:4077–4081, 2015.

[11] Thomas R. and A. J. MacLachlan, Brown M. D. and S. A. Haque. *"Structural, optical and charge generation properties of chalcostibite and tetrahedrite copper antimony sulfide thin films prepared from metal xanthates"* Journal of Materials Chemistry, 3, 24155, 2015.

[12] Rodrıguez-Lazcano Y., Nair M.T.S. and P.K. Nair. *"CuSbS2 thin film formed through annealing chemically deposited Sb2S3–CuS thin films".* Journal of Crystal Growth 223:399–406, 2001.

Influence of hetero-interfaces on photovoltaic performance in solar cells based on ZnSnP$_2$ bulk crystal

Shigeru Nakatsuka[1], Shunsuke Akari[2], Jakapan Chantana[2], Takashi Minemoto[2], and Yoshitaro Nose[1]

[1]Department of Materials Science and Engineering, Kyoto University, 606-8501, Japan
[2]College of Science and Engineering, Ritsumeikan University, 525-8577, Japan

Abstract — We fabricated solar cells using rare-metal-free ZnSnP$_2$ bulk crystals prepared by Sn flux method. It was clarified that the aqua regia etching of ZnSnP$_2$ bulk crystals before CBD process removed the defect layer in the crystal formed by mechanical polishing. Subsequently, we report that the conversion efficiency of approximately 2 % was achieved with the structure of Al/AZO/ZnO/CdS/ZnSnP$_2$/Cu, where the values of J_{SC}, V_{OC} and FF were 8.2 mA cm^{-2}, 0.452 V and 0.533, respectively. V_{OC} value was largely low, considering the bandgap value of ZnSnP$_2$, ~1.6 eV. In order to improve the high conversion efficiency, the optimization of buffer layer material is considered to be essential in the viewpoint of the band alignment.

I. INTRODUCTION

In recent years, solar cells using compound semiconductors have made large progress. Particularly, the solar cells based on CuIn$_{1-x}$Ga$_x$Se$_2$ (CIGS) and CdTe have achieved high conversion efficiencies of 22.6 [1] and 22.1% [2], respectively. However, the use of rare or toxic elements is not desirable in the viewpoint of widespread use of these materials.

In our research group, ZnSnP$_2$ has been focused on as a promising material for a solar absorber consisting of earth-abundant and safe elements. In the previous works, it was reported that ZnSnP$_2$ with a chalcopyrite structure showed a p-type conduction with the carrier concentration in a range of 10^{16}–10^{18}cm^{-3} [3] and has a direct bandgap of 1.6–1.7 eV [3]. According to Shockley–Queisser theory[4], the conversion efficiency over 30% was calculated in single-junction cells with ZnSnP$_2$ absorber under an AM 1.5G solar spectrum [5]. In addition, the absorption coefficient of ZnSnP$_2$ was reported to be about 10^5cm^{-1} in the visible light region, which is comparable to that of CIGS [6].

In our previous work, we fabricated ZnSnP$_2$ bulk crystal solar cells and confirmed that ZnSnP$_2$ acted as the light absorber. However, the conversion efficiency of the solar cells was less than 0.1 % [7]. In this work, we focused on the hetero-interface of CdS/ZnSnP$_2$ in order to improve photovoltaic performance.

III. EXPERIMENTAL PROCEDURE

ZnSnP$_2$ bulk crystals were grown by Sn flux method as reported in our previous work [8]. The raw materials, Zn shots (99.99%, Kojundo Chemical Laboratory), Sn shots (99.99%, Kojundo Chemical Laboratory) and red phosphorus flacks

(99.9999%, Kojundo Chemical Laboratory) were sealed in evacuated quartz ampule with the inner diameter of 10 mm under the pressure of 10^{-2} Pa. The nominal composition was controlled to be 92 mol%Sn in Sn-ZnP$_2$ pseudo binary diagram [8]. After that, the furnace was lifted up by the rate of 0.2 mm/h and the sample was unidirectionally solidified from the bottom. During crystal growth, the average cooling rate was approximately 0.7 °C/h. After the crystal growth, bulk crystals were cut into several wafers in perpendicular to the growth direction. The surface of each wafer was mechanically polished with a series of emery papers and finally with 1 μm diamond slurry on a buff sheet.

The solar cell structure of Al/AZO/ZnO/CdS/ZnSnP$_2$/Cu was adopted. First of all, Cu back electrode with the thickness of approximately 0.5 μm was prepared by direct current (DC) sputtering on the polished surface of ZnSnP$_2$ wafer with the diameter of about 8 mm and the thickness of about 200 μm. Before preparation of the electrodes, ZnSnP$_2$ bulk crystals were etched by immersing in 0.02 M Br$_2$ solution for 15 min. After the preparation of Cu back electrode, some samples were sealed in the evacuated quartz ampule below 10^{-2} Pa and annealed at 200 °C for 1 hour or 300 °C for 20 min. CdS with the thickness of about 50 nm was prepared on the opposite surface to back electrode by chemical bath deposition (CBD) method, where the chemical bath consisted of CdSO$_4$ (1.1 mM), ammonia (2.3 M) and thiourea (56 mM). The temperature and the deposition time were 80 °C and 11 min, respectively. Before preparation of CdS layer, the surface of ZnSnP$_2$ bulk crystals was etched by immersing in 1/2 or 1/4 diluted aqua regia solution for 1−15 min. Subsequently, ZnO and AZO films were formed by radio frequency (RF) magnetron sputtering at room temperature with the thickness of 50 and 300 nm, respectively. For the fabrication of ZnO and AZO films, ZnO (99.99 %, Furuuchi Chemical) and ZnO-2 wt. % Al$_2$O$_3$ (99.99 %, Furuuchi Chemical) were used as target materials. Finally, Al electrode with a grid pattern was fabricated by electron beam evaporation.

The current density−voltage (J−V) characteristics and the external quantum efficiency (EQE) of solar cells were investigated under the illumination conditions of 100 mW cm^{-2} and AM 1.5 G using the measurement system with solar simulator (Bunkoukeiki, CEP-25RR). The cross-sectional observation was performed using STEM-EDX (JEM-2100F, JEOL).

III. RESULTS AND DISCUSSION

The J–V characteristics of the solar cells with various etching conditions are shown in Figure 1. Table I also shows the solar cell parameters. These data indicates that aqua regia etching improved the photovoltaic performance such as short circuit current density, J_{SC}, open circuit voltage, V_{OC} and fill factor, FF. In the sample with 1/4 diluted aqua regia etching for 15 min, the significant improvement of photovoltaic performance was not observed compared with samples with 1/2 diluted aqua regia etching. Consequently, the conversion efficiency approximately 2 % was achieved in 1/2 diluted aqua regia etching for 15 min, where the values of J_{SC}, V_{OC} and FF were 8.2 mA cm^{-2}, 0.452 V and 0.533, respectively. Figure 2 shows the EQE spectra of the solar cells with and without 1/2 diluted aqua regia etching for 15 min, which also indicates the improvement of photovoltaic performance.

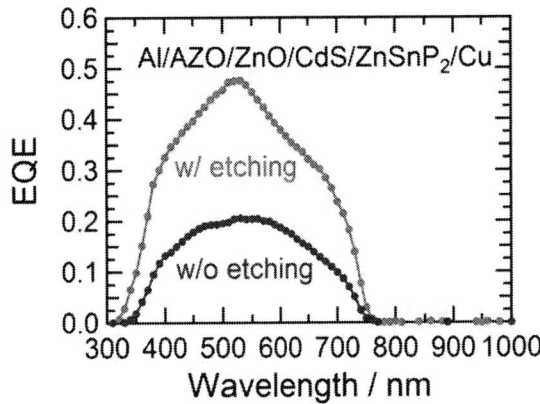

Fig. 2. EQE spectrum of the solar cell with 1/2 diluted aqua regia etching for 15 min.

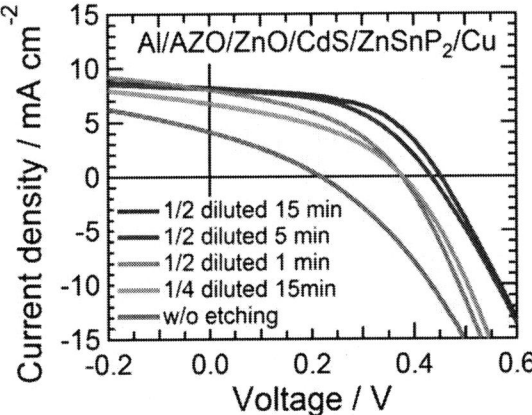

Fig. 1. J–V characteristics of ZnSnP$_2$ solar cells with and without aqua regia etching.

TABLE I
SOLAR CELL PARAMETERS OF ZnSnP$_2$ SOLAR CELLS

	J_{SC} / mA cm^{-2}	V_{OC} / V	FF	Efficiency / %
1/2diluted for 15 min	8.2	0.452	0.533	1.97
1/2 diluted for 5 min	8.1	0.433	0.496	1.73
1/2 diluted for 1 min	8.0	0.376	0.416	1.25
1/4 diluted for 15 min	6.7	0.378	0.396	1.00
w/o etching	4.1	0.217	0.308	0.27

In order to clarify the cause of these improvements, we carried out the TEM cross-sectional analysis. Figure 3 (a) shows a contrast due to lattice strain in the ZnSnP$_2$ bulk crystal near the interface of CdS/ZnSnP$_2$. When preparing ZnSnP$_2$ crystal solar cells, the surface was mechanically polished, which might introduce some lattice defects in crystals. A region including defects generally makes higher the recombination rate of minority carriers. Therefore, it is considered that the lower performance in the solar cells without etching might come from the region including defects. On the other hand, in the solar cells with 1/2 diluted aqua regia etching for 15 min, the strain contrast was not observed as shown in Figure 3 (b). Thus, it is considered that the above behavior in J–V characteristics and EQE spectra implies the improvement of carrier transport at the interface between CdS and ZnSnP$_2$ due to removing defect regions.

Although, the conversion efficiency of about 2 % was achieved in the ZnSnP$_2$ solar cells, more improvements of photovoltaic performance is essential for a practical use. In our previous work, XPS analysis suggested that the conduction band offset, ΔE_C, between CdS and ZnSnP$_2$ was about −1 eV, which is a large cliff and extremely limits the V_{OC} [9]. Therefore, the band alignment is needed between the buffer layer and ZnSnP$_2$ in the future work.

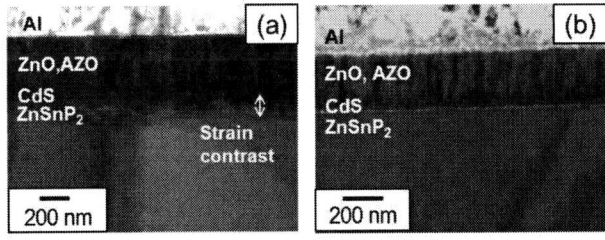

Fig. 3. Cross-sectional TEM images of the solar cells with the structure of Al/AZO/ZnO/CdS/ZnSnP$_2$/Cu (a) without and (b) with 1/2 diluted aqua regia etching for 15 min.

IV. CONCLUSION

Aqua regia etching was applied to $ZnSnP_2$ bulk crystals in order to remove the surface defect layer introduced by mechanical polishing. TEM cross-sectional observation clarified that the aqua regia etching was effective to remove the surface damaged layer. Consequently, the conversion efficiency of approximately 2 % was achieved with the structure of Al/AZO/ZnO/CdS/$ZnSnP_2$/Cu, where the values of J_{SC}, V_{OC} and FF were 8.2 mA cm^{-2}, 0.452 V and 0.533, respectively. In order to improve the photovoltaic performance of $ZnSnP_2$ solar cells, the optimization of buffer layer material is considered to be essential in the viewpoint of band alignment.

ACKNOWLEDGEMENT

This work was financially supported by Grant-in-Aid for JSPS Research Fellow Number 16J09443.

REFERENCES

[1] P. Jackson, R. Wuerz, D. Hariskos, E. Lotter, W. Witte, and M. Powalla, "Effects of heavy alkali elements in Cu(In,Ga)Se$_2$ solar cells with efficiencies up to 22.6%," *Phys. Status Solidi - Rapid Res. Lett.*, vol. 10, pp. 583–586, 2016.

[2] First Solar, "Fist Solar Hits Record 22.1% Conversion Efficiency for CdTe Solar Cell," 2016. .

[3] J. L. Shay and J. H. Wernick, *Chalcopyrite Semiconductors: Growth, Electronic Properties, and Applications*. Oxford: Pergamon Press, 1975.

[4] W. Shockley and H. J. Queisser, "Detailed balance limit of efficiency of p-n junction solar cells," *J. Appl. Phys.*, vol. 32, pp. 510–519, 1961.

[5] T. Yokoyama, F. Oba, A. Seko, H. Hayashi, Y. Nose, and I. Tanaka, "Theoretical Photovoltaic Conversion Efficiencies of $ZnSnP_2$, $CdSnP_2$, and $Zn_{1-x}Cd_xSnP_2$ Alloys," *Appl. Phys. Lett.*, vol. 6, pp. 61201, 2013.

[6] S. Minoura, K. Kodera, T. Maekawa, K. Miyazaki, S. Niki, and H. Fujiwara, "Dielectric function of Cu(In, Ga)Se$_2$-based polycrystalline materials," *J. Appl. Phys.*, vol. 113, pp. 63505, 2013.

[7] S. Nakatsuka, N. Yuzawa, J. Chantana, T. Minemoto, and Y. Nose, "Solar cells using bulk crystals of rare metal-free compound semiconductor $ZnSnP_2$," *Phys. Status Solidi*, vol. 214, pp. 1600650, 2016.

[8] S. Nakatsuka, H. Nakamoto, Y. Nose, T. Uda, and Y. Shirai, "Bulk crystal growth and carecterization of $ZnSnP_2$ copmpound semiconductor by flux method," *Phys. Status Solidi C*, vol. 12, pp. 520–523, 2015.

[9] S. Nakatsuka, Y. Nose, and Y. Shirai, "Band offset at the heterojunction interfaces of CdS/$ZnSnP_2$, ZnS/$ZnSnP_2$, and In$_2$S$_3$/$ZnSnP_2$," *J. Appl. Phys.*, vol. 119, pp. 1–6, 2016.

Junction by Diffusion of Elemental Sodium Alone into Bridgman Cu(In,Ga)Se$_2$

S. Park[1], C. H. Champness[1], S. Vanka[1], Z. Mi[1,2] and I. Shih[1]

[1]Electrical and Computer Engineering Department, McGill University, Montreal, Canada H3A 0E9
[2]Department of Electrical Engineering and Computer Science, Centre for Photonics and Multiscale Nanomaterials, University of Michigan, Ann Arbor, USA 48105

Abstract --The diffusion of elemental sodium into p-type Bridgman-grown CuInSe$_{2+x}$ has already been shown to create a pn-junction by its presence alone [1], provided the molecular amount of added [Na] exceeds the quantity $2x + \delta$ [2], (from the excess Se). Here, δ is the excess of Se found experimentally in the p-type but stoichiometrically-grown material (i.e.with x = 0). The present contribution demonstrates similar action in Bridgman-grown p-type quaternary Cu(In$_{1-y}$,Ga$_y$)Se$_2$, with y = 0.2 and 0.3. The existence of deep (more than 6 microns) junctions after the Na-diffusion, is demonstrated by sample lapping and hot probing with, photovoltaic and EBIC action. Information from XPS and XRD is also presented.

I. INTRODUCTION

It was reported earlier in the year 2,000 at the IEEE PVSC that p-type Bridgman CuInSe$_2$ was converted to n-type by the addition of sufficient elemental sodium, whether introduced into the melt before compound synthesis or by diffusion into existing solid p-type chalcopyrite material [3]. Later, it was shown that the atomic amount of sodium [Na] for type conversion was given by $[Na] > [Na]_{crit} = 2x + \delta$, where x is the stoichiometric excess of Se in the melt of CuInSe$_{2+x}$ and δ is the maximum sodium amount needed in nominal stoichiometric Bridgman material (i.e. with x = 0) to keep the ingot p-type [2]. However, in contrast to elemental Na, the ternary compound remains p-type with the addition of the *binary* compound Na$_2$Se, as in thin film CIS. The proposed interpretation of this action by elemental sodium is that the introduced Na atoms (electron affinity 0.55 [4]) are strongly attracted to the Se atoms (electron affinity 2.02 [4]) to form Na$_2$Se. It does so at the expense of the Se intended for the ternary chalcopyrite, which is then formed with a deficit of this element. In this laboratory, ingots prepared with a deficit of Se (i.e. x < 0) were always found to be n-type. Thus, the sodium, in elemental form, changes the conductivity type from p- to n-type, not by direct donor action, but by its effect on stoichiometry.

Arising from these considerations, the present paper shows that if sodium, in elemental form, is diffused into crystalline quaternary p-type CuIn$_{1-y}$ Ga$_y$Se$_2$ (y $\leq$ 0.3), at least prepared by the Bridgman method, a clear deep homojunction is formed.

II. CIGS SAMPLES

The samples for study were cut from p-type CIGS multi-crystalline ingots, grown previously in this laboratory by a Bridgman method with melt compositions CuIn$_{0.8}$Ga$_{0.2}$Se$_2$ and CuIn$_{0.7}$Ga$_{0.3}$Se$_2$. Results of XRD, cell photo-response and crystallographic examination of the material are given in reference [5]. For sodium diffusion, each sample was enclosed in aluminum foil, with a square opening of about 2 × 2 mm^2 in the top face and this was placed within a glass tube with one end closed. About 5 mg of elemental sodium of purity 99.99 %, was then positioned 10 cm from the sample in the tube, which was pumped out to a high vacuum and sealed off as an ampoule.

III. SODIUM IN-DIFFUSION

The prepared ampoule was positioned in a tube-furnace with a temperature gradient, so that the Na could be heated at a temperature 50 degrees above that of the sample. Typical sodium / sample temperatures were 300 °C / 250 °C and 280 °C / 230 °C for heating times of 5, 15, 20, 40 and 80 minutes. After this time and removal of the sample from the ampoule, the sample surface was found to be covered with a shiny orange-brown layer, which later was estimated to consist of metallic sodium mixed with copper and of thickness a few hundred nm. This shiny layer was removed with a cotton swab soaked in DI water to reveal the surface of the chalcopyrite. Thermoelectric probing with two metal points, one warm and the other at room temperature, indicated that the surface, originally checked to be p-type, was now n-type. The sample surface was then abrasively lapped, tested again with the hot probe and the resultant new thickness measured. This process of lap/hot probing/thickness measurement was repeated until p-type material was reached. From the beginning and final wafer thickness readings, the average junction depth could be determined.

978-1-5090-5606-4/17 $31.00 © 2017 IEEE

A. Hot Probe/Lap Method

Using the hot probe/lap method, the results of 4 different diffusion runs are shown schematically in Fig. 1 (a), (b), (c) and (d) on these wafers, typically about 5 mm thick. It is seen here that the n/p junctions are located tens of microns below the original surface. Taking the average junction depth from

Fig. 1(b) and 1(c) for 40 minutes diffusion, as 0.05 mm and that of Fig. 1(d) for 80 minutes as 0.06 mm, the average distance ratio of 0.06/0.05 is 1.2. This number is less than $\sqrt{80\ min/40\ min} = \sqrt{2} = 1.414$, which would be expected for a diffusion process. It is noted also that, in run Fig.1(a), the junction is much deeper for the sample with y = 0.3.

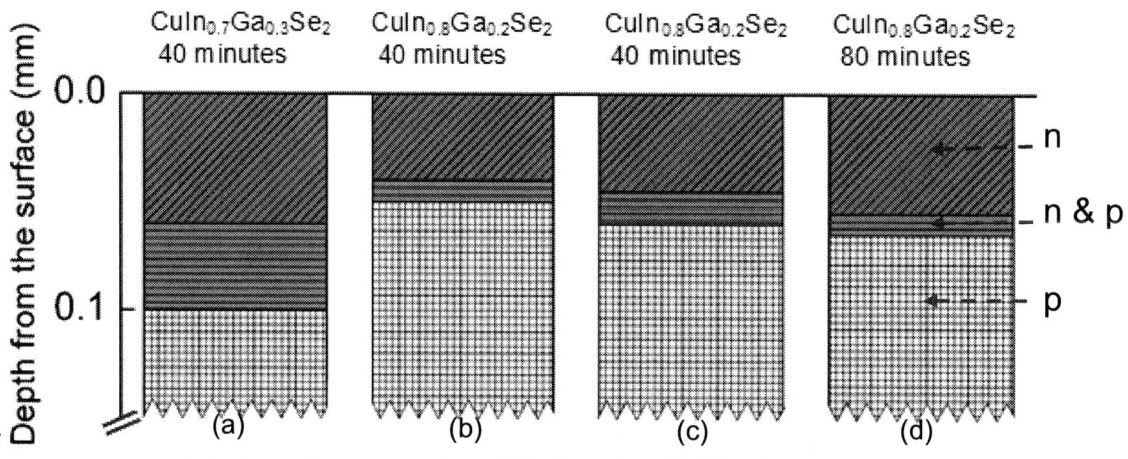

Fig. 1. Schematic n-type locations of 4 CIGS samples with diffused-in sodium.

B. EBIC Method

The electron beam induced current (EBIC) method of determining junction depths was applied to different CIGS samples. Fig. 2 shows EBIC traces for two samples of nominal composition $CuIn_{0.8}Ga_{0.2}Se_2$, where sodium was diffused in at 300 °C for 5 and 15 minutes. Here, the junction depths (with the shiny layer removed) are seen to be 6 and 10 microns

respectively. The square root of the diffusion time ratio is $\sqrt{15/5} = \sqrt{3} = 1.73$, which is close to the ration of the junction depths 10/6 = 1.67.

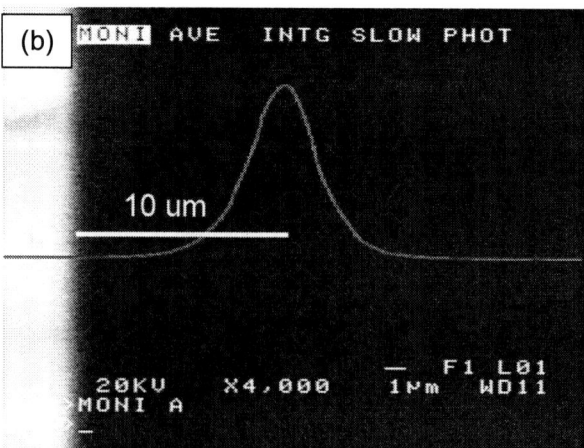

Fig. 2. EBIC currents superimposed on SEM images for two different $CuIn_{0.8}Ga_{0.2}Se_2$ samples, for which Na was diffused in at 300 °C for (a) 5 and (b) 15 minutes.

IV. PHOTOVOLTAIC ACTION

To check for photovoltaic action, a device structure was created from a CIGS sample of composition $CuIn_{0.8}Ga_{0.2}Se_2$, where the sodium was diffused in through a 0.04 cm^2 area aperture in surrounding aluminum foil. Following a chemical etch in chromic-sulphuric acid, gold was evaporated on the p-type bottom sample surface and a Wood's metal contact (low melting point eutectic alloy of 50% Bi, 26.7% Pb, 13.3% Sn, and 10% Cd by weight) was attached to the top surface. The structure was exposed to a xenon lamp with an approximate illumination intensity of "AM 1.5" or 100 mW/cm^2. The current density (j)-voltage (V) characteristic result for this structure is shown in Fig. 3. The photovoltaic characteristics are seen to be clearly observable but at lower levels compared to values from regular laboratory shallow junctions of CdS-CIGS solar cells. However, this is not unexpected for the present deep junction located more than 10 microns below the illuminated surface. Exposure to direct sunlight also showed significant photovoltaic sensitivity.

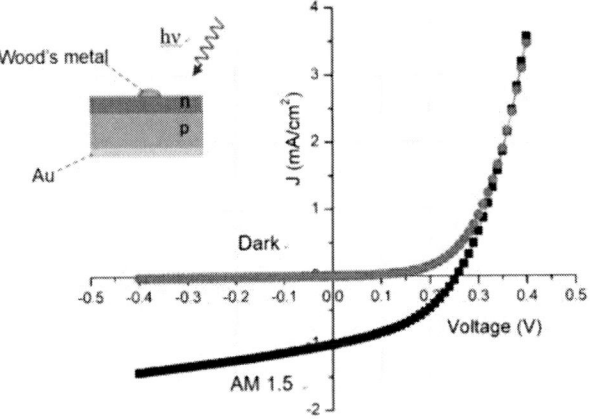

Fig. 3. Current density-voltage characteristic for sodium diffused-in sample $CuIn_{0.8}Ga_{0.2}Se_2$ under darkness and approximately "AM 1.5" illumination.

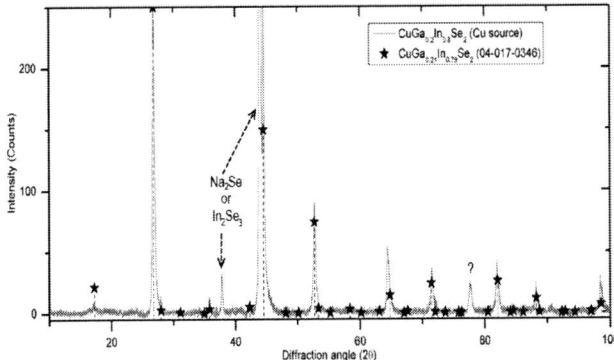

Fig. 4. X-ray diffractogram of powder scraped from n-type converted, Na-diffused, CIGS material, identified with PDF-4+ 2014 (The Powder Diffraction File) from ICDD (The International Centre for Diffraction Data) [11] * ★ :$CuGa_{0.21}In_{0.79}Se_2$ (Reference code:04-017-0346) [6]. Note the non-CIGS peaks at 2θ angles of about 37.3 °, 44 ° and 77.6 °

V. XRD RESULTS

Sufficient powder was collected from the n-type region of a sodium diffused-in sample of melt composition $CuIn_{0.8}Ga_{0.2}Se_2$ for XRD analysis, avoiding the p-type region. The result is shown in Fig. 4 for a copper X-ray target. From wider extensions of the 2θ axis (not shown), more accurate values of the chalcopyrite lattice parameter 'a' were determined from the (101), (112), (103), (211) and (220) peaks, yielding an average value of **a** = 5.7605 Å. This is to be compared with a value of 5.789 Å for the ternary $CuInSe_2$ [6], 5.744 for $CuIn_{0.8}Ga_{0.2}Se_2$ from reference [7] and 5.7478 for $CuIn_{0.79}Ga_{0.21}Se_2$ according to reference [8]. Close inspection of the spectrum of Fig. 4 indicates that all of the expected CIGS composition peaks are present but extra peaks exist at 2θ angles of 37.7 °, 44.0 ° and 77.6 °. The first two could correspond to Na_2Se (37.3 ° and 43.95 °) or In_2Se_3 (37.5 ° and 43.95 °). Fig. 5 shows an expanded 2θ scale in the range 43 ° to 45 °, separating one of these peaks from the CIGS (204, 220) chalcopyrite peak.

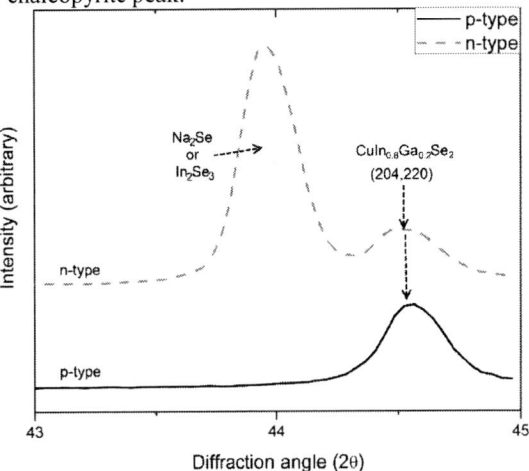

Fig. 5. Portion from XRD diffractogram of $CuIn_{0.8}Ga_{0.2}Se_2$ powder, taken from the n-region of Na diffused-in sample and from a p-type Na-free sample.

VI. XPS PROFILE AFTER Na DIFFUSION

Fig. 6 shows a deep XPS profile of the 5 elements Cu, In, Ga, Se and Na after a 40 minute, 300 °C sodium in-diffusion run into a sample of nominal composition $CuIn_{0.8}Ga_{0.2}Se_2$, with the initial shiny layer removed. The results indicate that the sodium content decreases monotonically with distance from the water-cleaned surface and, after 10^4 seconds of ion etching, amounts to about 2 at.% of the other elements, at a depth of about 6 microns. This is surely a proof that the sodium penetrates deeply into the chalcopyrite. The gallium percentage at this etching time is noted to be about 6 at.%, which is close to the amount expected from the nominal formula of $CuIn_{0.8}Ga_{0.2}Se_2$, yielding a [Ga]/[compound] ratio of 0.2/4 or 5 at.%.

Subsequent annealing in air of Na diffused-in $CuInSe_2$ samples have given rise to an extra XPS Se 3d peak at a binding energy of about 59 eV (not shown), indicating the presence of the ternary compound Na_2SeO_3 from oxidation of Na_2Se. This shows that the initial introduced elemental Na does not remain uncombined in the chalcopyrite.

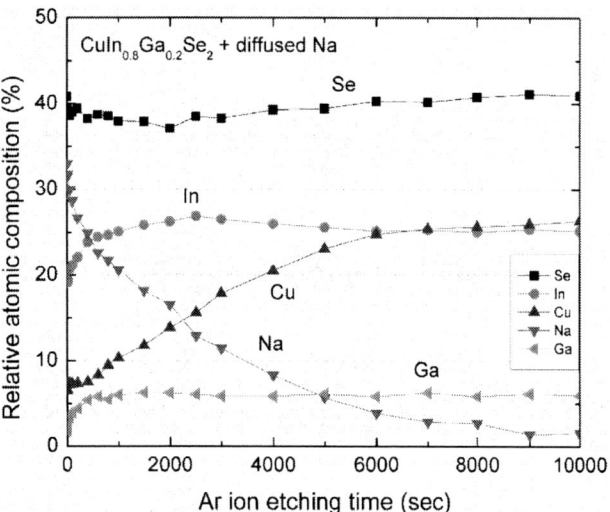

Fig. 6. XPS-determined variation of atomic percentages of diffused-in Na and other elements with Ar ion etching time (1 sec = 0.57 nm) in a diffused-in $CuIn_{0.8}Ga_{0.2}Se_2$ sample.

VII. DISCUSSION

The present results clearly show that, not only does added elemental sodium to CIGS (and CIS [1,9]) change p-type material to n-type, but it creates a pn-junction with observable photovoltaic action. As mentioned previously, it does this, not by direct donor action, but by the strong chemical attraction of sodium to selenium atoms already present, in the form of the chalcopyrite containing also Cu, In and Ga. This renders the created CIGS stoichiometrically deficient in Se and hence it becomes n-type unless x, the excess Se, is higher than $([Na] - \delta)/2$. If the added Na is in an already combined state, such as Na_2Se, the host p-type chalcopyrite stays p-type with a higher hole concentration and does not change type [10]. It also remains p-type with added elemental sodium if there is sufficient excess selenium present. Thus, 1 at.% of elemental Na added in p-type stoichiometrically-grown $CuInSe_2$ changes it to n-type but for the same added amount of 1 % in p-type $CuInSe_{2.2}$ (i.e. with 0.2 excess Se), the p-type character is maintained. The diffusion coefficient was determined to be in the range 10^{-10} to 10^{-8} cm^2/sec at 250°C in the Bridgman chalcopyrite materials.

REFERENCES

[1]. S.Park, C.H. Champness and I.Shih. *"Effect of sodium diffused into Bridgman CuInSe$_{2-x}$"*. J.Elect.Spect&Rel.Phen.**212**,21-27 2016.

[2]. H.F. Myers, C.H. Champness, I. Shih, "Electrical effect of introducing elemental sodium into the Bridgman melt of CuInSe$_{2+x}$ crystals", Journal of Crystal Growth, 387 (2014) 36-40.

[3]. H.P. Wang, I. Shih, C.H. Champness, "Characteristics of CuInSe2 Bridgman ingots grown with sodium", Proceedings of 28th IEEE Photovoltaic Specialists Conference, 15-22 Sept. 2000, IEEE, Anchorage, Alaska, 2000, pp. 642-645.

[4]. CRC Handbook of Chemistry and Physics, 88th edition, in: W.M. Haynes (Ed.), CRC Press, 2008, pp. 2704.

[5]. H. Du, C.H. Champness, I. Shih, T. Cheung, "Growth of Bridgman ingots of CuGa$_x$In$_{1-x}$Se$_2$ for solar cells", Thin Solid Films, 480 (2005) 42-45.

[6]. R.W. Birkmire, E. Eser, Polycrystalline thin film solar cells: Present status and future potential, in, Annual Reviews Inc, Palo Alto, CA, United States, 1997, pp. 625-653.

[7]. H. Du, C.H. Champness, I. Shih, T. Cheung, Growth of Bridgman ingots of CuGa$_x$In$_{1-x}$Se$_2$ for solar cells, Thin Solid Films, 480 (2005) 42-45.

[8]. M. Souilah, A. Lafond, C. Guillot-Deudon, S. Harel, M. Evain, Structural investigation of the Cu$_2$Se-In$_2$Se$_3$-Ga$_2$Se$_3$ phase diagram, X-ray photoemission and optical properties of the Cu$_{1-z}$(In$_{0.5}$Ga$_{0.5}$)$_{1+z/3}$Se$_2$ compounds, Journal of Solid State Chemistry, 183 (2010) 2274-2280.

[9]. C.H. Champness, H.F. Myers, I. Shih, Carrier polarity reversal with sodium addition in Bridgman-grown CuInSe2, Thin Solid Films, 519 (2011) 7337-7340.

[10]. H.F. Myers, Studies on the effect of sodium in Bridgman-grown copper indium selenide, M.Eng., McGill University (Canada), Montreal, 2008, pp. 136.

[11]. D.S. Kabekkodu, in, International Centre for Diffraction Data, Newtown Square, PA, USA., 2014.

Oxygen substitution and sulfur vacancies in NaBiS$_2$: a Pb-free candidate for solution processable solar cells

Robert J Patterson, Hongze Xia, Long Hu, Zhilong Zhang, Lin Yuan, Jianfeng Yang, Weijian Chen, Zihan Chen, Yijun Gao, Yicong Hu, Binesh Puthen Veettil, John A. Stride, Gavin Conibeer, Shujuan Huang

University of New South Wales, Sydney, NSW, 2033, Australia

Abstract — **High performance bismuth-based semiconductors are attractive large atom containing candidates needing further development. Theoretical investigations of defect mechanisms in NaBiS$_2$ have been performed using PBE GGA functionals. Effective masses are estimated at the band extrema and found to be 0.36m$_0$ or less. High levels of oxygen substitution and significant sulfur deficiency lead to a loss of the fundamental bandgap. Lattice parameters for the defective materials have been derived and provide an experimental signature for high levels of these defects. The work informs the development of promising but little explored lower symmetry bismuth compounds for solar energy applications.**

I. Introduction

The search for non-toxic "Pb-free" solution processable materials is proceeding at an accelerating rate. In this area, semiconductors based on bismuth are attractive since this large atom is a neighbor of lead, with Pb-based solution processed semiconductors known to out-perform others in colloidal quantum dot [1, 2] and halide perovskite solar cells [3]. Bismuth-based semiconductors continue to be developed. However, degenerate doped materials resulting from anion deficiency are often an issue [4]. However, some cases of bismuth incorporation into known high-efficiency materials have been explored, at least in theory, where substitution of Pb for Bi and Ag in perovskites has been proposed [5]. Other Ag and Bi compounds such as AgBiS$_2$ have also been proposed and these compounds have shown very strong absorption and promising efficiencies [5, 6], despite their having many polymorphs with complex unit cells [8]. From looking at the I-Bi-VII family of compounds, it is clear that the +1 cation (Cs, Rb, Na, Li, Ag, etc) has a strong influence on the bonding and the final crystal structure obtained [8]. Orthorhombic NaBiS$_2$ is unique among bismuth chalcogenides in that the bismuth atoms are octahedrally coordinated by sulfur anions. These interconnected octahedra appear to lead to a number of desirable properties, including overlapping conduction bands with many closely spaced minima that may lead to high absorption and small effective masses in both the conduction and valence bands. The structure is reminiscent of rocksalt [9] with octahedral coordination of the bismuth atom. This bonding arrangement is somewhat unique in this material family and suggests that charge donated by the electropositive Na ion, as well as its ionic radius, are important in determining the final crystal structure of the material.

Materials with reduced symmetry, such as orthorhombic Sb$_2$S$_3$, have been investigated previously with promising initial photovoltaic efficiencies of ~5% achieved [10]. The antimony compound NaSbS$_2$ has also been investigated with initial efficiencies of ~3% [11]. Whether or not a lower symmetry unit cell leads intrinsically to a higher probability for defects and thus low power conversion efficiencies (PCEs) in photovoltaic applications remains an open question.

Fabrication of NaBiS$_2$ has been attempted [12, 13] and these initial synthesis methods can continue to be built upon. Hydrothermal synthesis has been used, at relatively high pH, and appeared to yield highly crystalline material [12]. The formation energy of NaBiS$_2$ has been predicted to be particularly low [8] and high crystallinity from hydrothermal synthesis supports this, implying material stability. Additionally this synthesis was done in significant Na-excess, highlighting the importance of incorporating sufficient electropositive Na to prevent detrimental binary Bi-S phases from forming.

Ab-initio calculations can provide evidence as to whether oxygen incorporation or sulfur deficiency in the material creates undesirable mid-gap electronic states. Changes to the bandgap are often possible through alloying and material systems with tunable bandgaps are highly desirable, in tandem cell applications for example. For NaBiS$_2$, this would mean alloying with smaller anion, oxygen.

In this work, lattice parameters and electronic band-gaps are obtained from material's electronic bandstructure and density of states (DOS) for significant levels of oxygen substitution and sulfur vacancies. Effective masses are reported for the ideal material. The results inform the development of this particular bismuth compound and show that NaBiS$_2$ is a promising non-toxic candidate for photovoltaic cells where stoichiometry can be strictly maintained. This study also furthers investigations into semiconductors with low symmetry unit cells in the context of solar energy applications.

II. Computational Details

Ab-initio simulations were performed using the plane wave pseudopotential method that is part of the Quantum Espresso package [14]. To relax the bulk materials a 10x10x10 k-space grid was used as well as a plane wave cut off energy of 100 Ry. For the bandstructure calculations, the same energy cut off and 40 k-points per high symmetry direction in the Brillouin Zone (BZ) were computed. A 16x16x16 k-space grid was used for the DOS calculations.

The exchange functionals used for all calculations were PBE GGA functionals computed using the Martins-Troullier method. These functionals are very similar to the GGA method used in previous work [8] and the bulk band structure results appear to be quite similar, though the lattice parameters differ by a few percent.

The oxygen defect and sulfur vacancy were both introduced on the same atomic site. The sulfur at location k=[0.25, 0.25, 0.75] in crystal coordinates was either replaced by oxygen or removed to simulate a sulfur vacancy. After relaxation, subtle distortions in this ideal location are observed, with shifts on the order of ~0.01 evident (k=[0.26, 0.26, 0.74]).

TABLE I

LATTICE PARAMETERS FOR THE PRIMITIVE UNIT CELL AND ELECTRONIC BAND GAPS OBTAINED FOR BULK, OXYGEN SUBSTITUTED AND SULFUR DEFICIENT NaBiS2. DISTANCES IN ANGSTROMS, ANGLES IN DEGREES

	Bulk	O-sub	S-Def	Ref [8]
a	7.060	6.827	6.773	7.156
b	7.060	7.162	7.129	7.153
c	5.690	5.633	5.669	5.707
α	67.17	65.00	66.25	66.52
β	67.17	66.01	66.45	66.46
γ	46.98	48.06	48.13	47.01

III. Results & Discussion

Table 1 shows the crystalline parameters obtained in this work, with the parameters from Ref [8] given for comparison. The lattice parameters for the bulk material and that from Ref [8] differ by approximately 1.5%. The lattice parameters predicted for 25% atomic oxygen substitution on sulfur sites show an expected reduction in the lattice parameters relative to the fully sulfurized material, with the angles of the bonding also changing observably due to the reduced ionic radius of the oxygen. This suggests that this lattice structure is actually quite plastic, and very sensitive to changes in the electronic distribution. For sulfur vacancies the changes are even more dramatic, with reductions in the lattice constant up to 4% observed and similar changes in angle. These results provide a fingerprint for that can be compared straightforwardly to XRD data, showing whether material has significant oxygen incorporation at lattice sites, or where the sulfur content is sub-stoichiometric.

The bulk material shows highly symmetric bands about the conduction and valence band extrema (Figure 1). Estimated effective masses are shown in Table 2. These differ markedly

from those reported for the layered orthorhombic $RbBiS_2$ material tabulated in Ref [15].

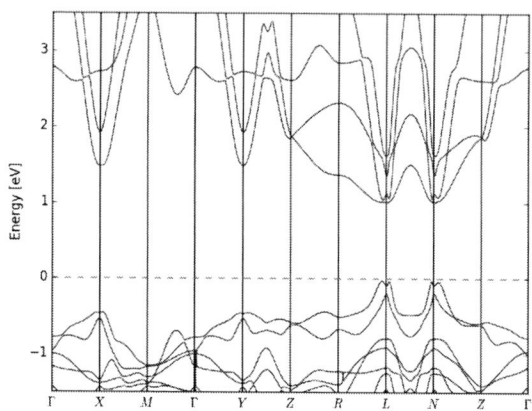

c)

Fig 1. Bulk band structure obtained for $NaBiS_2$ using PBE functionals. Band extrema are observed at the L and N points. The approximate bandgap is 1.01 eV. Effective masses at the band extrema are shown in TABLE 2.

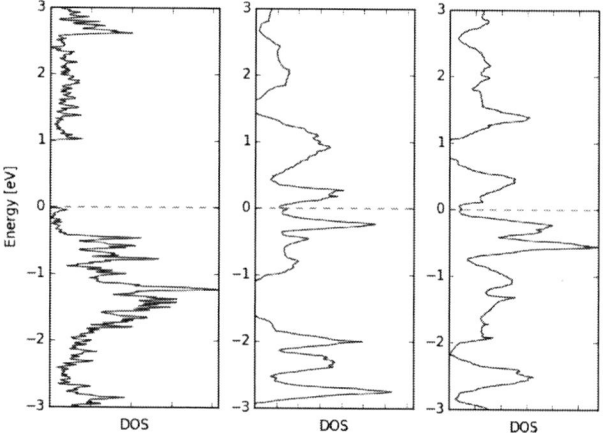

Figure 2 a) Electronic density of states (DOS) for bulk $NaBiS_2$, showing many peaks due to the many sharp extrema in the band structure. b) DOS for $NaBiS_{1.5}O_{0.5}$ showing bandgap closure, rather than widening, for oxygen incorporation, and c) DOS for $NaBiS_{2-x}$ (x=0.5) also showing bandgap closure due to lattice distortions and degenerate doping from the high level of sulfur vacancies.

As can be seen in Figure 2, significant oxygen incorporation to form an $NaBiS_{2-x}O_x$ alloy (x=0.5) reduces rather than enlarges the bandgap. Since bismuth oxides have wide bandgaps, this may run contrary to expectation. However, the distortions to the crystal lattice on incorporating the smaller oxygen anion appear significant, and these can lead to dramatic changes in the band structure. This is also consistent with experimental results where the material showed bandgaps between 1.3 and 1.5 eV [13], though underestimates of the bandgap are also expected using PBE GGA functionals. The results show that deep defects should be expected from oxygen substitution and from lattice distortions due to differing bond distances for Na, O and Bi.

Sulfur deficiency leads to deep electronic defects and severe bandgap narrowing, as shown in Figure 2. While the degree of sulfur deficiency in the material in this simulation is particularly high, at ~25% atomic, isolated vacancy defects likely cause similar deep electronic defects that act as recombination centers localized on particular unit cells. These deep electronic defects would also be expected to contribute significantly to degenerate doping of the material and reduced photovoltaic performance.

TABLE 2

EFFECTIVE MASS AND DIELECTRIC CONSTANT ESTIMATES OBTAINED FROM THE BULK STRUCTURE. THE STATIC DIELECTRIC CONSTANT FOR $NaBiS_2$ IS TAKEN FROM THE TRACE OF THE DIELECTRIC MATRIX

K-point {L/N}	Bulk (this work) [m_0]	$RbBiS_2$ (Ref[15]) [m_0]
m^*_e	0.36 @ 1.01 eV	0.20
m^*_h	0.25 @ 0.00 eV	10.96
ε_0	39.41	37.94

IV. Conclusions

The ab-initio results for the solution processable bismuth semiconductor $NaBiS_2$ have been found to be promising for photovoltaic applications. The bulk material has low charge carrier effective masses and an appropriate bandgap for solar energy harvesting. Incorporating oxygen into the material at lattice sites was found to be detrimental, which may run counter to intuition if lattice distortions from the smaller anion are not considered. Severe sulfur deficiency also leads to bandgap closing. Distortions to the crystal lattice have been quantified and this may assist in observing high levels of such defects experimentally. Further theoretical work will look at formation energies for S and Na vacancies and the inter-dependence of these two energies, since it is suspected that the electropositive Na is effectively "drawing in" the sulfur atoms due to its low electronegativity, in $NaBiS_2$ supercells.

More generally, this work provides insight into the beneficial effects of ionic, interconnected, large metal-anion octahedra as building blocks for high efficiency solution processable solar cells, even where the material has a unit cell with overall lower symmetry.

References

[1] Z. Zhang et al., "Significant Improvement in the Performance of PbSe Quantum Dot Solar Cell by Introducing a $CsPbBr_3$ Perovskite Colloidal Nanocrystal Back Layer," Adv. Energy Mater., vol. 7, no. 5, p. 1601773, Mar. 2017.

[2] M. Yuan, M. Liu, and E. H. Sargent, "Colloidal quantum dot solids for solution-processed solar cells," Nat. Energy, vol. 1, no. February, p. 16016, 2016.

[3] J. H. Heo et al., "Efficient inorganic–organic hybrid heterojunction solar cells containing perovskite compound and polymeric hole conductors," Nat. Photonics, vol. 7, no. 6, pp. 486–491, May 2013.

[4] H. Song et al., "Rapid thermal evaporation of Bi_2S_3 layer for thin film photovoltaics," Sol. Energy Mater. Sol. Cells, vol. 146, pp. 1–7, Mar. 2016.

[5] F. Giustino and H. J. Snaith, "Toward Lead-Free Perovskite Solar Cells," ACS Energy Lett., vol. 1, no. 6, pp. 1233–1240, Dec. 2016.

[6] M. Bernechea, N. C. Miller, G. Xercavins, D. So, A. Stavrinadis, and G. Konstantatos, "Solution-processed solar cells based on environmentally friendly $AgBiS_2$ nanocrystals," Nat. Photonics, vol. 10, no. 8, pp. 521–525, Jun. 2016.

[7] F. Viñes, M. Bernechea, G. Konstantatos, and F. Illas, "Matildite versus schapbachite: First-principles investigation of the origin of photoactivity in $AgBiS_2$," Phys. Rev. B, vol. 94, no. 23, p. 235203, Dec. 2016.

[8] G. L. W. Hart and R. W. Forcade, "Generating derivative structures from multilattices: Algorithm and application to hcp alloys," Phys. Rev. B, vol. 80, no. 1, p. 14120, Jul. 2009.; G. Bergerhoff, R. Hundt, R. Sievers, and I. D. Brown, "The inorganic crystal structure data base," J. Chem. Inf. Model., vol. 23, no. 2, pp. 66–69, May 1983.

[9] T. J. McCarthy et al., "Molten salt synthesis and properties of three new solid-state ternary bismuth chalcogenides, beta-$CsBiS_2$, gamma-$CsBiS_2$, and $K_2Bi_8Se_{13}$," Chem. Mater., vol. 5, no. 3, pp. 331–340, Mar. 1993.

[10] Y. Zhou et al., "Thin-film Sb_2Se_3 photovoltaics with oriented one-dimensional ribbons and benign grain boundaries," Nat. Photonics, vol. 9, no. 6, pp. 409–415, May 2015.

[11] S. U. Rahayu, C.-L. Chou, N. Suriyawong, B. A. Aragaw, J.-B. Shi, and M.-W. Lee, "Sodium antimony sulfide ($NaSbS_2$): Turning an unexpected impurity into a promising, environmentally friendly novel solar absorber material," APL Mater., vol. 4, no. 11, p. 116103, Nov. 2016.

[12] S. Kang, Y. Hong, and Y. Jeon, "A Facile Synthesis and Characterization of Sodium Bismuth Sulfide ($NaBiS_2$) under Hydrothermal Condition," Bull. Korean Chem. Soc., vol. 35, no. 6, pp. 1887–1890, Jun. 2014.

[13] I. Zumeta-Dubé, V.-F. Ruiz-Ruiz, D. Díaz, S. Rodil-Posadas, and A. Zeinert, "TiO_2 Sensitization with Bi_2S_3 Quantum Dots: The Inconvenience of Sodium Ions in the Deposition Procedure," J. Phys. Chem. C, vol. 118, no. 22, pp. 11495–11504, Jun. 2014.

[14] P. Giannozzi et al., "QUANTUM ESPRESSO: a modular and open-source software project for quantum simulations of materials," J. Phys. Condens. Matter, vol. 21, no. 39, p. 395502, Sep. 2009.

[15] R. E. Brandt, V. Stevanović, D. S. Ginley, and T. Buonassisi, "Identifying defect-tolerant semiconductors with high minority-carrier lifetimes: beyond hybrid lead halide perovskites," MRS Commun., vol. 5, no. 2, pp. 265–275, Jun. 2015.

Effect of Annealing on Performance of Solar Cells with New Oxide Absorber Mn₂V₂O₇

Pramod Ravindra[1], Eashwer Athresh[2], Rajeev Ranjan[3], Srinivasan Raghavan[1], Sushobhan Avasthi[1]

[1]Centre for Nanoscience and Engineering, Indian Institute of Science, Bangalore, India
[2]Interdisciplinary Centre for Energy Research, Indian Institute of Science, Bangalore, India
[3]Department of Materials Engineering, Indian Institute of Science, Bangalore, India

Abstract—All-oxide solar cells are attractive due to their stability, low cost and ease of fabrication. Very few oxide-absorbers are known and the efficiencies are limited by low mobility and lifetimes. In this work, for the first time, photovoltaic properties of a new oxide semiconductor, $Mn_2V_2O_7$ (MVO) are described. Optical measurements show that MVO has an indirect bandgap of 1.6 eV and a direct bandgap of 1.75 eV, which suggests that it absorbs efficiently in the visible region of the solar spectrum. The valence and conduction band positions of the intrinsically n-doped MVO film are determined to be at 5.5 eV and 3.9 eV. Schottky solar cells fabricated using Pt/MVO heterojunction show low short circuit current (J_{SC}) and open circuit voltage (V_{OC}). Annealing in nitrogen ambience results in a J_{SC} of 0.46 mA/cm² a 50x increase and of V_{OC} 0.21 V, almost 10x increase compared to un-annealed device. A maximum power of 24.1 μW/cm² is obtained, which is two orders of magnitude higher than un-annealed devices.

I. INTRODUCTION

Oxides are an attractive family of materials for solar absorption due to their abundance, chemical stability, non-toxicity, ease of fabrication and low cost of deposition. Oxide-based photovoltaics holds the promise of inexpensive and environmentally safe solar cells which is essential, besides high efficiency. Although oxides have been used as electrodes and carrier-selective contacts in solar cells, all-oxide photovoltaics has seen limited success, mainly due to low carrier lifetimes and low mobilities in the few known oxide-absorbers. Also, the best known absorbers have bandgaps that are not optimized for solar absorption [1, 2]

$Mn_2V_2O_7$ (MVO) has been studied widely for its magnetic[3], thermal properties [4] and as a pigment[5]. Recently, MVO has been demonstrated to be a candidate for solar water splitting [6]. In this work, we present, for the first time, MVO as an oxide-based solar absorber for thin film solar cells. Optical measurements show that MVO has a primary indirect bandgap of 1.6 eV and also a direct bandgap of 1.75 eV, which suggests that it absorbs efficiently in the visible region of the solar spectrum. The conduction band and valence band were measured to be at 3.9 eV and 5.5 eV respectively below vacuum level. Performance of solar cells improves dramatically after annealing in nitrogen, leading to a ≈400x increase in maximum generated power. V_{OC} of 0.21 V and a J_{SC}

of 0.46 mA/cm² are achieved, which suggests that MVO is a promising material for solar absorption in thin film solar cells.

II. EXPERIMENTAL DETAILS

All substrates were cleaned using acetone, followed by isopropanol in an ultrasonicating bath. The substrates were then rinsed in DI water and dried using a nitrogen jet. Thin films of MVO were deposited using a home-built PLD system whose base pressure is 5×10⁻⁵ mbar. A 248 nm KrF excimer laser was used to ablate the phase pure MVO target (Coherent ComPEX Pro). The target-substrate distance was maintained at 4 cm and the fluence was maintained at 1.5 J/cm². Deposition was carried out at a substrate temperature of 500 C at a frequency of 20 Hz. An oxygen pressure of 0.05 mbar was maintained during deposition. Annealing was carried out in a Tempress tube furnace. UV-visible spectroscopy was performed using a Shimadzu MPS3600 spectrophotometer with an integrating sphere attachment. UV-photoelectron spectroscopy was performed using a Kratos Ultra instrument/ Dark IV characteristics were measured using an Agilent B1500A semiconductor device analyzer. Light IV measurements were performed using a Keithley 2420 source meter and an Oriel solar simulator. The incident light was calibrated using a test cell provided by Oriel.

III. RESULTS AND DISCUSSION

A. Band-edge Determination of MVO

Absorption coefficient α of MVO was measured using UV-visible spectroscopy with integrating sphere attachment. Using α, a Tauc plot was constructed according to:

$$(\alpha h\vartheta) = A(h\vartheta - E_g)^{1/r} \qquad (1)$$

where h is the Planck's constant, υ is the frequency of light, and E_g is the band gap. When r = 2, the linear extrapolation of the plot reveals the indirect bandgap, whereas r = 0.5 reveals the direct bandgap. Tauc plots for extraction of both direct and indirect bandgap of MVO are shown in fig. 1(a). MVO has an indirect bandgap of 1.62 eV, and also a direct bandgap of 1.75 eV. This compares well with previously published data [6]. The

978-1-5090-5606-4/17 $31.00 © 2017 IEEE

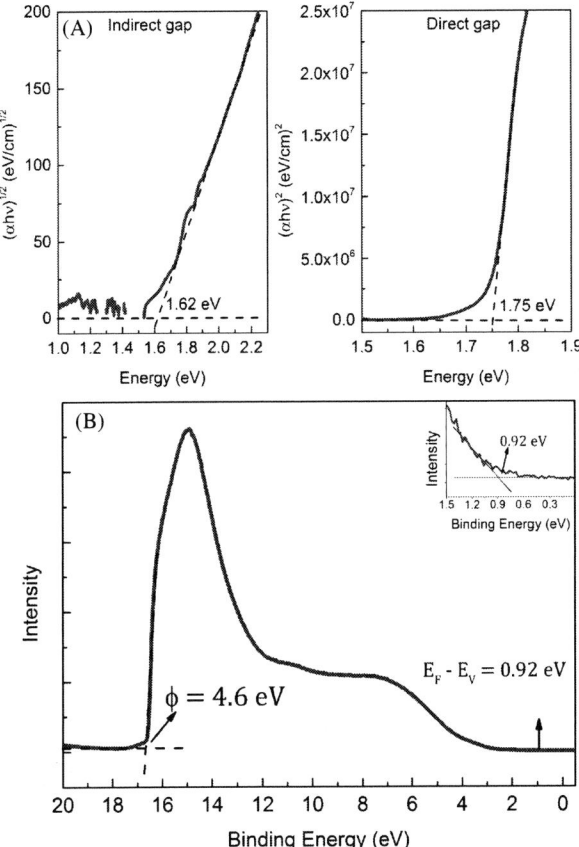

Fig 1. Determination of band-edges of MVO (a) Tauc plots showing both direct and indirect bandgap and (b)UPS spectrum for MVO

bandgap and the absorption coefficient (not shown) show that MVO absorbs strongly in the visible region below 700 nm, which constitutes a significant part of the solar spectrum.

UPS measurements using He-I line (22.12 eV) were used to determine the position of the valence-band maximum of MVO. Fig. 1(b) shows the UPS spectrum for MVO. From the onset at high binding energy, the work-function is determined to be 4.6 eV. From the cut-off at low binding energy, the valence band position (E_V) relative to the Fermi level (E_F) is found to be 0.9 eV. Using these, the position of the valence band is determined to be to be 5.5 eV below the vacuum level. Assuming the optical bandgap of 1.6 eV is equal to the electronic bandgap in MVO, the position of the conduction band (E_C) is calculated to be 3.9 eV below the vacuum level.

B. Pt/MVO Schottky Junctions

Using hot-probe method, MVO was determined to be an n-type semiconductor. This is unsurprising as oxides are typically n-

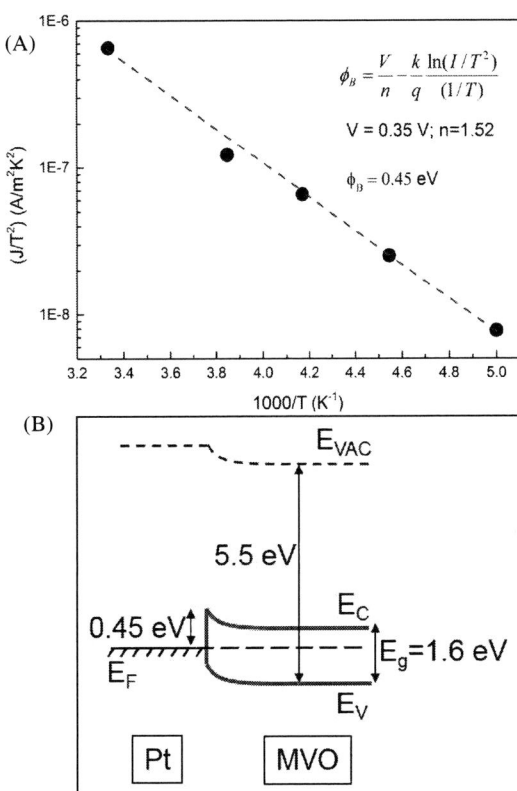

Fig. 2. (a) Richardson plot for Schottky barrier height measurement and (b) band diagram for Pt/MVO Schottky junction

type due to presence of oxygen vacancies. Platinum electrodes were sputtered on MVO to form a Schottky junction. To measure barrier height, temperature dependent I-V measurements were performed. Using the thermionic emission current model to describe carrier transport across the Pt/MVO junction, the barrier height was calculated according to

$$\varphi_B = \frac{V}{n} - \frac{k}{q} \frac{\ln(J/_{T2})}{(1/_T)} \qquad (2)$$

where φ_B is the barrier height, V is forward bias, n is the ideality factor, J the current density, T the temperature, k and q the Boltzmann constant and electron charge respectively. Fig. 2(a) shows the Richardson plot and the band-diagram. Current at a forward bias of 0.35 V was used for analysis. A Schottky barrier height of 0.45 eV was measured for the Pt/MVO junction.

C. Photovoltaic characteristics and effect of annealing

The Pt/MVO junction shows solar cell action upon illumination with A.M. 1.5 light. The as-fabricated device shows low short circuit current (J_{SC}) of 8 μA/cm^2 and an open

(A)

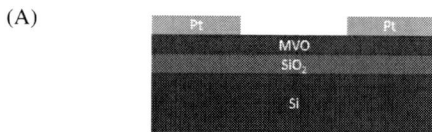

(B)

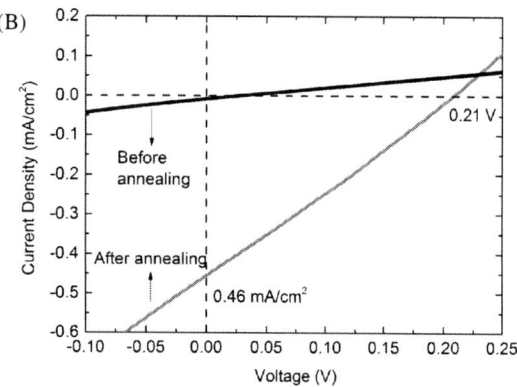

Fig. 3. (A) Device structure (B) I-V characteristics of Pt/MVO junction under A.M. 1.5 illumination before and after annealing, showing a dramatic improvement in performance

circuit voltage (V_OC) of 27 mV. The low values of J_{SC} and V_{OC} can be attributed to multiple factors. The structure consists of contacts that are 4 mm apart, which is probably much larger than the carrier diffusion length. So, photogenerated carriers are effectively collected from only a few diffusion lengths around the opaque platinum contacts which reduces the J_{SC}. The device does not employ a blocking layer, leading to high reverse currents. This, combined with interface and bulk defects acting as recombination centers, can result in a low value of V_{OC}. To increase grain size and reduce grain boundary recombination, the cell was annealed in a nitrogen ambient. Annealed devices show a dramatic improvement in the photovoltaic performance. The J_{SC} increases by 50x to 0.46 mA/cm^2, and the V_{OC} is 0.21 V, almost a 10x improvement compared to unannealed device (fig 3.). The ideality factor of

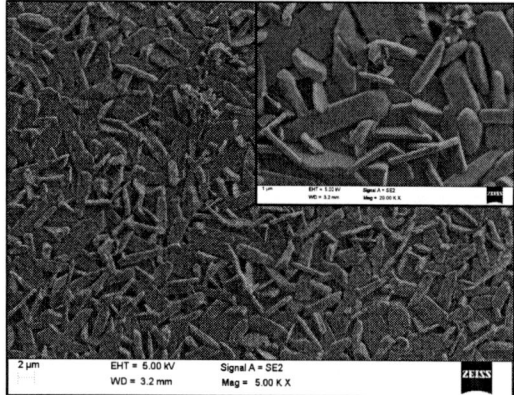

Fig. 4. SEM image of MVO film after N$_2$-anneailng. Inset shows the same film at a higher magnification, clearly showing grains > 1μm

the device after annealing is 2.4, and the reverse current is 4.2 × 10^{-6} A/cm^2.

Using

$$V_{OC} = \frac{nkT}{q} \ln\left(\frac{J_{SC}}{J_o}\right) \qquad (3)$$

V_{OC} is calculated to be 0.29 V, which matches well with the measured value of 0.21 V. The maximum power generated, 24.1 μW/cm^2, is comparable with the other early-stage oxide solar absorbers like Co$_3$O$_4$ (22.5 μW/cm^2)[7]. The large improvement in performance can be attributed to grain growth in the MVO film which results in grain sizes much large than 1μm (fig. 4.). This improves mobility and reduces grain boundary recombination, leading to increase in V_{OC}.

III. CONCLUSION

A new semiconducting oxide Mn$_2$V$_2$O$_7$ has been investigated for the first time as an absorber in thin film solar cells. The bandgap of the material is measured to be 1.6 eV with direct transitions at 1.75 eV which is desirable for a solar absorber. The band edges are determined to be at 3.9 eV and 5.5 eV, which makes it suitable for using common electron and hole blocking layers. Schottky solar cells formed using Pt/MVO junction after N$_2$-annealing show a J_{SC} 0.46 mA/cm^2 and a V_{OC} of 0.21 V, which is respectable for an early-stage oxide absorber. The maximum generated power is two orders of magnitude more than the un-annealed device. This improvement is attributed to grain growth in MVO. Our results show that Mn$_2$V$_2$O$_7$ is a promising candidate as an absorber for all-oxide solar cells.

REFERENCES

[1] T. Minami, Y. Nishi, and T. Miyata, "High-efficiency Cu2O-based heterojunction solar cells fabricated using a Ga2O3 thin film as an n-type layer," *Applied Physics Express,* vol. 6, p. 044101, 2013.

[2] D. Tiwari, D. J. Fermin, T. Chaudhuri, and A. Ray, "Solution-processed Bismuth ferrite thin films for all-oxide solar photovoltaics," *The Journal of Physical Chemistry C,* vol. 119, pp. 5872-5877, 2015.

[3] J.-H. Liao, F. Leroux, Y. Piffard, D. Guyomard, and C. Payen, "Synthesis, structures, magnetic properties, and phase transition of manganese (II) divanadate: Mn 2 V 2 O 7," *Journal of Solid State Chemistry,* vol. 121, pp. 214-224, 1996.

[4] T. Krasnenko and M. Rotermel, "Structural modification of Mn2V2O7: Thermal expansion and solid solutions," *Russian Journal of General Chemistry,* vol. 83, pp. 1640-1644, 2013.

[5] D. R. Swiler, "Manganese vanadium oxide pigments," ed: Google Patents, 2002.

[6] Q. Yan, G. Li, P. F. Newhouse, J. Yu, K. A. Persson, J. M. Gregoire, *et al.,* "Mn2V2O7: an earth abundant light absorber for solar water splitting," *Advanced Energy Materials,* vol. 5, 2015.

[7] B. Kupfer, K. Majhi, D. A. Keller, Y. Bouhadana, S. Rühle, H. N. Barad, *et al.*, "Thin Film Co3O4/TiO2 Heterojunction Solar Cells," *Advanced Energy Materials,* vol. 5, 2015.

[8] S. M. Sze and K. K. Ng, *Physics of semiconductor*

Electro-optical Properties of $Zn_2Mo_3O_8$ Thin-Films: A Novel Low-Bandgap Solar Absorber

Pramod Ravindra[1], Eashwer Athresh[2], Rajeev Ranjan[3], Srinivasan Raghavan[1], Sushobhan Avasthi[1]

[1]Centre for Nanoscience and Engineering, Indian Institute of Science, Bangalore, India
[2]Interdisciplanary Centre for Energy Research, Indian Institute of Science, Bangalore, India
[3]Department of Materials Engineering, Indian Institute of Science, Bangalore, India

Abstract- **Metal-oxide semiconductors are attractive as solar absorbers because they are abundant, stable, environmentally-safe, and low-cost. This work reports the electronic and optical properties of $Zn_2Mo_3O_8$ (ZMO), a novel oxide-based solar absorber. ZMO has a direct bandgap of 1.9-2.1 eV, so it can efficiently absorb visible photons. As-deposited films are poly-crystalline and unintentionally doped n-type with ~10^{17} cm^{-3} carriers. Electron mobility of 0.6 – 0.7 cm^2/V-s is comparable to other thin-film absorbers like CdTe. The conduction and valence band edges in ZMO are 4.4 eV and 6.3 eV below vacuum level, respectively. Preliminary devices show photoconductivity with a 1.5x increase in current upon illumination**

Index Terms — **photovoltaic cells, semiconductors, oxides, solar absorbers, thin-films**

I. INTRODUCTION

All-oxide solar cells, which employ metal-oxide absorbers, are interesting because typical oxides are earth-abundant, chemically stable, environmentally safe, and can be deposited by low-cost processes. However, oxide-based solar cells have not been very successful. The most successful demonstration of a solar cell shows an efficiency of 8.1%, where Bi_2FeCrO_6 was used as the absorber. However, these cells were made using epitaxial thin film layers on expensive Nb:STO substrates[1]. Cu_2O and $BiFeO_3$, the two most widely studied oxide absorbers in p-n heterojunction solar cells, have shown efficiencies of 5.38%[2] and 3.98%[3] respectively. These materials have bandgaps that are too large for optimal solar absorption.

$Zn_2Mo_3O_8$ (ZMO), a small bandgap oxide (E_G=1.9 eV), has been studied for its use in batteries[4] and photo-electrochemical cells[5]. A reported bandgap of 1.9 eV makes it suitable as a material for a top cell of a tandem cell with silicon. While bandgap has been reported, optical and electrical properties of ZMO thin films, and its use as a photovoltaic absorber has not been demonstrated. In this work, optical and electronic properties of ZMO are reported. We show using both UV-visible spectroscopy and spectroscopic ellipsometry that ZMOP has a bandgap of 1.9 – 2.1 eV and that it has a high absorption coefficient in the visible region. UV-Photoelectron Spectroscopy (UPS) is performed to measure the valence band

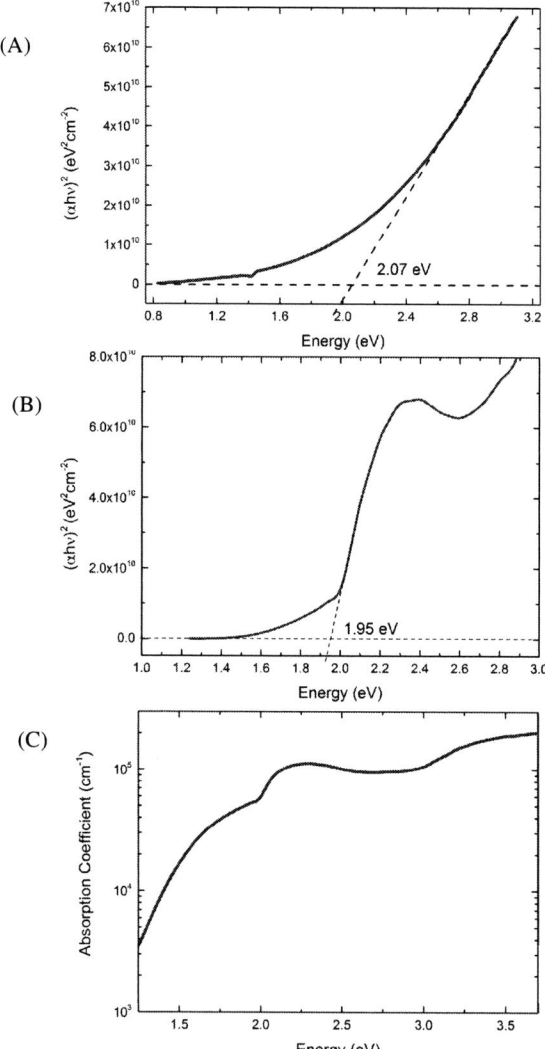

Fig. 1. Optical properties of ZMO (a) Tauc plot showing a direct bandgap of 2.07 eV obtained from UV-visible spectroscopy (b) Tauc plot using data from ellipsometry (c) absorption coefficient indicating the value at band edge

position. Room-temperature hall measurements reveal that ZMO is unintentionally n-doped with 10^{17} cm^{-3} electron concentration and a mobility of 0.6 – 0.7 cm^2/V-s. Finally, Schottky diodes using ZMO are shown to be photoactive, with a 1.5x increase in current upon illumination, suggesting that ZMO can be a promising solar absorber for thin-film solar cells.

II. EXPERIMENT DETAILS

Thin-films of ZMO were deposited at room temperature using pulsed laser deposition (PLD) on glass and fluorinated-tin oxide (FTO) glass substrates. Prior to deposition, the substrates were cleaned in acetone and isopropanol in an ultrasonic bath, followed by rinsing in DI water. Substrates were dried using a nitrogen jet.

PLD was carried out using a home-built system. The laser source was a comPEX pro KrF excimer laser (248 nm). Laser fluence was set at 2 J/cm^2. Deposition was done at a frequency of 5 Hz, at room temperature.

Thickness of the films was measured using a Dektak Surface Profiler. UV-visible spectroscopy was performed using a Shimadzu MPC3600 spectrometer, using integrating sphere attachment. SEM was done using a Zeiss Ultra 55 microscope. UV photoelectron spectroscopy were performed in a Kratos Axis Ultra using the He-1 excitation (21.22 eV). Electrical measurements were done using an Agilent B1500A semiconductor device analyser. Photocurrent measurements were done with a Keithley 2420 source meter and an Oriel solar simulator which was calibrated against a standard cell provided by Oriel.

III. RESULTS AND DISCUSSION

A. Optical and Electronic Properties of ZMO

Thin-films of ZMO were deposited at room temperature using pulsed laser deposition (PLD). XRD confirms that the

Fig. 2. SEM image of ZMO film showing particle sizes < 100 nm

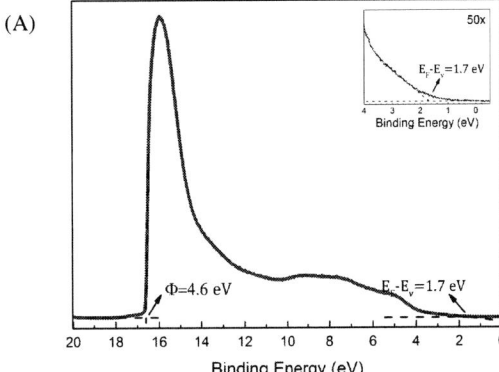

(A)

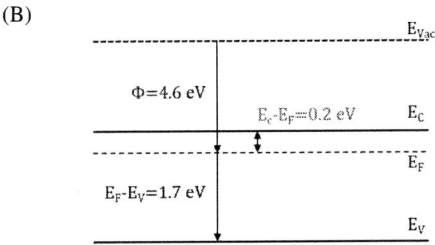

(B)

Fig. 3 (a) UPS spectrum of ZMO and (b) Band energy levels in ZMO film.

deposited films are polycrystalline ZMO (data not shown here).

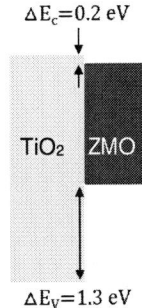

Fig. 4. Band alignment of ZMO with TiO$_2$ hole blocking layer

Bandgap and absorption coefficient of MVO were measured using UV-visible spectrophotometer with an integrating sphere. The absorption coefficient (α) was calculated from separately measured transmittance (T) and reflectance (R) using:

$$\alpha = -\frac{1}{d} \ln\left(\frac{T}{(1-R)}\right) \qquad (1)$$

Where d is the thickness of the film. Bandgap (E_G) is extracted from the Tauc plot, according to:

$$(\alpha h\vartheta) = A(h\vartheta - E_g)^{1/r} \qquad (2)$$

Where h is the Planck's constant, υ is the frequency of light and r = 2 for direct bandgap materials. For ZMO thin films, direct bandgap of 1.9 eV is extracted (Fig 1(a)), a relatively low value. The absorption coefficient at the band edge is ~2×10^4 cm^{-1}, so a 1.5μm thick film will absorb over 95% of the light incident on the film. Overall, the optical measurements show that ZMO can be an efficient solar absorber, suitable for thin-film solar cells.

B. Carrier concentration and Hall mobility

Results from Hall measurements on ZMO thin-films have been

Table 1. Summary of Hall experiment with ZMO

summarized in Table 1. The negative Hall voltage shows that electrons are the majority carriers, not surprising considering that oxygen vacancies in metal-oxide typically lead to n-type doping. Bulk mobility of 0.6-0.7 cm^2/V-s is measured, comparable to the reported mobility other thin-film solar absorbers such as CdTe (0.1 – 2.6 cm^2/V-s)[6]. The as-deposited films have small sized grains (Fig. 2). The average crystallite size extracted from XRD measurements is also only ≈25 nm. So grain boundary scattering is expected to be the limiting factor for mobility in these films.

Current (μA)	Field (T)	Hall voltage (mV)	Carrier concentration (cm^{-3})	Conductivity (S cm^{-1})	Mobility (cm^2/Vs)
5	2.5	-3.66	6.13×10^{17}	0.068	0.62
10	2.5	-8.15	6.83×10^{17}	0.067	0.69

C. Band-edge determination

UV photoelectron spectroscopy (UPS) was performed using He-I line (22.12 eV) to measure the position of the valence band maximum in ZMO films. From the onset at high binding-energy, the work function was found to be 4.6 eV (Fig 3(a)). From the cut-off at low-binding-energy, the position of the valence band (E_V) relative to the fermi level (E_F) was determined to be 1.7 eV. Using these two values, the position of the valence band can be calculated to be 6.3 eV below vacuum level. Assuming the optical bandgap to be equal to the electronic bandgap, we can calculate the position of the conduction band minimum to be at 4.4 eV. The extracted band energy levels are summarized in Fig 2(b). Comparing the energy levels of ZMO with TiO_2, a well-known electron-transport layer, it can be seen that the ZMO/TiO_2 interface will have a large valance band offset of 1.3 eV but a small conduction band offset of 0.2 eV (Fig 4). So TiO_2 is expected to function as a hole-blocking layer for ZMO.

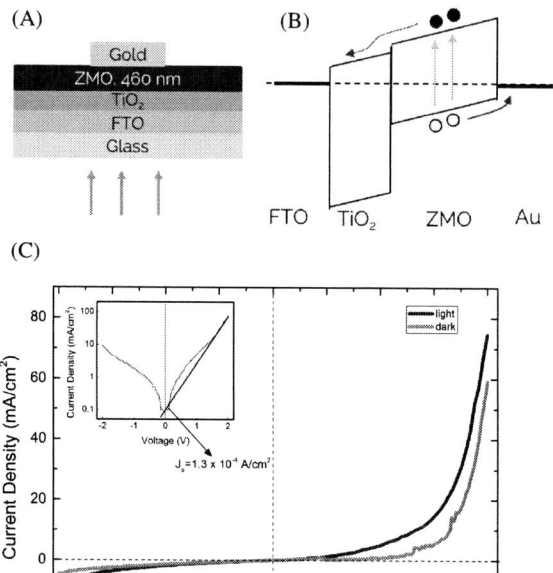

Fig. 5(a) Structure of test devices (b) band diagram of the one-sided device (c) I-V characteristics of the devices in dark and under A.M.1.5 illumination

D. Single-sided Solar Cells

Solar cells were fabricated by depositing ZMO on TiO_2-coated FTO substrates. Cells were illuminated through the transparent FTO electrode, with an evaporated Au film serving as the back electrode (Fig 4a). The expected band diagram for the structure shows that the device stack forms diode (Fig. 4b). ZMO doping is only ~10^{17} cm^{-3} (Table 1), so the 460 nm thick ZMO film is expected to be completely depleted. Under illumination, photogenerated electrons and holes are expected to collect at the FTO and the Au electrode, respectively.

The dark I-V characteristics of the devices show a high reverse current (Fig 5 (c)). We attribute this to shunts paths in the ZMO film caused by high surface roughness of PLD-deposited films. Under A.M. 1.5 illumination, the diode shows a 1.5x higher current, showing carrier are absorbed and separated in the ZMO diode. However, no measurable open-circuit voltage or short-circuit current is observed. Possibly because the carrier diffusion lengths is far smaller than thickness of the ZMO film.

IV. CONCLUSION

In conclusion, we report the electro-optical properties of $Zn_2Mo_3O_8$, a promising absorber for thin-film solar cells. The bandgap of ZMO is 1.9 eV, optimal for solar absorption. Hall measurements show that ZMO is n-type with 10^{17} cm^{-3} electrons with a reasonable mobility of $0.6 - 0.7$ cm^2/V-s. As-deposited films are polycrystalline with a crystallite size of ~25 nm. The conduction and valence bands edges in ZMO films are 4.4 eV and 6.3 eV below the vacuum level, respectively. TiO_2 is expected to form a hole-blocking heterojunction with ZMO. Preliminary FTP/TiO_2/ZMO/Au devices do show photoconductivity but no photovoltaic effect. Morphology and carrier diffusion length in the ZMO thin-film are the major challenges in ZMO thin-film solar cells.

REFERENCES

[1] R. Nechache, C. Harnagea, S. Li, L. Cardenas, W. Huang, J. Chakrabartty, *et al.*, "Bandgap tuning of multiferroic oxide solar cells," *Nature photonics,* vol. 9, pp. 61-67, 2015.

[2] T. Minami, Y. Nishi, and T. Miyata, "High-efficiency Cu2O-based heterojunction solar cells fabricated using a Ga2O3 thin film as n-type layer," *Applied Physics Express,* vol. 6, p. 044101, 2013.

[3] D. Tiwari, D. J. Fermin, T. Chaudhuri, and A. Ray, "Solution processed bismuth ferrite thin films for all-oxide solar photovoltaics," *The Journal of Physical Chemistry C,* vol. 119, pp. 5872-5877, 2015.

[4] S. Petnikota, S. K. Marka, V. V. Srikanth, M. V. Reddy, and B. V. Chowdari, "Elucidation of few layered graphene-complex metal oxide (A 2 Mo 3 O 8, A= Co, Mn and Zn) composites as robust anode materials in Li ion batteries," *Electrochimica Acta,* vol. 178, pp. 699-708, 2015.

[5] M. Paranthaman, G. Aravamudan, and G. S. Rao, "Photoelectrochemical properties of metal-cluster oxide compounds, A2Mo3O8 and (LiY) Mo3O8," *Bulletin of Materials Science,* vol. 10, pp. 313-322, 1988.

[6] Q. Long, S. A. Dinca, E. A. Schiff, M. Yu, and J. Theil, "Electron and hole drift mobility measurements on thin film CdTe solar cells," *Applied Physics Letters,* vol. 105, p. 042106, 2014.

Low temperature solution process for random high aspect ratio silver nanowire as promising transparent conductive layer

Arastoo Teymouri, Supriya Pillai, Zi Ouyang, Xiaojing Hao and Martin Green

School of Photovoltaic and Renewable Energy, University of NSW, Sydney, NSW, 2052, Australia

Abstract — **Indium Tin Oxide (ITO) is traditionally deposited at elevated temperatures to achieve the low resistivity needed for device applications. This work aims to replace (ITO) as transparent conductive (TC) film with high aspect ratio (length to diameter) silver nanowire (AgNW) in purpose of employing at heat-sensitive optoelectronic devices. Fused-joints of AgNW network required for good conductivity is normally achieved using high temperature annealing that can either damage heat-sensitive devices or cause difficulty in the fabrication process. Employing a low temperature post-treatment in deposition of high aspect ratio AgNWs showed a comparatively conductive AgNWs network with ITO. Conductive atomic force microscopic (C-AFM) and Four-point probe confirmed conductivity of the network for application as front electrode in heat-sensitive thin film solar cells. This cost-effective and solution-based process could be promising for third generation solar cells like Perovskite.**

I. INTRODUCTION

In today's optoelectronic device world, transparent conductive film plays one of the main roles to carry the charge in many applications, such as, solar cell, light emitting diode touch screens and flat panel displays. Silicon tandem is one of the highly efficient solar cells is seeking a high transmission front electrode in the large wavelengths from near ultraviolet (UV) to near infrared (NIR). For more than forty years, Indium Tin Oxide has been dominated as the most highly transparent conductive thin film. However, some intrinsic problems such as scarcity of Indium on the earth's crust, vulnerability to bending and high cost of production and the most important problem, free carrier absorption at longer wavelength have broadly driven scientific interests to find a replacement.

Metal nanowire network as TC film have shown comparable features with ITO, particularly Silver [1]. Many studies have been conducted in the last decade to prove the reliability of AgNWs as TC [1-3], but there are still some significant issues in applicability of AgNWs such as non-conductive areas between nanowires and the most important problem, supreme conductivity of AgNWs network is only enabled after a high temperature post-treatment [3]. Non-ballistic behaviour of Silver nanowires leads to high sheet resistance which means electrons cannot easily traverse from one wire to another [4]. To overcome the difficulty, annealing in high temperature (≥ 180 °C) has been applied as a standard treatment, so far [5]. At this temperature, a nanowire is able to make good contact between adjacent nanowires and AgNWs may gain a very low sheet resistance even below 10 $\Omega/\square$. On the other hand, high

temperature annealing could damage heat-sensitive devices. Emerging organic solar cells, perovskite and Copper Zinc Tin Sulfide (CZTS) solar cells are low temperature devices; the highest endurable temperature is preferably below 100°C where the structure of those solar cells remains unscathed [6]. In order to relax the compromise between the low sheet resistance and low annealing temperature, a new technique of solution-based coating has been applied in this study. In this study, we address this issue using longer and thinner (high aspect ratio) AgNWs with very simple and low temperature (60 °C) process. The aim was to achieve a comparable transparent conductive (TC) layer to Indium Tin Oxide (ITO) which suffers from parasitic absorption losses and reduced transmission at longer wavelengths that is not suited for the bottom cell in a tandem stack.

Studies put more efforts on replacing rather than the optimising annealing treatment, which makes the fabrication process more complicated. For instance, using physical pressing [7], coupling AgNWs with other conductive materials [8] and chemical treatment [9] of silver nanowires have been tried, but have not yet resulted in a desirable outcome, comparing to ITO. Furthermore, applying the above replacement methods could lower sheet resistance of AgNWs, but is not yet comparable to the annealing treatment in terms of higher sheet resistance and simplicity. In this work we aim to show that low temperature annealing is capable of providing low sheet resistance silver nanowire in order to use in heat-sensitive applications. This promising technique can alter the concept that high temperature annealing is the only method to have highly conductive and random nanowires that could be employed as TC layer.

II. EXPERIMENTAL AND THEORETICAL RESULTS

The high aspect ratio (length to diameter) nanowires employed in this study increase the probability of having more intimate contacts at the junctions, which causes nanowires to lay down perfectly on each other. Moreover, long nanowires can increase the chance of building an integrated conductive network; it results in that charges can find more unfailing paths to traverse to the next nanowire. This concept helps to meaningfully decrease the sheet resistance and improve transmittance.

Hence, the length of nanowire shall be considered in this matter. This length could be calculated from percolation theory [10]. According to the percolation theory; percolation length is directly and inversely proportional to number of bridging and length of silver nanowires, respectively [4].

$$I_C \propto \frac{N_C}{l} \qquad (1)$$

Which percolation length, number of nanowires and length of nanowire are respectively denoted by I_C, N_C and l. Based on this proportion, the longer nanowire leads to the better percolation; hence, 200um long silver nanowire selected to rise up the chance of lateral conductivity through the network, figure (1). The density of nanowires, N can be calculated by dividing the number of nanowires by the area, and is given by:

$$N = \frac{\dfrac{4CV}{D_{Ag}\Pi d^2 l}}{\Pi D^2 / 4} \qquad (2)$$

For a given volume of the nanowire solution V, with a concentration of C (0.2 mg/mL), D_{Ag} is the density of bulk silver (10.5 g/mL), d is the average diameter (70 nm), and l is the average length (200 μm) of the Ag nanowire applied in this work. Critical density of nanowires to start the percolation is given by:

$$l\sqrt{\Pi N_C} = 4.236 \qquad (3)$$

N_C for nanowires with length l is defined as the critical nanowire density required for the onset of conduction in a random network. Placing N_C in equation (2) the volume of solution with the certain concentration can be calculated.

Supplier (ACS materials) provides a concentration of 20mg/mL AgNWs in isopropyl alcohol, then diluted to a concentration of 0.2 mg/mL in Methanol.

Therefore, the main variation is the density of nanowires, which found to be 5% surface coverage (S) as a threshold to have an optimum transparency and conductivity.
In order to obtain minimum surface coverage (S) with the optimum length of AgNWs, experimental data were fitted based on percolation theory (Eqe 4).

Based on the percolation theory, sheet resistance (R_{sh}) is inversely proportional to surface coverage (S) of AgNWs (equation 4) [11].

$$R_{sh} \approx \alpha(S - S_0)^\beta \qquad (4)$$

where S_0 is the minimum surface coverage for percolation, parameter α is related to the total sheet resistance of AgNW

network and parameter β denotes junction resistivity of AgNWs [12].

To obtain a lower sheet resistance, transparency must be sacrificed which is not acceptable in many applications of TC layer; hence, optimizing sheet resistance of AgNWs has been accomplished with various density; Thereby, density is determined by optimum transmission as upper limit and percolation length (proportion to l) as lower limit. Based on the experiment results, having transmission (T) higher than 90% does not allow density of AgNWs to be larger than 20% of surface coverage (SC); higher SC drops T to below 90%. Therefore, employing AgNWs ratio as high as possible recommended in this paper. However, the diameter cannot be lower than the mean free path of AgNWs, which could be dramatically decrease conductivity [4]. Due to this fact, the lowest possible diameter of AgNWs 70nm has been selected and the results have been displayed on figure 3.

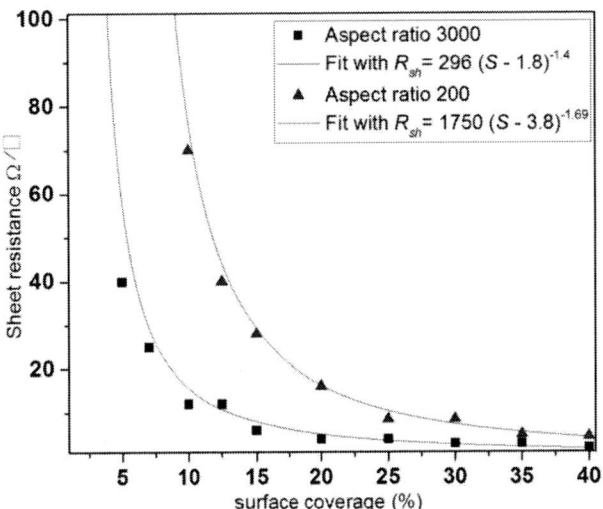

Fig. 1. Red line shows AgNWs (40um L, 130nm W), Blue line shows AgNWs (200um L, 70nm W)

Using longer nanowires expand the surface coverage of silver nanowires without increasing density (figure 3). Two different ratios of nanowires were tested; longer nanowires (ratio: 3000) have the superior conductivity without mitigating of transmission and enlarging surface coverage.

Figure 1 shows silver nanowire with higher ratio can compensate conductivity in lower density. As it can be seen, the optimum surface coverage of lower ratio AgNWs is larger (nearly 10%), while 10 order of magnitude higher ratio nanowire decreased optimum surface coverage to 5%.
AgNWs (aspect ratio 3000) with different annealing temperatures were imaged by Focused Ion Beam Scanning Electron Microscopy (FIB-SEM. Figure 2.a and 2.b show AgNWs annealed at 180 °C and 60 °C, respectively. The

samples were annealed immediately after deposition. The images were taken at an angle of 54° where the contact points of nanowires can be distinctly scrutinized by virtue of the side-view. In figure (2-a), AgNWs are closely laid down on each other and the melted joints are clear with many contact points between adjacent nanowires. After 60 °C annealing treatment, figure (2-b) shows AgNWs properly laying down on each other; however, no melted joints can be seen in entire network.

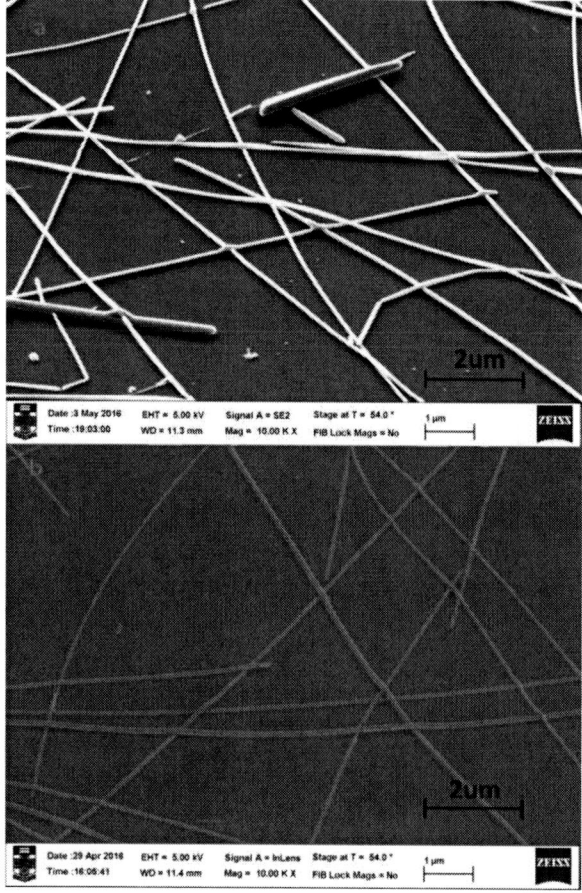

Figure 2.a) side-view (54°) SEM image of AgNWs sample annealed at 180 °C. b) side-view SEM image of AgNWs sample annealed at 60 °C.

Red line in figure 3 shows transmission of 90% in a wavelength range from 450 to 1000 nm wavelength after annealing treatment at 180 °C. In this work, post-treatment annealing at 60 °C has been applied to reach the excellent 23 Ω/□ sheet resistance with roughly 85% transmittance in the broad range of visible and NIR wavelengths, which is shown by the green plotline in figure 2. Black line is the transmittance of bare quartz used as reference.

Several samples have been tested to eventually find the best solution-based conditions of deposition on quartz sample with annealing post-treatment ≤100 °C. In order to compare the effect of annealing temperature on transmittance and conductance of AgNWs, two samples with quartz substrate were prepared. Both have the same sheet resistance of 23 Ω/□.

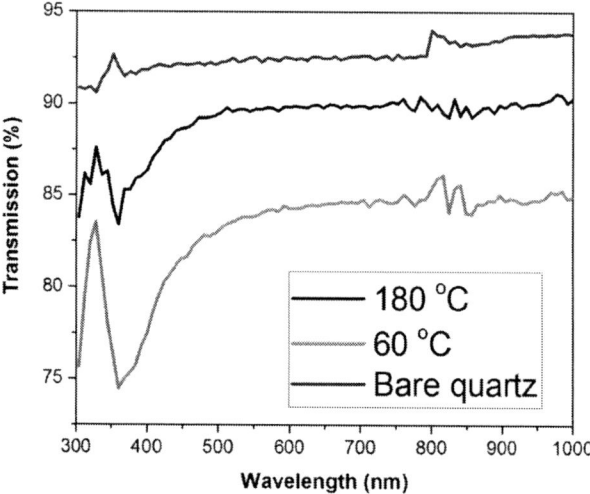

Figure 3. Effect of annealing temperature on transmission of AgNWs network with 23 Ω/□ sheet resistance; (Black line) bare quartz, (Red line) sample annealed in 180 °C and (Green line) sample annealed in 180 °C.

III. DATA ANALYSIS AND DISCUSSION

Comparing samples in the high and the low annealing temperatures proves that heat-treatment in high temperature provides enough activation energy to fuse adjacent silver nanowires in contact points. Thus, the network shows high conductivity to carry the charge. On the other hand, low temperature annealing makes just nano-contact between adjacent nanowires, which seems well enough to have a conductive network. Moreover, heat treatment immediately after deposition could remove organic materials like Polyvinylpyrrolidone (PVP) in between of the contact points and place nanowires closer to build a modest contact. Employing such high aspect ratio of nanowire (roughly 3000) enhances the chance of having more extensive contact in the junctions, which causes nanowires lay down perfectly on each other. It is obvious that higher annealing temperature propels nanowires to have better contacts with each other, which means lower density of nanowires gives higher conductivity. In lower temperature, using denser network of nanowires would relax this inevitable problem but at the cost of reduced transmission. Loss of transmission is in the range of 1/18 (roughly 5%).

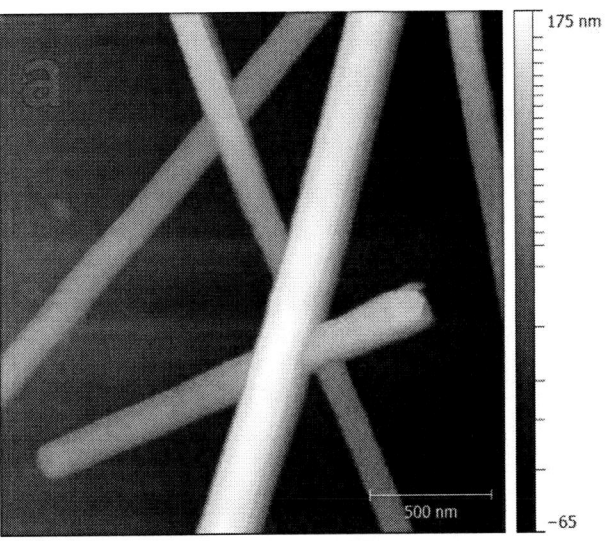

Figure 4.a) AFM image for Sample annealed in 60 °C, b) Conductive AFM image for the same area.

In order to prove the conductivity of AgNWs network using low temperature annealing, all samples had to be measured by Four-point probe. At least 10 points have been measured from each sample showing there are small conductive areas (between the probes) in every sample. In terms of collecting a reliable evidence to demonstrate that there is a connection between nanowire joints and nanowires perfectly connect to each other, C-AFM was employed. Figure 4.b, clearly shows the whole network shown in figure 4.a is highly conductive. Based on the fundamental principles of C-AFM, if any point is depicted in the current map, there is undoubtedly a conductive path that can drive electron from that point to the circuit.

IV. SUMMARY

We proposed that silver nanowire network can be used as transparent conductive layer with low temperature annealing. Owing to the cost-effectiveness and the simplicity of the method, this promising technique can be used not only for heat-sensitive emerging devices like Perovskite or CZTS cells, but also in many emerging devices. Eventually, AgNWs can be replaced with high-cost ITO as TC layer in numerous optoelectronic devices. Any type of solar cell is required its own optimization and work is underway to achieve this in the near future.

REFERENCES

[1] Hu. L., et al. (2010) "Scalable coating and properties of transparent, flexible, silver nanowire electrodes" ACS nano 4(5): 2955-2963.

[2] Leem, D. S., et al. (2011). "Efficient organic solar cells with solution-processed silver nanowire electrodes." Advanced Materials 23(38): 4371-4375.

[3] Mayoral, A., et al. (2011). "On the behavior of Ag nanowires under high temperature: in situ characterization by aberration-corrected STEM." Journal of Materials Chemistry 21(3): 893-898.

[4] Kumar, S., et al. (2005). "Percolating conduction in finite nanotube networks" Physical review letters 95(6): 066802.

[5] Madaria, A. R., et al. (2010). "Uniform, highly conductive, and patterned transparent films of a percolating silver nanowire network on rigid and flexible substrates using a dry transfer technique." Nano Research 3(8): 564-573.

[6] Shinde, N., et al. (2012). "Room temperature novel chemical synthesis of Cu 2 ZnSnS 4 (CZTS) absorbing layer for photovoltaic application" Materials Research Bulletin 47(2): 302-307.

[7] Yang, L., et al. (2011). "Solution-processed flexible polymer solar cells with silver nanowire electrodes.\" ACS applied materials & interfaces 3(10): 4075-4084.

[8] Liu, B.-T. and H.-L. Kuo (2013). "Graphene/silver nanowire sandwich structures for transparent conductive films." Carbon 63: 390-396.

[9] Tien, H.-W., et al. (2013). "Using self-assembly to prepare a graphene-silver nanowire hybrid film that is transparent and electrically conductive" Carbon 58: 198-207.

[10] C. Ruiyi, D. Suprem R, J. Changwook, K. M. Ryyan, J. David B and Alam MA, "Co-percolating Graphene Wrapped Silver Nanowire Network for High Performance, Highly Stable, Transparent Conductive Electrodes" *Nanoscale*, vol. 23(41), pp. 5150-5158, 2013.

[11] Mutiso, R. M., et al. (2013). "Integrating simulations and experiments to predict sheet resistance and optical transmittance in nanowire films for transparent conductors." ACS nano 7(9): 7654-7663.

[12] Jagota, M. and N. Tansu (2015). "Conductivity of nanowire arrays under random and ordered orientation configurations." Scientific reports 5.

Oxygen Incorporation into Si Nanocrystal/SiC Multilayers

Charlotte Weiss, Andreas Reichert, Johannes Hofmann, Stefan Janz

Fraunhofer Institute for Solar Energy Systems, Heidenhofstraße 2, 79110 Freiburg, Germany.

Abstract — We aimed to improve the properties of Si nanocrystal/SiC multilayers (ML) for the use as high bandgap solar cell absorber by the incorporation of oxygen (O). Therefore we compare the structural properties of Si nanocrystal/SiC ML with and without incorporated O by scanning electron microscopy, Raman spectroscopy and grazing incidence X-ray diffraction patterns. The O incorporation in the form of Si-O bonds was successful and a beneficial effect of O on the conservation of the ML structure during annealing and on the c-Si/c-SiC ratio is observed and discussed in detail.

Keywords: Si nanocrystals, multilayer, SiC, Si tandem

I. INTRODUCTION

A solar cell (SC) with a bottom cell of conventional crystalline bulk Si (c-Si) and a top cell with a higher bandgap of silicon nanocrystals (NC) embedded in a dielectric matrix represents a potential material system for a crystalline Si tandem SC. The ideal bandgap for a top cell on a c-Si (1.1 eV) bottom cell was calculated to be 1.7 eV [1]. This can be achieved in Si with the help of the quantum confinement effect [2], which means enlarging the Si bandgap by reduction of the Si crystals to the nano-scale.

The use of 3C-SiC as a matrix material for Si NC is interesting as it provides a small conduction band offset of 0.5 eV compared with other typical matrix materials (1.9 eV for Si_3N_4 or 3.2 eV for SiO_2 [2]), with the same trend expected for the valence band. A small band offset increases the tunneling probability from one Si NC to the other and hence the conductivity of the material, making transport less sensitive to variations in NC separation [2, 3].

The so-called multilayer (ML) approach has been developed to reach Si NC size control for the adjustment of the Si bandgap. Alternating single layers (SL) of stoichiometric SiC barriers (SiC) and Si rich SiC wells (SRC) with thicknesses in the nm range are deposited. Typically the deposition is performed by plasma enhanced chemical vapor deposition (PECVD), followed by an annealing step at 1100°C by furnace anneal [4–6].

During the annealing step, phase separation and Si NC formation are expected to occur in the SRC layers while the SiC layers should serve as diffusion barrier. Furthermore, the annealing causes hydrogen (H) effusion as the films are usually deposited from the precursor gases SiH_4 and CH_4. The ML approach is known to work very well for Si NC size control in SiO_2 [7], but is much more challenging in SiC because interdiffusion of the SiC and the SRC layer occurs [4, 6]. In addition, co-crystallization of Si and SiC NC was observed and no clear evidence of quantum confinement was adduced due to

the very weak photo-luminescence (PL) signal from Si_xC_{1-x} samples [6, 8, 9]. The absolute luminescence quantum yield of Si_xC_{1-x} samples was estimated by Schnabel *et al.* [10] to be very low ($<10^{-6}$), indicating a high density of non-radiative recombination-active defects, due to H effusion and co-crystallization during annealing [5]. The high defect densities must be strongly reduced in order to develop a good absorber material for SC.

This work proposes the incorporation of oxygen (O) into the Si NC/SiC ML as it is expected to (i) prevent the SiC crystallization to reduce the interfaces and stress in the material as reported by Kurokawa *et al.* [11], (ii) to passivate Si dangling bonds by the formation of Si-O bonds and (iii) to decrease the ML intermixing by behaving more like Si NC in SiO_2.

II. EXPERIMENTAL

Two types of substrates were used for layer deposition: (i) 250 µm thick p-doped float zone Si, (100)-oriented with a resistivity of 10 Ωcm. (ii) 1 mm thick fused silica (Suprasil®) substrates.

All ML and SL samples used in this work were deposited by PECVD conducted in a *Roth&Rau* AK400 reactor. The pressure during deposition was kept at 0.3 mbar and the substrate temperature at 270°C. The plasma power density was 100 mW/cm² at a frequency of 13.56 MHz. The variation of the gas fluxes of SiH_4, CH_4 and H_2 allows the deposition of hydrogenated a-SiC (a-Si_xC_{1-x}:H) with varying stoichiometry. After deposition, all samples were subjected to a furnace anneal step of 30 min@1100°C.

The samples on Si substrates were subsequently characterized by scanning electron microscopy (SEM), grazing incidence X-ray diffraction (GIXRD) and Fourier transformed infrared spectroscopy (FTIR) measurements. The samples on fused silica substrate were characterized by Raman spectroscopy.

In a first experiment, it was checked if the use of CO_2 as precursor gas leads to the desired O incorporation into the samples. Therefore, $Si_{0.50}C_{0.50}$ and $Si_{0.77}C_{0.23}$ SL were deposited under CO_2 fluxes varying between 0.0 and 10.0 sccm. To compensate a possible increase in C content in the samples due to the CO_2 flux, the CH_4 flux was reduced by the same amount as the CO_2 flux was increased. In a second experiment, the influence of O incorporation on ML stacks with 20 bilayers of $Si_{0.50}C_{0.50}/Si_{0.77}C_{0.23}$ (6 nm/9 nm) was examined. The choice of the sublayer thickness leads to an overall Si content of 63% and allows a direct comparison between ML and $Si_{0.63}C_{0.37}$ SL.

Three different types of ML and two different SL ($\approx$ 200 nm) for comparison were prepared:

- ML without CO_2 **ML0**
- ML with CO_2 flux during $Si_{0.77}C_{0.23}$ sublayer deposition **ML77**
- ML with CO_2 flux during $Si_{0.50}C_{0.50}$ sublayer deposition **ML50**
- SL $Si_{0.63}C_{0.37}$ with CO_2 **SL63**
- SL $Si_{0.77}C_{0.23}$ with CO_2 **SL77**

III. RESULTS AND DISCUSSION

The FTIR spectra in Fig. 1 show that the increase in CO_2 flux from 0.0 to 10.0 sccm leads to an increased vibration mode

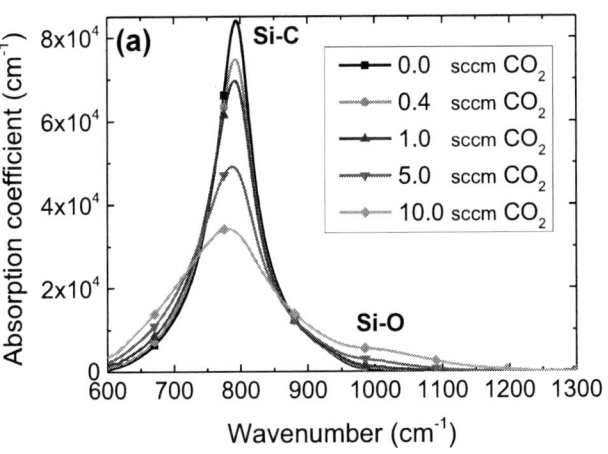

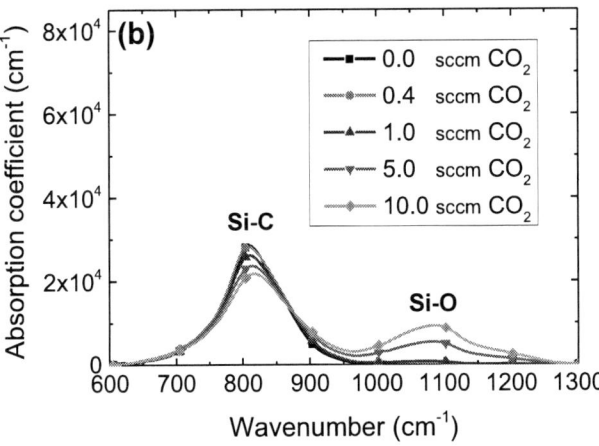

Fig. 1. FTIR spectra of (a) $Si_{0.50}C_{0.50}$ and (b) $Si_{0.77}C_{0.23}$ samples deposited with increasing amounts of CO_2 among the precursor gases and annealed for 30min@1100°C by furnace anneal. The increasing Si-O mode indicates the successful oxygen incorporation.

between 1000 and 1200 cm^{-1}. This result confirms the successful incorporation of O by CO_2 precursor gas as either Si-O or C-O vibrations are expected in this wavelength range

[12]. The increase of the vibration mode related signal for a certain CO_2 flux is stronger in the case of $Si_{0.77}C_{0.23}$ (Fig. 1(b)) than in $Si_{0.50}C_{0.50}$ samples (Fig. 1(a)). Therefore, it is assigned to Si-O vibrations as an increasing Si-O bond density with increasing Si content seems more obvious than an increasing C-O content with increasing Si content. It can be concluded from the SL experiments that O incorporation into the samples by adding CO_2 to the precursor gases was successful.

As a second step, the O-containing SL were combined to ML. In Fig. 2, SEM cross sections of the ML before and after annealing at 1100°C are shown.

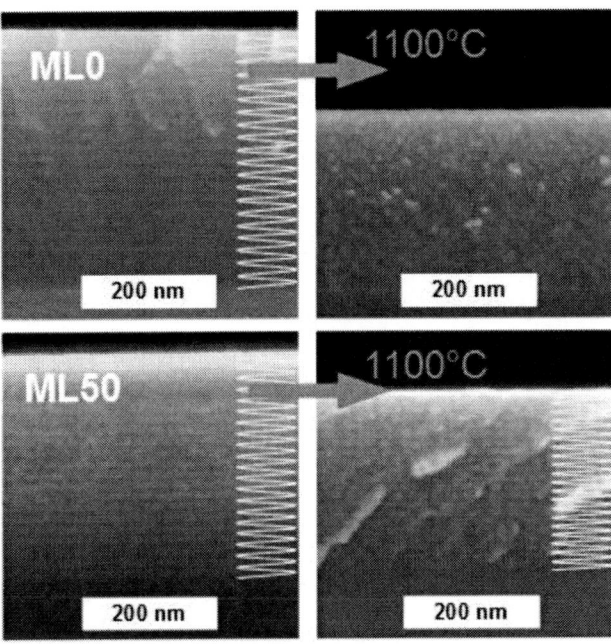

Fig. 2. SEM cross section images of ML deposited with and without CO_2 as precursor gas before and after annealing for 30min@1100°C by FA.

The ML with and without CO_2 show a layered structure in the SEM cross section before annealing. After annealing, the ML structure is lost in the ML0 samples as expected. However, in the case of ML with O, the ML structure is preserved during the annealing process as shown by the example of ML50 in Fig. 2. ML77 (not shown here) shows the same layered structure after annealing. For the first time, a ML structure with a Si content of only 77% in the Si-rich layers survived an annealing step of 30min@1100°C. It is concluded from this result that O in the samples hinders Si or C diffusion and therefore intermixing of the sublayers. FTIR spectra (not shown here) of as-deposited $Si_{0.50}C_{0.50}$ and $Si_{0.77}C_{0.23}$ samples prove the formation of Si-O bonds and show that approximately 60% of these bonds are already present after deposition. This is only a rough estimation as the Si-O vibration in the as-dep samples overlap strongly with the Si-H vibration. However, the large amount of Si-O bonds present prior to annealing are probably responsible for the reduced Si and C mobility during the annealing step. This interpretation is supported by the Raman and GIXRD results presented in Fig. 3(a) and (b), respectively. All normalized Raman spectra are plotted in Fig. 3(a). The black (◆) and the green (▨) spectra show SL63 and SL77, respectively. The brown (▲) spectrum represents ML0, and the orange (✳) and red (●) spectra represent the samples ML77 and ML50. Both ML have an overall Si content of 63%. The sample with the highest Si content (77%) clearly shows the most pronounced Raman c-Si peak (green spectra, ▨) and a c-Si peak position at higher wavenumbers than the other spectra. This can be explained by a larger fraction of crystalline Si present in the SL77 than in the

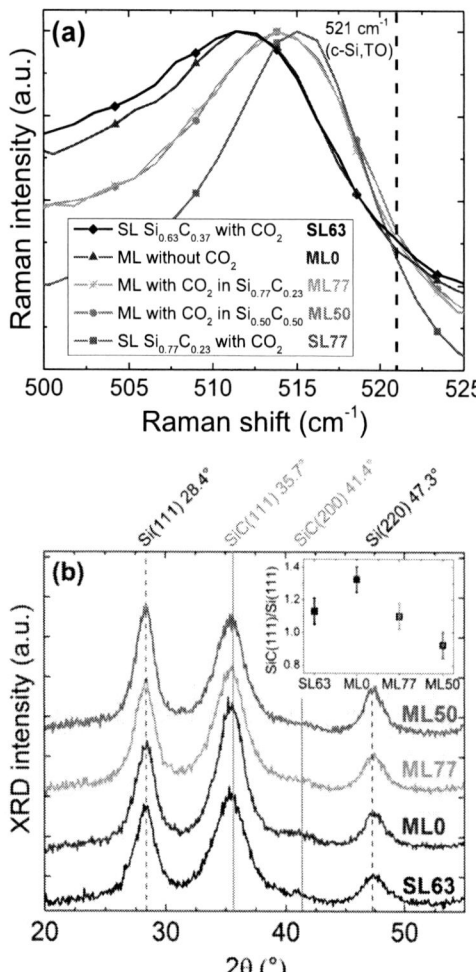

Fig. 3. Raman spectra (a) and GIXRD patterns (b) of ML with CO_2 (red) and without CO_2 (brown) annealed for 30min@1100°C by FA.

other samples. Raman results (not shown here) also prove that the Si crystallinity in $Si_{0.77}C_{0.23}$ SL samples is not influenced by

the O content in the layers. This is also true for the $Si_{0.63}C_{0.37}$ SL sample with and without CO_2. In Fig. 2(a) the SL63 sample clearly shows a smaller c-Si fraction than the SL77 sample. These results are not surprising so far. However, the comparison of ML0 and ML50/ML77 is still surprising as the ML0 shows a very similar c-Si peak at exactly the same peak position as the c-Si peak of SL63. This supports the assumption from Fig. 1 suggesting strong sublayers intermixing in the ML and therefore a behavior of ML0 after annealing which is quite similar to the properties of SL63 after annealing. Finally the Raman spectra of ML50 (red, ●) and ML77 (orange, ✳) are reviewed. The shape of the c-Si Raman modes and the peak positions lie between that of the ML0 and the SL77. This is an indication for a larger c-Si fraction in ML with CO_2 compared to ML without CO_2. A higher c-Si fraction requires a local Si concentration in the samples, which is higher than 63%. This strengthens the initial assumption that the O incorporation into the ML hinders sublayer intermixing.

In Fig. 2(b) the GIXRD pattern for all samples from Fig. 2(a) are depicted, except for SL77. At first sight, all GIXRD pattern look quite similar. However, the SiC(111)/Si(111) peak intensity ratios show significant differences as depicted in the inset of Fig. 2(b). The increase of SiC(111)/Si(111) for ML0 compared with SL63 could result from a slight difference in overall Si content in these two layers and will not be discussed further. Here the difference of the three ML will be examined. SiC(111)/Si(111) shows its highest value for ML0, decreases for ML77, and decreases further for ML50. This shows that O incorporation improved the c-Si to c-SiC ratio, either by the reduction of SiC crystallinity or by the increase of c-Si phase. It can be concluded that the Si-C bond density decreases starting with ML0 over ML77 down to the lowest value for ML50 from FTIR measurements (not shown here). As mentioned before, no clear statement about the SiC crystallinity is possible.

It seems to play a subordinated role if CO_2 is added to the $Si_{0.77}C_{0.23}$ or to the $Si_{0.50}C_{0.50}$ sublayers in the ML stack. It was just argued that the majority of the Si-O bonds are formed prior to annealing. Perhaps diffusion of the unbound O from the $Si_{0.50}C_{0.50}$ to the $Si_{0.77}C_{0.23}$ layers occurs during annealing. Di Ventra et al. [13] calculated that the activation energy for diffusion of a single O atom in 3C-SiC is only 1.7 eV, compared to 2.5 eV in bulk Si. However, it is doubtful if these values apply for $Si_{0.50}C_{0.50}$ compared to $Si_{0.77}C_{0.23}$ layers. The second possible reason is the deposition process itself. The regulation of the CO_2 flux at the PECVD tool is much slower than the regulation of the other precursor gases. As the deposition of a single sublayer takes less than 50 s, it is assumed that there is a certain amount of CO_2 molecules in the plasma during the whole process, leading to an incorporation of O in the whole ML samples.

Combining FTIR, GIXRD and Raman results, it can be concluded that from ML0 to ML77 the c-Si phase increases, whereas the c-SiC phase decreases due to the O incorporation.

For ML50, the c-SiC phase decreases further whereas the c-Si phase stays unchanged. The decrease in c-SiC phase is probably due to the competition between Si-O and Si-C bonds. The increase of c-Si phase due to O incorporation is probably due to an already mentioned hindering effect of O on Si diffusion. Therefore, the local Si density in the ML stays higher during annealing than without O and more c-Si phase forms - probably in the form of larger Si NC. However, this difference in Si NC size stays an assumption because it is too small to be proven experimentally. The second possible explanation for an increased c-Si phase could be a higher number of Si NC in ML with O, perhaps because the incorporated O acts as a nucleation seed.

The promising results of ML surviving the annealing step due to O incorporation will be investigated further by a sublayer thickness variation to examine if size control can be achieved by this method. Additionally, transmission electron microscopy measurements should be conducted to verify the maintenance of the ML and to show the Si NC size distribution.

IV. SUMMARY

Usually, Si NC/SiC ML structures show strong intermixing during the annealing step and are therefore not suitable for Si NC size control. In this work it is shown that the intermixing in the ML structure can be reduced by oxygen (O) incorporation during deposition. Furthermore, the formation of Si-O bonds and the increase of c-Si to c-SiC phase were observed. Although there is not enough data so far to prove that these changes of the ML with the help of O is accompanied by Si NC size control, by Si NC surface passivation, and by quantum confinement, this approach promises to fulfill those demands and will therefore be pursued further.

REFERENCES

[1] F. Meillaud, A. Shah, C. Droz, E. Vallat-Sauvain, and C. Miazza, "Efficiency limits for single-junction and tandem solar cells," *Sol. Energ. Mat. Sol. Cells*, vol. 90, no. 18-19, pp. 2952–2959, 2006.

[2] G. Conibeer *et al.,* "Silicon nanostructures for third generation photovoltaic solar cells," *Thin Solid Films*, vol. 511-2, pp. 654–662, 2006.

[3] C.-W. Jiang and M. A. Green, "Silicon quantum dot superlattices: Modeling of energy bands, densities of states, and mobilities for silicon tandem solar cell applications," *J. Appl. Phys.*, vol. 99, no. 11, p. 114902, 2006.

[4] C. Summonte *et al.,* "Silicon nanocrystals in carbide matrix," *Sol. Energ. Mat. Sol. Cells*, vol. 128, pp. 138–149, 2014.

[5] K. Ding *et al.,* "Annealing induced defects in SiC, SiO$_x$ single layers, and SiC/SiO$_x$ hetero-superlattices," *Phys. Status Solidi A*, published online, 2012.

[6] D. Song *et al.,* "Structural characterization of annealed Si(1-x)C(x)/SiC multilayers targeting formation of Si nanocrystals in a SiC matrix," *J. Appl. Phys.*, vol. 103, p. 83544, 2008.

[7] M. Zacharias *et al.,* "Thermal crystallization of amorphous Si/SiO2 superlattices," *Appl. Phys. Lett.*, vol. 74, no. 18, pp. 2614–2616, 1999.

[8] C. Weiss, M. Schnabel, A. Reichert, P. Löper, and S. Janz, "Structural and optical properties of silicon nanocrystals embedded in silicon carbide: Comparison of single layers and multilayer structures," *Appl. Surf. Sci.*, vol. 351, pp. 550–557, 2015.

[9] C. Summonte *et al.,* "Growth and characterization of Si nanodot multilayers in SiC matrix," in *23rd European Photovoltaic Solar Energy Conference: Proceedings*, 2008, pp. 730–733.

[10] M. Schnabel *et al.,* "Absorption and emission of silicon nanocrystals embedded in SiC: Eliminating Fabry-Pérot interference," *J. Appl. Phys.*, vol. 117, no. 4, p. 45307, 2015.

[11] Y. Kurokawa, S. Yamada, S. Miyajima, A. Yamada, and M. Konagai, "Effects of oxygen addition on electrical properties of silicon quantum dots/amorphous silicon carbide superlattice," *Curr. Appl. Phys.*, vol. 10, no. 3, S435–S438, 2010.

[12] H. Guenzler and H. M. Heise, *IR-Spektroskopie,* 3rd ed. Weinheim: VCH Verlagsgesellschaft, 1996.

[13] M. Di Ventra and S. T. Pantelides, "Atomic-Scale Mechanisms of Oxygen Precipitation and Thin-Film Oxidation of SiC," *Phys. Rev. Lett.*, vol. 83, no. 8, pp. 1624–1627, 1999.

Design of Cascaded Heterostructured p-i-i-n CdS/CdSe Low Cost Solar Cell

M. Zinaddinov, S. Mil'shtein

Advanced Electronics Technology Center, UMass Lowell, ECE Dept., MA, 01854, USA

Abstract — **Dominant presence of silicon in solar cells manufacturing is defined by maturity of its technology, and low cost of thin film solar cells made from this material. Better efficiency of cascaded solar cells compared to tandem devices performance was demonstrated in our previous work. It was previously demonstrated that some II-VI compounds are less costly than silicon and very efficient when used in solar cells. We present a novel p-i-i-n cell made of wide-gap, Eg=2.42eV, CdS and smaller energy gap, Eg=1.74eV, CdSe. The solar cell consists of top 0.1μm thick acceptor-doped ($2*10^{18}cm^{-3}$) CdS layer followed by 2μm of intrinsic CdS; next, CdSe intrinsic region of 2μm, followed by 1μm of donor-doped ($10^{16}cm^{-3}$) CdSe bottom layer. The energy diagram of cascaded solar cell illustrates favorable conditions for collection of both photo-electrons and photo-holes. Presence of wide-energy-gap CdS at the top of the device minimizes significantly the heat-up effect of following layers of CdSe. Solar cell demonstrated following performance characteristics: open-circuit voltage, Voc=1.45V; short-circuit current density, Jsc=23.36 mA/cm²; filling factor, FF=75%; efficiency equal 18.38%. The parameters listed are presenting operational of a very small segment of a solar cell. The surface area of the segment is 1μm². Our future study will be focused on design and production of CdS/CdSe solar cell deposited on thin CuO film.**

Index Terms — **heterostructure solar cell, cascaded structure, photovoltaic cells, cadmium sulfide, cadmium selenide.**

I. INTRODUCTION

Successful development of efficient solar cells is built nowadays on combination of harvesting of wider range of solar energy and design of multi semiconductor layers with viable energy gap. Thus, most of more sophisticated solar cells are using either tandem or hetero-structure designs. The success of different semiconductor materials is well reflected in recent Fraunhofer market study [1]. According to Fraunhofer report [1] the II-IV semiconductor compounds such as CdS, CdTe and/or CdSe started recently to prevail in competition with Silicon mostly due to the low cost of thin films of these compound semiconductors. The quantum dots made of CdS and CdSe already demonstrated 10- 17% efficiency [2]. The development of CdTe/CdS solar cells on flexible substrates with efficiency of 11% was reviewed recently [3]. Photovoltaic (PV) solar cells based on cadmium telluride (CdTe) represent the largest segment of commercial thin-film module production worldwide. Recent improvements have matched the efficiency of multicrystalline silicon while maintaining cost leadership. The United States is the leader in CdTe PV manufacturing, and NREL [4] has been at the forefront of research and development (R&D) in this area. In Europe, Fraunhofer research Institute demonstrated in joint German – French development of tandem SOI solar cells with concentrators efficiency of 46% [1]. The efforts to improve manufacturing technology of CdS and CdSe solar cells continue [5].

The deficiency of Si defined mostly by low absorption coefficient [6], what requires usage of thick layers of the material. The relatively small energy gap of Si leads to significant overheating [7] of the solar panels, which in turn decreases the operational efficiency of these widely used panels. Efficiency of silicon-based solar cells was significantly improved by design of a heterostructure of a-Si/c-Si. Research groups across the world [8-10] did prove that usage of wide gap amorphous silicon provides much higher efficiency compared to polysilicon or single crystal silicon solar cells. Our design of heterostructured thin-film silicon solar cells was different from any other design [11-12]. Offering the hetrostructure design with novel configuration p-i-i-n, we secured light absorption and generation of photo carriers in two intrinsic regions, where recombination of these carriers is significantly suppressed. The top p-region was designed to be very thin. Similar design ideas are explored in our novel CdS/CdSe p-i-i-n solar cell. Compering results of our recent design [13] of CdS/CdSe 15.6% efficient solar cell with cascaded p-i-n configuration we tend to believe that 18.5% efficiency of heterostructured CdS/CdSe p-i-i-n solar cell is achieved due to presence of double intrinsic layers. In current work we discuss the necessity to insert a buffer layer between the intrinsic regions to match lattice constants of CdS and CdSe. The cost efficiency of novel design is also assessed.

II. MODELING & DESIGN

A. Solar Spectrum Analysis

The structure consists of CdS and CdSe cascaded cell in a p-i-i-n configuration. Heterostructure solar cell with two intrinsic regions of different band gap allows to harvest photons more efficiently. Photons with shorter wavelengths are collected in the first CdS i- region with wide bandgap (Eg=2.42eV), and the remaining electrons are collected in the next intrinsic region made of CdSe (Eg= 1.74eV). This improves the thermal characteristics of the cell because high-energy photons do not generate much of heat up at the following layer of the semiconductor with a small energy band. Figure 1 shows the energy band diagram of the novel cell. Our calculations showed that 30% of the light within a spectrum of 0.4 - 0.71 μm. is collected by the first region and 70% by the bottom. The simulations also showed that the number of generated electron hole pairs was in the order of 10^{17} photons/s×cm^2 which corresponds to the number of photons in the given range.

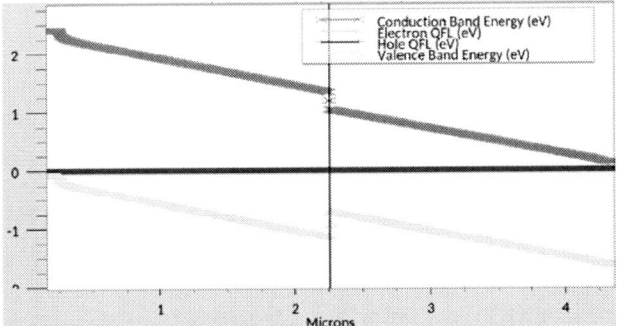

Fig. 1. Energy bands diagram of the CdS/CdSe cascaded p-i-i-n solar cell.

B. Layer Thicknesses and Doping

Design choices of the region thicknesses are described in this section. Optimal thicknesses configuration was expected to have a relatively smaller p-region and large i-regions. This is due to the fact that all electron-hole pairs created in that region will immediately recombine unlike inside of a p-i-i-n junction where carriers will be pulled apart by the build-in field. To maximize the light absorption, it is advantageous to have intrinsic regions relatively thick. If they are too thick, however, the strength of the build-in field decreases, in turn decreasing efficient harvesting of solar energy. Simulations showed that the optimal thicknesses for the regions are 0.1 μm for p-doped CdS, and 2 μm for both intrinsic CdS and CdSe. Our calculations with the extremely small lifetime of carriers in CdSe [14] motivated us to use the thickness of 0.1 μm for n-doped CdSe to decrease electron-hole recombination in that region and make the n- region smaller than the diffusion length, while keeping it thick enough for potential CVD manufacturing processes.

Design choices regarding the doping configuration where governed mainly by a need to have built-in potential across the junction as large as possible, while keeping the recombination of majority and minority carriers in p and n regions as small as possible to achieve maximum efficiency. Our modeling showed that having 2.3×10^{-18} cm^{-3} of acceptors in p-doped CdS region with 10^{-16} cm^{-3} donors in n-doped CdSe region provides optimal solar cell efficiency. In this simplified model, it was assumed that the regions are doped uniformly.

III. MODELING RESULTS

Figure 2 shows the I -V curve of the designed cell and characteristics of the simulated solar cell are shown in Table 1.

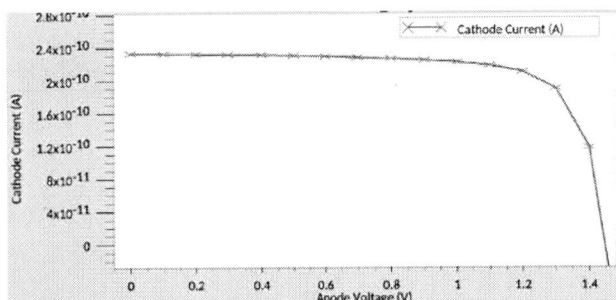

Fig. 2. IV relationship of the CdS/CdSe cascaded p-i-i-n solar cell.

TABLE I
CdS/CdSe SOLAR CELL CHARACTERISTICS

Jsc	23.36 mA*cm^{-2}
Voc	1.449 V
Pm	25.37 mW*cm^{-2}
Vm	1.200 V
Im	21.15 mA*cm^{-2}
FF	74.99%
Eff	18.38%

Notice that the modeled cross section of the device was set to be 1 μm^2, so the current values that are shown on the graph should be converted into current densities to be compared to other solar cells.

IV. ECONOMIC EVALUATION

The prices for raw materials being compared demonstrate that Si is about twice more expensive than CdS and one and a half times more expensive than CdSe [15-17]. Thin-film processing, however, depends on technology steps used, i.e. sputtering of Si is still less expensive than sputtering of CdS [18-19]. In the production of single crystals using epitaxial technology, Si appears to be more expensive [20-21].

V. CONCLUSIONS

The novel heterostructured CdS/CdSe p-i-i-n solar cell was designed to demonstrates the efficiency equal 18.4% with Voc=1.45V and short-circuit current density Jsc=23.36 mA/cm^2, which is better than designed by our group in 2012 solar cell from the same materials [13]. The low cost of thin films [20-21] of CdS and CdSe compared to Si is a major factor moving technology of II- IV compound semiconductors to the forefront of solar cells market. The complexity with p-type doping of CdS [22-24] was and still is a subject of studies, however it would be resolved with selection of epitaxial production technology. We are planning to select epitaxial technology for production of these solar cells in the nearest future. Depending on a selected technology we will design the buffer layer of graded CdS – CdSe, which should not exceed 30 – 40 Angstroms in thickness.

REFERENCES

[1] Fraunhofer Institute for Solar Energy Systems, ISE, "Photovoltaics Report", 17 November 2016

[2] Park H, Lee J, Park T, Lee S, and Yi W, "Enhancement of Photo-Current Conversion Efficiency in a CdS/CdSe Quantum-Dot-Sensitized Solar Cell Incorporated with Single-Walled Carbon Nanotubes" J Nanosci Nanotechnol. Vol15, 2, pp1614-7, 2015

[3] Xavier Mathew, J. Pantoja Enriquez, Alessandro Romeo, Ayodhya N. Tiwari, "CdTe/CdS solar cells on flexible substrates, Solar Energy", Volume 77, Issue 6, 2004, Pages 831-838

[4] National Renewable Research Laboratory, "Cadmium Telluride Solar Cells," Online, Accessed June 2017

[5] Wug-Dong Park, "Nanocrystalline CdS thin films prepared by chemical bath deposition," Nanotechnology Materials and Devices Conference, 2006. NMDC 2006. IEEE, vol.1, no., pp.460-461, 22-25 Oct. 2006

[6] S. M. Sze, "Physics of semiconductor devices"

[7] Joseph J. Wysoki, Paul Rappaport, "Effect of temperature on Photovoltaic Solar Energy conversion", Jour. Of App. Physics, Vol. 31, No. 3, March 1960.

[8] S. De Wolf , A. Descoeudres, Z. C. Holman, C. Ballif, "High-efficiency Silicon Heterojunction Solar Cells: A Review", Green, Vol. 2, Iss. 1, pp 7–24, 2012

[9] S. De Wolf, B. Demaurex, A. Descoeudres, and C. Ballif, "Very Fast Light-induced Degradation of a-Si:H/c-Si(100) Interfaces, Phys. Rev. B 83, 233301, 2011

[10] Z. C. Holman, A. Descoeudres, L. Barraud, F. Z. Fernandez and J. P. Seif et al. "Current Losses at the Front of Silicon Heterojunction Solar Cells", IEEE Jour. of Photovolt., vol. 2, # 1, p. 7-15, 2012

[11] S. Mil'shtein, M. Zinaddinov, N. Tokmoldin and S. Tokmoldin, "Design and Fabrication Steps of Silicon Heterostructured p-i-i-n Solar Cell with Corrugated Surface", Proceed. IEEE 43rd Intern. Confer. on Photovolt. Specialists, A26, 96, Aug., 2016.

[12] S. Mil'shtein, M. Zinaddinov, "Cascaded Heterostructured a-Si/c-Si Solar Cell with Increased Current Production"", Proceed. IEEE 43rd Intern. Confer. on Photovolt. Specialists, E 45,671, Aug., 2016.

[13] S. Mil'shtein, A. Pillai, S. Sharma, and G. Yessayan, "Design of Cascaded Low Cost Solar Cell with CuO Substrate",Intern. Confer. Phys. Semicond., ICPS-2012

[14] S.K. Tripathi, Alaa S. Al-Kabbi, Kriti Sharma, G.S.S. Saini, Mobility lifetime product in doped and undoped nanocrystalline CdSe, Thin Solid Films, Volume 548, 2013, Pages 406-410, ISSN 0040-6090, http://dx.doi.org/10.1016/j.tsf.2013.09.008.

[15] Sigma Alrich, "Silicon", Online, Accessed June 2017 [http://www.sigmaaldrich.com/catalog/product/aldrich/267414?l ang=en&region=US]

[16] Sigma Alrich, "Cadmium Selenide", Online, Accessed June 2017 [http://www.sigmaaldrich.com/catalog/product/aldrich/244600?l ang=en&region=US]

[17] Sigma Alrich, "Cadmium Sulfide", Online, Accessed June 2017 [http://www.sigmaaldrich.com/catalog/product/aldrich/217921?l ang=en&region=US&cm_sp=Insite-_-recent_fixed-_-recent5-2

[18] Kurt J. Lesker Company, "Cadmium Sulfide (CdS) Sputtering Targets", Online, Accessed June 2017 [http://www.lesker.com/newweb/deposition_materials/depositio nmaterials_sputtertargets_1.cfm?pgid=cd3]

[19] Kurt J. Lesker Company, "Silicon (Si) Sputtering Targets", Online, Accessed June 2017 [http://www.lesker.com/newweb/deposition_materials/depositio nmaterials_sputtertargets_1.cfm?pgid=si1]

[20] Barnett, A.M.; Rothwarf, A., "Thin-film solar cells: Aunified analysis of their potential", Electron Devices, IEEE Transactions on ,27, no.4, pp. 615- 630, Apr 1980

[21] H. Ullal,"CdTe PV", Report, National Solar Technology Roadmap, 2007

[22] R. Xie, J. Su, M. Li, L. Guo, "Structural and Photoelectrochemical Properties of Cu-Doped CdS Thin Films Prepared by Ultrasonic Spray Pyrolysis", Int. J. Photoenergy, Vol. 2013, 7 pages, Artile ID 620134, 2013

[23] K. Poornima, K. Gopala Krishnan, B. Lalitha, M. Raja, "CdS quantum dots sensitized Cu doped ZnO nanostructured thin films for solar cell applications, Superlattices and Microstructures", Volume 83, 2015, Pages 147-156, 2015

[24] M. Muthusamy, S. Muthukumaran, "Effect of Cu-doping on structural, optical and photoluminescence properties of CdS thin films, Optik - International Journal for Light and Electron Optics", Volume 126, Issue 24, 2015, Pages 5200-5206, 2015

Fast C-V method to mitigate effects of deep levels in CIGS doping profiles

P. K. Paul[1], J. Bailey[2], G. Zapalac[2], and A. R. Arehart[1]

[1]Electrical and Computer Engineering, The Ohio State University, Columbus, OH USA
[2]MiaSolé Hi-Tech Corp., Santa Clara, CA, USA

Abstract — **In this work, methods to determine more accurate doping profiles in semiconductors is explored where trap-induced artifacts such as hysteresis and doping artifacts are observed. Specifically in CIGS, it is shown that this fast capacitance-voltage (C-V) approach presented here allows for accurate doping profile measurement even at room temperature, which is typically not possible due to the large ratio of trap concentration to doping. Using deep level transient spectroscopy (DLTS) measurement, the deep trap responsible for the abnormal C-V measurement above 200 K is identified. Importantly, this fast C-V can be used for fast evaluation on the production line to monitor the true doping concentration, and even estimate the trap concentration. Additionally, the influence of high conductance on the apparent doping profile at different temperature is investigated.**

Index Terms — **CIGS, doping concentration, deep level, capacitance voltage measurement.**

INTRODUCTION

Accurate measurement of doping profiles is essential for accurate solar cell production, optimizing solar cell performance, and proper modeling and characterization. Typically, people use Hall measurement or capacitance-based approaches such as capacitance-voltage (C-V) or drive-level capacitance profiling (DLCP) to extract doping profiles [1-3]. However, the extracted doping can be influenced by deep levels and interface states[4-6]. In CIGS in particular, people use low temperature C-V or DLCP measurements. However, all of these techniques have their advantages and disadvantages [2,7,8]. For instance, Hall effect is a lateral technique, cannot be performed on actual solar cell structures [9]. The doping extracted from Hall measurement also does not provide any depth dependence, so non-uniformly doped samples can be problematic. Room temperature C-V measurements provides the depth dependant doping profile but the extracted doping profile can show large trap-induced hysteresis behavior [6,7]. Low temperature C-V measurements can successfully mitigate the hysteresis by freezing the effects of interface and bulk deep levels to measure the accurate doping concentration but it requires special equipment and longer total experiment times due to the cooling and heating [8]. Room temperature and low temperature DLCP measurements are used to eliminate the overestimation of doping concentration due to deep levels but DLCP measurements requires larger number of data acquisition and processing compared to C-V measurements and can not measure accurate doping profiles in non-uniform devices [10].

In CIGS solar cells, U-shaped doping profiles are commonly observed and suspected to be at least partially influenced by deep levels [7,8]. Some people use low temperature to slow down and avoid the trapping effects, and while this works it is more difficult and a priori knowledge of the defect time constants is required to ensure the trap emission is sufficiently slow at the measurement temperature [5]. Interpreting the actual doping profile from the temperature dependent extracted apparent U-shape doping profile, is a matter of debate [5,7]. Some groups consider the minimum point of the doping profile as the actual doping concentration [5] while others consider the highest reverse bias doping as the actual doping concentration [7]. Therefore, there is a strong need to understand the true doping profile.

In this paper, the deep trap responsible for the hysteresis in room temperature C-V is identified using deep level transient spectroscopy and a fast C-V measurement technique is proposed to avoid the influence of traps and accurately measure the doping profile even at room temperature. The trap-induced hysteresis and erroneous doping profiles is not limited to CIGS. Any semiconductor material system where the trap density is comparable to the doping density is subject to these issues, and the fast C-V approach is potentially applicable to all of these materials to achieve accurate doping profiles.

APPROACH

In this study, CIGS solar cells were grown by a roll-to-roll sputter deposition process on a flexible stainless steel substrate by MiaSolé [11]. First, the Mo metal back contact was deposited on the steel substrate followed by the sputter deposition of the CIGS absorber layer. Finally, the CdS buffer layer and transparent conducting oxide window layer were deposited. Then Ni/Al/Ni Ohmic top contacts were evaporated on the aluminum doped zinc oxide (AZO) and the devices were physically circumscribed to isolate approximately 2 mm^2 devices.

The fast C-V measurements were performed with an Agilent function generator and Boonton 7200 capacitance meter with a 100 kHz bandwidth. A triangular voltage ramp was applied to the device with variable ramp rates, and the capacitance was measured with a 1 MHz 30 mVp-p AC signal. DC voltage from -1.0 and up to 0.3 V were used during C-V measurements, and the capacitance, conductance, and voltage were simultaneously recorded with a National Instruments data acquisition card. Net doping profiles (N) were calculated using [1]

$$N = \frac{-c^3}{q\varepsilon_s A^2 \left(\frac{dC}{dV}\right)} \qquad (1)$$

978-1-5090-5606-4/17 $31.00 © 2017 IEEE

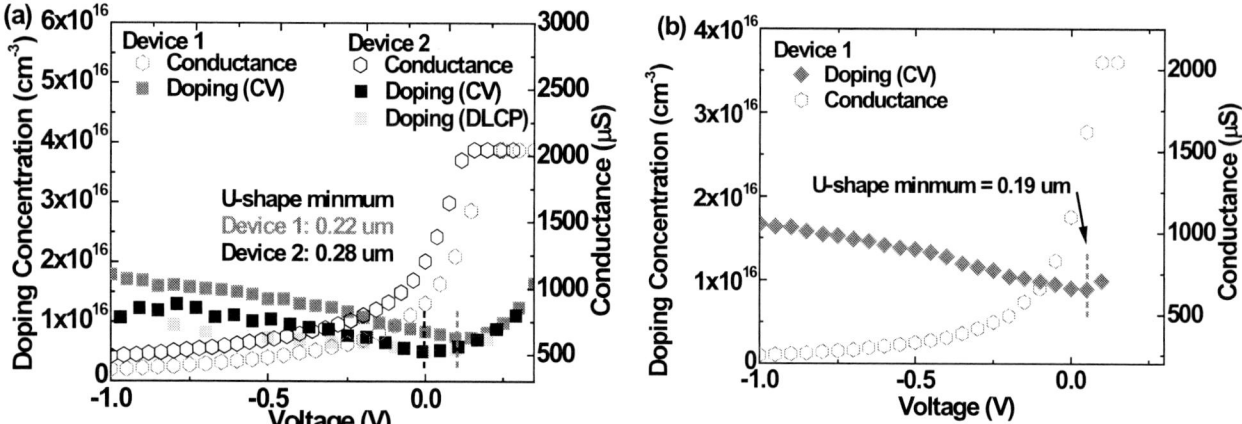

Figure 1: (a) Measured doping profiles by C-V using a 2 mV/s DC sweep rate and DLCP and conductance for two devices on the same sample at 110 K. The difference in devices is primarily in the amount of conductance. The DLCP and C-V extracted doping match quite well, but Devices 1 and 2 show different locations for the U-shaped doping, which likely arises in these devices because of the high conductance in forward bias causes artifacts in the measured capacitance – the doping begins to rise when the conductance approaches the maximum value of the meter (2047 µS). (b) Measured doping profiles by C-V and conductance for Device 1 at 210 K. the depletion depth at which the U-shaped minimum occurs has shifted with temperature suggesting it is not real. Like (a), the position of the U-shaped minimum is related to the conductance rise.

where A is the device area, ε_s is the permittivity, and q is the elementary charge.

To characterize the deep traps, deep level transient spectroscopy (DLTS) was performed from temperature range 80K to 325 K [12-14]. The DLTS transients were analyzed using the double boxcar method with rate windows from 0.8 to 2000 s^{-1}. For the DLTS measurement, traps were filled with a +0.2 V pulse for 10 ms and the trap emission was recorded in reverse bias with a -1.0 V applied. The activation energy of the deep trap was determined by Arrhenius analysis and the concentration was determined from DLTS signal peak height accounting the lamda effect [1].

RESULTS AND DISCUSSION

In Fig. 1, the CIGS solar cell doping profiles were extracted using C-V and DLCP measurements for two devices. These measurements were performed at 110 K and 210 K to avoid any hysteresis effects observed at higher temperatures. The extracted doping profile shows the typical U-shape doping concentration in the CIGS absorber layer. The depletion depth for minimum of the U-shape varies with both temperature and device suggesting it is the result of the measurement equipment namely the high conductance [6]. To explore this, the conductance was simultaneously recorded with the capacitance, which is also shown in Fig. 1. In Fig. 1(a), the doping profiles for two devices from the same sample were measured where the only difference was nominally the magnitude of the conductance. Both C-V and DLCP show very similar trends for the U-shaped minimum. However, the depletion depth at which the minimum occurs shifts from 0.22 to 0.28 µm as the conductance increases, which is likely because the high conductance in forward bias will at some point corrupt the

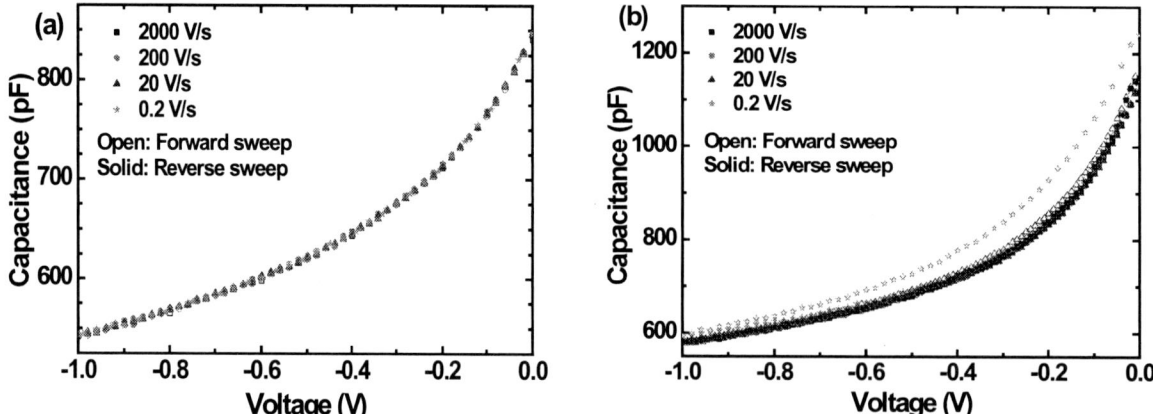

Figure 2: Measured C-V data measured at (a) 110 K and (b) 270 K with sweep rates of 2000 V/s, 200 V/s, 20V/s and 0.2 V/s where the reverse sweep was first followed by the forward sweep with no delay between the two directions. At 110 K no hysteresis is observed in the C-V measurements but at 270 K a signficant hysteresis is observed especially at the lowest sweep rate of 0.2 V/s.

978-1-5090-5606-4/17 $31.00 © 2017 IEEE

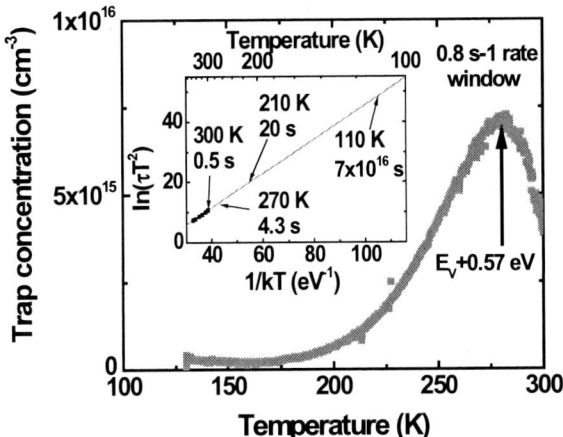

Figure 3: DLTS measurement on CIGS solar cell. DLTS spectra shows one dominant deep trap with activation energy Ev+0.57 eV. The inset shows the Arrhenius plot of the Ev+0.57 eV trap with estimated time constants for the various temperatures of measurement. The Ev+0.57 eV trap emission time constant τ is 0.5 s at 300 K, 4.3 s at 270 K, 20 s at 210 K, and $7x10^{16}$ s at 110 K. This confirms the traps are frozen at 110 K and can respond quickly to the slow C-V sweep rate at 300 K.

measured capacitance signal leading to erroneous extract doping profile.

To further explore this, Device 1 was measured at 210 K in Fig. 1(b) where the U-shaped minimum is reduced 0.03 μm compared with the 110 K case indicating again that conductance is likely playing a role in forward bias, but further work is needed to confirm this. Typically for high-fidelity capacitance measurements, it is desirable to have a Q factor (Q=ωC$_P$/G where ω is the AC angular velocity, C_P is the parallel capacitance, and G is the conductance) of 10, so as G increases Q decreases and the accuracy of the capacitance measurement decreases to the point where it cannot be trusted (Q < ~1) without additional verification. Forward bias is avoided in the rest of the C-V measurements to avoid these possible artifacts.

Usually, above 210 K the C-V measurements start to show hysteresis where the forward and reverse sweep capacitance do not match [8]. To understand how the hysteresis was affected by temperature and DC sweep rate, C-V measurements with forward and reverse sweeps were performed with different sweep rates at 110 K and 270 K, and the results are shown in Fig. 2. The 110 K C-V measurements in Fig. 2(a) show no distinguishable hysteresis at any sweep rate while the 270 K C-V measurements in Fig. 2(b) also show no hysteresis for DC sweep rates down to 20 V/s but shows large hysteresis of up to 100 pF at 0.2 V/s sweep rate. Traps were the most likely source of the hysteresis and the trap time constant depends exponentially on temperature. Therefore, defect spectroscopy was performed to identify the time constants and concentration of the deep levels.

Fig. 3 shows the DLTS spectra with one dominant trap with Ev+0.57 eV activation energy and a minimum trap concentration of $7x10^{15}$ cm^{-3}. Previously, with scanning-DLTS the Ev+0.57 eV was found to be spatially localized and located

only in specific intergrain regions [15-17]. The inset of Fig. 3 shows the Arrhenius plot of Ev+0.57 eV trap and the estimated trap emission time constants at several temperatures. The trap time constant τ_p follows [1]

$$\tau_p = \frac{1}{\sigma_p \langle v_{th} \rangle N_V} \exp\left(\frac{E_T - E_V}{kT}\right) \quad (2)$$

where σ_p is the hole capture cross section, N_v is the valence band density of states, v_{th} is the thermal velocity, $E_V - E_T$ is the trap energy relative to the valence band, T is the temperature, and k is Bolzmann's constant. The trap emission time constants were compared to the total C-V measurement time to determine if the traps had time to emit during the measurement. The total C-V measurement times were 1 ms, 10 ms, 100 ms, 10 s for the 2000, 200, 20 and 0.2 V/s sweep rates, respectively. At 110 K (Fig. 2a), the Ev+0.57 eV trap emission time constant is several orders of magnitude larger than the measurement time for all cases, so the traps could not response to the DC bias change and hence there was no observed hysteresis. In contrast, at 270 K (Fig. 2(b)) the trap emission time constant is 4.3 s so the 0.2 V/s sweep (10 s total measurement time) is longer then the trap time constant and therefore the trap cause hysteresis, which is also because the trap density is comparable to the doping density. Still, the 2000, 200, and 20 V/s sweep rates at 270 K were much faster than the trap time constant (4.3 s), so the traps could not respond and hence no hysteresis was observed Hence, we can conclude that the hysteresis behavior observed in the C-V is due to the Ev+0.57 eV trap.

Knowing the trap emission time constants from the Arrhenius plot, it is then possible to design C-Vs with fast and slow sweep rates as in Fig. 4, and then both the doping density and trap density can be measured. The total measurement time for the forward and reverse sweep t_m is,

$$t_m = \frac{2\Delta V}{r_{DC}} \quad (3)$$

where, r_{DC} is the DC bias sweep rate and ΔV is the measurement voltage. The maximum r_{DC} is limited by the capacitance meter bandwidth (BW) and is

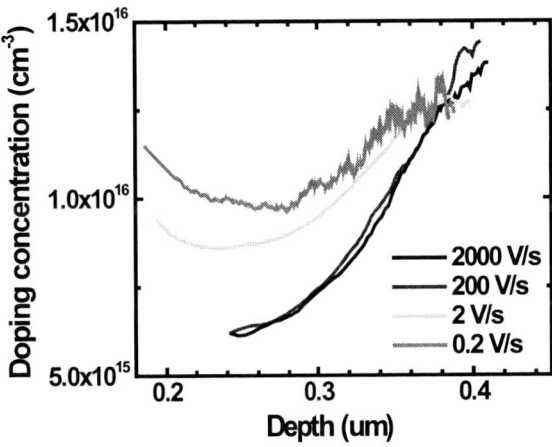

Figure 4: Measured doping profiles at 300 K for sweep rates listed in the legend.

$$r_{DC,max} = V_{res}BW \qquad (4)$$

where, V_{res} is the minimum voltage resolution and 20 mV was assumed here. With a bandwidth of 100 kHz. this gives an $r_{DC,max}$ of 2000 V/s for this setup. The maximum sweep rate can be experimentally confirmed as well by increasing r_{DC} until the C-V hysteresis begins to increase indicating a BW limited measurement. The minimum sweep rate ($r_{DC,min}$) is determined by the trap time constant. The total measurement time should be less than 10% of the trap time constant, so that most of the traps cannot emit during the measurement. This gives

$$r_{DC,min} = \frac{20\Delta V}{\tau_p} \qquad (5)$$

For the Ev+0.57 eV trap at 300 K, $r_{DC,min}$ is 40 V/s. For r_{DC} lower than 40 V/s, the trap-induced hysteresis would become visible and for r_{DC} higher than this the hysteresis would be negligible. Therefore,above 40 V/s traps cannot respond and the doping profiles will exclude any trapping effects (i.e. is only the doping) and would then overlap at all temperatures, which is shown experimentally in Fig. 4 confirming the theory. Finally, the frequency of the AC test bias, which measures the out-of-phase current to calculate the capacitance, should be much larger than $1/\tau_p$. Here, the AC frequency used is 1 MHz which is well above the minimum required frequency. By meeting all these requirements, a high-fidelity C-V measurement can be achieved and a doping profile without the effects of trapping can be measured.

Additionaly, it is possible with a fast and slow C-V to estimate the trap concentration. Using an r_{DC} less than $0.2\Delta V/\tau_p$ (i.e. 0.4 V/s at 300 K for the Ev+0.57 eV trap) will provide sufficiently slow rates such that the trap will stay in equilibrium with the applied bias, and the doping profile extracted will be the sum of the trap and doping concentrations. The fast C-V only measured the true doping profile, so the difference between the fast and slow C-V doping profiles is an estimate of the trap concentration. Comparing the 2000 V/s (fast) and 0.2 V/s (slow) doping profiles in Fig. 4, a trap density in the low- to mid-10^{15} cm^{-3} is estimated, which agrees well with the DLTS

result of 7×10^{15} cm^{-3} in Fig. 3. So, with a much faster and simpler measurement (fast and slow C-V) it is possible to get a quick and reasonably accurate estimates of both the trap concentration and doping profile at room temperature.

Finally, Fig. 5 shows the doping profile extracted from fast C-V using the highest 2000 V/s sweep rate for temperatures from 110 to 300 K. The extracted doping profiles are nominally identical over the whole temperature range indicating that traps are not influencing the doping profile at any of these temperatures. This indicates that the fast C-V approach to extract accurate doping profiles for CIGS or other material systems where the trap density is comparable with the doping density given the proper AC frequency and DC sweep rate. This also avoids the problem of needing to cool down thereby allowing for a simpler and cheaper setup to measure doping profiles.

CONCLUSIONS

Using this fast C-V method, more accurate doping profiles can be obtained at room temperature. The U-shaped doping profiles in these samples is observed using C-V and DLCP when the conductance reaches high values. These results suggest high device conductance negatively influences the accuracy of the extracted doping profile. It is demonstrated that comparing the high and low sweep rate doping profiles that the difference is comparable with the measured Ev+0.57 eV trap density suggesting this trap is primarily responsible for the difference in measured doping, and the time constants of the trap and total measurement time can explain the onset of the observed hysteresis. Finally, this simple method can be extended for quick measurements during production to monitor defect densities and doping.

Acknowledgements
The authors would like to thank the Department of Energy (Contract #DE-DD0007141) for financial support.

REFERENCES

1. D. K. Schroeder, *Semiconductor Material and Device Characterization*. New Jersey: John Wiley & Sons, Inc., 2006.
2. M. Islam, et al., "Effect of Se/(Ga+In) ratio on MBE grown Cu(In,Ga)Se2 thin film solar cell," *Journal of Crystal Growth*, vol. 311, pp. 2212-2214, 2009.
3. C. Michelson, A. Gelatos, and J. Cohen, "Drive-level capacitance profiling: Its application to determining gap state densities in hydrogenated amorphous silicon films," *Applied Physics Letters*, vol. 47, pp. 412-414, 1985.
4. L. Kimerling, "Influence of deep traps on the measurement of free-carrier distributions in semiconductors by junction capacitance techniques," *Journal of Applied Physics*, vol. 45, pp. 1839-1845, 1974.
5. G. Sozzi et al., "A numerical study of the use of C-V characteristics to extract the doping density of CIGS absorbers," 2016 IEEE 43rd Photovoltaic Specialists Conference (PVSC), Portland, OR, 2016, pp. 2283-2288.
6. S. Lany and A. Zunger, "Light- and bias-induced metastabilities in Cu(In,Ga)Se2 based solar cells caused by the (VSe-VCu) vacancy complex," *Journal of Applied Physics*, vol. 100, p. 113725, 2006.

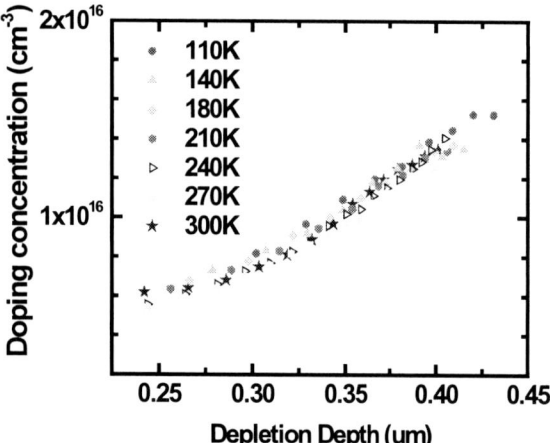

Figure 5: Measured doping profiles at different temperature with 2000 V/s sweep rate .

7. M. Cwil, M. Igalson, P. Zabierowski and S. Siebentritt, "Charge and doping distributions by capacitance profiling in Cu(In,Ga)Se$_2$ solar cells," *Journal of Applied Physics*, vol. 103, p. 063701, 2008.

8. J. Bailey, G. Zapalac, and D. Poplavskyy, "Metastable defect measurement from capacitance-voltage and admittance measurements in Cu(In, Ga)Se$_2$ Solar Cells," 2016 IEEE 43rd Photovoltaic Specialists Conference (PVSC), Portland, OR, 2016, pp. 2135-2140.

9. L.J. van der Pauw, "A method of measuring specific resistivity and Hall effect of discs of arbitrary shape," *Philips Res. Repts.*, vol. 13, pp. 1-9, Feb. 1958.

10. C. Warren, E. Roe, D. Miller, W. Shafarman and M. Lonergan, "An improved method for determining carrier densities via drive level capacitance profiling," *Applied Physics Letters*, vol. 110, p. 203901, 2017.

11. http://www.miasole.com.

12. A. R. Arehart, A. A. Allerman, and S. A. Ringel, "Electrical characterization of n-type Al$_{0.30}$Ga$_{0.70}$N Schottky diodes," Journal of Applied Physics, vol. 109, p. 114506, 2011.

13. D. V. Lang, "Deep-level transient spectroscopy: A new method to characterize traps in semiconductors," *Journal of Applied Physics*, vol. 45, pp. 3023-3032, 1974.

14. P. K. Paul, K. Aryal, S. Marsillac, S. A. Ringel, and A. R. Arehart, "Impact of the Ga/In ratio on defects in Cu (In, Ga) Se$_2$," 2016 IEEE 43rd Photovoltaic Specialists Conference (PVSC), Portland, OR, 2016, pp. 2246-2249.

15. P. K. Paul, et al., "Direct nm-Scale Spatial Mapping of Traps in CIGS," *IEEE Journal of Photovoltaics*, vol. 5, pp. 1482-1486, 2015.

16. P. K. Paul, K. Aryal, S. Marsillac, S. A. Ringel, and A. R. Arehart, "Identifying the source of reduced performance in 1-stage-grown Cu(In, Ga)Se$_2$ solar cells," 2016 IEEE 43rd Photovoltaic Specialists Conference (PVSC), Portland, OR, 2016, pp. 3641-3644.

17. S. Karki, et al., "In Situ and Ex Situ Investigations of KF Postdeposition Treatment Effects on CIGS Solar Cells," *IEEE Journal of Photovoltaics*, vol. 7, pp. 665-669, 2017.

18. F. Werner, T. Bertram, J. Mengozzi and S. Siebentritt, "What is the dopant concentration in polycrystalline thin-film Cu(In,Ga)Se$_2$?," *Thin Solid Films*, vol. 633, pp. 222-226, 2017.

Crystal Growth Phenomena in Polycrystalline (Cu)ZnTe/CdTe/CdS Via Molecular Dynamics

[1]Rodolfo Aguirre, [2]Jose J. Chavez, [2]Xiao W. Zhou, and [1]David Zubia

[1]The University of Texas at El Paso, El Paso, Texas, 79968, USA.

[2]Sandia National Laboratories, Livermore, CA 94550, USA.

Abstract — Molecular Dynamics (MD) simulations were applied to study the crystal growth phenomena in polycrystalline (Cu)ZnTe/CdTe/CdS heterostructures. Our results show that polycrystallinity, polytypism and Cu clustering are accurately predicted. The resulted films are zinc blende structure preferentially. The CdTe/CdS interface exhibits a high degree of disorder compared to the (Cu)ZnTe/CdTe interface. Stoichiometry plays an important role for the formation of Cu clusters and diffusion of Zn and Cu into the CdTe substrate. Dislocation motion is captured and analyzed.

I. INTRODUCTION

Material defects and stoichiometry variations impact the electronic properties and stability of the CdTe solar cell [1] [2]. Experimental techniques such as Time of Flight Secondary Ion Mass Spectroscopy (TOF-SIMS), Atomic Probe Tomography (APT), and Transmission Electron Microscope (TEM) have facilitated the study of defects and atomic distribution of the CdTe solar cell. For example, recent studies using TOF-SIMS have shown that the Phosphorous segregates into the CdTe grain boundaries [3]. In a separate study using APT, Cu has been found to form clusters at the ZnTe/CdTe interface [4]. Moreover, TEM micrographs have revealed some defects and dislocations at the CdTe [5]. To supplement the experimental work, we apply MD simulations to observe the formation and evolution of defects and growth of the (Cu)ZnTe/CdTe/CdS heterostructure. MD simulations offer a rapid, fast, inexpensive and unique solution to study the CdTe material at the atomic scale.

II. METHODOLOGY

The computationally intensive MD simulations were run using the Large-Scale Atomc/Molecular Massively Parallel Simulator (LAMMPS) [6] at the Gordon and Stampede computer clusters from the San Diego Supercomputer Center (SDSC) and the Texas Advanced Computing Center (TACC). A Stillinger-Weber potential [7] was used to define interatomic forces. Cu, Zn and Te were co-evaporated at vapor ratios of 0:1:1, 2:9:9 and 1:4:5 which nominally represent 0%, 10% and stoichiometric 10% Cu loadings, respectively. The polycrystalline CdTe substrate used for these simulations was grown on polycrystalline CdS. Prior to

that, the polycrystalline CdS was grown on an amorphous CdS substrate. Details of the preparation of the amorphous substrate and subsequent deposition of polycrystalline CdS and CdTe will be described elsewhere. RMS roughness of the polycrystalline CdTe surface was calculated to be 5.4 Å. Periodic boundary conditions were used in the plane direction (X and Z axes). Deposition temperature and rate were ~1200 K and ~0.002 A/ps, respectively. Total deposition time for the samples using 10% and stoichiometric 10% Cu loadings was 42.4 ns. In contrast, the total deposition time for the sample with 0% Cu loading was 21.2 ns.

Atomistic visualization, structural map and dislocation analysis was performed using the Open Visualization Tool (OVITO) [8].

III. RESULTS AND DISCUSSION

Cross-sectional atomic species and crystal structure maps of the three samples (0%, 10%, and S10%) are shown in Fig. 1. Figs. 1(a-c) show color coded dots indicating the position of the atoms whereas Figs. 1(d-f) show color coded dots indicating the crystalline structure associated with each atom. Green atoms are associated with a zinc blende (ZB) structure and red atoms are wurtzite. Atoms that were not determined to be either ZB or WZ are colored blue. In all three cases, the starting CdTe/CdS substrates were structurally identical but the (Cu)ZnTe depositions were distinct. However, the substrates were allowed to evolve distinctly during the (Cu)ZnTe deposition for each case.

Several phenomena are observed in the simulated films closely mimicking phenomena reported in experimental results. For example, polycrystallinity occurred remarkably similar to that observed in experimental films. This is indicated by grain boundary networks present in Figs. 1(d-f) which bound faulted grains that are oriented in a multitude of directions. The grains are predominantly zinc blende domains with stripped wurtzite regions denoting stacking faults.

Both disordered and high-symmetry (twinning) grain boundaries are present throughout the films. A preference towards disordered grain boundaries was detected at the CdTe/CdS interface which has a lattice mismatch of ~10%. This is indicated by the arrow in Fig. 1(f) emphasizing the blue (disordered) boundaries along the CdTe/CdS interface.

Moreover, there is minimal correlation between the orientations of the grains in CdS and CdTe suggesting a low degree of epitaxy. In contrast, a much higher degree of epitaxy is present at the (Cu)ZnTe/CdTe interface which has a lattice mismatch of ~6%. In this case, the grain domains extend beyond the (Cu)ZnTe/CdTe interface with minimal structural demarking of the metallurgical interface. White lines are used in Figs. 1(d-f) to mark the interface. We speculate that the nature of the interfaces will have significant ramifications on carrier transport across each junction.

Cu clustering occurred in the non-stoichiometric 10% sample which had an excess of group II atoms in the vapor flux consistent with experimental results in the literature [4]. A Cu cluster is indicated by an arrow in Fig. 1(b). Time evolution analysis of the data showed that Cu atoms start forming into clusters as early as 5.2 ns after the start of deposition. By 15.6 ns, the clusters increase in size starting with ~19 atoms and reaching a maximum of ~66 atoms. Thereafter, the cluster size remained constant until the end of the simulated deposition at 42.8 ns. In contrast, the Cu atoms were uniformly dispersed with no evidence of clustering in the stoichiometric case (Fig. 1. (c)) emphasizing the importance of using stoichiometric conditions during growth.

Intermixing and diffusion of the atoms occurred in all the samples. The arrows in Fig. 1(a) point to Zn and S atoms that have diffused into the CdTe. However a distinction is noted in that the S atoms diffused during the CdTe deposition whereas the Zn atoms diffused during the ZnTe deposition. This distinction is important because diffusion is generally greater in the growing film as compared to the substrate [9] [10]. This is corroborated in Fig. 1(a) which shows a greater number of S atoms in the CdTe compared to Zn.

Stoichiometry (and/or Cu clustering) and defects also appear to play important roles in the rate of diffusion. Although the CdTe/CdS substrates were structurally identical prior to (Cu)ZnTe growth, Fig. 1(b) clearly shows a greater degree of Zn diffusion into the CdTe in the non-stoichiometric 10% case compared to the 0% and stoichiometric 10% cases (Figs. 1(a) and (c)). Moreover, most of the diffusion occurred through grain boundaries and little to no diffusion in pristine areas.

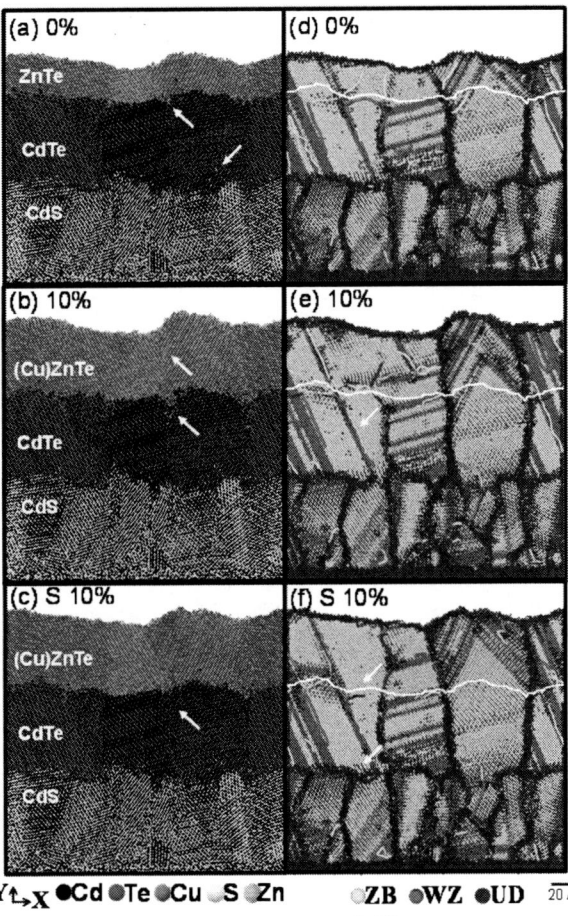

Fig. 1. Cross-sectional atomic species (a – c) and crystal structure maps (d – f) of (Cu)ZnTe deposited on polycrystalline CdTe/CdS using copper loading of 0%, 10%, and stoichiometric 10%, respectively.

Dislocation dynamics were also observed in the simulated films. The dislocations were usually associated with faults and interfaces in the films as observed in Figs. 1(d-f). In separate experiments, Shockley partials 1/6<112> were observed to be bounding stacking faults. For example, Fig. 1 (e) shows a Shockley partial dislocation located at a diagonal wurtzite strip as indicated by the arrow pointing to a green dislocation line. Moreover, ½<110> perfect dislocations were predominantly associated with interfaces and acted as misfit dislocations. 1/6<110> stair-rod and 1/3<111> Frank dislocations were also observed.

Fig. 2 shows the structure maps of the layers at different deposition times. Stacking faults projecting into the growth direction propagated into the growing (Cu)ZnTe layer as evidenced by the pair of staking faults highlighted by the arrow in Fig. 2(d) at 14 ns. However at ~21.1 ns (Fig. 2(f) the Shockley partials bounding the staking faults initiate glide in a manner that reduces the area of the stacking faults. The velocity of the Shockley partials is 0.2 Å/ps which is 2 orders of magnitude faster than the growth rate of the film. Another example of Shockley partial dislocation glide is observed to initiate at ~26 ns as indicated by the arrow in Fig. 2(g).

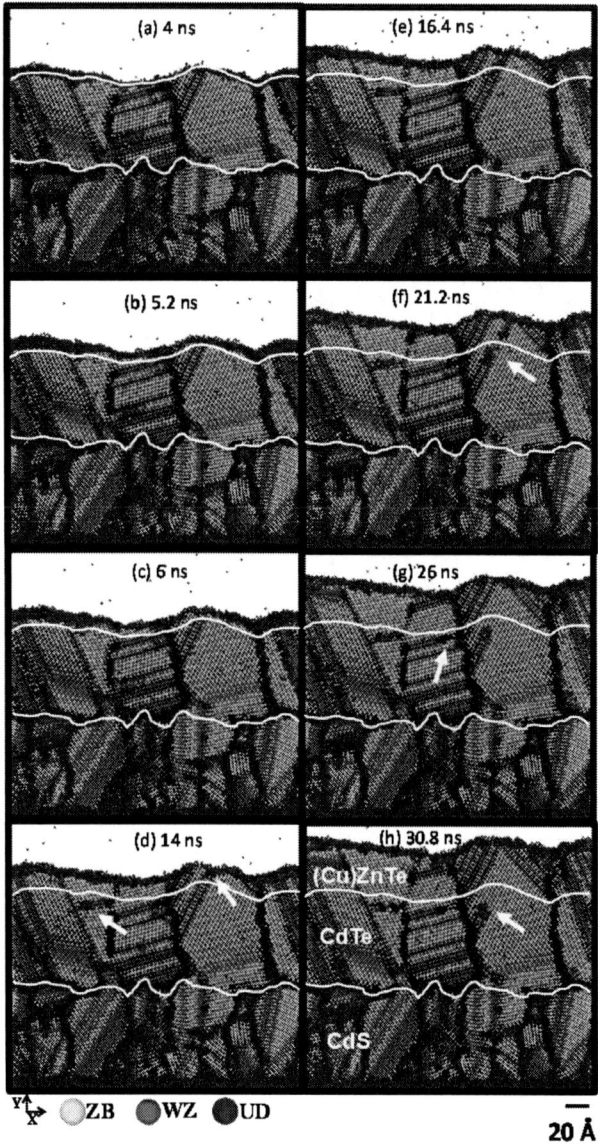

Fig. 2. Cross-sectional crystal structure maps of the films at several deposition times as indicated in the frames.

Finally dislocation dissociation was observed as shown in Fig. 3 which contains color coded lines indicating the location and Burgers vectors of the dislocations within the films. In Fig. 3(a) portions along a threading dislocation is observed to be dissociated into Shockley partials and another dislocation with undetermined Burgers vector as indicated by the arrow. However at 5.2 ns as shown in Fig. 3(b), the end of the dislocation dissociates into a stair-rod (pink) and perfect dislocation (blue). The perfect dislocation branch continues to

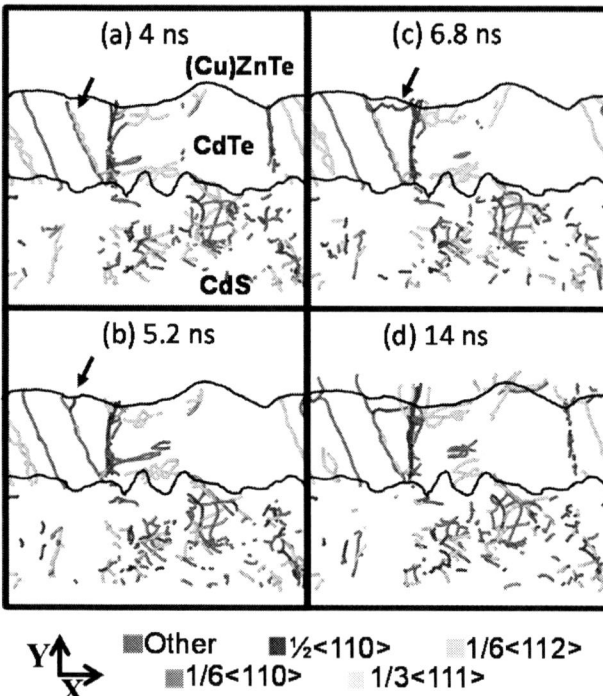

Fig. 3. Cross sectional view of the dislocation lines in the films. The color of the lines indicates the Burgers vectors.

grow and bends into the (Cu)ZnTe/CdTe interface as shown in Fig. 3(c). Presumably the dislocation was driven by strain to accommodate that lattice mismatch between the two layers. Thereafter the perfect dislocation remains stable and is also observed in Fig. 2(d) as indicated by the arrow.

IV. CONCLUSION

MD simulations were applied to study the crystal growth phenomena in polycrystalline CuZnTe/CdTe heterostructures. Generally speaking, the preferential structure is zinc blende in all the films. Extensive polycrystallinity, polytypism, and clustering is predicted in excellent agreement with reported experimental findings. The interfaces with the larger lattice mismatch (CdTe/CdS) exhibited much higher degrees of disorder compared to the (Cu)ZnTe/CdTe interfaces. A high degree of epitaxy was observed between (Cu)ZnTe and CdTe. Stoichiometry is an important growth condition as it affects clustering and diffusion into the substrate. Dislocation dynamics analysis indicates a dislocation mobility of ~100X faster than the growth rate. However suitable conditions are required to initiate glide in what otherwise might be pinned dislocations.

ACKNOWLEDGEMENT

This work was supported in part by the following grants: NSF/IGERT: DGE-0903670, DOE/BRIDGE:DE-EE0005958. Sandia National Laboratories is a multi-program laboratory managed and operated by Sandia Corporation, a whole owned subsidiary of Lockheed Martin Corporation, for the U. S. Department of Energy's National Security Administration under contract DE-AC094-94AL85000. This work used the computing resources provided by the Extreme Science and Engineering Discovery Environment (XSEDE) program, which is supported by National Science Foundation grant number ACI-1053575.

REFERENCES

[1] C. Li, "Understanding individual defects in CdTe thin-film solar cells via STEM: From atomic structure to electrical activity," *Materials Science in Semiconductor Processing,* 2017.

[2] M. Khan, "Stoichiometric effects in polycrystalline CdTe," 2014.

[3] E. Colegrove, "Phosphorous diffusion mechanisms and deep incorporation in polycrystalline and single-crystalline CdTe," *Physical Review Applied,* 2016.

[4] C. A. Wolden, "The roles of ZnTe buffer layers on CdTe solar cell performance," *Solar Energy Materials & Solar Cells,* vol. 147, pp. 203-210, 2016.

[5] Y. Yan, "Transmission electron microscopy study of dislocations and interfaces in CdTe solar cells," *Thin Solid Films,* pp. 7168-7172, 2011.

[6] S. Plimpton, "Fast parallel algorithms for short-range molecular dynamics," *Journal of Computational Physics,* vol. 117, pp. 1-19, 1995.

[7] T. A. W. Frank H. Stillinger, "Computer simulation of local order in condensed phases of silicon," *Physical Review B,* vol. 31, p. 1451, 1986.

[8] A. Stukowski, "Visualization and analysis of atomistic simulation data with OVITO - the Open Visualization Tool," *Modelling and simulation in materials science and engineering,* vol. 18, 2010.

[9] J. P. Enriquez, "S and Te inter-diffusion in CdTe/CdS hetero junction," *Solar Energy Materials & Solar Cells,* vol. 91, pp. 1392-1397, 2007.

[10] C. Li, "S-Te interdiffusion within grains and grain boundaries in CdTe solar cells," *Journal of Photovoltaics,* vol. 4, 2014.

Using High-Resolution Anomalous-Scattering X-Ray Diffraction to Observe Off-Stoichiometric Cu₂ZnSnS₄ Crystal Structures

Christopher J. Bosson[1], Max T. Birch[1], Douglas P. Halliday[1], Chiu C. Tang[2], and Peter D. Hatton[1]

[1]Department of Physics, The University of Durham, Durham, DH1 3LE, UK
[2]Diamond Light Source, Harwell Science and Innovation Campus, Didcot, OX11 0DE, UK

Abstract — Cu_2ZnSnS_4 (CZTS) is a promising material for the absorber layer in sustainable thin film solar cells, but its photovoltaic performance is currently limited by low open-circuit voltage, due in part to crystal structure disorder. High-resolution anomalous-scattering X-ray diffraction is shown in this investigation to be a very useful tool for elucidating disordered CZTS crystal structures. Three of five samples fabricated over a range of composition by solid state reaction displayed two distinct CZTS phases at room temperature, evident in minute splitting of some peaks due to different c/a lattice parameter ratios. These are attributed to different composition types of CZTS, defined by the prevalence of different charge-neutral defect complexes. This work is the first report of XRD from CZTS at high-enough resolution to distinguish these phases distinctly and the disorder type attribution is only possible because anomalous-scattering XRD can uniquely differentiate between copper and zinc site occupancy.

Index Terms — crystal structure, Cu_2ZnSnS_4, photovoltaic cells, X-Ray diffraction.

I. INTRODUCTION

Cu_2ZnSnS_4 (CZTS) is a promising solar cell absorber material with none of the problems suffered by more established materials, such as silicon, CdTe, and $Cu_2In_xGa_{1-x}Se_2$ (CIGS). Some such materials are expensive, for example silicon due to high manufacturing costs, and others due to element costs, particularly of Ga, In, and Te. Some contain elements not abundant enough to contribute electricity on the TW scale, such as Te, Se, and In. Some are toxic, such as Cd and Se. Some materials have a combination of these problems. [1]

CZTS, however, is composed of non-toxic, Earth-abundant, and low-cost elements. It is an intrinsically p-type direct gap semiconductor with near-optimal band gap of 1.4-1.5 eV and high absorption coefficient $>10^4$ cm^{-1}. It currently has a best efficiency of 9.2 %, [2] and 12.6 % as the toxic, more expensive, and less abundant selenium-containing CZTSSe. [3]

The efficiency of CZTS cells must be approximately doubled if they are to be adopted commercially. In order to do so, the low open-circuit voltage commonly reported must be increased. Disorder in the crystal structure and electrostatic potential fluctuations due to such disorder are likely to be limiting this. [4,5]

CZTS also has a complex phase diagram with a narrow region of stability, which is not yet fully correctly characterised. Therefore a better understanding of secondary phases, crystal structures, and crystal defects is needed.

This work aims to demonstrate the success of high-resolution anomalous-scattering powder XRD with thorough Rietveld refinement in improving the structural understanding of CZTS. Thorough Rietveld refinement includes using accurate experimentally measured compositions, secondary phase structures, antisite disorder

and vacancies on all cation lattice sites, and vacancies on the sulphur site. It follows a neutron diffraction study on two of the same samples that characterised the order-disorder and kesterite-sphalerite phase transitions CZTS undergoes around 500 K and 1250 K respectively. [6]

II. CZTS CRYSTAL STRUCTURES

CZTS is almost exclusively reported to form in the kesterite ($I\bar{4}$) crystal structure, [7,8] illustrated in Fig. 1, as this is its most stable configuration. [9] The alternative stannite structure is rarely reported experimentally. However, the kesterite structure often features sufficient Cu-Zn disorder to subtly change the structure to one with additional symmetry. This has historically been assumed to be the 'half-disordered kesterite' ($I\bar{4}2m$) structure illustrated in Fig. 1, in which the copper and zinc atoms in the z = ¼ and ¾ planes of the kesterite structure (the 2c and 2d Wyckoff positions) are randomly distributed, with full copper occupancy remaining in the z = 0 plane (the 2a Wyckoff position). Recently it has been confirmed that the 2a site shows equal or more disorder than the 2c and 2d sites, [6] so the 'fully disordered kesterite' ($I\bar{4}2m$) structure illustrated in Fig. 1 is more appropriate as an ideal model; although of course in real samples disorder will be less complete.

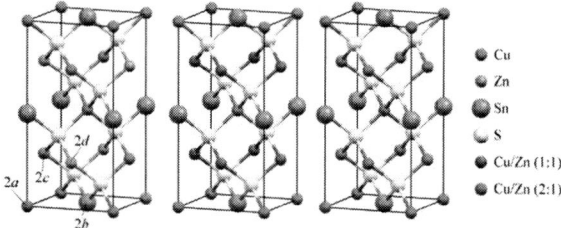

Fig. 1. kesterite (left), half-disordered kesterite (middle), and fully disordered kesterite (right) crystal structure of CZTS.

Cation disorder within the crystal structure is expected to be primarily only between copper and zinc, i.e. Cu_{Zn} and Zn_{Cu} antisite defects, because of the large chemical and size mismatch between tin and the other two cations. Ab initio calculations have indeed shown that Cu_{Zn} and Zn_{Cu} point defects have very low formation energies, [10] and the experimental prevalence of disordered kesterite structures confirms this. [8,11,12] The lack of tin disorder has also been confirmed experimentally for stoichiometric tin content. [13]

Cu-Zn disorder introduces antisite defect energy levels in the band gap (Cu_{Zn} at 0.15 eV above the valence band edge and Zn_{Cu} at 0.10 eV below the conduction band edge [14]), which act as recombination centres, reducing device efficiency. The Cu_{Zn} defect is thought to be the cause of the p-type behaviour in most CZTS, but it is replaced as the dominant acceptor by the copper vacancy for Cu-poor, Zn-rich compositions, which give the best device efficiencies.

Cu-Zn disorder has been shown to depend on the cooling rate after sample synthesis – 50 % antisite population at the 2c and 2d sites (i.e. 'complete' disorder) in water-quenched samples was reduced to 30 % in samples with a controlled cooling rate. [8] This is due to a second order phase transition to the fully disordered kesterite structure. The transition temperature was initially reported at 533 ± 10 K using Raman spectroscopy [11] and 552 ± 2 K using neutron diffraction, [12] and has recently been shown to depend on composition. [6] As usual device synthesis conditions involve annealing above the transition temperature, the disordered structure is formed initially and ordering occurs only during a lengthy cooling process below the critical temperature.

In addition to individual point defects, several charge-neutral defect complexes have been calculated to form in CZTS. Several of these stable defect complexes have been observed to form long-range order to the extent that they are responsible for grains of discrete CZTS phases, with particular stoichiometries. [15-17] These are described in Table . The stability provided by these off-stoichiometric defect complexes extends the composition region of phase-pure CZTS. [18] This is usually calculated as a very small region of the composition ternary phase diagram, with secondary phases also present in samples for most compositions. [19]

TABLE I. THE TYPES OF CZTS AS DEFINED BY LAFOND ET AL.[15], GURIEVA ET AL.[16], AND VALLE-RIOS ET AL.[17], WITH AN ADDITIONAL ENTRY FOR STOICHIOMETRIC, OR S-TYPE, CZTS.

CZTS type	Defect complex	Stoichiometry
S-type	$[Cu_{Zn}^- + Zn_{Cu}^+]$	Stoichiometric
A-type	$[V_{Cu}^- + Zn_{Cu}^+]$	Cu-poor, Zn-rich
B-type	$[Zn_{Sn}^{2-} + 2Zn_{Cu}^+]$	Cu-poor, Zn-rich
C-type	$[2Cu_{Zn}^- + Sn_{Zn}^{2+}]$	Cu-rich, Zn-poor
D-type	$[Cu_{Zn}^- + Cu_i^+]$	Cu-rich, Zn-poor
E-type	$[2V_{Cu}^- + Sn_{Zn}^{2+}]/[V_{Zn}^{2-} + V_{Cu}^- + Sn_{Cu}^{3+}]$	Cu-poor, Sn-rich
F-type	$[Zn_{Sn}^{2-} + 2Cu_i^+]/[Cu_{Sn}^{3-} + Zn_i^{2+} + Cu_i^+]$	Cu-rich, Sn-poor

Copper and zinc are isoelectronic in CZTS, so their X-ray scattering form factors, which are proportional to atomic number Z, are the same. It is therefore not possible to use conventional powder XRD to identify the exact structure. This can be overcome by using anomalous X-ray scattering using an absorption edge.

Anomalous scattering is the variation in atomic scattering factor for energies close to an absorption edge. It is accounted for by a correction of the form $f = f_0 + f' + i \cdot f''$, where f_0 is the uncorrected scattering factor, f' is the change in scattering factor magnitude and f'' is a phase shift. By comparing a non-resonant spectrum with one just below an absorption edge, where f' is large, scattering due to the respective element can be highlighted, revealing which planes it is present in and therefore its location within the crystal structure. Simultaneous Rietveld refinement using a non-resonant pattern and resonant ones for the Cu and Zn absorption edges can thus accurately determine the occupancy of each cation site.

III. EXPERIMENTAL DETAILS

Bulk polycrystalline samples with five different compositions were fabricated by solid state reaction. Finely ground Cu, Zn, and Sn powders were mixed in one alumina boat, S powder (with a 30 % excess to ensure full sulphurisation) was placed in another, and both were sealed together in an evacuated quartz ampoule. The ampoules were heated with a ramping rate of 5 K·min⁻¹ to 1073 K, at which they were kept for 24 hours and then left in the furnace to cool naturally back to room temperature (at ~0.5 K·min⁻¹).

An ampoule and sample are shown in Fig. 2. For diffraction experiments the samples were ground to a fine powder using an agate mortar.

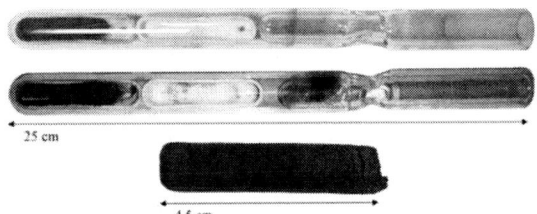

Fig. 2. Top to bottom: the elemental powders sealed in an ampoule pre-heat treatment, the ampoule post-heat treatment, and the final ingot of CZTS produced.

Inductively coupled plasma mass spectroscopy (ICPMS) was carried out using an Elan 6000 Perkin Elmer Sciex ICPMS to determine the post-fabrication elemental compositions of the samples.

SEM images were taken using an Hitachi SU-70 FEG SEM. EDX spectroscopy was carried out using the same SEM with an INCA x-act LN2-free analytical silicon drift detector and INCA software.

Anomalous scattering powder XRD experiments were carried out at the I11 beamline of Diamond Light Source. Resonant spectra were taken 10 eV below the Cu and Zn K absorption edges, 8.98 and 9.66 keV respectively, and non-resonant spectra at 15 keV.

Rietveld refinement was carried out using TOPAS v6.

IV. RESULTS AND ANALYSIS

A. Composition

The post-fabrication compositions, measured by ICPMS for copper, zinc, and tin, and EDX for sulphur, are plotted in Fig. 3 relative to the defect-complex-defined off-stoichiometric compositions.

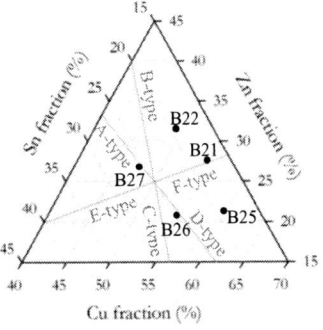

Fig. 3. The sample compositions measured by ICPMS, shown with the compositions of the defect-complex-defined CZTS types.

B. Crystal Structures

Due to the incredibly high resolution of the I11 beamline at Diamond, the crystal structures of the samples could be determined more accurately and precisely than yet reported. A typical non-resonant diffraction pattern is shown in Fig. 4.

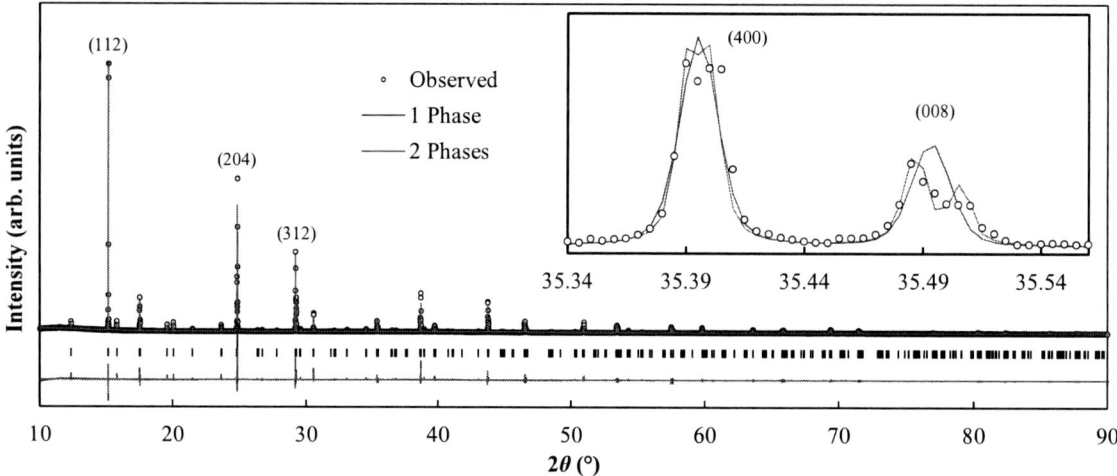

Fig. 4. X-ray diffraction pattern for B21 taken at room temperature using 15 keV X-rays at Diamond I11, showing models with a single kesterite phase and with two kesterite phases plotted over the data, the peak positions marked below the patterns, and the residual for each model at the bottom. Inset are the (400) and (008) peaks, showing peak splitting by ~0.1° due to the tetragonal ratio $a/2c$ being slightly less than one, and each individual peak showing further slight splitting of ~0.015° due to two nearly identical tetragonal phases being present.

It is clear from Rietveld refinement that all samples feature CZTS in the kesterite crystal structure, with significant disorder.

The tetragonal splitting of usually overlapping peaks due to the slightly non-unity $c/2a$ ratio of CZTS is easily discernable in these spectra; many such peaks are completely separately resolved. B21, B25, and B27 additionally exhibit even smaller splitting of individual peaks ($\Delta 2\theta \sim 0.02°$), which indicates two almost identical, but nevertheless distinct, phases present in the samples. This is the first report of diffraction at such high resolution from CZTS, meaning this is the first time such small splitting has been reported.

Pawley refinements reveal that both phases are tetragonal, ruling out the cubic ZnS secondary phase. The tetragonal Cu-Sn-S structures give a poor fit to the data, so are discounted. Secondary phases are additionally ruled out in all samples as the explanation for the peak splitting either by Raman, photoluminescence, and solid-state nuclear magnetic resonance (SSNMR) measurements (data not shown) discounting their presence, or by their additional presence in the XRD spectra at different positions. Close inspection of the refined models reveals that the stannite structure does not contribute intensity to several peaks, such as (123) and (231). This means that the best model is a mixture of kesterite types.

A future journal article currently in preparation will discuss in detail the identification of each phase, including their exact composition, structure, and the differences in disorder between them for each sample, and what this reveals about the effect of composition on CZTS.

For example, B21 has been found to consist of 56(1) % stoichiometric (S-type) and 44(1) % F-type CZTS, as illustrated in Fig. 5. The F-type phase shows almost complete order of the zinc atoms, i.e. they are almost all on the 2d site, with significant vacancy presence on all Cu and Zn sites. This implies that the presence of Cu_{Sn}, (and possibly Cu_i and Zn_i) defects discourages the formation of Cu_{Zn} and Zn_{Cu}. The S-type CZTS features equal disorder on the 2a and 2c sites, as recently confirmed. [6] The lattice parameters reveal a key difference between the two phases; the F-type phase has a larger a value, but a smaller c value than the S-type.

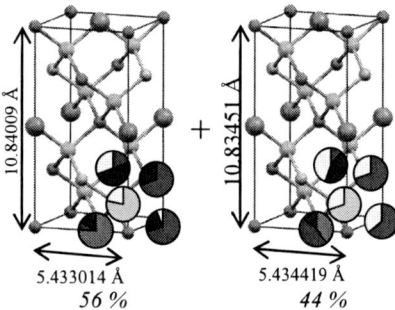

Fig. 5. The crystal structures, including cation disorder, of the two CZTS phases present in sample B21 at room temperature, S-type (left) and F-type (right).

These results confirm that disorder in CZTS is more complicated than simply Cu_{Zn} and Zn_{Cu} defects on the 2c and 2d sites, and so the ordering process is more complicated than a simple function of annealing time and temperature. The example of B21 discussed here shows the powerful ability of high-resolution anomalous-scattering XRD to resolve the structure, composition, and disorder of distinct CZTS phases within the same sample. The full journal article on this work currently in preparation will discuss these results for a full range of sample compositions. This will shed light on the effect of composition on CZTS device performance through a more detailed understanding of defect presence. It will also report how each of these structures changes over the order-disorder phase transition, and thus clarify the effect of this transition on CZTS fabrication.

V. CONCLUSION

High-resolution anomalous-scattering X-ray diffraction is shown to be a powerful tool to resolve the composition and disorder of distinct CZTS phases within the same sample.

Three of five bulk samples fabricated by solid state reaction over a range of compositions adopt two distinct kesterite phases with slightly different lattice parameters (Δ~0.004 Å). These are attributed to different types of CZTS, characterised by different charge-neutral defect complexes. The exact structure of each phase and the differences in disorder between them for each sample, as well as how they change over the order-disorder transition, will be discussed in detail in a future journal article currently under preparation.

It was only possible for these phases to be differentiated because of the incredibly high resolution of the I11 beamline at Diamond, the highest resolution yet used to study the structure of CZTS, and the unique ability of anomalous scattering to distinguish between isoelectronic copper and zinc ions. This is therefore the first time this two-phase structure has been observed using diffraction.

ACKNOWLEDGEMENTS

This work was financially supported by the UK Engineering and Physical Sciences Research Council (grant number 1335920), Diamond Light Source, and the University of Durham. It would not have been possible without the excellent technical staff and instrument scientists at the I11 and I15 beamlines at Diamond. ICPMS measurements were carried out by Dr Chris Ottley. Thanks also go to Jack Goodman, who assisted during some of the beamtime as part of an undergraduate project.

REFERENCES

1   A. Zuser & H. Rechberger. 'Considerations of resource availability in technology development strategies: The case study of photovoltaics'. *Resources, Conservation and Recycling* **56** (1), p.56-65, 2011.

2   T. Kato; H. Hiroi; N. Sakai; S. Muraoka; & H. Sugimoto. 'Characterisation of front and back interfaces on Cu_2ZnSnS_4 thin film solar cells'. *27th European Photovoltaic Solar Energy Conference*, p.2236-2239, 2012.

3   W. Wang et al. 'Device characteristics of CZTSSe thin film solar cells with 12.6% efficiency'. *Advanced Energy Materials* **4** (7), p.1301465.1-5, 2013.

4   J.J.S. Scragg et al. 'Cu–Zn disorder and band gap fluctuations in $Cu_2ZnSn(S,Se)_4$: Theoretical and experimental investigations'. *Physica Status Solidi B* **253** (2), p.247-254, 2016.

5   S.K. Wallace; D.B. Mitzi; & A. Walsh. 'The steady rise of kesterite solar cells'. *ACS Energy Letters* **2** (4), p.776-779, 2017.

6   C.J. Bosson; M.T. Birch; D.P. Halliday; K.S. Knight; & P.D. Hatton. 'Cation disorder and phase transitions in the structurally complex solar cell material Cu_2ZnSnS_4'. *(manuscript under review)*, 2017.

7   T. Washio et al. 'Analysis of lattice site occupancy in kesterite structure of Cu_2ZnSnS_4 films using synchrotron radiation X-ray diffraction'. *Journal of Applied Physics* **110** (7), p.074511.1-4, 2011.

8   S. Schorr. 'The crystal structure of kesterite type compounds: A neutron and X-ray diffraction study'. *Solar Energy Materials and Solar Cells* **95** (6), p.1482-1488, 2011.

9   S. Chen; X.G. Gong; A. Walsh; & S.-H. Wei. 'Crystal and electronic band structure of Cu_2ZnSnX_4 (X=S and Se) photovoltaic absorbers: First-principles insights'. *Applied Physics Letters* **94** (4), p.041903.1-3, 2009.

10 S. Chen; J.-H. Yang; X.G. Gong; A. Walsh; & S.-H. Wei. 'Intrinsic point defects and complexes in the quaternary kesterite semiconductor Cu_2ZnSnS_4'. *Physical Review B* **81** (24), p.245204.1-10, 2010.

11 J.J.S. Scragg; L. Choubrac; A. Lafond; T. Ericson; & C. Platzer-Björkman. 'A low-temperature order-disorder transition in Cu_2ZnSnS_4 thin films'. *Applied Physics Letters* **104** (4), p.041911.1-4, 2014.

12 A. Ritscher; M. Hoelzel; & M. Lerch. 'The order-disorder transition in Cu_2ZnSnS_4 – A neutron scattering investigation'. *Journal of Solid State Chemistry* **238**, p.68-73, 2016.

13 C. Malerba et al. 'Stoichiometry effect on Cu_2ZnSnS_4 thin films morphological and optical properties'. *Journal of Renewable and Sustainable Energy* **6** (1), p.011404.1-12, 2014.

14 S. Chen; A. Walsh; X.-G. Gong; & S.-H. Wei. 'Classification of lattice defects in the kesterite Cu_2ZnSnS_4 and $Cu_2ZnSnSe_4$ Earth-abundant solar cell absorbers'. *Advanced Materials* **25** (11), p.1522-1539, 2013.

15 A. Lafond; L. Choubrac; C. Guillot-Deudon; P. Deniard; & S. Jobic. 'Crystal structures of photovoltaic chalcogenides, an intricate puzzle to solve: The cases of CIGSe and CZTS materials'. *Zeitschrift für anorganische und allgemeine Chemie* **638** (15), p.2571-2577, 2012.

16 G. Gurieva et al. 'Structural characterisation of $Cu_{2.04}Zn_{0.91}Sn_{1.05}S_{2.08}Se_{1.92}$'. *Physica Status Solidi C* **12** (6), p.588-591, 2015.

17 L.E. Valle-Rios; K. Neldner; G. Gurieva; & S. Schorr. 'Existence of off-stoichiometric single phase kesterite'. *Journal of Alloys and Compounds* **657**, p.408-413, 2016.

18 A. Ritscher; A. Franz; S. Schorr; & M. Lerch. 'Off-stoichiometric CZTS: Neutron scattering investigations on mechanochemically synthesised powders'. *Journal of Alloys and Compounds* **689**, p.271-277, 2016.

19 I.D. Olekseyuk; I.V. Dudchak; & L.V. Piskach. 'Phase equilibria in the Cu_2S-ZnS-SnS_2 system'. *Journal of Alloys and Compounds* **368** (1–2), p.135-143, 2004.

Simulation of ZnMgO as the window layer for CdTe Solar Cells

Yunfei Chen[1], Shou Peng[2], Xin Cao[3], Alan E. Delahoy[1], Ken K. Chin[1]

[1]Department of Physics and CNBM New Energy Materials Research Center,
New Jersey Institute of Technology, Newark, NJ 07102, USA
[2]China Triumph International Engineering Co., Ltd, Shanghai, P.R. China 200063
[3]Bengbu Design & Research Institute for Glass Industry,
*TEL: 9734954666 Email: yc289@njit.edu

Abstract — CdS has been used as window layer for CdTe solar cells during the last decades of years. New wider band gap (E_g) materials is being found to replace CdS in order to remove blue loss so that short-circuit current (J_{SC}) of CdTe solar cells can be significantly increased. ZnMgO, synthesized by incorporating Mg content into ZnO, is one of the potential candidates. Some researchers have already successfully made high efficiency CdTe solar cells using ZnMgO. In this work, ZnMgO is used as the window layer and systematic computer simulation is finished to investigate the influence of the resistivity (R) of ZnMgO and the conduction band offset (ΔE_C) between ZnMgO layer and CdTe layer on the performance of CdTe solar cells. Simulation results indicate that low resistivity and high doping is needed for ZnMgO and the influence of conduction band offset is quite different when the doping concentration of ZnMgO varies. +0.2eV is an acceptable value for conduction band offset and the ideal Mg content in ZnMgO is around 10%. Finally, a new optimized CdTe solar cell with highest efficiency is designed and a flat band diagram is plotted.

Index Terms — CdTe solar cells, window layer, ZnMgO, resistivity, conduction band offset.

I. INTRODUCTION

CdTe solar cells are one of the three major branches of the thin film photovoltaic technology. It needs an n-type window layer to form a p-n junction with p-type CdTe layer. CdS has been used as the window layer during the last decades of years. However, with a relatively low band gap, the blue loss will happen that can significantly reduce J_{SC} of CdTe solar cells. The efficiency loss by this blue loss is estimated as approximately 2.3% [1]. Thus, it is helpful to find a new wider band gap material to replace CdS as the window layer in order to improve the efficiency of CdTe solar cells. Actually, the noticeable and dramatic efficiency improvement of CdTe solar cells recently is mainly originated from changing the window layer of CdTe solar cells. Many attempts have been made to find other proper window materials. ZnMgO is one of the potential candidates. Researchers from Colorado University successfully made a CdTe solar cell with efficiency reaching 18.3% by using ZnMgO directly as window layer [2]. However, deep discussion of ZnMgO as window layer is still required for the purpose of making best performance CdTe solar cells.

The first one is the resistivity of ZnMgO. As a window layer, ZnO should not be too resistive and should be highly doped. Otherwise, normal carrier transportation can be impeded, and thus decrease the short-circuit current of CdTe solar cells. On the other hand, window layer also has a function of 'buffer', which means they cannot be too conductive. Otherwise, contents in back contact material, such as Cu, can diffusion into TCO. CdTe solar cells then will be short-circuited and the open-circuit voltage (V_{OC}) of them can be extremely low.

The second one is the conduction band offset between ZnMgO layer and CdTe layer. The band gap of ZnMgO can range from 3.3eV (band gap of ZnO) to 7.8 eV (band gap of bulk MgO) or 6 to 6.5eV (band gap of nanoscale MgO) with the variation of Mg content [3, 4]. The electron affinity (χ) of ZnO is 4.3eV. According to common anion rule, materials with same anions should have very small conduction band offset. This means the change of band gap of ZnMgO will mainly leads to the change of conduction band offset. The conduction band offset has strong influence on the performance of CdTe solar cells. To acquire the highest efficiency CdTe solar cells, a certain range of conduction band offset is highly desired [5]. For the case of ZnMgO, the influence of conduction band offset on the CdTe solar cells still needs to be further investigated. Once the best value of conduction band offset is determined, the Mg content can be calculated from the equation (1) [6].

$$E_g=3.296+2.19x \quad (1)$$

where x is the atomic percent of Mg in the Zn/Mg alloy.

In this paper, SCAPS simulation is employed to find out the influence of resistivity of ZnMgO and conduction band offset on the performance of CdTe solar cells. Brief explanation of the simulation results is provided and a flat band diagram of the highest efficiency CdTe solar cell simulated is plotted.

II. METHOD

SCAPS (a solar cell capacitance simulator) is a one dimension computer program that is commonly applied in the research of solar cell areas [7]. Different kinds of solar cells with different layers including the interfaces between the layers can be built by the program. Then the program numerically solves equations, such as Poisson and continuity equations, for electrons and holes in one dimension to determine the band diagram of the solar cells devices and their response to illumination, voltage bias and temperature. By doing this, the performance of the solar cells can be simulated. Typical examples are J-V, C-V and QE.

A solar cell with the following orders is set up for the simulation. Back contact/ p-type CdTe/ (interface)/ n-type ZnMgO/ TCO / front contact. All the parameters in this simulation are shown in TABLE I. Part of the parameters are achieved from other publications [8, 9].

According to Equation (2), when the mobility (μ) of the ZnMgO is fixed, the resistivity of ZnMgO is only determined by the doping concentration (n) of it.

$$R = \frac{1}{\mu q n} \quad (2)$$

Thus, to simulate the influence of the resistivity of ZnMgO on the performance of CdTe solar cells, the simulation variable is the doping concentration of ZnMgO, while the other parameters are kept the same.

As mentioned above, the common anion rule tells us that materials with the same anions should have very small valence band offset. But it is not absolutely true in every case. In this paper, we assume by increasing the band gap of ZnMgO, 80% percent of the increment will go to conduction band, while the other 20% will go to valence band. For example, when the band gap of ZnMgO increases 0.1eV, the conduction band offset will increase 0.08eV and valence band offset will increase 0.02eV. From Anderson's rule, the conduction band offset is related to the electron affinity that forms the heterojunction. Accordingly, to simulate the influence of the conduction band offset between ZnMgO layer and CdTe layer on the performance of CdTe solar cells under the conditions of different doping concentration of ZnMgO, the variables of the simulation are the band gap of ZnMgO and its related conduction band offset. The other parameters are kept the same.

Once the ideal band gap of ZnMgO is determined, the Mg content for ZnMgO to yield highest efficiency CdTe solar cell will be calculated and the flat band diagram of the highest efficiency CdTe solar cell simulated is plotted.

III. RESULTS

A. The influence of resistivity of ZnMgO on the performance of CdTe solar cells

Figure 1 shows the efficiency of CdTe solar cells in terms with the doping concentration of ZnMgO. Information from the figure indicates that the doping concentration of ZnMgO should be as high as possible. However, this information could be wrong since the ZnMgO layer may also play a role of buffer in CdTe solar cells. It can help prevent the immigration of some harmful elements, for example Cu, into the front contact. If the doping concentration of ZnMgO is too high, the undesired immigration of the harmful elements maybe significantly enhanced. This kind of immigration will make the CdTe solar cells short-circuited, which should be

avoided. Since SCAPS is a one-dimension simulation program, the function of a buffer layer cannot be simulated. This means the information achieved from Figure 1 may be overthrown and the optimal doping concentration of ZnMgO should be further investigated by experiments.

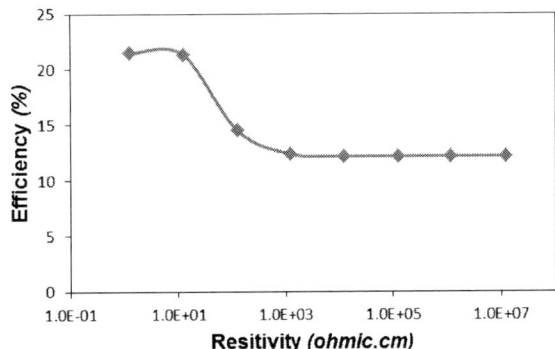

Fig.1. The efficiency of CdTe solar cells as a function of the resistivity of ZnMgO

B. The influence of the conduction band offset between ZnMgO layer and CdTe layer on the performance of CdTe solar cells

Figure 2 shows the influence of the conduction band offset on the performance of CdTe solar cells with the doping concentration of ZnMgO shifted.

The open-circuit voltage of CdTe solar cells increase slightly together with the conduction band offset. A proper explanation is the surface recombination rate is decreased when the conduction band offset is increased. This leads to the increase of carrier lifetime and further the decrease of dark current. Then, from either Shockley's model or Shockley-Read- Hall, the open-circuit voltage is increased because it is inverse proportional with the carrier lifetime. In addition, the increase of conduction band offset can also enhance the upper limit of the open-circuit voltage, which is the built-in voltage of CdTe solar cells, from Equation (3).

$$V_{OC} \le V_{bi} = 1/q \ (Egp + \Delta EC - \Delta En - \Delta Ep) \quad (3)$$

The short-circuit current and fill factor of CdTe solar cells decrease dramatically when the conduction band offset is too big. The reason is that a positive conduction band offset acts as a 'spike', which can prevent the movement of the photo-generated electrons from CdTe layer to ZnMgO layer. Then the short-circuit current will be subsequently reduced. However, if the positive conduction band offset is not too big, the electrons can use the thermal energy to climb through the 'spike' barrier and the solar cell can still work well.

It is also found that at higher doping concentration of ZnMgO, the threshold value of conduction band offset for short-circuit current to decrease is also higher. A possible explanation is a higher doping concentration of ZnMgO will decrease the value of ΔE_n, which means the the barrier the electrons has to climb through can be decreased a little bit.

The highest efficiencies of CdTe solar cells are obtained with the conduction band offsets of 0.1eV, 0.1eV, 0.18eV and 0.34eV for ZnMgO doping concentration of 10^{11} cm^{-3}, 10^{14} cm^{-3}, 10^{16} cm^{-3} and 10^{18} cm^{-3}, respectively. It looks that +0.2eV for the conduction band offset is always in the range of best performance CdTe solar cell no matter what the doping concentration of ZnMgO is. So the ideal Mg content for ZnMgO should be around 10% from Equation (1).

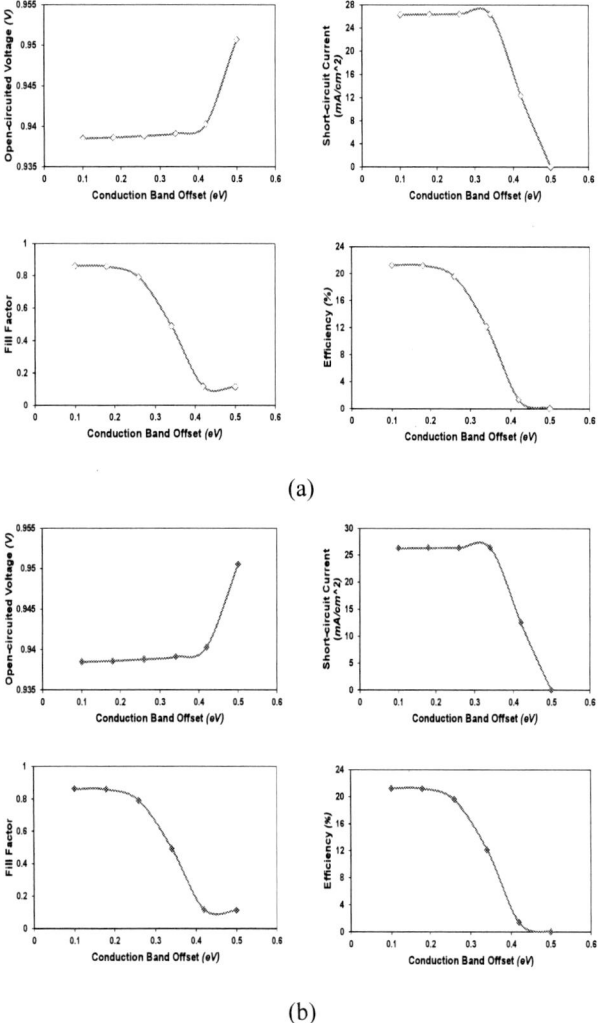

(a)

(b)

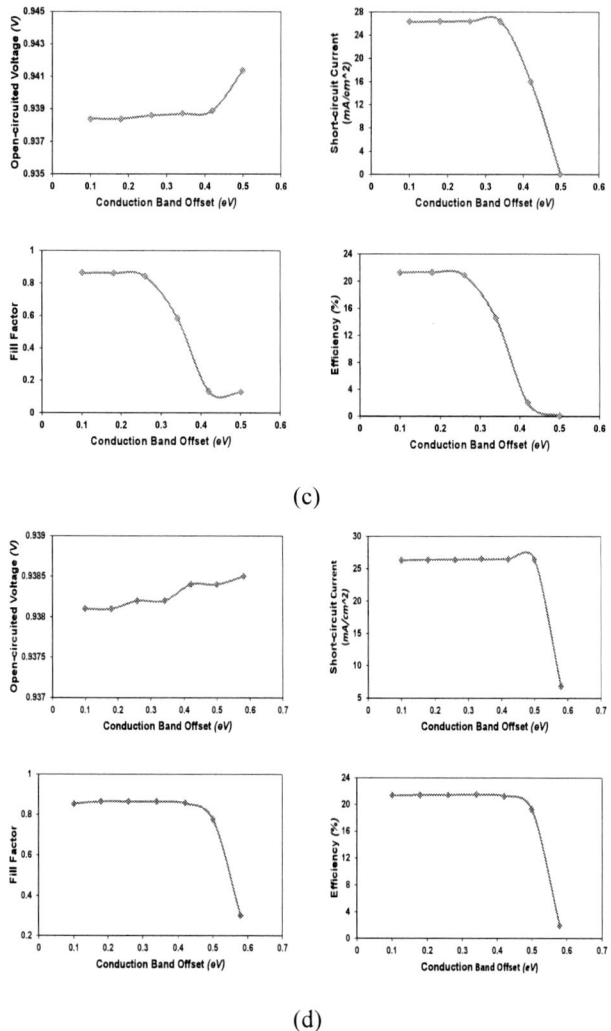

(c)

(d)

Fig.2. The performance of CdTe solar cell as a function of conduction band offset between ZnMgO layer and CdTe layer. (a) when the doping concentration of ZnMgO is 10^{11}cm^{-3}, (b) when the doping concentration of ZnMgO is 10^{14} cm^{-3}, (c) when the doping concentration of ZnMgO is 10^{16} cm^{-3}, (d) when the doping concentration of ZnMgO is 10^{18} cm^{-3}.

C.   The flat band diagram of the simulated highest efficiency CdTe solar cell

Figure 3 depicts the flat band diagram of the simulated highest efficiency CdTe solar cell. The doping concentration for the ZMO is 10^{18} cm^{-3}. The Mg content for ZnMgO is 10%, which forms a +0.2eV conduction band offset between CdTe layer and ZnMgO layer.

TABLE I
THE PARAMETERS OF THE SIMULATED SOLAR CELL

Layer	TCO	ZnMgO	Interface	p-CdTe
Thickness [nm]	500	70	-	2500
Band Gap [eV]	3.7	3.3-3.8	-	1.5
Electron Affinity [eV]	4.5	4.3-3.9	-	4.4
Relative Dielectric Permittivity	10	9	-	9.4
Density States of Conduction Band [cm^{-3}]	2.2×10^{18}	2.2×10^{18}	-	8.0×10^{17}
Density States of Valence Band [cm^{-3}]	1.8×10^{19}	1.8×10^{19}	-	1.8×10^{19}
Electron Mobility [cm^2/V.s]	32	5	-	320
Hall Mobility [cm^2/V.s]	30	1	-	40
Donor Concentration [cm^{-3}]	1.0×10^{19}	$1.0 \times 10^{11-18}$	-	-
Acceptor Concentration [cm^{-3}]	-	-	-	1.0×10^{15}
Dopant energy level [eV]	-	-	-	0.255
Defect Type	SA*	SD*	N*	SD*
Electron Cross Section [cm^{-2}]	1.0×10^{-15}	1.0×10^{-12}	1.0×10^{-15}	1.0×10^{-13}
Hole Cross Section [cm^{-2}]	1.0×10^{-12}	1.0×10^{-15}	1.0×10^{-15}	1.0×10^{-15}
Defect Concentration [cm^{-2}]	1.0×10^{16}	1.0×10^{18}	1.0×10^{13}	2.0×10^{13}
Defect energy level [eV]	2.0	1.65	0.6	0.585

* SA = single acceptor, N = Neutral, SD = single donor

IV. CONCLUSION

By increasing doping concentration to decrease resistivity of ZnMgO layer, simulation results indicates that the efficiency of CdTe solar cells improves and starts to reach the highest point when doping concentration is 10^{18} cm^{-3}. But this does not mean that higher doping concentration and lower resistivity of ZnMgO is desired. The reason is that SCAPS is a one-dimension simulation software and probably cannot reflect the role of ZnMgO as a buffer. The influence of conduction band offset between ZnMgO layer and CdTe layer is quite different when the doping concentration of ZnMgO varies. But +0.2eV is a value that is always in the range of best performance CdTe solar cells whenever the doping concentration of ZnMgO is. The ideal Mg content in ZnMgO is around 10%. Finally, a flat band diagram of the simulated highest efficiency CdTe solar cell is provided when the doping concentration of ZnMgO is 10^{18} cm^{-3} and the conduction band offset is +0.2eV.

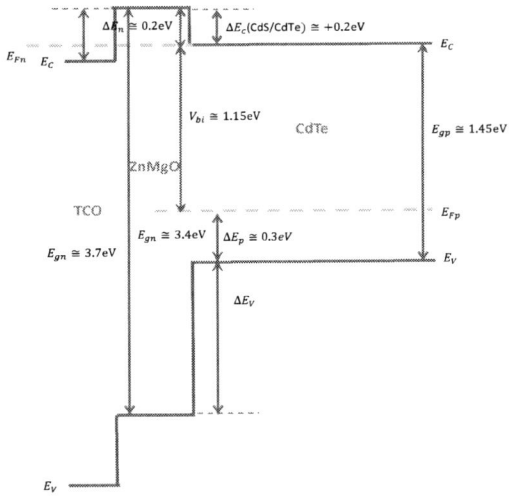

Fig.3. The flat band diagram of the simulated highest efficiency CdTe solar cell.

ACKNOWLEDGEMENT

The authors acknowledge the China Triumph International Engineering Co. (CTIEC), Shanghai, China and Bengbu

Design & Research Institute for Glass Industry, which offer generous financial support for this work.

REFERENCES

[1] J. Sites. Quantification of losses in thin-filmpolycrystalline solar cells. Solar Energy Materials & Solar Cells 75 (2003) 243–251

[2] J. Sites, A. Munshi, J. Kephart, D. Swanson and W. S. Sampath, Progress and challenges with CdTe cell efficiency, 2016 IEEE 43rd Photovoltaic Specialists Conference (PVSC), Portland, OR, 2016, pp. 3632-3635. doi: 10.1109/PVSC.2016.7750351

[3] C. Niedermeier, R. Rasander, S. Rhode, V. Kachkanov, B. Zou, N. Alford, M. Moram, Band gap bowing in $Ni_xMg_{1-x}O$, Scientific Reports 6, Article number: 31230 (2016) doi:10.1038/srep31230

[4] S. Heo, E. Cho, H. Lee, G. Park, H. Kang, T. Nagatomi, P. Choi, B. Choi, Band gap and defect states of MgO thin films investigated using reflection electron energy loss spectroscopy, AIP Advances 5, 077167 (2016); doi: http://dx.doi.org/10.1063/1.4927547

[5] Y. Chen et al., "An optimized structure for CdTe solar cells," 2016 IEEE 43rd Photovoltaic Specialists Conference (PVSC), Portland, OR, 2016, pp. 0423-0427. doi: 10.1109/PVSC.2016.7749625.

[6] M. Lorenz, E. M. Kaidashev, H. Von Wenckstern, V. Riede, C. Bundesmann, D. Spemann, G. Benndorf, H. Hochmuth, A. Rahm, H. C. Semmelhack, and M. Grundmann, "Optical and electrical properties of epitaxial (Mg,Cd)xZn1−xO, ZnO, and ZnO:(Ga, Al) thin films on c-plane sapphire grown by pulsed laser deposition," Solid-State Electronics 47, pp. 2205-2209, 2003.

[7] M. Gloeckler, "DEVICE PHYSICS OF Cu(In,Ga)Se2 THIN-FILM SOLAR CELLS," (2005).

[8] Y. Inoue, M. Hála, A. Steigert, R. Klenk and S. Siebentritt, "Optimization of buffer layer/i-layer band alignment," 2015 IEEE 42nd Photovoltaic Specialist Conference (PVSC), New Orleans, LA, 2015, pp. 1-5. doi: 10.1109/PVSC.2015.7355902

[9] J. Kephart, R. Geisthardt, W. Sampath. (2015), Optimization of CdTe thin-film solar cell efficiency using a sputtered, oxygenated CdS window layer, Prog. Photovolt: Res. Appl., doi: 10.1002/pip.2578

Modeling Effect of Defects on Efficiency of Nanowire CdS-CdTe Solar Cells

Hongmei Dang*, Esther Ososanya[1], Nian Zhang[1], Xiaohui Wang[2], Hojjatollah Sarvari[2] and Vijay P. Singh[2]

[1]Department of Electrical and Computer Engineering, University of the District of Columbia, Washington DC, 20008, U.S.A.

[2]Department of Electrical and Computer Engineering, University of Kentucky, Lexington, KY, 40506-0046, U.S.A.

ABSTRACT — **SCAPS simulation is conducted to model effect of defects on performance of the nanowire CdS-CdTe solar cells. In the simulation model, acceptor type traps with a liner distribution of $1.05*10^{18}/cm^3$ and $6.0*10^{17}/cm^3$ are introduced in CdS nanowires. Donor traps with a liner distribution of $2.5*10^{14}/cm^3$ and $8.0*10^{14}/cm^3$ are presented in p-CdTe layer. Acceptor type interface states are tuned to $1.2*10^{12}/cm^2$. The simulation models fit well with measured I-V, C-V and quantum efficiency characteristics of 12% nanowire solar cells. Simulation indicates that large energy bandgap in CdS layer reduces effects of defect traps of CdS layer on solar cell performance. In order to improve solar cell efficiency, it is necessary to search for approaches which increase shallow doping concentration and/or reduce donor trap concentration in the CdTe layer, and reduce interface states between CdS and CdTe.**

I. INTRODUCTION

Cadmium telluride (CdTe) becomes a leading thin film photovoltaic technology to supply low cost solar electricity which is comparable with conventional electricity[1,2]. Recently, we have developed nanowire CdS-CdTe solar cells where CdS nanowire array replaces planar CdS film as a window layer and CdTe maintains its polycrystalline structure on the top of the window layer[3-9]. The nanowire CdS-CdTe solar cells yielded power conversion efficiency of 12%. They shown a nearly ideal spectral response of quantum efficiency from 335nm to 850nm[7,9]. According to theory, they have potential to gain 27.8%-30% of power conversion efficiency[9].

To produce electricity near to theoretical efficiency, it is essential to understand the critical factors that create efficiency loss mechanism of the nanowire solar cells. Herein, we conducted numerical simulation of the 12% nanowire CdS/CdTe solar cells. Numerical simulations of current-voltage (I-V) and capacitance-voltage (C-V) characteristics of nanowire CdS-CdTe solar cells were performed by SCAPS software[10]. By fitting numerical I-V and CV simulations with measured I-V and C-V curves, we established numerical models of 12% nanowire CdS-CdTe solar cells. In the numerical models, we analyze effect of traps in nanowire CdS layer, traps in CdTe layer and interface states, traps on open circuit voltage, short circuit current, fill factor and efficiency. Because this work focused on device physics mechanism, investigation in this work will can give insight on the efficiency of planar CdS-CdTe solar cells.

Investigation of device physics mechanism plays a key role in further enhancing improve efficiency of the CdTe solar cells.

II. EXPERIMENTAL METHODS

Simulation of 12% nanowire CdS-CdTe solar cells was conducted by a SCAPS-1D simulator application. It estimates J-V curves, C-V curves and quantum efficiency[10]. In each layer, shallow donor or acceptor and its density are defined, and they are assumed to completely ionized and does not contribute to recombination[11,12]. Defect states like donor type defect states and acceptor type defect states are introduced in the SCAPS simulator. Interface defects between CdS nanowires and CdTe are also introduced in the SCAPS simulator. In the simulation, material parameters are chosen based on the reported values.

III. RESULT AND DISCUSSION

A. CdS Nanowire Defect States

Table1 and 2 show parameters which are used in CdS nanowires to simulate the measured I-V curves of the nanowire CdS-CdTe solar cells. In this simulation, thickness of CdS nanowires is 100nm, and energy band gap is 3.3eV, and shallow donor density of $1.10*10^{18}/cm^3$ is introduced in CdS nanowire layer, which fit experiment results of CdS nanowires.

Acceptor traps compensate free carriers and space charge in n-CdS. Simulation found that due to its large energy bandgap, the acceptor traps in the CdS nanowires have non-significant influence on quantum efficiency, photocurrent, open circuit voltage(Voc) and fill factor.

Herein, we introduce compensating acceptor type traps with density of $1.05*10^{18}/cm^3$ at region close to CdTe layer, and with density of $6.0*10^{17}/cm^3$ at region close to front contact side. The acceptor trap distribution is liner. Their energy level is in the middle gap (1.65eV) of CdS nanowires.

Capture cross section of electron and hole have been referred the literature [13,14]. CdS layer is responsible for quantum efficiency, photocurrent and fill factor, influence of CdS layer on Voc is slight. Therefore, these numerical values in the CdS nanowires provide fitted curves in quantum efficiency and photocurrent data, which are shown in figure 1 and 2.

Table 1 Parameter in the CdS Nanowire

Thickness(μm)	0.1
Energy bandgap(eV)	3.3
Shallow donor density Nd (1/cm^3)	$1.10*10^{18}$

Table 2 Acceptor Traps in the CdS Nanowires

Defect type	Acceptor
Capture cross section Electrons (cm^2)	$1.00*10^{-17}$
Capture cross section Holes (cm^2)	$1.00*10^{-12}$
Energy distribution	Single
Energy level (eV)	Ev+1.65
Nt distribution	Linear
Nt density at left (1/cm^3)	$6.00*10^{17}$
Nt density at right (1/cm^3)	$1.05*10^{18}$

B. Defect States in CdTe Layer

According to the simulation, shallow acceptor (doping) concentration of CdTe strongly affects open circuit voltage, and obviously impacts photocurrent and fill factor. For example, an increase in shallow acceptor density is made from $1*10^{15}$/cm^3 to $5*10^{15}$/cm^3, Voc increases by 90mV. In the CdTe layer, donor traps have similar effect on Voc, Jsc and fill factor as acceptor doping, However, their influence is less than influence of the acceptor doping on the photovoltaic parameters. Donor type traps in p-type CdTe compensate shallow acceptor doping concentration of CdTe. In CdTe layer, shallow acceptor doping concentration and donor traps are key parameters which significantly impact Voc.

In order to fit the IV and CV curves of the 12% CdS nanowire-CdTe solar cells, donor type traps are located in 0.28eV below conduction band of CdTe[13,14]. The donor type trap level, 0.28eV below conduction band of CdTe may be related to the ionized interstitial Cu$^+$ ion[13,14]. A liner

distribution of donor traps from $2.5*10^{14}$/cm^3 at region close to back contact side to $8.0*10^{14}$/cm^3 at region close to CdS side is introduced into p-CdTe layer, in order to obtain correspondence between simulated and measured Voc, Jsc and fill factor.

Observed from simulation, it is obvious that in order to make Voc higher than 900mV, it is necessary to increase acceptor doping concentration and reduce donor traps in CdTe.

Table 3 Donor Traps in the CdTe Layer

Defect type	Donor
Capture cross section Electrons (cm^2)	$3.1*10^{-13}$
Capture cross section Holes (cm^2)	$1.00*10^{-17}$
Energy distribution	Single
Energy level (eV)	Ec-0.28
Nt distribution	Linear
Nt density at left (1/cm^3)	$2.5*10^{14}$
Nt density at right (1/cm^3)	$8.0*10^{14}$

C. Interface States of CdS Nanowires and CdTe Layer

It is found that interface states are a key parameter which significantly impact solar cell performance. It is acceptor type interface states rather than donor type interface states that play a key role in solar cell efficiency. Table 4 shows acceptor type interface traps in the CdS nanowire-CdTe interface.

To realize a better correspondence between the simulated and measured curves, efficient interface recombination centers are introduced at the nanowire CdS-CdTe heterojunction. The efficient interface recombination centers are acceptor interface traps. The acceptor interface traps are uniform distribution, located in the 0.6eV above intrinsic fermi energy with characteristic energy of 0.1eV. The acceptor interface trap density is tuned to $1.2*10^{12}$/cm^2. Due to high interface state density, the recombination current is created by these acceptor type interface states, which reduces Voc.

Table 4 Acceptor Trap in the CdS nanowire-CdTe interface

Defect type	Acceptor
Capture cross section Electrons (cm^2)	$1.0*10^{-17}$
Capture cross section Holes (cm^2)	$1.0*10^{-12}$
Energy distribution	Uniform
Energy level (eV)	Ei+0.6
Characteristic energy (eV)	0.1
Total density (1/cm^2)	$1.2*10^{12}$
Density at peak energy (1/cm^2 .eV)	$1.2*10^{13}$

Voc and fill factor can be significantly impacted by interface states. As an example, Voc is remarkably increased by 80mV when acceptor type interface state density is reduced from to $1.2*10^{12}$/cm^2 to $1*10^{10}$/cm^2. Fill factor is obviously increased from 60% to 85%. Observed from simulation, influence of interface states on photocurrent is negligible until recombination rate via interface states is comparable with photogeneration rate. Then photocurrent is significantly reduced by the interface states. For an example, when interface state density reaches $1*10^{15}$/cm^2, photocurrent is reduced by 48%, indicating that junction has been deteriorated.

Simulation indicates that interface states have significantly impact on the Voc, Jsc, fill factor and efficiency. When interface has enough high acceptor traps, even photocurrent is remarkably reduced. Hence, it is crucial to reduce interface states in order to increase efficiency of the solar cells.

D. Simulation of CdS Nanowires-CdTe Solar Cells

Figure 1, 2 and 2 show the simulated and measured I-V, C-V and quantum efficiency curves of the 12% nanowire CdS-CdTe solar cells respectively. In these figures, dotted lines is measured curves and solid lines are simulated curves. Table 5 simulated and measured photovoltaics parameter.

By adopting defect in CdS nanowire and CdTe and interface states at CdS nanowires and CdTe heterojunction, simulated IV curve fits well with measured I-V curve. There is a slight different when supply voltage is larger than 0.85V.

It is caused by series resistance. Our solar cell sample has higher series resistance which reduces solar cell efficiency.

Simulated CV curve fits well with measured I-V curve when voltage is from -2V to -0.5V. When voltage is less than -2V or larger than -0.5V, simulated C-V curve deviates from measured C-V curve. In our simulated, we assume a linear distribution of donor traps below 0.28V conduction band of CdTe layer. However, our solar cell samples may have server types of traps [1] which are located forbidden band in the CdTe layer. These traps are not linear distribution along the thickness of CdTe layer. However other possible traps exist in the CdTe layer. We found that donor traps below 0.28V conduction band of CdTe layer significantly impact Voc.

Simulated quantum efficiency curve fits well with measured quantum efficiency from wavelength 300nm to 850nm. From 300 nm to 550 nm, the EQE response of the nanowire solar cells shows substantially high quantum efficiency. Simulation model indicates that quantum efficiency in CdS is mainly impacted by energy bandgap of CdS layer. When the CdS nanowire has large energy bandgap like 3.3eV, acceptor traps in the CdS nanowire have less influence on quantum efficiency, photocurrent, Voc and fill factor. It is considered that high energy bangap of CdS nanowires leads to strong quantum efficiency and high photocurrent. However, the CdS layer has small energy bandgap like 2.4eV. Then acceptor traps in the CdS layer will significantly reduce quantum efficiency, photocurrent and fill factor.

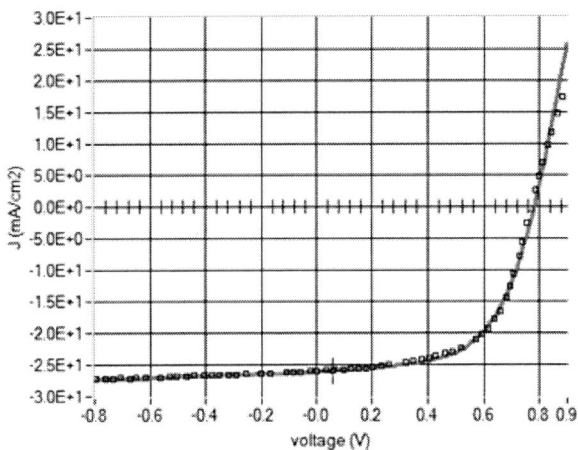

Fig.1. Simulated and measured I-V curves of 12% nanowire CdS-CdTe solar cells, where dotted line is measured I-V curve and solid line is simulated I-V curve.

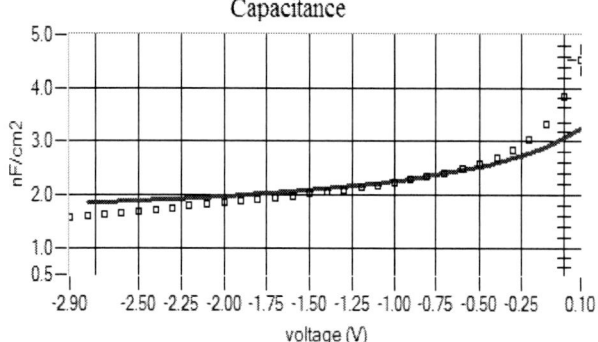

Fig.2. Simulated and measured C-V curves of 12% nanowire CdS-CdTe solar cells, where dotted line is measured C-V curve and solid line is simulated C-V curve.

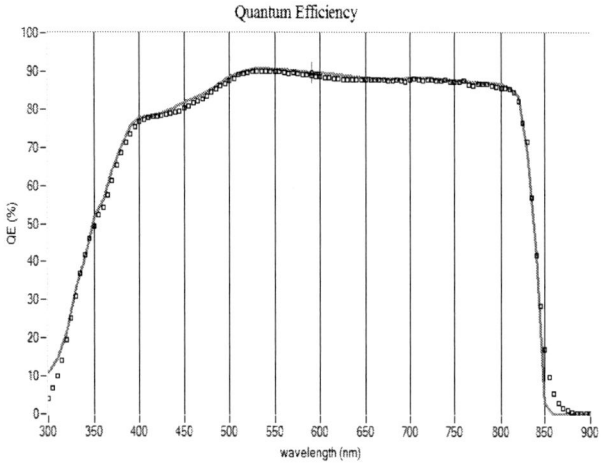

Fig.3. Simulated and measured quantum efficiency curves of 12% nanowire CdS-CdTe solar cells, where dotted line is measured quantum efficiency curve and solid line is simulated quantum efficiency curve.

Table 5 Simulated and Measured Photovoltaics Parameter of Nanowire CdS-CdTe Solar Cells.

Detail	Voc (mV)	Jsc(mA/cm²)	FF (%)	Efficiency (%)
Simulation	781	26.07	59.2	12.04
Experiment	770	26	59.8	12

E.. Predicted Efficiency in CdS Nanowires-CdTe Solar Cells

From simulation model of 12% CdS nanowire-CdTe solar cells, we have found that donor traps in CdTe layer, acceptor interface states are two important factors which impact Voc, photocurrent, fill factor and efficiency. Thus we predicted CdS nanowire CdTe solar cell performance when donor traps in CdTe layer and interface states in the heterojunction are reduced. When donor traps in the CdTe layer are reduced from average $5.25*10^{14}/cm^3$ to $1.0*10^{14}/cm^3$ and acceptor type interface states are $1.2*10^{12}/cm^2$ to $1.2*10^{10}/cm^2$, the nanowire CdS-CdTe solar cells will generate Voc of 907mV, photocurrent of 29 mA/cm², fill factor of 86% and power conversion efficiency of 22.7%.

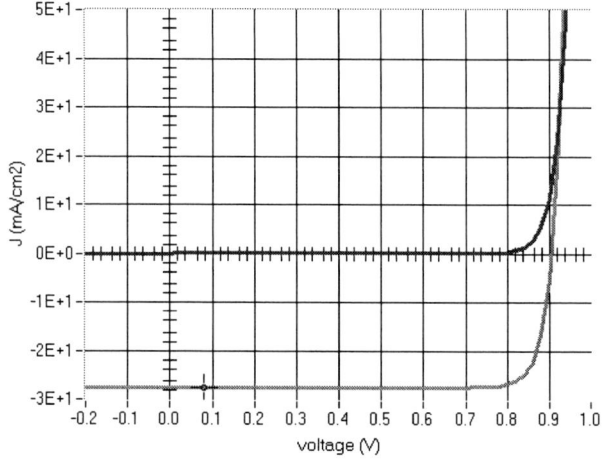

Fig.4. Simulated dark and light I-V curves of 22.7% nanowire CdS-CdTe solar cells.

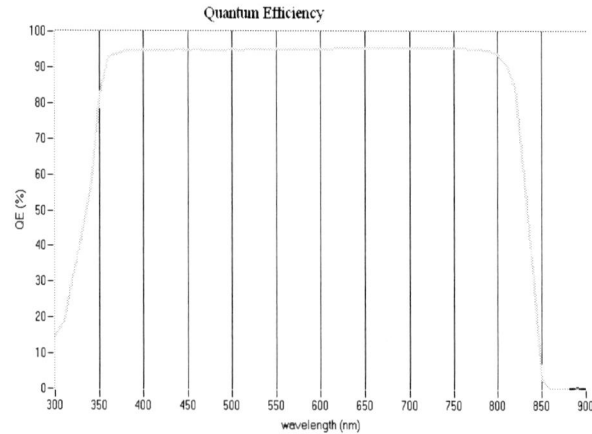

Fig.5. Simulated quantum efficiency of the nanowire CdS-CdTe solar cells.

Table 5 Simulated and Measured Photovoltaics Parameter of Nanowire CdS-CdTe Solar Cells.

Simulation	Voc(mV)	Jsc(mA/cm^2)	FF(%)	Efficiency (%)
Hit=$1*10^{10}$/cm^2 E1=$1*10^{14}$/cm^3	907	29	86	22.7

Figure 5 and 6 show simulated I-V and quantum efficiency of the nanowire CdS-CdTe solar cells with efficiency of 22.7%. Reducing donor traps in CdTe layer and acceptor type interface states significantly improve shunt resistance, series resistance, fill factor and Voc. Observed from quantum efficiency, the 22.7% of the nanowire CdS-CdTe solar cells show nearly ideal (95%) response spectrum from 350nm to 800nm. 5% spectrum response loss is caused by light reflection and scattering.

Thus, in order to achieve high efficiency CdTe solar cells, there are three important parameters which need to be considered. Firstly, CdS layer has large energy bandgap which provides high quantum efficiency and photocurrent, and also reduce effects of defect traps in the CdS layer on solar cell performance, especially photocurrent and quantum efficiency. Secondly, it is necessary to increase shallow doping concentration and/or reduce donor trap concentration in the CdTe layer. Thirdly, it is essential that interface between CdS and CdTe has to be significantly reduced. Therefore, an approaches has to be searched to reduce interface states by for example addressing lattice mismatch between CdS and CdTe.

IV. CONCLUSION

SCAPS simulation is conducted to establish effect of defects on performance of the nanowire solar cells. In the simulation model, compensating acceptor type traps with a liner distribution of $1.05*10^{18}$/cm^3 and $6.0*10^{17}$/cm^3 is introduced in CdS nanowires. Donor traps with a liner distribution of $2.5*10^{14}$/cm^3 and $8.0*10^{14}$/cm^3 are presented in p-CdTe layer. Acceptor type interface states are tuned to $1.2*10^{12}$/cm^2. The simulation models fit well with measured I-V, C-V and quantum efficiency characteristics of 12% nanowire solar cells. Simulation indicates that CdS layer with large energy bandgap reduce effects of defect traps in the CdS layer on solar cell performance. It is necessary to increase shallow doping concentration and/or reduce donor trap concentration in the CdTe layer. It is essential that acceptor type interface states between CdS and CdTe has to be significantly reduced.

ACKNOWLEDGMENT

This work was supported in part by grants from the National Science Foundation, (NSF-NIRT-ECS-0609064) and (NSF-EPCOR EPS-0447479), by grants from National Science Foundation (NSF-1505509), and by grants from the Kentucky Science and Engineering Foundation, (KSEF–148-502-02-27, KSEF- 148-502-03-68). Authors would like to acknowledge Dr. Sai Guduru for characterization support.

REFERENCES

1 Kranz, L. *et al.* Doping of polycrystalline CdTe for high-efficiency solar cells on flexible metal foil. *Nature communications* **4** (2013).

2 FirstSolar. *http://www.firstsolar.com/*.

3 Dang, H. *et al.* in *Photovoltaic Specialists Conference (PVSC), 2012 38th IEEE.* 002697-002701 (IEEE).

4 Dang, H. *et al.* Cadmium sulfide nanowire arrays for window layer applications in solar cells. *Solar Energy Materials and Solar Cells* **126**, 184-191 (2014).

5 Dang, H. & Singh, V. Effects of anodic aluminum oxide membrane on performance of nanostructured solar cells. *Materials Research Express* **2**, 055001 (2015).

6 Dang, H. & Singh, V. P. Nanowire CdS-CdTe solar cells with molybdenum oxide as contact. *Scientific reports* **5** (2015).

7 Dang, H., Singh, V. P. & Guduru, S. in *Photovoltaic Specialist Conference (PVSC), 2015 IEEE 42nd.* 1-6 (IEEE).

8 Dang, H., Singh, V. P., Guduru, S., Rajaputra, S. & Chen, Z. D. Nanotube photovoltaic configuration for enhancement of carrier generation and collection. *Nano Research* **8**, 3186-3196 (2015).

9 Dang, H., Singh, V. P., Guduru, S. & Hastings, J. T. Embedded nanowire window layers for enhanced quantum efficiency in window-absorber type solar cells like CdS/CdTe. *Solar Energy Materials and Solar Cells* **144**, 641-651 (2016).

10 Alex Niemegeers, M. B., Koen Decock, Stefaan Degrave, Johan Verschraegen. *http://scaps.elis.ugent.be/*.

11 Gloeckler, M., Fahrenbruch, A. & Sites, J. in *Photovoltaic Energy Conversion, 2003. Proceedings of 3rd World Conference on.* 491-494 (IEEE).

12 Nollet, P., Köntges, M., Burgelman, M., Degrave, S. & Reineke-Koch, R. Indications for presence and importance of interface states in CdTe/CdS solar cells. *Thin Solid Films* **431**, 414-420 (2003).

13   Seymour, F. H., Kaydanov, V., Ohno, T. R. & Albin, D. Cu and Cd Cl 2 influence on defects detected in CdTe solar cells with admittance spectroscopy. *Applied Physics Letters* **87**, 153507 (2005).

14   Balcioglu, A., Ahrenkiel, R. & Hasoon, F. Deep-level impurities in CdTe/CdS thin-film solar cells. *Journal of Applied Physics* **88**, 7175-7178 (2000).

Analytical description of charged grain boundary recombination in polycrystalline thin film solar cells

Benoit Gaury[1,2] and Paul M. Haney[1]

[1]Center for Nanoscale Science and Technology National Institute of Standards and Technology
Gaithersburg, MD, 20899, USA
[2]Maryland NanoCenter, University of Maryland College Park, MD 20742

Abstract—We present analytic expressions for the dark current-voltage relation of a pn^+ junction with a positively charged columnar grain boundary, containing a distribution of defect states in the band gap. A closed form relation for the open-circuit voltage V_{oc} is provided for an illuminated junction. These findings are verified by direct comparison with numerical simulations, and provide a quantitative understanding of the reduction of V_{oc} by grain boundaries in thin film photovoltaic materials.

Index Terms—grain boundary, thin films, recombination, open-circuit voltage.

I. INTRODUCTION

Despite years of research, precise guidelines for the improvement of chalcogenide materials, like CdTe, are still unavailable for photovoltaic applications [1]. In particular, the role of grain boundaries in these materials is not well understood. High densities of grain boundaries generally enhance recombination, hence reducing power conversion efficiency. However, recent development in thin-film photovoltaics based, for instance, on CdTe or Cu(In,Ga)Se$_2$, have led to unexpectedly high efficiencies despite large densities of grain boundaries [2].

Even though nanoscale measurements may be difficult to interpret [3], electron beam induced current [4]–[6] and Kelvin probe microscopy [7]–[9] experiments have revealed positively charged grain boundaries in these materials. So far, the corresponding body of theoretical studies has consisted of one-dimensional models [10]–[13], and more complex numerical simulations [14]–[17]. The latter revealed the paradoxical impact of grain boundaries on the efficiency of thin film solar cells. While grain boundaries might improve carrier collection, they also reduce the open-circuit voltage, leading to a more contrasted picture than often presented. Our recent analytical works Refs. [18], [19] are consistent with this finding.

In this work, we present new analytical expressions for the dark current-voltage $J(V)$ relation of a pn^+ junction with positively charged grain boundaries, containing many discrete defect states in the gap. We use physical descriptions of the electron and hole transport to create a simplified model which captures the essential features of the drift-diffusion-Poisson equations. The accuracy of this model and the corresponding

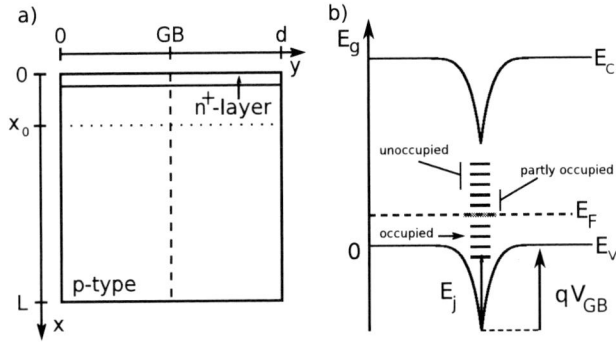

Fig. 1. (a) Two-dimensional model system of a pn^+ junction containing a single grain boundary. (b) Band structure in the neutral region of the p-doped semiconductor, orthogonal to the grain boundary. The dashed line is the thermal equilibrium Fermi level E_F, surrounded by the grain boundary defect states E_j. E_C and E_V are the maxima of the conduction and valence bands, E_g is the material bandgap energy, V_{GB} is the grain boundary built-in potential. The red line indicates the defect neutral energy level E_{GB}. We take the energy reference at the valence band edge in the bulk of the neutral region.

predictions for the grain boundary dark recombination current are verified numerically.

II. PHYSICAL MODEL OF A GRAIN BOUNDARY WITH MULTIPLE GAP STATES

Our model system, depicted in Fig. 1(a), is a pn^+ junction with a vertical grain boundary. We assume selective contacts such that the hole (electron) current vanishes at the $n(p)$-contact, and we use periodic boundary conditions in the y-direction. We note x_0 the position where electron and hole densities are equal in the grain interior.

Polycrystalline semiconductors possess a wide variety of defects, which all have specific formation energies as a function of the local environment and Fermi level. For a given Fermi level, we assume that the type of defect with the lowest formation energy is the most prevalent. Our model then considers a distribution of gap states that results from this particular defect. Because experimental evidence tend to indicate that grain boundaries are positively charged in materials like CdTe [4], [7], [8], we focus on distributions of gap states that provide positively charged defects. In order to conserve the electroneutrality of the device, the positive charges must be screened by nearby negative charges: free electrons in an n-type material or ionized acceptor dopants in a p-type material. In a p-type material, the system responds to the positive charge by developing an electric field around the grain boundary. This electric field repels holes from the grain boundary core, creating a depleted region around it. This depleted region is negatively charged because of the uncompensated ionized acceptors, and therefore compensates the positive charge of the defect. For the band structure across the grain boundary shown in Fig. 1(b), the electric field results in the formation of a built-in potential V_{GB} around the grain boundary. The amplitude of V_{GB} depends on the doping density, the distribution and density of the gap states.

The built-in potential and the Fermi level determine the type of the grain boundary, i.e. the majority carrier at the grain boundary core. In a p-type material, low values of built-in potentials create a hole depletion at the grain boundary, but holes remain majority carrier at the grain boundary core. Large V_{GB} values lead to type inversion at the grain boundary core, i.e. electrons become majority carrier. Each grain boundary type has a different carrier transport under nonequilibrium conditions. In this work we consider both inverted and non-inverted grain boundaries.

We model the grain boundary as a two-dimensional plane with amphoteric (acceptor and donor) states. The grain boundary charge reads

$$Q_{GB} = q\rho_{GB} \sum_{j=1}^{M} (1 - 2f_{GB}(E_j)) \qquad (1)$$

where ρ_{GB} is the 2D defect density of states, and M is the number of gap states. The occupancies of each state is given by

$$f_{GB}(E_j) = \frac{S_n n_{GB} + S_p \bar{p}_j}{S_n(n_{GB} + \bar{n}_j) + S_p(p_{GB} + \bar{p}_j)}, \qquad (2)$$

where n_{GB} (p_{GB}) is the electron (hole) carrier density at the grain boundary, S_n (S_p) is the electron (hole) surface recombination velocity, $\bar{n}_j$ and $\bar{p}_j$ are

$$\bar{n}_j = N_C e^{(-E_g + E_j)/k_B T} \qquad (3)$$

$$\bar{p}_j = N_V e^{-E_j/k_B T} \qquad (4)$$

where E_j is a defect energy level calculated from the valence band edge, N_C (N_V) is the conduction (valence) band effective density of states, E_g is the material bandgap, k_B is the Boltzmann constant and T is the temperature. At thermal equilibrium Eq. (2) reduces to the Fermi-Dirac distribution $f_{GB}(E) = (1 + \exp[(E - E_F)/k_B T])^{-1}$. We introduce E_{GB} as the neutral level of the distribution of states, such that the occupied and empty gap states (below and above E_{GB}) contribute an equal and opposite charge (red line in Fig. 1(b)). In thermal equilibrium, the problem of many defect states can therefore be mapped onto an effective *single* defect level

positioned at energy E_{GB}. The charge of the single effective level is then

$$Q_{GB} = qM\rho_{GB}(1 - 2f_0(E_{GB})), \qquad (5)$$

where f_0 is the effective occupancy of the effective state E_{GB}. In the limit of large defect density of states, we find that E_{GB} is given by the average of the gap state energies for an even number of states, and is equal to the gap state that has the same number of states above and under it for an odd number of states.

We consider large grain boundary defect densities such that the Fermi level is pinned at E_{GB} (see Fig. 1(b)). This situation occurs for densities above the critical value

$$\rho_{GB}^{crit} = \frac{1}{q} \frac{\sqrt{8\epsilon N_A(E_{GB} - E_F - k_B T)}}{\sum_{j=1}^{M} 1 - \frac{2}{1 + e^{(E_j - E_{GB} + k_B T)/k_B T}}}. \qquad (6)$$

Defining V_{GB}^0 as the equilibrium potential between the grain boundary and bulk of the neutral region, then assuming $\rho_{GB} > \rho_{GB}^{crit}$ leads to

$$qV_{GB}^0 \approx E_{GB} - E_F. \qquad (7)$$

The scope of this work is limited to built-in potentials such that $V_{GB}^0 \gg k_B T/q$.

III. ASSUMPTIONS AND GRAIN BOUNDARY PROPERTIES AWAY FROM EQUILIBRIUM

The full two-dimensional drift-diffusion-Poisson set of equations is not analytically solvable. We therefore focus on the solution along the grain boundary core, which reduces the problem to one dimension. This is made possible by the electrostatic confinement of electrons to the grain boundary core. Our analysis relies on assumptions that enable analytical solutions.

Our main assumption is that the hole quasi-Fermi level is flat along and across the grain boundary. Because electrons carry the current along the grain boundary, the corresponding hole current is negligible and so are the gradients of hole quasi-Fermi level. These gradients across the grain boundary are *a priori* not negligible. High surface recombination velocities at the grain boundary core can produce strong hole currents transverse to the grain boundary. In this case, our analysis relies on the assumption of high hole mobility, which enables strong hole currents without the need for gradients of hole quasi-Fermi level.

We further assume that the grain boundary charge does not change with applied voltage. This is justified by the limit of large defect density $Q_{GB}(V)/(q\rho_{GB}) \ll 1$. This assumption implies that f, the nonequilibrium counterpart to f_0 in Eq. (5), obeys $f = f_0 \approx 1/2$. This constraint leads to the relation

$$\sum_{j=1}^{M} f_{GB}(E_j) = M/2. \qquad (8)$$

Because Eq. (8) depends on the grain boundary carrier densities, the relative sizes of the terms in the occupancy Eq. (2) defines three regimes with distinct properties:

- "n-type" grain boundary: In this case, the grain boundary electron density determines the defect occupancy. f remains fixed because of the pinning of the *electron* quasi-Fermi level to the neutral point of the gap states E_{GB} for this case. We also assume that the electron quasi-Fermi level is relatively flat and equal to its bulk value. This is valid because the high electron density in the grain boundary core enables high currents with relatively small quasi-Fermi level gradients.

- "p-type" grain boundary: In this case, the grain boundary hole density determines the occupancy of the defect states. f remains fixed because of the pinning of the *hole* quasi-Fermi level to the neutral point of the gap states E_{GB} for this case. The relatively low electron density at the grain boundary core forces the electron quasi-Fermi level to develop gradients to drive the electron current along the grain boundary. In this case we solve a one-dimensional diffusion equation for the electron density along the grain boundary to obtain the carrier densities and recombination rate.

- High recombination: For sufficiently large applied voltages, the electron and hole carrier densities are the dominating terms in f and determine the defect states occupancies. There is no pinning of quasi-Fermi levels in this case, but we find that the grain boundary carrier densities satisfy

$$p_{GB} = \gamma(V) n_{GB} \qquad (9)$$

where the density ratio γ varies weakly with voltage. In this regime, Eq. (9) together with the assumption of flat hole quasi-Fermi level leads to a one-dimensional drift-diffusion equation for electrons confined to the grain boundary. Solving this equation leads to the carrier densities and recombination. In the case where each occupancy is independent of its gap state energy, one can show that $\gamma = 1$ and the carrier densities must be equal for the system to remain electrically neutral.

A final assumption is that the grains are not fully depleted. For doping densities on the order of 10^{15} cm^{-3}, this requirement implies grain size above 2 μm. For reference, a recent measurement [20] found that the average grain size in CdTe thin films (excluding twin boundaries) was 2.3 μm.

IV. GRAIN BOUNDARY DARK RECOMBINATION CURRENT

We provide expressions for the grain boundary dark recombination current. The general expression of the grain boundary dark current density reads

$$J_{GB}(V) = \frac{1}{d} \int_0^{L_{GB}} dx \, R_{GB}(x), \qquad (10)$$

with L_{GB} the length of the grain boundary. R_{GB} is the Schokley-Read-Hall recombination

$$R_{GB}(x) = \sum_{j=1}^M \frac{S_n S_p (n_{GB} p_{GB} - n_i^2)}{S_n (n_{GB} + \bar{n}_j) + S_p (p_{GB} + \bar{p}_j)}, \qquad (11)$$

with n_i the intrinsic carrier density. The complexity of many defect states, as opposed to a single one, will be incorporated into effective surface recombination velocities. The physical descriptions of the electron and hole transport are the same as in the single state problem, and the resulting grain boundary dark currents are formally identical to the ones presented in Ref. [18].

In the n-type grain boundary the recombination is determined by holes, which flow from the p-type grain interior into the grain boundary core. Because most of the absorber is p-type, the recombination is uniform along the entire grain boundary. The grain boundary carrier densities read

$$n_{GB}(x) = N_C e^{(-E_g + E_{GB})/k_B T} \qquad (12)$$

$$p_{GB}(x) = N_V e^{(-E_{GB} + qV)/k_B T}. \qquad (13)$$

Using the fact that $S_n n_{GB}$ is much larger than $S_p p_{GB}$ in the recombination Eq. (11) leads to the grain boundary dark current

$$J_{GB}(V) = \frac{S_p L_{GB}}{d} N_V e^{(-E_{GB} + qV)/k_B T}, \qquad (14)$$

where the effective surface recombination velocity $\mathcal{S}_p$ reads

$$\mathcal{S}_p = \sum_{j=1}^M \frac{1}{1 + \frac{\bar{n}_j}{\bar{n}_{GB}} + \frac{S_p \bar{p}_j}{S_n \bar{n}_{GB}}}. \qquad (15)$$

A close look at $\mathcal{S}_p$ shows that only states with energies $E_g - E_{GB} \lesssim E \lesssim E_{GB}$ contribute significantly to the recombination (blue region in Fig. 2(a)). The upper limit results from the fact that states above E_{GB} are empty of electrons. The lower limit is the energy at which holes are emitted from the defect state to the valence band faster than electrons relax from the conduction band to the defect state.

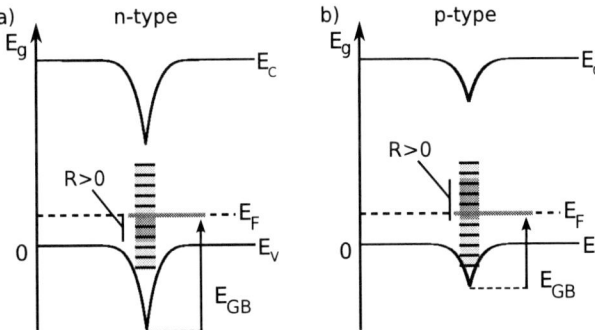

Fig. 2. Schematic of the states contributing to the recombination for the (a) n-type and (b) p-type grain boundary. E_{GB} is the grain boundary neutral energy level. States (not) contributing to the recombination are shaded in blue (grey).

The p-type grain boundary has a similar interpretation. The recombination is determined by electrons flowing into the grain boundary core from regions of the grain interior where $n > p$. This corresponds to $x < x_0$ in Fig. 1(a). The recombination is therefore mainly concentrated within the n-region of the pn junction depletion region, and is uniform for $x < x_0$. The grain boundary carrier densities read

$$n_{\mathrm{GB}}(x) = N_C e^{(-E_g + E_{\mathrm{GB}})/k_B T} e^{qV/k_B T} \qquad \text{for } x < x_0$$
$$= N_C e^{(-E_g + E_{\mathrm{GB}})/k_B T} e^{qV/k_B T} e^{-\frac{x - x_0}{L_n}} \quad \text{for } x > x_0 \tag{16}$$

$$p_{\mathrm{GB}}(x) = N_V e^{-E_{\mathrm{GB}}/k_B T}, \tag{17}$$

where $L_n = \sqrt{2 D_n L_{\mathcal{E}}/S_n}$ (D_n: electron diffusion coefficient) is the electron diffusion length, and $L_{\mathcal{E}} = k_B T/\mathcal{E}_y$ is the characteristic length of the electric field transverse to the grain boundary. $\mathcal{E}_y$. S_n is the effective surface recombination velocity in this case. Using the fact that $S_p p_{\mathrm{GB}}$ is much larger than $S_n n_{\mathrm{GB}}$ in the recombination Eq. (11) leads to the grain boundary dark recombination current

$$J_{\mathrm{GB}}(V) = \frac{S_n}{d} N_C e^{(-E_g + E_{\mathrm{GB}} + qV)/k_B T}$$
$$\times \left[x_0 + L_n \left(1 - e^{-\frac{L_{\mathrm{GB}} - x_0}{L_n}} \right) \right], \quad (18)$$

where S_n reads

$$S_n = \sum_{j=1}^{M} \frac{1}{1 + \frac{\bar{p}_j}{\bar{p}_{\mathrm{GB}}} + \frac{S_n \bar{n}_j}{S_p \bar{p}_{\mathrm{GB}}}}. \tag{19}$$

Equation (19) shows that only the states with energies $E_{\mathrm{GB}} \lesssim E \lesssim E_g - E_{\mathrm{GB}}$ contribute significantly to the recombination, as shown in Fig. 2(b). The lower limit results from the fact that states below E_{GB} are empty of holes. The upper limit is the energy at which electrons are emitted from the defect state to the conduction band faster than holes relax from the valence band to the defect state.

As voltage is increased, the grain boundary crosses over to the high-recombination regime. The carrier densities at the grain boundary are now constrained by Eqs (8) and (9). The charge carrier transport is analogous to the single state problem: holes flow towards the pn junction depletion region and the recombination is peaked at a hotspot there. The grain boundary carrier densities read

$$n_{\mathrm{GB}}(x) = \frac{1}{\sqrt{\gamma}} n_i e^{qV/(2k_B T)} e^{-\frac{x}{L'_n}} \tag{20}$$

$$p_{\mathrm{GB}}(x) = \sqrt{\gamma} n_i e^{qV/(2k_B T)} e^{-\frac{x}{L'_n}}, \tag{21}$$

where $L'_n = \sqrt{4 D_n L_{\mathcal{E}}/S}$ is the electron diffusion length, and $L_{\mathcal{E}}$ is the characteristic length of the electric field transverse to the grain boundary. S is the effective surface recombination velocity in this case

$$S = \frac{\gamma S_n S_p}{S_n + \gamma S_p} \sum_{j=1}^{M} \frac{1}{1 + \frac{S_n \bar{n}_j + S_p \bar{p}_j}{(S_n + \gamma S_p) \frac{n_i}{\gamma} e^{qV/(2k_B T)}}}. \tag{22}$$

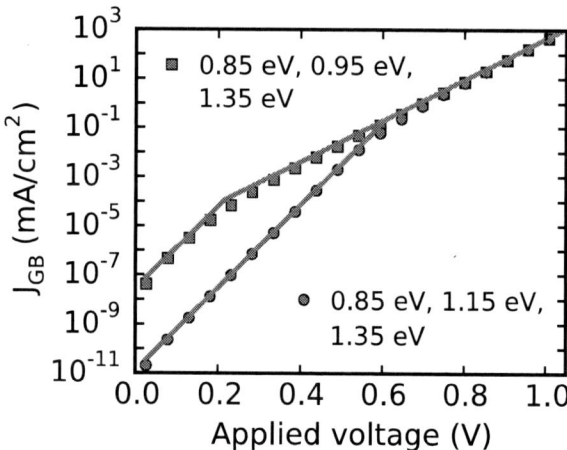

Fig. 3. Grain boundary dark recombination current as a function of voltage for two sets of defects energy levels. Symbols correspond to numerical data, full lines are analytic predictions. Parameters for the simulations are summarized in Table I.

TABLE I
LIST OF DEFAULT PARAMETERS FOR NUMERICAL SIMULATIONS.

Parameter	Value	Parameter	Value
L	$3\ \mu m$	$\mu_n = \mu_p$	$100\ \mathrm{cm^2/(V \cdot s)}$
d	$3\ \mu m$	$\tau_n = \tau_p$	10 ns
N_C	$8 \times 10^{17}\ \mathrm{cm^{-3}}$	$S_{n,p}$	10^5 cm/s
N_V	$1.8 \times 10^{19}\ \mathrm{cm^{-3}}$	ρ_{GB}	$10^{14}\ \mathrm{cm^{-2}}$
E_g	1.5 eV	ϵ	$9.4\ \epsilon_0$
N_A	$10^{15}\ \mathrm{cm^{-3}}$	N_D	$10^{17}\ \mathrm{cm^{-3}}$

The coefficient γ is found by solving Eq. (8) with the carrier densities Eqs. (20)-(21). The above results lead to the grain boundary dark recombination current

$$J_{\mathrm{GB}}(V) = \frac{S L'_n}{\sqrt{\gamma} d} n_i e^{V/(2V_T)} \left[1 - e^{-L_{\mathrm{GB}}/L'_n} \right]. \tag{23}$$

This result differs from the corresponding case in Ref. [18] by the voltage dependence of the effective surface recombination velocity.

To verify the accuracy of these expressions, we solved numerically the drift-diffusion-Poisson set of equations for the geometry presented in Fig. 1(a), and various distributions of defect states. Table I gives a list of the material parameters used for these calculations. We used periodic boundary conditions in the y-direction, and infinite (zero) surface recombination velocity for majority (minority) carriers at the contacts to simulate perfectly selective contacts. Fig. 3 shows the grain boundary recombination current for three gap states. The red (blue) curve corresponds to an n-type (p-type) grain boundary. Because we consider the limit of high defect density of states, the neutral point E_{GB} of each distribution of gap states corresponds to the intermediate energy of each case. A good agreement can be seen between our analytical model and the full numerical simulations.

V. OPEN-CIRCUIT VOLTAGE

Upon comparing the electron diffusion length to the length of the grain boundary in the above results, we find that limiting cases of Eqs. (14), (18) and (23) are of the general form

$$J_{\text{GB}}(V) = \lambda \frac{S}{d} N e^{-E_a/k_B T} e^{qV/(nk_B T)}, \quad (24)$$

where S is a surface recombination velocity, λ is a length characteristic of the recombination region, N is an effective density of states, E_a is an activation energy, n is the ideality factor and V is the applied voltage.

Equation (24) allows us to derive a closed form relation for the open-circuit voltage. It was shown in Ref. [18] that, around V_{oc}, the current-voltage relation of an illuminated junction is given by the sum of the short-circuit current J_{sc} and the dark current $J_{\text{dark}}(V)$. V_{oc} therefore satisfies $J_{\text{dark}}(V_{\text{oc}}) = J_{\text{sc}}$. Assuming large values of surface recombination velocities, the grain boundary recombination dominates over the bulk recombination and $J_{\text{dark}} = J_{\text{GB}}$. Solving for $J_{\text{GB}}(V_{\text{oc}}) = J_{\text{sc}}$ therefore leads to the general form of the open-circuit voltage

$$qV_{\text{oc}}^{\text{GB}} = nE_a + nk_B T \ln\left(\frac{dJ_{\text{sc}}}{S\lambda N}\right). \quad (25)$$

Equation (25) provides insight into how the parameters controlling the grain boundary and the distribution of gap states affect V_{oc}. For instance, the open-circuit voltage increases linearly with the defect activation energy. This energy corresponds to the neutral point of the distribution of gap states in the n-type and p-type regimes. However, $E_a = E_g/2$ in the high-recombination regime, so that the impact the initial distribution of gap states on V_{oc} is minimal (it remains present in the effective surface recombination velocity).

VI. CONCLUSION

We derived analytical relations for the grain boundary dark recombination current and the open-circuit voltage of a pn^+ junction with a positively charged columnar grain boundary. These new expressions account for a distribution of defect states in the gap of the absorber material. We showed that our simplified analytical model describes the essential features of the full numerical calculations.

ACKNOWLEDGMENT

B. G. acknowledges support under the Cooperative Research Agreement between the University of Maryland and the National Institute of Standards and Technology Center for Nanoscale Science and Technology, Award 70NANB14H209, through the University of Maryland.

REFERENCES

[1] S. G. Kumar and K. K. Rao, "Physics and chemistry of cdte/cds thin film heterojunction photovoltaic devices: fundamental and critical aspects," *Energy Environ. Sci.*, vol. 7, no. 1, pp. 45–102, 2014.

[2] M. A. Green, K. Emery, Y. Hishikawa, W. Warta, E. D. Dunlop, D. H. Levi, and A. W. Y. Ho-Baillie, "Solar cell efficiency tables (version 49)," *Prog. Photovolt. Res. Appl.*, vol. 25, no. 1, pp. 3–13, 2017.

[3] P. M. Haney, H. P. Yoon, B. Gaury, and N. B. Zhitenev, "Depletion region surface effects in electron beam induced current measurements," *J. Appl. Phys.*, vol. 120, p. 095702, 2016.

[4] H. P. Yoon, P. M. Haney, D. Ruzmetov, H. Xu, M. S. Leite, B. H. Hamadani, A. A. Talin, and N. B. Zhitenev, "Local electrical characterization of cadmium telluride solar cells using low-energy electron beam," *Sol. Energ. Mat. Sol. Cells*, vol. 117, pp. 499–504, 2013.

[5] O. Zywitzki, T. Modes, H. Morgner, C. Metzner, B. Siepchen, B. Späth, C. Drost, V. Krishnakumar, and S. Frauenstein, "Effect of chlorine activation treatment on electron beam induced current signal distribution of cadmium telluride thin film solar cells," *J. Appl. Phys.*, vol. 114, no. 16, p. 163518, 2013.

[6] C. Li, Y. Wu, J. Poplawsky, T. J. Pennycook, N. Paudel, W. Yin, S. J. Haigh, M. P. Oxley, A. R. Lupini, M. Al-Jassim, S. J. Pennycook, and Y. Yan, "Grain-boundary-enhanced carrier collection in cdte solar cells," *Phys. Rev. Lett.*, vol. 112, no. 15, p. 156103, 2014.

[7] I. Visoly-Fisher, S. R. Cohen, and D. Cahen, "Direct evidence for grain-boundary depletion in polycrystalline cdte from nanoscale-resolved measurements," *Appl. Phys. Lett.*, vol. 82, no. 4, pp. 556–558, 2003.

[8] H. Moutinho, R. Dhere, C.-S. Jiang, Y. Yan, D. Albin, and M. Al-Jassim, "Investigation of potential and electric field profiles in cross sections of CdTe/CdS solar cells using scanning kelvin probe microscopy," *J. Appl. Phys.*, vol. 108, no. 7, p. 074503, 2010.

[9] C.-S. Jiang, R. Noufi, J. AbuShama, K. Ramanathan, H. Moutinho, J. Pankow, and M. Al-Jassim, "Local built-in potential on grain boundary of Cu(In, Ga)Se$_2$ thin films," *Appl. Phys. Lett.*, vol. 84, no. 18, pp. 3477–3479, 2004.

[10] H. C. Card and E. S. Yang, "Electronic processes at grain boundaries in polycrystalline semiconductors under optical illumination," *IEEE Trans. Electron Devices*, vol. 24, no. 4, pp. 397–402, April 1977.

[11] J. G. Fossum and F. A. Lindholm, "Theory of grain-boundary and intragrain recombination currents in polysilicon pn-junction solar cells," *IEEE Trans. Electron Devices*, vol. 27, no. 4, pp. 692–700, 1980.

[12] M. A. Green, "Bounds upon grain boundary effects in minority carrier semiconductor devices: A rigorous perturbation approach with application to silicon solar cells," *J. Appl. Phys.*, vol. 80, no. 3, pp. 1515–1521, 1996.

[13] S. Edmiston, G. Heiser, A. Sproul, and M. Green, "Improved modeling of grain boundary recombination in bulk and p-n junction regions of polycrystalline silicon solar cells," *J. Appl. Phys.*, vol. 80, no. 12, pp. 6783–6795, 1996.

[14] M. Gloeckler, J. R. Sites, and W. K. Metzger, "Grain-boundary recombination in Cu(In,Ga)Se$_2$ solar cells," *J. Appl. Phys.*, vol. 98, no. 11, p. 113704, 2005.

[15] K. Taretto and U. Rau, "Numerical simulation of carrier collection and recombination at grain boundaries in Cu(In, Ga)Se$_2$ solar cells," *J. Appl. Phys.*, vol. 103, no. 9, p. 094523, 2008.

[16] U. Rau, K. Taretto, and S. Siebentritt, "Grain boundaries in Cu(In, Ga)(Se, S)$_2$ thin-film solar cells," *Appl. Phys. A*, vol. 96, no. 1, pp. 221–234, 2009.

[17] F. Troni, R. Menozzi, E. Colegrove, and C. Buurma, "Simulation of current transport in polycrystalline CdTe solar cells," *J. Electron. Mater.*, vol. 42, no. 11, pp. 3175–3180, 2013.

[18] B. Gaury and P. M. Haney, "Charged grain boundaries reduce the open-circuit voltage of polycrystalline solar cells–An analytical description," *J. Appl. Phys.*, vol. 120, p. 234503, 2016.

[19] B. Gaury and P. M. Haney, "Anatomy of charged grain boundaries in polycrystalline solar cells," *arXiv preprint arXiv:1704.04234*, 2017.

[20] J. Moseley, W. K. Metzger, H. R. Moutinho, N. Paudel, H. L. Guthrey, Y. Yan, R. K. Ahrenkiel, and M. M. Al-Jassim, "Recombination by grain-boundary type in cdte," *J. Appl. Phys.*, vol. 118, no. 2, 2015.

978-1-5090-5606-4/17 $31.00 © 2017 IEEE

Imaging the Effect of CdSe Window Layers in CdTe Photovoltaics

John M. Howard[1,2], Elizabeth M. Tennyson[1,2], William B. Gunnarsson[2,3],
Naba R. Paudel[4,5], Yanfa Yan[4,5], Marina S. Leite[1,2]

[1]Department of Materials Science and Engineering, Univ. of Maryland, College Park, MD, USA
[2]Institute for Research in Electronics and Applied Physics, Univ. of Maryland, College Park, MD, USA
[3]Department of Electrical and Computer Engineering, Univ. of Maryland, College Park, MD, USA
[4]Department of Physics and Astronomy, Univ. of Toledo, Toledo, OH, USA
[5]Wright Center for Photovoltaics Innovation and Commercialization, Univ. of Toledo, Toledo, OH, USA

Abstract — **Here we quantify and spatially resolve the carrier collection enhancement of CdSe window layers in CdTe photovoltaic devices by scanning photocurrent microscopy. We compare the photogenerated current of CdTe solar cells with both CdS and CdSe window layers. Our results show that, despite notable spatial variation in current in both cases, the device with the CdSe window layer outperforms the CdS one, in agreement with macroscopic analysis previously performed. The external quantum efficiency (EQE) enhancement occurs near the band edge of CdTe.**

Index Terms —CdTe, mesoscale, scanning photocurrent microscopy, window layers, CdSe, external quantum efficiency

I. INTRODUCTION

Polycrystalline photovoltaic (PV) devices are a promising solution to the need for low-cost renewable energy generation. Cadmium Telluride (CdTe) solar cells are particularly advantageous, as they have low energy payback times, a metric relating performance and system cost [1]. Nevertheless, CdTe-based devices are currently substantially below their theoretical maximum efficiency. The origin of their limited performance is related to their modest V_{oc}, which is likely caused by non-radiative recombination events that occur at the interfaces between the grains [2]-[3]. To improve device performance through bandgap engineering, CdSe has been evaluated as an alternative to the common CdS window layer. While the use of CdSe allows for higher short-circuit current (I_{SC}), it also reduces the open-circuit voltage (V_{OC}) of the devices [4]-[5]. This tradeoff underscores the need for functional imaging of such CdTe devices *in operando* to understand relationships between the morphology, composition, and electrical response of the device [6]. We implement scanning photocurrent microscopy (SPCM) [7]-[10] to perform a comparative analysis of two CdTe samples with CdS and CdSe window layers to measure the photocurrent response enhancement near the band edge of the latter.

II. DEVICE FABRICATION AND CHARACTERIZATION

The CdTe devices measured in this study are fabricated as follows. First, 100 nm of CdS or CdSe is deposited by radio frequency (RF) magnetron sputtering onto a fluorine-doped tin oxide (FTO) substrate at 250 °C under argon atmosphere. The ~4 μm CdTe layer is added *via* closed space sublimation (CSS) with the substrate heated to 610 °C. The CdTe layer is treated with CdCl$_2$ at 390 °C for 30 min. Back contacts consisting of 3.5 nm of Cu and 12 nm of Au, respectively, are added by thermal evaporation. Finally, the devices (Fig. 1 (a)) anneal under nitrogen gas at 200 °C for 20 minutes.

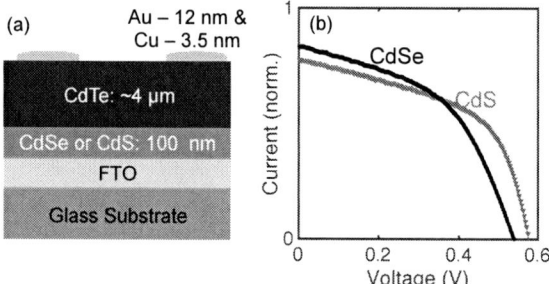

Fig. 1. (a) Schematic of CdTe devices with CdS and CdSe window layers. (b) Light *I-V* under AM1.5 global illumination for both samples.

Fig. 1(b) shows the macroscopic *I-V* characteristics for both devices. Here, the samples are illuminated through the CdTe side, allowing us to perform the photocurrent scans, as will be described in the next section. Due to the material absorption coefficient, there is a strong absorption near the surface of the CdTe layer for wavelengths below 800 nm. For such wavelengths, carrier generation occurs more than 3 μm from the *p-n* junction, dramatically lowering device performance. Nevertheless, for wavelengths > 800 nm we can probe the *p-n* junction of the solar cells, as desired.

III. SCANNING PHOTOCURRENT MICROSCOPY

Scanning photocurrent microscopy, as depicted in Fig. 2, is performed on the CdTe devices with both CdS and CdSe window layers. The photocurrent images are acquired through a confocal microscope using a 100x long working distance objective (NA 0.75) and a piezo scan stage with a 100x100 μm² maximum scan area. A supercontinuum laser fiber coupled into the microscope is used as the excitation source, where the wavelength of the laser is controlled through an

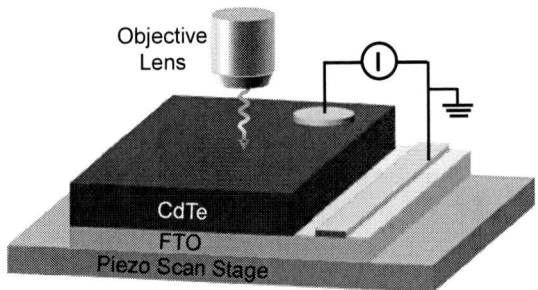

Fig. 2. Diagram of experimental setup for SPCM measurements.

acousto-optic tunable filter. Photocurrent images are obtained on both devices ~50 μm from the edge of the Au contact.

To compare the EQE of the different window layers in the NIR, we spatially resolve the photocurrent and perform a sequence of spectrally dependent measurements every 10 nm between 800 nm to 920 nm under constant incident photon flux (1.89×10^{23} photon s^{-1} m^{-2}). We monitor the laser power fluctuations through the use of a beam splitter and a photodiode. Image artifacts due to power fluctuations are excluded by subtracting the difference between the pixel-by-pixel incident power and the average incident power *via*:

$$I_{xy,cor} = I_{xy,meas} - C\left(P_{xy,diode} - \left\langle P_{xy,diode} \right\rangle\right) \quad (1)$$

where $I_{xy,cor}$ and $I_{xy,meas}$ are the corrected and measured

photocurrent scans, respectively, C is a proportionality constant, and $P_{xy,diode}$ is the power measured by the photodiode. In all cases, the average of the photocurrent scan changes less than 20 fA using this method. All SPCM images in this work have been corrected in this manner.

IV. RESULTS AND DISCUSSION

Optical micrographs of the 15×15 μm^2 representative regions imaged through SPCM reveal the CdTe grains for CdS and CdSe window layer devices, see Figs. 3 (a) and (f), respectively. Differences in the surface morphology are not expected, as the CdTe layers of both cells are deposited with the same method. The spectrally dependent photocurrent scans shown in Fig. 3 (b)-(e) and (f)-(j) exhibit maximum values at the nominal band gap of CdTe (~830 nm) for both CdS and CdSe window layer samples. This excitation wavelength marks a spatial transition in the photocurrent for both devices, as the features seen in the 800 nm scan expand into those at 880 nm. Uniquely, the CdSe window layer device displays a second spatial transition upon reaching 900 nm, as shown in Fig. 3 (j). The enhancement of the CdSe layer short-circuit current under these illumination conditions likely results from the increased external quantum efficiency (EQE) in the NIR, as previously reported [3]. The histograms of the photocurrent for each wavelength confirm how the charge carrier collection is more effective for the solar cells with the CdSe window

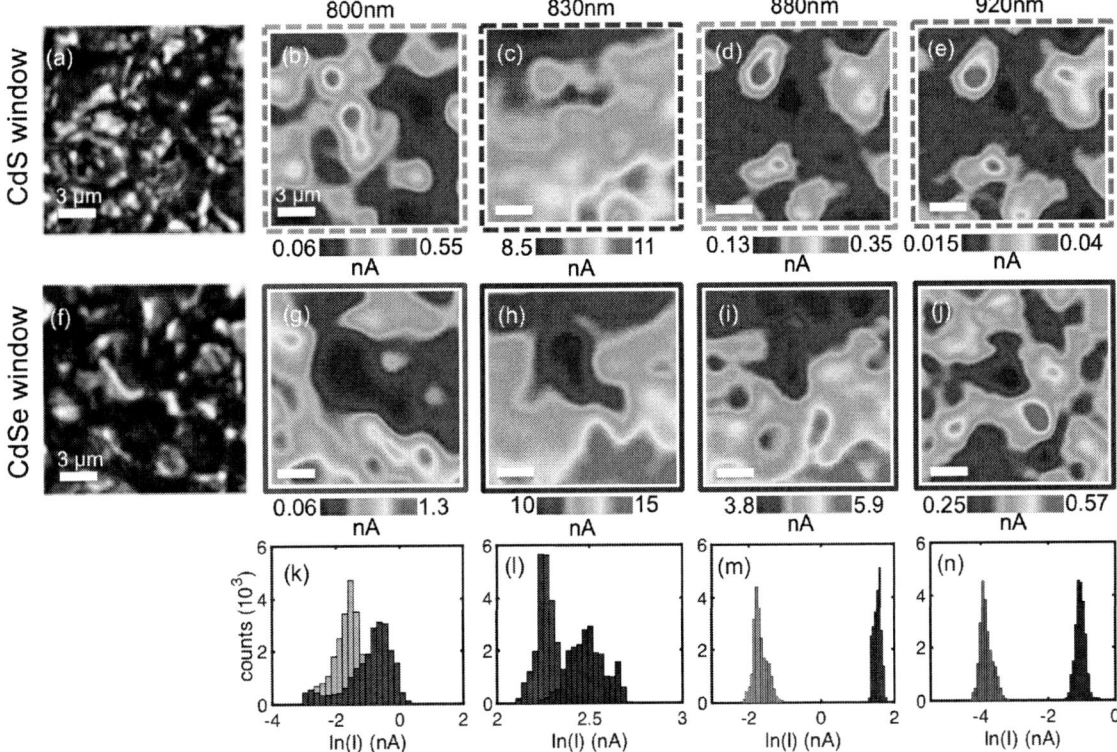

Fig. 3. (a, f) Optical micrographs showing CdTe grains. (b-e, g-j) SPCM images of CdS and CdSe devices at zero bias with varied excitation wavelength under equal incident photon flux. (k-n) Histograms of photocurrent (I) distributions for all scans, where the darker and lighter shades refer to the samples with CdSe and the CdS windows, respectively.

978-1-5090-5606-4/17 $31.00 © 2017 IEEE

layer for the whole wavelength under investigation here, see Fig. 3 (k)-(n). The difference between the two samples is even more evident for $\lambda > 880$ nm, where the photocurrent has a very narrow distribution, as shown by the histograms. We attribute this different electrical response to the fact that Te and Se interdiffuse more easily than Te and S, resulting in a CdTe$_x$Se$_{1-x}$ layer with a bandgap gradient between the p-doped and the n-doped materials that increases charge carrier collection [4], [11].

To better understand the effect of the different window layers on the carrier collection properties of the CdTe solar cells, we determine the EQE for each pixel of all photocurrent images according to:

$$EQE_{xy}(\lambda) = \frac{I_{xy}(\lambda)}{P_{inc}} \times \frac{hc}{\lambda q} \qquad (2)$$

where I_{xy} is the photocurrent measured via SPCM, P_{inc} is the incident power, λ is the wavelength of incident light, h is Planck's constant, c is the speed of light, and q is the elementary charge. To compare the carrier collection efficiency between the two cells, we calculate the average effective EQE of each scan, as shown in Fig. 4. The effective EQE is notably lower than that observed macroscopically, both due to the area and the direction of illumination. Nonetheless, the local EQE enhancement in the 850-920 nm range is substantial and in agreement with previous macroscopic results [4]-[5]. Between 860 and 900 nm, the EQE of the CdSe sample exceeds that of CdSe by > 7 percentage points.

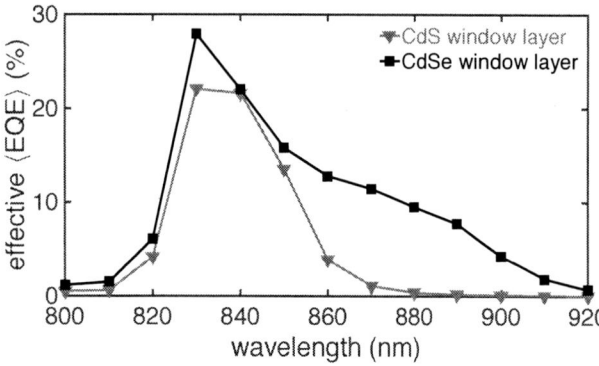

Fig. 4. Average effective EQE for CdSe and CdS window layer cells obtained from spectrally resolved scanning photocurrent microscopy measurements.

V. CONCLUSIONS

The photocurrent of two CdTe photovoltaic devices with CdSe and CdS window layers was spatially resolved with mesoscale resolution. By performing measurements between 800 nm and 920 nm we quantified the improved NIR carrier collection due to the CdSe window layer and showed that it occurs despite the local spatial variations. Our findings are an

important step towards understanding the electro-structural relationships governing performance in CdTe devices. Experiments combining electron back scattering diffraction (EBSD) and SPCM on the very same grains will enable us to determine the correlation between grain orientation and the electrical response of each interface and are planned for the near future.

ACKNOWLEDGEMENTS

The authors acknowledge the technical support from the Nanocenter, and the financial support from 2016 UMD Dean's Fellowship and Mtech - ASPIRE.

REFERENCES

[1] J. Peng, L. Lu, and H. Yang, "Review on life cycle assessment of energy payback and greenhouse gas emission of solar photovoltaic systems," *Renewable and Sustainable Energy Reviews*, vol. 19, pp. 255-274, 2013.

[2] M. A. Green and S. P. Bremner, "Energy conversion approaches and materials for high-efficiency photovoltaics," *Nature Materials*, 16, pp. 23-34, 2017.

[3] J. M. Burst, J. N. Duenow, D. S. Albin, E. Colegrove, M. O. Reese, J. A. Aguiar, C.-S. Jiang, M. K. Patel, M. M. Al-Jassim, D. Kuciauskas, S. Swain, T. Ablekim, K. G. Lynn and W. K. Metzger, "CdTe solar cells with open-circuit voltage breaking the 1 V barrier," *Nature Energy*, vol. 1, pp.16015-16021, 2016.

[4] N. R. Paudel and Y. Yan, "Enhancing the photo-currents of CdTe thin-film solar cells in both short and long wavelength regions," *Applied Physics Letters*, vol. 105, pp.183510-5, 2014.

[5] J. D. Poplawsky, W. Guo, N. R. Puadel, A. Ng, K. More, D. Leonard, Y. Yan, N. R. Paudel and Y. Yan, "Structural and compositional dependence of the CdTe$_x$Se$_{1-x}$ alloy layer photoactivity in CdTe-based solar cells," *Nature Communications*, vol. 7, pp. 12537-12546, 2016.

[6] E. M. Tennyson, J. M. Howard, M. S. Leite, "Mesoscale Functional Imaging of Materials for Photovoltaics," *ACS Energy Letters*, Invited Perspective, In Review, 2017.

[7] M. S. Leite, M. Abashin, H. J. Lezec, A. G. Gianfrancesco, A. A. Talin, and N. B. Zhitenev, "Nanoscale Imaging of Photocurrent and Efficiency in CdTe Solar Cells," *ACS Nano*, vol. 8, pp. 11883-11890, 2014.

[8] M. S. Leite, M. Abashin, H. J. Lezec, A. G. Gianfrancesco, A. A. Talin, and N. B. Zhitenev, "Mapping the Local Photoelectronic Properties of Polycrystalline Solar Cells Through High Resolution Laser-Beam-Induced Current Microscopy," *IEEE Journal of Photovoltaics*, vol. 4, pp. 311-316, 2014.

[9] E. M. Tennyson, J. A. Frantz, J. M. Howard, W. B. Gunnarsson, J. D. Myers, R. Y. Bekele, J. S. Sanghera, S. Na, M. S. Leite, "Photovoltage Tomography in Polycrystalline Solar Cells," *ACS Energy Letters*, vol. 1, pp. 899-905, 2016.

[10] S. L. Howell, S. Padalkar, K. Yoon, Q. Li, D. D. Koleske, J. J. Wierer, G. T. Wang, and L. J. Lauhon, "Spatial Mapping of Efficiency of GaN/InGaN Nanowire Array Solar Cells Using Scanning Photocurrent Microscopy," *Nano Letters*, vol. 13, pp. 5123-5128, 2013.

[11] D. W. Lane, "A review of the optical band gap of thin film CdS$_x$Te$_{1-x}$," *Solar Energy Materials and Solar Cells*, vol. 9, pp.1169-1175, 2006

Investigation of traps density and position in alkali treated Cu(In,Ga)Se$_2$ thin films and solar cells

Shankar Karki[1], Pran K. Paul[2], Grace Rajan[1], Chinedum Akwari[1], Angus Rockett[3], Steven A Ringel[2], Aaron R. Arehart[2], Sylvain Marsillac[1]

[1]Virginia Institute of Photovoltaics, Old Dominion University, Norfolk, VA 23529, USA
[2]Dept. of Electrical & Computer Engineering, The Ohio State University, Columbus, OH 43210, USA
[3]Dept. of Metallurgical and Materials Engineering, Colorado School of Mines, Golden, CO 80401, USA

Abstract — **This study contributes to the discussion on the deep traps in Cu(In,Ga)Se$_2$ material with KF post deposition treatment (KF PDT). The samples were prepared on nominally alkali-free alumina substrate. In order to investigate the deep traps, capacitance-based deep level transient and optical spectroscopies (DLTS/DLOS) were performed on the samples. Two defect levels were observed at Ev+0.59eV and Ev+0.98eV. Secondary ion mass spectrometry (SIMS) profile indicated the diffusion of potassium throughout the CIGS layer. Device measurements and simulation indicate that the Ev+0.98eV play an important role in reducing efficiency for devices without KF PDT.**

Index Terms — **Alkali PDT, CIGS, DLTS, DLOS, deep traps, thin-film solar cells.**

I. Introduction

With the introduction of alkali elements by post-deposition treatment, the efficiency of laboratory scale CIGS solar cells raised to record efficiency (22.6%) [1]. Studies show that K doping increases the open circuit voltage, the fill factor and also the net carrier density [2]. It has also been reported that KF treatment causes Cu and Ga depleted CIGS surface, widening of the surface band gap of the CIGS potentially due to the formation of K-In-Se compounds [3, 4]. Furthermore, surface modification has been linked to better CIGS/CdS hetero-junction. Usually, the CIGS solar cells are fabricated on soda lime glass (SLG) substrates, from which the beneficial Na impurity diffuses during high-temperature growth. In this study, in order to investigate the electrical properties of K-doped CIGS films, the films were grown on Mo-coated alumina substrate. The deep traps location and density was studied using a combination of capacitance-based deep level transient spectroscopy (DLTS) and deep level optical spectroscopy (DLOS) techniques.

II. Experiment

CIGS thin film about 2.3 μm in thickness were deposited on a Mo-coated polished alumina (Al$_2$O$_3$) substrate. Alkali-free alumina substrate, which has a comparable coefficient of thermal expansion to CIGS, has been demonstrated to produce above 15% efficient devices with appropriate Na doping [5]. The Mo back contact layer (800 nm) was deposited by dc magnetron sputtering using Ar gas. The CIGS absorber layers were grown by the three-stage co-evaporation process in a high vacuum chamber as described in our previous publication [6]. After the deposition of the CIGS film, the samples were cooled down. For KF PDT samples, 20 nm KF was deposited on top and subsequently the substrate temperature was raised to 350°C in order to allow KF diffusion through the film in presence of Se for 15 min. The solar cells were then completed by depositing 50 nm of CdS buffer layer by chemical bath deposition followed by rf sputtered i-ZnO (~75nm) and Al:ZnO (~275nm) transparent conducting oxide (TCO). The top contact (Ni/Al/Ni) grids were completed by electron beam evaporation. The KF PDT samples were rinsed in ammonia solution before deposition of the CdS layer in order to remove the residual K compounds on the surface. The average composition of the CIGS film as measured by X-ray fluorescence (XRF) show atomic fraction composition of 22.9% Cu, 17.6% In, 9.6% Ga and 49.9% Se with an average Ga/(Ga+In) of 0.35 and Cu/(Ga+In) of 0.85.

The current density-voltage (J-V) characteristics of the solar cells were measured under standard test condition using 100 mW/cm^2 (1.5G) spectrum. The external quantum efficiency (EQE) was measured using chopped monochromatic light. The experimental details of the DLTS and DLOS characterizations have been explained in our previous publications [7, 8].

III. Results and Discussion

Fig. 1 shows the depth profile of the Cu, In, Ga, Se and K elements in the CIGS film with and without KF PDT. Comparing the SIMS profile of the samples with and without KF PDT, the elemental distribution of the constituents did not show appreciable difference apart from the K depth profile. SIMS revealed that the K is distributed throughout the CIGS absorber layer in KF treated samples. The intensity of the K signal peaks at the surface at about 10% of the film depth from the surface; however, it remained fairly homogenous in the rest of the film thickness. After the validation of K incorporation from SIMS, defect spectroscopy measurements were performed on the CIGS films using DLTS and DLOS technique.

Fig. 2 shows the defect level distribution for the CIGS film from DLTS measurement in the temperature range from 150 K to 375 K. A positive peak in the DLTS spectra at ~365 K represents a majority carrier trap with a maximum density of ~

6.87×10^{13} cm^{-3} and 6.19×10^{13} cm^{-3} in KF untreated and KF treated samples respectively. The activation energy of this trap was determined to be approximately 0.59 eV above the valence band edge from the Arrhenius plot. This trap has been observed previously on CIGS samples with similar composition prepared on SLG substrate [7-9]. Furthermore, this defect was reported to be localized at some of the grain boundaries via scanning-DLTS maps superposed with the topography of the same area. The origin of this defect level has not been identified yet, although various sources of origin including In vacancies, anti-site defects and impurity contribution have been speculated [9].

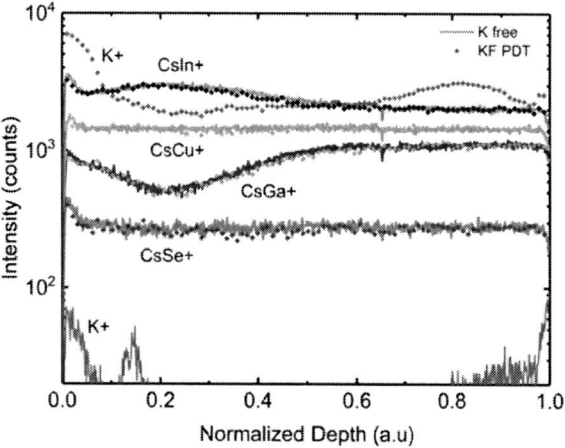

Fig. 1. SIMS depth profile of the samples with and without KF PDT

Fig. 3 shows the DLOS spectra of the CIGS sample. In this case, a deep level was observed at ~0.98 eV above the valence band edge with a density of ~3.0×10^{15} cm^{-3} and ~2.61×10^{15} cm^{-3} in KF untreated and KF treated samples respectively (~13% decrease). This trap has been previously observed in CIGS samples prepared on a SLG substrate [6] as well. The effect of varying concentration of this trap through SCAPS simulation has been reported previously [6], showing the decline in open circuit voltage and fill factor with increasing density of this particular deep center. Moreover, the reduction in the density of this trap was also observed for KF PDT CIGS samples with Na [6], indicating that KF plays the roles in passivating this defect irrespective of whether there is presence of Na or not.

Table 1 shows the average J-V parameters of the cells with and without KF PDT. It is evident that the KF treated samples have better power conversion efficiency, especially due to the

TABLE I.
THE AVERAGE CELL PARAMETERS OF THE CIGS CELLS STUDIED IN THIS WORK

Sample	Eff. (%)	V_{oc} (mV)	J_{sc} (mA/cm^2)	FF (%)
No KF	10.3	580	32.5	55.2
KF PDT	13.2	652	32.6	62.3

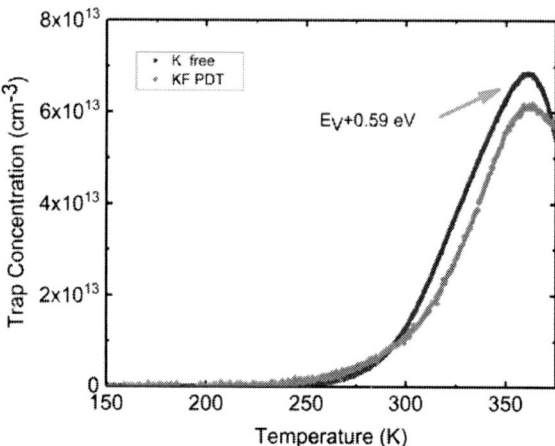

Fig. 2. DLTS spectra of CIGS sample with and without KF PDT

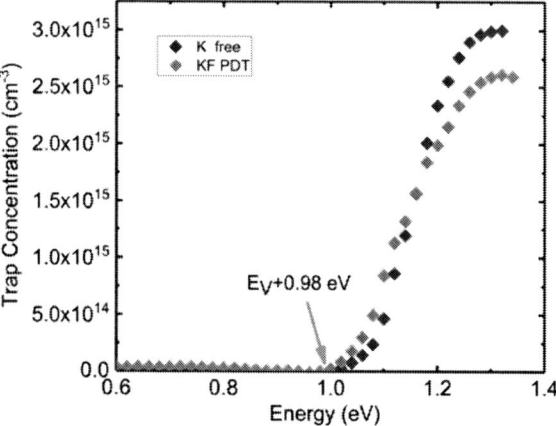

Fig. 3. DLOS spectra of CIGS samples with and without KF PDT

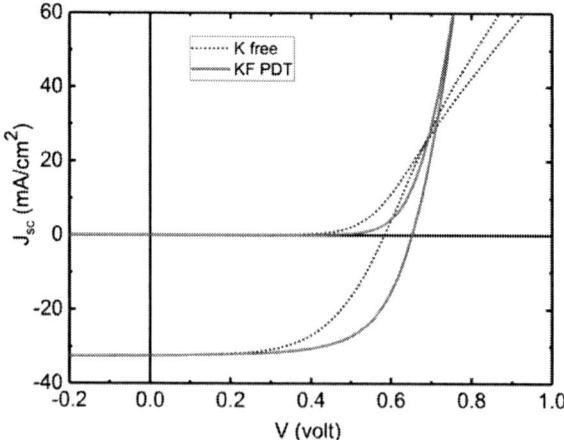

Fig. 4 J-V curve of the CIGS device with and without KF PDT

increase in open circuit voltage and fill factor. There is hardly any change in the short circuit current. Fig. 5 shows a representative J-V curve of the CIGS device with and without KF PDT. The K free sample showed a roll-over effect, which was eliminated in the KF treated sample. The roll-over effect is often observed in the alkali free (Na or K)[10, 11] samples which has been linked to the secondary diode at the Mo/CIGS back contact[10].

IV. CONCLUSION

To summarize, we have investigated the deep traps in KF treated CIGS thin film by DLTS and DLOS measurement through the bandgap and the K incorporation has been verified by SIMS in a nominally alkali free Alumina substrate. Two deep levels are found to be located at E_V+0.59 eV and E_V+0.98 eV. In particular, the concentration of E_V+0.98 eV trap was ~13% lower in the KF treated samples compared to untreated sample. The J-V measurements show that the power conversion efficiency improves, particularly due to enhanced open circuit voltage and fill factor in KF treated sample compared to the untreated sample and a roll over effect was observed in the alkali free samples which in turn was eliminated in the KF treated sample.

ACKNOWLEDGEMENT

This research was supported by the Department of Energy, under the Contract No. DE-EE0007141.

REFERENCES

[1] P. Jackson, R. Wuerz, D. Hariskos, E. Lotter, W. Witte, and M. Powalla, "Effects of heavy alkali elements in Cu(In,Ga)Se2 solar cells with efficiencies up to 22.6%," *physica status solidi (RRL) – Rapid Research Letters,* 2016.

[2] A. Laemmle, R. Wuerz, and M. Powalla, "Efficiency enhancement of Cu (In, Ga) Se2 thin‐film solar cells by a post‐deposition treatment with potassium fluoride," *physica status solidi (RRL)-Rapid Research Letters,* vol. 7, no. 9, pp. 631-634, 2013.

[3] A. Chirilă *et al.*, "Potassium-induced surface modification of Cu(In,Ga)Se2 thin films for high-efficiency solar cells," *Nat Mater,* Letter vol. 12, no. 12, pp. 1107-1111, 12//print 2013.

[4] E. Handick *et al.*, "Potassium Postdeposition Treatment-Induced Band Gap Widening at Cu (In, Ga) Se2 Surfaces–Reason for Performance Leap?," *ACS applied materials & interfaces,* vol. 7, no. 49, pp. 27414-27420, 2015.

[5] P. Salom, V. Fj?llstr?m, A. Hultqvist, and M. Edoff, "Na Doping of CIGS Solar Cells Using Low Sodium-Doped Mo Layer," *IEEE Journal of Photovoltaics,* vol. 3, no. 1, pp. 509-513, 2013.

[6] S. Karki *et al.*, "In Situ and Ex Situ Investigations of KF Postdeposition Treatment Effects on CIGS Solar Cells," *IEEE Journal of Photovoltaics,* vol. PP, no. 99, pp. 1-5, 2016.

[7] P. K. Paul, K. Aryal, S. Marsillac, S. A. Ringel, and A. R. Arehart, "Impact of the Ga/In ratio on defects in Cu(In, Ga)Se2," in *2016 IEEE 43rd Photovoltaic Specialists Conference (PVSC)*, 2016, pp. 2246-2249.

[8] P. K. Paul, K. Aryal, S. Marsillac, T. J. Grassman, S. A. Ringel, and A. R. Arehart, "Identifying the source of reduced performance in 1-stage-grown Cu(In, Ga)Se2 solar cells," in *2016 IEEE 43rd Photovoltaic Specialists Conference (PVSC)*, 2016, pp. 3641-3644.

[9] P. Paul *et al.*, "Direct nm-scale spatial mapping of traps in CIGS," *Photovoltaics, IEEE Journal of,* vol. 5, no. 5, pp. 1482-1486, 2015.

[10] R. Caballero *et al.*, "Influence of Na on Cu (In, Ga) Se 2 solar cells grown on polyimide substrates at low temperature: impact on the Cu (In, Ga) Se 2/Mo interface," *Applied Physics Letters,* vol. 96, no. 9, p. 092104, 2010.

[11] J. M. Raguse, C. P. Muzzillo, J. R. Sites, and L. Mansfield, "Effects of Sodium and Potassium on the Photovoltaic Performance of CIGS Solar Cells," *IEEE Journal of Photovoltaics,* vol. 7, no. 1, pp. 303-306, 2017.

The Effect of Deposition Stoichiometry and Post-deposition Treatments on Deep Defects in CdTe

Imran S. Khan, Vamsi Evani, Shamara Collins, Chih An Hsu, Vasilis Palekis and Chris Ferekides

University of South Florida, Tampa, Florida, USA

Abstract — Polycrystalline Cadmium Telluride (CdTe) with different gas phase stoichiometry was deposited by the Elemental Vapor Transport (EVT) technique. CdTe solar cells were fabricated with different post deposition processing, such as $CdCl_2$ heat treatment (HT) and Cu doping. The devices were characterized by standard Current-Voltage (JV) and Capacitance-Voltage (CV) measurements. Deep defect distribution was investigated with Deep Level Transient Spectroscopy (DLTS). Majority carrier traps with an activation energy of 0.4 eV were identified in devices at all deposition stoichiometry, and could be related to cadmium vacancy (V_{Cd}). Cl treated samples exhibited a deep minority carrier electron trap (~0.5 eV). Deep majority carrier traps were identified in various samples, possibly responsible for limiting the minority carrier lifetime. Samples with Cl and Cu showed a relatively shallower (~0.24 eV) electron trap with concentration up to 10^{14} cm^{-3}; this defect could be responsible for dopant compensation in CdTe.

Index Terms — Cadmium compound, copper, charge carrier lifetime, photovoltaic cells, semiconductor device doping.

I. INTRODUCTION

Polycrystalline CdTe as a thin film photovoltaic device absorber is one of the most cost effective semiconductor material to harness solar energy [1]. It has the lowest life cycle carbon footprint and is one of the most environment friendly photovoltaic technologies [2]. Although CdTe solar cells attained much improvement in efficiency over the last decade, the performance needs to be further enhanced to continue to be competitive with fossil fuel options. The efficiency of a solar cell is directly related to the doping concentration and carrier lifetime of the absorber material. Depending on the deposition techniques, deposition conditions, and post deposition

treatments shallow and deep defects can form in CdTe, which determine the carrier concentration and lifetime.

Shallow intrinsic defects such as Cd and Te vacancies (V_{Cd}, V_{Te}) can influence the doping concentration. The formation of compensating donor defects like Te_{Cd} and Cd_i can limit the p-type dopability of CdTe. The formation energy of these defect levels is a function of deposition stoichiometry and the Fermi energy [3]. Te-rich deposition condition enables formation of midgap defects (Te_{Cd} and Te_i) limiting the minority carrier lifetime. Another approach of attaining p-CdTe is to deposit in excess Cd condition and use group V materials as dopants. This method has the potential to attain better minority carrier lifetime by limiting midgap defects. Post deposition processing such as $CdCl_2$ HT and Cu doping is used respectively to improve grain boundary carrier collection and to improve doping along with back contact formation.

In this study, these two fundamental methods of manipulating the defect chemistry in CdTe– deposition stoichiometry and post-deposition treatment has been investigated using Deep Level Transient Spectroscopy (DLTS).

II. EXPERIMENTAL

Polycrystalline CdTe films were deposited by Elemental Vapor Transport (EVT) on glass substrates with ITO (RF Sputtering) as the transparent electrode and CdS (Chemical Bath Deposition) as the window layer. CdTe films were deposited under excess Te (Cd/Te vapor ratio 0.7), stoichiometric (ratio 1.0) and excess Cd (ratio 1.4) vapor conditions. All samples were $CdCl_2$ heat-treated under the same conditions. Graphite ink (no intentional Cu) was used as the

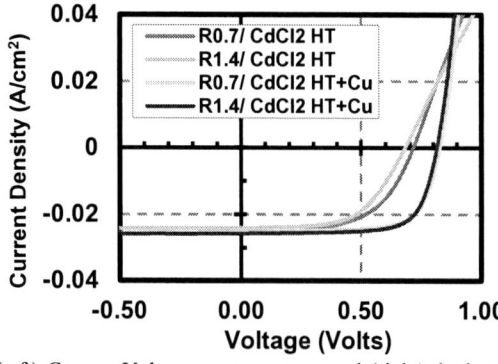 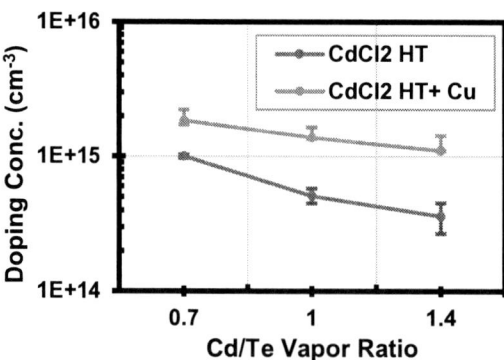

Fig. 1. (Left) Current-Voltage measurements and (right) doping concentration from Capacitance-Voltage measurements for CdTe devices with different Cd/Te vapor ratios and post deposition treatments. Excess Te deposited (ratio 0.7) devices exhibit better cell performance and doping concentration.

978-1-5090-5606-4/17 $31.00 © 2017 IEEE

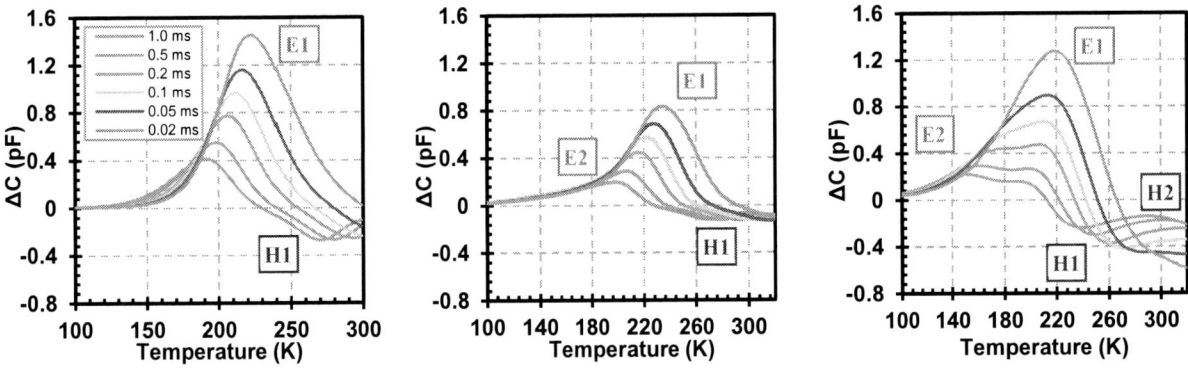

Fig. 2. DLTS spectra from CdS/CdTe heterojunction with CdCl₂ HT. Cd/Te deposition ratio 0.7 (left), 1.0 (center) and 1.4 (right).

back contact electrode for the samples without Cu doping. For Cu doped devices the back contact was Cu doped graphite ink. DLTS measurements were performed with a Sula Technologies Deep Level Spectrometer (Model DDS-12). The sample temperature was varied from 80K to 320K using liquid N_2 and heater inside a Janis VPF-100 cryostat. A bias pulse of -1V to 0V with a pulse width of 1ms was used. The resulting capacitance transients were analyzed with 6 different rate windows from 0.02 to 1 ms.

III. RESULTS AND DISCUSSION

A. Device Performance:

Photovoltaic device performance of the CdS/CdTe heterojunctions used in this study is shown in Fig. 1. Devices with CdTe deposited under excess Te vapor conditions (Cd/Te ratio 0.7) exhibited superior performance with 10~20 mV higher V_{OC}. Carrier concentrations obtained from capacitance-voltage measurements (Table 1) indicate that this improved performance is partly due to higher doping [4]. At this time it is speculated to be due to increased cadmium vacancy (V_{Cd}) concentration as a result of the Te-rich deposition conditions [5]. A doping concentration in the order of 10^{15} cm^{-3} is observed for Te-rich deposited samples. Inclusion of Cu in the back

contact improved the doping concentration (see Table 1), along with device performance (Fig. 1).

B. DLTS Spectra:

Figure 2 shows DLTS spectra for CdCl₂ HT CdTe devices with different vapor phase deposition stoichiometry. For the excess Te deposited (Cd/Te ratio 0.7) device two distinct peaks are observed. A positive peak E1 at 150~250K indicating a minority carrier trap and a negative peak H1 at 250K to room temperature due to a majority carrier trap. Both peaks E1 and H1 were also observed in stoichiometric and Cd-rich deposited films. A second positive peak (E2) at 100~180K indicates a shallower minority carrier trap. Moreover, ΔC continues to become increasingly more negative at higher temperatures in this case, indicating the presence of a deep majority carrier trap (H2).

The DLTS spectra for the devices with Cu doped back contact exhibit significant differences (Fig. 3). Peak E1 is no longer present and a shallower electron trap is observed for all devices regardless of stoichiometry. Based on its shape and temperature range, it is speculated to be the same minority trap E2 that appeared in Cd-rich deposited device without Cu (Fig. 3 right). A small negative peak at 120~140K indicates the presence of a shallow hole trap for Cd-rich deposited film with Cu (H3). For all Cu doped devices ΔC continues to greater

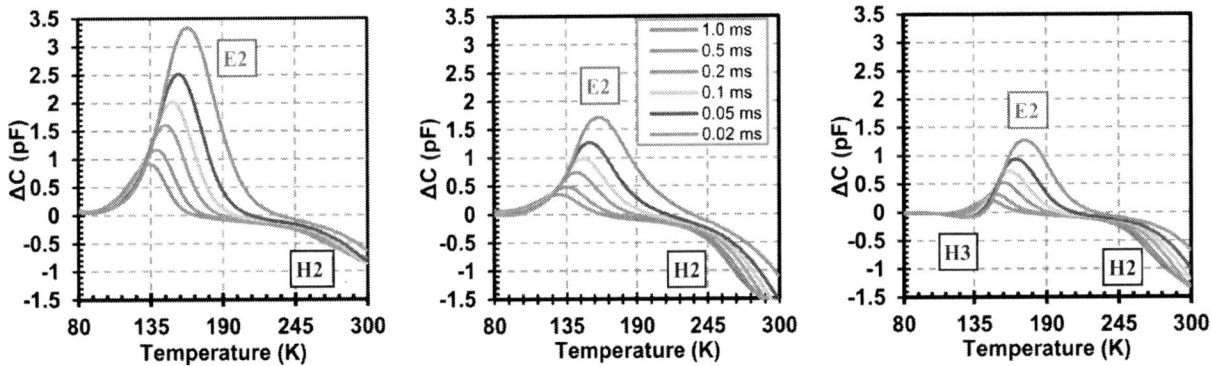

Fig. 3. DLTS spectra from CdS/CdTe heterojunction with CdCl₂ HT and Cu. Cd/Te deposition ratio 0.7 (left), 1.0 (center) and 1.4 (right).

TABLE I
TRAP CONCENTRATION AND ACTIVATION ENERGY OF DIFFERENT DEFECTS IDENTIFIED FROM DLTS MEASUREMENTS

Cd/Te Vapor Ratio	Post-deposition Treatment	Doping Concentration (cm^{-3})	E1		E2		H1	
			Activation Energy (eV)	Trap Conc. (cm^{-3})	Activation Energy (eV)	Trap Conc. (cm^{-3})	Activation Energy (eV)	Trap Conc. (cm^{-3})
0.7	CdCl$_2$	1.0E15	0.47	8E13			0.42	1E13
1.0		5.1E14	0.5	2E13			0.41	4E12
1.4		3.6E14	N/Q	N/Q	N/Q	N/Q	0.38	1E13
0.7	CdCl$_2$ + Cu	1.9E15			0.24	1E14		
1.0		1.4E15			0.22	5E13		
1.4		1.1E15			0.31	3E13		

negative values at high temperatures (>300K) suggesting the presence of a deeper majority carrier trap (labeled as H2 in the figure).

C. Defect Analysis:

The activation energy of H1 is calculated to be ~0.4 eV and is tentatively assigned to V_{Cd} [6]. This assignment is further corroborated by the observation that the peak disappeared in Cu doped samples, since Cu can occupy a cadmium vacancy in CdTe to form the acceptor defect Cu_{Cd} [7].

The electron trap E1 was previously reported on CSS deposited samples with and without Cl treatment [8]. It is speculated to be a native CdTe defect. The calculated activation energy of 0.4~0.5 eV suggests that it is either Cd_i or Te_{Cd}; or a combination of both [3].

The activation energy of E2 in devices deposited under stoichiometric and excess Cd conditions cannot be calculated due to its proximity to E1. However in Cu doped devices, the activation energy of the same is calculated to be 0.22~0.3 eV. Considering all the samples in this study are CdCl$_2$ heat treated, this defect is possibly related to Cl. Theoretical analysis anticipates Cl to occupy a Te site in CdTe. Cl_{Te} is a shallow donor defect with an activation energy of 0.19 eV. Cl_{Te} can also form a complex shallow acceptor defect (A-centers, V_{Cd}-Cl_{Te}) in conjunction with a cadmium vacancy [9]. The formation of A-centers are dependent upon the availability of V_{Cd}'s. The formation energy of Cu and Cl related defects in CdTe are the following [10].

$$Cu_i + V_{Cd} \rightarrow Cu_{Cd} \qquad \Delta H = -3.55eV$$
$$Cl_{Te} + V_{Cd} \rightarrow V_{Cd}\text{-}Cl_{Te} \qquad \Delta H = -2.02eV$$
$$Cu_i + V_{Cd}\text{-}Cl_{Te} \rightarrow Cu_{Cd} + Cl_{Te} \qquad \Delta H = -1.56eV$$

Negative energy of reaction indicates energetically favorable and exothermic reaction. A lower concentration of V_{Cd} is expected in Cd-rich deposited films. Cl_{Te} donor defects may not encounter an adequate amount of V_{Cd} to form A-centers, and remain as Cl_{Te}. If the V_{Cd}'s are all consumed, A-centers have higher formation energy than Cl_{Te}. When Cu is introduced in Cl treated film, it can occupy a cadmium vacancy site, or take the V_{Cd} site in an A-center. The former scenario is more favorable. However, when all the V_{Cd} sites are filled, Cu can occupy the A-centers. In such case, Cl_{Te} part of A-centers can emerge as

Cl_{Te} donors. The following two observations support this hypothesis - E2 appears in all Cu doped devices and in Cu-free devices it becomes more prominent as the films are grown in more Cd-rich conditions. This relatively shallow minority carrier defect with concentrations in the order of 10^{14} cm^{-3} possibly have a compensating effect in limiting the doping efficiency of polycrystalline CdTe.

Negative ΔC for all Cu doped devices at high temperatures (>300K) indicates the presence of a majority carrier tap H2 and is possibly a Cu related deep defect. Similar characteristic was also observed for CSS deposited Cu doped devices [8]. The presence of a deep defect in Cd-rich Cu free device, also is denoted H2, may or may not be the same defect that was observed in Cu doped sample. The presence of deep defects in Cd-rich deposited film is consistent with our previous minority carrier lifetime measurement on Cl treated Cd-rich samples; for samples with CdCl$_2$ HT, Te-rich deposited films demonstrated higher lifetime (7 ns) compared to Cd-rich deposited films (2 ns) [11]. For polycrystalline CdTe films, the grain boundary (GB) plays a significant role in determining the carrier lifetime. Cl incorporation is expected to passivate the grain boundary related deep states in CdTe. First principle studies suggest that Cl passivation is more effective for Te-core GB compared to Cd-core ones [12]. Hence, H2 in Cd-rich deposited Cu free device could be related to un-passivated Cd-core grain boundary states (possible SRH recombination center).

The presence of small negative peak H3 (possible shallow acceptor) in the Cd-rich Cu doped film cannot be explained/assigned at present and it is the subject of ongoing research. Table I lists the different traps that were identified in different samples along with their trap concentration and activation energy. The terms 'N/Q' used for some cases indicate the presence of the peak, however could not be quantified due to either low intensity or proximity to other peaks.

IV. CONCLUSION

The defect structure of EVT deposited polycrystalline CdTe with CdCl$_2$ treatment and Cu doping was investigated with Deep Level Transient Spectroscopy. Majority and minority carrier traps with different activation energies were identified. These traps were assigned to various native and extrinsic CdTe defects based on their activation energy and presence in specific devices. Cl related shallow minority trap in Cl and Cu treated devices indicated a possible dopant compensation mechanism. Lower minority carrier lifetime for Cd-rich deposited device with CdCl$_2$ HT was found to be due to deep majority carrier defects. Hence this paper provides a useful insight to the defect distribution of CdTe devices at different deposition stoichiometry, and how they are influenced by post deposition treatments such as CdCl$_2$ HT and Cu doping.

This work was supported by the DOE SunShot Initiative under the F-PACE program (DE-EE0005401) and the national Science Foundation (1144244).

REFERENCES

[1] P. Sinha, M. de Wild-Scholten, A. Wade, and C. Breyer, "Total cost electricity pricing of photovoltaics," *Proceedings of the 28th European Photovoltaic Solar Energy Conference*, 2013.

[2] V. M. Fthenakis, H. C. Kim, and E. Alsema. "Emissions from photovoltaic life cycles," *Environmental Science & Technology,* vol. 42(6), pp. 2168-2174, 2008.

[3] T. A. Gessert, et al. "Research strategies toward improving thin-film CdTe photovoltaic devices beyond 20% conversion efficiency." *Solar Energy Materials and Solar Cells,*" vol. 119, pp. 149-155, 2013.

[4] H. Zhao, et al. "The effect of impurities on the doping and V$_{OC}$ of CdTe/CdS thin film solar cells," *Thin Solid Films*, vol. 517(7), pp. 2365-2369, 2009.

[5] V. Evani, et al. "Effect of Cu and Cl on EVT-CdTe solar cells," *42nd IEEE Photovoltaic Specialist Conference (PVSC)*, 2015.

[6] Ji-Hui Yang, et al. "Review on first-principles study of defect properties of CdTe as a solar cell absorber." *Semiconductor Science and Technology,* vol. 31(8), pp. 083002, 2016.

[7] K. K. Chin, T. A. Gessert, and Su-Huai Wei, "The roles of Cu impurity states in CdTe thin film solar cells," *35th IEEE Photovoltaic Specialist Conference (PVSC)*, 2010.

[8] M. Khan, et al. "Study of defects in polycrystalline CdTe using DLTS." *43rd IEEE Photovoltaic Specialist Conference (PVSC)*, 2016.

[9] Ji-Hui Yang, et al. "First-principles study of roles of Cu and Cl in polycrystalline CdTe." *Journal of Applied Physics*, vol. 119(4), pp. 045104, 2016.

[10] Su-Huai Wei, and S. B. Zhang. "Chemical trends of defect formation and doping limit in II-VI semiconductors: The case of CdTe." *Physical Review B*, vol. 66(15), pp. 155211, 2002.

[11] V. Evani, et al. "Effect of Cu and Cl on EVT-CdTe solar cells." *42nd IEEE Photovoltaic Specialist Conference (PVSC)*, 2015.

[12] L. Zhang et al. "Effect of Co-passivation of Cl and Cu on CdTe Grain Boundaries," *Physical Review Letters*, vol. 101(15), pp. 122201, 2008

Testing the limits of mechanically-scribed CIGS microcells

Ombline Lafont[1], Nicolas Vandamme[1], Leia Ruffini[1], Jia Yu[1], Philip Jackson[2], Jose Alvarez[3], Daniel Lincot[1,4]

[1] Institut de Recherche et Développement sur l'Energie Photovoltaïque, IRDEP - CNRS/EDF/Chimie ParisTech - PSL Research University, 6 quai Watier, 78401 CHATOU Cedex, FRANCE
[2] Zentrum für Sonnenenergie- und Wasserstoff-Forschung Baden-Württemberg (ZSW), Meitnerstrasse 1, 70563 STUTTGART, GERMANY
[3] Laboratoire de Génie électrique et électronique de Paris, GeePs, UMR CNRS 8507, Centrale-Supélec, Univ. Paris-Sud, Université Paris-Saclay, Sorbonne Universités, UPMC Univ. Paris 06, 11 rue Joliot-Curie, Plateau de Moulon, 91192 GIF-SUR-YVETTE, FRANCE
[4] Institut Photovoltaïque d'Ile-de-France, IPVF, 8 rue de la Renaissance, 92160 ANTONY, FRANCE

Abstract — The performances of CIGS microcells of sizes ranging from 250μm×250μm to 500μm×500μm obtained by mechanical scribing are analyzed. Despite additional shunts due to the formation of cracks and spalling effects at the edges, the microcells reasonably preserve the efficiency of the standard cell (17.3%). The behavior of the cell edges is analyzed through SEM images, LBIC and PL measurements. Under 21 suns, an efficiency of 19.4% was obtained. This shows that the concept of concentration on mechanically-scribed CIGS microcells works well and could be applied on record efficiency CIGS cells (22%).

Index Terms — CIGS, high-efficiency solar cells, light concentration, mechanical scribing, microcell.

I. INTRODUCTION

The use of microcells is an interesting pathway to reduce balance of materials and tackle scarcity issues encountered in mainstream PV technologies. This is specifically the case for CIGS where Indium and Gallium contents are limiting components to their large-scale industrialization. Combined with the use of light concentration, devices with reduced dimensionality allow better performances along with improved possibilities of heat management and innovative architectures for solar panels [1]. Various strategies have been explored to define CIGS microcells: by photolithography [2], by chemical etching [3], or by electrodeposition [4]. The two first techniques are expensive and since they are top-down techniques, they do not allow material savings. Electrodeposition is a bottom-up technique that enables a precise deposition of CIGS on a patterned molybdenum surface. However, the state-of-the-art gives an efficiency below 10% under one sun [5]. Recently, Lotter *et al.* [6] have studied the performances of CIGS microcells using three different fabrication processes: mechanical scribing, laser scribing and chemical etching of the front window layers. On a 0.02 cm² scribed CIGS cell, they observe that the shunt resistance decreases with light concentration and that this effect is limited if edges are not illuminated. In this work, we pursue this discussion by studying tinier mechanically-scribed cells with sizes varying between less than 0.000625 cm² and

0.0025 cm². The large number of microcells studied allow to look at the statistics and conclude on the effect of the edges damaging under concentration.

II. EXPERIMENT

From an initial cell grown at ZSW with a conversion efficiency of 17.3% without anti-reflection coating, we fabricated microcells by mechanical scribing using an automatic scribing apparatus following a pattern with square and rectangle diodes which is imaged in Fig. 1. The initial cell with its front comb-shaped metal contact is also distinguishable in the same figure. Five diode series (A, B, C, D, and E) were fabricated corresponding respectively to the surface dimensions of 500×500 μm², 500×250 μm², 250×250 μm², 125×250 μm² and 125×125 μm². The grooves resulting from the mechanical scribing process show a width close to 70 μm. Due to lateral pressure, spalling effects and cracks are evidenced explaining the very low success rate for series D, and notably series E.

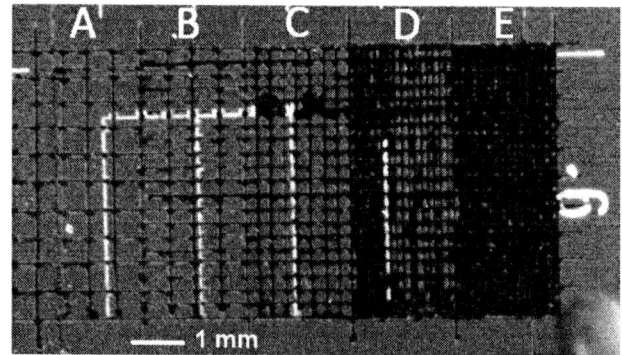

Fig. 1. Picture of the mechanically scribed sample. The letters A, B, C, D and E point out the different diode series with different surface dimensions.

Electrical Current-Voltage (IV) characteristics are measured under dark and one sun conditions with a two-probe setup. The front contact is obtained through a tungsten probe with a tip of 5 μm in diameter. In the following, we focus on A, B and C diodes where automatic IV measurements have been performed.

A. Dark characteristics

For each dark IV, we fit the electrical measurement with a two-diode model to differentiate contributions from an ideal diode and a recombination one [7]. For this purpose, the ideality factors are fixed to $n_1=1$ and $n_2=2$. This fit includes the evaluation of the resistances R_s and R_{sh}. The results of these fits are given in Table I. For simplicity, the cell surfaces are taken constant and equal to their nominal value (0.000625 cm² for C cells, 0.00125 for B cells and 0.0025 for A cells). This explains the large standard deviations given in Table I. J_{01} decreases with the cell size whereas J_{02} is unaffected. However, R_s stays constant whereas R_{sh} decreases by a factor 5 due to additional shunts induced by mechanical scribing, but does not depend on the cell size.

TABLE I
STATISTICAL ANALYSIS OF DARK MEASUREMENTS

	Area mm²	J_{01} (n=1) mA/cm²	J_{02} (n=2) mA/cm²	R_s Ω.cm²	R_{sh} Ω.cm²
Initial cell	50	2.7 10⁻¹²	5.2 10⁻⁶	1.9	5579
A	0.25	1.9 10⁻¹³	5.0 10⁻⁶	0.71	1497
		±1.4 10⁻¹³	±9.4 10⁻⁷	±0.28	±953
B	0.125	7.8 10⁻¹⁴	5.6 10⁻⁶	0.87	1227
		±8.9 10⁻¹⁴	±1.8 10⁻⁶	±0.92	±700
C	0.0625	1.1 10⁻¹⁴	5.4 10⁻⁶	0.85	1586
		±4.9 10⁻¹⁴	±2.2 10⁻⁶	±0.60	±981

Tab. 1. Average and standard deviation of electrical parameters extracted from the two-diode model fitting of I-V measurements on series A, B, and C. These figures are to be compared with values obtained for the initial cell reported in the first line of the table. Standard deviations are large, but partially due to the dispersion of diode areas that have not been corrected.

B. One sun illumination

We then measured the cell under one sun to check if the photovoltaic properties of the initial cell are conserved. The IV curves of the initial cell and of three cells in series A, B and C, which areas are respectively 0.00226, 0.00118 and 0.00033 cm² are analyzed. Jsc are respectively 25 mA/cm², 23.6 mA/cm² and 29.5 mA/cm². This dispersion is due to the shading by the probe needle, which we do not control and cannot measure under our solar simulator. To compare the results, the IV curves are translated so that short-circuit current matches the one of the initial cell (30.3 mA/cm²). This is

shown in Fig. 2. The open-circuit voltage is above 0.74 V for the three cells.

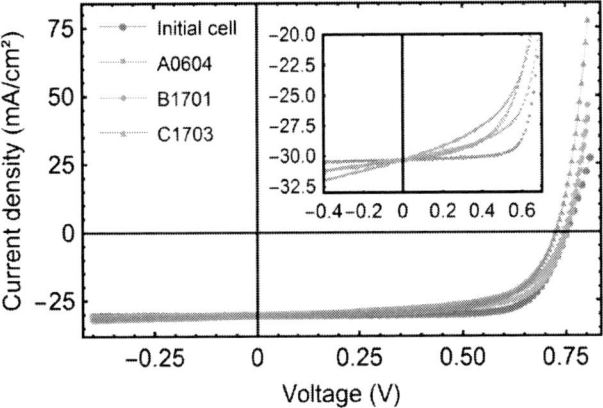

Fig. 2. Current-voltage curve of the initial cell (0.5cm²), and of three cells from series A, B and C. The curves are normalized in current to take into account the shading by the probe. The inset is a zoom in the region V=0.

However, the fill factor is much decreased (between 10 and 15%). This is linked to the decrease of R_{sh}, already observed under dark. Resulting efficiencies are 15.8% for cell A0604, 14.0% for cell B1701 and 14.3% for cell C1703.

III. ROLE OF EDGES

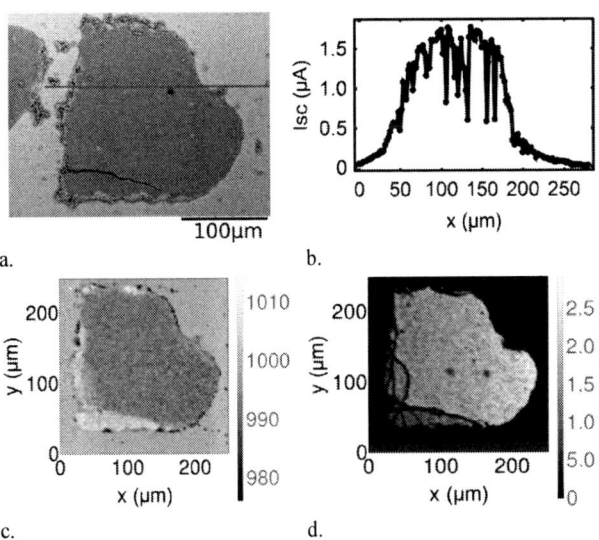

Fig. 3. a. SEM image of cell C1703. b. LBIC experiment on the red line indicated on a. c. PL intensity map, integrated between 900 and 1100nm. d. Map of the spectral center of gravity of the PL peak in nm (the information on PL intensity is lost).

To understand this loss of R_{sh} and probe the real role of the edges in mechanically scribed cells, we have performed LBIC, photoluminescence (PL) and SEM to analyze the morphology and composition of the edges. Fig. 3.a. shows the cell C1703 through SEM. A hole and a crack are visible. Some edges are neat, other exhibits flakes of probably CIGS (which means that the ZnO/ZnO:Al has been peeled off). However, those edges exhibit photoluminescence, as seen in Fig. 3.c. Therefore, they keep good photovoltaic properties. However, a more thorough analysis of the PL spectrum reveals at the edges that the PL peak is slightly shifted to longer wavelengths, as shown in Fig. 3.d. As we compare with Fig. 3.a. we conclude that this difference is not entirely due to the chemical composition. The displacement of the peak cannot be attributed to a temperature change either. This may be interpreted as the result of a partial peeling off the CIGS near the edges due to the scribing process which in turn affects the PL. In fact, this is probably due to additional back-reflections of PL photons on the back contact, caused by the reduction of the thickness of the cell and modification of the Mo-CIGS interface. Additionally, a LBIC measurement was performed with a 30 μm laser spot along the sample. The linear slope on the right lasts for only 20 μm, which is inferior to the spot size, which means that the right edge is very steep. On the left, the slope occurs during about 40 μm, which suggests an influence of the edge on a typical length of approximately 10 μm.

IV. CONCENTRATION

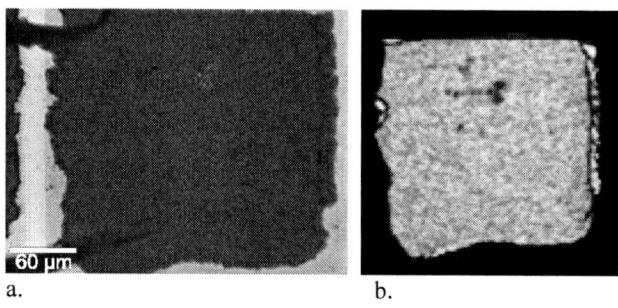

a.   b.

Fig. 4.   a. Optical microscope image of the cell F6 contacted using 200nm probe tips. b. F6 PL intensity map.

Measurements under concentration have been performed using a 532 nm laser. Close to one sun, there is no difference between two-probe and four-probe measurements since the contact resistance is negligible compared to the intrinsic series resistance. However, when the illumination is increased, the series resistance decreases [8], and the contact resistance starts to play an important role. It results in a diminution of the fill factor, which in turn decreases the efficiency. This interpretation is confirmed by a comparative measurement done on a cell crossed by a metallic grid (such as the third column of series A, see Fig. 1). The fill factor loss at concentrations over 10 suns is much decreased when the probe

is placed on the metallic grid. The shunt resistance also decreases with intensity, probably due to the photoconductivity of the damaged edges [6].

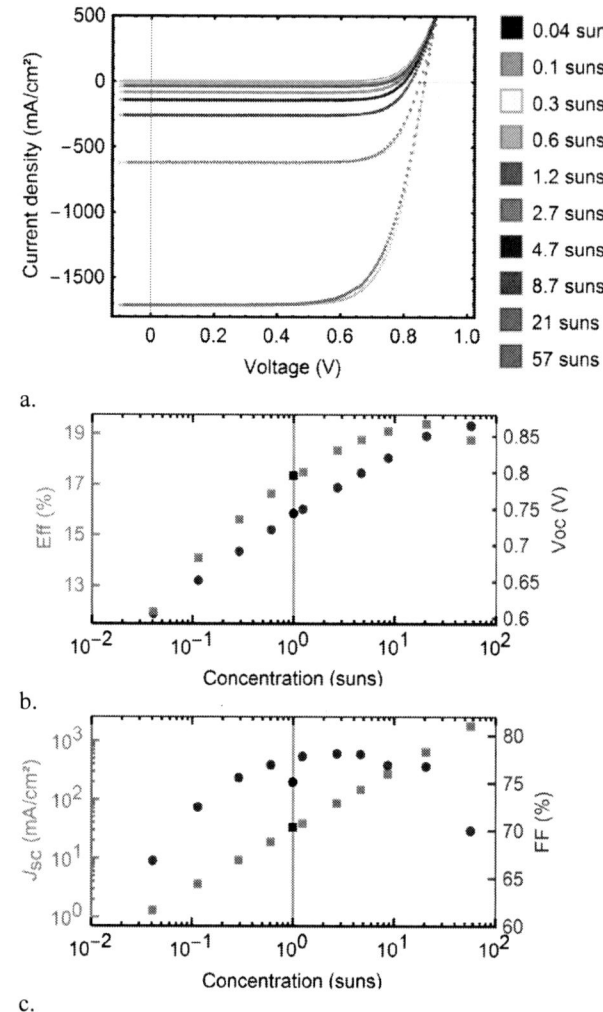

a.

b.

c.

Fig. 5.   a. Current-voltage curves at various laser concentrations. The measurements are done in forward (full disks) and reverse bias (open circles). b. Efficiencies and V_{oc} as a function of concentration. c. J_{sc} and FF as a function of concentration. The initial cell properties (before scribing) are shown as black squares (efficiency and Jsc) and circles (V_{oc} and FF).

Therefore, another part of the same sample was manually scribed by using a 5 μm diameter tungsten (W) tip. This approach enabled to realize microcells with cleaner edges (see Fig. 4.b.). Four-probe measurements were performed using 200 nm diameter W tips for the front contacts. The beam spot size is adjusted so as to optimize the V_{oc} and is here comparable to the microcell area (0.00064cm²). The concentration is obtained by dividing the short-circuit current by the one of the initial cell under one sun illumination (approximately 30mA/cm²). The shading by the contact probes is thus neglected, but it affects only the concentration values, not the obtained efficiencies. The results for F6 microcell are shown in Fig. 5. The maximum efficiency is 19.4% at 21 suns, which is an absolute 2.1% increase compared to the initial cell under one sun.

V. CONCLUSIONS

We demonstrate in this work that the concept of microcells for concentration applications holds even for low-cost and edge-damaging technique such as mechanical scribing. Potential shunting channels created by scribing are not detrimental to cell performances under concentrated light. This configuration is thus adapted to test the limits of microcell concept under high concentration using record cell plates (about 22% efficiency), without any new processing of the cell. These studies are the next step of this project in our laboratories. Indeed, preliminary measurements on a high-efficiency ZSW cell with alkali post deposition treatment (19% under one sun, no anti-reflection coating) show very promising results, with 22.5% efficiency achieved at about 50 suns using a 160µm FWHM laser spot size on a mechanically-scribed 0.0008cm² microcell.

ACKNOWLEDGEMENT

This work has received support under the program « post-doctorat innovation » from *PSL Research University* via Ombline Lafont fellowship.

REFERENCES

[1] M. Paire, L. Lombez, N. Péré-Laperne, S. Collin, J.-L Pelouard, J.-F Guillemoles, D. Lincot, "Microscale Cu(In,Ga)Se2 solar cells for concentration : a path to 30% efficiency – theoretical evaluation and first experimental results," *Proceedings of SPIE*, 2010

[2] M. Paire, A. Shams, L. Lombez, N. Pere-Laperne, S. Collin, J.-L. Pelouard, J.-F. Guillemoles, and D. Lincot, "Resistive and thermal scale effects for Cu(In,Ga)Se2 polycrystalline thin film microcells under concentration," *Energy Environ. Sci.*, vol. 4, no. 12, pp. 4972-4977, 2011.

[3] M. Paire, C. Jean, L. Lombez, S. Collin, J.-L Pelouard, Isabelle G., J.-F Guillemoles, and D. Lincot, "Cu(In,Ga)Se2 mesa diodes for the study of edge recombination," *Thin Solid Films*, vol. 582, pp. 258-262, 2015.

[4] M. Paire, C. Jean, L. Lombez, T. Sidali, A. Duchatelet, E. Chassaing, G. Savidand, F. Donsanti, M. Jubault, S. Collin, J.-L Pelouard, D. Lincot, and J.-F Guillemoles, "Characterization of Cu(In,Ga)Se2 Electrodeposited and Co-Evaporated Devices by Means of Concentrated Illumination," *IEEE Journal of Photovoltaics*, vol. 4, no. 2, pp. 693-696, March 2014.

[5] A. Duchatelet, K. Nguyen, P.-P. Grand, D. Lincot, and M. Paire, "Self-aligned growth of thin film Cu(In,Ga)Se2 solar cells on various micropatterns," *Appl. Phys. Lett.* 109, 253901 (2016).

[6] E. Lotter, P. Jackson, S. Paetel, and W. Wischmann, "Identification of loss mechanisms in cigs micro-cells for concentrator applications," in *32nd European Photovoltaic Solar Energy Conference and Exhibition*, 2016, pp. 1158-1160.

[7] A. Luque, Handbook of Photovoltaic Science and Engineering (John Wiley and Sons, 2003).

[8] M. Paire, L. Lombez, N. Pere-Laperne, S. Collin, J.-L. Pelouard, D. Lincot, and J.-F. Guillemoles, "Microscale solar cells for high concentration on polycrystalline Cu(InGa)Se2 thin films," *Appl. Phys. Lett.*, vol. 98, no. 26, pp. 264102, 2011.

[9] M. Paire, L. Lombez, F. Donsanti, M. Jubault, S. Collin, J.-L. Pelouard, J.-F. Guillemoles, and D. Lincot, "Cu(In,Ga)Se2 microcells: high efficiency and low material consumption," *JRSE* 5, p. 011202, 2013.

Photoluminescence imaging analysis of doping in thin film CdS and CdS/CdTe devices

C. Potamialis[1], F. Lisco[1], B. Maniscalco[1], M. Togay[1], A. Abbas[1], M. Bliss[1], J.W. Bowers[1], J.M. Walls[1]

I. Rimmaudo[2], R. Mis Fernandez[2], V. Rejon[2], J.L. Peña[2]

[1] CREST, Wolfson School of Mechanical, Electrical and Manufacturing Engineering, Loughborough University, Loughborough, UK

[2] Cinvestav, Unidad Merida, Merida, Yucatan 97310, Mexico

Abstract — The use of photoluminescence (PL) imaging analysis to assess the effectiveness of the passivation treatment due to the presence of chlorine in CdS thin films has been investigated. In this work, we show that the chlorine doping effect in the CdS window layer can be detected by PL imaging analysis, due to the formation of a defect complex of sulfur vacancy and Cl_S (V_S-Cl_S) and complexes between halogen ions and cadmium vacancies (V_{Cd}-Cl_S). CdTe devices with differently doped CdS layers were investigated. PL imaging, TEM, IV performance indicators and EQE analysis were performed to understand the effect of the different dopants on the electrical performances of CdTe devices.

I. INTRODUCTION

Thin film CdTe solar cells are already a commercially successful technology, due to a combination of low processing costs, and high module efficiency. The current efficiency record is 22.1 % by First Solar [1], where an important step in the device fabrication is the $CdCl_2$ post annealing treatment. The $CdCl_2$ treatment is necessary to produce high performing devices due to the beneficial structural and electrical impact on CdS/CdTe heterojunction devices. However, a weak $CdCl_2$ activation treatment results in poorly passivated grains, limiting the performance of the device, whilst an aggressive $CdCl_2$ activation process can cause excessive consumption of the CdS layer and may damage the junction with excessive chlorine build up at the interface [2], leading to delamination [3]. Usually optimization of this process consists of assessing the performance of completed devices which can be very time consuming and not cost effective.

Photoluminescence (PL) imaging is a powerful technique which allows fast, non-contact device analysis at each step of the fabrication process. This technique is suitable for large scale manufacturing use and assists the quest for further cost reduction of CdTe solar cells [4].

Commonly in CdS/CdTe solar cells, PL spectra peaks in the range of 1.3 – 1.5 eV originates from CdTe, while peaks in the range of 1.9 – 2.2 eV are related to CdS [7]. When the CdS is subjected to the $CdCl_2$ annealing treatment, a broad PL emission is generated at around 1.7 eV, which is associated with a defect complex of a sulfur vacancy and Cl_S (V_S-Cl_S) [5][6]. $CdCl_2$ activation introduces large amounts of Cl atoms which

may occupy sulfur vacancies and introduce the Cl_S states. Additionally, deep and shallow states are formed since complexes between halogen ions and cadmium vacancies are also likely to occur (V_{Cd}-Cl_S) [6]. The emission signal generated by the presence of chlorine in CdS films can be detected by PL imaging. This analysis can provide useful information about the effectiveness of the $CdCl_2$ treatment in CdS/CdTe devices even before the application of the back contact.

In this report, PL Imaging analysis was used to investigate the effect of different chlorine dopants in the CdS thin films used for CdTe solar cells. $CdCl_2$ treatment followed by annealing is known to improve the structural and electrical properties of CdS[6], as well as CdTe. Here we have used different chlorine dopants to understand whether the same improvements could be achieved on the as deposited CdS layer and on the electrical performance of the CdTe devices. The resulting electrical characteristics were studied against the presence of the various chlorine dopants.

II. METHODOLOGY AND EXPERIMENTAL PROCEDURES

A. PL imaging analysis

PL imaging was carried out using the system shown in Fig. 1. A 405nm LED was used as the excitation source, with a Si CCD camera fitted with a 720 nm long-pass filter used to detect the PL signal. The exposure time for all of the measured samples was kept at 10 seconds. All measurements were taken at room temperature and with the sample positioned with the glass facing the camera.

Fig. 2 shows the PL images of as-deposited CdS and CdS following a wet-$CdCl_2$ treatment. PL image acquisition shows the effect of the treatment on the film. The intensity of the signal (pixel counts) was used to compare untreated and treated material (Fig.2). When the Cl signal is not detectable the image appears completely dark as shown in the response from the untreated material. High intensity signals result in a brighter image, indicating the presence of chlorine complex emission from the $CdCl_2$ treated material. Using the scale bar as a reference, the intensity of ~1000 counts corresponds to untreated material and ~65,000 counts (detector saturation limit) is typical for a heavily treated layer.

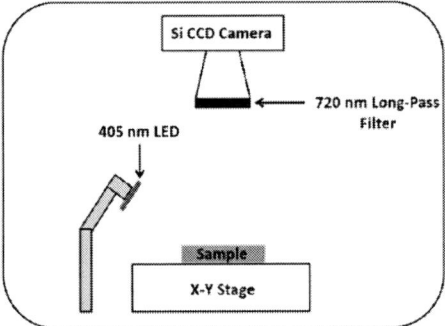

Figure 1: PL imaging system schematic diagram.

These values have been used to establish two extremes of the range and to investigate the performance and homogeneity of the passivation treatment.

During the CdCl$_2$ activation treatment, Te diffuses into the CdS to form a CdS$_{1-x}$Te$_x$ layer. This creates a broad PL emission in the range of 1.6-1.8 eV [5]. However, since the CdTe was not present, the contribution from Te diffusion into the CdS is removed and therefore the detected signal is due to the presence of chlorine in the CdS film. We believe that chlorine is responsible for the formation of sub band gap V$_s$-Cl$_s$ and V$_{Cd}$-Cl$_s$ complexes in the CdS [6].

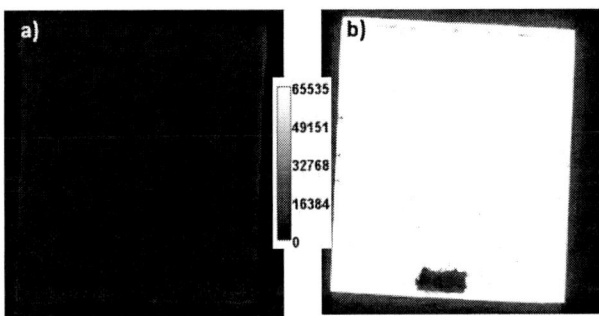

Figure 2: PL images of as deposited CdS (a) and CdCl$_2$ treated CdS (b) thin films.

Fig. 3a) shows the same sample as in Fig. 2, however after the deposition of the CdTe layer. The average PL signal decreased to 1.42×10^4, due to the formation of the p-n junction. This results in consumption of the CdS layer, therefore the detection of V$_s$-Cl$_s$ and V$_{Cd}$-Cl$_s$ complexes, by PL imaging, was limited and the PL emission showed a reduced signal. In Fig. 3b) The same sample after the CdCl$_2$ passivation treatment is illustrated. The average PL counts increased to 4.93×10^4 as a direct result of reintroduction of Cd and Cl in the CdTe device, which diffused into the CdS layer, due to the CdTe activation treatment by evaporation of CdCl$_2$.

PL imagining analysis can be related to the electrical properties of the devices. This investigation was assisted by measuring current density–voltage (J-V) curves, with the solar cell exposed to AM1.5G, 1000 W/m^2 illumination. EQE measurements were taken using an USHIO UXL-150so Xe Lamp with Newport 67005 lamp housing and H10 Horiba Jobin Yvon monochromator.

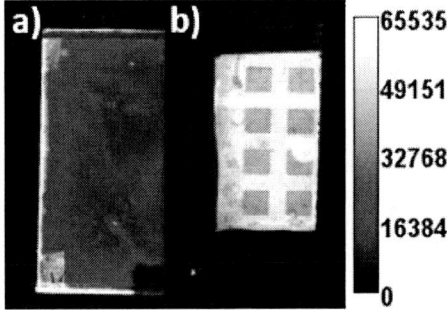

Figure 3: PL images of CdS: a) after CdTe deposition and b) after the second CdCl$_2$ activation treatment.

For a comprehensive analysis of on the effect of the dopants in CdS on the CdS/CdTe stack, structural investigations were also performed, using a Leo 1530 VP high-resolution field emission gun scanning electron microscope (FEGSEM).

B. CdS thin film deposition by sono-CBD

The substrates (50x50) mm^2 (Pilkington TEC 10, FTO) were ultrasonically cleaned before CdS deposition, in a de-ionised (DI) water solution containing 10% acetone and 10% IPA, for 1 hour at 60°C. They then received a 5-minute surface activation pre-treatment in an argon and oxygen plasma. CdS thin films were deposited using a sono-chemical bath (sono-CBD) [7], where an ultrasonic probe was used to replace a conventional magnetic stirrer to agitate the bath. 0.01M Cd(CH$_3$COO)$_2$, 0.1M CH$_4$N$_2$S, 25 wt % NH$_4$OH were used as precursors in 200 mL DI water in a preheated vessel at 70° C, resulting in films of ~150 nm thickness [4]. After CdS deposition, the substrates were rinsed with DI water and dried with dry compressed air. CdS doping was carried out using the same procedure, however in each case, alternative chlorine dopants were added to each bath (50% of the concentration of Cd source), with each deposition lasting for one hour. This resulted in different final thicknesses of the CdS films, according to the dopant added in the bath. In this work AlCl$_3$, and CdCl$_2$ were the investigated dopants. In the case of CdCl$_2$, this was used as a replacement cadmium source to the cadmium acetate used in the baseline process. The presence of AlCl$_3$ in the bath affected the deposition rate of the CdS layer significantly and resulted in an approximately 50 nm thick film.

978-1-5090-5606-4/17 $31.00 © 2017 IEEE   2458

C. CdTe film deposition by Close Space Sublimation and the CdCl₂ activation treatment

CdTe thin films were deposited in a home-made close-space sublimation system (CSS), using IR lamps, and graphite susceptors [8]. The prepared CdS film and source plate were placed in the vacuum chamber, and separated using quartz spacers. The deposition pressure during sublimation was 1 Torr, with a 6% O_2 in Ar gas mixture. The source and substrate temperatures were rapidly increased, to 630°C and 515°C, and the deposition time was fixed to 3 minutes, which results in 4 to 6 μm thick films. The CdCl₂ activation treatment was carried out by thermal evaporation. A quartz crucible was loaded with 0.5 g of CdCl₂, which was evaporated at ~1 x10⁻⁶ Torr. Post-annealing treatment was carried out after thermal evaporation. The samples were annealed on a hot plate at a dwell temperature of 415° C for 1 minute with ramping rate of 4°C/min starting from 370°C. Devices were completed with 80nm gold contacts deposited using thermal evaporation, with no intentional copper doping added in this study.

III. RESULTS

A. PL imaging analysis on CdS/CdTe devices

CdS/CdTe devices have been fabricated by varying the dopants in the CdS layer, as described in the experimental section. All samples underwent the same CdCl₂ post annealing treatment. Fig. 4 shows the PL image of CdTe devices with un-doped CdS, CdCl₂ wet- treated CdS, CdS deposited with CdCl₂ as Cd source and CdS:AlCl₃ .

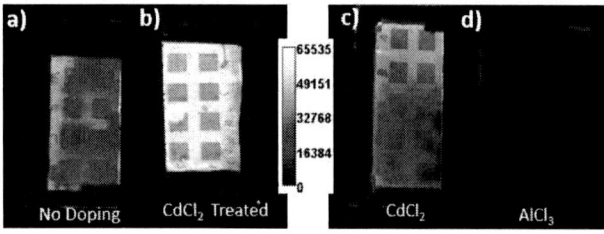

Figure 4: PL images of CdS/CdTe devices with differently doped CdS layers: a) un-doped CdS, b) wet treated CdS c) CdS with CdCl₂ as Cd Source and d) CdS:AlCl₃

The PL signal was averaged over the 8 individual cells on each sample (each cell has an area of 0.25 cm²). The baseline CdS/CdTe structure with no doping in the CdS layer, showed 2.36×10^4 average PL counts with a range of 8.7×10^3 to 4.08×10^4 counts. The CdS which has been CdCl₂ wet treated (Fig 4b) showed an average count intensity of 4.93×10^4 (as presented before in the methodology in Fig. 3b). Replacing Cd(CH₃COO)₂ with CdCl₂ as Cd source, for the deposition of the CdS layer, the average counts remain comparatively the same (2.32×10^4), whilst the maximum PL counts increased up to (4.36×10^4). This can be an indication of a favorable

incorporation of Cl inside the CdS layer. A low signal intensity was detected for CdS layers doped with AlCl₃ (2.2×10^3 average PL counts). This is an indication of either an under treated device, where the incorporation of CdCl₂ is almost undetectable or an over treated cell, which resulted in excessive consumption of CdS.

B. Electrical characterization of CdS/CdTe devices

Fig. 5 shows the J-V curves of CdS/CdTe devices with different CdS layers. The performance of the CdS/CdTe device where CdS layer has been deposited with no additional doping (considered as the baseline structure) gave 10.25 % efficiency with Voc, Jsc and FF of 811mV, 19.2 mA/cm² and 65% FF, respectively. Using CdCl₂ as the Cd source for the deposition of the CdS layer, the device parameters improved. Efficiency increased to 10.90 % due to an increase in the Jsc (19.9 mA/cm²) and FF (67%) while Voc remained constant (807 mV). This is in agreement with the PL imaging analysis, which showed an increase of the maximum PL counts.

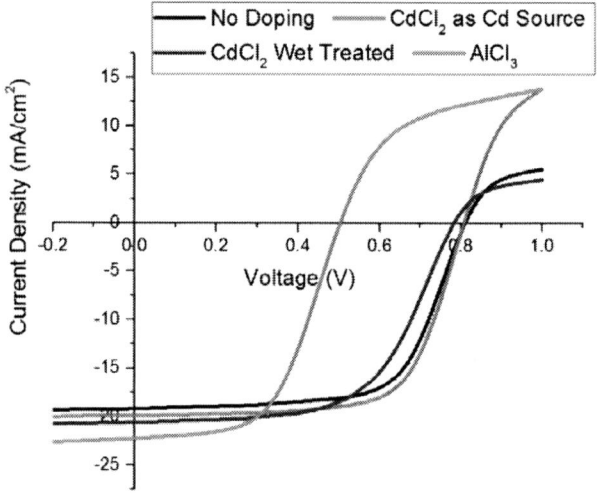

Figure 5: J-V curve of CdS/CdTe devices with different CdS doped layers (un-doped CdS, CdS with CdCl₂ as Cd Source, CdS: ZnCl₂, wet treated CdS and CdS:AlCl₃).

The CdTe device where the CdS layer was CdCl₂ wet-treated, prior to CdTe deposition, showed a reduction in electrical performance. The efficiency, Voc and FF decreased to 9.46%, 784 mV and 58%, respectively. There is a slight increase in current density (20.6 mA/cm²). PL imaging analysis showed an average of 4.93×10^4 counts, which was significantly higher in comparison to the other investigated devices. However, this sample cannot be directly compared, since it has been undergone a double CdCl₂ treatment (before and after the deposition of CdTe). Additionally, since the wet activation treatment performed on CdS films was not totally uniform, this could have affected the final performances of the device. In fact, in agreement with the PL imaging analysis, SEM results showed some areas completely depleted of CdS, as an effect of the over treatment. Doping CdS with $AlCl_3$ had a deleterious effect on the overall performance of the device. Efficiency, Voc, and FF decreased to 6.20%, 504 mV and 0.55, respectively. The significant decrease in voltage (~300 mV reduction compared with the baseline device) indicated a poor p-n junction formation. This was also confirmed by the PL image analysis where the PL emission showed 2.2×10^3 counts.

C. External Quantum Efficiency (EQE)

Fig. 6 shows the normalised external quantum efficiency (EQE) analysis performed on the investigated devices. The baseline CdTe structure with the un-doped CdS layer and the CdS layer deposited using $CdCl_2$ as source of cadmium showed comparable EQE responses. Only, in the range of 360 to 580 nm a slight decrease in EQE was observed for the un-doped CdS layer compared to the CdS deposited with $CdCl_2$ in the bath.

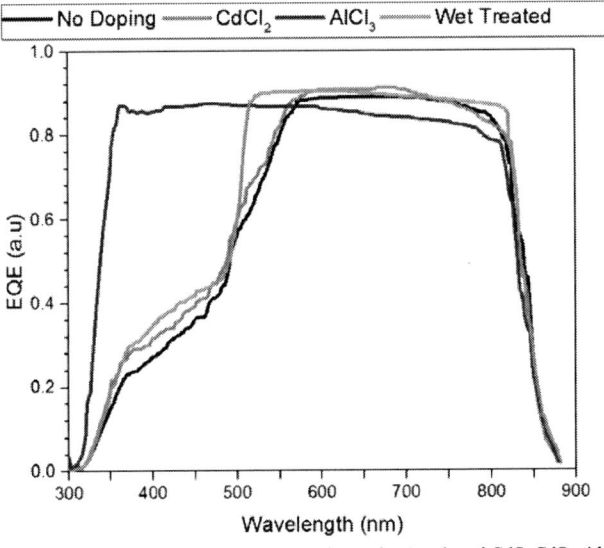

Figure 6: EQE spectra of the investigated samples (un-doped CdS, CdS with $CdCl_2$ as Cd Source, wet treated CdS and CdS:$AlCl_3$).

The CdTe device with the wet treated CdS layer showed a slightly higher EQE in the range of 450-600 nm. Additionally, the EQE response showed less back surface recombination (700-850 nm) in comparison with the un-doped and $CdCl_2$ doped devices[9]. The CdTe device with CdS:$AlCl_3$ doped layer showed an atypical EQE curve. The dramatic decrease in absorption in the range of 300-540 nm suggests significant consumption of CdS layer during the $CdCl_2$ activation treatment. It must be mentioned that the as deposited CdS layer doped with $AlCl_3$ was only 50 nm thick, with a consequence that the $CdCl_2$ activation treatment results in the CdS layer becoming fuly consumed. This is also in agreement with the PL imaging analysis, where the PL image gave a low signal of 2.2×10^3 counts.

D. SEM, EDX TEM and XRD analysis on CdS/CdTe Devices

To have a comprehensive understanding on how the dopants affect the CdS/CdTe interface after the $CdCl_2$ treatment, morphological and structural analyses were carried out. Fig. 7 shows the SEM cross section analysis of the investigated devices, focusing on the CdTe/CdS interface. The baseline structure with the un-doped CdS layer and the CdS deposited with $CdCl_2$ as the cadmium source exhibit uniform CdS layers with an average thickness of ~ 120 nm (Fig. 7a,b). Fig. 7c) shows the CdTe/CdS interface with the $CdCl_2$ wet treated CdS layer. This device showed a non-uniform CdS layer where some areas were completely depleted as an effect of the double treatment, causing shunt-paths in the device. SEM images of the CdTe device with the CdS:$AlCl_3$ doped window layer (Fig. 7e) showed that the CdS layer is barely visible, indicating total CdS consumption during the $CdCl_2$ activation treatment due to the insufficient initial thickness (~50 nm).

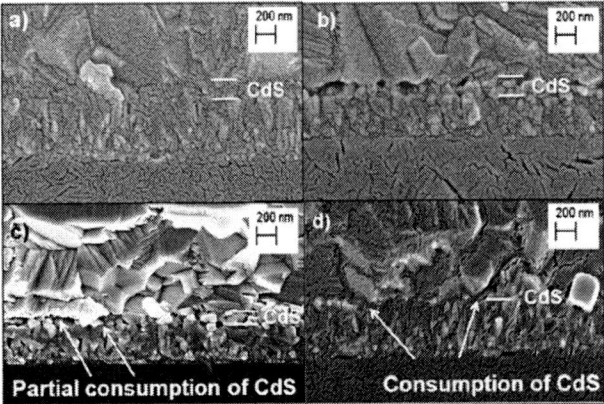

Figure 7: SEM cross section of CdS/CdTe interfaces with: a) un-doped CdS, b) CdS with $CdCl_2$ as Cd Source, c) wet treated CdS and d) CdS:$AlCl_3$.

These results agree with PL imaging analysis and electrical performance of the investigated devices.

TEM analysis was also performed on the device where $CdCl_2$ was used as Cd source for the deposition of the CdS layer. This showed the highest PL emission and electrical response. The TEM analysis confirmed the presence of chlorine at the CdTe/CdS interface (Fig. 8) after the $CdCl_2$ activation treatment. This is in agreement with the PL analysis where the chlorine can be detected due to the formation of V_s-Cl_s and V_{Cd}-Cl_s complexes which have energy levels in the band gap of the CdS layer.

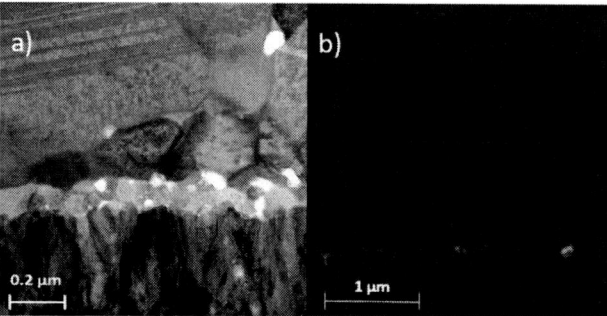

Figure 8: CdS/CdTe device with $CdCl_2$ as source of Cd during CdS deposition a) TEM image and b) chlorine map of the device.

It is known from literature that CdS films exhibit both zincblend (cubic) and wurzite (hexagonal) crystal structure, where the latter is the most stable one [6]. The introduction of dopants in CdS films might influence the crystal growth. Therefore, XRD was performed on all the investigated CdS layers to access any structural change. All the CdS films showed a mixed cubic-hexagonal structure, except the $CdCl_2$ wet treated layer, which showed the hexagonal phase. Furthermore optical characterisation showed a wider bandgap (2.4 eV) compared to the other investigated samples which can explain the reduction of absorption in the range of 450 nm – 550 nm.

IV. CONCLUSION

CdS thin films were deposited by CBD with a variety of different dopants. A comparison was made between CdTe devices where CdS layers were un-doped and doped with $AlCl_3$ and $CdCl_2$. These devices were subjected to the same $CdCl_2$ activation treatment. The presence of chlorine causes the formation of V_S-Cl_S and V_{Cd}-Cl_S complexes in the sub band gap of the CdS material. PL imaging analysis was used to detect the emission originated from these complexes. It has been verified that the PL signal is related to the presence of chlorine when the CdS film was treated with $CdCl_2$, prior to the deposition of CdTe.

It has been shown that the presence of different chorine dopants ($CdCl_2$, $AlCl_3$) in CdS films affected the PL emission signal, once the $CdCl_2$ treatment was performed on the final device.

Furthermore, PL emission was found to be useful to provide qualitative information about the effectiveness and the uniformity of the $CdCl_2$ treatment on the investigated CdTe devices.

Morphological and structural characterization has been also performed to have a comprehensive understanding on the role of the dopants on the final performance of the solar cells.

Further investigation using alternative dopants may lead to improvement of devices performances. Even though PL was proved to be a powerful tool in the assessment of different CdS dopants use, deeper characterization could enlarge the understanding on the process.

REFERENCES

[1] First Solar. Inc, "FIRST SOLAR ACHIEVES YET ANOTHER CELL CONVERSION EFFICIENCY WORLD RECORD," *February 23*, 2016. [Online]. Available: http://investor.firstsolar.com/releasedetail.cfm?ReleaseID=9 56479. [Accessed: 01-Jan-2001].

[2] L. D. L. Kranz, "Role of impurities and pn junction formation in CdTe thin film solar cells," no. 21826, 2014.

[3] D. R. Hodges, "Development of CdTe thin film solar cells on flexible foil substrates," University of South Florida, 2015.

[4] T. Trupke, B. Mitchell, J. W. Weber, W. McMillan, R. A. Bardos, and R. Kroeze, "Photoluminescence imaging for photovoltaic applications," *Energy Procedia*, vol. 15, no. 2011, pp. 135–146, 2012.

[5] A. E. Abken, D. P. Halliday, K. Durose, A. E. Abken, D. P. Halliday, and K. Durose, "Photoluminescence study of polycrystalline photovoltaic CdS thin film layers grown by close-spaced sublimation and chemical bath deposition Photoluminescence study of polycrystalline photovoltaic CdS thin film layers grown by close-spaced sublimation and," *J. Appl. Phys.*, vol. 64515, 2009.

[6] L. Wan, Z. Bai, Z. Hou, D. Wang, H. Sun, and L. Xiong, "Effect of CdCl2 annealing treatment on thin CdS films prepared by chemical bath deposition," *Thin Solid Films*, vol. 518, no. 23, pp. 6858–6865, 2010.

[7] F. Lisco, P. M. Kaminski, A. Abbas, K. Bass, J. W. Bowers, G. Claudio, M. Losurdo, and J. M. Walls, "The structural properties of CdS deposited by chemical bath deposition and pulsed direct current magnetron sputtering," *Thin Solid Films*, pp. 2–6, 2014.

[8] C. Potamialis, F. Lisco, J. W. Bowers, and J. M. Walls, "Fabrication of CdTe Thin Films by Close Space Sublimation," in *PVSAT-12 UoL*, 2016.

[9] J. Nelson, *The Physics of Solar Cells*, First Edit. London: Imperial Collage Press, 2003.

Application of Mapping Spectroscopic Ellipsometry for CdSe/CdTe Solar Cells: Optimization of Low-Temperature Processed Devices with All-Sputtered Semiconductors

Mohammed A. Razooqi, Adam B. Phillips, Geethika K. Liyanage, Fadhil K. Al-Fadhili, Maxwell M. Junda, Nikolas J. Podraza, Michael J. Heben, Robert W. Collins, and Prakash Koirala

Wright Center for Photovoltaics Innovation & Commercialization, and Department of Physics & Astronomy
University of Toledo, Toledo, Ohio, 43606, USA

Abstract — The incorporation of CdSe as a top component layer in superstrate CdTe photovoltaics (PV) technology has led to improvements in short-circuit current due to collection at both violet and near-infrared edges of the external quantum efficiency spectra. The achievable improvements are very sensitive to the details of device fabrication including the individual CdSe and CdTe layer thicknesses and the thermal profile of the entire process. In this research, mapping spectroscopic ellipsometry (M-SE) is being applied at different stages of the process, in conjunction with previously-established capabilities of through-the-glass SE (TG-SE) and variable-angle SE (VA-SE), all toward the development of an advanced metrology for this new absorber layer alloy system. In this report, correlations are presented between the effective thickness (or volume/area) of thin film CdSe, obtained by M-SE prior to CdTe deposition, and the ultimate device performance.

I. INTRODUCTION AND OVERVIEW

Increases in short-circuit current density have been the most significant contributors to the increases in record efficiency for laboratory CdTe solar cells over the last decade or more. Recent studies have demonstrated that the incorporation of CdSe as the top layer of the CdTe superstrate solar cell, or as an intermediate layer between thin CdS(:O) and thick CdTe, has resulted in improvements in current collection in both the violet-green and near-infrared ranges of the external quantum efficiency spectrum [1-4]. The improvement in the violet-green range is attributed to collection of electrons and holes generated within the topmost CdSe or II-VI alloy with Se. The improvement in the near-infrared is attributed to additional collection from a lower bandgap $CdSe_{1-x}Te_x$ alloy that arises due to a bandgap bowing effect in this alloy system [3].

Over the range of process temperatures used for superstrate solar cell fabrication, the top layer of CdSe exhibits a much stronger interaction with the overlying CdTe than does CdS, due to the higher solubility of Se in CdTe compared with S. Although this interaction enables variation of the bandgap and its profile through CdSe/CdTe alloying, the process is more challenging to optimize and control, also given the additional possibility of incorporating a very thin CdS or CdS:O layer at the interface with the transparent conductor [2]. As a result, metrology of the CdSe/CdTe solar cell structure is also more challenging than that of the CdS/CdTe structure. Similar

challenges have been met in the analyses of $CuIn_{1-x}Ga_xSe_2$ (CIGS) solar cells by applying ex situ variable-angle spectroscopic ellipsometry (VA-SE) performed on completed devices [5]. The structural information and bandgap profile for CIGS devices deduced by VA-SE are of sufficient quality for close simulation of the measured external quantum efficiency (EQE) spectrum without free parameters. The SE capability for CIGS has also been extended to mapping of structural and compositional parameters over large areas [6].

In this research, similar capabilities are being sought to characterize the structure and compositional profile of CdSe/CdTe solar cells. Here, we describe initial studies directed toward the optimization of CdSe/CdTe devices in which the two semiconductors are deposited by magnetron sputtering onto commercial SnO_2:F-coated glass with a top-most high resistivity transparent (HRT) layer. These initial studies do not use a CdS layer at the HRT interface, and the maximum processing temperature is 387°C which arises from the $CdCl_2$ treatment. Correlations have been developed between the CdSe/CdTe solar cell performance and the M-SE results for the CdSe effective thickness, the latter determined prior to CdTe deposition, while fixing all other device structure characteristics. Preliminary results are promising and suggest further applications of M-SE for analysis of the graded layers resulting from CdSe/CdTe interaction due to processing at elevated temperatures, as well as the more complex device structures incorporating CdS(:O)/CdSe/CdTe, tailored for maximum performance.

II. EXPERIMENTAL DETAILS

The superstrate used in this study was TEC™15/HRT, which is a soda-lime glass over-deposited with layers of $SnO_2/SiO_2/SnO_2$:F and a 100 nm HRT layer. These superstrates were cleaned ultrasonically with Micro-90™ (Int. Prod. Corp.) and with distilled water, each for 20 min. CdSe was deposited on the room temperature superstrates with nominal thicknesses of 80, 120, and 160 nm using an RF magnetron sputtering system. CdSe sputtering was performed in pure Ar at a pressure of 10 mTorr. The RF power was 25 W, yielding a deposition rate of ~ 10 nm/min. These CdSe samples were measured before further processing by mapping

978-1-5090-5606-4/17 $31.00 © 2017 IEEE

spectroscopic ellipsometry (M-SE; Accumap-SE, J. A. Woollam Co.) using a rotating-compensator multichannel instrument having a 0.75 to 6.5 eV photon energy range. The probe beam diameter is ~ 0.65 mm, and at the angle of incidence of 65.0°, an elliptical spot on the surface is generated with a major axis 1.5 mm in length. A total of 81 locations, corresponding to the positions of subsequently fabricated dot cells 0.125 cm^2 in area, were measured over each ~ 6.5 cm x 6.5 cm TEC™15/HRT/CdSe sample.

After the 80, 120, and 160 nm depositions of CdSe, the samples were moved to a second sputtering system, where the CdTe layer was deposited at a superstrate temperature of 250°C. Sputtering was performed in pure Ar at 10 mTorr pressure and a 23 standard cm^3/min flow rate. The RF sputtering power for CdTe was 200 W, yielding a deposition rate of 17 nm/min. After sputter deposition of a 2 μm thick layer of CdTe, the samples were each coated with a saturated solution of CdCl$_2$/methanol (0.04 mol/L), and annealed at 387°C for 30 min under dry air at atmospheric pressure and at a flow of 5 standard ft^3/hr. This is the maximum temperature step in the cell fabrication process. After cooling to room temperature, the sample surfaces were rinsed with methanol to remove residual CdCl$_2$. The back contacts consist of 3 nm Cu and 40 nm Au thermally evaporated through a shadow mask to form dot cells each with an area of 0.125 cm^2. A back contact anneal was performed at 150°C for 45 min in N$_2$.

For each of the three 6.5 cm x 6.5 cm samples having nominal CdSe thicknesses of 80, 120, and 160 nm, J-V measurements of the set of 81 dot cells were performed in the dark and under AM 1.5G illumination. EQE spectra were also measured at 0 V bias potential. The focus of this report is the M-SE, J-V, and EQE results which will be presented and discussed next. The goal of this study is to identify correlations for optimization purposes between the solar cell parameters deduced by J-V and the effective thickness of the CdSe film, measured by M-SE prior to CdTe deposition.

III. RESULTS AND DISCUSSION

A. Mapping spectroscopic ellipsometry (M-SE) of CdSe

Figure 1 shows typical ellipsometry spectra (ψ, Δ) for a single mapping point, measured from the film side after sputter deposition of the CdSe sample with 120 nm nominal thickness on TEC™15/HRT. In addition to the four layers that comprise the TEC™15/HRT structure, the model structure adopted in the simulation of the data incorporates a HRT/CdSe interface roughness layer that derives from the HRT surface roughness, as well as CdSe bulk and surface roughness layers. In spite of the complexity of the spectra, reasonably good agreement is obtained between the data and the best fit simulation. The real and imaginary parts (ε_1, ε_2) of the CdSe dielectric function used in the analysis were deduced via multi-thickness analysis from a deposition performed under identical conditions but on a Si wafer substrate. For multi-thickness analysis, multiple locations on the sample

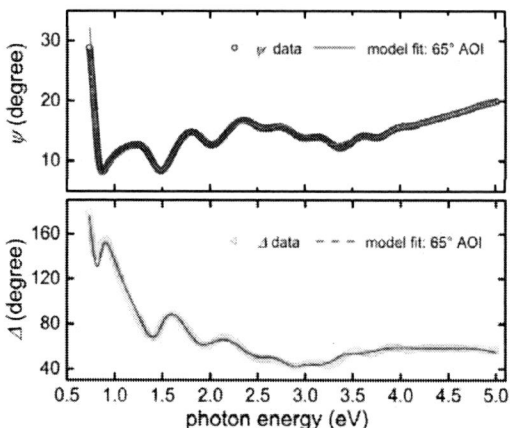

Fig. 1: Spectra in the experimental ellipsometry angles (ψ, Δ) (points) for TEC™15/HRT/CdSe plotted along with the best fit simulation (lines). The plot corresponds to one representative location on the map of Fig. 2 where the CdSe bulk layer and effective thicknesses are 67 and 94 nm, respectively.

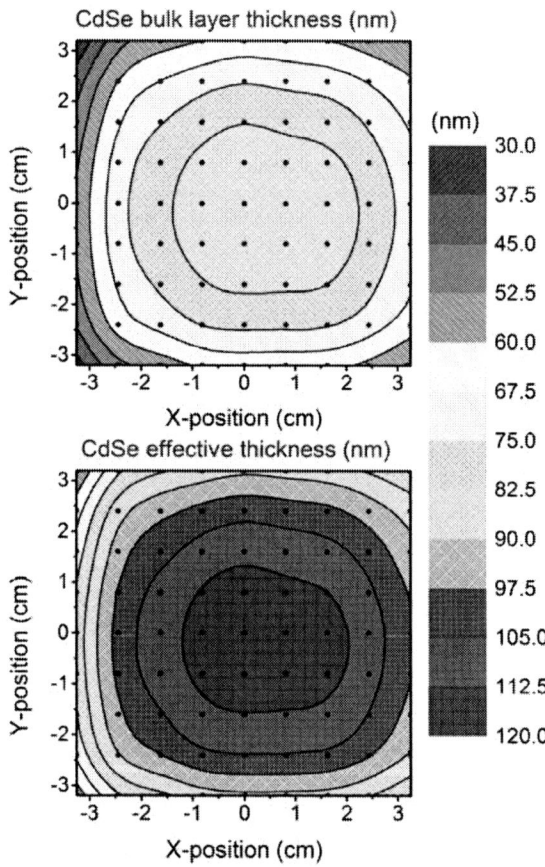

Fig. 2 Maps over an area of 6.5 cm x 6.5 cm for the bulk layer thickness (top) and effective thickness (bottom) in the analysis of a nominal 120 nm thick CdSe film deposited on a TEC™15/HRT superstrate. The effective thickness is a measure of CdSe material volume per unit area of superstrate, given as the product of the thickness and CdSe volume fraction summed over all the layers (interface/bulk/surface) that incorporate CdSe.

TABLE I. PARAMETERS FROM J-V MEASUREMENTS OF THE HIGHEST PERFORMING CdTe SOLAR CELLS INCORPORATING CdSe LAYERS OF DIFFERENT NOMINAL THICKNESSES, ALONG WITH THE RESULT FOR A STANDARD CdS/CdTe CELL. THE CdSe EFFECTIVE THICKNESS MEASURED BY M-SE AT THE SPECIFIC CELL LOCATION IS ALSO GIVEN.

Cell type with nominal thicknesses	CdSe eff. thickness (nm)	J_{SC} (mA/cm^2)	V_{OC} (V)	FF (%)	Eff. (%)
CdSe/CdTe: 160 nm/2 μm	148	26.0	0.753	67.0	13.1
CdSe/CdTe: 120 nm/2 μm	112	25.1	0.724	66.0	12.0
CdSe/CdTe: 80 nm/2 μm	77	24.7	0.657	63.8	10.4
CdS/CdTe: 120 nm/2 μm	0	22.1	0.832	72.7	13.4

were analyzed simultaneously to deduce bulk and surface roughness layer thicknesses along with (ε_1, ε_2). In all such analyses, the HRT/CdSe interface and CdSe surface roughness layers are modeled using the Bruggeman effective medium approximation as mixtures of (HRT + CdSe) and (CdSe + void), respectively [7]. The CdSe thickness sought for correlation with device performance is not the bulk layer value, but rather the effective value, given as the product of the layer thickness and CdSe volume fraction in the layer, summed over all three layers that incorporate the CdSe (interface/bulk/surface). As the CdSe effective thickness decreases to a value below the roughness thickness on the HRT layer, the bulk layer thickness tends to zero, and the film consists of only interface and surface roughness components. This does not imply that the film itself is discontinuous, but rather that the underlying roughness is greater than that which can be completely filled by the deposited CdSe. In general, the effective thickness is the material volume per area of substrate/superstrate and is appropriate in quantifying the accumulation of thin film material on rough surfaces.

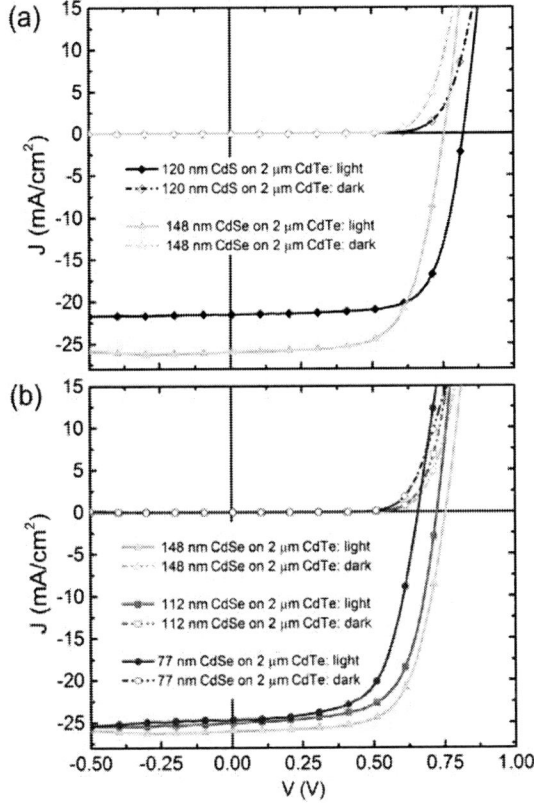

Fig. 3. Light and dark J-V characteristics for CdSe/CdTe solar cells having the CdSe effective thicknesses indicated in the legends, as measured at the cell location; (a) comparison of the highest efficiency CdSe/CdTe and CdS/CdTe cells; (b) comparison of the highest efficiency cells fabricated using CdSe nominal thicknesses of 80 nm (circles), 120 nm (squares), and 160 nm (triangles).

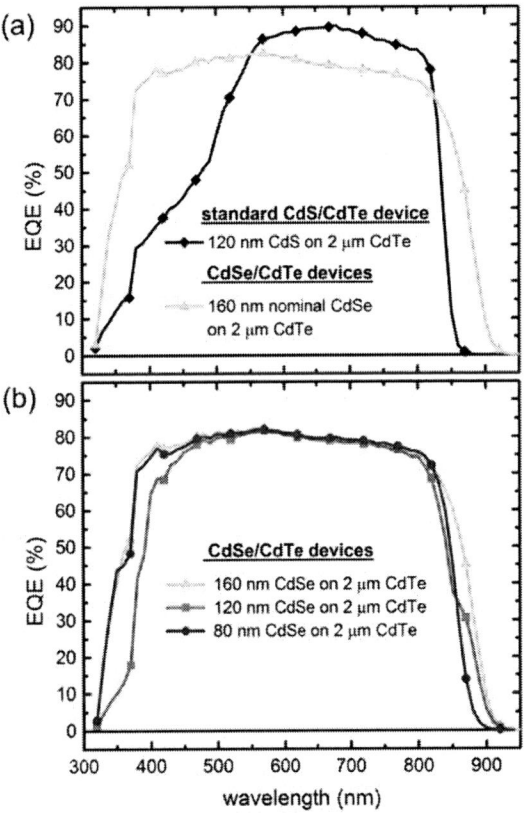

Fig. 4. External quantum efficiency spectra for representative CdSe/CdTe solar cells having the CdSe nominal thicknesses indicated in the legends; (a) comparison of the highest efficiency CdSe/CdTe and CdS/CdTe cells; (b) comparison of representative cells fabricated using the nominal thicknesses of 80 nm (circles), 120 nm (squares), and 160 nm (triangles).

978-1-5090-5606-4/17 $31.00 © 2017 IEEE

B. Effect of CdSe thickness on the photovoltaic performance

Table I shows the performance parameters for the highest efficiency CdSe/CdTe solar cells from the three samples fabricated with different nominal CdSe thicknesses of 80, 120, and 160 nm. No CdS layer was incorporated into these CdSe/CdTe devices. Figure 3 shows the light and dark (current density)-voltage (J-V) characteristics for the highest efficiency cells from the three samples, and Figure 4 shows EQE spectra representative of each of the three nominal CdSe thicknesses. Also shown in Table I and Figures 3(a) and 4(a) are the results for a standard CdS/CdTe solar cell of high efficiency used for comparison with the highest efficiency CdSe/CdTe cell with 148 nm CdSe effective thickness.

Table I shows that increases occur with increasing nominal and effective CdSe thicknesses in all solar cell performance parameters deduced from the J-V characteristics, including open-circuit voltage V_{OC}, short-circuit current J_{SC}, fill-factor (FF), and efficiency. The highest efficiency cells with 80 and 120 nm nominal thicknesses are located near the centers of the maps where the maximum thicknesses occur as in Fig. 2. The location of the highest efficiency cell with 160 nm nominal thickness appears displaced from the center to lower thickness, indicating an optimum CdSe effective thickness of

~ 150 nm. It is evident that J_{SC} is considerably higher for the CdSe/CdTe solar cell than for the standard CdS/CdTe cell due to the enhanced collection in the violet-green and the near-infrared ranges of the EQE spectra as shown in Figure 4(a). Both V_{OC} and FF are lower for the CdSe/CdTe solar cells than for the standard cell, however. The latter reductions lead to an efficiency for the best CdSe/CdTe cell deposited with 148 nm CdSe effective thickness that is only slightly lower (0.3% absolute) than that of the CdS/CdTe standard.

The EQE spectra show notable characteristics. First, at short wavelengths these spectra show surprisingly non-monotonic behavior with CdSe nominal thickness, meaning that representative cells from the 80 and 160 nm samples show a shorter wavelength onset compared to those with the intermediate nominal thickness of 120 nm. Such behavior could arise from effects having an optical or an electronic origin. Second, the EQE spectra for all three CdSe cells are relatively constant over the range of 500-800 nm, and are ~10% lower than that of the standard CdS/CdTe cell. This result may be consistent with a greater reflectance at the HRT/CdSe interface as compared to the HRT/CdS interface; optical modeling is needed to evaluate this suggestion. Third, at long wavelengths, an initial shift to lower energies in

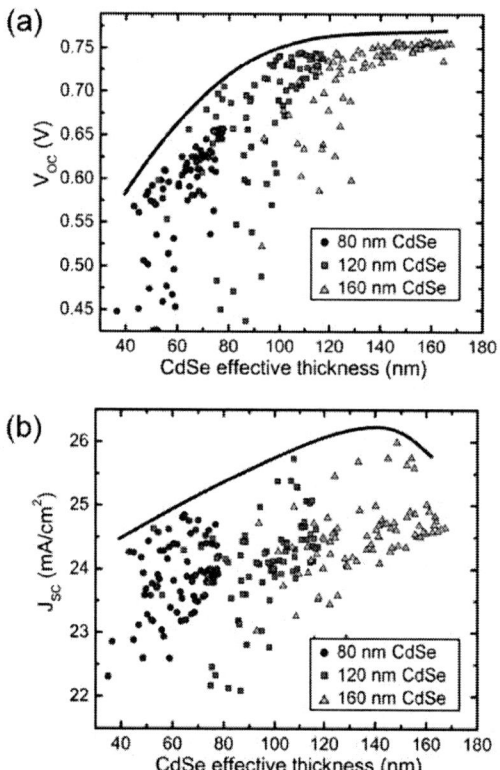

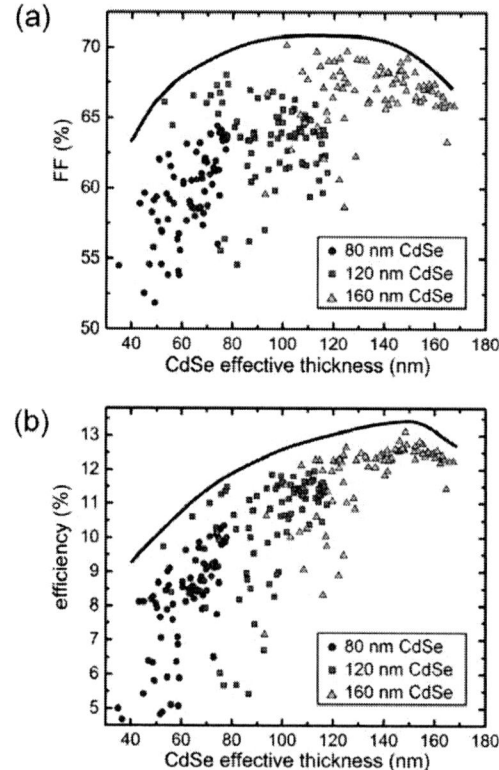

Fig. 5. Correlations between M-SE measurement of CdSe effective thickness at the precise solar cell location and the solar cell parameters of (a) open-circuit voltage V_{OC} and (b) short-circuit current J_{SC} for three sets of solar cells fabricated with nominal CdSe thicknesses of 80, 120, and 160 nm.

Fig. 6. Correlations between M-SE measurement of CdSe effective thickness at the precise solar cell location and the solar cell parameters of (a) fill-factor FF and (b) efficiency for three sets of solar cells fabricated with nominal CdSe thicknesses of 80, 120, and 160 nm.

the EQE-derived bandgap is observed followed by an apparent saturation with increasing CdSe thickness. This behavior is an indication that increased alloying of CdSe with CdTe is occurring with increasing CdSe thickness, thus reaching a maximum bowing effect [8]. Detailed device simulation, as has been performed for standard CdS/CdTe solar cells [7], would help to understand the differences and similarities among these CdSe/CdTe cells and between the CdSe/CdTe and CdS/CdTe cells.

In order to explore the optimization of CdSe/CdTe solar cells in greater detail, Figures 5 and 6 show the four solar cell parameters of V_{OC}, J_{SC}, FF, and efficiency correlated with CdSe effective thickness as obtained by M-SE. For these correlations, the M-SE measurement of effective thickness is performed prior to CdTe deposition at the same location as the solar cell fabricated subsequently. These correlations exploit the non-uniformity pattern as shown in Fig. 2 in order to generate essentially a continuous range of effective thicknesses from 40 to 170 nm based on three solar cell depositions with 80, 120, and 160 nm nominal thicknesses. For each CdSe effective thickness, however, non-uniformities occur in other characteristics of the solar cell such as variations in the application of CdCl$_2$ or in the thickness of the thin Cu layer forming the back contact, in addition to possible localized pinholes and contamination. These variations generate the considerable scatter at a given CdSe effective thickness. The scatter in Figures 5 and 6 appears to be greater for thinner CdSe.

Well-defined trends are observed in the maximum performance envelopes, shown as the solid lines in Figures 5 and 6. First, it is noted that V_{OC} shows a sharp increase with CdSe effective thickness from 40 to 100 nm and a more gradual increase above 100 nm, continuing to 170 nm, the maximum effective thickness for the sample with 160 nm nominal thickness. For each of the three samples defined by the nominal thickness, the maximum J_{SC} appears to oscillate with effective thickness, exhibiting maxima near 65, 105, and 150 nm. The behavior suggests an optical effect; however, such a conclusion must be verified through modeling. The FF correlation suggests a broad optimum near 110 nm. Finally, the efficiency exhibits an overall trend and an optimum versus CdSe effective thickness that reflects the behavior of J_{SC}, as the variations in V_{OC} and FF are relatively weak in the high efficiency range of effective thickness above 100 nm.

IV. SUMMARY

CdTe solar cells incorporating CdSe top layers in place of the traditional CdS or CdS:O have been deposited on transparent conducting oxide coated glass superstrates with overlying HRT layers. Both semiconductor layers of CdSe and CdTe are fabricated by magnetron sputtering for which the CdCl$_2$ treatment defines the maximum process temperature of 387°C. Mapping spectroscopic ellipsometry (M-SE) has been applied to the CdSe layer in order to identify accurate effective thicknesses (or volume/area) associated with the

CdSe in the starting film structure. The highest performance CdSe/CdTe solar cell incorporates ~150 nm effective thickness of CdSe, yielding a significantly higher current compared to standard CdS/CdTe cells due to contributions from the violet-to-green and near-infrared regions of the external quantum efficiency spectra. The reduced V_{OC} and FF of the CdSe/CdTe cells lead to similar maximum efficiencies for the CdSe/CdTe and CdS/CdTe solar cells. M-SE shows promise for further optimization of the CdSe/CdTe solar cell structure with the potential of mapping bandgap profiles as has been demonstrated in CIGS PV technology [6]. Correlation plots of the M-SE determined profile parameters will be used for a better understanding of the role of CdSe thickness in the fabrication process. M-SE is also expected to play an important role in more complicated CdSe/CdTe structures that incorporate CdS and CdS:O at the interface to the HRT.

REFERENCES

[1] N. R. Paudel and Y. Yan, "Enhancing the photo-currents of CdTe thin-film solar cells in both short and long wavelength regions," *Applied Physics Letters*, vol. 105, art. no. 183510 (2014).

[2] N. R. Paudel, J. D. Poplawsky, K. L. Moore, and Y. Yan, "Current enhancement of CdTe-based solar cells," *IEEE Journal of Photovoltaics*, vol. 5, pp. 1492-1496 (2015).

[3] J. D. Poplawsky, W. Guo, N. Paudel, A. Ng, K. More, D. Leonard, and Y. Yan. "Structural and compositional dependence of the CdTe$_x$Se$_{1-x}$ alloy layer photoactivity in CdTe-based solar cells," *Nature Communications*, vol. 7, art. no. 12537 (2016).

[4] X. Yang, Z. Bao, R. Luo, B. Liu, P. Tang, B. Li, J. Zhang, W. Li, L. Wu, and L. Feng, "Preparation and characterization of pulsed laser deposited CdS/CdSe bi-layer films for CdTe solar cell application," *Materials Science in Semiconductor Processing*, vol. 48, pp. 27-32 (2016).

[5] A.-R. Ibdah, P. Koirala, P. Aryal, P. Pradhan, S. Marsillac, A. A. Rockett, N. J. Podraza, and R. W. Collins, "Spectroscopic ellipsometry for analysis of polycrystalline thin-film photovoltaic devices and prediction of external quantum efficiency," *Applied Surface Science* (2017), in press. Online: doi.org/ 10.1016/j.apsusc.2016.12.236

[6] P. Aryal, A.-R. Ibdah, P. Pradhan, D. Attygalle, P. Koirala, N. J. Podraza, S. Marsillac, R. W. Collins, and J. Li, "Parameterized complex dielectric functions of CuIn$_{1-x}$Ga$_x$Se$_2$: applications in optical characterization of compositional non-uniformities and depth profiles in materials and solar cells," *Progress in Photovoltaics: Research and Applications* **24**, 1200-1213 (2016).

[7] P. Koirala, J. Li, H. P. Yoon, P. Aryal, S. Marsillac, A. A. Rockett, N. J. Podraza, and R. W. Collins, "Through-the-glass spectroscopic ellipsometry for analysis of CdTe thin-film solar cells in the superstrate configuration," *Progress in Photovoltaics: Research and Applications*, vol. 24, pp. 1055-1067 (2016).

[8] M. M. Junda, C. R. Grice, P. Koirala, R. W. Collins, Y. Yan, and N. J. Podraza, "Optical properties of CdSe$_{1-x}$S$_x$ and CdSe$_{1-y}$Te$_y$ alloys and their application for CdTe photovoltaics", these proceedings.

Assessing the Validity and Accuracy of Effective Electronic Materials: Can 1D Simulations Predict Polycrystalline Device Performance?

Yubo Sun[1], Allison Perna[1], Sudhajit Misra[2], Vasilios Palekis[3], Chris Ferekides[3], Jeffrey Aguiar[4], Peter Bermel and Michael A. Scarpulla[2]

[1] Purdue University, West Lafayette, IN, USA
[2] University of Utah, Salt Lake City, UT, USA
[3] University of South Florida, St. Petersburg, FL, USA
[4] Fuel Design and Development Department, Idaho National Laboratory, Idaho Falls, ID, 83404 USA

Abstract — General rules to reduce models of polycrystalline photovoltaic materials such as amorphous silicon (a-Si), cadmium telluride (CdTe), and Cu(In,Ga)Se$_2$ (CIGS) to 1D have been sought for decades. However, a key limitation associated with the 1D simulation of thin-film photovoltaics is the absence of grain boundaries (GBs), which are commonly found in polycrystalline solar cells. Depending on the orientation of growth, GBs are categorized as columnar GBs (roughly perpendicular to incident sunlight) and horizontal GBs (perpendicular to incident sunlight). In this work, we investigate whether polycrystalline grain materials can be appropriately treated as an effective medium, using 1D and 2D simulations of CdTe solar cells that are CSS grown and Br:MeOH+CdCl$_2$ treated. It is found that the grain boundary interface recombination has a marked effect in degrading Voc, which is consistently difficult to capture in a 1D model of an effective medium. In particular, we show that if the initial I-V curve of polycrystalline CdTe does not exhibit a low Schottky barrier, then Br:MeOH-etching the back will not necessarily improve device performance.

Index Terms — **cadmium telluride, photovoltaic cells, grain boundaries, polycrystalline materials**

I. INTRODUCTION

The drift-diffusion and charge neutrality equations are the foundation of semi-classical solar cell device physical analysis. A well-known analysis of a pn homojunction solar cell's operation under short circuit conditions in the light and dark was provided by Hovel, and forms the basis of the most common treatments in one dimension [1]. For device structures such as heterostructures, realistic absorption spectra, and defect recombination computer simulations are in standard use. While 2D and 3D simulation packages are available, few are affordable for many academic researchers, in terms of knowledge, purchase cost, and computational resources required. For widely-studied materials such as those used in Si devices, parameter values are well-known. In multicrystalline photovoltaic materials with very large grain sizes, often the effects of grain boundaries can be neglected. However, for polycrystalline compound semiconductor thin film devices still under development, validated material, interface, and boundary condition parameters may not be available. This state persists since compound semiconductor properties like grain boundary type, density, and recombination velocities depend sensitively on stoichiometry, native defects, and other growth parameters.

General rules for reducing 2D or 3D polycrystalline solar cell devices to 1D models without any parameters fitted to specific experiments have been sought for decades [1]. For specificity, we will assume a solar cell with lateral dimensions much greater than its thickness, that the top and bottom contacts are spatially uniform and iso-potential, and that the current and photovoltage occur in the thickness direction. From a device operation point of view, the lateral grain size should be compared to the minority carrier diffusion length, with depletion widths extending laterally from grain boundaries. Still, for polycrystalline devices, grain boundaries in principle cannot be ignored for any operating point of the solar cell. In practice, it is commonly assumed that this polycrystalline material may be regarded as an effective medium with some admixture of bulk and grain boundary properties. However, a simple *gendanken* experiment illustrates that a universal effective 1D medium cannot describe a true 3D structure with charged grain boundaries under all operating conditions.

For specificity, consider a p-type absorber layer of CdTe with columnar grains at zero bias and forward bias [2]. Assume that closed spaced sublimation (CSS) of these polycrystalline thin films is performed, followed by Br:MeOH etch and CdCl$_2$ treatment commonly used [3]. The GBs will be likely to have donor-like traps pinning the Fermi energy such that 100-300 mV potential difference exists with the grains. Under short circuit, electrons will be attracted into the GBs and holes repelled, thus the effective hole recombination velocity of the GBs will be reduced. At forward bias near Voc, the donor-like traps will have higher electron occupancy thus the induced band bending will be smaller and the effective recombination velocity for holes will be larger. It is not possible to accurately capture both of these conditions using a bulk homogeneous distribution of traps with fixed number and characteristics.

In this paper, we investigate the issues of defining an effective medium for 1D simulation, and apply it to recent experimental observations regarding the changes in grain boundary chemistry induced by bromine etching to prepare the back contact (Br:MeOH). Open questions still remain as to whether 1D simulations can correctly capture impact of these growth and treatment processes. In this manuscript, we describe the direct observations from TEM, which reveal that Br:MeOH etching increases the concentrations of cadmium vacancies, as well as removing Cl, which is known to passivate grain boundaries. From a device modelling standpoint, we interpret these chemical effects as increasing both the doping and number of recombination-active defects.

The structure of the remainder of this paper is as follows: next, we will discuss effective medium models in further detail; then, we will present our numerical modeling techniques for 1D and 2D problems; present the results, and compare them; and finally summarize and conclude this study.

II. EFFECTIVE MEDIUM MODELS

The key question in reducing 3D polycrystalline CdTe to 2D and/or 1D is how to represent the grains and grain boundaries as a single layer with fixed properties. This has been a long-standing problem and there is in fact no exact, fully general prescription for averaging the material properties of grains and grain boundaries. The primary reason is that band bending at grain boundaries is dynamic with regards to illumination and voltage bias; this in turn changes the effective recombination velocity of the GBs. Recognizing this limitation, we endeavor to evaluate the *accuracy* of two simple effective medium prescriptions. In other words we will assess the errors induced by treating the 3D case as 1D.

In the physical situation considered herein of Br:MeOH induced changes to GBs near the back contact, the changes to the material properties occur in the quasineutral region of the device. This reduces the complexity of the problem since the main depletion width is not involved – the band bending will dynamically change the free carrier density and defect occupancy both in the grain bulk and near the grain boundaries, whereas in the quasineutral region away from the grain boundaries, the defect occupancy and carrier density is fixed. Therefore, the accuracy of the 3D to 1D conversion is expected to scale with the relative volume of the GB-affected regions compared to the central regions of the grains – i.e. the grain size.

The main solar cell depletion width has much higher built-in voltage than typical grain boundaries and any such lateral depletion widths will be suppressed (screened) upon photogeneration near Voc. Thus the grain boundaries' built-in potential is of no concern and only the total concentration of doping and recombination-active defects will contribute to the material properties of the depletion width. In the quasineutral region (QNR), the averaging to form an effective medium requires more careful consideration.

Assuming a film thickness t and an array of columnar grains of square cross-section with edge length a, we propose two types of effective medium appropriate for averaging the lifetime and doping of the quasi-neutral region:

1) Conservation of the number of charged defects per film volume. In the 3D case charged defects (traps or dopants) are distributed only at the GBs while in the 1D case they are distributed randomly in the volume. Equating the numbers per volume yields $N_{3D}=2/a\ N_{2D}$. This type of averaging should always apply to the main junction's depletion region and may be accurate for the QNR.

2) Volume averaging the doping and minority carrier lifetime according to different characteristic distances from the GBs. This approach requires specification of the volumes affected by the GBs which will change with bias and illumination and of the appropriate lifetime, mobility, doping, and carrier densities.

In the first model, only the effective trap density is varied in order to handle cases of different grain size and the 1D simulation computes the traps' recombination behavior. The 2nd model requires somewhat more explanation and must be approximated for different bias and illumination limits. We can segment the grains into volumes that are bulk-like, that are within a diffusion length (L_n) of a depletion width edge or the GB itself, and that are within a depletion width of a charged GB. Grain boundaries in the quasi neutral region of columnar grains can form a depletion width W_d if they contain fixed charge. Per unit thickness, the bulk-like area in a columnar grain is $(a-2W_d-2L_n)^2$. The area contained in depletion widths would thus be $4aW_d-4W_d^2$. We retain the 2nd term since W_d may be a significant fraction of a. Similarly, rectangles of width L_n from the edge of the depletion width can be defined with area $4aL_d-8L_dW_d-4L_d^2$. Given these volume fractions, averaging can be carried out for materials properties in different physical cases.

We now consider a few cases of illumination and bias:
A) In the case of insignificant generation in the volume of interest (i.e. the QNR) and in the absence of bias through the cell, the depletion widths caused by fixed positive charge at GBs in the QNR will repel holes and thus prevent recombination within W_d for electrons. Within W_d, only the highest energy holes will be able to reach the GB to cause recombination. Thus the electron lifetime for the GB should be enhanced by a factor $\exp(V_{GB}/k_BT)$ above its value in the absence of V_{GB} where V_{GB} is the built-in voltage from the GB sheet charge. Both other geometric regions will exhibit the bulk lifetime τ_{bulk}.

B) If illumination causes significant generation in the QNR, V_{GB} and W_d will be reduced eventually to zero. Then, photogenerated carriers within L_d of the GBs will diffuse to the GB traps and recombine with a lifetime τ_{GB}. By definition, an electron that diffuses L_n and then recombines at the GB after τ_{GB} will have survived for $\tau_{bulk}+\tau_{GB}$ however carriers originating closer to the GB will diffuse for shorter times. Averaging the time for diffusion over the distances from L_n to 0 from the GB yields $\tau_{bulk}/3$. Thus the rectangles within L_n of the GBs are assigned $\tau_{bulk}/3 + \tau_{GB}$.

For a representative L_n of 1 μm (corresponding to 0.5 ns for electrons in CdTe), the variation volume fractions for different regions and effective lifetime assuming case B are plotted in Figure 1 as a function of lateral grain size. Note that typically grains in vapor deposited films will have lateral grain size within a factor of two of the film thickness.

II. NUMERICAL SIMULATION APPROACHES

The material properties for CdTe solar cells were based on the following considerations. First, we focus on closed space sublimation growth of polycrystalline CdTe, atop fluorinated tin oxide (FTO) glass, which is first covered by a layer of n-type CdS. Typical CdTe grain sizes range from sub-micron to several micron in value. These layers are contacted in the back by evaporated tellurium. KPFM experiments show 100-300 mV band bending between GBs and grains at back surface [3]. Second, TRPL has shown GB interface velocities from 10^4 to 10^5 cm/s when passivated with Cl [4]. These interface velocities are much worse (10^6 to 10^7 cm/s) when unpassivated. [5] Best CdTe has bulk recombination such that 2nd time constant in 2-photon pumped TRPL shows 10's of ns lifetime for p-doped (like GaAs) [4]. Most of the model parameters used in this study are taken from prior work by Colorado State [7]; other properties are taken from canonical work in the literature [8-11]. A full summary of all the materials properties used is provided in Table 1.

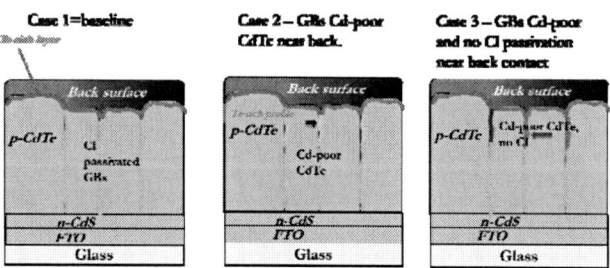

Figure 1: The three polycrystalline CdTe heterojunction geometries considered in this work. The first case experiences only CdCl₂ treatment, while the second and third cases are Br:MeOH etched, creating Cd-poor regions and a Te-rich skin layer in the back.

To fully understand the role of GBs in CdTe, a 2D TCAD simulation was performed using the Synopsys software tool "*Sentaurus*," which is a self-consistent electro-optical device solver [13]. It is designed to solve both drift-diffusion and Poisson equations in 1D, 2D, or 3D, with various extensions to incorporate physical effects such as thermionic emission [13]. As has been previously shown by Gloecker et. al [5], a

columnar GB in the absorber layer may lead to majority/minority carriers repulsive band bending, which amounts to downshifting of the valence band. Each case gives rise to completely different I-V characteristics. For example, majority carrier repulsive band bending at the GB is most likely due to minority carrier charged defects, which may create an effective minority carrier transport channel, thus boosting the short circuit current J_{SC}. In contrast, the majority carrier repulsive band bending will enhance bulk recombination in the grain interior (GI). There are generally two options to model columnar GBs in a 2D simulation. *First*, to create an intermediate layer between grains with grain width as a controllable parameter and by adding a defect profile (defect concentration, defect level, defect distribution, etc.) to vary the surface potential at the band bending. *Second*, to stack two grains together and specifying the interface recombination due to columnar GB by setting the interface recombination velocity for minority carriers. *Fig.1* below depicts the lateral energy band diagram (EBD) of donor-like charged GB in the P-type CdTe absorber under open circuit condition.

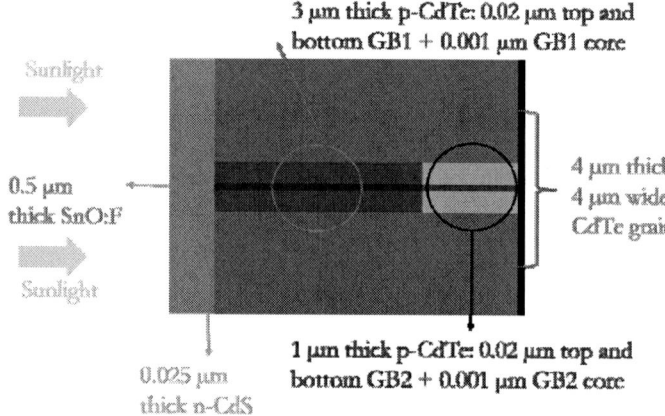

Figure 2: Periodic 2D geometry used to capture the 3 cases illustrated in Fig. 1 above using Sentaurus [13].

We utilize SCAPS 3.3.03 to solve the drift-diffusion and Poisson equations self-consistently in 1D [12]. At the FTO and CdTe edges of the simulation, 10^4 cm/s recombination velocity for electrons and holes were utilized and both contacts were set to flat bands for all conditions. The Te layer was omitted.

	FTO	CdS	Front CdTe	Back CdTe	Te
Thickness (μm)	0.5	0.025	3	1	0.05
E_g (eV)	3.6	2.4	1.5	1.5	0.33
χ (eV)	4	4	3.9	3.9	4.81
ε (relative)	9	10	9.4	9.4	9.4
N_c (#/cm³)	2.2×10^{18}	2.2×10^{18}	8×10^{17}	8×10^{17}	8×10^{17}
N_v (#/cm³)	1.8×10^{19}	1.8×10^{19}	1.8×10^{19}	1.8×10^{19}	1.8×10^{19}
μ_e (cm²/Vs)	100	100	320	320	320
μ_h (cm²/Vs)	25	25	40	40	40
N_d (#/cm³)	10^{17}	10^{17}	0	0	0
N_a (#/cm³)	0	0	varied	varied	10^{18}
$v_{th,e}$ (cm/s)	10^7	10^7	10^7	10^7	10^7
$v_{th,h}$ (cm/s)	10^7	10^7	10^7	10^7	10^7
$\beta_{radiative}$ (cm³/s)	-	-	2×10^{-10}	2×10^{-10}	2×10^{-10}
$C_{Auger,e}$ (cm⁶/s)	-	-	10^{-30}	10^{-30}	10^{-30}
$C_{Auger,h}$ (cm⁶/s)	-	-	10^{-30}	10^{-30}	10^{-30}
$\tau_{e,SRH}$ (ns)	100	10	varied	varied	varied
$\tau_{h,SRH}$ (ns)	0.1	10^{-4}	varied	varied	varied
N_t (#/cm³)	10^{15}	10^{18}	varied	varied	-
E_t (eV) & type	E_i (0/-)	E_i (0/-)	E_i (+/0)	E_i (+/0)	-
σ_e (cm²)	10^{-15}	10^{-17}	2×10^{-13}	2×10^{-13}	-
σ_h (cm²)	10^{-12}	10^{-12}	10^{-16}	10^{-16}	-

Table 1 – Material Properties Used in the Modelling, adapted from Refs. 8-11.

IV. RESULTS AND DISCUSSION

A. 1D Simulations

SCAPS simulations were performed to approximate Case 3 of Figure 1 by sweeping the back CdTe lifetimes and layer trap densities. This latter investigation corresponds to the 2D simulations described in the next subsection, with varying S_{gb} (in which the #/cm² of donor-like traps in the GB cores varies). Using the formula $N_{3D} = 2/a\, N_{2D}$, it is predicted that for recombination velocities of $10^4 - 10^7$ cm/s, randomly distributed 3D neutral trap densities will range from $2.5\times10^{13} - 2.5\times10^{16}$ /cm³. Obviously, if the bulk doping is 2×10^{14} /cm³, these cannot be treated as donor-like traps.

Within that effective medium model described in Section II, the 1D calculation predicts only about 5 mV of Voc loss from varying the back 1 μm of GB properties. In the next section, we will compare these results against a more realistic 2D model

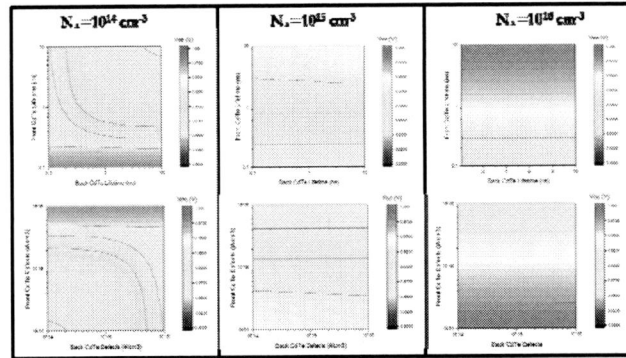

Figure 3: Illustration of the open circuit voltage obtained for a 1D SCAPS model of CdTe, using various levels of doping N_A with variable lifetime.

B. 2D Simulations

The grain boundaries formed in polycrystalline CdS/CdTe solar cells, as graphed in Figure 2, were explored in TCAD Sentaurus a device level drift-diffusion solver. Mid-gap traps were added in those grain boundary core layer to introduce lateral inhomogeneity at the grain boundary. Capture-cross section area of these midgap traps was fixed at 2×10^{13} /cm², the I-V characteristics of standard CdS/CdTe solar cell were plotted at various grain boundary recombination velocities enabled by tweaking the interface trap density. It was found that strong recombination activities at grain boundary could degrade open-circuit voltage of the solar cell.

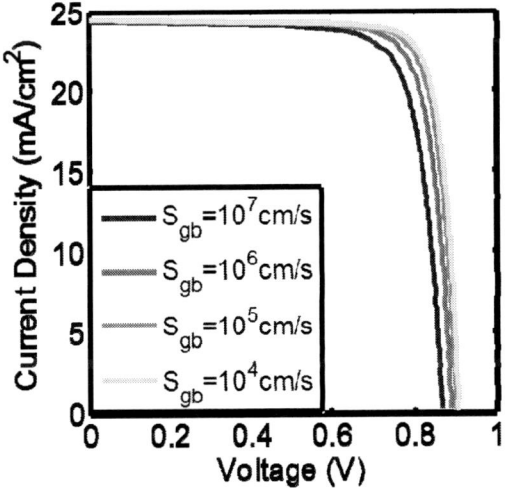

Figure 4: Current-voltage relation for a 2D Sentaurus model of polycrystalline CdTe with the geometry of Fig. 2.

The Br:MeOH etched CdS/CdTe sample was also studied in the 2D model. The etching would result in cadmium vacancies near the back contact, thus introduce shallow level acceptors to the core grain boundary level and form an ultrathin Te-rich layer between the back surface of the absorber and the back contact. As reported previously that Te layer has a 0.26eV valence band offset with CdTe layer, therefore the Schottky barrier Φ_B at the back can be mitigated. The I-V performance of standard CdTe and Br:MeOH etched CdTe were plotted and it was found that there is no substantial improvement for Br:MeOH etched CdTe. This is because standard CdTe sample is already with relatively low Schottky barrier of 0.3 V, non-ideal S-shape behavior was not observed in the output quadrant indicating that improving back contact quality by etching the absorber layer with Br:MeOH will not take effect. However, this treatment could be beneficial to CdTe samples with poor back contact quality.

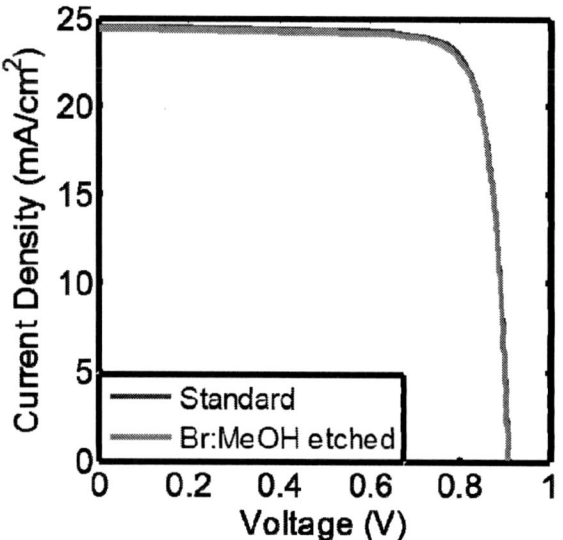

Figure 5: Current-voltage relation for a 2D Sentaurus model of polycrystalline CdTe with the geometry of Fig. 2, with etching from Br:MeOH. No appreciable improvement is predicted in this case.

These simulations predict about 40 mV loss. This contrasts significantly with the prior estimate of 5 mV loss from the effective medium model in 1D. As a result, we find that the effective medium model underestimates the Voc losses compared to a more realistic case, at least for the back 1 μm in a 4 μm device with 2×10^{14} cm^{-3} bulk doping and all the other relevant parameters.

V. CONCLUSIONS

In this paper, we considered the connection between the properties of polycrystalline materials commonly used in photovoltaics. This investigation began with a general hypothesis that polycrystalline grain material could be treated as an effective medium. To test this concept in detail, we then developed 1D and 2D simulation framework for modeling CdTe solar cells that are CSS grown and Br:MeOH+CdCl$_2$ treated. In the course of this study, we found that grain boundary interface recombination has a marked effect in degrading V$_{oc}$, which is hard to capture in a 1D model. In particular, we found that a 1D effective medium model only predicts approximately 5 mV out of a 40 mV loss in open circuit voltage associated with grain boundaries, in the case of a polycrystalline CdTe device etched in the back 1 μm of a 4 μm device with 2×10^{14} cm^{-3} bulk doping and all the other relevant parameters from Refs. 8-11. As a result, we also determined that if the initial I-V curve does not exhibit a low Schottky barrier, then Br:MeOH-etching the back will not necessarily improve device performance. Still, not all is lost. Alternatives to CdCl$_2$ treatment might reduce compensation, thus increasing doping and V$_{oc}$ [14].

VI. ACKNOWLEDGEMENTS

The authors thank Benoit H. Gaury, Paul Haney, and Muhammad Ashraful Alam for valuable discussions. Support was provided by the Department of Energy, under DOE Cooperative Agreement No. DE-EE0004946 (PVMI Bay Area PV Consortium), through the NCN-NEEDS program, which is funded by the National Science Foundation, contract 1227020-EEC, and through the CAREER award program, which is funded by the National Science Foundation under Award No. 1454315-EEC. B. G. acknowledges support under the Cooperative Research Agreement between the University of Maryland and the National Institute of Standards and Technology Center for Nanoscale Science and Technology, Award 70NANB14H209, through the University of Maryland.

VII. REFERENCES

1. Hovel, H. J. "The effect of depletion region recombination currents on the efficiencies of Si and GaAs solar cells." In *Photovoltaic Specialists Conference*, 10th, Palo Alto, Calif, pp. 34-39. 1974; Guo, Da, and Dragica Vasileska. "1D fast transient simulator for modeling CdS/CdTe solar cells." In *Photovoltaic Specialists Conference (PVSC)*, 2013 IEEE 39th, pp. 1961-1965.
2. Green, M.A. "Commercial progress and challenges for photovoltaics." *Nature Energy* 1 (2016): 15015.
3. Tiwari, A.N., G. Khrypunov, F. Kurdzesau, D. L. Bätzner, A. Romeo, and H. Zogg. "CdTe solar cell in a novel configuration." *Progress in Photovoltaics: Research and Applications* 12, no. 1 (2004): 33-38; Jiang, C-S., B. To, S. Glynn, H. Mahabaduge, T. Barnes, and M.M. Al-Jassim. "Recent progress in

nanoelectrical characterizations of CdTe and Cu (In, Ga) Se 2." In *Photovoltaic Specialists Conference (PVSC), 2016 IEEE 43rd*, pp. 3675-3680.

4. Johnston, S., K. Zaunbrecher, R. Ahrenkiel, D. Kuciauskas, D. Albin, and W. Metzger. "Simultaneous measurement of minority-carrier lifetime in single-crystal CdTe using three transient decay techniques." *IEEE Journal of Photovoltaics* 4, no. 5 (2014): 1295-1300.

5. Moore, J.E., X. Wang, E.K. Grubbs, J. Drayton, S. Johnston, D. Levi, M.S. Lundstrom, and P. Bermel. "Photoluminescence excitation spectroscopy characterization of cadmium telluride solar cells." In *Photovoltaic Specialists Conference (PVSC), 2016 IEEE 43rd*, (2016): 2223-2227.

6. Rios-Flores, A., J. L. Pena, V. Castro-Pena, O. Ares, R. Castro-Rodriguez, and A. Bosio. "A study of vapor CdCl2 treatment by CSS in CdS/CdTe solar cells." *Solar Energy* 84, no. 6 (2010): 1020-1026.

7. Gloeckler, M., A. L. Fahrenbruch, and J. R. Sites. "Numerical modeling of CIGS and CdTe solar cells: setting the baseline." In *Proceedings of 3rd World Conference on Photovoltaic Energy Conversion*, vol. 1 (2003): 491-494.

8. McMahon, T. J., and A. L. Fahrenbruch. "Insights into the nonideal behavior of CdS/CdTe solar cells." In Photovoltaic Specialists Conference, 2000. Conference Record of the Twenty-Eighth IEEE (2000): 539-542.

9. Bätzner, D. L., R. Wendt, A. Romeo, H. Zogg, and A. N. Tiwari. "A study of the back contacts on CdTe/CdS solar cells." Thin Solid Films 361 (2000): 463-467.

10. Balcioglu, A., R. K. Ahrenkiel, and F. Hasoon. "Deep-level impurities in CdTe/CdS thin-film solar cells." Journal of Applied Physics 88, no. 12 (2000): 7175-7178.

11. Castaldini, A., A. Cavallini, B. Fraboni, P. Fernandez, and J. Piqueras. "Midgap traps related to compensation processes in CdTe alloys." Physical Review B 56, no. 23 (1997): 14897.

12. Burgelman, M., P. Nollet, S. Degrave, "Modelling polycrystalline semiconductor solar cells", *Thin Solid Films* 361 (2000): 527-532; M. Burgelman, K. Decock, S. Khelifi and A. Abass, "Advanced electrical simulation of thin film solar cells", Thin Solid Films, **535** (2013) 296-301.

13. Wang, X., M.R. Khan, M. Lundstrom, and P. Bermel. "Performance-limiting factors for GaAs-based single nanowire photovoltaics." *Optics express* 22, no. 102 (2014): A344-A358.

14. Jensen, S.A., J.M. Burst, J.N. Duenow, H.L. Guthrey, J. Moseley, H.R. Moutinho, S.W. Johnston, A. Kanevce, M.M. Al-Jassim, and W.K. Metzger. "Long carrier lifetimes in large-grain polycrystalline CdTe without CdCl2." *Applied Physics Letters* 108, no. 26 (2016): 263903.

Characterizing recombination in CdTe-based solar cells by the temperature and excitation dependence of open-circuit voltage and photoluminescence

Craig H. Swartz[1,2], Sanjoy Paul[1], Corey R. Grice[3], Yanfa Yan[3], Lorelle Mansfield[4], Sachit Grover[5], Gang Xiong[5], and Jian V. Li[1,2]

[1]Department of Physics, Texas State University, San Marcos, TX, USA, [2]Materials Science, Engineering, and Commercialization Program, Texas State University, San Marcos, TX, USA, [3]Department of Physics, University of Toledo, Toledo, OH, USA, [4]National Renewable Energy Laboratory, Golden, CO, USA, [5]First Solar, Inc., Santa Clara, CA, USA

Abstract — **Further improvement of the open-circuit voltage (V_{oc}) of thin-film photovoltaic devices requires that the rate of non-radiative recombination be monitored and reduced. In this study, we investigate a method that attempts to account for the excitation and temperature dependence of V_{oc} and photoluminescence (PL), to extract detailed information about the spatial distribution of recombination and the effects of back contact offset.**

I. INTRODUCTION

To improve the efficiency of contemporary thin film photovoltaic devices, a central requirement will be an increase of the open-circuit voltage (V_{oc}). This entails control and monitoring of the rate of non-radiative recombination and the back contact offset [1], [2]. Band gap engineering, through the use of an alloy gradient, has been successfully employed in optimizing chalcogenide-based solar cells [3]. Gradients do, however, greatly complicate the metrology of recombination rates. Such rates are strongly spatially non-uniform, and recombination in the space charge region is not negligible in thin film photovoltaics.

Since V_{oc} and short circuit current (J_{sc}) are not sufficient guides to the level of recombination in the various interfaces and regions of a photovoltaic cell, a temperature-dependent and excitation-dependent analysis has been proposed [4]. Excitation-dependent photoluminescence (PL) has also proven to be an informative tool when separating the mechanisms of recombination [5], [6].

In this study, we investigate a method which attempts to fully account for the excitation and temperature dependence of V_{oc} and PL for extraction of more detailed information about the spatial distribution of recombination.

II. ANALYSIS

The recombination rates may be estimated with analytic expressions in the space charge region (SCR), in quasi-neutral region (QNR), and at the interface. We temporarily assume the quasi-equilibrium limit, i.e., that the hole and electron quasi-Fermi levels are constant, and assume that most recombination

in the SCR occurs at the point where the electron concentration is equal to the hole concentration.

It has been shown that V_{oc} can be expressed under these conditions as [7]

$$V_{oc} = 2V_{th} \ln\left(\frac{R_d/2}{R_i + R_b}\left(\sqrt{4IW\frac{R_i + R_b}{R_d^2} + 1} - 1 \right) \right) \quad (1)$$

where R_i and R_b refer to the interfacial and QNR recombination rates multiplied by $\exp(-V/V_{th})$, and R_d is the product of the depletion recombination rate and $\exp(-V/2V_{th})$. V_{th} is the thermal voltage, V is an applied voltage, I is the excitation intensity, and W is the depletion width.

The Rs have little dependence on V, but they have their own temperature dependence, in which we express the energy gap as $E_G \equiv qV_G$. The rate R_b is proportional to the absorber layer's intrinsic carrier concentration, or $\exp(V_G/(2V_{th}))$, and R_d is proportional to $\exp(V_G/V_{th})$. R_i is proportional to $\exp(-V_B/V_{th})$, where V_B is the interfacial potential barrier for holes. In the case of reasonably high injection I, the strong dependence of each R on temperature (by $V_{th} = k_B T/q$) allows an extrapolation of V_{oc} to a zero temperature limit, where

$$V_{oc} \rightarrow \frac{R_i V_B + R_b V_G}{R_i + R_b}. \quad (2)$$

If V_B and E_G are known, the ratio of the recombination rates R_i/R_b can be then found. The sum $R_i + R_b$ can be found from the I-dependence of V_{oc}, as shown in (1). The final rate R_d can be found from the fact that the summed rate of recombination is equal to the generation rate corresponding to intensity I in open circuit condition.

978-1-5090-5606-4/17 $31.00 © 2017 IEEE

Expanding this calculation to a wider array of situations may require abandoning a number of simplifications. If recombination in the SCR is sufficiently high, or if recombination in the QNR is sufficiently non-uniform, then one can no longer make the quasi-equilibrium assumption that the hole and electron quasi-Fermi levels are constant. In this case, or in the case of graded alloy films, or in the case of a back contact offset, more complete numerical simulations are needed that incorporate more fully the physical parameters of the situation including diverse recombination mechanisms, surface and interface phenomena, and contact offsets.

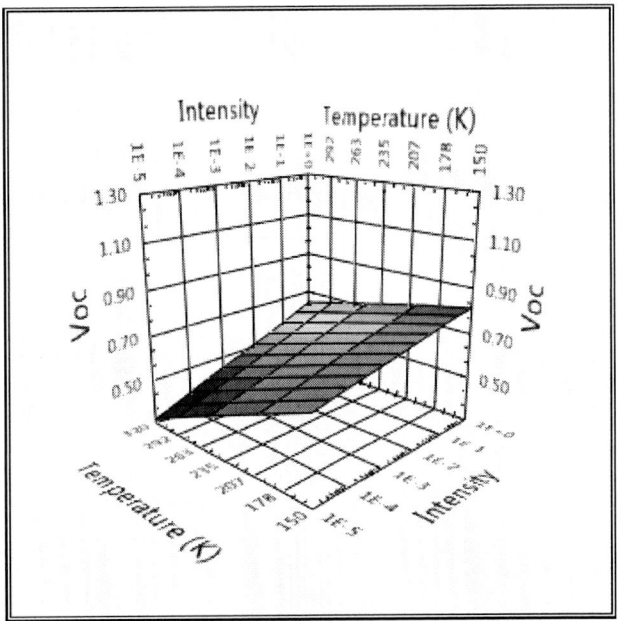

Fig. 2. Simulated intensity and temperature dependence of V_{oc} with a back contact offset of 0.3 eV.

An argon ion laser of 514 nm wavelength, modulated at 400 Hz, provided the excitation for PL versus I measurements (PL-I). A 6.5x objective lens focused the laser light to give a 120 μm FWHM Gaussian spot on the sample. The absolute excitation power was measured with a calibrated power meter. The luminescence was collected through the same objective, optically filtered to reject reflected laser light, and focused onto a photomultiplier tube. The PL intensity was recorded using a lock-in amplifier. The laser intensity was varied over six orders of magnitude by a series of calibrated neutral density filters.

Simulations were carried out using SCAPS software [9], in conjunction with external calculations of photon recycling [10]. The physical properties of the materials & traps were taken from the default simulation baseline given by Gloeckler [11].

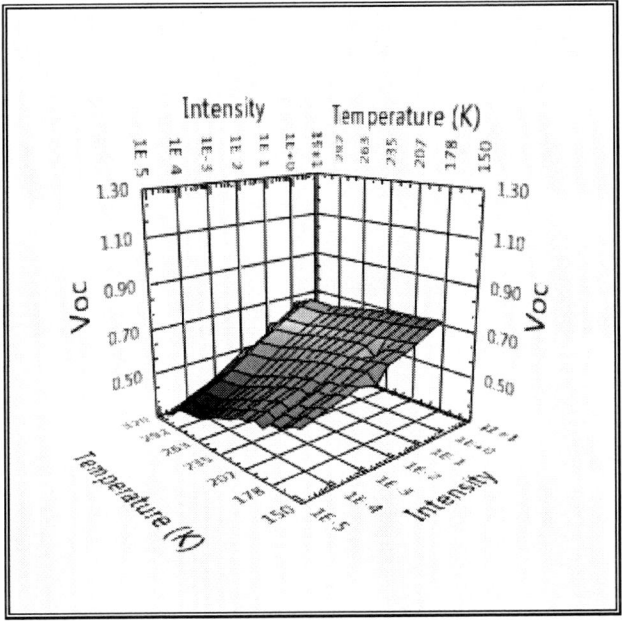

Fig. 1. Open circuit voltage of a CdTe-based solar cell from the University of Toledo with an unintentionally doped absorber layer as a function of both temperature and illumination intensity. The intensity is expressed in units of kW/m² to approximately correspond to 1 solar illumination.

III. EXPERIMENTAL DETAILS

Unintentionally doped CdTe-based solar cell devices were grown at the University of Toledo, and CdSe$_x$Te$_{1-x}$-based devices were grown at First Solar, by evaporation in the superstrate configuration on tin oxide-coated glass [8]. The absorbers in the CdSeTe devices exhibit a graded bandgap. The V_{oc} was measured while illuminated with a broad spectrum laser diode-driven light source (LDLS), and the intensity was then varied over several orders of magnitude by a series of neutral density filters. The temperature could also be varied from 20 to 330 K.

IV. RESULTS

A. Open circuit voltage

The measured relationship between I, T, and V_{oc} is shown in Figs. 1 and 2. The V_{oc} is related logarithmically to I, as expected. A low-temperature ceiling is visible in Fig. 1. Since carrier freeze-out can occur for p-type dopants in CdTe, the acceptors were assumed to have an ionization energy of 70 meV [12]. However, the amount of V_{oc} drop introduced by this freeze out was found insufficient to explain the ceiling effect.

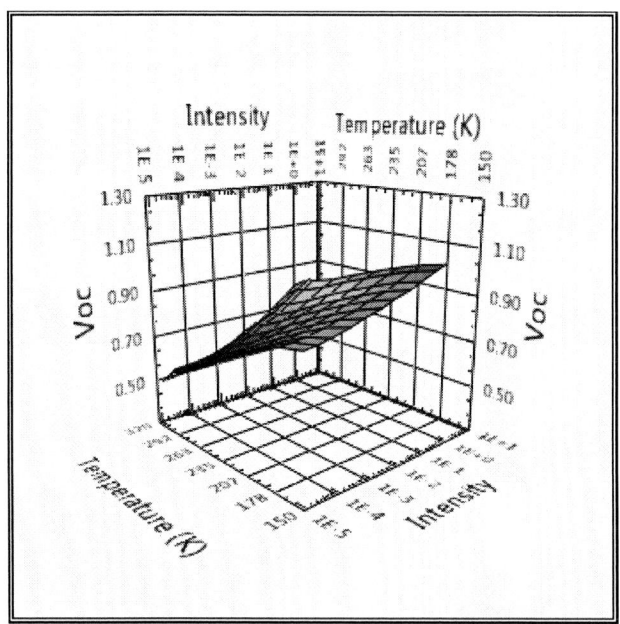

Fig. 3. Open circuit voltage of a CdTe-based solar cell with a doped absorber layer from First Solar, Inc. as a function of both temperature and illumination intensity.

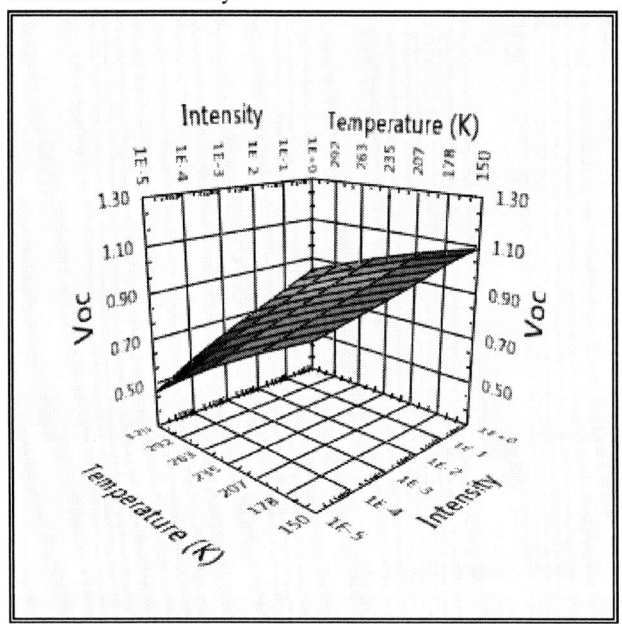

Fig. 4. Simulated intensity and temperature dependence of V_{oc}, without a back contact offset.

IV.

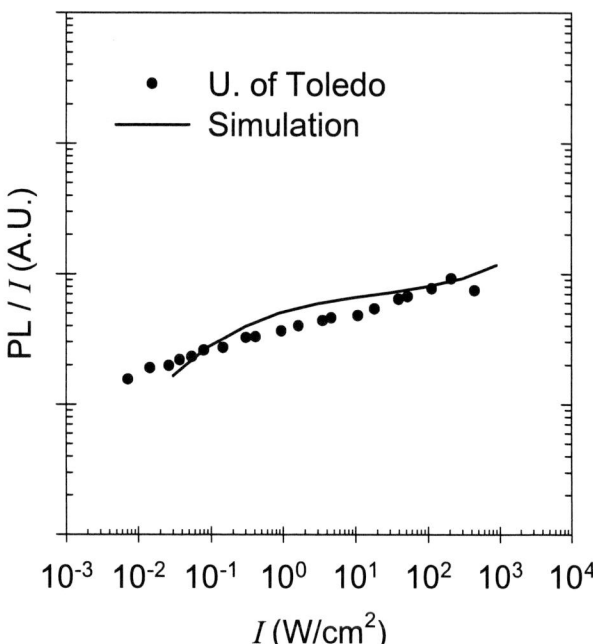

Fig. 5. PL-I results shown in terms of efficiency, or output intensity divided by the input excitation intensity.

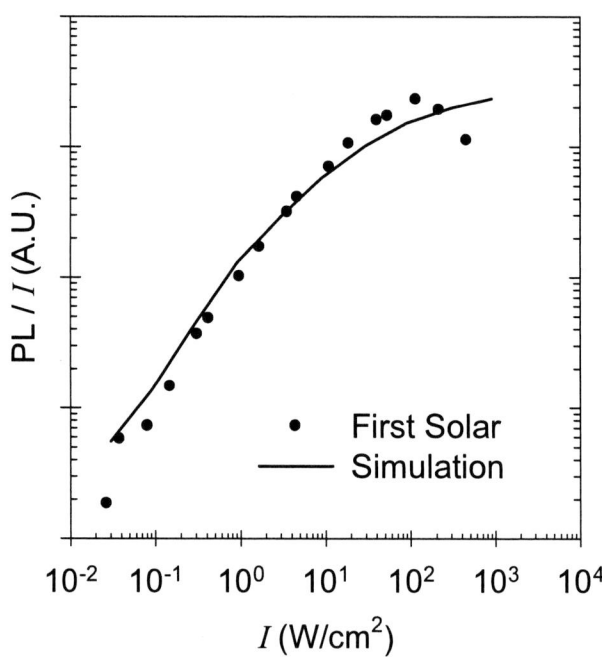

Fig. 6. PL-I results shown in terms of efficiency, or output intensity divided by the input excitation intensity.

Instead, a full numerical simulation shows that low temperature V_{oc} can be limited by a back contact offset. The results of the simulation can be seen in Figs. 3 and 4. A mid-gap trap was assumed with a density of $N_T = 10^{14}$ cm^{-3} and electron and hole capture cross sections of $\sigma_n = 5\times10^{-15}$, $\sigma_p = 5\times10^{-16}$ cm^{-2}, and a back contact recombination velocity of 10^4 cm/s [2]. In Fig. 3, a back contact offset was also included at 0.3 eV above the valence band. Note that the idealized, closed-form model in (1) deals only with non-radiative recombination rates, without including contact offset effects.

B. PL Intensity

The difference between the effect of non-radiative recombination centers versus the effect of back contact offset may be resolvable by examining the PL-I, as seen in Figs. 5 and 6. The PL-I is expressed in terms of efficiency, or output intensity divided by the input excitation intensity. Notably, the same parameters used for the V_{oc} calculation were also used to simulate the PL-I curves shown. The result of the PL-I measurement was more sensitive to bulk recombination centers and interfacial recombination than the V_{oc} measurement, and it was the comparison of PL-I data with simulation that led to the tentative conclusion that there is a higher front interfacial recombination in the unintentionally doped sample. The front interfacial recombination velocity was 10^4 cm/s, and this was also included in the V_{oc} calculation of Fig. 3, though its impact to V_{oc} turned out to be slight.

C. Thermal effects

The PL-I typically levels off at high intensities, approaching a limit of 100% radiative recombination efficiency (ε). A strong downturn is visible at the highest intensities in Figs. 5 and 6, and this appears on all evaporated polycrystalline CdTe-based solar cells so far tested. Furthermore, continuous measurement at the highest intensity causes a reduction in PL signal that nears 50% after 5 minutes, and this reduction appears to be permanent.

Such a feature is possible when the absorbed radiation is intense enough to significantly increase the temperature of the material at the laser spot and cause thermal damage. The amount of temperature shift ΔT caused by heating in a circular spot of radius r on the surface of a semi-infinite solid is [6]

$$\Delta T = I(1 - \varepsilon E_G/E_s)r/k \qquad (3)$$

where k is the thermal conductivity and E_G/E_s is the ratio of the band gap energy to the energy of an excitation photon, 1.5 eV / 2.4 eV in this case.

Previously reported experiments on single crystal CdTe structures did not exhibit such a change [5]. Our preliminary measurements investigating single crystal CdTe films using Raman scattering (514.5 nm) show that damage is induced by high intensities of 10^5 W/cm^2. The thermal conductivity of polycrystalline CdTe has been found to be close to single

crystal CdTe, about $k = 0.04$-0.07 W / (K cm) [13]. The glass substrate, however, may be five times lower. In the current context of multilayered structures, the situation is complicated by several factors. Numerous material interfaces are present in solar cells, prospectively producing thermal boundary resistances (TBR).

As ε approaches 100%, near the maximum intensity of 400 W/cm^2, (3) implies that ΔT should approach 60 K. This is reduced to $\Delta T = 30$ K because of the action of the chopper in halving the average absorbed power. A 30 K shift near room temperature does not typically cause a factor of two reduction in the PL signal, and this is also the case in our PL-I simulations. This is particularly true in the high intensity regime where ε approaches 100% and already has a weak dependence on changes in the carrier concentration or on thermal emission from gap levels. Therefore the drastic, and permanent, change seen in polycrystalline solar cell material is surprising. Either the interfacial thermal conductivity from the polycrystalline CdTe through the window layer into the glass substrate is smaller than expected, or visible light triggers diffusion to a degree that is out of proportion to its temperature rise.

This demonstrates that a more complete simulation, and a more complete dataset of V_{oc} and PL-I, particularly PL-I vs. temperature, would be very helpful in determining the mechanisms behind a junction's performance.

V. ACKNOWLEDGEMENTS

We acknowledge support by U.S. Department of Energy through Sunshot PVRD grant DE-EE-0007541 "Crosscutting recombination metrology for expediting Voc engineering".

REFERENCES

[1] R. M. Geisthardt, M. Topič, and J. R. Sites, "Status and Potential of CdTe Solar-Cell Efficiency," *IEEE Journal of Photovoltaics,* vol. 5, pp. 1217-1221, 2015.

[2] T. Song, A. Kanevce, and J. R. Sites, "Design of Epitaxial CdTe Solar Cells on InSb Substrates," *IEEE Journal of Photovoltaics,* vol. 5, pp. 1762-1768, 2015.

[3] I. Repins, L. Mansfield, A. Kanevce, S. A. Jensen, D. Kuciauskas, S. Glynn, *et al.,* "Wild band edges: The role of bandgap grading and band-edge fluctuations in high-efficiency chalcogenide devices," in *2016 IEEE 43rd Photovoltaic Specialists Conference (PVSC),* 2016, pp. 0309-0314.

[4] S. Grover, J. V. Li, D. L. Young, P. Stradins, and H. M. Branz, "Reformulation of solar cell physics to facilitate experimental separation of recombination pathways," *Applied Physics Letters,* vol. 103, p. 093502, 2013.

[5] C. H. Swartz, K. N. Zaunbrecher, S. Sohal, E. G. LeBlanc, M. Edirisooriya, O. S. Ogedengbe, *et al.*, "Factors influencing photoluminescence and photocarrier lifetime in CdSeTe/CdMgTe double heterostructures," *Journal of Applied Physics,* vol. 120, p. 165305, 2016.

[6] M. Passlack, R. N. Legge, D. Convey, Z. Y. Yu, and J. K. Abrokwah, "Optical measurement system for characterizing compound semiconductor interface and surface states," *IEEE Transactions on Instrumentation and Measurement,* vol. 47, pp. 1362-1366, 1998.

[7] J. V. Li, S. Grover, M. A. Contreras, K. Ramanathan, D. Kuciauskas, and R. Noufi, "A recombination analysis of Cu(In,Ga)Se2 solar cells with low and high Ga compositions," *Solar Energy Materials and Solar Cells,* vol. 124, pp. 143-149, 2014.

[8] N. R. Paudel, J. D. Poplawsky, K. L. Moore, and Y. Yan, "Current Enhancement of CdTe-Based Solar Cells," *IEEE Journal of Photovoltaics,* vol. 5, pp. 1492-1496, 2015.

[9] M. Burgelman, P. Nollet, and S. Degrave, "Modelling polycrystalline semiconductor solar cells," *Thin Solid Films,* vol. 361–362, pp. 527-532, 2000.

[10] M. A. Steiner, J. F. Geisz, I. García, D. J. Friedman, A. Duda, and S. R. Kurtz, "Optical enhancement of the open-circuit voltage in high quality GaAs solar cells," *Journal of Applied Physics,* vol. 113, p. 123109, 2013.

[11] M. Gloeckler, A. L. Fahrenbruch, and J. R. Sites, "Numerical modeling of CIGS and CdTe solar cells: setting the baseline," presented at the Proceedings of 3rd World Conference on Photovoltaic Energy Conversion, Osaka, Japan, 2003.

[12] V. Consonni, "Optical Properties of CdTe," in *CdTe and Related Compounds; Physics, Defects, Hetero- and Nano-structures, Crystal Growth, Surfaces and Applications.* vol. 1, R. Triboulet and P. Siffert, Eds., 1st ed., Amsterdam: Elsevier, 2010, p. 72.

[13] J. J. Alvarado-Gil, O. Zelaya-Angel, H. Vargas, and J. L. Lucio M, "Photoacoustic characterization of the thermal properties of a semiconductor-glass two-layer system," *Physical Review B,* vol. 50, pp. 14627-14630, 1994.

Experimental Evidence For CdS-related Transport Barrier in Thin Film Solar Cells and Its Impact on Admittance Spectroscopy

Florian Werner, Anastasiya Zelenina, and Susanne Siebentritt

Laboratory for Photovoltaics, Physics and Materials Science Research Unit, University of Luxembourg, 41, Rue du Brill, L-4422 Belvaux, Luxembourg

Abstract — We measure the temperature-dependent current–voltage characteristics of a Mo/CdS/i-ZnO/Al:ZnO thin film stack, which is conventionally used as a buffer/window stack at the front of chalcopyrite thin film solar cells. We observe a non-ohmic thermally-activated behavior, which suggests the presence of a transport barrier. We do not observe this barrier for stacks fabricated without the CdS layer, suggesting that conduction band offsets to the CdS or transport within the CdS are responsible for the reduced conductance. With regards to thermal admittance spectroscopy we show that the presence of such an additional element representing the buffer/window stack results in an additional step in the capacitance spectrum, which could easily be mistaken as a defect response.

I. INTRODUCTION

Thermal admittance spectroscopy is a widely used technique to characterize deep defects in Cu(In,Ga)(S,Se)$_2$ (CIGS) thin film solar cells. Such deep defects with transition levels within the bandgap of the absorber material can act as efficient recombination centers and thus limit the open-circuit voltage, and consequently the efficiency, of the solar cell. In thermal admittance spectroscopy, the recombination activity and energetic position of a defect level are derived from the temperature-dependence of steps observed in an experimental capacitance spectrum $C(\omega)$ [1]. While this relation between defect parameters and admittance spectrum is well known, it is often impossible to determine whether a given admittance signal is actually related to defects.

Recently, there has been an increasing consensus within the scientific community that transport barriers in the solar cell might play a crucial role in correctly interpreting the experimental admittance spectra [2]-[4]. In this contribution we study the current transport through the buffer/window stack in order to estimate its expected impact on the thermal admittance spectrum of a complete thin film solar cell.

II. CONDUCTANCE OF BUFFER/WINDOW LAYER STACK

In this study we measure the temperature-dependent current voltage characteristics (*IVT*) of two different buffer/window stacks deposited directly onto Mo-coated soda-lime glass (SLG):

A) SLG/Mo/**CdS**/i-ZnO/Al:ZnO
B) SLG/Mo/i-ZnO/Al:ZnO

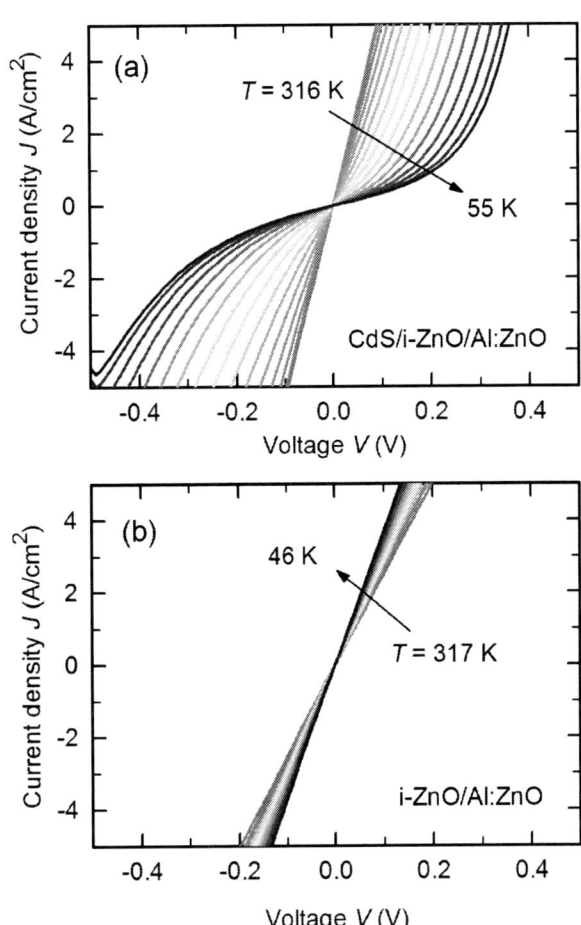

Fig. 1. Temperature-dependent *I–V* characteristics for the stack (a) with and (b) without CdS buffer layer for a range of set temperatures between 320 – 20 K (red to blue). Actual measured temperatures are indicated in the graphs. Some curves were omitted for clarity.

We use the same process parameters for the depositions as for standard CIGS solar cell fabrication, employing chemical bath deposition for CdS and sputtering for intrinsic and doped ZnO. Although the processing is nominally identical to standard solar cells, depositing directly on Mo rather than CIGS most likely will influence the transport properties, either due to different band offsets, differences in the Fermi level within the more lightly doped buffer layers, or simply due to a modified growth on the Mo surface. While the significance of any quantitative parameter is questionable, we assume that the qualitative transport behavior of a layer stack is at least indicative of its performance in the complete solar cell.

Figure 1 shows the *I–V* characteristics for stack A (with CdS) and stack B (without CdS buffer layer) in a temperature range between 320 – 20 K (setpoints, actual temperature ranges vary and are indicated in the graphs). The *I-V* characteristics for stack A are clearly non-ohmic, which becomes increasingly evident at low temperatures. In contrast, stack B without the CdS layer shows ohmic transport behavior down to low temperatures, and the conductance only weakly depends on temperature. This is a clear indication that the transport barrier in these buffer-only devices is caused by the CdS interlayer.

We obtain the conductance of each stack from the derivative $G_{bw} = dJ/dV$ around zero bias. For stack B without CdS, this value is between 20 – 40 S/cm^2 and is almost constant with respect to temperature. The conductance G_{bw} around zero bias for stack A including the CdS layer is shown as a function of inverse temperature $1000/T$ in Fig. 2. Starting from high temperatures, stack A first exhibits a thermally activated conductance with an activation energy around 70 meV, which then levels off to a virtually constant conductance at lower temperatures. For several samples (not shown here) we obtain activation energies in a range of 60 – 100 meV.

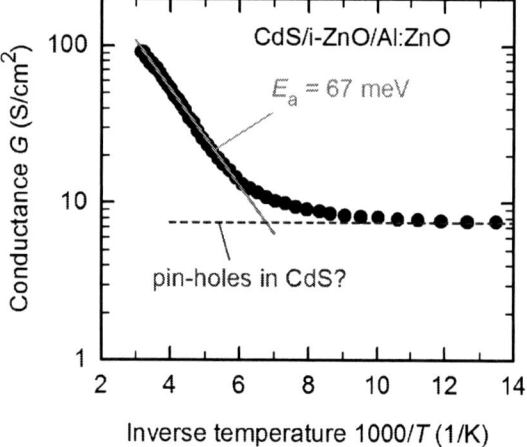

Fig. 2. Conductance *G* as a function of inverse temperature $1000/T$ for stack A (with CdS). The red solid line shows an exponential fit with activation energy $E_a = 67$ meV, the blue dashed line indicates a temperature-independent excess conductivity presumably due to pin-holes in the CdS layer.

The saturation of the conductance, rather than a continued thermally activated decrease, might be related to pin-holes in the CdS layer. Since the transport barrier only appears when the CdS layer is introduced into the layer stack, any surface area not covered by CdS is expected to be highly conductive

even at low temperatures, as corroborated by Fig. 1(b). Pin-holes in the CdS layer would thus allow current flow to bypass the more resistive CdS at low temperatures. This is represented in Fig. 2 by a constant conductivity corresponding to the blue dashed line. For a closed CdS layer, the temperature range where a thermally activated behavior of the buffer/window conductance is observed might thus in fact be larger than evident from Fig. 2. Taking into consideration the different growth surface (Mo instead of CIGS) during the chemical bath deposition, the CdS layer in our samples might indeed be less homogeneous than in a typical solar cell.

III. ADMITTANCE SPECTRUM OF A SERIAL ELECTRICAL EQUIVALENT CIRCUIT

As shown above, a thermally activated conductance through the buffer/window stack due to a transport barrier must be expected in a thin film solar cell, and its impact in thermal admittance spectroscopy must be considered. Determining the actual quantitative parameters of a transport barrier, even whether it is located at the front or back of the solar cell, is not an easy task. Therefore, we treat the barrier more generally as a "black box", which is defined by its conductance *G* and capacitance *C*. The most fundamental electrical equivalent circuit model for a solar cell with a transport barrier then consists of the series connection of two *GC* elements (more commonly referred to as *RC* elements, with $R = 1/G$), as sketched in Fig. 3. One of these elements represents the transport barrier (G_{bw} and C_{bw}), the other one originates from the space charge region in the *p/n* junction (G_j and C_j). If we assume an ideal solar cell free of deep defects, the junction capacitance is equal to the depletion capacitance of the space charge region, and the junction conductance is given by the voltage-derivative of the exponential diode current plus any shunt conductance bypassing the diode.

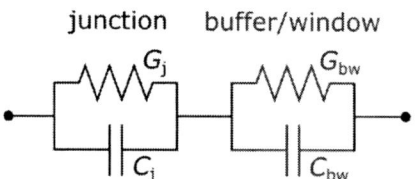

Fig. 3. Model electrical equivalent circuit showing the *p/n* junction (black) and the buffer/window – or any other – transport barrier (blue).

The admittance of the complete device containing the buffer/window stack and the junction is then fully determined by four frequency-independent parameters: conductance and capacitance of both *GC* elements, respectively. Note that the parameters of the space charge region forming the junction, although frequency-independent, are still functions of applied bias voltage.

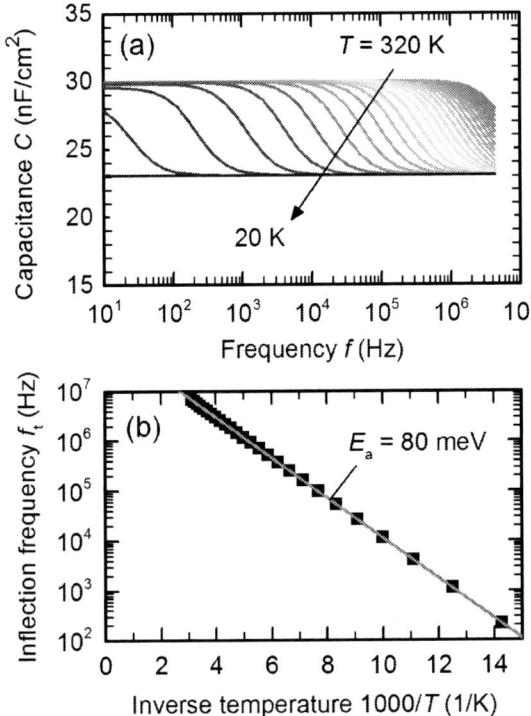

Fig. 4. (a) Simulated capacitance spectra for $T = 320 - 20$ K and representative parameters given in Table I. (b) Inflection frequencies f_t of the simulated capacitance spectra for each simulated temperature as a function of inverse temperature $1000/T$.

A. Dependence on measurement frequency

The admittance of a single GC element is $Y = G + i\omega C$. To calculate the admittance of the series connection of two GC elements, their respective admittances are inverted to impedances, added, and then inverted again to yield the total admittance of the full device. The total conductance and capacitance are then obtained from the real and imaginary part of the admittance, respectively. Performing this calculation for two frequency-independent GC elements, in our case for the buffer/window stack ($Y_{bw} = G_{bw} + i\omega C_{bw}$) and the solar cell junction ($Y_j = G_j + i\omega C_j$), yields a total capacitance of

$$C(\omega) = C_\infty + \frac{\Delta C}{1 + (\omega\tau)^2}, \qquad (1)$$

where $1/C_\infty = 1/C_j + 1/C_{bw}$ is the high-frequency limit, $\Delta C = \alpha(C_j + C_{bw})$ is the capacitance step height, $0 < \alpha < 1$ is a scaling factor depending on the relative magnitudes of G_{bw}, G_j, C_{bw}, and C_j, and the inflection frequency f_t is given by

$$f_t = \frac{1}{2\pi\tau} = \frac{1}{2\pi}\frac{G_{bw} + G_j}{C_{bw} + C_j}. \qquad (2)$$

Note that the capacitance spectrum described by (1) thus shows a step at the inflection frequency f_t, although all individual elements of the equivalent circuit were assumed to be independent of frequency.

The functional form of (1) is identical to the capacitance response expected for a deep defect in the space charge region, where C_∞ would be identified with the depletion capacitance of the space charge region and ΔC would be proportional to the defect concentration [1]. Accordingly, at a given temperature, the presence of a transport barrier in series with the space charge region results in an admittance spectrum which can easily be mistaken as a signature of deep defects.

B. Dependence on temperature

It is often found in experiment that the low- and high-frequency plateaus around a capacitance step do not vary much with changes in temperature, and that any significant temperature-dependence of the measured capacitance at a given frequency is only due to the presence of a capacitance step. Applied to our model in (1) this suggests that the individual capacitance values C_j and C_{bw} of junction and buffer/window stack are fairly constant with temperature. Note that this is only strictly valid for temperatures high enough to exclude carrier or mobility freeze-out, in which case the depletion capacitance of the space charge region would drop to the geometrical capacitance of the absorber layer.

If the capacitances are nearly constant, any temperature-dependence of the inflection frequency due to a transport barrier, given by (2), must be attributed to the temperature-dependence of the conductances G_j and G_{bw}. Without applying a significant forward bias voltage, the conductance of the p/n junction diode will be negligible compared to the conductance through the buffer/window stack for any reasonably good solar cell in a wide temperature range. For a thermally activated buffer/window conductance $G_{bw} = G_{bw.0} \exp(-E_a/kT)$ the inflection frequency of the capacitance spectrum is then approximated by

$$f_t(T) \approx \frac{1}{2\pi}\frac{G_{bw.0}}{C_{bw} + C_j}\exp\left(-E_a\Big/kT\right), \qquad (3)$$

and thus predominantly determined by the transport properties of the buffer/window stack.

C. Impact on thin film solar cell characterization

An exemplary capacitance spectrum is shown in Fig. 4(a) for a temperature range of $320 - 20$ K, calculated from (1) for the parameters given in Table I. Note that the parameters used for the buffer/window element are representative of values measured in Sec. II, while the junction capacitance of 30 nF/cm^2 corresponds to an absorber dopant concentration of the order of 10^{16} cm^{-3}. The conductance parameters of the junction do not significantly influence the qualitative

978-1-5090-5606-4/17 $31.00 © 2017 IEEE

capacitance spectrum as long as G_j is much smaller than G_{bw} at all temperatures. Figure 4(b) shows the inflection frequencies extracted from the simulated capacitance spectra in Fig. 4(a) at each temperature. As predicted by (3), these inflection frequencies indeed follow the same thermally activated behavior we have assumed for the buffer/window conductance.

TABLE I

DEVICE PARAMETERS FOR THE SIMULATED CAPACITANCE SPECTRA IN FIG. 4, ASSUMING THERMALLY ACTIVATED CONDUCTANCES $G = G_0 \exp(-E_a/kT)$.

Parameter:	Buffer/ window	Junction
Conductance prefactor G_0 (S/cm^2)	100	0.001
Conductance activation energy E_a (meV)	80	40
Capacitance C (nF/cm^2)	100	30

As shown in Sec. II, the conductance through the buffer/window stack, which determines the temperature-dependence of the inflection frequency, appears to be thermally activated with an activation energy of 60 – 100 meV, if the effect of pin-holes is ignored. In a complete solar cell we would thus expect to observe a temperature-dependent admittance spectrum, which according to (1) – (3) will be indistinguishable from a defect signature with a similar activation energy. Taking into account the uncertainty in obtaining the correct quantitative barrier height, for example due to CdS deposition on Mo rather than CIGS, this range of activation energies is indeed consistent with defect signatures commonly observed in CIGS solar cells [2].

So far we have assumed that the buffer/window conductance is thermally activated, although we could experimentally verify such behavior only in a fairly limited temperature range. Note, however, that the theory presented in this paper does not actually require a thermally activated behavior. As long as the buffer/window stack is much more conductive than the junction in a given temperature range, the temperature-dependence of the inflection frequencies will mimic the buffer/window conductance, no matter its actual functional form. In fact, different transport mechanisms through the CdS layer, which might deviate from an ideal thermally activated behavior, provide a compelling explanation for any deviations from a truly linear Arrhenius graph in the analysis of experimental admittance spectra of CIGS solar cells.

Identifying the origin of the capacitance step in admittance spectroscopy furthermore has severe implications for studying the absorber dopant concentration using capacitance-voltage profiling. If a capacitance step is due to the presence of a series GC element (the buffer/window stack in our model), the

simulated capacitance spectra shown in Fig. 4(a) demonstrate that the *low-frequency* capacitance approaches the space charge region capacitance, assumed to be $C_j = 30$ nF/cm^2 in our simulation. In contrast, the space charge region capacitance is given by the *high-frequency* capacitance limit if the capacitance step is caused by a deep defect [1].

IV. CONCLUSIONS

We have shown that the CdS buffer layer typically present in thin film solar cells introduces a transport barrier, which results in a thermally activated conductance through the buffer/window layer stack. This effect might also be present for alternative buffer layers with similar electronic properties to CdS. This transport barrier results in a capacitance step with a thermally activated inflection frequency, which could be misinterpreted as a defect signature in an admittance spectroscopy measurement of CIGS solar cells. The barrier height estimated for the transport barrier indeed agrees reasonably well with admittance signatures commonly observed for CIGS solar cells.

Our results emphasize that meticulous care must be taken in the interpretation of capacitance-based measurements. The buffer/window stack is an integral part of a thin film solar cell and its contribution to the device admittance cannot be ignored.

ACKNOWLEDGEMENTS

Parts of this study were funded by the Fonds National de la Recherche Luxembourg (FNR) in the project "Surface passivation for thin film photovoltaics" (SURPASS).

REFERENCES

[1] P. Blood and J. W. Orton, *The Electrical Characterization of Semiconductors*. London: Academic Press, 1992.

[2] T. Eisenbarth, T. Unold, R. Caballero, C. A. Kaufmann, and H.-W. Schock, "Interpretation of admittance, capacitance-voltage, and current-voltage signatures in Cu(In,Ga)Se$_2$ thin film solar cells," *Journal of Applied Physics*, vol. 107, 034509, 2010.

[3] J. Lauwaert, L. V. Puyvelde, J. Lauwaert, J. W. Thybaut, S. Khelifi, M. Burgelman, F. Pianezzi, A. N. Tiwari, and H. Vrielinck, "Assignment of capacitance spectroscopy signals of CIGS solar cells to effects of non-ohmic contacts," *Solar Energy Materials and Solar Cells*, vol. 112, pp. 78-83, 2013.

[4] G. Sozzi, S. Di Napoli, R. Menozzi, F. Werner, S. Siebentritt, P. Jackson, and W. Witte, "Influence of conduction band offsets at window/buffer and buffer/absorber interfaces on the roll-over of J-V curves of CIGS solar cells," accepted for the 44th IEEE Photovoltaics Specialists Conference (Washington, DC, 2017).

Transparent Conductive Adhesives for Tandem Solar Cells

Talysa R. Klein, Benjamin G. Lee, Manuel Schnabel, Emily L. Warren, Pauls Stradins, Adele C. Tamboli, Maikel F.A.M. van Hest

National Renewable Energy Laboratory, Golden, CO, 80401, USA

Abstract – **Methods of making silicon based tandem cells include wafer bonding, depositing absorber layers on the surface, and mechanical stacking. Of these, mechanically stacking after growth retains the individual cell quality during production, but typically requires a four-terminal device. We present a transparent conductive adhesive (TCA) interlayer, designed for transparency in wavelengths that Si absorbs, conductivity for out-of-plane conductivity between cells, and adhesive strength, enabling two-terminal stacked tandems. The developed TCA material is a polymer-particle blend of common transparent adhesives and metal-coated flexible microspheres, capable of bridging micro-scale gaps between uneven/textured surfaces. At 10% microsphere coverage, experimental and theoretical series resistance is compatible with high-efficiency tandem cells on silicon.**

Index Terms – **transparent conductive adhesive, polymer-particle blends, microspheres, lamination, tandem solar cells,**

I. INTRODUCTION

Although many photovoltaic (PV) materials have been discovered and explored [1], non-concentrated PV devices struggle with increasing efficiencies past 30% without the use of tandem solar cells [2]. Combining cells with complementary band gaps, such as silicon and III-V semiconductors, has been researched for many years [3]. In recent efforts, dual-junction GaAs/Si mechanically stacked tandem cell have reported efficiencies of 32.8% [4].

Methods of making tandem cells include wafer bonding, depositing absorber layers on Si substrates, and mechanical stacking [1,3-8]. Epitaxial growth on Si substrates is complicated due to mismatched lattice constants and thermal expansion coefficients. Wafer bonding requires extremely smooth surfaces, which adds an expensive polishing step [5]. Trials with depositing thin film PV materials on Si, such as perovskites [6] and OPV [1,7], show problems with film quality due to the texture of high efficiency Si cells. The simpler method of mechanically stacking solar cells after growth retains the quality of each cell individually during production [5]. Difficulties of mechanically stacked tandems are in designing the interlayer to connect the devices together both optically and electrically [3-8]. The simplest mechanical stacking method is mounting each cell to a common transparent substrate; this however produces 4-terminal devices. A 2-terminal device will be simpler for module integration and with current matching is capable of producing 37% devices [9]. The mechanical stacking option presented here is to stack the cells together using a transparent conductive adhesive (TCA) interlayer. This layer is designed to be acceptably transparent in wavelengths that Si absorbs, conductive to transfer current between the cells, and has adhesive strength to hold the cells together.

Prior research in developing a TCA has been primarily focused on replacing ITO in OLEDs and flexible electronics [10]. TCA materials studied are primarily nano-particle polymer blends of carbon nano-tubes [11], TCO nano-particles [12], or metal nano-wires [13]. A disadvantage to these nano-sized particle polymer blends is that their primary focus is for in-plane conductivity [10-13]. In the case of tandem solar cells, the conductivity needs to be in the out-of-plane, i.e perpendicular direction. Using nano-scale particles/wires for this application would require multiple layers to span the gap between cells on a micron-scale textured wafer, resulting in optical losses. The conductive polymer PEDOT:PSS when mixed with d-sorbitol [14] yields an adhesive-like substance that researchers have used in roll-to-roll lamination of solar cells [15]. This mixture has very little adhesive strength and is water-based, resulting in excess moisture within solar modules.

The purpose of this paper is to model a new TCA for mechanically stacked tandem cells using a polymer-particle blend of transparent adhesives and metal-coated flexible microspheres. The location in tandem cells and conductivity

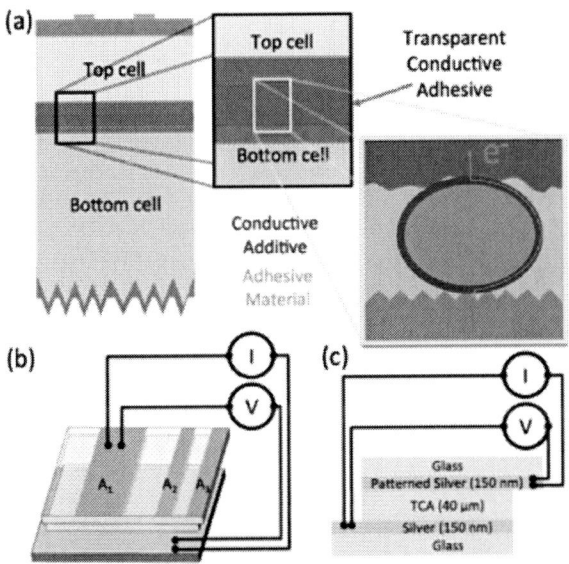

Fig. 1. (a) Schematic of TCA between top and bottom cells with micron-sized conductive particles in a transparent adhesive between textured substrates. Schematics of (b) the testing samples and (c) cross-section view used for taking series resistance measurements.

978-1-5090-5606-4/17 $31.00 © 2017 IEEE 2482

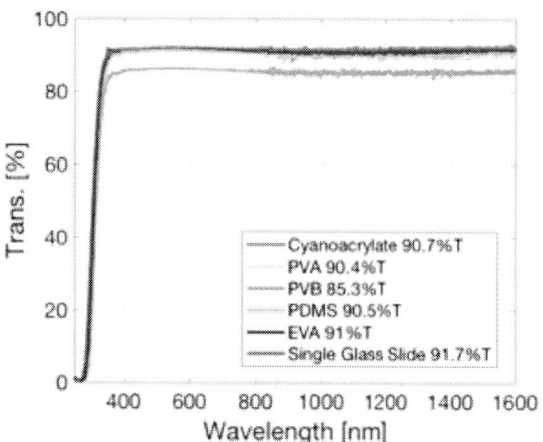

Fig. 2. Transmission measurements for the transparent adhesives and blank glass slide with the average percent transmission between 680-1130nm.

path for this TCA are shown in Fig. 1(a). A study of a variety of transparent adhesives with silver-coated poly(methyl methacrylate) (PMMA) microspheres was conducted, and a full description of the experimental results are given in ref. 16 A theoretical study of series resistance, transparency, and variable sensitivity was developed to determine the TCA's structure and requirements for tandem cell applications.

II. METHODS

Silver-coated microspheres with diameters ~40 μm were selected for their ability to deform and bridge uneven, textured surfaces. The diameter and flexibility of the spheres bridges the gaps between two independently grown cells and allows for increased contact area for individual spheres when deformed, as shown in Fig 1(a). Tailoring of the diameter of the spheres determines the thickness of the TCA layer. The transparent adhesives used were ethyl-vinyl acetate (EVA), cyanoacrylate, polydimethylsilozane (PDMS), polyvinyl alcohol (PVA), and polyvinyl butyral (PVB). The TCA was prepared according to ref. 16. Using Equation 1, the series resistance (R_s) was calculated using the current supplied (I), voltage measured (V), and sample area (A_1) shown in Fig. 1(e). Variation to the area using multiple samples allowed for statistical analysis of the measurements.

$$R_s = \frac{V}{I} A_1 = R * A_1 \qquad (1)$$

Theoretical modeling of polymer-particle blends was conducted in order to predict the series resistance through the system and determine the sensitivity of variables. This modeling provides the assessment of the sensitivity of the variables such as, size of the spheres, contacting area of spheres, metal coating on spheres, percent coverage of the spheres, contact resistance of the spheres, and material property of the spheres. These results are then compared to the experimental results for each polymer-particle blend and percent coverage.

III. RESULTS

A. Experimental Results

In Fig. 2, the transmission data from the transparent adhesives without microspheres between glass substrates are presented. From this data, it can be seen that the glass slide contributes the majority of losses while the transparent adhesives act as a index matching fluid between the substrates. With 91.7%T through a single glass slide minimal loses are seen from the adhesive layers and a secondary glass slide. With the addition of microspheres at various percent coverages, the transmission is expected to decrease proportional to the microsphere coverage. In Table I. a summary of the percent coverage, series resistance for the transparent material is shown. Demonstrating that the TCA particle polymer blends can be used for a variety of transparent adhesives.

TABLE I. SUMMARY OF PRELIMINARY TCA SERIES RESISTANCE TO PERCENT AREA COVERAGE OF MICROSPHERES.

Transparent Adhesive	% Microsphere Coverage	Series Resistance $\Omega\text{-}cm^2$
EVA	11%	0.38
PMMA	19%	0.46
Cyanoacrylate	3%	0.99
PVA	2%	1.5
PVB	4%	6.8

B. Theoretical Modeling

Calculations for a limiting case with the highest possible conductivity were first done assuming (1) the TCA is the same thickness as the average sphere, (2) the TCA is made of pure silver, and (3) no contact resistance between the surfaces. In this case the series resistance through the TCA $(R_{series,Ag})$ is given by equation (2), where R_{AgTCA} is the resistance of the TCA, calculated using ρ_{Ag}, ℓ_{TCA}, and A as the conductivity of silver, length of TCA, and the area of the sample, respectively. Therefore, the series resistance of this system is independent of sample size. Assuming $\rho_{Ag} = 1.59 * 10^{-6} \Omega cm$ and $\ell_{TCA} = 0.0042\ cm$ or the conductivity of silver and the average diameter of the microspheres, respectively. In the ideal case this results in a series resistance of $6.6 * 10^{-9}\ \Omega - cm^2$, an unrealistic lower bound due to the silver being a non-transparent material, which provides a starting point to introduce losses into the system.

978-1-5090-5606-4/17 $31.00 © 2017 IEEE

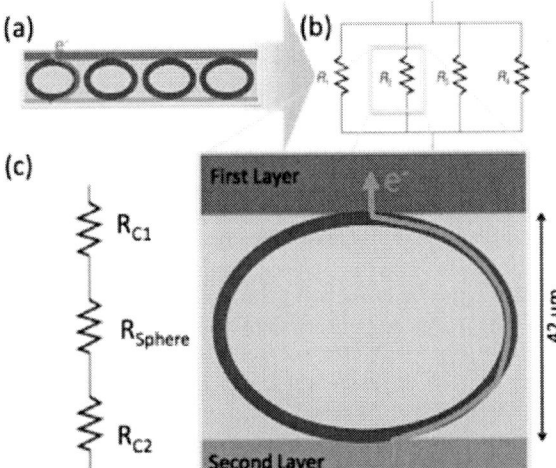

Fig. 3. Resistance model of the (a,b) microspheres in parallel and (c) each microsphere resistance is three resistors in series (top substrate contact resistance, resistance through the sphere, and bottom substrate contact resistance) with the representative schematic of these resistances.

$$R_{series,Ag} = \left(\frac{\rho_{Ag} * \ell_{TCA}}{A}\right) * A = \rho_{Ag} * \ell_{TCA} \quad (2)$$

Based on a simplified resistance model, the system was broken down into a series of parallel resistors, as seen in Fig 3a, where each resistor is representative of a single metal-coated PMMA microsphere. These resistors were then further broken down into three resistors in series, as shown in Fig. 3b. The three resistors represent R_{C1}, R_{Sphere}, and R_{C2}, which denote the contact resistance between the first cell and microsphere, resistance through the metal shell, and the contact resistance between the microsphere and second cell, respectively. For simplicity, all microspheres are assumed to be the size of an average sphere, $d = 42\mu m$, and same metal coating thickness, $t_{metal} = 250nm$, as provided by the manufacturer.

The first component of the TCA is determining the resistance through the microsphere, R_{Sphere}. This is

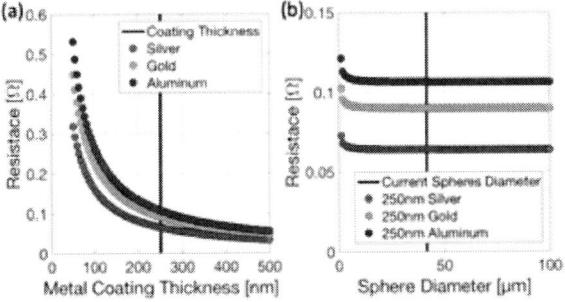

Fig. 4. (a) Resistance through a metal microsphere of 42 μm in diameter with varied metal coatings and metal thicknesses. (b) Resistance through a metal microsphere with 250nm of varied metal coating with varied microsphere diameter.

approximated by equation (3), with ρ as the conductivity of the metal coating, $\ell_{path} = \pi\frac{d}{2}$ is the length of the path around the sphere shown in Fig 3c., and A_{path} is the cross sectional area the electron will travel through the sphere. This cross sectional area of a coated sphere is calculated using equation (4).

$$R_{sphere} = \frac{\rho * \ell_{path}}{A_{path}} \quad (3)$$

$$A_{path} = \pi \left(\frac{d}{2}\right)^2 - \pi \left(\frac{d - 2(t_{metal})}{2}\right)^2 \quad (4)$$

This system is approximated using a two-dimensional approximation that doesn't account for the current limiting areas that could be present when small contact areas are present. Fig 4a plots the relationship of the metal coating thickness to resistance for various metals. Fig. 4b. shows the trend of the sphere diameter on various metal coatings of 250nm thickness.

The sensitivity to metal thickness decreases as the metal conductivity decreases, and with a 40% decrease in conductivity (from silver to gold) results in a 40% increase in resistance. The relationship to the diameter of the microsphere to resistance is much less sensitive, with the resistance bottoming out around 5 μm. For the microspheres used in this study, the resistance through a single microsphere is calculated to 0.06379 Ω per sphere when traveling top to bottom through a 250nm silver coated metal sphere of 42 μm in diameter.

The resistance of the contact starts with an approximation of the area of this contact. To determine this value the material properties of the spheres and substrates are used to find the deformation of the spheres. With each sphere the same size, it is first assumed that the force applied during pressing is evenly distributed over all microspheres, as represented by the schematic in Fig 5a. The force on a single sphere is calculated by equation (5), where F_{total} is the force applied over the sample during pressing. While the radius of the contact is calculated using calculations explained by J.A. Williams and R. S. Dwyer-Joyce [17] for contact stresses and deformations for two spheres in contact the contact radius (a), shown in Fig. 5b, is calculated by equation (6) where $R_1 (R_2)$ is the radius, $E_1 (E_2)$ is the moduli of elasticity, and $v_1 (v_2)$ is the Poisson's ratios for spheres 1 and 2, with sphere 2 as an infinitely large sphere to represent a flat plate. The deformation is assumed to be the same for both the top and bottom contacts. With material properties provided in a representative range, the contact area was iteratively compared to experimental results, summarized in Table II.

$$F = \frac{F_{total}}{Number\ of\ Spheres} \quad (5)$$

$$a = \left(\frac{3F\left(\left(\frac{1-v_1^2}{E_1}\right)+\left(\frac{1-v_2^2}{E_2}\right)\right)}{4\left(\frac{1}{R_1}+\frac{1}{R_2}\right)}\right)^{\frac{1}{3}} \quad (6)$$

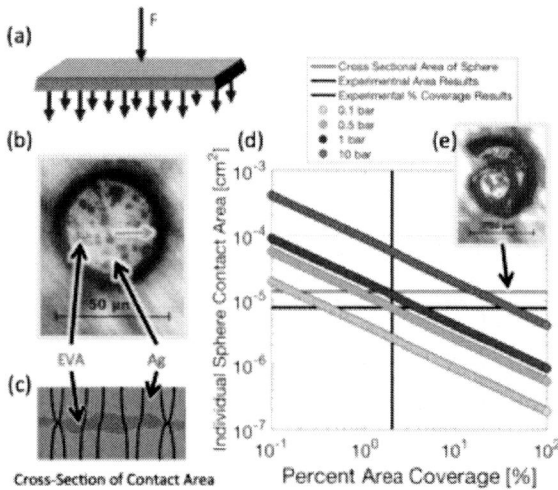

Fig. 5. (a) Schematic of the distribution of force, F, evenly over all spheres to calculate a, using equation (6), for the diameter of the contact area (b) shown on an experimentally evaluated contact area. (c) Representative contact between the sphere and silver coated glass with electron pathways concentrating on silver-silver contacts due to surface roughness and incomplete contacting, thus driving up the resistivity of the contact. (d) Calculated relationship between percent area coverage and individual sphere contact area for various processing pressures as well as (e) the microscope image of a broken sphere, represented by the red line where the pressure cause the spheres to rupture.

TABLE II. SUMMARY OF MATERIAL PROPERTIES USED IN THIS STUDY FOR MODEL INPUTS.

	Radius [$R_1(R_2)$]	Moduli of Elasticity [$E_1(E_2)$]	Poisson's ratios [$v_1(v_2)$]
Sphere 1 (Flexible Polymer Foam)	$42\mu m$	$4-12$ $*10^6 N/m^2$	$0.35-0.4$
Sphere 2 (Soda Lime Glass)	∞	$68-72$ $*10^9 N/m^2$	0.22

An individual microspheres contact area is shown in Fig. 5b, this image's area is calculated by measuring the diameter of the deformed section of the sphere. The area is assumed to be circular and shown with the particle percent area coverage of this sample in the plots as the black lines in Fig. 5d. This graph has the theoretical calculations of an individual spheres contact area as a function of particle density of the TCA when pressed at different pressures. Since the area of the contact is a function of force, the area will increase as the particle density decreases, since there are fewer particles to distribute the force while allowing the EVA to flow out through the sides. If the microsphere concentration is low enough, the spheres will no longer deform, but instead break. While it is difficult to model when the fracture will occur, the percent coverage below (and just before) where the contact area is equal to the cross-sectional area of the microspheres is to be assumed as broken.

The broken spheres are consistent with the experimental results, at 10% area coverage pressed at 10 bar will break all the spheres as shown in Fig. 5e.

The contact resistance, R_C, calculated using equation (7) where $A_{contact} = \pi a^2$ is the contact area, calculated using the equation (6), $\ell_{Contact}$ is the thickness of the contact, which is assumed to be 250 nm, and ρ_{mix} is the representative conductivity of the contact. This representative conductivity is what is used to increase the resistance of the contact artificially. However the representation of this can be viewed as the rough contacts between the sphere and the silver coated glass and polymer pockets trapped within the contacts. As shown in Fig. 5b,c, this contact has EVA pockets in both macro and micro sizes and well as metal contacting the substrate directly. To relate this more to the current contact resistance research the intermediate variable of specific contact resistance is calculated using equation (8) reducing the total contact resistance to equation (9). In Fig. 6a, the dependence of resistance of the contact on percent area coverage is shown for varied specific contact resistances. For reference, peak firing conditions for silicon top contacts see the specific contact resistance at 5 mΩ-cm^2 [18]. This relationship is a linear trend with increasing sphere coverage the more spheres to distribute the force, thus less contact area and increased total resistance of that contact. With both the top and bottom substrates in the test samples being of the sample material, the total contact resistance is assumed to be the equal.

$$R_C = \frac{\rho_{mix}*\ell_{Contact}}{A_{contact}} \tag{7}$$

$$r_c = \rho_{mix} * \ell_{Contact} \tag{8}$$

$$R_C = \frac{r_c}{A_{contact}} \tag{9}$$

Assuming that the contact resistance is the same for top (R_{C1}) and bottom (R_{C2}) and calculating the resistance through the sphere (R_{Sphere}), the total resistance of a single microsphere can be calculated with equation (10). This total resistance of a single microsphere, shown as a function of percent area coverage in Fig. 6b, can then be placed in parallel with the number of microspheres based on the percent coverage to calculate the total resistance of the system per cm^2 by equation (11), where n is the number of microspheres per cm^2, which is simply making an equivalent resistance for the diagram in Fig. 3. This resistance per total microsphere varies with percent coverage due to the distribution of force over the number of spheres within a sample will change the contact area of the spheres. The total resistance of the sample is displayed in Fig. 6d. Where the trend continues with the total resistance decrease as the number of microspheres increase, this is due the additional pathways through the TCA. However to make a direct comparison to the experimental TCA samples this resistance need to be need to be transformed into a series resistance. Using equation (12) the total sample resistance per unit area is transformed into a series resistance with the same

978-1-5090-5606-4/17 $31.00 © 2017 IEEE

area as an average TCA test sample made, $A_{average} = 0.6\ cm^2$. This plot is displayed in Fig. 6d.

$$R_{Sphere\ T} = R_{Sphere} + R_{C1} + R_{C2} = R_{Sphere} + 2R_C \quad (10)$$

$$\frac{1}{R_{sample\ T}} = \sum_{i=1}^{n} \frac{1}{R_i} = \frac{n}{R_{Sphere\ T}} \quad (11)$$

$$SR_{theoretical} = R_{Sample\ T}\big(A_{average}\big)^2 \quad (12)$$

To examine where the losses in the system are seen the ideal case with a pure silver TCA with a series resistance of $6.6 * 10^{-9}\ \Omega - cm^2$, an unrealistic lower bound due to the silver being a non-transparent material, with the addition of 100% area coverage of the sphere without the addition of contact resistances would result in $3.1 * 10^{-7}\ \Omega - cm^2$. This is two orders of magnitude loss due to the conduction through metal-coated microspheres instead of a flat sheet of metal. However, when you decrease these values to account for 10% transparency, assuming geometric shading, this value drops one more order of magnitude to $3.1 * 10^{-6}\ \Omega - cm^2$. Stating that the limiting factor of this system is not the spheres themselves. Comparisons to the lowest values recorded at this percent area coverage, $0.38\ \Omega - cm^2$, the additional five orders of magnitude are lost due to contact resistances between the spheres and the substrates. However, when moving to textured substrates have additional points of contacts due to the uneven nature will assist in mitigating these resistance losses.

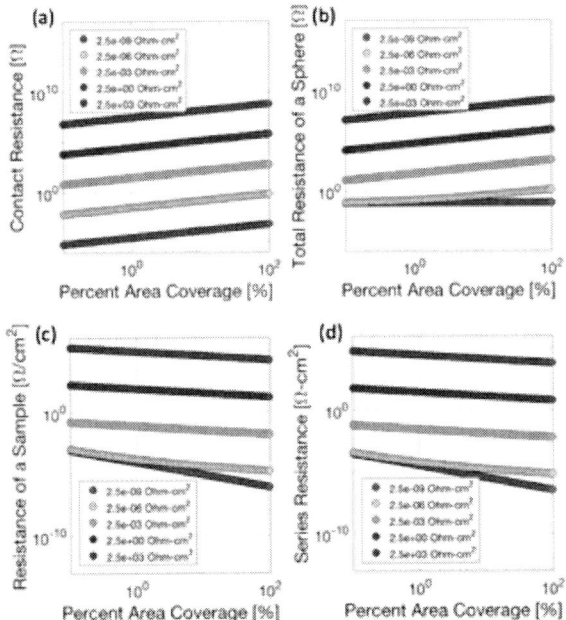

Fig. 6. (a) Calculated contact resistance of the sphere using equation (9) with (b) the corresponding total resistance of a sphere, (c) total resistance of a sample, and (d) series resistance of the TCA for varied percent area coverage and specific contact resistance.

IV. CONCLUSIONS

A new TCA material is modeled that consists of a polymer-particle blend comprised of transparent adhesives and metal-coated PMMA microspheres. This TCA results in series resistances of 0.38 Ω-cm², 0.46 Ω-cm², and 0.99 Ω-cm² for 11% microspheres in EVA, 19% microspheres in PMMA, and 3% microspheres in cyanoacrylate, respectively. The TCA material is made to be consistent with current manufacturing processes such as lamination and hot compression.

Theoretical modeling of the particle-polymer blends was conducted, demonstrating that the microsphere diameter and metal coatings have minimal impact on the series resistance of the system. This shows that tailoring of the microsphere diameter to dictate the thickness of the TCA can be done with minimal impact to the conductivity. Additionally, it was observed that the limiting factor of the system is not the resistance through the sphere but the contact resistance of the sphere to the substrate. This is indicated by comparisons made to an idealized TCA made of pure silver without contact resistance. When transferring the TCA from a solid conductive sheet to 100% coverage of metal-coated microspheres without contact resistance an increase of two orders of magnitude in series resistance is seen. Additionally when decreasing the percent area coverage of metal-coated spheres to 10% without contact resistance one additional order of magnitude is loss. Comparison to the lowest experimental result of 0.38 Ω-cm² the additional five orders of magnitude is loss from the contact resistance between the spheres and the substrates.

ACKNOWLEDGMENTS

Funding for this work at NREL was provided by DOE through EERE contract SETP DE-EE00030299 and under Contract No. DE- AC36-08GO28308. The United States Government retains and the publisher, by accepting the article for publication, acknowledges that the United States Government retains a non-exclusive, paid-up, irrevocable, world- wide license to publish or reproduce the published form of this manuscript, or allow others to do so, for United States Government purposes.

REFERENCES

[1] S.H. Park, I. Shin, K.H. Kim, R. Street, A. Roy, and A.J. Heeger, "Tandem Solar Cells Made form Amorphous Silicon and Polymer Bulk Heterojunction Sub-Cells," *Advanced Materials*, vol. 27(2), pp. 298-302, 2014

[2] National Renewable Energy Laboratory, "Best Research-Cell Efficiencies," http://www.nrel.gov/pv/assets/images/efficiency-chart.png; accessed 26 January 2017.

[3] S. Essig, J. Benick, M. Schachtner, A. Wekkeli, M. Hermle, and F. Dimroth, "Wafer-Bonded GaInP/GaAs//Si Solar Cells With 30% Efficiency Under Concentrated Sunlight," *IEEE journal of Photovoltaics*, vol. 5(3), pp. 977-981, 2015.

[4] S. Essig, C. Allebe, T. Remo, J.F. Geisz, M.A. Steiner, L. Barraud, J.S. Ward, M. Schnabel, K. Horowitz, A. Descoeudres, D.L. Young, M. Woodhouse, M. Despeisse, C. Ballif, and A.C. Tamboli, "32% efficient III-V/Si dual-junction solar cells and their challenging path towards cost competiveness," in *submitted to 44th IEEE PVSC, 2017.*

[5] A.C. Tamboli, M.F.A.M. van Hest, M.A. Steiner, S. Essig, E.E. Perl, A.G. Norman, N. Bosco, and P. Stradins, "III-V/Si wafer bonding using transparent, conductive oxide interlayers," *Applied Physics Letters*, vol. 106, 263901, 2015.

[6] D.P. McMeekin, G. Sadoughi, W. Rehman, G.E. Eperon, M. Saliba, M.T. Hörantner, A. Haghighirad, N. Sakai, L. Korte, B. Rech, M.B. Johnston, L.M. Herz, and H.J. Snaith, "A mixed-cation lead mixed-halide perovskite absorber for tandem solar cells," *Science*, vo. 351 (6269), pp. 151-155, 2016.

[7] S. Tanaka, K. Mielczarek, R. Ovalle-Robles, B. Wang, D. Hsu, and A.A. Zakhidov, "Monolithic parallel tandem organic photovoltaic cell with transparent carbon nanotube interlayer," *Applied Physics Letters*, vol. 94, 113506, 2009.

[8] M. Rohde, M. Zelt, O. Gabriel, S. Neubert. S. Kirner, D. Serverin, T. Stolley, B. Rau, B. Stannowski, and R. Schlatmann, "Plasma enhanced chemical vapor deposition process optimization for thin film silicon tandem junction solar cells, " *Thin Solid Films*, vol. 558, pp. 337-343, 2014.

[9] J.F. Geisz and D.J. Friedman, "III-N-V semiconductors for solar photovoltaic applications," *Semicond. Sci. Technol.*,17 No 8, 769-777, 2002.

[10] S. Yao and Y. Shu, "Nanomaterial-Enabled Stretchable Conductors: Strategies, Materials and Devices," *Advanced Materials*, vol. 27 (9), pp. 1480-1511, 2015.

[11] J. Du, S. Pei, L. Ma, and H. Cheng, "Carbon Nanotube- and Graphene-Based Transparent Conductive Films for Optoelectronic Devices," *Advanced Materials*, vol. 26(13). Pp. 1958-1991.

[12] S. Yoshidomi, M. Hasumi, and T. Sameshima, "Investigation of conductivity of adhesive layer including indium tin oxide particles fro multi-junction solar cells," *Applied Physics A*, vol. 116(4), pp. 2113-2118, 2014.

[13] D. Langley, G. Giusti, C. Mayousse, C. Celle, D. Bellet, and J. Simonato, "Flexible transparent conductive materials based on silver nanowire networks: a review," *Nanotechnology*, vol. 24(45), 452001, 2013.

[14] D. Angmo and G.C. Krebs, "Flexible ITO-Free Polymer Solar Cells," *Journal of Applied Polymer Science*, vol. 129, pp. 1-14, 2013.

[15] G.D. Spyropoulos, C.O.R. Quiroz, M. Salvador, Y. Hou, N. Gasparini, P. Schweizer, J. Adams, P. Kubis, N. Li, E. Spiecker, T. Ameri, H. Egelhaaf, and C.J. Brabec, "Organic and perovskite solar modules innovated by adhesive top electrode and depth-resolved laser patterning," *Energy & Environmental Science*, vol. 9, pp. 2302-2313, 2016.

[16] T.R. Klein, B.G. Lee, E.L. Warren, P. Stradins, A.C. Tamboli, M.F.A.M. van Hest. "Transparent Conductive Adhesives for Tandem Solar Cells Using Polymer-Particle Composites", In preparation.

[17] J. A. Williams and R.S. Dwyer-Joyce, "Chapter 3: Contact Between Solid Surfaces," Modern Tribology Handbook, CRC Press LLC, 2001.

[18] S.G. Kang, C.W. Lee, Y.J. Chung, C.G. Kim, S. Kim, D. Kim, C.J. Kim, and Y.K. Lee, "Nano-glass frit for inkjet printed front side metallization of silicon solar cells prepared by sol-gel process", *Phys. Status Solidi RRL*, vol 9, no. 5, 293-296, 2015.

Modeling three-terminal III-V/Si tandem solar cells

Emily L. Warren, Michael G. Deceglie, Paul Stradins, Adele C. Tamboli

National Renewable Energy Laboratory, 15013 Denver West Parkway, Golden, CO 80401, USA

Abstract — Three-terminal (3T) tandem cells fabricated by combining an interdigitated back contact (IBC) Si device with a wider bandgap top cell have the potential to provide a robust operating mechanism to efficiently capture the solar spectrum without the need to current match sub-cells or fabricate complicated metal interconnects between cells. Here we develop a two dimensional device physics model to study the behavior of IBC Si solar cells operated in a 3T configuration. We investigate how different cell designs impact device performance and discuss the analysis protocol used to understand and optimize power produced from a single junction, 3T device.

Index Terms — tandem solar cell, three-terminal, Sentaurus, TCAD, modeling

I. INTRODUCTION

There has been a great deal of recent interest in the development of III-V/Si tandem solar cells, with multiple new efficiency records set in the past year [1], [2]. Most tandem solar cells designs are either two-terminal (2T) devices, where the subcells are electrically connected in series, or 4 terminal (4T) devices, where each subcell is operated independently. While 2T devices are more common, recent modeling has shown that 4T devices are more resilient to variations in solar spectrum and can produce higher energy yield [3], [4]. An alternative cell configuration that has potential to combine the strengths of both is a three-terminal (3T) device consisting of a III-V top cell optically in series with a modified interdigitated back contact (IBC) Si cell featuring a top contact connected to the III-V cell as shown in Fig 1. Such a 3T tandem can be run either in a standard 2T configuration, or the second back contact can be used to extract excess current, making the device less sensitive to current matching between the subcells due to bandgap differences or spectral variation.

A 3T contacting geometry has the potential to benefit multiple approaches to creating tandem cells. These cells would not need lateral current extraction between the cells (which is required for 4T operation), enabling the use of transparent conductive adhesives (TCAs) or other low cost conductive layers [5]. However, there are unique design features of the 3T geometry that may detrimentally impact performance (such as the need for a conductive top surface in an IBC Si cell) that motivates the detailed modeling of this relatively unexplored tandem cell design.

The modeling of most tandem cells in the literature has been carried out by combining 1D solar cell models (e.g. PC1D) with separate circuit models or simply adding together the performance of the two subcells. While this is sufficient to capture the behavior of a 2T or 4T device, to accurately model the behavior of a 3T cell as described above requires at least a 2D model that is capable of handling device physics in two dimensions and with more than 2 contacts. There are no dedicated solar cell modeling software packages

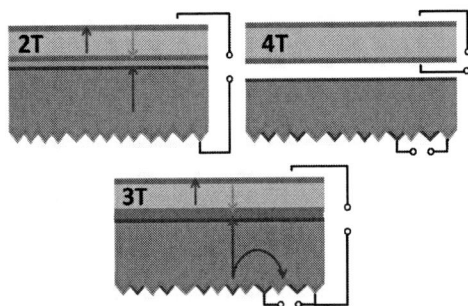

Fig. 1. Schematic comparing 2T, 3T, and 4T cell configuration. 2T devices are limited by current matching conditions, but offer simpler integration than 4T devices. 3T devices can leverage advantages of both approaches.

that meet this requirement, necessitating the use of a more powerful simulation environment. In this work, we have used a technology computer aided design (TCAD) software package to understand the performance of a 3T solar cell based on an IBC Si bottom cell.

II. SIMULATION METHODS AND PARAMETERS

The electrical simulations presented here were performed using Sentaurus TCAD (Synopsis). The two-dimensional nature of an IBC cell structure requires a 2D structure with a width equal to the pitch between the *p*- and *n*-type back contacts. The cell geometry was roughly based on the poly-silicon on oxide (POLO) IBC cell developed at the Institute for Solar Energy Research Hamelin (ISFH), although the device structure was not optimized to fully match experimental data [6], [7].

A. Optical Generation

Due to the vastly different length scales needed to capture the electrical and optical performance of an IBC device, the optical generation profile was calculated separately and then used to solve the device physics of the cell [8]. This enables the optical generation profile to guide the meshing of the device area, so that regions of high optical absorption are meshed more densely, which enhances the efficiency of convergence of the device physics model. Optical generation profiles were created within Sentaurus or using PV Lighthouse's module ray-tracing software [9]. All of the results presented here were calculated using the AM1.5G spectrum without any filtering from a top III-V cell or TCA layer. The optical stack included a standard 75 nm antireflective coating layer so that the simulated performance under standard illumination conditions can be more directly compared to standard single junction Si solar cells.

TABLE I
GEOMETRY OF SI CELL

Parameter	Value
Cell thickness	160 μm
Unit cell width	365 μm
poly-Si thickness	20 nm
Front contact parameters	
Tunnel oxide thickness	1.5 nm
P In-diffusion depth	300 nm
P In-diffusion peak	10^{19} cm^{-3}

TABLE II
DEVICE PARAMETERS FOR SI CELL

Parameter	Value
Temperature	300 K
Bulk doping	(P) 10^{15} cm^{-3}
Bulk lifetime	2 ms
SRV at Si/SiO$_2$ interface	10^3 cm/s
SRV at Si/SiN$_x$ interface	50 cm/s
SRV at metal contacts	10^7 cm/s
Tunneling model	Nonlocal tunneling
Tunneling effective mass (m$_{e,h}$)	0.4

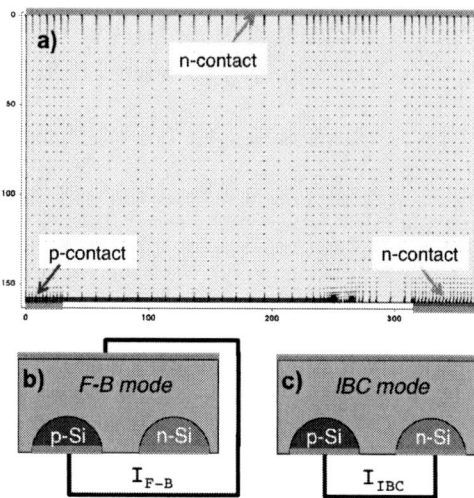

Fig. 2. a) TCAD model of a 3-terminal n-type Si cell with a n-type top contact and standard n and p IBC back contacts (units of x and y axes are in μm, and arrows indicate current flow across the device at max power point operation); simplified schematics of the two limiting operating modes of such a 3T cell: b) current extracted between *top-n* and *back-p* contact and c) current extracted between *back-n* contact and *back-p* contact.

B. Electrical Simulation

The geometry of the device is shown in Fig. 2a and consists of an *n*-type Si substrate with the back emitter/base defined by poly-Si *p* and *n* regions, respectively. The poly-Si regions are passivated with SiN$_x$ with smaller area metal openings that are contacted with Al. The entire front surface of the cell was coated with n-poly Si and then contacted with a transparent uniform contact to simulate a TCA layer. The basic geometric parameters of the cell are listed in Table I. For simplicity, the poly-Si layers are defined as separate c-Si regions with different doping densities [10], [11]. Gaussian doping profiles were used to simulating in-diffusion of dopants into the bulk. Different front surface conditions were investigated to understand the impact of passivation and recombination on 3T performance. A passivated front contact was modeled with a carrier selective contact using a thin tunneling SiO$_2$ layer, as has been previously demonstrated [11], [12]. An ohmic front contact (SRV = 10^7 cm/s) was modeled to provide an upper bound for the degradation of cell performance due to recombination at the front surface. The basic material parameters used to define the properties of the materials used are listed in Table II.

The device physics of the cell was solved using Sentaurus Device. To improve convergence of the model, the illumination was slowly turned on over multiple steps and then a quasi-stationary ramp was used to sweep the voltage of the *p* contact from 0V to 0.8V to extract the current voltage behavior of the cell. For 3T operation, another quasi-stationary ramp was used to set the current from one of the two *n* contacts prior to sweeping the voltage of the *back-p* contacts.

III. RESULTS AND DISCUSSION

For our proposed 3T tandem cell to work it must be possible to extract the excess current from an IBC Si cell that is operating in series with a wider bandgap top cell with minimal

losses to the overall performance of the device. We first examine each of the limiting 2T operating conditions for the simulated 3T Si cell.

There are two ways to extract power out a 3T Si cell, as shown in Fig 2. In the case of a cell with an n-type base, current can either be extracted between the front n-type contact of the cell and the back p-type contact (*F-B* mode, Fig 2b), or between the back n-type contact and the back p-type contact (*IBC* mode, Fig 2c).

We first investigate the limiting cases of performance for the device in the limiting 2T modes (*F-B* and *IBC*) under AM1.5G illumination to compare performance to standard 2T Si devices. It is well-documented that passivation of the front surface is critical to achieve high efficiencies for standard IBC cells, so it is important to determine whether a 3T device will suffer by having a conductive front surface. Figure 3 shows J-V data for cells operating in each mode with and without a passivating front contact (figure of merit data for each device is compiled in Table III.) When the poly-Si/SiO$_2$ passivated front contact structure is replaced with an ohmic contact (SRV = 10^7 cm/s), the performance of the cell drops dramatically, from >22% to <17% absolute, in both *IBC* mode and *F-B* mode. This demonstrates the importance of a well-passivated surface for the operation of a device with a large area contact. Interestingly, in both cases the cell performs slightly better in *F-B* mode than in *IBC* cell, for otherwise identical cell geometries. The difference in performance can be attributed to improved fill-factor (FF). It has previously been suggested that the direct current path between the *back-p* and full area *front-n* contact can minimize current bunching at localized contacts and improve FF, which also seems to improve performance in this simulation [11].

Although prior modeling of passivated contacts has been

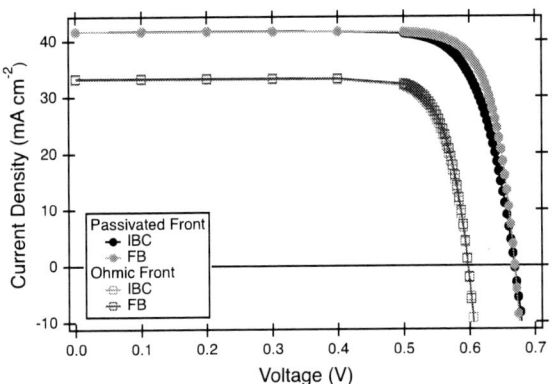

Fig. 3. J-V plots for a 3T Si cell, operated in either *IBC* mode or *F-B* mode. Solid markers are for a device with a passivated front contact, open squares for an ohmic front contact.

carried out by enabling tunneling across the thin SiO_2 region, recent work has shown that pinholes in the oxide play an important role in the operation of these contacts [13], [14]. In addition to the tunneling passivation, we also created an idealized front c-Si/poly-Si contact with a SRV of 10 cm/s ("front ideal c-Si/poly-S" in Table III), to represent an idealized passivated contact. The performance of these cells is very similar to those with an active tunneling layer, but the open circuit voltages (V_{oc}) of these devices were lower than when the SiO_2 was present. However this approach still provides a reasonable baseline device efficiency in both *IBC* and *F-B* mode.

In 3T mode, the cell's operating range has an extra degree of freedom, requiring two independent variables to specify the operating point of the cell. Since each n-type contact can create a different diode with the common p-type contact, each sub-circuit can sustain a different voltage. It is important to account for the power generated in each sub-circuit, not just the current-voltage behavior at any given operating condition of the cell. When the current at either n-type contact was set to a constant value (as might be the case for a tandem cell with the top cell connected in series to one sub-circuit), the total power out of both circuits matches or exceeds the 2T performance in *IBC* or *F-B* mode.

IV. CONCLUSION

We have developed a 2D TCAD model to investigate the performance of 3T Si IBC cells for integration into a 3T

TABLE III

FIGURES OF MERIT FOR OPERATION OF 3T SI CELL IN IBC AND F-B MODE UNDER DIFFERENT FRONT CONTACT CONDITIONS

Configuration	Voc (mV)	Jsc (mA cm^{-2})	FF	Eff (%)
IBC, front tunnel oxide	668	41.69	79.1	22.0
F-B, front tunnel oxide	668	41.69	82.2	22.9
IBC, ohmic front	597	33.14	81.2	16.1
F-B, ohmic front	597	33.32	81.7	16.3
IBC, front ideal c-Si/poly-Si	661	41.56	82.5	22.6
F-B, front ideal c-Si/poly-Si	662	41.56	82.7	22.7

tandem cell. Operating a tandem cell in such a configuration enables much simpler fabrication techniques than a 4T geometry, as no lateral conduction is needed between the top and bottom cells to extract current. It still enables more efficient utilization of all the available photocurrent than in a 2T configuration, where the total current is always limited by one of the subcells. We show that 3T cell designs utilizing a passivated front and two IBC back contacts enables Si devices to be operated in multiple modes with similar overall power conversion. This provides a pathway to new routes to high efficiency 3T Si-based tandem solar cells.

ACKNOWLEDGMENT

The authors thank Michael Rienacker (ISFH), Manuel Schnabel, Henning Schulte-Huxel, Robby Peibst (ISFH), and Ana Kavence for helpful discussions. Funding for this work at NREL was provided by DOE through EERE contract SETP DE-EE00030299 and under Contract No. DE- AC36-08GO28308. The United States Government retains and the publisher, by accepting the article for publication, acknowledges that the United States Government retains a non-exclusive, paid-up, irrevocable, world- wide license to publish or reproduce the published form of this manuscript, or allow others to do so, for United States Government purposes.

REFERENCES

[1] S. Essig, M. A. Steiner, C. Allebe, J. F. Geisz, B. Paviet-Salomon, A. Descoeudres, V. LaSalvia, L. Barraud, N. Badel, A. Faes, J. Levrat, M. Despeisse, C. Ballif, P. Stradins, and D. L. Young, "Realization of GaInP/Si Dual-Junction Solar Cells With 29.8% 1-Sun Efficiency," *J. of Photovoltaics*, vol. 6, no. 4, pp. 1012–1019, 2016.

[2] R. Cariou, J. Benick, P. Beutel, N. Razek, C. Fl, M. Hermle, D. Lackner, S. W. Glunz, S. Member, A. W. Bett, M. Wimplinger, and F. Dimroth, "Monolithic Two-Terminal III – V // Si Triple-Junction Solar Cells with 30.2% Efficiency under 1-Sun AM1.5G," *J. of Photovoltaics*, vol. 7, no. 1, pp. 367–373, 2017.

[3] H. Liu, Z. Ren, Z. Liu, A. G. Aberle, T. Buonassisi, and I. M. Peters, "The realistic energy yield potential of GaAs-on-Si tandem solar cells: a theoretical case study," *Optics Express*, vol. 23, no. 7, p. A382, 2015.

[4] S. Essig, S. Ward, M. A. Steiner, D. J. Friedman, J. F. Geisz, P. Stradins, and D. L. Young, "Progress Towards a 30% Efficient GaInP/Si Tandem Solar Cell," *Energy Procedia*, vol. 77, pp. 464–469, 2015.

[5] T. Klein, B. Lee, M. Schnabel, E. Warren, P. Stradins, A. Tamboli, and M. van Hest, "Transparent conductive adhesives for tandem solar cells," *Proceedings of the 44th IEEE PVSC*, 2016.

[6] M. Rienacker, M. Bossmeyer, A. Merkle, R. Udo, F. Haase, J. Kr, R. Brendel, and R. Peibst, "Junction Resistivity of Carrier-Selective Polysilicon on Oxide Junctions and Its Impact on Solar Cell Performance," *J. of Photovoltaics*, vol. 7, no. 1, pp. 11–18, 2017.

[7] F. Haase, F. Kiefer, S. Schäfer, C. Kruse, J. Krügener, R. Brendel, and R. Peibst, "Ibc solar cells with polycrystalline on oxide (polo) passivating contacts for both polarities," *Japanese Journal of Applied Physics*, 2017.

[8] "Optimization of Rear Contact Design in Monocrystalline Silicon Solar-Cell Using 3D TCAD Simulations," Synopsis, Tech. Rep., 2011.

[9] (2017, June) PV Lighthouse: Module Ray Tracer. [Online]. Available: https://www.pvlighthouse.com.au

[10] U. Römer, R. Peibst, T. Ohrdes, B. Lim, J. Krügener, E. Bugiel, T. Wietler, and R. Brendel, "Solar Energy Materials & Solar Cells Recombination behavior and contact resistance of n and p polycrystalline Si / mono-crystalline Si junctions," *Solar Energy Materials and Solar Cells*, vol. 131, pp. 85–91, 2014.

[11] H. Steinkemper, F. Feldmann, M. Bivour, and M. Hermle, "Numerical Simulation of Carrier-Selective Electron Contacts Featuring Tunnel Oxides," *IEEE Journal of Photovoltaics*, vol. 5, no. 5, pp. 1348–1356, 2015.

[12] H. Steinkemper, F. Feldmann, M. Bivour, and M. Hermle, "Theoretical Investigation of Carrier-selective Contacts Featuring Tunnel Oxides by Means of Numerical Device Simulation," *Energy Procedia*, vol. 77, no. 5, pp. 195–201, 2015.

[13] D. Tetzla, J. Krügener, Y. Larionova, S. Reiter, M. Turcu, F. Haase, R. Brendel, R. Peibst, U. Höhne, J. Kähler, and T. F. Wietler, "A simple method for pinhole detection in carrier selective POLO-junctions for high efficiency silicon solar cells," *Solar Energy Materials and Solar Cells*, 2017.

[14] R. Peibst, U. Römer, Y. Larionova, M. Rienäcker, A. Merkle, N. Folchert, and S. Reiter, "Working principle of carrier selective poly-Si / c-Si junctions : Is tunnelling the whole story ?" *Solar Energy Materials and Solar Cells*, pp. 1–8, 2016.

Wafer bonding approaches for III-V on Si multi-junction solar cells

Laura Vauche[1,2], Elias Veinberg-Vidal[1,2], Clément Weick[1,3], Christophe Morales[1,2], Vincent Larrey[1,2], Christophe Lecouvey[1,2], Mickaël Martin[1], Jérémy Da Fonseca[1,2], Christophe Jany[1,2], Thibaut Desrues[1,3], Céline Brughera[1,2], Philippe Voarino[1,3], Thierry Salvetat[1,2], Frank Fournel[1,2], Mathieu Baudrit[1,3] and Cécilia Dupré[1,2]

[1]Univ. Grenoble Alpes, Grenoble, 38000, France
[2]CEA, LETI, MINATEC Campus, Grenoble, 38054, France
[3]CEA, LITEN, INES, Le Bourget du Lac, 73375, France

ABSTRACT —**In this study, fabrication of monolithic triple junction (3J) GaInP/AlGaAs/Si solar cells is reported by different wafer bonding approaches: (i) the most straightforward, direct GaAs/Si wafer bonding; (ii) an innovative 2-step approach combining epitaxy of GaAs bonding layer on Si cell followed by GaAs/GaAs direct wafer bonding; and (iii) Surface-Activated GaAs/Si wafer Bonding (SAB). Bonding interfaces properties are comparatively studied, and the performance of resulting III-V on Si triple junction devices is examined. Open-circuit voltages close to 2.9V are obtained in all cases, but large differences in fill factors between the bonding methods are observed, probably due to the presence of an oxide layer.**

I. INTRODUCTION

Multi-Junction Solar Cells (MJSC) from III–V compound semiconductors have so far delivered the highest solar-electric conversion efficiencies. The broad spectrum of direct bandgap binary, ternary and quaternary III-V compounds, that can be grown lattice-matched on GaAs or Ge substrates, offers a wide range of possibilities for advanced device structures. However, GaAs or Ge substrates are expensive (typically more than 50% of the materials cost). III-V cells mechanically stacked, wafer bonded or epitaxially grown on much cheaper and widely available Si substrates, combined with III-V substrate reuse strategies in the case of bonding or stacking, could result in a significant cost reduction [1]. Therefore, the integration of III-V based materials on Si material may provide a cost breakthrough for PV technology, unifying the low-cost of crystalline silicon (c-Si) and the efficiency potential of III-V/c-Si multi-junction solar cells, allowing for a competitive Levelized Cost Of Electricity (LCOE) in one-sun or low-concentration terrestrial solar applications. Best power conversion efficiencies reported in literature for 2 terminal III-V/Si triple junction devices using different fabrication methods are reported in Table I. Direct epitaxy of III-V on Si is challenging, as GaAs and Si suffer from lattice and Coefficient of Thermal Expansion (CTE) mismatches, resulting in the presence of detrimental defects and lower device performance. Direct metal interconnect provides

bonding strength and electrical conductivity but may cause optical losses [2]. Direct wafer bonding is a well-established method in the semiconductors industry for Si-based materials and offers a great potential to combine III-V and Si materials through a permanent, electrically conductive and optically transparent interface. Although bonding of Si on III-V materials is challenging, the great potential of wafer bonding has been demonstrated with high device performances using Surface-Activated Bonding (SAB), as shown in Table I.

TABLE I: Power conversion efficiency of 2-terminal III-V on Si triple junction solar cells, as reported in literature

η (%) AM1.5G	V_{OC} (V)	3J materials	Fabrication method
30.2 [3]	3.05	GaInP/GaAs/Si GaInP/AlGaAs/Si	SAB
26.2 [3]	2.90		
25.2 [4]	2.88		
~26 [5]	2.85	GaInP/GaAs/Si	SAB
25.8 [2] 25.5 [6], [7]	2.73	GaInP/InGaAs/Si	Mechanical stack: direct metal interconnect
23.2 [8]	2.81	GaInP/GaAs/Si	Mechanical stack: Pd nanoparticles array
[9]	2.56	GaInP/GaAsP/Si	Epitaxy: GaAsP graded buffer layer

However, as SAB requires ultra-high vacuum, more straightforward bonding techniques could be more suitable for high throughput scenarios. In this study, we present III-V on Si triple junction solar cell fabrication processes including comparison of SAB, direct GaAs/Si wafer bonding and a novel strategy using GaAs/GaAs direct wafer bonding.

II. III-V ON Si SOLAR CELL FABRICATION PROCESS

III-V and Si cells preparation. n-on-p GaInP/AlGaAs cells are grown lattice matched by Fraunhofer-ISE institute with inverted layer configuration on 4-inch GaAs substrates using Metal-Organic Vapor Phase Epitaxy (MOVPE) including highly doped tunnel junctions and n-doped GaAs bonding layer, as reported in [10]. The n-on-p silicon bottom cells are manufactured independently by diffusions in p-type Cz (100) c-Si wafers resulting in the formation of P-doped n^+ emitter at the front surface; with a surface doping of 6.10^{20} at/cm^3 and a B-doped Back Surface Field (BSF).

Wafer surface preparation. To achieve good bonding mechanical quality, it is mandatory to bring together nearly perfect surfaces with low roughness and low particle contamination. A combination of cleaning treatments and Chemical Mechanical Polishing (CMP) steps is applied to GaAs and Si surfaces. Prior to bonding, a sulfur treatment followed by NH4OH scrub was implemented to GaAs surface, as sulfur treatments have been reported to improve bonding quality significantly with low-resistance interface [11]. On Si surface, after cleaning and CMP, two different finishing surface treatments were applied: UV/O$_3$ oxidizing treatment (A) or NH4OH scrub (B).

Bonding. **GaAs/Si:** Following the surface treatments, GaAs and Si surfaces are left oxidized. Hydrophilic n-GaAs/n-Si direct wafer bonding is performed under vacuum at room temperature. **GaAs/GaAs:** In this approach, a GaAs epitaxial layer is deposited on the Si cell in order to perform direct GaAs/GaAs bonding with the III-V cell. A 250 nm thick n-doped GaAs bonding epitaxial layer was deposited by MOVPE on the Si cell. Both GaAs surfaces (GaAs bonding layer on III-V cells, and GaAs epitaxial layer on Si cells) are then prepared with CMP followed by sulfur and NH4OH treatments. n-GaAs/n-GaAs hydrophilic bonding was performed under vacuum at room temperature. **SAB:** In a first step, GaAs surfaces are prepared using CMP followed by sulfur and NH4OH treatment, similarly than for GaAs/Si and GaAs/GaAs bondings. Si surfaces are prepared using CMP followed by NH4OH scrub. III-V and Si cells are loaded into EVG580 ComBond vacuum chamber. The oxide layers formed on GaAs and Si substrate surfaces are removed by Ar beam irradiation and then the surfaces are brought into contact under ultra-high vacuum.

Thermal treatments and GaAs substrate removal. For the bonding of hydrophilic surfaces, hydrogen bonds act at the contact points between the two wafers. Just after bonding at room temperature, bonding energies are still weak, and the bonding is reversible. To increase the bonding strength, a thermal treatment is usually performed on bonded structures [12], [13]. Therefore, after GaAs/Si or GaAs/GaAs bonding, annealing is performed under N$_2$ atmosphere for 1h in order to increase bonding strength and form covalent bonds. However, high temperatures can add additional stress due to CTE mismatch between III-V and Si materials. In this study, the annealing temperature after bonding was limited to 100°C or

200°C, because for higher temperatures the samples broke. In SAB bonding, covalent bonds are formed at room-temperature, which is convenient since it eliminates the need for annealing processes. Afterwards, the GaAs substrate is removed by wet chemical etching, leaving the few-micrometers-thick III-V active layers on the silicon bottom cell. In the case of GaAs/GaAs bonding, a 300°C annealing could be performed at this stage with reduced stress compared with the full structure with GaAs substrate. The resulting wafer-bonded GaInP/AlGaAs/Si 3J structures are presented in Fig 1, with process variations summarized in Table II.

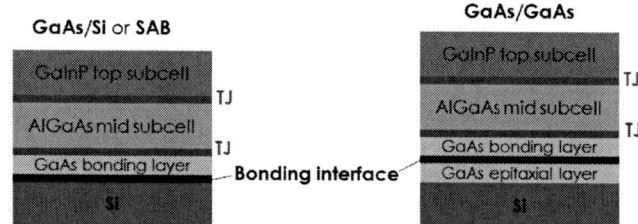

Fig. 1. 3J GaInP/AlGaAs/Si structures produced by different bonding approaches. Tunnel junctions (TJ) are represented.

Cell processing. Finally, front metal contacts designed for low concentration (10-20 suns, ~5% shading factor without bus bars [14]) and full sheet back metal contacts are deposited by evaporation. Mesa etching (trenches down to 2 μm inside Si) is performed to isolate individual cells of 1 and 2 cm², and a 65 nm Si$_3$N$_4$ Anti-Reflection Coating (ARC) is deposited by Plasma-Enhanced Chemical Vapor Deposition (PECVD).

TABLE II: Process variations.

Bonding method	Si finish Surface treatment	N$_2$ T°C	N$_2$ T°C after GaAs substrate removal.
GaAs/Si (A)	UV/O$_3$	200	-
GaAs/Si (B)	NH4OH scrub	100	-
GaAs/GaAs	-	200	300
SAB	NH4OH scrub + Ar irradiation	-	-

III. RESULTS AND DISCUSSION

Interface. The bond interface was analyzed by cross-sectional Transmission Electron Microscopy (TEM), as shown in Fig. 2. Bonding interfaces thinner than 5 nm were observed. Energy-dispersive X-ray spectroscopy (EDX) analysis revealed the presence of oxygen and sulfur at the GaAs/GaAs interface, as expected with the use of sulfur oxidizing surface treatments. Ar was detected at the interface of SAB-bonded device, as a result of Ar beam irradiation.

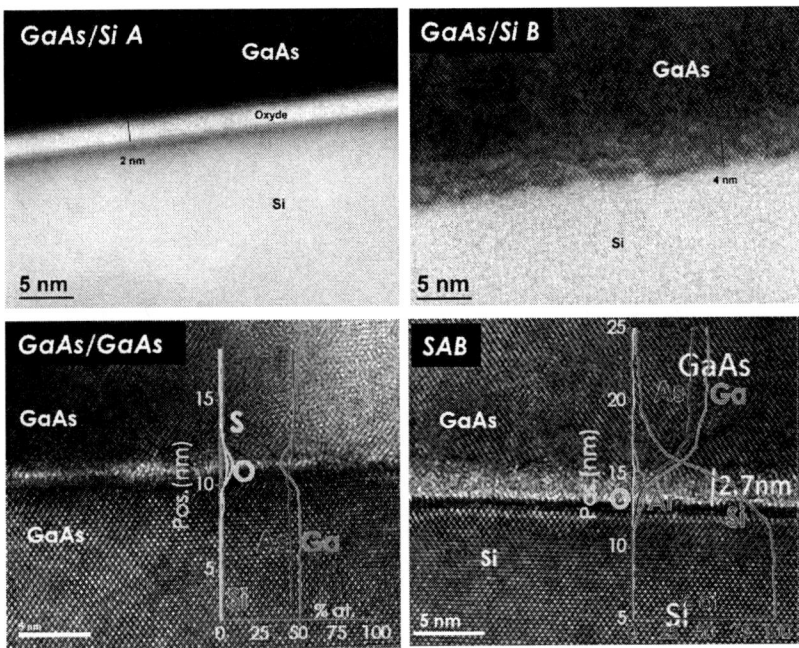

Fig. 2. TEM images of bonding interfaces. The EDX analysis was carried out on STEM images but EDX profiles were included in this Figure for clarity purposes.

The post-bonding quality is ensured through Scanning Acoustic Microscopy (SAM), employing ultrasound. This characterization technique has sub-mm resolution and hence allows to clearly show the areas where the bonding is complete (black) and the ones with voids (white).

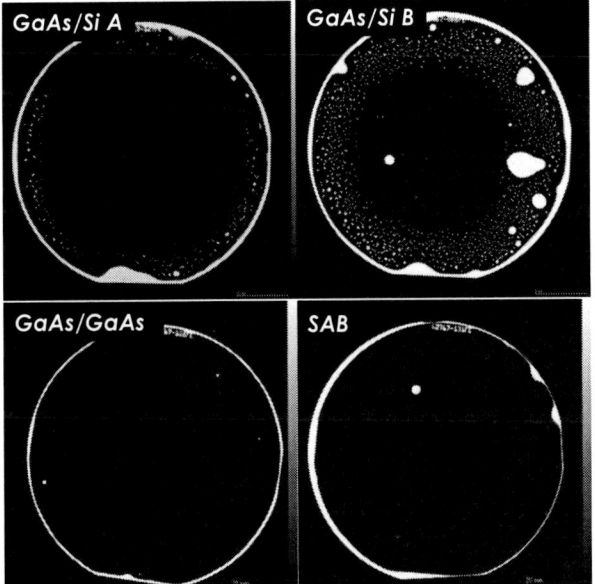

Fig. 3. SAM images of wafer-bonded 3J devices showing voids in white where bonding has not been successful. The voids may be caused by particle contamination on the surface of the wafer or by release of H_2 through diffusion of water trapped at the interface and

subsequent oxidation of Si with temperature [13]. All wafer-pairs are 100 mm diameter.

GaAs/Si interfaces bonded using surface finish B (NH_4OH scrub) are more defective than GaAs/Si interfaces with surface finish A (UV/O_3 treatment), as shown in Fig. 3, demonstrating the importance of wafer surface preparation. The probable origin of the defects in these hydrophilic GaAs/Si bondings is water. As the wafers are hydrophilic, water is adsorbed at the surface of the contacting wafers and thus trapped at the interface during bonding. This water plays a positive role at room temperature to enhance bonding thanks to hydrogen bonds but a deleterious one during the annealing phase as it may react with silicon, producing additional silicon oxide and gaseous hydrogen, therefore voids at the bond interface [12]. More voids are observed at GaAs/Si B interface, which is in agreement with larger production of additional silicon oxide and gaseous hydrogen at the interface, and with the observed thicker oxide layer, as observed by TEM (Fig. 2). SAB and GaAs/GaAs cells exhibit lower interface defectivity compared to GaAs/Si cells, as shown in Fig. 3. SAB bonding of GaAs/Si wafers has already been reported near defect-free in literature [15]. GaAs/GaAs bonding also provides a near defect-free interface (Fig. 3). This is a novel and original approach which shows a great potential, opening a new path for III-V on Si processing. In addition, cross-sectional TEM image of GaAs/GaAs device (Fig. 4) showed that heteroepitaxy induced defects (anti-phase domains, threading dislocations) were confined to the bonding layer and did not propagate in the III-V cells.

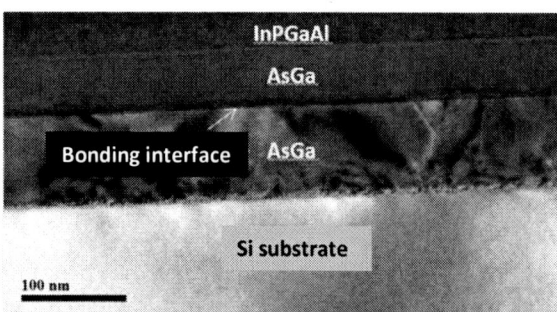

Fig. 4. TEM image of GaAs/GaAs bonding interface

Opto-electronic properties. I-V curves of the 3J devices were measured using a SpectraNova AM1.5g AAA solar simulator, as shown in Fig 5.

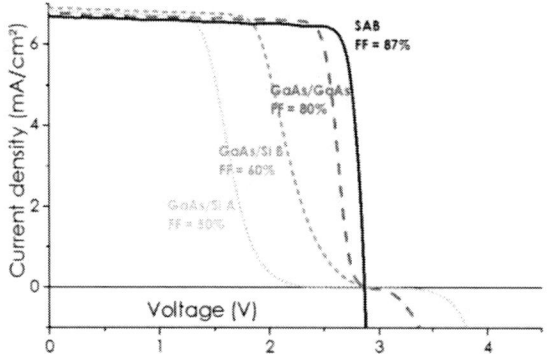

Fig. 5. I-V curves of GaInP/AlGaAs/Si devices with fill factors (FF). Process variations detailed in Table II.

The J_{SC} as measured in Fig. 5 is not significant because of spectral mismatch. Associated EQE measurements, shown in Fig. 6, revealed J_{SC} around 8 mA/cm² for the Si subcell, and J_{SC} higher than 10 mA/cm² for AlGaAs and GaInP subcells, which is much higher than ~ 6.5 mA/cm² measured by I-V in Fig. 5. In 2 terminal MJSC, sub-cells are connected in series and the lowest sub-cell current drives the total current of the MJSC. As a consequence, a small variation in the spectrum may lead to an underestimation or overestimation of one sub-cell current and therefore of the device J_{SC}. Ongoing work with flash solar simulator with component reference cells (isotypes) [16] will lead to more accurate performance evaluation.

Fill factor may also be impacted by spectral mismatch [17], but Fig. 5 gives a good qualitative estimation of the bond interface conductivity. GaAs/Si and GaAs/GaAs bonded cells show distorted I-V curves, with presence of a "S-kink", which may due to a potential barrier at the GaAs/Si or GaAs/GaAs

interface. By contrast, the I-V curve of SAB-bonded device did not show distortions, in accordance with SAB-bonded devices previously reported in literature [3], [5], which suggests that the S-kink most probable cause is the native oxide interlayers (removed in the case of SAB). Concerning direct GaAs/Si bondings, surface finish wet chemical treatments which would leave a very thin oxide layer at the Si surface are under development. Interestingly, GaAs/GaAs bonded devices exhibited superior performance, when compared to GaAs/Si bonded devices, as predicted by the lower defectivity at the interface (SAM analysis in Fig. 3), confirming the potential of this novel bonding approach. In addition, Si cell performance do not seem to be degraded by the epitaxy of GaAs, as shown by the EQE in Fig. 6.

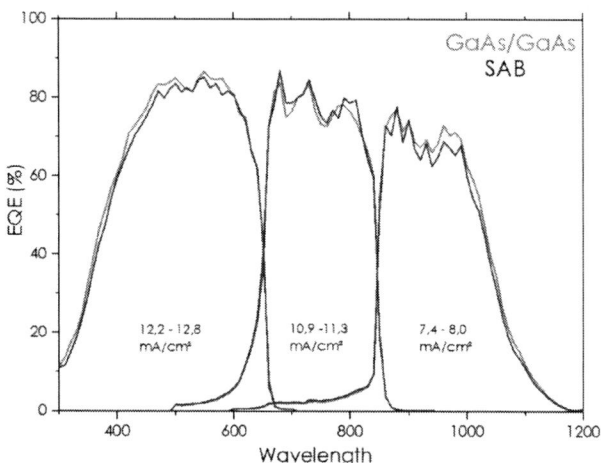

Fig. 6. External Quantum Efficiency (EQE) of GaInP/AlGaAs/Si devices. The calculated subcell short-circuit current densities (J_{SC} in mA/cm²) under AM1.5G at 1000 W/m² are shown for GaAs/GaAs and SAB devices. The extracted bandgaps from the EQE curves are 1.90 eV for the top sub-cell and 1.46 eV for the middle sub-cell.

Open-circuit voltages around 2.9V are reached, similarly to reported 1-sun 26.2% efficient device (see Table I).

Device optimization. The photocurrent mismatch between the top and the bottom sub-cells of the 3J devices is quite important (5 mA/cm² in the case of SAB device). Therefore, the J_{SC} of bottom Si sub-cell severely limits the performance of the 3J device. Implementation of a Double Layer Anti-Reflective Coating (DLARC) SiO_2/Si_3N_4 led to a 10% increase in the Si bottom subcell J_{SC}, as shown in Fig 7. Further 3J performance increase is expected from ongoing Si subcell optimization (minority carrier lifetime, doping profiles and passivation).

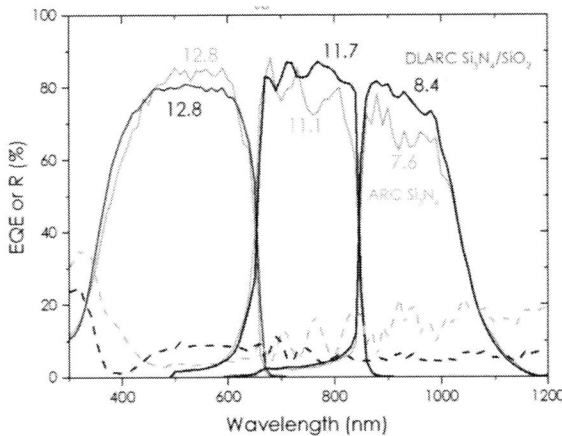

Fig. 7. EQE of SAB 3J solar cell with single layer ARC (green) and double layer ARC (dark blue). The calculated subcell short-circuit current densities (J_{SC} in mA/cm²) under AM1.5G at 1000 W/m² are shown for both devices. Respective reflection curves are shown in dashed lines.

IV. CONCLUSIONS AND FUTURE WORK

Different bonding approaches have been evaluated for the fabrication of III-V on Si solar cells. Best results were obtained using ultra-high vacuum SAB which gives low defective GaAs/Si bonding interfaces. Novel GaAs/GaAs bonding method also showed a great potential, opening a new path for III-V on Si processing.

V. ACKNOWLEDGEMENTS

The authors wish to acknowledge Anne-Marie Papon and SERMA Technologies for the TEM and EDX analysis, Pierre Mur, Marc Plissonnier and Hervé Ribot for their support.

REFERENCES

[1] J. S. Ward, T. Remo, K. Horowitz, M. Woodhouse, B. Sopori, K. VanSant, and P. Basore, "Techno-economic analysis of three different substrate removal and reuse strategies for III-V solar cells:," *Prog. Photovolt. Res. Appl.*, vol. 24, no. 9, pp. 1284–1292, 2016.

[2] J. Yang and R. Kleiman, "Optimization of bonded III-V on Si multi-junction solar cells," in *2013 IEEE 39th Photovoltaic Specialists Conference (PVSC)*, 2013, pp. 2151–2153.

[3] R. Cariou, J. Benick, P. Beutel, N. Razek, C. Flotgen, M. Hermle, D. Lackner, S. W. Glunz, A. W. Bett, M. Wimplinger, and F. Dimroth, "Monolithic Two-Terminal III–V//Si Triple-Junction Solar Cells With 30.2% Efficiency Under 1-Sun AM1.5g," *IEEE J. Photovolt.*, vol. 7, no. 1, pp. 367 – 373, 2017.

[4] S. Essig, J. Benick, M. Schachtner, A. Wekkeli, M. Hermle, and F. Dimroth, "Wafer-Bonded GaInP/GaAs//Si Solar Cells With 30% Efficiency Under Concentrated Sunlight," *IEEE J. Photovolt.*, vol. 5, no. 3, pp. 977–981, May 2015.

[5] N. Shigekawa, J. Liang, R. Onitsuka, T. Agui, H. Juso, and T. Takamoto, "Current–voltage and spectral-response characteristics of surface-activated-bonding-based InGaP/GaAs/Si hybrid triple-junction cells," *Jpn. J. Appl. Phys.*, vol. 54, no. 8S1, p. 08KE03, Aug. 2015.

[6] J. Yang, Z. Peng, D. Cheong, and R. Kleiman, "Fabrication of High-Efficiency III-V on Silicon Multijunction Solar Cells by Direct Metal Interconnect," *IEEE J. Photovolt.*, vol. 4, no. 4, pp. 1149–1155, Jul. 2014.

[7] J. Yang, Z. Peng, D. Cheong, and R. Kleiman, "III-V ON SILICON MULTI-JUNCTION SOLAR CELL WITH 25% 1-SUN EFFICIENCY VIA DIRECT METAL INTERCONNECT AND AREAL CURRENT MATCHING," in *EU PVSEC Proceedings*, 2012, pp. 160–163.

[8] H. Mizuno, K. Makita, T. Tayagaki, T. Mochizuki, H. Takato, T. Sugaya, H. Mehrvarz, M. Green, and A. Ho-Baillie, "A 'Smart Stack' Triple-Junction Cell Consisting of InGaP/GaAs and Crystalline Si," in *Photovoltaic Specialists Conference (PVSC), 2016 IEEE 43rd*, 2016.

[9] T. J. Grassman, D. J. Chmielewski, J. A. Carlin, and S. A. Ringel, "Development of Epitaxial 2- and 3-Junction III-V/Si Solar Cells," in *Photovoltaic Specialists Conference (PVSC), 2016 IEEE 43rd*, 2016.

[10] F. Dimroth, M. Grave, P. Beutel, U. Fiedeler, C. Karcher, T. N. D. Tibbits, E. Oliva, G. Siefer, M. Schachtner, A. Wekkeli, A. W. Bett, R. Krause, M. Piccin, N. Blanc, C. Drazek, E. Guiot, B. Ghyselen, T. Salvetat, A. Tauzin, T. Signamarcheix, A. Dobrich, T. Hannappel, and K. Schwarzburg, "Wafer bonded four-junction GaInP/GaAs//GaInAsP/GaInAs concentrator solar cells with 44.7% efficiency: Wafer bonded four-junction concentrator solar cells with 44.7% efficiency," *Prog. Photovolt. Res. Appl.*, vol. 22, no. 3, pp. 277–282, Mar. 2014.

[11] K. Nakayama, K. Tanabe, and H. A. Atwater, "Improved electrical properties of wafer-bonded p-GaAs/n-InP interfaces with sulfide passivation," *J. Appl. Phys.*, vol. 103, no. 9, p. 094503, 2008.

[12] H. Moriceau, F. Rieutord, F. Fournel, Y. Le Tiec, L. Di Cioccio, C. Morales, A. M. Charvet, and C. Deguet, "Overview of recent direct wafer bonding advances and applications," *Adv. Nat. Sci. Nanosci. Nanotechnol.*, vol. 1, no. 4, p. 043004, Feb. 2011.

[13] C. Ventosa, F. Rieutord, L. Libralesso, C. Morales, F. Fournel, and H. Moriceau, "Hydrophilic low-temperature direct wafer bonding," *J. Appl. Phys.*, vol. 104, no. 12, p. 123524, Dec. 2008.

[14] E. Veinberg-Vidal, C. Dupré, P. Garcia-Linares, C. Jany, R. Thibon, T. Card, T. Salvetat, P. Scheiblin, C. Brughera, F. Fournel, Y. Desieres, Y. Veschetti, V. Sanzone, P. Mur, J. Decobert, and A. Datas, "Manufacturing and Characterization of III-V on Silicon Multijunction Solar Cells," *Energy Procedia*, vol. 92, pp. 242–247, Aug. 2016.

[15] S. Essig, O. Moutanabbir, A. Wekkeli, H. Nahme, E. Oliva, A. W. Bett, and F. Dimroth, "Fast atom beam-activated n-Si/n-GaAs wafer bonding with high interfacial transparency and electrical conductivity," *J. Appl. Phys.*, vol. 113, no. 20, p. 203512, 2013.

[16] C. Domínguez, I. Antón, G. Sala, and S. Askins, "Current-matching estimation for multijunction cells within a CPV module by means of component cells," *Prog. Photovolt. Res. Appl.*, vol. 21, no. 7, pp. 1478–1488, Nov. 2013.

[17] G. Siefer, C. Baur, M. Meusel, F. Dimroth, A. W. Bett, and W. Warta, "Influence of the simulator spectrum on the calibration of multi-junction solar cells under concentration," in *Photovoltaic Specialists Conference, 2002. Conference Record of the Twenty-Ninth IEEE*, 2002.

Design Arithmetic of the Lateral III-V / Si Hybrid Module

Kenji Araki[1], Kyotaro Nakamura[2], Kan-Hua Lee[1], Takefumi Kamioka[1], Yu-Cian Wang[1], Nobuaki Kojima[1], Yoshio Ohshita[1], Masafumi Yamaguchi[1]

I1. Toyota Technological Institute, Nagoya, 468-8611 Japan,

2. Meiji University, Kawasaki, 214-8571 Japan

Abstract — **By stacking a III-V cell to the Si cell, it is expected to reach to more than 30 % of efficiency. However, the contribution of the Si bottom cell may be reduced to 5 to 6 % in maximum. Another problem is that the III-V cell cost is still expensive and high concentration without relying on tracking is expected. For solving these two problems, a new configuration that places small III-V multi-junction solar cells at four corners of a single-crystalline Si solar cell was invented.**

Index Terms — **Crystalline Si cell, III-V cell, Multi-junction cell, Mechanical stack cell, optics**

I. INTRODUCTION

To compete to the flat-plate crystalline Si cell module, various III-V cell technologies are developed. One of the candidates is III-V on Si structure. It is possible to improve the efficiency with more than 30 % without relying on concentration [1-2].

Because this is a combined technology, there may be wide variety of the design. Obviously, the common configuration is the tandem structure [1-2]. However, in this paper, we would like to propose different configuration, namely placing III-V and Si cell in lateral rather than stacking in vertical. A static concentrator technology is used to enhance the contribution from highly efficient III-V cells. Since its configuration is exotic, modern design arithmetic was needed to be developed. The new method is discussed in this paper.

II. WHAT IS A LATERAL HYBRID MODULE?

The Si bottom cell is filtered and its current output is significantly reduced. The measurement result of the InGaP/GaAs//Si mechanical stack cell suggested that the filtered current from Si cell was only 20 % and the gain to the cell efficiency was only 3.4 % (Fig. 1) [3].

Even though optical interface is improved, the best current ratio of the bottom Si cell will be 26 % weighted by AM1.5G spectrum and corresponding efficiency contribution from the bottom Si cell will be 5.2 % (Fig. 2).

It is true that the contribution from the Si bottom cell will be enhanced by the improvement of the response from longer wavelength region, however, it will be difficult to expect more than 6 % of the efficiency gain even from the ideal Si cell as far as we calculated (Fig. 3).

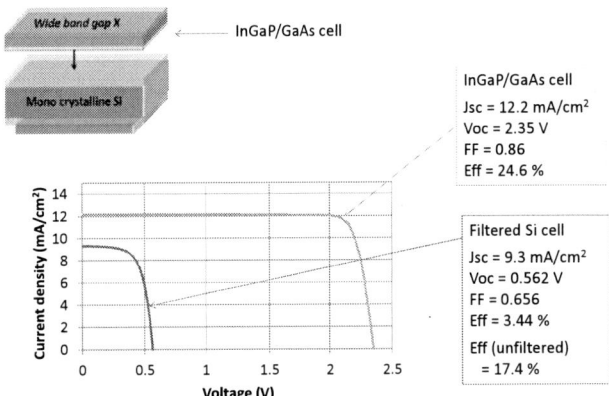

Fig. 1.   Measurement result of the III-V/Si tandem cell [3].

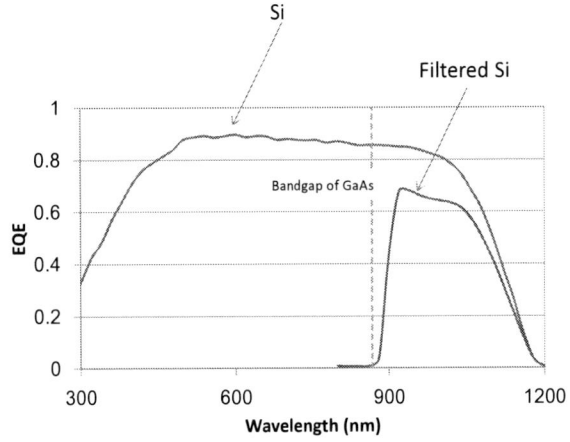

Fig. 2.   External quantum efficiency of the bottom Si cell [3].

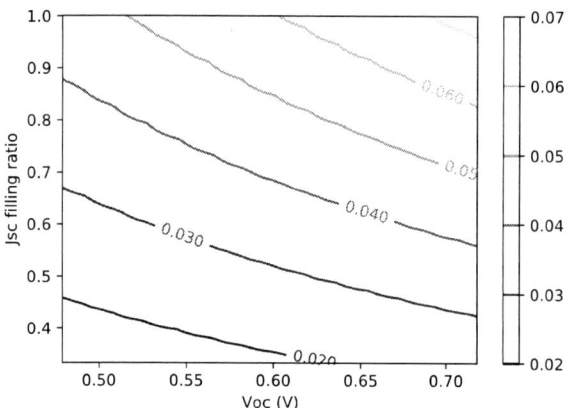

Fig. 3. Calculated contour plot of the gain by the bottom Si cell as a function of the Voc and Jsc filling ratio.

Such decline of the Si cell performance came from the filtered illumination. One possible idea is to place the III-V cell and Si cell laterally so that Si cell will not be filtered by III-V cell. We call it III-V/Si lateral hybrid. Because the current III-V cell is expensive, it may be wise to enhance it by static concentrator. Typical commercial crystalline Si cell was cut from the 200-mm diameter ingot to 156 mm cornered square shape. Considering 3 mm distance among electrically active cells, 19 mm square III-V multi-junction cells may be placed to the corners of 156 mm square Si cells (Fig. 4).

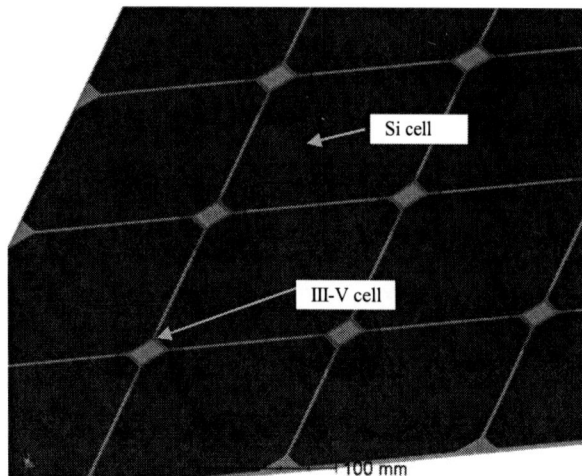

Fig. 4. Illustration of the III-V/Si lateral hybrid cell from the backside view. Optics is in the front side and it does not appear. It is discussed in the successive section.

Since the ratio of the III-V cell is only 1.5 % of the area of the Si cell, it is not sufficient to take the meaningful gain. Static concentrator on the III-V cells through 3 mm thick cover glass may be helpful.

III. DESIGN ARITHMETIC ON OPTICS

A. What is the difference from the conventional optics design?

The optics for the lateral hybrid is different in the following four points.

- Optical efficiency is not important
- Necessity of extremely wide acceptance angle
- Design by the synthetic light source rather than collimated sunlight.
- Consideration of shading from adjacent lenses

The first point looks strange. It is related to the two facts. One is that the ray that is not absorbed by the target of the III-V cell can be absorbed by the Si cells. Another is that the aperture of the optics is the entire area of the module. In other words, the geometrical concentration ratio in a narrow sense becomes extremely high.

All four points are out of the conditions that the conventional design method of the CPV optics can handle. The innovative design arithmetic is needed to be developed. Because it is necessary to handle complicated structure, for example, consideration of the shading loss from the adjacent lens peak due to handling non-collimated light, construction of the design equations [4-5] or variation method [6], combination of the Monte Carlo ray tracing and numerical optimization was useful.

B. Design arithmetic based on optimization of profile function.

The target of the design arithmetic is to obtain the optimum shape of the solid lens on the III-V cell. It was expressed by the linear combination of the series of the orthogonal functions with both radial and azimuthal parameters like Equation (1).

$$Z(r,\theta) = l + c_1 f_1(r,\theta) + c_2 f_2(r,\theta) + \cdots \quad (1)$$

where, $Z(r,\theta)$ is the profile function, $f_i(r,\theta)$ is the orthogonal function system, and l is the working distance from the origin of the lens profile to the cell.

TABLE I INDEX OF THE ORTHOGONAL FUNCTION SYSTEM $f_i(r,\theta)$

		Azimuthal degree																								
		0		1θ		2θ		3θ		4θ		5θ		6θ		7θ		8θ		9θ		10θ		11θ		12θ
odd/even		-		o	e	o	e	o	e	o	e	o	e	o	e	o	e	o	e	o	e	o	e	o	e
Radial degree	0																									
	1																									
	2	1																								
	3																									
	4	2								3																
	5																									
	6	4								5																
	7																									
	8	6								7							8									
	9																									
	10	9								10							11									
	11																									
	12	12								13							14									15

The target of the optimization is to find the best set of the vector $(l, c_1, c_2, \cdots)$ that maximizes the flux onto the III-V cell. Considering the III-V cell is square and has 4 symmetry planes, it can be expressed by sampling the Zernike's polynomial according to Table 1, index of the orthogonal function system used in Equation (1) structured by the matrix of the radial degree and azimuthal degree. Note that the shaded cells were not assigned in the Noll's sequence of the Zernike's polynomial [7]. l may be replaced by c_0, but assignment to the different variable was convenient to give additional geometrical constraints for the practical design.

The number of parameters was important for numerical optimization. Both the calculation time and the instability of convergence rapidly increases with the number of parameters. It was effective to select the effective parameters by the consideration of geometrical symmetry. Specifically, the index of both radial and azimuthal degree was selected to satisfy the following identities.

$$Z(r,\theta) \equiv Z(r,-\theta) \qquad (2)$$

$$Z(r,\theta) \equiv Z(r,\theta \pm \pi/2) \qquad (3)$$

The extracted result is shown in Table 1 with the function filtered by the identified equation (2) and (3).

Note that the equation (1) is no longer axially symmetrical after taking the series of functions by Table 1, but has 4-folded symmetrical planes, reflecting the fact that the cell aperture is square and on the same center point. This type of asymmetrical lens was known effective to the static low concentrator with wider acceptance half angle like the car-roof CPV [8].

Considering symmetry in the module, it may be convenient to generate the subset model. The minimum subset considering the shading effect is 3 x 3 arrays of the lens-cell pairs. The collection of the point source, typically 1,000,000 points placing on the square region of the square plane was generated by connecting the peaks of the 8 outer lenses. The direction of the ray was random but weighted by the distribution of the annually integrated incident angle so that the optimization results should be the result of maximizing the annually-integrated. In this design, it was given rom 47 representative sites in Japan and 411,720 sets of data consist of direct irradiance, diffused irradiance and the sun position [9]. At first, the histogram was generated using a rectangular window function. Then, a continuous distribution function was generated using a linear interpolation. Finally, a set of the incident angles was generated at 1,000,000 points using a uniform distribution ranging from 0°to 90°, weighted by the above-mentioned continuous distribution function.

The numerical optimization was done in ZEMAX ray-tracing software and its optimized result was output by CAD file so that the lens manufacture can directly form the lens and other manufacturer can check the assembly design rule and interference to another component by CAD environment. Most

importantly, this method is robust and capable of designing wide range of design parameters without worrying about asking reconstruction of the optical and geometrical equations to optical engineers.

The maximum number of the term of the equation (1) is important in the trade-off between the calculation time and accuracy. We found that azimuthal degree was important and the necessary value was 8 and preferably 12, indicating axially highly-ordered asymmetrical profile was effective to boost the efficiency.

It is useful to add some geographical constraints. Because we do not care about the optical efficiency, it is effective to increase the size of the unit lens for increasing the total flux to the III-V cells. Obviously, it is not practical in view of the material volume, weight that affects to structure cost and foundation cost, and thickness that affects to storage and transportation cost. Such constraints were included in the evaluation function in the optimization calculation.

IV. RESULTS – CALCULATION EXAMPLES

The lateral configuration has wide variety of the optimum solutions depending on initial values and geometrical constraints. Fig. 5 shows examples of the optimized lens profile solutions. In this calculation, Si cell was 156 mm x 156 mm and 20 % efficiency, III-V cell was 19 mm x 19 mm and 32 % efficiency. These cells were placed with 3 mm distance and were connected as the 4-terminal circuit. The entire cells were covered by 3 mm glass and additional bump structure was given coaxially on the center of the III-V cells. The lens bump was axially-asymmetrical optimized by Equation (1). The spectrum mismatching effect to III-V cells influenced by air mass fluctuation was considered in this calculation. In Fig. 5, the middle row corresponds to the annually averaged total flux density onto the III-V cell by the optimized sloped angle in the southern direction in Japan. The bottom row corresponds to the expected annually averaged module efficiency. Note that the module efficiency without lateral III-V cells and optics was 19.2 % using 20 % Si cells. In every case, the total flux of Si cell was more than 98 % and energy yield from Si cell was kept.

The module efficiency reached to 26.5 % by using 125 mm square (cut from 150 mm diameter Si ingot), 25 % Si cell, and 26 mm square, 35 % III-V cell, whereas the module efficiency without III-V cells was 22.7 %, namely 3.8 points of the gain thanks to the increased III-V and Si ratio to 4.5 % (Table 2).

Without geometrical constraints, both height and size of the lens bump tended to impractically big. However, even by the 40-mm bump height, the module efficiency gain was 1.5 points and considerably high.

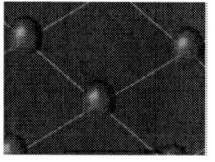 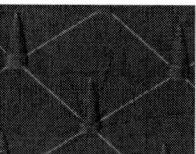 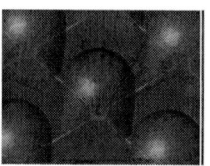

3.1 suns	3.8 suns	3.3 suns
20. 6 %	20.7 %	20.6 %

Fig. 5. Examples of optimized lens profile on III-V cells. The middle row corresponds to the annually averaged total flux density onto the III-V cell by the optimized sloped angle in the southern direction in Japan. The bottom row corresponds to the expected annually averaged module efficiency. Note that the module efficiency without lateral III-V cells and optics was 19.2 % using 20 % Si cells.

The module efficiency reached to 26.5 % by using 125 mm square (cut from 150 mm diameter Si ingot), 25 % Si cell, and 26 mm square, 35 % III-V cell, whereas the module efficiency without III-V cells was 22.7 %, namely 3.8 points of the gain thanks to the increased III-V and Si ratio to 4.5 %.

TABLE II DESIGN EXAMPLES TO 156-SQUARE SI CELL AND 120-SQUARE SI CELL

Si cell	III-V cell	Si Module efficiency	Lateral III-V/Si hybrid module efficiency
☐156 20%	☐19 32%	19.1%	20.6%
☐125 25%	☐26 35%	22.7%	26.6%

VI. CONCLUSION

III-V/Si cell has a potential of high efficiency but the contribution of Si cell dropped 5 to 6 % from 20 % efficiency. Alternatively, the lateral hybrid configuration kept the performance of Si cells.

The III-V cells were optically enhanced by the solid lens but the design requirements were completely different from conventional CPV. An innovative design arithmetic by numerically optimizing orthogonal function systems was developed.

The designed module efficiency was 20.6 to 20.7 % using 19.2 % Si module. The gain of the lateral hybrid was considerably high. The module efficiency reached to 26.5 % from 22.7 %, by using 125 mm square, 25 % Si cell, and 26 mm square, 35 % III-V cell, namely 3.8 points of the gain.

ACKNOWLEDGEMENT

This work has been partially supported by NEDO in Japan.

REFERENCES

[1] R. Cariou, J. Benick, M. Hermle, D. Lackner, S. Glunz, A. W. Bett, and F. Dimroth, "Development of Highly-Efficient III-V//Si Wafer-Bonded Triple-Junction Solar Cells", 43rd IEEE PVSC, 2016, Portland, USA

[2] S. Essig, C. Allebé, J. F. Geisz, M. A. Steiner, B. Paviet-Salomon, A. Descoeudres, A. Tamboli, L. Barraud, S. Ward, N. Badel, V. LaSalvia, J. Levrat, M. Despeisse, C. Ballif, P. Stradins, and D. L. Young, "Boosting the efficiency of III-V/Si tandem solar cells", 43rd IEEE PVSC, 2016, Portland, USA

[3] K. Araki et al., "Beyond the limit of Si solar cells – III-V on Si cell and its PCSC module concept", 26th PVSEC, 2016, Singapore.

[4] R. Leutz, A Suzuki, A. Akisawa, T. Kashiwagi, "Design of a Nonimaging Fresnel Lens for Solar Concentrators", Solar Energy, 65, 6, 199, pp. 379-388.

[5] K. Araki, H. Nagai, K. H. Lee, K. Ikeda, and M. Yamaguchi, "Design and Development of Dome-Shaped Fresnel Lens". IEEE Journal of Photovoltaics, 6(5), 2016, pp. 1339-1344.

[6] A. Cvetkovic, M. Hernandez, P. Benítez, J. C. Miñano, J. Schwartz, A. Plesniak, R. Jones, and D. Whelan, 2008. "The free form XR photovoltaic concentrator: a high performance SMS3D design," Proc. SPIE 7043, High and Low Concentration for Solar Electric Applications III, 70430E, 2008, September 09

[7] R. J. Noll, "Zernike polynomials and atmospheric turbulence". J Opt Soc Am 66, 1976, pp.207-211.

[8] T. Masuda, K. Araki, K. Okumura, S. Urabe, Y. Kudo, K. Kimura, ... & Yamaguchi, M. (2016, June). "Next environment-friendly cars: Application of solar power as automobile energy source". In Photovoltaic Specialists Conference (PVSC), 2016 IEEE 43rd (pp. 0580-0584). IEEE

[9] H. Itagaki, H. Okumura, and A. Yamada, "Preparation of meteorological data set throughout Japan for suitable design of PV systems", Proc. 3rd WCPEC, 2003, Volume: 2.

GaAsP Nanowire Solar Cell Development Towards Nanowire/Si Tandem Applications

Enrique Barrigon, Yang Chen, Gaute Otnes, Vilgaile Dagyte, Nicklas Anttu, Lars Samuelson, Magnus Borgström

Division of Solid State Physics and NanoLund, Lund University, Lund, Box 118, 22100, Sweden

Abstract — **III-V based nanowire solar cells are a promising candidate to be employed as the top cell in a III-V/Si tandem structure. Here, we report on the development of p/i/n GaAsP nanowire solar cells with the appropriate bandgap for such tandem structure. The performance of single NW devices is analyzed with current-voltage and electron beam induced current measurements.**

I. Introduction

The dominant photovoltaic (PV) technology is based on Si solar cells, with 1-sun world-record efficiencies higher than 26% for single junction Si cells [1]. However, Si PV-technology is considered to be mature and little efficiency improvement is expected at device level for a single junction cell. In order to increase the efficiency, a high energy bandgap subcell could be placed on top of the Si subcell creating a tandem cell. Such material for the top subcell could be made of a ternary III-V semiconductor such as GaInP or GaAsP. Creating such a tandem structure would mimic the approach followed by highly efficient III-V multijunction solar cells, but with the clear benefit of a reduced Si price. However, direct growth of III-V solar cells on Si is expensive and technologically difficult to achieve [2]. To circumvent direct growth, wafer bonding has been successfully employed with 1-sun efficiencies higher than 30% for a dual-junction cell [3].

Alternatively, the top solar cell could be made of III-V nanowires. The use of nanowires (NW) offers many advantages such as reduced material consumption and production cost [4], relatively high efficiencies [5] and even the possibility to grow them directly on Si, regardless of lattice mismatch [6]. NWs are typically grown in a well-defined array with appropriate dimensions to enhance light absorption due to resonant light trapping [7], yielding J_{sc} values similar to their planar counterparts. Current record efficiencies of single-junction III-V nanowire-based solar cells are 13.8% [8] and 15.3% [9] for epitaxially grown InP and GaAs nanowire arrays, respectively. A tandem cell consisting of a GaAs nanowire-array subcell grown epitaxially on top of a silicon subcell with an efficiency of 11.4% has been already demonstrated [10], but with suboptimal bandgap for the top cell.

In this line, this work deals with the development of GaAsP NW-based solar cell to be employed in a GaAsP/Si tandem solar cell. We first theoretically estimate the appropriate energy bandgap for the NWs given the experimental results achieved so far for NW solar cells, together with the array dimensions that optimize NW absorption to lead the way to a current matched tandem structure. Subsequently, we optimize the growth of GaAsP p/i/n NW structures and analyze the performance in terms of electron beam induced current (EBIC) and current-voltage (I-V) characteristics measured in single NW devices.

II. Experimental

$GaAs_{(1-x)}P_x$ NW arrays were grown on p-type GaAs (111)B substrates in a metal-organic vapor phase epitaxy (MOVPE) reactor with the vapour-liquid-solid (VLS) method. Reactor pressure was 100 mbar and total flow employed was 6 l/min. A hexagonal pattern of Au catalyst particles with a pitch of 500 nm was defined on the surface by nano imprint lithography (NIL) and a 70 nm thick SiNx growth mask was included for pattern preservation [11]. Trimethylgallium (TMGa), arsine (AsH_3) and phosphine (PH_3) were used for the NW growth at 450°C. The molar fraction of TMGa and AsH_3 employed was 1.60×10^{-5} and 1.67×10^{-3}, respectively. The PH_3 to AsH_3 ratio was varied from 0 to 30 (by keeping AsH_3 constant and using different PH_3 molar fractions) to control different $GaAs_{(1-x)}P_x$ composition values. Hydrogen bromide (HBr) was used to avoid radial overgrowth [12]. Diethylzinc (DEZn) and tetraethyltin (TESn) at a dopant/III ratio of 0.08 and 1.90, respectively, were used for p and n doping. For more general details on the NW growth, NIL and the benefits of using a SiNx mask, the reader is referred to [11].

High resolution X-ray diffraction (HRXRD) measurements were performed on a Bruker D8 system and SEM images were obtained in a Zeiss LEO 1560 SEM operated at 5 kV. EBIC measurements were performed in a Hitachi SU8010 SEM at room temperature operated at 5 kV and with a probe current in the tens of pA range, ensuring full excitation volume within the NWs and low excitation conditions. For the EBIC measurements, samples were cleaved to access NWs in the center of the sample. Samples were glued with silver paste to a metallic stub (back contact) and the Au particle on top of the NWs was contacted with a tungsten nanoprobe (top contact), controlled by a nano-probe system from Kleindiek Nanotechnik.

978-1-5090-5606-4/17 $31.00 © 2017 IEEE

III. ELECTROMAGNETIC MODELLING

A series-connected, 2-terminal NW/Si tandem solar cell implies some extra requirements of the NW material, NW array geometry (pitch) and NW dimensions (diameter and length) to produce a current matched device with the highest current possible [13]. To determine such requirements, we performed electromagnetic modelling of the optical response of a NW/Si tandem solar cell, including the full three-dimensional geometry of the NWs together with the necessary processing layers around them for large area devices (>1 mm^2). Such approach has already given good results in the past [8] and we use it here to determine the ideal characteristics of the NW array to optimize light absorption, to later try to create it as close as experimentally possible.

According to our modelling results, we firstly estimate that a bandgap of 1.64 eV gives a good current matching under the

TABLE I
SUMMARY OF OPTIMUM NW ARRAY AND DIMENSIONS

L (nm)	D (nm)	P (nm)	j$_{NWs}$(mA/cm^2)	j$_{Si}$(mA/cm^2)
2000	150	210	21.4	19.9
3000	150	230	22.1	19.2

AM1.5G spectrum, taking into account measured external quantum efficiency of NW solar cells and silicon cells [8], [9]. This value is slightly lower than the one obtained by Shockley-Queisser detailed balance analysis [14], which yields an optimum bandgap of 1.74 eV for the NWs assuming perfect absorption of above bandgap photons in both the NW array subcell and in the Si subcell.

With respect to the NW array, the optimized dimension values from the absorption standpoint are summarized in Table I for two typical values for the length of NWs, together with an equivalent short-circuit current density that takes into account only absorption, reflection and transmission in each subcell. For a bandgap of 1.64 eV, the ideal diameter is 150 nm with a pitch in the order of 200 nm. Putting these values into practice, it would be possible to grow NWs with a diameter close to 150 nm. However, a pitch of around 200 nm seems to be impractical for a NIL method. Alternatively, a process based on peel-off and NW re-arrangement with the desired pitch should be realized. Currently, we decided to proceed with the development of GaAsP NW solar cells with our standard pitch of 500 nm.

IV. RESULTS

A. GaAsP composition optimization

As seen in section III, we should aim to develop a GaAs$_{(1-x)}$P$_x$ with an energy bandgap of 1.64 eV. Table II summarizes the set of samples that have been employed to calibrate GaAsP NW material. The table contains the PH$_3$/AsH$_3$ ratio employed

to grow each sample, together with the difference found between the GaAs peak position of the substrate and the peak arising from the GaAsP NWs in HRXRD. The P composition varies between 7 and 31%, which covers the range between 1.50 and 1.80 eV according to Ref.[15].

TABLE II
SUMMARY OF THE SAMPLES GROWN

PH$_3$/AsH$_3$	Δθ (deg)	P(%)	E$_g$(eV)
5	0.0776	7	1.51
8.5	0.1282	12	1.57
10	0.1419	14	1.59
14	0.1946	19	1.65
20	0.2514	24	1.72
30	0.3195	31	1.80

B. GaAsP p/i/n growth and characterization

Fig.1 shows a SEM image (30° tilt) of a p/i/n GaAs$_{0.81}$P$_{0.19}$ NW array. The wires in this sample show a length of around 2.8 µm and a diameter of 160 nm. Generally speaking, there is a good pattern preservation of the stamp used in NIL, although the presence of some long, kinked wires (around 40 in this image, showing around 3000 wires in total) are evident. Strategies to avoid the growth of such kinked wires are underway. Inset of Fig.1 shows a zoomed in image of the sample, where the presence of the Au particle on top of the wires can be observed.

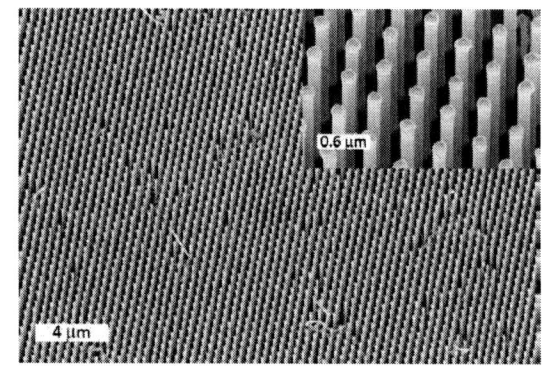

Fig. 1. SEM image of a GaAs$_{0.81}$P$_{0.19}$ NW array sample. Inset shows zoomed in image of the sample. A tilt of 30° has been employed during the measurements.

Fig.2(a) shows the SEM and EBIC response of a representative p/i/n GaAs$_{0.81}$P$_{0.19}$ wire. The Gaussian-like EBIC profile peaks at the interface between the nominally undoped and n-segment of the wire, indicating that the nominally undoped segment of the wire is actually p-type. An averaged effective diffusion length (L*) of minority carriers of

56±4 nm and 68±9 nm for the n- and p-segments, respectively, can be extracted by fitting the tails of the EBIC signal of several wires to a single exponential decay function [16]. The value obtained in the p-segment is similar to values reported for unpassivated p/n GaAs NWs with the same diameter. The reason for the lower value in the n-segment is not clear yet, but may be attributed to the presence of twin-planes typically observed when using high molar fractions of TESn, or a higher doping level achieved.

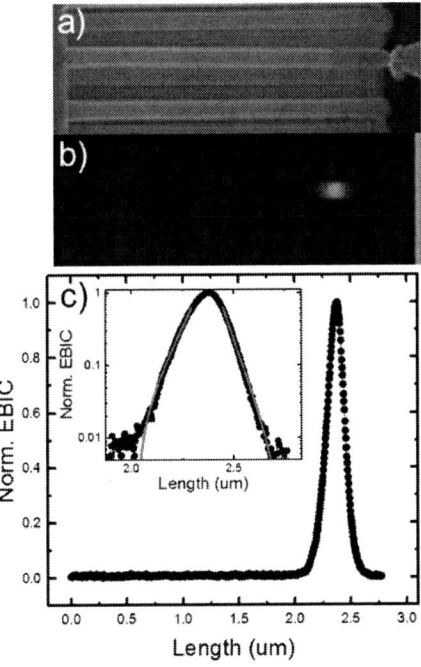

Fig. 2. SEM (a) and EBIC (b) of a single p/i/n $GaAs_{0.81}P_{0.19}$ NW solar cell. The EBIC line profile is plotted in (c), with an inset showing the extraction of the effective diffusion length.

The nano-probe system also allows for I-V measurements of single NWs. Fig. 3 shows the dark I-V characteristics plotted in semi-log scale of a representative wire of the same sample. It shows a diode characteristic with an ideality factor close to 2, which is generally observed in NW based solar cells. Rectification ratios of up to $\approx 10^5$ at ±1.5 V are achieved, indicating a good performing device.

In order to get an insight into the photogeneration properties of the wires, we repeated the I-V sweep with the e-gun on (25k magnification). In this way, instead of photogenerated carriers, we induce electron-generated carriers which allow to identify if the p/i/n diode in the NW shows well behaved I-V characteristics. Such characterization does not require time-consuming processing of the array into a full device. Indeed, from the semilog plot of absolute value of current vs voltage in Fig.3, we can confirm the presence of some I_{sc} and V_{oc} values.

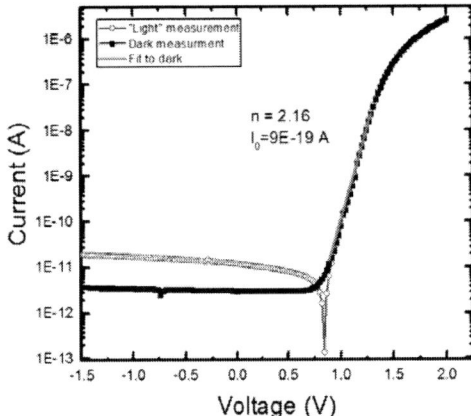

Fig. 3. Dark and e-light I-V characteristics of a single p/i/n $GaAs_{0.81}P_{0.19}$ NW

Further development of NW solar cells implies surface passivation, due to their high surface-to-volume ratio [9], [17]. In this way, surface recombination is reduced and higher V_{oc} values can be achieved. On the other hand, surface passivation may also increase I_{sc} [18]. Therefore, future passivation schemes such as epitaxial growth of high bandgap material [17], epitaxial growth of a thin layer of non-lattice matched materials such as GaP or InP [19] or wet chemical passivation with sulfides [16] will be tested.

IV. SUMMARY AND CONCLUSIONS

We have developed GaAsP NW solar cells with appropriate bandgap for NW/Si tandem structure. The wires show a diode-like I-V characteristic with an ideality factor of 2.16 and a Gaussian-like EBIC signal centered at the interface between the nominally undoped and the n-segment.

ACKNOWLEDGEMENTS

This work was performed within the NanoLund at Lund University, supported by the Swedish Research Council, the Swedish Energy Agency and the European Union's Horizon 2020 research and innovation programme under the Marie Sklodowska – Curie grant agreement No 656208 and under grant agreement No 641023 (NanoTandem). This article reflects only the author's view and the Funding Agency is not responsible for any use that may be made of the information it contains. Myfab – the Swedish research infrastructure for micro- and nanofabrication (http://www.myfab.se/) is also acknowledged.

REFERENCES

[1] K. Yoshikawa *et al.*, "Silicon heterojunction solar cell with interdigitated back contacts for a photoconversion efficiency over 26%," *Nat. Energy*, vol. 2, p. 17032, Mar. 2017.

[2] T. J. Grassman, J. A. Carlin, C. Ratcliff, D. J. Chmielewski, and S. A. Ringel, "Epitaxially-grown metamorphic GaAsP/Si dual-junction solar cells," in *2013 IEEE 39th Photovoltaic Specialists Conference (PVSC)*, 2013, pp. 0149–0153.

[3] R. Cariou *et al.*, "Monolithic Two-Terminal III-V//Si Triple-Junction Solar Cells With 30.2% Efficiency Under 1-Sun AM1.5g," *IEEE J. Photovolt.*, vol. 7, no. 1, pp. 367–373, Jan. 2017.

[4] M. Heurlin *et al.*, "Continuous gas-phase synthesis of nanowires with tunable properties," *Nature*, vol. 492, no. 7427, pp. 90–94, Nov. 2012.

[5] G. Otnes and M. T. Borgström, "Towards high efficiency nanowire solar cells," *Nano Today*, vol. 12, pp. 31–45, Feb. 2017.

[6] M. Borg *et al.*, "Vertical III–V Nanowire Device Integration on Si(100)," *Nano Lett.*, vol. 14, no. 4, pp. 1914–1920, Apr. 2014.

[7] N. Anttu and H. Q. Xu, "Efficient light management in vertical nanowire arrays for photovoltaics," *Opt. Express*, vol. 21, no. S3, p. A558, May 2013.

[8] J. Wallentin *et al.*, "InP Nanowire Array Solar Cells Achieving 13.8% Efficiency by Exceeding the Ray Optics Limit," *Science*, vol. 339, no. 6123, pp. 1057–1060, Mar. 2013.

[9] I. Aberg *et al.*, "A GaAs Nanowire Array Solar Cell With 15.3% Efficiency at 1 Sun," *IEEE J. Photovolt.*, vol. PP, no. 99, pp. 1–6, 2015.

[10] M. Yao *et al.*, "Tandem Solar Cells Using GaAs Nanowires on Si: Design, Fabrication, and Observation of Voltage Addition," *Nano Lett.*, vol. 15, no. 11, pp. 7217–7224, Nov. 2015.

[11] G. Otnes *et al.*, "Strategies to obtain pattern fidelity in nanowire growth from large-area surfaces patterned using nanoimprint lithography," *Nano Res.*, vol. 9, no. 10, pp. 2852–2861, Oct. 2016.

[12] A. Berg, K. Mergenthaler, M. Ek, M.-E. Pistol, L. R. Wallenberg, and M. T. Borgström, "In situ etching for control over axial and radial III-V nanowire growth rates using HBr," *Nanotechnology*, vol. 25, no. 50, p. 505601, 2014.

[13] N. Anttu, V. Dagytė, X. Zeng, G. Otnes, and M. Borgström, "Absorption and transmission of light in III–V nanowire arrays for tandem solar cell applications," *Nanotechnology*, vol. 28, no. 20, p. 205203, 2017.

[14] "Detailed Balance Limit of Efficiency of p-n Junction Solar Cells," *J. Appl. Phys.*, vol. 32, no. 3, pp. 510–519, Mar. 1961.

[15] V. Swaminathan and A. T. MacRander, *Material Aspects of Gaas and Inp Based Structures*. Englewood Cliffs, N.J: Prentice Hall PTR, 1991.

[16] C. Gutsche *et al.*, "Direct Determination of Minority Carrier Diffusion Lengths at Axial GaAs Nanowire p–n Junctions," *Nano Lett.*, vol. 12, no. 3, pp. 1453–1458, Mar. 2012.

[17] J. V. Holm, H. I. Jørgensen, P. Krogstrup, J. Nygård, H. Liu, and M. Aagesen, "Surface-passivated GaAsP single-nanowire solar cells exceeding 10% efficiency grown on silicon," *Nat. Commun.*, vol. 4, p. 1498, Feb. 2013.

[18] G. Mariani, A. C. Scofield, C.-H. Hung, and D. L. Huffaker, "GaAs nanopillar-array solar cells employing in situ surface passivation," *Nat. Commun.*, vol. 4, p. 1497, Feb. 2013.

[19] T. Haggren *et al.*, "Strong surface passivation of GaAs nanowires with ultrathin InP and GaP capping layers," *Appl. Phys. Lett.*, vol. 105, no. 3, p. 033114, Jul. 2014.

Demonstration of GaInP$_2$/Si Voltage Matched Tandem Solar Cells

David C. Bobela[1], Kenneth J. Schmieder[2], Matthew P. Lumb[2,3], James E. Moore[2], Robert J Walters[2], Eric A. Armour[4], Leo Matthew[5], Rajesh Rao[5], Angelo Mascarenhas[1], and Kirstin Alberi[1]

1. National Renewable Energy Laboratory, Golden, Colorado, 80403, USA
2. U.S. Naval Research Laboratory, Washington, DC, 20375, USA
3. George Washington University, Washington, DC, 200375, USA
4. Veeco MOCVD, Somerset, NJ, 08873, USA
5. Applied Novel Devices, Austin, Texas, 78717, USA

ABSTRACT — **Mechanically-stacked tandem solar cells combining existing photovoltaic technologies present a straightforward path toward higher conversion efficiencies. Voltage-matched configurations have been proposed as a simplified approach for fabricating two terminal modules, but they are subject to performance losses compared to four terminal configurations if voltage-matching conditions are not met. Here, we demonstrate a two terminal voltage-matched GaInP$_2$/Si tandem with 23% efficiency. Through a combination of experimental measurement and simulation, we explore the impact of voltage and fill factor mismatches between the top and bottom sub-cell strings. Our results show that efficiency losses are minimized when voltage-mismatches are accommodated by designing the bottom strings to have a higher voltage. Fill factor mismatches due to differences in series resistance also lead to changes in the optimal ratio of the number of cells in the top and bottom sub-cell strings.**

Index Terms — **mechanical stack, voltage matching, GaInP$_2$, silicon, photovoltaic cells**

I. INTRODUCTION

Decreasing the levelized cost of electricity (LCOE) generated by photovoltaics (PV) will require PV modules to become more efficient while maintaining or lowering production costs. Module cost reductions have been a substantial driver of PV adoption, while efforts in optimizing single junction solar cell performances toward the Shockley-Queisser limit have slowed[1]. Adding extra junctions, on the other hand, presents the possibilities of increasing the absolute module efficiency and lowering balance of systems costs. This is especially important for space-constrained applications where the total amount of energy generated per unit area is a key parameter. One avenue for accelerating the implementation of multiple junctions in flat plate modules is to combine existing, proven PV technologies (i.e. Si, CdTe, CIGS and III-Vs) into a hybrid architecture. In cases where processing incompatibilities prevent monolithic integration, mechanical stacking offers a straightforward fabrication alternative. Hybrid 4-terminal (4T) GaInP$_2$/Si, CdTe/CIGS and Perovskite/Si tandems have already been analyzed and even demonstrated to be attractive combinations [2]-[6].

The best approach for implementing mechanically stacked tandems at the module level continues to be an outstanding question. The stacked junctions could remain separate within a 4T module configuration. This design leaves the junctions unconstrained to operate at their optimal maximum power points. However, 4T tandem architectures may require additional module-level power electronics to prevent the build-up of large voltage differences between the closely spaced junctions or extra connections at every module. Two terminal (2T) configurations may still be more advantageous for reasons of safety and simplicity upon installation. The primary challenges for enabling 2T tandem modules are 1) minimizing the impact of performance mismatches between the top and bottom sub-cells and 2) simplifying the interconnection approach.

Voltage-matching (VM), rather than current-matching, may provide means to circumvent barriers to fabricating mechanically-stacked hybrid tandem solar cells [7],[8]. In general, strings of series-connected bottom cells and strings of series-connected top cells are connected in parallel to create a 2T device [9]. Selecting the number of sub-cells in each string, n_i, to produce the same maximum power point voltage, V_{mp}, for each string (i.e. $n_1 V_{mp,1} = n_2 V_{mp,2}$) ensures voltage-matching while relaxing the sub-cell bandgap energy and thickness restrictions for current matching. This architecture eliminates the need for tunnel junctions and allows the junctions to be fabricated separately in their preferred configurations by conventional manufacturing methods before being integrated using a compliant intermediate layer. The reduced sensitivity of the VM tandem architecture to diurnal variations in the solar spectrum relative to current-matched configurations (due to the diodes' logarithmic current-voltage relationship) also allows them to perform better under sub-optimal lighting conditions [9].

In this work, we demonstrate a 23% 2T VM GaInP$_2$/Si tandem. This material pairing was selected as a model system based on the near-ideal bandgap combination, potential for high efficiencies and well-behaved current-voltage, JV, characteristics [9]. While it is currently expensive to fabricate GaInP$_2$ cells, a number of innovative approaches related to lift-off and bonding processes are currently under exploration in order to re-use the GaAs substrate for multiple growths and incorporate layers grown on different substrates [10], [11]. Low cost alternatives to industry-standard metal-organic chemical vapor deposition (MOCVD) are also being developed for III-V epitaxy. This includes hydride vapor phase epitaxy (HVPE), which can be carried out at higher growth rates and utilizes cheaper precursor gases [12]. Future advances in each of these areas are anticipated to lower the cost of III-V cell production, and such tandems may eventually become competitive for space-constrained locations or for applications where energy production comes at a premium [2]. We use a combination of simulation and experimental measurements to understand the impact of voltage and fill factor, *FF*, mismatches on the tandem performance. Our results point to some design guidelines for 2T-VM modules.

II. EXPERIMENTAL

GaInP$_2$ cells were grown in an upright configuration in a Veeco K475i MOCVD reactor. Growth was carried out at 640 C on p-type (001) GaAs substrates (5 degree offcut toward <011>) using trimethylgallium, trimethylindium, trimethylaluminum, arsine, phosphine, disilane, diethyltellurium, dimethylzinc and carbon tetrabromide as precursors. The base layer was 900 nm thick. For each cell, a front gold contact grid was first deposited by electroplating, followed by the deposition of a ZnS/MgF$_2$ antireflection coating. The front surface of the cell was then attached to a 600 μm thick piece of Corning glass with transparent epoxy, and the GaAs substrate was chemically removed in an NH$_4$OH/H$_2$O$_2$ (1:1) solution. A back gold contact was added by electroplating, a cell area of 1 cm^2 was defined by wet chemical etching, and a second ZnS/MgF$_2$ antireflection coating was deposited. Within this construction, the cells were operated in a superstrate configuration. Typical efficiencies were 15%, with current-voltage, JV, characteristics of open circuit voltage, V_{OC} ~ 1.39 V, short circuit current, J_{SC} ~ 13 mA/cm^2 and fill factor, *FF* ~ 82%.

Interdigitated back contact (IBC) Si cells were fabricated from high lifetime silicon wafers. Heterojunctions were formed from p- and n-type backside patterns of amorphous silicon (a-Si) deposited over an optimized a-Si intrinsic layer. The intrinsic layers were chosen to be below 100A to improve passivation and the transport properties of the cell. Intrinsic and p-type a-Si layers were also deposited on the front side to form a front surface field. The cells were finished with silicon nitride deposited by low temperature plasma-enhanced chemical vapor deposition.

Individual GaInP$_2$ and Si cells were stacked together into units using transparent epoxy as the transparent interlayer, which provides light coupling and a mechanical bond between cells. These units were then combined to create 2x2 and 3x3 tandems. A schematic of the 2x2 tandem is shown in Fig. 1. The Si cells were connected in series to create either 2x1 or 3x1 2T bottom sub-cell strings (for example, string 1 in Fig. 1) with T_1 and T_2 terminals. Each of the GaInP$_2$ cells acted as a 1x1 string (for example, strings 2 and 3 in Fig. 1), and those cells were connected in parallel with T_3 and T_4 terminals to

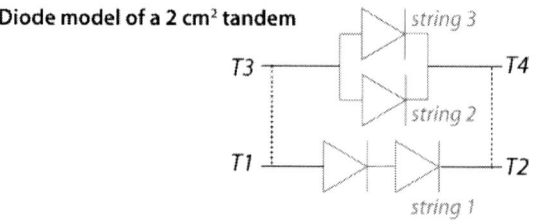

Diode model of a 2 cm² tandem

Schematic experimental 2 cm² tandem

Fig. 1 Schematic of our experimental GaInP$_2$/Si tandem solar cell. The stack can act as either a 4T or a 2T VM tandem based on whether parallel connections are made between terminals T1/T3 and T2/T4.

create either 1x2 or 1x3 2T top sub-cell strings. For 4T measurements, in which the top and bottom strings were not connected to each other, the series-connected Si sub-cell string was measured independently of the parallel-connected GaInP$_2$ top sub-cell string. For 2T-VM measurements, the Si sub-cell string was connected in parallel to the GaInP$_2$ strings (i.e. T_1 was connected to T_3 and T_2 was connected to T_4) to create a 2T device. JV measurements were carried out under 1-sun conditions at 25 °C.

III. RESULTS

JV curves of the independent 1x2 top and 2x1 bottom sub-cell strings of the 2x2 tandem, along with the 2T VM tandem are shown in Fig 2. For convenience the max power point, P$_{max}$, on each curve is annotated by a dot. The associated characteristics are presented in Table 1. Ideally, the sub-cell strings would be designed to have nearly equivalent V_{OC} and *FF* values so that their max power points occur at the same voltage. In these un-optimized test cells, the 2x1 Si sub-cell string V_{oc} is roughly 100 mV lower than that of each of the 1x1 GaInP$_2$ sub-cell strings. Additionally, the Si sub-cell string FFs are lower than those of the individual cells due to non-negligible series resistance in the cell interconnects. Overall, these two mismatches lead to a V$_{mp}$ mismatch of about 200 mV, as shown in Fig. 2. Leaving the two sets of strings separate, the 4T tandem has an efficiency of 24.9%. Connected in a 2T VM configuration, the tandem efficiency drops to 23.1%. Given the sizeable V_{mp} difference between the top and bottom strings (~17% relative to the GaInP$_2$ string V_{mp}), the efficiency drop (~8% relative to the 4T tandem) is modest.

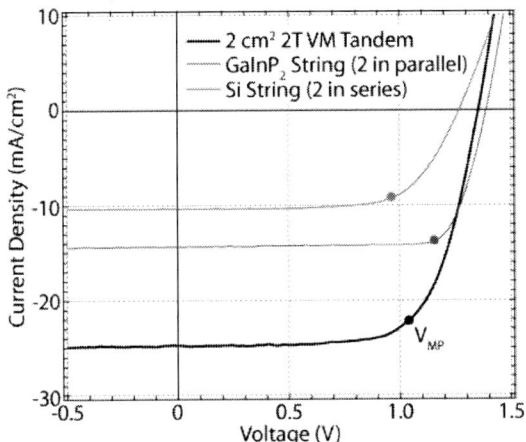

Fig. 2 Current-voltage curves for the top sub-cell string (two GaInP$_2$ cells in parallel), the bottom sub-cell string (two Si cells in series) and the 2T VM tandem for the 2x2 configuration. The dots mark the V_{mp} on each JV curve.

To better understand the impact these non-idealities have on the tandem performance, we computed the tandem efficiency as a function of V_{mp} mismatch defined as,

$$V_{mp} \, mismatch = \frac{V_{mp}^{GaInP} - V_{mp}^{Si}}{V_{mp}^{GaInP}} \quad (1)$$

where the V_{mp} components on the right hand side refer to the sub-cell string V_{mp} values. In the calculation, we construct the series-connected sub-cell string JV curves for the top and bottom junctions from identical individual cell JVs, ignoring cell to cell variations that may be present in real-world situations. We then computed the 2T tandem JV curves by adding those series connected sub-string curves in parallel. The resulting tandem efficiencies are plotted in Fig. 3 as a function of V_{mp} mismatch, which was determined from sub-cell string V_{mp} values according to Eq. 1. The 4T case for each simulation was also included as a horizontal line. No series resistances associated with the cell interconnects were included for the calculations presented in Fig. 3. In all calculations, we assume the total area of the top cells was equal to the total area of the bottom cells. We note that these calculations serve to illustrate general trends due to mismatching conditions, rather than be indicative of the absolute efficiency potential for GaInP$_2$/Si mechanical stacking technology.

The calculations were first performed for an "idealized" case (black solid and dashed lines), using JVs representative of best in class GaInP$_2$ and PERC Si cells, in order to set a baseline performance standard for voltage-matched GaInP$_2$/Si tandems [1], [2], [4]. The JV of the GaInP$_2$ top cell was generated with an analytical drift-diffusion model using values reported in the literature for III-V optical constants, band parameters and carrier mobilities [13]. The model also accounts for coherent reflections at interfaces within the cell

Table I. Experimentally-measured string and tandem characteristics for the 2x2 configuration

	GaInP$_2$ String (1x2 cm^2 in Parallel)	Si String (2x1 cm^2 in Series)	2T VM 2 cm^2 Tandem
V$_{OC}$ (V)	1.39	1.26	1.35
V$_{mp}$ (V)	1.16	0.96	1.04
J$_{SC}$ mA/cm^2)	14.36	10.3	24.9
FF (%)	80.12	68.2	68.7
Efficiency (%)	16.0	8.9	23.1

[14], [15]. The JV curve for a PERC Si bottom cell was generated with PC1D using an incident spectrum completely filtered by the bandgap of GaInP$_2$ [16].

The simulated 4T performance (horizontal black dashed line) is consistent with recent GaInP$_2$/Si tandem demonstrations [4]. The 2T VM efficiency (black solid line) matches that of the 4T when the V_{mp} mismatch approaches 0% but then drops with increasing V_{mp} mismatch. The magnitude of the drop is asymmetrical depending on whether the V_{mp} of the GaInP$_2$ sub-cell string is higher (positive V_{mp} mismatch value) or lower (positive V_{mp} mismatch value) than that of the Si sub-cell string. In either case, for very large V_{mp} mismatch, the 2T tandem efficiency asymptotes to the efficiency of the sub-cell string with the lower V_{mp}. For example, when V_{mp} mismatch approaches 100% (i.e the top GaInP$_2$ sub-cell string voltage is much greater than the Si sub-cell string voltage) the entire tandem V_{mp} is pinned by the Si sub-string voltage, effectively suppressing the power output of the GaInP$_2$ cells. This mismatch situation is analogous to current matched tandems, where the current produced by the tandem is dominated by the cell producing the least current. Vice versa, the tandem trends towards the GaInP$_2$ cell efficiency when the V_{mp} of the Si sub-string dominates. Since the GaInP$_2$ single cell efficiency is much greater, the tandem efficiency drop is less dramatic. *This asymmetry reveals that the tandem efficiency is least sensitive to V_{mp} mismatch when the mismatch is disproportionately placed on the lower performing junction.* Hence, if voltage mismatches must be accommodated, it is better to design the ratio of the number of sub-cells in the top and bottom strings such that the bottom sub-cell strings have a higher V_{mp} than the top sub-cell strings.

We also computed the 4T and 2T tandem performance from the experimental JV curves representative of our best Si and GaInP$_2$ cells (red solid and dashed lines). The same asymmetric trend is observed, although the absolute efficiency is lower due to the lower efficiencies of our cells compared to the ideal case. Finally, we included the experimental data points for the 2x2 and 3x3 tandems in 4T and 2T VM configurations as blue triangles and blue circles, respectively,

978-1-5090-5606-4/17 $31.00 © 2017 IEEE

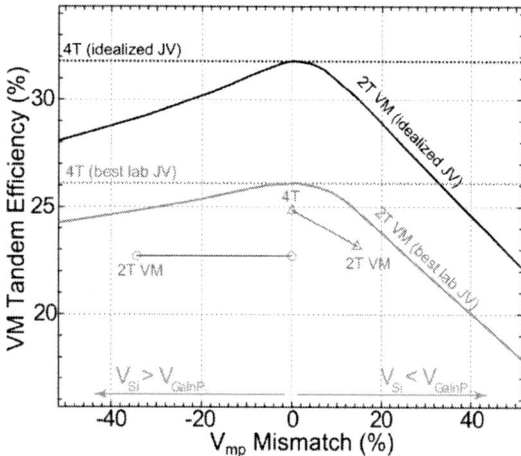

Fig. 3 Simulated (solid lines) and experimentally-measured (open symbols) tandem efficiencies as a function of V_{mp} mismatch between the GaInP$_2$ and Si strings. Note that V_{mp} Mismatch = 0% corresponds to the 4 terminal case. Open triangles and open circles represent 2x2 and 3x3 tandem configurations, respectively.

in Fig. 3. The 2T VM 2x2 tandem represents the case where the V_{mp} of the GaInP$_2$ top sub-cell string is greater than that of the Si bottom sub-cell string, while the 2T VM 3x3 tandem represents the opposite case. The data points for the 4T tandem efficiencies are placed at 0% V_{mp} mismatch to represent that they are unconstrained. The degradation of the tandem efficiencies follows the computed trend. Specifically, the efficiency loss due to voltage-mismatch in the 2T VM 2x2 tandem is much greater than that observed in the 2T VM 3x3 tandem. The additional efficiency loss between the values computed with the experimental single cell JV curves and our demonstrated tandems likely stems from 1) performance mismatch between cells in each sub-string (i.e. some cells perform worse than the best performances used in the simulation) and 2) additional series resistance in the cell interconnects. The first factor is particularly acute in the 3x3 tandem, where there is a greater spread in cell performances.

To explore the impact of parasitic resistances on the tandem performance, we have also calculated the tandem efficiency composed of series or shunt resistance-degraded cells. The general trends were captured for four test cases listed in Table II. We artificially reduced the *FF* by adding either a series resistor or an internal shunt resistor to the idealized JV model. Resistance values were adjusted until the *FF* dropped from high (~85% typical) to the "degraded case" of ~65%. Doing so had negligible impact on the J_{SC} and V_{OC}. Results from these calculations are shown in Fig. 4. Since the reduced *FF* affects the overall efficiency of the sub-cell strings as well as their V_{mp}, we expect that it will ultimately alter the optimal ratio of cells in the bottom sub-cell strings to cells in the top sub-strings that will produce voltage-matching conditions. We therefore have plotted the 2T tandem efficiency in Fig. 4 as a function of the bottom/top cell ratio to examine the overall effect of parasitic resistance on the optimal ratio.

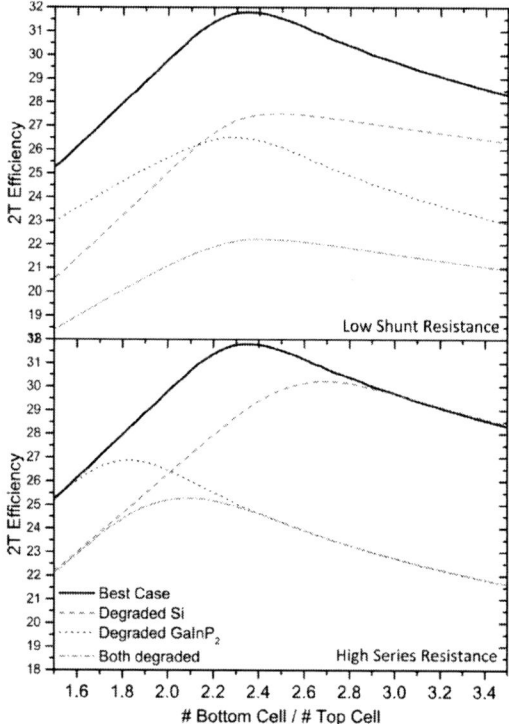

Figure 4: Simulations of 2 terminal efficiency for cases where cell FF have been artificially degraded, by either an internal shunt resistor (top panel) or series resistor (bottom panel). "Best case" corresponds to cells having performance metrics listed in Table 1.

In general, the optimum ratio for voltage matching fluctuates, depending on the case, but the fluctuations are small when low shunt resistance dominates the degradation mechanism. In either case, the shift in optimum ratio is a result of V_{mp} becoming smaller in the presence of parasitic resistance. The trend is most pronounced in the high series resistance scenario. When just the Si cell is degraded, the sub-string V_{mp} decreases, requiring more bottom cells (higher optimum ratio) to achieve V_{mp} to the GaInP$_2$ top string (red curve). By the same effect, fewer GaInP$_2$ are needed to match the Si string (lower optimum ratio), when the high quality GaInP$_2$ cell is degraded (blue curve).

From Fig. 4, it is clear that any degraded sub-cell string reduces the tandem efficiency most dramatically for the low shunt resistance scenario. Hence the tandem tolerates series resistance issues better than shunt resistance issues. This observation is a consequence of increased characteristic resistance, R_c ($R_c = V_{oc}/I_{sc}$) that triples when the shaded Si sub-cell string produces 3x less current. As a result, the sensitivity of efficiency to low shunt resistance increases. For example, the degraded Si *FF* under 1-sun will degrade more when under low-light conditions. In the high series resistance scenario, the larger R_c actually reduces sensitivity to series resistance, causing less of a reduction in tandem efficiency (degraded Si *FF* under 1-sun actually *increases* under low light conditions).

Table II. Cases investigated in *FF* degradation simulations

Case	GaInP$_2$ Cell FF (%)	Si Cell FF (%)	R_{Series} (Ω)	R_{Shunt} (Ω)
Best Case	88	82	0	10000
Degraded Si	88	65	3.5	67
Degraded GaInP$_2$	65	82	21	307
Both Degraded	65	65	Values listed above	

For either scenario, degrading the GaInP$_2$ cells causes the most harm, since the top cell usually produces about 2/3 of the total efficiency to the tandem [17].

IV. CONCLUSION

We have experimentally demonstrated a 23% 2T VM GaInP$_2$/Si tandem solar cell. Our 2x2 and 3x3 tandems closely followed simulated performance trends, which highlight some basic design guidelines for VM tandems. If any voltage mismatch must be accommodated between the top and bottom sub-cell strings, the bottom strings should be designed to have a higher V_{mp}. Degradation of the *FF* through a reduction in the shunt resistance is expected to decrease the tandem performance more than through an increase in the series resistance, although a change in the series resistance can lead to a change in the optimum ratio of the number of cells in the top and bottom sub-cell strings. This initial VM tandem demonstration had a rather modest performance, and there is much room for improvement. Higher efficiency cells, more uniformity in the performance of those cell, and larger module-sizes (needed to obtain ideal bottom/top cell ratios) will all help to improve the efficiency.

ACKNOWLEDGEMENT

The work was supported by the U.S. Department of Energy EERE contract SETP DE-EE00025786. The United States Government retains and the publisher, by accepting the article for publication, acknowledges that the United States Government retains a non-exclusive, paid-up, irrevocable, world-wide license to publish or reproduce the published form of this manuscript, or allow others to do so, for United States Government purposes.

REFERENCES

[1] M.A. Green, *et al.*, Solar cell efficiency tables (version 49), *Prog. Photovolt: Res. Appl.*, **25**, 3 (2017).

[2] D.C. Bobela, L. Gedvilas, M. Woodhouse, K.A.W. Horowitz and P.A. Basore, Economic competitiveness of III-V on silicon tandem one-sun photovoltaic solar modules in favorable future scenarios, *Prog. Photovolt: Res Appl.*, **25**, 41 (2017).

[3] S.U. Nanayakkara, K. Horowitz, A. Kanevce, M. Woodhouse and P. Basore, Evaluating the economic viability of CdTe/CIS and CIGS/CIS tandem photovoltaic modules, *Prog. Photovolt. Res. Appl.*, **25**, 271 (2017).

[4] S. Essig, *et al.*, Progress toward a 30% efficient GaInP/Si tandem solar cell, *Energy Procedia*, **77**, 464 (2015).

[5] K. Tanabe, K. Watanabe and Y. Arakawa, III-V/Si hybrid photonic devices by direct fusion bonding, *Scientific Reports*, **2**, 349 (2012).

[6] P. Loper, *et al.*, Organic-inorganic halide perovskite/crystalline silicon four-terminal tandem solar cells, *Phys. Chem. Chem. Phys.*, **17**, 1619 (2015).

[7] A.L. Lentine, *et al.*, Optimal cell connections for improved shading, reliability and spectral performance of microsystem enabled photovoltaics (MEPV) modules, 35th IEEE Photovoltaic Specialists Conference, PVSC 2010; Honolulu, HI; United States; 20 June 2010 through 25 June 2010, pp. 3048.

[8] R. Strandberg, Detailed balance analysis of area de-coupled double tandem photovoltaic modules, *Appl. Phys. Lett.*, **106**, 033902 (2015).

[9] J.M. Gee, A comparison of different module configurations for multi-band-gap solar cells, *Solar Cells*, **24**, 147 (1988).

[10] F. Dimroth, *et al.*, Wafer bonded four-junction GaInP/GaAs/GaInAsP/GaInAs concentrator solar cells with 44.7% efficiency, *Prog. Photovolt: Res. Appl.*, **22**, 277 (2014).

[11] D. Gomez, *et al.*, Process Capability and Elastomer Stamp Lifetime in Micro Transfer Printing, 66th IEEE Electronic Components and Technology Conference, ECTC 2016; The Cosmopolitan of Las VegasLas Vegas; United States; 31 May 2016 through 3 June 2016, pp. 680

[12] J. Simon, K.L. Schulte, N. Jain, S. Johnston, M. Young, M.R. Young, D.L Young and A. J. Ptak, Upright and inverted single-junction GaAs solar cells grown by hydride vapor phase epitaxy, *IEEE. J. Photovolt.*, **7**, 157 (2017).

[13] J.F. Geisz, M.A. Steiner, I. Garcia, S.R. Kurtz and D.J. Friedman, Enhanced external radiative efficiency for 20.8% efficient single-junction GaInP solar cells, *Appl. Phys. Lett.*, **103**, 041118 (2013).

[14] M.P. Lumb, *et al.*, Extending the one-dimensional hovel model for coherent and incoherent back reflections in homojunction solar cells, *IEEE J. Quantum Electron.*, **49**, 462 (2013).

[15] M.P. Lumb, M.A. Steiner, R.J. Walters and J.F. Geisz, Incorporating photon recycling into the analytical drift-diffusion model of high efficiency solar cells, *J. Appl. Phys.*, **116**, 194504 (2014).

[16] D.A. Clugston and P.A. Basore, PC1D version 5: 32-bit solar cell modeling on personal computers, in *Conference Record of the Twenty-Sixth IEEE Photovoltaic Specialists Conference, 1997*, Sep. 1997, pp. 207.

[17] T.J. Coutts, J.S. Ward, D.L. Young, K.A. Emery, T.A. Gessert and R. Noufi, Critical Issues in the Design of Polycrystalline, Thin-Film Tandem Solar cells, *Prog. Photovolt: Res. Appl. 2003;* **11**:359-375.

Wafer Bonded III-V on Silicon Multi-Junction Cell with Efficiency beyond 31%

Romain Cariou[1], Jan Benick[1], Paul Beutel[1], Nico Tucher[1], Martin Graf[1], David Lackner[1], Martin Hermle[1], Stefan W. Glunz[1,2], Andreas W. Bett[1], and Frank Dimroth[1]

[1]Fraunhofer Institute for Solar Energy Systems, Heidenhofstraße 2, 79110 Freiburg, Germany
[2]Laboratory for Photovoltaic Energy Conversion, University of Freiburg, 79110 Freiburg, Germany

Abstract — **Silicon-based multi-junction solar cells are under the spotlight as it is foreseen to be the next step for highest efficiency terrestrial photovoltaic flat panels. Indeed, recently it was experimentally shown that adding III-V top cells on a silicon bottom cell allowed exceeding the intrinsic Si single junction efficiency limit of 29.4% set by Auger recombination. In this study we focus on monolithic III-V//Si multi-junction two-terminal devices, as this architecture is directly compatible with actual module standards. We use surface activated wafer bonding to join the III-V and Si cells. As a result of material quality and process improvement, we present here a 2-terminal, 4 cm² GaInP/AlGaAs//Si triple-junction solar cell reaching 31.3% under AM1.5g 1-sun conditions.**

I. Introduction

In the quest for better photovoltaic system energy yield and lower levelized cost of electricity, improving the cell efficiency is an important research direction. A silicon single-junction solar cells have recently reached the remarkable efficiency value of 26.6% [1]. While being a huge step forward, this result is also closing the gap towards the practical upper efficiency limit for this material (~27%) [2]. However, Si-based multi-junction is a promising way to unlock this limit and go well beyond 30% efficiency. According to the 2016 international technology roadmap for photovoltaics report (ITRPV) it could represent 5% of the technology market share in 2026. Among the high bandgap candidates, III-V semiconductors offer attractive top cell options for future highly efficient 1-sun solar modules (tuneable bandgap, stability, etc.). Indeed, it was recently shown that surface activated wafer bonding technology enables to experimentally reach a 1-sun AM1.5g efficiency of 30.2% by permanently joining GaInP/GaAs top cells with an active Si bottom cell [3]. While being an important milestone for 2-terminal III-V//Si multi-junction devices, this approach did not reach its potential yet. We present here our latest achievements for this technology.

II. Si-based Multi-junction Solar Cells

Literature results [4], [5] for the best 1-sun AM1.5g single-junctions and Si-based multi-junction solar cells are summarized in Fig. 1. Silicon and gallium arsenide record

single junction devices reach 26.6% and 28.8% efficiency respectively. For the Si based multi-junctions, two main material systems are leading the competition: perovskites and III-V semiconductors. In addition, we distinguish here the results depending on the number of junctions (2J and 3J), and the interconnection configuration (2-terminal or 4-terminal). The perovskite on silicon actual status is 23.6% and 25.2% efficiency in 2- and 4-terminal respectively. In the case of III-V on silicon, efficiencies of 30.5% have been published for a 2J 4-terminal mechanically stacked device, and we present here our 3J 2-terminal solar cell reaching 31.3% efficiency (star in Fig. 1).

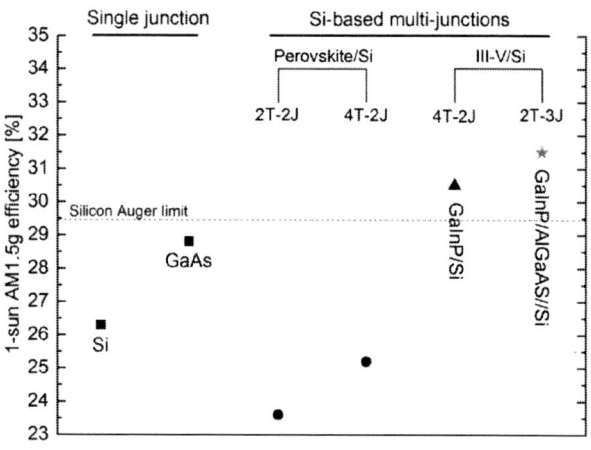

Fig. 1. AM1.5g record efficiency values from literature for single-junctions and for Si-based 2- and 3-junctions (2J and 3J), in 2-terminal or 4-terminal configuration (2T an 4T) [4], [5]. The red star shows the 31.3% wafer bonded GaInP/AlGaAs//Si device presented in this study.

III. Wafer Bonding Process

In order to achieve monolithic integration of III-V top cells on silicon, we use the surface activated wafer bonding approach. An overview of the fabrication process is presented in Fig. 2: The $Ga_{0.51}In_{0.49}P/Al_{0.04}Ga_{0.96}As$ top tandem cells were grown lattice matched on 4-inch GaAs substrates using metal-organic vapor phase epitaxy [6]. The solar cell layer stack was grown in inverted direction whereas the silicon bottom cells were prepared independently by implanting phosphorous atoms in p-type float-zone (FZ) c-Si wafers.

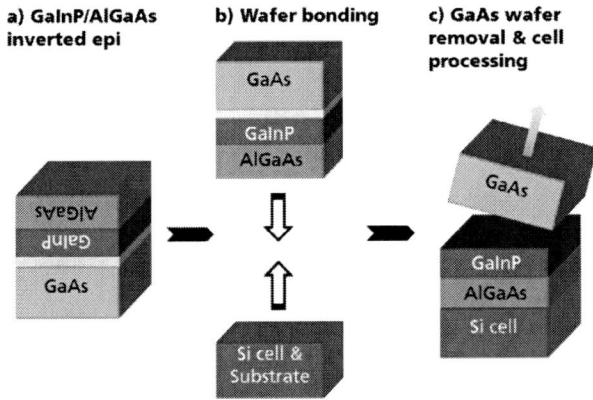

Fig. 2. Fabrication process flow for GaInP/AlGaAs//Si 3J wafer-bonded solar cells.

GaAs and Si wafers were then loaded on top and bottom facing chucks, heated to 120°C in an Ayumi SAB200 bond chamber. After reaching a base pressure $<5 \times 10^{-8}$ mbar, the Si and GaAs wafer surfaces were exposed to an argon fast atom beam [7] for deoxidization (Ar kinetic energy ~ 0.4 keV). The wafers were then pressed together with a force of 10 kN for a few minutes.

The GaAs growth substrate is etched away, leaving the few-micrometers-thick GaInP/AlGaAs top cells permanently bonded to the silicon bottom p-n diode. After this etching step, the silicon rear side was passivated using an Al_2O_3/SiN_x stack, followed by Al evaporation and local rear contact formation using the laser-fired contact process [8]. Finally, front metal contacts and anti-reflection coatings were evaporated, and mesa trenches etched (down to ~5μm in Si) to isolate the individual cells with a 4 cm^2 area.

IV. SOLAR CELL RESULTS

In our previous study we demonstrated a process route leading to an efficiency of 30.2% for a 4 cm^2 wafer bonded GaInP/GaAs//Si triple-junction solar cells [3]; the key points to achieve this result were: i) GaAs//Si bond interface offering both bulk-like mechanical strength and low contact resistance, ii) an improved passivation quality at the silicon back side, and iii) a relatively high open circuit voltage for the GaAs middle cell. However, our device performance was limited by a low parallel resistance and by imperfect current matching. In this study, we have performed incremental improvements of the cell structure to address those main issues. Our new generation of III-V//Si bonded solar cell shows a net increase in

efficiency, with the best 4 cm^2 device reaching 31.3% efficiency under 1-sun AM1.5g.

The corresponding solar cell results are presented in Fig. 3. The EQE of the device is shown in Fig. 3a). The GaInP top cell produces 13.4 mA/cm^2 (blue curve). The AlGaAs middle cell (4% Al, green curve), with a higher band gap and thicker layers than the previous GaAs middle cell, produces 12.6 mA/cm^2. And the silicon (red curve) limits here the overall current to 11.7 mA/cm^2. Thus, current matching between the subcells is not yet achieved with GaInP and AlGaAs producing excess current. The sum of subcells current density is 37.7 mA/cm^2 which can potentially lead to a current-matched 12.6 mA/cm^2 by thinning down a bit the top cells. The oscillations visible in total EQE (sum of subcells, black curve) are linked to the layer stack reflectivity. The front metal grid was adapted to current densities of 3J 1-sun devices, thus reducing the front metal coverage compared to our previous solar cell generation.

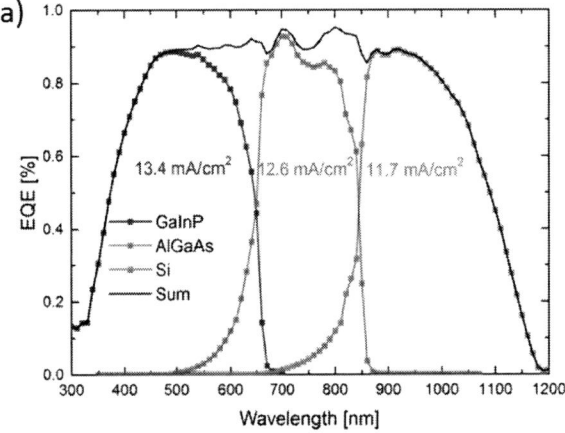

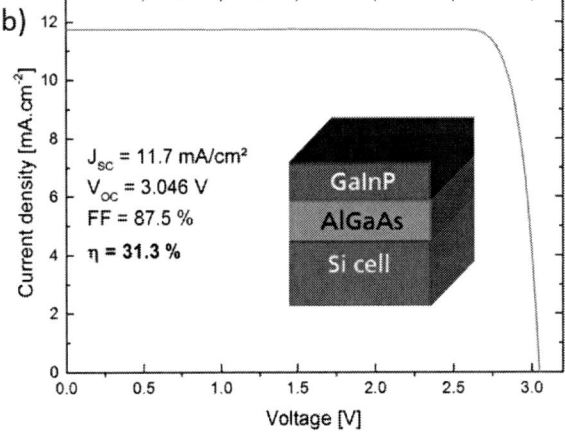

Fig. 3. a) External Quantum Efficiency of GaInP/AlGaAs//Si subcells as well as the sum of the three quantum efficiencies. b) Corresponding AM1.5g I–V characteristics measured under a spectrally adjustable solar simulator at Fraunhofer ISE calibration laboratory with an aperture of 3.984 cm^2. Device parameters are listed in the inset.

The corresponding AM1.5g 1-sun IV-curve measured with an aperture mask of 3.984 cm^2 (including fingers and bus bars) is shown in Fig. 3b): it reaches a V_{oc} of 3.046V, a FF of 87.5% and a short-circuit current density of 11.7 mA/cm^2 resulting in an efficiency of 31.3%. The limitation due to shunts of the previous device is not visible any more (excellent FF), mainly because of the change from current limiting subcell from GaAs to Silicon. The new solar cell reaches a record efficiency for a 1-sun 2-terminal III-V on silicon device.

V. CONCLUSION AND PERSPECTIVE

In summary, we have demonstrated a 4 cm^2 2-terminal GaInP/AlGaAs//Si triple junction solar cell reaching 31.3% efficiency under the AM1.5g spectrum. However this III-V on silicon approach did not reach its full potential yet. The current density can be significantly enhanced by current matching the device. In addition, while the Si front side has to remain perfectly flat to enable wafer bonding, light trapping features, such as random pyramids or more advanced structures [9], [10], can be implemented on the back side to boost even further the current. Other pathways towards higher performances will be explored in the future: higher bandgap top and middle cells, material quality improvements and enhanced passivation quality in the silicon bottom cell.

ACKNOWLEDGMENT

The authors would like to acknowledge V. Klinger, E. Oliva, S. Stättner, K. Wagner, R. Marlene da Silva Freitas, A. Henkel, and R. Koch for depositions and processing, M. Graf for LFC process, and G. Siefer, M. Schachtner, A. Wekkeli, E. Schäffer, and E. Fehrenbacher for solar cell characterization. This project has received funding from the European Union's Horizon 2020 research & innovation program under the Marie Sklodowska-Curie grant agreement No 655272.

REFERENCES

[1] K. Yoshikawa et al., "Silicon heterojunction solar cell with interdigitated back contacts for a photoconversion efficiency over 26%," Nat. Energy, vol. 2, p. 17032, Mar. 2017.

[2] D. D. Smith, P. Cousins, S. Westerberg, R. De Jesus-Tabajonda, G. Aniero, and Y.-C. Shen, "Toward the Practical Limits of Silicon Solar Cells," IEEE J. Photovolt., vol. 4, no. 6, pp. 1465–1469, Nov. 2014.

[3] R. Cariou et al., "Monolithic Two-Terminal III-V//Si Triple-Junction Solar Cells With 30.2% Efficiency Under 1-Sun AM1.5g," IEEE J. Photovolt., vol. 7, no. 1, pp. 367–373, Jan. 2017.

[4] M. A. Green et al., "Solar cell efficiency tables (version 49)," Prog. Photovolt. Res. Appl., vol. 25, no. 1, pp. 3–13, Jan. 2017.

[5] J. Werner et al., "Efficient Near-Infrared-Transparent Perovskite Solar Cells Enabling Direct Comparison of 4-Terminal and Monolithic Perovskite/Silicon Tandem Cells," ACS Energy Lett., vol. 1, no. 2, pp. 474–480, Aug. 2016.

[6] F. Dimroth et al., "Comparison of Direct Growth and Wafer Bonding for the Fabrication of GaInP/GaAs Dual-Junction Solar Cells on Silicon," IEEE J. Photovolt., vol. 4, no. 2, pp. 620–625, Mar. 2014.

[7] S. Essig and F. Dimroth, "Fast Atom Beam Activated Wafer Bonds between n-Si and n-GaAs with Low Resistance," ECS J. Solid State Sci. Technol., vol. 2, no. 9, pp. Q178–Q181, Jan. 2013.

[8] E. Schneiderlöchner, R. Preu, R. Lüdemann, and S. W. Glunz, "Laser-fired rear contacts for crystalline silicon solar cells," Prog. Photovolt. Res. Appl., vol. 10, no. 1, pp. 29–34, Jan. 2002.

[9] J. Eisenlohr et al., "Rear side sphere gratings for improved light trapping in crystalline silicon single junction and silicon-based tandem solar cells," Sol. Energy Mater. Sol. Cells, vol. 142, pp. 60–65, Nov. 2015.

[10] N. Tucher et al., "Crystalline Silicon Solar Cells with Enhanced Light Trapping via Rear Side Diffraction Grating," Energy Procedia, vol. 77, pp. 253–262, Aug. 2015.

[11] N. Tucher et al., "Optical simulation of photovoltaic modules with multiple textured interfaces using the matrix-based formalism OPTOS," Opt. Express, vol. 24, no. 14, pp. A1083–A1093, Jul. 2016.

[12] F. Feldmann, C. Reichel, R. Müller, and M. Hermle, "The application of poly-Si/SiOx contacts as passivated top/rear contacts in Si solar cells," Sol. Energy Mater. Sol. Cells, vol. 159, pp. 265–271, Jan. 2017.

Integration of thin Al films on $In_{0.18}Ga_{0.82}As$ metamorphic grade structures for low-cost III-V photovoltaics

Alessandro Giussani, Michael A. Slocum, and Seth M. Hubbard

Rochester Institute of Technology, Rochester, NY, 14623

Nathan Smaglik, Nikhil Pokharel, and S. Phillip Ahrenkiel

South Dakota School of Mines and Technology, Rapid City, SD, 57701

Abstract — $In_{0.18}Ga_{0.82}As$ has a bandgap of 1.16eV, ideal for single-junction devices under moderate solar concentration, and is lattice-matched to Al, an earth-abundant and widely available metal which may potentially be employed as substrate or buffer layer for low-cost III-V photovoltaics. In this paper the feasibility of an heteroepitaxial approach to integrate Al on GaAs via metamorphic $In_{0.18}Ga_{0.82}As$ is discussed. Metal organic chemical vapor deposition delivered unstrained $In_{0.18}Ga_{0.82}As$ templates of high quality but proved to be inadequate to fabricate pure Al layers at least at the temperatures typically used for III-V growth, owing to incorporation of considerable amounts of carbon from the precursors. *Ex situ* thermal evaporation of the metal was attempted, yielding polycrystalline material without epitaxial registry to the $In_{0.18}Ga_{0.82}As$ template. Detailed characterization of the grown specimens and discussion of the issues faced are presented.

Index Terms — InGaAs, metamorphic grade buffer, aluminum, heteroepitaxy, lattice matching, MOCVD.

I. INTRODUCTION

Metamorphic epitaxial growth is the growth of fully relaxed material on a substrate with considerable structural difference, typically a large lattice mismatch. Misfit strain is deliberately relieved by the formation of dislocations at the interface or at multiple interfaces in a graded structure, away from the active areas of the device. Since any generated threading segment contributes only marginally to strain relaxation but detrimentally acts as a non-radiative recombination center, the density of threading dislocations is to be minimized by eliminating kinetic barriers to dislocation glide, e.g., compositional inhomogeneity and high surface roughness. In photovoltaics, metamorphic epitaxy enables the integration of absorbing materials with a wider energy bandgap range than lattice-matched systems offer, allowing for a more optimized partitioning of the solar spectrum [1], [2].

In the first part of this paper, the optimization of the Metal Organic Chemical Vapor Deposition (MOCVD) of fully relaxed $In_{18}Ga_{0.82}As(001)$ layers on GaAs(001) via metamorphic grade (MMG) heterostructures is presented. This alloy composition features a bandgap of 1.16 eV, ideal for reaching the maximum Shockley-Queisser efficiency limit of single-junction devices under moderate solar concentration,

and a lattice parameter that matches Al upon mutual in-plane rotation by 45° of the two (001)-oriented unit cells. The integration of Al with III-V materials is intriguing. It could open avenues for cost-effective photovoltaics and optoelectronics, as suggested by the pioneering work of *Ptak et al.* on the coincident site lattice-matched epitaxy of $In_{0.18}Ga_{0.82}As$ on Al on $MgAl_2O_4$ spinel substrates [3]. The Al buffer was selectively etched in HF to lift off and layer-transfer the III-V heterostructure, and recycle the spinel substrate [4]. The proof-of-principle integration on GaAs(001) substrates of pure Al via metamorphic $In_{0.18}Ga_{0.82}As$ is the subject of the second section of this manuscript.

II. EXPERIMENTAL

$In_{0.18}Ga_{0.82}As$ / $In_{0.66}Ga_{0.34}P$ double heterostructures (DHs) lattice-matched to metamorphic $In_{0.18}Ga_{0.82}As$ on GaAs(001) were fabricated by MOCVD using an Aixtron Closed Coupled Showerhead® 3x2" reactor. In some samples a metallic Al interlayer was inserted in between the $In_{0.18}Ga_{0.82}As$ template and the DH. III-V layers were grown using trimethylgallium (TMGa) and trimethylindium (TMIn) as group III precursors, and either arsine (AsH_3) or phosphine (PH_3) as group V precursors. The deposition of Al was attempted on metamorphic $In_{0.18}Ga_{0.82}As$ by MOCVD, employing both trimethylaluminum (TMAl) and tris(tertiarybutyl)aluminium (TTBAl), and by thermal evaporation using a PVD75C evaporator and 99.99% pure Al pellets as source material, both from Kurt J. Lesker.

Five types of GaAs(001) substrates were explored aiming at achieving the highest quality $In_{18}Ga_{0.82}As$ template for the subsequent integration of Al: on-axis, 2° to [110], 2° to <111>A, 2° to <111>B, and 6° to <111>A. Prior to growth the substrates were annealed for 5 min under AsH_3 at 700°C to desorb the native oxide. For each substrate type, 5 x 300 nm, 10 x 300 nm and 5 x 600 nm $In_xGa_{1-x}As$ MMGs with x increasing in identical steps up to 22 at% were grown. The grading rate was 1.05% misfit / μm for the first structure and 0.53% misfit / μm for the other two. The top $In_{0.22}Ga_{0.78}As$ film, which is +1.58% lattice-mismatched to GaAs, overshoots the target $In_{0.18}Ga_{0.72}As$ composition to compensate for residual coherency strain [5]. The growth temperature was 600°C for the 5 x 300 nm and 10 x 300 nm series, and 600 and 650°C for the 5 x 600 nm specimens. MMG heterostructures were overgrown with an unstrained $In_{0.18}Ga_{0.82}As$ Fall Back Layer (FBL) in the 1-3 μm range at 650°C to isolate active regions of the device from defects. The DH design, starting from the FBL, consists of 20 nm $In_{0.66}Ga_{0.34}P$ / 2μm Si-doped $In_{0.18}Ga_{0.82}As$ / 20 nm $In_{0.66}Ga_{0.34}P$ / 10 nm $In_{0.18}Ga_{0.82}As$. The active layer was grown at 650°C, whereas barriers and cap were grown at 575°C. The Al MOCVD growth temperature was either 600 or 650°C, below the Al melting point. Real temperature, as well as reflectivity at 405, 633 and 950 nm, and curvature were monitored *in situ* with a LayTec EpiCurve® TT sensor. The reactor pressure was 100 mbar throughout the growth process. During thermal evaporation of Al, the specimen was not directly heated and its temperature was not monitored. The deposition rate was 0.3 nm / s.

The degree of relaxation of the heteroepitaxial films was studied by means of high resolution x-ray diffraction (HRXRD) reciprocal space maps (RSMs), carried out in grazing incidence geometry with a Bruker D8 four-circle diffractometer operated at 40 kV / 40 mA and equipped with Cu Kα₁ radiation, a Ge(022) four-bounce monochromator, and a Ge(220) three-bounce analyzer crystal. Mapping (004) and {224} reflections at a given substrate azimuth is necessary to correctly calculate alloy composition and strain in presence of epilayer tilt. A second analysis with the x-ray beam impinging along a 90°-rotated azimuth to the first measurements was performed, given that $In_xGa_{1-x}As$ is reported to relax asymmetrically along <110> and <1-10> [6]. Powder XRD by means of a Bruker D2 Phaser working at 30 kV / 10 mA was utilized to study the crystallinity of the Al films and the DHs grown on the Al buffer. The surface morphology at various stages of the growth process was examined using a Veeco atomic force microscope (AFM) in tapping mode. MMGs, FBLs and DHs were also studied by means of a photoluminescense (PL) setup consisting of a 532 nm laser source with 256 W·cm⁻² power density, a Princeton Instruments Acton SpectraPro SP-2300 spectrometer and an InGaAs detector. Finally, transmission electron microscopy (TEM) analysis was performed at 200 KV on a JEM-2100 LaB₆ with scanning TEM (STEM) and secondary electron microscopy (SEM) capabilities. Energy-dispersive x-ray (EDX) spectra were collected by means of an Oxford Inca system with a silicon drift detector.

III. RESULTS AND DISCUSSION

The top row of Fig. 1 shows the (004) RSM and the AFM of a 5 x 300 nm MMG grown at 600°C on a non-miscut substrate. The diffraction peaks are broad and poorly defined. The AFM image exhibits a short crosshatch arising from strain fields around the misfit dislocations formed along <1-10> [7]. The root-mean-square (rms) roughness over (40x40) μm² is ~ 12 nm. On the bottom row of Fig. 1 the 5 x 600 nm MMG grown

at 650°C on GaAs(001) offcut by 2° towards <111>B instead presents neatly distinguished reflections from each layer of the grade, and a long, uninterrupted crosshatch resulting in a rms roughness of ~ 6 nm.

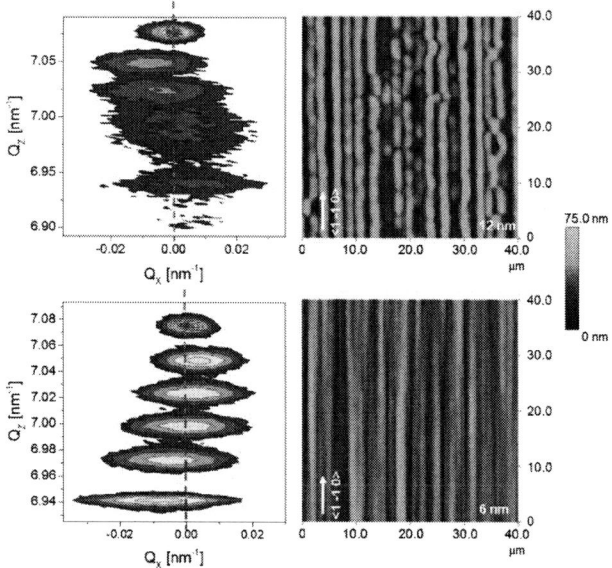

Fig. 1. (Left) (004) RSMs with the x-ray beam in-plane projection along <1-10> and (right) AFM (40x40)μm² images of (top) a 5 x 300 nm MMG grown at 600°C on on-axis GaAs(001) and (bottom) a 5 x 600 nm MMG grown at 650°C on 2°-offcut-to-<111>B GaAs(001). The red dashed lines in the RSMs highlight the presence of lattice tilt in the epilayers. The height scale bar is the same for both AFM images. Rms roughness is indicated on the bottom right corner of each image.

Reducing the grading rate of the 5 x 300 nm structure, by either increasing the layer thickness (5 x 600 nm) or the number of layers (10 x 300 nm), yields well defined reflections in the RSMs even on-axis substrates, although the sample surface still features interrupted crosshatch and approximately double rms roughness than on any offcut GaAs(001). The importance of lowering the misfit change rate is further consolidated by PL measurements. Fig. 2(a) presents PL spectra of different MMG structures for a fixed miscut, 6° to <111>A. Results agree with theory and experiments published in the literature [8]. Raising the growth temperature for a given heterostack favors dislocation gliding, which leads to higher quality material and hence brighter PL. The most intense PL signal is detected for the 10 x 300 nm heterostructure at 600°C though, suggesting that doubling the number of layers may be more beneficial than growing 50°C hotter. The reason could be that a larger number of interfaces provide more grounds for relaxation via misfit dislocations and more efficiently filter threading dislocation segments. The lowest PL peak is recorded for the 5 x 300 nm grade, confirming the AFM and HRXRD findings. PL on the 10 x 300 nm buffers grown at 600°C shows that 6° to <111>A yields the brightest PL among the investigated miscuts, followed by 2° to <111>B, and reinforces that on-axis growth

978-1-5090-5606-4/17 $31.00 © 2017 IEEE   2515

is of inferior quality, the signal being the weakest and at least twice as broad as the features for the offcut depositions (Fig. 2(b)).

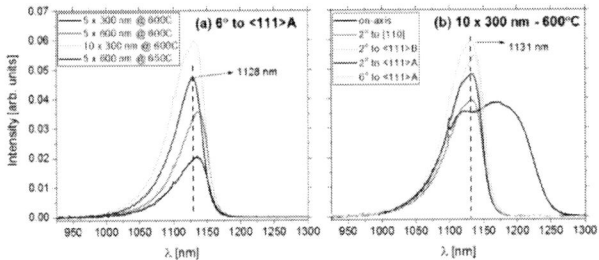

Fig. 2. PL of MMGs as a function of (a) the stack and the growth temperature for the 6° to <111>A offcut; (b) the offcut for the 10 x 300 nm stack at 600°C.

Fig. 3 illustrates for the 5 x 600 nm MMG at 650°C on 2°-offcut-to-<111>B GaAs(001) the trend of the XRD-calculated in-plane lattice constant (black data-points) across the heterostructure as the indium concentration is increased by ~ 4.5 at% from layer to layer. The increment rate of the lattice parameter is nearly constant up to the fourth film and slows down for the OSL. This appears to be due to a much steeper increase in magnitude of the XRD-derived negative (compressive) in-plane strain between the fourth layer and the OSL compared to the underlying films (blue data-points). This phenomenon is recurring in all the samples analyzed and is currently under study, as it may suggest avenues to improve the quality of the MMG.

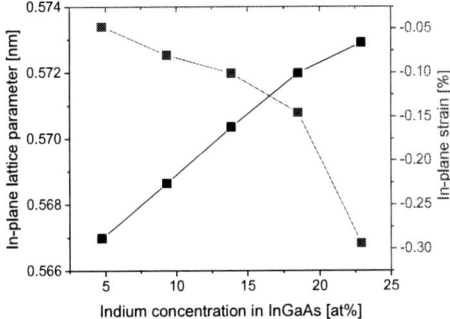

Fig. 3. XRD-derived (black) in-plane lattice parameter and (blue) in-plane strain as a function of the indium concentration for the 5 x 600 nm MMG at 650°C on 2°-offcut-to-<111>B GaAs(001). Values are averaged between <110> and <1-10> to account for anisotropic relaxation.

Fig. 4 summarizes the behavior of the indium concentration (top figure) and the in-plane relaxation (bottom figure) of the OSL, calculated from RSMs, as a function of the substrate offcut and the crystal orientation. It emerges that the amount of indium in the alloy is slightly above target (22 at%) for all the samples. It is also larger for the measurements taken with the x-ray beam impinging along <110> than <1-10> except for the MMG grown on on-axis GaAs(001). Similarly, the in-plane

relaxation is different along the two orthogonal in-plane directions, namely, it is smaller along <1-10> for all the specimens except the growth on the on-axis substrate. This means that the $In_xGa_{1-x}As$ layers relax anisotropically. Misfit dislocations are of the 60° $a/2$<110>{111} type and occur along two different dislocation lines in this material system. α dislocations, characterized by As-terminated cores and [1-10] line direction, have typically lower formation energy and higher glide velocity than β dislocations with Ga-terminated cores and [110] line direction. Since strain is relieved perpendicularly to the dislocation line direction, it is speculated that more readily formed and gliding α dislocations are responsible for the higher relaxation observed along <110> [6]. This hypothesis is in agreement with the AFM finding that, for all the MMGs grown on offcut substrate, there are more surface striations along <1-10> than <110> (see bottom image of Fig. 1). The specimen grown on on-axis GaAs represents an exception, with a larger relaxation along <1-10>. It is noted that a number of factors are reported to influence dislocation formation and kinetics, among them temperature, substrate offcut, surface morphology and group V partial pressure [6], [9]. It could be that the different relaxation behavior of the on-axis growth is related to the short and interrupted crosshatch of Fig. 1.

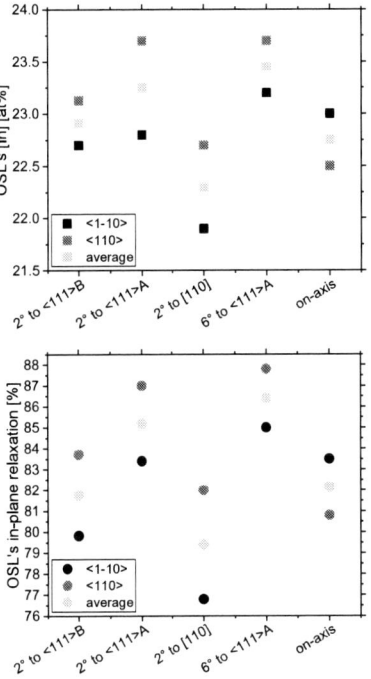

Fig. 4. XRD-derived (top) indium concentration and (bottom) in-plane relaxation of the OSL as a function of the substrate offcut for the 5 x 600 nm MMG series at 650°C.

Based on the AFM, HRXRD and PL data described above, 2° towards <111>B and 6° towards <111>A substrates, and 5 x 600 nm and 10 x 300 nm grade structures were selected for the overgrowth of the $In_{0.18}Ga_{0.82}As$ FBL. The FBL resulted in in-plane tensely strained material, independent of the growth temperature in the (600-650)°C range and of the thickness in the (1-3)μm range, as revealed by *in situ* curvature analysis (curvature slope ~ +0.03 km^{-1}·s^{-1}; stress ~ +0.2 GPa) and *ex situ* RSMs (in-plane relaxation ~ 110%). The tensile nature of the strain of the FBL and the excess of indium in the OSL with respect to the targeted structure, indicated by Fig. 3, suggested reducing the overshoot's composition by nominal 1 at% [In] [5]. Such change led to an *in situ* curvature slope of -0.001 km^{-1}·s^{-1} (stress ~ -0.006 GPa) compared to the previous + 0.03 km^{-1}·s^{-1}, pointing to a compressive strain of lesser magnitude. In-plane relaxation from XRD turned out to be ~ 96% along <110> and ~ 104% along <1-10>, averaging ~ 100%. It is noted that the *in situ* and *ex situ* strain measurements are not contradictory since the former is taken at growth temperature and the latter after the specimen has cooled down to room temperature. Also, the presence of a minor compressive strain during the deposition is not critical according to the literature, as it was reported that inverted $In_{0.27}Ga_{0.73}As$ solar cells with in-plane compressive strain as large as -0.2 % still feature an open-circuit voltage approaching ideality, whereas a similar magnitude of tensile strain results in significant device degradation [10]. Cross-sectional TEM shows a neat interface between OSL and FBL with no new defects forming in the latter nor propagating up there from the bottom layers (Fig. 4).

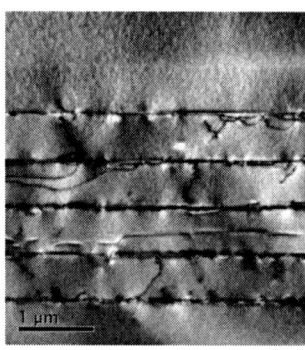

Fig. 4. Dark field, g = {220} TEM cross-section of a 5 x 600 nm MMG at 650°C on 2°-offcut-to-<111>B GaAs(001).

To further test the quality of the grown material, a DH was deposited on top of the FBL / MMG as detailed in the experimental section. It is noted that for some samples the growth process was interrupted after the FBL to allow for *ex situ* cleaving of the 2 inch wafer into quarters to carry out DH splits. In this case, a 200 nm $In_{0.18}Ga_{0.72}As$ homo-epitaxial layer preceded the first $In_{0.66}Ga_{0.34}P$ barrier to avoid growing the DH directly on a surface exposed to the atmosphere. PL proves that DHs with growth interruption exhibit nearly identical spectra to uninterrupted growths. RSMs confirm the DH is exactly lattice-matched to the underlying FBL, as demonstrated by the presence of a single Bragg peak for all the layers following the OSL (Fig. 5).

Next, MOCVD deposition of metallic Al on the $In_{0.18}Ga_{0.72}As$(001) FBL was attempted. Both TMAl and

TTBAl precursors at 600°C and 650°C produced layers which do not exhibit any peak in powder diffraction. As an alternative method for the growth of Al epitaxial layers on lattice-matched $In_{0.18}Ga_{0.72}As$, thermal evaporation was pursued. This implies interrupting the MOCVD growth and dealing with a number of challenges, among which exposing to air the $In_{0.18}Ga_{0.72}As$ template prior to evaporation and then the Al surface before recommencing the MOCVD growth of the DH.

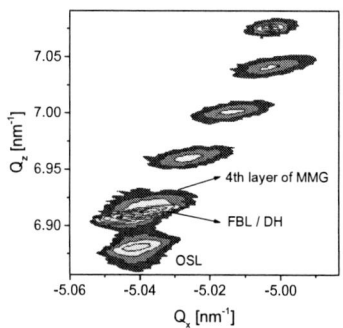

Fig. 5. (224) RSM of a DH on 2°-offcut-to-<111>B GaAs(001), with the x-rays aligned along GaAs<1-10>.

Fig. 6 presents an AFM study through the Al deposition process. Fig. 6(a) shows the typical crosshatch of the surface of a 3 μm FBL / 10 x 300 nm MMG structure on a 6° to <111>A substrate over (40 x 40)μm². The crosshatch appears more complex than for a 2° to <111>B substrate (Fig. 1, bottom right). In addition to wide ridges and troughs running parallel to <1-10> and alternating along <110> with peak-to-valley distances as large as 15 nm, perpendendicularly there is a superimposed finer structure contributing features slightly tilted around <110>. The evaporation of 250 nm Al results in a coating of the surface with grains in the 25 nm range. The crosshatch of the underlying FBL is still visible and the roughness does not increase (Fig. 6(b-d)).

No Al-related reflections could be detected by means of either an HRXRD wide range specular scan or an RSM around the bulk position of Al(002), which is expected in case in-plane 45°-rotated coincidence lattice epitaxy of Al on $In_{0.18}Ga_{0.72}As$(001) has occurred. Measurements were carried out in triple axis and double axis configuration with increasing detector slit size. Powder diffraction was then utilized to verify whether the Al film is crystalline in any form. Fig. 7 proves that Al is polycrystalline with (111) and (001) orientations along the growth direction (red curve). A piece of such an Al template was transferred within minutes from the evaporator into the MOCVD to minimize atmospheric contamination, i.e., oxidation of the metal surface. Once in the reactor it was ramped up in AsH$_3$ to 700°C, temperature at which it was baked for 5 min, in the very same way any MOCVD growth on GaAs substrates or on lattice-matched $In_{0.18}Ga_{0.72}As$ commences. The idea behind the bake at 700°C, which exceeds the melting point of Al (660°C), is to melt the metal and let it re-crystallize upon cooling down to 650°C, aiming at inducing solid phase epitaxy from the underlying FBL. The DH was subsequently deposited starting with a 200 nm $In_{0.18}Ga_{0.72}As$ film as for the interrupted growth of DHs without Al buffer. Powder diffraction revealed the disappearance of metallic Al features and the onset of multiple reflections from $In_{0.66}Ga_{0.34}P$ and $In_{0.18}Ga_{0.72}As$, which are labeled as III-V peaks (Fig. 7 – black curve). This indicates that the DH grows polycrystalline. Interestingly, to the right of the III-V peaks some additional reflections show up (marked by arrows), suggesting the presence of material with a smaller lattice constant than the one of the ideal lattice-matched DH.

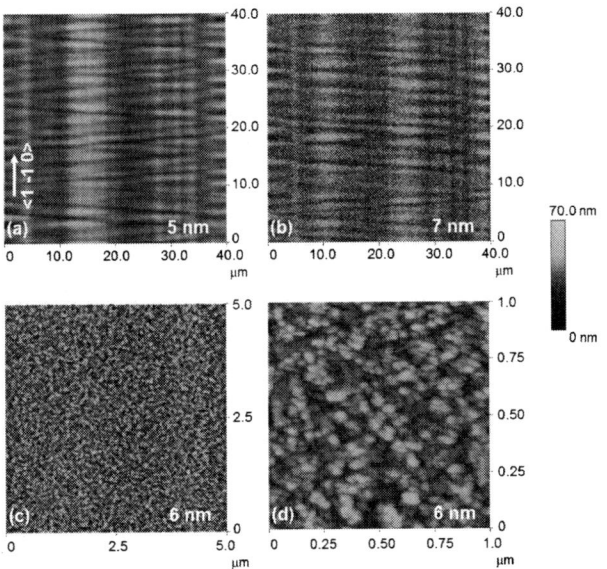

Fig. 6. AFM of a 3 μm FBL / 10 x 300 nm MMG on 6° to <111>A GaAs(001) (a) pre-evaporation of Al over (40 x 40) μm²; (b-d) post-evaporation of 250 nm Al over (40 x 40), (5 x 5), (1 x 1) μm², respectively. Rms roughness is reported on the bottom right corner. Vertical side of all images is parallel to GaAs<1-10>. Height scale bar is the same for all images.

The specimen was analyzed by TEM and EDS. Fig. 8(a) shows a cross-sectional zoom-in on the evaporated film, which points out that severe roughening occurred during the DH growth (the rms roughness of the final surface is ~ 150 nm over (40 x 40)μm² as opposed to the few nm post-Al evaporation – see Fig. 6(b)). The EDS elemental maps of Fig. 8(b-c) show that Al has fully converted into AlAs, probably owing to AsH_3 being flown during the bake at 700°C. Fig. 8(d) reveals also undesired oxidation of the metal caused by the MOCVD-to-evaporator-to-MOCVD transfers. In spite of the pronounced roughness, the two $In_{0.66}Ga_{0.34}P$ barriers appear continuous and sandwich the active layer.

Another piece from the same Al template was overgrown by a DH for which the maximum temperature during the process was lowered to 600°C (bake temperature in AsH_3 and $In_{0.18}Ga_{0.72}As$ growth temperature). 600°C is safely below the melting point of Al. Yet the powder diffraction scan (green curve of Fig. 7) is very similar to the one of the DH grown at higher temperature. Only a very weak Al(111) reflection

remains compared to the scan of the as-evaporated metal (Fig 7(b)). Peaks to higher angles with respect to the lattice-matched DH's ones, compatible with the presence of AlAs, are again present. This suggests that, even below the melting point of Al, AsH_3 arsenizes the metal throughout the entire thickness, probably by diffusing through the numerous grain boundaries provided by the polycrystalline nature and the small grain size of the evaporated Al. In fact it is possible that the Al layer has not melted even when the maximum temperature of the MOCVD process was 700°C. It could have converted to AlAs before the melting point of Al was reached. The impact of the oxygen contamination on the layer crystallization has to be thoroughly investigated too.

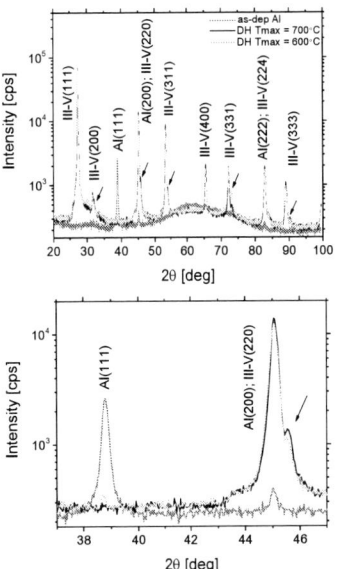

Fig. 7. (a) Powder diffraction of (red curve) as deposited 250 nm Al of Fig. 6; DH on Al using (black curve) 700°C and (green curve) 600°C as maximum MOCVD temperature. (b) zoom-in on 37°< 2θ < 47° to highlight the Al(111) behavior post-DH growth.

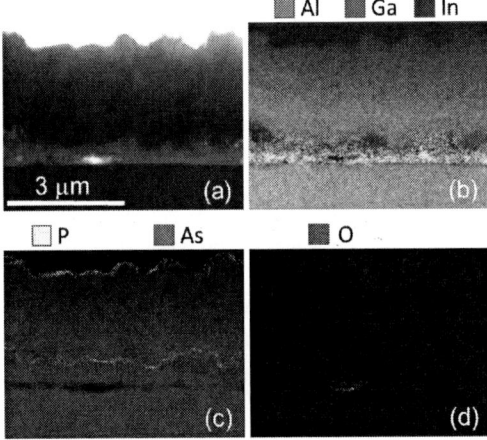

Fig. 8. (a) Bright field STEM cross section of FBL / Al / DH. EDS mapping of (b) Al, Ga and In; (c) P and As; (d) O.

978-1-5090-5606-4/17 $31.00 © 2017 IEEE   2518

In spite of the observed difficulties with the integration of Al into the DH in these preliminary experiments, PL emission from the $In_{0.18}Ga_{0.72}As$ active layer could be measured. Peaks from the DHs with Al are red-shifted by ~ 20-50 nm, ~ 3 orders of magnitude less intense and ~ 3 times broader than for the DH baseline without Al (Fig. 9). The difference in peak intensity and shift between high and low temperature growth could be due to the Al layer of the latter having been exposed to air for considerably longer.

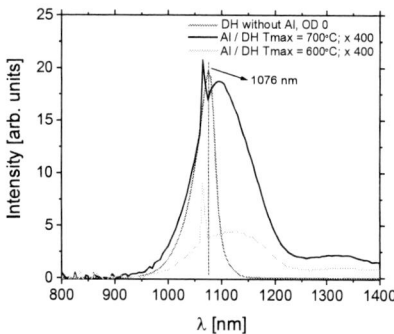

Fig. 9. PL of the samples of Fig. 7. (Red) control DH without Al interlayer; DH on evaporated Al with (black curve) 700°C and (green curve) 600°C maximum temperature during the MOCVD process. An optical density filter of 1 was used for the red curve; the other two signals are multiplied by a factor of 400. The narrow feature visible for the low intensity curves stems from the laser (2 x 532 nm).

IV. CONCLUSIONS

Fully relaxed $In_{0.18}Ga_{0.72}As$ templates, theoretically lattice-matched to Al via coincidence lattice epitaxy, were fabricated by MOCVD on GaAs(001) using $In_xGa_{1-x}As$ metamorphic buffers featuring an overshoot layer to x = 21 at%. Different grade designs, different substrate offcut angles / directions, and two growth temperatures, 600°C and 650°C, were studied. By combining HRXRD, AFM and PL, it was inferred that, within the space of parameters explored, accommodating the 1.3% misfit to GaAs in either 10 layers of 300 nm or 5 layers of 600 nm and employing either 2° to <111>B or 6° to <111>A substrates yield MMGs of the highest quality. The 3 μm unstrained fall back layer exhibits rms roughness of a few nm over (40 x 40) μm². The integration of Al on such a template via MOCVD was not successful. Nor TMAl nor TTBAl produced crystalline Al films in the temperature range (600-650)°C, owing to undesired incorporation of carbon from the metal-organic precursor, which does not fully dissociate in absence of AsH₃. To overcome this issue, thermal evaporation of Al was attempted at the cost of exposing to air first the $In_{0.18}Ga_{0.72}As$ fall back layer and then the metal film. No epitaxial registry of the evaporated Al to the template was observed. The metal grows polycrystalline with (111)- and (001)-oriented domains. Upon re-introduction into the MOCVD reactor for the growth of $In_{0.66}Ga_{0.34}P$ / $In_{0.18}Ga_{0.72}As$ / $In_{0.66}Ga_{0.34}P$ DH, the Al film converts to AlAs during the ramp up and bake in AsH₃. Compared to the control DH without Al interlayer, the DH on Al is ~ 10 times rougher and emits red-shifted PL with ~ 10^3 times lower intensity.

ACKNOWLEDGEMENT

This work was funded by the U. S. Department of Energy under Award Number DE-EE0007363.

REFERENCES

[1] C. J. K. Richardson and M. L. Lee, "Metamorphic Epitaxial Materials," *MRS Bulletin*, vol. 41, no. 3, pp. 193–197, 2016.

[2] T. J. Grassman *et al.*, "Expanding the palette: Metamorphic strategies over multiple lattice constant ranges for extending the spectrum of accessible photovoltaic materials," in *Photovoltaic Specialists Conference (PVSC), 2011 37th IEEE*, 2011, pp. 3375–3380.

[3] Y. Lin, A. G. Norman, W. E. McMahon, H. R. Moutinho, C.-S. Jiang, and A. J. Ptak, "Single-crystalline aluminum grown on MgAl2O4 spinel using molecular-beam epitaxy," *J. Vac. Sci. Technol. B*, vol. 29, no. 3, p. 03C128, 2011.

[4] A. J. Ptak, Y. Lin, A. Norman, and K. Alberi, "Methods of producing free-standing semiconductors using sacrificial buffer layers and recyclable substrates," US 9,041,027 B2, 2015.

[5] S. P. Ahrenkiel, M. W. Wanlass, J. J. Carapella, R. K. Ahrenkiel, S. W. Johnston, and L. M. Gedvilas, "Optimization of buffer layers for lattice-mismatched epitaxy of $Ga_xIn_{1-x}As/InAs_yP_{1-y}$ double-heterostructures on InP," *Sol. Energy Mater. Sol. Cells*, vol. 91, no. 10, pp. 908–918, 2007.

[6] R. France, A. J. Ptak, C.-S. Jiang, and S. P. Ahrenkiel, "Control of asymmetric strain relaxation in InGaAs grown by molecular beam epitaxy," *J. Appl. Phys.*, vol. 107, no. 10, p. 103530, 2010.

[7] E. A. Fitzgerald, S. B. Samavedam, Y. H. Xie, and L. M. Giovane, "Influence of strain on semiconductor thin film epitaxy," *J. Vac. Sci. Technol. Vac. Surf. Films*, vol. 15, no. 3, p. 1048, 1997.

[8] E. A. Fitzgerald, A. Y. Kim, M. T. Currie, T. A. Langdo, G. Taraschi, and M. T. Bulsara, "Dislocation dynamics in relaxed graded composition semiconductors," *Mater. Sci. Eng. B*, vol. 67, no. 1, pp. 53–61, 1999.

[9] R. M. France *et al.*, "Reduction of crosshatch roughness and threading dislocation density in metamorphic GaInP buffers and GaInAs solar cells," *J. Appl. Phys.*, vol. 111, no. 10, p. 103528, 2012.

[10] J. F. Geisz, A. X. Levander, A. G. Norman, K. M. Jones, and M. J. Romero, "In situ stress measurement for MOVPE growth of high efficiency lattice-mismatched solar cells," *J. Cryst. Growth*, vol. 310, no. 7–9, pp. 2339–2344, 2008.

Temperature dependent characteristics of GaInP/GaAs/GaInNAsSb solar cell under simulated AM0 spectra

Riku Isoaho, Arto Aho, Antti Tukiainen, and Mircea Guina

Optoelectronics Research Centre, Tampere University of Technology, Tampere, FI-33720, Finland

Abstract — **We report on the temperature characteristics of GaInP/GaAs/GaInNAsSb triple junction solar cell monolithically grown by molecular beam epitaxy. In particular, we have compared the temperature dependent light-biased current-voltage characteristics of the cell at simulated AM0 spectral conditions produced by two solar simulators: a customized three band solar simulator and a Xenon simulator equipped with an AM0 filter. For the three band simulator, the temperature coefficients corresponding to short-circuit current density and open-circuit voltage were found to be 5.3 μA/cm^2/°C and -6.8 mV/°C, respectively. These values are in agreement with literature reports for GaInP/GaAs/Ge solar cells. Illumination using a filtered single Xenon lamp leads to an erroneously high temperature coefficient value for short-circuit current density.**

Index Terms — **AM0, dilute nitride, molecular beam epitaxy, multijunction solar cell, solar simulator, temperature coefficient.**

I. INTRODUCTION

Multijunction solar cells (MJSC) based on III–V compounds are an attractive choice for concentrated photovoltaics (CPV) and space applications due to the versatility of III–V materials enabling high conversion efficiency. For example, with four junction III–V solar cells, a conversion efficiency of 46% has been demonstrated [1]. By increasing the number of junctions, the conversion efficiency of MJSCs is projected to surpass 50% in CPV applications [2].

The operating conditions (i.e. illumination conditions and temperature) have a great impact on the solar cell performance. For example, in near space the illumination intensity is higher compared to one-sun terrestrial spectral conditions where the illumination intensity is attenuated by the atmosphere. The atmosphere also affects the solar spectrum as radiation at different wavelengths are not absorbed uniformly causing variation of the terrestrial spectrum. Secondly, the operating temperatures of the cells can vary substantially depending on the application; in satellite orbit the cell temperatures can vary between -160°C and 100°C [3] whereas in typical CPV operation the cell temperature can realistically rise up to 50–60°C above ambient temperature [4]-[5].

For optimal energy harvesting, the solar cells should be tailored to fit their corresponding application, and from the design point-of-view it is vital to understand the effects of temperature and illumination conditions on the cells. In this paper, we report on the temperature characteristics of triple junction (3J) GaInP/GaAs/GaInNAsSb solar cell grown by plasma-assisted molecular beam epitaxy (PAMBE). We have measured the light-biased current-voltage (LIV) characteristics

for the cell under AM0 spectral conditions. We have compared the temperature characteristics of the cell when measured with a customized three band solar simulator and a single Xenon lamp simulator.

II. EXPERIMENTAL

A. Solar cell fabrication

The cell was monolithically grown on 2" p-GaAs(100) substrate using Veeco GEN20 PAMBE reactor. The epitaxy system was equipped with SUMO type effusion cells for group III materials, valved cracker sources for As, P and Sb and a RF plasma source for introduction of atomic N. The bandgap energies of the GaInP, GaAs and GaInNAsSb sub-cells were 1.9 eV, 1.4 eV and 1.0 eV, respectively. For lattice-matching the dilute nitride bottom cell, an In concentration of ~2.7 times the N concentration was used [6]. The wafer was processed into cells with size of 4×4 mm^2 with an active area of 0.1175 cm^2, and the cells were coated with a TiO$_2$–SiO$_2$ antireflective coating.

B. Light-biased current-voltage characterization

For the LIV characterization, we used primarily a three band solar simulator comprising individually adjustable Xenon and two halogen light sources that were optically filtered to be suitable for the characterization of 3J cells. The short wavelength band for the GaInP sub-cell is produced using the Xenon lamp due to its better spectral characteristics in the UV range compared to halogen light sources. Different filter configurations are needed for the simulation of AM0 and AM1.5 spectral conditions.

The spectral bands of the illumination were generally designed such that their edges were close to the bandgaps of the sub-cells (i.e. each sub-cell would be illuminated with only one spectral band) and thus, allowing independent adjustment of intensity incident on each sub-cell. For the AM0 spectral conditions a compromise between the sharp spectral bands and UV illumination had to be made, which causes coupling between spectral bands intended for GaInP and GaAs sub-cells (see Fig. 1a). For the AM1.5 configuration, the transitions between spectral bands near the band edges of the sub-cells are sharp (see Fig. 1b), which allows the individual sub-cell currents to be determined by altering the bias light intensity incident on each sub-cell.

978-1-5090-5606-4/17 $31.00 © 2017 IEEE

The intensity calibration of the three band simulator was done with single junction GaInP, GaAs, and GaInNAsSb solar cells that were calibrated by external quantum efficiency measurements. The beam size and spatial uniformity of the three band solar simulator limited the cell size to 1×1 cm^2. The long term temporal instability of the three band simulator was $\pm 2\%$. The sample stage allowed heating with a resistive element and the stage temperature was monitored with an AD590 temperature transducer. The temperature could be determined with an accuracy of $0.1°C$ with relative error of $\pm 2.5°C$.

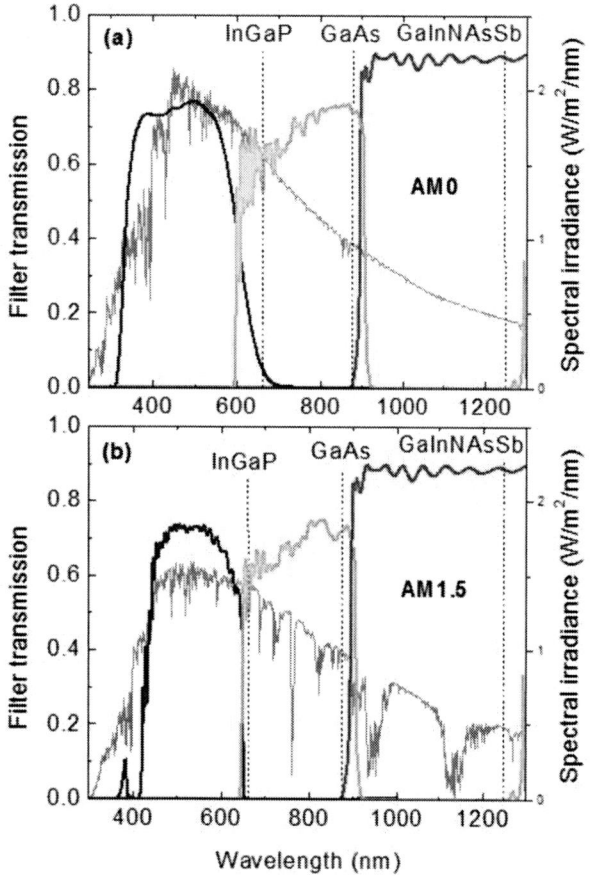

Fig. 1. (a) The filter transmissions of AM0 configuration with ASTM E490-00a reference spectrum [7] and (b) AM1.5 configuration with ASTM G173-3 AM1.5D reference spectrum (1000 W/m^2 normalization) [8]. The absorption edges of the sub-cells are indicated with dotted lines.

For comparison, the temperature dependent LIV characteristics of the 3J cell were also measured by using a 1000 W single source Xenon solar simulator manufactured by Oriel. The Xenon simulator could be equipped with AM0 or AM1.5G air mass filters. The AM0 illumination intensity of the Xenon simulator was calibrated by using a dual junction

solar cell calibrated at Fraunhofer ISE. The same sample stage that was used for the three band simulator measurements was also used for the Xenon simulator measurements.

III. RESULTS

The LIV characteristics of the 3J sample were measured in the temperature range of 25–90°C with the three band simulator under one-sun AM0 spectral conditions utilizing ASTM E490-00a AM0 [7] as the reference spectrum. The measurements were repeated by using the Xenon simulator calibrated to one-sun AM0 spectral conditions. The LIV curves measured from the 3J cell with both simulator systems are presented in Fig. 2.

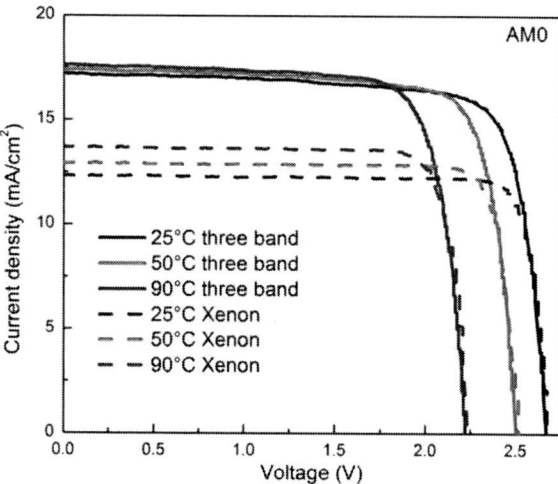

Fig. 2. Temperature dependent LIV curves measured for the 3J at AM0 spectral conditions. The solid and dashed lines present the curves measured with the three band simulator and the Xenon simulator, respectively.

From Fig. 2. we can observe a clear difference between the behavior measured with the two simulators; approximately 30% lower short-circuit current density (J_{sc}) values were measured with the Xenon simulator. Under illumination from the three band simulator, J_{sc} of the 3J cell was determined to be 17.2 mA/cm^2 at 25°C. The corresponding value under Xenon source illumination was 12.3 mA/cm^2. The large variation in the measured J_{sc} values also affects the conversion efficiency significantly; with the three band simulator the active area efficiency (η_{active}) was 26.7% whereas with illumination from the Xenon simulator η_{active} was only 21.1% at 25°C. The deviations in the J_{sc} values suggests that the spectra produced by the Xenon simulator does not sufficiently match the reference spectra and causes imbalance in the current generation between the sub-cells, thus, resulting in distorted J_{sc} values. The spectrum of the Xenon source deviates from the reference AM0 spectrum below 300 nm range as well as in the 750–1000 nm range, where the spectrum exhibits several peaks and dips with a major dip around 800–850 nm [9]. Due to the significant differences in the measured J_{sc} values, the Xenon simulator should be used with caution for MJSC characterization.

In addition, LIVs in Fig. 2. indicate that the cell current increases slightly with increasing temperature, regardless of the source of illumination, as the bandgap energies of the sub-cells red-shift with temperature causing changes in the absorption bands of the sub-cells. Fig. 3. illustrates the change in the J_{sc} when the cell temperature is varied.

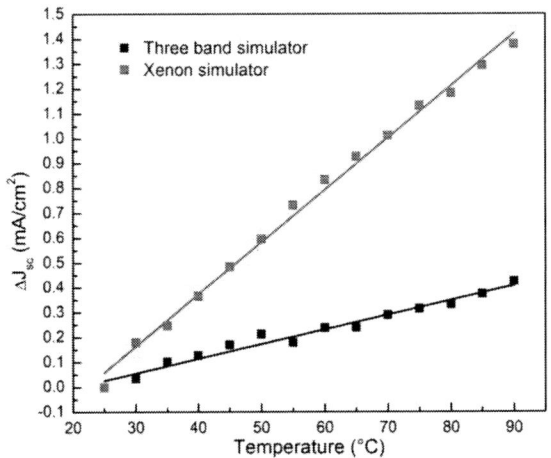

Fig. 3. ΔJ_{sc} values measured from the 3J cell between 25–90°C under AM0 illumination.

In Fig. 2. the open-circuit voltage (V_{oc}) exhibits a negative temperature dependence that is caused by increased dark current due to increased thermal generation of carriers [10]. The fill factor (FF) also decreases as function of temperature. The negative temperature dependencies of V_{oc} and FF overcome the positive change in cell current, thus, causing the temperature dependence of conversion efficiency to become negative.

All temperature induced changes are close to linear. The temperature coefficients for J_{sc} were determined from the slopes of linear fits of the experimental data. With illumination from the three band simulator a temperature coefficient of 5.3 µA/cm²/°C was determined. The temperature coefficient is in good accordance with the value of 5 µA/cm²/°C reported for GaInP/GaAs/Ge cell under AM0 illumination [11]. The coefficient value for J_{sc} extracted from the Xenon simulator measurements was 20.1 µA/cm²/°C, which is significantly higher than the value presented in the literature. This supports the statement that the Xenon source is not adequate for MJSC characterization.

The temperature coefficients for V_{oc} were also extracted from the LIV data. The analysis yielded -6.8 mV/°C and -6.6 mV/°C with illumination for the three band simulator and the

Xenon simulator, respectively. The measured coefficients are reasonably close to each other regardless of the source of illumination, and are also close to values that have been previously published for other III–V MJSCs [11]-[12]. The temperature dependent behavior of V_{oc} is illustrated in Fig. 4.

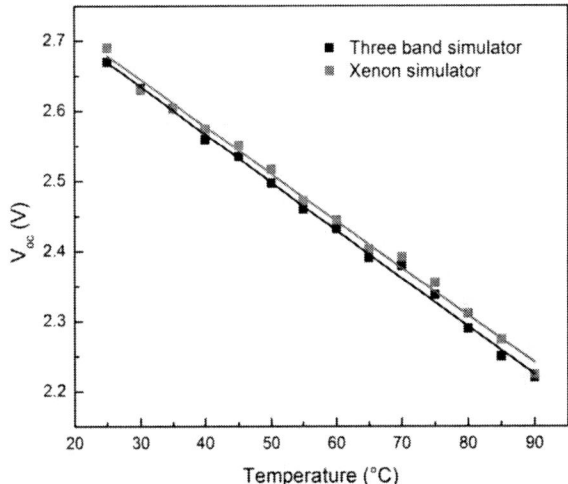

Fig. 4. Temperature dependence of V_{oc} in between 25–90°C under AM0 illumination.

The conversion efficiency of the cell decreases with increasing temperature. The temperature dependence of η_{active} for the 3J cell is illustrated on Fig. 5.

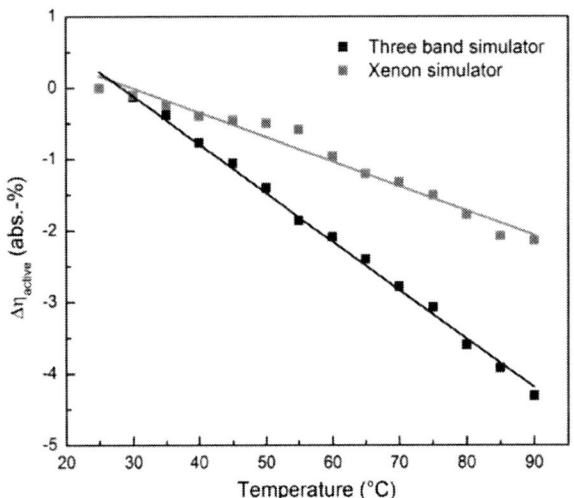

Fig. 5. $\Delta\eta_{active}$ values determined for the 3J cell in the temperature range of 25–90°C with AM0 spectral conditions.

The change in the η_{active} is significantly higher for measurements made with the three band simulator compared to the Xenon simulator. This difference in the temperature behavior of the conversion efficiency is a direct consequence from the difference in the temperature coefficients for J_{sc} obtained for the cell with different illumination (see Fig. 3.).

With illumination from the three band simulator the η_{active} decreased with a slope of -0.068 abs.-%/°C. With illumination from the Xenon simulator the temperature dependency of conversion efficiency was significantly smaller with a corresponding rate of -0.034 abs.-%/°C. The temperature coefficient for η_{active} measured with the three band solar simulator is reasonably close to temperature coefficient value of -0.057 abs.-%/°C reported for GaInP/InGaAs/Ge 3J cell [13].

IV. CONCLUSIONS

The temperature dependent behavior of a MBE-grown GaInP/GaAs/GaInNAsSb solar cell was studied at AM0 spectral conditions. The solar simulation for the experiments was realized with a single source Xenon simulator and a customized three band solar simulator. With illumination from the Xenon simulator the cell current was observed to be underestimated by ~30% compared to the three band simulator measurements. Temperature coefficients for J_{sc}, V_{oc} and η_{active} were extracted from the LIV data. In general, the determined temperature coefficient values were found to be in a reasonably good agreement with temperature coefficient values published for other 3J cells but the temperature coefficient for J_{sc} measured with the Xenon simulator was found to deviate significantly from the literature value. The large difference in the coefficient for J_{sc} is was attributed to insufficient spectral matching of the Xenon solar simulator that causes imbalance in the current generation between the sub-cells.

ACKNOWLEDGEMENT

This work was financially supported by European Research Council within the AMETIST project (ERC-2015-AdG, action #695116). In addition, R.I. acknowledges the financial support from Doctoral Training Network in Condensed Matter and Material Physics (CMMP).

REFERENCES

[1] M.A. Green, K. Emery, Y. Hishikawa, W. Warta and E.D. Dunlop, "Solar cell efficiency tables (version 47)", *Progress in Photovoltaics: Research and Applications*, vol. 24, no. 1, pp. 3-11, 2016.

[2] R.R. King, D., Bhusari, A. Boca, D. Larrabee, X.-Q. Liu, W. Hong, C.M. Fetzer, D.C. Law, and N.H. Karam, N. H., "Band gap-voltage offset and energy production in next-generation multijunction solar cells", *Progress in Photovoltaics: Research and Applications*, vol. 19, pp. 797–812, 2011.

[3] S.H. Liu, E.J. Simburger, J. Matsumoto, A. Garcia, J. Ross and J. Nocerino, "Evaluation of thin-film solar cell temperature coefficients for space applications", *Progress in Photovoltaics: Research and Applications*, vol. 13, no. 2, pp 149-156, 2005.

[4] G. Siefer and A.W. Bett, "Calibration of III-V Concentrator Cells and Modules", *2006 IEEE 4th World Conference on Photovoltaic Energy Conference*, Waikoloa, HI, pp. 745-748, 2006.

[5] A.W. Walker, J.F. Wheeldon, O. Theriault, M.D. Yandt and K. Hinzer, "Temperature dependent external quantum efficiency simulations and experimental measurement of lattice matched quantum dot enhanced multi-junction solar cells", *2011 37th IEEE Photovoltaic Specialists Conference*, Seattle, WA, pp. 000564-000569, 2011.

[6] A. Aho, A. Tukiainen, V. Polojärvi and M. Guina, "Performance assessment of multijunction solar cells incorporating GaInNAsSb", *Nanoscale Research Letters*, vol. 9, p. 61, 2014.

[7] ASTM E490-00a, *Standard Solar Constant and Zero Air Mass Solar Spectral Irradiance Tables*, ASTM International, West Conshohocken, PA, 2000.

[8] ASTM G173-03, "Standard Tables for Reference Solar Spectral Irradiances: Direct Normal and Hemispherical on 37° Tilted Surface", ASTM International, West Conshohocken, PA, 2003.

[9] Newport Corporation, "Curve Normalization", http://www.newport.com/Curve-Normalization/412214/1033/content.aspx, January 2017.

[10] E. Radziemska and E. Klugmann, "Thermally affected parameters of the current–voltage characteristics of silicon photocell", *Energy Conversion and Management*, vo. 43, no. 14 pp. 1889–1900, 2002.

[11] Spectrolab, "28.3% Ultra Triple Junction (UTJ) Solar Cells Datasheet", http://www.spectrolab.com/DataSheets/TNJCell/utj3.pdf, January 2017.

[12] A. Braun, E.A. Katz, J.M. Gordon, "Basic aspects of the temperature coefficients of concentrator solar cell performance parameters", *Progress in Photovoltaics: Research and Applications*, vol. 21, no. 5, pp. 1087-1094, 2013.

[13] B. Cho, J. Davis, L. Hise, A. Korostyshevsky, G. Smith, A. V. Ley, P. Sharps, T. Varghese, M. Stan, "Qualification testing of the ZTJ GaInP2/GaInAs/Ge solar cell to the AIAA S-111 standard", *Photovoltaic Specialists Conference (PVSC)*, 2009 34th IEEE, pp. 001009-001014, 2009.

Efficiency of GaAs P/Si Two-junction Solar Cells with Multi-Quantum Wells: a Realistic Modeling with Carrier Collection Efficiency

Boram Kim[1], Kasidit Toprasertpong[1], Oliver Supplie[2], Agnieszka Paszuk[2], Thomas Hannappel[2], Yoshiaki Nakano[1] and Masakazu Sugiyama[1]

[1]School of Engineering, the University of Tokyo, Bunkyo-ku, Tokyo, 113-8656, Japan
[2]Institute of Physics, Technical University Ilmenau 98683, Germany

Abstract — A new structure for III-V on Si two-junction solar cells is proposed with the use of multi-quantum wells (MQWs), reducing the lattice constant of the GaAsP-based top cell and minimizing the thickness of a metamorphic buffer layer between the top cell and the GaP seed layer on Si. The strain-balanced MQWs helps the reduction of As content in the top-cell matrix while extending the absorption edge to a longer wavelength by narrow-gap quantum wells. The efficiency is predicted to be as large as 40% with an entire MQW thinner than 800 nm. The model takes into account the drawbacks of MQWs such as limited light absorption and the bottleneck of carrier collection from the confinement states. This work addresses a new direction toward high-efficiency III-V on Si cells.

I. INTRODUCTION

III-V compound semiconductors are promising materials for achieving high efficiency solar cells, but its high cost is still a bottleneck of application for solar cells. While, Si is the most general material of solar cells for its low cost and matured technologies for mass production, the energy conversion efficiency of a Si solar cell has been saturated, and the best score of Si solar cell increased by less than 1% in recent 10 years. For these reasons, solar cells with III-V material integrated on Si substrate has been researched widely for lowering fabrication cost and enhancing efficiency [1].

In III-V semiconductors, GaAs or its alloys are the most required material for high efficiency multi-junction solar cells [2]. To integrate III-V materials such as GaAs on a Si substrate, buffer layers are needed to suppress defects coming from differences of material properties - such as differences of lattice constant, polarization and thermal expansion coefficient - between Si and III-V materials. The combination of a GaP seed layer and a GaAsP graded buffer layer is a good candidate for the integration of Si and III-V alloys because GaP is optically transparent to a Si bottom cell and its lattice constant is matched with Si [1, 3-5]. For these reasons, we here consider about only GaAs or GaAsP as a top cell material.

II. CALCULATION

A. The structure of two-junction cells with MQWs

The theoretical limit (by Shockley-Quessier modeling) of Si single junction solar cell is 30% under AM1.5 [6]. But, if Si can be used as a bottom cell of a two-junction solar cell, its efficiency can reach up to 45%. And the maximum efficiency can be achieved when the top cell energy bandgap is 1.73eV as shown in Fig1. For GaAsP alloy to achieve this bandgap, As content must be as high as 75% and a thick buffer layer (4~5μm) is required in order to manage a large lattice mismatch of 2.8% between a GaAs$_{0.75}$P$_{0.25}$ top cell and a GaP seed layer [7].

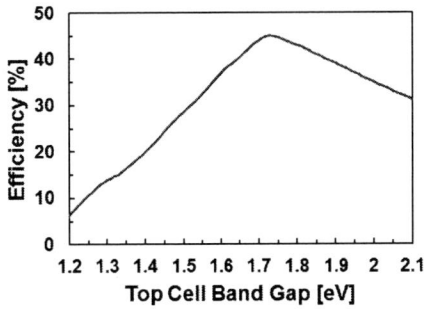

Fig.1 Efficiency of two-junction solar cell with Si bottom cell calculated by detailed balance modeling (AM1.5)

To reduce the thickness of buffer layer, it is mandatory to use a GaAsP top cell with a smaller As content so that we can reduce the lattice mismatch between the top cell and the GaP seed layer, which inevitably results in the current mismatch between subcells and the efficiency of the two-junction cell is degraded. To solve this trade-off, we here propose a new structure: a GaAsP/Si two-junction solar cell with multi-quantum wells (MQWs) in the top cell, as shown in Fig.2. As an example, GaAs$_{0.5}$P$_{0.5}$ which has lower arsenic content than GaAs$_{0.75}$P$_{0.75}$ is adopted as a host material of the top cell, the bandgap of which is 2.05eV and is not suitable for current-matching with the Si bottom cell. However, by inserting QW layers in the top cell, the output current can be enhanced by extending the absorption edge to longer wavelengths [8]-[10], and the efficiency degradation due to current mismatch can be mitigated. GaAs$_{0.5}$P$_{0.5}$ can reduce lattice mismatch versus GaP to ~1.8% compared with 2.8% for GaAs$_{0.75}$P$_{0.25}$ and we can expect substantial decrease in the thickness of the metamorphic buffer layer.

978-1-5090-5606-4/17 $31.00 © 2017 IEEE

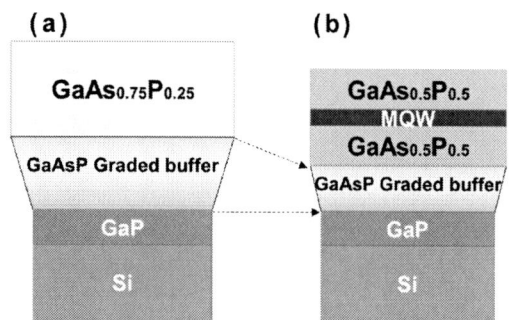

Fig.2 Schematic of GaAsP/Si 2-junction solar cell with (a)As 75% top cell (b)As 50% top cell with multi-quantum wells (MQWs)

B. Assumptions in the model

Based on the proposed structure of two-junction solar cell with MQWs, the expected maximum efficiency η (%) was calculated by incorporating the degradation in carrier collection efficiency (CCE) due to carrier confinement inside the QWs [11-12], as will be described in the next section.

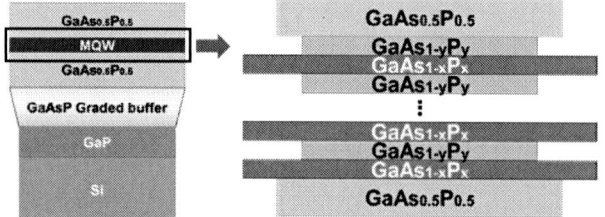

Fig.3 Schematic diagram of GaAs$_{0.5}$P$_{0.5}$ with MQWs on Si 2-junction solar cell and the modeling of multi-quantum wells

Fig.3 shows the structure of two-junction solar cell with MQWs in the top cell on the basis of the following assumptions. For the MQWs embedded in GaAs$_{0.5}$P$_{0.5}$ and composing the top cell, the content of wells and barriers are set to GaAs$_{1-x}$P$_x$ and GaAs$_{1-y}$P$_y$, and the thicknesses were set to L_w and L_b, respectively (here x < 0.5 < y). The phosphorous content in the barrier was decided to satisfy the strain-balance condition as in (1),

$$A_{bulk} = \frac{L_w A_w + L_b A_b}{L_w + L_b} \qquad (1)$$

Here, A_{bulk}, A_w and A_b represents a lattice constant of bulk material (GaAs$_{0.5}$P$_{0.5}$), the well and the barrier, respectively. The well thickness was fixed at 5nm.

The bandgaps of the well and barrier were calculated taking into account the impact of strain such as elastic stiffness, shear modulus and deformation potential. And there are several additional assumptions in the simulation:

- No optical loss in the metamorphic buffer layer, and the GaP seed layer is ideal so that there is no performance degradation in those layers.
- 100% light absorption in Si and GaAs$_{0.5}$P$_{0.5}$ matrix.
- Absorption coefficient in quantum well was set to 32000 cm^{-1} (~80% absorbed in 100-period MQWs)
- No efficiency loss associated with non-radiative recombination

All the calculation was done with AM1.5 spectrum.

C. Carrier collection efficiency in relation to effective mobility

Behavior of carriers in MQWs is inferior to the one in a bulk material due to quantum confinement effect. But numerical simulation including all the carrier transport processes is time consuming and is not suitable for the efficient structural optimization. Here, we approximated MQWs as a quasi-bulk material and introduced a concept of effective carrier mobility μ_{eff} for the entire MQWs [12]. It is a sum of thermal mobility μ_{th}, tunneling mobility μ_{tun} and thermal-assisted tunneling mobility μ_{tun_th}, as in (2),

$$\mu_{eff} = \mu_{th} + \mu_{tun} + \sum_{i \geq 2} \mu_{tun\_th,i} \qquad (2)$$

where, each component can be obtained using the formulae which are the functions of the barrier height from the confined energy levels in a well numbered as i. Due to the space limitation, the detailed description on these formulae will be given in another publication. The effective mobility then gives CCE as

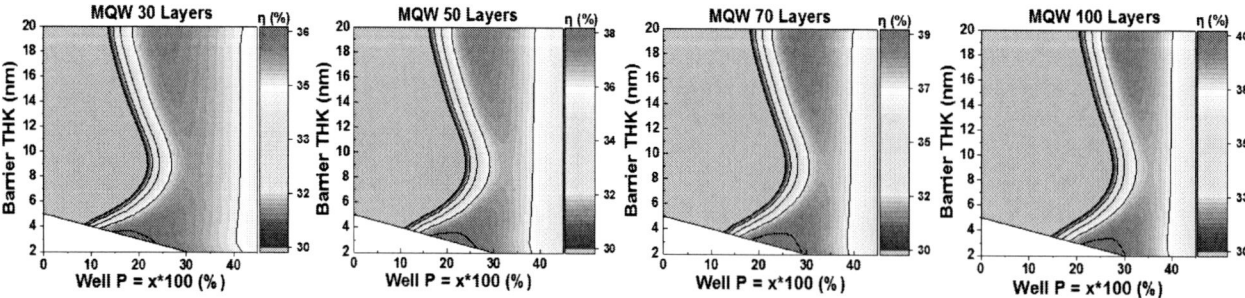

Fig.4 Expected efficiency from GaAs$_{0.5}$P$_{0.5}$ with MQWs on Si two-junction solar cell.

TABLE I

MAXIMUM EFFICIENCY RESULT FROM CARRIER ESCAPE CONSIDERED CALCULATION

Barrier Thickness	MQW = 30Layers			MQW = 50Layers			MQW = 70Layers			MQW = 100Layers		
	η (%)	x	y	η (%)	x	y	η (%)	x	y	η (%)	x	y
2 nm	35.3	0.3	1	37.2	0.3	1	38.6	0.3	1	39.8	0.3	1
3 nm	35.8	0.2	1	38.1	0.2	1	39.2	0.25	0.92	40.1	0.25	0.92
4 nm	35.6	0.25	0.81	37.6	0.25	0.81	38.7	0.25	0.81	39.5	0.3	0.75
5 nm	35.4	0.25	0.75	37.1	0.3	0.7	38.3	0.3	0.7	39.0	0.3	0.7
10 nm	35.1	0.3	0.6	36.5	0.35	0.58	37.4	0.35	0.58	38.2	0.35	0.58
15 nm	35.3	0.3	0.57	37.1	0.3	0.57	38.1	0.3	0.57	38.6	0.3	0.57
20 nm	35.5	0.25	0.56	37.5	0.25	0.56	38.6	0.3	0.55	39.5	0.3	0.55

$$CCE = \exp\left(-\frac{L}{4L_d}\right) \qquad (3)$$

where L_d is the average drift length of electrons and holes in the MQW region and L is the entire thickness of MOWs.

III. RESULTS AND DISCUSSION

The calculated efficiency of the two-junction cells are shown in table I and Fig. 4. The variables were 1) phosphorous content in well x in the range of 0 to 0.45, 2) barrier thickness L_b in the range of 2 to 20 nm, and 3) the period of quantum wells from 1 to 100.

In the contour plot Fig.4, the region filled with gray has no advantage over the theoretical efficiency 30% for Si single-junction cell. As for the impact of the barrier thickness, high efficiency can be achieved with either very thin barriers (under 4 nm with high P content) or very thick barriers (over 20 nm with low P content). For thin and high barriers, carrier transport by tunneling is dominant. For thick and low barriers, on the other hand, thermal escape is dominant.

According to effective mobility calculated in this simulation, effective tunneling mobility is less than 0.1 cm^2/Vs when its barrier thickness is over 5 nm. And effective thermal mobility can exceed 1 cm^2/Vs if barrier thickness is over 30 nm accompanied by reduced P content.

For each MQW period, η takes the maximum when the barrier thickness was 3 nm. Thin barrier is also beneficial for the purpose of reducing the total thickness of MQWs.

This simulation confirmed that 40% efficiency can be achieved by the two-junction solar cell with the GaAs$_{0.5}$P$_{0.5}$ top cell including 100-period MQWs and Si as a bottom cell.

IV. CONCLUSION

We proposed a new structure for high efficiency and low cost III-V/Si two-junction solar cell in which GaAs$_{0.5}$P$_{0.5}$ top cell include strain-balanced MQWs composed of GaAsP for the purpose of extending the absorption edge to longer wavelengths and improving the balance of subcell currents. The thickness of the metamorphic buffer layer can be reduced by a smaller lattice mismatch between GaAs$_{0.5}$P$_{0.5}$ and GaP (~1.8%) compared with 2.8% for a GaAs$_{0.75}$P$_{0.25}$ bulk, which is necessary for current matching without MQWs. Degradation of carrier collection efficiency (CCE) from the MQWs due to carrier confinement in the wells was incorporated in the efficiency estimation for the first time. According to the calculation, very thin (under 4nm) or very thick (over 30nm) barrier is suitable for enhancing CCE, and the efficiency can reach 40% if 100-periods MQWs are implemented. Even 50-periods MQWs can lead to 38% in efficiency. For the purpose of cost reduction, we need to reduce the thickness of entire III-V layers and the thicknesses of 3nm for the barriers and 5 nm for the wells can suppress the total thickness of MQWs under 800nm even 100-periods MQWs are implemented.

REFERENCES

[1] N. Jain and M. K. Hudait, "III–V Multijunction Solar Cell Integration with Silicon: Present Status, Challenges and Future Outlook," *Energy Harvesting and Systems*, 1(3-4), p. 121-145, 2014.

[2] K. Derendorf, S. Essig, E. Oliva, V. Klinger, T. Roesener, S. P. Philipps, J. Benick, M. Hermle, M. Schachtner, G. Siefer, W. Jager and F. Dimroth, "Fabrication of GaInP/GaAs//Si Solar Cells by Surface Activated Direct Wafer Bonding, " *IEEE Journal of Photovoltaics*, vol. 3, p. 1423-1428, 2013

[3] S. A. Ringel, J. A. Carlin, T.J. Grassman, B. Galiana, A.M. Carlin, A.M. Carlin, D. Chmielewski, L. Yang, M.J. Mills, Al Mansouri, S. P. Bremner, A. Ho-Baillie, X. Hao, H. Mehrvarz, G. Conibeer and M. A. Green, "Ideal GaP/Si Heterostructures Grown by MOCVD: III-V/Active-Si Subcells, Multijuntions, and MBE-to-MOCVD III-V/Si Interface Science," *39th IEEE Photovoltaic Specialists Conference*, 2013

[4] T. J. Grassman, M. R. Brenner, S. Rajagopalan, R. Unocic, R. Dehoff, M. Mills, H. Fraser, and S. A. Ringel, "Control and elimination of nucleation-related defects in GaP/Si(001) heteroepitaxy," Applied Physics Letter, vol. 94, 232106, 2009

[5] T. J. Grassman, J. A. Carlin, C. Ratcliff, D. J. Chmielewski, S. A. Ringel, "Epitaxially-Grown Metamorphic GaAsP/Si Dual-Junction Solar Cells," *39th IEEE Photovoltaic Specialists Conference*, 2013

[6] W. Shockley and H. J. Queisser, "Detailed Balance Limit of Efficiency of p-n Junction Solar Cells," *Journal of Applied Physics*, vol. 32, p. 510.

[7] K. N. Yaung, M. Vaisman, J. Lang, and M. L. Lee, "GaAsP solar cells on GaP/Si with low threading dislocation density," *Applied Physics Letters*, vol. 109, 032107, 2016

[8] K. W. J. Barnham and G. Duggan, "A new approach to high-efficiency multi-band-gap solar cells," *Journal of Applied Physics*, vol. 67, p. 3490-3493, 1990

[9] M. Sugiyama, Y. Wang, H. Fujii, H. Sodabanlu, K. Watanabe and Y. Nakano, "A quantum-well superlattice solar cell for enhanced current output and minimized drop in open-circuit voltage under sunlight concentration," *Journal of Physics D: Applied Physics*, vol. 46, 024001, 2013

[10] H. Fujii, K. Toprasertpong, Y. Wang, K. Watanabe, M. Sugiyama and Y. Nakano, "100-period, 1.23-eV bandgap InGaAs/GaAsP quantum wells for high-efficiency GaAs solar cells: toward current-matched Ge-based tandem cells," *Pogress in Photovoltaics: Research and Applications*, vol. 22, p. 784–795, 2014

[11] H. Fujii, K. Toprasertpong, K. Watanabe, M. Sugiyama and Y. Nakano, "Evaluation of Carrier Collection Efficiency in Multiple Quantum Well Solar Cells," *IEEE Journal of Photovoltaics*, vol. 4, p. 237-243, 2014

[12] K. Toprasertpong, T. Tanibuchi, H. Fujii, T. Kada, S. Asahi, K. Watanabe, M. Sugiyama, T. Kita and Y. Nakano, "Comparison of Electron and Hole Mobilities in Multiple Quantum Well Solar Cells Using a Time-of-Flight Technique," *42nd IEEE Photovoltaic Specialists Conference*, 2015

Inverse Metamorphic III-V/epi-SiGe Tandem Solar Cell Performance Assessed by Optical and Electrical Modeling

Raphaël Lachaume[1,2], Martin Foldyna[3], Gwénaëlle Hamon[3,4], Nicolas Vaissière[3], Jean Decobert[5], Romain Cariou[3,5], Pere Roca i Cabarrocas[3], José Alvarez[1] and Jean-Paul Kleider[1]

[1]GeePs; CNRS UMR 8507; CentraleSupélec; Univ Paris-Sud ; Sorbonne Universités-UPMC Univ Paris 06; 11 rue Joliot-Curie, Plateau de Moulon, F-91192 Gif-sur-Yvette Cedex, France

[2]Institut Photovoltaïque d'Ile-de-France (IPVF), Antony, France

[3]LPICM, CNRS, Ecole Polytechnique, Université Paris-Saclay, 91128 Palaiseau, France

[4]TOTAL New Energies, 24 cours Michelet, 92069 Paris La Défense Cedex, France

[5]III-V Lab, 1 av. Augustin Fresnel, 91767 Palaiseau, France

Abstract — **Recent developments have unlocked the main issues arising from the combination of III-V and silicon and have opened a new way to fabricate tandem solar cells. We here propose to evaluate such tandem concept based on inverse metamorphic growth of c-Si(Ge) on GaAs by means of numerical simulation. Electrical and optical models are first faced to experimental realizations of single junction cells to calibrate material parameters and to assess the electrical quality of the epi-SiGe layer. Then the tandem structure is optimized, current matching conditions are given and the benefit of using a 2D grating at the back-side is studied.**

I. INTRODUCTION

The growing interest in new concepts of III-V/Si tandem solar cells arises from the need to reduce the cost of high efficiency III-V based multijunctions by using low cost substrates such as silicon [1]. Because it is a challenge to grow III-V materials directly onto Si wafers due to thermal expansion coefficient and lattice parameter mismatches but also polar on non-polar material growth issues, other ways of combining III-V compounds and Si have been developed. Among them is the inverse metamorphic concept recently proposed by Cariou et al. [2]. In the latter approach, the crystalline silicon bottom cell is deposited at low temperature (<200°C) by plasma-enhanced chemical deposition (PECVD) directly on the III-V top-cell, as shown in Fig. 1, preventing the degradation of the electrical properties of the underlying III-V layers grown by metal organic chemical deposition (MOCVD). It has also been demonstrated that it is possible to grow crystalline silicon-germanium on III-V as an alternative to silicon using such low temperature PECVD process [2], [3]. SiGe should be preferred to Si to benefit from the higher absorption coefficient of the SiGe alloy, allowing for sufficient photogeneration in thin epitaxial absorber layers [4]. The tandem cell is then transferred to a low cost carrier and the GaAs substrate can be reclaimed.

Although the possibility to achieve a high crystalline quality for the epitaxial Si or SiGe alloys has already been demonstrated [2], precise evaluations of the electrical transport properties and the optical absorption in the epi-Si(Ge) absorber are mandatory, as well as the calculation of the theoretical efficiency of such a tandem cell. For this purpose, an in-depth electrical and optical modeling study is proposed in order to analyze this novel concept in terms of performance and thus to provide guidelines for the design and fabrication of these tandem solar cells.

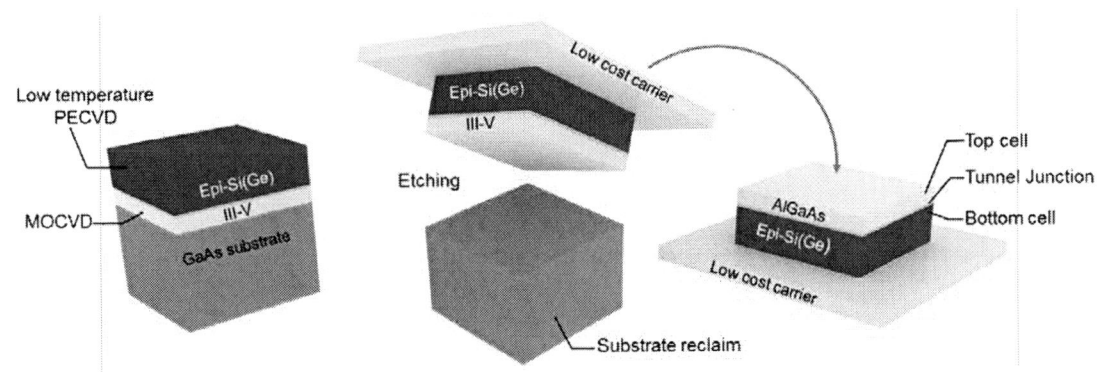

Fig. 1.   Principle of the IMPETUS approach: inverted metamorphic growth of Si(Ge) on III-V.

II. RESULTS AND DISCUSSION

Simulations of the tandem cell electrical performance have been made using a commercial TCAD numerical tool. It has been coupled with an in-house advanced optical model to take into account diffraction of light in case a nanostructured light trapping scheme is present in the tandem design. Electrical material parameters have been either extracted from literature or thoroughly calibrated by fitting experimental data obtained from previously fabricated single junction solar cells. Indeed, we have successfully reproduced GaAs and AlGaAs experimental cells characteristics such as AM1.5G current-voltage (I-V) or internal quantum efficiency curves published in [5] and [6]. We have also successfully reproduced the I-V curves of previously fabricated epi-Si(Ge) heterojunction solar cells [3], [7] obtained for different absorber thicknesses. Namely, the latter simulation-experiment comparison has allowed us to retrieve the interface and bulk defects density inside the epitaxial Si and SiGe layers and hence to estimate the effective diffusion length (L_d) in such absorbers. We have shown that L_d was mainly limited by the interface quality rather than by the bulk. Bulk defects density has actually been found to be exponentially decaying with increasing thickness of epitaxial absorber, reaching values below 2×10^{14} cm^{-3} for a capture cross section fixed to 10^{-16} cm^2. Resulting L_d values of ~ 16 μm and ~ 9 μm could be obtained for 4.2 μm thick epi-Si and 1.9 μm thick epi-SiGe, respectively, i.e. L_d is ~ 4 times larger than the thickness of the absorber. This ensures a very good carrier collection in thin epitaxial SiGe solar cells and demonstrates the feasibility of using these thin epitaxial layers as high quality absorbers in tandems.

Once we have verified the suitability of the top and bottom models for reproducing fabricated cells characteristics, we have optimized the tandem cell first considering no defective layers, then introducing bulk and interface defects densities previously extracted in the epi-SiGe absorber.

The AlGaAs/epi-SiGe tandem structure that we have simulated and optimized is described in Fig. 2. The Anti-Reflective Coating (ARC) composition and thickness have been adjusted to minimize the cell reflectivity. The Al$_x$Ga$_{1-x}$As top cell absorber composition x must be tuned to adjust the band gap to have the top and bottom cells short circuit current density (J_{sc}) matched for given top and bottom absorbers thicknesses. The optimized emitter thickness is 100 nm while the top absorber thickness is fixed to 1 μm, ensuring a sufficient light absorption in the top cell [4]. The tunnel junction shown here is based on GaAs, which doping and thicknesses are optimized to ensure a sufficient tunneling current density as well as to provide an efficient internal electric field inside the bottom absorber made out of epi-Si$_{0.73}$Ge$_{0.27}$. The amorphous silicon bilayer (i/n) a-Si:H plays the combined roles of passivation layer and back surface field for the bottom cell. The thickness of the bottom epitaxial absorber achievable experimentally is variable from few microns to tens of microns. For this range of

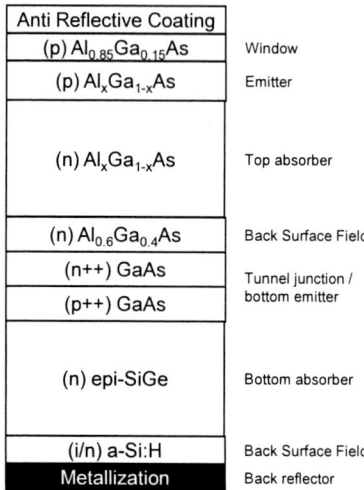

Anti Reflective Coating	
(p) Al$_{0.85}$Ga$_{0.15}$As	Window
(p) Al$_x$Ga$_{1-x}$As	Emitter
(n) Al$_x$Ga$_{1-x}$As	Top absorber
(n) Al$_{0.6}$Ga$_{0.4}$As	Back Surface Field
(n++) GaAs	Tunnel junction /
(p++) GaAs	bottom emitter
(n) epi-SiGe	Bottom absorber
(i/n) a-Si:H	Back Surface Field
Metallization	Back reflector

Fig. 2.   Simulated AlGaAs/SiGe tandem solar cell structure.

thickness, it is mandatory to enhance the light absorption, by increasing the light path inside this layer *e.g.* using an efficient back reflector. In this study we compared different back metallization schemes on the amorphous silicon layer: a simple aluminum layer, a bilayer of ZnO (25 nm) on silver and a 2D grid with a pitch of 750 nm on ZnO and Ag substrate.

An example of our tandem simulation results is given in Fig. 3. We here compare simulated tandem J_{sc} contour maps obtained for the Al reflector and the ZnO/Ag reflector cases, without introducing defects. This figure shows the best Al$_x$Ga$_{1-x}$As composition x to reach the highest J_{sc} value for epi-SiGe thicknesses ranging from 5 μm to 40 μm. The resulting best (x,y) pairs are indicated by a black dotted line, also giving the top and bottom cells current matching condition. The ZnO/Ag being more efficient than a single Al metallization as a reflector, higher J_{sc} values can be obtained. However, the current matching condition changes and therefore it is important to take this into account in the design of the tandem cell.

The analysis of balancing currents in the tandem device has also been done including the 2D grating to emphasize the benefits in terms of the efficiency gain or epi-SiGe thickness reduction. With a grating, the optimum Si$_{0.73}$Ge$_{0.27}$ thickness is 8.5 μm for a fixed 17% Al content whereas for the flat device and using the same optical model, we have achieved the same current matching for a SiGe thickness of 17 μm.

We have also simulated the effect of interface and bulk defects inside the epi-SiGe bottom cell on the tandem solar cell I-V characteristics. Given the density values that were previously extracted, it has been observed that despite open-circuit voltage (V_{oc}) and fill factor (FF) losses inherent to increased recombinations, the J_{sc} parameter remains unaffected. We thus can claim that (i) the electrical quality of the epi-SiGe material ensures a very high carrier collection efficiency in the bottom cell and (ii) current matching conditions previously calculated without defects remain very suitable.

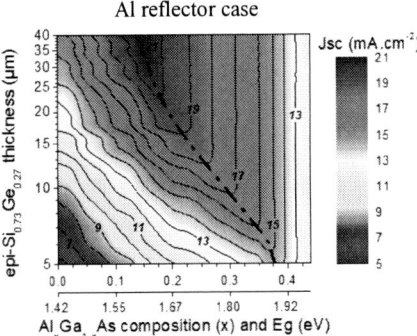

Al reflector case

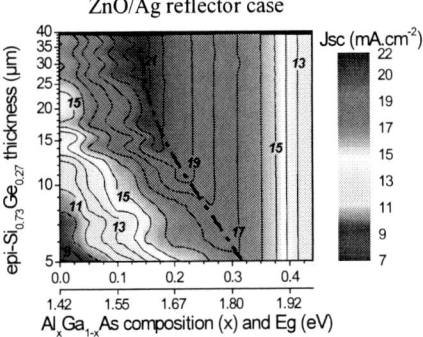

ZnO/Ag reflector case

Fig. 3. 2D contour maps of simulated tandem cell J$_{sc}$ as a function of bottom cell absorber thickness and top cell absorber Al composition x obtained for two different back reflectors.

Finally, as shown in Fig. 4, we have simulated the efficiency of the optimized tandem solar cell as a function of the thickness of the epi-SiGe for three different scenarios corresponding to the ideal case, interface defects only and both interface and bulk defects in epi-SiGe, respectively. Performance is mainly affected by interface defects for epi-SiGe thicknesses below 9 μm and by bulk defects above this value. Therefore, the interface quality is a key parameter that requires specific attention during the device fabrication,

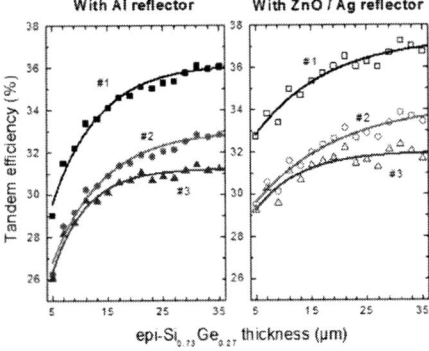

Fig. 4. Simulated tandem cell efficiency as a function of bottom cell absorber thickness for three different scenarios #1, #2 and #3 and for two different back reflectors: Al reflector (left) and ZnO / Ag reflector (right). #1: no defects; #2: interface defects only; #3 interface and bulk defects.

especially if light trapping schemes are used to reduce the required bottom absorber thickness. Finally, we have shown that over 30 % efficiency could be achieved for the tandem with only ~ 7 μm of epi-SiGe and ZnO/Ag back metallization in the most pessimistic scenario, which represents an absolute efficiency difference of ~ 3.5 % compared to the ideal case. In the latter ideal scenario, the highest efficiency achievable is ~ 37 % with 1.1 μm thick Al$_{0.15}$Ga$_{0.85}$As top cell and epi-Si$_{0.73}$Ge$_{0.27}$ bottom cell less than ~ 30 μm thick.

III. SUMMARY

We present an opto-electrical modeling of the recently proposed inverse metamorphic III-V/epi-SiGe tandem solar cell concept. Model parameters are rigorously calibrated and compared to previous experimental realizations of single junction devices. The optimal current matching conditions are found for a very wide range of epi-SiGe thickness and Al content in the AlGaAs top cell absorber. The detailed influence of the bulk and interface electrical quality in the epi-SiGe bottom cell is also assessed. Finally, the prediction of the tandem device performance according to different realistic scenarios is presented. Advanced optical models are used to investigate the impact of using back side light trapping schemes. Using a periodic grating allows us to reduce by 50 % the SiGe film thickness for the same cell performance, which is very important in a thin-film approach and has a tremendous impact on the cost of the device as short deposition time reduces the fabrication cost.

ACKNOWLEDGEMENTS

This work has been carried out within the IMPETUS project N°ANR-13-PRGE-0009-03.

REFERENCES

[1] J. P. Connolly, D. Mencaraglia, C. Renard, and D. Bouchier, "Designing III–V multijunction solar cells on silicon," *Prog. Photovolt. Res. Appl.*, Jan. 2014.

[2] R. Cariou *et al.*, "Low temperature plasma enhanced CVD epitaxial growth of silicon on GaAs: a new paradigm for III-V/Si integration," *Sci. Rep.*, vol. 6, p. 25674, May 2016.

[3] R. Cariou, J. Tang, N. Ramay, R. Ruggeri, and P. Roca i Cabarrocas, "Low temperature epitaxial growth of SiGe absorber for thin film heterojunction solar cells," *Sol. Energy Mater. Sol. Cells*, vol. 134, pp. 15–21, Mar. 2015.

[4] R. Lachaume *et al.*, "Performance Analysis of AlxGa1-xAs/epi-Si(Ge) Tandem Solar Cells: A Simulation Study," *Energy Procedia*, vol. 84, pp. 41–46, Dec. 2015.

[5] A. van Geelen, P. R. Hageman, G. J. Bauhuis, P. C. van Rijsingen, P. Schmidt, and L. J. Giling, "Epitaxial lift-off GaAs solar cell from a reusable GaAs substrate," *Mater. Sci. Eng. B*, vol. 45, no. 1, pp. 162–171, Mar. 1997.

[6] S. Heckelmann, D. Lackner, C. Karcher, F. Dimroth, and A. W. Bett, "Investigations on AlxGa1-xAs Solar Cells Grown by MOVPE," *Ieee J. Photovolt.*, vol. 5, no. 1, pp. 446–453, Jan. 2015.

[7] R. Cariou, "Epitaxial growth of Si(Ge) materials on Si and GaAs by low temperature PECVD: towards tandem devices," phdthesis, Ecole Polytechnique, 2014.

Towards Monolithically Integrated GaAs on Si Tandem Solar Cell

Zhen Liu[1,2], Zekun Ren[3], Haohui Liu[1], Tonio Buonassisi[2,3], Ian Marius Peters[2,3]

[1] Solar Energy Research Institute of Singapore, Singapore, 7 Engineering Drive 1, 117574, Singapore
[2] Massachusetts Institute of Technology, 77 Massachusetts Avenue, Cambridge, MA 02139, USA
[3] Singapore MIT Alliance for Research and Technology, 1 CREATE way, 138602, Singapore

Abstract — GaAs and Si solar cells are the technologies that achieve the highest single-junction efficiencies. Due to the non-ideal bandgap combination, incorporating those two materials into a monolithic integrated tandem is, however, not straightforward. In this paper, we discuss the potential of GaAs/GaAs/Si triple-junction architecture. As the reachable short-circuit current in a GaAs solar cell is about two thirds of that reachable with Si, current matching can be achieved by combining a double-junction GaAs/GaAs top cell with a Si bottom cell. As the first result, we fabricated a prototype of GaAs/GaAs dual-junction top cell with an efficiency of 17.8% and demonstrated a four-terminal tandem solar cell with an efficiency of 20.4%. However, with state-of-the-art material properties, the integrated GaAs/GaAs/Si concept is envisioned to reach a standard-testing-condition efficiency of 33.0% under AM1.5G.

I. INTRODUCTION

Single junction solar cells made from Si and III-V compound semiconductors have each achieved very high efficiencies and have demonstrated reliable long-term performance during outdoor operation. Hence, combining these two technologies to realize a high-efficiency one-sun tandem solar cell is an attractive perspective [1]–[3]. The efficiency potential of GaAs on Si solar cells is among the highest achievable for any Si based tandem solar cell with technology available today [4]–[6]. However, the ideal bandgap partner for silicon is in the range of 1.75 to 1.8 eV, which can be realized by $GaIn_{(1-x)}$. GaAs, on the other hand, has a band gap of 1.42 eV, making it a far less ideal choice. Correspondingly, the radiative efficiency limit for the combination $GaIn_xP_{(1-x)}/Si$ is 45%, and only 41.7% for GaAs/Si, as shown in Fig. 1.

The radiative limit, however, stands in contrast to the efficiencies that were actually achieved using those materials. GaAs stands out as the material that has achieved the highest single-junction one-sun conversion efficiency at 28.8% [7], marking approx. 90% of its radiative limit. Efficiencies achieved with $GaIn_xP_{(1-x)}$ are far lower. The current record for $GaIn_xP_{(1-x)}$ is 20.8% [7] which is approx. 75% with respect to the radiative limit. Si has most recently achieved 26.3% efficiency [8]. Taking these realistic values into consideration, Fig. 1 shows that the efficiency potentials of integrated GaAs/Si tandem solar cells are very attractive.

From the techno-economic perspective, GaAs and Si combination is also attractive because both solar cells can achieve similar efficiencies, which have shown to be an important metric for tandem integration to be economically preferable [9], In addition, metal-organic chemical vapour deposition (MOCVD) of GaAs absorber is much cheaper than deposition of $GaIn_xP_{(1-x)}$ absorber due to higher growth rate –

approximately four times higher according to the cost analysis by Woodhouse and Goodrich [10].

In this paper, we discuss a novel concept of the integrated GaAs/GaAs/Si architecture. A sketch of this architecture is shown in Fig. 2. This architecture solves the current matching issue in an integrated tandem for GaAs and Si combination. We investigate both efficiency and energy yield potential for this architecture and present the first results for a prototype tandem, consisting of a GaAs/GaAs double-junction top cell and a silicon bottom cell in a 4T configuration.

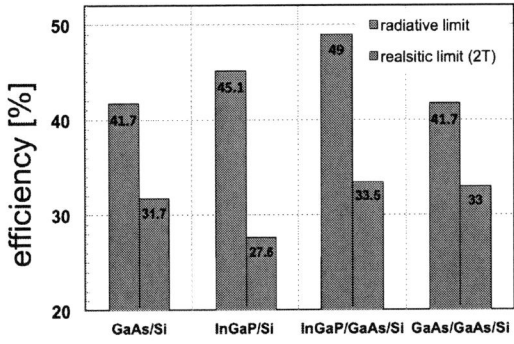

Fig. 1. Comparison of efficiencies between different material combinations of III-V and silicon.

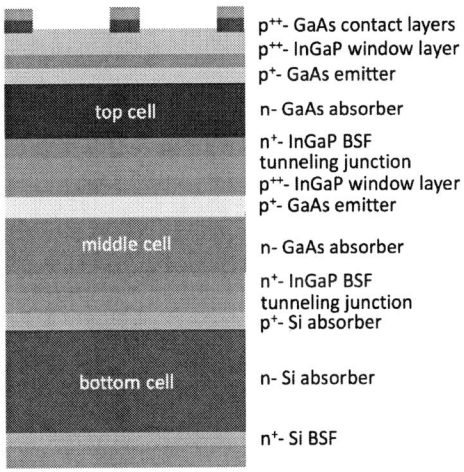

Fig. 2. Schematic diagram of the architecture of the GaAs/GaAs/Si tandem solar cell concept.

Particular attention is dedicated to energy yield as a figure of merit, as tandem solar cells have a different sensitivity to meteorological influences like solar spectrum or temperature than single junction solar cells. A comparison of standard testing condition efficiencies is, therefore, not sufficient to determine the performance potential increase when transferring from a single-junction to a tandem technology.

II. ENERGY YIELD POTENTIAL

Varying operating conditions have an impact on solar cell performance. The most dominant ones are variations in illumination intensity and operating temperature. Integrated tandem solar cells are additionally affected by variations in the spectral composition, as these variations have an impact on the distribution of current generation in top- and bottom cell [11]–[13]. Because of the current-matching requirement, any such variation will affect an integrated tandem negatively. To determine the magnitude of this effect, we have calculated the harvesting efficiency (the ratio of electrical energy generated and available solar energy) for different integrated tandem solar cells for the year of 2014 for conditions in Denver, using best-in-class solar cell parameters. Apart from the GaAs/GaAs/Si architecture that this paper focuses on, we have also considered other III-V on Si tandem solar cell architectures for comparison.

The results of these calculations are summarized in table I. Due to the mentioned variations in insolation, temperature and spectrum, harvesting efficiencies are generally lower than standard testing condition (STC) efficiencies. Triple-junction cells are typically suffering a higher loss, as the current matching effect is more pronounced if the total current is smaller. The InGaP on Si result marks an exception. Due to the lower material quality of InGaP, the top cell is thinner than ideal current generation would dictate, and variations in the spectrum can improve the overall current.

TABLE I: Comparison of standard testing condition (STC) and harvesting efficiencies for different monolithic, two terminal tandem cell architectures.

Architecture	η_{STC}	$\eta_{harvest}$ (Denver)
GaAs/Si	31.7%	31.2%
InGaP/Si	27.6%	27.6%
InGaP/GaAs/Si	33.5%	32.6%
GaAs/GaAs/Si	33.0%	32.1%

Note that, despite the band-gap mismatch, it is possible to approach current matching in an integrated GaAs/Si tandem by thinning down the top cell. We calculated that a top-cell thickness of 240 nm would be required to achieve current matching. The efficiency potential of the GaAs/Si architecture using best-in-class cell parameters would be 31.7%. One limitation of this architecture is that, because of the current

matching requirement, photons are utilized in silicon that could more efficiently be used in GaAs. This limitation can be addressed by replacing the GaAs top cell with a GaAs/GaAs double junction. We estimate the potential of the triple junction GaAs/GaAs/Si architecture with best-in-class parameters at 33.0%, similar to that of InGaP/GaAs/Si at 33.5%. Although the efficiency potential of the GaAs/GaAs/Si tandem is slightly lower than the Inga/GaAs/Si tandem, the optimum absorber thickness of the top GaAs cell is much thinner than the top InGaP cell and the throughput of growing GaAs films is also much faster than growing InGaP films.

III. PROTOTYPE RESULTS

We fabricated a prototype GaAs/GaAs double-junction solar cell with an area of 1×1 cm^2 and semi-transparent rear-side contacts [14]. The double-junction solar cell was deposited with an AIXTRON Crius MOCVD reactor on epi-ready <100> oriented GaAs on-axis wafer substrates. The growth temperature was 630 °C under a reactor pressure of 100 mbar. TMIn, TMGa, AsH$_3$ and PH$_3$ were used as precursor gasses and H$_2$ at 32 standard liters per minute was used as carrier gas. A tunnel junction consisting of a 20 nm heavily doped GaAs layer doped with Te at 2×10^{19} cm^{-3} and C at 3×10^{19} cm^{-3} was grown subsequently. The base of the top GaAs was 0.12 μm thick an n-doped with Si at, 1×10^{17} cm^{-3}. The emitter was 0.1 μm thick and p$^+$-doped GaAs with Zn at 2×10^{18} cm^{-3}. The base of the middle cell was 1.4 μm thick an n-doped with Si at 1×10^{17} cm^{-3}. The emitter was 0.1 μm thick and p$^+$-doped with Zn at 2×10^{18} cm^{-3}. The window layer consisted of InGaP, highly doped with Zn at 2×10^{18} cm^{-3}, the BSF layer was made of the same material composition but doped with Si at 2×10^{18} cm^{-3}.

This double-junction GaAs/GaAs solar cell had a one-sun efficiency of 17.8% under AM1.5G. The current-voltage curve is shown in Fig. 3a. The efficiency of this dual-junction GaAs/GaAs cell is mainly limited by the greater-than-ideal thickness (220 nm) of the GaAs top cell. Developing a thinner top cell with a thickness below 150 nm is in progress, but these thin cells are challenging due to shunting and defects during growth.

The bottom cell was a p-type multicrystalline silicon wafer solar cell fabricated by the Renewable Energy Corporation (REC). The passivated emitter rear contact (PERC) device was manufactured according to standard industrial practice and had a 17.9% one-sun efficiency under AM1.5G. The device was laser cut into 1×1 cm^2 cells to match the size of the top cell.

We have stacked the prototype double-junction GaAs/GaAs solar cell mechanically to the bottom cell. Without the GaAs substrate removal, the Si bottom cell received very low illumination due to the parasitic absorption in the doped GaAs wafers [15]–[17]. After substrate removal, the silicon cell achieved an efficiency of 2.6%, resulting in a mathematically combined efficiency of the tandem of 20.4%. Measured current voltage characteristics and quantum efficiencies are shown in

Fig. 3 and Fig. 4 respectively. We notice that the short-circuit current in the Si bottom cell is still much lower than the GaAs/GaAs top cell even though the GaAs substrate is removed. The low photogeneration in the Si bottom cell also suggests that the thicknesses of the top and middle GaAs cells have to be further reduced in order to allow monolithic integration. Our previous optical simulation indicates that the ideal thickness for the top and middle GaAs cell is around 105 nm and 500 nm respectively in order to achieve current matching with a Si solar cell. This current-matched ideal condition could possibly improve the GaAs/GaAs/Si efficiency to 33% as shown in Fig. 1.

(a)

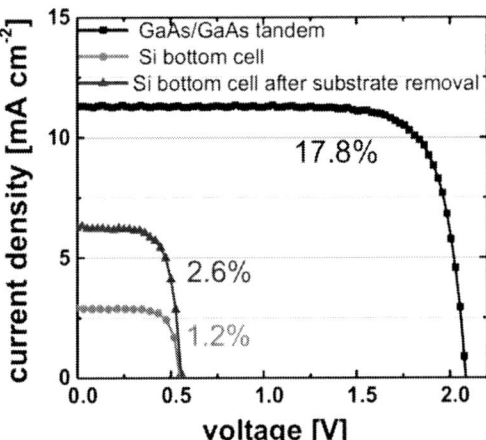

(b)

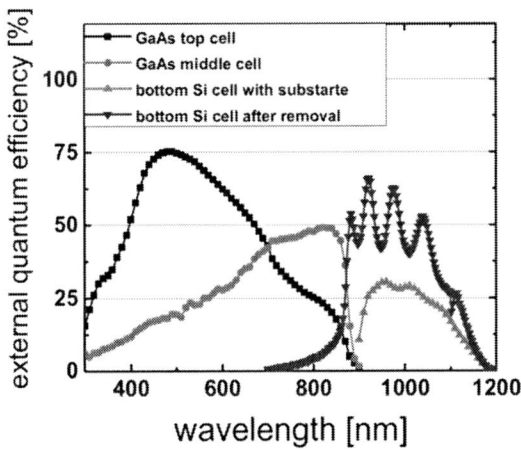

Fig. 3. (a) Current-voltage characteristics of the GaAs/GaAs tandem (black) and the silicon bottom cell before (red) and after (blue) GaAs substrate removal. (b) Quantum efficiencies of the GaAs top (black) – and middle (green) cells, as well as for the silicon bottom cell.

IV. SUMMARY AND OUTLOOK

In this paper, we discussed the GaAs/GaAs/Si architecture. This architecture aims to solve the current matching issue in integrated two-terminal tandems of GaAs and Si, which forms a non-ideal band-gap pair. Yet, GaAs and Si are the materials with the highest achieved single-junction efficiencies. The efficiency potential for best-in-class cell parameters reveals that this combination can generate world-record efficiencies for one-sun tandems.

We estimated the realistic STC efficiency potential of the GaAs/GaAs/Si tandem to be 33.0%. This efficiency was calculated by taking into account cell parameters derived from world record cells. We also calculated the harvesting efficiency using the actual operating conditions for Denver in the year 2014 for the suggested architecture and compared it to values obtained for other double- and triple junction solar cells. Monolithically integrated tandem solar cells generally show a higher sensitivity to actual operating conditions. For the GaAs/GaAs/Si architecture we obtain a harvesting efficiency potential of 32.1%.

We fabricated a prototype GaAs/GaAs tandem solar cell with 17.8% efficiency. To generate a tandem with silicon, we removed the substrate of the GaAs/GaAs double-junction solar cell and stacked it mechanically onto a silicon bottom solar cell The shaded silicon cell delivered an efficiency of 2.6%, resulting in a mathematically combined value of 20.4% for the GaAs/GaAs//Si tandem.

We are currently improving the efficiency of the GaAs/GaAs sub-cell by fabricating a very thin top cell (<150 nm) in order to achieve an current-matched integration into a Si-based tandem solar cell. Furthermore, we envision the monolithically integrated GaAs/GaAs/Si tandem fabricated by layer transferring and wafer bonding technologies will be a good candidate for high-efficiency terrestrial applications.

ACKNOWLEDGEMENT

This work was supported by funding from Singapore's National Research Foundation through the Singapore MIT Alliance for Research and Technology's "Low energy electronic systems (LEES) IRG" and by the U.S. Department of Energy (DOE) under Contract No. DEEE0006707. SERIS is sponsored by the National University of Singapore (NUS) and Singapore's National Research Foundation (NRF) through the Singapore Economic Development Board (EDB).

REFERENCES

[1] S. Essig, J. Benick, M. Schachtner, A. Wekkeli, M. Hermle, and F. Dimroth, "Wafer-bonded GaInP/GaAs/Si solar cells with 30% efficiency under concentrated sunlight," *IEEE Journal of Photovoltaics*, vol. 5, no. 3, pp. 977–981, 2015.

[2] S. Essig, M. A. Steiner, C. Allebé, J. F. Geisz, B. Paviet-Salomon, S. Ward, A. Descoeudres, V. LaSalvia, L. Barraud, N. Badel, A.

Faes, J. Levrat, M. Despeisse, C. Ballif, P. Stradins, and D. L. Young, "Realization of GaInP/Si dual-junction solar cells with 29.8% 1-sun efficiency," *IEEE Journal of Photovoltaics*, vol. 6, no. 4, pp. 1012–1019, 2016.

[3] R. Cariou, J. Benick, P. Beutel, N. Razek, C. Flötgen, M. Hermle, D. Lackner, S. W. Glunz, A. W. Bett, M. Wimplinger, and F. Dimroth, "Monolithic Two-Terminal III-V//Si triple-junction solar cells with 30.2% efficiency under 1-sun AM1.5g," *IEEE Journal of Photovoltaics*, vol. 7, no. 1, pp. 367–373, 2017.

[4] Z. Ren, J. P. Mailoa, Z. Liu, H. Liu, S. C. Siah, T. Buonassisi, and I. M. Peters, "Numerical analysis of radiative recombination and reabsorption in GaAs/Si tandem," *IEEE Journal of Photovoltaics*, vol. 5, no. 4, pp. 1079–1086, 2015.

[5] Z. Ren, N. Sahraei, Z. Liu, Y. Zhu, H. Hou, H. Liu et al., "Performance potential analysis of a 21.3% GaAs on industrial c-Si tandem solar cell," published online, 2017. DOI:10.13140/RG.2.2.10021.91363.

[6] Z. Yu, M. Leilaeioun, and Z. Holman, "Selecting tandem partners for silicon solar cells," *Nature Energy*, vol. 1, p. 16137, 2016.

[7] "Research cell record efficiency chart (Rev. 12-02-2016)," National Renewable Energy Laborary, online: www.nrel.gov/pv/, 2016.

[8] K. Yoshikawa, H. Kawasaki, W. Yoshida, T. Irie, K. Konishi, K. Nakano, T. Uto, D. Adachi, M. Kanematsu, H. Uzu, and K. Yamamoto, "Silicon heterojunction solar cell with interdigitated back contacts for a photoconversion efficiency over 26%," *Nature Energy*, vol. 2, no. 5, p. 17032, Mar. 2017.

[9] I. M. Peters, S. E. Sofia, J. P. Mailoa, and T. Buonassisi, "Techno-economic analysis of tandem photovoltaic systems," *RSC Advances*, vol. 6, pp. 66911–66923, 2016.

[10] M. Woodhouse and A. Goodrich, "A Manufacturing Cost Analysis Relevant to Single- and Dual-Junction Photovoltaic Cells Fabricated with III-Vs and III-Vs Grown on Czochralski Silicon," National Renewable Energy Labooratory, 2015.

[11] H. Liu, Z. Ren, Z. Liu, A. G. Aberle, T. Buonassisi, and I. M. Peters, "The realistic energy yield potential of GaAs-on-Si tandem solar cells: a theoretical case study," *Optics Express*, vol. 23, no. 7, pp. A382–A390, 2015.

[12] H. Liu, A. G. Aberle, T. Buonassisi, and I. M. Peters, "On the methodology of energy yield assessment for one-Sun tandem solar cells," *Solar Energy*, vol. 135, pp. 598–604, 2016.

[13] H. Liu, Z. Ren, Z. Liu, A. G. Aberle, T. Buonassisi, and I. M. Peters, "Predicting the outdoor performance of flat-plate III–V/Si tandem solar cells," *Solar Energy*, vol. 149, pp. 77–84, 2017.

[14] Z. Liu, Z. Ren, H. Liu, J. P. Mailoa, N. Sahraei, S.-C. Siah, S. E. Sofia, F. Lin, T. Buonassisi, and I. M. Peters, "Light management in mechanically-stacked GaAs/Si tandem solar cells: optical design of the Si bottom cell," in *Proc. of the 42nd IEEE Photovoltaic Specialists Conference (PVSC)*, 2015, pp. 1–4.

[15] Z. Liu, Z. Ren, H. Liu, N. Sahraei, F. Lin, R. Stangl, T. Buonassisi, and I. M. Peters, "Optical loss analysis of four-terminal GaAs/Si tandem solar cells," in *Proc. of the 43rd IEEE Photovoltaic Specialists Conference (PVSC)*, 2016, pp. 1914–1917.

[16] Z. Liu, Z. Ren, H. Liu, N. Sahraei, F. Lin, R. Stangl, A. G. Aberle, T. Buonassisi, and I. M. Peters, "A Modeling Framework for Optimizing Current Density in Four-Terminal Tandem Solar Cells: A Case Study on GaAs/Si Tandem," *Solar Energy Materials and Solar Cells*, in press, 2017.

$ZnSiP_2$ Thin Film Growth for Si-Based Tandem Photovoltaics

Aaron D. Martinez*, Elisa M. Miller[†], Andrew G. Norman[†], Paul Stradins*[†],
Eric S. Toberer*[†], and Adele C. Tamboli*[†]

*Colorado School of Mines, Golden, Colorado, 80401, USA
[†]National Renewable Energy Laboratory, Golden Colorado, 80401, USA

Abstract—$ZnSiP_2$ is a ternary III-V analog with 0.5% lattice mismatch with Si and a 2.1 eV band gap, in the appropriate range for a top cell on a Si-based tandem device. We have previously shown that $ZnSiP_2$ has many properties suitable for applications to Si-based tandem photovoltaics using bulk single crystals grown in a Zn flux. The favorable results obtained from characterization of bulk material encourage the development of $ZnSiP_2$ as a photovoltaic absorber. To pursue this development, we have constructed a thin film growth reactor. This reactor employs a combination of chemical vapor deposition, using silane and phosphine as precursor gases, and physical vapor deposition, using an effusion cell to evaporate elemental Zn. We will present the results of $ZnSiP_2$ film growth on (100) Si substrates. The composition, structure, and morphology of these films have been characterized by energy dispersive X-ray spectroscopy and X-ray photoelectron spectroscopy, X-ray diffraction and transmission electron diffraction, and electron microscopy, respectively. These promising results represent significant advancement towards implementing $ZnSiP_2$ as a top cell material on Si-based tandem photovoltaics.

Index Terms—$ZnSiP_2$, silicon, thin films, photovoltaic cells, tandem photovoltaics.

I. SUMMARY

There has been a longstanding need for optically-active materials that can be integrated with silicon, both for tandem photovoltaics and for other optoelectronic applications. Typically, lattice-mismatched III-V materials are combined with silicon, either by growth that can result in a high defect density or by methods such as wafer bonding or mechanical stacking. However, the II-IV-V_2 materials provide an alternative pathway via growth of lattice-matched materials with wide band gaps, i.e. $ZnSiP_2$ and $ZnGeP_2$. This class of materials is structurally and chemically similar to the III-Vs, but their ternary chemistry enables wider tunability of properties, and they have received considerably less study for photovoltaic applications. We focus on $ZnSiP_2$, a material with 0.5% lattice mismatch with Si and a 2.1 eV band gap, in the relevant range for a top cell in a Si-based tandem.

We have previously demonstrated that $ZnSiP_2$ (0.5% lattice mismatch with Si and a 2.1 eV band gap) has many properties suitable for applications to silicon-based tandem photovoltaics. [1]–[3] Using bulk single crystals grown in a Zn flux, we have studied the fundamental properties of the material. As a III-V analog, $ZnSiP_2$ may be expected to have optoelectronic quality similar to the III-V's. In bulk form we have found this material to be stable up to 800C in vacuum, and that it forms in an ordered chalcopyrite structure with few defects. [3] These

defects form shallow energy levels, resulting in a minority carrier lifetime of 7 ns that is comparable to CdTe (20 ns). [3], [4] We characterized the photoresponse of $ZnSiP_2$ in a photoelectrochemical configuration and observed a high open circuit voltage of 1.3 V and an energy conversion efficiency of 0.9%. [3] Finally, we will discuss progress on epitaxial growth of $ZnSiP_2$ films on Si for solid-state devices.

The favorable results obtained from characterization of bulk material encourage the development of $ZnSiP_2$ as a photovoltaic absorber, but thin film studies have been sparse. [5], [6]. To pursue this development, we have constructed a thin film growth reactor that employs a combination of chemical vapor deposition, using silane and phosphine as precursor gases, and physical vapor deposition, using an effusion cell to evaporate elemental Zn. We will present the results of $ZnSiP_2$ film growth on (100) Si substrates. The composition, structure, and morphology of these films have been characterized by energy dispersive X-ray spectroscopy and X-ray photoelectron spectroscopy, X-ray diffraction and transmission electron diffraction, and electron microscopy, respectively. These promising results represent significant advancement towards implementing $ZnSiP_2$ as a top cell material on Si-based tandem photovoltaics.

ACKNOWLEDGEMENTS

A. Tamboli was supported by the U.S. Department of Energy (DOE), Office of Science, Basic Energy Sciences, Materials Sciences and Engineering Division. A. Martinez was supported by the U.S. DOE, Solar Energy Technologies Office (SETO) under contract # SETP DE-EE00025786. Work at NREL is covered under DOE EERE contract DE-AC36-08GO28308. The United States Government retains and the publisher, by accepting the article for publication, acknowledges that the United States Government retains a nonexclusive, paid-up, irrevocable, world-wide license to publish or reproduce the published form of this manuscript, or allow others to do so, for United States Government purposes.

REFERENCES

[1] A. D. Martinez, B. R. Ortiz, N. E. Johnson, L. L. Baranowski, L. Krishna, S. Choi, P. C. Dippo, B. To, A. G. Norman, P. Stradins, V. Stevanovic, E. S. Toberer, and A. C. Tamboli, "Development of $ZnSiP_2$ for Si-Based Tandem Solar Cells," *IEEE J. Photovolt.*, vol. 5, no. 1, pp. 17–21, 2015.

[2] A. D. Martinez, E. L. Warren, P. C. Dippo, D. Kuciauskas, B. R. Ortiz, H. Guthrey, A. Duda, A. G. Norman, E. S. Toberer, and A. C. Tamboli, "Single Crystal Growth and Phase Stability of Photovoltaic Grade $ZnSiP_2$ by Flux Technique," *Proc. 41st IEEE PVSC*, 2015.

[3] A. D. Martinez, E. L. Warren, P. Gorai, K. A. Borup, D. Kuciauskas, P. C. Dippo, B. R. Ortiz, R. T. Macaluso, S. D. Nguyen, A. L. Greenaway, S. W. Boettcher, A. G. Norman, V. Stevanovic, E. S. Toberer, and A. C. Tamboli, "Solar energy conversion properties and defect physics of $ZnSiP_2$," *Energy Environ. Sci.*, vol. 9, no. 3, pp. 1031–1041, 2016.

[4] J. Ma, D. Kuciauskas, D. Albin, R. Bhattacharya, M. Reese, T. Barnes, J. V. Li, T. Gessert, and S.-H. Wei, "Dependence of the minority-carrier lifetime on the stoichiometry of CdTe using time-resolved photoluminescence and first-principles calculations," *Physical Review Letters*, vol. 111, no. 6, pp. 067 402–1–067 402–5, 2013.

[5] B. Curtis and P. Wild, "The preparation and growth of polycrystalline layers of $ZnSiP_2$ in an open flow system," *Materials Research Bulletin*, vol. 5, no. 2, pp. 69–72, 1970.

[6] V. P. Popov and B. R. Pamplin, "Epitaxial growth of solid solutions of $ZnSiP_2$ in Si," *Journal of Crystal Growth*, vol. 15, no. 2, pp. 129–132, 1972.

In situ control over the sublattice orientation of GaP/Si(100):As virtual substrates for tandem absorbers

Agnieszka Paszuk,[1] Oliver Supplie,[1] Sebastian Brückner,[1] Matthias M. May,[2] Anja Dobrich,[1] Andreas Nägelein,[1] Boram Kim,[3] Yoshiaki Nakano[3], Masakazu Sugiyama,[3] Peter Kleinschmidt,[1] and Thomas Hannappel[1]

[1] Institute of Physics, Department for Photovoltaics, Ilmenau University of Technology, Ilmenau, Germany
[2] Chemistry Department, University of Cambridge, Cambridge, United Kingdom
[3] School of Engineering, University of Tokyo, Bunkyo-ku, Tokyo, Japan

Presenting author: Thomas Hannappel, thomas.hannappel@tu-ilmenau.de

Abstract — **III-V integration on Si processed in MOCVD ambient which contains As opens up new opportunities for high-efficiency multi-junction solar cells. Here, we study the interaction of As with vicinal Si(100) surfaces, the formation of atomically well-ordered, As-modified Si(100) surfaces and its impact on subsequently grown GaP epilayers. We combine optical *in situ* spectroscopy with surface science techniques in ultra-high vacuum to understand the As-modified Si(100) surface and the III-V/Si interface at atomic scale. We demonstrate that depending on dimer orientation on the Si(100) surface, we are able to control the sublattice orientation of subsequently grown GaP.**

I. INTRODUCTION

Integration of III-V semiconductors and Si is highly desired in opto- and in microelectronics and holds perspectives for photovoltaic [1,2] and water splitting devices [3,4] with high conversion efficiencies. Major advantages of Si are its low cost and its suitable bandgap, which enables photovoltaic conversion efficiencies close to optimum in a tandem absorber device [2,5,6]. Metalorganic chemical vapor deposition (MOCVD) allows for device manufacturing at industrially relevant scale. Growth of III-V on Si in MOCVD ambient, however, is highly complex due to interaction of grown surfaces with the carrier gas, residuals from previous processes and the competition between kinetically and energetically driven processes. A further challenge is related to polar-on-nonpolar epitaxy [7]: To avoid anti-phase domains (APDs) in the subsequently grown III-V epilayers, it is necessary to prepare single domain Si(100) surfaces with double-layer steps. Commonly, a thin GaP buffer layer is applied as a transition layer between nonpolar Si substrates and polar III-V heterostructures, due to its close lattice matching to Si. Single domain Si(100) surface preparation [8–13] and subsequent defect-free GaP(100) growth [14–18] in As-free MOCVD ambient have been studied in great detail. For most realistic device structures, however, As plays an important role, for example in As-based graded buffer layers [19] or dilute nitride compounds [4,20]. Moreover, As facilitates Si deoxidation [21,22] and reduces out-diffusion of

Si into the GaP epilayers [23]. Recent studies revealed that Si preparation in As-rich MOCVD ambient leads to (almost) single-domain Si:As surfaces [22,24]. These were prevalently (1×2) reconstructed [22,24] and the GaP/Si:As heterointerface was suggested to be less abrupt than in the As-free case [17]. The sublattice orientation of the GaP epilayer was shown to be inverted when grown on As-modified Si(100) (Si:As) in comparison to monohydride-terminated Si(100) (Si:H) [24]. Control over the GaP sublattice orientation has not yet been shown for the practically more relevant entirely As-rich ambient. Also, the influence of MOCVD process routes on the atomic structure of the Si(100):As surface and the corresponding effect on the formation of the GaP/Si:As heterointerface has only been studied little so far.

Here, we study the dimer orientation on the vicinal Si(100) surface in dependence on the preparation conditions in As-rich MOCVD ambient with optical *in situ* spectroscopy and benchmarking to surface science techniques applied in UHV. We show that specific processing routes enable to prepare either prevalently (1×2) or (2×1) reconstructed Si(100):As surfaces. The surface formation can be tuned *in situ* due to characteristic reflection anisotropy spectra. Thereby, we obtain control over the sublattice orientation of the subsequently grown GaP epilayer and can prepare virtual substrates, which are free of anti-phase disorder for both sublattice orientations. Further, we present important information on the microscopic origin of the *in situ* spectra.

II. EXPERIMENTAL

All samples were grown in a horizontal AIX-200 MOCVD reactor (Aixtron) with H_2 as a process gas. We employed vicinal Si(100) substrates with 6° offcut towards [011] direction, which enable the possibility of *in situ* studies during the entire Si preparation process, including high temperatures [9]. Si(100) substrates were thermally deoxidized in 950 mbar H_2 ambient for 30 min at 1000°C. In order to prepare the (1×2) reconstructed As-modified surfaces, the substrates were annealed first with tertiarybutylarsine (TBAs) and secondly without TBAs in H_2 and background As_x at 850°C (before cooling to room temperature). GaP was nucleated by pulses of tertiarybutylphosphine and triethylgallium at 420°C prior to GaP growth of about 40 nm above 570°C [25,16]. Subsequently, the P-rich, (2×2)/c(4×2) reconstructed surface was prepared [17]. The entire process was controlled *in situ* by reflection anisotropy spectroscopy (RAS, LayTec EpiRAS-200) [26]. RAS measures the difference in reflection of linearly polarized light between [011] and [0$\bar{1}$1] directions in the surface plane, normalized to the total reflection. The RA spectra can have different microscopic origin and are thus often difficult to interpret. Therefore, we benchmark them with the atomic surface structure obtained from electron-based surface sensitive analysis tools in UHV, such as low energy electron diffraction (LEED, Specs ErLEED 100-A) and X-ray photoelectron spectroscopy (XPS, Specs Focus 500, Phoibos 150).

III. RESULTS AND DISCUSSION

Figure 1 shows the RA spectrum of the As-modified Si(100) 6° "A-type" surface (orange line), its corresponding LEED pattern and, as a reference, the RA spectrum of an As-modified, double-layer stepped Si(100) 2° surface (gray line), which is discussed in detail the *Ref.* [24]. Here, the term A-type refers to a notation of Chadi [27] and corresponds to (1×2) domain with dimers oriented perpendicularly to the step edge. The corresponding LEED pattern shows a clear (1×2) surface reconstruction, without presence of the spots at half order from the minority (2×1) domain.

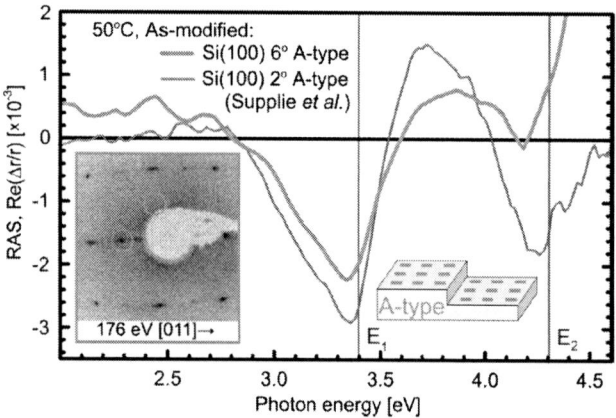

Fig. 1. RA spectra (measured at 50°C) of As-modified Si(100) 6° with predominant A-type majority domains (orange) and, as a reference, of the As-modified Si(100) 2° surface (grey) [24]. The LEED pattern of the Si(100) 6° sample (inset) confirms (1×2) surface reconstruction. The sketch indicates the prevalent dimer orientation at the surfaces.

The RA spectrum of the As-modified Si(100) 6° surfaces exhibits a very similar line shape compared to the Si(100) 2° reference. Both spectra exhibit two characteristic minima

close to the E_1 and E_2 critical points of bulk Si and a maximum at around 3.7 eV. This similar RAS line shape indicates that the RAS signal is mostly terrace related; however we cannot entirely exclude other microscopic origins which also contribute to the spectra as suggested in *Ref.* [24]. We combined the RAS measurements with XPS after contamination-free transfer to UHV [28]. A quantitative XPS analysis reveals similar intensity ratios of the As $2p_{3/2}$ to the Si 2p photoemission (PE) line for the As-modified Si(100) 6° and the Si(100) 2° A-type surfaces [24]. While this indicates a similar As-coverage on both samples, XPS measurements varying the photoelectron take-off angle indicate less intermixing in case of the vicinal A-type Si(100) 6° surface.

To elaborate on the influence of the As coverage on the line shape of the RA spectrum, we performed additional annealing steps in TBAs and background As$_x$ at two different temperatures with cooling to 420°C, in order to measure RAS at rather low T where the spectral features are sharper. First, the A-type domain surface was prepared as described above and cooled down to 420°C. In the next step, the sample was heated to 830°C, where it was annealed first with, and secondly without TBAs (1st annealing, about one minute each), and then cooled down to 420°C to measure RAS (Fig. 2, red line). Subsequently, this annealing procedure was repeated on the identical sample at 670°C before cooling again to 420°C (2nd annealing, Fig. 2, blue line). The entire process was controlled *in situ* with RAS and at each step the RAS signal from the surface was stable before going to the next step.

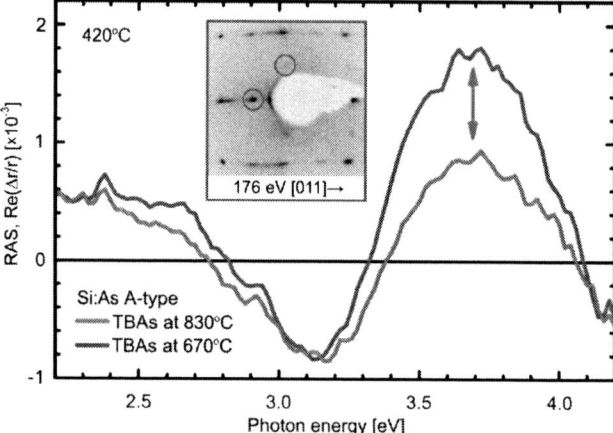

Fig. 2. RA spectra (measured at 420°C) of As-modified Si(100):As 6° A-type surface after 1st (red) and 2nd (blue) additional annealing with and without TBAs supply at 830°C and 670°C, respectively. The inset shows LEED pattern corresponding to the Si(100):As 6° A-type sample after the additional annealing in TBAs at 670°C (blue frame).

The RA spectrum of the As-modified Si(100) 6° sample after 1st annealing did not differ from the one taken right after the surface preparation and first cooling down to 420°C (not shown here). The RA spectra of the Si(100) sample after 1st and 2nd additional annealing have a very similar shape and identical characteristic peak positions. However, the amplitude of the peak at 3.7 eV is significantly different. The LEED pattern of the sample after 2nd annealing (at 670°C, Fig. 2. inset, blue frame) exhibits a strong prevalence of (1×2) domain on the surface. A higher As coverage on this surface compared to the surface prepared as in Fig. 1. (orange line), can be confirmed by XPS: The intensity ratio As $2p_{3/2}$/Si $2p$ PE lines is larger for the sample annealed at 670°C in comparison to the sample prepared as on Fig. 1 (orange line) by a factor of 1.3. At 830°C, the desorption rate of As from the Si surface is higher than at 670°C, which is in line with the XPS data. Accordingly, the RA intensity of the peak at 3.7 eV corresponds to a lower As coverage on the Si surface. The lack of change in amplitude of the peak at 3.2 eV hints to a different microscopic origin, as also suggested in [24]. A modification of the cooling procedure (after the deoxidation step) was done to prepare Si(100):As 6° surface with a prevalence of (2×1) domain on the surface (to be published elsewhere).

RA spectra of heteroepitaxially grown GaP layers on the vicinal A- and B-type As-modified Si(100) surfaces are shown in Fig. 3 (orange and violet line, respectively). The line shape and sign of the RA signal of the GaP/Si(100):As 6° B-type surface (violet line) is similar to the RA spectrum of the P-rich GaP(100) [29]. The signal has a characteristic negative peak at 2.35 eV and a positive peak at about 3.4 eV, which both correspond to a B-type (2×2)/c(4×2) reconstructed P-rich GaP(100) surface, on which buckled P-dimers are aligned (2×1)-like on the surface (see violet inset) [17,30]. Small differences of the line shape between the RAS signal of the P-rich GaP(100) [29] and GaP/Si(100):As 6° B-type surfaces are due to internal reflection and anisotropy of the buried heterointerface. RA signal of the GaP(100) grown on Si(100):As 6° A-type surface (orange line) is very similar in the line shape to the one grown on the Si:As B-type but with a flipped sign, which implies a rotation of majority P-dimers on the surface by 90° (see orange inset). Due to the tetrahedral coordination in the crystal lattice, the orientation of the P-dimers at the P-rich GaP surface corresponds to the GaP sublattice orientation [17]. The sublattice orientation of the two GaP layers thus is inverted. Both GaP/Si(100):As surfaces were measured by LEED. The spots at half order in both LEED patterns verify that the P-dimers are aligned (2×1)- and (1×2)-like, respectively on the surfaces. In addition, in both LEED patterns we did not observe half order spots corresponding to the minority domain, and a rough quantification based on the amplitude of the RAS signal amplitude [31,32] indicates that the APD concentration at the surface is lower than 1.5% and 7% for GaP grown on Si(100):As 6° A- and B-type respectively.

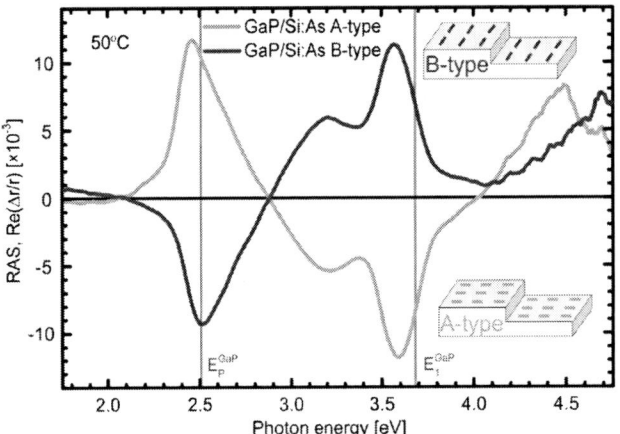

Fig. 3. RA spectra of P-rich, (2×2)/c(4×2) reconstructed GaP layers grown on As-modified Si(100) 6° A-type (orange) and B-type (violet). Grey vertical lines indicate energies of the surface state of P-rich reconstruction (E_P^{GaP}) [33] and interband transition of GaP (E_1^{GaP}) at 50°C. Sketches indicate the prevalent dimer orientation at the surfaces.

LEED and RAS confirm that the orientation of the P-dimers at the GaP/Si(100) surface (sublattice orientation) depends on the dimer orientation on the Si(100):As substrate: GaP grown on the As-modified Si A-type (B-type) surface results in A-type (B-type) polarity. This is the opposite relation as observed for GaP heteroepitaxy in As-free ambient. We are thus able to prepare GaP epilayers almost free of antiphase disorder with adjustable sublattice orientation in As-rich MOCVD ambient by controlling the dimer orientation on the Si:As (100) surface prior heteroepitaxy. These virtual GaP/Si:As (100) substrates are promising for subsequent III-V integration with low-defect density.

IV. CONCLUSION

We are able to control the prevalence of the majority domain on the Si(100) 6° surface in As-rich MOCVD ambient by specific process routes. We showed that the RAS signal contains information not only from the terrace-related structure, such as dimers, but also from the As coverage. Based on XPS measurement of the As-modified Si(100) 6° A-type surfaces we find indications for intermixing of Si and As atoms within near-surface layers. GaP grown on As-modified Si(100) with 6° offcut and A-type surfaces results in almost single-domain epilayers with an inverted sublattice compared to the one grown on B-type surfaces. All surfaces involved exhibit characteristic RA spectra. *In situ* RAS thus allows for identifying the optimum process parameters and conditions and to control the surface formation also in As-rich ambience.

ACKNOWLEDGEMENT

The authors are grateful for experimental support by Antonio Müller and Mathias Biester. M.M.M., A.N., and A.P. acknowledge scholarships of the German National Academy of Sciences Leopoldina, Carl Zeiss Stiftung, and Landesgraduiertenschule PhotoGrad, respectively. This work was financially supported by the German Federal Ministry for Education and Research (BMBF, project no. 03SF0525B).

REFERENCES

[1] F. Dimroth, M. Grave, P. Beutel, U. Fiedeler, C. Karcher, T. N. D. Tibbits, E. Oliva, G. Siefer, M. Schachtner, A. Wekkeli, A. W. Bett, R. Krause, M. Piccin, N. Blanc, C. Drazek, E. Guiot, B. Ghyselen, T. Salvetat, A. Tauzin, T. Signamarcheix, A. Dobrich, T. Hannappel, and K. Schwarzburg, "Wafer bonded four-junction GaInP / GaAs // GaInAsP / GaInAs concentrator solar cells with 44 . 7 % efficiency," *Prog. Photovolt Res. Appl.*, vol. 22, no. January, p. 277, 2014.

[2] M. Feifel, J. Ohlmann, J. Benick, T. Rachow, S. Janz, M. Hermle, F. Dimroth, A. Beyer, K. Volz, and D. Lackner, "MOVPE Grown Gallium Phosphide – Silicon Heterojunction Solar Cells," *IEEE J. photovoltaics*, vol. 7, no. 2, p. 502, 2017.

[3] J. L. Young, M. A. Steiner, H. Döscher, R. M. France, J. A. Turner, and T. G. Deutsch, "Direct solar-to-hydrogen conversion via inverted metamorphic multi-junction semiconductor architectures," *Nat. Energy*, vol. 2, no. March, p. 1, 2017.

[4] O. Supplie, M. M. May, H. Stange, C. Höhn, H.-J. Lewerenz, and T. Hannappel, "Materials for light-induced water splitting: In situ controlled surface preparation of GaPN epilayers grown lattice-matched on Si(100)," *J. Appl. Phys.*, vol. 115, no. 11, p. 113509, 2014.

[5] H. Döscher, O. Supplie, M. M. May, P. Sippel, C. Heine, A. G. Muñoz, R. Eichberger, H.-J. Lewerenz, and T. Hannappel, "Epitaxial III-V films and surfaces for photoelectrocatalysis.," *Chemphyschem*, vol. 13, no. 12, p. 2899, 2012.

[6] S. Essig, S. Ward, M. A. Steiner, D. J. Friedman, J. F. Geisz, P. Stradins, and D. L. Young, "Progress towards a 30% efficient GaInP/Si tandem solar cell," *Energy Procedia*, vol. 77, p. 464, 2015.

[7] H. Kroemer, "Polar-on-nonpolar epitaxy," *J. Cryst. Growth*, vol. 81, p. 193, 1987.

[8] A. Dobrich, P. Kleinschmidt, H. Döscher, and T. Hannappel, "Quantitative investigation of hydrogen bonds on Si(100) surfaces prepared by vapor phase epitaxy," *J. Vac. Sci. Technol. B Microelectron. Nanom. Struct.*, vol. 29, no. 4, p. 04D114, 2011.

[9] S. Brückner, H. Döscher, P. Kleinschmidt, and T. Hannappel, "*In situ* investigation of hydrogen interacting with Si(100)," *Appl. Phys. Lett.*, vol. 98, no. 21, p. 98, 2011.

[10] S. Brückner, H. Döscher, P. Kleinschmidt, O. Supplie, A. Dobrich, and T. Hannappel, "Anomalous double-layer step formation on Si(100) in hydrogen process ambient," *Phys. Rev. B*, vol. 86, no. 19, p. 195310, 2012.

[11] S. Brückner, P. Kleinschmidt, O. Supplie, H. Döscher, and T. Hannappel, "Domain-sensitive *in situ* observation of layer-by-layer removal at Si(100) in H_2 ambient," *New J. Phys.*, vol. 15, no. 11, p. 113049, 2013.

[12] S. Brückner, O. Supplie, A. Dobrich, P. Kleinschmidt, A. Paszuk, and T. Hannappel, "Control over dimer orientations on vicinal Si(100) surfaces in H_2 ambient: Kinetics vs. energetics," *to be submitted*, 2017.

[13] B. Kunert, I. Németh, S. Reinhard, K. Volz, and W. Stolz, "Si (001) surface preparation for the antiphase domain free heteroepitaxial growth of GaP on Si substrate," *Thin Solid Films*, vol. 517, no. 1, p. 140, 2008.

[14] I. Németh, B. Kunert, W. Stolz, and K. Volz, "Heteroepitaxy of GaP on Si: Correlation of morphology, anti-phase-domain structure and MOVPE growth conditions," *J. Cryst. Growth*, vol. 310, no. 7, p. 1595, 2008.

[15] A. Beyer, J. Ohlmann, S. Liebich, H. Heim, G. Witte, W. Stolz, and K. Volz, "GaP heteroepitaxy on Si(001): Correlation of Si-surface structure, GaP growth conditions, and Si-III/V interface structure," *J. Appl. Phys.*, vol. 111, no. 8, p. 083534, 2012.

[16] K. Volz, A. Beyer, W. Witte, J. Ohlmann, I. Németh, B. Kunert, and W. Stolz, "GaP-nucleation on exact Si (001) substrates for III/V device integration," *J. Cryst. Growth*, vol. 315, no. 1, p. 37, 2011.

[17] O. Supplie, S. Brückner, O. Romanyuk, H. Döscher, C. Höhn, M. M. May, P. Kleinschmidt, F. Grosse, and T. Hannappel, "Atomic scale analysis of the GaP/Si(100) heterointerface by *in situ* reflection anisotropy spectroscopy and *ab initio* density functional theory," *Phys. Rev. B*, vol. 90, no. 23, p. 235301, 2014.

[18] O. Supplie, M. M. May, S. Brückner, A. Nägelein, P. Kleinschmidt, and T. Hannappel, "Time-resolved optical in situ spectroscopy during formation of the GaP / Si (100) interface and benchmarking to photoelectron spectroscopy," no. Ewmovpe Xvi, p. 2, 2015.

[19] J. F. Geisz, J. M. Olson, M. J. Romero, C. S. Jiang, and A. G. Norman, "Lattice-mismatched GaAsP solar cells grown on silicon by OMVPE," *IEEE 4th World Conf. Photovolt. Energy Conf.*, vol. 1, p. 772, 2006.

[20] S. R. Kurtz, A. A. Allerman, E. D. Jones, J. M. Gee, and J. J. Banas, "InGaAsN solar cells with 1.0 eV band gap, lattice matched to GaAs," *Appl. Phys. Lett.*, vol. 74, no. 5, p. 729, 1999.

[21] T. Hannappel, W. E. McMahon, and J. M. Olson, "An RDS, LEED, and STM study of MOCVD-prepared Si(100) surfaces," *J. Cryst. Growth*, vol. 272, p. 24, 2004.

[22] E. L. Warren, A. E. Kibbler, R. M. France, A. G. Norman, P. Stradins, and W. E. McMahon, "Growth of antiphase-domain-free GaP on Si substrates by metalorganic chemical

vapor deposition using an in situ AsH$_3$ surface preparation," *Appl. Phys. Lett.*, vol. 107, no. 8, p. 082109, 2015.

[23] Y. Kohama, K. Uchida, T. Soga, T. Jimbo, and M. Umeno, "Quality improvement of metalorganic chemical vapor deposition grown GaP on Si by AsH3 preflow," *Appl. Phys. Lett.*, vol. 53, no. 10, p. 862, 1988.

[24] O. Supplie, M. M. May, P. Kleinschmidt, A. Nägelein, A. Paszuk, S. Brückner, and T. Hannappel, "In situ controlled heteroepitaxy of single-domain GaP on As-modified Si(100)," *APL Mater.*, vol. 3, no. 12, p. 126110, 2015.

[25] O. Supplie, M. M. May, G. Steinbach, O. Romanyuk, F. Grosse, A. Nägelein, P. Kleinschmidt, S. Brückner, and T. Hannappel, "Time-resolved in situ spectroscopy during formation of the GaP/Si(100) heterointerface," *Phys. Chem. Lett.*, vol. 6, no. 3, p. 464, 2015.

[26] D. E. Aspnes and A. A. Studna, "Anisotropies in the above-band-gap optical spectra of cubic semiconductors," *Phys. Rev. Lett.*, vol. 54, no. 17, p. 1956, 1985.

[27] D. J. Chadi, "Stabilities of single-layer and bilayer steps on Si(001) surfaces," *Phys. Rev. Lett.*, vol. 59, no. 15, p. 1691, 1987.

[28] T. Hannappel, S. Visbeck, L. Töben, and F. Willig, "Apparatus for investigating metalorganic chemical vapor deposition-grown semiconductors with ultrahigh-vacuum

based techniques," *Rev. Sci. Instrum.*, vol. 75, no. 5, p. 1297, 2004.

[29] T. Löben, T. Hannappel, K. Möller, H.-J. Crawack, C. Pettenkofer, and F. Willig, "RDS , LEED and STM of the P-rich and Ga-rich surfaces of GaP (100)," *Surf. Sci. Lett.*, vol. 494, no. 1, p. L755, 2001.

[30] P. H. Hahn, W. G. Schmidt, F. Bechstedt, O. Pulci, and R. Del Sole, "P-rich GaP(001) (2x1)/(1x2) surface: A hydrogen-adsorbate structure determined from first-principles calculations," *Phys. Rev. B*, vol. 68, no. 3, p. 033311, 2003.

[31] H. Döscher and T. Hannappel, "In situ reflection anisotropy spectroscopy analysis of heteroepitaxial GaP films grown on Si(100)," *J. Appl. Phys.*, vol. 107, no. 12, p. 123523, 2010.

[32] H. Döscher, T. Hannappel, B. Kunert, A. Beyer, K. Volz, W. Stolz, H. Döscher, T. Hannappel, B. Kunert, A. Beyer, K. Volz, and W. Stolz, "In situ verification of single-domain III-V on Si(100) growth via metal-organic vapor phase epitaxy," *Appl. Phys. Lett.*, vol. 93, no. 17, p. 172110, 2008.

[33] P. Sippel, O. Supplie, M. M. May, R. Eichberger, and T. Hannappel, "Electronic structures of GaP(100) surface reconstructions probed with two-photon photoemission spectroscopy," *Phys. Rev. B*, vol. 89, no. 16, p. 165312, 2014.

III-V/Si tandem cell to module interconnection – comparison between different operation modes

Henning Schulte-Huxel, Emily L. Warren, Manuel Schnabel, Paul Stradins, Daniel Friedman, Adele C. Tamboli

National Renewable Energy Laboratory, 15013 Denver West Parkway, Golden, CO 80401, USA

Abstract — III-V/Si tandem solar cells have the potential to surpass the theoretical efficiency limit of silicon solar cells. On the cell level it was shown at this conference that three-terminal tandem (3T) devices consisting of III-V top and Si bottom cell perform as well as operating the subcells independently. However, integrating these 3T devices in a module requires voltage matching of the top and the bottom cell. Here we investigate the robustness of 3T III-V/Si tandem devices in comparison with independently operated and current matched two terminal devices with respect to spectral and thermal effects. Under most conditions, interconnected voltage matched devices are able to perform as well as those with independent operation of the top and bottom cell, and prove that 3T devices significantly outperform current matched devices.

I. INTRODUCTION

Due to their potential to surpass the theoretical efficiency limit of single junction silicon solar cells, III-V/Si tandem solar cells have been a focus of research in the recent years and have reached efficiency records of over 32% [1]. These results were achieved using GaAs or GaInP top cells featuring bandgaps of 1.42 eV and 1.81 eV, respectively. Depending on the design of the subcells and the bonding technique, different configurations of the tandem cell are possible, .i.e., a tandem cell featuring two, three, or four terminals [2].

Two terminal (2T) devices have the advantage that for module integration they can be interconnected in series as single junction devices, see Fig. 1 a). However, the series interconnection requires a matching of their currents. This current matching results in a strong restriction of the bandgap of the top cell and causes significant performance losses under varying spectra [2–4]. Operating the two subcells independently as in the case of four terminal (4T) devices [2, 3, 5], circumvents the current matching, see Fig. 1 c). However, this results in a more complex system integration.

At this conference, it was shown that three terminal (3T) devices perform as well as 4T devices [6–8] and can be beneficial with respect to cell processing and interconnection. Nevertheless, the parallel/series interconnection of 3T devices in modules requires that the voltages of series connected bottom cells (e.g. 2) is matched with the voltage of one top cell, see Fig. 1 b). This requirement of voltage matching applies also to the 4T devices integrated into modules featuring only two external contacts as for common single-junction modules [2]. However, in past studies, only the performance of individual devices was analyzed. There are

few works that compared interconnected tandem cells [2] focused mainly on special applications [9, 10].

In this work we investigate the robustness of parallel/series interconnection 3T III-V/Si tandem devices in comparison with series interconnected 2T devices and independent operated 4T devices with respect to spectral and thermal effects.

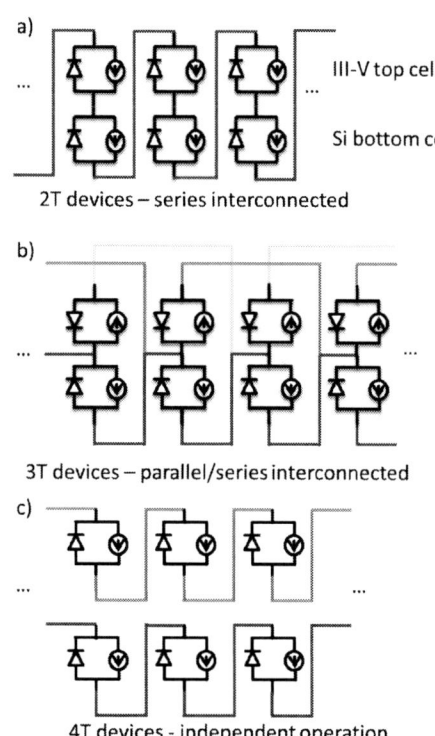

Fig. 1 Schemes of the interconnection of a) 2T, b) 3T, and c) 4T tandem cells. 2T devices are limited by the current matching of the subcells and interconnected 3T devices by the matching of the voltages in the substrings of top and bottom cells.

II. SIMULATION MODEL

We use the 1D ideal diode model of a solar cell in order to model the theoretical efficiency for the III-V/Si tandem cell with varying bandgap E_g of the top cell. As input parameters for the top and the bottom cell, we use the dark saturation current density J_0 and the short circuit current density J_{sc}.

978-1-5090-5606-4/17 $31.00 © 2017 IEEE

J_0 of the top and bottom is based on an empirically determined difference between E_g and the open circuit voltage V_{oc} of 0.4 eV at 25°C [11]

$$J_0 = J_s \exp(-qV_{oc}/nkT), \qquad (1)$$

with a standard current density J_s of 16 mA/cm² [11] and an ideality factor n of unity. We implement the temperature dependence of J_0 through [12]

$$J_0(T) \sim T^3 \exp(-E_g(T)/kT), \qquad (2)$$

using J_0 at 25°C as reference. The temperature dependence of $E_g(T)$ is calculated using the Varshni coefficients for Si [13] and for the top cell of GaAs [14], which is used as representative material for a direct bandgap III-V semiconductor of the top cell. For simplicity, we use the linear interpolation of the Varshni coefficients of GaAs between -40°C and 80°C.

The short circuit current density J_{sc} of the top cell is determined by using the Beer–Lambert law and the spectral dependent absorption coefficient of GaAs [15] shifted by the difference of the bandgap edge with respect to GaAs at 25°C. J_{sc} for the bottom cell is determined by using the spectral dependent absorption coefficient of Si [16] and the transmission of the top cell assuming no reflection at the interface. The thickness of the top cells is assumed to be 1 μm and for the bottom cell we assume slight light trapping resulting in an effective optical thickness of 1 mm. We use the AM 1.5G spectrum (ASTM G-173-03) and the AM1G to AM10G spectra normalized to 1000 W/m² calculated using SMARTS 2.95 [17, 18] (average photon energy of the used spectra of 1.84 – 1.58 eV for 280 – 1200 nm). Losses due to parasitic absorption, reflection, shading, and resistances are not taken into account unless stated otherwise. Luminescent coupling of the junctions is taken to be negligible.

In case of the interconnection of 3T devices, the power of individual cells at the string ends is partially lost [2]. Depending on the cell design the power loss is on the order of the power of one tandem cell, i.e., for 60 cell string this corresponds to a 1.6% loss. For simplicity these losses are neglected in this work assuming an infinite string.

III. RESULTS AND DISCUSSION

Fig. 2 shows the evolution of the efficiency for III-V/Si tandem solar cell under standard testing conditions (STC, 1000 W/m², AM1.5G, 25°C) operating the top and bottom cells independently, series interconnected (2T devices), or parallel/series interconnected (3T devices). In the case of parallel/series interconnected 3T devices we assume one top cell interconnected in parallel to a series of one, two, or three bottom cells, denoted as 1:1, 1:2, and 1:3, respectively.

Operating the cells independently (blue curve) leads to the highest efficiency for all bandgap energies of the top cell with a maximum efficiency η_{max} of 37.8% at $E_g = 1.81$ eV, which corresponds to the bandgap of GaInP cells [19]. In case of the series interconnected 2T devices (green curve) η_{max} is 37.4% at 1.67 eV. The optimal bandgap depends on the cell thicknesses and other material properties [11]. For parallel/series interconnected 3T devices, the ratio of 1:1 leads to $\eta_{max} = 26.9\%$, which is approximately the same as the efficiency of the bottom Si cell. This is caused by the fact that the voltage of the Si cell limits the device voltage and in our model every absorbed photon leads to an electron-hole pair. In case of a ratio of 1:3, an efficiency of 34.1% is reached at $E_g = 2.36$ eV. This is well below the efficiency of the parallel/series interconnected 3T devices with a ratio of 1:2 with $\eta_{max} = 37.8\%$ at $E_g = 1.80$ eV, which is almost identical to the results for independent operation. This shows that the results presented at this conference for individual 3T devices operating as well as 4T devices [6–8] also hold for GaInP top cells in a 3T configuration, interconnected in a module with two resulting terminals.

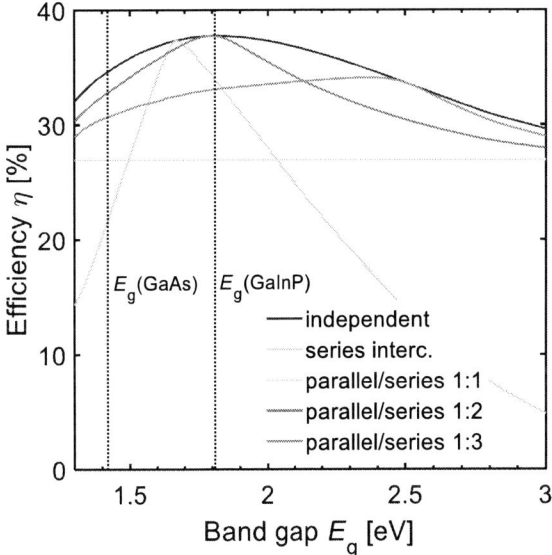

Fig. 2 Efficiency under STC for III-V/Si tandem solar cells using different operation modes, independent operation, series interconnection, and parallel/series intereconnections, as a function bandgaps of the top cells. Additionally, the bandgap energies of GaAs and GaInP are indicated as dashed lines.

Changes in the illumination spectra have significant impact on the performance of tandem solar cells, since the ratio of currents generated in the top and bottom cell is affected and the two cells operate at different voltages at the maximum power point.

Fig. 3 show the effect of the illumination of the tandem devices under AM1G to AM10G spectra normalized to 1000 W/m². The shape of the functions is the same as under

STC. For series interconnected 2T devices (green solid lines) we observe a significant decrease in the optimal bandgap from 1.70 eV to 1.46 eV when changing from AM1G to AM10G.

In contrast, for parallel/series interconnected 3T devices (red solid lines) the optimal bandgap is almost unaffected by spectral changes. For the 1:2 configuration the optimum bandgap is constant at 1.80 eV (red dashed line).

The consequence is that parallel/series interconnected 3T devices are much more robust against spectral changes compared to series interconnected 2T devices, when keeping the bandgaps for the two configurations constant at the optimum value for AM1.5G spectra (dashed lines). Also, the parallel/series interconnected 3T devices perform as well as operating the subcells independently (blue solid lines) at optimal bandgap under AM1.5G.

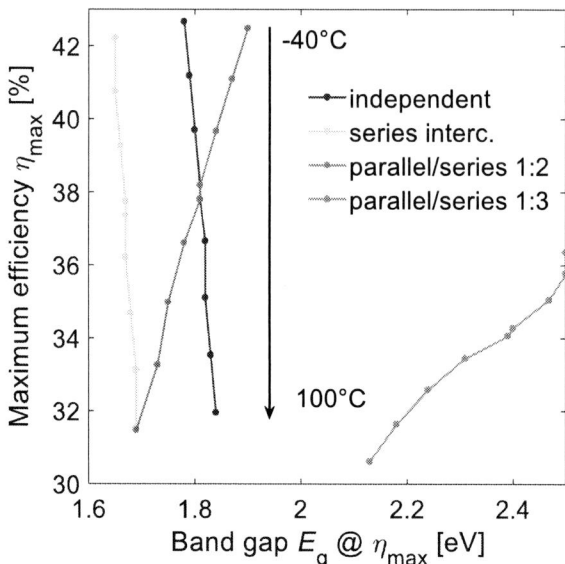

Fig. 4 Optimum bandgap of the top cell for temperatures between -40°C and 100°C and the corresponding maximum efficiency η_{max} for tandem solar cells using different operation modes.

As indicated in Fig. 5, the parallel/series interconnected devices 1:2 perform as well as operating the subcells independently only at a certain temperature. For instance, in Fig. 5 the band gap is optimized for 25°C. Nevertheless, for the majority of realistic operation temperatures (up to 60°C) [20], the parallel/series interconnected device with ratio 1:2 performs better than series interconnected tandems.

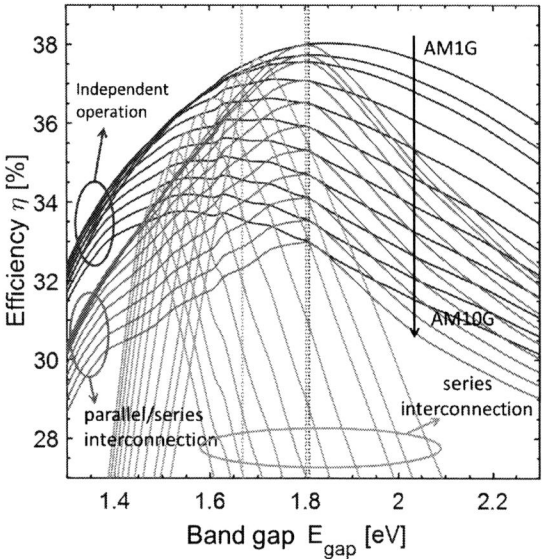

Fig. 3 Efficiency under under AM1G to AM10G spectra normalized to 1000 W/m² for III-V/Si tandem solar cells using different operation modes, independent operation (blue), series interconnection (green), and parallel/series interconnection 1:2 (red), as a function bandgaps of the top cells. The dashed lines indicate the optimal bandgaps under AM1.5G spectra.

In contrast, the temperature dependence is more pronounced for parallel/series interconnected 3T devices compared to series interconnected 2T devices. Fig. 4 shows that for the independent operation and series interconnection the increase of the optimal bandgap with temperature is about 0.05 eV. At higher temperatures the bandgap is decreased and thus for the optimal splitting of the spectra and current matching a larger bandgap is required. However, for the parallel/series interconnected devices the optimal bandgap decreases by 0.21 eV between -40°C and 100°C for a ratio 1:2 and more than 0.4 eV for a ratio 1:3, since the voltage of the two/three bottom cells decreases more than of the top cell.

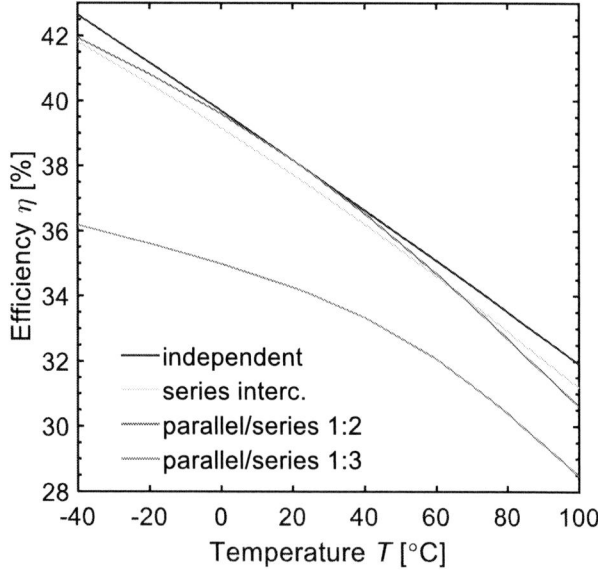

Fig. 5 Efficiency for tandem solar cell as a function of operation temperature. The bandgap of the top cells is the optimum bandgap under STC for each interconnection case.

IV. CONCLUSION

We showed that in the case of top cells with an bandgap of $E_g = 1.81$ eV (GaInP) on silicon bottom cells ($E_g = 1.12$ eV) parallel/series interconnected 3T (1:2) devices perform as well as operating the top and bottom cell individually. As a consequence, modules with 3T devices integrated in one circuit have the potential to reach the same power as modules with separate circuits for the top and bottom cells of 4T devices. This holds also under varying spectra between AM1G to AM10G. Depending on the spectra, the advantage of parallel/series interconnected 3T devices compared to series interconnected 2T devices is up to $8\%_{abs}$ when choosing the bandgap of the top cell to be the optimum under STC (parallel/series interconnected 3T devices (1:2): 1.8 eV, series interconnected 2T devices: 1.7 eV) under the assumption of no interconnection losses and infinite strings. Only in the case of high operation temperatures ($> 60°C$) do series interconnected devices perform better than parallel/series interconnected ones, when both topcells optimized for STC. However, this temperature limit can be shifted to higher values by optimizing the bandgap of the top cell for higher temperatures. Therefore, we conclude that modules with parallel/series interconnected tandem devices are much more robust against changes to the operation conditions than series interconnected devices for a wide range of top cell's bandgaps including GaAs and GaInP.

ACKNOWLEDGEMENT

H. Schulte-Huxel acknowledges support for the Research Fellowship by Deutsche Forschungsgemeinschaft (DFG) (grant agreement No: SCHU 3206/1). Funding for this work at NREL was provided by DOE through EERE contract SETP DE-EE00030299 and under Contract No. DE-AC36-08GO28308. The authors thank M. Rienecker (ISFH), R. Witteck (ISFH), S. McAlpine (NREL) and U. Römer (UNSW) for the fruitful discussion.

REFERENCES

[1] S. Essig, C. Allebe, T. Remo, J. F. Geisz, M. A. Steiner, L. Barraud, J. S. Ward, M. Schnabel, K. A. W. Horowitz, A. Descoeudres, D. L. Young, M. Woodhouse, M. Despeisse, C. Ballif, and A. C. Tamboli, "32% efficient III-V/Si dual-junction solar cells and their challenging path towards cost competiveness," *To be presented at this conference.*

[2] J. M. Gee, "A comparison of different module configurations for multi-band-gap solar cells," *Solar Cells*, vol. 24, no. 1-2, pp. 147–155, 1988.

[3] I. Almansouri, A. Ho-Baillie, S. P. Bremner, and M. A. Green, "Supercharging Silicon Solar Cell Performance by Means of Multijunction Concept," *IEEE J. Photovoltaics*, vol. 5, no. 3, pp. 968–976, 2015.

[4] T. Trupke and P. Würfel, "Improved spectral robustness of triple tandem solar cells by combined series/parallel interconnection," *J. Appl. Phys.*, vol. 96, no. 4, p. 2347, 2004.

[5] S. Essig, J. F. Geisz, M. A. Steiner, A. Merkle, R. Peibst, J. Schmidt, R. Brendel, S. Ward, D. J. Friedman, P. Stradins, and D. L. Young, "Development of highly-efficient GaInP/Si Tandem Solar Cells," in *Proceedings of the 42nd IEEE Photovoltaic Specialists Conference*, 2015, pp. 1–4.

[6] M. Rienäcker, E. L. Warren, M. Schnabel, S. Kajari-Schröder, R. Niepelt, R. Brendel, P. Stradins, A. C. Tamboli, and R. Peibst, "Current-matching of monolithic tandem cells by using interdigitated back contact bottom cells with three terminals," *To be presented at the 33rd European Photovoltaic Solar Energy Conference and Exhibition*, 2017.

[7] M. Schnabel, M. Rienäcker, A. Merkle, T. R. Klein, N. Jain, S. Essig, van Hest,Maikel F. A. M., J. Geisz, J. Schmidt, R. Brendel, R. Peibst, P. Stradins, and A. C. Tamboli, "III - V/Si Tandem Cells Utilizing Interdigitated Back Contact Si Cells and Varying Terminal Configurations," *To be presented at this conference.*

[8] E. L. Warren, M. G. Deceglie, P. Stradins, and A. C. Tamboli, "Modeling three-terminal III-V/Si tandem solar cells," *To be presented at this conference.*

[9] A. L. Lentine, G. N. Nielson, M. Okandan, W. C. Sweatt, J. L. Cruz-Campa, and V. Gupta, "Optimal cell connections for improved shading, reliability, and spectral performance of microsystem enabled photovoltaic (MEPV) modules," in *Proceedings of the 35th IEEE Photovoltaic Specialists Conference*, 2010, pp. 3048–3054.

[10] R. Strandberg, "Detailed balance analysis of area de-coupled double tandem photovoltaic modules," *Appl. Phys. Lett.*, vol. 106, no. 3, p. 33902, 2015.

[11] R. M. France, J. F. Geisz, M. A. Steiner, D. J. Friedman, J. S. Ward, J. M. Olson, W. Olavarria, M. Young, and A. Duda, "Pushing Inverted Metamorphic Multijunction Solar Cells Toward Higher Efficiency at Realistic Operating Conditions," *IEEE J. Photovoltaics*, vol. 3, no. 2, pp. 893–898, 2013.

[12] J. C. Fan, "Theoretical temperature dependence of solar cell parameters," *Solar Cells*, vol. 17, no. 2-3, pp. 309–315, 1986.

[13] V. Alex, S. Finkbeiner, and J. Weber, "Temperature dependence of the indirect energy gap in crystalline silicon," *Journal of Applied Physics*, vol. 79, no. 9, pp. 6943–6946, 1996.

[14] I. Vurgaftman, J. R. Meyer, and L. R. Ram-Mohan, "Band parameters for III–V compound semiconductors and their alloys," *Journal of Applied Physics*, vol. 89, no. 11, pp. 5815–5875, 2001.

[15] B. Schumann, *Properties of Gallium Arsenide. EMIS Datareviews Series no. 2, second edition. INSPEC, The Institute of Electric Engineering, London and New York*, 1991.

[16] C. Schinke, P. Christian Peest, J. Schmidt, R. Brendel, K. Bothe, M. R. Vogt, I. Kröger, S. Winter, A. Schirmacher, S. Lim, H. T. Nguyen, and D. MacDonald, "Uncertainty analysis for the coefficient of band-to-band absorption of crystalline silicon," *AIP Advances*, vol. 5, no. 6, p. 67168, 2015.

[17] C. A. Gueymard, "SMARTS, A Simple Model of the Atmospheric Radiative Transfer of Sunshine: Algorithms and Performance Assessment," *Professional Paper FSEC-PF-270-95. Florida Solar Energy Center, 1679 Clearlake Rd., Cocoa, FL 32922.*

[18] C. A. Gueymard, "Parameterized transmittance model for direct beam and circumsolar spectral irradiance," *Solar Energy*, vol. 71, no. 5, pp. 325–346, 2001.

[19] J. F. Geisz, M. A. Steiner, I. García, S. R. Kurtz, and D. J. Friedman, "Enhanced external radiative efficiency for 20.8%

efficient single-junction GaInP solar cells," *Appl. Phys. Lett.*, vol. 103, no. 4, p. 41118, 2013.

[20] J. Haschke, J. P. Seif, Y. Riesen, A. Tomasi, J. Cattin, L. Tous, P. Choulat, M. Aleman, E. Cornagliotti, A. Uruena, R. Russell, F. Duerinckx, J. Champliaud, J. Levrat, A. A. Abdallah, B. Aïssa, N. Tabet, N. Wyrsch, M. Despeisse, J. Szlufcik, S. de Wolf, and C. Ballif, "The impact of silicon solar cell architecture and cell interconnection on energy yield in hot & sunny climates," *Energy Environ. Sci.*, vol. 9, p. 2122, 2017.

InGaP/GaAs/ITO/Si Hybrid Triple-Junction Cells with GaAs/ITO Bonding Interfaces

Naoteru Shigekawa, Tomoya Hara, Tomoki Ogawa and Jianbo Liang
Graduate School of Engineering, Osaka City University, Osaka 558-8585, Japan

Takefumi Kamioka, Kenji Araki and Masafumi Yamaguchi
Toyota Technological Institute, Nagoya 468-8511, Japan

Abstract—Using surface-activated bonding technologies we fabricate InGaP/GaAs/ITO/Si hybrid triple-junction (3J) cells with p^+-GaAs/ITO and those with n^+-GaAs/ITO bonding interfaces. ITO films deposited on the emitter of Si bottom cells work as intermediate layers between III-V and Si sub cells. The samples are not heated during the bonding process. The photovoltaic characteristics of the fabricated 3J cells are compared with characteristics of conventional 3J cells without intermediate layers. The InGaP/GaAs/ITO/Si 3J cells with n^+-GaAs/ITO bonding interfaces reveal the highest conversion efficiency and the lowest differential resistance among the investigated 3J cells, which implies the potential of ITO-based intermediate layers for achieving more excellent performances of hybrid multi-junction cells.

I. INTRODUCTION

The surface activated bonding (SAB) technologies [1] have successfully been applied for fabricating III-V/Si multi-junction solar cells [2] since dissimilar semiconductor materials with different lattice constants and thermal expansion coefficients can be bonded to each other at low temperatures. It is well understood that the electrical properties of bonding interfaces are influenced by the interface states introduced during the Ar beam irradiation in the SAB process. The performances of hybrid multi-junction cells are, consequently, likely to be limited by the resistance across the bonding interfaces. Although comparatively low interface resistances (~ 0.1 Ωcm^2) were obtained for junctions made of heavily-doped substrates or epitaxially-grown layers [3], the interface resistance in the actual hybrid 3J cells was much higher due to the limited thicknesses ($\sim$several nm) of bonding layers, which also worked as emitters, of Si bottom cells [4].

More importantly, in general, surfaces with the averaged roughness of <1 nm are required so as to achieve firmly-bonded junctions using SAB. This means that the surfaces of bonding layers of the respective sub cells must be flat, i.e., artifacts for enhancing the efficiencies such as textures and passivation films can not be incorporated to the sub cells in fabricating hybrid multi-junction cells.

Here we expect that better performances could be realized in hybrid multi-junction cells by replacing their SAB-based semiconductor/semiconductor junctions by junctions with optically and electrically transparent intermediate layers since such intermediate layers are assumed to play a role of lowering the interface resistance as well as flattening the surfaces of sub cells. In this work, we examine the feasibility of indium-tin-oxide (ITO) films, which are deposited on the emitters of Si bottom cells, as practical candidates for such intermediate

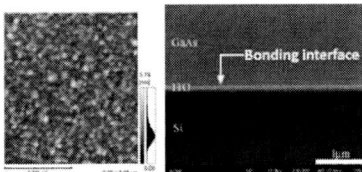

Fig. 1. An AFM image of ITO films deposited on Si substrates as well as a SEM image of GaAs/ITO/Si junctions.

layers. All of the investigated junctions and cells are fabricated without heating samples.

II. EXPERIMENTS

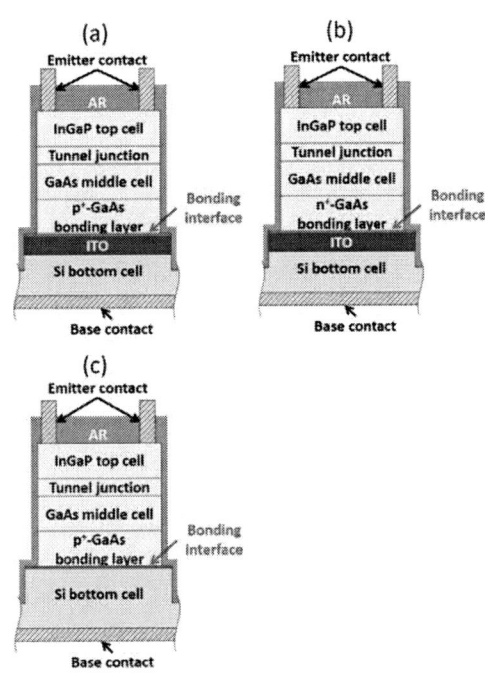

Fig. 2. Schematic cross sections of (a) InGaP/GaAs/ITO/Si 3J cells with p^+-GaAs/ITO bonding interfaces (3J(a)), (b) InGaP/GaAs/ITO/Si 3J cells with n^+-GaAs/ITO bonding interfaces (3J(b)), and (c) InGaP/GaAs/Si 3J cells with p^+-GaAs/n^+-Si bonding interfaces (3J(c)).

In a preparatory study, we performed AFM observations of surfaces of ITO films deposited on Si substrates. We found that the averaged roughness in surfaces of ITO films was ≈ 0.3 nm as is shown in Fig. 1, which was small enough for the films to be firmly bonded. We actually confirmed that ITO films deposited on Si substrates were bonded to other Si substrates and the obtained Si/ITO/Si junctions revealed excellent electrical characteristics [5]. In Fig. 1 is also shown an SEM image of GaAs/ITO/Si junctions, which demonstrated that interfaces with no voids were achieved. Furthermore we observed that the transmittance of ITO films was not deteriorated by irradiating Ar beams in the SAB process.

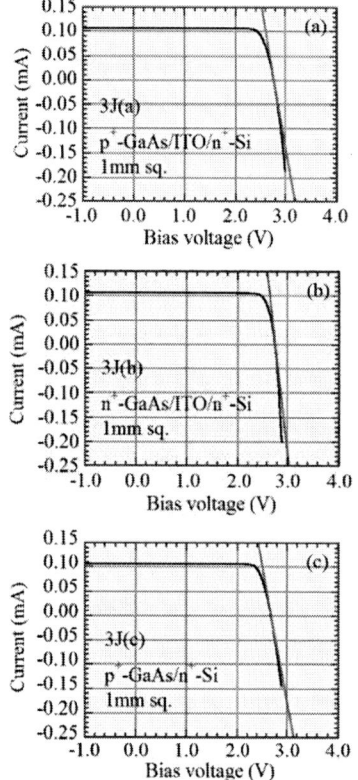

Fig. 3. $I-V$ characteristics of (a) 3J(a), (b) 3J(b), and (c) 3J(c) cells.

TABLE I
CHARACTERISTIC PARAMETERS OF THE HYBRID 3J CELLS.

3J type	3J(a)	3J(b)	3J(c)
short-circuit current (mA)	0.106	0.107	0.106
open-circuit voltage (V)	2.77	2.76	2.67
FF (%)	83.4	85.6	84.1
-dV/dI @0 mA (Ω)	1700	1130	1800
as-measured efficiency (%)	24.6	25.1	24.0

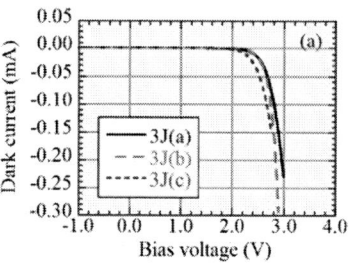

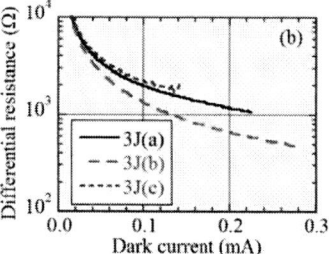

Fig. 4. (a) $I-V$ characteristics of the 3J cells in the dark. (b) The relationships between the differential resistance and the dark current of the 3J cells in the dark.

90-nm-thick ITO films were deposited on the several-nm-thick n^+-Si emitters of Si bottom cells. The emitters were fabricated by the ion implantation and annealing. We prepared n-on-p InGaP/GaAs 2J heterostructures with p^+-GaAs bonding layers and those with n^+-GaAs bonding layers. We bonded each of the 2J heterostructures to the Si bottom cells with the capping ITO films. By using previously-reported process steps [2], we fabricated two types of InGaP/GaAs/ITO/Si hybrid 3J cells, or 3J(a) and 3J(b), which are equipped with p^+-GaAs/ITO and n^+-GaAs/ITO bonding interfaces, respectively. In addition we prepared 3J cells with p^+-GaAs/n^+-Si bonding interfaces (3J(c)). The schematic cross sections of the 3J(a), 3J(b), and 3J(c) cells are shown in Figs. 2(a), 2(b), and 2(c), respectively. The mesa area of the respective 3J cells was 1 mm^2.

The current-voltage (I-V) characteristics of the 3J cells under the air mass 1.5G/one sun irradiance, which were measured using in-house facilities, are shown in Figs. 3(a), 3(b), and 3(c) for 3J(a), 3J(b), and 3J(c), respectively. Values of parameters characterizing the respective curves are summarized in Table I. The conversion efficiencies of the respective cells were 24.6 (3J(a)), 25.1 (3J(b)), and 24.0% (3J(c)). Their differential resistance at $I = 0$ mA, which we obtained by the least-square fitting, was 1700, 1130, and 1800 Ω for 3J(a), 3J(b), and 3J(c), respectively, as is seen from the slopes of the red straight lines in the figures. The intrinsic conversion efficiency of the 3J(b) cell was estimated to be $\approx 27\%$ by compensating the shadow loss of the emitter contacts and the estimated contribution of the 10-μm Si periphery.

The I-V characteristics of the 3J cells measured in the dark are shown in Fig. 4(a). The relationships between the differen-

978-1-5090-5606-4/17 $31.00 © 2017 IEEE

tial resistance and the dark current, which were obtained from the respective I-V curves, are shown in Fig. 4(b). We find that the differential resistance was lowered by employing the ITO intermediate layers. The lowest resistance was obtained in 3J(b), or 3J cells with n^+-GaAs/ITO bonding interfaces.

The difference in resistance in the dark among the three 3J cells is consistent with the results under the solar irradiance. The difference between 3J(a) and 3J(b) is explained by the n-type features of ITO films. The obtained results indicate that the ITO intermediate layers should play a role of lowering the series resistance in hybrid multijunciton cells and are useful in achieving higher efficiencies in hybrid multi-junction cells.

III. Conclusion

We successfully fabricated InGaP/GaAs/ITO/Si hybrid 3J cells using the SAB technologies. The 3J cells with n^+-GaAs/ITO bonding interfaces revealed higher conversion efficiencies and lower interface resistances in comparison with the 3J cells with p^+-GaAs/ITO or p^+-GaAs/n^+-Si bonding interfaces. The results imply that the ITO films should be applicable as intermediate layers in hybrid multi-junction cells.

Acknowledgment

Heterostructures for InGaP/GaAs 2J cells were prepared by Sharp Corporation. This work was supported by the "Research and Development of ultra-high efficiency and low-cost III-V compound semiconductor solar cell modules (High efficiency and low-cost III-V/Si tandem)"project of New Energy and Industrial Technology Development Organization (NEDO).

References

[1] H. Takagi, K. Kikuchi, R. Maeda, T. R. Chung, and T. Suga, Appl. Phys. Lett. **68**, 2222 (1996).

[2] N. Shigekawa, J. Liang, R. Onitsuka, T. Agui, H. Juso, and T. Takamoto, J. Appl. Phys. **54**, 08KE03 (2015).

[3] J. Liang, S. Nishida, M. Morimoto, and N. Shigekawa, Elec. Lett. **49**, 830 (2013).

[4] N. Shigekawa and J. Liang, to be presented in 2017 IEEE PVSC-44.

[5] J. Liang, T. Ogawa, K. Araki, T. Kamioka, and N. Shigekawa, to be presented in LTB-3D 2017.

Measurements of Potentials at Tap Contacts and Estimation of Resistance across Bonding Interfaces in InGaP/GaAs/Si Hybrid Triple-Junction Cells

Naoteru Shigekawa and Jianbo Liang

Graduate School of Engineering, Osaka City University, Osaka 558-8585, Japan

Abstract—InGaP/GaAs/Si hybrid triple-junction (3J) cells equipped with tap contacts are fabricated. The difference in potentials between bases of GaAs middle cells and emitters of Si bottom cells is measured while the 3J cells are illuminated. The resistance across the p-GaAs/n-Si bonding interfaces is estimated to be ~ 4 Ωcm^2, which is 30 times higher than the resistance measured for interfaces of heavily-doped p-GaAs and n-Si substrates. Such high resistances in bonding interfaces in the 3J cells are likely to be attributed to their thin (several-nm-thick) n^+-Si emitters that also work as bonding layers.

I. INTRODUCTION

InGaP/GaAs/Si triple-junction (3J) cells are promising candidates for high-efficiency and low-cost photovoltaics from the practical viewpoints [1]. Due to the difficulties in the epitaxial growth of III-V layers on Si substrates, such multi-junction cells are mostly fabricated by using surface-activated bonding (SAB) technologies [2], or using hybrid approaches. In the SAB process, the sample surfaces are cleaned prior to bonding by using fast atom beams of noble gas species such as Ar.

Although the electrical properties of bonding interfaces might be deteriorated by damages due to the Ar beam irradiation [3], preparatory studies with doped semiconductor substrates revealed that the resistance across the bonding interfaces is decreased to practically-applicable values by increasing concentrations of impurities in bonding layers [4] or by annealing the bonding interfaces [5]. Actually we previously achieved p^+-GaAs substrate/n^+-Si substrate junctions with an interface resistance of 0.13 Ωcm^2 [4]. The resistance across the bonding interfaces in actual hybrid multi-junction cells, however, have not yet been fully investigated although such resistance could contribute to the series resistance (R_s) in hybrid multi-junction cells. In this work, we directly measure potentials at bases of GaAs middle cells and emitters of Si bottom cells, or those at bonding layers, in hybrid 3J cells with additional tap contacts and estimate the resistance across the bonding interfaces.

II. EXPERIMENTS

Si-based bottom cell structures were fabricated by the implantation of phosphor (P) and boron (B) ions and the rapid thermal annealing. The dose of P atoms for forming emitters was 4.3×10^{14} cm^{-2}. The thickness and sheet resistance of emitter layers were estimated to be several nm and ≈ 800 Ω/sq. by SIMS and TLM measurements, respectively.

We separately prepared heterostructures for InGaP/GaAs double-junction (2J) cells by the epitaxial growth on GaAs substrates. We fabricated n-on-p InGaP/GaAs/Si 3J cells with a nominal top-cell mesa area of 0.041 cm^2 (2.02 mm by 2.02

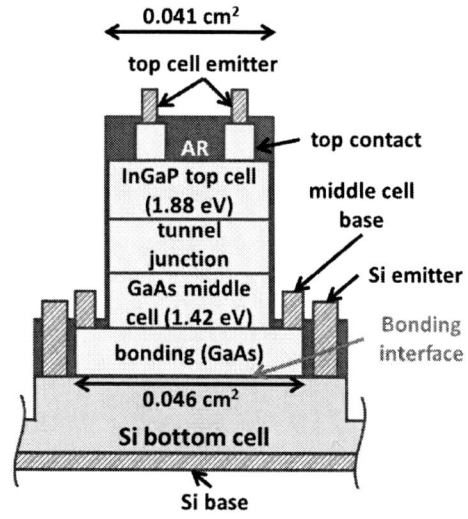

Fig. 1. A schematic cross section of 2.02-mm-by-2.02-mm InGaP/GaAs/Si 3J cells with tap contacts.

mm) by surface activation bonding (SAB) of the heterostructures and Si-based bottom cells and using the conventional device process. We formed tap contacts on bases of GaAs middle cells and emitters of Si bottom cells. The other steps for fabricating cells were the same as previous reports. The schematic cross section of 3J cells is shown in Fig. 1. It is notable that a part of the surface of Si bottom cells is exposed to the air so that their effective area is larger than that of top cells. The area of p-GaAs/n-Si bonding interfaces is 0.046 cm^2.

The current-voltage ($I - V$) characteristics of the 3J cell illuminated under the air mass 1.5G/one sun condition are shown in Fig. 2. The $I - V$ characteristics of the InGaP/GaAs 2J sub cell and those of the Si bottom cell, which were measured by using tap contacts, are also shown in this figure. The open-circuit voltage (V_{OC}) of the 3J cell (2.84 V) almost agreed with the sum of V_{OC} of the 2J sub cell (2.31 V) and Si bottom cell (0.49 V), i.e., the additivity of V_{OC} was confirmed. The short-circuit current (I_{SC}), maximum output power (P_{max}), resistance at the open-circuit-voltage condition ($-dV/dI|_{V=V_{OC}}$) of the 3J cell are 0.448 mA, 1.041 mW, and 464 Ω, respectively. We normalized the achieved I_{SC} and P_{max} by the area of top cells and obtained the short-circuit current density of ≈ 11 mA/cm^2 and the conversion

978-1-5090-5606-4/17 $31.00 © 2017 IEEE

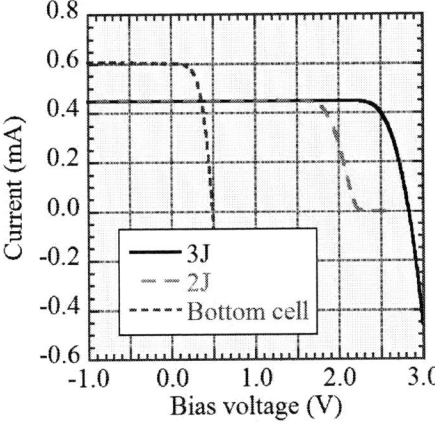

Fig. 2. $I - V$ characteristics of InGaP/GaAs/Si 3J cells, InGaP/GaAs 2J sub cells, and Si bottom cells illuminated under the condition of air mass of 1.5G and one sun.

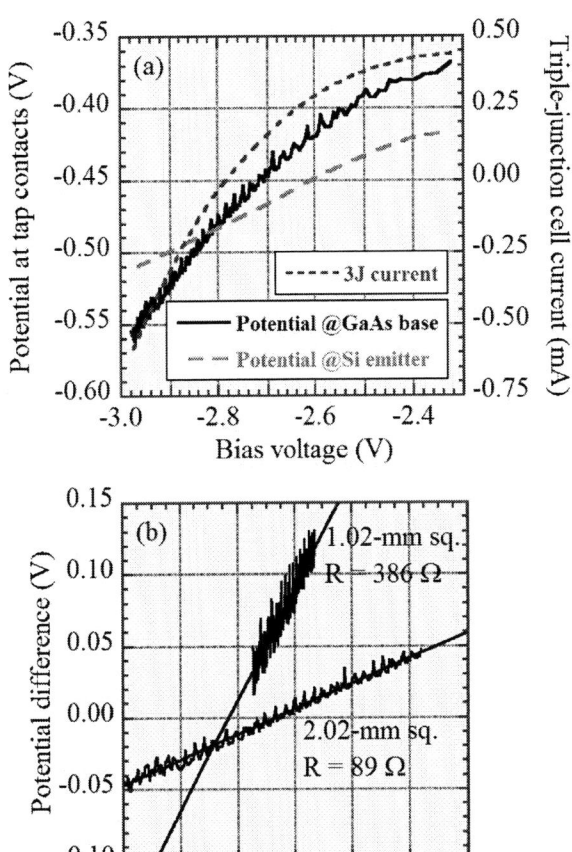

Fig. 3. (a) The dependencies of the potentials at the base of GaAs middle cell and the emitter of Si bottom cell and the current across the 2.02-mm-by-2.02-mm 3J cell on the bias voltage. (b) The relationships between the difference between bases of middle cells and emitters of bottom cells and the 3J-cell current for the 2.02-mm-by-2.02-mm and 1.02-mm-by-1.02-mm 3J cells. The results of least-square fitting to straight lines and the estimated resistances across the bonding interfaces are also shown.

efficiency of 25.5%, which are larger than our previous report [1]. In addition, I_{SC} of the Si bottom cell (0.601 mA) is larger than that of the 2J sub cell (0.447 mA). These results are attributable to uncovered parts of surfaces of the Si bottom cell. We also find that the series resistance for the 2J sub cell is higher than that for the 3J cell from Fig. 2, which is due to a large resistance at the tap contact on the GaAs base.

Potentials at the two tap contacts were measured while the 3J cell was illuminated and bias voltages were applied. The base of the 3J cell was grounded during measurements. The dependencies of the potentials at tap contacts and the current across the 3J cell on the potential at the emitter of 3J cell, or the bias voltage, are shown in Fig. 3(a). The oscillation in the potential at the contact on the GaAs base is likely to be due to the high contact resistance. V_{OC} of the 3J cell, 2J sub cell, and Si bottom cell was ≈ 2.78, 2.3, and 0.48 V, respectively. A slight discrepancy between these V_{OC} values and the results shown in Fig. 2 might be due to the difference in conditions for measurements.

Figure 3(b) gives the relationship between the difference in potentials at the tap contacts, which should be the bias voltage applied to the bonding interfaces, and the current in the 3J cell. The resistance across the bonding interface $R_{bonding}$ was estimated to be 89 Ω from the slope of the curve obtained by the least-square fitting. We measured potentials at tap contacts in 1.02-mm-by-1.02-mm 3J cells by using the same method. We found that the resistance was 386 Ω for bonding interfaces with an area of 0.013 cm^2, as is also shown in Fig. 3(b) .

III. DISCUSSIONS

The difference between the slope of $I - V$ characteristics at the open-circuit voltage ($-dV/dI|_{V=V_{OC}}$) and the estimated resistance at the bonding interface, which is $464 - 89 = 375$ Ω for the 2.02-mm-by-2.02-mm 3J cell, is likely to be attributed to the diode resistance, the substrate (bulk) resistance, and the contact resistances. Based on the scheme of differential

resistance of $p-n$ diodes, the diode resistance is approximately given by

$$(n_1 + n_2 + n_3)k_BT/(qI_{SC}) = 3 \times nk_BT/(qI_{SC}), \quad (1)$$

where k_BT/q is the thermal voltage (25.9 mV at $T = 300$ K) and n_1, n_2, and n_3 are the ideality factors of the respective sub cells. The right-hand side in the above equation is based on the assumption that the ideality factors of the respective sub cells are equal to n. Ignoring the contribution of the substrate resistance and the contact resistance, we obtained

$$3 \times nk_BT/(qI_{SC}) = 375 \ \Omega, \quad (2)$$

which corresponded to $n = 2.2$. This value was likely to disagree with the requirement of the physics of $p - n$ diodes that $n \leq 2$. The disagreement might be due to the resistances that we ignore. Similar analyses were performed for the 1.02-mm-by-1.02-mm 3J cell so that we obtained $n = 1.9$.

The estimated resistances across the bonding interfaces corresponded to ~ 4 Ωcm^2 for both 1.02-mm-by-1.02-mm and 2.02-mm-by-2.02-mm 3J cells, which was approximately 30 times higher than a previously-reported resistance for junctions made of p^+-GaAs and n^+-Si substrates. The discrepancy was likely to be due to thin (several-nm thick) n^+-Si emitter layers in bottom cells. The obtained results suggested that emitter structures in bottom cells should be revised for achieving bonding interfaces with lower resistances in 3J hybrid tandem cells.

IV. CONCLUSION

The resistance across the p-GaAs/n-Si bonding interfaces in InGaP/GaAs/Si 3J hybrid tandem cells were estimated by measuring potentials at tap contacts on bases of GaAs middle cells and emitters of Si bottom cells. The resistance amounted to ~ 4 Ωcm^2, which was several ten times higher than that observed for junctions of heavily-doped substrates. Such high resistance in the bonding interfaces of 3J cells was likely to be attributed to thin n^+ emitter layers of bottom cells.

ACKNOWLEDGMENT

Heterostructures for InGaP/GaAs 2J cells were prepared by Sharp Corporation. This work was supported by the "Research and Development of ultra-high efficiency and low-cost III-V compound semiconductor solar cell modules (High efficiency and low-cost III-V/Si tandem)" Project of New Energy and Industrial Technology Development Organization (NEDO).

REFERENCES

[1] N. Shigekawa, J. Liang, R. Onitsuka, T. Agui, H. Juso, and T. Takamoto, J. Appl. Phys. **54**, 08KE03 (2015).
[2] H. Takagi, K. Kikuchi, R. Maeda, T. R. Chung, and T. Suga, Appl. Phys. Lett. **68**, 2222 (1996).
[3] M. Morimoto, J. Liang, S. Nishida and N. Shigekawa, Jpn. J. Appl. Phys. **54**, 030212 (2015).
[4] J. Liang, S. Nishida, M. Morimoto, and N. Shigekawa, Elec. Lett. **49**, 830 (2013).
[5] J. Liang, L. Chai, S. Nishida, M. Morimoto, and N. Shigekawa, Jpn. J. Appl. Phys. **54**, 030211 (2015).

Optimization of a GaAsP Top Cell for Implementation in a III-V/Si Tandem Structure

Amber C. Silvaggio, Daniel L. Lepkowski, Daniel J. Chmielewski, Jacob T. Boyer, Steven A. Ringel, Tyler J. Grassman

The Ohio State University, Columbus, Ohio, 43210, USA

Abstract — **Significant performance improvement in a top cell isotype for a GaAs$_{0.75}$P$_{0.25}$/Si tandem solar cell was achieved using a model-driven optimization process. Physics-based TCAD software was used to extract key materials parameters via quantum efficiency curve fitting against a reference prototype device. These parameters were then used to guide the optimization of three key aspects of the top cell device structure: emitter doping, emitter thickness, and base thickness. An isotype cell following this new design was then grown via MOCVD, processed, and characterized. While the new design did not perform up to the predicted level, it was nonetheless found to provide an absolute increase in AM1.5G conversion efficiency of 2.3% (from 12.7% to 15.0%, without antireflection coating) due to significant improvements in all relevant metrics, most notably J$_{SC}$ (+12.7% relative) and fill factor (+3.3% relative). This initial modeling-guided effort has thus demonstrated a high degree of efficacy, requiring only a single iteration to yield significant results.**

Index Terms —**III-V semiconductor materials, solar panels, electronic device modeling.**

I. INTRODUCTION

III-V/Si multijunction solar cells are being explored as a potentially high-efficiency, low-cost avenue for terrestrial power generation. There are two main approaches for device fabrication currently being explored toward this end: bonding of exfoliated III-V device structures with a Si cell [1-3] and direct epitaxy of III-V materials/devices onto the Si cell [4-6]. We focus on the latter here, based on our previous demonstration of an all-epitaxial, metalorganic chemical vapor deposition (MOCVD)-grown, series-connected GaAs$_{0.75}$P$_{0.25}$/Si

dual-junction (tandem) solar cell [4], depicted in Figure 1. The promising initial results from this prototype device revealed a clear pathway for achieving high conversion efficiencies (i.e. >30% AM1.5G), including several aspects of design optimization that are needed to maximize the performance of the GaAs$_{0.75}$P$_{0.25}$/Si tandem cell. This work focuses on the top cell design.

Within the GaAs$_{0.75}$P$_{0.25}$/Si tandem device structure, the most fundamental limiter of performance of the III-V top cell is expected to be excessive threading dislocation density (TDD) resulting from the ~3% lattice mismatch with Si. Significant progress has been made in this respect, with recent demonstrations of TDD down to the mid-10^6 cm^{-2} range [7]. Nonetheless, if actual materials properties are taken into account—namely, minority carrier lifetime as dictated by residual TDD—it is possible the optimize device performance for the given achievable material quality [8]. To this end, the work described herein targets future device designs following an assumed reduction in residual TDD to the low 10^6 cm^{-2} range. The ultimate achievable top cell performance at this design point should serve as a conservative milestone for realistically achievable performance.

II. APPROACH

Baseline reference and semi-optimized test n$^+$/p GaAs$_{0.75}$P$_{0.25}$ cells, with an n$^+$-Al$_{0.64}$In$_{0.36}$P window layer and a p$^+$-Ga$_{0.64}$In$_{0.36}$P back surface field (BSF), as schematically shown in Figure 2, were grown on p-type GaAs$_y$P$_{1-y}$/GaAs tensile-graded virtual substrates by metal-organic chemical vapor deposition (MOCVD). 2 mm × 2 mm solar cell test

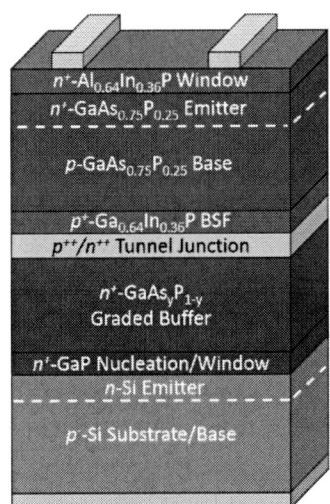

Figure 1: GaAs$_{0.75}$P$_{0.25}$/Si dual-junction solar cell structure [4].

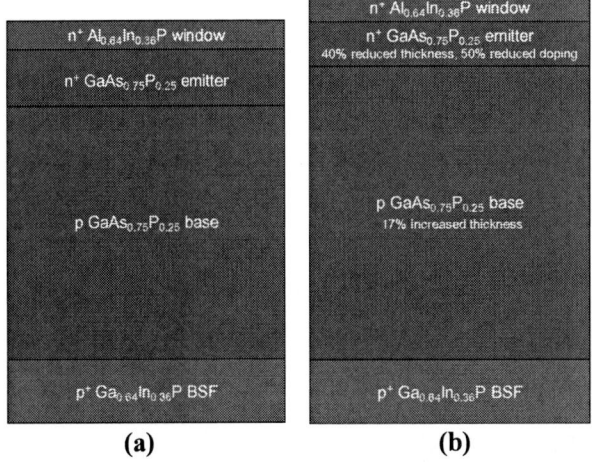

Figure 2: (a) Original reference and (b) semi-optimized cell designs.

978-1-5090-5606-4/17 $31.00 © 2017 IEEE

devices were processed using standard photolithographic and both wet and dry etching methods. Top (n-type) Ohmic contacts were made using a Ni/GeAu metal stack deposited on a highly-doped GaAs$_{0.75}$P$_{0.25}$ surface contact layer and were annealed at 390°C; rear (p-type) contact, made to the bottom of the GaAs substrate, was provided via full coverage Ti/Au. The surface metal grid coverage was 8.6% and no antireflection coating was applied. The cells were characterized by external quantum efficiency (EQE) using a custom-built small spot system and illuminated current-voltage (LIV) under simulated AM1.5G (OAI TriSol). Lamp intensity drift between the measurements of both characterized samples was measured to be < 1.5%. TDD in the processed cells were measured by electron beam induced current (EBIC), revealing a residual density of ~2×10^6 cm^{-2}. Further TDD reduction is expected to be possible within this grading system [9], but this value was considered adequate to serve as a conservative analog to the projected virtual substrate material quality for a >30% efficient tandem structure.

Silvaco Atlas technology computer aided drafting (TCAD) software was used to model the GaAs$_{0.75}$P$_{0.25}$ reference cell in order to extract unknown materials properties (e.g. minority carrier diffusion length, interface recombination velocity, etc.). The simulated device structure—layer thicknesses, doping concentrations, and materials compositions—was set to the best known values based upon empirical data or nominal growth targets. EQE curves were then calculated, with optical effects included using the transfer matrix method. The minority carrier diffusion length (L$_D$) and interface recombination velocities (IRV) were adjusted for each region/interface until the EQE of the model fit the measured EQE data from the reference cell. Simulated LIV measurements were used for secondary verification of model accuracy. The material parameters of each region were then extracted from this fit model and kept constant for optimization simulations.

Optimization was performed by maximizing the simulated EQE. Design aspects that were targeted for this work, based upon analysis of the experimental EQE data suggesting room for the most improvement (in the short- and mid-wavelength ranges), included emitter layer thickness, emitter layer doping, and base layer thickness. These values were manually varied in a systematic manner using a sequential "hill climb" optimization methodology. Note that this initial optimization round was relatively coarse since it was performed manually, but future work will examine the same and additional variables on a finer scale. The new design that resulted from the optimization modeling was grown under nominally identical conditions on a nominally identical virtual substrate as the reference cell. Test devices were processed and characterized in the same manner to enable direct comparison.

III. RESULTS AND DISCUSSION

A schematic comparison of the initial and final cell designs are provided in Figure 2. Upon analyzing the simulated EQE results from the various structures, a semi-optimized structure was created. A 50% reduced emitter doping was selected following the observation that the modeled designs with lower emitter doping consistently yielded a higher short wavelength response versus the original design. This result thus suggested

an excessive Shockley-Read-Hall recombination in the original emitter design due to excessive doping. Reduction in doping thus should improve the diffusion length in the emitter allowing for a higher collection probability.

For emitter thickness, the model indicated continuous EQE improvement with thicknesses approaching zero. This, too, is likely a result of SRH-based carrier loss, but also suggests that perhaps the Al$_{0.64}$In$_{0.36}$P/GaAs$_{0.75}$P$_{0.25}$ IRV was being underestimated. Additionally, some amount of emitter thickness is needed to ensure a sufficiently low surface sheet resistance to avoid fill factor (FF) loss, since window layer thickness was not considered in this optimization round. Thus, as a reasonable compromise based on expected growth uniformity control and use of the lightly cross-hatched metamorphic virtual substrates, an emitter thickness reduction of 40% was chosen.

Finally, a base thickness increase of 17% was selected. This change was in response to the indication of a slight near-bandgap EQE increase owing to higher total absorption with a thicker base layer. However, this improvement was found to saturate at only about 15-20% increase versus the original design, ostensibly due to being still somewhat limited by the minority carrier diffusion length.

Figure 3 presents the EQE results from the simulated and experimental studies from both reference and semi-optimized cells. We can see here that the improvement in performance of the new cell design is effectively limited to the short wavelength (λ ≤ 500 nm). This can be straightforwardly assigned to the changes made to the emitter. The reduced thickness helps to reduce the total amount of photogeneration within the (still) relatively highly doped (and thus still relatively low lifetime) emitter, ensuring that absorption occurs in a region with high collection probability (i.e. the depletion region or base). The reduced doping provides a slight increase in depletion width within the emitter, which combined with the thinner overall layer helps to ensure better collection altogether, likely assisted somewhat by slightly longer lifetime due to the reduced doping. The combination of the thinner emitter and wider depletion region creates a region with significantly higher carrier collection probability much closer to the emitter/window

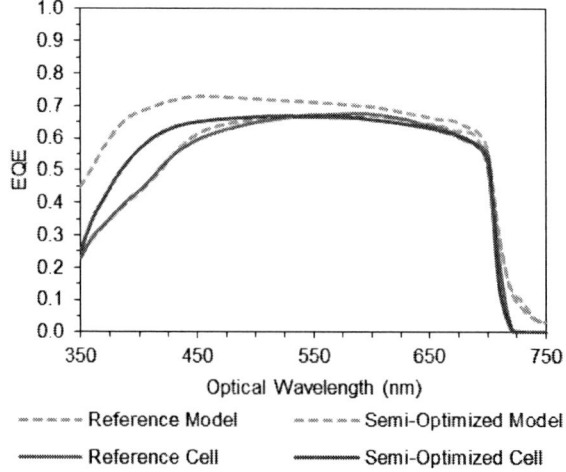

Figure 3: Simulated and measured EQE data from the two non-antireflection-coated GaAsP cell designs.

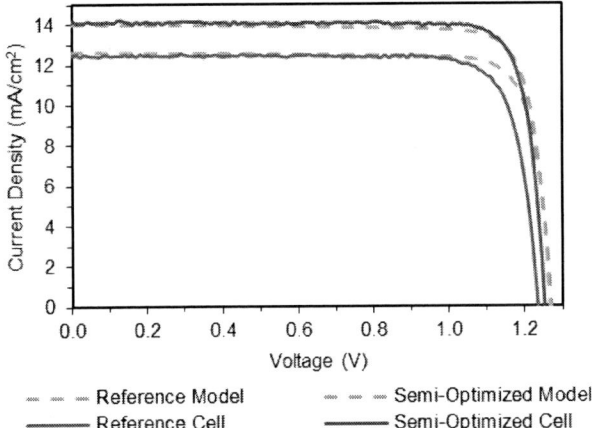

Figure 4: LIV data from the two fabricated GaAsP cells under a simulated AM1.5G spectrum.

interface, which is where the maximum photogeneration for sub 500 nm photons lies.

The semi-optimized and reference cells were found to be nearly identical in their longer wavelength responses. Based on available optical data, the model did suggest that there was some transmission loss in the reference structure indicated by the projected improvement in long wavelength response with increased base thickness, but the experimental results indicate that this was not a significant performance limiter at present. Most likely the long wavelength performance is being limited by non-ideal minority carrier lifetime due to dislocation content. However, since there has also not yet been any substantial effort undertaken to optimize the $GaAs_{0.75}P_{0.25}$ growth conditions, the overall material quality (e.g. point defects) could also be playing a role.

Figure 4 presents the simulated and experimental LIV results from representative examples of the reference and semi-optimized cells under simulated AM1.5G illumination. The simulated LIV curves were corrected to account for front-side grid shadowing, which was not explicitly included within the present TCAD model. A significant improvement in operating characteristics of the semi-optimized cell over the reference can clearly be seen. Even with the improvement in carrier collection (EQE) limited to the short wavelength regime, the largest increase was still found in the J_{SC}: from 12.5 mA/cm^2 for the reference cell to 14.1 mA/cm^2 for the new, semi-optimized design, a relative improvement of +12.7%. The fill factor also experienced a significant increase, from 82.7% (reference) to 85.4% (semi-optimized), a +3.3% relative improvement. The V_{OC} saw the smallest increase, only +1.6% relative (1.23 V to 1.25 V). It may be that there was some improvement due to the reduced recombination in the emitter, which was somewhat cancelled out by the reduced emitter doping. Nonetheless, altogether these improved metrics result in a total increase in absolute AM1.5G efficiency of +2.3% (12.7% to 15.0%, not third-party verified). Addition of a good quality antireflection coating is expected to further increase this toward 20%, which based on minimum requirements modeling [10], should be sufficient to support achievement of a ≥ 30% $GaAs_{0.75}P_{0.25}$/Si tandem cell.

While this is a promising result, it is clear from Figure 3 that

the actual fabricated cell underperformed versus the simulated prediction of EQE. One immediate point of discrepancy is likely related to optical aspects. Slight differences were observed between measured specular reflectance and simulated reflectance. While these differences were relatively small, they will nonetheless propagate into the materials properties extraction effort, and thus into the device design. Improved empirical data and optical model approach should help minimize this source of uncertainty. From a practical standpoint, it is also likely that there were slight variations between the reference and semi-optimized growths (i.e. TDD, doping, interface quality, etc.), which would yield some degree of unpredictable variability.

It is also highly likely that there are significant inaccuracies within the model due to incorrect materials properties either supplied or extracted by the operator or calculated internally within the software. Indeed, it is our experience that many materials properties within the built-in library are incorrect, and where known they have been replaced, but many of the values and interpolation models used are not easily verified or updated. Further, the extracted minority carrier diffusion length, ostensibly the most important property, depends upon both minority carrier lifetime and minority carrier mobility, neither of which are currently known for $GaAs_{0.75}P_{0.25}$ at this TDD. The minority carrier mobility was estimated via extrapolation from tabulated GaAs data and our own measured $GaAs_{0.75}P_{0.25}$ majority carrier mobilities. However, if the mobilities used in the simulation were incorrect, so too would be the extracted minority carrier lifetime (due to their mutual relationship with diffusion length). This potential discrepancy could explain the relative underperformance of the semi-optimized cell design. e.

Finally, possibly the largest source of uncertainty at the present is the relative values of, and interplay between, the window/emitter IRV and the emitter bulk minority carrier lifetime, which both impact the emitter minority carrier diffusion length. The best fit was found when the input parameters indicated that collection in the emitter is bulk transport limited. As such, any IRV effect would have been wrapped up into an effective lifetime, which while enabling a good fit for the original reference cell, is not necessarily transferrable to a different design (e.g. thinner emitter in the semi-optimized cell). This underestimation of the window/emitter IRV may also explain the slight V_{OC} overestimation in the modeled LIV curves. The relationship between interface recombination and bulk diffusion length is currently being explored with more detailed modeling approaches.

IV. CONCLUSION

A semi-optimized $GaAs_{0.75}P_{0.25}$ top cell isotype was designed through the application of TCAD modeling and predictive simulation. The resultant new cell design was fabricated, characterized, and compared to the reference cell used to develop the initial model. Following only a single modeling-experiment iteration, a significant improvement in cell performance was observed, with a total absolute AM1.5G efficiency improvement of +2.3% (+28% relative). While the achieved performance level was not as high as expected based upon the modeling, likely due to growth variations and discrepancies in extracted and assumed materials properties, this result nonetheless resulted in a top cell design capable of supporting $\geq$ 30% GaAs$_{0.75}$P$_{0.25}$/Si tandem performance, with still plenty of room for further improvement.

ACKNOWLEDGMENT

This material is based upon work supported by the Department of Energy, Office of Energy Efficiency and Renewable Energy (EERE), under Award Number DE-EE0007539 and The Ohio State University College of Engineering Undergraduate Honors Research Distinction scholarship program.

REFERENCES

[1] R. Cariou, J. Benick, P. Beutel, N. Razek, C. Flötgen, M. Hermle, D. Lackner, S. W. Glunz, A. W. Bett, M. Wimplinger and F. Dimroth, "Monolithic Two-Terminal III-V//Si Triple-Junction Solar Cells With 30.2% Efficiency Under 1-Sun AM1.5g," *IEEE Journal of Photovoltaics*, vol. 7, pp. 367-373, 2017.

[2] S. Essig, M. A. Steiner, C. Alleb, J. F. Geisz, B. Paviet-Salomon, S. Ward, A. Descoeudres, V. LaSalvia, L. Barraud, N. Badel, A. Faes, J. Levrat, M. Despeisse, C. Ballif, P. Stradins and D. L. Young, "Realization of GaInP/Si Dual-Junction Solar Cells With 29.8% 1-Sun Efficiency," *IEEE Journal of Photovoltaics*, vol. 6, pp. 1012-1019, 2016.

[3] A. C. Tamboli, M. F. A. M. v. Hest, M. A. Steiner, S. Essig, E. E. Perl, A. G. Norman, N. Bosco and P. Stradins, "III-V/Si wafer bonding using transparent, conductive oxide interlayers," *Applied Physics Letters*, vol. 106, p. 263904, 2015.

[4] T. J. Grassman, D. J. Chmielewski, S. D. Carnevale, J. A. Carlin, and S. A. Ringel, "GaAs$_{0.75}$P$_{0.25}$/Si Dual-Junction Solar Cells Grown by MBE and MOCVD," *IEEE Journal of Photovoltaics*, vol. 6, pp. 326-331, 2016.

[5] T. J. Grassman, D. J. Chmielewski, J. A. Carlin and S. A. Ringel, "Development of epitaxial 2- and 3-junction III-V/Si solar cells," *Proc. 43rd IEEE Photovoltaic Specialists Conference*, pp. 2036-2039, 2016.

[6] M. Vaisman, K. N. Yaung, Y. Sun and M. L. Lee, "GaAsP/Si solar cells and tunnel junctions for III-V/Si tandem devices," *Proc. 43rd IEEE Photovoltaic Specialists Conference*, pp. 2043-2047, 2016.

[7] A. Mehrotra, W. Wang, and A. Freundlich, "Modeling and fabrication of GaAs solar cells with high dislocation tolerance," *Proc. 40th IEEE Photovoltaic Specialist Conference*, pp. 0514-0519, 2014.

[8] K. N. Yaung, J. R. Lang, and M. L. Lee, "Towards high efficiency GaAsP solar cells on (001) GaP/Si," *Proc. 40th IEEE Photovoltaic Specialist Conference*, pp. 0831-0835, 2014.

[9] M. J. Mori, S. T. Boles, and E. A. Fitzgerald, "Comparison of compressive and tensile relaxed composition-graded GaAsP and (Al)InGaP substrates," *Journal of Vacuum Science and Technology A*, vol. 28, pp. 182-188, 2010.

[10] T. P. White, N. N. Lal, and K. R. Catchpole, "Tandem Solar Cells Based on High-Efficiency c-Si Bottom Cells: Top Cell Requirements for >30% Efficiency," *IEEE Journal of Photovoltaics*, vol. 4, pp. 208-214, 2014.

[11] Atlas ver. 5.19.20.R, Silvaco Inc., Santa Clara, CA, 2013.

Theoretical Design of Perovskite/CdTe Four-terminal Tandem Solar Cells

Tao Tang[1], Huan Zhang[1], Xingzhi Du[1], Yiming Liu[2], Hang Zhou[1]*

1.The School of Electronic and Computer Engineering, Peking University Shenzhen Graduate School, Peking University, Shenzhen, 518055, China

2.NanoSYD, Mads Clausen Institute, University of Southern Denmark, Alsion 2, DK-6400 sønderborg, Denmark

Abstract — In high efficiency CdTe solar cells, a thin compact CdS window layer is usually required to ensure the formation of good heterojunction and to allow sufficient sunlight being absorbed by the CdTe, which inevitably adds to the manufacturing difficulties. Here, a four-terminal tandem solar cell with perovskite top cell and CdTe bottom cell is designed and simulated by wxAMPS. The Br content (*x*) in MAPb(I$_{1-x}$Br$_x$)$_3$ perovskite is tuned to match the bandgap of CdS/CdTe bottom cells. In such configuration, the adverse effect of thick CdS can be obviated.

Index Terms — wxAMPS, perovskite, CdTe, tandem solar cells.

I. INTRODUCTION

Polycrystalline cadmium sulfide/cadmium telluride (CdS/CdTe) is one of the most well-known thin-film solar cells. Loferski first proposed the use CdTe in photovoltaic device in 1956 [1], and three years later Rappaport fabricated the first single crystal homojunction CdTe solar cell with a efficiency of 2% [2]. The research of CdTe photovoltaics has received considerable attention in the past sixty years, and in 2016, a certified 22.1% power conversion efficiency on CdTe solar cells was achieved by First Solar [3]. CdTe photovoltaics is the second largest commercial photovoltaics technology after conventional crystalline silicon solar cells. One way to achieve high photocurrent in the CdS/CdTe solar cell is to reduce the window (CdS) layer thickness. Theoretically, CdS layer thickness should be less than 100nm. However, pinholes in CdS and non-conformal coverage of CdS on transparent conducting oxide layer will cause leakage current, leading to the deterioration of device performance [4]. In addition, to prepare leakage-free CdS film with a thickness less than 100nm requires precise fabrication conditions. In this paper, a four terminal tandem cell approach of using a higher bandgap top cell is introduced. In such configuration, the overall power conversion efficiency can be improved without reducing the thickness of CdS layer in the bottom CdTe solar cell.

Perovskite materials are ideal light harvesting candidates for solar spectrum between 400 and 550nm [5]. The bandgap of inorganic-organic hybrid perovskite, such as methylammonium-lead(II)-halogen with the chemical formula CH$_3$NH$_3$PbX$_3$ (MAPbX$_3$), can be continuously tuned up from 1.57eV to 2.30 eV by substituting Br for I to make MAPb(I$_{1-}$ $_x$Br$_x$)$_3$ (0≤*x*≤1) , which makes perovskite solar cells especially attractive for tandem applications [6]. It has been reported that mechanically-stacked tandems using hybrid perovskites as the top cell absorber onto copper indium gallium diselenide (CIGS) and low-quality multicrystalline silicon (Si) has improved the utilization of sunlight and the efficiency of tandem cell. However, reports on the Perovskite/CdTe tandem cell are limited. Here, we perform device modelling on the MAPb(I$_{1-x}$Br$_x$)$_3$, CdTe, and MAPb(I$_{1-x}$Br$_x$)$_3$/CdTe four-terminal tandem solar cells using wxAMPS, respectively. In our device model, we simulate perovskite with different Br content to match the absorption spectrum of CdS/CdTe bottom cell, aiming to alleviate the strict requirements of CdS layer.

II. DEVICE SIMULATION PARAMETERS

A. Theoretical Model

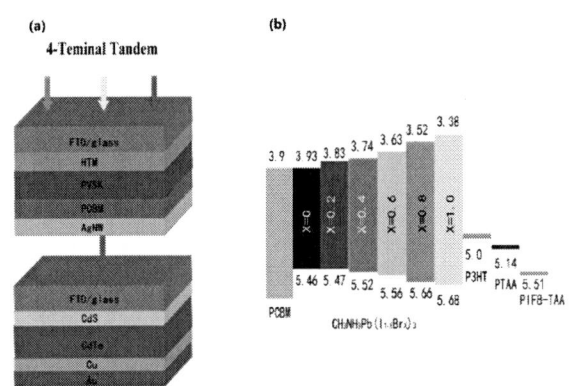

Fig. 1. (a)Proposed structure of tandem cells; (b)Energy level diagram of MAPb(I$_{1-x}$Br$_x$)$_3$ perovskite solar cell.

The proposed MAPb(I$_{1-x}$Br$_x$)$_3$/CdTe four-terminal tandem cells are schematically shown in Fig.1(a). When the light is coming through the MAPb(I$_{1-x}$Br$_x$)$_3$ top cell, almost all of short wavelength is absorbed by the top cell, the filtered transmitted light is then passed to the CdTe bottom cell. In this way the tandem cells achieve complementary absorption spectrum and improve the utilization of light.

978-1-5090-5606-4/17 $31.00 © 2017 IEEE

TABLE 1
SIMULATION PARAMETERS OF MAPb($I_{1-x}Br_x$)$_3$/ CdTe TAMEND CELLS [8-10]

	PCBM	MAPb($I_{1-x}Br_x$)$_3$	PIF8-TAA	PTAA	P3HT	SnO$_2$	CdS	CdTe
ε_r	3.9	30	3	3	3	9	10	9.4
E_g (eV)	2.0	1.57~2.30	2.88	3.3	2.0	3.6	2.4	1.5
χ (eV)	3.9	3.38~3.93	2.63	1.8	3.0	4	4	3.9
Thickness (nm)	50	300	50	50	50	500	25	4000
μ_n or μ_p (cm^{2}V^{-1}S^{-1})	0.2	14	4	0.4	0.1	100	100	320
N_A (cm^{-3})	0	6×10^{14}	2×10^{16}	2×10^{16}	2×10^{16}	0	0	2×10^{14}
N_D (cm^{-3})	2.93×10^{17}	0	0	0	0	1×10^{17}	1.1×10^{18}	0
N_C (cm^{-3})	2.5×10^{21}	2.5×10^{20}	1×10^{22}	1×10^{22}	1×10^{22}	2.2×10^{18}	2.2×10^{18}	8×10^{17}
N_V (cm^{-3})	2.5×10^{21}	2.5×10^{20}	1×10^{22}	1×10^{22}	1×10^{22}	1.8×10^{19}	1.8×10^{19}	1.8×10^{19}

For the top perovskite cell, silver nanowires (AgNW), which have good conductivity and high optical transmission, are selected as the electrode in order to improve transmittance of the top cell [7]. Based on the previous reports, PIF8-TAA, instead of PTAA and P3HT was chosen as hole conductors in later device modellings.

For the bottom CdTe solar cell, we referred to the simulation report by Gloeckler M [8], and simulated the output characteristics and EQE spectra of CdS/CdTe cells with different thickness of CdS layers.

B. Parameter Simulation

In the simulation, temperature is set at 300K, and AM 1.5 solar radiation spectrum is selected as the illuminating source. The material parameters are summarized in Table 1, most of which are selected from the reported papers [8-10]. For the organic hole transporting materials, we assume that the hole mobility is increased by two orders of magnitude and the doping concentration is set to be 2×10^{16} cm^{-3} by doping Lithium bis(trifluoromethylsulfonyl)-imide (Li-TFSI) and tert-Butyl pyridine (TBP).

For the perovskite materials MAPb($I_{1-x}Br_x$)$_3$, their bandgap, lowest unoccupied molecular orbital (LUMO) and highest occupied molecular orbital (HOMO) of the film are adapted with different Br composition (x), as shown in Fig.1(b) [11].Meanwhile, their absorption coefficients also change with x, according to a recent report [12]. Models of perovskite solar cell device were constructed with thin film stacks of PIF8-TAA/ MAPb($I_{1-x}Br_x$)$_3$/PCBM, where p-type PIF8-TAA and n-type PCBM thin film was used as HTM and ETM, respectively. The light reflection of the top and the bottom contacts were set to be 0 and 1. We simulated the output characteristics and corresponding external quantum efficiency (EQE) spectra of MAPb($I_{1-x}Br_x$)$_3$ cells with x progressionally increasing from 0 to 1 by an interval of 0.2.

III. RESULTS AND DISCUSSION

A. Simulation of MAPb($I_{1-x}Br_x$)$_3$ and CdTe solar cells

According to our simulation results，for the perovskite solar cell, the open-circuit voltage (V_{oc}) increases with x increasing; whereas the short-circuit current (J_{sc}) decrease significantly with x, as shown in Fig 2 (a). The reduction of J_{sc} with increasing x is directly related to the blue-shift of absorption onset and the reduced absorption coefficient of the perovskite materials. The improvement of Voc is attributed to the widening of the bandgap with increasing x in MAPb($I_{1-x}Br_x$)$_3$. The EQE spectra of the devices are shown in Fig 2(b), an average EQE approaching 80% in the range of 450-600 nm was observed, and the integrated photocurrent densities from EQE spectra were also well matched with the Jsc.

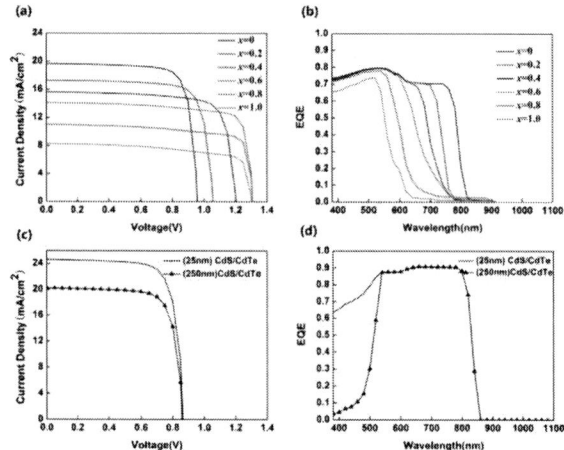

Fig. 2. I-V curves (a) and EQE spectra (b) of MAPb($I_{1-x}Br_x$)$_3$ solar cell with different x values, I-V curves (c) and EQE spectra (d) of CdTe solar cell with 25nm and 250nm CdS.

For the CdTe solar cell, as the thickness of CdS is increased tenfold from 25nm to 250nm, the device efficiency decreases from 16.3% to 13.2%, shown in Fig. 2(c). EQE spectra show that light absorption declined substantially from 400 to 550nm, which matches well with the I-V curve.

B. Simulation of MAPb(I_{1-x}Br_x)_3/ CdTe Tandem Solar Cells

The light absorption of MAPb(I_{1-x}Br_x)_3 top cell varies with x so that the filtered light that reaches the CdS/CdTe bottom cell and the final efficiency are affected by the x value. The spectral intensities irradiated on the surface of the CdTe cell are related to the transmittance of each layer of the top cell. Considering the thickness of PIF8-TAA and PCBM thin film were far less than perovskite layer, we assumed the transmittance of both HTM and ETM are 1. The transmittance of top cell is equal to the product of the transmission coefficient of the perovskite layer and the transmission coefficient of the AgNW electrode. The transmittance of the AgNW transparent electrode was set at 0.9 [7].

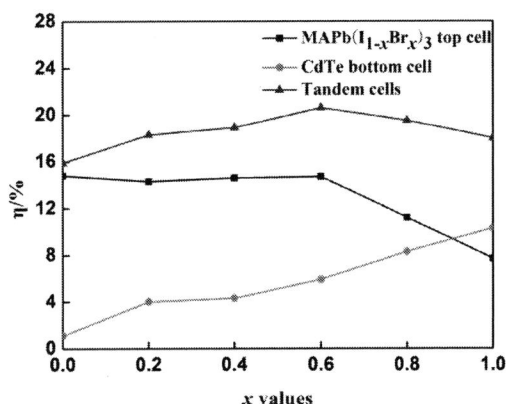

Fig. 3. The efficiencies of MAPb(I_{1-x}Br_x)_3 top cells, CdTe bottom cells and MAPb(I_{1-x}Br_x)_3/ CdTe tandem cells with different x values.

Due to the light absorption of MAPb(I_{1-x}Br_x)_3 cell, the Jsc and efficiency of CdTe cell increases with x increasing. The efficiency of tandem cells are determined by the sum of the efficiency of the bottom cell and top cell. As shown in Fig 3, efficiency of MAPb(I_{1-x}Br_x)_3/ CdTe cells changes with x and tandem cells efficengcy reaches a maximum of 20.6% when x equals 0.6. We picked x=0.6 to simulate the effect of thick CdS layer on tandem cell in the following section.

C. MAPb(I_{0.4}Br_{0.6})_3/ CdTe and CdS Thickness

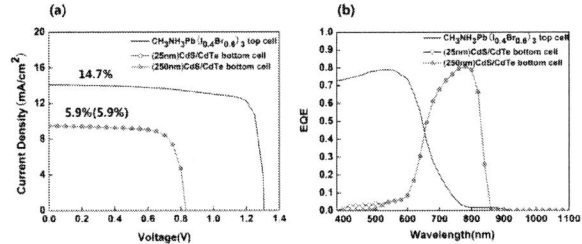

Fig. 4. I-V (a) and EQE (b) of CH_3NH_3Pb(I_{0.4}Br_{0.6})_3/CdTe solar cells with 25nm and 250nm CdS.

We further simulated MAPb(I_{0.4}Br_{0.6})_3/ CdTe tandem cell with different CdS layer thickness-----25nm and 250nm, and compared I-V and EQE spectra. As shown in Fig.4, when we changed the CdS layer thickness from 25nm to 250nm, the I-V and EQE curves of the CdTe cell with 250nm CdS are substantially almost overlapped with those from the counterpart of 25 nm CdS in MAPb(I_{0.4}Br_{0.6})_3/ CdTe tandem cells. The efficiency of tandem cells and CdTe bottom cell remain unchanged with thicker CdS layer, which can be ascribed to the aborption of the short wavelength of light (400-550 nm) by the MAPb(I_{0.4}Br_{0.6})_3 top cell. Therefore, the strict requirement on the thickness of the CdS layer is alleviated.

IV. CONCLUSION

We performed MAPb(I_{1-x}Br_x)_3, CdTe, MAPb(I_{1-x}Br_x)_3/ CdTe four-terminal tandem solar cells using wxAMPS simulation tool. The tandem cell efficiency reaches a maximum of 20.6% when x equals 0.6. The theoretical design of perovskite/ CdTe tandem cell is helpful to alter the strict requirements of CdS window layer. As the thickness of CdS is increased tenfold from 25nm to 250nm, the efficiency of MAPb(I_{0.4}Br_{0.6})_3/ CdTe tandem cell and CdTe bottom cell remain unchanged. The simulation could be useful to advancing experimental progress of the perovskite/ CdTe tandem cell.

ACKNOWLEDGEMENT

Hang Zhou would like to acknowledge the Shenzhen Science and Technology Innovation Fund under Grant No. JCYJ20160229122349365，and JCYJ20150629144006876.

REFERENCES

[1] J. J. Loferski, "Theoretical Considerations Governing the Choice of the Optimum Semiconductor for Photovoltaic Solar Energy Conversion," *Journal of Applied Physics,* vol. 27, pp. 777-784, 1956.

[2] P. Rappaport, "The photovoltaic effect and its utilization," *Solar Energy,* vol. 3, pp. 8-18, 1959.

[3] Research Cell Efficiency Records. NREL (http://www.nrel.gov/pv/assets/images/efficiency chart.jpg).

[4] A. Waleed, Q. Zhang, M. M. Tavakoli, S.-F. Leung, L. Gu, J. He, X. Mo, and Z. Fan, "Performance improvement of solution-processed CdS/CdTe solar cells with a thin compact TiO2 buffer layer," *Science Bulletin,* vol. 61, pp. 86-91, 2016.

[5] C. D. Bailie, M. G. Christoforo, J. P. Mailoa, A. R. Bowring, E. L. Unger, W. H. Nguyen, J. Burschka, N. Pellet, J. Z. Lee, and M. Grätzel, "Semi-transparent perovskite solar cells for tandems with silicon and CIGS," *Energy & Environmental Science,* vol. 8, pp. 956-963, 2015.

[6] J. H. Noh, S. H. Im, J. H. Heo, T. N. Mandal, and S. I. Seok, "Chemical management for colorful, efficient, and stable inorganic–organic hybrid nanostructured solar cells," *Nano letters,* vol. 13, pp. 1764-1769, 2013.

[7] G. Y. Margulis, M. G. Christoforo, D. Lam, Z. M. Beiley, A. R. Bowring, C. D. Bailie, A. Salleo, and M. D. McGehee, "Spray Deposition of Silver Nanowire Electrodes for Semitransparent Solid - State Dye - Sensitized Solar Cells," *Advanced Energy Materials,* vol. 3, pp. 1657-1663, 2013.

[8] M. Gloeckler, A. Fahrenbruch, and J. Sites, "Numerical modeling of CIGS and CdTe solar cells: setting the baseline," in *Photovoltaic Energy Conversion, 2003. Proceedings of 3rd World Conference on,* 2003, pp. 491-494.

[9] Y. Wang, Z. Xia, J. Liang, X. Wang, Y. Liu, C. Liu, S. Zhang, and H. Zhou, "Towards printed perovskite solar cells with cuprous oxide hole transporting layers: a theoretical design," *Semiconductor Science and Technology,* vol. 30, p. 054004, 2015.

[10] J. H. Heo, D. H. Song, and S. H. Im, "Planar CH3NH3PbBr3 Hybrid Solar Cells with 10.4% Power Conversion Efficiency, Fabricated by Controlled Crystallization in the Spin - Coating Process," *Advanced Materials,* vol. 26, pp. 8179-8183, 2014.

[11] B.-w. Park, B. Philippe, S. M. Jain, X. Zhang, T. Edvinsson, H. Rensmo, B. Zietz, and G. Boschloo, "Chemical engineering of methylammonium lead iodide/bromide perovskites: tuning of opto-electronic properties and photovoltaic performance," *Journal of Materials Chemistry A,* vol. 3, pp. 21760-21771, 2015.

[12] A. Sadhanala, F. Deschler, T. H. Thomas, S. n. E. Dutton, K. C. Goedel, F. C. Hanusch, M. L. Lai, U. Steiner, T. Bein, and P. Docampo, "Preparation of single-phase films of CH3NH3Pb (I1–x Br x) 3 with sharp optical band edges," *The journal of physical chemistry letters,* vol. 5, pp. 2501-2505, 2014.

Wafer-Bonded AlGaAs//Si Dual-Junction Solar Cells

Elias Veinberg-Vidal[1,2], Laura Vauche[1,2], Clément Weick[1,3], Jérémy Da Fonseca[1,2], Christophe Jany[1,2], Christophe Morales[1,2], Christophe Lecouvey[1,2], Thibaut Desrues[1,3], Philippe Voarino[1,3], Frank Fournel[1,2], Anne Kaminski-Cachopo[1], Alejandro Datas[4], Pablo Garcia-Linares[4], Mathieu Baudrit[1,3], Pierre Mur[1,2] and Cécilia Dupré[1,2]

[1] Université Grenoble Alpes, Grenoble, 38000, France
[2] CEA, LETI, MINATEC Campus, Grenoble, 38054, France
[3] CEA, LITEN, INES, Le Bourget du Lac, 73375, France
[4] IES-UPM, 28040 Madrid, Spain

Abstract — Monolithic two-terminal III-V on Si dual-junction (2J) solar cells were fabricated by means of Surface-Activated direct wafer Bonding (SAB). $Al_{0.2}Ga_{0.8}As$ single-junction cells are grown on GaAs substrate by Metal-Organic Vapor Phase Epitaxy (MOVPE) and bonded at room temperature to independently fabricated Si solar cells. The n^+-GaAs//n^+-Si bonding interface is characterized by Transmission Electron Microscopy (TEM) revealing a 2-3 nm thick amorphous interlayer. The performance of the 1 cm² tandem cells, designed for low concentration applications, was studied by External Quantum Efficiency (EQE) and J-V measurements showing a power conversion efficiency of 17% under one-sun AM1.5G spectrum. To our knowledge, this is the highest efficiency ever reported for a wafer-bonded 2J III-V on Si solar cell. Limitations to performance have been identified and therefore higher efficiencies are expected.

Index Terms — photovoltaic solar cells, multijunction, III-V on silicon, surface-activated direct wafer bonding.

I. INTRODUCTION

Today, Si based solar cells is the dominant photovoltaic technology with more than 90% of the market share. However, Si cell improvement gradually slows down as the power conversion efficiency record, 26.3% at this time [1], approaches the theoretical efficiency limit for Si of 29.4% [2]. Stacking highly efficient III–V junctions on top of Si solar cells is a promising way to overcome the Si single-junction efficiency limit. In addition, unlike III-V multijunction solar cells (MJSC) grown on expensive GaAs or Ge substrates, III-V cells mechanically stacked or wafer bonded to much cheaper and widely available Si wafers, combined with III-V substrate reuse strategies, could result in a significant cost reduction, thus potentially allowing for a competitive Levelized Cost Of Electricity (LCOE) in one-sun or low-concentration terrestrial solar applications [3].

In the '80s and '90s, much effort was invested in the epitaxial growth of (Al)GaAs on Si. However, mismatches in lattice parameter and in coefficient of thermal expansion, polar/nonpolar materials and AlGaAs minority carrier lifetime sensitivity to oxygen contamination have limited the efficiency to 21.4%, as reported by Soga et al. back in 1997 for a tandem AlGaAs/Si solar cell [4].

An attractive approach to overcome heteroepitaxy issues consists of joining independently processed Si and III–V solar cells by direct wafer bonding or mechanical stack techniques, as recently demonstrated by Cariou et al. with a triple junction GaInP/GaAs//Si solar cell using SAB technique, showing 30.2% one-sun power conversion efficiency [5].

In the case of a III-V on Si 2J solar cell, detailed balance modeling shows a Shockley-Queisser limit conversion efficiency beyond 40% for a top-cell band gap of about 1.7 eV under the AM1.5G spectrum [6]. Therefore, a 2J III-V on Si solar cell has the potential to reach very high efficiencies and it could be more cost competitive compared to the 3J-MJSC due to the high cost of III-V additional epitaxial layers needed [7]. To our knowledge, the best efficiency reported so far for a wafer-bonded 2J III-V on Si solar cell under AM1.5G standard spectrum is 11.1% achieved by Shigekawa et al. with a GaInP//Si cell structure [8].

In this work, the fabrication process of monolithic two-terminal $Al_{0.2}Ga_{0.8}As$//Si tandem solar cells using SAB is presented along with EQE and current density-voltage (*J-V*) characteristics under one-sun AM1.5G spectrum.

II. EXPERIMENTAL

A. Sample Preparation

Wafer-bonded $Al_{0.2}Ga_{0.8}As$//Si 2J solar cells were fabricated according to the process described hereunder and the resulting cell structure is presented in Fig. 1.

1) $Al_{0.2}Ga_{0.8}As$ top cells (1.7 eV) are grown lattice matched with inverted layer configuration on 4-inch GaAs substrates by MOVPE, including a highly doped GaAs tunnel junction and a 500 nm thick bonding layer.

2) The Si bottom cells are manufactured independently by thermal diffusion processes on p-type Czochralski (CZ) c-Si wafers (14-22 Ω·cm, 525 µm), resulting in the formation of P-doped n^+ emitter and a B-doped p^+ back surface field (BSF).

3) Si and GaAs wafer surfaces are cleaned using wet chemical treatments and planarized by Chemical-Mechanical Polishing (CMP), resulting in a bonding layer thickness of 170 nm.

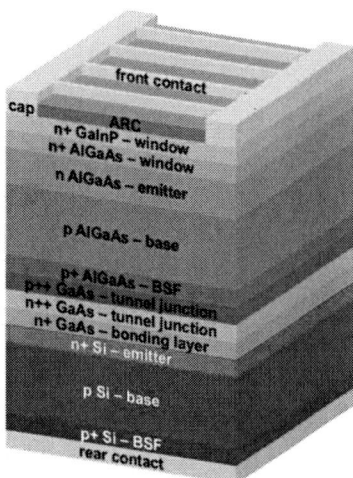

Fig. 1.   AlGaAs//Si 2J solar cell structure scheme (not to scale).

Afterwards, the wafers are loaded into an EVG580 ComBond vacuum chamber and then exposed to an Ar ion beam of few hundreds of eV that removes the native oxide layers via sputtering and leaves very reactive surfaces with dangling bonds. The wafers are then brought into contact and covalent bonds are formed at room temperature without the need of annealing.

4)   The GaAs substrate and GaInP etch-stop layer (not represented in Fig. 1) are etched away respectively by $H_2O_2+NH_4OH$ and HCl/H_3PO_4 solutions, leaving the few-micrometers-thick III-V active layers on the Si bottom cell.

5)   Finally, front metal contacts designed for low concentration (~20 suns, ~5% shading factor without bus bars [9]) and full sheet back metal contacts are deposited by evaporation. Mesa etching (trenches down to 2 μm inside Si) is performed to partially isolate individual cells of 1 or 2 cm² and a 65 nm Si_3N_4 Anti-Reflection Coating (ARC) is deposited by PECVD.

Single-junction $Al_{0.2}Ga_{0.8}As$ and Si stand-alone solar cells were also manufactured, performing step 1 (in upright configuration) and 5 for the AlGaAs solar cell and 2 and 5 for the Si solar cell.

B. Characterization

Spectral response and reflectivity measurements were performed using a conventional EQE system as previously reported in [10]. J-V measurements were carried out using a Spectra-Nova one-sun AM1.5G class AAA solar simulator. A shadow mask, designated area of 0.68 cm², was used to avoid potential contributions from the Si wafer outside the cell area.

III. Results and Discussion

The implementation of the SAB technique using Ar fast atom beam sputtering results in the formation of a ~3 nm thick

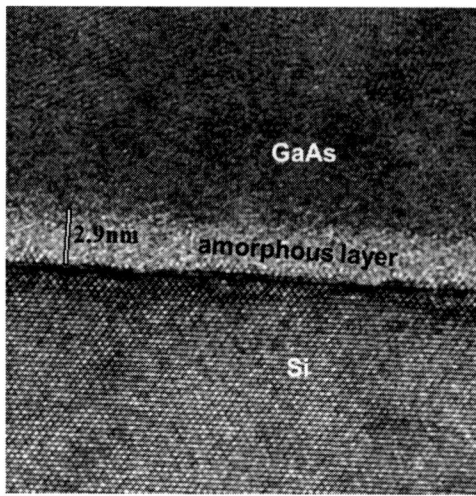

Fig. 2.   TEM image taken under Si orientation showing the ~3 nm thick amorphous layer between the GaAs-Si bonding interfaces.

amorphous interlayer, as revealed by TEM analysis shown in Fig. 2. No voids or crystal dislocations could be found.

In Fig. 3 and Fig. 4, EQE and J-V measurements of the tandem AlGaAs//Si subcells are presented along with the AlGaAs and Si single-junction stand-alone reference cells. J_{SC} can be calculated from EQE by convolution with the AM1.5G spectrum obtaining 13.5 and 12.7 mA/cm² for top and bottom subcells respectively. This suggests that the Si bottom junction limits the current of the series connected tandem device, although the subcells are not far from being current matched. However, both top and bottom junctions could produce much higher currents as detailed hereunder:

a)   Anti-Reflection Coating. The reflectivity curve at normal incidence (black dotted line in Fig. 3) shows that the single-layer Si_3N_4 ARC is adapted for AlGaAs top subcell (~550 nm) but is not very performant for the Si bottom subcell spectral absorption region (reflectivity from 10 to 20% between 730 to 1200 nm). A more adapted bilayer ARC with optimum materials and layer thicknesses will be used to significantly increase the EQE, especially in the bottom subcell region.

b)   Material quality. $Al_{0.2}Ga_{0.8}As$ stand-alone cell suffers from a high band gap-voltage offset $W_{OC} = E_g/q - V_{OC} = 0.51$ V, calculated using the V_{OC} shown in the inset table of Fig. 4 and the bandgap energy of 1.7 eV, suggesting low crystal quality probably due to oxygen contamination during epitaxy caused by the strong bond between oxygen and aluminum [11]. Si stand-alone high $W_{OC} = 0.54$ V could be explained by the low crystal quality of the CZ substrate that is not adapted to the 525 μm thickness. PC1D simulations (not shown here) indicate that the use of float-zone (FZ) Si substrates, with higher minority carrier lifetime, coupled with an improved emitter profile and rear surface passivation will allow to improve the Si bottom subcell V_{OC} and J_{SC}.

978-1-5090-5606-4/17 $31.00 © 2017 IEEE   2563

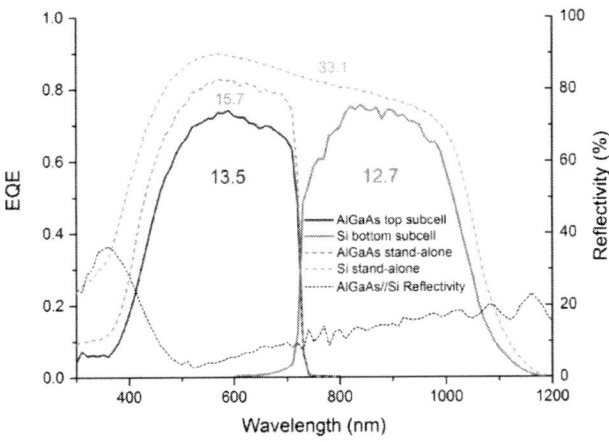

Fig. 3. EQE and reflectivity of tandem AlGaAs//Si subcells along with the AlGaAs and Si single-junction stand-alone cells used as reference. J_{SC} (mA/cm²) calculated by convolution with the AM1.5G spectrum is also shown.

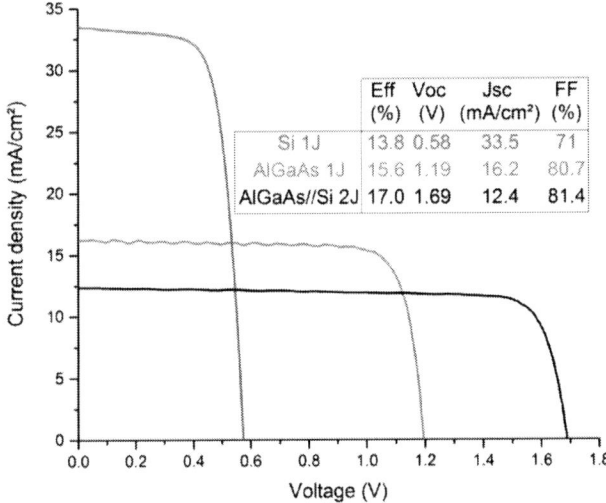

	Eff (%)	Voc (V)	Jsc (mA/cm²)	FF (%)
Si 1J	13.8	0.58	33.5	71
AlGaAs 1J	15.6	1.19	16.2	80.7
AlGaAs//Si 2J	17.0	1.69	12.4	81.4

Fig. 4. J-V curve at one sun AM1.5G spectrum of tandem AlGaAs//Si subcells along with the AlGaAs and Si single-junction stand-alone cells used as reference.

c)   Front contact processing and inverted growth issues.
$Al_{0.2}Ga_{0.8}As$ top subcell shows degraded performance compared to the stand-alone single-junction cell (J_{SC} = 15.7 mA/cm², dashed blue line from Fig. 3). TEM and Energy-Dispersive X-ray spectroscopy (EDX) analysis shown in Fig. 5) suggest different hypotheses that could explain the lower EQE of the grown inverted top subcell respect to the upright 1J cell. In the inverted growth structure, a ~5 nm GaAs layer was found at the interface between the GaInP barrier and the AlGaAs window layers. This may be caused by dopant diffusion due to the different growth sequence in the inverted configuration [12]. In addition, TEM analysis also revealed the presence of gold particles in the front surface of the tandem cell, probably

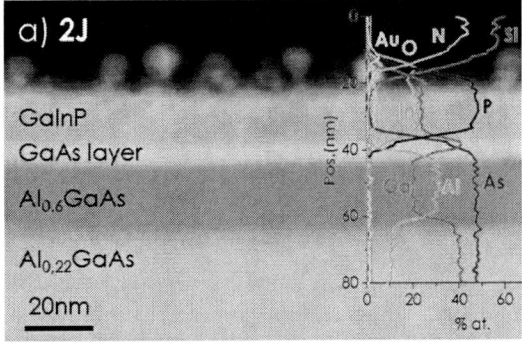

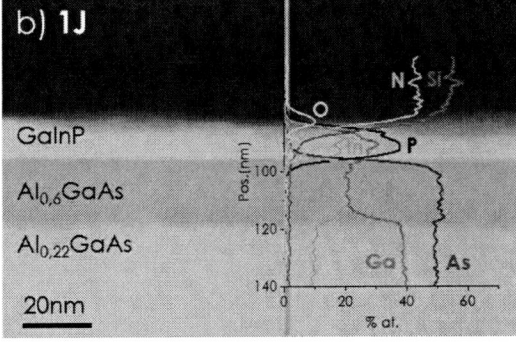

Fig. 5. TEM images and EDX profiles of the 2J and 1J upper cell layers.

due to some difficulties during the tandem cell front contact processing steps. In contrast, gold particles were not found in the stand-alone cell surface and the GaInP barrier layer was ~10 nm thinner, which could reduce losses due to absorption. Solving these epitaxy and front contact processing issues a J_{SC} greater than 15 mA/cm² should be attainable in the top subcell, as already demonstrated for the stand-alone cell.

d)   GaAs absorption. The bottom Si subcell shows low EQE from 730 to 800 nm due to optical absorption in the 170 nm of the GaAs bonding layer and in the 50 nm of the GaAs tunnel junction. These GaAs layers have indeed a 1.4 eV bandgap that is lower than that of the top junction and hence they block part of the light that should be absorbed by the Si bottom subcell. This will be partially mitigated in future work by thickness reduction of the bonding layer to around 50 nm.

As the subcells are connected in series, a small variation in one region of the spectrum may lead to an underestimation or overestimation of one of the subcells current and hence to the J_{SC} of the total device. For this reason, despite of the solar simulator being AAA class, it is not perfectly tuned for MJSC characterization. Therefore, the possible spectral imbalance produced by the solar simulator could be responsible for the increase in the current mismatch between the subcells and hence explain the slight higher fill factor (FF) of the tandem cell [13]. Nonetheless, very little difference could be observed between the J_{SC} calculated from EQE and the one extracted from the J-V curves. Measurements under concentrated light from 1 to 30 suns are currently being done using a flash solar simulator calibrated for MJSC accurate characterization [14]. Similar cells with the same front metal contact designed, optimized for ~15 suns, have shown a boost of around 3% in efficiency when measured under the same concentrations with respect to the one-sun performance.

IV. CONCLUSIONS AND FUTURE WORK

In conclusion, $Al_{0.2}Ga_{0.8}As$//Si tandem solar cells has been demonstrated using SAB showing improved efficiency compared to its AlGaAs and Si stand-alone counterparts. A power conversion efficiency of 17% under one-sun AM1.5G spectrum was measured, which is higher than the present record for a wafer-bonded 2J III-V on Si solar cell [8]. Moreover, this performance level can be further improved by solving the following issues: a) non-optimum ARC, b) Si and AlGaAs material quality, c) front contact processing and inverted growth difficulties and d) over-thick GaAs bonding layer. Ongoing work on these issues is expected to bring large performance improvements.

V. ACKNOWLEDGEMENTS

The authors wish to acknowledge Anne-Marie Papon and SERMA Technologies for the TEM and EDX analysis and Marc Plissonnier and Hervé Ribot for their support.

REFERENCES

[1] "News Release: World's Highest Conversion Efficiency of 26.33% Achieved in a Crystalline Silicon Solar Cell — A World First in a Practical Cell Size —," NEDO New Energy Technology Department, Kaneka Corporation, Sep. 2016.

[2] A. Richter, M. Hermle, and S. W. Glunz, "Reassessment of the Limiting Efficiency for Crystalline Silicon Solar Cells," *IEEE J. Photovolt.*, vol. 3, no. 4, pp. 1184–1191, Oct. 2013.

[3] J. S. Ward *et al.*, "Techno-economic analysis of three different substrate removal and reuse strategies for III-V solar cells: Techno-economic analysis for III-V solar cells," *Prog. Photovolt. Res. Appl.*, 2016.

[4] T. Soga, K. Baskar, T. Kato, T. Jimbo, and M. Umeno, "MOCVD growth of high efficiency current-matched AlGaAsSi tandem solar cell," *J. Cryst. Growth*, vol. 174, no. 1–4, pp. 579–584, Apr. 1997.

[5] R. Cariou *et al.*, "Monolithic Two-Terminal III-V//Si Triple-Junction Solar Cells With 30.2% Efficiency Under 1-Sun AM1.5g," *IEEE J. Photovolt.*, vol. 7, no. 1, pp. 367–373, Jan. 2017.

[6] J. P. Connolly, D. Mencaraglia, C. Renard, and D. Bouchier, "Designing III–V multijunction solar cells on silicon," *Prog. Photovolt. Res. Appl.*, vol. 22, no. 7, pp. 810–820, Jul. 2014.

[7] M. Woodhouse and A. Goodrich, "Manufacturing Cost Analysis Relevant to Single-and Dual-Junction Photovoltaic Cells Fabricated with III-Vs and III-Vs Grown on Czochralski Silicon (Presentation)," National Renewable Energy Laboratory (NREL), Golden, CO., 2014.

[8] N. Shigekawa, M. Morimoto, S. Nishida, and J. Liang, "Surface-activated-bonding-based InGaP-on-Si double-junction cells," *Jpn. J. Appl. Phys.*, vol. 53, no. 4S, p. 04ER05, Jan. 2014.

[9] E. Veinberg-Vidal *et al.*, "Manufacturing and Characterization of III-V on Silicon Multijunction Solar Cells," *Energy Procedia*, vol. 92, pp. 242–247, Aug. 2016.

[10] P. García-Linares, C. Domínguez, P. Voarino, P. Besson, and M. Baudrit, "Effect of the encapsulant temperature on the angular and spectral response of multi-junction solar cells," in *40th IEEE Photovoltaic Specialist Conference*, 2014, pp. 3298–3303.

[11] S. Heckelmann, D. Lackner, F. Dimroth, and A. W. Bett, "Material quality frontiers of MOVPE grown AlGaAs for minority carrier devices," *J. Cryst. Growth*, vol. 464, pp. 49–53, Apr. 2017.

[12] M. A. Steiner, J. F. Geisz, R. C. Reedy, and S. Kurtz, "A direct comparison of inverted and non-inverted growths of GaInP solar cells," in *33rd IEEE Photovoltaic Specialists Conference*, 2008, pp. 1–6.

[13] G. Siefer, C. Baur, M. Meusel, F. Dimroth, A. W. Bett, and W. Warta, "Influence of the simulator spectrum on the calibration of multi-junction solar cells under concentration," in *29th IEEE Photovoltaic Specialists Conference*, 2002, pp. 836–839.

[14] C. Domínguez, I. Antón, G. Sala, and S. Askins, "Current-matching estimation for multijunction cells within a CPV module by means of component cells," *Prog. Photovolt. Res. Appl.*, vol. 21, no. 7, pp. 1478–1488, Nov. 2013.

Enhancement of Si Photovoltaic Module by Introducing III-V/Si Hybrid Configurations and Cost Evaluations under Various Cost Ratios of III-V/Si Photovoltaics

Yu-Cian Wang[1], Kenji Araki[1], Kyotaro Nakamura[2], Kan-Hua Lee[1], Takefumi Kamioka[1], Nobuaki Kojima[1], Yoshio Ohshita[1], Masafumi Yamaguchi[1]

1 Toyota Technological Institute, Nagoya, 468-8511, Japan

2 Meiji University, Kawasaki, 214-8571 Japan

ABSTRACT — Combining Si PV module technology with an efficient III-V CPVs top PV into a tandem device is an attractive approach to enhance the efficiency of PV modules. Here, a method called III-V/Si hybrid configurations for enhancing the Si module efficiency is placing III-V static CPVs at the edge gap of each Si wafer rather than stacking in the top of Si. The prototype PV module shows the enhancement of efficiency. The cost evaluations of the III-V/Si hybrid configurations compared with Si module is discussed under various presumed cost ratio of III-V/Si PVs.

Index Terms — III-V/Si hybrid configurations, static concentrator photovoltaic (static CPV), geometric concentration ratio of III-V cell (GCR)

I. INTRODUCTION

Wafer-based crystalline Si photovoltaics (PVs) have developed for several decades and occupy a PV market share of over 90%. PV systems based on Si PVs have been shown to offer high reliability and limited efficiency degradation over a period (>25 years) and continued reduction in manufacturing costs at the cells and modules. A lot of efforts have been contributed to non-concentrating, flat-plate Si PV technologies over decades. However, single-junction technologies are reaching to their practical limits; crystalline silicon which is with the practical limit of ~29% [1] has been accomplished 26.3 % efficiency [2]. The present highest single junction PVs is held by GaAs with 28.8% [3]. Single junction technology cannot realistically conquer the Shockley–Queisser limit of about 33% [4] with lots of delicate techniques. The use of multi-junction PVs and concentrator PVs (CPVs) has been introduced to improve the PV efficiencies. Multi-junction PVs provide one effective solution to gain current efficiencies. However, III-V type semiconductors served as top absorber in multi-junction PVs confront the challenge of the high fabrication cost, mostly due to the limited throughput of the used deposition reactors and expensive substrates to grow cells on [5]. Fortunately, multi-junction PVs with CPV system offer a comparable cost because of relatively high efficiency and reduce the consumption of expensive III-V semiconductor materials such as gallium which is rarer than gold. Hence, the savings in material enable the technology to achieve great

decrease in unit price of electricity produced from the cell and lower the module price. Although CPV techniques can provide a method for reducing the cost III-V PVs, the practical cost of III-V PVs still hardly compete with the Si PVs.

In the study, a method for enhancing the Si module efficiency is placing III-V static CPVs at the edge gap of each Si wafer rather than stacking in the top of Si. A static concentrator technology is used to enhance the contribution from highly efficient III-V cells. III-V static CPVs integrating at the vacancies of Si modules is proposed and the prototype module is also demonstrated. The cost of various geometric concentration ratio of III-V cell (GCR) is also discussed.

II. FABRICATION AND RESULTS OF PRELIMINARY PHOTOVOLTAIC MODULES

As the concept of III-V/Si hybrid systems is proposed, the availability for applying the structure to the present Si PV module should be evaluated. It is known that the III-V on Si two-junction PVs serves as a filter which the incident photos decrease for the bottom Si solar cells [6]. As shown in Fig. 1, the incident photos declined to less than 50% for a bottom Si PV compared to the totally covered top III-V PV (A point). Under the optimization of static concentrator design on the top of III-V PVs with a practicable means, the bottom Si solar cells reaches to almost 1 sun with shrinkage area of the III-V CPVs (B point). The schematic of the prototype III-V/Si hybrid configurations is depicted in Fig. 2. The III-V concentrating solar cells are piled at the blank area of the conventional Si solar cell modules. The dimensions of Si and III-V solar cells are 156x156 mm^2 and 19x19 mm^2, respectively. The spacing between each solar cell is 3 mm. The area ratio of the III-V solar cells compared to the Si is approximately 1.5%. These cells were connected as the 4-terminal circuit. The entire cells were covered by 3 mm glass and bump structures served as concentration lens were given coaxially on the center of the III-V cells.

978-1-5090-5606-4/17 $31.00 © 2017 IEEE

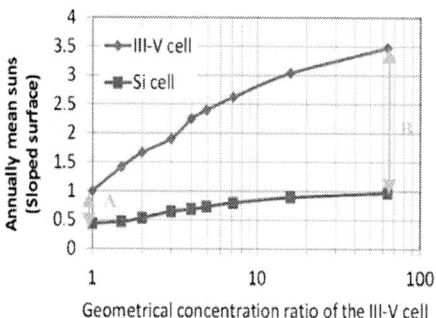

Fig. 1. The output of III-V/Si with various geometric concentration ratios

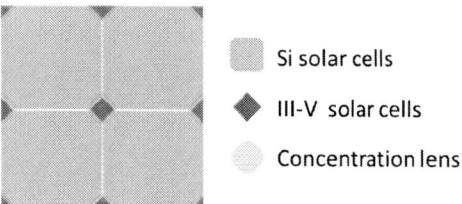

Fig. 2. The schematic picture of the III-V/Si hybrid configurations

The Si efficiency of 19% was implemented in the prototype III-V/Si hybrid module. The efficiency of the III-V/Si hybrid module increased to 20.1%. Compared to other efforts to increase the efficiency of Si PV modules, the III-V/Si hybrid configuration is a possible alternative to simplify the process. However, it is expected to reach more than 30 % of efficiency by stacking a III-V cell to the Si cell, especially for the high-efficiency PV panel on the roof of the car. According to a statistics by Ministry of Japan [7, 8], two thirds of family car runs less than 30 km for a day. The necessary electricity of average annual energy will be 642 kWh per year for cars with the lighting of the weight. It provides a promising solution to meet the power requirement for majority of the family cars run by the sun without supplying gasoline that the use of more than 30 % of high-efficiency PV in a limited installation space like a car roof. Based on the above-mentioned structure, the increment of the efficiency is not high enough for some applications but the enhancement of efficiency can be achieved by increasing the area ratio of the III-V solar cells compared to the Si in a limited space. From the results, there exists a common trade-off issue between the cost main caused by the extra III-V CPVs and the enhancement of the efficiency of the module with III-V/Si hybrid configurations. Many studies are attributed to decrease the production cost by increasing high throughputs and improving production yields. Hence, it is expected to reduce the III-V production cost to a competitive cost range or an acceptable price with higher module efficiency to meet some specific purposes.

III. THE COST EVALUATION UNDER VARIOUS COST RATIO OF III-V/SI

The cost evaluation includes some assumptions. In the cost calculations, there are two kinds of cost in a PV module; one is the cost with area dependence and the other is the cost with area independence. The cost fractions of area dependence and area independence are 70% and 30% respectively. The cost with area independence is assumed to equal to the cost of PVs themselves. The cost ratio of 3 is anticipated that III-V materials grown by hydride vapor phase epitaxy (HVPE), which can obtain higher growth rate, shorter process time and lower source cost compared to metal-organic chemical vapor deposition (MOCVD) [9]. Cost/Watt and the efficiency under various GCR are plotted in Fig. 3. Referring to the cost ratio of 3 and 6, the Cost/Watt can be reduced to less than the Si PV modules and this result shows the positive potential for the requirement of high efficiency III-V/Si (> 30%) with high coverage of III-V CPVs on Si PVs. Until the improvement of the III-V growth process, the cost ratio of III-V of 10 higher is hardly inevitable. The profitable cost (< Si cost) under the gain of PV module efficiency can be achieved with those with high GCR in view of the current state of III-V/Si cost ratio.

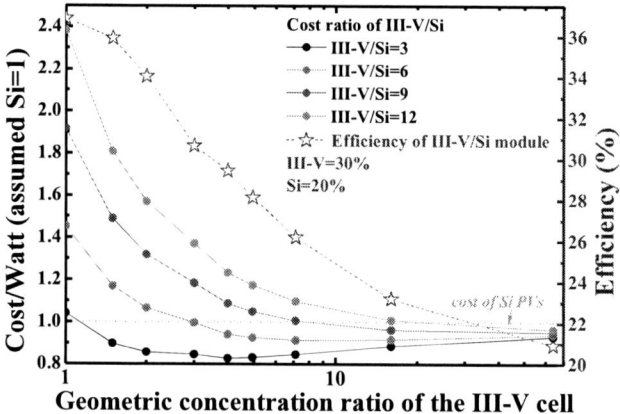

Fig. 3. The Cost/Watt and efficiency v.s Geometric concentration ratio of the III-V cell under different cost ratio of III-V/Si

In the relatively high GCR region, the configuration is to place small III-V static CPVs on four corners of the single crystalline Si PV. There are two sizes III-V CPVs can be integrated in the existing two sizes of Si wafer, 125x125 and 156x156 mm^2 with spaces for III-V CPV area with 19x19 and 26x26 mm^2. Table 1 shows the efficiency of III-V/Si PV modules with high GCR with two sizes of Si PVs. The inset figure is the one of designs of the III-V static CPVs which is considered as reducing transportation costs because of easy stacking without ruining the lens. In the calculation of PV performance, Si efficiency of 25% and III-V efficiency of 35% are taken into account for some occasions demanded of high efficiency modules. The module efficiency gain was considerably high in both cases. For the requirement of high PV efficiency (>30%), the cost/watt of III-V/Si hybrid configurations with relatively low GCR can compete with that of Si PV is expected to the cost ratio of III-V/Si less than 6.

Table 1 Efficiency and cost/watt of III-V CPVs/Si PV modules integrated with the conventional sizes of Si PV wafers.

	Efficiency w/ III-V/Si hybrid	Si module Efficiency	Cost/Watt	
			Cost ratio of III-V/Si	
			3	12
III-V(26x26) /Si(125x125)	25.4 %	23.4 %	0.98	1.08
III-V(19x19) /Si(156x156)	25.5 %	23.9 %	0.97	1.01

IV. CONCLUSIONS

A simple method to increase the Si module is achieved. The gain of efficiency for the prototype of PV module with III-V/Si hybrid configurations is 1.1 point (from 19% to 20.1%) and III-V/Si hybrid configurations hold the performance of Si PVs. From the view of cost competing with present Si PV, the cost/watt of PV modules with III-V/Si hybrid configurations can be reached with PV module with high GCR; namely, the configurations of small of III-V static CPVs without shading the Si PVs can provide both increasing the module efficiency and comparable costs. As soon as the cost ratio of III-V/Si is below 6, it is supposed to meet the high efficiency over 30% with market potential costs for the PV module with III-V/Si hybrid configurations at relatively low GCR.

ACKNOWLEDGMENTS

This work was supported by the New Energy and Industrial Technology Development Organization (NEDO) under the Ministry of Economy, Trade and Industry (METI), Japan.

REFERENCES

[1] A. Richter, et al., "Reassessment of the Limiting Efficiency for Crystalline Silicon Solar Cells" IEEE Journal of Photovoltaics vol 3, pp. 1184–1191, 2013.
[2] K. Yoshikawa, et al., "Silicon heterojunction solar cell with interdigitated back contacts for a photoconversion efficiency over 26%," Nature Energy, vol. 2, 17032 (8), 2017.
[3] B. M. Kayes, et al., "27.6% Conversion efficiency, a new record for single-junction solar cells under 1 sun illumination," in 37th IEEE Photovoltaic Specialists Conference, 2011, p. 4.
[4] W. Shockley, H. J. Queisser, "Detailed balance limit of efficiency of p-n junction solar cells," Journal of Applied Physics, vol. 32, pp. 510–519, 1961.
[5] M. Woodhouse, A. Goodrich, "A Manufacturing Cost Analysis Relevant to Single-and Dual-Junction Photovoltaic Cells Fabricated with III-Vs and III-Vs Grown on Czochralski Silicon," National Renewable Energy Laboratory, Golden, Colorado, 2013.
[6] K. Araki et al., "Beyond the limit of Si solar cells – III-V on Si cell and its PCSC module concept", in 26th Photovoltaic Science and Engineering Conference, 2016, Singapore
[7] EPA report, "Light-Duty Automotive Technology, Carbon Dioxide Emissions, and Fuel Economy Trends: 1975 through 2015", EPA-420-R-15-016, Dec. 2015
[8] Japanese Ministry of Land, Infrastructure, Transport and Tourism, "Road Traffic Census" (in Japanese, http://www.mlit.go.jp/road/ir/ir-data/ir-data.html, Sep. 2015.
[9] J. Simon, D. Young, A. Ptak, "Low-cost III–V solar cells grown by hydride vapor-phase epitaxy," in 42th IEEE Photovoltaic Specialists Conference, 2014, p.

Numerical Simulation of p-type front junction PERL Silicon cell for III-V/Si Tandem Devices

Chuqi Yi, Fa-Jun Ma, Anita Ho-Baillie and Stephen Bremner

School of Photovoltaic and Renewable Energy Engineering, University of New South Wales, Sydney 2052, Australia

Abstract — **Extensive modelling work is required to guide the optimization of prototype epitaxial grown $GaAs_{0.75}P_{0.25}$/Si dual junction solar cell for achieving > 30% efficiency. In this work, Sentaurus TCAD is used to model p-type PERL silicon cell under $GaAs_{0.75}P_{0.25}$ truncated one-sun AM1.5G spectrum. The PERL cell J_{sc} is found to be greatly dependent on GaP/Si Interface Recombination Velocity (IRV), while FF and V_{oc} remain largely unaffected. Moreover, further increase of minority carrier lifetime above 500μs in p-type silicon cell does not enhance its performance significantly, providing the base doping density is lower than 1E17cm⁻³. Loss analysis shows that heavily-doped deep junction emitters should be avoided in order to get optimum bottom cell J_{sc} when GaP/Si IRV is low.**

I. INTRODUCTION

Owing to reduced thermalization loss, multi-junction solar cell (MJSC), amongst many other photovoltaic (PV) techniques, is the only one that has surpassed the Shockley-Queisser efficiency limit so far [1]. State-of-the-art MJSCs involve the integration of comprehensive III-V semiconductor materials on conventional substrates such as Gallium Arsenide (GaAs) or Germanium (Ge), which are more expensive compared to silicon. This cost issue has been the main driving force for researching the integration of cheap, robust and scalable silicon material for tandem devices [2], leveraging the long history of silicon PV research and current existing manufacturing infrastructures.

Successful elimination of heterovalent nucleation-related defects by employing a Gallium Phosphide (GaP) nucleation layer on offcut silicon (001) [3] has re-invigorated III-V/Si integration research. Subsequent metamorphic step graded buffer (SGB) layers allow the fabrication of III-V materials with desirable lattice constants and band gap values.

All these technical advances have enabled new designs such as epitaxially grown monolithic $GaAs_{0.75}P_{0.25}$/Si dual junction solar cells [4], corresponding to the optimal band gap combination of 1.7/1.12 eV, with a maximum attainable 1 sun AM1.5G efficiency of 37% [5]. Despite the successful transition of epitaxy techniques from Molecular Beam Epitaxy (MBE) to the more industrially compatible Metal Organic Chemical Vapor Deposition (MOCVD) [6], much work remains to ensure $GaAs_{0.75}P_{0.25}$/Si solar cells have the performance and cost compatible with industry. To this end, computer aided simulation as a cost and time efficient approach, is necessary to guide III-V/Si cell design and

optimization for achieving more than 30% AM1.5G efficiency.

The spectrum seen by the bottom silicon cell in III-V/Si tandems is different from single junction silicon solar cells, since most of the high energy photons are absorbed by top III-V materials. In order to achieve better current matching and higher tandem efficiency, silicon bottom cell must be redesigned for better long-wavelength response. In this paper, Synopsys Sentaurus Technology Computer Aided Design (TCAD) software is employed to simulate $GaAs_{0.75}P_{25}$/Si dual junction solar cell, with a focus on optimizing silicon cell performance under this III-V truncated spectrum.

II. SIMULATION DETAILS

The presence of a small conduction band and large valence band offset between GaP and silicon dictates an n-GaP/n-Si interface in the $GaAs_{0.75}P_{0.25}$/Si architecture. Therefore, the most relevant configurations for the silicon bottom cell are a p-type base with a front junction or n-type base with a rear junction. Both configurations have been analyzed with the results presented here being for a conventional p-type front junction PERL (Passivated Emitter, Rear locally-diffused) silicon cell. The PERL was initially designed for better long-wavelength performance, and is deployed with a planar front surface as shown in Fig. 1. All input parameters for the PERL cell are adopted from [7], except the cell thickness which is 280μm.

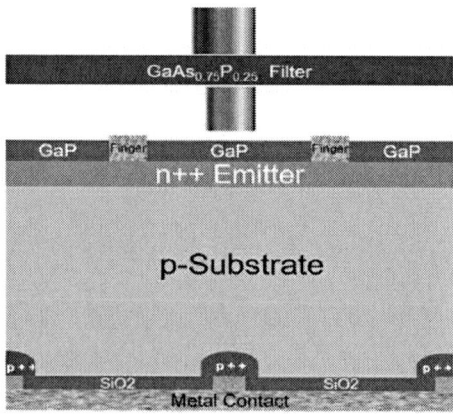

Fig. 1. Schematic of simulated cell structure in this work.

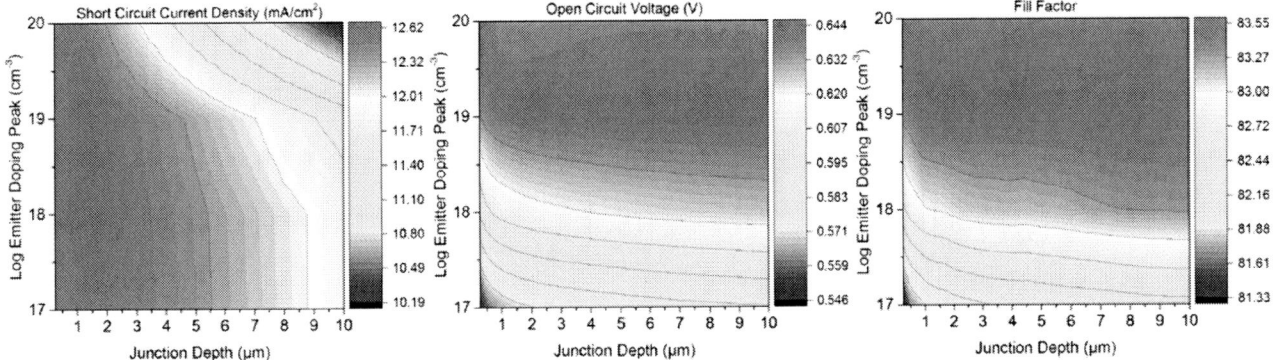

Fig. 2. Performance of PERL silicon bottom cell under $GaAs_{0.75}P_{0.25}$ truncated 1-sun AM1.5G spectrum, as a function of front emitter peak doping density and junction depth. GaP/Si IRV=1E4 cm/s and without ARC. Left: J_{sc}. Center: V_{oc}. Right: FF.

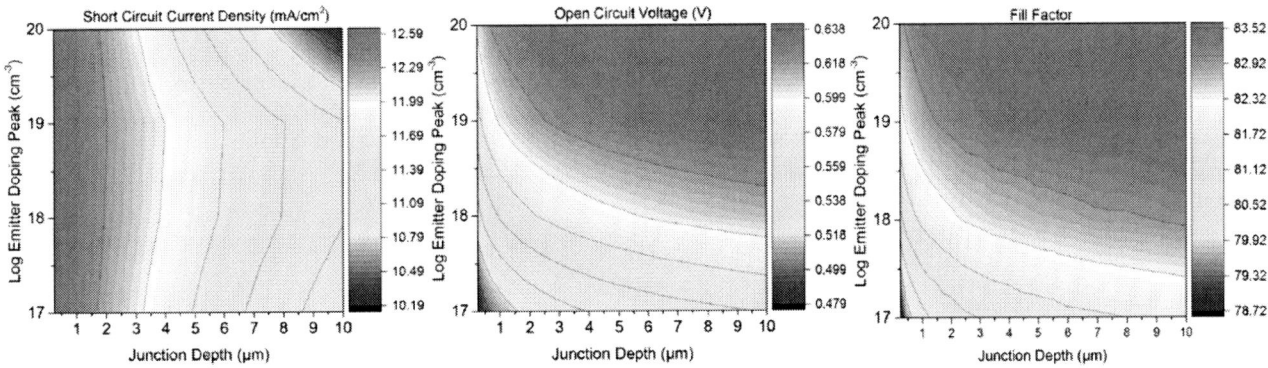

Fig. 3. Performance of PERL silicon bottom cell under $GaAs_{0.75}P_{0.25}$ truncated 1-sun AM1.5G spectrum, as a function of front emitter peak doping density and junction depth. GaP/Si IRV=1E6 cm/s and without ARC. Left: J_{sc}. Center: V_{oc}. Right: FF.

For simplicity, top $GaAs_{0.75}P_{0.25}$ cell and GaP nucleation layer are simulated as 2μm and 200nm filters, respectively, along with bottom silicon cell in 3 dimensional (3D) optical simulations, where the transfer matrix method is used to account for light interference across layers. The generated 3D optical profile is then converted to 1D for electrical simulations.

A Varshni model of temperature dependent band gaps [8] is used for silicon with temperature set at 300K, together with Schenk model to account for band gap narrowing effect [9]. Klaassen model [10] of temperature dependent lattice mobility is employed, as shown in (1), where room temperature mobility and the exponent values are adopted from [11].

$$\mu_{v}^{L} = \mu_{v,300}^{L} \cdot \left(\frac{T_L}{300K} \right)^{\gamma_{0,v}} \quad (1)$$

For the emitter doping profiles, the net ionized doping concentration N(z) as a function of depth is calculated using Gaussian approximation as defined in PC1D [12]:

$$N(z) = N_p \cdot \exp[-(z - z_p)^2 / z_f^2] \quad (2)$$

Where Np is the peak doping concentration, z_p the depth of the peak concentration and z_f the depth factor. The emitter peak doping concentration is varied between 1E17 to 1E20 cm-3, while for each concentration the depth factor is used to adjust the junction depth between 0.25 and 10μm.

III. EMITTER DOPING PROFILE SIMULATION FOR P TYPE BASE PERL DESIGN

As indicated in [13], GaP/Si IRV greatly affects silicon bottom cell performance, as do the emitter doping profiles. Fig. 2 and 3 show simulated output data of PERL cell under various emitter doping conditions for both low and high interface recombination velocity (IRV). The results show that J_{sc} is very sensitive to the GaP/Si IRV whereas V_{oc} and FF are not greatly affected, as when the IRV is high (above 1E6 cm/s), J_{sc} higher than 12.3mA/cm^2 can only be achieved with shallow junctions (within 3.5μm depth). In both cases, for a given emitter junction depth, the short circuit current density remains largely unaffected while changing peak doping density, especially when the junction depth is small. When emitter doping level increases, cell dark currents decreases and therefore higher V_{oc} is observed. Cell FF keeps relatively

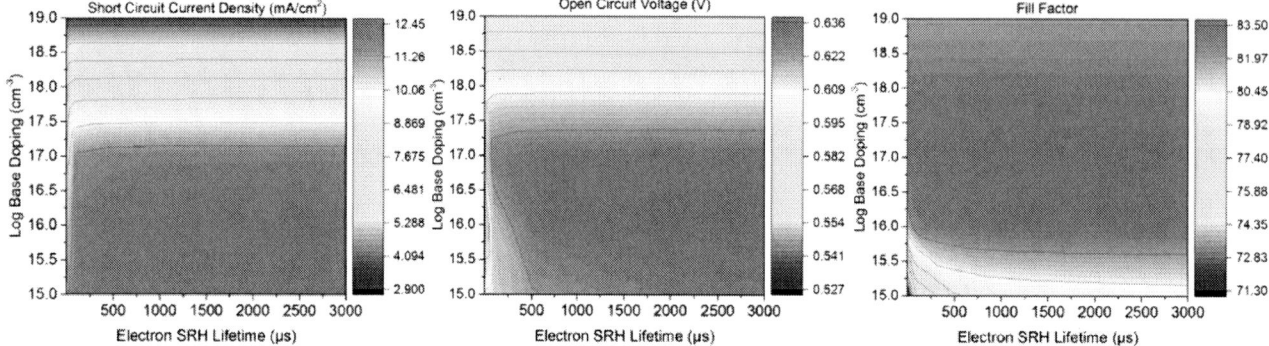

Fig. 4. Performance of PERL silicon bottom cell under GaAs0.75P0.25 truncated 1-sun AM1.5G spectrum, as a function of base peak doping density and electron SRH lifetime. GaP/Si IRV=1E4 cm/s and without ARC. Left: J_{sc}. Center: V_{oc}. Right: FF.

stable at above 80 for most doping profiles, except for extremely low peak concentration and shallow junction depths, where high sheet resistance leads to higher overall series resistance. It should be noted that, in tunnel junction interconnected GaAs0.75P0.25/Si dual junction cell, no current lateral flow is required in the emitter and therefore FF is not expected to be significantly affected by emitter doping profiles.

IV. EFFECTS OF BASE DOPING LEVEL AND MINORITY CARRIER LIFETIME

In practice, it is the first essential step to know the optimal parameter window of choosing silicon wafers so that it will not impose performance limitation on the final III-V/Si tandem cells. Two parameters, namely the base doing density and minority carrier lifetime of the silicon wafers, are most significant in affecting the final cell performance. Therefore, effects of those two parameters on the final cell performance are simulated and the results are shown in Fig. 4. For the simulated p-type front junction PERL silicon cells, bulk doping density is varied between 1E15 and 1E19 cm^{-3}, while electron Shockley-Read-Hall (SRH) lifetime is altered between 0.1 and 3000μs. All other parameters, including the emitter and rear localized doping profiles, are adopted from [7], except for the cell thickness which is at 280μm and GaP/Si IRV is at 1e4cm/s.

Results from Fig.4 show that the prerequisite of attaining high Jsc and Voc output for the p-type PERL silicon bottom cell in III-V/Si architecture is that a threshold background doping density of 1E17 cm^{-3} should not be exceeded. Above 1E17 cm^{-3} base doping density, both Jsc and Voc of the silicon bottom cell drop dramatically due to the significant rise in carrier recombination rate in the bulk. On the other hand, for a given background doping level, silicon cell performance remains largely unaffected while minority carrier lifetime is varied. Yet when minority carrier lifetime is

extremely low, cell performance degradation is still observed; particularly when electron SRH lifetime is lower than 500μs cell open circuit voltage is greatly affected. Therefore, silicon cell lifetime degradation during MBE or MOCVD heat up [14, 15] needs to be tackled before high efficiency III-V/Si tandem solar cells can be ensured.

V. CURRENT LOSS ANALYSIS

Light-generated current density losses in different parts of the PERL cell with a typical heavily doped emitter were calculated, with results summarized in Fig. 4. It is clear that under the truncated spectrum, for a PERL cell with this exact type of configuration, the cell rear surface is losing more current than emitter itself, followed by the base. Therefore, the most efficient step to further improve this cell performance is via optimization of its rear surface, including a better passivation scheme, or a reduced local diffusion area to bring down rear surface recombination velocity (RSRV).

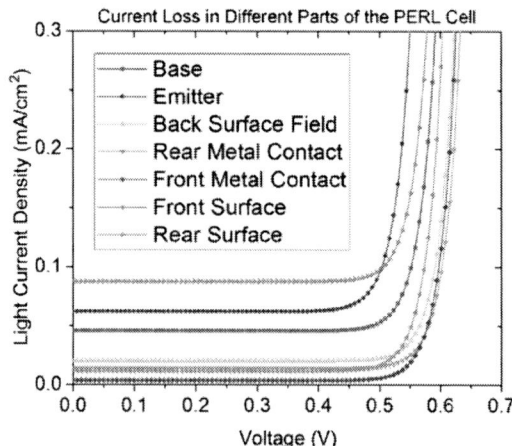

Fig. 5. Light-generated current density loss in various parts of the PERL cell with a typical 1E20 cm^{-3} Peak and 2 μm deep n++ emitter. GaP/Si IRV=1E4 cm/s.

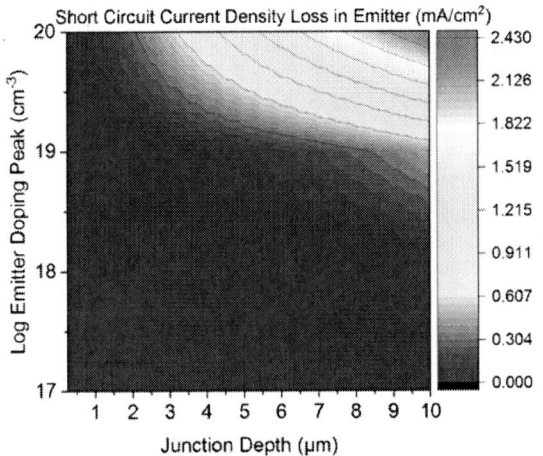

Fig. 6. J_{sc} loss in emitter with various emitter doping profiles. GaP/Si IRV=1E4 cm/s. All other parameters are as specified in [7].

When voltage equals zero, the current losses shown in Fig. 4 become the losses of short circuit current density for different parts of the cell. Fig. 5 illustrates J_{sc} losses in the emitter with various doping profiles. It shows that J_{sc} losses in emitter remain relatively low except for heavily doped and deep junction emitters. Therefore, to improve current output of silicon bottom cell and thus a better current matching, heavily doped deep junction emitters should be avoided, which is in line with the results shown in Fig. 2 and 3. Results suggest for a GaP/Si interface with an IRV = 1E4cm/s, the emitter profile should not deeper than 2μm but should have peak doping density at 1E20/cm3.

VI. SUMMARY

A P-type front junction PERL silicon solar cell under GaAs$_{0.75}$P$_{0.25}$ truncated AM1.5G spectrum has been simulated using Sentaurus TCAD. Front emitter designs have been explored by simulating the cell performance with different emitter doping profiles. It was found that J_{sc} is sensitive to the GaP/Si IRV while V_{oc} and FF remain largely unaffected. Effects of base doping and minority carrier lifetime on cell performance are also investigated. Results indicate silicon base doping concentration should not exceed 1E17cm^{-3}, whilst the effects of minority carrier lifetime is not as pronounced, providing it is greater than 500μs. Loss analysis indicates heavily doped deep junction emitters should be avoided as far as silicon bottom cell J_{sc} is concerned. Future work will include the simulation of n-type rear junction configuration and textured silicon rear surface for better current matching between silicon and III-V sub-cells.

REFERENCES

[1]   G. Conibeer, "Third-generation photovoltaics," *Materials today,* vol. 10, no. 11, pp. 42-50, 2007.

[2]   I. Almansouri, A. Ho-Baillie, S. P. Bremner, and M. A. Green, "Supercharging silicon solar cell performance by means of multijunction concept," *IEEE Journal of Photovoltaics,* vol. 5, no. 3, pp. 968-976, 2015.

[3]   T. Grassman *et al.,* "Control and elimination of nucleation-related defects in GaP/Si (001) heteroepitaxy," *Applied Physics Letters,* vol. 94, no. 23, p. 232106, 2009.

[4]   T. J. Grassman, D. J. Chmielewski, J. A. Carlin, and S. A. Ringel, "Development of epitaxial 2-and 3-junction III-V/Si solar cells," in *Photovoltaic Specialists Conference (PVSC), 2016 IEEE 43rd,* 2016, pp. 2036-2039: IEEE.

[5]   J. Geisz and D. Friedman, "III–N–V semiconductors for solar photovoltaic applications," *Semiconductor Science and Technology,* vol. 17, no. 8, p. 769, 2002.

[6]   T. Grassman, J. Carlin, B. Galiana, F. Yang, M. Mills, and S. Ringel, "MOCVD-grown GaP/Si subcells for integrated III–V/Si multijunction photovoltaics," *IEEE Journal of Photovoltaics,* vol. 4, no. 3, pp. 972-980, 2014.

[7]   A. Fell *et al.,* "Input Parameters for the Simulation of Silicon Solar Cells in 2014," *IEEE Journal of Photovoltaics,* vol. 5, no. 4, pp. 1250-1263, 2015.

[8]   Y. P. Varshni, "Temperature dependence of the energy gap in semiconductors," *Physica,* vol. 34, no. 1, pp. 149-154, // 1967.

[9]   A. Schenk, "Finite-temperature full random-phase approximation model of band gap narrowing for silicon device simulation," *Journal of Applied Physics,* vol. 84, no. 7, pp. 3684-3695, 1998.

[10]   D. Klaassen, "A unified mobility model for device simulation," in *Electron Devices Meeting, 1990. IEDM'90. Technical Digest., International,* 1990, pp. 357-360: IEEE.

[11]   S. M. Sze and K. K. Ng, *Physics of semiconductor devices.* John wiley & sons, 2006.

[12]   P. Basore and P. Clugston, "PC1D v. 5.0," *University of New South Wales,* 1997.

[13]   I. Almansouri *et al.,* "Designing bottom silicon solar cells for multijunction devices," *IEEE Journal of Photovoltaics,* vol. 5, no. 2, pp. 683-690, 2015.

[14]   E. García-Tabarés, I. García, J.-F. Lelièvre, and I. Rey-Stolle, "Impact of a Metal–Organic Vapor Phase Epitaxy Environment on Silicon Substrates for III–V-on-Si Multijunction Solar Cells," *Japanese Journal of Applied Physics,* vol. 51, no. 10S, p. 10ND05, 2012.

[15]   L. Ding, C. Zhang, T. U. Nærland, N. Faleev, C. Honsberg, and M. I. Bertoni, "Silicon Minority-Carrier Lifetime Degradation During Molecular Beam Heteroepitaxial III-V Material Growth," *Energy Procedia,* vol. 92, pp. 617-623, 2016.

Epitaxial GaP Layers Grown on Si Substrates using Migration Enhanced and Molecular Beam Epitaxy

Chaomin Zhang[1], Allison Boley[2], Nikolai Faleev[1], David J. Smith[2], Christiana B. Honsberg[1]

[1] Arizona State University, Ira Fulton School, Solar Power Lab, Tempe, AZ, U.S.A.

[2] Arizona State University, Physics Department, Tempe, AZ, U.S.A.

Abstract — This study compares the microstructure of epitaxial GaP/Si heterostructures grown using the migration-enhanced epitaxy (MEE) and molecular beam epitaxy (MBE) techniques. High-resolution X-ray diffraction and cross-section transmission electron microscopy for thin (~50 nm) GaP layers grown by MEE indicated greatly improved crystallinity and much reduced defect density for off-cut Si wafers compared with precisely oriented Si wafers. Observations of MBE-grown GaP/Si samples grown with thicknesses ranging from ~35 nm to 2 microns revealed very low defect densities for thin layers but substantial threading defects and high Stacking Fault density predominated in the thicker, almost fully relaxed structures.

I. INTRODUCTION

GaP has attracted attention as a possible nucleation layer for the growth of III–V solar cells on Si wafers due to the small lattice mismatch of ~0.37% between GaP and Si. However, the performance of GaP/Si solar cells are still limited by the quality of the epi-layers [1]. Anti-phase domains (APDs) are typically created during the heteroepitaxial growth of polar material (GaP) on non-polar material (Si), and crystal defects related to the APDs are liable to act as deep levels in the forbidden gap of the material. Furthermore, crystalline defects such as stacking faults and dislocations with threading segments created during the heteroepitaxial growth are likely to decrease the minority-carrier lifetime of III–V solar cells and thereby reduce their open-circuit voltage (V_{oc}). However, the degraded crystal quality of GaP grown on Si maybe improved by introducing migration-enhanced epitaxy (MEE) technique and post-growth annealing [2-5]. The off-cut ($\geq 4°$) Si wafers have been demonstrated to reduce anti-phase domain (APD) by self-annihilation [2-5]. However, the Si substrates with the exact (001) orientation are more common

Here, we present the results of high-resolution X-ray diffraction (XRD) and transmission electron microscopy (TEM) structural investigations of Si(001)-GaP epitaxial structures grown on nominal and 4 degree offcut oriented substrates using MEE and molecular beam epitaxy (MBE) techniques. Structural investigation has revealed correlations between the growth conditions and the crystal perfection of the epitaxial structures, related to the wafer orientation (MEE-grown structures) and the GaP layer thickness (MBE-grown structures). The P-polarity of GaP layer grown by MEE on

offcut wafers was directly confirmed by aberration-corrected TEM micrographs.

II. EXPERIMENTAL DETAILS

A) Epitaxial Growth of GaP-Si(001) structures

The GaP epitaxial layers were grown on Si(001) substrates using a solid-source Veeco GEN III MBE system with a phosphorus valved cracker. The Si wafers were *n*-type float-zone material, precisely oriented (001) or with ~4 degrees offcut towards the [011] direction. This wafer misorientation should help to form double steps on the substrate surface during epitaxial growth and possibly minimize APD creation during GaP epitaxial growth.

Prior to deposition, the Si wafers were chemically cleaned using the standard RCA solution, which was finally combined with 10% hydrogen fluoride (HF) for surface refresh. The GaP layers were grown on the Si substrates following preheating at 820°C for 5 min to fully remove the residual native silicon oxide layer. Surface reconstructions during the preheating and deposition were monitored *in situ* by reflection-high-energy electron diffraction (RHEED). Initially, under the presence of the native oxide, mild (1x1) RHEED patterns were observed, while clear and streaky (2x1) patterns, indicative of native-oxide removal and perfect surface reconstruction, were observed after the substrate preheating. After annealing, the substrate temperature was decreased to the growth temperature for further deposition.

Two MEE-grown GaP-Si(001) heterostructures of 50-nm GaP layer thickness were deposited at 440°C onto precisely oriented (MEE-I structure) and off-cut (MEE-II structure) Si wafers. The deposition loop consisted of a sequence of 5s Ga deposition, 1s pause with closed Ga and P sources, 8s exposure under P flux, and 5s pause, with the loop being repeated for a total of 184 cycles. Before commencing MEE growth, the P shutter was opened for 30 s and then closed for 12s before the initial Ga deposition. The P and Ga flux ratio was ~5.

Several GaP-Si(001) epitaxial structures with GaP layer thicknesses ranging from ~35 nm up to ~2 μm were epitaxially grown by MBE on precisely oriented wafers at 580°C using a P/Ga ratio of ~ 4.5. At the initiation of growth, the P shutter

978-1-5090-5606-4/17 $31.00 © 2017 IEEE

was open for 20 s for P deposition, then 10 loops of short-period quasi-SL with 5s GaP deposition and 5s pause under P-flux were applied to improve the planarity of the growth front and to accelerate GaP surface reconstruction to (2×1), and then the main MBE growth process was started. During the entire deposition period, the RHEED pattern was 2×1, corresponding to the perfectly reconstructed growth surface.

B. High-Resolution XRD and TEM Investigations

High-resolution XRD studies were performed using an X'Pert MRD diffractometer with a multilayer focusing mirror under double- and triple-axis alignment. Hybrid Ge(220) monochromator ensured 18 arc.sec collimated and almost monochromatic CuKα-1 incident radiation, while a Ge(220) analyzer with 12 arc.sec acceptance angle allowed separate coherent and diffuse-scattered radiation in the vicinity of the (004) and (224) reflections. Coherent double - (DC) and triple-crystal (TC) ω-2θ and ω rocking curves (RCs) and Reciprocal Space Maps (RSM) were used to determine stress relaxation and specify preferred type, spatial distribution, and density of crystal defects in the structures.

Specimens suitable for TEM observation were prepared using focused-ion-beam milling as well as additional argon-ion-milling to remove some surface milling artefacts. Observations were made with a FEI-Philips CM200 high-resolution electron microscope, equipped with a double-tilt specimen holder, and an aberration-corrected ARM-200F microscope. Cross-section observations were made along {110}-type zone axes so that the surface normal would be perpendicular to the incident-beam direction.

III. EXPERIMENTAL RESULTS

A) Investigation of MEE-grown structures

Double-crystal ω-2θ and TC ω RCs shown in Fig.1 revealed that the crystal perfection of 50-nm GaP layer grown by MEE under the same growth conditions on the precisely oriented wafer was substantially lower than the crystal perfection of the GaP layer grown on the offcut wafer.

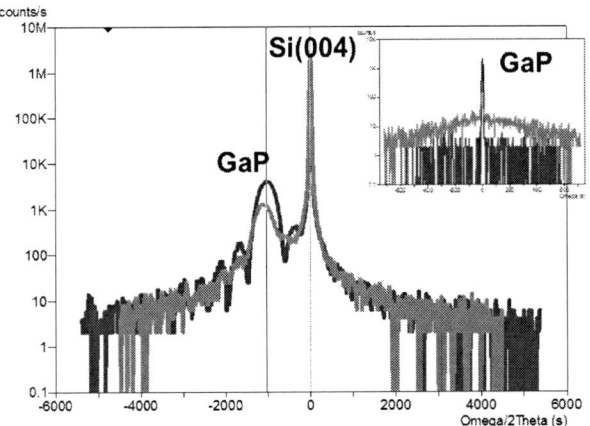

Fig. 1, DC coherent ω-2θ RCs measured in the vicinity of Si-GaP (004) reflections. Red curve corresponds to the precise MEE-I structure, blue one to offcut MEE-II structure. Insert shows the corresponding GaP TC ω RCs.

Diffusion of interference fringes for the red ω-2θ RC means the diminution of x-ray vertical scattering coherence of the GaP layer, caused by crystal defects with edge segment(s) in the volume of the epitaxial layer created during epitaxial growth. The wide diffuse base and diminished intensity of the central narrow peak of the MEE-I ω RC (red) indirectly confirm this suggestion: most probably, these defects will be identified as closed dislocation loops or stacking faults (SFs). Narrow central peak reveals that bending of the structure, caused by these defects, is rather small, while their density is ~1.5-2×10^9 cm^{-2}. On other hand, MEE-II epitaxial structure and GaP layer are almost perfectly crystalline, and the volumetric density of crystal defects is ~2×10^5 cm^{-2}.

TEM cross-section micrographs (Fig. 2) visually confirm the trends of the XRD results. The MEE-I GaP sample (Fig. 2a) revealed many stacking faults and multiple twins (red circles) in the volume of the GaP layer, some of them intersected in the volume. By comparison, the MEE-II structure (Fig. 2b) showed no extended crystal defects over the same lateral distance. Offcut angle significantly diminishes symmetry related structural deteriorations on the growth front, resulted in SFs and microtwins creation.

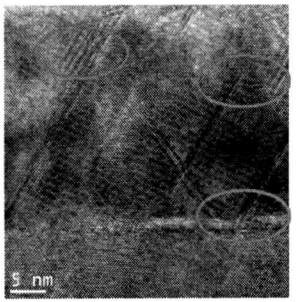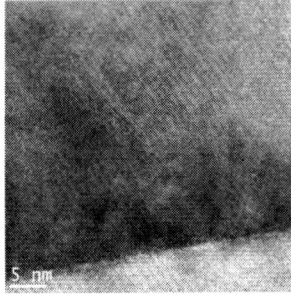

Fig. 2. TEM cross-section micrographs of MEE grown structures: a) GaP grown on precise Si wafer, b) GaP grown on offcut Si wafer.

B) Investigation of MBE-grown structures

The MBE-grown structures exhibited high crystal perfection for the thin (~35 nm) GaP layer, while the thicker (≥250 nm) layers demonstrated diffusion of interference fringes in the ω-2θ DC RC due to deterioration of the vertical coherence of the GaP layer caused by edge segments created by the intersected SFs. Relaxation of the initial elastic stress demonstrates a gradual increase for GaP layers thicker than 250 nm (Fig. 3, DC ω-2θ RCs). The relaxation for 500-nm thickness is ~ 9%, rising successively up to 40% and 75%, respectively, in the 1 and 2 μm GaP layer structures. Supposedly, the relaxation in these structures commences at ≤ 250 nm through the creation of 60° dislocations at the GaP-Si interface, and is followed by their gradual density increase and structural transformation to Lomer dislocations at the bottom interface, which are typical for low initially deteriorated epitaxial structures. [6].

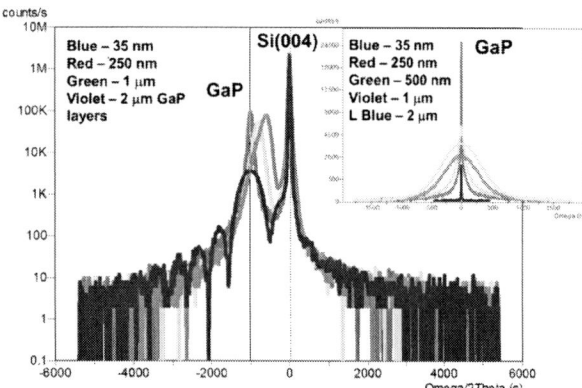

Fig. 3. DC coherent ω-2θ RCs measured in the of vicinity of Si-GaP (004) reflections. Black and violet bars specify the angle position of fully elastically stressed GaP layer and Si(004) substrate peaks. TC ω RCs shown insert.

On the other hand, bending evolution of Si-GaP epitaxial structures, correlated with the type and density of volumetric threading defects, is opposite to partially relaxed low deteriorated epitaxial structures without effect of symmetry change [5]. Density of threading defects, evaluated by the FWHM of the ω RCs, is steadily increasing ($\sim 2.2 \times 10^7$ cm^{-2} in 500 nm, $\sim 8 \times 10^8$ cm^{-2} in 1μm, and $\sim 1.1 \times 10^9$ cm^{-2} in 2μm GaP layer), as well as the bending elastic stress, induced by them.

The electron microscopy observations (see Fig. 4) confirm the trends in defect development, predicted from the X-ray data. The thinnest GaP layer (~ 35 nm) shows an abrupt and coherent GaP/Si interface and although the top surface is slightly uneven there are virtually no structural defects visible in the GaP layer across large lateral distances. Conversely, the thick (~ 2 μm) GaP layer contains a high density of threading dislocations, as well as considerable {111}-type inclined stacking faults.

IV. CONCLUDING REMARKS

Crystal perfection of MEE and MBE grown Si(001)-GaP structures were investigated by high-resolution XRD and TEM. MEE structures grown under the same growth conditions revealed that 4 degrees offcut angle significantly affects crystal perfection of GaP. Higher perfection of offcut structure most probably correlates to improved atomic migration and atomic incorporation on the growth front. MBE grown structures revealed high crystal perfection of the thin GaP layer, then gradual relaxation of the initial elastic stress in

thicker GaP layers, and increasing bending of the thick GaP structures caused by threading dislocations, created in these layers.

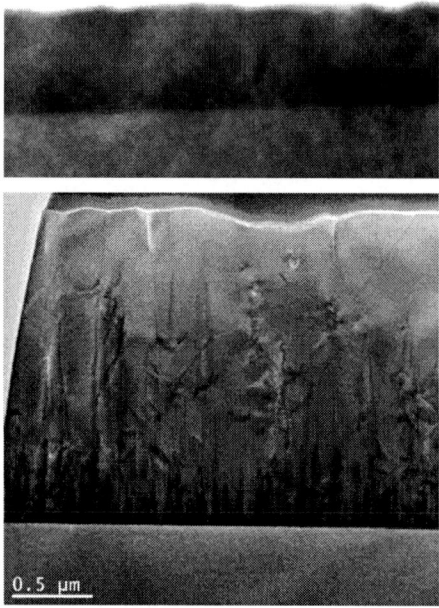

Fig.4. TEM cross-section micrographs of MBE-grown GaP/Si structures grown: top) thin (~35 nm) GaP layer; bottom) thick (~ 2 μm) GaP

REFERENCES

[1] C. Zhang, N. N. Faleev, L. Ding, M. Boccard, M. Bertoni, Z. Holman, R. R. King, and C. B. Honsberg, "Hetero-emitter GaP/Si solar cells with high Si bulk lifetime," in *2016 IEEE 43rd Photovoltaic Specialists Conference (PVSC)*, 2016, pp. 1950–1953.

[2] T. J. Grassman, J. A. Carlin, B. Galiana, F. Yang, M. J. Mills, and S. A. Ringel, "MOCVD-Grown GaP/Si Subcells for Integrated III–V/Si Multijunction Photovoltaics," *IEEE J. Photovoltaics*, vol. 4, pp. 972–980, 2014.

[3] J. R. Lang, J. Faucher, S. Tomasulo, K. N. Yaung, and M. L. Lee, "Comparison of GaAsP solar cells on GaP and GaP/Si," *Appl. Phys. Lett.*, vol. 103, pp. 092102–1–092102–5, 2013.

[4] T. J. Grassman, M. R. Brenner, M. Gonzalez, A. M. Carlin, R. R. Unocic, R. R. Dehoff, M. J. Mills, and S. A. Ringel, "Characterization of Metamorphic GaAsP/Si Materials and Devices for Photovoltaic Applications," *IEEE Trans. Electron Devices*, vol. 57, pp. 3361–3369, 2010.

[5] K. Yamane, T. Kawai, Y. Furukawa, H. Okada, and A. Wakahara, "Growth of low defect density GaP layers on Si substrates within the critical thickness by optimized shutter sequence and post-growth annealing," *J. Cryst. Growth*, vol. 312, pp. 2179–2184, 2010.

[6] A. Maros, N. Faleev, R. R. King, C. B. Honsberg, D. Convey, H. Xie, F. A. Ponce, "Critical thickness investigation of MBE-grown GaInAs/GaAs and GaAsSb/GaAs heterostructures", *J. Vac. Sci. Technol.* B vol. 34, 02L113, 2016.

Investigation of carrier-induced defect behavior in p-type multi-crystalline silicon

Catherine E. Chan, Tsun H. Fung, David N. R. Payne, Daniel Chen, Malcolm D. Abbott, Alison M. Ciesla, Ran Chen, Brett J. Hallam, and Stuart R. Wenham

School of Photovoltaic and Renewable Energy Engineering, University of New South Wales, Sydney, NSW 2052, Australia

Abstract — The underlying mechanism behind carrier-induced degradation in multi-crystalline silicon (mc-Si) solar cells remains unknown, however clues can be gained through studies of the behavior of the defect/s in response to different processing steps. Recently, surprisingly significant changes to the kinetics of the defect/s after low-temperature dark annealing have been observed for mc-Si PERC cells. In this study we apply the same processes to symmetrical lifetime test structures to validate that these changes to the kinetics are indeed caused by changes in the bulk of the wafers.

I. INTRODUCTION

Multi-crystalline silicon (mc-Si) suffers from severe carrier-induced degradation (CID) effects, particularly when used in passivated emitter and rear cells (PERC) [1]. There have been numerous studies carried out in recent years investigating this carrier-induced defect's behavior [2]–[17], characterizing its recombination properties [5], [7], [17]–[20] and investigating mitigation processes to reduce its impact on solar cell performance [4], [11], [12]. Mc-Si CID (also broadly known as light and elevated temperature induced degradation – LeTID) is a serious issue for the photovoltaic (PV) industry as it shifts towards PERC production on mc-Si substrates [21]. In an extensive study performed by Hanwha Q-Cells of mc-Si wafers from seven different suppliers, no wafer was found to be immune to the degradation [22]. A comprehensive review of what is known about mc-Si CID was given in references [23] and [24]. Despite the significant research efforts, the root cause behind the degradation is yet to be conclusively identified.

We recently demonstrated that thermal history plays a significant role in altering the kinetics of degradation and regeneration of mc-Si PERC cells [16]. PERC cells that had a 2.5 hour dark anneal prior to light soaking showed very different degradation and regeneration behavior depending on the dark anneal temperature, with increments of only 25 °C having dramatic effects on the kinetics. Figure 1 shows the degradation and regeneration curves for four of these PERC cells, highlighting four different modes we defined:

1. 'Standard' degradation and regeneration (control)

2. Degradation is accelerated and the extent of degradation is increased, followed by an accelerated regeneration (DA <= 200 °C)

3. Degradation is accelerated and the extent of degradation is increased, followed by an extremely slowed regeneration (DA = 225 °C)

4. Degradation is suppressed but is slowed significantly - no regeneration apparent after 1000 hours of light soaking (DA> 225 °C).

From these results, it is apparent that dark annealing processes can alter the state of the mc-Si CID defect system. However in that work it was not clear whether the significant changes observed were linked to the front or rear surface of the PERC cells, or whether this was indeed a true bulk effect. Here we build upon the previous findings by investigating degradation and regeneration behaviour for a variety of different symmetrical substrates.

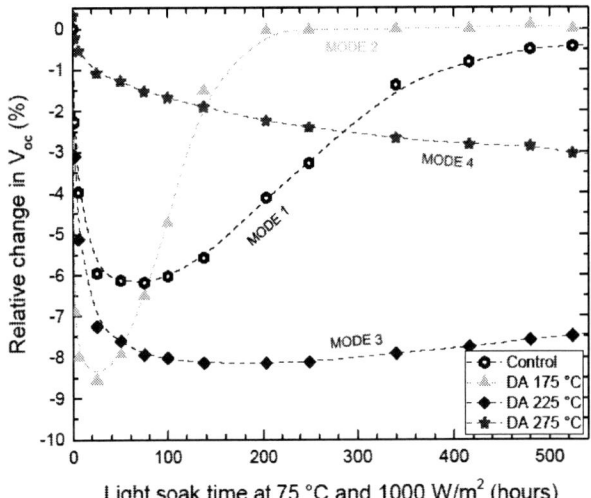

Fig. 1.   Degradation and regeneration curves highlighting the four distinct modes of the defect evolution during light soaking. The dashed lines are B-splines serving as guides to the eye. Data is taken from [16].

978-1-5090-5606-4/17 $31.00 © 2017 IEEE

II. EXPERIMENTAL METHOD

In order to separate any front and rear surface effects, symmetrical lifetime test structures were fabricated. The lifetime test structures investigated were:

1. p-type mc-Si wafer with phosphorus diffusion and SiN_x passivation to represent the front surface of the solar cell
2. p-type mc-Si wafer with $AlOx/SiN_x$ stack to represent the rear surface of the solar cell
3. p-type mc-Si wafer with only SiN_x passivation (no phosphorus diffusion) to investigate whether the diffused layer has any effect on the defect kinetics

To minimize effects of different fabrication steps and thermal history on the results, all samples underwent as near-identical processing as possible. The process flow for all three structures is shown in Figure 2. Firstly, all samples (12 full size 156 mm × 156 mm 1.6 ohm.cm boron-doped neighboring 'sister' wafers, where 4 sister wafers were used per structure) were etched in an alkaline saw-damage etch solution to remove surface damage (acidic texturing was unavailable at the time of this experiment). Following RCA cleaning all samples underwent $POCl_3$ diffusion and PSG removal. For sample structures 2 and 3, the diffused layers were then removed using an identical saw-damage etch solution as previous to ensure that the surfaces remained similarly textured to sample structure 1. Note that sample structure 1 had double the amount of silicon removed in the first etch step to account for the extra etch step on structures 2 and 3. Samples then received PECVD layers on both sides (SiN_x only for structures 1 and 3, and $AlOx/SiN_x$ stack for structure 2). All samples then underwent firing at set peak temperature of 835 °C in a Sierratherm infra-red fast firing furnace, corresponding to an actual sample temperature of approximately 740 °C.

Samples were then cleaved into smaller 39 mm × 39 mm tokens for further processing. Sister tokens underwent dark annealing at three different temperatures: 175 °C, 225 °C and 275 °C for 2.5 hours, which should each activate a different mode of degradation according to Figure 1. One sister token was also kept as a control with no dark anneal treatment. Injection-dependent effective lifetime measurements were performed using a photo-conductance lifetime tester (WCT-120, Sinton Instruments [25]) on tokens after firing, at incremental steps during dark annealing up to 2.5 hours and then at incremental steps during subsequent stability testing. The effective lifetime was extracted at an excess minority carrier density (Δn) of 9.1×10^{14} cm^{-3} (corresponding to 10% of the bulk doping density). Accelerated stability testing was carried out using a laser based setup as detailed in reference [6]. The sample temperature was 130 °C under an illumination intensity of 34.6 kW/m^2.

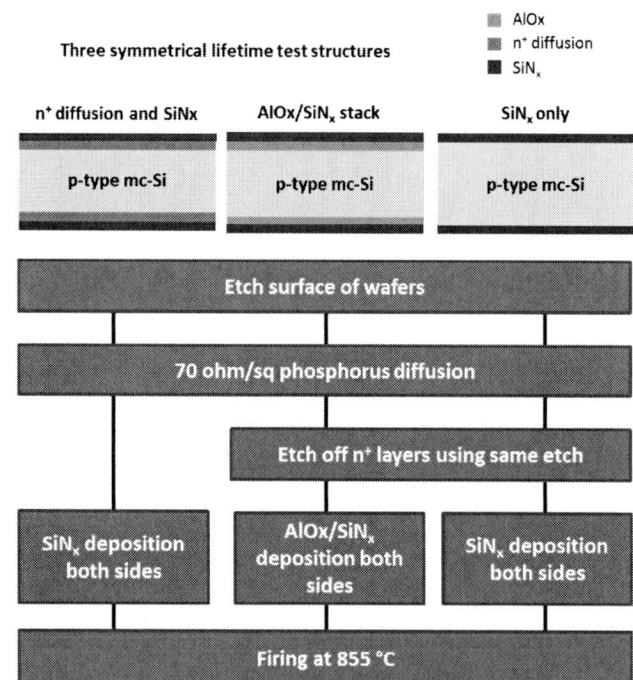

Fig. 2.   Process flow for symmetrical lifetime test structures

III. RESULTS AND DISCUSSION

A. Evolution of effective lifetime during dark annealing

Figures 3 (a)-(c) show the effective lifetime at incremental points during the dark anneal process for each dark anneal temperature. For the 175 °C case, all samples degraded in lifetime over the 2.5 hours of annealing. At 225 °C, all samples at first degraded in lifetime, reaching a minimum after 15-30 mins after which the lifetime then improved again. At 275 °C, structures 1 and 3 initially showed very little change, and then a lifetime improvement, whilst structure 2 showed a clear degradation at first followed by significant lifetime improvement and then a gradual decline. Analysis of the injection-dependent lifetime curves indicated that these changes during dark annealing were predominantly bulk effects rather than changes to the surface passivation. However it should be noted that the AlOx/SiNx layer did not seem to be stable, with the effective lifetime of these samples changing significantly after successive flashes with the Sinton tester (unrelated to FeB pair dissociation). Care was taken to ensure the lifetimes plotted here were taken after only a single flash after each dark anneal increment.

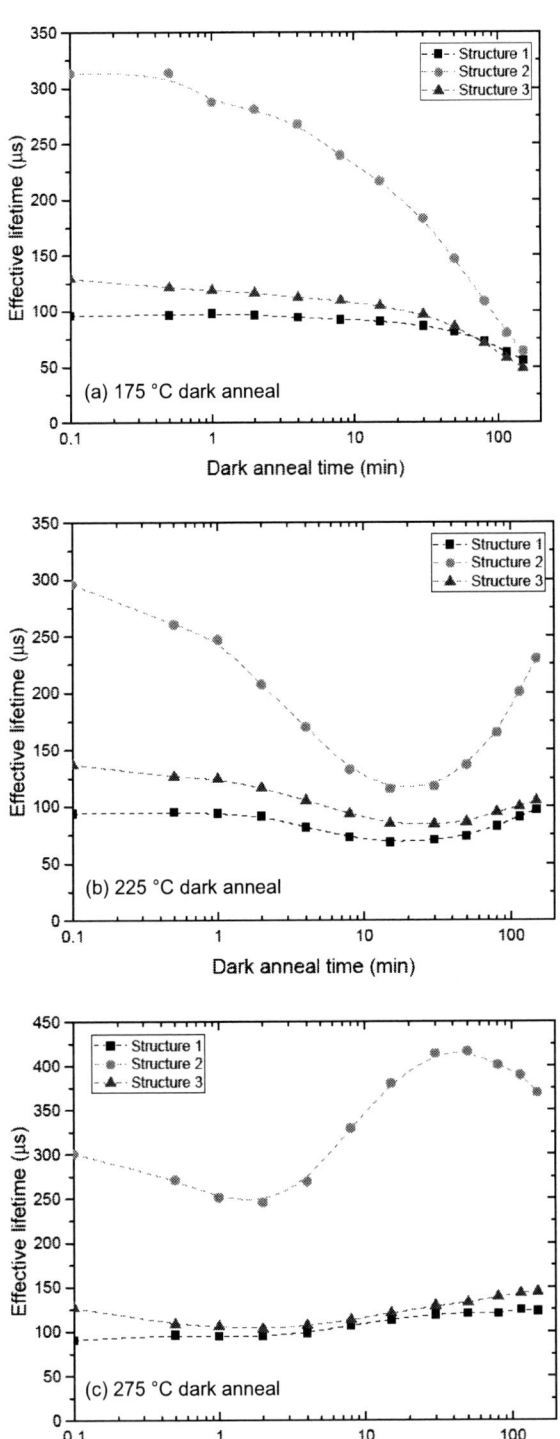

Fig. 3. Evolution of effective lifetime during dark annealing for the three lifetime test structures at (a) 175 °C, (b) 225 °C and (c) 275 °C.

B. Effect of pre-dark anneal temperature on subsequent degradation and regeneration of lifetime test structures

The effective lifetime of the three structures as a function of laser illumination time are shown in Figure 4 (a)-(c). Regardless of the structure type, each dark anneal treatment altered the kinetics of degradation and regeneration in a similar way. Relative to the case of there being no pre dark anneal treatment, a 175 °C dark anneal prior to light soaking accelerated and enhanced the degradation extent, and also accelerated the regeneration process. A dark anneal at 225 °C also accelerated the degradation and enhanced the degradation extent, but somewhat slowed the regeneration rate compared to the control case. Finally, the 275 °C dark anneal appeared to suppress the degradation considerably and even further suppress the regeneration rate, consistent with the results in [16]. Note that while initial lifetimes were different, and the extent of degradation for each of the three structures varied, the fact that all three structures responded in a similar way to the dark anneal treatments prior to light soaking reveals that the changes to the carrier-induced defect kinetics of the PERC cells reported in [16] are not significantly tied to the front or rear surface layers and are most likely a bulk effect.

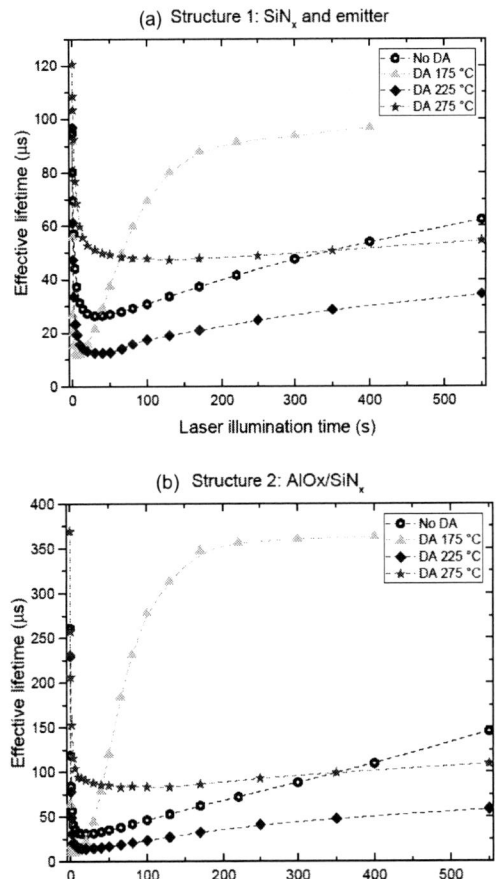

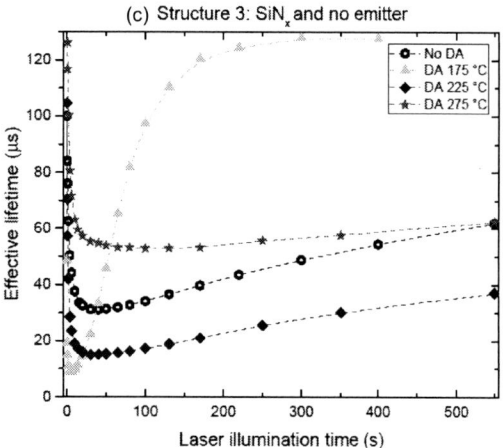

(c) Structure 3: SiN$_x$ and no emitter

Legend:
- No DA
- DA 175 °C
- DA 225 °C
- DA 275 °C

Fig. 4. Evolution of effective lifetime as a function of laser illumination time after dark annealing for 2.5 hours at several temperatures for (a) structure 1, (b) structure 2 and (c) structure 3.

C. Shockley-Read-Hall analysis at the degraded state for each mode of degradation

Injection-dependent Shockley-Read-Hall (SRH) lifetimes of the defect were realized by the difference between inverse carrier lifetimes before and after degradation. By using this approach, we assume that other recombination mechanisms, bulk carrier lifetimes and any defects intrinsic to the samples remain identical in both conditions. Therefore, any changes in the lifetime are assumed to be caused only by the defect under investigation. To minimize error, the dark saturation current densities (J_0) of both states were quantified and monitored for changes. No significant change in the extracted J_0 values were observed from the initial to degraded state. A Chi-square analysis was undertaken to evaluate the quality of fit and determine the range for the electron to hole capture cross section ratios (k-values) for the defects [26]. The extracted k-values (assuming a mid-band-gap defect level) are shown in Table 1 for each of the structures and dark anneal conditions. Note that the extracted k-values for the AlO$_x$/SiN$_x$ passivated samples (structure 2) are likely to have an increased (unquantified) error since the passivation layer was not stable during illumination with the Sinton flash lamp (as discussed earlier). For the SiN$_x$ passivated samples (with and without emitter), the extracted k-values agreed well with each other for the case of no dark anneal, 175 °C and 225 °C dark anneals, implying that the defect induced under these conditions was likely the same. Interestingly, an increased k-value was observed for the samples that underwent the higher temperature 275 °C dark anneal prior to light soaking. This could hint at the possibility that there is a different type of defect precursor activated at this higher temperature, or that another defect with a higher k-value is also activated.

TABLE I

EXTRACTED K-VALUES FOR THE DEFECT AT THE MOST DEGRADED POINT FOR EACH STRUCTURE AND DARK ANNEAL CONDITION

	SiNx + emitter	AlOx/SiNx no emitter	SiNx no emitter
No DA	32.1 ± 0.9	40 +/- 2.2	32.7 +/- 1.3
175 °C	32.1 ± 1.0	32.7 +/- 0.8	29.4 +/- 1.4
225 °C	30.1 ± 0.9	37.7 +/- 1.6	31.2 +/- 0.8
275 °C	38.8 ± 0.8	43.4 +/- 2.9	38.1 +/- 1.7

Note that, the k-values extracted could be strongly affected by the limited range of injection level and the assumption for the energy level of the defect (mid-gap). To eliminate the need to make this assumption and to obtain a unique solution, further analysis must include temperature and injection dependent lifetime spectroscopy (TIDLS) such as that presented in [19] and a larger number of samples.

IV. CONCLUSION

Dark annealing can alter the state of the carrier-induced defect system in mc-Si PERC cells, with relatively small changes in temperature leading to significant changes to the degradation and regeneration kinetics. This is a point that is particularly important for studies where dark annealing is performed on samples prior to studying the mc-Si defect, or for cases where samples to be compared undergo different thermal processing steps. Here it was shown that these changes to the kinetics do not appear to be dependent on the front or rear surface of a PERC cell through analysis of symmetrical lifetime test structures. SRH fitting of the injection-dependent lifetime curves in their most degraded state indicates that the defect is likely the same in the control samples with no dark anneal, and the samples that received 175° C and 225 °C anneals. The samples annealed at 275 °C prior to light soaking showed increased k-values, indicating that the defect may be different or that an additional defect has been introduced with a k-value higher than the one under investigation.

REFERENCES

[1] K. Ramspeck, S. Zimmermann, H. Nagel, A. Metz, Y. Gassenbauer, B. Birkmann, and A. Seidl, "Light Induced Degradation of Rear Passivated Mc-Si Solar Cells," *27th Eur. Photovolt. Sol. Energy Conf. Exhib.*, vol. 1, pp. 861–865, 2012.

[2] F. Fertig, K. Krauß, and S. Rein, "Light-induced degradation of PECVD aluminium oxide passivated silicon solar cells," *Phys. status solidi - Rapid Res. Lett.*, vol. 9, no. 1, pp. 41–46, 2015.

[3] F. Kersten, P. Engelhart, H. Ploigt, A. Stekolnikov, T. Lindner, F. Stenzel, M. Bartzsch, A. Szpeth, K. Petter, J. Heitmann, and J. W. Müller, "A New mc-Si Degradation Effect called LeTID," in *IEEE Photovoltaics Specialists Conference*, 2015.

[4] K. Krauss, A. A. Brand, F. Fertig, S. Rein, and J. Nekarda, "Fast Regeneration Processes to Avoid Light-Induced Degradation in Multicrystalline Silicon Solar Cells," *IEEE J. Photovoltaics*, vol. 6, no. 6, pp. 1427–1431, Nov. 2016.

[5] D. Bredemeier, D. Walter, S. Herlufsen, and J. Schmidt, "Lifetime degradation and regeneration in multicrystalline silicon under illumination at elevated temperature," *AIP Adv.*, vol. 6, no. 3, p. 35119, Mar. 2016.

[6] D. N. R. Payne, C. E. Chan, B. J. Hallam, B. Hoex, M. D. Abbott, S. R. Wenham, and D. M. Bagnall, "Acceleration and mitigation of carrier-induced degradation in p-type multi-crystalline silicon," *Phys. status solidi - Rapid Res. Lett.*, vol. 10, no. 3, pp. 237–241, 2016.

[7] K. Nakayashiki, J. Hofstetter, A. E. Morishige, T.-T. A. Li, D. B. Needleman, M. A. Jensen, and T. Buonassisi, "Engineering Solutions and Root-Cause Analysis for Light-Induced Degradation in p-Type Multicrystalline Silicon PERC Modules," *IEEE J. Photovoltaics*, vol. 6, no. 4, pp. 860–868, Jul. 2016.

[8] T. Luka, S. Großer, C. Hagendorf, K. Ramspeck, and M. Turek, "Intra-grain versus grain boundary degradation due to illumination and annealing behavior of multi-crystalline solar cells," *Sol. Energy Mater. Sol. Cells*, vol. 158, pp. 43–49, 2016.

[9] J. Lindroos, S. Dubois, N. Enjalbert, and M. Rinio, "Light Beam Induced Current of light-induced degradation in multicrystalline silicon solar cells," in *26th NREL Workshop on Crystalline Silicon Solar Cells & Modules*, 2016.

[10] D. Skorka, A. Zuschlag, and G. Hahn, "Spatially Resolved Degradation and Regeneration Kinetics in mc-Si," *32nd Eur. Photovolt. Sol. Energy Conf. Exhib.*, pp. 643–646, 2016.

[11] D. N. R. Payne, C. E. Chan, B. J. Hallam, M. D. Abbott, S. R. Wenham, and D. M. Bagnall, "Rapid Passivation of Carrier-induced Defects in p-type Multi-crystalline Silicon," *Sol. Energy Mater. Sol. Cells*, vol. 158, no. 1, pp. 102–106, 2016.

[12] C. E. Chan, D. N. R. Payne, B. J. Hallam, M. D. Abbott, T. H. Fung, A. M. Wenham, B. S. Tjahjono, and S. R. Wenham, "Rapid Stabilization of High-Performance Multicrystalline P-type Silicon PERC Cells," *IEEE J. Photovoltaics*, vol. 6, no. 6, pp. 1473–1479, Nov. 2016.

[13] A. Zuschlag, D. Skorka, and G. Hahn, "Degradation and regeneration in mc-Si after different gettering steps," *Prog. Photovoltaics Res. Appl.*, 2016.

[14] R. Eberle, W. Kwapil, F. Schindler, M. C. Schubert, and S. W. Glunz, "Impact of the firing temperature profile on light induced degradation of multicrystalline silicon," *Phys. status solidi - Rapid Res. Lett.*, vol. 10, no. 12, pp. 861–865, Nov. 2016.

[15] M. Selinger, W. Kwapil, F. Schindler, K. Krauß, F. Fertig, B. Michl, W. Warta, and M. C. Schubert, "Spatially resolved analysis of light induced degradation of multicrystalline PERC solar cells," *Energy Procedia*, vol. 92, pp. 867–872, 2016.

[16] C. Chan, T. H. Fung, M. Abbott, D. Payne, A. Wenham, B. Hallam, R. Chen, and S. Wenham, "Modulation of Carrier-Induced Defect Kinetics in Multi-Crystalline Silicon PERC Cells Through Dark Annealing," *Sol. RRL*, vol. 1, no. 2, p. 1600028, Feb. 2017.

[17] T. H. Fung, C. E. Chan, B. J. Hallam, D. N. R. Payne, M. D. Abbott, and S. R. Wenham, "Impact of annealing on the formation and mitigation of carrier-induced defects in multi-crystalline silicon," in *7th International Conference on Silicon Photovoltaics, SiliconPV*, 2017.

[18] A. E. Morishige, M. A. Jensen, D. B. Needleman, K. Nakayashiki, J. Hofstetter, T.-T. A. Li, and T. Buonassisi, "Lifetime Spectroscopy Investigation of Light-Induced Degradation in p-type Multicrystalline Silicon PERC," *IEEE J. Photovoltaics*, vol. 6, no. 6, pp. 1466–1472, Nov. 2016.

[19] C. Vargas, Y. Zhu, G. Coletti, C. Chan, D. Payne, M. Jensen, and Z. Hameiri, "Recombination parameters of lifetime-limiting carrier-induced defects in multicrystalline silicon for solar cells," *Appl. Phys. Lett.*, vol. 110, no. 9, p. 92106, Feb. 2017.

[20] M. A. Jensen, A. E. Morishige, J. Hofstetter, D. B. Needleman, and T. Buonassisi, "Evolution of LeTID Defects in p-Type Multicrystalline Silicon During Degradation and Regeneration," *IEEE J. Photovoltaics*, pp. 1–8, 2017.

[21] ITRPV Seventh Edition, "International Technology Roadmap for Photovoltaic Results 2015," 2016.

[22] K. Petter, K. Hubener, F. Kersten, M. Bartzsch, F. Fertig, B. Kloter, and J. Muller, "Dependence of LeTID on brick height for different wafer suppliers with several resistivities and dopants," in *9th International Workshop on Crystalline Silicon for Solar Cells*, 2016.

[23] T. Luka, C. Hagendorf, and M. Turek, "Multicrystalline PERC solar cells: Is light-induced degradation challenging the efficiency gain of rear passivation?," *Photovoltaics International Volume 32*, pp. 37–44, 2016.

[24] F. Fertig, F. Kersten, K. Petter, M. Bartzsch, F. Stenzel, A. Mette, B. Kloter, and J. Mueller, "Light and Elevated Temperature Induced Degradation of Multicrystalline Silicon Solar Cells and Modules," in

26th NREL Workshop on Crystalline Silicon Solar Cells & Modules, 2016.

[25] R. A. Sinton and A. Cuevas, "Contactless determination of current–voltage characteristics and minority-carrier lifetimes in semiconductors from quasi-steady-state photoconductance data," *Appl. Phys. Lett.*, vol. 69, no. 17, p. 2510, Oct. 1996.

[26] N. Nampalli, T. H. Fung, S. Wenham, B. Hallam, and M. Abbott, "Statistical analysis of recombination properties of the boron-oxygen defect in p-type Czochralski silicon," *Front. Energy*, vol. 11, no. 1, pp. 4–22, 2017.

Magnetron Sputtered Hydrogenated Silicon Thin Films: Assessment for Application in Photovoltaics

Dipendra Adhikari, Maxwell M. Junda, Sylvain X. Marsillac*, Robert W. Collins, and Nikolas J. Podraza

Wright Center for Photovoltaics Innovation and Commercialization & Department of Physics and Astronomy, University of Toledo, Toledo, Ohio, 43606, USA

*Virginia Institute of Photovoltaics, Old Dominion University, Norfolk, Virginia, 23529, USA

Abstract — **Hydrogenated amorphous and nanocrystalline silicon (a-Si:H, nc-Si:H) thin films have been prepared by radio frequency magnetron sputtering with process parameters used to manipulate optical properties, structure, and growth evolution. *In-situ* real time spectroscopic ellipsometry is applied to construct growth evolution diagrams by tracking nucleation and coalescence of crystallites from the amorphous phase, as well as surface roughening within the amorphous growth regime, as functions of hydrogen-to-argon gas ratio. Infrared spectroscopic ellipsometry determines Si-H$_n$ vibrational modes, and grazing incidence x-ray diffraction provides crystallite orientation. The suitability of these materials for photovoltaics is assessed.**

Index Terms — **ellipsometry, hydrogenated silicon, virtual interface analysis, sputtering.**

I. INTRODUCTION

Thin film hydrogenated silicon (Si:H) is a useful material in photovoltaics (PV) because of its higher absorption coefficient and lower cost fabrication methods compared to crystalline (c-Si) as well as its passivation ability. Nanocrystalline Si:H (nc-Si:H) has low photo-induced degradation with enhanced infrared (IR) absorption compared to amorphous Si:H (a-Si:H).

Radio frequency (RF) and very high frequency (VHF) plasma enhanced chemical vapor deposition (PECVD) as well as hot-wire chemical vapor deposition have been extensively studied for deposition of Si:H for solar cells. Here RF magnetron sputtering, a simple and cost effective deposition technique without the need for Si-carrying source gases, has been explored for Si:H. Microstructural properties such as crystallinity, hydrogen incorporation, and surface morphology can be effectively controlled during sputtering [1]. The microstructural evolution of PECVD Si:H films has been investigated [2]-[4] and similar techniques are now applied for sputtered material. Two series of Si:H samples with different growth rates (low rate: ~0.15 Å/s and high rate: ~1.3 Å/s) are fabricated as functions of hydrogen-to-total gas flow ratio, p_{H2} = [H$_2$] / {[H$_2$] + [Ar]} × 100%. Using systematic process parameter variations, we have produced growth evolution diagrams as functions of p_{H2} during sputtering and studied the effects of deposition conditions on resultant film properties as steps toward achieving high quality a-Si:H and nc-Si:H material for PV. Initial assessments are made for a-Si:H and nc-Si:H absorber layers in single and tandem junction PV as well as for a-Si:H to passivate crystalline Si in HIT cells.

II. EXPERIMENTAL DETAILS

Si:H films have been RF sputter deposited on native oxide covered c-Si wafer substrates held at 200°C using a 3-inch diameter undoped Si target powered at 100 to 250 W and a 5.3-inch target-to-substrate distance. Hydrogen is added to argon gas during sputtering to control film microstructure with total pressures ranging from 10 to 30 mTorr. Two series of films have been prepared under conditions yielding low deposition rates of ~ 0.15 Å/s (RF power = 100 W, pressure = 30 mTorr) and high deposition rates of ~ 1.3 Å/s (RF power = 250 W, pressure = 10 mTorr). Growth evolution diagrams for each series have been developed as functions of varying p_{H2}. *In situ* real time spectroscopic ellipsometry (RTSE) data are collected using a dual rotating compensator ellipsometer (RC2, J. A. Woollam Co.) during growth to detect changes in structure (film and surface roughness thickness) and near IR to ultraviolet (UV) optical properties for a-Si:H and nc-Si:H.

IR optical properties of films pertaining to silicon-hydrogen (SiH$_n$; n = 1, 2, 3) bonding modes have been obtained from ellipsometric spectra (IR-VASE, J. A. Woollam Co.). Grazing incidence x-ray diffraction (GIXRD) measurements have been performed with a Rigaku/Altima-III x-ray diffractometer using Cu-K$_\alpha$ radiation (λ = 1.54059 Å) to confirm the presence of crystallinity and identify crystallite orientation.

III. RESULTS AND DISCUSSION

Studies of growth evolution for Si:H identify the transition from amorphous to nanocrystalline material, reflected in the amorphous-to-mixed-phase [a→(a+nc)] and mixed-phase-to-single-phase [(a+nc)→nc] structural transitions corresponding to the nucleation of crystallites from the amorphous phase followed by their subsequent coalescence when covering the film surface (Fig. 1). An additional transition is observed for a-Si:H in which a smooth stable surface is reached then followed by roughening dictated by atomic scale self-shadowing [a→a]. The thickness at which these transitions occur are functions of process parameters as shown in growth evolution diagrams (Fig. 2). The initial appearance of crystallites is accompanied by an abrupt increase in the surface roughness thickness with a maximum roughness thickness reached at crystallite coalescence. Surface roughness thickness as a function of bulk film thickness and these structural transitions are obtained from analysis of RTSE data as has been done previously for PECVD Si:H [4]. The a→(a+nc) and (a+nc)→nc transitions, as well as the volume fraction of crystallites at the top film surface, are also identified by contrast in near IR to UV complex dielectric function ($\varepsilon = \varepsilon_1 + i\varepsilon_2$) spectra between a-Si:H and nc-Si:H and virtual interface analysis (VIA) of RTSE data [5]-[7].

Fig. 1 shows an example of surface roughness and crystallite fraction as functions of bulk film thickness for high deposition rate p_{H2} = 80% material. The deposition rate of sputtered Si:H increases with increasing RF power and decreases with increasing total pressure (Fig. 3). Fig. 2 shows the growth evolution diagrams of sputtered Si:H for low and high deposition rates, respectively. The same a→a, a→(a+nc), and (a+nc)→nc structural transitions observed for PECVD Si:H are observed for the sputtered material. At lower p_{H2}, the films are amorphous. Nanocrystallites nucleate from the amorphous phase at moderate p_{H2} and the initial appearance of crystallites indicated by the a→(a+nc) transition shifts to lower bulk film thickness with increasing p_{H2}. With sufficient accumulated thickness, crystallites coalesce to form a single phase nc-Si:H layer on the surface indicated by the (a+nc)→nc transition. This transition also shifts towards lower thickness with increasing p_{H2}. Crystallite nucleation shifts to higher p_{H2} with

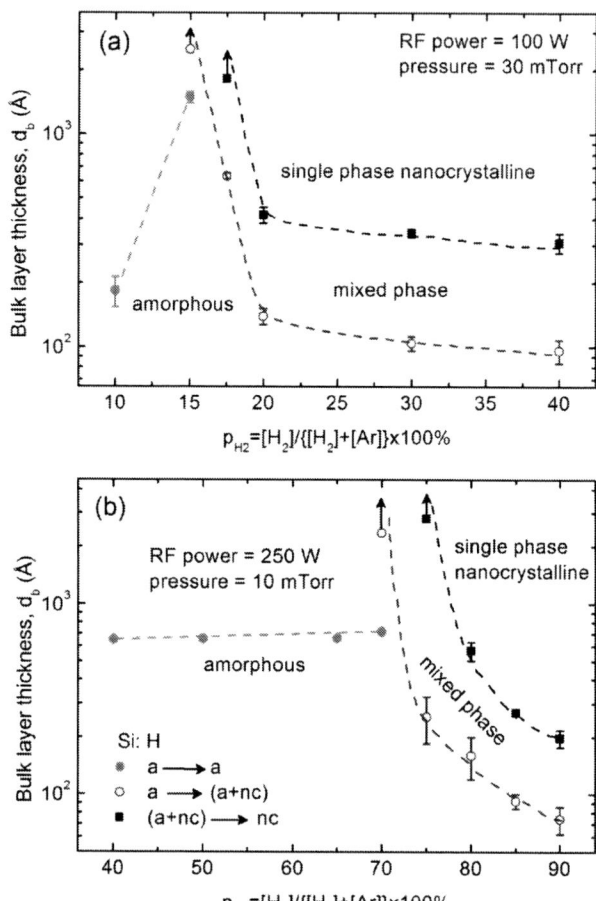

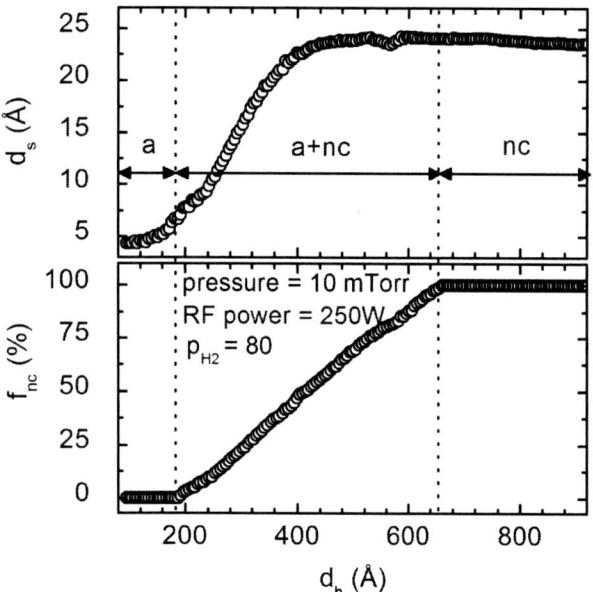

Fig. 1. Surface roughness thickness (d_s) and nanocrystallite fraction (f_{nc}) as functions of bulk layer thickness (d_b) obtained from RTSE for a high rate series p_{H2} = 80% Si:H film.

Fig. 2. Growth evolution diagrams of Si:H obtained from real time spectroscopic ellipsometry (RTSE) depicting thicknesses of roughening in the amorphous phase [a→a], initial appearance of crystallites [a→(a+nc)], and crystallite coalescence [(a+nc)→nc] for (a) low rate (top) and (b) high rate (bottom) Si:H as functions of hydrogen-to-argon ratio, p_{H2}. The upward arrows indicate transitions occur beyond the maximum thickness measured after Ref. [7].

increasing deposition rate.

For films remaining in the amorphous phase, the deposition rate is relatively stable for both series (Fig. 3). This rate continues to remain stable for slightly higher p_{H2} after crystallites first nucleate, however further increases in p_{H2} cause a reduction in growth rate. This reduction occurs with crystallite nucleation at < 150 Å and coalescence at < 400 Å for both series. At these higher p_{H2} conditions, hydrogen may not be as effectively incorporated into crystallites compared to a-Si:H and grain boundary material produced at lower p_{H2}. This additional unincorporated hydrogen may lead to etching of weakly bound material lowering the overall deposition rate.

As depicted in Fig. 4, the GIXRD patterns of the high and low deposition rate series clearly show (111) and (220) peaks

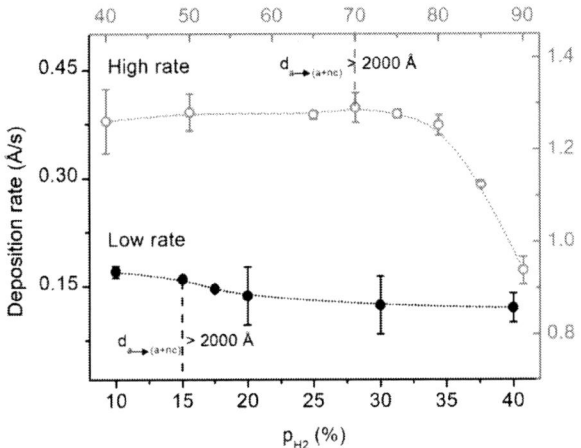

Fig. 3. Deposition rates as functions of p_{H2} for low (lower horizontal axis) and high (upper horizontal axis) rate sputtered Si:H.

of Si at $2\theta \sim 28.4°$ and $\sim 47.3°$ respectively. The (111) peak at $28.4°$ is more intense than (220) peak at $47.3°$ indicating preferential crystallite orientation along the (111) direction. For both peaks, the broadening decreases and the amplitude increases with increasing p_{H2}, indicating increases in size and volume fraction of crystallites.

Optical properties in the form of the complex dielectric function ($\varepsilon = \varepsilon_1 + i\varepsilon_2$) spectra describing a-Si:H are obtained from fitting RTSE measurements at a bulk layer thickness (d_b) of ~200 Å before nanocrystallite nucleation. The spectra in ε that best represent the optical properties of nc-Si:H are obtained from VIA [5],[7] by fitting to RTSE measurements collected when at least ~200 Å of fully nc-Si:H material has accumulated after the (a+nc)→nc transition. Figs. 5 and 6 depicts spectra in ε for relatively dense and less dense a-Si:H as well as for

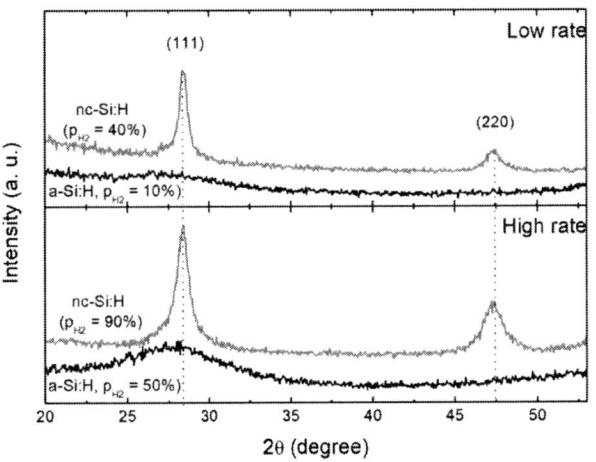

Fig. 4. Representative grazing incidence x-ray diffraction (GIXRD) patterns of high and low deposition rate a-Si:H and nc-Si:H.

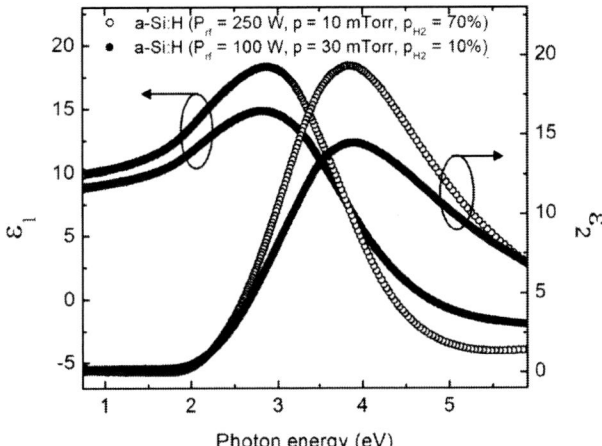

Fig. 5. Complex dielectric function ($\varepsilon = \varepsilon_1 + i\varepsilon_2$) spectra for two a-Si:H films of different relative density. The denser film is prepared at $p_{H2} = 70\%$ (high rate) and the less dense film at $p_{H2} = 10\%$ (low rate).

nc-Si:H prepared under different conditions. Differences in ε for a-Si:H and nc-Si:H are consistent with [2]-[6], with a-Si:H exhibiting a single broad absorption feature and nc-Si:H exhibiting dampened and broadened critical point transitions at energies similar to those found in c-Si. For the nc-Si:H films in Fig. 6, the lower magnitude of features in ε may arise from poor grain boundary passivation, i.e. voids between the crystalline grains. In both a-Si:H and nc-Si:H, defects such as dangling bonds or grain boundaries are passivated by hydrogen to enhance electronic performance.

Si-H_n IR vibrational modes for Si:H are observed in the IR spectra range ε with Fig. 7 showing ε_2 for high rate $p_{H2} = 50\%$ a-Si:H as a representative example. The absorption feature

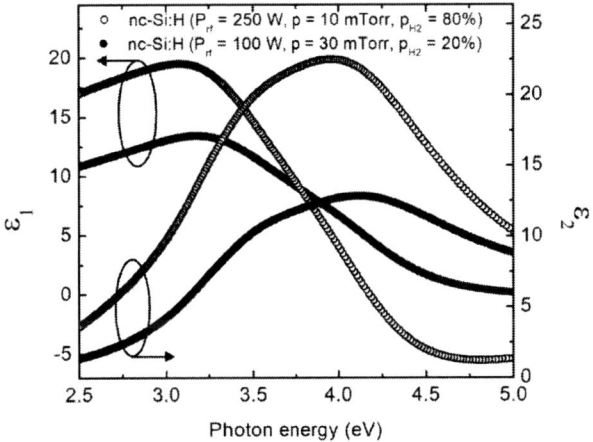

Fig. 6. Spectra in ε for two nc-Si:H films of different relative density. The denser film is prepared at $p_{H2} = 80\%$ (high rate) and the less dense film at $p_{H2} = 20\%$ (low rate).

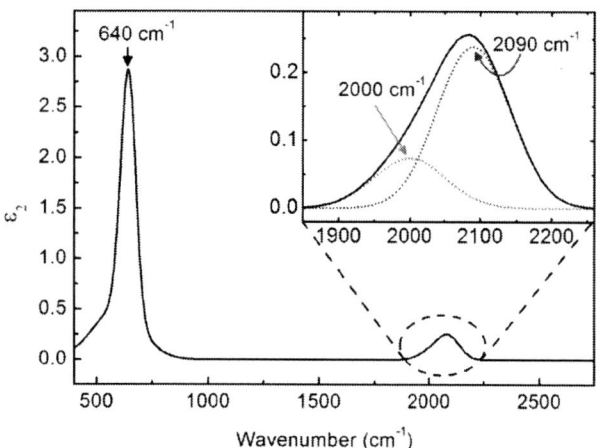

Fig. 7. Representative IR spectra in ε_2, shown for high rate $p_{H2} = 50\%$ sputtered Si:H.

centered at 590 cm^{-1} represents Si-H$_2$ rocking mode [8], 640 cm^{-1} is attributed to Si-H$_n$ wagging modes [4],[8]-[10], and high and low stretching modes due to Si-H$_2$ and Si-H are seen at 2090 and 2000 cm^{-1}, respectively [8]-[10]. From the wagging mode absorption peak at 640 cm^{-1}, the hydrogen content, C_H (at. %), is calculated using [8],[9]

$$C_H = \left[A_\omega \left(\int \frac{\alpha(\omega)}{\omega} \frac{d\omega}{N_{Si}} \right) \right] \times 100\% \qquad (1)$$

where $\alpha(\omega)$ is the absorption coefficient, A_w is the oscillation strength ($A_{640} = 2.1 \times 10^{19}$ cm^{-2} for 640 cm^{-1} wagging mode), ω is the frequency, and N_{Si} is the atomic density of c-Si (5×10^{22} cm^{-3}). Fig. 8 shows the calculated C_H as a function of p_{H2} for

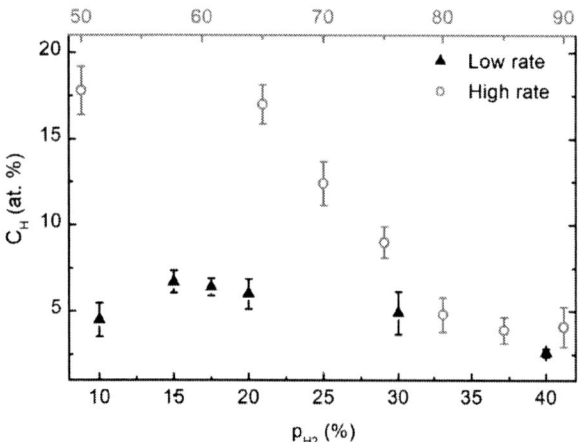

Fig. 8. Hydrogen content C_H (at. %) as a function of p_{H2} comparison for low (lower horizontal axis) and high (upper horizontal axis) rate Si:H.

both series which are comparable to reported values [9]. C_H decreases at higher p_{H2} for both series of samples, and the higher rate samples have significantly higher values of C_H compared to low rate samples. All low rate samples have $C_H \leq$ 7%, suggesting that hydrogen is not effectively incorporated into the low rate samples, even those which are amorphous. Similarly, the area under the deconvoluted absorption peaks at 2000 and 2090 cm^{-1} associated to Si-H (monohydride) and Si-H$_2$ (dihydride) stretching modes as a function of p_{H2} is shown in Fig. 9 for both series. The result shows more pronounced 2090 cm^{-1} features relative to 2000 cm^{-1} in a-Si:H.

The relative void fraction of all samples with respect to the optically densest film (high rate, $p_{H2} = 90\%$ for both series) is determined by comparing IR ε at $\omega = 5000$ cm^{-1} where the material is non-absorbing. Specifically, a Bruggeman effective medium approximation [11] containing the variable fractions of optically densest film and void is applied to fit ε describing each film with relative void fractions presented in Table I. Some samples of the low rate series are relatively less dense compared to those of the high rate series. Combined with the lower values of C_H, this lower density implies that the low rate samples may be more porous with less hydrogen passivation compared to the high rate samples.

TABLE I
RELATIVE VOID FRACTIONS IN SPUTTERED SI:H FILMS

Low rate series		High rate series	
p_{H2} (%)	Void fraction	p_{H2} (%)	Void fraction
10	0.009	50	0.061
15	0.090	65	0.053
17.5	0.003	70	0.016
20	0.119	75	0.008
30	0.004	80	0.003
40	0.083	85	0.016
		90	0

External quantum efficiency (EQE) simulations are applied to determine if the spectra in ε for sputtered a-Si:H and nc-Si:H may provide suitable absorption in single junction solar cells. EQE is simulated using a-Si:H ($p_{H2} = 70\%$, from high rate series) and nc-Si:H ($p_{H2} = 80\%$, from high rate series) as intrinsic absorber layers in p-i-n superstrate configuration solar cells. For each simulation, ε is temperature corrected [12],[13] such that the optical properties are those of the film at room temperature, rather than at the deposition temperature as initially measured by RTSE. Thicknesses and ε describing all layers other than the intrinsic layer are taken from a previously fabricated ~9% efficient solar cell [14] and are used as input for simulating EQE of these hypothetical devices incorporating sputtered intrinsic absorber layers [15],[16]. The simulated EQE spectra for 3000 Å thick a-Si:H and 1.8 μm thick nc-Si:H absorbers are plotted in Fig. 10. At short wavelengths, EQE of simulated cells using both sputtered a-Si:H and nc-Si:H

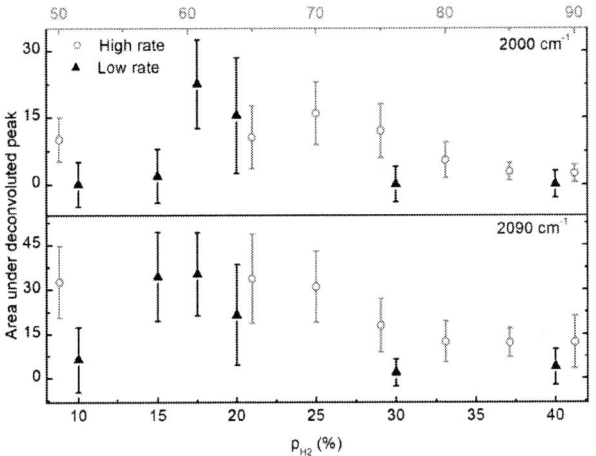

Fig. 9. Area under deconvoluted 2000 and 2090 cm^{-1} absorption peaks as a function of p_{H2} for low (lower horizontal axis) and high (upper horizontal axis) rate series.

absorber layers converges. The simulation incorporating sputtered nc-Si:H has increased EQE at longer wavelengths, due to the much lower bandgap energy of nc-Si:H compared to a-Si:H. As in the case of PECVD materials, sputtered a-Si:H and nc-Si:H could make suitable tandem junction PV device partners.

The EQE simulations indicate that optical performance of PV devices using sputtered Si:H are comparable to those traditionally made by PECVD [17]-[20], although the electronic quality of these sputtered Si:H films have not been evaluated. When assessed in aggregate, the results presented indicate that the lower rate samples are of considerably less practical interest for single and tandem junction PV due to both

poor material quality and deposition rates not practical for fabricating 0.3 to 1.8 μm thick absorber layers. Predominant Si-H$_2$ bonding seen in both series for a-Si:H is also not desirable for absorber layers [21],[22]. However, sputtered a-Si:H may still be suitable for grain boundary and surface passivation in nc-Si:H and c-Si. In particular, sputtering at low power and high pressure, such as the conditions used in the low rate series, may be well suited for passivating c-Si in HIT cells due to the lower ion bombardment damage of the underlying crystalline surface. For HIT structures, the total thickness of the undoped + doped a-Si:H layers are typically ≤ 100 Å thick so even these low deposition rates do not preclude inclusion in devices. Spectra in ε for low and high rate sputtered a-Si:H are similar to that reported for HIT cell passivation [23], indicating that similar density material may be device relevant, although the quality of the material for electronic passivation has not yet been assessed.

IV. CONCLUSION

RF sputtered Si:H undergoes phase transformation from amorphous to mixed-phase and then to single-phase nanocrystalline during growth depending on the deposition parameters. From the growth evolution diagrams, deposition rates do not decrease at the lowest p_{H2} at which crystallites nucleate. Si-H bonding is identified from IR spectra in ε, with higher rate material exhibiting greater C_H indicating more effective hydrogen incorporation relative to the low rate samples. All a-Si:H films have more pronounced features at 2090 cm^{-1} relative to that observed at 2000 cm^{-1} which is considered undesirable for absorber layers in PV devices. EQE simulations of hypothetical devices with sputtered a-Si:H and nc-Si:H have optical performance comparable to traditionally produced PV. Low deposition rate films, with low ion bombardment of the underlying surfaces, may also be applicable for depositing a-Si:H in HIT cells. If of suitable passivation quality for HIT cells and overall electronic quality for absorbers, sputter deposition of Si:H could provide a pathway to lowering fabrication costs by eliminating the need for hazardous Si-carrying precursor gases.

ACKNOWLEDGEMENT

This work was supported by University of Toledo start-up funds, the Ohio Department of Development (ODOD) Ohio Research Scholar Program (Northwest Ohio Innovators in Thin Film Photovoltaics, Grant No. TECH 09-025), and the Office of Naval Research (Grant No. 11847944).

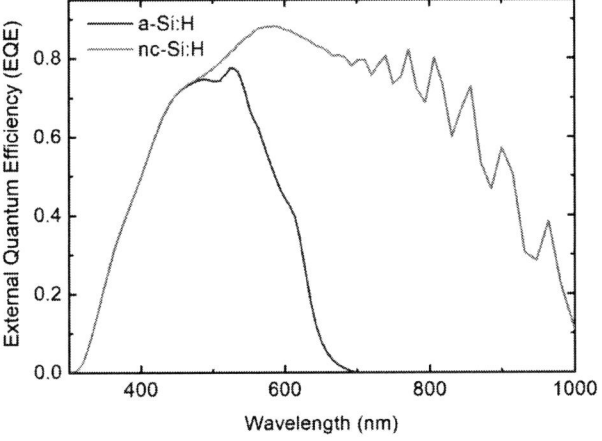

Fig. 10. External quantum efficiency (EQE) simulated for p-i-n single junction devices using room temperature spectra in ε for 3000 Å thick a-Si:H ($p_{H2} = 70\%$) and 1.8 μm thick nc-Si:H ($p_{H2} = 80\%$) absorbers.

REFERENCES

[1] P. Dutta, S. Paul, S. Tripathi, Y. Chen, S. Chatterjee, V. Bommisetty, D. Galipeau, and A. Liu, "Comparative study of nc-Si:H deposited by reactive sputtering using crystalline and

978-1-5090-5606-4/17 $31.00 © 2017 IEEE   2586

polycrystalline silicon targets," in *33rd IEEE Photovoltaic Specialist Conference*, 2008, p. 4.

[2] R. W. Collins, A. S. Ferlauto, G. M. Ferreira, C. Chen, J. Koh, R. J. Koval, Y. Lee, J. M. Pearce, and C. R. Wronski, "Evolution of microstructure and phase in amorphous, protocrystalline, and microcrystalline silicon studied by real time spectroscopic ellipsometry," *Solar Energy Materials & Solar Cells,* vol. 78, pp. 143-180, 2003.

[3] L. R. Dahal, J. Li, J. A. Stoke, Z. Huang, A. Shan, A. S. Ferlauto, C. R. Wronski, R. W. Collins, and N. J. Podraza, "Applications of real-time and mapping spectroscopic ellipsometry for process development and optimization in hydrogenated silicon thin-film photovoltaics technology," *Solar Energy Materials & Solar Cells*, vol. 129, pp. 32-56, 2014.

[4] L. K. Gautam, M. M. Junda, H. F. Haneef, R. W. Collins, and N. J. Podraza, "Spectroscopic ellipsometry studies of n-i-p hydrogenated amorphous silicon based photovoltaic devices," *Materials*, vol. 9, pp. 1-23, 2016.

[5] A. Ferlauto, G. Ferreira, R. Koval, J. Pearce, C. Wronski, R. Collins, M. Al-Jassim, and K. Jones, "Evaluation of compositional depth profiles in mixed-phase (amorphous + crystalline) silicon films from real time spectroscopic ellipsometry," *Thin Solid Films,* vol. 455, pp. 665-669, 2004.

[6] N. Podraza, J. Li, C. Wronski, E. Dickey, M. Horn, and R. Collins, "Analysis of $Si_{1-x}Ge_x$:H thin films with graded composition and structure by real time spectroscopic ellipsometry," *Phys. Status Solidi,* vol. 205, pp. 892-895, 2008.

[7] D. Adhikari, M. M. Junda, S. X. Marsillac, R. W. Collins, and N. J. Podraza, "Nanostructure evolution of magnetron sputtered hydrogenated silicon thin films," *Journal of Applied Physics,* 2017 (Submitted).

[8] M. H. Brodsky, M. Cardona, and J. J. Cuomo, "Infrared and Raman spectra of the silicon-hydrogen bonds in amorphous silicon prepared by glow discharge and sputtering," *Physical Review B,* vol. 16, pp. 3556-3571, 1977.

[9] A. A. Langford, M. L. Fleet, B. P. Nelson, W. A. Lanford, and N. Maley, "Infrared absorption strength and hydrogen content of hydrogenated amorphous silicon," *Physical Review B,* vol. 45, pp. 13367-13377, 1992.

[10] E. C. Freeman and W. Paul, "Infrared vibrational spectra of rf-sputtered hydrogenated amorphous silicon," *Physical Review B,* vol. 18, pp. 4288-4300, 1978.

[11] H. Fujiwara, J. Koh, P. I. Rovira, and R. W. Collins, "Assessment of effective-medium theories in the analysis of nucleation and microscopic surface roughness evolution for semiconductor thin films," *Physical Review B* 61, p. 10832, 2000.

[12] N. J. Podraza, C. R. Wronski, M. W. Horn, and R. W. Collins, "Dielectric functions of a-$Si_{1-x}Ge_x$:H versus Ge Content, temperature, and processing: Advances in optical function parameterization," *Materials Research Society Symposium Proceedings.* Vol. 910, 2006.

[13] R. W. Collins and A. S. Ferlauto, in: *Handbook of Ellipsometry*, edited by H. G. Tomkins and E. A. Irene (William Andrew, Norwich, NY), pp. 92-235, 2005.

[14] M. M. Junda, A. Shan, P. Koirala, R. W. Collins, and N. J. Podraza, "Spectroscopic ellipsometry applied in the full p-i-n a-Si:H solar cell device configuration," *IEEE Journal of Photovoltaics*, vol. 5, pp. 307-312, 2015.

[15] F. Abeles, "Recherche sur la propagation des ondes electromagnetiques sinusoidales dans les milieux stratifies. Applications aux couches minces," *Annales de Physique (Paris)*, vol. 5, pp. 596-640, 1950.

[16] F. Leblanc, J. Perrin, and J. Schmitt, "Numerical modeling of the optical properties of hydrogenated amorphous silicon based pin solar cells deposited on rough transparent conducting oxide substrates," *Journal of Applied Physics*, vol. 75, pp. 1074-1087, 1994.

[17] F. Leblanc, J. Perrin, and J. Schmitt, "Numerical modeling of the optical properties of hydrogenated amorphous silicon based pin solar cells deposited on rough transparent conducting oxide substrates," *Journal of Applied Physics*, vol. 75, pp. 1074-1087, 1994.

[18] P. Aryal, J. Chen, z. Huang, L. R. Dahal, M. N. Sestak, D. Attygalle, R. Jacobs, V. Ranjan, S. Marsillac, and R. W. Collins, "Quantum efficiency simulations from on-line compatible mapping of thin-film solar cells," *Proc. 37th IEEE Photovoltaics Specialists Conference*, 2011, p. 6.

[19] S. N. Agbo, S. Dobrovolskiy, G. Wegh, R. A. C. M. M. van Swaaij, F. D. Tichelaar, P. Sutta, and M. Zeman, "Structural analyses of seeded thin film microcrystalline silicon solar cell," *Progress in Photovoltaocs: Research and Applications*, vol. 22, pp. 346-355, 2012.

[20] A. V. Shah, H. Schade, M. Vanecek, J. Meier, E. Vallat-Sauvain, N. Wyrsch, U. Kroll, C. Droz, and J. Bailat, "Thin-film silicon solar cell technology," *Progress in Photovoltaocs: Research and Applications*, vol. 12, pp. 113-142, 2004.

[21] A. H. M. Smets, W. M. M. Kessels, and M. C. M. van de Sanden, "Vacancies and voids in hydrogenated amorphous silicon," *Applied Physics Letters*, vol. 82, pp. 1547-1549, 2003.

[22] J. Melskens, A. H. M. Smets, M. Schouten, S. W. H. Eijt, H. Schut, and M. Zeman, "New insights in the nanostructure and defect states of hydrogenated amorphous silicon obtained by annealing," *IEEE Journal of Photovoltaics*, vol. 3, pp. 65-71, 2013.

[23] H. Fujiwara, T. Kaneko, and M. Kondo, "Application of hydrogenated silicon oxide layers to c-Si heterojunction solar cells," *Applied Physics Letters*, vol. 91, pp. 133508-1-3, 2007.

High Quality and Thin Silicon Wafer for Next Generation Solar Cells

Yoshio Ohshita[1], Takuto Kojima[2], Ryota Suzuki[2], Kosuke Kinoshita[2], Tomoyuki Kawatsu[3], Kyotaro Nakamura[2], and Atsushi Ogura[2]

[1]Toyota Technological Institute, Nagoya, Aichi, 468-8511, Japan
[2]Meiji University, Kawasaki 214-8571, Japan
[3]Komatsu NTC Ltd., Nanto, Toyama 939-1595, Japan

Abstract — The high quality and thin Si wafer technology for the future higher conversion efficiency and lower cost crystalline silicon solar cells are realized. The high minority carrier lifetimes even after the processes are obtained by controlling the Czochralski growth condition, which prevents the interstitial oxygen segregation enhanced by the substitutional carbon. The Cz ingot is sliced by the advanced diamond multi-wire saw technology and the thin wafers with relatively thin damaged layer and 100μm kerf-loss are realized. The thin wafer with low kerf-loss decreases the wafer cost and improve the cell performance. The thin wafer solar cells are fabricated using the PERC processes, and these technologies are evaluated from the view point of photovoltaic.

Index Terms — crystal silicon wafer, crystal growth, thin wafer, low kerf loss, photovoltaic cells, minority carrier lifetime.

I. INTRODUCTION

The mono- and multi- crystalline silicon (Si) solar cells have dominated the present photovoltaic market, and they will be expected as the future workhorse of main PV power generation. The next generation crystalline silicon solar cells require the higher conversion efficiency over 23%, lower cost as $0.5/W and higher reliability. To realize the higher efficiencies with lower production cost and higher reliability, we are developing many innovative manufacturing technologies for crystalline silicon solar cells and evaluating new materials and processes under NEDO project (Fig. 1). We will establish intelligent technologies for realizing the future wafer-based Si solar cells in corporation of universities.

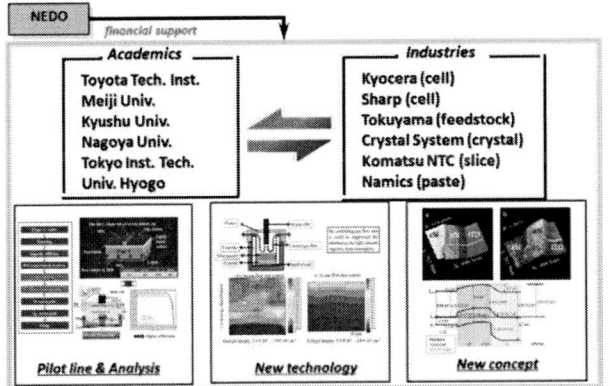

Fig. 1 NEDO project: Development of high performance and reliable PV modules to redcuce levellized cost of energy.

One of our targets is the realize of the high quality Si wafer with low cost. Recently, the conversion efficiencies of crystalline Si solar cells, such as, back contact, TOPcon, and hetero-back contact, are exceeding 25% by several research institutes [1-5]. To realize such high conversion efficiency solar cells, the high quality Si crystals, which have long minority carrier lifetime after the fabrication processes, must be used. Nowadays, the Czochralski grown (Cz) wafers are widely used, because a relatively high conversion efficiency can be obtained with low cost. However, there are many residual oxygen atoms in the crystal, and some of them will segregate as SiO_2 during the solar cell fabrication. The residual impurities such as carbon also enhance the SiO_2 precipitates formation. Thus, the degradation of solar cell performance due to this segregation must be suppressed. Another important issue is the slicing technology. Now, the diamond multi-wire slicing is the main stream for obtaining the single crystal Si wafers. To improve the conversion efficiency of solar cells and to decrease the slice cost, thin wafer with the lower damage and lower kerf-loss are required.

In this paper, the high quality and thin Si wafer technologies for the future higher conversion efficiency and lower cost crystalline Si solar cells will be discussed. To realize the uniform profile of minority carrier lifetime in the ingot grown by Cz method and the high minority carrier lifetime after the processes, the effects of the residual carbon and growth condition on the interstitial O segregation are studied. The photoluminescence (PL) mapping is used for the evaluation of the damages at the surface of wafers sliced using the diamond multi-wire saw. 100μm thin wafer and 100μm kerf-loss are realized by the advanced technology. The thin wafer solar cells are fabricated to evaluate the Si wafer quality from the view point of solar cells, and to develop the manufacturing technologies.

II. HIGH QUALITY CRYSTAL

A. Low degradation after processes

Cz grown Si wafers contain a large amount of residual oxygen (O_i) incorporated during the crystal growth, which exists as interstitial ones. Some of these O_i precipitate as silicon dioxide (SiO_2) due to the high temperature processes during the cell fabrication. This precipitation generates crystal defects. Those defects act as recombination centers, which

978-1-5090-5606-4/17 $31.00 © 2017 IEEE

decrease the short circuit current (I_{sc}) and the open circuit voltage (V_{oc}) in solar cell performances. Therefore, it is important to prevent this degradation of solar cell properties inducted by the precipitation during the cell fabrication.

However, it is difficult to decrease the concentration of O_i in the Cz crystal. But, the SiO_2 precipitation can be suppressed by controlling the Si crystal growth conditions, although there are some amount of residual O_i and substitutional carbon (C_s) atoms in the crystal. The effects of new growth condition were evaluated by the implied Voc changes after the thermal treatment corresponding to the n-type bifacial solar cell as shown in Fig. 2. Several types of phosphorous (P) doped Si wafers were used. After the texturing process, the boron (B) was dopes to fabricate the emitter layer using the boron − silicate glass (BSG) deposited by the atmosphere pressure chemical vapor deposition (APCVD). Then, P was diffused from the phosphorus-silicate glass (PSG). As the passivation layers, SiO_2 was formed by the thermal oxidation and SiNx was deposited using the plasma enhanced chemical vapor deposition (PECVD) using the SiH_4 and NH_3 gas system. The O_i concentrations in these wafers were almost constant around 20 ~ 40 ppma, and the C_s were changed from ~0.04 to ~1 ppma.

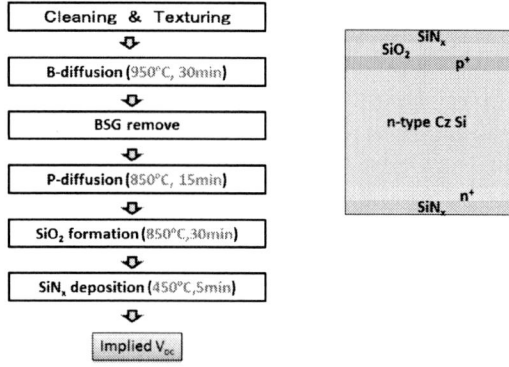

Fig. 2 Thermal budget for fabricating n-type bi-facial solar cell.

The relationships between the implied V_{oc} and the C_s concentration are shown in Fig. 3. The implied V_{oc} dramatically decreases with increasing C_s concentration. The implied V_{oc} is determined by the number of recombination centers (N) as follows,

$$\text{Implied } V_{oc} \propto \ln(\tau_{eff}) \propto \ln(1/N)$$

Therefore, with increasing C_s concentration, the effective minority carrier lifetime was decreased and the performance of solar cell was degraded. This means that a SiO_2 precipitation during the cell fabrication processes was enhanced by C_s and that they increased the recombination centers. However, although the O_i concentrations are almost the same, the absolute values of the implied V_{oc} are increased depending on the crystal growth conditions in this experiment. Ingots A and B were grown under the different conditions, while the residual O_i concentrations were almost the same in these crystals. In the case of ingot A, the higher Voc was obtained, which indicates that the precipitations of oxygen impurities were suppressed compared with that in another ingot even with the same thermal budget. The high minority carrier lifetime after the solar cell fabrication processes can be realized by the growth condition optimization, along with decreasing the Cs concentration.

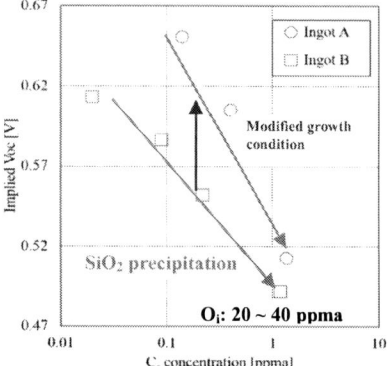

Fig. 3 Relationships between the C_s and implied V_{oc}. The implied V_{oc} is improved by controlling the growth condition and decreasing the C_s concentration.

B. High uniformity of minority carrier lifetime

To decrease the cost of Si ingot, the uniformity of lifetime in the grown one is another important issue. Because of the non-uniformity of electrical properties, some part of ingot cannot be used for the cell production. The minority carrier lifetime and resistivity profiles in the n-type Si crystal grown by the advanced Cz method are shown in Fig. 4. The uniform lifetime from the top to bottom was realized in Cz-A by controlling the growth condition, based on the above mentioned results. The resistivity uniformity is also high. They decrease the waste of ingot, which reduces the production cost.

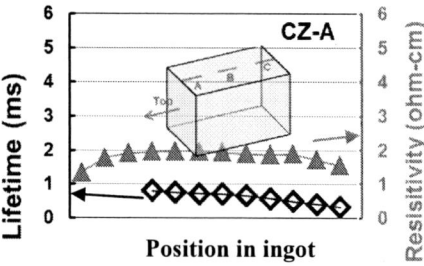

Fig. 4 Uniform profiles of minority carrier lifetime and resistivity in the ingot grown by the advanced Cz method.

III. THIN WAFER AND LOW KERF-LOSS DIAMOND MULTI-WIRE SAW

The next step for crystal silicon solar cell fabrication is the slicing and the diamond multi-wire slicing technology is widely used (Fig. 5). KOMATSU NTC Ltd is developing the advanced crystal Si ingot slicing equipment with the diamond-coated multi-wire saw. By decreasing the diameter of the wires and controlling of slicing condition, such as, the wire speed and the tension, it realizes the lower damage, thinner Si wafers, and thinner kerf-loss.

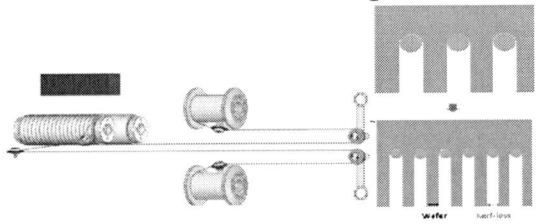

Fig. 5 Schematic image of diamond multi-wire saw equipment. By decreasing the wire diameter, the low kerf-loss and thin wafer are realized.

A. High quality slicing of Si ingot

After slicing, there were damaged areas near the surface of Si wafer. Most of them can be removed by the damage and texturing etching. Although the fixed abrasive sawing reduces the surface damages as compared with the slurry based sawing, some of them remained after the etching and they deteriorate the solar cell performances. Thus, these damage layers, which remain after the etching processes, must not be generated during the slicing processes, in order to fabricate the high conversion efficiency crystalline Si solar cells. However, the detecting of these remained damages is difficult by using the conventional optical microscope. Therefore, in this study, photo-luminescence (PL) imaging technology was adopted to evaluate these damages, and the slicing processes were improved based on the obtained results. The InGaAs imager was used at -80°C. The detection area was about 900-1700nm square and the pixel size and number were1.7μm and 640x512. Therefore, the spatial resolution was around 10nm.

The PL image obtained with 532nm laser excitation is shown in Fig. 6. These wafers were etched using the conventional etching process. They are the maps of band edge emission intensity, which correspond to the minority carrier lifetime in the crystal. The dark area means the short minority carrier lifetime region, where there are relatively larger amount of recombination centers. The damage areas are observed in the edge region of the wafers. These damages are not bulk defects, and they were introduced during the slicing process. The appropriate slice condition suppressed these damage area generation. The sawing process causes the generation of dislocations, which act as active recombination centers of minority carriers. They may deteriorate the solar cell

performances of the higher conversion efficiency such as the bifacial n-type single crystalline Si solar cell.

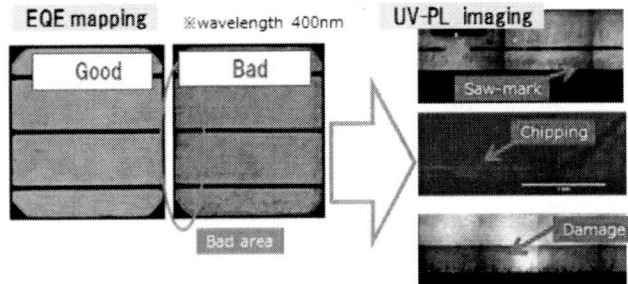

Fig. 6 PL images using 532nm laser. There are saw mark, chipping, and damage layers.

To evaluate the effect of these damages and the impact of advanced slicing technology on the solar cell properties, the n-type wafer bifacial cells were fabricated. The schematic cell structure and the external quantum efficiency (EQE) of fabricated solar cells is shown in Fig. 7. When there were remained damages after the damage etching, the EQE was dramatically degraded in the short wave length region, indicating that these defects must be removed.

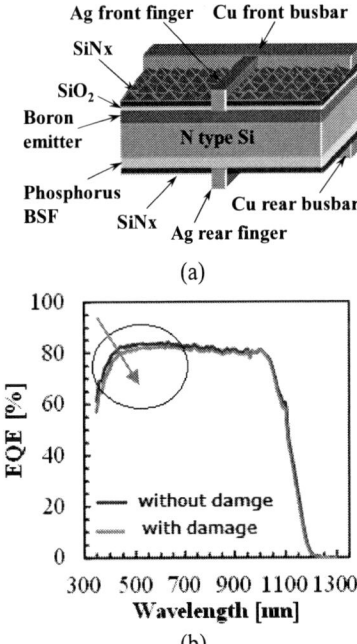

Fig. 7 (a): N-type bifacial solar cell structure. (b): External quantum efficiency of fabricated solar cells. The EQE in short wave length decreases due to the damaged layers which remian after the etching process. On the other hand, EQE increase due to the improved slice technology.

The solar cell performances are shown in Fig. 8. When there are damage in the crystal after the etching process, the open ciruicut voltage is decreased and the relativel low conversion

978-1-5090-5606-4/17 $31.00 © 2017 IEEE   2590

efficiency is obtained. The higher conversion efficinecy is realized when the damage was suppressed slicing the wafer using the improved conditions.

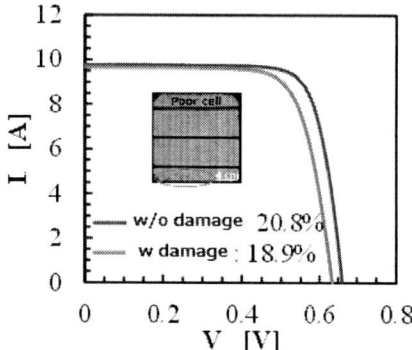

Fig. 8 Current-Voltage relationships of n-type bifacial solar cells, fabricated using Si wafers with and without the remained saw damages confirmed by PL. The higher open circuit voltage is obtained due to the lower damage area.

The relationship between the slicing condition and the damage area was clarified and the slice technology was improved. This allows the high conversion efficiency solar cells.

B. Thin wafer slicing with low kerf loss

When the wafer thickness becomes thinner, the higher Voc is expected as well as the cost reduction of crystalline Si solar cells. The 156 x 156 mm^2 size wafers with 100μm thickness were sliced from the bulk of Cz crystal by the multi wire slicing equipment. To realize such thin wafers with low kerf-loss, the coolant supply, angle processing in main roller groove shape were improved, the appropriate thickness of the plating and the concertation of diamond (Fig. 9) were determined.

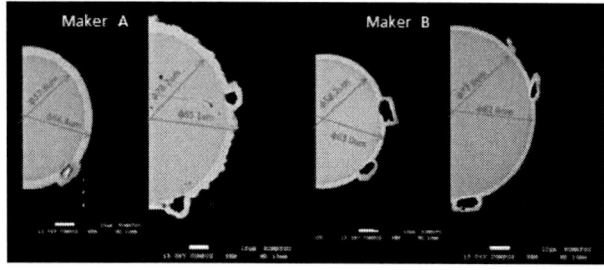

Fig. 9 Cross-sectional images of diamond wire. The thickness of plating and the concertation of diamond on the surface are controlled.

They decrease the kerf-loss. The present kerf-loss is around 100μm, and our next targets are 75μm and then 60μm (Fig. 10). This lower kerf-loss means the ability of thin wafer

fabrications. 100μm kerf-loss thickness slicing technology realized the 100μm thin wafer as shown in Fig. 11. As decreasing the kerf-loss, the wafer thickness which can be produced, becomes thinner as 75μm thickness and then 50μm.

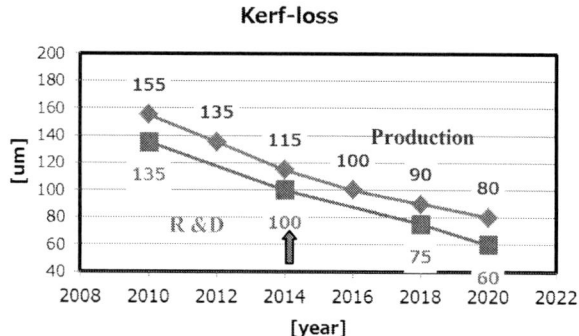

Fig. 10 Technology trend of kerf-loss by our diamond multi-wire saw system. The present kerf-loss is around 100μm. The next target is 75μm.

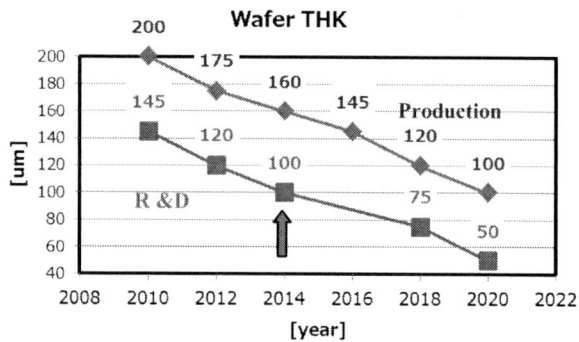

Fig. 11 Technology target for wafer thickness obtained by the diamond multi-wire slicing. Now, the 100μm thickness slicing is realized. The next challenge is 75μm.

To evaluate the wafer quality of this thin wafer and to develop the thin wafer solar cell fabrication processes, we fabricated PERC solar cells using ultra-thin wafer, as shown in Fig. 12, whose thickness was 105μm after slicing and 88μm after texturing, and confirmed that the cell efficiency of ultra-thin PERC was almost equal to normal thick cell. The conversion efficiency of the thin cell was around 19% [6]. This result proves the feasibility of our developed ultra-thin PERC to realize high efficiency and low cost.

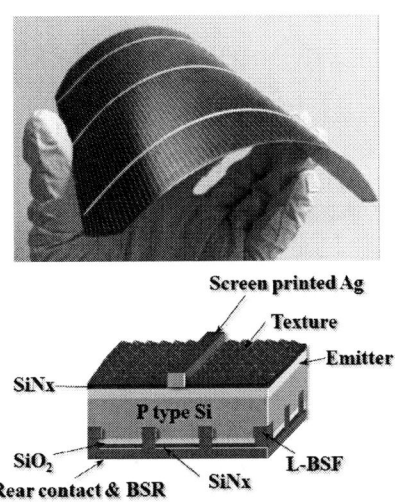

Screen printed Ag
Texture
Emitter
SiNx
P type Si
SiO₂
Rear contact & BSR
SiNx
L-BSF

Fig. 12 PERC solar cell using ultra-thin wafer, whose thickness was 105μm after slicing and 88μm after texturing.

VI. CONCLUSION

The high quality and thin Si wafer technologies for the future higher conversion efficiency and lower cost crystalline Si solar cells were realized. The effects of the residual carbon and growth condition on the interstitial O segregation were studied. Based on the obtained results, the uniform profile of minority carrier lifetime in the ingot grown by the advanced Cz method was obtained, and the high minority carrier lifetime in the bulk after the processes was realized. PL image showed the damage area induced by the slicing, which cannot be removed after the damage etching process. The improved slicing condition prevented the formation of these defects and realized the higher conversion efficiency. By decreasing the wire diameter and controlling the slice condition, the 100μm thin wafer with 100μm kerf-loss was realized by the diamond multi-wire slicing technology, and the next targets are 75μm thickness and kerf-loss. These advanced crystalline Si wafer technologies will reduce the wafer cost and improve the cell performances.

ACKNOWLEDGEMENT

The authors would like to thank sincerely co-workers for their good results and fruitful discussion. This work was supported by New Energy and Industrial Technology Development Organization (NEDO) under the Ministry of Economy, Trade and Industry (METI). We also would like to thank Motoo Morimura, Norihiko Maeda, Shimako Naito, Takefumi Kamioka and Mari Aoki for the cell fabrications.

REFERENCES

[1] K. Masuko, M. Shigematsu, T. Hashiguchi, D. Fujishima, M. Kai, N. Yoshimura, T. Yamaguchi, Y. Ichihashi, T. Mishima, N. Matubara, T. Yamanishi, T. Takahama, M. Taguchi, E. Maruyama, and S. Okamoto, IEEE J-PV 4 (2014) 1433.

[2] D. D. Smith, P. Cousins, S. Westerberg, R. De Jesus-Tabajonda, G. Aniero, and Yu-Chen Shen, IEEE J-PV 4 (2014) 1465.

[3] J. Nakamaura, N. Asano, T. Hieda, C. Okamoto, H. Katayama, and K. Nakamura, IEEE J-PV 4 (2014) 1491.

[4] K. Yamamoto, D. Adachi, H. Uzu, T. Uto, T. Irie, M. Hino, M. Kanematsu, H. Kawasaki, K. Konishi, R. Mishima, K. Nakano, T. Terashita, K. Yoshikawa, M. Ichikawa, T. Kuchiyama, T. Suezaki, T. Meguro, N. Nakanishi, M. Yoshimi, D. Schroos, N. Valckx, N. Menou, J. L. Hernández, Proc. 31st EU PVSEC, (2015) 1003.

[5] Kaneka Corporation News Release. World's Highest Conversion Efficiency of 26.33% Achieved in a Crystalline Silicon Solar Cell http://www.kaneka.co.jp/kaneka-e/images/topics/1473811995/1473811995_101.pdf (accessed 2017-01-26).

[6] K. Nakamura, A. Tanizaki, K. Okamoto, Y. Kawamoto, Y. Ohshita, presented at 25th International Photovoltaic Science & Engineering Conference, 2015.

First Demonstration of Radial Junction Silicon Nanowire Solar Mini-Modules Prepared by PECVD and Laser Scribing

Mutaz Al-Ghzaiwat[a], Martin Foldyna[a], Takashi Fuyuki[a], Wanghua Chen[a], Erik V. Johnson[a], Jacques Meot[b], and Pere Roca i Cabarrocas[a]

[a] LPICM, CNRS, Ecole Polytechnique, Université Paris-Saclay, 91128 Palaiseau, France.
[b] SOLEMS, 3 rue Léon Blum, 91120 Palaiseau, France.

Abstract — Based on recent advancements of radial junction silicon nanowire (RJ SiNWs) solar cells, a demonstration of 5x5 cm² RJ SiNW solar mini-module is presented in this work. The SiNW devices were grown by plasma-assisted vapor-liquid-solid technique at low temperature in a plasma-enhanced chemical vapor deposition reactor. The 5x5 cm² mini-modules have been obtained using a commercial laser scribing apparatus. The laser scribing insures a monolithic integration of electrically separated cells. We have obtained a power generation of 10 mW from 5 individual cells of total active area of 8.6 cm². The mini-module has an open-circuit voltage of 3.85 V. The performance was evaluated using solar simulator and short comings (high series resistance) were analyzed using a home-made electroluminescence setup.

Index Terms: Mini-modules, Radial junction, a-Si:H, silicon nanowire, VLS method, PECVD, laser scribing.

I. INTRODUCTION

Solar cells based on silicon thin film have been under development to achieve high energy generation, low material consumption and low fabrication cost [1]. Moreover, a promising research field based on radial junction silicon nanowires (RJ SiNWs) brings several advantages over the traditional planar junction solar cells. High built-in electrical field due to the ultra-thin absorber layer (~100 nm), enhanced light trapping and anti-reflection properties, the efficient absorption of light and the photo-generated carriers collection being decoupled due to the unique geometry of NWs [2].

Recently, the energy conversion efficiency (η) of SiNW solar cells has reached ~9.2 % for 0.126 cm² area [3]. Based on that, upscaling and extending compatibility with industrial processes are required to meet demands of PV market.

Laser scribing of thin film solar devices is a common approach towards large-area fabrication [4]. It is a non-contact process where a selective and precise removal of thin film materials can be achieved. In addition, a monolithically integration of a series connected and isolated solar cells can be realized. The laser scribing process consists of three main steps: (i) P1, a selective laser scribing of the back contact to insure electrical isolation between the separated TCO segments. (ii) P2, a selective laser scribing of PIN junction material to make a contact for the top and back contacts. (iii) P3, a selective scribing of the top contact to insure isolation of each individual solar cell. Technical parameters of laser scribing (e.g. wavelength and pulse duration) control the quality of the scribed area are described elsewhere [5].

In this work, we have developed SiNW mini-modules based on plasma-assisted chemical vapor deposition (PECVD) grown RJ and laser scribing cell separation.

II. SAMPLE FABRICATION

Silicon NWs were grown on the top of SnO_2:F (FTO) substrates using plasma-assisted vapor-liquid-solid (VLS) growth process [2], where 600 nm thick FTO was deposited on 2.2 mm glass substrates using chemical vapor deposition (CVD) technique. After loading the sample inside the plasma enhanced chemical vapor deposition (PECVD) reactor, a hydrogen (H_2) plasma was ignited for 2 min at RF power of 5 W, the H_2 pressure was fixed at 600 mTorr at flow rate of 100 standard cubic centimeters per minute (sccm) and the substrate temperature was 200 °C.

Major et al. have shown that FTO surfaces get reduced (under H_2 plasma treatment) to yield elemental tin (Sn) [6]. Banerjeet et al. have studied the degradation of FTO film under H_2 plasma treatment; they have reported that under temperatures higher than 180 °C, surface change occur on FTO films. However, the sheet resistance increases significantly when the H_2 plasma treatment temperature increases [7]. For the work presented in this paper, we have grown SiNWs from the produced metallic Sn while keeping the sheet resistance of FTO ~12 Ω/□.

As shown in Fig. 1, the growth of SiNWs was initiated by introducing a gas mix of silane (SiH_4), H_2, and trimethylboron (TMB). Typically, the length of the p-type self-assembled SiNWs is around 600-800 nm, and the diameters of top and bottom are ~20 and ~50 nm, respectively. Subsequently, a 100 nm of intrinsic a-Si:H absorber layer was deposited [8]. Afterwards, n-type μc-SiO_x:H layer was deposited on top of the absorber layer in the presence of SiH_4, H_2, Phosphine (PH_3) and CO_2.

To build mini-modules based on RJ SiNWs, substrates of 5x5 cm² (600 nm FTO on 2.2 mm thick glass) have been used. Prior to loading the substrates into the PECVD reactor, laser scribing using YAG (1.064 μm) nanosecond pulsed laser with Q-switch was conducted on the FTO back contact layer (step

P1). The scribed width on the FTO layer was around 120 μm as shown in Fig. 2.a. After loading the samples in the PECVD reactor, a PIN RJ was deposited using the previously discussed process. Afterwards, the substrates were scribed using a YAG (532 nm) nanosecond pulsed laser with Q-switch (step P2). Laser spots on PIN RJ layer were obtained with ~50 μm width as shown in Fig. 2(b). Finally, an indium-tin-oxide (ITO) was sputtered with nominal thickness of 300 nm (measured on flat surface) on top of RJ SiNWs. The pattern of the ITO was defined using lift off technique (step P3), where a paste of titanium dioxide (TiO_2) was spread (width is around 450 μm) on the surface of the PIN RJs, and followed by sputtering of ITO layer. Afterwards, the paste is removed by rinsing the samples in alcohol. The deposition of patterned ITO on top of the PIN RJ is shown in Fig. 2(c).

A reference PIN solar cell has been fabricated to compare the performance with the fabricated mini-modules. We have used the same substrates (5x5 cm^2) without using P2 and P3 steps. Instead, ~ 280 nm of ITO was sputtered on the top of PIN RJs through a shadow mask with circular pads of 2 mm and 4 mm diameters.

Prior the characterization, all samples have been annealed for 1 hour at 180 ˚C in air.

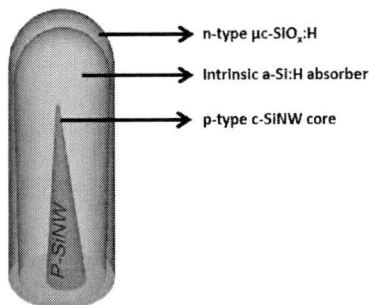

Fig. 1.   Schematic showing the plasma-assisted VLS grown p-type SiNW, coated with 100 nm intrinsic a-Si:H absorber layer and n-type μcSiOx layer.

III. RESULTS AND DISCUSSION

The electrical performance of the fabricated SiNW solar mini-modules is shown in the Fig. 4. The active area of the mini-module is defined by 5 interconnected cells with dimensions of 4.3x0.4 cm^2 (on 5x5 cm^2 substrates) as shown in Fig. 3. The first results show high open-circuit voltage of 3.85 V (0.77 per cell) indicating a working electrical interconnection between the individual cells. In addition, the Voc for each cell is in good agreement with the reference sample that has been discussed previously. However, due to the presence of a relatively low shunt and high series resistances, a poor fill factor (FF) of 27.56 % and relatively low efficiency of 1.17 % was obtained.

High series resistance of these devices originates (mainly) from the insufficient removal of PIN RJ SiNWs as shown in the SEM micrographs in Fig. 2(b). The SEM micrograph

shows clear separation and non-uniformity of areas defined by pulsed laser during scribing of PIN RJ.

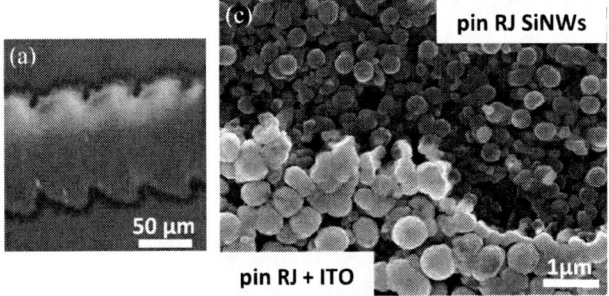

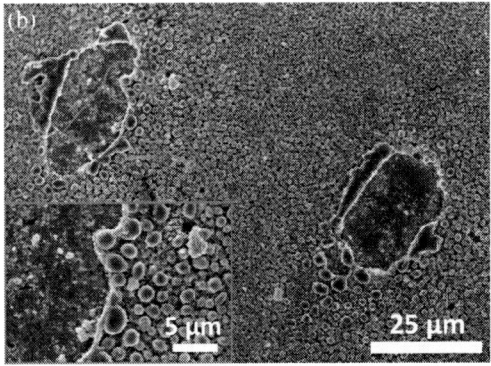

Fig. 2.   SEM micrographs of laser scribing based fabrication process. (a) Scribes on FTO layer before depositing RJ SiNWs using (1.064μm) nanosecond pulsed laser. (b) Scribes on RJ SiNWs using (532nm) nanosecond pulsed laser, inset shows zooming on the scribed area. (c) The sputtered ITO on top of SiNWs and SINWs covered by TiO_2.

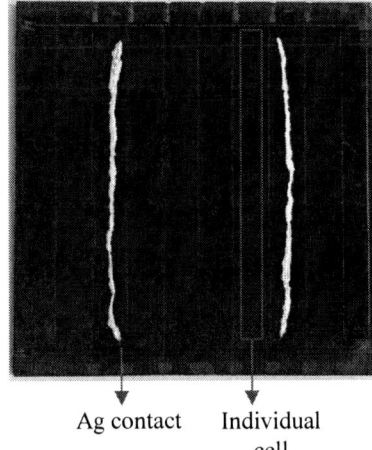

Fig. 3. RJ SiNW solar mini-module fabricated by PECVD using plasma-assisted VLS technique, with assistance of laser scribing for cells separation.

Tani et al. have studied the presence of series resistance in the top contact (TCO) of thin film modules and its role on collection of photo-generated carriers using

978-1-5090-5606-4/17 $31.00 © 2017 IEEE

electroluminescence technique (EL) [9]. Fig. 5(a) shows EL image of a RJ SiNW mini-module obtained using a charge-coupled device (CCD) camera having Si detector with a bandpass filter removing wavelengths below 900 nm, taken with an applied voltage of 13 V. Fig. 5(b) shows the EL intensity plotted along the horizontal distance of the interconnected cells [yellow dashed line in Fig. 5(a)]. At the injection edge of each individual cell, a high EL emission can be observed. However, the EL emission was decreased at the collection edge which is mainly due to the high resistive losses in the top ITO contact.

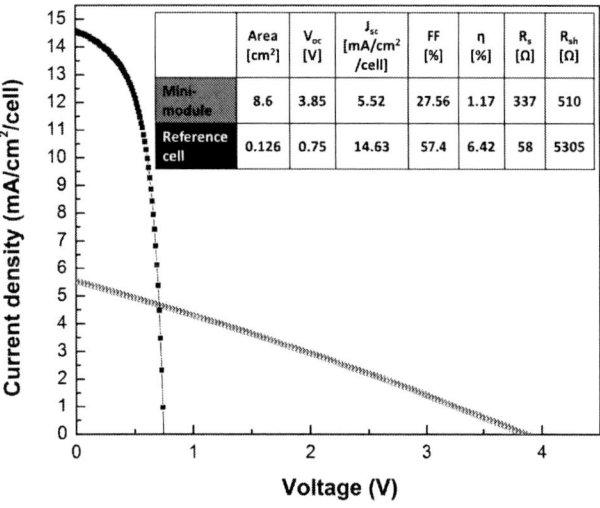

Fig. 4. J-V characteristics of mini-module based on RJ SiNWs (black curve) in comparison with reference SiNW cell (red curve). The electrical parameters are included in the graph.

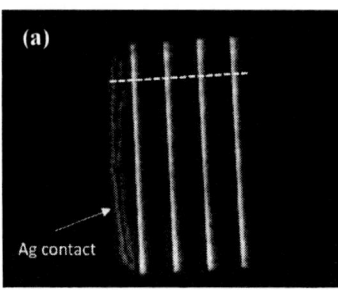

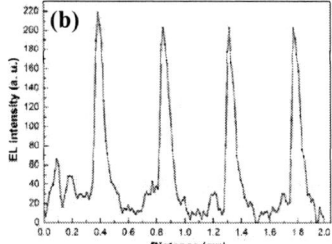

Fig. 5. (a) EL image obtained after removing the background light using ImageJ software. (b) EL profile along the yellow dashed line. The lowering of EL intensity away from the injecting part indicates high resistive losses due to ITO.

IV. CONCLUSION

We have successfully fabricated mini-modules based on RJ SiNWs. In addition, a high open-circuit voltage of 3.85 V has been obtained, thanks to laser scribing technique which allows the interconnection between SiNW solar cells. Moreover, the results demonstrate reliability of the large-area fabrication based on RJ SiNWs. However, new designs of mini-modules are being developed as well as optimizing laser scribing parameters to improve mini-modules performance. Moreover, for better carriers collection an optimized silver grid will be utilized.

ACKNOWLEDGMENT

The authors would like to thanks Solems company for its precious help with laser scribing. We acknowledge financial support of French National Research Agency within SOLARIUM project N° ANR-14-CE05-0025. The PhD thesis of Mutaz Al-Ghzaiwat has been funded by the French government through Campus France. T. Fuyuki acknowledges funding from Total via the chaire CAP.

REFERENCES

[1] F. Meillaud, M. Boccard, G. Bugnon, M. Despeisse, S. Hänni, F.-J. Haug, J. Persoz, J.-W. Schüttauf, M. Stuckelberger, and C. Ballif, "Recent advances and remaining challenges in thin-film silicon photovoltaic technology," *Materials Today*, vol. 18, issue 7, pp. 378-384, 2015.

[2] S. Misra, L. Yu, W. Chen, M. Foldyna, and P. Roca i Cabarrocas, "A review on plasma-assisted VLS synthesis of silicon nanowires and radial junction solar cells," *Journal of Physics D: Applied Physics*, vol. 47, pp. 393001, 2014.

[3] S. Misra, L. Yu, M. Foldyna, and P. Roca i Cabarrocas, "New approaches to improve the performance of thin-film radial junction solar cells built over silicon nanowire arrays," *IEEE Journal of Photovoltaicss*, vol. 5, pp. 40-45, 2015.A.

[4] Bubenzer, P. Lechner, H. Schade, and H. Rübel, "Process technology for mass production of large-area a-Si solar modules," *Solar energy materials and solar cells*, vol. 34, issues 1-4, pp. 347-358, 1994.

[5] A. Schoonderbeek, V. Schuetz, O. Haupt, and U. Stute, "Laser processing of thin films for photovoltaic applications," *Journal of Laser Micro/Nanoengineering*, vol. 5, pp. 248-255, 2010.

[6] S. Major, S. Kumar, M. Bhatnagar, and K. L. Chopra, "Effect of hydrogen plasma treatment on transparent conducting oxides," *Applied Physics Letters*, vol. 49, pp. 394-396, 1986.

[7] R. Banerjee, A. De, S. Ray, A.K. Barua, and S.R. Reddy, "Hydrogen plasma degradation of SnO2: F films prepared by the APCVD method," *Journal of Physics D: Applied Physics*, vol. 26, pp. 2144-2147, 1993.

[8] S. Misra, L. Yu, M. Foldyna, and P. Roca i Cabarrocas, "High efficiency and stable hydrogenated amorphous silicon radial junction solar cells built on VLS-grown silicon nanowires," *Solar Energy Materials and Solar Cells*, vol. 118, pp. 90-95, 2013.

[9] A. Tani, and T. Fuyuki, "Direct assessment of series resistance in thin film solar cells utilizing electroluminescence," *35th IEEE PVSC*, pp. 001689-001691, 2010.

Impact of induced defects on device performance in silicon heterojunction solar cells

Pradeep Balaji, André Augusto and Stuart G. Bowden

Arizona State University, PO Box 875706, Tempe, AZ 85287, United States

Abstract — Recent results show the potential of silicon heterojunction solar cells working as a bottom cell in tandem with wide band gap materials. The surface quality of the bottom cell is critical to reduce the recombination current and promote a good interface between the top and bottom cell. In this paper, we investigate the impact on performance of surface defects that are present after the i-p/i-n junction formation in silicon heterojunction cells. We induce controlled distribution of defects after the junction formation using a laser, which amounts to 1% of the total area of the cell. We measured an 8% relative power loss between samples with and without the induced damages. The most affected parameter is the voltage at the maximum power point that shows a drop of 13 mV after the induced damages.

Index Terms — photovoltaics, silicon heterojunction, induced defects.

I. INTRODUCTION

Silicon heterojunction solar cells are now at efficiencies over 26% [1]. Parameters such as doping concentration, thickness of passivating layers, mobility of carriers in the transparent conducting oxide and defect density play a crucial role in the heterojunction device capabilities [2], [3]. The high-quality interface between amorphous silicon (a-Si) and crystalline silicon (c-Si) leads to higher voltages compared to other silicon solar cell architectures [4]. SHJ cells are highly sensitive to defects/contaminants on the surface. Hence quality control of these devices is essential for mass production. Wafer sawing process can introduce surface defects and contaminants on c-Si. Such defects can act as nucleation sites for metal/impurity precipitation, leading to areas of high recombination which affects the overall performance of the device [5]. The minority carrier recombination at lower injection levels in these devices increase significantly due to surface contamination. It is common to find regions/spots of high recombination in these high-performance heterojunction cells [6]-[8].

In this paper, we investigate the impact of non-uniform passivation in SHJ cells on final solar cell efficiency. We induce controlled distribution of defects on the solar cell after i-p/i-n junction formation, and compare the performance of devices with and without induced damages.

II. EXPERIMENT

SHJ cells (Figure 1) were fabricated using a 6-inch n-type c-Si wafer with an area of 239 cm^2, starting thickness of ~200 μm, and bulk resistivity of ~2-4 Ω cm. Saw damage removal and texturing was performed using a mixture of KOH and GP solar

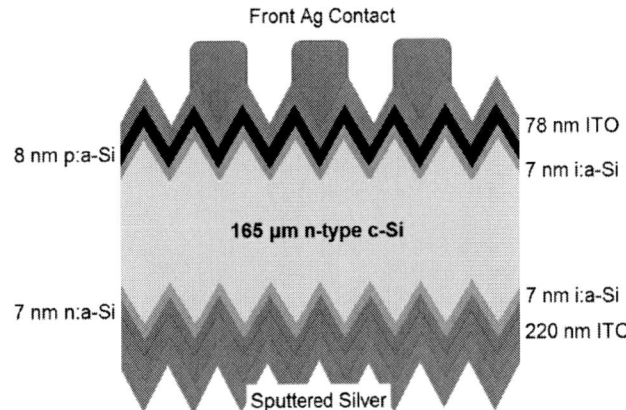

Figure 1: Schematic diagram of the heterojunction solar cells used in this study

additive, followed by wet chemical cleaning. Average wafer thickness was ~165 ± 5 μm after this process. The i-p/i-n stack was deposited via plasma enhanced chemical vapor deposition (PECVD). A nanosecond pulsed solid state laser, with a wavelength of 1064 nm was then used to induce surface damage on half of these passivated wafers to recreate areas of high recombination generally seen on silicon heterojunction solar cells. Spot sizes 0.7 mm in diameter were created on these wafers using 65% of maximum power density of the laser, 18 kHz pulse frequency with a writing speed of 70 mm/s. A total of 168 such spots were created on a specific area of these wafers as shown in figure 2, which corresponds to a total area of 1% of the solar cell. Indium tin oxide (ITO) was then sputtered on both sides and silver on the rear surface. Finally, front surface metallization was carried by screen printing silver paste. All the samples were then annealed for 2 hours at 200 °C.

Sinton photoconductance lifetime tester, steady-state photoluminescence (PL), scanning electron microscopy (SEM) were used for characterization of defects in these cells. PL intensity is proportional to the minority carrier density (Δn) and lifetime at any point in the solar cell processing [9]. Lifetime is directly correlated to the rate of recombination, which in turn is proportional to cell efficiency. The final device performance was characterized using Sinton FCT-400 I-V tester.

III. RESULTS AND DISCUSSION

Characteristic and recurring defects are observed in PL images of wafers that have undergone saw damage removal and

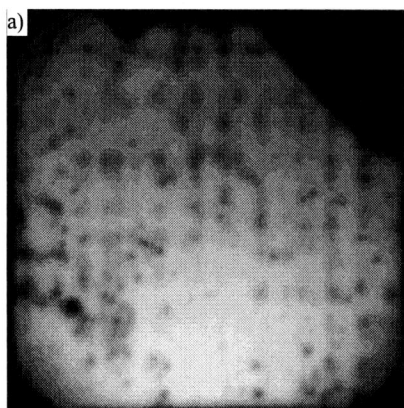

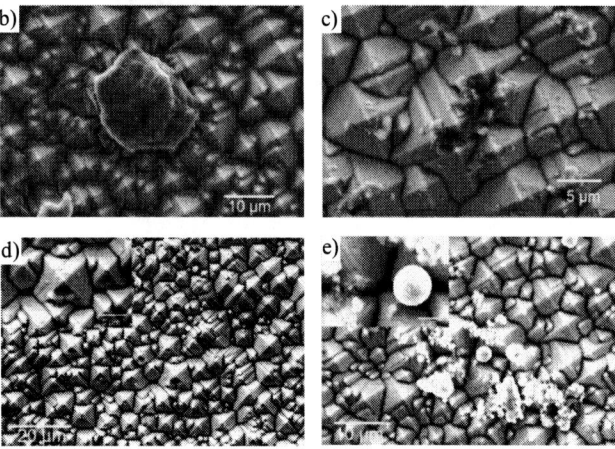

Figure 2: a) Photoluminescence image of defect rich sample used for SEM analysis. b) Non-etched silicon particles. c) Damage on passivating layer. d) Damaged pyramids seen on c-Si. e) Various debris particles.

and texturing, even after subsequent RCA-B and Piranha wet chemical cleaning. Various kind of defects (Figure 2) are observed on c-Si wafers as seen in the SEM images. User handling of the wafers, the use of carriers for deposition of a-Si layers in the PECVD tool and other uncontrollable processing variables play a major role in creating these areas of high

recombination. The PECVD deposition chambers must be cleaned and conditioned prior to the deposition of the a-Si layers to avoid debris particles on the surface of the wafer.

To study the effect of these high recombination areas on device performance, a quantitative investigation was carried out by comparing the characteristics of devices with no induced defects and devices with controlled distribution of laser induced defects. Seven samples went through the standard silicon heterojunction solar cell process as described above and seven other wafers went through the process with laser induced defects on their surfaces. The effective minority carrier lifetime and implied open circuit voltage (iV_{OC}) of these set of wafers are all the same order (Figure 4) of 4.5 ms and 735 mV respectively before damage at an injection level of 10^{15} cm^{-3}. The average lifetime and iV_{OC} of the wafers decreases to 2.8 ms and 731 mV after laser induced damage. A change in 1.7 ms lifetime corresponds to almost 37% change from the initial value. The average emitter saturation current density (J_{0e}) increased from 1.8 fA/cm^2 to 3.1 fA/cm^2. Sputtering of ITO is known to induce damage on the surface of wafers reducing the carrier lifetimes [10]. Minority carrier lifetimes were recovered by annealing the samples at 200 °C for 40 minutes. Areas of high recombination, when quantified using PL images, tend to reduce significantly after the deposition of ITO. Lateral transport of photogenerated carriers in ITO can be the reason for this [11].

After metallization, parameters such as short circuit current density J_{SC}, open circuit voltage V_{OC}, voltage and current at maximum power point, V_{MP} and J_{MP}, fill factor FF, were extracted for each cell. Figure 5 shows a comparison of these parameters for control and damaged devices. The average efficiency of the control samples was at 18.6% and the average of the damaged ones was at 18.1%, i.e. a relative decrease of 2% in the efficiency. It is evident from the I-V characteristics that the induced defects didn't have any considerable change on the short circuit current density of the wafers, which can be expected since the damaged area is around 1% of the total area. There is a relative change of only 0.5% or 4mV on the V_{OC}. The biggest impact of the damages was seen at V_{MP}, with a

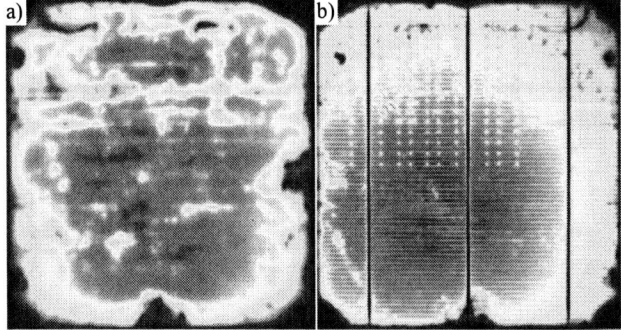

Figure 3: Representative PL images of the samples that were used for this study a) passivated wafer b) metalized wafer with laser induced defects

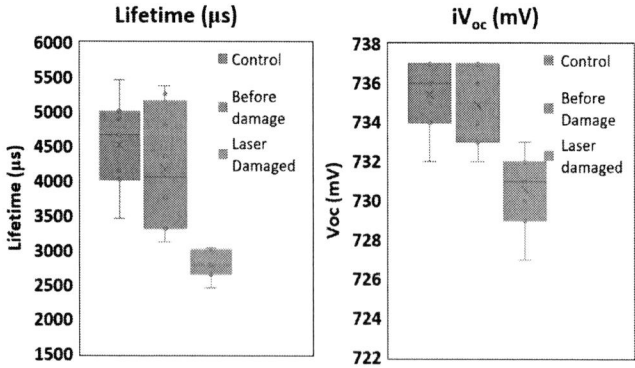

Figure 4: Lifetime and iV_{OC} of wafers before metallization

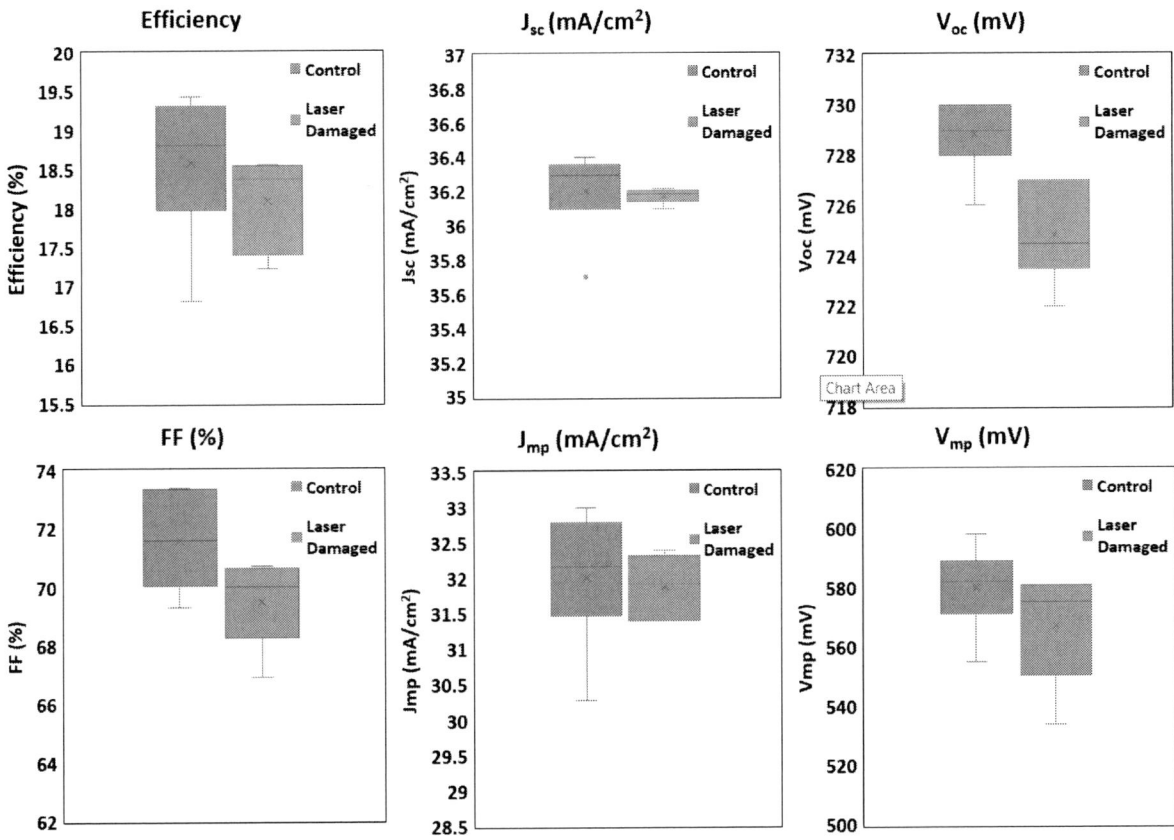

Figure 2: I-V characteristics of the metallized sample

difference of 13 mV, i.e. a relative change of 2%, although there was no appreciable change in the J_{MP} of the cell. There was an absolute 2% loss in the fill factor between them. These factors correspond to an absolute loss of 0.12 W per cell, a relative loss of 8% per cell. Using calibrated photoluminescence images we can spatially resolve lifetime, V_{OC} and J_{0e} images of cells at various stages of processing, which can be further used to model and predict the efficiencies of wafers before metallization [9],[10].

IV. CONCLUSION

It was shown that areas of high recombination are detrimental to device performance. An area of 1% of defects on a silicon wafer can lead to an 8% relative loss in power generated from the device. Also, V_{MP} shows a drop of 13 mV, and negligible change in J_{SC} after the induced damages. These results indicate that each damaged area can be treated as a diode that brings down the voltage across the device. Additional investigations are underway to understand the impact of induced damages on V_{MP} when compared with other parameters of the solar cell.

V. ACKNOWLEDGEMENTS

This material is based upon work primarily supported by the Engineering Research Center Program of the National Science Foundation and the Office of Energy Efficiency and Renewable Energy of the Department of Energy under NSF Cooperative Agreement No. EEC ‐ 1041895. Any opinions, findings and conclusions or recommendations expressed in this material are those of the author(s) and do not necessarily reflect those of the National Science Foundation or Department of Energy.

VI. REFERENCES

[1] Yoshikawa, K., Kawasaki, H., Yoshida, W., Irie, T., Konishi, K., Nakano, K., ... & Yamamoto, K. (2017). Silicon heterojunction solar cell with interdigitated back contacts for a photoconversion efficiency over 26%. Nature Energy, 2, 17032.

[2] De Wolf, S., Descoeudres, A., Holman, Z. C., & Ballif, C. (2012). High-efficiency silicon heterojunction solar cells: A review. green, 2(1), 7-24.

[3] Werner, J., Weng, C. H., Walter, A., Fesquet, L., Seif, J. P., De Wolf, S., ... & Ballif, C. (2015). Efficient monolithic perovskite/silicon tandem solar cell with cell area> 1 cm2. The journal of physical chemistry letters, 7(1), 161-166.

[4] Augusto, A., Herasimenka, S. Y., King, R. R., Bowden, S. G., & Honsberg, C. (2017). Analysis of the recombination mechanisms

of a silicon solar cell with low bandgap-voltage offset. Journal of Applied Physics, 121(20), 205704.

[5] B. L. Sopori, "Defect Clusters in Silicon: Impact on the Performance of Large-Area Devices", Materials Science Forum, Vols. 258-263, pp. 527-534, 1997

[6] Herasimenka, S. Y., Tracy, C. J., Sharma, V., Vulic, N., Dauksher, W. J., & Bowden, S. G. (2013). Surface passivation of n-type c-Si wafers by a-Si/SiO2/SiNx stack with< 1 cm/s effective surface recombination velocity. Applied Physics Letters, 103(18), 183903.

[7] Wang, F., Zhang, X., Wang, L., Jiang, Y., Wei, C., Sun, J., & Zhao, Y. (2014). Role of hydrogen plasma pretreatment in improving passivation of the silicon surface for solar cells applications. ACS applied materials & interfaces, 6(17), 15098-15104.

[8] Herasimenka, S. Y., Dauksher, W. J., & Bowden, S. G. (2013). > 750 mV open circuit voltage measured on 50 μ m thick silicon heterojunction solar cell. Applied Physics Letters, 103(5), 053511.

[9] Trupke, T., Mitchell, B., Weber, J. W., McMillan, W., Bardos, R. A., & Kroeze, R. (2012). Photoluminescence imaging for photovoltaic applications. Energy Procedia, 15, 135-146.

[10] Kuwano, K., & Ashok, S. (1997). Investigation of sputtered indium-tin oxide/silicon interfaces: ion damage, hydrogen passivation and low-temperature anneal. Applied surface science, 117, 629-633.

[11] Filipič, M., Holman, Z. C., Smole, F., De Wolf, S., Ballif, C., & Topič, M. (2013). Analysis of lateral transport through the inversion layer in amorphous silicon/crystalline silicon heterojunction solar cells. Journal of Applied Physics, 114(7), 074504.

[12] Karsten Bothe, David Hinken, "Quantitative Luminescence Characterization of Crystalline Silicon Solar Cells. In Gerhard P. Willeke, and Eicke R. Weber, editors: Semiconductors and Semimetals," Vol. 89, Burlington: Academic Press, 2013, pp. 259-339.

[13] B. Hallam, Y. Augarten, B. Tjahjono, T. Trupke, and S. Wenham, "Photoluminescence imaging for determining the spatially resolved implied open circuit voltage of silicon solar cells," Journal of Applied Physics, vol. 115, no. 4, pp. 044901, 2014..

Laser Hydrogenation on Heavily Dislocated Cast-Mono Silicon Cells

Alison M. Ciesla, Catherine E. Chan, Sisi, Wang, Malcolm D. Abbott, CheeMun Chong and Stuart R. Wenham

The University of New South Wales, Sydney, NSW, SPREE TETB 2052, Australia.

Abstract — Annealing processes both with and without the use of illumination have previously been shown to significantly improve the performance of cells made on highly dislocated cast-mono silicon material. This work further investigates this process, and finds that at temperatures <300 °C, whether illuminated or not, significant changes occur in the first second of processing. However, there is still a clear dependence on the illumination intensity, with different trends appearing for cells processed in the dark, ~2 suns or ~33 suns, indicating the charge state of the hydrogen is likely still important. While the exact mechanisms are not yet understood, this work should provide further insight into the behavior of hydrogen and passivation in heavily dislocated silicon.

I. INTRODUCTION

Hydrogen is known to be able to passivate a wide range of defects in silicon [1]. Recent efforts have focused on understanding the behavior of hydrogen in silicon and improving hydrogenation processes [2]-[6]. It is now understood that hydrogen can exist in three charge states in silicon, which is dependent upon the position of the Fermi-level [7]. The authors of this work have been particularly interested in understanding the ability of hydrogen to passivate defects in poorer quality silicon to enable the production of solar cells on cheaper forms of silicon without sacrificing electrical performance. Cast-mono silicon is one material of interest that is susceptible to a high density of dislocations in parts of the ingot [8]. Developing hydrogenation processes to passivate such dislocations is of increasing importance given recent announcements by GCL, the world's largest silicon and wafer producer for photovoltaics, that the main focus of their product in the future is planned to be the cast-mono material [9].

Previous results have shown that the use of laser illumination to help control the charge states of the hydrogen atoms (laser hydrogenation) during a thermal anneal can improve heavily dislocated cast-mono silicon [10][11]. This was attributed to the high illumination intensities acting to increase the minority charge state fractions which is beneficial for improved passivation. More recently, and somewhat surprisingly, it was found that similar results were achieved simply through low temperature thermal processes for 1-4 minutes, without illumination [12].

This work follows on from these previous studies. Since previous results showed such significant improvement after the first minute of treatment, but yet still improving at 4 minutes [12], this work investigates the cell performance after hydrogenation treatments as short as 1 s, and extending to

periods longer than 4 mins. The previous studies also only used illumination intensities corresponding to photon counts above the equivalent of ~30 suns. In this study, the optics for the laser are altered to enable much lower illumination intensity (~2 suns). Further, the large time steps between measurements in the previous studies made it difficult to ascertain if any trends existed in passivation rates. In this study, the time intervals are reduced to enable identification of any trends that may exist due to different rates of passivation under varied illumination.

II. EXPERIMENTAL DETAILS

The substrates used in this work are 156 x156 mm 1 Ω.cm heavily dislocated cast-mono crystalline silicon wafers with alkaline texturing. After a standard RCA clean and HF dip, the samples were phosphorous diffused in a Tempress Systems POCl$_3$ tube furnace (795°C pre-deposition for 28 min, 885°C drive-in for 30 min) to achieve a phosphorous emitter with a final sheet resistance of ~65 Ω /□. After diffusion, 75 nm of SiNx:H is deposited on the front surface in a Roth and Rau MAiA PECVD tool [13]. The wafers were cleaved into 39 x 39 mm tokens, which were screen-printed with a full rear aluminium contact and a silver front contact grid. Cells were fired in an industrial Sierratherm infrared fast firing furnace with a peak temperature of ~800 °C and power to the lamps controlled for increased illumination of the wafer top surface and improved hydrogenation [6]. The cells were laser edge isolated for a final cell area of 7.84 cm^2. The cells were characterized using a custom built 1-Sun I-V tester, photoluminescence (PL) imaging in a BT Imaging tool and suns-Voc on a Sinton Instruments tool.

In an initial hydrogenation test (*A*), a cell was illuminated by 5.2 x 10^{17} photons/cm^2/s (~2 suns, where 1 sun is ~2.5 x 10^{17} photons/cm^2/s [14]) using a 938 nm laser on a hotplate with set point 321 °C, resulting in an actual cell temperature of 292 °C. The cell was measured before annealing, and at increasing intervals of annealing time (starting at 1s), up to 16 mins total anneal time.

Following the interesting results of (*A*) (shown in the next section), in (*B*) a set of 3 sister tokens/cells was selected for hydrogenation treatments (from neighboring positions in the ingot with aligned crystallographic defects) to ensure the most similar electrical characteristics and defects for passivation. All three cells were annealed on a hotplate set to 321 °C, using optical lenses to vary the illumination intensity of the 938 nm laser illumination.

The conditions seen by the sister cells were:

- ~33 suns (8.3 x 10^{18} photons/cm^2/s), ~302 °C
- ~2 suns (5.2 x 10^{17} photons/cm^2/s), ~292 °C
- 0 suns (dark anneal), ~292 °C

The ~2 sun illumination had negligible impact on the temperature, while the ~33 suns caused ~10 °C temperature increase. A separate experiment has shown that the ~10 °C temperature variation does not measurably affect results.

According to published theory [7][15], the absence of illumination should cause virtually all the interstitial hydrogen within the p-type silicon to be in the positive charge state (H$^+$), while at the other extreme, the highest illumination should give increased concentrations of the minority charge states H^0 and H$^-$. Converting the hydrogen into the minority charge states could potentially have multiple benefits; improved reactivity due to the presence of electrons could potentially yield rapid improvements, whereas increased diffusivity (particularly for neutral H^0) would likely lead to changes over longer time frames as the hydrogen relocates to regions where the hydrogenation takes place.

The cells were each annealed for 8 mins total, with intermittent measurements of increasing time increments, taken at cumulative times of 0 s, 1 s, 2 s, 4 s, 8 s, 16 s, 32 s, 1 min, 2 mins, 4 mins, 8 mins. At each point, light IV and suns-V_{OC} measurements are taken.

III. RESULTS AND DISCUSSION

A. Incremental laser hydrogenation at ~2 suns

Fig. 1. shows the impact of the initial ~2 suns illuminated anneal on the test cell's performance parameters over a period of 16 mins, plotted on a linear time scale. The absolute performance parameters for the heavily dislocated cast-mono cell before and after 16 min of illuminated anneal can be seen in Table 1, with the "start" values clearly highlighting the need for hydrogen passivation.

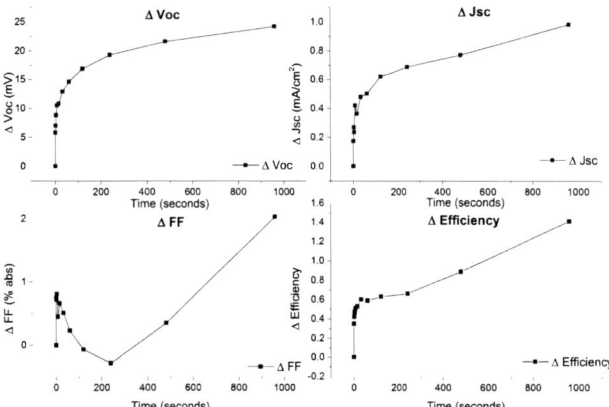

Fig. 1. Changes in cell performance due to increasing anneal times with illumination at ~2 suns.

Table. 1. Performance of a heavily dislocated cast-mono cell before and after ~2 suns hydrogenation anneal:

	Start	After 16 mins
V_{OC}	592 mV	617 mV
J_{SC}	34.7 mA/cm^2	35.7 mA/cm^2
FF	67.2 %	69.2 %
Efficiency	13.8 %	15.2 %

In Fig. 1, a significant improvement can be seen after just 1 second of processing. All parameters improve, with the short circuit current increasing 0.2 mA/cm^2, FF nearly 1% absolute and V_{OC} over 5 mV, resulting in an efficiency improvement of ~0.4% absolute. This perhaps indicates improved hydrogen reactivity due to charge state control, which could have almost instantaneous impact provided the atomic hydrogen is already well distributed throughout the device as may be expected following the previous high temperature firing.

The results here under ~2 suns illumination are similar to that of previous studies in the dark and with ~33 suns, in that the voltage increased ~20 mV over the first 4 minutes. This indicates that the improvements are independent of light intensity, and also not highly temperature sensitive since cell temperatures across the experiments range from 292-313 °C. However, it is now clear with the higher resolution of data points that the current also increased with time. This was not discernable in the previous work.

Previous results also indicated that the fill factor increased in a similar manner to the other parameters. Here however, some strange effects are seen. Initially the fill factor improved, but then at 4 mins/240 s (the time the previous experiments completed) it reached a minimum, worse than its starting point. It was initially considered that the differing FF effects might be due to the continual heating and cooling of the sample as the process was broken down into many smaller steps, however the next experiment proved this to be an unlikely cause. Other possibilities are that the low light intensity caused the variation (as opposed to very high intensity or dark, which are tested in the next section (*B*)), or else different defects exist in this sample that responded differently.

With all effects combined, the efficiency still appeared to improve in a similar manner to previous experiments for the first 4 minutes (240 s) due to the increasing V_{OC} and J_{SC}, despite the decreasing FF.

After the first 4 minutes, the V_{OC} and J_{SC} continued to improve but at a much slower rate. At this time, the fill factor started to increase relatively quickly causing the efficiency to also start to increase faster at 4 mins.

B. Incremental laser hydrogenation with varied intensity

The changes in V_{OC} with processing time (relative to starting V_{OC}s) are plotted in Fig. 2. on a ~log$_2$ time scale.

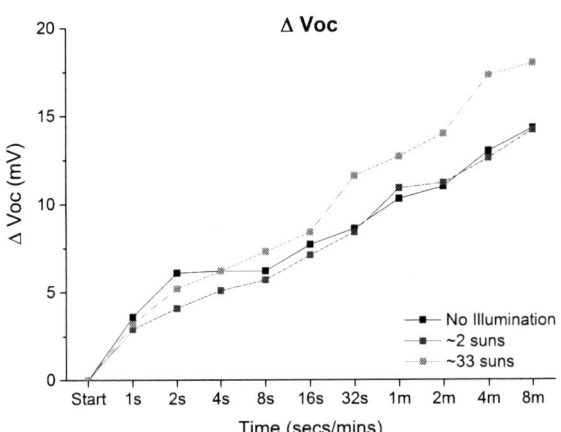

Fig. 2. : Changes in Voc of cast-mono cells due to increasing laser hydrogenation processing times, plotted on a ~log$_2$ scale.

The cell exposed to the maximum ~33 suns illumination intensity shows the largest improvement, with >25% larger increase in V_{OC} after 8 mins. All cells still appeared to be improving, albeit at a slower rate; in the last 4 mins of processing, the cells still improved >1 mV on average, while initially the cells improved at a rate of >3 mV/sec on average.

Such significant changes in the first second again perhaps indicate passivation reactions occurring almost instantaneously such as due to altered hydrogen charge states, however the same improvements also occurred in the dark, which is somewhat surprising given that the atomic hydrogen in the p-type region should remain almost entirely H$^+$. In [12], it was hypothesised that the large improvements due to annealing even after the firing hydrogenation process was due to the back-surface field (not yet present during firing) to allow hydrogen to build up in the bulk allowing improved passivation. However, a time interval of only 1 s is not sufficient for hydrogen of any charge state to move significantly for this to be the case. It is unclear at this stage what enabled such a rapid improvement in cell performance due to a simple 1 s thermal process even after the firing hydrogenation process.

PL images of the 3 cells before and after the hydrogenation anneals are shown in Fig. 3. to give a visual indication of how the cells improved with laser processing. Before annealing, the dislocations were severe, giving almost no PL signal and appearing almost black. The benefit of the high ~33 suns illumination is clearly evident as the best of the three cells, with the dislocations still visible but to a much lesser degree. The cells processed in the dark and low light experienced similar levels of improvement (14.2 and 14.3 mV respectively); although the cell processed in the dark appears better than that exposed to low light, it also started with a higher V_{OC} which may not be obvious in the PL images.

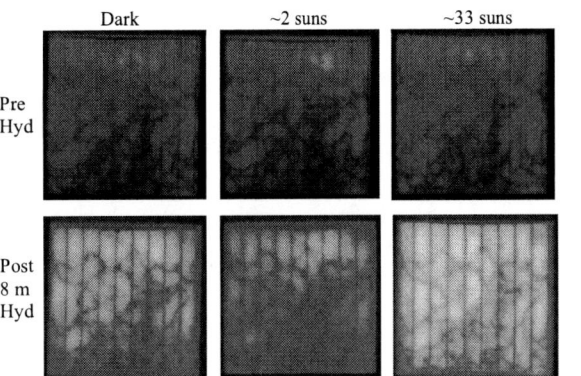

Fig. 3. PL images of cells before and after laser hydrogenation with varied illumination intensity

The changes in J_{SC} relative to starting J_{SC} with processing time are plotted in Fig. 4 on a ~log$_2$ time scale.

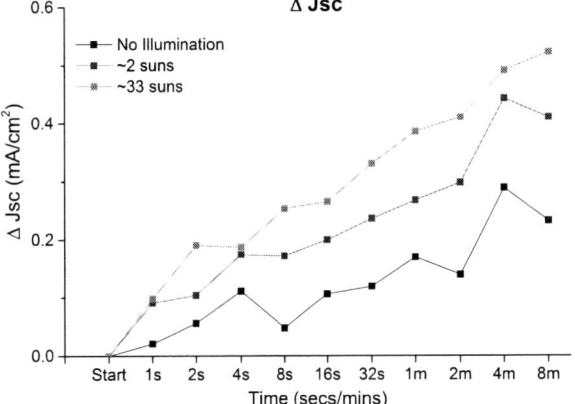

Fig. 4. Changes in J_{SC} of cast mono cells due to increasing laser hydrogenation processing times on a ~log$_2$ scale?

It must be noted that the J_{SC} is the parameter most prone to error during measurement and the changes in J_{SC} are not as significant as the changes in V_{OC}, however there is a clear improvement trend. This plot again indicates that stronger illumination yields increased improvement; the cell annealed in the dark improved least, while the cell exposed to ~33 suns showed the most improvement. In the last 4 mins, the difference became more pronounced as all the samples processed in the dark or with low light (~2 suns) appeared to decrease, while the cell under the ~33 suns light intensity continued to improve.

The changes in fill factor (FF) due to the hydrogenation anneals are significant and particularly interesting. Fig. 5. compares the changes in FF measured on the IV tester, with the pseudo-FF (pFF) measured on the Sinton suns-V_{OC} tool that is free from effects of Rs.

Firstly, it is important to note the poor starting FFs of these heavily dislocated cast-mono cells; similar to previous works, the FFs were all initially less than 66%. When comparing this with the starting pseudo fill factors which are all ~80% or higher, the poor fill factor can be mostly attributed to series resistance effects, which seem to improve with hydrogenation.

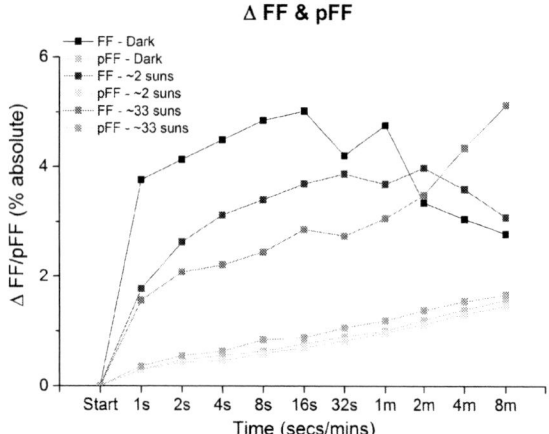

Fig. 5. Changes in FF and pFF with different illumination intensities as a function of increasing laser hydrogenation processing times plotted on a ~$\log_2$ scale.

The suns-V_{OC} measurements show clear and similar improvements in pFF independent of illumination, as well as a reduction in ideality factor (not shown here) indicating reduced junction recombination. The improvements in pFF are significant (~1.5% abs), however do not account for the up to 5% absolute improvement in FF. Similar to previous results, this suggests that the largest changes in the FF are a result of changes in series resistance [12]. Here, the FF of all cells improved well over 1% absolute in just the first second alone. 1s at ~300 °C is unlikely to cause any significant structural change in metal contact formation as previously proposed in [12] and could again suggest some involvement with hydrogen and perhaps the charge state of the hydrogen atoms. Further investigation is required to ascertain what could be causing such dramatic changes so rapidly.

The changes in FF also present other unexplained effects, which appear to be series resistance related. The lower the light intensity, the faster the rate of FF improvement initially; until a switching point, when the FF of the cells under low light intensities degraded again (with the cell in dark degrading the most severely), while the cell under the highest light intensity continued to improve and was still improving after 8 mins. This implies the series resistance effects are highly illumination dependent, however it is unclear exactly what is causing these changes.

It is interesting that the ~2 suns anneal did not cause initial degradation of the FF as it did in (A); since the sister wafers of B were not crystallographically matched to that of A, this may indicate that different dislocations respond differently to the passivation processes.

The overall effect of the hydrogenation anneals on efficiency is shown in Fig. 6. The trends in efficiency changes seem to largely mimic the FF trends indicating that despite the improvement in all parameters, the FF is having the dominant

impact. The lower light intensities show more improvement than the highest intensity up to 1 min, but at 8 mins are degrading again due to the degradation in FF. The highest ~33 suns light intensity provides the best overall improvement of 1.7% absolute, from 13.4% to 15.1% and the cell is still improving after 8 mins.

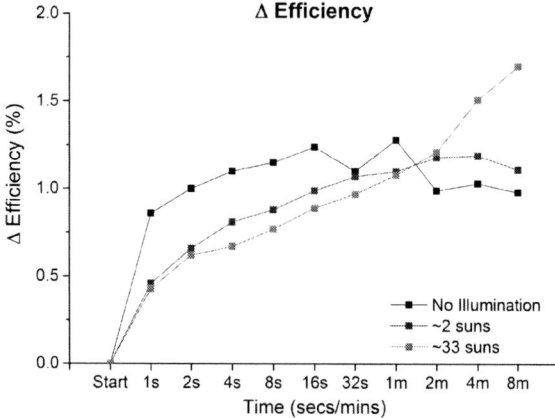

Fig. 6. Changes in efficiency due to different illumination intensity laser hydrogenation anneals plotted on a ~$\log_2$ scale.

III. SUMMARY AND CONCLUSIONS

This work demonstrates the ability of low temperature (~300 °C) anneals to significantly improve cells on heavily dislocated cast mono silicon in as little time as 1s. Improvements are seen in all performance parameters; most significantly in the first few seconds, with V_{OC} and J_{SC} still improving after 8 minutes of processing time. Significant changes in FF are seen due to series resistance changes, which seem highly dependent on the illumination intensity. Rapid improvements occur irrespective of illumination during the first second, however at 8 mins only the highest light intensity ~33 suns continues improving, while the lower light intensities degrade again. The differing results and particularly the extremely rapid changes appear to show some dependence on hydrogen and the charge state of the hydrogen atoms, but also seem heavily dependent on the sample dislocations. While the exact mechanisms causing the changes are not well understood, this work will hopefully provide insight and motivation into better understanding the behavior of hydrogen in heavily defected silicon.

REFERENCES

[1] J. I. Pankove and N. M. Johnson, "Hydrogen in semiconductors," *Boston, MA : Academic Press*, 1991.
[2] B. Sopori, "Silicon Solar-Cell Processing for Minimizing he Influence of Impurities and Defects," *J. of Elec. Mat*, vol. 31, p. 972-980, 2002.
[3] S. Martinuzzi, "Hydrogen passivation of defects in multicrystalline silicon solar cells," *SOLMAT*, vol. 80, p. 343-353, 2003.

[4] M. Gläser and D. Lausch, "Towards a quantitative model for BO regeneration by means of charge state control of hydrogen," *Energy Procedia*, vol. 77, p.592-598, 2015.

[5] C.Sun, F. Rougieux & D. Macdonald, "A unified approach to modelling the charge state of monatomic hydrogen and other defects in crystalline silicon," *Journal of Applied PV*, vol.117 p.. 2015.

[6] B. J. Hallam, P. G. Hamer, S. R. Wenham, *et al.*, "Advanced Bulk Defect Passivation for Silicon Solar Cells', *IEEE Journal of Photovoltaics*, vol. 4, no. 1, 2014.

[7] C. Herring, N. M. Johnson and C. G. Van de Walle, "Energy levels of isolated interstitial hydrogen in silicon," *Phys. Rev. B*, vol. 64, 2001.

[8] I. Guerrero, V. Parra, T. Carballo *et al.*, "About the origin of low wafer performance and crystal defect generation on seed-cast growth of industrial mono-like silicon ingots,", *Progress in PV: Res. And Appl.*, vol. 22 p. 923-932, 2014

[9] Yuepeng Wan, CTO of GCL Holdings Ltd., *CSPV Conference*, Shanghai, China, Nov. 2016.

[10] L. Song, A. Wenham, S. Wang *et al.*, "Laser Enhanced Hydrogen Passivation of Silicon Wafers," *Int. J. of Photoenergy*, vol. 2015.

[11] A. Wenham, B. Hallam, L. Song *et al.*, "Efficiency enhancement for screen printed solar cells on quasi-mono wafers through hydrogen passivation of dislocations", *Eu PVSEC*, 2015.

[12] A. Wenham, L. Song, M. Abbott *et al.*, "Defect passivation on cast-mono crystalline screen-printed cells," *Front. Energy*, vol. 11, p. 60-66, 2017.

[13] Z. Hameiri, N. Borojevic, L. Mai *et al.*, "Low-Absorbing and Thermally Stable Industrial Silicon Nitride Films With Very Low Surface Recombination," *IEEE Journal of Photovoltaics*, DOI: 10.1109/JPHOTOV.2017.2706424, 2017

[14] T. Trupke, R. A. Bardos, M. C. Schubert & W. Warta, 'Photoluminescence imagine of silicon wafers', *Applied Physics Letters*, vol. 89, pp. 044107, 2006

[15] B. J. Hallam, P. G. Hamer, S. Wang, *et al.*, "Advanced Hydrogenation of Dislocation Clusters and Boron-oxygen Defects in Silicon Solar Cells," *Energy Procedia*, vol.77 p. 799-809, 2015.

Performance Optimization of Semi-Transparent Thin-Film Amorphous Silicon Solar Cells

Yuan Gao[1,2,3], Fai Tong Si[1], Olindo Isabella[1], Rudi Santbergen[1], Guangtao Yang[1], Jianfei Dong[2,4], Guoqi Zhang[5], and Miro Zeman[1]

[1]Photovoltaic Materials and Devices, Delft University of Technology, Delft, 2628 CD, The Netherlands
[2]Beijing Research Center, Delft University of Technology, Beijing, China
[3]State Key Laboratory of Solid State Lighting, Changzhou, China
[4]Suzhou Institute of Biomedical Engineering and Technology, Chinese Academy of Sciences, Suzhou, China
[5]Department of Microelectronics, Delft University of Technology, Delft, 2628 CD, The Netherlands

Abstract — **Semi-transparent solar cells possess tremendous potential in glass-based PV applications. Using an optical model, GenPro4, we provide with a simulation method to optimize the configuration of such solar cells. For a single-junction amorphous silicon solar cell, the optimized thickness of the absorber layer is obtained at 170 nm to realize a target average transmittance of 20% in the visible range of the sunlight spectrum. A sample cell was fabricated accordingly to verify the proposed method. Measurement results show that an average transmittance of 20.04% is achieved with the conversion efficiency of 6.94%. The presented optimization method can also be applied to the estimation of color appearance of semi-transparent solar cells.**

Index Terms — **semi-transparent solar cells, amorphous silicon, GenPro4, optical simulation, BIPV.**

I. INTRODUCTION

Semi-transparent photovoltaics (STPVs) are drawing more attention due to the prospect of PV applications integrated with architecture, agriculture, automobile, etc. [1, 2]. STPVs not only contribute to the reduction of greenhouse gas emission, but also to the esthetic design of PV systems because of the colorful and translucent appearance [3]. Comparing with other promising candidates for STPV (such as Perovskite [4], organic photovoltaics [5], etc.), thin-film silicon technology reveals its advantages in the respect of stability, abundant raw material, capability of mass production and very large area deposition [6]. Currently, commercially available STPV products applying thin-film silicon technology vary in module efficiency, transmittance, and color appearance [3]. Customized modules can be installed as the windows in buildings [1], the envelope of glass greenhouse [2], and even automotive sunroof and windows [5]. One of the challenges for promoting this technology is to improve the efficiency of energy conversion while preserving the transparency of target surfaces.

Unlike conventional opaque solar cells, the back contact in semi-transparent solar cells acts as only a conductive layer, instead of a back reflector. To achieve desired transparency, most previous research was conducted by replacing the metal back contact with transparent back contact (TBC) and tuning the thickness of certain layers based on existing structures of opaque solar cells. However, experimental results of those research were presented by different criteria on the transparency of STPV, such as the average transmittance over a certain spectral band, the transmittance at a certain wavelength, etc. Therefore, here relevant literatures are reviewed individually without concluding the best candidate among them. Yeop Myong *et al.* invented and investigated three approaches to fabricating proper back contacts for colored semi-transparent PV modules, i.e. TBC type with green color, hybrid type with additional laser-scribed patterns and blue encapsulating film, and opaque back contact (OBC) type with additional laser-scribed patterns. Module-level values of average total transmittance in the wavelength range of 360-750 nm (T_{360_750}) and the nominal conversion efficiency (η) were shown in TBC type ($T_{360\_750} = 14.0\%$, $\eta = 5.6\%$), hybrid type ($T_{360\_750} = 30\%$, $\eta = 3.5\%$), and OBC type ($T_{360\_750} = 10\%$, $\eta = 6.4\%$), respectively [3]. Lim *et al.* conducted a series of experiments to obtain the optimal thickness of transparent conductive oxide (TCO) [7], high-bandgap (HB) n/i-interface layers [8], and p-Si layer [9], respectively. In [7], the thickness of TCO was taken as the variable, and hydrogenated amorphous silicon (a-Si:H) with TCO thickness of 300 nm exhibited an outstanding performance ($T_{400\_800} = 21.6\%$, $\eta = 5.6\%$). In [8], HB layers were introduced to decrease shunt loss and increase carrier collection. Among all HB-layer structures, the cell fabricated with triple HB layers showed the best optoelectronic performance ($T_{400\_800} = 23.6\%$, $\eta = 6.9\%$). In [9], the thin p-Si layer was employed to obtain high transparency and high current density while a buffer layer was used to enhance V_{oc}, which was reduced by the thin p-Si layer. The best performance ($T_{500\_800} = 30.7\%$, $\eta = 5.36\%$) was found when the thickness of p-Si layer was 7.5 nm and the thickness of buffer layer was 0.5 nm. Lee *et al.* investigated ultra-thin intrinsic a-Si/organic hybrid structures which are able to produce transmissive or reflective colors (red, green, and blue). It was claimed that the power efficiency of the hybrid cell was obtained up to 3%. Transmission spectra were plotted in curves, without calculating the values of average transmittances [10].

Previous research shares a common problem that the variable thicknesses of target layers were chosen manually

978-1-5090-5606-4/17 $31.00 © 2017 IEEE

based on the experience of researchers. The interval of chosen variables cannot be too small since fabricating cell samples costs time and efforts. Therefore, precise simulation plays an important role in optimizing the configurations and structures of semi-transparent solar cells quickly. However, details of simulation methods were rarely mentioned in previous research. In this study, GenPro4, an optical model developed at the PVMD group, is applied to the design and optimization of semi-transparent single-junction a-Si:H solar cells [11, 12]. GenPro4 represents the solar cell as a multilayer structure and calculates the fraction of incident light absorbed in each layer, taking into account scattering of light at the interfaces. Such model and its previous versions have been validated for a wide variety of wafer-based and thin-film solar cells [11-14]. The photocurrent density associated to the absorptance in the absorbing i-layer, to the reflection and to the transmittance of the solar cell can be accurately calculated with proper input data, such as surface morphologies, layers thickness and refractive indexes. GenPro4 supports our efforts in fabricating, testing and measuring solar cells with various configurations to obtain solar cells with balanced optical and electrical performance.

In this paper, an optimization method is proposed to design and fabricate semi-transparent solar cells. At first, the device structure was modified based on the structure of a state-of-the-art opaque solar cell. Then, GenPro4 was deployed to optimize the thickness of the target i-layer to obtain desired transmittance. According to the simulation results, a sample was fabricated and measured to compare with our estimation. Conclusions were drawn at the end, along with necessary analysis and discussion.

II. DESIGN AND OPTIMIZATION

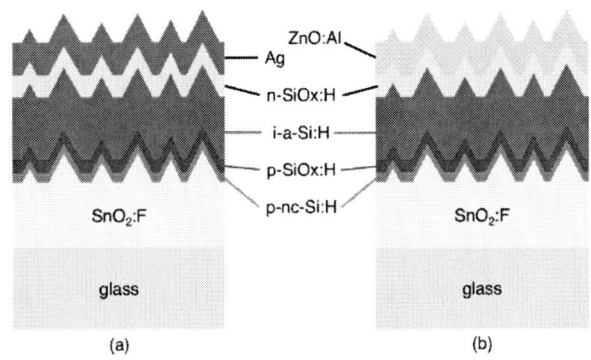

Fig. 1. The schematic structures of single-junction a-Si:H solar cells with (a) Ag and (b) ZnO:Al as back contact, respectively. The direction of light incidence is from bottom to top.

State-of-the-art opaque single-junction a-Si:H cells have a device structure as shown in Fig. 1 (a). To obtain a semi-transparent structure, the Ag back reflector is replaced with ZnO:Al transparent conductive oxide (TCO), which enables

the transmittance of un-absorbed light through the solar cell, as shown in Fig. 1 (b). Since GenPro4 is a purely optical model, which does not consider the electrical characteristics of the solar cell, we assume that the open circuit voltage (V_{oc}) and the fill factor (FF) maintain typical values when we adjust the thickness of each layer within reasonable ranges. With GenPro4, the implied photocurrent can be calculated quickly as a function of the thickness of absorber layer. In this way, we can possibly estimate the conversion efficiency of the solar cell in terms of simulation variables, i.e. the thickness of crucial layers.

Theoretically, the thickness of all the seven layers should be taken as input variables. However, relatively thin layers, such as p-layer and n-layer, contribute to a relatively small portion of light absorption, but have critical impact on electrical characteristics. Therefore, the thickness of p-nc-Si:H, p-SiO$_x$:H, and n-SiO$_x$:H is kept constant in the optical simulation. Meanwhile, the thickness of glass and fluorine-doped tin oxide (SnO$_2$:F) is determined by the nanotextured Asahi-VU substrate, and the thickness of ZnO:Al is chosen to be 300 nm, ensuring both high transmittance and sufficiently low sheet resistance. Thus, the variable layers are narrowed down to the i-layer only.

To estimate the conversion efficiency (η) of the cell, we assume that V_{oc} stays at 0.85 V, and FF at 0.7. The *implied* short circuit current density (J_{sc}) can be obtained from the optical simulation. Then, as the power input (I_{in}) is the standard 100 mW/cm^2, $\eta = J_{sc} \cdot V_{oc} \cdot FF / I_{in}$.

Fig. 2 presents the efficiency and transmittance with varying thickness of i-layer from 50 to 500 nm. As both curves reveal distinct trends, a trade-off must be made to achieve decent efficiency and satisfying transmittance, which is set at 20% for STPV applications [15]. The corresponding thickness of the i-layer is in this way determined.

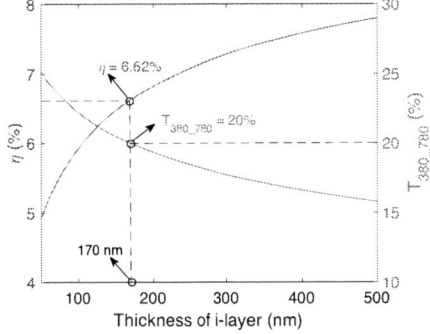

Fig. 2. The estimated efficiency (η) and the average transmittance between 380 nm and 780 nm (T$_{380\_780}$) of the semi-transparent a-Si:H solar cell as a function of the thickness of i-layer.

The detailed structure of semi-transparent device is shown in Table I. Owing to GenPro4 model, experiments with layer thickness as variables can be avoided. Also, GenPro4 can generate the reflectance, absorptance and transmittance of

978-1-5090-5606-4/17 $31.00 © 2017 IEEE

each layer at different wavelength as shown in Fig. 3. The average transmittance of the solar cell is calculated according to the transmitted spectrum in the wavelength between 380 and 780 nm. The simulation tool can also help us investigating the optical contribution of each layer for further improvement.

TABLE I
LAYERS OF SEMI-TRANSPARENT SOLAR CELL WITH OPTIMAL THICKNESS

Layer	Thickness (nm)
glass	7×10^5
SnO$_2$:F	700
p-nc-Si:H	4
p-SiO$_x$:H	15
i-a-Si:H	170
n-SiOx:H	30
AZO	300

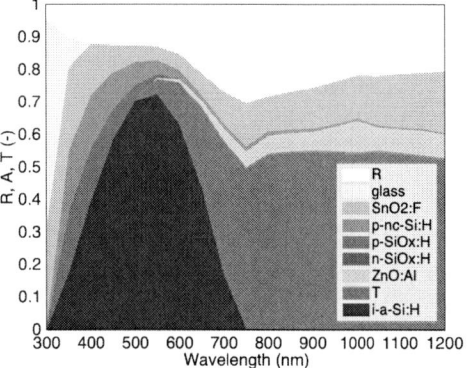

Fig. 3.   Area plot of reflectance, absorptance and transmittance of each layer at different wavelength.

III. EXPERIMENTAL PROCESS AND RESULTS

Fig. 4.   Semi-transparent cell sample viewed at the rear with the window as background.

STPV cells were fabricated on Asahi-VU type substrates, which are glass coated with nanotextured SnO$_2$:F serving as front TCO and light-scattering surface. The solar cells use a-Si:H as the absorber and have a structure of glass/700 nm SnO$_2$:F /4 nm p-nc-Si:H/20 nm p-SiO$_x$:H/170 nm i-a-Si:H/60 nm n-SiO$_x$:H/300 nm ZnO:Al. The thin-film silicon alloy materials were deposited in a multi-chamber system using plasma-enhanced chemical vapor deposition (PECVD) at radio frequency (RF) of 13.56 MHz. ZnO:Al is used as the rear transparent electrode and was deposited by RF magnetron

sputtering with a shadow mask, which also defines the area of the solar cells. The semi-transparent sample fabricated as above is presented in Fig. 4, in which the rear of the cell is facing the camera with the window as background.

The performance of the solar cells was examined by current-voltage (*I-V*) and external quantum efficiency (EQE) measurements. In specific, *I-V* measurement was conducted under AM1.5G solar spectrum with an irradiance of 1000 W/m^2 with a dual-lamp continuous solar simulator (WACOM WXS-90S-L2, class AAA) at a controlled sample temperature of 25 °C. The sample was tested in open rear configuration (i.e. no reflective chuck). The EQE measurement was performed using an in-house system. V_{oc} and FF were determined by the *I-V* measurement. J_{sc} was obtained by weighting the measured EQE with the AM1.5G solar spectrum.

Table II presents the measured parameters comparing with those obtained from the optical simulation. It turns out that the best measured efficiency is 0.32% higher than the simulation results due to the underestimation of J_{sc}. The unexpected good performance of J_{sc} might come from the deviation in fabrication or measurement process.

TABLE II
SOLAR CELL PARAMETERS FROM SIMULATION AND MEASUREMENT

Parameters	Simulation	Measurement
V_{oc} (V)	0.85	0.83
J_{sc} (mA/cm^2)	11.13	13.24
FF	0.70	0.63
η (%)	6.62	6.94

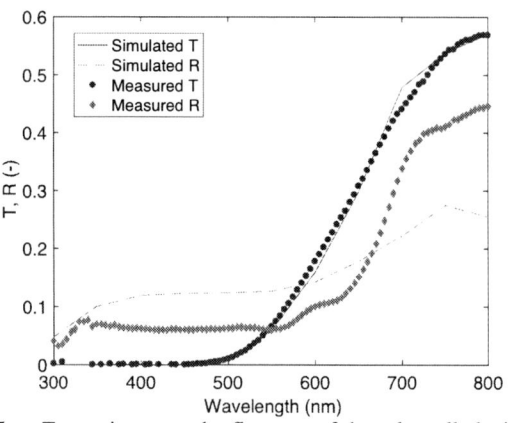

Fig. 5.   Transmittance and reflectance of the solar cell obtained by measurement and simulation.

The reflectance and transmittance of the sample are also measured with a Lambda 950 spectrophotometer. Measurement and simulation results are shown in Fig. 5. The measured reflectance is lower than that in simulation in the range between 300 to 660 nm. On the other hand, the measured transmittance fits well with the simulation data. It means that more light is absorbed by the solar cell than expected, which could explain why J_{sc} is higher than the estimated data. As for the transmittance, the measurement result is 20.04%, which is close to the desired value 20%.

IV. CONCLUSION AND DISCUSSION

The optical model GenPro4 was applied to the optimization of STPV cells based on thin-film silicon technology. The simulation method was proven to be able to find out the optimal thickness of the variable layer according to the target transmittance of solar cells. A sample cell was fabricated accordingly and measured in experimental environment. Results show that proposed method can deliver precise estimation of transmittance. Electrical parameters are estimated by empirical data, whose accuracy cannot be ensured. The optimal semi-transparent cell sample shows competitive optoelectronic performance ($T_{380\_780} = 20.04\%$, $\eta = 6.94\%$) comparing with the results in previous research.

Literature review on semi-transparent solar cells reveals that it is difficult to evaluate the optoelectronic performance since various criteria were used in the results. Here we suggest that three types of transmittance should be calculated based on the transmittance spectrum, i.e. average transmittance over a certain spectral band, energy-weighted transmittance, and vision-weighted transmittance [16]. Details on the transmittance calculations shall be discussed in the future to standardize the evaluation method of semi-transparent solar cells, which benefits both research and industrial community.

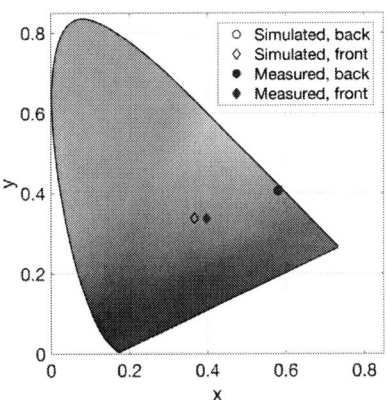

Fig. 6. Color appearances of the semi-transparent solar cell marked in the CIE 1931 color space chromaticity diagram. Circles represent the color appearance viewed from the side of the back contact, i.e. inside the window, and diamonds represent the color appearance viewed from the side of the front glass, i.e. outside the window. Hollow markers represent simulated results, and solid markers represent measured results.

Further research on STPV can also be conducted by the proposed method. One of the critical parameters of semi-transparent solar cells is, for instance, the color appearance, which can be calculated according to the transmission and reflection spectra obtained by GenPro4 [17]. Fig. 6 presents the color appearances of the semi-transparent solar cell viewed from both front and back sides. The xy coordinates of simulated and measured colors are marked in the CIE 1931 color space chromaticity diagram. It's obvious that the back color obtained from simulation is basically overlapped with the simulated result, which is calculated depending on the spectral transmission yielded by GenPro4. It implies that tedious experimental process can be replaced by programming in GenPro4 by varying materials and tuning the thickness of each layer in order to obtain the desired color appearance. The simulated front color drifts from the measured one due to the deviation of spectral reflectance.

Also, optical characteristics under different light sources or acting on the angle of incidence and polarization can be simulated in GenPro4 by replacing the source files. For example, the color appearance at night can be obtained by replacing the standard solar light source files with indoor LED light source files while reversing the sequence of layers. The design process of STPV cell can therefore be dramatically accelerated by the proposed optimization method.

ACKNOWLEDGEMENT

The authors thank process engineer Martijn Tijssen and technician Stefaan Heirman for providing with experimental materials and helping with measurement procedures.

This work is supported by a joint project cooperated by the group of Photovoltaic Materials and Devices at Delft University of Technology, State Key Laboratory of Solid State Lighting at Changzhou Base in China, and TU Delft Beijing Research Center.

REFERENCES

[1] G. Quesada, D. Rousse, Y. Dutil, M. Badache, and S. Hallé, "A comprehensive review of solar facades. Transparent and translucent solar facades," *Renew. Sustain. Energy Rev.*, vol. 16, no. 5, pp. 2643–2651, Jun. 2012.

[2] A. Yano, M. Onoe, and J. Nakata, "Prototype semi-transparent photovoltaic modules for greenhouse roof applications," *Biosyst. Eng.*, vol. 122, pp. 62–73, Jun. 2014.

[3] S. Yeop Myong and S. Won Jeon, "Design of esthetic color for thin-film silicon semi-transparent photovoltaic modules," *Sol. Energy Mater. Sol. Cells*, vol. 143, pp. 442–449, Dec. 2015.

[4] K. T. Lee, L. Guo, and H. Park, "Neutral- and Multi-Colored Semitransparent Perovskite Solar Cells," *Molecules*, vol. 21, no. 4, p. 475, 2016.

[5] K. S. Chen, J. F. Salinas, H. L. Yip, L. Huo, J. Hou, and A. K. Y. Jen, "Semi-transparent polymer solar cells with 6% PCE, 25% average visible transmittance and a color rendering index close to 100 for power generating window applications," *Energy Environ. Sci.*, vol. 5, no. 11, p. 9551, Nov. 2012.

[6] F. J. Haug and C. Ballif, "Light management in thin film silicon solar cells," *Energy Environ. Sci.*, vol. 8, no. 3, pp. 824–837, 2015.

[7] J. W. Lim, S. H. Lee, D. J. Lee, Y. J. Lee, and S. J. Yun, "Performances of amorphous silicon and silicon germanium semi-transparent solar cells," *Thin Solid Films*, vol. 547, pp. 212–215, Nov. 2013.

[8] J. W. Lim, M. Shin, D. J. Lee, S. H. Lee, and S. Jin Yun, "Highly transparent amorphous silicon solar cells fabricated using thin absorber and high-bandgap-energy n/i-interface layers," *Sol. Energy Mater. Sol. Cells*, vol. 128, pp. 301–306, 2014.

[9] J. W. Lim, D. J. Lee, and S. J. Yun, "Semi-Transparent Amorphous Silicon Solar Cells Using a Thin p-Si Layer and a Buffer Layer," *ECS Solid State Lett.*, vol. 2, no. 6, pp. Q47–Q49, 2013.

[10] J. Y. Lee, K. T. Lee, S. Seo, and L. J. Guo, "Ultra-thin intrinsic amorphous silicon/organic hybrid structure for decorative photovoltaic applications," *2014 IEEE 40th Photovolt. Spec. Conf. PVSC 2014*, pp. 956–958, 2014.

[11] M. Zeman, O. Isabella, S. Solntsev, and K. Jäger, "Modelling of thin-film silicon solar cells," *Sol. Energy Mater. Sol. Cells*, vol. 119, pp. 94–111, 2013.

[12] R. Santbergen, T. Meguro, T. Suezaki, G. Koizumi, K. Yamamoto, and M. Zeman, "GenPro4 Optical Model for Solar Cell Simulation and Its Application to Multijunction Solar Cells," *IEEE J. Photovoltaics*, vol. 7, no. 3, pp. 919–926, 2017.

[13] R. Santbergen and R. J. C. van Zolingen, "The absorption factor of crystalline silicon PV cells: A numerical and experimental study," *Sol. Energy Mater. Sol. Cells*, vol. 92, no. 4, pp. 432–444, 2008.

[14] A. Ingenito, O. Isabella, S. Solntsev, and M. Zeman, "Accurate opto-electrical modeling of multi-crystalline silicon wafer-based solar cells," *Sol. Energy Mater. Sol. Cells*, vol. 123, pp. 17–29, 2014.

[15] C. Tsai and C. Tsai, "Development of Tandem Amorphous/ Microcrystalline Silicon Thin-Film Large-Area See-Through Color Solar Panels with Reflective Layer and 4-Step Laser Scribing for Building-Integrated Photovoltaic Applications," *Hindawi Publ. Corp. J. Nanomater.*, vol. 2014, 2014.

[16] ISO 9050:2003, "Glass in Building-Determination of Light Transmittance, Solar Direct Transmittance, Total Solar Energy Transmittance, Ultraviolet Transmittance and Related Glazing Factors." *International Organization for Standardization*, Geneva, 2003.

[17] R. W. G. Hunt and M. R. Pointer, "A colour-appearance transform for the CIE 1931 standard colorimetric observer," *Color Res. Appl.*, vol. 10, no. 3, pp. 165–179, 1985.

Low temperature spalling of silicon: A crack propagation study

Pablo Guimera Coll[*,1], Tine Uberg Nærland[1], Nathan Stoddard[2], Michael Stuckelberger[1], and Mariana Bertoni[1]

[1] Ira A. Fulton Schools of Engineering, Arizona State University, Tempe, AZ 85287
[2] SolarWorld Industries America, Hillsboro, OR 97124

*Email: pguimera@asu.edu

ABSTRACT — **Spalling is a promising kerfless method for cutting thin silicon wafers while doubling the yield of a silicon ingot. The main obstacle in this technology is the high total thickness variation of the spalled wafers, often as high as 100% of the wafer thickness. It has been suggested before that a strong correlation exists between low crack velocities and a smooth surface, but this correlation has never been shown during a spalling process in silicon. The reason lies in the challenge associated to measuring such velocities. In this contribution, we present a new approach to assess, in real time, the crack velocity as it propagates during a low temperature spalling process. Understanding the relationship between crack velocity and surface roughness during spalling can pave the way to attain full control on the surface quality of the spalled wafer.**

Index Terms — **Spalling, silicon, kerfless, wafering, crack velocity**

I. INTRODUCTION

The goal of kerfless technologies is to develop new methods for cutting silicon that eliminates slurry and wires while doubling the yield of a silicon ingot. One of the challenges is to do so repeatedly through the ingot without losses in lifetime. Spalling has been shown to be a promising technology, which capitalizes on the crystalline and mechanical properties of silicon. However, previous implementations of this technology, especially those using high temperature with metal foils, carry several challenges for PV application [1]. In some cases, the substrate needs to be heated to temperatures higher than 600°C in order to sufficiently activate the spalling mechanism that kicks in upon the subsequent cooling [2], [3]. This range of temperatures combined with the presence of metals from the stressor layers substantially degrades the electronic properties of the substrate.

The only way to avoid this degradation is to perform the spalling at lower temperatures (<150°C) but even experiments performed at these temperatures show poor wafer performance [4]. The main cause for this low performance is that the surface of the cleaved wafers is not smooth enough and often present a total thickness variation as high as 100% of the wafer thickness [5].

The root cause of this issue relies on the velocity at which the crack propagates during spalling. At high velocities, typical of spontaneous spalling, the crack becomes unstable. This instability leads to critical variations in the crack trajectory with a consequent branching and corrugation of the surface [6].

The high velocities add another problem; when the crack propagates above a certain range of velocities, it emits a wave [7]. This emitted wave then reaches the edges of the sample and reflects back into the material, interfering with the crack tip, causing substantial ripples on the surface [8]. This process can drastically change the velocity and the direction of the propagating crack and, thus, the uniformity of the cleaved surface.

A clear correlation between surface roughness and crack velocity was presented by Arakawa during a study on brittle polymers [9]. What is more, he demonstrated experimentally, that the roughness of a cleaved wafer depends on the product of the crack velocity and the stress state at the crack tip, also known as the stress intensity factor, K. This *K-value* depends mainly on the material, the sample geometry, and the applied load to the system.

By controlling the loading conditions, the roughness of the spalled wafer can be confined. In order to do so, a greater understanding of the crack velocity and dynamics must be achieved. However, while crack velocities have been measured during traditional tension tests, no reliable method to measure the crack propagation velocity for spalling of silicon substrates has ever been presented. In the following paper, we report on a novel method to measure in real time the velocity of a dynamic crack during spalling of silicon wafers.

II. BACKGROUND

In fracture mechanics, several measurement techniques have been developed to measure crack velocities for a range of materials during tension tests. In general, three different techniques have been implemented to obtain the valuable information of a crack's velocity:

(a) A common technique takes advantage of the interaction of the crack and its front waves after the crack has propagated fully through the material. This interaction locally induces a perturbation causing the crack front to curve, called Wallner lines. From a geometry analysis, the crack velocity can be calculated for different points in the material [10]. The complexity of observing and differentiating the crack front from the Wallner lines in rough surfaces, as those produced in spalling, makes this technique highly non-reliable.

(b) A second method is high-speed photography. As seen by Bellanger *et al.* this technique has several limitations on obtaining information about the crack tip during spalling [4]. The cameras cannot determine the exact initiation time of the crack propagation and cannot precisely provide information of the crack position with time, something crucial to establish a correlation with roughness.

(c) Finally, the potential drop technique has been a successful way used to calculate the crack velocity during tension tests in insulating and conductive materials [11]–[13]. A strip of a thin conductive layer is coated along the predicted crack path. After fracture initiation in a pre-made notch, a dynamic crack will break the resistive coating thereby varying the sample's resistance as shown in Fig. 1 redrawn based on Fineberg *et al.* [13]. The instantaneous resistance of the conductive layer is calculated by measuring the voltage drop across it.

With this system geometry, a nearly linear relation between the crack length and the resistance of the coated plate can be obtained. This method requires a conductive strip along the predicted crack path, and for current spalling methods this would be extremely complicated, as the strip and electrodes needed to measure the voltage drop would have to be applied on the side of the entire thin substrate (~700μm).

In this paper, we revolutionize the concept of the potential drop technique to facilitate the measurement during spalling of conductor substrates. The modification overcomes the difficulty associated with the geometric constraints of the low temperature spalling setup and it allows, for the first time, to track the position and velocity of a dynamic crack with high spatial and temporal resolution during initiation, propagation and arrest. We refer to this technique as "In-plane Crack Dynamics" (ICD).

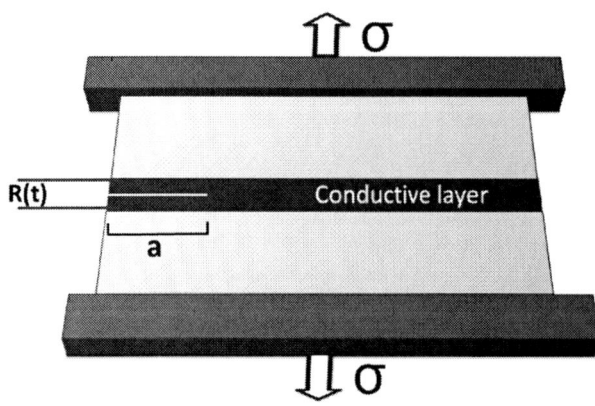

Figure 1. Schematic view of a traditional voltage drop technique system in a tension test with an applied load σ, crack length a, and the resistance of the material $R(t)$. Redrawn based on Fineberg *et al.*[13]

III. IN-PLANE CRACK DYNAMICS

The main characteristic in the traditional potential drop techniques is the conductor strip located at the top, bottom or side of the substrate that is ruptured during fracture. As the fracture advances, the resistance of the strip changes, and provides exact information on the position and speed of the crack. However, for any spalling technique in silicon or any other brittle material, a stressor layer is bonded to the bottom of a thin substrate through deposition or epoxy [14]–[16]. Thus, it is not possible to successfully use the bottom or side of the substrate to deposit the conductive strip. However, the benefit of using a semiconductor as silicon allows the implementation of a more advantageous configuration.

In this novel ICD technique, two electrode strips made of silver are deposited by thermal evaporation at 10^{-7} Torr on the surface of the silicon substrate, separated by 5mm, with a thickness of approximately 200nm. A low silver strip to substrate thickness ratio is important to assure that the silver strips do not affect the spalling process.

The silver strips are then connected to a Wheatstone bridge as shown in Fig.2. The Wheatstone bridge is powered by batteries for noise reduction in the measurements with a voltage V_{in} applied. The voltage output of the bridge (V_b) is digitized at a rate of 2 MHz by a PXIe 12-bit Analog to Digital converter from National Instruments. The resistance of the substrate between electrodes, R_x, is obtained by measuring V_b, and using the known values for the bridge resistors R_a, R_b, R_c and V_{in} following Eq. 1.

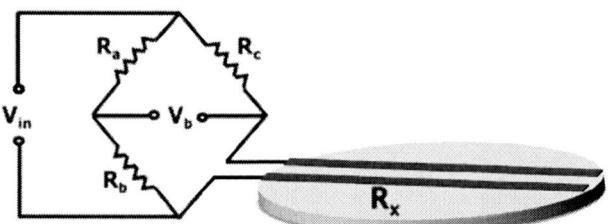

Figure 2. The diagram shows the system composed by the Wheatstone bridge and silicon substrate with two silver strips deposited on top as electrodes. It consists of three known resistors (R_a,R_b,R_c), a silicon substrate with resistance R_x, a battery (V_{in}) and the measured voltage output of the bridge (V_b).

$$R_x = \frac{\dfrac{R_b}{R_a + R_b} + \dfrac{V_b}{V_{in}}}{1 - \left(\dfrac{R_b}{R_a + R_b} + \dfrac{V_b}{V_{in}} \right)} \cdot R_c \qquad (1)$$

The resistance R_x is a function of the crack length and is sensitive to the geometry of the silver strips and the substrate. The relations are explained in the following:

The resistance, R_x, across the two metal strips depends on three factors, the resistivity of the substrate, ρ, the distance between both metal strips, d, and the cross sectional area, A,

$$R_x = \frac{\rho \cdot d}{A} = \frac{\rho \cdot d}{L \cdot T} \qquad (2)$$

where A can be expressed as product of the length of the strips, L and the thickness of the wafer, T, as seen in Fig.3.

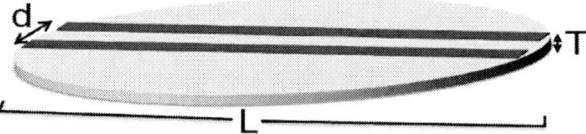

Figure 3. Schematic of the silicon substrate with the most relevant dimensions.

As the crack propagates from a pre-made notch on the side of the substrate, the thickness of the measured substrate varies as depicted in Fig. 4. When the crack is propagating, there are two distinct sections that contribute to the resistance between the strips; one related to the section that has been spalled and; one other section that has not been spalled yet. The spalled section of the substrate will have a thickness $t = T\text{-}w$, where T is the entire substrate thickness and w the thickness of the final spalled wafer. The length of the spalled section, $a(t)$, is time-dependent and defined by the position of the crack front.

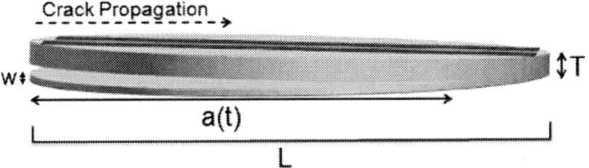

Figure 4. Schematics of system configuration as the crack propagates from left to right.

The un-spalled part of the substrate has a thickness, T, and a time-dependent length of $l(t) = L - a(t)$. Therefore, the resistance of the substrate changes with time following the equation:

$$R_x(t) = \rho d \left(\frac{1}{a(t)(T-w) + \big(L-a(t)\big)T} \right) \qquad (3)$$

Rearranging this equation, the crack length is:

$$a(t) = \frac{1}{w} \left(LT - \frac{\rho d}{R_x(t)} \right) \qquad (4)$$

Thus, by simply differentiating the crack position data for every point in time, the crack velocity can be calculated at any point in distance or time.

Using this methodology to investigate the crack propagation velocity we would typically observe that the velocity profile would start at zero as long as the stress field at the crack tip, K, is below a critical value, K_{IC} [17]. Once the crack starts propagating, the velocity will increase abruptly and, after that, the crack could keep accelerating, slow down or arrest if the stress field at the tip is below the critical value, K_{IC}. In Fig. 5 we show preliminary data of the crack position and velocity versus time using In-plane Crack Dynamics technique.

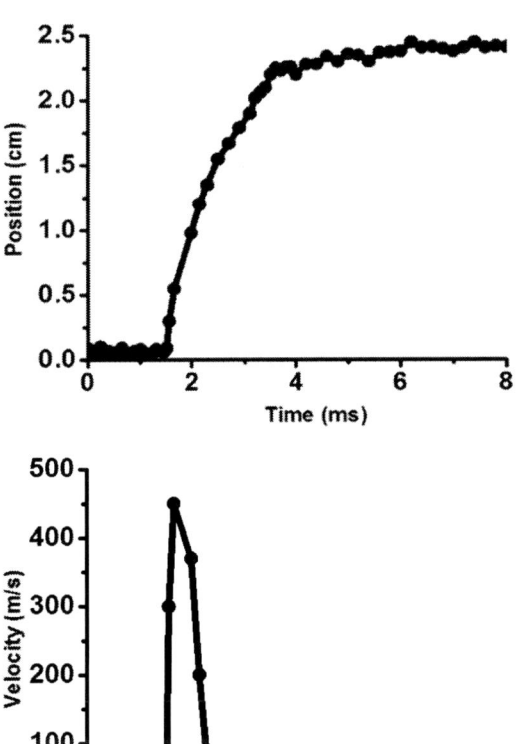

Figure 5. Illustration of position and velocity calculation for a crack during spalling. The graph ranges are approximate but representative for the measurement setup.

The correlation of this crack velocity to the stress applied to the system and to the surface roughness of the spalled wafer is ongoing work and will be presented in a future report. These new insights into the dynamics of propagating cracks are crucial for the further understanding of the relationship between surface quality and crack velocity, with the ultimate goal of being able to control crack motion to achieve a desired roughness.

978-1-5090-5606-4/17 $31.00 © 2017 IEEE   2612

IV. CONCLUSION

In this paper, we have introduced a new technique for measuring the velocity of a propagating crack during spalling denominated "In-plane Crack Dynamics" (ICD).

To the best of our knowledge, this technique is the first one capable to provide a direct measurement of crack velocity during the spalling of silicon wafers or any other semiconductor or conductor material. The developed technique grants full information on the entire fracture process, including initiation, spalling, and arrest with high spatial and temporal precision.

One of the main issues with spalling is the control of the crack propagation and the resulting poor surface quality of the spalled wafers. This new technique enables the collection of significantly more information collected about the crack velocity and in overcoming the main barrier of spalling, this new crack measurement technique plays a key role in gaining full control over the crack dynamics and surface roughness.

REFERENCES

[1]   P. Bellanger et al., "New stress activation method for kerfless silicon wafering using Ag/Al and epoxy stress-inducing layers," *IEEE J. Photovoltaics*, vol. 4, no. 5, pp. 1228–1234, 2014.

[2]   F. Dross et al., "Stress-induced large-area lift-off of crystalline Si films," *Appl. Phys. A*, vol. 89, no. 1, pp. 149–152, 2007.

[3]   J. Vaes, A. Masolin, a. Pesquera, and F. Dross, "SLiM-cut thin silicon wafering with enhanced crack and stress control," *Soc. Photo-Optical Instrum. Eng. Conf. Ser.*, vol. 7772, p. 777212, 2010.

[4]   P. Bellanger, A. Slaoui, A. Minj, R. Martini, M. Debucquoy, and J. M. Serra, "First Solar Cells on Exfoliated Silicon Foils Obtained at Room Temperature by the SLIM-Cut Technique Using an Epoxy Layer," *IEEE J. Photovoltaics*, vol. 6, no. 5, pp. 1115–1122, 2016.

[5]   A. Masolin, "Fabrication and Characterization of Ultra-Thin Silicon Crystalline Wafers for Photovoltaic Applications using a Stress-Induced Lift-Off Method," KU Leuven, 2012.

[6]   D. Sherman, "Velocity dependent crack deflection in single crystal silicon," *Scr. Mater.*, vol. 49, no. 6, pp. 551–555, 2003.

[7]   J. Boudet and S. Ciliberto, "Interaction of Sound with Fast Crack Propagation," *Phys. Rev. Lett.*, vol. 80, no. 2, pp. 341–344, 1998.

[8]   H. Wallner, "Linienstrukturen an Bruchflachen," *Zeitschrift fur Phys.*, vol. 114, no. 5–6, pp. 368–378, 1939.

[9]   K. Arakawa, "Relationships between fracture parameters and fracture surface roughness of brittle polymers," *Int. J. Fract.*, vol. 48, no. 2, pp. 103–114, 1991.

[10]   A. Rabinovitch, V. Frid, and D. Bahat, "Wallner lines revisited," *J. Appl. Phys.*, vol. 99, no. 7, pp. 1–4, 2006.

[11]   J. A. Hauch and M. P. Marder, "Energy Balance in Dynamic Fracture, Investigated by a Potential Drop Technique," *Int. J. Fract.*, vol. 90, no. 2, pp. 133–151, 1998.

[12]   T. Cramer, A. Wanner, and P. Gumbsch, "Crack velocities during dynamic fracture of glass and single crystalline silicon," *Phys. Stat. Sol.*, vol. 164, p. R5, 1997.

[13]   E. Sharon and J. Fineberg, "Microbranching instability and the dynamic fracture of brittle materials," *Phys. Rev. B*, vol. 54, no. 10, pp. 7128–7139, 1996.

[14]   K. Lobato et al., "Comparative study of stress inducing layers to produce kerfless thin wafers by the Slim-cut technique," *2013 IEEE 39th Photovolt. Spec. Conf.*, pp. 0177–01180, 2013.

[15]   R. Niepelt, J. Hensen, V. Steckenreiter, R. Brendel, and S. Kajari-Schöder, "Kerfless exfoliated thin crystalline Si wafers with Al metallization layers for solar cells," *J. Mater. Res.*, vol. 30, no. 21, pp. 3227–3240, 2015.

[16]   S. W. Bedell et al., "Kerf-less removal of Si, Ge, and III-V layers by controlled spalling to enable low-cost PV technologies," *IEEE J. Photovoltaics*, vol. 2, no. 2, pp. 141–147, 2012.

[17]   R. C. Petersen, "Accurate Critical Stress Intensity Factor Griffith Crack Theory Measurements by Numerical Techniques," *Sample J.*, pp. 737–752, 2013.

New Findings of Thermal Effect on pm-Si:H Solar Cells Optoelectronic Properties

L. Hamui[1,2*], L. A. Gómez-González[2], G. Santana[2]

[1] Facultad de Ingeniería, Universidad Anáhuac, Av. Universidad Anáhuac 46, Col. Lomas Anáhuac, Huixquilucan, Estado de México, México, C.P. 52786.

[2] Instituto de Investigaciones en Materiales, Universidad Nacional Autónoma de México. A.P. 70-360, Coyoacán, C.P. 04510, México, D.F.

Abstract — **This work describes the effect of temperature on pm-Si:H PIN and NIP structures deposited by PECVD. Temperature dependent measurements were conducted to understand hydrogen diffusion and cell performance by analyzing structural and optical properties. Hydrogen rearrangement on films and microstructure changes were observed. The total hydrogen that effuses from PIN structure is lower than for the NIP related to defects creation after induced hydrogen diffusion, break of Si-H bonds and the generation of dangling bonds. These could give significant evidence for the development of a more stable device during operation besides an enhanced functional performance of this type of structure.**

Index Terms — **polymorphous silicon, photovoltaic cell, optical properties, diffusion**

I. INTRODUCTION

Pm-Si:H is a material composed by silicon nanocrystals in an a-Si:H matrix and suitable for solar cell devices. The latter is because it exhibits better electronic transport and stability properties compared to those of the conventional a-Si:H [1]-[3]. Most of the improved properties related to these materials are a consequence of the hydrogen ability to passivate defects and impurities in the Si network. Stable devices and higher efficiencies for solar energy conversion are the goals of the solar cell industry. Nevertheless, a-Si:H has a light-induced degradation effect referred to as the Staebler–Wronski Effect (SWE) [3]. Which then act as recombination centers, greatly reducing the efficiency during device operation [4], [5]. The latter indicates that pm-Si:H could be an important materials due to its higher stability. Furthermore, the change on the temperature of the device may induce a similar effect and a reduction on the solar cell efficiency may be presented. Bonded hydrogen atoms may break with temperature and diffuse through the amorphous network generating an increase of the dangling bonds. These atoms may be trapped by the dangling bonds within the film or may exodiffuse leading to a film richer in defects [6]. The trapping probability of free hydrogen atoms increases with temperature due to a higher dangling bonds formation and it is more pronounced closer to the film surface, which shortens the diffusion length [6]. So

due to its importance, hydrogen bonding and stability should be analyzed, in order to determine the hydrogen bonding environment and configuration within the dependent device structure. At moderate temperatures (100-300 °C), a slow diffusion was observed while at higher temperatures (400-800°C) fast hydrogen interstitial diffusion is present [7]-[8]. During hydrogen diffusion pm-Si:H can be crystallized due to subsurface reactions of bond breaking taking place as part of the exodiffusion process [8]. However, in order to understand the temperature effect on pm-Si:H solar cells optoelectronic properties, in this work we examined the PIN and NIP structures properties and correlate them with exodiffusion experiments. Their structural and optical properties were analyzed in terms of the different hydrogen diffusion processes for samples exposed to heat. A comparison of both structures was done in order to determine which one is less affected. Furthermore, the last could give significant evidence for the development of a more stable device during operation along with an enhanced functional performance of this type of structure.

II. MAIN EXPERIMENTAL

Pm-Si:H thin films PIN and NIP structures were grown in a conventional plasma enhanced chemical vapor deposition (PECVD) in ARCAM system with parallel plates, activated by a RF signal of 13.56 MHz [9]. The samples were deposited on ITO and on corning glass #7059 substrates in both configurations for the different characterization techniques. The depositions were done with a chamber pressure of 3.3 Torr, substrate temperature of 275 °C and a RF power of 30 W in order to obtain good quality films and relatively high deposition rate (5.2-6.3 Å/s). Hydrogen (H_2) and silane (SiH_4) were used as precursor gases with mass flow rates of 200 and 40 sccm, respectively. Doped layers were grown using conditions similar to previous works [7]. Doping is obtained by adding 5 sccm of a doping gas (trimethylboron for p-doping and phosphine for n-doping) diluted at 2% in hydrogen. To avoid any oxidation effect, the films were grown in a single process without breaking vacuum. The p and n type layers were around 20-30 nm of thickness, while the intrinsic layer was ~200 nm thick. The parameters for both structures were maintained constant and only the order of deposition of the layers was varied. Raman spectroscopy was used to investigate the nanostructure of the film via a Jobin-Yvon Horiba triple monochromator T6400 micro-Raman spectrometer in backscattering configuration equipped with a He–Ne laser (632 nm) excitation. All the measurements were performed at room temperature in open air. A low incident power (2 mW) was selected in order to avoid any beam-induced structural changes during measurements. The spectrometer is equipped with a CCD detector and a confocal microscope with a 10 x objective. Thermal annealing during exodiffusion experiments was done with an incremental ramp of 10°C/min starting at room temperature and up to 500 °C for hydrogen evolution studies in all the pm-Si:H structures. The temperature limit of 500 °C was selected because it has been shown on previous works that this temperature represents the maximum of the exodiffusion, indicating that the major part of the hydrogen content already effused from the structures. These experiments were performed in a vacuum chamber with a controlled heating system which increases the temperature and the effusing hydrogen was detected with a mass spectrometer. Dark and illuminated IV curves were recorded on a Keithley 2400 programmable electrometer on planar geometry using aluminum electrodes evaporated in a vacuum lower than 10^{-5} Torr. A halogen lamp with an intensity of 100 mW/cm^2 and similar spectrum to that of the solar spectrum was selected for illuminated electrical measurements.

III. RESULTS AND DISCUSSION

TABLE I
DESCRIPTION OF THE PIN AND NIP STRUCTURES.

Structure	Total thickness (nm)	Substrate
PIN	243.7	glass/ SnO$_2$
NIP	248.3	glass/ SnO$_2$

Table I shows a general description of the PIN and NIP structures studied on this work. Both structures were deposited on the following order glass/ SnO$_2$ / structure (PIN or NIP). The total thickness (~246 nm) is similar for both structures and the growth parameters were maintained constant to be able to compare these devices. Both structures were heated to 500 °C, during exodiffusion, in order to understand temperature effect on the optoelectronic and structural properties of these devices.

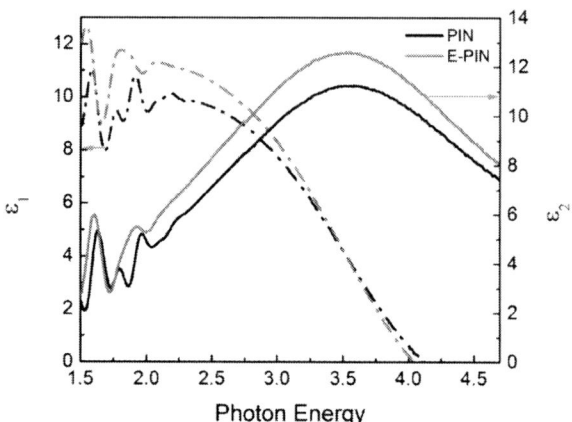

Fig. 1. Real and imaginary part of the pseudo-dielectric function for the pm-Si:H PIN structure as grown and after exodiffusion.

The real and imaginary part of the pseudo-dielectric function for the pm-Si:H PIN and NIP structure as grown and after exodiffusion are shown on figure 1 and 2. It can be observed that these spectra are very different indicating that the structure order affects the properties of the device. The film structural and optical properties varied with slightly with the thermal annealing. In figure 1 it can be observed that the PIN structure after exodiffusion presents a shift to higher values for ε_2 and it is observed as well for e1 for values of energy lower than 3 eV. The last could be related to a film structure order change due to hydrogen diffusion during annealing. In figure 2, there is an important change on both spectras and it is more pronounced on imaginary part of the pseudodielectric function maybe related to a variation on interface quality. On the other hand, it can be appreciated that the exodiffussion experiment decrease the value of ε_1, contradictory to what is observed on the PIN structure (figure 1). The last indicates that there is a different diffusion process depending on the film order (device architecture). To understand these effects during annealing, exodiffusion experiments were conducted.

The hydrogen exodiffusion spectra is shown on Figure 3 for both pm-Si:H structures. First, it can be observed that the hydrogen exodiffusion is more intense for the NIP and also it means that a larger amount of hydrogen effuses from the sample. The hydrogen exodiffusion depends on the device structure. While hydrogen exodiffusion is more pronounced for the NIP structure also a shoulder ~380 °C is observed. The lack of appearance of the shoulder on the PIN structure is dependent of the hydrogen diffusion mechanism which are different from the NIP structure that require less energy for these process.

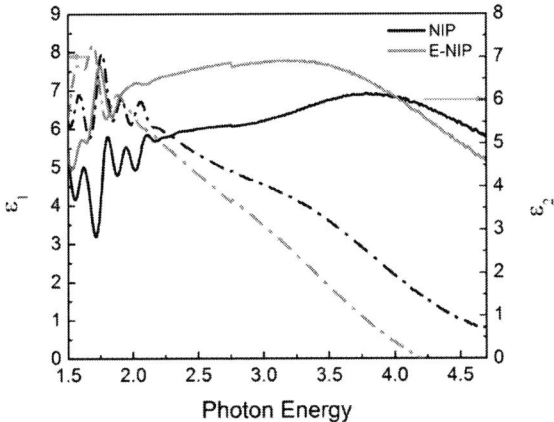

Fig. 2. Real and imaginary part of the pseudo-dielectric function for the pm-Si:H NIP structure as grown and after exodiffusion.

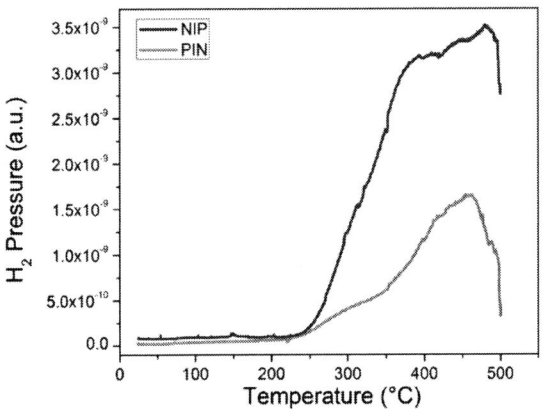

Fig. 3. Exodiffusion spectra for the pm-Si:H PIN and NIP structures as grown.

Figure 4 shows the evolution of the Raman spectra for the NIP and PIN structures before and after exodiffusion. First, it is observed that both structures present similar Raman spectra. Moreover, it can be observed a blueshift after exodiffusion for both structures but more pronounced for the PIN. The fact that a blueshift occurs could be related to an increase of desorder on the films similar to the observation made on SE. It is reasonable that during exodiffusion a large amount of hydrogen is concentrated on the surface of the PIN top film generating a change on the microstructure that can be apreciated with Raman Spectroscopy. Nevertheless, this observation is not valid for the NIP structure due to the large effusion presented during annealing compared to the PIN structure.

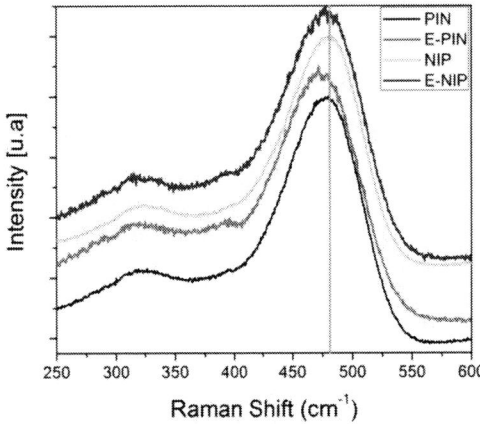

Fig. 4. Evolution of the Raman spectra for the NIP and PIN structures before and after exodiffusion.

IV curves were conducted for the PIN and NIP strustures as observed on figure 5 and 6 respectively. In both figures it is shown dark IV curves compared to the iluminated IV Curves. First it can be observed a big change due to ilumination and a variation of the iluminated curve shape between the structures. Both structures present an ohmic behavior on the dark-IV curve. The PIN structure IV curve under illumination is more affected than the NIP structure. Second, the Voc is larger for the NIP structure 0.15 V compared to the PIN structure 0.04 V. Moreover, the PIN Isc is larger (0.12 mA) compared to the NIP structure (0.08 mA)

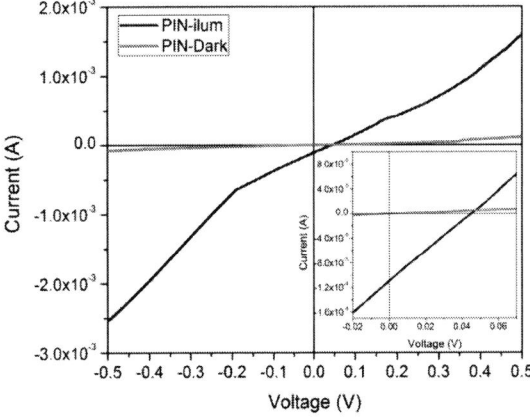

Fig. 5. Dark IV curves for the pm-Si:H PIN structure and evolution of the IV curve as a consequence of light exposition.

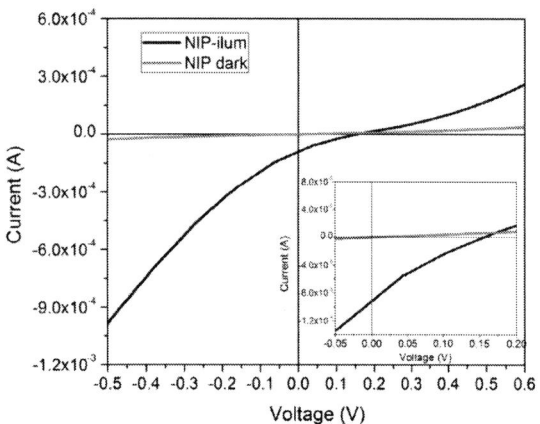

Fig. 6. Dark IV curves for the pm-Si:H NIP structure and evolution of the IV curve as a consequence of light exposition.

These structures Dark-IV curves show an ohmic behavior with similar resistivity. For illuminated IV curves it can be observed a diode effect which is shifted from zero to negative values due to the light exposition effect. This shift for both structures present very small Voc and Isc which is not very desirable on solar cell devices but may be improved.

IV. CONCLUSIONS

The film structural and optical properties varied with slightly with the thermal annealing. The crystalline fraction of these structures is observed to be maintained almost constant. A peak appearance around 400 cm-1 is observed on Raman spectra with the thermal annealing. The real and imaginary part of the pseudo-dielectric function is dependent of the device structure and is augmented with the thermal annealing for the PIN structure. Hydrogen exodiffusion is more pronounced for the NIP structure and it presents a shoulder ~380 °C. Both structures present an ohmic behavior on the dark-IV curve. The PIN structure IV curve under illumination is more affected than the NIP structure.

ACKNOWLEDGMENTS

We acknowledge partial financial support for this work from DGAPA-UNAM PAPIIT Projects IB101612 and IN100914, CONACyT Mexico under projects 153948 and 179632, SENER-CONACyT project 151076 and Universidad Anáhuac.

REFERENCES

[1] R. Butté, S. Vignoli, M. Meaudre, R. Meaudre, O. Marty, L. Saviot, P. Roca i Cabarrocas, "Structural, optical and electronic properties of hydrogenated polymorphous silicon films deposited at 150°C", J. Non-Cryst. Solids Vol. 266-269, pp. 263–268, 2000.

[2] M.Y. Soro, M.E. Gueunier-Farret, J.P. Kleider, "Structural and electronic properties of hydrogenated polymorphous silicon films deposited at high rate", J. Appl. Phys. Vol. 109 pp. 023713, 2011.

[3] L. Hamui, A. Remolina, M.F. Garcia-Sanchez, A. Ponce, M. Picquart, M. López-López, B.M. Monroy, G. Santana, "Deposition, opto-electronic and structural characterization of polymorphous silicon thin films to be applied in a solar cell structure", Materials Science in Semiconductor Processing Vol. 30 pp. 85–91, 2015.

[4] D.L. Staebler, C.R. Wronski, "Reversible conductivity changes in discharge produced amorphous Si", Appl. Phys. Lett. Vol. 31, pp. 292-294, 1977.

[5] H.M. Branz, "Hydrogen collision model: quantitative description of metastability in amorphous silicon", Phys. Rev. B Vol. 59, pp. 5498-5512, 1999.

[6] D. Smeets, B.C. Johnson, J.C. McCallum, C.M. Comrie, "Real-time in situ study of hydrogen diffusion in amorphous Si formed by ion implantation", Nucl. Instr. and Meth. in Phys. Res. B Vol. 269, pp. 2657-2661, 2011.

[7] N. Pham, A. Hadjadj, P. Roca i Cabarrocas, O. Jbara, F. Kail, "Interpretation of the hydrogen evolution during deposition of microcrystalline silicon by chemical transport", Thin Solid Films Vol. 517, pp. 6225-6229, 2009.

[8] L. Hamui, B.M. Monroy, P. Roca i Cabarrocas, G. Santana, "Effect of light-soaking on the hydrogen effusion mechanisms in polymorphous silicon thin film structures", Materials Chemistry and Physics Vol. 163, p.p. 311-316, 2015.

[9] K.H. Kim, E.V. Johnson, P. Roca i Cabarrocas, "Irreversible light-induced degradation and stabilization of hydrogenated polymorphous silicon solar cells", Solar Ener. Mater. & Sol. Cells, Vol. 105, pp. 208–212, 2012.

Study of PV module degradation rate prediction through correlation of field-aged and accelerated-aged module degradation data

Babak T. Hamzavy[1], William J. Grieco[1], Brian J. Fields[1], Cara S. Libby[2], William B. Hobbs[3], Olga Lavrova[4], and C. Birk Jones[4]

(1) Southern Research Institute; (2) Electric Power Research Institute; (3) Southern Company Services, Inc.; (4) Sandia National Laboratories

Abstract — The Electric Power Research Institute (EPRI) is leading a DOE-funded project to study photovoltaic (PV) module degradation rates and mechanisms. The project is intended to advance the state of the art in terms of module certification and wear-out degradation prediction certainty with an emphasis on the correlation of field-aged and accelerated-aged module degradation data. The methodologies described are applied to fielded PV modules at a utility-scale plant with multiple years of on-sun exposure and reserve modules from the same plant. Accelerated aging of the reserve modules will be performed with two common qualification standards: IEC 61215 and Qualification Plus. Interim results are reported on the module selection approach, variability encountered vis-à-vis bill of materials, electroluminescent and thermal imaging evaluation, visual inspection methodology, manufacturer and lab-measured power rating comparison, and selection of the utility-scale project for this study.

Keywords: LCOE, Degradation, Qualification, Electroluminescent, Reliability, Mechanism, BOM, and Backsheet.

I. INTRODUCTION

The remarkable growth of the solar industry within the last 15 years, along with innovations in module materials and fabrication methods, has driven down module costs. As a result, the 3¢/kWh LCOE (levelized cost of electricity) target set by the U.S. Department of Energy's (DOE) SunShot Program is becoming a more achievable goal, and solar is already competitive with conventional energy resources in some markets. Accurate assessment of PV module degradation rates is still a difficult task. Additionally, it is unclear whether the degradation rate is linear or non-linear [1]. This is partly due to the difficulty in establishing feasible time scale conditions in laboratory testing that mimic the service environment. The difficulty in assessing degradation rates accurately is compounded by an often evolving bill of materials in PV module manufacturing. Many component materials have changed significantly over the last decade, a time span considerably shorter than the typical 25-year module life expectancy [2].

The EPRI project is funded through the DOE's Physics of Reliability: Evaluating Design Insights for Component Technologies in Solar 2 (PREDICTS2) funding opportunity. It is a collaborative study with Southern Research (SR), Southern Company Services, Inc. (SCS), and Sandia National Laboratories (SNL). Testing and non-destructive evaluation are conducted at the Southeastern Solar Research Center (SSRC) and its PV laboratory on SR's campus (Fig.1), and destructive evaluation will take place at SNL. The study aims to compare degradation rates in reserve modules exposed to qualification

Fig.1: SSRC and PV laboratory

test conditions in IEC 61215 and Qualification Plus (Qual Plus) and correlate them with degradation rates observed in field-aged modules. This paper provides a description of the project approach and interim results.

II. SELECTION OF UTILITY SCALE PF PV SYSTEM.

A survey was distributed to industry contacts at more than 80 organizations to identify commercial PV plants with sufficient reserve modules for testing. Required criteria included a minimum of 40 reserve multi-crystalline modules in pristine condition and of the same make, model and vintage as the fielded modules. A plant that met the required conditions was identified, and reserve modules were shipped to the SSRC test facility.

The preliminary visual assessment results revealed differences in the bill of material, power rating, and unique details associated with PV cell binning within a module. These observations compelled the team to compare the quality and consistency of the field and reserve samples.

The selected plant provided 57 modules of the same model number, a large enough pool to allow for the selection of the desired sample count (36 modules) for this study. The modules have a 12 x 6 cells layout -, based on 72 polycrystalline cells with 3 bus bars. Of the 57 modules, 16 have a power rating of 285 W and 41 have a rating of 290 W. All modules meet the manufacturer-specified rating within ±3%. Two distinct backsheet types were identified generically as A and B, with 21 of the former and 36 of the latter.

978-1-5090-5606-4/17 $31.00 © 2017 IEEE

A. Characterization approach

A visual assessment was performed using infrared (IR) and electroluminescent (EL) imaging. First, IR aerial imaging (Fig.2) of the fielded modules was performed to identify areas with minimal hot spots.

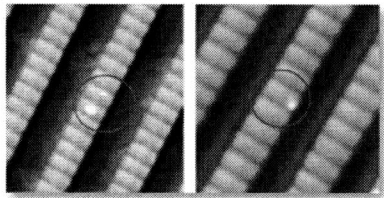

Fig.2: Aerial IR images from the utility system; hotspots identified (imaging by Heliolytics Inc- www.heliolytics.com .

In total, 72 modules from these defect-free areas of the plant were subjected to *in situ* EL imaging (Fig. 3) at night using a portable power supply [3]. Similarly, reserve modules were subjected to EL imaging in the laboratory dark room at SSRC. A side by side comparison of EL images for field and reserve module samples (Fig. 3) showed that the reserve modules selected for the experiment have similar visual characteristics as those on sun in the service environment. This visual

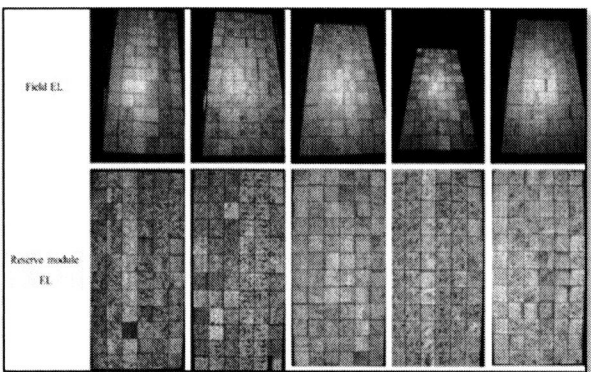

Figure 3: EL images of fielded and reserve module samples

assessment was performed in an effort to ensure that the reserve modules are of comparable quality to those currently in the service environment.

Two observers compared the EL images for the reserve and fielded modules, counting the defects in each image. Gage R&R was used to measure consistency between the defect counts in the reserve and fielded modules (Fig. 4). The classification of defects consisted of cell damage in the form of fractures and dark areas observed along the edge of the cell as shown (Fig. 6). One observer identified an average of 0.9 defects per reserve module and 1.4 defects per fielded module, and the second observer identified 2.1 and 2.8 defects per module for reserve and fielded modules, respectively. There is a measurable difference between the average counts for reserve and fielded modules, but there is not enough information to

connect the relatively low defect counts to long-term performance outcomes; therefore, it was decided not to use the defect count as a tool for selecting reserve modules for testing

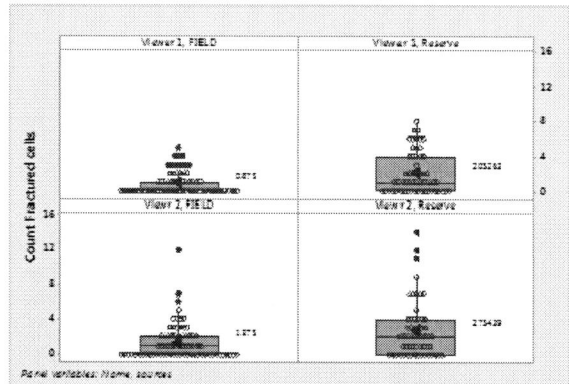

Fig.4: Cell fracture count comparison for 2 observers.

in the study. Instead, a randomized sample of 36 modules was selected using serial numbers from the 57 available reserve modules. The sample population was divided into three batches of 12 modules (one set of 12 for IEC-61215 and two sets of 12 for Qual Plus). Both backsheet types (set A and B) and both power ratings (285 and 290 W) are represented in the set of 36 but no effort was made to intentionally select for either characteristic.

The power characteristics of all 36 samples were measured using a Pasan solar simulator. The distribution of the measured max output power as compared to the module's nameplate power rating is shown (Fig. 5). The differences seen in power rating within the same module design are likely a function of cell binning, which impacts the power and other electrical characteristics of the module. These differences are believed to

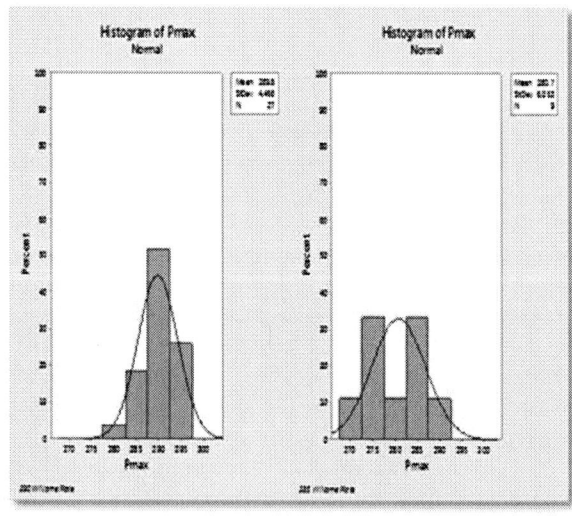

Figure 5: Reserve sample power distribution; Left module rated at 290 W, Right module rated at 285 W.

occur frequently and may provide an additional perspective for better understanding cell binning during manufacturing and its influence on long-term degradation rate.

B. Observed module defects

The visual assessment technique utilized EL imaging at the Isc (short circuit current) specified on the module's nameplate. The field imaging, however, used a lower current level than module Isc, limited by the availability of a portable power supply with Isc level output. Several types of defects were identified in reserve samples and fielded samples. Despite the difference in the amount of forward bias current supplied to the reserve and fielded modules, the image quality did not appear to be impacted. M. Köntges *et al.* [4] suggests that long-term performance is most strongly associated with cell fractures proximate and parallel to bus bars and fractures that impact a large area of the cell. IR imaging of all modules was performed on sun under ≥ 800 Wm^{-2} solar irradiance in short circuit configuration. The cell temperature under this outdoor condition showed an inverse relation with the magnitude of the EL signal. While the correlation is not exact, the worst EL cell images (i.e., cell uniformly lit in forward bias) matched the hottest cells in the IR images (Fig.6A). A glossary of the defect types seen among the field and reserve modules is shown in Fig.6B. Cells that are dim in EL are more susceptible to reverse bias and therefore exhibit higher resistive behavior vis-à-vis hot spots. Fig. 6C shows I-V (current-voltage) curve traces for cells with low and high temperatures, as revealed by IR images, in a module [5]. The I-V trace was obtained using a Pasan solar simulator where a low temperature cell and a high temperature cell were alternately shaded and flashed. When the higher temperature cell experiences reverse bias (when shaded) it displays a considerably higher series resistance that results in a high operating temperature impacting the module performance overall. Such a cell(s) can also impact the bypass diode [5]. The latter combined with dirt, bird droppings, and other natural occurrences of shading can influence the string performance.

C. Test procedure

IEC 61215 and Qualification Plus (Qual Plus) are qualification standards that have been applied in the solar industry as a means of evaluating the durability of new module designs and bill of materials [6]. The standards are not used to assess PV module reliability in the service environment; rather, they are effective in identifying infant mortality issues as a result of poor module design and bill of material. Despite a lack of correlation with service life and accurate estimation of degradation rate, both methods are widely used by manufacturers to assess the robustness of their module design. In many cases, manufacturers subject modules to repeated cycles of IEC 61215, with a focus on testing to module failure.

In this study, reserve modules will be exposed to two important stress factors present in the service environment. The relevant stress factors are thermal cycling (10.11) and humidity

freeze (10.12) per IEC 61215. These conditions test the integrity of a PV module and the resiliency of the circuitry. These stresses combined with dynamic mechanical loading, as defined in the Qual Plus standard will be used to stress the reserve module samples and induce degradation. I-V sweeps of the reserve modules will be measured at multiple points in the testing protocols and compared to the I-V sweeps for fielded modules as well as module performance at the test onset. The planned testing sequence for this study is shown in Fig.7A. The pre-conditioning and the UV pre-exposure is specified by IEC 61215. The team has developed several methods, with corresponding algorithms, for quantifiable detection of degradation changes of all PV modules. I-V sweeps will be collected for individual PV modules and at the string-level, and time series measurements will be obtained for the duration of the project to detect changes in performance relative to baseline measurements. The degradation detection methods will follow the multi-step process shown in Fig. 7B as well as using EL imaging technique described in [3]. These methods have been validated by applying them to fault detection and classification in the context of performance of the module in situ on sun [7,8].

Automatic fault detection and review of in-situ I-V curves was performed using a Support Vector Machine (SVM) and Genetic Programming (GP) algorithm to first classify the existing condition, and then to estimate ideal I-V curve behavior at a given plane-of-array (POA) irradiance and module temperature value. Classification was used to determine whether a PV string was performing well or experiencing fault behavior. The estimate of normal behavior was performed such that a potential loss of electrical power caused by a fault condition could be calculated. The classifier determined whether the particular I-V curve was a fault or not. If a fault was not detected, the curve was determined to be normal and the process ended. However, if a fault was found then the GP regression algorithm estimated the potential I-V curve under normal operating conditions [7,8]. Based on this estimate, the lost power was calculated by comparing it to the

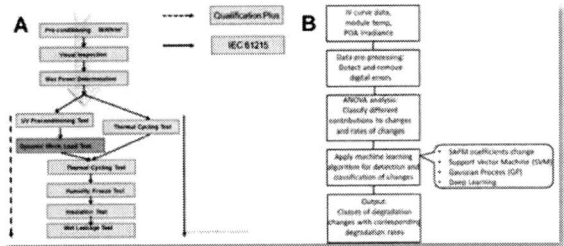

Fig.7: (A) IEC 61215 and Qual Plus flow charts; (B) Multi step process to measure degradation rate.

actual I-V curve data for the particular instance. Furthermore, to detect individual module degradation, individual I-V curves from Qual Plus samples will be correlated with fielded samples at the utility plant. Identical individual module I-V tracers will be installed on reserve modules tested on sun at SR and on select fielded modules at the utility plant from which the reserve modules were supplied. Real-time I-V sweeps will be collected for the project duration, after IEC 61215 and Qual Plus testing are completed. Qualitative and quantitative changes in I-V curves between lab-aged reserve modules and fielded modules will then be compared and correlated on a continuous basis, which will provide information from which to derive degradation rates. Historical trending information of all other relevant variables will be used to provide further correlation between performance changes and degradation of field-aged and lab-aged modules, which will also provide information with which to derive degradation rates.

III. CONCLUSION

IEC 61215 and Qualification Plus test methodologies are used at SR to assess the performance of reserve PV modules from a utility scale PV power plant. The module selection approach involved a visual assessment of lab-based EL images for the experimental samples compared to in-situ EL images for modules of identical design in the service environment. A measurable difference between the average defect counts for reserve and fielded modules was detected, but there was not enough information to connect the relatively low defect counts to long-term performance outcomes. This provided the necessary justification to randomly choose 3 sets of 12 modules for this study; 12 modules for IEC 61215, 2 sets of 12 each for Qualification Plus. It is also expected that the differences noted in the nameplate power rating will provide additional insights on module performance as a function of cell binning that may manifest in the visual differences seen in the EL images as well as power output. The test of IEC 61215 set has commenced. The Qual Plus samples are currently being exposed to UV preconditioning (15 kWhm^{-2}), and this is followed by dynamic mechanical load testing (Fig. 7A).

IV. ACKNOWLEDGEMENT

The information, data, or work presented herein was funded in part by the Office of Energy Efficiency and Renewable Energy (EERE), U.S. Department of Energy, under Award Number DE-EE0007137. Sandia National Laboratories is a multi-mission laboratory managed and operated by National Technology and Engineering Solutions of Sandia, LLC., a wholly owned subsidiary of Honeywell International, Inc., for the U.S. Department of Energy's National Nuclear Security Administration under contract DE-NA0003525.

REFERENCES

[1] Manuel Va´zquez, and Ignacio Rey-Stolle, "Photovoltaic Module Reliability Model Based on Field Degradation Studies," PROGRESS IN PHOTOVOLTAICS: RESEARCH AND APPLICATIONS 2008; 16:419–433 .

[2] Alexander Z. Bradley, "Materials can be key to differences in module durability", http://www.dupont.com/content/dam/dupont/products-and-services/solar-photovoltaic-materials/solar-photovoltaic-materials-landing/documents/PV-Tech-article.pdf

[3] Lockridge, Britny P., Olga Lavrova, and William J. Hobbs. "Comparison of electroluminescence image capture methods." Photovoltaic Specialists Conference (PVSC), 2016 IEEE 43rd.

[4] M. Köntges, S. Kajari-Schröder, I. Kunze, U. Jahn, "CRACK STATISTIC OF CRYSTALLINE SILICON PHOTOVOLTAIC MODULES," 26th European Photovoltaic Solar Energy Conference and Exhibition, 2011.

[5] Babak Hamzavy, Alexander Z. Bradley, Safety and Performance Analysis of a Commercial Photovoltaic Installation. SPIE, 2013.

[6] Soh Suzuki [1,2], Tadanori Tanahashi[2], Takuya Doi[3] and Atsushi Masuda[3], "An examination of the acceleration method of thermal cycling test for crystalline silicon PV modules."
[1]Photovoltaic Power Generation Technology Research Association (PVTEC), Japan
[2]ESPEC, Japan, [3]National Institute of Advanced Industrial Science and Technology (AIST), Japan (NREL PV reliability Workshop, February 2017)

[7] Jones, C. B., Martínez-Ramón, M., Smith, R., Carmignani, C. K., Lavrova, O., Robinson, C., & Stein, J. S. (2016, June). "Automatic fault classification of photovoltaic strings based on an in situ IV characterization system and a Gaussian process algorithm." In Photovoltaic Specialists Conference (PVSC), 2016 IEEE 43rd (pp. 1708-1713). IEEE.

[8] Jones, C. B., King, B.H., Stein, J. S., Lavrova, O., "Quantify PV Module Degradation based on Model Coefficients Extracted from an Embedded Data Acquisition Devices", In Photovoltaic Specialists Conference (PVSC), 2017 IEEE 44th

Advanced Analysis of Multi Wire Wafering Processes

Ringo Koepge[1], Samuel Brinnig[1], Felix Kaule[1], Hartmut Schwabe[1] S, Stephan Schoenfelder[1,2]

[1]Fraunhofer Center for Silicon Photovoltaics CSP, Otto-Eissfeldt-Strasse 12, 06120 Halle (Saale), Germany
[2]Leipzig University of Applied Sciences, Karl-Liebknecht-Strasse 134, 04277 Leipzig, Germany

Abstract — **The improvement of each single process step is the major focus of all photovoltaic companies in order to decrease costs and provide high quality products. Within the entire solar cell module costs the wafer manufacturing is still about 16 percent. In this work a new wafering process analysis based on big data sets from machine and wafer characterization is presented. A correlation between given process recipe and the resulting wafer geometry are shown. Therefore new parameters are defined which are indicating directly process variations. The wafering experiments show that the variation of the ratio between feed speed and wire speed causes a change of wire usage that is correlated to a change of wafer thickness distribution. In addition, wire types are varied to observe the impact on the wafer quality and wire performance. Thus, the analysis of wafer thickness distribution was improved to create a multi wire process benchmark to address process issues and wire performance.**

Index Terms — **Wafering, Multi Wire, Process Analysis, Silicon Wafer, Wafer Strength, Wafer Geometry**

I. INTRODUCTION

The major targets of all technological optimizations are an improvement of the material and consumable consumption by keeping or increasing the quality of the product, in order to reduce the manufacturing costs. In the crystalline silicon photovoltaic industry the manufacturing costs for silicon wafers is about 16 percent of the entire cost of a solar module today. Thus, there is still a reason for process development [1]. Nowadays the wafering process technology is subdivided into the diamond wire technology and the conventional slurry process. In both cases specific subtopics are currently under investigation [2, 3, 4, 5, 6, 7]. Huge steps of process innovation in wafering technology were the change of wire types, e.g. from straight to structured wire and from straight wire to diamond wire [7, 8]. In the case of structured wire versus straight wire the change to structured wire type leads to an increase of the feed speed by a factor of two and a significant decrease of wire and slurry consumption [7].

In this work a process analysis is presented which can be adapted to any multi wire wafering process. Wafering runs with different wire technologies are performed and analyzed by improving existing methods and combing different subtopics of previous research. The wire contact length is an indirect parameter for wire wear and was introduced by Sunder et.al [9, 10]. Here, the wire contact length is replaced by a more suitable parameter called specific silicon removal area. Furthermore, the correlation of the specific silicon removal area to the local thickness variation of the wafers was determined, instead of conventional TTV analysis [9]. Additional to the wafer thickness variation other quality

parameters like the mechanical strength are used for correlation to the wire usage. Finally, wire performance parameter can be determined and a new process time dependent analysis is established.

II. MATERIALS AND METHODS

The process analysis includes the results of four wafering experiments. That wafering runs are performed on a Meyer Burger DS264 multi wire saw. Structured wire and straight wire were used in a slurry based process.

Table 1: Main machine settings and parameters for multi wire sawing runs

	Run 1	Run 2	Run 3	Run 4
Wire Type	structured	structured	structured	straight
Wire Diameter	115 µm	115 µm	115 µm	120 µm
Slurry Type	F600	F600	F600	F600
Material Type	mc Si	mc Si	mc Si	mc Si
Total Brick Length [mm]	801	760	761	760
Feed Rate [mm/min]	0.60	0.5-0.9	0.75	0.33-0.5
Wire Speed [m/s]	12	12	12-15	14

The main process parameters are summarized in Table 1. Wafering runs were performed with a stepwise increase or decrease of process velocities in order to evaluate the maximum wire performance. Run 1 represents the reference process with constant wire speed and feed rate during the process. An increase of process velocities was realized in Run 2 and Run 3 to see the influence of changing machine settings and a higher wire usage. The wire performance of a straight wire was analyzed in Run 4 for comparison to the structured wire performance. After the wafering process the wafers are separated from the glass beams, cleaned and a geometric analysis was performed. The wafer thickness distribution was determined by Hennecke Metrology Systems in a line scan from the bottom to the top of the wafer. Figure 1 shows the approach for advanced data analysis. A schematic view of a silicon brick that was sliced into wafers is shown on the left side of the figure. The first wire web contact occurs at the bottom of the brick.

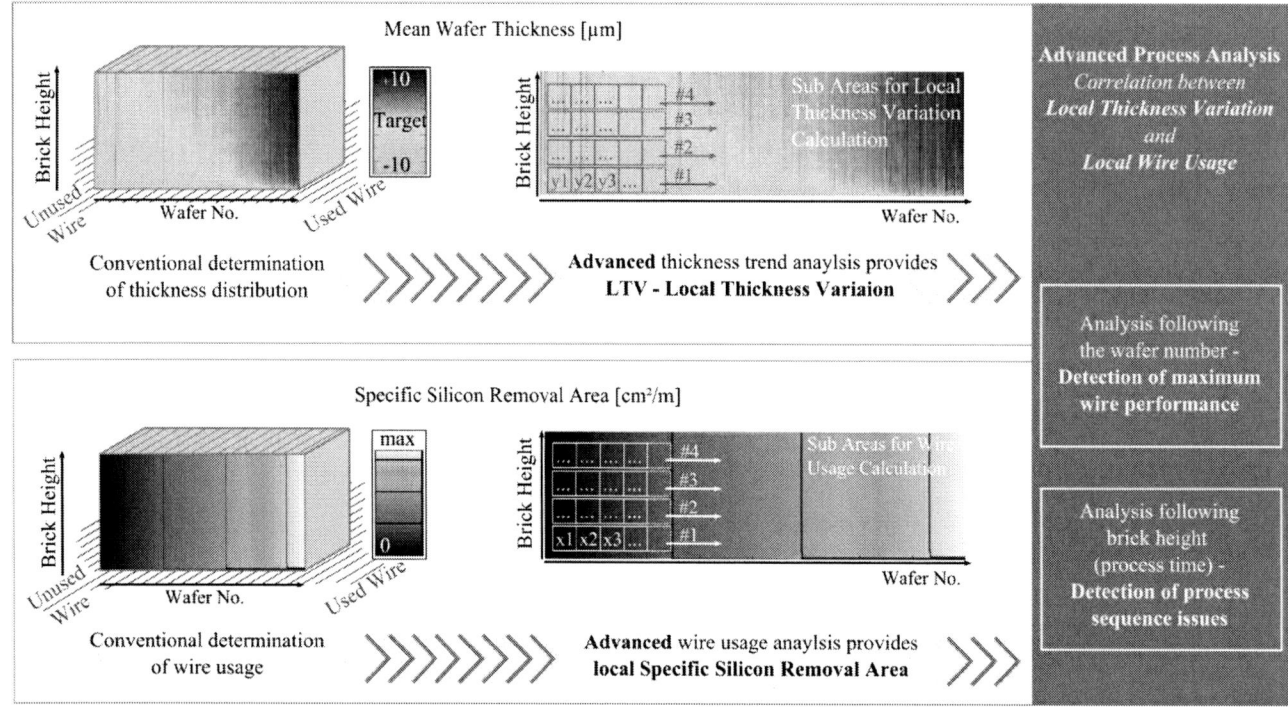

Figure 1: Approach for advanced process analysis, local analysis of thickness variation and wire uses provides wire performance parameter and critical process sequences.

The wafer thickness was measured capacitively by three line scans and an average thickness distribution for each wafer was calculated. To evaluate the wire performance a second data set is needed. The wire wear cannot directly be measured. Thus, the usage was calculated for each wafering run. The wire usage describes the amount of new silicon surface that was created by one meter of the sawing wire. Thus, the wire usage is normalized to the wire length and it is called specific silicon removal area. At the bottom left in Figure 1 the variation of the specific silicon removal area over the block is shown. In that case the wire usage increases with an increase of the wafer number which is common for slurry based processes. In a second step the measured thickness distribution und the calculated wire usage were put together. In the center of Figure 1 the approach for advanced data analysis is illustrated. Subareas are created at equal positions in both data sets to calculate a local thickness variation (difference of maximum and minimum wafer thickness in that area of interest) out of the thickness data and a mean specific silicon removal area. Additionally, the wafer strength was determined by performing ball-on-ring tests [11] at small samples out of different extraction positions, not shown here. Silicon chips with a dimension of 20 x 20 mm² were prepared by a LASER out of the silicon wafers at a wire usage state of 0.1 cm²/m to 3.0 cm²/m. The wire performance can be defined as quantitative correlation between the wire usage and the wafer properties like variation of thickness or wafer strength. With use of the described method a maximum of wire performance will be detected as well as process time dependent issues, like the loss of wafer quality. Performance parameters can be defined which indicates the amount of wire performance with respect to the wafer properties. Thus, different wafering technologies can be benchmarked directly by performing a small number of wafering runs and an advanced local analysis.

III. RESULTS

A. Wire Usage vs. Wafer Geometry

The thickness distribution of all wafers of run one to four is shown in Figure 2. For run number one there is a slight trend in wafer thickness but the average of the entire wafer stack is almost at 160 μm. For run number two and three the distributions are quite equal to each other. That similar pattern correlates to the equal wire usage that was calculated previously. Run number four shows a very different wafer thickness that is caused by another wire type (straight wire), another wire thickness and different process settings. Furthermore the crystal orientation seems to influence the thickness trend dramatically. Each single silicon brick (four bricks in total) shows a unique thickness distribution. In Figure 3 the specific silicon removal area is presented. Run number one shows a linear increase of the wire use due to an increase of the wafer number. Run number two was divided into four process segments. The wire speed was kept constant and the feed speed was increased. Thus, the wire usage was increased in comparison to the reference run number one.

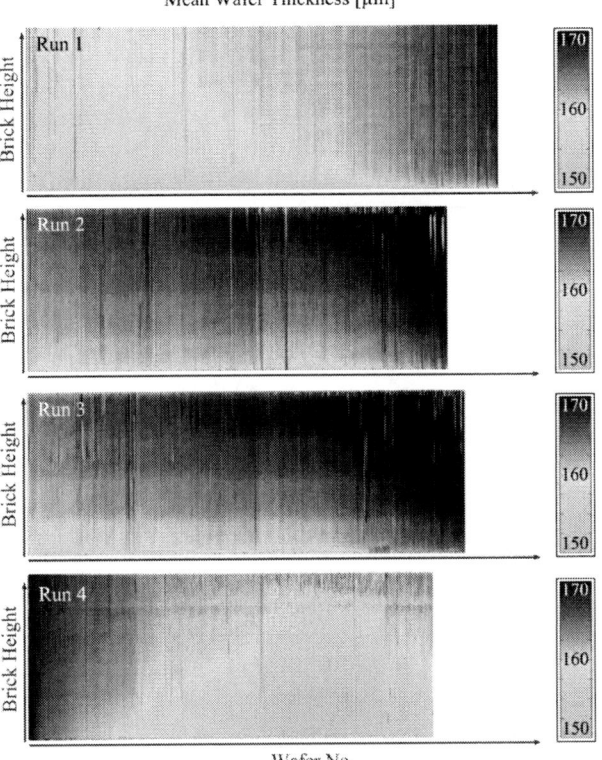

Figure 2: Thickness distributions of all wafering runs

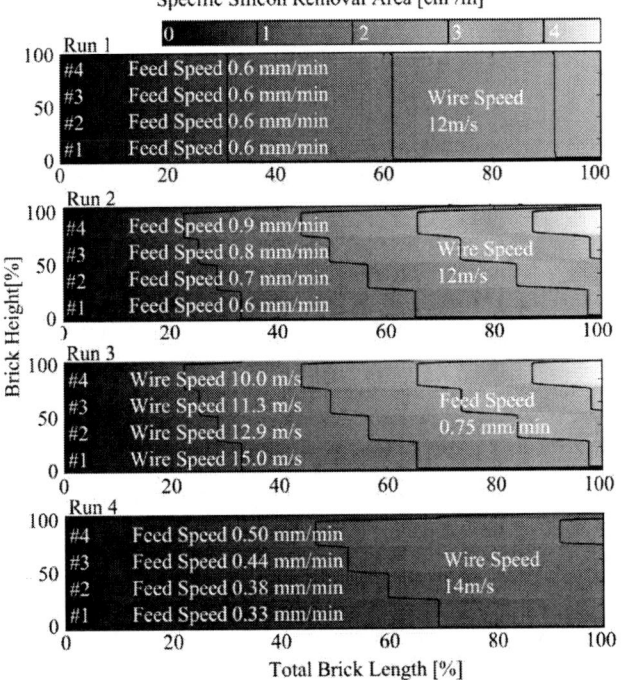

Figure 3: Specific silicon removal area of all wafering runs. Run two, three and four are subdivided in four several process segments.

The wire usage of run number three was equal to run two that was expected in spite of very different machine settings. For run three the feed speed was kept constant and the wire speed was decreased, so the impact of the process velocities can be analyzed. Regarding to a less wire performance of a straight steel wire the settings for run four are not equal to the other wafering experiments. Run four was performed to determine the wire performance for a different wire type than structured wire.

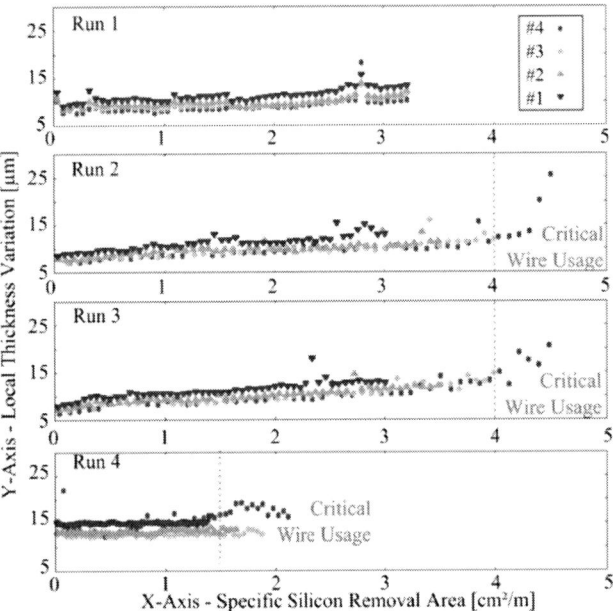

Figure 4: Specific silicon removal area of all wafering runs. Run two, three and four are subdivided in four several process segments.

The correlation between the thickness variation and the wire usage can finally be seen in Figure 4. The reference run shows a slight increase of the local thickness variation due to the increase of the specific silicon removal area. Run one was analyzed by dividing the process in four segments. Thus, a change of wafer properties regarding to the process sequences (increase of brick height) can be determined by the sequential analysis. The results of run one show a higher level of thickness variation at segment #1 that could be correlated to any process issues at the starting sequence.

Run number two and three show an equal result in comparison to each other and also in comparison to run number one. By an increase of the wire usage the local thickness variation also slightly increases. The absolute values for all three runs are at the same level and in all three cases any issue to the starting sequence were observed. In addition, for run two and three a critical point of wire usage was detected. For both runs an increase of the local thickness variation occurs by passing a specific silicon removal area of 4 cm² /m. Moreover, it can be seen that wire usage is dominating the wafer properties, the change of feed speed and wire speed show no significant influence in that case.

Results for Run 4 are quite different. There was no increase of the LTV detected with an increase of the specific silicon removal area and the specific wire usage of run 4 is low in comparison to the structured wire process. A critical wire usage can be assumed by passing 1.5 cm² /m. It has to kept in mind that all results are based on single wafering runs, which should be validated by further wafering runs.

B. Wire Usage vs. Wafer Strength

In a second investigation the mechanical properties of the wafers were determined. Wafer chips of run number three were extracted and the wafer strength was determined and correlated to the specific silicon removal area. Figure 5 shows the fracture stresses of all silicon chips. The detailed statistical Weibull analysis shows no significant difference in strength Though, as a trend the material strength slightly increases by an increase of the specific silicon removal area. Thus, a decrease of the wafer damaging could correlate with an increase of wire usage.

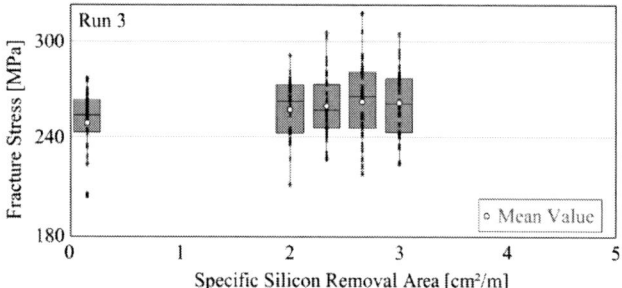

Figure 5: Boxplots of fracture stresses that were determined by ball-on-ring test.

IV. DISCUSSION

A. Impact of Crystal Deviation

The results of run one, two and three show clearly a correlation between the wire usage and the new defined local thickness variation, but that kind of analysis should be validated by a sufficient number of wafering runs. Nevertheless, a minimal number of runs can be used because of defined wire usage parameters in one sawing run. A strong impact of the multi crystalline structure was observed for the wafering run number four (see Figure 2). Each silicon brick shows a unique thickness distribution that is potentially caused by a change of the crystal structure. For getting more significant correlation between wire usage and local wafer geometry in multi crystalline silicon a higher number of wafering runs is needed.

B. Considering the Loading Forces

There is a correlation between the process velocity and the applied contact pressure that was published in the beginning of the 20th century by Preston [12]. A material removal rate

was described. Besides the machine data protocol and the wafer geometry also the forces in wire direction and feed direction were determined for this investigation (not shown here). The specific silicon removal area is calculated based on the feed rate in the silicon brick and the wire speed. By considering the normal forces a process coefficient can be calculated similar to the Preston coefficient to quantify the material removal more suitable.

C. Definition of Performance Parameter

If the observed correlation between the specific silicon removal area and the local thickness variation is validated by a higher number of wafering runs, new parameters can be defined to classify the wire performance. A linear behavior at the very beginning of the data points can be assumed and the slop of a linear fit can quantify the intensity of the loss of wafer quality (Quality Loss Factor). In addition at some point the local thickness variation will strongly increase and the data points will leave the linear behavior. That silicon removal area describes the performance limit of the wire and can also be defined as characteristic value (Critical Wire Usage). Both wire performance parameters are illustrated in Figure 6.

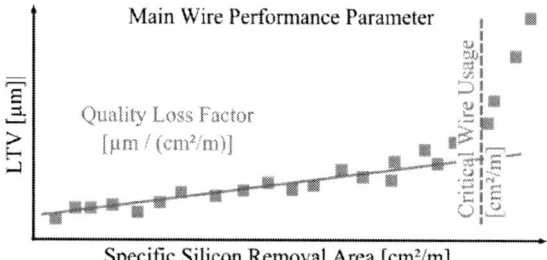

Figure 6: Schematic view of the correlation between specific silicon removal area and local thickness variation. The slope represents the loss of wafer quality due to wire usage. The wire cutting performance is reached by leaving the linear correlation.

V. SUMMARY AND CONCLUSION

An advanced process analysis was performed by the calculation and determination of local quality parameters and local wire usage parameters in wire sawing processes. The presented analysis method enables a comparison of different wire types and multi wire wafering processes regarding their wire performance and the resulting wafer properties. For structured wire a wire usage of 4 cm²/ m was observed until the local thickness variation reaches a defined quality value of 15 µm. In comparison the straight wire reaches that value at 1.5 cm² /m. The wire performance limit defines the maximum cutting capability and can be converted into wire cost per wafer. Wire costs can be compared directly regarding their performance. In a second step the wafer strength was observed. There were no significant differences based on Weibull analysis. The fracture stresses show a slightly increase by an increase of the specific silicon removal area.

Current standard wafer specification analysis was enhanced by calculation of local thickness parameters instead of the conventional TTV value and a trend analysis in wire usage direction and process time direction. Thus, issues in several process sequences can be observed by a brick height dependent (vertical) analysis of the wafer thickness distribution. The method can be used for quality control of the wafering process which cannot be seen by conventional wafer specification analysis.

ACKNOWLEDGEMENT

The financial support of Federal Ministry of Education and Research within the projects "MechSi" (contract no. 03IPT607X) and Project "DiaCell" (contract no. 03IPT607A) is gratefully acknowledged.

REFERENCES

[1] International Technology Roadmap for Photovoltaik (ITRPV), Eighth Edition, March 2017

[2] H. Wu, "Wire Sawing Technology: A State-of-the-Art Review", Precision Engineering, 2016

[3] Möller, H.J., "Chapter Two - Wafering of Silicon", Semiconductors and Semimetals Vol. 92, 2015

[4] A. Bidiville, K. Wasmer, M. Van der Meer, C. Ballif, "Wire-sawing processes: parametrical study and modeling", Solar Energy Materials and Solar Cells, 2015

[5] K. Sunder, O. Anspach, "On the Parameters that Impact the Performance of Diamond Wire in the Production of Silicon Wafers", 32nd European Photovoltaic Solar Energy Conference and Exhibition, 2016

[6] Liedke, T. and Kuna, M., "Discrete element simulation of micromechanical removal processes during wire sawing", WEAR, 2013

[7] O. Anspach, B. Hurka, K. Sunder, "Structured wire: From single wire experiments to multi-crystalline silicon wafer mass production", Solar Energy Materials and Solar Cells, 2014

[8] J- I. Bye, L. Norheim, B. Holme, Ø. Nielsen, S. Steinsvik, S. A. Jensen, G. Fragiacomo, I. Lombardi, "Industrialised Diamond Wire Wafer Slicing for High Efficiency Solar Cells", 26th European Photovoltaic Solar Energy Conference and Exhibition, 2011

[9] K. Sunder, H. Uhle, S. Knöppel, O. Anspach, "Prediction of Wire Wear, Wire Bow and Total Thickness Variation in the Diamond Wire Wafering Process", 28th European Photovoltaic Solar Energy Conference and Exhibition, 2013

[10] R. Koepge, C. Klute, F. Kaule, K. Sunder, O. Anspach, S. Schoenfelder, "Mechanical and Microstructural Wire Sawn Wire Sawn Wafers Considering the Wire Wear in Sawing Process." 31st European Photovoltaic Solar Energy Conference and Exhibition, 2015

[11] Schoenfelder, S.; Bagdahn, J. & Petzold, M. , Burghartz, J. (Ed.), "Mechanical Characterisation and Modelling of Thin Chips", 17, Ultra-thin Chip Technology and Applications. Springer New York , 195-218, 2011

[12] F. W. Preston, "The Theory and Design of Plate Glass Polishing Machines", Journal of the Society of Glass Technology, 1927

Consideration on open-circuit voltage of
Si heterojunction solar cells under low concentration condition

Makoto Konagai

Tokyo City University, 8-15-1 Todoroki, Setagaya-ku, Tokyo, JAPAN 158-0082

Abstract — **We conducted a one-dimensional device simulation of Si heterojunction low-concentrator solar cells using AFORS-HET focusing on the influence of Auger recombination on the open-circuit voltage. It was found that the influence of Auger recombination becomes apparent at approximately 10-30 suns. We measured the properties under concentration of prototype heterojunction solar cells and compared the theoretical and experimental results. The open-circuit voltage at 1 sun was 0.72 V, but increased to 0.77 V at 12 suns. In addition to the characterization with solar simulators, the relation between incident light intensity and Voc was measured using a Suns-Voc apparatus. As a result, a Voc of 0.81-0.82 V was obtained at around 30 suns.**

Index Terms — **amorphous silicon, Auger recombination, concentrator solar cells, heterojunction solar cells, silicon, Suns-Voc.**

I. INTRODUCTION

Attempts to use silicon solar cells at low concentration ratios have been made since the 1970s. Subsequently, improvements in the performance and efficiency of flat-plate Si solar cells caused research and development of low-concentration technology to slow; however, a definite trend has recently emerged toward further reducing the levelized cost of electricity by means of low concentration ratios. In terms of improving conversion efficiency by means of low concentration ratios, in 2005, A. Slade et al. reported using a conventional p-n junction Si solar cell to obtain an open-circuit voltage (V_{oc}) of 0.83 V and an energy conversion efficiency of 26.3% at a concentration of 260 suns[1]. Also, Kaneka Corporation has recently reported the properties of its heterojunction solar cells under low-concentration ratios, including the achievement of open-circuit voltages between approximately 0.77 V and 0.78 V under low-concentration ratios[2].

The open-circuit voltage of conventional p-n junction Si solar cells has always been considered to increase monotonically with concentration ratio (suns). Also, in conventional p-n solar cells, recombination of minority carriers via recombination centers and surface states dominates, and a carrier concentration at which Auger recombination dominates is not reached. However, heterojunction solar cells use Si wafers that have extremely high carrier lifetimes, and so defect densities are low and Auger recombination is expected to have a large influence on the device properties. Based on

the above considerations, we carried out theoretical analyses centered on the influence of Auger recombination on solar cell properties. Furthermore, the effect of Auger recombination on V_{oc} was experimentally verified.

II. SIMULATION OF PROPERITES UNDER CONCENTRATION USING AFORS-HET

We conducted a one-dimensional device simulation using AFORS-HET. The structure used in the device analysis is a heterojunction using an n-type Si wafer, with a p-type a-Si or n-type a-Si layer formed on each side of the wafer.

First, the simulation was carried out at various doping concentrations of the 5 nm thick p-type amorphous layer. At 10 suns, characteristics vary greatly depending on acceptor concentration. A substantial deterioration in characteristics can be observed just as a result of taking the acceptor concentration to be 6×10^{19} cm^{-3}, which is slightly lower than its reference value (7.47×10^{19} cm^{-3}). At Na = 6×10^{19} cm^{-3}, a so-called S curve becomes apparent, which is commonly observed experimentally in the vicinity of the open-circuit voltage. These analysis results show that the doped amorphous p-layer is one factor that greatly limits the properties of the Si heterojunction solar cell under concentration. This is because, even if the doping concentration in the amorphous Si layer is high, this layer contains many defects (1×10^{20} cm^{-3}) and tail states exist, which means that its hole concentration is extremely low. In this analysis, Na was taken as 9×10^{19} cm^{-3} in order to sufficiently increase the hole concentration.

We simulated the influence of bulk defect density N_D on solar cell properties. In AFORS–HET, when bulk defect density N_D is 1×10^{10} cm^{-3}, the parameters are adjusted so that carrier lifetime becomes 1 msec. Accordingly, when bulk defect density N_D is 1×10^9 cm^{-3}, carrier lifetime becomes 10 msec. When bulk defect density is lower than 1 msec, solar cell properties deteriorate rapidly in both the 1 sun and 10 suns cases. Also, in the case of 10 suns, when the bulk defect density N_D is below 1×10^{10} cm^{-3}, an open-circuit voltage of 0.8 V is obtained.

Based on the above device simulation, we investigated the level of conversion efficiency that could realistically be obtained when the heterojunction Si solar cell being developed in this study is applied to a low concentration system. The results showed that if the solar cell is operated at around 10 suns using a wafer with a carrier lifetime of 10 msec, a conversion efficiency of approximately 27% (Voc=0.810V, FF=0.867, Isc= 0.384A/cm²) can be expected as shown in Fig.1.

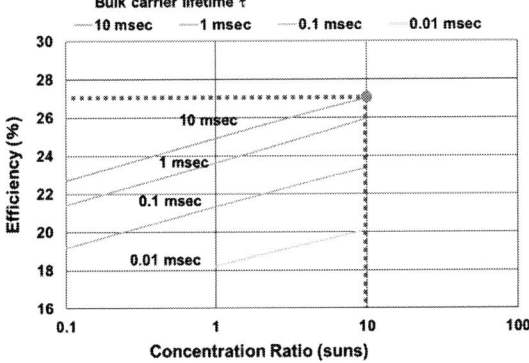

Fig.1 Simulation results for efficiency and concentration ratio with the bulk carrier lifetime as a parameter.

III. DARK I-V AND I_{sc}-V_{oc}

Generally, in an ideal p-n solar cell, current–voltage (I–V) characteristics under dark conditions and Isc-Voc characteristics under illuminated conditions correspond exactly. However, in heterojunction solar cells, the ratio of recombination current via recombination centers and surface states is decreased because wafers that originally have extremely low bulk defects are used, and this means that high open-circuit voltages are obtained even at a concentration ratio of 1 sun. Therefore, we determined the dark I-V and Isc-Voc characteristics of the heterojunction solar cell using a simulation and performed a comparison.

Figure 2 shows the dark I-V and Isc-Voc of the heterojunction Si solar cell with a wafer thickness of 200 μm.

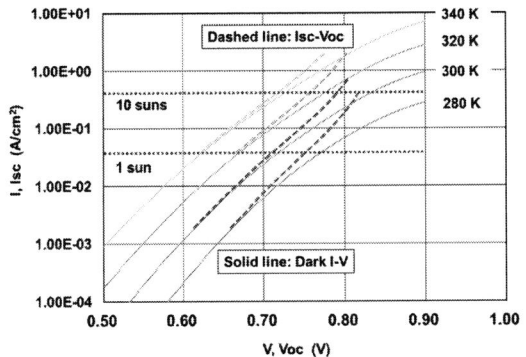

Fig.2 Simulation results for dark I-V and Isc-Voc of heterojunction Si solar cell of 200 μm thickness.

Looking first at dark I-V, the influence of series resistance is apparent after the current level exceeds about 1 sun. Whereas, looking at the Isc-Voc characteristics, the increase in open-circuit voltage with increasing injection level is smaller than expected, and a slight tendency towards saturation is present. This phenomenon is considered to be a characteristic of heterojunction solar cells.

IV. INFLUENCE OF AUGER RECOMBINATION

The results of experiments using an ordinary p-n solar cell under concentration show that even if the open-circuit voltage increases due to concentration, when it is approximately 0.8 V, the relationship of log(Isc)–Voc characteristics becomes almost linear. However, in the case of heterojunction solar cells, high open-circuit voltages exceeding 0.75 V are obtained even at 1 sun because wafers with extremely high carrier lifetimes are used and surface recombination is kept very low due to the incorporation of amorphous passivation layers. Therefore, even at a low concentration ratios of about 10 suns, the influence of Auger recombination is thought to become apparent and the growth of Voc to slow. Accordingly, in our simulation, we set the Auger constant to zero. The results showed that the log(Isc)-Voc characteristics became almost linear(Fig.3). The above simulation showed that when a heterojunction solar cell is used as a low-concentrator cell, reduction due to Auger recombination must be taken into account in the design of the device.

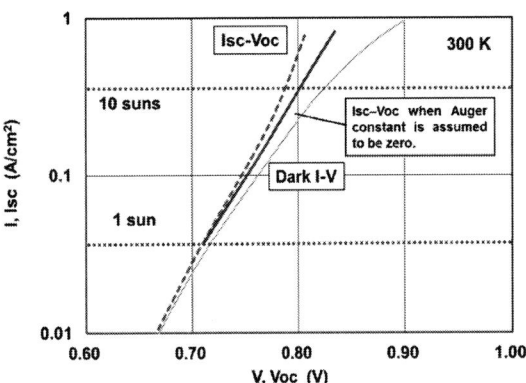

Fig.3 Analysis results under conditions shown in Fig.2 with Auger constant taken as zero.

The most dominant way to avoid the decline of Voc due to Auger recombination is to make the thickness of Si wafer extremely thin. If solar cells are made using a very thin wafer, the volume of recombination of carriers will decrease, so Auger recombination will decrease and Voc will improve.

978-1-5090-5606-4/17 $31.00 © 2017 IEEE   2628

Figure 4 is a simulation of the relationship between the thickness of Si wafer and Voc. The thickness of Si was assumed to be 0.1 μm, 1 μm, 10 μm, and 100 μm.

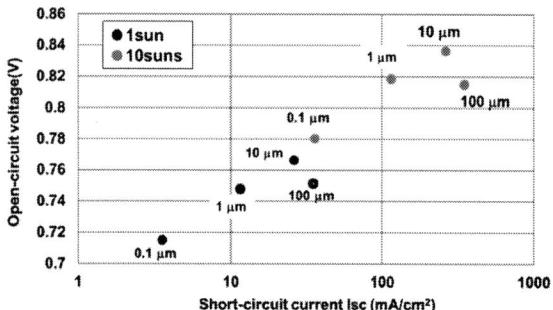

Fig.4 Relationship between the short-circuit current Isc and the open circuit voltageVoc. The thickness of Si is assumed to be 0.1 μm, 1 μm, 10 μm, and 100 μm.

Needless to say, the Voc of solar cell changes greatly by photocurrent. Especially when Si wafer becomes thin, enough photocurrent can not be obtained. Therefore, in Fig.4, the performance at 1sun and 10suns is simulated and the relationship between Voc and photocurrent is drawn. The simulation results show that Voc of 0.76 V or more can be obtained even in 1 sun condition even if the thickness of the wafer is 10 μm. In addition, Voc is improved to 0.84 V by concentrating at 10 suns. Of course, if the Si wafer is thin, optical confinement is important to improve the photocurrent, and it is necessary to develop a technology that further evolves the texture structure used in common Si solar cells.

Figure 5 shows the theoretical analysis of the relationship between the thickness of the n - Si wafer and Voc. In the figure, the effect of Auger recombination on Voc is analyzed under the conditions of irradiation intensity of 1 sun and 20 suns.

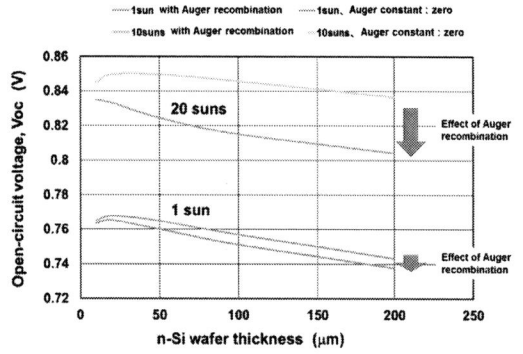

Fig.5 Relationship between n-Si wafer thickness and the open-circuit voltage Voc.

At 20 suns when carrier's injection level gets higher, the effect of Auger recombination can be confirmed more clearly. From this figure, it can be seen that Voc exceeding 0.82 V can be obtained at 20 suns if the thickness of the n - Si wafer can be reduced to 50 μm.

In general, when solar cells are operated at low temperatures, open-circuit voltage increases monotonically. This is because the dark current is determined by processes that are related to thermal energy, whether the injection current or the recombination current in the depletion layer, and these vary exponentially with absolute temperature.

In this analysis, we made large changes to the physical parameters such as defect density and surface recombination velocity, as well as to the cell structure parameters such as each layer's thickness, so that the open-circuit voltage varied between 0.5 V and 0.75 V at 1 sun. We also varied AM 1.5 illumination intensity between 0.0001 and 1 sun.

Figure 6 shows the theoretical analysis of the temperature dependence of Voc. The results demonstrate that when the operating temperature is extended to 0 K, Voc is asymptotic to the Si energy bandgap of 1.1 eV in all cases. This analysis does not take components related to the tunnel effect into account as components of solar cell current. In general, carrier transport limited by the tunnel effect is not a thermal effect, and so it is not applicable.

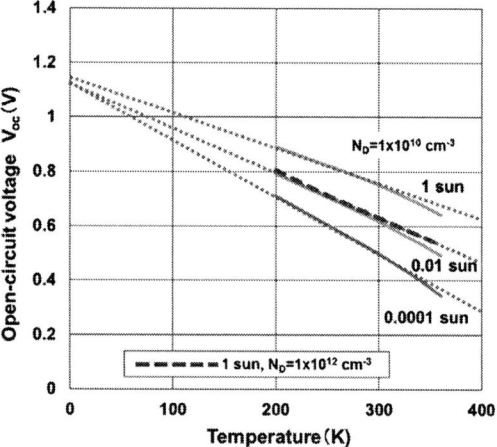

Fig.6 Theoretical analysis of Voc–T characteristics for Si hetero junction solar cell. Voc is asymptotic to Si energy bandgap at 0 K, irrespective of conversion efficiency at room temperature.
The parameters used are as follow,
    p-a-Si, thickness: 5nm, Na: 9×10^{19} cm^{-3},
    n-Si, N$_D$: 1×10^{10} cm^{-3}, carrier lifetime: 1msec.

V. EXPERIMENTAL RESULTS

We measured the properties under concentration of a prototype heterojunction solar cell with a conversion efficiency of approximately 17-20% built to 1-sun specifications, and compared the theoretical and experimental results. Basically, the heterojunction structure comprises an n-type wafer with amorphous Si(O):H incorporated into the interface layers[3]. The cell area is 1 cm². Due to the 1-sun specification, the series resistance of the electrode and TCO was large, and the fill factor of the prototype sample decreased substantially with concentration ratio; however, we obtained important results regarding open-circuit voltage. The open-circuit voltage of this cell at 1 sun was 0.72 V, but increased to 0.77 V at 12 suns. These results show that if a cell displaying an open-circuit voltage of approximately 0.75 V at 1 sun can be produced, an open-circuit voltage of approximately 0.8 V can be obtained at 10 suns[4].

In addition to the characterization with solar simulators, the relation between incident intensity and Voc was measured using a Suns-Voc apparatus. As a result, a Voc of 0.81-0.82 V was obtained at 30 suns although the influence due to some temperature rise was observed.

According to the simulation results, it seems that the influence of Auger recombination already appears in the vicinity of Voc = 0.82 V. At the next stage of simulation , we will also discuss the degree of influence while comparing it with the carrier lifetime measurement result by Quasi-Steady-State Photoconductance (QSSPC) lifetime measurement .

Figure 7 shows the current-voltage characteristics of the heterojunction solar cell under dark conditions, as well as the temperature characteristics of the Isc-Voc plot measured when varying the illumination intensity. The measurement temperature was varied within a range from 5°C to 25°C. For this sample, it was found that a plot of log(Isc)-Voc varies linearly up to approximately 10 suns, and when the sample is cooled to 5°C, the open-circuit voltage increases up to 0.80 V. This study is the first in the world to measure an open-circuit voltage of 0.8 V in a heterojunction solar cell, albeit at a low temperature of 5°C.

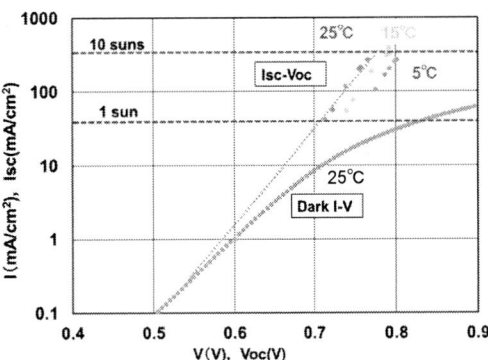

Fig.7 Plot of measured dark I-V and Isc-Voc of a-SiO/Si heterojunction solar cell. The performance of this cell at 1 sun is Isc = 36.7 mA/cm² and Voc = 0.717 V.

VI. CONCLUSIONS

The influence of Auger recombination becomes apparent in heterojunction solar cells at concentrations of approximately 10-30 suns, and so it is necessary to take measures such as making the Si wafer thin. Innovations in device design are expected to result in open-circuit voltages above 0.8 V at low concentrations of approximately 10 suns.

ACKNOWLEDGEMENT

This study was supported in part by the Ministry of Education, Culture, Sports, Science and Technology of Japan (MEXT), FUTURE−PV Innovation Project(FY2012-2016).

REFERENCES

[1] A.Slade and V.Garboushian, *PVSEC-15*, Shanghai (2005)

[2] J.L.Hernandez et.al, *28th EUPVSEC*, Paris (2013)

[3] K.Nakada, J.Irikawa, S. Miyajima, and M. Konagai, *Japanese Journal of Applied Physics* ,54, 052303 (2015) doi:10.7567/JJAP.54.052303

[4] R.Sato, T.Kubota, K.Sawano, P.Sichanugrist, H.Zhang, K.Nakada, M.Konagai , *26th International Photovoltaic Science and Engineering Conf.(PVSEC-26)*, 24-28 October, 2016, Singapore

Characterization of Microcrystalline Silicon Thin Film Solar Cells Prepared by High Working Pressure Plasma-enhanced Chemical Vapor Deposition

Jung-Dae Kwon[1], Dong-Ho Kim[1], Ji-Hoon Lee[1], Myungkwan Song[1], Myunghun Shin[2]

[1]Advanced Functional Thin Films Dept., Surface Technology Division, Korea Institute of Materials Science, Changwon 51508, Republic of Korea

[2]School of Electronics and Information Engineeringg, Korea Aerospace University, Goyang-city, Gyeonggi-do 412-791, South Korea

Abstract — **Using the high working pressure plasma-enhanced chemical vapor deposition (HWP-PECVD) technique, the hydrogenated microcrystalline silicon (μc-Si:H) films for photovoltaic layers of thin film solar cells was investigated. The μc-Si:H films were deposited on surface textured fluorine-doped tin oxide (FTO) glass substrates at 100 Torr in a 100 MHz very high frequency (VHF) plasma of gas mixtures containing He, H₂, and SiH₄. It was found that an optimum ratio of the H₂/SiH₄ flow-rate existed for growing a homogenous microcrystalline through the whole film without amorphous incubation layer. When an intrinsic μc-Si:H thin film was deposited at n-i-p single junction solar cell, the cell performances were dependent on with or without an amorphous incubation layer. With an amorphous incubation layer, the open circuit voltage (V$_{oc}$) of cell was 0.8 V, which was typical cell property of hydrogenated amorphous silicon (a-Si:H). On the other hand, at the optimum ratio of the H₂/SiH₄ flow-rate, μc-Si:H single cell responding an infrared light showed the Voc of 0.49 V. Intrinsic hydrogenated microcrystalline silicon (μc-Si:H) thin film, exhibiting the photovoltaic performance (Eff: 7.2%, V$_{oc}$: 0.49V, J$_{sc}$: 23mA/cm², FF:64%) was able to be successfully fabricated.**

I. INTRODUCTION

The thin-film silicon solar cell has been attracting considerable attention as high conversion efficiency could be achieved at relatively low manufacturing costs.[1] Hydrogenated microcrystalline silicon (μc-Si:H) cell responds mainly from red to infrared light, and the optical absorption coefficients of μc-Si:H at long wavelengths are almost identical to those of the Si single crystal.[2] Because absorption coefficient of μc-Si:H is relatively low, a thickness of 1~2 μm is usually required for μc-Si:H photovoltaic-active layer. Therefore, for achieving a short production time, or for industrializing the solar cell, researches about plasma-enhanced chemical vapor deposition (PECVD) process using very high frequency (VHF, 50~150 MHz) at high pressure (10~750 Torr) have been reported. Meanwhile, control of the crystalline volume is crucial because the Si crystallinity has a significant influence on the optical gap, defect density, and conductivity of the materials. The commonest method for controlling the crystalline volume fraction (X$_c$) of μc-Si is to

adjust the dilution ratio of silane (SiH₄) and hydrogen (H₂) gases, or the plasma power. Here, we fabricated the single junction μc-Si:H thin film solar cells by using a high working pressure plasma-enhanced chemical vapor deposition (HWP-PECVD) system at 100 Torr with a cylindrical rotary electrode; this system is superior to conventional PECVD because it has the following features: a high deposition rate as a result of the high partial pressure of the reactive gas and a high plasma density by the very high frequency of 100 MHz; the ability to control the film uniformity because of the homogeneous distribution of reactants by the rotary electrode system; and low bombardment damage because of the lower kinetic energy. When an intrinsic μc-Si:H thin film was deposited at n-i-p single junction solar cell, the cell performances were dependent on with or without an amorphous incubation layer. With an amorphous incubation layer, the cell did not respond an infrared light. On the other hand, when an intrinsic Si film was homogeneous microcrystalline through the whole film, the cell responded an infrared light, and its performance was a behavior of μc-Si:H thin film solar cell.

II. Experimental

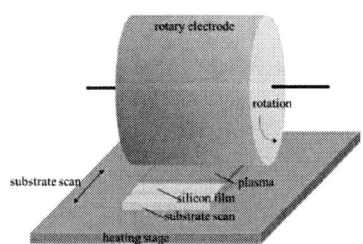

Fig. 1.   Schematic illustration of the experimental setup.

The HWP-PECVD system used to produce μc-Si:H was based on previously reported designs (see Fig. 1). The diameter of the cylindrical rotary electrode was 300 mm and

978-1-5090-5606-4/17 $31.00 © 2017 IEEE

the width was 240 mm. The HWP-PECVD of μc-Si:H was performed on 1.8-mm-thick soda-lime glass (200 × 200 mm) substrates at 200 °C, with a deposition pressure of 100 Torr.

Before deposition, the base pressure of the chamber was reduced to 2×10^{-6} Torr, using dry and turbo molecular pumps. After closing the main valve connected to the turbo molecular pump, helium (He), H_2, and SiH_4 gases were simultaneously injected into the chamber until a pressure of 100 Torr was attained, and continuously supplied into the reaction chamber during Si film deposition. The purity of the gases was 99.999%. The SiH_4+H_2 concentration was 4% and the ratios of the H_2/SiH_4 flow-rates was 13~35. The electrode rotation speed was 1000 rpm, and the deposition gap between the electrode and the substrate was 0.5 mm. The substrate scan distance was 150 mm and the scan speed was fixed at 10 mm/s. An impedance matching unit supplied 100 MHz VHF power of 20 W/cm^2 to the electrode. The Si film thickness on surface textured fluorine-doped tin oxide (FTO) glass [NSG TECTM 8 of PILKINGTON] was measured by α-step (TENCOR P-11). For confirmation of the Si crystalline volume fraction, Raman spectra were measured using a Jobin Yvon LabRam HR800 (Horibo, Ltd., Kyoto, Japan) UV/micro-Raman spectrometer at room temperature. The measurements were carried out at 632.8 nm using a HeNe laser, below 50 mW to avoid thermally induced crystallization. The microstructures of the films were observed by transmission electron microscopy (TEM: JEM-2100F, JEOL, Tokyo, Japan) at 200 kV. The TEM specimens for cross-sectional observations were prepared by using Ar ion source of 3.2 keV with precision ion polishing system (PIPS). For the process of manufacturing the solar cells, the cell structure was glass substrate/textured FTO/aluminum-doped zinc oxide (AZO)/p-i-n/metal electrode. The 100 nm thick AZO film was deposited by sputter system to protect FTO from hydrogen plasma. Only the i-layer with thickness of 1 μm was deposited by HWP-PECVD, and the p- and n-layers with thickness of about 30 nm were formed by conventional low-pressure PECVD. Thus, the p/i and i/n interfaces of those cells were exposed to the air during the specimen transfer. For the metal electrode, silver was thermally deposited in a high vacuum chamber (~ 2×10^{-6} Torr) using a shadow mask to define a cell active area of 0.25 cm^2. The FTO glass contained nine cells and the cell performance was considered to be the average value determined after measuring the performances of nine cells. Current density-voltage (J-V) characteristics were measured using a Keithley 2400 source meter under 100 mW/cm^2 (AM 1.5G) irradiation from a solar simulator (Pecell Technologies Inc., PEC-L11). In addition to the photovoltaic performance, external quantum efficiency (EQE) of each photovoltaic device was obtained by using a 200 W Xe lamp and a grating monochromator, and the light intensity was measured by a calibrated Si solar cell (PV measurement).

III. RESULTS AND DISCUSSION

The crystallinity of the Si films was confirmed by Raman spectroscopy, as shown in Fig. 2. All the about 1 μm thick Si films were deposited on the textured FTO glass substrate because the films peeled off on bare glass substrate owing to the residual stress of μc-Si. The Si transverse optical (TO) peaks were deconvoluted into their integrated crystalline Gaussian peak (I_c, ~520 cm^{-1}), amorphous Gaussian peak (I_a, ~480 cm^{-1}), and intermediate Gaussian peak (I_m, ~510 cm^{-1}). Following this, the crystalline volume fraction (X_c) was calculated from the simple equation, $X_c = (I_c + I_m)/(I_c + I_m + I_a)$. The X_c increased from 39 to 76% when the ratios of H_2/SiH_4 flow-rates (R) increased from 15 to 20.

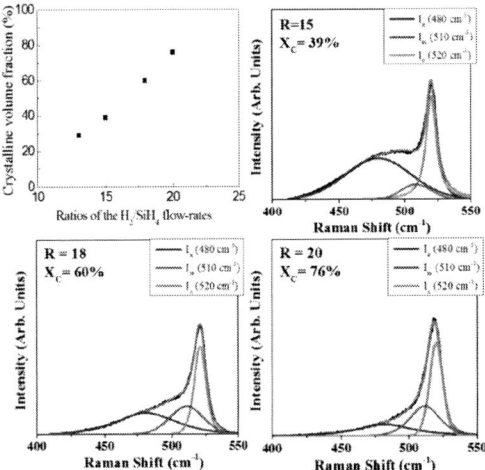

Fig. 2. Raman spectra of μc-Si:H films grown at the ratio of H_2/SiH_4 flow-rates (R)

In order to fabricate the μc-Si:H n-i-p single junction solar cell, intrinsic μc-Si layer was deposited by HWP-PECVD at R = 18. The thickness of i-layer was about 1 μm by regulating the number of substrate scanning on a large area (200 × 200 mm). We measured the cell performances of the J-V characteristic, including short-circuit current (J_{sc}), open-circuit voltage (V_{oc}), and fill factor (FF), as shown in Fig. 3(a). The conversion efficiency of 5.6% (Jsc: 13 mA/cm^2; V_{oc}: 0.8V; and FF: 54%) has been achieved. Specially, the open circuit voltage (0.8 V) of the cell is higher than that of previous reports (0.5 V) in μc-Si thin film solar cell. The V_{oc} of hydrogenated amorphous silicon (a-Si:H) single cell is usually 0.8 ~ 0.9 V. T. Matsui et. al. reported that V_{oc} of μc-Si single solar cell increased from 0.4 to 0.5 V as the crystalline volume fraction in intrinsic Si layer decreased from 80 to 50 %.[3] However, the X_c of HWP-PECVD μc-Si:H single cell with V_{oc} (0.8 V) was 60 %. To investigate the microstructure of the deposited HWP-PECVD μc-Si thin film, cross-sectional TEM observations was performed as shown in Fig. 3 (b). The film exhibited that 800 nm thick microcrystalline Si grew on the initial 200 nm thick amorphous incubation layer on the textured FTO glass. The inset shows the diffraction patterns of

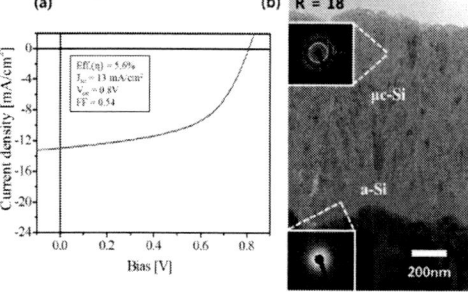

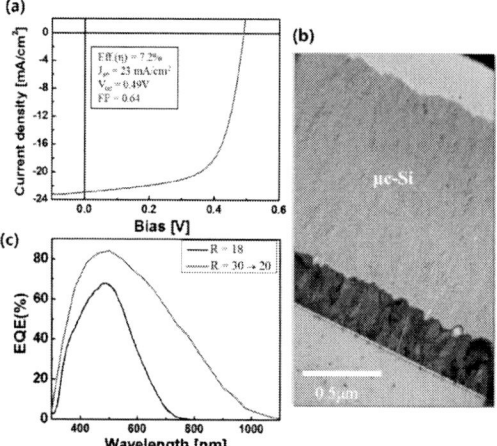

Fig. 3. (a) Photocurrent vs voltage characteristic of solar cell with μc-Si:H i-layer deposited at the ratio of H_2/SiH_4 flow-rate of 18. (b) Cross-sectional image of the solar cell with μc-Si:H i-layer deposited by HWP-PECVD at the ratio of H_2/SiH_4 flow-rates of 18.

each layer. The diffraction pattern in the μc-Si:H layer indicated peaks corresponding to cubic polycrystalline phase Si in the (111), (220), and (311) lattice planes, however the diffraction pattern for a-Si:H layer exhibited fuzzy rings and was shadowy. According to the growth dynamics of μc-Si:H, it nucleates from within the growing a-Si:H phase after a critical phase-transition (amorphous to crystalline) thickness that decreases with increasing H_2 dilution ratio. Thus, it seemed that the high V_{oc} of 0.8 V was generated by intrinsic amorphous incubation layer.

In order to fabricate the μc-Si:H n-i-p single junction solar cell without the amorphous incubation Si layer, we performed the two-step intrinsic Si deposition process in which a 200 nm thick μc-Si:H film was deposited at R = 30, and then the ratio of H_2/SiH_4 flow-rates was changed to 20. The two-step process was to prevent excessive Si grain growth which may induce the increase of defect density and decrease of photosensitivity. Figure 4 (a) shows the two-step intrinsic μc-Si:H cell performances of the *J-V* characteristic. The open circuit voltage was 0.49 V, which was typical cell property of μc-Si:H n-i-p single junction solar cell. By injecting a high hydrogen gas at the initial Si film growth, the amorphous incubation layer was not observed through the cross-sectional TEM image, as shown in Fig. 4 (b). Figure 4 (c) shows the EQE spectra for two μc-Si:H single junction solar cells, which were intrinsic Si thin films deposited by R=18, and two-step intrinsic Si deposition process by R=30→20. The cell prepared at R=18 shows a typical spectral response for a-Si:H cells to absorb in spectral range (350 ~ 700nm). Thus, it was confirmed that the μc-Si:H solar cell fabricated by HWP-PECVD at a high pressure of 100 Torr operated well in the long wavelength range.

Fig. 4. (a) Photocurrent vs voltage characteristic of solar cell with μc-Si:H i-layer deposited by two-step process at the ratios of H_2/SiH_4 flow-rates of 30→20. (b) Cross-sectional image of the solar cell with μc-Si:H i-layer deposited by HWP-PECVD at the ratio of H2/SiH4 flow-rates of 30→20. (c) External quantum efficiency spectra of solar cell with μc-Si:H i-layer deposited at the ratios of H_2/SiH_4 flow-rates of 18, 30→20.

IV. CITING PREVIOUS WORK

The intrinsic μc-Si:H films for photovoltaic layers of thin film solar cells successfully deposited by using the high working pressure plasma-enhanced chemical vapor deposition (HWP-PECVD) technique. The crystalline volume fraction of Si could be controlled by changing the ratio of H_2/SiH_4 flow-rate. When the μc-Si:H n-i-p single junction solar cell was fabricated at the ratio of 18, the intrinsic μc-Si:H films contained the 200nm thick amorphous Si incubation layer, that resulted in the V_{oc} (0.8 V) of a-Si:H single cell performance. On the other hand, when the two-step intrinsic Si deposition process by R=30→20 was carried out, homogenous μc-Si:H films responding from red to infrared light through the whole film without amorphous incubation layer could be obtained.

REFERENCES

[1] H. Keppner, J. Meier, P. Torres, D. Fischer and, A. Shah, "Microcrystalline silicon and micromorph tandem solar cell," *Appl. Phys. A*, vol. 69, pp. 169-177, 1999.

[2] A. Poruba, A. Fejfar, Z. Remeš, J. Špringer, M. Vaněček, J. Kočka, J. Meier, P. Torres, and A. Shah, "Optical absorption and light scattering in microcrystalline silicon thin films and solar cells," *J. Appl. Phys.*, vol. 88, no. 1, pp. 148-160, 2001.

[3] T. Matsui, M. Tsukiji, H. Saika, T. Toyama, and H. Okamoto, " Correlation between microstructure and photovoltaic performance of polycrystalline silicon thin film solar cells," *Jpn. J. Appl. Phys.*, vol. 41, pp. 20-27, 2002.

978-1-5090-5606-4/17 $31.00 © 2017 IEEE

Atomic-layer-deposited V_2O_{5-x} Films as a Highly-efficient p-type Layer for Thin Film a-Si Solar Cells

Ji-Hoon Lee, Myungkwan Song, Dong-Ho Kim, and Jung-Dae Kwon[*]

Advanced Functional Thin Films Department, Surface Technology Division, Korea Institute of Materials Science (KIMS), Changwon, Gyeongnam 641-831, Republic of Korea

Abstract — V_2O_{5-x} thin films were fabricated by using atomic-layer-deposition to be applied as a p-type layers for thin film amorphous Si solar cells for the first time. The film growth characteristics of V_2O_{5-x} including growth rate, crystallinity, and surface roughness are related to the exposure time of ozone (O_3), which is an oxidant. The effect of O_3 exposure time on photovoltaic performances of thin film solar cell constructed with p-type V_2O_{5-x} films were systematically investigated. Under the O_3 pulse time of 1 s, the best power conversion efficiency (PCE) of 5.35 % (J_{SC}: 15.76 mAcm^{-2}, V_{OC}: 0.6 V, FF: 0.564) was obtained, while that under pulse time of 5 s was slightly reduced to 4.18 %. It was revealed that open-circuit voltage (V_{OC}), which is strongly dependent on O_3 pulse time, is the major factor for determining PCE. From the ultraviolet photoelectron spectroscopy (UPS) and X-ray photoelectron spectroscopy (XPS) analyses, it was confirmed that increased proportion of V^{4+} ions in V_2O_{5-x} films with increase in O_3 pulse time lead to the reduction in work function (and V_{OC}) and resulted PCE.

I. INTRODUCTION

Considering the environmental pollution crisis caused by fossil fuel-based electric power production and the limited supply of these fuels, developing alternative power stations is strongly urged. In this context, solar energy harvesting have been intensively researched. Among various types of solar cells (SCs), thin film SCs based on a-Si:H are expected as the most promising one with a view point of abundance in nature, stable performance, and possibility to expand application field into the windows for building integrated photovoltaic system (BIPV) [1].

To date, most of the p-type layers in p-i-n (or n-i-p) multiple stacks in thin film SCs have been constructed with diborane (B_2H_6), boron trifluoride (BF_3), or trimethylborane (TMB)-doped a-Si:H, which lead to severe health and environmental issues. Therefore, development of alternative P-layer materials showing the features of environmental friendly as well as high power conversion efficiency (PCE) has been strongly urged.

Recently, it has been proved that un-doped metal oxides including WO_3, MoO_3, and V_2O_{5-x} can replace the p-type a-Si:H layers because of large difference in work function between metal oxides and n-type Si [2, 3]. V_2O_5 has received particular attention, which stems from its outstanding physical properties, e.g., excellent electrical properties, a high optical band gap of 2.8 eV, and a high work function of 6.85 eV [4-6].

In this study, high-quality V_2O_{5-x} thin films were fabricated by using atomic-layer-deposition (ALD) by means of self-limiting reaction and their use in the thin film SCs as a p-type layer were investigated for the first time. This study provided the major key for optimizing the ALD process for V_2O_{5-x} films and systematic investigation on the relationship between physical properties of the V_2O_{5-x} and photovoltaic parameters of open-circuit voltage (V_{OC}), current density (J_{SC}), and fill factor (FF).

II. EXPERIMENTAL

A. Atomic-layer-deposition (ALD) process for V_2O_{5-x} films

V_2O_{5-x} films were grown via the ALD process (ALD system Lucida M100, NCD) at a low temperature of 180 °C. Vanadium triisopropoxide (VO(OC$_3$H$_7$)$_3$, VTOP) and ozone gas were used as the source of vanadium and oxygen, respectively. An ozone delivery system (OWG-15A-UP, OZONEWORKS) was employed to supply a stable O_3 flow of 15 g per h. The temperature of the VTOP source was maintained at 40 °C using a heating circulator, and the stainless-steel feed line was heated to 80 °C to prevent condensation of the precursor in the line. During the ALD process, argon gas was consistently supplied to the chamber at a flow rate of 100 sccm. Each deposition cycle of the V_2O_{5-x} consisted of a pulse and purge step of the VTOP and ozone, respectively. The ozone pulse time was varied from 1 to 8 s, for a fixed flow rate of 25 sccm. The pulse time of the VTOP precursor was set to 3 s and the purge time was fixed at 9 s, when argon gas was used for both the VTOP and the ozone.

B. Fabrication of thin film SCs

solar cells were manufactured under vacuum using a cluster system consisting of three-chambers (one for ALD and the other two for plasma-enhanced chemical vapor deposition (PECVD)). To construct a-Si:H thin-film solar cells, the V_2O_{5-x} film was first prepared on the front electrode, which was a fluorine-doped tin oxide (FTO, SnO$_2$:F)-coated glass (5 cm × 5 cm). The intrinsic and n-type a-Si:H films were subsequently prepared via the PECVD. The PECVD process was performed at a respective substrate temperature, RF power, and working pressure of 250 °C, 50 W, and 0.5 Torr, for both the intrinsic and n-type layers. The thickness of the intrinsic and n-type a-Si:H films was adjusted to 400 nm and 25 nm, respectively. In

978-1-5090-5606-4/17 $31.00 © 2017 IEEE

the case of the n-type layer, the PH₃ was used as a dopant gas and the total flow rate was maintained at 180 sccm, i.e., H₂: 120 sccm, SiH₄: 30 sccm, and PH₃: 30 sccm). As a back electrode, 200-nm-thick silver (Ag) was thermally deposited in a high-vacuum (~2 × 10⁻⁶ Torr) chamber, by using a shadow mask to define each of the eighteen 0.25 cm² (area) active cells.

C. Characterization

Optical properties and crystallinity of structures were determined via spectroscopic ellipsometry (SE) and X-ray diffraction (XRD), respectively. In addition, the work function and bonding state of the films were analyzed using ultraviolet photoelectron spectroscopy (UPS) and X-ray photoelectron spectroscopy (XPS), respectively. The average photovoltaic performance was determined after measuring the performance of each of these eighteen cells of the a- Si:H thin-film solar cells. The corresponding current density voltage characteristics were measured by using a Keithley 2400 source meter under 100 mWcm⁻² (AM 1.5G) irradiation, from a solar simulator (Pecell Technologies Inc., PEC-L11). In addition, the external quantum efficiency (EQE) of each photovoltaic device was obtained by using a 200 W Xe lamp and a grating monochromator; the light intensity was measured by using a calibrated Si solar cell (Photovoltaic measurement).

III. RESULTS AND DISCUSSION

A. Growth behavior of ALD-V₂O₅₋ₓ films

The growth rate of V₂O₅₋ₓ films was monitored as shown in Fig. 1 (a). The rate decreased from 0.75 to 0.6 Å/cycle at the short period of O₃ pulse time, and then saturated with increasing pulse time of 3 ~ 8 sec. It was founded that the crystallinity and preferred orientation of the V₂O₅₋ₓ films are strongly related to O₃ pulse time (Fig. 1 (b)). At the O₃ pulse time of 1 s, films show typical amorphous structure indicating the presence of organic ligand impurities resulting from incomplete reaction between VTOP precursor and O₃. On the other hand, clear diffraction peaks from orthorhombic V₂O₅₋ₓ films were noticed with increase in O₃ pulse time. Comparing the relative ratio of (001) to (101) planes, V₂O₅₋ₓ films may experience further crystallization, and the plane perpendicular to the z-axis preferentially formed with increasing O₃ pulse time. Increase in crystallization was also supported by surface morphology and roughness as shown in Fig. 1 (c) and (d). This morphology may be influenced by the degree of crystallization, which in turn varies with the pulse time, i.e., amorphous and crystalline films have smooth and rough surfaces, respectively. The surface of the amorphous V₂O₅₋ₓ (at 1 s) was clean and smooth. The surface was roughened, however, with increasing pulse time, indicating that crystallization may be induced by long exposure to ozone. The crystal size of the V₂O₅₋ₓ film increased and the corresponding roughness increased gradually with increasing pulse time. As the figure shows, the roughness increased from ~ 0.15 to 1.98 nm in pulse times ranging from 1 to 8 s.

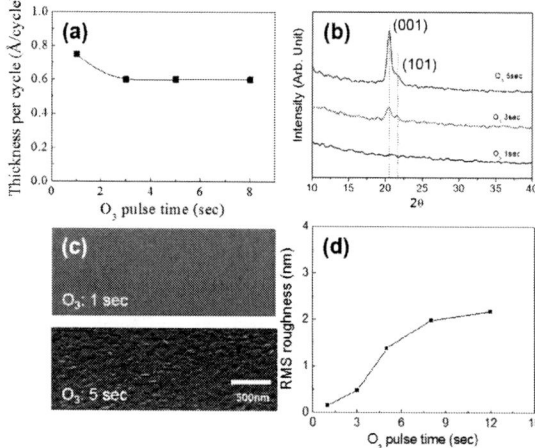

Fig. 1 Thickness per cycle (a) and XRD patterns (b) of ALD-V₂O₅₋ₓ film as a function of O₃ pulse time. (c) SEM surface images of V₂O₅₋ₓ film with O₃ pulse time of 1 (top) and 3 (bottom) sec. (d) RMS roughness as a function of O₃ pulse time.

B. Photovoltaic performances

Fig. 2 shows the typical J-V characteristics of fabricated thin films SCs with a multilayered structure of Ag/n-a-Si:H/i-a-Si:H/ALD-V₂O₅₋ₓ/FTO as a function of O₃ pulse time. The photovoltaic parameters including PCE, J_SC, FF, and V_OC were calculated and the results are shown in Fig. 3. From the result of significant improvement in PCE by insertion of ALD-V₂O₅₋ₓ, it was confirmed that V₂O₅₋ₓ films successfully acted as a hole-transporting layer. Among the various O₃ pulse time, exposure time of 1 s shows the highest value of PCE (5.35 %). The PCE was mainly determined by the variation in V_OC as shown in Fig. 3. V_OC was abruptly increased upon exposure to O₃ gas, and then slightly reduced with O₃ pulse time increased.

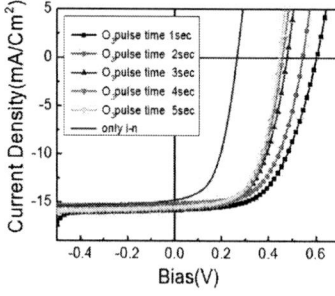

Fig. 2 Typical J-V characteristics of ALD-V₂O₅₋ₓ p-type layers grown at various O₃ pulse times.

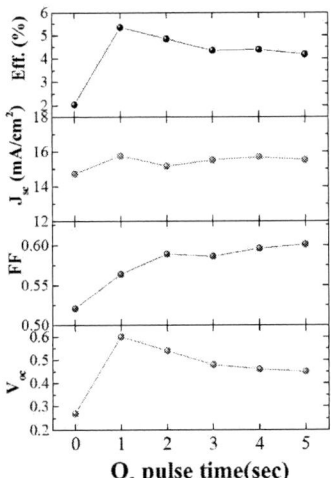

Fig. 3 Photovoltaic parameters of PCE, J_{SC}, FF, and V_{OC} as a function of O_3 pulse time.

V_{OC} indicates the difference in work function between p- and n-type layers, thus V_{OC} were varied proportionally with the work function of p-type V_2O_{5-x} layers. The work function is determined by photon energy – the binding energy of the secondary edge observed in UPS spectra (Fig. 4). The work function of V_2O_{5-x} films with O_3 pulse times of 1. 3, and 5 s were calculated as 5.31, 5.11, and 4.92 eV, respectively, indicating the work function was decreased with increasing pulse time.

For identifying correlation between work function and chemical bonding state of V_2O_{5-x} films, XPS analysis was performed as shown in Fig. 5. For all films, V^{4+} ions were coexisted with V^{5+} ions, and portion of low valence V^{4+} is increased with increase in O_3 pulse time. According to the previous study on the effect of vanadium valence state on work function [7], the work function of V^{4+} (VO_2) is lower than that of V^{5+} (V_2O_5). Consequently, reductions in the V_{OC} with increasing ozone pulse time may have resulted from an increase in the amount of VO_2 phase present in the film.

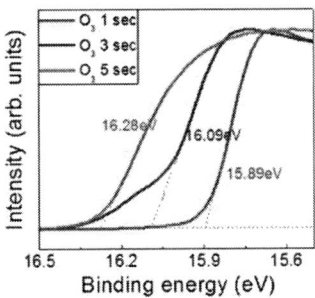

Fig. 4 UPS spectra of V_2O_{5-x} thin films

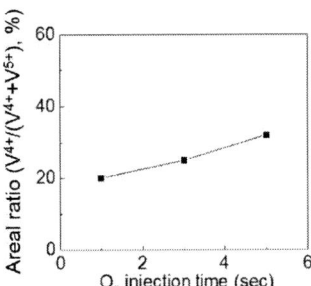

Fig. 5 The areal ratio of $V^{4+}/(V^{4+}+V^{5+})$ determined from XPS V core level $2p_{3/2}$ spctra

IV. SUMMARY

A p-doped layer-free a-Si:H thin film SCs constructed with a V_2O_{5-x} layers were successfully demonstrated by using the VTOP precursor and O_3 as reactants in the ALD system. At a O_3 pulse time of 1 s, the best PCE (i.e., 5.35%), was achieved because of the highest fraction of V_2O_5 phase, which induced increase in the work function (5.31 eV). This study provides an alternate path for the development of an optimized ozone pulse time of the V_2O_5, by ALD, for a thin-film a-Si:H SCs. More importantly, the hole-transporting layer is fabricated with environmental-friendly process (without using toxic gases).

REFERENCES

[1] S.Y. Myong, K. Sriprapha, Y. Yashiki, S. Miyajima, A. Yamada, and M. Konagai, "Silicon-based thin-film solar cells fabricated near the phase boundary by VHF PECVD technique," *Sol. Ener. Mater. Sol. Cells*, vol. 92, pp. 639-645, 2008.

[2] J.S. Lee, I.H. Jang, and N.G. Park, "Effects of oxidation state and crystallinity of tungsten oxide interlayer on photovoltaic property in bulk hetero-junction solar cell," *J. Phys. Chem. C*, vol. 116, pp. 13480-13487, 2012.

[3] H.H. Jung, J.D. Kwon, S. Lee, C.S. Kim, K.S. Nam, Y. Jeong, K.-B. Chung, S.Y. Ryu, T. Ocak, A. Eray, D.-H. Kim, "Doping-free silicon thin film solar cells using a vanadium pentoxide window layer and a LiF/Al back electrode," *Appl. Phys. Lett.*, vol. 103, pp. 073903, 2013.

[4] M.T. Greiner, M.G. Helander, W.M. Tang, Z.B. Wang, J. Qiu, and Z.H. Lu, "Universal energy-level alignment of molecules on metal oxides," *Nat. Mater.*, vol. 11, pp. 76-81, 2011.

[5] K. Takanezawa, K. Tajima, and K. Hashimoto, "Efficiency enhancement of polymer photovoltaic devices hybridized with ZnO nanorod arrays by the introduction of a vanadium oxide buffer layer," *Appl. Phys. Lett.*, vol. 93, 063308, 2008.

[6] M. Benmoussa, E. Ibnouelghazi, A. Bennouna, and E.L. Ameziane, "Structural, electrical and optical properties of sputtered vanadium pentoxide thin films," *Thin Solid Films*, vol. 265, pp. 22-28, 1995.

[7] M.T. Greiner, L. Chai, M.G. Helander, W.M. Tang, and Z.H. Lu, "Transition metal oxide work functions: the influence of cation oxidation state and oxygen vacancies," *Adv. Func. Mater.*, vol. 22, pp. 4557-4568, 2012.

A Novel Defect Passivation Method for Multicrystalline Si Wafer by H$_2$S Reaction

Hsiang-Yu Liu, Ujjwal K. Das, and Robert W. Birkmire

Institute of Energy Conversion, University of Delaware, Newark, DE 19716, USA

Abstract — The passivation of multicrystalline silicon (mc-Si) wafers by reaction in H$_2$S is discussed where the effective minority carrier lifetime (τ_{eff}) was measured to evaluate the improvements of the bulk quality. Mc-Si τ_{eff} improved to >200 μsec due to bulk and surface passivation by H$_2$S reaction, compared to 40~60 μsec with only surface passivation by quinhydrone-methanol (QHY/ME) or hydrogenated amorphous Si (a-Si:H), two well-developed surface passivation methods for monocrystalline Si. Injection level dependent τ_{eff} showed the reduction of both trap and recombination states for mc-Si passivation in H$_2$S. τ_{eff} > 250 μsec was yielded through surface repassivation in QHY/ME after removal of H$_2$S reaction-formed surface layer by wet chemical etching. This indicates the passivated bulk quality by H$_2$S reaction was not affected even after a series of wet chemistry treatments.

Index Terms — bulk passivation, H$_2$S reaction, minority carrier effective lifetime, multicrystalline Si, passivation stability, trap effect

I. INTRODUCTION

Deployment of multicrystalline silicon (mc-Si) wafer solar cells has expanded in recent years due to much lower cost compared to monocrystalline silicon wafer solar cells. The greatest challenge in mc-Si wafer solar cell is to improve the device performance that is primarily limited by traps and recombination states along the grain boundaries. These recombination defects are detrimental to the minority carrier lifetime (τ_{eff}) and cell efficiency[1].

Hydrogen passivation of mc-Si bulk defects at the grain boundaries is the most common method to improve mc-Si solar cell performance[2]. Several methods have been used to diffuse hydrogen into silicon bulk for passivation such as: hydrogen plasma[3], low energy hydrogen implantation[4], and plasma-enhanced chemical vapor deposition (PECVD) of hydrogenated silicon nitride (a-SiN$_x$:H)[5]. Hydrogen plasma and hydrogen implantation have inherent limitation due to ion damage, making a-SiN$_x$:H the most common method for hydrogen diffusion to reduce bulk defects in mc-Si and therefore increase the cell efficiency.

In this work, a new approach for mc-Si bulk defect passivation is presented using H$_2$S reaction. We have previously reported [6] that sulfur can passivate silicon surface defects and improve minority carrier lifetime in monocrystalline Si through reaction with H$_2$S. The concept of Si surface passivation by H$_2$S is based on theoretical arguments that sulfur can be a suitable valence mending adsorbate to restore silicon surface[7]. Besides being a proper candidate for Si surface restoration, sulfur has the highest diffusivity in Si among all group VI elements and the potential for bulk defect passivation. Han et.al.[8] reported that reaction of H$_2$S with Si (100) surface consisted of H$_2$ desorption accompanied by diffusion of sulfur into bulk silicon at elevated temperature. Comparing to hydrogen, the double valency of sulfur could passivate double dangling bond sites along grain boundaries, while hydrogen can only passivate single dangling bond sites.

Saha et al. reported the passivation of mc-Si bulk using H$_2$S reaction[9], with the best passivation results leading to τ_{eff} ~ 25 μsec at H$_2$S reaction temperature 525°C, followed by post-annealing at 500°C. However, the apparent τ_{eff} curve as a function of minority carrier density (Δn) displayed an obvious trap effect at low injection level, which implies presence of significant number of traps in the bulk.

II. EXPERIMENTAL

N-type 190 μm mc-Si wafers with resistivity 3.0 Ω-cm were cleaned using the following sequential steps: 1) solvent cleaning with ultrasonic agitation; 2) piranha oxidation in a mixture of sulfuric acid and hydrogen peroxide (3:1); 3) removal of surface oxide layer in 10% hydrofluoric acid (HF); 4) 2-3 μm of Si etching in an acid mixture of HF : nitric acid = 1 : 100 (HNA); and 5) formation of a H-terminated surface using 10% HF dip for 60 sec just prior to loading into the H$_2$S reactor. N-type 145 μm monocrystalline Cz wafers with resistivity 5.0 Ω-cm were used as control samples, where the cleaning procedure was same as mc-Si wafers.

The H$_2$S reaction was carried out in a custom-built quartz tube chemical vapor deposition (CVD) reactor, where the reaction temperature was controlled by a heating jacket and monitored by K-type thermocouple positioned beneath the sample holder. The tube reactor was evacuated to 10^{-6} Torr after loading the H-terminated silicon samples, and then pressurized to 520 Torr with H$_2$S gas at room temperature, followed by temperature ramp-up to the set point. All reactions were done in a static H$_2$S environment at temperature between 600°C - 675°C for 30 minutes. After the reaction was completed, the system was cooled down in static H$_2$S to 250°C, and then to room temperature in flowing Ar.

Si defect passivation was also studied by two other methods for comparison using quinhydrone-methanol (QHY/ME) and hydrogenated amorphous Si (a-Si:H). H-terminated Si wafers after the last cleaning (step 5) were immersed in QHY/ME solution that is known to passivate surface defects by organic

978-1-5090-5606-4/17 $31.00 © 2017 IEEE

adsorbates[10]. In another method, a thin (10 nm) a-Si:H layer were deposited on both sides of clean wafers at 200°C by PECVD as discussed elsewhere [11].

The passivation quality was evaluated by measuring injection level dependent τ_{eff} using a Sinton tester by photoconductance decay (PCD) method. The τ_{eff} values reported in the text are at a minority carrier density of 1×10^{15} cm^{-3}. Photoluminescence (PL) images were used to evaluate the uniformity of surface passivation where the image was captured using silicon germanium detector. X-ray photoelectron spectroscopy (XPS) characterization was performed on control monocrystalline Cz wafer with its surface passivated in H$_2$S reaction where the samples were transferred to XPS without exposing to air using a vacuum-sealed transport vessel.

III. RESULTS & DISCUSSION

Fig. 1(a) and (b) shows the τ_{eff} as a function of Δn of monocrystalline Cz wafer and mc-Si wafer, respectively with three different passivation approaches. Excellent defect passivation for Cz wafers were achieved using all three methods: QHY/ME, deposited a-Si:H, and H$_2$S reaction, where the $\tau_{eff} > 2000 \mu$sec were achieved as shown in Fig 1(a). Monocrystalline Cz wafers used in this work have high bulk quality, therefore the passivation on Cz wafers can be considered as surface passivation only. The Fig. 1(b), however, shows that the defect passivation levels are different for three different methods. Moreover, the shape of the τ_{eff} curve for QHY/ME and a-Si:H passivated wafers are very different from the H$_2$S passivated wafer. The anomalous increase in τ_{eff} at low injection level ($<1\times10^{15}$ cm^{-3}) for QHY/ME and a-Si:H passivated samples indicate presence of trap states as commonly observed in poorly passivated mc-Si due to presence of dislocation and grain boundaries [12,13]. At a low injection level, the excess carrier density becomes comparable to the trap states density. This increases the photo conductance and hence the carrier lifetime estimated from PCD[13]. The H$_2$S passivated wafer shows significant suppression of this apparent increase in τ_{eff} at low injection level due to efficient passivation of trap states. It also improves τ_{eff} at high injection ($>1\times10^{15}$ cm^{-3}) indicative of efficient passivation of recombination states. Therefore we conclude that QHY/ME and deposited a-Si:H layer can not effectively passivate the trap and recombination states in mc-Si, while H$_2$S reaction reduces both types of defects.

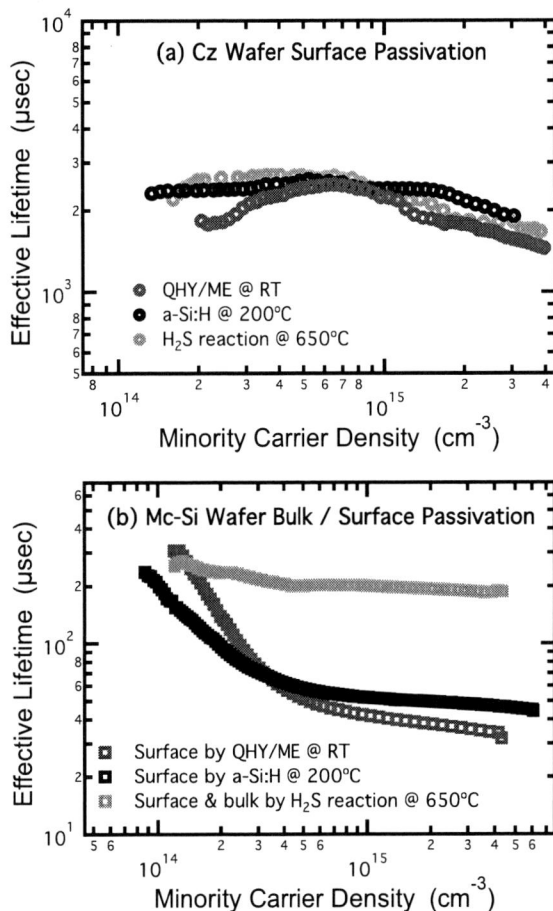

Fig. 1. (a) τ_{eff} as a function of Δn of monocrystalline Cz wafers comparing three different Si surface passivation methods; (b) τ_{eff} as a function of Δn of mc-Si wafers with the same passivation methods as with Cz wafers demonstrating QHY/ME and a-Si:H can not passivate trap and recombination states in mc-Si effectively, while H$_2$S reaction passivates both types of defects.

The stability of the τ_{eff} for the Cz wafers after reaction with H$_2$S is inferior compared to the mc-Si wafers. Unstable surface passivation level is within expectation since the surface passivating silicon sulfide compound is unstable in air due to ease of hydrolysis and oxidation[14]. Fig. 2 shows the degradation of τ_{eff} of a Cz wafer after passivation in H$_2$S, where the initial τ_{eff} is ~2000 μsec and degraded to below 100 μsec within 70 minutes. The mc-Si wafer has relatively better stability comparing to Cz wafers, as seen in Fig. 2. The slightly degraded τ_{eff} on mc-Si is most likely caused by the instability of surface passivation level, which can be further understood in the next surface removal and repassivation work.

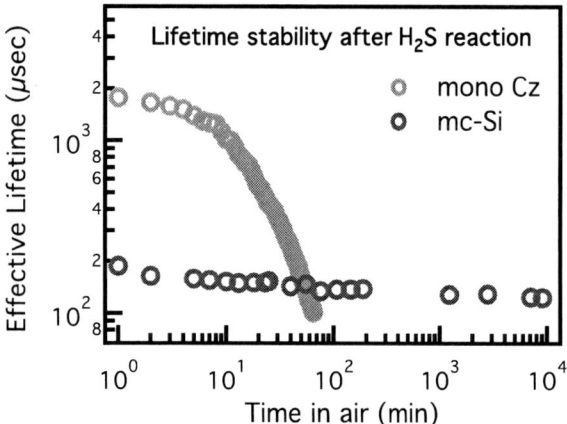

Fig. 2. Comparison of the passivation stability after H$_2$S reaction: The Cz wafer τ_{eff} degraded from initial ~2000 μsec to below 100 μsec within 70 min with storage in air; the mc-Si wafer was relatively stable over time.

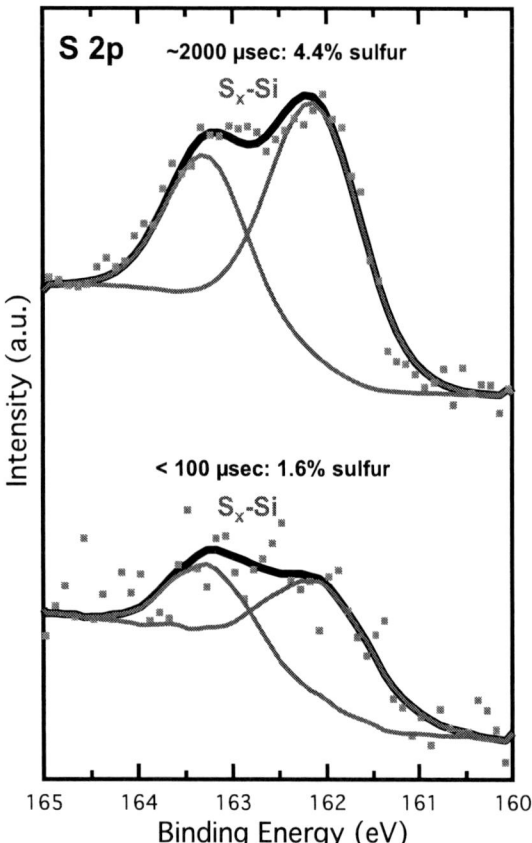

Fig. 3. XPS spectra of S 2p on a Cz Si surface after reaction with H$_2$S at 650°C: comparison of the sulfur concentration before and after degradation where the initial τ_{eff} is ~2000 μsec and the degraded τ_{eff} is below 100 μsec.

XPS analysis of S 2p peak was performed on Cz wafers after reaction with H$_2$S at 650°C to compare the sulfur concentration and τ_{eff} before and after degradation is shown in Fig. 3. The concentration of sulfur on Si surface decreased from 4.4% to 1.6% concurrent with the change in τ_{eff} from ~2000 μsec to below 100 μsec.

To further separate the effect of surface and bulk passivation and instability in τ_{eff}, a series of wet chemistry treatments for surface removal and repassivation was studied as shown in Fig. 4(a). A mc-Si wafer cleaned in HF had an initial τ_{eff} of only 25 μsec before H$_2$S reaction, which increased to 208 μsec after reaction in H$_2$S at 650°C for 30 minutes and is attributed to reduction of both traps and recombination states. The passivated mc-Si wafer was then etched in HNA solution to remove ~ 2 μm Si layer from both surfaces and τ_{eff} after surface removal dropped down to 62 μsec. QHY/ME was then used to repassivate the surface, resulting a τ_{eff} of 252 μsec. This indicates H$_2$S reaction passivates bulk defects in mc-Si and the bulk passivation remains unaffected even after a series of wet chemical etching and cleaning.

Fig. 4(b) exhibits the injection level dependent τ_{eff} after each processing step. Removal of surface layer on both sides after H$_2$S reaction (blue curve) resulted in unpassivated surfaces, but the bulk passivation effect still can be seen when comparing with clean bare mc-Si wafer (black curve). The reduction in trap states and recombination states after H$_2$S reaction (red curve) and after surface repassivation in QHY/ME (green curve) demonstrates the bulk passivation quality is unaffected and preserved during a series of wet chemistry treatments. Therefore the results in Fig. 4 indicate: i) H$_2$S reaction provides both bulk and surface defect passivation effectively; ii) H$_2$S-passivated bulk quality is stable over time, and the unstable τ_{eff} is due to degradation of surface passivation, similar to Cz wafer; and iii) the passivated surface can be restored and repassivated with few simple wet chemical steps without affecting the bulk passivation provided by H$_2$S. This developed wet chemical process will allow the fabrication of surface sensitive high efficiency Si solar cell structures, such as a-Si/c-Si heterojunction (HJ) on mc-Si wafer that has H$_2$S-passivated bulk using a-Si:H as a surface passivation layer. The completion of device will provide approaches to investigate other electrical properties, such as diffusion length, emitter saturation current density and suitability for high efficiency solar cell structure on mc-Si with bulk passivated by H$_2$S reaction.

(a)

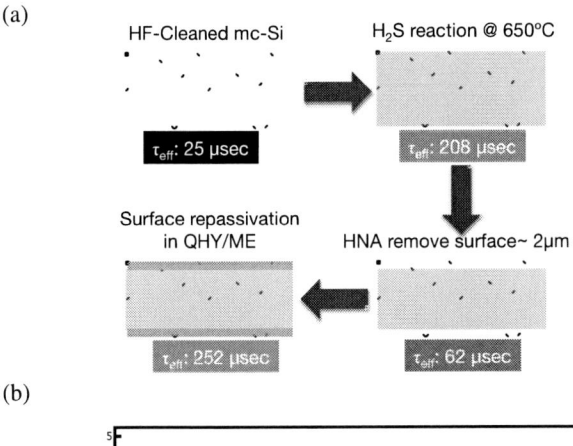

(b)

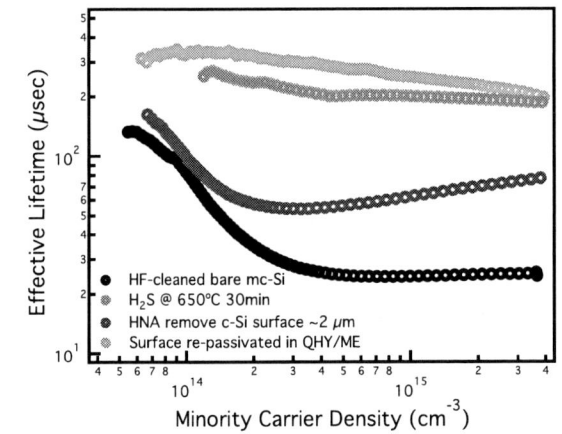

Fig. 4. (a) Graphic depiction of processing on mc-Si starting from HF-cleaned bare wafer, followed by H_2S reaction for bulk and surface passivation. After H_2S reaction, surface was removed by HNA and then repassivated by QHY/ME. Each step has τ_{eff} reported at 10^{15} cm^{-3}. (b) Injection level dependent τ_{eff} after each step of processing, verifying the bulk passivation quality can be maintained with trap effect eliminated during surface removal and repassivation in wet chemistry.

The PL images of the unreacted and H_2S reacted mc-Si are shown in Fig. 5. The bare mc-Si with τ_{eff} 7 μsec does not display any luminescence. However, for the mc-Si wafer processed in H_2S at 650°C with an initial $\tau_{eff} = 208$ μsec, there is a well-defined PL image, demonstrating the improved bulk and surface quality after reaction in H_2S.

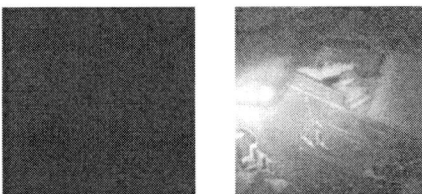

Fig. 5. PL images on clean bare mc-Si wafer (left) shows no luminescence and (right) 650°C H_2S-passivated mc-Si wafer with a clear defined PL image.

IV. CONCLUSION

H_2S reaction with mc-Si provides both surface and bulk passivation where the τ_{eff} increased from 7 μsec to > 200 μsec and was directly observed by PL luminescence imaging. Mc-Si had the anomalous increase of τ_{eff} at low injection level caused by trap effect no longer exist after passivation in H_2S. Surface passivation using H_2S reaction degrades due to the unstable Si-S compound in air, which was confirmed by evaluating the H_2S passivation on Cz wafers and XPS analysis. The H_2S-passivated surface layer on mc-Si wafer can be completely removed without damaging the bulk passivation, and surface passivation can be restored using a-Si:H or QHY/ME solution. This approach confirms the stability of bulk passivation over time and not affected by wet chemical processing, and therefore provides a pathway to fabricate HJ solar cells on mc-Si wafers with surface passivated by a-Si:H and bulk passivated by H_2S.

ACKNOWLEDGEMENT

Authors thank Gowri Sriramagiri for helping on sample processing, Nuha Ahmed for photoluminescence characterization at IEC, and Dr. Thomas Beebe and Zachary Voras from Department of Chemistry and Biochemistry for help in XPS measurements and many helpful discussions. XPS characterization in this work was supported by US National Science Foundation award number 1428149.

REFERENCES

[1] A. K. Ghosh, C. Fishman, and T. Feng, "Theory of the electrical and photovoltaic properties of polycrystalline silicon," *J. Appl. Phys.*, vol. 51, no. 1, pp. 446–454, 1980.

[2] C. E. Dube and J. I. Hanoka, "Hydrogen passivation of multicrystalline silicon," *Photovoltaic Specialists Conference,* 2005.

[3] S. Rattanapan, H. Yamamoto, S. Miyajima, T. Sato, and M. Konagai, "Hydrogen plasma treatment for improving bulk passivation quality of c-Si solar cells," *Current Applied Physics*, vol. 10, no. S, pp. S215–S217, 2010.

[4] S. Ashok and S. A. Ringel, "Low-energy hydrogen implantation for silicon Schottky barrier modification," *Vacuum*, vol. 36, no. 11, pp. 917–920, 1986.

[5] F. Duerinckx and J. Szlufcik, "Defect passivation of industrial multicrystalline solar cells based on PECVD silicon nitride," *Solar Energy Materials and Solar Cells*, vol. 72, no. 1, pp. 231–246, 2002.

[6] H.-Y. Liu, U. K. Das, S. Hegedus, Z. E. Voras, T. P. Beebe Jr, T. P. Beebe, and R. Birkmire, "Si surface passivation by H_2S reaction for c-Si solar cell ," *Electronic Materials Conference,* 2016.

[7] E. Kaxiras, "Semiconductor-surface restoration by valence-mending adsorbates: Application to Si(100):S and Si(100):Se," *Phys. Rev. B*, vol. 43, no. 8, pp. 6824–6827, 1991.

[8] M. Han, Y. Luo, N. Camillone, and R. M. Osgood, "Reaction of H 2S with Si(100)," *J. Phys. Chem. B*, vol. 104, no. 28, pp. 6576–6583, 2000.

[9] A. Saha, H. Zhang, and W. C. Sun, "A new method for bulk passivation in multicrystalline-Si by sulfur," *42th IEEE Photovoltaic Specialists Conference*, 2015.

[10] B. Chhabra, S. Bowden, R. L. Opila, and C. B. Honsberg, "High effective minority carrier lifetime on silicon substrates using quinhydrone-methanol passivation," *Appl. Phys. Lett.*, vol. 96, no. 6, pp. 063502, 2010.

[11] U. K. Das, M. Z. Burrows, M. Lu, and S. Bowden, "Surface passivation and heterojunction cells on Si (100) and (111) wafers using dc and rf plasma deposited Si: H thin films," *Appl. Phys. Lett.*, vol. 92, no. 6, pp. 063504, 2008.

[12] J. R. Haynes and J. A. Hornbeck, "Trapping of Minority Carriers in Silicon. II. n-Type Silicon," *Phys. Rev.*, vol. 100, no. 2, pp. 606–615, Oct. 1955.

[13] D. Macdonald and A. Cuevas, "Trapping of minority carriers in multicrystalline silicon," *Appl. Phys. Lett.*, vol. 74, no. 12, pp. 1710–1712, Mar. 1999.

[14] A. Haas, "The Chemistry of Silicon-Sulfur Compounds," *Angewandte Chemie International Edition*, vol. 4, no. 12, pp. 1014–1023, Dec. 1965.

Carrier Transportation at Novel Silver Paste Contact

Takefumi Kamioka[1], Satoshi Kameyama[1], Kazuo Muramatsu[2], Aki Tanaka[2], Naotaka Iwata1, Kyotaro Nakamura[3], Atsushi Ogura[3], and Yoshio Ohshita[1]

[1]Toyota Technological Institute, Nagoya, Aichi 468-8511, Japan

[2]NAMICS Corporation, Nigorigawa, Niigata 650-3131, Japan

[3]Meiji University, Kawasaki, Kanagawa 214-8571, Japan

Abstract — **When our novel Ag paste is used, the high solar cell performance is obtained. A thick SiON layer is formed between Ag line and Si substrate after the firing process. To understand the effect of this layer on the electrical properties, the current-voltage analysis of the contact is carried out. The obtained results indicate that there might be locally contacted areas between Ag and Si, and these regions are distributed in laterally, which determined the carrier transportation at the contact. This SiON layer with local contacts prevents the Ag diffusion into Si crystal during the firing process, and decreases the minority carrier recombination, resulting in the relative high performances.**

Index Terms — **contact, diode, screen printing, silicon, silver paste, solar cell.**

I. INTRODUCTION

To realize the high conversion efficiency of crystalline silicon (Si) solar cells, the higher performance of metallization is required. The front and back metals are widely formed by the screen printing processes. However, there are three problems which deteriorate the solar cell properties. One is the line metal contamination. During the metallization process, silver (Ag) atoms diffuse into the semiconductor material due to the high firing temperature. The Ag diffusion decrease the minority carrier lifetime and increase the p-n junction leakage current. The second one is the recombination at the metal and Si interface. The Ag/Si interface acts as an active recombination center, which increases the reverse saturation current density (J_0), resulting in the decrease of the open circuit voltage (V_{OC}). The third issue is the contact resistance to the high resistivity emitter. In the conventional crystalline Si solar cells, the high concentration phosphorus (P) atoms, which exceed to 10^{20} atoms/cm^3, are doped to form an n-type layer as an emitter on p-type Si substrate. Because of this high doped impurities, the minority carrier lifetime in the emitter region decreases through the Auger process. This causes the lower V_{OC}. However, when the conventional Ag paste is used to the high resistivity emitter, the contact resistance increases, and the fill factor (FF) is decreased.

To solve these problems, the new Ag paste has been developed [1]. It demonstrates the low contact resistance to the relatively low dosed n-type layer, and the high FF and V_{OC}

were obtained. By using this paste, the n-type bifacial PERT cells were realized with the relatively high conversion efficiency without the selective emitter structure (Fig. 1) [2]. There exists a SiON layer, which contains fine Ag particles between the Ag lines and Si substrate [1]. This layer may play an important role in the realization of these unique properties. However, the reasons for the obtained results are not clear yet. Especially, the layer seems to be too thick to allow enough current for the solar cell performance and the carrier transportation mechanism is not clear neither.

In this paper, the carrier transportation at the interface between the Ag and Si substrate is studied based on the current–voltage analysis. The Ag metal contact is formed on the p- or n-type Si substrate using the novel or conventional pastes. Although there exists a thick SiON layer between Ag and Si, the high current is obtained. The obtained results indicate that the current and voltage relation is determined as a Schottky contact and that the Ag directly contacted to Si substrate determines the carrier transportation.

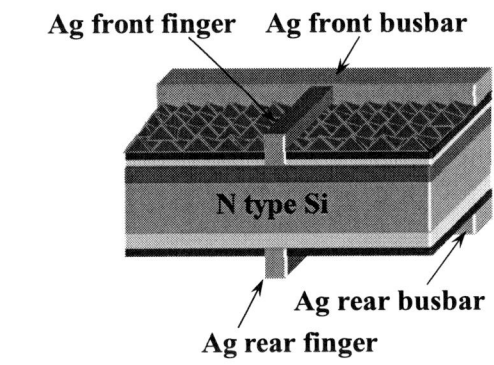

J_{SC} (mA/cm^2)	V_{OC} (V)	FF	Eff. (%)
40.6	0.656	0.743	19.8

Fig. 1 Schematic image of the fabricated bifacial solar cell. Both front and rear fingers were formed by our novel Ag paste [2].

978-1-5090-5606-4/17 $31.00 © 2017 IEEE

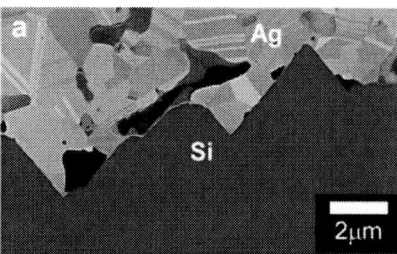

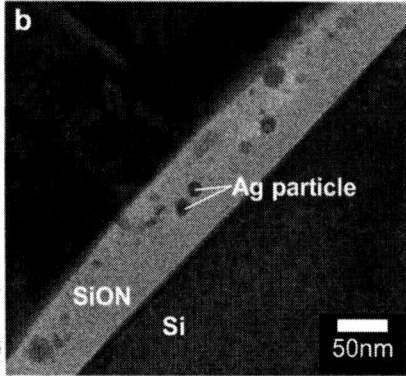

Fig. 2 (a) Cross-sectional SEM image of the contact layer. (b) Relatively thick SiON layer is formed at the interface between the front Ag line and Si substrate. Fine Ag particles exist in this layer.

II. EXPERIMENT

The current-voltage (I–V) characteristic was obtained using a simple structure device, as shown in Fig. 3. The (001) n-type (arsenic or phosphorus doped) and p-type (boron doped) Si wafers were used as substrates. The silicon-nitride (SiN) film was deposited on the wafer by the parallel-plate plasma-enhanced chemical vapour deposition using the silane (SiH$_4$) and ammonia (NH$_3$) gas system. The film thickness was 80 nm. The deposition condition was same as the solar cell fabrication. The Ag electrode was printed by the screen printing method. Here, our novel and the conventional Ag pastes were used. The Ag electrode was obtained by the firing process. The firing condition was same as the cell production. The Ag/Al was deposited for the back-side contact formation.

The I–V relationships were obtained at room temperature and also the effect of measurement temperature dependences were evaluated. Based on these relationships, the carrier transportation properties at the contact were analyzed.

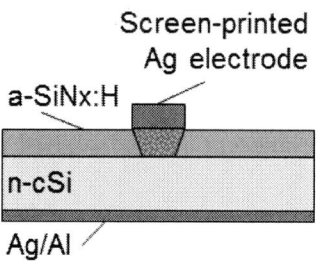

Fig. 3 Schematic of the test device used for evaluating carrier transportation properties between Ag and n-type Si substrate.

III. RESULTS AND DISCUSSION

The current–voltage (I–V) relationship for Ag/n-Si contact at room temperature is shown in Fig 4. It shows that the contact has the rectification properties and is not Ohmic.

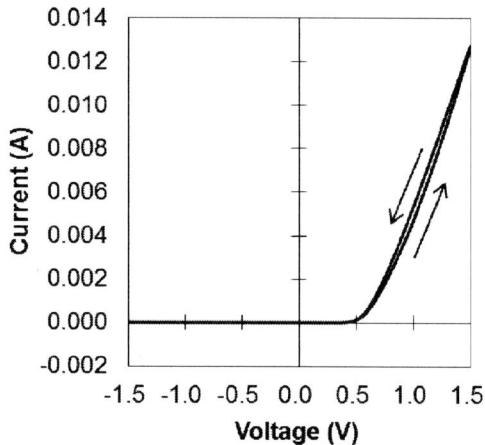

Fig. 4 Current–voltage relationship. This shows the rectification properties.

The temperature dependence of I-V curve is shown in Fig. 5. At a small forward voltage (< 0.5 V), the current exhibits the exponential behavior and has a strong temperature dependence. This result suggests that the current is limited by the thermionic-emission at the interface, and not by the tunneling phenomena.

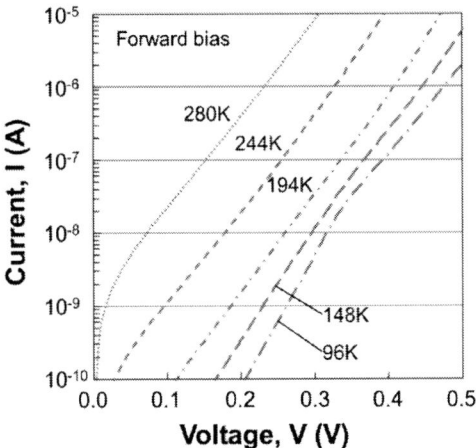

Fig. 5 Current–voltage relationships as a function of measurement temperature.

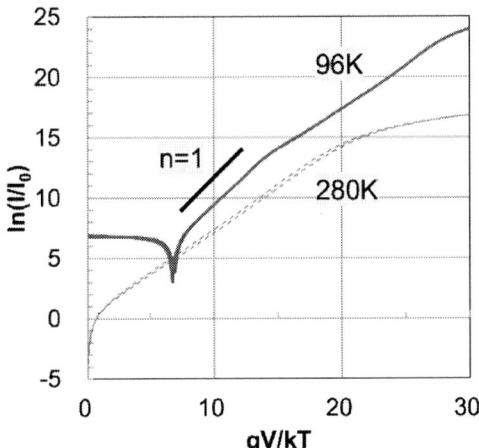

Fig. 6 Temperature dependence of current-voltage properties at 96 and 280 K.

Figure 6 suggests that there is a Schottky barrier at the interface. The current density–voltage (J–V) characteristic of a metal-semiconductor Schottky contact, where the current is determined by the thermionic emission, is expressed as follows [4].

$$J = A^* T^2 \exp\left(-\frac{\Phi_B}{kT}\right)\left[\exp\left(\frac{qV}{nkT}\right) - 1\right]. \quad (1)$$

Here, k is the Boltzmann constant, q the elemental charge, T the temperature, A^* the Richardson constant and n the ideal factor. Schottky barrier height, Φ_B, is the difference between the metal workfunction and Si semiconductor electron affinity. The experiment result is well explained by this equation. Since the doping concentration of Si substrate is low, this barrier is formed.

By fitting the experimental I–V curve to the ideal Schottky contact diode model, the extracted Schottky barrier height

(Φ_B) were obtained about 0.63 eV for n-type and 0.28 for p-type values, respectively. These values are explained by using the Fermi level of Si crystal and Ag metal workfunction [5]. These results suggested that a direct contact between Ag and Si was formed at the interface during the metallization process, and that the dominant carrier transport mechanism in this range can be explained by the Schottky contact model. The secondary electron microscopy (SEM) image of the Ag/Si interface showed that almost all the Si surface was covered with a SiON layer [1]. Combined with the present I–V analysis, there might be the locally contacted areas between Ag and Si, and these regions are distributed in laterally, which determine the carrier transportation. In a solar cell structure, the contacts have high doping concentration at the emitter region and this barrier width becomes very narrow. Therefore, the current is limited by the tunneling at the Schottky barrier. On the other hand, this SiON layer decreases the minority carrier recombination at the contact areas and prevents the Ag diffusion from the electrodes to the semiconductor.

VI. CONCLUSION

Our new paste showed the low contact resistance to the relatively low dosed n-layer and the high FF and V_{OC}. There was a uniform SiON layer between the Ag lines and Si substrate. Despite of this thick SiON layer, the high current was obtained. To understand the carrier transportation mechanism at the contact layer, the Ag contact was formed on the Si substrate without p-n junction by the firing method and the current-voltage was analyzed. The dominant carrier transport mechanism in this range was well explained by the Schottky contact model. This indicated that there might be the locally contacted areas between Ag and Si and these regions were distributed in laterally, which determined the carrier transportation.

ACKNOWLEDGEMENT

This work was partly supported by the New Energy and Industrial Technology Development Organization (NEDO) under the Ministry of Economy, Trade and Industry of Japan.

REFERENCES

[1] T. Takahashi, "Contact formation issues for high efficiency crystalline silicon solar cells," *Proc. Workshop on Crystalline Silicon Solar Cells & Modules: Materials and Processes*, 2015, p. 56.
[2] K. Nakamura, T. Takahashi, and Y. Ohshita, "Novel silver and copper pastes for n-type bi-facial PERT cell," *Proc. of European Photovoltaic Science and Engineering Conference*, 2015, pp. 536.
[3] T. Kamioka, Tetsu Takahashi, Kazuo Muramatsu, Aki Tanaka, Naotaka Iwata, Kyotaro Nakamura, Atsushi Ogura, and Yoshio Ohshita, "Novel silver paste to n- and p-layers for fabricating high efficiency crystalline Si solar cells," in *26the Photovoltaic Science and Engineering Conference*, 2016.

[4] B. L. Anderson and R. L. Anderson, *Fundamentals of semiconductor Devices 1st ed*. New York, New York: McGraw-Hill, 2005.

[5] A. M. Cowley and S. M. Sze, "Surface states and barrier height of metal-semiconductor systems," *J. Appl. Phys.*, vol. 36, pp. 3212-3220, 1965.

Influence of Deposition Parameters on Silicon Thin Films Deposited by Magnetron Sputtering

Grace Rajan[1], Tejaswini Miryala[1], Shankar Karki[1], Robert W. Collins[2], Nikolas Podraza[2], Sylvain Marsillac[1]

[1] Virginia Institute of Photovoltaics, Old Dominion University, Norfolk, VA, USA
[2] Department of Physics and Astronomy, The University of Toledo, Toledo, OH, USA

Abstract — **Amorphous and crystalline silicon films are widely used for microelectronic and photovoltaic devices. Here, silicon thin films were fabricated by magnetron sputtering and the deposition parameters were modified to enhance crystallinity of the films. The sputtering deposition was notably assisted by a low energy argon ion beam. The substrate temperature and sputtering power, as well as the energy of the ion beam, were used as deposition parameters. Finally, the optical and electrical properties of the deposited thin films were studied and demonstrate the possibility of using several deposition parameters to enhance the films crystallinity.**

Keywords: Silicon, Ion beam, Sputtering, Thin film properties

I. INTRODUCTION

Micro/nanocrystalline silicon has attracted considerable attention as an innovative, low-cost and stable material for thin films in solar cells and thin film transistors. They are mixed phase materials typically containing both a crystalline silicon phase and an amorphous silicon phase [1]. These films have improved carrier mobility, lifetime, diffusion length, and conductivity, as well as higher doping efficiency and stability against light soaking under illumination in comparison with hydrogenated amorphous silicon films (a-Si:H) [1, 2]. The crystalline films can be fabricated directly in deposition processes by different methods such as low pressure chemical vapor deposition and plasma enhanced chemical vapor deposition, or they can be formed from a-Si(:H) by solid phase crystallization, excimer laser annealing or metal induced crystallization [3]. Magnetron sputtering is also a promising method for the fabrication of silicon thin films, as the crystallinity can be controlled through a variety of deposition conditions. One approach in alternative for using higher growth temperature for controlling the growth kinetics is by increasing the momentum of the arriving atoms (to enhance the surface diffusion). This can be achieved by employing ion beam deposition with low energy ions, or ion beam assisted deposition (IBAD) (25-300 eV) [4]. In this paper, the different deposition parameters in magnetron sputtering with IBAD capability are varied to study the optical and electrical properties of the deposited silicon thin films. The impact of deposition power, temperature and energies of sputtered atoms on the crystallinity of the deposited films is studied.

II. EXPERIMENTAL DETAILS

P-type doped silicon thin films were prepared by RF sputtering from a 3" boron doped p-type silicon target in an argon atmosphere. The deposition process was carried out at a base pressure of 4×10^{-7} Torr to reduce the contamination from the atmospheric gas. The argon pressure was maintained at 0.5 mTorr for all depositions. A Kaufman KDC-40 4 cm gridded ion source aimed at the growing film was used along with the RF sputtering process to assist in the deposition process. Soda lime glass, quartz, fused silica, and silicon wafers with native and thermal (500 nm) oxide layers were used as substrates. The depositions were carried out at temperatures varying from room temperature to 750°C. The highest temperatures were restricted to quartz, fused silica, and silicon wafer substrates. *Ex-situ* spectroscopic ellipsometry (SE) data were acquired after film growth using a multichannel ellipsometer with a photon energy range from 0.75 to 6.5 eV at angles of incidence of 55°, 65° and 75°. Complementary characterization of the as-deposited films included Hall Effect and resistivity measurements (4-point probe).

III. RESULTS AND DISCUSSION

A. Effect of substrate temperature

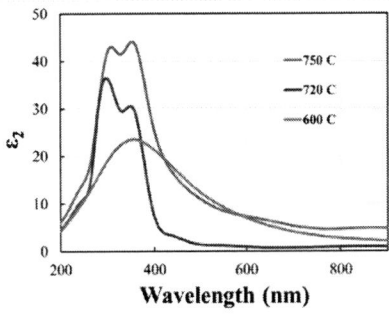

Fig. 1: Imaginary part ε_2 of the complex dielectric function extracted by spectroscopic ellipsometry for silicon films deposited at different substrate temperatures on top of silicon wafers with native oxides.

Silicon thin films were deposited by sputtering on different substrates at temperatures ranging from room temperature to 750°C. The complex dielectric functions extracted from SE on silicon wafer substrates with native oxide are illustrated in Figure 1. It can be seen that at temperatures of 600°C below, the silicon layer does not show any features, indicative of an amorphous layer. For deposition temperatures above 700°C, two key features appear notably between 300-400 nm, indicating that the samples are nanocrystalline.

B. Effect of substrate

The films were deposited on different substrates – namely soda lime glass, quartz, fused silica, and silicon wafers with native and thermal oxides. Quartz and fused silica substrates were chosen for Hall Effect measurements of films deposited at higher temperatures. The electrical properties were measured primarily for the Si films on quartz and fused silica substrates, to avoid the possibility of probing through the oxides to the underlying silicon substrates.

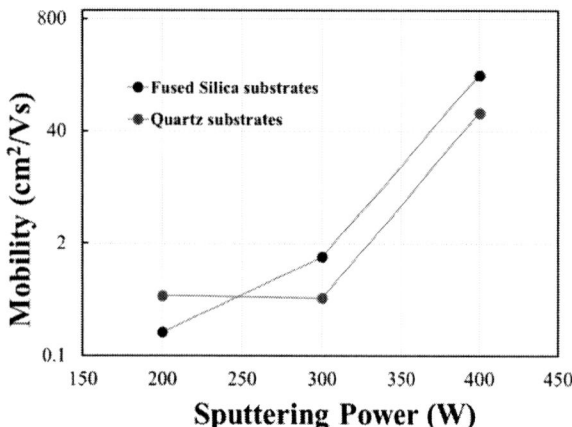

Fig. 2: Influence of sputtering power on the mobility of thin film silicon deposited on different substrates.

Figure 2 compares the variation in the mobility with sputtering power for the two different substrates. As one can see, there is not a significant difference between the results for the two substrate types.

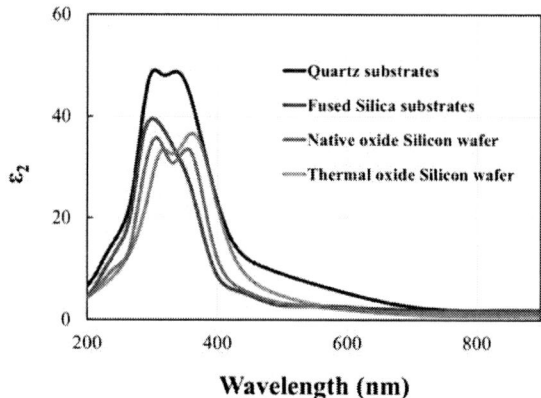

Fig. 3: Imaginary part ε_2 of the complex dielectric function extracted by spectroscopic ellipsometry for different substrates.

The imaginary part ε_2 of the complex dielectric function as extracted from ex-situ SE is shown in Figure 3 for different substrates. It is observed that the films deposited on silicon wafers (both native oxide and thermal oxide covered) have better crystalline quality, as indicated by the best resolved features. The films deposited on quartz and fused silica substrates, however, do not have well resolved peaks at 295 nm and 364 nm (4.2 eV and 3.4 eV) and the films are likely to incorporate smaller grains. In some cases, one peak is enhanced relative to the other. In fact, the peaks for the Si film on quartz are very strong and characteristic of crystalline silicon.

C. Effect of Ion beam deposition

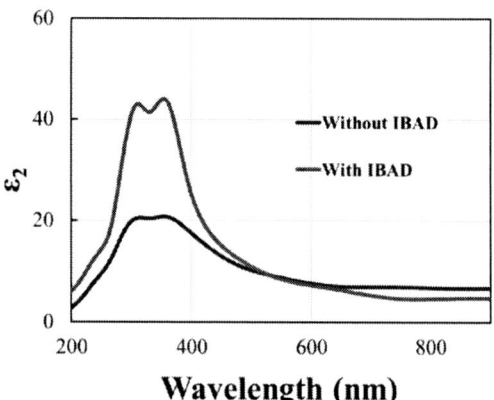

Fig. 4: Influence of an ion beam assist on the imaginary part ε_2 of the complex dielectric function of the sputtered silicon films extracted by spectroscopic ellipsometry.

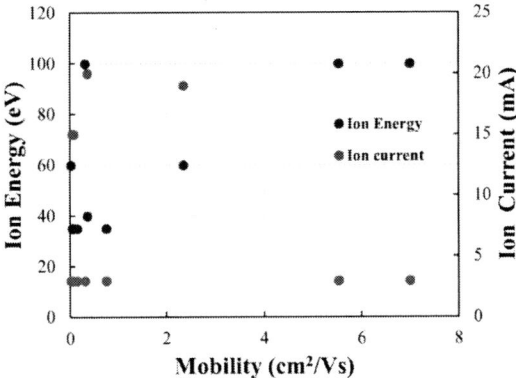

Fig. 5: Influence of ion beam parameters on the mobility for the silicon films.

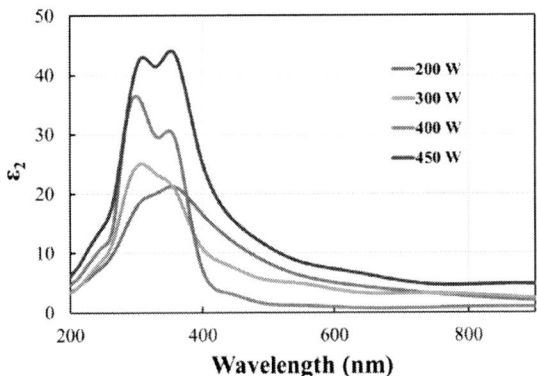

Fig. 6: Imaginary part ε2 of the complex dielectric function spectra for silicon films deposited with IBAD at different sputtering powers.

Figure 4 illustrates the influence of the ion beam on the deposition of thin film silicon, by inspecting the changes in the dielectric functions. The other sputtering deposition parameters were kept constant and the temperature was maintained at 700 °C. A clear increase in the density of nanocrystalline films is observed with IBAD. The mobility of surface and near surface atoms can be controlled during the growth using a low energy ion beam. The beam voltage was varied from 35 V to 400 V and the maximum ion current was fixed for a constant beam voltage to avoid direct impingement of the energetic ions. As shown in Figure 5, higher mobilities were obtained at a low ion energy (100 eV) and an ion beam current of 3 mA. An increase in the ion energy above 250 – 300 eV resulted in the etching of the films.

D. Effect of Sputtering power

The sputtering power during deposition was varied from 200 W to 450 W in order to explore its effect on the crystallinity of the silicon films. The other deposition parameters such as deposition pressure, substrate temperature and ion beam parameters were maintained constant. Figure 6 shows the variation of ε2 as obtained by SE for different sputtering power levels. There is a clear increase in the density and crystallinity of the film with the increase in the power. All the films show distinct critical point transactions, so are likely nanocrystalline.

The influence of the RF power on the electrical properties of the silicon film is shown in Figure 7. The mobility of the films tends to improve with the increase in sputtering power, indicating higher quality of the films. Overall, the higher sputtering power provides greater momentum per arriving adatom, improving the surface mobility and resulting in larger crystallite sizes, which in turns enhances the electrical properties of the films.

With the increase in the sputtering power along with IBAD, the surface adatoms are provided with higher momentum and the nucleation and growth of the film is promoted. In the low energy regime of the ion beam, the grain boundaries of the films can be easily moved and thus the films show larger crystallite sizes with higher RF power and low ion beam energy. The silicon film deposited at 450 W can be clearly identified as a nanocrystalline thin film.

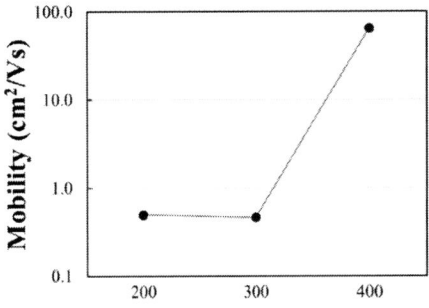

Fig. 7: Influence of sputtering power on the mobility for the silicon films.

IV. CONCLUSIONS

The effect of substrate temperature, sputtering power, and substrate on the crystallinity of sputtered silicon films was observed. A low energy ion beam was used to assist the sputter deposition and a clear increase in the density of the nanocrystalline phase was observed. Future work includes implementing this deposition process while passivating grain boundaries with hydrogen, producing thin films at reasonably high deposition rates, and also implementing these layers effectively into PV test structures.

ACKNOWLEDGEMENTS

This work was supported by the Office of Naval Research under contract #15-SN-0002.

REFERENCES

[1] J.-D. Kwon, K.-S. Nam, Y. Jeong, D.-H. Kim, S.-G. Park, S.-Y. Choi, "Control of crystallinity in nanocrystalline silicon prepared by high working pressure plasma-enhanced chemical vapor deposition", Adv. Mat. Sci. Engg., vol. 2012, 2012.

[2] A. Parashar, S. Kumar, J. Gope, C.M.S. Rauthan, S.A. Hashmi, P.N. Dixit, "RF power density dependent phase information in

hydrogenated silicon films", J. Non-crystal. Solids, vol. 356, pp. 1774-1778, 2010.

[3] S. Gall, C. Becker, E. Conrad, P. Dogan, F. Fenske, B. Gorka, K.Y. Lee, B. Rau, F. Ruske and B. Rech, "Polycrystalline silicon thin-film solar cells on glass", Sol. Energy Mat. Sol. Cells, vol. 93, pp. 1004-1008, 2009.

[4] F.A. Smidt, "Use of ion beam assisted deposition to modify the microstructure and properties of thin films", Int. Mater. Rev., vol. 35, pp. 61-128, 1990.

Minority Carrier Lifetime Variations in Multicrystalline Silicon Wafers with Temperature and Ingot Position

Sissel Tind Søndergaard[1], Jan Ove Odden[2], and Rune Strandberg[1]

[1]University of Agder, Department of Engineering Sciences, P.O. Box 509, NO-4898 Grimstad, Norway
[2]Elkem Solar AS, P.O. Box 8040 Vaagsbygd, NO-4675 Kristiansand S, Norway

Abstract—The minority carrier lifetimes of multicrystalline silicon wafers are mapped using microwave photoconductive decay for different temperatures and ingot positions. Wafers from the top of the ingot display larger areas with lower lifetimes compared to wafers from the bottom. The lifetimes of low-lifetime areas are found to increase with the temperature, while the lifetimes of some high-lifetime areas decrease or remain unchanged. The relative improvement of the low-lifetime areas is considerably larger than the relative change in the high-lifetime areas. We suggest that the above-mentioned observations explain, at least partially, why previous studies have found the relative temperature coefficients of mc-Si cells to improve towards the top of the ingot.

Index Terms—Charge carrier lifetime, multicrystalline silicon, silicon ingot, temperature, microwave photoconductive decay (μ-PCD).

I. Introduction

The performance of a silicon solar cell depends on the minority carrier lifetime, which is affected by several parameters such as temperature, injection level and the type and concentration of impurities in the cell [1] - [4]. The temperature dependence of the lifetime should be of special interest for industrial purposes since field operation temperatures can be relatively high, negatively affecting the power output from the cell [5].

Multicrystalline (mc-Si) wafers made from a polycrystalline feedstock (poly-Si) are widely used for industrial production of silicon solar cells due to the cost-effectiveness of the fabrication process. However, a relatively large and uneven distribution of impurities and crystal defects can be found throughout the ingot because of the quality of the feedstock and the solidification process [6], [7]. Several papers have shown that the minority carrier lifetime of mc-Si wafers varies with the wafer position in the ingot and this effect is considered to be impurity related [3], [8]. In recent work, it has been shown that the temperature coefficients of solar cells made from mc-Si vary along the height of an ingot [9]. A natural question to be raised is thus whether temperature variations affect the minority carrier lifetime differently throughout the ingot. Investigating this might provide information about how to optimize cells originating from different positions in an ingot as well as improve our understanding of how silicon cells perform in the field.

The present work is a study of the variations in minority carrier lifetime of mc-Si wafers, combining the effect of temperature and ingot position using microwave photoconductive decay (μ-PCD). The lifetimes are mapped at room temperature and at an elevated temperature of 56 °C, which is close to temperatures often encountered in the field.

II. Experimental Details

The studied samples were mc-Si wafers made from directional solidification of a poly-Si feedstock, produced industrially and passivated with a layer of 43 nm a-Si:H in a research laboratory. The lifetimes were measured using μ-PCD in a Semilab WT-2000PVN. To heat the wafers, a heat plate with a top section consisting of a thick slab of sintered aluminum oxide (alsint) was used. This material is an electrical insulator and the thickness of the slab ensures that underlying metallic parts do not interfere with the lifetime measurements. To secure exactly similar conditions, the wafers were placed on the heat plate also when mapping lifetimes at room temperature.

To control the repeatability of the lifetime measurements, one particular wafer was mapped at several occasions throughout the period through which the series of measurements was conducted. The results of these repeated control measurements are shown in Fig. 1 where the shortest measured lifetime is presented as a function of measurement number. The error bars show the standard deviation of the 23 °C and the 56 °C measurements, respectively. The measured lifetimes change slightly during the period considered, however within the statistical fluctuations.

III. Results

The average lifetime of each wafer is shown in Fig. 2 as a function of the position of the wafer in the ingot. Position 1 indicates the wafer closest to the bottom whereas 600 is close to the top. The average lifetime is found to increase by 3.4 % on average when the temperature is increased from 23 °C to 56 °C. However, conducting a Student's t-test, the difference in lifetimes with temperature is not found to be statistically significant. Furthermore, we observe a tendency that the average lifetime decreases throughout the ingot, which matches results in the literature [3], [8], [10].

The 0.1 % pixels with the longest measured lifetime on each wafer is shown in Fig. 3 as a function of the wafer position. The longest lifetime decreases by 6.5 % on average when the temperature is increased, however this decrease is

978-1-5090-5606-4/17 $31.00 © 2017 IEEE

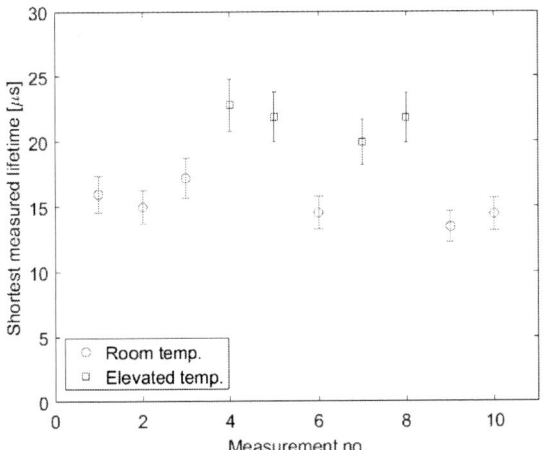

Fig. 1. Repeated control measurements of the shortest measured lifetime on a wafer. The error bars show the standard deviation of the 23 °C and the 56 °C measurements, respectively.

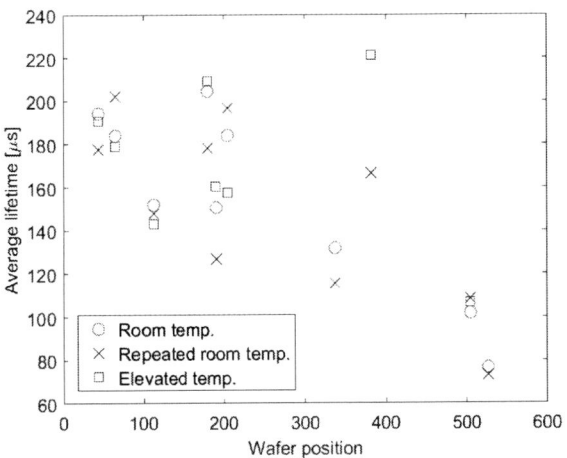

Fig. 2. Average lifetime of each wafer as a function of wafer position.

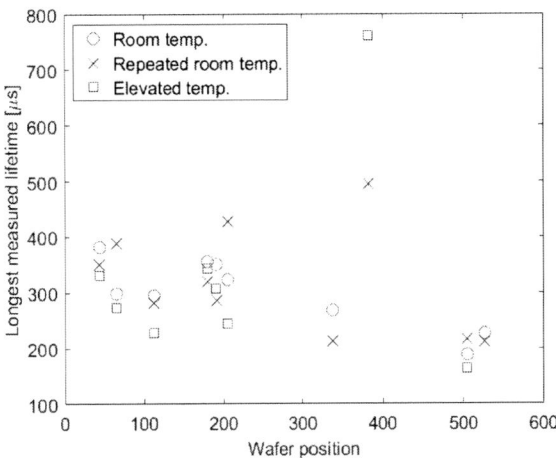

Fig. 3. 0.1 % pixels with longest measured lifetime as a function of wafer position.

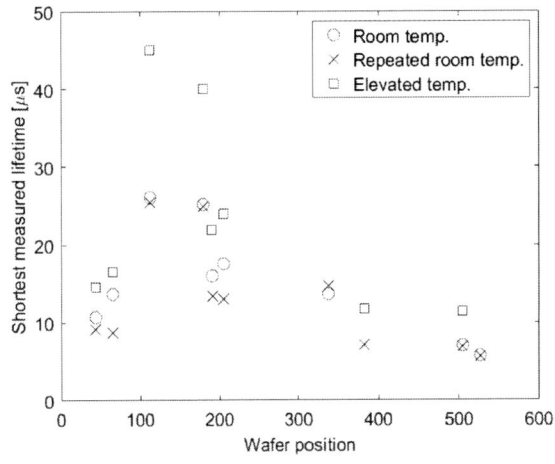

Fig. 4. 0.1 % pixels with shortest measured lifetime as a function of wafer position.

also not found to be statistically significant. Furthermore, we observe the tendency that the longest measured lifetime is lowest towards the top of the ingot.

The 0.1 % pixels with the shortest measured lifetime on each wafer is shown in Fig. 4 as a function of the wafer position. The shortest lifetime is found to increase by 47.3 % on average when the temperature is increased. Furthermore, we observe a tendency that the shortest measured lifetime takes the shortest value at the top of the ingot.

A consistent increase in lifetime with temperature was found for areas on the wafers with lifetimes below 50 µs. The

fraction of pixels on each wafer with lifetimes below 50 µs is shown in Fig. 5 as a function of the wafer position. The fraction decreases by 55.3 % on average when the temperature is increased. In addition, Fig. 5 shows the tendency that the wafers at the top of the ingot consist of a larger number of pixels with lifetimes below 50 µs. This matches the tendency found in Fig. 4 where the shortest measured lifetime was shortest at the top of the ingot.

To visualize the effect of the temperature change, some additional data of the < 50 µs areas are plotted in Fig. 6. It shows the ratio of the fraction of pixels with lifetimes below 50 µs at 56 °C to the same fraction at 23 °C. Hence a low ratio corresponds to a drastic decrease in the number of low

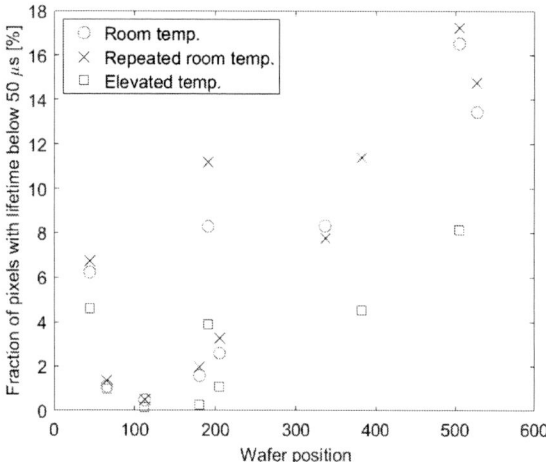

Fig. 5. Fraction of pixels with lifetime below $50\,\mu s$ as a function of wafer position.

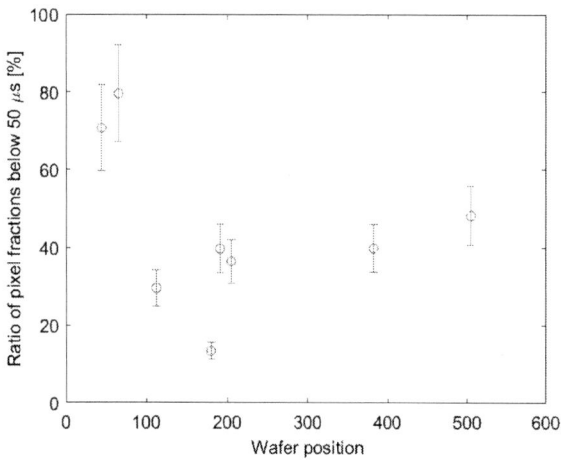

Fig. 6. Ratio of fractions of pixels with lifetime below $50\,\mu s$ at $56\,°C$ to $23\,°C$ as a function of wafer position. The error bars are estimated from the standard deviation of repeated control measurements.

lifetime pixels when the temperature is increased. The error bars are estimated from the standard deviation of repeated control measurements of the lifetimes below $< 50\,\mu s$. This was calculated for one particular wafer and generalized to the remaining wafers.

We observe that the number of low-lifetime pixels decreases for all wafers when the temperature is increased. In addition, we see a tendency that the wafers from the middle and the top of the ingot seem to benefit considerably from the temperature increase, whereas the effect is relatively small for wafers from the bottom. Because of the low number of mea-

surements this should only be interpreted as a possible trend. From Fig. 6, we would expect a constant ratio throughout the ingot. However, the change in composition and distribution of impurities throughout the ingot might cause the wafers to respond differently to temperature changes, as will be discussed later.

A further investigation of the temperature effect can be seen in Fig. 7, which shows the fraction of pixels in different lifetime intervals for two bottom and two top wafers at the relevant temperatures. The top wafers are found to consist of a larger fraction of pixels with lifetimes below $50\,\mu s$ compared to bottom wafers, consistent with the trend observed in Fig. 5. This fraction decreases with temperature for all lifetime intervals below $50\,\mu s$ for top wafers, while this is not always the case for bottom wafers. Furthermore, the decrease seems to be larger for top wafers. The number of pixels with lifetimes above $150\,\mu s$ responds differently to the temperature increase for the different wafers as shown in Fig. 8. Some of the wafers experience a shift of the pixels towards lower lifetimes, whereas one of the top wafers display the opposite trend (note that the lifetime intervals for "top wafer (position 382)" differ from the other wafers). This shows that an increase in temperature in general is beneficial for areas with low lifetime but not necessarily for areas with a high lifetime.

A PC1D simulation is shown in Fig. 9 relating the cell efficiency to the carrier lifetime at the two relevant temperatures. The simulation is based on standard cell parameters. For cells with lifetimes above $50\,\mu s$, a temperature increase results in a relatively large efficiency reduction regardless of how the lifetime changes. For cells with lifetimes below $50\,\mu s$, an increase in temperature results in a significantly smaller efficiency reduction if the lifetime increases with the temperature, as is the case for the low-lifetime areas in this study. Since wafers from the top of the ingot in general show larger areas with low lifetimes, the relative performance of such cells is expected to decrease less with temperature compared to cells originating from the bottom of the ingot. Consequently, we expect the top cells to show better temperature coefficients, which agrees with the results in Ref. [9].

During the directional solidification of mc-Si, the majority of impurities will segregate towards the top of the ingot leaving a higher concentration of impurities in this area [8]. In addition, back diffusion from the crucible into the ingot can result in a higher concentration of impurities with large diffusion coefficients right at the bottom of the ingot [6]. An uneven distribution and composition of impurities might cause the wafers to respond differently to temperature changes and the wafers from the top of the ingot to display better temperature coefficients. This seems plausible since the capture cross sections of different impurities depend differently on temperature and might decrease, as observed for iron, copper, molybdenum and titanium [4], [11] - [13].

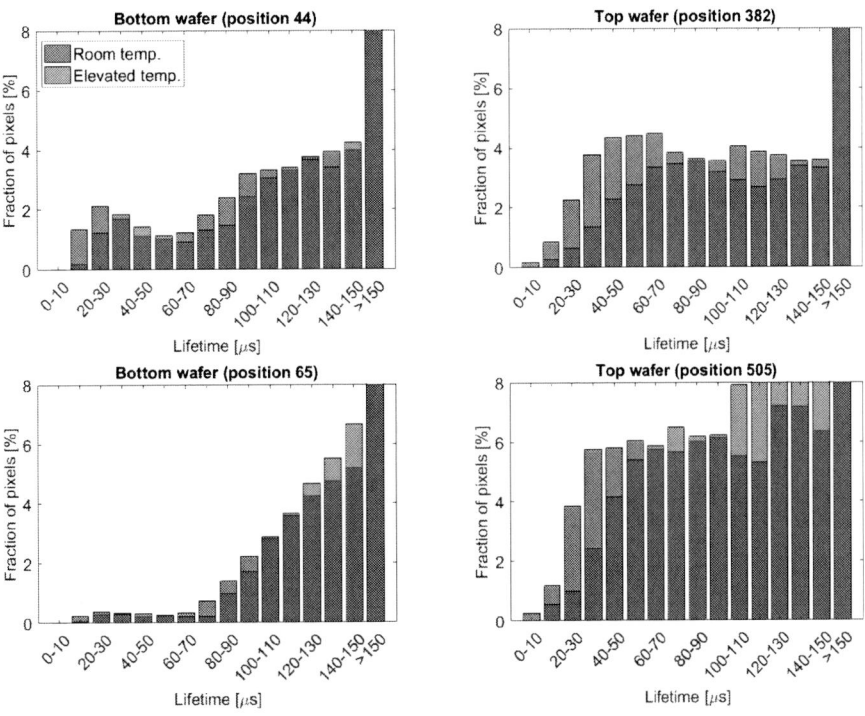

Fig. 7. Fraction of pixels of different low-lifetime intervals for bottom and top wafers.

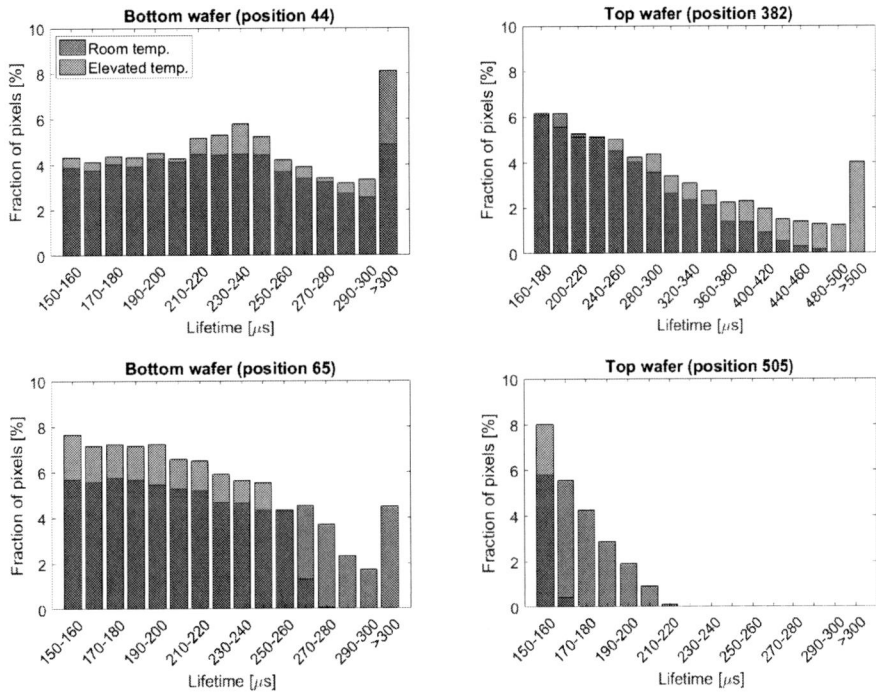

Fig. 8. Fraction of pixels of different high-lifetime intervals for bottom and top wafers. Note that the intervals for top wafer (position 382) differ from the other wafers.

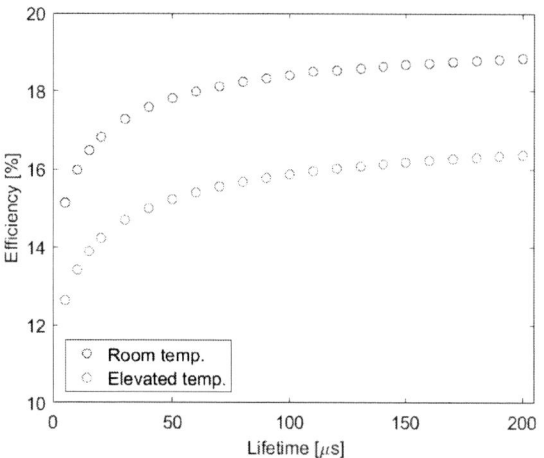

Fig. 9. PC1D simulation of cell efficiency as a function of lifetime.

IV. CONCLUSION

The minority carrier lifetimes of mc-Si wafers have been studied for different temperatures and ingot positions. We found that wafers from the top of the ingot contained larger areas with lifetimes below $50\,\mu s$ compared to wafers from the bottom. Areas with lifetimes below $50\,\mu s$ in general experienced an increased lifetime when the temperature was increased. In high-lifetime areas, the lifetime decreased with temperature for some of the wafers, whereas others increased or remained constant.

These findings are believed to be impurity related, since the composition and distribution of impurities change throughout the ingot and in general is found to be larger at the top. Since the capture cross sections of some impurities decrease with temperature, this could explain why temperature coefficients previously have been found to be better for cells originating from the top of an ingot.

ACKNOWLEDGMENT

We would like to thank Gaute Stokkan at SINTEF, for generously sharing the heat plate with us, and Elin Stubhaug and Ragnar Ekker, who designed and made the heat plate as a part of their Masters project at NTNU.

REFERENCES

[1] Y. Hayamizu, T. Hamaguchi, S. Ushio, and T. Abe, "Temperature dependence of minority-carrier lifetime in iron-diffused p-type silicon wafers", *Journal of Applied Physics*, vol. 69, pp. 3077, 1991.

[2] D. Macdonald, A. Cuevas, S. Rein, P. Lichtner, and A. W. Glunz, "Temperature- and injection-dependent lifetime spectroscopy of copper related defects in silicon", in *3th World Conference on Photovoltaic Energy Conversion*, 2003, pp. 87-90.

[3] G. Coletti, R. Kvande, V. D. Mihailetchi, L. J. Geerligs, L. Arnberg, and E. J. Øvrelid, "Effect of iron in silicon feedstock on p- and n-type multicrystalline silicon solar cells", *Journal of Applied Physics*, vol. 104, pp. 104913, 2008.

[4] B. B. Paudyal, K. R. McIntoch, and D. H. Macdonald, "Temperature dependent carrier lifetime studies of Ti in multicrystalline silicon", *Journal of Applied Physics*, vol. 105, pp. 124510, 2009.

[5] M. A. Green, *Solar Cells: Operating Principles, Technology, and System Applications*, Englewood Cliffs, NJ: Prentice-Hall, 1982.

[6] M. Di Sabatino, G. Tranell, and E. J. Øvrelid, "Impurities through the silicon solar cell value chain", in *39th IEEE Photovoltaic Specialists Conference*, 2013, pp. 001-006.

[7] D. Macdonald, A. Cuevas, A. Kinomura, Y. Nakano, and L. J. Geerligs, "Transition-metal profiles in a multicrystalline silicon ingot", *Journal of Applied Physics*, vol. 97 (3), pp. 033523, 2005.

[8] A. Bentzen, H. Tathgar, R. Kopecek, R. Sinton, and A. Holt, "Recombination lifetime and trap density variations in multicrystalline silicon wafers through the block", in *31th IEEE Photovoltaic Specialists Conference*, 2005, pp. 1074-1077.

[9] C. Berthod, R. Strandberg, and J. O. Odden, "Temperature coefficients of compensated silicon solar cells-influence of ingot position and blend-in-ratio", *Energy Procedia*, vol. 77, pp. 15-20, 2015.

[10] K. Lauer, M. Ghosh, A. Lawerenz, and S. Dauwe, "Minority carrier lifetime, trap density and interstitial iron content in multicrystalline silicon raw wafers versus ingot position", in *Proceedings of the 21st European PV SEC*, 2006, pp. 1362.

[11] B. B. Paudyal, K. R. McIntoch, and D. H. Macdonald, "Temperature dependent electron and hole capture cross sections of iron-contaminated boron-doped silicon", in *34th IEEE Photovoltaic Specialists Conference*, 2009, pp. 1588-1593.

[12] A. Inglese, J. Lindroos, H. Vahlman, and H. Savin, "Recombination activity of light-activated copper defects in p-type silicon studied by injection- and temperature-dependent lifetime spectroscopy", *Journal of Applied Physics*, vol. 120, pp. 125703, 2016.

[13] B. B. Paudyal, K. R. McIntoch, D. H. Macdonald, and G. Coletti, "Temperature dependent carrier lifetime studies of Mo in crystalline silicon", *Journal of Applied Physics*, vol. 107 (5), pp. 124510, 2010.

CuO nanowires-based Radial hetero-junction thin film silicon solar cells with a high open-circuit voltage

Xiaolin Sun[1,2], Jiawen Lu[1], FanYang[1], Linwei Yu[1*], Jun Xu[1], Ling Xu[1], Kunji Chen[1]

[1] National Laboratory of Solid State Microstructures and School of Electronics Science and Engineering/Collaborative Innovation Center of Advanced Microstructures, Nanjing University, Nanjing 210093, China

[2] College of Electronical and Information Engineering, Sanjiang University, Nanjing 210012, China

Abstract — **In this article, we demonstrated the fabrication of 3D p-type CuO nanowires (NWs)/intrinsic hydrogenated amorphous silicon(i-a-Si:H) Ridial Hetero-juntion Solar Cells(RHSC) for the first time. we also obtained different composition of oxide NWs and different lengths of NWs by controlling the growing time and temperature upon low-cost stainless steel substrates. A highly p-doped a-Si:H thin film was introduced as a passivation layer to establish more depletion and decrease the leakage at hetero-interface. Photovoltaic(PV) property with a high open-circuit voltage of 740mV has been reached for the cell, which is higher than that of solar cells fabricated with CuO-based planner structure.**

Key Words — **CuO nanowires, RHSC, I-V character, Raman spectra, solar cells.**

I. INTRODUCTION

Radial Hetero-junction Solar Cells constructed over a matrix of nanowires (NWs) have been widely investigated as a promising 3D architecture that will not only boost the light harvesting performance of thin film solar cells, but also allow a new control dimension to optimize the junction design for achieving a fast carrier separation and better stability against lasting sunlight exposure. This concept has been explored and proven effective in our previous works on hydrogenated amorphous Si (a-Si:H) thin film RHSC deposited over p-type doped SiNWs, where a high open circuit voltage of Voc=0.92 V and a power conversion efficiency of 9.2% [1, 2] were achieved.

In parallel, p-type copper oxide(CuO), has many interesting properties with a narrow energy gap of 1.24eV, which is close to the energy gap of c-Si. CuO is also a relatively nontoxic, environment-friendly, and abundantly available semiconductor material, and has been mainly investigated as a kind of light-absorbing material for exploring a new type of c-Si/CuO heterojunction cells[3, 4]. Many studies suggested that copper oxide could be used not only as an absorption layer but also as a junction layer that forms a high built-in voltage in copper oxide and Si-based heterojunction thin-film solar cells [CuO_x]. But, the potential of a radial heterojunction thin film Si/CuO solar cell has not been explored so far. Actually, the p-type CuO nanowires could serve as both an ideal 3D framework and a p-type electrode in a p-i-n Si RHSC.

In this study, we chosen stainless steel as the substrate, which is low-cost and can be used as electrode directly, then deposited Copper film using thermal evaporation system upon the substrate, the next step is to anneal CuO NWs by the method of thermal oxidation, followed by a highly conformal coating over the long CuO NWs by PECVD thin film deposition, which is full compatibility with the conventional a-Si:H thin film technology. We have analyzed both the components of the oxide by different annealing temperatures and the length of CuO NWs by controlling the annealing time. Based on our understanding, it may be the first report of fabricating p-type CuO NWs/a-Si:H RHSC. Furthermore, the p-type CuO NWs /i-c-Si:H radial junction can increase the light-trapping effect and suppress the recombination of the carriers and in the end improve the open circuit voltage of RHSC, which is mainly attributed to the radial junction structure of solar cells. More importantly, CuO NWs growth exempts the use of extremely toxic borane gas, and without H_2 plasma treatment process for metal catalyst, leading to significant procedure simplification and cost reduction for building RHSC.

II. EXPERIMENT PROCEDURES

Fig.1(a) shows the 3D structure of p-CuO NWs/a-Si:H/n-a-Si:H RHSC. First of all, we prepared 1μm Cu films by thermal evaporation system upon stainless steel substrate. Then we putted the substrate in annealing furnace at different temperatures (250℃, 300℃, 350℃, 450℃, 550℃) for 3hours, meanwhile, gaseous mixture containing 30% oxygen and 70% argon were admitted in the silica tube. We also obtained different lengths of CuO NWs by controlling the annealing time (1h, 3h and 5h) at 450℃. Then the samples were loaded into a PECVD (Plasma-enhanced chemical vapor deposition) chamber for depositing intrinsic hydrogenated amorphous Si (a-Si:H) thin film, followed by a n+ doped window layer, which was introduced by our previous work[1]. During the deposition process, the gasses SiH4 and PH3 (1.5 diluted in H_2) were used. The i-layer was deposited for 1h, with typical chamber pressure, RF power, H_2 flow, SiH4 flow of 30Pa, 20 W, 20sccm and 6sccm respectively; the n-layer was deposited for 6min at RF power of 20W, chamber pressure at 30Pa with a supply of H_2 at 20sccm, SiH4 at 6sccm, and PH3 at 1.7sccm.

978-1-5090-5606-4/17 $31.00 © 2017 IEEE

At last, ITO (Indium Tin Oxides) thin film and metal electrode were prepared by Radio Frequency Magnetron Sputtering.

For characterization of the CuO NWs, we observed the morphology using a scanning electron microscope (Fig.1), measured composition of the NWs at different temperatures using Raman spectroscopy(Fig.2).

The current-voltage (I-V) performances were characterized by the J-V curves using a solar simulator with illumination of air mass(AM) 1.5 at room temperature (as show in Fig.3), the main photovoltaic (PV) parameters are summarized in the table of Fig.3.

III. RESULTS AND DISCUSSION

Prior to applying the CuO NWs to fabricate the solar cells, we investigated the formation the CuO NWs by the method of thermal oxidation. There are two mechanisms: vapor-liquid-solid(VLS) and vapor-solid(VS) have been most widely adopted for the growth of nanowires in the surrounding gas. VS mechanism seems to be responsible for the growth of CuO NWs[5,6], and has been applied frequently to explain the formation of nanowires recently. The two steps of reactions can be summarized as the following functions [7].

$$4Cu + O_2 \rightarrow 2Cu_2O \quad (1)$$
$$2Cu_2O + O_2 \rightarrow 4CuO \quad (2)$$

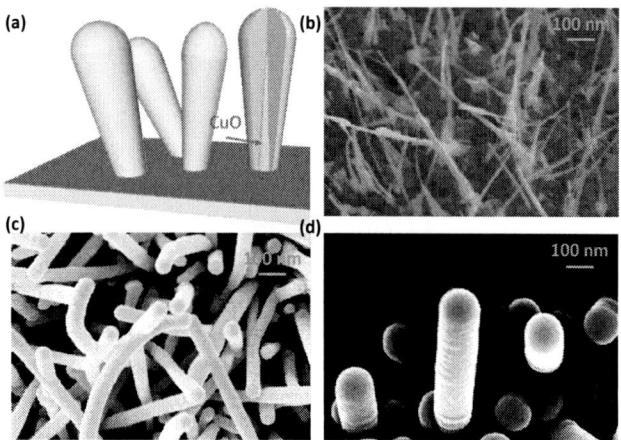

Fig. 1. (a) illustrates the 3D multilayer structure deposited around the CuO NWs core. (b)The SEM image of the CuO NWs（c,d）The SEM images of i-layer, n-type a-Si:H and final top ITO layer coating over CuO NWs matrix.

Cu film deposited on the stainless steel substrates was thermal oxidated in gaseous mixture of Ar_2 and O_2 and CuO NWs were formed. Fig.2 shows the Raman spectra of oxide at different temperatures. The samples were oxided for 3h at different temperatures for studying the oxide components. When annealing temperature is between 250℃ and 350℃, it presents mixture phase of Cu_2O and CuO; when annealing temperature is above 350℃, Cu_2O phase is

less and less, and CuO becomes the dominate phase; when annealing temperature reached up to 400℃, there is only CuO left. Because Cu_2O has a wide band gap of 2.0eV, which is not good to absorb visible light for solar cells, we calcined CuO NWs at high temperature (normally higher than 400℃) to fabricated 3D RHSC.

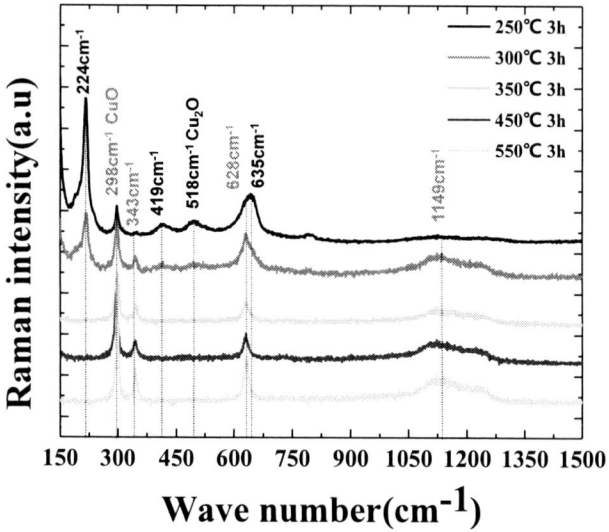

Fig. 2. Raman spectra of oxide growing at different temperatures (250℃, 300℃, 350℃, 450℃, 550℃) for 3 hour

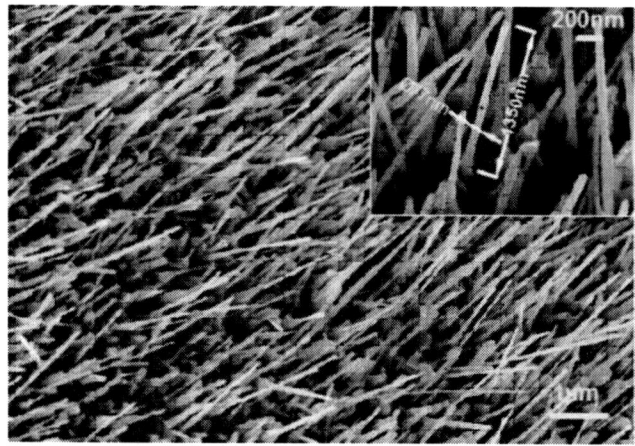

Fig. 3. SEM image of the density and morphology of CuO NWs

Fig.1 shows the SEM the structure of CuO NWs-basted solar cells. Fig.1 (a) shows the 3D structure of CuO NWs-basted solar cells. Fig.1 (b) is the SEM image of CuO NWs. As can be seen from the Fig.1 (b), CuO NWs are straight and slender, even if coated with Si:H and ITO. Fig.2 shows the Raman spectra of oxide growing at different temperatures. As mentioned before, Cu_2O NWs will be formed at lower temperature (lower than 350 ℃), In order to remove the

978-1-5090-5606-4/17 $31.00 © 2017 IEEE

interference of Cu_2O totally, we chose 450℃ to grow CuO NWs for further CuO NWs-basted solar cells. Fig.1 (c,d) is the SEM images of CuO NWs coating with i-Si:H layer and n-a-Si:H, and with the final top ITO contact. The structure and the highly conformal coating over the CuO NWs by PECVD thin film deposition are exactly same as the SiNWs-based RHSC, which is introduced in detail in our previous work[1].

Fig.3 shows the density and morphology of CuO NWs. As show in Fig.3, CuO NWs are much more straight and slender than Si NWs, and the density is around $2*10^{14}$ /m^2. The CuO NWs in this study were grown to a length of ~1350 nm with diameters in the range of ~70 nm When we growing CuO NWs at 450℃ for 3 hours.

What we pay the most attention to is the Optical-Electrical Characteristic of 3D CuO NWs-based RHSC. In order to analyze the effect of the length of CuO NWs on characteristic of solar cells, we tried to grow different lengths of CuO NWs by controlling the growing time (1h, 3h and 5h) at 450℃, the corresponding J-V curves measured under standard AM1.5 irradiation are shown in Fig.4, and the extracted photovoltaic parameters are summarized in the table in Fig.4, which shows J-V characteristics of p-CuO NWs/c-Si:H/n-Si:H RHSC with the Voc of 557mV and Jsc of 7.74mA/cm^2 for the CuO NWs annealed at 450°C for 3h, which higher than the photovoltaic parameters of solar cells whose CuO NWs annealed for 1h and 5h. We assumed that it is mainly due to the CuO NWs of variance. However, the measured Voc of 557mV was lower than we expected, which was probably due to both insufficient built-in potential caused by a low doping level of the p-CuO NWs and high leakage at the hetero-interface.

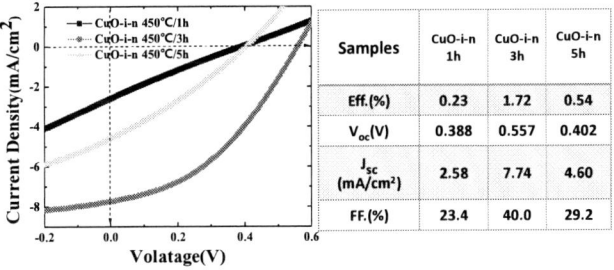

Fig. 4. The J-V curves of the corresponding CuO NWs (with different growning period) thin film solar cells, as well as the extracted parameters under AM1.5 illumination are presented in the table.

In order to further improve the performance of the CuO NWs-based RHSC, we deposited a highly doped p-type a-Si:H passivation layer (5nm) between the p-CuO NWs and i-a-Si:H layers during the fabricating process. Fig.5 exhibits an unexpected high open-circuit voltage of Voc=740 mV, it is even higher than the highest value(715mV) reported so far for CuO_x/a-Si:H planner heterojunction solar cells [7]. It indicated that the p-type a-Si:H passivation layer supplied more depletion at the heterojunction and mainly contributed to the

increase in Voc at the hetero-interface, at the same time decreased the leakage through the heterojunction. However, because the p-type a-Si:H passivation layer will introduce toxic gasses during the process, it would be better to replace it with an oxide-based passivation layer to improve the performance of RHSC and to process safety in the future.

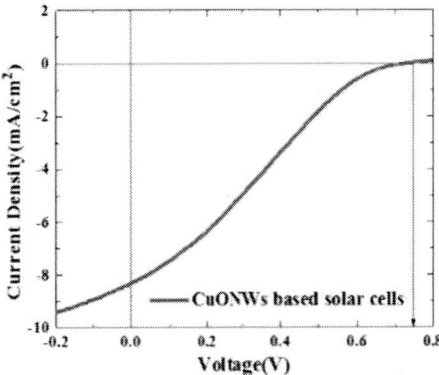

Fig. 5. Current density-voltage(J-V) curve of radial junction a-Si:H/CuO NWs solar cells after depositing a thin film of p-type passivation layer.

IV. CONCLUSION

In summary, CuO NWs growing on the stainless steel substrate through the process of thermal annealing of Cu film has been demonstrated to have the capability for fabricating 3D radial hetero-junction solar cells. We annealed and characterized p-CuO NWs, and based on which we demonstrated successfully fabrication of p-CuO x /i-a-Si:H 3D RHSCs for the first time, and a maximum Voc of 740mV was obtained with a diameter of 77nm p-CuO NWs. We showed great potential for low-cost, high-open circuit voltage radial junciton solar cells, expanded the application of CuO NWs for Si solar cells and provided a potential new photovoltaic technology.

ACKNOWLEDGEMENT

We acknowledge the financial supports from the National Natural Science Foundation of China under Nos. 61674075, 11274155 and 51572120, the National Basic Research 973 Program under Grant Nos. 2014CB921101, 2013CB632101, and 2013CB932900, the Jiangsu Excellent Young Scholar Program under No. BK20160020, the Scientifc and Technological Support Program in Jiangsu province under No. BE2014147-2,Jiangsu Shuangchuang Program, the Fundamental Research Funds for Central Universities, and the support from the Key Laboratory of Advanced Photonic and Electronic Materials.

REFERENCES

[1] S. Misra, L. Yu, M. Foldyna, P. Roca i Cabarrocas, High efficiency and stable hydrogenated amorphous silicon radial junction solar cells built on VLS-grown silicon nanowires, *SOL. ENERG. MAT. SOL. C.*, 118 (2013) 9095.

[2] S. Misra, L. Yu, M. Foldyna, P. Roca i Cabarrocas, New Approaches to Improve the Performance of Thin-Film Radial Junction Solar Cells Built Over Silicon Nanowire Arrays, *IEEE Journal of Photovoltaics*, 5 (2015) 40-45.

[3] S.H. Lee, M. Shin, S.J. Yun, J.W. Lim, CuOx/a-Si:H heterojunction thin-film solar cell with an n-type μc-Si:H depletion-assisting layer, *Photovoltaics: Research and Applications*, (2015).

[4] p-CuO/n-Si heterojunction solar cells with high open circuit voltage and photocurrent through interfacial engineering, *Photovoltaics: Research and Applications*, (2014).

[5] X. Jiang, T. Herricks, Y. Xia, CuO Nanowires Can Be Synthesized by Heating Copper Substrates in Air, *Nano. Lett.* 2 (2002) 1333.

[6] L.S. Huang, S.G. Yang, T. Li, B.X. Gu, Y.W. Du, Y.N. Lu, S.Z. Shi J.Crystal , Microwave-assisted Synthesis of Single-crystalline CuO Nanoleaves, *Growth* 260 (2004) 130.

[7] Adegboyega, G. A. Niger. J. Preparation and characterization of thermally oxidized copper substrates for photothermal and photovoltaic energy conversion *Renewable Energy*, 1990, 1, 21.

The Effect of Chemical Composition on Porous Etching for Epi and Lift-off Wafer Process

Teng-Yu Wang[1,2*], Peng-Wei Chen[1,3], and Han-Wen Liu[3]

[1]Green Energy and Environment Research Labs, Industrial Technology Research Institute, Hsinchu, Taiwan, R.O.C.

[2]Material and Chemical Research Labs, Industrial Technology Research Institute, Hsinchu, Taiwan, R.O.C.

[3]Graduate Institute of Optoelectronic Engineering, National Chung Hsing University, Taichung, Taiwan, R.O.C.

Abstract — **Epi and lift-off process can create single crystalline silicon wafers without kerf-loss. There are three steps in epi and lift-off process: porous layer formation, epitaxial growth, and exfoliation. The porous layer is created by electrochemical etching in hydrofluoric acid solution. The ethanol is usually added into the etching solution to enhance the contact between substrate surface and solution. Because the ethanol is volatile, the ethanol content inside the etching solution is adjusted to stabilize the solution composition in this study. A relatively solvent, isopropanol, was also used to replace the ethanol to control evaporation rate of the solution.**

I. INTRODUCTION

Silicon solar cells is the main products in PV. In the beginning, solar technology was expensive and the efficiency was not acceptable. With the progress of technology, the energy conversion efficiency nowadays is over 20%, the cost is also decreased. The methods of cost reduction include thinning wafer, reduce the silver cost, and the mass production technique. A new technology is needed to continue to reduce the cost in the future. Kerf-free process is a low cost technique. The cost reduced because no material waste in wafering process. There are several kinds of kerf-free technology, such as smart-cut [1], ribbon growth [2], Slim-cut (stress-induced lift-off method) [3], and ELO (epitaxial and lift-off) [4].

The first step in ELO process is electrochemical etching silicon substrate to create a porous release layer in hydrofluoric acid solution. The second step is the epitaxial growth of single-crystalline silicon layer above the porous layer. Finally, the silicon wafer is lift-off from the mother substrate. R. Hao et. al fabricated ELO wafer and 22.5% PERC solar cell was obtained [5]. Y. Woon et al. transfer 18-μm-thick ELO wafer on a steel substrate and fabricated a flexible solar cell with 16.8% efficiency [6]. The ELO wafer had been proved that even with conventional solar cell structure, 20% energy conversion efficiency could be obtained [7]. Most of the study used ethanol and hydrofluoric acid as the porous etching solution. In this study, the effect of composition of etching solution for porous etching was discussed.

II. EXPERIMENTAL

Figure 1 and figure 2 are the etching device and the process flow diagram for ELO process, respectively. The boron-doped p-type silicon wafer and phosphorous-doped n-type silicon wafer with a resistivity of 0.01-0.05 Ω-cm was used as the substrate. The electrochemical etching was carry on in a Teflon tank with platinum electrode and the silicon substrate was fixed on the anode. Two step etching with different current density and etching time were used to obtain double layer of porous silicon. The current density for first step etching was 2 mA/cm^2 and 240 sec. Which produced the low porosity layer with a thickness about 2~3 μm. The current density for second step was adjusted depending on the type of substrate. For P-type substrate, the current density was 30-75 mA/cm^2. For N-type substrate, the current density was adjusted to 75-125 mA/cm^2. The thickness of high porosity layer produced by second step etching is about 0.2-1 μm. The etching solution is hydrofluoric acid solution mixing with ethanol (C_2H_5OH, 95%). The composition is varied from HF:C_2H_5OH:H_2O =1:1:1 to 1:1:8. Another experiment replaced ethanol with isopropanol (C_3H_7OH, IPA, 99.5%) in this study.

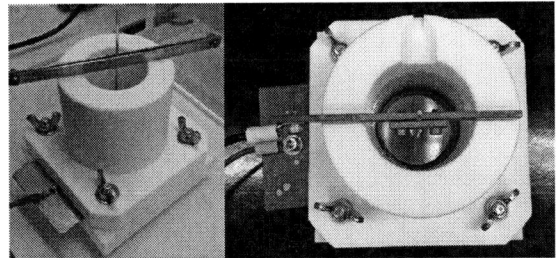

Fig. 1.   The Teflon device for porous etching.

978-1-5090-5606-4/17 $31.00 © 2017 IEEE

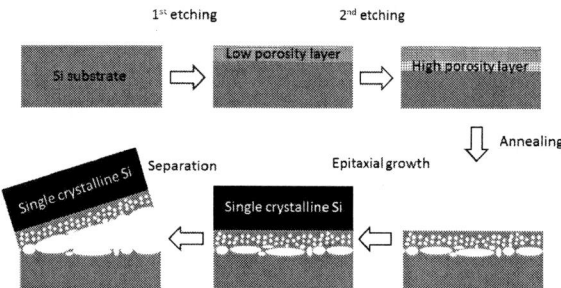

Fig. 2.   Process flow diagram of ELO process in this study.

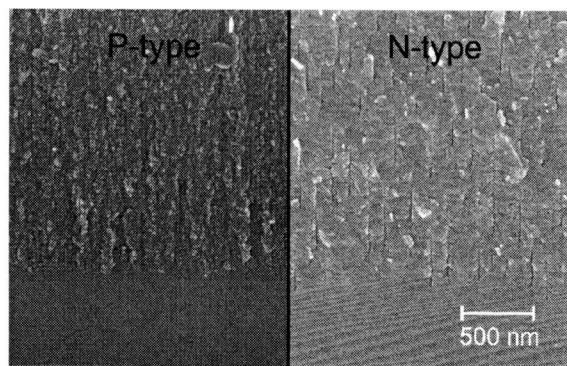

Fig. 3.   SEM cross section of porous structure on P-type and N-type substrate

III. DISCUSSIONS

Hydrofluoric is the main agent in etching reaction. When the oxide layer on the silicon surface dissolved in hydrofluoric acid, the silicon wafer will become hydrophobic and affect the etching reaction. Therefore, ethanol is added into the etching solution to improve the solution characteristic. In literatures, the familiar composition of the etching solution were $HF:C_2H_5OH=1:1$ or $HF:C_2H_5OH:H_2O =1:1:1$. When the ethanol content in the solution reaches 30%, the composition of solution is not easy to control. This is because the ethanol is a volatilizable solution. In this study, the composition of etching solution was adjusted to reduce the volatile solution content. A low-volatile solutions, such as IPA, are also used to replace ethanol to change the characteristics. Table I is the comparison of the chemicals added in the etching solution. IPA is widely used in the PV industry. Which is usually added into the lye for silicon texturization process to improve the uniformity of etching. The IPA has lower boiling temperature and vapor pressure than ethanol, the evaporating rate is slower. It is believed that the IPA is suitable for porous etching process.

TABLE I COMPARISON OF SOLUTION PROPERSITES

Solution	Ethanol	IPA
Boiling point	78.2°C	82.6°C
Vapor pressure(at 20°C)	43.7 mmHg	18 mmHg
Concentration	95%	99.5%

Figure 3 is the cross section of porous structure obtained from electrochemical etching in ethanol solution. The etching time is 120 sec with the current density of 50 mA/cm². The etching velocity in N-type substrate is faster than P-type substrate. The holes in N-type substrate have bigger radius and deeper depth. However, the distance between each hole is longer in N-type substrate than P-type substrate. Therefore, the optimum etching condition will different in N-type substrate and P-type substrate.

The composition effect of the solution was observed in the study. Table II is the experimental parameters of the etching solution. The ratio of solution is varied from 1:1:1 to 1:1:8. In order to reduce the evaporation rate of etching solution, the water content of the solution is increased. For P-type substrate, the low porosity layer peeled when the water content in the etching solution is higher than 76% ($HF:C_2H_5OH:H_2O=1:1:4$). It is assumed that when the ethanol content is low, the contact between substrate and solution became weak. The current distribution in the solution became non-uniform. Parts of the surface on the substrate with high current concentrated was peeled-off. Therefore, the suitable composition for ethanol solution is $HF:C_2H_5OH:H_2O =1:1:1$ to 1:1:3. Figure 4 is the SEM cross section of porous structure on P-type substrate obtained from $HF:C_2H_5OH:H_2O =1:1:3$ etching solution. The current density of first step and second step etching is 2 mA/cm² and 75 mA/cm², respectively. The thickness of low porosity layer and high porosity layer are 1.29 µm and 0.30 µm, respectively. The low porosity layer is stable during etching and can be used as the epitaxial growth surface.

TABLE II COMPOSITION OF ETHANOL SOLUTION

Solution	$HF:C_2H_5OH:H_2O$	Water content
EtOH-1	1:1:1	52.0%
EtOH-2	1:1:2	64.0%
EtOH-3	1:1:3	71.2%
EtOH-4	1:1:4	76.0%
EtOH-5	1:1:5	79.4%
EtOH-6	1:1:8	85.6%

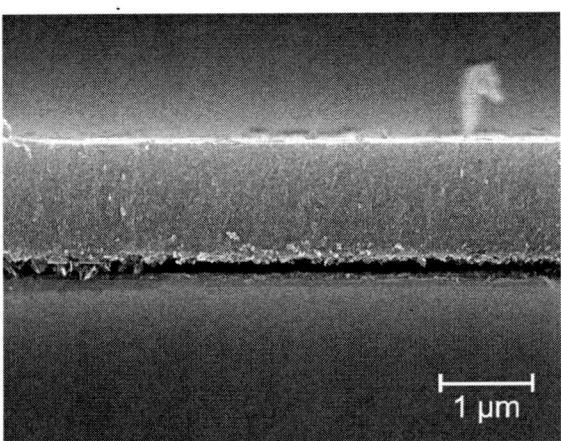

Fig. 4. SEM cross section of porous structure on P-type substrate (HF:C_2H_5OH:H_2O =1:1:3).

Fig. 5. SEM cross section of porous structure on P-type substrate (HF:IPA:H_2O =1:1:3).

When the ethanol was replaced with IPA, the etching result was similar. Table III is the experimental parameters of the etching solution. The ratio of solution is varied from HF: IPA:H_2O=1:1:1 to 1:1:8. Similarly, the substrate surface was partially peeled-off when the water contact was higher than 75% (HF: IPA:H_2O=1:1:4). The acceptable etching solution concentration of IPA solution observed in the experiments was close to ethanol solutions. Figure 5 is the SEM cross section of porous structure on P-type substrate obtained from HF:IPA:H_2O =1:1:3 etching solution. The current density of first step and second step etching is 2 mA/cm^2 and 75 mA/cm^2, respectively. The thickness of low porosity layer and high porosity layer are 1.10 μm and 0.38 μm, respectively. The low porosity layer is stable during etching and can be used as the epitaxial growth surface. It is verify that the IPA can be used to replace the ethanol in electrochemical etching to obtain double porous layer.

When the substrate was an N-type wafer, the etching result was different. If the etching carried out in the ethanol solution, only the composition of HF:C_2H_5OH:H_2O =1:1:1 can obtain stable porous layer. The low porosity layer peeled partially when the water content in the etching solution is higher than 64% (HF:C_2H_5OH:H_2O=1:1:2). If the etching carried out in the IPA solution, the composition could be adjusted to HF:IPA:H_2O =1:1:3. Figure 5 is the SEM cross section of porous structure on N-type substrate. Figure 6(a) is obtained from ethanol solution. The current density of first step and second step etching is 2 mA/cm^2 and 125 mA/cm^2, respectively. The thickness of low porosity layer and high porosity layer are 2.53 μm and 0.63 μm, respectively. Figure 6(b) is obtained from IPA solution. The current density of first step and second step etching is 2 mA/cm^2 and 125 mA/cm^2, respectively. The thickness of low porosity layer and high porosity layer are 2.13 μm and 0.74 μm, respectively.

TABLE III COMPOSITION OF IPA SOLUTION

Solution	HF: IPA:H_2O	Water content
IPA-1	1:1:1	50.5%
IPA -2	1:1:2	62.9%
IPA -3	1:1:3	70.3%
IPA -4	1:1:4	75.3%
IPA -5	1:1:5	78.8%
IPA -6	1:1:8	85.2%

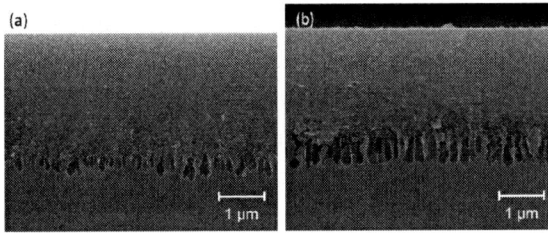

Fig. 6. SEM cross section of porous structure on N-type substrate (a) ethanol solution; (b) IPA solution.

Figure 7 is the ELO wafer obtained in this study. The silicon wafer is phosphorous doped N-type wafer with thickness of 107 μm. The resistivity is 2.36 Ω-cm, which is suitable to use in solar cell fabrication.

Fig. 7. The silicon ELO wafer with thickness of 107 μm.

IV. CONCLUSIONS

The characteristics of double porous layer for epi and lift-off process were study. In order to prevent the evaporation of solution and have stable etching, the composition of etching solution was tested. The composition of etching solution could be adjust to 1:1:3. If the solution is thinner, the current distribution will become non-uniform and the surface will partially peel-off. The IPA was confirmed to be used to replace ethanol, which can maintain the etching quality and reduce the evaporation rate of the solution.

ACKNOWLEDGMENT

The financial support provided by Bureau of Energy is gratefully acknowledged.

REFERENCES

[1] F. Henley, A. Lamm, S. Kang, Z. Liu, and L. Tian, "Direct film transfer (DFT) technology for kerf-free silicon wafering", 23rd EUPVSEC (2008).

[2] M. J. McCann, K. R. Catchpole, K. J. Weber, and A. W. Blakers, "A review of thin-film crystalline silicon for solar cell applications. Part 1: Native substrates", Solar Energy Materials & Solar Cells 68 (2001) 135-171.

[3] F. Dross, A. Milhe, J. Robbelin, I. Gordon, P. O. Bouchard, G. Beaucarne, and J. Poortmans, "SLIM-cut: a kerf-loss-free method for wafering 50-μm-thick crystalline Si wafers based on stress-induced lift-off", 23rd EUPVSEC (2008).

[4] R. Brendel, K. Feldrapp, R. Horbelt, and R. Auer, " 15.4%-efficient and 25 μm-thin crystalline Si solar cell from layer transfer using porous silicon", Physical state sol. 197 (2003) pp.497–501.

[5] R. Hao, T. S. Ravi, V. Siva, J. Vatus, I. Kuzma-Filipek, F. Duerinckx, M. Recaman-Payo, M. Aleman, E. Cornagliotti, P. Choulat, R. Russell, A. Sharma, L. Tous, A. Uruena, J. Szlufcik, and J. Poortmans, "Kerfless epitaxial mono crystalline Si wafers with built-in junction and from reused substrates for high-efficiency PERx cells", 43th IEEE PVSC (2016).

[6] W. Yoon, A. Lochtefeld, N. Kotulak, D. Scheiman, A. Barnett, P. Jenkins, and R. Walters, "Enhanced surface passivation of epitaxially grown emitters for high efficiency ultrathin crystalline Si solar cells", 43th IEEE PVSC (2016).

[7] N. Milenkovic, M. Driessen, B. Steinhauser, J. Benick, S. Lindekugel, M. Hermle, S. Janz, and S. Reber, "20% efficient solar cells fabricated from epitaxially grown and freestanding n-type wafers", Solar Energy Materials & Solar Cells 159 (2017) 570-575.

Electrical and Optical Performance of Silicon Solar Cells Using Plasmonics Indium Nanoparticles Layer Embedded in SiO$_2$ Antireflective Coating

Hao-Yu Yang, Wen-Jeng Ho*, Sheng-Kai Feng, Jheng-Jie Liu, Ta-Wei Chuang, Guan-Yi Li, Yun-Chie Yang, Cho-Chun Chiang, and Yao-Hui Chen

Department of Electro-Optical Engineering, National Taipei University of Technology, Taipei 10608, Taiwan, R.O.C. *: wjho@ntut.edu.tw

Abstract — In this study, we demonstrate the photovoltaic performance enhancement of silicon solar cell by means of plasmonics indium nanoparticles (In-NPs) layer embedded in SiO$_2$ antireflective coating (ARC). The optical reflectance, external quantum efficiency, and photovoltaic current-voltage are measured and compared. Impressive conversion efficiency enhancement of 35.94% for the cell with double In-NPs layers ARC, 34.77% for the cell with single In-NPs layer ARC, and 26.67% for the cell with a pure SiO$_2$ ARC were obtained which is compared to the reference cells. Besides, the gain in absolute efficiency of 1.26-1.52% for the cells with In-NPs ARC was higher than that of the cell with a pure SiO$_2$ ARC, due to the contribution of plasmonics scattering of In-NPs.

I. INTRODUCTION

The metallic nanoparticles (NPs), such as Au NPs and Ag NPs, have attracted considerable attention in photovoltaic device applications for their ability to resonantly couple with incident light and scatter incident light into the active absorption layer of the solar cell [1-3]. In general, the optical properties of metallic particles are depended on the particles size, shape, spacing, and surrounding material [4]. The plasmonic effects of metallic particles also changed with the optical properties of NPs [5]. The applications of metallic NPs on the photovoltaic devices have been wildly researched, however, only a few of papers have reported on the use of indium (In) NPs, which can conduct great plasmonic light scattering effect to improve the photovoltaic device performance [6].

In this study, the plasmonic light scattering effects of single or double In-NPs layer embedded in a SiO$_2$ antireflective coating (ARC) with a thickness of 90 nm on silicon solar cells are studied. The optical and electrical properties of the cells are examined by the measurements of optical reflectance, photovoltaic current-voltage (I–V), external quantum efficiency (EQE). The novelty of this study is the near-field and far-field plasmonic effects on the cell with single In-NPs layer as well as double In-NPs layers embedded in the SiO$_2$ ARC that can be examined by using optical reflectivity spectrum and EQE response.

II. EXPERIMENT

This study employed a 525-μm-thick born-doped (p-type) crystalline silicon(c-Si) wafer with orientation (100), polished on one side with resistivity of 10 Ω-cm and it was cut into small sample (1 cm × 1 cm). For device processing, all c-Si samples were first cleaned in acetone and iso-propyl alcohol solutions, then rinsed with de-ionized water and finally removed the oxide by using diluted HF solution. Next, a liquid-type phosphorus source (Phosphorosilicafilm; Emulsitone Co., USA) was spun on the clean front-side of c-Si samples using a spin-on film process followed by a baking process to remove the solvents and promoting cross-linking. After baking, a rapid thermal annealed (RTA) treatment at 900˚C for 1 min was used in order to diffuse the phosphorus and resulting in an n$^+$-Si emitter of approximately 0.4-μm in thickness. Next, a 450-nm thick aluminum (Al) and a 20-nm Ti/250-nm Al film were evaporated on the rear side p-Si and front side n$^+$-Si and f subsequently annealed in RTA chamber, respectively. For further comparing, the samples were isolation etched in HF and HNO$_3$ mixed solution using a photolithography process to obtain individual areas of 4 × 4 mm^2 served as the bare-type reference solar cell. The fabricated bare silicon solar cell is shown in Fig. 1(a).

To investigate the enhancement of the photovoltaic performance achieved by the light scattering of the In NPs, the silicon solar cells consisting following structure: (A) a cell with a 90-nm thick SiO$_2$ ARC layer (Fig. 1(b)), (B) a cell with single In-NPs layer embedded in the SiO$_2$ ARC (Fig. 1(c)), and, (C) a cell with double In-NPs layer embedded in the SiO$_2$ ARC (Fig. 1(d)) were proposed. For the cell with single In-NPs layer, an 18-nm thick SiO$_2$ space layer was deposited firstly on the bare silicon solar cell using e-beam evaporation. Then, a 3.8 nm thick indium film was deposited over the SiO$_2$ space layer and subsequently annealed in RTA chamber at 200˚C for 30 min in ambient H$_2$, forming In NPs. For double In-NPs layers, additional 18-nm thick SiO$_2$ layer and 3.8 nm thick indium films were formed as the same the first time processing. In this work, we maintained constant conditions during In NP deposition and RTA annealing. Finally, a SiO$_2$ film was deposited over In-NPs layer until the total thickness of SiO$_2$ layer that was reached a thickness of 90 nm (be called In-NPs ARC).

The optical reflectance, EQE response (Enli Technology Co., Ltd., Kaohsiung, Taiwan), and photovoltaic current density-voltage (J-V) of the cells under AM 1.5 G illumination

978-1-5090-5606-4/17 $31.00 © 2017 IEEE

with and without In-NPs layer embedded in SiO$_2$ ARC are measured and compared.

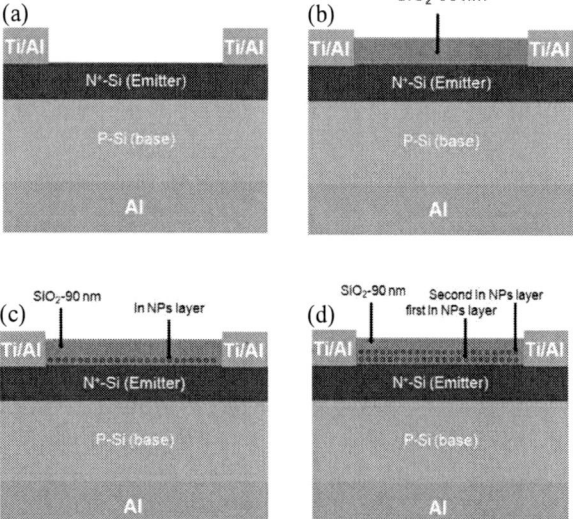

Fig. 1. Schematic diagram of the proposed silicon solar cells: (a) bare silicon solar cell as the reference cell, (b) cell with a 90-nm SiO$_2$ ARC. (c) Single In-NPs layer embedded in SiO$_2$ ARC, (d) double In-NPs layers embedded in SiO$_2$ ARC.

III. RESULT AND DISCUSSION

To examine the sizes and profiles of the In NPs, the field-emission electron scanning microscopy (FE-SEM) was conducted, as shown in Fig. 2(a). The diameter of the In NPs was ranged mainly from 10 to 30 nm and the coverage of In NPs was approximately 32.87%, which was calculated by analyzing the SEM image shown in Fig. 2(b) using J-image software.

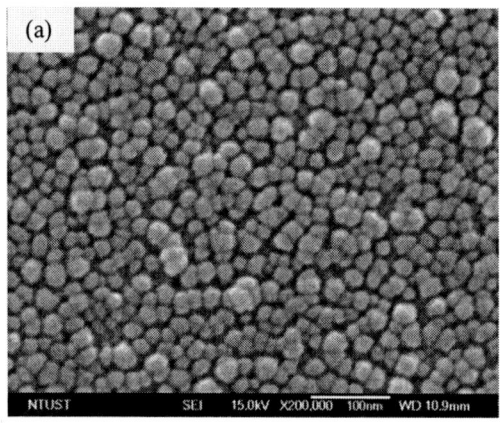

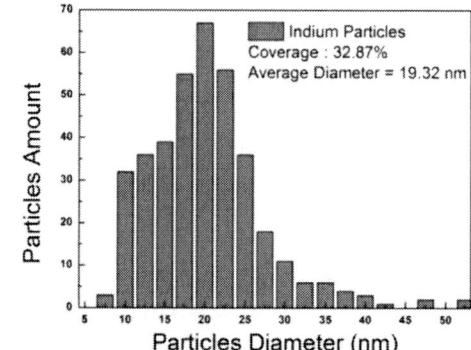

Fig. 2. (a) Electron scanning microscopy (SEM) image of In NPs sizes and profiles. (b) Using J-image software to calculate particle diameter and coverage by analyzing the SEM image.

Fig. 3. shows the optical reflectance of a bare solar cell, a cell with a 90-nm SiO$_2$ layer, and cells with single and double In NPs layers embedded in the 90-nm SiO$_2$ layer. The reflectance of the cell with a 90-nm SiO$_2$ layer was lower than that of the bare cell due to antireflection of SiO$_2$ layer. However, the reflectance of the cell with single In-NPs layer embedded in 90-nm SiO$_2$ layer further decreased across the entire wavelength range to a level below that of the cell with only a 90-nm SiO$_2$ layer, this is due to the ARC and plasmonic forward scattering of photons by the In-NPs layer embedded in SiO$_2$ layer. Furthermore, the reflectance of the cell with double In-NPs embedded in SiO$_2$ layer was further decreased at the wavelengths of 350-650 nm due to much plasmonic light scattering induced by the more In NPs.

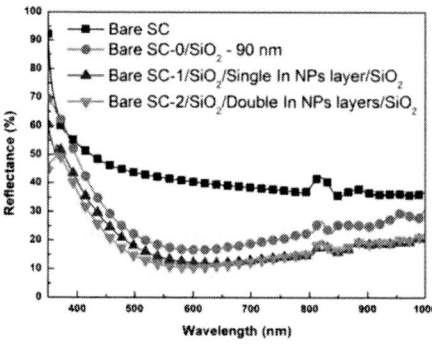

Fig. 3. Reflectance spectrum of the proposed solar cells.

Fig. 4. shows the EQE and ΔEQE response of all proposed solar cells in this work. The EQE values of the cell with single In-NPs ARC were higher than that of the cell with only a 90-nm SiO$_2$ ARC at wavelengths of 350-1000 nm, which is in agreement with the results of optical reflectance obtained at wavelengths of 350-1000 nm due to the plasmonic scattering of In NPs. The EQE values of the cell with double In-NPs layers were higher than that of the cell with single In-NPs layer at wavelength of 350-650 nm, which well agreement with the results of optical reflectance, due to much plasmonic light scattering induced by the more In NPs, The ΔEQE, which was calculated by [(EQE of the cell with SiO$_2$ ARC or

978-1-5090-5606-4/17 $31.00 © 2017 IEEE

the cell with In-NPs ARC) — (EQE of bare cell) / (EQE of bare cell)]. It shows that the EQE enhancement of the cell with In-NPs ARC due to plasmonic scattering of In-NPs was higher than that of the cell with a SiO_2 ARC in full wavelength range. In addition, the EQE enhancement of the cell with double In-NPs layers was higher than that of the cell with single In-NPs layer at the wavelength range of 350-650 nm because of plasmonics effects of In-NPs enhancing at short wavelengths by much more In-NPs particles.

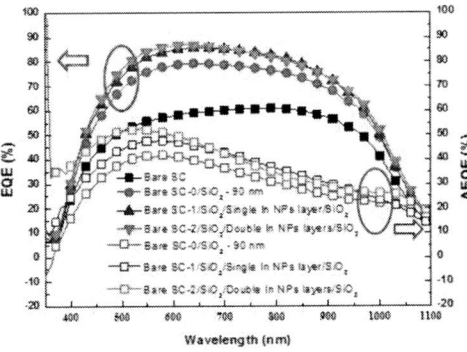

Fig. 4. EQE and ΔEQE spectra of the proposed solar cells.

The photovoltaic J-V characteristics of the cells under one-sun AM 1.5G illumination are presented in Fig. 5 and the photovoltaic performances are summarized in Table I. The short-circuit current density (J_{sc}) and open-circuit voltage (V_{oc}) as well as the conversion efficiency (η) are increased when a SiO_2 ARC or In-NPs ARC was applied on the bare solar cells. In general, J_{sc} is proportion to the EQE and η is proportion to J_{sc}. In this work, the η enhancement was well agreed to the increased in J_{sc} and V_{oc}. Another important finding in this study indicated that the J_{sc}, V_{oc} and η of the cells with In-NPs ARC were higher than that of the cell with a conventional SiO_2 ARC. Thus, the plasmonics effects of In NPs layer embedded in SiO_2 ARC can be additional increased the absolute efficiency of 1.26-1.52%.

TABLE I

PHOTOVOLTAIC PERFORMANCES OF THE PROPOSED SOLAR CELLS

	J_{sc} (mA/ cm^2)	V_{oc} (mV)	F.F. (%)	η (%)	ΔJ_{sc} (%)	Δη (%)
Bare SC-0	26.48	526.2	74.0	10.31	-	-
Bare SC-0 with SiO_2 ARC	33.41	528.4	74.1	13.08	26.41	26.86
Bare SC-1	26.89	533.3	74.2	10.64	-	-
Bare SC-1 with single In-NPs layer-ARC	35.84	536.4	74.6	14.34	33.28	34.77
Bare SC-2	26.33	535.3	76.2	10.74	-	-
Bare SC-2 with double In-NPs layers ARC	35.88	541.1	75.2	14.60	36.16	35.94

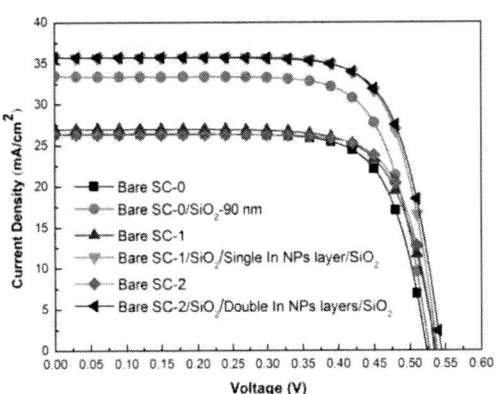

Fig. 5. Photovoltaic J-V of the proposed solar cells.

IV. CONCLUSION

The photovoltaic performance enhancement of silicon solar cell using plasmonics In-NPs layer embedded in SiO_2 ARC was demonstrated. The optical reflectance, EQE response, and photovoltaic J-V were used to examine the contribution of In-NPs plasmonics effects. The η enhancement of 35.94% for the cell with double In-NPs layer was higher than that of the cell with single In-NP ARC of 34.77% and the cell with a pure SiO_2 ARC of 26.86%, compare to the reference bare cells.

ACKNOLOWLEDGEMENTS

The authors would like to thank the Ministry of Science and Technology (MOST) of the Republic of China for financial support under Grant MOST 103-2221-E-027-049-MY.

REFERENCES

[1] R. J. Mukti, A. Islam, "An enhanced efficient thin film silicon solar cell design based on silver nanoparticle," *5th International Conference on Informatics, Electronic and Vision (ICIEV 2016)*, 7760156, pp. 1052-1056.

[2] H. Nourolahi, A. Behjat, S. M. M. Hosseini Zarch, M. M. Bororizadeh, "Silver nanoparticle plasmonic effects on hole-transport material-free mesoporous heterojunction perovskite solar cells," *Solar Energy*, vol. 139, pp. 475–483, 2016.

[3] N. F. Fahim, Z. Ouyang, B. Jia, Y. Zhang, Z. Shi, M. Gu, "Enhanced photocurrent in crystalline silicon solar cells by hybrid plasminic antireflection coating," *Applied Physics Letter*, vol. 101, p. 261102, 2012.

[4] K. L. Kelly, E. Coronado, L. L. Zhao, G. C. Schatz, "The optical properties of metal nanoparticles: The influence of size, shape, and dielectric environment," *The Journal of Physical Chemistry B*, vol. 107, pp. 668–677, 2003.

[5] M. Schmid, P. Andare, P. Manley, "Plasmonic and photonic scattering and near fields of nanoparticles," *Nanoscale Research Letter*, vol. 9:50, pp. 1-11, 2015.

[6] W. J. Ho, Y. Y. Lee, S, Y, Su, "External quantum efficiency response of thin silicon solar cell based on plasmonic scattering of indium and silver nanoparticles," *Nanoscale Research Letter*, vol. 9:483, pp. 1-8, 2014.

Electroluminescence Analysis For Separation of Series Resistance From Recombination Effects in Silicon Solar Cells with Interdigitated Back Contact Design

Nuha Ahmed[1,2], Lei Zhang[1,2], Ujjwal Das[1], and Steven Hegedus[1,2]

[1]Institute of Energy Conversion, University of Delaware, Newark, DE 19716, USA

[2]Department of Electrical and Computer Engineering, University of Delaware, Newark, DE 19716, USA

Abstract — **Electroluminescence (EL) is used to separate series resistance (R_s) from junction recombination effects in interdigitated back contact (IBC) Si solar cells. We applied iterative techniques using EL images at different injection bias to derive spatial maps of R_s and dark saturation current density (J_0) in two devices with different processing. Higher R_s regions were around emitter strips while higher J_0 regions were around base strips, some regions showed both effects. Mean values of R_s and J_0 from EL agreed very well with lumped values from dark IV diode analysis with mean R_s and J_0 values of 0.5 Ohms-cm², 5.3*10⁻⁹A/cm²,1.1 Ohms-cm² and 4*10⁻¹⁰A/cm² for both cells respectively.**

Index Terms — **electroluminescence, heterojunction, interdigitated back contacts, silicon solar cells, laser fired contacts, junction recombination, series resistance, diode voltage.**

I. INTRODUCTION

EL is a quick non-destructive characterization tool that has been used to map solar cell parameters such as the minority carrier effective diffusion length [1], local junction voltage, series resistance [2], and the dark saturation current density. Most of the EL analysis in the literature focuses on one dimensional (1D) front junction (FJ) silicon solar cells [3] and a few papers focus on 2 dimensional (2D) diffused junction silicon solar cells [4]. There has been little analysis done on 2D heterojunction IBC cells (IBC-HJ) or cells with complex 2D emitter or contact structures. In this work, EL is applied to IBC-HJ solar cells which have complex 2D lateral current transport. We analyze EL to produce spatial maps of R_s and J_0. Both loss mechanisms degrade the device performance but have different origins and solutions. J_0 represents the junction minority carrier recombination while R_s stems from the majority carriesr bulk properties and lateral geometries. However, IBC cells operate at medium-high level injection where R_s effects are due to both carriers-electrons and holes. Spatial mapping of the loss mechanisms helps in analyzing the more complex 2D lateral current transport as occurs in IBC-HJ solar cells presented here. Our method provides insight into the design and operational features of the device as well as random process-related non-uniformities. We analyze two IBC-SHJ devices with different back patterning processes, back metal, gap dimensions and JV parameters.

II. EXPERIMENTAL

A. Sample Preparation

IBC cells were prepared from textured n-type CZ wafers with a resistivity of ~8.5 Ohms-cm. The wafers were randomly textured and passivated with an intrinsic a-Si:H layer followed by antireflection coating of ~75 nm a-SiN$_x$:H and ~20 nm a-SiC:H. All a-Si:H layers were deposited by plasma-enhanced chemical vapor deposition (PECVD). All metal layers were deposited by metal beam evaporation. The two devices discussed in this paper differ in their processing and their gap dimensions. The first device was made using standard 3 steps photolithography (P-Lith) to pattern the back surface and has a gap width of 25 μm and Al metal as back contacts. The second device differs in that the contacts to the n base region were laser fired (LFC) through a back metal stack of Titanium/Antimony and Aluminum [5] and has a gap width of 100 μm. The gap region provides isolation between the 1300 μm width p emitter strip and the 300 μm n region base strip. Complete details of the sample preparation is discussed in [6]. The JV curves and parameters of the IBC devices are shown in Fig. 1 and Table 1 below. They have very comparable V_{oc} and I_{sc} but significantly different FF.

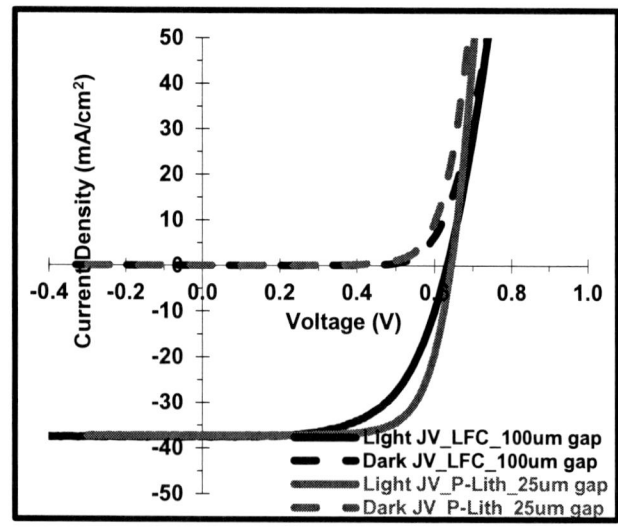

978-1-5090-5606-4/17 $31.00 © 2017 IEEE

Fig.1.   JV curves and parameters for two complete IBC devices with front textured and p-emitter and n-base contacts in the back and different back patterning processes, back metal and gap width.

TABLE I
JV PARAMETERS OF TWO DIFFERENT IBC DEVICES: 1. WITH THREE STEP PHOTOLITHOGRAPHY AND 2. LASER FIRED CONTACTS IN THE BACK

Device	V_{oc} (V)	J_{sc} (mA/cm)	FF (%)	Eff (%)	Gap Width (µm)
P-Lith	0.644	37.35	71.9	17.3	25
LFC	0.649	37.6	60.2	14.3	100

B. EL System

The EL measurement system used a Si-Ge CMOS imaging detector equipped with a thermo -electric cooler at -80° C. It is enclosed in a light-tight dark box. To provide the injection current we used a Keithley source meter in a 4 wire mode. In addition to the dark frame subtraction provided by the camera, we subtracted a dark current image at the same EL conditions from each image to eliminate the dark current signal. The detector array size is 480 by 640 pixels and the final resolution of the system was ~25 um. All images have an exposure time of 0.533 seconds. We used the program "FIJI" developed by the National Institute of Health (NIH) and based on ImageJ Java open source software for image processing [7].

III. VARIABLE INTENSITY METHODOLOGY AND RESULTS

A. Analysis of Intensity Dependent EL

There are several methods to obtain a lumped global R_s, however, methods like EL allow the spatially resolved mapping of localized R_s. We applied an iterative technique similar to the method proposed by Breitenstein [2] where two EL images obtained at different voltage biases are used to derive a spatial map of R_s in the device. This method is based on an area related resistance that is connected in series to the local diode. The relation between EL emission intensity at a given pixel i and the local junction voltage V_{di} can be expressed by (1) below.

$$\emptyset_i(intensity) = C_i exp(\frac{V_{di}}{V_t}) \qquad (1)$$

Where C_i is a calibration constant and V_t is the thermal voltage. There are two key assumptions [2]. One is that at the lowest bias condition there are negligible Ohmic losses due to the low current. A second is that C_i is independent of injection level. We have plotted C_i at different injection levels and verified that there is no dependency since EL is only exponentially dependent on the diode voltage. To begin the iterative analysis, we assume at first that R_s effects are negligible at the lower bias voltage and resulting injection current: 0.64 V and 25 mA for the 3 step P-Lith device and

0.63V and 25.5 mA for the LFC device shown in Fig. 2(a) and Fig.2(c) below, respectively such that all the applied voltage is available to bias the junction.

$$V_{app} = V_{d,i1} \qquad (2)$$

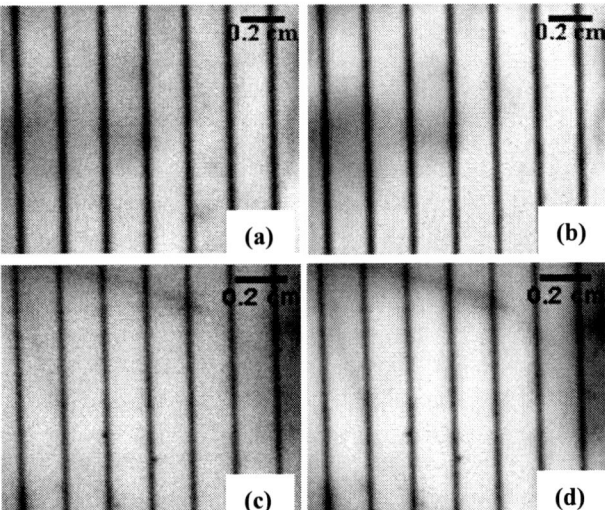

Fig.2.   EL images of two IBC device (a) 3 step P-Lith: Low bias image 0.64V, 25 mA (b) 3 step P-Lith: High bias image 0.69V, 50 mA (c) LFC-Low bias image 0.63 V, 25.5 mA (d) LFC-High bias image 0.68V,42.9 mA.

This allowed us to obtain a C_i for each pixel. We then use this C_i to obtain a spatial map of the internal voltage losses in the higher voltage bias image of 0.69 V and 50 mA for the 3 step (P-Lith) device and 0.68 V and 42.9 mA injection current for the LFC contact device as shown in Fig. 2(b) and Fig. 2(d) respectively above. This voltage loss is due to the combined effects of R_s and J_0 according to (3) below.

$$V_{di} = V_{app} - R_{si}J_{0i}exp(\frac{V_{di}}{nKT}) \qquad (3)$$

Where R_{si} is the localized spatially dependent R_s, J_{0i} is the localized spatially dependent J_0 and n is the diode ideality factor. To isolate R_s and J_0 effects, we assume C_i is inversely related to J_0 through a scaling factor f such that

$$C_i = \frac{f}{J0i} \qquad (4)$$

We apply standard dark diode analysis techniques [8] to obtain global lumped values for R_s, n and J_0. The last two (J_0 and n) are used directly as initial values in the first iteration to get the first R_s values from the second image at a higher bias. The scaling factor f is chosen such that the mean values of J_{0i} correspond to the global value of J_0 from diode analysis. This scaling factor is also used to calculate the diode voltage for the second (high voltage) image. The first R_s and J_0 values are then used to account for the losses in the low voltage image and new values of J_0 are obtained and used in the high voltage image to get new values of R_s. These new values of R_s and J_0 are then compared to the values at the beginning of the step. These iterations are repeated until good agreement is achieved between values assumed at the beginning of the step and those obtained at the end. We typically find agreement within ±1% within 4 iterations and values in very good agreement with global R_s and J_0 values. The final and fourth iteration for the internal higher voltage distribution at 50 mA and 0.69 V for the 3 step P-Lith device and at 42.5 mA and 0.68V for the LFC device are shown in Fig. 3(a) and 3(b) respectively below with voltage values ranging from approximately 0.649 V to 0.66 V and from 0.643 V to 0.655 V.

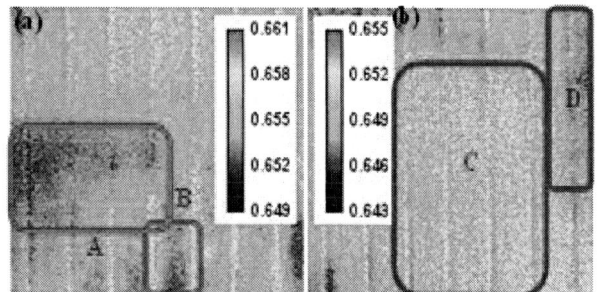

Fig.3. Diode voltage images (a) 3 step P-Lith device at 0.69 V and 50 mA (b) LFC device at 0.68 V and 42.5 mA.

We then use this internal voltage distribution with (1)-(4) above to derive R_s and J_0 maps as shown in Fig. 4(a), Fig. 4(b) Fig. 4(c) and Fig. 4(d) below respectively.

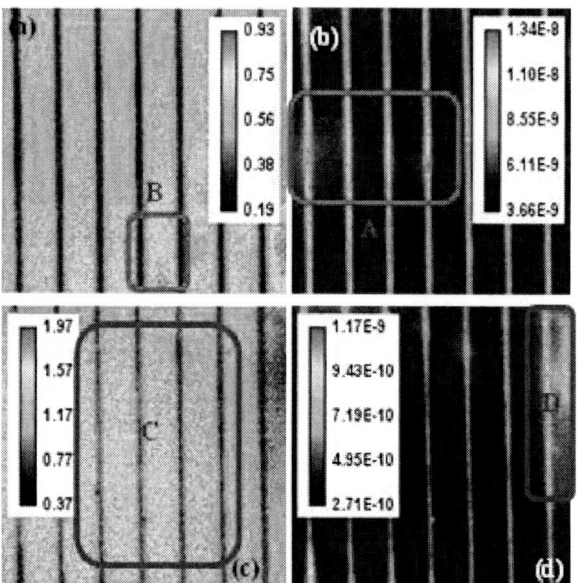

Fig.4. R_s and J_0 for IBC devices (a) R_s : 3 step P-Lith spatial EL map-color bar gives R_s in Ohms-cm^2 (b) J_0 : 3 step P-Lith Spatial EL map-color bar gives J_0 in A/cm^2(c) R_s: LFC- color bar gives R_s in Ohms-cm^2 (d) J_0: LFC device color bar gives J_0 in A/cm^2.

B. Minority Carrier Concentration

First we will discuss the spatial variations due to the IBC design and operation. The EL image shown in Figs. 2(a)-(d), have a characteristic pattern of higher emission over the wider p strip and lower emission over the narrower n strip and gap. We attribute this to the minority carrier concentration in each region of the device as the cell is forward biased. We have found this to be the case for IBC cells as well as for test structures with either front or back emitter with IBC patterns in the back. When minority carriers (holes) are injected from the p strips, they will recombine with electrons in the base of the device above the strip and a photon will be emitted for each electron-hole pair that recombines radiatively. The voltage images shown in Fig. 3(a) and Fig. 3(b) for both devices show an overall higher internal voltage over the n strips compared to over the p strips which supports our analysis that the lower emission over the n strips shown in the EL images in Figs. 2(a)-(d) is not due to voltage losses. Note that the higher voltage values in Fig. 3(a) and Fig. 3(b) go up to 0.661 V and 0.655 V respectively since for the EL analysis we are only using (~1.2 cm X 1.2 cm) of the whole imaged device area (~1.5 cm X 1.65 cm) and excluding the bus bars which would have higher internal voltage values. Thus for both devices the characteristic patterns in the EL images correspond to the minority carrier concentration in each region.

IV. DISCUSSION OF SPATIAL DEPENDENCE OF R_s AND J_0

The R_s map in Fig. 4(a) and Fig. 4(c) show higher R_s values over the wide p strips and lower R_s values over the n strips. This R_s pattern for IBC cells has also been observed by other groups [4] and can be attributed to the lower lumped contribution of the series resistance for the electron current (shorter path) over the n strips versus the hole current over the p strips (prolonged path) as each carrier type experiences a different resistance path. The electron current density over the n strip is ~4X higher compared to the hole current density at the same voltage due to the difference between the contact area for each region: the p strips are ~ 4X wider than the n strips. This results in a considerable decrease of R_s over the n strips. The fact that this is consistent for both sets of devices further validates this assumption.

The J_0 distribution in Fig. 4(b) and Fig. 4(d) shows a complementary pattern to the R_s map in Fig. 4(a) and Fig. 4(c) with higher J_0 values obtained over the narrower n strips and lower J_0 values obtained over the wider p strips. This is slightly in disagreement: we find the implied V_{oc} and lifetimes for the p strips are lower compared to the n strips indicating lower passivation yet in this paper the J_0 EL analysis show better passivation for the p strips. We can attribute this to the spatial difference in the ideality factor n for both strips from the EL analysis of test structures with IBC patterns and back emitter from which we get lower n values for the p strip (1.42) and higher n values for the n strip (1.57). However, please note that the magnitude of the difference between high and low values for J_0 in both devices are relatively small (about 2X). This is surprising given that one injects majority carriers while the other injects minority carriers, and that one cell has n-type HJ contact and the other a localized LFC contact

To analyze the EL analysis based on the device processing, region C in the in Fig.3 (b) for the LFC device shows higher voltage values over both the n and the p strips. Fig. 4(c) which is the R_s map obtained from the diode voltage map shows higher R_s values over the p strips and lower R_s over the n strips. Fig 4. (d) shows the final J_0 map in which slightly higher J_0 values are obtained over the n strips and very low J_0 values are obtained for the p strips. The high voltage values in the diode voltage in Fig.3 (b) are due to the very low J_0 values over the p strips and low J_0 values over the n strips which indicates that the laser firing process does not degrade the device passivation. It is reasonable that the slightly higher J_0 values are obtained over the n strips given that these are the strips that had the local LFC's. Comparing the J_0 maps in both the 3 step P-Lith device and the LFC device, there is a smaller contrast in the J_0 values for the LFC compared to the 3 steps P-Lith which have higher contrast between the n and p strips. The JV parameters of both devices are very similar except that the LFC has a lower fill factor that might be due to the laser firing processing or to the gap parameters.

Some regions such as the center-left region in Fig. 3(a) red rectangle (A) show lower voltage values spanning both strips.

This corresponds to the dark regions shown in the EL images in Fig. 2(a) and Fig. 2(b). This voltage loss is due to both effects R_s and J_0. Decoupling that leads to most of this voltage drop to be attributed to the higher J_0 values as shown in the J_0 image in Fig. 4(b).The other dark region shown in the center lower voltage image in Fig. 3(a) (red smaller rectangle (B)) is mainly due to R_s as shown in Fig.4 (a). Other regions for the LFC device is shown in Fig.3 (b) with the purple narrower rectangle (D) which shows lower internal voltage values. Fig.4 (d) shows the J_0 map image for the LFC device with the narrow purple rectangle region (D) showing higher J_0 values leading to a loss in the diode voltage. Without this R_s and J_0 analysis we would not have been able to isolate the source of these voltage losses which then allows for specific targeting of solutions to the random defects.

The R_s intensity histogram distribution for the 3 step P-Lith device in Fig. 5(a) below shows 2 regions: lower R_s (~0.2 Ohms-cm^2) over the narrower n strips and higher R_s (~0.6 Ohms-cm^2) over wider p strip with a mean R_s value of 0.5 Ohms-cm^2 which is in good agreement with the global lumped value of R_s obtained from IV analysis as shown in Table II. The mean value of J_0 obtained from EL ($5.3*10^{-9}$ A/cm^2) from Fig. 4(b) is also in good agreement with the global value from IV analysis ($9*10^{-9}$ A/cm^2). Similarly the R_s intensity histogram distribution for the LFC device shown in Fig. 5(b) shows 2 regions: lower R_s (~0.4 Ohms-cm^2) over the narrower n strips and higher R_s (~1.6 Ohms-cm^2) over the wider p strips. The close agreement of the key parameters R_s and J_0 from EL mapping and JV analysis shown in Table II below further validates applying the EL technique to IBC-HJ cells with different back pattering process.

Note that on the second or LFC device (100 µm gap) all three parameter maps - V_{di}, R_{si} and J_{0i}- show no evidence of the localized LFC which is the only place the majority current actually flows from and is known to have high recombination. This uniformity of response over the LFC strips is consistent with the photocurrent mapping via laser beam induced current (LBIC) maps on these same samples.

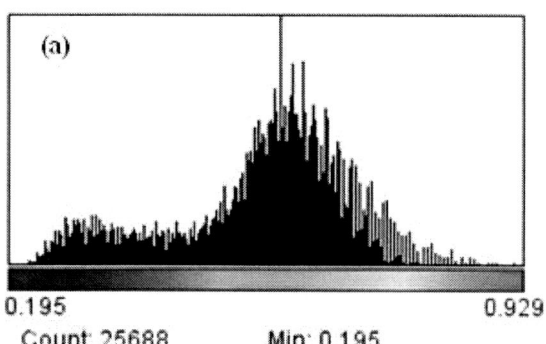

0.195   0.929
Count: 25688 Min: 0.195
Mean: 0.553   Max: 0.929

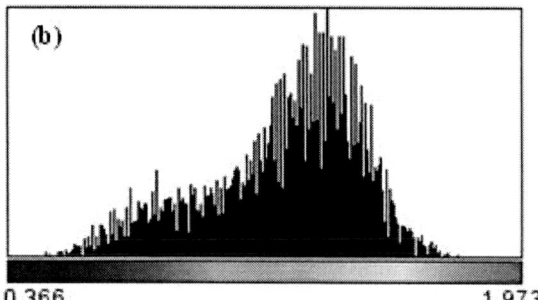

0.366   1.973
Count: 25688 Min: 0.366
Mean: 1.221   Max: 1.973

Fig.5. R_s Intensity Histogram (a) R_s Intensity Histogram: 3 step P-Lith device (b) R_s Intensity Histogram: LFC device

TABLE II
R_S AND J_0 OF TWO IBC DEVICES OBTAINED FROM STANDARD DARK DIODE ANALYSIS AND FROM EL ITERATIVE ANALYSIS

Device	Method	Values	R_s $\Omega\text{-cm}^2$	J_0 $A\text{-cm}^2$
IBC-3steps P-Lith	Standard Diode Analysis	Lumped Sum	0.45	$9*10^{-9}$
IBC-3steps P-Lith	EL Analysis	Average	0.55	$5.3*10^{-9}$
IBC-LFC	Standard Diode Analysis	Lumped Sum	1.10	$6*10^{-10}$
IBC-LFC	EL Analysis	Average	1.22	$4*10^{-10}$

V. CONCLUSIONS

Using an iterative approach to analyze EL images at two intensities, quantitative spatial maps were derived for R_s and J_0 for IBC cells that differ in their processing. The method separates the impact of series resistance from junction recombination and spatially maps regions affected by these two loss mechanisms. R_s and J_0 maps correlated with the devices parameters and further validated that there are no significant losses in the IBC devices with LFC. Values obtained from the EL iterative technique and values obtained from dark diode analysis are in very good agreement with each other which validates the analysis. The EL technique discussed in this paper is not only applicable to IBC-HJ solar cells but could be applied to 1D devices with ID HJ emitter (or contact) and 2D patterned back contact or emitters to gain further insight into the minority and majority carrier effects.

REFERENCES

[1] P.Wurfel, T.Trupke and T.Puzzer, "Diffusion length of silicon solar cells from luminescence images", *Journal of Applied Physics*", vol.101 no 123110, pp1-10, 2007

[2] O. Breitenstein, A. Khanna, Y. Augarten, J. Bauer, J. M. Wagner, and K. Iwig, "Quantitative evaluation of electroluminescence images of solar cells", *Physics Status Solidi Rapid Research Letters*, vol. 4 no. 1-2, pp 7-9, 2010.

[3] J. Giesecke, M. Kasemann, and W. Warta, "Determination of local minority carrier diffusion length in crystal silicon from luminescence images", *Journal of Applied Physics,* vol. 106, no.1, 2009

[4] M. Padilla, B. Michl, N. Hagedorn, C. Reichel, S. Kluska, A. Fell, M. Kasemann, W. Warta, and M. C. Schubert,"Local series resistance imaging of silicon solar cells with complex current path", *IEEE Journal of Photovoltaics,* vol. 5, no. 3, 2015

[5] J.He, S.Hegedus, U.Das, Z.Shu, M.Bennett, L.Zhang and R.Birkmire, "Laser-fired contact for n-type crystalline Si solar cells", *Progress in Photovoltaics: Research and Applications,* vol.23, pp 1091–1099, 2015

[6] U. Das, H. Liu, J. He, and S. Hegedus, "The role of back contact patterning on stability and performance of Si IBC HJ solar cells", *40th IEEE PVSC*, pp 0590-0593, 2014

[7] ImageJ, https://imagej.nih.gov/ij/

[8] S.Hegedus and W. Shafarman,"Thin-film solar cells: device measurements and analysis", *Progress in Photovoltaics: Research and Applications,* vol.12, pp 155-176, 2004

Indoor Measurement of Angle Resolved Light Absorption by Black Silicon

Mekbib W. Amdemeskel[1], Beniamino Iandolo[2], Rasmus S. Davidsen[2], Ole Hansen[2], Gisele A. dos Reis Benatto[1], Nicholas Riedel[1], Peter B. Poulsen[1], Sune Thorsteinsson[1], Anders Thorseth[1], Carsten Dam-Hansen[1]

[1] Department of Photonics Engineering, Technical University of Denmark, Frederiksborgvej 399, 4000 Roskilde, Denmark, mekwub@fotonik.dtu.dk
[2] Department of Micro- and Nanotechnology, Technical University of Denmark, Ørsteds Pl., 2800 Kongens Lyngby, Denmark

Abstract — **Angle resolved optical spectroscopy of photovoltaic (PV) samples gives crucial information on PV panels under realistic working conditions. Here, we introduce measurements of angle resolved light absorption by PV cells, performed indoors using a collimated high radiance broadband light source. Our indoor method offers a significant simplification as compared to measurements by solar trackers. As a proof-of-concept demonstration, we show characterization of black silicon solar cells. The experimental results showed stable and reliable optical responses that makes our setup suitable for indoor, angle resolved characterization of solar cells.**

I. INTRODUCTION

The short circuit current output of solar cells under working conditions is affected by several factors. In particular, the effect of angle of incidence (AOI) on the optical properties, and therefore on the short circuit current, is considerable for AOI beyond 45° and needs to be taken into account when assessing performance of solar cells. Reindl et al. investigated the effect of diffuse irradiation as function of irradiation angle using four different models [1]. The effect of angle of incidence (AOI) on performance of crystalline silicon (c-Si) solar cells was studied as a function of irradiation angle by King et al. [2]. A comparison of angle resolved optical characterization of black Si solar cells with conventionally textured Si solar cells was recently presented by Davidsen et al. [3], however this work presented data on raw cells. In this work, we take a further step towards characterizing black Si based PV panels in working conditions by measuring short circuit current as function of incidence angle on encapsulated, one cell mini modules, black Si multi crystalline solar cell and a reference cell without black Si. In a more general perspective, we have built a set-up that allows automatized, reliable measurements of the short circuit current of solar cells as function of incidence angle with collimated (with an angular divergence of about 0.1°), simulated sunlight. This indoor set-up has several advantages over measuring the AOI effect outdoors on a tracker: (i) the measurements are not compromised by the effects of atmospheric diffuse or ground reflected light; (ii) the spectrum and intensity of the light source remain constant throughout the test; and (iii) the

rotation stage can move easily and accurately within -90° and + 90° AOI. Furthermore, indoor measurements using light sources such as flash lights are highly affected by the divergence of the beam. In contrast, the light source used here provides bright illumination across the UV-VIS-NIR range together with high spatial and power stability. In addition the light source is well collimated by collection optics to give a stable and reliable power measurement. PV simulation tools such as the familiar PVsyst describe the optical losses from increased AOI as an "incidence angle modifier" (IAM). It is also possible to use our set up to measure IAM data and use it in PVsyst simulations and estimate energy production solar cells across different locations.

II. EXPERIMENTAL METHOD

Black Si based solar cells were fabricated on p-type wafers as previously described [3], using a maskless reactive ion etch (RIE) process to texture the Si surface. In brief, the following steps were followed: (1) saw damage removal by etching in 30 % KOH at 75 °C for 2 min followed by cleaning in 20 % HCl at room temperature for 5 min and rinsing using deionized water; (2) Maskless RIE at room temperature in a plasma with $O_2 : SF_6 \approx$ 1:1 gas flow ratio, chamber pressure of 28 mTorr, 13.56 MHz radio-frequency and platen power of 30 W, (STS RIE); (3) Emitter formation using a tube furnace (Tempress Systems) with liquid $POCl_3$ as dopant source and N_2 as carrier gas at 840 °C and atmospheric pressure for 50 min, followed by removal of phosphor-silicate glass (PSG) in 5% hydrofluoric acid (HF); (4) Plasma enhanced chemical vapor deposition (PECVD) of 60 nm hydrogenated amorphous silicon nitride (SiNx:H) anti- reflective coating at 400 °C using a PlasmaLab System 133 (Oxford Instruments); (5) Screen-printing of Ag front and Al rear contacts with standard Ag and Al pastes using an EkraX5-STS screen printer, followed by co-firing of the front and rear contacts at 800 °C using a RTC Model LA-309 belt furnace; (6) Edge isolation by laser ablation using a J-1030-515-343FS System (Oxford Lasers Ltd). For the encapsulation, ribbons were soldered on the solar cell and a standard lamination procedure was performed at 140 °C following the encapsulants

978-1-5090-5606-4/17 $31.00 © 2017 IEEE

manufacturer's recommendations, with a stack consisting of 4 mm window glass, 450 μm TPO (Thermo Plastic Poly olefin), the cell, 450 μm TPT (Tedlar Polyster Tedlar) backsheet, and a black colored TPT back-sheet.

Angle resolved absorption measurements of the black silicon cells were performed using the set-up sketched in Fig. 1(a). A laser driven light source (LDLS) EQ-99FXC (Energetiq Technology, Inc.) with spectral emission between 190 and 2100 nm, collimated using an off-axis parabolic mirror, was used together with a UV filter, a sample holder stepper motor (THORLABS-NR360S) and a LABVIEW controlled short circuit current (I_{sc}) measuring transducer with an impedance of 0.670 ohms. The UV filter was positioned immediately after the light source in order to remove the UV-C part. The spectrum of the LDLS with and without filter in comparison with the solar spectrum is shown in Fig. 2. The measurement room was kept at a temperature of 21°C using an air conditioning system.

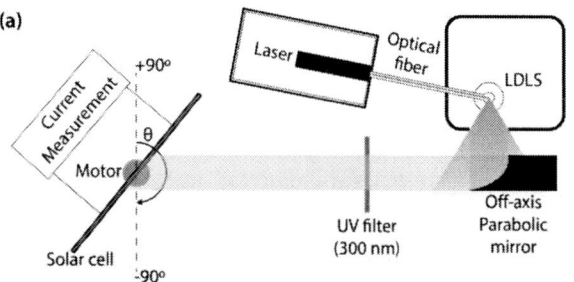

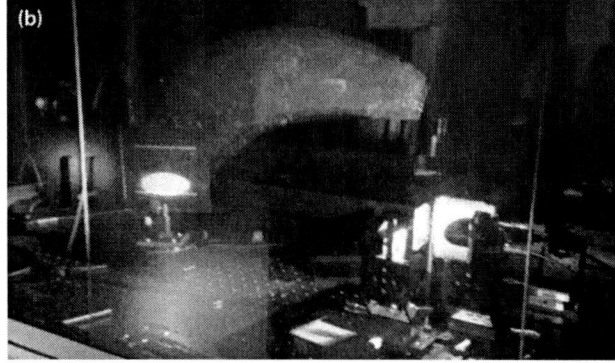

Fig. 1 Top view Schematic (a) and photograph (b) of the setup for measuring angle resolved absorption.

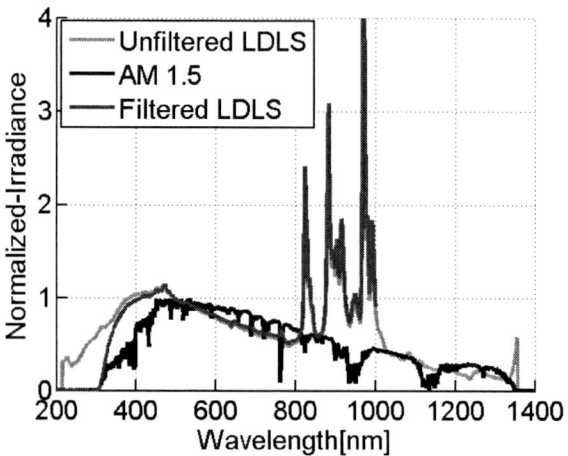

Fig. 2 The sun, filtered and unfiltered LDLS spectrum normalized to sun peak wavelength.

The AOI was changed by rotating the solar cells using a stepper motor around the vertical axis. The AOI are indicated with respect to normal incidence. The electrical transducer together with LABVIEW station was used to measure the I_{sc} for AOI between 90° to -90° with a 5° step size.

III. RESULTS AND DISCUSSION

Short circuit current versus angle of incidence (I_{sc}-AOI, or I_{sc}-θ) curves were obtained for two different TPO encapsulated black Si based samples, black Si multi crystalline solar cell and a reference cell without black Si Fig. 5 (a). We have performed an initial repeatability test by acquiring I_{sc}-AOI curves 10 times for the same samples. This repeatability of the set up was checked by calculating the error-I_{sc} and relative standard deviation (RSD) of the individual angular measurements. Results are summarized in Fig. 3 and Fig. 4. In particular, we determined the highest relative standard deviation of I_{sc} to be 0.75 % for AOI larger than + 80°. Hence, for all of our calculations we have used the average of the measurements at a given angle. From these calculations we can conclude that our measurement set up is highly repeatable.

978-1-5090-5606-4/17 $31.00 © 2017 IEEE   2673

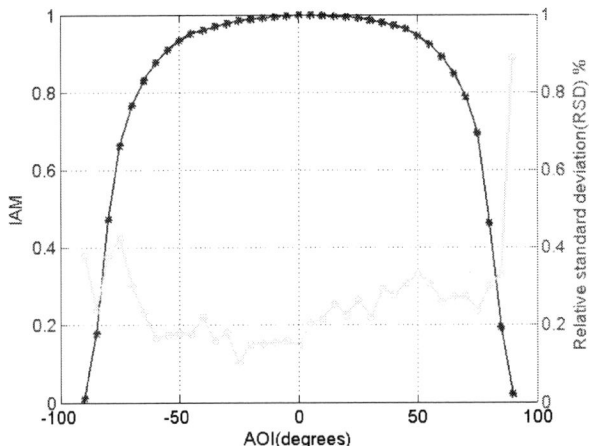

Fig. 3 AOI-IAM and AOI –RSD plot for reproducibility of our measurement setup.

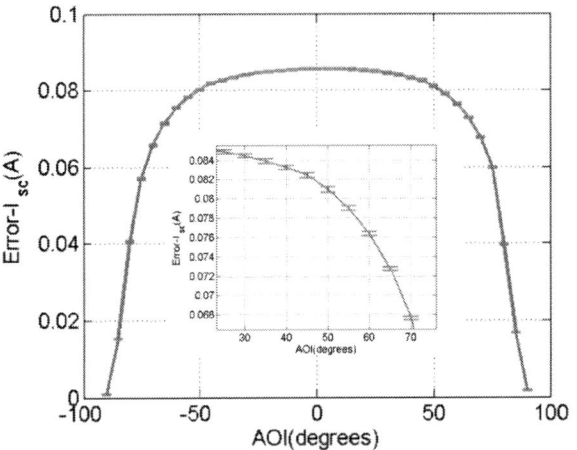

Fig. 4 AOI-Isc error plot for reproducibility of our measurement setup.

It is worth mentioning that, for AOI larger than ± 75°, not all the light beam hits the solar cell, as can be seen in Fig. 5(b) and (c). Therefore, we have computed an area i.e. power correction factor (it is calculated by dividing the partial area of the beam on the cell by the full elliptical area of the beam as shown in Fig. 5 (b) and (c)) for the partial beam spot to recalculate the actual reduced power reaching the cell from the total power (full spot area). This area correction for large angles is similar to the standard cosine correction for measurements setups where the spot size is larger than the cell.

Once corrected by the area factor, I_{sc} was normalized to the value at the normal angle incidence, i.e. for AOI = 0°. Since we are using a collimated light source, we have neglected the diffuse component. Thus for this setup the normalized and area corrected current is in fact the Incident Angle Modifier (IAM) used in e.g. PV-syst. We characterized two black

silicon cells, and the IAM for these samples is shown in Fig. 6. The graphs look very similar, with only a very slightly faster decay at larger AOI for sample 1. Furthermore, for a better comparison, we have compared two other samples, named black Si multi crystalline solar cell and a reference cell without black Si. In the same way as for the two black silicon measurements, we have carried out I_{sc} versus AOI measurement. After area correction, the IAM was computed and plotted for these samples as shown in Fig. 7.

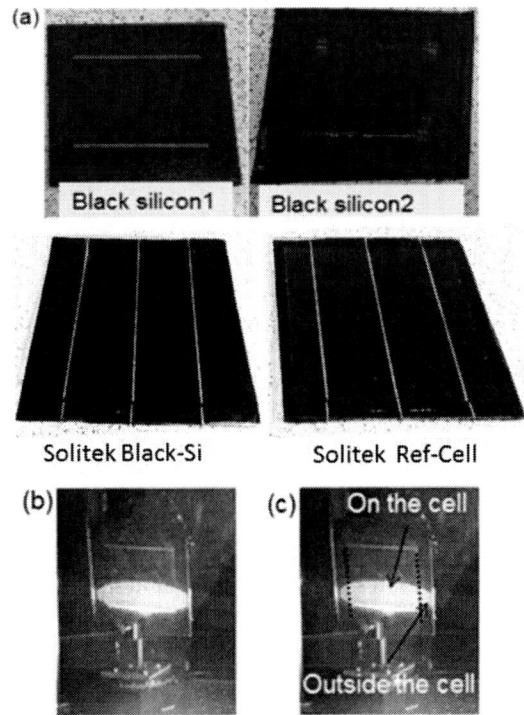

Fig. 5 Photograph showing the three black silicon solar cells and reference cell without black Si (a) and illustration of portion of the beam missing the solar cell for AOI larger than ± 75°.

978-1-5090-5606-4/17 $31.00 © 2017 IEEE

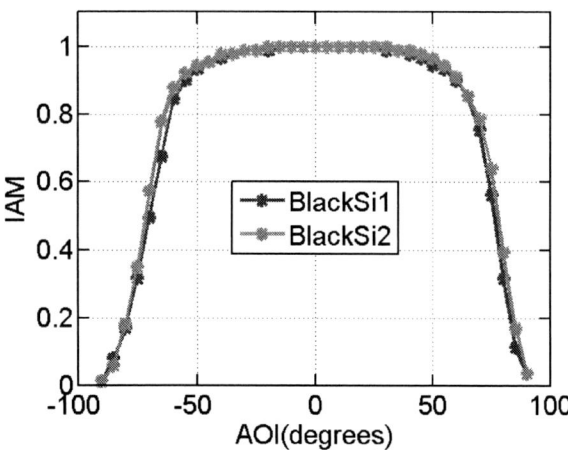

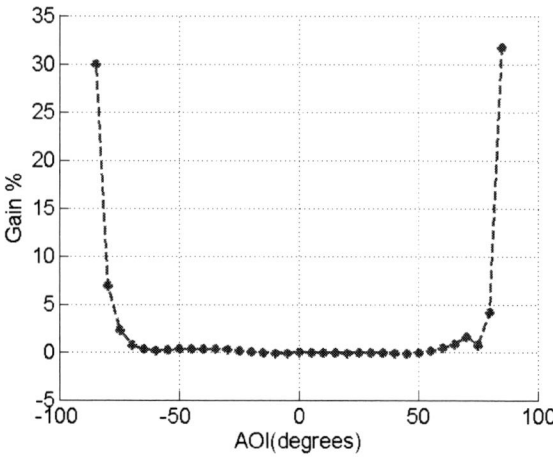

Fig. 6 IAM-AOI plot for the two black silicon solar cells.

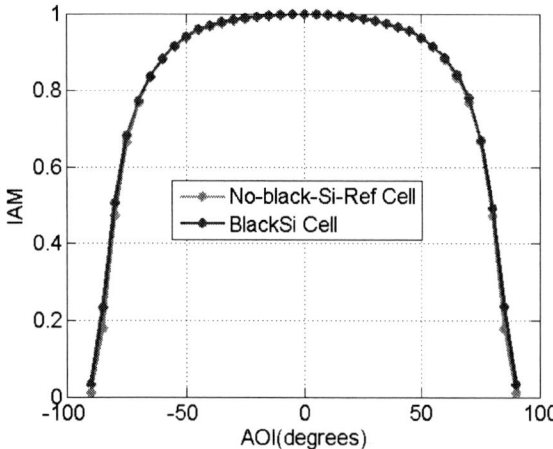

Fig. 8 A plot showing the gain of the black silicon solar cell over a reference cell without black silicon.

IV. CONCLUSIONS

In conclusion, we presented here an indoor study on the effect of the angle of incidence on the short circuit current of encapsulated, black silicon based solar cells, using a broadband, collimated light source. In a wider perspective, our setup allows for automatized, reliable measurements of short circuit current for AOI between +90° and -90° and is not restricted to measurements on black silicon. Results indicate the setup to be a viable method for determining the IAM. Importantly, reproducibility on the AOI was very high with a maximum average relative standard deviation of 0.75%. Moreover, from our measurement of black Si and reference cell without black cell; we have shown that the black silicon yields more significant gain in the output than the latter standard cell at a higher angle from the normal angle of incidence (which is 32% for 85°). In the future we have planned to do more measurements on standard cells for validation of the method and systematic tests allowing for determination of the angular performance for black silicon with EVA encapsulated cells in combination with different types of glasses (ARC coated and structured). The setup presented here is a valuable tool for indoor measuring of the IAM i.e. the angular performance on solar cells and mini modules.

Fig. 7 AOI-IAM plot for black Si multi crystalline solar cell and a reference cell without black Si.

From the measurement between black Si multi crystalline solar cell and a reference cell without black Si, we have calculated the gain obtained by black Si in comparison with the reference cell without black Si. The AOI versus gain plot is shown in the Fig. 8

978-1-5090-5606-4/17 $31.00 © 2017 IEEE

REFERENCES

[1] D. T. Reindl, W. A. Beckman, and J. A. Duffie, "Evaluation of hourly tilted surface radiation models," *Sol. Energy*, vol. 45, no. 1, pp. 9–17, 1990.

[2] D. L. King, J. A. Kratochvil, and W. E. Boyson, "Measuring solar spectral and angle-of-incidence effects on photovoltaic modules and solar irradiance sensors," *Conf. Rec. Twenty Sixth IEEE Photovolt. Spec. Conf. - 1997*, no. September,

pp. 1113–1116, 1997.

[3] R. S. Davidsen, J. Ormstrup, M. L. Ommen, P. E. Larsen, M. S. Schmidt, A. Boisen, Ø. Nordseth, and O. Hansen, "Angle resolved characterization of nanostructured and conventionally textured silicon solar cells," *Sol. Energy Mater. Sol. Cells*, vol. 140, pp. 134–140, 2015.

Impact of non-flat photogeneration and carrier profiles on the luminescent emission and detection of silicon solar cells

Nekane Azkona, Federico Recart, Pedro Rodríguez, Vanesa Fano, Aloña Otaegi and Juan Carlos Jimeno

Institute of Microelectronic Technology – UPV/EHU, 48013 Bilbao, Bizkaia, Spain

Photoluminescence is a powerful technique for the characterization of solar cells, with applications including diffusion length [1] and minority carrier lifetime extraction [2]. However, the analysis of the luminescent signal usually neglects the impact of non-flat carrier profiles in the measurement, which is minimized using infrared excitation (800 or 850 nm wavelengths) and detecting the emission from the illuminated side [3-6]. In the analysis of the signal or the image, both the generation profile and the resulting excess carrier concentration are usually assumed to be flat and the emitted photon reabsorption is neglected. This work analyses the impact on the photoluminescence emission of non uniform carrier profiles, which can be modulated by using shorter illumination wavelengths, but are unavoidable even if infrared excitation is used, in cases of non-uniform recombination. The study includes a quantification of the effect of photon reabsorption for steep profiles, which modifies the spectral distribution of the detected signal and makes the detection from the illuminated side almost 50% greater than from the opposite one.

Index Terms — crystalline silicone, solar cells, luminescence, photon reabsorption.

I. INTRODUCTION

The use of the luminescence emission as a tool for silicon solar cell characterization has gained widespread in the last decade. Interest has focused mainly on the photoluminescence (PL), as it has two characteristics that make it very attractive: fast and contactless measurement of any type of sample; the extra being the possibility of different detections, including a two-dimensional image acquisition.

The implementation of characterization techniques based on photoluminescence has been quite challenging, due to the difficulties in the calibration and interpretation of the luminescent emission. Even if at the beginning photoluminescence imaging (PLI) was mainly used to get qualitative information, several methods have been presented which allow the extraction of a variety of quantitative data, such as lifetime [7], series resistance [8] or diffusion length [1]. Besides, calibration methods based on the use of two images taken under different polarization, illumination or detection conditions were introduced to overcome the measurement uncertainty [1, 7-11].

Besides all the advances on the application of the luminescent emission for solar cell characterization purposes, some approximations and simplifications are usually used, such as uniform generation and flat excess carrier profile. This assumption is justified by a near infrared excitation (800 nm - 850 nm wavelength [3-6]). With regard to the photon reabsorption, this is most of the times simply disregarded,

given that the induced error is considered within acceptable limits [13].

The aim of this work is to quantify the impact of the excess carrier profile on the detected luminescence. IR light is weakly absorbed, giving a quite uniform excess carrier profile, but shorter wavelength being more energetic, will be more rapidly absorbed within the sample, generating less as we go deep in it. Depending on the sample's lifetime related parameters, the obtained profile can show asymmetries even under IR excitation.

The sample will emit photons from any surface, and detection can, in principle, be carried out from the illuminated side or from the contrary. Under the flat excess carrier concentration assumption, both detections will be equal.

In this work, extreme scenarios for the carrier excess profiles are proposed to delimit the photon reabsorption induced differences that can be obtained when measuring front emission (in reference to the illuminated side) and rear.

II. ANALYSIS AND RESULTS

A. Influence of the generated carrier profile in the PL measurement

In the presence of a junction, characterization methods like suns-PL or QSS-PL use the luminescence intensity to extract the minority carrier excess at the junction (Δn_j) and a voltage using (1).

$$V_{impl} = V_T \cdot \ln\left(\frac{n \cdot p}{n_i^2}\right) \approx V_T \cdot \ln\left(\frac{R_{RAD}}{k \cdot B \cdot n_i^2}\right) \quad (1)$$

Where V_T is the thermal voltage, n_i the intrinsic carrier concentration, n the electron concentration, p the hole concentration, B the radiative recombination coefficient, R_{RAD} the emitted radiation and k a calibration constant.

If low level injection condition apply, with the minority carrier excess profile uniform along the sample volume, from x=0 to x=W, the approximation in (2) is valid, and the average minority carrier excess in the sample equals the excess at the junction.

$$\int_x n(x) \cdot p(x) dx \approx N_B \cdot \Delta n_j \cdot W \quad (2)$$

Where Δnj is the excess at the junction, N_B the base doping and W the sample thickness. In this case, approximation in (1) is valid and it may be used to extract the voltage, but it will not in a general case, as samples with identical Δn_j will give different minority carrier averages if they have different carrier profiles. If the average carrier excess is assumed to equal the carrier excess at the junction in the case of a non-uniform profile, there is an underestimation of the real Δn_j.

Assuming low injection level and a carrier profile changing only in the sample depth, the overall volume radiative recombination can be approximated by (3).

$$R_{RAD} = \iint\limits_{\lambda \ x} B \cdot n(x) \cdot p(x) dx d\lambda \approx B \cdot N_B \cdot W \cdot \Delta n_j \cdot fp \quad (3)$$

Here $fp = \int fp(x)dx$, is a factor, $0 < fp \leq 1$, related to the shape excess carrier profile. Figure 1 shows some limit cases: very high diffusion length (L) combined with low/high back surface recombination velocity (Sr) and very low diffusion length.

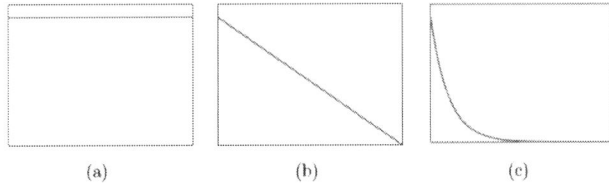

(a)   (b)   (c)

Fig. 1.   Excess profiles in the volume for three extreme cases: (a)P1: L>>W and low Sr, (b) P2: L>>W and high Sr and (c) P3: L<<W.

Table I gathers numerically what can be seen in figure 1, where the evolution of fp in the exponentially decaying profile (P3 profile) depends on the particular values of the diffusion length and sample width, but it always lie between 0 and 1.

TABLE I
VOLUME EXCESS CARRIER PROFILE INTEGRAL
APPROXIMATION (LOW INJECTION)

L (cm)	Sr (cm/s)	$\int n(x) \cdot p(x) \cdot dx$	f_p
high	low	$N_B \cdot n_j \cdot W$	1
high	high	$N_B \cdot n_j \cdot W/2$	1/2
low	$\forall$	$N_B \cdot n_j \cdot L_n[1 - exp(-W/L_n)]$	f=f(L,W)

B. Influence of the excitation wavelength on the PL measurement

As indicated in the previous section, the estimation of the union voltage from the luminescent intensity in EL and PL measurements may be a poor approximation in some cases.

Particularly in the PL case, the excess carrier profile will be conditioned, not only by the characteristics of the samples, but also by the excitation wavelength.

In figure 2 the resulting excess carrier profiles for three different cells under the same excitation wavelength are shown. Simulations have been carried using PC1D [14] and the lifetime and surface recombination values have been chosen to follow the previously described three scenarios. Other characteristics like doping or width are kept the same in all three cells.

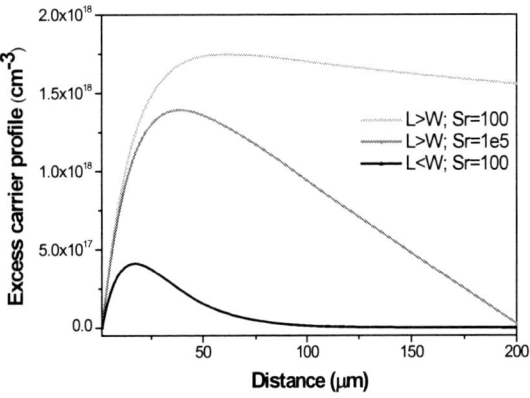

Fig. 2.   Different minority excess carrier concentration profiles due to the sample characteristics for 850 nm excitation.

Figure 3 shows how, for the same cell characteristics, absorption, and thus, excess profile, is illumination wavelength dependent. In this case a cell with good lifetime and surface recombination characteristics (P1 profile type cell) have been simulated, and exposed to four wavelengths in the VIS range.

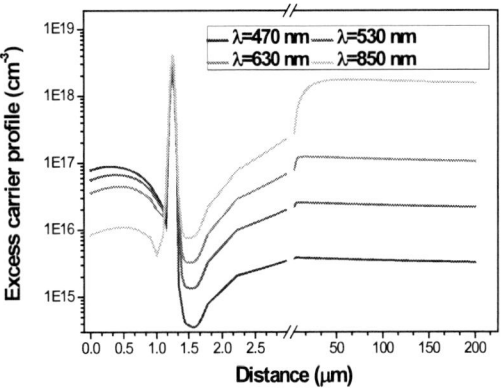

Fig. 3.   Photogeneration for four excitation wavelengths (470 nm, 530 nm, 630 nm, 850 nm) in a sample with L>W and Sr=100 (cm/s).

The cell response to different wavelengths may allow for quantum efficiency measurements. In the particular case of tandem cells, where different semiconductor layers allow a better use of the solar spectrum, the possibility of generating at different depths permits to infer pseudo IV curves of the superimposed subcells [12].

C. Quantification of the impact of the reabsorption in the front and rear PL detection (in relation to the illuminated side)

In the case of single semiconductor based cells, band to band recombination will emit photons of a determined energy, approximately equal to the semiconductor band gap. In the case of silicon, the luminescent emission is comprised between 900 nm and 1200 nm. Consequently, these photons have a poor reabsorption probability, making this effect negligible in general, with a low error [13].

This can be considered an adverse effect, but is not necessarily so: as in LBIC, generation at different depths and detection of the generated photocurrent allows for the determination of the diffusion length, is interesting how Würfel eta al. turn the concept on its head to determine the diffusion length from the reabsorption detected in the luminescence signal [1].

The reabsorption effect has been applied to the silicon emission spectrum, taking axis x starting from the illuminated side: $fa(\lambda,x)=exp(-\alpha(\lambda)\cdot x)$ for detection from the illuminated surface (front) and $fa(\lambda,x)=exp(-\alpha(\lambda)\cdot(W-x))$ for rear side detection.

Figure 4 shows the expected spectral distribution of the photons reaching the front and the rear surfaces for a P3 type profile. It is evident that the shorter the photon wavelength the greater the suffered reabsorption, as more energetic photons have more probability to be absorbed. Photodiodes or silicon CCD cameras are commonly used for luminescence detection, what can increase the aforementioned effect as their spectral response shows higher sensitivity to shorter wavelengths. Figure 5 the response of a silicon photodiode has been applied to the distributions in figure 4. The result shows the expected spectral distribution of the detected luminescence from the front and rear surfaces.

The impact of the photon reabsorption in the measurement is analyzed using the three extreme cases (excess carrier profiles presented in section A) and the following equation:

$$N_{fDET} = \iint_{\lambda\ x} B(\lambda)\cdot p(x)\cdot n(x)\cdot fa(\lambda,x)dxd\lambda \quad (4)$$

Where N_{fDET} is the number of detected photons, and $fa(\lambda, x)$ represents the absorption. Using the approximation for the profiles gathered in table I, equation (4) turns into (5).

$$N_{fDET} \approx N_B\cdot\Delta n_j\iint_{\lambda\ x} B(\lambda)\cdot fp(x)\cdot fa(\lambda,x)dxd\lambda \quad (5)$$

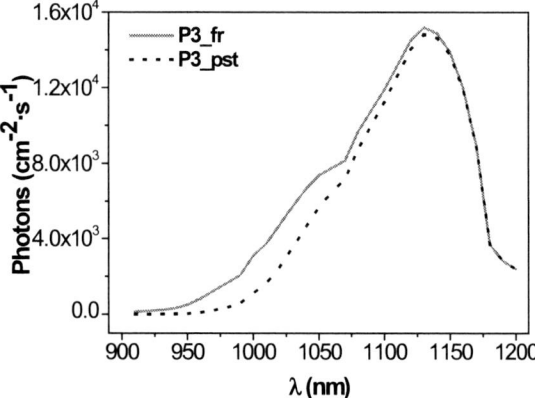

Fig. 4. Expected spectral distribution of photons reaching the front (fr) and rear (pst) surface of a W=200 μm sample with $\Delta n_j=10^{14}$ cm^{-3} for L<<W profile (P3).

To delimit the effect of photon reabsorption on the detection, the spectral sensitivity of the detector is added to the previous calculations. Figure 5 shows the result for the profile in which the effect is more evident.

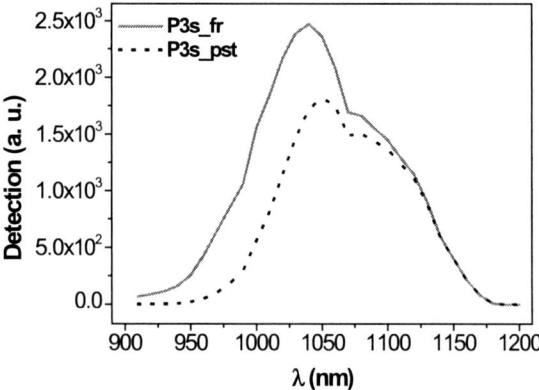

Fig. 5. Spectral distribution of photons detected with a silicon photodiode from the front (fr) and rear (pst) surface of a W=200 μm sample with $\Delta n_j=10^{14}$ cm^{-3} for the L<<W profile (P3).

The scenario in which the effect of photon reabsorption is more evident is the one shown in figure 5. For this extreme case the overall effect of the exponentially decaying profile and the detectors response, lead to a front side detection nearly 50% greater than rear side one.

Table II shows the values obtained for the relation between front and rear sides, with and without including the detectors response.

TABLE II

IMPACT OF PHOTON REABSORPTION IN THE RELATION BETWEEN FRONT AND REAR DETECTION

Front/rear	P1	P2	P3
Outgoing photons	1	1.06	1.15
Detected photons	1	1.17	1.46

III. ANALYSIS AND RELEVANCE OF THE RESULTS

In general, for luminescence based measurements, flat carrier excess profiles are considered. This approximation makes it easier to obtain the obtaining implied voltage maps from luminescence intensity images. But it has been shown that, due to the sample characteristics or to the excitation wavelength, profiles far from flat can result, and in the case of very steep profiles this approximation is not valid, as it underestimates the junction voltage.

On the other hand, the effect of photon reabsorption has been delimited by applying it to limit cases for the carrier excess profiles and calculating the front side (illuminated one) and rear side detection. The calculations show that, for the flat profile case, the detection side does not affect the result. Besides, in this case the averaged excess carrier equals the excess at the junction, making the approximations valid. But in the other two cases, the combined effect of the non flat profile and the photon reabsortion, can lead to a front side detection 46% higher from the illuminated surface compared to the rear one.

IV. SUMMARY

Being diverse the uses of the photoluminescence detection for the solar cell characterization, almost all of them have something in common: the measurement setup and conditions. On one side the excitation: a long wavelength, near IR, excitation is used, making more suitable the flat profile approximation. The detected luminescence gives information on the average of the excess carrier concentration, which, under these conditions, is supposed to equal the excess at the junction. Shorter wavelengths will lead to steeper profiles, but even with a near IR excitation the resulting profile can be rather far from flat, depending on the emitter recombination, bulk lifetime and surface recombination velocities of the sample. In these cases the flat profile assumption would underestimate the junction carrier excess concentration and the voltage.

This non flat profile would make it necessary to take into account the usually neglected photon reabsorption. The calculations indicate that in the case of a steep carrier excess profile, the impact of this reabsorption can lead to differences up to 46% between the front and the back side detection.

ACKNOWLEDGEMENT

This work has been supported by the MINECO/FEDER within the framework of the project ENE2014-56069-CA-1-R.

REFERENCES

[1] P. Würfel, T. Trupke, T. Puzzer, E. Schäffer, W. Warta, & SW Glunz, "Diffusion lengths of silicon solar cells from luminescence images" *Journalof applied physics*, 101:123110, 2007.

[2] T. Trupke, R. A. Bardos, M. D. Abbott, P. Würfel, E. Pink, Y. Augarten, F.W. Chen, K. Fisher, J.E. Cotter, M. Kasemann, et al. "Progress with luminescence imaging for the characterisation of silicon wafers and solar Cells" *Proceedings of the 22nd European Photovoltaic Solar Energy Conference*, 2007, pp. 22-31.

[3] T. Trupke, R. A. Bardos, M..C. Schubert, & W. Warta, "Photoluminescence imaging of silicon wafers" Applied Physics Letters, 89(4):044107, 2006.

[4] T. Trupke, & RA Bardos, "Photoluminescence: a surprisingly sensitive lifetime technique" in 31st IEEE Photovoltaic Specialists Conference, 2005, pp. 903-906.

[5] T. Trupke, R. A. Bardos, M. D. Abbott, & J. E. Cotter. "Suns-photoluminescence: Contactless determination of current-voltage characteristics of silicon wafers" Applied Physics Letters, 87:093503, 2005.

[6] T. Trupke, R. A. Bardos, M. D. Abbott, F. W. Chen, K. Fisher, & J. E. Cotter, "Luminescence imaging: an ideal characterization tool for silicon" in 16th Workshop on Crystalline Silicon Solar Cell Materials and Processes: Extended Abstracts and Papers, 2006, pp. 50-57.

[7] T. Trupke, B. Mitchell, J. W. Weber, & J. Nyhus, "Bulk minority carrier lifetime from luminescence intensity ratios measured on silicon bricks" in 25th European Photovoltaic Solar Energy Conference, 2010, pp. 1307-1311.

[8] T. Trupke, E. Pink, R. A. Bardos, & M. D. Abbott. "Spatially resolved series resistance of silicon solar cells obtained from luminescence imaging" Applied Physics Letters, 90:093506, 2007.

[9] T. Trupke, R. A. Bardos, & M. D. Abbott, "Self-consistent calibration of photoluminescence and photoconductance lifetime measurements" Applied Physics Letters, 87(18):184102, 2005.

[10] J. A. Giesecke, M. Kasemann, & W. Warta, "Determination of local minority carrier diffusion lengths in crystalline silicon from luminescence images" Journal of Applied Physics, 106(1):014907, 2009.

[11] S. Herlufsen, K. Ramspeck, D. Hinken, A. Schmidt, J. Müuller, K. Bothe, J. Schmidt, & R. Brendel, "Dynamic lifetime imaging based on photoluminescence measurements" in 25th European Photovoltaic Solar Energy Conference, 2010, pp. 2369-2373.

[12] X. Bubnova, "Contactless characterisation of III-V multi-junction solar cells using laser spectroscopy". in 25th 25th European Photovoltaic Solar Energy Conference, 2010, pp. 505-507.

[13] T. Trupke, "Influence of photon reabsorption on quasi-steady-state photoluminescence measurements on crystalline silicon". *Journal of applied physics*, 100:063531, 2006.

[14] D.A. Clugston and P. A. Basore. "PC1D version 5: 32-bit solar cell modeling on personal computers" in 26th IEEE Photovoltaic Specialists Conference, 1997, pp.207-210.

Development of outdoor luminescence imaging for drone-based PV array inspection

Gisele A. dos Reis Benatto[1], Nicholas Riedel[1], Sune Thorsteinsson[1], Peter B. Poulsen[1], Anders Thorseth[1], Carsten Dam-Hansen[1], Claire Mantel[1], Søren Forchhammer[1], Kenn H. B. Frederiksen[2], Jan Vedde[3], Michael Petersen[4], Henrik Voss[5], Michael Messerschmidt[5], Harsh Parikh[6], Sergiu Spataru[6] and Dezso Sera[6]

[1]Department of Photonics Engineering, Technical University of Denmark, Frederiksborgvej 399, 4000, Roskilde, Denmark
[2]Kenergy, Grønningen 43, 8700, Horsens, Denmark
[3]SiCon Silicon & PV consulting, J N Vinthersvej 5, 3460, Birkerød, Denmark
[4]Skive Kommune, Torvegade 10, 7800 Skive, Denmark
[5]Sky-Watch A/S, Østre Alle 6 Støvring, Nordjylland, 9530, Denmark
[6]Aalborg University, Aalborg, 9220, Denmark

Abstract — **In this work we investigate and present preliminary results for two methods for luminescence imaging of photovoltaic (PV) modules in outdoor conditions, with the aim of choosing the most suitable method for implementation on a drone PV plant inspection system. We examined experimentally both electroluminescence (EL) and photoluminescence (PL) PV module imaging methods under natural light conditions, and determined that fast pulsed EL imaging with InGaAs detector cameras can yield reasonably accurate results under daylight conditions. Moreover, we formulated the necessary requirement for a PL light source, which would allow PL imaging of modules under daylight conditions.**

Index Terms — **drone-based PV inspection, electroluminescence imaging, image processing, outdoor defect detection, photoluminescence imaging.**

I. Introduction

In order to ensure expected return on investment (ROI) of small and large-scale photovoltaic (PV) installations, regular fault detection for effective maintenance, is highly important. Present day PV panels are designed to operate for 25-30 years, however field experience shows that after 11-12 years of operation 2% or more of all PV panels fail [1]. However, the failure rate is even higher for older installations, especially those manufactured before the year 2000 [2].

In practice, the frequency and inspection detail level is often limited by manpower and cost. Presently, drone-based infrared (IR) thermography inspection of solar plants is a reality, and the technology is expected to develop further into automated solar plant inspection [3]–[5]. The accuracy of thermographic fault detection though, presents limitations – primarily related to deconvoluting the failure signature into failure type and severity, which can be overcome when performed in combination to electro-(EL) or photo-(PL) luminescence imaging of the panels. The combination of defect detection techniques has been already tested in laboratory [1], [6], although many limitations still need to be addressed in order to obtain image acquisition outdoors and integrate, automatize and optimize the imaging system in a drone.

In this work, we investigate and present preliminary results for two methods for luminescence imaging of PV modules in outdoor conditions, with the aim of choosing the most suitable method for implementation on a drone PV plant luminescence inspection system. First, we investigate a pulsed EL imaging method under daylight conditions, to determine the necessary camera and measurement parameters. In the second part, we examine a PL imaging method under natural low light conditions, do determine if PL imaging would be feasible for outdoor PL imaging, along with the necessary light source requirements. The concept of PL/EL in a drone is illustrated in Fig. 1.

Fig. 1.   Sketch of the concept of automatized drone inspection.

II. Experiment and Methodology

The experimental tests performed in this work are focused on investigating EL and PL imaging techniques that are suitable for implementation into a drone-based inspection system. The PL technique avoids the need for electrical contact into the solar panels, which is a time limiting factor for drone-based inspection, especially in large-scale solar plants.

978-1-5090-5606-4/17 $31.00 © 2017 IEEE

The luminescence emission peak for silicon-based solar cells at ambient temperature is at 1150 nm [7], near a water absorption band in the solar spectrum (AM 1.5), as illustrated in Fig. 2. In the same figure, the quantum efficiencies of two camera detectors: i) a cooled Si charge-coupled device (CCD) ii) and short-wave infrared (SWIR) InGaAs, able to detect the emission peak, are plotted. Comparing the curves with the peak positions it is clear that a CCD camera can acquire only a small portion of the emission peak. At the same time, the SWIR InGaAs can detect the emission entirely, providing fast integration times, ideal for drone operation. Additionally, the InGaAs detector avoids the most intense section of the solar spectrum. Even though, to avoid the relatively intense sunlight, a sharp optical band-pass filter is used, with the transmission wavelength illustrated as the white area in Fig. 2.

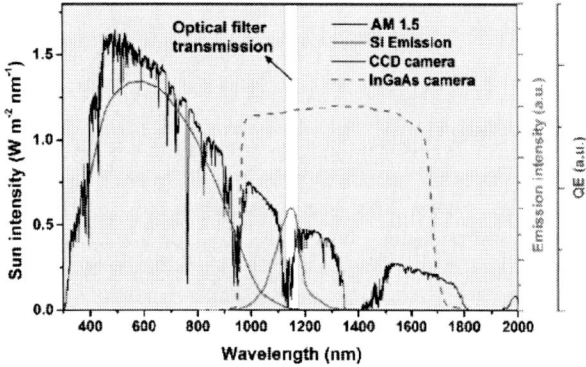

Fig. 2.   AM 1.5 sun spectrum (black line), CCD (solid blue line) and InGaAs (dashed blue line) camera QE curves, and silicon emission peak (red line). The grey areas show wavelengths that are cut off with the use of an optical filter in order to avoid detection of the sunlight.

A. Electroluminescence

The EL images shown in this paper are acquired from a mechanically stressed 36 cell multicrystalline silicon solar panel with 1 x 1 meter dimension. An InGaAs camera from Hamamatsu model C12741-03, and an OD>4.0 1150nm band-pass filter with 50nm FWHM was used to obtain all EL images.

A sequential image acquisition system was implemented in order to enhance the quality of the images obtained at high noise level during the day. Such system synchronizes the image acquisition with an electrical forward bias applied by a DC power supply. Fig. 3 illustrates the synchronizing circuit, driven by an Arduino logic controller, and the pulse width modulated (PWM) waveforms applied to the PV panel and camera. The exposure time is established separately in the camera software.

To estimate the effect of the sun on the imaging process and to better understand the noise characteristics towards an InGaAs detector and develop image processing strategies, we acquired sequences of 100 images (50 under forward bias (signal) and 50 as background images for subtraction) at 6 Hz with 20 ms exposure time, under 300, 500, 600, and 800 W/m²

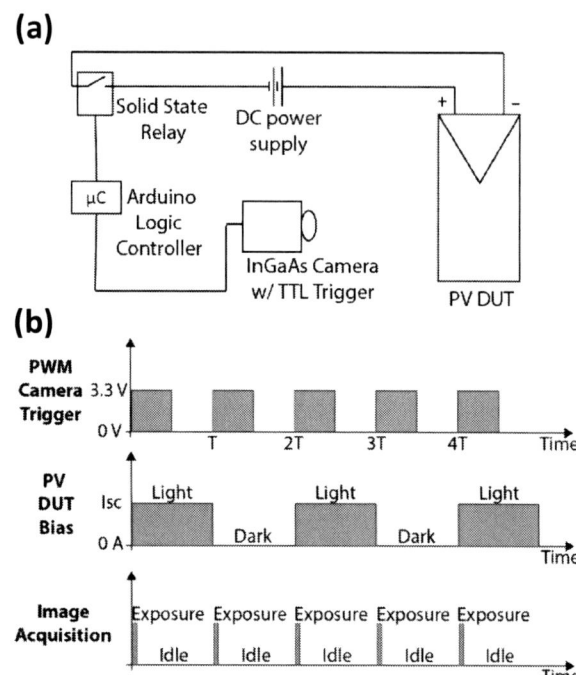

Fig. 3.   Basic trigger circuit that synchronizes the imaging and forward bias from a DC power supply (a); and the PWM waveforms driven by the Arduino logic controller (b).

global horizontal irradiance (GHI). The solar irradiance was acquired by a weather station [8] located just few meters from the imaging position. The iris aperture was f4 for 300 W/m² and f8 for the remaining. The images were acquired at around 2.6 meters from the panel.

As averaging several pictures is the most common way the minimize noise from images, we perform it the enhance image quality. The image processing included taking the difference between the average of light and dark images, automatic stretching of the initial dynamic range (the source images are 16bits) to 8 bits for display and cropping of the whole image to the region of interest. T-tests of the images pixel values were performed for better understanding of signal the noise ratio (SNR).

B. Photoluminescence

PL images indoors were acquired using a laser diode at 800 nm with capacity of 13 W maximum optical power, while the camera and the laser were circa 0.5 meter far from the sample. The sample consisted of a multicrystalline silicon cracked PV cell. The same camera and filter used for EL was used to obtain PL images. The images were acquired at 700 ms exposure time for 54, 71, 87, 104 and 120 W/m² light intensity on the plan of the cell, which corresponded respectively to 3, 4, 5, 6 and 7 W of optical power from the laser.

III. RESULTS AND DISCUSSION

A. Electroluminescence

EL imaging during the night is comparable to indoors EL imaging, where normally there is no significant ambient light noise levels. For comparison, Fig. 4 shows the indoors EL image of the mechanically stressed module. However, during the day such images are surrounded by high, and very often variable, light noise levels, primarily from the sunlight itself.

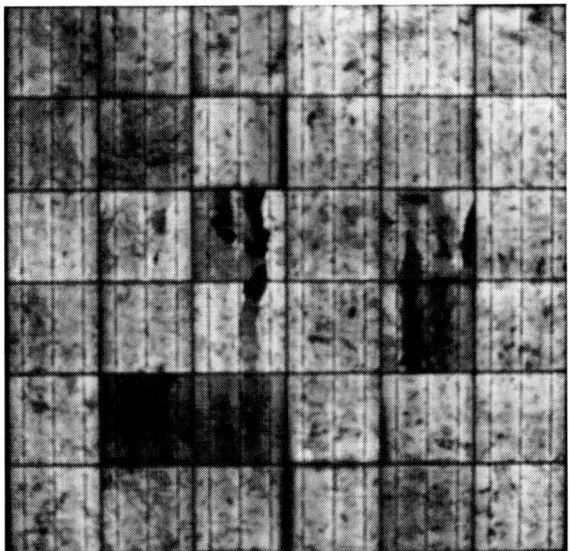

Fig. 4.   Indoors EL image of mechanically stressed PV module, showing cracks and disconnected cell areas.

Fig 5a shows the EL images acquired under different sun irradiation levels, after averaging, subtracting, and image processing. The irradiance on the plane of the module array (POA) was 117.6, 332.3, 467.8, and 714.5 W/m^2 respectively; using the Hay Davie's model for diffuse irradiance on a tilted surface [9]. At 800 W/m^2 for example, there was a strong direct beam irradiance when the image was taken, which did not limit the EL image quality after processing. All the final images present roughly similar information for broken cells and shunt defects present in the module and detected in Fig. 4.

The plots in Fig. 5b show the two sample t-tests that compare the image pixel values under dark (0 current bias) and light (I_{SC}

bias) conditions for each test irradiance. The y-axis show the summation of pixel values in the InGaAs detector (512x640), where the value of each pixel is a value between 0.0 (i.e. completely dark) and 1.0 (i.e. completely saturated).

Each t-test shows data from 50 light and 50 dark images at a given irradiance. The green diamonds show the 95% confidence interval of the mean wherein the horizontal green line shows the mean and the top and bottom corners show the confidence interval. The dark and light pixel distributions are considered significantly different when the confidence intervals do not overlap. The t-tests show that the dark and light pixel distributions are significantly different ($p < 0.05$) for all irradiances except for 500 W/m^2 ($p = 0.06$). The difference between dark (background) and light (I_{SC} bias) image means illustrates how it was possible to obtain the images in Fig. 5a. As the difference decreases with the irradiation level, the SNR was lower, yet with significant difference.

The sun intensity though constantly varies during the image acquisition, dependent on to the time of the day and cloud cover. Fig. 5c show the time series of light and dark images sequences, as their pixel values changed during the circa 17 seconds that sequence took to acquire. Such variation is directly related to the ambient illumination (sun intensity) variation as it equally affects both the light and background reference (dark) images. Even though irradiance fluctuations bring challenges in particular situations, such variations did not impose visual limitations in the daylight outdoor EL imaging after processing. Nevertheless, automatic aperture adjustment will be required to avoid image saturation during drone inspection.

Averaging several pictures is the most common way the minimize noise from images. Although, to take several pictures of the same scene, if the system camera-power supply is not fast enough, it can be limiting for the drone movement. Fig. 6 shows the resulting averaged and subtracted images when different amounts of light and dark images are used under. All images in Fig. 6 were taken under 300 W/m^2 illumination in natural sunlight. The average of 8 light and dark images (16 in total images), show a good level of noise removal. For this, it is required that the drone keeps position for 0.32s, with a power supply as fast as the camera triggering for 20ms exposure time. This is achievable if the drone is equipped with an appropriate camera stabilization gimbal, in addition to performing digital imaging stabilization on the acquired images. However, fewer averaged pictures does allow the detection of major defects in this example.

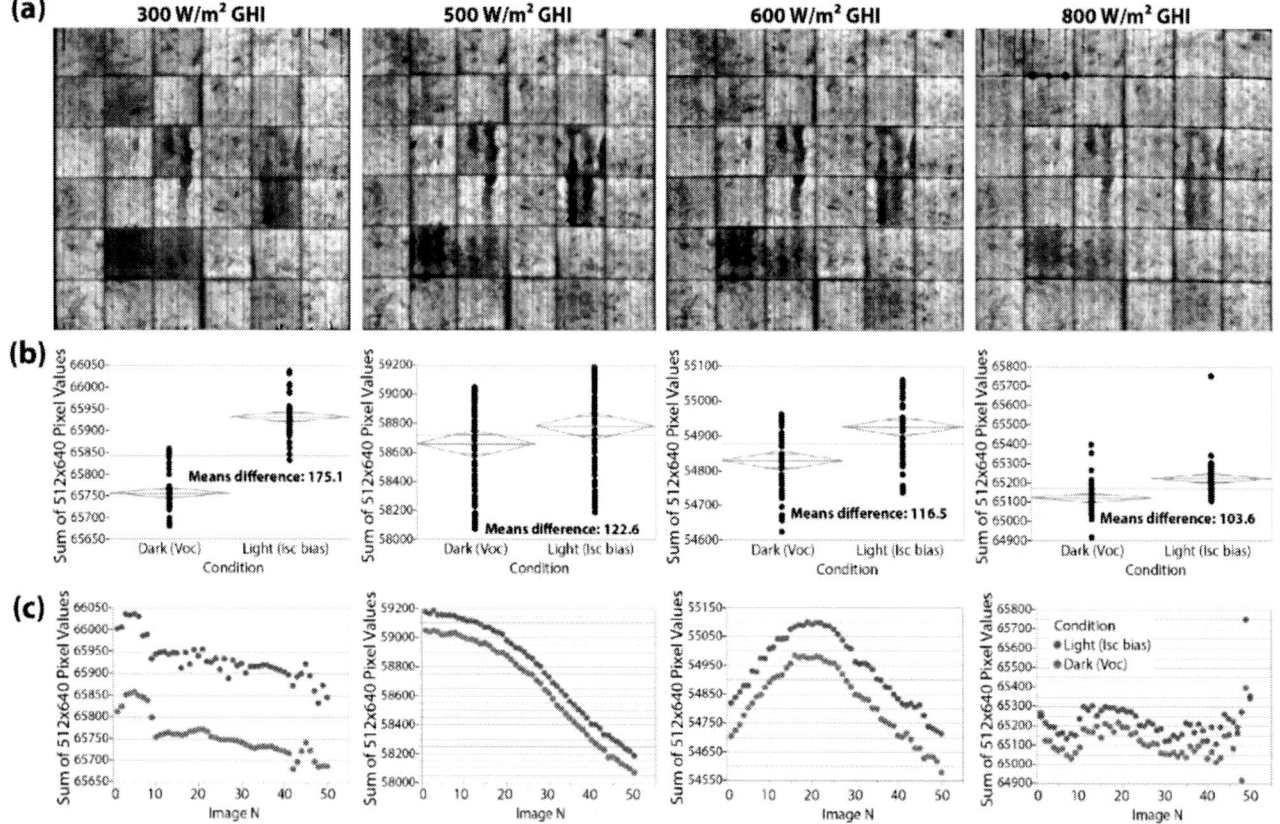

Fig. 5. EL images obtained under different sun irradiation levels, after image processing, automatic cropping and contrast correction (a); sum of pixel values of dark and light images of the correspondent image above, with the mean and 95% confidence interval shown as green diamonds (b); correspondent time series of light (blue dots) and dark (red dots) images, showing variations of light intensity during the acquisition of the sequence of images (c).

B. Photoluminescence

In addition to the outdoors EL imaging tests, laboratory PL tests were performed. Fig. 7 show the images acquired at 700 ms exposure time for 54, 71, 87, 104 and 120 W/m² light intensity on the plan of the cell. Therefore, long exposure times were required for such low illumination rates. As the intensity of a light source complies with the inverse-square law, even a powerful light source will have limited maximum distance from the panels. At the same time, such powerful light source will require cooling, which poses size and weight challenges for the drone.

For the development of the best strategy of minimum weight and best light source for PL outdoor imaging, a measurement modeling was developed, in order to correlate the relevant variables of the system. Such model for the PL image signal generated by an artificial light source (S) in arbitrary units can be expressed as the equation below:

$$S = \frac{1}{D^2} \frac{P_{optical}}{A} \tau \, \eta_{PL}(\lambda_{Light}) \, \rho_{Camera}(\lambda_{PL}) \qquad (1)$$

Where D is the distance from sample (panel) to the light source; $P_{optical}$ is the optical power output of the light source which is related to the electrical power input and device

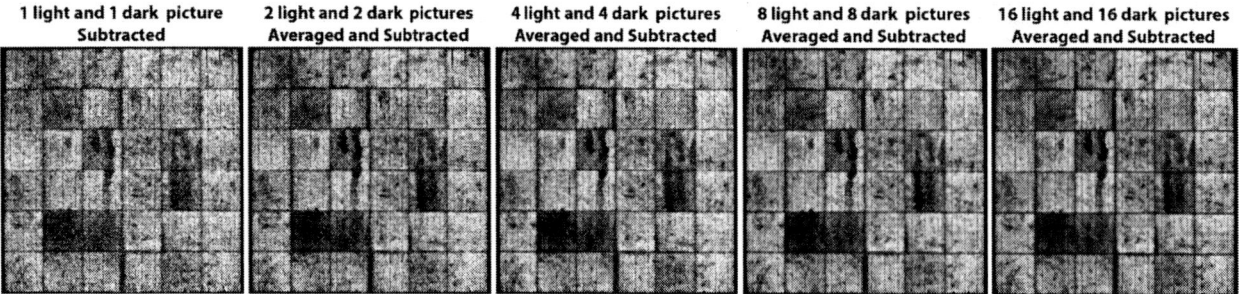

Fig. 6. Averaged and subtracted EL pictures of a solar panel acquired outdoors under 300 W/m² sun illumination.

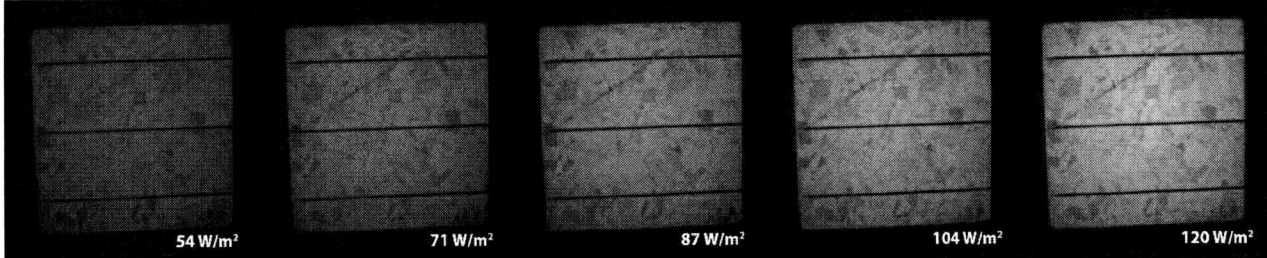

Fig. 7. PL images taken using 800nm laser diode, under 700ms exposure time, from a polycrystalline silicon wafer under different illumination.

efficiency, leading to known heating and consequentially the need for certain cooling. A is the area to be illuminated, contained by designed optics in order to avoid light loss. τ is the exposure time or the time required for the light to be on, here correspondent also to the camera exposure time. $\eta_{PL}(\lambda_{Light})$ is the PL quantum yield, related to the absorbed photons in the PV module (PV/silicon quantum efficiency) at the given wavelength of the light source and reemitted in the same direction as the incident light. $\rho_{Camera}(\lambda_{PL})$ is the camera acquisition factor that correlates quantum efficiency, dynamic range, among other sensor and camera designed features, at a certain wavelength, in this case correspondent to the luminescence signal wavelength (centred at 1150 nm).

With such vision of the system, it was possible to correlate the signal response measured in the laboratory with the needed requirements to build a drone integrated light source that will allow outdoor PL imaging. Taking into consideration the available technologies, a laser line scan following the drone movement is the one that complies sufficiently with optical power and current image acquisition requirements. In this case, the area needed to be illuminated is smaller, making the $\frac{P_{optical}}{A}$ factor lower. Fig. 8 illustrates the above described approach.

According to (1), the laser line scan approach with current available technology will able to acquire the similar signal presented in Fig. 7 (71 W/m^2) at three meters distance of a full size module with 20 ms exposure time. Another example would be a fast pulsing high power laser, supplied by capacitors. The approaches as pulsing laser and LEDs so far presented too low signal intensity for the current setup, but as the whole system is in development, they are not discarded.

V. CONCLUSION

The EL imaging performed for PV inspection during the day under high sun intensity address the possibility of performing EL imaging inspection with more freedom, during more hours of the day, and simultaneous IR and EL. In a drone system, the compatibility will remain the same for either daylight or nighttime EL. In future work the forward bias and camera triggering will be carried out via wireless communication. In addition, we will focus efforts on improving imaging processing, which can be done automatically and without losing flight time.

The next step for this analysis is to develop a controllable test bed with a moving camera and check the limit of frames from a video taken at certain speeds. The camera used in this work has the maximum frame rate of 60 fps, which is relatively low for this application. Consequently, a faster camera will allow more pictures to be taken in a shorter period, and the future tests will define how fast the drone can move while it takes different amounts of pictures for the image processing.

The PL indoor measurement parameters permitted the measurement modeling for the light source development, which indicates a line laser scan as the most promising light source for outdoor PL and drone integration.

ACKNOWLEDGEMENT

The authors acknowledge the financial support from Innovation Fund Denmark for the project 6154-00012B DronEL – Fast and accurate inspection of large photovoltaic plants using aerial drone imaging.

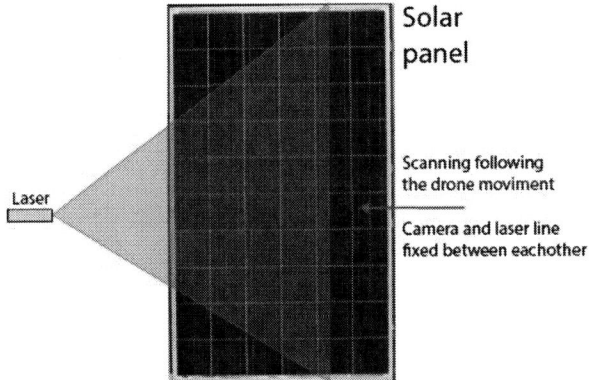

Fig. 8. Laser line scan PL approach.

REFERENCES

[1] M. Köntges *et al.*, "Review of Failures of Photovoltaic Modules," 2014.

[2] D. C. Jordan and S. R. Kurtz, "Photovoltaic degradation rates - An Analytical Review," *Prog. Photovoltaics Res. Appl.*, vol.

21, no. 1, pp. 12–29, 2013.

[3] S. Dotenco *et al.*, "Automatic detection and analysis of photovoltaic modules in aerial infrared imagery," *2016 IEEE Winter Conf. Appl. Comput. Vis.*, pp. 1–9, 2016.

[4] "The drones that inspect and maintain photovoltaic power plants," *sUAS News - The Business of Drones*. [Online]. Available: https://www.suasnews.com/2015/04/the-drones-that-inspect-and-maintain-photovoltaic-power-plants/. [Accessed: 25-Jan-2017].

[5] "Drones to cut plant inspection costs as South Africa eyes quality," *PV Insider*. [Online]. Available: http://analysis.pv-insider.com/drones-cut-plant-inspection-costs-south-africa-eyes-quality. [Accessed: 25-Jan-2017].

[6] S. Johnston and T. Silverman, "Photoluminescence and Electroluminescence Imaging Workstation," *NREL*, 2015.

[7] W. S. Yoo, K. Kang, G. Murai, and M. Yoshimoto, "Temperature Dependence of Photoluminescence Spectra from Crystalline Silicon," *ECS J. Solid State Sci. Technol.*, vol. 4, no. 12, pp. P456–P461, 2015.

[8] DTU Fotonik, "DTU Risø PV Weather Data." [Online]. Available: http://ftnk-psolmaal.win.dtu.dk/B130_PV_Weather/index.html. [Accessed: 18-May-2017].

[9] J. E. Hay and J. A. Davies, "Calculations of the solar radiation incident on an inclined surface," in *Proc. of First Canadian Solar Radiation Data Workshop*, 1980, p. 59.

Climbing Drum Peel (CDP) Test Method for Characterizing Adhesion in Flexible PV Modules

Venkata Bheemreddy and Kedar Hardikar

MiaSole Hi-Tech Corp., Santa Clara, CA, 95051, USA

Abstract — Characterization of adhesion in flexible PV modules during the product development and field exposure is critical to ensure long-term adhesion in the field over a typical warranty period of 25 years. Currently, no established standard test is available for a quantitative evaluation of adhesion in flexible PV modules. In this work, a Climbing Drum Peel (CDP) test is proposed as a preferred method for this purpose. The proposed CDP test method is demonstrated on a flexible PV module laminate configuration subjected to varying levels of environmental exposure including field exposure. With a simpler sample preparation and fracture mechanics based analysis of test results, CDP can potentially become a preferred test standard for emerging flexible PV products.

Index Terms — adhesion, climbing drum peel, environmental exposure, flexible PV, fracture mechanics.

I. INTRODUCTION

Flexible PV modules rely on packaging materials to ensure their durability in harsh operating environments [1]-[2]. This packaging provides protection for the solar cells and other electrical components from the operating environment thereby enabling long service life. With ever growing choices of packaging materials available, conducting rigorous experimentation for evaluating these materials during the product development cycle is critical. In building a flexible PV module, several thermoplastic polymer layers are laminated together, encapsulating the solar cell and other electric components. During lamination, all the layers in the stack are adhesively bonded to each other. The quality of this bond is crucial for ensuring the module performance. Delamination of the packaging materials is a major safety concern which can lead to exposing the solar cell and electrical circuitry to the outside environment. Delamination can be a potential reason for field failures – more than 90% of the customer-returns of conventional PV modules are related to delamination failures based on a study conducted by Kleiss et al. [3]. The fact that such a failure mode was not detected during the product development cycle indicates that current testing methods are not rigorous or the requirements are not adequately setup. For emerging flexible PV products the challenge is even more pronounced. Since the products are recently introduced, sufficient field data is not expected to be available. This is complicated further by the fact that no established standard test method is agreed upon for characterizing adhesion in flexible PV modules while some effort is underway for conventional PV modules [4]-[5].

Peel tests such as 180° peel, 90° peel and T-peel are typically used for characterizing adhesion of packaging materials in conventional PV modules. However these methods are considered qualitative in nature attributing to their unrealistic mode of failures compared to field failures. Peel tests typically use peak load and average load per unit width of bond as measures of bond strength. It is to be noted here that these measures are only relative measures of bond strength and do not correspond to an intrinsic material value. On the contrary, fracture mechanics based quantity such as the critical strain energy release rate (G_{IC}) is a characteristic measure of the bond strength. It is well known that the peel tests have to conform to specific conditions including the elastic deformation hypothesis for the adherend such that the fracture energy can be estimated from the measured loads [6]. However, this is particularly not the case when polymeric materials are involved in which, such methods suffer from plastic deformation of the adherend and severe plastic deformation at the leading delamination front that dominates the fracture behavior and the associated characterization of adhesion.

Double Cantilever Beam (DCB) test has been widely used as a suitable test method for measuring G_{IC}. One of the difficulties here is that the G_{IC} computation is a function of crack length and as a result a specialized setup is required to monitor the delamination crack front during the test [7]. While secondary approaches such as compliance based techniques exist which eliminate the need to measure the crack length during testing, it is desired to have a test method that is mechanics based and is independent of crack length for measuring G_{IC}. Methods such as corner adhesion test and tapered beam method have been proposed to eliminate the need for measuring the crack length for computing G_{IC} [5] however these methods including DCB do not enable exposing the samples as-is in a consistent manner during environmental testing thereby evaluating adhesion as a function of environmental exposure (temperature, moisture and UV) is not trivial. Considering all these factors, the above mentioned techniques are not preferred for characterizing adhesion in flexible PV modules. Desired characteristics of an adhesion test method for flexible PV products are:

1. Fracture mechanics based test that characterizes G_{IC}
2. Simple sample preparation
3. Enables environmental exposure in a consistent manner

4. Enables testing of indoor test samples and field exposed samples in a consistent manner.

In this work, a Climbing Drum Peel (CDP) test [7]-[9] is proposed as a preferred method for characterizing adhesion in flexible PV modules and measuring G_{IC}. CDP is a popular test method in the aerospace and adhesive communities. Unlike the DCB test, CDP does not require crack length measurements for computing the G_{IC} and enables exposing the samples in environmental testing where primary moisture and light paths are not altered. In CDP test the crack propagation is proportional to the crosshead displacement which is not the case with the DCB test. Sample preparation and testing are relatively simple and testing can be conducted in a similar manner on coupons and at module level. In this work, the proposed CDP test is demonstrated on flexible PV module specimens subjected to different levels of environmental exposure which includes the field exposure.

The paper is organized as follows. In section 2, the CDP test method is described. Section 3 discusses the samples for adhesion measurement using CDP test and associated results. Section 4 concludes the paper with a viable outlook.

II. CLIMBING DRUM PEEL (CDP) TEST METHOD

A representative schematic of the CDP test is shown in Fig. 1. The setup used in this work consists of a flanged drum, a flexible adherend (specimen peel arm), a rigid adherend, flexible cables, suitable clamps and a testing machine capable of applying tensile loads. The peel arm is secured to the drum using a drum clamp and the specimen is glued to the rigid adherend which provides the required stiffness to prevent the specimen from bending during the test. The top of the rigid adherend and the bottom of the flexible cables are clamped and connected to an Instron testing machine. The bottom clamp is constrained from moving while a constant displacement rate is applied on the top clamp and the corresponding force is measured using the Instron data acquisition system. When the external load is applied, the drum tends to climb along the specimen due to the inherent nature of the CDP setup and thereby propagating the debond. It is assumed here that the peel arm conforms well to the radius of the drum.

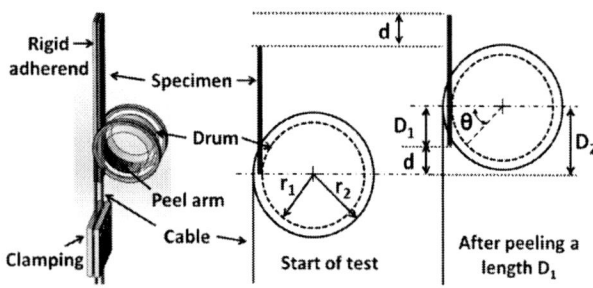

Fig. 1. Representative schematic of the CDP test setup.

Fig. 2 shows a typical load-displacement plot for the CDP test. It is characterized by:

(a) a load P_w which corresponds to the load required to overcome the drum weight and to wind the peel arm around the drum.

(b) a load P_d which includes the load contribution due to debond propagation.

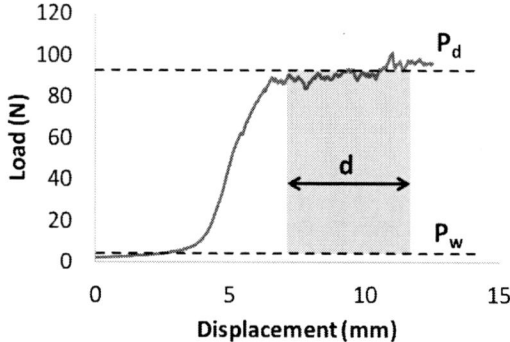

Fig. 2. Typical load-displacement plot for a CDP test.

For the CDP test, ASTM D1781 suggests bond strength evaluation using 'peel torque per unit width', given by:

$$T = \frac{(P_d - P_w)(r_2 - r_1)}{w} \qquad (1)$$

Where 'T' is the peel torque per unit width, 'w' is the width of the specimen, 'r_1' is the radius of the drum + one half the peel arm thickness, and 'r_2' is the radius of the flange. Although this measure enables evaluating bond strength of different adhesive interfaces, it is only a relative measure and does not correspond to a characteristic material value. An alternative is to measure the mode I critical strain energy release rate (G_{IC}) which is an intrinsic material property. The energy needed to debond the interface of interest can be estimated by measuring the shaded area in Fig. 2. G_{IC} can be computed from this by dividing this energy by the surface area of the delaminated interface on the specimen.

$$G_{IC} = \frac{(P_d - P_w) \times d}{wD_1} \qquad (2)$$

Where 'd' is the Instron frame displacement and 'D_1' is the increase in debond length. From Fig. 1, the total drum displacement (D_2) can be expressed in terms of 'D_1' and 'd' as follows:

$$D_2 = D_1 + d \qquad (3)$$

Here both 'D_1' and 'D_2' can be expressed in terms of the angle of rotation of the drum (θ) as follows:

$$D_1 = r_1\theta \qquad (4)$$

$$D_2 = r_2\theta \qquad (5)$$

From Eq. 4 and Eq. 5,

$$D_2 = \frac{r_2}{r_1} D_1 \qquad (6)$$

Substituting Eq. 6 in Eq. 3 and rewriting,

$$D_1 = \frac{r_1}{r_2 - r_1} d \qquad (7)$$

Combining Eq. 2 and Eq. 7 gives:

$$G_{IC} = \frac{(P_d - P_w) \times (r_2 - r_1)}{w r_1} \qquad (8)$$

As shown in Eq. 8, computing G_{IC} requires measuring the loads P_w, P_d, and knowing the geometry specifications of the specimen and CDP setup. Unlike the DCB test, no secondary measurements such as the progressive crack length are required here.

III. ADHESION MEASUREMENT AND ANALYSIS

Fig. 3a shows a representative flexible PV module, and the adhesive interface explored during product development for adhesion characterization using CDP test is shown in Fig. 3b. As mentioned earlier, the PV module materials need to survive harsh climates, ensuring reliable performance over a typical warranty period of 25 years. While the highlighted region (tab/front barrier interface) in Fig. 3b poses no concern related to leakage current based on rigorous reliability testing conducted in-house, it is of interest to quantify the adhesion at the candidate material interface due to varying levels of environmental exposure.

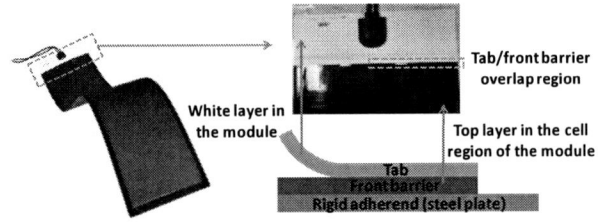

(a) Representative flexible PV module   (b) CDP sample configuration

Fig. 3.   CDP test sample configuration.

Four sets of test samples were chosen for this study. Table 1 provides the environmental conditioning history in each of these test sample sets. Test specimens are evaluated for adhesion using the CDP test with a cross-head displacement rate of 1 mm/min. The measured peel loads were found to be repeatable for different levels of environmental exposure. Using the average peel load, the values of G_{IC} for all the test samples were computed using Eq. 8 and compared as shown in Fig. 4.

TABLE I
SAMPLE SETS FOR CDP TEST

Sample Set	Environmental Conditioning
Set 1	No environmental exposure
Set 2	Damp Heat 1000 hours, 85°C/85% RH, IEC 61646
Set 3	2 years field exposure Santa Clara, CA, USA
Set 4	Damp Heat 3000 hours, 85°C/85% RH, 3X IEC 61646

Set 1 samples with no environmental exposure were chosen as the reference for adhesion evaluation. As shown in Fig. 4, Set 2 samples with Damp Heat 1000 hours (DH1000) exposure have shown some drop in adhesion relative to Set 1, which was measured to be 4%. Set 3 samples with 2 years field exposure however, have not shown any signs of degradation in adhesion relative to Set 1. Finally, Set 4 samples with Damp Heat 3000 hours (DH3000) exposure have shown ~69% degradation in adhesion relative to Set 1. The results indicate that laminate interface used in this study was robust for DH1000 exposure which is the industry standard for evaluating resistance to long-term penetration of humidity as defined by International Electrotechnical Commission (IEC)-61646. However, by extended exposure of the sample to DH3000 which is 3X industry standard stress level adhesion level had dropped significantly. Additional testing of samples with temperature conditioning and UV is an ongoing study. This data, when available, can be compared with the data presented in this study to understand the sensitivity of the candidate interface adhesion to temperature, humidity and UV.

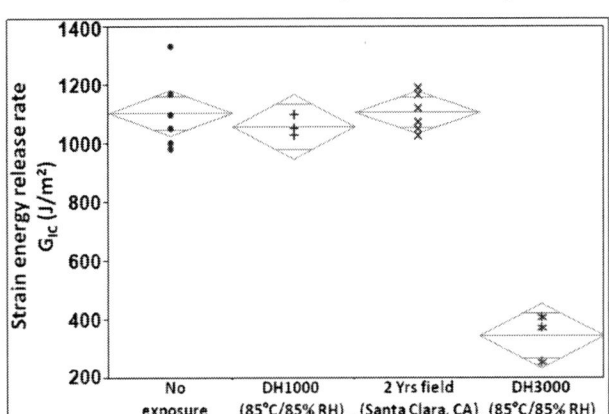

Fig. 4.   Computed strain energy release rates using CDP test for different levels of environmental exposure.

IV. CONCLUSIONS

In this work, a Climbing Drum Peel (CDP) test was proposed as a preferred method for characterizing adhesion in flexible PV modules and measuring G_{IC}. This method, unlike the DCB test, does not require crack length measurements for computing the G_{IC}; and a major advantage over other adhesion test methods is that CDP enables exposing the samples as-is in a consistent manner during environmental testing. The method involves a simple sample preparation and testing can be conducted on coupons and at module level in a consistent manner. Testing is straight forward and can be done using common mechanical testing equipment. Flexible PV module samples with different levels of environmental exposure, including the field exposure, were evaluated for adhesion using the proposed CDP test and the measurements were repeatable. Additional testing is ongoing for characterizing adhesion in samples exposed to temperature conditioning and UV. Overall, the proposed CDP test can bring value to the emerging flexible PV products by enabling characterization of adhesion during the product development stage and field exposures, thereby ensuring long-term adhesion over a typical warranty period of 25 years.

REFERENCES

[1] J. Zhu, G. Surier, D. Wu, D. Montiel-Chicharro, T. R. Betts, and R. Gottschalg, "Adhesion requirements for photovoltaic modules of polymeric encapsulation," in *Proceedings of 2016 IEEE International Reliability Physics Symposium*, 2016, pp. 756-760.

[2] G. J. Jorgensen, K. M. Terwilliger, M. D. Kempe, and T. J. McMahon, "Testing of packaging materials for improved PV module reliability," in *31st IEEE Photovoltaic Specialists Conference*, 2005, pp. 1-4.

[3] G. Kleiss, J. Kirchner, and K. Reichart, "Quality and reliability - sometimes the customer wants more," in *Proceedings of PV Module Reliability Workshop*, 2015, p. 1.

[4] N. Bosco, S. Kurtz, and R. H. Dauskardt, "A fracture mechanics based approach for adhesion testing in the PV module laminate," in *Proceedings of PV Module Reliability Workshop*, 2015, p. 1.

[5] N. Bosco, "Moving the PV industry to a quantitative adhesion test method," in *3rd Atlas/NIST Workshop on PV Materials Durability*, 2015, p. 1.

[6] K. S. Kim and N. Aravas, "Elasto-plastic analysis of the peel test," *International Journal of Solids and Structures*, vol. 24, pp. 417-435, 1988.

[7] F. Daghia and C. Cluzel, "The Climbing Drum Peel test: an alternative to the Double Cantilever Beam for the determination of fracture toughness of monolithic laminates," *Composites: Part A*, vol. 78, pp. 70-83, 2015.

[8] ASTM D1781, "Standard test method for Climbing Drum Peel for adhesives," *ASTM International*, 2004.

[9] A. Nettles, E. D. Gregory, and J. R. Jackson, "Using the Climbing Drum Peel (CDP) test to obtain a GIC value for core/face sheet bonds," *Journal of Composite Materials*, vol. 41, pp. 2863-2876, 2007.

Accuracy of Solar Simulator Spectral Determination Using Band-Pass Filtering Method

Weston Dobson[1], Harrison Wilterdink[1], Cassidy Sainsbury[1], Adrienne Blum[1], Justin Dinger[1], Ronald A. Sinton[1], Karsten Bothe[2], David Hinken[2], Martin Wolf[2]

[1]Sinton Instruments, Boulder, CO, USA [2]Institute for Solar Energy Research Hamelin (ISFH), Germany

Abstract — CCD spectrometers have become the industry-standard tools for classifying solar simulator light spectrums, but they have drawbacks. A simpler method is to measure the amount of light present in each relevant band defined by the classification standards *directly*, by filtering the incident light appropriately and measuring its magnitude with a basic detector. This is much less expensive, and also has some other surprising benefits. The difficult aspect of this method is quantifying and correcting for the uncertainties introduced in the process. These will be outlined in detail in order to build confidence in the accuracy of this approach.

Index Terms — Band-pass filters, CCD Image Sensors, Characterization, Error Correction, Standards, System Testing

I. INTRODUCTION

Characterizing the light spectrum of solar simulators has become a very common task in the solar industry. The goal is typically to determine the spectral class as defined by the IEC 60904-9 solar simulator standard [1]. These measurements are most frequently done using CCD spectrometers which measure the spectrum in high-resolution wavelength increments. Typically corrections for nonlinearity, dark count generation, and variations in the CCD quantum efficiency at different wavelengths are applied to convert the raw counts output by the CCD pixels into meaningful irradiance metrics. Software is then used to perform numerical integrals on this data to determine the cumulative irradiance in each of the wavelength bands defined by the standard and determine if these magnitudes are within the specified limits. A simpler, more direct method of performing these measurements is to filter the spectrum using precision, high performance optical filters to match the defined radiation bands and measure the magnitude of these bands *directly* with an appropriate detector. This method is not novel [2] and is even described as an acceptable approach by the IEC 60904-9 standard [1] itself, but its use, historically, has been infrequent. The method is attractive for many reasons [3], including:

- Significantly reduced cost
- Can be made to be very rugged, compact and portable
- High signal-to-noise ratio, which allows nearly instantaneous measurements unlike the longer integration times required by CCDs. This enables:
  - Easier measurement of very short or rapidly changing flash pulses

  - Characterization of spectral classification as a function of intensity in a single measurement

Despite these advantages, this method does require some careful considerations in its application in order to obtain accurate, useful results. This paper will detail the sources of uncertainty that arise from this approach and how to correct them. In addition, it will draw attention to a nuance present in the IEC spectral characterization standard which can complicate the understanding of the impact of spectral mismatch for all spectral characterization methods, and can be very easy to overlook.

II. METHOD

The concept behind characterizing a solar simulator using the band-pass filtering method is simple. The first step is to obtain band-pass filters having pass bands that correspond to each of the wavelength bands defined by the IEC standard [1]. Then, a light detector (in our implementation a high-efficiency silicon reference cell) can be exposed to a light source while being obscured by each of the band-pass filters in sequence. For each filter, the magnitude of the signal produced at each step is measured and logged. A practical implementation of this concept is shown in Fig. 1.

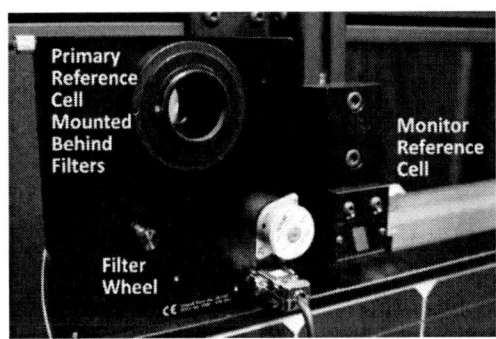

Fig. 1   A "Filter Wheel Spectrometer"

If this system were ideal, the bandpass filters would have perfect cut-on and cut-off edges at the intended wavelengths and the light detector would have a completely flat spectral response. Is so, the relative magnitudes measured by this system for each of the spectral bands would be directly comparable to the total amount of irradiance present in each of these bands in the reference spectrum of the IEC 60904-9 standard [1]. However, in a real system there are technical challenges that must be addressed before these comparisons can be made.

III. ERROR CORRECTION AND UNCERTAINTY

There are three sources of error that appear when attempting to characterize a spectrum using the 'filter/detector' method:

1. Imperfect response of the optical filters (filter effects)
2. Non-ideal light detector quantum efficiency (QE effects)
3. The complication that practical light detectors will actually measure photon flux and not irradiance

A. Filter and QE Effects

In order to account for these effects, the calibration and test laboratory at ISFH (ISFH CalTeC) collaborated with Sinton Instruments to measure spectral response curves for each of the filter/detector combinations of the design, and exposed the unit to well-characterized xenon-halogen and blackbody light sources to evaluate the effectiveness of the corrections in practice. To visualize these imperfections, Fig. 2 shows two of the actual spectral response curves that were measured for this system. These two example spectral response curves are shown alongside a scaled ideal spectral responsivity curve for reference (shown as a diagonal dashed line), and the gray areas

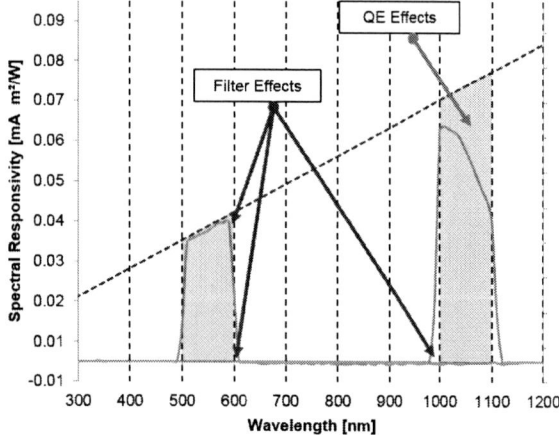

represent ideal spectral responses for these individual bands.
Fig. 2   Spectral response curves for two different spectral bands vs. their ideal counterparts

Any spectral response that fails to follow the border of these gray regions is non-ideal. The filter effects are associated more specifically with the accuracy and sharpness of the cut-on/cut-off wavelengths of the optical filters, while the QE effects are due to the non-flat QE response of the silicon reference cell inherent to the device itself. To account for these effects, we determine correction factors ($ß_i$) for each of the spectral

response curves to force them to match the results that would be obtained with ideal spectral response. These correction factors are all interdependent though, because what is relevant for classification purposes is the *ratio* of each spectral response to the sum of all of the spectral responses. If we also calculate the ratio of the ideal spectral response for a particular band to the total of all of the ideal spectral responses, then the correction factors can be represented by the ratio of these two ratios. Expressed mathematically, this can be represented as shown in equation (1) to calculate a correction factor ($ß_i$) for a particular spectral response band (SR_i) out of a set number (n) of spectral response curves, over a wavelength range defined between wavelengths λ_{min} to λ_{max}.

$$ß_i = \frac{\int_{\lambda_{ilower}}^{\lambda_{iupper}} SR_{ideal}(\lambda)d\lambda}{\int_{\lambda_{min}}^{\lambda_{max}} SR_{ideal}(\lambda)d\lambda} \Bigg/ \frac{\int_0^\infty SR_i(\lambda)d\lambda}{\sum_{i=1}^n \int_0^\infty SR_i(\lambda)d\lambda} \quad (1)$$

As a concrete example using the IEC standard [1], imagine that we are trying to calculate the correction factor for the 400-500 nm spectral response curve. We need to take the *ideal* spectral response from 400-500 nm, divide that by the ideal spectral response from 400-1100 nm, and then divide that entire ratio by the ratio of the integral of the 400 nm spectral response curve to the sum of all of our spectral response curves, as seen in equation (2).

$$ß_i = \frac{\int_{400}^{500} SR_{ideal}(\lambda)d\lambda}{\int_{400}^{1100} SR_{ideal}(\lambda)d\lambda} \Bigg/ \frac{\int_0^\infty SR_{400}(\lambda)d\lambda}{\sum_{i=1}^n \int_0^\infty SR_i(\lambda)d\lambda} \quad (2)$$

Calculating correction factors this way will produce generic correction factors that account for the filter and QE effects, but *only in the special case of the system being exposed to a completely flat irradiance spectrum!* These correction factors can account for effects such as a band-pass filter "turning on" a couple nanometers too early. However, if we imagine that the spectrum that we're trying to measure happens to have an unusual feature right in that same region, it will affect the result more than we would expect from just calculating the increase in signal from the additional bandwidth of that filter alone. At first it seems almost impossible to have any confidence in the ultimate accuracy of this approach, because if we're measuring an unknown spectrum, we don't know what that spectrum is doing in the regions where our spectral responses are non-ideal, so we don't know what the impact on the accuracy will ultimately be. In theory, we could even have an unlimited amount of uncertainty if we were to imagine a spectrum that was simply a delta function of photons right in one of those trouble regions.

The good news is that we *do* have some knowledge of the types of spectrums that we will potentially be measuring with this tool, *if* the goal is not to build a completely general-purpose spectrometer but to build a purpose-built instrument designed specifically for characterizing xenon flashes, sunlight, and possibly LED-based sources. For example, *all* xenon light sources will have peaks at 827 nm, 764nm, etc. because those peaks correspond to the atomic emission lines for xenon. We can always count on those being there if it's a xenon source. The specific shape of different xenon spectrums will depend on any number of factors, but we can still do much better than 'flat' correction factors for our spectral response curves, especially in regions that contain known features. With this knowledge, we can recalculate the correction factors by weighting the correction calculations with a reference spectrum.

This reference should represent the typical spectrum of the light source that we're attempting to characterize. This will make the errors small when characterizing a spectrum that is close to that reference. To calculate these correction factors correctly, we should recalculate them to normalize the amount of *current* that would be produced by each filter/detector combination compared to the ideal case. This is done simply by replacing each spectral response integral with an integral of the *product* of spectral response and the weighting spectrum $S(\lambda)$:

$$\text{ß}_i = \frac{\int_{\lambda_{ilower}}^{\lambda_{iupper}} SR_{ideal}(\lambda)S(\lambda)d\lambda}{\int_{\lambda_{min}}^{\lambda_{max}} SR_{ideal}(\lambda)S(\lambda)d\lambda} \Bigg/ \frac{\int_0^\infty SR_i(\lambda)S(\lambda)d\lambda}{\sum_{i=1}^n \int_0^\infty SR_i(\lambda)S(\lambda)d\lambda} \quad (3)$$

Correction factors of this type can be calculated for every type of 'target' spectrum that could be anticipated for a particular design of solar simulator. See Table I for a list of example correction factors that were calculated for this device using different weighting spectrums.

TABLE I: CORRECTION FACTORS CALCULATED FOR VARIOUS WEIGHTING SPECTRUMS

λ Range (nm)	ß$_i$ ISFH Class "A+" Xe + Ha	ß$_i$ Sinton Class A Xenon	ß$_i$ Sinton Class C Xenon	ß$_i$ AM1.5 Global
400-500	1.079	1.109	1.106	1.084
500-600	1.045	1.074	1.101	1.045
600-700	0.982	0.999	1.011	0.979
700-800	0.941	0.954	0.955	0.941
800-900	0.991	1.032	1.051	0.979
900-1000	0.928	0.929	0.935	0.961
1000-1100	1.042	0.921	0.951	1.040

Notice that these correction factors are not large. Even across a wide range of spectral types, none of the correction factors increase or decrease the signal produced by each filter by much more than 10%. This is an encouraging result, because it suggests that even without any correction, we were not far from an accurate result even *with* the imperfections in the filter responses and the imperfect QE of the reference cell. One of the advantages of this method is that even without any correction at all, it's still possible to be "in the ballpark."

However, even with these correction factors, the measurements will still not be corrected perfectly because the correction factors themselves were determined by spectral responsivity measurements which have their own uncertainties.

Determining the uncertainties of these correction factors involves propagating these uncertainty sources out through the full calculation shown in equation (3) for each correction factor separately. Determined with a coverage factor of k = 2, (~95% confidence interval) the calculated uncertainty values for the ISFH CalTeC hybrid xenon/halogen light source correction factors are shown in Table II.

TABLE II: UNCERTAINTY VALUES FOR EACH OF THE CORRECTION FACTORS BY WAVELENGTH RANGE

λ range (nm)	400-500	500-600	600-700	700-800	800-900	900-1000	1000-1100
U (%)	7.29%	5.65%	5.03%	5.34%	4.82%	6.00%	3.87%

These uncertainties contribute directly to the uncertainties of the final bin ratios that are reported for the letter-grade assignment of spectral class, because converting to the final bin ratios from here is simply a series of multiplication and division operations by constants, which does not affect the proportional uncertainties.

We have now determined part of the final uncertainty of our measurement using this method, but we have to remember that our correction factors were calculated in the first place for a reference "target" spectrum which should represent a spectrum very close to what we expect to be measuring. If we happen to be measuring a spectrum that is significantly different from that reference spectrum, it will contribute additional uncertainty. To get an idea of how much uncertainty this could cause, it's helpful to look once again at the values shown in Table 1. When comparing the correction factors that were calculated for completely different types of spectrums, there can be some significant differences between the different correction factors. However, what is interesting is that if we look only at the two pure xenon light sources: one graded class A and the other class C, the correction factors are not as different as one might expect. The correction factors and the percentage difference between them are shown in Table III.

TABLE III: CORRECTION FACTORS FOR TWO PURE XENON LIGHT SOURCES CLASSIFIED AS A AND C ACCORDING TO IEC 60904-9 [1]

λ range (nm)	400-500	500-600	600-700	700-800	800-900	900-1000	1000-1100
Class A	1.109	1.074	0.999	0.954	1.032	0.929	0.921
Class C	1.106	1.101	1.011	0.955	1.051	0.935	0.951
Diff. (%)	0.3%	-2.5%	-1.2%	-0.1%	-1.8%	-0.6%	-3.3%

These correction factors are very similar. The biggest difference between them in in the final band (3.3%). That is a surprisingly small deviation when considering how quantitatively different these spectrums are. From a distance, a class C xenon spectrum has a different shape than a class A xenon spectrum, but on a local scale the features are very similar, but with different magnitudes. Only local features cause problems because these correction factors are concerned with errors introduced by unexpected spikes or drops in a measured spectrum only in regions where the filter and QE performance is non-ideal, which are small and localized. If we were to actually measure a class C spectrum while using the class A correction factors, the results we obtained would be off by no more than an additional 3.3%, and even then only in that final band, where the non-ideal QE behavior dominates. This is exceptional considering that the class C spectrum represents a completely unfiltered worst case. As such, the 3.3% represents a *practical* upper limit of additional uncertainty that we would expect when measuring xenon simulator spectrums, because we can make the reasonable assumption that any of the xenon spectrums we would be attempting to qualify would fall, in terms of uncertainty, well between the reference class A spectrum and the worst case.

Even more commonly we would be evaluating spectrums that are much closer to the reference class A spectrum, and the additional uncertainty would be much less than 3.3%. Even so, we should perform one final error propagation calculation to determine the total uncertainty of the final reported ratios which takes into account the contributions from the correction factors as well as the uncertainty due to the measured spectrum potentially not matching the reference spectrum. Because the cause of these uncertainties is independent from the uncertainties in the correction factors themselves, we can simply use the root sum of squares rule for error propagation, and these final, combined values are shown in Table IV.

TABLE IV: FINAL UNCERTAINTIES OF THE REPORTED BIN RATIOS

λ range (nm)	400-500	500-600	600-700	700-800	800-900	900-1000	1000-1100
U (%)	7.30%	6.18%	5.18%	5.34%	5.14%	6.02%	5.09%

The largest of these values is 7.3%, and represents a practical worst-case uncertainty with 95% confidence. It should be noted that this is not a general result for all 'filter wheel spectrometers' and is only valid for this particular device construction and the particular spectral responsivity tools used to characterize it. A similarly constructed device could have better or worse uncertainty depending on a number of factors, but as a proof of concept, this result shows that this method is completely usable for confidently assigning spectral classifications to xenon spectrums, considering that the specification limits for class A in the IEC standard [1] are currently ±25%. Also of note, these uncertainty values are almost entirely based on invariant factors like the spectral responses of each filter, which do not typically change. What this means is that while overall *accuracy* is limited by the uncertainties outlined in this section, the repeatability of the tool is not, making it an extremely reliable choice for monitoring spectral changes over time.

B. Photon Flux vs. Irradiance

Finally, we have to consider that what this type of spectrometer measures is actually photon flux and not irradiance. This isn't necessarily an uncertainty, but it is a complication. First of all, it means that instead of comparing irradiance totals to what is specified in the IEC standard [1], we have to calculate photon flux totals from the AM 1.5 reference spectrum and compare our measurements to those instead. This is simple, but there's still one final problem, and that is that after performing our measurements, when we assign a class to the spectrum, those classes are explicitly defined in the standard by *irradiance* norms only. While this distinction makes no difference in terms of determining how well a spectrum matches the AM 1.5 standard for *device performance*, because the letter of the standard does not allow photon flux comparisons as an alternative to irradiance for class assignment purposes, there will be a small uncertainty that depends on the band.

We can calculate the limit of this uncertainty by imagining an absolute worst case for the IEC 60904-9 standard [1]. If we imagine two spectrums, one of which is a delta function of photons right at 400 nm, and another which has a delta function of exactly the same number of photons at 500 nm, then the two functions will have equivalent photon flux in the 400-500 nm band. However, the spectrum containing entirely 400 nm photons will actually contain 25% more energy than the spectrum with 500 nm photons, simply because 400 nm photons are more energetic and their energy content is linearly dependent on their wavelength. The 100 nm difference between them represents 25% of the starting wavelength in the band, which is 400 nm, hence 25% more energy. As the wavelengths increase for each band, this worst-case uncertainty decreases, because the 100 nm gap represents a smaller proportional difference between the starting and ending points of the bands. The trend continues up until the final band of the IEC standard, which is a double-width band of 200 nm. This maximum uncertainty between irradiance and photon flux is summarized in Table V.

TABLE V: WORST-CASE UNCERTAINTIES FOR BAND RATIOS OBTAINED BY PHOTON FLUX INSTEAD OF IRRADIANCE

λ range (nm)	400-500	500-600	600-700	700-800	800-900	900-1100
U (%)	25%	20%	16.7%	14.3%	12.5%	22.2%

These uncertainties might seem quite large and unmanageable at first, but we must recall that these uncertainties were obtained by calculating them for the most extreme laser line spectrums imaginable. No broadband spectrum that would ever be used for testing photovoltaic devices and also attempting to conform to the IEC standard would have uncertainties anywhere remotely near this level. As a more practical example of realistic uncertainties that could be expected by classifying spectrums using photon flux instead of irradiance, the following table lists the discrepancies observed between the ratios reported when a representative class A xenon spectrum was measuring using a calibrated CCD spectrometer in both photon flux (ratio$_p$) and irradiance (ratio$_i$) modes. These results are shown in Table VI.

TABLE VI: RESULTS REPORTED MEASURING A CLASS A XENON SPECTRUM BY PHOTON FLUX AND IRRADIANCE COMPARISONS

λ range (nm)	400-500	500-600	600-700	700-800	800-900	900-1100
Ratio$_p$	0.982	1.029	1.063	0.978	1.047	0.927
Ratio$_i$	0.969	1.020	1.053	0.977	1.038	0.942
Diff. (%)	1.32%	0.87%	0.94%	0.10%	0.86%	-1.62%

When comparing results obtained from photon flux totals against those obtained from irradiance totals on a realistic example spectrum, the difference was 1.62% in the worst bin, which is an order of magnitude lower than the theoretical worst case uncertainty. For the purpose of determining absolute uncertainty when using this tool to classify spectrums according to the IEC standard [1], these uncertainties should technically be rolled into the total uncertainty values obtained in the previous section, but since cells "see" photons and not irradiance, we would argue that classifying spectrums by photon flux is actually superior to classifying spectrums by irradiance, and that these are not errors in the traditional sense. We would also argue that the IEC standard [1] should allow for comparisons to photon flux totals for class assignment, especially since that is *more* relevant to spectral matching for characterization purposes of photovoltaic devices than irradiance matching.

IV. PRACTICAL TESTING

As a bit of a sanity check that all of our corrections and uncertainty levels were reasonable, a second collaboration was done between Sinton Instruments and ISFH CalTeC to test a second separately assembled unit against ISFH CalTeC's well-characterized hybrid xenon/halogen light source. It was calibrated in this case simply by adjusting each set of correction factors until the results matched the first calibrated unit at Sinton Instruments. ISFH CalTeC performed measurements of their light source and reported the measured bin ratios using their internal CCD spectral radiometer (Ratio$_{CCD}$) as well as this filter-wheel radiometer (Ratio$_{FW}$) These are shown in Table VII.

TABLE VII: RESULTS FROM THE SAME HYBRID XE-HA LIGHT SOURCE CHARACTERIZED BY A CCD SPECTROMETER AND THE "FILTER WHEEL" SPECTROMETER

λ range (nm)	400-500	500-600	600-700	700-800	800-900	900-1100
Ratio$_{CCD}$	0.98	1.06	1.02	0.93	0.96	0.99
Ratio$_{FW}$	1.00	1.09	1.02	0.94	0.95	1.00
Diff. (%)	2%	2.7%	0%	1.1%	-1.1%	1%

These figures show excellent agreement between the two spectral characterization methods, demonstrating once again as a proof of concept the power of the band-pass filtering approach when properly handled. In addition, the values are all significantly less than the calculated uncertainties, as expected.

IV. CONCLUSIONS

Characterizing solar simulator spectrums using a band-pass filter approach can have impressive accuracy if done correctly. In this study, absolute uncertainty had a *practical* upper limit of no more than 7.5% using k=2, which is more than accurate enough for confident spectral class assignment, pass/fail evaluations, and precise drift tracking. That combined with its low cost, ruggedness, excellent measurement speed, and exceptional repeatability makes it a method that warrants serious consideration for spectral evaluation.

REFERENCES

[1] IEC 60904-9:2007, 2007
[2] Correction Procedures for the Flasher Calibration of PV Devices Resulting in Reduced Restrictions and Uncertainties, 2nd World Conference on PV Solar Energy Conversion – Wien 1998
[3] Low-Cost, High-Speed Spectral Characterization Using Edge Filters, Sinton, NREL Silicon Workshop

Correlation of I-V Curve Parameters with Module-Level Electroluminescent Image Data Over 3000 Hours Damp-Heat Exposure

Justin S. Fada *, Andrew J. Loach *, Alan J. Curran *, Jennifer L. Braid *, Shuying Yang [†], Timothy J. Peshek *,
Roger H. French *
* Solar Durability and Lifetime Extension (SDLE) Research Center,
Case Western Reserve University, 10900 Euclid Ave., Cleveland, Ohio 44106, USA
Email: see sdle.case.edu
[†] SunEdison Inc., 600 Clipper Drive, Belmont, California 94002, USA

Abstract—**Four module brands of three samples each were exposed to 3000 hours of accelerated damp-heat testing. Current voltage (I-V) curve tracing and electroluminescent (EL) imaging were performed at 500 hour steps. The I-V curves were analyzed to get 8 common I-V parameters: maximum power (P_{mp}), fill factor (FF), current at maximum power (I_{mp}), voltage at maximum power (V_{mp}), current at short circuit (I_{sc}), voltage at open circuit (V_{oc}), series resistance (R_s), and shunt resistance (R_{sh}). The EL images were processed with a data processing pipeline using the open source coding language Python, employing techniques such as filtering, thresholding, convex Hull, regression fitting, and perspective transformation. This pre-processing was necessary to re-orient the images in a uniform manner as there was variance involved in the alignment of the modules during the measurement process. With the pre-processed data in hand, module-level parameters of median, mean, and standard deviation were calculated from the image pixel intensity distributions. These image data parameters were plotted against all the I-V curve parameters to identify trends in these data types. Correlation heat maps were generated depicting the relationships between data type parameters from IV and EL measurements.**

Index Terms—**I-V, electroluminescent imaging, accelerated exposure, degradation.**

I. INTRODUCTION

In recent years, global photovoltaic (PV) installation has seen a significant increase with the U.S. solar market alone growing 97% by installing 14.76 GW of solar PV in 2016 [1] [2]. With continued adoption of PV systems intended to produce power for decades after installation comes a need to further understand module lifetime performance and gain insights into the mechanisms causing module degradation.

Global installation of PV causes modules to experience a wide range of environmental conditions with variable doses of exposure. Under each one of those settings a particular module design is expected to perform reliably and last for the duration of the warranty provided by the manufacturer, which is typically ~ 25 years. Common stressors include temperature, humidity, and ultra-violet irradiance, which lead to the emergence of defective behavior through degradation modes that include cell cracking, series resistance increases, discoloration, hot spots, encapsulant chemical changes, and many others [3], [4]. Under real-world outdoor conditions diverse stressor combinations occur leading to complex degradation signatures from which to deduce important mechanistic

information [5]. To focus mechanistic research, lab-based accelerated testing is performed to control module exposure precisely and reduce the time of experimentation.

Possible measurement techniques for solar modules are numerous, spanning a range of dimensionally, density, ease of acquisition, and potential usefulness in deducing lifetime behavior [6]. The most common of these are maximum power point tracking and current-voltage ($I - V$) curve tracing, which provide point-in-time data regarding the electrical state of the test module [3], [7]. Of these, $I - V$ tracing performs a complete electrical sweep from short-circuit current (I_{sc}) to open-circuit voltage (V_{oc}) of an illuminated PV module. From the current and voltage measurements, a variety of useful module-level parameters can be derived. These include maximum power (P_{mp}), fill factor (FF), current at maximum power (I_{mp}), voltage at maximum power (V_{mp}), current at short circuit (I_{sc}), voltage at open circuit (V_{oc}), series resistance (R_s), and shunt resistance (R_{sh}).

Research has shown that changes in these parameters over exposure time are related to particular module phenomena, such as I_{sc} relating to cracked cells and V_{oc} relating to the short-circuit bypass diode [3], [8]. However, these parameters yield only a coarse measure of performance in heterogeneous systems as they measure the input/output electrical signal of the full module and cannot provide information on variability due to cracking or shunting.

Visual inspection of test modules can give greater insights, however only qualitatively. Higher order quantitative data is possible to obtain using image measurements, such as electroluminescence (EL), photoluminescence (PL), and thermography.

Electroluminescent imaging uses an applied forward-biased voltage to cause charge carrier recombination in semiconductor wafer materials leading to photon emission (for silicon this emission occurs in the near-infrared light range). Each pixel represents a spatially resolvable data point as a function of local photon emission registered on the camera sensor, typically generating millions of data points with each measurement [3], [9]. High density data with spatial information about emission intensity variables on the module surface provides many avenues for study [10], [11]. Most simply by providing additional metrics pertaining to module performance including mean, median and standard deviation of pixel values at the module-level, sub-string and cell-level by slicing apart the module image after image processing.

Our aim is to demonstrate correlation between common $I-V$ parameters and EL derived intensity parameters. This method will be refined and explored further as a means of studying degradation patterns in modules under more diverse exposure types with the desire to obtain mechanistic insights through integration of measurement techniques.

The data set used in this work consists of 4 brands of commercial 60-cell modules each consisting of 3 samples. These modules span 3 cell material types: mono-crystalline silicon Al-BSF (Aluminum Back Surface Field), multi-crystalline silicon Al-BSF, and mono-crystalline passivated emitter rear contact (PERC). The modules were exposed to standard IEC61215 damp heat test conditions and were measured using I-V curve tracing and EL imaging at 500 hour intervals from 500 to 3000 hours [12].

II. DATA PROCESSING

Due to the process by which the EL images are captured, there is module orientation variability present between images. To ensure the images were uniformly oriented for analysis, image processing techniques are implemented to create a data processing pipeline using the open source language Python [13], [14]. Filtering and thresholding methods are used to initially pre-process the data to reduce noise and remove unimportant background data. With a noise reduced image, a convex Hull algorithm is used to identify the cell areas and mark them as a "1" (white pixel) while every other pixel was assigned a "0" (black pixel). A series of 1-dimensional x-axis and y-axis parallel slices are taken through this array of 1's and 0's to identify the steps up (0 to 1) and steps down (1 to 0) across the slice. These steps correspond to the module edge. A linear regression line is fit to these edge points and the intersections of the edge lines found. These 4 intersection points identify the corners of the variably oriented PV module. A perspective transformation is then applied to uniformly orient the module. An example of the original and transformed image can as seen in Figure 1 [14].

After processing, each EL image is 2500 x 1500 pixels, or 3,750,000 individual data points existing as gray-scale floating point values between 0 (black) and 1 (white). These data points were used to obtain the median, mean and standard deviation of all image pixels per module. As an additional step to the data processing pipeline, the images can be sliced into substring areas and cell areas and the median, mean, and standard deviation of pixel values calculated.

Using the full $I-V$ curve, the measurement device calculated the aforementioned 8 parameters under test conditions. Using the language R, the $I-V$ and EL parameters were merged into a single data-frame for analysis [15].

III. ANALYSIS

As an initial probe of the data set, a correlation matrix of 8 $I-V$ parameters (maximum power (P_{mp}), fill factor (FF), current at maximum power (I_{mp}), voltage at maximum

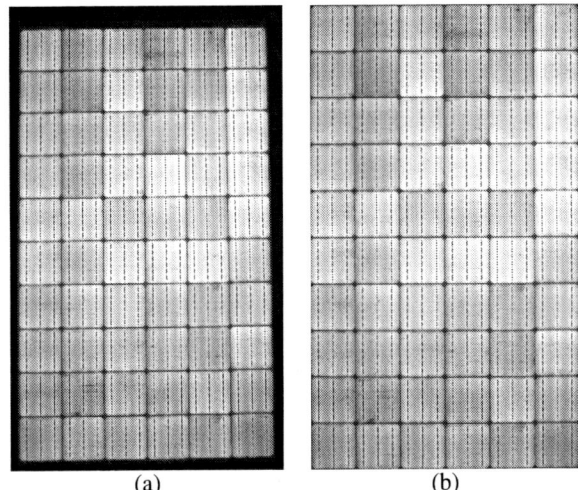

Fig. 1. (a) EL image captured after 500 hours of damp heat exposure. (b) EL image in Figure 1 (a) after image processing from its non-uniformly oriented raw state into a uniformly oriented version.

power (V_{mp}), short circuit current (I_{sc}), open circuit voltage (V_{oc}), series resistance (R_s), and shunt resistance (R_{sh})) and 3 EL module parameters (median pixel intensity, mean pixel intensity and standard deviation of pixel intensity) was generated (seen in Figure 2). The correlation values are color coded from 1 (dark blue) for positive correlation to -1 (dark red) for negative correlation. Additionally, the parameters were clustered using hierarchical clustering to group variables with similar correlations. Using this matrix, it can be seen that EL parameters have high correlation with some I-V parameters, such as R_s, V_{mp}, I_{mp}, and P_{mp}.

EL imaging provides a means of observing local electrical characteristics and therefore the modules resistances (R_s, and R_{sh}) along with P_{mp} and FF from $I-V$ measurements will be analyzed further against EL parameters.

A. Maximum Power

Maximum power is a critical value of interest in degradation analysis as this directly relates to the module's energy production. In the above correlation matrix, the correlation coefficient of P_{mp} to EL median is 0.785, indicating moderate to strong correlation. As mentioned, this data contains three wafer types. Visualizing the $I-V$ P_{mp} and EL median intensity (values were centered and scaled using the base R function "scale" to properly overlay data of differing magnitude [16]) over 3000 hours of exposure, clear groupings of the data is seen (Figure 3). The $I-V$ and EL data of each wafer type trend together exhibiting the same slope direction between each exposure step.

With clearly differing performance based on wafer material types, the data is subset by wafer material and the same correlation calculated. It is seen that the correlation is highly dependent on the material type, with mono-Si Al-BSF and mono-Si PERC cells showing very high correlation and multi-

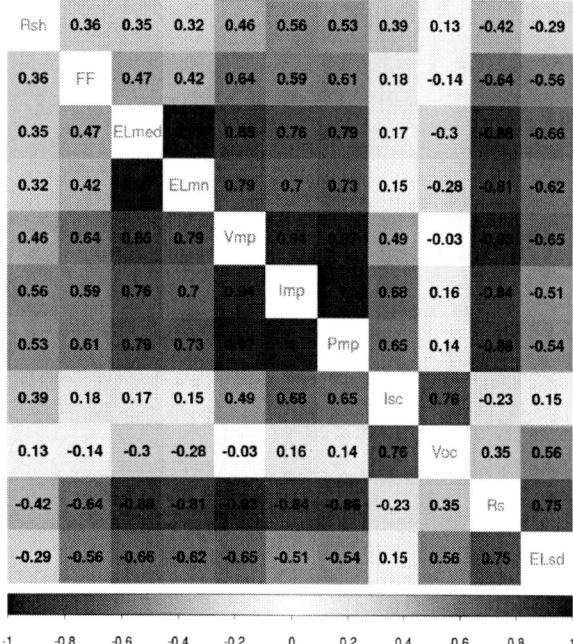

Fig. 2. A correlation heat map each IV parameter and EL intensity parameter. Where the diagonal variables in the figure represent: Pmp - P_{mp}, FF- FF, Imp - I_{mp}, Vmp - V_{mp}, Isc - I_{sc}, Voc - V_{oc}, Rs - R_s, Rsh - R_{sh}, ELmed - median EL pixel intensity, ELmn - mean EL pixel intensity, and ELsd - standard deviation of EL pixel intensity

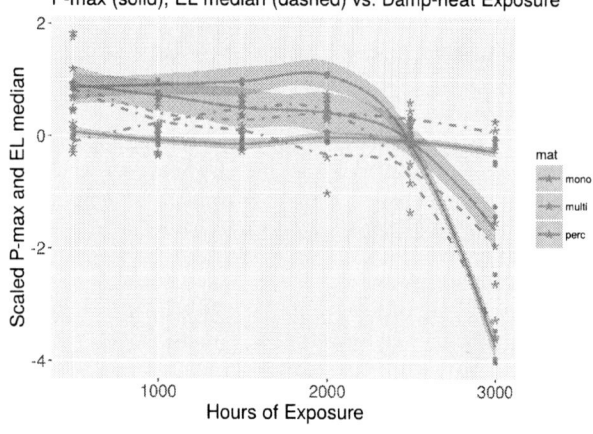

Fig. 3. The solid lines depict the $I - V$ P_{mp} while the dashed lines depict the EL median values. Standard error for P_{mp} shown only (for plot clarity). Scaling was used to overlay the data [16].

Si Al-BSF showing a low correlation, as seen in Table I. This low correlation for the multi-Si Al-BSF modules is likely due to the lack of degradation observed and therefore lack of a mechanism driven trend. Table II shows the percent degradation from 500 to 3000 hours for each brand alongside wafer materials. For multi-Si brands A and D the percent degradation over 3000 hours of damp-heat exposure is only

5.94% and 2.77%, respectively, and therefore lack significant trends that would yield higher correlation of $I - V$ and EL parameters.

TABLE I
RESULTS OF CORRELATIONS OF P_{mp} AND EL MEDIAN SUBSET BY WAFER MATERIAL TYPE.

Material	$P_{max} \sim EL_{med}$ Correlation
Mono-Si AL-BSF	0.966
Mult-Si AL-BSF	0.386
PERC	0.896
All	0.785

TABLE II
PERCENT CHANGE IN P_{mp} FROM 500 TO 3000 HOURS OF DAMP-HEAT EXPOSURE.

Brand	Material	Percent P_{mp} Loss (%)
A	Mult-Si AL-BSF	5.94
B	PERC	29.99
C	Mono-Si AL-BSF	54.18
D	Multi-Si AL-BSF	2.77

B. Series Resistance

Module-level series resistance is a composite measure of all cell series resistances and interconnect resistances. It would be expected that the brightness of the modules as a function of exposure duration will trend with the series resistance as defects appear. Additionally, as defects appear the module intensity distribution would be expected to become more heterogeneous.

Using the entire data set, the correlation between R_s and median EL intensity is -0.878 indicating that as series resistance increases the modules are not emitting as many photons. Sub-setting the data by wafer material the behavior appears similar to that in Table I where the multi-Si AL-BSF modules have low correlation and the remaining with high correlation.

TABLE III
RESULTS OF CORRELATIONS OF R_s AND EL MEDIAN SUBSET BY WAFER MATERIAL TYPE.

Material	$R_s \sim EL_{med}$	$R_s \sim EL_{sd}$
Mono-Si AL-BSF	-0.96	0.75
Mult-Si AL-BSF	-0.49	0.58
PERC	-0.78	0.68

Visualizing the relationships of R_s with step-wise EL median intensity in Figure 4 it is clear that these correlation values are accurate. Additionally it is seen that the standard deviation of pixel intensities increase as the series resistance increases.

Calculating the same intensity values for each of the 3 substrings we see similar trends. However, to gain a measure of how different each cell in a module is when series resistance changes, the median value of each cell region was calculated and the standard deviation of those 60 values per module was calculated. This measure shows that even though the standard deviation of all module pixels increases with increased series

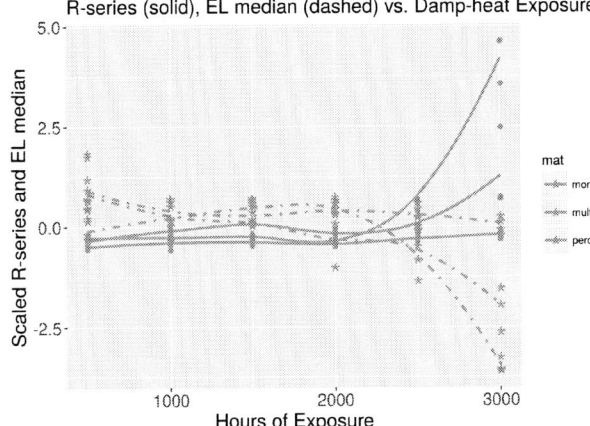

Fig. 4. The Solid lines depict the $I - V\ R_s$ while the dashed lines depict the EL median values.

TABLE IV

CORRELATION OF $I - V\ R_s$ WITH THE STANDARD DEVIATION OF PIXEL VALUES IN EACH OF THE 3 MODULE SUB-STRING REGIONS AND WITH THE STANDARD DEVIATION OF THE MEDIAN VALUE IN EACH OF THE 60 CELL REGIONS OF EACH MODULE.

Material	Sub-Str. 1	Sub-Str. 2	Sub-Str. 3	Cell
Mono-Si AL-BSF	0.77	0.76	0.72	-0.73
Multi-Si AL-BSF	0.60	0.58	0.50	-0.21
PERC	0.64	0.67	0.69	-0.56

resistance (i.e. the difference from light to dark pixels is larger), over increased exposure the cells in each module tend to become more similar as they degrade. This can be seen in Table IV.

C. Fill Factor

Fill factor (FF) is a measure of module quality calculated as the ratio of P_{mp} to the product of I_{sc} and V_{oc}). The correlation of FF and median EL intensity for all data is 0.916 (this was after removal of a single anomalous "0" logged by the $I - V$ tracing unit causing a correlation of 0.475). Sub-setting by wafer material, module EL mean and standard deviation along with standard deviation of the median value of each cell was calculated (Table V).

TABLE V

CORRELATION OF $I - V\ FF$ WITH MODULE EL MEDIAN, MEAN, AND STANDARD DEVIATION. CORRELATION OF $I - V\ FF$ WITH THE STANDARD DEVIATION OF THE MEDIAN VALUE IN EACH 60 CELL REGION OF EACH MODULE.

Material	EL Med.	EL Mean	EL S.D.	Cell S.D.
Mono-Si AL-BSF	0.96	0.94	-0.78	0.73
Multi-Si AL-BSF	0.46	0.57	-0.86	0.27
PERC	0.87	0.87	-0.81	0.56

These values indicate a similar trend to the other $I - V$ parameters discussed above, however with stronger correlations across all EL variables. As fill factor decreases, the standard deviation of pixel values is seen to increase. Additionally, a

similar trend to series resistance is present where the decrease in FF results is a reduction in the standard deviation of the median values of each cell.

D. Shunt Resistance

The shunt resistance is a measure of current bypassing in the active cell area. Cell level shunting typically does not manifest itself in module $I - V\ R_{sh}$ as the other cells in series mask the local effect [3]. However, if enough shunting occurs, this value will decrease noticeably and correlate with parameters such as FF and P_{mp} [3].

In these data, R_{sh} does not have a strong correlation with any of the EL parameters when all the data are used. When the data are subset by wafer materials there is moderate agreement between R_{sh} and EL parameters. Regardless of the magnitude, the trend sign indicates that when shunting occurs the overall brightness of the module decreases as current is carried away by local shunts and that decreased shunt resistance indicates a weak to moderate increase in pixel variability (Table VI).

TABLE VI

R_{sh} CORRELATED WITH EL MEDIAN , MEAN AND STANDARD DEVIATION SUBSET BY WAFER MATERIAL.

Material	$R_{sh} \sim EL_{med}$	$R_{sh} \sim EL_{mean}$	$R_{sh} \sim EL_{sd}$
mono-Si	0.652	0.638	-0.458
multi-Si	0.146	0.123	-0.210
PERC	0.435	0.410	-0.552
All	0.354	0.322	-0.291

In the literature, it is noted that the R_{sh} does not indicate local effects at low degradation, but when the degradation gets high and significant shunting occurs then the module $I - V$ R_{sh} will reduce [3].

To test this, the data was split by hours of exposure such that 500, 1000, and 1500 hours of exposure were in one group and the remaining long exposure modules were placed in the second group. When the correlation of shunting with EL parameters were calculated for the low exposure group, the values were small and without a significant trend (Table VII). However, when the second group was studied median EL had high correlation, aside from the multi-Si material. The other EL parameters did not yield conclusive results.

TABLE VII

R_{sh} CORRELATED WITH EL MEDIAN INTENSITY SUBSET BY WAFER MATERIAL WITH DATA SPLIT AT 1500 HOURS OF EXPOSURE. THE 1500 HOUR STEP IS INCLUDED WITH THE LOWER EXPOSURE GROUP.

Material	$R_{sh} \sim EL_{med} \leqslant 1500 Hrs.$	$R_{sh} \sim EL_{med} > 1500 Hrs.$
mono-Si	-0.299	0.932
multi-Si	0.120	-0.103
PERC	-0.487	0.784
All	-0.212	0.469

IV. CONCLUSION

It has been shown that EL image parameters correlate well with trends seen in data collected via I-V curve measurements. Evaluation of the relationship of series resistance,

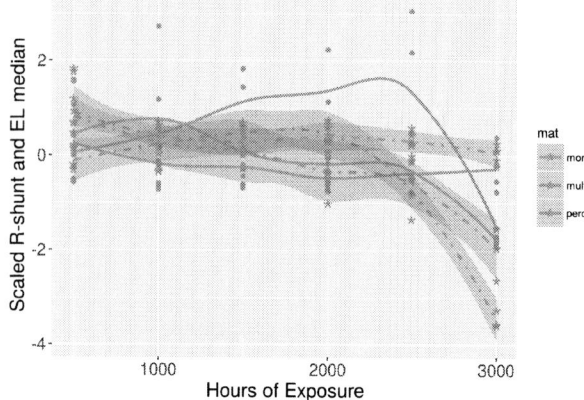

Fig. 5. The Solid lines depict the $I - V\ R_{sh}$ while the dashed lines depict the EL median values. Standard error for EL median is shown only (to reduce plot clutter). Scaling was used to overlay the data [16].

shunt resistance, fill factor and maximum power were examined using calculated median, mean, and standard deviation of pixel intensity values at the module, sub-string, and cell levels. Of the results demonstrated, as the modules studied degraded there was a trend indicating that the between cell variation in a module reduces with increased degradation manifested in common $I - V$ parameters. This initial probing of this data set provides validation to continue exploration of data integration between EL and $I - V$ measurements with refined feature analysis and modeling. This we hope will provide additional insights into the performance of PV systems and their degradation processes.

ACKNOWLEDGMENT

The authors are grateful for funding for this work from the Department of Energy (Award No. DE-EE-0007140). The authors are also grateful for the SDLE Research Center.

REFERENCES

[1] "SEIA/GTM Research U.S. Solar Market Insight," Tech. Rep., 2017.
[2] "IEA Key World Energy Statistics 2015," 2015.
[3] M. Kontges, S. Kurtz, C. Packard, U. Jahn, K. Berger, K. Kato, T. Friesen, H. Liu, and M. Van Isehegam, "IEA-PVPS {Task 13}: Review of Failures of PV Modules," Tech. Rep., May 2014. [Online]. Available: http://iea-pvps.org/index.php?id=275
[4] D. C. Jordan, T. J. Silverman, J. H. Wohlgemuth, S. R. Kurtz, and K. T. VanSant, "Photovoltaic failure and degradation modes," *Progress in Photovoltaics: Research and Applications*, vol. 25, no. 4, pp. 318–326, Apr. 2017. [Online]. Available: http://onlinelibrary.wiley.com/doi/10.1002/pip.2866/abstract
[5] R. H. French, R. Podgornik, T. J. Peshek, L. S. Bruckman, Y. Xu, N. R. Wheeler, A. Gok, Y. Hu, M. A. Hossain, D. A. Gordon, P. Zhao, J. Sun, and G.-Q. Zhang, "Degradation science: Mesoscopic evolution and temporal analytics of photovoltaic energy materials," *Current Opinion in Solid State and Materials Science*, vol. 19, no. 4, pp. 212–226, Aug. 2015. [Online]. Available: http://www.sciencedirect.com/science/article/pii/S1359028614000989

[6] M. A. Hossain,, Y. Hu,, N. R. Wheeler,, A. Gok,, P. Zhao,, W. Du,, M. Randall,, E. Weiss,, C. Fagerholm,, R. Kidwell,, J .Fada,, Y. Xu,, L.S. Bruckman,, T. J. Peshek,, GQ Zhang,, J. Sun,, and R. H. French,, "Lifetime and Degradation Science Approach towards Photovoltaic System Reliability," Sep. 2014. [Online]. Available: http://www.ieee.org/conferences_events/conferences/conferencedetails/index.html?Conf_ID=33864
[7] C. B. Jones, M. Martnez-Ramn, C. Carmignani, J. S. Stein, and B. H. King, "Wondering what to blame? Turn PV performance assessments into maintenance action items through the deployment of learning algorithms embedded in a Raspberry Pi device," in *2016 IEEE 43rd Photovoltaic Specialists Conference (PVSC)*, Jun. 2016, pp. 0261–0266.
[8] T. J. Peshek, J. S. Fada, Y. Hu, Y. Xu, M. A. Elsaeiti, E. Schnabel, M. Khl, and R. H. French, "Insights into metastability of photovoltaic materials at the mesoscale through massive IV analytics," *Journal of Vacuum Science & Technology B*, vol. 34, no. 5, p. 050801, Sep. 2016. [Online]. Available: http://scitation.aip.org/content/avs/journal/jvstb/34/5/10.1116/1.4960628
[9] Justin S. Fada, Nicholas R. Wheeler, Davis Zabiyaka, Nikhil Goel, Timothy J. Peshek, and Roger H. French, "Democratizing an electroluminescence imaging apparatus and analytics project for widespread data acquisition in photovoltaic materials," *Review of Scientific Instruments*, vol. 87, no. 8, p. 085109, Aug. 2016. [Online]. Available: http://scitation.aip.org/content/aip/journal/rsi/87/8/10.1063/1.4960180
[10] T. Trupke, J. Nyhus, and J. Haunschild, "Luminescence imaging for inline characterisation in silicon photovoltaics," *physica status solidi (RRL)-Rapid Research Letters*, vol. 5, no. 4, pp. 131–137, 2011. [Online]. Available: http://onlinelibrary.wiley.com/doi/10.1002/pssr.201084028/full
[11] D. Hinken, K. Ramspeck, K. Bothe, B. Fischer, and R. Brendel, "Series resistance imaging of solar cells by voltage dependent electroluminescence," *Applied Physics Letters*, vol. 91, no. 18, p. 182104, 2007. [Online]. Available: http://scitation.aip.org/content/aip/journal/apl/91/18/10.1063/1.2804562
[12] IEC, "IEC61215 ed2.0 - Excerpts- Crystalline silicon terrestrial photovoltaic (PV) modules - Design qualification and type approval | IEC Webstore | Publication Abstract, Preview, Scope," Tech. Rep., 2005. [Online]. Available: http://webstore.iec.ch/webstore/webstore.nsf/Artnum_PK/34077
[13] Python, "Python.org," 2013. [Online]. Available: https://www.python.org/
[14] J. S. Fada, M. A. Hossain, J. L. Braid, S. Yang, T. J. Peshek, and R. H. French, "Electroluminescent Image Processing and Cell Degradation Type Classification via Computer Vision and Statistical Learning Methodologies." Washington, D.C.: IEEE PVSC 44, Jun. 2017.
[15] "R: A Language and Environment for Statistical Computing," Vienna, Austria, 2016, 00061. [Online]. Available: http://www.R-project.org/
[16] Becker, R. A., Chambers, J. M., and Wilks, A. R., "The New S Language." 1988.

A Novel Method to Investigate Stoichiometry and Performance of Buried Passivated Contacts Utilizing Time-of-Flight SIMS

Steven P. Harvey, William Nemeth, Jeff Aguiar, Craig Perkins, Pauls Stradins

National Renewable Energy Laboratory, Golden, CO, 80403, USA

Abstract — We report on methods using time-of-flight secondary-ion mass spectrometry (TOF-SIMS) to elucidate critical information about the buried interfaces in interdigitated back-contact silicon solar cells. We examined the stoichiometry of the tunneling oxide with TOF-SIMS and X-ray photoelectron spectroscopy (XPS). Interestingly, the data indicates that the stoichiometry of the tunneling oxide moves away from SiO_2 towards lower stoichiometry (SiO_x) upon the crystallization annealing treatment, and towards SiO_2 again upon Al_2O_3 deposition and a subsequent forming gas anneal. We also report on an important artifact for samples with an Al_2O_3 passivation layer related to the oxygen content of the matrix.

Index Terms — TOF-SIMS, passivated contact, IBC cell, XPS.

I. INTRODUCTION

Secondary-ion mass spectrometry (SIMS) is a powerful analytical technique for determining elemental and isotopic distributions in solids. SIMS is well known in the photovoltaic industry for generating 1-D dopant distribution profiles due to its great sensitivity. However, with TOF-SIMS, one can look in much more detail by examining the many different secondary-ion species that are all simultaneously detected during a measurement. Specifically, TOF-SIMS is capable of resolving an interfacial SiO_x layer in the polySi/$SiO_{(x)}$/Si stack and its stoichiometry by following the secondary-ion complex Si_3O (as well as others) as described by Chiba [1, 2]. In addition, TOF-SIMS is an ideal tool for investigating buried interfaces with high depth resolution because the two-beam nature of the measurement allows the sputter conditions to be set up independent of the analysis conditions; this is unlike dynamic SIMS, where there is only one ion beam for both analysis and sputtering. In this work, we adapt the TOF-SIMS methodology developed in [1] to assess performance of the cells with passivated surfaces and contacts, and to provide secondary confirmation of the stoichiometry of the buried oxide via X-ray photoelectron spectroscopy (XPS). If successful, this methodology would allow for a high-throughput, < 1-nm depth resolution of the near-interface oxide stoichiometry that governs contact and surface (interface) passivation. Thus, TOF-SIMS could provide information typically obtained from a combination of XPS and EELS measurements. Furthermore, we could apply that knowledge to the performance of the passivated contacts and gain valuable insight into the contact performance with a high-throughput characterization method.

II. EXPERIMENTAL

Single-side polished n-Cz (22 Ω–cm) from WRS, double-side polished Topsil p-FZ (4 Ω–cm), and KOH-etched Norsun n-Cz (4 Ω–cm) KOH saw-damage-removed wafers were RCA cleaned (SC-1, SC-2), and then subjected to thin (~1.5 nm as verified by ellipsometry) SiO_x growth thermally, as a low-temperature oxide in a quartz tube furnace with a 6:1 N_2/O_2 mixture at around 700°C as described in detail previously [3]. On this tunneling oxide, a-Si:H layers were then grown via 13.56-MHz plasma-enhanced chemical vapor deposition (PECVD) in a multi-chamber cluster tool manufactured by MVSystems, Inc., at a substrate temperature of about 300°C and at 1.0 torr and H_2-to-SiH_4 ratio of R=100, on both sides of the wafer. The a-Si:H was thermally solid-phase crystallized (SPC) into poly-Si in a tube furnace with N_2 flow at 850°C. It should be noted that the adhesion of as-grown a-Si:H as well as crystallized poly-Si to SiO_x depends on the H_2/SiH_4 ratio as well as the SiO_x preparation. After that, samples received atomic-layer-deposited (ALD) Al_2O_3 film grown in a Beneq reactor using trimethlyaluminum and H_2O precursors at 200°C. Samples were then annealed in a tube furnace with forming gas flow at 400°C for 20 min.

For poly-Si structures, the Al_2O_3 was removed with dilute HF, and the poly-Si was etched to a thickness of 1–3 nm using a solution of 5% TMAH in deionized water at 80°C to prepare the samples for XPS analysis of the buried oxide. XPS measurements were completed using a Physical Electronics 5600 system. Monochromated AlKα radiation was used as the excitation source. The binding energy scale of the spectrometer was calibrated by measuring sputter-cleaned foils of copper, gold, and molybdenum.

TOF-SIMS was completed using an ION-TOF TOF SIMS V instrument. Secondary ions for analysis were created by a 3-lens 30-keV BiMn ion gun. The primary beam was operated in bunched mode, with an 11-ns pulse width and a pulsed beam current of 1 pA. A cesium or oxygen ion beam with an energy of 1 keV was used as the sputtering beam (sputtering current 5–10 nA). These conditions allow for sub-nm depth resolution when profiling.

978-1-5090-5606-4/17 $31.00 © 2017 IEEE

III. RESULTS

In a preliminary study reported elsewhere, we presented results where we resolved the buried tunneling oxide by TOF-SIMS for both pre- and post-SPC structures.[4] We noted that we saw similar depth profiles both for the Si_3O signal, indicating the presence of a sub-stoichiometric oxide at the cSi/SiO_x interface,[1,2] and for the more "intense" Si_2O cluster. We tracked the species SiH as a marker for the hydrogen content and the species SiO_2 as a marker for oxygen content. The SiH signal was seen in the amorphous silicon layer with a pile-up in the SiO_x layer, which returns to baseline levels with dehydrogenation via the SPC annealing treatment. This is indicative of a defective, passivated interface at the buried oxide becoming less defective after the SPC thermal treatment.

In this follow up study we investigated a set of samples through various steps of our standard device processing: 1) after deposition of the doped polysilicon, 2) after a subsequent solid-phase crystallization of the polysilicon, and 3) after a subsequent Al_2O_3 deposition and hydrogenation process (forming gas anneal). The Al_2O_3 layer was subsequently removed and the samples were subjected to TOF-SIMS analysis, the results for which are show in Figure 1.

The TOF-SIMS hydrogen profiles in Fig. 1A shows as expected a large hydrogen signal though the a-Si layer. Interestingly, the SPC anneal does not remove all the hydrogen from the now polysilicon layer, and the remaining hydrogen seems to have accumulated at the wafer side of the tunneling oxide. After FGA with Al_2O_3 the hydrogen is very much localized in the tunneling oxide, with an additional accumulation on each side of the tunneling oxide, indicative of passivated interfaces.

We can examine the TOF-SIMS Si_2O signal in Fig 1B to investigate the nature of any substoichiometric oxide present at the buried interface. [1, 2] Interestingly, it appears as if sub-stoichiometric oxide signal 'spreads out' after the SPC treatment, indicating a more diffuse interface and a shift from SiO_2 towards more SiO_x. After the FGA the Si_2O signal significantly sharpens, and shifts to the poly interface, indicating a shift away from SiO_x towards SiO_2.

XPS results for the silicon oxide binding energy region of the same samples are presented in Fig. 2. The data was fit with three components to the silicon-oxide spectrum, and the FWHM for each component was held constant for all three samples. We neglected to include a fourth Si^0 component as prior work indicated it did not improve the fitting.[2] It is assumed the highest binding energy component at ~104.1eV (orange in all the plots) corresponds to SiO_2, and those at lower BE (blue and red curves) correspond to substoichiometric oxides (e.g. Si^{3+} and Si^{2+}). Interestingly, Fig 2 shows that the buried oxide has the largest fraction of SiO_2 just after polysilicon deposition (Fig. 2A), and the oxide trends more towards Si_2O_3 than SiO_2 after the SPC treatment (Fig 2B), which is consistent with TOF SIMS results in Fig 1B.

After the Al_2O_3 deposition and FG anneal the amount of SiO_2 increases, but a surprisingly large amount of substoichiometric oxide persists, which is also consistent with the relative amounts of Si_2O signal observed in the TOF-SIMS data (Fig 2B). The results from XPS and TOF SIMS tell a consistent story; that there is a more diffuse tunneling oxide with a larger substoichiometric component after SPC than is present after polysilicon deposition. The correlation between the two techniques, along with the prior work in this area [1, 2] suggests analysis of TOF-SIMS secondary ion clusters could be used to reliably investigate the stoichiometry of buried interfaces and passivated contacts and relate this information to the performance of the contacts.

During the course of this study we noted an important artifact for samples analyzed without etching off the Al_2O_3 capping layer (structure $AlO_x/SiO_x/pFZ$). Figure 3A shows a standard SIMS analysis of the matrix species, Al, O, and Si; and, in addition, the secondary-ion complex Si_3O is overlaid on this graph to show the presence of the sub-stoichiometric SiO_x at the tunneling oxide/wafer interface. Several secondary-ion species associated with the alumina layer are shown in Fig. 2B. These data were collected under conditions where the depth resolution of the measurement is on the order of one nanometer. Surprisingly, Fig. 3A suggests that the thickness of the alumina layer extracted from the signals for Al, AlO, and Al_2O_3 appears to vary. In particular, the AlO signal appears to extend several nanometers past the buried SiO layer (which is where the Al^- signal begins to decrease).

This apparently anomalous AlO signal might be the result of different secondary-ion yields for the species Al, AlO, and Al_2O_3. In particular, the high oxygen content in the sample in and around the buried silicon oxide layer can influence the secondary-ion yield (and thus, the intensity) for AlO species more strongly than for the Al_2O_3 species. To test this hypothesis, we performed additional measurements under conditions that reduce these matrix effects on the resultant secondary-ion signals. First, we introduce a controlled oxygen leak during the measurement under conditions otherwise identical to Fig. 3A. Even with the enhanced oxygen content in the ambient (base pressure 1×10^{-5} mbar instead of the typical 1×10^{-9} mbar), a strong difference is still observed in the depth at which the Al, AlO, and Al_2O_3 signals start to decrease when entering the buried oxide and silicon wafer. Finally, TOF-SIMS data was collected under special conditions known to eliminate matrix-related artifacts. The data taken in Fig. 3A used cesium as a sputter source, which is common for measurements done in negative SIMS polarity. However, when Cs is used as a sputter source, one can perform what is known as an "MCs^+ analysis," where, for example, instead of following Al or Si, one follows the secondary-ion complex of $CsAl^+$ or $CsSi^+$, and conducts the measurement in positive SIMS polarity. This is known to mitigate matrix-related effects on the yield of a secondary-ion signal. The data collected in this manner is shown in Fig. 3B. In this case, one can clearly see that the Al^+, $CsAlO^+$, and AlO^+ signals start to drop off at identical depths (there was no Al_2O_2 or Al_2O_3 or signal measureable under these conditions). This indicates that the high-AlO^- signal noted in Fig 3A was due to the complex interplay between secondary-ion yield and matrix oxygen content. These observations suggest that the TOF-SIMS ion-complex profiles might sensitively depend on their bonding structure and neighboring environment (e.g., SiO_x next to Al_2O_3). These interactions might provide a valuable link to the device performance.

IV. CONCLUSIONS

We have reported on a case-study to use TOF-SIMS to investigate the stoichiometry of buried oxides in passivated contact silicon solar cells. A substoichiometric oxide is present in the buried oxide for polySi/SiO(x)/Si with increased hydrogen content, indicative of a passivated, defective interface. It appears as the substoichiometric oxide becomes more diffuse after SPC treatment, and sharpens significantly after Al_2O_3 deposition and subsequent FGA. These results were consistent for analysis of both TOF-SIMS secondary ion clusters related to substoichiometric SiO_x, and XPS data taken on the buried oxide of the same samples. This work suggests a novel TOF-SIMS methodology as a new, high-throughput method to investigate passivated contacts in silicon solar cells.

ACKNOWLEDGMENTS

Funding for this work was provided by the United Sates Department of Energy EERE contract SETP DE-EE00030301 (SuNLaMP) and under Contract No. DE-AC36-08GO28308. The U.S. Government retains and the publisher, by accepting the article for publication, acknowledges that the U.S. Government retains a nonexclusive, paid up, irrevocable, worldwide license to publish or reproduce the published form of this work, or allow others to do so, for U.S. Government purposes.

REFERENCES

[1] K. Chiba and S. Nakamura, "Characterization of ion species of silicon oxide films using positive and negative secondary ion mass spectra," *Applied Surface Science,* vol. 253, pp. 412-416, 2006.

[2] K. Chiba and Y. Takenaka, "Embedded structure of silicon monoxide in SiO2 films," *Applied Surface Science,* vol. 254, pp. 2534-2539, 2008.

[3] B. Nemeth, D. L. Young, M. R. Page, V. LaSalvia, S. Johnston, R. Reedy, and P. Stradins, "Polycrystalline silicon passivated tunneling contacts for high efficiency silicon solar cells," *Journal of Materials Research,* vol. 31, pp. 671-681, 2016/003/28 2016.

[4] B. Nemeth, S. P. Harvey, J. Li, D. L. Young, A. Upadhyaya, V. LaSalvia, B. G. Lee, M. R. Page, and P. Stradins, "Effect of the SiO2 interlayer properties with solid-source hydrogenation on passivated contact performance and surface passivation," *Energy Procedia,* vol. Submitted, 2017.

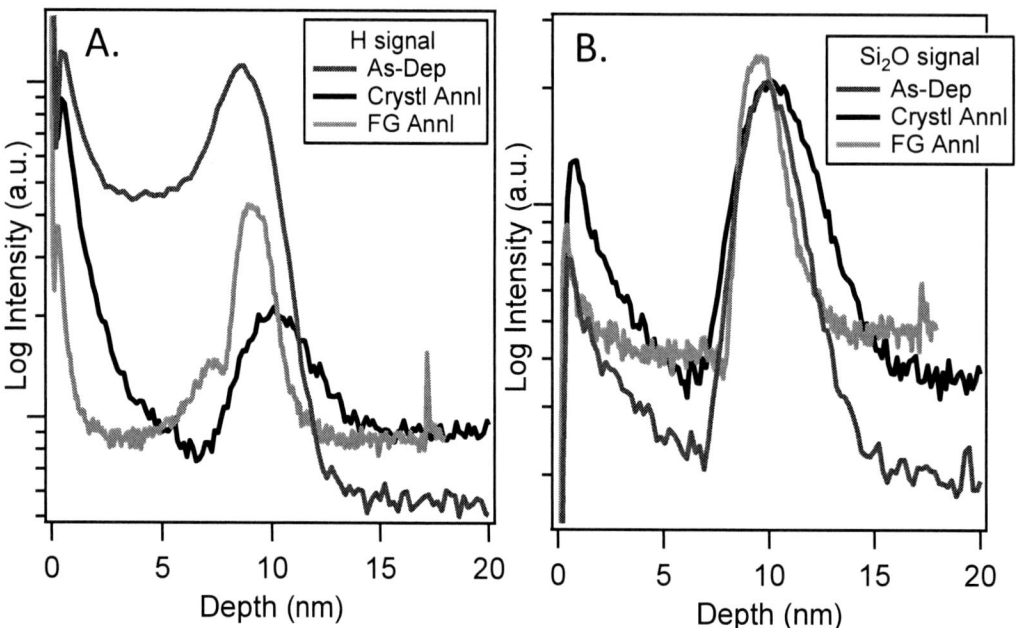

Figure 1. (a) TOF-SIMS hydrogen profiles for samples after polysilicon deposition (As-Dep), after SPC treatment (Crystl Annl), and after forming gas anneal (FG Annl). (b) TOF-SIMS Si_2O profiles for samples after polysilicon deposition (As-Dep), after SPC treatment (Crystl Annl), and after forming gas anneal (FG Annl). The profiles in B suggest the oxide becomes more diffuse and with a larger SiO_x component after SPC.

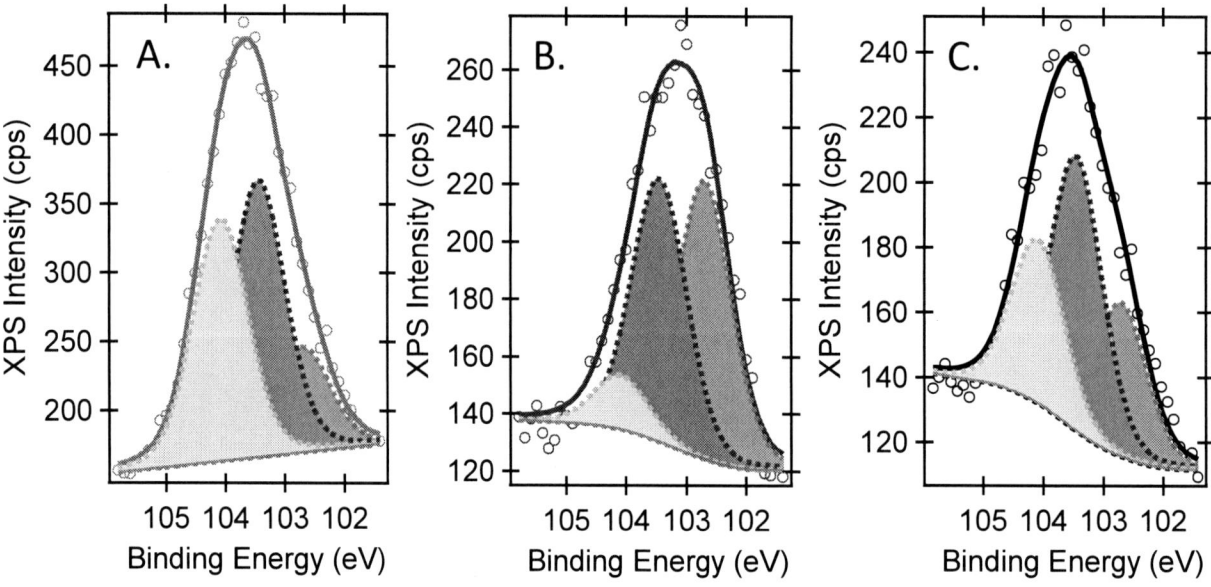

Figure 2. XPS data showing the silicon oxide portion of the BE spectrum, included peak-fitting results. The largest BE component at 104.1eV is assumed to be SiO_2, the blue and red components at lower BE are assumed to be substoichiometric oxides, e.g. Si_3O_4 and SiO (a) Tunneling oxide after polysilicon deposition. (b) Tunneling oxide after SPC treatment. (c) Tunneling oxide after Al_2O_3 deposition and forming gas annealing treatment.

978-1-5090-5606-4/17 $31.00 © 2017 IEEE   2705

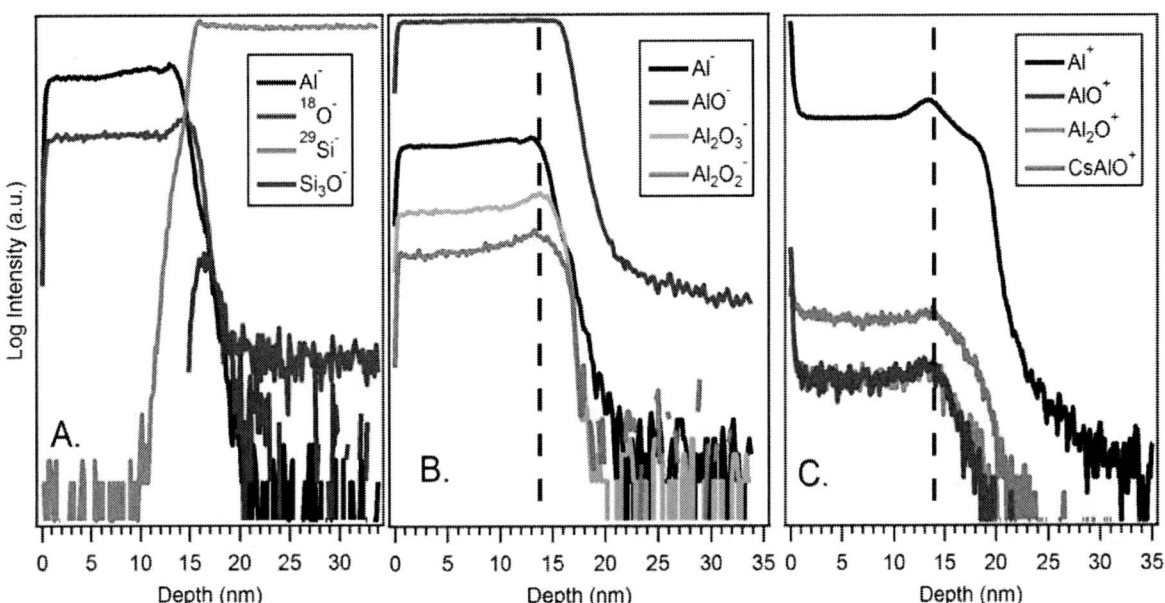

Figure 3. TOF-SIMS profile data for an $Al_2O_3/SiO_{(x)}/Si$ structure. A) Standard negative polarity measurement showing matrix signals and the Si_3O signal indicative of substoichiometric SiO_x. B) Standard negative polarity measurement showing signals associated with the alumina layer. C) MCs^+ positive polarity measurement, which can mitigate the effect of matrix oxygen content on secondary-ion yield.

A comparison between quasi-steady state and transient photoconductance lifetimes in silicon ingots: simulations and measurements

[1]Mohsen Goodarzi, [2]Ronald Sinton, [3]Daniel Chung, [3]Bernhard Mitchell, [3]Thorsten Trupke and [1]Daniel Macdonald

[1]Research School of Engineering, The Australian National University, Canberra ACT 2601, Australia

[2]Sinton Instruments Inc., Boulder CO USA. http://www.sintoninstruments.com

[3]School of Photovoltaic and Renewable Energy Engineering, UNSW, Sydney, NSW 2052, Australia

Abstract — **We present and compare numerical simulations and experimental data for Quasi-Steady State (QSS) and transient photoconductance lifetime measurements on silicon ingots. The simulation results show that the QSS method is generally more accurate for lifetimes below 150 μs, whereas transient measurements are more accurate above this value. However, transient measurements require sufficient time to have elapsed after the flash is terminated, to ensure that the impact of the unpassivated ingot surface is reduced. The results also show that the surface recombination velocity has a slightly reduced impact on n-type material in comparison with p-type material, due to the reduced minority carrier mobility. The simulation results are also compared with measured QSS and transient lifetimes on a standard p-type monocrystalline block.**

Index Terms — **Carrier lifetime, QSSPC, Silicon ingot, Surface recombination, Transient lifetime.**

I. INTRODUCTION

The carrier lifetime is one of the key parameters for the performance of silicon solar cells, and is invaluable for process control. The transient Photoconductance Decay (PCD) and Quasi-Steady State Photoconductance (QSSPC) techniques have been developed and are widely used for bulk lifetime measurements on silicon wafers and ingots [1]. However, the effective lifetimes measured on ingots with the QSS and transient methods may differ noticeably, depending on the carrier lifetime itself [2]. To investigate the reasons behind this discrepancy, we simulate the carrier dynamics of the QSS and the transient modes for both p-type and n-type ingot measurements. The p-type simulation results are compared to measurements with a BCT-400 boule tester from Sinton Instruments on a monocrystalline p-type Cz-grown silicon block.

II. SIMULATION AND EXPERIMENT DETAILS

The QSS and the transient modes are simulated in this work to compare the results measured by the two methods when applied to ingots. In both cases, the simulation of the excess carrier profiles is based on the continuity equation [3]:

$$\frac{\partial \Delta n}{\partial t} = G(x,t) - U(x,t) + \frac{1}{q}\frac{dJ_n}{dx} \qquad (1)$$

where G and U are the generation and the recombination rates respectively, and the last term is a diffusion term caused by non-uniform carrier densities. The excess carrier density $\Delta n(x,t)$ is then a function of both depth x and time t. (1) was numerically solved for the QSS mode (when $\frac{d\Delta n(t)}{dt} = 0$) and the transient mode (when $G(t) = 0$). The simulations were based on a finite element approach, with the depth and time intervals Δx and Δt chosen to be small enough to ensure the local changes in the excess carrier density are approximately linear in x and t.

For both the QSS and transient modes, the simulated excess carrier densities are modulated by the depth sensitivity of the coil [2], and then used to calculate an average excess carrier density, Δn_{avg} (weighted average) as explained in detail in [4], where the effective thickness (w_{eff}) is also defined.

For the QSS mode, the effective QSS lifetime ($\tau_{QSS\text{-}eff}$) based on the average excess carrier density is then used to estimate the QSS bulk lifetime ($\tau_{QSS\text{-}bulk}$), via a transfer function, as proposed by Bowden and Sinton [4]. This transfer function is intended to remove the effect of surface recombination arising at the unpassivated ingot surface.

For the transient mode simulation, the first stage is a QSS mode simulation, in order to build up the initial carrier concentration before the flash is terminated and the transient decay commences. This results in a $\Delta n_{avg}(t)$ at time t elapsed after terminating the light source, with the transient lifetime $\tau_{transient}$ calculated from the slope of the average excess carrier density decay. No transfer function is applied to the transient analysis.

In the simulations presented here, the true bulk lifetime τ_{Bulk} is defined as an input parameter, and is injection-independent, for simplicity.

To allow comparison with the simulation results, both QSSPC and transient lifetime measurements were performed with a BCT-400 boule tester from Sinton Instruments on a p-type monocrystalline block cut from the central region of a commercially grown Cz ingot. For the QSS mode, a standard flash decay time of 5 ms was used with an IR 1000 filter. In this work, measured QSS and transient lifetimes are reported at $\Delta n_{avg} = 10^{15}$ cm^{-3}.

III. RESULTS AND DISCUSSION

Fig. 1. shows the simulated excess carrier density profiles during a transient photoconductance decay, over a depth of 1 cm, at different times after the light source was turned off. In this case $\tau_{bulk} = 300$ μs and the peak flash intensity was 400 suns. It shows that the high surface recombination velocity significantly impacts the decay during the early parts of the measurement, which will cause the transient lifetime to be artificially low. However, this effect is reduced over time, and eventually the bulk recombination process becomes the major influence in the reduction of the excess carrier density.

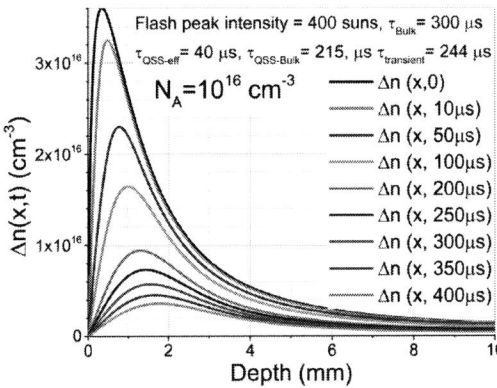

Fig. 1.   Excess carrier density profile vs depth for different times after the light source is turned off, for p-type silicon. $\tau_{QSS-eff}$ is the simulated effective QSS lifetime and $\tau_{QSS-Bulk}$ is the QSS lifetime after applying the transfer function. All lifetimes are determined at $\Delta n_{avg} = 10^{15}$cm^{-3}.

Fig. 2 shows the simulated average excess carrier density as a function of time for transient measurements with different bulk lifetimes. The transient lifetimes extracted at $\Delta n_{avg} = 10^{15}$ cm^{-3}, are also shown, as are the corresponding QSS bulk lifetimes. In this case, which represents quite typical measurement conditions, the QSS results are closer to the true bulk lifetime values for lifetimes lower than 150 μs, whereas the transient lifetimes are more accurate above this value.

In general however, the transient lifetime can always provide a more accurate result, if a sufficiently long time is elapsed before the slope is extracted. However, this is often not practical, as the peak sun intensity may need to be

unreasonably high to allow the lifetime to be extracted at the desired excess carrier density after a sufficient elapsed time. Furthermore, the decay time constant of most flashes does not allow accurate transient measurements below 100 microseconds in any case.

As the bulk lifetime increases above 150 μs, the QSS-bulk lifetime is increasingly inaccurate in the simulation results presented here. This is due to the transfer function implemented in the Sinton Instruments tool being optimized for lower lifetimes. In a real QSS measurement, additional uncertainties arise in high lifetime samples in the QSS mode, due to dynamic affects occurring when the carrier lifetime becomes comparable to the flash decay time, meaning that the assumption of Quasi-Steady State conditions becomes increasingly invalid [5, 6].

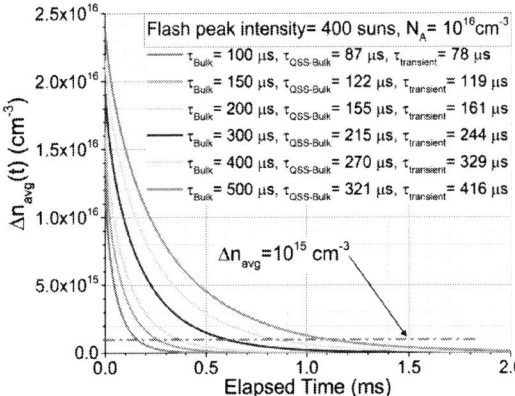

Fig. 2.   Simulated excess carrier density as a function of time in transient lifetime measurement, lifetimes extracted at $\Delta n_{avg}=10^{15}$cm^{-3}.

As mentioned above, and as discussed in [7], the unpassivated surface induces a large carrier diffusion current towards the surface, placing another constraint on the transient measurements. These surface recombination effects are largely removed from the carrier density decay over time. Consequently, the accuracy of the transient lifetime measurement depends on the time elapsed after which the decay rate is measured, as illustrated in Fig 3. The figure shows that if a sufficiently long time has elapsed, the transient method can in principle provide a very accurate result, at least for an injection-independent bulk lifetime, as assumed here. In practice, the achievable delay time is constrained by the initial flash intensity. Figure 3 also indicates the required delay time after the flash is terminated for the simulated bulk lifetime to approach the true bulk lifetime with 95% accuracy.

Fig. 3 also shows the elapsed time required for 95% accuracy of the transient lifetime with a peak flash intensity of 400 suns. For lower lifetimes, the delay time should be about 10 times the bulk lifetime, reducing to 6 times for lifetimes around 500 microseconds. In general, a conservative approach is to allow an elapsed time of at least ten times the

978-1-5090-5606-4/17 $31.00 © 2017 IEEE   2708

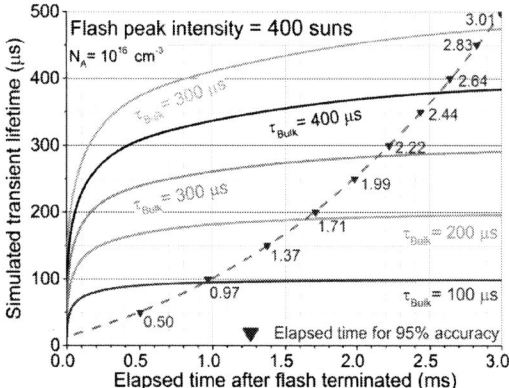

Fig. 3. Simulated transient lifetime value vs the time of measurement elapsed after the light source is terminated. The labelled data are the required elapsed time to report the lifetime with 95% accuracy. The dashed line is fitted to present the required elapsed time for 95% accuracy as a function of the bulk lifetime.

carrier lifetime, to ensure accurate results.

The surface recombination effect on both QSS and transient lifetimes also depends on the minority carrier mobility. The minority carrier electrons in p-type silicon material have significantly higher mobility (larger effective diffusion coefficient, D_{eff}) than the holes in n-type material. Consequently, minority carriers in n-type silicon diffuse towards the surface more slowly than electrons in p-type material, resulting in a higher excess carrier lifetime in n-type material with the same doping, lifetime and illumination intensity.

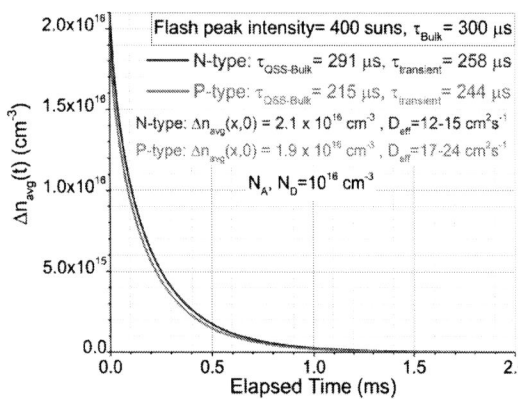

Fig. 4. Δn_{avg} vs the time of measurement elapsed after the light source is terminated for p-type (red) and n-type (blue) silicon.

This is illustrated in Fig. 4 for a transient decay, where the same illumination intensity is applied to p- and n-type doped material. Subsequently, the rate at which the carrier density profile decays is quicker in p-type material resulting in a smaller extracted transient lifetime.

The Klaassen [8, 9] mobility model is used to calculate the effective diffusion coefficient in this work. The value of D_{eff} is calculated as explained in ref [10] and is injection level dependent, resulting in a range of diffusion coefficients during the decays, as shown in Fig 4.

The initial excess carrier density in a sample directly depends on the illumination source intensity, as well as the carrier lifetime. Higher initial injection levels allow a longer delay time until the injection level decays to $\Delta n_{avg} = 10^{15}$ cm^{-3}. According to Fig. 3, a longer delay should result in a higher lifetime closer to the actual bulk lifetime. Fig. 5. shows the results for simulated transient lifetimes with different flash peak intensities, as well as the QSS-bulk lifetime, all reported at $\Delta n_{avg} = 10^{15}$ cm^{-3}. The transient lifetime values with 400 suns flash peak intensity are labeled (blue) in Fig. 5. This intensity is commonly used in the BCT-400 boule tester in transient mode.

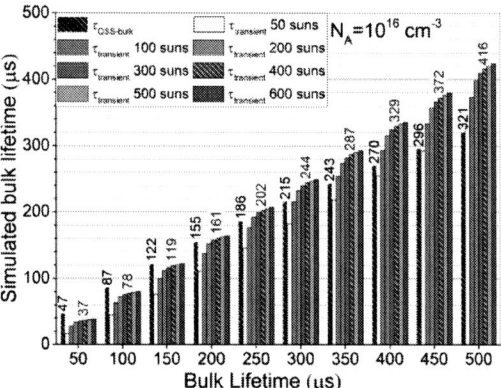

Fig. 5. Simulated transient lifetime with different flash peak intensities compared with the QSS-bulk lifetime (black data).

As can be seen, the transient measurement results with the peak intensities below 200 suns report noticeably lower lifetimes than the bulk lifetime. In contrast, the reported lifetime with 500 and 600 suns are only slightly higher than the results with 400 suns peak intensity. Therefore, the flash with 400 suns peak intensity appears to be suitable for transient measurements at $\Delta n_{avg} = 10^{15}$ cm^{-3} in this lifetime range.

Several transient measurements were performed on a monocrystalline p-type silicon block applying different flash peak intensities to find the peak at which the measured transient lifetime saturated. The results are shown in Fig. 6. The transient lifetime results are generally higher than the QSS-bulk results, as expected based on the simulations. There are significant increases between the measured transient lifetimes when the peak intensity increased from 180 to 413 suns. However, the lifetime is more stable when the peak intensity increased from 413 to 475 and 584 suns. This is also qualitatively consistent with the simulation results above.

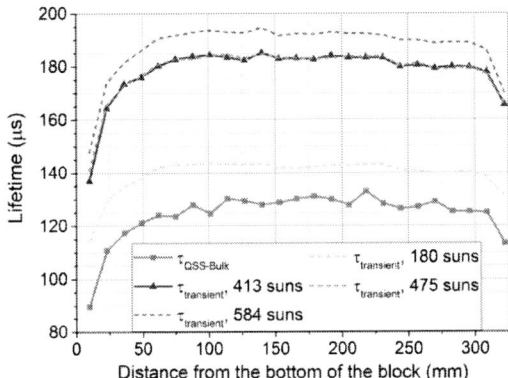

Fig. 6. The lifetime measurement results on a standard p-type monocrystalline block at $\Delta n_{avg}=10^{15}$ cm^{-3}. The transient measurements were performed with flash peak intensities of 180, 413, 475 and 584 suns.

IV. CONCLUSIONS

Numerical simulations show that the QSSPC technique is accurate for measuring bulk lifetimes below 150 µs on ingots, for which the transfer function eliminates the surface recombination effects efficiently, resulting in good agreement with the actual bulk lifetime. However, above 150 µs, the transfer function is less accurate. Hence, the transient measurement is the preferred technique in this range, provided the lifetime is extracted after a sufficiently long time to ensure that the bulk recombination is the dominant recombination process in the carrier decay. A conservative approach to satisfying this requirement is to allow a delay time of 10 times the carrier lifetime. Measured lifetime data on a p-type monocrystalline silicon block were qualitatively consistent with these simulation results.

ACKNOWLEDGEMENT

This work has been supported by the Australian Renewable Energy Agency (ARENA) through research grant "RND009".

REFERENCES

[1] D. Klein, F. Wuensch, and M. Kunst, "The determination of charge-carrier lifetime in silicon," *physica status solidi (b),* vol. 245, pp. 1865-1876, 2008.

[2] J. S. Swirhun, R. A. Sinton, M. K. Forsyth, and T. Mankad, "Contactless measurement of minority carrier lifetime in silicon ingots and bricks," *Progress in Photovoltaics: Research and Applications,* vol. 19, pp. 313-319, 2011.

[3] H. Nagel, C. Berge, and A. G. Aberle, "Generalized analysis of quasi-steady-state and quasi-transient measurements of carrier lifetimes in semiconductors," *Journal of Applied Physics,* vol. 86, p. 6218, 1999.

[4] S. Bowden and R. A. Sinton, "Determining lifetime in silicon blocks and wafers with accurate expressions for carrier density," *Journal of Applied Physics,* vol. 102, p. 124501, 2007.

[5] J. A. Giesecke, M. C. Schubert, F. Schindler, and W. Warta, "Harmonically Modulated Luminescence: Bridging Gaps in Carrier Lifetime Metrology Across the PV Processing Chain," *IEEE Journal of Photovoltaics,* vol. 5, pp. 313-319, 2015.

[6] R. Sinton, T. Mankad, S. Bowden, and N. Enjalbert, "Evaluating silicon blocks and ingots with quasi-steady-state lifetime measurements," in *Proceedings of the 19th European Photovoltaic Solar Energy Conference, Paris, France,* 2004, pp. 520-523.

[7] Mohsen Goodarzi, Ronald A Sinton, Hao Jin, Peiting Zheng, Wei Chen, Quanzhi Wang, *et al.,* "Accuracy of interstitial iron measurements on p-type multicrystalline silicon blocks by quasi-steady-state photoconductance," *Submitted to IEEE Journal of Photovoltaics,* (2017).

[8] D. Klaassen, "A unified mobility model for device simulation—I. Model equations and concentration dependence," *Solid-State Electronics,* vol. 35, pp. 953-959, 1992.

[9] D. Klaassen, "A unified mobility model for device simulation—II. Temperature dependence of carrier mobility and lifetime," *Solid-State Electronics,* vol. 35, pp. 961-967, 1992.

[10] A. Cuevas, "Modelling silicon characterisation," *Energy Procedia,* vol. 8, pp. 94-99, 2011.

New development in Glow Discharge Optical Emission Spectrometry for the characterization and the thickness measurement of layers for photovoltaic applications

Philippe Hunault[1], Matthieu Chausseau[1], Patrick Chapon[2], Sofia Gaiaschi[2], Anaïs Loubat[3,4], Muriel Bouttemy[3], Arnaud Etcheberry[3]

[1]HORIBA Scientific, Edison, NJ, 3880 Park Avenue, 08820, USA

[2]HORIBA Scientific, 16-18 rue du Canal, Longjumeau, 91165, France

[3]Institut Lavoisier de Versailles (ILV), UMR 8180 CNRS-UVSQ, 45 av. des Etats-Unis, Versailles, 78035, France.

[4]Institut Photovoltaïque d'Ile-de-France (IPVF), 8 rue de la Renaissance, Antony, 92160, France.

Abstract — Glow Discharge Optical Emission Spectrometry (GD-OES) is an elemental depth profiling technique, considered as complementary to Auger Spectroscopy, Secondary Ion Mass Spectrometry or X-ray Photoelectron Spectroscopy for the characterization of thin films.

GD-OES has already been widely used for the characterization of materials used for photovoltaic applications and results obtained for CIGS and Perovskites based technologies will be presented.

With its most recent development, allowing for the direct measurement of layer thicknesses, GD-OES has become even more attractive for materials characterization. This innovative feature called Differential Interferometry Profiling (DiP) will be described and illustrated by examples on CIGS solar cells.

I. INTRODUCTION

Pulsed Radio Frequency Glow Discharge Optical Emission Spectrometry (RF GD-OES) is becoming a powerful technique for the analysis of photovoltaic materials and devices. It relies on the very fast sputtering (typically µm/min) of a representative area of a sample by a high density and low energy plasma, to provide the direct measurement of the elemental depth profile, with nanometric resolution.

While the use of a RF source allows the analysis of conductive, non-conductive or hybrid materials, patented features such as the Ultra-Fast Sputtering (UFS) allow the characterization of metallic layers embedded behind polymeric ones, which are of particular interest for hybrid photovoltaic devices. Moreover, the pulsed source allows the efficient analysis of fragile materials (e.g. organics samples and coatings on glass) as it reduces the thermal load on the sample, avoiding an excessive heating which could cause either the breaking of the sample or the migration of some elements, such as sodium or calcium. In addition, the pulsed mode ensures also an improved depth resolution.

RF GD-OES offers the possibility to measure in few minutes all elements of interest from H to U, including light and alkali ones. Thanks to its sputtering rate, it allows the fast control over all steps of a fabrication process as well as the study of complete photovoltaic devices.

Figure 1 shows a schematic representation of a GD-OES plasma source. The sample is mounted outside the plasma chamber and pressed against an o-ring in order to assure its sealing. No Ultra High Vacuum is need and this makes GD-OES a fast and easy to operate technique and allows access to the sample for specific in-operando studies.

Fig. 1.   Schematic of RF Glow Discharge Source

A schematic representation of the Glow Discharge is presented in Figure 2. In GD-OES an Ar plasma is used to sputter a significant area of the sample. When the discharge breaks down, the Ar ions are accelerated towards the surface of the sample, which is grounded, and they bombard it. Then, the sputtered species move inside the glow discharge where they are excited through collisions with Ar ions. The de-excitation of the excited species causes the emission of light at characteristic wavelengths depending on elements. By using a polychromator, the emitted light is measured, allowing for a time resolved elemental analysis and therefore a depth profile analysis.

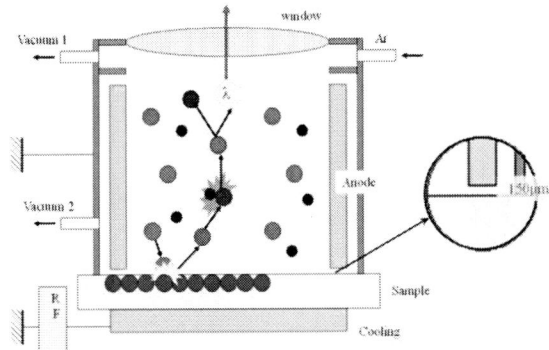

Fig. 2.   Schematic of RF Glow Discharge principle

With the recent development of a Differential Interferometry Profiling (DiP) accessory, which can be mounted inside the GD-OES instrument (Figure 3), it is now possible to obtain the direct measurement of the crater depth during the analysis. GD-OES is a comparative technique providing the evolution of the elemental intensity as a function of the sputtering time. One of the challenges of this technique is the quantification step. Indeed, by establishing a set of calibration curves it is possible to obtain quantitative information: concentration as a function of depth. However, particular care has to be paid for the time-to-depth conversion. Until now in GD-OES such conversion was the last step of the quantification process and was based on the assumption of the material density. The big advantage of DiP is that the crater depth and the sputtering rate are now directly measured and no more assumption on the material density is required

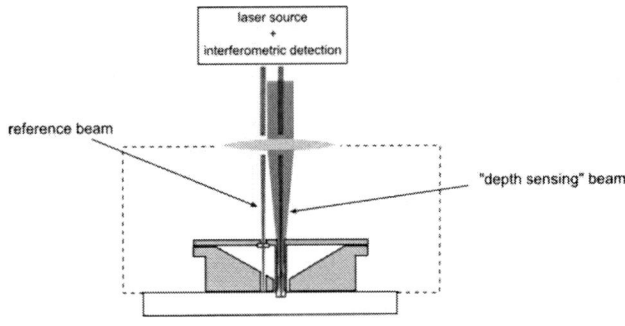

Fig. 3.   Schematic of Differential Interferometry Profiling Module with reference and probe beams displayed

II. RESULTS AND DISCUSSION

A. Study of Sodium diffusion in interfaces

N. Naghavi used GD-OES to study the diffusion of Sodium in the Zn(S,O) layer, which is known to lead to meta-stability problems [1]. Results were presented during the French GD users meeting in 2016. Figure 5a shows the depth profile of a CIGS solar cell exhibiting a low amount of Sodium at the

interface, whereas Figure 5b displays the GD-OES results obtained on a different CIGS solar cell showing a significantly higher amount of Sodium.

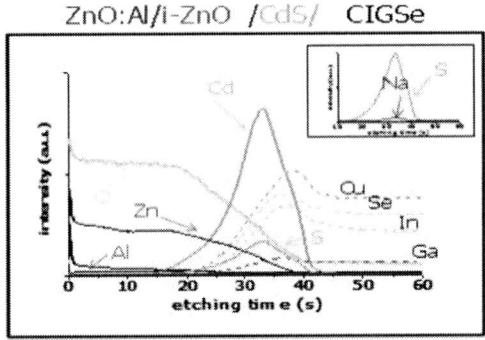

Fig. 5a   Depth profiles of CIGS solar cell exhibiting low amount of Sodium (courtesy of N. Naghavi [1])

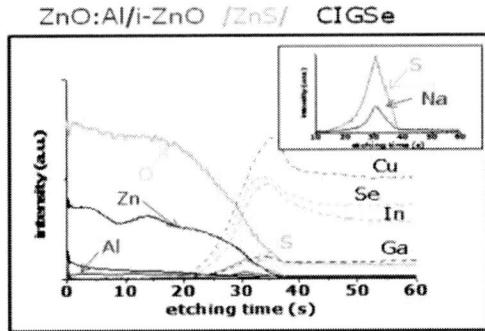

Fig. 5b   Depth profiles of CIGS solar cell exhibiting high amount of Sodium (courtesy of N. Naghavi [1])

This study could be performed thanks to the use of the pulsed source as the reduction of the thermal load on the sample allows to avoid the unwanted diffusion of elements during the analysis, revealing the real diffusion due to fabrication processes.

B. Elemental Depth Profiling of Perovskites

GD-OES is also efficient technique for the analysis of Perovskites based cells.

Using GD-OES, Cojocaru et al. [2] were able to study the structure of the devices and they correlated these results with EDS-STEM measurement (Figures 6a and 6b)

Thanks to its design and to the absence of ultra-high vacuum, in-operando characterizations are possible during a GD-OES analysis. Indeed, Lee *et al.* [3] published recently a paper on the study of Halide Ionic Migration under Bias in $CH_3NH_3PbI_{3-x}Cl_x$-Based Perovskite Solar Cells Using GD-OES Analysis.

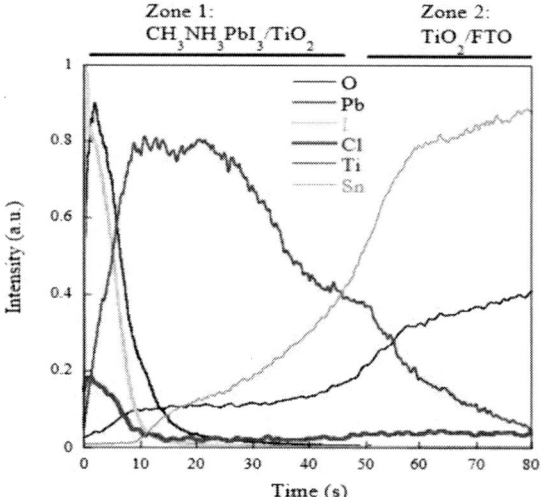

Fig. 6a GD-OES profiles of perovskite films $CH_3NH_3PbI_{3-x}Cl_x$ deposited on 50 nm TiO_2 on the FTO substrates (Courtesy of Cojocaru et al. [2])

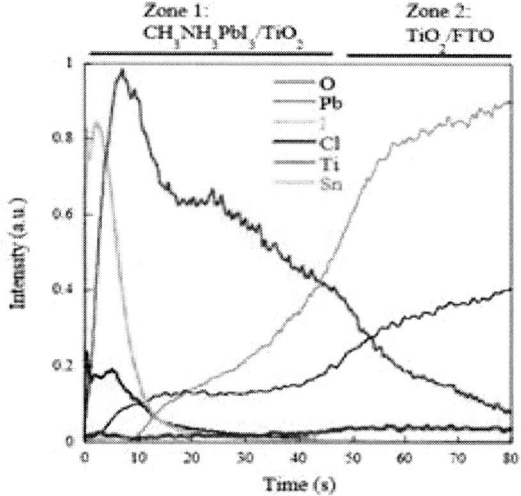

Fig. 6b GD-OES profiles of perovskite films $CH_3NH_3PbI_3$ deposited on 50 nm TiO_2 on the FTO substrates (Courtesy of Cojocaru et al. [2])

Depth profiles obtained as a function of the negative or positive applied biases are shown in Figures 7a and 7b, respectively, proving the mobility of the I ions.

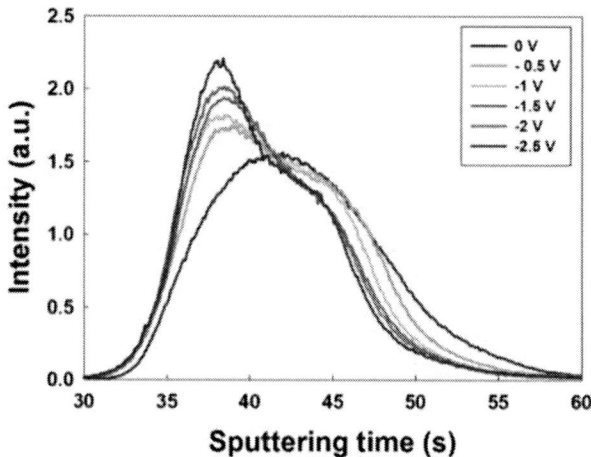

Fig. 7a GD-OES profile lines of iodine species in Perovskite films depending on negative applied biases (Courtesy of Lee et al. [3])

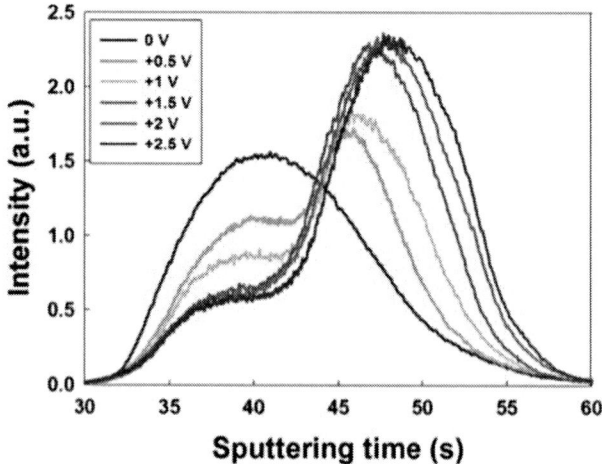

Fig. 7b GD-OES profile lines of iodine species in Perovskite films depending on positive applied biases (Courtesy of Lee et al. [3])

C. Elemental Depth Profile and layer thickness measurement on CIGS solar cells

The most recent development for GD-OES is the Differential Interferometry Profiling module that allows to get real time depth measurement as a function of time, for the first time in Glow Discharge.

Thanks to the DiP module, layer thicknesses of CIGS solar cells were measured along with the elemental composition. CIGS samples are particularly challenging for DiP and XPS as they are often rough, partially transparent and oxidized at the surface. For this reason CIGS/Mo/glass samples were preliminary etched using a $HBr:Br_2 = 0.2:0.02$ mol/L solution [4] to facilitate the study at first. Indeed, this chemical etching leads to a progressive decrease of the roughness while maintaining the surface composition of the CIGS absorber. By varying the etching time, a fine modulation of the roughness can be obtained. For this study, 2 samples etched 2 and 5 min were prepared, and their corresponding roughnesses have been checked by AFM measurements.

The elemental depth profile for the CIGS/Mo/glass sample before etching (RMS 277nm) is shown in Figure 8. Figures 9 and 10 are showing the elemental depth profiles after 2 minutes (RMS 65 nm) and 5 minutes (RMS 37 nm) of etching time, respectively. These results were obtained in the frame of a collaborative work between HORIBA FRANCE and the Lavoisier Institute of Versailles and were presented at the 2016 EMRS-Spring meeting and at the 8th GD Day.

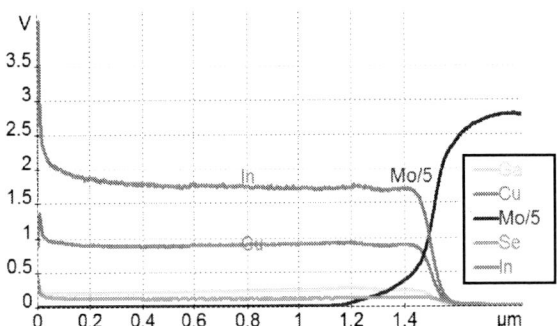

etching. 3 different samples have been measured and all values obtained are in perfect agreement, showing the capability of the Differential Interferometry Profiling module to provide accurate measurement of the depth and thus of the layer thicknesses.

III. CONCLUSION

GD-OES is an ideal tool for the analysis of materials used for photovoltaic applications. It can be used to optimize and monitor processes for films depositions and also to characterize solar cells for research purposes. Examples have been shown here for CIGS and Perosvkites based solar cells but other results have been published on CdTe or Thin Film Si technologies [5-8]. GD-OES can perform elemental depth profiling, giving the elemental distribution of the elements to check for efficiency of processes but, thanks to the pulsed RF source, it also allow studying diffusion effects of elements at the interfaces. Thanks to the innovative Differential Interferometry Profiling module, the GD-OES is now able to provide the layer thickness in real time for samples that are characterized as well as erosion rate and full crater depth. The use of DiP can also help to identify variations in density within a layer as the sputtering rate should be affected.

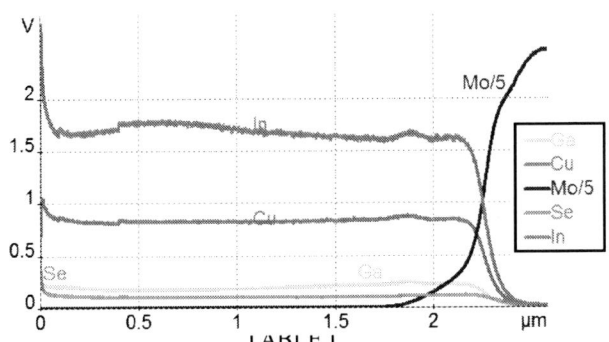

TABLE I
COMPARISON OF DEPTH VALUES ON CIGS SOLAR CELL

Etching time	Sample	DiP μm	Profilometer μm
2 minutes	A	2.30	2.15
	B	2.05	2.11
	C	2.35	2.10
5 minutes	A	1.80	1.55
	B	1.81	1.71
	C	1.70	1.61

Fig. 8   Elemental depth profile as a function of depth in μm for CIGS/Mo/glass solar cell layers before etching (RMS 277 nm)

Fig. 9   Elemental depth profile as a function of depth in μm for CIGS/Mo/glass solar cell layers after 2 minutes etching time (RMS 65 nm)

Fig. 10   Elemental depth profile as a function of depth in μm for CIGS/Mo/glass solar cell layers after 5 minutes etching time (RMS 37 nm)

To validate the accuracy of depth measurements using DiP, measurements were also done using a profilometer and values were compared. Depth values obtained by DiP and by Profilometer are compared in Table I for CIGS solar cells after

References

[1] N. Naghavi, GD Users Meeting, 2016
[2] L. Cojocaru, S. Uchida, D. Matsubara, H. Matsumoto, K. Ito, Y. Otsu, P. Chapon, J. Nakazaki, T. Kubo, H. Segawa, Direct Confirmation of Distribution for Cl- in $CH_3NH_3PbI_{3-x}Cl_x$ Layer of Perovskite Solar Cells , Chem. Lett. 2016, 45, 884–886
[3] H. Lee, S. Gaiaschi, P. Chapon, A. Marronnier, H. Lee, J.C. Vanel, D. Tondelier, J.E. Bourée, Y. Bonnassieux, B. Geffroy, Direct Experimental Evidence of Halide Ionic Migration under Bias in $CH_3NH_3PbI_{3-x}Cl_x$-Based Perovskite Solar Cells Using GD-OES Analysis, ACS Energy Lett. 2017, 2, 943−949
[4] M. Bouttemy, P. Tran-Van, I. Gerard, T. Hildebrandt, A. Causier, J. L. Pelouard, G. Dagher, Z. Jehl, N. Naghavi, G. Voorwinden, B. Dimmler, M. Powalla, J.F. Guillemoles, D. Lincot, A. Etcheberry, Thinning of CIGS solar cells: Part I: Chemical

processing in acidic bromine solutions, Thin Solid Film 519 (2011) 7207-7210

[5] P. Sanchez, D. Alberts, B. Fernandez, A. Menendez, R. Pereiro, A. Sanz-Medel, Endogenous and exogenous hydrogen influence on amorphous silicon thin films analysis by pulsed radiofrequency glow discharge optical emission spectrometry, Anal. Chim. Acta (2011)

[6] P. Sanchez, O. Lorenzo, A. Menendez, J.L. Menendez, D. Gomez, R. Pereiro, B. Fernandez, Characterization of Doped Amorphous Silicon Thin Films through the Investigation of Dopant Elements by Glow Discharge Spectrometry: A Correlation of Conductivity and Bandgap Energy Measurements, International Journal of Molecular Sciences 12(4):2200-15 (2011)

[7] P. Sanchez, B. Fernandez, A. Menendez, D. Gomez, R. Pereiro, A. Sanz-Medel, A path towards a better characterisation of silicon thin-film solar cells: depth profile analysis by pulsed radiofrequency glow discharge optical emission spectrometry, Prog. Photovolt: Res. Appl. (2013)

[8] O. Zywitzki, T. Modes, M. Dienel, H. Morgner, C. Metzner, E. Schwuchow, Effect of Microstructure on Chlorine Activation of CdTe Thin Film Solar Cells, 31st European Photovoltaic Solar Energy Conference and Exhibition, 1210-1215 (2015)

Deep level transient spectroscopy measurements of silicon heterojunction cells

Sanchit Khatavkar[1], C. V. Kannan[2], Vijay Kumar[2], P. R. Nair[1], and B. M. Arora[1]

[1]Dept. of Electrical Engineering, IIT Bombay, Mumbai India; [2]Moser Baer Photovoltaic Pvt. Ltd., Greater Noida, U.P.-201306, India;

Abstract — **Excellent open circuit voltage (V_{oc}) reported by Si Heterojunction (Si-HJ) solar cells has been a topic of considerable interest among the PV community. One of the reasons attributed to this large V_{oc} is the reduction of interface recombination due to the presence of an inversion layer at a-Si:H/cSi interface. Here, we employ deep level transient spectroscopy (DLTS) measurements of silicon heterojunction cells (Si-HJ) to probe process induced interface traps at the a-Si:H/cSi interface. Interestingly, contrary to literature reports, we find both majority and minority peaks in differently processed Si-HJ solar cells – a result which could have interesting implications towards performance optimization.**

Index Terms — **silicon heterojunction cells, rate window, polarity of DLTS, solar cell processing.**

I. INTRODUCTION

Silicon heterojunction cells (Si-HJ) have the potential to achieve high efficiency, which has been amply demonstrated by Panasonic[1]. The main attraction in these cells is the high open circuit voltage (V_{oc}), with Panasonic demonstrating a value of 750mV[1]. A lot of work has gone into finding out the reason for this high V_{oc} and all of the groups attribute this to the low recombination rate between hydrogenated amorphous silicon (a-Si:H) and crystalline silicon (c-Si) [2]–[4]. After the existence of an inversion layer at this heterointerface was proven by Maslova et al.[5], the reason for low recombination rate at the interface was found out to be the repulsion of majority carriers due to this inversion layer. Jian Li et al.[3] has done deep level transient spectroscopy (DLTS) of Si-HJ solar cells and has shown that the DLTS peaks are due to the inversion charge at the heterointerface. The potential of DLTS technique to understand this phenomenon has been harnessed in this work, where DLTS results of four Si-HJ cells having differences in processing conditions are reported and it is interesting to note how the differences in processing lead to difference in DLTS signatures.

II. EXPERIMENTAL

The silicon heterojunction cells investigated in this work were fabricated from CZ n-type monocrystalline silicon substrates of resistivity of about 4 Ω-cm. After the saw damage removal, wafers were textured, and surfaces were passivated. This was followed by deposition of about 3-5 nm i-layer a-Si:H on both sides and subsequent 6-12 nm of p layer

a-Si:H deposition on the front side and about 50 nm n-layer a-Si:H on the back side by PECVD for solar cells of Type (A). 100 nm thick TCO layers were then deposited on both, the front and the back sides, followed by the deposition of front silver contact grid and a large area back aluminium contact. For solar cells of Type (B), the sequence of depostion was the same but for two differences, a) the thicknesses of the amorphous layers, viz. 8nm for i-layer a-Si:H, 8nm for p-layer a-Si:H and 24nm n-layer a-Si:H at the backside and b) back contact is screen printed silver.

Deep level transient spectroscopy measurements were done using SULA Technologies DLTS setup. The solar cells were cleaved to an area of ~ 1.5mm^2 so that the capacitance is less than 1nF, the measurement limit of the SULA setup. A capacitance voltage sweep was performed for each sample prior to the DLTS experiment and the expected C-V behaviour was seen. The DLTS measurements were conducted at a quiescent voltage of -2V and pulse voltage height of 2V, in other words, the solar cells were pulsed from reverse bias to zero bias. The filling pulse width was kept at 300μs and period was 5ms. The DLTS Spectra were taken at rate window values of 0.02ms, 0.05ms, 0.1ms, 0.2ms and 1ms. The temperature was varied from 77K to 350K.

III. POLARITY OF DLTS PEAKS

The solar cells used for DLTS measurements are of n-type substrate, so the inversion layer will contain holes, which are the minority carriers in the solar cells. It is well known that the interface of a-Si:H/cSi heterojunctions contain traps. The steady state electron occupany of traps is given by Eq. 1.[6]

$$n_T = \frac{c_n n}{c_n n + c_p p} N_T \tag{1}$$

where N_T is trap density, n is number of electrons in conduction band of the semiconductor, p is the number of holes in the valence band of the semiconductor, c_n is the capture coefficient of electrons and c_p is the capture coefficient of holes.

If we assume $c_p \gg c_n$ and $p \approx n$, then most traps will be occupied by holes. At t=0, the traps are neutral with $n_T \approx 0$ and space charge region density $N_{scr} = N_D$, the background doping density. When the device is pulsed from zero to negative bias in a DLTS experiment, minority holes are emitted from the traps, the charge of the traps changes from neutral to negative, and $N_{scr} \sim N_D - n_T$ for $t \to \infty$. The space charge region density decreases, space charge region width increases and the capacitance decreases with time[6]. The difference signal δC which is measured by DLTS, is given as Eq. 2.

$$\delta C = C(t_1) - C(t_2) \qquad (2)$$

where $C(t_1)$ and $C(t_2)$ are the capacitance values at times t_1 and t_2 respectively ($t_2 > t_1$). If the polarity of the DLTS peaks is positive, then it is due to emission of holes, from the above explanation, $C(t_1) > C(t_2)$, giving the positive sign to the peaks and thus these peaks are called minority peaks. Converse is true for majority peaks, which will be due emission of electrons in our case and $C(t_1) < C(t_2)$, giving the negative sign to the DLTS peaks.

IV. RESULTS AND DISCUSSION

Fig. 1 shows the DLTS Spectra of a solar cell of Type (A) taken at rate window values 0.02ms, 0.05ms, 0.1ms and 0.2ms. It can be seen from Fig.1 that the polarity of the peaks is positive and the peaks are seen at low temperature (77K - 200K). The polarity of DLTS peaks suggests hole trapping and can be attributed to the inversion charge at the heterointerface. The activation energy for Type (A) solar cells was 0.14 - 0.15eV which can be seen in Fig. 2. It matches with the activation energy of admittance spectroscopy measurements (0.16eV) and is in close agreement with the activation energy of DLTS measurements (0.227±0.15eV) reported by Jian Li et al[3].

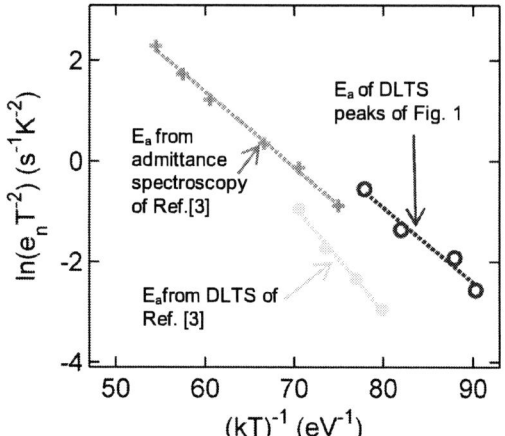

Fig.2. Activation energy plot from DLTS of type (A) solar cells (blue circles) compared with activation energy plots from DLTS (green solid circles) and from admittance spectroscopy (magenta stars) reported by Jian Li et al[3].

The DLTS spectra of a solar cell of Type (B) taken at rate window values 0.02ms, 0.1ms, 0.2ms and 1ms, are shown in Fig. 3. The DLTS peaks of Type (B) cells are negative, implying majority carrier emission, which is that of electrons in our case. These DLTS signatures are completely different from what is seen in Type (A) solar cells and no hole trapping is observed. There are two possibilities which can give rise to the observed behavior: a) Fermi energy at the interface is getting pinned, preventing hole capture or emission and /or b) the thickness of i-aSi:H layer for Type (B) is 8nm, which is twice that of Type (A), inducing some electron trapping defects at the interface.

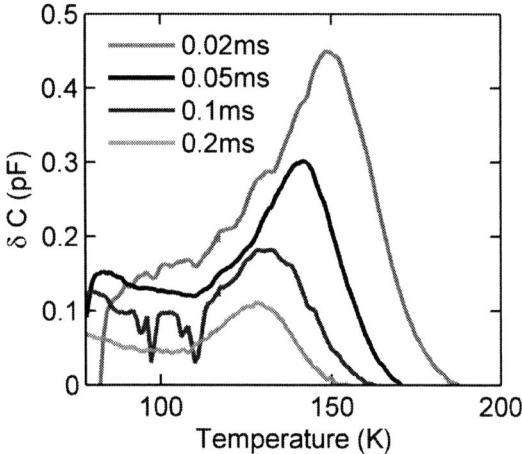

Fig.1. DLTS Spectra of one solar cell of Type (A), taken at rate window values of 0.02ms, 0.05ms, 0.1ms and 0.2ms.

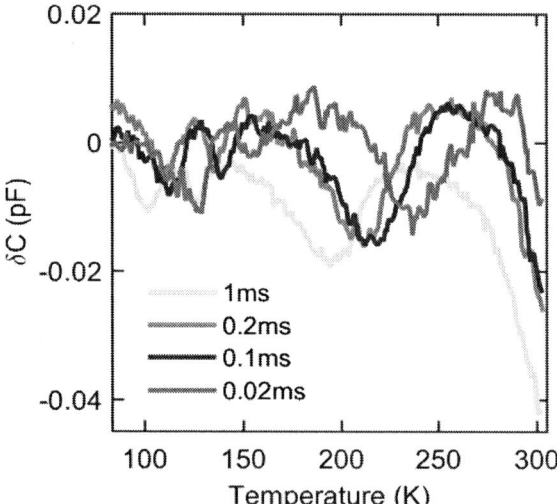

Fig.3. DLTS Spectra of solar cell of Type (B) taken at rate window values of 0.02ms, 0.1ms, 0.2ms and 1ms.

V. SUMMARY

We have thus reported the results of DLTS measurements of silicon heterojuntion cells fabricated under different processing conditions. It is interesting to see that this difference can give rise to different DLTS signatures, viz. the polarity of DLTS peaks. Even though the presence of the inversion layer is confirmed in all reports of silicon heterojunction cells, we cannot ascertain its presence in all types of silicon heterojunctions, atleast in n-type silicon heterojunctions. A more detailed analysis of these cells needs to be done in order to fully understand the origin of the different DLTS signatures and also the role of DLTS in understanding the relation of inversion charge to the performance of silicon heterojuntion cells.

ACKNOWLEDGEMENTS

This paper was based upon work supported in part by the Solar Energy Research Institute for India and the United States (SERIIUS), funded jointly by the U.S. Department of Energy (under Subcontract No. DE-AC36-08GO28308) and the Government of India's Department of Science and Technology (under Subcontract No. IUSSTF / JCERDC SERIIUS/2012). The authors also acknowledge the National Centre for Photovoltaic Research and Education (NCPRE), IITBombay for providing the characterization facility and the financial support during the course of this work.

REFERENCES

[1] K. Masuko *et al.*, "Achievement of More Than 25% Conversion Efficiency With Crystalline Silicon Heterojunction Solar Cell," *IEEE J. Photovoltaics*, vol. 4, no. 6, pp. 1433–1435, Nov. 2014.

[2] U. Rau, V. X. Nguyen, J. Mattheis, M. Rakhlin, and J. H. Werner, "Recombination at a-Si:H/c-Si heterointerfaces and in a-Si:H/c-Si heterojunction solar cells," *3rd World Conf. onPhotovoltaic Energy Conversion, 2003. Proc.*, vol. 2, pp. 1124–1127, 2003.

[3] J. V. Li, R. S. Crandall, D. L. Young, M. R. Page, E. Iwaniczko, and Q. Wang, "Capacitance study of inversion at the amorphous-crystalline interface of n-type silicon heterojunction solar cells," *J. Appl. Phys.*, vol. 110, no. 11, p. 114502, 2011.

[4] O. Maslova *et al.*, "Understanding inversion layers and band discontinuities in hydrogenated amorphous silicon/crystalline silicon heterojunctions from the temperature dependence of the capacitance," *Appl. Phys. Lett.*, vol. 103, no. 18, p. 183907, 2013.

[5] O. a. Maslova *et al.*, "Observation by conductive-probe atomic force microscopy of strongly inverted surface layers at the hydrogenated amorphous silicon/crystalline silicon heterojunctions," *Appl. Phys. Lett.*, vol. 97, no. 25, p.

[6] 252110, 2010.

D. Shroder, *Semiconductor Material and Device Characterization*. John Wiley & Sons, Inc., 2006.

Characterization of Modules and Arrays with SunsVoc

Alex Killam and Stuart Bowden

Arizona State University, Tempe, AZ, 85283, USA, Solar Power Labs, QESST

Abstract — Suns-V_{OC} is currently used as a characterization technique for cell level testing. The idea of Suns-V_{OC} can be applied using natural light to modules and arrays to characterize the performance, degradation, and diagnose failures. Modeling simulations are show as evidence that Suns-V_{OC} can be applied as a characterization technique. Experimental results begin to show how the technique is applied in field application. The characterization technique has proven evidence that more in depth analysis can be performed to provide as an affordable characterization technique for modules and arrays.

I. INTRODUCTION

The concept of Suns-V_{OC} (more properly termed illumination-V_{OC}) is used extensively in the characterization of silicon solar cells. Suns-V_{OC} relies on the principle of superposition; in that a solar cell illuminated IV curve is the dark IV curve shifted by the short circuit current. The basic theory and techniques for the concept of Suns-V_{OC} and the related technique of J_{SC}-V_{OC} were first outlined in 1963 [1] and used with various implementations [2]. Despite the very early publication, Suns-V_{OC} wasn't widely adopted until Sinton et al [3] in 2000 and the subsequent availability of a commercial tool. Suns-V_{OC} is now a standard cell measurement and the concept of pseudo-FF is now widely used.

The idea of Suns-V_{OC} used at the cell level was applied to large scale modules and arrays. Instead of using an artificial light source (as done in typical cell testing) the use of natural variation of the solar illumination at sunrise and sunset was used to provide the necessary changes in light intensity. The use of the sun as a light source has significant advantages: the entire array can be uniformly illuminated, the cost of the data acquisition is simplified to array V_{OC} and illumination level, and there is minimal interruption to the array operation as the measurements can be done at sunrise and sunset when the system is producing little power. While a seemingly simple technique, Suns-V_{OC} potentially provides a wealth of information through measurements such as potential module defects. A challenge is to interpret the measurements for real world applications at the large-scale module and array size. The most widely predicted application is the ability of examining the Suns-V_{OC} measurements for its ability to predict and interpret a variety of array failure modes. The most common failure modes found in modules and arrays are delamination, cracked cells, degraded junction boxes, and failed interconnects [4].

II. IMPLEMENTING SUNS-V_{OC}

Using an artificial variable light source to measure and test an entire array is impractical. Instead, the sun itself is used as the variable light source - particularly during sunrise and sunset. The use of the sun greatly simplifies the system. The light intensity is monitored with the use of a smaller reference solar cell next to the module/array. Companion cells have been shown to provide equivalent or superior performance to pyranometers [5-6] particularly when temperature correction is applied. The use of the Suns-V_{OC} curve for arrays was first described in the literature by Sinton instruments [7].

The Suns-V_{OC} technique uses the principle that the illuminated IV curve is just the ideal IV curve shifted by the short circuit current. A Suns-V_{OC} trace is developed by measuring the cell V_{OC} at varying illumination levels. The result is that an IV curve that can be constructed from a series of illumination-V_{OC} measurements. By plotting the curve as one minus measured amount of sun intensity, a curve similar to the standard power-IV curve can be plotted. The term power-IV is used to refer to the standard current voltage curve that would be measured with a variable load. The pseudo-IV curve is the curve derived from the Suns-V_{OC} using the 1–suns transformation. Comparing the power-IV curve to the pseudo-IV enables the measurement of the cell series resistance. Alternatively, the Suns-V_{OC} curve can be plotted on a semi log plot to enable advanced characterization such as J_{01}, J_{02} and ideality factors.

J_{SC}-V_{OC} curves are closely related to the concept of Suns-V_{OC}. The only difference is that the cell itself is used to monitor light intensity in the form of the short circuit current density, J_{SC}. J_{SC}-V_{OC} has the practicality and advantage of not requiring a separate device to monitor the light intensity, unlike most current module testers. Suns-V_{OC} is faster and easier to implement since the cell does not need to be switched back and forth between V_{OC} and J_{SC} for each data point.

III. SUNS-V_{OC} MODELING APPLICATION TO MODULE AND ARRAYS

Circuit modeling is used to show how the Suns-V_{OC} technique would be applied to a real-life application. LT Spice was the modeling software of choice. A 60-cell module was constructed using the cell characteristics found in table 1. The cell was constructed to represent a typical production monocrystalline silicon solar cell. The 60-cell module was

978-1-5090-5606-4/17 $31.00 © 2017 IEEE

interconnected as shown below in figure 1. This configuration was also constructed to represent a typical production module. Common modules utilize bypass diodes roughly every 20 cells to minimize power loss and hot spots due to mismatch effects. A 10-module array was also modeled using the configuration shown below in figure 2 and the specifications found in table 2. The 10-module array was used as a proof of concept because modules can be connected in much more varying ways depending on the application. This array utilized blocking diodes, which are placed in arrays to keep current from flowing back into the cells.

Individual Cell Characteristics	
Cell ISC	8.5 A
Cell VOC	715 mV
Bypass Diode IAVG	10 A
Bypass Diode VBRK	100 V
Resistor Value	100 kΩ

Table 1 – 60-cell module characteristics

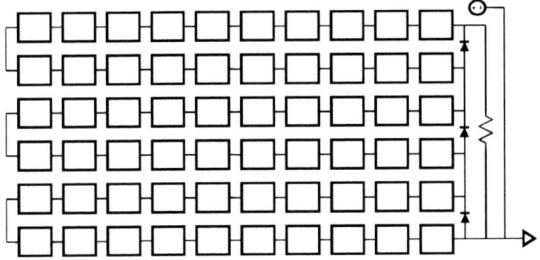

Figure 1 – 60-cell module layout

10 Module Array Characteristics	
Module ISC	8.5 A
Module VOC	42.9 V
Blocking Diode IAVG	10 A
Blocking Diode VBRK	350 V
Resistor Value	100 kΩ

Table 2 – 10-module array characteristics

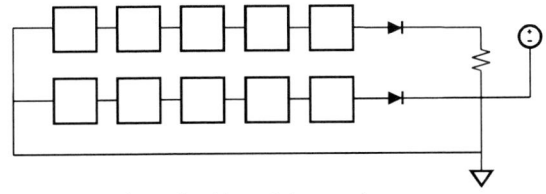

Figure 2 – 10-module array layout

"1 sun" measurements were performed on the 60-cell module using Suns-V_{OC} against the familiar power-IV curve. Figure 3 shows the results of this modeling. The Suns-V_{OC} curve

represents a pseudo IV curve as the technique does not account for the module's series resistance. Therefore, the series resistance can be calculated by the delta of the two curves. Series resistance can be used as a key factor for determining a module's health. The same can also be applied to arrays as shown below in figure 4.

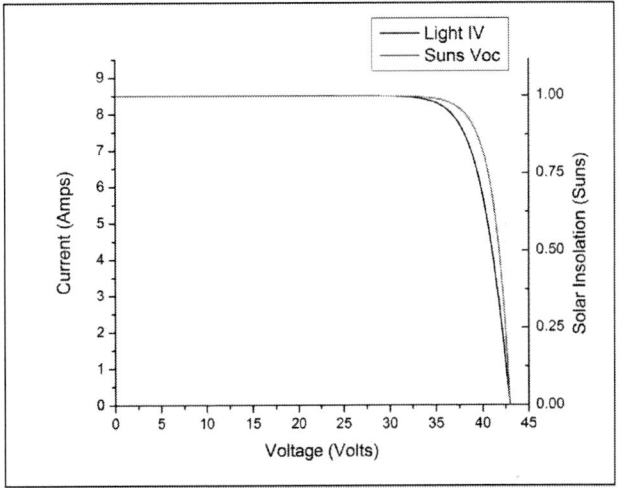

Figure 3 – 60-cell module IV curve

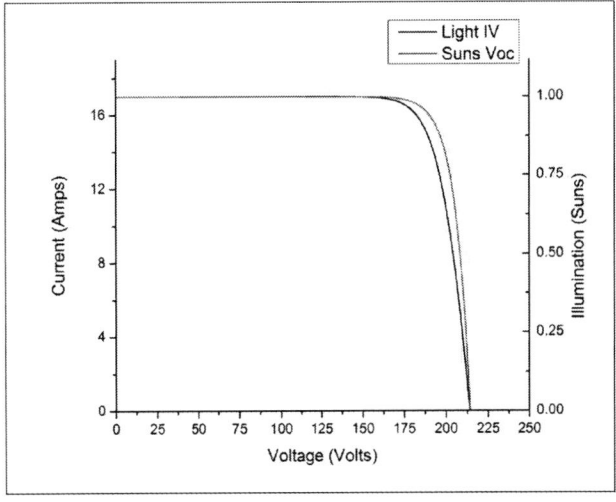

Figure 4 – 10-module array IV curve

In order to use Suns-V_{OC} as a valuable module and array characterization method, shading must not register as a defect. Shading almost has not effect on the open circuit voltage, meaning it does not impact the measurements. To justify this, modeling was conducted on the same module in array with SunsVoc against the power-IV curve. Respectively, figure 5 and 6 show the IV curves for the module with 1 cell shaded 50% and the array with 1 module shaded 50%. As displayed, the SunsVoc technique is not effected by the shading whereas the power-IV curve has drastic effects.

$$V_{OC} \equiv -2.2 \frac{mV}{°C} \qquad (1)$$

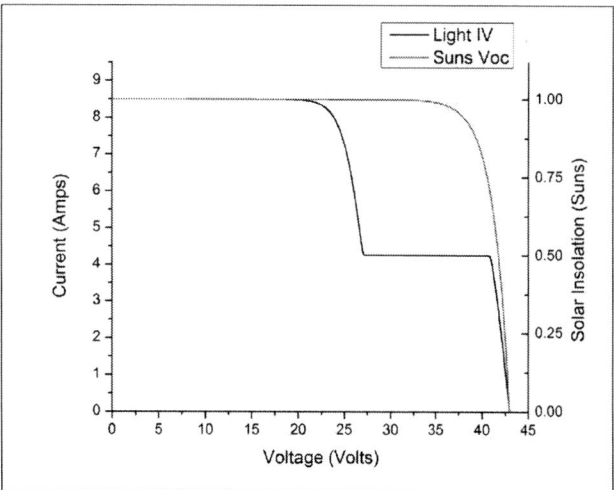

Figure 5 – 60-cell module with 1 shaded cell

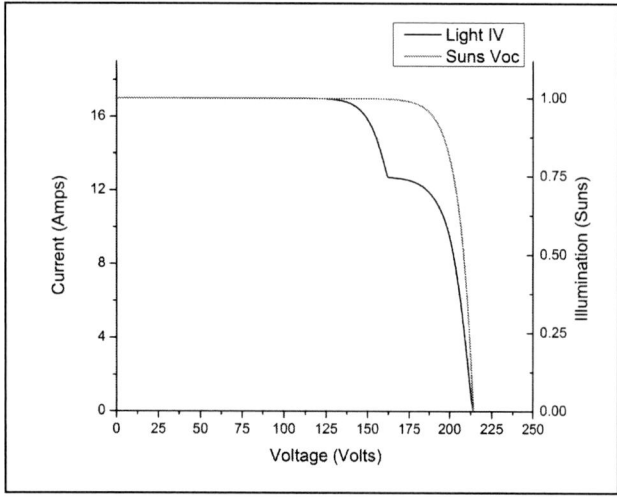

Figure 6 – 10-module array with 1 shaded module

IV. SUNS-V$_{OC}$ EXPERIMENTAL APPLICATION TO MODULE AND ARRAYS

Experimental results were performed on a 96-cell module composed of typical 6" crystalline silicon solar cells. The results of the measurements are presented below in figures 7-10. The raw Suns-V$_{OC}$ data is the data displayed in figure 6 in green. As the data shows, there is clearly a large amount of noise with data on different days giving quite different results. Using equation 1, the data was corrected for temperature giving the corresponding red curve where all the data subsequently flows.

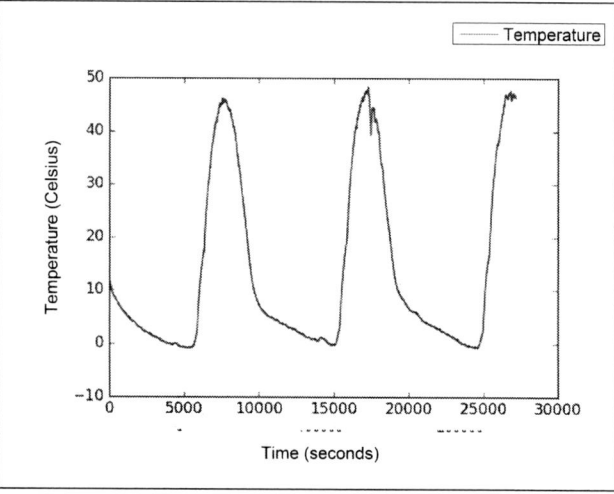

Figure 7 - Module temperature over 3-day testing period in Dec 16'

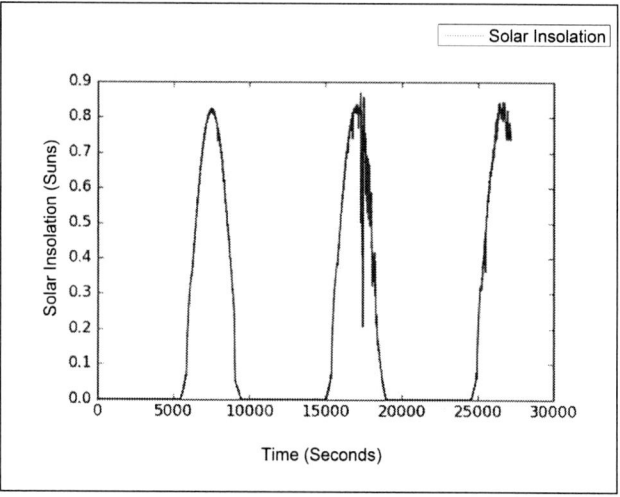

Figure 8 - Number of suns over 3-day testing period in Dec 16'

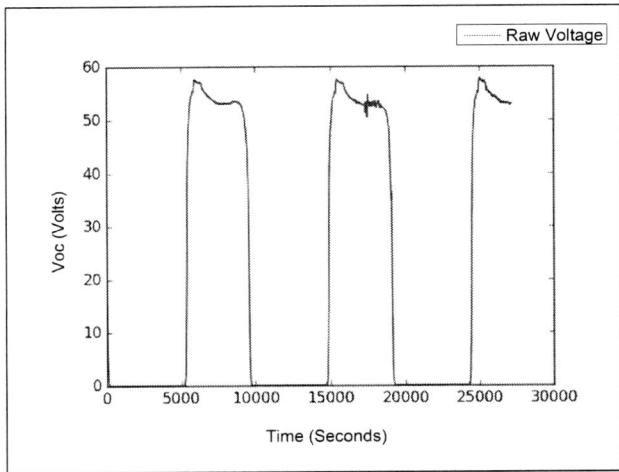

Figure 9 - Raw Voc of module over 3-day testing period in Dec 16'

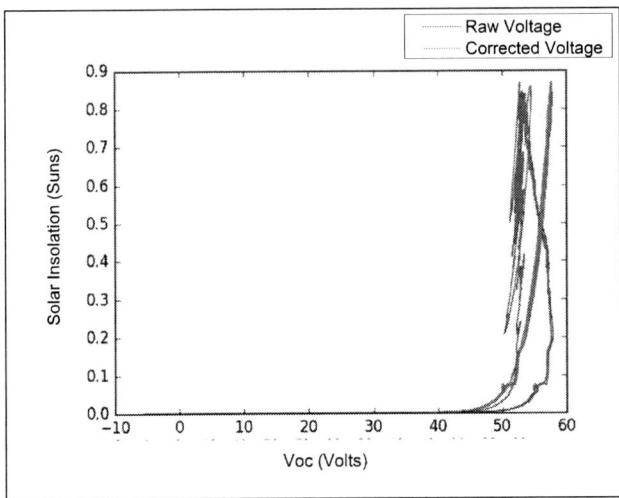

Figure 10 – Suns-V$_{OC}$ of module over 3-day testing period in Dec 16'

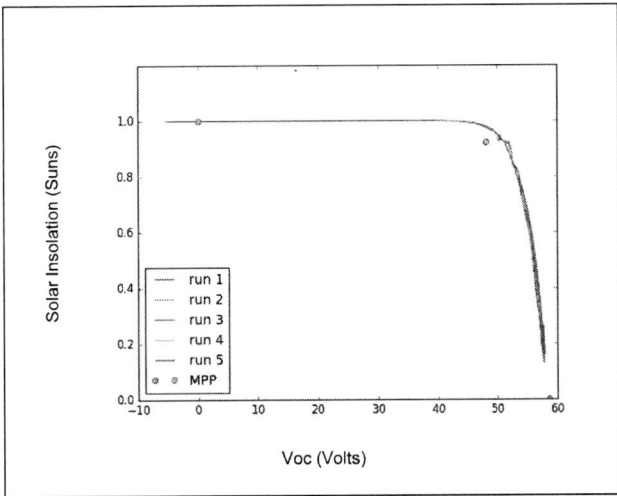

Figure 11 – Pseudo-IV curve from module over 3-day testing period in Dec 16'. Each sunrise/sunset is denoted as an independent run.

Once the Suns-Voc data is taken, it is compared to the power-IV curve as shown in figure 11. In this case, the power-IV fill factor is 75.5%. From the Suns V$_{OC}$ curve, we can measure fill factor of the module without the effect of series resistance (often termed pseudo-FF) as 80.5%. In a system, the inverter typically only gives the maximum power point, but the addition of the Suns V$_{OC}$ measurements provides a wealth of extra information at almost no extra cost. Figure 11 shows the ability to measure the pseudo IV curve across a range of light intensities.

IV. CONCLUSION

Suns-Voc has a significant role in characterization at the solar cell level. These preliminary results show that there is a potential role for Suns-Voc to be used as characterization at the module and array level. These techniques potentially can be used to identify defects in modules such as cell cracks, bypass diode failures, hot spots, etc. It can also be used to study the degradation rate of modules as the understanding of reliability is becoming more important.

REFERENCES

[1] M. Wolf and H. Rauschenbach, "Series Resistance Effects on Solar Cell Measurements," *Advanced Energy Conversion*, vol. 3, 1963.

[2] S. Bowden, V. Yelundur, and A. Rohatgi, "Implied-Voc And Suns-Voc Measurements In Multicrystalline Solar Cells," in *Proceedings of the 29th IEEE Photovoltaic Specialists Conference*, New Orleans, LA, USA, 2002.

[3] R.A. Sinton and A. Cuevas, "A Quasi-Steady-State Open-Circuit Voltage Method for Solar Cell Characterization" In: *16th European Photovoltaic Solar Energy Conference*. Glasgow, Scotland; 2000. pp. 1152–1155.

[4] E. D. Dunlop, D. Halton, H. A. Ossenbrink, "20 years of life and more: Where is the end of life of a PV module?", Proc. 31st IEEE Photovolt. Spec. Conf., pp. 1593-1596, 2005-Jan.-37.

[5] L. Dunn, M. Gostein, and K. Emery, "Comparison of pyranometers vs. PV reference cells for evaluation of PV array performance," in *Photovoltaic Specialists Conference (PVSC), 2012 38th IEEE*, 2012, pp. 002899–002904.

[6] J. Meydbray, E. Riley, L. Dunn, K. Emery, and S. Kurtz, Pyranometers and Reference Cells: Part 2: What Makes the Most Sense for PV Power Plants? Oct, 2012.

[7] M. K. Forsyth, M. Mahaffey, A. L. Blum, W. A. Dobson, and R. A. Sinton, "Use of the Suns-Voc for diagnosing outdoor arrays amp; modules," in *2014 IEEE 40th Photovoltaic Specialist Conference (PVSC)*, 2014, pp. 1928–1931

A Study of Performance Characterization with Rear Light Source in Conventional Bifacial Solar Cells

Soo Min Kim, Sang Hoon Jung, Hae-won Choi, Yong Bae Kim, Min Gu Kang, Hee-eun Song, Gyu-seok Choi

Gumi Electronics & Information Technology Research Institute, Gumi, 39171, South Korea
Korea Institute of Energy Research, 152 Gajeong-ro, Yuseong-gu, Daejeon 305-343, South Korea

Abstract—We compared bifacial solar cell measurement results between bifacial irradiation and theoretically calculated values. Conventional solar simulator was modified to measure bifacial solar cell with rear side LED light source. Bifacial solar cells which were used to measure were from 2 companies mass production. In this paper, each side of bifacial cells was measured under standard test condition as first time action. The other cell side which is opposite side of light irradiation was covered with nonreflecting velvet plate to prevent light reentry from cell penetration or unintended directions to cell. So, we expected that measurement data could be exacted by excluding unnecessary factors. These measurement results that each side of cell, front and rear side, are used to calculate expected bifacial cell characteristics. Secondly, we measured bifacial solar cells characteristics with bifacial irradiation with newly installed bifacial solar simulator. The measurement result with bifacial irradiation is compared with theoretical bifacial cell values which were calculated with bifacial cell formula. As a result, we can verify difference values of efficiency and power were within $\pm 3\%$. So, it can be possible that conventional solar simulator could be used to measure bifacial solar cell if there are expected diode ideality factor (n) and front/rear side of cell measurement result under the standard test condition (STC).

Index Terms—bifacial solar cell, bifacial illumination, bifacial efficiency, solar simulator, currentvoltage characterization.

I. INTRODUCTION

Recently crystalline silicon solar cell has 23% of market price in solar module industry, but it should be reduced gradually. [1], [2] Thus many companies have been concentrated on developing highly efficient cells to reduce the levelized cost of electricity(LCOE). Since bifacial approach has been investigated at 1960s, the bifacial solar cells have been developed with various different types and attracted particular interests in recent years. [3] The bifacial cell can produce electron-hole pair bifacially by illuminating light to both sides of the cell simultaneously. Because of this characteristic, bifacial module can generate more power than conventional type of solar module such as monofacial p–type PERC or PERL modules and so on for the same installation configurations(area, albedo, etc.) and climate conditions. Nowadays, there are so many cell types and companies which are manufacturing bifacial solar cells and one of the bifacial cells such as Pananonic HIT^{TM}(Heterojuction with Intrinsic Thin layer) has been reached to 23.0% in conversion efficiency. [4]

To evaluate the performance and quality, the solar cells must be measured in an indoor circumstances under standard test conditions(STC, 1000W/m^2, AM1.5G and 25°C) as defined by the International Electrotechnical Commission(IEC). For conventional type of solar cells, standards such as IEC 60904 have been adopted to assess the cells. [5], [6] Since bifacial cells operates under simultaneous front and rear side irradiance in outdoor condition with non-zero albedo, it is desirable to characterize these cell types for bifacial irradiance. Bifacial cell manufacturers suffer with the problem of standard indoor measurements and announcing the conversion efficiency of the bifacial solar cells depending on albedo or rear side irradiance ratio.

In the absence of standards, most bifacial cell manufacturers report the front side monofacial electrical parameters under STC and tabulate the efficiency/power with a simple linear addition of front and rear side efficiencies for particular rear side irradiance conditions. [7], [8] The Panasonic HITTM Double 225 datasheet says that the rated power of the cell is measured under Standard Test Conditions (STC) and it does not account for power produced from the back face of panels. [7] However, since bifacial cell efficiency and power does not vary linearly with the illumination, the simple addition of front and rear side efficiencies will lead into inaccuracy in power assessment. As a possible solution for accurately measuring the current–voltage(I–V) parameters under bifacial irradiance, some researchers presented reported solar simulators with simultaneous bifacial irradiance capability. [9], [10] In addition, theoretical method was demonstrated to predict the performance of bifacial solar cell to within 1% variation with actual measurements by using one-diode model. [9]–[11] However, these methods are neither practical nor feasible to characterize conventional bifacial cells experimentally and theoretically.

In this paper, we compared bifacial solar cell characteristics between bifacial irradiation and theoretically calculated bifacial cell result values by using bifacial solar simulator which is modified from conventional solar simulator with rear side LED light source.

II. EXPERIMENT

A. Sample preparation and apparatus

For this experiment, we modified conventional solar simulator(NEWPORT 94063A: class AAA) to characterize a bifacial cell. We installed newly designed LED lamp for the rear side irradiance. The LED lamp is classified to CBA and

TABLE 1
LED LAMP SPECTRUM CLASS ANALYSIS RESULTS.

Wavelength(nm)	Energy ratio	AM 1.5	Matching	Class[a]
400-499	21.66	18.4	1.18, A	
500-599	29.65	19.9	1.49, C	
600-699	17.22	18.4	0.94, A	C
700-799	9.24	14.9	0.62, B	
800-899	7.67	12.5	0.61, B	
900-1100	15.57	15.9	0.92, A	

[a] IEC 60904 spectrum matching class standard: A (0.75-1.25), B (0.6-1.4), C (0.4-2.0)

Fig. 1. Experimental apparatus for measurement of bifacial solar cell, (a) solar simulator with two light sources, (b) front light source of short ARC 1kW Xenon lamp – Class AAA, (c) rear light source of eight set LED – Class CBA.

its intensity can vary from zero to one sun as shown in Fig. 1. Table 1 shows spectral match of the LED source. The spectrum only in the wavelength from 500 to 599 nm range was classified to C. Various kinds of LED chips(white, 450nm, 650nm, 750nm, 850nm) are arranged in the LED lamp and the intensity of each LED chip can be controlled to reach real sun light. Therefore, we can characterize the bifacial cell with two light sources to both sides simultaneously.

We used conventional bifacial cells two company's with 6 inches scale and p-type Cz-Si (100) for this measurement test. Each cell was measured front and rear side of cell individually under STM condition. The other cell side which is opposite side of light irradiation was covered with nonreflecting velvet plate to prevent light reentry from cell penetration or unintended directions to cell in order to take an accurate measured data. These results were utilized to derive theoretically calculated bifacial solar cell characteristics.

Bifacial cell was exposed bifacial light source to measure real bifacial characteristics. The front side of cell was exposed under STM condition and rear side of cell was exposed from 0 to 1sun as unit of 0.1sun at the same time. To prevent rising temperature of cell during test, simulator was modified flash type. A JIG for the bifacial cell was designed to block any other unintended light source except front and rear side illumination.

B. Theoretical model for bifacial solar cell

Singh et al. suggested I–V characterization parameters theoretically to estimate the bifacial cell performance under bifacial illumination. [11], [12] They used one-diode model to derive bifacial parameters and assumed linear current response with light intensities. However, since bifacial cell efficiency and power does not vary linearly with the illumination, the simple addition of front and rear side efficiencies will lead into inaccuracy in power assessment. We cited bifacial solar cell theoretical formula to verify our newly installed bifacial solar simulator and check result conformity between real measurement data and theoretical data. The formula

was derived from diode equation and additional rear side light irradiation and we modified partial formula from quoted formula to make more precise result in real measurement.

To calculate bifacial solar cell performance, firstly two short-circuit current was defined with bifacial illumination such as I_{sc-f} means short-circuit current measured for front side illumination of the cell at STC (A), I_{sc-r} is for rear side. Therefore, bifacial short-circuit current can be defined

$$I_{sc-bi} = I_{sc-f} + xI_{sc-r} = R_{I_{sc}}I_{sc-r} \qquad (1)$$

where $R_{I_{sc}}$ is the relative current gain which is dimensionless value. Dimensionless irradiance ratio represents x that is consist of (G_r/G_f), irradiance on the front and rear side of cell are G_r and G_f with W/m² unit. Relative current gain with dimensionless unit can be written as follows:

$$R_{I_{sc}} = \frac{I_{sc-f} + xI_{sc-r}}{I_{sc-f}} = 1 + x\frac{I_{sc-r}}{I_{sc-f}} \qquad (2)$$

Photovoltaic potential equation in bifacial illumination was improved using one–diode model for single solar cell like this

$$V_{oc-bi} = V_{oc-f}\frac{ln\left(R_{I_{sc}}\frac{I_{sc-f}}{I_0}\right)}{ln\left(\frac{I_{sc-f}}{I_0}\right)} \qquad (3)$$

where, V_{oc-f} is the open-circuit voltage measured for front side illumination of cell at STC, and I_0 is the saturation current of solar cell in dark state. The saturation current that represents recombination of solar cell diode is defined as follows:

$$I_0 = \frac{I_{sc-f}}{exp\left(\frac{qV_{oc-f}}{nkT}\right)} \qquad (4)$$

where q and k are the elementary charge constants and the boltzmann constants. Non-ideality of diode represent as n that means diode ideality factor. This saturation current concepts can help to reduce variation of theory model results.

TABLE 2
EACH CELL FRONT AND REAR MEASURED DATA.

	Cell[a]	V_{oc} (V)	I_{sc} (A)	F.F. (%)	P_m (W)	Efficiency (%)	Bifaciality (%)[b]	Diode ideality factor (n)
A-1	(Front)	0.634	9.53	74.81	4.52	18.92	96.0	1.21
	(Rear)	0.631	9.17	75.02	4.34	18.17		
A-2	(Front)	0.628	9.46	74.89	4.45	18.62	96.3	1.27
	(Rear)	0.625	9.1	75.34	4.29	17.94		
A-3	(Front)	0.629	9.47	75.63	4.51	18.86	96.2	1.17
	(Rear)	0.629	9.16	75.2	4.33	18.14		
B-1	(Front)	0.659	9.92	79.1	5.17	21.65	86.5	1.08
	(Rear)	0.659	8.58	79.07	4.47	18.72		
B-2	(Front)	0.659	10.04	76.55	5.07	21.2	89.0	1.20
	(Rear)	0.659	8.91	76.72	4.51	18.86		
B-3	(Front)	0.659	10.04	76.67	5.07	21.24	90.0	1.28
	(Rear)	0.659	8.9	77.63	4.55	19.06		

[a]Each side of cell was measured under STM condition and the other cell side which is opposite side of light irradiation was covered with nonreflecting velvet plate.

[b]Power, rear/front

Typically used pseudo fill factor equation was modified on bifacial illumination using bifacial open-circuit voltage and ideality factor.

$$_pFF_{bi} = \frac{\left(\frac{qV_{oc-bi}}{nkT}\right) - ln\left(\frac{qV_{oc-bi}}{nkT} + 0.72\right)}{\left(\frac{qV_{oc-bi}}{nkT} + 1\right)} \qquad (5)$$

This bifacial pseudo fill factor can be reflected real experimental data with empirical assumption. The bifacial fill factor can be defined as fallow

$$FF_{bi} = _pFF_{bi} - \frac{R_{I_{sc}}\left(_pFF_f - FF_f\right)V_{oc-f}}{V_{oc-bi}} \qquad (6)$$

From fill factor definition, bifacial maximum power can be represented like this equation,

$$P_{bi} = I_{sc-bi}V_{oc-bi}FF_{bi} \qquad (7)$$

Finally, we can defined power conversion efficiency on bifacial illumination state as follows:

$$\eta_{bi} = \frac{I_{sc-bi}V_{oc-bi}FF_{bi}}{A_{cell}\left(G_f + G_r\right)} \qquad (8)$$

To use this bifacial cell formula, measurement data from front/rear side of bifacial solar cell and n are necessary.

III. RESULTS AND DISCUSSIONS

Bifacial cells from A and B companies were measured front and rear side individually. As a result, we can see the different cell characteristics between two companies' cell. As shown in Table 2, each company's cell has different measurement result. and bifaciality also different. A company cells were good bifaciality as 96% but other characteristics were worse than B company. On the contrary, B company cell were better characteristics than A company but bifaciality was lower as 88%. It can be expected that each company cell has different manufacturing technology. Results in Table 2 would be used to theoretically calculated bifacial solar cell parameters. And we need to investigate effect of bifacial solar cell characteristics later.

Theoretically calculated bifacial solar cell parameters can be worked out from theoretical formula with front and rear side measurement results and n. Measurement results which are front and rear side of cell are easy to know. However, n is not simple to get in order to calculate bifacial cell parameters. n is a fitting parameter that describes how closely the diodes behavior matches that predicted by theory. [13] To get n we used simulation fitting program which we made algorithm. n was derived from comparison between real measured data with bifacial irradiation and calculated parameters with bifacial formula. Diode ideality factor of each cell was decided when the difference of results between actuality and theory is got the lowest error. Even though each cell was manufactured from other company and has different cell parameters, results of diode ideality factor are in $1.08 < n < 1.27$ range. Table 2 is shown each cell's diode ideality factor from simulation fitting program and real measured data. Result is presented that normal diode ideality factor (n) is about n=1.1 or 1.2 from 2 companies 6 cells. It means that if the cell has not special structure or problem, typical diode ideality factor (n) of cell could be about n=1.1 or 1.2. So, when characteristics of bifacial solar cell is needed to know with conventional solar simulator, we think carefully that this result of n=1.1 or 1.2 can be applied when bifacial solar cell parameters are calculated.

Every cell was measured within variation of irradiation ratio (G_r/G_f, G_f=1) 0~1. As light intensity is higher, bifacial cell parameters, V_{oc-bi}, I_{sc-bi}, P_{bi} and η_{bi}, are increased

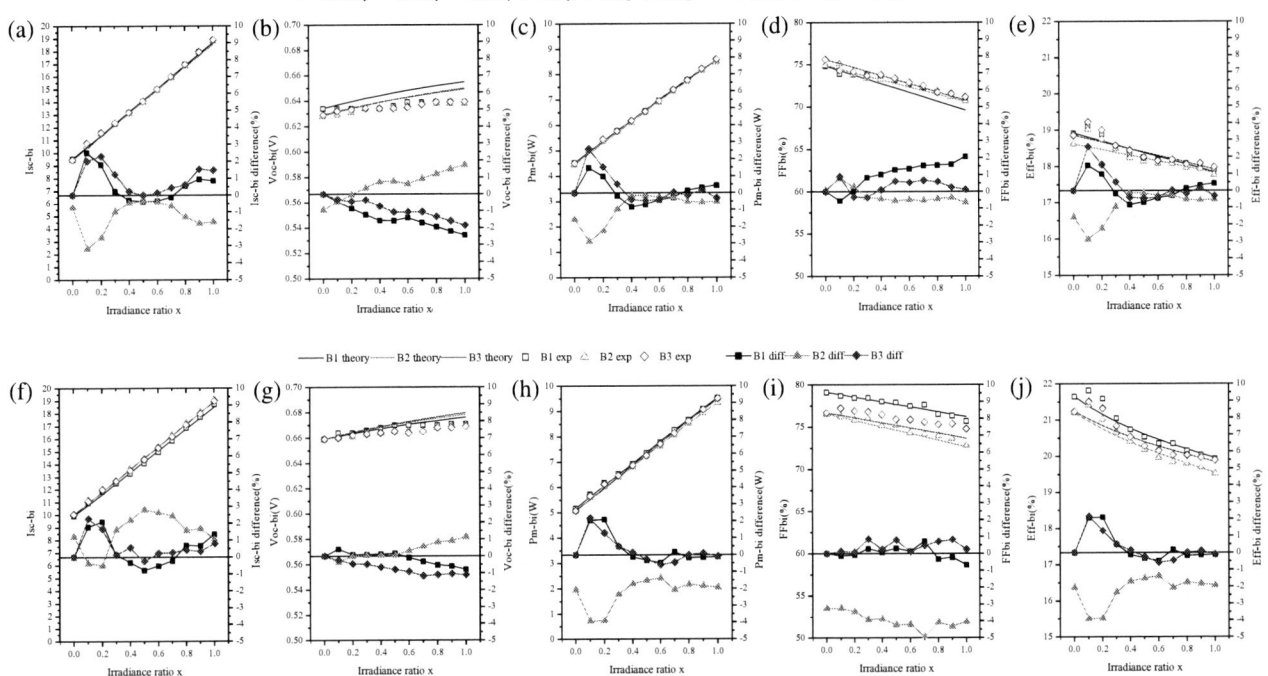

Fig. 2. Bifacial solar cell performance results compare with theory and experiment, and their difference percentage, (a–e) A type solar cell results, (f–j) B type results (a),(f) short-circuit current I_{sc-bi}, (b),(g) open-circuit voltage V_{oc-bi}, (c),(h) max power P_{m-bi}, (d),(i) fill factor FF_{bi}, (e),(i) efficiency η_{bi}.

because of induced current. But, FF_{bi} is decreased because of power loss ($P = I^2 R$) according to resistance factor and induced current in cell. Even though I_{sc-bi} is shown linear increase by light intensity, final result that power is not shown linear increase because of FF_{bi}. So, it is not simple to calculate by summation of front and rear cell characteristic to estimate bifacial cell characteristics. Each cell characteristic was compared with real bifacial irradiation result and calculated bifacial solar cell parameters with diode ideality factor value from simulation fitting program. Parameters that V_{oc-bi}, I_{sc-bi}, P_{bi} and η_{bi} were confirmed and the difference of each parameter value between bifacial irradiation and calculated parameter was within ±3% in fig. 2. We can identify that newly modified simulator performance and accuracy. From this result, it is confirmed that bifacial solar cell parameters can be estimated from conventional solar simulator with accurate front/back measured data and predicted diode ideality factor.

IV. CONCLUSION

The bifacial solar simulator is modified from conventional solar simulator with rear side light source and bifacial cell JIG. Tested cells are 6 from 2 company's commercialized bifacial solar cell. Bifacial cell was measured with bifacial irradiation and front/rear side of cell under STM condition individually. And then theoretical bifacial parameters are calculated by front and rear side measured data. Each company's

cell has different characteristics which are bifaciality, V_{oc}, I_{sc}, FF, p_m and η. It seems that these results are came from different cell structure and manufacture process. Each cell has n that range is $1.08 < n < 1.27$ according to simulation results. The result with bifacial irradiation is not same with simple summation of front and rear characteristics because of resistance factor and power loss by induced current in cell. Theoretical bifacial characteristics are calculated from formula model with front and rear side measured data and calculated n using (1) – (8). The best way to get accurate bifacial solar parameter is real measurement with bifacial irradiation solar simulator. However, we can confirmed that real bifacial irradiation result and theoretically calculated result are different within ±3. In this present experiment, it is shown our newly installed bifacial solar simulators performance and accuracy. And conventional solar simulator also can measure bifacial cell characteristics by using bifacial theoretical formula with accurate front/back side measured data and predicted n.

ACKNOWLEDGMENT

This work was conducted under the framework of the research and development program of the Korea Institute of Energy Research(KIER) (B7-2426) and supported by the New & Renewable Energy of the Korea Institute of Energy Technology Evaluation and Planning(KETEP) grant funded

by the Korea government Ministry of Knowledge Economy
(No.20163010012230)

REFERENCES

[1] Renewable energy technology: Cost analysis series – solar photovoltaic. Technical report, International Renewable Energy Agency, June 2012.

[2] International Technology Roadmap for Photovoltaic (ITRPV). Technical report, March 2017.

[3] H.U.S. Mori. patent no. 3,278,811, October 1966.

[4] Mikio Taguchi, Yasufumi Tsunomura, Hirotada Inoue, Shigeharu Taira, Takeshi Nakashima, Toshiaki Baba, Hitoshi Sakata, and Eiji Maruyama. High-efficiency hit solar cell on thin (<100 μm) silicon wafer. In *Proceedings of the 24th European Photovoltaic Solar Energy Conference*, pages 1690–1693, 2009.

[5] International Electrotechnical Commission. *IEC Standard 60904-3: Photovoltaic Devices, Part 3: Measurement Principles for Terrestrial Photovoltaic (PV) Solar Devices with Reference Spectral Irradiance Data*, 2008.

[6] Antonio Luque and Steven Hegedus. Handbook of photovoltaic science and engineering, 2003.

[7] Panasonic Corporation. Bifacial photovoltaic module bifacial photovoltaic module vbhn225dj06 hit double 225, 2014.

[8] Prism Solar Technologies. *Prism Solar Installation Manual, B265(L), B260, B255, B250, B245, B200, B150, HB180*, December 2015.

[9] H Ohtsuka, M Sakamoto, M Koyama, K Tsutsui, T Uematsu, and Y Yazawa. Characteristics of bifacial solar cells under bifacial illumination with various intensity levels. *Progress in Photovoltaics: Research and Applications*, 9(1):1–13, 2001.

[10] A Edler, M Schlemmer, J Ranzmeyer, and R Harney. Flasher setup for bifacial measurements. In *présenté à BIFIPV workshop, Konstanz*, 2012.

[11] Jai Prakash Singh, Armin G Aberle, and Timothy M Walsh. Electrical characterization method for bifacial photovoltaic modules. *Solar Energy Materials and Solar Cells*, 127:136–142, 2014.

[12] Jai Prakash Singh, Timothy M Walsh, and Armin G Aberle. A new method to characterize bifacial solar cells. *Progress in Photovoltaics: Research and Applications*, 22(8):903–909, 2014.

[13] Robert F Pierret. *Semiconductor Device Fundamentals*. Pearson Education India, September 1996.

Electrical characterization of the carrier transport properties in a Cu(In,Ga)Se₂ solar cell

Roberto Lopez[1], Sanjoy Paul[1], Ingrid Repins[2], and Jian V. Li[1,3]

1. Department of Physics, Texas State University, San Marcos, TX, 78666 United States

2. National Renewable Energy Lab, Golden, CO, 80401 United States

3. Materials Science, Engineering and Commercialization (MSEC), Texas State University, San Marcos, TX, 78666 United States

Abstract — **This work uses electrical characterization techniques to study the transport properties of carriers in the Cu(In,Ga)Se₂ (CIGS) absorber of a thin-film photovoltaic device. The majority carrier (hole) density, mobility and resistivity are measured in the dark using a coordinated admittance spectroscopy and capacitance-voltage technique that takes advantage of the bias dependence of the modified dielectric relaxation in Cu(In,Ga)Se₂ solar cell. The bulk of the data were taken at low temperatures where the temperature dependencies of the above carrier transport parameters are also measured.**

I. INTRODUCTION

One of the most promising thin-film photovoltaic materials being used to elaborate solar cells at the moment is Cu(In,Ga)Se₂ (CIGS). Its direct band gap (1.0-1.4 eV) [1], allows this material to absorb a significant portion of the solar spectrum, achieving a very high efficiency and allowing the possibility of tandem CIGS devices.

The hole mobility in CIGS materials has been studied by different groups over the years, most of them using Hall measurement [2] – [6]. Only one group has used capacitance-based methods [6]. The biggest advantage of this technique is that it can be applied to finished solar cell devices, unlike Hall measurements which require specially prepared thin-film test samples. In this work, we use capacitance-voltage (C-V) and admittance spectroscopy techniques to extract the hole mobility.

II. METHODS

apacitance measurements are commonly conducted by applying a sinusoidal voltage stimulus to the device under test [7]. If the signal is of frequency ω and amplitude V_0 then $V(t) = V_0 \sin(\omega t)$ describes the voltages as a function of time to the device under test. The capacitance $C=dQ/dV=I/(dV/dt)$ is calculated from the current $I(t) = C\frac{dV}{dt} = C\omega V_0 \cos(\omega t)$, which is 90-degree out of phase with respect to voltage.

In a p-n junction, the depletion, or space-charge, region is fully depleted of free carriers, and the bulk region is neutral. When we apply a bias, the only change in charge occurs at the edge of the depletion zone (Fig. 1), giving rise to the junction capacitance.

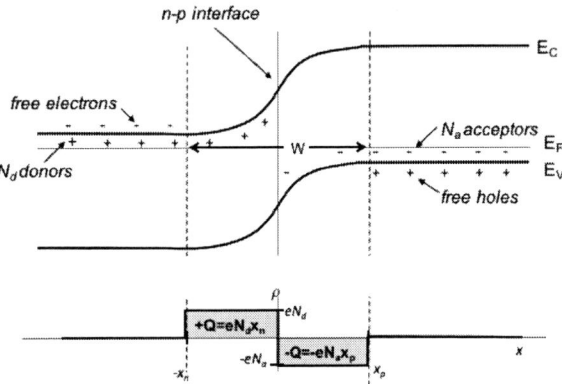

Figure 1 n-p junction band diagram and charge distribution

The junction capacitance can be described in terms of other junction properties [8]. For an asymmetrical p-n⁺ junction, which is a good approximation for the CdS/CIGS device, the depletion region width is $W \approx \sqrt{\frac{2\varepsilon(V_{bi}-V)}{qN_a}}$ where ε is the dielectric constant, V_{bi} the built-in voltage, and N_a the hole density in the p layer. From the depletion capacitance $C=dQ/dV=\varepsilon A/W$, one can extract N_a by $N_a = -\frac{2}{q\varepsilon A^2}\left(\frac{\Delta(1/C^2)}{\Delta V}\right)^{-1}$ where A is the device area.

Strictly speaking, the above CV analysis using capacitance and its voltage dependence to obtain the carrier concentration is only valid for the frequencies at which the majority carriers in the neutral region can respond to the ac bias modulation. Under such circumstances, the ac capacitance measures the mobile charge density at the edge of the depletion width. Under other circumstances, for example, trap charges may respond to ac modulation at low frequencies

978-1-5090-5606-4/17 $31.00 © 2017 IEEE

but not at high frequencies, where the demarcation point between high and low frequencies is determined by the trap's capture/emission time. Because of this, traps contribute to the low frequency capacitance (but not to the high frequency) and trap properties can be extracted from the frequency dependence of capacitance, which is the basis of admittance spectroscopy technique [9] when used for characterizing defect levels from capacitance measured at varying frequencies and temperatures.

Some properties of the absorber can be derived from further analysis of the bias and frequency dependence of the capacitance data. [10] The equivalent circuit of the absorber in a typical thin-film solar cell device (Fig. 2) can be modeled as the series connection of contribution from the depletion region (represented by the depletion capacitance C_d) and that from the quasi-neutral region (represented by the parallel connection of bulk conductance G_g' and capacitance C_g'). Ordinarily, the quasi-neutral CIGS region, characterized by resistivity ρ and permittivity ε, exhibits dielectric relaxation at frequency $\omega_{dr}=1/\rho\varepsilon$. Below ω_{dr}, the neutral region behaves like a conductor with $G_g'=A/(t-W)/\rho$, where t is the thickness of the absorber. Above ω_{dr}, the neutral region behaves like a capacitor $C_g'=\varepsilon A/(t-W)$.

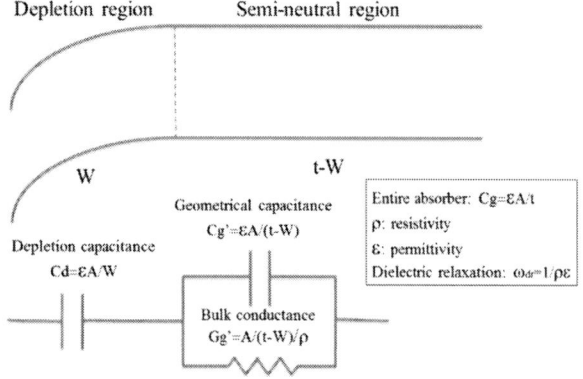

Fig. 2. The band diagram and equivalent-circuit model for the absorber in a typical thin-film solar cell device. The depletion region is modeled as depletion capacitance C_d. The semi-neutral region is modeled as the parallel connection of lump-parameter geometrical capacitance C_g' and conductance G_g'.

Let the total admittance of the circuit in Fig. 2 be Y_{total}. A straightforward circuit analysis yields

$$Y_{total} = \frac{\omega^2 C_d \rho \varepsilon \left(\frac{C_d}{C_g}-1\right)+j\omega C_d (\omega^2 \frac{C_d}{C_g}\rho^2 \varepsilon^2 +1)}{\omega^2 \frac{C_d^2}{C_g^2}\rho^2 \varepsilon^2 +1} \quad (1)$$

Now:

$$(Y_{total})_{Imaginary} = \omega C = \frac{\omega C_d (\omega^2 \frac{C_d}{C_g}\rho^2 \varepsilon^2 +1)}{\omega^2 \frac{C_d^2}{C_g^2}\rho^2 \varepsilon^2 +1} \quad (2)$$

The equivalent capacitance C of the circuit as a function of frequency is therefore:

$$C = C_d \frac{\omega^2 \rho^2 \varepsilon^2 \frac{C_d}{C_g}+1}{\omega^2 \rho^2 \varepsilon^2 \frac{C_d^2}{C_g^2}+1} \quad (3)$$

where ω is the frequency. The capacitance C versus frequency ω plot thus exhibits a step transition around the inflection point ω_p, below which C approaches C_d and above which C approaches C_g, where $C_g=\varepsilon A/t$. Recall the dielectric relaxation frequency $\omega_d=1/\rho\varepsilon$, one can arrive at the inflection frequency

$$\omega_p = \frac{C_g}{C_d \rho \varepsilon} = \frac{W}{t}\left(\frac{1}{\rho\varepsilon}\right) \quad (4)$$

Note that the depletion width W depends on the bias V, one can take advantage of the bias voltage dependence of the ω_p to extract the resistivity ρ and mobility μ. First, the thickness of the thin film solar cell t is measured from the depletion width at low temperatures when full depletion of the absorber occurs or, when full depletion is not achievable, via other thickness measurements such as through spectroscopic ellipsometry and a profilometer. The electrically active absorber thickness t equals $\varepsilon A/C$. Then, the resistivity ρ and mobility μ are extracted from the bias dependence of the inflection frequency ω_p expressed as:

$$\omega_p^2 = \frac{W^2}{t^2}\left(\frac{1}{\rho\varepsilon}\right)^2 = \frac{2}{q\varepsilon N_a \rho^2 t^2}(V_{bi} - V) \quad (5)$$

If one plots ω_p^2 against V, the slope of this plot can be used to determine the resistivity:

$$\rho = \sqrt{\frac{2}{q\varepsilon N_a t^2 Slope}} \quad (6)$$

or equivalently, the mobility:

$$\mu = \frac{1}{qN_a\rho} = \sqrt{\frac{\varepsilon t^2 Slope}{2qN_a}} \quad (7)$$

where the carrier density N_a is determined from capacitance-voltage measurement.

III. Experiment

The CIGS film (M3083-21) was deposited by co-evaporation utilizing the three-stage recipe. The substrate temperature of the sample (M3083-21) was 435 C.

To perform the CV measurement and admittance spectroscopy, an Agilent 4294 Impedance Analyzer was used to collect the raw data. For the capacitance-voltage measurement, the DC voltage was varied from -1.5 to 0.6 V while keeping an AC modulation amplitude of 50 mV$_{p-p}$ at 10 kHz. For admittance spectroscopy, the DC bias was varied from -1.5 to 0.6 V, the temperature from 14 to 350 K with a step size of 7 K in a Janis cryostat, and the frequency from 10^3 to 10^8 Hz.

IV. Results

Using data measured at low frequencies, we calculate the hole density N$_a$ from the local slope of the Mott-Schottky plot of C^2 vs bias N$_a$=-2/slope/qεA^2 and depletion region W=εA/C. N$_a$ is determined to be 2.2x10^{14} cm^{-3} at 150 K. Note this value is lower than that in the high-efficiency devices (~ 10^{16} cm^{-3}), which allows high-quality data presented in this work be measured from instrument at hand but should not be interpreted as an inherent limit of the technique. At lower temperatures, the entire CIGS absorber becomes depleted and the thickness of the absorber is determined to be 3.3 μm using a dielectric constant value of 13.6 (Fig. 3).

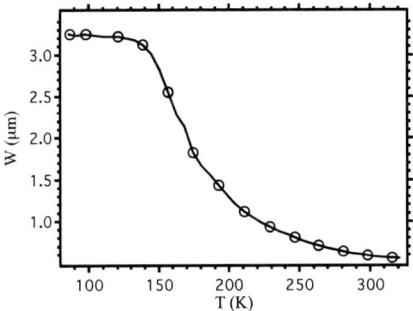

Fig. 3 At lower temperatures, the entire CIGS absorber becomes depleted and the thickness of the absorber is determined to be 3.3 μm using a dielectric constant value of 13.6.

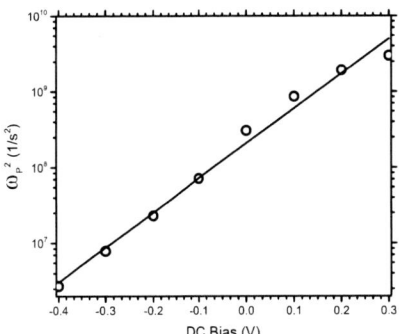

Fig. 4. The square of ω_p exhibits a linear dependence on the bias voltage as described by Eq. 5, from which a slope is determined and in turn ρ and μ extracted according to Eq. 6 and Eq. 7.

We then inspect the frequency dependence of C at a fixed temperature. The C vs ω spectrum shows a step transition separating two plateaus (data not shown): C$_d$ at $\omega< \omega_p$ and C$_g$ at $\omega> \omega_p$. We identify ω_p from the negative peak in the ωdC/dω vs ω spectrum. The square of ω_p exhibits a linear dependence on the bias voltage as described by Eq. 5, from which a slope is determined and in turn ρ and μ extracted according to Eq. 6 and Eq. 7 (Fig. 4). With measured values of t and N_a, we determine the resistivity of the film as 2.2×10^7 Ωcm and mobility as 0.005 cm^2/V/s at 150 K.

We also performed measurement of the majority carrier transport characteristics at room temperature and determined $N_a = 4.0 \times 10^{14}$ cm^{-3}, $\rho = 2.4 \times 10^4$ Ωcm, and $\mu = 0.66$ cm^2/V/s. This result compare favorably to previous reports on CIGS with a hole mobility ranging from 0.02 to 0.7 cm^2/V/s..[2]

V. Conclusions

We report the usage of the admittance spectroscopy method to measure resistivity and hole mobility in CIGS photovoltaic devices. This method produces results consistent with other techniques yet can be applied to complete solar cell devices instead of the specially prepared Hall test structure...

VI. Acknowledgements

We acknowledge support by U.S. Department of Energy through Sunshot PVRD grant DE-EE-0007541 "Crosscutting recombination metrology for expediting Voc engineering".

References

[1] U. Rau, H.W. Schock, Electronic properties of ZnO/CdS/Cu(In,Ga)Se2 solar cells – recent achievements, current understanding, and future challenges, Appl. Phys. A 69, 131-147 (1999).

[2] S. A. Dinca, E. A. Schiff, B. Egaas, R. Noufi, D. L. Young, and W. N. Shafarman, "Hole drift mobility measurements in polycrystalline $CuIn_{1-x}Ga_xSe_2$", *Physical Review B* 80, 2009, pp. 235201-1-12.

[3] M. Rusu, P. Gashin, and A. Simashkevich, "Electrical and luminescent properties of $CuGaSe_2$ crystals and thin films", *Solar Energy Materials and Solar Cells* 70, 2001, pp. 175-186.

[4] S. Schuler, S. Nishiwaki, J. Beckmann, N. Rega, S. Brehme, S. Siebentritt, and M. Ch. Lux-Steiner, "Charge carrier transport in polycrystalline $CuGaSe_2$ thin films", *IEEE Photovoltaics Specialists Conference*, 2002, pp. 504–507.

[5] D. J. Schroeder, J. L. Hernandez, G. D. Berry, and A. A. Rockett, "Hole transport and doping states in epitaxial $CuIn_{1-x}Ga_xSe_2$", *Journal of Applied Physics* 83, 1998, pp. 1519-1526.

[6] J.W. Lee, J.D. Cohen, and W.N. Shararman, The determination of carrier mobilities in CIGS photovoltaic devices using high-frequency admittance measurements, *Thin Solid Films* 480–481 (2005) 336–340.

[7] P.H. Mauk, H. Tavakolian, J.R. Sites, "Interpretation of Thin-Film Polycrystalline Solar Cell Capacitance," IEEE Transactions on Electron Devices, 37 (2), 1990, 422-427.

[8] R.S. Muller, T.I. Kamins, and M. Chan, "Device Electronics for Integrated Circuits", 3rd ed., John Wiley and Sons, Inc., 2003.

[9] T. Walter, R. Herberholz, C. Muller, H. W. Schock "Determination of defect distributions from admittance measurementsand application to $Cu(In,Ga)Se_2$ based heterojunctions," J. Appl. Phys. 80 (8), 15 October 1996, 4411-4420

[10] J.V. Li, X. Li, D.S. Albin, and D.H. Levi, "A Method to Measure Resistivity, Mobility, and Absorber Thickness in Thin-Film Solar Cells with Application to CdTe Devices," Sol. Energy Mat. Sol. Cells 94, 2010, pp. 2073-2077

Systematic Thermalphotovoltaic Solar Cell Optimization

Zheng Lyu[1], Muyu Xue[2], Junyan Chen[3], Jieyang Jia[1], Shanhui Fan[1], James Harris[1,2]

[1]Department of Electrical Engineering, Stanford University, Stanford, CA, 94305, US

[2]Department of Material Science and Engineering, Stanford University, Stanford, CA, 94305, US

[3]School of Physics, Peking University, Beijng, 100871, China

Abstract— In this work, we present a systematical simulation study on narrow band gap $In_{0.53}Ga_{0.47}As$ single junction solar cell, latticed-matched to InP substrate, for the application of thermophotovoltaic systems. The $In_{0.53}Ga_{0.47}As$ solar cell performance is simulated based on multiple parameters including: thermal emitter emission spectrum of the TPV system, epitaxial design of cell structure and cell temperature. This study demonstrates a pathway to optimize the TPV cell structure, and analyzes the effect of tailoring emitter spectrum and cell temperature.

Index terms- Photovoltaic cells, simulation, optimization methods

I. INTRODUCTION

The concept of thermophotovoltaics (TPV) [1-4] has been proposed as a potential candidate for overcoming the Shockley-Queisser limit for single junction solar cell. Currently although the TPV systems which incorporate thermal absorber/emitter and narrow band gap solar cells have been built experimentally, the measured system conversion efficiencies are still below typical single junction solar cells [6-7]. One of the biggest efficiency limiting factor for the whole TPV system is the narrow band gap TPV cell, which requires the accurate match of the cell bandgap and the emitter spectrum, good material quality and low cell operating temperature. The InGaAs single junction cell lattice-matched to InP substrate can be good candidate for a TPV system. InGaAs cell in this study has a band gap of 0.75 eV and a composition as $In_{0.53}Ga_{0.47}As$.

In this study, a systematical simulation of $In_{0.53}Ga_{0.47}As$ TPV cell was conducted using Rsoft Lasermod under certain thermal emission spectrum from the thermal emitter. The tailored emitter spectrum with the peak energy aligned to the band gap of the $In_{0.53}Ga_{0.47}As$ was first calculated and used as the incident light spectrum. The cell performance was simulated with the variation of (a) the window/back surface field (BSF) layer doping, (b) the emitter/base thickness and (c) the cell temperature. The original cell epitaxial design is based on the structure reported in [8]. In the optimization process, the cell temperature was first set at 300 K, the doping levels of the window and BSF were adjusted. In the next step, the thicknesses of the emitter and base layer were varied respectively to obtain an optimized V_{OC} and cell efficiency. With the optimized structure at temperature of 300 K, the temperature effect on the cell efficiency was further analyzed and discussed. Our simulation results showed that the $In_{0.53}Ga_{0.47}As$ cell is able to reach conversion efficiency of 49% at 300 K. However, as temperature increases to 500 K, the simulated efficiency drops to 33%. This work demonstrates a systematic simulation method for TPV cell structure optimization and reveals the importance of heat management for building a high-efficiency TPV system.

II. CELL STRUCTURE AND EMITTER SPECTRUM DESIGN

A. $In_{0.53}Ga_{0.47}As$ cell structure design

The $In_{0.53}Ga_{0.47}As$ cell structure, as illustrated in Fig 1, was designed based on the similar design previous reported in [8]. The doping concentration and thickness of each layer are listed in Table I.

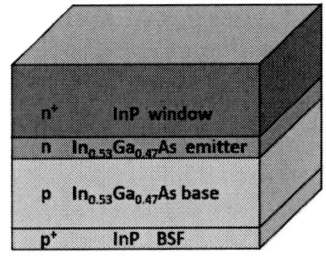

Fig 1. Schematic of the structure of the simulated $In_{0.53}Ga_{0.47}As$ cell

The 0.75 eV band gap $In_{0.53}Ga_{0.47}As$ cell is lattice matched to InP substrate.

TABLE I
ORIGIN THICKNESS AND DOPING CONCENTRATION

	Window (n-type)	Emitter (n-type)	Base (p-type)	BSF (p-type)
Thickness (μm)	1.5	0.2	2.5	0.3
Doping (cm⁻³)	1×10^{18}	1×10^{18}	1×10^{18}	1×10^{18}

Adjustments on the structure were conducted to optimize the cell efficiency. Doping concentrations of the window layer and BSF layer were changed to value of 1×10^{18} cm⁻³, 1×10^{19} cm⁻³ and 1×10^{20} cm⁻³ respectively. By obtaining the best doping level for the structure, the optimal thicknesses of the emitter and base layer were subsequently decided. The thickness of the emitter layer was varied to values of 0.1 μm, 0.2 μm to 0.5μm.

978-1-5090-5606-4/17 $31.00 © 2017 IEEE

Similarly, the thickness of the base layer was changed to 0.5 µm, 2.5 µm, 3.5 µm.

B. Calculation of the incident light spectrum

In a TPV system, the incident light collected by the cell is a tailored narrow band width spectrum from the emitter's blackbody irradiation. In our study, to get the highest efficiency, the incident spectrum was tailored to match the absorption of the $In_{0.53}Ga_{0.47}As$ cell. The band gap (E_g) of the $In_{0.53}Ga_{0.47}As$ cell corresponds to the wavelength (λ) of 1.65 µm calculated with the relationship of $E_G = \frac{hc}{\lambda}$, where h is the Planck constant, c is the light speed. The relationship between the peak of the emission spectrum and the temperature follows the equation:

$$\lambda_{peak} = \frac{b}{T} \qquad (1)$$

where the constant b is Wien's displacement constant, which is equals to 2.898×10^{-3} K m. λ_{peak} is the wavelength of peak of the spectrum, and T is the temperature. Substituting the wavelength corresponding to the bandgap of the InGaAs (0.75 eV), the corresponding temperature is T=1751 K. The black body irradiance was calculated based on the Planck's Law:

$$I(\lambda, T) = \frac{2hc^2}{\lambda^5} \frac{1}{e^{\frac{hc}{\lambda kT}} - 1} \qquad (2)$$

where $I(\lambda, T)$ is the blackbody irradiance, k is the Boltzmann constant, T is the temperature.

The calculated spectrum is shown in Fig 2. Noted that in order to conduct the simulation, the irradiance was taken the solid angle as 1 sr. The incident light was selected with the wavelength range from 1.47 µm to 1.65 µm (corresponding to the band gap of 0.85 eV and 0.75 eV).

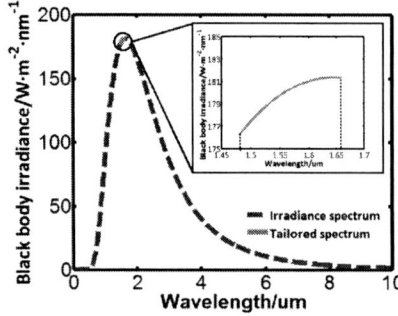

Fig 2. Calculated blackbody irradiance spectrum

III. SIMULATION RESULTS AND DISCUSSION

A. Window and BSF layer doping level

First, doping concentration of the Window layer and BSF layer was changed to value of 1×10^{18} cm^{-3}, 1×10^{19} cm^{-3} and 1×10^{20} cm^{-3} respectively. The J-V curves were simulated and are shown in Fig 3 and the detailed simulation result was shown in Table II.

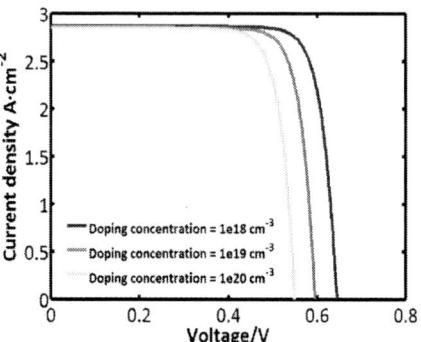

Fig 3. Simulation results of the InGaAs cell with different Window & BSF doping concentrations.

The short circuit current J_{sc} remains relatively constant while the open circuit voltage V_{oc} decreases from 0.65 V to 0.55 V as the doping level increases from 1×10^{18} cm^{-3} to 1×10^{20} cm^{-3}, indicating more recombination in the Window and BSF layer due to higher doping level. Such effect decreases the cell efficiency from 43.1% to 35.5%.

TABLE II
SIMULATION RESULTS WITH DOPING LEVEL

Doping/cm^{-3}	1×10^{18}	1×10^{19}	1×10^{20}
Efficiency /%	43.1	39.3	35.5
$J_{sc}/A \cdot cm^{-2}$	2.87	2.86	2.85
V_{oc} /V	0.650	0.600	0.550
FF /%	82.1	81.3	79.9

B. Thickness of Emitter & base layer

With the optimized window and BSF doping level of 1×10^{18} cm^{-3}, the thickness of the emitter layer was changed to values of 0.1 µm, 0.2 µm and 0.5 µm. The simulation result is shown in Fig 4 with details shown in Table III.

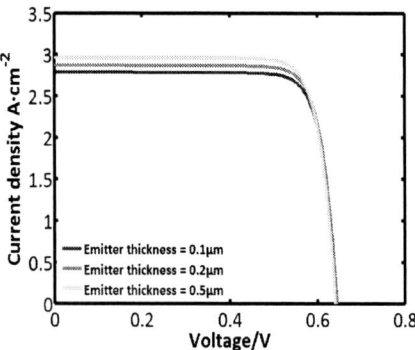

Fig 4. Simulation results of the InGaAs cell with different emitter thicknesses of 0.1 µm, 0.2 µm, 0.5 µm.

TABLE III
SIMULATION RESULTS WITH EMITTER THICKNESS

Thickness/μm	0.1	0.2	0.5
Efficiency /%	41.9	43.1	44.0
$J_{sc}/A \cdot cm^{-2}$	2.79	2.86	2.96
V_{oc} /V	0.650	0.650	0.650
FF / %	81.1	82.1	81.4

The V_{oc} remains relatively the same at 0.65 V, with a slight increase of J_{sc} from 2.79 A/cm^2 to 2.96 A/cm^2, indicating addition light absorption in the thicker Emitter layer, which leads to the increase of cell efficiency from 41.92% to 44.04%.

The thickness of the base layer was further optimized with the values of 0.5 μm, 2.5 μm and 3.5 μm. The result is shown in Fig 5 with the details in Table IV.

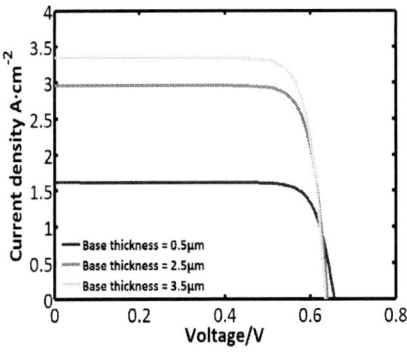

Fig 5. Simulation results of the InGaAs cell with different base thickness of 0.5 μm, 2.5 μm, 3.5 μm.

TABLE IV
SIMULATION RESULTS WITH EMITTER THICKNESS

Thickness/μm	0.5	2.5	3.5
Efficiency /%	24.7	44.0	49.3
$J_{sc}/A \cdot cm^{-2}$	1.62	2.96	3.34
V_{oc} /V	0.660	0.650	0.640
FF / %	81.6	82.1	81.4

While the V_{oc} slightly decreases from 0.65 V to 0.64 V, the J_{sc} doubles from 1.62 A/cm^2 to 3.34 A/cm^2, which leads to a significant increase in the efficiency from 24.7% to 49.3%. Such efficiency also reaches the best simulation result, with the parameters listed in Table V.

TABLE V
OPTIMIZED PARAMETERS OF THE CELL

	Doping /cm^{-3}	Emitter /μm	Base /μm	Efficiency /%
Value	1×10^{18}	0.5	3.5	49.3

C. Temperature effect

Since the emitter in the TPV system would be at a high temperature (1751 K in this study), it is important to evaluate the cell performance at temperatures higher than 300 K. The

temperature effect on the In$_{0.53}$Ga$_{0.47}$As cell was further explored, with the result shown in Fig 6. The details were shown in Table VI.

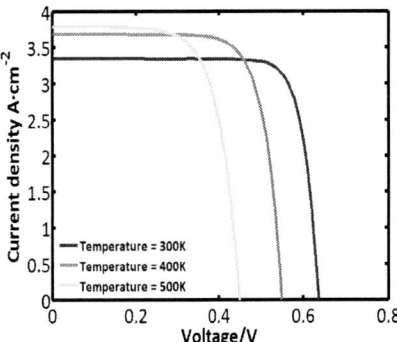

Fig 6. Simulation results of the optimized cell at different cell temperature of 300 K, 400 K, 500 K.

TABLE VI
SIMULATION RESULTS WITH TEMPERATURE

Temperature/K	300	400	500
Efficiency /%	49.3	43.7	33.4
$J_{sc}/A \cdot cm^{-2}$	3.34	3.68	3.79
V_{oc} /V	0.64	0.55	0.46
FF / %	81.4	76.0	67.6

The higher temperature results in a slightly higher J_{sc}, but the corresponding V_{oc} drops rapidly resulting in the decrease of the cell efficiency from 49.3% to 33.4%. The result shows the importance of heat management to maintain the low cell operating temperature in the TPV system.

IV. CONCLUSION

In this study, a systematic simulation method for the InGaAs based TPV cell was demonstrated. By optimizing the doping level and the thickness of specific layers, a significant efficiency increase was observed from simulation. Such method can potentially facilitate the design of high-efficiency TPV system.

ACKNOWLEDGEMENT

The authors would like to acknowledge the GCEP and BAPVC for financial support.

REFERENCES

[1] Swanson R M. A proposed thermophotovoltaic solar energy conversion system[J]. Proceedings of the IEEE, 1979, 67(3): 446-447..

[2] Harder N P, Würfel P. Theoretical limits of thermophotovoltaic solar energy conversion[J]. Semiconductor Science and Technology, 2003, 18(5): S151..

[3] Narayanaswamy A, Chen G. Surface modes for near field thermophotovoltaics[J]. Applied Physics Letters, 2003, 82(20): 3544-3546.

[4] Rephaeli E, Fan S. Absorber and emitter for solar thermo-photovoltaic systems to achieve efficiency exceeding the Shockley-Queisser limit[J]. Optics express, 2009, 17(17): 15145-15159.

[5] Shockley W, Queisser H J. Detailed balance limit of efficiency of p - n junction solar cells[J]. Journal of applied physics, 1961, 32(3): 510-519.

[6] Lenert A, Bierman D M, Nam Y, et al. A nanophotonic solar thermophotovoltaic device[J]. Nature nanotechnology, 2014, 9(2): 126-130.

[7] Bierman D M, Lenert A, Chan W R, et al. Enhanced photovoltaic energy conversion using thermally based spectral shaping[J]. Nature Energy, 2016, 1: 16068.

[8] Wilt D M, Fatemi S, Hoffman Jr R W, et al. InGaAs PV device development for TPV power systems[C]//AIP Conference Proceedings. AIP, 1995, 321(1): 210-220.

Characterization of Tellurium as a Back Contact for CdTe Solar Cells

C.E. Moffett and W.S. Sampath

Department of Mechanical Engineering, Colorado State University, Fort Collins, CO 80523 USA

Abstract—CdTe solar cells currently use a hetero-junction design allowing for ideal band alignment engineering. The back contact has often been a source of CdTe performance limitations due to the formation of a barrier at the back interface. Tellurium has been studied as a possible high quality back contact to CdTe. Temperature dependent current density-voltage measurements and x-ray and ultraviolet photoelectron spectroscopy were carried out in order to approximate offsets at the CdTe/Te interface.

Index Terms — back contact, cadmium telluride, photovoltaic cells, spectroscopy, tellurium, thin-films.

I. INTRODUCTION

THE growing PV market requires technologies that utilize economic large scale manufacturing with high device efficiency. Cadmium Telluride (CdTe) has been expanding in the thin film photovoltaic market due to low manufacturing costs and recent record efficiencies. CdTe has recently used band gap grading to develop more ideal band gap engineering, ideal for optimizing to the Shockley-Queisser limit [1]. CdTe's high absorption coefficient allows for absorption of nearly all incident photons at only 1 μm [1]. First Solar Inc., a leading developer of CdTe thin film solar devices, has demonstrated record efficiencies of small area CdTe devices at 22.1. % [2].

The development of high quality contacts is essential to improving cell performance and stability [3]. The back contact has long been a source of uncertainty and instability in CdTe technology [3, 4]. The back contact barrier formed at the interface impedes carrier transport, which has a current limiting effect on devices [5]. Tellurium (Te) has been proposed to make a passivated, low barrier back contact to CdTe [6]. This is typically achieved through an etching process of the CdTe absorber, leaving a Te rich layer [7, 8]. However Te deposited directly onto CdTe has also demonstrated an improvement to overall efficiency [9, 10].

In this work, a Te layer was evaporated on thin film CdTe devices fabricated at Colorado State University (CSU). Te was deposited at 2nm to simulate an etched surface and 8nm was used to represent a contact used in devices. The various thicknesses and various configurations were used to study the material and electrical properties of the contact.

II. MATERIALS AND METHODS

Small area CdTe solar devices were fabricated at CSU utilizing a fully-automated, single vacuum deposition chamber with integrated close space sublimation sources as described by Swanson *et. al.* [11]. The CSU baseline device structure with a Te back contact is shown in Fig. 1. All samples fabricated use commercially available SnO_2:F coated 3 mm Pilkington TEC12D soda lime glass.

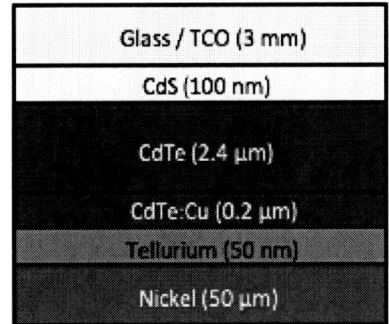

Fig. 1. Diagram of CdS/CdTe device with tellurium back contact fabricated at CSU.

The Te layer was deposited using thermal evaporation in a Cooke evaporator with the substrate held at room temperature. Te is deposited at 10^{-5} Torr in an environment that is flushed with argon to minimize atmospherics gases. The thickness and deposition rate were measured using a R.D. Mathis quartz crystal monitoring instrument. The deposition rate was held constant at 1 nm/s and time is used to vary the Te thickness. Cross sectional TEM and EDS of evaporated Te on CdTe indicated uniform and conformal coverage of the Te.

XPS and UPS were used to characterize the material properties of the Te film. The methods used are highly surface sensitive processes [6]. Practices were performed to limit oxidation and contamination of the samples. Samples were prepared within 24 hours of analysis and kept in a Ar vacuum atmosphere glove box with an environment with < 0.3 ppm of oxygen prior to measurement.

XPS was performed in the Central Instruments Facility (CIF) at CSU using a Phi 5800 surface analysis system which utilizes a monochromatic Al Kα anode (hν=1486.6 eV). A pass energy of 187.85 eV for survey scans and 5.85 eV for high-resolution scans was used. The base pressure of the chamber is maintained at 10^{-9} Torr during data acquisition.

The UPS system utilizes a Prevac UVS 40A2 source with He I (hν=21.2 eV) discharge gas. A pass energy of 2.95 eV for valence band edge scans was used. The base pressure of the chamber for UPS data acquisition is held at <10^{-7} Torr.

III. RESULTS

A. Electrical Characterization

Electrical characterization carried out at CSU suggests a low barrier formation at the CdTe/Te interface. Fig. 2 presents temperature-dependent-current vs. voltage measurements (J-V-T) of a CdS/CdTe cell with a Te back contact as shown in Fig. 1. As the temperature was reduced from 25 to -75°C, the diode behavior changed significantly, as the contact barrier has a progressively larger impact on carrier transport with decreasing temperature. The barrier height was calculated experimentally using the procedure defined by Koishiyev *et. al.* [12], where curves displaying roll-over were used for fit and calculation of the barrier height. The average barrier height was determined as 0.39 ± .02 eV also shown in Table III.

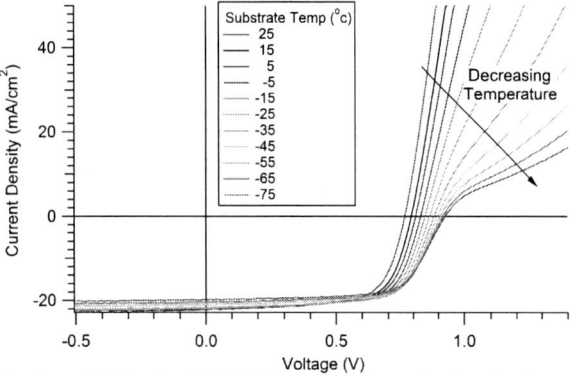

Fig. 2.   J-V-T of a CdS/CdTe cells with a tellurium back contact displaying a barrier behavior at lower substrate temperatures.

B. Material Characterization

The band offsets at the CdTe/Te interface were approximated by measuring the binding energy of the core levels (CL) of Te and Cd and valence band maxima (VBM) of Te and CdTe separately. Fig. 3 shows three separate film stacks used for characterizing the Te and CdTe films. The CdTe film stacks were not passivated with CdCl$_2$ and did not receive any intentional copper doping. The structure shown in Fig. 3(a) was used to approximate the CL binding energy and the VBM of Te; Fig. 3(b), to measure the CL binding energy of Cd and the VBM of CdTe; and Fig. 3(c) to measure the CL binding energies of Cd and Te at the interface. Several Te thicknesses were investigated to determine the properties of the CdTe/Te interface samples and to discern if bulk

properties were affected by Te thickness. The Te layer was kept thin enough to measure the Cd CL emission.

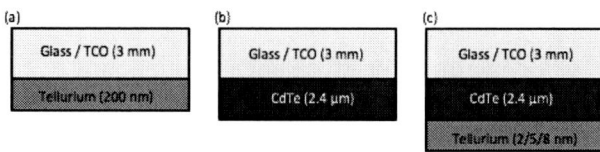

Fig. 3.   Schematics of sample structures utilized for XPS/UPS measurement. (a) Bulk Te, (b) bulk CdTe and (c) CdTe/Te interface sample with varying Te thickness.

Using software packages Fityk and Multipak, the XPS data were analyzed to fit CL peaks and extrapolate the valence band edge. Core level peaks were fit with Voigt functions, which are commonly used for peak fit in XPS [6]. UPS data was used for analysis of the valence band edge due to a much smaller energy range, allowing for a higher resolution [13]. An Au sample was measured and the Fermi level determined and used as a reference. The VBM of structures shown in Fig. 3(a) and (b) were determined by linearly extrapolating the leading edge of the valence band [14]. The valence band offset can be determined using (1) [15],

$$E_{VBO} = (E_{CL}^{Te} - E_{VBM}^{Te}) - (E_{CL}^{Cd} - E_{VBM}^{CdTe}) - \Delta E_{CL} \quad (1)$$

where E$_{CL}$ is the core level binding energy and E$_{VBM}$ is the valence band maxima for both the Te and CdTe films. ΔE$_{CL}$ is defined by (2) as the difference in core levels of Cd and Te, measured from the interface of CdTe/Te, as shown in Fig. 3(c) [15].

$$\Delta E_{CL} = E_{CL}^{Te} - E_{CL}^{Cd} \quad (2)$$

The XPS measured Te3d$_{5/2}$ and Cd3d$_{5/2}$ CL peaks exhibited well-defined behavior and were used for measurement of E_{CL}^{Te} and E_{CL}^{Cd} values, respectively. E_{VBM}^{Te} and E_{VBM}^{Cd} were determined from UPS analysis of the valence band edge. Fig. 4(a) displays the Te3d$_{5/2}$ peak and Fig. 4(b) displays the Te valence band edge, both analyzed from the 200 nm Te sample shown in Fig. 3(a). A similar analysis was performed for the Cd3d$_{5/2}$ peak of the 2.4 μm CdTe sample described in Fig. 3(b). The average of several measured CL and VBM binding energy values with the associated tool error are summarized Table I.

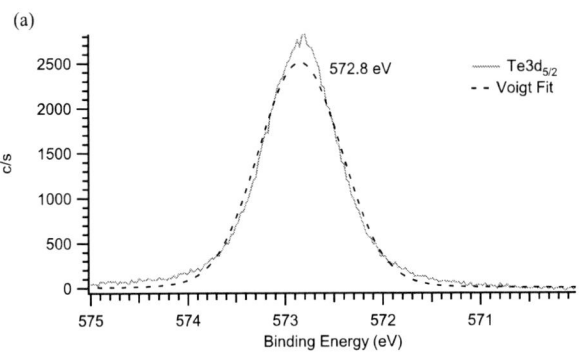

(a)

(b)

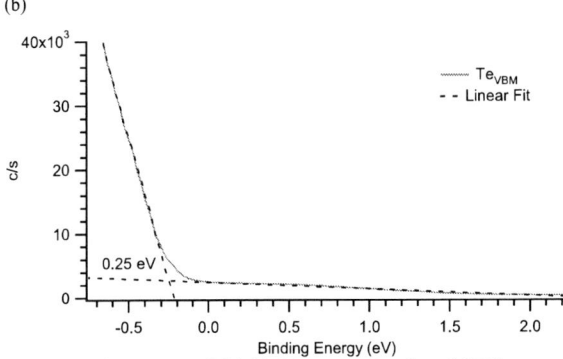

Fig. 4. XPS spectra of (a) Te3d$_{5/2}$ core level and UPS spectra of (b) Te valence band edge measured both from the 200 nm Te sample, shown in red. The core level peak was fit with a Voigt function, shown in black. The valence band edge was linearly extrapolated, shown in black.

TABLE I
CORE LEVEL AND VALENCE BAND MAXIMA VALUES FROM
STRUCTURES SHOWN IN FIGURE 3

Energy Levels	Binding Energy (eV)
E_{CL}^{Te}	573.1 ± 0.1
E_{CL}^{Cd}	405.1 ± 0.1
E_{VBM}^{Te}	0.26 ± 0.05
E_{VBM}^{CdTe}	0.88 ± 0.05

The XPS measured Te3d$_{5/2}$ and Cd3d$_{5/2}$ CL peaks determined from the structures shown in Fig. 3(c) (2.4 μm CdTe/ 2-8 nm Te) are shown in Fig. 5. A decreasing intensity of the Te3d$_{5/2}$ emission with decreasing Te thickness is observed in Fig. 5. A similar effect was observed with the Cd3d$_{5/2}$ emission peak as Te thickness increased. The CL values were determined and the averages of several peaks are summarized in Table II. A slight shift among the binding energies with the varying thicknesses is apparent. ΔE_{CL} was calculated using (2) and used for calculation of VBO values for the varying CdTe/Te interface samples. (1) was used to

calculate the VBO values and these are summarized in Table III.

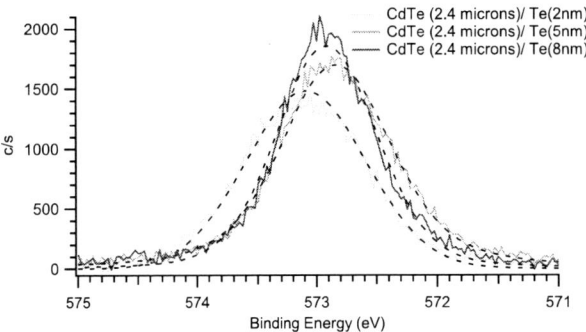

Fig. 5. XPS spectra of Te3d$_{5/2}$ core level peaks from 2.4 μm CdTe/ 2-8 nm Te interface samples, shown in blue. The core level peaks were fit with a Voigt function, shown in black.

TABLE II
CORE LEVEL VALUES FROM CDTE/TE INTERFACE STRUCTURES
SHOWN IN FIGURE 3

Cell Structure	E_{CL}^{Te} (eV)	E_{CL}^{Cd} (eV)
2.4 μm CdTe/ 2 nm Te	573.0±0.1	405.3±0.1
2.4 μm CdTe/ 5 nm Te	573.1±0.1	405.2± 0.1
2.4 μm CdTe/ 8 nm Te	573.1±0.1	405.2±0.1

TABLE III
CALCULATED VALENCE BAND OFFSET VALUES FROM J-V-T
AND XPS/UPS ANALYSIS

Cell Structure	E_{VBO} (eV)
2.4 μm CdTe/ 2 nm Te	0.87± 0.2
2.4 μm CdTe/ 8 nm Te	0.63± 0.2
Device shown in Fig. 1 & 2	0.39± 0.02

IV. DISCUSSION

The 0.88 eV calculated VBM of CdTe suggests that CdTe exhibits slightly n-type behavior as this would put the approximate Fermi level slightly above the middle of the gap. This is inconsistent with hot probe tests, which probe bulk properties and indicate a slightly p-type doping under typical CSS CdTe deposition conditions. Kraft *et.* al [6] suggests that the surface material characteristics of CdTe vary from those found in bulk CdTe.

XPS and UPS analysis was performed to investigate the barrier at the CdTe/ Te interface and determine the VBO. The VBO values at the interface were calculated and listed in Table III. Below a certain thickness it has been shown the bulk Te properties are not fully developed [8]. The VBO value decreases as the Te thickness increases from 2 to 8 nm. The VBO value further decreases as the CdTe device is CdCl$_2$ passivated, copper doped, and 50 nm Te is used.

Niles *et al.* [8] performed similar experiments studying the CdTe/Te interface and reported a valence band offset value of 0.26 eV for Te. This is significantly less than all listed values

978-1-5090-5606-4/17 $31.00 © 2017 IEEE

in Table III. The VBO calculated from the JVT of the passivated, copper doped, 50 nm Te device presented in Fig. 2 lies between the values presented by Niles and our calculated values. Passivation and doping used in finished devices appear to have significant effects on the VBO.

V. CONCLUSION AND ADDITIONAL WORK

Tellurium has been characterized as a back contact for CdTe solar cells. XPS and UPS characterization were performed to approximate the VBO at the un-passivated, un-doped CdTe/Te interface. We observed that the offset decreases as the tellurium is thickened from 2 to 8 nm, indicating the tellurium thickness impacts the VBO offset at the interface. In addition, JVT of $CdCl_2$ passivated, copper doped, 50 nm Te CdTe devices show a further reduction in the VBO. This implies that tellurium thickness and device processing affect the barrier developed at the back contact. Additional XPS and UPS characterization of $CdCl_2$ passivated and copper doped CdTe with Te contacts would give further insight into the VBO at the back of the device.

ACKNOWLEDGMENTS

The authors are thankful for funding support from NSF AIR, and assistance from Drew Swanson, Jennifer Drayton, Andrew Moore, Jason Kephart, James Sites, Patrick McCurdy, Marina D'Ambrosio, Carey Reich, Kurt Barth, and Kevan Cameron.

REFERENCES

[1] Shockley, William and Queisser," Detailed Balance Limit of Efficiency of p-n Junction Solar Cells," *Journal of Applied Physics*, vol. 32, pp. 510-519, 1961.

[2] S. Kurtz and D Levi, "Best Research-Cell Efficiencies," (2017), available at:https://commons.wikimedia.org/wiki/File:PVeff (rev170117).png

[3] Kevin D. Dobson, Iris Visoly-Fisher, Gary Hodes, David Cahen, "Stability of CdTe/CdS thin-film solar cells," *Solar Energy Materials and Solar Cells*, vol. 62, pp. 295-325, 2000.

[4] Andrew Moore, Tian Fang, James Sites, Cu Profiles in CdTe Solar Cells, *in 42nd IEEE Photovoltaic Specialist Conference*, 2015.

[5] S. H. Demtsu and J. R. Sites, "Effect of back-contact barrier on thin-film CdTe solar cells," *Thin Solid Films*, vol. 510, no. 1–2, pp. 320–324, 2006.

[6] Kraft, D. and Thissen, A. and Broetz, J. and Flege, S. and Campo, M. and Klein, A. and Jaegermann, W., Characterization of tellurium layers for back contact formation on close to technology treated CdTe surfaces, *Journal of Applied Physics*, vol. 94, pp. 3589-3598, 2003.

[7] Brian E. McCandless, Kevin D. Dobson, "Processing options for CdTe thin film solar cells", *Solar Energy*, vo. 77, pp. 839-856, 2004.

[8] Niles, David W. and Li, Xiaonan and Sheldon, Peter and Höchst, Hartmut, "A photoemission determination of the band diagram of the Te/CdTe interface," *Journal of Applied Physics*, vol. 77, pp. 4489-4493, 1995.

[9] Wei Xia, Hao Lin, Hsiang Ning Wu, Ching W. Tang, Irfan Irfan, Chenggong Wang, Yongli Gao, "Te/Cu bi-layer: A low-resistance back contact buffer for thin film CdS/CdTe solar cells," *Solar Energy Materials and Solar Cells*, vol. 128, pp. 411-420, 2014.

[10] Niles, D. W., Li, X., Albin, D., Rose, D., Gessert, T. and Sheldon, P., "Evaporated Te on CdTe: A vacuum-compatible approach to making back contact to CdTe solar cell devices," *Prog. Photovolt: Res. Appl.*, vol. 4, pp. 225–229, 1996.

[11] D. E. Swanson, J. Kephart, P. Kobyakov, K. Walters, K. Cameron, J. Drayton, J. R. Sites, K. Barth, and W.S. Sampath,, "A single-vacuum Closed-Space-Sublimation Chamber to Fabricate CdTe solar cells," JVST, 2015.

[12] G.T. Koishiyev, J.R. Sites, S.S. Kulkami, N.G. Dhere, "Determination of back contact barrier height in Cu(In,Ga)(Se,S)$_2$and CdTe solar cells," *in 33rd IEEE Photovoltaic Specialists Conference*, 2008.

[13] C.L. Perkins, F.S. Hasoon, H.A. Al-Thani, and S.E. Asher. "XPS and UPS Studies of Thin Film PV Materials Modified by Reactions in Liquids," *DOE Solar Tech. Program Review Meeting*. 2004

[14] Y. Zhu, N. Jain, and M.K.Hudait, "X-ray photoelectron spectroscopy analysis and band offset determination of CeO_2 deposited on epitaxial (100),(110), and (111)Ge," *Journal of Vacuum Science & Technology*, vol. 32, 011217, 2014.

[15] E. A. Kraut, R. W. Grant, J. R. Waldrop, and S. P. Kowalczyk, "Precise determination of the valence-band edge in X-Ray photoemission spectra: Application to measurement of semiconductor interface potentials," *Phys. Rev. Lett.*, vol. 44, pp. 1620–1623, 1980

On the Different Explanations of the Recombination Currents with High Ideality Factor in Silicon Solar Cells

A. Otaegi, V. Fano, N. Azkona, J. R. Gutiérrez and J. C. Jimeno

Technological Institute of Microelectronics (TiM), University of the Basque Country, 48013 Bilbao, Spain

Abstract — The recombination of the neutral regions of the solar cell is well modeled by means of an exponential term with ideality factor close to 1 in the low injection regime but non ideal performances evidenced in the *I-V* characteristic curves lead to diverse theories to explain the apparent high values of the ideality factor. This work analyzes a critical review of the existing explanations and clarifies their validity areas. Our experimental results indicate the evidence of high current areas that could be due to localized defects connected to the rest of the circuitry by a high connecting resistance.

Index Terms — Ideality factor, modeling, recombination.

I. Introduction

The classic theory of Shockley-Read-Hall, SRH, on the analysis of solar cell performance assumes an ideality factor $m=1$ to the neutral areas and $m=2$ to the depletion region, with uniformly distributed traps [1-3]. Performances different from the ideal are soon evidenced as high recombination currents are observed, leading to assume the existence of ideality factors different from the ideal. M. Wolf introduced high values of the ideality factor ($m>2$) as a result of undesirable shapes (shoulders, kinks, humps) in the non ideal performances of the solar cells in the 0.1-0.5 Volt range of the current-voltage curve [4], [5]. In the same year, H. J. Queisser found out an $m \approx 3$ performance, outcome from localized regions of the junction area due to recombination centers [6]. S. C. Choo theoretically verified that the recombination in the depletion region for asymmetrical *p-n* junctions ($\zeta_{n0} \neq \zeta_{p0}$) do not follow $m=2$ [7] and E. A. Faulkner et al. proposed a modified theory, for silicon *p-n* junctions, assuming traps in a localized region of the depletion region; as a result, m can take values from 1 to 2 [8].

Non ideal *I-V* curves (with $m \neq 2$) do not follow the double-diode model and a proper quantification of the characteristic parameters (J_{01}, J_{02}, R_S, R_P and m) may be difficult. Diverse models and explanations have tried to clarify the phenomenon according to different theories, beginning from recombination via tunneling of carriers, trap-assisted tunneling, intercenter charge transfer, edge defects or shunt resistance effects [9-13].

Discussion but caution found in the explanations of the non ideality phenomena are significant since different mechanisms are involved in the *I-V* curves performance. Regarding to different values of m, two main postulates are found in literature, those that justify m-s varying from 1 to 2 [14, 15], and those that justify apparent high values (with $m>2$) which do not follow the classic SRH recombination model [16].

First explanations about apparently high m factors were introduced by S. J. Robinson et al. [17]. As they indicate,

unequal electron and hole capture rates induce a high recombination region situated in different areas of the solar cell: space charge region, cell substrate or back surface. For every case, there is a transition from a high to a low recombination area, due maybe to a transition from low to high injection in the area beneath the damaged region. The transition into the two areas with different current performance implies a region with high m. If the damage is situated in the neutral region (the substrate or the rear surface), the two transition regions are $m=1$ but if the damage is in the space charge region, the two transition regions are $m>1$ (close to 2). The intermediate region forces the high m. This explanation hints the source of high m-s: a transition between two mechanisms or regions with different recombination, on which one of them saturates and stops acting. In the case explained by S. J. Robinson et al., the recombination in the transition from low to high injection involves also the recombination of the majorities; for conventional solar cells (with N_B approx. 10^{16} cm^{-3}) and high bias voltages implies that the high m-s are noticed for high voltages (close to 500 mV).

The explanation introduced by Hernando et al. [18] indicates that high values of the ideality factor are due to localized defects (broken textured pyramids). The localized defects induce regions with high recombination; these regions are connected through a high connecting resistance (which is indeed the factor that saturates in this model) to the rest of the structure; this is to say that in the low current regime there is a high recombination due to the punctual effect of the recombination, but as long as the current through the defect increases, it is limited due to the ohmic losses that connect the defect to the rest of the structure. This effect is superposed to the rest of the circuitry evidencing a hump *I-V* performance [19]. The key factor is then the interconnecting resistance.

K. R. McIntosh et al. follow the explanation introduced by Hernando et al. to justify the recombination in the edges of the cell [20, 21]. They analyzed the effect according to different doping densities in the emitter and the substrate. They limited the connecting resistance to the resistance obtained due to the geometry of the emitter and its distance from the defect. As a results, they maximum value of m in the space charge region was found to vary from 1.7 to 2. This value is low enough to explain the deviations of the ideality factors except if the damage performs as $m>2$, which was not considered in previous theories.

An explanation to high ideality factors ($m>2$) is introduced by O. Breitenstein et al. [22]. They assert that the recombination is multilevel, i.e. via coupled pair of defects in

978-1-5090-5606-4/17 $31.00 © 2017 IEEE

heavily damaged *p-n* junction regions or via deep donor acceptor pair recombination (DAP). The coupled defect level recombination (CDL), with two mutually coupled defect levels acting as trapping centers, has its origin in trapping and recombination of carriers directly from intercenters [23-25], while the DAP recombination via deep levels is outlined to be the responsible of *m>2* performances when going from low to high defect densities [26].

Another explanation to the problem would involve the single trap defect consideration, for simplicity centered in E_g and $\zeta_{n0} = \zeta_{p0}$, but with a defect small enough to have limited the spreading resistance (inversely proportional to the size of the defect) and justifying resistances greater than the geometric resistance. Our experimental results indicate that high values of current of *m* could be related to localized defects with high connecting resistance values [27].

II. ONE-DIMENSIONAL DAMAGE MODELS

Non ideal behavior of solar cells is analyzed in the nineties [17], inducing a defected region inside the substrate of the solar cells, as it is shown in Figure 1.

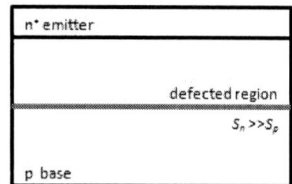

Fig. 1. Structure with a defected region for the one-dimensional model, with different capture cross sections.

As they indicate, unequal electron and hole capture rates induce a high recombination region situated in different areas of the solar cell: space charge region, cell substrate or back surface. For every case, there is a transition from a high to a low recombination area, due maybe to a transition from low to high injection in the area beneath the damaged region. The transition into the two areas with different current performance implies a region with high *m*. If the damage is situated in the neutral region (the substrate or the rear surface), as in the figure, the two transition regions are *m=1* but if the damage is in the space charge region, the two transition regions are *m>1* (close to 2). The intermediate region forces the high *m*. This explanation hints the source of high *m*-s: a transition between two mechanisms or regions with different recombination, on which one of them saturates and stops acting. In the case explained by S. J. Robinson et al., the recombination in the transition from low to high injection involves also the recombination of the majorities; for conventional solar cells (with N_B approx. 10^{16} cm^{-3}) and high bias voltages implies that the high *m*-s are noticed for high voltages (close to 500 mV).

In general, the mechanisms of *m* larger than 2, involve a saturation effect of a high recombination term.

In 1995, epitaxially grown diodes with large ideality factors were analyzed [23]. In that work, ideality factors larger than 2 were explained by means of the coupled defect level recombination. The trapping and recombination of carriers directly from intercenters, as it is shown in Figure 2, lead to high ideality factors only if the defect levels are shallow, close to the conduction band and the valence band.

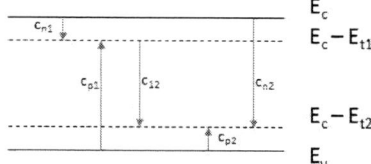

Fig. 2. Capture processes in the CDL model.

Another authors, following CDL model, explained the damage in solar cells (by laser grooving, scratching, etc) based on a large number of defects, about 10^{20} cm^{-3}, between recombination centers in the midgap proximities, but requiering a difference in the capture cross sections, up to 10^{12}, from one to the other recombination center, which enables a good connection from one recombination center to the conduction band and from the other recombination center to the valence band, in spite of being in the middle of the bandgap [22, 26].

On the other hand, the CDL the model explains the reverse characteristics of the diode.

III. TWO-DIMENSIONAL DAMAGE MODELS

The explanation introduced by Hernando et al. [18] indicates that high values of the ideality factor are due to localized defects. These defects induce regions with high recombination, which are connected through a high connecting resistance to the rest of the structure; this is to say that in the low current regime there is a high recombination due to the punctual effect of the recombination, but as long as the current through the defect increases, it is limited due to the ohmic losses that connect the defect to the rest of the structure. The high connecting resistance, is indeed, the factor that saturates in this model.

Fig. 3 shows the electronic circuit for the damage, an induced recombination diode with *m=2*, connected to the depletion region diode by the connecting resistance R_c. In Fig. 4, the effect of the damage is observed in the dark *I-V* characteristic curve of the cells, as a transition, from an *m=2* region to an *m=2* region for middle voltage ranges, made by the R_c [27].

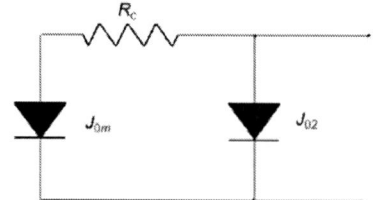

Fig. 3. Electronic circuit for localized defects and depletion region behavior.

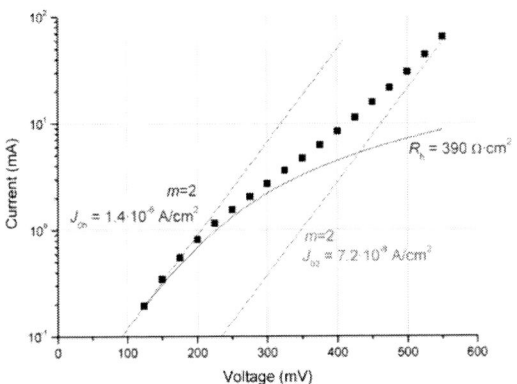

Fig. 4. Transition from low current regime to high current regime through the connecting resistance.

The behavior of the ideality factor is analyzed if Fig. 5, for different values of R_c. As it can be shown in the figure, the determinant effect on the behavior of the ideality factor is the connecting resistance, whose effect, a hump, would be masked by the J_{02} component if this resistance had low values; but this hump would be clearly observed for high values of R_c. This connecting resistance has been considered low up to now [20], so its effect has been ignored. Even at low voltages, for high connecting resistances, the ideality factor could reach to 4 at voltages as low as 100 mV.

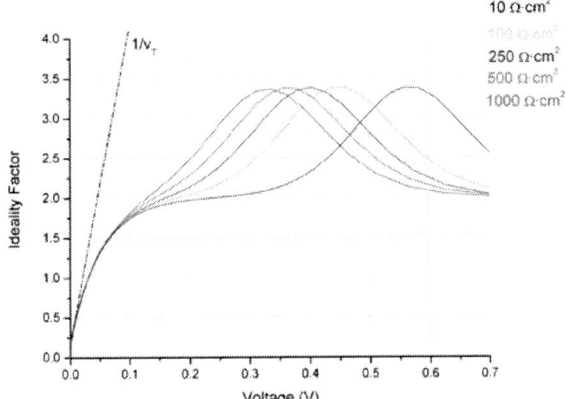

Fig. 5. Curves of ideality factors for several R_c.

It should be pointed out that for given values of the J_{0m} and R_c, $5 \cdot 10^{-8}$ A/cm^2 and 1000 Ωcm^2 for the curves shown in Fig. 6, the maximum slope of the ideality factor is determined by $1/v_T$, but this value is modulated by the distance between the J_{0m} of the damage and the J_{02} of the solar cell. Reducing this distance, the ideality factor can be reduced, influencing also in the maximum value of the ideality factor.

In some cases, as in the CDL model, ideality factor values higher than two remain not very variable with respect to the

voltage, but a distributed model of diodes for the damage, see Fig. 7, would diminish the slopes as it is shown in Fig. 8.

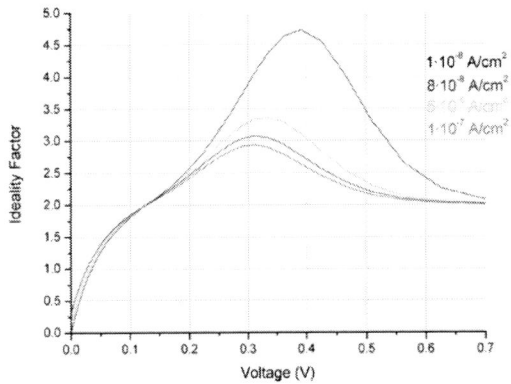

Fig. 6. Curves of ideality factors for several J_{02}.

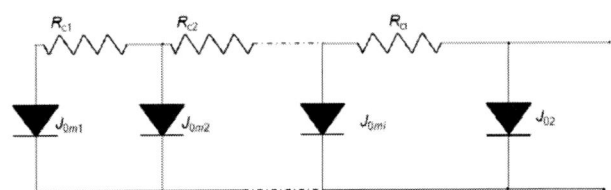

Fig. 7. Distributed circuit model for the damage in the solar cell.

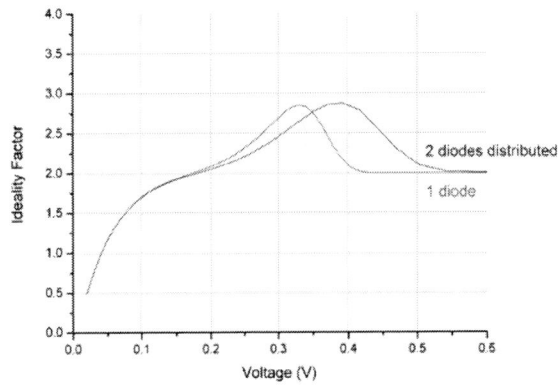

Fig. 8. Comparison of the ideality factor for a distributed and a localized defect.

IV. CONCLUSIONS

All the models that justify ideality factors larger than 2 are based in the Shockley Read Hall model. To justify such large values, they are based on high recombination models that suffer a saturation effect. It is in the transition between two mechanisms or regions with different recombination, on which one of them saturates and stops acting, where ideality factors larger than 2 are found.

978-1-5090-5606-4/17 $31.00 © 2017 IEEE

Two different models explain the large m values; in the one-dimensional models, the large m values at low currents are explained by the appearance of large number of defects, above 10^{20} cm^{-3}, that lead to recombination centers with particular characteristics that tend to saturate as long as they are filling.

Two-dimensional models refer to conventional recombinations with high values. The access to reach that recombination is made through a resistive path, which for certain currents separates electrically the damage from the structure.

Both models, the one-dimensional and the two-dimensional, characterize the damage in the solar cells; the simulation of the CDL model explains the characteristic of reverse in the solar cells but the two-dimensional model is thermally verified, although only for the forward characteristic.

ACKNOWLEDGMENTS

This work has been supported by the MINECO/FEDER within the framework of the project ENE2014-56069-CA-1-R.

REFERENCES

[1] W. Shockley. "The theory of *p-n* junctions in semiconductors and *p-n* junction transistors". Bell Sys. Tech. J. 8:425-489 (1949).

[2] W. Shockley, W. J. Read. "Statistics of the recombinations of holes and electrons." Phys. Rev. 87: 835 (1952).

[3] C-T. Sah, R. N. Noyce, W. Shockley. "Carrier generation and recombination in *p-n* junctions and *p-n* junction characteristics". Proceedings of the IRE, 1228-1257 (1957).

[4] M. Wolf. "Advances in silicon solar cell development". Energy conversion for space power. 231-261 (1961).

[5] M. Wolf. "Series resistance effects on solar cell measurements". Advanced energy conversion. 3:455-479 (1961).

[6] H. J. Queisser. "Forward characteristics and efficiencies of silicon solar cells" Solid State Electronics, 5:1-10 (1962).

[7] S. C. Choo. "Carrier generation-recombination in the space-charge region of an asymmetrical *p-n* junction." Solid-State Electronics, 11: 1069-1077 (1968).

[8] E. A. Faulkner, M. J. Buckingham. "Modified theory of the current/voltage relation in silicon *p-n* junctions". Electronics Letters, 4:17; 359-360 (1968).

[9] S. M. Sze, "Physics of semiconductor devices"2nd edition. New York, Wiley.

[10] A. Schenk. "An improved approach to the Shockley Read Hall recombination in inhomogeneous fields of space-charge-region" J. Appl. Phys. Vol. 18. Pp. 2261-2268. 1985.

[11] W. M. Chen, B. Monemar, E. Janzén, J. L. Lindström. "Direct observation of intercenter charge transfer in dominant nonradiative recombination channels in silicon." Physical review letters 67: 1914-1917 (1991).

[12] M.A. Green, A. W. Blakers, C.R. Osterwald. "Characterization of high efficiency solar cells" J. Appld. Phys. Vol 58, pp. 4402-4408. 1985.

[13] PP Altermatt, G. Heiser, X. Dai, J.J. Armin, G. Aberle, S.J. Robinson, T. Young, S.R. Wenham, M.A. Green. "Rear surface passivation of high efficiency silicon solar cells by a floating junction". Journal of Applied Physics, 1996.

[14] P. Ashburn, D. V. Morgan, M. J. Howes. "A theoretical and experimental study of recombination in silicon p-n junctions." Solid-State Electronics, 18: 569-577 (1975).

[15] J. Pallarés, L. F. Marshall, X. Correig, J. Calderer, R. Alcubilla. "Space charge recombination in *p-n* junctions with discrete and continuous trap distribution." Solid State Electronics 41: 17-23 (1975).

[16] A. Kaminski, J. J. Marchand, A. Laugier. "Non ideal dark I-V curves behavior of silicon solar cells." Solar Energy Materials & Solar Cells, 51: 221-231 (1998).

[17] S. J. Robinson, A. G. Aberle, M. A. Green. "Recombination saturation effects in silicon solar cells." IEEE Transactions on Electron Devices, 41: 1556-1569 (1994).

[18] F. Hernando, R. Gutiérrez, G. Bueno, F. Recart, V. Rodríguez. "Humps, a surface damage explanation." 2nd WC PSEC, 1321-1324 (1998).

[19] J. Beier, B. Voß. "Humps in dark I-V curves – Analysis and explanation." 23rd IxEEE PSC, 321-326 (1993).

[20] K. R. McIntosh, C. B. Honsberg. "The influence of edge recombination on a solar cell's *I-V* curve." 17th EPVSEC, 1578-1581 (2001).

[21] K. R. McIntosh, P. P. Altermatt, G. Heiser. "Depletion region recombination in silicon solar cells: when does $m_{DR}=2$?" 17th EPVSEC, 1578-1581 (2001).

[22] O. Breitenstein, P. P. Altermatt, K. Ramspeck, M. A. Green, J. Zhao, A. Schenk. "Interpretation of commonly observed *I-V* characteristics of C-Si cells having ideality factor larger than two." Proceedings on the 4th WC on PEC IEEE, 879-884 (2006).

[23] A. Schenk, U. Krumbein. "Coupled defect-level recombination: Theory and application to anomalous diode characteristics." Journal of Applied Physics, 78:3185-3192 (1995).

[24] W. M. Chen, B. Monemar, E. Janzén, J. L. Lindström. "Direct observation of intercenter charge transfer in dominant nonradiative recombination channels in silicon." Phys. Rev. Lett. 67: 1914-1917.

[25] A. M. Frens, M. T. Bennebroek, A. Zakrzewski, J. Schmidt, W. M. Chen, E. Janzén, J. L. Lindström, B. Monemar. "Observation of rapid direct charge transfer between deep defects in silicon." Phys. Rev. Lett. 72: 2939-2942.

[26] S. Steingrube, O. Breitenstein, K. Ramspeck, S. Glunz, A. Schenk, P. P. Altermatt. "Explanation of commonly observed shunt current in c-Si solar cells by means of recombination statistics beyond the Shockley-Read-Hall approximation." Journal of Applied Physics, 110:014515 (2011).

[27] A. Otaegi, V. Fano, M. A. Rasool, J. R. Gutiérrez, J. C. Jimeno, N. Azkona, E. Cereceda. "Progress on the explanation and modeling of the laser-induced damage." Submitted to journal.

Identification of shunts in a monolithic multijunction GaAs/GaAs device by spectrometric characterization

Felipe Oviedo[1,2], Liu Zhe[3], Zekun Ren[2], Kevin Nay Yaung[2], Maung Thway[3,4], Liu Haohui[3],
Tonio Buonassisi[1,2], Ian Marius Peters[1,2]

[1]Massachusetts Institute of Technology (MIT), 77 Massachusetts Avenue, Cambridge, MA 02139, United States
[2]Singapore-MIT Alliance for Research and Technology (SMART), 1 CREATE Way, 138602, Singapore
[3]Solar Energy Research Institute of Singapore (SERIS), National University of Singapore, 7 Engineering Drive 1, 117574, Singapore
[4]Department of Electrical and Computer Engineering, National University of Singapore, 4 Engineering Drive 3, 117583, Singapore

Abstract. **Monolithic multijunction solar cells are one of the most promising concepts for future, high-efficiency commercial solar cells. However, the characterization of this type of device, and the identification of detrimental mechanisms such as shunts, tends to be complex compared to single junction devices. In this work, we propose a methodology to identify the shunted subcell in monolithic multijunction devices by spectrometric characterization and equivalent circuit modelling. Current-voltage (*J-V*) characteristics of a GaAs/GaAs dual junction solar cell are measured under different spectra and constant intensity. A series-connected, two-diode equivalent circuit model is developed, and the shunted subcell in the tandem stack is identified by fitting the measured data. The *J-V* characteristics for different spectra and the effect of shunts on them are explained.**

I. INTRODUCTION

High-efficiency solar cells are necessary to reduce the price of solar-generated electricity and allow significant growth in installed photovoltaic capacity [1], [2]. Multijunction solar cells are among the most promising technologies for achieving substantial efficiency gains in comparison with state-of-the-art single junction devices. The higher complexity of multijuction device architectures and the various operating conditions (spectrum shape, illumination intensity, current-matching conditions, etc.) define specific fabrication, characterization and operation constraints [3].

Two-terminal monolithic multijunction devices are comprised of two or more subcells electrically connected in series by tunneling junctions. The current density of the device is usually limited by the lowest current density of the subcell stack. Thus, certain electrical and optical constraints are required for the subcell integration. Due to these constraints, certain physical conditions and irregularities such as current-mismatch and shunts have different effects on the multijunction solar compared to the well-studied single junction case. For instance, when current-mismatch and significant shunts in either subcell exist, it is possible that the short-circuit current density (j_{sc}) of the multijunction device surpasses that of the current-limiting device [4]. In consequence, shunts affect the *J-V* characteristics of a monolithic multijunction device in different ways according to the specific current matching conditions, shunt magnitude and shunt location.

Conventionally, in order to maximize the power output of the device, the individual currents of the subcells under standard AM1.5G illumination conditions are matched by adjusting the thicknesses or band gaps of the subcells. Nevertheless, spectral variations during solar cell operation could affect significantly the currents and *J-V* characteristics of the subcells [5]. If one of the subcells has a shunt, the maximum power output of the device might not correspond to the current-matched case due to changes in the *J-V* characteristics of the device. Furthermore, a *J-V* measurement under standard AM1.5G laboratory conditions may not reveal the presence of shunts which affect the device efficiency under real operating conditions.

The individual subcells of a monolithic multijunction device are usually difficult to access and characterize independently. In this case, spectrometric characterization is a common technique used to study the subcell *J-V* parameters. The technique consists in systematically varying the spectral distribution of the illumination over a multijunction solar cell. For a dual junction device, given that the blue part of the spectrum is mainly absorbed by the top sub cell and the infrared part of the spectrum is absorbed by the bottom sub cell, the spectral variations cause different current matching conditions, and allow to analyze each subcell quasi-independently, optimize current-matching conditions and determine spectrum-induced voltage shifting. Spectrometric characterization has been performed before, with various degrees of accuracy, employing filters in a single-source solar simulator or Xenon lamp, or using LED solar simulators.

In this work, spectrometric characterization is used to identify the shunted subcell in a monolithic GaAs/GaAs device that presents a shunted *J-V* curve, and quantify the shunt effects under various current-matching conditions. The resulting *J-V* curves are fitted through an equivalent-circuit model with the aim of interpreting the observed effects over *J-V* characteristics. This methodology allows the identification and characterization of shunts in multijunction devices for a wide range of operating conditions.

II. METHODOLOGY

A schematic structure of the studied GaAs/GaAs device is shown in Figure 1. The GaAs/GaAs cell was fabricated on an epi-ready <100> oriented *n*-type GaAs on-axis wafers using a Metal Organic Chemical Vapour Deposition (MOCVD) reactor. The MOCVD growth was performed under a pressure of 100 mbar using TMGa, TMIn, AsH$_3$ and PH$_3$ as precursors

978-1-5090-5606-4/17 $31.00 © 2017 IEEE

and H_2 as carrier gas. Heavily doped $n++$ and $p++$ 20 nm GaAs layers form the tunnel junction, allowing electron-hole recombination in the interface and, thus, connecting top and bottom subcell electrically. In the absence of significant shunts or major quality issues, our baseline dual junction cell for this process is over 20% [6]. For this study, the efficiency of the test GaAs/GaAs device, without an antireflective coating, is 13.9%. A priori, the device is believed to be shunted, but the actual location and shunt characteristics are unknown.

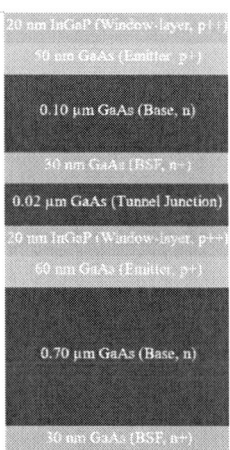

Figure 1: Schematic of studied GaAs/GaAs monolithic solar cell

The spectrometric characterization procedure was performed with a Wavelabs, Sinus-220 LED-based I-V tester. The tester is comprised of 21 LEDs of different wavelengths that allow approximating the distribution of any arbitrary solar spectrum. By varying the spectrum shape it is possible to excite either the front or the bottom subcell of the device. The average photon energy (APE), which represents the average energy per photon in a given spectrum, is used as a proxy for spectrum shape. A higher APE value corresponds to a blue-rich spectrum and, in consequence, a higher photogenerated current in the top GaAs subcell. In the same way, a lower APE value corresponds to spectral distribution shifted to the infrared [7]. For this study, APE values in the range of 1.70 eV – 1.95 eV were approximated, corresponding to real spectra measured in Singapore. To avoid variations in irradiance level, all the spectra were scaled to match the power of the AM1.5G spectrum in the 300-1100 nm wavelength range. Figure 2 shows the LED-approximated spectra with different APE values. The APE value of the standard AM1.5G spectrum is 1.83. Higher values correspond to blue-rich spectra, and lower values to red-rich spectra as evidenced in Figure 2, as explained before.

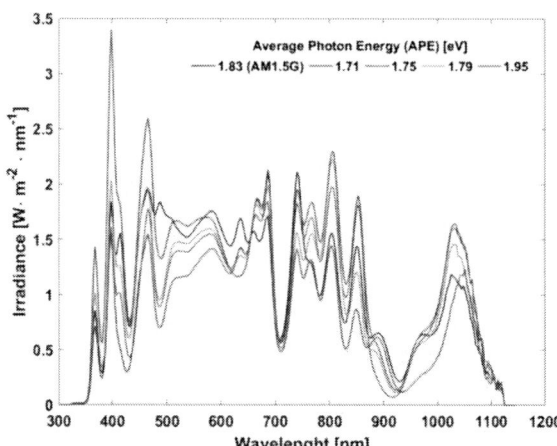

Figure 2: LED spectra of the equal irradiance level and different APE values, approximating real spectra measured in Singapore.

The *J-V* characteristics of the multijunction device were measured for each spectrum. To analyze the observed trends, an equivalent-circuit model composed of two double-diode equivalent-circuit models connected in series, representing respectively the top and bottom subcells in a tandem device, was used. A model schematic is shown in Figure 3. The shunt and series resistance are considered to be fully ohmic. The standard double-diode equation was modified to account for photoluminescence coupling and photon recycling [6]. In this case, each subcell can be described by:

$$
J(V) = (1 - \eta) J_L - J_{01} \left\{ \exp\left[\frac{qV + JR_s}{kT} \right] - 1 \right\} - \\
J_{02} \left\{ \exp\left[\frac{qV + JR_s}{mkT} \right] - 1 \right\} - \frac{V + JR_s}{R_{shunt}} - \frac{K}{\beta} J_{L'} \ (1)
$$

In equation 1, J_{01} and J_{02} correspond to the ideal a non-ideal reserve saturation currents of the diode respectively, m is the diode ideality factor (assumed to be 2, in this case), R_{shunt} is the equivalent shunt resistance, R_s is the equivalent series resistance, J_L is the photo-generated current of the subcell of interest, $J_{L'}$ is the photo-generated current of the complementary subcell, k is the Boltzmann constant, T is the absolute temperature of the junction, η is the internal electroluminescence efficiency accounting for photon recycling and $\frac{K}{\beta}$ is the photoluminescence efficiency factor from the complementary junction to the junction of interest. The coupling factors η and $\frac{K}{\beta}$ are based on the geometry of the structure and can be obtained from [6], [8]–[10].

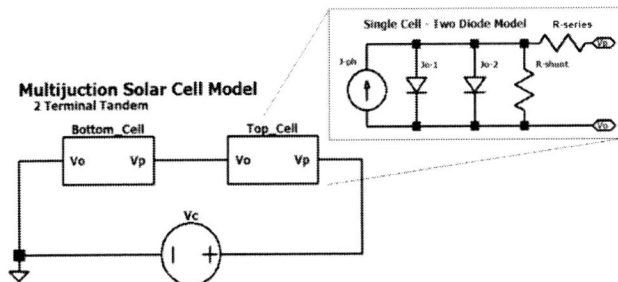

Figure 3: Equivalent circuit model of GaAs/GaAs solar cell

The dark saturation current and the short-circuit current density (j_{sc}), equal to the photogenerated current density J_L of each subcell, are required for the equivalent circuit model. These inputs parameters are not readily available by direct measurements. For that reason, the input parameters were found by fitting the J-V characteristics of the device under AM1.5G illumination (APE 1.83 eV) through a Sentaurus TCAD model. A two-dimensional 650 µm symmetry element of the solar cell was modelled, corresponding to the spacing between metal fingers. The GaAs/GaAs tunnel junction, consisting of two 10 nm heavily doped GaAs layers was simulated through an interband tunnelling model, accounting for dynamic tunnelling [11]. The calculated input parameters for the equivalent-circuit model are shown in Table 1.

Table 1: Input parameters and characteristics of GaAs/GaAs device at standard conditions (AM 1.5G).

GaAs top subcell	
Cell parameter	Parameter value
Jo1, n=1	1.36-17 mA/cm^2
Jo2, n=2	1.11e-11 mA/cm^2
Jsc	8.98 mA/cm^2
Voc	1.028 V
GaAs bottom subcell	
Cell parameter	Parameter value
Jo1, n=1	8.48-17mA/cm^2
Jo2, n=2	1.33e-11 mA/cm^2
Jsc	8.82 mA/cm^2
Voc	0.994 V
GaAs/GaAs monolithic tandem device	
Cell parameter	Parameter value
Jsc	8.82 mA/cm^2
Voc	2.022 V
FF	0.8425
Efficiency	13.9%

III. RESULTS AND DISCUSSION

The GaAs/GaAs tandem cell was measured under the specified spectrum conditions at constant intensity. The measured J-V curves are show in Figure 4. The j_{sc} of the tandem

device and the cell acting as the current-limiting subcell change as the spectrum is shifted. For APE values over the AM 1.5G condition (1.83 eV), the j_{sc} of the tandem device is reduced as the bottom cell is depleted of light. For APE values over the AM 1.5G condition, a higher shunt resistance starts to be observed in the J-V characteristics. As the top cell becomes current limiting, for the APE value of 1.71 and 1.75 eV, the equivalent shunt resistance of the J-V curve is reduced, revealing the presence of a shunt, or a defect causing significantly low shunt resistance. The shunt resistance seems to increase as the top cell become current-limiting.

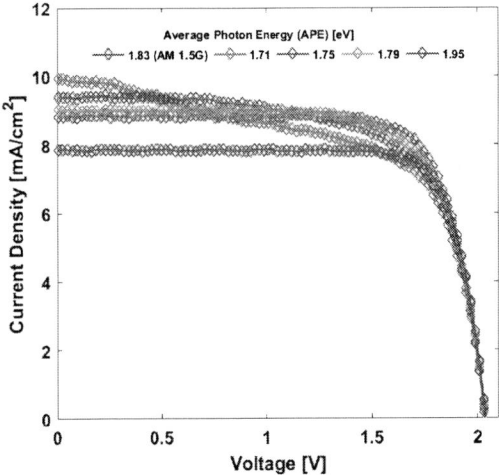

Figure 4: J-V characteristics of GaAs/GaAs dual junction structure under varying APE spectra.

At first glance, is not obvious whether the observed shunt is present in the top or in the bottom subcell. The equivalent circuit model of Figure 3 is helpful to determine the shunt location. Initially, the generated current density of each subcell under the extreme APE values of 1.71 and 1.95 eV was found. Subsequently, the shunt and series resistance of each top and bottom subcell were varied to fit the experimental J-V curves at the corresponding APE values. Two cases were investigated: 1.) Shunt in the bottom subcell, and 2) Shunt in the top subcell. The type of shunt was supposed to be ohmic, which refers to a fully linear contribution to the reverse saturation current.

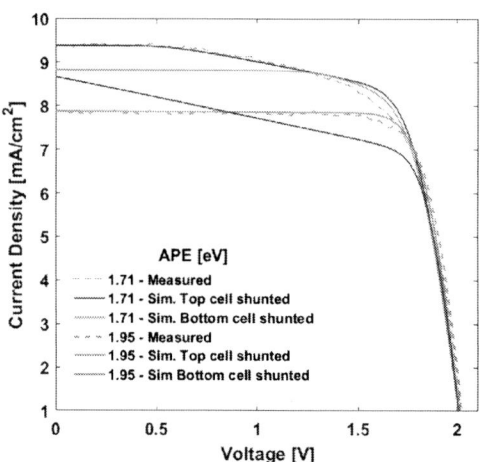

Figure 5: Measured and simulated J-V curves of cell at APE values of 1.71 and 1.95 eV

Figure 5 compares the measured and simulated *J-V* curves at the two APE values. Both simulated curves of top and bottom shunted cell have the same equivalent shunt resistance. It is possible to identify the shunted cell as the top cell, because it approximates with good agreement the measured *J-V* curves. For the 1.71 eV APE condition, the top cell is the current-limiting cell. However, the final j_{sc} of the simulated cell varies according to which subcell is shunted. In the case in which the shunt is occurring in the current-limiting cell, the j_{sc} of the tandem device is observed to be higher than that j_{sc} of the current-limiting subcell . The effect of the non-zero slope close to the shunted subcell's j_{sc} explains the observed j_{sc} transition.

We know that for short-circuit conditions the voltages of both subcells connected by a working tunnel junction, V_{top} and V_{bot}, are subjected to $V_{top} + V_{bot} = 0$ [12]. If the slope of the shunted subcell close to its j_{sc} is non-horizontal, the cell will operate in reverse bias. In that case, the diode equation can only be satisfied by a j_{sc} that is higher than the j_{sc} of the current-limiting cell, explaining the current transition. In the context of the equivalent circuit representation of the tandem device, this trend could be interpreted as the reduced shunt resistance electrically isolating the rectifying characteristics of the current-limiting cell and, consequently, reducing its limiting effect over the whole tandem device. Consistently, this effect is not observed in the case of the shunting in the bottom subcell, which is the non-limiting subcell for lower APE values. In the opposite case, for the 1.95 APE value condition, the bottom subcell acts as the current-limiting subcell. In this scenario, the shunt in the top subcell fits adequately the measured data, because the measured j_{sc} of the device does not surpass the j_{sc} of the current-limiting subcell.

The identification of the top shunt in the top subcell provides important fabrication information, and contributes to the isolation of the shunting mechanism. A limitation of this method is the consideration of the shunt as fully linear (ohmic). The most common types of shunts usually have a non-linear contribution to the recombination current that is not adequately captured by the model in certain cases. Furthermore, the diode ideality factor, assumed to be 2, might differ from that value for different subcell types and architectures.

The methodology could be combined with other characterization methodologies that estimate the *J-V* characteristics of the device, or a more elaborate two-diode fitting procedure to avoid the use of more complex Sentaurus TCAD simulations.

IV. CONCLUSION

A new methodology for identifying the shunted subcell under varying spectrum conditions in a monolithic multijunction devices was developed. Spectrometric characterization was used to measure the *J-V* characteristics of a GaAs/GaAs device under various current-limiting conditions. By using an equivalent circuit model, the shunted subcell was identified and the observed *J-V* characteristics were explained. Photon recycling and luminescence coupling are considered.

The proposed methodology can be transferred to any monolithic multijunction device and could be used in the development process of the new multijunction architectures with multiple integrated layers of material, such as perovskite/Si tandem devices. Future work should use the variations in *J-V* curves under different spectra to better characterize the shunt types and relate them to observable quality defects in the subcells.

REFERENCES

[1] D. B. Needleman, J. R. Poindexter, R. C. Kurchin, I. M. Peters, G. Wilson, and T. Buonassisi, "Economically Sustainable Scaling of Photovoltaics to Meet Climate Targets," *Energy Environ. Sci.*, vol. 9, pp. 2122–2129, 2016.

[2] D. M. Powell, M. T. Winkler, A. Goodrich, and T. Buonassisi, "Modeling the cost and minimum sustainable price of crystalline silicon photovoltaic manufacturing in the united states," *IEEE J. Photovoltaics*, vol. 3, no. 2, pp. 662–668, 2013.

[3] J. P. Mailoa, C. D. Bailie, E. C. Johlin, E. T. Hoke, A. J. Akey, W. H. Nguyen, M. D. McGehee, and T. Buonassisi, "A 2-terminal perovskite/silicon multijunction solar cell enabled by a silicon tunnel junction," *Appl. Phys. Lett.*, vol. 106, no. 12, p. 121105, 2015.

[4] A. Braun, N. Szabó, K. Schwarzburg, T. Hannappel, E. A. Katz, J. M. Gordon, A. Braun, N. Szabó, K. Schwarzburg, T. Hannappel, E. A. Katz, and J. M. Gordon, "Current-limiting behavior in multijunction solar cells Current-limiting behavior in multijunction solar cells," vol. 223506,

no. 2011, pp. 19–22, 2015.

[5] H. Liu, Z. Ren, Z. Liu, A. G. Aberle, T. Buonassisi, and I. M. Peters, "The realistic energy yield potential of GaAs-on-Si tandem solar cells: a theoretical case study," *Opt. Express*, vol. 23, no. 7, p. A382, 2015.

[6] Z. Ren, H. Liu, Z. Liu, C. S. Tan, A. G. Aberle, T. Buonassisi, and I. M. Peters, "The GaAs/GaAs/Si solar cell ??? Towards current matching in an integrated two terminal tandem," *Sol. Energy Mater. Sol. Cells*, vol. 160, no. August 2016, pp. 94–100, 2017.

[7] T. Minemoto, Y. Nakada, H. Takahashi, and H. Takakura, "Uniqueness verification of solar spectrum index of average photon energy for evaluating outdoor performance of photovoltaic modules," *Sol. Energy*, vol. 83, no. 8, pp. 1294–1299, 2009.

[8] J. F. Geisz, M. A. Steiner, I. Garcia, R. M. France, W. E. McMahon, C. R. Osterwald, and D. J. Friedman, "Generalized Optoelectronic Model of Series-Connected Multijunction Solar Cells," *IEEE J. Photovoltaics*, vol. 5, no. 6, pp. 1827–1839, 2015.

[9] D. J. Friedman, J. F. Geisz, and M. A. Steiner, "Effect of luminescent coupling on the optimal design of multijunction solar cells," *IEEE J. Photovoltaics*, vol. 4, no. 3, pp. 986–990, 2014.

[10] Z. Ren, J. P. Mailoa, Z. Liu, H. Liu, S. C. Siah, T. Buonassisi, and I. M. Peters, "Numerical Analysis of Radiative Recombination and Reabsorption in GaAs/Si Tandem," *IEEE J. Photovoltaics*, vol. 5, no. 4, pp. 1079–1086, 2015.

[11] M. Hermle, G. Létay, S. P. Philipps, and A. W. Bett, "Numerical Simulation of Tunnel Diodes for Multi-junction Solar Cells," *Prog. Photovolt Res. Appl.*, vol. 15, no. February 2013, pp. 409–418, 2008.

[12] A. Braun, N. Szabó, K. Schwarzburg, T. Hannappel, E. A. Katz, and J. M. Gordon, "Current-limiting behavior in multijunction solar cells," *Appl. Phys. Lett.*, vol. 98, no. 22, 2011.

A Simulation Study on Radiative Recombination Analysis in CIGS Solar Cell

Sanjoy Paul [1,2], Roberto Lopez [1], Md Dalim Mia [2], Craig H. Swartz [2], and Jian V. Li [1,2]

[1] Physics Department, Texas State University, San Marcos, TX 78666, USA

[2] Materials Science, Engineering, and Commercialization Program, Texas State University, San Marcos, TX 78666, USA

Abstract — **Recombination is an important issue to be addressed in order to develop high-efficiency solar cells. This report presents numerical simulations based on SCAPS-1D program to analyze recombination in thin film CIGS solar cells. The temperature and intensity dependent recombination analysis gives insights and is useful to analyze experimental observation.**

Index Terms — **Recombination, open-circuit voltage, photoluminescence efficiency, SCAPS-1D.**

I. INTRODUCTION

$CuIn_{1-x}Ga_xSe_2$ (CIGS, where x = Ga content) absorber materials is one of the best thin film PV materials with record efficiency about 22% at lab scale and about 16% for commercial modules [1]. Recombination phenomenon is a critical issue in further development of PV materials and devices. There are three major types of recombinations: Shockley-Read-Hall (SRH), radiative, and Auger recombination [2]. In the case of CIGS solar cell, the SRH and radiative recombinations dominate. Auger recombination is sufficiently low unless in the case of concentrator solar cell [3]. The total recombination rate in the solar cell is given by the equation

$$R_{Total} = \frac{(np - n_i^2)}{\tau_n(p + p_1) + \tau_p(n + n_1)}$$
$$+ B(np - n_i^2)$$
$$+ (np - n_i^2)(A_n n + A_p p)$$

(1)

Here, the first, second, and third terms are due to SRH, Radiative, and Auger recombination respectively. The parameters are n/p: electron/hole density, n_i: intrinsic carrier concentration, n_1/p_1: electron/hole occupancy at the defect level, τ_n/τ_p: lifetime of electron/hole, B is the radiative recombination coefficient, and A_n/A_p are the Auger recombination coefficients of electron/hole respectively.

It is truly challenging to identify experimentally the spatial distribution of recombination. Experimentally identifying the strength of recombination (basically SRH recombination) in few regions (space-charge, quasi-neutral, and interfaces) was attempted in CIGS solar cell *via* temperature and light intensity dependent open-circuit voltage analysis [4], [5]. However, the effect of radiative recombination was totally omitted.

Here in this study we focus on the radiative recombination analysis in the solar cell structure *via* simulation using SCAPS-1D platform. Temperature and intensity dependent radiative recombination simulations have been performed to investigate the effects of various electronic parameters such as surface recombination velocities of electrons and holes at front and back contact, capture cross section of electrons and holes, defect density and its energetic location, etc.

II. NUMERICAL SIMULATION

The SCAPS (a Solar Cell Capacitance Simulator), a one-dimensional solar cell simulator program, was particularly developed for CIGS and CdTe solar cell and a useful platform to simulate any solar cell devices and its properties [6], [7]. One major advantages is that all layer properties in the cell structure can be graded [8]. Besides, the advantage of batch calculation, scripting, and recording of solar cell parameters as function of any semiconductor parameters is a great convenience.

In this simulation, we have used the modified definition file ("Numos CIGS baseline.def") whose solar cell structure is [n-ZnO (50 nm)/n-CdS (50 nm)/p-CIGS (2 μm)]. The simulated energy band profile at T = 300K and one sun illumination is shown in Fig. 1. The bandgap of CIGS can be tuned from 1.04 eV (pure $CuInSe_2$) to 1.65 eV (pure $CuGaSe_2$) depending on the Ga content (x) according to the relation $E_g = 1.04 + 0.67x + bx(x - 1)$. Here, b is the optical bowing coefficient and its value is between 0.11 and 0.24 eV [9]. The band gap of CIGS in this simulation was double graded (as shown in the inset of Fig. 1) *via* compositional based grading.

978-1-5090-5606-4/17 $31.00 © 2017 IEEE

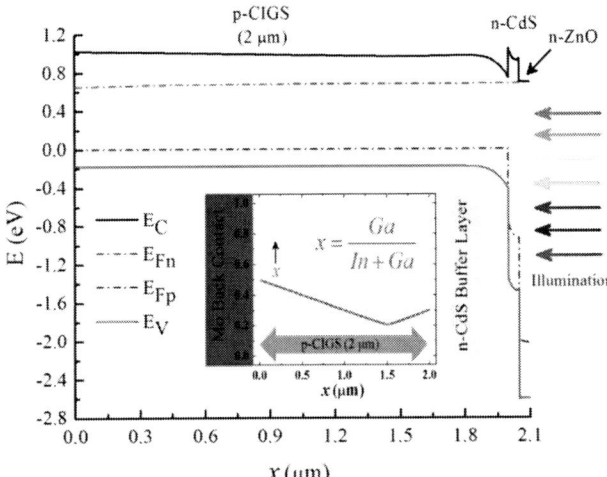

Fig. 1. The band diagram of CIGS solar cell at T = 300K using one sun (AM1.5G) illumination. The inset shows the band gap grading in CIGS absorber.

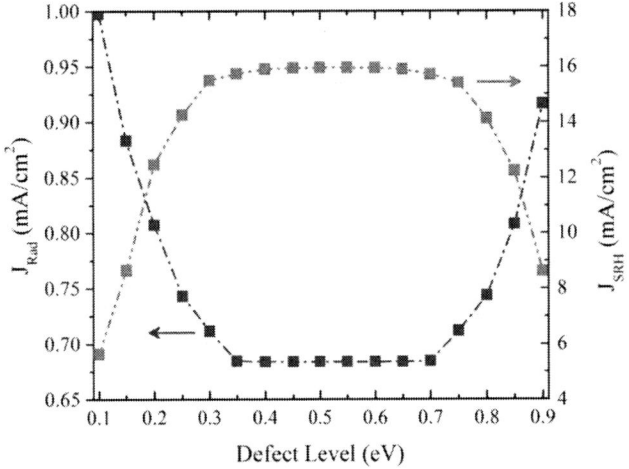

Fig. 2. Radiative and SRH recombination current density (at V = V_{OC}) as a function of the defect level in the CIGS absorber above valance band.

The uniform mid-gap defect level (of defect density 1.24×10^{13} cm^{-3}) was considered. In the range of level $0.35 - 0.7$ eV, constant radiative and SRH recombination currents (at T = 300K, and one sun illumination) were observed (as shown in Fig. 2). We considered the radiative recombination coefficient as 1×10^{-10} cm^3/s. Surface recombination velocity (for electrons and holes) at the front and back contact were chosen as 1×10^5 cm/s. The current-voltage characteristics, SRH and Radiative recombination profile, and open-circuit voltage (V_{OC}) were recorded as a function of temperature ($200 - 400$K) and illumination intensity with white light (AM1.5G) by changing the ND filter (of density between -2.0 to 2.0). As in most cases the radiative recombination study (via photoluminescence measurement) has been performed using monochromatic laser light, so we have performed some simulation (temperature and intensity dependent) using monochromatic light (540 nm) as well.

In order to record V_{OC}, recombination profile and recombination currents at open circuit condition ($V = V_{OC}$), the voltage was swept with very small step size (0.5 mV) and stopped at V_{OC}. Fig. 3 shows the generated recombination (SRH, radiative, and total) current-voltage characteristics at room temperature under one sun condition. The radiative recombination current used here is simulated at $V = V_{OC}$. Radiative recombination efficiency (i.e. photoluminescence efficiency) is the ratio of radiative current density (J_{Rad}) to the incident power density (in W/cm^2).

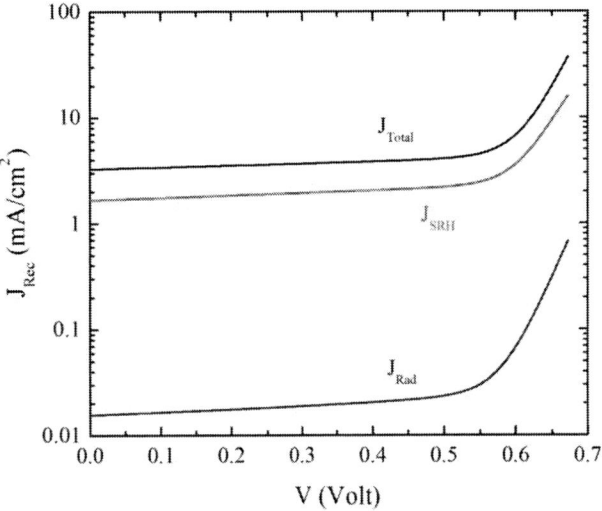

Fig. 3. The recombination current-voltage characteristics at T = 300K under one sun (AM1.5) condition.

III. RESULTS AND DISCUSSION

The SCAPS-1D simulated temperature and intensity dependent V_{OC} surface plot is shown in the Fig. 4. This figure shows the V_{OC} is sensitive to temperature and intensity and informative for analyzing SRH recombination [10]. Corresponding temperature and intensity dependent surface plot of radiative recombination efficiency (PL efficiency) is shown in Fig. 5.

978-1-5090-5606-4/17 $31.00 © 2017 IEEE

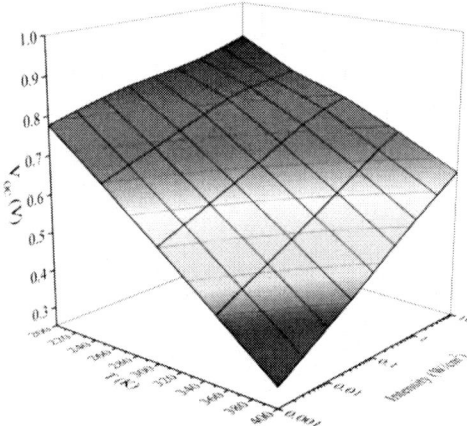

Fig. 4. SCAPS-1D generated temperature and intensity dependent V_{OC} data with CIGS solar cell structure (as shown in Fig. 1) using white light illumination (AM1.5).

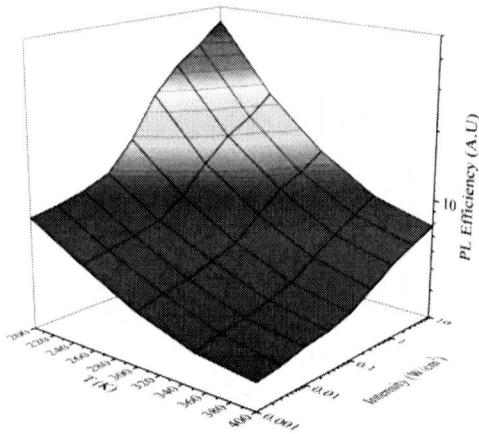

Fig. 5. Temperature and intensity dependent surface plot of radiative recombination efficiency (corresponds to Fig. 4) using white light illumination (AM1.5).

SCAPS-1D generated SRH and radiative recombination profiles (for T = 200K and 300K) is shown Fig. 6 at V = V_{OC}. This shows both that recombination is temperature dependent and that most of the recombination occurs near the buffer/absorber interface and in the depletion region. The SRH recombination is highly dominating over radiative one in CIGS solar cells which strongly depend on the defect density and its energetic location above the valence band of CIGS absorber. At lower temperature (T = 200K) radiative recombination is higher than at T = 300K. This is because of the electron and hole accumulation at the minima of conduction band and maxima of valance band respectively which enhances the radiative recombination [11]. At higher temperature, the carriers spread over because of thermal energy which reduces the radiative recombination [12]. Radiative recombination is a strong

function of light intensity as recombination yield is proportional to the product of generated electrons and holes. Increased light intensity increases the proportion of radiative recombination (as shown in Fig. 7) in comparison to SRH. At very high light level (1000 suns in Fig. 7) radiative recombination dominates over SRH, which is commonly seen in concentrator solar cells.

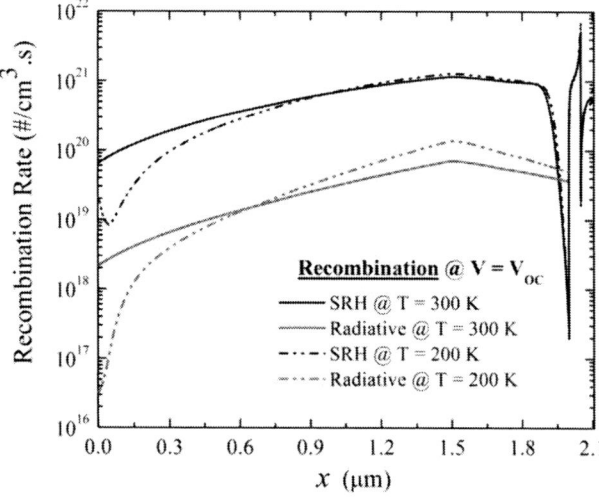

Fig. 6. SRH and radiative recombination profile at V = V_{OC} (T = 200K and 300K) under one sun illumination (AM1.5).

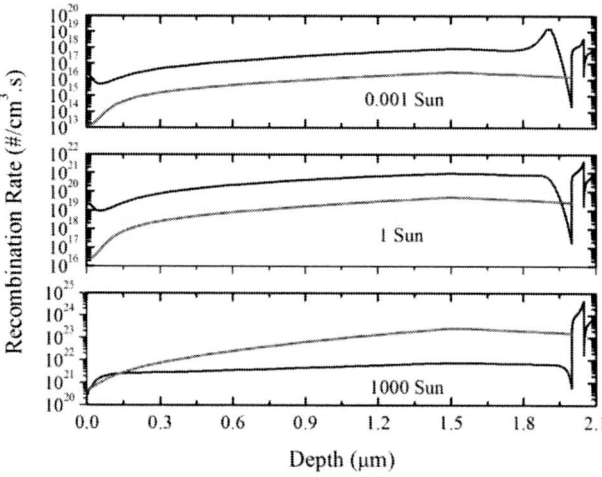

Fig. 7. Depth dependent Radiative (red) and SRH (black) recombination rate at T = 300K with different intensities [0.001 sun (top), 1 sun (middle), and 1000 suns (bottom)].

Photoluminescence study is the experimental technique to probe radiative recombination in PV materials and devices. The experimental photoluminescence study is commonly done with monochromatic laser light focused onto the sample to achieve high intensity. Because of this reason, we present SCAPS-1D simulation to see the shapes of two-dimensional V_{OC} and PL efficiency profile using 540 nm excitation. Fig. 8 (a) shows the V_{OC} profile and Fig. 8 (b) the corresponding PL efficiency profile. The capture cross-section for electron (σ_n) and hole (σ_p) used in this simulation is $\sigma_n = 5 \times 10^{-13}$ cm^2 and $\sigma_p = 1 \times 10^{-15}$ cm^2 respectively. The surface recombination velocity for electron (S_e) and hole (S_h) used in this simulation are (back-contact: $S_e = 10$ cm/s, and $S_h = 10$ cm/s; and front-contact: $S_e = 1 \times 10^5$ cm/s, and $S_h = 1 \times 10^5$ cm/s). Note that there is similarity in V_{OC} and PL efficiency surface plots with the white light excitation as expected.

semiconductor surface promote recombination and surface passivation scheme is required in order to reduce recombination [13]. The capture cross-section represents the probability of capturing free carriers via trapping. During the movement of carriers, the attraction between the trap and the carrier increases capture cross-section and have impact on recombination [14]. For capture cross-section $\sigma_n = 1 \times 10^{-12}$ cm^2, and $\sigma_p = 1 \times 10^{-16}$ cm^2 (with same surface recombination velocities as in case of Fig. 8(b)), the change in PL profile shape is shown in Fig. 9.

With $\sigma_n = 5 \times 10^{-13}$ cm^2, and $\sigma_p = 1 \times 10^{-15}$ cm^2 if we change the surface recombination velocities of electron and holes (front-contact: $S_e = 10$ cm/s, and $S_h = 10$ cm/s; back-contact: $S_e = 1 \times 10^5$ cm/s, and $S_h = 1 \times 10^5$ cm/s) we observe a significant change in PL efficiency profile (Fig. 10).

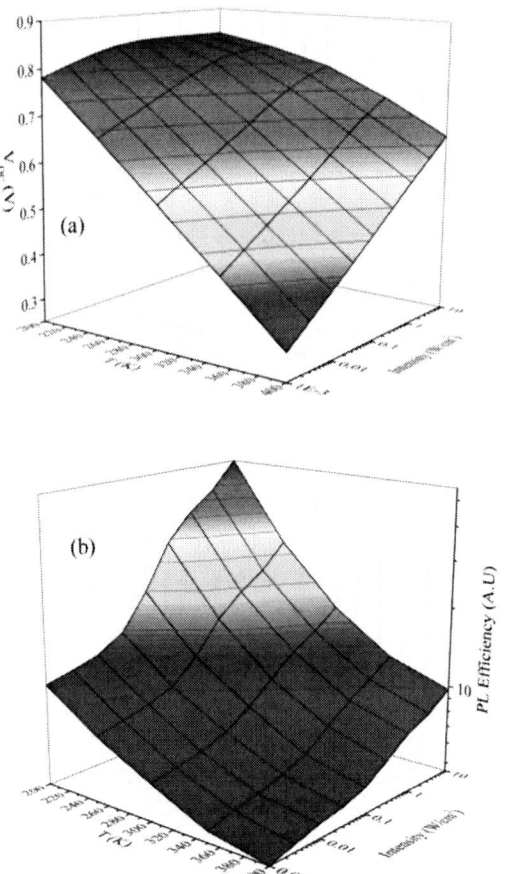

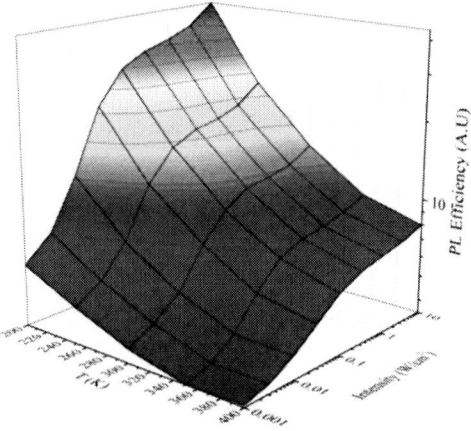

Fig. 9. PL efficiency profile depends on the capture cross-section ($\sigma_n = 1 \times 10^{-12}$ cm^2, $\sigma_p = 1 \times 10^{-16}$ cm^2) using 540 nm light.

Fig. 8. Temperature (K) and intensity (W/cm^2) dependent V_{OC} (a) and PL efficiency (b) profile in CIGS solar cell. The wavelength used in this simulation is 540 nm.

It was found that the shapes of PL efficiency profile are strong function of capture cross-section and surface recombination velocities of electrons and holes. At the surface of a semiconductor, the semiconductor atomic lattice is abruptly interrupted. The defects or impurities at the

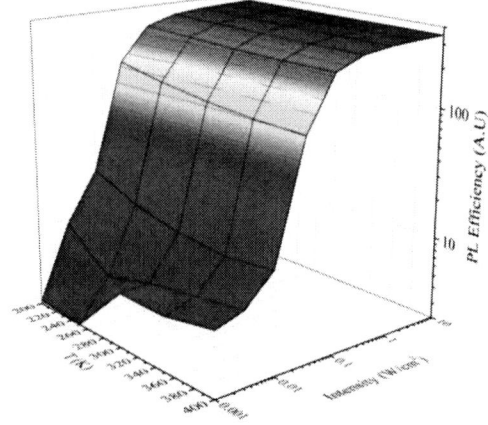

Fig. 10. PL efficiency profile depends on the surface recombination velocities (front-contact: $S_e = 10$ cm/s, and $S_h = 10$ cm/s; back-contact: $S_e = 1 \times 10^5$ cm/s, and $S_h = 1 \times 10^5$ cm/s) using 540 nm light.

The effect of trap grading on V_{OC} and PL efficiency using 540 nm excitation is shown in Fig. 11. Here the same amount of trap (of peak density 1×10^{13} cm^{-3}) graded exponentially in the defect menu of SCAPS-1D. Fig. 11(a) and (b) shows the V_{OC} and PL efficiency surface plot due to exponential decay of peak defect density 1×10^{13} cm^{-3} (at $x = 0$ μm) with the characteristics length 0.1 μm (left). Fig. 11(c) and (d) shows the V_{OC} and PL efficiency surface plot due to defect density grading with the characteristics length 0.5 μm (both left and right). Fig. 11 (e) and (f) shows the V_{OC} and PL efficiency surface plot due to exponential decay of peak defect density 1×10^{13} cm^{-3} (at $x = 2.0$ μm) with the characteristics length 0.1 μm (right). Note that all surface plots (V_{OC} and PL efficiency) has same scale. The change observed in V_{OC} and PL efficiency surface plots is due to the defect density and its energetic location.

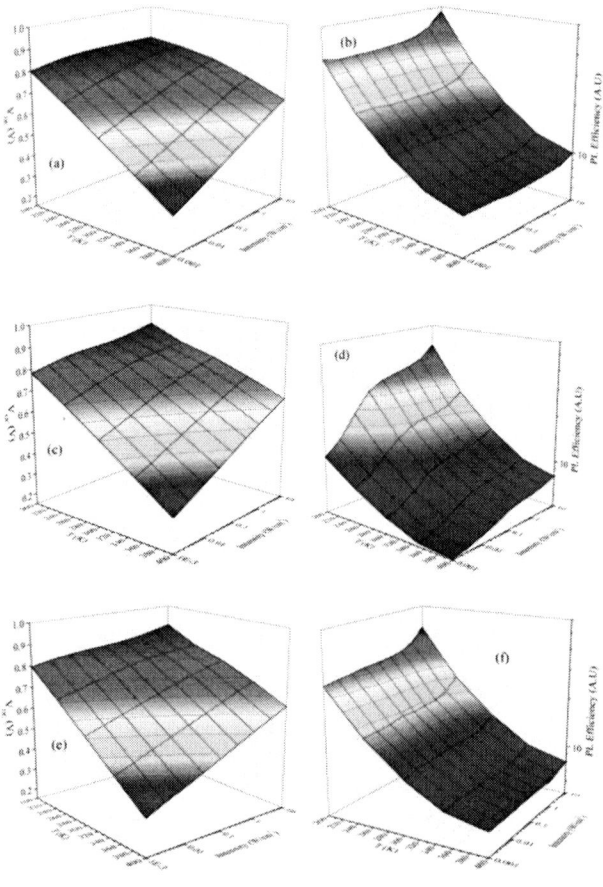

Fig. 11. Surface plot of V_{OC} and PL efficiency with defect level located at the left [V_{OC} (a), PL efficiency (b)], at the middle [V_{OC} (c), PL efficiency (d)], and right [V_{OC} (e), PL efficiency (f)] using 540 nm excitation.

There are various other electronic and device parameters which may be responsible for different shapes in PL efficiency. Here we only presented a few as examples. SCAPS-1D simulation is a tool to assist understanding of the solar cell devices and recombination phenomena. The validity of the radiative recombination simulation studies required standard characterization techniques (for example photoluminescence study). To compare the simulation results with the experimental one need calibrated PL set up that can give absolute photoluminescence data. Combined temperature and intensity dependent electrical (for SRH recombination) and optical (for radiative recombination) experimental analysis can give real picture on recombination – which is particularly of future interest.

With absolute spectral PL measurement, the quasi-fermi level splitting can be estimated and compare to V_{OC} in considering the generalized Plank's law proposed by P. Würfel [15]–[17]. There is another approach to analyze radiative recombination *via* PL-I method (intensity dependent PL efficiency measurement) [18]–[20].

V. CONCLUSIONS

SCAPS-1D is a suitable platform to simulate solar cell characteristics including recombination analysis. Graded bandgap CIGS solar cell was simulated using SCAPS-1D to visualize and understand the recombination as a function of temperature and light intensity. Numerical results could be potentially helpful in developing high efficiency solar cells and understand experimental results.

ACKNOWLEDGEMENT

Authors would like to acknowledge Prof. Marc Burgelman and coworkers, ELIS, University of Gent, Belgium, for providing us the SCAPS-1D program used in this work. We acknowledge support by U.S. Department of Energy through Sunshot PVRD grant DE-EE-0007541 "Crosscutting recombination metrology for expediting V_{OC} engineering".

REFERENCES

[1] K. Rui *et al.*, "New World Record Cu(In,Ga)(Se,S)2 Thin Film Solar Cell Efficiency Beyond 22%," in *Proceedings of the 43rd IEEE PVSC*, 2016, pp. 3–7.

[2] S. M. Sze and K. K. Ng, *Physics of Semiconductor Devices*. John Wiley & Sons, Inc., 2006.

[3] N. E. Gorji, U. Reggiani, and L. Sandrolini, "Auger generation effect on the thermodynamic efficiency of Cu(In,Ga)Se2 thin film solar cells," *Thin Solid Films*, vol. 537, pp. 285–290, Jun. 2013.

[4] S. Grover, J. V. Li, D. L. Young, P. Stradins, and H. M. Branz, "Reformulation of solar cell physics to facilitate experimental separation of recombination pathways," *Appl. Phys. Lett.*, vol. 103, no. 9, p. 93502, Aug. 2013.

[5] J. V. Li, S. Grover, M. A. Contreras, K. Ramanathan, D. Kuciauskas, and R. Noufi, "A recombination analysis of Cu(In,Ga)Se2 solar cells with low and high Ga compositions," *Sol. Energy Mater. Sol. Cells*, vol. 124, pp. 143–149, May 2014.

[6] A. Niemegeers, M. Burgelman, and K. Decock, "Alex Niemegeers, Marc Burgelman, Koen Decock, Johan Verschraegen, Stefaan Degrave, 'SCAPS Manual (Source: http://scaps.elis.ugent.be/)', Version: 27 june 2015."

[7] A. Niemegeers, M. Burgelman, R. Herberholz, U. Rau, D. Hariskos, and H.-W. Schock, "Model for electronic transport in Cu(In,Ga)Se2 solar cells," *Prog. Photovoltaics Res. Appl.*, vol. 6, no. 6, pp. 407–421, Nov. 1998.

[8] K. Kim *et al.*, "Simulations of chalcopyrite/c-Si tandem cells using SCAPS-1D," *Sol. Energy*, vol. 145, pp. 52–58, 2017.

[9] O. Lundberg, M. Edoff, and L. Stolt, "The effect of Ga-grading in CIGS thin film solar cells," *Thin Solid Films*, vol. 480, pp. 520–525, 2005.

[10] R. E. Brandt, N. M. Mangan, J. V Li, Y. S. Lee, and T. Buonassisi, "Determining interface properties limiting open-circuit voltage in heterojunction solar cells High spectral selectivity for solar absorbers using a monolayer transparent conductive oxide coated on a metal Determining interface properties limiting open-circ," *J. Appl. Phys.*, no. 10, pp. 185301–184502, 2017.

[11] D. Abou-Ras, T. Kirchart, and Rau Uwe, *Advanced Characterization Techniques for Thin Film Solar Cells*. Wiley- VCH Verlag GmbH & Co. KGaA, 2016.

[12] Y.-K. Liao *et al.*, "Observation of unusual optical transitions in thin-film Cu(In,Ga)Se_2 solar cells," *Opt. Express*, vol. 20, no. S6, p. A836, Nov. 2012.

[13] O. J. Sandberg, A. Sundqvist, M. Nyman, and R. Osterbacka, "Relating Charge Transport, Contact Properties, and Recombination to Open-Circuit Voltage in Sandwich-Type Thin-Film Solar Cells," *Phys. Rev. Appl.*, vol. 5, no. 4, p. 44005, Apr. 2016.

[14] F.-C. Chiu and Fu-Chien, "Surface State Capture Cross-Section at the Interface between Silicon and Hafnium Oxide," *Adv. Mater. Sci. Eng.*, vol. 2013, pp. 1–5, 2013.

[15] P. Würfel, *Physics of Solar Cells: From Basic Principles to Advanced Concepts*. Wiley, 2016.

[16] F. Babbe, L. Choubrac, and S. Siebentritt, "Quasi Fermi level splitting of Cu-rich and Cu-poor Cu(In,Ga)Se 2 absorber layers," *Appl. Phys. Lett.*, vol. 109, no. 8, p. 82105, Aug. 2016.

[17] P. Wurfel, "The chemical potential of radiation," *J. Phys. C Solid State Phys.*, vol. 15, no. 18, pp. 3967–3985, Jun. 1982.

[18] C. H. Swartz *et al.*, "Study of defects in CdTe heterostructures using imaging confocal photoluminescence and photoluminescence intensity measurements," in *2014 IEEE 40th Photovoltaic Specialist Conference (PVSC)*, 2014, pp. 2419–2424.

[19] A. W. Walker *et al.*, "Nonradiative lifetime extraction using power-dependent relative photoluminescence of III-V semiconductor double-heterostructures," *J. Appl. Phys.*, vol. 119, no. 15, p. 155702, Apr. 2016.

[20] C. H. Swartz *et al.*, "Radiative and interfacial recombination in CdTe heterostructures," *Appl. Phys. Lett.*, vol. 105, no. 22, p. 222107, Dec. 2014.

Simulation and Spectroscopy of Carrier Relaxation in GaSb and GaAs

A.C. Scofield[a], A.I. Hudson[a], B.L. Liang[b], B.C. Juang[b], D.L. Huffaker[b], S.M. Hubbard[c], W.T. Lotshaw[a]

[a]The Aerospace Corporation, El Segundo, CA, USA, [b]University of California Los Angeles, Los Angeles, CA, [c]Rochester Institute of Technology, Rochester, NY

Abstract — **Numerical simulation of carrier dynamics in GaAs and GaSb double heterostructures is used in combination with steady-state and time-resolved spectroscopy to extract quantitative material properties in order to predict the end performance of photovoltaic devices. We find that precise details of even simple double heterostructures, including barrier composition, surface recombination, substrate doping and lifetime must be taken into account in order to accurately extract this information.**

I. INTRODUCTION

The end performance of III-V photovoltaic cells is largely determined by the carrier dynamics of their constituent materials. Luminescence spectroscopy is widely used as a rapid feedback qualitative diagnostic to assess material quality. However, extracting quantitative measures of material quality either by time-resolved or steady-state luminescence spectroscopy can be challenging, often yielding ambiguous results.

Time-resolved PL (TRPL) can provide direct measurements of carrier lifetimes as shown in Figure 1. The analysis of such data, however, can be complicated by the presence of non-radiative recombination through Shockley-Read-Hall (SRH) centers [1,2] and background carrier concentration. In this case, both the majority and minority carrier lifetimes must be considered. Another significant complication is the dependence of the carrier lifetimes on the carrier density resulting from optical injection. In many cases, the lifetime can be

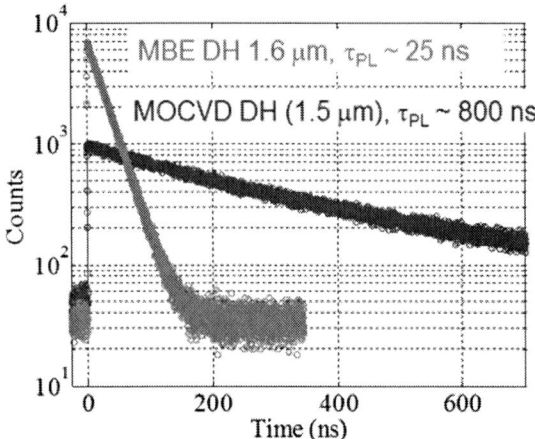

Figure 1. TRPL measurements of MOCVD and MBE grown GaAs double-heterostructures at room temperature, where the MOCVD grown GaAs shows a lifetime of ~800 ns while the MBE grown GaAs shows only ~25 ns.

significantly extended by saturation of SRH centers [2] and photon recycling [3], or reduced due to the presence of bimolecular recombination of optically generated carriers [4]. In these cases, the TRPL data become complex and require analysis within the constraints of physics-based relaxation models to extract meaningful information on the carrier recombination rates and mechanisms [1]. These considerations are especially important where PL is used as a diagnostic of material quality or performance in optoelectronic and photovoltaic device development.

Because carrier recombination rates can depend nonlinearly on the optically injected carrier density, an examination of the steady-state PL (SSPL) intensity as a function of excitation power can be a useful tool for analysis of carrier recombination mechanisms in semiconductors. We apply a model of SSPL derived from a generalized rate equation to show that both the background carrier concentration (Nd) and the non-radiative recombination rate (knr) factor strongly into the measured excitation dependence of SSPL intensity from GaAs double-heterostructures, and alternative interpretations of the causal relaxation mechanisms.

Finally, we describe the results of numerical modeling and simulations of the heterostructures using Silvaco TCAD to elucidate limiting factors such as carrier leakage (across barriers) and substrate/surface recombination.

II. EXPERIMENTAL DETAILS

In order to simplify measurements and analysis, the samples tested in this work consist of GaAs and GaSb double-heterostructures (DHs) where a GaAs or GaSb active region is clad with AlGaAs or AlGaSb barriers to prevent surface and substrate recombination. A series of GaAs and GaSb DHs included in this study were grown by both molecular-beam epitaxy (MBE) and by metal-organic chemical vapor deposition (MOCVD). All samples are grown on semi-insulating (SI) (100) substrates with a 200 nm buffer layer, an AlxGa1-xAs(Sb)/GaAs(Sb)/ AlxGa1-xAs(Sb) DH, and a 20 nm GaAs(Sb) capping layer, where the Al mole fraction x is varied between 0.25 and 0.6 and the active layer thickness is varied between 0.9 and 1.6 um.

For SSPL experiments the DH samples were excited with a Ti:Sapphire laser at 780 nm that was operated in the continuous wave (CW) mode. Steady-state (SS) photoluminescence measurements were made using a home built spectrometer consisting of a 0.5 m monochromator with a 600 g/mm grating blazed at 1000 nm and a PMT with spectral coverage from 300

978-1-5090-5606-4/17 $31.00 © 2017 IEEE

to 1700 nm (Hamamatsu R5509-72). The laser was mechanically chopped at 300-400 Hz and the PMT current was detected with a lock-in amplifier. Measurements were made between 4K and 300K using a closed-loop He cryostat (Montana Instruments). The PL intensity was measured at the emission peak, by recording the lock-in signal for ~100 s and averaging the data in that interval. The laser average power was adjusted using a waveplate-polarizer pair and measured after a focusing lens with a calibrated Si photo-diode power meter. The focused spot was elliptical with major and minor axes of 359 um and 258 um respectively as measured with a CCD camera. Time resolved PL was recorded using time correlated single photon counting (TCSPC). The Ti:Sapphire was operated in a short (sub 200 fs) pulse mode with a selectable repetition rate using an acousto-optic pulse picker. The PL transient was collected with the same spectrometer except that a hybrid-PMT (Becker & Hickl HPM-100-50) with an impulse response function of ~120 ps was used as the detector.

III. BARRIER LEAKAGE IMPACT ON LIFETIME

TRPL measurements were taken as a function of temperature to determine carrier lifetimes and identify recombination mechanisms. The MOCVD-grown GaAs DH lifetimes have a clear excitation dependence from 25-295K indicating the carrier recombination is radiative limited up to room temperature. The MBE-grown GaAs DH lifetimes have excitation dependence only up to 200K, and at higher temperatures the lifetime is independent of pulse energy, indicating a transition from radiative limited lifetime to non-radiative limited lifetime.

Analysis of the temperature dependent lifetime of the MBE-grown GaAs DH from 200-295K reveals that the non-radiative lifetime does not follow the expected $\tau_{nr} \, \alpha \, T^{-1/2}$ for SRH recombination: exhibiting instead an activation energy of ~121 meV. This result indicates that the mechanism limiting the lifetime of the MBE-grown material is thermally activated with a potential barrier equal to the activation energy.

In order to elucidate the mechanisms limiting the PL rate of the MBE grown samples identified by the TRPL and SSPL analysis, Silvaco TCAD software was employed to solve a complete drift-diffusion model of the sample structure. The MBE grown samples are limited by a thermally activated process with an energy of ~121 meV. Curiously, this activation energy is nearly equal to the valence band offset between the GaAs active layer and the AlxGa1-xAs barrier in this sample (where x=0.25). Using Silvaco, both the transient decay and the excitation dependence of the radiative recombination rate are simulated as a function of Al mole fraction in the AlxGa1-xAs barriers while assuming a constant lifetime of 1 μs for the active layer, 1 ns for the substrate, and a surface recombination velocity of 5x105 cm/s.

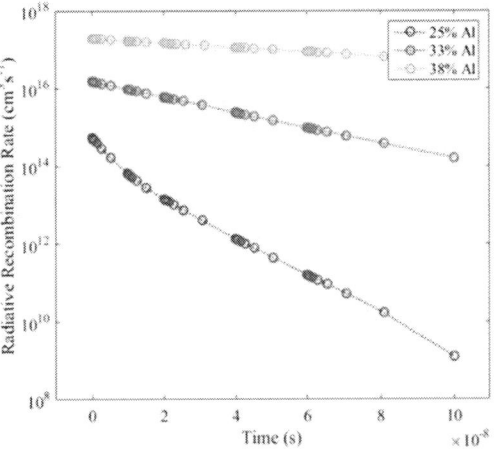

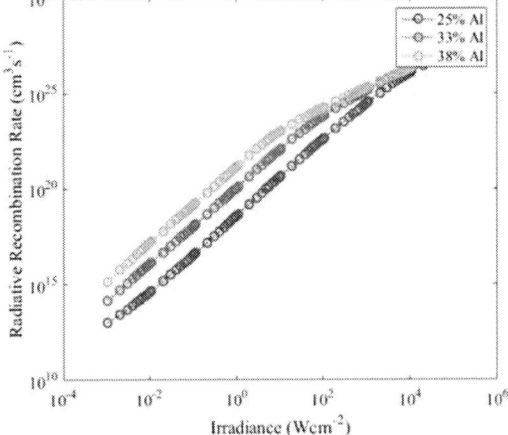

Figure 2. Silvaco simulation results for GaAs double-heterostructures with differing Al composition in the AlGaAs barriers. (Top) Transient decary of radiative recombination in the GaAs active layer, and (bottom) excitation irradiance dependence of SSPL.

The results of the transient and steady-state simulations are shown in Figure 2. While all material parameters except for the AlxGa1-xAs composition are kept constant, the lifetime changes dramatically as the Al fraction is varied from x=0.25 to x=0.38. At x=0.25, the PL decay shows a lifetime of 3 ns, while at x=0.38, the decay shows a lifetime of 77 ns. Similarly, the excitation irradiance simulation of SSPL shows that at low levels of irradiance, the radiative recombination rate differs by greater than two orders of magnitude as the Al fraction of the barrier increases from 0.25 to 0.38.

In order to verify the predictions of the Silvaco simulations, a series of GaAs and GaSb DHs were grown by MBE with different AlxGa1-xAs compositions ranging from x=0.25 to x=0.6. As shown in Figure 3, the room-temperature TRPL measurements show lifetimes that increase by over an order of magnitude with increased Al fraction.

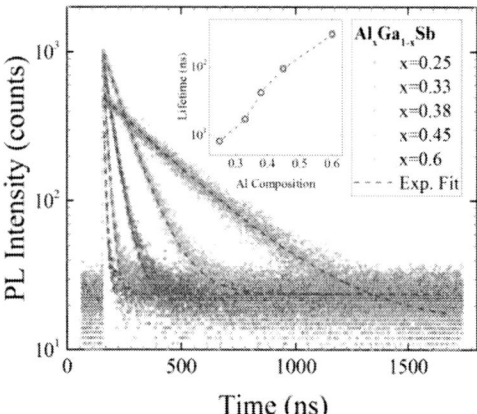

Figure 3. Room-temperature TRPL measurements of MBE grown GaSb double-heterostructures with varying Al fraction in the AlGaSb barriers.

IV. MODELING OF BARRIER LEAKAGE

The effect of radiative and non-radiative recombination along with the loss of carriers due to barrier leakage is modeled by steady-state solutions of carrier rate equations:

$$G = \frac{2D_n}{WL_B} n_{lk} + \frac{N_D n_{act} + n_{act}{}^2}{\tau_p[N_D + n_{act}] + \tau_n n_{act}} + B[N_D n_{act} + n_{act}{}^2]$$

$$R = B[N_D n_{act} + n_{act}{}^2]$$

where n_{act} is the carrier concentration in the active region and n_{lk} is the carrier concentration in the barriers contributing to leakage current. These two carrier concentrations are related by the continuity of the quasi-fermi level at the active/barrier interface:

$$n_{act} = \frac{1}{2\pi} \left(\frac{2m_e}{\hbar^2}\right)^{3/2} \int_{E_C}^{\infty} \frac{\sqrt{E - E_C}}{1 + e^{(E - E_{Fc})/kT}} dE$$

$$n_{lk} = \frac{1}{2\pi} \left(\frac{2m_e}{\hbar^2}\right)^{3/2} \int_{E_B}^{\infty} \frac{\sqrt{E - E_B}}{1 + e^{(E - E_{Fc})/kT}} dE$$

These equations can be calculated numerical or solved analytically using the Boltzman approximation at low carrier injection. A comparison between measured SSPL curves and calculated SSPL curves is shown in Figure 4.

V. CONCLUSION

A combination of TRPL, SSPL, and numerical simulations have been employed to investigate the mechanisms limiting non-radiative recombination in GaAs and GaSb double heterostructures. The results show that it is necessary to account

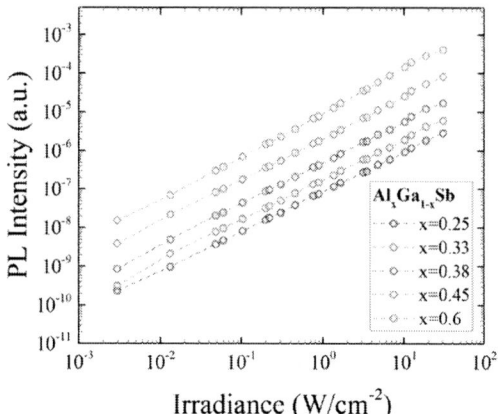

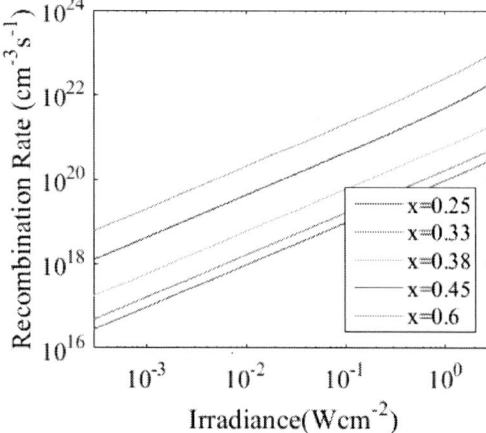

Figure 4. Comparison of experimental (top) and numerical calculations (bottom) of SSPL of GaSb double-heterostructures with varying Al fraction.

for all carrier relaxation mechanisms throughout the entire structure, including carrier leakage over the barriers in addition to Shockley-Read-Hall and surface recombination.

REFERENCES

[1] Marvin, D. C.; Moss S. C.; and Halle, L. F., "Analysis of transient photoluminescence measurements on GaAs and AlGaAs double heterostructures", J. Appl. Phys. 72, 1970-1984 (1992)

[2] Ahrenkiel, R. K.; Keyes, B. M.; and Dunlavy, D. J., "Intensity-dependent minority-carrier lifetime in III-V semiconductors due to saturation of recombination centers", J. Appl. Phys. 70, 225-231 (1991)

[3] Ahrenkiel, R. K.; Dunlavy, D. J.; Keyes, B. M.; Vernon, S. M.; Dixon, T. M.; Tobin, S. P.; Miller, K. L.; and Hayes, R. E. "Ultralong minority-carrier lifetime epitaxial GaAs by photon recycling", Appl. Phys. Lett. 55, 1088-1090 (1989)

[4] Ahrenkiel, R. K. "Minority Carrier Lifetime in III-V Semiconductors", <u>Semiconductors and Semimetals</u> 39, 40-145

This Page Intentionally Left Blank.

Computational Design of Dopants in CdTe Grain Boundaries for Efficient Photovoltaics

Fatih G. Sen[1], Tadas Paulauskas[2], Ce Sun[3], Moon Kim[3], Robert F. Klie[2], Maria K.Y. Chan[1]

[1] Argonne National Laboratory, Lemont, IL, 60439, U.S.A.
[2] University of Illinois at Chicago, Chicago, IL, 60607, U.S.A.
[3] University of Texas at Dallas, Richardson, TX, 75080, U.S.A.

Abstract — A fundamental understanding of the role of dopants in electronic structure of polycrystalline CdTe may lead to efficiency improvements. In the present work, we investigated effect of Cl, P and S doping on CdTe grain boundaries using first principles density functional theory (DFT) calculations. In addition to already known Cl, P and S can segregate in grain boundaries and incorporation of these elements can effectively reduce midgap states to increase the photovoltaic efficiency of CdTe. The methodology we presented can be used to design other alloying elements to CdTe to improve photovoltaic efficiency towards reaching the theoretical limit.

Index Terms — CdTe, photovoltaics, grain boundary, density functional theory, doping

I. INTRODUCTION

High efficiency, and low manufacturing costs make CdTe as one of the common photovoltaic material used in solar cells. The practical efficiencies of polycrystalline CdTe photovoltaic cells of 22.1% are still well below the theoretical limit (30%), indicating possible room for improvement [1-3]. The most recent improvements in the efficiency are generally achieved by overcoming the difficulties in forming ohmic and transparent contacts. Another photovoltaic efficiency limiting factor is due to Shockley-Read-Hall (SRH) [4, 5] recombination at grain boundaries and within grains. Recent studies has shown that $CdCl_2$ treated polycrystalline CdTe films had much larger efficiency after segregation of Cl in the grain boundaries by substituting Te, and passivation of dangling bonds [6-8].

The efficiency enhancement mechanisms by Cl incorporation to CdTe grain boundaries are still not completely known. A systematic understanding of atomic structure of CdTe grain boundaries can be gained by combining density functional theory (DFT) calculations with high resolution electron microscopy experiments [9]. When the underlying mechanisms of Cl doping to grain boundaeis revealed, new dopants can be designed reduce recombination rates, and allow for high-performance CdTe devices reaching the theoretical photovoltaic efficiency [10].

Twin boundaries that are frequently observed in poly-CdTe have mild effects on photovoltaic properties and do not change recombination rates [11]. Doping effects on the optoelectronic properties of CdTe are more pronounced in high and low angle grain boundaries, and dislocation cores. In the present study, we first doped 4.8° (110)‖(110) CdTe grain boundary dislocation cores that has been previously shown to have midgap states with Cl, and revealed the electronic structure changes that yielded an efficiency increase upon doping with different concentrations. We have also studied the effect of S and P doping first time on the photovoltaic properties regarding its efficiency in solar cells.

II. METHODOLOGY

The atomic structure of 4.8° (110)‖(110) CdTe grain boundaries were created using the crystallographic information obtained from STEM images. STEM images of grain boundaries are processed with image analysis methods and subsequently atomistic simulations were carried out with the molecular dynamics program LAMMPS [12] using available empirical potentials for CdTe, namely Stillinger-Weber (SW) and bond-order potential (BOP). The atomic structure consisted of 1440 atoms. The details of the construction of atomic structure is given in Ref. [13]

The doped structures of 4.8° (110)‖(110) grain boundary was created by substituting the Te atoms at the dislocation core shown in Fig. 1 with Cl, S and P. There are total 6 Te sites at the dislocation core that can be substituted. We studied the amount of doping by varying the number of Te atoms substituted with 1, 2, and 6 dopants. To identify if the substitution of dopants are thermodynamically feasible, we calculated the segregation energy (E_{seg}) for each dopant using the relation:

$$E_{seg}^X = (E_{GB}^X - E_{GB}) - (E_{bulk}^X - E_{bulk}) \quad (1)$$

where E_{GB}^X and E_{GB} are the total energies of the grain boundary structure with and without dopant atoms, respectively. E_{bulk}^X and E_{bulk} are the total energies of the bulk CdTe 3x3x3 supercell with and without dopant atoms, respectively.

All calculations are carried out using density functional theory (DFT). The changes in the electronic structure are investigated via the density of states (DOS). All DFT calculations were carried out using Vienna Ab initio Simulation Package (VASP) [14, 15]. Projector augmented wave (PAW) potentials [16] was used, and the exchange correlation energy was estimated using the generalized gradient approximation (GGA) parameterized by Perdew-Burke-Ernzerhof (PBE) [17]. We used a kinetic energy cutoff

TABLE I
Segregation Energies, E_{SEG}, (eV/atom) of Cl, S and P at 4.8° (110)∥(110) Grain Boundary

Dopant	Coverage	
	1/6	6/6
Cl	-2.2	-0.8
S	-1.2	0.4
P	-2.4	-0.6

of 274 eV, and all atomic positions are relaxed to give an energy convergence of 10^{-5} eV/atom.

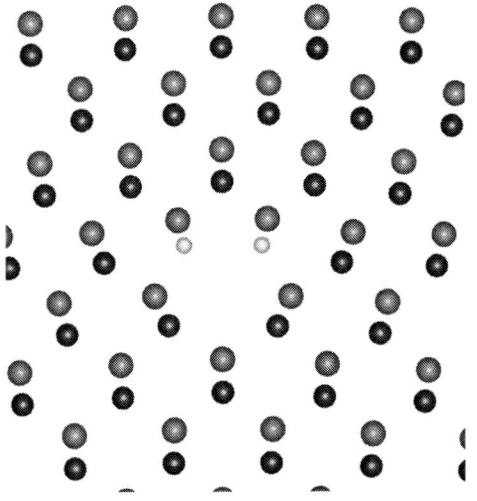

Fig. 1.   Atomic structure of the dislocation core at 4.8° (110)∥(110) grain boundary. The Te atoms that were substituted by Cl, P and S are colored in green.

III. RESULTS AND DISCUSSION

A. Segregation Energies of Dopants

Segregation energies (E_{seg}) of Cl, P and S at 4.8° (110)∥(110) grain boundary are given in Table I. Table I shows that at 1/6 concentration, E_{seg} of all Cl, S and P are negative indicating that segregation is thermodynamically feasible. If all Te atoms at the dislocation core are substituted corresponding to concentration of 6/6, E_{seg} of Cl and P at the grain boundary are still negative, while E_{seg} of S is slightly positive so it is not thermodynamically feasible to have all dislocation core is doped with S.

B. Electronic Structure

Electronic structure changes in the 4.8° (110)∥(110) grain boundary with incorporation of dopants are studied by means of density of states (DOS). Fig. 2 shows the DOS of Cl doped dislocation core at different concentrations namely, 1/6, 2/6

and 6/6, and compared with undoped dislocation core. The large peak near the Fermi level (E_F) in the undoped dislocation core indicates the formation of midgap states due to the dislocation core. When Cl is incorporated, substantial decrease in the midgap states are observed. At higher Cl concentrations of 6/6, the midgap states at the E_F are almost disappeared. Thus, Cl incorporation is expected to increase the photovoltaic efficiency by reducing the recombination, which is in accordance with the experimental observations.

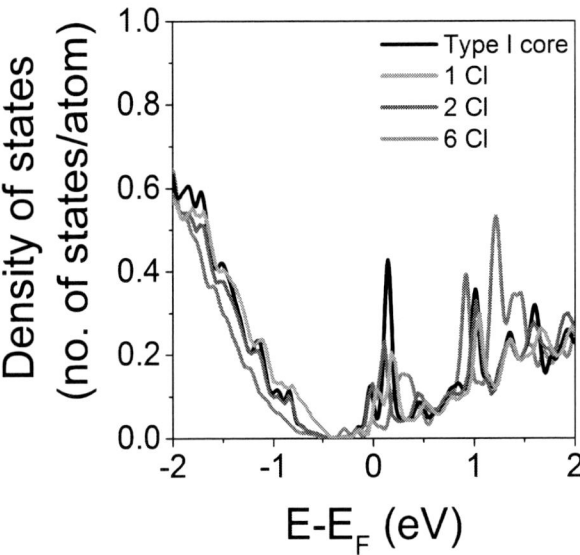

Fig. 2.   Total density of states (DOS) of Cl doped dislocation core at different concentrations compared to un-doped Type I core.

Since, the largest decrease in the midgap states with Cl doping is achieved with the highest dopant concentration, we investigated the DOS of the 6/6 concentration of P and S. Fig. 3 shows the DOS of P doped dislocation core in comparison with the un-doped core. A drastic decrease in the midgap states is noted when P is incorporated in Fig 3. Similarly when CdTe grain boundary is doped with S, midgap states are reduced significantly in Fig. 4. The electrostatic potential profiles averaged around the dislocation core was also calculated and showed that upon Cl, S and P doping the grain boundary is still attracting holes, which will enable charge separation. Consequently, it is predicted that both P and S can be promising alloying elements to polycrystalline CdTe to reduce recombination rates.

IV. CONCLUSIONS

We successfully designed alloying elements to CdTe using computational tools to increase the photovoltaic efficiencies. Incorporation of dopants to CdTe grain boundaries and their effect on photovoltaic properties are investigated using density functional theory methods. In addition to Cl, S and P are identified to be promising candidates to reduce recombination rates in CdTe solar cells. The methodology we followed can be further applied to search other doping elements to increase photovoltaic efficiency by engineering the grain boundaries.

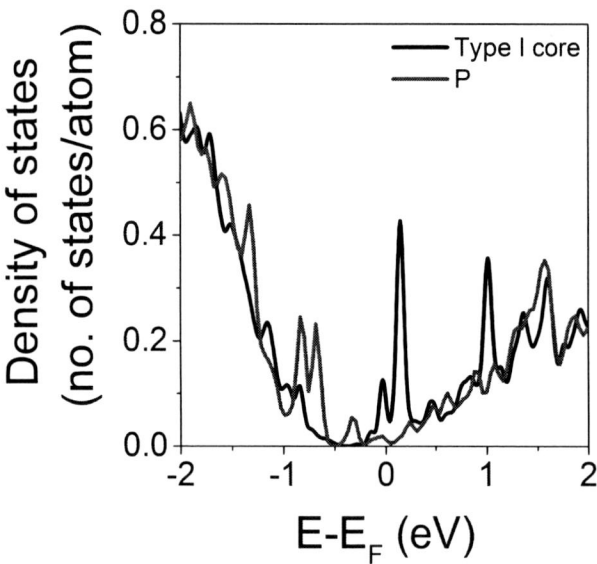

Fig. 3. Total density of states (DOS) of P doped dislocation core at concentration of 6/6 compared to un-doped Type I core.

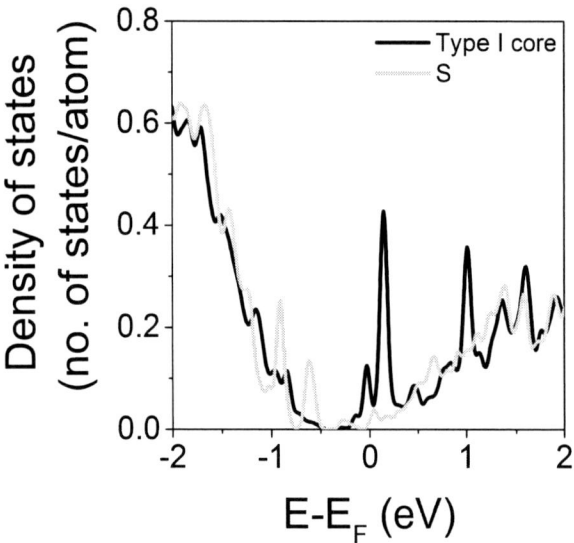

Fig. 4. Total density of states (DOS) of S doped dislocation core at concentration of 6/6 compared to un-doped Type I core.

ACKNOWLEDGEMENT

We acknowledge funding from the DoE Sunshot program under contract # DOE DEEE005956. Use of the Center for Nanoscale Materials, an Office of Science user facility, was supported by the U.

S. Department of Energy, Office of Science, Office of Basic Energy Sciences, under Contract No. DE-AC02-06CH11357. This research used resources of the National Energy Research Scientific Computing Center, a DOE Office of Science User Facility supported by the Office of Science of the U.S. Department of Energy under Contract No. DE-AC02-05CH11231.

REFERENCES

[1] C. Ferekides and J. Britt, "CdTe solar cells with efficiencies over 15%," Solar Energy Materials and Solar Cells, vol. 35, no. 1-4, pp. 255-262, Sep 1994.

[2] X. Wu, "High-efficiency polycrystalline CdTe thin-film solar cells," (in English), Solar Energy, vol. 77, no. 6, pp. 803-814, Dec. 2004.

[3] W. Shockley and H. J. Queisser, "Detailed balance limit of efficiency of p-n junction solar cells," Journal of Applied Physics, vol. 32, no. 3, pp. 510-519, 03 1961.

[4] W. Shockley and W. T. Read, Jr., "Statistics of the recombinations of holes and electrons," Physical Review, vol. 87, pp. 835-842, 09 01 1952.

[5] R. N. Hall, "Electron-hole recombination in germanium," Physical Review, vol. 87, pp. 387-387, 07 15 1952.

[6] M. Tuteja et al., "Direct observation of CdCl2 treatment induced grain boundary carrier depletion in CdTe solar cells using scanning probe microwave reflectivity based capacitance measurements," The Journal of Physical Chemistry C, vol. 120, no. 13, pp. 7020-7024, 2016.

[7] J. Major et al., "In-depth analysis of chloride treatments for thin-film CdTe solar cells," Nature Communications, vol. 7, 2016.

[8] C. Li et al., "Grain-boundary-enhanced carrier collection in CdTe solar cells," Physical review letters, vol. 112, no. 15, p. 156103, 2014.

[9] T. Paulauskas et al., "Atomic-resolution characterization of the effects of CdCl2 treatment on poly-crystalline CdTe thin films," Applied Physics Letters, vol. 105, no. 7, p. 071910, 2014.

[10] A. P. Sutton and R. W. Balluffi, "Overview .61. On geometric criteria for low interfacial energy," Acta Metallurgica, vol. 35, no. 9, pp. 2177-2201, Sep 1987.

[11] S.-H. Yoo, K. T. Butler, A. Soon, A. Abbas, J. M. Walls, and A. Walsh, "Identification of critical stacking faults in thin-film CdTe solar cells," Applied Physics Letters, vol. 105, no. 6, Aug 11 2014, Art. no. 062104.

[12] S. Plimpton, "Fast parallel algorithms for short-range molecular dynamics," Journal of computational physics, vol. 117, no. 1, pp. 1-19, 1995.

[13] C. Sun et al., "Atomic and electronic structure of Lomer dislocations at CdTe bicrystal interface," Scientific reports, vol. 6, 2016.

[14] G. Kresse and J. Hafner, "Ab-initio molecular-dynamics simulation of the liquid-metal amorphous-semiconductor transition in germanium," Physical Review B, vol. 49, no. 20, pp. 14251-14269, May 15 1994.

[15] G. Kresse and J. Furthmuller, "Efficiency of ab-initio total energy calculations for metals and semiconductors using a plane-wave basis set," Computational Materials Science, vol. 6, no. 1, pp. 15-50, Jul 1996.

[16] G. Kresse and D. Joubert, "From ultrasoft pseudopotentials to the projector augmented-wave method," (in English), Physical Review B, vol. 59, no. 3, pp. 1758-1775, Jan 15 1999.

[17] J. P. Perdew, K. Burke, and M. Ernzerhof, "Generalized gradient approximation made simple," (in English), Physical Review Letters, vol. 77, no. 18, pp. 3865-3868, Oct 28 1996.

Analyses of Photovoltaic Power Plant Performance Estimates Based on Detailed Laboratory Module Characterizations and Typical Real-World Input Data Sources

Rajeev Singh[1], John L. R. Watts[1],
and Kellen Gillispie[2]

DNV GL PVEL, Berkeley, CA, 94710, USA[1]
Stratasense, Oakland, CA, 94608, USA[2]

Abstract — Accurate performance estimation of a PV power plant based on detailed laboratory performance characterizations of a target PV module is a critical component of project development. We utilized state of the art laboratory-based PV module performance measurements to extract module performance characteristics over wide operating ranges to generate detailed performance models. The characterized PV modules were then deployed in a real-world operating environment along with I-V curve tracers and incident irradiance and module temperature sensors. PV module power output calculations of the deployed modules using the generated models and input data from different incident irradiance and module temperature sensors are compared to module power output data acquired using I-V curve tracers. The differences between the calculated and measured power output values for different operating parameter inputs to the performance model are analyzed and discussed. These analyses can inform realistic best-case uncertainties of PV power plant performance estimates.

Index Terms — Cell temperature, energy yield, photovoltaic, performance estimate, pyranometer irradiance, photovoltaic sensor irradiance.

I. INTRODUCTION

A photovoltaic (PV) module type considered as a component in a modern PV power plant is typically characterized for performance, reliability, and durability during the project development phase to enable informed product selection decisions. The accuracy and applicability of such testing are crucial to project finance. It is therefore important to compare laboratory characterization data to real-world operational data to validate laboratory methods and quantify differences in measured performances between the two settings. The differences between laboratory and outdoor measurement results can be used to estimate the uncertainties in module real-world performance predictions that are based on laboratory module characterizations.

The standard IEC 61853-1:2011 is commonly applied to the characterization of PV modules for energy yield estimates. The standard specifies module performance measurements over a set of operating conditions that encompasses typical operating ranges of PV modules at many locations throughout the globe. The accuracies of PV power plant energy yield estimates are limited by the generalized measurement conditions specified in the standard, the PV module performance measurement accuracies, and the accuracies of the modelling tools and associated data utilized for the performance predictions.

Additionally, when validating the accuracies of any performance estimate using operational plant data, one must consider the accuracies of the measurements of the real-world operating conditions and PV power plant output used to evaluate the performance model.

In this work, semi-empirical models of PV modules generated with isothermal PV module flash-test data are evaluated using different real-world irradiance and module temperature data sources. We compare these detailed performance predictions of PV modules to their measured real-world performances and demonstrate the range of predicted PV module performance values possible under even the most favorable of experimental and analytical conditions. Such evaluations can be utilized to establish best-case uncertainties for PV plant energy yield estimates.

II. EXPERIMENT

We utilized state of the art laboratory performance characterization data (per IEC 61853-1) of two commercial PV modules in conjunction with the Sandia Array Performance Model (SAPM) to develop performance models that are specific to each module [1]. The two modules were then deployed in a real-world operating environment and connected to current-versus-voltage (I-V) curve tracers with integrated thermocouple meters manufactured by Stratasense to monitor the electrical operating parameters and temperatures of the modules in real-time [2]. In addition, an industry-standard CMP11 pyranometer and a temperature-corrected PV irradiance sensor manufactured in the European Solar Test Installation (ESTI) configuration were installed adjacent to and in the plane of the modules to monitor incident irradiance [3]. The CMP11 pyranometer was chosen due to its industry-standard designation. The operating temperatures of the modules were monitored by thermocouples affixed to the backsheets of the modules and via the open-circuit voltages (V_{OC}) of the modules as measured by the I-V curve tracers. Since the open-circuit voltages of the modules were characterized with respect to both cell-temperature and incident irradiance in great detail in a laboratory setting, average cell-temperatures of the modules are calculated to an estimated accuracy of $\pm$ 2 °C. Estimated measurement uncertainties of the ESTI PV irradiance sensor and the laboratory-based PV module measurements under standard test conditions (STC) are

± 2 % in irradiance and maximum power (P_{MAX}), respectively. The manufacturer-calibrated CMP11 is specified to an overall accuracy of ± 2 % in irradiance. Stratasense I-V curve tracers are specified to accuracies of ± 1 % in voltage and ± 2 % in P_{MAX}. Because of varying real-world measurement parasitics, the accuracies of module cell-temperature measurements using backsheet-affixed thermocouples are not clear. The PV modules and irradiance sensors were cleaned of any soil weekly. Data presented were acquired between August and December of 2016.

III. RESULTS AND DISCUSSION

Fig. 1 below is a plot of the ratios of measured real-world module performance ratios (PR) as referenced to the irradiance reported by the PV sensor to the PR as referenced to the irradiance reported by the pyranometer over the course of a typical summer day. Because the spectral response and incidence-angle modifier (IAM) characteristics of the PV irradiance sensor and the PV modules are closely matched, the measured PRs referenced to the PV sensor are larger during the morning and evening relative to the PRs referenced to the pyranometer. The magnitude of this effect will depend on location and time of year increasingly as location moves away from the equator. The asymmetry in the PR ratio profile can be

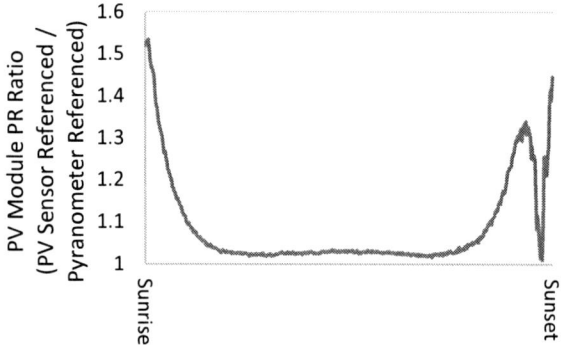

Fig. 1. Ratios of PV module performance ratio referenced to a PV irradiance sensor to performance ratio referenced to a pyranometer

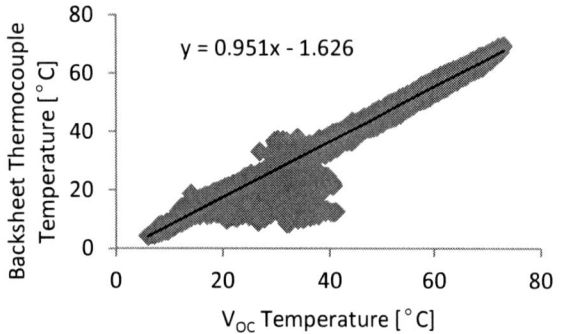

Fig. 2. Backsheet temperature as measured by an affixed thermocouple versus V_{OC}-derived average module cell-temperature

due to differences in atmospheric conditions between the solar angles at sunrise and sunset. For example, atmospheric pollution levels are increased at sunset relative to sunrise, and therefore the spectral irradiance incident on a PV plant will differ between those periods [4]. Since the CMP11 pyranometer is not temperature-corrected while the PV irradiance sensor is temperature-corrected, the PR referenced to the PV irradiance sensor increases slightly toward noon relative to the PR referenced to the pyranometer. The two PRs begin to converge after noon. This is because the pyranometer is reporting a relatively higher irradiance during periods of higher ambient temperatures and irradiances. For the data in Fig 1., the magnitude of the PR ratio increase at noon relative to the minimum PR ratio values is approximately 0.09. A component of the PR ratio values is simply due to calibration offset (e.g. at STC) between the two irradiance sensors.

Fig. 2 above is a plot of real-world module temperatures as measured using backsheet-affixed thermocouples (one per module on center) and as calculated from the measured V_{OC} values. The laboratory-measured relationships between V_{OC}, temperature, and incident irradiance of the PV modules were used along with the real-world irradiance values reported by the plane-of-array (PoA) PV irradiance sensor to calculate the average operating cell-temperatures. The V_{OC} versus cell-temperature relationships were calibrated in an isothermal chamber using Type-T thermocouples, which is the same thermocouple type utilized for the real-world backsheet temperature measurements. The thermocouple reports temperatures that are typically lower than the average module cell-temperatures calculated from V_{OC}. This is consistent with the fact that the cells are the hottest components of a PV module under operating conditions. Although there are points at which the thermocouple readings are higher than the V_{OC}-derived readings, the majority of readings represent the case for which the V_{OC}-derived values are significantly larger than the thermocouple readings. Accurate measurements of the operating temperatures of a PV module using one or more external temperature sensors are extremely difficult due to the varying temperature differences (that increase with irradiance) between the cells and the backsheet, non-ideal thermal contact between the sensor and the backsheet, the perturbation of the thermal characteristics of the module at the locations of the affixed sensors, convective cooling of the temperature sensors from the side opposite the measurement interface, and in-plane module temperature gradients [5]. In contrast, utilizing measured module V_{OC} that is calibrated for both cell-temperature and irradiance is an excellent method for determining average cell-temperature in a PV module.

Fig. 3 below is a plot of the measured P_{MAX} values of the PV modules versus the P_{MAX} values calculated using the SAPM and the following four input datasets: PoA irradiance as measured by the PV irradiance sensor (PVS), PoA irradiance as measured by the pyranometer (PY), module temperature measured using a backsheet-affixed thermocouple (T), and measured V_{OC}-derived average cell-temperature (V). The standard deviation of the ratios of the directly measured P_{MAX} values to the calculated values referenced to the pyranometer is 0.105 while

TABLE II. CALCULATED PV MODULE MAXIMUM POWER VALUES USING THE LINEAR REGRESSION COEFFICIENTS OF TABLE I

Measured Power [W]	Calculated Power [W]				% Difference (Calculated - Measured)			
	PVS, V	PY, V	PVS, T	PY, T	PVS, V	PY, V	PVS, T	PY, T
50	50.4	51.9	51.0	52.2	0.9	3.9	2.1	4.4
100	98.9	101.4	100.8	103.2	-1.1	1.4	0.8	3.2
150	147.4	150.8	150.6	154.3	-1.8	0.6	0.4	2.8
200	195.8	200.3	200.3	205.3	-2.1	0.1	0.2	2.6
250	244.3	249.7	250.1	256.3	-2.3	-0.1	0.0	2.5
300	292.8	299.2	299.8	307.3	-2.4	-0.3	-0.1	2.4

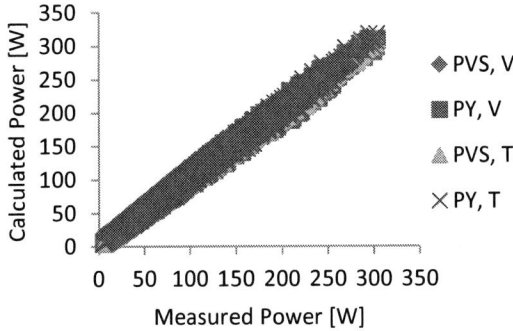

Fig. 3. Measured real-world PV module maximum power versus calculated maximum power using the SAPM and different irradiance and module temperature measurement devices

the standard deviation of the ratios referenced to the PV irradiance sensor is 0.083. Those values are indicative of the effects of spectral and IAM mismatches between the PV modules and the pyranometer. Table I below lists the coefficients of the linear regressions of the datasets of Fig. 3. As expected, because the PV sensor measures lower irradiance values relative to the pyranometer and the V_{OC}-derived cell-temperatures are typically higher than the temperatures measured by the backsheet-affixed thermocouples, the calculated power values that utilize those inputs are lowest. Table II above lists calculated power values using the linear regression coefficients of Table I at several input measured-power values throughout the range. The table includes the percent differences of the calculated power values relative to the measured power values. The PV module output power values calculated using the different input data sources can vary as widely as approximately 5 %.

TABLE I. LINEAR REGRESSION COEFFICIENTS FOR EACH OF THE DATASETS

	Slope	y-intercept
PVS, V	0.969	1.976
PY, V	0.989	2.490
PVS, T	0.995	1.275
PY, T	1.021	1.186

IV. CONCLUSION

The results of this investigation suggest that even when using detailed semi-empirical models based on state of the art laboratory measurements of the specific PV modules under test in conjunction with real-world sensor data taken at/of the same operating PV modules, the differences between calculated performance values that utilize input values provided by typical sensors and the actual measured performance values are large. There is also a large range of calculated performance values derived from various sensor datasets. The approximately 5 % range in calculated performance values does not include the P_{MAX} measurement errors of either the laboratory flash-test measurements of the modules or the real-world I-V curve traces because those errors are common to the four calculated datasets. Instead, those two error sources only shift the results of the calculated datasets relative to the real-world measured performance values. Additionally, errors due to the utilization of lower-quality (e.g. historical, satellite) weather data commonly used for PV plant performance predictions, PV module soiling assumptions, PV module temperature estimations, and estimations of system losses (e.g. module mismatch, conductor, inverter, transformer) will further degrade the accuracies of PV plant energy-yield estimates.

REFERENCES

[1] King, David L., Jay A. Kratochvil, and William Earl Boyson. *Photovoltaic array performance model.* United States. Department of Energy, 2004.

[2] Quiroz, Jimmy E., et al. "In-situ module-level I–V tracers for novel PV monitoring." *Photovoltaic Specialist Conference (PVSC), 2015 IEEE 42nd.* IEEE, 2015.

[3] Ossenbrink, H., and K. A. Münzer. "The ESTIsensor–A New Reference Cell for Monitoring of PV Plant Performance." *11th European Photovoltaic Solar Energy Conference and Exhibition, Montreux, Switzerland.* 1992.

[4] Lynch, David K., and William Charles Livingston. *Color and light in nature.* Cambridge University Press, 2001.

[5] Singh, Rajeev, and John L. R. Watts. "Irradiance- and temperature-dependent PV module performance measurement." *3rd PV Performance Modeling Workshop.* 2014.

Critical Evaluation of the Foundations of Solar Simulator Standards

Ronald A. Sinton, Harrison Wilterdink, Justin Dinger, Adrienne L. Blum,

Weston Dobson, and Cassidy Sainsbury

Sinton Instruments, Boulder, CO, 80301, USA

Abstract — Simulator classifications are often cited for performance measurement of silicon solar cells and modules. While there is significant technical work investigating these classifications (spectral classification in particular), the implications of these results have not been clearly articulated in practical terms. This paper reviews the classification metrics, and concludes that there are many misconceptions based on interpretations of what the classification means. In particular, it is found that the classifications in the standards do not necessarily predict simulator accuracy, and can promote the opposite.

Index Terms—Silicon, measurement, photovoltaic cells, photovoltaic modules.

I. INTRODUCTION

Simulator standards, in particular IEC 60904-9 Ed. 2 (2007), classify solar simulators for cell and module measurement in terms of spectrum, uniformity, and light and data acquisition parameters [1]. The result is a classification for each category, varying from A to C. It is commonly presumed that an AAA simulator is inherently more accurate than a CCC simulator. In recent years, an additional informal classification "A+" has arisen, with parameters closer to the IEC metrics than the AAA classification, which leads to the presumption that this will be better (in some way) than the AAA classification.

There is extensive work indicating that the classifications do not in fact correlate directly to measurement accuracy [2 – 5]. This creates confusion within industry, especially when "suppliers and users of solar simulators refer to these classes in speaking in terms of technical performance quality" [3]. The problems arising from this perception include:

- False ranking of value between instrument types
- Simulator products that do not meet expectations based on classification
- Perverse incentives to optimize simulators towards irrelevant metrics
- Barriers to innovations towards better, more cost-effective tools due to arbitrary fixed requirements
- A misplaced confidence in simulator classification as a proxy for measurement accuracy

This paper discusses the three classifications, to probe for weaknesses in the classification methods and recommend potential alternatives. The authors are relatively new to this discipline of simulator and spectral response standards, and hope to encourage discussion to further educate us on the parts that we may not yet understand. For specificity and brevity, we focus on the special case of silicon module testing.

II. SPECTRAL CLASSIFICATION

The first classification category of IEC 60904-9 is spectral match. The metric for classifying simulators is to compare the integrated energy within six wavelength "bins" between the simulator spectrum and AM1.5G. A letter grade is given according to how closely the simulator matches these energy bins: Class A is ±25% match, Class B ±40%, Class C is +100/-60%; an unofficial class "A+" was created by TUV Rheinland, corresponding to ±12.5%. The overall letter grade is determined by lowest letter of the six bins. The six bins are:

400-500nm	500-600nm	600-700nm	700-800nm	800-900nm	900-1100nm

This metric brings us to the first question: why match energy in each bin? Solar cells do not integrate *energy*, they integrate *photons*. Each photogenerated electron relaxes to the band edge and contributes equally to photocurrent, independent of the energy of the initial photon. The methodology of classifying energy leads to significant ambiguities in the total number of photons available. As an illustrative thought experiment, one can construct two spectra *perfectly matched in energy* ("Class A+++…") from 900 nm – 1100 nm but *differing by 20% in photon count*, simply by using light at only one of the limits of this wavelength range.

The situation can become worse if we consider the spectral response (or EQE) of the device under test, as shown for some Si devices in Fig. 1. For our hypothetical spectrum with photons concentrated near 900 nm, Si will convert almost all of these into photocurrent (since EQE $\approx$ 100%). Conversely, for the spectrum with photons concentrated near 1100 nm, Si will convert less than half of these into photocurrent (since EQE $\approx$ 40%). Using the numbers quoted above, these two theoretical "A+++…" simulators would generate photocurrents differing by about 50% for the range 900 nm – 1100 nm. For perspective, this would be a discrepancy of about *10% in the total photocurrent* (generated across the whole range where Si is spectrally active: 300 nm – 1200 nm).

The IEC 60904-9 spectral classes have significant ambiguity in photon count, since the method compares integrated energy (not photons) using relatively wide wavelength ranges. Additionally, the wavelength-dependent spectral response of the actual device under test can amplify the effect of this ambiguity, as we have shown for Si.

Another example can be used to illustrate some implications of the IEC 60904-9 methodology. The example is as follows:

1) Take some module EQE curves from the literature [6], as shown in Fig. 1.

2) Take some example simulator spectrums. In this case, we use two Xe spectra and an LED simulator comprised of 5 LEDs [4]. The spectral class of these simulators is shown in Table II, and the spectral irradiance of the Xe spectra are shown in Fig. 2.

3) Calculate the spectral mismatch factor (MMF, per IEC 60904-7) for a "perfect" reference device (EQE = 1 for all wavelengths) in the range of 300 nm – 1100 nm [7]. The "perfect" reference evaluates the MMF relative to the case of using every photon available in the AM1.5 spectrum in the 300-1100nm range.

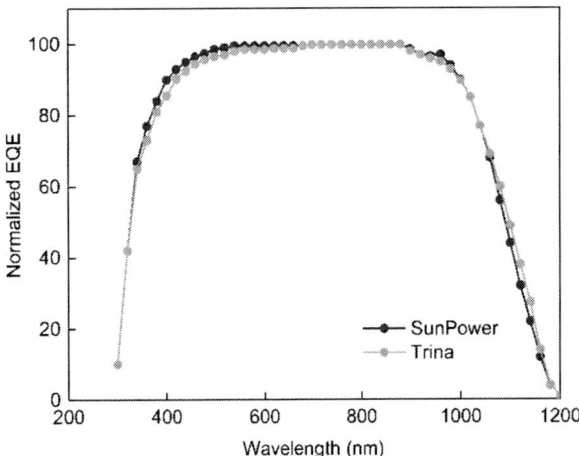

Fig. 1. EQE of mono- and multi-crystalline record modules [6].

TABLE II
SUMMARY RESULTS (MISMATCH PERFECT REFERENCE)

Type	Class	SunPower Mono	Trina Multi
Xenon	C	1.009	1.008
Xenon	A	1.027	1.025
5-LED	"A+"	1.044	1.041

4) Compare the resulting cumulative photocurrent to what would result from an AM1.5G illumination.

The resulting photocurrent curves from step 4) are shown in Fig. 3. The ratio of each curve's cumulative photocurrent at 1100 nm to that of AM1.5G is the spectral mismatch factor (MMF). These values are summarized in Table II. Counterintuitively, the Class C spectrum has very low spectral mismatch, yielding a photocurrent very similar to the AM1.5G spectrum, and significantly better than both the Class "A+" LED spectrum and Class A filtered Xe spectrum. How can this be? Examining Fig. 1 and 3, the key is that in the wavelength range where the EQE curves are flat and near-unity (roughly 480 nm – 950 nm), the wavelength of the incoming photons does not matter—they will all be converted to current equally. Therefore an exact match to the AM1.5G photon distribution in this region is relatively unimportant for Si. In the wavelength range where the EQE curves are rapidly changing and non-unity, an exact match to AM1.5G is much more important. Largely through coincidence, the Class C Xe spectrum meets these criteria—it does not perfectly match AM1.5G in the relatively unimportant range 480 nm – 950 nm, but it does match AM1.5G very well in the regions of rapidly-changing EQE. Our examination of Fig. 3 suggests the Class A Xe and "A+" LED simulators were optimized for the IEC 60904-9

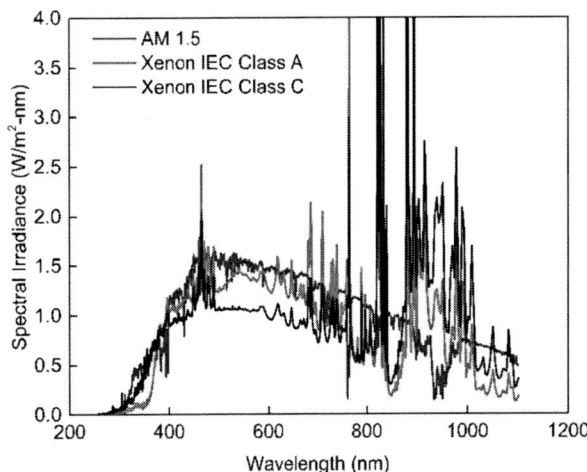

Fig. 2. Examples of a Class C and Class A Xe flash spectrum. For the Class A spectrum, where the strong lines from 800 nm – 1100 nm have been attenuated to meet the Class A metric.

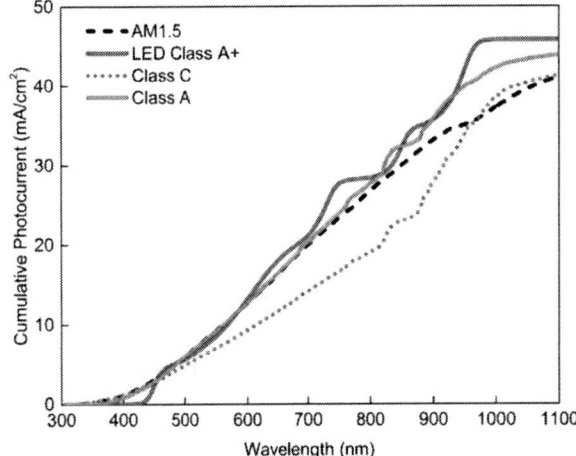

Fig. 3. The cumulative photocurrent for 4 cases of spectrum for the SunPower module with EQE shown in Fig. 1.

metric across six wavelength ranges, and as a result, do not match AM1.5G very well in the critical regions of rapidly-changing EQE.

The major errors in photocurrent occur in the 900 nm – 1100 nm range. In this range, the LED simulator generates most of its photons near 900 nm, and the Si module collects them at very high EQE resulting in over-reporting of current and a significant MMF. Additionally, no light is generated from the LED simulator for wavelengths > 950 nm; this is despite the "A+" rating. We can qualitatively explain the Class A Xe spectrum in the same way. Aggressive filtering of the Xe spectral lines at wavelengths > 800 nm shifted the center of the photon distribution closer to 900 nm, resulting in over-reporting current and a high MMF. Similar to the LED spectrum, there are too few photons from 1000 nm – 1100 nm. A major conclusion is that by filtering the Class C spectrum, a higher classification of "A" was achieved at the expense of measurement accuracy, especially for light trapping in the module that determines the performance from 1000 nm – 1100 nm. Based on published data [3], we believe this is a common phenomenon for Class A filtered Xe simulators in general. Fig. 3 indicates that the Class C unfiltered Xe spectrum is superior in emulating AM1.5G in the critical spectral ranges that matter for Si.

For industrial measurements employing best practices, a spectrally-matched device is used as a reference to set the simulator irradiance, resulting in no spectral mismatch error for any spectral class of tester. In this case, the point is to have measurement sensitivity in the wavelength ranges where process variation may alter the device performance (e.g., glass/encapsulant effects on UV absorption, and back-surface reflection effects on IR absorption). These are the same wavelength ranges identified above where the Si EQE is rapidly changing, and therefore, these are the regions where the simulator should have close match to the AM1.5G. In regions where the spectral response is constant, or is not strongly influenced by process variation, the precise photon distribution is not important—only the integral matters over this region. A purpose-built simulator designed according to these principles does not necessarily require a perfect spectral match to AM1.5G across the entire spectral range (or a high classification per IEC 60904-9) in order to attain high measurement accuracy.

III. SPATIAL NON-UNIFORMITY OF IRRADIANCE

The measurement of irradiance non-uniformity consists of dividing the test plane into roughly 64 sections, and measuring the irradiance of each section with a reference cell. Non-uniformity is reported using the maximum and minimum measured irradiance values [1]:

$$\frac{I_{max} - I_{min}}{I_{max} + I_{min}} \times 100\%. \tag{1}$$

In this case, Class A is ±2%, Class B is ±5%, and Class C is ±10 %. The informal Class "A+" is defined as ±1% by TUV Rheinland.

One issue is immediately apparent. Why measure 64 points and report a result based only on the minimum and maximum? Standard statistics texts frown on metrics based on this difference (often called the range), because the result depends critically on the random error for these two points. There may be some historical significance to (1), since a series string of solar cells is limited at short-circuit current by the lowest current (lowest illumination × I_{sc} product for a single cell), which could correspond to the minimum in the intensity plane. The actual case of measuring *power* is more interesting, since the measurement of power is very insensitive to non-uniformity. It is useful to clarify two separate components of measurement error that result from spatial non-uniformity of irradiance.

The first error component is the deviation of the *average incident power* from STC (1000 W m⁻², or "one sun"). Since the power from most PV devices is approximately linear in irradiance, the error due to measuring at higher or lower irradiance is a straightforward two-sided uncertainty (±1% error in incident power yields ±1% error in measured power output from the PV device). Most simulators have a procedure for adjusting the average incident power, or may also provide methods for correcting measured data to STC based on irradiance recorded by a reference cell during the measurement. In production environments, average incident power is set by adjusting the simulator irradiance to match the known power output of a spectrally-matched, traceable reference module. In test laboratories the average incident power can be measured using various methods, for example, using a reference cell mapped across the module plane.

The second error component is a loss of power (relative to the case of perfectly uniform irradiance), caused by each cell operating off of its own maximum power point. This is a current mismatch effect, and it results in one-sided uncertainty, since current mismatch always decreases the module's maximum power. The maximum power point is determined by:

$$Power = I_{mp} \times V_{mp}. \tag{2}$$

At the module maximum power point, the effect of the lowest-current cell is to decrease module I_{mp}. However, the effect on I_{mp} is much less pronounced than for the short-circuit case. A rule of thumb is that low-current cells will affect the maximum power negligibly, as long as the lowest-current cell is within ±4% of the average module current [8 – 10]. Qualitatively, this is because the I-V curve of each cell is relatively *compliant* near its maximum power point [9]. When current *increases*, voltage *decreases* (and vice versa), so power is relatively constant over a range of cell operating currents. Cells can operate off of their individual maximum power points in order to be at the module operating maximum power current without much loss in power. Contrast this to the short-circuit portion of a cell I-V curve, where a small change in current causes cells and strings to go into reverse bias, and the module current to be limited by a single cell. Measured I_{sc} can be a relatively strong function of non-uniformity, while power is not.

There is a common misperception within the industry that the two error components described above are large, and that their net effect on the power-measurement error is equal to the non-uniformity calculated from (1) [11]. A straightforward analysis shows otherwise: here, we show a simulated plot of error in power as a function of non-uniformity (Fig. 4) for several distributions of irradiance across a 60-cell PERC module [9]. The distributions of light all have the *same average incident power*, therefore eliminating the first error component discussed above and focusing on the effects of the second error component. In order of increasing severity, the illumination patterns are:

1) *Best case*—one cell at each extreme (minimum and maximum) and the rest at the average irradiance.
2) *Normal distribution*—two cells at each extreme irradiance corresponding to ±3 standard deviations of a normal distribution.
3) *Uniform staircase*—each cell at one unique irradiance, spread uniformly from lowest to highest.
4) *Worst case*—half of the cells at the minimum irradiance, half at the maximum, and no cells operating at the average irradiance.

Fig. 4 shows that for non-uniformity < 5%, there is no case where the error in power is as great as the non-uniformity itself. For the normal case (perhaps most typical of actual data distributions), a non-uniformity of 5% yields an error in power less than 0.5%. The non-uniformity defined by (1) has no information about how many cells are operating off of their maximum power point current, or by how much. Therefore, there is a different curve for each distribution of intensities across the module. Contrast this to a plot of the standard deviation of incident irradiance across the module for the same simulation (Fig. 5). The standard deviation *does* contain information about how many cells operate off of their maximum power point, and at what distance from I_{mp}. For relatively small non-uniformity (< 4%), all of the data falls onto a universal curve! This result is extremely useful and relevant for estimating measurement error due to non-uniformity effects, though it is also somewhat coincidental.

The data in Fig. 4 and 5 are numerically calculated, but can be conceptually understood by inverting the problem statement: instead of calculating power from many cells running at the same current (but different irradiance), instead calculate the power from many cells running at the same irradiance (but different I_{mp} on a universal one-sun power curve. Then, the problem is to evaluate the power loss from spreading points across the universal power vs. current curve due to variations in cell current [12]. For example, a cell in a module receiving 1.03 times the average irradiance will transform to a cell at one sun operating at 0.97 of the nominal I_{mp}. These problems are

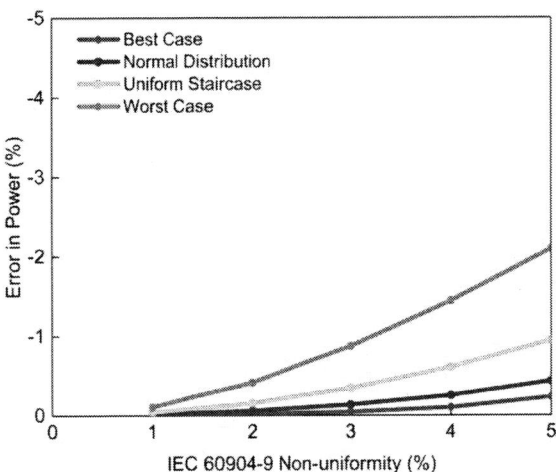

Fig. 4. Error in power measurement due to non-uniformity calculated by using (1) for four cases of distribution of the non-uniformity.

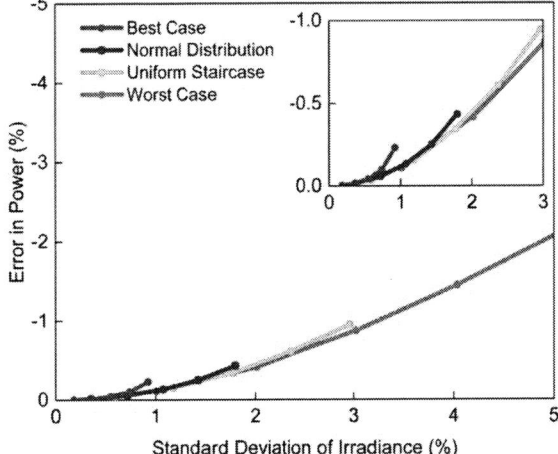

Fig. 5. The error in power measurement as a function of the standard deviation of the illumination distribution. The points on each curve correspond to 1, 2, 3, 4, and 5% non-uniformity as defined by (1). For non-uniformity < 4%, all data falls on a single curve.

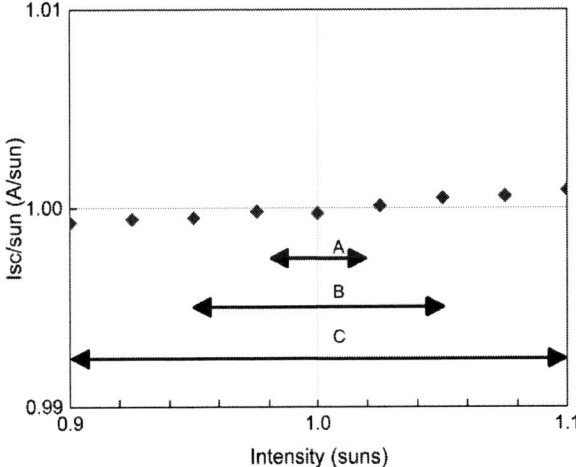

Fig. 6. Short-circuit current per sun, measured for a 96-cell high-efficiency silicon module. Black arrows indicate the corresponding IEC 60904-9 letter grade for long-term instability (LTI) of irradiance over the measured irradiance ranges.

equivalent for small differences in intensities since the trends in voltage and series resistance loss with intensity are opposing, small, and similar in magnitude. Numerical or analytical calculations (not presented here) indicate the transformation between these two problems is especially applicable in the specific case of cells with an ideality factor of 1 and series resistance of $0.67\ \Omega\ cm^2$. In this special case the voltage effects cancel out for small intensity variations.

Since the maximum power point on the current vs. power curve is defined by a maximum at $(dP/dJ = 0)$, we know that the Taylor series expansion around $J = J_{mp}$ will be dominated by a quadratic term for small deviations from J_{mp}. This means that summing up the deviations in power for each cell is proportional to the sum of squares in $(J-J_{mp})$. Therefore the power deviation due to non-uniformity is roughly linear in the variance of $(J-J_{mp})$, which is the square of the standard deviation in current density. This is the exact functional relationship illustrated in Fig. 5. This error is not a random uncertainty. It is a single-sided power loss due to cells forced to run off of their maximum power point to match the common module current.

This lucky coincidence, illustrated in Fig. 5, means that the power uncertainty due to non-uniformity can be calculated based standard deviations of the incident irradiance, as observed in Fig. 5, for any distribution of irradiance across the module. Taking the case of the normal distribution, we see a smooth function with less than 0.4% error for non-uniformity up to 5%. There is no reason to assign special significance (Class A) to 2%; the whole range of values between 0% and 5% non-uniformity result in relatively small measurement errors.

The key data that enables calculation of the power uncertainty are the incident irradiance for each cell in the module plane, from which we calculate the average irradiance, the standard deviation of irradiance, and the minimum irradiance. The simple analysis will hold until the difference between the average irradiance and the minimum irradiance is greater than the difference between I_{mp} and I_{sc} for the cells. At this point (about 4% – 5% of I_{sc}), the lowest intensity point will force the average cell off of the maximum power point to a higher-than-optimal voltage. The maximum irradiance cell, on the other hand, will have a very weak effect on the module power. More details on these topics will be available in future publications [12].

It is quite clear that the standard deviation of irradiance is a much more relevant parameter than the non-uniformity as defined by (1). Errors are monotonic in standard deviation for cases where the difference between average and minimum illumination is less than 4% (Fig. 5). In contrast, a non-uniformity of 5% does not discriminate between a case where the error would be 0.2% and a case where the error would be 2%, as can be seen from Fig. 4.

The discussion of errors due to non-uniformity here contrasts with the recommendation to use I_{sc} calibration for module testers [11] but is in agreement with the consensus in the literature [8, 10]. We have shown here that the uncertainties in

power are very weakly dependent on the non-uniformity (for non-uniformity < 5%) when calibration by power is used in preference to short-circuit current. Another simple observation is relevant here: at I_{sc} in the presence of ANY cell mismatch or non-uniformity, the typical Si module is bypassing 2 of its 3 strings through diodes. Therefore, calibration at I_{sc} is using only 1/3 of the module area to estimate the power incident upon the module under test. Additionally, this single string is likely to be current-limited by a single cell. All of this suggests calibration of simulator irradiance using module I_{sc} will be extremely sensitive to irradiance changes over a relatively small area of the module test plane (and therefore *insensitive* to the average irradiance over the whole plane). Calibration of simulator irradiance using module power is tolerant to non-uniformity, as shown in Fig. 4, and uses the entire module area during this calibration. Careful attention to module temperature and electrical connection insure minimal errors using calibration to a reference module of the same size, shape, and spectral response as the module under test [10].

IV. INTENSITY VS. TIME AND DATA ACQUISITION

The third letter in the IEC 60904-9 classification is based on the light being at the same intensity for all data points, and the data acquisition being simultaneous for measurements of light, current, and voltage. Given the variety of analysis and corrections for light intensity and data acquisition available, the limits ±2%, ±5%, and ±10% for A, B, and C classification (respectively) may be completely arbitrary with respect to predicting accuracy or repeatability of the simulator.

As an example, we show a plot of I_{sc}/sun vs. suns to illustrate the long term instability (LTI), the effect of taking data at other than exactly one sun. We plot I_{sc}/sun (a simple correction for short-circuit current taken near one sun) for a range of intensities spanning Class A (1 sun ±2%) to Class C (±10%) in Fig. 6. No significant difference is seen, indicating that for this Si module this specification is irrelevant. We know of no data indicating that LTI is a critical parameter determining accuracy of module power measurements, and the authors are interested to know where it came from. If there is no supporting data then it should be discarded or disregarded. A specification that requires constant light intensity but permits time-varying voltage and current is arbitrary, and unnecessarily limiting with respect to instrument design and implementation.

IV. CAPACITANCE AND OTHER ISSUES

The largest uncertainty in measurement of high-efficiency silicon modules is often the capacitance, requiring careful attention to voltage-sweep characteristics. Simulator classification does not account for capacitance measurement issues at all. Other issues such as angular acceptance of the module and choice of reference cell can also have more effect on accuracy than the classification parameters in the standard.

IV. CONCLUSION

Simulator classifications, especially IEC 60904-9, are given a great deal of deference in industry. However, the results of the metrics in IEC 60904-9 for spectrum and illumination uniformity do not reliably assign higher letter-grade classifications to more accurate simulators. Additionally, we find no purpose for the long-term-instability classification.

Direct calculations of measurement uncertainty using the underlying data are straightforward and much more relevant in each of the three cases. As a prime example, simulators with Class C spectral classification can often be more accurate for measuring silicon modules than Class A, or unofficial Class "A+."

Our analysis indicates that the following procedures could be used to assess the accuracy of different simulators for any given application, in preference to the sometimes misleading letter-grade classifications from the simulator standard.

1) Use the spectral irradiance curve of the simulator to calculate the mismatch factor of the simulator for the technology to be measured. This calculation can use an ideal reference cell integrated over the range of wavelengths utilized by the target technology, as in this paper, or the spectral response of the actual reference cell that will be used in testing.

2) Use maximum power for a reference module to calibrate the power on the module area, or use a reference cell map to determine the average power incident upon the module.

3) To report illumination non-uniformity, report the minimum, the average, and the standard deviation of intensity incident on all of the cells on the module. Reporting only the minimum and maximum (as "non-uniformity") is statistically weak and is not sufficient information for a meaningful evaluation of errors. For a specific technology, the errors can be calculated from the information on standard deviation of illumination.

4) No significance should be assigned to the LTI classification without knowledge of the analysis and data acquisition details.

The first and second points are well supported by the technical consensus from the literature. The third and fourth points here appear to be new or perhaps not widely appreciated.

REFERENCES

[1] IEC 60904-9:2007 (Ed. 2) §5.1

[2] W. Herrmann et al., "Advanced Intercomparison Testing of Testing of PV Modules in European Test Laboratories," in *22nd EU PVSEC* (Milan, 2007).

[3] W. Herrmann and L. Rimmelspacher, "Uncertainty of Solar Simulator Spectral Irradiance Data and Problems with Spectral Match Classification," in *27th EU PVSEC* (Frankfurt, 2012).

[4] W. Herrmann, "Advances in Spectral Irradiance Analysis of Solar Simulators," in *26th PVSEC* (Singapore, 2016).

[5] C. Monokroussoss et al., "IEC 60904-9 Spectral Classification and Impact on Industrial Rating of c-Si Devices," WCPEC-6, 23rd-27th November 2014, Kyoto, Japan

[6] M. Green et al., "Solar Cell Efficiency Tables," in *Progress in Photovoltaics* Vol. 24, Issue 1, pp. 3 – 11, November 2015

[7] IEC 60904-7:2008 (Ed. 3)

[8] W. Herrmann and W. Wiesner, "Modelling of PV Modules – The Effects of Non-Uniform Irradiance on Performance Measurements with Solar Simulators," 16th EU PVSEC (Glasgow, 2000).

[9] H. Wilterdink et al., "Spatial non-uniformity of irradiance and uncertainty in power rating for c-Si modules," NREL PV Module Reliability Workshop (PVMRW), 2016

[10] C. Monokroussos et al., "Impact of Calibration Methodology into the Power Rating of c-Si PV Modules Under Industrial Conditions," 28th EU PVSEC (Paris, 2013), pp. 2926 – 2934.

[11] V. Fakhfouri et al., "Uncertainty assessment of PV Power measurement in industrial environments," 26th EU PVSEC (Hamburg, 2011), pp. 3408-3412

[12] R. Sinton et al., Submitted for publication

Impact of Infrared Optical Properties on Crystalline Si and Thin Film CdTe Solar Cells

Indra Subedi[1], Timothy J Silverman[2], Michael Deceglie[2], and Nikolas J. Podraza[1]

[1]Department of Physics & Astronomy and Wright Center for Photovoltaics Innovation & Commercialization, University of Toledo, Toledo, OH 43606, USA

[2]National Renewable Energy Laboratory, Golden, CO 80401, USA

Abstract — Photons with energies below the photovoltaic (PV) absorber band gap do not generate current and adversely impact performance when absorbed in other solar cell components to produce heat. Here we incorporate infrared (IR) optical response in simulations for understanding thermal losses. Spectroscopic ellipsometry is used to measure Al-Si interface optical properties in Si PV. Reflectance of an Si module has been analyzed to account for encapsulant and Al-Si interface contributions. IR-extended quantum efficiency simulations calculate efficiency gains / losses arising from variations in current generated and total reflectance as functions of transparent front contact material in thin film CdTe PV.

Index Terms — Al-Si interface, CdTe, optical properties, ray tracing, silicon solar cell, transparent conducting oxide.

I. INTRODUCTION

When the temperature of a solar cell increases, the charge carrier recombination rate also increases. This increase directly affects power conversion efficiency by decreasing open circuit voltage (V_{oc}) and fill factor. Temperature can be managed by rejecting unwanted heat, increasing in-the-field efficiency of solar modules. For every 1 K rise in temperature there is about 0.3-0.5 % [1] decrease in relative efficiency for wafer based silicon (Si) photovoltaic (PV) modules. The devices are heated by light absorbed in other component layers of the solar cell—those not generating electrical current. Therefore, reduction in absorbance within the other component layers of the PV device may be used to lower operating temperature. The vast majority (~99 %) of solar irradiance reaching the solar cell lies in the 300-2500 nm spectral range with a subset absorbed in the active layer of the PV device. Typically, absorption of longer wavelength photons with energy lower than the active layer band gap are mainly responsible for heating and increasing the temperature of modules. For example, the aluminum-silicon (Al-Si) interface at the back contact of Si solar cell absorbs sub Si band gap infrared (IR) photons that may heat the device and are accounted in simulations here.

Similarly, transparent conducting oxides (TCOs) used as the front electrical contact in CdTe solar cells also absorb IR photons. The interfaces between different component layers in thin film CdTe solar cells also affect the total reflectance of devices. Earlier studies on optical losses on thin film CdS / CdTe solar cells has been limited at long wavelengths to near the CdTe band gap [2]. To account for the full reflectance of the CdTe device and absorption in the TCO including that in the IR region, our simulations span from 300 to 2500 nm.

To reduce heating, it may be desirable to make devices more IR reflective and with less parasitic absorption in the near-IR and visible spectral range. We have obtained Al-Si interface optical properties below the Si band gap for Si device reflectance simulation. This simulation is consistent with experimental data for a commercial multicrystalline Si module. Additionally, we have compared external quantum efficiency (QE), total reflectance, and parasitic absorption with associated efficiency gains / losses for thin film CdTe PV with different TCO structures by using IR-extended optical simulations. From the IR-extended simulations, we see that changing the TCO may be a pathway to improving the operating efficiency by rejecting more light in the IR region or heat. As a case study, variations in carrier concentration (N) and mobility (μ) for aluminum doped zinc oxide (AZO) are used to control the IR optical response and electrical properties of that TCO.

II. SIMULATION DETAILS

Two simulation approaches have been applied. Packaged Si solar cell performance and total reflectance from the device stack is simulated using the Module Ray Tracer (MRT) from PV Lighthouse [3], which includes contributions from the Al-Si interface and ethylene-vinyl acetate (EVA) encapsulant layers. This MRT software combines Monte Carlo ray tracing and thin film optics making it suitable to account for multilayer stack and periodic texture on the Si. Each ray tracing simulation by MRT is performed with 5×10^5 rays with 5000 rays per run at 20 nm wavelength intervals. To simulate the real irradiance, AM 1.5G [4] solar irradiance spectrum is used as input. The Si module structure used in this simulation is glass / EVA / anti-reflection coating (ARC)

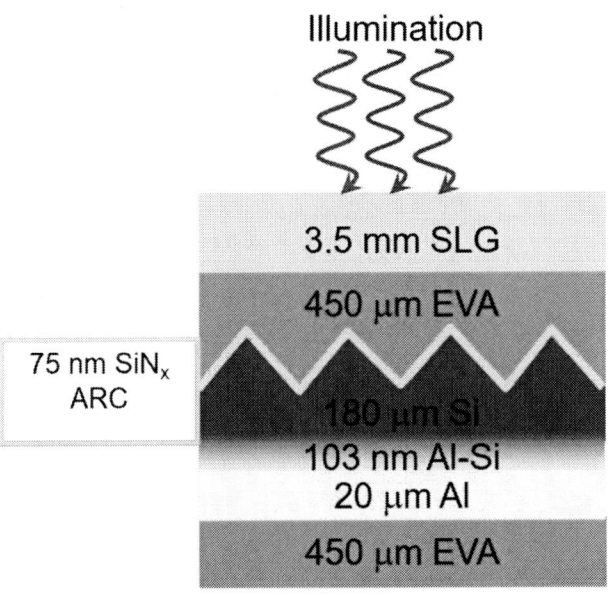

Fig.1. Schematic of encapsulated Si solar cell structure used in the Module Ray Tracer (MRT) simulation.

pyramidal structure / Si PV device / EVA with nominal thicknesses as shown in Fig. 1. The full details of this approach are available elsewhere [5]. The optical properties used in MRT are from literature [6]–[11] except for the Al-Si interface, which has not been well-characterized optically to date. For the Al-Si interface, we fabricated a test sample for characterization of interfacial optical response particularly below the Si band gap. An optically opaque Al film is thermally evaporated onto one side of a double side polished single crystal Si (c-Si) wafer which is then annealed at 450 °C for 15 min in a helium ambient to promote interdiffusion. Spectroscopic ellipsometry measurements are performed in a through-the-Si configuration [5] at 55° angles of incidence on the c-Si side, with the intention that the beam from the initial ambient / c-Si interface is excluded while reflection from the back c-Si / Al interface is collected. This approach yields Al-Si complex optical properties over the 1128-2500 nm wavelength range. Shorter wavelengths than the c-Si band gap (>1.1 eV or <1128 nm) are not accessible in this configuration as the c-Si wafer is heavily absorbing and sufficiently thick. The experimental ellipsometric spectra are fitted to a structural model including c-Si / finite thickness Al-Si interface / Al in a least squares regression analysis.

We also conducted a series of QE simulations using our own software to deduce the potential for increased efficiency in CdTe thin film solar cells by balancing loss in current generated with improved IR and total reflectance. The QE simulation software used here applies Fresnel's equations and multilayer matrix methods, but does not consider texture [12]. The baseline CdTe simulation applied is that obtained to match the QE spectrum of a functional CdTe solar cell fabricated at University of Toledo [13]. The QE and reflectance are simulated for CdTe solar cells with different TCOs and compared relative to commercial SnO_2:F (FTO). Optical and structural parameters obtained from spectroscopic ellipsometry serve as an input for simulation. The thickness of the TCO structure is constrained to values which yield the same sheet resistance as the commercial FTO, nominally 15 Ω / sq. Short-circuit current density (J_{sc}) is evaluated from QE by assuming 100% collection of photo-generated carriers. Power gains (losses) due to cooling (heating) are quantified from the integral of the product of total reflectance and the AM 1.5 spectrum.

III. RESULTS AND DISCUSSION

A. Si Module Reflectance

Wavelength dependent optical properties in terms of complex index of refraction, $N (\lambda) = n (\lambda) + ik (\lambda)$, for the Al-Si interfacial region obtained from through-the-Si ellipsometry is shown in Fig. 2 [5]. The Al-Si interface is significantly more absorbing than c-Si in this spectral range, as evidenced by $k (\lambda)$. $N (\lambda)$ of the Al-Si interface serves as input for the discrete 103 nm interface layer of the structure in Fig. 1. Using that relatively simple interfacial structure and reference N for each component, the simulated total normal incidence reflectance is in good agreement with measured reflectance of a commercial multicrystalline Si module, particularly at wavelengths below the Si band gap to 1700 nm. Fig. 3 shows the results from the simulation compared to experimental measurement. For wavelengths near 600-800 and 1700-2300 nm, there is greater deviation between the simulation and experiment. These deviations may arise from differences in the assumed and actual texturing uniformity of the Si surface; variations in $N (\lambda)$ for the Si absorber, EVA, covering glass, SiN_x; and texturing in the glass.

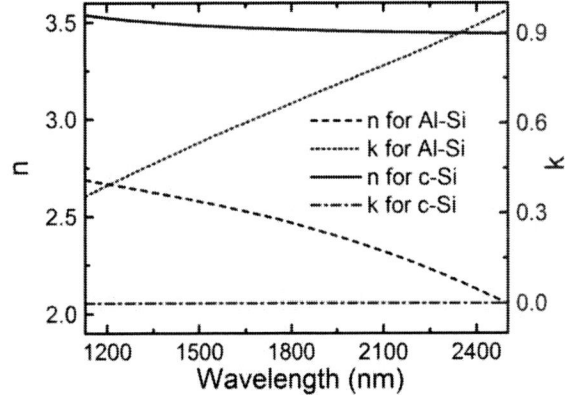

Fig. 2. Complex index of refraction, $N (\lambda) = n (\lambda) + ik (\lambda)$, spectra obtained for the Al-Si interface by through-the-Si spectroscopic ellipsometry and reference spectra for crystalline Si [6].

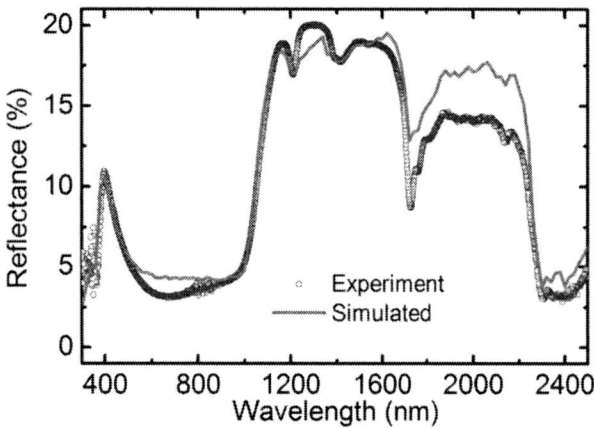

Fig. 3. Simulated and measured total reflectance of a Si module as a function of wavelength.

B. Gains / Losses in CdTe Thin Film Solar Cells

The comparison of QE and total reflectance for different TCOs, FTO [13] and two types of AZO [14]–[15], is shown in Fig. 4. Optical properties for AZO (1) are directly measured in [14], while those for AZO (5) use the parametric model for AZO (1) but with N and μ from [15]. QE in the CdTe absorbers are relatively similar, but the device with AZO (1) has significantly lower long-wavelength reflectance while device with AZO (5) has higher IR reflectance in comparison with FTO. Similar curves have been calculated for additional TCOs such as indium tin oxide (ITO), differently sourced FTO, AZO [16], and SrVO₃ [17] Relative gain or loss in J_{sc} of devices with different TCOs relative to the baseline response for a CdTe device with a FTO TCO are indicated by the vertical position of data points in Fig. 5. Relative gain in efficiency from increased total and IR reflection is calculated assuming 30 W/m² in reflected power

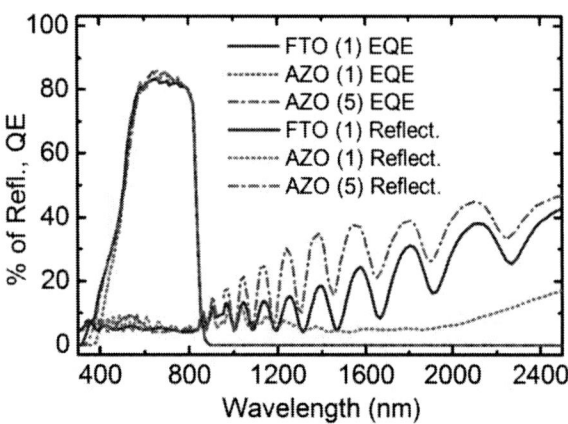

Fig. 4. Quantum efficiency (QE) and total reflectance for CdTe solar cells with SnO₂:F (FTO) and aluminum doped zinc oxide (AZO) transparent conducting oxide (TCO) front contacts.

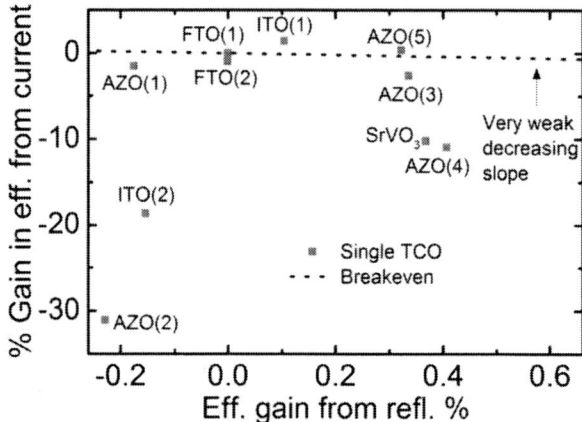

Fig. 5. Comparison of gains / losses in efficiency (%) from current generation and reflectance management for reduced heating in CdTe solar cells relative to using a FTO TCO. The breakeven line is 1 to -1 line having very small decreasing slope.

results in a 1 K temperature reduction with a 0.34 % relative reduction [18] in efficiency expected for every 1 K rise in temperature.

A device with ITO (1) has a small gain in J_{sc} and reflection in comparison with FTO, while that with AZO (5) also has a small J_{sc} gain but a more substantial gain from reflectance. Higher value of N and μ in AZO improves gain from both J_{sc} and total reflection relative to FTO. We also simulated AZO with varying μ keeping conductivity and thickness same as for AZO (1), denoted by S(n). When μ is increased, gains from J_{sc} increase and a small gain from total reflection relative to FTO is observed as shown in Fig. 6. Table I shows summary of N, μ, and thickness for different simulated and experimentally determined AZO TCOs.

Transparency in near IR to visible range for combinations of N and μ must be ensured to obtain high J_{sc}. Stability of

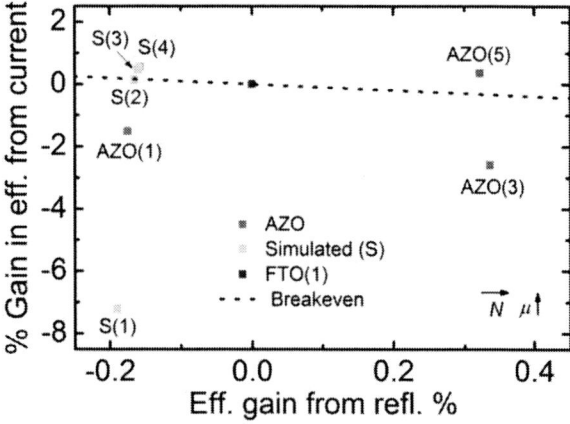

Fig. 6. Gains / losses in efficiency (%) from current generation and total reflectance for various AZO TCOs with varying mobility (μ) and carrier concentration (N) as listed in Table I.

TABLE I
CARRIER CONCENTRATION, MOBILITY, AND THICKNESS OF AZO USED IN CDTE SIMULATIONS

	N (x 10^{20} 1/cm^3)	μ (cm^2/Vs)	Thickness (nm)
AZO (1) [Ref. 14]	1.33	21	1500
AZO (3) [Ref. 16]	6.81	24	253
AZO (5) [Ref. 15]	6.88	37	165
S (1)	1.84	15	1500
S (2)	1.10	25	1500
S (3)	0.92	30	1500
S (4)	0.79	35	1500

both ITO and AZO under the full CdTe fabrication procedure makes these candidates challenging to those devices. Detrimental changes in the properties of ITO occur at the processing temperatures (~ 600 °C) required for CdTe PV. An issue for implementing AZO as a TCO is it can also react with CdS in the CdTe device because fabrication is done in higher thermal budget about 600 °C. Implementation of a highly resistive buffer layer between AZO and CdS has been proposed and tested [19]. In our simulations, an undoped tin oxide layer is applied as a transparent highly resistive buffer layer between AZO and CdS, demonstrating that the presence of this layer does not negatively impact performance. From these cumulative results, improvements in efficiency from reflectance management from different TCOs are obtainable but near IR to visible range transparency must be assured to prevent losses from reduced J$_{sc}$.

IV. CONCLUSION

We have incorporated IR optical properties in simulations of a wafer Si module and thin film CdTe solar cells. Agreement between simulated and experimental reflectance from 300 to 2500 nm for a commercial Si module is demonstrated using an optical model incorporating contributions from the Al-Si interface. CdTe device simulations incorporating different TCOs are used to quantify gains and losses due to both total reflectance and J$_{sc}$ variations. In particular, AZO TCOs are evaluated and indicate that gains in both relative J$_{sc}$ and cooling from improved IR and total reflectance are possible with increasing N and μ, which are within the range of those reported in literature. Application of these modeling techniques will be helpful in assessing device designs to enhance efficiency by reducing parasitic absorption in near IR to visible spectral range and maximizing IR reflection to reduce heating.

ACKNOWLEDGEMENT

This work was supported by the U.S. Department of Energy under Contract No. DE-AC36-08GO28308 with the National Renewable Energy Laboratory. Funding provided by U.S. Department of Energy Office of Energy Efficiency and Renewable Energy Solar Energy Technologies Office.

The U.S. Government retains and the publisher, by accepting the article for publication, acknowledges that the U.S. Government retains a nonexclusive, paid-up, irrevocable, worldwide license to publish or reproduce the published form of this work, or allow others to do so, for U.S. Government purposes.

The authors would like to thank Dr. Prakash Koirala for providing the baseline model of CdTe with FTO.

REFERENCES

[1] E. Skoplaki and J. Palyvos, "On the temperature dependence of photovoltaic module electrical performance: A review of efficiency/power correlations," *Solar Energy,* vol. 83, pp. 614-624, 2009.

[2] L. Kosyachenko, E. Grushko, and X. Mathew, "Quantitative assessment of optical losses in thin-film CdS/CdTe solar cells," *Solar Energy Materials and Solar Cells,* vol. 96, pp. 231-237, 2012.

[3] PV Lighthouse: Module Ray Tracer, accessed 24 April 2017, https://www.pvlighthouse.com.au.

[4] C. Gueymard, *SMARTS2: A Simple Model of the Atmospheric Radiative Transfer of Sunshine: Algorithms and Performance Assessment*: Florida Solar Energy Center Cocoa, FL, 1995.

[5] I Subedi, T. J Silverman, M. Deceglie, and N. J. Podraza, "Al+Si interface optical properties obtained in the Si solar cell configuration", Manuscript submitted to *Physica Status Solidi (a),* 2017.

[6] C. Herzinger, B. Johs, W. McGahan, J. A. Woollam, and W. Paulson, "Ellipsometric determination of optical constants for silicon and thermally grown silicon dioxide via a multi-sample, multi-wavelength, multi-angle investigation," *Journal of Applied Physics,* vol. 83, pp. 3323-3336, 1998.

[7] C. Schinke, P. Christian Peest, J. Schmidt, R. Brendel, K. Bothe, M. R. Vogt, I. Kröger, S. Winter, A. Schirmacher, and S. Lim, "Uncertainty analysis for the coefficient of band-to-band absorption of crystalline silicon," *AIP Advances,* vol. 5, p. 067168, 2015.

[8] A. D. Rakić, A. B. Djurišić, J. M. Elazar, and M. L. Majewski, "Optical properties of metallic films for vertical-cavity optoelectronic devices," *Applied Optics,* vol. 37, pp. 5271-5283, 1998.

[9] L. K. Gautam, L. Ye, and N. J. Podraza, "LPCVD SiN$_x$ thin film on c-Si wafer by spectroscopic ellipsometry," *Surface Science Spectra,* vol. 23, pp. 51-54, 2016.

[10] M. R. Vogt, "Development of physical models for the simulation of optical properties of solar cell modules," *submitted PhD thesis, Leibniz University of Hannover,* pp. 21-24, 2015.

[11] M. Rubin, "Optical properties of soda lime silica glasses," *Solar Energy Materials,* vol. 12, pp. 275-288, 1985.

[12] P. Aryal, J. Chen, Z. Huang, L. R. Dahal, M. N. Sestak, D. Attygalle, R. Jacobs, V. Ranjan, S. Marsillac, and R. Collins, "Quantum efficiency simulations from on-line compatible mapping of thin-film solar cells," in *Photovoltaic Specialists*

Conference (PVSC), 2011 37th IEEE, 2011, pp. 002241-002246.

[13] P. Koirala, J. Li, H. P. Yoon, P. Aryal, S. Marsillac, A. A. Rockett, N.J. Podraza, and R.W. Collins, "Through-the-glass spectroscopic ellipsometry for analysis of CdTe thin-film solar cells in the superstrate configuration," *Progress in Photovoltaics: Research and Applications,* vol. 24, pp. 1055-1067, 2016.

[14] P. Uprety, M. M. Junda, K. Ghimire, D. Adhikari, C. R. Grice, and N. J. Podraza, "Spectroscopic ellipsometry determination of optical and electrical properties of aluminum doped zinc oxide," *Applied Surface Science,* 2017.

[15] J. Nomoto, H. Makino, and T. Yamamoto, "Carrier mobility of highly transparent conductive Al-doped ZnO polycrystalline films deposited by radio-frequency, direct-current, and radio-frequency-superimposed direct-current magnetron sputtering: grain boundary effect and scattering in the grain bulk," *Journal of Applied Physics,* vol. 117, p. 045304, 2015.

[16] N. Ehrmann and R. Reineke-Koch, "Ellipsometric studies on ZnO: Al thin films: Refinement of dispersion theories," *Thin Solid Films,* vol. 519, pp. 1475-1485, 2010.

[17] L. Zhang, Y. Zhou, L. Guo, W. Zhao, A. Barnes, H.-T. Zhang, C. Eaton, Y. Zheng, M. Brahlek, H. F. Haneef, N. J. Podraza, M. H. W. Chan, V. Gopalan, K. M. Rabe, and R. Engel-Herbert, "Correlated metals as transparent conductors," *Nature Materials,* vol. 15, pp. 204-210, 2016.

[18] First Solar Series 4™ PV Module - Advanced Datasheet, accessed 8 June 2017, http://www.firstsolar.com/-/media/First-Solar/Technical-Documents/Series-4-Datasheets/Series-4V2-Datasheet.ashx.

[19] J. Perrenoud, L. Kranz, S. Buecheler, F. Pianezzi, and A. Tiwari, "The use of aluminium doped ZnO as transparent conductive oxide for CdS/CdTe solar cells," *Thin Solid Films,* vol. 519, pp. 7444-7448, 2011.

The Impact of Impurities on the Relative Efficiencies of Solar Cells from different Silicon Feedstocks

Muhammad Tayyib[1], Aleksandr Dobroliubov[1], Zekija Ramic[1], Muhammad Nadeem Akarm[1], Jan Ove Odden[2]

[1]University College of Southeast Norway, P.O. Box 235, NO-3603 Kongsberg, Norway,
[2]Elkem Solar AS, P.O. Box 8040 Vaagsbygd, NO-4675 Kristiansand S, Norway,

Abstract — **To enhances the efficiency of silicon feedstock, it is important to fully understand the performance behavior of solar cells made from those, especially in the low efficiency sites. Solar cells made from two different feedstocks have been investigate using number of instruments to find the effect of efficiency drop on impurities and imperfections in silicon crystal. It is found that cell regions consisting of dislocation clusters along with number of metallic impurities exhibit low performance to the incident light. Similar finding are obtained from both feedstock materials — a relatively cheap Elkem Solar grade silicon and polysilicon made through Silane process.**

Index Terms — **Elkem Solar, impurities, EDS, EBSD, LBIC**

I. INTRODUCTION

Crystalline silicon covers over 90% of global market share for the production of solar cells from the photovoltaic (PV) industry [1]. Multi crystalline (mc) silicon contains some concentration of impurity elements trapped during comparatively fast solidification of ingot comparing to single crystalline silicon produced by Czochralski or float zone method. Mc-Silicon manufactured by directional solidification contains many crystal defects such as grain boundaries (GBs) and dislocations. These sites act as recombination centres for the generated charge carriers causing a significant negative effect on the conversion efficiency of mc solar cells [2]. According to the literature [3], the solidification process induces mechanical stresses at high temperature which act as driving forces for the generation of dislocations. It is often believed that dislocations are primarily created in the lower part and subsequently multiplies towards the top of the ingot. As a result, high dislocation densities are observed in the top part of the silicon bricks. The existence of the dislocations and other crystallographic defects creates shunt regions for majority carriers [e] which results in low current in these regions.

In order to improve the quality of the solar cells, detailed studies have to be undertaken in laboratories for identifying the physics behind and the causes of defective regions and their impact on the performance. In the present paper, we have studied defective regions of solar cells by using various tools to identify root causes of the charge recombination centers.

II. EXPERIMENTAL SETUP

Micro defects and impurities analyses have been performed on multi-crystalline silicon solar cells. Cell materials manufactured from two different methods 1) the Siemens process (Silane production method) and 2) a metallurgical route – the Elkem Solar Silicon (ESS®) were compared.

Both materials are solidified in same G5 size furnace. To study the additional effect of ingot lining, corner bricks from both ingots have been selected for current experiment. Ingot sawing until passivated emitter rear contact (PERC) solar cells fabrication have been carried out on same production line in order to track down differences between both initial feedstock silicon materials. A number of wafers along known ingot height have been selected and processed to solar cells. Following sequential steps are carried out to find impact of impurities on the performance of the PERC solar cells.

- IV measurements using AAA sun simulator.
- Light beam induced current (LBIC) mapping of all solar cells using Semilab SDI WT-2000 instrument equipped with four different wavelength lasers (655, 854, 950, 974 nm).
- Identifying bad responding areas on solar cells and cutting using wafer dicing instrument.
- Electron beam induced current (EBIC) using FE-SEM for detailed study of charge recombination sites at region of interest (ROI).
- Grinding down the front contact and silicon nitride layer using 2000 grid paper and subsequently polishing with 1 µm diamond paste.
- Chemical polishing using HF:HNO3 (1:6).
- Defect site decoration using HF:1.5MCrO$_3$ (1:1) solution.
- The distribution of impurities and crystal structure analyses on ROI using SEM equipped with Energy-dispersive X-ray spectroscopy (EDS) and crystal orientation using Electron backscatter diffraction (EBSD).

Hitachi cold FE-SEM (model 8230) equipped with EBIC detector (from Hitachi) and EDS, EBSD detectors (from Oxford) is used to perform micro analyses of solar cells before and after polishing/etching.

III. RESULTS AND DISCUSSION

All the PERC solar cells for the present study had conversion efficiencies above 18%. The dislocation distribution was very inhomogeneous over the silicon wafer / solar cell surface. It is also well known that the impurities are the last to solidify during directional solidification [4]. The present research cells have therefore been selected both from top and bottom part of the silicon brick. The LBIC and reflection maps were measured at different laser wavelengths; i.e. 655, 854, 950, 974 nm. In the present paper, only the 974 nm maps are presented due to high penetration depth of the near infra-red laser, which leads to more detailed information of bulk recombination farther below the PN junction in the solar cell.

Figure 1 shows such LBIC map of an ESS® solar cell near the bottom of the ingot. Two sides of the wafer show relatively low output current due to contribution of impurities from the ingot lining. The concentration of these impurities were very low which could not be measured using SEM-EDS.

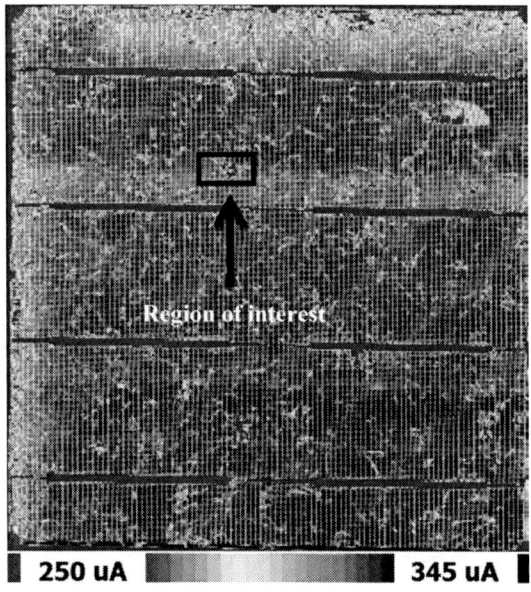

Fig. 1. LBIC map of an ESS® solar cell (near bottom of the ingot) measured at 974 nm wavelength with average 309μA current output. The conversion efficiency is 18.5% at STC.

Figure 2 shows a close up SEM image of an area of the interest whereas the EBIC contrast reveals the recombination sites of produced charges. EBSD map further discloses the fact that such recombinations are more pronounced at certain grain boundaries (GB) and dislocation clusters. The presence of the impurities in silicon material introduces many small / low angle grain boundaries leading almost similar crystal orientations on both sides of GBs. While EDS analyses of one of the dislocation cluster confirms the presence of metal impurities (Fe and Cr) c.f. Fig. 3.

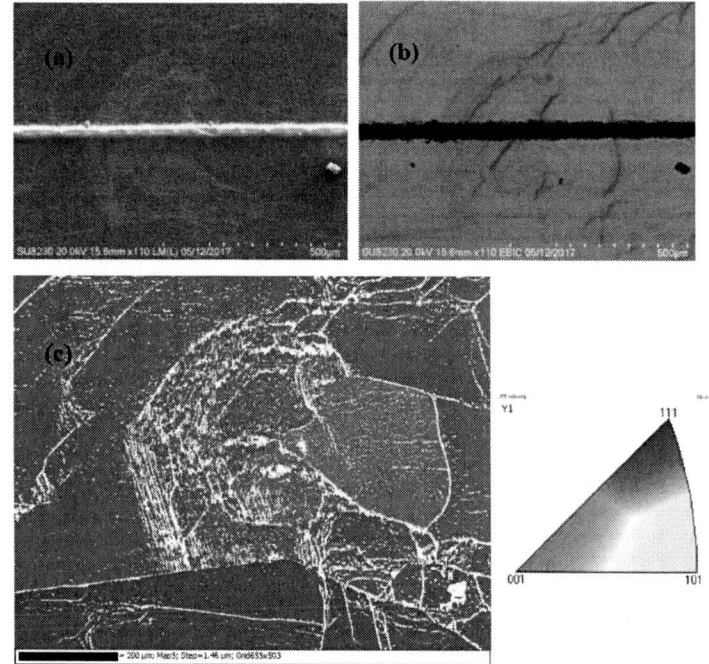

Fig. 2. SEM image (a), EBIC (b) and EBSD map (c) of the region of interest showing crystal structure with different orientation. The inverse pole figure shows the color scheme with orientations.

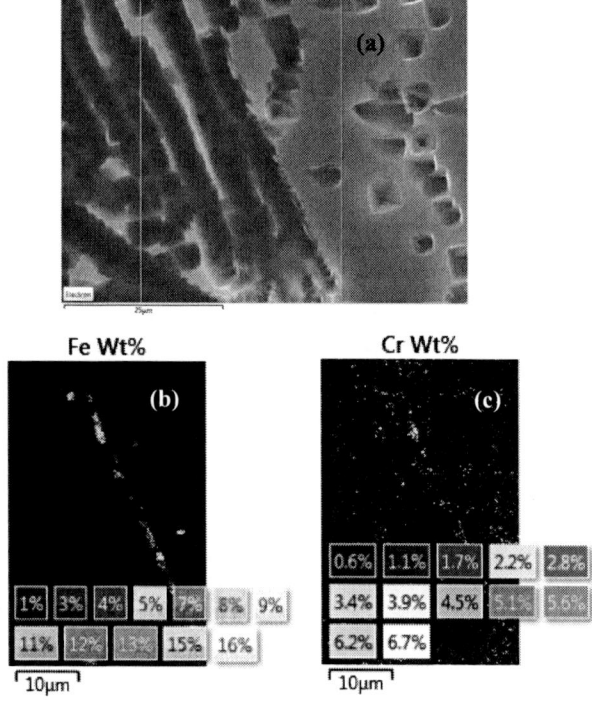

Fig. 3. SEM image (a) and EDS maps of the selected area from ROI, the presence of the Fe (b) and Cr (c) at dislocation cluster show charge recombination correlation due to metal impurities.

Figure 4 shows LBIC map of a polysilicon solar cells near the bottom of the ingot. Two sides of the wafer show a uniform border of relatively higher charge recombination due to contribution of impurities like carbon and oxygen from the ingot lining. The area of interest inspected in SEM using EBIC shows the electrical recombinations spread over the whole ROI c.f Fig 5(a). EBSD map of the same ROI exposes the crystal orientations and grain boundaries in the same area. Although recombinations sites are present over all types of grain boundaries. However, low-angle grain boundaries (LAGBs) shows relatively intense charge trapping sites. Hence, one of such LAGB was examined under EDS and we found the segregation of Cu and W on such site along with extremely small amount of Fe and Cr.

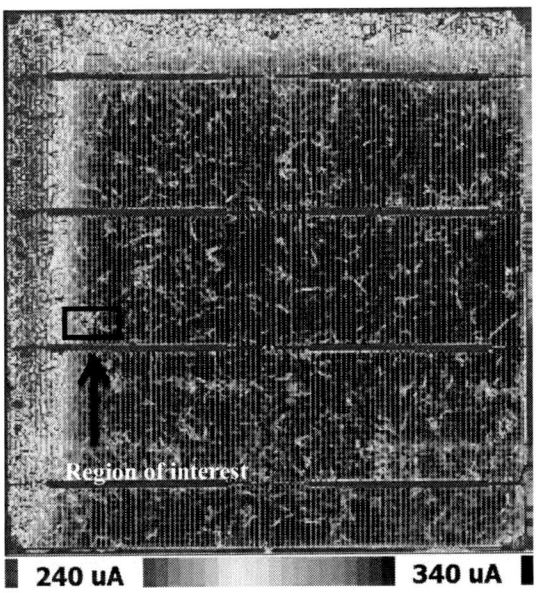

Fig. 4. LBIC map of a Polysilicon solar cell (near bottom of the ingot) measured at 974 nm wavelength with average 303μA current output. The conversion efficiency is 18.03% at STC.

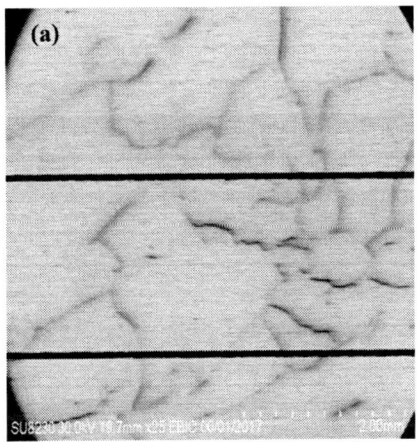

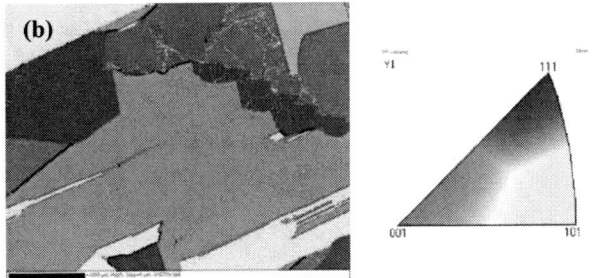

Fig. 5. EBIC (a) and EBSD map (b) of region of interest showing crystal structure with different orientation. The inverse pole figure shows the color scheme with orientations. The red color on grain indicates Σ3 grain boundaries.

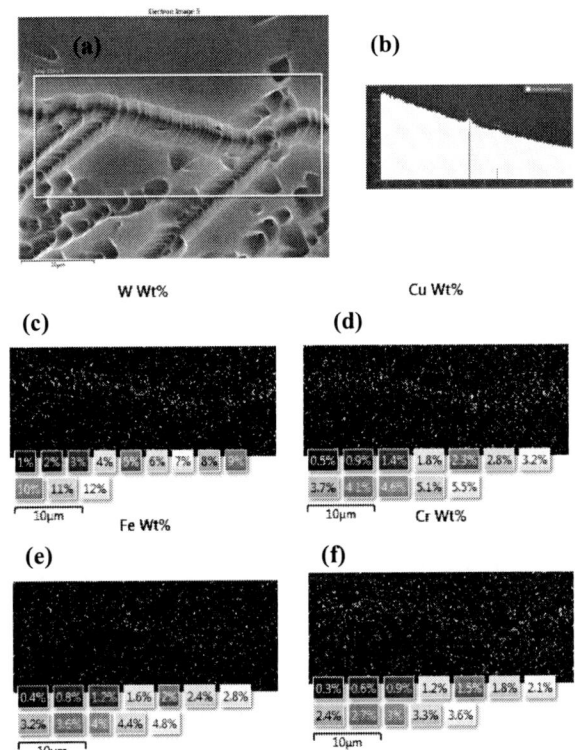

Fig. 6. SEM image (a) and EDS maps of the selected dislocation region from ROI, the presence of the W (b), Cu (c), Fe (e) and Cr (f) at dislocation cluster show charge recombination correlation with metal impurities.

LBIC result from one ESS® solar cell near the top of the ingot is presented in Fig. 7. As the dislocations and impurities tend to pile up towards the top of the ingot hence relatively large number of bad responding areas can be noticed on the surface of the solar cell. Regardless of this fact, the efficiency of the solar cell at STC is above 18%. One ROI is selected for further analyses. EBIC shows number of charge recombination sites on the ROI. However, further investigation with EBSD revealed that majority of such charge trapping sites are mostly LAGB. EDS results reveal the presence of metal impurities at these LAGBs c.f. Fig 9.

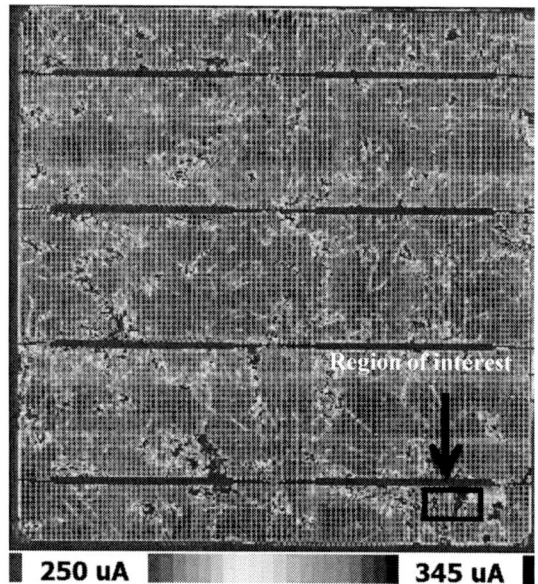

Fig. 7. LBIC map of an ESS® solar cell (near top of the ingot) measured at 974 nm wavelength with average 304μA current output. The conversion efficiency is 18.25% at STC.

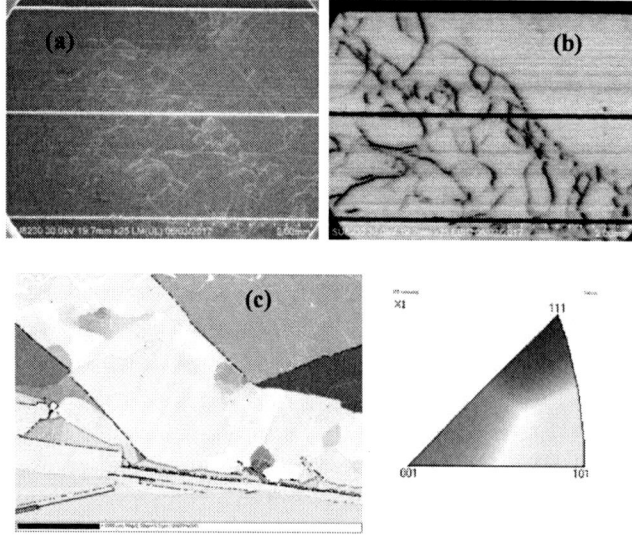

Fig. 8. SEM (a), EBIC (b) and EBSD map (c) of the region of interest showing crystal structure with different orientation. The inverse pole figure shows the color scheme with orientations. The black color on grain indicates Σ3 GB while pink color boundaries are Σ9 GB.

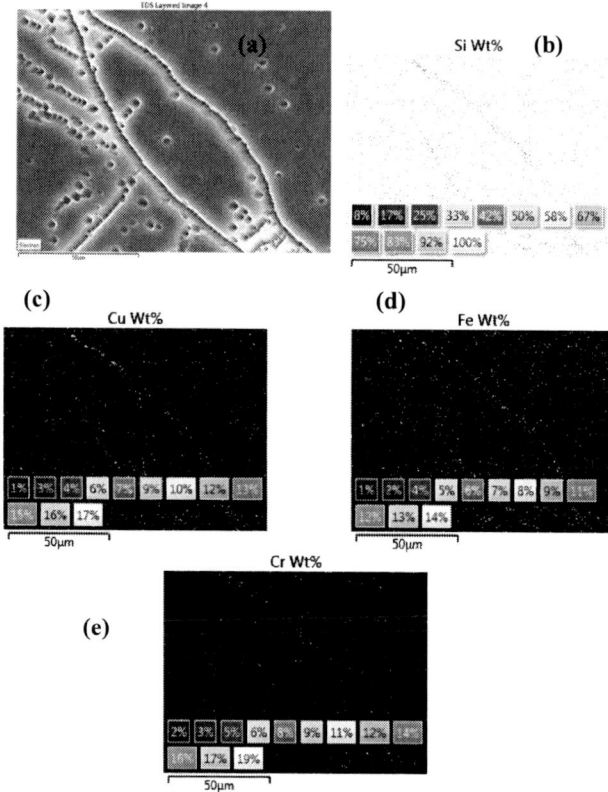

Fig. 9. SEM image (a) and EDS maps of the selected dislocation region from ROI, the presence of the Si (b) Cu (c), Fe (d) and Cr (e) at LAGB show charge recombination correlation with metal impurities.

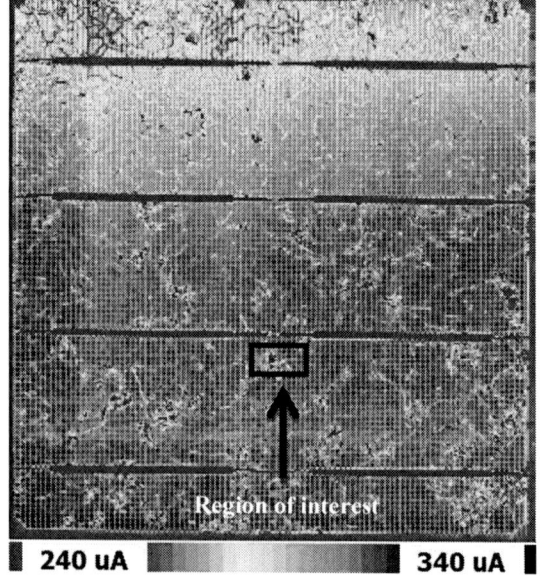

Fig. 10. LBIC map of a polysilicon solar cell (near top of the ingot) measured at 974 nm wavelength with average 295μA current output. The conversion efficiency is 17.88% at STC.

One polysilicon solar cell is selected from similar position as ESS® near the top of the ingot and we studied the LBIC response c.f. Fig. 10. The efficiency of this cell is little lower than ESS®. Similarly, this top cell also shows number of small areas with high recombination sites. One ROI is selected for further investigation of the root cause of low responding area. Here again EBIC and EBSD results reveals similar relationship of LAGBs with charge trapping sites. EDS mapping confirms the presence of the small amount of metallic impurities like W, Cu and Fe on such GBs c.f. Fig 12.

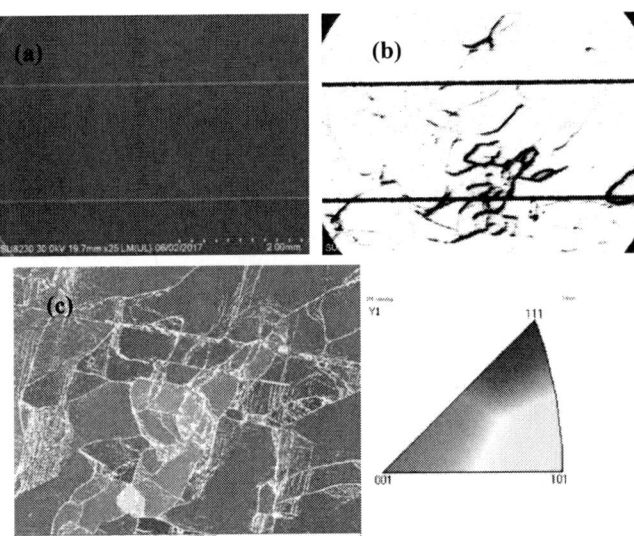

Fig. 11. SEM (a), EBIC (b) and EBSD map (c) of the region of interest showing crystal structure with different orientation. The inverse pole figure shows the color scheme with orientations. The whole surface is made of different size of grains separated by LAGB.

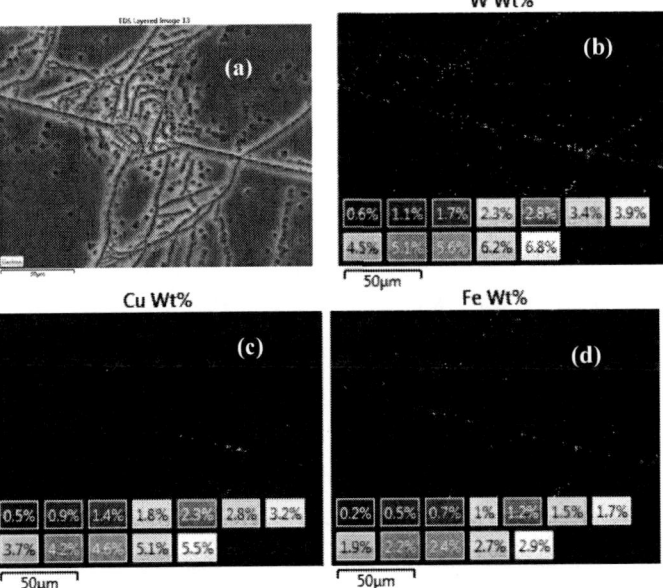

Fig. 12. SEM image (a) and EDS maps of the selected dislocation

region from ROI, the presence of the W (b) and Cu (c) at dislocation cluster show charge recombination correlation with metal impurities.

Impurity induced performance loss is primarily due to a reduction of the base diffusion length in the solar cell [5]. Bottom solar cells are relatively pure, as the impurity tend to solidify towards the top of the ingot. Hence, only few low responding sites have been observed during LBIC mapping for bottom solar cells from both feedstocks. On the other hand, the top cells contain somewhat higher concentration of dislocation clusters that lead to distribution of relatively large number of smaller areas of low efficiency sites. Different sigma boundaries are available on silicon wafer but it is observed that the impurities tend to get trapped mostly at low angle grain boundaries and dislocation clusters. Sigma-3 grain boundary is inactive at room temperature. At few places sigma-9 and sigma-27A grain boundaries had acted as local shunts for induced charges. EDS analysis have some measurement limitation due to less penetration depth (few micrometer) even at 30kV acceleration voltage. Hence, actual concentration of impurity atoms cannot be determined with this method. Instead, it gives an indication of the presence of certain kind of elements at the defect sites.

IV. CONCLUSION

Both EBIC and EBSD results reveal a direct relationship between low responding areas of solar cells and the crystallographic defects in the silicon wafer. EDS analyses confirms the presence of metallic impurities like W, Cu, Cr and Fe on charge trapping sites. Such findings are the same for solar cells coming from two different feedstocks. Hence, cost effective ESS® feedstock exhibit similar performance as polysilicon-feedstock from Siemens process. This study is limited to few cells, which cannot necessarily represent the whole production line of both feedstocks.

REFERENCES

[1] Corsin Battaglia et al., High-efficiency crystalline silicon solar cells: status and perspectives, Energy Environ. Sci., 2016,9, page 1552-1576.

[2] A. D. Kurtz, et al., Effect of Dislocations on the Minority Carrier Lifetime in Semiconductors, Phys. Rev. Vol. 101, Iss. 4, Feb 1956.

[3] Isao Takahashi et al., "Generation mechanism of dislocations during directional solidification of multicrystalline silicon using artificially designed seed", Journal of Crystal Growth vol 312 pages 897–901, 2010.

[4] M.P. Bellmann et al., "Impurity segregation in directional solidified multi-crystalline silicon", Journal of Crystal Growth, Volume 312, Issue 21, 2010, Pages 3091-3095,

[5] J. R. Davis et al., "Impurities in silicon solar cells," in IEEE Transactions on Electron Devices, vol. 27, no. 4, pp. 677-687, April 1980.

Accuracy Evaluation of Absolute Electroluminescence-Efficiency Measurements of Solar Cells Using a Sensitivity-Calibrated-Photodetector Contact Method

Masahiro Yoshita[1]*, Yoshihiro Hishikawa[1], Yoshihiko Kanemitsu[2], and Hidefumi Akiyama[3]

1) National Institute of Advanced Industrial Science and Technology (AIST),
Tsukuba, Ibaraki 305-8568, Japan
2) Institute for Chemical Research (ICR), Kyoto Univ., Uji, Kyoto 611-0011, Japan
3) Institute for Solid-State Physics (ISSP), Univ. of Tokyo, Kashiwa, Chiba 277-8581, Japan

Abstract — **Absolute electroluminescence (EL) measurements have been becoming essential methods not only to evaluate internal properties of individual subcells in multi-junction solar cells but also to characterize performance of series-connected solar-cell arrays. A photodetector-contact method, in which a spectral-sensitivity-calibrated photodetector is proximately placed on a sample surface in face-to-face geometry, is a simple and useful quantitative method for characterizing the absolute EL intensities of the solar cells. In this work, we investigated collection efficiencies of emission for this method by numerical calculations based on a ray-tracing calculation model and evaluated accuracy in the absolute EL measurements of the solar-cell devices using this method.**

Index Terms — **solar cells, electroluminescence, quantitative measurement, electroluminescence quantum efficiency, photodetector contact method.**

I. INTRODUCTION

Precise and accurate performance characterization is essential for development of high-efficiency solar cells and modules. Recently, quantitative luminescence measurements have been becoming powerful methods to investigate not only structural [1,2] but also internal electrical and optical properties of solar cells and modules [3-5]. Quantitative electroluminescence- (EL-) intensity measurements combined with emission-efficiency analyses can thoroughly reveal internal properties of individual subcells in multi-junction (MJ) solar cells [6-11] and can also be applicable to the performance characterization of series-connected solar-cell arrays or solar modules [12].

In a photodetector-contact method, which is one of the useful quantitative methods for characterizing the absolute EL intensity of the solar cells [10,13], a spectral-sensitivity-calibrated photodetector such as a photodiode is placed proximately on a sample surface in face-to-face geometry and EL from the sample under forward bias is quantitatively measured. However, in this method, the collection efficiency of the emission from the sample is sensitive to the proximity of the detector to the cell surface, which significantly affects accuracy in the estimation of the EL efficiency of the devices. To establish the contact method as a useful and more reliable one, proximity-distance dependence of the collection

efficiency of the emission from the sample should be quantitatively evaluated.

In this work, we present numerical calculations of the collection efficiencies of the emission in the photodetector contact method. We developed a calculation model for the collection efficiency of the emission from the light emitter by means of ray-tracing calculations including effects of multiple reflections between surfaces of the sample and the photodetector and calculated the collection efficiencies of the emissions from a point light source and a square one as functions of various setting parameters such as a sample-detector separation distance etc. Furthermore, to confirm the validity of the calculation model developed, we performed collection-efficiency measurements for a light source using

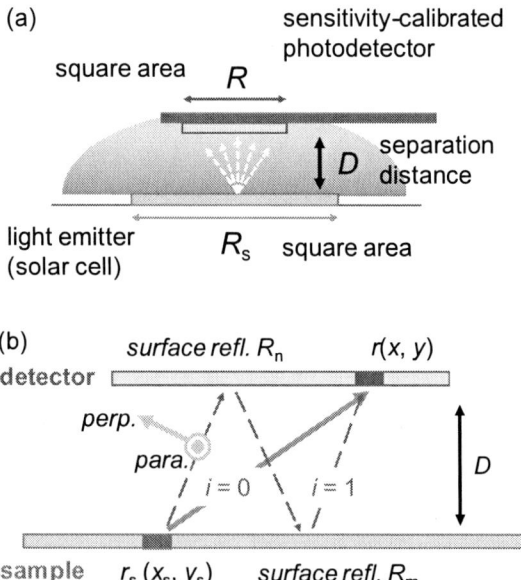

Fig. 1 (a) Schematic of the quantitative emission efficiency measurement using a photodetector-contact method in face-to-face geometry. (b) Simulation model based on a ray-tracing calculation for the photodetector contact method.

$$P_{total} = \iint_{r\ r_s} P_{partial}(r,r_s)\,dx\,dy\,dx_s\,dy_s\,, \qquad (1)$$

where

$$P_{partial}(r,r_s) = \sum_i \frac{P_t}{\pi} M_i(r,r_s)$$

$$\left[\frac{(2i+1)D}{(r-r_s)^2 + \{(2i+1)D\}^2} \right]^2 ,$$
$$(2)$$

and

$$M_i(r,r_s) = \frac{M_i^{para}(r,r_s) + M_i^{perp}(r,r_s)}{2} R_m^2. \qquad (3)$$

photodetector contact method.

II. PHOTODETECTOR CONTACT METHOD AND MODELLING

Figure 1(a) shows schematic configuration of the photodetector-contact method for measuring absolute EL-efficiency of the solar cells and modules. A spectral-sensitivity-calibrated photodetector is closely placed on a solar cell or module surface in face-to-face geometry and EL from the sample operated in forward bias is detected by the photodetector. Emission intensities detected by the photodetector might be dependent on the separation distance between the solar cell surface and the detector and also on the area sizes of the solar cell and the photodetector. In addition, if the surface reflectivities of the sample and the photodetector are not zero, multiple reflections of the emission between the surfaces would cause overestimation of the total emission intensity from the light emitter.

To simulate parameter dependence of the collection efficiency of the emission from the light source, we considered a calculation model that consists of a planar square light emitter with a side of R_s and a surface reflectivity of R_m and a photodetector with a side of R and a surface reflectivity of R_n placed in parallel with a separation distance (D), as shown in Fig. 1(b). Equations (1) – (3) describe total intensities of the emission incident on the detector surface derived by means of a lay-tracing calculation. The integrand $P_{partial}\ (r,r_s)$ in Eq. (1) represents an incident intensity at a position $r(x,\ y)$ of the photodetector for an emission from a position $r_s\ (x_s,\ y_s)$ of the sample. P_t is the total emission intensity of the light source. In this calculation, the light source is assumed to be a Lambertian light emitter, whose emission pattern follows a cosine's law. This assumption is mostly valid for semiconductor crystalline solar cells and planar LEDs [12]. In Eqs. (1) - (3), the effect of the multiple reflections of emission are considered and index i represents number of reflections at the sample surface. The component of $i = 0$ in Eqs. (2) and (3) corresponds to the light component directly incident to the detector without multi-reflections. In the calculation, non-polarized emission from the

light source was assumed. M_i^{para} and M_i^{perp} in Eq. (3) are multiple-reflection factors corresponding to parallel and perpendicular components of the polarization direction of the emission, respectively.

III. CALCULATION RESULTS

A. Collection efficiency of emission from point sources

Figure 2 shows calculation results of the collection efficiencies of the emission from a point light source as a function of the separation distance (D) between the sample and the detector. The surface reflectivity (R_n) of the detector for normal incident light was set to 0.2, which approximately corresponds to that of the photodiode used in the previous absolute EL experiment [13]. The vertical axis in Fig. 2 indicates the relative collection efficiency normalized with the light intensity collected by the detector via a single path under the hypothetical condition that all the emissions from the light source are vertically incident on the detector. The curve at $R_m = 0$ corresponds to the result in case of the single-path detection ignoring multiple-reflection effects.

The collection efficiency is largely dependent on both D and R_m as seen in Fig. 2. As the surface reflectivity R_m of the sample is 1.0, the collection efficiency exceeds 1.0 as D is decreased below 2 mm. This overestimation is caused by the effects of the multiple reflection of the emission between the surfaces of the sample and detector [14]. On the other hand, as the sample surface reflectivity is zero ($R_m = 0$), the relative collection efficiency becomes close to 1.0 at D smaller than 2 mm and less sensitive to the separation distance. However, it should be note that as D becomes zero, the collection efficiency approaches to 0.96 not to 1.0. This deviation from 1.0 at D = 0 comes from the incident angle dependence of the

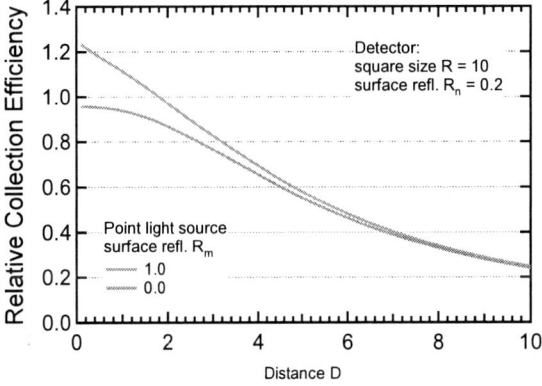

Fig. 2. Collection efficiency of the emission from a point light source as a function of the separation distance (D) between the sample and the detector. The surface reflectivity R_n is set to 0.2 corresponding to that of a photodetector used in the previous experiments[13]. The unit of D and R is millimeter (mm).

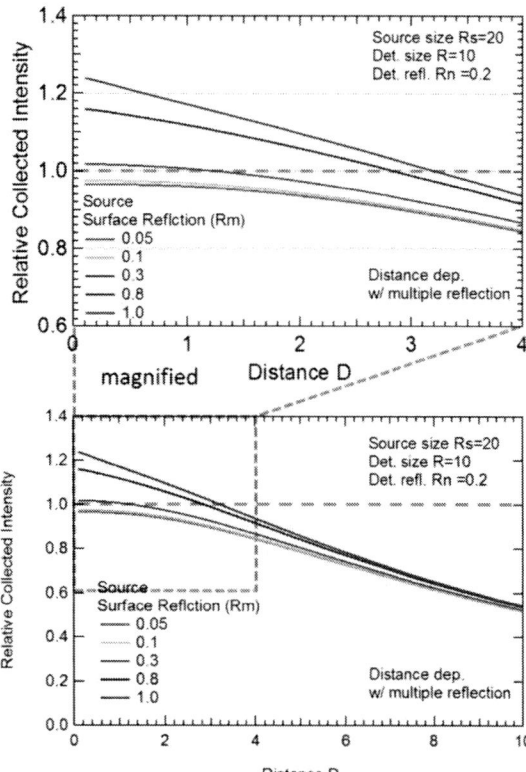

Fig. 3. Collection efficiencies of the emission as a function of the separation distance (D) between the sample and the detector at various surface reflectivities (R_m) of the light source. The surface reflectivity R_n is set to 0.2 corresponding to that of a photodetector used in the previous experiments[13]. Top figure is the magnified one of the bottom figure. The unit of D, R, R_s is millimeter (mm).

Fresnel reflections at the surface of the detector. These results indicate that (1) the effects of multiple reflection should be taken into account in the case of high R_m and that (2) for high-precision quantitative analysis of the collection efficiency with the estimation deviation less 5%, the surface reflection of the detector should also be treated exactly in the calculations.

B. Collection efficiency of emission from square sources

To evaluate collection efficiencies of the emission from a solar cell, we performed numerical simulations for a square light source. Figure 3 shows calculation results of the collection efficiencies for a square light source with the side R_s as a function of D at various R_m. The parameters of the detector are the same as those in Fig. 2. The vertical axis indicates the relative collection efficiency as well as that in Fig. 2.

The results are strongly dependent on both D and R_m as well as those for the point-light source. In the cases of R_m larger than 0.3, the calculation results exceeded unity as D approached to zero as seen in Fig. 3. This behavior indicates

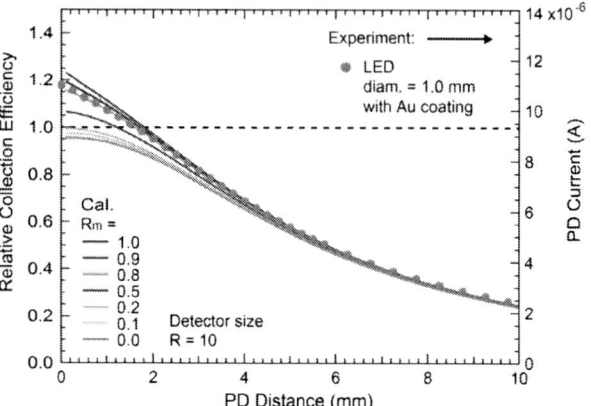

Fig. 4 Separation-distance dependence of the collection efficiency of the emission.

that due to effects of the multiple reflections of the emission, the collection efficiency is overestimated. In the cases of R_m less than 0.3, the collection efficiencies at D smaller than 2 mm are close to 1.0 with deviations less than 0.05 in Fig. 3. However, it should be again noted that the collection efficiency does not approach to 1.0.

The results obtained from the numerical simulations suggest that, in the photodetector-contact method, we should take the effects of the multiple reflection into account in the analysis of the absolute EL measurements for the solar cells. The results in Fig. 3 also suggest that for the solar cells with an anti-reflection coat of $R_m < 0.3$, the collection-efficiency of the emission can be accurately measured with a deviation as small as 5 % if the detector is placed at D closer than 2 mm. To realize high-precision measurements of the collection efficiencies, the calculated collection efficiency can also be used as a calibration factor.

IV. EXPERIMENTAL RESULTS

To confirm the validity of the calculation model, we performed collection-efficiency measurements for the contact method. As a preliminary experiment, we used a planar light-emitting diode (LED) having a circular emitting aperture (diameter of 1mm) surrounded by a gold contact as a point light source.

Figure 4 shows separation-distance (D) dependence of the collection efficiency measured for the test LED. In the experiment, a 10-mm-square photodiode whose spectral sensitivity was separately calibrated was used. For comparison, calculation results of the collection efficiency for the light emitter with various R_m from 0.0 to 1.0 are also shown in Fig. 4. The obtained experimental result agreed well with the calculation curve for a point emitter with a surface reflectivity of around 0.8 - 0.9. This quantitative agreement indicates the validity of the calculation model to evaluate the collection efficiency of the planar light source in the contact method.

Moreover, the calculation results can be used as calibration factors to estimate total luminescence intensities from solar-cell devices in the absolute EL measurements.

IV. SUMMARY

We developed the calculation model based on the ray tracing to simulate collection efficiencies of the light emission from the solar cells in the absolute EL measurements by means of the photodetector contact method. The calculation results obtained indicate that in the photodetector-contact method, we should take the effects of the multiple reflection into account in the analysis of the absolute EL measurements. We also performed collection-efficiency measurements of the emission from a LED by using the contact method to confirm the validity of the developed calculation model. The experimental results for the LED with a circular aperture as a point light source agreed well with the calculation result.

ACKNOWLEDGEMENT

This work was partly supported by KAKENHI No.26390075 from JSPS in Japan. The work in part at University of Tokyo and Kyoto University was also supported by KAKENHI No.15H03968 from JSPS, the Photon Frontier Network Program of MEXT, JST-SENTAN, JST-CREST, and NEDO in Japan.

REFERENCES

[1] T. Fuyuki, H. Kondo, T. Yamazaki, Y. Takahashi, and Y. Uraoka, Applied Physics Letters, 86, 262108, 2005.

[2] C. G. Zimmermann, Journal of Applied Physics, 100, 023714, 2006.

[3] U. Rau, *Physical Review B* 76, 085303, 2007.

[4] T. Kirchartz, U. Rau, M. hermle, A. W. Bett, A. Helbig, and J. H. Werner, *Applied Physics Letters*, 92, 123502, 2008.

[5] S. Roensch, R. Hoheisel, F. Dimroth, and A. W. Bett, *Applied Physics Letters*, 98, 251113, 2011.

[6] J. F. Geisz, M. A. Steiner, I. Garcia, S. R. Kurtz, and D. J. Friedman, *Applied Physics Letters*, 103, 041118, 2013.

[7] J. F. Geisz, M. A. Steiner, I. Garcia, M. France, W. E. McMahon, C. R. Osterwald, and D. J. Friedman, *IEEE J. Photovolt*, vol. 5, pp. 1827-1839, 2015.

[8] A. Delamarre, M. Paire, J.-F. Guillemoles, and L. Lombez, *Prog. Photovolt: Res. Appl.*, 23, 1305-1312, 2014.

[9] S.-Q. Chen, L. Zhu, M. Yoshita, T. Mochizuki, C. Kim, H. Akiyama, M. Imaizumi, Y. Kanemitsu, *The 40th IEEE Photovoltaic Specialist Conference*, 2014, p.1780 − 1783.

[10] S.-Q. Chen, L. Zhu, M. Yoshita, T. Mochizuki, C. Kim, H. Akiyama, M. Imaizumi, and Y. Kanemitsu, *Scientific Reports*, 5, 7836, 2015.

[11] L. Zhu, M. Yoshita, S.-Q. Chen, T. Nakamura, T Mochizuki, C. Kim, M. Imaizumi, Y. Kanemitsu, and H. Akiyama, *J. Photovoltaics.*, 6, 777-782, 2016.

[12] T. Mochizuki, C. Kim, M. Yoshita, J. Mitchell, Z. Lin, S.-Q. Chen, H. Takato, Y. Kanemitsu, and H. Akiyama, *Journal of Applied Physics*, 119, 034501, 2016.

[13] M. Yoshita, L. Zhu, C. Kim, H. Kubota, T. Nakamura, M. Imaizumi, Y. Kanemitsu, and H. Akiyama, The *43rd IEEE Photovoltaic Specialist Conference*, 2016, p.3570 − 3573.

[14] M. Yoshita *et al.* (in preparation).

Nanometer-Scale Carrier Imaging of Potential-Induced Degradation in c-Si Solar Cells

C.-S. Jiang,[1] C. Xiao,[1] H.R. Moutinho,[1] S. Johnston,[1] M.M. Al-Jassim,[1] X. Yang,[2] and Y. Chen,[2] and J. Ye[3]

[1] National Renewable Energy Laboratory (NREL), Golden, CO 80401, USA
[2] Trina Solar Inc., Changzhou, P.R. China

[3] Ningbo Institute of Industrial Technology, Chinese Academy of Science, Ningbo, P.R. China

Abstract — We report on nm-resolution imaging of charge-carrier distribution around local potential-induced degradation (PID) defects using scanning capacitance microscopy. We imaged cross sections of heavily field-degraded module areas as cored out and selected by mm-scale photoluminescence imaging. Localized areas with abnormal carrier behavior or junction damage were found: the apparent n-type carrier extends vertically into the absorber to ~1–2 μm from the cell surface, and laterally in similar lengths; in defect-free areas, the n-type carrier extends ~0.5 μm, which is consistent with the junction depth. For comparison, we also investigated areas of the same module exhibiting less PID stress, and did not find any such heavily damaged junction area. Instead, we found slightly abnormal carrier behavior, where the carrier-type inversion in the absorber did not occur, but the p-type carrier concentration changed slightly in a much smaller lateral length of ~300 nm. These nanoelectrical findings suggest that the existing extended defects, which may not be significantly harmful to cell performance, were changed by PID to heavily damaged junction areas.

Index Terms — c-Si solar cell, potential-induced degradation, scanning capacitance microscopy (SCM), nanoelectrical property.

I. INTRODUCTION

Potential-induced degradation (PID) is an important degradation component in all the major solar panels, and it increases in severity with increasing system voltage of a solar array. To date, a widely reported mechanism of PID in c-Si solar cells is local shunts caused by Na drift to extended defects—specifically, stacking faults [1]. A theoretical model proposes an interstitial Na plane at the stacking faults, which results in half-filled Si anti-bonding states deep in the Si bandgap [2]. Electrical conduction through the gap states can cause local shunting if the stacking fault cuts through the p-n junction. This atomic structure/chemistry of the defects affects the macroscopic current-voltage (I-V) because of the current flowing through the defects, but also, flowing through the surrounding materials by changing the surrounding electrical behaviors. However, still lacking is how the atomic defect affects the surrounding materials in nm–μm scales.

In this paper, we report on nm-resolution imaging of charge-carrier distribution around local PID defects using scanning capacitance microscopy (SCM)—which is, to our knowledge, the first nanoelectrical characterization of

reliability-related defects. We found abnormal carrier areas with type inversion, likely changed from existing extended defects (which are not significantly harmful for PV) by the PID stressing.

II. EXPERIMENTAL

Areas in a field-failed multi-Si module heavily degraded by PID were selected by photoluminescence (PL) imaging and cored out [3] for SCM investigation [4–6]. The cored Si cell was cleaved through a dark PL area (Fig. 1(a)), and the cross section was polished using diamond pads with particle sizes down to 100 nm and finally using 50-nm colloidal silica. The sample was then annealed at 300°C for 30 min with ultraviolet illumination to form a high-quality SiO_2 thin layer (~10 nm) on the cross-sectional surface for SCM measurement [4]. SCM images the carrier polarity and concentration in sub-10-nm resolutions by measuring the local dC/dV signal, where C and V are the metal/insulator/semiconductor (MIS) capacitance formed by the probe/oxide/sample, and the bias voltage between the probe and sample, respectively [4]. SCM is based on an atomic force microscope (AFM) (Veeco D3100 and Nanoscope IIIa). A whole metal probe (Rocky Mountain Nanotechnology, 25Pt300B) was used to collect the capacitance signal by enhancing microwave conduction. An AC voltage (Vac) with an amplitude of 1 V and frequency of 90 kHz was applied to the sample to obtain the dC/dV SCM signal by a lock-in-amplifier. No DC voltage (Vdc=0) was applied to the sample.

III. RESULTS AND DISCUSSIONS

Figures 1(b) and 1(c) show a SCM and the corresponding contact-mode AFM images. The images are constructed of three adjacent images in 5×5 μm² scan size. We took images one by one along the cross section for more than a 1-mm length (200 images) to statistically find areas with damaged junctions. In good junction areas, the n-type emitter in SCM images appears as negative dC/dV (dark) and the p-type absorber as positive (bright), as shown in Figs. 1(b) and 1(d). Around the electrical junction where electrons and holes have the same concentration, the dC/dV has a small flat region with dC/dV ~0 if the oxide on the cross section is high quality or does not trap charges in the oxide and at the oxide/Si interface [4,5]. The junction depth or emitter width is ~500 nm. This normal junction was imaged in most areas along the junction. However, abnormal junction areas such as the one shown in Figs. 1(b) and 1(e) were also found, where the apparent n/p carrier interface is at a depth of ~2 μm from the cell surface. This junction-damaged area likely has two defects paired, and they are separated by ~1 μm. The lateral width of the affected area is ~2 μm around the defect pair. Inside the affected area, the apparent n-type carrier concentration or dC/dV is highly nonuniform. Although the n/p carrier interface exists, we believe that such a junction change or damage from a normal junction should largely change the junction electronic behavior, and further affect the macroscopic cell I-V output. If the n/p interface has significant trap states, the junction cannot be well formed and local I-V through the damaged junction should not behave as a normal diode. The apparent carrier type-inversion around a defect and across a junction is caused

by the total effects of two electric fields of corresponding defect-induced charge-trapping and built-in field of dopant ions. In addition, this apparent appearance of inversion in the SCM image is affected by the two electric fields and the third field applied to the MIS structure between the probe and sample (or an equivalent field from trapped charges in the oxide or at the Si/oxide interface). All the electric fields complicate analysis of carrier distribution [4]. Because of low carrier density in the depletion region, the carrier inversion may not need an electric field around the defect as strong as in a charge-neutral region. Or, in other words, the appearance of carrier inversion, its apparent boundary, and dC/dV amplitudes should be sensitive to the electric fields across the defect area.

Figures 2(a) and 2(b) show an example of the junction damage with an individual defect and with a similar size of ~2 μm deep and ~1 μm wide. Figures 2(c) and 2(d) show another paired defect, which appears smaller than the pair in Fig. 1(b), but with similar separation distance. We randomly cut through

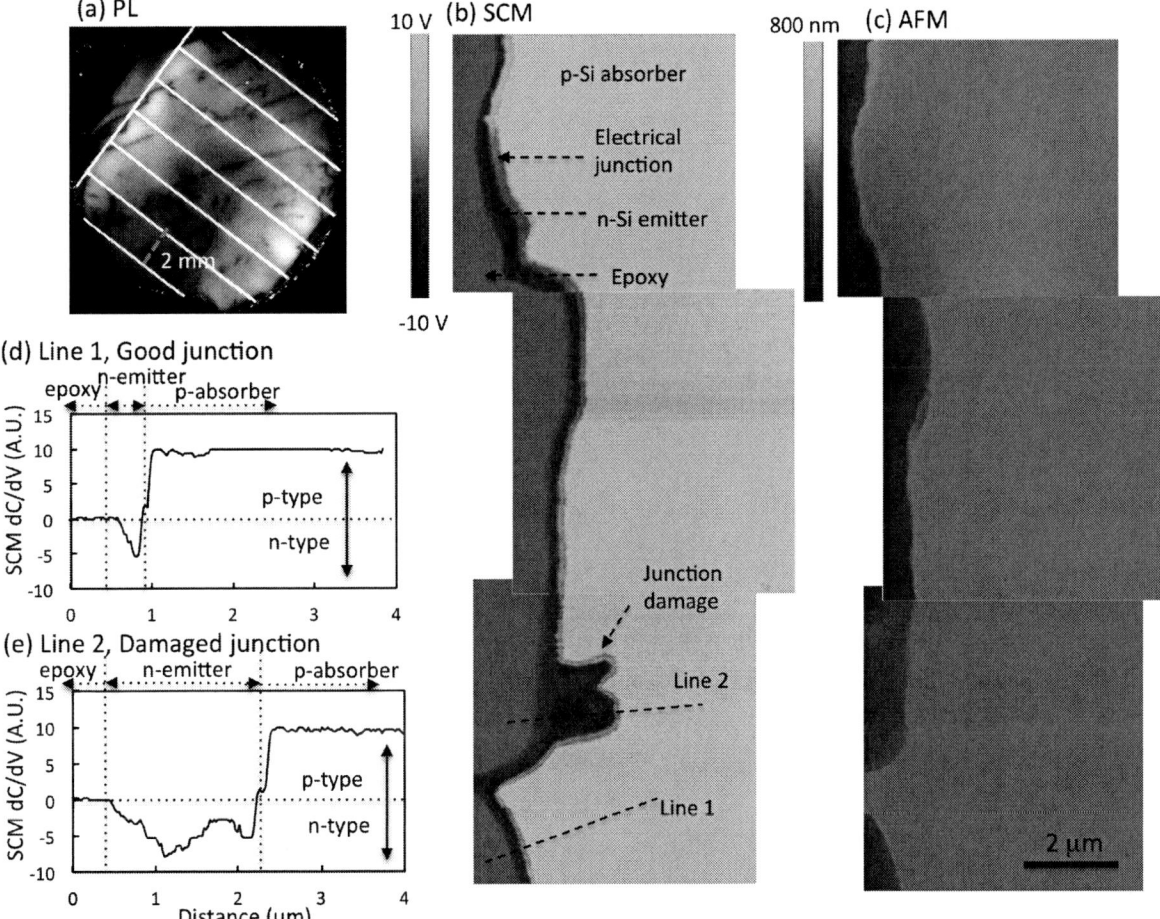

Fig. 1(a) A PL image taken on a highly degraded area cored out from a field-failed module. The red dashed line indicates the location where SCM imaging was performed. (b) and (c) show SCM and the corresponding AFM images taken on the cross sections along the junction. (d) and (e) are SCM line profiles along the dashed lines in (b).

the dark PL area when making the cross-sectional sample, and mechanically polished away many μm. Defects should appear randomly on the cross-sectional surface. If a plane defect intersects on the cross section at its edge, the defect size on the cross section can be smaller than the case of cutting through its center. However, the separation of two defects should not depend on the location that the defect planes cut through. These junction damages have a concentration of $\sim 10^5/\text{cm}^2$, as estimated by $C=N/2\mu m \times L$, where N is the number of defects found, and L is the length along the junction imaged by SCM. We note that this estimate is very rough because probing a defect needs to make a large number of scans (hundreds) so that we can only find a few defects, making the concentration estimation inaccurate.

To compare with the PID effect, we have also performed SCM imaging on less PID-stressed areas as identified and cored out by the bright PL on the same module. Interestingly, we also found abnormal carrier or defect areas, as shown in Fig. 3. However, the SCM images around the defects are much different from the defects in heavily PID-affected areas. Although the depth of the defect is similar to Figs. 1 and 2, the lateral affected size is much smaller (~300 nm). Unlike the PID defects, the carrier in the absorber does not show the apparent type-inversion, but shows only a change in carrier

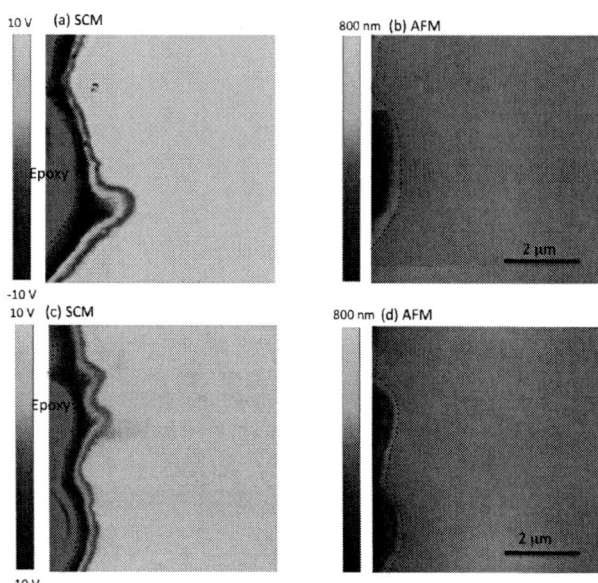

Fig. 2(a) and (b) A SCM and the corresponding AFM images showing an individual defect; (c) and (d) The images showing a paired defect.

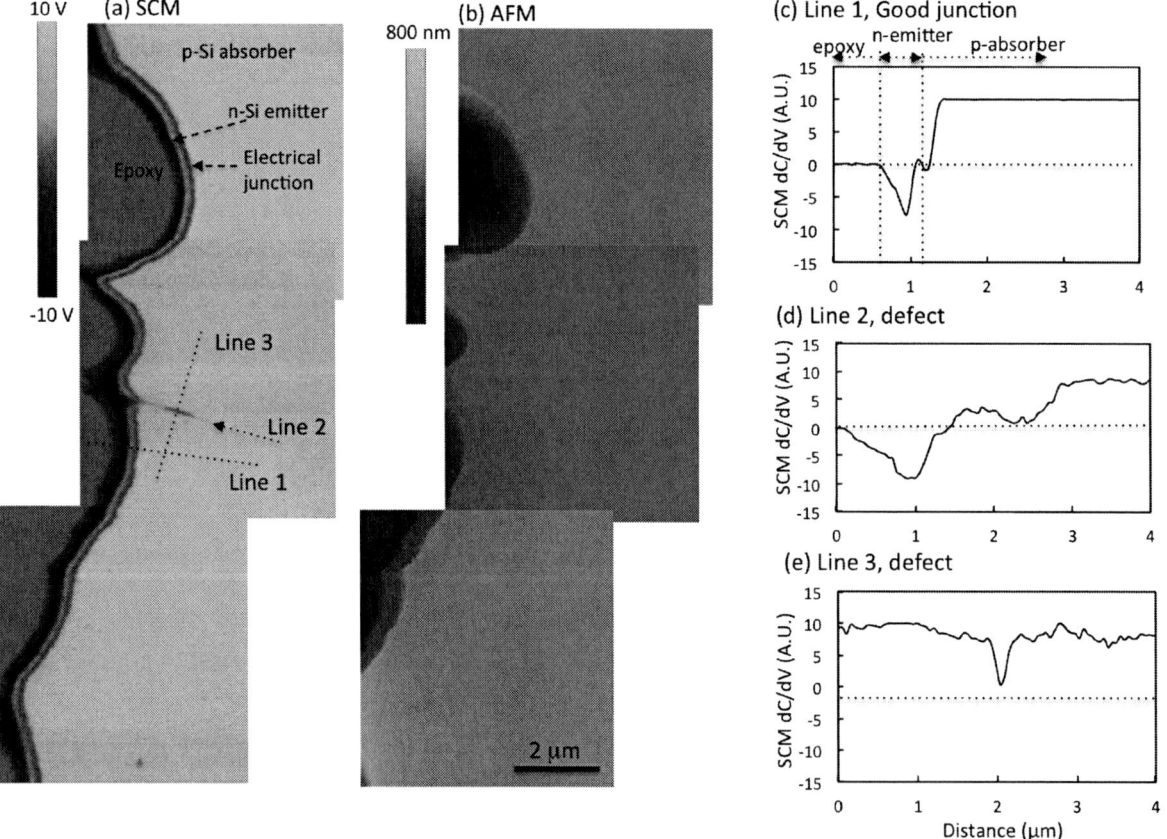

Fig. 3. (a) and (b) SCM and the corresponding images taken on the cross section of a piece cored out from a less PID-stressed area. (c), (d), and (e) show SCM line profiles along the dashed lines in (a).

concentration around the defect. Because the defect states do not affect the surrounding carrier distribution as much as the PID defects, this defect may not have gap states as deep and/or as dense as the PID defects. Therefore, we believe that this defect is not as harmful to the device performance as the heavily PID-stressed ones.

Extended defects such as dislocations, stacking faults, and grain boundaries are formed in casting of multicrystalline-Si ingots. The density depends on individual lots; a $\sim 10^5/cm^2$ density is in the reasonable range. With this density of extended defects, solar modules exhibit their usual efficiency, indicating that the modules should be tolerable with the defects. However, our SCM results suggest a significant change in a module's electronic behavior by PID. These defects may become harmful to the device performance by the larger size of the electrical effect. Na decoration of the defects can be a plausible reason for the defect change by PID. Because SCM is a 2-dimensional imaging technique, it is usually easy to image a defect with a 1-dimensional curve or line, but not with a 0-dimensional point, unless the point size is adequately large. A line appearance on the cross-sectional surface should be a plane defect and a point is a line defect in the bulk. Therefore, line defects in the bulk such as dislocations may be overlooked by the SCM imaging.

IV. CONCLUSION

We have imaged carrier distribution around PID defects in a c-Si cell module and found defect areas 2 μm deep and 1–2 μm wide with apparent nonuniform carrier inversion. By comparing with less-stressed pieces of the module, the damaged junction is discussed in terms of changes by the PID—from existing extended defects that may not be significantly detrimental to the device performance before the PID. These nanoelectrical findings are expected to provide knowledge to bridge the gaps between atomic-size electronic structure and cell-level macroscopic I-V output.

ACKNOWLEDGMENT

This work was supported by the U.S. Department of Energy under Contract No. DE-AC36-08GO28308 with the National Renewable Energy Laboratory.

The U.S. Government retains and the publisher, by accepting the article for publication, acknowledges that the U.S. Government retains a nonexclusive, paid-up, irrevocable, worldwide license to publish or reproduce the published form of this work, or allow others to do so, for U.S. Government purposes.

REFERENCES

[1] V. Naumann, D. Lausch, A. Hahnel, J. Bauer, O. Breitenstein, A. Graff, M. Werner, S. Swatek, S. Groper, J. Bagdahn, and C. Hagendorf, "Explanation of potential-induced degradation of the shunting type by Na decoration of stacking faults in Si solar cells," *Solar Energy Materials & Solar Cells* 120, 383–389, 2014.

[2] B. Ziebarth, M. Mrovec, C. Elsasser, and P. Gumbsch, "Potential-induced degradation in solar cells: Electronic structure and diffusion mechanism of sodium in stacking faults of silicon," *J. Appl. Phys.* 116, 093510, 2014.

[3] H.R. Moutinho et al., to be published

[4] C.C. Williams, "Two-dimensional dopant profiling by scanning capacitance microscopy," *Annu. Rev. Mater. Sci.* 471–504, 1999.

[5] V.V. Zavyalov, J.S. McMurray, and C.C. Williams, "Scanning capacitance microscope methodology for quantitative analysis of p-n junction," *J. Appl. Phys.* 85, 7774–7783, 1999.

[6] C.-S. Jiang, J.T. Heath, H.R. Moutinho, and M.M. Al-Jassim, "Scanning capacitance spectroscopy on n^+-p asymmetrical junction in multicrystalline Si solar cells," *J. Appl. Phys.* 110, 014514 (2011).

NREL Efforts to Address Soiling on PV Modules

Lin J. Simpson,[1] Matthew Muller,[1] Michael Deceglie,[1] Helio Moutinho,[1] Craig Perkins,[1] C. S. Jiang,[1] David C. Miller,[1] Leonardo Micheli,[1,2] Govindasamy Tamizhmani,[3] Sai Ravi Vasista Tatapudi,[3] and Mowafak Al-Jassim[1]

(1) National Renewable Energy Laboratory, Golden, CO 80401, USA

(2) Colorado School of Mines, Golden, CO 80401

(3) Arizona State University, Tempe, AR 85281

Abstract — **Natural soiling has reduced the energy output of PV systems since the technology was first used. Projecting even a small ~4% average annual soiling loss (found in some places in the U.S.), translates to ~10 GW of power loss worldwide, which correlates today to ~$2 billion in lost revenue annually, worldwide. Production losses due to soiling may be even higher in high soiling environments, substantially increasing the levelized cost of electricity (LCOE). Furthermore, while soiling has been discussed in the literature for more than 70 years, solutions to many problems are still needed. NREL is working with the PV industry to develop the tools/knowledge so that the effects of soiling can be predicted for different environmental conditions and cost effective mitigation can be implemented. For this paper, we will describe our efforts to (1) predict PV module soiling rates based on environmental factors at a PV installation and from its energy production data, (2) quantitatively measure the adhesion forces to understand the physics enabling soiling, and (3) develop related standards on PV module coatings and artificial soiling. For example, NREL has used soiling station data to identify the most important environmental factors that are correlated with average soiling losses, and is using this information to develop models that accurately predict soiling losses at prospective sites without the need for local soiling stations. Furthermore, by providing a detailed understanding of soiling mechanisms and the corresponding requirements for PV module coatings, we are providing valuable information about what improvements in performance may be possible at a given PV site. Ultimately, this effort will help reduce the uncertainty in PV plant power output and maintenance requirements, and thus reduce costs.**

Index Terms — **photovoltaic cells, electronic packaging.**

I. Introduction/Overview

NREL's recent comprehensive review of solar energy soiling[1] identifies that soiling has been discussed in the literature for more than 70 years, and yet "the fundamental properties of dust and its effect on energy transfer are still not fully understood, nor is there a clear solution to the problem." NREL is working with the PV industry to address several of the most important problems that include predicting soiling losses, identifying and quantifying soiling adhesion mechanisms, and developing standards on PV coatings and artificial soiling (Fig. 1). These efforts focus on developing the appropriate tools and knowledge to devise cost effective mitigation that ultimately reduces the levelized cost of electricity (LCOE) by increasing power output, reducing cleaning costs, and reducing finance costs. Recent evaluations of large PV installations found that soiling rates within the U.S. range from 0% to 0.3% per day, with annual average losses of ~4% for many southwestern U.S. sites.[2] Investigations also found that a substantial amount of rain is often required to remove the contamination, and that without active washing, the PV energy output can be reduced 20% to 40% between individual weather events.[2] Furthermore, the most insolation rich regions in the world can have higher soiling rates. For example, some solar thermal plants need to be cleaned twice weekly, while some Middle East located PV installations are cleaned daily. Even a 4% loss in energy collection results in an ~10 GW reduction (from the >270 GW total) in peak power worldwide (~1 GW loss in the U.S.) which correlates to ~$2 billion of lost revenue annually, worldwide. Production loss and thus LCOE will be even more in higher soiling environments.

II. Method

To have maximum impact on the soiling problem, NREL has fully engaged U.S. PV and coating manufacturers to identify the most important problems that can be quickly addressed. The overall goal of our efforts is to reduce LCOE by developing models that more accurately predict annualized soiling losses at PV installations, and by providing the PV industry with the tools and knowledge needed for successful long-term use of modules with value-added coatings (e.g., antireflection and/or anti-soiling coatings). Such coatings target increased energy yield, leading to lower LCOE. However, this can only be realized if the coating is appropriate for specific environment at the PV site and has the necessary long-term durability. Accurate soiling rate predictions reduce performance uncertainty, leading to lower finance rates and thus lower LCOE.

III. Results

Soiling Rate Predictions: Past attempts to model soiling rates (primarily trying to use first principals) have been largely unsuccessful, and thus the soiling rate, presently, must be directly measured using specialized soiling station equipment at every installation (Fig. 2). However, adapting the soiling station results to understand PV power loss is not always direct; ultimately better and less expensive methods are needed to lower maintenance costs. NREL has developed automated routines that extract the soiling losses directly from PV production data provided by several of our industrial partners (Fig. 1).[3] Initial evaluation of this approach relative to soiling losses measured from soiling stations at the same sites indicates that the methodology has potential. We are therefore expanding our efforts from the initial 10 sites, to look at production data from hundreds of sites.

preliminary, due to the limited number of sites examined, the results suggest soiling risk analysis may be possible using existing environmental data sets. For example, Environmental Protection Agency (EPA) particulate data is available for much of the U.S., (Fig. 1), and PRISM rainfall data is available for the continental U.S. Although the results are not statistical, individual examination of the 20 sites showed that the worst-case locations were all in desert climates with irrigated agriculture. For example, an unirrigated Mojave Desert location had an average soiling loss less than 2 % while a site within an irrigated agriculture region in the San Joaquin Valley had losses near 6%. NREL is expanding the pool of

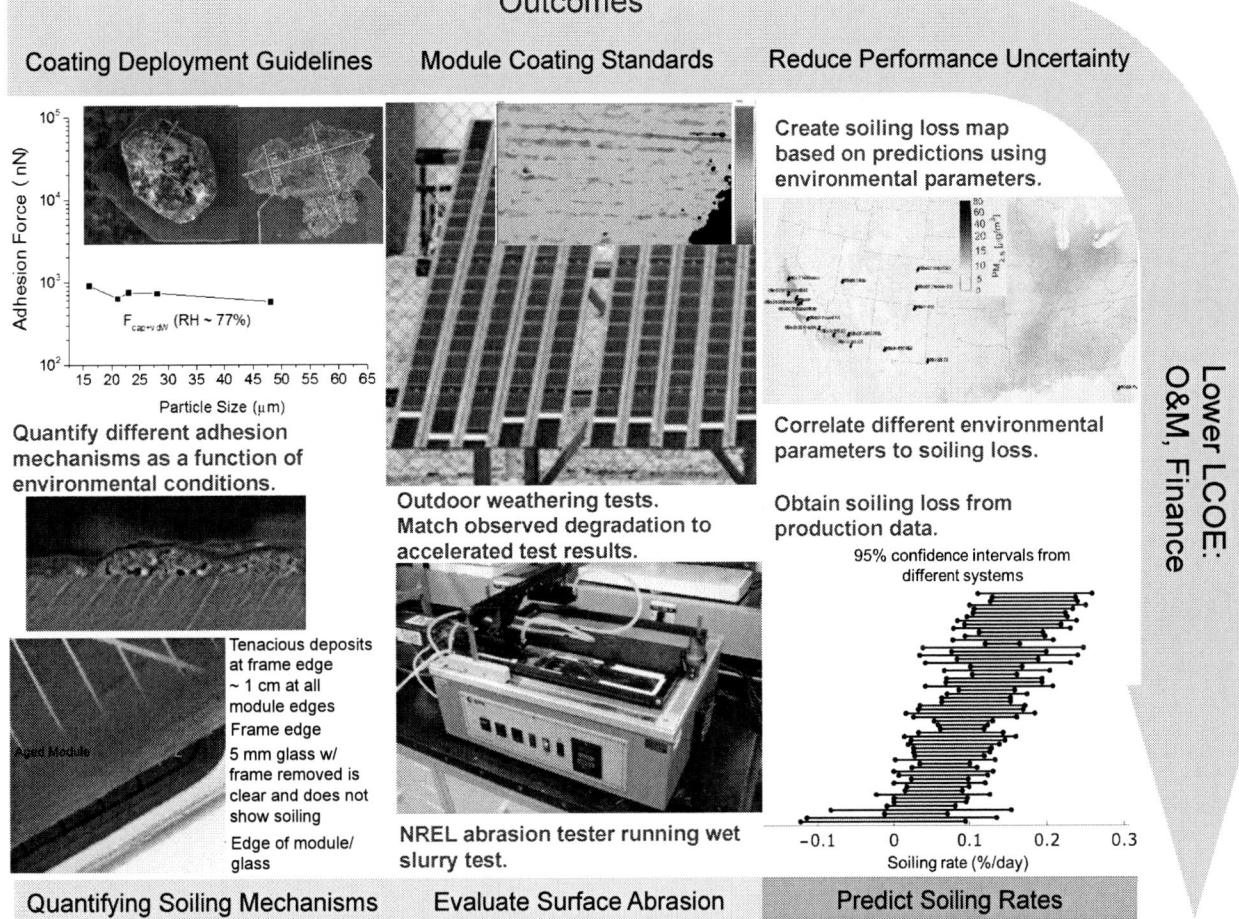

Fig. 1. Illustration showing how the complementary focus areas at NREL work together to directly address the DOE goal of reducing LCOE. These focus areas are considered high priority by the PV industry.

Because soiling station results are often not available at all PV installations, NREL has also started development efforts to predict annual soiling losses using just environmental parameters. For example, our initial efforts using a two-variable regression found an adjusted R^2 of 0.90 for a combination of PM2.5 (particulate matter smaller than 2.5 microns) and a binary classification for the average length of dry days between rain events.[4] Although this work is

data used to further validate the initial findings and to develop a more sophisticated model for predicting variation in PV soiling losses strictly from the site characteristics. Such a model can then be applied to prospect low cost PV installations. Engineers can use this specific site enabled model to better design systems to minimize soiling, and to improve operation and maintenance based on expected soiling losses and system economics. Financiers are expected to require lower interest rates as soiling derates become driven by precise data rather than speculative overestimates.

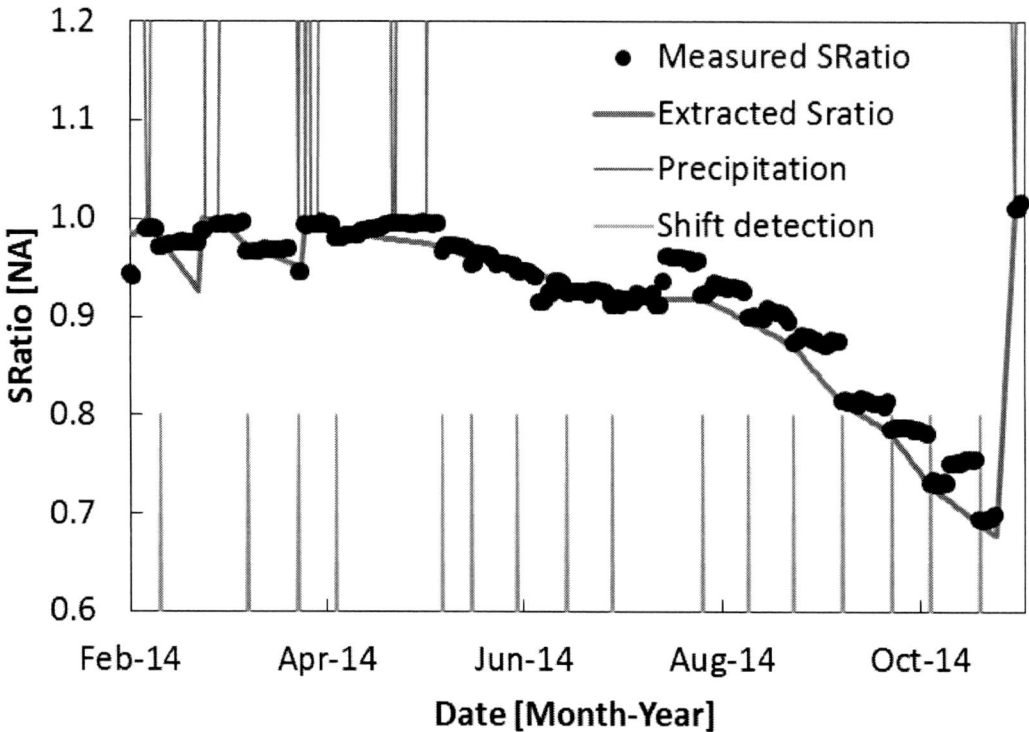

Fig. 3. An example of soiling station data (black dots) from one site in the southwest U.S. and how a SRatio correction algorithm is applied; red lines connect each set of data points that the algorithm accepts. The SRatio correction accounts for days with precipitation and downward shift detection that are believed to be cell cleanings about every two weeks. The results indicate that daily reference cell cleaning will reduce the uncertainty the soiling station data.

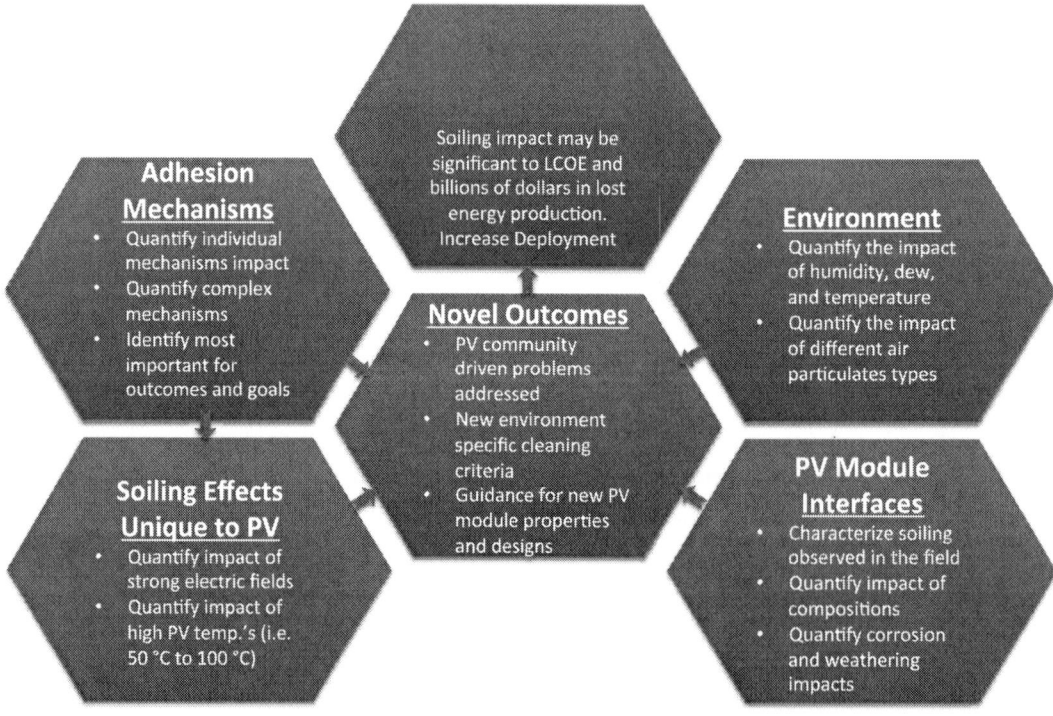

Fig. 2. Different interactions that are being quantified at NREL to better understand soiling. Several key results will impact mitigation strategies.

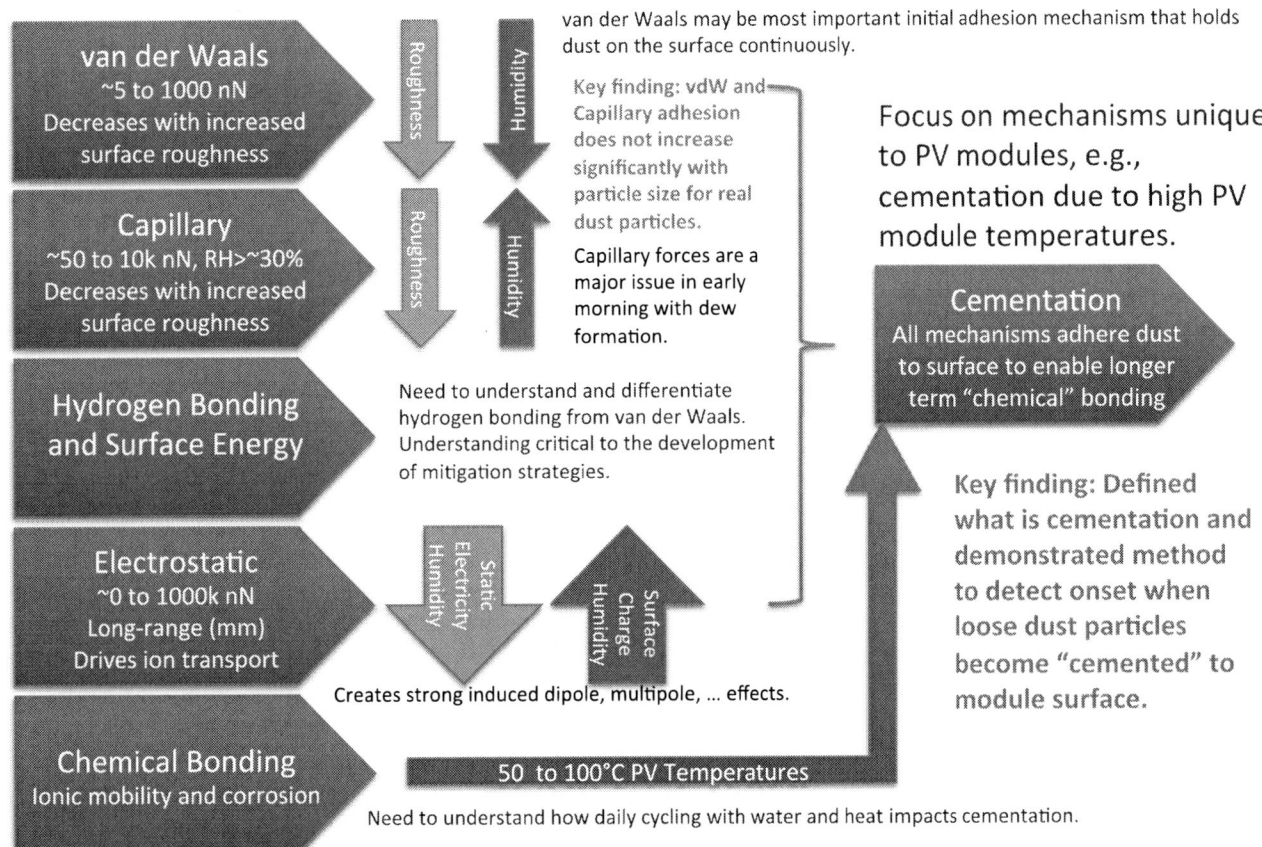

Fig. 4. Different interactions that are being quantified at NREL to better understand soiling. Several key results will impact mitigation strategies.

Soiling Mechanisms: One of the main factors missing in the soiling literature is a detailed evaluation of the specific adhesion mechanisms and their corresponding significance to soiling. As shown in Fig. 3, NREL is evaluating a wide variety of factors that govern the adhesion of dust particles to the PV surfaces as well as the processes involved to keep the particles there and to perhaps transform their interaction. The factors are being used to help evaluate different novel outcomes including developing guidance for cleaning and new PV interface designs. At a more basic level, as shown in Fig. 4, the focus of our efforts is to identify and quantify the specific adhesion mechanisms associated with the operating conditions specific to PV modules (e.g., cyclic dew formation and much higher daytime temperatures than attained for windows or concentrated solar mirrors). As shown in Fig. 4, one key finding to date is that standard adhesion mechanisms like van der Waals (vdW) and capillary forces do not have a significant dependence on real dust particle size, but rather on contact area (Fig. 1). Another key finding is the need to develop a consensus definition for cementation and then using that definition to measure the onset of cementation of real dust particles. Our results indicate that while vdW and capillary

forces account for much of the initial accumulation of particulate matter, the subsequent adhesion forces are much smaller than expected for large particles. Furthermore, it may be possible to substantially reduce the capillary adhesion force with the appropriate surface properties. Other less understood mechanisms like cementation may facilitate stronger adhesion that requires enhanced cleaning methods.

Recent work using a quartz crystal microbalance, demonstrates that cementation can be defined as particulates transforming to "solid" materials that transmit shear waves. In this way, NREL has demonstrated the ability to apply humid and dry air at ambient temperature to induce and determine the onset of cementation of loose particles. These initial results indicate that some materials like halite, kaolinite, and palygorskite quickly forms cement with humidity. However, other materials like silica particles do not at ambient temperatures. In addition to decreasing light transmission across the entire array, cementation may cause non-uniform soiling that may induce additional PV module degradation. For example, sufficient shading of a single cell may induce a hot spot where the cell disproportionally heats up higher than other cells in the string. This causes the cell to become more resistive, and thus the cell heats even more as electricity from other cells in the string effectively become bottle necked at that cell. Depending on the string configuration and the protective diodes in place, this single "hot cell" can cause the entire string to become bypassed, and/or the higher temperature may induce faster degradation of cell components or module protective materials.

Industry Standards: In addition to understanding the appropriate interface properties needed to decrease dust adhesion and soiling rates, any PV module surface treatment or coating must be durable enough to provide sustained performance improvement in the field, where weathering can occur from corrosion, impacting matter, and cleaning. NREL is working with the international PV community to determine the amount of abrasion damage that occurs in the field and correlating these results to accelerated laboratory tests (see Fig. 1) that can be used to qualify a coating for use in PV. Initial work in this area has demonstrated that the typical brushes used for PV cleaning in the field do not themselves induce damage to PV glass, but small amounts of dust contaminants in the slurry causes some haze after an appropriate amount of brush cycles. However, an existing sand drop test induced a substantial amount of damage to PV glass, such that the method will have to be tailored to the PV industry. These results will ultimately be compared with acquired field-aged modules as well as our own set of glass coupons with and without coatings that have been deployed in harsh soiling environments. An initial set of these samples are being evaluated by NREL after a one year deployment in the field. Interestingly, the growth of organics on the glass surface is an interesting and unexpected finding. Finally, we are developing and have initial results for an inexpensive artificial soiling test method that can be used to simulate PV module dew and heat cycles, to test the efficacy of different coatings for soiling.

REFERENCES

[1] Sarver et. al., Renewable and Sustainable Energy Reviews, 2013, vol. 22, issue C, pages 698-733

[2] Kimber et al. Proceedings of the world conference on photovoltaic energy conversion, 2, p. 2391 (2006)

[3] *"A Scalable Method for Extracting Soiling Rates from PV Production Data"*, Deceglie et. al., IEEE PVSC 2016

[4] *"An Investigation of the Key Parameters for Predicting PV Soiling Losses"*, Micheli et al. Progress in PV, *http://onlinelibrary.wiley.com/doi/10.1002/pip.2860/epdf.*

IV. CONCLUSIONS

NREL is performing a set of tasks to address several important issues associated with soiling of PV modules. The results from these tasks indicate that it may be possible to determine soiling losses from production data of present PV systems, as well as from environmental conditions; such that a soiling station may not be necessary for every site. Furthermore, we have found that van der Waals and capillary forces on real dust particles are much lower than expected based on the particle size, and are more related to the much smaller contact areas associated with rough surfaces. In addition, defining cementation as a process where loose particles agglomerate to the point that they can transmit shear waves have provided a good working definition that has enable us to use a QCM to detect and quantify the onset and buildup of cementation with wet and dry air cycling. Finally, we have been developing and evaluating the types of accelerated tests, as well as collecting the field data needed to assemble an IEC standard for PV module coatings that may be used for antireflection and/or anti-soiling purposes. The initial results indicate that many test procedures used in other industries are too aggressive for PV module testing, and that an unique combination of factors that include higher heating, ultraviolet exposure, humidity, and soiling, must be considered for the standard.

ACKNOWLEDGMENTS

The authors would like to acknowledge and thank all those who contributed to this project including Sarah Kurtz and all of our industrial and academic collaborators. This work was supported by the U.S. Department of Energy under Contract No. DE-AC36-08GO28308 to the National Renewable Energy Laboratory.

The U.S. Government retains and the publisher, by accepting the article for publication, acknowledges that the U.S. Government retains a nonexclusive, paid up, irrevocable, worldwide license to publish or reproduce the published form of this work, or allow others to do so, for U.S. Government purposes.

Modeling Potential-Induced Degradation (PID) of Field-Exposed Crystalline Silicon Solar PV Modules: Focus on a Regeneration Term

Eleonora Annigoni[1], Alessandro Virtuani[1], Fanny Sculati-Meillaud[1], Christophe Ballif[1,2]

[1] École Polytechnique Fédérale de Lausanne (EPFL), Institute of Microengineering (IMT), Photovoltaics and Thin Film Electronics Laboratory, Neuchâtel, CH-2000, Switzerland
[2] CSEM, PV-Center, Neuchâtel, CH-2000, Switzerland

Abstract — **In this work, we further elaborate a model that was previously presented to simulate the effect of potential-induced degradation (PID) on the performance of field-exposed modules. We add to the predictive model a term aimed at describing the effect of regeneration during the night. We also investigate the impact of irradiance, temperature, bias, and load conditions on power regeneration. Laboratory PID-degraded mini-modules were subjected to indoor regeneration tests at different stress levels in order to model the contribution of each factor.**
This allows us to implement into our PID predictive model, besides a degradation term, a second term describing the regeneration. Finally, we apply this model with outdoor climatic data to obtain a more realistic simulation of the power evolution over time of PID-affected PV modules in different climatic conditions.

I. INTRODUCTION

In this work, we aim at developing a comprehensive model to predict the performance of solar photovoltaic (PV) modules in operation as a consequence of the Potential-Induced Degradation (PID) effect (see e.g. [1] and [2]). We consider crystalline silicon (c-Si) PV modules with conventional p-type cells. In a recent contribution, we presented a model for PID power degradation (in [3]) based on laboratory accelerated ageing tests.

With the aim of simulating the actual performance of field-exposed modules, we complement the model by including the contribution of a regeneration term. Power regeneration can occur at night or under illumination during hot and dry periods. In particular, we study the impact of irradiance, temperature, bias and load conditions on the power regeneration of PID degraded PV modules.

First we performed indoor tests on pre-degraded mini-modules at different regenerating conditions. From the results of such tests we then develop a mathematical description of PID recovery (as can be found in other contributions e.g., [4] and [5]).

This allows us to express a more comprehensive equation modeling PID, with the inclusion of a regeneration term besides a degradation one [3]. This general expression can then be used, in combination with a set of input meteorological data for a specific location, to model the evolution of PID on PV modules in operation. As a first attempt, we apply the model using the parameters obtained for our test devices and by selecting three locations with different climatic conditions.

II. PREVIOUS CONTRIBUTION: DEGRADATION

In [3] we presented a mathematical model for maximum power degradation, which we briefly recall. We subjected two-cell mini-modules to accelerated aging in climatic chamber at different levels of the main stress factors (which, for degradation, are: temperature, T, relative humidity, RH, and voltage, V). The maximum power was measured at specified times during the tests. The final equation we obtained to describe the time-evolution of power degradation as function of the stress factors is the following:

$$\frac{P_{max}(t)}{P_{max,0}} = 1 - C_V \cdot V \qquad (1)$$

The quantity $P_{max}(t)$ is the mini-module's power after t hours of exposure to constant stress levels, and $P_{max,0}$ is the initial power. Such equation applies for modules exposed to negative voltage with respect to the ground, and V is the absolute value of this potential difference. The coefficient C_V contains the effects of T, RH, and time t (see (2)), and is in agreement with an equation previously proposed by Hacke et al. in [5] (κ is the Boltzmann constant).

$$C_V(T, RH, t) := A \cdot e^{-\frac{Ea}{\kappa \cdot T}} \cdot RH^B \cdot t^2 \qquad (2)$$

The values we obtained for the parameters by fitting our experimental results with equation 1 are: $A = 6.0188 \cdot 10^{-15} \, h^{-2}$, $B = 4.43 \, [-]$, and $E_a = 0.86 \, eV$.

III. EXPERIMENTAL WORK: REGENERATION

It is well known from field experience that the PID mechanism is reversible: modules performance can show some level of regeneration, for example when modules are not exposed to voltage (at nighttime) or under illumination during hot and dry periods [2]. The recovery process was subject of some recent studies. For instance, in [4] an equation for the regeneration in dark as function of the temperature was

978-1-5090-5606-4/17 $31.00 © 2017 IEEE

proposed. Authors of ref. [6] show that the recovery mechanism can be affected by irradiance too. However, to our knowledge, a clear relation describing the combined effect of temperature, irradiance, and voltage in the regeneration is still missing.

In this work we improve the model in [3] by adding a term that describes the recovery process.

We laminated at PV-Lab two-cell mini-modules (size 20 cm x 40 cm), all samples featuring the same cell type (commercial mono crystalline silicon p-type cells). The modules are laminated in a glass/backsheet configuration, with soda-lime glass at the front, a polyethylene-based backsheet, and a commercial ethylene-vinyl acetate (EVA) as encapsulant material. All samples were prepared with the same processes, the same bill of material, and in a single batch. We simulated the presence of a metallic frame by means of an electrically conductive aluminum tape on the four sides. To degrade their performance, we subjected the mini-modules to PID tests in climatic chamber according to the procedure described in the proposed IEC Technical Specification for PID in c-Si modules (IEC TS 62804-1:2015-08). Mini-module's leads are short-circuited and a negative voltage is applied between the cells and the grounded frame. The temperature was increased to 85°C instead of the specified 60°C in order to accelerate degradation. Electrical parameters of the mini-modules were determined by means of IV measurements at standard test conditions (STC: AM1.5G, 25°C, 1000 W/m²) using a LED-halogen based sun simulator. The mini-modules were characterized initially and at the end of the degradation tests (96h).

After the degradation, the same mini-modules were subjected to regeneration tests. On the one hand, to investigate the regeneration occurring at night, tests were performed at different temperatures in dark and with no applied voltage. On the other hand, to reproduce the daily regeneration (that may occur during hot and dry periods), the samples were exposed to different irradiance levels. We also investigated for the first time the effect of (1) bias voltage during regeneration under irradiance, by applying a bias to some of the samples, and (2) of load conditions.

A. Regeneration in the dark

For the study of nighttime recovery, we subjected pre-degraded mini-modules to recovery in dark. The pre-degradation tests were performed for all samples in a climatic chamber at 85°C / 85% RH / -1000 V for 96h. Regeneration tests were then performed in an oven, at different temperatures and under dry conditions. Samples were kept in open-circuit conditions.

By defining the relative recovery at time t as:

$$RR(t) := \frac{P_{max}(t) - P_{min}}{P_{ini} - P_{min}} \qquad (3)$$

where P_{ini} and P_{min} are the mini-module maximum power at, respectively, the beginning and the end of the degradation test, we obtained (see Fig. 1) the following relation for the relative recovery as function of the temperature, in agreement with [4]:

$$RR(t) = 1 - \exp\left(-\left(\frac{t}{\tau(T)}\right)^{\beta}\right) \qquad (4)$$

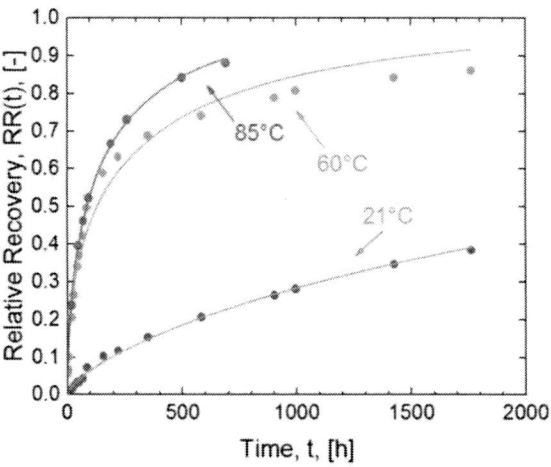

Fig. 1. Relative recovery over time for mini-modules during indoor testing in dark at different temperatures, with the fitting curves according to Eq. (4).

In Eq. (4), the parameter β seems to be independent on temperature and has an average value of 0.55 [-]. The parameter τ [h] follows an Arrhenius law with respect to temperature, with an activation energy of 0.52 eV.

The fact that higher temperatures induce a faster power recovery is consistent with the most accredited theory for the regeneration process, which attributes it to an out-diffusion of Na atoms from stacking faults in the pn junction [7]. As previously suggested in [4], this indicates that – under dry conditions - power regeneration might be more pronounced during daytime, with higher temperatures, than during nighttime, even in the presence of a negative applied bias to the modules. This is the main motivation behind this work.

B. Regeneration during light exposure

Here we aim at modeling the effect of irradiance on power recovery. As for regeneration in dark, all the samples were pre-degraded in climatic chamber at 85°C / 85% RH / -1000 V for 96h. The recovery was then performed at different irradiance levels, in dry conditions (relative humidity around 15%). A solar simulator system of class C was used that comprises a chamber with temperature controlled in a range of ±5°C. While the samples regenerating in the dark were kept at open-circuit (OC) conditions, here we wanted to reproduce the conditions at which fielded modules are kept during daylight. For this reason, we applied to some of the samples a negative voltage (simulating modules at the negative extreme of a string). In order to assess the impact on regeneration of irradiance only, other samples were left with no voltage applied (see Table I).

TABLE I
EXPERIMENTAL MATRIX OF TEST CONDITIONS TO
INVESTIGATE THE REGENERATION MECHANISM

T [°C]	Irradiance [W/m^2]	Load conditions	Voltage [V]
60	1000	PID (SC)	-1000
		SC	/
		OC	/
		Mpp	/
	800	PID (SC)	-1000
		Mpp	/
	640	PID (SC)	-1000
		Mpp	/

For a specific set of conditions (1000 W/m2, 60°C, - 1000 V), we investigated the effect of load conditions as well, by exposing the modules, respectively, in short-circuit (SC), open-circuit (OC), and, by applying a resistive load, maximum power point (Mpp).

Fig. 2 shows the relative power over time of the samples regenerating under 1000 W/m2 at different load conditions. Two samples were tested per each condition.

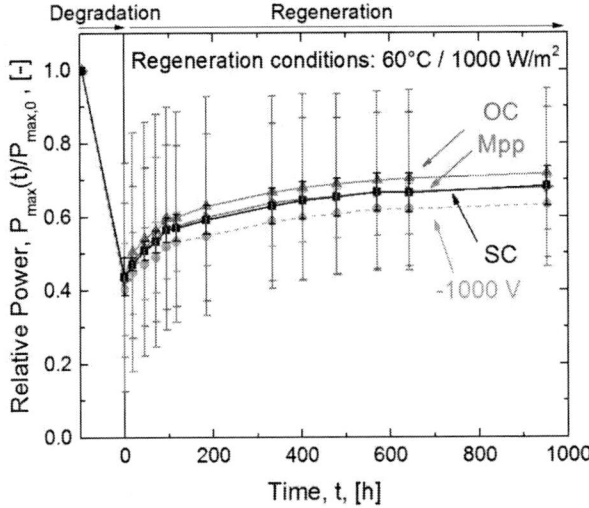

Fig. 2. Power evolution of mini-modules normalized by their initial power. The pre-degradation is performed at 85°C / 85% RH / -1000V for 96h. Regeneration starts at time 0 h, with samples exposed to 60°C, 1000 W/m^2 and at different load conditions. Relative humidity was low to enhance the regeneration (around 15%). Per each load conditions, two samples were tested. Lines are a guide to the eye.

A slight difference can be seen in the regeneration speed and in the extent of maximum recovery. For instance, negatively-biased (-1000 V) mini-modules regenerate slower than the others, while samples in OC conditions show the strongest recovery, which we tend to ascribe to the slightly higher temperatures (a few degrees) of the devices held in OC. However, what needs to be highlighted is that, under these test conditions, even samples with a high negative voltage applied are regenerating. This means that daytime regeneration can occur in the field even for modules in strings exposed to high negative voltages towards ground, at least when humidity is low and/or in the absence of rain.

Results of regeneration under different irradiance levels are shown in Fig. 3. Dotted lines are for samples with -1000 V applied, while straight lines represent samples in Mpp conditions. The effect of irradiance on the power regeneration seems not to be very pronounced, whereas a stronger impact is rather given by the load conditions.

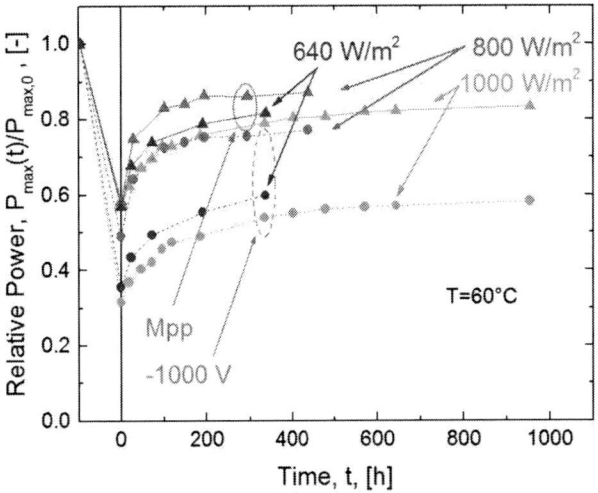

Fig. 3. Power evolution of mini-modules normalized to their initial power (before degradation). PID degradation is performed at 85°C / 85% RH / -1000V for 96h. Regeneration starts at time 0 h, with samples exposed to 60°C and different irradiance levels. Relative humidity was low to enhance the regeneration (around 15%). Lines are a guide to the eye. Dotted lines are for samples with a -1000 V bias voltage applied, straight lines are for samples in Mpp conditions.

For the same samples, also the values of relative recovery (as defined in Eq. (3)) suggest that irradiance may have a secondary relevance in promoting the regeneration compared to load conditions (see Fig. 4). Here, however, the "800 W/m^2 / -1000 V" data-set contradicts partly these observations. One explanation for this may lie in possible temperature variations from the set value of 60°C in the different tests: samples regenerating faster might actually be exposed to slightly higher temperatures than the others. We then speculate that irradiance might have only a secondary relevance in promoting regeneration, whereas temperature plays a larger role. Further tests are ongoing to definitely assess the contribution of irradiance on power recovery.

In addition, even if - in order to avoid contradictory results - particular care has been dedicated in sample processing (same bill of material, processes, and processing time), the samples show different level of degradation (i.e. not all samples degrade by the same amount), as can be seen in Fig 3. This variability then translates into different levels of recovery or of relative recovery (see Fig. 4), which may have a considerable impact on our findings/conclusions. Further, it is widely observed (and reported in the literature) that the samples undergoing a more severe degradation tend to recover less.

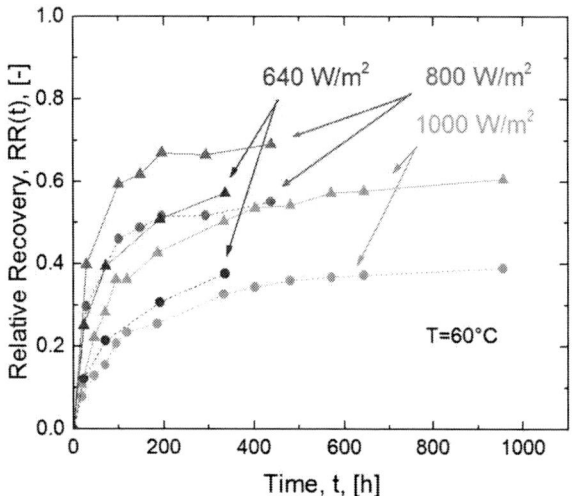

Fig. 4. Relative recovery (defined in Eq. (3)) referred to the same samples as in Fig. 3.

In order to complete the model for daytime power recovery, on-going work (not presented in this paper) is investigating, for a fixed irradiance level (i.e. 1000 W/m²), the effects that (1) temperature and (2) bias voltage have on the regeneration. The latter to assess the distribution of the recovery process as a function of the position of the module in a string.

IV. SIMULATING OUTDOOR PID

The final goal of this work is to model the evolution of the PID phenomenon and the impact on module performance as a function of different climatic conditions. To do this we applied the equations determined by means of indoor accelerated tests (Eq. (1) and (4)), with Typical Meteorological Year data as input parameters for the simulations.

Applying these equations, determined at constant stress laboratory conditions, to outdoor weather data is not straightforward. Indeed, indoors module's power undergoes a continuous degradation described by (1) and by a fixed initial value $P_{max,0}$ (and analogously for regeneration), whereas outdoors $P_{max}(t)$ fluctuates continuously depending on the time-varying temperature, irradiance, and other environmental parameters. Our approach is described in [3]; in particular we

employ the concept of equivalent time to make the implementation with varying stress conditions consistent.

Here we complete the model by adding the contribution of the regeneration process.

In Fig. 5, by using the indoor-determined parameters of our test devices, we simulate the power of a 2-cell mini-module exposed to -1000 V in three different climates. We assume, as a first approximation, that degradation occurs during daytime, according to Eq. (1), and thermally-driven recovery during nighttime, Eq. (4). Degradation is supposed to occur for any level of relative humidity (according to (4)), differently from other works, such as [8], where it is supposed to apply only when relative humidity is above a certain threshold level. Simulation results show a clear impact of relative humidity on the different degradation levels reached in the three climates. One can observe that the effect of seasonality is also evident, e.g. in Miami the trade-off between degradation and regeneration is well in correspondence with the rain and the dry season respectively of this location. A similar seasonal trend was observed in [2] for modules installed in PID conditions in Florida.

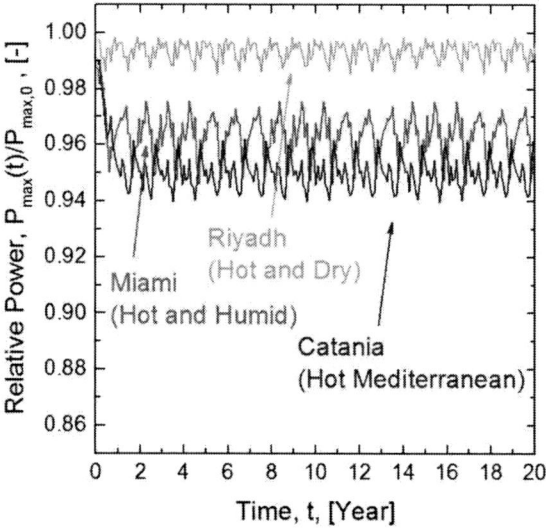

Fig. 5. Simulated time-evolution of a 2-cell mini-module's power as consequence of daytime PID and nighttime regeneration in three different climates.

The planned experimental work described in the previous section should clarify more the dependence of power regeneration on irradiance. We will then integrate the term for daytime regeneration in the simulations.

As the model presented in this work is obtained for 2-cell mini-modules, it also needs to be adapted to full-size modules taking into account the different voltage that affects each cell. This extension of the model will be based on previous works, such as [6] and [8].

V. SUMMARY

We present an improved model for PID prediction, based on a study of the regeneration mechanism. For the first time, the relationship between power recovery and irradiance, temperature, voltage, and load conditions was investigated, by means of regeneration tests performed at different stress levels.

In dark conditions (replicating nighttime recovery), the power regeneration is promoted by higher temperatures according to an Arrhenius behavior, in agreement with what found in [4].

Under illumination, what needs to be highlighted is that, under these specific test conditions (illumination, 60°C, low humidity), regeneration clearly takes place as well for the samples exposed to a considerable high negative bias-voltage (-1000 V). The results obtained so far suggest that different irradiance levels do not have a strong effect on the power regeneration, whereas a clear impact is given by the load conditions. However, in spite of the accuracy with which we manufactured the samples in order to guarantee the repeatability of the results (all samples were laminated on the same day and with the same roll of encapsulant and of backsheet), we still obtain a quite high variability in the results. One possible reason could be that the temperature in the different tests had some variations from the set value of 60°C, so that samples exposed to a higher temperature could regenerate faster than the others. In view of these preliminary results we speculate that irradiance may play a secondary role in the recovery mechanism, whereas temperature is a more relevant factor.

The results of regeneration tests performed so far allow us modeling the term for nighttime regeneration, which we included in our previous predictive model for degradation, [3]. Finally, we apply this PID model with outdoor climatic data to obtain a realistic simulation of modules performance in different climatic conditions, see Fig. 5. A trade-off between power degradation and regeneration is reproduced, and it matches the yearly alternation of the humid and the dry season. Such seasonal effect in the case of Miami climate is well in correspondence with that observed in [2] for modules exposed outdoors in Florida.

ACKNOWLEDGMENTS

We gratefully acknowledge the financial support of EOS Holding and of the Swiss National Science Foundation (SNSF) through the National Research Program (NRP) "Active Interfaces" (NRP 70). We are grateful to Xavier Niquille for support in the experimental work and to the all PV-lab team.

REFERENCES

[1] S. Pingel, O. Frank, M. Winkler, S. Daryan, T. Geipel and H. B. J. Hoehne, "Potential Induced Degradation of Solar Cells and Panels," 2010.

[2] P. Hacke, R. Smith, K. Terwilliger, G. Perrin, B. Sekulic and S. Kurtz, "Development of an IEC test for crystalline silicon modules to qualify their resistance to system voltage stress," *Progress in Photovoltaics: Research and Applications,* vol. 22, pp. 775-783, 2014.

[3] E. Annigoni, M. Jankovec, F. Galliano, H. Li, L. Perret-Aebi, M. Topic, F. Sculati-Meillaud, A. Virtuani and C. Ballif, "Modeling potential-induced degradation (PID) in crystalline silicon solar cells: from accelerated-aging laboratory testing to outdoor prediction," in *32nd EU PVSEC*, Munich, 2016.

[4] P. Lechner, S. Hummel and J. Schnepf, "Evaluation of Recovery Methods after Potential Induced Degradation of PV Modules," in *31st European Photovoltaic Solar Energy Conference and Exhibition*, Hamburg, 2015.

[5] P. Hacke, S. Spataru, K. Terwilliger, G. Perrin, S. Glick, S. Kurtz and J. Wohlgemuth, "Accelerated Testing and Modeling of Potential-Induced Degradation as a Function of Temperature and Relative Humidity," *IEEE Journal of Photovoltaics,* vol. 5, no. 6, pp. 1549-1553, 2015.

[6] S. Koch, J. Berghold, C. Hinz, S. Krauter and P. Grunow, "Improvement of a Prediction Model for Potential Induced Degradation by Better Understanding the Regeneration Mechanism," in *31st European Photovoltaic Solar Energy Conference and Exhibition*, Hamburg, 2015.

[7] D. Lausch, V. Naumann, A. Graff, A. Hähnel, O. Breitenstein, C. Hagendorf and J. Bagdahn, "Sodium outdiffusion from stacking faults as root cause for the recovery process of potential-induced degradation (PID)," *Energy Procedia ,* vol. 55, pp. 486-493, 2014.

[8] J. Hattendorf, W.-M. Gnehr, R. Loew, T. Roth , D. Koshnicharov and M. Zentgraf, "Potential-induced degradation and temperature-driven regeneration: a realistic simulation," Paris, 2013.

Soiling loss on PV modules at two locations in India studied using a water based artificial soiling method

Sonali Bhaduri[1], Sachin Zachariah[2], Lawrence L. Kazmerski[2,3], Balasubramaniam Kavaipatti[1], and Anil Kottantharayil[2]

[1]Department of Energy Science and Engineering, [2]Department of Electrical Engineering,
[1,2]Indian Institute of Technology Bombay, Mumbai, Maharashtra, 400076, India.
[3]National Renewable Energy Laboratory, Golden, Colorado.

Abstract – **India, being a country in the sun-belt region, has an abundance of solar energy. However, it is seen that photovoltaic (PV) modules in India suffer from more power loss than many other parts of the world due to soiling. In our study, we had developed a water - based artificial dust deposition chamber inside a laboratory environment, by which we can deposit dust with a uniformity of 93% on a glass substrate for dust densities upto 0.6mg/cm² which signifies approximately 30% of J_{sc} loss due to the effect of soiling on PV modules. The dust samples were collected from the surface of PV modules deployed in the field. We have compared soiling losses from two locations, namely Mumbai and Gandhinagar, in India by water - based dust deposition method. We have also investigated the dust sample collected in Mumbai in 2013 and 2016. It is seen that the soil composition changes with time. This would be the major challenge to find an effective method of mitigating dust deposition on PV modules.**

Index Terms — **Soiling loss, photovoltaic modules, dust, J_{sc} loss.**

I. INTRODUCTION

India has an abundance of solar energy which makes PV sector one of the highest growth sectors in the energy industry. However environmental factors like high temperature and humidity have a negative effect on the module performance and durability [1]. India also has a high concentration of particulate matter (PM10) in the air [2], which makes soiling a major reason for power losses in PV modules. Dust in every climate will have different chemical composition and particle size distributions leading to differences in soiling losses [3, 4].

Previously we had reported on artificial soiling experiments using acetonitrile and DI water as carrier solvents [5]. We had reported that acetonitrile modifies the composition of the dust. Also, the soiling losses were overestimated when acetonitrile was used. Hence, we used DI (Deionized) water as a carrier solvent for our further studies. In the natural environment, dust deposition on the PV modules is non - uniform. However, for laboratory studies, high uniformity is desirable for meaningful comparison between different sources of dust. We studied the soiling loss for Gandhinagar and Mumbai and found that the losses vary significantly with location. Also, dust collected at 3 years apart are significantly different in composition.

II. EXPERIMENTAL METHODS AND RESULTS

A. Preparation of soil suspension

The dust was collected from the surface of PV modules from Mumbai and Gandhinagar. Suspension of 3 g of dry dust in 100 ml of DI water was made as it gave approximately 0.2 mg/cm² dust with each spray. The glass substrate was from Borosil Glass Works Limited, India and had a dimension of 2 cm × 2 cm. Before the dust deposition, the glass substrate was cleaned with IPA and was dried under ambient conditions.

B. Dust deposition

The methodology for the dust deposition was same as reported previously [5]. Deposition chamber design was improved to increase the uniformity to 93%. We introduced a time control switch that enables us to control the spray duration in the range of milli seconds. For the experiments reported in this paper, we used 0.6 sec of spray duration for achieving a high uniformity over 2 cm x 2 cm samples. The spray gun was placed vertically, and the glass substrate was heated continuously and was maintained at 70ºC. Commercial grade N_2 was kept at 20 psi which was used as carrier gas. The distance between the spray head and glass substrate was 25 cm.

C. Measurement of the density of the dust.

The glass substrates were weighed using a micro balance (Citizen CX85S) of resolution 0.00001 g before ($M_{cleaned}$) and after (M_{soiled}) the dust deposition and the weight difference is divided by the area of the glass substrate (A), to obtain the areal density of the dust. Calibration of the weighing balance was done before every measurement to ensure precise results. The soil gravimetric density (SGD) is calculated as follows

$$\text{SGD (g/m}^2) = \frac{M\,soiled - M\,cleaned}{A} \qquad (1)$$

D. Characterization

D.I. Uniformity of the deposited dust

Uniformity was the key challenge for water based dust deposition method. We achieved high uniformity by

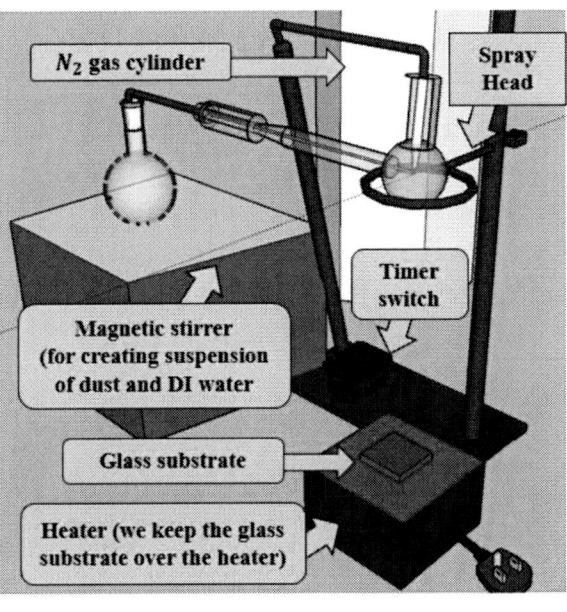

(a)

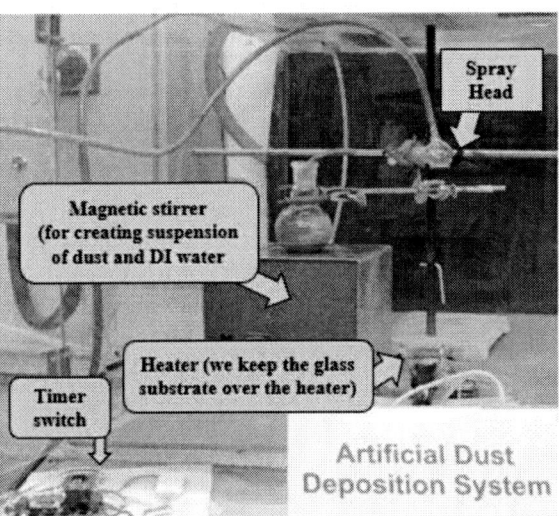

(b)

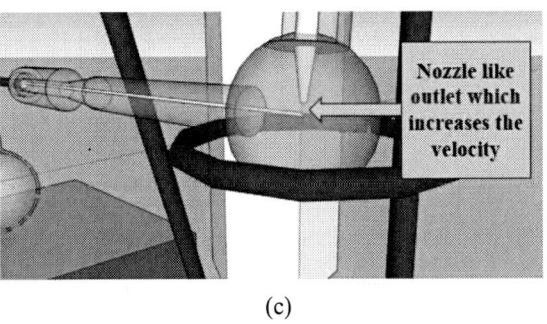

(c)

Fig. 1: (a) Schematic design of modified dust deposition chamber. (b) Real dust deposition set up. (c) Magnified view of the spray head

optimizing the parameters like spray duration, carrier gas pressure, and the position of the glass substrate. Water has a tendency of forming spherical droplets of non-uniform radius in a free fall condition. To overcome this, we increased the pressure of the carrier gas to get finer water droplets. The time duration of the spray was also decreased to milliseconds so that the fine droplets do not combine with each other forming spherical drop like structure on the glass substrate. The pressure of the carrier gas in a 6 mm diameter tube was fixed to 20psi and was connected to a solenoid switch along with a timer circuit which enabled automated on and off spray duration and allowed us to maintain consistency of spray period.

A low-cost nozzle was designed for this dust deposition chamber. An atomizer that allows uniform spray in larger area could not be used because our sample was not a solution but a suspension of dust in DI water. The spray head had two ends (shown in Fig. 1: (c)), which had a nozzle like shape that helped us to increase the velocity at the outgoing tip of the spray.

The optical images of the low (0.15 mg/cm^2), medium (0.61 mg/cm^2) and high (1.5 mg/cm^2) density of dust depositions are shown in Fig. 2. The uniformity was measured by the quantum efficiency (QE) measurements of a multi Si solar cell which was overlaid by the dust deposited glass substrate. We used Bentham PVE300 quantum efficiency (QE) measurement system. QE scans were taken at 5-nm increments from 300 to1100 nm wavelengths. A light beam of 1.5 mm X 5 mm was used for the QE measurements. The slit width was choosen such that it can be focused on the active area between the metallic fingers, this helped us to eliminate the losses due to metallic fingers. First, the QE of the cell, overlaid with clean glass substrate, was measured, and this was considered as QE_{clean}. This was considered as the reference for measurements of QE for the glass substrates with dust. Then, QE was measured with substrates of different soil gravimetric densities overlaid on the cell and was labeled as QE_{soiled}. QE loss due deposition of the dust layer was defined as the QE_{clean} - QE_{soiled} (Fig. 4).

J_{SC} was estimated by integrating the QE curve (soiled sample). J_{SC} was measured at 5 different positions as shown in Fig. 2. The J_{SC} values are listed in Table I, and it is seen that, within the sample non-uniformity is about 7% for dust densities upto 0.6mg/cm^2 which signifies approximately 30% of J_{sc} loss due to the effect of soiling.

Fig. 2. Dust deposited samples with low (0.15mg/cm^2), medium (0.61 mg/cm^2) and high (1.5 mg/cm^2) densities.

978-1-5090-5606-4/17 $31.00 © 2017 IEEE

TABLE I: Uniformity of low, medium and high density of dust deposited by water based dust deposition.

Location of measurement on the sample (see Fig. 2)	J_{SC} value estimated from QE measurement (mA/cm²)		
	S1-Low density (0.15 mg/cm²)	S2-Medium density (0.61 mg/cm²)	S3-High Density (1.5 mg/cm²)
1	25.05	22.44	9.54
2	25.01	21.21	9.85
3	25.32	20.86	10.39
4	26.02	22.79	10.81
5	26.08	22.93	11.90

At very high densities like 1.5mg/cm² which results in about 70% J_{sc} loss, the non-uniformity increased to 14%. This condition of very high soiling is not yet observed by us in the actual field scenario. The uniformity of the dust deposition also depends on the size of the dust particle. Our dust particles (taken from PV modules located in Mumbai) size was majorly between 16 µm to 31 µm (shown in Table II).

D.II. Comparison of dust samples from Gandhinagar and Mumbai

We collected dust samples from Gandhinagar (Hot and Dry Climate) and Mumbai (Hot and Humid Climate) and analyzed the soiling loss based on J_{SC} for various dust densities as shown in Fig. 3. These dust samples were collected in 2016. QE loss for various dust densities for both Gandhinagar and Mumbai are shown in Fig. 4. The J_{SC} loss was calculated by the following equation.

$$J_{SC} \text{ (loss)} = 1 - \frac{J_{SC\_soiled}}{J_{SC\_cleaned}} \quad (2)$$

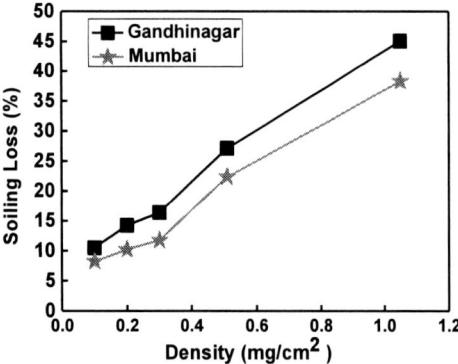

Fig. 3: Soiling Loss based on J_{SC} of Gandhinagar and Mumbai at various dust densities.

Gandhinagar shows more J_{SC} and QE loss than Mumbai (shown in Fig. 3, 4). In our previous work [3], we had

reported the highest soiling loss in Mumbai among Gurgaon, Hanle, Jodhpur and Agra, Pondicherry and Mumbai. It was argued that the reasons for this are the differences in the mineral composition and the particle size distribution in the soil [3, 6].

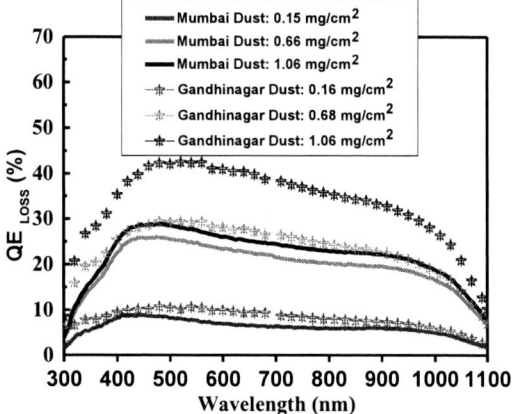

Fig. 4: QE loss of Gandhinagar and Mumbai dust at various dust densities.

TABLE II – The particle size distribution for Gandhinagar and Mumbai dust.

D (µm)	% of Total Samples		Sediment Type
	Gandhi-nagar	Mumbai	
0-4	4.97	1.10	Clay
4-8	18.77	4.37	Very Fine Silt
8-16	70.06	27.72	Fine Silt
16-31	6.20	48.77	Medium Silt
31-63	0.00	18.04	Coarse Silt
63-125	0.00	0.00	Very Fine Grained

The sediment type analysis of the two dust samples is shown in Table II. This analysis was done using Beckman Coulter Laser Diffraction Particle Size Analyzer. This system has the ability to measure particle sizes from 0.017 to 2000 µm. By this analysis we observed that, Gandhinagar dust is much finer than the Mumbai dust. This supports the argument that the finer the dust, higher the losses. We also notice that the particle distribution of Mumbai dust shown in Table II is different from our previous study on Mumbai dust collected in 2013 [3].

We also carried out XRD analysis for both Gandhinagar and Mumbai dust to see the difference in mineralogy as shown in Fig. 5. The XRD analysis was done using the PANalytical Empyrean system. The XRD pattern was analyzed to identify the minerals using the PANalytical's XRD software suite. The mineral content of Mumbai and Gandhinagar's dust samples differs significantly as expected.

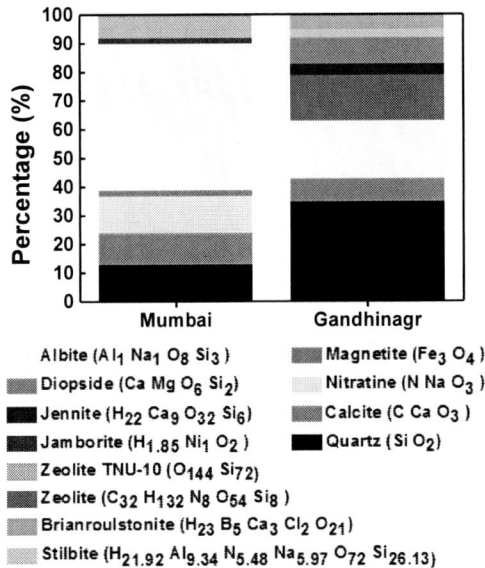

Fig. 5: XRD of dry dust of Gandhinagar and Mumbai. Both dust samples were collected in 2016.

D.III. Variation of dust composition with time

Mineral composition of dust can vary with time. Fig. 6 shows the XRD analysis of the Mumbai dry dust from 2013 and 2016. It is seen that the quartz and calcite were the main components in 2013, and the proportion of these had dropped substantially in 2016.

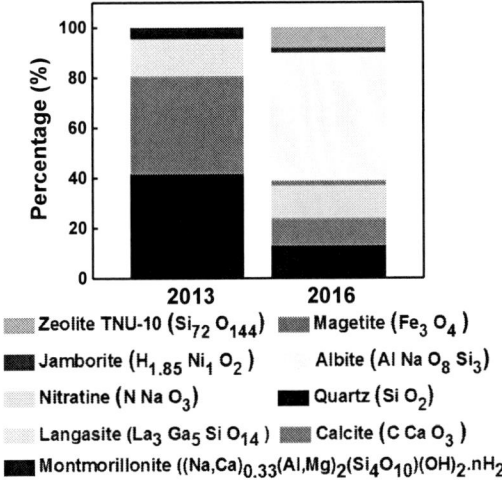

Fig. 6: XRD analysis of dust collected in 2013 and 2016 from Mumbai.

The Mumbai site had several construction activities around the site during 2013. However, these activities had subsided by 2016. This could partly explain the reduction in calcite, which is a component in cement. In the dust composition of 2016, we found high content of albite which has sodium. Sodium can come in the dust by the formation of bonds with the glass surface over the years. This is known as cementation effect which alters the surface of the modules [7].

III. CONCLUSIONS

We have developed a dust deposition technique that uses DI water as the carrier solvent, and the within sample uniformity is 93% on a glass substrate for dust densities upto 0.6 mg/cm^2 which signifies 30% J_{sc} loss . We used DI water as it is present in form of humidity and dew in the natural enviornment and also is a key factor that influences the adhesion of dust on PV modules. Gandhinagar dust showed higher QE and J_{sc} loss than Mumbai dust, as the dust particle of Gandhinagar was finer than Mumbai. The chemical composition of the Mumbai dust which was deposited on the PV modules has also changed with time, one of the factors can be cementation where the dust reacts with the module glass surface and thus resulted in high content of sodium. The other factor could be the dust generating activities around the PV installation. These variations would pose a challenge to the development of effective anti-soiling coatings.

ACKNOWLEDGEMENTS

This research is based upon work supported by the Solar Energy Research Institute for India and the U.S. (SERIIUS) funded jointly by the U.S. Department of Energy subcontract DE AC36-08G028308 (Office of Science, Office of Basic Energy Sciences, and Energy Efficiency and Renewable Energy, Solar Energy Technology Program, with support from the Office of International Affairs) and the Government of India subcontract IUSSTF/JCERDC-SERIIUS/2012 dated 22nd Nov. 2012. The authors also acknowledge the National Centre for Photovoltaic Research and Education (NCPRE), Civil Engineering Department and Earth Science Department at IIT Bombay for allowing us to use their facilities for QE measurement, sediment analysis and XRD respectively.

References

[1] R. Dubey et al., "Correlation of electrical and visual degradation seen in field survey in India", Proceedings of the 43rd IEEE *Photovoltaic Specialists Conference (PVSC)*, pp. 1692-1696, 2016.

[2] World Health Organization Report on "Exposure to particulate matter with an aerodynamic diameter of 10um or less (pm10) in 1600 urban areas, 2008-2013". Health Statistics and Information Systems(HSI). [Online]-http://www.who.int/gho/phe/outdoor_air_pollution/exposure/en/

[3] J. J. John, S. Warade, G. TamizhMani, and A. Kottantharayil, "Study of Soiling Loss on Photovoltaic Modules with Artificially Deposited Dust of Different Gravimetric Densities and Compositions collected from Different Locations in India", IEEE Journal of Photovoltaics, 6(1), 236 – 243, 2016.

[4] T. Sarver, A. AI-Qaraghuli and L. L. Kazmerski, "A Comprehensive Review of the Impact of Dust on the Use of Solar Energy: History, Investigations, Results, Literature, and Mitigation Approaches," *Renewable and Sustainable Energy Reviews,* vol. 22, p. 689-733, 2013.

[5] S. Bhaduri, S. Warade, J. J. John. B. Kavaipatti, and A. Kottantharayil, "Artificial Dust Deposition Using Water as Carrier Solvent for Investigation of Soiling Losses in Photovoltaic Modules", Proceedings of the 43rd IEEE *Photovoltaic Specialists Conference (PVSC)*, pp. 2076-2079, 2016.

[6] M. S. El-Shobokshy and F. M. Hussein, "Effect of dust with different physical properties on the performance of photovoltaic cells," *Solar Energy*, vol. 51, pp. 505–511, 1993.

[7] K. Ilse et al., "Microstructural analysis of the cementation process during soiling on glass surfaces in arid and semi-arid climates", *Physica Status Solidi (RRL) - Rapid Research Letters*, vol. 10, pp. 525 - 529, 2016.

Quantifying Year-to-Year Variations in Solar Panel Soiling from PV Energy-Production Data

Michael G. Deceglie[a], Leonardo Micheli[a,b], and Matthew Muller[a]

[a] National Renewable Energy Laboratory, Golden, CO 80401, USA
[b] Department of Chemistry, Colorado School of Mines, Golden, CO 80401, USA

Abstract — We present a method for quantifying solar panel soiling loss from PV energy-production data and show how it can be used to quantify year-to-year variations in soiling loss. Cleaning events are automatically detected, eliminating the need for precipitation data or assumptions about cleaning. The method also calculates the reduction of energy yield due to soiling, not just a soiling rate. We apply the method annually and also to multi-year datasets. We find significant year-to-year variation in soiling loss, a result with important implications for site planning and financing. The method presented here can leverage existing datasets of daily energy yield to quantify annual and total soiling risk without the need for new hardware and serves as a supplement to standard plant planning practices.

I. INTRODUCTION

Deposition of dirt and dust on the surface of solar panels, known as soiling, is an important factor affecting energy yield of photovoltaic (PV) systems. It is useful to quantify the impact of soiling from both an operations and maintenance (O&M) and a system-planning perspective. Here we present a method, which builds upon prior work [1], to quantify the impact of soiling using only PV energy-production data. The method enables quantification of year-to-year variations in soiling loss.

The method presented here improves upon prior efforts in several ways. First, cleaning events are automatically detected, eliminating the need for any assumptions about links between cleaning events and precipitation. It thus eliminates the need for precipitation data. Another advantage of the method we describe is that it quantifies the impact of soiling on *energy yield*. We achieve this by calculating a daily soling derate to be applied to daily insolation. In contrast, many prior studies focus on measuring the soiling *rate* and pairing that rate with precipitation data [1 – 4]. The approach we describe is designed to find the soiling signal in an otherwise complex daily time series of PV energy-production that could be affected by many non-soiling operational factors such as faulty equipment, measurement errors, or permanent degradation. This approach also naturally captures the different impacts of soiling during high- or low- irradiance parts of the year at a given site.

We begin by explaining the method. Then we present results from several sites and highlight important features of the method itself and the insights gained about the nature of soiling and year-to-year variations in soiling loss. Finally, we present a validation of the method against soiling loss calculated from soiling stations.

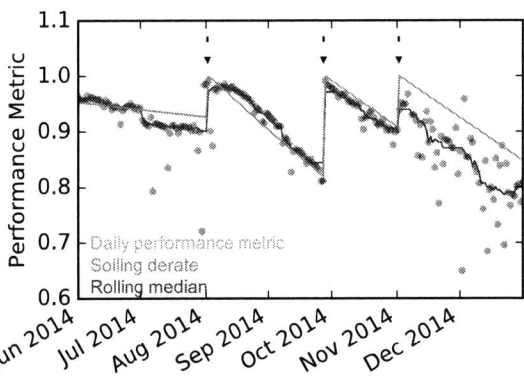

Fig. 1. Example soiling calculation. A rolling median filter is applied to daily performance metric to detect positive shifts, interpreted as cleaning events and marked by arrows. The slopes of the periods between these events are fit with a linearly decreasing soiling derate.

II. SOILING LOSS CALCULATIONS

The method presented here is performed on a time series of the daily performance metric from a PV array and solar resource (plane-of-array insolation). We use the term daily performance metric to refer to any quantification of the daily energy production of a PV system, for example performance index. The soiling loss calculation can, in principle, use various performance metrics calculated in different ways.

Here (and as previously described [1]), we use a temperature-corrected and normalized performance, which we calculate for each system from 15-minute AC power, irradiance, and ambient temperature. The raw data are filtered to eliminate points with irradiance <100 W/m^2, affected by clipping, and with low power (less than 1% of the maximum). Then, each AC power point is temperature corrected based on measured module temperature. The temperature-corrected powers are then integrated daily to calculate a temperature-corrected daily energy yield, which is normalized by daily plane-of-array insolation. Finally these daily values are normalized to the 95th percentile to obtain a dimensionless performance metric.

The daily performance metric is used in the following steps to extract soiling loss:

1. Detect positive steps in the data indicating cleaning events. This is achieved by forward filling missing data then calculating an 11-day moving median. When the moving median increases by more than 0.015 in a single day, the day is flagged as a cleaning event.

978-1-5090-5606-4/17 $31.00 © 2017 IEEE   2804

2. Use the Theil-Sen method [5] to calculate the slope of performance metric vs. time between each pair of cleaning events. The 95% confidence interval in each slope is also calculated and recorded.

3. Calculate a daily derate due to soiling (shown by the red line in Fig. 1) based on the slope of each interval and assuming the soiling derate starts at one for each interval (perfectly clean).

4. Multiply the daily soiling derate by plane-of-array insolation to obtain a soiling-derated insolation.

5. Compare the raw and derated insolation to quantify loss due to soiling.

6. Perform Monte Carlo simulation to estimate uncertainty by regenerating the daily derate series based on randomly drawn slopes from the 95% confidence intervals of each soiling period.

The first step is to detect positive shifts in the daily performance metric. These sudden increases in array output are interpreted as cleaning events. In contrast with previous work [1–4], there is no need for rainfall data or cleaning logs, the cleanings are automatically detected. Fig. 1 shows examples of these positive shifts. We detect these shifts by applying an eleven-day rolling median filter to the daily performance metric data. The rolling median filter is particularly useful in this application because it reduces the influence of noise, but preserves step-like shifts in the data. When the rolling median increases by more than 0.015, the day is flagged as a cleaning event. The 0.015 value was found to work well for the data considered here, but sensitivity to this threshold is an important consideration for future work.

The intervals between cleaning events can be understood as soiling intervals. Note that these intervals may have negligible soiling and thus negligible slope. This is accounted for in the method for accumulating losses. Each interval is fit with the Theil-Sen method, a regression method that is robust to outliers [5]. The Theil-Sen method also yields a confidence interval for the slope of the regressed data. Here we use the 95% confidence interval in the slope of each interval for Monte Carlo simulations described below.

A time series of daily soiling derates is then calculated for each soiling interval assuming a linear change based on the fitted slope and a starting derate value of one for the interval. The derate can be interpreted as the expected reduction in resource due to soiling. This calculation is designed to isolate the soiling loss from other factors that may affect daily yield. Intervals containing negative steps of greater than 5% (indicative of other system problems) or with positive slopes, are assigned a slope of 0, and a derate of one (no loss). An example derate time series is plotted in Fig. 1.

The daily soiling derate is then multiplied by daily plane-of-array insolation. This daily-derated insolation, shown in the bottom panel of Fig. 2, is integrated over the time period of interest and compared with the full insolation to calculate a soiling loss. A key feature of this technique is that the soiling loss is weighted by

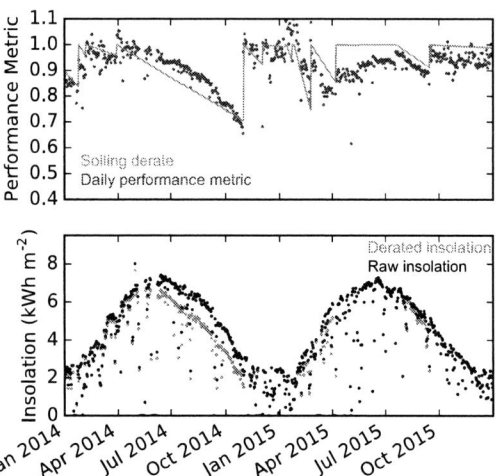

Fig. 2 The soiling calculation shown for site A over a two-year period. The top panel shows the time series of performance metric and soiling derate. The bottom panel shows both the raw and derated insolation. More soiling loss is observed in 2014 than 2015.

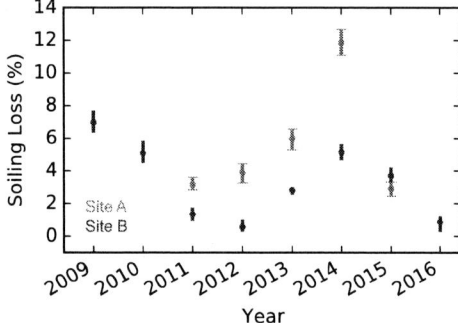

Fig. 3 Annual soiling loss at two different sites in California. The error bars indicate the 95% confidence interval of the loss. Notably, significant variation is observed from year to year.

daily insolation, so that soiling during low-irradiance periods is less impactful than during high-irradiance periods.

To estimate the uncertainty associated with the interval regressions, a Monte Carlo simulation is performed using the confidence intervals of the slope for each interval. New slopes are randomly drawn for each soiling interval from a uniform distribution spanning the 95% confidence interval of the respective regressions. In this way, a new series of daily soiling derates is calculated. This process is repeated 1000 times to build up a sample of integrated soiling losses representative of the underlying uncertainty in each interval's regression. The difference between the 2.5 and 97.5 percentiles of this sample are taken as the 95% confidence interval for the integrated soiling loss.

The entire process can be applied to different time periods. Here, we have applied it to individual calendar years, and also to full multi-year datasets. This provides information on the year-to-year variations in soiling loss and also in the long-term average.

III. RESULTS AND DISCUSSION

We performed the soiling calculations on production data from ten inverters at nine different sites in the United States, for both individual calendar years, and for the entire multi-year datasets. All sites were utility-scale power plants. Fig. 3 shows the yearly soiling loss calculated for the two sites found to have the most severe soiling. Note the year-to-year variations can be quite high, and are larger than the confidence intervals of the individual annual calculations. This suggests that the largest source of uncertainty for the soiling that a plant may experience in a given year comes not from our ability to quantify the soiling loss, but from the variation in climatic conditions.

An example of the year-to-year variations is more closely examined in Fig. 2. The top panel shows the soiling derate and performance metric time series. The performance metric exhibits a period of slow decline followed by rapid recovery in the summer of 2014, a pattern not repeated in 2015. This pattern is consistent with a lengthy soiling interval. The bottom panel shows the impact of the soiling derate on daily insolation. This illustrates a key feature of the method: the interaction between variations in soiling throughout the year and the solar resource is captured by the method.

The method described here represents a significant advancement beyond methods that simply yield a soiling rate. Comparing sites solely on the soiling rate can be misleading as sites with high soiling rates can actually have low soiling energy losses due to the complex interaction between the rate, the precipitation patterns, and the available irradiance. Here, the impact on solar resource is explicitly considered. This incorporates information about whether soiling occurs during high-irradiance times or low-irradiance periods. The method also incorporates information about the length of the soiling runs, which indirectly incorporates information about the precipitation patterns.

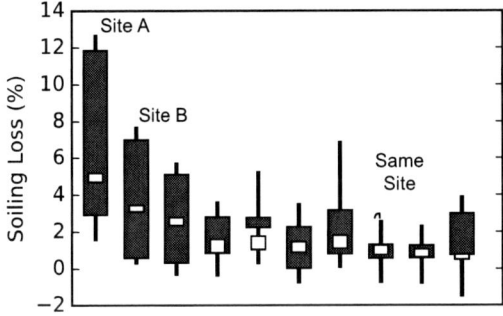

Fig. 4. Soiling loss calculated for ten inverters at nine different sites. The white boxes indicate the 95% confidence interval for the soiling loss over the entire multi-year dataset. The blue boxes indicate the range of single-year median soiling loss values found for the dataset. The black lines indicate the union of all the single-year 95% confidence intervals.

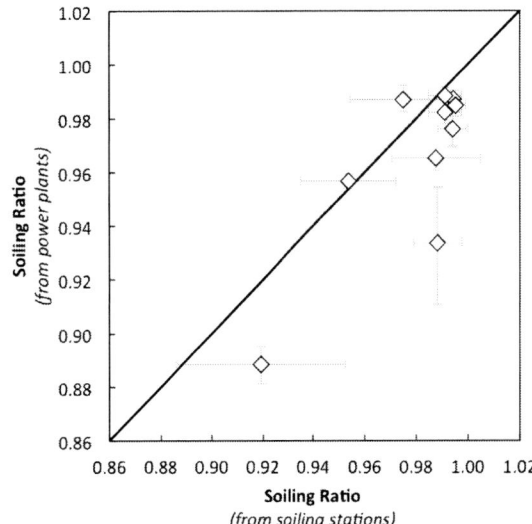

Fig. 5 Comparison between soiling ratio (ratio between soiled and ideal energy yield) obtained using the method described here (vertical axis) and using soiling station data [6]. Confidence intervals are indicated. The methods rely on different assumptions about precipitation and cleaning, but general agreement is observed. The line indicates perfect agreement.

Fig. 4 compares soiling losses calculated for individual years and multi-year datasets for each inverter considered. The white boxes indicate the 95% confidence interval for the soiling loss calculated over the entire dataset. These are observed to be small relative to the year-to-year variation represented by the blue boxes (median values) and black bars (union of all the single-year 95% confidence intervals). This observation supports the conclusion that year-to-year variation in soiling loss is a large source of uncertainty when predicting annual losses.

Another important feature of the method illustrated in Fig. 2 is that the soiling derate does not perfectly track the performance metric. This is because the derate is designed to isolate the soiling loss from other factors affecting daily energy yield. To confirm that this method is successful in isolating the soiling loss, we validated it against soiling stations deployed at the same sites. The details of the soiling station analysis are presented elsewhere [6]. The comparison between soiling loss calculated from soiling stations and from production data, during the same time periods, is shown in Fig. 5. While general agreement is observed, the extraction method presented here tends to yield lower soiling ratios than the soiling station analysis. This is likely due to the use of precipitation data in the soiling station analysis; if a cleaning event was detected without a precipitation event, that cleaning event was assumed artificial and the recovery not included in the soiling ratio calculation. In contrast the extraction method presented here does not utilize precipitation data to make such inferences.

IV. CONCLUSION

We have presented a method for extracting solar-resource-weighted soiling loss from PV production data and demonstrated its use in quantifying year-to-year variations in soiling loss. In future work, we plan to explore the uncertainty associated with the magnitude of cleaning events and clipping due to DC/AC ratio; we will also investigate the use of satellite-based irradiance data.

In contrast to methods that focus on soiling rate, the method presented here captures the interplay between soiling, precipitation, and solar resource patterns to quantify energy loss due to soiling. Cleaning events are automatically detected, thus the method does not require any precipitation data or assumptions about links between precipitation and cleaning. When combined with cleaning logs for a PV system, the method can be extended to calculate historical soiling loss under various hypothetical O&M cleaning schedules, from only natural cleanings to regular manual or robotic cleanings.

Another important feature of this method is that it can be applied to individual years or to multi-year datasets. Our results indicate that year-to-year variations in soiling loss are larger than the uncertainty in long-term average loss. This has important implications for planning and financing of PV systems. It also points to risk of using a single year of prospecting data to plan for deployment at a particular site.

Since this method uses production data, it can leverage the large amounts of data already existing in the industry, without the need for new hardware. Because the method is designed to isolate soiling losses from other factors affecting daily energy yield it is robust when applied to different performance metrics from differently modeled PV systems. This enables consistent soiling quantification at large-scale across diverse datasets. When scaled and deployed on fleet-level data, this method will provide historical statistics on annual losses and the year-to-year variation that can be expected in different areas. These long-term soiling loss statistics can supplement industry standard site prospecting practices and reduce financial risk.

ACKNOWLEDGMENT

The authors thank Sarah Kurtz (NREL) and Greg Kimball (SunPower) for insightful discussions.

This work was supported by the U.S. Department of Energy under Contract No. DE-AC36-08GO28308 with the National Renewable Energy Laboratory. Funding provided by U.S. Department of Energy Office of Energy Efficiency and Renewable Energy Solar Energy Technologies Office.

The U.S. Government retains and the publisher, by accepting the article for publication, acknowledges that the U.S. Government retains a nonexclusive, paid-up, irrevocable, worldwide license to publish or reproduce the published form of this work, or allow others to do so, for U.S. Government purposes.

REFERENCES

[1] M. Deceglie, M. Muller, Z. Defreitas, and S. Kurtz, *Photovoltaic Specialists Conference (PVSC)*, 2016, pp. 2061–20165.

[2] A. Kimber, L. Mitchell, S. Nogradi, and H. Wenger, *Photovoltaic Energy Conversion*, vol. 2, May 2006, pp. 2391–2395.

[3] F. A. Mejia and J. Kleissl, "Soiling losses for solar photovoltaic systems in California," Solar Energy, vol. 95, pp. 357–363, 2013.

[4] J. Caron and B. Littmann, "Direct monitoring of energy lost due to soiling on First Solar modules in California," Photovoltaics, IEEE Journal of, vol. 3, no. 1, pp. 336–340, Jan 2013.

[5] P. K. Sen, "Estimates of the regression coefficient based on Kendall's tau," Journal of the American Statistical Association, vol. 63, no. 324, pp. 1379–1389, 1968.

[6] M. Muller, L. Micheli, and A. A. Martinez-Morales, "A Method to extract soiling loss data from soiling stations with imperfect cleaning schedules" *Photovoltaic Specialists Conference (PVSC)*, 2017, submitted.

Accurately Measuring PV Soiling Losses with Soiling Station Employing PV Module Power Measurements

Michael Gostein[1], Bill Stueve[1], Mandy Chan[2]

[1]Atonometrics, 8900 Shoal Creek Blvd., Ste. 116, Austin, TX 78757, USA

[2]E.ON Climate & Renewables North America, 20 California St., Ste. 500, San Francisco, CA 94111, USA

Abstract — Monitoring of PV soiling losses has become increasingly important for both operational PV power plants and pre-construction site surveys. Therefore, dedicated soiling measurement stations are being routinely deployed at many sites. Such stations typically compare the output of a "soiled" PV device with that of a "clean" device. While many stations measure only the short-circuit current of the soiled & clean devices, using the assumption that power output is proportional to short-circuit current, for accurate characterization it is often necessary to measure the actual power output in addition to the short-circuit current, especially when using crystalline silicon modules. This is essential in cases where soiling accumulates non-uniformly, such as along the edges or sides of the modules. Small amounts of non-uniform soiling may cause significant power losses that cannot be detected through measurements of short-circuit current alone. We present a case study from a pre-construction site-survey soiling station in the US Southwest region. Results presented here are an update to a previous publication about this site, with data now covering a 2-year period. The results illustrate examples of both uniform and non-uniform soiling, and demonstrate the importance of using module power measurements to get accurate results.

Index Terms — photovoltaic systems, performance analysis, solar power generation, solar energy.

I. INTRODUCTION

The effects of dust, dirt, pollen, and other contaminant accumulation on PV modules, commonly referred to as soiling, can cause significant reductions in PV power plant energy generation [1], [2]. Therefore, accurate prediction and monitoring of soiling losses has become increasingly important. As a result, dedicated soiling measurement stations are being routinely deployed at many sites, including both pre-construction site-survey locations and operational PV power plants, especially for utility-scale projects [2].

Soiling measurement stations typically employ a pair of PV reference devices, one of which is kept continuously clean and the other of which is allowed to be soiled at the natural rate [2]. The effect of soiling is then determined by comparing the output of the soiled reference device with that of the clean one. The soiled reference device is a PV module identical to those in use (or to be used) at the site, while the clean reference device may be either a reference cell or another PV module. The clean device is maintained in a clean state through either manual or automatic washing.

In the simplest method, the output of the soiled and clean reference devices is compared on the basis of the short-circuit current of each device. As a PV cell becomes soiled, the irradiance reaching the cell is reduced, and the short-circuit current of the cell is reduced proportionally. When all the cells in a module are uniformly soiled, the reduction in the module's short-circuit current is a good measure for estimating the reduction in the module's total output power.

Often, however, soiling accumulates on modules in a spatially non-uniform manner [3], [4]. One driver for this is the effect of gravity, which results in a tendency for particles to accumulate near the bottom of a PV module under the influence of small amounts of precipitation sufficient to move particles but insufficient to produce total cleaning [3]. Framed modules are typically susceptible to soiling accumulation near their edges, typically at the bottom but also possibly along the sides, due to the effects of gravity, precipitation, and wind [4].

Such spatially non-uniform soiling can have a significant effect on module power output. Due to the way in which shading affects a module's I-V curve, soiling particles accumulated in one area of a module can reduce power output by an amount much greater than expected if the same number of particles were distributed uniformly over the module, particularly for typical crystalline silicon modules [5]. In these cases the reduction in module short-circuit current due to soiling is no longer a good predictor of the reduction in module output power [5],[6][7].

Therefore, due to the possible influence of spatially non-uniform soiling, soiling measurement stations should typically measure both the short-circuit current and the power output of their soiled reference module, in order to produce the most accurate measurement [5].

In this work we present a case study from a pre-construction site-survey soiling station in the US Southwest region, as an update to a previous publication on this site [8] with data now covering a 2-year period. The results show examples of the effects of both uniform and non-uniform soiling with greater clarity than in the initial work, and demonstrate the importance of using module power measurements to get accurate soiling assessment.

II. DATA COLLECTION

The data collection station was installed in an undeveloped area with potential for future solar projects, and has been online since August of 2014. The station includes typical meteorological instruments as well as a dedicated soiling measurement system. The soiling measurement system is

978-1-5090-5606-4/17 $31.00 © 2017 IEEE

implemented using two identical 295 W framed polycrystalline-silicon PV modules, one of which is designated the clean reference and the other of which is designated the soiled reference. The clean reference is manually cleaned 2-3 times per week, and at each cleaning both modules are photographed. The modules are installed co-planar on a fixed rack, with a tilt angle of 5°, chosen to represent the average position of a potential single-axis tracker. Fig. 1 shows photographs of the two reference modules under various conditions which will be described below. The I-V curves of both modules are measured once per minute by a control unit (Atonometrics RDE300 series), which also fits the curves and calculates I-V parameters. The results are logged and periodically uploaded via a cell modem for analysis.

From the results, we calculate two soiling ratio (SR) metrics. The first, the soiling ratio Isc index, is defined as:

$$SR^{Isc} = \frac{I_{sc}^{soiled}}{I_0^{soiled} \cdot \left(1 + \alpha \cdot \left(T^{soiled} - T_0\right)\right) \cdot \left(G/G_0\right)} \quad (1)$$

where I_{sc}^{soiled} is the measured short-circuit current of the soiled reference module, I_0^{soiled} is the module's short-circuit current at a reference condition (e.g. STC), α is the module's temperature coefficient of short-circuit current, T^{soiled} is the measured module temperature, T_0 is the module temperature at the reference condition (e.g. 25° C at STC), G is the measured plane-of-array irradiance determined from the clean module short-circuit current, and G_0 is the irradiance at the reference condition (e.g. 1000 W/m² at STC). The denominator in this equation is the "expected" short-circuit current in the absence of soiling and the numerator is the actual measured short-circuit current.

The second metric, the soiling ratio Pmax index, is defined similarly as:

$$SR^{Pmax} = \frac{P_{max}^{soiled}}{P_0^{soiled} \cdot \left(1 + \gamma \cdot \left(T^{soiled} - T_0\right)\right) \cdot \left(G/G_0\right)} \quad (2)$$

where P_{max}^{soiled} is the measured maximum power of the soiled module, P_0^{soiled} is the module maximum power at the reference condition, γ is the temperature coefficient of maximum power, and the other terms are as in eq. (1). In eq. (2), the denominator is the "expected" maximum power in the absence of soiling and the numerator is the actual measured maximum power.

In the absence of soiling, both SR^{Isc} and SR^{Pmax} should be unity.

We integrate the once-per-minute readings of the two metrics to calculate daily average values for each, denoted $<SR^{Isc}>$ and $<SR^{Pmax}>$. Following the method described in [2], these daily values are calculated as the irradiance-weighted averages of those measurements which are performed within ±2 hours of solar noon on each day.

(a) 02/22/2016, uniform soiling:

(b) 02/29/2016, non-uniform soiling:

(c) 05/06/2016, bottom right still soiled:

(d) 05/13/2016, both modules clean:

Fig. 1. Photographs of the clean (left) and soiled (right) reference module pair on different dates, corresponding to the lettered annotations in Fig. 2, illustrating various conditions.

III. RESULTS AND DISCUSSION

Fig. 2 shows the $<SR^{Isc}>$ and $<SR^{Pmax}>$ daily average soiling ratio metrics over a 24-month period, together with the measured daily rainfall. Initially the soiled reference module is clean and both soiling metrics are close to unity. However as soiling accumulates the metrics fall indicating lower output of the soiled module compared to the clean one. During many periods the Isc-based metric well represents the actual power loss measured by the Pmax-based metric. However during other periods the two metrics diverge, for example in early 2015 (as previously reported [8]) and then more dramatically in mid-2016.

A close examination of the mid-2016 period is instructive. On Feb. 22, 2016, as shown in Fig. 1 (a), the soiling is fairly uniform and the $<SR^{Isc}>$ and $<SR^{Pmax}>$ values, as shown in Fig. 2 at point (a), are about equal. Following a very slight rain

978-1-5090-5606-4/17 $31.00 © 2017 IEEE   2809

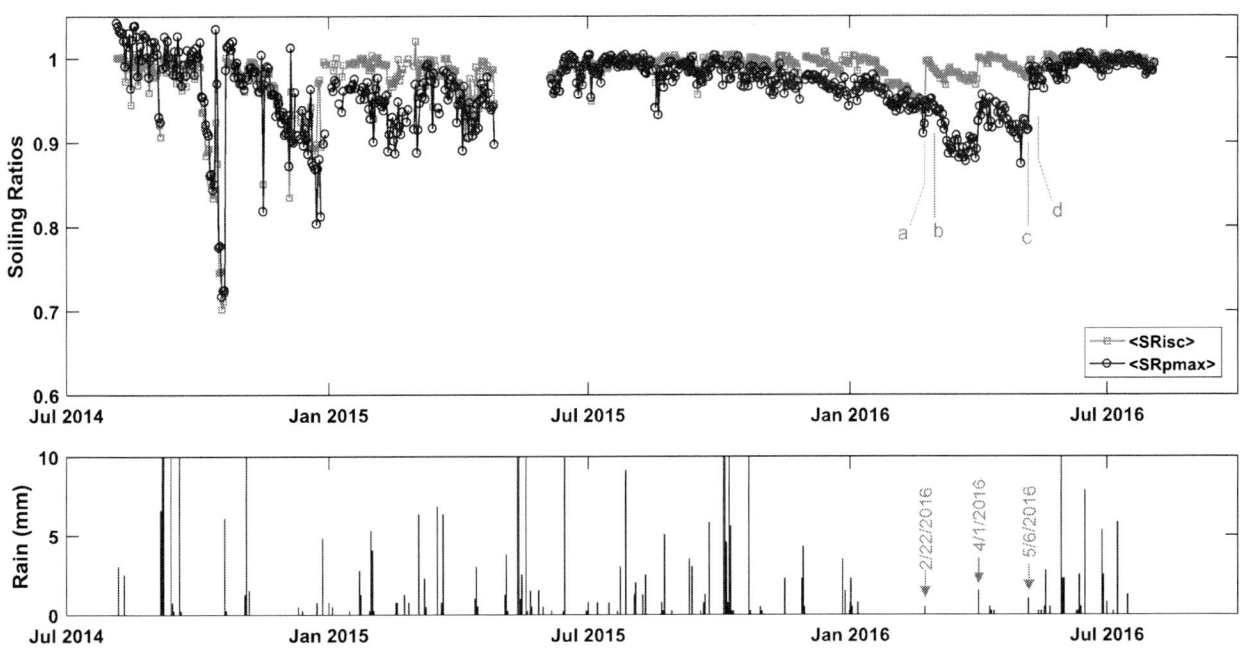

Fig. 2. Measured daily average values of the SR^{Isc} and SR^{Pmax} metrics (top) along with daily rainfall (bottom, y-axis limited to 10 mm) over a 24-month period. The four lettered arrows (a, b, c, d) indicate the days shown in the module photographs in Fig. 1.

(0.5 mm) late in the day on Feb. 22 the $<SR^{Isc}>$ value recovers to unity, while the $<SR^{Pmax}>$ is hardly affected. From the photo in Fig. 1 (b) taken on Feb. 29 we see that the rain has only cleaned the top and left portions of the soiled module, allowing the short-circuit current to recover but with power still suppressed, as shown by point (b) in Fig. 2. A similar event occurs following the 1.5 mm rain on Apr. 1, 2016, which causes a complete recovery of the $<SR^{Isc}>$ but only a partial recovery of $<SR^{Pmax}>$. The photo in Fig. 1 (c) taken early on May 6, 2016 shows that soiling has accumulated in the corner of the module, and Fig. 2 (c) shows that this causes the $<SR^{Pmax}>$ to be suppressed to 0.90 (a 10% loss of power) while $<SR^{Isc}>$ is still 0.97 (only a 3% loss). Finally, the 1.0 mm rain late in the day of May 6 completely cleans the module, as shown in Fig. 1 (d), causing both $<SR^{Isc}>$ and $<SR^{Pmax}>$ to recover as shown in Fig. 2 at point (d).

For the entire 24-month period, the average loss in current $(1 - <SR^{Isc}>)$ was only 1.9% while the average loss in power $(1 - <SR^{Pmax}>)$ was 3.7%.

These results demonstrate several important points. First, it is important to measure the actual module power in order to determine the true soiling loss, which can be significantly higher than the loss in short-circuit current. In addition, we see that small rain events of approximately 1 mm can easily result in redistributing dust on PV modules into a non-uniform pattern, which results in disproportionate power loss [5].

The results in this example illustrate the benefits of incorporating module power measurements into soiling

measurement stations in order to detect non-uniform as well as uniform soiling events, providing for an accurate overall measurement of soiling-related power losses.

REFERENCES

[1] S. Canada, "Quality Assurance: Impacts of Soiling on Utility-Scale PV System Performance," *SolarPro Magazine*, no. 6.3, pp. 14–20, May 2013.

[2] M. Gostein, J. R. Caron, B. Littmann, "Measuring soiling losses at utility-scale PV power plants," in *40th IEEE Photovoltaic Specialists Conference*, Denver, CO, 2014.

[3] H. Qasem, et al, "Dust-induced shading on photovoltaic modules," *Progress in Photovoltaics: Research and Applications*, vol. 22(2), pp. 218-226, February 2014.

[4] E. Lorenzo, R. Moretón, and I. Luque, "Dust effects on PV array performance: in-field observations with non-uniform patterns," in *Progress in Photovoltaics: Research and Applications*, vol. 22(6), pp. 666-670, June 2014.

[5] M. Gostein, B. Littmann, J. R. Caron, L. Dunn, "Comparing PV Power Plant Soiling Measurements Extracted from PV Module Irradiance and Power Measurements," in *39th IEEE Photovoltaic Specialists Conference*, Tampa, FL, 2013.

[6] C. Schill, S. Brachmann, M. Koehl, "Impact of soiling on IV-curves and efficiency of PV-modules," *Solar Energy*, 2015.

[7] J. Lopez-Garcia, A. Pozza, T. Sample, "Long-term soiling of silicon PV modules in a moderate subtropical climate," *Solar Energy*, 2016.

[8] M. Gostein, T. Düster, and C. Thuman, "Accurately Measuring PV Soiling Losses with Soiling Station Employing Module Power Measurements," in *42nd IEEE Photovoltaic Specialists Conference*, New Orleans, LA, 2015.

Performance of Monocrystalline Silicon solar cell – Influence of dust on Ultra-Violet and Visible region during early stage of deposition

Hemaprabha Elangovan[1], Upasna Ranjan[1], Jagdish A K[2], Praveen C. Ramamurthy[1,2],

Kamanio Chattopadhyay[1,2]

[1]Interdisciplinary Centre for Energy Research, Indian Institute of Science, Bangalore 560012, India

[2]Department of Materials Engineering, Indian Institute of Science, Bangalore 560012, India

Abstract — In general, solar irradiance reaching the photovoltaic (PV) panel decides the power output. But the dust deposition on the panel decreases the energy conversion efficiency over time. Dust affects several parameters and in this study, the focus is given on the light absorbance during initial stages of dust deposition. It is seen that the dust deposition has deteriorated the irradiance reaching the panel. Particles of various size has hindered different wavelength region of the solar spectrum, which has been understood by simulation. Pronounced effect is seen in Ultra-Violet (UV) range light during early stage of dust deposition. Hindrance in visible region of the light is dominated by larger particles.

Index terms – Dust deposition, monocrystalline silicon, UV, Visible, Absorbance, Solar cell performance.

I. INTRODUCTION

Solar energy represents the dominant source for renewable energy and harnessing is primarily carried out through solar Photovoltaic (PV) panels. Therefore, the nature of deployment and its performance are subject of major research and development efforts [1], [2].

Efficient use of solar energy resource can be achieved by optimizing the property of solar light absorber materials, the architecture of the PV module and through efficient maintenance protocols. It is well-known that efficient performance of PV panels is affected by external factors like environmental and climatic conditions [3].

Fig. 1. Experimental setup showing the solar panel and dust collection glass coupon arrangement.

Dust is one such factor, which can cause adverse effect on the PV performance [4]. Dust particles can originate from both earth and atmosphere, vehicular exhaust, storms, agricultural and industrial effects based on the local environment [5], [6]. Dust directly interacts with the solar irradiance and thereby affecting the light available to the panels by attenuating and scattering the incident light [9]. The presence of dust also increases the panel temperature, degrading the quality of the panel [7], [8].

Several reports are found on the nature of light interaction by dust particles on the decrease in panel efficiency [10], [11]. The interaction of light depend on the nature of particle size and shape [12]. However, the effect of particle size on the light scattering has not been reported so far. Understanding the effect of particle size on light scattering can assist in deciding the cleaning methods and schedule to be followed for PV panel maintenance [13], [14].

In this study, decoupling of UV and visible range interaction by the dust particles is carried out on monocrystalline silicon solar panel. In addition, the real-time condition is simulated and the scattering exhibited by the various particle sizes has been studied.

II. EXPERIMENTAL DETAILS

Performance of the monocrystalline silicon solar panels installed on the rooftop of ICER building in the premises of Indian Institute of Science, Bengaluru, India (13° 29′ North, 77° 56′ East), from 01/March/2016 to 01/April/2016 is monitored. The locality has pleasant climate unlike desert regions and hence the major cause of concern here is pollution and dust generated in the urban condition. Monocrystalline panel (Vikram Solar Pvt. Ltd, India) was used in the study and the specifications are listed in Table I.

TABLE I
SPECIFICATIONS OF PV PANEL

Rated Voltage, V_{mp}	32.1 V
Rated Current, I_{mp}	8.42 A
Peak Power, P_{peak}	270 W
Fill Factor, FF	79
Efficiency	15.94 %

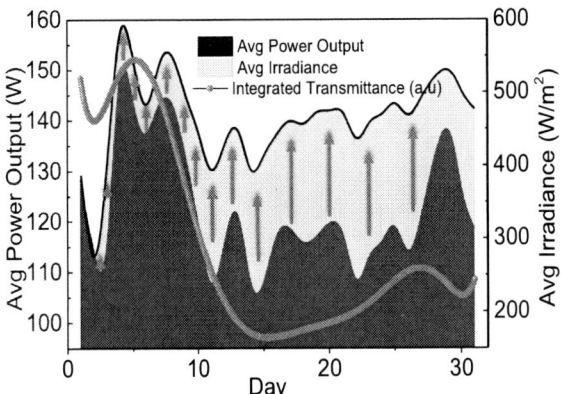

Fig. 2. Average power output with respect to the average irradiance over days.

The panel was placed at 0-degree inclination, with respect to the ground. Surfaces of thirty glass slides of dimension 75 mm*25 mm*1.2 mm were cleaned by sonicating in acetone, dried, weighed and placed on the side of the panels like an array in same tilt angle as that of the panel (i.e., 0-degree inclination), as shown in Fig. 1. One glass slide was taken every day for further dust characterization.

Copper substrates were placed near the glass slides that were used for studying the deposits using Scanning Electron Microscope (SEM).

To provide the meteorological data, weather station and pyranometer were installed in the rooftop where solar modules are installed. These are connected to Solar resources and

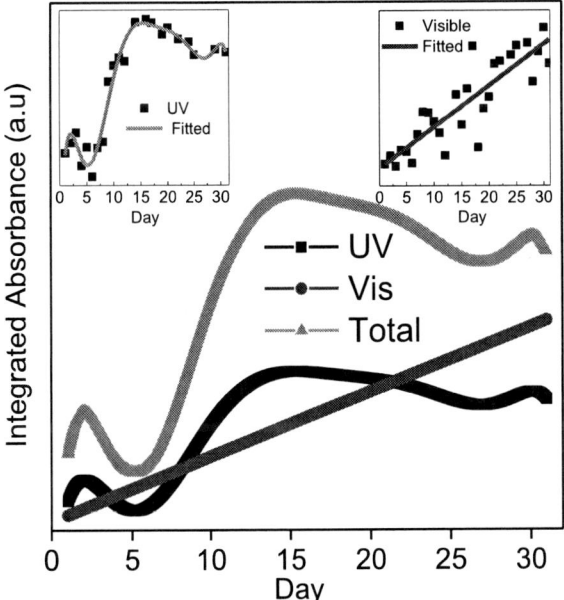

Fig. 3. The effect of dust deposition in the absorbance spectra in UV and visible regions. Inset shows the curve fitting

reliability dashboard, which is an indigenously designed software setup to integrate the panel parameters such as energy and power output. This software – hardware integration has helped in collecting the power output and the irradiance data with time resolution of a second.

Composition analysis and particle size were studied using Energy Dispersive Spectroscopy (EDS) attached with the SEM (ESEM Quanta 30 kV equipped with thermionic emission gun) and the Dektak make profilometer.

The absorbance spectra were recorded with PerkinElmer UV/Vis/Near IR spectrophotometer (LAMBDA 750). The transmittance of light through the dust layer to the glass from air is computed using frequency domain electromagnetic simulations. Finite element simulations were employed for this purpose using COMSOL-5. For simplicity, a monolayer of equi-sized and equidistant particles of SiO_2 extending in periodically in two dimensions is assumed. A minimum element size of 5 nm is employed, whereas the minimum wavelength for which the computation is carried out is 280 nm.

III. RESULTS AND DISCUSSIONS

Relation between the power output and the irradiance is given by

$$\eta = \frac{P_{max}}{A \times G} \times 100\% \qquad (1)$$

where
P_{max} is the maximum rated power of PV module
A is the area of the PV module
G is the inclined irradiance on PV module
η is the efficiency of PV module

The change in average power output (P_{avg}) with corresponding average irradiance (G_{avg}) is shown in Fig. 2. The trend of P_{avg} follows that of the G_{avg}. During the initial days, the difference between the average irradiance and P_{avg} (light gray region) is small, and increases rapidly from 5th day to 15th day. Beyond this, the difference remains constant. Ideally, without any dust deposition, the difference need not increase, as the power output is always directly proportional to the irradiance, based on Eqn. 1. With increasing days, the difference increases due to the dust deposition, as dust particles can interfere with the irradiance incident on the panel. This was studied by subjecting the experimentally collected glass slides to obtain the change in the absorbance data.

The absorbance measurement was performed from wavelength of mid UV region (280 nm – 315 nm) through near UV region (315 nm – 380 nm) to visible range (380 nm – 800 nm). The effect was considered in these wavelength regions as most of the solar spectrum falls in this range. Fig. 3 shows the trend of UV and visible range integrated absorbance with respect to time. Inset shows the graphs fitted using the measured values, with $R^2 = 0.91$ for UV curve and $R^2 = 0.96$ for visible curve. The curve for UV absorbance shows that during initial days of dust deposition, (from 5th day to 15th day) the absorbance

978-1-5090-5606-4/17 $31.00 © 2017 IEEE

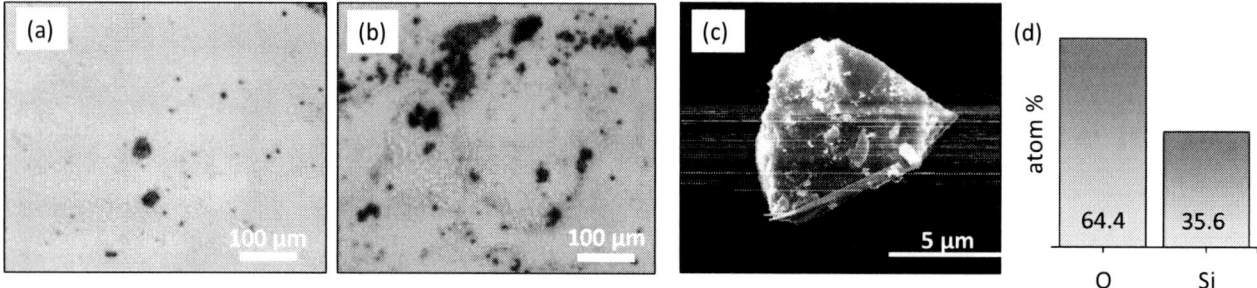

Fig. 4. Optical micrographs of the glass slides in transmission mode a) 2nd day b) 10th day of dust accumulation. c) SEM micrograph of the deposited dust particle and the d) composition corresponds to SiO2.

rapidly increases and over time the curve becomes flat. But the visible range absorbance increases linearly over the period of dust accumulation.

Light absorbance by the dust particles was experimentally observed using the optical microscopy images taken in the transmission mode, as given in Fig. 4. Black region possesses the particles, which absorb the incident visible light during imaging. From second day of experiment (Fig. 4a) to 10th day (Fig. 4b) the number of black particles increases. This can be due to the agglomeration of several smaller particles over days. SEM micrograph of the deposited dust particle (Fig. 4c) on copper substrate and the corresponding EDS atomic composition is equivalent to that of SiO2.

Knowing the particle coverage per area based on the Dektak results (Fig. 5) the total contribution of light absorbance was calculated using Eqn. 2. and the results obtained are shown in Fig. 6.

$$A_{UV} = \Sigma (A_i)_{UV} * n_i \; ; A_{Vis} = \Sigma (A_i)_{Vis} * n_i \qquad (2)$$

where,
A_{UV} = UV range absorbance, A_{Vis} = Visible range absorbance, i = particle size, A_i = Absorbance value estimated by simulation for each particle size, n_i = Area covered by each particle size.

Based on the particle size, light scattering can vary in different wavelength region [15]. To understand the effect, a trend for particle size vs absorbance (Fig. 6) was obtained using simulation. It can be seen from Fig.3, that for smaller particles (<=1 µm), the absorbance in UV range is higher than visible range. However, for larger particles (>2 µm), a huge increment in absorbance in both UV and visible region is observed and the absorbance in visible range dominates over UV range.

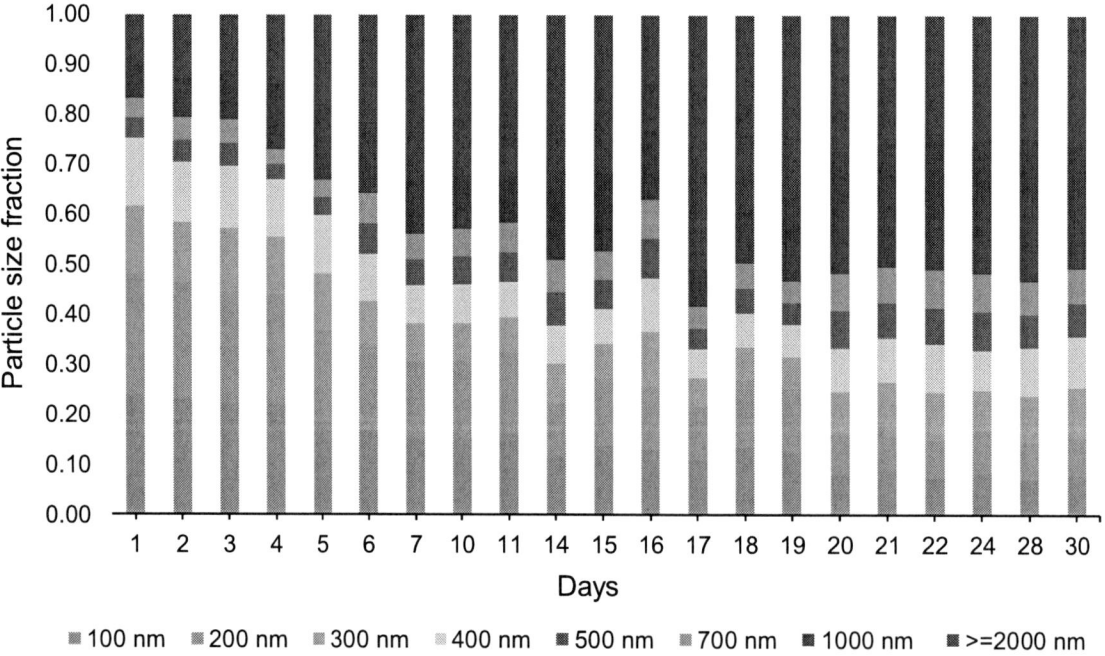

Fig. 5. Particle size fraction of the deposited dust obtained using Dektak profilometer.

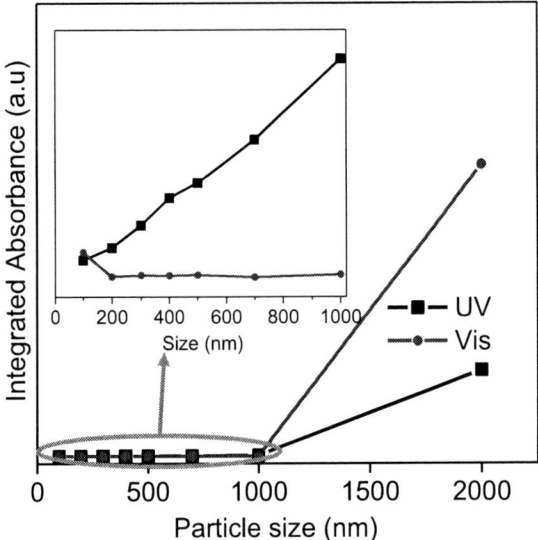

Fig. 6. Integrated absorbance as a function of particle size obtained through simulation.

The estimated trend of both UV and visible curves is same in Fig. 7. This is because, the contribution from larger particles is dominating, based on the estimated absorbance for larger particles (>2 μm), as seen in Fig. 5. Also, A_{Vis} by these particles is higher than A_{UV}. Hence in the estimated curve spectra, irrespective of the day, the visible curve is always higher than the UV curve.

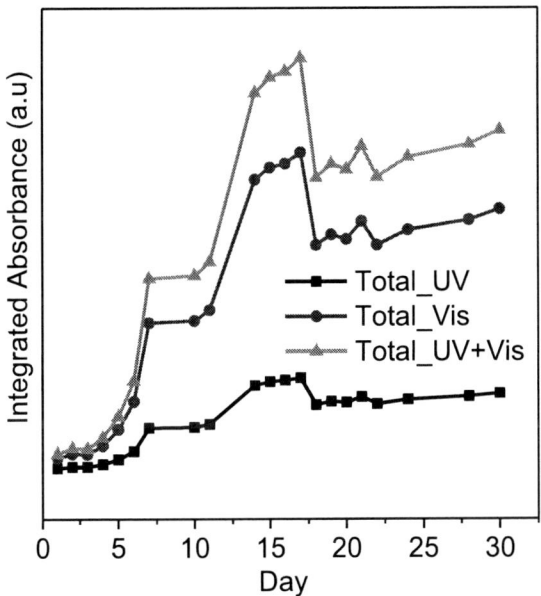

Fig. 7. Estimated absorbance in UV and Visible wavelength region

However, the experimental data (Fig. 3) shows that during initial days, the UV range absorbance dominates, and after that period absorbance is influenced by visible range.

The particle size fraction shown in Fig. 5, suggests that during initial days, the fraction of smaller particle is high. Thereafter, significant increase of larger particles fraction is observed. As a result, during initial days of dust accumulation after cleaning, smaller particles hinder significant amount of UV irradiance. As the larger particle size fraction increases, interaction with visible range predominates.

IV. CONCLUSIONS

(i) Major decrease in power output from the solar panel, happens during 5 – 15 days of dust deposition.

(ii) During early stage of dust deposition, drop in power output is due to the UV range absorbance, due to the absorbance of irradiance by smaller particles.

ACKNOWLEDGEMENT

This work is based upon work supported in part under the US– India Partnership to Advance Clean Energy-Research (PACE-R) for the Solar Energy Research Institute for India and the United States (SERIIUS), funded jointly by the U.S. Department of Energy (Office of Science, Office of Basic Energy Sciences, and Energy Efficiency and Renewable Energy, Solar Energy Technology Program, under Subcontract DE-AC36-08GO28308 to the National Renewable Energy Laboratory, Golden, Colorado) and the Government of India, through the Department of Science and Technology under Subcontract IUSSTF/JCERDC-SERIIUS/ 2012 dated 22nd Nov. 2012.

REFERENCES

[1] W. Grossmann, K. W. Steininger, C. Schmid, and I. Grossmann, "Investment and employment from large-scale photovoltaics up to 2050," *Empirica*, vol. 39, no. 2, pp. 165–189, May 2012.

[2] S. K. Sahoo, "Renewable and sustainable energy reviews solar photovoltaic energy progress in India: A review," *Renew. Sustain. Energy Rev.*, vol. 59, pp. 927–939, Jun. 2016.

[3] B. R. Paudyal and S. R. Shakya, "Dust accumulation effects on efficiency of solar PV modules for off grid purpose: A case study of Kathmandu," *Sol. Energy*, vol. 135, pp. 103–110, Oct. 2016.

[4] F. M. Zaihidee, S. Mekhilef, M. Seyedmahmoudian, and B. Horan, "Dust as an unalterable deteriorative factor affecting PV panel's efficiency: Why and how," *Renew. Sustain. Energy Rev.*, vol. 65, pp. 1267–1278, Nov. 2016.

[5] N. Beattie, R. Moir, S. Roberts, G. Buffoni, P. Graham, and N. Pearsall, "Sand and dust accumulation on photovoltaic modules in dry regions," 2011.

[6] J. K. Kaldellis and A. Kokala, "Quantifying the decrease of the photovoltaic panels' energy yield due to phenomena of natural air pollution disposal," *Energy*, vol. 35, no. 12, pp. 4862–4869, Dec. 2010.

[7] M. J. Adinoyi and S. A. M. Said, "Effect of dust accumulation on the power outputs of solar photovoltaic modules," *Renew. Energy*, vol. 60, pp. 633–636, Dec. 2013.

[8] Z. A. Darwish, H. A. Kazem, K. Sopian, M. A. Al-Goul, and H. Alawadhi, "Effect of dust pollutant type on photovoltaic

performance," *Renew. Sustain. Energy Rev.*, vol. 41, pp. 735–744, Jan. 2015.

[9]   L. Cristaldi *et al.*, "Simplified method for evaluating the effects of dust and aging on photovoltaic panels," *Measurement*, vol. 54, pp. 207–214, Aug. 2014.

[10]   B. Aïssa, R. J. Isaifan, V. E. Madhavan, and A. A. Abdallah, "Structural and physical properties of the dust particles in Qatar and their influence on the PV panel performance," *Sci. Rep.*, vol. 6, p. 31467, Aug. 2016.

[11]   B. S. Yilbas, H. Ali, M. M. Khaled, N. Al-Aqeeli, N. Abu-Dheir, and K. K. Varanasi, "Influence of dust and mud on the optical, chemical, and mechanical properties of a pv protective glass," *Sci. Rep.*, vol. 5, p. 15833, Oct. 2015.

[12]   M. a. C. Potenza *et al.*, "Shape and size constraints on dust optical properties from the Dome C ice core, Antarctica," *Sci. Rep.*, vol. 6, p. 28162, Jun. 2016.

[13]   A. Al Shehri, B. Parrott, P. Carrasco, H. Al Saiari, and I. Taie, "Impact of dust deposition and brush-based dry cleaning on glass transmittance for PV modules applications," *Sol. Energy*, vol. 135, pp. 317–324, Oct. 2016.

[14]   A. Rifai, N. Abu Dheir, B. S. Yilbas, and M. Khaled, "Mechanics of dust removal from rotating disk in relation to self-cleaning applications of PV protective cover," *Sol. Energy*, vol. 130, pp. 193–206, Jun. 2016.

[15]   C. F. Bohren and D. R. Huffman, "Absorption and Scattering by a Sphere," in *Absorption and Scattering of Light by Small Particles*, Wiley-VCH Verlag GmbH, 1998, pp. 82–129.

A Comprehensive Study of Light Soaking Effect in CdTe Solar Cells

D. Guo[1], A. Moore[2], D. Krasikov[3], I. Sankin[3] and D. Vasileska[1]

[1]School of Electrical, Computer and Energy Engineering, Arizona State University, Tempe, AZ, USA
[2]Department of Physics, Colorado State University, Fort Collins, CO, USA
[3]First Solar Inc., Perrysburg, OH, USA

Abstract — **In this work, both experiments and simulations were employed to study Cu ion's role in light soaking effect of CdTe solar cells. Both experiment and simulation show Cu ion's migration under forward-bias could cause device performance enhancement. As consistent results are achieved from both aspects, we conclude that migration of Cu ions or other charged species could be the cause of device performance enhancement observed in light soaking experiments.**

Index Terms – **CdTe, Photovoltaic cells, numerical simulation, semiconductor device reliability.**

I. INTRODUCTION

Nearly all PV technologies exhibit device performance changes under extended illumination, or "light soaking", despite the fact that both the trend and the magnitude of such changes are not always the same among different solar cell technologies. Both commercial modules and research cells based on CdTe technology have shown improvement of cell performance under light soaking conditions for up to 20 hours [1]. Many have accredited such phenomena to the passivation of traps and migration of Cu ions [2], [3]. In this work, combined experimental and theoretical effort is being pursued for a comprehensive view of this issue. In the experiment, we observed that device performance enhancement was initiated by the voltage bias applied to the cells rather than the light soaking itself. Simulation result suggests that Cu ion migration under forward-bias can be used to explain such performance enhancement.

As consistent results are achieved from both aspects, we may conclude that the migration of Cu ions or other charged species could be the main cause of the device performance enhancement observed in the light soaking experiments.

II. EXPERIMENTAL SETUP AND RESULTS

Since our primary interest is to find out whether trap passivation or Cu ion migration is the main cause of light soaking effect, the following experiment was performed to test whether electric-field drifted mobile ions or light-generated excess carriers are responsible for the light soaking effect: four "CSU" CdTe solar cells with Te/Ni back contact [4] were dark soaked at 60°C for up to 17 hours. Afterwards, two samples were light soaked at the same temperature with AM1.5 solar irradiance, while the other two samples were kept in dark at 60°C. In both groups, one device was short-circuited and the other one was forward-biased: open-circuited in the light soaking group or 0.7V forward-biased in the dark group. In the

biased case, smaller internal field is expected, comparing to the short-circuited cells. Several in-situ light J-V measurement were performed for each cells during these soaking experiments at the same stress temperature of 60°C. Hence, temperature-dependent behvior of solar cells is avoided in this study.

Figure 1 shows the changes in device performance of all four devices as a function of time, under different light and voltage bias conditions as indicated. It is clear that performance enhancement is presented in both forward-biased devices (marbles), regardless of the light conditions. On the other hand, no significant increment were observed in both short-circuited devices (squares). However, steady decrease in efficiency and fill factors were presented in the short-circuited and light-soaked device (red solid squares), which could be an indication of commonly observed light-induced degradation in CdTe solar cells [5], [6]. Stable performance was observed in the short-circuited device under dark soaking (blue open squares), as neither light nor voltage bias was applied to it. No significant changes were presented in J_{SC} of all devices thus they are neglected in this discussion.

Between the forward-biased (performance enhanced) solar cells, V_{OC} curves share similar path, which might suggest that the V_{OC} growth is caused by the reduced internal field in these devices. However, the FF curves do not follow with each other: gradual increase was presented in the dark soaked device (green open marbles) with minor fluctuations after 10 hour of stress while an initial increment followed by steady reduction was observed in the Light & Biased cell (black solid marbles). Since the other light-soaked but short-circuited cell (red solid squares)

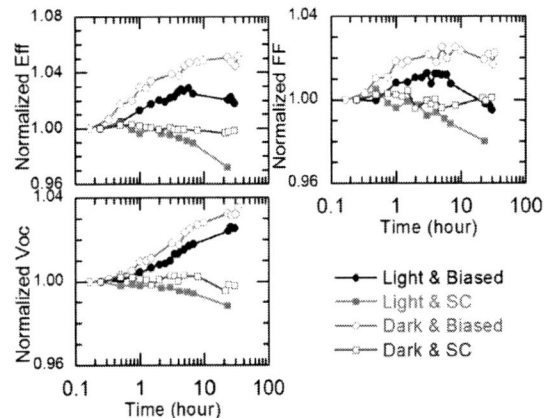

Fig. 1.   Device performance under different soaking conditions.

shows similar degradation of FF, especially after 6 hour of light soaking, it is highly possible that the drop in FF is related to light soaking itself. Such reduction in FF also limited the efficiency growth of the Light & Biased device.

In summary, experiment shows that light soaking (light-generated excess carrier) decreases FF hence overall device performance while forward-bias helped to boost V_{OC}, FF and overall conversion efficiency.

III. SIMULATION RESULTS AND DISCUSSION

We employed a 1D diffusion-reaction simulator to investigate Cu migration and dopant ionization changes under different light soaking conditions [7]. The simulator solves diffusion-reaction equations for free carriers and point defects in time-space domain self-consistently with global Poisson equation [8], thus real-time electrostatic potential and device performance can be extracted from simulations. The diffusion-reaction for a species X (target defect) is of the form

$$\frac{dX}{dt} = -\frac{dJ_X}{dx} - R_X \qquad (1)$$

where R_X is the net reaction rate of defect X and J_X, the flux of defect X, is calculated with the drift-diffusion equation:

$$J_X = D_X\left(\frac{d[X]}{dx} + \frac{[X]}{kT}\frac{d[q\varphi + G]}{dx}\right) \qquad (2)$$

where D_X is the diffusivity, φ is the electrostatic potential and G is the spatially dependent formation energy of the defect. In this work, we are primarily interested in the interactions between free carriers and major Cu dopants such as $Cu_i(+)$ donors and $Cu_{Cd}(-)$ acceptors. Hence, besides the drift-diffusion of Cu dopants, ionization reactions are also included. For example, Equation 3a and 3b are listed for $Cu_{Cd}(-)$ acceptors.

$$Cu_{Cd}^- + h_V^+ \underset{K^b}{\overset{K^f}{\rightleftarrows}} Cu_{Cd}^0 \qquad (3a)$$

$$Cu_{Cd}^0 + e_C^- \underset{K^b}{\overset{K^f}{\rightleftarrows}} Cu_{Cd}^- \qquad (3b)$$

Rate law is applied to calculate the reaction rate. Details of the numerical solution technique can be found in Ref [8], [9].

In this research, a standard ZnTe/CdTe/CdS structure is employed with simplified dopant compensation picture. Namely, 10^{16} cm^{-3} $Cu_{Cd}(-)$, 0.4×10^{16} cm^{-3} $Cu_i(+)$ and 0.5×10^{16} cm^{-3} background donor concentration are uniformly distributed in CdTe as the initial defect distribution, resulting in 10^{15} cm^{-3} hole density in the CdTe absorber layer.

Since solar cells were dark soaked for 17 hours prior to any soaking experiment, we first simulate the equilibrium of the defect system in CdTe cells under dark without any voltage applied. Figure 2 shows the equilibrium distribution of Cu dopants as well as the band diagram of the solar cell under dark at 60°C. Due to built-in potential of the p-n junction, most of $Cu_i(+)$ is pushed away from the depletion region, resulting in higher net acceptor concentrations in the junction area.

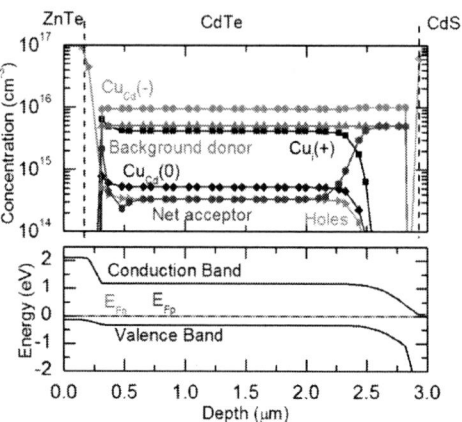

Fig. 2.   Equilibrium of the solar cells under dark conditions.

Figure 3 gives the equilibrium defect distribution the cells under 1 Sun illumination with 0.7V forward bias (or V_{OC}). In contrast to the dark equilibrium case, under light illumination and forward bias, $Cu_i(+)$ moves deeper into the depletion region due to reduced potential difference across the junction, further decreasing the net acceptor density near the junction. As we have discussed in the experimental section (Section II), expected behavior of Cu dopants is observed for the dark-soaked and forward-biased cell. For the case of dark-soaked and short-circuited device, no significant migration was observed. Based on the differences in doping profiles, both our 1D solver and Silvaco Atlas predict an increment in FF for the forward-biased device, as shown in Figure 4.

As a time-dependent solver was employed in this work, continuous migration of the defects and transient behavior of device performance were simulated as well. Similarly to Figure 1, Figure 5 shows simulated device performance as a function of time, under different conditions. Again, forward-biased "devices" show performance enhancement while no change is observed in the zero-biased "devices". But, neither does the short-circuit condition decrease FF nor does the forward bias

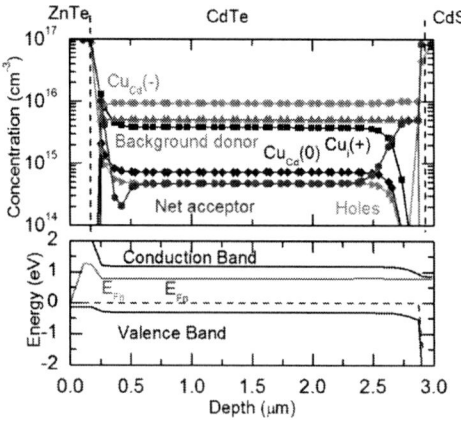

Fig. 3.   Equilibrium of the light-soaked and open-circuited cell.

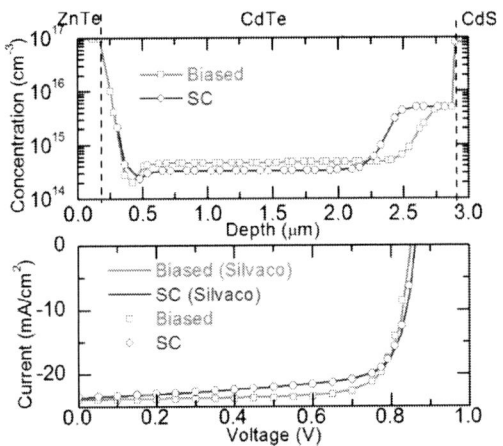

Fig. 4. Comparison of simulated net acceptor distributions under different bias conditions and their corresponding IV curves.

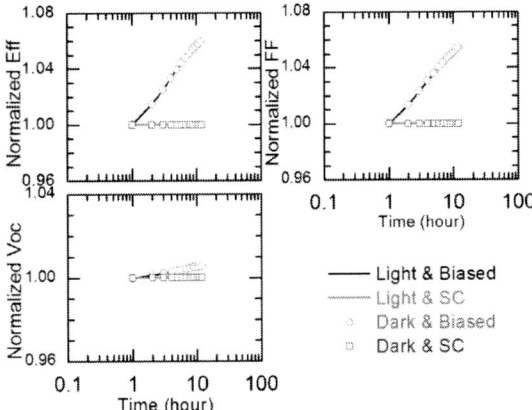

Fig. 5. Simulated device performance under different soaking conditions. IV curves were calculated by employed 1D solver.

considerably boost Voc. However, both phenomena were presented in our previous work, in which $Cu_{Cd}(-)$ and $Cu_i(+)$ interactions were investigated in the context of long term stability of CdTe solar cells [10]. This might indicate that the inclusion of defect reactions in this research could help to further understand light soaking effect in CdTe solar cells. E.g. interaction of Cu and Cl in real CdTe absorbers can lead to formation of pair complexes, introducing new mechanisms of slow defect migration and new instability mechanisms related to association/dissociation of complexes [11].

IV. CONCLUSION

To conclude, we have investigated the light soaking effect in CdTe solar cells both experimentally and via numerical simulations. Forward-bias condition is identified as the main reason of device performance enhancement in experiment. Our simulation results further show that Cu ion migration could be responsible for such metastable behavior. Inclusion of Cu interactions could lead to better understand of light soaking effects in CdTe solar cells.

ACKNOWLEDGEMENT

The information, data or work presented herein was funded in part by the Office of Energy Efficiency and Renewable Energy (EERE), U.S. Department of Energy, under award number DE-EE-0006344 and DE-EE-0007536.

REFERENCES

[1] M. Gostein and L. Dunn, "Light soaking effects on photovoltaic modules: Overview and literature review," in *Photovoltaic Specialists Conference (PVSC), 2011 37th IEEE*, 2011, pp. 3126–3131.

[2] J. A. del Cueto and B. von Roedern, "Long-term transient and metastable effects in cadmium telluride photovoltaic modules," *Prog. Photovoltaics Res. Appl.*, vol. 14, no. 7, pp. 615–628, Nov. 2006.

[3] T. J. Silverman, M. G. Deceglie, B. Marion, and S. R. Kurtz, "Performance Stabilization of CdTe PV Modules Using Bias and Light," *IEEE J. Photovoltaics*, vol. 5, no. 1, pp. 344–349, Jan. 2015.

[4] A. Moore and J. Sites, "Stability of CdTe solar cells with various back contact methods," in *2016 IEEE 43rd Photovoltaic Specialists Conference (PVSC)*, 2016, pp. 2218–2222.

[5] S. Demtsu, S. Bansal, and D. Albin, "Intrinsic stability of thin-film CdS/CdTe modules," in *2010 35th IEEE Photovoltaic Specialists Conference*, 2010, pp. 001161–001165.

[6] A. Moore and J. Sites, "Cu profiles in CdTe solar cells," in *2015 IEEE 42nd Photovoltaic Specialist Conference (PVSC)*, 2015, pp. 1–5.

[7] D. Guo and D. Vasileska, "Modeling of light soaking effect in CdTe solar cells," in *2016 International Conference on Numerical Simulation of Optoelectronic Devices (NUSOD)*, 2016, pp. 181–182.

[8] D. Brinkman, D. Guo, R. Akis, C. Ringhofer, I. Sankin, T. Fang, and D. Vasileska, "Self-consistent simulation of CdTe solar cells with active defects," *J. Appl. Phys.*, vol. 118, no. 3, p. 35704, 2015.

[9] D. Guo, T. Fang, A. Moore, D. Brinkman, R. Akis, D. Krasikov, I. Sankin, C. Ringhofer, and D. Vasileska, "Numerical Simulation of Copper Migration in Single Crystal CdTe," *IEEE J. Photovoltaics*, vol. 6, no. 5, pp. 1286–1291, Sep. 2016.

[10] D. Guo, R. Akis, D. Brinkman, A. Moore, T. Fang, I. Sankin, D. Vasileska, and C. Ringhofer, "Cu Migration and its Impact on the Metastable Behavior of CdTe Solar Cells," in *Proceedings of the 42th IEEE Photovoltaic Specialists Conference*, 2015.

[11] D. Krasikov and I. Sankin, "Defect interactions and the role of complexes in CdTe solar cell absorber," *J. Mater. Chem. A*, 2017, DOI: 10.1039/C6TA09155E

Correction for Metastability in the Quantification of PID in Thin-Film Module Testing

Peter Hacke,[1] Sergiu Spataru,[2] and Steve Johnston[1]

[1] National Renewable Energy Laboratory, Golden, CO 80401 USA

[2] Aalborg University, Aalborg, 9220, Denmark

Abstract — **A fundamental change in the analysis for the accelerated stress testing of thin-film modules is proposed, whereby power changes due to metastability and other effects that may occur due to the thermal history are removed from the power measurement that we obtain as a function of the applied stress factor. In this work, initial thermal treatment of the module is performed before application of the independent variable stress of system voltage so that any temperature–dependent processes (*e.g.*, diffusion) that affect the module power are largely activated beforehand. Secondly, the power of reference modules normalized to an initial state—undergoing the same thermal and light exposure history but *without* the applied stress factor such as humidity or voltage bias—is subtracted from that of the stressed modules. For better understanding and appropriate application in standardized tests, the method is demonstrated and discussed for potential-induced degradation testing in view of the parallel-occurring but unrelated physical mechanisms that can lead to confounding power changes in the module.**

Index Terms — **potential-induced degradation, PID, CdTe, thin-film modules, high voltage**

I. INTRODUCTION

A frequently sought-after goal when testing thin-film modules is achieving some common comparison point for evaluation. This is usually standard test conditions (STC, 25°C, 1000 W/m^2 irradiance), achieved by a conditioning or "stabilization" step [1]. IEC 61215 ed. 3 Module Quality Test (MQT) 19 describes applying 800 W/m^2 to 1000 W/m^2 with the module temperature at 50 °C ± 10 °C until the module power is deemed not to change within 2% [2]. Electrical characteristics of CIGS and CdTe modules are observed to undergo performance changes known as metastability, which have been associated with charging-discharging of defects states [3,4]. Additionally, power changes due to ion transport based on the stresses applied to the module may occur. Cu at the back contact of the CdTe module can improve the contact with the wide-bandgap p-CdTe layer and it is known to form Cu_{Cd} acceptor-state levels, which increase the carrier density such that higher open-circuit voltage (V_{oc}) is obtained [5]. However, stability or power performance can also degrade as Cu diffuses through grain boundaries to the CdTe/CdS interface [6]. It may be desired to separate such mechanisms when one is trying to study mechanisms specific to the applied stress factors, such as with the effect of damp heat or system voltage stress.

Additionally, it has been shown that the open-circuit condition can accelerate positive Cu^+ ion migration toward the junction, leading to power loss [6,7]. Efficiency loss is less for CdTe cells held at 100°C biased to maximum power (P_{max}) or in short circuit than in open circuit [7]. Similar trends have been reported in fielded modules [6]. Therefore, the state of loading or bias on the module can influence how ions migrate into regions of the absorber layer.

The MQT 19 stabilization process may not sufficiently recover the degradation from the dark heat soaks that accompany many chamber stress tests. Such power changes particular to the specimen and its history make it difficult to measure the effect of the specific environmental stress factor applied. For example, it has been shown that there is degradation of STC power in CIGS modules after placing them unbiased in 85°C dark heat that the MQT 19 stabilization process cannot recover [8]. It was found that the degradation might be mistakenly assigned to the humidity in damp heat testing if the effect of the heat is not specifically controlled and consideration to the junction bias is not given.

Basic experimental procedure involves (1) an *independent variable* that is deliberately changed; (2) a *dependent variable* whose outcome is measured as a function of the independent variable; and (3) *control variables*, which can affect the outcome and need to be kept constant if possible, or else carefully monitored so that their influence on the dependent variable can be quantified [9]. In this vein, MQT 19 performed before and after stress testing presents at least two concerns. First, the MQT 19 temperature of 50°C ± 10°C is insufficient to activate processes that may occur at a subsequent 85°C temperature level occurring in (for example) the following IEC 61215 stress tests. Consequently, any control variables that are activated at 85°C will not be maintained as constant during the stress tests. Second, after potential-induced degradation (PID) stress tests, slow thermal recovery has been observed in some CIGS modules [10]. Application of IEC 61215 MQT 19 to regenerate power loss associated with metastability may also recover losses due to PID to some extent making it difficult to evaluate the PID independently from the metastability.

MQT 19 represents a relatively high insolation and temperate module field condition, omitting real stress factors that exist in fielded modules such as system voltage that drives PID. Many environments can be damp, cold, and have relatively low insolation, not at all represented by MQT 19 conditions. It is of interest to understand the extent of PID power loss resulting from application of a PID stress test before recovery, as well as the potential to recover. Therefore, deconvolution of PID recovery from metastability recovery during application of MQT 19 requires attention.

In this work, we discuss procedures to address the above concerns for the stress testing of thin-film modules. To better isolate the control variables, unbiased modules as references or controls are placed alongside and compared with high-voltage biased modules for characterizing PID degradation. The extent of PID exhibited and any PID recovery is clearly quantified by separation from other effects in CdTe modules including metastabilities and copper diffusion. The concepts presented here are anticipated to be applicable to standardized testing, including for quantification of PID in thin-film modules.

II. EXPERIMENT

A commercial CdTe thin-film module type of double-glass and edge-seal construction was used. Three modules were processed according to the sequence shown in Fig. 1. At the damp heat stage, one module was stressed with -1000 V applied to the shorted modules leads, another with +1000 V, and the third in open circuit without any voltage applied. Stabilization was performed with a light chamber according to IEC 61215-2 MQT 19. Subsequently, dark dry-heat soaks were performed in environmental chambers at 55°C [11] and less than 5% relative humidity (RH). Flash testing was performed at STC conditions. PID stress tests were performed in damp heat at 85°C and 85% RH with an apparatus previously described [12]. Key points in the process indicated in Fig. 1 include L_0, the flash tester-determined power measurement after the initial light-stabilization procedure (MQT 19) with temperature 50°C ± 10°C; D_0, after the 55°C dry dark soak; D_n, post-stress dark-state measurements; D_r, after a thermal-recovery step; and L_n, module power measurement after light stabilization. Flash testing during environmental chamber procedures was performed in stages at about one-week intervals, with some exceptions.

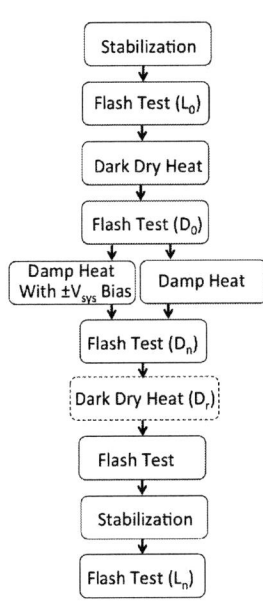

Fig. 1. Process sequence.

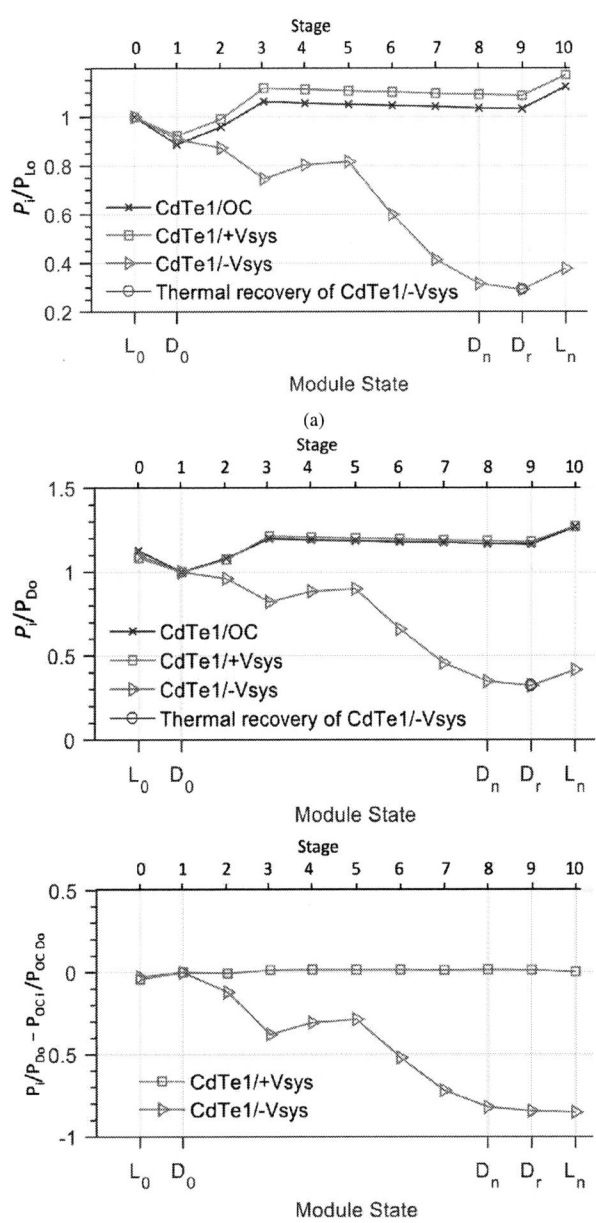

Fig. 2. Two CdTe modules undergoing PID stress testing and one in-chamber unbiased reference module in open circuit at various testing states. The indicated stage numbers represent approximately weekly flash test measurements. The x-axis labels L_0, D_0, D_n, D_r, and L_n correspond to steps in the process as indicated in Fig. 1. System voltage, V_{sys}, (+ or - 1000 V) was applied at D_0 and released at D_n. (a) Module power normalized to their value after initial stabilization L_0, (b) module power normalized to their value after dark dry-heat D_0, and (c) fraction power change of the PID-stressed modules relative to the unbiased in-chamber reference module based the D_0-normalized data in Fig 2(b).

III. RESULTS AND DISCUSSION

Figure 2 shows the results of the sequence in Fig. 1 applied to CdTe modules. The three samples indicated are either in open circuit (OC) or have positive or negative 1000 V bias applied to the module leads, which are respectively labeled as $+V_{sys}$ and $-V_{sys}$. Figure 2(a) shows the results for the module power normalized to their values at L_0, whereas Fig. 2(b) shows the results normalized at D_0. It can be seen in both the OC sample and the sample with $+V_{sys}$ stress applied (a non-PID-sensitive configuration that closely follows the power of the unbiased OC module) that there is an initial decrease in power during the dark dry heat process associated with metastability followed by a net increase in power up to the measurement at stage 3. The extent of the power decrease from L_0 to D_0 due to metastability appears to vary in Fig. 2(a).

To isolate this control variable associated with metastability for the study of the subsequent PID, it is beneficial to view the data normalized to each module's power at point D_0 after the dark dry heat soak as shown in Fig. 2(b). As discussed in the introduction, the initial net increase in power seen in the OC and $+V_{sys}$ samples up to stage 3 in the testing may be attributed to net acceptor formation. V_{oc} was measured to increase about 4% with markedly increased electroluminescence intensity attributable to the increased net carrier density [5]. This is followed in these two samples by a slow, steady decline in power—degradation associated with a fill factor drop of about 2.6% to the D_n point is seen, attributable to Cu diffusion to the junction. After the application of 1000 V to the $+V_{sys}$ and $-V_{sys}$ samples just after D_0, only the $-V_{sys}$ sample appears to degrade by PID.

It is critical to recognize that the V_{oc} increase, attributed to the Cu incorporation to form additional carriers in the p-CdTe layer, was not achieved with the initial light stabilization at the L_0 point, likely due to the insufficient high temperature (50°C ± 10°C). On the other hand, we can see the relative stability in the OC and $+V_{sys}$ samples after stage 3 at 85°C, whereby the copper incorporation to the acceptor state, a control variable in this case, has been isolated. In this experiment, we applied bias before completely maximizing the copper incorporation to the acceptor state. For better isolation of this control variable, application of the bias and placing D_0 at stage 3 would be more appropriate.

After the PID stress testing, attempts for thermal recovery (85°C) of the PID in the $-V_{sys}$ sample between D_n and D_r did not result in increased power; but instead, additional degradation occurred, attributable to continued Cu diffusion to the junction. For the final IEC 61215-2 MQT 19 stabilization step between D_r and L_n it can be seen that the increase in power of the PID-degraded sample under $-V_{sys}$ bias and the non-PID producing conditions, $+V_{sys}$ and OC, are all of about the same magnitude. Also taking into account that thermal PID recovery did not occur between D_n and D_r, it is likely that the power increase with stabilization at L_n is associated with metastability rather than PID recovery in this module type.

A solution for dealing with these unisolated control variables for determining the actual power loss due to PID is to measure the power change relative to the in-chamber reference module undergoing damp heat without applied bias. The change in power of the CdTe modules in the $+V_{sys}$ and $-V_{sys}$ configurations relative to the unbiased modules is shown in Fig. 2(c). The $-V_{sys}$ module data, for example, are calculated at each stage i as $P_{-Vsys, i}/P_{-Vsys, Do} - P_{OC, i}/P_{OC, Do}$. Viewing the data in this way, we first see the fractional power loss in the PID-stressed modules devoid of the influence of the control variables, such as the effect of Cu_{Cd} acceptor-state formation and Cu diffusion into the junction. The $+V_{sys}$ biased module, despite the varying power measured over the course of the PID stress test, is absolutely flat in Fig. 2(c) after the D_0 normalization point where the voltage bias was applied. This clarifies that the $+V_{sys}$ polarity does not produce PID. This further gives us confidence that the signal we are seeing in the $-V_{sys}$ biased module is free of effects of control variables and representative of the actual PID. Secondly, we can see that there is no PID recovery in the PID-affected module after the adjustment that was made in Fig. 2(c)—the increase in power we see in Figs. 2(a) and 2(b) at the L_n stage is consistent with the other samples, and therefore, attributable largely to reversible components of metastability.

Based on previously published results [7], isolation of the control variable associated with copper diffusion into the junction over the course of testing leading to fill factor degradation may also be addressed to an extent by applying forward bias voltage or light during the course of the PID stress test. These possibilities are included in the draft standard IEC 62804-2 "Test methods for the detection of potential-induced degradation – Part 2: Thin film." However, these require careful consideration of current and load because placing the cells in forward bias during 85°C tests has been shown to cause additional degradation in some cases; e.g., with the module maintained at maximum power [13] or illuminated and in short circuit [8]. Further, application of light while maintaining the prescribed RH uniformly on the module surface during a damp-heat test may be experimentally challenging because of the required irradiance and temperature uniformity. Further studies on the behavior of light and voltage bias over the junction during PID stress tests are under way.

III SUMMARY AND CONCLUSIONS

There are multiple factors affecting the power that we measure from thin-film modules over the course of stress tests. To better isolate control variables such as metastability, contact annealing, and Cu migration from an independent variable such as system voltage stress, this paper proposes the following:

1) Insert a dry-heat step at or above the stress temperature before application of the intended stress factor or independent

variable (e.g., humidity, voltage bias) to precipitate beforehand processes that will occur at the eventual stress temperature.

2) Because of unisolated control variables leading to, for example, power increases during testing, use reference modules (*in-situ* controls) that follow the samples through the test process omitting only the key independent variable (e.g., system voltage), so that the effect of the intended stress can be better gauged.

3) As necessary, include light or forward-bias voltage during the stress test to maintain the junction of the cells in a field-representative configuration and to isolate controlled variables such as Cu ion migration to the junction in CdTe modules. However, careful analysis is required so that degradation or enhancement of power is not additionally introduced.

ACKNOWLEDGEMENT

This work was supported by the U.S. Department of Energy under Contract No. DE-AC36-08GO28308 with the National Renewable Energy Laboratory. The U.S. Government retains and the publisher, by accepting the article for publication, acknowledges that the U.S. Government retains a nonexclusive, paid up, irrevocable, worldwide license to publish or reproduce the published form of this work, or allow others to do so, for U.S. Government purposes.

REFERENCES

[1] C.A. Deline, J.A. del Cueto, D.S. Albin, S.R. Rummel, J. *Photon. Energy* 2(1) 022001 (2012).

[2] Terrestrial photovoltaic (PV) modules – Design qualification and type approval – Part 2: Test procedures, International Electrotechnical Commission, Geneva (2016).

[3] M. Igalson, M. Cwil, M. Edoff, *Thin Solid Films* 515, 6142–6146 (2007).

[4] P. Zabierowski, U. Rau, M. Igalson, *Thin Solid Films* 387, 147–150 (2001).

[5] S.H. Demtsu, D.S. Albin, J.R. Sites, W.K. Metzger, A. Duda, *Thin Solid Films* 516 (8), 2251–2254 (2008).

[6] N. Strevel, L. Trippel, M. Gloeckler, *Photovol. International* 17, 1–6, (2012).

[7] C.R. Corwine, A.O. Pudov, M. Gloeckler, S.H. Demtsu, J.R. Sites, *Solar Energy Materials & Solar Cells* 82(4), 481–489 (2004).

[8] K. Sakurai, K. Ogawa, H. Shibata, A. Masuda, EU PVSEC 2016, 20–24 June 2016, Munich (5DO.10.3)

[9] A. Rubin, *Statistics for Evidence-Based Practice and Evaluation*, Brooks/Cole Cengage Learning, 2013.

[10] S. Yamaguchi, S. Jonai, K. Hara, H. Komaki, Y. Shimizu-Kamikawa, H. Shibata, S. Niki, Y. Kawakami, A. Masuda, *Jap. J. Appl. Physics* 54, 08KC13 (2015).

[11] The 55°C dark heat soak was performed at a lower temperature than the subsequent PID stress temperature of 85°C. As it will be shown, the dark heat soak should have preferably been performed at or above the eventual stress temperature to activate any temperature-dependent processes before the subsequent stress test.

[12] P. Hacke, K. Terwilliger, R. Smith, S. Glick, J. Pankow, M. Kempe, S. Kurtz, 37th IEEE PV Specialists Conference, Seattle, WA, 814–820 (2011).

[13] P. Hacke, K. Terwilliger, S.H. Glick, G. Perrin, J. Wohlgemuth, S. Kurtz, K. Showalter, J. Sherwin, E. Schneller, S. Barkaszi, R. Smith, *J. Photonics Energy* 5, 053083 (2015).

978-1-5090-5606-4/17 $31.00 © 2017 IEEE

A Fine Model of Power Degradation for Crystalline Silicon Solar Modules

Wenshuang He[a], Baosong Duan[b], Fumei Wang[a], Ao Wang[a], Jipeng Chang[a], He Wang[a], Hong Yang[a,*], Jie Ding[c], Junjun Zhang[c], Jingsheng Huang[c]

[a]MOE Key laboratory for Nonequilibrim Synthese and Modulation of Condensed Matter, School of Science,

Xi'an Jiaotong University, Xi'an 710049, People's Republic of China

[b]Xi'an Communications Institute, Xi'an 710106, People's Republic of China

[c]China electric power research institute, Nanjing, 210003, People's Republic of China

*Corresponding author: Hong Yang, hongy126@126.com

Abstract — In this paper, a detailed model of power degradation was established by tracking the Photovoltaic (PV) power plant installed in Jinchang, China. And using this model predicts that the mean lifetime of PV modules in this plant could be up to 30 years. In addition, we found that the power distribution of solar modules in this plant follows a Gaussian distribution. Finally, the accelerated stress tests were used to assess reliability of the PV modules and verify the accuracy of model that we established.

Index Terms — photovoltaic module, power distribution, detailed model, reliability.

I. INTRODUCTION

Reliability evaluation based on degradation models is commonly applied in highly reliable products as a cost effective and confident way of evaluating their reliability [1]. In the whole photovoltaic industry, PV modules are often considered to be the most reliable component of a PV system. At present, PV module manufacturers provide a warranty of 25 years with 20% reduction in its power output over this period [2-4]. In order to achieve the goal of 25 years lifespan, a lot of work on how to improve the reliability of PV modules had been done over the past 50 years [5]. But there's no systematic and comprehensive research about the reliability of PV modules. In recent years, due to the stimulation of market and shortage of energy, the application of PV modules rose sharply [6]. And the reliability of PV modules has been greatly improved along with the improvement of packaging materials and packaging equipment. Therefore, it is high time to carry out in-depth research about outdoor test of PV modules in order to improve the reliability further, reduce the power degradation, and prolong the service lifetime of the PV modules.

In general, the degradation of photovoltaic module is assessed by the output power, and the power loss during its lifetime compared to its initial power. Crystalline silicon solar modules are exposed to outdoor with harsh conditions. And environmental conditions strongly influence the performance

and reliability of PV modules. From all of the degradation, the main influence factors are: ultraviolet radiation, temperature and humidity [7-8]. People cannot wait 20 or 25 years to see what happens to the PV modules. So it's necessary to establish the degradation model of PV modules. And the accelerated reliability tests are necessary to assess whether a module type will survive 25 years anywhere it is deployed. We strongly believe that in this new scenario the aid of reliability models based on degradation may become of utmost importance as a tool to quantify reliability parameters and thereby to establish realistic (for the customer) and defensible (for the manufacturer) power warranties.

Although the reliability evaluation based on the degradation model is an important method to evaluate the reliability of the product, there are few studies of the reliability models in photovoltaic. In this paper, a fine degradation model was established by tracking the PV plant installed in Jinchang, China. And we found that the power distribution of solar modules in this plant follows a Gaussian distribution. Then the accelerated tests are used to verify the accuracy of the model, and to assess the reliability of the PV modules that installed in Jinchang of China. Finally, the model we established was used to predict that the mean lifetime of PV modules in this plant could be up to 30 years.

II. EXPERIMENTAL RESULTS AND DISCUSSION

The data is collected from the PV power plant located in Jinchang of Gansu province, the northwest of China. And Jinchang is located at 37.47 degrees North latitude and 102.43 degrees East longitude with a temperate continental climate where is dry, hot and rainless all the year round. The scene of this PV power plant is showed in Fig.1.

Fig. 1.   The field scene of PV power plant

A. System description

The crystalline silicon solar cells used in this study is 156 mm×156 mm polycrystalline silicon solar cells. The tabbing and stringing are carried out by manual welding respectively by using a 1.6 mm width tinned copper ribbon. The polycrystalline silicon solar module consists of a glass superstrate, polycrystalline silicon solar cell, polymer layers and tinned copper ribbons. The back side of the module is a white TPA with a stronger anti-ultraviolet aging ability. The circuit diagram of solar module connected by 60 solar cells (3 substring) in series and each substring contains 20 solar cells. The capacity of the plant is 50 MWp which divided into 50 PV units. Two modules were randomly selected in each unit, so we got tracking test data of 100 modules.

The average value of the important electrical performance parameters for the 100 PV modules are shown in the following table. Voc is the average open circuit voltage, Isc is the average short circuit current, Vm is the average maximum voltage, Im is the average maximum current, and Pm is the average maximum power.

Table I
THE PARAMETERS OF PV MODULES

	Voc (V)	Isc (A)	Vm (V)	Im (A)	Pm (W)
Before running	37.54	8.65	29.92	8.25	246.87
After running 1 year	37.41	8.61	29.85	8.14	242.90
After running 2 years	37.39	8.59	29.77	8.05	239.72

With handing the data of 100 modules by mathematical statistical method, we found that the power distribution of the whole plant follows the Gauss distribution before and after two years outdoor test, the curves of power distribution of 100 modules were obtained by software SPSS as shown in Fig.2 and Fig.3. And the probability density is given by [9]:

$$f(P) = \frac{1}{\sigma \cdot \sqrt{2\pi}} \exp(-\frac{1}{2}(\frac{P - P(t)}{\sigma})^2) \quad (1)$$

Where P(t) is the mean power value of the whole plant at a given time and σ is its standard deviation.

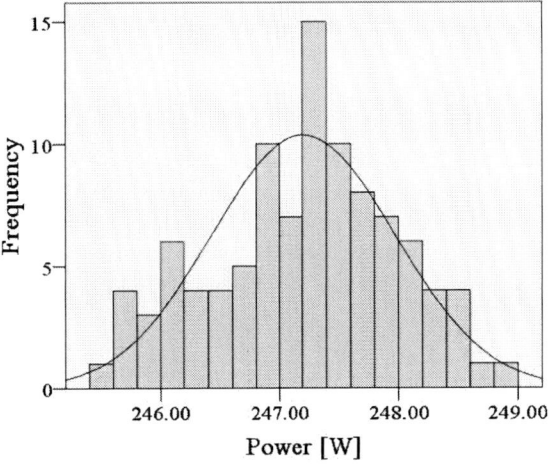

Fig. 2.   The power distribution of 100 modules before outdoor test

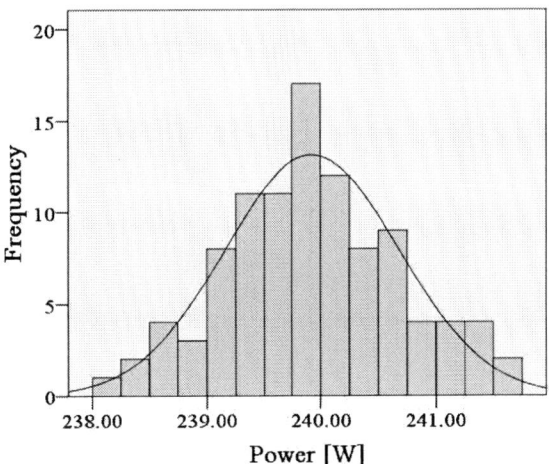

Fig.3.   The power distribution of 100 modules after 2 years outdoor test

B. The degradation model

From the regular of power degradation of the PV modules, we found that average power degradation of modules over time is not linear. In the first few months of operation, the rate of power degradation is relatively high due to the light induced degradation. And the rate of power degradation decreases gradually and tends to be stable in the later running time. So we propose a non-linear model for power degradation of PV modules output power given by [9]:

$$P(t) = P_0 \times [1 - \exp(-\alpha \cdot \beta)] \quad (2)$$

Where P_0 is the power of the PV modules before the outdoor test, α and β are parameters of the degradation model.

Concretely, P_0 is the average power of 100 modules before the outdoor test, α and β can be received by the data of the first two years and the curve fitting of the custom meaning function, so we got the following specific model expression.

$$P(t) = 246.87 \times [1 - \exp(-4.17 \times t^{-0.28})] \quad (3)$$

Fig.4. shows the variation of the power degradation with time. There are three red triangle symbols, and they represent the actual measured data in the past two years. The black smooth line is the function model of the power degradation. Through the following figure, we found that after 30 years of operation, the average power of PV modules is about 200W, and the rate power degradation has not up to 20%. It means that the working lifetime of PV modules could be up to 30 years.

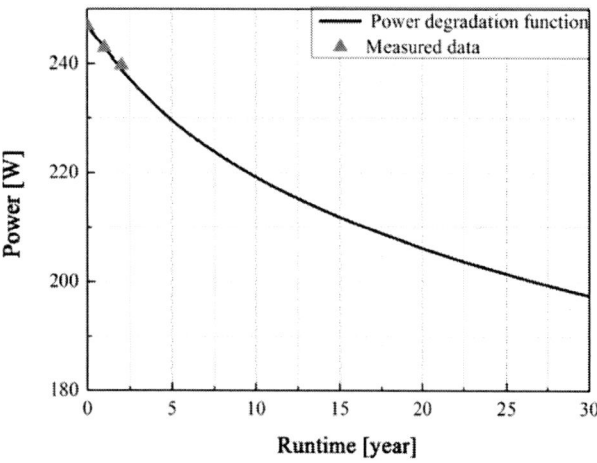

Fig. 4.   The relation of the power of PV modules to the time

C. Qualification testing

Most commercial crystalline silicon PV module manufactures qualify their modules with standard IEC 61215 (crystalline silicon terrestrial PV modules—design qualification and type approval) [10]. In this section, 4 PV modules were used as the experimental samples. These modules have been running outdoors for two years, and numbered A, B, C, D. IEC 61215 tests are not reliability tests but can indeed provide some information on reliability, which include the many stress tests, the specific tests utilized in this paper are as follows [11-13].

• Damp heat (DH) exposure at 85 ℃ and 85% relative humidity for 1000 hours.

• 500 Thermal cycles (TC) from -40 ℃ to +85 ℃.

Under 85 ℃, 85% relative humidity, A and B were tested under 1000 hours, 1500 hours and 2000 hours. C and D were tested 50, 200 and 500 thermal cycles from -40 ℃ to +85 ℃. The experimental results take shown in the Table II and III.

Table II
DAMP HEAT TESTING

Module A		
Test	Pmax (W)	Power Loss (%)
Initial	240.50	--
DH1000 h	237.92	1.08%
DH1500 h	233.26	3.01%
DH2000 h	228.43	5.02%
Module B		
Test	Pmax (W)	Power Loss (%)
Initial	238.30	--
DH1000 h	236.03	0.95%
DH1500 h	231.25	2.96%
DH2000 h	226.40	4.98%

Table III
LONG TERM THERMAL CYCLING TESTING

Module C		
Test	Pmax (W)	Power Loss (%)
Initial	241.70	--
50TC	240.78	0.38%
200TC	240.66	0.43%
500TC	240.39	0.54%
Module D		
Test	Pmax (W)	Power Loss (%)
Initial	240.00	--
50TC	239.09	0.38%
200TC	238.96	0.43%
500TC	238.68	0.55%

Formal qualification tests have strict pass criteria. For example, in IEC 61215 a power loss of more than 5% in any one test is grounds for failure. According to the results of accelerated life test, after 1000 hours Damp heat of 85/85%, the maximum power loss of PV modules is 2.6 W, which only account for 1.08% of the initial power. After 500 thermal cycles from -40 ℃ to +85 ℃, the maximum power loss of PV modules is 1.31 W, which only account for 0.54% of the initial power. It is obvious that these modules had lost less that 5% of their initial power, which pass the test easily.

Fig.5. and Fig.6. show the relationship between power degradation rate with Damp heat time and Thermal cycles respectively. It is easy to calculate that after 5750 hours of Damp heat test, the power degradation rate of PV module will up to 20%, it means that the PV module will failure. And 54150 Thermal cycles could achieve the same effect. In addition, it indicates that the influence of damp heat on power degradation of PV modules is much greater than that of thermal cycles. Compared with the data from outdoor testing, we found that the PV modules can run 10 years under the influence of damp and temperature, but can run 40 years under the single factor of temperature. It is undeniable that under the conditions of actual outdoor operation, there will not be a single factor to influence the power degradation of PV modules, and it must be the result of the interaction of many factors. This means that it is not accurate to use the accelerated life test in predicting the life of PV modules, but at least it reflects the reliability of modules in some ways.

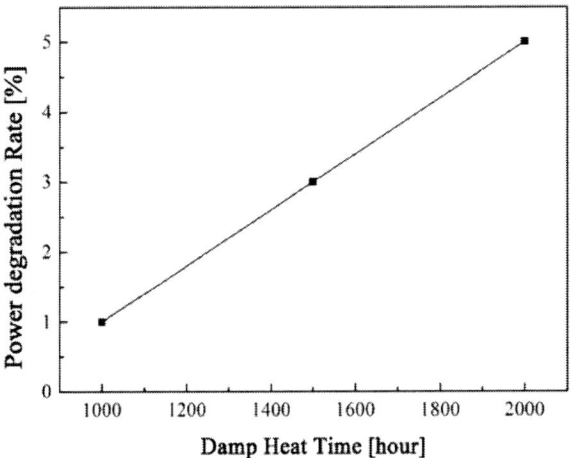

Fig. 5.   Relationship between degradation rate and Damp heat time.

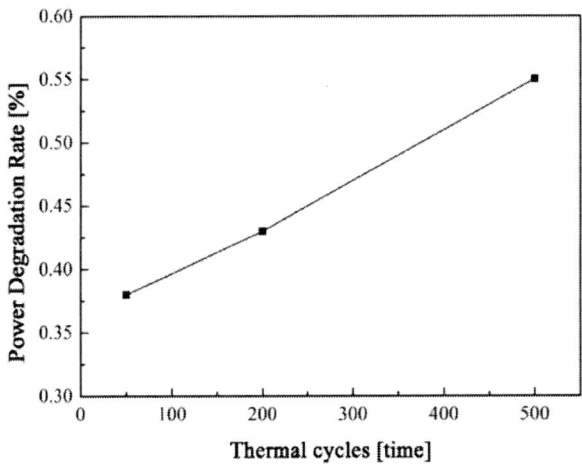

Fig. 6.   Relationship between degradation rate and Thermal cycles

III. CONCLUSION

In this paper, we made a complete and systemic analysis of the power degradation. Then a non-linear model for power degradation of PV modules was established which dismissed the simplified liner model. Using this model, the mean working lifetime of the plant installed in Jinchang was predicted that could be up to 30 years. In addition, we found that the power distribution of the plant follows a Gaussian distribution at a given time. As a result, the power distribution

of a plant and the power degradation of a PV module can be predicted. Finally, the accelerated stress tests were used to assess long reliability of the PV modules installed in Jinchang.

ACKNOWLEDGEMENT

The authors would like to thank the support of Natural Science Foundation of China (Grant No. 61376067 and 61274050). This study was also supported by the National Key Technology Research and Development Program of the Ministry of Science and Technology of China (Grant No. 2015BAA09B01).

REFERENCES

[1] Vázquez M, Rey-Stolle I. Photovoltaic module reliability model based on field degradation studies[J]. Progress in Photovoltaics Research & Applications, 2008, 16(5):419-433.
[2] Conibeer G J, Richards B S. A comparison of PV/electrolyser and photoelectrolytic technologies for use in solar to hydrogen energy storage systems[J]. International Journal of Hydrogen Energy, 2007, 32(14):2703-2711.
[3] Kuitche J M. A Statistical Approach to Solar Photovoltaic Module Lifetime Prediction[J]. Dissertations & Theses - Gradworks, 2014.
[4] Alshushan M A S, Saleh I M. Power degradation and performance evaluation of PV modules after 31 years of work[C]// Photovoltaic Specialists Conference. IEEE, 2013:2977-2982.
[5] John H. Wohlgemuth, Daniel W. Cunningham, Paul Monus, Jay Miller, and Andy Nguyen "Long Term Reliability of Photovoltaic Modules", 4th WPEC, 2006
[6] Wohlgemuth, J., et al. "Failure modes of crystalline Si modules." PV Module Reliability Workshop 2010.
[7] Sharma V, Chandel S S. Performance and degradation analysis for long term reliability of solar photovoltaic systems: A review[J]. Renewable & Sustainable Energy Reviews, 2013, 27(6):753-767.
[8] Wang H, Wang A, Yang H, et al. Performance degradation of crystalline silicon solar module in various ultraviolet radiation area[C]// IEEE, Photovoltaic Specialists Conference. IEEE, 2016:1757-1760.
[9] Munoz M A, Alonso-Garca M C, Vela N, et al. Early degradation of silicon PV modules and guaranty conditions[J]. Solar Energy, 2011, 85(9):2264-2274.
[10] International Electrotechnical Commission (IEC) 61215: 2nd edn, 2005. Crystalline silicon terrestrial photovoltaic modules—Design qualification and type approval.
[11] Ndiaye A, Charki A, Kobi A, et al. Degradations of silicon photovoltaic modules: A literature review[J]. Solar Energy, 2013, 96(2013):140-151.
[12] IEC 61215 "Crystalline silicon terrestrial photovoltaic (PV) modules –Design qualification and type approval."
[13] John Wohlgemuth, "Overview of Failure Mechanisms and PV Qualification Tests", Proceedings of PV Module Reliability Workshop,http://www1.eere.energy.gov/solar/pv_module_reliab-ility_workshop_2010.html,2010.

Test Methods for Hydrophobic Coatings on Solar Cover Glass

Kenan Isbilir, Biancamaria Maniscalco, Ralph Gottschalg and John Michael Walls

CREST, Wolfson School of Mechanical, Electrical and Manufacturing Engineering, Loughborough
University, Loughborough, Leicestershire, LE11 3TU, UK

Abstract — The world market for solar energy continues to expand. However, to be competitive with traditional energy sources, photovoltaic (PV) modules must be capable of continuous and reliable high performance. Performance losses occur due to the soiling of the cover glass on modules. Soiling can be reduced by using hydrophobic coatings. These decrease surface energy and thus minimize adhesion to soiling. These coatings can help reduce maintenance and retain consistent electrical output. It is not yet clear, how hydrophobic coatings can be assessed and compared. In this paper, test methods to simulate the stresses that coatings experience in their life-time are assessed. These test methods help to predict the durability and useful lifetime of the coatings when applied to solar cover glass. Various test methods from different standards have been applied to hydrophobic coated glass surfaces and optimized to simulate real-outdoor conditions. A sand impact test and a water drop simulation test have been devised to study the effect of sand and rain on hydrophobic performance and durability.

Index Terms — hydrophobic coating performance, photovoltaic (PV) anti-soiling, coating durability tests.

I. INTRODUCTION

Solar Photovoltaic (PV) modules need to be highly efficient and reliable to compete with conventional energy sources. Currently they are produced with a manufacturer's warranty period of 25 years [1]. It is crucial to maintain the maximum performance throughout their life-time.

Soiling on PV cover glass is a significant problem that affects performance and requires costly maintenance. Dust and grime accumulation is a complex phenomenon and is influenced by diverse site-specific environmental and weather conditions. Accumulation of debris such as dust, sand, bird-droppings and water-stains (salts) cause a reduction of the incident solar irradiance to the PV absorber and degrade the efficiency of the module [2]. The presence of soiling causes a reduction in transmitted light into the module. It also leads to inhomogeneous shading with an increased possibility of triggering hot spots [3]. It causes a reduction in the power output, thus increasing the cost of energy production. The solar industry schedules periodic cleaning of the PV arrays in solar farms to prevent the build-up of soiling on the module cover glass. However, the ongoing cost of maintenance is an additional financial burden for the operator.

The surface of the PV cover glass can be coated with a thin layer that acts as an anti-soiling coating. This layer can either be hydrophilic [4], that has high surface energy which attracts sufficient water to clean the surface or hydrophobic that has low surface energy which repels the water and washes the

soiling away. These coatings reduce soiling and minimise the frequency of the cleaning task. The coating must be transparent otherwise it would cause optical losses.

The cleaning action of hydrophobic coatings stems from their high water contact angle. When water impinges on the hydrophobic surface, droplets of almost spherical shape form. These then roll away, carrying away the dust and dirt [5], [6]. Hydrophobic coatings have the potential to play an important role in reducing the level of soiling by presenting a low surface energy for reduced adhesion. They also make cleaning and maintenance easier for retaining performance [7].

Figure 1-Measurement of the water contact angle on a glass surface with a hydrophobic coating

The life-time of hydrophobic coatings is crucial to sustain performance and ideally should match the life-time of the module. The coatings should resist the harshest outdoor conditions. The potential of hydrophobic coatings has already been recognized commercially and they are beginning to be deployed in the field. However, standards do not yet exist for performance and durability assessment. It is important that this is addressed to enable a fair and unbiased assessment of different anti-soiling coatings. In this work, methods to test and predict the durability and performance of hydrophobic coatings for PV cover glass are assessed. The results presented are obtained from hydrophobic coatings under development for application to solar cover glass. The formulation of the coatings was changed during these trials. However, the purpose of this study was to develop testing techniques for performance and durability. The results of tests on specific formulations will be published elsewhere.

978-1-5090-5606-4/17 $31.00 © 2017 IEEE

II. EXPERIMENTAL

Tests to assess the performance of hydrophobic coatings include optical transmission, water contact angle and roll-off angle. The durability tests include damp heat, ultraviolet (UV) exposure, thermal cycling, adhesion, abrasion resistance, solubility, sand impact and rain drop simulation, all of which simulate outdoor conditions or exposure to cleaning and maintenance. These tests have been performed on a variety of hydrophobic formulations. The test results reported are on hydrophobic coatings under development for commercial use by the PV industry. The focus has been on performance and not cost-effectiveness.

A. Performance of Hydrophobic Coatings for PV Cover Glass

The water contact angle (WCA) and the roll-off angle at a certain surface inclination determines the performance of hydrophobic coatings. In PV applications, the optical measurements also play an important role to avoid any absorption/reflection losses at the air-glass interface. Thus, the transmittance (T) and reflectance (R) measurements are considered important measures of coating performance. For simplicity, transmittance and reflectance measurements at 550 nm are used.

B. Durability

Durability is defined as the ability to withstand mechanical and environmental stresses. Hydrophobic coated cover glass surfaces were exposed to environmental stress tests such as damp heat, thermal cycling and UV exposure tests. These tests were carried out in accordance with IEC 61646:2008 to evaluate the scale and causes of degradation [8]. The surfaces were also exposed to various types of mechanical and chemical stress tests, including adhesion, resistance to abrasion, and water solubility.

The coated glass surface was visually inspected before and after each test. The optical measurements were obtained using a Varian Cary® UV 5000 spectrophotometer. The water contact angle was measured using a Dataphysics OCA 20 system after coating and at increasing time intervals during each of the accelerated durability tests. Optical and Scanning Electron Microscopy (SEM) images (Jeol® 7100F) were obtained to characterize the severity of the degradation. Roll-off angle was measured before and after each test with an in-house developed measurement system, using between 20-40 μl water drops.

A sand impact test (SIT) was devised to evaluate durability to sand erosion. A new water drop test was developed to assess the impact of rain. In both of these devised tests, the samples were positioned on a platform at a 45° angle, which is adjustable.

III. RESULTS

A. Damp Heat Test

The hydrophobic coatings were exposed to a damp heat test at 85°C and 85% relative humidity (RH). The surface was exposed for a total of 1000 hours with intermediate optical and water contact angle measurements made at 100, 250 and 500 hours. The performance of the hydrophobic coating was severely reduced by the damp heat test after 1000 hours of exposure. Table I shows T, R, roll-off angle and WCA measurements for initial, post 500 hours and final measurements until the surface lost the hydrophobic property which is defined when WCA is measured below 90° [5]. The increase in roll-off angle shows that the damp heat exposure also increases the adhesion of the water drops. Despite degradation in hydrophobic performance, the optical performance showed a slight improvement associated with the effects of heat treatment on the coating.

TABLE I
DAMP HEAT TEST RESULTS

	Initial	500 hours	1000 hours
WCA (°)	106.3	90.7	83.1
Roll-off angle (°)	73.0	-	78.2
R (%)	7.3	7.0	6.3
T (%)	92.9	92.8	93.1

B. Thermal Cycling Test

The hydrophobic coatings were stressed using a thermal cycling test. One cycle comprises of 10 minutes of exposure to 85°C followed by 10 minutes at -40°C with no humidity control according to the testing protocol IEC 61646:2008. The sample was exposed to a total of 200 cycles. The optical and hydrophobic measurement results in Table II shows that the coating was resilient to thermal cycling. The WCA measurements reveal a gradual degradation with increasing number of cycles with a reduction of 4.0% after 200 cycles. Transmittance and reflectance measurements show no significant change. However, an improvement in the roll-off angle is observed on exposure to thermal cycling.

TABLE II
THERMAL CYCLING TEST RESULTS

	Initial	100 cycles	200 cycles
WCA (°)	106.3	104.5	102.1
Roll-off angle (°)	72.9	-	50.1
R (%)	7.0	6.8	6.7
T (%)	93.1	92.9	92.9

C. UV Exposure

The UV Exposure test was applied to hydrophobic coated surfaces for 500 hours with an intermediate check at 250 hours to determine its effect. The test was conducted with a minimum of 15 kWh/m^2 of UV light with a 3% to 10 % of the total energy in the UVB light range. The WCA measurement result shows that the UV exposure affects the hydrophobic performance of the coatings noticeably, but does not affect the optical performance significantly. The roll-off angle is increased slightly with UV exposure. This shows that the effect of UV exposure on roll-off angle and WCA can vary separately depending on the coating chemistry, as the two are not related [9]. The T, R, roll-off and WCA results are shown in Table III.

TABLE III
UV EXPOSURE TEST RESULTS

	Initial	250 hours	500 hours
WCA (°)	106.3	100.7	99.1
Roll-off angle (°)	72.5	-	79.4
R (%)	7.8	7.6	7.3
T (%)	92.7	92.3	92.4

D. Adhesion Test

A tape adhesion test [10] was used to determine the adhesion of the coatings to the glass substrate. The surface was tested according to Mil-C-675C with minor changes. A 1inch long and 1inch wide adhesive tape was pressed firmly on the coated area for 10 seconds with an extra 1inch to hold the tape. The tape is then removed rapidly within 1s. The results revealed no peeling or detachment of the coating. No change was observed to the hydrophobic and optical performance. The results shown in Table IV are consistent with good adhesion.

TABLE IV
ADHESION TEST RESULTS

	Initial	Post exposure
WCA (°)	106.3	105.9
Roll-off angle (°)	71.0	72.5
R (%)	7.9	8.0
T (%)	92.6	92.6

E. Abrasion Resistance Test

The abrasion resistance test is important to assess the durability of hydrophobic coatings when exposed to maintenance or cleaning cycles. A test method similar to BS EN 1096-2 [11] was performed using a reciprocating abraser. A CS-10 grade abrader material [12] was used to simulate the degradation caused by particles. This provided a 'medium to hard' abrasion of the hydrophobic coating. Figure 2 shows an optical microscope image of an abraded area on the coating tested. The optical results are similar to the uncoated glass suggesting that in some cases most of the coating was removed during the test.

Figure 2-Optical Image of the coating after Abrasion Resistance Test

The initial and final results are shown in Table V. The WCA was reduced dramatically to below 90°. A significant increase in the roll-off angle of the coating could be caused by the increased roughness or hydrophobic coating removal in the abraded area..

Figure 3 shows the transmittance measurements over the 350 to 800 nm wavelength range for comparison with an uncoated glass sample. As seen on the graph, the transmittance of the coating (green) reduced to the level of the uncoated glass (amber) after the abrasion test.

TABLE V
ABRASION TEST RESULTS

	Initial	100 cycles	200 cycles
WCA (°)	104.8	76.5	70.2
Roll-off angle (°)	44.3	70.2	89.1
R (%)	7.2	8.3	8.3
T (%)	92.9	92.0	92.0

F. Solubility Test

Hydrophobic coated samples were used to evaluate the applicability of a solubility test according to BS ISO 9211-4 [13]. There are 12 levels of severity for these tests. The severity can be increased gradually. The least severe exposure is immersion in DI water and the most severe exposure consists of immersion of the coating in boiling DI water for 2 minutes followed by immersion in DI water at room temperature for 1 minute. The coatings were exposed gradually to the most severe condition. Table VI shows the measurements after 10 cycles at the most severe exposure level.

Optical measurement results indicate that the coating had slight degradation in transmittance and reflectance. However a lower WCA was measured at the edges of the coated samples after tests at severity level 3. This is a total 96 hours of immersion in de-ionized (DI) water. Nevertheless, the coating preserved its hydrophobicity until testing at severity level 12. The post-exposure WCA results shown in Table VI are average values over the surface. The lowest was 99.6° and highest was 118.5° for the sample tested.

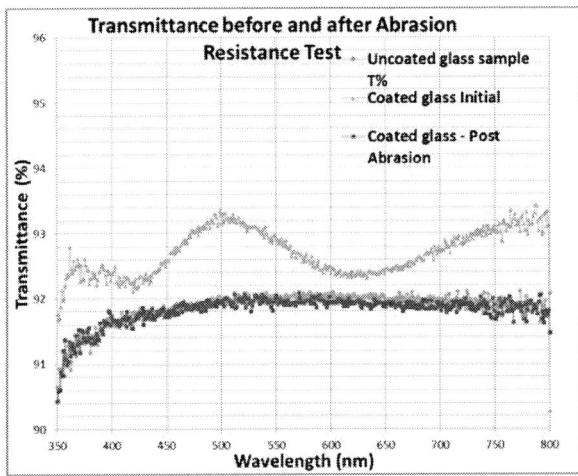

Figure 3-Measured transmittance before and after the Abrasion Resistance Test

TABLE VI
SOLUBILITY TEST RESULTS

	Initial	Post exposure
WCA (°)	118.6	104.4
Roll-off angle (°)	47.6	75.3
R (%)	4.7	5.3
T (%)	95.1	94.7

G. Sand Impact Test (SIT)

Hydrophobic coated samples were exposed to 500g of sand impacting the surface in 90s. The system was adopted from ASTM D968-15 [14]. The size of the sand particles used was ~0.25 mm. The results shown in Table VII correspond to the coating performance after sand exposure. The optical performance is measured approximately on the same point on each of the samples. The WCA results shown are the average of measured values over the coated area. The hydrophobic performance was not measured after the SIT without cleaning to avoid misleading results due to the retention of sand particles. The results after cleaning with pressurized air were compared with surfaces cleaned with IPA and DI water.

The SEM images in Figure 4 – A and 4 – B show the effect of abrasion on the surface of the coating. Some areas were only slightly damaged while on others the coating was completely removed. The optical and hydrophobic measurement results confirm the areas damaged by different degrees results in different levels of performance degradation. The lowest transmittance and reflectance measured over the coated area was 94.4% and 5.2%, respectively. The lowest WCA measured post SIT after cleaning with air and with IPA/DI water was 62.5° and 70.0°, respectively.

TABLE VII
SAND IMPACT TEST RESULTS

	Initial	Post SIT- no cleaning	Post SIT- cleaning with air	Post SIT- cleaning with IPA and DI water
WCA (°)	119.5	-	84.6	92.8
Roll-off angle (°)	76.8	-	-	>90.0
R (%)	2.8	4.4	3.4	4.2
T (%)	96.8	96.5	96.0	96.1

Figure 4-A - SEM image of a damaged area on the coating after Sand Impact Test

H. Rain Drop Simulation Test (RDST)

Deionized (DI), rain and tap water are used to simulate the effect of rain. The test was carried out for a continuous period of 1 minute to evaluate the coating performance in a controlled laboratory environment. In this way, outdoor factors such as temperature, dust etc. that may affect performance could be controlled. The diameter of the simulated drops was measured to be between 2 to 3 mm on a flat glass substrate. After exposure, the sample was left to dry under atmospheric conditions. The results show that the optical and hydrophobic performance was reduced mainly by water stains remaining on the surface after drying.

The results obtained using tap water are shown in Table VIII. The results indicate that the effect of water stains is highly detrimental to the coating performance. A slight increase in the WCA is possibly due to additional roughness caused by the exposure. However this additional roughness also caused an increase in roll-off angle. The water stains obstruct the movement of the water drop at roll-off. Figure – 5 is the image of a coating surface after drying under atmospheric conditions.

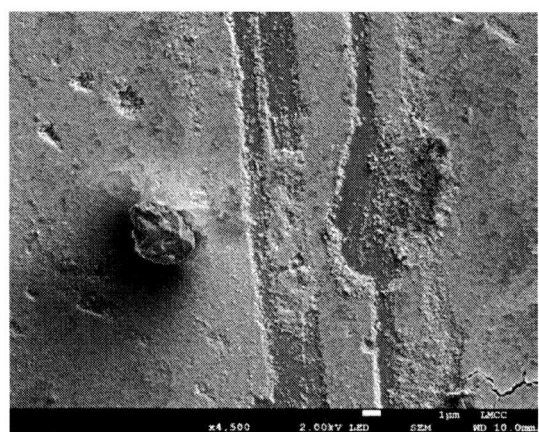

Figure 4-B - SEM image of partially damaged area after Sand Impact Test

TABLE VIII
RAIN DROP SIMULATION TEST (RDST) RESULTS

	Initial	Post RDST-highest	Post RDST-lowest	Post RDST-after cleaning	Post RDST-stain after cleaning
WCA (°)	118.9	128.0	52.0	125.0	90.0
Roll-off angle (°)	61.1	-	>90.0	83.0	>90
R (%)	1.3	1.1	5.4	1.5	1.4
T (%)	98.6	98.2	96.0	98.5	98.4

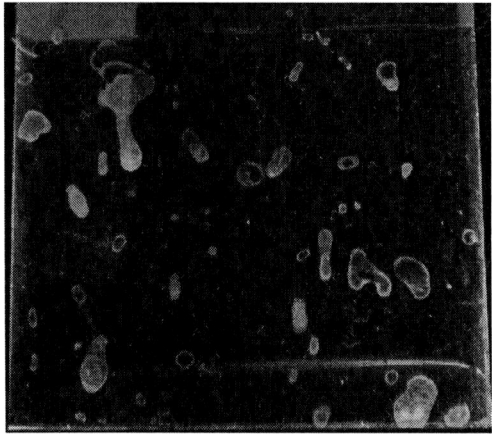

Figure 5-Surface of the coating post-RDST after drying at room temperature

IV. CONCLUSION

Hydrophobic coatings on solar cover glass have the potential to minimize the accumulation of soiling and maintain module performance. The coatings are already becoming available commercially. Durability is a key concern. It will determine if the coatings will need to be re-applied and at what time interval. It is clear that durability will be dependent on the local environment. It is vital that the optical transmission of the coating is maintained. This paper considers the tests and standards that should be applied to hydrophobic coatings to ensure they are fit for purpose.

The environmental stress tests which are adopted from PV qualification standards are used to evaluate the degradation of hydrophobic coatings for PV modules. The test results show that the hydrophobic performance is degraded at different levels for different environmental stresses. Increase in optical performances was observed if heat is involved which acts as heat treatment of the coating surface.

Sand Impact Tests show how both optical and hydrophobic performance of coatings can suffer during the exposure. The coatings can be partially or completely removed. Optical measurements show that the point where the coating first suffered from sand impact has highest degradation. Variation between the WCA results over the coated surface was investigated using SEM. The images confirm the different effects of the sand impact at different locations.

The results from the rain drop simulation test emphasize that the roll-off angle should be below a 'critical value', so that the water drops clean the surface and reduce the risk of staining on the surface. Water stains are detrimental for the hydrophobic properties of coatings. Ideally, the roll-off angle should be below the installation angle of the PV modules.

V. SUMMARY

In this paper, the testing methods and performance requirements for hydrophobic coatings used on solar cover glass are compiled and assessed. The results of different environmental and mechanical stresses on hydrophobic coatings under development for commercial use by the PV industry are analysed. It is clear that the performance of the coatings may vary according to local environmental conditions. However, hydrophobic and optical performance should be consistent. The challenges for durability and performance of hydrophobic coatings in the field are highlighted.

ACKNOWLEDGEMENT

The authors are grateful to EPSRC for support of the SOLplus project funded through EP/N510014/1. The authors acknowledge use of facilities within the Loughborough Materials Characterisation Centre (LMCC).

REFERENCES

[1] A. Ndiaye, A. Charki, A. Kobi, C. M. F. Kebe, P. A. Ndiaye, and V. Sambou, "Degradations of silicon photovoltaic modules: A literature review," *Sol. Energy*, vol. 96, pp. 140–151, 2013.

[2] J. K. Kaldellis and P. Fragos, "Ash deposition impact on the energy performance of photovoltaic generators," *J. Clean. Prod.*, vol. 19, no. 4, pp. 311–317, 2011.

[3] H. Qasem, T. R. Betts, H. Müllejans, H. Albusairi, and R. Gottschalg, "Dust-induced shading on photovoltaic modules," *Prog. Photovoltaics Res. Appl.*, vol. 22, no. 2, pp. 218–226, 2014.

[4] T. Lorenz, E. Klimm, and K. A. Weiss, "Soiling and anti-soiling coatings on surfaces of solar thermal systems - Featuring an economic feasibility analysis," *Energy Procedia*, vol. 48, pp. 749–756, 2014.

[5] H. Dodiuk, P. F. Rios, A. Dotan, and S. Kenig, "Hydrophobic and self-cleaning coatings," *Polym. Adv. Technol.*, vol. 18, no. April, pp. 746–750, 2007.

[6] I. P. Parkin and R. G. Palgrave, "Self-cleaning coatings," *J. Mater. Chem.*, vol. 15, no. 17, pp. 1689–1695, 2005.

[7] B. Brophy, Z. R. Abrams, P. Gonsalves, and K. Christy, "Field Performance and Persistence of Anti-Soiling Coatings on Photovoltaic Glass," in *31st European Photovoltaic Solar Energy Conference and Exhibition*, 2015, pp. 2598–2602.

[8] IEC, "Thin-film terrestrial photovoltaic (PV) modules — Design qualification and type approval," 61646, 2008.

[9] N. K. Mandsberg and R. Taboryski, "The rose petal effect and the role of advancing water contact angles for drop confinement."

[10] Military Specification, "Coating of Glass Optical Elements (Anti-Reflection)," MIL–C–675C, 1986.

[11] BS EN, "Glass in building. Coated glass. Requirements and test methods for class A, B and S coatings," 1096-2, 2012.

[12] ASTM, "Standard Test Method for Abrasion Resistance of Organic Coatings by the Taber Abraser," D4060-14, 2014.

[13] BS ISO, "Optics and photonics - Optical coatings, Part 4: Specific test methods," 9211–4, 2012.

[14] ASTM, "Standard Test Methods for Abrasion Resistance of Organic Coatings by Falling Abrasive," D968-15, 2015.

AUTHOR INDEX

Aaditya, Gayathri 604
Abbas, A........................... 1691, 2457, 3430
Abbas, Ahmed E. 1888
Abbas, Ali 186, 752, 1674
Abbott, Malcolm D............. 1322, 2576, 2600
Abdalla, L.B 1245
Abdallah, Amir A....................................3435
Abdallah, Shaimaa A. 219
Abdellaoui, Imane...................................900
Abdullah, Ahmad...................................2128
Aberle, Armin..2318
Aberle, Armin G. 284, 496, 499, 1922
Ablekim, Tursun 3422
Aboubakr, Benazzouz...............................487
Abouelkhair, Hussain M..........................2324
Abtahi, Amir 638
Abudayyeh, Omar K..................................88
Acebo, Laura 155
Addamane, S. J......................................281
Adewoyin, Adeyinka 2381
Adhikari, Dipendra 2582
Affouda, Chaffra A. 259
Agarwal, Mohit 2330
Agarwal, Sumit......................................1777
Agarwal, Vivek.............. 2952, 2981, 2986, 3050
Agbo, Solomon N....................................2114
Ager, Joel 3410
Agrawal, Rakesh 1449
Aguiar, Jeff 2702
Aguiar, Jeffrey 2467
Aguirre, Rodolfo 2419
Ahamioje, Joseph A. 2931
Ahanzhamejhad, Ramez H..........................170
Ahlswede, E. 3260
Ahlswede, Erik 791
Ahmad, Jawad.......................................3096
Ahmed, Benlarabi 487
Ahmed, Nuha.................................658, 2667
Aho, Arto 297, 2520
Aho, T. 1189
Aho, Timo 297
Ahrenkiel, P..869
Ahrenkiel, Phil 206, 831
Ahrenkiel, Richard K. 3448
Ahrenkiel, S. Phillip 2514
Ahsan, Nazmul 2334
Aierken, Abuduwayiti...............................226
Aindow, Mark 1522
Aïssa, Brahim 3435
Akaki, Yoji...2338
Akari, Shunsuke 2385
Akarm, Muhammad Nadeem 2776
Ake-Sultan, Bernt 2864
Akiki, Tilda...1968
Akimoto, Katsuhiro 33, 160, 900

Akimoto, Naoki...................................... 712
Akiyama, Hidefumi 721, 2781, 3528
Akwari, Chinedum 735, 2446
Al Mahmud, Abdullah 1067
Alahmed, Ahmed.................................. 1110
Alam, Giri Wahyu 1498
Alam, Muhammad A................... 1055, 1259
Alam, Muhammad Ashraful....................... 1904
Alberi, Kirstin 2506
Albin, David....................1196, 3305, 3319
Alcubilla, R. 1781
Alcubilla, Ramón 944
Aleman, Monica 2227, 3435
Alexander, Jessica A. 966
Alfadhili, Fadhil K. 730, 815
Al-Fadhili, Fadhil K. 2462
Al-Ghzaiwat, Mutaz 2593
Algora, C. 1210
Algora, Carlos....................................... 1204
Alharbi, Fahhad H. 963
Alharthi, Yahya Z. 1018, 1110
Ali, Asad 1228
Ali, Jaffar Moideen Yacob....................... 2318
Ali, Waqar... 1228
Alivisatos, A. Paul................................. 1737
Al-Jassim, M. 1196
Al-Jassim, M.M. 1312, 2280, 2785
Al-Jassim, Mowafak.................. 62, 1371, 1381,
  1400, 2789, 2887, 3305, 3319
Al-Jassim, Mowafak M. 3147
Aljaziri, Marwa 2011
Alkhayat, Rabee B. 815
Allebé, C. 50, 2073
Allebé, Christophe..................... 3254, 3256
Allen, Thomas....................................... 2076
Almheiri, Anwar 1946
Almonacid, Florencia 2858
Alrashidi, Hameed.................................. 2858
Altermatt, Pietro P. 1922, 2220, 3304
Alvarez, Diego Alonso 1339
Alvarez, Genesis 2941
Alvarez, José 2453, 2528
Aly, Shahzada P..................................... 963
Alzahmi, Wadhah 1946
Amdemeskel, Mekbib W........................... 2672
Anctil, Annick 2124
Anderberg, A. 467
Andler, Joseph 1449
Ando, Daisuke 931
Ando, Yasutaka 970
Ando, Yuta 192
Andreani, Lucio 290
Angeles-Ordóñez, G. 142
Annigoni, Eleonora.................... 1395, 2794
Anoma, Marc Abou.................................. 1549

AUTHOR INDEX

Anselmo, Andrew..................74, 2839, 2897
Antony, Aldrin 1755
Anttu, Nicklas..................2502
Anyadiegwu, Ifeanacho 970
Anyanwu, Uchechi..................319
Araki, Hideaki..................2338
Araki, Kenji..................359, 412, 1479, 1711, 1714, 1743, 2498, 2548, 2566
Aranguren, G. 643
Archer, Alexander 771
Arehart, A. R...................30, 2414
Arehart, Aaron R...................215, 2446, 3139
Arinze, Ebuka S...................667
Armour, Eric..................827
Armour, Eric A...................210, 2506
Armoush, Maher..................1058
Arnold, Daniel B...................3002
Arnou, Panagiota..................146, 186
Arora, B. M...................396, 1995, 2716
Arora, Brij M...................3478
Arp, Juergen..................1411
Arredondo, C. A...................2031
Artegiani, Elisa..................752, 1669, 2372
Aryal, Krishna..................182
Asadirad, Mojtaba..................866
Asahi, Shigeo..................23
Asgharzadeh, Amir..................1537, 1543, 3333
Ashrafee, Tasnuva..................735
Aslam, Aasma..................2355
Asomoza, René..................632
Astakhov, Oleksandr..................2114
Aswani, U...................1898
Athresh, Eashwer..................2395, 2399
Atia, Adam A...................3230
Atkins, R...................229
Atlan, Olivier..................626
Atwater, Harry A...................512, 521, 558, 572, 1248, 1589, 1737, 2236
Augarten, Yael..................1651
Augusto, André 1589, 2596
Avasthi, Sushobhan........ 251, 837, 841, 986, 2395, 2399
Avenet, Julien..................1933
Avery, J. E...................1863
Awadallah, Osama..................3473
Awasthi, Vishnu..................2345
Ayala, Orlando..................735
Azkona, N...................2740
Azkona, Nekane..................2677
Azzolini, Joseph A...................608
Baba, Masaaki..................1724
Babbe, Finn..................151, 2054
Babcock, Sean J...................2298
Bachman, Benjamin F. 3381
Badel, N...................50
Badosa, Jordi..................626

Badr, Ikken 487
Bae, Soohyun 935
Baggu, Murali 2991
Baik, Sungsun 2242
Bailey, C. 845
Bailey, Christopher G. 2298
Bailey, J. 2414
Bailey, Jeff. 1686, 3327
Baines, Tom 742, 1445
Baka, Maro 3343
Baker, Rupesh 3172
Bakhshi, Sara 322
Bakker, Klaas 2875
Balaji, Pradeep 2596
Balakrishnan, G. 281
Balasubramaniam, Kavaipatti R. 1704
Baldus-Jeursen, Christopher..................1908
Ball, Greg 2263
Ballif, C. 50, 2073
Ballif, Christophe 55, 1220, 1395, 2104, 2794, 3254, 3256, 3435
Baloch, Ahmer A.B...................963, 1058
Banda, Pedro 1946
Banerje, Rangan 1151
Banerjee, Sanjay K. 363
Barahman, Gil 2285
Barakel, Damien 2255
Barnes, T. M. 138
Barnes, Teresa M...................3422
Barnett, Allen 315
Barraud, L. 50
Barraud, Loris 3254, 3256
Barrigon, Enrique 2502
Barth, Kurt 424
Bartolo, Robert E. 195
Bartsch, J. 884
Basore, Paul A. 2163
Bastide, Stéphane 3402
Bastola, Ebin 738, 781
Basu, Prabir K. 396
Battaglia, A. 1747
Baudrit, Mathieu 2492, 2562
Bauer, Andreas..................791, 2058
Bauer, Jan 1376
Bauhuis, G. 1189
Baumann, Thomas 1077
Baumgartner, Franz..................1077
Baur, Carsten..................541, 2087
Baxter, Jason B...................3143
Bearda, Twan..................1233
Beauchemin, Ryan D. 102
Becerril-Romero, Ignacio 155
Becker, Jacob J. 3366, 3410
Bedair, Salah M...................2195
Belanger, Ted..................1427

AUTHOR INDEX

Belletête, Marc 1579
Belluardo, Giorgio 3360, 3482
Bemrrr, Andreas...3500
Benamara, Mourad...3370
Benatto, Gisele A. Dos Reis 2672, 2682
Benick, J..2064
Benick, Jan 2511
Bennett, Dirk..2042
Bennett, Mitchell...247
Bennett, Mitchell F...............210, 259, 873, 2091
Berardone, Irene 402
Berg, Alexander 1773
Berg, Morgann 3417
Bermel, Peter...1904, 2467
Bernard, Annie...................................2870, 2891
Berry, Joseph J...2176
Bert, J..1733
Bertoni, Mariana.....................944, 2610, 2854
Bertoni, Mariana I. 2179, 3309
Besanger, Yvon...3102
Bett, Alexander J..1253
Bett, Andreas W..2511
Bettenwort, Gerd 1965
Beutel, Paul..2511
Beutner, Volker...1855
Bhaduri, Sonali...................................2799, 3478
Bhan, Mohan Krishan 496
Bhandari, Khagendra P..............738, 748, 781, 815
Bhatia, A. 1656
Bhatia, Swasti...1755
Bhattacharya, Indranil 3083
Bhattacharya, Sitangshu 2376
Bheemreddy, Venkata 2688
Bialek, Tom 2991
Bidiville, Adrien 1333
Biedenham, Richard E.......................................3245
Biegelsen, D. K. 1733
Biiss, M. 2457
Binetti, Simona..1669
Birch, Max T. 2423
Birkmirc, Robert..726
Birkmire, Robert W...2637
Bishop, Doug 3275
Bishop, Douglas...726
Bishop, Douglas M. 1441
Bissels, G...1189
Biswas, R...1350
Bittau, F. 3430
Bittau, Francesco 752
Bittner, Zachary...677
Bittner, Zachary S................18, 202, 2084
Bivour, M...2064
Bivour, Martin..1253
Blakely, Logan...1573
Blanche, Pierre-Alexandre 1147

Bläsi, Benedikt 352
Blasi, David 1531
Blum, Adrienne 2692
Blum, Adrienne L...2765
Bob, Brion 2258
Bobela, David C..2506
Bobyl, A.V. 1025, 1811
Boca, Andreea..2099
Boccard, Mathieu 55, 1220, 1317, 1790, 3366
Boeck, Torta..3396
Bohra, Rakesh 1912
Boizot, Bruno 83, 2087
Bolaji, Adewumi...2381
Boley, Allison 2573
Bolke, J. G. 1656
Bonnassieux, Yvan 626
Bonomo, Pierluigi...2118
Book, Felix 1824
Bora, Birinchi 3478
Borgers, Tom 3343
Borgström, Magnus 2502
Borgström, Magnus T 1286
Borland, John 2947
Borne, Axel 2864
Borowik, Lukasz...1516
Bosco, Nick...3190, 3200
Bosco, Nick S...2864
Bosson, Christopher J...2423
Bostock, Peter..2267
Bothe, Karsten...2692
Bourcois, Jérôme..2087
Bourdin, Vincent 626
Bourgoin, Jacques C. 2087
Bourne, Ben C..1549
Bousselham, Abdelkader 1058
Bouttcmy, Muriel..2711
Bowden, Stuart...................240, 925, 1797, 2719
Bowden, Stuart G...................................1589, 2596
Bowen, Leon..1445
Bowers, J.W. 2457, 3430
Bowers, Jake W..................146, 186, 752, 2349
Boyce, Ken 2000
Boyce, Kenneth...1933
Boyd, Matthew..1933
Boyer, Jacob...215
Boyer, Jacob T. 2079, 2554
Boyer-Richard, Soline 2192
Brabec, Christoph J. 1346, 3500
Bradshaw, Geoffrey K. 88, 301, 531
Brady, Brendan 3388
Braga, Daniel Sena..2307
Braid, Jennifer L. 1927, 2697, 3456
Brammertz, G...3260
Brand, A.A. 884
Brates, Nanu 1728

AUTHOR INDEX

Bräuninger, Matthias 3256
Breitenstein, Otwin 1376
Breitwieser, M. 3135
Bremner, S. P. 953
Bremner, Stephen 858, 1215, 1845, 2186, 2569
Bremner, Stephen P. 948
Brendel, Rolf 1366, 3371
Breus, V. 1752
Bright, Jamie M. 1405
Brinnig, Samuel....................................2622
Brito, Pedro P.....................................2307
Britt, Jeffrey 1455
Brittman, Sarah 2245
Broderick, Robert.................................3008
Broderick, Robert J. 1435, 1555, 1567, 1573, 3025, 3031
Brolo, Alexander G. 3388
Bruckman, Laura....................................1933
Bruckman, Laura S. 2000
Brückner, Sebastian................................2538
Brughera, Céline 2492
Brule, Carlton.....................................1728
Brulo, Gregory S. 1469
Bryan, Jonathan 1317
Buchanan, Wayne 1196
Büchler, A. 884, 3135
Buckner, Jessica...................................537
Buerhop, Claudia 3500
Bukowsky, Colton R. 1737
Bulkin, P. 1781
Bulkin, Pavel 1237
Bullock, James.............................59, 2076
Buonassisi, T. 648, 1140, 3295
Buonassisi, Tonio 284, 1264, 1491, 2242, 2532, 2744, 3236, 3290, 3300
Burgers, A.R. 3150
Burgers, Antonius R. 917
Burnham, Laurie 1435
Burroughs, Scott 272, 1469
Busquet, Severine 1061
Butt, Isaac..182
Cabarrocas, Pere Roca I............464, 1237, 2528, 2593
Cachet-Vivier, Christine...........................3402
Caffy, Florent 1516
Calderón-Obaldía, Fausto 626
Calle, Eric 944
Calvo-Barrio, Lorenzo..............................3285
Campa, Andrej......................................1346
Campanelli, Mark...................................437
Campbell, Calli M..........................3366, 3410
Campesato, Roberta 76, 541, 545
Campos, Cláudio Dias 2307
Camus, Christian 3500
Cañadillas, David 429, 1116

Canino, A. 1747
Caño, P. 1210
Cao, Huihui..1619
Cao, Wenkai..696
Cao, Xin...2427
Cao, Yunxue....................392, 1430, 1873, 2918
Cappelluti, F. 1189
Cardona, Dagoberto.................................670
Cardwell, D. 3511
Cariou, Romain 2511, 2528
Carlin, John A. 215
Carlson, David E. 3442
Carlson, Emily 701
Carneiro, Lucas M. 417
Carolus, Jome 2875
Carpenter, Bernard 537
Carr, Anna J. 1081
Carriere, Jarrett 2833
Carruthers, Steve 3514
Carter, Catrice M. 3393
Carter, Cedric....................................2135
Carter, Sam 3514
Casale, Mariacristina 76, 541, 545
Casallas-Moreno, Y. L. 670
Casper, Chadwick 1476
Cassini, Denio A. 1917
Castañeda, Carlos A. Rodríguez 1858
Catthoor, Francky 3343
Cattin, Jean......................................3435
Cattoni, Andrea 1289
Cavani, Olivier....................................2087
Cédola, A. P. 1189
Cendagorta-Galarza, Manuel.........................429
Cepeda, Kyle 876
Cesar, I. 3150
Cesar, Ilkay 917
Chai, Gaoda 976
Chai, Jing 1922
Chakraborty, Sagnik................................3300
Chamarthi, Phani Kumar.............................2952
Chamberlin, Charles 1271
Champliaud, J. 50
Champliaud, Jonathan 3435
Champness, C. H. 2388
Chan, Calvin 3417
Chan, Catherine E. 2576, 2600
Chan, Mandy 2808
Chan, Maria K.Y. 6, 1256, 2759
Chan, R. 3511
Chandralal, Sreeram 1674
Chandran, Deepak 2986
Chang, Jipeng 1873, 2823
Chang, Sheng-Hao...................................1051
Chang, Via-Chung...................................1051
Chantana, Jakapan 757, 2385

AUTHOR INDEX

Chaporr, Patrick...2711
Chapuis, Valentin..2104
Chattopadhyay, Kamanio 2811
Chattopadhyay, Shashwata....... 1850, 1858, 2849, 3478
Chaudhry, Ghulam M. 1018, 1110
Chaujar, Rishu..377
Chaurasia, Saloni..................................837, 841
Chausseau, Matthieu 2711
Chavali, Raghu Vamsi Krishna 1904
Chavez, Jose J. 2419
Chemisana, Daniel 1339
Chen, Benjamin....................................2358
Chen, Chien-Hsun....................................911
Chen, Chun-Chi....................................1635
Chen, Daniel.......................................2576
Chen, Eric Y. 1598, 3384
Chen, Haiyan.......................................2220
Chen, Hung-Ling 1289
Chen, Junyan............................1835, 2732
Chen, Kaifeng......................................2185
Chen, Kunji.......................................2656
Chen, Lung-Chien...................................367
Chen, Meixi...........................326, 999, 2035
Chen, Peng-Wei....................................2660
Chen, Ran...2576
Chen, Renfang......................................1241
Chen, Shi-Wei.......................................1627
Chen, Sung-Yu......................................911
Chen, Tsung-Cheng 329
Chen, Tzu-Yu..1627
Chen, Wanghua.....................................2593
Chen, Weijian......................................2392
Chen, Y. 2785
Chen, Yang...2502
Chen, Yao- Hui..............................893, 2664
Chen, Yifeng...........................1922, 2220
Chen, Yunfei........................761, 2427
Chen, Yusi...1835
Chen, Zhi David....................................1044
Chen, Zihan..2392
Chendo, Michael...................................2381
Cheng, Y. 14
Cheng, Yan..667
Cheng, Yuh-Jen...................................1610
Cheng, Zhe...3473
Cheng, Zhongkai..................................3393
Chenna, Shiva Tarun.............................1674
Chiang, Cho-Chun.........................893, 2664
Chiang, Fu-Kuo....................................198
Chikhalkar, Abhinav....................823, 827
Chin, Ken K. 761, 2427
Chinnusamy, Saravanan 980
Chiu, Chun-Yu....................................1169
Chiu, P..2094
Chiu, Philip......................................2099

Chmielewski, Daniel J............215, 2079, 2554
Cho, Eunhwan 333, 1824, 1838
Cho, Junsik 810
Cho, Yasuo 3323
Choi, Gyu-Seok................................. 2723
Choi, J. -K 2019
Choi, Rae-Won 2723
Choi, Seungkeun 1037
Choi, Sungjin 1758
Chong, Cheemun 2600
Choubisa, Hitarth 1022
Choudhury, K. R. 2312
Chouhan, Arun Singh 986
Choulat, Patrick 2227, 3435
Chow, E.M. 1733
Chowdhury, Ahrar Ahmed 888
Christians, Jeffrey A. 2176
Christmann, G. 50
Chu, Chi-Wei 1051
Chu, Haifeng 1222
Chu, Sheng 1299
Chua, Soo Jin 284
Chuang, Ta-Wei 343, 367, 893, 2664
Chung, Daniel 2707, 3304
Chung, Haejun 1904
Chung, Simon 696, 2186
Ciesla, Alison M. 2576, 2600
Cifuentes, L. 1210
Cifuentes, Luis 1204
Ciocia, Alessandro 3096
Cirino, Daniel A. Merced.................... 3044
Clayton-Warwick, D. 138
Cleveland, Erin 247, 2091
Clinton, Evan A. 305
Cobo-Yepes, Nicolás.......................... 2963
Codd, Daniel S. 3245
Cohen, Bat-El.................................. 2170
Cole, Wesley J................................. 2163
Colegrove, E. 1312
Colegrove, Eric 3147, 3319
Coll, Pablo Guimera........................... 2610
Collin, Stéphane................... 1289, 3147
Collins, Robert W.............807, 2462, 2582, 2646, 3426
Collins, Shamara 802, 1638, 2449, 3413
Comagliotti, Emanuele....................... 3435
Condorelli, G. 1747
Conibeer, Gavin.....................696, 2186, 2392
Conlon, Benjamin P. 219
Conrad, Brianna 315
Cordeiro, Patricia............................ 2135
Cordova, Adam 1965
Cornagliotti, Emanuele................ 1804, 2227
Cornaro, Cristina 3482
Cornell, Robert................................ 1275
Correa, J.M. 433

AUTHOR INDEX

Correa-Baena, Juan-Pablo 3300
Cossio, Gabriel 1181
Costa, Sara...............1979
Costa, Suellen C. 2307
Côté, Alexandre...............1908
Cousar, Larry C. 921
Cravens, R. 2094
Crawford, L. 1733
Crupi, F...............2073
Cruz, José Ortega 1959, 1990
Cruz, Leila R. De Oliveira...............2307
Cruz-Campa, Jose Luis 337
Cuevas, Andres...............2076
Cui, Jie 2076
Cui, Min...............1765
Cunningham, Daniel W. 1463
Cunningham, Joseph 3161
Cur, Jie...............517
Curran, Alan J. 1927, 2697, 3488
Curvat, L. 50
Cushing, Scott K. 417
Da Fonseca, Jérémy 2492, 2562
Dabney, M. S. 138
D'Abrigeon, Laurent...............545
Daenen, Michael...............2875
Dagenais, Mario 195, 1048
Dagyte, Vilgaile...............2502
Dahal, Saroj...............309, 3123
Dahal, Som...............240
Dai, Yushuai 18, 222, 677, 1184
Dalal, V L...............1350
Dalal, Vikram...............2247
Dalpian, G...............1245
Dam-Hansen, Carsten...............2672, 2682
Danel, A. 1747
Dang, Hongmei...............2432
Dangate, Milind S. 980
Daniil, Andreana...............944
Danzl, F.J.K...............3150
Darbali-Zamora, Rachid 2957, 2963
Das, Ujjwal 408, 1473, 1761, 1828, 2667
Das, Ujjwal K. 2637
Datas, Alejandro 2562
Dauskardt, Reinhold...............3190, 3200
Davidsen, Rasmus S...............2672
Davies, J. I. 1210
Davis, Kristopher O. 74, 322, 1804, 3448
Davis, Tracy...............537
De Coux, Patricia 464
De Melo, O...............2342
De Nicolas, S. Martin 50
De Oliveira, Michele C. C...............1917
De Villers, Bertrand J. Tremolet 1354
De Wolf, Stefaan 55, 3256, 3435
De, F. C. Lins Vanessa 1917

Debnath, M. C. 14
Debnath, Tanmoy...............1067
Deboever, Jeremiah...............1555, 1567
Debrot, F. 50
Debucquoy, Maarten...............1233
Debusschere, Vincent 3102
Deceglie, Michael 2771, 2789
Deceglie, Michael G...............2488, 2804, 3452
Deckerl, D. 1752
Decobert, Jean 2528
Deer, Tanya...............1908
Deitz, Julia I...............3139
Delahoy, Alan E. 761, 2427
Delhotal, J. 3224
Deligiannls, D. 3150
Deline, Chris 116, 1537, 1922, 3184, 3333
Demadrille, Renaud 1516
Demirkan, Korhan 820
Deng, Changhong 1158
Deng, Weiwei 2220
Denk, Patrick 1360
Descoeudres, A...............50
Descoeudres, Antoine 3254
Despeisse, M. 50, 2073
Despeisse, Matthieu...............3254, 3256, 3435
Desrues, Thibaut 2492, 2562
Deutsch, Todd G...............47
Devos, Arnaud 464
Dewitt, Daniel 1835
Dey, Anamika 1034
Dhakal, Tara P. 989
Dhere, N. 389, 1701
Di Leo, Paolo 3096
Di Mare, Simone 2372
Di Napoli, Simone...............2205
Diaz, Liliana Ruiz...............1147
Diercks, David R...............46
Dimitrievska, Mirjana...............3285
Dimopoulos, Theodoros 178
Dimroth, Frank 2511
Ding, Jie 1937, 2823
Dinger, Justin...............2692, 2765
Diniz, Antonia Sônia A. C...............1917, 2307
Dirriwachter, Antonius B. 3448
Dise, John 132, 1104
Dise, Skip...............1427
Dobrich, Anja 2538
Dobroliubov, Aleksandr...............2776
Dobson, Kevin...............315
Dobson, Kevin D. 658
Dobson, Weston 2692, 2765
Dogan, Yusuf 229
Doi, T. 441
Dominguez, A...............2342
Dong, Jianfei 2605

AUTHOR INDEX

Doolittle, William A. 305
Dooraghi, Michael...................................1169
Döscher, Henning...47
Doty, Matthew F. 1598, 3384
Dougher, Chris...3245
Dougherty, Brian 1933
Drahi, Etienne 464
Drayton, Jennifer A. 164
Drees, M. 3511
Dréon, Julie 55
D'Rozario, Julia 18
Drummy, Lawrence F. 966
Du, Chen-Hsun..911
Du, Xingzhi...2558
Du, Zhongming.................198, 767, 1707
Duan, Baosong......................................392, 2823
Duan, Wenqi...346
Dubey, R. 1995
Dubey, Rajiv 1704, 2849, 3478
Dubois, Anne Migan 626
Duenow, J.N. 1312
Duenow, Joel 1196, 3147
Duerinckx, Filip 2227, 3435
Dugan, Roger C. 3055
Dugdill, Brian 2014
Dumbrell, Robert.................................420, 3315
Dunham, Scott T. 3119
Dupré, Cécilia.....................................2492, 2562
Durand, Olivier 2192
Durose, Ken 742, 1445
Durstock, Michael F. 966
Durygin, Andriy 3473
Dusane, Rajiv O 2330
Dussarrat, Christian 326
Dutt, A. 370
Dutt, Ateet 2342
Dutta, P. 869
Dutta, Pavel 866, 2368
Duttagupta, S.P. 1898
Eafanti, Joshua...3190
Ebe, Falko 2996
Ebert, Matthieu 1531
Ebong, Abasifreke..888
Ediger, E. 2364
Edinger, Stefan 178
Edoff, Marika 796
Edwards, Daniel J 3514
Eeles, Alexander.................................146, 186
Efthymiou, Venizelos 3107
Egbe, Daniel Ayuk Mbi 1360
Eggink, Wouter 2109
Ekins-Daukes, Ned 1339
El Assimi, Taha 3402
Elangovan, Hemaprabha...........................2811
Elanzeery, Hossam.................................151, 2054

Eldho, T.I. 1898
El-Henawey, Mohamed.............................2247
Elkhatib, Mohamed..........................2141, 2969
Elleuch, Omar..359
Ellibee, Donald...1543
Ellingson, Randall 2926
Ellingson, Randall J. 1030
Ellingson, Randy J.............738, 748, 781, 815
Ellis, Chase T. 873
Elnosh, Ammar...1946
Elsehrawy, Farid 1189
Emery, K.A. 490
Engerer, Nicholas A. 1405
Eriksen, Ryan 2870
Eriksen, Ryan S. 2891
Ermer, J. 2094
Ermer, James 2099
Ermer, Jim H. 37
Escarra, Matthew D. 37, 3245
Escobar, D. Martínez 1959, 1990
Esfandiari, Parichehr 178
Espinct-Gonzalez, Pilar 558
Espíndola-Rodríguez, Moisés 155, 512, 572, 3265
Espinet-González, Pilar..........................521, 1248
Essa, Gharibah 2011
Essig, Stephanie 55, 3254, 3371
Etcheberry, Arnaud 2711
Etgar, Lioz 2170
Eugen, Rene 2864
Evani, Vamsi.........................802, 1638, 2449
Evans, Garrett Z...921
Evstigneev, M.........................663, 1025, 1811
Evstigneev, Mykhaylo A. 690
Eylers, Katharina 3396
Fada, Justin S. 2697, 3456, 3488
Faes, A. 50
Fairbrother, Andrew.............1933, 2000, 3204
Faleev, Nikolai 1215, 2573
Fan, S. 3376
Fan, Shanhui 2185, 2732
Fang, Liang 226
Fang, Y. 1603
Fang, Yi 305
Fano, V. 2740
Fano, Vanesa 2677
Faraj, Abudul 2014
Farnung, Boris...2267
Farré, Laia Arqués 3285
Farshchi, Rouin 1459, 1686
Faur, Maria...896
Faur, Orry...896
Favre, W. 1747
Fedina, Maria 2070
Fejfar, A. 2073
Felder, T. 2312

AUTHOR INDEX

Feldmann, F. 2064
Feng, Sheng-Kai 343, 367, 893, 2664
Feng, Shien-Ping 1012
Feng, Zhiqiang 1922, 2220
Fenning, David P. 1494, 2245
Ferekides, Chris 802, 1638, 2449, 2467, 3413
Ferekides, Chris S. 1511
Ferekides, Christos 175
Ferguson, Andrew J. 1354
Ferguson, L. 1863
Fernández, Eduardo F. 2858
Fernandez, R. Mis 1691, 2457
Fetzer, C. 2094
Fiducia, Thomas 424
Fields, Brian J. 2618
Filipic, Miha 1233
Filonovich, Sergej 464, 1237
Firth, Peter 1317
Fischer, A. 1603
Fischer, Alec 823
Fischer, Alec M. 305
Fisher, Brent 210, 272, 1469
Fisher, Dallas 989
Fitzgerald, Eugene A. 213
Fleming, Robert A. 1869
Flicker, J. D. 3224
Flicker, Jack 1280
Florides, Michalis 1941
Foldyna, Martin 2528, 2593
Forberich, Karen 1346
Forbes, David V. 3468
Forchhammer, Soren 2682
Forsh, P.A. 1811
Foster, Robert 2014
Fouchier, Marin 3402
Fournel, Frank 2492, 2562
Fraas, L. M. 1863
Fraas, Lewis 2042
France, Ryan M. 47, 232
Fraser, Ray 337
Frederiksen, Kenn H. B. 2682
Freeman, Janine M. 3494
Freiburger, Brennen M. 1869
French, Roger 1933
French, Roger H. 1927, 2000, 2697, 3456, 3488
Freundlich, Alexandre 236, 673, 1452
Fridman, Lucas 2000
Friedman, Daniel 549, 2543
Friedman, Daniel J. 42, 268, 1201
Friend, Mari Paz 429
Fritzsche, M. 1752
Frontini, Francesco 2118
Fthenakis, V. 2019
Fthenakis, V. M. 3230
Fthenakis, Vasilis 3077

Fu, Ran 1259, 1463
Fuhrich, Alexander 3396
Fuhrmann, Bianca 83
Fujiwara, Koji 1973
Fukuda, Tetsuya 931
Funabiki, Shigeyuki 2906
Fung, Tsun H. 2576
Fuyuki, Takashi 2593
Gabetta, Giuseppe 76, 545
Gabor, Andrew M. 74, 2839, 2897
Gaddy, Edward 585
Gahr, Stefan 178
Gai, Boju 549, 2291
Gaiaschi, Sofia 2711
Gallon, Joshua B. 3448
Galtieri, Jason 2975, 3214
Gambogi, W. 2312
Gao, Hui 226
Gao, Peng 1648
Gao, Wei 226
Gao, Y. 869
Gao, Yijun 2392
Gao, Ying 2368
Gao, Yuan 2048, 2605
Gao, Yujie 2870, 2891
García, I. 1210
Garcia, Iván 1204
Garcia, Juan Lopez 402
Garcia-Linares, Pablo 2562
Garg, Vivek 2345
Garner, Sean 2870
Garner, Sean M. 2891
Garnett, Erik C. 2245
Garreau-Iles, L. 2312
Garrillo, Pablo A. Fernández 1516
Garuz, Richard 2255
Gaury, Benoit 1303, 2438
Gdoutos, Eleftherios E. 558
Geelan-Small, Peter 3304
Gehre, Simon 3500
Geissbiihler, J. 50
Geissbuehler, Jonas 3256
Geisz, John 549, 3371
Geisz, John F. 232, 268, 1737, 2195, 3254
Georghiou, George E. 276, 1163, 1941, 1954, 3107
Geraghty, Paul 1342
Gerardi, C. 1747
Gerber, Andreas 1400, 1651
Gerdimenes, Anne 619
Gervasi, Massimo 541
Ghaisas, S.V 389, 1701
Ghimire, Kiran 993
Ghosh, Kunal 716
Gibbs, Jacob M. 730
Gibelli, François 2192

AUTHOR INDEX

Giebink, Noel C. 1469
Giguère, Jean-Benoit 1360
Gilchrist, James B. 966
Gillispie, Kellen 2762
Giordano, Francesco.................3096
Giraldo, Sergio 3265, 3285
Giussani, A..............................845
Giussani, Alessandro........206, 831, 2514
Givot, Bradley L.....................2864
Gladden, Christopher 1476
Glasgow, Nate........................1427
Glatthaar, M...................884, 3135
Gloeckler, Markus 1193
Glunz, S...............................3135
Glunz, S.W............................2064
Glunz, Stefan W. 1253, 2511
Gokkaya, Huseyin Cem 958
Goldschmidt, Jan Christoph 1253
Golembeski, Andrew A...............3143
Goma, Elias Garcia...................3462
Gombia, Enos 541
Gona, Michael N. 2349
Gong, Chen............................1585
Gong, Jue.............................2251
Gonzálcz-Díaz, Benjamín............3240
Gonzalez, Maria 259
Gonzalez, S............................3224
Gonzalez, Sigifredo 2147, 3002, 3020
Gonzalez-Díaz, Benjamín 429, 1116
Goodarzi, Mohsen 2707
Gooding, Renee 1280, 1543
Goodnick, S. M.......................1603
Goodnick, Stephen 1790
Goodnick, Stephen M. 305, 582, 1797
Gordillo, G.......................433, 503
Gordon, Ivan..........................1233
Gori, Gabriele..........................76
Gorman, Brian 62, 1371, 1381
Górnez-González, L. A.................2614
Gostein, Michael................2808, 2923
Goswamy, Naveen 1908
Gotoh, Kazuhiro 1765, 1794
Gottschalg, Ralph 1411, 2827, 3208
Govaerts, Jonathan 3343
Goverde, Hans........................3343
Gowda, Ramesh Rame...............1912
Graf, Martin...........................2511
Graham, Kenneth 1044
Grandidier, Jonathan 2099
Grassman, Tyler J............182, 215, 2079, 2554, 3139
Greco, Erminio 76, 541
Grede, Alex J.........................1469
Green, Martin...................2213, 2403
Green, Martin A........................858
Green, Michael.......................2926

Greenhalgh, R.C......................3430
Gregory, Geoffrey 74
Grévin, Benjamin 1516
Grice, Corey R...........771, 1643, 2473, 3426
Grieco, William J......................2618
Griffin, Alecia 2870
Griffin, Alecia C. 2891
Grijalva, Santiago................1555, 1567
Grini, S...............................3269
Große, T.............................1752
Großer, Stephan 2232
Grossklaus, Kevin 701
Großschädl, Bettina 1329
Grovenor, Chris.......................424
Grover, Sachit...................1193, 2473
Grübel, B.............................884
Gu, Fei...............................1346
Gu, Tian.............................1473
Gu, Tingyi...........................1828
Gu, Xiaohong 1933, 2000, 2844, 3195, 3204
Guarracino, Ilaria 1339
Guay, Nathan.........................1543
Gudla, Sushanth 1389
Guerrero-Lemus, Ricardo 429
Guillemoles, Jean-François........1289, 2192
Guillevin, N..........................3150
Guillevin, Nicolas 917
Guina, M.............................1189
Guina, Mircea 297, 2520
Guischard, Felix 2836
Gunawan, Oki...................1441, 3275
Gunnarsson, William B................2443
Guo, D.........................1603, 2816
Guo, Hong...........................1299
Guo, Q.................................3
Guo, Qi..............................226
Guo, Shuwen 1430, 1873, 2918
Guo, Yongjie.........................1719
Gupta, Amit Kumar 2952, 2981, 2986, 3050
Gupta, Mool C. 937
Gupta, Neeti.........................696
Gupta, Ritesh Kant 1034
Gupta, Shivam........................377
Gupta, V.............................1733
Gustafsson, Mattias 2025
Guthrey, Harvey................1400, 2887
Gutiérrez, J. R. 2740
Gutiérrez, R. 643
Gutscher, S..........................884
Guwaeder, Abdulmunim 1122
Gwak, Jihye..........................810
Ha, Dongheon 1585
Habermann, D. 1752
Hack, James.........................999
Hack, James H. 326

AUTHOR INDEX

Hacke, Peter 1371, 1381, 1421, 1922, 2819, 2854, 3305

Hackl, Wolfeanz 178

Haddad, M. 2094

Haddadian, Rojiar 1927

Hadi, Sabina Abdul...........................213, 1741

Hadjipanayi, Maria............................276

Hadke, Shreyash 986

Hadley, Wendy......................................2014

Haegel, Nancy M. 62

Hagendorf, Christian.................1376, 2232

Hägglund, Carl.......................................796

Hahn, Carina E. 175

Hai, Hoang Tri 931

Haight, Richard.....................................1441

Hajimiri, Ali.................................521, 558

Hajizadeh, Amin 3092

Halbwax, Mathieu.................................3402

Hall, Allen...1511

Hallam, Brett J.......................................2576

Halliday, Douglas P...............................2423

Hamadani, Behrang H..............263, 437, 508

Hameiri, Ziv 66, 420, 3290, 3315

Hamon, Gwénaëlle.................................2528

Hamui, L. 2614

Hamzaoui, Saad 900

Hamzavy, Babak T.................................2618

Han, Sang M. 88

Han, Xinyue 1719

Han, Youngsik 2242

Hanada, Toru 940

Handwerker, Carol A............................1449

Haney, Paul 1303

Haney, Paul M.......................................2438

Hanley, J. 2094

Hanna, Amir 1055

Hannappel, Thomas................2524, 2538, 2538

Hanriot, Sergio De Morais 2307

Hansen, Clifford............ 1127, 1537, 3184, 3333, 3348

Hansen, Clifford W...............110, 1543, 1549

Hansen, Ole 2672

Hansen, Richard....................................2042

Hansen, Shirley 2042

Hao, Xia...160

Hao, Xiaojing.....................858, 2213, 2403

Haohui, L. 1140

Haohui, Liu 2744

Haque, K A S M Ehteshamul 346

Haque, M. D. 552

Hara, Shigeomi...........................1950, 3339

Hara, Tomoya.......................................2548

Harari, Joseph.......................................3402

Hardikar, Kedar 2688

Häring, Adrian 2263

Hariskos, Dimitrios...............................2058

Harmand, Jean-Christophe 1289

Harris, Christian 319

Harris, James............................... 1835, 2732

Harris, Tom 2991

Harvey, Steven......................................2887

Harvey, Steven P. 1371, 1381, 2702, 3305, 3319

Haschke, Jan 3435

Haslinger, Michael 1804

Hassan, Ibrahim A. I. 2858

Hatch, S. 14

Hatton, Peter D. 2423

Hauch, Jens 3500

Haug, F.-J. 2073

Hausgen, Paul E. 102

Hausmann, J. 1752

Havu, Ville 2070

Haysom, Joan E. 1094

He, Junwen 1469, 1737

He, Qiuxiang 3304

He, Wenshuang......................................392

Hea, Wenshuang 2823

Heben, Michael......................................2926

Heben, Michael J. 170, 730, 748, 815, 1030, 2462

Hegedus, Steven.................408, 1473, 1761, 1828, 2667

Hegedus, Steven S.................................658

Heidmann, Berit 3396

Heilbrunner, Herwig 1360

Heilscher, Gerd.....................................2996

Heinz, F. D. 3135

Heinze, Matthias 2263

Heller, Dominic.....................................1077

Henes, Dan 1094

Hentz, Sandrine 966

Hermle, M. 2064

Hermle, Martin...........................1253, 2511

Hernandez, J. A. 2031

Hernández, Johan..................................1143

Hernandez, Joseph 3473

Hernandez-Alvidrez, Javier..................2153

Hernández-Gutiérrez, C. A. 670

Hernández-Rodríguez, Cecilio 429

Herrera, Daniel J...................................219

Herrmann, W. 107

Herz, Magnus..3360

Heta, Y. 2312

Hetterich, Michael 1682, 2216

Hettick, Mark 59, 823, 2076

Heurlin, Magnus 1286

Hickey, Benjamin 1459

Hidaka, Kazuyuki 1973

Higa, M. 441

Hilfiker, M. 2364

Hill, Alex 1893

Himwas, Chalermchai...........................1289

Hindi, Basel...1058

AUTHOR INDEX

Hinken, David 2692
Hinojosa, M. 1210
Hinzer, Karin 1094
Hirai, Masakazu 1769
Hirata, Yoichi 613
Hirose, Kotaro 3323
Hirstl, Louise C. 2091
Hishikawa, Y. 441, 1003
Hishikawa, Yoshihiro 480, 2781
Ho, Jian Wei 496
Ho, Wen-Jeng 343, 367, 893, 2664
Hoang, Bao 96
Ho-Baillie, Anita...................858, 1845, 2569
Hobbs, William B. 2618
Hoerteis, Matthias 914
Hoex, Bram 517
Hoff, Thomas 132, 1104
Hofmann, Johannes 2407
Hoheisel, Raymond 247, 3514
Höhn, Oliver 352
Holman, Z. C. 3376
Holman, Zachary 1790, 3366
Holman, Zachary C. 1220, 1228, 1317,
      1322, 1820, 3250
Holmgren, William F. 110, 1127
Holzmann, Daniel...914
Hong, Chung-Yu 294
Hong, Keunkee 399
Honsberg, Christiana...........................827, 3088
Honsberg, Christiana B. 240, 305, 582,
      681, 1215, 1841, 2573
Hopf, Markus 1965
Horenstein, Mark..2870
Horenstein, N Mark 2891
Horner, Greg S. 3448
Horowitz, Kelsey A.W. 1259, 1463
Horzel, J. 50, 2073
Hoshii, Takuya 2334
Hosokawa, Kazuya 613
Hossain, Istiaque 2247
Hossain, Mohammad A. 3456
Hossain, Mohammad I....................................963
Howard, John M. 2443
Hsi, Edward 1275
Hsu, Chia-Jhe 1623
Hsu, Chih An 1638, 2449, 3413
Hsu, Lung-Hsing 1610
Hsu, Shun-Chieh 1606, 1623
Hsu, Shu-Tsung.......................445, 448, 476
Hsu, Wei-Lun 1048
Hsu, Yu -Chen 888
Hu, Chehao 229
Hu, Cheng-Shun 329
Hu, Hailin 1858
Hu, Juejun 1473

Hu, Lilei...3129
Hu, Long 2392
Hu, Yang 1927
Hu, Yicong...2392
Huang, Jialiang 2213
Huang, Jingsheng 1937, 2823
Huang, Jing-Shun.................512, 521, 558, 572, 1248
Huang, Shujuan 2392
Huang, Vi-Wen 1631
Huang, Weijing...................................1873, 2918
Huang, Wei-Ming 1627, 1631
Huang, Wen-Hsi 385
Huang, Ying-Yuan 1807
Huang, Yi-Wen 1627
Huang, Yu-Ming 1606
Huang, Yu-Ting 1012
Huang, Z. 3260
Huayamave, Victor..2839
Hubbard, S. M. 552, 845, 2755
Hubbard, Seth 677
Hubbard, Seth M....................18, 202, 206, 222,
      831, 1184, 2084, 2298, 2514
Huber, Christian 2216
Hudson, A.I. 2755
Huey, Bryan D. 1522
Huffaker, D.L. 2755
Huffaker, Diana 202
Huhn, Vito...1651
Huld, Thomas 2167
Hunault, Philippe 2711
Hung, Yung-Jr 1606
Huo, Yijie 1835
Husein, Sebastian 944
Huss, Alexandra M...164
Hussain, Babar 451, 2355
Hussain, Muhammad M. 1055
Hutchings, Douglas...1869
Hutchings, Douglas A. 921
Hutter, Oliver S. 1445
Hwang, James...333
Hyvl, M. 2073
Iandolo, Beniamino 2672
Ianno, N.J. 2364
Ichikawa, Yukimi 1769
Idlbi, Basem 2996
Ikki, Osamu 2159
Ilic, Ognjen 1737
Imai, Jun 2906
Imaizumi, Mitsuru 567, 3506
Imtiaz, Syed N. 1067
Ingenhoven, Philip...3482
Ingenito, A...2073
Inns, Daniel 3113
Isabella, Olindo 2605
Isbilir, Kenan 2827

AUTHOR INDEX

Isherwood, Patrick J. M. 2349
Ishii, Tomoaki 455
Ishino, Yuya...757
Ishizuka, Shogo 33
Islam, Kazi..37
Islam, Muhammad Monirul.....................33, 900
Islam, Raisul...1835
Isoaho, Riku...2520
Iwasaki, Kazuya...2338
Iwata, Naotaka...2642
Iwuoha, Emmanuel.......................................1360
Iyer, Abhishek....................................326, 999, 2035
Iyer, Parameswar K.......................................1034
Izquierdo-Roca, Victor........................3265, 3285
Jackson, Christine...215
Jackson, Philip 2205, 2453
Jacob, David...1549
Jacobson, Arne...1271
Jadkar, S.R..1701
Jaeckel, Bengt..1411
Jae-Yun, Fa-Jun Ma, ...1845
Jagdish, A K...2811
Jäger-Waldau, Arnulf.....................................2167
Jahn, Ulrike..3360
Jain, Aditi..333
Jain, Nikhil 42, 46, 232, 578, 2195, 3371
Janoch, Rob...............................74, 2839, 2897
Jansen, Mark J..1081
Jany, Christophe..................................2492, 2562
Janz, Stefan...83, 2407
Jaramillo, Adolfo..1143
Jared, Bradley...1473
Jarmar, T..30
Jasti, Naga Prathibha 986
Jaswal, Rohit..3172
Javed, Mehwish Azher.....................................1317
Javey, Ali...............................59, 823, 2076
Jeangros, Q...2073
Jenkins, P. P...845
Jenkins, Phillip P...........247, 373, 1838, 2091, 3514
Jensen, Brian..2014
Jensen, M. A...3295
Jensen, Mallory A. 1491, 3290, 3300
Jensen, Soren..1196, 3147
Jeong, Woo-Lim.....................................777, 1665
Jhaveri, Janam..1773
Ji, Liang..1933, 2000
Ji, Yaping..37
Ji, Yaping Vera..3245
Jia, Jieyang...1835, 2732
Jian, Ding-Rung.....................................1627, 1631
Jiang, C. S...1312, 2789
Jiang, C.-S...2280, 2785
Jiang, Chun-Sheng...................................62, 1371
Jiang, Lian L...589

Jiang, Lian Lian 120
Jiang, Xuefang..1937
Jiang, Yu...3220
Jimeno, J. C.......................................643, 2740
Jimeno, Juan Carlos.......................................2677
Jin, C..1781
Jin, Yu..3119
John, Jim J...1946
John, Joachim..1804, 2227
John, Suru Vivian...1360
Johnson, A. D..1210
Johnson, E.V...1781
Johnson, Erik V...2593
Johnson, J. L.......................................1656, 1661
Johnson, Jay 2135, 2141, 2153, 2969, 3002, 3008
Johnston, S..2785
Johnston, Steve.....................62, 202, 459, 1371,
    1381, 1400, 2213, 2819, 2887, 3305, 3452
Jones, C. Birk....................2618, 3008, 3155, 3488
Jones, David...1342
Joonwichien, Supawan......................................904
Joshi, Madhuwanti S................2952, 2981, 2986, 3050
Joshi, Pranav..2247
Jošt, Marko...1346
Jovanovic, Raka...963
Juang, B.C..2755
Juang, Bor-Chau...202
Juárez, A. Sánchez..1990
Juárez, Aarón Sánchez......................................1959
Juhl, Mattias K......................................420, 3315
Juhl, Mattias Klaus..66
Julien, Scott..1933, 2000
Junci, Wang..496
Junda, Maxwell 771
Junda, Maxwell M.............2462, 2582, 3426, 3468
Jung, Jae Hak..487
Jung, Jiirgen...2864
Jung, Sang Hoon...2723
Jung, Sang Hyun...244
Juso, Hiroyuki...3506
Kabalan, Amal...2358
Kaczynski, Ryan...1455
Kaizu, Toshiyuki 23
Kaizuka, Izumi...2159
Kakosimos, Konstantinos E.................................1888
Kalainatharr, Sivaperuman..................................2334
Kalb, J..1733
Kale, Abhijit...1801
Kale, Abhijit S...1777
Kallickal, Johnson...................................1543, 3348
Kalt, Heinz...1682, 2216
Kamata, N..552
Kamevama, Satoshi..2642
Kamino, Brett..3256
Kamins, Ted..1835

AUTHOR INDEX

Kaminski, P.M. 3430

Kaminski, Piotr 1674

Kaminski-Cachopo, Anne 2562

Kamioka, Takefumi 2498, 2548, 2566, 2642

Kanemitsu, Yoshihiko.................................721, 2781

Kanevce, A. 1312

Kanevce, Ana 3147

Kang, Ho Kwan 244

Kang, Min Gu..................................356, 1758, 2723

Kang, Yoonmook 935

Kankiewicz, Adam 132, 1104, 1132, 1427

Kannan, C. V. 2716

Kao, Ming-Hsuan...1627

Kaplan, Stephen..600, 1071

Kaplar, R. 3224

Karas, Joseph 925

Karki, Shankar... 182, 735, 807, 2298, 2446, 2646, 3139

Karmarkar, M...1661

Karpowich, Lindsey...914

Karthik, Shravan 3172

Kashkoush, Ismail...322

Kaslin, Remo 1077

Kasry, Amal..2858

Kasu, Makoto 1950, 3339

Kato, Takekazu 1175

Kato, Takuya 160

Katsube, Ryoji 2361

Kaule, Felix..2622

Kausika, Bala Bhavya...................................3014, 3167

Kavaipatti, Balasubramaniam 2799

Kawatsu, Tomoyuki...................................381, 2588

Kazmcrski, Lawrence L. 2799

Kazmerski, L.L. 1245

Kazmerski, Lawrence L. 1917, 2307

Kazumi, Kenji 2361

Keeler, Gordon..1473

Keller, Nico 1077

Kelly, George...1275, 2263

Kelly, Matthew..3514

Kelzenberg, Michael D. 512, 521, 558, 572, 1248

Kempe, M.D. 138

Kempe, Michael 1933, 2000

Kempe, Michael D..3208

Kephart, Jason 785

Kephart, Jason M. 3417

Kern, Gregory 2147

Kern, Gregory A. 3020

Kesavan, Arul Varman..1614

Kessels, Wilhelmus M.M. 1817

Kessler, Emily..206

Khadimallah, A. 869

Khalili, A. 1189

Khan, Imran 802, 1638, 3413

Khan, Imran S. 2449

Khan, Mohammad R. 1055

Khan, Taj M. 451

Khanna, Raghav..2926

Kharait, Rounak A. 2833

Kharel, Khim 236, 673, 1452

Khatavkar, Sanchit 2716

Khatiwada, D. 869

Khatiwada, Devendra 866

Khatri, Ishwor 192

Khatri, Trijul 377

Khomcnko, Denis V..690

Khoo, Yong Sheng 1922

Khor, Alan 3172

Khoram, Parisa 2245

Khorenko, Victor...83, 2087

Khoury, R. 1781

Kiefer, Fabian 1366

Killam, Alex 2719

Killinger, Sven 126, 1405

Kim, Boram 2201, 2524, 2538

Kim, Chang Zoo 244

Kim, D. 1189

Kim, Dae Young 626

Kim, Dong Seop 399

Kim, Dong-Ho...2631, 2634

Kim, Donghwan 935, 1758

Kim, Hae-Sun 777, 1665

Kim, Hyo Jin 849

Kim, In-Young 777

Kim, Jae Hyun.............................363, 2844, 3195, 3204

Kim, Jin-Hyeok 777

Kim, Jisun 399

Kim, Ka-Hyun 1758

Kim, Kangho 244

Kim, Kihwan 810

Kim, Kyoung- Tae 1037

Kim, Min-Soo...487

Kim, Moon 2759

Kim, Sangpyeong...240

Kim, Soo Min 2723

Kim, Woo Kyoung 487

Kim, Yeongho 827

Kim, Yong Bae 2723

Kim, Yong Whan 849

Kim, Youngjo 244

Kimbal, Gregory M. 110

Kimura, Daiki 854

Kindole, Dickson 970

Kindvall, Anna..785

King, Bruce H. 3155, 3488

King, Richard 827

King, Richard R. 301, 823, 1215, 1841

Kingma, Aldo 541

Kini, Roshan 2926

Kinoshita, Kosuke 1504, 2588

Kirk, A..3511

AUTHOR INDEX

Kita, Takashi 23
Kleider, Jean-Paul 2528
Klein, Talysa R. 2482, 3371, 3439
Kleinschmidt, Peter 2538
Klemm, Hagen W. 3396
Klenk, Markus 1077
Klie, Robert F. 2759
Klimm, Elisabeth 2836
Klise, Katherine A. 3161
Klisel, Geoffrey T. 3494
Kluska, S. 884, 3135
Knight, Bruce 2014
Knopf, Hannes 1965
Ko, Changhee 326
Kobayashi, Jonathan 1061
Koehl, Michael 3488
Koepgel, Ringo 2622
Kogler, Willi 791
Kohlstädt, Markus 1253
Koike, Junichi 931
Koirala, Prakash 2462, 3426
Kojima, Nobuaki 359, 2498, 2566
Kojima, Takuto 1504, 2588
Komsa, Hannu-Pekka 2070
Konagai, Makoto 1769, 2627
König, M. 1752
Konstantinou, Georgios 1941
Kontges, Marc 1366
Kopecek, Radovan 1222
Koschny, T. 1350
Kostylyov, V.P. 1025, 1811
Kostylyov, Vitaliy P. 690
Kotipalli, Ratan 2209
Kottantharayil, A. 1995
Kottantharayil, Anil 396, 716, 1850, 2799, 3478
Kottokkaran, Ranjith 2247
Kotulak, Nicole 999, 1838
Kotulak, Nicole A. 247
Koyama, Koichi 1765, 1787
Kozodoy, Peter 1476
Krabb, Peter 178
Krantz, Patrick W. 730
Krasikov, D. 2816
Krc, Janez 1346
Krein, Philip T. 3214
Krich, Jacob J. 1294
Krishnan, Mani R. 1912
Krishnan, Sheeja 76
Krishnaswami, Hariharan 2931, 2936
Krogen, J. 2094
Krügener, Jan 1494
Krut, Dimitri D. 37
Ku, Chen-Hao 329
Kubiniec, Alex 132, 1132, 1427
Kuciauskas, Darius 1679

Kudriavtsev, Yu. 670
Kuitche, Joseph 1877, 1883
Kulish, Mykola R. 690
Kum, Hyun 18, 222, 677, 2084
Kumar, Rajesh 3478
Kumar, Shailendra 2345
Kumar, Sukanya Santhosh 980
Kumar, Vijay 2716
Kumari, Khushboo 251
Kuo, Hao-Chung 1610, 1627
Kuo, Po-Tsun 1006
Kuo, Ting-Wei 329
Kurdgelashvili, Lado 2035
Kurihara, Risa 2159
Kurimoto, Yuji 931
Kurokawa, Yasuyoshi 1765
Kurstjens, Rufi 83
Kurtz, Sarah 1275, 1922, 2263, 3190
Kusaki, Kazuki 23
Kuthanazhi, Vivek 3478
Kwon, Jung-Dae 2631, 2634
Kwon, Sang Jik 195
Kyureghian, H. 2364
La Centra, Ricci 2870, 2891
Lachaurne, Raphaël 2528, 3402
Lachowicz, A. 50
Lackner, David 2511
Lacroix, Jean-Sébastien 1579
Lafleur-Lambert, Antoine 1360
Lafont, Ombline 2453
Lagumavarapu, Ramesh B. 202
Lai, B. 3295
Lai, Barry 1494, 2170, 2179, 2245, 3300, 3309
Lai, Yi 1009
Laine, Hannu S. 1491, 1494, 3236
Lakshmanan, Ramakrishnan 2870, 2891
Landgraf, D. 1752
Lang, Mario 2216
Lapierre, Ray R. 1294
Larrey, Vincent 2492
Larsen, Ross E. 1354
Larson, Bryon W. 1354
Lasalvia, Vincenzo 881, 1491, 1801, 2242, 3439
Laschinski, Joachim 1965
Lassise, Maxwell 3410
Latham, Joseph 1086
Lau, Derwin 3220
Lau, Kei May 578
Lave, Matthew 1435, 3008, 3025, 3031, 3184, 3348
Lavrova, Olga 1280, 2618, 3488, 3494
Law, D. 2094
Lazarou, Constantinos 276
Le Corre, Alain 2192
Le Donne, Alessia 1669
Le Gall, Sylvain 3402

AUTHOR INDEX

Le Guen, Vincent 70
Le Rouzo, Judikaël 2255
Lebreton, Fabien 464, 1237
Leclerc, Christophe....................................558, 572
Lecouvey, Christophe 2492, 2562
Ledinek, Dorothea 796
Ledinsky, M...2073
Lee, Angela 417
Lee, Benjamin G. 881, 1737, 1801, 1832, 2482, 3439
Lee, Calvin 1342
Lee, Dong-Seon 777, 1665
Lee, Eunjoo...399
Lee, Eunsang..2124
Lee, Hae-Seok..935
Lee, Hyeonseok...1012
Lee, Jaejin 244
Lee, Jeong In 356, 1758
Lee, Ji-Hoon 2631, 2634
Lee, Jihwan 1181
Lee, Jinwoo 1455
Lee, Jongwon 681, 1215
Lee, Kan-Hua 359, 412, 1479, 1711,
 1714, 1743, 2498, 2566
Lee, Kyumin 1526
Lee, Kyu-Tae 1469
Lee, M. L. 3376
Lee, Minjoo...2291
Lee, Minjoo L. 42
Lee, Mitch 600
Lee, Mitchell...595
Lee, Seunghun 1253
Lee, Soonil 363
Lee, Yeonbae 1204
Lee, Yun Seog 1441
Lefebvre, Amy 1933
Lefebvre, Amy L. 2000
Lehman, Peter 1271
Lehr, J. 3224
Leilaeioun, M. 3376
Leilaeioun, Mehdi 1322, 1790
Leite, Marina S....................... 1508, 1585, 2443
Lekx, David 1094
Lemaître, Aristide 1289
Lemus, Ricardo Guerrero 1116, 3240
Lennon, Alison 3220
Lennon, Kyle 3384
Lennon, Kyle R. 1598
Leone, Stephen R. 417
Leonhardt, M. 1752
Leow, Shin Woei 3275
Lepkowski, Daniel..215
Lepkowski, Daniel L. 2079, 2554
Lester, Luke F. 219
Leto, Riccardo 1728
Leu, S. 1752

Levcenco, Sergiu...3396
Levi, D.H. 467, 490
Levi, Dean 483
Levrat, J. 50
Levrat, Jacques 3435
Levy, David H. 3442
Li, Chu Tu 1094
Li, Duanhui 1473
Li, Guan-Yi..........................343, 367, 893, 2664
Li, Jian 771, 1643
Li, Jian V..........................2473, 2728, 2749
Li, Joel B. 3300
Li, Kexue 424
Li, L. 3295
Li, Lan 1473
Li, Li 1175
Li, Lu 1619
Li, Mengjie 3315
Li, Qiang 578
Li, Rui..1094
Li, Siming 3143
Li, Wenjie 3275
Li, Xiaoping 1193
Li, Xinyi..255
Li, Xueying 2170
Li, Y. 869
Li, Yongkuan 2368
Li, Yunjun 907
Li, Yunpeng 2220
Li, Zhanhang 226
Li, Zhuohui 1598, 3384
Liang, B.L. 2755
Liang, Jianbo 2548, 2551
Liao, Anqi 948
Liao, Yuanxun 696
Liao, Yuaxun 2186
Libby, Cara S..2618
Licht, Abigail 701
Lichty, Marlene L..2298
Lie, Stener 3275
Lim, Bianca 2318
Lin, Albert 294, 1631
Lin, Albert S. 1627
Lin, Cheng-Shian..1006
Lin, Chien-Chung....................1606, 1610, 1623, 1627
Lin, Ching-Fuh 1006, 1009
Lin, Fen 284
Lin, Ming-Yi 1051
Lin, Shang-Pang 1006
Lin, Yan 1100
Lin, Yandan 2048
Lin, Yan-Zhang 1623
Lin, Yida 667
Lin, Yu-Hsuan 911
Lin, Yung-Sheng...329

AUTHOR INDEX

Lin, Zong-Xian 367
Lincoln, Jason 2897
Lincoln, Jason L. 2839
Lincot, Daniel 2453
Linton, John 337
Lipovšek, Benjamin 1346
Lipski, Michael V. 1469
Lisbona, Emilio Fernandez 545
Lisco, F. 1691, 2457
Litjens, Geert 3014
Liu, A. Y. 1485
Liu, Chenxi 3172
Liu, Fang Fang 1648
Liu, Fangyang 2213
Liu, H. 1189
Liu, H.Y. 14
Liu, Haitao 472
Liu, Han-Wen 2660
Liu, Haohui 284, 2532
Liu, Hsiang-Yu 2637
Liu, Huiyun 3370
Liu, Jheng-Jie 343, 893, 2664
Liu, Kanglin 1100
Liu, Mengxia 3129
Liu, Qihang 1245
Liu, Ruimin 2220
Liu, Simon H. 93
Liu, X.Q. 2094
Liu, Xiangxin 198, 767, 1707
Liu, Xinbing 1728
Liu, Xing-Quan 2099
Liu, Zhe 284
Liu, Zhen 2532
Liu, Zhengjun 1494
Liu, Zhengxin 1241
Livera, Andreas 276, 1954
Liyanage, Geethika K. 170, 730, 815, 2462
Llin, Lourdes Ferre 1339
Lloyd, Alexis 2870
Lloyd, Michael A. 726, 3143
Lnr, Yiming 2558
Loach, Andrew J. 2697
Lodha, Saurabh 716
Lokanath, Sumanth 1275
Loke, Samuel P. 512, 521, 558
Loke, W.K. 1210
Lombardero, I. 1210
Lombez, Laurent 70, 1289, 2192
Lonergan, Mark 802
Long, Yean-San 448, 476
Looney, Erin E. 1491, 3236, 3290, 3300
Löper, P. 2073
Löper, Philipp 55
Lopez, Cristina S. Polo 2118
López, G. 1781

Lopez, Roberto 2728, 2749
López-González, J.M. 1781
López-López, M. 670
Lopez-Marino, Simón 155
Lorentzen, Justin 3514
Lorenzo, Antonio T. 1127
Loser, Ulrich 2272
Lossen, Jan 1222
Lotshaw, W.T. 2755
Lou, Chaogang 1619
Loubar, Anais 2711
Loyer, Camille 2000
Lu, Ching-Ying 1835
Lu, Hongbo 255
Lu, J.P. 1733
Lu, Jiawen 2656
Lu, Kyle B. 3448
Lu, Zhou 1728
Lubenow, Tomas 3333
Lujan, R. 1733
Luka, Tabea 2232
Lumb, Matthew P. 210, 247, 259, 272, 873, 2506
Luna, Miguel A. 632
Lunacek, Monte 3008
Lunt, Richard R. 2124
Luo, Shiqiang 976
Luo, Wei 1922
Luo, Yanqi 2170, 2245
Luria, Justin L. 1522
Luther, Joseph M. 2176
Lynn, Kelvin G. 3422
Lyons, Alan 2285
Lyu, Yadong 1933, 2844, 3195, 3204
Lyu, Zheng 1835, 2732
Ma, D. 229
Ma, Fa-Jun 2569
Ma, Xiaokun 1469
Macalpine, Sara 1537
Macco, Bart 1817
Macdonald, D. 1485, 3295
Macdonald, Daniel 2707, 3300
Mack, C. 490
Mack, I. 2073
Mack, Shawn 259, 873
Mackie, Neil 820
Maclaren, Scott 1511
Macmaster, Steven W. 2864
Madani, Keeya 940, 1824
Madsen, C. K. 229
Maeda, P.Y. 1733
Magaña, Ernesto 1494
Magdaleno, R. Santos 1990
Magdaleno, Rocío De La Luz Santos 1959
Magnin, Vincent 3402
Magnone, Lydie 1415

AUTHOR INDEX

Mahadik, N. A. 845
Mahapatra, Chiranjibi 2849
Maia, Cristiana Brasil 2307
Maidaniuk, Yurii 3370
Mailoal, Jonathan 1264
Major, Jonathan D. 742, 1445
Makita, Kikuo 861, 1724
Makoutz, Emily A. 3381
Makrides, George 1163, 1941, 1954, 3107
Malhotra, Raghav 3172
Malik, Roger 1193
Maliya, Heini 226
Malkov, Andrei V. 146, 186
Mallick, Tapas K. 2858
Manda, Surya 761
Mandelis, Andreas 3129
Manganiello, Patrizio 3343
Mangelinck-Noël, Nathalie 1498
Mani, Monto 604
Maniscalco, B. 2457
Maniscalco, Biancamaria 2827
Mann, Colin 1248
Mann, Colin J. 93, 512
Mansfield, Lorelle 1400, 2473
Mansoori, A. 281
Mantel, Claire 2682
Manzoor, Salman 1228, 1322
Marie, Benoit 1498
Marion, Bill 1134, 1537, 1543, 3333, 3348
Markevich, V. P. 1485
Markides, Christos N. 1339
Maros, Aymeric 1215
Marsh, Brett M. 417
Marsillac, Sylvain 182, 735, 807, 2298, 2446, 2646, 3139
Marsillac, Sylvain X. 2582
Marti, Shilpa 2936
Martín, I. 1781
Martin, Mickaël 2492
Martinez, Aaron D. 2536, 3406
Martinez-Morales, Alfredo A. 2881
Martínez-Pérez, Alejandro 3285
Martín-Martín, D. 3376
Martins, Ana C. 2104
Martinson, Alex B.F. 6, 1256
Masada, Isao 1504
Mascarenhas, Angelo 2506
Maser, Jörg 3309
Maser, Jörg M. 2179
Maskell, Douglas L. 120, 589
Mastroianni, Simone 1253
Masuda, Atsushi 1268
Masuda, Shota 1794
Masutomi, Yasuki 3339
Matei, I. 1733

Mather, Barry 1561
Mathew, Leo 363
Mathew, X. 142
Mathews, N. R. 142
Matsubara, Koji 381, 1333
Matsui, Takuya 381, 1333
Matsumoto, Yasuhiro 632
Matsumura, Hideki 1765, 1787
Matsuo, K. 3
Matthew, Leo 2506
Maximenko, S. I. 845
Maximenko, Sergey 2091
Maximenko, Sergey I. 873
May, Matthias M. 2538
Mayberry, Ryan 914
Mazumder, Malay 2870
Mazumder, Malay K. 2891
Mazur, Yuriy I. 3370
Mccandless, Brian 1196, 3319
Mccandless, Brian E. 726
Mcclary, Scott A. 1449
Mcclung, Larry 2833
Mcclure, E. L. 845
Mcclure, Elisabeth L. 2298
Mcclure, Harumi 2947
Mccndless, Brian E. 3143
Mccomb, David W. 966, 3139
Mcdanal, A.J. 525
Mcdanold, Byron K. 2864
Mcfavilen, Heather 305, 582
Mcintosh, Keith R. 1322
Mcintyre, Maxwell 1040
Mcintyre, Michael 1086
Mckenna, Russell 126
Mcmahon, William E. 268, 3381, 3406
Mcmeans, Philip A. 921
Mcpheeters, Claiborne 42, 525, 2099
Meakin, David 1927
Medic, V. 2364
Medici, Vasco 2118
Meeker, Michael A. 873
Mehlich, H. 1752
Mehta, Hitesh K. 3038
Meier, Florian 2996
Meissner, Dieter 178
Meitl, Matt 272
Meitl, Matthew 873
Melamed, Celeste L. 3406
Melchiorre, Michele 151, 2054
Meleco, A. J. 14
Mellor, Alexander 1339
Melnikov, Alexander 3129
Melvin, Andrew 1
Méndez, Juan A. 3240
Meng, Fanying 1241

AUTHOR INDEX

Meng, Hsin-Fei 1635
Meng, Xiaodong 2854
Menossi, Daniele 752, 1669, 2372
Menozzi, Roberto 2205
Men-Pérez, E. 370
Meot, Jacques 2593
Merdzhanova, Tsvetelina 2114
Merghcim, Julia 3500
Merkle, Agnes 3371
Merz, Christopher 1965
Merzlic, Sebastien 1933, 2000
Messer, Alexander 512
Messer, Alexander J. 521, 558
Messerschmidt, Michael 2682
Messmer, C. 2064
Metzger, W.K. 1312
Metzger, Wyatt K. 1196, 3147, 3305, 3319
Meuris, M. 3260
Mewe, Agnes A. 917
Meyers, Bennet 3354
Mi, Z. 2388
Mi, Zetian 1299
Mia, Md Dalim 2749
Michaelson, Lynne 925
Micheli, Leonardo 2301, 2789, 2804, 2858, 2881
Michl, Bernhard 1329
Mihailetchi, Valentin D. 1222
Mihaylov, Blagovest 1411
Mikofski, Mark 3354
Mikofski, Mark M. 110
Milakovich, Timothy 213
Miller, Bill 1473
Miller, David C. 2789, 2864, 3195, 3208
Miller, Elisa M. 2536
Milleville, Christopher C. 1598, 3384
Mil'shtein, S. 2411
Min, Jung-Hong 777
Minemoto, Takashi 455, 757, 2385
Minkin, L. 1863
Mints, Paula 2039
Miryala, Tejaswini 2646
Mishima, T. D. 14
Mishra, Himani 2376
Misra, Sudhajit 175, 802, 2467
Mitchell, Bernhard 2707, 3304
Mittag, Max 1531
Miyajima, Sakutaro 480
Miyashita, Naoya 854, 2334
Mizuno, Hidenori 1724
Moffett, C.E. 2736
Mohammed, Khaja H. 921
Mohapatra, Soumya Ranjan 3050
Mohr, Christian 83
Monnard, Raphäel 3256
Montenegro, Davis 3055

Montes, Carlos 429
Montgomery, Kyle H. 531
Montiel-Chicharro, Daniel 3208
Moon, Soo-Jin 3256
Moore, A. 2816
Moore, Andrew 1522
Moore, James 1838, 2091
Moore, James E. 259, 272, 373, 2506
Moore, Jay 2947
Moosa, Hassa 2011
Moosa, Maitha 2011
Moradi, Hadis 638, 2941
Moraitis, Panagiotis 3167
Morales, Christophe 2492, 2562
Morales, Cristian 2870, 2891
Morales-Acevedo, A. 670
Morel, Don 802, 1638, 3413
Morgado-Dias, F. 3178
Moriarty, T. 490
Moriarty, Tom 483
Moriki, Akinori 2906
Morin, Jean-Francois 1360
Morishige, Ashley E. 1494, 3236, 3290, 3300
Morita, Hiroshi 1973
Morral, Anna Fontcuberta I. 944
Morris, Jeromie 2996
Morrison, Matthew 229
Mortazavi, Soheyl 2875
Moseley, J. 1312
Moseley, John 62, 1196, 1381, 2887, 3123, 3147
Moser, David 3360, 3482
Moustafa, A. 1747
Moutinho, H.R. 1312, 2280, 2785
Moutinho, Helio 62, 2789, 3305
M'sirdi, Nacer K. 1968
Muaddi, Saad 1110
Mueller, Thomas 496, 2318
Mukherjee, Shaibal 2345
Mulder, P. 1189
Muller, Bjorn 126, 2267
Muller, M. 2280
Muller, Matthew 2294, 2301, 2789, 2804, 2858, 2881
Müller, R. 2064
Munasinghe, Anjali 2124
Munday, Jeremy N. 1585
Mundt, Laura 1253
Mundus, Markus 1253
Munkhammar, Joakim 3067
Muñoz, D. 1747
Munoz, Krystal 925
Munshi, Amit 1674
Munshi, Amita 980
Mur, Pierre 2562
Muralidharan, Pradyumna 1790, 1797
Muramatsu, Kazuo 2642

AUTHOR INDEX

Murphy, J. D. 1485
Murugesan, Arumugam 2172
Muskovin, Eric.................537
Mutitu, James.................315
Mwove, Johnson Kyalo.................2014
Myers, Matt.................525
Nærland, Tine Uberg.................2610
Nagaoka, Akira 1679
Nagarajan, Adarsh.................2991
Nage, M..................3150
Nagel, H. 3135
Nägelein, Andreas 2538
Nair, P. R..................2716
Nair, Pradeep R..................1015, 1022, 1755
Nakada, Tokio 192
Nakamur, Tetsuya 567
Nakamura, Kyotaro.................1504, 1794, 2498, 2566, 2588, 2642
Nakamura, Shigeyuki 2338
Nakamura, Tetsuya.................562, 3506
Nakano, Yoshiaki 854, 2201, 2524, 2538, 3528
Nakata, Tatsuya 854
Nakatsuka, Shigeru.................2385
Nam, Wooseok.................2242
Nanda, A..................229
Nandal, Vikas 1015
Nanduri, Sai Naga Raghuram.................1018
Naqavi, Ali.................512, 521, 558, 572, 1248
Narasimhan, K.L.................396, 1850, 1995, 3478
Nardone, Marco.................309, 3123
Naseem, Hameed A..................921
Natsheh, Ammar.................2011
Naumann, Volker.................1376
Nawara, Witek.................3462
Nawaz, Syed F..................1067
Nayfeh, Ammar.................213, 1741
Naylor, Mark.................914
Nayshevsky, Illya.................2285
Ndione, Paul.................1253
Needell, David R..................1737
Neely, J..................3224
Neely, Jason.................2141
Neergat, Manoj.................1704
Nehme, Bechara.................1968
Nelson, George T.................202, 206, 222, 1184, 2084
Nemeth, William 1777, 1801, 1817, 1832, 2242, 2702, 3439
Nespoli, Lorenzo.................2118
Nett, Zach 1737
Neuschitzer, Markus.................155
Neuwirth, Markus.................1682
Newlands, Allan.................2042
Ng, Annie 958
Ngan, Lauren.................600
Nguven, Dac-Trung.................2192

Nguyen, H. T..................3295
Nguyen, Tinh 3204
Nickel, Benedikt.................3388
Nicolay, S..................50
Nicolay, Sylvain 3256
Niemi, T..................1189
Niesen, Bjoern.................3256
Nietzold, Tara.................944, 2179, 3309
Nii, Kohdai.................85
Niki, Shigeru 33
Nilsson, Ulf H..................2864
Nishikawa, Naoyuki 1385
Nishio, M..................3
Nishioka, Kensuke.................480, 1479
Noack, Max.................2247
Nobre, André M..................3172
Nobuhara, Shohei.................1175
Nocerino, John.................93
Noda, Naoto 326
Noda, Yoshimasa 970
Nofuentes, Gustavo.................2858
Nogay, G..................2073
Noh, Shinyoung 858
Nonnenmacher, H. J..................1752
Norman, Andrew.................1381, 2887
Norman, Andrew G..................2536, 3406
Norwood, Robert A..................1147
Nose, Yoshitaro.................1679, 2361, 2385
Nowakowski, Marilyn L..................3524
Nsofor, Ugochukwu 1828
Nukala, Tejeswar.................3061
Nunomura, Shota 381
Nurdin, Muhammad 3102
Nussbaumer, Hartmut.................1077
Nuzzo, Ralph G..................1469, 1737
Nyirjesy, Gabrielle 667
Oberbeck, Lars.................3370
O'Brien, Greg 1933
O'Brien, Gregory.................2000
Ocaña, Luis 429
O'Carroll, Deirdre M..................3393
Ochoa, M..................1210
Odden, Jan Ove 2651, 2776
Oehler, Fabrice 1289
Ogawa, Tomoki 2548
Ogura, Atsushi.................1504, 2588, 2642
Ogutman, Kortan 1804, 3448
Oh, Jaewon 1858, 1877, 1883, 2912
Oh, Seung Kyu.................866
Oh, Soo-Young.................487
Ohdaira, Keisuke.................1385, 1787
Ohigashi, Takashi.................2159
Ohshima, H..................441
Ohshima, Takeshi.................562, 567
Ohshita, Yoshio 1504, 1794, 2498, 2566, 2588, 2642

AUTHOR INDEX

Ohta, Taisuke 3417
Ok, Young-Woo 333, 1807, 1838
Oka, Naotaka...1973
Okada, Yoshitaka.................. 10, 85, 854, 2334
Okafor, Jonathan O. 219
Okano, Y. 3
Okel, Lars A.G. 1081
Oliva, Florian 3265, 3285
Olopade, Muteeu......................................2381
Olvera, María De La Luz 632
O'Neill, Mark 525
Oney, Michael F. T 2176
Onno, Arthur...3370
Onunkwo, Ifeoma.....................................2135
Oo, W.M. Hlaing.....................................1661
Opila, Robert...................................999, 2035
Opila, Robert L. 315, 326
Oreski, Gemot 178
Orlovskaya, Nina A. 2324
Ortega, E. 643
Ortega, Pablo...944
Ortiz, Brenden R. 3406
Ortiz-Rivera, Eduardo I. 2957, 2963
Ory, Daniel 70
Oshima, Ryuji...861
Ososanya, Esther.....................................2432
Osowski, M. 3511
Osterwald, C.R. 467, 490
Ota, Yasuyuki...1479
Otaegi, A. 2740
Otaegi, Aloña...2677
Otnes, Gaute 1286, 2502
Ottoson, L. 467, 490
Ouyang, Zi......................................2403, 3220
Oviedo, Felipe...2744
Ozanne, A. -S. 1747
Paap, Scott 1473
Packard, Corinne E. 46
Page, Matthew.............................1777, 2242
Paggi, Marco 402
Palekis, Vasilios.......... 175, 802, 1511, 1638, 2467, 3413
Palekis, Vasilis.......................................2449
Palitzsch, Wolfram...................................2272
Palmer, Evan 496
Palmiotti, Elizabeth.................................1400
Palmquist, Nathan 667
Pan, Hui..1100
Pan, N...3511
Pan, Zhen..226
Panchal, A. K. 3061
Panchal, Ashish K. 3038
Pandey, Rahul..377
Paolone, Mario..1415
Parashar, Parag...............................1627, 1631
Paraskeva, Vasiliki....................................276

Parenti, Robert C. 3520
Parikh, Anuja V. 3123
Parikh, Harsh 2682
Park, Chinho 487
Park, Ji-Sang 6, 1256
Park, Joo Hyung 810
Park, Kyung Ho 244
Park, S. 2388
Park, Seonyong 2087
Park, Somin 1044
Park, Sungeun 935
Park, Won-Kyu 244
Partain, Larry...2042
Passow, Kendra 595, 600, 1071
Paszuk, Agnieszka 2524, 2538
Patra, Payal...761
Patterson, Robert J 2392
Paudel, Naba R. 2443
Paul, Douglas 1339
Paul, Nicolas..70
Paul, P. K. 30
Paul, Pran K. 2446, 3139
Paul, Sanjoy 2473, 2749
Paulauskas, Tadas 2759
Paull, P. K. 2414
Paull, Sanjoy 2728
Paulsen, Andrew 3514
Pavgi, Ashwini 1877, 1883
Paviet-Salomon, B. 50
Paviet-Salomon, Bertrand 3256
Pavilonis, Michael...................................1476
Pavlov, Marko 626
Pavlovsky, Igor 907
Pawar, Vaibhav 2986
Payne, David 315
Payne, David N.R. 2576
Peaker, A. R. 1485
Peale, Robert E. 2324
Peharz, Gerhard 178, 1329
Peibst, Robby 1366, 3371
Pellegrino, Sergio..............512, 521, 558, 572
Peña, J.L. 1691, 2457
Pena, Juan Luis 1669, 2372
Peña, Ramón 632
Peng, Jun..2076
Peng, Shou 761, 2427
Penning, David P 2170
Peppanen, Jouni 3025
Pera, David 1979
Peraca, Nicolás Márquez 263
Perez, Richard 132, 1104
Pérez-Rodríguez, Alejandro 3265, 3285
Perez-Wurfl, Ivan 315
Perkins, C. 2280
Perkins, Craig 2702, 2789

AUTHOR INDEX

Perkins, Craig L. 2294
Perl, E. 3376
Perl, Emmett E. 42, 1201
Perna, Allison 2467
Pesala, Bala 2858
Peschel, Gina 3396
Peshek, Timothy J. 1927, 2697, 3456
Peters, I. M. 648, 1140
Peters, Ian Marius.............. 284, 1264, 2532, 2744
Peters, Marius 3236
Petersen, Michael 2682
Peterson, Chris 512
Peterson, Josh 1169
Petoukhoff, Christopher E. 3393
Pfiester, Nicole.....................................701
Phillips, Adam 748
Phillips, Adam B. 170, 730, 815, 1030, 2462
Phillips, Laurie J. 1445
Phillips, Nancy H.................................2864
Phinikarides, Alexander 1954
Picard, Sandrine..................................2087
Piccinelli, Fabio 1669, 2372
Pickel, Tobias 3500
Pierro, Marco 3482
Pieters, Bart E. 1651
Pihan, Etienne 1498
Pillai, Supriya 2403
Pistor, Paul 155, 3285
Piszczor, Michael 525
Pitalúa, Nun..632
Platzer-Björkman, C.............................3269
Plessing, Lukas 178
Pleus, Albert 1835
Plochowietz, A. 1733
Podraza, Nikolas 2646
Podraza, Nikolas J. 2462, 2582, 2771, 3426, 3468
Poindexter, Jeremy 3300
Poissant, Yves.....................................1908
Pokharel, Nikhil 831, 2514
Polojärvi, Ville.....................................297
Poncho, Corpuz....................................2947
Poortmans, J. 3260
Poortmans, Jef....................................1233
Pop, Sergiu C. 921, 1869
Poplavskyy, Dmitry.....................1459, 1686
Porter, Ilana J. 417
Potamialis, C. 2457
Pötz, Sandra 178
Pouladi, S. 869
Pouladi, Sara 866
Poulsen, Peter B. 2672, 2682
Powalla, Michael 791
Previtali, Jonathan 1275
Price, Jared S......................................1469
Prietl, Christine 1329

Printraza, Nikolas J................................993
Procel, P. 2073
Ptak, Aaron J. 46, 62, 2275
Puska, Martti J. 2070
Puthanveettil, Suresh E............................76
Qazi, Farah 1317
Qian, Gary 667
Qian, Shen 958
Qin, Xuefei 1594
Qiu, Botong 667
Qudsia, Syeda 1317
Quinto, Carlos 429
Quiroz, Jimmy E. 1280
Rada, Jacob..1271
Radhakrishnan, Hariharsudan
Sivaramakrishnan 1233
Raghavan, Srinivasan 837, 841, 986, 2395, 2399
Ragunathan, Gautham...........................1181
Rahman, Mosaddequr 1067
Rahn, Christopher D...............................1469
Raiker, Gautam A. 3073
Raj, Samuel......................... 284, 496, 499
Rajan, Grace................182, 735, 807, 2298, 2446, 2646
Rajbhandari, Pravakar P. 989
Rajput, Amit Singh 499
Raju, T. Bhim 1034
Raker, David 2926
Rale, Pierre 1289, 3147
Ramakumar, Rama...............................1122
Ramamurthy, Praveen C. 1614, 2811
Rambabu, Sugguna 3478
Ramic, Zekija 2776
Ramírez, A. 503
Ramirez, A.A. 433
Ramírez, E. A. 433, 503
Ramos, Helena Geirinhas 3178
Ramos, Javier 2255
Ramprasad, Sumukh 496
Ramu, Govind..................... 1275, 2263
Rancoita, P.G. 541
Rand, James...925
Ranjan, Rajeev 2395, 2399
Ranjan, Upasna 2811
Ranjbar, S. 3260
Ransome, Steve 652
Rao, Arun D 1614
Rao, B.V. 1898
Rao, Rajesh 363, 2506
Rao, Roshan R 604
Raorane, Neha 1755
Raote, Yojak 1022
Rashkin, L. 3224
Rastogi, A.C. 3279
Rathi, M. 869
Rathi, Monika............................. 866, 2368

AUTHOR INDEX

Rathore, Sudharm 2902
Rau, Uwe...1651, 2114
Raupp, Christopher 1984
Ravindra, M. 76
Ravindra, Pramod 2395, 2399
Raychaudhuri, S..1733
Razooqi, Mohammed A. 2462
Recart, Federico...2677
Reddy, Anurag 3528
Reddy, K.S..2858
Reddy, Rekha 3524
Reed, S. 869
Reedy, Robert C..881
Reese, M. O. 138
Regalado-Pérez, E. 142
Reichel, C. 2064
Reichert, Andreas 2407
Reinders, Angèle 2109
Reindl, T. 1140
Reise, Christian 2267
Rejon, V. 1691, 2457
Ren, Zekun 284, 2532, 2744
Ren, Zhiwei..958
Reno, Matthew J. 1555, 1567, 1573,
    1579, 2975, 3025, 3031, 3055
Renteria, E. J. 281
Repins, Ingrid...2728, 3452
Reusser, Jean 2255
Reyes-Banda, M.G. 142
Rey-Stolle, I. 1210
Rey-Stolle, Ignacio 1204
Rhodes, Christopher 1476
Riaz, Hiba 1741
Ribeyron, P. -J...1747
Ricardo, Julian Do Nascimento 3077
Rich, Geoffrey 600
Richards, J...3224
Richardson, Walter...1116
Richter, A...1752, 2064
Richter, Mauricio...3360
Riedel, Nicholas...2672, 2682
Rienacker, Michael 3371
Riesen, Yannick..3435
Rigdon, Terry B. 3448
Riggs, Brian 37
Riggs, Brian C..3245
Riley, Daniel 1537, 3155, 3184, 3348
Riley, Daniel M.......................................1543, 1549
Rimmaudo, I. 1691, 2457
Rimmaudo, Ivan..1669, 2372
Rincon-Charris, Amilcar A. 2963
Ringel, Steven A. 215, 2079, 2446, 2554
Ringleb, Franziska 3396
Rivera, Eduardo I. Ortiz 3044
Riverola, Alberto...1339

Robert, Sofie 1804
Roberts, Jesse...3083
Robertson, John..37
Robertson, Kyle W. 1294
Robinson, Charles D. 3155
Rochat, Raphael 326
Rocheleau, Richard E...1061
Rockett, A. 30
Rockett, Angus......................182, 1400, 2446
Rockett, Angus A. 1511
Rodrigues, Sandy..3178
Rodriguez, D. J. 2031
Rodríguez, Diego J. 1143
Rodríguez, Pedro 2677
Rodríguez-Gallegos, Carlos D. 2318
Roest, Stefan 3462
Roeth, A. J. 14
Rogers, John A. 1469
Rohatgi, Ajeet 333, 940, 1807, 1824, 1838
Roland, Paul J...1030
Roller, John...508
Romanin, Vince 37, 3245
Romeo, Alessandro..................752, 1669, 2372
Ronoh, Geoffrey Kibiegon 970
Rooijakkers, Tom T.H. 1081
Ropp, Michael...2147, 3020
Rosales-Ascensio, Enrique....................................3240
Rose, Volker 2179, 3300
Ross, N. 3269
Rotoli, P. 1747
Rounsaville, Brian 940, 1807, 1824
Routhier, Alexander F. 3088
Rowell, David..3524
Roy, Sam 2358
Roy, Tatiana A. 521, 558
Royer, Fabien..558
Rozza, Davide 541
Rubbard, Seth M. 3468
Ruffini, Leia 2453
Ruiz, Carmen M. 2255
Ruiz, E. O. Ángel 1990
Rummel, S. 467
Rupp, B. 1733
Ruppalt, Laura B...873
Russell, Annie 1094
Russell, Richard..2227, 3435
Russell, Thomas C.R. 2236
Ruth, Daniel 2301
Ryou, J. 869
Ryou, Jae-Hyun 866, 2368
Saavedra, Michael 1473
Sablon, Kimberly 1181
Sabnis, Sanjeev 2849
Sacchetto, Davide 3256
Sachenko, A.V...663, 1025, 1811

AUTHOR INDEX

Sachenko, Anatoliy V 690
Sáenz, M.J 643
Saetre, Tor Oskar 685
Sahayaraj, S 3260
Sahli, Florent 3256
Sahraei, N. 648
Sai, Hitoshi 381, 1333
Saifullah, Muhammad 810
Sainsbury, Cassidy 2692, 2765
Saito, K. 3
Saito, Tomohiro 931
Saive, Rebecca 1589, 2236
Sakamoto, Katsuyoshi 85
Sakamoto, Norihiko 1268
Sakurai, Takeaki 33, 160, 900
Salamo, Gregory J. 3370
Salavei, Andrei 2372
Salazar, J. 370
Salazar-Duque, John E. 2963
Salo, Kristian 1494
Salome, Pedro 796
Salpakari, Jyri 3236
Salvetat, Thierry 2492
Samoilenko, Yegor 1697
Sampath, W.S. 980, 2736
Sampath, Walajabad 424, 785, 1674
Sampath, Walajabad S. 3417
Sample, Tony 1275
Sampson, Matthew D. 6, 1256
Samuelson, Lars 2502
Samundsett, C. 3295
Samundsett, Chris 2076
Sánchez, Yudania 155
Sánchez-Pérez, P. A. 1959, 1990
Sanchiz, Joaquín 429
Sandeep, K. 396
Sang, Baosheng 1455
Sang, Shiyu 472, 1430, 2918
Sangjeong, Myeong 356
Sankaran, M. 76
Sankin, I. 2816
Santana, G. 370, 2342, 2614
Santana-Rodríguez, G. 670
Santbergen, Rudi 2605
Santhanam, Parthiban 2185
Santos, M. B. 14
Santoyo-Salazar, J. 370
Saraf, Akash 761
Saraswat, Krishna 1835
Sargent, Edward H. 3129
Sarmah, Nabin 2858
Sarvari, Hojjatollah 1044, 2432
Sarwar, Jawad 1888
Sasaki, A. 1003
Sastry, O. S. 3478

Sato, Daisuke 1743
Sato, Shin-Ichiro 562
Sato, S-I. 552
Satzinger, Valentin 178
Saucedo, Edgardo 155, 3265, 3285
Savin, Hele 944, 1494, 3236
Sawallich, S. 3150
Sayed, Islam E.H. 2195
Sayyah, Arash 2891
Scaccabarozzi, Andrea 1289
Scarpulla, M.A. 1656, 1661
Scarpulla, Michael A. 175, 802, 1679, 2467
Schäfer, Nicolas 2216
Schaller, Richard D. 6
Scheiman, David 1838, 3514
Schelhasl, Laura T. 2176
Scheltens, Frank J. 966
Schenller, E.J. 1701
Schermer, J. 1189
Schindler, F. 2064
Schitthelm, F. 1752
Schlemmer, James 1104
Schmid, Martina 3396
Schmidt, Jan 3371
Schmidt, Thomas 3396
Schmieder, Kenneth J. ...210, 259, 272, 873, 2091, 2506
Schnabe, Thomas 2216
Schnabel, Erdmut 3488
Schnabel, Manuel ...1817, 2482, 2543, 3254, 3371, 3439
Schnabel, T. 3260
Schnabel, Thomas 791
Schneider, Kevin 1476
Schneller, Ej 389
Schneller, Eric J. 2839, 2897, 3448
Schoenfeld, Winston 2839
Schoenfeld, Winston V. 322, 1804
Schoenfelder, Stephan 2622
Schoenwald, David 2969
Scholl, Jonathan A. 1549
Schoop, Urs 1455
Schorch, M. 1752
Schriemer, Henry P. 1094
Schubert, M. C. 3135
Schubert, Martin C. 1329
Schulte, Kevin 62
Schulte, Kevin L. 46, 232, 2275
Schulte-Huxel, Henning 1366, 2543, 3371
Schulz, Gerd 914
Schulze, Patricia S.C. 1253
Schwabe, Hartmut 2622
Schweiger, M. 107
Sclj, Josefine 619
Scofield, A.C. 2755
Scolari, Enrica 1415
Sculati-Meillaud, Fanny 2794

AUTHOR INDEX

Seif, Johannes P. 3435
Seigneur, Hubert 2839, 2897
Sellami, Nazmi 2858
Sellers, Andrew 2926
Sellers, Diane G. 3384
Sellers, I. R. 14
Selvamanickam, V. 869
Selvamanickam, Venkat 866, 2368
Semichaevsky, Andrey 319
Sen, Fatih G. 2759
Senaud, L.-L. 50
Sengar, Brajendra S. 2345
Sengupta, Manajit 116, 1169
Senthilarasu, S. 2858
Sepeher, Mohsen M. 1094
Sera, Dezso 1421, 2682
Serra, João M. 1979
Sethia, Saurabh 2902
Seydel, Elisabeth 1682
Shafarman, William N. 26
Shah, S. 1350
Shahirinia, Amir 3092
Shanmugam, Vinodh 2318
Sharma, Ashok K. 396
Sharma, Romika 3300
Sharma, S. 2094
Sharps, Paul 42, 525, 2099
She, Hui 1863
Shen, Chang-Hong 1627
Shen, Zeqing 3393
Shephard, Les E. 1116
Shervin, Kaveh 1452
Shervin, Shahab 866
Shetty, Nishit 876
Shi, Jianwei 1820
Shi, Jiatiwei 1322
Shi, Xuanyi 3220
Shi, Zhan 1037
Shibata, Hajime 33, 1268
Shieh, Jia-Min 1627
Shigekawa, Naoteru 2548, 2551
Shih, Cheng-Hao 2035
Shih, I. 2388
Shih, Ishiang 1299
Shima, D. M. 281
Shimura, H. 1003
Shin, Hyun-Beom 244
Shin, Myunghun 2631
Shin, Seunghyun 935
Shin, Woo Jung 385
Shinde, O.S. 389, 1701
Shirasawa, Katsuhiko 904, 931
Shkrebtii, Anatoli I. 690
Shoji, Yasushi 10
Shore, Andrew 437

Shrestha, Niraj 1030
Shrestha, Santosh 696, 2186
Shu, Chia-Jhe.................................... 1606
Shu, Jinn-Kong 1606
Shubhrant, Abhishek 2902
Si, Fai Tong 2605
Siddiki, Mahbube K. 1018, 1110
Sidhu, Navjot Kaur 3279
Siebentritt, Susanne 151, 2054, 2205, 2478
Siepchen, Bastian............................... 761
Sikchang, Hyo 356
Silva, Francois 464, 1237
Silva, José A. 1979
Silvaggio, Amber C. 2554
Silverman, Timothy............................ 1259, 1893
Silverman, Timothy J. 1400, 2771, 3452
Simon, John 42, 46, 62, 1201, 2275
Simon, Kirby...................................... 876
Simpson, L. 2280
Simpson, Lin 2294
Simpson, Lin J. 1893, 2789
Sinapis, Kostas............................ 1081, 1090
Singh, Aparna 2902
Singh, Ashish 1034
Singh, Ashish K. 1704
Singh, Hemant K. 1995, 3478
Singh, Rajeev 2762
Singh, Rhythm 1151
Singh, Rubina 1855
Singh, Sukvhinder 2227
Singlr, Vijay P. 2432
Sinha, Archana 3478
Sinha, Parikhit 2005
Sinisuka, Ngapuli I 3102
Sink, Joseph 3333
Sinton, Ronald 2707
Sinton, Ronald A. 2692, 2765
Sio, H. C. 3295
Sio, Hang Cheong 3300
Sites, James R. 164, 1308
Slocum, Michael 677
Slocum, Michael A. 18, 202, 206, 222, 831, 1184, 2084, 2514, 3468
Slooff, Lenneke H. 1081
Smaglik, Nathan 831, 2514
Smestad, Greg P. 2858
Smith, Benjamin 1134
Smith, Brittany L. 18, 1184
Smith, David J. 2573
Smith, Mathew................................... 2941
So, Won-Shup 487
Soares, Gabriela De Amorim................ 2875
Sodabanlu, Hassanet 854
Söderström, T. 1752
Sofia, Sarah E. 1264

AUTHOR INDEX

Sogabe, Tomah 85, 712
Sokolovskyi, I.O. 663, 1025, 1811
Sokolovskyi, Igor O. 690
Solanki, Chetan S. 3478
Solanki, Chetan Singh 1850
Soltanmohammad, Sina...................................26
Soman, Anishkumar 1828
Søndergaard, Sissel Tind 2651
Song, Dengyuan...1430
Song, Hee-Eun.......................................356, 1758
Song, Myungkwan 2631, 2634
Song, Tao...1308
Song, Zhaoning................... 170, 730, 748, 815, 1030
Sonp, Hee-Eun..2723
Sood, Neeru 2858
Sossan, Fabrizio 1415
Soudachanh, A. L...281
Sozzi, Giovanna..2205
Spandana, B...396
Spataru, Sergiu................... 1421, 2682, 2819
Spaulding, David.............................820, 1686
Spertino, Filippo...3096
Spiering, Stefanie 791
Spinelli, P. 3150
Spooner, Ted.......................................1275, 2263
Sreekumar, Nimisha 1755
Sridharan, Akirt...999
Sriramagiri, Gowri 658, 1196
Srivatsan, R. 120, 589
Stark, Cameron..1855
Starkl, Hannes...178
Steeman, Rob..337
Steenhoff, Volker 3388
Stefancich, Marco 1946
Steijvers, Henk...2875
Stein, Joshua....................... 1537, 3333, 3348
Stein, Joshua S............... 1543, 3155, 3161, 3184, 3488
Steiner, Myles A. 42, 47, 232, 1201, 2195, 3254
Steinfedt, Jeff...525
Stender, C. 3511
Stender, Christopher L.................................3524
Stephan, Jack...2124
Stevens, Margaret...701
Steward, Malia...1037
Stewart, J. 3224
Stika, K. 2312
Stiles, Phil 2833
Stoddard, Nathan...2610
Stokes, Adam.......................................1381, 2887
Stolt, L. 30
Stone, Kevin H...2176
Stradins, Paul881, 1491, 1777, 1801, 1817, 1832, 2242, 24
Stradins, Pauls.....................................2482, 2702
Strandberg, Rune..............................706, 2651
Stride, John A...2392

Stuart, Thomas 2926
Stuckelberger, J...2073
Stuckelberger, Michael 2610, 2854, 3309
Stuckelberger, Michael E..............................2179
Stueve, Bill 2808, 2923
Sturm, James C. 1773
Stutz, Elias Z. 944
Su, Bojie 392, 1430, 1873, 2918
Su, Chengfeng.............................. 392, 1873
Subbiah, Jegadesan 1342
Subedi, Indra 2771, 3468
Subedi, Kamala Khanal 781
Sudbury, Benjamin A. 1322
Suga, Mitsunobu 567
Sugaya, Takeyoshi.......................... 562, 861, 1724
Sugimoto, Hiroki 160
Sugiyama, Masakazu 854, 2201, 2524, 2538, 3528
Sugiyama, Mutsumi 192
Sugiyama, Ryo 712
Suhana, Hadi 3102
Sumita, Taishi 3506
Sun, C. 1485
Sun, Ce 2759
Sun, Chang..3300
Sun, Chenguang..1241
Sun, Kaiwen 2213
Sun, Qiang...1648
Sun, Qiming 3129
Sun, S. 869
Sun, Sicong..2368
Sun, Wen-Cheng...2227
Sun, Xiaolin...2656
Sun, Xingshu...................1055, 1259, 1904
Sun, Yaojie 2048
Sun, Yubo...2467
Sun, Yukun 2291
Sun, Zeming 937
Supplie, Oliver..............................2524, 2538
Surya, Charles 958
Sutou, Yuji 931
Sutterlueti, Juergen 652
Suzuki, Ryota 1504, 2588
Swain, Santosh K..3422
Swartz, Craig H. 2473, 2749
Sweatt, William 1473
Syu, Hong-Jhang 1009
Szabo, Sandor 2167
Szlufcik, Jozef.....................1233, 2227, 3435
Tabet, Nouar.......................963, 1058, 3435
Tacconi, Mauro 541
Tachibana, Shoji 1504
Tadese, Alemu 1104
Tadesse, Alemu.....................132, 1132, 1427
Tae, Christian 1835
Taekjeong, Kyung 356

AUTHOR INDEX

Takahashi, Akiko 2906
Takahashi, Isao 1765, 1794
Takahashi, Takuji 455
Takahashi, Yasuhito 1973
Takamoto, Tatsuya 3506
Takato, Hidetaka 381, 904, 1724, 3323
Takenouchi, T. 441
Tamaki, Ryo 10
Tamboli, Adele 3254, 3371
Tamboli, Adele C. 578, 2482, 2488,
    2536, 2543, 3381, 3406
Tamizhmani, Govindasamy 1389, 1850,
    1858, 1877, 1883, 1959, 1984, 2789, 2912
Tan, Jin 1158
Tan, Joel M. R. 3275
Tan, K.H. 1210
Tan, Xuehai 761
Tanahashi, Katsuto 3323
Tanahashi, Tadanori.............................1268
Tanaka, Aki 2642
Tanaka, T. 3
Tanaka, Takahiro 2947
Tang, Chiu C. 2423
Tang, Houjun 1100
Tang, Mingchu 3370
Tang, Tao 2558
Tanke-Pedretti, Anna 1473
Tao, Meng 385, 608
Tao, Yuguo 1824
Tappan, Ian A. 2864
Tassone, Christopher J..........................2176
Tatapudi, Sai 1850, 1858, 1877, 1959, 2912
Tatapudi, Sai Ravi Vasista 2789
Tatavarti, Rao 1184, 2084
Tatavarti, Sudersena Rao........................1181
Tate, John Keith...................................333
Tayagaki, T. 3
Tayyib, Muhammad 2776
Tchemycheva, Maria 1289
Tedeschi, Giampiero 2372
Teena, Percis 3113
Tennyson, Elizabeth M. 1508, 2443
Terheiden, Barbara..............................1824
Terukov, E.I. 1025, 1811
Teubner, Thomas................................3396
Teymouri, Arastoo 2403
Thanh, Nguyen Cong 1765
Thankalekshmi, Ratheesh R. 3279
Theelen, Mirjam.................................2875
Theigi, San 881
Theingi, San......................................1832
Theocharides, Spyros....................1163, 3107
Therrien, Francis 1579
Thibeault, Brian..................................315
Thimsen, Elijah..................................876

Thompson, Christopher.........................1196
Thompson, Corey S. 1869
Thon, Susanna M..................................667
Thorseth, Anders........................2672, 2682
Thorsteinsson, Sune 2672, 2682
Thway, Maung 284, 2744
Tidwell, Steven 1086
Timò, Gianluca 290
Tirumalai, Tejas 2923
Tischler, Joseph G. 873
Titus, Jochen 820
To, Alexander 517
To, B. 2280
Toberer, Eric S. 2536, 3406
Todorov, Teodor 1441
Togay, M. 2457
Togay, Mustafa 146, 186
Tomasi, A. 50
Tomasi, Andrea 3435
Tomasulo, Stephanie 2091
Tonic, Marko 1346
Toor, Fatima 346, 1537, 1543, 3184, 3333, 3348
Toprasertpong, Kasidit..................2201, 2524
Torralba, Encarnacion...........................3402
Tous, Loïc 2227, 3435
Tracy, Jared..............................3190, 3200
Traverse, Christopher...........................2124
Trout, T. John....................................2312
Trupke, Thorsten 66, 420, 2707, 3304, 3315
Tsafarakis, Odysseas............................1090
Tsai, Cheng- Ying 3366
Tsai, Jia-Lin 1606
Tsai, Jia-Ling 294
Tseng, Zong-Liang 367
Tsutsumi, S. 3
Tu, Wei-Chen 1051
Tucher, Nico....................352, 1253, 2511
Tukiainen, Antti 297, 2520
Tuminello, F. 3511
Tummala, Abhishiktha 2912
Turek, Marko 2232
Turner, John A. 47
Tuteja, Mohit 1511
Tyagi, Astha 716
Tyler, Kevin 301
Tyson, Tom 925
Tzolov, Marian 1040
Ubukata, Akinori..................................861
Ueda, Kohsuke 3506
Ueda, T. 1003
Uematsu, Takumi 1950
Ulbricht, Christoph 1360
Ulicná, Sona 146, 186
Uma, B. R. 76
Umishio, Hiroshi.................................381

AUTHOR INDEX

Unold, Thomas 3396
Unsur, Veysel...888
Upadhyaya, Ajay D.........................940, 1807
Upadhyaya, Vijay D.......................................333
Upadhyaya, Vijaykumar 940, 1807, 1824
Uprety, Prakash 3468
Urbano, J. Antonio..632
Uruena, Angel...............................2227, 3435
Usami, Noritaka 1765, 1794
Utsunomiya, Satoshi 904
Vadiee, E...1603
Vadiee, Ehsan.....................305, 827, 1841
Vagidov, Nizami Z. 531
Vähänissi, Ville...1494
Vaida, Mihai E...417
Vaidya, Nina............................512, 521, 558, 572, 1248
Vaisman, M. 3376
Vaisman, Michelle.......................578, 3381
Vaissiére, Nicolas 2528
Valderrama, Nicolas 1893
Valdivia, Christopher E.................................1094
Van Aken, Bas B. 3462
Van Alsburg, Jane...1455
Van De Loo, Bas W.H....................................1817
Van Der Heide, Arvid....................................3343
Van Hest, Maikel F.A.M...............2482, 3371, 3439
Van Sark, Wilfried...3014
Van Sark, Wilfried G.J.H.M. 1090, 3167
Vandamme, Nicolas 2453
Vandervelde, Thomas E. 701
Vanka, S..2388
Vanka, Srinivas 1299
Vansant, Kaitlyn 1922, 3452
Vargas, Carlos...3290
Vasi, J..1995
Vasi, Juzer 1850, 3478
Vasileska, D. 1603, 2816
Vasileska, Dragica..........................1790, 1797
Vasilyev, Leonid A. 3448
Vasudevan, Saravanan 2172
Vauche, Laura.................................2492, 2562
Vedde, Jan 2682
Veettil, Binesh Puthen 2392
Vehse, Martin 3388
Veinberg-Vidal, Elias.....................2492, 2562
Veith-Wolf, Boris 1366
Velappan, Krishnakumar................................761
Venizelou, Venizelos 276, 1163, 1941, 3107
Verbitskiy, V.N..1811
Verlinden, Pierre J.........................1922, 2220
Vermang, B..3260
Vermang, Bart...2209
Verschac, Rodrigo...1175
Vetter, E..1752
Viana, Marcelo Machado 1917, 2307

Vignola, Frank..1169
Vijh, Aarohi 3520
Vilcot, Jean-Pierre..3402
Vincent, Nina 1893
Vines, L. 3269
Vinogradova, Tatiana 512
Vinogradova, Tatiana G. 521, 558, 572
Virtuani, Alessandro 1395, 2104, 2794
Vlasiuk, V.M...1025
Vlasyuk, V.M..1811
Vleugels, J. 3260
Voarino, Philippe............................2492, 2562
Vogt, Malte Ruben 1366
Von Gastrow, Guillaume 944
Voroshazi, Eszter..3343
Voss, Henrik..2682
Waddle, John M. 309, 3123
Wade, Andreas 2005
Wagner, Sigurd 1773
Waiis, J.M..2457
Waldhauser, Wolfgang 1329
Walker, Don............................93, 512, 1248
Walls, J.M..1691, 3430
Walls, John 1674
Walls, John M.................146, 186, 752, 2349
Walls, John Michael 2827
Walls, Michael 424
Walter, Arnaud 3256
Walters, Joseph 2839, 2897
Walters, R. J. 845
Walters, Robert 3514
Walters, Robert J. 210, 247, 259, 272, 373, 873, 1838, 2091, 2506
Waltmger, A. 1752
Walukiewicz, W...3
Walukiewicz, Wladek....................................1204
Wan, Kai-Tak 1933, 2000
Wan, Ronghua 226
Wan, Yimao 59, 2076
Wang, Ao...1937, 2823
Wang, Baomin 1469
Wang, Changlei 993
Wang, Da-Wei 3220
Wang, Deng..2048
Wang, Feng 1044
Wang, Fumei.....................392, 1937, 2823
Wang, Haotian 1342
Wang, He. 226, 392, 1430, 1648, 1873, 1937, 2823, 2918
Wang, Hongfeng 1215
Wang, Laidong 385
Wang, Mu 3370
Wang, Q. 1733
Wang, Rui 1100
Wang, Shenghao 160
Wang, Shizhen 976

AUTHOR INDEX

Wang, Sisi 2600
Wang, Teng-Yu 2660
Wang, Xiaohui.......................................2432
Wang, Y. 1733
Wang, Y. D. 1733
Wang, Yan 1922
Wang, Yichen 1299
Wang, Yiwang 1100
Wang, Yongqian 2220
Wang, Yu 1933, 2000
Wang, Yu-Cian 2498, 2566
Wang, Zigang 1922
Ward, J. Scott..3254
Warmann, Emily......................................1248
Warmann, Emily C. 512, 521, 558, 572
Warner, Jeffery. H.................................2091
Warren, Emily.......................................3371
Warren, Emily L. 578, 2482, 2488, 2543, 3381
Washington, Lori 3520
Washio, Hidetoshi 3506
Watanabe, Kentaroh..........................854, 3528
Watanabe, Yasuyuki................................613
Waters, Martin.......................................2923
Watson, S..648
Watthage, Suneth C. 170, 730, 748, 815, 1030
Watts, John L.R. 2762
Weeber, Arthur.......................................2875
Weick, Clément...............................2492, 2562
Weigand, William 1790
Weiss, Charlotte...............................83, 2407
Weiss, Dirk...1264
Weiss, Karl-Anders..................................2836
Wen, Ching-Chang 329
Wen, Xiaoming..696
Wenham, Stuart R. 2576, 2600
Werner, Florian...............................2205, 2478
Werner, Jérémie.............................55, 3256
West, Bradley M. 2179, 3309
Western, N. J...953
Western, Ned J.......................................948
Wheeler, Tobias 1476
Whipple, Steven 88
Whiteside, V. R......................................14
Wibowo, A. 3511
Wibowo, Andre 1181, 1184, 2084
Wicaksono, S. 1210
Widén, Joakim 3067
Wieghold, Sarah 3300
Wienands, Karl 1253
Wiese, Martin..1531
Wille-Haussmann, Bernhard 126
Williams, J. 1603
Williams, Joshua J. 305, 582
Williams, R. 490
Wilson, Gregory 3236

Wilson, Marshall 322
Wilt, David M. 88, 102, 301, 531
Wilt, Sam... 301
Wilterdink, Harrison 2692, 2765
Winkler, Kristina 1253
Winkler, Thilo 3500
Wirsching, Sven 3500
Wirth, Harry.. 1531
Wischmann, Wiltraud 2058
Wissen, A.. 1752
Witte, Wolfram 2205
Witteck, Robert 1366
Wohlgemuth, John 1275
Wojtowicz, Anna 164
Wolden, C. A. 138
Wolden, Colin A. 1697
Wolf, Martin 2692
Wolffersdorff, Paul 595
Wong, Johnson 499, 3113
Wong, Johnson Kai Chi 496
Wong, Lydia H. 3275
Woodhouse, Michael 1259
Woods, Jason 1893
Woods-Robinson, Rachel 3410
Worrell, Ernst 3014
Wright, Lewis D. 146, 186
Wu, Gordon 96
Wu, J. 1189
Wu, Jiang 3370
Wu, Kuen-Yi 911
Wu, Po-Ching 1623
Wu, Ruei-Ying 1635
Wu, Shang-Hsuan 1051
Wu, Teng-Chun 448, 476
Wu, Yonggang 1594
Wu, Yuh-Renn 294
Wu, Zhuopeng 1241
Würfel, Uli 1253
Wyrsch, Nicolas 3435
Wyss, P. 2073
Xia, Hongze 2392
Xia, Zihuan 1594
Xiao, C. 1312, 2785
Xiao, Chuanxiao 62, 1371
Xiao, T. Patrick 2185
Xiao, Zhi Bin 1648
Xie, Yu 116
Xiong, Gang 1193, 2473
Xiong, Zhen 2220, 3304
Xu, Jun 2656
Xu, Ling 2656
Xu, Lu 1737
Xu, Menglei 1233
Xu, Qi 37, 3245
Xu, Qianfeng 2285

AUTHOR INDEX

Xu, Tao 2251
Xu, Xiaojie 3410
Xu, Zhaoran 59
Xue, Muyu 1835, 2732
Yablonovitch, Eli...................................2185
Yachi, Toshiaki.......................................613
Yadav, Karan Shishir 2902
Yadav, Tarun S. 396
Yakes, Michael K. 873, 2091
Yamada, Noboru.............................1724, 1743
Yamada, Nobuyuki 2906
Yamagami, Takeru 192
Yamagoe, K. 441
Yamaguchi, Hiroshi 3506
Yamaguchi, Koichi...................................712
Yamaguchi, Masafumi......................359, 412, 1479,
        1711, 1714, 1743, 2498, 2548, 2566
Yamaguchi, Seira 1385
Yamamichi, Masaaki 1275, 2263
Yamaya, Haruki 2159
Yan, Chang 2213
Yan, Di 2076
Yan, Yanfa 771, 993, 1643, 2443, 2473, 3426
Yancey, Billy...2128
Yanchilin, Anton 585
Yang, Fan 2656
Yang, Guangtao 2605
Yang, Hao-Yu................... 343, 367, 893, 2664
Yang, Hong............. 392, 1430, 1873, 1937, 2823, 2918
Yang, Jianfeng..2392
Yang, Mohshi..907
Yang, Peter 1100
Yang, Shuying 2697, 3456
Yang, X. 2785
Yang, Yang 2220
Yang, Yi Tong 1648
Yang, Yun-Chie 893, 2664
Yang, Zhihao 74
Yao, Li You..1648
Yao, Y. 869, 1752
Yao, Yangyi 1048
Yao, Yao 866, 2368
Yarnaguchi, Koichi.....................................85
Yates, Peter..1445
Yaung, K. Nay 3376
Yaung, Kevin Nay 284, 2744
Ye, Feng 2220
Ye, J. 2785
Ye, Qilin 948
Yeh, Chun-Ming.......................................911
Yellowhair, Julius 2870
Yellowhair, Julius E.................................2891
Yi, Chuqi...2569
Yilmaz, S...3430
Yoo, Chang Youn 399

Yoon, Howard W.......................................437
Yoon, Jongseung 549, 2291
Yoon, S. F. 1210
Yoon, Woojun 373, 1838
Yoshiba, Shuhei 1769
Yoshino, Kenji 1679
Yoshita, M. 1003
Yoshita, Masahiro 2781
You, Bang-Jin 367
You, Liang-Chian 1635
Young, David 46
Young, David L.............1817, 1832, 2275, 3254
Young, James L..47
Young, Steven 582
Youssef, Amanda................1491, 2242, 3300
Youtsey, Christopher 3524
Yu, Edward T. 363, 1181
Yu, Jia 2453
Yu, K. M. 3
Yu, Kin Man 1204
Yu, Li-Chieh 3204
Yu, Linwei 2656
Yu, Ming 1193
Yu, Peichen 294, 1610, 1635
Yu, Pei-Chen 1606
Yu, Sun 1522
Yu, Zhengshan J..............1228, 1317, 1322, 2039, 3250
Yuan, Bo 315
Yuan, Lin 2392
Yue, Yao 93
Yun, Jae Ho 810
Zachariah, S. 1995
Zachariah, Sachin 2799, 2849, 3478
Zahler, James..1463
Zahler, James M.....................................3245
Zakaria, Naimi 487
Zamora, Rachid Darbali 3044
Zang, Kai 1835
Zapalac, G. 2414
Zapalac, Geordie 820, 3327
Zauner, Andy...1237
Zech, Tobias 1531
Zelenina, Anastasiya 2054, 2478
Zeman, Miro 2605
Zeng, Guoping 907
Zeng, Xulu 1286
Zeyu, L. 1781
Zhai, Yonghui 472
Zhan, Tien-Chien 294
Zhang, Bao 226
Zhang, C. 1603
Zhang, Chaomin.............240, 827, 1215, 1841, 2573
Zhang, Guoqi 2605
Zhang, Hua 3304
Zhang, Huan 2558

AUTHOR INDEX

Zhang, Jili 1100
Zhang, Jing 3384
Zhang, Junjun 1937, 2823
Zhang, Lei 408, 1761, 1828, 2667
Zhang, Liang 2247
Zhang, Liping 1241
Zhang, Nian 2432
Zhang, Qiming 226
Zhang, Wei 255, 1193
Zhang, Weijie 820
Zhang, X. 2094
Zhang, Xiaochen 1567
Zhang, Xue 392, 1430, 1873, 2918
Zhang, Yang 195
Zhang, Yi 696, 2186
Zhang, Yong-Hang 3366, 3410
Zhang, Yufeng 198, 767, 1707
Zhang, Z. 2019
Zhang, Zhilong 2392
Zhang, Zongyi 1594
Zhangl, Xiaochen 1555
Zhao, Dewei 993
Zhao, Hui 392, 1430, 1873, 2918
Zhao, J. 1752
Zhao, Jing 1845
Zhao, Pan 1430, 1873, 2918
Zhao, Xin-Hao 3366
Zhao, Yuan 3366
Zhao, Yuetao 1044
Zhe, Liu 2744
Zheng, N. 869
Zhigunov, D.M. 1811
Zhongbiao, Ye 1044
Zhou, Guomin 472
Zhou, Hang 976, 2558
Zhou, Jian 1594
Zhou, Xiao W. 2419
Zhu, Jiang 3208
Zhu, Lin 721, 3528
Zhu, Yan 66, 3290
Zhu, Ziyao 198
Zide, Joshua M. O. 3384
Zielnik, Allen 3208
Zilles, Roberto 1917
Zilouchian, Ali 638, 2941
Zimmerman, Jeramy D. 3381
Zin, Ngwe 322
Zinaddinov, M. 2411
Zoppi, Guillaume. 742
Zubia, David 2419
Zunger, Alex 1245

IEEE
445 Hoes Lane
Piscataway, NJ 08854-4141

ISBN 978-1-5090-5606-4

9 781509 056064